ICF	intracellular fluid
ICN	intensive care nursery
ICU	intensive care unit
ID	intradermal
IF	intrinsic factor
IFN	interferon
Ig	immunoglobin
IM	intramuscular; infectious mononucleosis
IN	internist
IOP	intraocular pressure
IPPA	inspection, palpation, percussion, auscultation
IUD	intrauterine device
IV	intravenous
IVC	inferior vena cava
IVF	in vitro fertilization
IVP	intravenous pyelogram
IVT	intravenous transfusion
JGA	juxtaglomerular apparatus
KS	Kaposi's sarcoma
KUB	kidneys, ureters, bladder
LBB	left breast biopsy
LDL	low-density lipoprotein
LFT	liver function test
LG	laryngectomy
LH	luteinizing hormone
LLQ	left lower quadrant
LMP	last menstrual period
LOC	loss of consciousness
LP	lumbar puncture
LPN	licensed practical nurse
LRI	lower respiratory infection
LUQ	left upper quadrant
LVAD	left ventricular assist device
MAb	monoclonal antibody
mEq/l	milliequivalents per liter
MG	myasthenia gravis
MI	myocardial infarction
MLT	medical laboratory technologist
mm^3	cubic millimeter
mm Hg	millimeters of mercury
MOA	medical office assistant
MRI	magnetic resonance imaging
MS	multiple sclerosis
MSH	melanocyte-stimulating hormone
MSOF	multisystem organ failure
MVP	mitral valve prolapse
NBM	nothing by mouth
ND	natural death
NE	norepinephrine
NGU	nongonococcal urethritis
NLMC	nocturnal leg muscle cramping
NMJ	neuromuscular junction
NREM	nonrapid eye movement
NSAID	nonsteroidal antiinflammatory drug
NSU	nonspecific urethritis
NTG	nitroglycerin
NTP	normal temperature and pressure
OB/GYN	obstetrician-gynecologist; obstetrics-gynecology
OC	oral contraceptive
OD	overdose; right eye
OHS	open heart surgery
OI	opportunistic infection
OR	operating room
ORT	operating room technician
OT	oxytocin
OTC	over-the-counter
OV	office visit
P	pressure
PABA	para-aminobenzoic acid
PCP	*Pneumocystis carinii* pneumonia
PD	Parkinson's disease
PE	pulmonary embolism; physical examination
PED	pediatrics; pediatrician
PEG	pneumoencephalogram
PEMF	pulsating electromagnetic field
PET	positron emission tomography
PG	prostaglandin
pH	hydrogen-ion concentration
PID	pelvic inflammatory disease
PKU	phenylketonuria
PMH	past medical history
PMN	polymorphonuclear leucocyte
PMP	plasma membrane protein
PMS	premenstrual syndrome
PNS	peripheral nervous system
PRL	prolactin
PROG	progesterone
PT	prothrombin time; physical therapist
PTCA	percutaneous transluminal coronary angioplasty
PTD	permanent and total disability
PTH	parathyroid hormone
PTX	pneumothorax
PUBS	percutaneous umbilical blood sampling
PUL	percutaneous ultrasonic lithotripsy
Px	prognosis; pneumothorax
PX	physical examination
R	roentgen (unit of x-radiation)
RA	rheumatoid arthritis
RAD	radiation absorbed dose
RAS	reticular activating system
RBB	right breast biopsy
RBC	red blood cell; red blood count
RBOW	rupture of bag of waters
RDA	recommended daily allowance
RDS	respiratory distress syndrome
REM	rapid eye movement
Rh	*Rhesus*
RHC	respirations have ceased
RK	radial keratotomy
RLQ	right lower quadrant
RLX	relaxin
RM	radical mastectomy
RN	registered nurse
RNA	ribonucleic acid
ROS	review of symptoms
RR	respiratory rate
RRR	regular rate and rhythm (heart)
RS	Reye's syndrome
RT	radiotherapy; radiologic technologist
RUQ	right upper quadrant
Rx	prescription
SA	sinoatrial (sinuatrial)
SBP	systolic blood pressure
SC	subcutaneous
SCA	sickle-cell anemia
SCD	sudden cardiac death
SCID	severe combined immunodeficiency syndrome
SDS	same-day surgery
SF	synovial fluid
SG	skin graft; specific gravity
SH	social history
SIDS	sudden infant death syndrome
SIG	sigmoidoscopy; sigmoidoscope
SIW	self-inflicted wound
SLE	systemic lupus erythematosus
SMD	senile macular degeneration
SNS	somatic nervous system
SOB	shortness of breath
SPF	sun protection factor
S/S (Sx)	signs and symptoms
STD	sexually transmitted disease
SubQ or SQ	subcutaneous
SV	stroke volume
SVC	superior vena cava
T	temperature
TAH	total artificial heart
TB	tuberculosis
TIA	transient ischemic attack
TM	transcendental meditation
TMJ	temporomandibular joint
TND	transient neurologic deficit
TOP	termination of pregnancy
TPE	therapeutic plasma exchange
TPN	total parenteral nutrition
TPR	temperature, pulse, and respiration
TSH	thyroid-stimulating hormone
TSS	toxic shock syndrome
Tx	treatment
UA	urinalysis
UDO	undetermined origin
UG	urogenital
URI	upper respiratory infection
US	ultrasound; ultrasonography
UTI	urinary tract infection
UV	ultraviolet
VD	venereal disease
VDRL	venereal disease research laboratory test (blood test for syphilis)
VF	ventricular fibrillation
VPC	ventricular premature contraction
VS	vital signs
VT	ventricular tachycardia
VV	varicose veins; vulva and vagina
WBC	white blood cell; white blood count
WNL	within normal limits
X-match	cross-match
XRT	x-ray therapy

Principles of Human Anatomy

Principles of Human Anatomy

SIXTH EDITION

Gerard J. Tortora
BERGEN COMMUNITY COLLEGE

HarperCollinsPublishers

Sponsoring Editor: **Bonnie Roesch**
Developmental Editor: **Meryl R. G. Muskin**
Project Editor: **Thomas R. Farrell**
Art Director: **Teresa J. Delgado**
Art Coordinator: **Claudia Durrell**
Text and Cover Design: **Circa 86, Inc.**
Cover Illustration: **Study of nude for *The Battle of Cascina,* Michelangelo. Teyler Museum, Haarlem.**
Photo Researcher: **Mira Schachne**
Production: **Jimmy Spillane**
Compositor, Printer, and Binder: **Arcata Graphics/Kingsport**
Cover Printer: **Lehigh Press**

Principles of Human Anatomy, Sixth Edition

Library of Congress Cataloging-in-Publication Data

Tortora, Gerard J.
Principles of human anatomy / Gerard J. Tortora. — 6th ed.
p. cm.
Includes bibliographical references and index.
ISBN 0-06-500243-1—ISBN 0-06-500907-X (Teacher Edition)
1. Human physiology. 2. Human anatomy. I. Title.
[DNLM: 1. Anatomy. 2. Physiology. QS 4 T712pa]
QP34.5.T68 1992
612—dc20
DNLM/DLC
for Library of Congress 91-20795
CIP

92 93 94 9 8 7 6 5 4 3 2

The sixth edition of Principles of Human Anatomy *is dedicated to Geraldine C. Tortora, my wife, whose love, support, encouragement, and assistance have contributed so much to my success as an author.*

Contents in Brief

Contents in Detail vii

Preface xv

Note to the Student xix

1 *An Introduction to the Human Body 1*

2 *Cells 28*

3 *Tissues 62*

4 *The Integumentary System 94*

5 *Bone Tissue 110*

6 *The Skeletal System: The Axial Skeleton 128*

7 *The Skeletal System: The Appendicular Skeleton 161*

8 *Joints 181*

9 *Muscle Tissue 212*

10 *The Muscular System 236*

11 *Surface Anatomy 317*

12 *The Cardiovascular System: Blood 333*

13 *The Cardiovascular System: The Heart 351*

14 *The Cardiovascular System: Blood Vessels 377*

15 *The Lymphatic System 427*

16 *Nervous Tissue 455*

17 *The Spinal Cord and the Spinal Nerves 471*

18 *The Brain and the Cranial Nerves 497*

19 *The Autonomic Nervous System 546*

20 *Sensory and Motor Systems 561*

21 *The Endocrine System 603*

22 *The Respiratory System 628*

23 *The Digestive System 652*

24 *The Urinary System 694*

25 *The Reproductive Systems 717*

26 *Developmental Anatomy 768*

Appendix: Answers to Self Quizzes A-1

Glossary of Combining Forms, Word Roots, Prefixes, and Suffixes G-1

Glossary of Terms G-6

Bibliography B-1

Index I-1

Preface xv
Note to the Student xix

1 An Introduction to the Human Body 1

Anatomy Defined 2
Levels of Structural Organization 2
Life Processes 5
Overview of the Human Body 8
Structural Plan 8
Anatomical Position and Anatomical Names 8
Directional Terms 9
Planes and Sections 11
Body Cavities 000 12
Abdominopelvic Regions 15
Abdominopelvic Quadrants 15
Medical Imaging 21
Measuring the Human Body 24
Study Outline 25
Review Questions 26
Self Quiz 26

2 Cells 28

Generalized Animal Cell 29
Plasma (Cell) Membrane 30
Chemistry and Structure 30
Functions 30
Movement of Materials Across Plasma Membranes 32
Passive Processes 33 Active Processes 33
Cytosol 34
Organelles 34
Nucleus 36
Ribosomes 39
Endoplasmic Reticulum (ER) 39
Golgi Complex 39
Lysosomes 39
Peroxisomes 42
Mitochondria 42
The Cytoskeleton 44
Centrosome and Centrioles 45
Flagella and Cilia 45
Cell Inclusions 46
Normal Cell Division 47
Somatic Cell Division 47
Interphase 48 Mitosis 48 Cytokinesis 50 Time Required 50
Reproductive Cell Division 51
Meiosis 52
Abnormal Cell Division: Cancer (CA) 53
Definition 53
Types 53
Growth and Spread 54
Possible Causes 55
Treatment 56
Cells and Aging 56
Key Medical Terms Associated with Cells 57
Study Outline 57
Review Questions 59
Self Quiz 60

3 Tissues 62

Types of Tissues and Their Origins 63
Epithelial Tissue 63
General Features 63
Covering and Lining Epithelium 64
Arrangement of Layers 64 Cell Shapes 64 Cell Junctions 64 Classification 66 Simple Epithelium 72 Stratified Epithelium 72 Pseudostratified Columnar Epithelium 73
Glandular Epithelium 73
Structural Classification of Exocrine Glands 73 Functional Classification of Exocrine Glands 74
Connective Tissue 75
General Features 75
Basic Elements 75
Cells 75 Ground Substance 76 Fibers 76
Classification 77
Embryonic Connective Tissue 84

Mature Connective Tissue 84
Loose Connective Tissue 84 Dense Connective Tissue 85 Cartilage 85 Bone (Osseous) Tissue 86 Blood (Vascular) Tissue 86
Membranes 86
Mucous Membranes 86
Serous Membranes 87
Cutaneous Membrane 87
Synovial Membranes 87
Muscle Tissue 87
Nervous Tissue 89
Study Outline 90
Review Questions 92
Self Quiz 92

4 | *The Integumentary System 94*

Skin 95
Anatomy 95
Epidermis 95
Dermis 97
Functions 98
Skin Color 98
Epidermal Ridges and Grooves 99
Blood Supply 100
Epidermal Derivatives 100
Hair 100
Anatomy 100 Color 102
Glands 102
Sebaceous (Oil) Glands 102 Sudoriferous (Sweat) Glands 103 Ceruminous Glands 103
Nails 103
Aging and the Integumentary System 104
Developmental Anatomy of the Integumentary System 104
Applications to Health 105
Key Medical Terms Associated with the Integumentary System 106
Study Outline 107
Review Questions 108
Self Quiz 108

5 | *Bone Tissue 110*

Functions 111
Histology 111
Compact Bone 113
Spongy Bone 114
Ossification: Bone Formation 115
Intramembranous Ossification 116
Endochondral Ossification 116
Bone Growth 118
Bone Replacement 119
Blood and Nerve Supply 119
Exercise and the Skeletal System 120
Aging and the Skeletal System 120
Developmental Anatomy of the Skeletal System 120
Applications to Health 122
Key Medical Terms Associated with Bone Tissue 125
Study Outline 125
Review Questions 126
Self Quiz 126

6 | *The Skeletal System: The Axial Skeleton 128*

Types of Bones 129
Surface Markings 129
Divisions of the Skeletal System 129
Skull 130
Sutures 131
Fontanels 131
Cranial Bones 133
Frontal Bone 133 Parietal Bones 134 Temporal Bones 134 Occipital Bone 136 Sphenoid Bone 137 Ethmoid Bone 139
Cranial Fossae 139
Facial Bones 139
Nasal Bones 139 Maxillae 140 Paranasal Sinuses 141 Zygomatic Bones 141 Mandible 142 Lacrimal Bones 143 Palatine Bones 143 Inferior Nasal Conchae 143 Vomer 143
Orbits 144
Foramina 144
Hyoid Bone 144
Vertebral Column 145
Divisions 145
Normal Curves 146
Typical Vertebra 146
Cervical Region 148
Thoracic Region 151
Lumbar Region 152
Sacrum and Coccyx 152
Thorax 154
Sternum 154
Ribs 155
Applications to Health 156
Study Outline 157
Review Questions 158
Self Quiz 158

7 | *The Skeletal System: The Appendicular Skeleton 161*

Pectoral (Shoulder) Girdle 162
Clavicle 162
Scapula 162
Upper Extremity 162
Humerus 162
Ulna and Radius 164
Carpals, Metacarpals, and Phalanges 166
Pelvic (Hip) Girdle 167
Lower Extremity 171
Femur 171
Patella 174
Tibia and Fibula 174
Tarsals, Metatarsals, and Phalanges 175
Arches of the Foot 176
Female and Male Skeletons 178
Study Outline 179
Review Questions 180
Self Quiz 180

8 | *Joints 181*

Classification of Joints 182
Structural Classification 182
Functional Classification 182
Fibrous Joints 182
Sutures 182
Syndesmoses 182
Gomphoses 182
Cartilaginous Joints 183
Synchondroses 183
Symphyses 183
Synovial Joints 183
Structure 183
Contact and Movement at Synovial Joints 185
Types of Movements at Synovial Joints 186
Gliding 186 Angular 186 Rotation 187 Circumduction 188 Special 188
Types of Synovial Joints 190
Gliding Joint 190 Hinge Joint 190 Pivot Joint 192 Ellipsoidal Joint 192 Saddle Joint 192 Ball-and-Socket Joint 192
Summary of Joints 192
Selected Joints of the Body 192
Applications to Health 207
Key Medical Terms Associated with Joints 209
Study Outline 209
Review Questions 210
Self Quiz 210

9 | *Muscle Tissue 212*

Characteristics 213
Functions 213
Types 213
Skeletal Muscle Tissue 213
Connective Tissue Components 213
Nerve and Blood Supply 215
Histology 217
Contraction 218
Sliding-Filament Mechanism 218
Neuromuscular Junction 218
Motor Unit 219
Mechanism 220
All-or-None Principle 221
Muscle Length and Force of Contraction 222
Muscle Tone 222
Muscular Atrophy and Hypertrophy 223
Types of Skeletal Muscle Fibers 223
Cardiac Muscle Tissue 224
Smooth (Visceral) Muscle Tissue 225
Regeneration of Muscle Tissue 228
Aging and Muscle Tissue 229
Developmental Anatomy of the Muscular System 229
Applications to Health 230
Key Medical Terms Associated with the Muscular System 231
Study Outline 231
Review Questions 233
Self Quiz 234

10 | *The Muscular System 236*

How Skeletal Muscles Produce Movement 237
Origin and Insertion 237
Lever Systems and Leverage 238
Arrangement of Fasciculi 239
Group Actions 239
Naming Skeletal Muscles 240
Principal Skeletal Muscles 240
Study Outline 314
Review Questions 314
Self Quiz 315

11 | *Surface Anatomy 317*

Head 318
Neck 319

Trunk 320
Upper Extremity 323
Lower Extremity 328
STUDY OUTLINE 331
REVIEW QUESTIONS 331
SELF QUIZ 331

12 The Cardiovascular System: Blood 333

Functions of Blood 334
Physical Characteristics of Blood 334
Components of Blood 335
Formed Elements 336
Plasma 336
Origin of Blood Cells 339
Erythrocytes (Red Blood Cells) 339
Structure 339
Functions 339
Life Span and Number 340
Production 340
Blood Group Systems 341
Leukocytes (White Blood Cells) 341
Structure and Types 341
Functions 342
Neutrophils and Monocytes 342 Eosinophils 342
Basophils 342 Lymphocytes 342
Life Span and Number 343
Production 344
Thrombocytes (Platelets) 344
Structure 344
Function 344
Life Span and Number 344
Production 344
Applications to Health 345
KEY MEDICAL TERMS ASSOCIATED WITH BLOOD 347
STUDY OUTLINE 348
REVIEW QUESTIONS 349
SELF QUIZ 350

13 The Cardiovascular System: The Heart 351

Location 352
Pericardium 353
Heart Wall 354
Chambers of the Heart 355
Great Vessels of the Heart 358
Valves of the Heart 358
Atrioventricular (AV) Valves 358
Semilunar Valves 358
Skeleton of the Heart 358
Surface Projection 360
Blood Supply 360
Conduction System 362
Electrocardiogram (ECG or EKG) 364
Cardiac Cycle 364
Autonomic Control 365
Artificial Heart 365
Heart–Lung Machine 366
Risk Factors in Heart Disease 366
Developmental Anatomy of the Heart 367
Applications to Health 368
KEY MEDICAL TERMS ASSOCIATED WITH THE HEART 373
STUDY OUTLINE 373
REVIEW QUESTIONS 374
SELF QUIZ 375

14 The Cardiovascular System: Blood Vessels 377

Arteries 378
Elastic (Conducting) Arteries 378
Muscular (Distributing) Arteries 378
Anastomoses 378
Arterioles 380
Capillaries 380
Venules 381
Veins 381
Blood Reservoirs 382
Circulatory Routes 383
Systemic Circulation 384
Hepatic Portal Circulation 416
Pulmonary Circulation 417
Fetal Circulation 417
Aging and the Cardiovascular System 421
Developmental Anatomy of Blood Vessels and Blood 421
Applications to Health 421
KEY MEDICAL TERMS ASSOCIATED WITH BLOOD VESSELS 423
STUDY OUTLINE 423
REVIEW QUESTIONS 424
SELF QUIZ 425

15 The Lymphatic System 427

Lymphatic Vessels 428
Lymphatic Tissue 430
Lymph Nodes 430
Tonsils 432
Spleen 432

Thymus Gland 434
Lymph Circulation 435
Route 435
Thoracic (Left Lymphatic) Duct 435 Right Lymphatic Duct 436
Maintenance 437
Principal Groups of Lymph Nodes 437
Aging and the Immune System 445
Developmental Anatomy of the Lymphatic System 445
Applications to Health 446
KEY MEDICAL TERMS ASSOCIATED WITH THE LYMPHATIC SYSTEM 451
STUDY OUTLINE 452
REVIEW QUESTIONS 452
SELF QUIZ 453

16 | *Nervous Tissue 455*

Organization 456
Histology 456
Neuroglia 456
Neurons 458
Structure 458 Structural Variation 461 Classification 462
Nerve Impulse 463
Speed 463
Synapses 464
Neurotransmitters 465
Regeneration 466
Chromatolysis 466
Wallerian Degeneration 467
Retrograde Degeneration 467
Repair 467
Organization of Neurons 467
STUDY OUTLINE 468
REVIEW QUESTIONS 469
SELF QUIZ 470

17 | *The Spinal Cord and the Spinal Nerves 471*

Grouping of Neural Tissue 472
Spinal Cord 472
Protection and Coverings 472
Vertebral Canal 472 Meninges 472
General Features 474
Structure in Cross Section 474
Functions 478
Impulse Conduction 478 Reflex Center 479
Spinal Nerves 481
Names 481
Composition and Coverings 481
Distribution 483
Branches 483 Plexuses 483 Intercostal (Thoracic) Nerves 489
Dermatomes 489
Applications to Health 490
STUDY OUTLINE 493
REVIEW QUESTIONS 494
SELF QUIZ 494

18 | *The Brain and the Cranial Nerves 497*

Brain 498
Principal Parts 498
Protection and Coverings 498
Cerebrospinal Fluid (CSF) 499
Blood Supply 502
Brain Stem 503
Medulla Oblongata 503 Pons 508 Midbrain 508
Diencephalon 509
Thalamus 509 Hypothalamus 511 Epithalamus 513 Subthalamus 513
Cerebrum 513
Lobes 516 White Matter 516 Basal Ganglia (Cerebral Nuclei) 516 Limbic System 518 Functional Areas of Cerebral Cortex 519 Electroencephalogram (EEG) 521
Brain Lateralization (Split-Brain Concept) 521
Cerebellum 521
Structure 522 Functions 522
Cranial Nerves 523
Olfactory (I) 524
Optic (II) 525
Oculomotor (III) 526
Trochlear (IV) 527
Trigeminal (V) 527
Abducens (VI) 527
Facial (VII) 527
Vestibulocochlear (VIII) 528
Glassopharyngeal (IX) 529
Vagus (X) 529
Accessory (XI) 530
Hypoglossal (XII) 530
Aging and the Nervous System 530
Developmental Anatomy of the Nervous System 531
Applications to Health 536
KEY MEDICAL TERMS ASSOCIATED WITH THE CENTRAL NERVOUS SYSTEM 541
STUDY OUTLINE 542
REVIEW QUESTIONS 543
SELF QUIZ 544

19 | The Autonomic Nervous System 546

Somatic Efferent and Autonomic Nervous Systems 547
Structure of the Autonomic Nervous System 547
Visceral Efferent Pathways 547
Preganglionic Neurons 548 Autonomic Ganglia 548 Postganglionic Neurons 550
Sympathetic Division 551
Parasympathetic Division 552
Physiology of the Autonomic Nervous System 553
Neurotransmitters 553
Receptors 554
Activities 556
Visceral Autonomic Reflexes 556
Control by Higher Centers 557
Biofeedback 557
Meditation 558
STUDY OUTLINE 558
REVIEW QUESTIONS 559
SELF QUIZ 559

20 | Sensory and Motor Systems 561

Sensations 562
Definition 562
Characteristics 562
Classification of Receptors 563
Location 563 Stimulus Detected 563 Simplicity or Complexity 563
General Senses 563
Cutaneous Sensations 563
Tactile Sensations 563 Thermoreceptive Sensations 565 Pain Sensations 565
Proprioceptive Sensations 567
Receptors 567
Levels of Sensation 569
Sensory Pathways 569
Somatosensory Cortex 569
Proprioception, Discriminative Touch, Two-Point Discrimination, and Vibration 570
Pain, Temperature, Light Touch, and Pressure 570
Cerebellar Tracts 570
Motor Pathways 571
Linkage of Sensory Input and Motor Responses 571
Motor Cortex 572
Pyramidal Pathways 572
Extrapyramidal Pathways 572
Special Senses 573
Olfactory Sensations 573
Structure of Receptors 573 Olfactory Pathway 574
Gustatory Sensations 574
Structure of Receptors 574 Gustatory Pathway 576
Visual Sensations 576
Accessory Structures of Eye 576 Structure of Eyeball 578 Visual Pathway 583
Auditory Sensations and Equilibrium 583
External (Outer) Ear 583 Middle Ear 587 Internal (Inner) Ear 588 Auditory Pathway 588 Mechanism of Equilibrium 593
Aging and the Special Senses 595
Applications to Health 595
KEY MEDICAL TERMS ASSOCIATED WITH SENSORY STRUCTURES 597
STUDY OUTLINE 598
REVIEW QUESTIONS 600
SELF QUIZ 600

21 | The Endocrine System 603

Endocrine Glands 604
Pituitary (Hypophysis) 605
Adenohypophysis 606
Neurohypophysis 607
Thyroid 608
Parathyroids 612
Adrenals (Suprarenals) 614
Adrenal Cortex 614
Adrenal Medulla 617
Pancreas 617
Ovaries and Testes 621
Pineal Gland (Epiphysis Cerebri) 621
Thymus 622
Aging and the Endocrine System 622
Developmental Anatomy of the Endocrine System 622
Other Endocrine Tissues 624
KEY MEDICAL TERMS ASSOCIATED WITH THE ENDOCRINE SYSTEM 624
STUDY OUTLINE 624
REVIEW QUESTIONS 626
SELF QUIZ 626

22 | The Respiratory System 628

Organs 630
Nose 630
Pharynx 632
Larynx 634
Trachea 635
Bronchi 638

Lungs 639
Gross Anatomy 640 Lobes and Fissures 640 Lobules 641 Alveolar–Capillary (Respiratory) Membrane 644 Blood and Nerve Supply 644
Aging and the Respiratory System 645
Developmental Anatomy of the Respiratory System 645
Applications to Health 646
KEY MEDICAL TERMS ASSOCIATED WITH THE RESPIRATORY SYSTEM 648
STUDY OUTLINE 649
REVIEW QUESTIONS 650
SELF QUIZ 650

23 | *The Digestive System 652*

Digestive Processes 653
Organization 653
General Histology 653
Mucosa 653 Submucosa 654 Muscularis 654 Serosa 654
Peritoneum 655
Mouth (Oral Cavity) 656
Tongue 657
Salivary Glands 658
Teeth 661
Dental Terminology 662 Dentitions 662 Blood and Nerve Supply 662
Pharynx 663
Esophagus 663
Histology 664
Activities 664
Blood and Nerve Supply 665
Stomach 665
Anatomy 665
Histology 666
Activities 668
Blood and Nerve Supply 669
Pancreas 669
Anatomy 669
Histology 670
Activities 670
Blood and Nerve Supply 671
Liver 671
Anatomy 671
Histology 673
Activities 674
Blood and Nerve Supply 674
Gallbladder (GB) 675
Histology 675
Activities 675
Blood and Nerve Supply 675
Small Intestine 675
Anatomy 675
Histology 676
Activities 677
Blood and Nerve Supply 679
Large Intestine 679
Anatomy 679
Histology 681
Activities 681
Blood and Nerve Supply 683
Aging and the Digestive System 684
Developmental Anatomy of the Digestive System 684
Applications to Health 684
KEY MEDICAL TERMS ASSOCIATED WITH THE DIGESTIVE SYSTEM 689
STUDY OUTLINE 690
REVIEW QUESTIONS 691
SELF QUIZ 692

24 | *The Urinary System 694*

Kidneys 695
External Anatomy 695
Internal Anatomy 698
Nephron 698
Blood and Nerve Supply 702
Juxtaglomerular Apparatus (JGA) 704
Physiology 704
Hemodialysis Therapy 705
Ureters 707
Structure 707
Histology 707
Physiology 707
Blood and Nerve Supply 707
Urinary Bladder 707
Structure 708
Histology 708
Physiology 708
Blood and Nerve Supply 710
Urethra 710
Histology 710
Physiology 711
Aging and the Urinary System 711
Developmental Anatomy of the Urinary System 711
Applications to Health 711
KEY MEDICAL TERMS ASSOCIATED WITH THE URINARY SYSTEM 713
STUDY OUTLINE 714
REVIEW QUESTIONS 715
SELF QUIZ 715

25 | *The Reproductive Systems* 717

Male Reproductive System 718
Scrotum 718
Testes 718
Spermatogenesis 723 Spermatozoa 725
Ducts 725
Ducts of the Testis 725 Epididymis 726 Ductus (Vas) Deferens 726 Ejaculatory Ducts 729 Urethra 729
Accessory Sex Glands 730
Semen (Seminal Fluid) 731
Penis 731
Female Reproductive System 733
Ovaries 734
Oogenesis 738
Uterine (Fallopian) Tubes 739
Uterus 740
Menstrual Cycle 742
Hormonal Control 743 Menstrual Phase (Menstruation) 744 Preovulatory Phase 744 Ovulation 745 Postovulatory Phase 745 Menarche and Menopause 746
Vagina 746
Vulva 747
Perineum 748
Mammary Glands 749
Structure 749 Breast Cancer 749
Birth Control (BC) 752
Sterilization 752
Hormonal 752
Intrauterine Devices (IUDs) 752
Barrier 753
Chemical 753
Physiologic 753
Coitus Interruptus (Withdrawal) 753
Induced Abortion 753
Male Contraception 754
Aging and the Reproductive Systems 754
Developmental Anatomy of the Reproductive Systems 754
Applications to Health 757
KEY MEDICAL TERMS ASSOCIATED WITH THE REPRODUCTIVE SYSTEMS 763
STUDY OUTLINE 763
REVIEW QUESTIONS 765
SELF QUIZ 766

26 | *Developmental Anatomy* 768

Development During Pregnancy 769
Fertilization and Implantation 769
Fertilization 769 Formation of the Morula 771 Development of the Blastocyst 771 Implantation 771
In Vitro Fertilization (IVF) 773
Embryonic Development 773
Beginnings of Organ Systems 774
Embryonic Membranes 774
Placenta and Umbilical Cord 777
Fetal Growth 779
Gestation 780
Structural and Functional Changes 780
Exercise and Pregnancy 782
Prenatal Diagnostic Tests 782
Amniocentesis 782
Chorionic Villus Sampling (CVS) 783
Fetal Ultrasonography 783
Electronic Fetal Monitoring (EFM) 784
Alphafetoprotein (AFP) Test 784
Parturition and Labor 784
KEY MEDICAL TERMS ASSOCIATED WITH DEVELOPMENTAL ANATOMY 786
STUDY OUTLINE 786
REVIEW QUESTIONS 787
SELF QUIZ 788
Appendix: Answers to Self Quizzes A-1
Glossary of Combining Forms, Word Roots, Prefixes, and Suffixes G-1
Glossary of Terms G-6
Bibliography B-1
Index I-1

Preface

AUDIENCE

Designed for the introductory course in human anatomy, the sixth edition of *Principles of Human Anatomy* assumes no previous study of the human body. The text is geared to students in health-oriented, medical, and biological programs. Among the students specifically served by this text are those pursuing careers as nurses, medical assistants, physicians' assistants, medical laboratory technologists and technicians, perfusionists, radiation therapy technologists and radiographers, respiratory therapists, dental hygienists, physical and occupational therapists, surgical assistants and technologists, diagnostic medical sonographers, cytotechnologists and histologic technologists, electroencephalographic (EEG) technologists, emergency medical technicians-paramedics, nuclear medicine technologists, morticians, and medical record administrators and technicians. Because of the scope of the book, however, *Principles of Human Anatomy,* Sixth Edition, is also useful for students in the biological sciences, science technology, liberal arts, physical education, and premedical, predental, and prechiropractic programs.

OBJECTIVES

The objectives of *Principles of Human Anatomy* remain unchanged in this sixth edition. Because human anatomy is such a large and complex body of knowledge to present in an introductory course, the first objective is to concentrate on data and unified concepts that contribute to a basic understanding of the structure of the human body. Data not essential to this objective have been minimized. The second objective is to present essential technical vocabulary and important, but difficult, concepts at a reading level that can be handled by the average student. Easy-to-comprehend explanations of terms and step-by-step development of concepts are presented to meet this objective.

THEMES

The sixth edition of this textbook gives some emphasis to physiology, disorders, and clinical applications. The basic content of the book is anatomy, but *because structure and function are so inseparably related,* it makes little sense to present students with anatomical detail without relating anatomy to some function. The discussion of function gives students a better understanding of anatomical concepts.

An understanding of structure and function may be enhanced by considering variations from the normal, or the defects and disorders observed in clinical situations. In turn, such conditions are better understood once the context of normal anatomy has been established. Disorders are treated under the special heading **Applications to Health** at the ends of chapters, and **Clinical Applications** integrated throughout the text.

NEW TO THIS EDITION

ORGANIZATION AND CONTENT

The book is organized by systems rather than by regions. In the sixth edition, each chapter has been carefully revised.

Chapter 1 introduces the levels of structural organization, organ systems, structural plan, anatomical position, regional names, directional terms, planes and sections, cavities of the human body, and units of measurement. The chapter has been improved by the inclusion of two exhibits, one that summarizes body cavities and one that presents medical imaging techniques.

Chapter 2, which deals with the cellular level of organization, describes a generalized animal cell to demonstrate the basic structural features and functions of cells. It includes updated material on the nucleus, cell division, and cancer. New to the chapter are discussions on bulk flow and an exhibit on processes by which substances move across plasma membranes.

The tissue level of organization is presented in Chapter 3 through descriptions of the structure, functions, and locations of the principal kinds of tissues. New to the chapter is a reclassification of connective tissues. The discussions of epithelial and connective tissues have also been revised. New line art has been placed adjacent to the epithelial tissue and areolar tissue photomicrographs. Several new color photomicrographs have been added as replacements.

The discussion of the organ and organ-system levels of organization begins with Chapter 4 on the integumentary system. This chapter includes revised material on aging and the integumentary system.

In Chapter 5, there is a revised section on ossification and a new section on exercise and the skeletal system.

Chapter 6, on the axial skeleton, has been updated to include a new section on the orbits and a new summary exhibit on comparing vertebrae in different regions.

In Chapter 7, which covers the appendicular skeleton, there is a revised discussion of the pelvis. New to Chapter 8 on articulations is a section on arthroplasty.

The muscular system is analyzed through a study of muscle tissue and the locations and actions of the principal skeletal muscles. Chapter 9 contains a new section on the regeneration of muscle tissue.

Chapter 10, on skeletal muscles, has been revised to include use of a new technique for illustrating skeletal muscles. Several new color photographs and cross-sectional illustrations have also been added.

As in the previous edition, the chapter on surface anatomy has been placed immediately following the muscular system, rather than at the end of the book. However, because of the way Chapter 11 is structured, that is, separate exhibits of various regions of the body with accompanying art, surface anatomy can be studied at *any* time during the course. The approach used to study surface anatomy is left to the option of the instructor and student. Many new surface anatomy features have been added.

The student is next introduced to the cardiovascular and lymphatic systems. Chapter 12, on blood, contains newly added material on blood doping and blood groups and a revised discussion of red blood cell structure and function.

Chapter 13, on the heart, contains new sections on the heart–lung machine and revised discussions on the artificial heart and atherosclerosis.

New to Chapter 14, on blood vessels, are overview sections added to blood vessel exhibits that focus on the blood vessels under consideration.

Chapter 15, on the lymphatic system, contains new sections on aging and the lymphatic system and types of transplants. The section on acquired immune deficiency syndrome (AIDS) has been expanded and revised.

The next major area of emphasis is the nervous system. Students are introduced to the structure and function of nervous tissue, the spinal cord and spinal nerves, the brain and cranial nerves, and the autonomic nervous system. Chapter 16, on nervous tissue, contains a new section on synapses. The section on nerve damage and repair has been moved from Chapter 17 to Chapter 16. Chapter 18, which deals with the brain, has revised sections on the hypothalamus, diencephalon, and limbic system and a new section on delirium. In Chapter 19, on the autonomic nervous system, the exhibit on the activities of the system has been revised.

Sensory structures are considered next. New to Chapter 20 is a section on anesthesia. Also, the motor pathways have been moved to follow immediately the sensory pathways. Attention is then turned to the endocrine system in Chapter 21.

Chapter 22, on the respiratory system, contains a revised section on the nasal cavity.

New topics added to Chapter 23, on the digestive system, include cystic fibrosis (CF), obesity, and dietary fiber. In addition, the section on activities of the liver has been expanded.

Chapter 24, on the urinary system, has been expanded to include a revised discussion of the blood supply to the kidneys.

Chapter 25, on the reproductive systems, has further been strengthened by the addition of material on ovarian and testicular hormones and cancer of the prostate gland. The section on birth control has been completely revised.

Finally, Chapter 26, on developmental anatomy, now contains a new section on exercise and pregnancy and a revised discussion of in vitro fertilization (IVF).

THE VISUAL TEXT

The new design for this edition has been developed to present both the words and art more effectively, making the text easier for the student to use. The illustration program has been greatly enhanced in the sixth edition.

■ ***Line Art.*** The anatomically accurate line drawings in the book are large, so that details are easily seen. In the sixth edition, many new illustrations have been added, and full color is used throughout to differentiate structures and regions. There are over 40 brand-new illustrations, and approximately 200 other illustrations have been redrawn for greater clarity and effectiveness.

■ ***Photographs.*** The photographs amplify the narrative and the line drawings. Numerous photomicrographs (in color), newly added scanning electron micrographs, and transmission electron micrographs enhance the histological discussions. Color photographs of specimens clarify gross anatomy discussions and have been added throughout. Photographs of regional dissections, most in full color, have also been added to the sixth edition.

SPECIAL FEATURES

As in previous editions, the book contains numerous learning aids. Users of the book have cited the pedagogical aids as one of the book's many strengths. All the tested and successful learning aids of previous editions have been kept in the sixth edition, and several new ones have been added. These special features include the following:

1. ***Student Objectives.*** Each chapter opens with a list of student objectives. Each numbered objective describes a knowledge or skill students should acquire while studying the chapter. (See ***Note to the Student*** for an explanation of how the objectives can be used.)

2. ***Chapter Outline.*** Each chapter contains an outline of its contents to help students overview the sequence of topics.
3. ***Exhibits.*** Health-science students are generally expected to learn a great deal about the anatomy of certain organ systems, specifically, skeletal muscles, articulations, blood vessels, lymph nodes, and nerves. To avoid interrupting the discussion of concepts and to organize the data, anatomical details have been presented in tabular form in exhibits, most of which are accompanied by illustrations. This method allows a clearer statement of general concepts in the narrative and organizes the specific details to be learned. New summary exhibits have been added to the sixth edition.
4. ***Phonetic Pronunciations.*** Throughout the text, phonetic pronunciations are provided in parentheses for selected terms. These pronunciations are given at the point where the terms are introduced and are repeated in the ''Glossary of Terms.'' The ***Note to the Student*** explains the pronunciation key. Many new phonetic pronunciations have been added in the sixth edition.
5. ***Clinical Applications.*** Throughout the text, clinical applications are integrated within the narrative. In the sixth edition, the title appears at the beginning of each application. Many new clinical applications have been added.
6. ***Applications to Health.*** Abnormalities of structure or function are grouped at the end of appropriate chapters in sections entitled ''Applications to Health.'' These sections provide a review of normal body processes as well as demonstrations of the importance of the study of anatomy to a career in any of the health fields. All disorders have been updated, and many new ones have been added.
7. ***Key Medical Terms.*** Glossaries of selected medical terms appear at the end of appropriate chapters; these listings are entitled ''Key Medical Terms.'' All glossaries have been revised for the sixth edition. In addition, phonetic pronunciations have been retained in the medical terminology lists in the sixth edition.
8. ***Study Outline.*** A study outline at the end of each chapter provides a brief summary of major topics. This section consolidates the essential points covered in the chapter so that students can recall and relate the points to one another. Page numbers added to the outline make it easier to locate topics within chapters.
9. ***Review Questions.*** Review questions at the end of each chapter provide a check to see if the objectives stated at the beginning of the chapter have been mastered. Page numbers have been added to the questions in the sixth edition to help students locate the answers. After answering the questions, students should reread the objectives to determine whether they have met the goals.
10. ***Self Quizzes.*** These quizzes are designed to evaluate for an understanding of principles and concepts discussed in the chapter. Answers to the questions are provided at the back of the book in the Appendix.
11. ***Appendix.*** The Appendix, ''Answers to Self Quizzes,'' contains the correct responses to the self-quiz tests for each chapter.
12. ***Glossaries.*** Two glossaries appear at the end of the book. The first, which is new to the sixth edition, deals with combining forms, word roots, prefixes, and suffixes. The second is a comprehensive glossary of terms and has been greatly updated and expanded.
13. ***Bibliography.*** The revised bibliography lists current references for instructors and students.
14. ***Inside Front and Back Covers.*** Three helpful listings have been placed on the inside of the front and back covers. *Symbols and Abbreviations* (formerly Appendix A) is an alphabetical list of commonly encountered medical abbreviations, and it has been expanded for the sixth edition. *Eponyms Used in This Text* (formerly Appendix B) is an alphabetical list of commonly encountered eponyms and the corresponding current terminology. (Eponyms are cited in the text in parentheses immediately following the preferred current terms.) *Measurements* consists of three tables that deal with metric units of length, mass, and volume along with some U.S. equivalents.

SUPPLEMENTARY MATERIALS

The following supplementary items are available from HarperCollins to accompany the sixth edition of *Principles of Human Anatomy*.

1. ***Instructor's Manual.*** Each chapter in the complimentary *Instructor's Manual*, prepared by Gerard J. Tortora, consists of a chapter overview, a list of instructional concepts relating to the chapter, and lists of audiovisual materials relating to the chapter topic.
2. ***Test Bank.*** A complimentary test bank, prepared by Gerard J. Tortora, contains a variable number of multiple-choice questions for each chapter in the textbook. The test bank is also available on computer disks (Testmaster) and may be secured from the publisher.
3. ***Laboratory Manual for Human Anatomy: With Cat Dissections,* Second Edition,** by Patricia J. Donnelly, is newly revised. Each muscle dissection is accompanied by a photo, line illustration, and simple step-by-step instructions. Reference is made to comparable human organs. Newly developed laboratory report sheets at the end of each lab have been included.
4. ***Laboratory Manual for Human Anatomy Using Cadavers,*** by Victor Eroschenoko, is presented primarily as a syllabus for use with prosected human cadavers. The manual can be equally successful with the use of models. The manual contains detailed descriptions and illustrations of all major systems of the body. The

illustrations have been prepared not only to maximize the recognition of anatomical parts in the prepared human cadaver, but also to allow optional coloring of the described anatomical parts.

5. ***Transparencies.*** One hundred seventy-five full-color transparencies are available. Illustrations from the text that are often shown and discussed in class are the subjects for the transparencies, and special care has been taken to make them clear and usable as overhead projections.
6. ***The Anatomy Coloring Book.*** This popular study aid, written by Wynn Kapit and Larry Elson, helps students memorize structures of the body through coloring in line drawings.
7. ***Software Programs.*** Numerous software programs are available to enhance student understanding of anatomy. Varying programs cover such activities as identifying key structures, pronunciation, spelling, vocabulary dictionaries, and self quizzes.
8. ***Videos.*** A number of videos are available to enhance your classroom and laboratory presentations. *BodyWorks* is a video specifically created by HarperCollins and takes your students through a complete presentation of the nervous, circulatory, and muscular systems using a prosected cadaver, moving footage, and animated art. Videos of cat musculature dissection and fetal pig dissection are also available.
9. ***Laser Disk.*** *Laster-Touch Anatomy,* a computer-controlled interactive videodisk, is available exclusively through HarperCollins. Available are programs for the skeletal system and the muscular system. The audiovisual tutorial shows diagrams and actual body parts, along with narration and captions to reinforce spelling and pronunciation of terms. Quizzes and exercises are included throughout the presentation, with a tutorial section for wrong answers. This program uses the best of laser-disk technology with touch-screen access on the computer screen.

ACKNOWLEDGMENTS

I wish to thank the following people for their pointed comments and thoughtful review for the sixth edition:

Annalisa Berta, *San Diego State University*
Paul Biersuck, *Nassau Community College*
Stephen L. Dodd, *Louisiana State University*
Larry Ganion, *Ball State University*
Anne Johnson, *Southern Illinois University School of Medicine*
L. Henry Kermott, *St. Olaf College*
Audrey L. Mackey, *Austin Community College*
John K. McDonald, *Emory University School of Medicine*
Lowell D. Neudeck, *Northern Michigan University*
Izak Paul, *Mount Royal College*
Patricia Placke Munn, *Longview Community College*
William Presch, *California State University, Fullerton*
James H. Sheetz, *Birmingham Southern College*

Special thanks to Dr. Michael Kennedy of Hahnemann Medical College for his expertise in reviewing the illustration program.

Continuing thanks go to the many instructors and students who were kind enough to send me their suggestions for improvement and to those people who worked on the five previous editions. Gratitude is also extended for the contributions of the individuals and organizations whose names appear with the photographs in the text.

Finally, for typing drafts of the manuscript and numerous other duties associated with the task of putting together a textbook, thanks to Geraldine C. Tortora.

The first edition of this textbook was published in 1977 by Harper & Row, Publishers. Since that time, I have received the highest level of support and encouragement from all personnel involved in the development of all editions of this text. For this edition, I extend my very special thanks to the following individuals at HarperCollins Publishers for their continued support and encouragement: Susan Katz (Publisher), Marianne Russell (Vice President and Editorial Director), Bonnie Roesch (Allied Health Sciences Editor), Meryl R. G. Muskin (Developmental Editor), Thomas R. Farrell (Project Editor), Teresa J. Delgado (Art Director), Claudia Durrell (Art Coordinator), Mira Schachne (Photo Researcher), and Kewal K. Sharma (Production Manager).

As the acknowledgments indicate, the participation of many individuals of diverse talent and expertise is required in the production of a textbook of this scope and complexity. For this reason, readers and users of the sixth edition are invited to send their reactions and suggestions to me so that plans can be formulated for subsequent editions.

GERARD J. TORTORA
Natural Sciences and Mathematics S229
Bergen Community College
400 Paramus Road
Paramus, NJ 07652

Note to the Student

At the beginning of each chapter is a listing of ***Student Objectives.*** Before you read the chapter, please read the objectives carefully. Each objective is a statement of a skill or knowledge that you should acquire. To meet these objectives, you will have to perform several activities. Obviously, you must read the chapter carefully. If there are sections of the chapter that you do not understand after one reading, you should reread those sections before continuing. In conjunction with your reading, pay particular attention to the figures and exhibits; they have been carefully coordinated with the textual narrative. Also, at the beginning of each chapter is a ***Chapter Outline,*** designed to provide you with an overview of the sequence of topics discussed in each chapter.

At the end of each chapter are several other learning guides that you may find useful. The ***Study Outline*** is a concise summary of important topics discussed in the chapter. This section is designed to consolidate the essential points covered in the chapter, so that you may recall and relate them to one another. Page numbers have been added to the outline so that you can locate topics in the chapter more easily. The ***Review Questions*** are a series of questions designed specifically to help you master the objectives. After you have answered the review questions, you should return to the beginning of the chapter and reread the objectives to determine whether you have achieved the goals. ***Key Medical Terms*** appear in some chapters. This is a listing of terms, with phonetic pronunciations and definitions, designed to build your medical vocabulary. New to the sixth edition are ***Self Quizzes*** for each chapter. The questions are designed to evaluate for an understanding of principles and concepts discussed in the chapter. Answers to the questions are provided in the Appendix.

As a further aid, we have included pronunciations for many terms that may be new to you. These appear in parentheses immediately following the new words, and they are repeated in the glossary of terms at the back of the book. (Of course, since there will always be some conflict among medical personnel and dictionaries about pronunciation, you will come across variations in different sources.) Look at the words carefully and say them out loud several times. Learning to pronounce a new word will help you remember it and make it a useful part of your medical vocabulary. Take a few minutes now to read the following pronunciation key, so it will be familiar as you encounter new words.

PRONUNCIATION KEY

1. The strongest accented syllable appears in capital letters, for example, bilateral (bī-LAT-er-al) and diagnosis (dī-ag-NŌ-sis).
2. If there is a secondary accent, it is noted by a prime mark (′), for example, constitution (kon′-sti-TOO-shun) and physiology (fiz′-ē-OL-ō-jē). Any additional secondary accents are also noted by a prime mark, for example, decarboxylation (dē′-kar-bok′-si-LĀ-shun).
3. Vowels marked with a line above the letter are pronounced with the long sound, as in the following common words:

   ā as in *māke*
   ē as in *bē*
   ī as in *īvy*
   ō as in *pōle*

4. Vowels not so marked are pronounced with the short sound as in the following words:

   e as in *bet*
   i as in *sip*
   o as in *not*
   u as in *bud*

5. Other phonetic symbols are used to indicate the following sounds:

   a as in *above*
   oo as in *sue*
   yoo as in *cute*
   oy as in *oil*

An Introduction to the Human Body

1

STUDENT OBJECTIVES

1. Define anatomy, with its subdivisions, and physiology.
2. Define each of the following levels of structural organization that make up the human body: chemical, cellular, tissue, organ, system, and organismic.
3. Identify the principal systems of the human body, list representative organs of each system, and describe the function of each system.
4. List and define several important life processes of humans.
5. Define the anatomical position and compare common and anatomical terms used to describe various regions of the human body.
6. Define several directional terms used in association with the human body.
7. Define the common anatomical planes that may be passed through the human body and distinguish a cross section, frontal section, and midsagittal section.
8. List by name and location the principal body cavities and the organs contained within them.
9. Describe how the abdominopelvic cavity is divided into nine regions and into quadrants.
10. Contrast the principles employed in several medical imaging techniques in the diagnosis of disease.

CHAPTER OUTLINE

- **Anatomy Defined**
- **Levels of Structural Organization**
- **Life Processes**
- **Overview of the Human Body**

Structural Plan
Anatomical Position and Anatomical Names
Directional Terms
Planes and Sections
Body Cavities
Abdominopelvic Regions
Abdominopelvic Quadrants

- **Medical Imaging**
- **Measuring the Human Body**

You are about to begin a study of the human body in order to learn how it is organized and how it functions. This study involves many branches of science. Each contributes to an understanding of how your body normally works and what happens when it is injured, diseased, or placed under stress. You will also learn how your body attempts to maintain ***homeostasis,*** that is, keeping its internal environment within certain limits even though internal and external conditions are constantly changing.

ANATOMY DEFINED

Anatomy (a-NAT-o-me; *anatome* = to dissect) refers to the study of *structure* and the relationships among structures. Several subdivisions of anatomy are described in Exhibit 1-1.

Whereas anatomy and its branches deal with structures of the body, ***physiology*** (fiz′-ē-OL-ō-jē) deals with *functions* of the body parts, that is, how the body parts work. Since function cannot be completely separated from structure, you will learn about the human body by studying first its anatomy and then the necessary physiology. The structure of a part often determines the functions it will perform. For example, fused bones of the skull protect the brain, whereas spaces between long bones in the limbs permit various types of movements. The chambers of the heart that pump blood greater distances have thicker walls than those that pump blood shorter distances. The external part of the ear is specifically shaped for the efficient collection of sound waves that normally result in hearing.

EXHIBIT 1-1

Subdivisions of Anatomy

SUBDIVISION	DESCRIPTION
Surface Anatomy	Study of the form (morphology) and markings of the surface of the body.
Gross (Macroscopic) Anatomy	Study of structures that can be examined without the use of a microscope.
Systemic (Systematic) Anatomy	Study of specific systems of the body such as the nervous system or respiratory system.
Regional Anatomy	Study of a specific region of the body such as the head or chest.
Radiographic (rā′-dē-ō-GRAF-ik; *radio* = ray; *graph* = to write) **Anatomy**	Study of the structure of the body that includes the use of x-rays.
Developmental Anatomy	Study of development from the fertilized egg to adult form.
Embryology (em′-brē-OL-ō-jē; *logos* = study of)	Study of development from the fertilized egg through the eighth week in utero.
Histology (hiss′-TOL-ō-jē; *histio* = tissue)	Microscopic study of the structure of tissues.
Cytology (sī-TOL-ō-jē; *ctyo* = cell)	Microscopic study of the structure of cells.
Pathological (path′-ō-LOJ-i-kal; *patho* = disease) **Anatomy**	Study of structural changes associated with disease.

LEVELS OF STRUCTURAL ORGANIZATION

The human body consists of several levels of structural organization that are associated with one another in various ways (Figure 1-1). The lowest level of organization, the ***chemical level,*** includes all chemical substances essential for maintaining life. The chemicals are made up of atoms such as carbon (C), hydrogen (H), oxygen (O), nitrogen (N), calcium (Ca), potassium (K), and sodium (Na). The atoms combine to form molecules. Familiar examples of molecules are proteins, carbohydrates, fats, and vitamins.

Chemicals, in turn, combine to form the next higher level of organization: the ***cellular level. Cells*** are the basic structural and functional units of an organism. As you will see in Chapter 2, cells contain specialized structures called *organelles,* such as a nucleus and mitochondria, that perform specific functions. Among the many kinds of cells in your body are muscle cells, nerve cells, and blood cells. Figure 1-1 shows several isolated cells from the lining of the stomach. Each has a different structure, and each performs a different function.

The next higher level of structural organization is the ***tissue level. Tissues*** are groups of similar cells that together with their intercellular material (substance between cells) usually have a similar origin in an embryo and perform special functions. When the cells shown in Figure 1-1 are joined together, they form a tissue called *epithelium,* which lines the stomach. Each type of cell in the tissue has a specific function. *Parietal cells* produce hydrochloric acid (HCl), which is found in the stomach's gastric juice. *Mucous cells* produce mucus, a slippery, thick secretion that protects the stomach lining from damage by its own juices. *Zymogenic cells* produce an enzyme that starts the digestion of proteins. Other examples of tissues in your body are muscle tissue, connective tissue, and nervous tissue.

When different kinds of tissues are joined together they form an even higher level of organization: the ***organ level. Organs*** are structures that are composed of two or more different tissues, have specific functions, and usually have recognizable shapes. Examples of organs are the heart, liver, lungs, brain, and stomach. Figure 1-1 shows several tissues that make up the stomach. Working from the outside inward, the *serosa* is a layer of epithelial tissue and

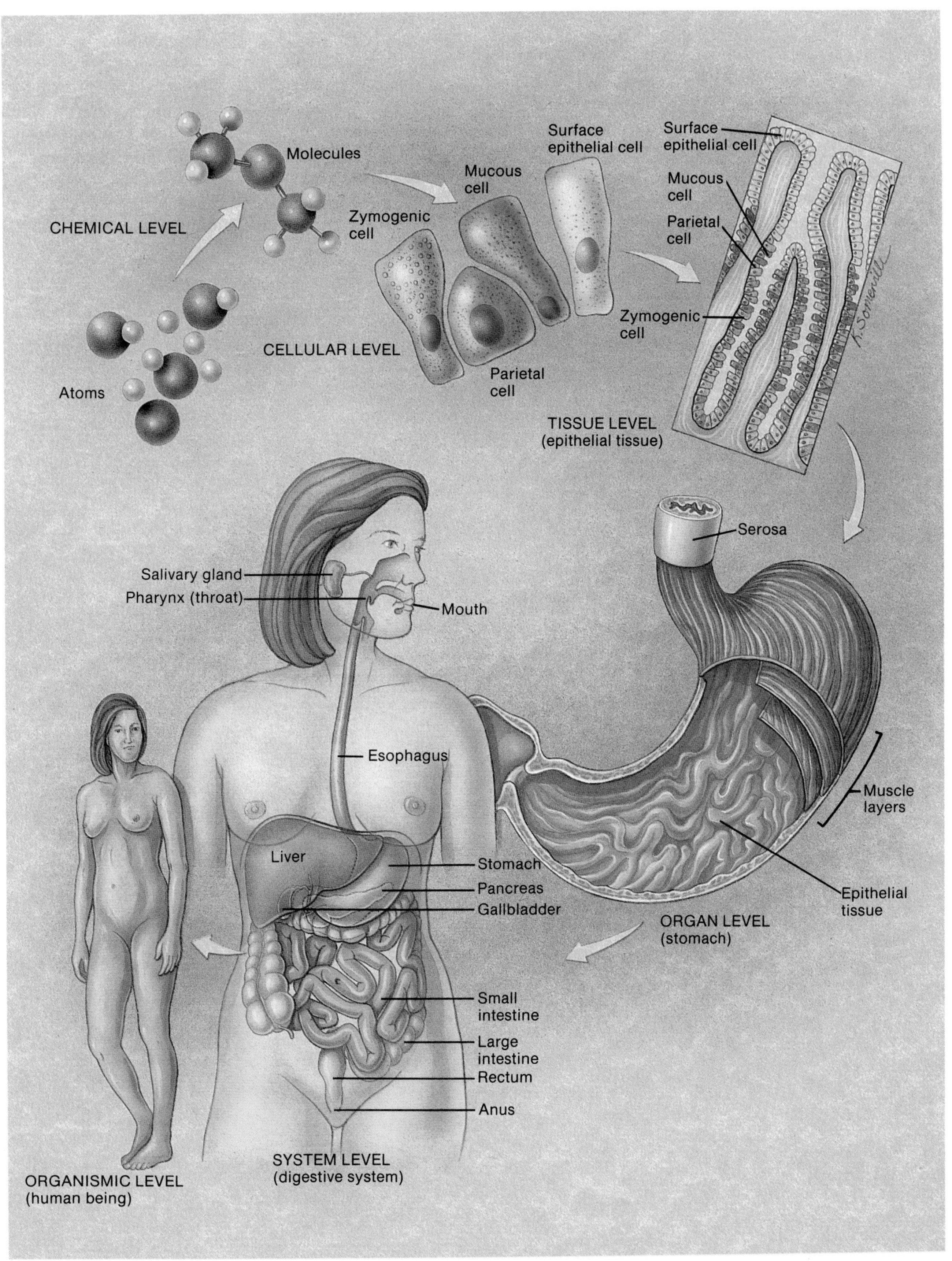

FIGURE 1-1 Levels of structural organization that compose the human body.

connective tissue that protects the stomach and reduces friction when the stomach moves and rubs against other organs. The *muscle tissue layers* of the stomach contract to mix food and pass it on to the next digestive organ (small intestine). As noted earlier, the *epithelial tissue* layer lining the stomach produces acid, mucus, and an enzyme.

The next higher level of structural organization in the body is the ***system level.*** A ***system*** consists of related organs that have a common function. The digestive system, which functions in the breakdown and absorption of food, is composed of these organs: mouth, saliva-producing glands called salivary glands, pharynx (throat), esophagus, stomach, small intestine, large intestine, rectum, liver, gallbladder, and pancreas. Sometimes an organ is part of more than one system. The pancreas, for example, is part of both the digestive system and the hormone-producing endocrine system.

The highest level is the ***organismic level.*** All the systems of the body functioning with one another constitute the total ***organism***—one living individual.

In the chapters that follow, you will examine the anatomy and physiology of the major body systems. Exhibit 1-2

EXHIBIT 1-2

Principal Systems of Human Body: Representative Organs and Functions

1. INTEGUMENTARY
Definition: The skin and structures derived from it, such as hair, nails, and sweat and oil glands.
Function: Helps regulate body temperature, protects the body, eliminates wastes, helps synthesize vitamin D, and receives certain stimuli such as temperature, pressure, and pain.

2. SKELETAL
Definition: All the bones of the body, their associated cartilages, and the joints of the body.
Function: Supports and protects the body, assists in body movements, houses cells that produce blood cells, and stores minerals.

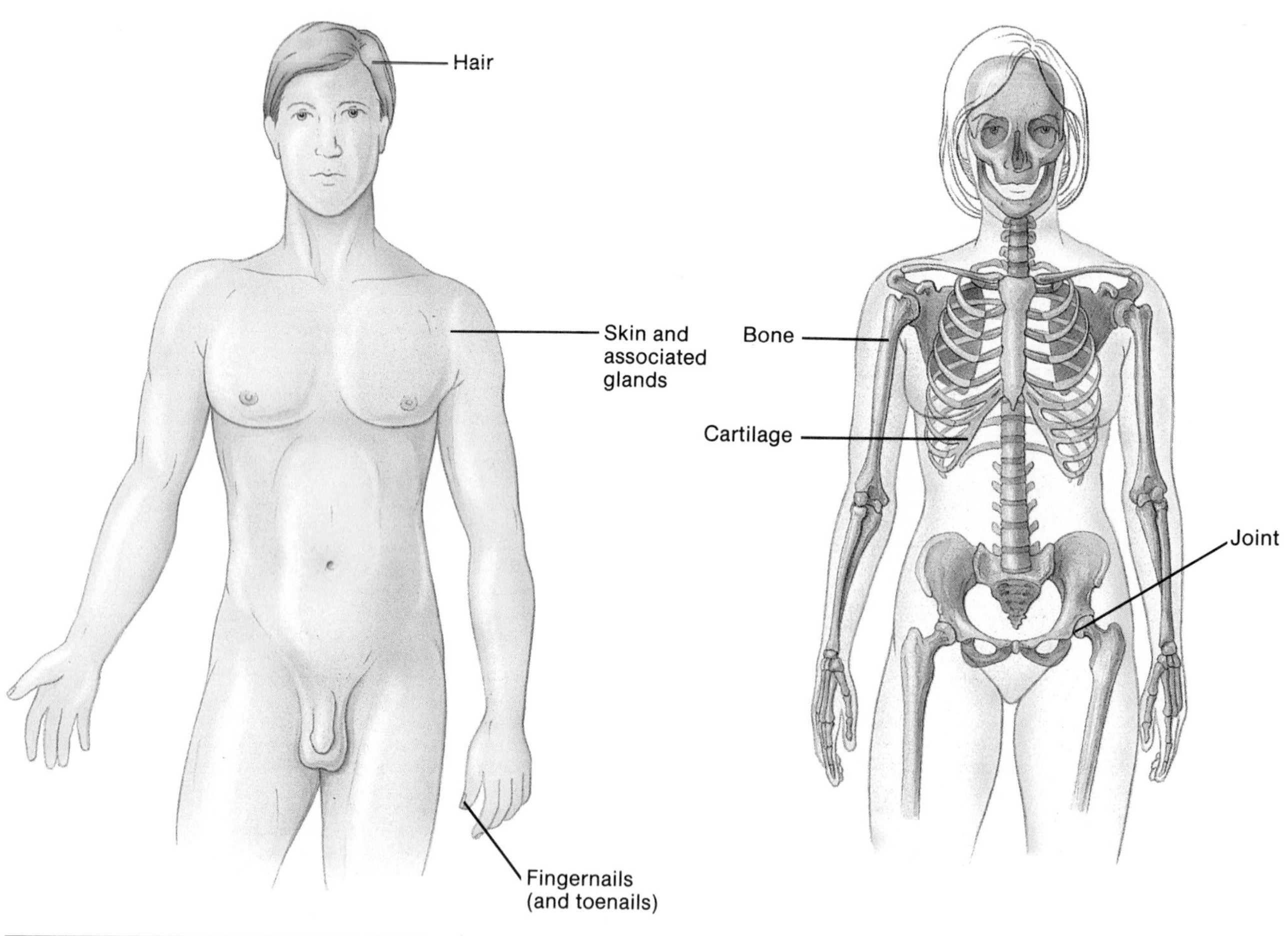

describe these systems in terms of their representative organs and their general functions. The systems are presented in the exhibit in the order in which they are discussed in later chapters.

LIFE PROCESSES

All living organisms carry on certain processes that set them apart from nonliving things. Following are several of the more important life processes of humans:

1. ***Metabolism*** is the sum of all the chemical reactions that occur in the body. One phase of metabolism, called ***catabolism,*** provides us with energy needed to sustain life—by breaking down food molecules, for example. The other phase of metabolism, called ***anabolism,*** uses the energy from catabolism to make various substances that form the body's structural and functional components.
2. ***Responsiveness*** is the ability to detect and react to changes in the external or internal environment. Different cells detect different sorts of changes and respond in

3. MUSCULAR

Definition: Specifically refers to skeletal muscle tissue (shown in the illustration); other muscle tissues include smooth (visceral) and cardiac.

Function: Participates in bringing about movement, maintains posture, and produces heat.

4. CARDIOVASCULAR

Definition: Blood, heart, and blood vessels.

Function: Distributes oxygen and nutrients to cells, carries carbon dioxide and wastes from cells, helps maintain the acid–base balance of the body, protects against disease, prevents hemorrhage by forming blood clots, and helps regulate body temperature.

Continued

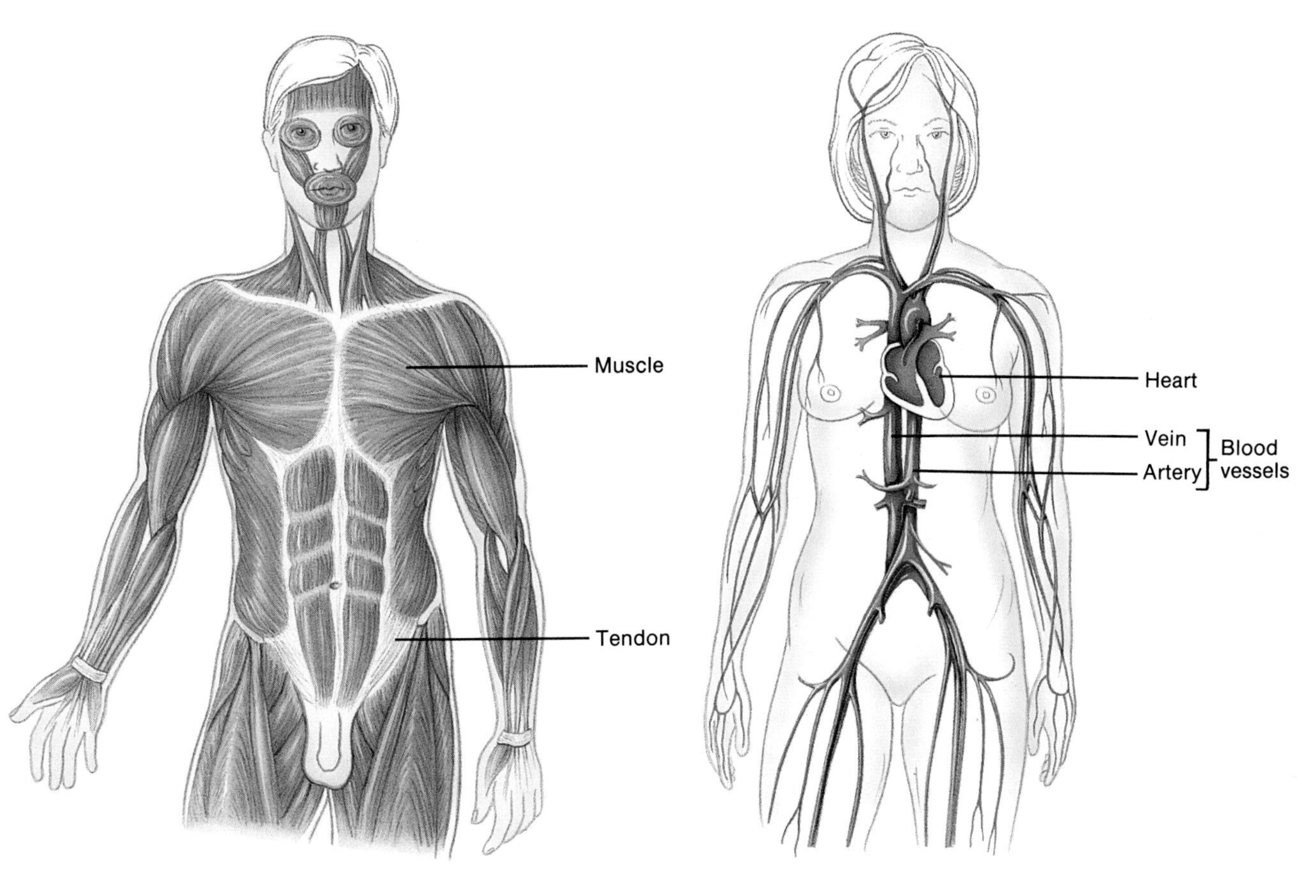

characteristic ways. For example, nerve cells can respond by generating signals, known as action potentials (nerve impulses). Action potentials may carry messages over long distances, such as between your brain and your big toe. Muscle fibers (cells) respond by contracting—that is, becoming shorter—and thus generating forces that move body parts. Endocrine cells in the pancreas respond to elevated blood glucose (sugar) level by secreting the hormone insulin to try to lower blood sugar level to normal.

3. ***Movement*** includes motion of the whole body, individual organs, single cells, or even organelles inside cells. For example, the coordinated contraction of several leg muscles moves the whole body from one place to another, as when walking or running. When we eat a meal, the gallbladder contracts as a single organ and releases bile that helps in digestion of fats. When a body tissue is damaged or infected, certain white blood cells move from the blood into the tissue to help clean up and repair the area. At times in the life of a cell, certain

EXHIBIT 1-2 *(Cont.)*

5. LYMPHATIC

Definition: Lymph, lymphatic vessels, and structures or organs containing lymphatic tissue (large numbers of white blood cells called lymphocytes) such as lymph nodes, the spleen, thymus gland, and tonsils.

Function: Returns proteins and plasma to the cardiovascular system, transports fats from the gastrointestinal tract to the cardiovascular system, filters body fluid, produces white blood cells, and protects against disease.

6. NERVOUS

Definition: Brain, spinal cord, nerves, and sense organs, such as the eye and ear.

Function: Controls and integrates body activities by sending nerve action potentials (nerve impulses) to effectors of the body (muscles and glands).

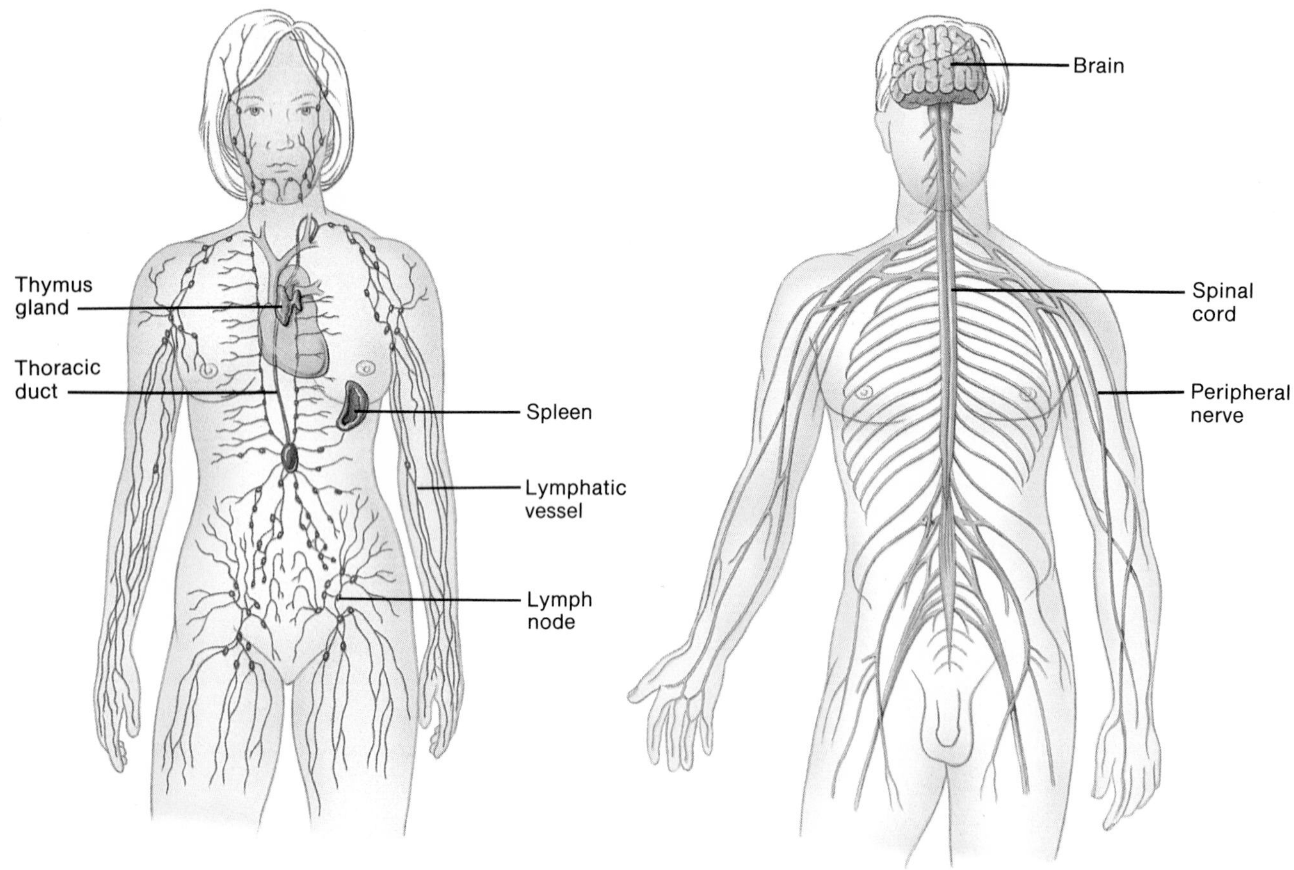

organelles move from one position to another to help ingest, digest, and eliminate various substances.

4. ***Growth*** refers to an increase in size. It is due to an increase in the number or size of cells or both. Sometimes, a tissue increases in size because the substance between cells increases in amount. In growing bone, for example, mineral deposits accumulate around the bone cells. ***Differentiation*** is the change that a cell undergoes from an unspecialized one to a specialized one. Specialized cells have structural and functional characteristics that differ from their undifferentiated ancestor cells. Through differentiation, a fertilized egg normally develops into an embryo, and then into a fetus, infant, child, and finally an adult. At each stage there is a diversity of differentiated cells. A high level of differentiation is accompanied by a loss of a cell's ability to reproduce.
5. ***Reproduction*** refers to either the formation of new cells for growth, repair, or replacement, or the production of a new individual. Through reproduction, life continues from one generation to the next.

7. ENDOCRINE
Definition: All glands and tissues that produce hormones.
Function: Controls and integrates body activities by sending chemical signals (hormones) via the blood to effectors of the body (muscles and glands).

8. RESPIRATORY
Definition: The lungs and associated passageways such as the pharynx (throat), larynx (voice box), trachea (windpipe), and bronchial tubes leading into and out of them.
Function: Supplies oxygen, eliminates carbon dioxide, and helps regulate the acid–base balance of the body.

Continued

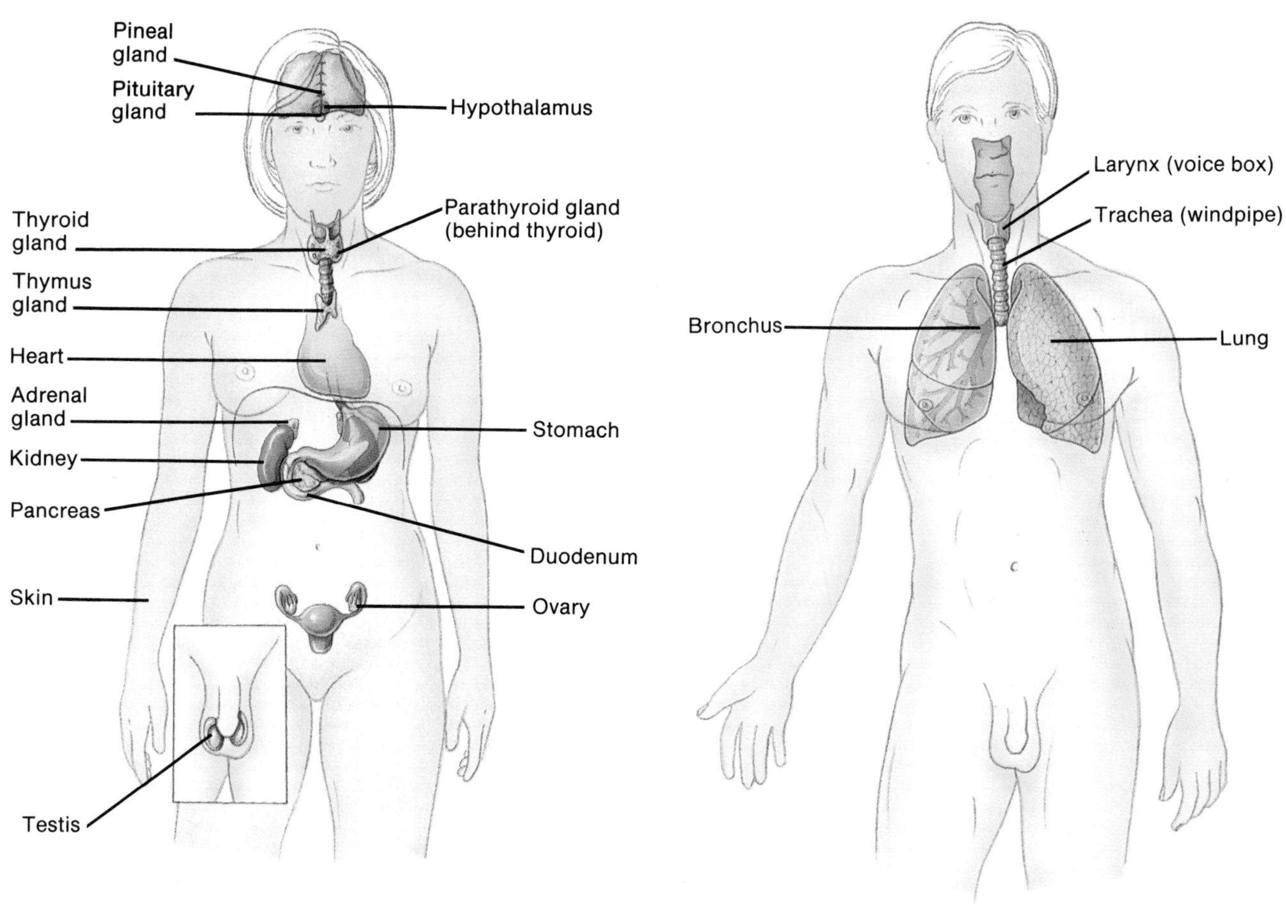

OVERVIEW OF THE HUMAN BODY

STRUCTURAL PLAN

The human body has certain general ***anatomical characteristics*** that will help you to understand its overall structural plan. For example, humans have a ***backbone*** **(*vertebral column*)**, a characteristic that places them in a large group of organisms called ***vertebrates.*** Another characteristic is the body's ***tube-within-a-tube*** construction. The outer tube is formed by the body wall; the inner tube is part of the gastrointestinal tract. Moreover, humans are for the most part ***bilaterally symmetrical;*** that is, essentially the left and right sides of the body are mirror images.

ANATOMICAL POSITION AND ANATOMICAL NAMES

In anatomy, there is universal agreement that descriptions

EXHIBIT 1-2 *(Cont.)*

9. DIGESTIVE
Definition: A long tube called the gastrointestinal (GI) tract and associated organs such as the salivary glands, liver, gallbladder, and pancreas.
Function: Breaks down and absorbs food for use by cells and eliminates solid and other wastes.

10. URINARY
Definition: Organs such as the kidneys, ureters, urinary bladder, and urethra that together produce, collect, and eliminate urine.
Function: Regulates the chemical composition of blood, eliminates wastes, regulates fluid and electrolyte balance and volume, helps maintain the acid–base balance of the body, and helps regulate red blood cell count.

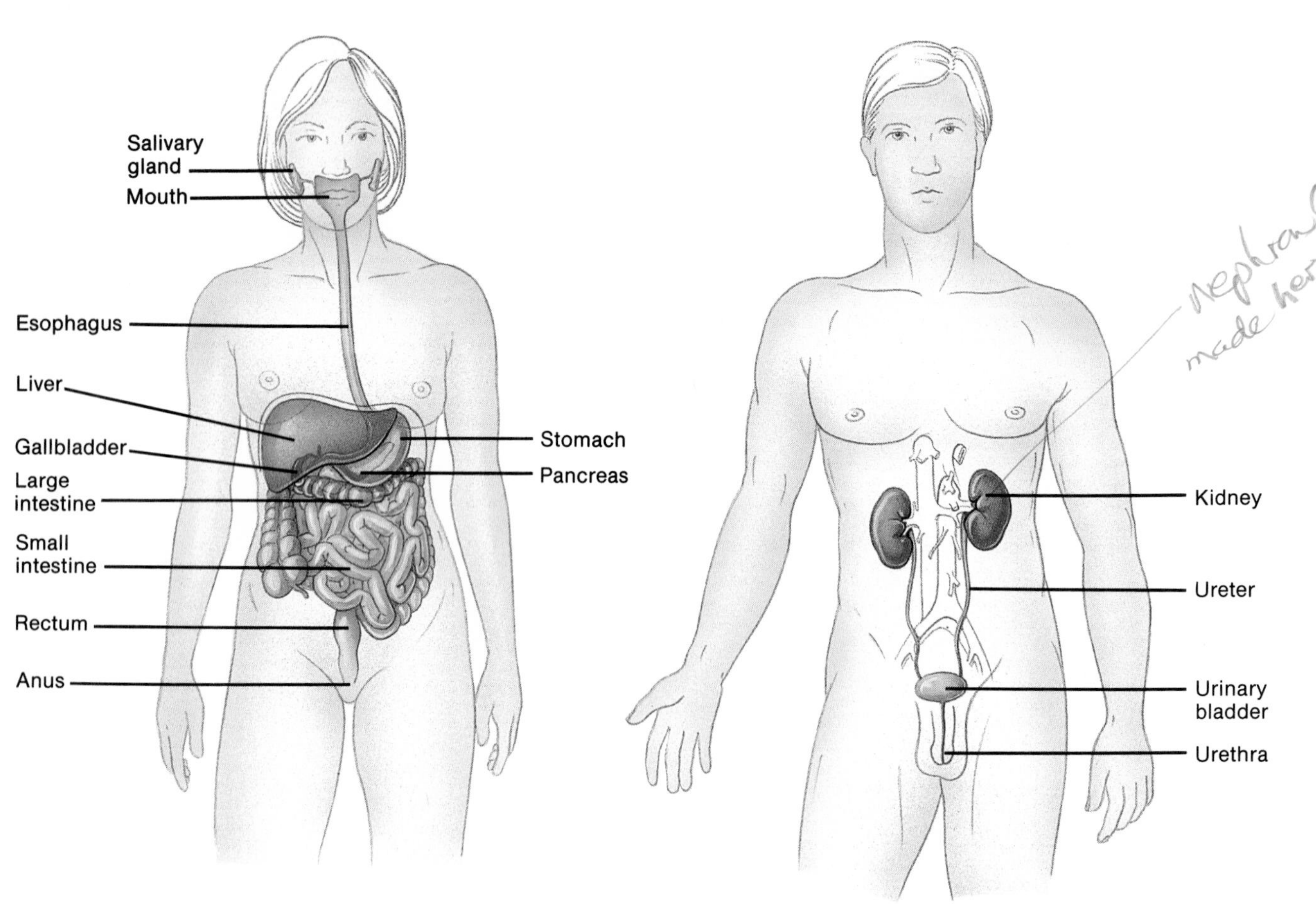

of any region or part of the human body assume that the body is in a specific position, called the ***anatomical position,*** so that directional terms are clear and any part can be related to any other part. In the anatomical position, the subject is standing erect (upright position) facing the observer, the upper extremities (limbs) are placed at the sides, the palms are turned forward, and the feet are flat on the floor (Figure 1-2). The common and anatomical names of the principal body regions are also presented in Figure 1-2.

DIRECTIONAL TERMS

To explain exactly where various body structures are located in relation to each other, anatomists use certain ***directional terms.*** Such terms are precise and avoid the use of unnecessary words. Many directional terms are defined in Exhibit 1-3, and the parts of the body referred to in the examples are labeled in Figure 1-3. Studying the exhibit and the figure together should make clear to you many of the directional relationships among various body parts.

11. REPRODUCTIVE
Definition: Organs (testes and ovaries) that produce reproductive cells (sperm and ova) and other organs that transport and store reproductive cells, such as the uterine (Fallopian) tubes and uterus in females and the ductus (vas) deferens and penis in males.
Function: Reproduces the organism.

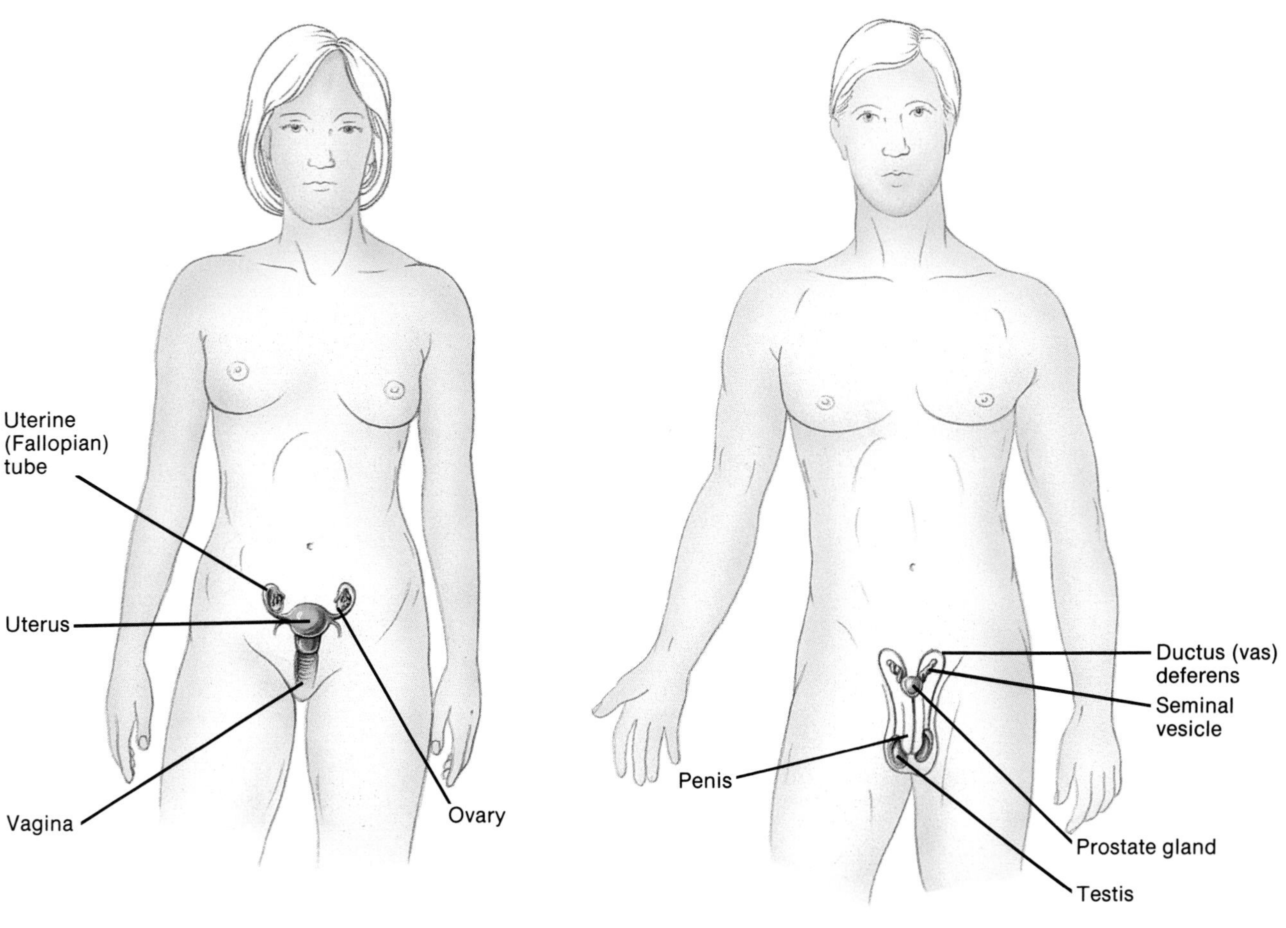

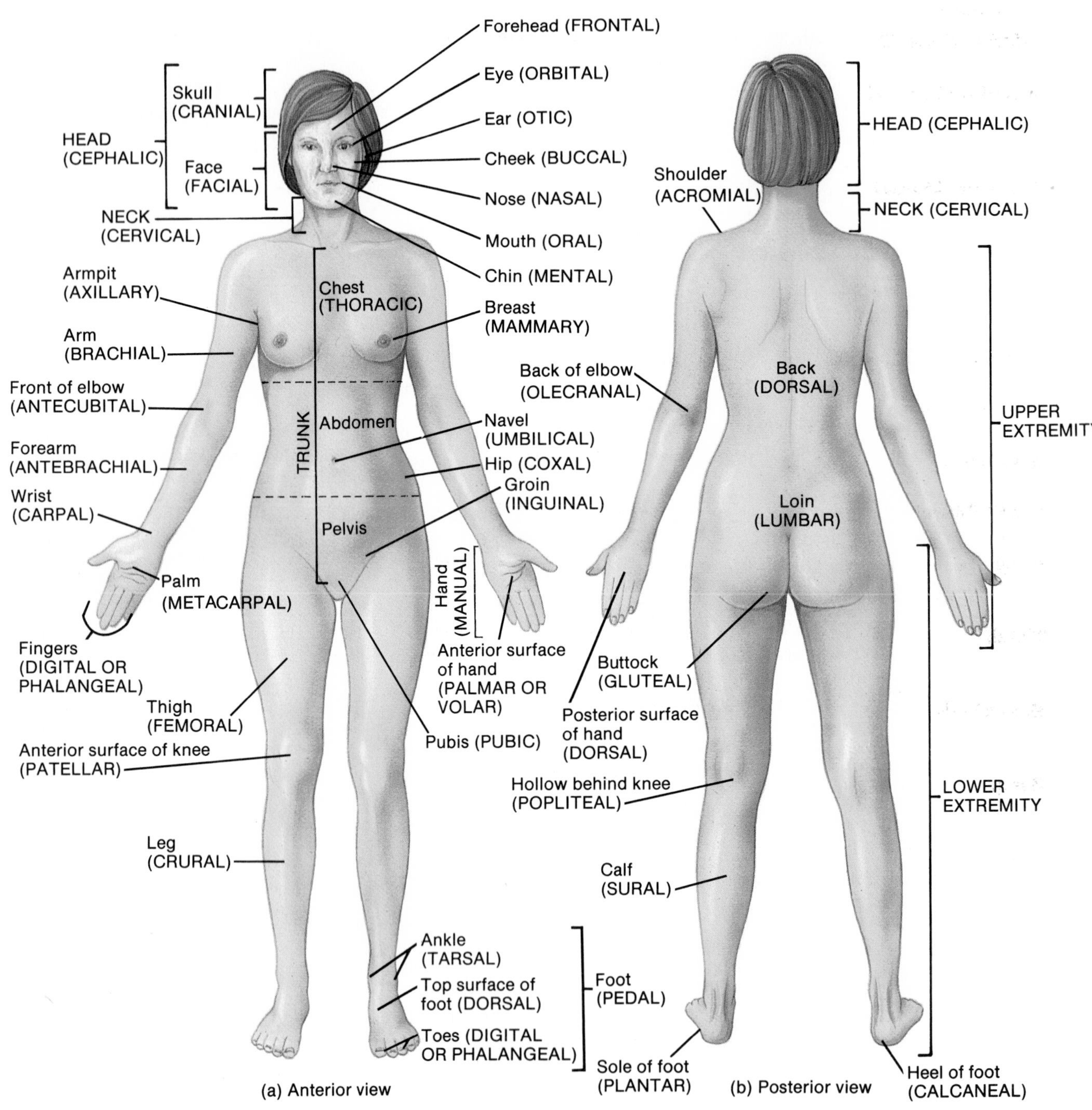

FIGURE 1-2 Anatomical position. The common names and anatomical terms, in parentheses, are indicated for many of the regions of the body. For example, the chest is the thoracic region.

EXHIBIT 1-3

Directional Terms[a]

TERM	DEFINITION	EXAMPLE
Superior (Cephalic or Cranial)	Toward the head or the upper part of a structure.	The heart is superior to the liver.
Inferior (Caudal)	Away from the head or toward the lower part of a structure.	The stomach is inferior to the lungs.
Anterior (Ventral)	Nearer to or at the front of the body. In the ***prone position,*** the body lies anterior side down. In the ***supine position,*** the body lies anterior side up.	The sternum is anterior to the heart.
Posterior (Dorsal)	Nearer to or at the back of the body.	The esophagus is posterior to the trachea.
Medial (Mesial)	Nearer to the midline of the body or a structure. The ***midline*** is an imaginary vertical line that divides the body into equal left and right sides. The *anterior midline* is on the front surface of the body, and the *posterior midline* is on the back surface.	The ulna is on the medial side of the forearm.
Lateral	Farther from the midline of the body or a structure.	The lungs are lateral to the heart.
Intermediate	Between two structures.	The ring finger is intermediate between the little and middle fingers.
Ipsilateral	On the same side of the body.	The gallbladder and ascending colon of the large intestine are ipsilateral.
Contralateral	On the opposite side of the body.	The ascending and descending colons of the large intestine are contralateral.
Proximal	Nearer to the attachment of an extremity (limb) to the trunk or a structure; nearer to the point of origin.	The humerus is proximal to the radius.
Distal	Farther from the attachment of an extremity (limb) to the trunk or a structure; farther from the point of origin.	The phalanges are distal to the carpals (wrist bones).
Superficial	Toward or on the surface of the body.	The muscles of the thoracic wall are superficial to the viscera in the thoracic cavity. (See Figure 1-7C.)
Deep	Away from the surface of the body.	The ribs are deep to the skin of the chest. (See Figure 1-7C.)

[a] Study this exhibit with Figures 1-3 and 1-7 in order to visualize the examples given.

PLANES AND SECTIONS

The human body may also be studied with respect to ***planes*** (imaginary flat surfaces) that pass through it (Figure 1-4). A ***sagittal*** (SAJ-i-tal) ***plane*** is a vertical plane that divides the body or organs into right and left sides. Such a plane may be midsagittal or parasagittal. A ***midsagittal*** (***median***) ***plane*** passes through the midline of the body and divides the body or organs into *equal* right and left sides. A ***parasagittal*** (*para* = near) ***plane*** does not pass through the midline of the body and therefore divides the body or organs into *unequal* left and right sides. A ***frontal*** (***coronal;*** kō-RŌ-nal) ***plane*** divides the body or organs into anterior (front) and posterior (back) portions. Finally, a ***horizontal*** (***transverse***) ***plane*** divides the body or organs into superior (upper) and inferior (lower) portions.

When you study a body structure, you will often view it in section, meaning that you look at only one surface

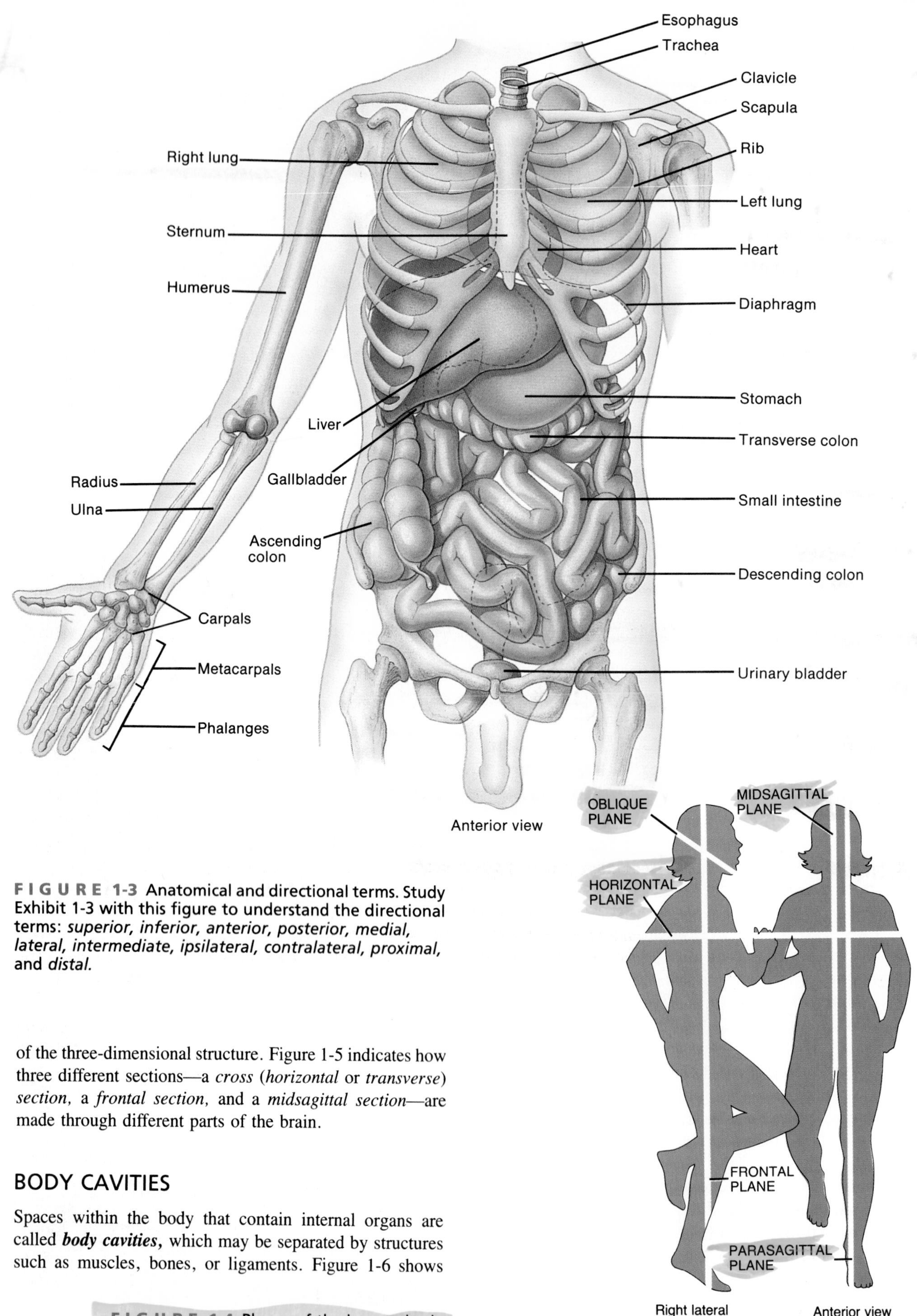

FIGURE 1-3 Anatomical and directional terms. Study Exhibit 1-3 with this figure to understand the directional terms: *superior, inferior, anterior, posterior, medial, lateral, intermediate, ipsilateral, contralateral, proximal,* and *distal.*

of the three-dimensional structure. Figure 1-5 indicates how three different sections—a *cross* (*horizontal* or *transverse*) *section,* a *frontal section,* and a *midsagittal section*—are made through different parts of the brain.

BODY CAVITIES

Spaces within the body that contain internal organs are called ***body cavities,*** which may be separated by structures such as muscles, bones, or ligaments. Figure 1-6 shows

FIGURE 1-4 Planes of the human body.

FIGURE 1-5 Planes and sections through different parts of the brain. The planes are shown in the diagrams on the left and the resulting sections are shown in the photographs on the right. (a) Courtesy of Stephen A. Kieffer and E. Robert Heitzman, *An Atlas of Cross-Sectional Anatomy,* Harper & Row, Publishers, Inc., New York, 1979. (b) Courtesy of Lester Bergman and Associates. (c) © 1988, Rotker, Phototake.

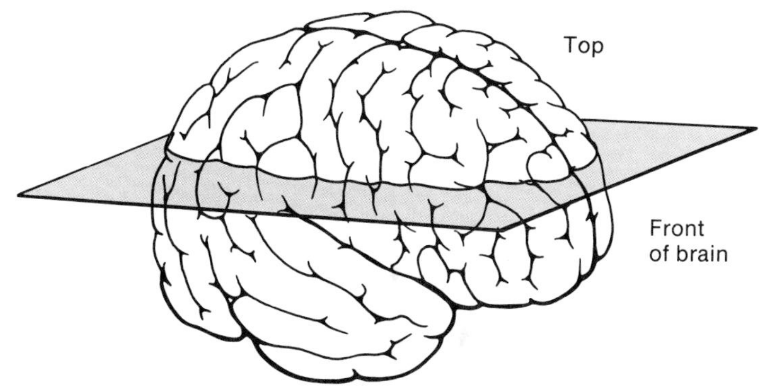

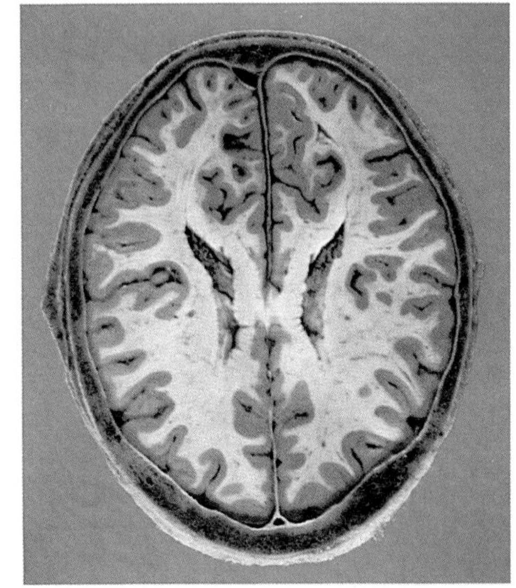

(a) Cross section of brain

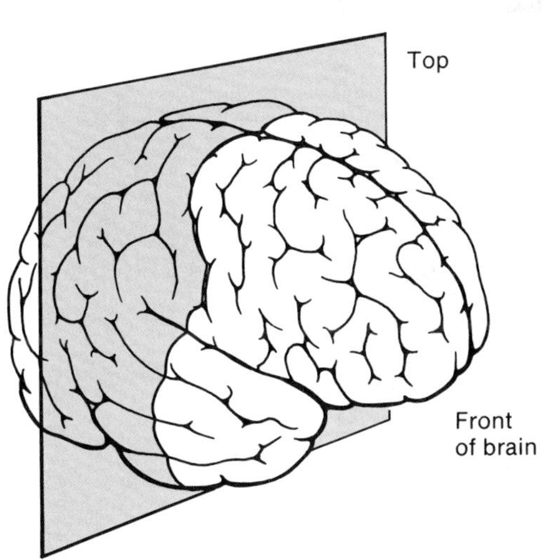

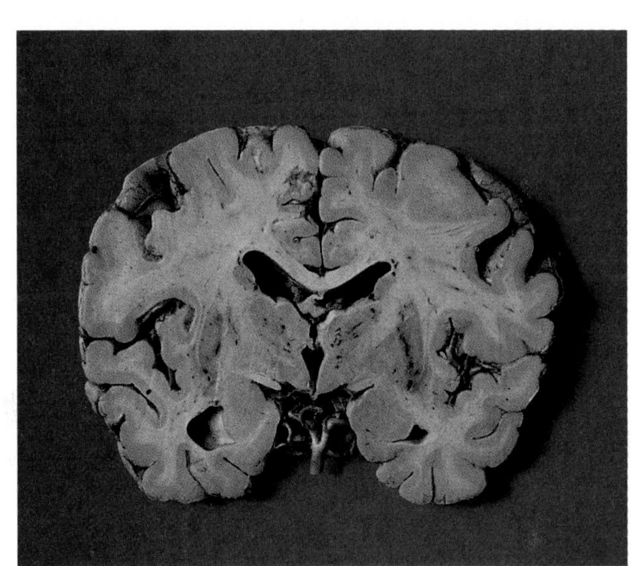

(b) Frontal section of brain

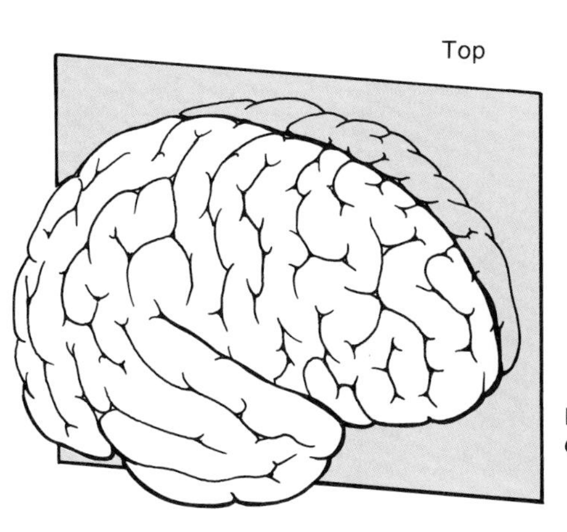

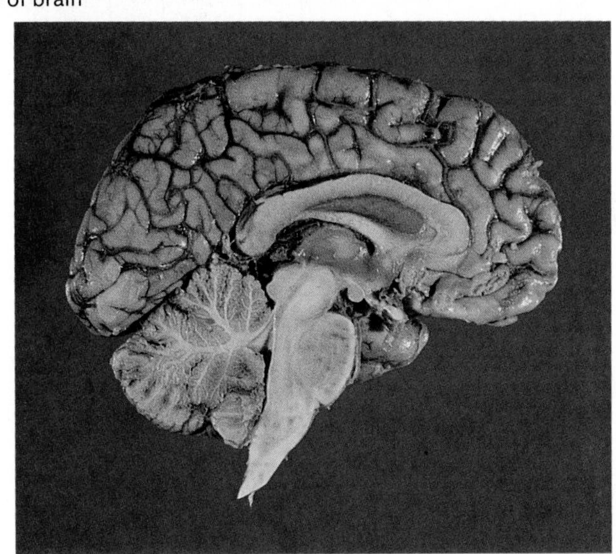

(c) Midsagittal section of brain

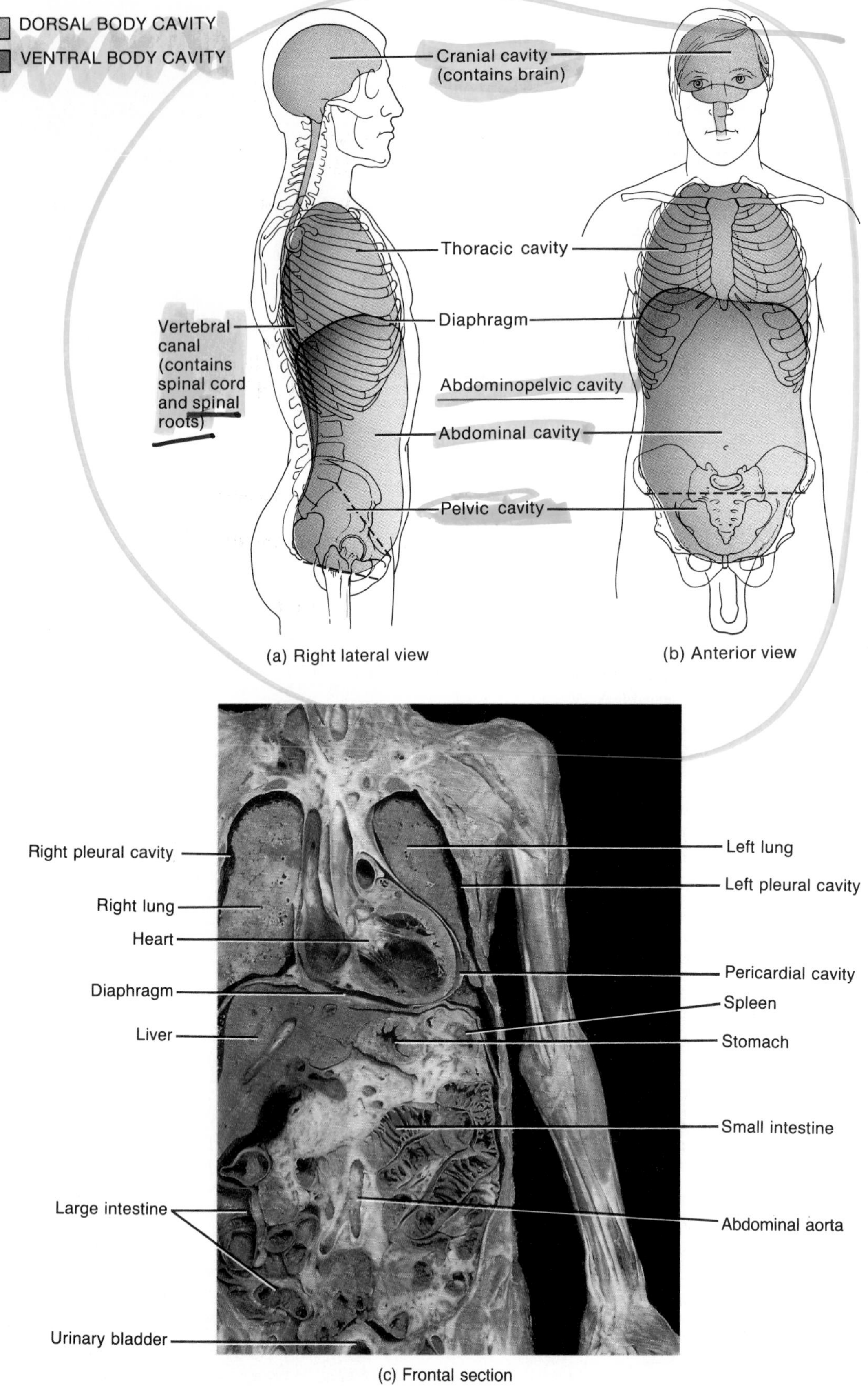

FIGURE 1-6 Dorsal and ventral body cavities. (c) Courtesy of J. A. Gosling, P. F. Harris, *et al.*, *Atlas of Human Anatomy,* Gower Medical Publishing Ltd., 2nd ed., 1991.

the two principal body cavities: dorsal and ventral. The ***dorsal body cavity*** is located near the posterior (dorsal) surface of the body. It is further subdivided into a ***cranial cavity,*** which is a bony cavity formed by the cranial (skull) bones and containing the brain, and a ***vertebral (spinal) canal,*** a bony cavity formed by the vertebrae of the backbone and containing the spinal cord and the beginnings (roots) of spinal nerves.

The other principal body cavity is the ***ventral body cavity,*** which is located on the anterior (ventral) aspect of the body. Its walls are composed of skin, connective tissue, bone, muscle, and a *serous membrane,* a thin, slippery membrane that lines the wall of the cavity and covers the organs within it. The organs inside the ventral body cavity are called ***viscera*** (VIS-er-a). Like the dorsal body cavity, the ventral body cavity has two principal subdivisions—an upper portion, called the ***thoracic*** (thō-RAS-ik) ***cavity*** (or chest cavity), and a lower portion, called the ***abdomino-pelvic*** (ab-dom′-i-nō-PEL-vik) ***cavity.*** The structure that divides the ventral body into the thoracic and abdominopelvic cavities is the diaphragm (DĪ-a-fram; *diaphragma* = partition or wall), an important muscle for breathing.

The thoracic cavity has three compartments. First, there are two ***pleural cavities,*** each surrounding a lung (Figure 1-7c). Each pleural cavity is a small, fluid-filled space between the part of the serous membrane that covers the lung and the part that lines the chest cavity. The serous membrane associated with the lungs is called the ***pleura.*** The ***pericardial*** (per′-i-KAR-dē-al; *peri* = around; *cardi* = heart) ***cavity*** is a fluid-filled space between the part of the serous membrane that covers the heart and the part that lines the chest cavity. The serous membrane associated with the heart is called the ***pericardium.*** The ***mediastinum*** (mē′-dē-as-TĪ-num; *media* = middle; *stare* = stand in) is a region between the lungs, extending from the sternum (breastbone) to the vertebral column or backbone (Figure 1-7). The mediastinum includes all of the contents of the thoracic cavity except the lungs themselves. Among the structures in the mediastinum are the thymus gland, heart, esophagus, trachea, and many large blood and lymphatic vessels.

The abdominopelvic cavity, as the name suggests, is divided into two portions, although no wall separates them (see Figure 1-6). The serous membrane that lines the abdominopelvic cavity and covers the organs within it is called the ***peritoneum.*** The upper portion of the abdominopelvic cavity, the ***abdominal*** (*abdere* means to hide, because it hides the viscera) ***cavity,*** contains the stomach, spleen, liver, gallbladder, pancreas, small intestine, and most of the large intestine. The lower portion, the ***pelvic cavity,*** contains the urinary bladder, portions of the large intestine, and the internal organs of reproduction. The pelvic cavity is the region between two imaginary planes, shown by dashed lines in Figure 1-6a.

The various body cavities are summarized in Exhibit 1-4.

EXHIBIT 1-4

Summary of Body Cavities

CAVITY	COMMENTS
DORSAL	
Cranial	Formed by cranial bones and contains brain.
Vertebral	Formed by vertebrae and contains spinal cord and beginnings of spinal nerves.
VENTRAL	
Thoracic	Chest cavity; separated from abdominal cavity by diaphragm.
Pleural	Contains lungs.
Pericardial	Contains heart.
Mediastinum	Region between the lungs from the breastbone to backbone that contains heart, thymus gland, esophagus, trachea, and many large blood and lymphatic vessels.
Abdominopelvic	Subdivided into abdominal and pelvic cavities.
Abdominal	Contains stomach, spleen, liver, gallbladder, pancreas, small intestine, and most of large intestine.
Pelvic	Contains urinary bladder, portions of large intestine, and internal reproductive organs.

ABDOMINOPELVIC REGIONS

To describe the location of organs easily, the abdominopelvic cavity may be divided into the ***nine regions*** shown in Figure 1-8a. Note which organs and parts of organs are in the different regions by carefully examining Figure 1-8b-d and Exhibit 1-5. Although some parts of the body in the illustrations and exhibit may be unfamiliar to you at this point, they will be discussed in detail in later chapters.

ABDOMINOPELVIC QUADRANTS

The abdominopelvic cavity may be divided more simply into ***quadrants*** (*quad* = four). These are shown in Figure 1-9. In this method, frequently used by clinical personnel, a horizontal line and a vertical line are passed through the umbilicus (navel). These two lines divide the abdomen into a ***right upper quadrant (RUQ), left upper quadrant (LUQ), right lower quadrant (RLQ),*** and ***left lower quadrant (LLQ).*** Whereas the nine-region division is more widely used for anatomical studies, the quadrant division is more commonly used for locating the site of an abdominopelvic pain, tumor, or other abnormality.

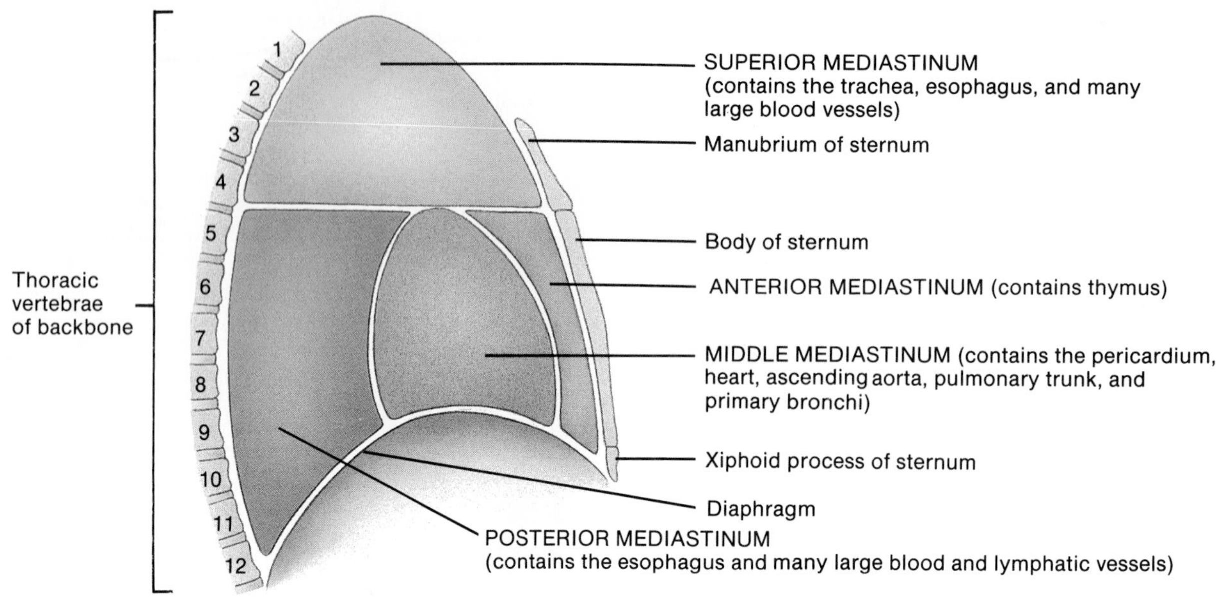

(a) Right lateral view

Esophagus
Trachea
Sternum
Right pulmonary artery
Aorta
Thoracic vertebrae of backbone
Left atrium
Right ventricle
Heart
Diaphragm
Liver

(b) Near midline sagittal section

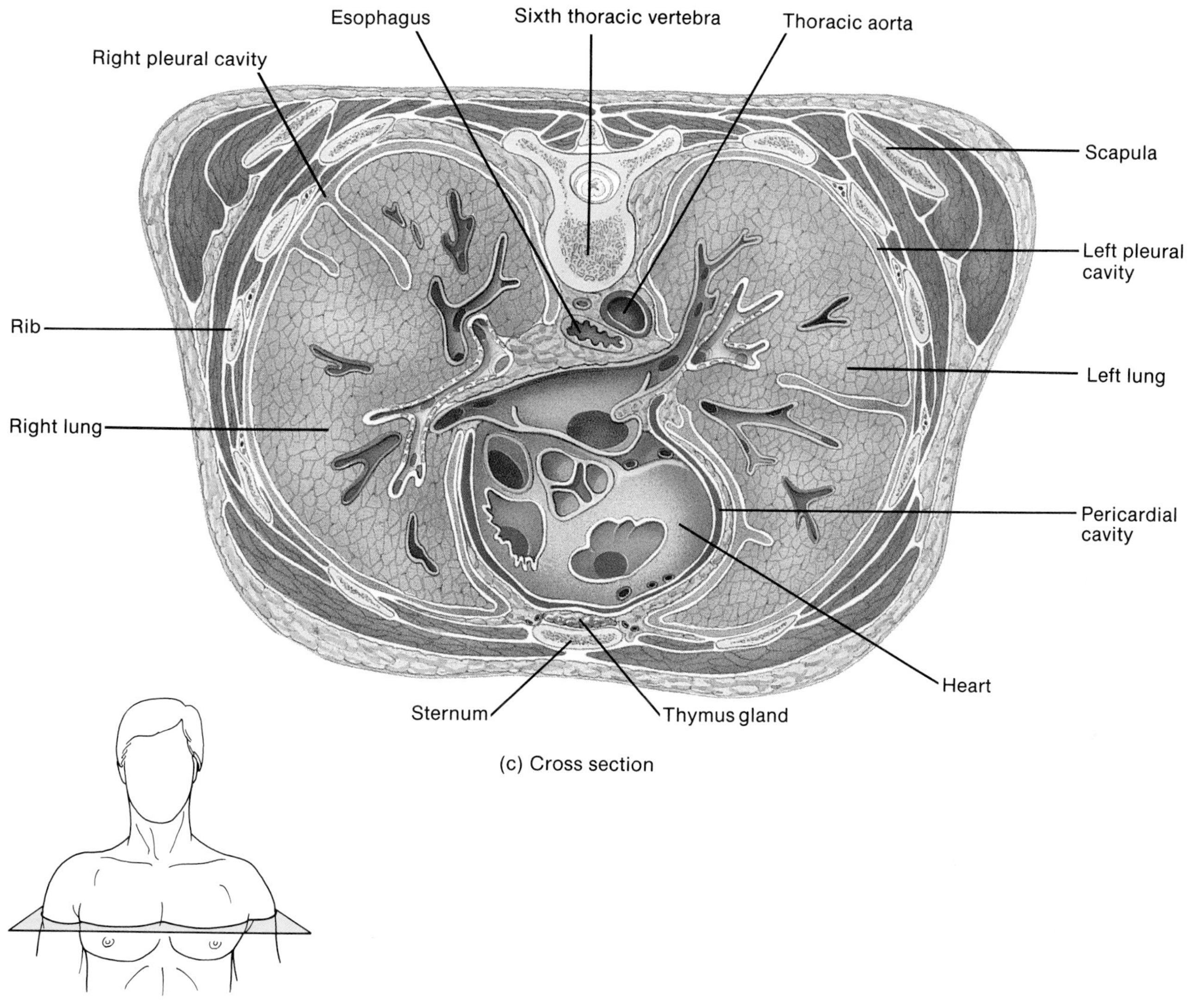

(c) Cross section

Pronounce media stin

FIGURE 1-7 Mediastinum. (a) Diagram of subdivisions of the mediastinum and some of its contents. (b) Photograph through the thorax showing some mediastinal structures. Courtesy of J. A. Gosling, P. F. Harris, *et al.*, *Atlas of Human Anatomy,* Gower Medical Publishing Ltd., 2nd ed., 1991. (c) Diagram of mediastinum. Some of the structures shown and labeled may be unfamiliar to you now. However, they are discussed in detail in later chapters.

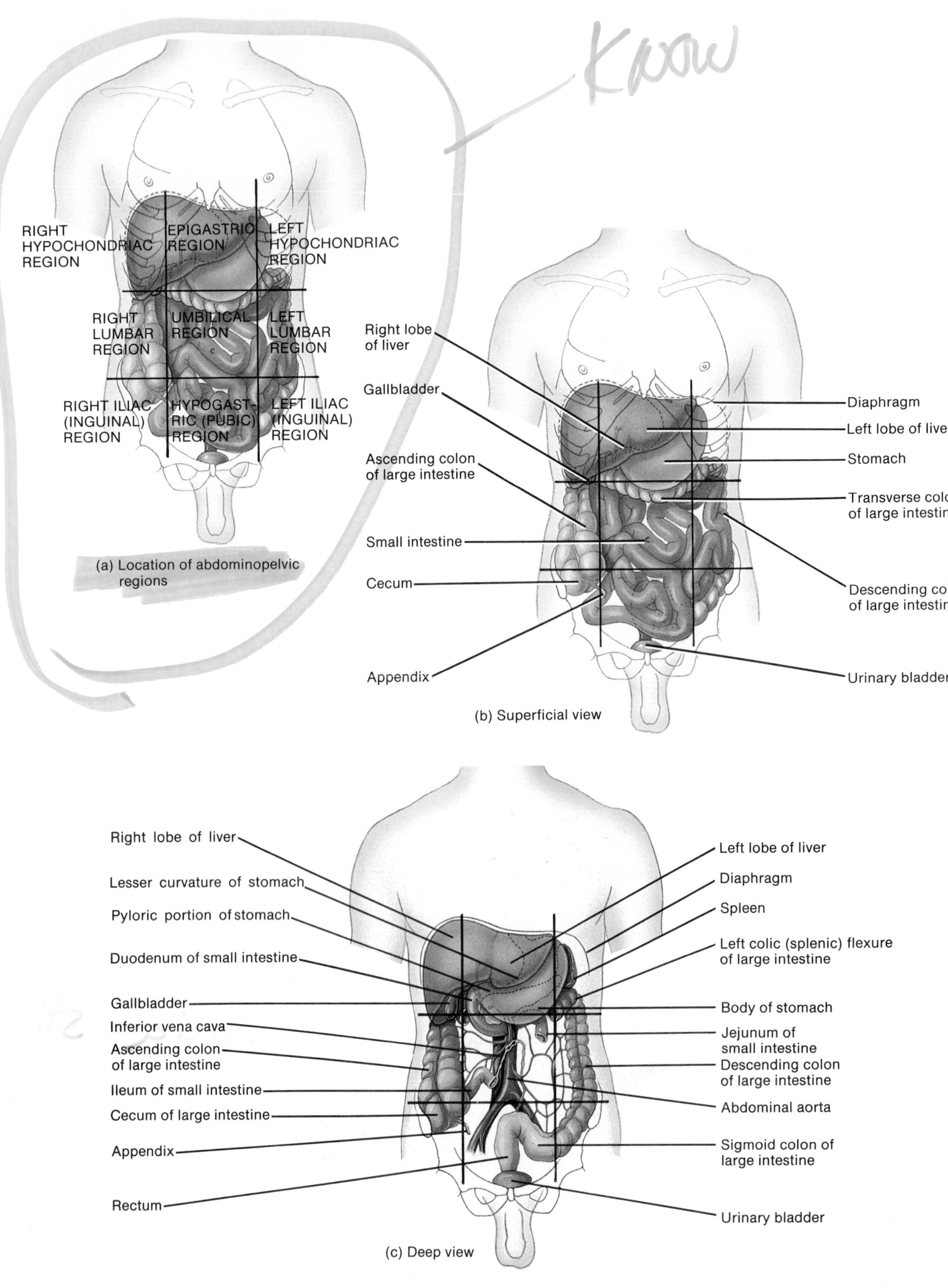

(a) Location of abdominopelvic regions

(b) Superficial view

(c) Deep view

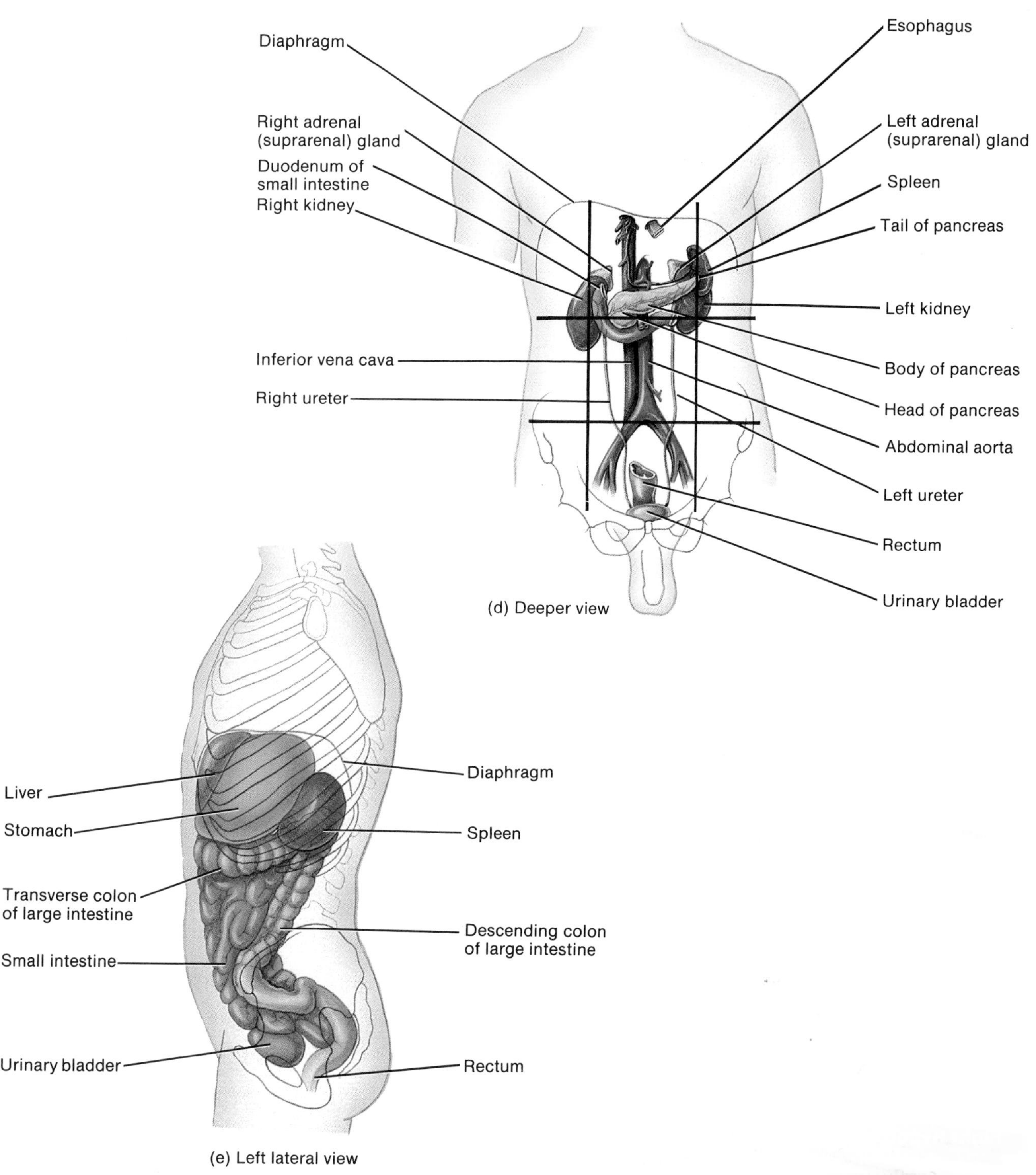

FIGURE 1-8 Abdominopelvic cavity. (a) The nine regions. The ***subcostal*** (top horizontal) ***line*** is drawn just inferior to the bottom of the rib cage, across the lower portion of the stomach. The ***transtubercular*** (bottom horizontal) ***line*** is drawn just inferior to the tops of the hipbones. The ***left*** and ***right midclavicular*** (two vertical) ***lines*** are drawn through the midpoints of the clavicles, just medial to the nipples. The two horizontal and two vertical lines divide the abdominopelvic cavity into a larger middle section and smaller left and right sections. (b) The greater omentum has been removed. (c) Most of the small intestine and transverse colon have been removed to expose deeper structures. (d) Many organs have been removed, exposing the posterior structures. The internal reproductive organs in the pelvic cavity are shown in Figures 25-1 and 25-11 and may be reviewed in Exhibit 1-2, Reproductive System.

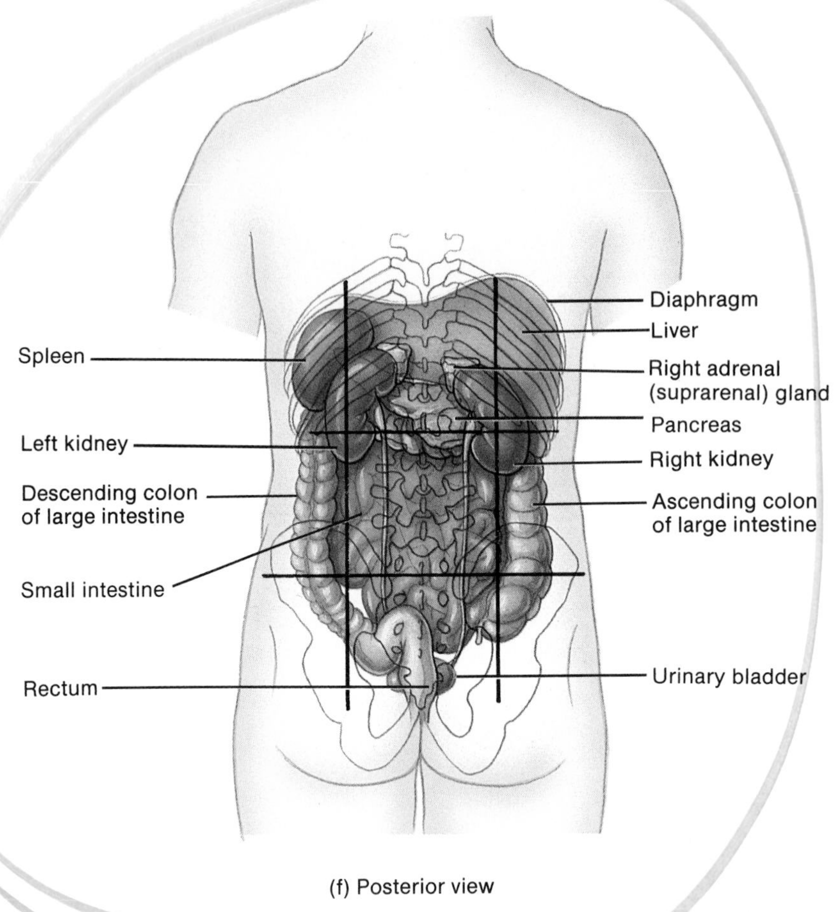

(f) Posterior view

EXHIBIT 1-5

Representative Structures Found in the Abdominopelvic Regions

REGION	REPRESENTATIVE STRUCTURES
Epigastric (ep-i-GAS-trik; *epi* = above; *gaster* = stomach)	Left lobe and medial part of right lobe of liver, pyloric portion and lesser curvature of stomach, superior and descending portions of duodenum, body and superior part of head of pancreas, and right and left adrenal (suprarenal) glands.
Right Hypochondriac (hī-pō-KON-drē-ak; *hypo* = under; *chondro* = cartilage)	Right lobe of liver, gallbladder, and upper superior third of right kidney.
Left Hypochondriac	Body and fundus of stomach, spleen, left colic (splenic) flexure, superior two-thirds of left kidney, and tail of pancreas.
Umbilical (um-BIL-i-kul)	Middle portion of transverse colon, inferior part of duodenum, jejunum, ileum, and bifurcations (branching) of abdominal aorta and inferior vena cava.
Right Lumbar (*lumbus* = loin)	Superior part of cecum, ascending colon, right colic (hepatic) flexure, inferior lateral portion of right kidney, and small intestine.
Left Lumbar	Descending colon, inferior third of left kidney, and small intestine.
Hypogastric (Pubic)	Urinary bladder when full, small intestine, and part of sigmoid colon.
Right Iliac (Inguinal) (IL-ē-ak; iliac refers to superior part of hipbone)	Lower end of cecum, appendix, and small intestine.
Left Iliac (Inguinal)	Junction of descending and sigmoid parts of colon and small intestine.

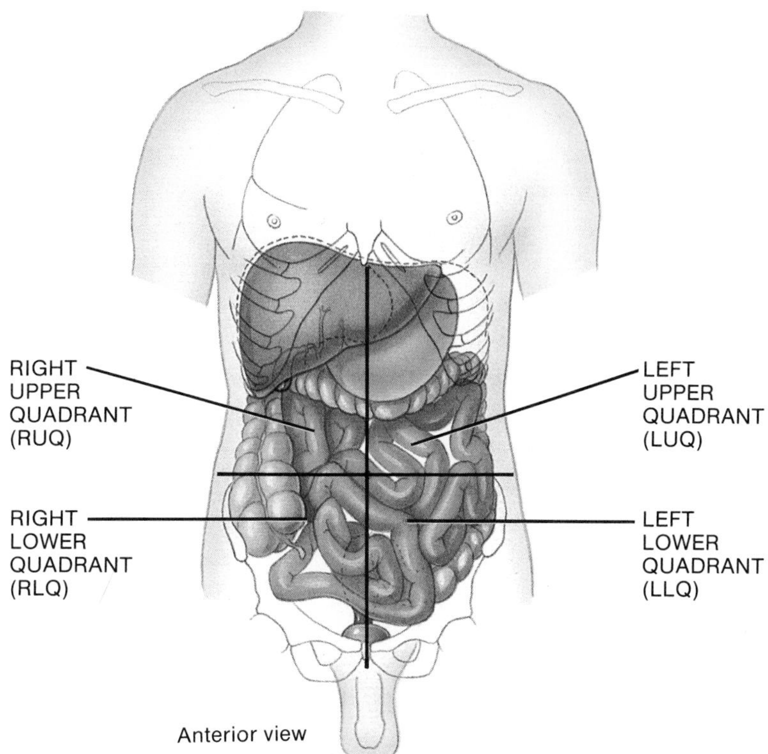

FIGURE 1-9 Quadrants of the abdominopelvic cavity. The two lines intersect at right angles at the umbilicus to define the quadrants.

CLINICAL APPLICATION

Autopsy

To determine the cause of death accurately, it is necessary to perform an ***autopsy*** (AW-top-sē; *auto* = self; *opsis* = to see with one's own eyes), that is, a postmortem examination of the body. In addition, an autopsy can also be used to uncover the existence of diseases not detected during life, support the accuracy of diagnostic tests, determine the effectiveness and side effects of drugs, analyze the effects of environmental influences on the body, and educate health-care students. Moreover, an autopsy can reveal conditions that may affect offspring or siblings (such as congenital heart defects) and can help resolve disputes and affect benefits that survivors receive as part of insurance settlements.

A typical autopsy consists of three principal phases of examination. The first phase is examination of the exterior of the body for the presence of wounds, scars, tumors, or other abnormalities. The second phase includes the dissection and gross examination of the major body organs, looking for pathological changes or evidence of violent destruction. The third phase of autopsy consists of microscopic examination of tissues to ascertain any pathology. Depending on the circumstances, techniques may also be used to detect and recover microbes and toxic chemicals and to determine the presence of foreign substances in the body.

A complete autopsy commonly begins with a Y-shaped incision to expose the thoracic, abdominal, and pelvic viscera. The upper parts of the Y begin in front of each shoulder, extend inferior to the nipples, and join just inferior to the sternum. The incision is then continued down the middle of the abdominopelvic wall to the pubic symphysis. The sternum is removed by cutting the ribs to expose the thoracic viscera, and the tissues of the abdominopelvic wall are folded back to expose the contents of the abdominopelvic cavity. Unless the information is of special clinical significance, the face and extremities are not usually dissected.

The first and second phases of an autopsy usually last two to four hours, depending on their comprehensiveness. When the procedure is completed, the organs are returned to the body, except for those that are donated or used to extract a particular pharmacologic substance, and all incisions are sutured.

MEDICAL IMAGING

In recent years, there has been an explosion in the development of various kinds of ***medical imaging*** techniques. These techniques are essential for diagnosing a wide range of disorders and are summarized in Exhibit 1-6.

EXHIBIT 1-6

Summary of Medical Imaging Procedures

PROCEDURE	DESCRIPTION AND EXAMPLE
Conventional Radiography	A single barrage of x-rays passes through the body and produces a two-dimensional image of the interior of the body, called a ***radiograph*** (RĀ-dē-ō-graf′). *Comment:* Overlap of structures can make diagnosis difficult and subtle differences in tissue density cannot always be differentiated.

Left clavicle

Rib

Left lung

Heart

Diaphragm

Radiograph of chest (© Biophoto, SPL, Photo Researchers)

PROCEDURE	DESCRIPTION AND EXAMPLE
Computed Tomography (CT) Scanning [formerly computerized axial tomography (CAT) scanning]	An x-ray source moves in an arc around the body, and x-rays are converted to electronic signals to produce a cross-sectional picture, called a ***CT scan,*** onto a video monitor; CT scans can now be converted to three-dimensional images that can be rotated and used in plastic surgery and radiology. *Comment:* Quick, painless, very detailed images; excellent for detecting tumors, aneurysms (bulges in blood vessels), kidney stones, gallstones, infections, tissue damage, deformities.

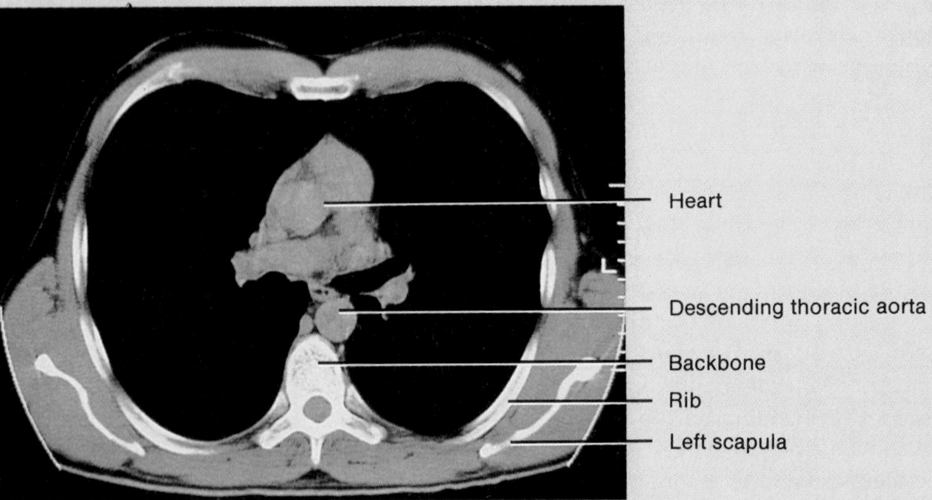

CT scan of chest (Courtesy of Fraser, SPL, Photo Researchers)

PROCEDURES	DESCRIPTION AND EXAMPLE	
Dynamic Spatial Reconstruction (DSR)	A highly sophisticated x-ray machine produces moving, three-dimensional, life-size images from any view. *Comment:* Image can be rotated, tipped, "sliced open," enlarged, replayed, and viewed in slow motion or at high speeds; a good procedure for heart, lung, and blood vessel imaging, measuring movements and volumes, and assessing tissue damage.	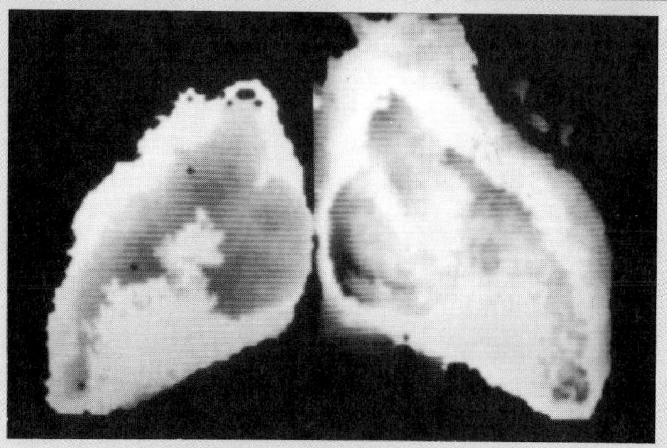 DSR Image of "opened" heart (Courtesy of Dr. Richard A. Robb, Mayo Foundation, Rochester, MN)
Magnetic Resonance Imaging (MRI) [formerly nuclear magnetic resonance (NMR) imaging]	Protons of hydrogen atoms in tissues are used and their response to magnetism is determined. The image produced is a color, two- or three-dimensional biochemical blueprint of cellular activity. *Comment:* Noninvasive and uses no radiation, but is not indicated for pregnant women or persons with artificial pacemakers or metal joints. Useful in detecting tumors and artery-clogging fatty plaques, assessing mental disorders, revealing brain changes, measuring blood flow, and studying metabolism.	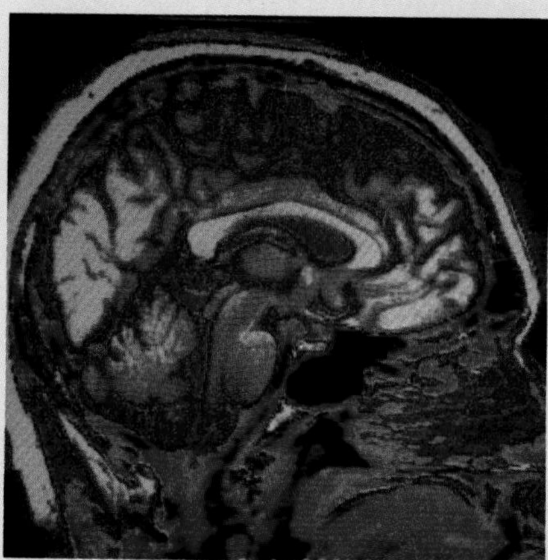 MRI of brain (Courtesy of Technicare Corporation, Cleveland, OH)
Ultrasound (US)	High-frequency sound waves produced by a hand-held transducer bounce back to produce an image on a video monitor called a ***sonogram.*** The image may be still or moving. *Comment:* Used to study abdominal and pelvic organs, blood flow (Doppler ultrasound), the heart (echocardiography), a developing fetus (fetal ultrasound), and guide amniocentesis. Procedure uses no harmful radiation, is noninvasive and painless, and uses no dyes. Obesity and scars interfere with the procedure.	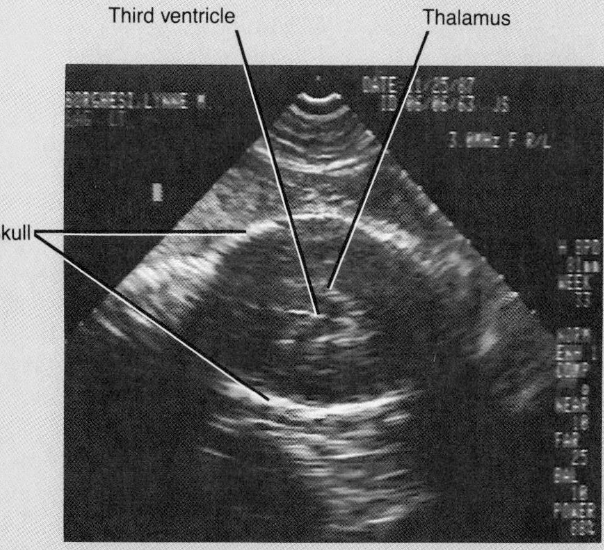 Sonogram of fetal head at 33 weeks (Courtesy of Lynne, James, and James Gerard Borghesi. Labels provided by Professor Carole Schanel, Bergen Community College)

Continued

EXHIBIT 1-6 *(Continued)*

Summary of Medical Imaging Procedures

PROCEDURES	DESCRIPTION AND EXAMPLE	
Positron Emission Tomography (PET)	Radioisotopes injected into the body emit electronlike particles called positrons that collide with electrons in body tissues and release gamma rays, which are similar to x-rays. The gamma rays are relayed to a computer, which constructs a ***PET scan,*** a colored image that shows where the radioisotopes are being used. *Comment:* PET scans provide information on function rather than structure and are very useful in detecting chemical changes associated with normal organs and dysfunctional organs such as the brain and heart.	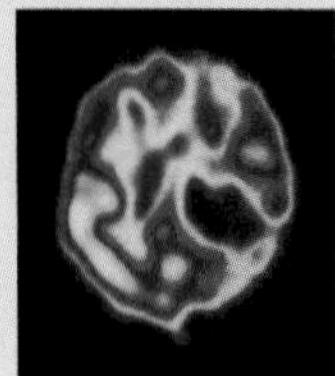 PET showing blood flow through brain (Dr. Michel M. TerPogossian, Washington University Medical Center)
Digital Subtraction Angiography (DSA)	A computer compares an x-ray of a region of the body before and after a contrast dye has been introduced. A blood vessel that shows up in the first image can be subtracted (erased) from the second, leaving an unobstructed image of the first. *Comment:* Used primarily to study blood vessels in the brain and heart.	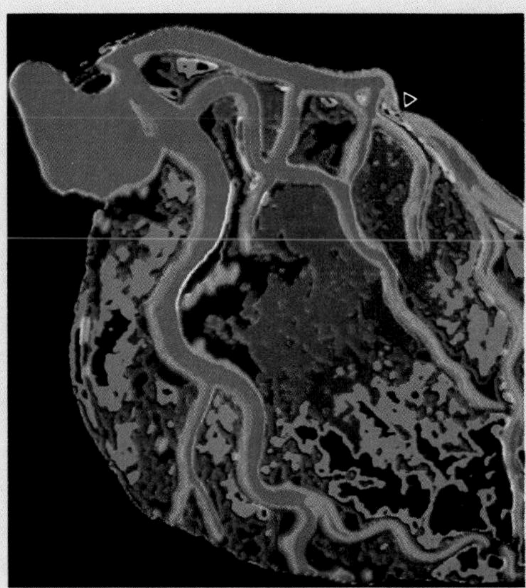 DSA image of heart showing narrowed blood vessel (triangle) (Courtesy of The Bayer Company)

MEASURING THE HUMAN BODY

An important aspect of describing the body and understanding how it works is ***measurement***—what the dimensions of an organ are, how much it weighs, how long it takes for a physiological event to occur. Such measurements also have clinical importance, for example, in determining how much of a given medication should be administered. As you will see, measurements involving time, weight, temperature, size, length, and volume are a routine part of your studies in a medical science program.

Whenever you come across a measurement in the text, the measurement will be given in metric units. The metric system is standardly used in sciences. To help you compare the metric unit to a familiar unit, the approximate U.S. equivalent will also be given in parenthesis directly after the metric unit. For example, you might be told that the length of a particular part of the body is 2.54 cm (1 in.).

As a first step in helping you understand the correlation between the metric system and the U.S. system of measurement, three exhibits have been prepared and are located inside the back cover: (1) metric units of length and some U.S. equivalents, (2) metric units of mass and some U.S. equivalents, and (3) metric units of volume and some U.S. equivalents.

STUDY OUTLINE

Anatomy Defined (p. 2)

1. Anatomy is the study of structure and the relation among structures.
2. Subdivisions of anatomy include surface anatomy (form and markings of surface features), gross anatomy (macroscopic), systemic, or systematic, anatomy (systems), regional anatomy (regions), radiographic anatomy (x-rays), developmental anatomy (development from fertilization to adulthood), embryology (development from fertilized egg through eighth week in utero), histology (tissues), cytology (cells), and pathological anatomy (disease).
3. Physiology is the study of how body structures function.

Levels of Structural Organization (p. 2)

1. The human body consists of several levels of structural organization; among these are the chemical, cellular, tissue, organ, system, and organismic levels.
2. Cells are the basic structural and functional units of an organism.
3. Tissues consist of groups of similarly specialized cells and their intercellular material that usually have a similar embryological origin and that perform certain special functions.
4. Organs are structures of definite form that are composed of two or more different tissues and have specific functions.
5. Systems consist of related organs that have a common function.
6. The human organism is a collection of structurally and functionally integrated systems.
7. The systems of the human body are the integumentary, skeletal, muscular, nervous, endocrine, cardiovascular, lymphatic, respiratory, digestive, urinary, and reproductive sytems (see Exhibit 1-2).

Life Processes (p. 5)

1. All living forms have certain processes that distinguish them from nonliving things.
2. Among the life processes in humans are metabolism, responsiveness, movement, growth, differentiation, and reproduction.

Overview of the Human Body (p. 8)

Structural Plan (p. 8)

1. The human body has certain general characteristics.
2. Among the characteristics are a backbone, a tube-within-a-tube organization, and bilateral symmetry.

Anatomical Position and Anatomical Names (p. 8)

1. When in the anatomical position, the subject stands erect facing the observer, the upper extremities are placed at the sides, the palms of the hands are turned forward, and the feet are flat on the floor.
2. Regional names are terms given to specific regions of the body for reference. Examples of regional names include cranial (skull), thoracic (chest), brachial (arm), patellar (knee), cephalic (head), and gluteal (buttock).

Directional Terms (p. 9)

1. Directional terms indicate the relation of one part of the body to another.
2. Commonly used directional terms are superior (toward the head or upper part of a structure), inferior (away from the head or toward the lower part of a structure), anterior (near or at the front of the body), posterior (near or at the back of the body), medial (nearer the midline of the body or a structure), lateral (farther from the midline of the body or a structure), intermediate (between a medial and lateral structure), ipsilateral (on the same side of the body), contralateral (on the opposite side of the body), proximal (nearer the attachment of an extremity to the trunk or a structure), distal (farther from the attachment of an extremity to the trunk or a structure), superficial (toward or on the surface of the body), and deep (away from the surface of the body).

Planes and Sections (p. 11)

1. Planes are imaginary flat surfaces that are used to divide the body or organs into definite areas. A midsagittal (median) plane divides the body or organs into equal right and left sides; a parasagittal plane divides the body or organs into unequal right and left sides; a frontal (coronal) plane divides the body or organs into anterior and posterior portions; and a horizontal (transverse) plane divides the body or organs into superior and inferior portions.
2. Sections are flat surfaces resulting from cuts through body structures. They are named according to the plane on which the cut is made and include cross sections, frontal sections, and midsagittal sections.

Body Cavities (p. 12)

1. Spaces in the body that contain internal organs are called cavities.
2. The dorsal and ventral cavities are the two principal body cavities. The dorsal cavity contains the brain and spinal cord. The organs of the ventral cavity are collectively called viscera.
3. The dorsal cavity is subdivided into the cranial cavity, which contains the brain, and the vertebral (or spinal) canal, which contains the spinal cord and beginnings of spinal nerves.
4. The ventral body cavity is subdivided by the diaphragm into an upper thoracic cavity and a lower abdominopelvic cavity and is lined by a serous membrane.
5. The thoracic cavity contains two pleural cavities, a pericardial cavity, and mediastinum.
6. The mediastinum is a region between the lungs that extends from the sternum to the vertebral column; it contains all contents of the thoracic cavity, except the lungs.
7. The abdominopelvic cavity is divided into a superior abdominal and an inferior pelvic cavity.
8. Viscera of the abdominal cavity include the stomach, spleen, pancreas, liver, gallbladder, small intestine, and most of the large intestine.
9. Viscera of the pelvic cavity include the urinary bladder, sigmoid colon, rectum, and internal female and male reproductive structures.

Abdominopelvic Regions (p. 15)

1. To describe the location of organs easily, the abdominopelvic cavity may be divided into nine regions by drawing four

imaginary lines (left midclavicular, right midclavicular, subcostal, and transtubercular).
2. The names of the nine abdominopelvic regions are epigastric, right hypochondriac, left hypochondriac, umbilical, right lumbar, left lumbar, hypogastric (pubic), right iliac (inguinal), and left iliac (inguinal).

Abdominopelvic Quadrants (*p. 15*)

1. To locate the site of an abdominopelvic abnormality in clinical studies, the abdominopelvic cavity may be divided into quadrants by passing imaginary horizontal and vertical lines through the umbilicus.
2. The names of the abdominopelvic quadrants are right upper quadrant (RUQ), left upper quadrant (LUQ), right lower quadrant (RLQ), and left lower quadrant (LLQ).

Medical Imaging (p. 21)

1. Conventional radiography uses a single barrage of x-rays. The photographic two-dimensional image produced is called a radiograph.
2. Computed tomography (CT) scanning combines the principles of x-ray and advanced computer technology. The image produced, called a CT scan, provides a very accurate cross-sectional picture of any area of the body.
3. The dynamic spatial reconstructor (DSR) is a highly sophisticated x-ray machine that can produce moving, three-dimensional images of different organs of the body.
4. Magnetic resonance imaging (MRI) is based on the reaction of protons of hydrogen atoms to magnetism.
5. Ultrasound (US) is based on high-frequency sound waves that are reflected and translated into images.
6. Positron emission tomography (PET) is a form of radioisotope scanning based on the emission of positrons from radioisotopes.
7. Digital subtraction angiography (DSA) compares a blood vessel before and after a contrast medium is introduced.

REVIEW QUESTIONS

1. Define anatomy. List and define the various subdivisions of anatomy. Define physiology. (p. 2).
2. Give several examples of how structure and function are related. (p. 2).
3. Define each of the following terms: cell, tissue, organ, system, and organism. (p. 2).
4. Using Exhibit 1-2 as a guide, outline the functions of each system of the body, then list several organs that compose each system. (p. 4).
5. What is a vertebrate? Why is the body considered to be a tube-within-a-tube? What does bilateral symmetry mean? (p. 8).
6. List and define the life processes of humans. (p. 5).
7. Define the anatomical position. Why is the anatomical position used? (p. 5).
8. Review Figure 1-2. See whether you can locate each region on your own body and name each by its common and anatomical term. (p. 10).
9. What is a directional term? Why are these terms important? Use each of the directional terms listed in Exhibit 1-3 in a complete sentence. (p. 9).
10. Define the various planes that may be passed through the body. Explain how each plane divides the body. Describe the meaning of cross section, frontal section, and midsagittal section. (p. 11).
11. Define a body cavity. List the body cavities discussed and tell which major organs are located in each. What is the mediastinum? (p. 12).
12. Describe how the abdominopelvic area is subdivided into nine regions. Name and locate each region and list the organs, or parts of organs, in each. (p. 15).
13. Describe how the abdominopelvic cavity is divided into quadrants and name each quadrant. (p. 15).
14. Why is an autopsy performed? Describe the basic procedure. (p. 21).
15. Explain the principle and clinical application of conventional radiography, computed tomography (CT) scanning, dynamic spatial reconstruction (DSR), magnetic resonance imaging (MRI), ultrasound (US), positron emission tomography (PET), and digital subtraction angiography (DSA). (p. 21)

SELF QUIZ

1. Complete the following table relating common terms to anatomical terms.

COMMON TERM	ANATOMICAL TERM
a.	Axillary
b. Arm	
c.	Cephalic
d. Chest	
e. Neck	

Choose the one best answer to these questions.

___ 2. The microscopic study of cells is called
A. histology; B. embryology; C. cytology; D. gross anatomy; E. surface anatomy.

___ 3. Which of the following statements about the function of the respiratory system is *not* true?
A. it supplies oxygen; B. it eliminates carbon dioxide; C. it helps regulate acid–base balance in the body; D. it filters blood; E. it includes the lungs.

___ **4.** Which is most inferiorly located?
A. abdominal cavity; B. pelvic cavity; C. mediastinum; D. diaphragm; E. pleural cavity.

___ **5.** The spleen, tonsils, and thymus are all organs in which system?
A. nervous; B. lymphatic; C. cardiovascular; D. digestive; E. endocrine.

___ **6.** The system responsible for providing support, protection, assisting in movement, storage of minerals, and housing cells that produce blood cells is
A. nervous; B. lymphatic; C. cardiovascular; D. digestive; E. skeletal.

___ **7.** In the anatomical position
A. the body is in a supine position; B. the body is in a prone position; C. the palmar surface of the hand is anterior; D. the dorsal surface of the hand is anterior; E. the body is prone and the palmar surface of the hand is anterior.

___ **8.** The general anatomical characteristics of the human body include
(1) a backbone (vertebral column).
(2) a tube-within-a-tube construction.
(3) bilateral symmetry.
A. (1) only; B. (2) only; C. (3) only; D. (1) and (2); E. all of the above.

___ **9.** Which word describes the location of the stomach with reference to the pancreas?
A. anterior; B. distal; C. dorsal; D. proximal; E. sagittal.

___ **10.** Any part of the body that is away from the midline is said to be
A. medial; B. lateral; C. distal; D. superior; E. posterior.

___ **11.** Where does the knee joint lie in reference to the leg bones?
A. anterior; B. medial; C. inferior; D. lateral; E. proximal.

___ **12.** The microscopic examination of tissues is referred to as
A. embryology; B. physiology; C. histology; D. gross anatomy; E. surface anatomy.

___ **13.** The development of unspecialized cells into specialized ones is called
A. contractility; B. responsiveness; C. metabolism; D. conductivity; E. differentiation.

___ **14.** Which of the following statements is/are true of a plane that divides the body into superior and inferior parts?
(1) This plane must pass through the umbilicus (navel).
(2) This plane is a transverse (horizontal) plane.
(3) This plane is a midsagittal (median) plane.
(4) This plane is called a coronal (frontal) plane.
A. (1) only; B. (2) only; C. (3) only; D. (4) only; E. (1) and (3).

___ **15.** Which of the following is/are true?
(1) The thoracic cavity is the same as the pleural cavity.
(2) The diaphragm divides the abdominopelvic cavity into abdominal and pelvic portions.
(3) The organs of importance in the ventral body cavity are the brain and spinal cord.
A. (1) only; B. (2) only; C. (3) only; D. (1) and (2); E. none of the statements is true.

___ **16.** You would *not* look in the pelvic cavity to find the
A. rectum; B. sigmoid colon; C. thymus gland; D. uterus; E. urinary bladder.

Complete the following.

17. From superior to inferior, the three abdominal regions on the right side are right hypochondriac, right lumbar, and ________.

18. When you lie face down in a pool, as if to do the "dead man's float," you are lying on your ________ surface.

19. Since the stomach and the spleen are both located on the left side of the abdomen, they could be described as ________-lateral.

20. A ________ plane divides the brain into equal left and right sides.

Cells

2

STUDENT OBJECTIVES

1. Define a cell and list its generalized parts.
2. Explain the chemistry, structure, and functions of the plasma membrane.
3. Describe how materials move across plasma membranes by diffusion, facilitated diffusion, osmosis, filtration, active transport, and endocytosis.
4. Describe the chemical composition and functions of the cytosol.
5. Describe the structure and functions of the following: nucleus, ribosomes, endoplasmic reticulum (ER), Golgi complex, lysosomes, peroxisomes, mitochondria, cytoskeleton, centrioles, cilia, and flagella.
6. Define a cell inclusion and give several examples.
7. Describe cancer (CA) as a type of abnormal cell division.
8. Explain the relation between aging and cells.
9. Define key medical terms associated with cells.

CHAPTER OUTLINE

■ **Generalized Animal Cell**
■ **Plasma (Cell) Membrane**
Chemistry and Structure
Functions
Movement of Materials Across Plasma Membranes
Passive Processes
Active Processes
■ **Cytosol**
■ **Organelles**
Nucleus
Ribosomes
Endoplasmic Reticulum (ER)
Golgi Complex
Lysosomes
Peroxisomes
Mitochondria
The Cytoskeleton
Centrosome and Centrioles
Flagella and Cilia
■ **Cell Inclusions**
■ **Normal Cell Division**
Somatic Cell Division
Mitosis
Cytokinesis
Time Required
Reproductive Cell Division
Meiosis
■ **Abnormal Cell Division: Cancer (CA)**
Definition
Types
Growth and Spread
Possible Causes
Treatment
■ **Cells and Aging**
■ **Key Medical Terms Associated with Cells**

The study of the body at the cellular level of organization is important because activities essential to life occur in cells and disease processes originate there. A ***cell*** may be defined as the basic, living, structural, and functional unit of the body and, in fact, of all organisms. ***Cytology*** (sī-TOL-ō-jē; *cyt* = cell; *logos* = study of) is the branch of science concerned with the study of cells. This chapter concentrates on the structure, functions, and reproduction of cells.

A series of illustrations accompanies each cell structure that you study. A diagram of a generalized animal cell shows the location of the structure within the cell. An electron micrograph shows the actual appearance of the structure. An ***electron micrograph (EM)*** is a photograph taken with an electron microscope. Some electron microscopes can magnify objects up to 1 million times. In comparison, the light microscope that you probably use in your laboratory magnifies objects up to 1000 times their size. A diagram of the electron micrograph clarifies some of the small details by exaggerating their outlines. Finally, an enlarged diagram of the structure shows its details.

GENERALIZED ANIMAL CELL

A ***generalized animal cell*** is a composite of many different cells in the body. Examine the generalized cell illustrated in Figure 2-1, but keep in mind that no such single cell actually exists.

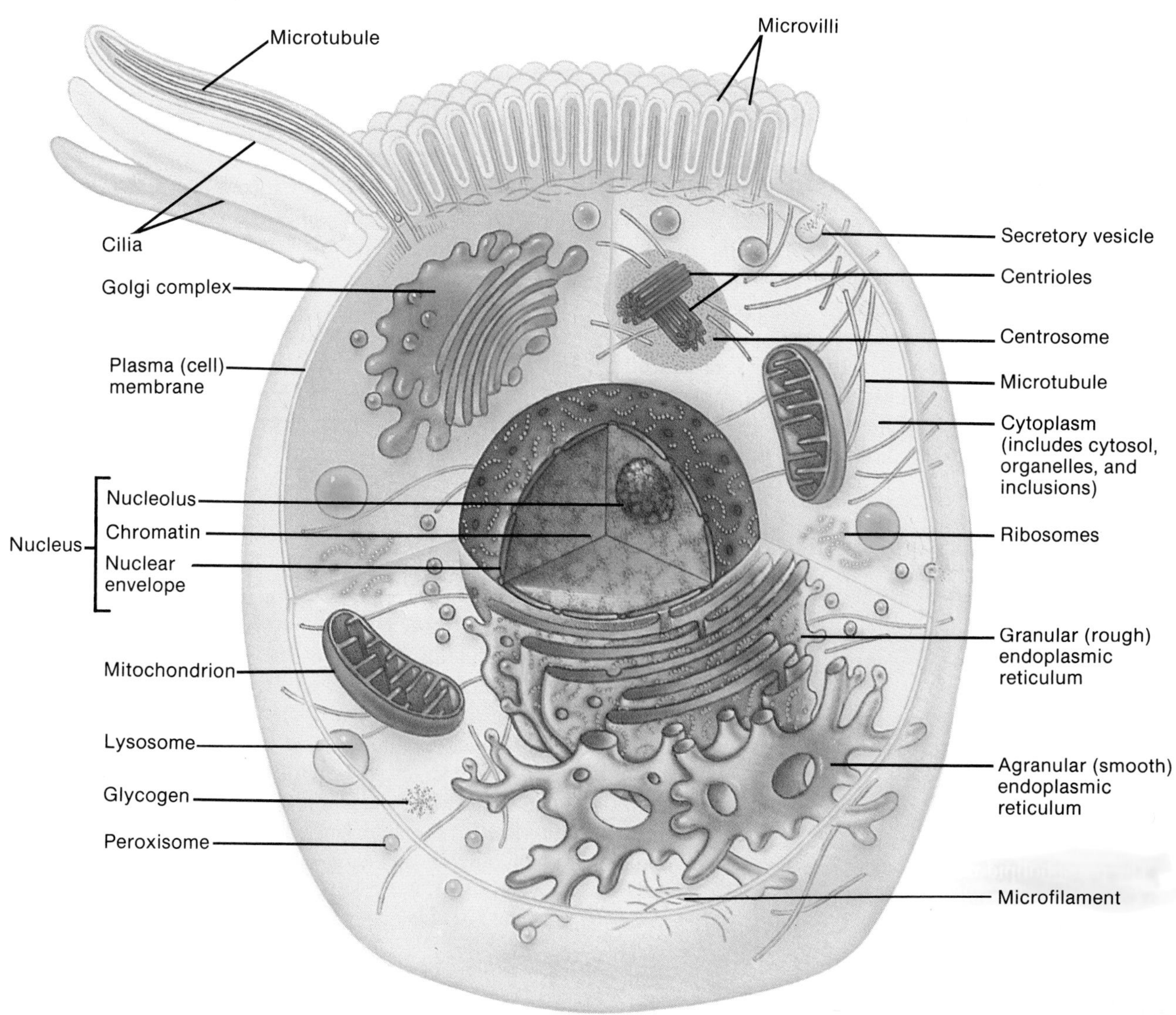

FIGURE 2-1 Generalized animal cell based on electron microscope studies.

For convenience, we can divide the generalized cell into four principal parts:

1. ***Plasma (cell) membrane.*** The outer, limiting membrane separating the cell's internal parts from the extracellular materials and external environment. *Extracellular materials,* which are substances external to the cell surface, will be examined in connection with tissues in Chapter 3.
2. ***Cytosol.*** The term ***cytoplasm*** refers to all of the cellular contents located between the plasma membrane and nucleus. The term ***cytosol*** (SĪ-tō-sol) refers to the thick semifluid portion of cytoplasm in which organelles and inclusions are suspended and solutes are dissolved. The cytosol contains many soluble proteins and enzymes, nutrients, ions, and other small molecules, all of which participate in various phases of metabolism.
3. ***Organelles.*** Permanent highly organized structures with characteristic shapes that are highly specialized for specific cellular activities.
4. ***Inclusions.*** Temporary structures that contain secretions and storage products of cells.

PLASMA (CELL) MEMBRANE

The exceedingly thin structure that separates the internal components of a cell from the external environment is called the ***plasma (cell) membrane*** (Figure 2-2a). The membrane measures from 4.5 nm at the phospholipid bilayer regions up to 10 nm in regions where membrane proteins are present. Formerly, dimensions of cells and parts of cells were given in angstroms (Å). 1 Å = 0.1 nm.

CHEMISTRY AND STRUCTURE

Plasma membranes consist primarily of phospholipids (lipids that contain phosphorus), the most abundant chemicals, and proteins. Other chemicals in lesser amounts include *cholesterol* (a lipid), *glycolipids* (combinations of carbohydrates and lipids), and carbohydrates called *oligosaccharides* (Figure 2-2b).

The currently accepted model of plasma membrane structure is known as the ***fluid mosaic model.*** A mosaic is a pattern of many small pieces fitted together. According to this concept, the membrane is a mosaic of proteins floating like icebergs in a sea of lipids.

The phospholipid molecules are arranged in two parallel rows, forming a ***phospholipid bilayer.*** A phospholipid molecule consists of a polar, phosphate-containing "head" that mixes with water (*hydrophilic*) and nonpolar fatty acid "tails" that do not mix with water (*hydrophobic*). The molecules are oriented in the bilayer so that "heads" face outward on either side, and the "tails" face each other in the membrane's interior. The phospholipid bilayer is dynamic in that the phospholipid molecules can move sideways and exchange places in their own row; however, movement of phospholipid molecules between rows rarely occurs. The bilayer is also self-sealing; if a needle is pushed through it and pulled out, the puncture site will seal automatically. The phospholipid bilayer forms the basic framework of plasma membranes.

The plasma membrane proteins (PMPs) are classified into two categories: integral and peripheral. ***Integral proteins*** are embedded in the phospholipid bilayer among the fatty acid "tails." Some of the integral proteins lie at or near the inner and outer membrane surfaces, whereas others penetrate the membrane completely. Since the phospholipid bilayer is somewhat fluid and flexible and the integral proteins have been observed moving from one location to another in the membrane, the relationship has been compared to icebergs (proteins) floating in the sea (phospholipid bilayer).

The subunits of some integral proteins form minute channels through which substances can be transported into and out of the cell (described shortly). Other integral proteins are bound to branching chains of carbohydrates called *oligosaccharides.* The oligosaccharides and integral proteins together provide receptor sites that enable a cell to recognize other cells of its own kind so that they can associate to form a tissue, to recognize and respond to foreign cells that might be potentially dangerous, and to recognize and attach to hormones, nutrients, and other chemicals. Red blood cells also have receptors that prevent them from clumping and producing unwanted clumps. Type II diabetes is a disease believed to be caused by faulty cell receptor sites.

Peripheral proteins are loosely bound to the interior and exterior surfaces of the membrane and are easily separated from it. Far less is known about them than integral proteins, and their functions are not yet completely understood. For example, some peripheral proteins are believed to serve as enzymes that catalyze certain cellular reactions. An example is cytochrome *c,* involved in cellular respiration. Other peripheral proteins, such as spectrin of red blood cells, are believed to have a mechanical function by serving as a scaffolding to support the plasma membrane. It is also believed that peripheral proteins may assume a role in changes in membrane shape during such processes as cell division, locomotion, and ingestion.

The presence of cholesterol molecules makes the membrane less flexible and less permeable. Glycolipids mediate cell-to-cell recognition and communication, participate in cellular growth and development, and may be infection sites for several kinds of viruses and bacteria.

FUNCTIONS

Based upon the discussion of the chemistry and structure of the plasma membrane, we can now describe its several important functions.

1. It provides a flexible boundary that ***gives shape*** to a cell and ***encloses*** and ***protects*** the cellular contents. Some peripheral proteins help to ***support*** the membrane.
2. It ***separates*** a cell from the external environment.
3. Some integral proteins and oligosaccharides provide ***receptor sites*** (that is, binding sites) that enable a cell to recognize cells of its own kind to form a tissue; respond to foreign cells that might be potentially dangerous; and attach to hormones, nutrients, antibodies, neurotransmitters, and other chemicals.
4. Some peripheral proteins function as ***enzymes*** that catalyze various chemical reactions.

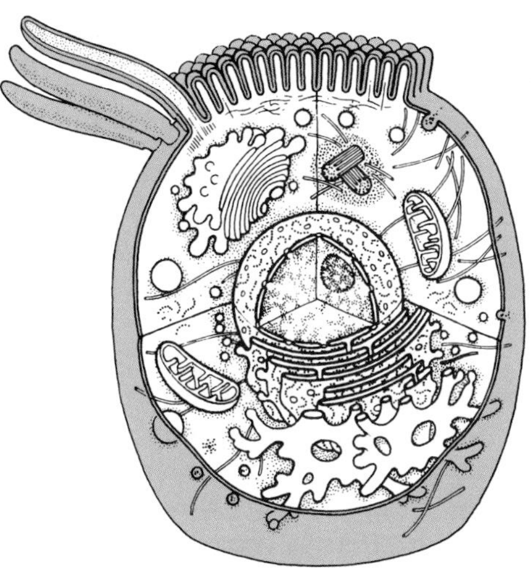

Generalized cell showing location of plasma membrane

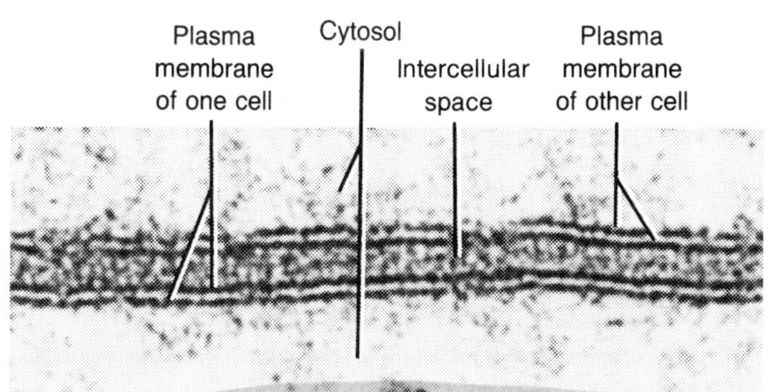

(a) Electron micrograph of two plasma membranes, 200,000×

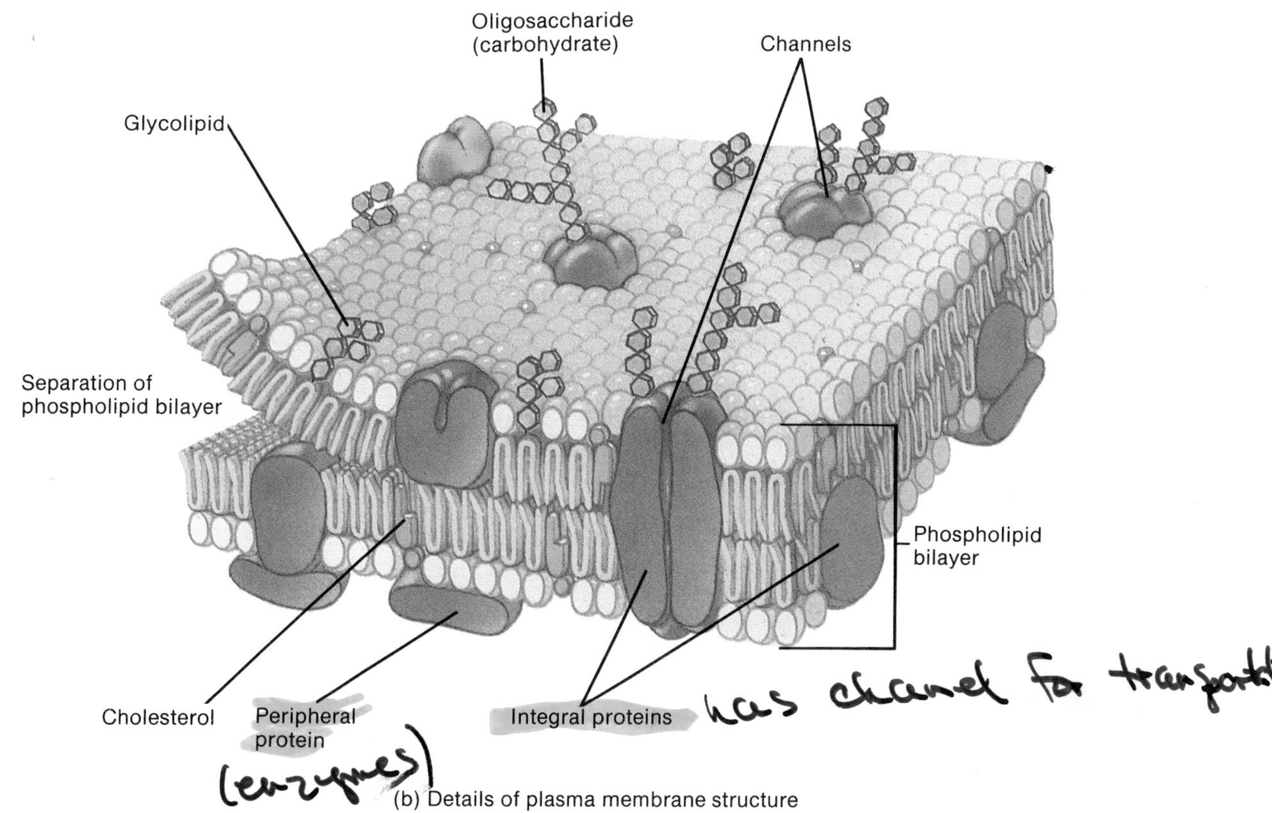

(b) Details of plasma membrane structure

FIGURE 2-2 Plasma membrane. (a) Copyright © Dr. Donald Fawcett, Science Source/Photo Researchers. (b) The separation of the phospholipid bilayer is for illustrative purposes only.

5. Some peripheral proteins help ***change membrane shape*** during processes such as cell division, locomotion, and ingestion.
6. It regulates the entrance and exit of materials. The ability of a plasma membrane to permit certain substances to enter and exit, but to restrict the passage of others, is called ***selective permeability***.

MOVEMENT OF MATERIALS ACROSS PLASMA MEMBRANES

Before actually discussing how materials move into and out of cells, we will first describe the location of the various fluids through which the substances move. Fluid outside body cells is called ***extracellular*** (*extra* = outside) ***fluid*** (***ECF***) and is found in two principal places. The fluid filling the microscopic spaces between the cells of tissues is called ***interstitial*** (in′-ter-STISH-al) ***fluid*** (*inter* = between) or ***intercellular fluid***. The extracellular fluid in blood vessels is termed ***plasma*** (Figure 2-3) and in lymphatic vessels is called ***lymph***. Fluid within cells is called ***intracellular*** (*intra* = within, inside) ***fluid*** (***ICF***). Among the substances in extracellular fluid are gases, nutrients, and electrically charged chemical particles called *ions*—all needed for the maintenance of life. Extracellular fluid circulates through the blood and lymphatic vessels and from there moves into the spaces between the tissue cells. Thus, it is in constant motion throughout the body. Essentially, all body cells are surrounded by the same fluid environment. The movement of substances across a plasma membrane, or across membranes within cells, is essential to the life of the cell. Certain substances, for example, must move into the cell to support life, whereas waste materials or harmful substances must be moved out. Plasma membranes regulate the movements of such materials.

The processes involved in the movement of substances across plasma membranes are classified as either passive or active. In ***passive processes***, substances move across plasma membranes without the use of energy—from the breakdown of ATP (adenosine triphosphate) by the cell. ATP is a molecule that stores energy for various cellular uses. The movement of substances in passive processes involves the *kinetic energy* (the energy of motion) of individual molecules or ions. The substances move on their own down a concentration gradient—that is, from an area where their concentration is high to an area where their concentration is low. The substances may also be forced across the plasma membrane by pressure from an area where the pressure is high to an area where it is low. In ***active processes*** the cell uses energy (from the breakdown of ATP) in moving the substance across the membrane since the substance typically moves against a concentration gradient—that is, from an area where its concentration is low to an area where its concentration is high.

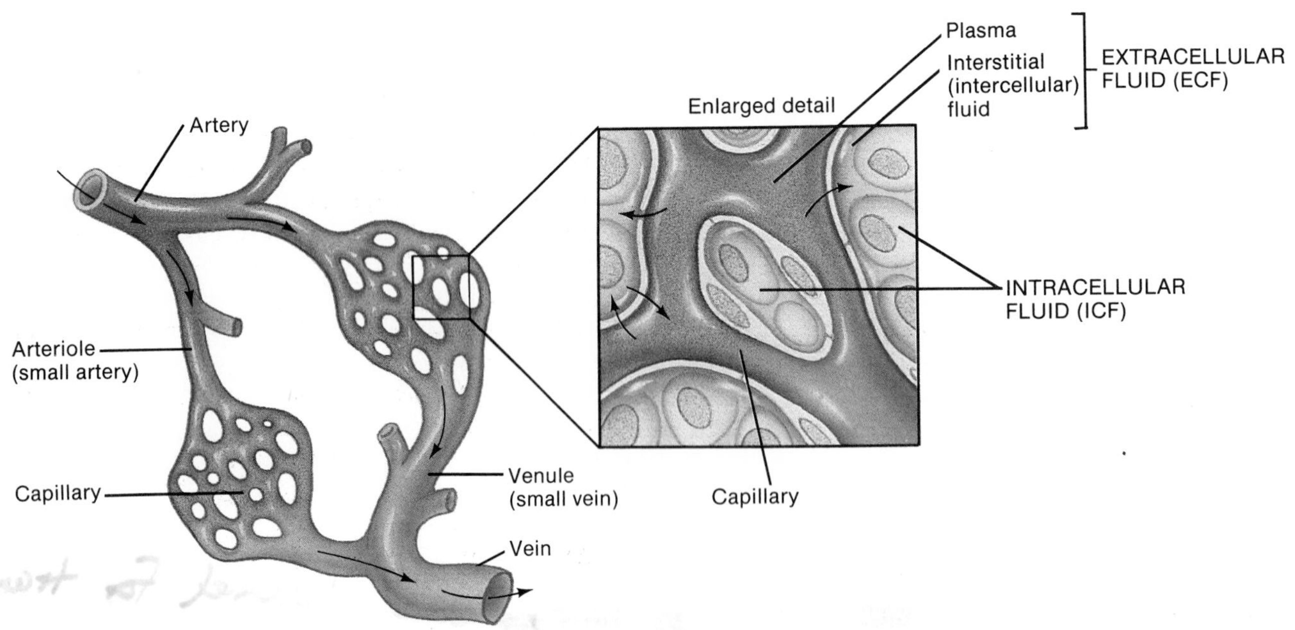

FIGURE 2-3 Principal fluid compartments. Extracellular fluid is found in two principal places: in blood vessels as plasma and between cells as interstitial fluid. Plasma circulates through arteries and arterioles and then into microscopic blood vessels called capillaries. From there, certain components of plasma move into the spaces between body cells, where it is called interstitial fluid. Some of this fluid then returns to capillaries as plasma and passes through the venules, then into the veins.

Passive Processes

■ ***Diffusion*** A passive process called ***diffusion*** (*dif-fus* = spreading) occurs when there is a *net* (greater) movement of molecules or ions from a region of their higher concentration to a region of their lower concentration—that is, when the molecules move from an area where there are more of them to an area where there are fewer of them. The movement from high to low concentration continues until the molecules are evenly distributed. At this point they move in both directions at an *equal* rate; there is no net diffusion. This point of even distribution is called equilibrium. The difference between high and low concentrations is called the *concentration gradient.* Molecules moving from the high-concentration area to the low-concentration area are said to move *down* or *with* the concentration gradient.

One of the most important factors that determines whether a substance moves across a membrane by diffusion is its ability to dissolve in lipid. A good example of diffusion in the body is the movement of oxygen from the blood into the cells and the movement of carbon dioxide from the cells back into the blood. This movement is essential in order for body cells to maintain homeostasis. It ensures that cells receive adequate amounts of oxygen and eliminate carbon dioxide as part of their normal metabolism. Small molecules that are not lipid-soluble, such as certain ions (sodium, potassium, chloride), are able to diffuse through channels formed by integral proteins in the membrane.

■ ***Facilitated Diffusion*** Another type of diffusion through a selectively permeable membrane occurs by a process called ***facilitated diffusion.*** This process is accomplished with the assistance of integral proteins in the membrane that serve as carriers. Through this process, some chemical substances that are large molecules and insoluble in lipids can still pass through the plasma membrane. Among these are various sugars, especially glucose. In facilitated diffusion it is believed that the integral protein (carrier) attaches to glucose, making it more soluble. Then, by altering its shape the integral protein transports the glucose across the plasma membrane. In facilitated diffusion the cell does not use energy from the breakdown of ATP, and the movement of large and lipid-soluble substances is from a region of their higher concentration to a region of their lower concentration.

■ ***Osmosis*** Another passive process by which materials move across membranes is ***osmosis.*** It is the net movement of water molecules through a selectively permeable membrane from an area of high water concentration to an area of low water concentration. The water molecules pass through channels in integrated proteins in the membrane and the net movement continues until equilibrium is reached. Water moves between various compartments of the body by osmosis.

■ ***Bulk Flow*** The movement of *large* numbers of ions, molecules, or particles in the same direction as a result of forces that push them is referred to as ***bulk flow.*** The substances move in unison in response to forces such as osmotic or hydrostatic (water) pressure at rates far greater than can be accounted for by diffusion or osmosis alone. Examples of bulk flow in the body are movements of substances through blood capillary membranes, blood flow within vessels, and movement of air into and out of the lungs.

■ ***Filtration*** Another passive process involved in moving materials in and out of cells is ***filtration.*** This process involves the movements of solvents such as water and dissolved substances such as sugar across a selectively permeable membrane by gravity or mechanical pressure, usually hydrostatic (water) pressure. Such a movement is always from an area of high pressure to an area of lower pressure and continues as long as a pressure difference exists. Most small- to medium-sized molecules can be pushed through a cell membrane, whereas large molecules or aggregates cannot.

An example of filtration occurs in the kidneys, where the blood pressure supplied by the heart forces water and small molecules like urea through thin cell membranes of tiny blood vessels and into the kidney tubules. In this basic process, protein molecules are retained in the blood since they are too large to be forced through the cell membranes of the blood vessels. The molecules of harmful substances such as urea are small enough to be forced through and eliminated in the urine, however.

Active Processes

When cells actively participate in moving substances across membranes, they must expend energy by the breakdown of ATP. Cells can even move substances against a concentration gradient. The active processes considered here are active transport, endocytosis (phagocytosis, pinocytosis, and receptor-mediated endocytosis), and exocytosis.

■ ***Active Transport*** The process by which substances are transported across plasma membranes typically from an area of their low concentration to an area of their high concentration is called ***active transport.*** In order to transport a substance against a concentration gradient, the membrane uses energy supplied by the breakdown of ATP. In fact, a typical body cell probably expends up to 40 percent of its ATP for active transport. In active transport the substance being moved enters a channel in an integral membrane protein and makes contact with a specific site in the channel. Then the ATP splits, and the energy from the breakdown of ATP causes a change in the shape of the integral membrane protein that expels the substance on the opposite side of the membrane.

As just noted, glucose can be transported across cell membranes via facilitated diffusion from areas of high concentration to areas of low concentration. Glucose can also be moved by the cells lining the gastrointestinal tract from the cavity of the tract into the blood, even though blood concentration of glucose is higher. This movement involves active transport. As you will see later, kidney cells also have the ability to actively transport glucose back into the blood so that it is not lost in the urine. Active transport is also an important process in maintaining higher concentrations of some ions inside body cells and higher concentrations of other ions outside body cells.

■ ***Endocytosis*** Large molecules and particles pass through plasma membranes by a process called ***endocytosis,*** in which a segment of the plasma membrane surrounds the substance, encloses it, and brings it into the cell. The export of substances from the cell occurs by a reverse process called ***exocytosis,*** a very important mechanism for secretory cells. There are three basic kinds of endocytosis: phagocytosis, pinocytosis, and receptor-mediated endocytosis.

In ***phagocytosis*** (fag′-ō-sī-TŌ-sis) or "cell eating," projections of cytoplasm, called ***pseudopodia*** (soo′-dō-PŌ-dē-a), engulf large solid particles external to the cell (Figure 2-4a,b). Once the particle is surrounded, the membrane folds inwardly, forming a membrane sac around the particle. This newly formed sac, called a ***phagocytic vesicle,*** breaks off from the plasma membrane, and the solid material inside the vesicle is digested by enzymes provided by lysosomes (discussed shortly). Indigestible particles and cell products are removed from the cell by exocytosis. Phagocytosis and pinocytosis are important because molecules and particles of material that would normally be restricted from crossing the plasma membrane because of their large size can be brought into or removed from the cell. Through phagocytosis, the white blood cells engulf and destroy bacteria and other foreign substances. The phagocytic white blood cells of the body constitute a vital defense mechanism that helps protect us against disease.

In ***pinocytosis*** (pi′-nō-sī-TŌ-sis), or "cell drinking," the engulfed material consists of an extracellular liquid rather than a solid (Figure 2-4c). Moreover, no cytoplasmic projections are formed. Instead, a minute droplet of liquid is attracted to the surface of the membrane. The membrane folds inwardly, forms a ***pinocytic vesicle*** that surrounds the liquid, and detaches from the rest of the intact membrane. Whereas few cells are capable of phagocytosis, many cells carry on pinocytosis.

Receptor-mediated endocytosis is a highly selective process in which cells can take up large molecules or particles. The interstitial fluid that bathes cells contains a large number of chemicals, most of which are in concentrations lower than in the cells themselves. These chemicals, called ***ligands,*** serve a variety of functions. For example, some of the substances are nutrients—like amino acids, iron, and vitamins—needed for various chemical reactions that sustain life. Some are hormones that deliver messages to cells so that cells can carry on specific physiological responses. Other substances are waste products or poisonous materials that certain cells have the ability to break down so that they do not interfere with normal functioning of cells.

The plasma membrane contains protein receptors that have binding sites for ligands (Figure 2-4d). The binding between them causes the plasma membrane to fold inward, forming a ***vesicle*** around the ligand. As a vesicle moves inward from the plasma membrane, it fuses with another vesicle to form a larger structure called an ***endosome.*** Each endosome develops into an even larger structure called a ***CURL (compartment of uncoupling of receptor and ligand).*** Within the CURL, the ligands separate from the receptors. Ligands have different destinations in the cell. However, in most instances following this separation, the vesicular portion fuses with a lysosome, where the ligands are broken down by powerful digestive enzymes. The tubular portion recycles receptors to the plasma membrane for reuse.

The various passive and active processes by which substances move across plasma membranes are summarized in Exhibit 2-1.

CYTOSOL

The cellular content inside the cell's plasma membrane and external to the nucleus is called ***cytoplasm*** (SĪ-tō-plazm′) (Figure 2-5a,b). The semifluid portion of cytoplasm in which organelles and inclusions are suspended and solutes are dissolved is called the ***cytosol,*** which is actually intracellular fluid (see Figure 2-1). Physically, the cytosol is described as a thick, semitransparent, elastic fluid.

Chemically, the cytosol is 75 to 90 percent water plus solid components. Proteins, carbohydrates, lipids, and inorganic substances (substances that usually lack carbon) compose the bulk of the solid components. The inorganic substances and most carbohydrates, amino acids, and peptides are soluble in water and are present as a solution. The majority of organic compounds, however, are found as ***colloids***—particles that remain suspended in the surrounding medium. Since the particles of a colloid bear electrical charges that repel each other, they remain suspended and separated from each other.

Functionally, the cytosol is the substance in which some chemical reactions occur. The cytosol receives raw materials from the external environment by way of extracellular fluid and converts them into usable energy by decomposition reactions. The cytosol is also the site where new substances are synthesized for cellular use.

ORGANELLES

Despite the numerous chemical activities occurring simultaneously in the cell, there is little interference of one reaction with another. This is because the cell has a system of

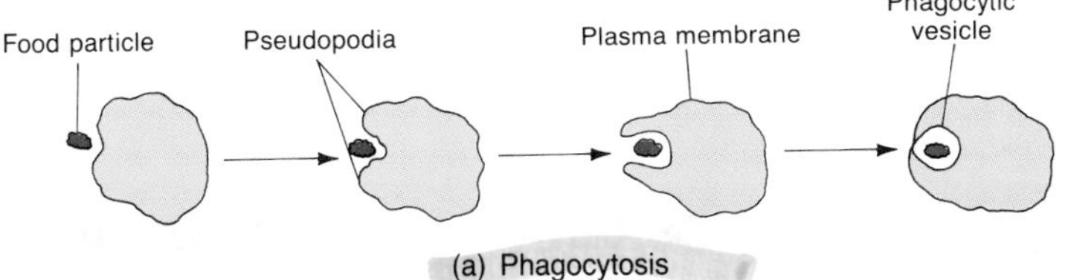

(a) Phagocytosis

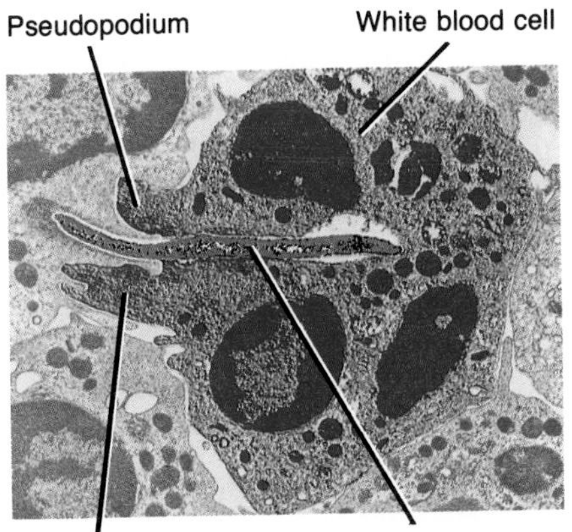

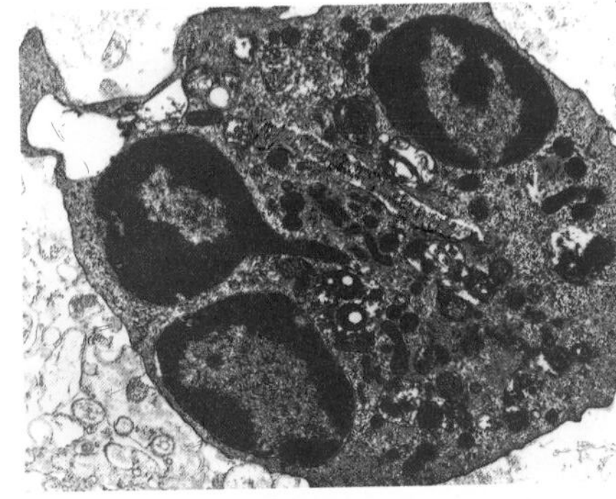

(b) Color-enhanced photomicrographs of phagocytosis

(c) Pinocytosis

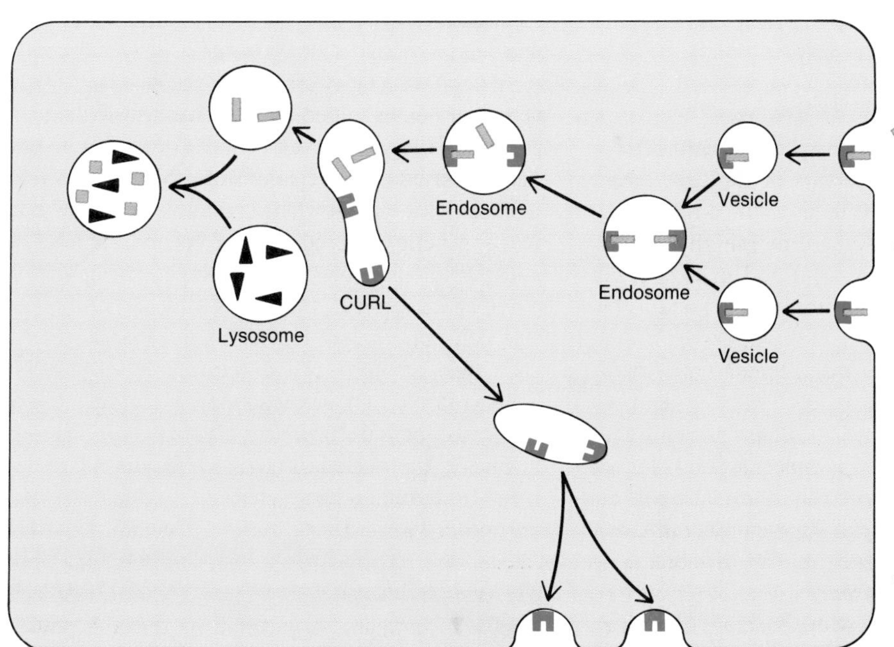

(d) Receptor-mediated endocytosis

FIGURE 2-4
Endocytosis. The photomicrograph in part (b) on the left shows a human white blood cell (neutrophil) engulfing a microbe, and the photomicrograph on the right shows a later stage of phagocytosis in which the engulfed microbe is being destroyed. (Courtesy of Abbott Laboratories.) In the variation of pinocytosis shown on the left in (c), the ingested substance enters a channel formed by the plasma membrane and becomes enclosed in a pinocytic vesicle at the base of the channel. In the variation on the right, the ingested substance becomes enclosed in a pinocytic vesicle that forms at the surface of the cell and detaches.

EXHIBIT 2-1
Summary of Processes by Which Substances Move Across Plasma Membranes

PROCESS	DESCRIPTION
PASSIVE PROCESSES	Substances move on their own down a concentration gradient from an area of higher to lower concentration or pressure; cell does not expend energy by the breakdown of ATP.
Diffusion	Net movement of molecules or ions due to their kinetic energy from an area of higher to lower concentration until an equilibrium is reached.
Facilitated Diffusion	Diffusion of larger molecules across a selectively permeable membrane with the assistance of integral proteins in the membrane that serve as carriers.
Osmosis	Net movement of water molecules due to kinetic energy across a selectively permeable membrane from an area of higher to lower concentration of water until an equilibrium is reached.
Bulk Flow	Movement of large numbers of ions, molecules, or particles in the same direction as a result of forces that push them.
Filtration	Movement of solvents (such as water) and solutes (such as glucose) across a selectively permeable membrane as a result of gravity or hydrostatic (water) pressure from an area of higher to lower pressure.
ACTIVE PROCESSES	Substances move against a concentration gradient from an area of lower to higher concentration; cell must expend energy released by the breakdown of ATP.
Active Transport	Movement of substances, usually ions, across a selectively permeable membrane from a region of lower to higher concentration by an interaction with integral proteins in the membrane.
Endocytosis	Movement of large molecules and particles through plasma membranes in which the membrane surrounds the substance, encloses it, and brings it into the cell. Examples include phagocytosis ("cell eating"), pinocytosis ("cell drinking"), and receptor-mediated endocytosis.
Exocytosis	Export of substances from the cell by reverse endocytosis.

compartmentalization provided by structures called ***organelles*** ("little organs"). These structures are permanent specialized portions of the cell with characteristic shapes that assume specific roles in growth, maintenance, repair, and control. The number and types of organelles vary among different cells, depending on their functions.

NUCLEUS

The ***nucleus*** (NOO-klē-us) is usually a spherical or oval organelle and is the largest structure in the cell (Figure 2-5). It contains the hereditary factors of the cell, called *genes*, which control cellular structure and direct many cellular activities. Most body cells contain a single nucleus, although some, such as mature red blood cells, do not contain a nucleus. Skeletal muscle fibers (cells) and a few other cells contain several nuclei.

The nucleus is enclosed by a ***nuclear envelope (membrane)***, a structure that consists of the following components: (1) inner nuclear membrane, (2) outer nuclear membrane, (3) perinuclear cistern, and (4) nuclear pores (Figure 2-5c). The ***inner*** and ***outer nuclear membranes*** are parallel to each other and resemble the plasma membrane in structure. The space between the two nuclear membranes is called the ***perinuclear cistern (space)***. The outer surface of the nuclear envelope is studded with ribosomes (described next) and is continuous at certain points with granular (rough) endoplasmic reticulum, also described shortly. Spaced at relatively equal distances throughout, the inner and outer nuclear membranes curve together and join to form openings called ***nuclear pores***, which provide a means of communication between the nucleus and cytoplasm. The pores themselves are not actually unrestricted channels since a thin protein diaphragm, which forms a pluglike structure, stretches across the pore. It appears that nuclear pores permit the passage of large molecules such as RNA and various proteins.

Several structures are inside the nucleus. One or more spherical bodies called the ***nucleoli*** (nu-KLĒ-ō-lī) may be present. These structures do not contain a membrane and are composed of aggregations of protein, DNA (deoxyribonucleic acid), and RNA (ribonucleic acid). Whereas DNA is the genetic material (makes up genes), RNA, together with DNA, is involved in protein synthesis. Nucleoli disperse and disappear during cell division and re-form once new cells are formed. Nucleoli are the sites of the synthesis of a type of RNA called ribosomal RNA (rRNA). The RNA is stored in nucleoli and assumes a function in protein synthesis (manufacture). In cells actively engaged in protein synthesis, the nucleolus tends to be quite large and prominent. In a nondividing (interphase) cell, DNA and protein are loosely packed and form a network collectively called ***chromatin***. (During cell division, DNA and protein are condensed and coiled into rod-shaped bodies called ***chromosomes***.) Chromatin may be dispersed and poorly stained (*euchromatin*), which is associated with a high degree of synthetic activity, or clumped and densely stained (*heterochromatin*), which is associated with a low degree of

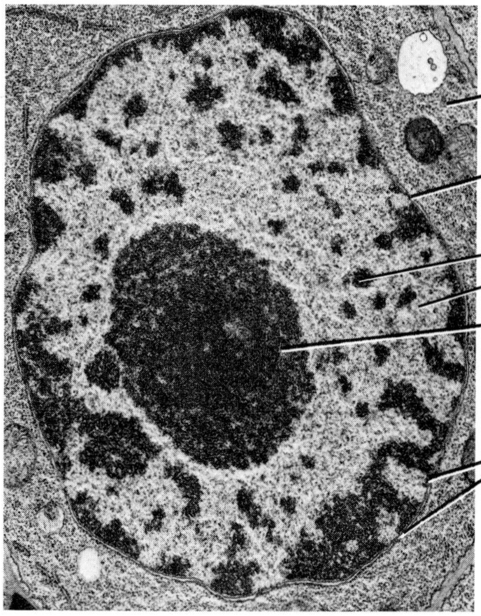

Cytosol
Nuclear pore
Chromatin
Heterochromatin
Euchromatin
Nucleolus
Nuclear envelope

(a) Electron micrograph of cytosol and nucleus, 31,600×

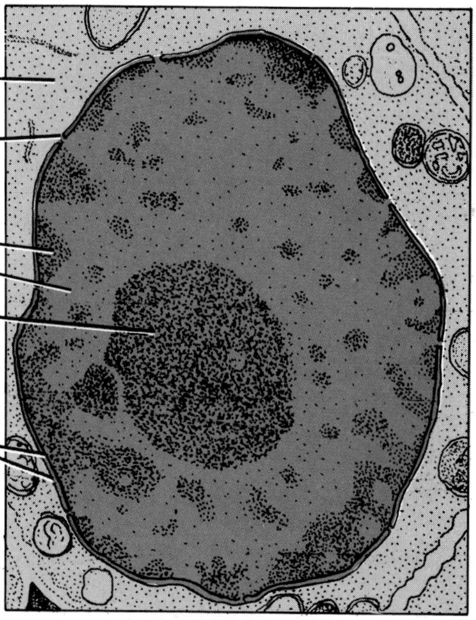

(b) Diagram of Electron Micrograph

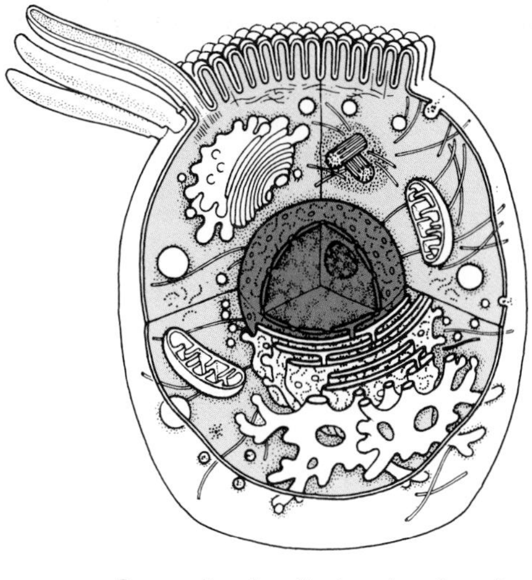

Generalized cell showing location of cytosol and nucleus

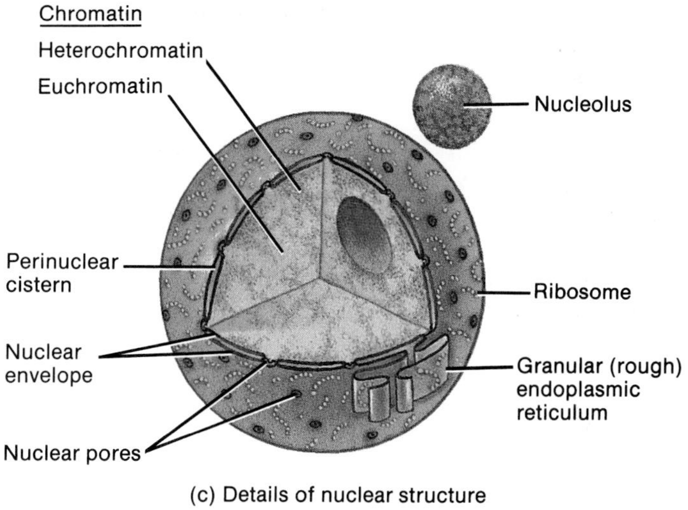

Chromatin
Heterochromatin
Euchromatin
Nucleolus
Perinuclear cistern
Ribosome
Nuclear envelope
Granular (rough) endoplasmic reticulum
Nuclear pores

(c) Details of nuclear structure

Mitochondrion
Nucleus
Plasma membrane

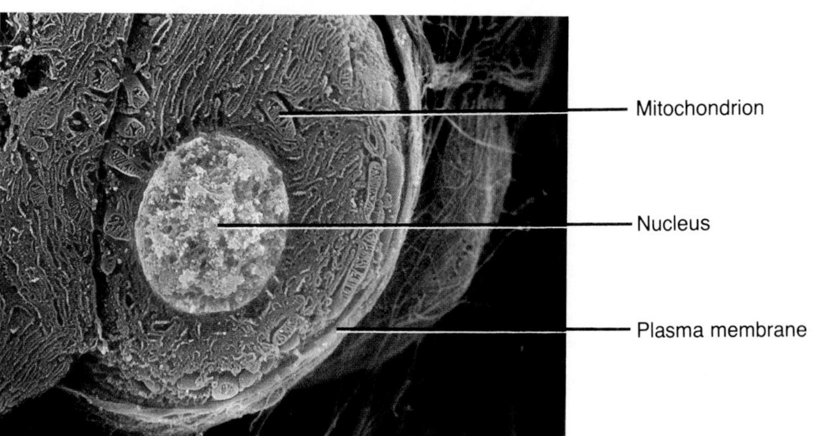

(d) Scanning electron micrograph of a nucleus, 7800×

FIGURE 2-5 Cytosol and nucleus. (a) Copyright © Dr. Myron C. Ledbetter, Biophoto Associates/ Photo Researchers. In (c) the nucleolus has been "lifted" out of the nucleus so that its shape can be seen. (d) Courtesy of L. Nilsson, *The Body Victorious,* Delacorte Press.

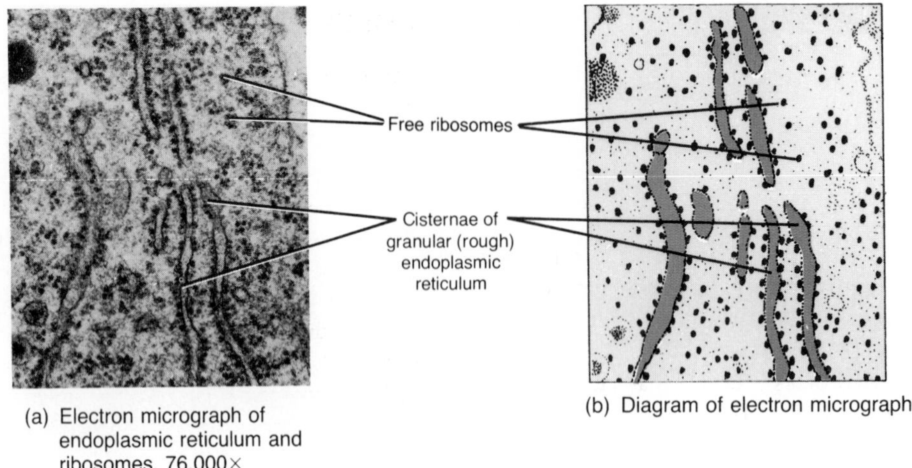

(a) Electron micrograph of endoplasmic reticulum and ribosomes, 76,000×

(b) Diagram of electron micrograph

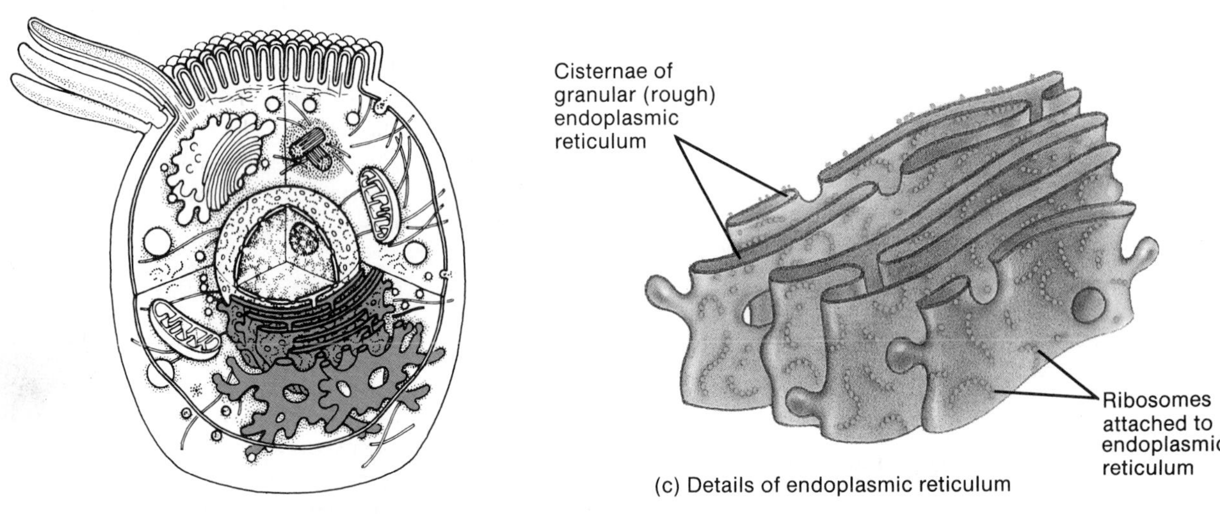

Generalized cell showing location of endoplasmic reticulum and ribosomes

(c) Details of endoplasmic reticulum

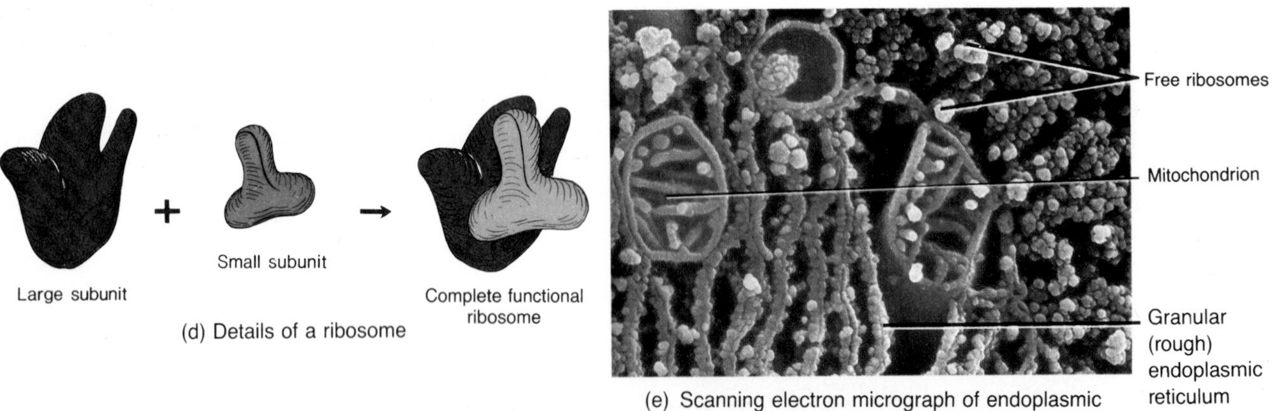

(d) Details of a ribosome

(e) Scanning electron micrograph of endoplasmic reticulum and ribosomes, 60,000×

FIGURE 2-6 Endoplasmic reticulum and ribosomes. (a) Copyright © Dr. Myron C. Ledbetter, Biophoto Associates/Photo Researchers. See if you can find the agranular (smooth) endoplasmic reticulum in Figure 2-7a. (e) Courtesy of L. Nilsson, *The Body Victorious,* Delacorte Press.

synthetic activity. The ***nucleoplasm*** is the semifluid substance in which the chromatin and nucleoli are suspended and serves as a medium for the movement of substances within the nucleus.

RIBOSOMES

Ribosomes (RĪ-bō-sōms) are tiny granules, 25 nm at their largest dimension, that are composed of a type of RNA called ***ribosomal RNA (rRNA)*** and a number of specific ribosomal proteins. The rRNA is manufactured by DNA in the nucleolus. Ribosomes were so named because of their high content of RNA. Structurally, a ribosome consists of two subunits, one about half the size of the other. Scientists have provided three-dimensional models of the structure of a ribosome (Figure 2-6d). Functionally, ribosomes are the sites of protein synthesis: they receive genetic instructions and use them to produce protein. Amino acids are joined one at a time on ribosomes into a protein chain. The completed chain then may fold into a protein molecule that can serve as part of the cell's structure or as an enzyme.

Some ribosomes, called ***free ribosomes,*** are scattered in the cytosol; they have no attachments to other parts of the cell. The free ribosomes occur singly or in clusters, and they are concerned primarily with synthesizing proteins for use inside the cell. Other ribosomes are attached to a cellular structure called the endoplasmic reticulum (ER). These ribosomes are involved in the synthesis of proteins for export from the cell.

ENDOPLASMIC RETICULUM (ER)

Within the cell, there is a system of double membranous channels, called ***cisternae,*** of varying shapes. This system is known as the ***endoplasmic reticulum*** (en′-dō-PLAS-mik re-TIK-yoo-lum), or ***ER*** (Figure 2-6). The channels are continuous with the nuclear envelope.

On the basis of its association with ribosomes, the ER is divided into two types. ***Granular (rough) ER*** is studded with ribosomes; ***agranular (smooth) ER*** is free of ribosomes. Agranular ER is synthesized from granular ER.

Numerous functions are attributed to the ER. It contributes to the mechanical support and distribution of the cytosol. The ER is also involved in the intracellular exchange of materials with the cytosol and provides a surface area for chemical reactions. Various products are transported from one portion of the cell to another via the ER, so the ER is considered an intracellular transportation system. The ER also serves as a storage area for synthesized molecules. Together with a cellular structure called the Golgi complex, the ER assumes a role in the synthesis and packaging of molecules. Ribosomes associated with granular ER synthesize proteins. Agranular ER is associated with lipid synthesis, inactivation or detoxification of certain molecules, and release of calcium ions involved with muscle contraction.

GOLGI COMPLEX

An organelle called the ***Golgi*** (GOL-jē) ***complex*** is generally near the nucleus. In cells with high secretory activity, the Golgi complex is extensive. It consists of four to six flattened membranous sacs, called ***cisternae,*** stacked upon each other like a pile of dishes. Associated with the cisternae are small ***vesicles*** which cluster along the expanded ends of the cisternae (Figure 2-7). On the basis of function, cisternae are designated as *cis, medial,* and *trans* (described shortly).

The principal function of the Golgi complex is to process, sort, package, and deliver proteins to various parts of the cell. Proteins synthesized at ribosomes associated with granular ER are transported into the granular ER (Figure 2-8). While still in the granular ER, sugar molecules are added to the proteins if needed (glycoproteins). Next, the proteins become surrounded by a ***transfer vesicle*** formed by a piece of the granular ER membrane, and the transfer vesicle buds off and fuses with the *cis* cisternae of the Golgi complex. The *cis* cisternae are the ones closest to the granular ER. As a result of this fusion, the proteins enter the Golgi complex. Once inside the Golgi complex, the proteins are transported by other vesicles formed by the Golgi complex from the *cis* cisternae to the *medial* cisternae to the *trans* cisternae, the ones farthest from the granular ER. As the proteins pass in succession through the Golgi cisternae, they become modified in various ways depending on their function and destination. The proteins are sorted and packaged (in vesicles) in the *trans* cisternae. Some vesicles become ***secretory vesicles,*** which move toward the surface of the cell where the protein is released (secretion). The contents of the secretory vesicles are discharged into the extracellular fluid, and the vesicle membrane is incorporated into the plasma membrane. Cells of the gastrointestinal tract that secrete enzymes utilize this mechanism. The secretory vesicle prevents "digestion" of the cytoplasm of the cells by the enzymes as it moves toward the cell surface. Other vesicles that pinch off from the Golgi complex are loaded with special digestive enzymes and remain within the cell. They become cellular structures called lysosomes (described shortly).

The Golgi complex is also associated with lipid secretion. Lipids synthesized by the agranular ER pass through the ER into the Golgi complex. Packaged lipids are discharged at the surface of the cell and enter extracellular fluid. In the course of moving through the cytosol, the vesicle may release lipids into the cytoplasm before being discharged from the cell. These appear in the cytosol as lipid droplets. Among the lipids secreted in this manner are steroids.

LYSOSOMES

When viewed under an electron microscope, ***lysosomes*** (LĪ-so-sōms; *lysis* = dissolution; *soma* = body) appear as membrane-enclosed spheres (Figure 2-9). They are formed from Golgi complexes and have a single membrane. They contain

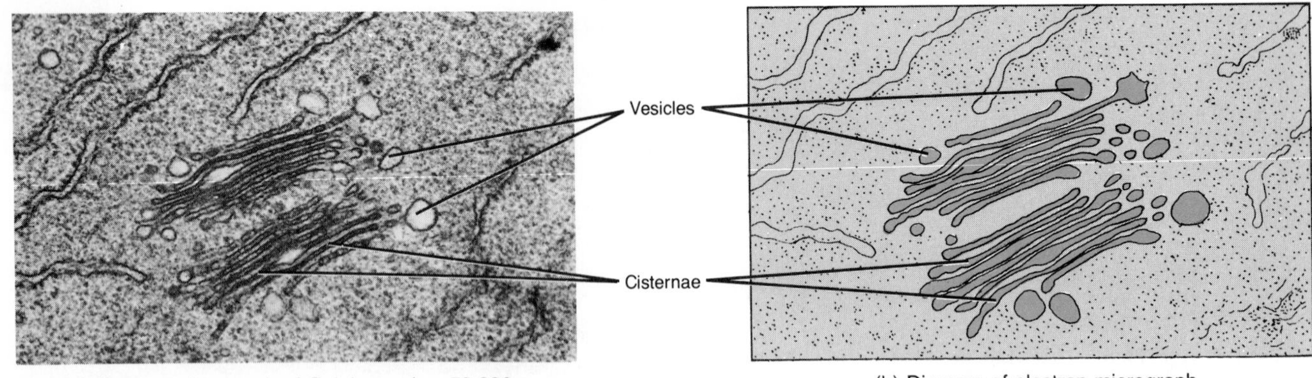

(a) Electron micrograph of Golgi complex, 78,000×

(b) Diagram of electron micrograph

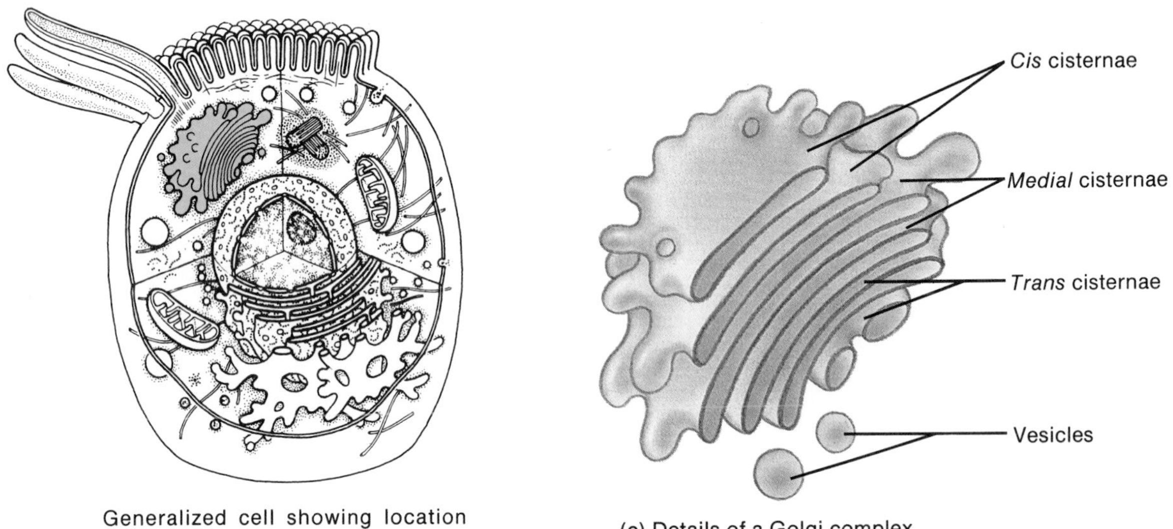

Generalized cell showing location of Golgi complex

(c) Details of a Golgi complex

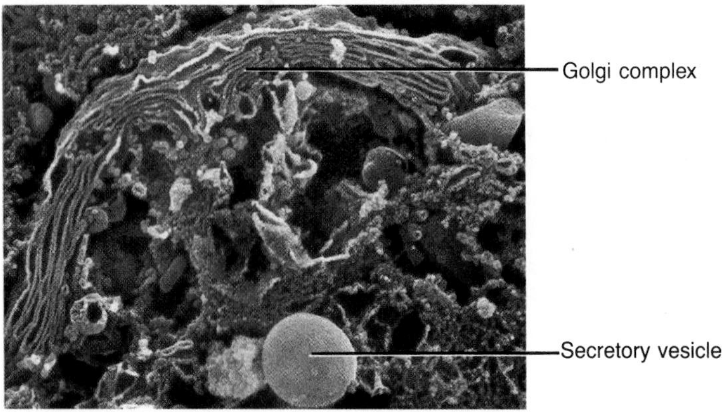

(d) Scanning electron micrograph of Golgi complex, 20,000×

FIGURE 2-7 Golgi complex. (a) Copyright © Dr. Myron C. Ledbetter, Biophoto Associates/Photo Researchers. (d) Courtesy of L. Nilsson, *The Body Victorious,* Delacorte Press.

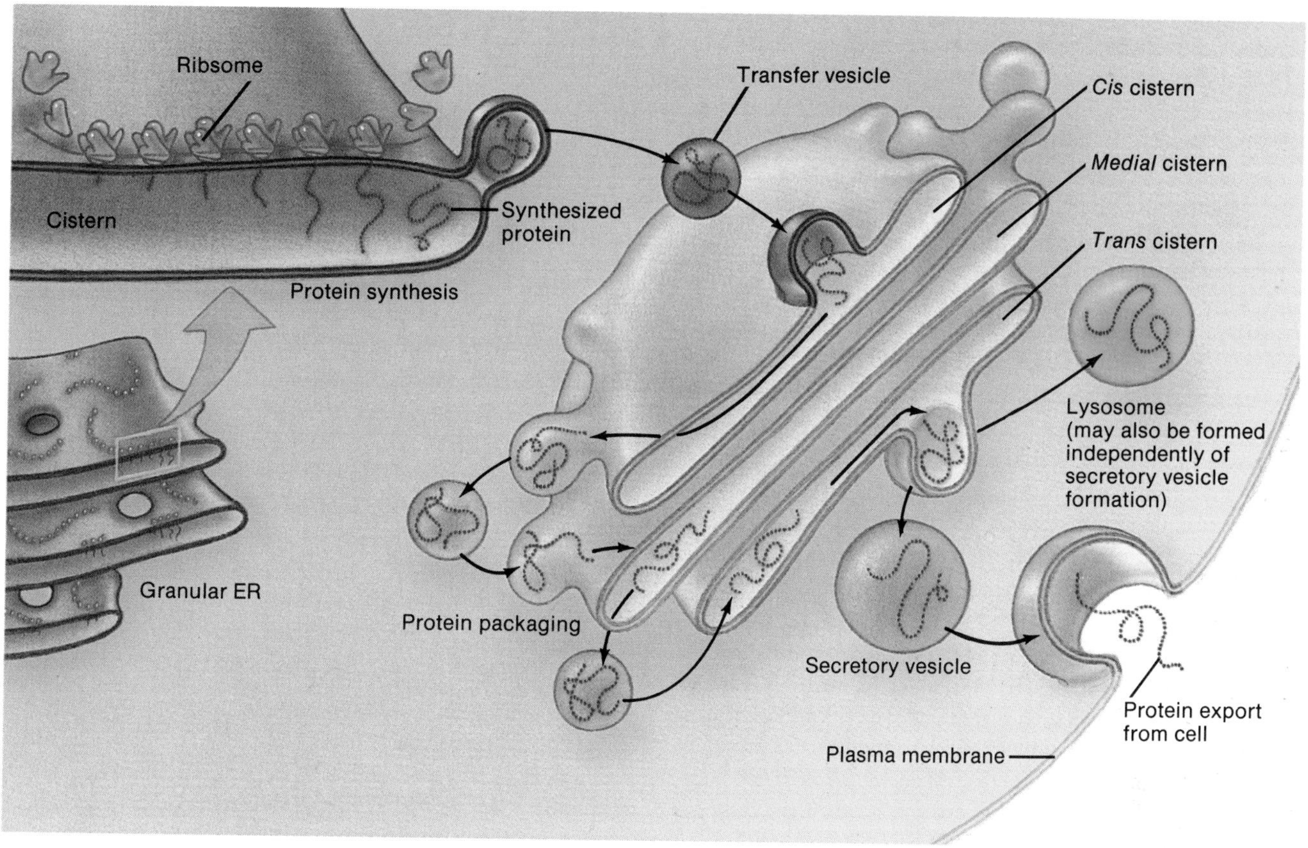

Ribsome
Transfer vesicle
Cis cistern
Medial cistern
Cistern
Synthesized protein
Trans cistern
Protein synthesis
Lysosome (may also be formed independently of secretory vesicle formation)
Granular ER
Protein packaging
Secretory vesicle
Protein export from cell
Plasma membrane

FIGURE 2-8 Packaging of synthesized protein for export from the cell.

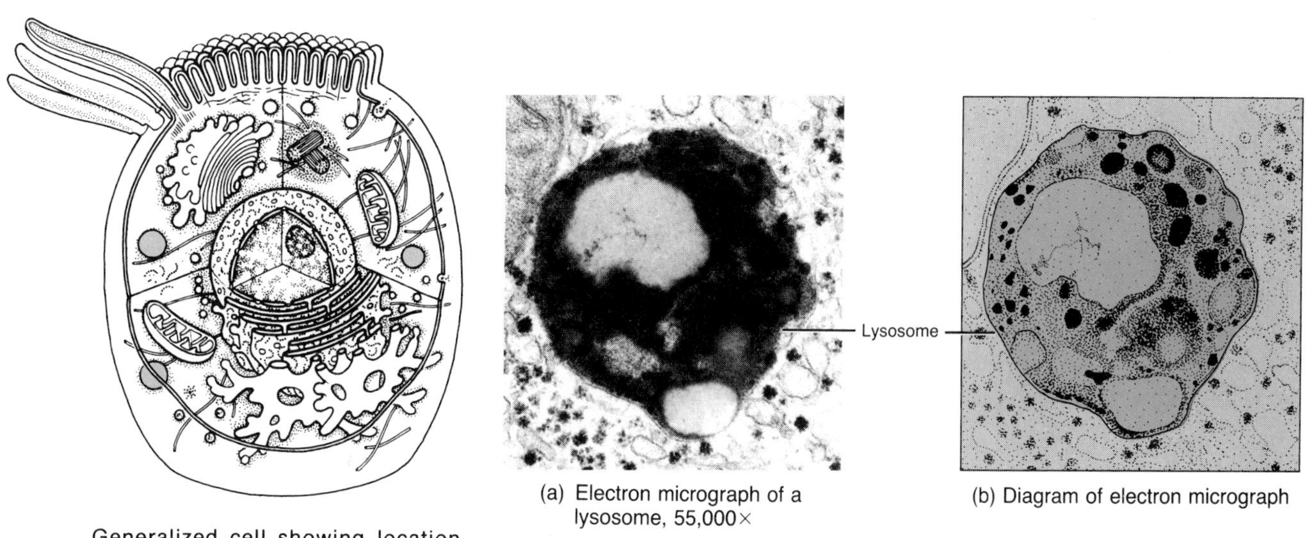

Generalized cell showing location of lysosomes

(a) Electron micrograph of a lysosome, 55,000×

(b) Diagram of electron micrograph

FIGURE 2-9 Lysosome. (a) Courtesy of F. Van Hoof, Université Catholique de Louvain.

powerful digestive (hydrolytic) enzymes capable of breaking down many kinds of molecules. You will see in Chapter 18 that Tay-Sachs disease results from a deficiency of a lysosomal enzyme. These enzymes are also capable of digesting bacteria and other substances that enter the cell in phagocytic vesicles. White blood cells, which ingest bacteria by phagocytosis, contain large numbers of lysosomes.

Lysosomal enzymes are believed to be synthesized on the granular ER and then transported to the Golgi complex, probably in an inactive form to prevent the unwanted digestion of structures through which they pass. Lysosomes develop as buds that pinch off the ends of Golgi cisternae. When first formed, the lysosome is referred to as a ***primary lysosome,*** meaning that it contains enzymes but is not yet engaged in digestive activity. In order to participate in digestion, a primary lysosome must fuse with a membrane-bound vacuole or other organelles containing food particles. A ***vacuole*** is formed from the plasma membrane and, in animal cells, frequently functions in temporary storage or transportation. As a result of this fusion, the primary lysosome becomes known as a ***secondary lysosome***—that is, a lysosome engaged in digestive activity. In the digestive process, the contents of the vacuole are broken down into smaller and smaller components. Eventually, the products of digestion are small enough to pass out of the lysosome into the cytosol of the cell to be recycled for the synthesis of various molecules needed by the cell.

Lysosomes function in intracellular digestion in a number of ways. For example, during phagocytosis lysosomal enzymes digest the solid material contained in phagocytic vesicles. A similar process occurs in pinocytosis and receptor-mediated endocytosis. Lysosomes also use their enzymes to recycle the cell's own molecules. A lysosome can engulf another organelle, digest it, and return the digested components to the cytosol for reuse. In this way, worn-out cellular structures are continually renewed. The process by which worn-out organelles are digested is called ***autophagy*** (aw-TOF-a-jē; *auto* = self; *phagio* = to eat). A human liver cell can recycle about half its contents in a week. Under normal conditions the lysosomal membrane is impermeable to the passage of enzymes into the cytosol, thereby preventing digestion of the cellular contents. However, under certain conditions, ***autolysis*** or self-destruction by lysosomes occurs. For example, during human embryological development, the fingers and toes are webbed. The normal development of individual fingers and toes requires the selective removal of the cells of the web between the digits. This process involves autolysis, in which lysosomal enzymes digest the tissue between the digits. It is actually a programmed destruction of cells. Because of this activity, lysosomes are sometimes referred to as "suicide packets."

Lysosomes also function in extracellular digestion. Lysosomal enzymes released at sites of injury help to digest away cellular debris and prepare the injured area for effective repair. Prior to fertilization, the head of a sperm cell releases lysosomal enzymes capable of digesting a barrier around the egg so that the sperm cell can penetrate it and fertilization can take place. Also, release of lysosomal enzymes from a cell may be the process responsible for bone removal. In bone reshaping, especially during the growth process, special bone-destroying cells, called ***osteoclasts,*** secrete extracellular enzymes that dissolve bone. Bone tissue cultures given excess amounts of vitamin A remove bone apparently through an activation process involving lysosomes.

CLINICAL APPLICATION

Lysosomes and Steroids

Although an adequate intake of vitamin A is required for lysosomal function, animals overfed on vitamin A develop spontaneous fractures, suggesting greatly increased lysosomal activity. On the other hand, cortisone and hydrocortisone, ***steroid hormones*** produced by the adrenal gland, ***have a stabilizing effect on lysosomal membranes.*** The steroid hormones are well known for their antiinflammatory properties, which suggests that they reduce destructive cellular activity by lysosomes.

PEROXISOMES

Organelles similar in structure to lysosomes, but smaller, are called ***peroxisomes*** (pe-ROKS-i-sōms). See Figure 2-1. They are so named because they usually contain one or more enzymes that use molecular oxygen to oxidize (remove hydrogen atoms from) various organic substances. Such reactions produce hydrogen peroxide (H_2O_2). One of the enzymes in peroxisomes, called *catalase,* uses the H_2O_2 generated by other enzymes to oxidize a variety of other substances, including phenol, formic acid, formaldehyde, and alcohol, toxic substances that may enter the bloodstream. This type of oxidation is especially important in liver and kidney cells whose peroxisomes detoxify the potentially harmful substances.

MITOCHONDRIA

Small, rod-shaped, or filamentous structures called ***mitochondria*** (mī-tō-KON-drē-a) appear throughout the cell. Because of their function in generating energy, they are referred to as the "powerhouses" of the cell. Within mitochondria, energy is transferred from carbon compounds, such as glucose, to ATP. When sectioned and viewed under an electron microscope, each reveals an elaborate internal organization (Figure 2-10). A mitochondrion consists of two membranes, each of which is similar in structure to the plasma membrane. The outer mitochondrial membrane is smooth, but the inner membrane is arranged in a series of folds called ***cristae.*** The central cavity of a mitochondrion enclosed by the inner membrane and cristae is called the ***matrix.*** Within the matrix are ***matrix granules,*** which are probably accumulations of calcium phosphate. Their

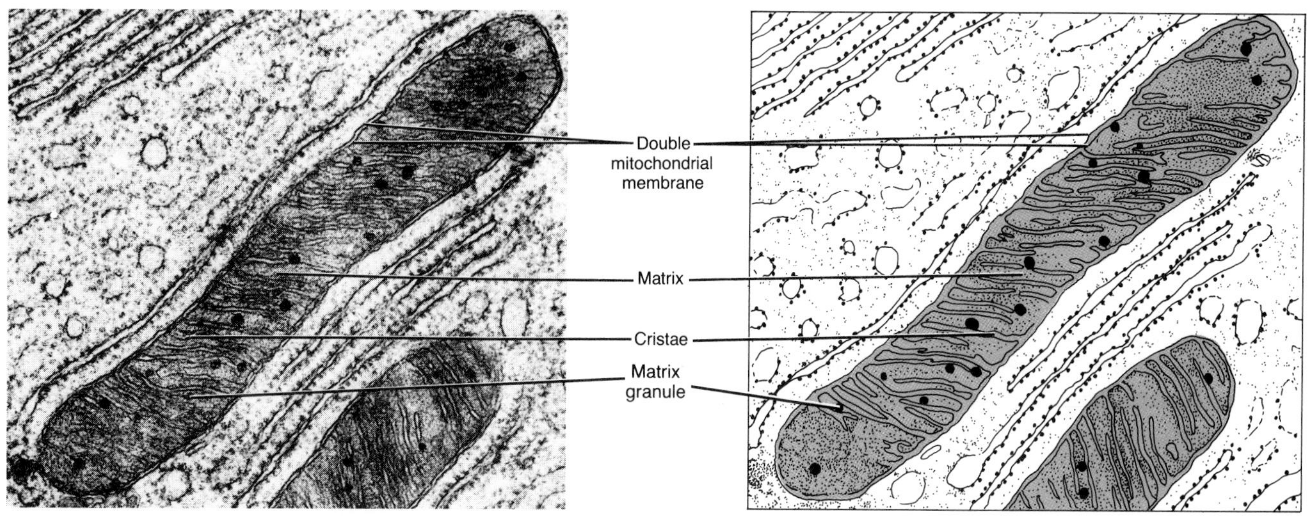

(a) Electron micrograph

(b) Diagram of electron micrograph

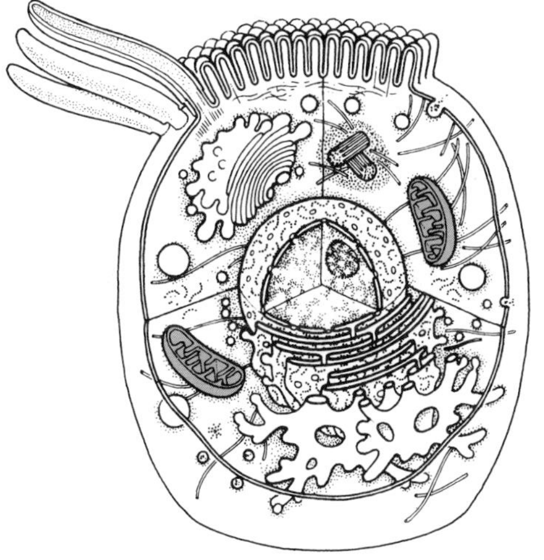

Generalized cell showing location of mitochondria

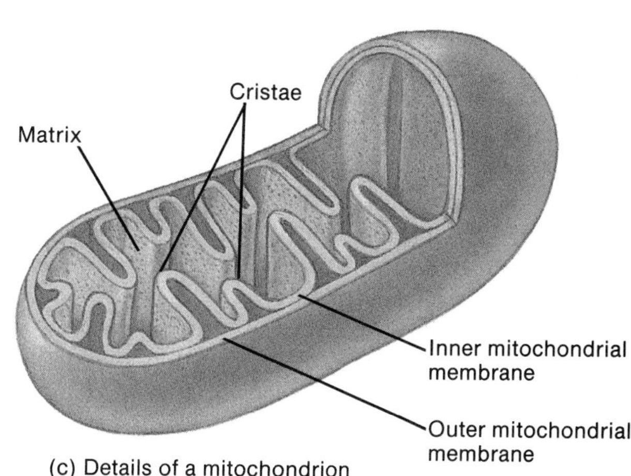

(c) Details of a mitochondrion

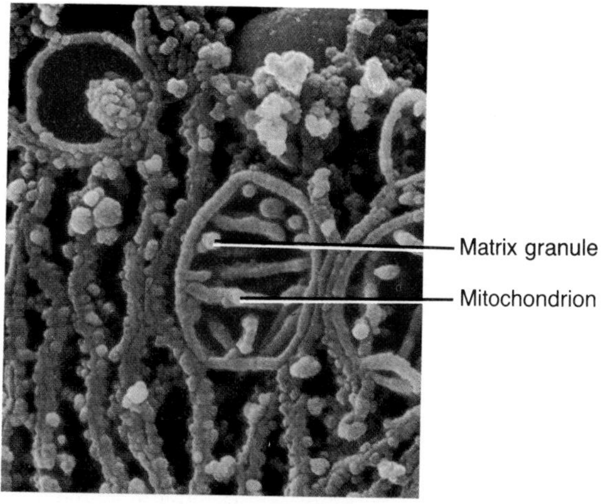

(d) Scanning electron micrograph of a mitochondrion, 30,000×

FIGURE 2-10 Mitochondria. (a) Courtesy of Lester V. Bergman & Associates, Inc. (d) Courtesy of L. Nilsson, *The Body Victorious,* Delacorte Press.

presence helps to maintain a typically low level of calcium ions in the cell.

Because of the nature and arrangement of the cristae, the inner membrane provides an enormous surface area for chemical reactions, collectively referred to as ***cellular respiration.*** Enzymes involved in energy-releasing reactions that form ATP are located on the cristae. Active cells, such as muscle, liver, and kidney tubule cells, have a large number of mitochondria because of their high energy expenditure.

Mitochondria are self-replicative—that is, they can divide to form new ones. The replication process is controlled by DNA that is incorporated into the mitochondrial structure. Self-replication usually occurs in response to increased cellular need for ATP. The energy-releasing reactions only occur if oxygen (O_2) is present and result in the catabolism of nutrient molecules. Some of the energy released as these nutrients are broken down is used to re-form the high-energy molecule ATP.

THE CYTOSKELETON

The cytosol has a complex internal structure, containing a series of exceedingly small microfilaments, microtubules, and intermediate filaments, together referred to as the ***cytoskeleton*** (Figure 2-11).

Microfilaments are rodlike structures that are 6 nm in diameter and are of variable length. Microfilaments consist of a protein called ***actin.*** In muscle tissue, actin microfilaments (*thin myofilaments*) and myosin microfilaments (*thick myofilaments*) are involved in the contraction of muscle fibers (cells). This mechanism is described in Chapter 9. In nonmuscle cells, microfilaments help to provide support and shape and assist in the movement of entire cells (phagocytes and cells of developing embryos) and movements within cells (secretion, phagocytosis, pinocytosis).

Microtubules are relatively straight, slender, cylindrical structures that range in diameter from 18 to 30 nm, usually averaging about 24 nm. They consist of a protein called ***tubulin.*** Microtubules and microfilaments help to provide support and shape for cells. Microtubules may also form conducting channels through which various substances can move throughout the cytosol. This mechanism has been studied extensively in nerve cells. Microtubules also assist in the movement of pseudopodia that are characteristic of phagocytes. As you will see shortly, microtubules form the structure of flagella and cilia (cellular appendages involved in motility), centrioles, and the mitotic spindle (structures that are involved in cell division).

Intermediate filaments range between 8 and 12 nm in diameter. An example is neurofilaments found in nerve cells (Chapter 16). Although the functions of intermediate filaments are not completely understood, it appears that they provide structural reinforcement in some cells and assist in contraction of others.

Some investigators believe that the microfilaments, microtubules, and intermediate filaments, as well as other cytoplasmic components, are held together by a three-dimensional meshwork of fine filaments called ***microtrabeculae*** (mī′-krō-tra-BEK-yoo-lē), which are about 10–15 nm thick. Together, all the microtrabeculae constitute the ***microtrabecular lattice*** (Figure 2-11). This lattice is be-

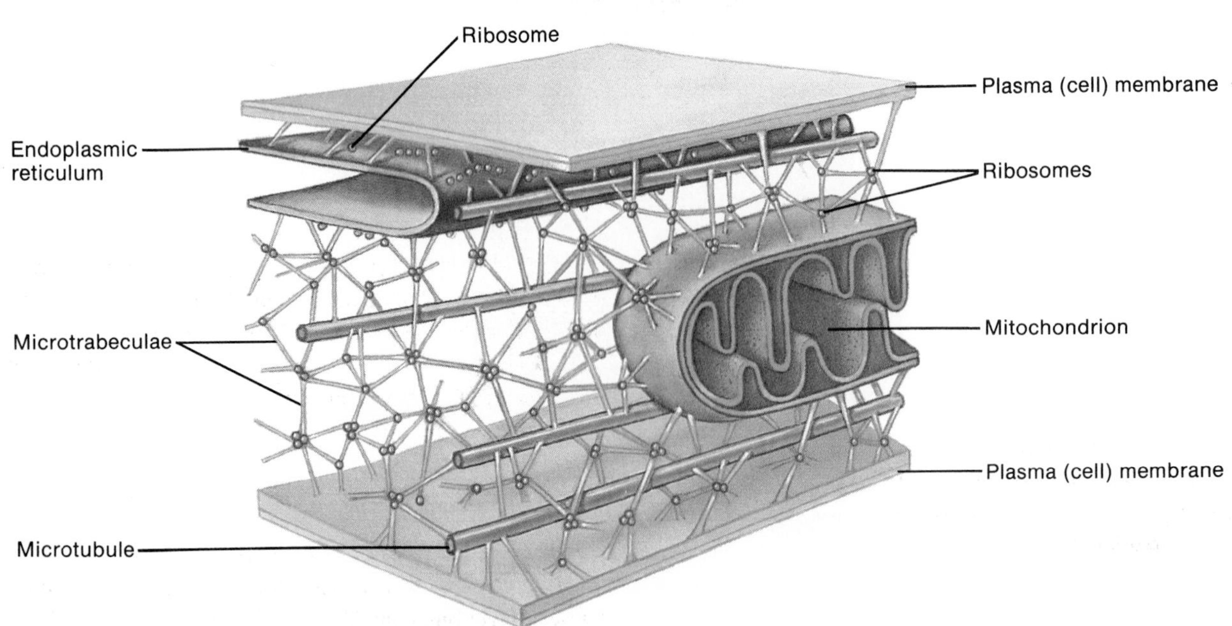

FIGURE 2-11 Microtrabecular lattice. (After Keith R. Porter and Jonathan B. Tucker).

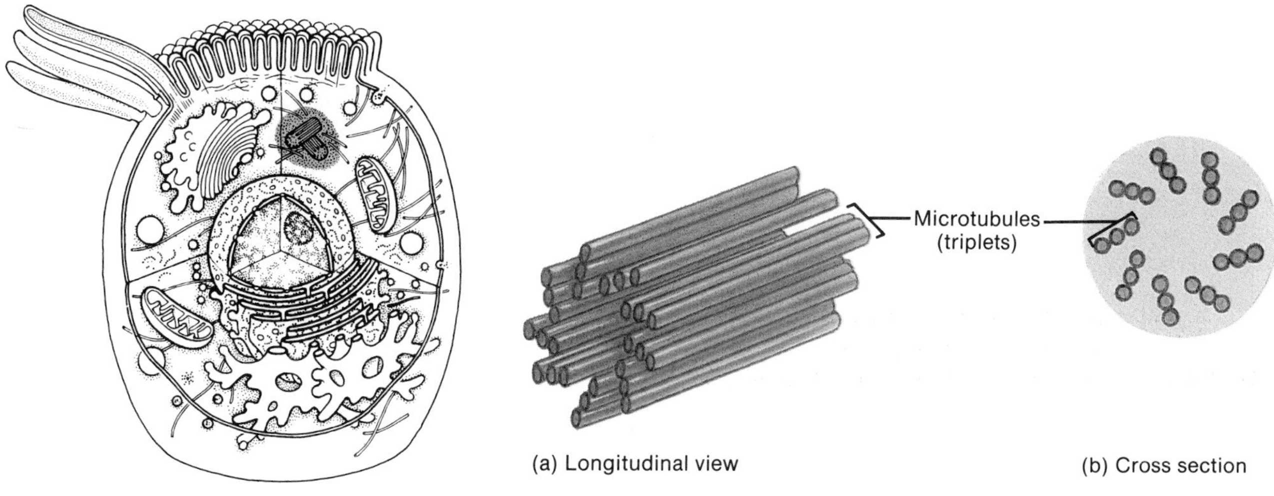

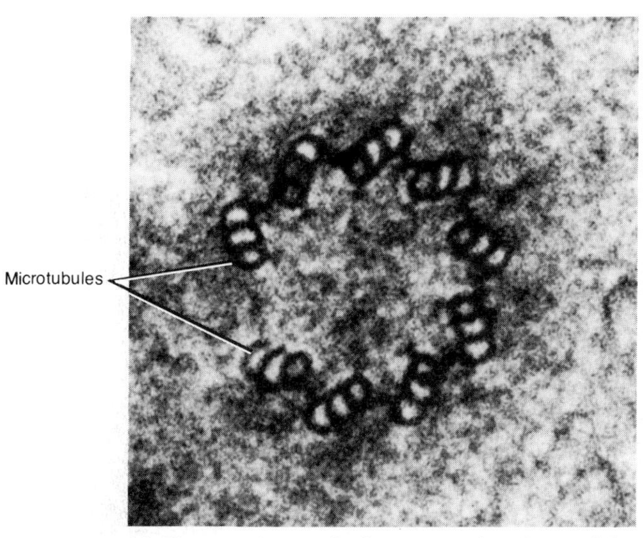

(c) Electron micrograph of a cross section of a centriole, 67,000×

FIGURE 2-12 Centrosome and centrioles. (c) Courtesy of Biophoto Associates/Photo Researchers.

lieved to provide organization for chemical reactions that occur within the cytoplasm and to assist in the transport of substances through the cell.

CENTROSOME AND CENTRIOLES

A dense area of cytosol, generally spherical and located near the nucleus, is called the ***centrosome*** (***centrosphere***). Within the centrosome is a pair of cylindrical structures called ***centrioles*** (Figure 2-12). Each centriole is composed of nine triplet clusters of microtubules arranged in a circular pattern. Centrioles lack the two central single microtubules found in flagella and cilia. The two centrioles are situated so that the long axis of one is at right angles to the long axis of the other. Centrosomes assume a role in cell reproduction by serving as centers about which the mitotic spindle involved in chromosome movement is organized. This role will be described shortly as part of cell division. Like mitochondria, centrioles contain DNA that controls their self-replication.

FLAGELLA AND CILIA

Some body cells possess projections for moving the entire cell or for moving substances along the surface of the cell. These projections contain cytosol and are covered by the plasma membrane. If the projections are few and long in proportion to the size of the cell (typically occurring singly or in pairs), they are called ***flagella*** (fla-JEL-a). The only example of a flagellum in the human body is the tail of a sperm cell, used for locomotion (see Figure 25-5). If the projections are numerous and short, resembling many hairs, they are called ***cilia*** (SIL-ē-a). In humans, ciliated cells of the respiratory tract move mucus that has trapped foreign particles over the surface of the tissue (see Figure 22-5b). Electron microscopy has revealed no fundamental structural difference between cilia and flagella. Both consist of nine pairs of microtubules that form a ring around two single microtubules in the center.

CELL INCLUSIONS

Cell inclusions are a large and diverse group of substances produced by cells, some of which have recognizable shapes. These products are principally organic and may appear or disappear at various times in the life of the cell. Examples include melanin, glycogen, and lipids. ***Melanin*** is a pigment stored in certain cells of the skin, hair, and eyes. It protects the body by screening out harmful ultraviolet rays from the sun. ***Glycogen*** is a polysaccharide that is stored in the

EXHIBIT 2-2

Cell Parts and Their Functions

PART	FUNCTIONS
PLASMA MEMBRANE	Protects cellular contents; makes contact with other cells; provides receptors for hormones, enzymes, and antibodies; mediates the entrance and exit of materials.
CYTOSOL	Serves as the ground substance in which chemical reactions occur.
ORGANELLES	
Nucleus	Contains genes and nucleoli and controls cellular activities.
Ribosomes	Sites of protein synthesis.
Endoplasmic Reticulum (ER)	Contributes to mechanical support; facilitates intracellular exchange of materials with cytosol; provides a surface area for chemical reactions; provides a pathway for transporting chemicals; serves as a storage area; together with Golgi complex synthesizes and packages molecules for export; ribosomes associated with granular ER synthesize proteins; agranular ER synthesizes lipids, detoxifies certain molecules, and releases calcium ions involved with muscle contraction.
Golgi Complex	Packages synthesized proteins for secretion in conjunction with endoplasmic reticulum; forms lysosomes; secretes lipids; combines carbohydrates with proteins to form glycoproteins for secretion.
Lysosomes	Digest substances and foreign microbes; may be involved in bone removal.
Peroxisomes	Contain several enzymes related to hydrogen peroxide formation and disposal.
Mitochondria	Sites of production of ATP.
Microfilaments	Form part of cytoskeleton; involved in muscle fiber (cell) contraction; provide support and shape; assist in cellular and intracellular movement.
Microtubules	Form part of cytoskeleton; provide support and shape; form intracellular conducting channels; assist in cellular movement; form the structure of flagella, cilia, centrioles, and mitotic spindle.
Intermediate Filaments	Form part of cytoskeleton; probably provide structural reinforcement in some cells.
Centrosomes	Help organize mitotic spindle during cell division.
Flagella and Cilia	Allow movement of entire cell (flagella) or movement of particles trapped in mucus along surface of cell (cilia).
INCLUSIONS	Assorted functions: melanin (pigment in skin, hair, eyes) screens out ultraviolet rays; glycogen (stored glucose) can be decomposed to provide energy; lipids (stored in fat cells) can be decomposed to produce energy.

liver, skeletal muscle fibers (cells), and the vaginal mucosa. When the body requires quick energy, liver cells can break down the glycogen into glucose and release it into the blood. ***Lipids,*** which are stored in adipocytes (fat cells), may be decomposed for producing energy.

The major parts of the cell and their functions are summarized in Exhibit 2-2.

NORMAL CELL DIVISION

Most of the cell activities discussed thus far maintain the life of the cell on a day-to-day basis. However, cells become damaged, diseased, or worn out—and die. New cells must be produced as replacements and for growth. In addition, sperm and egg cells must be produced by cell division. In a 24-hour period, the average adult loses billions of cells from different parts of the body. Obviously, these cells must be replaced. Cells that have a short life span, such as the cells of the outer layer of skin, are continually being replaced.

Cell division is the process by which cells reproduce themselves. It consists of a nuclear division and a cytoplasmic division. Because nuclear division can be of two types, two kinds of cell division are recognized.

In the first kind of division, often called ***somatic cell division,*** a single starting cell called a ***parent cell*** divides to produce two identical cells called ***daughter cells.*** This process consists of a nuclear division called ***mitosis*** and a cytoplasmic division called ***cytokinesis*** (*kinesis* = motion). The process ensures that each daughter cell has the same *number* and *kind* of chromosomes as the original parent cell. After the process is complete, the two daughter cells have the same hereditary material and genetic potential as the parent cell. This kind of cell division results in an increase in the number of body cells and is the means by which dead or injured cells are replaced and new ones are added for body growth.

The second type of cell division is called ***reproductive cell division*** and is the mechanism by which sperm and egg cells are produced, cells required to form a new organism. The process consists of a nuclear division called ***meiosis*** plus ***cytokinesis.*** We will first discuss somatic cell division.

SOMATIC CELL DIVISION

When a cell reproduces, it must replicate (produce duplicates of) its chromosomes so that its hereditary traits may be passed on to succeeding generations of cells. A ***chromosome*** (*chromo* = colored) is a highly coiled DNA molecule that is partly covered by protein. The protein causes changes in the length and thickness of the chromosome. Hereditary information is contained in the DNA portion of the chromosome in units called ***genes.*** Humans have about 100,000 functional genes. However, at any given time, only a fraction of the genes are operating.

Before taking a look at the relation between chromosomes and cell division, it is necessary to examine briefly the structure of DNA, the basic component of chromosomes.

A molecule of DNA is a chain composed of repeating units called ***nucleotides.*** Each nucleotide of DNA consists of three basic parts (Figure 2-13): (1) It contains one of four possible ***nitrogenous bases,*** which are ring-shaped structures containing atoms of C, H, O, and N. The nitrogenous bases found in DNA are named adenine, thymine,

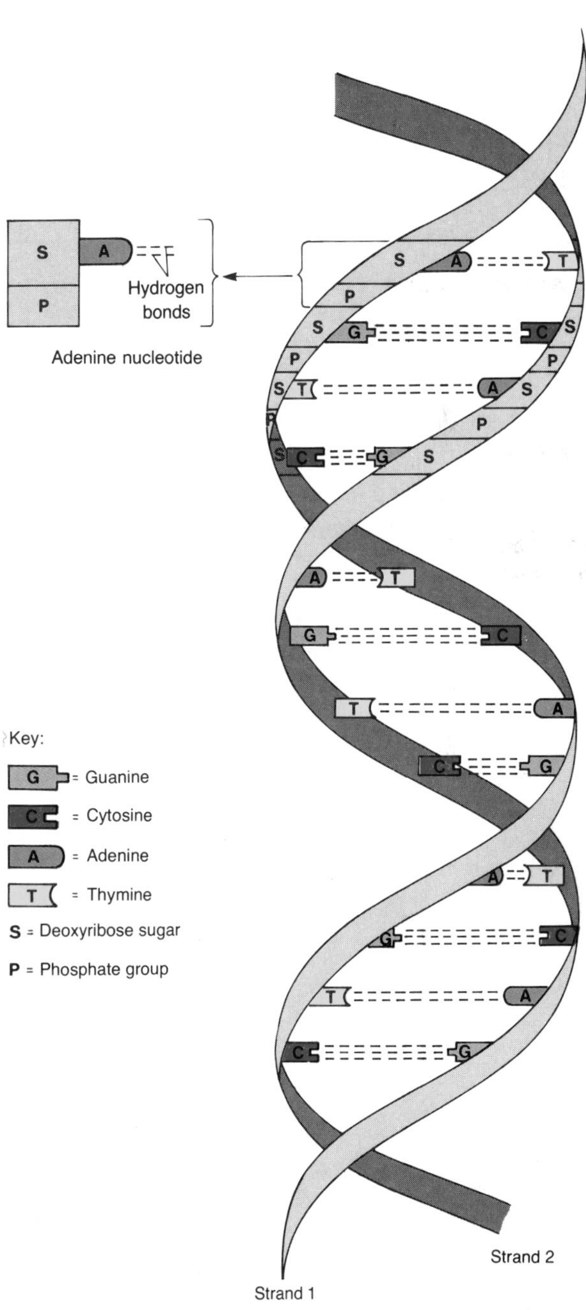

FIGURE 2-13 DNA molecule.

cytosine, and guanine. (2) It contains a sugar called ***deoxyribose.*** (3) It has a phosphoric acid called the ***phosphate group.*** The nucleotides are named according to the nitrogenous base that is present. Thus, a nucleotide containing thymine is called a ***thymine nucleotide,*** one containing adenine is called an ***adenine nucleotide,*** and so on.

The chemical composition of the DNA molecule was known before 1900, but it was not until 1953 that a model of the organization of the chemicals was constructed. This model was proposed by J. D. Watson and F. H. C. Crick on the basis of data from many investigations. Figure 2-13b shows the following structural characteristics of the DNA molecule: (1) The molecule consists of two strands with crossbars. The strands twist about each other in the form of a ***double helix*** so that the shape resembles a twisted ladder. (2) The uprights of the DNA ladder consist of alternating phosphate groups and the deoxyribose portions of the nucleotides. (3) The rungs of the ladder contain paired nitrogenous bases. As shown, adenine always pairs off with thymine, and cytosine always pairs off with guanine. It is estimated that the nucleus of a human diploid cell (cell with two sets of chromosomes) contains about three billion nitrogenous base pairs.

Interphase

When a cell is between divisions it is said to be in ***interphase (metabolic phase).*** It is during this stage that the replication (synthesis) of chromosomes, centrosomes, and centrioles occurs, and the RNA and protein needed to produce structures required for doubling all cellular components are manufactured. The period of interphase during which chromosomes are replicated is referred to as the ***S*** (for synthesis) ***period.***

The S period is preceded by a ***G_1*** (for gap or growth) ***period,*** during which cells are engaged in growth, metabolism, and the production of substances required for division. Following chromosomal replication in the S period, there is another G period called the ***G_2 period.*** During this gap period there is activity similar to that occurring in the G_1 period. Since the G periods are times during which there are no events related to chromosomal replication, they are thought of as gaps in the divisional cycle (see Figure 2-16).

When DNA replicates, its helical structure partially uncoils (Figure 2-14). Those portions of DNA that remain coiled stain darker than the uncoiled portions. This unequal distribution of stain causes the DNA to appear as a granular mass called ***chromatin*** (Figure 2-15a), which consists of euchromatin and heterochromatin. During uncoiling, DNA separates at the points where the nitrogenous bases are connected. In the presence of an enzyme, each exposed nitrogenous base then picks up a complementary nitrogenous base (with associated sugar and phosphate group). This uncoiling and complementary base pairing continues until each of the two original DNA strands is matched and joined with one newly formed DNA strand. The original DNA molecule has become two DNA molecules.

A microscopic view of a cell during interphase shows a clearly defined nuclear membrane, nucleoli, nucleoplasm, and chromatin. Once a cell completes its replication of DNA, centrosomes, and centrioles and its production of RNA and proteins during interphase, mitosis begins.

Mitosis

The events that take place during mitosis and cytokinesis are plainly visible under a microscope after the cells have been stained in the laboratory.

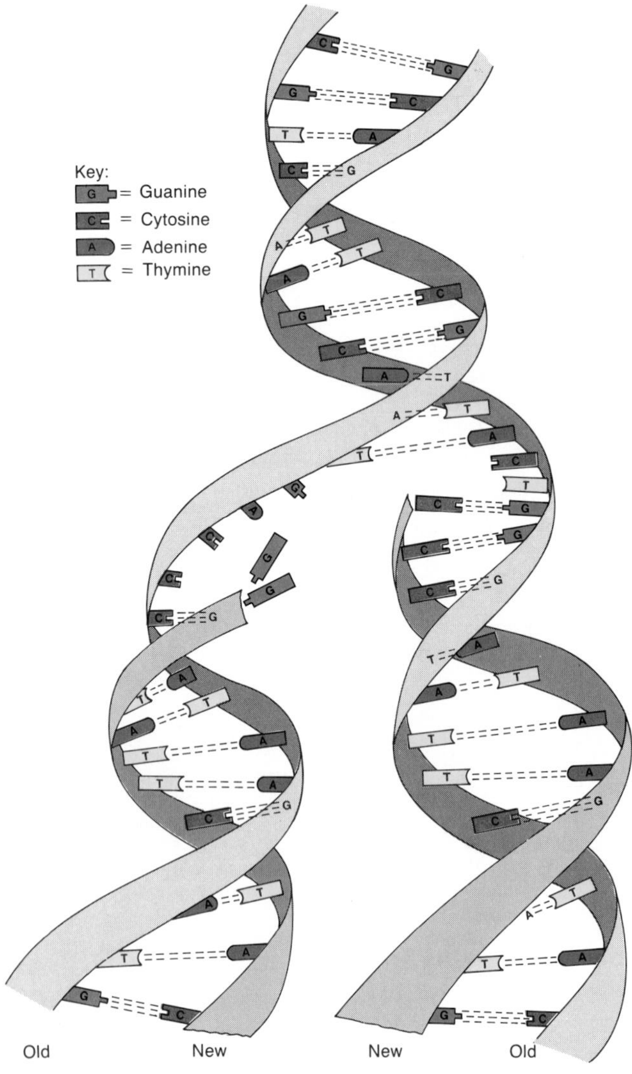

FIGURE 2-14 Replication of DNA. The two strands of the double helix separate by breaking the hydrogen bonds between nucleotides. New nucleotides attach at the proper sites, and a new strand of DNA is paired off with each of the original strands. After replication, the two DNA molecules, each consisting of a new and an old strand, return to their helical structure.

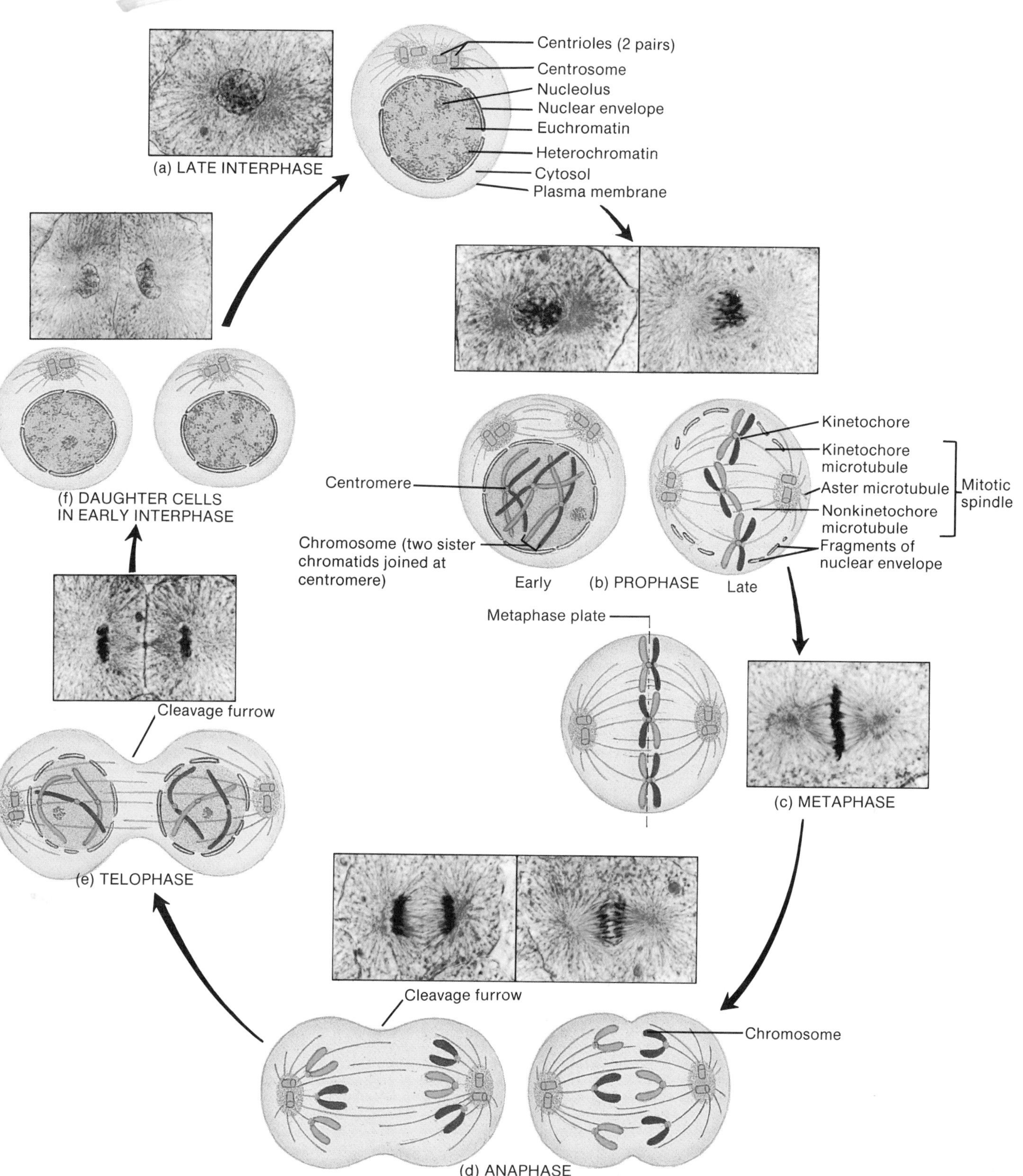

FIGURE 2-15 Cell division: mitosis and cytokinesis. Photomicrographs and diagrammatic representations of the various stages of cell division in whitefish eggs. Read the sequence starting at (a) and move clockwise until you complete the cycle. (Photographs by Carolina Biological Supply Company.)

The process called ***mitosis*** is the distribution of the two sets of chromosomes into two separate and equal nuclei following the replication of the chromosomes of the parent nucleus. For convenience, biologists divide the process into four stages: prophase, metaphase, anaphase, and telophase. These are arbitrary classifications. Mitosis is actually a continuous process, one stage merging imperceptibly into the next.

■ ***Prophase*** The first stage of mitosis is called ***prophase*** (*pro* = before) (Figure 2-15b). During early prophase, the chromatin condenses and shortens into chromosomes. This condensation is necessary because long strands would become tangled and this would interfere with their movement. Since DNA replication took place during interphase, each prophase chromosome is actually composed of a pair of identical double-stranded DNA molecules called ***chromatids.*** Each chromatid pair is held together by a small spherical body called a ***centromere,*** which is required for the proper segregation of chromosomes. Attached to the outside of each centromere is a protein complex referred to as a ***kinetochore*** (ki-NET-ō-kor), whose function will be described shortly.

Later in prophase, the nucleoli disperse and the nuclear envelope breaks up. In addition, a pair of centrioles and its centrosome each move to opposite poles (ends) of the cell. As they do so, the centrosomes start to form the ***mitotic spindle,*** a football-shaped assembly of microtubules that are responsible for the movement of chromosomes. The lengthening microtubules between centrosomes pushes the centrosomes to the poles of the cell so that the spindle extends from pole to pole. As the mitotic spindle continues to develop, three types of microtubules are recognized on the basis of where their ends terminate: (1) ***nonkinetochore microtubules*** grow from the centrosomes, extend inward, but do not bind to kinetochores; (2) ***kinetochore microtubules*** grow from centrosomes, extend inward, and attach to kinetochores; and (3) ***aster microtubules,*** which grow out of centrosomes but radiate outward from the mitotic spindles. Overall, the spindle is an attachment site for chromosomes and it also distributes chromosomes to opposite poles of the cell.

■ ***Metaphase*** During ***metaphase*** (*meta* = after), the second stage of mitosis, the centromeres of the chromatid pairs are lined up by the kinetochore microtubules at the exact center of the mitotic spindle. This midpoint region is called the ***metaphase plate*** (Figure 2-15c).

■ ***Anaphase*** The third stage of mitosis, ***anaphase*** (*ana* = upward), is characterized by the splitting and separation of centromeres (and kinetochores) and the movement of the two sister chromatids of each pair toward opposite poles of the cell (Figure 2-15d). Once separated, the sister chromatids are referred to as chromosomes. The movement of chromosomes is due to shortening of kinetochore microtubules and elongation of the nonkinetochore microtubules, processes that increase the distance between separated chromosomes. As the chromosomes move during anaphase, they appear V-shaped.

■ ***Telophase*** The final stage of mitosis, ***telophase*** (*telo* = far or end), begins as soon as chromosomal movement stops. Telophase is essentially the opposite of prophase. During telophase, the identical sets of chromosomes at opposite poles of the cell uncoil and revert to their threadlike chromatin form; kinetochore microtubules disappear; nonkinetochore microtubules elongate even more; a new nuclear envelope reforms around each chromatin mass; new nucleoli reappear in the daughter nuclei; and eventually the mitotic spindle breaks up.

Cytokinesis

Division of a parent cell's cytoplasm and organelles is called ***cytokinesis*** (sī-tō-ki-NĒ-sis). It usually starts during late anaphase and terminates during telophase (Figure 2-15e,f). Cytokinesis begins with the formation of a ***cleavage furrow,*** a slight indentation of the plasma membrane that extends around the center of the cell. The furrow gradually deepens until opposite surfaces of the cell make contact and the cell is split in two. The result is two separated daughter cells, each with separate portions of cytoplasm and organelles and its own set of identical chromosomes.

If we consider the cell cycle in its entirety, the sequence of events can be summarized as follows: G_1 period → S period → G_2 period → mitosis → cytokinesis (Figure 2-16).

A summary of the events that occur during the interphase and cell division is presented in Exhibit 2-3.

Time Required

The time required for mitosis varies with the kind of cell, its location, and the influence of factors such as temperature. Furthermore, the different stages of mitosis are not equal in duration. However, in order to give you some idea of the length of a cell cycle, mammalian cells in culture have been studied and often have the following time intervals: The G_1 period is highly variable, ranging from almost nonexistent in rapidly dividing cells to days, weeks, or years. However, it typically takes about 8–10 hours. The S period takes about 6–8 hours, the G_2 period about 4–6 hours, and mitosis and cytokinesis about 30–45 minutes. Within the mitosis and cytokinesis time interval, prophase takes longest and anaphase is shortest. As you can see, mitosis and cytokinesis represent only a small part of the life cycle of a cell. Together, the various phases of the cell cycle require about 18–24 hours in many cultured mammalian cells.

REPRODUCTIVE CELL DIVISION

In sexual reproduction, each new organism is produced by the union and fusion of two different sex cells, one produced by each parent. The sex cells, called ***gametes,*** are the ovum produced in the female gonads (ovaries) and the sperm produced in the male gonads (testes). The union and fusion of gametes is called ***fertilization,*** and the cell thus produced is known as a ***zygote.*** The zygote contains a mixture of chromosomes (DNA) from the two parents and, through its repeated mitotic division, develops into a new organism.

Gametes differ from all other body cells (somatic cells) with respect to the number of chromosomes in their nuclei. Somatic cells, such as brain cells, stomach cells, kidney cells, and all other uninucleated somatic cells, contain 46 chromosomes in their nuclei. Some somatic cells, such as skeletal muscle fibers (cells), are multinucleated and thus contain more than 46 chromosomes. However, since most somatic cells are uninucleated, these are the cells to which we will refer in the following discussion. Of the 46 chromosomes, 23 are a complete set that contain one copy of all the genes necessary for carrying out the activities of the cell. In a sense, the other 23 chromosomes are a duplicate set. The symbol n is used to designate the number of different chromosomes within the nucleus. Since somatic cells contain two sets of chromosomes, they are referred to as ***diploid*** (DIP-loyd; *di* = two) ***cells,*** symbolized as $2n$. In a diploid cell, two chromosomes that belong to a pair are called ***homologous*** (hō-MOL-ō-gus) ***chromosomes,*** or ***homologues.*** In human diploid cells, 22 of the 23 pairs of chromosomes are morphologically similar and are called ***autosomes.*** The other pair is called the ***sex chromosomes*** and designated as X and Y. In females, the homologous pair of sex chromosomes consists of two X chromosomes; in males, the pair consists of an X and a Y chromosome.

If gametes had the same number of chromosomes as somatic cells, the zygote formed from their fusion would have double the number. The somatic cells of the resulting individual would have twice the number of chromosomes ($4n$) as the somatic cells of the parents, and with every succeeding generation, the number of chromosomes would double. The chromosome number does not double with each generation because of a special nuclear division called ***meiosis.*** Meiosis occurs only in the development of gametes, and it results in the production of cells that contain only 23 chromosomes. Thus, gametes are ***haploid*** (HAP-loyd) ***cells,*** meaning "one-half," and are symbolized as n.

EXHIBIT 2-3

Summary of Events Associated with Interphase and Cell Division

PERIOD OR STAGE	ACTIVITY
INTERPHASE	Cell is between divisions.
G_1 Period	Cell engages in growth, metabolism, and production of substances required for division; no chromosomal replication.
S Period	Chromosomal replication occurs.
G_2 Period	Same as for G_1 period.
CELL DIVISION	Single parent cell produces two identical daughter cells.
Prophase	Chromatin shortens and coils into chromosomes (chromatids), nucleoli and nuclear envelope become less distinct, and a pair of centrioles and its centrosome each move to opposite poles of cell and centrosomes form mitotic spindle.
Metaphase	Centromeres of chromatid pairs line up on metaphase plate of cell.
Anaphase	Centromeres divide and identical sets of chromosomes move to opposite poles of cell.
Telophase	Nuclear envelope reappears and encloses chromosomes, chromosomes resume chromatin form, nucleoli reappear, and mitotic spindle disappears.
Cytokinesis	Cleavage furrow forms around center of cell, progresses inward, and separates cytosol and organelles into two separate and equal portions.

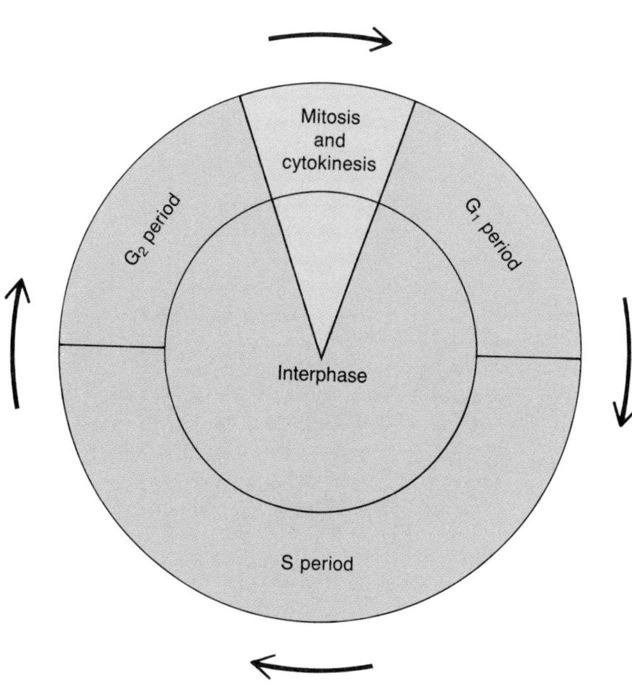

FIGURE 2-16 Various phases and periods in a cell cycle. Relative amounts of time are indicated by the size of the compartments.

Meiosis

The formation of haploid sperm cells in the testes of the male consists of several phases and is called ***spermatogenesis.*** One of the phases involves meiosis. The formation of haploid ova (eggs) in the ovaries of the female also involves several phases and is referred to as ***oogenesis.*** It, too, involves meiosis. Both spermatogenesis and oogenesis are discussed in detail in Chapter 25. At this point, we will examine only the essentials of meiosis.

Meiosis occurs in two successive nuclear divisions referred to as ***reduction division (meiosis I)*** and ***equatorial division (meiosis II).*** During the interphase that precedes reduction division of meiosis, the chromosomes replicate themselves. This replication is similar to that in the interphase preceding the mitosis of somatic cell division. Once chromosomal replication is complete, reduction division begins. It consists of four phases referred to as prophase I, metaphase I, anaphase I, and telophase I (Figure 2-17).

Prophase I is an extended phase in which the chromosomes shorten and thicken, the nuclear envelope and nucleoli disappear, the centrioles replicate, and the mitotic spindle appears. Unlike the prophase of mitosis, however, a unique event occurs in prophase I of meiosis. The chromosomes become arranged in homologous pairs. The pairing is called ***synapsis.*** The four chromatids of each homologous pair are referred to as a ***tetrad.*** Another unique event of meiosis occurs within a tetrad. Portions of one chromatid may be exchanged with portions of another, a process called ***crossing-over*** (Figure 2-18). This process, among others, permits an exchange of genes among chromatids so that subsequent daughter cells produced are unlike each other genetically and unlike the parent cell that produced them. This phenome-

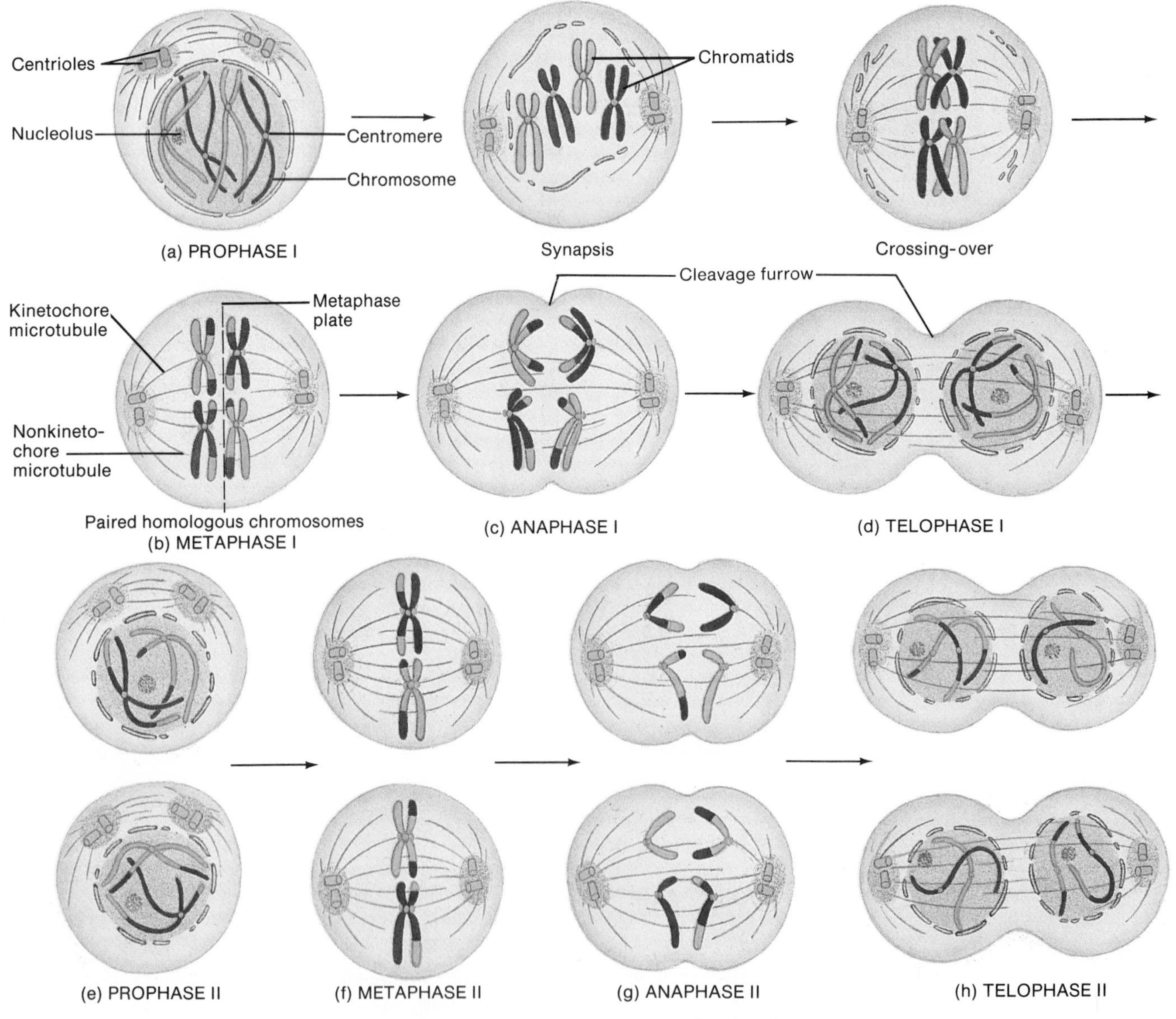

FIGURE 2-17 Meiosis. See text for details.

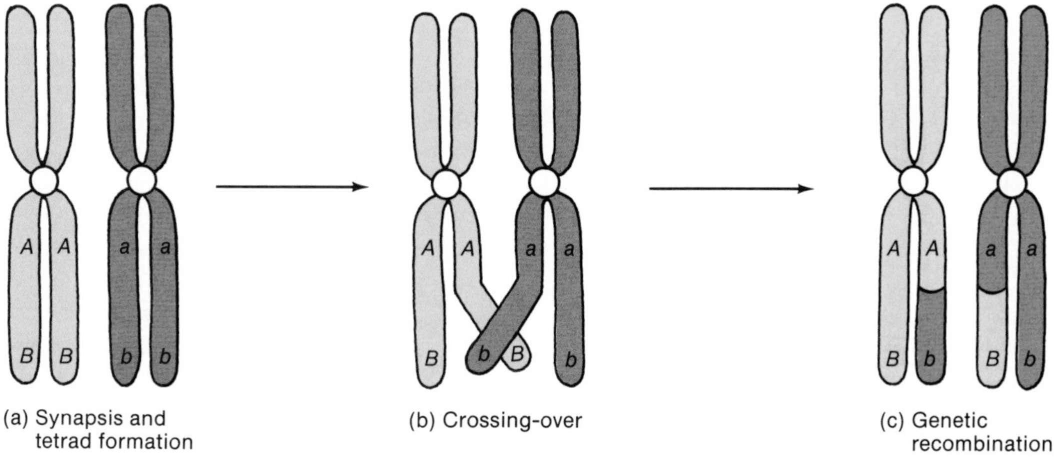

FIGURE 2-18 Crossing-over within a tetrad.

non accounts for part of the great genetic variation among humans and other organisms that form gametes by meiosis.

In metaphase I, the homologous chromosomes line up along the metaphase plate of the cell, with the homologues side by side. Recall that there is no pairing of homologous chromosomes during the metaphase of mitosis. The homologues randomly attach to the mitotic spindle. The centromeres of each chromatid pair form kinetochore microtubules that attach the centromeres to opposite poles of the cell.

Anaphase I is characterized by separation of the members of each homologous pair, with one member of each pair moving to an opposite pole of the cell. During anaphase I the centromeres do not split and the paired chromatids, held by a centromere, remain together. During the anaphase of mitosis the centromeres split and the sister chromatids separate.

Telophase I and cytokinesis are similar to telophase and cytokinesis of mitosis. The net effect of reduction division is that each resulting daughter cell contains the haploid number of chromosomes; each cell contains only one member of each pair of the original homologous chromosomes in the starting parent cell.

The interphase between reduction division and equatorial division is either brief or lacking altogether. It does differ from the interphase preceding reduction division in that there is no replication of DNA between the reduction and equatorial divisions.

The equatorial division of meiosis consists of four phases, referred to as prophase II, metaphase II, anaphase II, and telophase II. These phases are essentially similar to those that occur during mitosis since the centromeres split and the sister chromatids separate and move toward opposite poles of the cell.

In reviewing the overall process, note that during reduction division we start with a parent cell with the diploid number and end up with two daughter cells, each with the haploid number. During equatorial division, each haploid cell formed during reduction division divides, and the net result is four haploid cells that are all genetically different. As you will see later, all four haploid cells develop into sperm cells in the testes of the male, but only one of the haploid cells has the potential to develop into an ovum in the female. The other three become structures called *polar bodies* that do not function as gametes.

A very simplified comparison of mitosis and meiosis is illustrated in Figure 2-19.

ABNORMAL CELL DIVISION: CANCER (CA)

DEFINITION

When cells in some area of the body duplicate without control, the excess tissue that develops is called a ***tumor, growth,*** or ***neoplasm.*** The study of tumors is called ***oncology*** (*onco* = swelling or mass; *logos* = study of), and a physician who specializes in this field is called an ***oncologist.*** Tumors may be cancerous and sometimes fatal, or they may be quite harmless. A cancerous growth is called a ***malignant tumor,*** or ***malignancy.*** A noncancerous growth is called a ***benign tumor.*** Benign tumors do not spread to other parts of the body, but they may be removed if they interfere with a normal body function or are disfiguring.

TYPES

At present, cancers are classified by their microscopic appearance and the body site from which they arise. At least 100 different cancers have been identified in this way. If finer details of appearance are taken into consideration, the number can be increased to 200 or more. The name of the cancer is derived from the type of tissue in which it develops. ***Carcinoma*** (*carc* = cancer; *oma* = tumor) refers to a malignant tumor consisting of epithelial cells. An example is a ***melanoma,*** cancer of melanocytes, skin

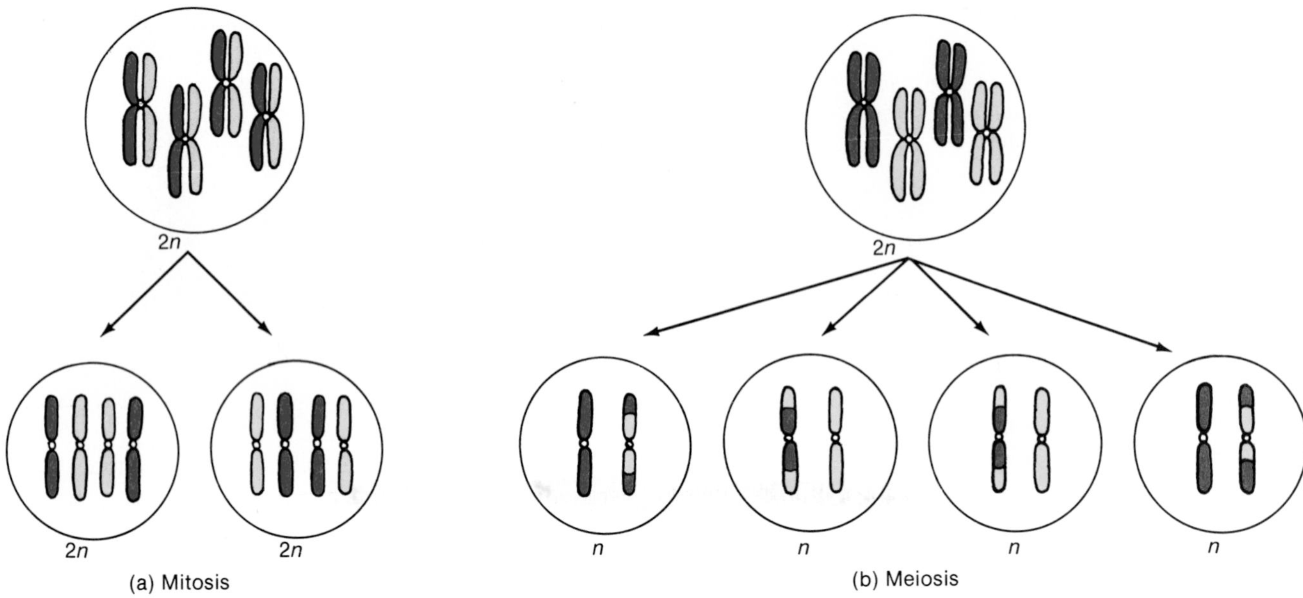

FIGURE 2-19 Very simplified comparison between mitosis and meiosis.

cells that produce the pigment melanin. A tumor that develops from a gland is called an ***adenocarcinoma*** (*adeno* = gland). ***Sarcoma*** is a general term for any cancer arising from connective tissue. ***Osteogenic sarcomas*** (*osteo* = bone; *genic* = origin), the most frequent type of childhood cancer, destroy normal bone tissue and eventually spread to other areas of the body. ***Myelomas*** (*myelos* = marrow) are malignant tumors, usually occurring in middle-aged and older people, that interfere with the blood cell–producing function of bone marrow and cause anemia. ***Chondrosarcomas*** (*chondro* = cartilage) are cancerous growths of cartilage.

CLINICAL APPLICATION

Grading and Staging Tumors

Two methods are used to categorize malignant tumors: grading and staging. In ***grading,*** pathologists classify tumors into four categories based on the degree to which cells are altered in size, shape, and organization as seen microscopically. Grade I tumors closely resemble normal cells; grade IV tumors are altered dramatically from normal cells. Grades II and III tumors are intermediate in appearance between grades I and IV. Grading aids physicians in cancer prognosis in that grade I tumors have the highest survival rates, whereas grade IV tumors have the lowest. Grading is also used to evaluate precancerous cells such as those of the uterine cervix (Pap smear).

Staging is based on the progression of tumor growth and development. One example of a staging system is the most frequently used TNM system that is usually applied to breast and lung cancers as well as others. *T* stands for tumor (T1–T4 defines increasing extent of tumor size), *N* refers to regional lymph nodes affected by the malignancy (N1–N3 indicates advancing nodal disease), and *M* refers to metastasis (M0 indicates no metastasis and M1–M3 indicates advancing degrees of metastasis).

GROWTH AND SPREAD

Cells of malignant tumors duplicate continuously and very often quickly and without control. The majority of cancer patients who die are killed not by the ***primary tumor*** that develops but by secondary infections of bacteria and viruses due to lowered resistance as a result of ***metastasis*** (me-TAS-ta-sis), the spread of cancerous cells to other parts of the body. One of the unique properties of a malignant tumor is its ability to metastasize. Cancer cells secrete a protein called ***autocrine motility factor (AMF)*** that enables them to metastasize. Metastatic groups of cells are more difficult to detect and eliminate than primary tumors.

In the process of metastasis, there is an initial invasion of the malignant cells into surrounding tissues. As the cancer grows, it expands and begins to compete with normal tissues for space and nutrients. Eventually, the normal tissue atrophies and dies. The invasiveness of the malignant cells may be related to mechanical pressure of the growing tumor, motility of the malignant cells, and enzymes produced by the malignant cells. When normal cells of the body divide and migrate (for example, skin cells that multiply to heal a superficial cut), their further migration is inhibited by contact on all sides with other skin cells—called ***contact inhibition.*** Unfortunately, malignant cells do not conform to the rules of contact inhibition; they have the ability to invade healthy body tissues with very few restrictions.

Following invasion, some of the malignant cells may detach from the primary tumor and invade a body cavity (abdominal or thoracic) or enter the blood or lymph. This latter condition can lead to widespread metastasis. In the next step in metastasis, those malignant cells that survive in the blood or lymph invade adjacent body tissues and establish ***secondary tumors.*** It is believed that some of the invading cells involved in metastasis have properties different from those of the primary tumor that enhance metastasis. These include appropriate mechanical, enzymatic, and surface properties. In the final stage of metastasis, the secondary tumors become vascularized; that is, they take on new networks of blood vessels that provide nutrients for their further growth. Any new tissue, whether it results from repair (wound repair, menstruation, ovulation), normal growth (tissue growth, placenta formation), or tumors, requires a blood supply. Proteins that serve as chemical triggers for blood vessel growth are called ***tumor angiogenesis factors (TAFs),*** which have been isolated from human colon tumors. In all stages of metastasis, the malignant cells resist the antitumor defenses of the body. The pain associated with cancer develops when the growth puts pressure on nerves or blocks a passageway so that secretions build up pressure.

POSSIBLE CAUSES

What triggers a perfectly normal cell into losing control and becoming abnormal? Scientists are uncertain. First, there are environmental agents: substances in the air we breathe, the water we drink, the food we eat. A chemical or other environmental agent that produces cancer is called a ***carcinogen.*** The World Health Organization estimates that carcinogens may be associated with 60 to 90 percent of all human cancers. Examples of carcinogens are the hydrocarbons found in cigarette tar. Ninety percent of all lung cancer patients are smokers. Another environmental factor is radiation. Ultraviolet (UV) light from the sun, for example, may cause genetic mutations in exposed skin cells and lead to skin cancer, especially among light-skinned people.

Viruses, a second cause of cancer, are tiny packages of nucleic acids, either DNA or RNA, that are capable of infecting cells and converting them to virus-producers. With over 100 separate viruses identified as carcinogens in many species and tissues of animals, it is also probable that at least some cancers in humans are due to viruses. For example, *human T-cell leukemia-lymphoma virus-1 (HTLV-1)* is strongly associated with *leukemia,* a malignant disease of blood-forming tissues, and *lymphoma,* a cancer of lymphoid tissue. A variant of HTLV-1 known as *human immunodeficiency virus (HIV)* is the causative agent of acquired immune deficiency syndrome (AIDS). People with AIDS often develop *Kaposi's sarcoma,* cancer of blood vessels. The *Epstein-Barr virus (EBV)*, the causative agent of infectious mononucleosis, has been linked as the causative agent of several human cancers—*Burkitt's lymphoma* (a cancer of white blood cells called B-cells), *nasopharyngeal carcinoma* (common in Chinese males), and *Hodgkin's disease* (a cancer of the lymphatic system). The *hepatitis B virus (HBV)* has been associated with cancer of the liver. Also, *type 2 herpes simplex virus,* the causative agent of genital herpes, has been implicated in cancer of the cervix of the uterus, and the *papilloma virus,* a virus that causes warts, has been associated with cancer of the cervix, vagina, vulva, and penis, and cancer of the colon.

A great deal of cancer research is now directed toward studying ***oncogenes*** (ONG-kō-jēnz)—genes that have the ability to transform a normal cell into a cancerous cell when they are *inappropriately activated.* Oncogenes are derived from normal genes that regulate growth and development, called ***proto-oncogenes.*** These genes may undergo some change that either causes them to produce an abnormal product or disrupts their control so that they are expressed inappropriately, making their products in excessive amounts or at the wrong time. It is believed that some oncogenes cause extra production of growth factors, chemicals that stimulate cell growth. Other oncogenes may cause changes in a surface receptor, causing it to send signals as though it were being activated by a growth factor. As a result, the growth pattern of the cell becomes abnormal.

Every human cell contains oncogenes. In fact, oncogenes apparently carry out normal cellular functions until a malignant change occurs. It appears that some proto-oncogenes are activated to oncogenes by various types of mutations in which the DNA of the proto-oncogenes is altered. Such mutations are induced by carcinogens. Other proto-oncogenes are activated by viruses. Some oncogenes can also be activated by rearrangement of a cell's chromosomes in which segments of DNA are exchanged. This rearrangement is sufficient to activate oncogenes by placing them near genes that enhance their activity. Burkitt's lymphoma, malignant tumors of the colon and rectum, and one type of lung cancer are linked to oncogenes.

Researchers have also determined that some cancers are not caused by oncogenes but may be caused by genes called ***anti-oncogenes*** or ***tumor-suppressing genes.*** These genes may function to oppose the action of the oncogene and can cause cancer when they are *inappropriately inactivated.* The prototype for a cancer caused by an anti-oncogene is a rare, inherited childhood cancer of the eye called ***retinoblastoma.*** Such inappropriate inactivation may also be involved in some types of breast cancer and one type of lung cancer.

Tumorigenesis (too′-mor-i-JEN-e-sis), the process by which tumors develop, seems to involve not only oncogenes, which promote excessive cell proliferation, but also anti-oncogenes, which suppress proliferation. Tumorigenesis is a multistep process that may begin with an oncogene but may also be influenced by anti-oncogenes.

Currently, scientists are also trying to establish a relationship between stress and cancer. Some believe that stress may play a role not only in the development but also in the metastasis of cancer.

There is also great interest in determining the effects of alterations of the immune system and nutrition in the development of cancer. In 1982 the National Research Council issued a series of guidelines related to diet and cancer. The main recommendations were: (1) to reduce fat intake; (2) to increase consumption of fiber, fruits, and vegetables; (3) to increase intake of complex carbohydrates (potatoes, pasta); and (4) to reduce consumption of salted, smoked, and pickled foods, as well as simple carbohydrates (refined sugars).

TREATMENT

Treating cancer is difficult because it is not a single disease and because all the cells in a single population (tumor) do not behave in the same way. The same cancer may contain a diverse population of cells by the time it reaches a clinically detectable size. Although tumor cells look alike when stained and viewed under the microscope, they do not necessarily behave in the same manner in the body. For example, some metastasize and others do not. Some divide and others do not. Some are sensitive to drugs and some are resistant. As a consequence of differences in drug resistance, a single chemotherapeutic drug may destroy susceptible cells but permit resistant cells to proliferate. This is probably one of the reasons that combination chemotherapy is usually more successful. Certain tumor cells may be simultaneously resistant to completely unrelated drugs, a phenomenon called ***cross-resistance,*** and this may result from a single mutation. It has recently been learned that resistance to chemotherapeutic drugs is due to a mechanism in which a plasma membrane protein called P-glycoprotein pumps drugs out of tumor cells. Research is concentrating on ways to inhibit the functioning of P-glycoprotein so that anticancer drugs can concentrate in tumor cells to kill them.

Added to this problem is the fact that solid tumors, such as those that arise in the breasts, lungs, colon, and other organs, develop barriers to anticancer drugs delivered by the blood. Within such tumors are high-pressure areas located deep inside them that collapse blood vessels in the tumor. This makes it difficult, if not impossible, for blood-borne anticancer agents to penetrate the tumor.

In addition to chemotherapy, radiation therapy, surgery, hyperthermia (abnormally high temperatures), and immunotherapy (bolstering the body's own defenses) may be used alone or in combination to treat tumors.

Scientists are moving closer to developing a vaccine for cancer. What happens in cancer is that the immune system fails to protect the body. Accordingly, the goal of a cancer vaccine is to stimulate the immune system into marshaling a successful attack against the cancer cells. Tumor cells prepared from a person's own cancer are used in vaccines to elicit a specific immune response to the particular tumor from which the patient suffers. In early chemical trials, tumor cell vaccines appear to help patients with melanoma and colon and kidney cancer fight metastases by stimulating immune responses.

Another new treatment for certain types of cancer is called ***photodynamic therapy (PDT).*** In this procedure, used primarily for esophageal, lung, and urinary bladder tumors, patients are given an intravenous injection of a chemical that accumulates in cancer cells and becomes lethal to the cells when exposed to red light.

CELLS AND AGING

Aging is a normal process accompanied by a progressive alteration of the body's adaptive responses that attempt to maintain health. It is a general response that produces observable changes in structure and function and increased vulnerability to environmental stress and disease. Disease and aging probably accelerate each other. The specialized branch of medicine that deals with the medical problems and care of elderly persons is called ***geriatrics*** (jer′-ē-AT-riks; *geras* = old age; *iatrike* = surgery, medicine).

The obvious characteristics of aging are well known: graying and loss of hair, loss of teeth, wrinkling of skin, decreased muscle mass, and increased fat deposits. The physiological signs of aging are gradual deterioration in function and capacity to respond to environmental stress. Thus, basic kidney and digestive metabolic rates decrease, as does the ability to respond effectively to changes in temperature, diet, and oxygen supply in order to maintain a constant internal environment. These manifestations of aging are related to a net decrease in the number of cells in the body (thousands of brain cells are lost each day) and to the dysfunctioning of the cells that remain.

The extracellular components of tissues also change with age. Collagen fibers, responsible for the strength in tendons, increase in number and change in quality with aging. These changes in the collagen of arterial walls are as much responsible for their loss of extensibility as are the deposits associated with *atherosclerosis,* the deposition of fatty materials in arterial walls. Elastin, another extracellular component, is responsible for the elasticity of blood vessels and skin. It thickens, fragments, and acquires a greater affinity for calcium with age—changes that may also be associated with the development of atherosclerosis.

Several kinds of cells in the body—heart cells, skeletal muscle fibers (cells), nerve cells—are incapable of replacement. Experiments have proved that many other cell types are limited when it comes to cell division. Cells grown outside the body divide only a certain number of times and then stop. The number of divisions correlates with the donor's age and with the normal life span of the different species from which the cells are obtained—strong evidence for the hypothesis that cessation of mitosis is a normal, genetically programmed event. According to this view, an "aging" gene is part of the genetic blueprint at birth, and it turns on at a preprogrammed time, slowing down or halting processes vital to life.

Another theory of aging is the ***free radical theory.*** Free radicals, or oxygen radicals, are oxygen molecules that bear free electrons, are highly reactive, and can easily tie up and weaken proteins. As a result, cells grow rigid as nutrients are excluded and wastes are locked in. The effects exhibited include wrinkled skin, stiff joints, and hardened arteries. Free radicals may also cause damage to DNA. Among the factors that produce free radicals are pollution, radiation, and certain foods we eat. Other substances in the diet such as vitamin E, vitamin C, beta-carotene, and selenium, are antioxidants and inhibit free radical formation.

Recently, it has been learned that glucose, the most abundant sugar in the body, may play a role in the aging process. According to one hypothesis, glucose is added, haphazardly, to proteins, forming irreversible cross-links between adjacent protein molecules. As a person ages, the formation of more cross-links probably contributes to the stiffening and loss of elasticity that occurs in aging tissues.

Whereas some theories of aging explain the process at the cellular level, others concentrate on regulatory mechanisms operating within the entire organism. For example, one such theory holds that the immune system, which manufactures antibodies against foreign invaders, turns on its own cells. This autoimmune response might be caused by changes in the surfaces of cells, causing antibodies to attack the body's own cells. As surface changes in cells increase, the autoimmune response intensifies, producing the well-known signs of aging. Another organismic theory suggests that aging is programmed in the pituitary gland, a gland that produces and stores hormones and is attached to the undersurface of the brain. Supposedly, at a set time in life, the gland releases a hormome that triggers age-associated disruptions.

The effects of aging on the various body systems are discussed in their respective chapters.

KEY MEDICAL TERMS ASSOCIATED WITH CELLS

NOTE TO THE STUDENT

Each chapter in this text that discusses a major system of the body is followed by a glossary of ***key medical terms.*** Both normal and pathological conditions of the system are included in these glossaries. You should familiarize yourself with the terms, since they will play an essential role in your medical vocabulary.

Some of these disorders, as well as disorders discussed in the text, are referred to as local or systemic. A ***local disease*** is one that affects one part or a limited area of the body. A ***systemic disease*** affects either the entire body or several parts.

The science that deals with why, when, and where diseases occur and how they are transmitted in a human community is known as ***epidemiology*** (ep′-i-dē′-mē-OL-ō-jē; *epidemios* = prevalent; *logos* = study of). The science that deals with the effects and uses of drugs in the treatment of disease is called ***pharmacology*** (far′-ma-KOL-ō-jē; *pharmakon* = medicine; *logos* = study of).

Atrophy (AT-rō-fē; *a* = without; *tropho* = nourish) A decrease in the size of cells with subsequent decrease in the size of the affected tissue or organ; wasting away.

Biopsy (BĪ-op-sē; *bio* = life; *opsis* = vision) The removal and microscopic examination of tissue from the living body for diagnosis.

Deterioration (de-te′-rē-ō-RĀ-shun; *deterior* = worse or poorer) The process or state of growing worse; disintegration or wearing away.

Dysplasia (dis-PLĀ-zē-a; *dys* = abnormal; *plas* = to grow) Alteration in the size, shape, and organization of cells due to chronic irritation or inflammation; may progress to neoplasia (tumor formation, usually malignant) or revert to normal if the stress is removed.

Hyperplasia (hī′-per-PLĀ-zē-a; *hyper* = over) Increase in the number of cells due to an increase in the frequency of cell division.

Hypertrophy (hī-PER-trō-fē) Increase in the size of cells without cell division.

Insidious (in-SID-ē-us) Hidden, not apparent, as a disease that does not exhibit distinct symptoms of its arrival.

Metaplasia (met′-a-PLĀ-zē-a; *meta* = change) The transformation of one cell into another.

Metastasis (me-TAS-ta-sis; *stasis* = standing still) The transfer of disease from one part of the body to another that is not directly connected with it.

Necrosis (ne-KRŌ-sis; *necros* = death; *osis* = condition) Death of a group of cells.

Neoplasm (NĒ-ō-plazm; *neo* = new) Any abnormal formation or growth, usually a malignant tumor.

Progeny (PROJ-e-nē; *progignere* = to bring forth) Offspring or descendants.

Senescence (se-NES-ens) The process of growing old.

STUDY OUTLINE

Generalized Animal Cell (p. 29)

1. A cell is the basic, living, structural and functional unit of the body.
2. A generalized cell is a composite that represents various cells of the body.
3. Cytology is the science concerned with the study of cells.
4. The principal parts of a cell are the plasma (cell) membrane, cytosol, organelles, and inclusions.

Plasma (Cell) Membrane (p. 30)

Chemistry and Structure (p. 30)

1. The plasma (cell) membrane surrounds the cell and separates it from other cells and the external environment.
2. It is composed primarily of phospholipids and proteins. According to the fluid mosaic model, the membrane consists of a phospholipid bilayer with integral and peripheral proteins.

Functions (p. 30)

1. Functionally, the plasma membrane provides shape; encloses and protects cells; separates cells from the external environment; provides receptor sites; and regulates the passage of materials.
2. The membrane's selectively permeable nature restricts the passage of certain substances.

Movement of Materials Across Plasma Membranes (p. 32)

1. Passive (physical) processes depend on the kinetic energy of individual molecules.
2. Diffusion is the net movement of molecules or ions from an area of higher concentration to an area of lower concentration until an equilibrium is reached.
3. In facilitated diffusion, certain molecules, such as glucose, combine with a carrier to become soluble in the phospholipid portion of the membrane.
4. Osmosis is the movement of water through a selectively permeable membrane from an area of higher water concentration to an area of lower water concentration.
5. Bulk flow is the movement of large numbers of ions, molecules, or particles in the same direction as a result of forces (hydrostatic, osmotic) that push them.
6. Filtration is the movement of water and dissolved substances across a selectively permeable membrane by pressure.
7. Active (physiological) processes depend on the use of ATP by the cell.
8. Active transport is the movement of ions across a cell membrane from lower to higher concentration.
9. Endocytosis (phagocytosis, pinocytosis, receptor-mediated endocytosis) is the movement of substances through plasma membranes in which the membrane surrounds the substance, encloses it, and brings it into the cell.
10. Phagocytosis is the ingestion of solid particles by pseudopodia. It is an important process used by white blood cells to destroy bacteria that enter the body.
11. Pinocytosis is the ingestion of a liquid by the plasma membrane. In this process the liquid becomes surrounded by a vacuole.
12. Receptor-mediated endocytosis is the selective uptake of large molecules by cells.

Cytosol (p. 34)

1. Cytoplasm contains all of the cellular contents between the plasma membrane and nucleus. Cytosol is the semifluid portion of cytoplasm in which cellular organelles and inclusions are suspended and solutes are dissolved.
2. Cytosol is composed mostly of water plus proteins, carbohydrates, lipids, and inorganic substances. The chemicals in the cytosol are either in solution or in a colloid (suspended) form.
3. Functionally, cytosol is the medium in which chemical reactions occur.

Organelles (p. 34)

1. Organelles ("little organs") are specialized portions of the cell with characteristic morphology that carry on specific activities.
2. They assume specific roles in cellular growth, maintenance, repair, and control.

Nucleus (p. 36)

1. Usually the largest organelle, the nucleus controls cellular activities and contains the genetic information.
2. Most body cells have a single nucleus; some (red blood cells) have none, whereas others (skeletal muscle fibers) have several.
3. The parts of the nucleus include the inner and outer nuclear membranes, perinuclear cistern, nuclear pores, nucleoli, genetic material (DNA), and nucleoplasm.
4. Chromosomes are composed of DNA and histones.

Ribosomes (p. 39)

1. Ribosomes are granular structures consisting of ribosomal RNA and ribosomal proteins.
2. They occur as free ribosomes in the cytosol (singly or in clusters) or in conjunction with endoplasmic reticulum.
3. Functionally, ribosomes are the sites of protein synthesis.

Endoplasmic Reticulum (ER) (p. 39)

1. The ER is a network of parallel membranes continuous with the plasma membrane and nuclear membrane.
2. Granular or rough ER has ribosomes attached to it. Agranular or smooth ER does not contain ribosomes.
3. The ER provides mechanical support, releases calcium ions involved in muscle contraction, conducts intracellular exchange of materials with the cytosol, transports substances intracellularly, synthesizes lipids and proteins, stores synthesized molecules, and helps export chemicals from the cell.

Golgi Complex (p. 39)

1. The Golgi complex consists of four to eight stacked, flattened membranous sacs (cisternae) referred to as *cis, medial,* and *trans*.
2. The principal function of the Golgi complex is to process, sort, and deliver proteins within the cell. It also secretes proteins and lipids and forms lysosomes.

Lysosomes (p. 39)

1. Lysosomes are spherical structures that contain digestive enzymes. They are formed from Golgi complexes.
2. They are found in large numbers in white blood cells, which carry on phagocytosis.
3. Lysosomes function in intracellular digestion.
4. If the cell is injured, lysosomes release enzymes and digest the cell. Thus, they are called "suicide packets," and the process is called autolysis.
5. Lysosomes may be involved in bone removal and play a role in embryonic development and the destruction of phagocytized microorganisms.

Peroxisomes (p. 42)

1. Peroxisomes are similar to lysosomes but smaller.
2. They contain enzymes (e.g., catalase) involved in the metabolism of hydrogen peroxide.

Mitochondria (p. 42)

1. Mitochondria consist of a smooth outer membrane and a folded inner membrane surrounding the interior matrix. The inner folds are called cristae.
2. The mitochondria are called "powerhouses" of the cell because ATP is produced in them.

The Cytoskeleton (p. 44)

1. Together microfilaments, microtubules, and intermediate filaments form the cytoskeleton.
2. Microfilaments are rodlike structures consisting of the protein actin or myosin. They are involved in muscular contraction, support, and movement.
3. Microtubules are cylindrical structures consisting of the protein tubulin. They support, provide movement, and form the structure of flagella, cilia, centrioles, and the mitotic spindle.
4. Intermediate filaments appear to provide structural reinforcement in some cells.
5. The cytoskeleton and other cytoplasmic components are held together by microtrabeculae; interconnecting microtrabeculae form the microtrabecular lattice.

Centrosome and Centrioles (p. 45)

1. The dense area of cytosol containing the centrioles is called a centrosome. It assumes an important role in cell reproduction by helping to organize the mitotic spindle.
2. Centrioles are paired cylinders arranged at right angles to one another.

Flagella and Cilia (p. 45)

1. These cellular projections have the same basic structure and are used in movement.
2. If projections are few (typically occurring singly or in pairs) and long, they are called flagella. If they are numerous and hairlike, they are called cilia.
3. The flagellum on a sperm cell moves the entire cell. The cilia on cells of the respiratory tract move foreign matter trapped in mucus along the cell surfaces toward the throat for elimination.

Cell Inclusions (p. 46)

1. Cell inclusions are chemical substances produced by cells. They are usually organic and may have recognizable shapes.
2. Examples are melanin, glycogen, and lipids.

Normal Cell Division (p. 47)

1. Cell division is the process by which cells reproduce themselves. It consists of nuclear division (mitosis or meiosis) and cytoplasmic division (cytokinesis).
2. Cell division that results in an increase in body cells is called somatic cell division and involves a nuclear division called mitosis plus cytokinesis.
3. Cell division that results in the production of sperm and eggs is called reproductive cell division and consists of a nuclear division called meiosis plus cytokinesis.

Somatic Cell Division (p. 47)

1. Prior to mitosis and cytokinesis, the DNA molecules, or chromosomes, replicate themselves so that the same chromosomal complement can be passed on to future generations of cells.
2. A cell carrying on every life process except division is said to be in interphase (metabolic phase).
3. Mitosis is the distribution of two sets of chromosomes into separate and equal nuclei following their replication.
4. It consists of prophase, metaphase, anaphase, and telophase.
5. Cytokinesis usually begins in late anaphase and terminates in telophase.
6. A cleavage furrow forms at the cell's center and progresses inward, cutting through the cell to form two separate portions of cytoplasm.

Reproductive Cell Division (p. 51)

1. Gametes contain the haploid (n) chromosome number, and uninucleated somatic cells contain the diploid ($2n$) chromosome number.
2. Meiosis is the process that produces haploid gametes. It consists of two successive nuclear divisions called reduction division (meiosis I) and equatorial division (meiosis II).
3. During reduction division, homologous chromosomes undergo synapsis and crossing-over; the net result is two haploid daughter cells.
4. During equatorial division, the two haploid daughter cells undergo mitosis, and the net result is four haploid cells.

Abnormal Cell Division: Cancer (CA) (p. 53)

1. Cancerous tumors are referred to as malignant; noncancerous tumors are called benign; the study of tumors is called oncology.
2. The spread of cancer from its primary site is called metastasis.
3. Carcinogens are chemicals or environmental agents that can produce cancer.
4. Oncogenes are genes that can transform normal cells into cancerous cells; their normal counterparts are called proto-oncogenes. Anti-oncogenes are genes that suppress cell proliferation and may cause cancer by their inactivation.
5. Treating cancer is difficult because all the cells in a single population do not behave in the same way.

Cells and Aging (p. 56)

1. Aging is a normal process accompanied by progressive alteration of the body's homeostatic adaptive responses.
2. Many theories of aging have been proposed, including genetically programmed cessation of cell division and excessive immune responses, but none successfully answers all the experimental objections.
3. All the various body systems exhibit definitive and sometimes extensive changes with aging.

REVIEW QUESTIONS

1. Define a cell. What are the four principal portions of a cell? What is meant by a generalized cell? (p. 29)
2. Discuss the chemistry and structure of the plasma membrane with respect to the fluid mosaic model. (p. 30)

3. How do integral and peripheral membrane proteins differ in function? (p. 30)
4. Describe the various functions of the plasma membrane. What determines selective permeability? (p. 30)
5. What are the major differences between passive (physical) processes and active (physiological) processes in moving substances across plasma membranes? (p. 32)
6. Define and give an example of each of the following: diffusion, facilitated diffusion, osmosis, bulk flow, filtration, active transport, phagocytosis, pinocytosis, and receptor-mediated endocytosis. (p. 33)
7. What is cytoplasm? Discuss the chemical composition and physical nature of the cytosol. What is its function? (p. 34)
8. What is an organelle? (p. 36) By means of a labeled diagram, indicate the parts of a generalized animal cell.
9. Describe the structure and functions of the nucleus of a cell. (p. 39)
10. Discuss the distribution of ribosomes. What is their function? (p. 39)
11. Distinguish between granular (rough) and agranular (smooth) endoplasmic reticulum (ER). What are the functions of ER? (p. 39)
12. Describe the structure and functions of the Golgi complex. (p. 39)
13. List and describe the various functions of lysosomes by contrasting autophagy and autolysis. (p. 39)
14. Why are mitochondria referred to as ''powerhouses'' of the cell? (p. 42)
15. What is the importance of peroxisomes? (p. 42)
16. Contrast the structure and functions of microfilaments, microtubules, and intermediate filaments. What is the microtrabecular lattice? (p. 44)
17. Describe the structure and function of centrosomes. (p. 45)
18. How are flagella and cilia distinguished on the basis of structure and function? (p. 45)
19. Define a cell inclusion. Provide examples and indicate their functions. (p. 46)
20. Distinguish between the two types of cell division. Why is each important? (p. 47)
21. Define interphase. What happens during the G_1, S, and G_2 periods of interphase? (p. 48)
22. Describe the principal events of each stage of mitosis. (p. 48)
23. Distinguish between haploid (n) and diploid ($2n$). (p. 51)
24. Define meiosis and contrast the principal events of reduction division and equatorial division. (p. 52)
25. What is a tumor? Distinguish between malignant and benign tumors. Describe the principal types of malignant tumors. How are malignant tumors graded and staged? (p. 53)
26. Define metastasis. What factors contribute to metastasis? (p. 54)
27. Discuss several possible causes of cancer (CA). Distinguish oncogenes, proto-oncogenes, and anti-oncogenes. (p. 55)
28. How is cancer treated? What are some of the problems with respect to treating cancer? (p. 56)
29. What is aging? List some of the characteristics of aging. (p. 56)
30. Briefly describe the various theories regarding aging. (p. 56)
31. Refer to the glossary of key medical terms associated with cells. Be sure that you can define each term. (p. 57)

SELF QUIZ

Choose the one best answer to these questions.

___ **1.** A membrane that permits the passage of only certain materials is described as
A. impermeable; B. freely permeable; C. selectively permeable; D. endocytic; E. macromolecular.

___ **2.** As a result of somatic cell division, each daughter cell has
A. half as many chromosomes as its parent cell; B. twice as many chromosomes as its parent cell; C. exactly the same number of chromosomes as its parent cell; D. one-quarter as many chromosomes as its parent cell; E. three-quarters as many chromosomes as its parent cell.

___ **3.** Air moves into and out of the lung as a result of
A. bulk flow; B. osmosis; C. active transport; D. phagocytosis; E. pinocytosis.

___ **4.** The process by which worn-out organelles are digested is called
A. hemolysis; B. exocytosis; C. autolysis; D. endocytosis; E. pinocytosis.

___ **5.** The important difference between meiosis and mitosis is that
A. in meiosis the number of chromosomes in the resulting cells is reduced by one-half; B. fewer sperm cells are produced, as compared with the number of egg cells; C. meiosis occurs only in somatic cells; D. abnormalities seldom occur in mitosis but are fairly common in meiosis; E. egg cells are produced by meiosis, sperm cells by mitosis.

___ **6.** Which process does not belong with the others?
A. phagocytosis; B. osmosis; C. diffusion; D. bulk flow; E. filtration.

___ **7.** Which is not a cell inclusion?
A. lipid; B. glycogen; C. melanin; D. lysosome.

___ **8.** Cytokinesis is the division of the cytoplasm of a cell. A similar equal division of the nucleus is a process called
A. synapsis; B. pinocytosis; C. involution; D. diplopia; E. mitosis.

9. Match the following:

___**a.** site of direction of cellular activities by means of genes located here	**A.** centrosome
___**b.** sites of protein synthesis; may occur attached to ER or scattered freely in cytoplasm	**B.** cilia
	C. endoplasmic reticulum (ER)
___**c.** system of membranous channels providing pathways for transport within cell and surface areas for chemical reactions; may be granular or agranular	**D.** flagella
	E. Golgi complex
	F. lysosomes
	G. microfilaments
___**d.** stacks of cisternae; involved in pack-	**H.** microtubules

aging and secretion of proteins and lipids and synthesis of carbohydrates
___e. called "suicide packets" since they may release enzymes that lead to autolysis of the cell
___f. similar to lysosomes, but smaller; contain the enzyme catalase
___g. cristae-containing structures, called "powerhouses of the cell" since ATP production occurs here
___h. form part of cytoskeleton; involved with cell movement and contraction
___i. part of cytoskeleton, provide support and give shape to cell; form flagellae, cilia, centrioles, and mitotic spindle
___j. helps organize mitotic spindle used in cell division
___k. long, hairlike structures that help move entire cell, for example, sperm cell
___l. short, hairlike structures that move particles over cell surface

I. mitochondria
J. nucleus
K. peroxisomes
L. ribosomes

Complete the following.

10. Phagocytosis is the process of cell (eating? drinking?).

11. The plasma membrane is composed of two main chemical components: a bilayer of ________ embedded with protein.

12. In the formation of mature sperm and egg cells, the nuclear division is known as ________.

13. The period of interphase when chromosomes are replicated is the ________ period.

14. By the time that human cells complete meiosis, they will contain (46? 23?) chromosomes, the (2*n*? *n*?) number.

15. Match the following:

___ **a.** chromosomes move toward opposite poles of the cell
___ **b.** nuclear envelope disappears; chromatin thickens into distinct chromosomes (chromatids); centrosomes move to opposite ends of cell and mitotic spindle forms
___ **c.** the series of events is essentially the reverse of prophase; cytokinesis occurs
___ **d.** chromatids line up on the metaphase plate
___ **e.** the cell is not involved in cell division; it is in a metabolic phase that follows telophase

A. anaphase
B. interphase
C. metaphase
D. prophase
E. telophase

Tissues

3

STUDENT OBJECTIVES

1. Describe the principal types of extracellular materials.
2. Define tissues, compare their origins, and classify the tissues of the body into four major types.
3. Discuss the distinguishing characteristics of epithelial tissue.
4. List the structure, location, and function for the basic types of epithelial tissue.
5. Define a gland and distinguish between exocrine and endocrine glands.
6. Identify the distinguishing characteristics of connective tissue.
7. Discuss the cells, ground substance, and fibers that constitute connective tissue.
8. List the structure, function, and location of the basic types of connective tissue.
9. Define an epithelial membrane and list the location and function of mucous, serous, cutaneous, and synovial membranes.
10. Contrast the three types of muscle tissue with regard to structure and location.
11. Describe the structural features and functions of nervous tissue.

CHAPTER OUTLINE

- **Types of Tissues and Their Origins**
- **Epithelial Tissue**

General Features
Covering and Lining Epithelium
Arrangement of Layers
Cell Shapes
Cell Junctions
Classification
Simple Epithelium
Stratified Epithelium
Pseudostratified Columnar Epithelium
Glandular Epithelium
Structural Classification of Exocrine Glands
Functional Classification of Exocrine Glands

- **Connective Tissue**

General Features
Basic Elements
Cells
Ground Substance
Fibers
Classification
Embryonic Connective Tissue
Mature Connective Tissue
Loose Connective Tissue
Dense Connective Tissue
Cartilage
Bone (Osseous Tissue)
Blood (Vascular Tissue)

- **Membranes**

Mucous Membranes
Serous Membranes
Cutaneous Membrane
Synovial Membranes

- **Muscle Tissue**
- **Nervous Tissue**

Cells are highly organized units, but they do not function in isolation. They work together in groups of similar cells called tissues. Outside the cells of a tissue are extracellular materials, which, among other functions, support and bind cells together in a tissue. As noted earlier, the substances that lie outside cells are called ***extracellular materials.*** They include body fluids, which provide a medium for dissolving, mixing, and transporting substances. Among the body fluids are interstitial (intercellular) fluid, the fluid that fills the microscopic spaces (interstitial spaces), and plasma, the liquid portion of blood, found in blood vessels. Extracellular materials also include secretions like mucus and special substances that form the matrix (substance between cells) in which cells of a tissue called connective tissue are embedded.

TYPES OF TISSUES AND THEIR ORIGINS

A ***tissue*** is a group of similar cells and their intercellular substance that have a similar origin in an embryo and function together to perform a specialized activity. The science that deals with the study of tissues is called ***histology*** (hiss′-TOL-ō-jē; *histio* = tissue; *logos* = study of).

CLINICAL APPLICATION

Pathologists and Biopsies

A ***pathologist*** (pa-THOL-ō-gist, *pathos* = disease) is a physician who specializes in laboratory studies of cells and tissues to help other physicians reach accurate diagnoses. Pathologists also conduct autopsies (see Chapter 1) and supervise other laboratory personnel who test and microscopically examine blood and other body fluids.

A ***biopsy*** (BĪ-op′-sē) is the removal of a sample of living tissue for microscopic examination. It is used to help diagnose many disorders, especially cancer, to determine whether a cancer has metastasized, and to determine the cause of unexplained infections and inflammations. Once the tissue sample is removed, it may be embedded in wax and then cut into very thin sections for microscopic analysis. One of the principal functions of a pathologist is to examine the tissue for any distortions or alterations that might indicate disease.

The various tissues of the body are classified into four principal types according to their function and structure:

1. Epithelial (ep′-i-THĒ-lē-al) tissue, which covers body surfaces, lines body cavities and ducts, and forms glands.
2. Connective tissue, which protects and supports the body and its organs, binds organs together, and stores energy reserves.
3. Muscular tissue, which is responsible for movement.
4. Nervous tissue, which initiates and transmits action potentials (nerve impulses) that coordinate body activities.

About eight days after a sperm fertilizes an egg, the mass of cells that has resulted embeds in the lining of the uterus and begins to form three ***primary germ layers: ectoderm, endoderm,*** and ***mesoderm.*** These are the embryonic tissues from which all tissues and organs of the body develop. Epithelial tissues develop from all three germ layers. Many epithelial tissues derive from endoderm, for example, the epithelial lining of the gastrointestinal and respiratory systems, and several internal organs, such as the thyroid gland and the liver. Mesoderm gives rise to epithelial tissues in much of the urogenital system (kidneys and gonads and their ducts). Ectoderm is the source of epithelial tissues in some endocrine glands, the skin, and portions of the eyes, ears, nose, and mouth. Connective tissues and muscle tissues all derive from mesoderm. Ectoderm gives rise to nervous tissue. (See Exhibit 26-1 for a more detailed list of structures derived from the primary germ layers.)

Epithelial tissue and connective tissue, except for bone and blood, will be discussed in detail in this chapter. The general features of bone tissue and blood will be introduced here, but their detailed discussion occurs later in the book. Similarly, the detailed discussion of muscle tissue and nervous tissue will be taken up later in Chapters 9 and 16, respectively.

EPITHELIAL TISSUE

Epithelial tissue (or, more simply, ***epithelium***) may be divided into two subtypes: (1) ***covering and lining epithelium*** and (2) ***glandular epithelium.*** Covering and lining epithelium forms the outer covering of external body surfaces and the outer covering of some internal organs. It lines body cavities and the interiors of the respiratory, digestive, urinary, and reproductive systems, as well as blood vessels and ducts. Epithelium is also combined with nervous tissue to make up special sense organs for smell, hearing, vision, and touch, which respond to stimuli. And it is the tissue from which gametes (sperm and eggs) develop. Glandular epithelium constitutes the secreting portions of glands such as sweat glands and the thyroid gland.

GENERAL FEATURES

1. Epithelium consists largely or entirely of closely packed cells with little extracellular substance between adjacent cells.
2. Epithelial cells are attached to each other and other structures by points of attachment called cell junctions (described shortly).
3. Epithelial cells are arranged in continuous sheets that may be single or multilayered.

4. Epithelia overlie and adhere firmly to connective tissue, which holds the epithelium in position and prevents it from being torn.
5. Epithelia are ***avascular*** (*a* = without; *vascular* = blood vessels). The blood vessels that supply oxygen and nutrients and remove carbon dioxide and wastes are located in the underlying connective tissue. Materials diffuse through the connective tissue to be delivered to the epithelium.
6. Epithelial cells rest on an extracellular structure called the ***basement membrane,*** located between the epithelium and underlying connective tissue. The membrane is composed of the ***basal lamina*** (collagen and glycoproteins) secreted by the epithelium, and the ***reticular lamina*** (reticular fibers and glycoproteins) produced by the connective tissue. The basement membrane provides physical support for epithelium, provides for cell attachment, serves as a filter in the kidneys, and guides cell migration during development and repair.
7. Epithelia have a nerve supply.
8. Since epithelium is subject to a certain amount of wear and tear and injury, it has a high capacity for renewal. Examining epithelial cells that are sloughed off provides the basis for the Pap smear, a test for precancer and cancer diagnosis of the uterus, cervix, and vagina (Chapter 25).
9. Epithelia are diverse in origin. They are derived from all three primary germ layers (ectoderm, mesoderm, and endoderm).
10. Functionally, epithelia participate in protection, lubrication, secretion, digestion, absorption, transportation, excretion, sensory reception, and reproduction.

CLINICAL APPLICATION

Epithelial Tissue Regeneration

The ***ability of epithelial tissue to regenerate*** is of clinical significance following injury and surgery. On the basis of regenerative capacity, epithelial cells (as well as other body cells) are distinguished into two principal types: labile cells and stable cells. *Labile* (LĀ-bīl) *cells* continue to regenerate throughout life under normal conditions. Examples include epidermal cells of the skin and epithelial cells of mucous membranes of the gastrointestinal (GI), urinary, and respiratory tracts. *Stable cells* cease regeneration at about the time of puberty but may resume regeneration under special conditions, such as injury. Epithelial cells of the liver, pancreas, kidneys, and thyroid glands are examples.

There are also cells of the body called *permanent cells* that cease to regenerate after birth and remain that way throughout life. An example is a nerve cell.

COVERING AND LINING EPITHELIUM

Arrangement of Layers

Covering and lining epithelium is arranged in several different ways related to location and function. If the epithelium is specialized for absorption or filtration and is in an area that has minimal wear and tear, the cells of the tissue are arranged in a single layer. Such an arrangement is called ***simple epithelium.*** If the epithelium is found in an area with a high degree of wear and tear, then the cells are stacked in several layers. This tissue is referred to as ***stratified epithelium.*** A third, less common arrangement of epithelium is called ***pseudostratified columnar.*** Like simple epithelium, pseudostratified epithelium has only one layer of cells. However, some of the cells do not reach the surface—an arrangement that gives the tissue a multilayered, or stratified, appearance.

Cell Shapes

In addition to classifying covering and lining epithelium according to the number of its layers, we may also categorize it by cell shape. The cells may be flat, cubelike, columnar, or a combination of shapes. ***Squamous*** (SKWĀ-mus) cells are flattened and scalelike. They are attached to each other and form a mosaic pattern. ***Cuboidal*** cells are usually cube-shaped in cross section. They sometimes appear as hexagons when viewed from above. ***Columnar*** cells are tall and cylindrical, appearing as somewhat rectangular in shape when set on end. ***Transitional*** cells have a variety of shapes and are found where there is a great degree of distention or expansion in the body. Transitional cells in the bottom layer of an epithelial tissue may range in shape from cuboidal or polyhedral (having many surfaces). In the superficial layer, they may range from cuboidal to squamous, depending on how much they are pulled out of shape during certain body functions.

Cell Junctions

Epithelial cells, unlike connective tissue cells, are tightly joined to form a close functional unit. The points of attachment between adjacent plasma membranes of epithelial cells (and a few other types of cells as well) are called ***cell junctions.*** In addition to providing cell-to-cell attachments, cell junctions prevent the movement of materials between certain cells and provide channels for communication between other cells. Four categories of cell junctions are recognized: (1) tight junction, (2) intermediate junction, (3) desmosome, and (4) gap junction. These cell junctions are illustrated in Figure 3-1.

■ ***Tight Junction*** A ***tight junction*** is typically located along the lateral surfaces of adjacent plasma membranes, closest to the side facing the lumen. In a tight junction, the outer layers of adjacent plasma membranes are fused together at various points, enabling the cells to adhere firmly to each other. Tight junctions are common in epithelial cells lining the small intestine. Here, they prevent the movement of substances via the intercellular route. Instead, selected substances are directed to enter the cells of the small intestine by absorption through fingerlike projections called microvilli.

■ ***Intermediate Junction*** An ***intermediate junction*** is located just below the tight junction between the epithelial cells of the small intestine. Intermediate junctions are believed to form firm connections between adjacent cells. In an intermediate junction, the adjacent plasma membranes are separated by an intercellular space measuring about 20 nm. The inner surfaces of the neighboring plasma membranes contain a moderately dense area of cytoplasmic material in which actin-containing ***microfilaments*** are embedded. Collectively, the microfilaments constitute the ***terminal web.***

■ ***Desmosome*** One of the most frequently encountered types of cell junction is called a ***desmosome*** (*desmos* = bond; *soma* = body). Desmosomes are numerous in stratified epithelium, especially the epidermis of the skin. They are also found in the epithelial lining of the small intestine. Desmosomes are scattered over adjacent cell surfaces somewhat like spot welds and form firm intercellular attachments between cells. The adjacent plasma membranes at a desmosome appear thickened because there is a dense proteinaceous ***plaque*** of cytoplasmic material on the inner surfaces of the neighboring membranes. Cytoplasmic filaments called ***tonofilaments*** converge in the region of the plaques and make a U-turn back into the cytoplasm. The plasma membranes of the adjacent cells at a desmosome are separated by an intercellular space measuring about 22–24 nm. In addition to serving as firm sites of adhesion between adjacent cells, desmosomes also help to support the entire layer of epithelial tissue and function as sites of attachment for the cytoskeleton to the cell surface.

In the basal cells of epithelium, the plasma membranes are in contact with the underlying basement membrane rather than adjacent plasma membranes. Here, there are structures that look like half a desmosome, called ***hemidesmosomes*** (*hemi* = half). Hemidesmosomes are assumed to provide anchorage for the basal epithelial cell plasma membranes to the basement membrane.

■ ***Gap Junction*** In a ***gap junction,*** the outer layers of adjacent plasma membranes approach each other, leaving a gap of about 2 nm between them. The gap is bridged

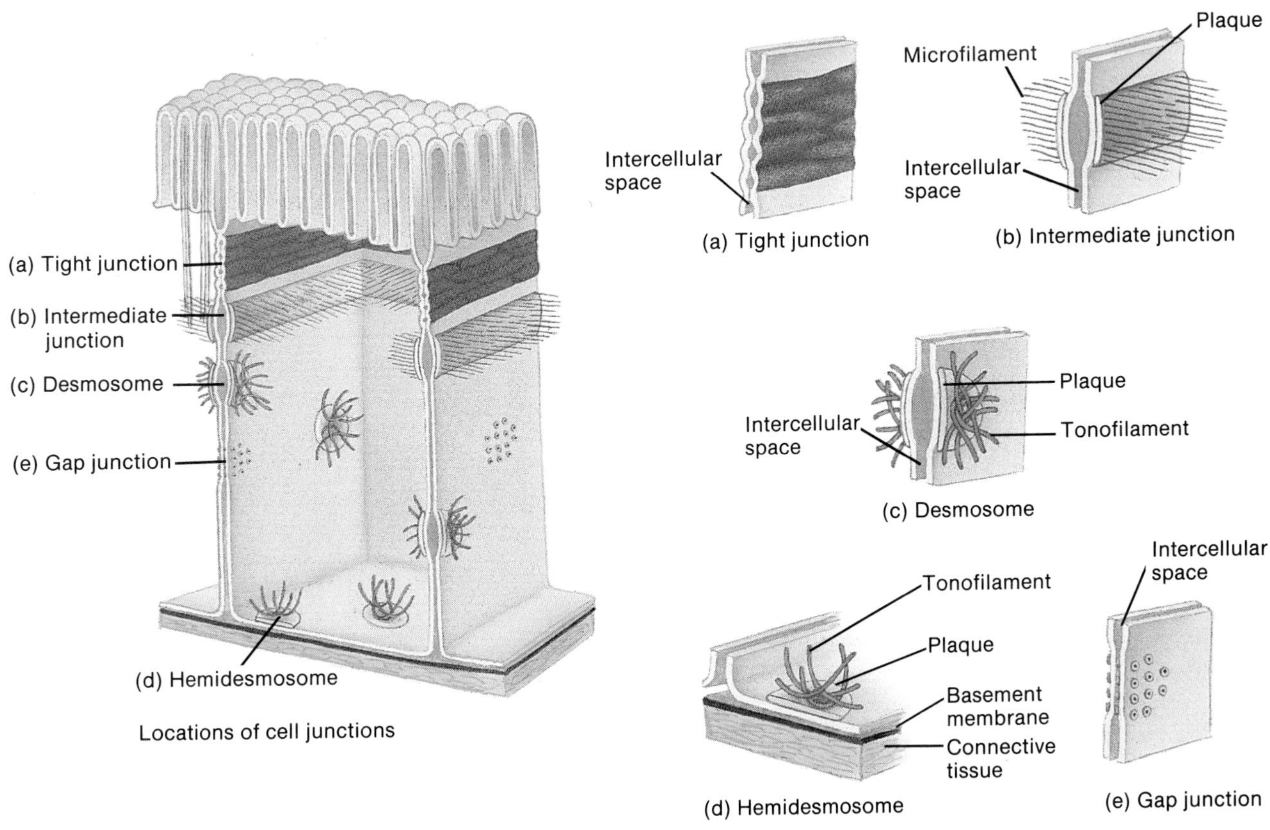

FIGURE 3-1 Cell junctions.

by structures in the membrane that link the cells together into units called ***connexons.*** These structures may be minute channels formed by membrane proteins that enable adjacent cells to communicate. It is postulated that ions and small molecules pass rapidly through gap junctions from one cell to the next, resulting in spontaneous action potential conduction. Gap junctions are found between smooth and cardiac muscle fibers (cells), as well as between epithelial cells. The presence of gap junctions between muscle fibers presumably enables the cells to conduct the action potentials rapidly and contract in an orderly and coordinated manner.

Classification

Considering layers and cell shapes in combination, we may classify covering and lining epithelium as follows:

Simple
1. Squamous
2. Cuboidal
3. Columnar

Stratified
1. Squamous
2. Cuboidal
3. Columnar
4. Transitional

Pseudostratified columnar

Each of the epithelial tissues described in the following sections is illustrated in Exhibit 3-1.

EXHIBIT 3-1
Epithelial Tissues

COVERING AND LINING EPITHELIUM

Simple Squamous Epithelium

Description: Single layer of flat, scalelike cells; centrally located nuclei.
Location: Lines air sacs of lungs, glomerular (Bowman's) capsule of kidneys, and inner surface of the tympanic membrane (eardrum) of ear. Called endothelium when it lines heart, blood, and lymphatic vessels, and forms capillaries. Called mesothelium when it lines the ventral body cavity and covers viscera as part of a serous membrane.
Function: Filtration, absorption, exchange, and secretion in serous membranes.

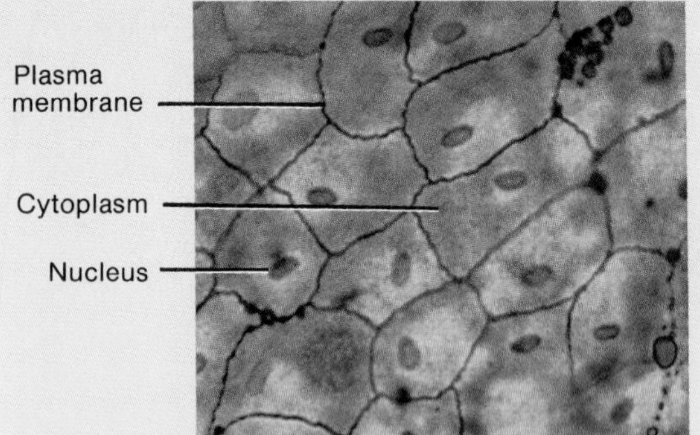

Surface view of mesothelial lining of peritoneal cavity (243 x)

(© Biophoto, Photo Researchers)

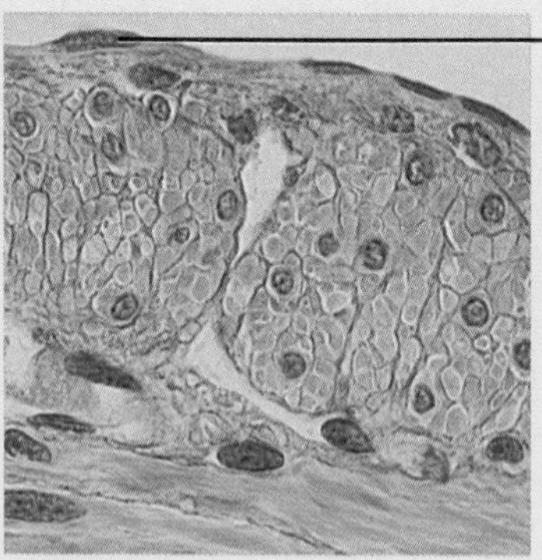

Sectional view of intestinal serosa (245 ×)

(Courtesy of Douglas Merrill)

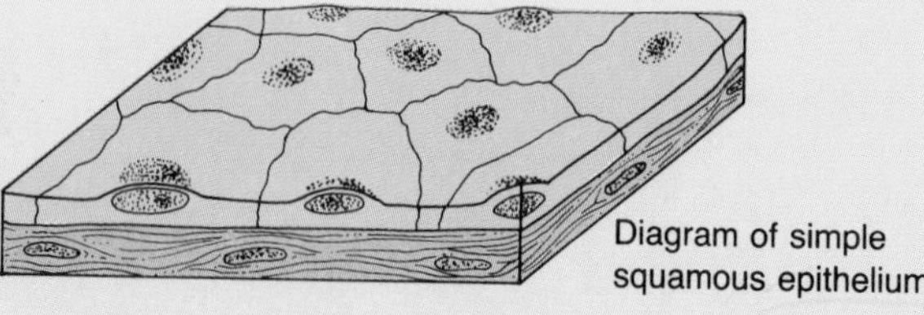

KNOW & RECOGNIZE

Simple Cuboidal Epithelium

Description: Single layer of cube-shaped cells; centrally located nuclei.
Location: Covers surface of ovary, lines anterior surface of capsule of the lens of eyes, forms pigmented epithelium of retina of eye, and lines kidney tubules and smaller ducts of many glands.
Function: Secretion and absorption.

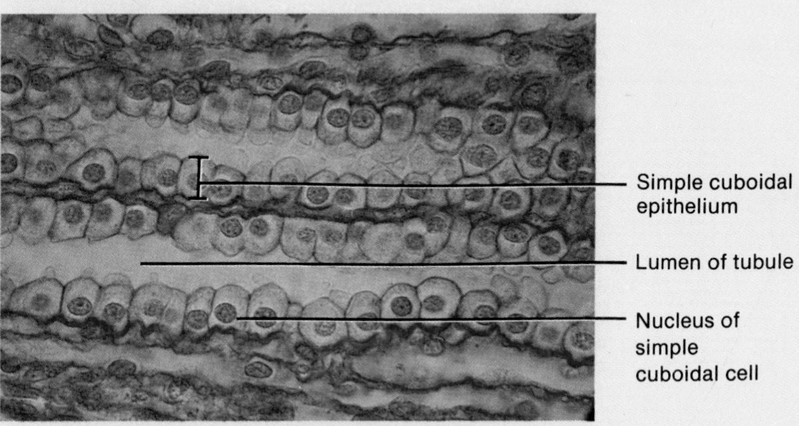

Sectional view of kidney tubules, (575 ×)

(Courtesy of Ed Reschke)

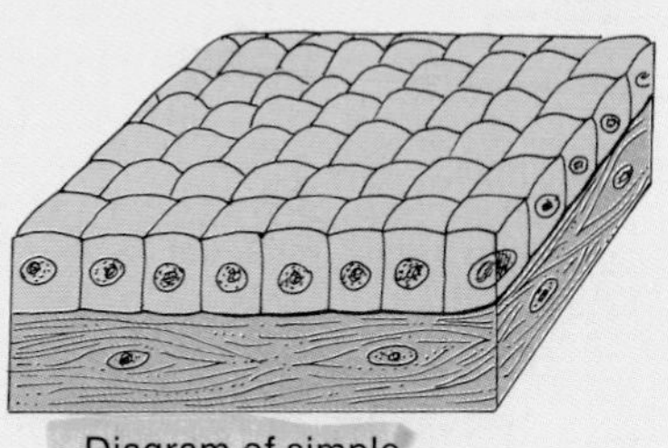

Diagram of simple cuboidal epithelium

Simple Columnar (Nonciliated) Epithelium

Description: Single layer of nonciliated rectangular cells; nuclei at bases of cells; contains goblet cells in some locations.
Location: Lines the gastrointestinal tract from the stomach to the anus, excretory ducts of many glands, and gallbladder.
Function: Secretion and absorption.

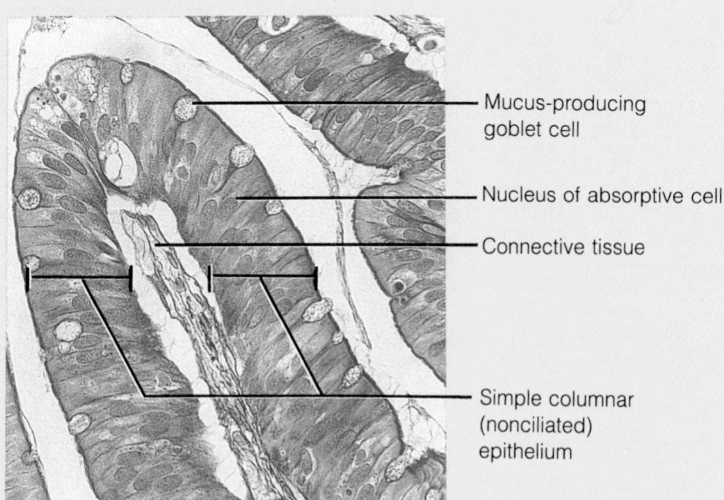

Sectional view of goblet cells from the small intestine (140x)

(Courtesy of Lester Bergman and Associates)

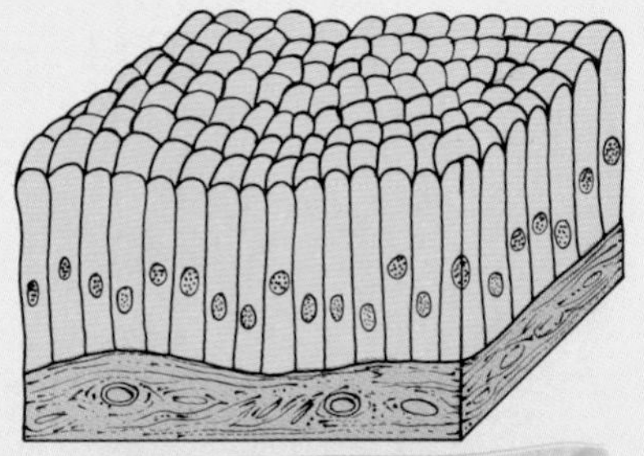

Diagram of simple columnar (nonciliated) epithelium

Continued

EXHIBIT 3-1 (*Continued*)

Simple Columnar (Ciliated) Epithelium

Description: Single layer of ciliated columnar cells; nuclei at bases of cells; contains goblet cells in some locations.
Location: Lines a few portions of upper respiratory tract, uterine (Fallopian) tubes, uterus, some paranasal sinuses, and central canal of spinal cord.
Function: Moves fluids such as mucus by ciliary action.

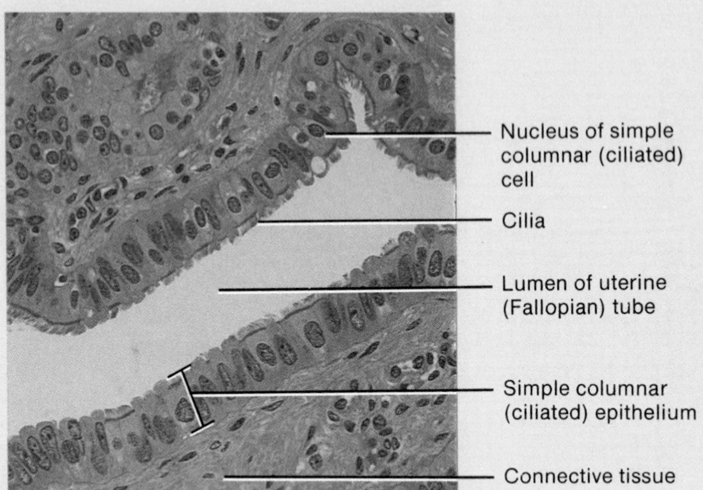

Sectional view of uterine (Fallopian) tube
(100 ×)
(Courtesy of Douglas Merrill)

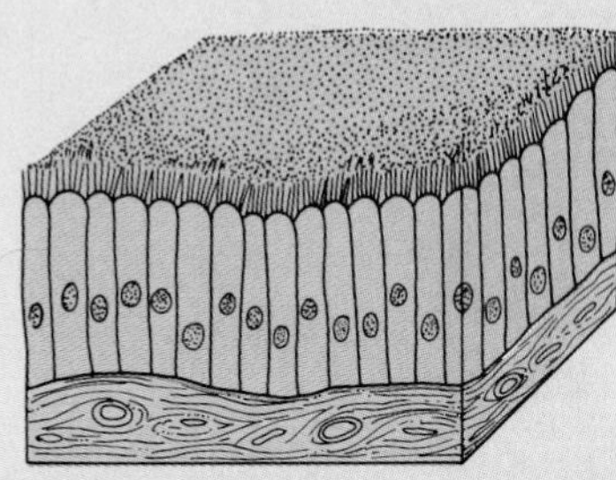

Diagram of simple columnar (ciliated) epithelium

Stratified Squamous Epithelium

Description: Several layers of cells; cuboidal to columnar shape in deep layers; squamous cells in superficial layers; basal cells replace surface cells as they are lost.
Location: Nonkeratinized variety lines wet surfaces such as lining of the mouth, tongue, esophagus, part of epiglottis, vagina, and covering of tongue; keratinized variety forms outer layer of skin.
Function: Protection.

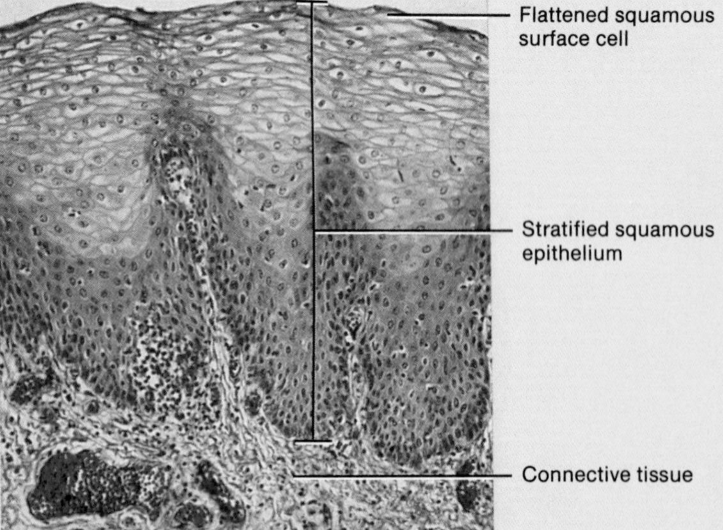

Sectional view of vagina (200 x)
(Courtesy of Biophoto, SPL, Photo Researchers)

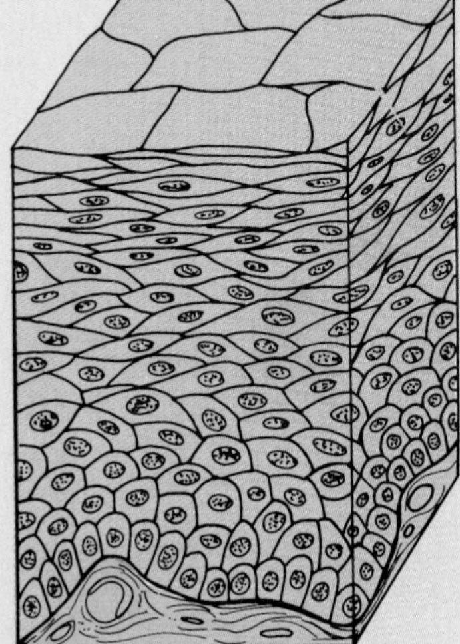

Diagram of stratified squamous epithelium

Stratified Cuboidal Epithelium

Description: Two or more layers of cells in which the surface cells are cube-shaped.
Location: Ducts of adult sweat glands, portion of conjunctiva of eye, cavernous urethra of male urogenital system, pharynx, and epiglottis.
Function: Protection.

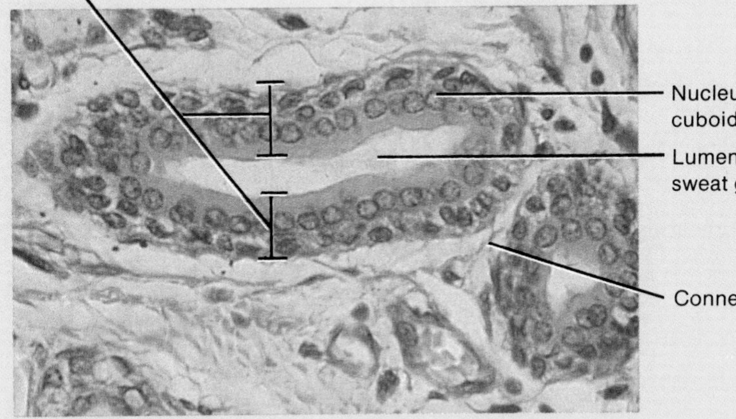

Sectional view of the duct of a sweat gland (450x)
(Courtesy of Biological Photo Service)

Diagram of stratified cuboidal epithelium

Stratified Columnar Epithelium

Description: Several layers of polyhedral cells; columnar cells only in superficial layer.
Location: Lines part of male urethra, large excretory ducts of some glands, and small areas in anal mucous membrane.
Function: Protection and secretion.

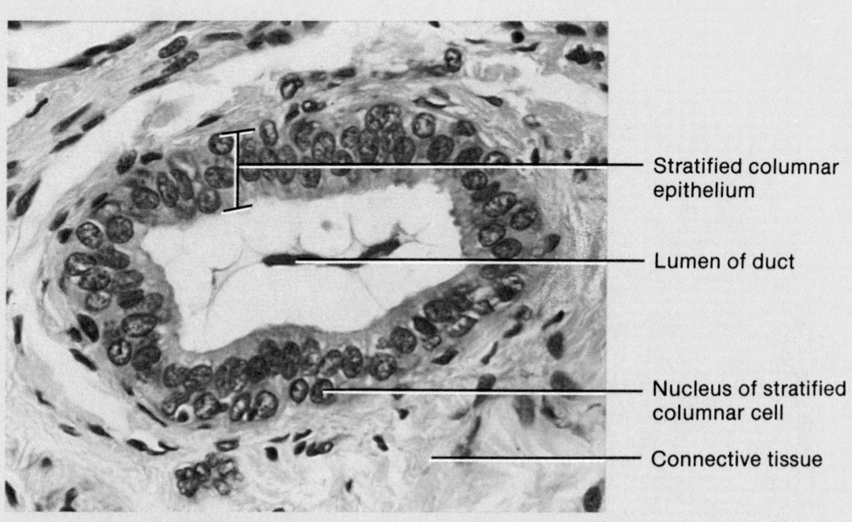

Sectional view of the duct of the submandibular salivary gland (495 x)
(Courtesy of Biophoto, SPL, Photo Researchers)

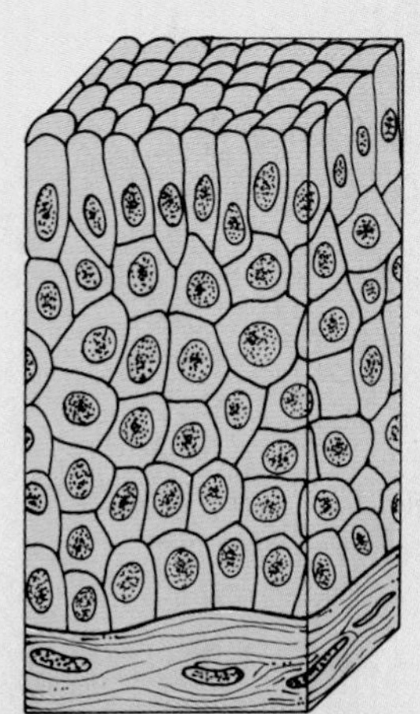

Diagram of stratified columnar epithelium

Continued

EXHIBIT 3-1 (*Continued*)

Transitional Epithelium

Description: Resembles nonkeratinized stratified squamous tissue, except that superficial cells are larger and have a rounded free surface.
Location: Lines urinary bladder and portions of ureters and urethra.
Function: Permits distention.

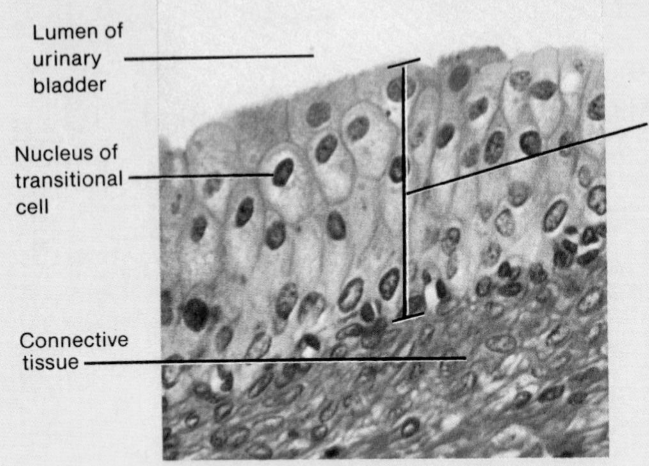

Sectional view of urinary bladder in relaxed state (240x)
(Biophoto, Photo Researchers)

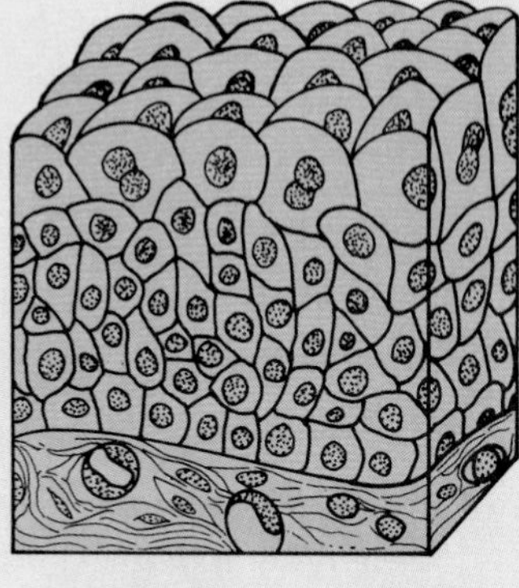

Diagram of transitional epithelium

Pseudostratified Columnar Epithelium

Description: Not a true stratified tissue; nuclei of cells at different levels; all cells attached to basement membrane, but not all reach surface.
Location: Lines larger excretory ducts of many large glands, epididymis, male urethra, and auditory (Eustachian) tubes; ciliated variety with goblet cells lines most of the upper respiratory tract and some ducts of male reproductive system.
Function: Secretion and movement of mucus and sperm cells by ciliary action.

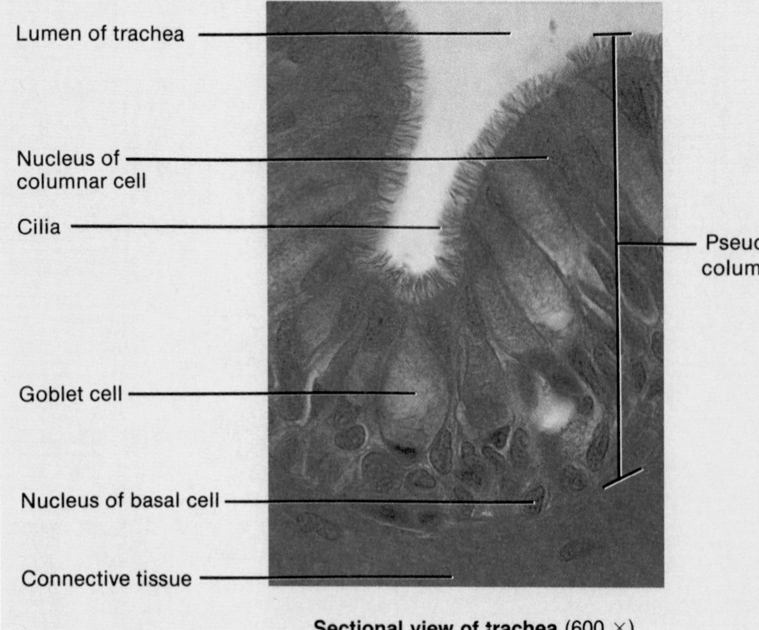

Sectional view of trachea (600 ×)
(Courtesy of Ed Reschke)

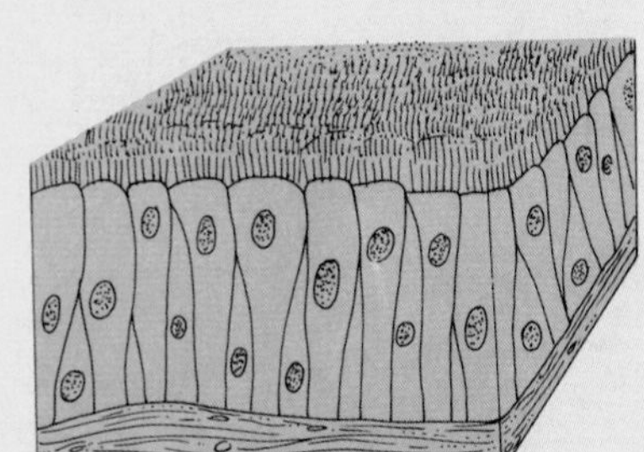

Diagram of pseudostratified columnar epithelium

GLANDULAR EPITHELIUM

Exocrine Gland

Description: Secretes products into ducts.
Location: Sweat, oil, wax, and mammary glands of the skin; digestive glands such as salivary glands that secrete into mouth cavity and pancreas that secretes into the small intestine.
Function: Produces mucus, perspiration, oil, wax, milk, or digestive enzymes.

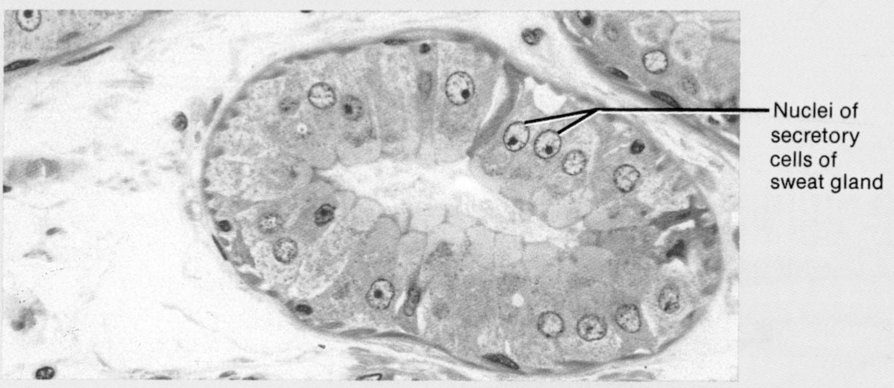

Sectional view of the secretory portion of a sweat gland (1532×)

(© Bruce Iverson)

Endocrine Gland

Description: Secretes hormones into blood.
Location: Pituitary at base of brain, thyroid and parathyroids near larynx, adrenals (suprarenals) above kidneys, ovaries in pelvic cavity, testes in scrotum, and thymus in the thoracic cavity.
Function: Produces hormones that regulate various body activities.

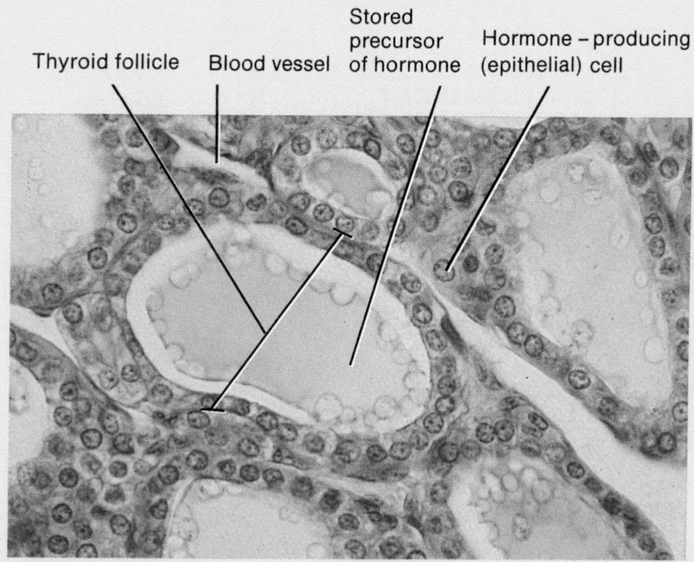

Sectional view of thyroid gland (500 ×)

(Courtesy of Lester Bergman and Associates)

Simple Epithelium

■ ***Simple Squamous Epithelium*** This type of simple epithelium consists of a single layer of flat, scalelike cells. Its surface resembles a tiled floor. The nucleus of each cell is centrally located and oval or spherical. Since simple squamous epithelium has only one layer of cells, it is highly adapted to diffusion, osmosis, and filtration. Thus, it lines the air sacs of the lungs, where respiratory gases (oxygen and carbon dioxide) are exchanged between air spaces and blood. It is present in the part of the kidney that filters blood. It also lines the inner surface of the tympanic membrane (eardrum) of the ear. Simple squamous epithelium is found in body parts that are subject to little wear and tear.

Simple squamous epithelium that lines the heart, blood vessels, and lymphatic vessels and forms the walls of capillaries is known as ***endothelium.*** Simple squamous epithelium that forms the epithelial layer of serous membranes is called ***mesothelium.*** Serous membranes line the thoracic and abdominopelvic cavities and cover the viscera within them. Both endothelium and mesothelium are derived from mesoderm.

■ ***Simple Cuboidal Epithelium*** Viewed from above, the cells of simple cuboidal epithelium appear as closely fitted polygons. The cuboidal nature of the cells is obvious only when the tissue is sectioned at right angles. Like simple squamous epithelium, these cells possess a central nucleus that is usually round. Simple cuboidal epithelium covers the surface of the ovaries, lines the anterior surface of the capsule of the lens of the eye, and forms the pigmented epithelium of the retina of the eye. In the kidneys, where it forms the kidney tubules and contains microvilli, it functions in water reabsorption. It also lines the smaller ducts of some glands and the secreting units of glands, such as the thyroid.

Simple cuboidal epithelium performs the functions of secretion and absorption. ***Secretion*** is the production and release by cells of a fluid that may contain a variety of substances such as mucus, perspiration, or enzymes. ***Absorption*** is the intake of fluids or other substances by cells of the skin or mucous membranes.

■ ***Simple Columnar Epithelium*** The surface view of simple columnar epithelium is similar to that of simple cuboidal tissue. When sectioned at right angles to the surface, however, the cells appear somewhat rectangular. The nuclei, usually located near the bases of the cells, are commonly oval.

The luminal surfaces (surfaces adjacent to the lumen, or cavity of a hollow organ, vessel, or duct) of simple columnar epithelial cells are modified in several ways, depending on location and function. Simple columnar epithelium lines the gastrointestinal tract from the stomach to the anus, the gallbladder, and excretory ducts of many glands. In such sites, the cells protect the underlying tissues. Many of the cells are also modified to aid in the digestive process. In the small intestine especially, the plasma membranes of the columnar cells have microscopic fingerlike projections called ***microvilli*** (see Figure 2-1). These structures serve to increase the surface area of the plasma membrane and thereby allow larger amounts of digested nutrients and fluids to be absorbed into the body.

Interspersed among the typical columnar cells of the intestine are other modified columnar cells called ***goblet cells.*** These cells, which secrete mucus, are so named because the mucus accumulates in the upper half of the cell, causing the area to bulge out. The whole cell resembles a goblet or wine glass. The secreted mucus serves as a lubricant and protects the walls of the gastrointestinal tract.

Another modification of columnar epithelium is found in cells with hairlike processes called ***cilia*** (*cillio* = to move). In a few portions of the upper respiratory tract, ciliated columnar cells are interspersed with goblet cells. Mucus secreted by the goblet cells forms a film over the respiratory surface. This film traps foreign particles that are inhaled. The cilia wave in unison and move the mucus, with any foreign particles, toward the throat, where it can be swallowed or eliminated. Ciliated columnar epithelium is also found in the uterus and uterine (Fallopian) tubes of the female reproductive system, some paranasal sinuses, and the central canal of the spinal cord.

Stratified Epithelium

In contrast to simple epithelium, stratified epithelium consists of at least two layers of cells. Thus, it is more durable and can protect underlying tissues from the external environment and from wear and tear. Some stratified epithelial cells also produce secretions. The name of the specific kind of stratified epithelium depends on the shape of the *surface* cells.

■ ***Stratified Squamous Epithelium*** In the more superficial layers of this type of epithelium, the cells are flat, whereas in the deep layers, cells vary in shape from cuboidal to columnar. The basal, or bottom, cells continually replicate by cell division. As new cells grow, they push the surface cells outward. The cells of the base layer continually shift upward and outward. As they move farther from the deep layer and their blood supply, they become dehydrated, shrink, and become harder. At the surface, the cells are rubbed off. New cells continually emerge, are sloughed off, and replaced.

One form of stratified squamous epithelium is called ***nonkeratinized stratified squamous epithelium.*** This tissue is found on wet surfaces that are subjected to considerable wear and tear—such as the lining of the mouth, the covering of the tongue, the esophagus, and the vagina. Another form of stratified squamous epithelium is called ***keratinized stratified squamous epithelium.*** The surface cells of this type form a tough layer of material containing ***keratin,*** a protein that is waterproof and resistant to friction and helps to resist bacterial invasion. The outer layer of skin, the epidermis, consists of keratinized stratified squamous epithelium.

■ ***Stratified Cuboidal Epithelium*** This relatively rare type of epithelium is found in the ducts of the sweat glands of adults, a portion of the conjunctiva of the eye, cavernous urethra of the male urogenital system, pharynx, and epiglottis. It sometimes consists of more than two layers of cells. Its function is mainly protective.

■ ***Stratified Columnar Epithelium*** Like stratified cuboidal epithelium, this type of tissue is also uncommon in the body. Usually the basal layer or layers consist of shortened, irregularly polyhedral cells. Only the superficial cells are columnar in form. This kind of epithelium lines part of the male urethra, some larger excretory ducts such as lactiferous (milk) ducts in the mammary glands, and small areas in the anal mucous membrane. It functions in protection and secretion.

■ ***Transitional Epithelium*** This kind of epithelium is very much like nonkeratinized stratified squamous epithelium. The distinction is that cells of the outer layer in transitional epithelium tend to be large and rounded rather than flat. This feature allows the tissue to be stretched (distended) without the outer cells breaking apart from one another. When stretched, they are drawn out into squamous-like cells. Because of this arrangement, transitional epithelium lines hollow structures that are subjected to expansion from within, such as the urinary bladder, and parts of the ureters and urethra. Its function is to help prevent a rupture of the organ.

Pseudostratified Columnar Epithelium

The third category of covering and lining epithelium consists of columnar cells and is called pseudostratified columnar epithelium. The nuclei of the cells are at varying depths. Even though all the cells are attached to the basement membrane in a single layer, some do not reach the surface. This feature gives the impression of a multilayered tissue, the reason for the designation *pseudo*stratified epithelium. The cells that do reach the surface either secrete mucus or contain cilia that move mucus and foreign particles for eventual elimination from the body. It lines the larger excretory ducts of many glands, epididymis, parts of the male urethra, and the auditory (Eustachian) tubes, the tubes that connect the middle ear cavity and upper part of the throat. Pseudostratified columnar epithelium may contain cilia and goblet cells. It lines most of the upper respiratory tract and certain ducts of the male reproductive system.

GLANDULAR EPITHELIUM

The function of glandular epithelium is secretion, accomplished by glandular cells that lie in clusters deep to the covering and lining epithelium. A ***gland*** may consist of one cell or a group of highly specialized epithelial cells that secrete substances into ducts, onto a surface, or into the blood. The protection of such substances always requires active work by the cells and results in an expenditure of energy.

All glands of the body are classified as exocrine or endocrine according to whether they secrete substances into ducts (or directly onto a free surface) or into the blood. ***Exocrine glands*** secrete their products into ducts (tubes) that empty at the surface of covering and lining epithelium or directly onto a free surface. The product of an exocrine gland may be released at the skin surface or into the lumen of a hollow organ. The secretions of exocrine glands include mucus, perspiration, oil, wax, and digestive enzymes. Examples of exocrine glands are sweat glands, which eliminate perspiration to cool the skin and salivary glands, which secrete a digestive enzyme. ***Endocrine glands*** are ductless and ultimately secrete their products into the blood. The secretions of endocrine glands are always hormones, chemicals that regulate various physiological activities. The pituitary, thyroid, and adrenal (suprarenal) glands are examples of endocrine glands.

Structural Classification of Exocrine Glands

Exocrine glands are classified into two structural types: unicellular and multicellular. ***Unicellular glands*** are single-celled. A good example of a unicellular gland is a goblet cell (see Exhibit 3-1, simple columnar, nonciliated). Although goblet cells do not contain ducts, they are often classified as unicellular, mucus-secreting exocrine glands. Goblet cells are found in the epithelial lining of the digestive, respiratory, urinary, and reproductive systems. They produce mucus to lubricate the free surfaces of these membranes.

Multicellular glands occur in several different forms (Figure 3-2). If the secretory portions of a gland are tubular, it is referred to as a ***tubular gland.*** If they are flasklike, it is called an ***acinar*** (AS-i-nar) ***gland.*** If the gland contains both tubular and flasklike secretory portions, it is called a ***tubuloacinar gland.*** Further, if the duct of the gland does not branch, it is referred to as a ***simple gland;*** if the duct does branch, it is called a ***compound gland.*** By combining the shape of the secretory portion with the degree of branching of the duct, we arrive at the following structural classification for exocrine glands:

I. ***Unicellular.*** Single-celled gland that secretes mucus. Example: goblet cell of the digestive and respiratory systems.

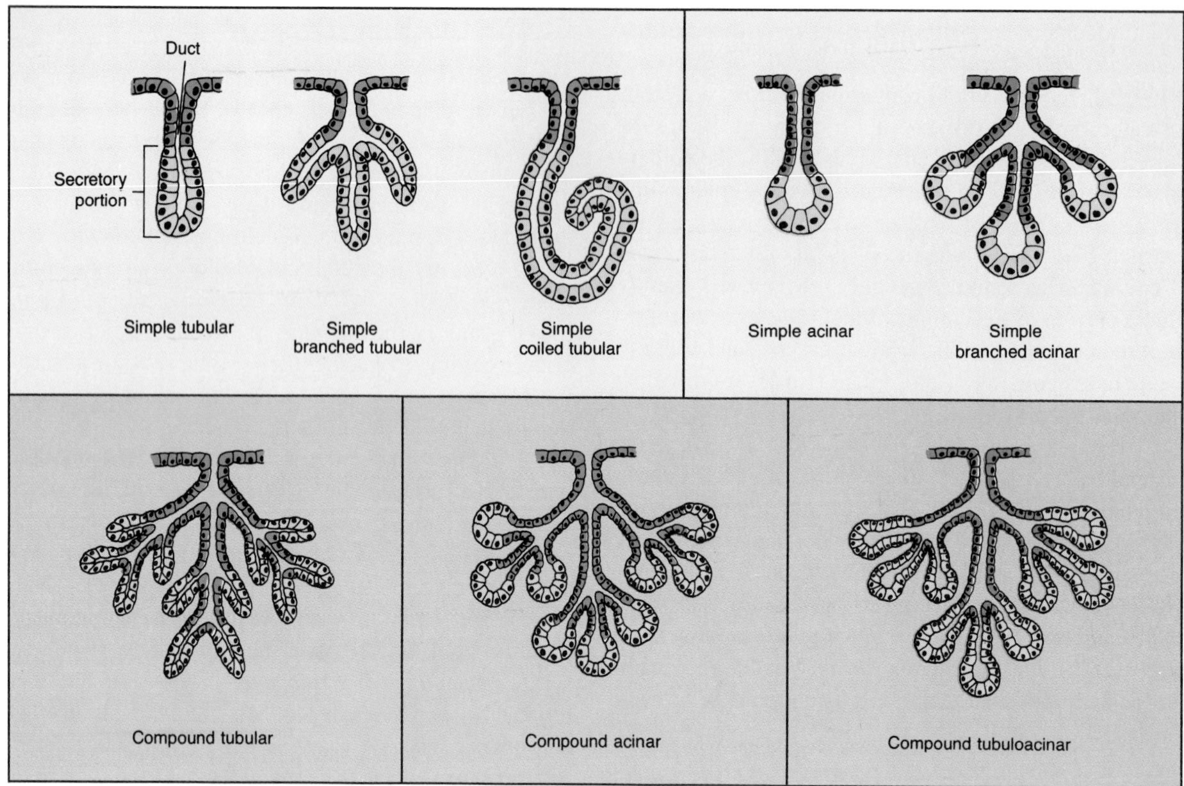

FIGURE 3-2 Structural types of multicellular exocrine glands. The secretory portions of the glands are indicated in gold. The blue areas represent the ducts of the glands.

II. ***Multicellular.*** Many-celled glands.
   A. ***Simple.*** Single, nonbranched duct.
      1. ***Tubular.*** The secretory portion is straight and tubular. Example: intestinal glands.
      2. ***Branched tubular.*** The secretory portion is branched and tubular. Examples: gastric and uterine glands.
      3. ***Coiled tubular.*** The secretory portion is coiled. Example: eccrine sudoriferous (sweat) glands.
      4. ***Acinar.*** The secretory portion is flasklike. Example: seminal vesicle glands.
      5. ***Branched acinar.*** The secretory portion is branched and flasklike. Example: sebaceous (oil) glands.
   B. ***Compound.*** Branched duct.
      1. ***Tubular.*** The secretory portion is tubular. Examples: bulbourethral (Cowper's) glands, testes, and liver.
      2. ***Acinar.*** The secretory portion is flasklike. Example: salivary glands (submandibular).
      3. ***Tubuloacinar.*** The secretory portion is both tubular and flasklike. Examples: salivary glands (parotid) and pancreas.

Functional Classification of Exocrine Glands

The functional classification of exocrine glands is based on whether a secretion is a product of a cell or consists of entire or partial glandular cells themselves. The three recognized categories are holocrine, merocrine, and apocrine glands. ***Holocrine glands*** accumulate a secretory product in their cytoplasm. The cell then dies and is discharged with its contents as the glandular secretion (Figure 3-3a). The discharged cell is replaced by a new cell. One example of a holocrine gland is a sebaceous (oil) gland of the skin. ***Merocrine (eccrine) glands*** simply form the secretory product and discharge it from the cell (Figure 3-3b). Most exocrine glands of the body are merocrine. Examples of merocrine glands are the salivary glands and pancreas. ***Apocrine glands*** accumulate their secretory product at the apical (outer) margin of the secreting cell. That portion of the cell pinches off from the rest of the cell to form the secretion (Figure 3-3c). The remaining part of the cell repairs itself and repeats the process. Examples of apocrine glands are large, modified sweat glands found in the axillary, anal, and genital glands.

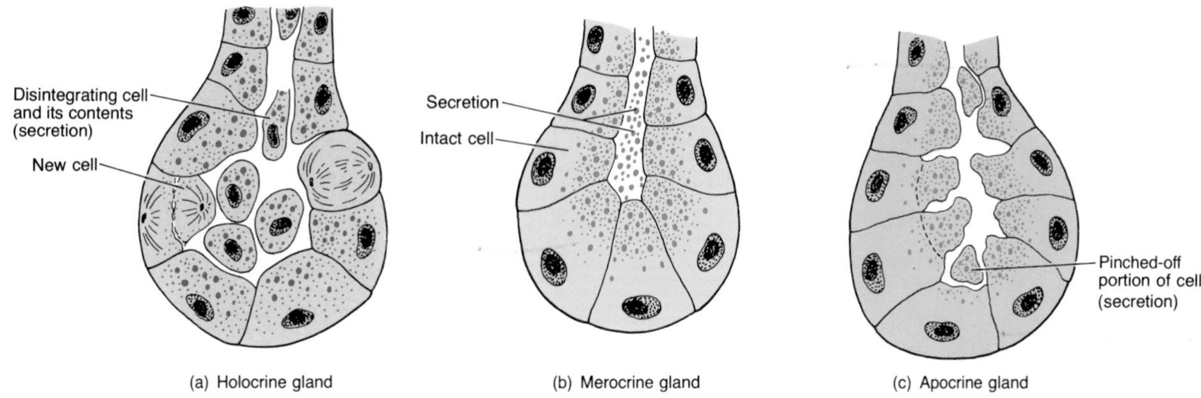

FIGURE 3-3 Functional classification of multicellular exocrine glands.

CONNECTIVE TISSUE

Connective tissue is found throughout the body. In fact, it is the most abundant and widely distributed of the four basic tissues.

GENERAL FEATURES

1. Connective tissue consists of three basic elements: cells, ground substance, and fibers. Together the ground substance and fibers, which are outside the cells, are referred to as the ***matrix.*** The three elements vary, depending on the tissue and its functions. Unlike epithelium, connective tissue cells are widely scattered with considerable matrix.
2. Unlike epithelium, connective tissue does not occur on free surfaces, such as the surface of a body cavity or the external surface of the body.
3. Except for cartilage, connective tissue, like epithelium, has a nerve supply.
4. Unlike epithelium, connective tissue usually is highly vascular (rich blood supply). Whereas tendons have a scanty blood supply, cartilage is avascular.
5. The matrix of a connective tissue—which may be fluid, semifluid, gellike, or fibrous—largely determines its qualities. In blood, the matrix is fluid. In cartilage, it is firm but pliable. In bone, it is considerably harder and not pliable.
6. Connective tissue with a good blood supply, such as bone, regenerates better than one with a poor blood supply, such as dense connective tissue, which makes up tendons.
7. All connective tissue develops from mesenchyme, a tissue derived from the mesoderm germ layer.
8. Functionally, connective tissue connects various body parts, supports, provides strength, protects, insulates, separates structures such as skeletal muscles, and functions in transportation (blood).

BASIC ELEMENTS

As noted earlier, connective tissue consists of three basic elements: cells, ground substance, and fibers.

Cells

The cells in connective tissue are derived from embryonic stem cells called mesenchymal cells. Each of the major groups of connective tissue contains an undifferentiated cell type that ends in *-blast.* This cell retains its capacity for division and secretes the matrix that is characteristic of the tissue. In most connective tissues, once the matrix is produced, the undifferentiated cells lose their capacity for cell division, maintain the matrix, and become mature cells whose name ends in *-cyte.* For example, the undifferentiated cell in connective tissue proper is called a fibroblast, whereas the mature cell is known as a fibrocyte. In cartilage the respective cells are chondroblasts and chondrocytes, and in bone they are osteoblasts and osteocytes.

Following are some of the cells contained in various types of connective tissue. The specific tissues to which they belong will be described shortly.

1. ***Fibroblasts*** (FĪ-brō-blasts) are large, flat cells with branching processes that produce matrix.
2. ***Fixed macrophages*** (MAK-rō-fā-jēz; *macro* = large; *phagein* = to eat), or ***histiocytes,*** are derived from ***monocytes,*** a type of white blood cell. Macrophages are irregular in shape with short branching projections and are capable of engulfing bacteria and cellular debris by phagocytosis. Thus, they provide a vital defense for the body.
3. ***Plasma cells*** are small and either round or irregular and develop from types of white blood cells called ***lymphocytes (B cells).*** Plasma cells produce antibodies and, accordingly, provide a defense mechanism through immunity. They are found in many places in the body, but most are found in connective tissue, especially that of the gastrointestinal tract and the mammary glands.

4. ***Mast cells*** are found in abundance along blood vessels. They form heparin, an anticoagulant that prevents blood from clotting in the vessels. Mast cells are also believed to produce histamine and serotonin, chemicals that dilate small blood vessels during inflammation.
5. Other cells in connective tissue include ***adipocytes*** **(*fat cells*)** and ***leucocytes*** **(*white blood cells*).**

Ground Substance

The ***ground substance*** is amorphous; that is, it has no specific shape. It may range from a fluid to a gel to a solid. It is produced by connective tissue cells and deposited in the space between the cells. Several examples of chemicals contained in ground substance are as follows: ***Hyaluronic*** (hī′-a-loo-RON-ik) ***acid*** is a viscous, fluidlike substance that binds cells together, lubricates joints, and maintains the shape of the eyeballs. ***Chondroitin*** (kon-DROY-tin) ***sulfate*** is a jellylike substance that provides support and adhesiveness in cartilage, bone, and blood vessels. ***Dermatan sulfate*** is found in the skin, tendons, and heart valves, and ***keratan sulfate*** is found in the cornea of the eye and bone.

The ground substance supports cells, binds them together, and provides a medium through which substances are exchanged between the blood and cells.

Fibers

Three types of fibers are embedded in the matrix between the cells of connective tissue: collagenous, elastic, and reticular fibers.

1. ***Collagenous*** (*kolla* = glue) ***fibers*** are very tough and resistant to a pulling force, yet allow some flexibility in the tissue because they are not taut. These fibers often occur in bundles. They are composed of many minute fibers called *fibrils* lying parallel to one another. The bundle arrangement affords a great deal of strength. Chemically, collagenous fibers consist of the protein *collagen*. Collagenous fibers are found in most types of connective tissue, especially bone, cartilage, tendons, and ligaments.

EXHIBIT 3-2

Connective Tissues

EMBRYONIC CONNECTIVE TISSUE

Mesenchyme

Description: Consists of star-shaped mesenchymal cells embedded in a fluid ground substance that contains delicate reticular fibers.

Location: Under skin and along developing bones of embryo; some mesenchymal cells found in adult connective tissue, especially along blood vessels.

Function: Forms all other kinds of connective tissue.

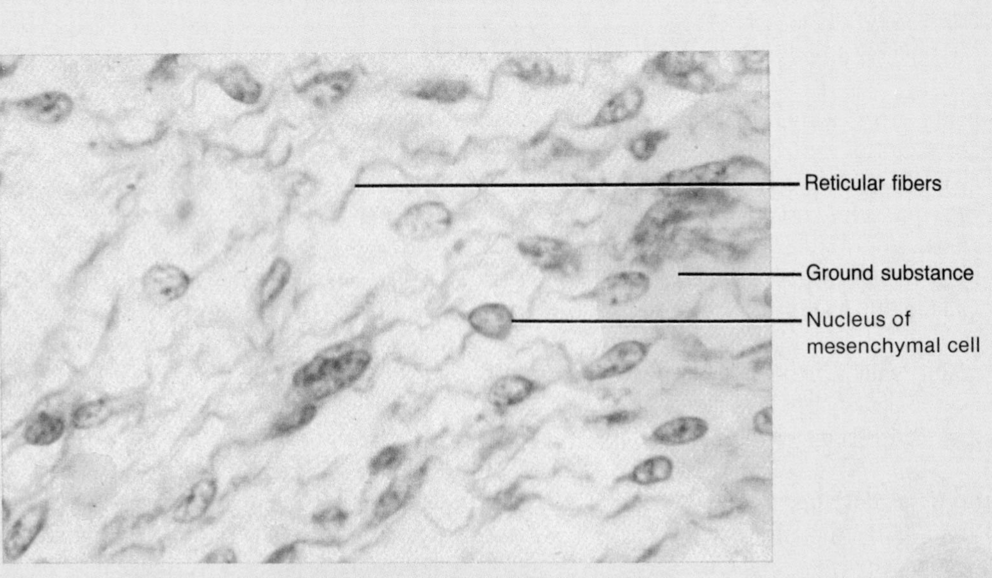

Sectional view of mesenchyme from a developing fetus (1008x)

(Courtesy of Andrew J. Kuntzman)

2. ***Elastic fibers,*** by contrast, are smaller than collagenous fibers and freely branch and rejoin one another. Elastic fibers consist of a protein called *elastin*. These fibers also provide strength and have great elasticity, up to 50 percent of their length. Elastic fibers are found in the skin, blood vessels, and lungs.
3. ***Reticular*** (*rete* = net) ***fibers*** also consist of collagen, plus some *glycoprotein*. They are very thin fibers that form branching networks. Like collagenous fibers, they provide support and strength and also form the ***stroma*** (framework) of many soft organs, such as the spleen and lymph nodes. The fibers also are found around fat, nerve and muscle cells, and in the basement membrane.

CLASSIFICATION

The classification of connective tissues is difficult because of the diversity of cells, ground substance, and fibers, and their relative proportions. Thus, the separation of connective tissues into categories is not always clear-cut. With this in mind, following is one of several classifications that might be used.

I. Embryonic connective tissue
- **A.** Mesenchyme
- **B.** Mucous connective tissue

II. Mature connective tissue
- **A.** Loose connective tissue
  - **1.** Areolar connective tissue
  - **2.** Adipose tissue
  - **3.** Reticular connective tissue
- **B.** Dense connective tissue
  - **1.** Dense regular
  - **2.** Dense irregular
  - **3.** Elastic
- **C.** Cartilage
  - **1.** Hyaline cartilage
  - **2.** Fibrocartilage
  - **3.** Elastic cartilage
- **D.** Bone (osseous) tissue
- **E.** Blood (vascular) tissue

Each of the connective tissues described in the following sections is illustrated in Exhibit 3-2.

Mucous Connective Tissue

Description: Consists of star-shaped cells embedded in a viscous ground substance that contains collagenous fibers.
Location: Umbilical cord of fetus.
Function: Support.

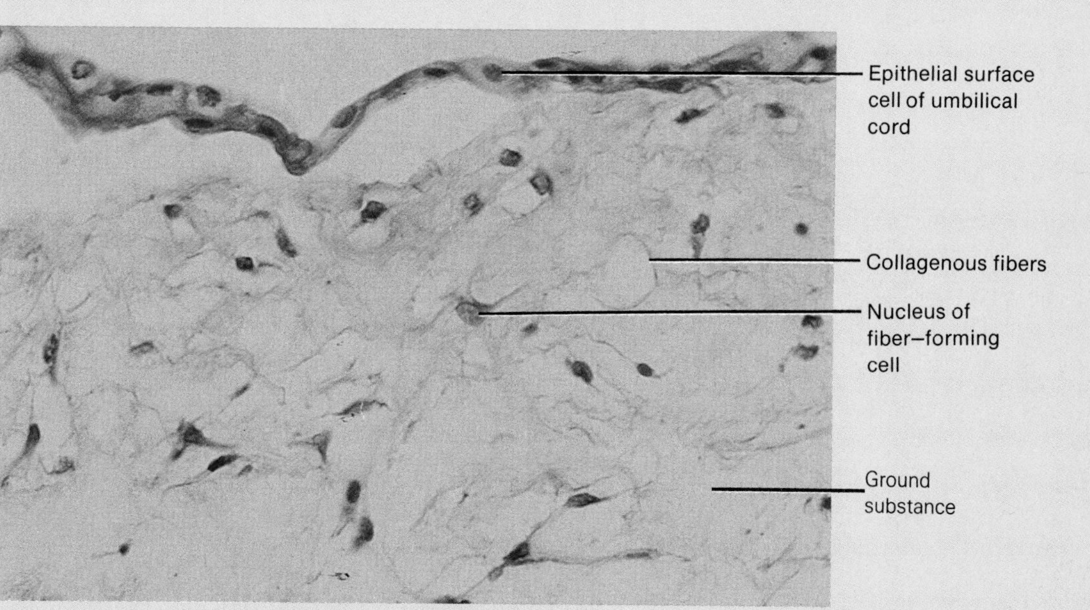

Sectional view of the umbilical cord (457 x)

(Courtesy of Lester Bergman and Associates)

Continued

EXHIBIT 3-2 (*Continued*)

MATURE CONNECTIVE TISSUE
Loose Connective Tissue
Areolar Connective Tissue

Description: Consists of fibers (collagenous, elastic, and reticular) and several kinds of cells—such as fibroblasts, fixed macrophages, plasma cells, and mast cells—embedded in a semifluid ground substance.
Location: Subcutaneous layer of skin, mucous membranes, blood vessels, nerves, and body organs.
Function: Strength, elasticity, and support.

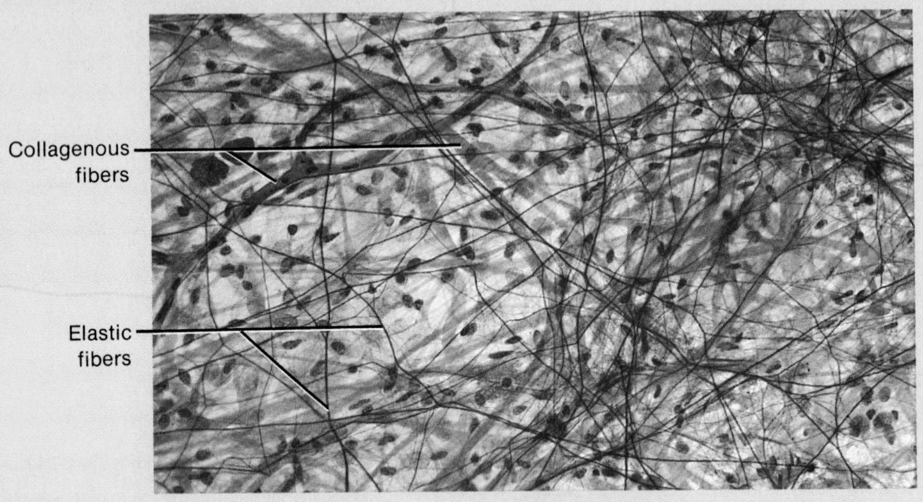

Sectional view of subcutaneous tissue (224x)
(Courtesy of Biophoto, Photo Researchers)

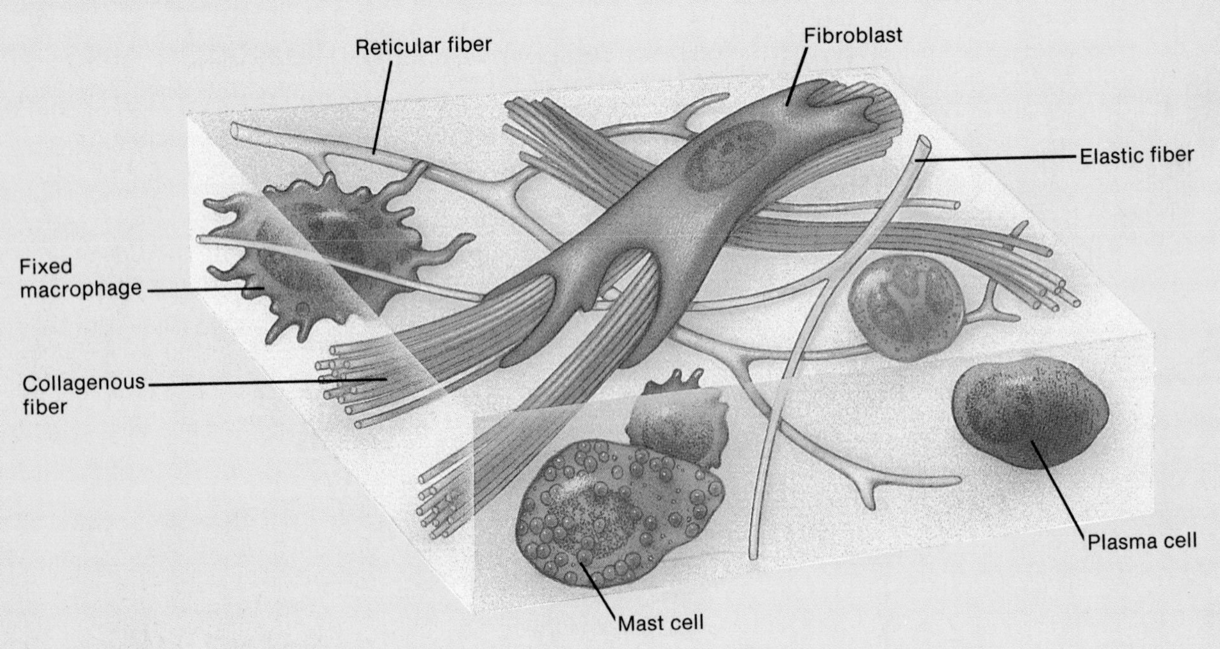

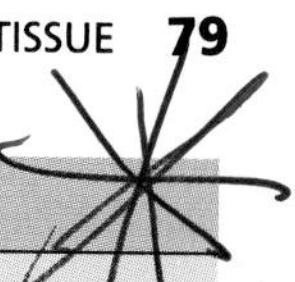

Adipose Tissue

Description: Consists of adipocytes, "signet ring"–shaped cells with peripheral nuclei, that are specialized for fat storage.
Location: Subcutaneous layer of skin, around heart and kidneys, marrow of long bones, and padding around joints.
Function: Reduces heat loss through skin, serves as an energy reserve, supports, and protects.

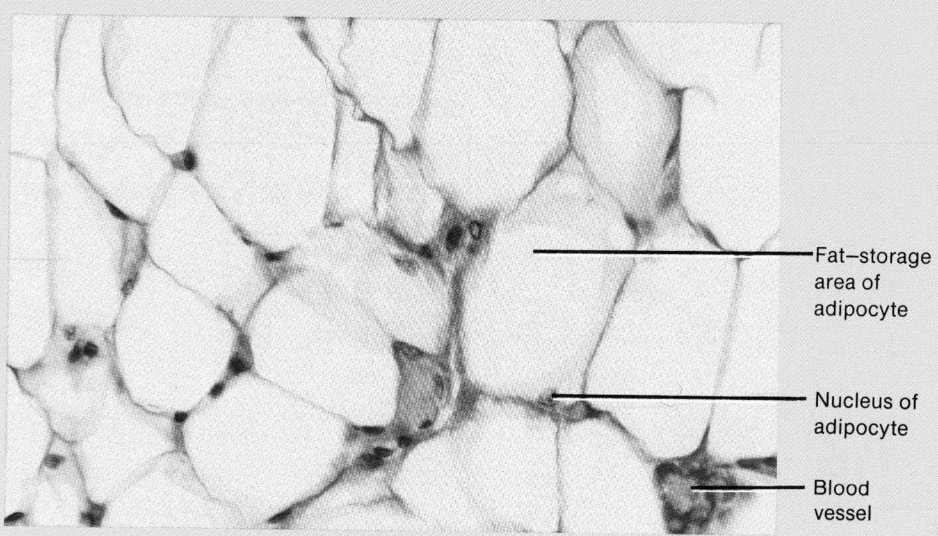

Sectional view of adipocytes of white fat of the pancreas (1720×)

(© Bruce Iverson)

Reticular Connective Tissue

Description: Consists of a network of interlacing reticular fibers and reticular cells.
Location: Liver, spleen, lymph nodes, and basal lamina of basement membranes.
Function: Forms stroma of organs, binds together smooth muscle tissue cells.

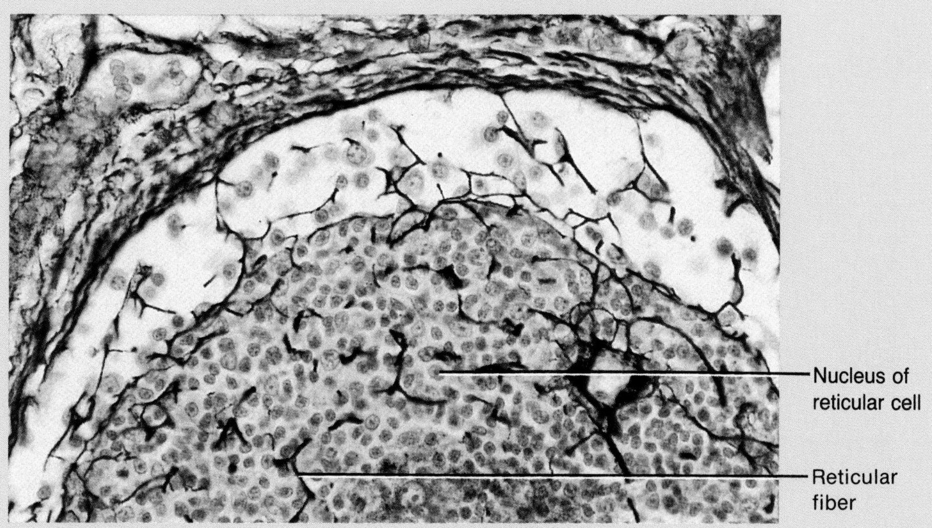

Sectional view of lymph node (496x)

(Courtesy of Biophoto, Photo Researchers)

Continued

EXHIBIT 3-2 (Continued)

Dense Connective Tissue

Dense Regular Connective Tissue

Description: Consists of predominantly collagenous fibers arranged in bundles; fibroblasts present in rows between bundles.
Location: Forms tendons, ligaments, and aponeuroses.
Function: Provides strong attachment between various structures.

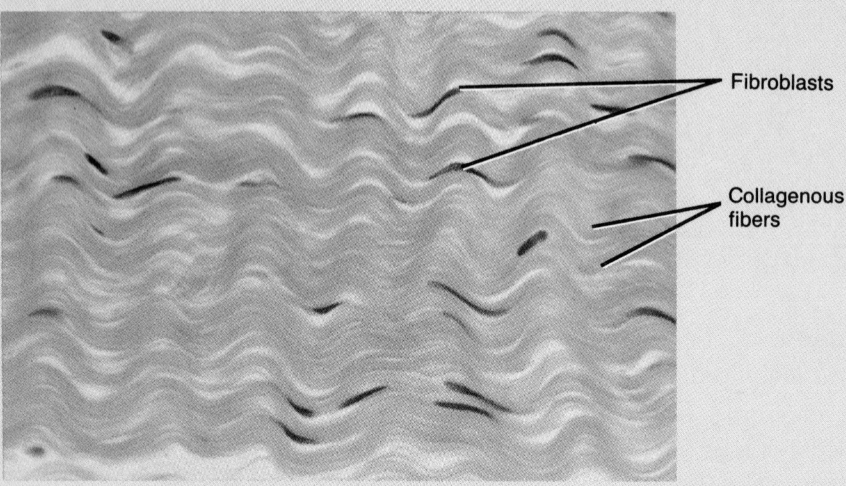

Sectional view of a tendon (250 ×)

(Courtesy of Andrew J. Kuntzman)

Dense Irregular Connective Tissue

Description: Consists of predominantly collagenous fibers randomly arranged.
Location: Fasciae, dermis of skin, and membranes around various organs.
Function: Provides strength.

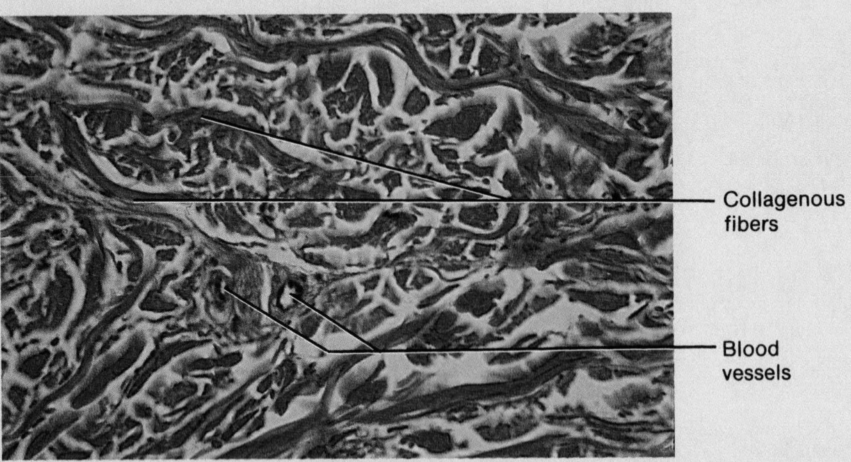

Sectional view of dermis of skin (275 ×)

(Courtesy of Ed Reschke)

Elastic Connective Tissue

Description: Consists of predominantly freely branching elastic fibers; fibroblasts present in spaces between fibers.
Location: Lung tissue, wall of arteries, trachea, bronchial tubes, true vocal cords, and ligmenta flava of vertebrae.
Function: Allows stretching of various organs.

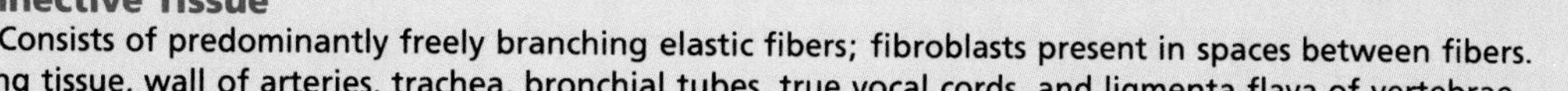

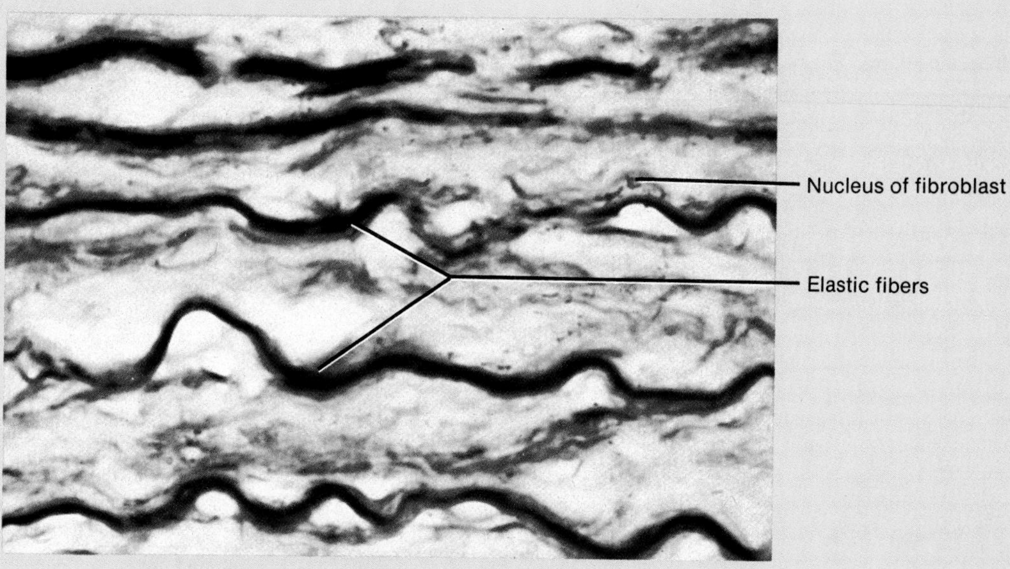

Sectional view of the aorta (largest artery in the body) (300x)

(Courtesy of Biophoto, Photo Researchers)

Cartilage

Hyaline Cartilage

Description: Also called gristle; ground substance appears as a bluish white, glossy mass; contains fine collagenous fibers and numerous chondrocytes; is the most abundant type of cartilage.
Location: Ends of long bones, ends of ribs, nose, parts of larynx, trachea, bronchi, bronchial tubes, and embryonic skeleton.
Function: Provides movement at joints, flexibility, and support.

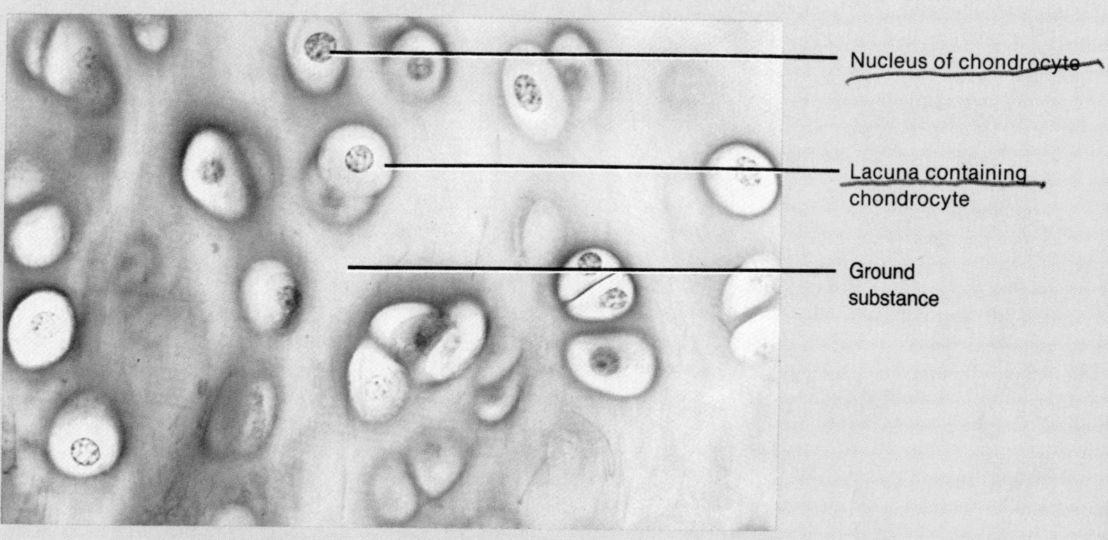

Sectional view of hyaline cartilage from trachea (512 x)

(Courtesy of Biophoto, Photo Researchers)

Continued

EXHIBIT 3-2 (*Continued*)

Fibrocartilage

Description: Consists of chondrocytes scattered among bundles of collagenous fibers.
Location: Pubic symphysis, intervertebral discs, and menisci of knee.
Function: Support and fusion.

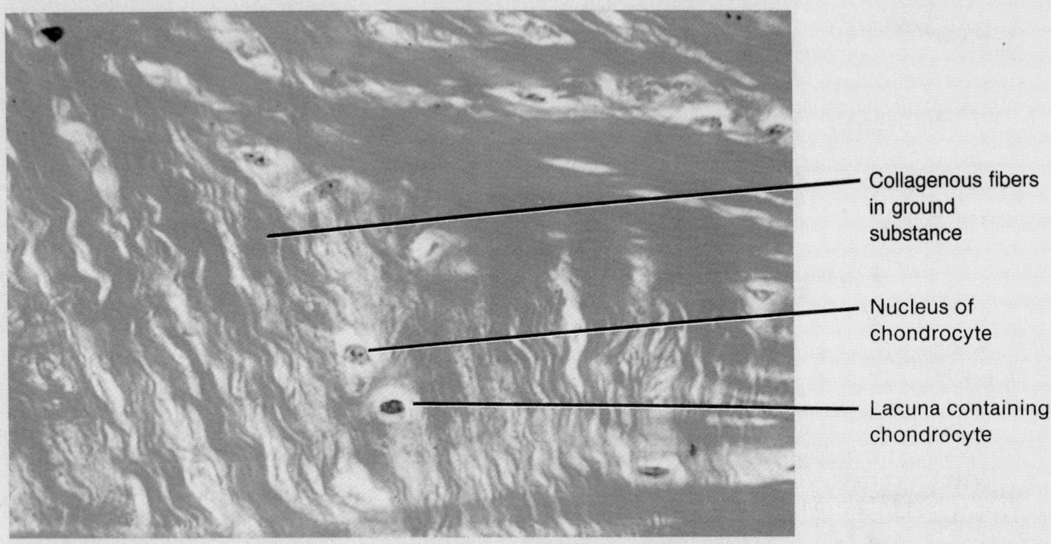

Sectional view of fibrocartilage from the patellar tendon insertion (288×)

(Courtesy of Biophoto, Photo Researchers)

Elastic Cartilage

Description: Consists of chondrocytes located in a threadlike network of elastic fibers.
Location: Epiglottis of larynx, external ear, and auditory (Eustachian) tubes.
Function: Gives support and maintains shape.

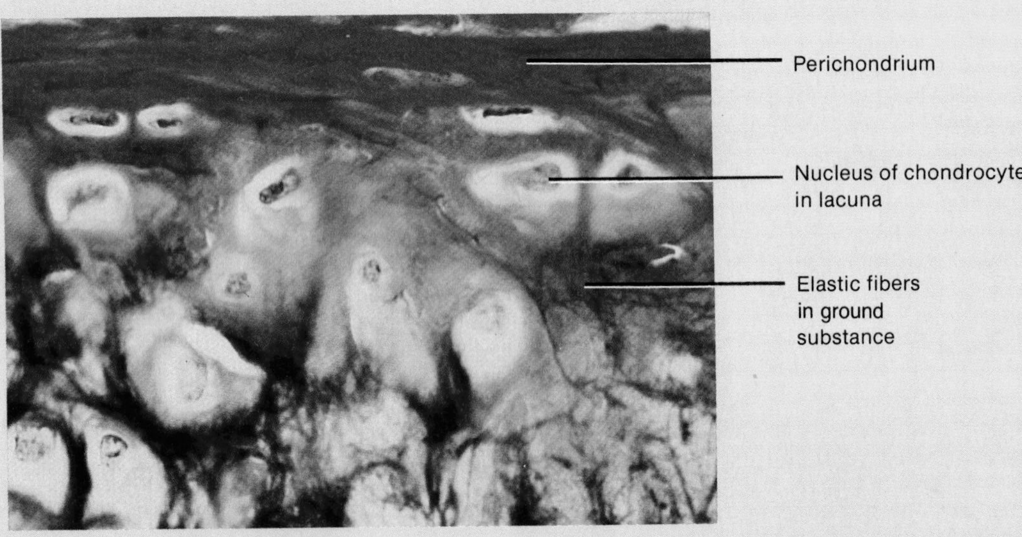

Sectional view of elastic cartilage from pinna of ear (288x)

(Courtesy of Biophoto, Photo Researchers)

Bone (Osseous)

Description: Compact bone consists of osteons (Haversian systems) that contain lamellae, lacunae, osteocytes, canaliculi, and central (Haversian) canals. See also Figure 5-3.
Location: Both compact and spongy bone comprise the various bones of the body.
Function: Support, protection, storage, houses blood-forming tissue, and serves as levers that act in conjunction with muscle tissue to provide movement.

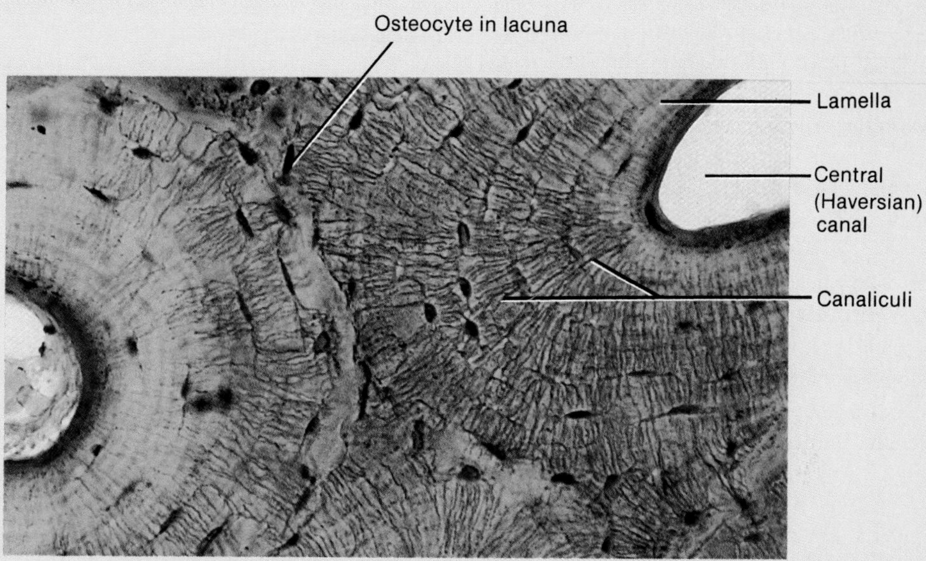

Sectional view of portions of two osteons (Haversian systems) from the femur (150x)

(Courtesy of Biophoto, Photo Researchers)

Blood (Vascular)

Description: Consists of plasma and formed elements (erythrocytes, leukocytes, and thrombocytes).
Location: Within blood vessels (arteries, arterioles, capillaries, venules, and veins).
Function: Erythrocytes transport oxygen and carbon dioxide, leucocytes carry on phagocytosis and are involved in allergic reactions and immunity, and thrombocytes are essential for the clotting of blood.

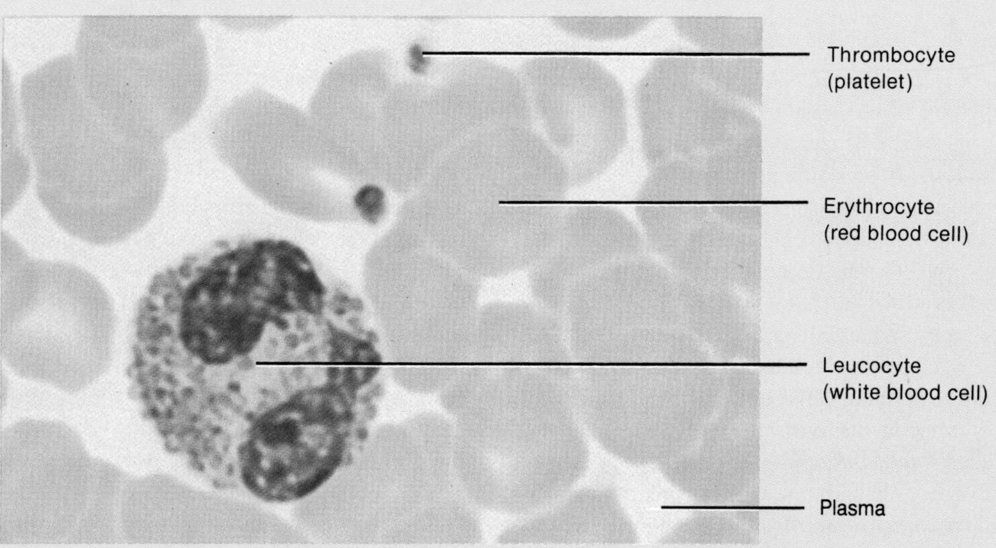

Blood smear (2350x)

(Courtesy of Lester Bergman and Associates)

EMBRYONIC CONNECTIVE TISSUE

Connective tissue that is present primarily in the embryo or fetus is called ***embryonic connective tissue.*** The term *embryo* refers to a developing human from fertilization through the first two months of pregnancy; a *fetus* refers to a developing human from the third month of pregnancy to birth.

One example of embryonic connective tissue found almost exclusively in the embryo is ***mesenchyme*** (MEZ-en-kīm)—the tissue from which all other connective tissues eventually arise. It is composed of star-shaped mesenchymal cells, a fluid ground substance, and delicate reticular fibers. Mesenchyme is located beneath the skin and along the developing bones of the embryo. Some mesenchymal cells are scattered irregularly throughout mature connective tissue, most frequently around blood vessels. Here mesenchymal cells differentiate into fibroblasts that assist in wound healing.

Another kind of embryonic connective tissue is ***mucous connective tissue* (*Wharton's jelly*),** found primarily in the fetus. It is a form of mesenchyme that contains star-shaped cells, a more mucous ground substance, and collagenous fibers. This tissue is located in the umbilical cord of the fetus, where it supports the wall of the cord.

MATURE CONNECTIVE TISSUE

Mature connective tissue exists in the newborn, has differentiated from the mesenchyme, and does not change after birth. It is subdivided into several kinds.

Loose Connective Tissue

Loose connective tissue contains groups of loosely arranged fibers and many cells. The types are areolar connective tissue, adipose tissue, and reticular connective tissue.

■ ***Areolar Connective Tissue*** Areolar (a-RĒ-ō-lar) connective tissue is one of the most widely distributed connective tissues in the body. It contains several types of cells, including fibroblasts, fixed macrophages, plasma cells, mast cells, adipocytes, and a few white blood cells. The ground substance contains hyaluronic acid, chondroitin sulfate, dermatan sulfate, and keratan sulfate. The fibers present are collagenous, elastic, and reticular.

The matrix materials normally aid the passage of nutrients from the blood vessels of the connective tissue into adjacent cells and tissues. The thick consistency of hyaluronic acid, however, may impede the movement of some drugs. If an enzyme called ***hyaluronidase*** is injected into the tissue, the ground substance changes to a watery consistency. This feature is of clinical importance because the reduced viscosity hastens the absorption and diffusion of injected drugs and fluids through the tissue and thus can lessen tension and pain. White blood cells, sperm, and some bacteria produce hyaluronidase.

Areolar connective tissue is continuous throughout the body. It is present in all mucous membranes and around all blood vessels and nerves. It also occurs around body organs and in the papillary (superficial) region of the dermis of the skin. Combined with adipose tissue, it forms the ***subcutaneous*** (sub′-kyoo-TĀ-nē-us; *sub* = under, *cut* = skin) ***layer,*** the layer of tissue that attaches the skin to underlying tissues and organs. The subcutaneous layer is also referred to as the ***superficial fascia*** (FASH-ē-a) or ***hypodermis.***

CLINICAL APPLICATION

Marfan Syndrome

Marfan* (*mar-FAN*) *syndrome is an inherited disorder that affects about 1 person in 10,000. It results in abnormalities of connective tissue, most likely a defect in collagen or elastin, especially in the skeleton, eyes, and cardiovascular system. Marfan victims tend to be tall, with disproportionately long arms, an unusually long lower half of the body, very long fingers and toes, an especially elongated thumb, an overly curved backbone, a malformed breastbone that either curves outward or inward, a backward curve of the legs, flat feet, and loose joints. Other common, visible signs of Marfan syndrome include leanness, small muscle mass, crowded teeth, and nearsightedness. Among the cardiovascular problems—the most serious ones—are a weakened area in the aorta (main artery that emerges from the heart), weakened heart valves, and heart murmurs. Flo Hyman, an outstanding American female volleyball player, died of Marfan syndrome in January 1986.

■ ***Adipose Tissue*** Adipose tissue is basically a form of loose connective tissue in which the cells, called ***adipocytes,*** are specialized for fat storage. Adipocytes are derived from fibroblasts, and the cells have the shape of a "signet ring" because the cytoplasm and nucleus are pushed to the edge of the cell by a large droplet of fat. The thin rim of cytoplasm represents the circular part of the ring, whereas the nucleus represents the oval signet. Adipose tissue is found wherever areolar connective tissue is located. Specifically, it is in the subcutaneous layer, around the kidneys, and in the marrow of long bones, as a padding around joints, and behind the eyeball in the orbit. Adipose tissue is a poor conductor of heat and therefore reduces heat loss through the skin. It is a major energy reserve and generally supports and protects various organs.

Most of the fat in adults is white fat, the type just described. Another type, called ***brown fat,*** differs from white fat in that the adipocytes are smaller, the fat is stored in several smaller droplets rather than a single large one, and the fat is brown in color due to a very rich blood supply and numerous mitochondria (the colored cytochromes in mitochondria help impart the brown color). Also, although brown fat is widespread in the fetus and infant, in adults only small amounts are found in places such as in the subcutaneous tissue between the shoulder blades, mediastinum, and inguinal region. Brown fat is concerned primarily with heat production and thus helps to regulate body temperature in the newborn.

CLINICAL APPLICATION

Suction Lipectomy (Liposuction)

A surgical procedure, called ***suction lipectomy*** (*lipo* = fat; *ectomy* = removal of) or ***liposuction,*** involves suctioning out small amounts of fat from certain areas of the body. The technique can be used in a number of areas of the body in which fat accumulates, such as the inner and outer thighs, buttocks, area behind the knee, underside of the arms, breasts, and abdomen. However, suction lipectomy is not a treatment for obesity, since it will not result in a permanent reduction in fat in the area. Rather, it is a body-contouring procedure. Moreover, certain complications may develop, including fat emboli (clots), infection, fluid depletion, and injury to internal structures.

■ ***Reticular Connective Tissue*** Reticular connective tissue consists of interlacing reticular fibers and reticular cells. It helps to form a delicate supporting stroma (framework) for many organs, including the liver, spleen, and lymph nodes. It is also found in the basal lamina of the basement membrane and around blood vessels and muscle. Reticular connective tissue also helps to bind together the fibers (cells) of smooth muscle tissue.

Dense Connective Tissue

Dense connective tissue contains more numerous and thicker fibers but considerably fewer cells than loose connective tissue. The types are dense regular connective tissue, dense irregular connective tissue, and elastic connective tissue.

■ ***Dense Regular Connective Tissue*** In this tissue, bundles of collagenous fibers have an orderly, parallel arrangement that resists pulling. The tissue is adapted for tension in one direction.

Fibroblasts, which produce the fibers and ground substance, are placed in rows between the fibers. The tissue is silvery white, tough, yet somewhat pliable. Because of its great strength, it is the principal component of ***tendons,*** which attach muscles to bones; ***aponeuroses*** (ap′-ō-noo-RŌ-sēz), which are sheetlike tendons connecting one muscle with another or with bone; and many ***ligaments*** (collagenous ligaments), which hold bones together at joints.

■ ***Dense Irregular Connective Tissue*** This tissue contains collagenous fibers that are arranged without regular orientation and is found in parts of the body where tensions are exerted in various directions. The tissue usually occurs in sheets. It forms some fasciae, the reticular (deeper) region of the dermis of the skin, the periosteum of bone, the perichondrium of cartilage, and the ***membrane (fibrous) capsules*** around organs, such as the kidneys, liver, testes, and lymph nodes.

■ ***Elastic Connective Tissue*** Elastic connective tissue has a predominance of freely branching elastic fibers, which give the unstained tissue a yellowish color. Fibroblasts are present only in the spaces between the fibers. Elastic connective tissue can be stretched and will snap back into shape. It is a component of the walls of elastic arteries, the trachea, bronchial tubes to the lungs, and the lungs themselves. Elastic connective tissue provides stretch and strength, allowing structures to perform their functions efficiently. Yellow elastic ligaments, as contrasted with collagenous ligaments, are composed mostly of elastic fibers; they form the *ligamenta flava* of the vertebrae (ligaments between successive vertebrae), the suspensory ligament of the penis, and the true vocal cords.

Cartilage

Cartilage is capable of enduring considerably more stress than the tissues just discussed. Unlike other connective tissues, cartilage has no blood vessels or nerves, except for those in the perichondrium (membranous covering). Cartilage consists of a dense network of collagenous fibers and elastic fibers firmly embedded in chondroitin sulfate, a jellylike ground substance. Whereas the strength of cartilage is due to its collagenous fibers, its resilience (ability to assume its original shape after deformation) is due to chondroitin sulfate. The cells of mature cartilage, called ***chondrocytes*** (KON-drō-sīts), occur singly or in groups within spaces called ***lacunae*** (la-KOO-nē) in the matrix. The surface of cartilage is surrounded by irregularly arranged dense connective tissue called the ***perichondrium*** (per′-i-KON-drē-um; *peri* = around; *chondro* = cartilage). Three kinds of cartilage are recognized: hyaline cartilage, fibrocartilage, and elastic cartilage (Exhibit 3-2).

■ ***Hyaline Cartilage*** This cartilage, also called ***gristle,*** contains a resilient gel as its ground substance that appears in the body as a bluish-white, shiny substance. The fine collagenous fibers, though present, are not visible with ordinary staining techniques, and the prominent chondrocytes are found in lacunae. Hyaline cartilage is the most abundant kind of cartilage in the body. It is found at joints over the ends of the long bones (where it is called ***articular cartilage***) and forms the ***costal cartilages*** at the ventral ends of the ribs. Hyaline cartilage also helps to form the nose, larynx, trachea, bronchi, and bronchial tubes leading to the lungs. Most of the embryonic skeleton consists of hyaline cartilage. Hyaline cartilage affords flexibility and support and, as articular cartilage, reduces friction and absorbs shock.

■ ***Fibrocartilage*** Chondrocytes scattered through many bundles of visible collagenous fibers are found in this type of cartilage. Fibrocartilage is found at the pubic symphysis, the point where the hipbones fuse anteriorly at the midline. It is also found in the intervertebral discs between vertebrae and the menisci of the knee. This tissue combines strength and rigidity.

■ ***Elastic Cartilage*** In this tissue, chondrocytes are located in a threadlike network of elastic fibers. Elastic cartilage provides strength and elasticity and maintains the shape of certain organs—the epiglottis of the larynx, the external part of the ear (auricle), and the auditory (Eustachian) tubes.

Bone (Osseous) Tissue

Together, cartilage, joints, and ***bone*** or ***osseous*** (OS-ē-us) ***tissue*** comprise the skeletal system. Mature bone cells are called ***osteocytes***. The matrix consists of mineral salts, primarily calcium phosphate and calcium carbonate, and collagenous fibers. The salts are responsible for the hardness of bone.

Bone tissue is classified as either compact (dense) or spongy (cancellous), depending on how the matrix and cells are organized. At this point, we will discuss compact bone only. The basic unit of compact bone is called an ***osteon (Haversian system)***. Each osteon consists of ***lamellae***, concentric rings of hard, intercellular substance; ***lacunae***, small spaces between lamellae that contain osteocytes; ***canaliculi***, minute canals that project from lacunae and provide numerous routes so that nutrients can reach osteocytes and wastes can be removed from them; and a ***central (Haversian) canal*** that contains blood vessels and nerves. Notice that whereas bone tissue is vascular, cartilage is not. Also, the lacunae of bone tissue are interconnected by canaliculi; those of cartilage are not.

Functionally, the skeletal system supports soft tissues, protects delicate structures, works with skeletal muscles to facilitate movement, stores calcium and phosphorus, houses red marrow, which produces several kinds of blood cells, and houses yellow marrow, which contains lipids as an energy source.

The details of compact and spongy bones are discussed in Chapter 5.

Blood (Vascular) Tissue

Blood (vascular tissue) is a liquid connective tissue that consists of a matrix called plasma and formed elements (cells and cell-like structures). ***Plasma*** is a straw-colored liquid that consists mostly of water with a wide variety of dissolved substances (nutrients, enzymes, hormones, respiratory gases, and ions). The formed elements are erythrocytes (red blood cells), leukocytes (white blood cells), and thrombocytes (platelets). The fibers characteristic of all connective tissues are present in blood only when it is clotted.

Erythrocytes function in transporting oxygen to body cells and removing carbon dioxide from them. ***Leukocytes*** (also spelled ***leucocytes***) are involved in phagocytosis, immunity, and allergic reactions. The five basic types of leukocytes are called neutrophils, eosinophils, basophils, lymphocytes, and monocytes. ***Thrombocytes*** function in blood clotting.

The details of blood are considered in Chapter 12.

MEMBRANES

The combination of an epithelial layer and an underlying connective tissue layer constitutes an ***epithelial membrane***. The principal epithelial membranes of the body are mucous membranes, serous membranes, and the cutaneous membrane, or skin. Another kind of membrane, a ***synovial membrane***, does not contain epithelium.

MUCOUS MEMBRANES

A ***mucous membrane***, or ***mucosa***, lines a body cavity that opens directly to the exterior. Mucous membranes line the entire gastrointestinal, respiratory, and reproductive tracts and most of the urinary system (see Figure 23-2). They consist of a lining layer of epithelium and an underlying layer of connective tissue. In addition, most mucous membranes also contain a layer of smooth muscle called the ***muscularis mucosae***.

The epithelial layer of a mucous membrane secretes mucus, which prevents the cavities from drying out. It also traps particles in the respiratory passageways and lubricates and absorbs food as it moves through the gastrointestinal tract. In addition, the epithelial layer is responsible for the secretion of digestive enzymes.

The connective tissue layer of a mucous membrane is called the ***lamina propria***. The lamina propria is so named because it belongs to the mucous membrane (*proprius* = one's own). It binds the epithelium to the underlying structures and allows some flexibility of the membrane. It also holds the blood vessels in place, protects underlying muscles from abrasion or puncture, and allows the diffusion of oxygen and nutrients to the epithelium covering it and the diffusion of carbon dioxide and wastes from the epithelium covering it.

The muscularis mucosae contains smooth muscle fibers (cells) and separates the mucosa from the submucosa beneath. Its role in the gastrointestinal tract is considered in Chapter 23.

SEROUS MEMBRANES

A ***serous membrane,*** or ***serosa,*** lines a body cavity that does not open directly to the exterior, and it covers the organs that lie within the cavity. Serous membranes consist of thin layers of areolar connective tissue covered by a layer of mesothelium, and they are composed of two portions. The part attached to the cavity wall is called the ***parietal*** (pa-RĪ-e-tal) ***portion.*** The part that covers and attaches to the organs inside these cavities is the ***visceral portion.*** The serous membrane lining the thoracic cavity and covering the lungs is called the ***pleura*** (see Figure 1-7). The serous membrane lining the heart cavity and covering the heart is the ***pericardium*** (*cardio* = heart). The serous membrane lining the abdominal cavity and covering the abdominal organs and some pelvic organs is called the ***peritoneum.***

The epithelial layer of a serous membrane secretes a lubricating fluid, called ***serous fluid,*** that allows the organs to glide easily against one another or against the walls of the cavities. The connective tissue layer of the serous membrane consists of a relatively thin layer of areolar connective tissue.

CUTANEOUS MEMBRANE

The ***cutaneous membrane,*** or skin, constitutes an organ of the integumentary system and is discussed in the next chapter.

SYNOVIAL MEMBRANES

Synovial membranes line the cavities of the freely movable joints (see Figure 8-1a). Like serous membranes, they line structures that do not open to the exterior. Unlike mucous, serous, and cutaneous membranes, they do not contain epithelium and are therefore not epithelial membranes. They are composed of areolar connective tissue with elastic fibers and varying amounts of fat. Synovial membranes secrete ***synovial fluid,*** which lubricates the articular cartilage at the ends of bones as they move at joints and nourishes the articular cartilage covering the bones that form the joints. These are articular synovial membranes. Other synovial membranes line cushioning sacs, called bursae, and tendon sheaths in our hands and feet that facilitate the movement of muscle tendons.

MUSCLE TISSUE

Muscle tissue consists of fibers (cells) that are highly specialized for the active generation of force for contraction. As a result of this characteristic, muscle tissue provides motion, maintenance of posture, and heat production. On the basis of certain structural and functional characteristics, muscle tissue is classified into three types: skeletal, cardiac, and smooth (Exhibit 3-3).

Skeletal muscle tissue is named for its location—attached to bones. It is also ***striated;*** that is, the fibers (cells) contain alternating light and dark bands (striations) that are perpendicular to the long axes of the fibers. The striations are visible under a microscope. Skeletal muscle tissue is also ***voluntary*** because it can be made to contract or relax by conscious control. A single skeletal muscle fiber is cylindrical, and the fibers are parallel to each other in a tissue. Each muscle fiber contains a plasma membrane, the ***sarcolemma,*** surrounding the cytoplasm, or ***sarcoplasm.*** Skeletal muscle fibers are multinucleate (they have more than one nucleus), and the nuclei lie close to the sarcolemma. The contractile elements of skeletal muscle fibers are proteins that are assembled into ***myofilaments.*** They contain wide, transverse, dark bands and narrow light ones that give the fibers the striated appearance.

Cardiac muscle tissue forms the bulk of the wall of the heart and its contraction results in pumping blood to all parts of the body. Like skeletal muscle tissue, it is striated. However, unlike skeletal muscle tissue, it is usually involuntary; its contraction is usually not under conscious control. Cardiac muscle fibers are roughly quadrangular and branch to form networks throughout the tissue. The fibers usually have only one nucleus that is centrally located. Cardiac muscle fibers are bound to each other by transverse thickenings of the sarcolemma called ***intercalated discs,*** which contain both gap junctions and desmosomes. These are unique to cardiac muscle. Desmosomes serve to strengthen the tissue and gap junctions aid muscle action potential (impulse) conduction.

Smooth (visceral) muscle tissue is located in the walls of hollow internal structures such as blood vessels, the stomach, intestines, and urinary bladder. Its contraction helps break down food, move food and fluids through the body, and evacuate water. Smooth muscle fibers are usually involuntary, and they are ***nonstriated (smooth).*** Each smooth muscle fiber is thickest in the midregion, with either end tapering to a point, and contains a single, centrally located nucleus. Many individual fibers in some smooth muscle, for example in the wall of the intestines, are connected by gap junctions and thus contract as a single unit. In other locations—for example, the iris of the eye—smooth muscle fibers contract individually, like skeletal muscle fibers, because gap junctions are absent.

EXHIBIT 3-3

Muscle Tissue

Skeletal Muscle Tissue

Description: Cylindrical, striated fibers with several peripheral nuclei; voluntary.
Location: Usually attached to bones.
Function: Motion, posture, heat production.

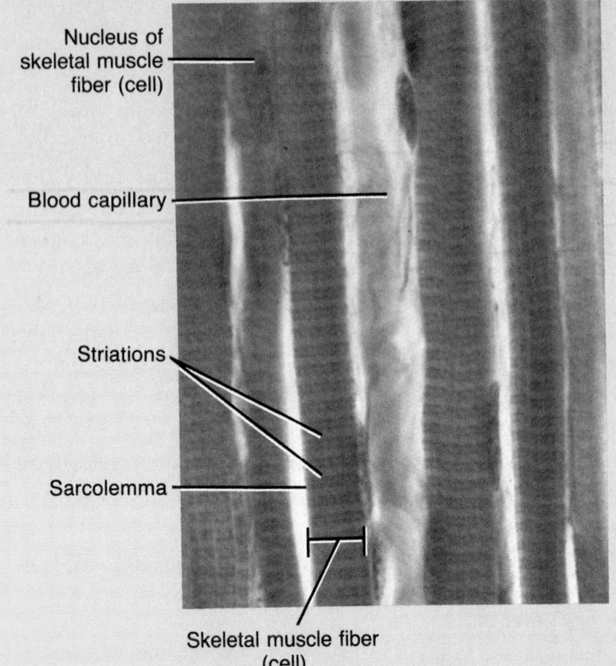

Section of skeletal muscle (800×)

(Courtesy of Andrew J. Kuntzman)

Cardiac Muscle Tissue

Description: Quadrangular, branching, striated fibers with one centrally located nucleus; contains intercalated discs; usually involuntary.
Location: Heart wall.
Function: Motion (contraction of heart that pumps blood to all parts of body).

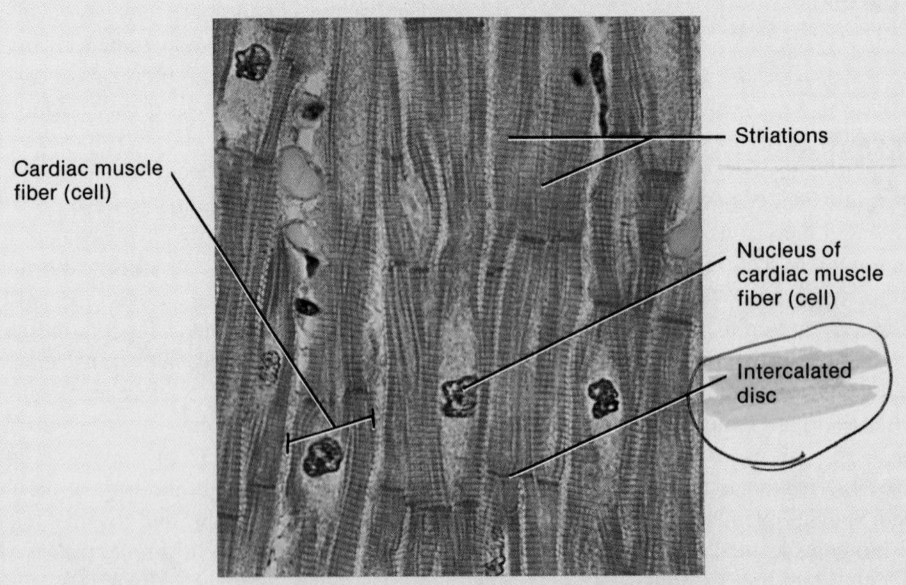

Section of cardiac muscle (700 ×)

(Courtesy of Ed Reschke)

Smooth (Visceral) Muscle Tissue

Description: Spindle-shaped, nonstriated fibers with one centrally located nucleus; usually involuntary.

Location: Walls of hollow internal structures such as blood vessels, stomach, intestines, gallbladder, and urinary bladder.

Function: Motion (constriction of blood vessels, propulsion of foods through gastrointestinal tract; contraction of gallbladder).

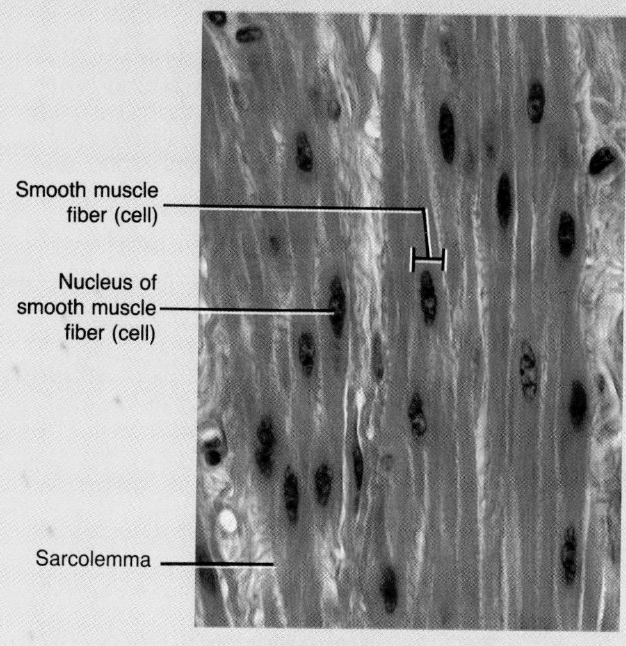

Section of smooth muscle (840x)

(Courtesy of Andrew J. Kuntzman)

A more detailed discussion of muscle tissue is considered in Chapter 9.

NERVOUS TISSUE

Despite the tremendous complexity of the nervous system, it consists of only two principal kinds of cells: neurons and neuroglia. ***Neurons,*** or nerve cells, are highly specialized cells that are sensitive to various stimuli; converting stimuli to action potentials (nerve impulses); and conducting action potentials to other neurons, muscle fibers, or glands. Neurons are the structural and functional units of the nervous system. Most consist of three basic portions: cell body and two kinds of processes called dendrites and axons (Exhibit 3-4). The ***cell body*** contains the nucleus and other organelles. ***Dendrites*** are highly branched processes of the cell body. ***Axons*** are single, long processes of the cell body that conduct action potentials away from the cell body.

Neuroglia are cells that protect and support neurons (see Figure 16-2). They are of clinical interest because they are frequently the sites of tumors of the nervous system.

The detailed structure and function of neurons and neuroglia are considered in Chapter 16.

A summary of the principal tissues discussed in this chapter is presented in Exhibit 3-5.

EXHIBIT 3-4

Nervous Tissue

Description: Neurons (nerve cells) consist of a cell body and processes extending from the cell body called dendrites (usually conduct action potentials toward cell body) or axons (usually conduct action potentials away from cell body). Neurons are supported by neuroglia.

Location: Nervous system.

Function: Exhibits sensitivity to various types of stimuli, converts stimuli to action potentials, and conducts action potentials to other neurons, muscle fibers, or glands.

Nucleolus

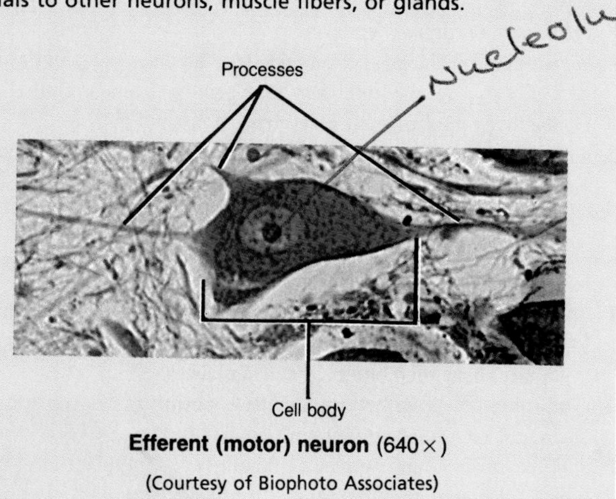

Efferent (motor) neuron (640 ×)

(Courtesy of Biophoto Associates)

EXHIBIT 3-5
Summary of Tissues

TISSUE	COMMENT
EPITHELIUM	
Covering and Lining	Forms outer covering of body and some viscera; lines body cavities, some viscera, blood vessels, and ducts; makes up parts of sense organs; and produces gametes.
Simple	Single layer of cells.
Squamous	Flat, scalelike cells.
Cuboidal	Cube-shaped cells.
Columnar	Rectangular-shaped cells.
Stratified	Two or more layers of cells.
Squamous	Flat, scalelike cells in top layer.
Cuboidal	Cube-shaped cells in top layer.
Columnar	Rectangular-shaped cells in top layer.
Transitional	Cells variable in shape.
Pseudostratified	Single layer of cells that appears to be stratified.
Glandular	Forms secretory portions of glands.
Exocrine	Secrete products into ducts.
Endocrine	Secrete hormones into the blood.
CONNECTIVE	
Embryonic	Present primarily in embryo and fetus.
Mesenchyme	Embryonic tissue from which all other connective tissues develop.
Mucous	Fetal tissue found in umbilical cord.
Mature	Found in newborn and does not change after birth.
Loose	
Areolar	Most abundant connective tissue.
Adipose	Specialized for fat storage.
Reticular	Forms stroma (framework) of many organs.
Dense	
Dense Regular	Forms tendons and ligaments.
Dense Irregular	Found in dermis of skin, fasciae, and membrane around various structures.
Elastic	Provides stretch and strength.
Cartilage	Has no blood or nerve supply.
Hyaline	Found at ends of long bones and ribs.
Fibrocartilage	Found in pubic symphysis and intervertebral discs.
Elastic	Found in larynx and ear.
Bone (osseous)	Contains very rigid intercellular substance and is classified as compact or spongy.
Blood (vascular)	Liquid connective tissue consisting of plasma and formed elements (erythrocytes, leukocytes, and thrombocytes).
MUSCULAR	Highly specialized for contraction.
Skeletal	Found attached to bones, striated, voluntary.
Cardiac	Found in the heart, striated, involuntary.
Smooth (visceral)	Found in viscera and blood vessels, nonstriated, involuntary.
NERVOUS	
Neurons	Specialized for detecting stimuli, converting them into action potentials, and conducting action potentials.
Neuroglia	Protect and support neurons.

STUDY OUTLINE

Types of Tissues and Their Origins (p. 63)

1. A tissue is a group of similar cells and their intercellular substance that have a similar embryological origin and are specialized for a particular function.
2. Depending on their function and structure, the various tissues of the body are classified into four principal types: epithelial, connective, muscular, and nervous.
3. Ectoderm, mesoderm, and endoderm all give rise to epithelium; mesoderm gives rise to connective tissues and muscle tissue; and ectoderm gives rise to nervous tissue.

Epithelial Tissue (p. 63)

1. Epithelium has many cells, little intercellular material, and no blood vessels (avascular). It is attached to connective tissue by a basement membrane. It can replace itself.
2. The subtypes of epithelium include covering and lining epithelium and glandular epithelium.
3. Cell junctions found between epithelial cells include tight junctions, intermediate junctions, desmosomes, and gap junctions.

Covering and Lining Epithelium (p. 64)

1. Layers are arranged as simple (one layer), stratified (several layers), and pseudostratified (one layer that appears as several); cell shapes include squamous (flat), cuboidal (cubelike), columnar (rectangular), and transitional (variable).
2. Simple squamous epithelium is adapted for diffusion and filtration and is found in lungs and kidneys. Endothelium lines the heart and blood vessels. Mesothelium lines the thoracic and abdominopelvic cavities and covers the organs within them.
3. Simple cuboidal epithelium is adapted for secretion and absorption. It is found covering ovaries, in kidneys and eyes, and lining some glandular ducts.
4. Nonciliated simple columnar epithelium lines most of the gastrointestinal tract. Goblet cells secrete mucus. In a few portions of the respiratory tract, the cells are ciliated to move foreign particles trapped in mucus out of the body.
5. Stratified squamous epithelium is protective. It lines the upper gastrointestinal tract and vagina and forms the outer layer of skin.

6. Stratified cuboidal epithelium is found in adult sweat glands, pharynx, epiglottis, and portions of the urethra.
7. Stratified columnar epithelium protects and secretes. It is found in the male urethra and large excretory ducts.
8. Transitional epithelium lines the urinary bladder and is capable of stretching.
9. Pseudostratified epithelium has only one layer but gives the appearance of many. It lines larger excretory ducts, parts of the urethra, auditory (Eustachian) tubes, and most upper respiratory structures, where it protects and secretes.

Glandular Epithelium (p. 73)

1. A gland is a single cell or a mass of epithelial cells adapted for secretion.
2. Exocrine glands (sweat, oil, and digestive glands) secrete into ducts or directly onto a free surface.
3. Structural classification includes unicellular and multicellular glands; multicellular glands are further classified as tubular, acinar, tubuloacinar, simple, and compound.
4. Functional classification includes holocrine, merocrine, and apocrine glands.
5. Endocrine glands secrete hormones into the blood.

Connective Tissue (p. 75)

1. Connective tissue is the most abundant body tissue. It has few cells, an extensive matrix, and a rich blood supply (vascular), except for cartilage. It does not occur on free surfaces.
2. The basic elements in connective tissue are cells (fibroblasts, fixed macrophages, plasma cells, mast cells), ground substance (hyaluronic acid, chondroitin sulfate, dermatan sulfate, keratan sulfate), and fibers (collagenous, elastic, reticular).
3. Connective tissue is classified into two principal types: embryonic and mature.

Embryonic Connective Tissue (p. 84)

1. Mesenchyme forms all other connective tissue.
2. Mucous connective tissue is found in the umbilical cord of the fetus, where it gives support.

Mature Connective Tissue (p. 84)

1. Mature connective tissue exists in the newborn and does not change after birth. It is subdivided into several kinds: loose (areolar, adipose, reticular), dense (regular, irregular, elastic), cartilage, bone tissue, and blood tissue.
2. Areolar connective tissue is one of the most widely distributed connective tissues in the body. Its ground substance contains fibers (collagenous, elastic, and reticular) and various cells (fibroblasts, macrophages, plasma, and mast). Areolar connective tissue is found in all mucous membranes, around body organs, and in the subcutaneous layer.
3. Adipose tissue is a form of loose connective tissue in which the cells, called adipocytes, are specialized for fat storage. It is found in the subcutaneous layer and around various organs.
4. Reticular connective tissue consists of interlacing reticular fibers and forms the stroma of the liver, spleen, and lymph nodes.
5. Dense (regular) connective tissue has a close packing of collagenous fibers in parallel bundles. It is found in tendons, ligaments, and aponeuroses.
6. Dense irregular connective tissue has irregularly arranged collagenous fibers. It is found in fasciae, dermis of skin, and membranes around organs.
7. Elastic connective tissue has a predominance of freely branching elastic fibers that give it a yellow color. It is found in elastic arteries, trachea, bronchial tubes, and true vocal cords.
8. Cartilage has a jellylike matrix containing collagenous and elastic fibers and chondrocytes.
9. Hyaline cartilage is found in the embryonic skeleton, at the ends of bones, in the nose, and in respiratory structures. It is flexible, allows movement, and provides support.
10. Fibrocartilage connects the hipbones and the vertebrae. It provides strength.
11. Elastic cartilage maintains the shape of organs such as the larynx, auditory (Eustachian) tubes, and external ear.
12. Bone (osseous) tissue consists of mineral salts and collagenous fibers that contribute to the hardness of bone and of cells called osteocytes. It supports, protects, helps provide movement, stores minerals, and houses red marrow.
13. Blood (vascular) tissue consists of plasma and formed elements (erythrocytes, leukocytes, and thrombocytes). Functionally, its cells transport, carry on phagocytosis, participate in allergic reactions, provide immunity, and bring about blood clotting.

Membranes (p. 86)

1. An epithelial membrane is an epithelial layer overlying a connective tissue layer. Examples are mucous, serous, and cutaneous membranes.
2. Mucous membranes line cavities that open to the exterior, such as the gastrointestinal tract.
3. Serous membranes (pleura, pericardium, peritoneum) line closed cavities and cover the organs in the cavities. These membranes consist of parietal and visceral portions.
4. The cutaneous membrane is the skin.
5. Synovial membranes line joint cavities and do not contain epithelium.

Muscle Tissue (p. 87)

1. Muscle tissue is modified for contraction and thus provides motion, maintenance of posture, and heat production.
2. Skeletal muscle tissue is attached to bones, is striated, and is voluntary.
3. Cardiac muscle tissue forms most of the heart wall, is striated, and is usually involuntary.
4. Smooth (visceral) muscle tissue is found in the walls of hollow internal structures (blood vessels and viscera), is nonstriated, and is usually involuntary.

Nervous Tissue (p. 89)

1. The nervous system is composed of neurons (nerve cells) and neuroglia (protective and support cells).
2. Most neurons consist of a cell body and two types of processes called dendrites and axons.
3. Neurons are specialized to pick up stimuli, convert stimuli into nerve impulses, and conduct nerve impulses.

REVIEW QUESTIONS

1. What are extracellular materials? Give several examples. (p. 63)
2. Define a tissue. What are the four basic types of human tissue? From which primary germ layer do they develop? (p. 63)
3. Distinguish covering and lining epithelium from glandular epithelium. (p. 63)
4. What are the general features of epithelium? (p. 63)
5. Describe the various layering arrangements and cell shapes of epithelium. (p. 64)
6. Define a cell junction. List several types and their functions. (p. 64)
7. How is epithelium classified? List the various types. (p. 66)
8. For each of the following kinds of epithelium, briefly describe the microscopic appearance, location in the body, and functions: simple squamous, simple cuboidal, simple columnar (nonciliated and ciliated), stratified squamous, stratified cuboidal, stratified columnar, transitional, and pseudostratified columnar. (p. 72)
9. Define the following terms: endothelium, mesothelium, secretion, absorption, goblet cell, and keratin. (p. 72)
10. What is a gland? Distinguish between endocrine and exocrine glands. (p. 73)
11. Describe the classification of exocrine glands according to structure and function and give at least one example of each. (p. 73)
12. What are the general features of connective tissue? (p. 75)
13. Describe the basic elements of connective tissue. (p. 75)
14. How are connective tissues classified? List the various types. (p. 77)
15. How are embryonic connective tissue and mature connective tissue distinguished? (p. 84)
16. Describe the following connective tissues with regard to microscopic appearance, location in the body, and function: areolar, adipose, reticular, dense regular, dense irregular, elastic, hyaline cartilage, fibrocartilage, elastic cartilage, bone (osseous) tissue, and blood (vascular) tissue. (p. 85)
17. Define the following kinds of membranes: mucous, serous, cutaneous, and synovial. Where is each located in the body? What are their functions? (p. 86)
18. Describe the histology of muscle tissue. How is it classified? What are its functions? (p. 87)
19. Distinguish between neurons and neuroglia. Describe the structure and function of neurons. (p. 89)
20. Following are some descriptions of various tissues of the body. For each description, name the tissue described. (p. 73)
    - **a.** An epithelium that permits distention (stretching).
    - **b.** A single layer of flat cells concerned with filtration and absorption.
    - **c.** Forms all other kinds of connective tissue.
    - **d.** Specialized for fat storage.
    - **e.** An epithelium with waterproofing qualities.
    - **f.** Forms the framework of many organs.
    - **g.** Produces perspiration, wax, oil, or digestive enzymes.
    - **h.** Cartilage that shapes the external ear.
    - **i.** Contains goblet cells and lines the intestine.
    - **j.** The most widely distributed connective tissue.
    - **k.** Forms tendons, ligaments, and aponeuroses.
    - **l.** Specialized for the secretion of hormones.
    - **m.** Provides support in the umbilical cord.
    - **n.** Lines kidney tubules and is specialized for absorption and secretion.
    - **o.** Permits extensibility of lung tissue.
    - **p.** Stores red marrow, protects, supports.
    - **q.** Nonstriated, usually involuntary muscle tissue.
    - **r.** Consists of leukocytes, erythrocytes, and thrombocytes.
    - **s.** Composed of a cell body, dendrites, and axon.

SELF QUIZ

Choose the one best answer to these questions.

___ **1.** A surgeon performing abdominal surgery will pass through the skin, then subcutaneous tissue, then muscle, to reach the ________ membrane lining the inside wall of the abdomen:
A. parietal pleura; B. parietal pericardium; C. parietal peritoneum; D. visceral pleura; E. visceral pericardium; F. visceral peritoneum.

___ **2.** Which statement about connective tissue is false?
A. cells are very closely packed together; B. connective tissue has an abundant blood supply; C. matrix is present in large amounts; D. it is the most abundant tissue in the body; E. it does not cover or line body surfaces.

___ **3.** Modified columnar cells that are unicellular glands secreting mucus are known as
A. cilia; B. microvilli; C. goblet cells; D. branched tubular glands; E. basal cells.

___ **4.** A group of similar cells and their intercellular substance operating together to perform a specialized activity is called a(n)
A. organ; B. tissue; C. system; D. organ system; E. organism.

___ **5.** Which statement best describes covering and lining epithelium?
A. it is always arranged in a single layer of cells; B. it contains large amounts of intercellular substance; C. it has an abundant blood supply; D. its free surface is exposed to the exterior of the body or to the interior of a hollow structure; E. its cells are widely scattered.

___ **6.** Which statement best describes connective tissue?
A. usually contains a large amount of matrix; B. always arranged in a single layer of cells; C. primarily concerned with secretion; D. usually lines a body cavity; E. is avascular.

___ **7.** A gland
A. is either exocrine or endocrine; B. may be single celled or muticellular; C. with a duct system may be classed as simple or compound; D. forms and produces secretions; E. is described by all of the above.

___ **8.** Mucous membranes
A. line cavities of the body that are not open to the outside; B. secrete a thin watery serous fluid; C. cover the outside of such organs as the kidney and stomach; D. are found lining the respiratory and urinary passages; E. are described by none of the above.

___ **9.** Which of the following statements are correct?
(1) Simple squamous epithelium lines blood vessels.
(2) Endothelium is composed of cuboidal cells.
(3) Ciliated epithelium is found only in the respiratory system.
(4) Stratified epithelium is found on the surface of the skin.
(5) Transitional epithelium is found in the urinary tract.
A. (1), (2), and (3); B. (1), (3), and (5); C. (1), (4), and (5); D. (1), (2), (3), and (4); E. (2), (4), and (5).

___ **10.** Which type of cell junction assumes a role in action potential (impulse) conduction between muscle fibers?
A. desmosome; B. intermediate junction; C. tight junction; D. hemidesmosome; E. gap junction.

___ **11.** Since the seminal vesicles contain single, nonbranched ducts and flasklike secretory portions, they are classified as
A. compound tubular; B. simple branched acinar; C. simple tubular; D. compound tubuloacinar; E. simple acinar.

___ **12.** Which of the following is involuntary and striated?
A. skeletal muscle tissue; B. cardiac muscle tissue; C. smooth (visceral) muscle tissue; D. neuroglial tissue; E. neural tissue.

___ **13.** Which tissue is characterized by the presence of cell bodies, dendrites, and axons?
A. muscle; B. vascular; C. nervous; D. epithelial; E. osseous.

14. Match the following:

___ **a.** lines the inner surface of the stomach and intestine.
___ **b.** lines urinary tract, as in bladder, permitting distension
___ **c.** lines mouth; present on outer surface of skin
___ **d.** single layer of cube-shaped cells; found in kidney tubules and ducts of some glands
___ **e.** lines air sacs of lungs where thin cells are required for diffusion of gases into blood
___ **f.** not a true stratified; all cells on basement membrane, but some do not reach surface of tissue
___ **g.** derived from lymphocyte, gives rise to antibodies and so is helpful in defense
___ **h.** phagocytic cell; engulfs bacteria and cleans up debris; important during infection
___ **i.** believed to form collagenous and elastic fibers in injured tissue
___ **j.** abundant along walls of blood vessels; believed to produce heparin, an anticoagulant, as well as histamine, which dilates blood vessels
___ **k.** tissue forming most of heart wall
___ **l.** contains lacunae and chondrocytes
___ **m.** forms fasciae and dermis of skin
___ **n.** stores fat and provides insulation

A. cardiac muscle tissue
B. transitional epithelium
C. fibroblast
D. pseudostratified columnar epithelium
E. dense irregular connective tissue
F. simple columnar epithelium
G. macrophage
H. stratified squamous epithelium
I. adipose
J. simple cuboidal epithelium
K. plasma cell
L. simple squamous epithelium
M. mast cell
N. cartilage

Complete the following.

15. Bone tissue is also known as ________ tissue. Compact bone consists of concentric rings, or (lamellae? canaliculi?) with bone cells, called (chondrocytes? osteocytes?), located in tiny spaces called ________.

16. The portion of serous membranes that covers organs (viscera) is called the ________ layer; that portion lining the cavity is named the ________ layer.

17. The embryonic tissue from which all other connective tissues arise is called ________.

18. The structure that attaches epithelium to underlying connective tissue is called the ________.

19. Blood, or ________ tissue, consists of a fluid called ________ containing three types of formed elements.

20. A gland that secretes its product into a duct is referred to as a(n) ________ gland.

The Integumentary System

STUDENT OBJECTIVES

1. Define the integumentary system.
2. List the various layers of the epidermis and describe their structure and functions.
3. Describe the composition and functions of the dermis.
4. Explain the basis for skin color.
5. Describe the blood supply of the skin.
6. Compare the structure, distribution, and functions of hair, skin glands, and nails.
7. Describe the effects of aging on the integumentary system.
8. Describe the development of the epidermis, its derivatives, and the dermis.
9. Describe the causes and effects for the following skin disorders: skin cancer, burns, acne, systemic lupus erythematosus (SLE), psoriasis, decubitus ulcers, and sunburn.
10. Define key medical terms associated with the integumentary system.

CHAPTER OUTLINE

- **Skin**

Anatomy
Epidermis
Dermis
Functions
Skin Color
Epidermal Ridges and Grooves
Blood Supply

- **Epidermal Derivatives**

Hair
  Anatomy
  Color
Glands
  Sebaceous (Oil) Glands
  Sudoriferous (Sweat) Glands
  Ceruminous Glands
Nails

- **Aging and the Integumentary System**
- **Developmental Anatomy of the Integumentary System**
- **Applications to Health**
- **Key Medical Terms Associated with the Integumentary System**

A group of tissues that performs a specific function is an ***organ.*** The next higher level of organization is a ***system***—a group of organs operating together to perform specialized functions. The skin and organs derived from it, such as hair, nails, glands, and several specialized receptors, constitute the ***integumentary*** (in′-teg-yoo-MEN-tar-ē) ***system*** of the body. The developmental anatomy of the integumentary system is considered at the end of the chapter.

Of all the body's organs, none is more exposed to inspection, disease, and injury than the skin. Because of its visibility, skin is a reflection of emotions as evidenced by frowning, blushing, and perspiring. Diseases of internal organs may be revealed by changes in the skin such as color changes (paleness, redness, yellow coloration, or bluish coloration) or abnormal eruptions or rashes (chickenpox, cold sores, or measles). The skin is also subject to a number of disorders that involve just itself, such as warts, age spots, or pimples. The skin's exposure to the environment makes it susceptible to damage from trauma, sunlight, and microbes.

Many interrelated factors may affect both the appearance and health of the skin, including nutrition, hygiene, circulation, age, immunity, genetic factors, psychological state, and drugs. So important is the skin to body image that people spend a great deal of time and money, and may even undergo surgery, to restore skin to a more normal or youthful appearance.

SKIN

The ***skin*** is an organ because it consists of different tissues joined together to perform specific activities. It is one of the larger organs of the body in terms of surface area. For the average adult, the skin occupies a surface area of approximately 2 sq m (22 sq ft). The skin is not just a simple thin covering that keeps the body together and gives it protection. The skin is quite complex in structure and performs several functions essential for survival. ***Dermatology*** (der′-ma-TOL-ō-jē; *dermato* = skin; *logos* = study of) is the medical specialty that deals with the diagnosis and treatment of skin disorders.

ANATOMY

Structurally, the skin consists of two principal parts (Figure 4-1). The outer, thinner portion, which is composed of *epithelium,* is called the ***epidermis.*** The epidermis is cemented to the inner, thicker, *connective tissue* part called the ***dermis.*** Thick skin has a relatively thick epidermis, whereas thin skin has a relatively thin epidermis. Beneath the dermis is a ***subcutaneous (SC) layer.*** This layer, also called the ***superficial fascia*** or ***hypodermis,*** consists of areolar and adipose tissues. Fibers from the dermis extend down into the subcutaneous layer and anchor the skin to it. The subcutaneous layer, in turn, attaches to underlying tissues and organs.

EPIDERMIS

The ***epidermis*** is composed of stratified squamous epithelium and contains four distinct types of cells (Figure 4-2). The most numerous is known as a ***keratinocyte*** (ker-i-TIN-ō-sīt; *kerato* = horny). Keratinocytes produce the protein keratin, which helps waterproof and protect the skin and underlying tissues, and participate in immunity. A second type of cell is called a ***melanocyte*** (MEL-a-nō-sīt), which can also be found in the dermis. It produces melanin, one of the pigments responsible for skin color, and absorbs ultraviolet (UV) light. The third and fourth types of cells in the epidermis are called ***nonpigmented granular dendrocytes,*** two distinct cell types, also known as ***Langerhans' cells*** and ***Granstein cells.*** These cells differ both in their sensitivity to damage by UV radiation and their functions in immunity. Langerhans' cells are a small population of dendrocytes that arise from bone marrow and migrate to the epidermis and other areas of the body that contain stratified squamous epithelial tissue. These cells are sensitive to UV radiation and lie superficial to the basal layer of keratinocytes. Langerhans' cells interact with cells called helper T cells to assist in the immune response. Granstein cells are dendrocytes that are more resistant to UV radiation and interact with cells called suppressor T cells to assist in the immune response.

The keratinocytes of the epidermis are held together by cell junctions called desmosomes and are organized into four or five layers, depending on location in the body (see Figures 4-1 and 4-2). Where exposure to friction is greatest, such as in the palms and soles, the epidermis has five layers. In all other parts of the body it has four layers. The names of the five layers from the deepest to the most superficial are as follows:

1. ***Stratum basale.*** This single layer of cuboidal to columnar cells contains stem cells that are capable of continued cell division and melanocytes. As the stem cells multiply, they push up toward the surface and become part of the layers to be described next. As the cells move upward, their nuclei degenerate, and the cells die. Eventually, the cells are shed when they occupy the top layer of the epidermis. Other cells in the stratum basale grow into the dermis and give rise to sweat and oil glands and hair follicles. The stratum basale is sometimes referred to as the ***stratum germinativum*** (jer′-mi-na-TĒ-vum) to indicate its role in germinating new cells. The stratum basale of hairless skin contains nerve endings sensitive to touch called ***tactile (Merkel's) discs.***

   Between or just below cells of the stratum basale are *melanocytes,* cells that produce melanin, a pale yellow to black pigment that contributes to skin color.
2. ***Stratum spinosum.*** This layer of the epidermis contains 8 to 10 rows of polyhedral (many-sided) cells that fit closely together. The surfaces of these cells contain spinelike projections (*spinosum* = prickly) that join the cells together. Melanin is also found in this layer.

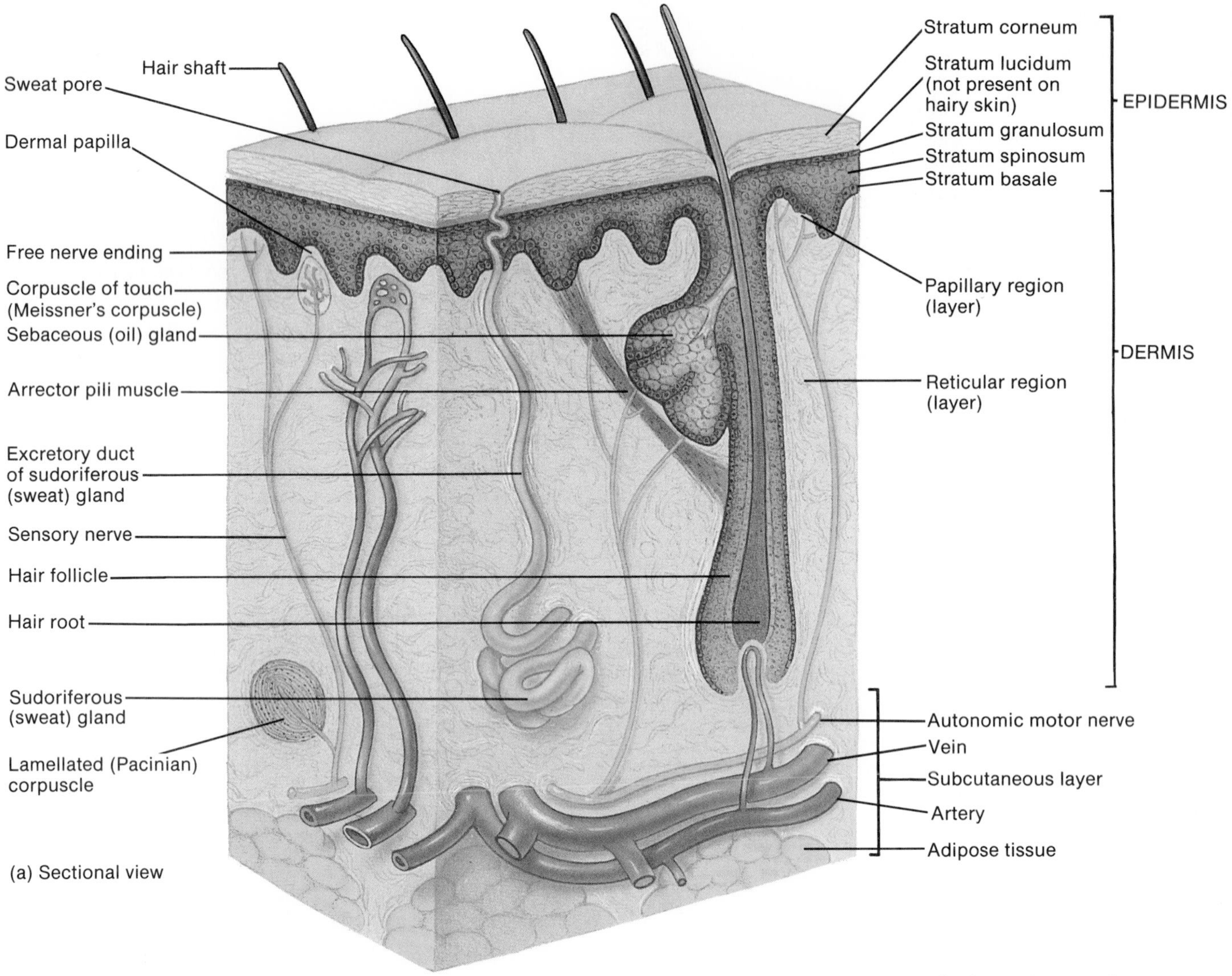

Hair shaft
Sweat pore
Dermal papilla
Free nerve ending
Corpuscle of touch (Meissner's corpuscle)
Sebaceous (oil) gland
Arrector pili muscle
Excretory duct of sudoriferous (sweat) gland
Sensory nerve
Hair follicle
Hair root
Sudoriferous (sweat) gland
Lamellated (Pacinian) corpuscle
Stratum corneum
Stratum lucidum (not present on hairy skin)
Stratum granulosum
Stratum spinosum
Stratum basale
EPIDERMIS
Papillary region (layer)
DERMIS
Reticular region (layer)
Autonomic motor nerve
Vein
Subcutaneous layer
Artery
Adipose tissue

(a) Sectional view

Hair shafts
EPIDERMIS
DERMIS
Blood vessels

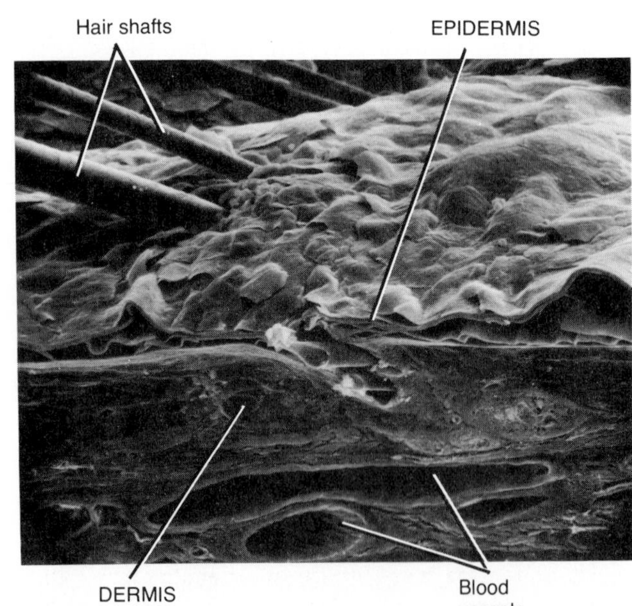

(b) Scanning electron micrograph of skin, 260×

Sweat pore
Surface epithelial cell

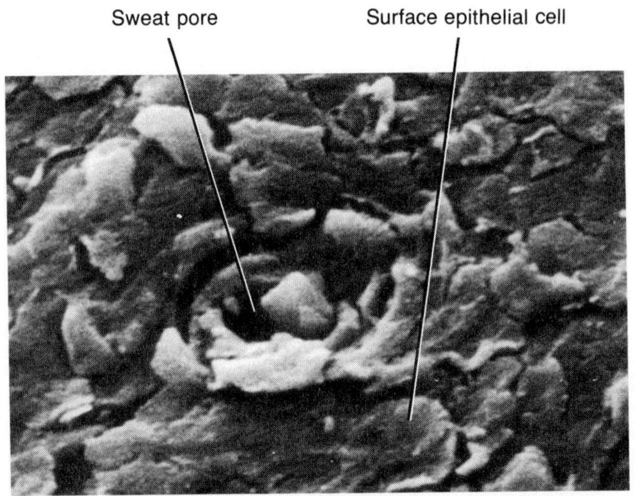

(c) Scanning electron micrograph of a sweat pore, 110×

FIGURE 4-1 Skin. (b) Courtesy of Richard K. Kessel and Randy H. Kardon, *Tissues and Organs: A Text-Atlas of Scanning Electron Microscopy.* Copyright © 1979 by W. H. Freeman & Co. (c) Courtesy of Brain/Science Photo Library/Photo Researchers.

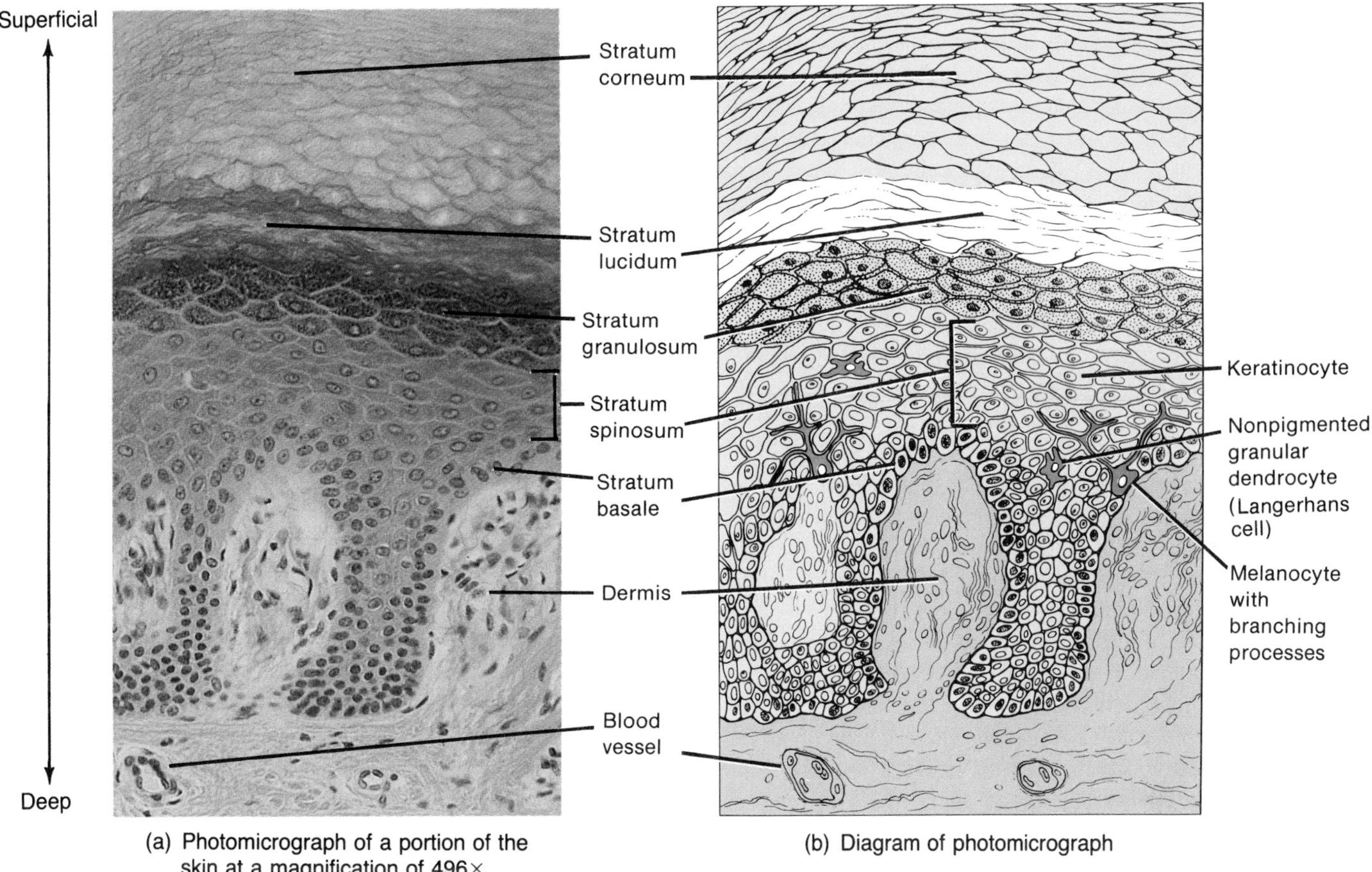

(a) Photomicrograph of a portion of the skin at a magnification of 496×

(b) Diagram of photomicrograph

FIGURE 4-2 Structure of the epidermis. (a) Courtesy of Lester Bergman and Associates.

3. ***Stratum granulosum.*** The third layer of the epidermis consists of three to five rows of flattened cells that contain darkly staining granules of a substance called ***keratohyalin*** (ker′-a-tō-HĪ-a-lin). This compound is involved in the first step of keratin formation. ***Keratin*** is a protein found in the top layer of the epidermis that protects the skin from injury and microbial invasion and makes it waterproof. The nuclei of the cells in the stratum granulosum are in various stages of degeneration. As these nuclei break down, the cells are no longer capable of carrying out vital metabolic reactions and die.
4. ***Stratum lucidum.*** This layer is normally found only in the thick skin of the palms and soles and is absent in thin skin. It consists of three to five rows of clear, flat, dead cells that contain droplets of a substance called ***eleidin*** (el-Ē-i-din). The layer is so named because eleidin is translucent (*lucidum* = clear). Eleidin is formed from keratohyalin and is eventually transformed to keratin.
5. ***Stratum corneum.*** This layer consists of 25 to 30 rows of flat, dead cells completely filled with keratin. These cells are continuously shed and replaced by cells from deeper strata. The stratum corneum serves as an effective barrier against light and heat waves, bacteria, and many chemicals.

In the process of ***keratinization,*** newly formed cells produced in the basal layers are pushed up to the surface. As the cells move upward, the cytoplasm, nucleus, and other organelles are replaced by keratohyalin, then eleidin, and finally keratin, and the cells die. Eventually, the keratinized cells are sloughed off and replaced by underlying cells that, in turn, become keratinized. The whole process by which a cell forms in the basal layers, rises to the surface, becomes keratinized, and sloughs off takes about two to four weeks.

DERMIS

The second principal part of the skin, the ***dermis,*** is composed of connective tissue containing collagenous and elastic fibers (see Figure 4-1). The cells of the dermis are relatively few and consist of fibroblasts, macrophages, and adipocytes. The dermis is very thick in the palms and soles and very thin in the eyelids, penis, and scrotum. It also tends to be thicker on the dorsal aspects of the body than the ventral and thicker on the lateral aspects of extremities than medial aspects. Numerous blood vessels, nerves, glands, and hair follicles are embedded in the dermis.

The upper region of the dermis is named the ***papillary region*** or ***layer.*** It consists of areolar connective tissue containing fine elastic fibers. Its surface area is greatly increased by small, fingerlike projections called ***dermal papillae*** (pa-PIL-ē). These structures project into the epidermis, and many contain loops of capillaries. Some dermal papillae also contain tactile receptors called ***corpuscles of touch*** or ***Meissner's*** (MĪS-nerz) ***corpuscles,*** nerve endings that are sensitive to touch. Dermal papillae cause ridges in the overlying epidermis. It is these ridges that leave fingerprints on objects that are handled.

The remaining portion of the dermis is called the ***reticular region*** or ***layer.*** It consists of dense, irregular connective tissue containing interlacing bundles of collagenous and coarse elastic fibers. It is named the reticular (*rete* = net) region because the bundles of collagenous fibers interlace in a netlike manner. Spaces between the fibers are occupied by a small quantity of adipose tissue, hair follicles, nerves, oil glands, and the ducts of sweat glands. Varying thicknesses of the reticular region, among other factors, are responsible for differences in the thickness of the skin.

The combination of collagenous and elastic fibers in the reticular region provides the skin with strength, extensibility, and elasticity. (***Extensibility*** is the ability to stretch; ***elasticity*** is the ability to return to original shape after extension or contraction.) The ability of the skin to stretch can readily be seen during conditions of pregnancy, obesity, and edema. The small tears that occur during extreme stretching are initially red and remain visible afterward as silvery white streaks called ***striae*** (STRĪ-ē).

The reticular region is attached to underlying organs, such as bone and muscle, by the subcutaneous layer. The subcutaneous layer also contains nerve endings called ***lamellated*** or ***Pacinian*** (pa-SIN-ē-an) ***corpuscles*** that are sensitive to pressure (see Figure 20-1).

Whereas cold receptors, which are probably free nerve endings, are found in and just below the dermis, warmth receptors, also probably free nerve endings, are located in the upper and middle dermis.

CLINICAL APPLICATION

Lines of Cleavage and Surgery

The collagenous fibers in the dermis extend in all directions, but in particular regions of the body they tend to be oriented more in one direction than another. The predominant direction of the underlying collagenous fibers is indicated in the skin by ***lines of cleavage (tension lines).*** The lines are especially evident on the palmar surfaces of the fingers, where they run parallel to the long axis of the digit. Lines of cleavage are of particular interest to a surgeon because an incision running parallel to the collagenous fibers will heal with only a fine scar. An incision made across the rows of fibers disrupts the collagen, and the wound tends to gape open and heal in a broad, thick scar.

FUNCTIONS

The numerous functions of the skin are as follows:

1. ***Regulation of body temperature.*** In response to high environmental temperature or strenuous exercise, the production of perspiration by sudoriferous (sweat) glands helps to lower body temperature back to normal. Changes in the flow of blood to the skin also alter its insulating properties and help to adjust body temperature.
2. ***Protection.*** The skin covers the body and provides a physical barrier that protects underlying tissues from physical abrasion, bacterial invasion, dehydration, and ultraviolet (UV) radiation. Hair and nails also have protective functions, as described shortly.
3. ***Reception of stimuli.*** The skin contains numerous nerve endings and receptors that detect stimuli related to temperature, touch, pressure, and pain.
4. ***Excretion.*** Not only does perspiration assume a role in helping to regulate normal body temperature, it also assists in the excretion of small amounts of water, salts, and several organic compounds.
5. ***Synthesis of vitamin D.*** The term ***vitamin D*** actually refers to a group of closely related compounds synthesized naturally from a precursor molecule present in the skin upon exposure to ultraviolet (UV) radiation. The precursor is converted to other compounds in the liver and kidneys and ultimately to vitamin D, which stimulates the absorption of calcium and phosphorus from dietary foods. Vitamin D is actually a hormone, since it is produced in one location in the body, transported by the blood, and then exerts its effect in another location.
6. ***Immunity.*** Certain cells of the epidermis are important components of immunity, your ability to fight disease by producing antibodies (Chapter 15).

SKIN COLOR

The color of the skin is due to ***melanin,*** a pigment in the epidermis; carotene, a pigment mostly in the dermis; and hemoglobin, a pigment in red blood cells in capillaries in the dermis. The amount of melanin varies the skin color from pale yellow to black. This pigment is found primarily in the basale and spinosum layers. Melanin is synthesized in melanocytes, located either just beneath or between cells of the stratum basale. Melanocytes are produced from ***melanoblasts*** (MEL-a-nō-blasts), precursor cells that vary in number from 800 to 2000 cells per cubic millimeter (mm^3). Maximum numbers of melanoblasts are found in mucous membranes, the penis, face, and extremities. Since the number of melanocytes is about the same in all races, differences in skin color are due to the amount of pigment the melanocytes produce and distribute. An inherited inability of an individual in any race to produce melanin results in ***albinism*** (AL-bin-izm). In albinism, the pigment is absent in the hair and eyes as well as the skin. An individual affected with albinism is called an ***albino*** (al-BĪ-no). The partial or complete loss of melanocytes from areas of skin produces patchy, white spots, and the condition is called ***vitiligo*** (vit-i-LĪ-gō). In some people, melanin tends to form in patches called ***freckles.***

Melanocytes synthesize melanin from the amino acid *tyrosine* in the presence of an enzyme called *tyrosinase.* Exposure to ultraviolet radiation increases the enzymatic activity of melanocytes and leads to increased melanin production. The cell bodies of melanocytes send out long processes between epidermal cells (see Figure 4-2b). Upon contact with the processes, epidermal cells take up the melanin by phagocytosis. When the skin is again exposed to ultraviolet radiation, both the amount and the darkness of melanin increase, tanning and further protecting the body against radiation. Thus, melanin serves a vital protective function.

CLINICAL APPLICATION

Malignant Melanoma

Overexposure of the skin to the ultraviolet light of the sun may lead to skin cancer. Among the most lethal skin cancers is ***malignant melanoma*** (*melano* = dark-colored; *oma* = tumor), cancer of the melanocytes. Fortunately, most skin cancers involve basal and squamous cells and can be treated by surgical excision.

It is a good practice to examine one's skin periodically for ***moles*** that may develop highly irregular borders, uneven surfaces, or a mixture of colors, especially red, white, and blue. These signs of changing appearance plus changes in size and bleeding may indicate that a melanoma is developing.

Some individuals, usually over 55 years of age, may develop ***liver spots*** (***age spots***). These are flat skin patches that look like freckles and range in color from light brown to black. Liver spots are clusters of melanocytes and are medically insignificant; they do not become cancerous.

Another pigment, called ***carotene*** (KAR-o-tēn), is found in the stratum corneum and fatty areas of the dermis and subcutaneous layer in people of Asian origin. Together, carotene and melanin account for the yellowish hue of their skin.

The pink to red color of Caucasian skin is due to ***hemoglobin,*** a red pigment in red blood cells that carries oxygen in blood. The pink to red color of the skin depends on the amount and quality of the blood moving through capillaries in the dermis. The redness of the vessels is not heavily masked by pigment because there are lower amounts of melanin in Caucasian skin. The epidermis has no blood vessels, a characteristic of all epithelia.

EPIDERMAL RIDGES AND GROOVES

The outer surface of the skin of the palms and fingers and soles and toes is marked by a series of ridges and grooves that appear either as fairly straight lines or as a pattern of loops and whorls, as on the tips of the digits.

The ***epidermal ridges*** develop during the third and fourth fetal months as the epidermis conforms to the contours of the underlying dermal papillae (Figure 4-3). The function of the ridges is to increase the grip of the hand or foot by increasing friction and acting like tiny suction cups. Since the ducts of sweat glands open on the summits of the epidermal ridges, fingerprints (or footprints) are left when a smooth object is touched. The ridge pattern, which is genetically determined, is unique for each individual. Normally, it does not change throughout life, except to enlarge, and thus can serve as the basis for identification through fingerprints or footprints.

Epidermal grooves on other parts of the skin divide the surface into a number of diamond-shaped areas. Examine the dorsum of the hand as an example. Note that hairs typically emerge at the points of intersection of the grooves. Note also that the grooves increase in frequency and depth in regions where your fingers bend.

CLINICAL APPLICATION

Treating Scars and Wrinkles

It is possible to improve in many cases the appearance of scars from deep acne, chickenpox, burns, cleft lip, and other disorders and to make age-related wrinkles disappear. The procedure is called a ***collagen implant.*** Collagen is the body's principal structural material. It is found in skin, bone, cartilage, tendons, ligaments, and various viscera and accounts for almost one-third of the total protein of the body.

The collagen implant is prepared from cattle collagen. Once injected into the skin, the collagen becomes a stationary fleshlike substance that is incorporated into the skin and smoothes it out. The collagen implant becomes colonized by blood vessels and cells and acts as a natural structural framework in the skin.

A recent treatment for wrinkles and brown spots related to sun-damaged (photodamaged) skin is the drug ***tretinoin*** (***Retin-A***), a derivative of vitamin A that has been used to treat acne since the 1960s. Among the reported benefits of using tretinoin are increased cell proliferation and thickening of the epidermis, smoother stratum corneum, diminished number and size of melanocytes, increased production of collagen and elastin, dilation of blood vessels in the dermis, and regression of precancerous lesions. Dermatologists caution that tretinoin does not reverse the aging process or improve advanced changes associated with aging, and it may increase the risk of cancers caused by UV light exposure.

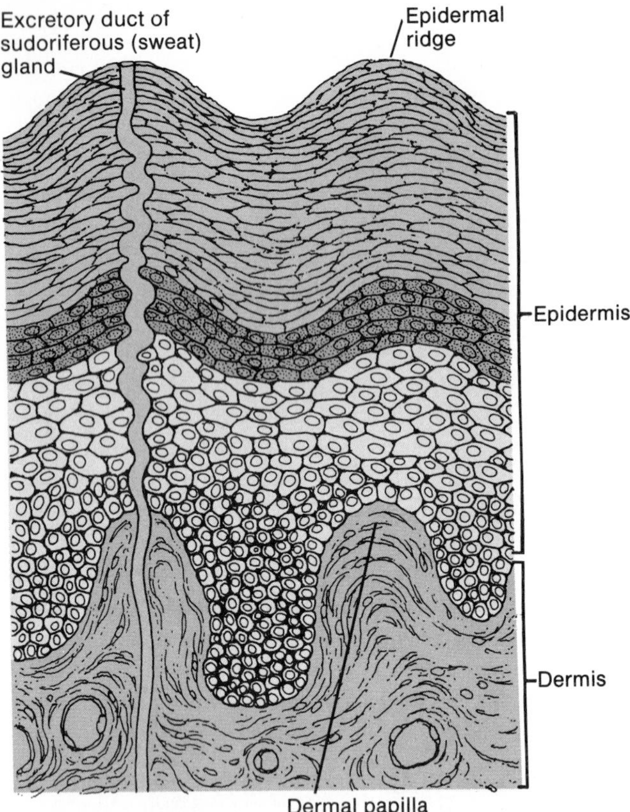

FIGURE 4-3 Relation between epidermal ridges and dermal papillae.

BLOOD SUPPLY

Although the epidermis is avascular, the dermis is well supplied with blood (see Figure 4-1a). The arteries supplying the dermis are generally derived from branches of arteries supplying skeletal muscles in a particular region. Some arteries supply the skin directly. One plexus of arteries, the ***cutaneous plexus,*** is located at the junction of the dermis and subcutaneous layer and sends branches that supply the sebaceous (oil) and sudoriferous (sweat) glands, the deep portions of hair follicles, and adipose tissue. The ***papillary plexus,*** formed at the level of the papillary layer, sends branches that supply the capillary loops in the dermal papillae, sebaceous glands, and superficial portions of hair follicles. The arterial plexuses are accompanied by venous plexuses that drain blood from the dermis into larger subcutaneous veins.

EPIDERMAL DERIVATIVES

Organs developed from the embryonic epidermis—hair, glands, nails—perform functions that are necessary and sometimes vital. Hair and nails protect the body. The sweat glands help regulate body temperature. The enamel of teeth is also an epidermal derivative and is discussed in Chapter 23.

HAIR

Growths of the epidermis variously distributed over the body are ***hairs,*** or ***pili*** (PI-lē). Their primary function is protection. Although the protection is limited, hair on the head guards the scalp from injury and the sun's rays. It also helps prevent heat loss. Eyebrows and eyelashes protect the eyes from foreign particles. Hair in the external ear canal and nostrils protects these structures from insects and numerous airborne particles. Touch receptors associated with hair follicles are activated when hair is even slightly moved.

Hairs are distributed on nearly all parts of the body. It has been estimated that an average adult has about five million hairs, of which about 100,000 are in the scalp. They are absent from places such as the palms of the hands, soles of the feet, dorsal surfaces of the distal phalanges, lips, nipples, clitoris, glans penis, and umbilicus.

Normal hair loss in an adult scalp is about 70–100 hairs per day. Both the rate of growth and the replacement cycle may be altered by illness, diet, and other factors. For example, high fever, major illness, major surgery, blood loss, or severe emotional stress may increase the rate of shedding. Rapid weight-loss diets involving severe restriction of calories or protein also increase hair loss. An increase in the rate of shedding can also occur for three to four months after childbirth. Certain drugs and radiation therapy are also factors in increasing hair loss.

Anatomy

Each hair is a thread of fused, keratinized cells that consists of a shaft and a root (see Figure 4-1a). The ***shaft*** is the superficial portion, most of which projects above the surface of the skin. The shaft of coarse hairs consists of three principal parts (Figure 4-4). The inner ***medulla*** is composed of rows of polyhedral cells containing granules of eleidin and air spaces. The medulla is poorly developed or not present at all in fine hairs. The second principal part of the shaft is the middle ***cortex.*** It forms the major part of the shaft and consists of elongated cells that contain pigment granules in dark hair, but mostly air in white hair. The ***cuticle of the hair,*** the outermost layer, consists of a single layer of thin, flat, scalelike cells that are the most heavily keratinized. They are arranged like shingles on the side of a house, but the free edges of the cuticle cells point upward rather than downward like shingles. The ***root*** is the portion below the surface that penetrates into the dermis and even into the subcutaneous layer and, like the shaft, contains a medulla, cortex, and cuticle.

Surrounding the root is the ***hair follicle,*** which is made up of an external zone of epithelium (the external root sheath), an internal zone of epithelium (the internal root

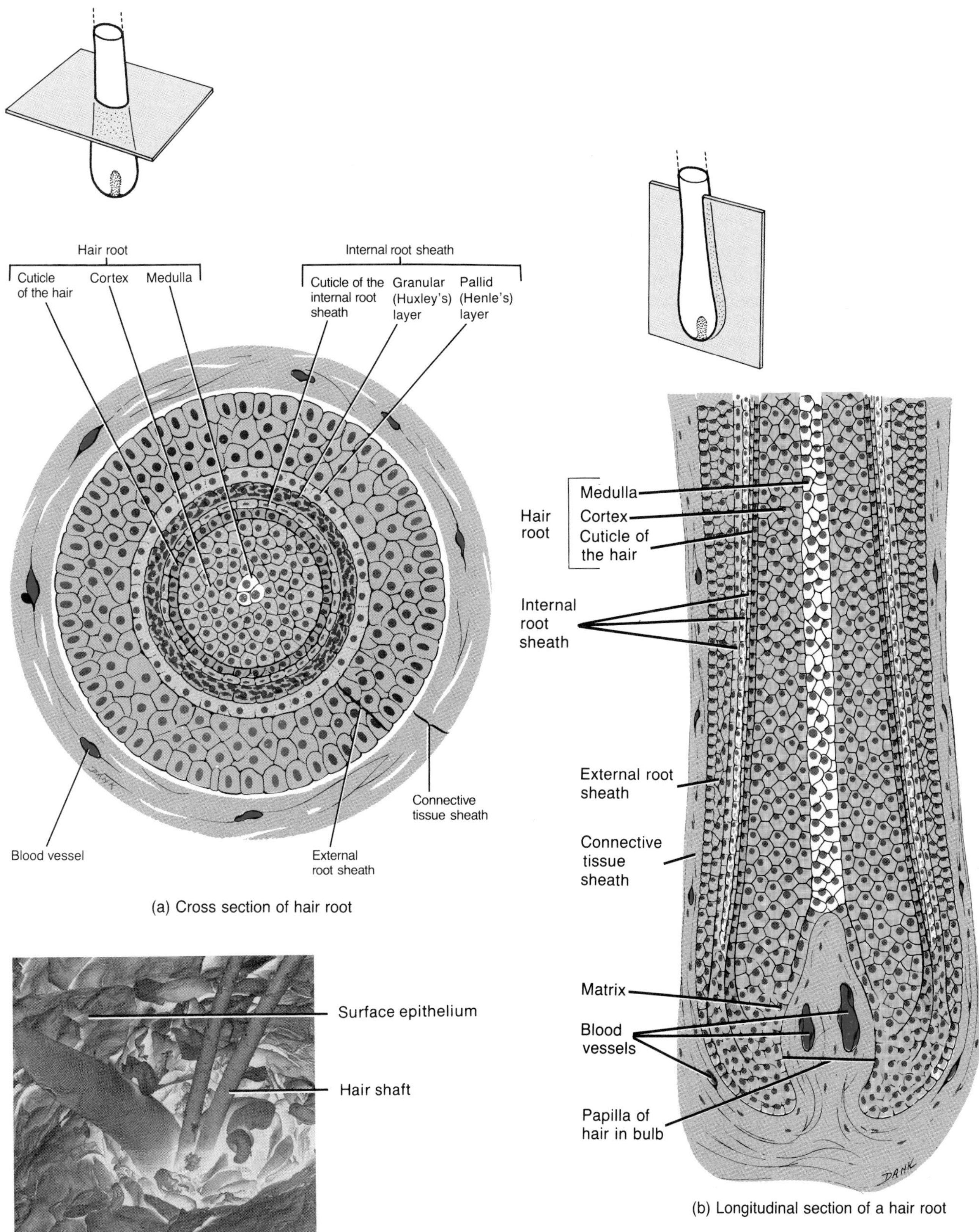

(a) Cross section of hair root

(b) Longitudinal section of a hair root

(c) Scanning electron micrograph of several hair shafts (400 x)

FIGURE 4-4 Hair. Note the shinglelike cuticular scales of the hair shafts. (c) © David Scharf, 1986, All Rights Reserved. Peter Arnold, Inc.

sheath), and a connective tissue sheath, derived from the dermis. The ***external root sheath*** is a downward continuation of the basale and spinosum layers of the epidermis. Near the surface it contains all the epidermal layers. As it descends, it does not exhibit the superficial epidermal layers. At the bottom of the hair follicle, the external root sheath contains only the stratum basale. The ***internal root sheath*** is formed from proliferating cells of the matrix (described shortly) and takes the form of a cellular tubular sheath deep to the external root sheath. The internal root sheath extends only partway up the follicle and consists of (1) an inner layer, the ***cuticle of the internal root sheath,*** which is a single layer of flattened cells with atrophied nuclei, (2) a middle ***granular (Huxley's) layer,*** which is one to three layers of flattened nucleated cells, and (3) an outer ***pallid (Henle's) layer,*** which is a single layer of cuboidal cells with flattened nuclei.

The base of each follicle is enlarged into an onion-shaped structure, the ***bulb.*** This structure contains an indentation, the ***papilla of the hair,*** filled with areolar connective tissue. The papilla of the hair contains many blood vessels and provides nourishment for the growing hair. The bulb also contains a region of germinal cells called the ***matrix,*** derived from the stratum basale layer. The cells of the matrix produce new hairs by cell division when older hairs are shed. This replacement occurs within the same follicle.

CLINICAL APPLICATION

Hair Removal

A substance that removes superfluous hair is called a ***depilatory.*** It dissolves the protein in the hair shaft, turning it into a gelatinous mass that can be wiped away. Since the hair root is not affected, regrowth of the hair occurs. In ***electrolysis,*** the hair bulb is destroyed by an electric current so that the hair cannot regrow.

Sebaceous (oil) glands and a bundle of smooth muscle are also associated with hair. Details of the sebaceous glands will be discussed shortly. The smooth muscle is called ***arrector pili;*** it extends from the dermis of the skin to the side of the hair follicle (Figure 4-1a). In its normal position hair emerges at an angle to the surface of the skin. The arrectores pilorum muscles contract under stresses of fright, cold, and emotions and pull the hairs into a vertical position. This contraction results in "goosebumps" or "gooseflesh" because the skin around the shaft forms slight elevations.

Around each hair follicle are nerve endings, called ***hair root plexuses,*** that are sensitive to touch (see Figure 20-1). They respond if a hair shaft is moved.

Color

The color of hair is due primarily to melanin. It is formed by melanocytes distributed in the matrix of the bulb of the follicle. There are two basic classes of melanin: eumelanin (brown-black) and pheomelanin (yellow to reddish). One distinguishing characteristic is that pheomelanin has a higher content of sulfur. The many variations in hair color are combinations of different amounts of the pigments. Light-colored hair has a predominance of pheomelanin, as does red hair. Dark-colored hair contains mostly eumelanin. Graying of hair is the loss of pigment believed to be the result of a progressive inability of the melanocytes to make tyrosinase, the enzyme necessary for the synthesis of melanin. White hair results from air in the medullary shaft.

CLINICAL APPLICATION

Hair and Hormones

At puberty, when their testes begin secreting significant quantities of male sex hormones (androgens), males develop the typical male pattern of hair growth, including a beard and a hairy chest. In females, both the ovaries and the adrenal glands produce small quantities of androgens. Occasionally, a tumor of one of these glands oversecretes androgens and causes *hirsutism* (hur-SOO-tiz-um), excessive hairiness of the upper lip, chin, chest, inner thighs, and abdomen in females or prepuberal males.

Surprisingly, androgens also must be present for the most common form of baldness, *male-pattern baldness,* to occur. In genetically predisposed males, androgens somehow inhibit hair growth. A chance observation reported in 1979 that a new blood pressure medicine had the side effect of causing hair growth caused an immediate stampede to see whether it could be used to treat baldness. (Pharmaceutical companies knew it would be an overnight best-seller if it worked.) The drug is minoxidil (Rogaine), a potent vasodilator. When applied topically, it does stimulate some hair regrowth in some persons with thinning hair due to male-pattern baldness. For many, however, the hair growth is meager, and the drug does not help individuals who already are bald.

GLANDS

Several kinds of glands are associated with the skin: sebaceous glands, sudoriferous glands, ceruminous glands and mammary glands. The mammary glands will be discussed in Chapter 25 as a component of the female reproductive system.

Sebaceous (Oil) Glands

Sebaceous (se-BĀ-shus), or ***oil glands,*** with few exceptions, are connected to hair follicles (see Figure 4-1a). The secreting portions of the glands lie in the dermis, and those glands associated with hairs open into the necks of hair follicles. Sebaceous glands not associated with hair follicles open directly onto the surface of the skin (lips, glans penis, labia minora, and tarsal glands of the eyelids). Sebaceous glands are simple branched acinar glands. Absent in the palms and soles, sebaceous glands vary in size and shape in other regions of the body. For example, they are small in most areas of the trunk and extremities but are large in the skin of the breasts, face, neck, and upper chest.

The sebaceous glands secrete an oily substance called ***sebum*** (SĒ-bum), a mixture of fats, cholesterol, proteins, and inorganic salts. Sebum helps keep hair from drying and becoming brittle, forms a protective film that prevents excessive evaporation of water from the skin, keeps the skin soft and pliable, and inhibits the growth of certain bacteria. When sebaceous glands of the face become enlarged because of accumulated sebum, acne lesions called ***blackheads*** develop. Since sebum is nutritive to certain bacteria, ***pimples*** or ***boils*** often result. The color of blackheads is due to melanin and oxidized oil, not dirt. Although sebaceous gland activity increases during adolescence, it decreases with aging. This decrease contributes to wrinkles and increased fragility of the skin.

Sudoriferous (Sweat) Glands

Sudoriferous (soo′-dor-IF-er-us; *sudor* = sweat; *ferre* = to bear), or ***sweat glands*** are divided into two principal types on the basis of structure and location: eccrine and apocrine.

Eccrine sweat glands are much more common than apocrine sweat glands and are simple, coiled tubular glands. They are distributed throughout the skin except for the margins of the lips, nail beds of the fingers and toes, glans penis, glans clitoris, labia minora, and eardrums. Eccrine sweat glands are more numerous in the skin of the palms and the soles; their density can be as high as 3000 per square inch in the palms. The secretory portion of eccrine sweat glands is located in the subcutaneous layer, and the excretory duct projects upward through the dermis and epidermis to terminate at a pore at the surface of the epidermis (see Figure 4-1). Eccrine sweat glands function throughout life and produce a secretion that is more watery than that of apocrine sweat glands.

Apocrine sweat glands are simple, branched tubular glands. Their distribution is limited primarily to the skin of the axilla, pubic region, and pigmented areas (areolae) of the breasts. The secretory portion of apocrine sweat glands is located in the dermis or subcutaneous layer and the excretory duct opens into hair follicles. Apocrine sweat glands begin to function at puberty and produce a more viscous secretion than eccrine sweat glands. They are responsive during times of emotional stress.

Perspiration, or ***sweat,*** is the substance produced by sudoriferous glands, mainly eccrine sweat glands because they are so numerous. Perspiration is a mixture of water, salts (mostly NaCl), urea, uric acid, amino acids, ammonia, sugar, lactic acid, and ascorbic acid. Its principal function is to help regulate body temperature by way of evaporation of water in perspiration, which carries off large quantities of heat energy from the body surface. It also helps to eliminate wastes.

Since the mammary glands are actually modified sudoriferous glands, they could be discussed here. However, because of their relationship to the reproductive system, they will be considered in Chapter 25.

Ceruminous Glands

In certain parts of the skin, sudoriferous glands are modified as ***ceruminous*** (se-ROO-mi-nus) ***glands.*** Such modified glands are simple, coiled tubular glands present in the external auditory meatus (canal). The secretory portions of ceruminous glands lie in the subcutaneous layer, deep to sebaceous glands, and the excretory ducts open either directly onto the surface of the external auditory meatus or into ducts of sebaceous glands. The combined secretion of the ceruminous and sebaceous glands is called ***cerumen*** (*cera* = wax). Cerumen, together with hairs in the external auditory meatus, provides a sticky barrier that prevents the entrance of foreign bodies.

CLINICAL APPLICATION

Impacted Cerumen (Earwax)

Some people produce an abnormal amount of cerumen, or earwax, in the external auditory meatus. It then becomes impacted and prevents sound waves from reaching the tympanic membrane (eardrum). The treatment for ***impacted cerumen*** is usually periodic ear irrigation or removal of wax with a blunt instrument by trained medical personnel.

NAILS

Plates of tightly packed, hard, keratinized cells of the epidermis are referred to as ***nails.*** The cells form a clear, solid covering over the dorsal surfaces of the terminal portions of the fingers and toes. Each nail (Figure 4-5) consists of a nail body, a free edge, and a nail root. The ***nail body*** is the portion of the nail that is visible, the ***free edge*** is the part that may project beyond the distal end of the digit, and the ***nail root*** is the portion that is hidden in the proximal nail groove. Most of the nail body is pink because of the underlying vascular tissue. The whitish semilunar area of the proximal end of the body is called the ***lunula*** (LOO-nyoo-la), meaning *little moon.* It appears whitish because the vascular tissue underneath does not show through owing to the thickened stratum basale in the area. The ***eponychium*** (ep′-ō-NIK-ē-um) or ***cuticle*** is a narrow band of epidermis that extends from the margin of the nail wall (lateral border), adhering to it. It occupies the proximal border of the nail and consists of stratum corneum.

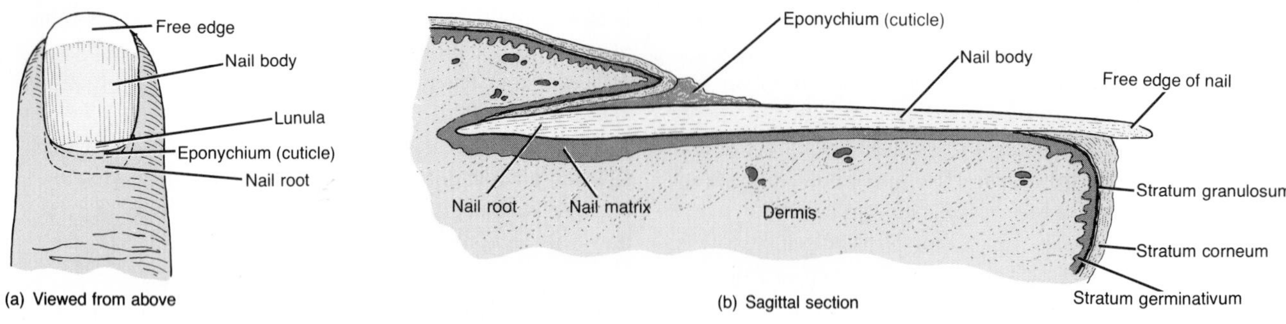

FIGURE 4-5 Structure of fingernails.

The epithelium of the proximal part of the nail bed is known as the ***nail matrix.*** Its function is to bring about the growth of nails. Essentially, growth occurs by the transformation of superficial cells of the matrix into nail cells. In the process, the outer, harder layer is pushed forward over the stratum germinativum. The average growth in the length of fingernails is about 1 mm (0.04 in.) per week. The growth rate is somewhat slower in toenails. For some reason, the longer the digit, the faster the nail grows. Adding supplements such as gelatin to an otherwise healthy diet has no effect on making nails grow faster or stronger.

Functionally, nails help us to grasp and manipulate small objects in various ways and provide protection against trauma to the ends of the digits.

AGING AND THE INTEGUMENTARY SYSTEM

Although skin is constantly aging, the pronounced effects do not occur until a person reaches the late forties. Collagen fibers decrease in number, stiffen, break apart, and form into a shapeless, matted tangle. Elastic fibers lose some of their elasticity, thicken into clumps and fray, and as a result, the skin wrinkles. Fibroblasts, which produce both collagenous and elastic fibers, decrease in number, and macrophages become less efficient phagocytes. Nonpigmented granular dendrocytes (Langerhans' cells) decrease in number, thus decreasing immune responsiveness of older skin. In addition, blood vessels in the dermis become thick-walled and less permeable. Other effects include a loss of subcutaneous fat; atrophy of sebaceous (oil) glands, producing dry and broken skin that is susceptible to infection; a decrease in perspiration production, which probably contributes to the increased incidence of heat stroke in the elderly; a decrease in the number of functioning melanocytes, resulting in gray hair and atypical skin pigmentation; and an increase in the size of some melanocytes, producing pigmented blotching (liver spots).

In general, aged skin, especially the dermis, is thinner than young skin, and migration of cells from the basal layer to the superficial epidermal layers slows down considerably. Aged skin also heals poorly and becomes more susceptible to pathological conditions such as skin cancer, senile pruritis (itching), decubitus ulcers (bedsores), and herpes zoster (shingles). Prolonged exposure to the UV rays of sunlight accelerates the aging of skin and results in considerable damage (photodamage). With increased age, the growth rates of hair and nails decrease.

DEVELOPMENTAL ANATOMY OF THE INTEGUMENTARY SYSTEM

In this and subsequent chapters, the developmental anatomy of the systems of the body will be discussed at the end of each appropriate chapter. Since the principal features of embryonic development are not treated in detail until Chapter 26, it will be necessary to explain and review a few terms at this point so that you can follow the development of organ systems.

As part of the early development of a fertilized egg, a portion of the developing embryo differentiates into three layers of tissue called ***primary germ layers.*** On the basis of position, the primary germ layers are referred to as ***ectoderm, mesoderm,*** and ***endoderm.*** They are the embryonic tissues from which all tissues and organs of the body will eventually develop (see Exhibit 26-1).

The *epidermis* is derived from the ***ectoderm,*** the outermost primary germ layer. At the beginning of the second month the ectoderm consists of simple cuboidal epithelium. These cells become flattened and are known as the ***periderm.*** By the fourth month all layers of the epidermis are formed and each layer assumes its characteristic structure.

Nails are developed during the third month. Initially, they consist of a thick layer of epithelium called the ***primary nail field.*** The nail itself is keratinized epithelium and grows forward from its base. It is not until the ninth month that the nails actually reach the tips of the digits.

Hair follicles develop between the third and fourth months as downgrowths of the stratum basale of the epidermis into the dermis below. The downgrowths soon differentiate into the bulb, papilla of the hair, beginnings of the epithelial portions of sebaceous glands, and other structures associated with hair follicles. By the fifth or sixth month, the follicles produce ***lanugo*** (delicate fetal hair), first on the head and then on other parts of the body. The lanugo is usually shed prior to birth.

The epithelial (secretory) portions of *sebaceous (oil) glands* develop from the sides of the hair follicles and remain connected to the follicles.

The epithelial portions of *sudoriferous (sweat) glands* are also derived from downgrowths of the stratum basale of the epidermis into the dermis. They appear during the fourth month on the palms and soles and a little later in other regions. The connective tissue and blood vessels associated with the glands develop from ***mesoderm,*** the middle primary germ layer.

The *dermis* is derived from wandering ***mesenchymal (mesodermal) cells.*** The mesenchyme becomes arranged in a zone beneath the ectoderm and there undergoes changes into the connective tissues that form the dermis.

APPLICATIONS TO HEALTH

SKIN CANCER

Excessive sun exposure can result in ***skin cancer,*** the most common cancer in Caucasians. However, everyone, regardless of skin pigmentation, is a potential victim of skin cancer if exposure to sunlight is sufficiently intense and continuous. Natural skin pigment can never give complete protection.

The three most common forms of skin cancer are basal cell carcinoma (BCC), squamous cell carcinoma (SCC), and malignant melanoma. ***Basal cell carcinomas (BCCs)*** account for over 75 percent of all skin cancers. The tumors arise from the epidermis and rarely metastasize. They are believed to be caused by years of chronic sun exposure. ***Squamous cell carcinomas (SCCs)*** also arise from the epidermis, and though less common than BCCs, they have a variable tendency to metastasize. Most SCCs arise from preexisting lesions on sun-exposed skin. ***Malignant melanomas,*** as noted earlier, arise from melanocytes and are the leading cause of death from all diseases arising in the skin since they metastasize rapidly. Fortunately, malignant melanomas are the least common, accounting for only about 3 percent of all skin cancers. The principal cause is chronic sun exposure. A clinical trial is under way to test an antimelanoma vaccine made from human melanoma cells. The vaccine apparently sensitizes the patients so that they are more receptive to other anticancer substances, such as interleukin-2, or chemotherapy. The antimelanoma vaccine is designed to stop continued tumor growth rather than prevent tumor formation. Some studies also show that interferon may be active against malignant melanoma.

The treatment of most skin cancers consists of curettage, excision, electrodesiccation, cryosurgery, laser surgery, and radiation therapy.

Among the risk factors for skin cancer are:

1. ***Skin type.*** Persons with fair skin who never tan but always burn are at high risk.
2. ***Geographic location.*** Areas with many days of sunlight per year and high-altitude locations show high incidences of skin cancer.
3. ***Age.*** Older people are more prone to skin cancer owing to greater exposure to sunlight.
4. ***Personal habits.*** Individuals engaged in outdoor occupations have a higher risk.
5. ***Immunologic status.*** Persons who are immunosuppressed have a higher incidence of skin cancer.

In recent years the increasing popularity of ***suntanning salons*** has become a concern to physicians. Many salons claim to use "safe" wavelengths of UV light—that is, longer (A) portions of the UV spectrum (UVA). According to most medical authorities, the radiation emitted by sunlight and by lamps in many suntanning parlors is dangerous. Among the harmful effects are increased risk for skin cancer, premature aging of the skin, suppression of the immune system (changes in and decreased numbers of Langerhans' cells), retinal damage or cataracts, and chemical photosensitivity (adverse skin reactions that occur when certain chemicals, including some drugs, are applied topically or taken internally at the same time an individual is exposed to UV radiation).

If you must be in direct sunlight for long periods of time, use a suitable sunscreen.

BURNS

Tissues may be damaged by excessive heat or by electrical, radioactive, or corrosive chemical agents, all of which can destroy (denature) the proteins in the exposed cells and cause cell injury or death. Such damage is a ***burn.*** Burns disrupt homeostasis because they destroy the protection afforded by the skin, allowing microbial invasion and infection, loss of fluid, and loss of temperature control. The injury to tissues directly or indirectly in contact with the damaging agent, such as the skin or the linings of the respiratory and gastrointestinal tracts, is the local effect of a burn. Generally, however, the systemic effects of a burn are a greater threat to life than the local effects. The systemic effects of a burn may include (1) a large loss of water, plasma, and plasma proteins, which causes shock; (2) bacterial infection; (3) reduced circulation of blood; and (4) decreased production of urine.

ACNE

Acne is an inflammation of sebaceous (oil) glands that usually begins at puberty when the sebaceous glands, under the influence of androgens (male hormones), increase production of sebum. The condition occurs when bacteria grow in the sebum.

SYSTEMIC LUPUS ERYTHEMATOSUS (SLE)

Systemic lupus erythematosus (er-i′-them-a-TŌ-sus), ***SLE***, or ***lupus*** is an autoimmune, inflammatory disease, occurring mostly in young women in their reproductive years. An *autoimmune disease* is one in which the body attacks its own tissues, failing to differentiate between what is foreign and what is not. In SLE, damage to blood vessel walls releases chemicals that cause inflammation. The blood vessel damage then affects virtually every body system. One of the symptoms of SLE is a rash across the bridge of the nose and cheeks called a "butterfly rash."

PSORIASIS

Psoriasis (sō-RĪ-a-sis) is a chronic, occasionally acute, noncontagious relapsing skin disease characterized by distinct, small, reddish, round skin elevations covered with scales. Psoriasis ordinarily involves the scalp, elbows, knees, back, and buttocks.

The cause of psoriasis is an abnormally high rate of mitosis in epidermal cells that may be related to a substance carried in the blood, a defect in the immune system, or a virus. Triggering factors such as trauma, infections, certain drugs (beta blockers and lithium), seasonal and hormonal changes, and emotional stress can initiate and intensify the skin eruptions.

DECUBITUS ULCERS

Decubitus (dē-KYOO-bi-tus) ***ulcers,*** also known as ***bedsores*** or ***pressure sores,*** are caused by a constant deficiency of blood to tissues overlying a bony projection that has been subjected to prolonged pressure against an object such as a bed, cast, or splint. The deficiency results in tissue ulceration. Small breaks in the epidermis become infected, and the sensitive subcutaneous and deeper tissues are damaged. Eventually the tissue is destroyed.

SUNBURN

Sunburn is injury to the skin as a result of acute, prolonged exposure to the ultraviolet (UV) rays of sunlight. UV rays inhibit DNA and RNA synthesis and thus lead to cell death and possibly damage to blood vessels and other structures. Overexposure over a period of years results in a leathery skin texture, wrinkles, skin folds, sagging skin, warty growths called keratoses, freckling, a yellow discoloration due to abnormal elastic tissue, premature aging of the skin, and skin cancer.

To retain a youthful appearance of your skin and to decrease your chance of skin cancer, decrease photodamage by limiting exposure to UV light. Covering the body remains the best protection against UV exposure, and a topical sunscreen will help protect uncovered parts. One of the most effective agents for protection against overexposure to ultraviolet rays of the sun is para-aminobenzoic acid (PABA). The alcohol preparations of PABA are best because the active ingredient binds to the stratum corneum of the skin. The degree of protection is indicated by a numerical rating, the ***sun protection factor*** **(*SPF*),** on sunscreen products. The SPF is the time required to produce erythema when a product is used divided by the time required to produce erythema when the product is not used. For example, if you can remain exposed to the sun *with protection* for 75 minutes before your skin gets red, but you can remain exposed to the sun *without protection* for only 5 minutes, then the SPF of the product you are using is 15. In other words, the product allows you to stay in the sun 15 times as long.

KEY MEDICAL TERMS ASSOCIATED WITH THE INTEGUMENTARY SYSTEM

Abrasion (a-BRĀ-shun; *ab* = away; *rasion* = scraped) A portion of the skin that has been scraped away.

Athlete's (ATH-lēts) **foot** A superficial fungus infection of the skin of the foot.

Chickenpox Highly contagious disease that initiates in the respiratory system and is caused by the varicella-zoster virus and characterized by vesicular eruptions on the skin that fill with pus, rupture, and form a scab before healing. Also called **varicella** (var′-i-SEL-a). Shingles is caused by the latent chickenpox virus.

Cold sore (KŌLD sor) A lesion, usually in oral mucous membrane, caused by type 1 herpes simplex virus (HSV), transmitted by oral or repiratory routes. Triggering factors include UV radiation, hormonal changes, and emotional stress. Also called a **fever blister.**

Contusion (kon-TOO-shun; *contundere* = to bruise) Condition in which tissue below the skin is damaged, but the skin is not broken.

Corn (KORN) A painful conical thickening of the skin found principally over toe joints and between the toes. It may be hard or soft, depending on the location. Hard corns are usually found over toe joints, and soft corns are usually found between the fourth and fifth toes.

Cyst (SIST; *cyst* = sac containing fluid) A sac with a distinct connective tissue wall, containing a fluid or other material.

Eczema (EK-ze-ma; *ekzein* = to boil out) An acute or chronic superficial inflammation of the skin, characterized by redness, oozing, crusting, and scaling. Also called **chronic dermatitis.**

Erythema (er′-e-THĒ-ma; *erythema* = redness) Redness of the skin caused by an engorgement of capillaries in lower layers of the skin. Erythema occurs with any skin injury, infection, or inflammation.

German measles Highly contagious disease that initiates in the respiratory system and is caused by the rubella virus and characterized by a rash of small red spots on the skin. Also called **rubella** (roo-BEL-a).

Hemangioma (hē-man′-jē-Ō-ma; *hemo* = blood; *angio* = blood vessel; *oma* = tumor) Localized tumor of the skin and subcutaneous layer that results from an abnormal increase in blood vessels; one type is a **port-wine stain,** a flat, pink, red, or purple lesion present at birth, usually at the nape of the neck.

Hirsutism (HER-soot-izm) Excessive growth of hair in females and children, with a distribution similar to that of adult males, resulting from higher than normal levels of androgens.

Hives (HĪVZ) Condition of the skin marked by reddened elevated patches that are often itchy. Most commonly caused by infections, physical trauma, medications, emotional stress, food additives, and certain foods. Also called **urticaria** (yoor-ti-KAR-ē-a).

Hypodermic (hī-pō-DER-mik; *hypo* = under) Relating to the area beneath the skin. Also called **subcutaneous.**

Impetigo (im′-pe-TĪ-go) Superficial skin infection caused by staphylococci or streptococci; most common in children.

Intradermal (in′-tra-DER-mal; *intra* = within) Within the skin. Also called **intracutaneous.**

Laceration (las′-er-Ā-shun; *lacerare* = to tear) Wound or irregular tear of the skin.

Measles Highly contagious disease that initiates in the respiratory system and is caused by the measles virus and characterized by a papular rash on the skin. Also called **rubeola** (roo-bē-Ō-la).

Nevus (NE-vus) A round, pigmented, flat, or raised skin area that may be present at birth or develop later. Varying in color from yellow-brown to black. Also called a **mole** or **birthmark.**

Polyp (POL-ip) A tumor on a stem found especially on mucous membranes.

Pruritus (proo-RĪ-tus; *pruire* = to itch) Itching, one of the most common dermatological disorders. It may be caused by skin disorders (infections), systemic disorders (cancer, kidney failure), or psychogenic factors (emotional stress).

Subcutaneous (sub′-kyoo-TĀ-nē-us) Beneath the skin. Also called **hypodermis.**

Topical (TOP-i-kal) Pertaining to a definite area; local. Also in reference to a medication, applied to the surface rather than ingested or injected.

Wart (WORT) Mass produced by uncontrolled growth of epithelial skin cells; caused by a virus (papillomavirus). Most warts are noncancerous. Treatment may involve cryosurgery with liquid nitrogen, electrosurgery, chemical destruction, injections, carbon dioxide, laser surgery, surgical excision, and immunotherapy.

STUDY OUTLINE

Skin (p. 95)

1. The skin and organs derived from it (hair, glands, and nails) constitute the integumentary system.
2. The principal parts of the skin are the outer epidermis and inner dermis. The dermis overlies the subcutaneous layer.
3. The epidermal layers, from deepest to most superficial, are the stratum basale, spinosum, granulosum, lucidum, and corneum. The basale undergoes continuous cell division and produces all other layers. Epidermal cells include keratinocytes, melanocytes, and nonpigmented granular dendrocytes (Langerhans' and Granstein cells).
4. The dermis consists of a papillary region and a reticular region. The papillary region is areolar connective tissue containing blood vessels, nerves, hair follicles, dermal papillae, and corpuscles of touch (Meissner's). The reticular region is dense irregular connective tissue containing adipose tissue, hair follicles, nerves, sebaceous (oil) glands, and ducts of sudoriferous (sweat) glands.
5. Lines of cleavage indicate the direction of collagenous fiber bundles in the dermis and are considered during surgery.
6. The skin is one of the larger organs of the body. It performs the functions of regulating body temperature; protection; receiving stimuli; excretion of water, salts, and several organic compounds; synthesis of vitamin D; and providing immunity.
7. The color of the skin is due to melanin, carotene, and blood in capillaries in the dermis.
8. Epidermal ridges increase friction for better grasping ability and provide the basis for fingerprints and footprints.
9. The epidermis is avascular; the dermis is abundantly supplied by the cutaneous plexus and papillary plexus.

Epidermal Derivatives (p. 100)

1. Among the organs derived from the epidermis are hair, skin glands (sebaceous, sudoriferous, and ceruminous), and nails.

Hair (p. 100)

1. Hairs are epidermal growths that function in protection.
2. Hair consists of a shaft above the surface, a root that penetrates the dermis and subcutaneous layer, and a hair follicle.
3. Associated with hairs are sebaceous (oil) glands, arrectores pilorum muscles, and hair root plexuses.
4. Hair color is due to combinations of various amounts of two basic classes of melanin: eumelanin (brown-black) and pheomelanin (yellow to reddish). Graying is due to the loss of melanin.
5. New hairs develop from cell division of the matrix in the bulb. "Male-pattern" baldness is caused by androgens and heredity.

Glands (p. 102)

1. Sebaceous (oil) glands are usually connected to hair follicles; they are absent in the palms and soles. Sebaceous glands produce sebum, which moistens hairs and waterproofs the skin. Enlarged sebaceous glands may produce blackheads, pimples, and boils.
2. Sudoriferous (sweat) glands are divided into apocrine and eccrine. Eccrine sweat glands have an extensive distribution; their ducts terminate at pores at the surface of the epidermis. Apocrine sweat glands are limited in distribution to the skin of the axilla, pubis, and areolae; their ducts open into hair follicles. Sudoriferous glands produce perspiration, which carries small amounts of wastes to the surface and assists in maintaining body temperature.
3. Ceruminous glands are modified sudoriferous glands that secrete cerumen. They are found in the external auditory meatus.

Nails (p. 103)

1. Nails are plates of tightly packed, hard, keratinized epidermal cells over the dorsal surfaces of the terminal portions of the fingers and toes.

2. The principal parts of a nail are the body, free edge, root, lunula, eponychium, and matrix. Cell division of the matrix cells produces new nails.

Aging and the Integumentary System (p. 104)

1. Most effects of aging occur when an individual reaches the late forties.
2. Among the effects of aging are wrinkling, loss of subcutaneous fat, atrophy of sebaceous glands, and decrease in the number of melanocytes.

Developmental Anatomy of the Integumentary System (p. 104)

1. The epidermis is derived from ectoderm. Hair, nails, and skin glands are epidermal derivatives.
2. The dermis is derived from wandering mesodermal cells.

Applications to Health (p. 105)

1. Skin cancer can be caused by excessive exposure to sunlight. Types include basal cell carcinoma (BCC), squamous cell carcinoma (SCC), and malignant melanoma.
2. Tissue damage that destroys protein is called a burn.
3. Acne is an inflammation of sebaceous (oil) glands.
4. Systemic lupus erythematosus (SLE) is an autoimmune inflammatory disease of connective tissue.
5. Psoriasis is a chronic skin disease characterized by reddish, raised plaques or papules.
6. Decubitus ulcers are caused by a chronic deficiency of blood to tissues subjected to prolonged pressure.
7. Sunburn is a skin injury resulting from prolonged exposure to the UV rays of sunlight.

REVIEW QUESTIONS

1. What is the integumentary system? (p. 95)
2. Compare the structure of epidermis and dermis. What is the subcutaneous (SC) layer? (p. 95)
3. List and describe the epidermal layers from the deepest outward. What is the importance of each layer? Describe the various cells that comprise the epidermis. (p. 95)
4. Contrast the structural differences between the papillary and reticular regions of the dermis. (p. 98)
5. What are lines of cleavage? What is their importance during surgery? (p. 98)
6. List the six principal functions of the skin. (p. 98)
7. Explain the factors that produce skin color. What is an albino? (p. 99)
8. Describe how melanin is synthesized and distributed to epidermal cells. (p. 99)
9. How are epidermal ridges formed? Why are they important? (p. 99)
10. List the receptors in the epidermis, dermis, and subcutaneous (SC) layer and indicate the location and role of each. (pp. 95, 98)
11. Describe the blood supply of the skin. (p. 100)
12. Describe the structure of a hair. How are hairs moistened? What produces "goose bumps" or "gooseflesh"? (p. 100)
13. Contrast the locations and functions of sebaceous (oil) glands, sudoriferous (sweat) glands, and ceruminous glands. (p. 102)
14. Distinguish between eccrine and apocrine sweat glands. (p. 103)
15. From what layer of the skin do nails form? Describe the principal parts of a nail. (p. 103)
16. Describe the effects of aging on the integumentary system (p. 104)
17. Describe the origin of the epidermis, organs derived from it, and the dermis. (p. 104)
18. Define each of the following disorders of the integumentary system: skin cancer, burn, acne, systemic lupus erythematosus (SLE), psoriasis, decubitus ulcers, and sunburn. (p. 105)
19. Refer to the glossary of key medical terms associated with the integumentary system. Be sure that you can define each term. (p. 106)

SELF QUIZ

Choose the one best answer to these questions.

___ 1. Which of the following statements about the function of skin is *not* true?
A. it helps regulate body temperature; B. it protects against bacterial invasion and dehydration; C. it absorbs water and salts; D. it participates in the synthesis of vitamin D; E. it detects stimuli related to temperature and pain.

___ 2. The layer of the dermis that is in direct contact with the epidermis is the
A. papillary region; B. stratum corneum; C. stratum basale; D. reticular region; E. stratum granulosum.

___ 3. In that part of a course in anatomy dealing with the integumentary system, one is concerned with
A. mucous membranes; B. the viscera of the abdominal and thoracic cavities; C. the skin and related structures; D. bones and joints; E. lymphatic tissues.

___ 4. The layer of the skin from which new epidermal cells are derived is the
A. stratum corneum; B. stratum basale; C. stratum lucidum; D. stratum granulosum; E. reticular layer.

___ 5. "Goose bumps" occur as a result of
A. contraction of arrector pili muscles; B. secretion of sebum; C. contraction of elastic fibers in the bulb of the hair follicle; D. contraction of dermal papillae; E. secretion of perspiration.

6. Match the following:

___ **a.** deepest layer of epithelium, consisting of a single layer of cuboidal to columnar cells
___ **b.** eight to ten rows of polyhedral cells that contain spinelike projections
___ **c.** third layer of epidermis, consisting of three to five rows of flattened, degenerating cells
___ **d.** flat, dead cells that are clear due to presence of eleiden; normally found only in thick skin of palms and soles
___ **e.** most superficial layer of skin, consisting of 25 to 30 rows of flat, dead cells

A. basale
B. corneum
C. granulosum
D. lucidum
E. spinosum

Complete the following:

7. This type of cell produces the pigments that give skin its color: ________.

8. The epidermis is derived from the (ecto-? meso-? endo-?) derm. All its layers are formed by the (second? fourth?) month. The dermis arises from the ________ -derm.

9. The ________ is the part of a hair follicle where cells undergo mitosis permitting growth of a new hair.

10. Present in fingerlike projections known as dermal ________ are corpuscles of touch (Meissner's corpuscles), sense receptors sensitive to (pressure? touch?).

11. The outer layer of skin is named the ________. It is composed of (connective tissue? epithelium?).

12. The inner portion of skin, called the ________, is made of (connective tissue? epithelium?).

13. Arrange the parts of a hair from superficial to deep:
A. shaft; ___ B. bulb; ___ C. root ___

14. Arrange the layers of a hair from external to internal:
A. medulla; ___ B. cuticle; ___ C. cortex ___

15. State whether the following descriptions refer to sebaceous, sudoriferous, or ceruminous glands:
A. sweat glands: ________
B. simple branched acinar glands leading directly to hair follicle: secrete sebum that keeps hair and skin from drying out: ________
C. line the external auditory meatus: secrete ear wax: ________

Circle T (true) or F (false) for the following:

T F 16. Eccrine sweat glands are more numerous than apocrine sweat glands and are especially dense on palms and soles.

T F 17. The dermis consists of two regions; the papillary region is most superficial, and the reticular region is deeper.

T F 18. The internal root sheath is a downward continuation of the epidermis.

T F 19. Both epidermis and dermis contain blood vessels (are vascular).

T F 20. The brachial plexus forms part of the blood supply of the skin.

Bone Tissue

5

STUDENT OBJECTIVES

1. Discuss the components and functions of the skeletal system.
2. List and describe the gross features of a long bone.
3. Describe the histological features of compact and spongy bone tissue.
4. Contrast the steps involved in intramembranous and endochondral ossification.
5. Describe the processes of bone construction and destruction involved in bone remodeling.
6. Describe the conditions necessary for normal bone growth and replacement.
7. Describe the blood and nerve supply of bone tissue.
8. Explain the effects of aging on the skeletal system.
9. Describe the development of the skeletal system.
10. Define the following: osteoporosis, rickets, osteomalacia, Paget's disease, and osteomyelitis.
11. Define a fracture, describe several common kinds of fractures, and describe the sequence of events involved in fracture repair.
12. Define key medical terms associated with bone tissue.

CHAPTER OUTLINE

- **Functions**
- **Histology**
  - Compact Bone
  - Spongy Bone
- **Ossification: Bone Formation**
  - Intramembranous Ossification
  - Endochondral Ossification
- **Bone Growth**
- **Bone Replacement**
- **Blood and Nerve Supply**
- **Exercise and the Skeletal System**
- **Aging and the Skeletal System**
- **Developmental Anatomy of the Skeletal System**
- **Applications to Health**
- **Key Medical Terms Associated with Bone Tissue**

The framework of bones and cartilage that provides support, protects our organs, and allows us to move is called the ***skeletal*** (*skeletos* = dried up) ***system.*** The specialized branch of medicine that deals with the preservation and restoration of the skeletal system, articulations (joints), and associated structures is called ***orthopedics*** (or′-thō-PĒ-diks; *ortho* = correct or straighten; *pais* = child).

The developmental anatomy of the skeletal system is considered at the end of the chapter.

FUNCTIONS

The skeletal system performs several basic functions.

1. ***Support.*** The skeleton provides a framework for the body, and as such, it supports soft tissues and provides points of attachment for many skeletal muscles.
2. ***Protection.*** Many internal organs are protected from injury by the skeleton. For example, the brain is protected by the cranial bones, the spinal cord by the vertebrae, the heart and lungs by the rib cage, and internal reproductive organs by the hipbones.
3. ***Movement.*** Skeletal muscles are attached to bones and when the muscles contract they pull on bones and produce movement, as discussed in detail in Chapter 10.
4. ***Mineral storage.*** Bones store several minerals (especially calcium and phosphorus) that can be distributed to other parts of the body upon demand.
5. ***Storage of blood cell–producing cells.*** In certain bones a connective tissue called red marrow produces blood cells, a process called ***hematopoiesis*** (hēm′-a-tō-poy-Ē-sis) or ***hemopoiesis. Red marrow*** consists of blood cells in immature stages, fat cells, and macrophages. Red marrow produces red blood cells, some white blood cells, and platelets. Hematopoiesis is discussed in detail in Chapter 12.
6. ***Storage of energy.*** Fats stored in cells of yellow marrow are an important chemical energy reserve.

HISTOLOGY

Structurally, the skeletal system consists of several types of connective tissue: cartilage, bone, bone marrow, and the periosteum, the membrane around bones. We described the microscopic structure of cartilage and dense connective tissue in Chapter 3. Here, our attention will be directed to a detailed discussion of the microscopic structure of bone tissue.

Like other connective tissues, ***bone,*** or ***osseous*** (OS-ē-us) ***tissue,*** contains a great deal of matrix surrounding widely separated cells. The matrix is about 67 percent mineral salts (inorganic) and 33 percent protein fibers, mainly collagen (organic). Four types of cells are characteristic of bone tissue: osteoprogenitor (osteogenic) cells, osteoblasts, osteocytes, and osteoclasts. ***Osteoprogenitor*** (os′-tē-ō-prō-JEN-i-tor; *osteo* = bone; *pro* = precursor; *gen* = to produce) ***cells*** are unspecialized cells derived from mesenchyme. They possess mitotic potential and have the ability to differentiate into osteoblasts. ***Osteoblasts*** (OS-tē-ō-blasts′; *blast* = germ or bud) do not have mitotic potential and are associated with bone formation. They secrete some of the organic components and mineral salts involved in bone formation. ***Osteocytes*** (OS-tē-ō-sīts′; *cyte* = cell), or mature bone cells, are the principal cells of bone tissue. Like osteoblasts, osteocytes have no mitotic potential. Osteocytes are actually osteoblasts that become isolated within the bony matrix that they deposit around themselves and whose structure changes. Whereas osteoblasts initially form bone, osteocytes maintain daily cellular activities of bone tissue. ***Osteoclasts*** (OS-tē-o-clasts′; *clast* = to break) develop from circulating monocytes (one type of white blood cell). They function in bone removal by resorption (degradation), which is important in the development, growth, maintenance, and repair of bone.

Unlike other connective tissues, the matrix of bone contains abundant mineral salts, primarily tricalcium phosphate $[Ca_3(PO_4)_2 \cdot (OH)_2]$, called *hydroxyapatite,* and some calcium carbonate ($CaCO_3$). In addition, there are small amounts of magnesium hydroxide, fluoride, and sulfate. As these salts are deposited in the framework formed by the collagenous fibers of the matrix, crystallization occurs and the tissue hardens; that is, it becomes ***calcified.*** Mineral salts (inorganic) compose about 67 percent of the weight of bone, and mostly collagenous fibers (organic) make up the remaining 33 percent or so. The collagenous fibers contribute resiliency and pliability to bone, making it less brittle than other calcium-based products, such as egg shells and oyster shells.

At one time, it was thought that calcification simply occurred when enough mineral salts were present to form crystals. Now, however, it is known that the process occurs only in the presence of collagen. Mineral salts accumulate in microscopic spaces between collagenous fibers. There the salts crystallize and become hardened. Then, after the spaces are filled, mineral salts are deposited around collagenous fibers, where the salts again crystallize and harden. The combination of crystallized salts and collagen produces the characteristic properties of bone—hardness due to crystallized mineral salts and great strength due to collagen.

The microscopic structure of bone may be analyzed by first considering the anatomy of a long bone such as the humerus (arm bone) shown in Figure 5-1a. A typical long bone consists of the following parts:

1. ***Diaphysis*** (dī-AF-i-sis; *dia* = through; *physis* = growth). The shaft or long, main portion of the bone.
2. ***Epiphyses*** (ē-PIF-i-sēz; *epi* = above; *physis* = growth). The extremities or ends of the bone (singular is ***epiphysis***).
3. ***Metaphysis*** (me-TAF-i-sis). The region in a mature bone where the diaphysis joins the epiphysis. In a growing

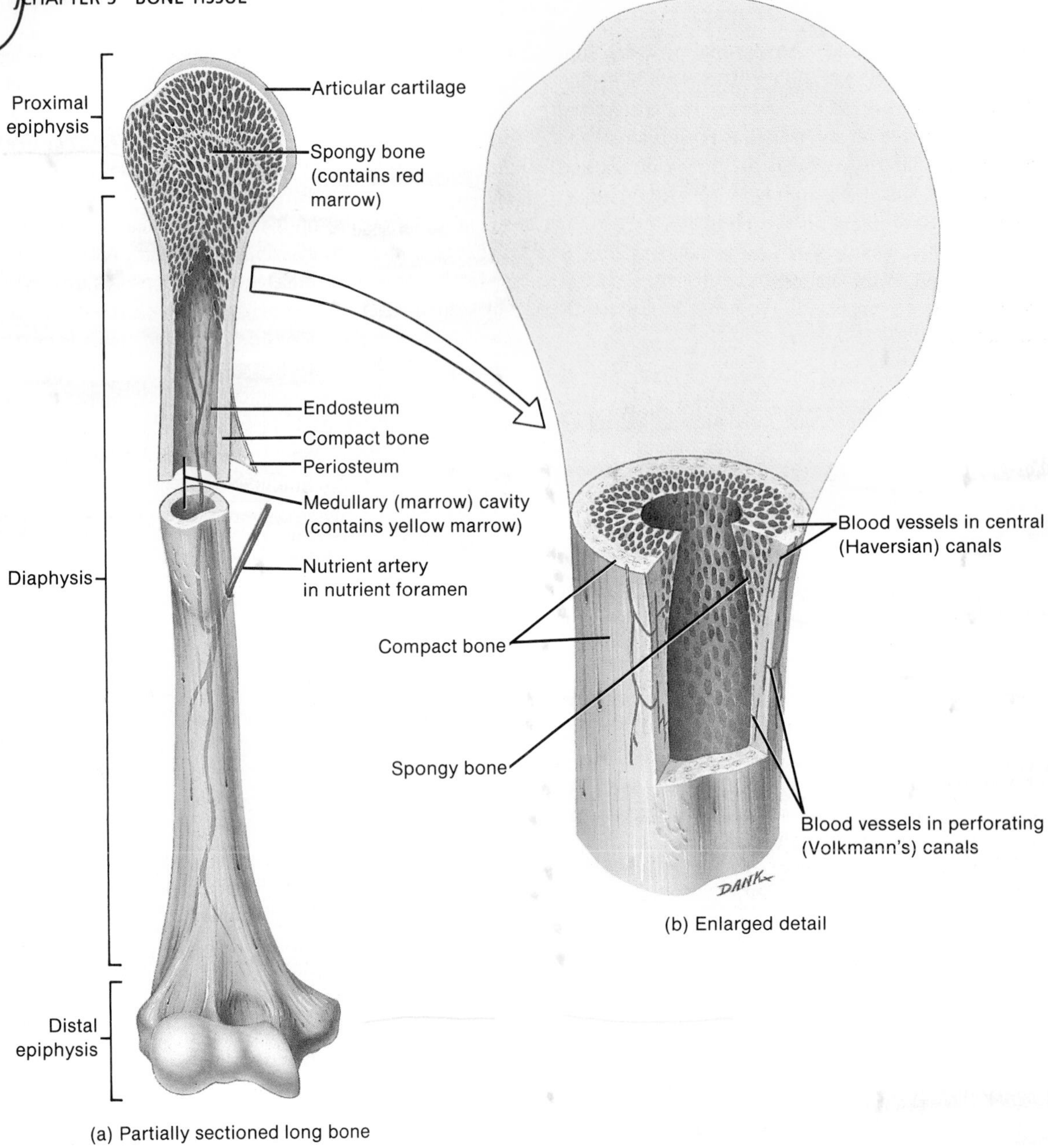

FIGURE 5-1 Bone tissue.

bone, it is the region including the epiphyseal plate where cartilage is reinforced and then replaced by bone (described later in the chapter).

4. ***Articular cartilage.*** A thin layer of hyaline cartilage covering the epiphysis where the bone forms a joint with another bone. The cartilage reduces friction and absorbs shock at freely movable joints.
5. ***Periosteum*** (per′-ē-OS-tē-um). The periosteum (*peri* = around; *osteo* = bone) is a dense, white, fibrous covering around the surface of the bone not covered by articular cartilage. It consists of two layers. The outer ***fibrous layer*** is composed of connective tissue containing blood vessels, lymphatic vessels, and nerves that pass into the bone. The inner ***osteogenic*** (os′-tē-ō-JEN-ik) ***layer*** contains elastic fibers, blood vessels, osteoprogenitor cells, osteoclasts, and osteoblasts. The periosteum is essential for bone growth, repair, and nutrition. It also serves as a point of attachment for ligaments and tendons.
6. ***Medullary*** (MED-yoo-lar′-ē) or **marrow cavity.** The space within the diaphysis that contains the fatty ***yellow marrow*** in adults. Yellow marrow consists primarily of fat cells and a few scattered blood cells. Thus, yellow marrow functions in fat (energy) storage.
7. ***Endosteum*** (end-OS-tē-um). The lining of the medullary cavity that consists of osteoprogenitor cells and osteoblasts and scattered osteoclasts.

Bone is not completely solid. In fact, all bone has some spaces between its hard components. The spaces provide channels for blood vessels that supply bone cells with nutrients. The spaces also make bones lighter. Depending on the size and distribution of the spaces, bone may be categorized as compact or spongy (Figure 5-2; see Figure 5-1a,b also).

Compact (dense) bone tissue contains few spaces. It forms a layer over the spongy bone tissue. The layer of compact bone is thicker in the diaphysis than the epiphyses. Compact bone tissue provides protection and support and helps the long bones resist the stress of weight placed on them. ***Spongy (cancellous) bone tissue,*** by contrast, contains many larger spaces filled with red marrow. It makes up most of the bone tissue of short, flat, and irregularly shaped bones and most of the epiphyses of long bones. Spongy bone tissue in the hipbones, ribs, breastbone, skull, and ends of some long bones is the only site of red marrow in adults.

COMPACT BONE

We can compare the differences between spongy and compact bone tissues by looking at the highly magnified section in Figure 5-3a. (See Figure 5-2 also.) One main difference is that adult compact bone has a concentric-ring structure, whereas spongy bone appears as an irregular latticework. Nutrient arteries and nerves from the periosteum penetrate the compact bone through ***perforating (Volkmann's) canals.*** The blood vessels of these canals connect with blood vessels and nerves of the medullary cavity and those of the ***central (Haversian) canals.*** The central canals run longitudinally through the bone. Around the canals are ***concentric lamellae*** (la-MEL-ē)—rings of hard, calcified matrix. Between the lamellae are small spaces called ***lacunae*** (la-KOO-nē; *lacuna* = little lake) which contain osteocytes. ***Osteocytes,*** as noted earlier, are mature osteoblasts that no longer secrete matrix materials, but instead support daily cellular activities of bone tissue. Radiating in all directions from the lacunae are minute canals called ***canaliculi*** (kan′-a-LIK-yoo-lī), which contain slender processes of osteocytes surrounded by extracellular fluid (Figure 5-3b). The canaliculi connect with one another and, eventually, with the central canals. Thus an intricate network is formed through-

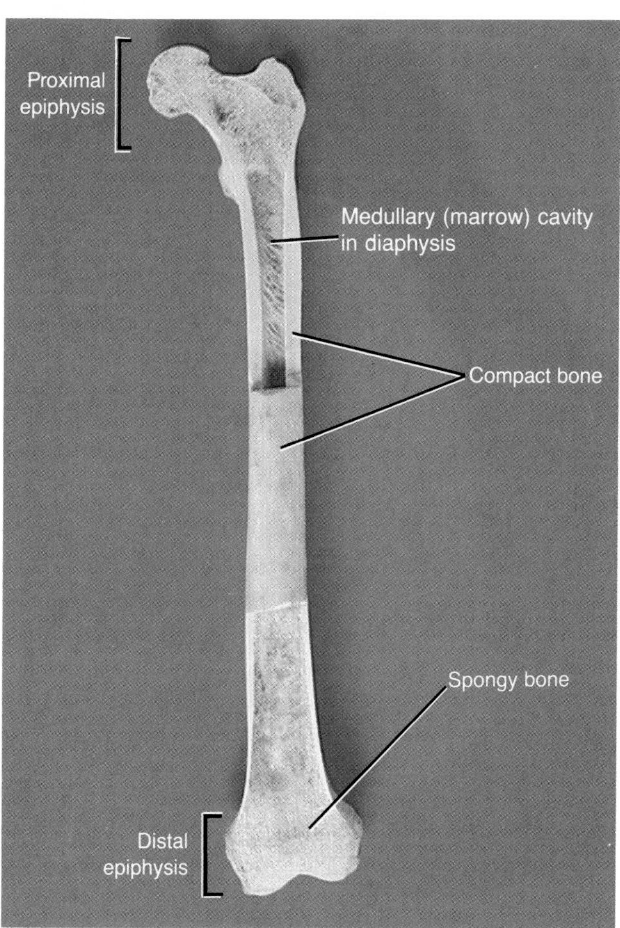

(a) Photograph of partially sectioned femur

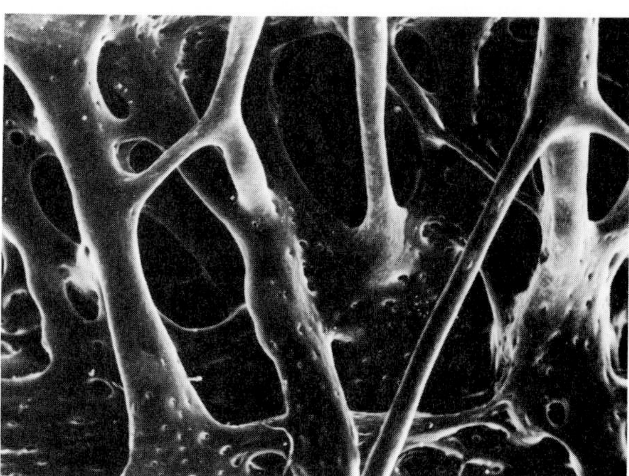

(b) Scanning electron micrograph of spongy bone trabeculae, 25×

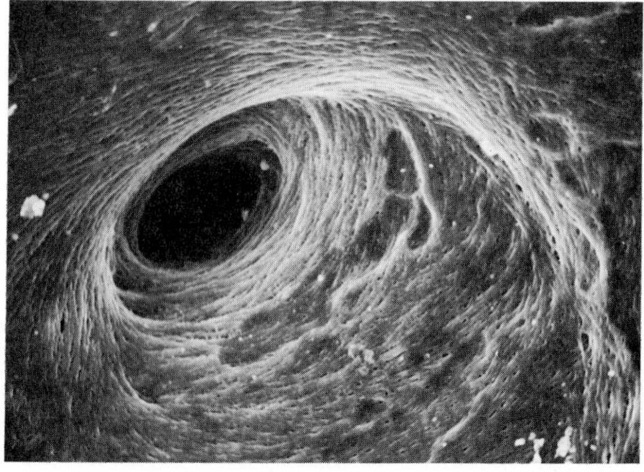

(c) Scanning electron micrograph of a central (Haversian) canal of compact bone, 250×

FIGURE 5-2 Spongy and compact bone. (a) Courtesy of Lester V. Bergman & Associates, Inc. (b) and (c) Courtesy of Fisher Scientific Company and S.T.E.M. Laboratories, Inc. Copyright © 1975.

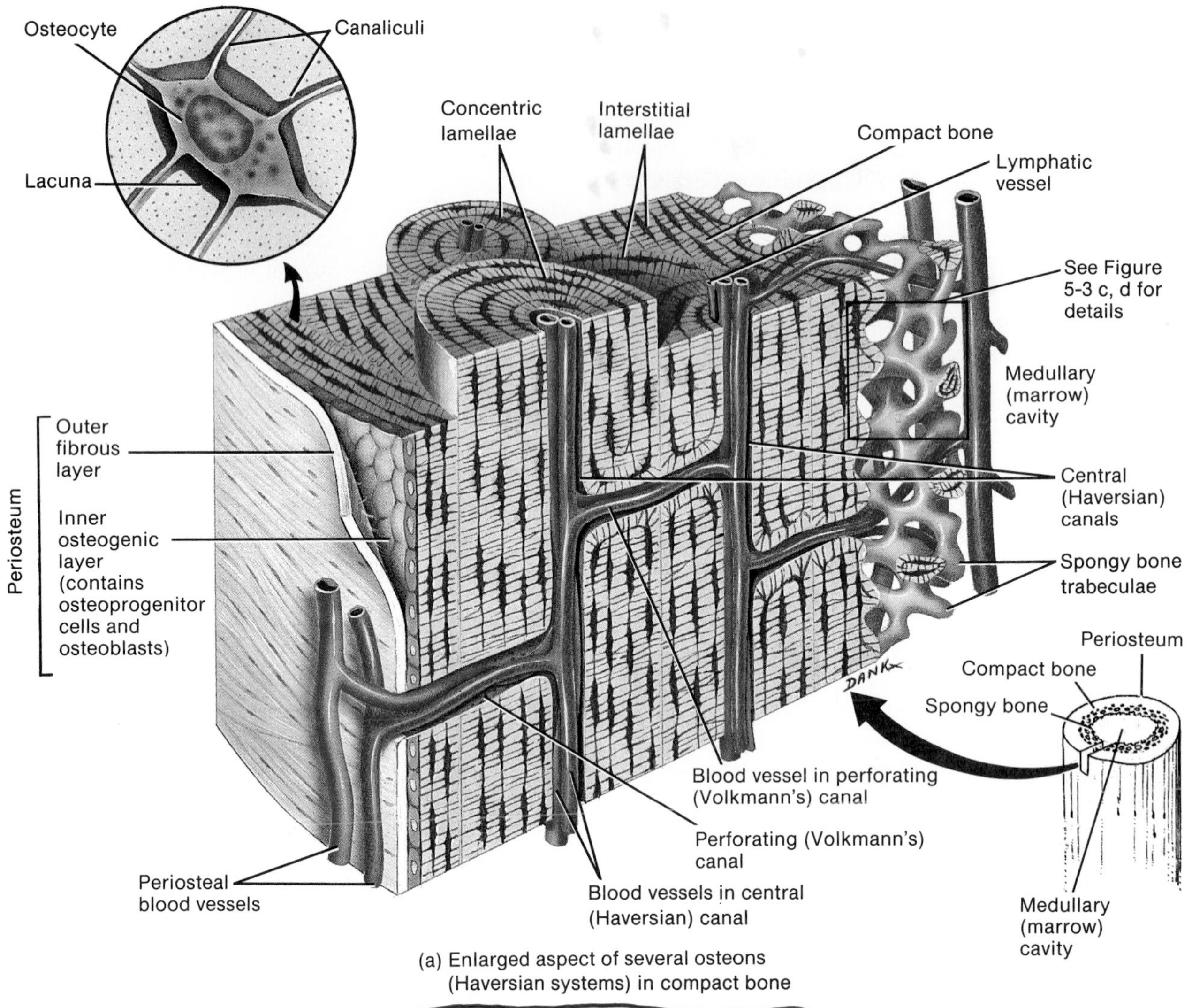

FIGURE 5-3 Histology of bone tissue. A photomicrograph of compact bone is shown in Exhibit 3-2. (b) Courtesy of Biophoto Associates/Photo Researchers.

out the bone. This branching network of canaliculi provides numerous routes so that nutrients and oxygen can reach the osteocytes and wastes to diffuse away. Osteocytes from neighboring lacunae form gap junctions with each other, facilitating easy movement of materials from cell to cell. Each central canal, with its surrounding lamellae, lacunae, osteocytes, and canaliculi, is called an ***osteon* (*Haversian system*).** Osteons are characteristic of adult compact bone. The areas between osteons contain ***interstitial lamellae.*** These areas also possess lacunae with osteocytes and canaliculi, but their lamellae are usually not connected to the osteons. Interstitial lamellae are fragments of older osteons that have been partially destroyed during bone replacement.

SPONGY BONE

In contrast to compact bone, spongy bone does not contain true osteons (Figure 5-3a,c). It consists of an irregular latticework of thin plates of bone called ***trabeculae*** (tra-BEK-yoo-lē). See Figure 5-2 also. The spaces between the trabeculae of some bones are filled with red marrow, which produces blood cells. Within the trabeculae lie lacunae, which contain osteocytes. Blood vessels from the periosteum penetrate through to the spongy bone. Osteocytes in the trabeculae are nourished directly from the blood circulating through the medullary cavities. Osteons are necessary in spongy bone because osteocytes are not deeply buried, as in compact bone, and have access to nutrients directly from the blood.

Most people think all bone is very hard and rigid; however, the bones of an infant are quite soft and become completely rigid only after growth stops during late adolescence. Even then, bone is constantly broken down and rebuilt. It is a dynamic, living tissue. Let us now see how bones are formed and how they grow.

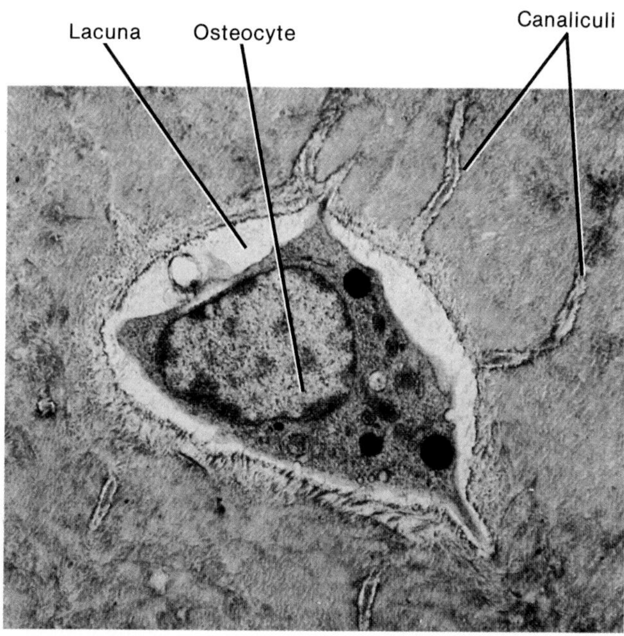

(b) Electron micrograph of an osteocyte, 10,000×

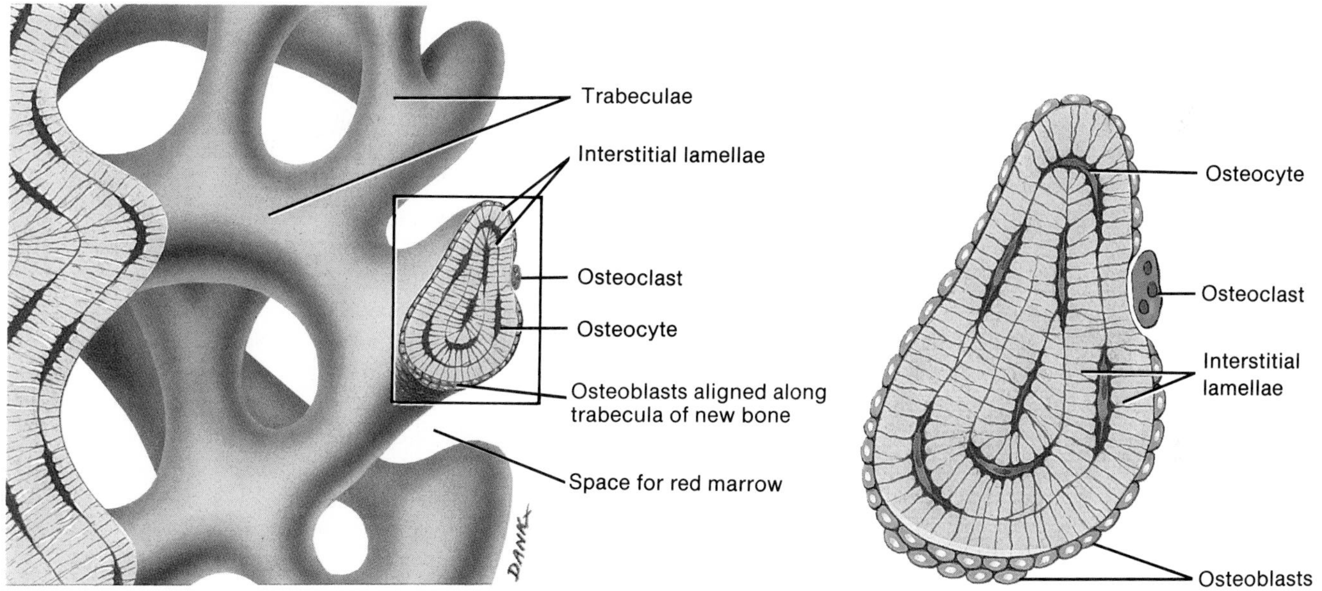

(c) Enlarged aspect of spongy bone trabeculae

(d) Details of a section of a trabecula

OSSIFICATION: BONE FORMATION

The process by which bone forms is called ***ossification*** (os′-i-fi-KĀ-shun) or ***osteogenesis.*** The "skeleton" of a human embryo is composed of fibrous connective tissue membranes and hyaline cartilage. Both are loosely shaped like bones and provide the supporting structures for ossification. Ossification begins around the sixth or seventh week of embryonic life and continues throughout adulthood. Bone formation follows one of two patterns. The first is called ***intramembranous*** (in′-tra-MEM-bra-nus; *intra* = within; *membranous* = membrane) ***ossification.*** This term refers to the formation of bone directly on or within the fibrous connective tissue membranes. The second kind, ***endochondral*** (en′-dō-KON-dral; *endo* = within; *chondro* = cartilage) ***ossification,*** refers to the formation of bone in hyaline cartilage. These two kinds of ossification do *not* lead to differences in the structure of mature bones. They are simply different methods of bone formation. Both mechanisms involve the replacement of a preexisting connective tissue with bone.

The first stage in the development of bone is the migration of embryonic mesenchymal connective tissue cells into the area where bone formation is about to begin. These cells increase in number and size and become osteoprogenitor cells. In some skeletal structures where capillaries are lacking, they become chondroblasts; in others where capillaries are present, they become osteoblasts. The chondroblasts are responsible for cartilage formation. Osteoblasts produce bone tissue by either intramembranous or endochondral ossification.

INTRAMEMBRANOUS OSSIFICATION

The flat bones of the skull, mandible (lower jawbone), and clavicles (collarbones), develop directly on or within fibrous connective tissue membranes formed by mesenchymal cells. This is called ***intramembranous ossification.*** The essentials of the process are as follows:

1. At the site where the bone will develop, mesenchymal cells become vascularized, cluster, and differentiate into osteoprogenitor cells and then osteoblasts. The site of such a cluster is called a ***center of ossification.*** Osteoblasts secrete the organic matrix of bone, called the ***osteoid*** (OS-tē-oid). As soon as the cells become enclosed within matrix and have no room to form additional matrix, they are called osteocytes. Osteocyte processes are located within canaliculi. Later, calcium salts are deposited and the tissue becomes calcified.
2. As the bone matrix forms, it develops into ***trabeculae.*** As trabeculae develop in various ossification centers, they fuse with one another to create the open latticework appearance of spongy bone. The spaces between trabeculae are filled with vascularized connective tissue, which differentiates into red bone marrow.
3. On the outside of the bone, vascularized mesenchyme in the fibrous connective tissue membrane develops into the periosteum. Eventually, some of the surface layers of spongy bone are replaced by compact bone. Much of the newly formed bone will be remodeled (destroyed and reformed) so that the bone may reach its final adult size.

ENDOCHONDRAL OSSIFICATION

The replacement of hyaline cartilage by bone is called ***endochondral (intracartilagenous) ossification.*** Most bones of the body are formed in this way. This type of ossification is best observed in a long bone. The essentials of the process are as follows:

1. ***Development of the cartilage model.*** At the site where the bone is going to form, mesenchymal cells crowd together in the shape of the future bone. The mesenchymal cells differentiate into ***chondroblasts*** that produce cartilage matrix, and therefore the model consists of hyaline cartilage (Figure 5-4a). In addition, a ***perichondrium*** (per-i-KON-drē-um), a membrane around the cartilage model, develops.
2. ***Growth of the cartilage model.*** The cartilage model grows in length by continual cell division of chondrocytes accompanied by further secretion of cartilage matrix by the daughter cells. This pattern of growth, which results in an increase in length, is called ***interstitial*** (in′-ter-STISH-al) ***growth***—that is, growth from within. Growth of the cartilage in thickness is mainly due to the addition of more matrix to its periphery by new chondroblasts that develop from the perichondrium. This growth pattern of cartilage in which matrix is deposited on its surface is called ***appositional*** (a-pō-ZISH-a-nal) ***growth.***

   As the cartilage model continues to grow, chondrocytes in its mid-region hypertrophy (increase in size), probably because they accumulate glycogen for energy and produce enzymes to catalyze further chemical reactions. The cells burst, resulting in a change in pH in the matrix, which triggers calcification. Once the cartilage becomes calcified, nutrients required by the cartilage cells no longer diffuse through the matrix, and the cartilage cells die. The lacunae of the cells that have been killed are now empty, and the thin partitions between them break down (Figure 5-4b).

   In the meantime, a nutrient artery penetrates the perichondrium in the mid-region of the model, stimulating osteoprogenitor cells in the perichondrium to differentiate into osteoblasts and lay down a thin shell of compact bone under the perichondrium called the ***periosteal bone collar.*** Once the perichondrium starts to form bone, it is called the ***periosteum.***
3. ***Development of the primary ossification center.*** Near the middle of the model, capillaries of the periosteum grow into the disintegrating calcified cartilage. These vessels, and the associated osteoblasts, osteoclasts, and red marrow cells, are known as the ***periosteal bud.*** On growing into the cartilage model, the capillaries produce a ***primary ossification center,*** a region where bone tissue will replace most of the cartilage (Figure 5-4c). In the center, osteoblasts begin to deposit bone matrix over the remnants of calcified cartilage, forming spongy bone trabeculae. As the center enlarges toward the ends of the bone, osteoclasts break down the newly formed spongy bone trabeculae, leaving the medullary (marrow) cavity in the center of the model. The cavity then fills with red bone marrow.
4. ***Development of the diaphysis and epiphysis.*** The diaphysis (shaft), which was once a solid mass of hyaline cartilage, is replaced by compact bone, the central part of which contains a red marrow–filled medullary (marrow) cavity. When blood vessels (epiphyseal arteries) enter the epiphyses, ***secondary ossification centers*** develop, usually after birth (Figure 5-4d). In these centers, bone formation is similar to that in the primary ossification with a few exceptions: (1) spongy bone is retained in the interior of the epiphyses, (2) no medullary cavities are formed in the epiphyses; and (3) hyaline cartilage remains covering the epiphyses (***articular cartilage***) and

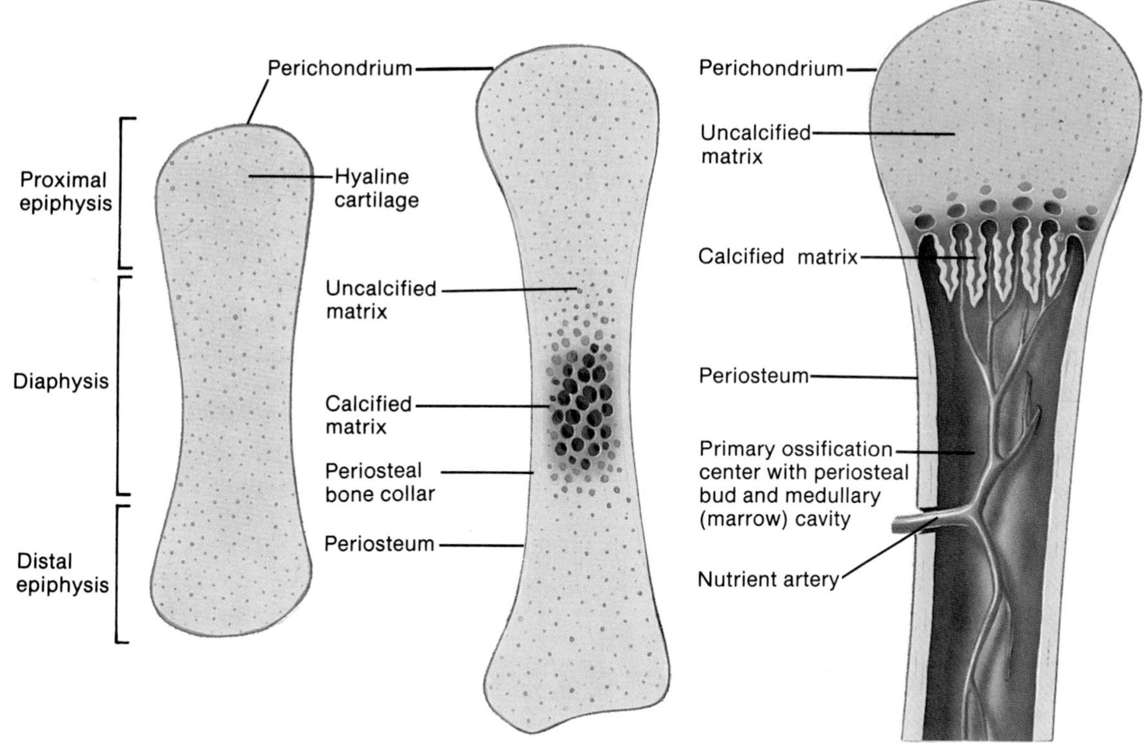

(a) Mesenchymal cells differentiate into chondroblasts which form the hyaline cartilage model

(b) Cartilage model grows by interstitial and appositional growth, chondrocytes in midregion calcify the matrix, vacated lacunae form small cavities, osteoblasts in perichondrium produce periosteal bone collar

(c) With development of periosteal bud, primary ossification center forms and medullary cavity forms

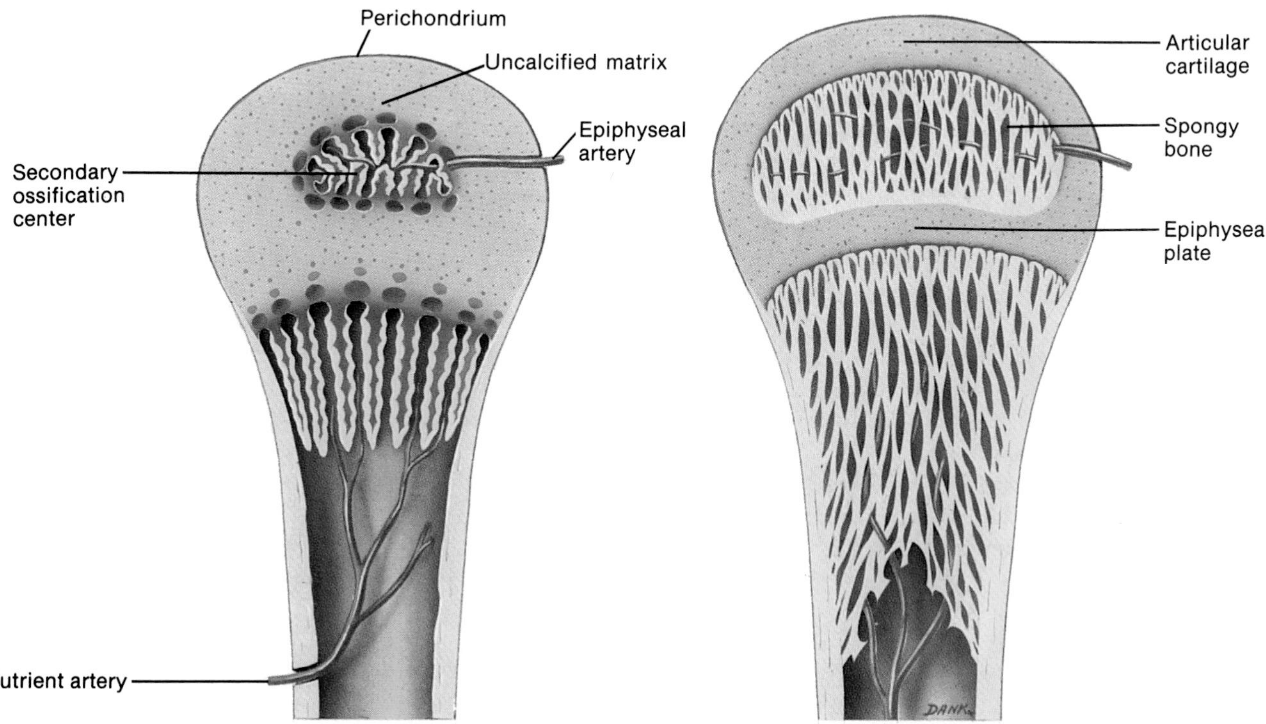

(d) Postnatal development of secondary ossification center in epiphysis. On a long bone, a secondary ossification center also develops in the distal epiphysis

(e) Remnants of hyaline cartilage as articular cartilage and epiphyseal plate

FIGURE 5-4 Endochondral ossification of a long bone (tibia).

between the diaphysis and epiphysis (***epiphyseal plate***), which is responsible for the lengthwise growth of long bones (Figure 5-4e).

BONE GROWTH

In order to understand how a bone grows in length, you will need to know some of the details of the structure of the epiphyseal plate.

The epiphyseal (ep′-i-FIZ-ē-al) plate consists of four zones (Figure 5-5). The ***zone of reserve cartilage*** is adjacent to the epiphysis and consists of small chondrocytes that are scattered irregularly throughout the matrix. The cells of this zone do not function in bone growth; they anchor the epiphyseal plate to the bone of the epiphysis.

The ***zone of proliferating cartilage*** consists of slightly larger chondrocytes arranged like stacks of coins. The function of this zone is to make new chondrocytes by cell division to replace those that die at the diaphyseal surface of the epiphyseal plate.

The ***zone of hypertrophic*** (hī-per-TRŌF-ik) ***cartilage*** consists of even larger chondrocytes that are also arranged in columns, with the more mature cells closer to the diaphysis. The lengthwise expansion of the epiphyseal plate is the result of cell divisions at the zone of proliferating cartilage and maturation of the cells in the zone of hypertrophic cartilage.

The ***zone of calcified matrix*** is only a few cells thick and consists mostly of dead cells because the matrix around them has calcified. The calcified matrix is taken up by osteoclasts, and the area is invaded by osteoblasts and capillaries from the bone in the diaphysis. These cells lay down bone on the calcified cartilage that persists. As a result, the diaphyseal border of the epiphyseal plate is firmly cemented to the bone of the diaphysis.

The region between the diaphysis and epiphysis of a bone where the calcified matrix is replaced by bone is called the ***metaphysis*** (me-TAF-i-sis). The activity of the epiphyseal plate is the only mechanism by which the diaphysis can increase in length. Unlike cartilage, which can grow by both interstitial and appositional growth, bone can grow only by appositional growth.

The epiphyseal plate allows the diaphysis of the bone to increase in length until early adulthood. It also shapes the articular surfaces. The rate of growth is controlled by hormones such as human growth hormone (hGH) produced by the pituitary gland and sex hormones produced by the ovaries and testes. As the child grows, cartilage cells are produced by mitosis on the epiphyseal side of the plate. They are then destroyed, and the cartilage is replaced by bone on the diaphyseal side of the plate. In this way, the thickness of the epiphyseal plate remains almost constant, but the bone on the diaphyseal side increases in length.

Growth in diameter occurs along with growth in length. In this process, the bone lining the medullary cavity is

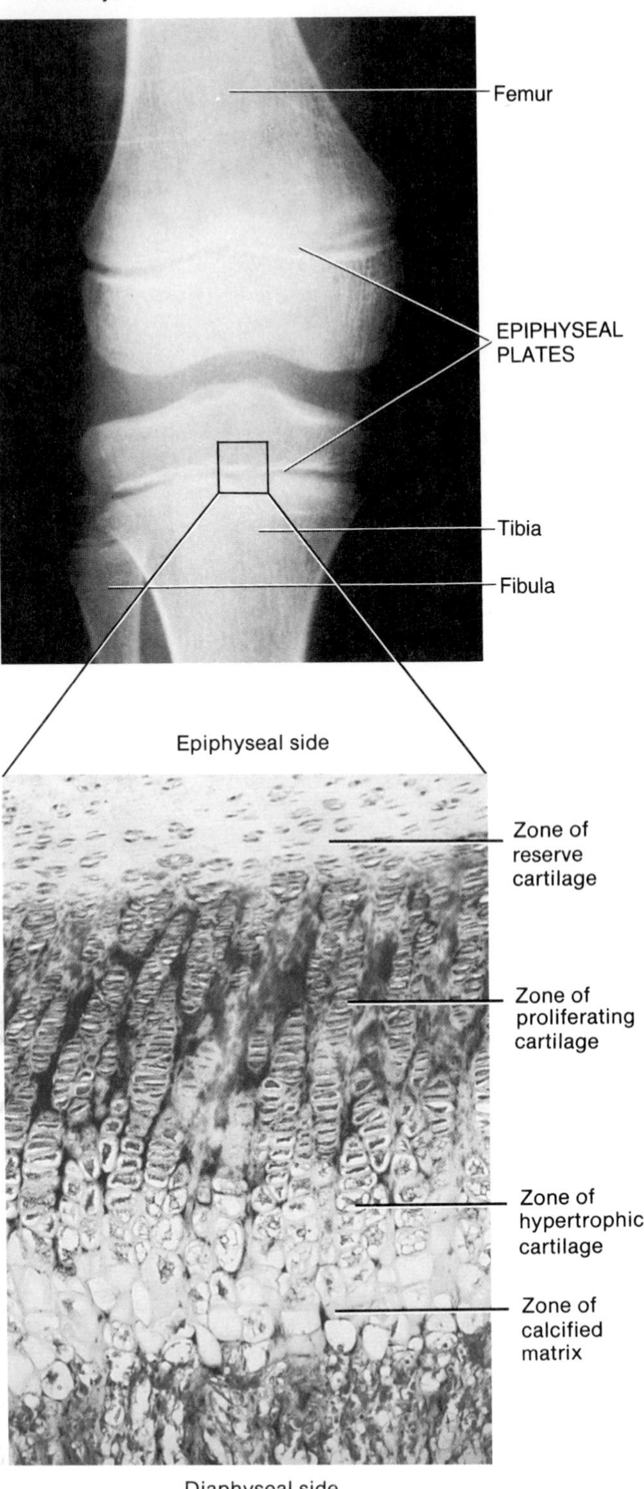

FIGURE 5-5 Epiphyseal plate. (a) Courtesy of Arthur Provost, R.T. (b) Shown are the four zones of the plate. © Biophoto, SPL, Photo Researchers.

destroyed by osteoclasts, so the cavity increases in diameter. At the same time, osteoblasts from the periosteum add new bone tissue around the outer surface of the bone. Initially, diaphyseal and epiphyseal ossification produce only spongy bone. Later, by remodeling, the outer region of spongy bone is reorganized into compact bone.

Eventually, the epiphyseal cartilage cells stop dividing, and the cartilage is replaced by bone. The newly formed bony structure is called the ***epiphyseal line,*** a remnant of the once active epiphyseal plate. With the appearance of the epiphyseal line, bone growth in length stops. The clavicle is the last bone to stop growing. Ossification of most bones is usually completed by age 25. In general, lengthwise growth in bones in females is completed before that in males.

BONE REPLACEMENT

Bone, like skin, continually replaces itself throughout adult life. ***Remodeling*** is the ongoing replacement of old bone tissue by new bone tissue. Bone is never metabolically at rest; it constantly remodels and reappropriates its matrix and minerals along lines of mechanical stress. Compact bone is formed from spongy bone. The diameter of a long bone is increased by the destruction of the bone closest to the medullary cavity and the construction of new bone around the outside of the diaphysis. However, even after bones have reached their adult shapes and sizes, old bone is continually destroyed and new bone tissue is formed in its place.

Remodeling takes place at different rates in various body regions. The distal portion of the femur (thighbone) is replaced about every four months. By contrast, bone in certain areas of the shaft will not be completely replaced during the individual's life. Remodeling allows worn or injured bone to be removed and replaced with new tissue. It also allows bone to serve as the body's storage area for calcium. Many other tissues in the body need calcium in order to perform their functions. For example, a nerve cell needs calcium for nerve impulse conduction, muscle needs calcium to contract, and blood needs calcium to clot. The blood continually trades off calcium with the bones, removing calcium when it and other tissues are not receiving enough of this element and resupplying the bones with dietary calcium to keep them from losing too much bone mass.

The cells believed to be responsible for the resorption (loss of a substance through a physiological or pathological process) of bone tissue are osteoclasts. In the healthy adult, a delicate homeostasis is maintained between the action of the osteoclasts in removing calcium salts and collagen and the action of the bone-making osteoblasts in depositing calcium salts and collagen. Should too much new tissue be formed, the bones become abnormally thick and heavy. If too much calcium is deposited in the bone, the surplus may form thick bumps (spurs) on the bone that interfere with movement at joints. A loss of too much tissue or calcium weakens the bones and allows them to break easily or to become very flexible. As you will see later, a greatly accelerated remodeling process results in a condition called Paget's disease.

In the process of resorption, osteoclasts send out projections that secrete protein-digesting enzymes released from lysosomes and several acids (lactic, carbonic, and citric). The enzymes may function by digesting the collagen and other organic substances, whereas the acids may cause the bone salts to dissolve. It is also presumed that the osteoclastic projections may phagocytize whole fragments of collagen and bone salts. Magnesium deficiency inhibits the activity of osteoblasts.

Normal bone growth in the young and bone replacement in the adult depend on several factors. First, sufficient quantities of calcium and phosphorus, components of the primary salt (hydroxyapatite) that makes the matrix of bone hard, must be included in the diet. Recent studies also suggest that boron may be a factor in bone growth by inhibiting calcium loss and increasing levels of estrogen. Manganese may also be important in bone growth. It has been shown that manganese deficiency significantly inhibits laying down of new bone tissue.

Second, the individual must obtain sufficient amounts of vitamins, particularly vitamin D, which participates in the absorption of calcium from the gastrointestinal tract into the blood, calcium removal from bone, and kidney reabsorption of calcium that might otherwise be lost in urine. Vitamin C helps to maintain the matrix of bone and other connective tissues. Vitamin C deficiency leads to decreased production of collagen and bone matrix, thereby causing retardation of bone growth and delayed healing of fractures. Vitamin A helps to control the activity, distribution, and coordination of osteoblasts and osteoclasts during development. Its deficiency results in a decreased rate of growth in the skeleton. Vitamin B_{12} may also play a role in osteoblast activity.

Third, the body must manufacture the proper amounts of the hormones responsible for bone tissue activity (Chapter 21). Human growth hormone (hGH), secreted by the pituitary gland, is responsible for the general growth of bones. Calcitonin (CT), produced by the thyroid gland, promotes bone formation. The parathyroid glands secrete parathyroid hormone (PTH), which promotes bone resorption. Other hormones, especially the sex hormones (estrogen and testosterone), promote the growth of new bone. Insulin and thyroid hormones are also important for normal bone growth and maturity.

BLOOD AND NERVE SUPPLY

Bone is richly supplied with blood, and blood vessels are especially abundant in portions of bone containing red bone marrow. Blood vessels pass into bones from the periosteum.

Although the blood supply to bones varies according to their shape, here we shall consider the blood supply to a long bone only.

Because the artery to the diaphysis of a long bone is usually the largest, it is referred to as the ***nutrient artery.*** The artery first enters the bone during the early development and passes through an opening in the diaphysis, the ***nutrient foramen*** (see Figure 5-1a), which leads into a ***nutrient canal.*** On entering the medullary cavity, the nutrient artery divides into a proximal and a distal branch, each supplying most of the marrow, inner portion of compact bone of the diaphysis, and metaphysis. As the nutrient artery passes through compact bone on its way to the medullary cavity, it sends branches into the central (Haversian) canals. Branches of the ***articular arteries*** enter many vascular foramina in the epiphysis and supply the marrow, metaphysis, and bony tissue of the epiphysis. ***Periosteal arteries,*** accompanied by nerves, enter the diaphysis at numerous points through perforating (Volkmann's) canals (see Figure 5-3a). These blood vessels run in the central canals and supply the outer part of the compact bone of the diaphysis. Veins accompany the several types of arteries; the principal veins leave the bone by numerous vascular foramina at the epiphyses of the bone.

Although the nerve supply to bone is not extensive, some nerves (vasomotor) accompany blood vessels and some (sensory) occur in the periosteum. The nerves of the periosteum are primarily concerned with pain, as might be associated with a fracture or tumor.

EXERCISE AND THE SKELETAL SYSTEM

Within limits, bone has the ability to alter its strength in response to mechanical stress. When placed under mechanical stress, bone tissue increases its deposition of mineral salts and production of collagenous fibers. Removal of mechanical stress induces removal of mineral salts and collagenous fibers. Among the mechanical stresses to which bone is subjected are those that result from the pull of skeletal muscles on bones and supporting the weight of the body against the pull of gravity. Bones of athletes, which are stressed to a high degree, become considerably thicker than those of nonathletes. (It appears that activities with high loads and stress, such as weight lifting, build more bone mass than activities with lots of repetitions, such as jogging or swimming). In the absence of such stresses, bone weakens through demineralization (loss of bone minerals) and collagen reduction. If a person has a fractured leg bone in a cast but continues to walk on the opposite leg, it can be noted that the fractured bone becomes decalcified (loses calcium) within a few weeks, from lack of mechanical stress, whereas the opposite bone remains normally calcified. Also, bones of astronauts show some degree of loss of mass as a result of the weightless environment.

In response to mechanical stress, bone (mostly collagen) produces very minute currents of electricity. It is believed that stress deforms crystals in collagen, causing a separation of electrical positive and negative charges (that is, polarization), which will make an electric current flow. The production of a current in this manner is called the ***piezoelectric*** (pē-e-zō-e-LEK-trik) ***effect,*** believed to stimulate formation of more osteoblasts in the negatively charged area, which make additional bone matrix. One effect of regular exercise in which bones are stressed, such as walking or running, is to stimulate bone growth. Another effect is to increase the production of calcitonin (CT) by the thyroid gland, a hormone that inhibits the activity of bone-destroying osteoclasts and thereby inhibits bone resorption.

AGING AND THE SKELETAL SYSTEM

There are two principal effects of aging on the skeletal system. The first effect is decalcification. This loss usually begins after age 30 in females, accelerates greatly around age 40–45 as estrogen levels decrease, and continues until as much as 30 percent of the calcium in bones is lost by age 70. In males, calcium loss typically does not begin until after age 60. The loss of calcium from bones is one of the factors related to a condition called osteoporosis, which will be described shortly.

The second principal effect of aging on the skeletal system is a decrease in the rate of protein formation, which results in a decreased ability to produce the organic portion of bone matrix. As a consequence, bone matrix accumulates a lesser proportion of organic matrix and a greater proportion of inorganic matrix. In some elderly individuals this process can cause their bones to become quite brittle and more susceptible to fracture.

DEVELOPMENTAL ANATOMY OF THE SKELETAL SYSTEM

Bone forms about the sixth or seventh week of embryonic development by either of two processes: ***intramembranous ossification*** or ***endochondral ossification.*** Both processes begin when ***mesenchymal (mesodermal) cells*** migrate into the area where bone formation will occur. In some skeletal structures, the mesenchymal cells develop into ***chrondroblasts*** that form *cartilage*. In other skeletal structures, the mesenchymal cells develop into ***osteoblasts*** that form *bone tissue* by intramembranous or endochondral ossification. (The details of ossification were discussed earlier in the chapter.)

Discussion of the development of the skeletal system provides us with an excellent opportunity to trace the development of the extremities. The *extremities* make their appearance about the fifth week as small elevations at the sides of the trunk called ***limb buds*** (Figure 5-6). They

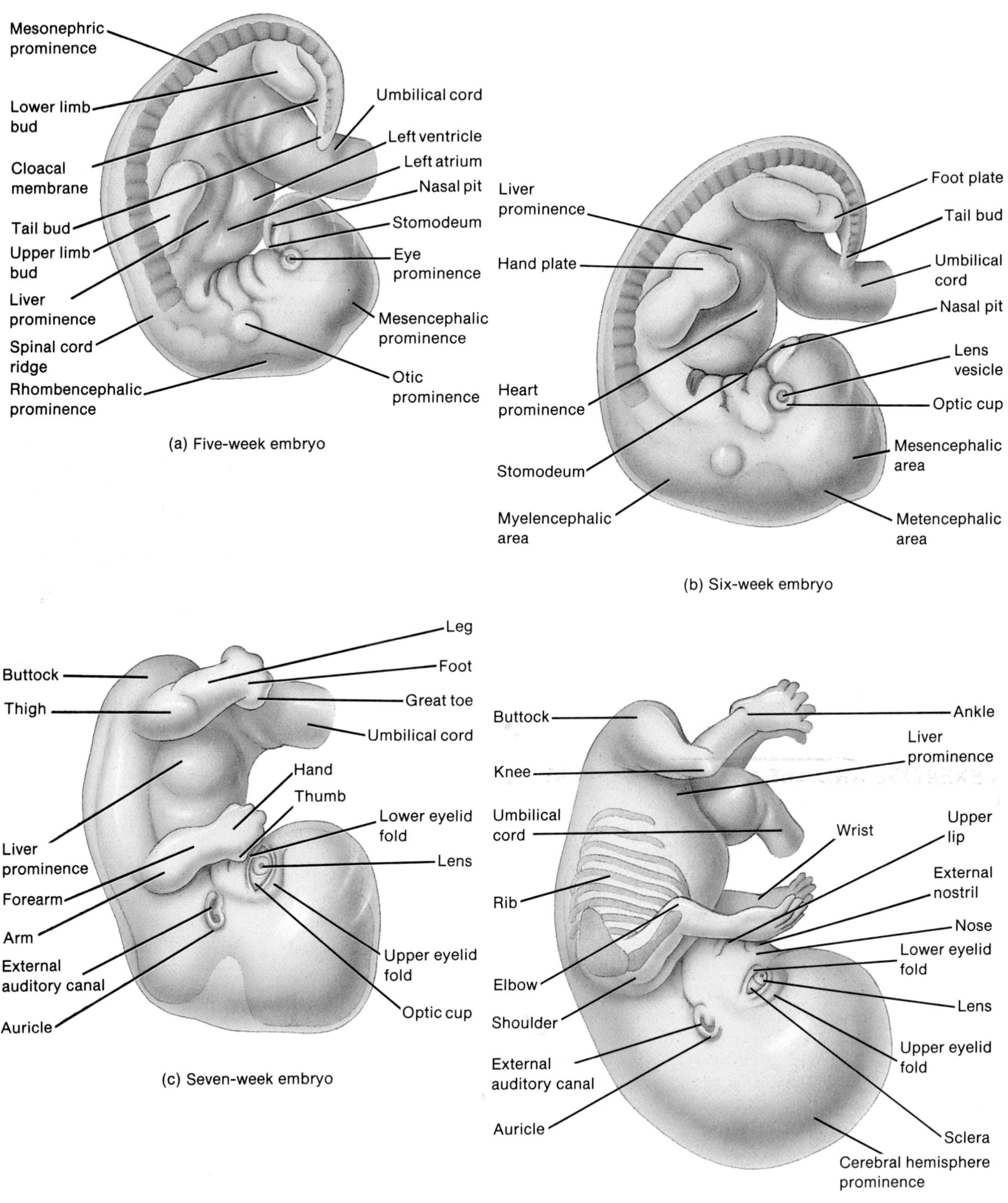

FIGURE 5-6 External features of a human embryo at various stages of development. Many of the labeled structures are discussed in later chapters.

consist of masses of general ***mesoderm*** covered by ***ectoderm.*** At this point, a mesenchymal skeleton exists in the limbs; some of the masses of mesoderm surrounding the developing bones will become the skeletal muscles of the extremities. By the sixth week, the limb buds develop a constriction around the middle portion. The constriction produces distal segments of the upper buds called ***hand plates*** and distal segments of the lower buds called ***foot plates.*** These plates represent the beginnings of the *hands* and *feet*, respectively. At this stage of limb development, a cartilaginous skeleton is present. By the seventh week, the *arm, forearm,* and *hand* are evident in the upper limb bud, and the *thigh, leg,* and *foot* appear in the lower limb bud. Endochondral ossification has begun. By the eighth week, the upper limb bud is appropriately called the *upper extremity* as the *shoulder, elbow,* and *wrist* areas become apparent, and the lower limb bud is referred to as the *lower extremity* with the appearance of the *knee* and *ankle* areas.

The ***notochord*** is a flexible rod of tissue that lies in a position where the future vertebral column will develop (see Figure 9-12b). As the vertebrae develop, the notochord becomes surrounded by the developing vertebral bodies, and the notochord eventually disappears except for remnants that persist as the *nucleus pulposus* of the intervertebral discs (see Figure 6-20).

APPLICATIONS TO HEALTH

OSTEOPOROSIS

Osteoporosis (os′-tē-ō-pō-RŌ-sis) is an age-related disorder characterized by decreased bone mass and increased susceptibility to fractures as a result of decreased levels of estrogens. The disorder primarily affects middle-aged and elderly people—women more than men and whites more than blacks. Between puberty and the middle years, sex hormones, especially estrogen, maintain osseous tissue by stimulating the osteoblasts to form new bone. Osteoblasts contain estrogen receptors in their nuclei. Women produce smaller amounts of sex hormones, especially estrogens, after menopause, and both men and women produce smaller amounts during old age. As a result, the osteoblasts become less active and there is a decrease in bone mass. Aging is also directly responsible for an increased use of parathyroid hormone (PTH), which increases osteoclastic activity, decreased levels of vitamin D, and a decline in vitamin D absorption from the gastrointestinal tract, which inhibits bone formation. Osteoporosis can also occur during pregnancy and nursing and in individuals exposed to prolonged treatment with cortisone or high levels of thyroid hormones. The first symptom of osteoporosis occurs when bone mass is so depleted that the skeleton can no longer withstand the mechanical stresses of everyday living, and fractures result. Osteoporosis is responsible for shrinkage of the backbone and height loss, hunched backs, hip fractures, and considerable pain. Osteoporosis affects the entire skeletal system, especially the vertebral bodies, ribs, proximal femur, hip, humerus, and distal radius.

Among the factors implicated in osteoporosis, besides race and gender, are (1) body build (short females are at greater risk since they have less total bone mass; (2) weight (thin females and those who overdo exercise are at a greater risk since adipose tissue is a great source of estrone, an estrogen that retards bone loss); (3) smoking (smoking decreases blood estrogen levels); (4) calcium deficiency and malabsorption; (5) vitamin D deficiency; (6) exercise (sedentary people are more likely to develop bone loss); (7) certain drugs (alcohol, some diuretics, cortisone, and tetracycline promote bone loss); (8) premature menopause (which may be due to excessive exercise); and (9) family history of osteoporosis (daughters of women with osteoporosis have reduced bone mass, especially in the lumbar spine and perhaps in the neck of the femur).

Estrogen replacement therapy (ERT), calcium supplements, and weight-bearing exercise are prescribed to prevent or retard the development of osteoporosis in postmenopausal women. Adequate diet and exercise are the mainstays of preventing osteoporosis in middle-aged and older men. Among treatments for established osteoporosis are the preventive measures given above as well as anabolic steroids and calcitonin (CT). The most important aspect of treatment is prevention, and many physicians now urge women in their twenties and thirties to pay attention to exercise and adequate calcium intake in advance of the perimenopausal years when, without ERT and calcium supplements, a negative calcium balance favors the development of osteoporosis.

VITAMIN DEFICIENCIES

Rickets

A deficiency of vitamin D in children results in ***rickets.*** It is characterized by an inability of the body to transport calcium and phosphorus from the gastrointestinal tract into the blood for utilization by bones. As a result, epiphyseal cartilage cells cease to degenerate, and new cartilage, rather than bone tissue, is produced. Epiphyseal cartilage thus becomes wider than normal. At the same time, the soft matrix laid down by the osteoblasts in the diaphysis fails to calcify, and thus bones stay soft. When the child walks, the weight of the body causes the bones in the legs to bow.

Osteomalacia

Deficiency of vitamin D in adults causes the bones to lose excessive amounts of calcium and phosphorus. This loss, called ***demineralization,*** occurs mainly in the bones of the pelvis, legs, and spine. Demineralization caused by vitamin D deficiency is called ***osteomalacia*** (os′-tē-ō-ma-LĀ-shē-a; *malacia* = softness). With demineralization, the weight of the body produces bowing of the leg bones, shortening of the backbone due to compression, and flattening of the hipbones.

Osteomalacia may also result from failure to absorb fats (steatorrhea) because vitamin D is soluble in fats and calcium combines with fats. As a result, vitamin D and calcium remain with the unabsorbed fat and are lost in the feces.

PAGET'S DISEASE

Paget's disease is characterized by a greatly accelerated remodeling process in which osteoclastic resorption is massive and new bone formation by osteoblasts is extensive. As a result, there is an irregular thickening and softening of the bones and greatly increased vascularity, especially in bones of the skull, pelvis, and extremities.

OSTEOMYELITIS

The term ***osteomyelitis*** (os′-tē-ō-mī-i-LĪ-tis) includes all infectious diseases of bone. These diseases may be localized or widespread and may also involve the periosteum, marrow, and cartilage.

FRACTURES

A ***fracture*** is any break in a bone. Usually, the fracture is restored to normal position by manipulation without surgery. This procedure of setting a fracture is called ***closed reduction.*** In other cases, the fracture must be exposed by surgery before the break is rejoined. This procedure is known as ***open reduction.***

Although fractures may be classified in several different ways, the following scheme is useful (Figure 5-7):

1. ***Partial (incomplete).*** A fracture in which the break across the bone is incomplete.
2. ***Complete.*** A fracture in which the break across the bone is complete, and therefore the bone is broken into two or more pieces.
3. ***Closed (simple).*** A fracture in which the bone does not break through the skin.
4. ***Open (compound).*** A fracture in which the broken ends of the bone protrude through the skin.
5. ***Comminuted*** (KOM-i-nyoo′-ted). A fracture in which the bone is splintered at the site of impact and smaller fragments of bone are found between the two main fragments.
6. ***Greenstick.*** A partial fracture in which one side of the bone is broken and the other side bends; occurs only in children.
7. ***Spiral.*** A fracture in which the bone is usually twisted apart.
8. ***Transverse.*** A fracture at right angles to the long axis of the bone.
9. ***Impacted.*** A fracture in which one fragment is firmly driven into the other.
10. ***Pott's.*** A fracture of the distal end of the fibula, with serious injury of the distal tibial articulation.
11. ***Colles'*** (KOL-ēz). A fracture of the distal end of the radius in which the distal fragment is displaced posteriorly.
12. ***Displaced.*** A fracture in which the anatomical alignment of the bone fragments is not preserved.
13. ***Nondisplaced.*** A fracture in which the anatomical alignment of the bone fragments is preserved.

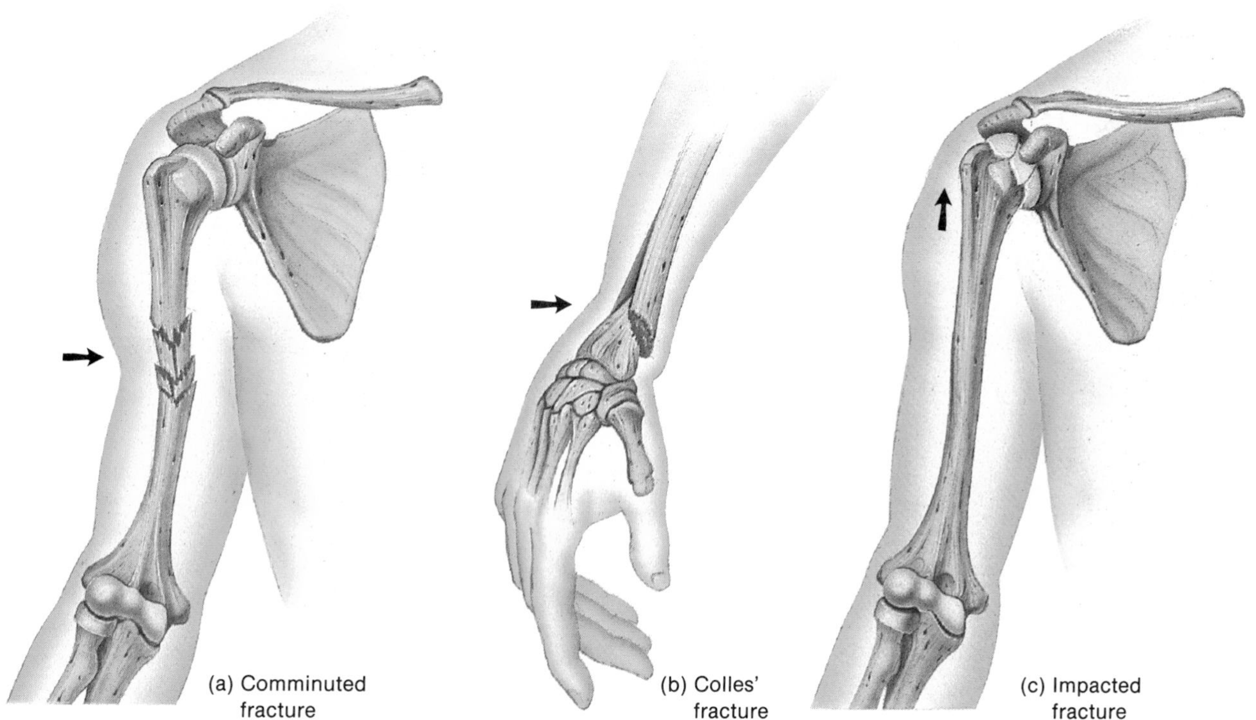

FIGURE 5-7 Types of fractures. (g) Courtesy of Lester Bergman & Associates.

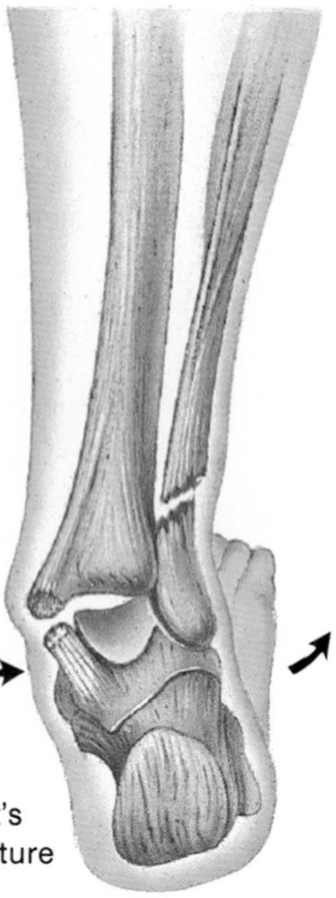

(d) Pott's fracture

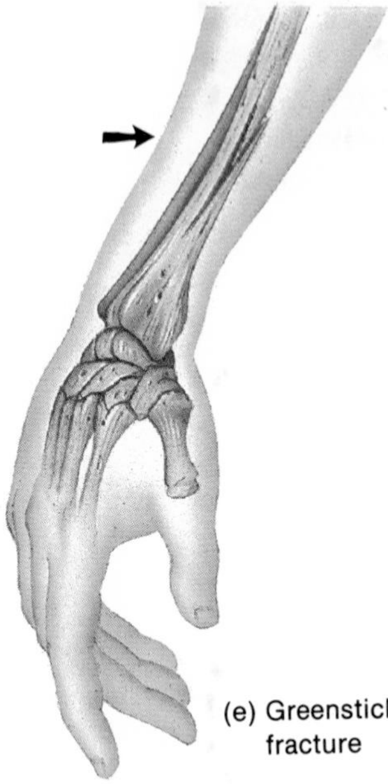

(e) Greenstick fracture

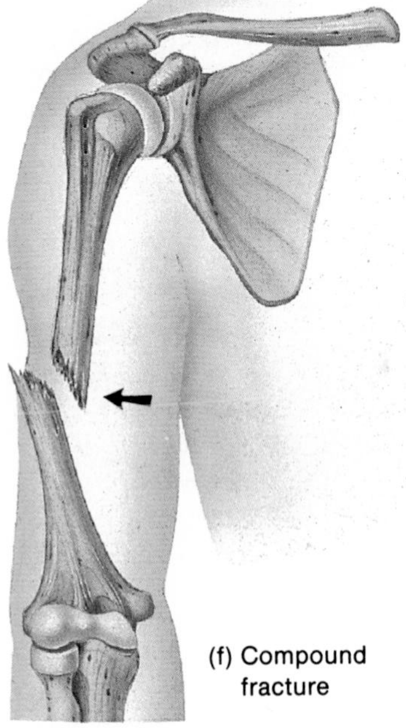

(f) Compound fracture

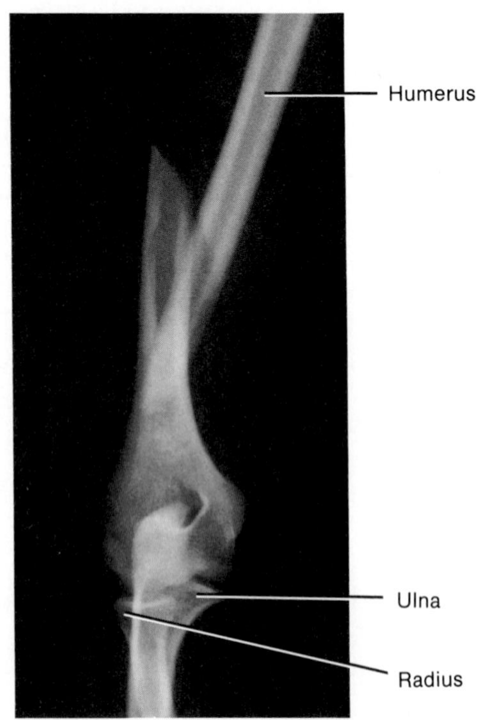

(g) X-ray of compound fracture of the humerus

14. ***Stress.*** A partial fracture resulting from inability to withstand repeated stress due to a change in training, harder surfaces, longer distances, and greater speed. About 25 percent of all stress fractures involve the fibula, typically the distal third.
15. ***Pathologic.*** A fracture due to weakening of a bone caused by disease processes such as neoplasia, osteomyelitis, osteoporosis, or osteomalacia.

Bone sometimes requires months to heal. A fractured femur, for example, may take six months to heal. Sufficient calcium to strengthen and harden new bone is deposited only gradually. Bone cells also grow and reproduce slowly. Moreover, in a fractured bone the blood supply is decreased, which helps to explain the difficulty in the healing of an infected bone.

CLINICAL APPLICATION

Pulsating Electromagnetic Fields

In the past when a fracture failed to unite, the patient could either wait, hoping that nature and time would solve the problem, or choose surgery. Now there is another alternative, called ***pulsating electromagnetic fields (PEMFs)***, that involves electrotherapy to stimulate bone repair.

Essentially, the fracture is exposed to a weak electric current generated from coils that are fastened around the cast. Osteoblasts adjacent to the fracture site become more active metabolically in response to the stimulation. The increased activity apparently causes an acceleration of calcification, vascularization, and endochondral ossification, resulting in acceleration of fracture repair. It was noted earlier that parathyroid hormone (PTH) increases osteoclastic activity and thus stimulates bone destruction. One hypothesis suggests that electricity also heals fractures by keeping PTH from acting on osteoclasts, thus increasing bone formation and repair.

KEY MEDICAL TERMS ASSOCIATED WITH BONE TISSUE

Achondroplasia (a-kon′-drō-PLĀ-zē-a; *a* = without; *chondro* = cartilage; *plasia* = growth) Imperfect ossification within cartilage of long bones during fetal life; also called **fetal rickets.** It causes a form of dwarfism.

Craniotomy (krā′-nē-OT-ō-mē; *cranium* = skull; *tome* = a cutting) Any surgery that requires cutting through the bones surrounding the brain.

Osteitis (os′-tē-Ī-tis; *itis* = inflammation of) Inflammation or infection of bone.

Osteoarthritis (os′-tē-ō-ar-THRĪ-tis; *arthro* = joint) The degeneration of cartilage, allowing the bony ends to touch and, from the friction of bone against bone, creating a bony reaction that worsens the friction and worsens the condition; usually associated with the elderly.

Osteochondroma (os′-tē-ō-kon-DRŌ-ma; *chondro* = cartilage) A benign tumor of bone and cartilage.

Osteopenia (os′-tē-ō-PĒ-nē-a; *osteo* = bone; *penia* = poverty) Reduced bone mass due to a decrease in the rate of bone synthesis to a level insufficient to compensate for normal bone breakdown; any decrease in bone mass below normal. Examples include osteoporosis and osteomalacia.

Osteosarcoma (os′-tē-ō-sar-KŌ-ma; *sarcoma* = connective tissue tumor) A malignant tumor composed of osseous tissue.

STUDY OUTLINE

Functions (p. 111)

1. The skeletal system consists of all bones attached at joints and cartilage between joints.
2. The functions of the skeletal system include support, protection, movement, mineral storage, and housing blood-forming tissue.

Histology (p. 111)

1. Bone tissue consists of widely separated cells surrounded by large amounts of matrix. The four principal types of cells are osteoprogenitor cells, osteoblasts, osteocytes, and osteoclasts. The matrix contains collagenous fibers and abundant hydroxyapatites (mineral salts), consisting mainly of calcium phosphate salts.
2. Parts of a typical long bone are the diaphysis (shaft), epiphyses (ends), metaphysis, articular cartilage, periosteum, medullary (marrow) cavity, and endosteum.
3. Compact (dense) bone consists of osteons (Haversian systems) with little space between them. Compact bone lies over spongy bone and composes most of the bone tissue of the diaphysis. Functionally, compact bone protects, supports, and resists stress.
4. Spongy (cancellous) bone consists of trabeculae surrounding many red marrow–filled spaces. It forms most of the structure of short, flat, and irregular bones, and the epiphyses of long bones. Functionally, spongy bone stores some red and yellow marrow and provides some support.

Ossification: Bone Formation (p. 111)

1. Bone forms by a process called ossification (osteogenesis) that begins when mesenchymal cells become transformed into osteoprogenitor cells, which undergo cell division giving rise to cells that differentiate into osteoblasts and chondroblasts.
2. The process begins during the sixth or seventh week of embryonic life and continues throughout adulthood. The two types of ossification—intramembranous and endochondral—involve the replacement of a preexisting connective tissue with bone.
3. Intramembranous ossification occurs within fibrous connective tissue membranes.
4. Endochondral ossification occurs within a hyaline cartilage model. The primary ossification center of a long bone is in the diaphysis. Cartilage degenerates, leaving cavities that merge to form the medullary (marrow) cavity. Osteoblasts lay down bone. Next, ossification occurs in the epiphyses, where bone replaces cartilage except for the articular cartilage and epiphyseal plate.

Bone Growth (p. 118)

1. The anatomical zones of the epiphyseal plate are the zones of reserve cartilage, proliferating cartilage, hypertrophic cartilage, and calcified matrix.
2. Because of the activity of the epiphyseal plate, the diaphysis of a bone increases in length by appositional growth.
3. Bone grows in diameter as a result of the addition of new bone tissue by periosteal osteoblasts around the outer surface of the bone.

Bone Replacement (p. 119)

1. The homeostasis of bone growth and development depends on a balance between bone formation and resorption.
2. Old bone is constantly destroyed by osteoclasts, while new bone is constructed by osteoblasts. This process is called remodeling.
3. Normal bone growth depends on calcium, phosphorus, and vitamins, especially vitamin D, and is controlled by hormones that are responsible for bone mineralization and resorption.

Blood and Nerve Supply (p. 119)

1. Long bones are supplied by nutrient, articular, and periosteal arteries; veins accompany the arteries.
2. The nerve supply to bones consists of vasomotor and sensory nerves.

Exercise and the Skeletal System (p. 120)

1. Bone can alter its strength in response to mechanical stress.
2. Bone that is stressed produces a minute electric current, by way of its mineral salt crystals (piezoelectric effect), that stimulates osteoblastic activity.
3. Regular exercise can stimulate osteoblasts and inhibit osteoclasts.

Aging and the Skeletal System (p. 120)

1. The principal effect of aging is a loss of calcium from bones, which may result in osteoporosis.
2. Another effect is a decreased production of organic matrix, which makes bones more susceptible to fracture.

Developmental Anatomy of the Skeletal System (p. 120)

1. Bone forms from mesoderm by intramembranous or endochondral ossification.
2. Extremities develop from limb buds, which consist of mesoderm and ectoderm.

Applications to Health (p. 122)

1. Osteoporosis is a decrease in the amount and strength of bone tissue owing to decreases in hormone output.
2. Rickets is a vitamin D deficiency in children in which the body does not absorb calcium and phosphorus. The bones soften and bend under the body's weight.
3. Demineralization caused by vitamin D deficiency in adults results in osteomalacia.
4. Paget's disease is the irregular thickening and softening of bones, related to a greatly accelerated remodeling process.
5. Osteomyelitis is a term for the infectious diseases of bones, marrow, and periosteum.
6. A fracture is any break in a bone. The types of fractures include partial, complete, closed (simple), open (compound), comminuted, greenstick, spiral, transverse, impacted, Pott's, Colles', displaced, nondisplaced, and stress.
7. Treatment by pulsating electromagnetic fields (PEMFs) has provided dramatic results in healing fractures that would otherwise not have mended properly. Its application for limb regeneration and stopping the growth of tumor cells is being investigated.

REVIEW QUESTIONS

1. Define the skeletal system. What are its six principal functions? (p. 111)
2. Why is bone tissue considered a connective tissue? Describe the cells present and the composition of the matrix. (p. 111)
3. Diagram the parts of a long bone and list the functions of each part. (p. 111)
4. Distinguish between spongy and compact bone in terms of microscopic appearance, location, and function. (p. 113)
5. Diagram the microscopic appearance of compact bone and indicate the functions of the various components. (p. 113)
6. What is meant by ossification? Describe the initial events of ossification. (p. 115)
7. Outline the major events involved in intramembranous and endochondral ossification and explain the main differences. (p. 117)
8. Describe the histology of the various zones of the epiphyseal plate. How does the plate grow? What is the significance of the epiphyseal line? (p. 118)
9. Define remodeling. How does balance between osteoblast activity and osteoclast activity demonstrate the homeostasis of bone? (p. 119)
10. Describe the blood and nerve supplies of a long bone. (p. 119)
11. Explain the effects of exercise and aging on the skeletal system. (p. 120)
12. Describe the development of the skeletal system. (p. 120)
13. Define the following: osteoporosis, rickets, osteomalacia, Paget's disease, and osteomyelitis. (p. 122)
14. What is a fracture? Distinguish several principal kinds. (p. 123)
15. Refer to the glossary of key medical terms associated with the bone tissue. Be sure that you can define each term. (p. 125)

SELF QUIZ

1. Match the following:

___ **a.** thin layer of hyaline cartilage at end of long bone
___ **b.** region of mature bone where diaphysis joins epiphysis
___ **c.** covering over bone to which ligaments and tendons attach
___ **d.** inner layer of covering over bone; contains osteoprogenitor cells and osteoblasts

A. articular cartilage
B. endosteum
C. fibrous periosteum
D. metaphysis
E. osteogenic periosteum

___ **e.** layer of osteoprogenitor cells and osteoblasts lining the medullary cavity

Choose the one best answer to these questions.

___ **2.** The cells responsible for the resorption (destruction) of bone tissue are
A. osteocytes; B. osteoblasts; C. osteoclasts; D. chondrocytes; E. chondroblasts.

___ **3.** You have four sets of vitamins and minerals: Set I—vitamins B and K; Set II—calcium and phosphorus; Set III—vitamins A, C, and D; Set IV—iron and sodium. Which would you recommend in the diet of a child for the growth of his or her bones?
A. all of the above; B. Sets I, II, and IV; C. Sets II, III, and IV; D. Sets II and III; E. Sets I, II, and III.

___ **4.** Which of the following is *not* considered a function of bone or of bony tissue?
A. attachment of muscles, tendons, and ligaments; B. secretion of calcitonin (CT) that affects calcium levels in the blood; C. formation of blood cells and fat storage; D. support and protection of soft organs and tissues; E. depot for calcium and phosphorus.

___ **5.** The periosteum of a bone
(1) is the membrane lining the medullary (marrow) cavity of the bone.
(2) is a dense, fibrous membrane covering the outside of most of a long bone, but not the articular ends.
(3) is the name of the outer layer of compact bone.
(4) contains nerves and blood vessels.
A. (1) only; B. (2) only; C. (3) only; D. (4) only; E. (2) and (4).

___ **6.** Which of the following is *not* true regarding bone types?
A. compact bone tissue is dense and strong and is located on the outside of bones; B. compact bone tissue is also called spongy bone; C. spongy bone has many open (marrow) spaces between the bony tissue; D. the microscopic structure of compact bone consists of osteons (Haversian systems); E. the plates of bone forming an open network pattern in spongy bone are called trabeculae.

___ **7.** Which of the following is/are true?
(1) Bone maturation and modeling go on continually during the life of the individual.
(2) The surface skull bones develop via intramembranous ossification.
(3) Most bones are laid down as cartilage, which is replaced by bone.
A. (1) only; B. (2) only; C. (3) only; D. (1) and (2); E. all of the above

___ **8.** Which of the following statements is false?
A. the outer connective tissue covering of a bone is the periosteum; B. porous, spongelike bone is called spongy bone; C. the medullary (marrow) cavity of a long bone is lined by periosteum; D. the central cavity within the shaft of a long bone is called the medullary cavity; E. the tissue occupying the medullary cavity is yellow bone marrow.

___ **9.** Which of the following is/are true.
(1) Spongy bone is found in the ends of long bones and inside flat and irregular bones.
(2) The microscopic structure of compact bone shows osteons (Haversian systems).
(3) Short bones consist entirely of compact bone, which is composed of osteons (Haversian systems).
(4) Spongy bone consists of interlacing plates and bars of bony material with many spaces between.
A. (1) only; B. (2) only; C. (3) only; D. (4) only; E. (1), (2), and (4).

Complete the following.

10. Typical of all connective tissues, bone consists mainly of (cells? matrix?).

11. Bones and cartilage are formed from (ecto-, meso-, endo-) derm that later differentiates into the embryonic connective tissue called ________.

12. The two main kinds of tissue that compose the "skeleton" of a developing embryo or fetus are ________ and ________.

13. A. Spaces containing bone cells: ________
B. Bone cells: ________
C. Concentrically arranged calcified layers of bone: ________
D. Small (microscopic) canals running longitudinally through bone and containing blood vessels: ________
E. Large (macroscopic) cavity filled with yellow marrow: ________
F. Minute canals, radially arranged between lacunae: ________
G. Horizontal canals carrying blood vessels from periosteum of osteons (Haversian systems) and marrow: ________

Arrange the answers in correct sequence.

___ ___ ___ **14. a.** From most superficial to deepest:
A. endosteum
B. periosteum
C. compact bone

___ ___ ___ **b.** Phases in formation of bone in embryonic life, in chronological order:
A. mesenchyme cells
B. osteocytes
C. osteoblasts

Circle T (true) or F (false) for the following.

T F 15. Osteons (Haversian systems) are found in compact bone but not in spongy bone.

T F 16. Another name for the epiphysis is the shaft of the bone.

T F 17. Osteons (Haversian canals) run longitudinally (lengthwise) through bone, but perforating (Volkmann's) canals run horizontally across bone.

T F 18. In a long bone the primary ossification center is located in the diaphysis, whereas the secondary center of ossification is in the epiphysis.

T F 19. The epiphyseal plate appears earlier in life than the epiphyseal line.

T F 20. Compact bone that is of intramembranous origin differs structurally from compact bone developed from cartilage.

T F 21. The organic matrix of bone secreted by osteoblasts is called the osteoid.

T F 22. Growth of cartilage in diameter is due to interstitial growth.

The Skeletal System: The Axial Skeleton

6

STUDENT OBJECTIVES

1. Define the four principal types of bones on the basis of shape.
2. Describe the various markings on the surfaces of bones.
3. List the components of the axial and appendicular skeleton.
4. Identify the bones of the skull and the major markings associated with each.
5. Identify the principal sutures, fontanels, paranasal sinuses, and foramina of the skull.
6. Identify the bones of the vertebral column and their principal markings.
7. List the defining characteristics and normal curves of each region of the vertebral column.
8. Identify the bones of the thorax and their principal markings.
9. Contrast herniated (slipped) disc, abnormal curves, spina bifida, and fractures of the vertebral column as disorders associated with the skeletal system.

CHAPTER OUTLINE

- **Types of Bones**
- **Surface Markings**
- **Divisions of the Skeletal System**
- **Skull**

Sutures
Fontanels
Cranial Bones
Frontal Bone
Parietal Bones
Temporal Bones
Occipital Bone
Sphenoid Bone
Ethmoid Bone
Cranial Fossae
Facial Bones
Nasal Bones
Maxillae
Paranasal Sinuses
Zygomatic Bones
Mandible
Lacrimal Bones
Palatine Bones
Inferior Nasal Conchae
Vomer
Orbits
Foramina

- **Hyoid Bone**
- **Vertebral Column**

Divisions
Normal Curves
Typical Vertebra
Cervical Region
Thoracic Region
Lumbar Region
Sacrum and Coccyx

- **Thorax**

Sternum
Ribs

- **Applications to Health**

The skeletal system forms the framework of the body. For this reason, a familiarity with the names, shapes, and positions of individual bones will help you understand some of the other organ systems. For example, movements such as throwing a ball, typing, and walking require the coordinated use of bones and muscles. To understand how muscles produce different movements, you need to learn the parts of the bones to which the muscles attach and the types of joints acted upon by the contracting muscles. The respiratory system is also highly dependent on bone structure. The bones in the nasal cavity form a series of passageways that help clean, moisten, and warm inhaled air. Furthermore, the bones of the thorax are specially shaped and positioned so that the chest can expand during inhalation. Many bones also serve as landmarks to students of anatomy as well as to surgeons. As you will see, bony landmarks can be used to locate the outlines of the lungs and heart, abdominal and pelvic viscera, and structures within the skull. Blood vessels and nerves often run parallel to bones. These structures can be located more easily if the bone is identified first.

CLINICAL APPLICATIONS

Why Scientist Study Bones

Skeletal remains may persist for thousands of years after a person has died. They make it possible to trace patterns of disease and nutrition, evaluate the effects of certain social and economic changes, and deduce patterns of reproduction and mortality. Skeletal remains also may reveal an individual's sex, age, height, and race.

Many disorders can leave permanent effects on skeletal material. Three of the many common causes of bone pathologies are malnutrition, tumors, and infections. Each may cause specific changes in bone that permit diagnosis from skeletal remains.

TYPES OF BONES

Almost all the bones of the body may be classified into four principal types on the basis of shape: long, short, flat, and irregular. ***Long bones*** have greater length than width and consist of a diaphysis and a variable number of epiphyses. For example, metacarpals, metatarsals, and phalanges have only one epiphysis. The femur actually has four. Other long bones have two. Long bones are slightly curved for strength. A curved bone is structurally designed to absorb the stress of the body weight at several different points so the stress is evenly distributed. If such bones were straight, the weight of the body would be unevenly distributed and the bone would easily fracture. Examples of long bones include bones of the thighs, legs, toes, arms, forearms, and fingers. Figure 5–1a shows the parts of a long bone.

Short bones are somewhat cube-shaped and nearly equal in length and width. Their texture is spongy except at the surface, where there is a thin layer of compact bone. Examples of short bones are the wrist and ankle bones.

Flat bones are generally thin and composed of two more or less parallel plates of compact bone enclosing a layer of spongy bone. Flat bones afford considerable protection and provide extensive areas for muscle attachment. Examples of flat bones include the cranial bones (which protect the brain), the sternum and ribs (which protect organs in the thorax), and the scapulas. The spongy bone found between plates of compact bone in skull bones is called the ***diploë*** (DIP-lō-ē).

Irregular bones have complex shapes and cannot be grouped into any of the three categories just described. They also vary in the amount of spongy and compact bone present. Irregular bones include those of the vertebral column (backbone) and certain facial bones.

There are two additional types of bones that are not included in this classification by shape, but instead are classified by location. ***Sutural*** (SOO-chur-al) ***bones*** are small bones between the joints (sutures) of certain cranial bones (see Figure 6-2d). Their number varies greatly from person to person. ***Sesamoid bones*** are small bones in tendons where considerable pressure develops—for instance, in the wrist. These, like sutural bones, are also variable in number. Two sesamoid bones, the patellas (kneecaps), are present in all individuals.

SURFACE MARKINGS

The surfaces of bones have various structural features adapted to specific functions. These features are called ***surface markings.*** Long bones that bear a great deal of weight have large, rounded ends that can form sturdy joints and provide adequate surface area for the attachment of ligaments and muscles. Other bones have depressions that receive the rounded ends. Rough areas serve as points of attachment for muscles, tendons, and ligaments. Grooves in the surfaces of bones provide for the passage of blood vessels. Openings occur where blood vessels and nerves pass through the bone. Exhibit 6-1 describes the different markings and their functions.

DIVISIONS OF THE SKELETAL SYSTEM

The adult human skeleton consists of 206 named bones grouped in two principal divisions: the ***axial skeleton*** and the ***appendicular skeleton.*** The longitudinal ***axis,*** or center, of the human body is a straight line that runs through the body's center of gravity. This imaginary line extends through the head and down to the space between the feet. The midsagittal plane and midline are drawn through this line. The axial division consists of the bones that lie around

EXHIBIT 6-1
Bone Markings

MARKINGS	DESCRIPTION	EXAMPLE
DEPRESSIONS AND OPENINGS		
Fissure (FISH-ur)	A narrow, cleftlike opening between adjacent parts of bones through which blood vessels or nerves pass.	Superior orbital fissure of the sphenoid bone (Figure 6-2).
Foramen (fō-RĀ-men; *foramen* = hole)	An opening through which blood vessels, nerves, or ligaments pass.	Infraorbital foramen of the maxilla (Figure 6-2).
Meatus (mē-Ā-tus; *meatus* = canal)	A tubelike passageway running within a bone.	External auditory meatus of the temporal bone (Figure 6-2).
Paranasal Sinus (*sin* = cavity)	An air-filled cavity within a bone connected to the nasal cavity.	Frontal sinus of the frontal bone (Figure 6-8).
Groove or Sulcus (*sulcus* = ditchlike groove)	A furrow or depression that accommodates a soft structure such as a blood vessel, nerve, or tendon.	Intertubercular sulcus of the humerus (Figure 7-4).
Fossa (*fossa* = basinlike depression)	A depression in or on a bone.	Mandibular fossa of the temporal bone (Figure 6-4).
Fontanel (fon'-ta-NEL; = little fountain)	Dense connective tissue-filled space between skull bones at birth.	Anterior fontanel between frontal and parietal bones (Figure 6-3).
PROCESSES	Any prominent projections.	Mastoid process of the temporal bone (Figure 6-2).
Processes That Form Joints		
Condyle (KON-dīl; *condylus* = knucklelike process)	A large, rounded articular prominence.	Medial condyle of the femur (Figure 7-11).
Head	A rounded, articular projection supported on the constricted portion (neck) of a bone.	Head of the femur (Figure 7-11).
Facet	A smooth, flat surface.	Articular facet for the tubercle of rib on a vertebra (Figure 6-15).
Processes to Which Tendons, Ligaments, and Other Connective Tissues Attach		
Tubercle (TOO-ber-kul; *tuber* = knob)	A small, rounded process.	Greater tubercle of the humerus (Figure 7-4).
Tuberosity	A large, rounded, usually roughened process.	Ischial tuberosity of the hipbone (Figure 7-8).
Trochanter (trō-KAN-ter)	A large, blunt projection found only on the femur.	Greater trochanter of the femur (Figure 7-11).
Crest	A prominent border or ridge on a bone.	Iliac crest of the hipbone (Figure 7-7).
Line	A less prominent ridge than a crest.	Linea aspera of the femur (Figure 7-11).
Spinous Process or Spine	A sharp, slender process.	Spinous process of a vertebra (Figure 6-13).
Epicondyle (*epi* = above)	A prominence above a condyle.	Medial epicondyle of the femur (Figure 7-11).

the axis: bones of the skull, auditory ossicles, hyoid bone, ribs, sternum (breastbone), and vertebral column (backbone).

The appendicular division contains the bones of the ***upper*** and ***lower extremities*** (***limbs***), plus the bones called ***girdles,*** that connect the extremities to the axial skeleton.

The 80 bones of the axial division and the 126 bones of the appendicular division are typically grouped as shown in Exhibit 6-2.

Refer to Figure 6-1 to see how the two divisions are joined to form the complete skeleton. The bones of the axial skeleton are shown in gold.

SKULL

The ***skull,*** which contains 22 bones, rests on the superior end of the vertebral column and includes two sets of bones:

EXHIBIT 6-2

Divisions of the Skeletal System

REGIONS OF THE SKELETON	NUMBER OF BONES
AXIAL SKELETON	
Skull	
Cranium	8
Face	14
Hyoid	1
Auditory Ossicles[a] **(3 in Each Ear)**	6
Vertebral Column	26
Thorax	
Sternum	1
Ribs	24
	80
APPENDICULAR SKELETON	
Pectoral (Shoulder) Girdles	
Clavicle	2
Scapula	2
Upper Extremities	
Humerus	2
Ulna	2
Radius	2
Carpals	16
Metacarpals	10
Phalanges	28
Pelvic (Hip) Girdle	
Coxal, Pelvic, or Hipbone	2
Lower Extremities	
Femur	2
Fibula	2
Tibia	2
Patella	2
Tarsals	14
Metatarsals	10
Phalanges	28
	126
	Total = 206

[a] Although the auditory ossicles are not considered part of the axial or appendicular skeleton, but rather as a separate group of bones, they are placed with the axial skeleton for convenience. The auditory ossicles are exceedingly small bones named for their shapes. Their names are the malleus, incus, and stapes, commonly called the hammer, anvil, and stirrup, respectively. The middle portion of each ear contains three auditory ossicles held together by a series of ligaments. The auditory ossicles vibrate in response to sound waves that strike the eardrum and assume a key function in the mechanism involved in hearing. This is described in detail in Chapter 20.

cranial bones and facial bones. The ***cranial bones*** enclose and protect the brain. The eight cranial bones are the frontal bone, parietal bones (2), temporal bones (2), occipital bone, sphenoid bone, and ethmoid bone. There are 14 ***facial bones:*** nasal bones (2), maxillae (2), zygomatic bones (2), mandible, lacrimal bones (2), palatine bones (2), inferior nasal conchae (2), and vomer. Be sure you can locate all the skull bones in the various views of the skull (Figure 6-2a-d).

SUTURES

A ***suture*** (SOO-chur; *sutura* = seam) is an immovable joint found only between skull bones. Very little connective tissue is found in a suture between the bones. Sutures begin to fuse between ages 20 and 30 and by age 60 almost all sutures are united. Four prominent sutures, which are shown in Figures 6-2 and 6-3, are:

1. ***Coronal suture*** between the frontal bone and the two parietal bones.
2. ***Sagittal suture*** between the two parietal bones.
3. ***Lambdoid*** (LAM-doyd) ***suture*** between the parietal bones and the occipital bone.
4. ***Squamosal*** (skwa-MŌ-sal) ***suture*** between the parietal bones and the temporal bones.

FONTANELS

The "skeleton" of a newly formed embryo consists of cartilage or fibrous membrane structures shaped like bones. Gradually, the cartilage or fibrous membrane is replaced by bone, a process called ossification. At birth, membrane-filled spaces called ***fontanels*** (fon′-ta-NELZ; = little fountains) are found between cranial bones (Figure 6-3). These "soft spots" are areas of dense connective tissue that will eventually be replaced with bone through intramembranous ossification. They (1) enable the fetal skull to compress as it passes through the birth canal, (2) permit rapid growth of the brain during infancy, (3) help determine the degree of brain development by their state of closure, (4) serve as landmarks (anterior fontanel) for withdrawal of blood from the superior sagittal sinus, and (5) aid in determining the position of the fetal head before birth.

Although an infant may have many fontanels at birth, the form and location of several are fairly constant.

1. The ***anterior (frontal) fontanel*** is located between the angles of the two parietal bones and the two segments of the frontal bone. This fontanel is roughly diamond-shaped, and it is the largest of the fontanels. It usually closes 18 to 24 months after birth.
2. The ***posterior*** (***occipital;*** ok-SIP-i-tal) ***fontanel*** is situated between the two parietal bones and the occipital bones. This diamond-shaped fontanel is considerably smaller than the anterior fontanel. It generally closes about two months after birth.
3. The ***anterolateral*** (***sphenoidal;*** sfē-NOY-dal) ***fontanels*** are paired. One is located on each side of the skull at the junction of the frontal, parietal, temporal, and sphenoid bones. These fontanels are small and irregular in shape. They normally close about three months after birth.

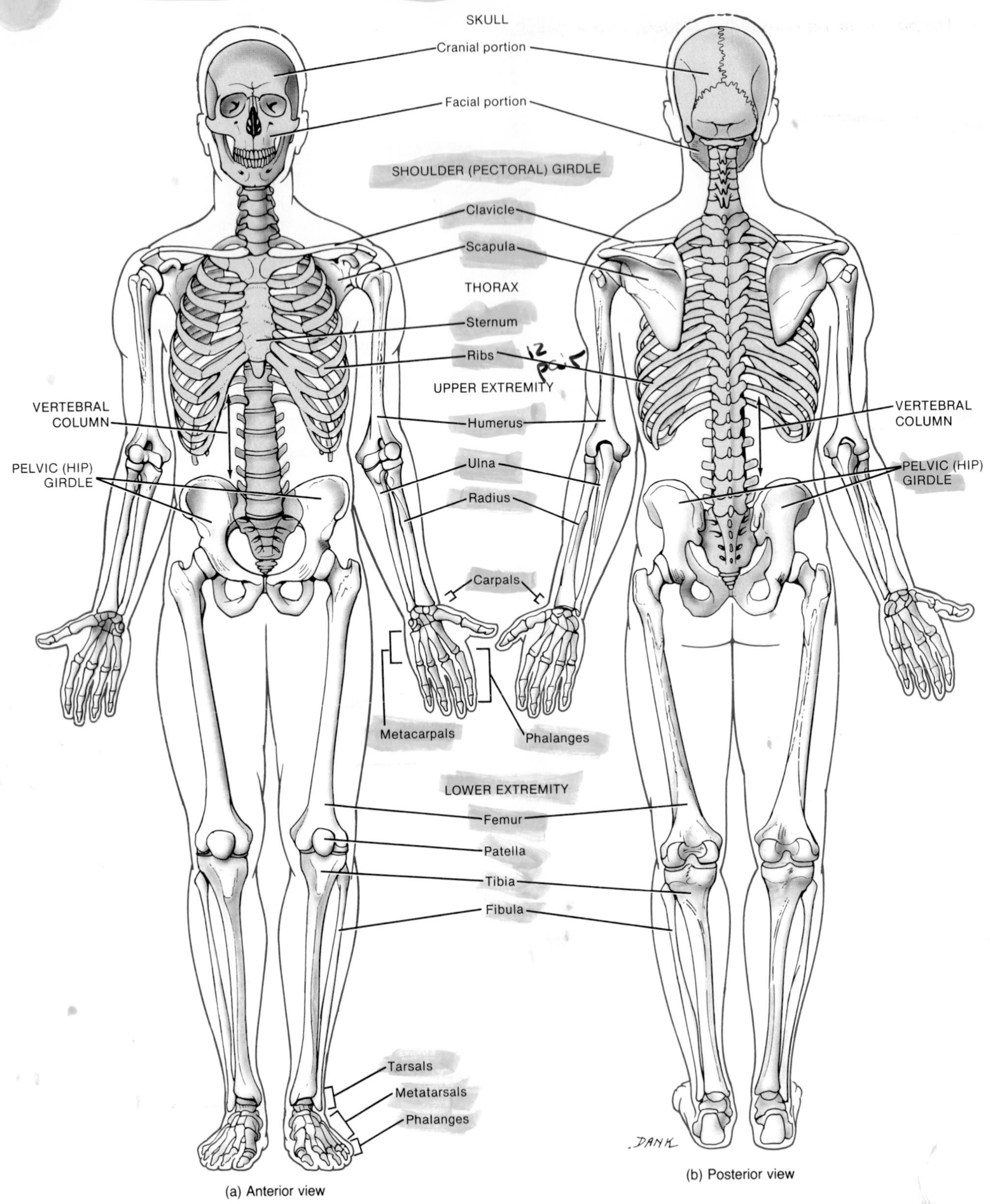

(a) Anterior view

(b) Posterior view

FIGURE 6-1 Divisions of the skeletal system. The axial skeleton is indicated in gold and the appendicular skeleton is indicated in light brown. Note the position of the hyoid bone in Figure 6-2c.

4. The ***posterolateral (mastoid) fontanels*** are also paired. One is situated on each side of the skull at the junction of the parietal, occipital, and temporal bones. These fontanels are irregularly shaped. They begin to close one or two months after birth, but closure is not generally complete until the age of 12 months.

CRANIAL BONES

Frontal Bone

The ***frontal bone*** forms the forehead (the anterior part of the cranium), the roofs of the orbits (eye sockets), and most of the anterior part of the cranial floor. Soon after birth, the left and right parts of the frontal bone are united by a suture, which usually disappears by age 6.

If you examine the anterior and lateral views of the skull in Figure 6-2, you will note the ***frontal squama*** (SKWĀ-ma; *squam* = scale) which corresponds to the forehead. It projects slightly above its lower edge on either side of the midline to form the ***frontal eminences.*** Inferior to each eminence is a horizontal ridge, the ***superciliary arch,*** caused by the projection of the frontal sinuses posterior to the eyebrow.

Between the eminences and the arches just superior to the nose is a flattened area, the ***glabella.*** A thickening of the frontal bone is called the ***supraorbital margin.*** From this margin the frontal bone extends posteriorly to form the roof of the orbit and part of the floor of the cranial cavity.

Within the supraorbital margin, slightly medial to its midpoint, is a hole (frequently a notch) called the ***supraorbital foramen (notch).*** As the various foramina associated with cranial bones are discussed, refer to Exhibit 6-5 to note which structures pass through them. The ***frontal sinuses***

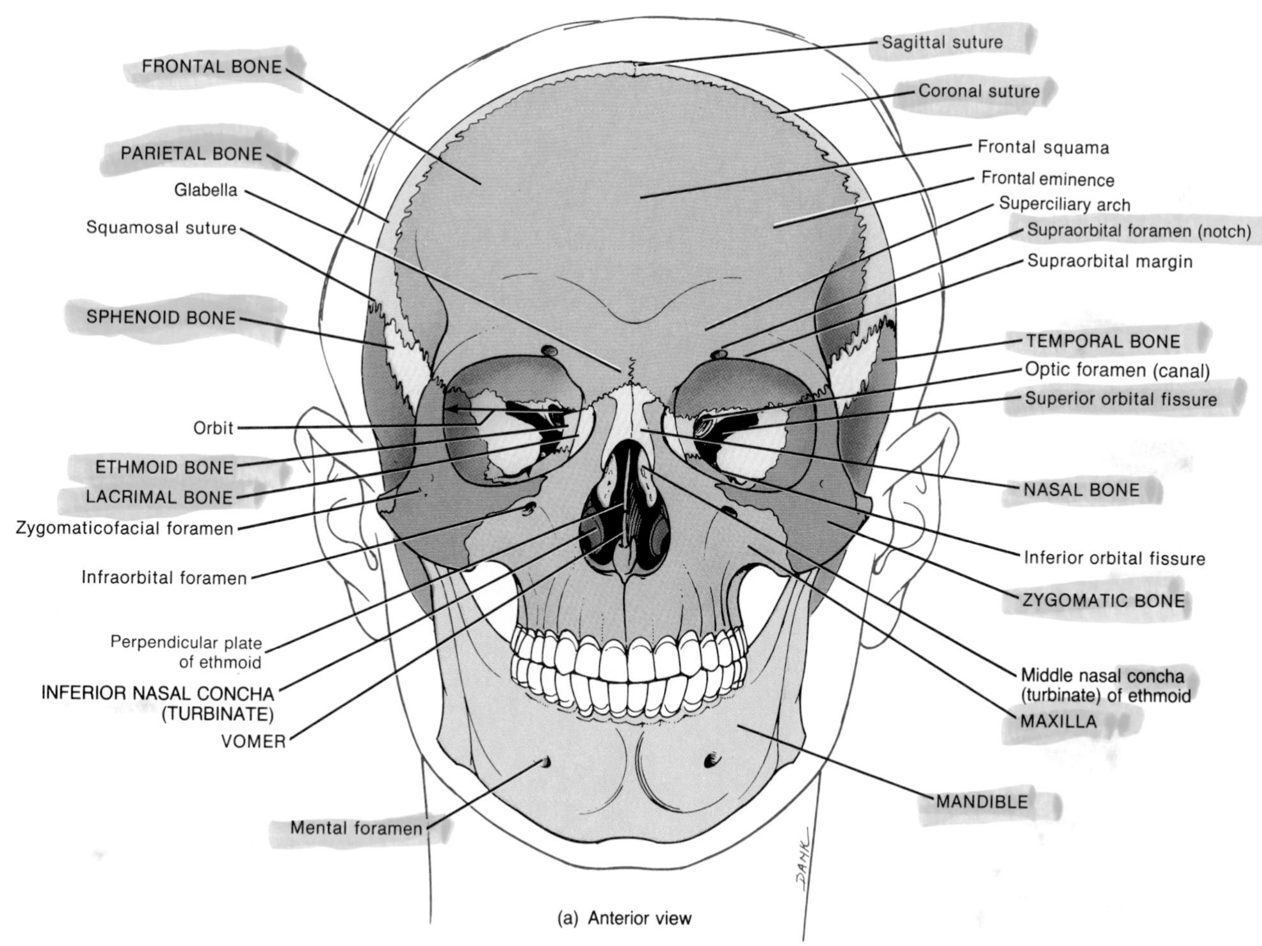

(a) Anterior view

FIGURE 6-2 Skull. (b) Although the hyoid bone is not part of the skull, it is included in the illustration for reference. The sutural bones in (d) are exaggerated for emphasis.

lie deep to the frontal squama. These mucous membrane–lined cavities act as sound chambers that give the voice resonance.

CLINICAL APPLICATION

Black Eye

Just above the supraorbital margin is a sharp ridge that overlies the frontal sinus. A blow to the ridge frequently lacerates the skin over it, resulting in bleeding. Bruising of the skin over the ridge causes tissue fluid and blood to accumulate in the surrounding connective tissue and gravitate into the upper eyelid. The resulting swelling and discoloration is called a ***black eye.***

Parietal Bones

The two ***parietal bones*** (pa-RĪ-e-tal; *paries* = wall) form the greater portion of the sides and roof of the cranial cavity. The external surface contains two slight ridges that may be observed by looking at the lateral view of the skull in Figure 6-2. These are the ***superior temporal line*** and a less conspicuous ***inferior temporal line*** below it. The internal surfaces of the parietal bones contain many protrusions and depressions that accommodate the blood vessels supplying the outer meninx (covering) of the brain called the ***dura mater.***

Temporal Bones

The two ***temporal*** (*tempora* = temples) ***bones*** form the inferior sides of the cranium and part of the cranial floor.

In the lateral view of the skull in Figure 6-2b, note the ***temporal squama***—a thin, large, expanded area that forms the anterior and superior part of the temple. Projecting from the inferior portion of the temporal squama is the ***zygomatic process,*** which articulates with the temporal process of the zygomatic bone. The zygomatic process of the

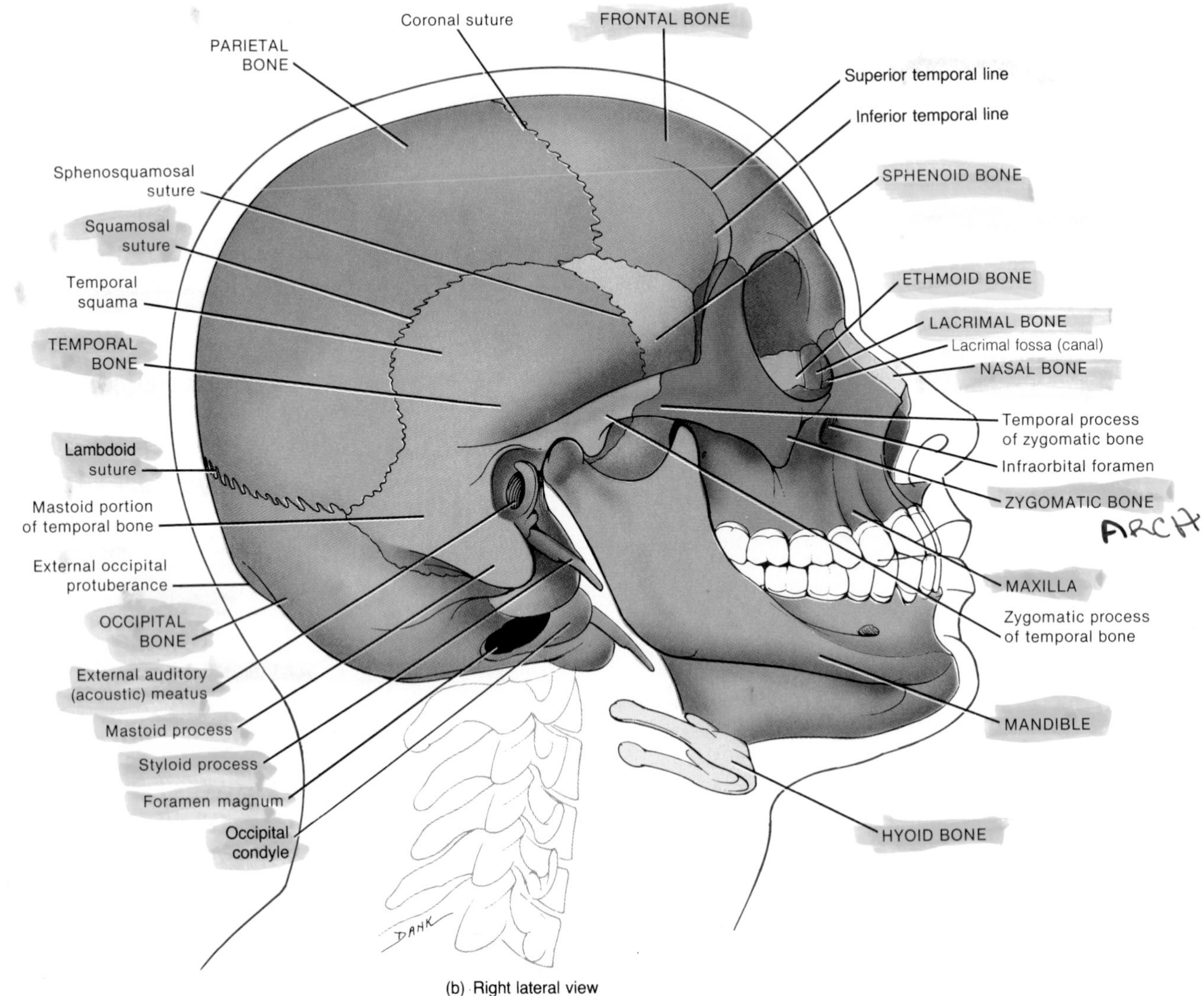

(b) Right lateral view

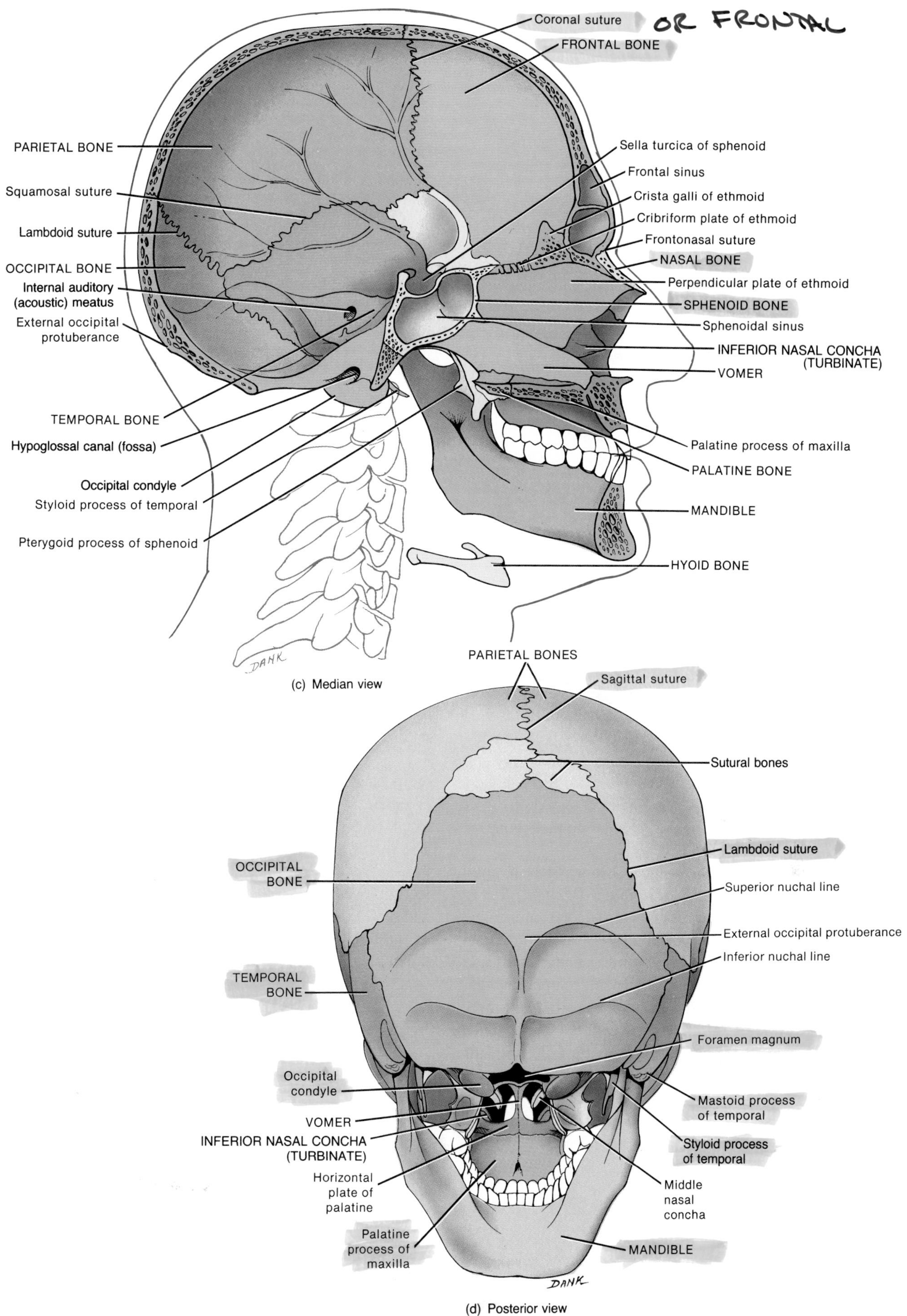

(c) Median view

(d) Posterior view

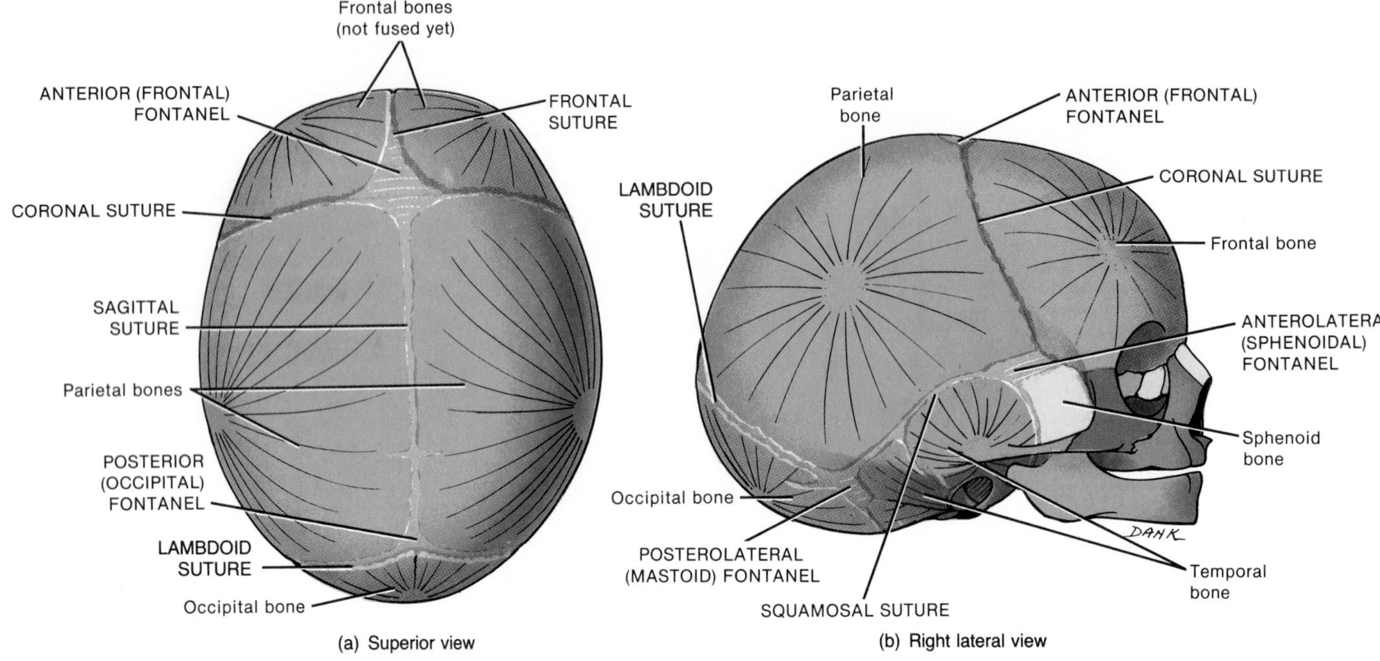

FIGURE 6-3 Fontanels of the skull at birth.

temporal bone and the temporal process of the zygomatic bone constitute the ***zygomatic arch***.

At the floor of the cranial cavity, shown in Figure 6-5, is the ***petrous portion*** of the temporal bone. This portion is triangular and located at the base of the skull between the sphenoid and occipital bones. The petrous portion contains the internal and middle ear structures which are involved in hearing and equilibrium (balance). It also contains the ***carotid foramen (canal)*** (see Figure 6-4). Posterior to the carotid foramen and anterior to the occipital bone is the ***jugular foramen (fossa)***.

Between the squamous and petrous portions of the temporal bone is a socket called the ***mandibular (glenoid) fossa***. Anterior to the mandibular fossa is a rounded eminence, the ***articular tubercle***. The mandibular fossa and articular tubercle articulate (form a joint) with the condylar process of the mandible (lower jawbone) to form the temporomandibular joint (TMJ). The mandibular fossa and articular tubercle are seen best in Figure 6-4.

In the lateral view of the skull in Figure 6-2, you will see the ***mastoid portion*** of the temporal bone, located posterior and inferior to the external auditory meatus, or ear canal. In the adult, this portion of the bone contains several ***mastoid air "cells."*** These are air spaces separated from the brain only by thin bony partitions.

CLINICAL APPLICATION

Mastoiditis

If ***mastoiditis,*** the inflammation of the mastoid air cells occurs, the infection may spread to the brain or its outer covering. These mastoid air cells do not drain as do the paranasal sinuses.

The ***mastoid process*** is a rounded projection of the temporal bone posterior to the external auditory meatus. It serves as a point of attachment for several neck muscles. Near the posterior border of the mastoid process is the ***mastoid foramen***. The ***external auditory (acoustic) meatus*** is the canal in the temporal bone that leads to the middle ear. The ***internal auditory (acoustic) meatus*** is superior to the jugular foramen (see Figure 6-5a). The ***styloid process*** projects downward from the undersurface of the temporal bone and serves as a point of attachment for muscles and ligaments of the tongue and neck. Between the styloid process and the mastoid process is the ***stylomastoid foramen*** (see Figure 6-4).

Occipital Bone

The ***occipital*** (ok-SIP-i-tal) ***bone*** forms the posterior part and most of the base of the cranium (Figure 6-4).

The ***foramen magnum*** is a large hole in the inferior part of the bone through which pass the medulla oblongata (part of the brain that joins with the spinal cord) and its membranes, the spinal portion of the accessory nerve, the vertebral and spinal arteries, and the meninges.

The ***occipital condyles*** are oval processes with convex surfaces, one on either side of the foramen magnum, that articulate with depressions on the first cervical vertebra. Extending laterally from the posterior portion of the condyles

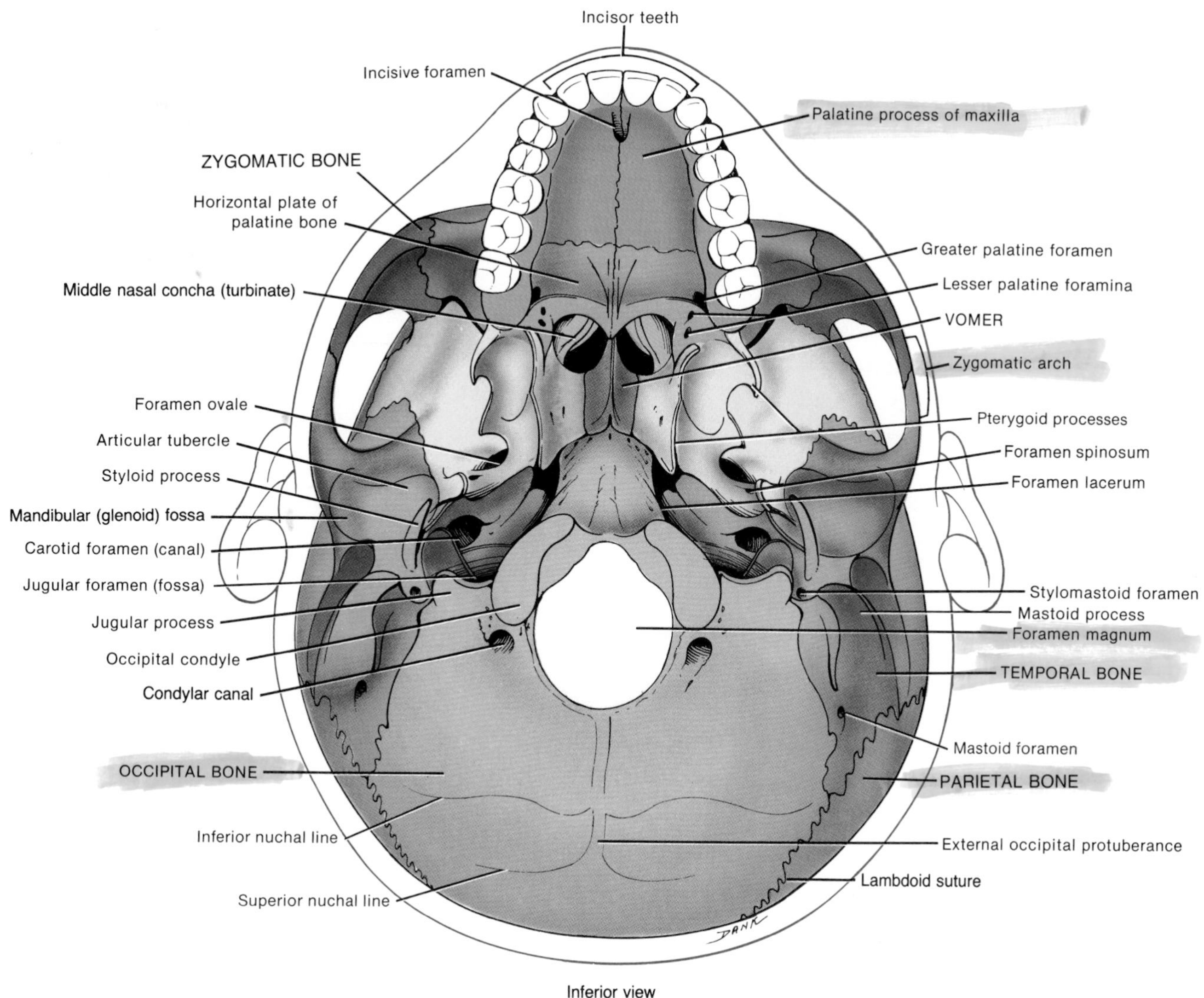

FIGURE 6-4 Skull.

are plates of bone called the ***jugular processes.*** Superior to the condyles is the ***hypoglossal canal*** (***fossa***) (see Figure 6-5).

The ***external occipital protuberance*** is a prominent projection on the posterior surface of the bone just superior to the foramen magnum. You can feel this structure as a definite bump on the back of your head, just above your neck. The protuberance is also visible in Figure 6-2c. Extending laterally from the protuberance are two curved lines, the ***superior nuchal lines,*** and below these two ***inferior nuchal lines,*** which are areas of muscle attachment (see Figure 6-2e).

Sphenoid Bone

The ***sphenoid*** (SFĒ-noyd) ***bone*** is situated at the middle part of the base of the skull. *Spheno* means wedge. This bone is referred to as the keystone of the cranial floor because it articulates with all the other cranial bones. If you view the floor of the cranium from above in Figure 6-5, you will note that the sphenoid articulates with the frontal bones anteriorly, the temporal bones laterally, and the occipital bone posteriorly. It lies posterior and slightly superior to the nasal cavities and forms part of the floor and sidewalls of the orbit (eye socket). The shape of the sphenoid resembles a bat with outstretched wings.

The ***body*** of the sphenoid is the cubelike central portion between the ethmoid and occipital bones. It contains the ***sphenoidal sinuses,*** which drain into the nasal cavity (see Figure 6-8). On the superior surface of the sphenoid body is a depression called the ***sella turcica*** (SEL-a TUR-sika = Turk's saddle). The deepest part of this depression houses the pituitary gland. Posteriorly, the sella turcica is bounded by a ridge that projects upward, the ***dorsum sellae.***

The ***greater wings*** of the sphenoid are lateral projections from the body and form the anterolateral floor of the cranium.

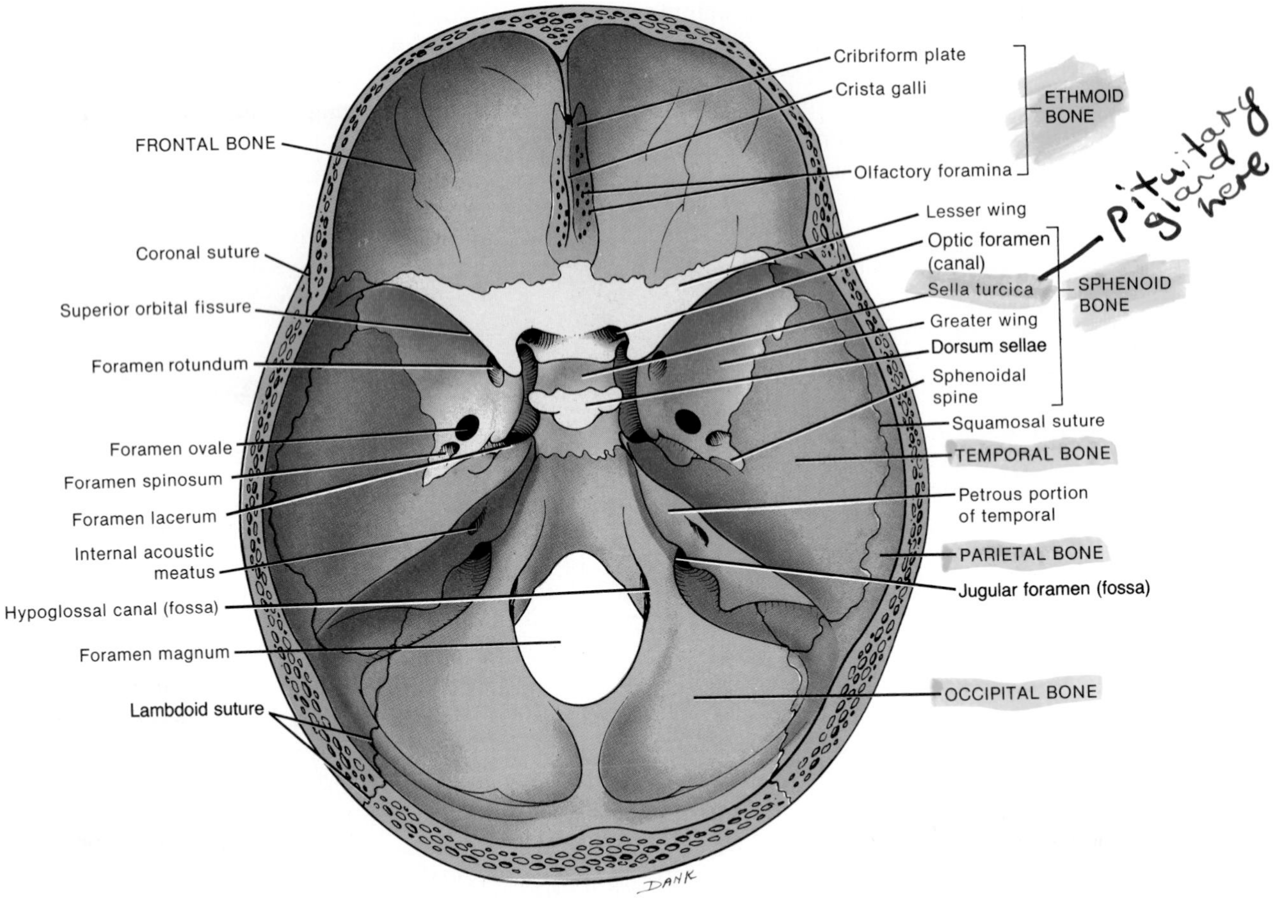

Cribriform plate
Crista galli
ETHMOID BONE
FRONTAL BONE
Olfactory foramina
Lesser wing
Optic foramen (canal)
Sella turcica
SPHENOID BONE
Greater wing
Dorsum sellae
Sphenoidal spine
Coronal suture
Superior orbital fissure
Foramen rotundum
Squamosal suture
TEMPORAL BONE
Foramen ovale
Foramen spinosum
Foramen lacerum
Petrous portion of temporal
Internal acoustic meatus
PARIETAL BONE
Jugular foramen (fossa)
Hypoglossal canal (fossa)
Foramen magnum
OCCIPITAL BONE
Lambdoid suture
DANK
pituitary gland here

(a) Viewed from above in floor of cranium

Lesser wings
Optic foramen (canal)
Superior orbital fissure
Sphenoidal sinuses
Foramen rotundum
Pterygoid canal
Body
Sphenoidal spine
Greater wings
Pterygoid process
DANK

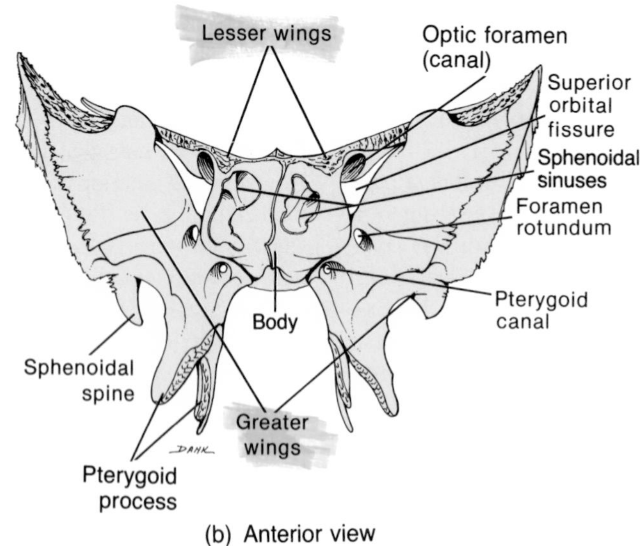

(b) Anterior view

FIGURE 6-5 Sphenoid bone.

The greater wings also form part of the lateral wall of the skull just anterior to the temporal bone. The posterior portion of each greater wing contains a triangular projection, the ***sphenoidal spine,*** that fits into the angle between the squama and petrous portion of the temporal bone.

The ***lesser wings*** are anterior and superior to the greater wings. They form part of the floor of the cranium and the posterior part of the orbit.

Between the body and lesser wing, you can locate the ***optic foramen (canal).*** Lateral to the body between the greater and lesser wings is a somewhat triangular slit called the ***superior orbital fissure.*** This fissure may also be seen in the anterior view of the skull in Figure 6-2a.

On the inferior part of the sphenoid bone are the ***pterygoid*** (TER-i-goyd) ***processes.*** These structures project inferiorly from the points where the body and greater wings unite. Attached to the pterygoid processes are some of the muscles that move the mandible.

At the base of the lateral pterygoid process in the greater wing is the ***foramen ovale.*** Another foramen, the ***foramen spinosum,*** lies at the posterior angle of the sphenoid. The ***foramen lacerum*** is bounded anteriorly by the sphenoid bone and medially by the sphenoid and occipital bones. The foramen is covered in part by a layer of fibrocartilage in living subjects. Another foramen associated with the sphenoid bone is the ***foramen rotundum,*** located at the junction of the anterior and medial parts of the sphenoid bone.

Ethmoid Bone

The ***ethmoid bone*** is a light, spongy bone located in the anterior part of the floor of the cranium between the orbits. It is anterior to the sphenoid and posterior to the nasal bones (Figure 6-6). The ethmoid bone forms part of the anterior portion of the cranial floor, the medial wall of the orbits, the superior portions of the nasal septum, or partition, and most of the sidewalls of the nasal roof. The ethmoid is the principal supporting structure of the nasal cavities.

Its ***lateral masses (labyrinths)*** compose most of the wall between the nasal cavities and the orbits. They contain several air spaces, or "cells," ranging in number from 3 to 18. It is from these "cells" that the bone derives its name (*ethmos* = sieve). The ethmoid "cells" together form the ***ethmoidal sinuses.*** The sinuses are shown in Figure 6-8. The ***perpendicular plate*** forms the superior portion of the nasal septum (see Figure 6-7). The ***cribriform (horizontal) plate*** lies in the anterior floor of the cranium and forms the roof of the nasal cavity. The cribriform plate contains the ***olfactory foramina.*** Projecting upward from the cribriform plate is a triangular process called the ***crista galli*** (= cock's comb). This structure serves as a point of attachment for the membranes (meninges) that cover the brain.

The lateral masses contain two thin, scroll-shaped projections on either side of the nasal septum. These are called the ***superior nasal concha*** (KONG-ka; *concha* = shell) or ***turbinate*** and the ***middle nasal concha (turbinate).*** The conchae cause inhaled air to whirl around and, as a result, many inhaled particles strike and become trapped in the mucus found on the lining of the nasal passageways. This turbulence thus cleanses inhaled air before it passes into the trachea, bronchi, and lungs.

CRANIAL FOSSAE

The floor of the cranium contains three distinct levels, from anterior to posterior, called ***cranial fossae*** (see Figure 6-5a). The fossae contain depressions for the various brain convolutions, grooves for cranial blood vessels, and numerous foramina. The highest level, the ***anterior cranial fossa,*** is formed largely by the portion of the frontal bone that constitutes the roof of the orbits and nasal cavity, the crista galli and cribriform plate of the ethmoid bone, and the lesser wings and part of the body of the sphenoid bone. This fossa houses the frontal lobes of the cerebral hemisphere. The rough surface of the frontal bone can lead to tearing of the frontal lobes of the cerebral hemispheres during head trauma. The next fossa is below the level of the anterior cranial fossa, is posterior, and is called the ***middle cranial fossa.*** It is shaped like a butterfly and has a small median portion and two expanded lateral portions. The median portion is formed by part of the body of the sphenoid bone, and the lateral portions are formed by the greater wings of the sphenoid bone, temporal squama, and parietal bone. The middle cranial fossa cradles the temporal lobes of the cerebral hemispheres. The last fossa, at the lowest level, is the ***posterior cranial fossa,*** the largest of the fossae. It is formed largely by the occipital bone and the petrous and mastoid portions of the temporal bone. It is a very deep fossa that accommodates the cerebellum, pons, and medulla.

FACIAL BONES

The shape of the face changes dramatically during the first two postnatal years, owing in part to the expanding brain and cranial bones, the formation and eruption of teeth, and the increase in size of the paranasal sinuses. Growth of the face ceases at approximately 16 years of age.

Nasal Bones

The paired ***nasal bones*** meet at the middle and superior part of the face (see Figures 6-2 and 6-7) and form part of the bridge of the nose. The inferior portion of the nose, indeed the major portion, consists of cartilage.

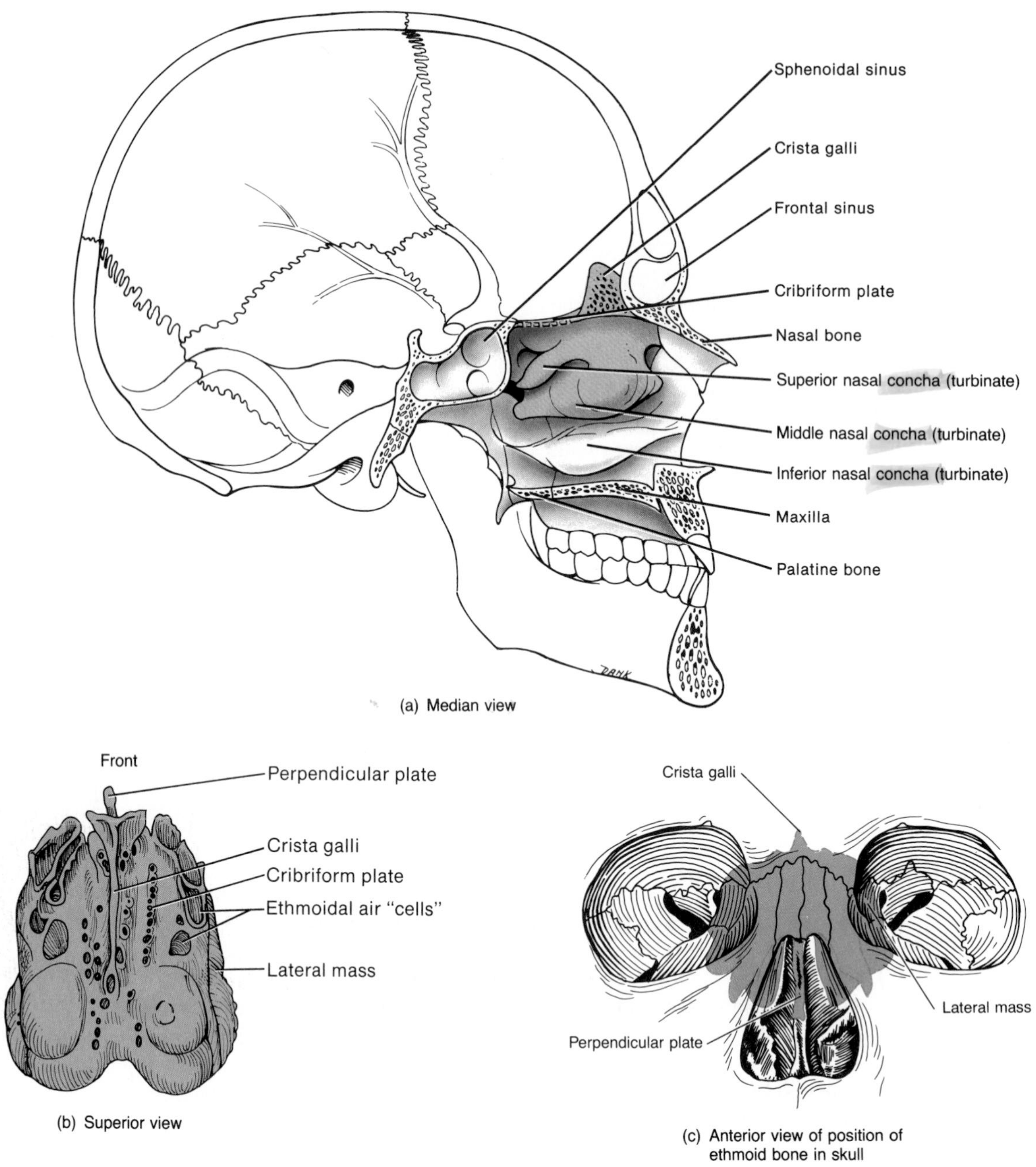

FIGURE 6-6 Ethmoid bone. (c) is a highly diagrammatic representation showing the approximate position of the ethmoid bone in the skull.

Maxillae

The paired ***maxillae*** (mak-SIL-ē) unite to form the upper jawbone (Figure 6-7) and articulate with every bone of the face except the mandible, or lower jawbone. They form part of the floors of the orbits, part of the roof of the mouth (most of the hard palate), and part of the lateral walls and floor of the nasal cavity.

Each maxillary bone contains a ***maxillary sinus*** that empties into the nasal cavity (see Figure 6-8). The ***alveolar*** (al-VĒ-ō-lar) ***process*** (*alveolus* = hollow) contains the ***alveoli*** (bony sockets) into which the maxillary (upper) teeth are set. The ***palatine process*** is a horizontal projection of the maxilla that forms the anterior three-fourths of the hard palate, or anterior portion of the roof of the oral cavity. The left and right maxillary bones unite, and the fusion is normally completed before birth.

CLINICAL APPLICATION

Cleft Palate and Cleft Lip

If the palatine processes of the maxillary bones do not unite before birth, a condition called ***cleft palate*** results.

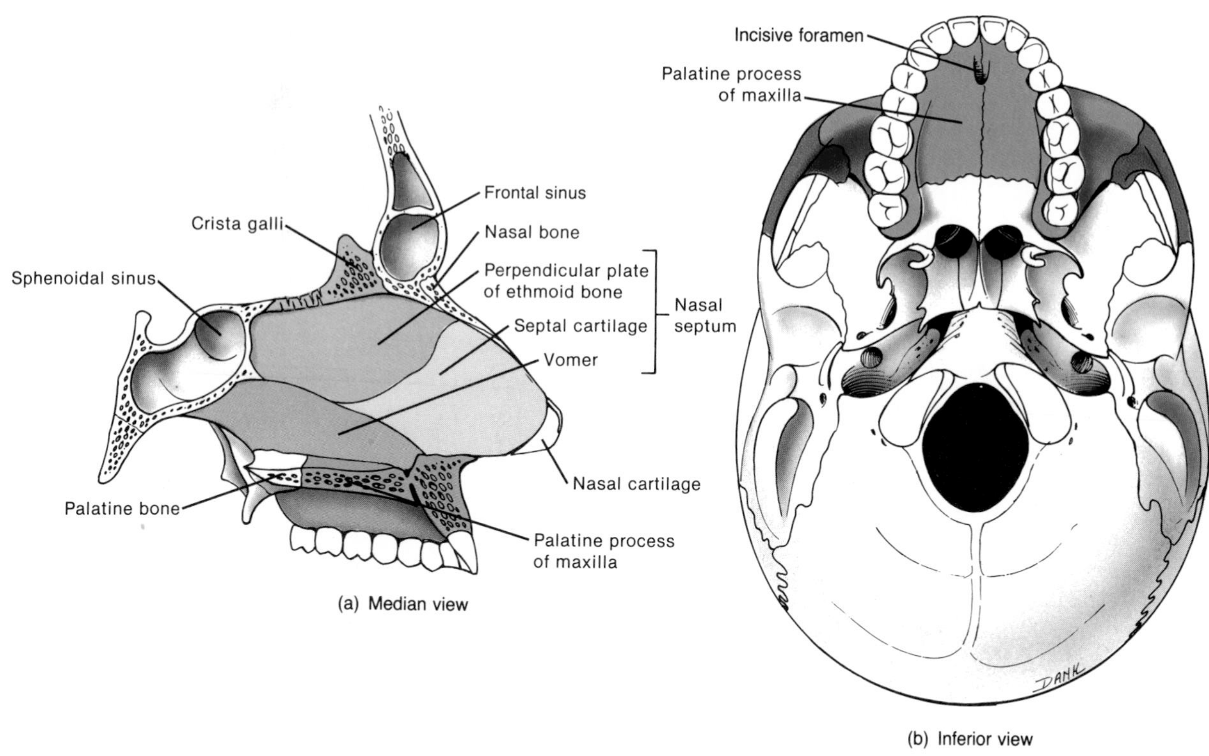

FIGURE 6-7 Maxillae.

The condition may also involve incomplete fusion of the palatine bones (see Figure 6-4). Another form of this condition, called ***cleft lip,*** involves a split in the upper lip. Cleft lip is often associated with cleft palate. Depending on the extent and position of the cleft, speech and swallowing may be affected. Facial and oral surgeons recommend closure of cleft lip during the first year of life, and surgical results are excellent. Repair of cleft palate is done between the first and second year of life, before the child begins to talk. Here again, results are usually excellent, and orthodontic therapy may be valuable in aligning the teeth.

The ***infraorbital foramen,*** which can be seen in the anterior view of the skull in Figure 6-2, is an opening in the maxilla inferior to the orbit. Another prominent foramen in the maxilla is the ***incisive foramen*** just posterior to the incisor teeth. A final structure associated with the maxilla and sphenoid bone is the ***inferior orbital fissure.*** It is located between the greater wing of the sphenoid and the maxilla (see Figure 6-4).

Paranasal Sinuses

Although they are not cranial or facial bones, this is an appropriate point to discuss paired cavities, called ***paranasal sinuses,*** that are located in certain cranial and facial bones near the nasal cavity (Figure 6-8). The paranasal sinuses are lined with mucous membranes that are continuous with the lining of the nasal cavity. Cranial bones containing paranasal sinuses are the frontal, sphenoid, ethmoid, and maxillae. (The paranasal sinuses were described in the discussion of each of these bones.) Besides producing mucus, the paranasal sinuses lighten the skull bones and serve as resonating chambers for sound as we speak or sing.

CLINICAL APPLICATION

Sinusitis

Secretions produced by the mucous membranes of the paranasal sinuses drain into the nasal cavity. An inflammation of the membranes due to an allergic reaction or infection is called ***sinusitis.*** If the membranes swell enough to block drainage into the nasal cavity, fluid pressure builds up in the paranasal sinuses, and a sinus headache results.

Zygomatic Bones

The two ***zygomatic bones,*** commonly referred to as the cheekbones, form the prominences of the cheeks and part of the outer wall and floor of the orbits (see Figure 6-10). The zygomatic bones articulate with the frontal, maxilla, sphenoid, and temporal bones.

The ***temporal process*** of the zygomatic bone projects posteriorly and articulates with the zygomatic process of

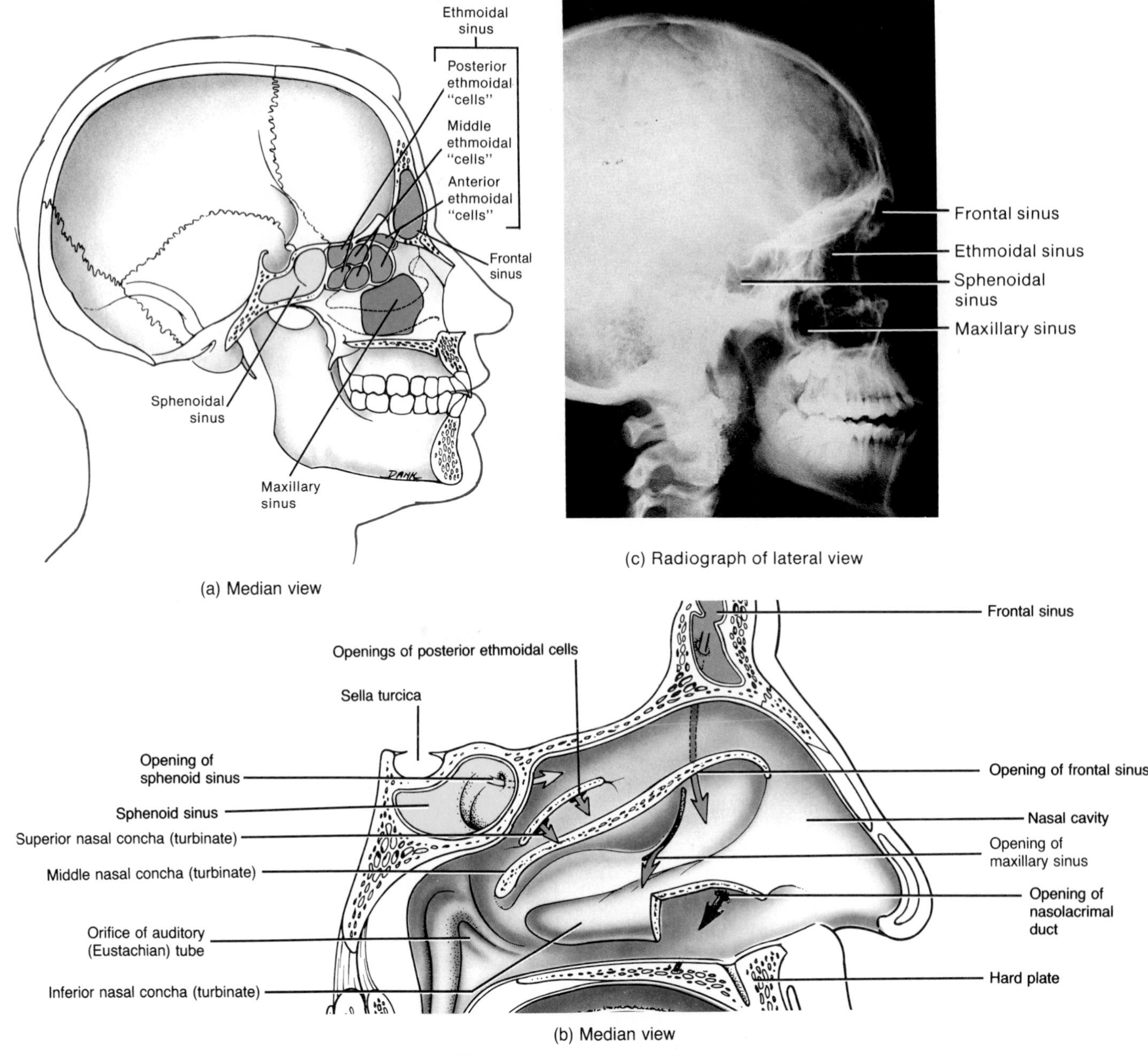

FIGURE 6-8 Paranasal sinuses. (a) Location of the sinuses. (b) Openings of the paranasal sinuses into the nasal cavity. Portions of the conchae have been sectioned. (c) © Omikron Photo Researchers.

the temporal bone. These two processes form the ***zygomatic arch*** (see Figure 6-4). A foramen associated with the zygomatic bone is the ***zygomaticofacial foramen*** near the center of the bone (see Figure 6-2).

Mandible

The ***mandible,*** or lower jawbone, is the largest, strongest facial bone (Figure 6-9). It is the only movable skull bone.

In the lateral view you can see that the mandible consists of a curved, horizontal portion called the ***body*** and two perpendicular portions called the ***rami*** (RĀ-mī). The ***angle*** of the mandible is the area where each ramus meets the body. Each ramus has a posterior ***condylar*** (KON-di-lar) ***process*** that articulates with the mandibular (glenoid) fossa and articular tubercle of the temporal bone to form the temporomandibular joint (TMJ). It also has an anterior ***coronoid*** (KOR-ō-noyd) ***process*** to which the temporalis muscle attaches. The depression between the coronoid and condylar processes is called the ***mandibular notch.*** The ***alveolar process*** is an arch containing the ***alveoli*** (sockets) for the mandibular (lower) teeth.

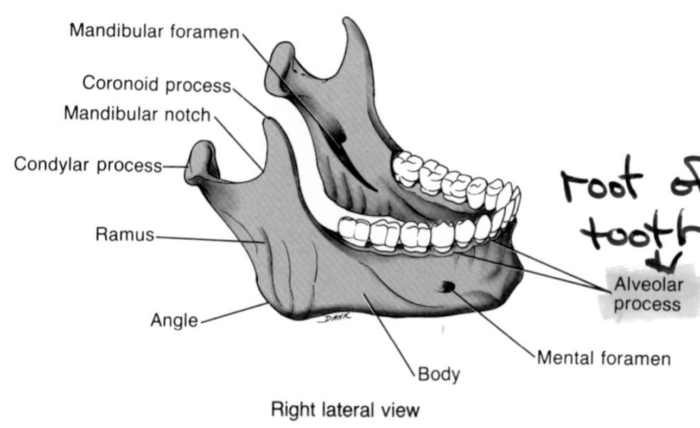

FIGURE 6-9 Mandible.

The *mental foramen* (*mentum* = chin) is approximately below the second premolar tooth. It is through this foramen that dentists sometimes reach the nerve when injecting anesthetics. Another foramen associated with the mandible is the *mandibular foramen* on the medial surface of the ramus, another site frequently used by dentists to inject anesthetics. The mandibular foramen is the beginning of the *mandibular canal,* which runs forward in the ramus deep to the roots of the teeth. The canal carries branches of the inferior alveolar nerve and blood vessels to the teeth. Parts of these nerves and vessels emerge through the mental foramen.

CLINICAL APPLICATION

Temporomandibular Joint (TMJ) Syndrome

One problem associated with the temporomandibular joint (TMJ) is ***TMJ syndrome.*** It is characterized by dull pain around the ear, tenderness of the jaw muscles, clicking or popping noise when opening or closing the mouth, limited or abnormal opening of the mouth, headache, tooth sensitivity, and abnormal wearing of the teeth. TMJ syndrome might be caused by improperly aligned teeth, grinding or clenching the teeth, trauma to the jaw, or arthritis. Treatment may consist of application of moist heat or ice, keeping lips together and teeth apart, a soft diet, taking aspirin, muscle retraining, use of an occlusal splint, adjusting or reshaping the teeth, orthodontic treatment, or surgery.

Lacrimal Bones

The paired *lacrimal bones* (LAK-ri-mal; *lacrima* = tear) are thin bones roughly resembling a fingernail in size and shape. They are the smallest bones of the face. These bones are posterior and lateral to the nasal bones in the medial wall of the orbit. They can be seen in the anterior and lateral views of the skull in Figure 6-2.

The lacrimal bones form a part of the medial wall of the orbit. They also contain the *lacrimal fossae (canals)* through which the tear ducts pass into the nasal cavity (see Figure 6-2).

Palatine Bones

The two *palatine* (PAL-a-tīn) *bones* are L-shaped and form the posterior portion of the hard palate, part of the floor and lateral wall of the nasal cavity, and a small portion of the floors of the orbits. The posterior portion of the hard palate, which separates the nasal cavity from the oral cavity, is formed by the *horizontal plates* of the palatine bones. These can be seen in Figure 6-4.

Two foramina (plural) associated with the palatine bones are the greater and lesser palatine foramina (see Figure 6-4). The *greater palatine foramen* is at the posterior angle of the hard palate. The *lesser palatine foramina,* usually two or more on each side, are posterior to the greater palatine foramina.

Inferior Nasal Conchae

Refer to the views of the skull in Figures 6-2a and 6-6a. The two *inferior nasal conchae* (KONG-kē) or *turbinates* are scroll-like bones that form a part of the lateral wall of the nasal cavity and project into the nasal cavity inferior to the superior and middle nasal conchae of the ethmoid bone. They serve the same function as the superior and middle nasal conchae of the ethmoid bone; that is, they promote the turbulent filtration and circulation of air before it passes into the lungs. The inferior nasal conchae are separate bones and not part of the ethmoid.

Vomer

The *vomer* (= plowshare) is roughly triangular bone that forms the inferior and posterior part of the nasal septum. It is clearly seen in the anterior view of the skull in Figure 6-2a and the inferior view in Figure 6-4.

The inferior border of the vomer articulates with the cartilage septum that divides the external nose into a right and left nostril. Its superior border articulates with the perpendicular plate of the ethmoid bone. The structures that form the *nasal septum,* or partition, are the perpendicular plate of the ethmoid, septal cartilage, vomer, and parts of the palatine bones and maxillae (see Figure 6-7a).

CLINICAL APPLICATION

Deviated Nasal Septum (DNS)

A ***deviated nasal septum (DNS)*** is deflected laterally from the midline of the nose. The deviation usually occurs at the junction of bone with the septal cartilage. If the deviation is severe, it may entirely block the nasal passageway. Even a partial blockage may lead to infection and inflammation may develop and cause nasal congestion, blockage of the paranasal sinus openings, chronic sinusitis, headache, and nosebleeds.

A summary of bones of the skull is presented in Exhibit 6-3.

ORBITS

Each ***orbit*** (eye socket) is a pyramid-shaped space that contains the eyeball and associated structures. It is formed by seven bones of the skull (Figure 6-10) and has four walls and an apex (back end). The roof of the orbit consists of parts of the frontal and sphenoid bones. The lateral wall is formed by portions of the zygomatic and sphenoid bones. The floor of the orbit is formed by parts of the maxilla, zygomatic, and palatine bones. The medial wall of the orbit is formed from portions of the maxilla, lacrimal, ethmoid, and sphenoid bones.

The principal openings of each orbit are the following:

1. ***Optic foramen (canal)*** at the junction of the roof and medial wall.
2. ***Superior orbital fissure*** at the upper lateral angle of the apex.
3. ***Inferior orbital fissure*** at the junction of the lateral wall and floor.
4. ***Supraorbital foramen (notch)*** on the medial side of the supraorbital margins of the frontal bone.
5. ***Canal for nasolacrimal duct*** in the nasal bone.

FORAMINA

Some ***foramina*** (singular = foramen) of the skull were mentioned along with the descriptions of the cranial and facial bones with which they are associated. As preparation for studying other systems of the body, especially the nervous and cardiovascular systems, these foramina, as well as some additional ones, and the structures passing through them are listed in Exhibit 6-4. For your convenience and for future reference, the foramina are listed alphabetically.

EXHIBIT 6-3

Summary of Bones of the Adult Skull[a]

CRANIAL BONES	FACIAL BONES
Frontal (1)	Nasal (2)
Parietal (2)	Maxillae (2)
Temporal (2)	Zygomatic (2)
Occipital (1)	Mandible (1)
Sphenoid (1)	Lacrimal (2)
Ethmoid (1)	Palatine (2)
	Inferior nasal conchae (2)
	Vomer (1)

[a] The numbers in parentheses indicate how many of each bone are present.

HYOID BONE

The single ***hyoid bone*** (*hyoedes* = U-shaped) is a unique component of the axial skeleton because it does not articulate with any other bone. Rather, it is suspended from the styloid process of the temporal bone by ligaments and muscles. The hyoid is located in the neck between the mandible and larynx. It supports the tongue and provides attachment for some of its muscles and for muscles of the neck and pharynx. Refer to the median and lateral views of the skull in Figure 6-2b,c to see the position of the hyoid bone.

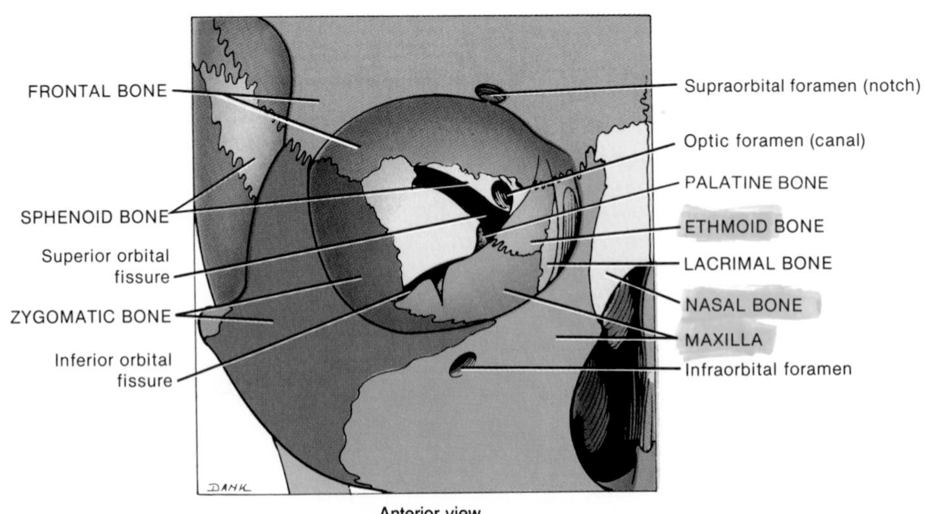

FIGURE 6-10 Right orbit.

EXHIBIT 6-4

Summary of Foramina of the Skull

FORAMEN	LOCATION	STRUCTURES PASSING THROUGH
Carotid (Figure 6-4)	Petrous portion of temporal.	Internal carotid artery.
Greater Palatine (Figure 6-4)	Posterior angle of hard palate.	Greater palatine nerve and greater palatine vessels.
Hypoglossal (Figure 6-5)	Superior to base of occipital condyles.	Hypoglossal (XII) nerve[a] and branch of ascending pharyngeal artery.
Incisive (Figure 6-7)	Posterior to incisor teeth.	Branches of greater palatine vessels and nasopalatine nerve.
Inferior Orbital (Figure 6-4)	Between greater wing of sphenoid and maxilla.	Maxillary branch of trigeminal (V) nerve and infraorbital vessels.
Infraorbital (Figure 6-2a)	Inferior to orbit in maxilla.	Infraorbital nerve and artery.
Jugular (Figure 6-4)	Posterior to carotid canal between petrous portion of temporal and occipital.	Internal jugular vein, glossopharyngeal (IX) nerve, vagus nerve (X), and accessory (XI) nerve.
Lacerum (Figure 6-5)	Bounded anteriorly by sphenoid, posteriorly by petrous portion of temporal, and medially by the sphenoid and occipital.	Branch of ascending pharyngeal artery.
Lesser Palatine (Figure 6-4)	Posterior to greater palatine foramen.	Lesser palatine nerves and artery.
Magnum (Figure 6-4)	Occipital bone.	Medulla oblongata and its membranes, accessory (XI) nerve, and the vertebral and spinal arteries and meninges.
Mandibular (Figure 6-9)	Medial surface of ramus of mandible.	Inferior alveolar nerve and vessels.
Mastoid	Posterior border of mastoid process of temporal bone.	Emissary vein to transverse sinus and branch of occipital artery to dura mater.
Mental (Figure 6-9)	Inferior to second premolar tooth in mandible.	Mental nerve and vessels.
Olfactory (Figure 6-5)	Cribriform plate of ethmoid.	Olfactory (I) nerve.
Optic (Figure 6-5)	Between upper and lower portions of small wing of sphenoid.	Optic (II) nerve and ophthalmic artery.
Ovale (Figure 6-5)	Greater wing of sphenoid.	Mandibular branch of trigeminal (V) nerve.
Rotundum (Figure 6-5)	Junction of anterior and medial parts of sphenoid.	Maxillary branch of trigeminal (V) nerve.
Spinosum (Figure 6-5)	Posterior angle of sphenoid.	Middle meningeal vessels.
Stylomastoid (Figure 6-4)	Between styloid and mastoid processes of temporal.	Facial (VII) nerve and stylomastoid artery.
Superior Orbital (Figure 6-5)	Between greater and lesser wings of sphenoid.	Oculomotor (III) nerve, trochlear (IV) nerve, ophthalmic branch of trigeminal (V) nerve, and abducens (VI) nerve.
Supraorbital (Figure 6-2a)	Supraorbital margin of orbit.	Supraorbital nerve and vein.
Zygomaticofacial (Figure 6-2a)	Zygomatic bone.	Zygomaticofacial nerve and vessels.

[a] The Roman numerals associated with cranial nerves indicate the order in which the nerves arise from the brain, from front to back.

The hyoid consists of a horizontal ***body*** and paired projections called the ***lesser*** and the ***greater horns*** (Figure 6-11). Muscles and ligaments attach to these paired projections.

The hyoid bone is frequently fractured during strangulation. As a result, it is carefully examined in an autopsy when strangulation is suspected.

VERTEBRAL COLUMN

DIVISIONS

The ***vertebral column*** (***spine***), together with the sternum and ribs, forms the skeleton of the ***trunk*** of the body.

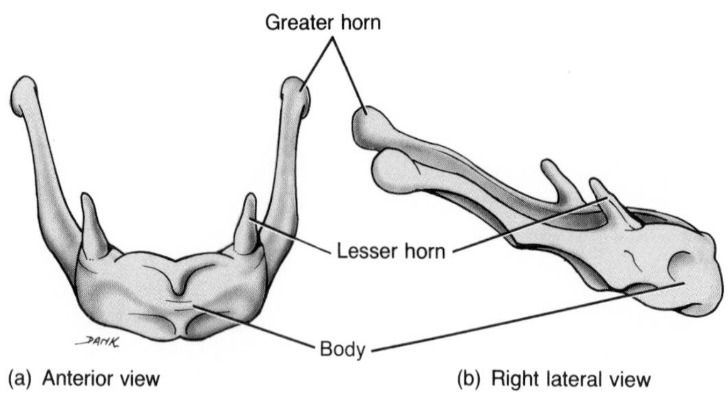

FIGURE 6-11 Hyoid bone.

The vertebral column makes up about two-fifths of the total height of the body and is composed of a series of bones called ***vertebrae.*** In an average adult male, the column measures about 71 cm (28 in.) in length; in an average adult female, it measures about 61 cm (24 in.) in length. In effect, the vertebral column is a strong, flexible rod that moves anteriorly, posteriorly, and laterally and rotates. It encloses and protects the spinal cord, supports the head, and serves as a point of attachment for the ribs and the muscles of the back. Between vertebrae are openings called ***intervertebral foramina.*** The nerves that connect the spinal cord to various parts of the body pass through these openings.

The adult vertebral column typically contains 26 vertebrae (Figure 6-12a,b). These are distributed as follows: 7 ***cervical vertebrae*** (*cervix* = neck) in the neck region; 12 ***thoracic vertebrae*** (*thorax* = chest) posterior to the thoracic cavity; 5 ***lumbar vertebrae*** (*lumbus* = loin) supporting the lower back; 5 ***sacral vertebrae*** fused into one bone called the ***sacrum*** (SĀ-krum); and usually 4 ***coccygeal*** (kok-SIJ-ē-al) ***vertebrae*** fused into one or two bones called the ***coccyx*** (KOK-six). Prior to the fusion of the sacral and coccygeal vertebrae, the total number of vertebrae is 33. Whereas the cervical, thoracic, and lumbar vertebrae are movable, the sacral and coccygeal vertebrae are immovable.

Between adjacent vertebrae from the first vertebra (atlas) to the sacrum are fibrocartilaginous ***intervertebral discs.*** Each disc is composed of an outer fibrous ring consisting of fibrocartilage called the ***annulus fibrosus*** and an inner soft, pulpy, highly elastic structure called the ***nucleus pulposus*** (Figure 6-12c). The discs form strong joints, permit various movements in portions of the vertebral column, and absorb vertical shock. Under compression, they flatten, broaden, and bulge from their intervertebral spaces (Figure 6-12c). Above the level of the sacrum, the intervertebral discs comprise about one-fourth the length of the vertebral column.

NORMAL CURVES

When viewed from the side, the vertebral column shows four ***normal curves*** (Figure 6-12b). Two of them, called cervical and lumbar curves, are anteriorly convex [ )], and two, called thoracic and sacral, are anteriorly concave [( ]. The curves of the column, like the curves in a long bone, are important because they increase its strength. The curves also help maintain balance in the upright position, absorb shocks from walking, and help protect the column from fracture.

In the fetus, there is only a single anteriorly concave curve. At approximately the third month after birth, when an infant begins to hold its head erect, the ***cervical curve*** develops. Later, when the child stands and walks, the ***lumbar curve*** develops. Because these curves are modifications of the fetal positions, they are called ***secondary curves.*** Because the other two curves, the ***thoracic curve*** and the ***sacral curve,*** retain the anterior concavity of the fetus, they are referred to as ***primary curves.***

TYPICAL VERTEBRA

Although there are variations in size, shape, and detail in the vertebrae in different regions of the column, all the vertebrae are basically similar in structure (Figure 6-13). A typical vertebra consists of the following components.

1. The ***body*** is the thick, disc-shaped anterior portion that is the weight-bearing part of a vertebra. Its superior and inferior surfaces are roughened for the attachment of intervertebral discs. The anterior and lateral surfaces contain nutrient foramina for blood vessels.
2. The ***vertebral (neural) arch*** extends posteriorly from the body of the vertebra. With the body of the vertebra, it surrounds the spinal cord. It is formed by two short, thick processes, the ***pedicles*** (PED-i-kuls), which project posteriorly from the body to unite with the laminae. The ***laminae*** (LAM-i-nē) are the flat parts that join to form the posterior portion of the vertebral arch. The space that lies between the vertebral arch and body contains the spinal cord. This space is known as the ***vertebral foramen.*** The vertebral foramina of all vertebrae together form the ***vertebral (spinal) canal.*** The pedicles are notched superiorly and inferiorly in such a way that

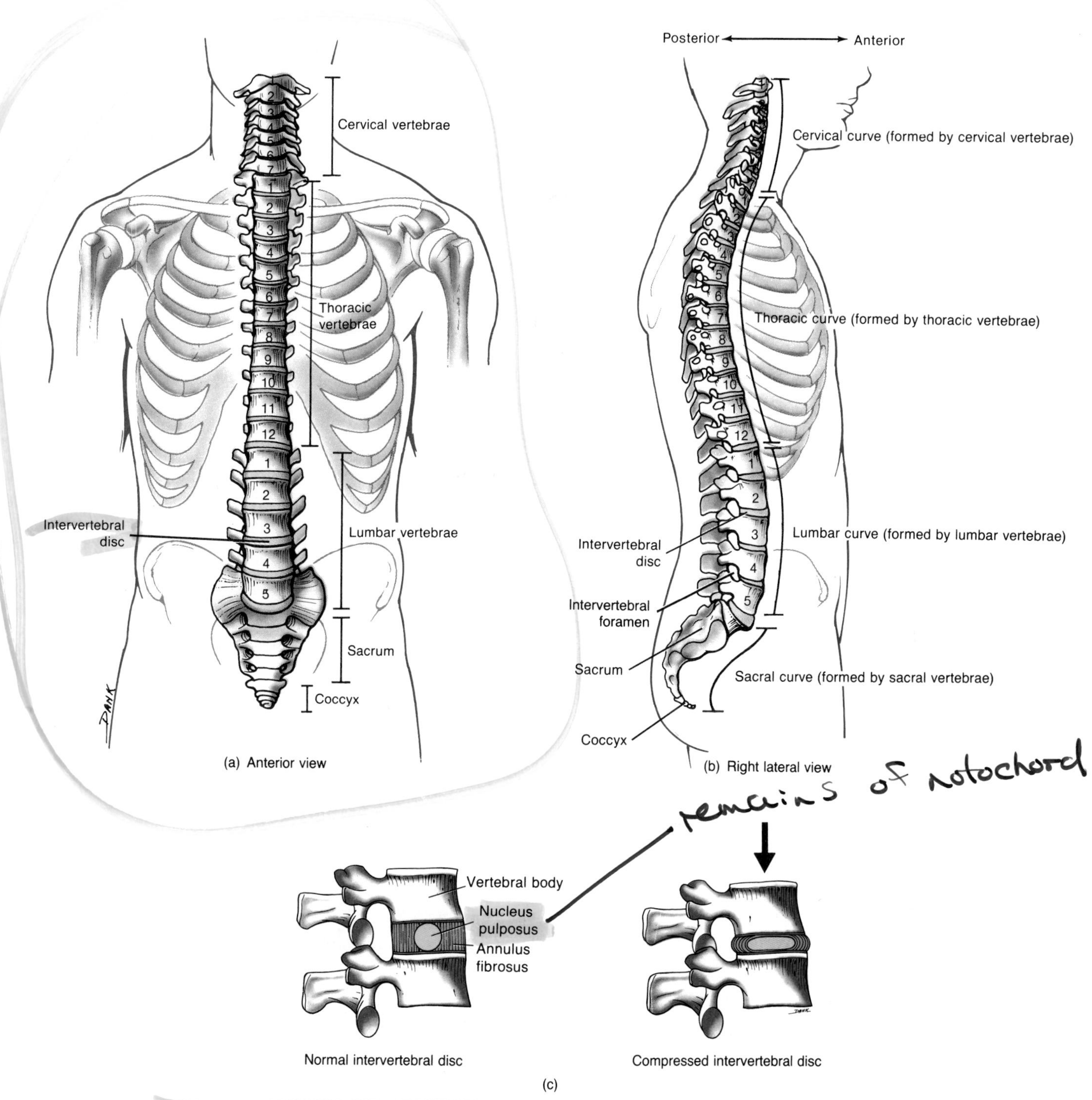

FIGURE 6-12 Vertebral column. In (c), the relative size of the disc has been enlarged for emphasis. A "window" has been cut in the annulus fibrosus so that the nucleus pulposus can be seen.

when they are arranged in the column, there is an opening between vertebrae on each side of the column. This opening, the ***intervertebral foramen,*** permits the passage of a single spinal nerve.

3. Seven ***processes*** arise from the vertebral arch. At the point where a lamina and pedicle join, a ***transverse process*** extends laterally on each side. A single ***spinous process*** (***spine***) projects posteriorly and inferiorly from the junction of the laminae. These three processes serve as points of attachment for muscles. The remaining four processes form joints with other vertebrae. The two ***superior articular processes*** of a vertebra articulate with the vertebra immediately superior to them. The two ***inferior articular processes*** of a vertebra articulate with the vertebra inferior to them. The articulating surfaces of the articular processes are referred to as ***facets.***

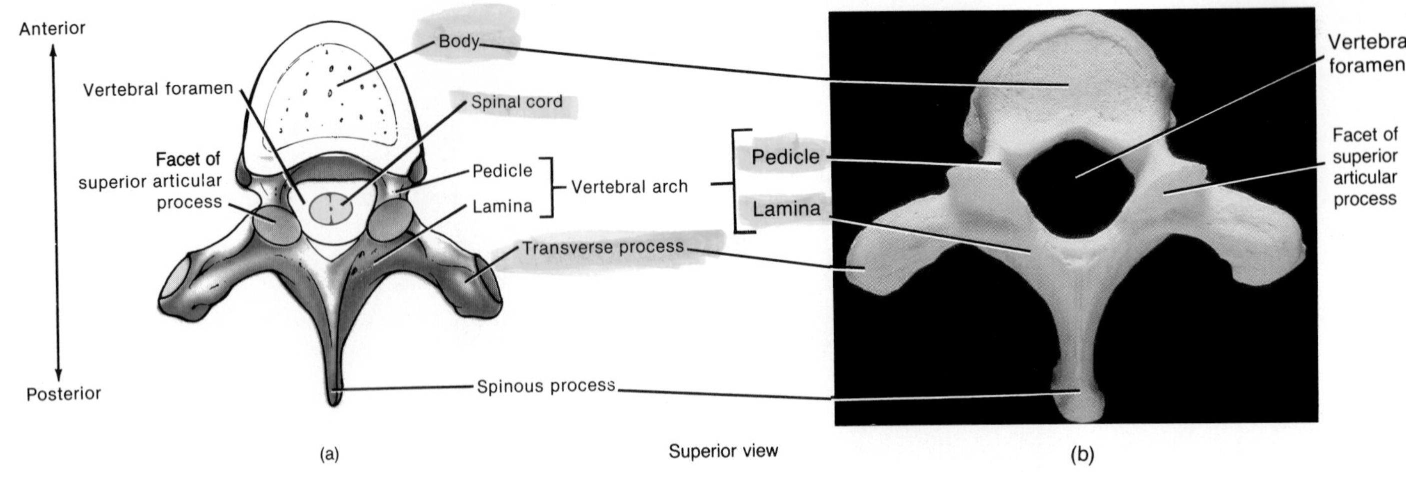

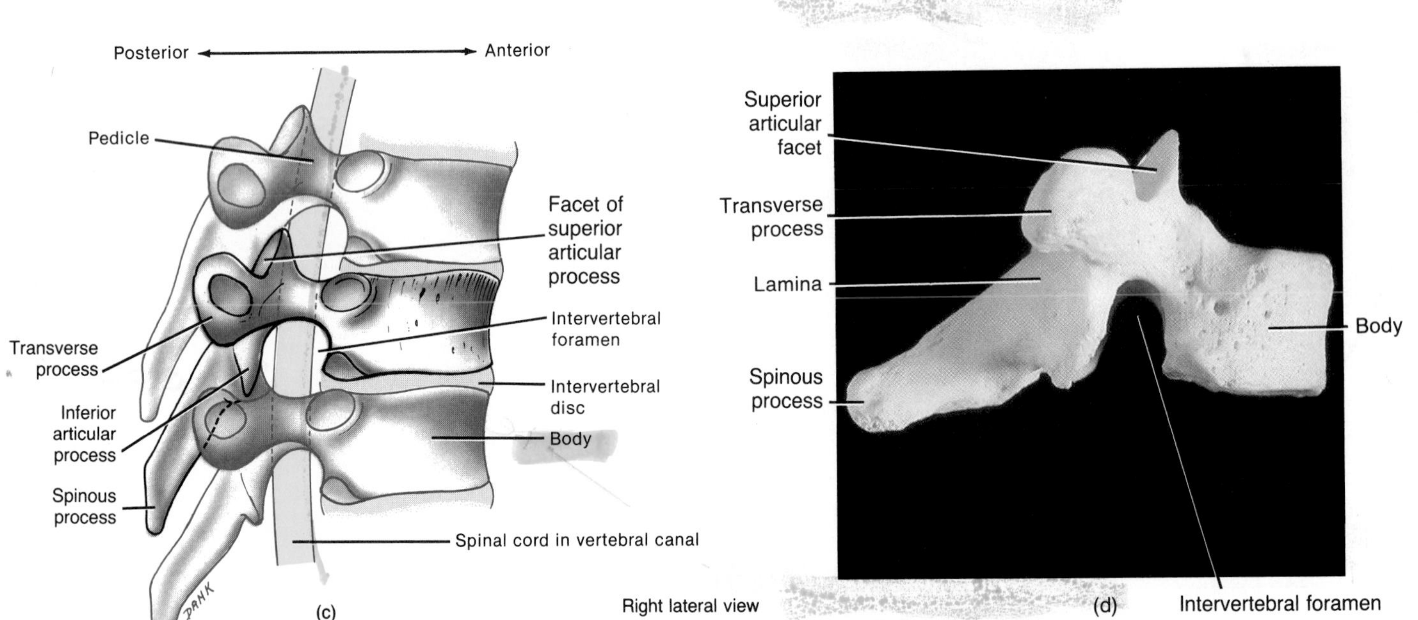

FIGURE 6-13 Typical vertebra as illustrated by a thoracic vertebra. (b) and (d) Courtesy of J. A. Gosling, P. F. Harris, *et al.*, *Atlas of Human Anatomy,* Gower Medical Publishing Ltd., 2nd ed., 1991.

CERVICAL REGION

The bodies of ***cervical vertebrae*** are smaller than those of thoracic vertebrae (Figure 6-14). The vertebral arches, however, are larger. The spinous processes of the second through sixth cervical vertebrae are often *bifid*—that is, having a cleft. All cervical vertebrae have three foramina: the vertebral foramen and two transverse foramina. The vertebral foramina of cervical vertebrae are the largest in the vertebral column because they house the cervical enlargement of the spinal cord. Each cervical transverse process contains a ***transverse foramen*** through which the vertebral artery and its accompanying vein and nerve fibers pass.

The first two cervical vertebrae differ considerably from the others. The first cervical vertebra (C1), the ***atlas,*** is so named because of its support of the head. Essentially, the atlas is a ring of bone with ***anterior*** and ***posterior arches*** and large ***lateral masses.*** It lacks a body and a spinous process. The superior surfaces of the lateral masses, called ***superior articular facets,*** are concave and articulate with the occipital condyles of the occipital bone. This articulation permits the movement seen when nodding the head. The inferior surfaces of the lateral masses, the ***inferior articular facets,*** articulate with the second cervical vertebra. The transverse processes and transverse foramina of the atlas are quite large.

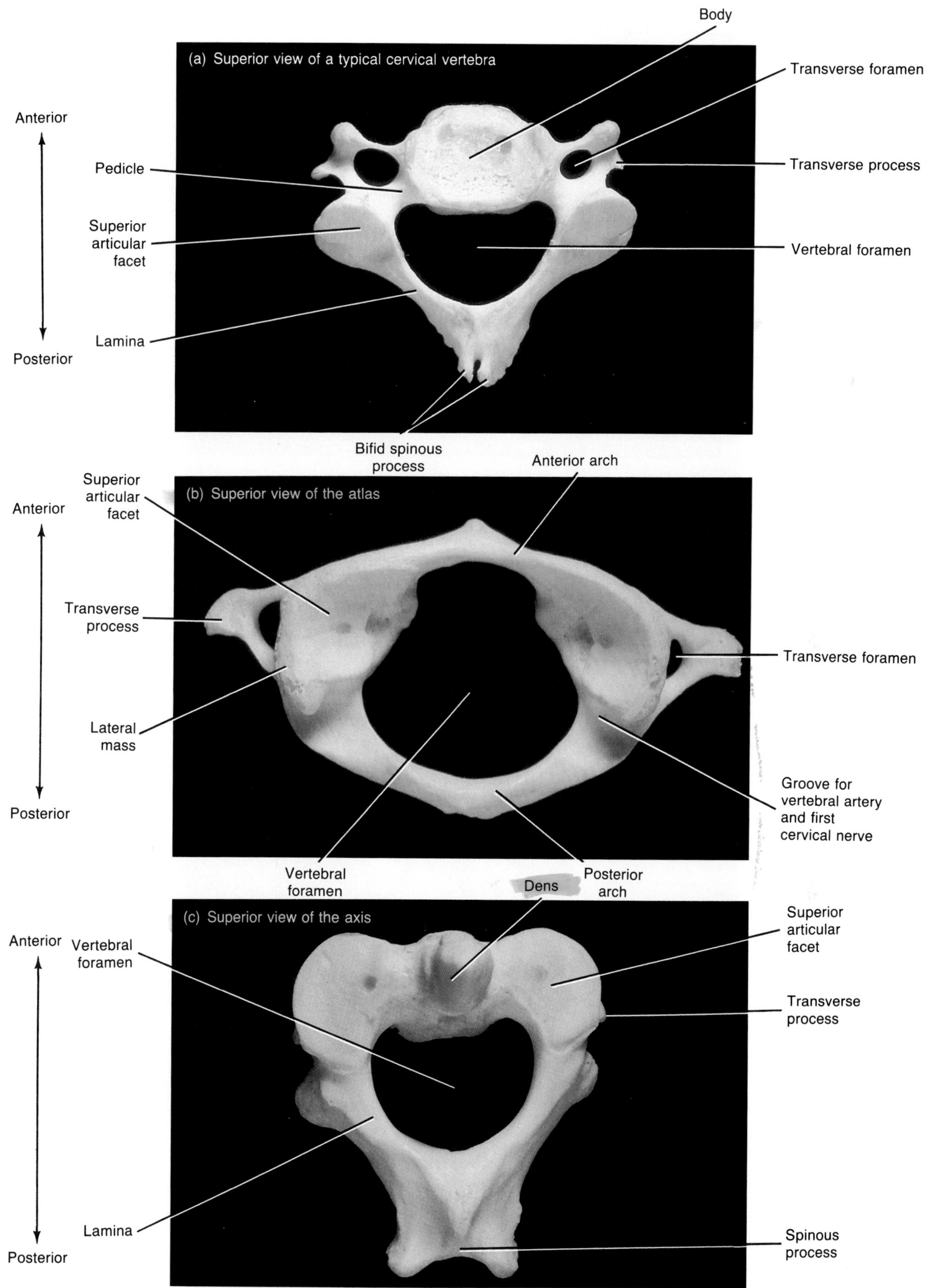

FIGURE 6-14 Cervical vertebrae. (Photographs courtesy of J. A. Gosling, P. F. Harris, *et al. Atlas of Human Anatomy,* Gower Medical Publishing Ltd., 2nd ed., 1991.)

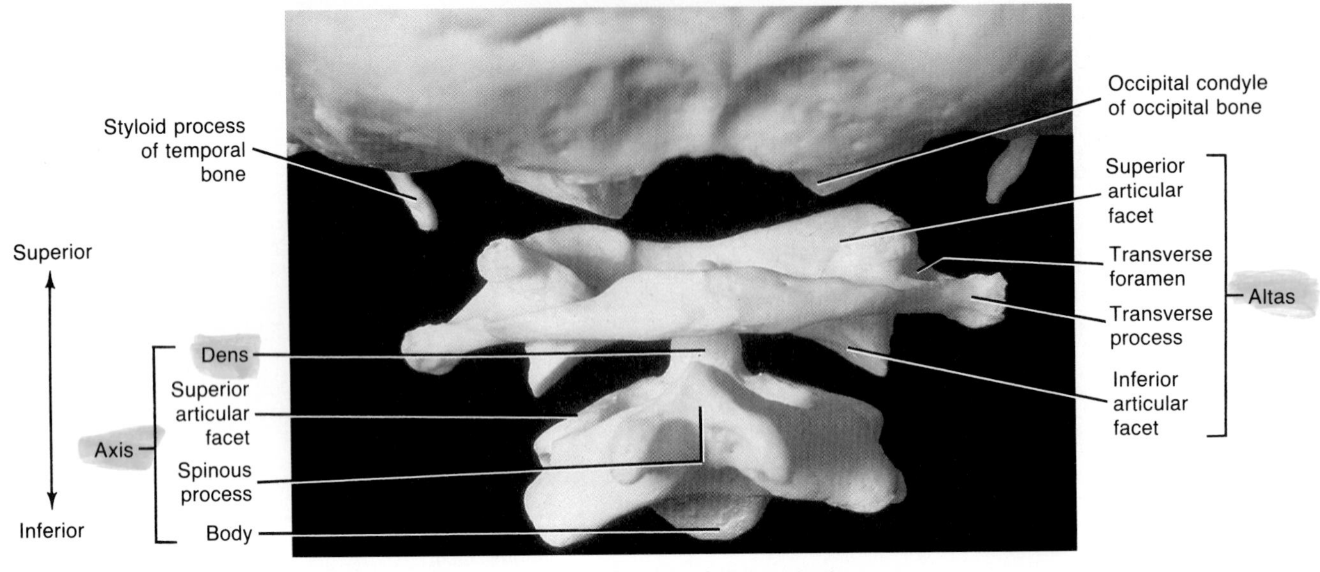

(d) Posterior view of atlas and axis

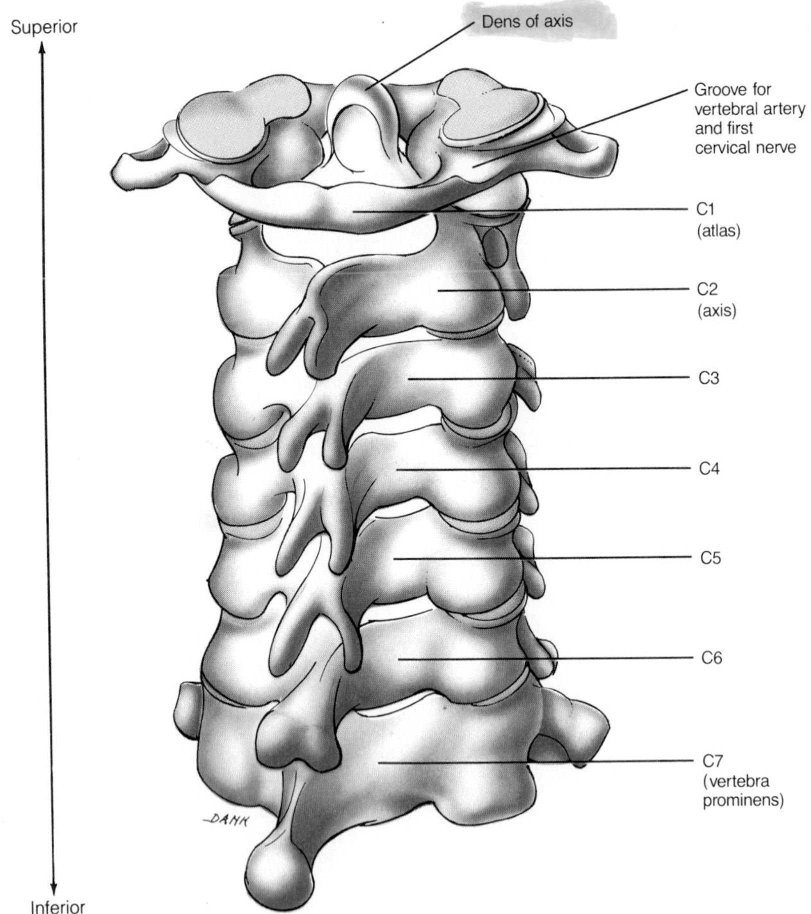

(e) Posterior view of articulated cervical vertebrae

The second cervical vertebra (C2), the ***axis,*** does have a body. A peglike process called the ***dens*** (*dens* = tooth) projects up through the ring of the atlas. The dens makes a pivot on which the atlas and head rotate. This arrangement permits side-to-side rotation of the head. In various instances of trauma, the dens of the axis may be driven into the medulla oblongata of the brain. This injury is the usual cause of fatality in ***whiplash injuries*** that result in death.

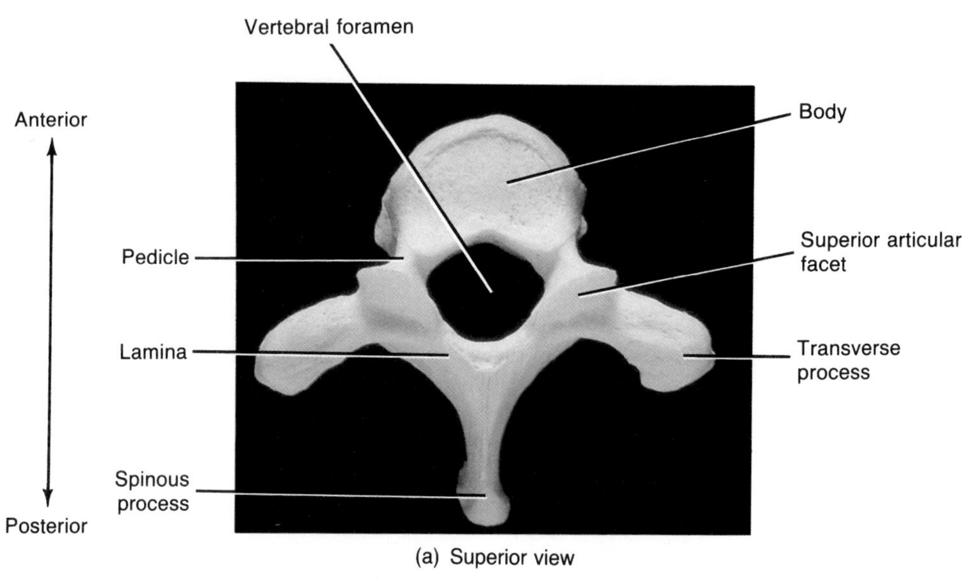

(a) Superior view

FIGURE 6-15 Thoracic vertebrae. (Photographs courtesy of J. A. Gosling, P. F. Harris, et al., *Atlas of Human Anatomy,* Gower Medical Publishing Ltd., 2nd ed., 1991.)

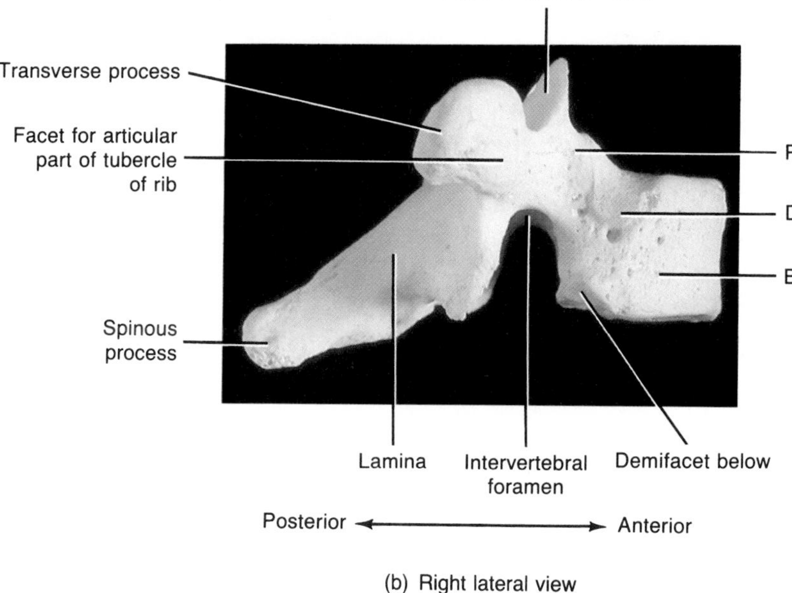

(b) Right lateral view

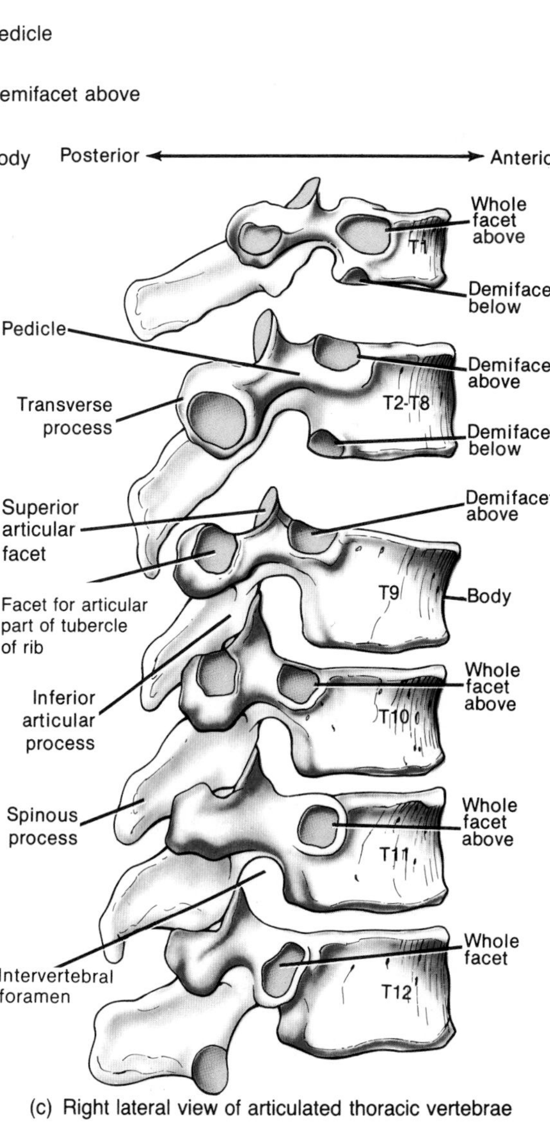

(c) Right lateral view of articulated thoracic vertebrae

The third through sixth cervical vertebrae (C3–C6) correspond to the structural pattern of the typical cervical vertebra previously described.

The seventh cervical vertebra (C7), called the ***vertebra prominens,*** is somewhat different. It is marked by a large, nonbifid spinous process that may be seen and felt at the base of the neck.

THORACIC REGION

Thoracic vertebrae (T1–T12) are considerably larger and stronger than cervical vertebrae (Figure 6-15). In addition, the spinous process on each vertebra is long, laterally flattened, and directed inferiorly. Thoracic vertebrae also have longer and heavier transverse processes than cervical vertebrae.

Except for the eleventh and twelfth thoracic vertebrae, the transverse processes have ***facets*** for articulating with the tubercles of the ribs. The bodies of thoracic vertebrae also have whole facets or half-facets, called ***demifacets,*** for articulation with the heads of the ribs. Movements of the thoracic region are limited by thin intervertebral discs and the attachment of ribs to the sternum.

LUMBAR REGION

The ***lumbar vertebrae*** (L1–L5) are the largest and strongest in the column (Figure 6-16). Their various projections are short and thick. The superior articular processes are directed medially instead of superiorly. The inferior articular processes are directed laterally instead of inferiorly. The spinous processes are quadrilateral in shape, thick, and broad and project nearly straight posteriorly. The spinous processes are well adapted for the attachment of the large back muscles.

A summary of the major structural differences among cervical, thoracic, and lumbar vertebrae is presented in Exhibit 6-5.

SACRUM AND COCCYX

The ***sacrum*** (= sacred or holy bone) is a triangular bone formed by the union of five sacral vertebrae. These are indicated in Figure 6-17 as S1–S5. Fusion begins between 16 and 18 years of age and is usually completed by the midtwenties. The sacrum serves as a strong foundation for the pelvic girdle. It is positioned at the posterior portion of the pelvic cavity between the two hipbones.

The concave anterior side of the sacrum faces the pelvic cavity. It is smooth and contains four ***transverse lines (ridges)*** that mark the joining of the four pairs of ***anterior sacral (pelvic) foramina.*** The lateral portion of the superior surface contains a smooth surface called the ***ala*** (wing), which is formed by the transverse process of the first sacral vertebra (S1).

The convex, posterior surface of the sacrum is irregular. It contains a ***median sacral crest,*** the fused spinous processes of the upper sacral vertebrae; a ***lateral sacral crest,*** the transverse processes of the sacral vertebrae; and four pairs of ***posterior (dorsal) sacral foramina.*** These foramina communicate with the anterior sacral foramina through which nerves and blood vessels pass. The ***sacral canal*** is a continuation of the vertebral canal. The laminae of the fifth sacral vertebra, and sometimes the fourth, fail to meet, thereby leaving an inferior entrance to the vertebral canal called the ***sacral hiatus*** (hī-Ā-tus). On either side of the sacral hiatus are the ***sacral cornua,*** the inferior articular processes of the fifth sacral vertebra. They are connected by ligaments to the coccygeal cornua of the coccyx.

CLINICAL APPLICATION

Caudal Anesthesia

Anesthetic agents that act on the sacral and coccygeal nerves are sometimes injected through the sacral hiatus, a procedure called ***caudal anesthesia*** that is used most frequently in obstetrics. Since the sacral hiatus is between the sacral cornua, the cornua are important bony landmarks for locating the hiatus. Anesthetic agents may also be injected through the posterior (dorsal) sacral foramina.

The superior border of the sacrum exhibits an anteriorly projecting border, the ***sacral promontory*** (PROM-on-tō′-rē). It is an obstetrical landmark for measurements of the pelvis. Laterally, the sacrum has a large ***auricular surface*** for articulating with the ilium of the hipbone. Posterior to the auricular surface is a roughened surface, the ***sacral tuberosity,*** that contains depressions for the attachment of ligaments. The sacral tuberosity is another surface of the sacrum that unites with the ilium of the hipbone to form the sacroiliac joint. The ***superior articular processes*** of the sacrum articulate with the fifth lumbar vertebra.

The ***coccyx*** is also triangular in shape and is formed by the fusion of the coccygeal vertebrae, usually the last

EXHIBIT 6-5
Comparison of Principal Structural Features of Cervical, Thoracic, and Lumbar Vertebrae

CHARACTERISTIC	CERVICAL	THORACIC	LUMBAR
Size	Small	Larger	Largest
Foramina	3	1	1
Spinous Processes	Slender and bifid	Long and fairly thick	Short and blunt
Transverse Processes	Small	Fairly large	Large and blunt
Articular Facets for Ribs	No	Yes	No
Direction of Articular Facets			
Superior	Posterosuperior	Posterolateral	Medial
Inferior	Anteroinferior	Anteromedial	Lateral
Size of Intervertebral Discs	Thick relative to size of vertebral bodies	Thin relative to vertebral bodies	Massive

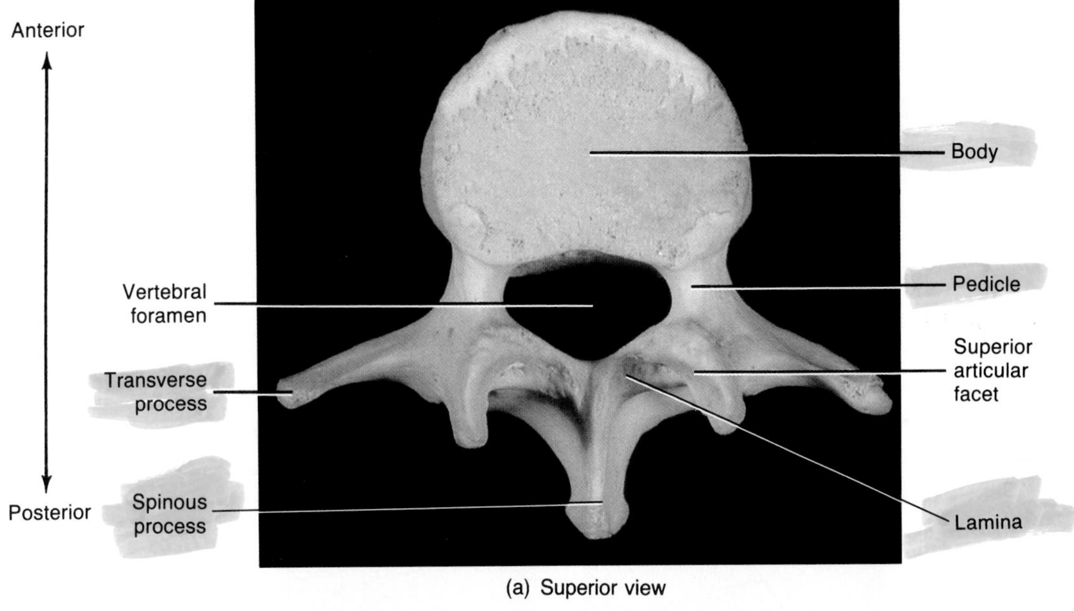

(a) Superior view

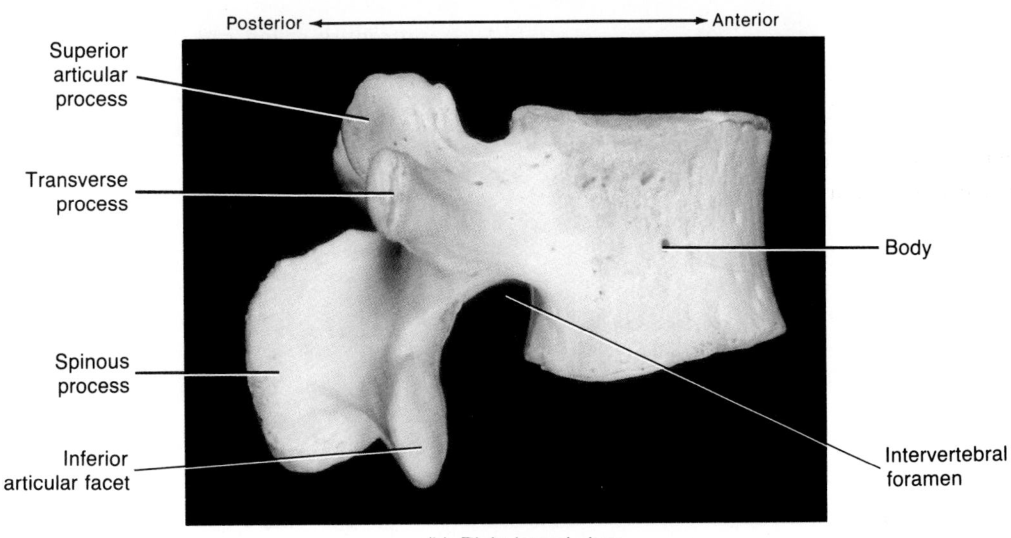

(b) Right lateral view

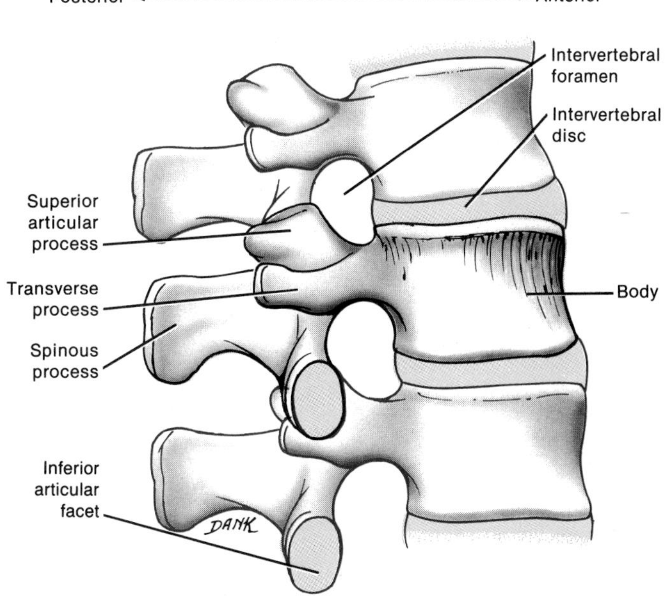

(c) Right lateral view of articulated lumbar vertebrae

FIGURE 6-16 Lumbar vertebrae. Photographs courtesy of J. A. Gosling, P. F. Harris, et al., *Atlas of Human Anatomy,* Gower Medical Publishing Ltd., 2nd ed., 1991.)

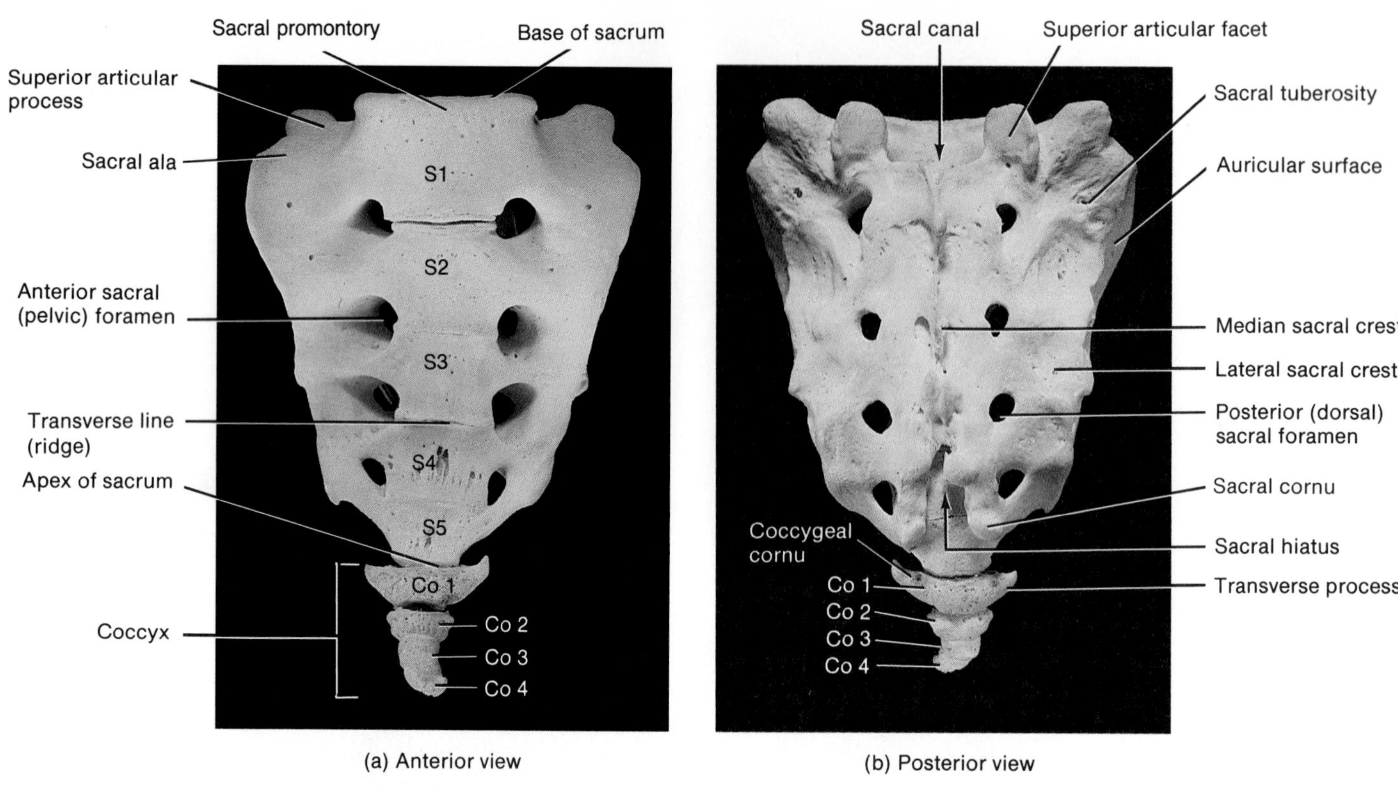

FIGURE 6-17 Sacrum and coccxy. Courtesy of Lester Bergman and Associates.

four. These are indicated in Figure 6-17 as Co1–Co4. Fusion generally occurs between 20 and 30 years. The dorsal surface of the body of the coccyx contains two long ***coccygeal cornua*** that are connected by ligaments to the sacral cornua. The coccygeal cornua are the pedicles and superior articular processes of the first coccygeal vertebra. On the lateral surfaces of the body of the coccyx are a series of ***transverse processes,*** the first pair being the largest. The coccyx articulates superiorly with the sacrum.

THORAX

The term ***thorax*** refers to the entire chest. The skeletal portion of the thorax is a bony cage formed by the sternum, costal cartilages, ribs, and the bodies of the thoracic vertebrae (Figure 6-18).

The thoracic cage is roughly cone-shaped, the narrow portion being superior and the broad portion inferior. It is flattened from front to back. The thoracic cage encloses and protects the organs in the thoracic cavity and upper abdomen. It also provides support for the bones of the shoulder girdle and upper extremities.

STERNUM

The ***sternum,*** or breastbone, is a flat, narrow bone measuring about 15 cm (6 in.) in length. It is located in the median line of the anterior thoracic wall. The sternum may be split in the midsagittal plane to allow surgeons access to structures in the thoracic cavity such as the thymus gland, heart, and great vessels of the heart.

The sternum (Figure 6-18) consists of three basic portions: the ***manubrium*** (ma-NOO-brē-um), the superior portion; the ***body,*** the middle, largest portion; and the ***xiphoid*** (ZĪ-foyd) ***process,*** the inferior, smallest portion. The junction of the manubrium and body forms the ***sternal angle.*** The manubrium has a depression on its superior surface called the ***suprasternal notch.*** On each side of the suprasternal notch are ***clavicular notches*** that articulate with the medial ends of the clavicles. The manubrium also articulates with the first and second ribs. The body of the sternum articulates directly or indirectly with the second through tenth ribs. The xiphoid process has no ribs attached to it but provides attachment for some abdominal muscles. The xiphoid process consists of hyaline cartilage during infancy and childhood and does not ossify completely until about age 40. If the hands of a rescuer are mispositioned during cardiopulmonary resuscitation (CPR), there is danger of fracturing the ossified xiphoid process, separating it from the body and driving it into the liver.

CLINICAL APPLICATION

Sternal Puncture

Since the sternum possesses red bone marrow throughout life and because it is readily accessible and has a thin

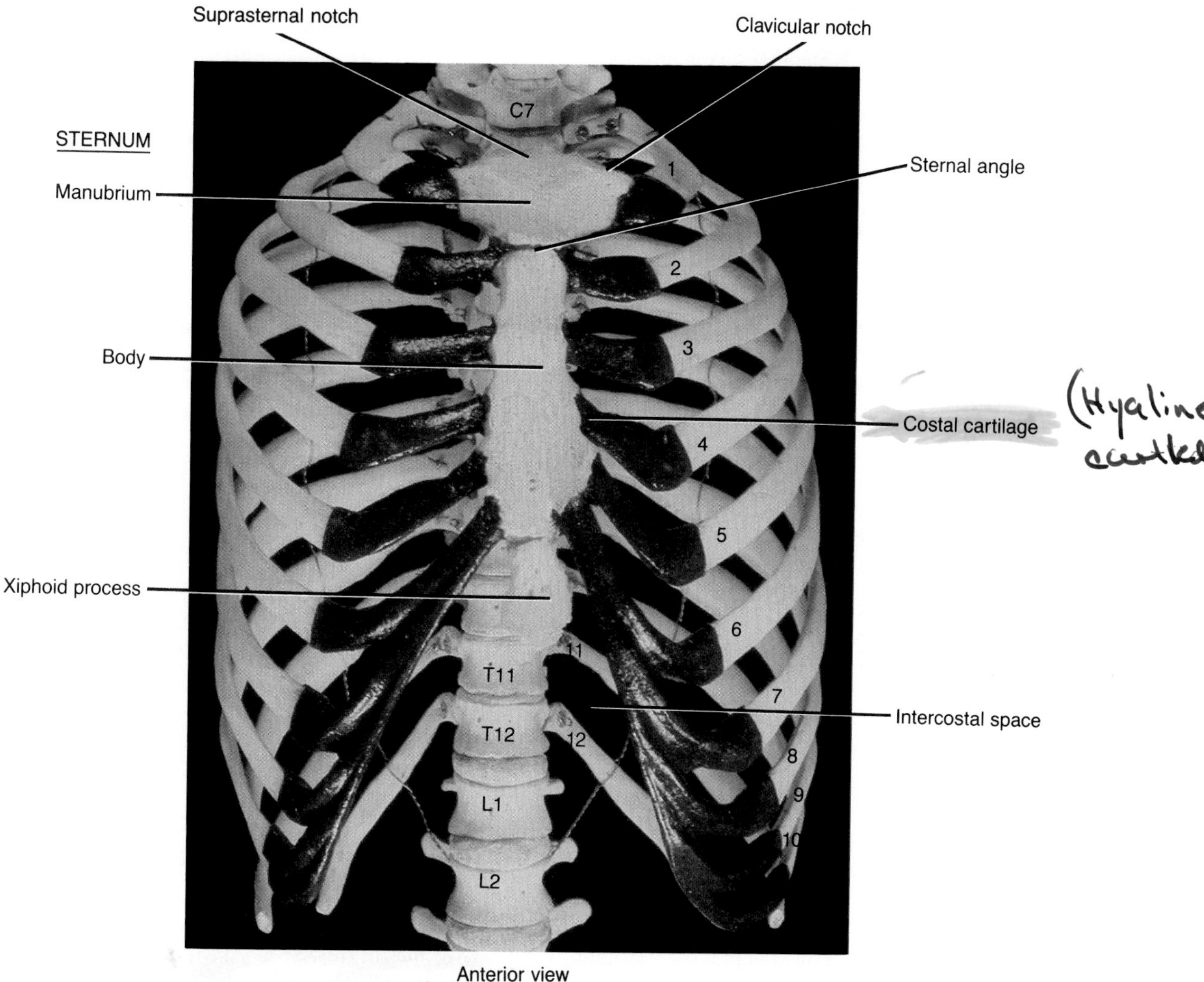

FIGURE 6-18 Skeleton of the thorax. Courtesy of J. A. Gosling, P. F. Harris, et al., *Atlas of Human Anatomy,* Gower Medical Publishing Ltd., 2nd ed., 1991.

external surface compact bone, it is a common site for withdrawal of marrow for biopsy **(*marrow aspiration*).** Under a local anesthetic, a wide-bore needle is introduced into the marrow cavity of the sternum for aspiration of a sample of red bone marrow. This procedure is called a ***sternal puncture.***

RIBS

Twelve pairs of ***ribs*** make up the sides of the thoracic cavity (Figure 6-18). The ribs increase in length from the first through seventh. Then they decrease in length to the twelfth rib. Each rib articulates posteriorly with its corresponding thoracic vertebra. In order to number the ribs anteriorly, count downward from the second costal cartilage, which articulates with the sternum at the sternal angle.

The first through seventh ribs have a direct anterior attachment to the sternum by a strip of hyaline cartilage called ***costal cartilage*** (*costa* = rib). These ribs are called ***true*** (***vertebrosternal***) ***ribs.*** The remaining five pairs of ribs are referred to as ***false ribs*** because their costal cartilages do not attach directly to the sternum. The cartilages of the eighth, ninth, and tenth ribs attach to each other and then to the cartilage of the seventh rib. These false ribs are called ***vertebrochondral ribs.*** The eleventh and twelfth false ribs are designated as ***floating*** (***vertebral***) ***ribs*** because their anterior ends do not attach at all to the sternum. They attach only posteriorly to the thoracic vertebrae.

Although there is some variation in rib structure, we will examine the parts of a typical (third through ninth) rib (Figure 6-19). The ***head*** of a typical rib is a projection at the posterior end of the rib. It is wedge-shaped and consists of one or two ***facets*** that articulate with facets on the bodies of adjacent thoracic vertebrae. The inferior facet is larger than the superior facet. The ***neck*** is a constricted portion just lateral to the head. A knoblike structure on the posterior surface where the neck joins the body is called

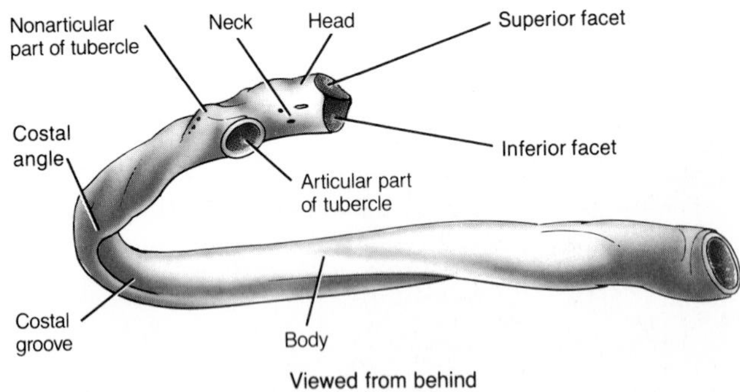

FIGURE 6-19 Typical rib. Shown is a left rib.

a *tubercle* (TOO-ber-kul). It consists of a ***nonarticular part*** that affords attachment to the ligament of the tubercle and an ***articular part*** that articulates with the facet of a transverse process of the inferior of the two vertebrae to which the head of the rib is connected. The ***body (shaft)*** is the main part of the rib. A short distance beyond the tubercle, there is an abrupt change in the curvature of the shaft. This point is called the ***costal angle.*** The inner surface of the rib has a ***costal groove*** that protects blood vessels and a small nerve.

The posterior portion of the rib is connected to a thoracic vertebra by its head and articular part of a tubercle. The facet of the head fits into a facet on the body of a vertebra, and the articular part of the tubercle articulates with the facet of the transverse process of the vertebra.

Spaces between ribs, called ***intercostal spaces,*** are occupied by intercostal muscles, blood vessels, and nerves (see Figure 6-18a). Surgical access to the lungs or other structures in the thoracic cavity is commonly undertaken through an intercostal space. Special rib retractors are used to create a wide separation between ribs. The costal cartilages are sufficiently elastic to permit considerable bending.

CLINICAL APPLICATION

Rib Fractures

Rib fractures represent the most common chest injuries and usually result from direct blows, most commonly from steering wheel impact, falls, and crushing injuries to the chest. Ribs tend to break at the point where greatest force is applied, but may also break at the weakest point (that is, the site of greatest curvature), which is just anterior to the costal angle. In children the ribs are highly elastic, and fractures are less frequent than in adults. Since the first two ribs are protected by the clavicle and pectoralis major muscle, and the last two ribs are mobile, they are the least commonly injured. The middle ribs are the ones most commonly fractured. In some cases, fractured ribs may cause damage to the heart, great vessels of the heart, lungs, trachea, bronchi, esophagus, spleen, liver, and kidneys.

APPLICATIONS TO HEALTH

HERNIATED (SLIPPED) DISC

If the anterior and posterior ligaments of the intervertebral discs become injured or weakened, the pressure developed in the nucleus pulposus may be great enough to rupture the surrounding fibrocartilage. As a result, the nucleus pulposus may herniate (protrude) posteriorly or into one of the adjacent vertebral bodies. This condition is called a ***herniated (slipped) disc.***

Most often the nucleus pulposus slips posteriorly toward the spinal cord and spinal nerves (Figure 6-20). This movement exerts pressure on the spinal nerves, causing considerable, sometimes very acute, pain. If the roots of the sciatic nerve, which passes from the spinal cord to the foot, are pressured, the pain radiates down the back of the thigh, through the calf, and occasionally into the foot. If pressure is exerted on the spinal cord itself, nervous tissue may be destroyed.

Traction, bed rest, and analgesics usually relieve the pain. If such treatment is ineffective, surgical decompression of the spinal nerves or removal of some of the nucleus pulposus may be necessary.

ABNORMAL CURVES

As a result of various conditions, the normal curves of the vertebral column may become exaggerated, or the column may acquire a lateral bend, resulting in ***abnormal curves*** of the spine.

Scoliosis (skō′-lē-Ō-sis; *scolio* = bent), a lateral bending of the vertebral column, usually in the thoracic region, is the most common of the abnormal curves. It may be congenital (present at birth), in which vertebrae are malformed, or it may result from chronic sciatica, paralysis of muscles

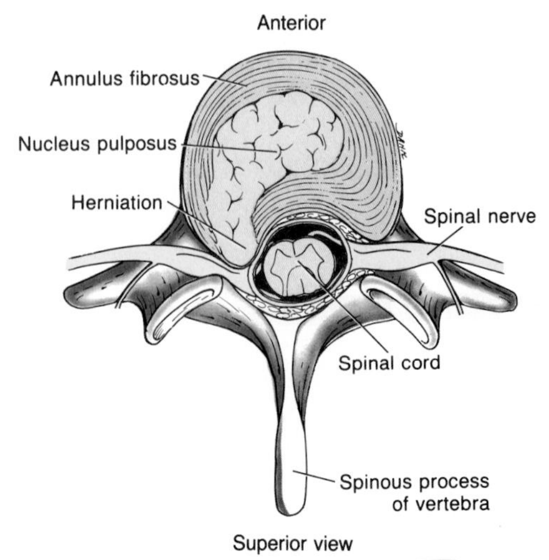

FIGURE 6-20 Herniated (slipped) disc.

on one side of the vertebral column, poor posture, or one leg being shorter than the other.

Kyphosis (kī-FŌ-sis; *kypho* = hunchback) is an exaggeration of the thoracic curve of the vertebral column. In tuberculosis of the spine, vertebral bodies may partially collapse, causing an acute angular bending of the vertebral column. In the elderly, degeneration of the intervertebral discs leads to kyphosis. Kyphosis may also be caused by rickets or poor posture. The term *round-shouldered* is an expression for mild kyphosis. Electrical muscle stimulation is being studied to assess its effects on kyphosis as well as on scoliosis.

Lordosis (lor-DŌ-sis; *lordo* = swayback) is an exaggeration of the lumbar curve of the vertebral column. It may result from increased weight of abdominal contents as in pregnancy or extreme obesity, poor posture, rickets, or tuberculosis of the spine.

SPINA BIFIDA

Spina bifida (SPĪ-na BIF-i-da) is a congenital defect of the vertebral column in which laminae fail to unite at the midline. In serious cases, protrusion of the membranes (meninges) around the spinal cord or the spinal cord itself produce serious problems, such as partial or complete paralysis, partial or complete loss of urinary bladder control, and the absence of reflexes. Spina bifida may be diagnosed prenatally by a test of the mother's blood, sonography, and/or amniocentesis.

FRACTURES OF THE VERTEBRAL COLUMN

Fractures of the vertebral column most commonly involve T5, T6, and T9–T12. Thoracic fractures usually result from a flexion-compression type of injury such as might be sustained in landing on the feet or buttocks after a fall from a height or having a heavy weight fall on the shoulders. Cervical vertebrae may be fractured or, more commonly, dislodged by a fall on the head with acute flexion of the neck, as might happen on diving into shallow water. Dislocation may even result from the sudden forward then backward jerk (''whiplash'') that may occur when an automobile or airplane crashes. Spinal nerve damage may occur as a result of fractures of the vertebral column.

STUDY OUTLINE

Types of Bones (p. 129)

1. On the basis of shape, bones are classified as long, short, flat, or irregular.
2. Sutural bones are found between the sutures of certain cranial bones. Sesamoid bones develop in tendons or ligaments.

Surface Markings (p. 129)

1. Surface markings are structural features visible on the surfaces of bones.
2. Each marking is structured for a specific function—joint formation, muscle attachment, or passage of nerves and blood vessels.
3. Terms that describe markings include fissure, foramen, meatus, fossa, process, condyle, head, facet, tuberosity, crest, and spine. See Exhibit 6-1.

Divisions of the Skeletal System (p. 129)

1. The axial skeleton consists of bones arranged along the longitudinal axis. The parts of the axial skeleton are the skull, hyoid bone, vertebral column, sternum, and ribs.
2. The appendicular skeleton consists of the bones of the girdles and the upper and lower extremities. The parts of the appendicular skeleton are the pectoral (shoulder) girdles, bones of the upper extremities, pelvic (hip) girdle, and bones of the lower extremities.

Skull (p. 130)

1. The skull consists of the cranium and the face. It is composed of 22 bones.
2. Sutures are immovable joints between bones of the skull. Examples are coronal, sagittal, lambdoid, and squamosal sutures.
3. Fontanels are dense connective tissue membrane-filled spaces between the cranial bones of fetuses and infants. The major fontanels are the anterior, posterior, anterolaterals, and posterolaterals.
4. The eight cranial bones include the frontal, parietal (2), temporal (2), occipital, sphenoid, and ethmoid.
5. The 14 facial bones are the nasal (2), maxillae (2), zygomatic (2), mandible, lacrimal (2), palatine (2), inferior nasal conchae (2), and vomer.
6. Paranasal sinuses are cavities in bones of the skull that communicate with the nasal cavity. They are lined by mucous membranes. The cranial bones containing the paranasal sinuses are the frontal, sphenoid, ethmoid, and maxillae.
7. The orbits (eye sockets) are formed by seven bones of the skull.
8. The foramina of the skull bones provide passages for nerves and blood vessels.

Hyoid Bone (p. 144)

1. The hyoid bone is a U-shaped bone that does not articulate with any other bone.
2. It supports the tongue and provides attachment for some of its muscles as well as some neck muscles and muscles of the pharynx.

Vertebral Column (p. 145)

1. The vertebral column, sternum, and ribs constitute the skeleton of the trunk.
2. The bones of the adult vertebral column are the cervical vertebrae (7), thoracic vertebrae (12), lumbar vertebrae (5), the sacrum (5, fused), and the coccyx (4, fused).
3. The vertebral column contains normal primary curves (thoracic and sacral) and normal secondary curves (cervical and lumbar). These curves give strength, support, and balance.

4. The vertebrae are similar in structure, each consisting of a body, vertebral arch, and seven processes. Vertebrae in the different regions of the column vary in size, shape, and detail.

Thorax (p. 154)

1. The thoracic skeleton consists of the sternum, ribs and costal cartilages, and thoracic vertebrae.
2. The thoracic cage protects vital organs in the chest area and upper abdomen.

Applications to Health (p. 156)

1. Protrusion of the nucleus pulposus of an intervertebral disc posteriorly or into an adjacent vertebral body is called a herniated (slipped) disc.
2. Exaggeration of a normal curve or lateral bending of the vertebral column results in an abnormal curve. Examples include scoliosis, kyphosis, and lordosis.
3. The imperfect union of the vertebral laminae at the midline, a congenital defect, is referred to as spina bifida.
4. Fractures of the vertebral column most often involve T5, T6, and T9–T12.

REVIEW QUESTIONS

1. What are the four principal types of bones? Give an example of each. Distinguish between a sutural (Wormian) and a sesamoid bone. (p. 129)
2. What are surface markings? Describe and give an example of each. (p. 129)
3. Distinguish between the axial and appendicular skeletons. What subdivisions and bones are contained in each? (p. 129)
4. What are the bones that compose the skull? The cranium? The face? (p. 130)
5. Define a suture. What are the four prominent sutures of the skull? Where are they located? (p. 131)
6. What is a fontanel? Describe the location of the six fairly constant fontanels. (p. 131)
7. What is a paranasal sinus? What cranial bones contain paranasal sinuses? (p. 141)
8. Describe the components of each orbit. (p. 144)
9. What bones form the skeleton of the trunk? Distinguish between the number of nonfused vertebrae found in the adult vertebral column and in that of a child. (p. 145)
10. What are the normal curves in the vertebral column? How are primary and secondary curves differentiated? What are the functions of the curves? (p. 146)
11. What are the principal distinguishing characteristics of the bones of the various regions of the vertebral column? (p. 146)
12. What bones form the skeleton of the thorax? What are the functions of the thoracic skeleton? (p. 154)
13. How are ribs classified on the basis of their attachment to the sternum? (p. 155)
14. What is a herniated (slipped) disc? Why does it cause pain? How is it treated? (p. 156)
15. Define the following: scoliosis, kyphosis, lordosis, and spina bifida. (p. 156)
16. Define the following: black eye (p. 134), cleft palate and cleft lip (p. 140), sinusitis (p. 141), temporomandibular joint (TMJ) syndrome (p. 143), deviated nasal septum (DNS) (p. 143), caudal anesthesia (p. 152), sternal puncture (p. 154), and fractured rib (p. 156).

SELF QUIZ

1. Match the following:

___ **a.** this bone forms the lower jaw, including the chin
___ **b.** these are the cheekbones; they also form lateral walls of the orbit of the eye
___ **c.** tear ducts pass through tiny foramina in these bones; they are the smallest bones in the face
___ **d.** the bridge of the nose is formed by these bones
___ **e.** organs of hearing (internal part of ears) are located in and protected by these bones
___ **f.** this bone sits directly over the spinal column; it contains the hole through which the spinal cord connects to the brain

A. ethmoid
B. frontal
C. inferior nasal concha
D. lacrimal
E. mandible
F. maxilla
G. nasal
H. occipital
I. palatine
J. parietal
K. sphenoid
L. temporal
M. vomer
N. zygomatic

___ **g.** the name means "wall"; the bones form most of the roof and much of the side walls of the skull
___ **h.** these bones form most of the roof of the mouth (hard palate) and contain the sockets into which upper teeth are set
___ **i.** L-shaped bones form the posterior parts of the hard palate and nose
___ **j.** commonly called the forehead, it provides protection for the anterior portion of the brain
___ **k.** a light spongy bone, it forms much of the roof and internal structure of the nose
___ **l.** it serves as a "keystone," since it binds together many of the other bones of the skull; it is shaped like

a bat, with the wings forming part of the sides of the skull and the legs at the back of the nose

___ **m.** this bone forms the inferior part of the septum dividing the external nose into two sides, left and right.

___ **n.** two delicate bones form the lower parts of the side walls of the nose

Choose the one best answer to these questions.

___ **2.** Which of the following statements is/are true?
(1) The metatarsals are examples of long bones.
(2) Flat bones have a middle layer of spongy bone between two layers of compact bone.
(3) Irregular bones consist of compact bone only.
A. (1) only; B. (2) only; C. (3) only; D. (1) and (2); E. All of the above.

___ **3.** A foramen is
A. a cavity within a bone; B. a depression; C. a hole for blood vessels and nerves; D. a ridge; E. for muscle attachment.

___ **4.** Which of the following are bones found in the axial skeleton?
A. ribs, vertebrae, and radius; B. skull, sternum, and vertebral column; C. costal cartilages, clavicle, and coccyx; D. ilium, ischium, and sacrum; E. parietal, pelvic, and sesamoid bones.

___ **5.** Which of the following is/are true?
(1) The squamosal suture is where the temporal bone articulates with the sphenoid and parietal bones.
(2) The coronal suture is associated with the posterior fontanel of the fetal skull.
(3) The parietal and temporal bones are separated by the lambdoid suture.
A. (1) only; B. (2) only; C. (3) only; D. none of the above; E. all of the above.

___ **6.** Which of the following pairs of terms are properly matched, that is, have the closest relation?
A. mastoid air cells–paranasal sinuses; B. temporal bone–infraorbital foramen; C. vomer–anterior cranial fossa; D. temporal bone–styloid process; E. palatine bone–crista galli.

___ **7.** Which of the following structures is part of the frontal bone?
A. infraorbital foramen; B. crista galli; C. supraorbital foramen (notch); D. optic foramen; E. petrous portion.

___ **8.** The opening in the base of the skull through which the spinal cord, as a continuation of the brain stem, passes is the
A. transverse fissure; B. intervertebral foramen; C. vertebral canal; D. sphenoid sinus; E. foramen magnum.

___ **9.** The sphenoid bone
(1) lies posterior to the ethmoid bone.
(2) is anterior to the occipital bone.
(3) has part (the greater wing) visible on the outside of the skull.
(4) articulates with the mandible.
A. (1) only; B. (2) only; C. (3) only; D. (4) only; E. (1), (2), and (3).

___ **10.** The sella turcica
(1) is a depression in the ethmoid bone.
(2) is located in approximately the center of the floor of the cranium.
(3) contains the pineal gland.
(4) identifies the location of the auditory ossicles.
A. (1) only; B. (2) only; C. (3) only; D. (4) only; E. (1), (2), and (3).

___ **11.** Which of the following is *not* part of the ethmoid bone?
A. cribriform plate; B. crista galli; C. middle concha; D. sella turcica; E. superior nasal concha.

___ **12.** All of the following articulate with the maxilla *except* the
A. nasal bone; B. frontal bone; C. mandible; D. palatine bone; E. zygomatic bone.

___ **13.** Which of the following bones is correctly matched with the process?
A. zygomatic bone–temporal process; B. maxilla–pterygoid process; C. vomer–mandibular process; D. temporal–occipital process; E. sphenoid–styloid process.

___ **14.** Which bone contributes to the formation of three cranial cavities (the floor and lateral wall of the nasal cavity, the roof of the mouth, and the floor of the orbit) but does not contain a sinus?
A. palatine; B. sphenoid; C. ethmoid; D. vomer; E. none of the above.

___ **15.** The second cervical vertebra
(1) has a characteristic feature, the dens.
(2) articulates with the first cervical vertebra in a pivot joint.
(3) is called the vertebra prominens because of its long spinous process.
A. (1) only; B. (2) only; C. both (1) and (2); D. (3) only; E. none of the above.

___ **16.** The skeleton of the thorax
A. is formed by 12 pairs of ribs and costal cartilages, the sternum, and the 12 thoracic vertebrae; B. protects the internal chest organs, as well as the liver and stomach (partially); C. is roughly cone-shaped; D. aids in supporting the bones of the shoulder girdle; E. is described by all of the above.

___ **17.** The sterum
A. is classed as a sesamoid bone; B. articulates with all the "true" ribs by means of costal cartilages; C. consists of two parts: a manubrium and a xiphoid process; D. forms part of the pectoral girdle; E. is described by none of the above.

___ **18.** Ribs
A. always articulate with the vertebrae at the sternum; B. always articulate with the sternum by means of cartilages; C. may be classified as true, false, or floating; D. may be considered part of the appendicular skeleton; E. do not support any parts of the body.

Arrange the answers in correct sequence.

___ ___ ___ **19. a.** From anterior to posterior:
A. ethmoid bone
B. sphenoid bone
C. occipital bone

___ ___ ___ **b.** Vertebral regions, from superior to inferior:
A. lumbar
B. thoracic
C. cervical

___ ___ ___ **c.** From superior to inferior:
A. atlas
B. manubrium of sternum
C. hyoid

20. Match the following:

___ **a.** small body, foramina for vertebral blood vessels in transverse processes
___ **b.** atlas and axis
___ **c.** five in an adult
___ **d.** only vertebrae that articulate with ribs
___ **e.** massive body, blunt spinous process and articular processes directed medially or laterally
___ **f.** long spinous processes that point inferiorly
___ **g.** most inferior part of vertebral column
___ **h.** articulate with the two hipbones

A. cervical
B. coccyx
C. lumbar
D. sacral
E. thoracic

The Skeletal System: The Appendicular Skeleton

STUDENT OBJECTIVES

1. Identify the bones of the pectoral (shoulder) girdle and their major markings.
2. Identify the upper extremity, its component bones, and their markings.
3. Identify the components of the pelvic (hip) girdle and their principal markings.
4. Identify the lower extremity, its component bones, and their markings.
5. Define the structural features and importance of the arches of the foot.
6. Compare some structural differences between female and male skeletons, especially those that pertain to the pelvis.

CHAPTER OUTLINE

- **Pectoral (Shoulder) Girdle**
  Clavicle
  Scapula
- **Upper Extremity**
  Humerus
  Ulna and Radius
  Carpals, Metacarpals, and Phalanges
- **Pelvic (Hip) Girdle**
- **Lower Extremity**
  Femur
  Patella
  Tibia and Fibula
  Tarsals, Metatarsals, and Phalanges
  Arches of the Foot
- **Female and Male Skeletons**

This chapter discusses the bones of the appendicular skeleton, that is, the bones of the pectoral (shoulder) and pelvic (hip) girdles and extremities. The differences between female and male skeletons are also compared.

PECTORAL (SHOULDER) GIRDLE

The ***pectoral*** (PEK-tō-ral) or ***shoulder girdles*** attach the bones of the upper extremities to the axial skeleton (Figure 7-1). Each of the two pectoral girdles consists of two bones: a clavicle and a scapula. The clavicle is the anterior component and articulates with the sternum at the sternoclavicular joint. The posterior component, the scapula, which is held in position by complex muscle attachments, articulates with the clavicle and humerus. The pectoral girdles have no articulation with the vertebral column. Thus, the shoulder joints are not very stable, but they are freely movable and thus allow movement in many directions.

CLAVICLE

The ***clavicles*** (KLAV-i-kuls), or collarbones, are long, slender bones with a double curvature (Figure 7-2). The medial one-third of the clavicle is convex anteriorly, whereas the lateral one-third is concave anteriorly. Since the junction of two curves is the weakest point of a structure, this is the most frequent site of clavicular fractures. The clavicles lie horizontally in the superior and anterior part of the thorax superior to the first rib.

The medial end of the clavicle, the ***sternal extremity,*** is rounded and articulates with the sternum. The broad, flat, lateral end, the ***acromial*** (a-KRŌ-mē-al) ***extremity,*** articulates with the acromion of the scapula. This joint is called the ***acromioclavicular joint.*** (Refer to Figure 7-1 for a view of these articulations.) The ***conoid tubercle*** on the inferior surface of the lateral end of the bone serves as a point of attachment for a ligament. The ***costal tuberosity*** on the inferior surface of the medial end also serves as a point of attachment for a ligament.

CLINICAL APPLICATION

Fractured Clavicle

Because of its position, the clavicle transmits forces from the upper extremity to the trunk. If such forces are excessive, as in falling on one's outstretched arm, a ***fractured clavicle*** may result. In fact, it is the most frequently broken bone in the body.

SCAPULA

The ***scapulae*** (SCAP-yoo-lē), or shoulder blades, are large, triangular, flat bones situated in the posterior part of the thorax between the levels of the second and seventh ribs (Figure 7-3). Their medial borders are located about 5 cm (2 in.) from the vertebral column.

A sharp ridge, the ***spine,*** runs diagonally across the posterior surface of the flattened, triangular ***body.*** The end of the spine projects as a flattened, expanded process called the ***acromion*** (a-KRŌ-mē-on), easily felt as the high point of the shoulder. The length of the upper limb is measured from the acromion, even by tailors. This process articulates with the clavicle. Inferior to the acromion is a depression called the ***glenoid cavity.*** This cavity articulates with the head of the humerus to form the shoulder joint.

The thin edge of the body near the vertebral column is the ***medial (vertebral) border.*** The thick edge closer to the arm is the ***lateral (axillary) border.*** The medial and lateral borders join at the ***inferior angle.*** The superior edge of the scapular body, called the ***superior border,*** joins the vertebral border at the ***superior angle.*** The ***scapular notch*** is a prominent indentation along the superior border near the coracoid process; through it passes the suprascapular nerve.

At the lateral end of the superior border is a projection of the anterior surface called the ***coracoid*** (KOR-a-koyd) ***process*** to which muscles attach. Above and below the spine are two fossae: the ***supraspinous*** (soo′-pra-SPĪ-nus) ***fossa*** and the ***infraspinous fossa,*** respectively. Both serve as surfaces of attachment for shoulder muscles. On the ventral (costal) surface is a lightly hollowed-out area called the ***subscapular fossa,*** also a surface of attachment for shoulder muscles.

UPPER EXTREMITY

The ***upper extremities*** consist of 60 bones. The skeleton of the right upper extremity is shown in Figure 7-1. Each upper extremity includes a humerus in the arm, ulna and radius in the forearm, carpals (wrist bones), metacarpals (palm bones), and phalanges in the fingers of the hand.

HUMERUS

The ***humerus*** (HYOO-mer-us), or arm bone, is the longest and largest bone in the upper extremity (Figure 7-4). It articulates proximally with the scapula and distally at the elbow with both ulna and radius.

The proximal end of the humerus consists of a ***head*** that articulates with the glenoid cavity of the scapula. It also has an ***anatomical neck,*** the former site of the epiphyseal plate, which is an oblique groove just distal to the head. The ***greater tubercle*** is a lateral projection distal to the neck. It is the most laterally palpable bony landmark of the shoulder region. The ***lesser tubercle*** is an anterior projection. Between these tubercles runs an ***intertubercular sulcus (bicipital groove)*** which contains a tendon from the long head of the biceps brachii muscle. The ***surgical neck***

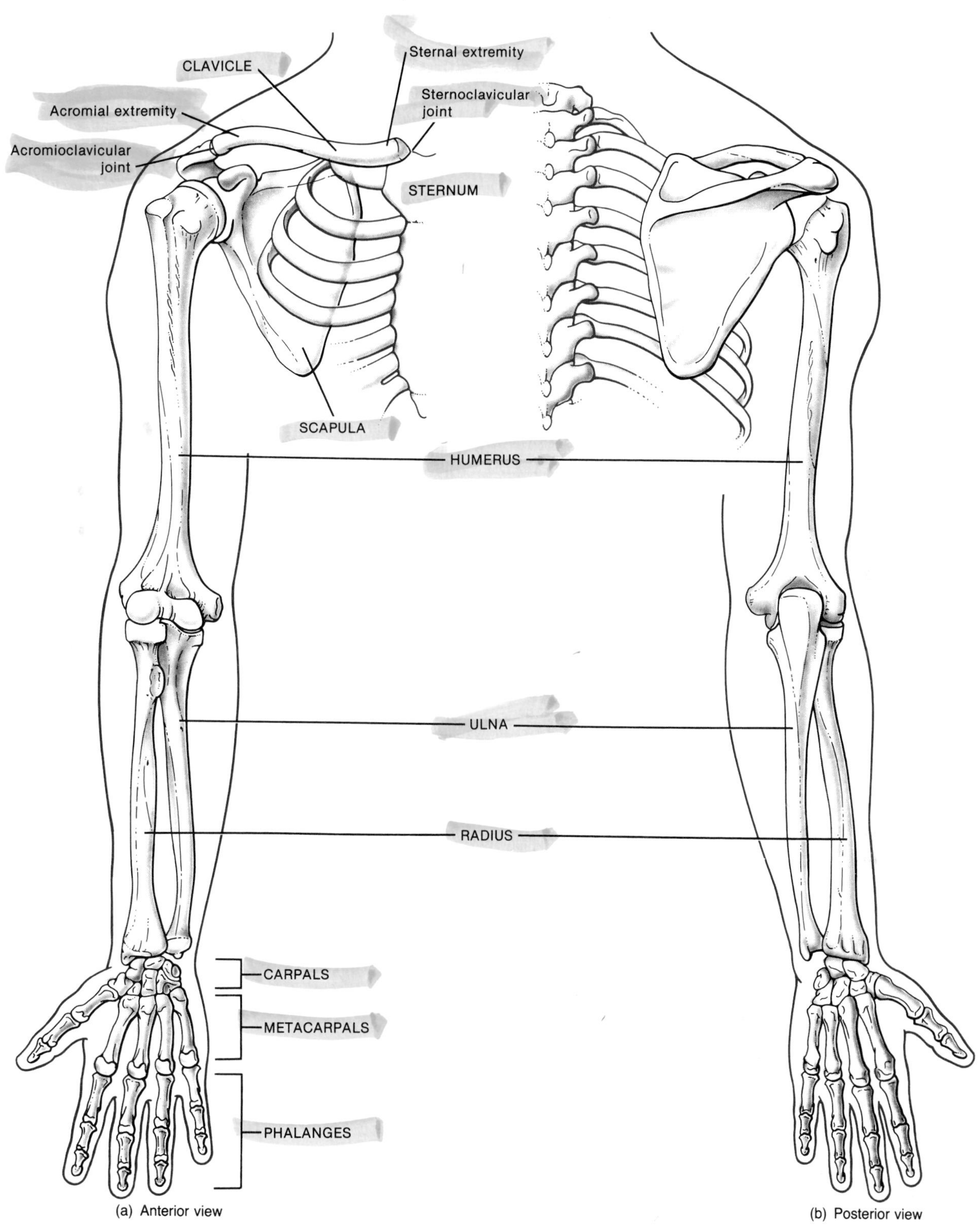

FIGURE 7-1 Right pectoral (shoulder) girdle and upper extremity.

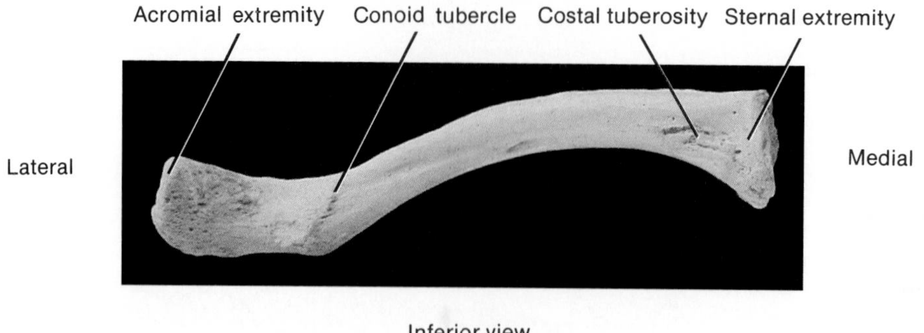

FIGURE 7-2 Right clavicle.

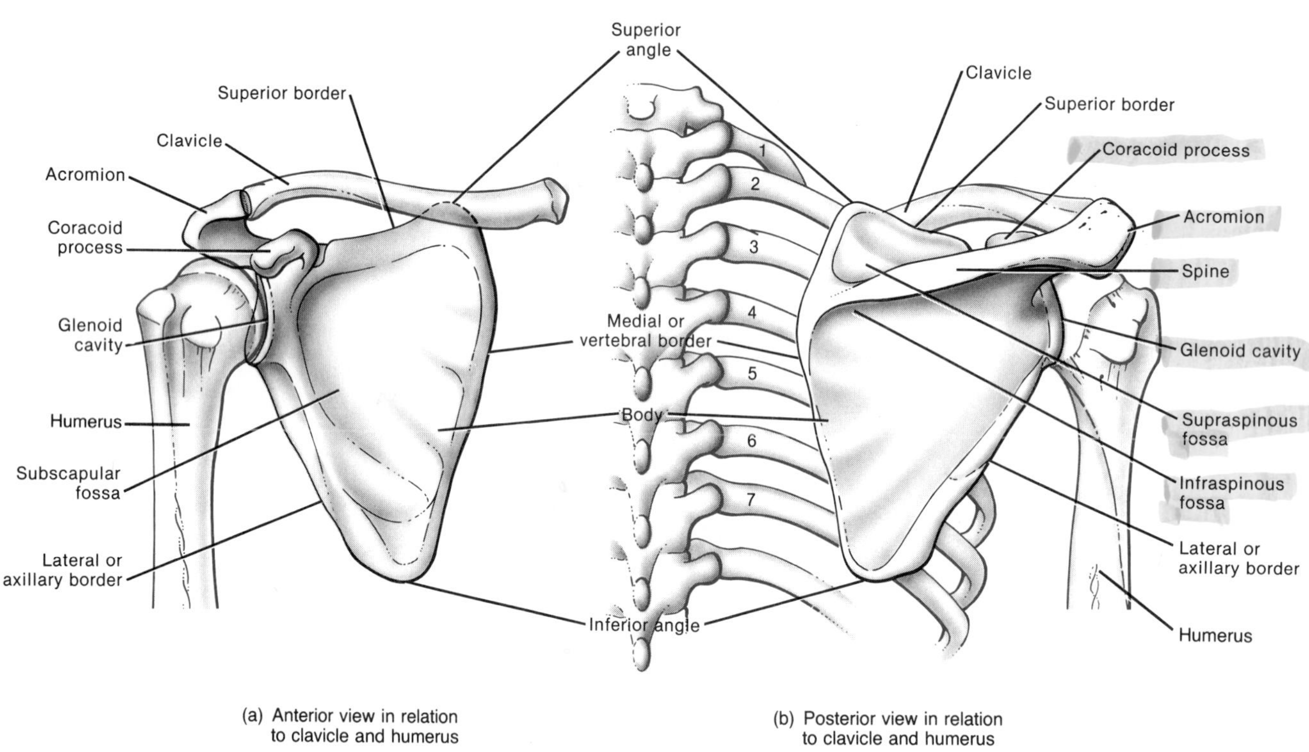

FIGURE 7-3 Right scapula. (c)–(e) Courtesy of Lester Bergman and Associates.

is a constricted portion just distal to the tubercles. It is so named because of its liability to fracture.

The ***body (shaft)*** of the humerus is cylindrical at its proximal end. It gradually becomes triangular and is flattened and broad at its distal end. Along the middle portion of the shaft, there is a roughened, V-shaped area called the ***deltoid tuberosity.*** This area serves as a point of attachment for the deltoid muscle.

The following parts are found at the distal end of the humerus. The ***capitulum*** (ka-PIT-yoo-lum) is a rounded knob that articulates with the head of the radius. The ***radial fossa*** is a depression that receives the head of the radius when the forearm is flexed. The ***trochlea*** (TRŌK-lē-a) is a pulleylike surface that articulates with the ulna. The ***coronoid fossa*** is an anterior depression that receives part of the ulna when the forearm is flexed. The ***olecranon*** (ō-LEK-ra-non) ***fossa*** is a posterior depression that receives the olecranon of the ulna when the forearm is extended. The ***medial epicondyle*** and ***lateral epicondyle*** are rough projections on either side of the distal end to which most muscles of the forearm are attached. The ulnar nerve lies on the posterior surface of the medial epicondyle and may easily be rolled between the finger and the medial epicondyle.

ULNA AND RADIUS

The ***ulna*** is the medial and longer bone of the forearm (Figure 7-5). In other words, it is located at the little finger

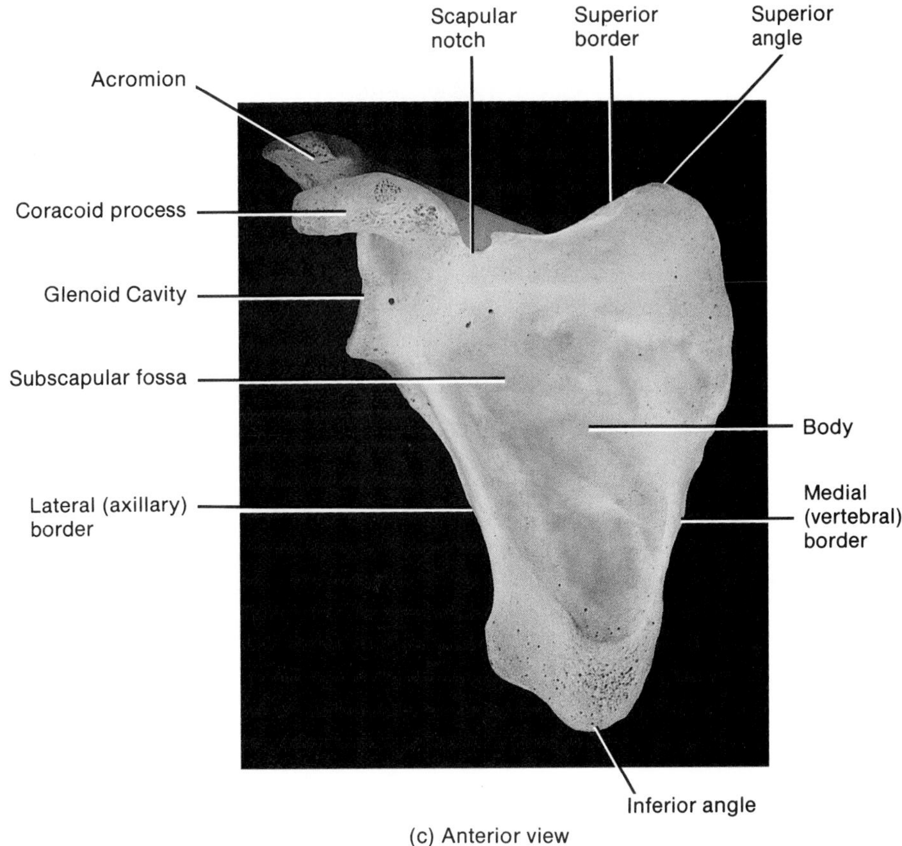

(c) Anterior view

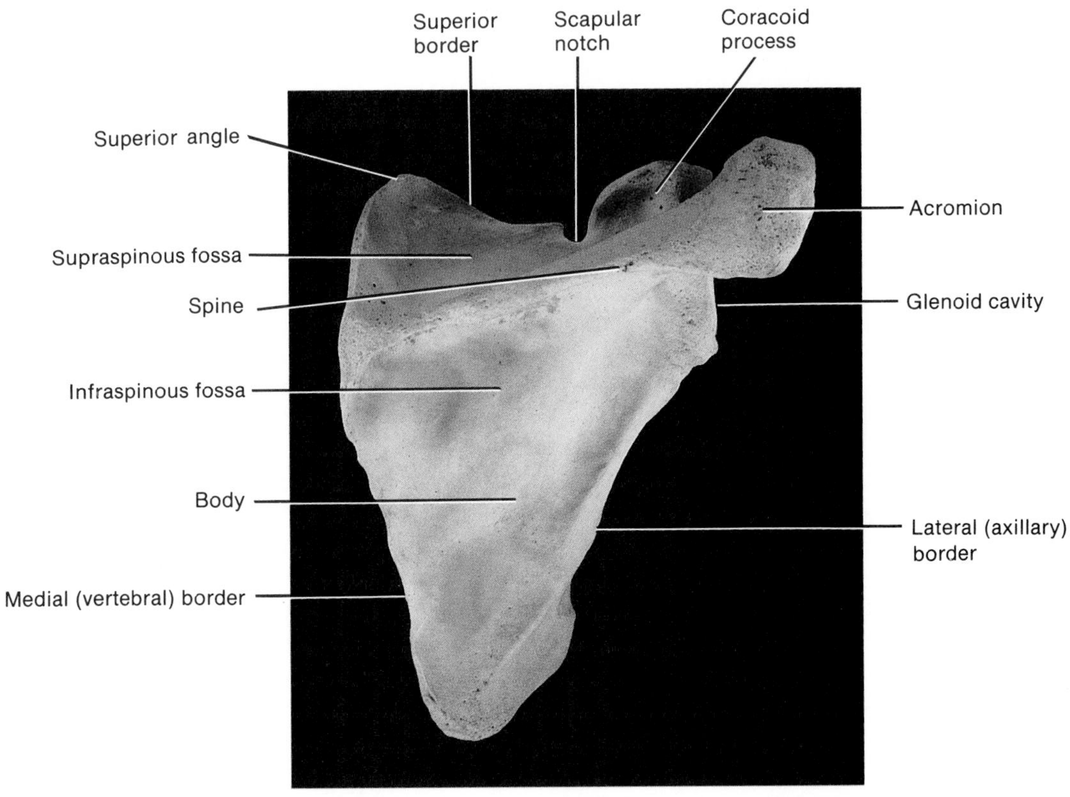

(d) Posterior view

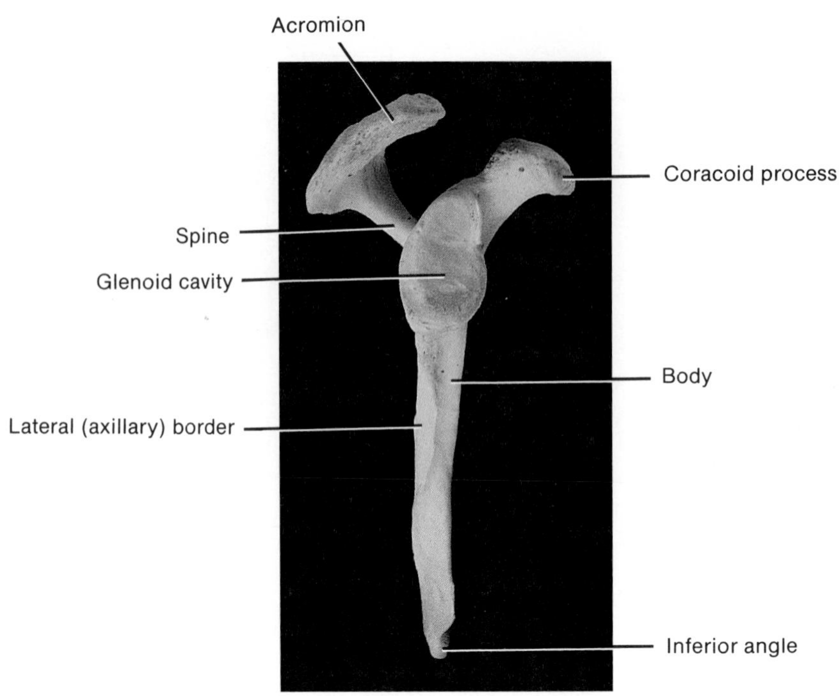

(e) Lateral border view

side. At the proximal end of the ulna is the ***olecranon*** **(*olecranon process*),** which forms the prominence of the elbow. The ***coronoid process*** is an anterior projection that, together with the olecranon, receives the trochlea of the humerus. The ***trochlear (semilunar) notch*** is a curved area between the olecranon and the coronoid process. The trochea of the humerus fits into this notch. The ***radial notch*** is a depression located laterally and inferior to the trochlear notch. It receives the head of the radius. The distal end of the ulna consists of a ***head*** that is separated from the wrist by a fribrocartilage disc. A ***styloid process*** is on the posterior side of the distal end.

The ***radius*** is the lateral and shorter bone of the forearm; that is, it is situated on the thumb side. The proximal end of the radius has a disc-shaped ***head*** that articulates with the capitulum of the humerus and radial notch of the ulna. It also has a raised, roughened area on the medial side called the ***radial tuberosity.*** This is a point of attachment for the biceps brachii muscle. The shaft of the radius widens distally to form a concave inferior surface that articulates with two bones of the wrist called the lunate and scaphoid bones. Also at the distal end is a ***styloid process*** on the lateral side and a medial, concave ***ulnar notch*** for articulation with the distal end of the ulna.

CARPALS, METACARPALS, AND PHALANGES

The skeleton of the hand is composed of three regions: (1) proximal carpus, (2) intermediate metacarpus, and (3) distal phalanges. The ***carpus*** (or wrist) consists of eight small bones, the ***carpals,*** united to each other by ligaments (Figure 7-6). The palmar surface of the carpals is concave and forms the carpal tunnel (described shortly).

The bones of the carpus are arranged in two transverse rows, with four bones in each row, and they are named for their shapes. In the anatomical position, the proximal row of carpals, from the lateral to medial position, consists of the ***scaphoid*** (resembles a boat), ***lunate*** (resembles a crescent moon in its anteroposterior aspect), ***triquetrum*** (has three articular surfaces), and ***pisiform*** (pea-shaped). In about 70 percent of cases involving carpal fractures, only the scaphoid is involved. The distal row of carpals, from the lateral to medial position, consists of the ***trapezium*** (four-sided), ***trapezoid*** (also four-sided), ***capitate*** (its rounded projection, the head, articulates with the lunate), and ***hamate*** (named for a large hook-shaped projection on its anterior surface).

Together, the concavity formed by the pisiform and hamate (on the ulnar side) and scaphoid and trapezium (on the radial side) plus the flexor retinaculum (deep fascia) constitute a space called the ***carpal tunnel.*** Through it pass the long flexor tendons of the digits and thumb and the median nerve. As you will see in Chapter 17, a condition called carpal tunnel syndrome is associated with the carpal tunnel.

The five bones of the ***metacarpus*** (*meta* = after) constitute the palm of the hand. Each metacarpal bone consists of a proximal ***base,*** a ***shaft,*** and a distal ***head.*** The metacarpal bones are numbered I to V, starting with the one in

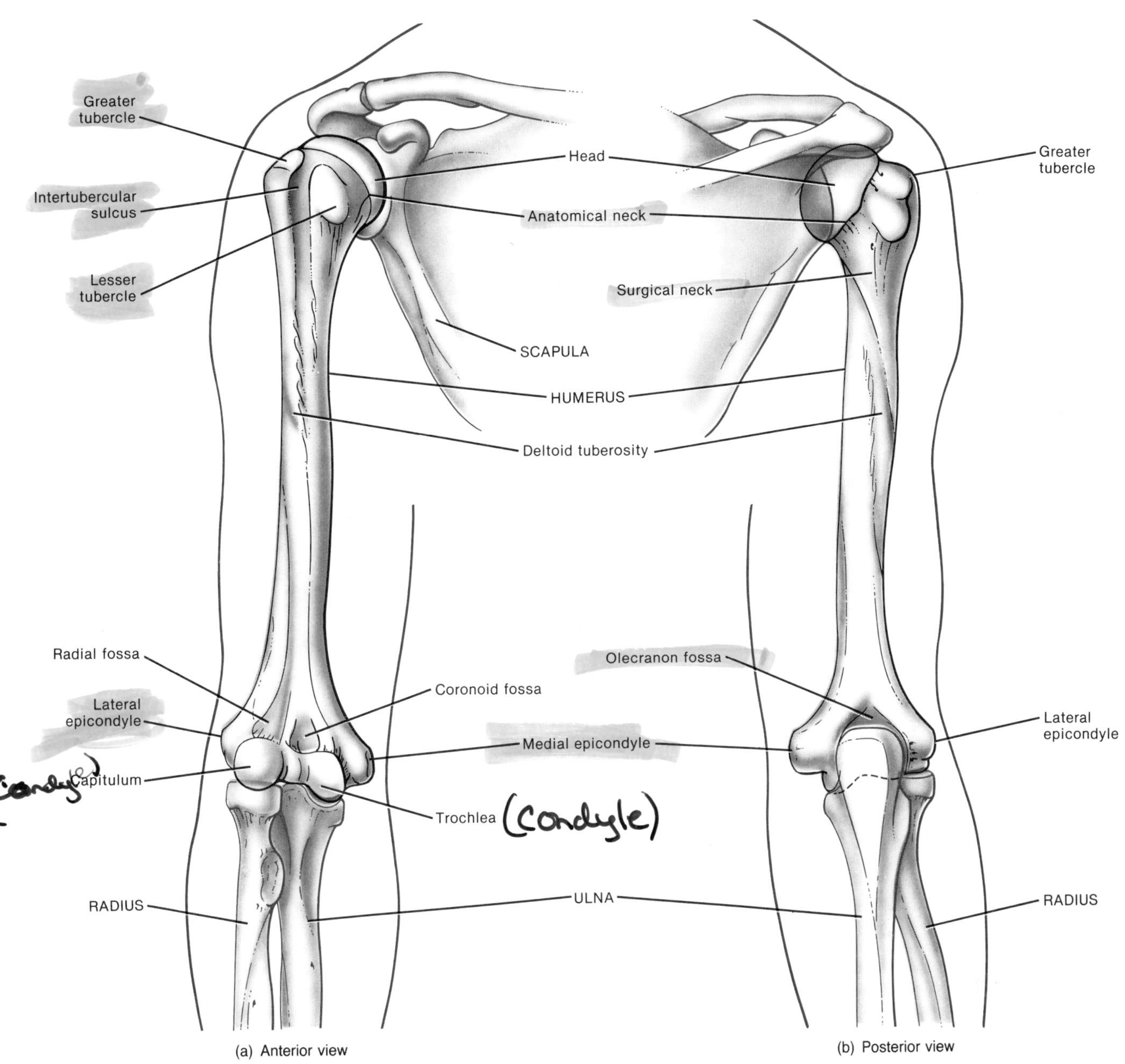

FIGURE 7-4 Right humerus in relation to scapula, ulna, and radius.

the thumb. The bases articulate with the distal row of carpal bones. The heads articulate with the proximal phalanges of the fingers. The heads of the metacarpals are commonly called the "knuckles" and are readily visible when the fist is clenched.

The ***phalanges*** (fa-LAN-jēz), or bones of the fingers, number 14 in each hand. A single bone of the finger (or toe) is referred to as a ***phalanx*** (FĀ-lanks). Each phalanx consists of a proximal ***base,*** a ***shaft,*** and a distal ***head.*** There are two phalanges in the first digit, called the ***pollex (thumb),*** and three phalanges in each of the remaining four digits. These digits, moving medially from the thumb, are commonly referred to as the index finger, middle finger, ring finger, and little finger. The first row of phalanges, the ***proximal row,*** articulates with the metacarpal bones and second row of phalanges. The second row of phalanges, the ***middle row,*** articulates with the proximal row and the third row. The third row of phalanges, the ***distal row,*** articulates with the middle row. The thumb has no middle phalanx.

PELVIC (HIP) GIRDLE

The ***pelvic (hip) girdle*** consists of the two ***hipbones*** or ***coxal*** (KOK-sal) ***bones*** (Figure 7-7). The pelvic girdle provides a strong and stable support for the lower extremities

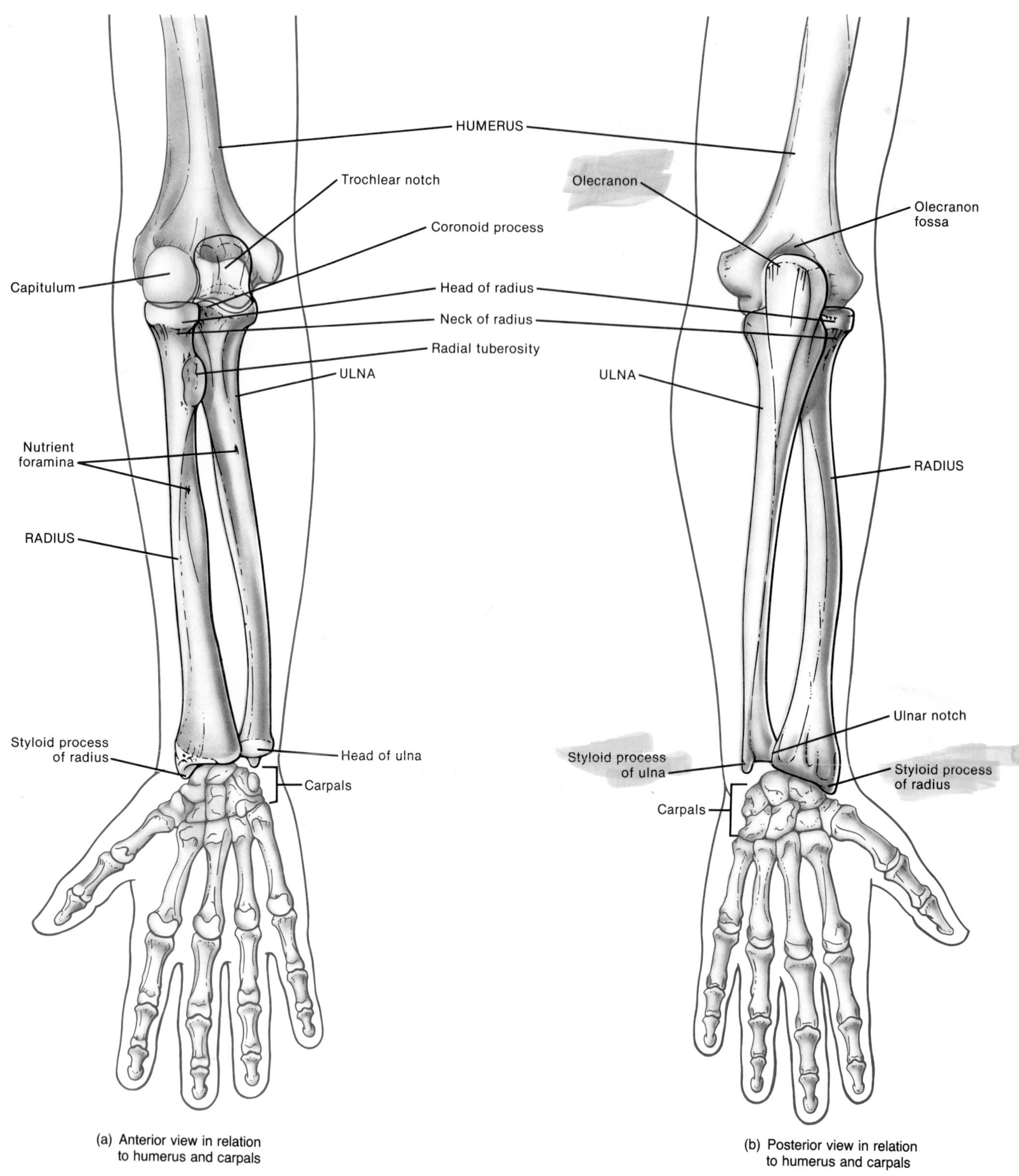

FIGURE 7-5 Right ulna and radius. (c) Courtesy of Lester Bergman and Associates.

on which the weight of the body is carried. The hipbones are united to each other anteriorly at the ***pubic symphysis*** (SIM-fi-sis). They unite posteriorly to the sacrum.

Each of the two ***hipbones*** of a newborn consists of three components: a superior ***ilium,*** an inferior and anterior ***pubis,*** and an inferior and posterior ***ischium*** (Figure 7-8). Eventually, the three separate bones fuse into one. The area of fusion is a deep, lateral fossa called the ***acetabulum*** (as′-e-TAB-yoo-lum), which serves as the socket for the head of the femur. Although the adult hipbones are both single bones, it is common to discuss the bones as if they still consisted of three portions.

The ilium is the largest of the three subdivisions of the hipbone. Its superior border, the ***iliac crest,*** ends anteriorly

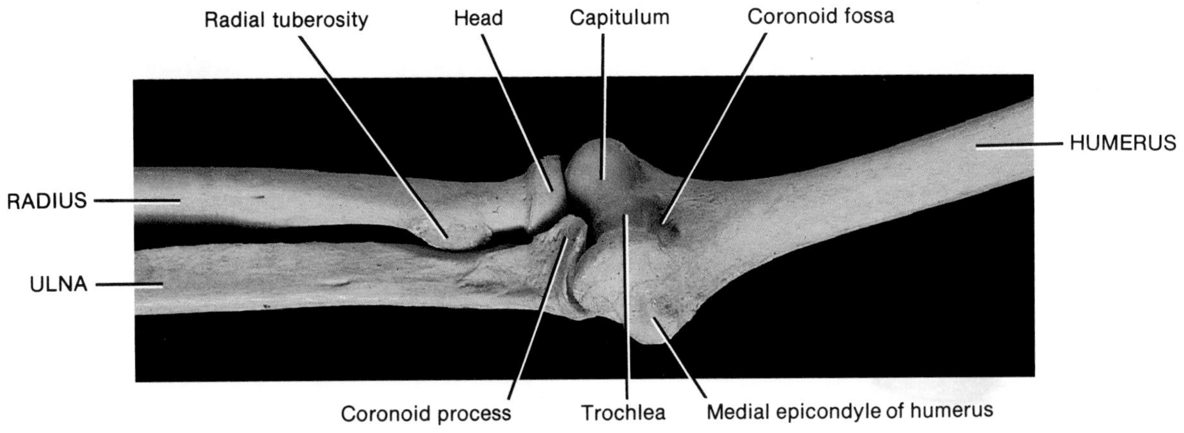

(c) Medial view in relation to humerus

FIGURE 7-5 Continued.

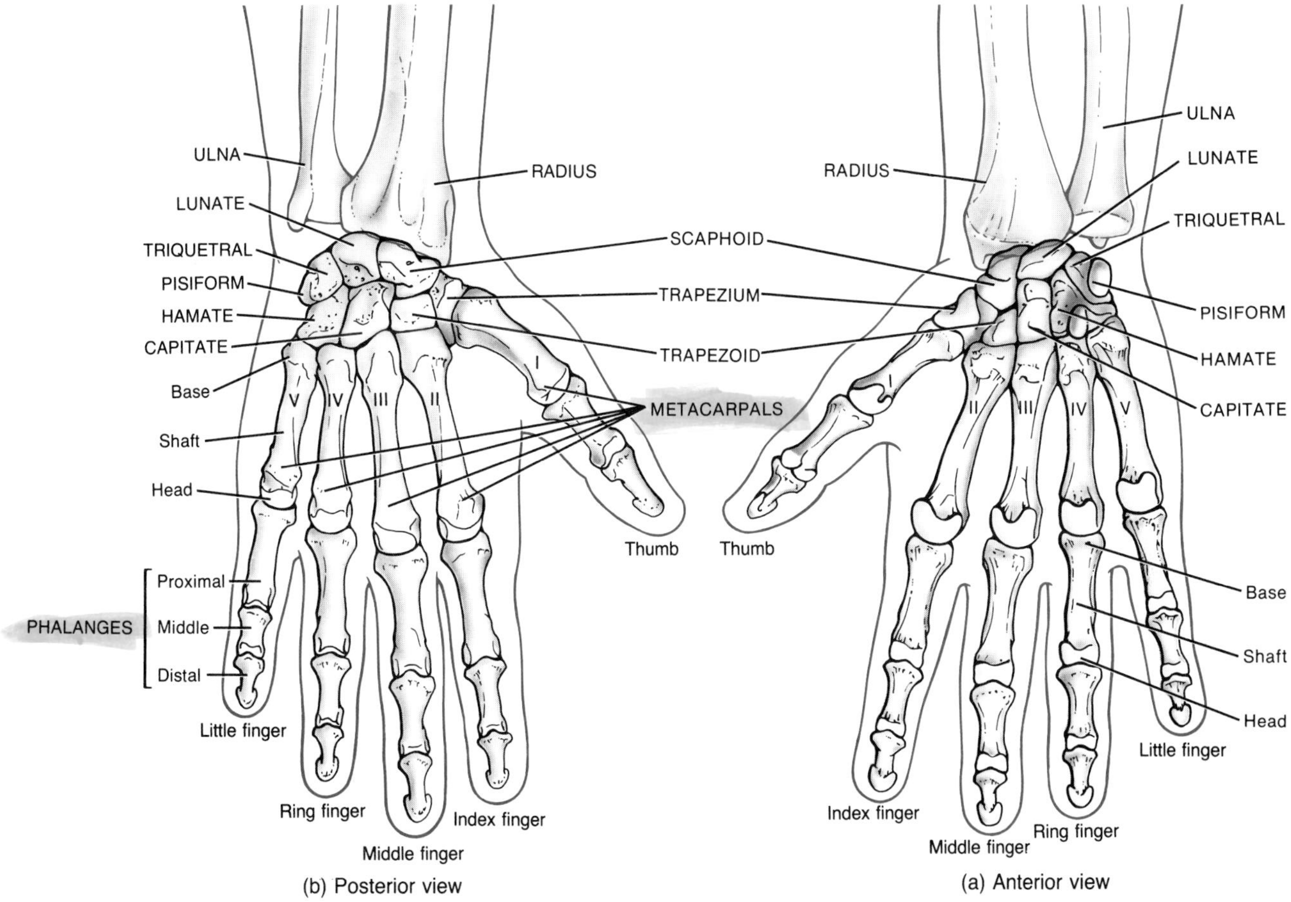

(b) Posterior view

(a) Anterior view

FIGURE 7-6 Right wrist and hand in relation to ulna and radius.

in the ***anterior superior iliac spine.*** Posteriorly, the iliac crest ends in the ***posterior superior iliac spine.*** The ***posterior inferior iliac spine*** is just inferior to it. The spines serve as points of attachment for muscles of the abdominal wall. Inferior to the posterior inferior iliac spine is the ***greater sciatic*** (sī-AT-ik) ***notch.*** The internal surface of the ilium seen from the medial side is the ***iliac fossa.*** It is a concavity where the iliacus muscle attaches. Posterior to this fossa are the ***iliac tuberosity,*** a point of attachment for the sacroiliac ligament, and the ***auricular surface,*** which articulates with the sacrum to form the ***sacroiliac joint.*** The other conspicuous markings of the ilium are three arched lines

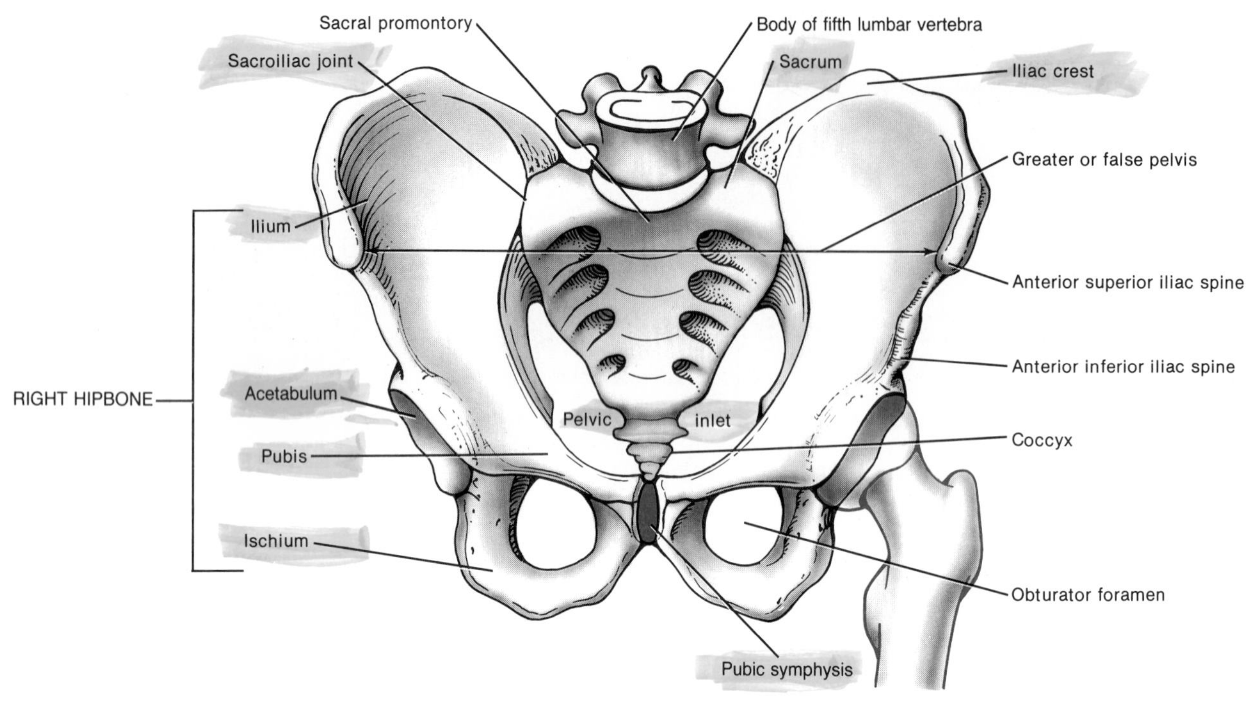

FIGURE 7-7 Pelvic (hip) girdle of a male in relation to sacrum and femur.

on its gluteal (buttock) surface called the ***posterior gluteal line,*** the ***anterior gluteal line,*** and the ***inferior gluteal line.*** The gluteal muscles attach to the ilium between these lines.

The ischium is the inferior, posterior portion of the hipbone. It contains a prominent ***ischial spine,*** a ***lesser sciatic notch*** below the spine, and an ***ischial tuberosity.*** The rest of the ischium, the ***ramus,*** joins with the pubis, and together they surround the ***obturator*** (OB-too-rā′-ter) ***foramen.***

The pubis is the anterior and inferior part of the hipbone. It consists of a ***superior ramus,*** an ***inferior ramus,*** and a ***body*** that contributes to the formation of the pubic symphysis. The anterior border of the body is known as the ***pubic crest*** and at its lateral end is a projection, the ***pubic tubercle.***

As noted previously, the pubic symphysis is the anterior joint between the two hipbones (see Figure 7-7). It contains a pad of fibrocartilage. The acetabulum is the fossa formed by the ilium, ischium, and pubis. It is the socket for the head of the femur. Two-fifths of the acetabulum is formed by the ilium, two-fifths by the ischium, and one-fifth by the pubis. On the inferior portion of the acetabulum is a deep indentation, the ***acetabular notch.*** It forms a foramen through which nutrient vessels and nerves enter the joint and serves as a point of attachment for the ligament of the head of the femur.

Together, the sacrum, coccyx, and two hipbones form a basinlike structure called the ***pelvis.*** The superior and inferior portions of the pelvis are separated from each other by an oblique plane that connects the sacral promontory (posterior), pubic symphysis (anterior), and several joints laterally. The circumference of this plane is called the ***pelvic brim*** (Figure 7-9a,b).

The portion of the pelvis above the pelvic brim is called the ***greater (false) pelvis.*** It is bordered by the lumbar vertebrae (posteriorly), the upper portions of the hipbones (laterally), and abdominal wall (anteriorly). The greater pelvis is actually part of the abdomen and does not contain any pelvic organs, except for the urinary bladder when full and the uterus in pregnancy. The greater pelvis is not an obstacle to childbirth.

The portion of the pelvis below the pelvic brim is called the ***lesser (true) pelvis.*** It is bounded by the sacrum and coccyx (posteriorly), inferior portions of the ilia (laterally), and pubis (anteriorly). The lesser pelvis surrounds the pelvic cavity (see Figure 1-6a). The superior opening of the lesser pelvis, actually the pelvic brim, is called the ***pelvic inlet;*** the inferior opening is called the ***pelvic outlet*** (Figure 7-9b). Whereas the pelvic inlet in females is wider and oval-shaped, in males it is more narrow and heart-shaped. The ***pelvic axis*** is an imaginary curved line passing through the lesser pelvis at right angles to the center of the planes of the pelvic inlet and outlet. It is the course taken by the baby's head as it descends through the pelvis.

CLINICAL APPLICATION

Pelvimetry

Pelvimetry is the measurement, usually manually, of the size of the pelvic inlet and outlet. This measurement is

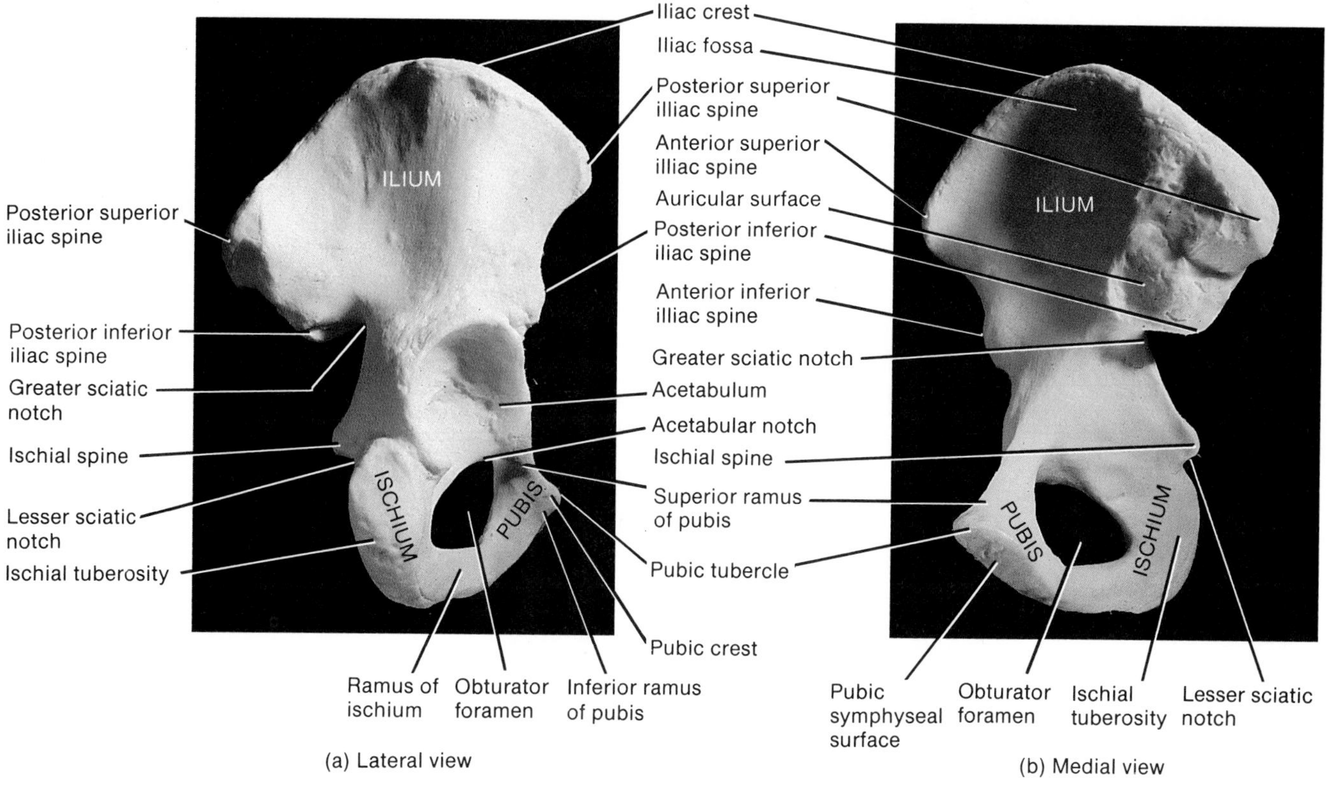

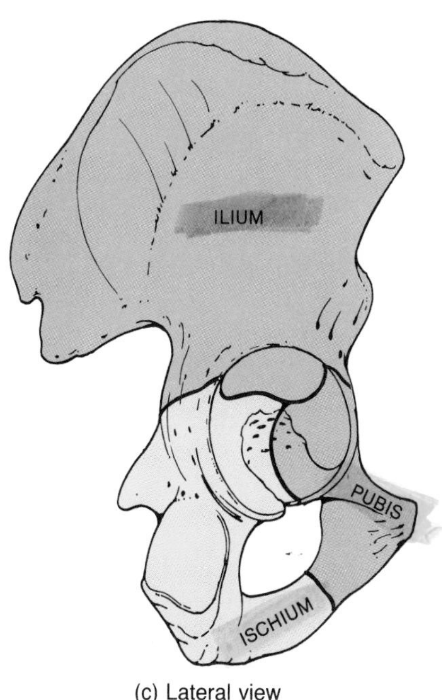

FIGURE 7-8 Right hipbone. (a) and (b) Courtesy of Lester Bergman and Associates. In the three divisions of the hipbone shown in (c), the lines of fusion of the ilium, ischium, and pubis are not always visible in an adult.

important because the fetus must pass through the narrower lesser (true) pelvis at birth.

LOWER EXTREMITY

The ***lower extremities*** are composed of 60 bones (Figure 7-10). Each extremity includes the femur in the thigh, patella (kneecap), fibula and tibia in the leg, tarsals (anklebones), metatarsals, and phalanges in the toes.

FEMUR

The ***femur,*** or thighbone, is the longest and heaviest bone in the body (Figure 7-11). Its proximal end articulates with the hipbone. Its distal end articulates with the tibia. The body (shaft) of the femur angles medially as it approaches the femur of the opposite thigh. As a result, the knee joints are brought nearer at the midline. The degree of convergence is greater in the female because the female pelvis is broader.

The proximal end of the femur consists of a rounded ***head*** that articulates with the acetabulum of the hipbone. The ***neck*** of the femur is a constricted region distal to the head. A fairly common fracture in the elderly occurs at the neck of the femur. Apparently, the neck becomes so weak that it fails to support the body. The ***greater trochanter*** (trō-KAN-ter) and ***lesser trochanter*** are projections that serve as points of attachment for some of the thigh and

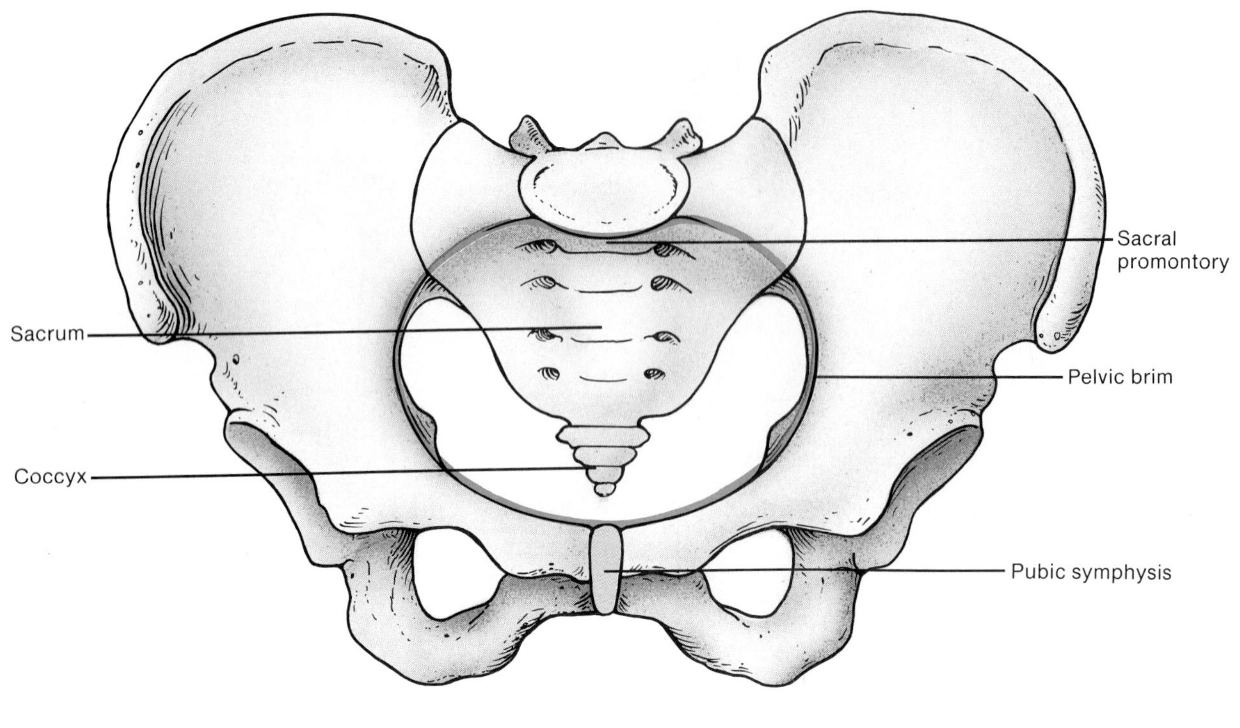

(a) Anterior view

Sacral promontory
Sacrum
Plane of pelvic inlet (pelvic brim)
Lesser (true) pelvis
Pelvic axis
Coccyx
Posterior
Anterior
Plane of pelvic outlet
Pubic symphysis

(b) Sagittal section

FIGURE 7-9 Female pelvis.

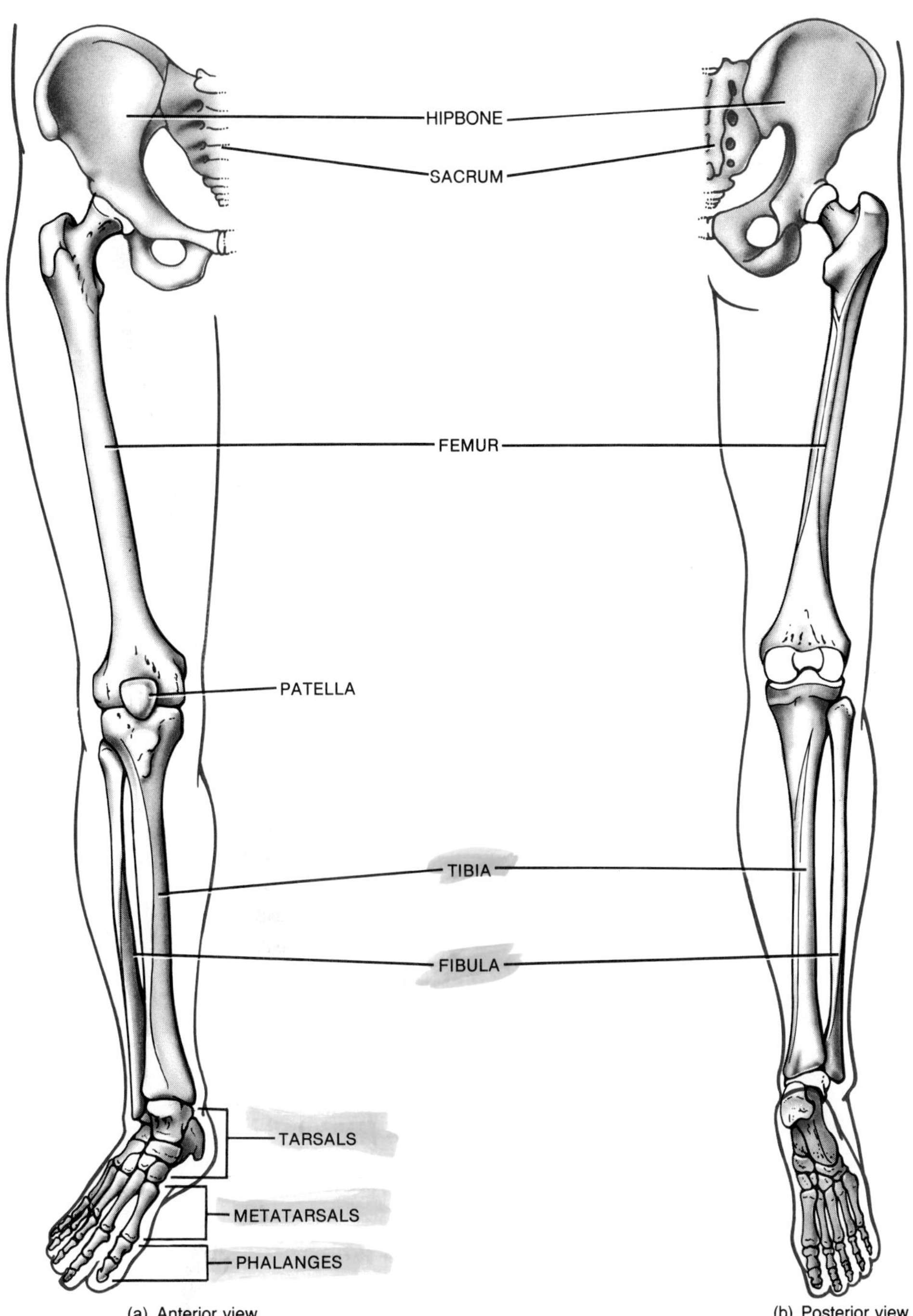

FIGURE 7-10 Right pelvic (hip) girdle and lower extremity.

buttock muscles. The greater trochanter is the prominence felt and seen anterior to the hollow on the side of the hip. The lesser trochanter is inferior and medial to the greater trochanter. Between the trochanters on the anterior surface is a narrow ***intertrochanteric line.*** Between the trochanters on the posterior surface is an ***intertrochanteric crest.***

The diaphysis (shaft) of the femur contains a rough vertical ridge on its posterior surface called the ***linea aspera.*** This ridge serves for the attachment of several thigh muscles.

The distal end of the femur is expanded and includes the ***medial condyle*** and ***lateral condyle.*** These articulate with the tibia. Superior to the condyles are the ***medial epicondyle*** and ***lateral epicondyle.*** A depressed area between the condyles on the posterior surface is called the ***intercondylar*** (in′-ter-KON-di-lar) ***fossa.*** The ***patellar surface*** is located between the condyles on the anterior surface.

Pathologic changes in the angle of the neck of the femur result in abnormal posture of the lower limbs. A decreased

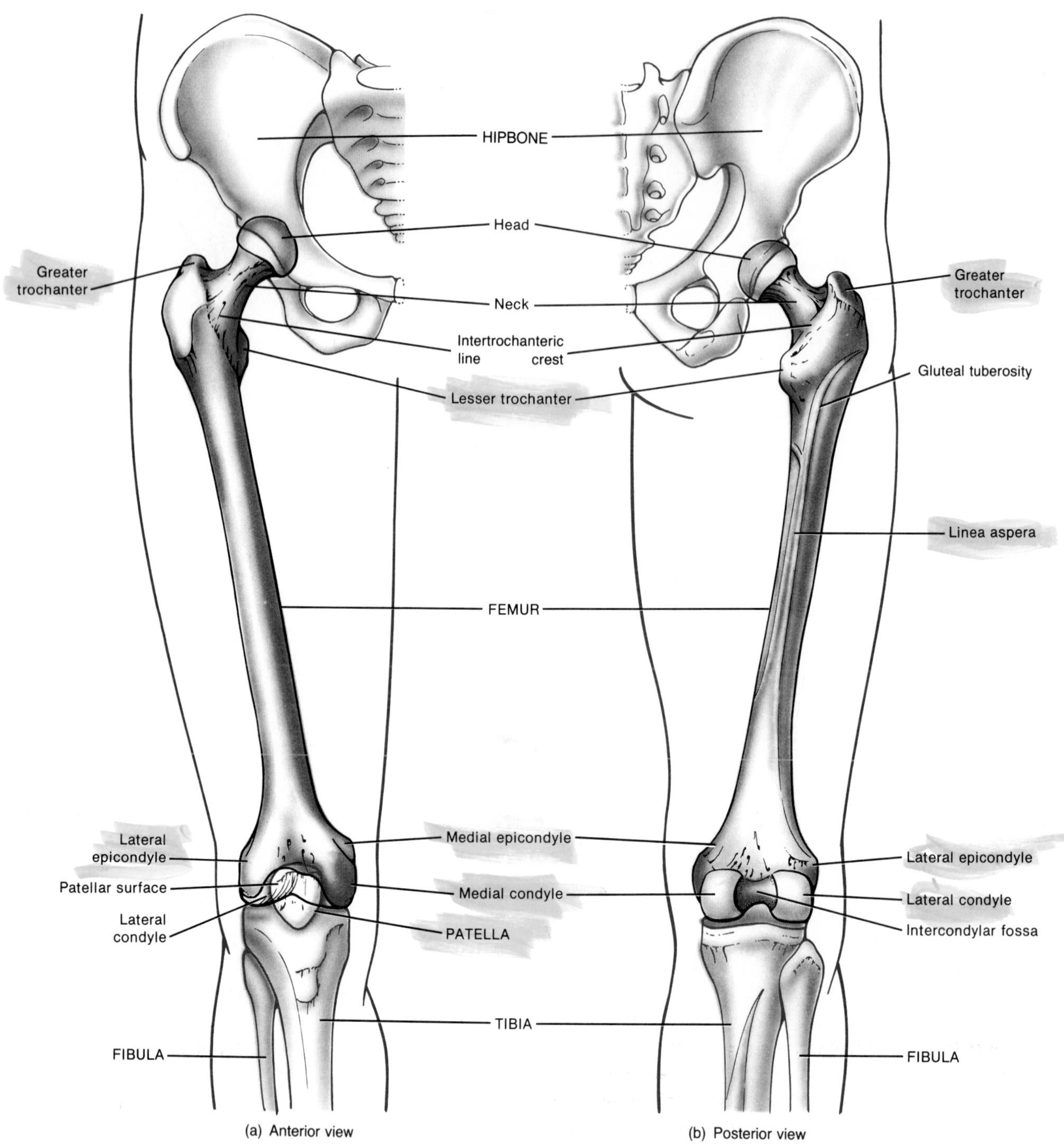

FIGURE 7-11 Right femur in relation to the hipbone, patella, tibia, and fibula.

angle produces ''knock-knee'' condition. An abnormally large angle produces ''bowleg'' condition. Either condition places an abnormal strain on the knee joints.

PATELLA

The ***patella,*** or kneecap, is a small, triangular bone anterior to the knee joint (Figure 7-12). It is a sesamoid bone that develops in the tendon of the quadriceps femoris muscle. The broad superior end of the patella is called the ***base.*** The pointed inferior end is the ***apex.*** The posterior surface contains two ***articular facets,*** one for the medial condyle and the other for the lateral condyle of the femur.

TIBIA AND FIBULA

The ***tibia,*** or shinbone, which is the larger medial bone of the leg (Figure 7-13), bears the major portion of the

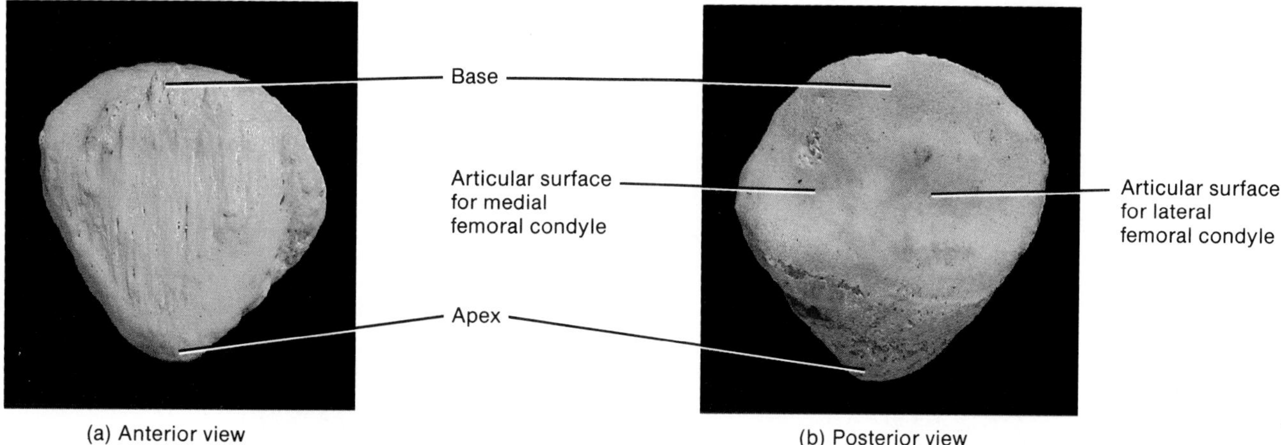

FIGURE 7-12 Right patella. Courtesy of Lester Bergman and Associates.

weight of the leg. The tibia articulates at its proximal end with the femur and fibula and at its distal end with the fibula of the leg and talus bone of the ankle.

The proximal end of the tibia is expanded into a ***lateral condyle*** and a ***medial condyle,*** which articulate with the condyles of the femur. The inferior surface of the lateral condyle articulates with the head of the fibula. The slightly concave condyles are separated by an upward projection called the ***intercondylar eminence.*** The ***tibial tuberosity*** on the anterior surface is a point of attachment for the patellar ligament.

The medial surface of the distal end of the tibia forms the ***medial malleolus*** (mal-LĒ-ō-lus). This structure articulates with the talus bone of the ankle and forms the prominence that can be felt on the medial surface of your ankle. The ***fibular notch*** articulates with the fibula.

CLINICAL APPLICATION

Shinsplints

Shinsplints (tibia stress syndrome) refers to soreness or pain along the tibia, probably caused by inflammation of the periosteum (periostitis) brought on by the repeated tugging of the muscles and tendons attached to the periosteum or by compression of the muscle mass by tissues enclosing the muscles. The condition may occur as a result of walking or running up and down hills or by vigorous activity of the legs following a period of relative inactivity. Rest usually alleviates the pain. Patients who do not respond to rest may be given local injections of cortisonelike steroid drugs or may have to undergo minor surgery to release pressure in the soft tissues around the bone.

The ***fibula*** is parallel and lateral to the tibia. It is considerably smaller than the tibia and bears less weight. The ***head*** of the fibula, the proximal end, articulates with the inferior surface of the lateral condyle of the tibia below the level of the knee joint. The distal end has a projection called the ***lateral malleolus*** that articulates with the talus bone of the ankle, forming the prominence on the lateral surface of the ankle. The inferior portion of the fibula also articulates with the tibia at the fibular notch. A fracture of the lower end of the fibula with injury to the tibial articulation is called a ***Pott's fracture*** (see Figure 5-8d).

TARSALS, METATARSALS, AND PHALANGES

The skeleton of the foot is composed of three regions: (1) proximal tarsus, (2) intermediate metatarsus, and (3) distal phalanges. The ***tarsus*** is a collective designation for the seven bones of the ankle called ***tarsals*** (Figure 7-14). The term *tarsos* pertains to a broad, flat surface. The ***talus*** (TĀ-lus) and ***calcaneus*** (kal-KĀ-nē-us) are located on the posterior part of the foot. The anterior part contains the ***cuboid, navicular,*** and three ***cuneiform*** (*cuneiform* = wedge-shaped) ***bones*** called the ***first (medial), second (intermediate),*** and ***third (lateral) cuneiform.*** The talus, the uppermost tarsal bone, is the only bone of the foot that articulates with the fibula and tibia and has no muscle attachments. It is surrounded on one side by the medial malleolus of the tibia and on the other side by the lateral malleolus of the fibula. During walking, the talus initially bears the entire weight of the body. About half the weight is then transmitted to the calcaneus. The remainder is transmitted to the other tarsal bones. The calcaneus, or heel bone, is the largest and strongest tarsal bone.

The ***metatarsus*** consists of five metatarsal bones numbered I to V from the medial to lateral position. Like the metacarpals of the palm of the hand, each metatarsal consists of a proximal ***base,*** a ***shaft,*** and a distal ***head.*** The metatarsals articulate proximally with the first, second, and third cuneiform bones and with the cuboid. Distally, they articulate with the proximal row of phalanges. The first metatarsal is thicker than the others because it bears more weight.

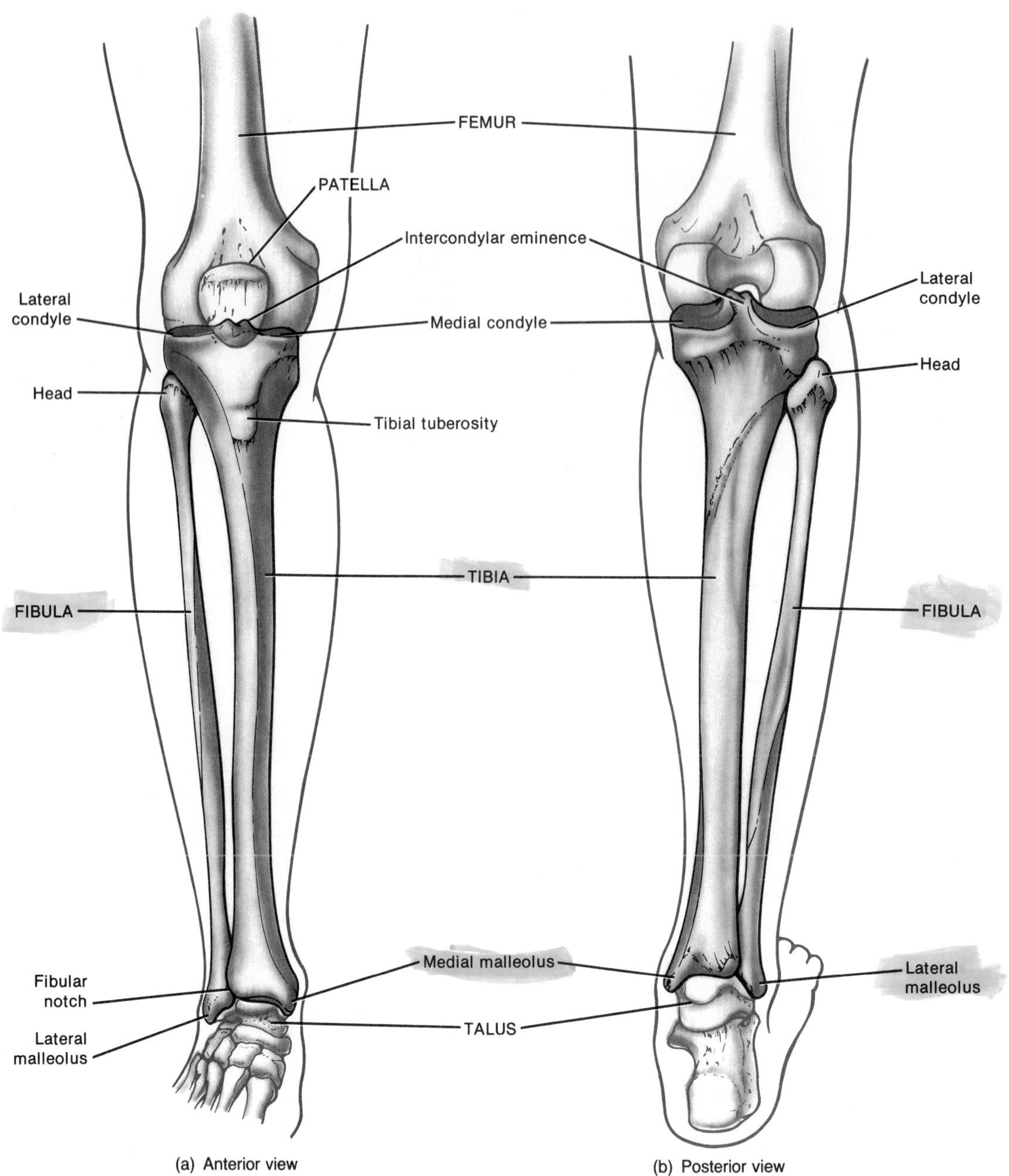

FIGURE 7-13 Right tibia and fibula in relation to femur, patella, and talus.

The ***phalanges*** of the foot resemble those of the hand both in number and arrangement. Each also consists of a proximal ***base,*** a middle ***shaft,*** and a distal ***head.*** The ***hallux*** (great or big toe) has two large, heavy phalanges called proximal and distal phalanges. The other four toes each have three phalanges—proximal, middle, and distal.

ARCHES OF THE FOOT

The bones of the foot are arranged in two ***arches*** (Figure 7-15). These arches enable the foot to support the weight of the body, provide an ideal distribution of body weight over the hard and soft tissues of the foot, and provide leverage while walking. The arches are not rigid. They yield as weight is applied and spring back when the weight is lifted, thus helping to absorb shocks. In most cases, arches are fully developed by age 12 or 13.

The ***longitudinal arch*** has two parts. Both consist of tarsal and metatarsal bones arranged to form an arch from the anterior to the posterior part of the foot. The ***medial*** (inner) ***part*** of the longitudinal arch originates at the calcaneus. It rises to the talus and descends through the navicular, the three cuneiforms, and the heads of the three medial metatarsals. The talus is the keystone of this arch. The

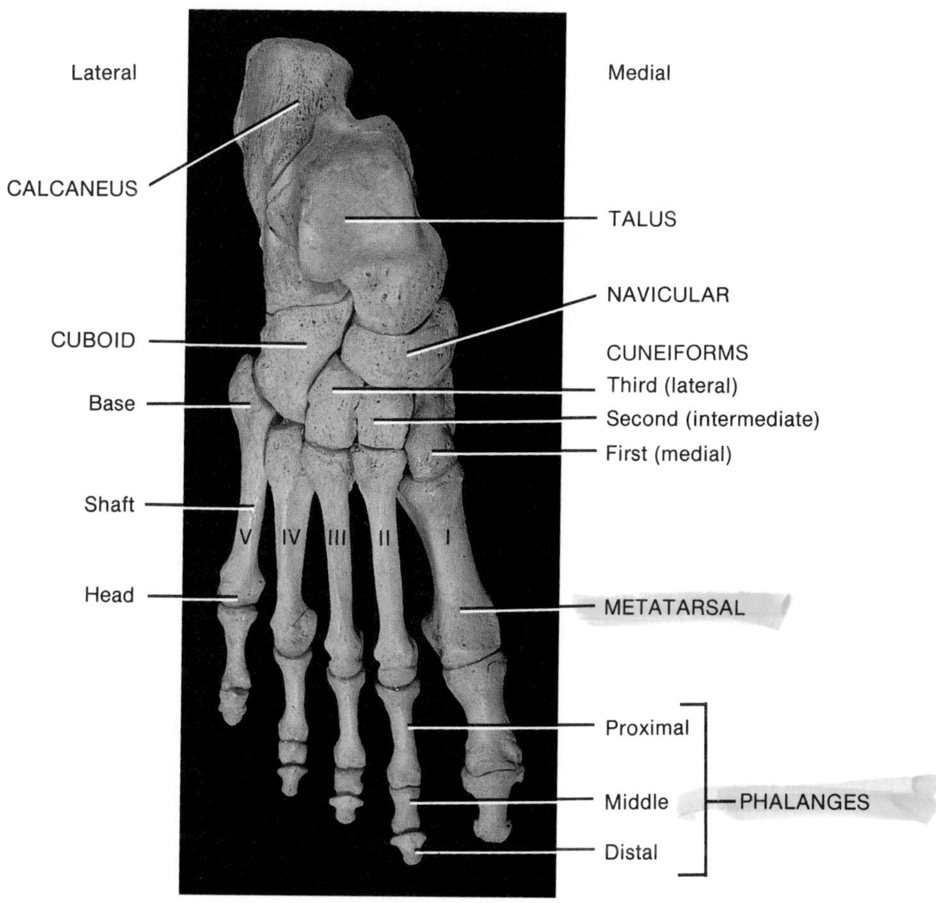

FIGURE 7-14 Right foot. Courtesy of Lester Bergman and Associates.

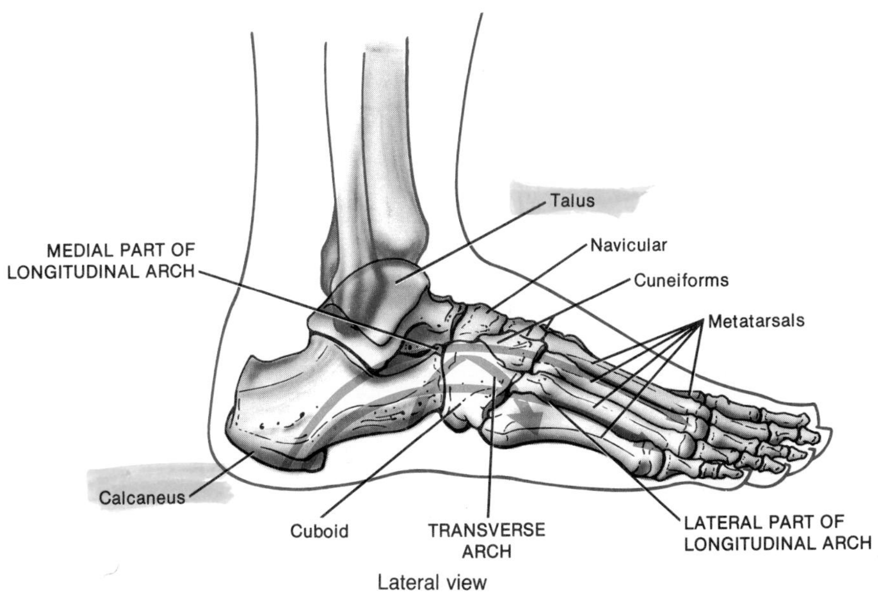

FIGURE 7-15 Arches of the right foot.

lateral (outer) ***part*** of the longitudinal arch also begins at the calcaneus. It rises at the cuboid and descends to the heads of the two lateral metatarsals. The cuboid is the keystone of this arch.

The ***transverse arch*** is formed by the navicular, three cuneiforms, and the bases of the five metatarsals.

CLINICAL APPLICATION

Flatfoot, Clawfoot, and Bunions

The bones composing the arches are held in position by ligaments and tendons. If these ligaments and tendons are weakened, the height of the medial longitudinal arch may decrease or "fall." The result is ***flatfoot.*** Among the causes are excessive weight, postural abnormalities, weakened supporting tissues, and severe pounding on an unyielding surface, which sometimes occurs among athletes. Flatfoot may lead to plantar fasciitis (inflammation in the ligaments in the bottoms of the foot), shinsplints, stress fractures, Achilles tendonitis, bunions, and calluses. Flatfoot is frequently treated by use of a custom-designed arch support called an orthotic.

Clawfoot is a condition in which the medial longitudinal arch is abnormally elevated. It is frequently caused by muscle imbalance, such as that resulting from poliomyelitis.

A ***bunion*** (***hallux valgus;*** *valgus* = bent outward) is a deformity of the great toe. Although the condition may be inherited, it is typically caused by wearing tightly fitting shoes. It is characterized by lateral deviation of the proximal phalanx of the great toe and medial displacement of metatarsal I. Arthritis of the first metatarsophalangeal joint may also be a predisposing factor. The condition produces inflammation of bursae (fluid-filled sacs at the joint), bone spurs, and calluses.

FEMALE AND MALE SKELETONS

The bones of the male are generally larger and heavier than those of the female. The articular ends are thicker in relation to the shafts. In addition, since certain muscles of the male are larger than those of the female, the points of attachment—tuberosities, lines, ridges—are larger in the male skeleton.

Many significant structural differences between female and male skeletons are noted in the pelvis; most are related to pregnancy and childbirth. The typical differences are listed in Exhibit 7-1 and illustrated in Figure 7-16.

EXHIBIT 7-1

Comparison of Typical Female and Male Pelvis

POINT OF COMPARISON	FEMALE	MALE
General Structure	Light and thin	Heavy and thick
Joint Surfaces	Small	Large
Muscle Attachments	Rather indistinct	Well marked
Greater (False) Pelvis	Shallow	Deep
Pelvic Inlet (Pelvic Brim)	Larger and more oval	Heart-shaped
Pelvic Outlet	Comparatively large	Comparatively small
Pubic Arch	Greater than a 90° angle	Less than a 90° angle
Pubic Symphysis	Less deep	More deep
Ischial Tuberosity	Turned outward	Turned inward
Ilium	Less vertical	More vertical
Acetabulum	Small	Large
Obturator Foramen	Oval	Round
Greater Sciatic Notch	Wide	Narrow

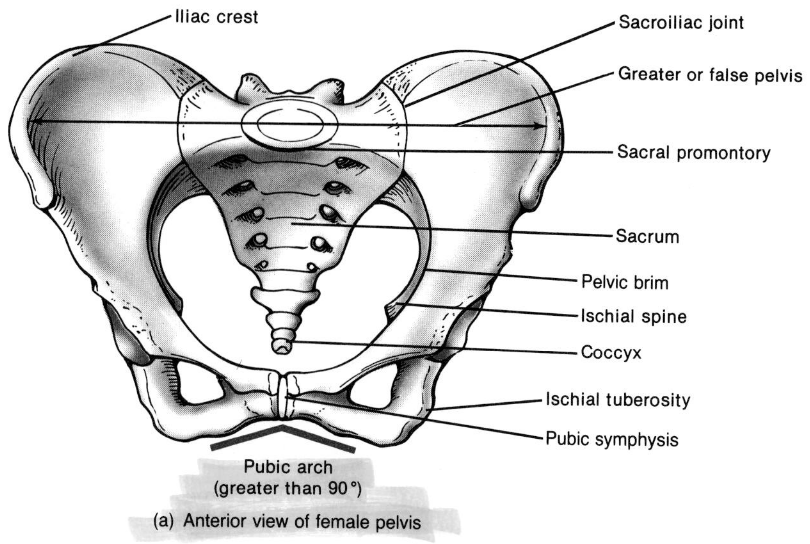

(a) Anterior view of female pelvis

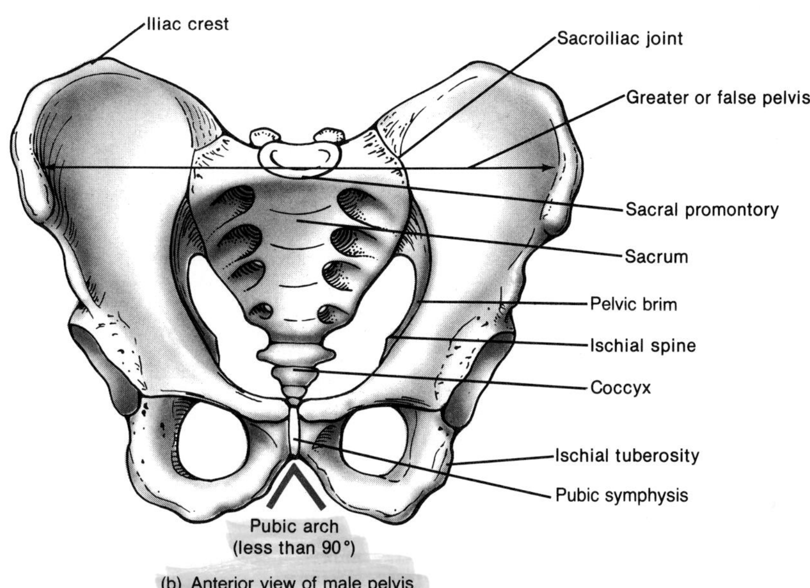

(b) Anterior view of male pelvis

FIGURE 7-16 Comparison of female and male pelvis.

STUDY OUTLINE

Pectoral (Shoulder) Girdle (p. 162)

1. Each pectoral (shoulder) girdle consists of a clavicle and scapula.
2. Each attaches an upper extremity to the trunk.

Upper Extremity (p. 162)

1. The bones of each upper extremity include the humerus, ulna, radius, carpals, metacarpals, and phalanges.

Pelvic (Hip) Girdle (p. 167)

1. The pelvic (hip) girdle consists of two hipbones.
2. It attaches the lower extremities to the trunk at the sacrum.
3. Each hipbone consists of three fused components—ilium, pubis, and ischium.

Lower Extremity (p. 171)

1. The bones of each lower extremity include the femur, patella, tibia, fibula, tarsals, metatarsals, and phalanges.
2. The bones of the foot are arranged in two arches—the longitudinal arch and the transverse arch—to provide support and leverage.

Female and Male Skeletons (p. 178)

1. The female pelvis is adapted for pregnancy and childbirth. Some differences in pelvic structure are listed in Exhibit 7-1.
2. Male bones are generally larger and heavier than female bones and have more prominent markings for muscle attachment.

REVIEW QUESTIONS

1. What is the pectoral (shoulder) girdle? Why is it important? (p. 162)
2. What are the bones of the upper extremity? (p. 162)
3. What is the pelvic (hip) girdle? Why is it important? (p. 167)
4. What are the bones of the lower extremity?
5. What is pelvimetry? What is the clinical importance of pelvimetry? (p. 170)
6. In what ways do the upper extremity and lower extremity differ structurally?
7. Describe the structure of the longitudinal and transverse arches of the foot. What is the function of an arch? (p. 176)
8. What are the principal structural differences between typical female and male skeletons? Use Exhibit 7-1 as a guide in formulating your response. (p. 178)
9. Define the following: fractured clavicle (p. 162); pelvimetry (p. 170); shinsplints (p. 175); flatfoot, clawfoot, and bunion (p. 178).

SELF QUIZ

Choose the one best answer to these questions.

___ **1.** Which structures are on the posterior surface of the upper extremity (in the anatomical position)?
A. radial fossa and radial notch; B. trochlea and capitulum; C. coronoid process and coronoid fossa; D. olecranon process and olecranon fossa; E. lesser tubercle and intertubercular sulcus.

___ **2.** Choose the false statement:
A. the capitulum articulates with the head of the radius; B. the medial and lateral epicondyles are located at the distal ends of the tibia and fibula; C. the coronoid fossa articulates with the ulna when the ulna is flexed; D. the trochlea articulates with the trochlear notch of the ulna; E. the ulna is the medial bone of the forearm.

___ **3.** The bones of the pectoral (shoulder) girdle
A. articulate with the sternum anteriorly and the vertebrae posteriorly; B. include both clavicles and scapulae and the upper part of the sternum; C. are considered to be part of the axial skeleton; D. both articulate with the head of the humerus, forming the shoulder joint; E. are described by none of the above.

___ **4.** Which of the following pairs of terms is correctly matched or related?
A. clavicle–breastbone; B. scapula–shoulder blade; C. sternum–wrist; D. coronoid bone–collarbone; E. pelvis–backbone.

___ **5.** The anatomical name for the socket that the humerus fits into is the
A. acetabulum; B. subscapular fossa; C. glenoid cavity; D. sella turcica; E. iliac fossa.

___ **6.** Which two structures (anatomical features) are directly anterior/posterior to each other?
A. the two tubercles of the humerus; B. the coronoid and olecranon fossae of the humerus; C. the two epicondyles of the humerus; D. the anatomical and surgical necks of the humerus; E. the trochlea and capitulum of the humerus.

___ **7.** The radius
A. has a tuberosity on its distal end; B. has an acromial process at its proximal end; C. is longer and larger than the ulna; D. has a styloid process on its distal end; E. and the ulna are called carpal bones.

___ **8.** The portion of the pelvis above the pelvic brim is the
A. lesser pelvis; B. pelvic inlet; C. pelvic axis; D. greater pelvis; E. true pelvis.

___ **9.** The greater trochanter is a large bony prominence located
A. on the humerus; B. on the proximal part of the femur; C. near the tuberosity of the tibia; D. at the base of the skull; E. near the medial (vertebral) border of the scapula.

___ **10.** Which of the following is a part of the femur?
A. obturator foramen; B. trochlea; C. tibial tuberosity; D. medial malleolus; E. none of the above.

___ **11.** Which of the following statements regarding the male pelvis is *not* true?
A. the bones are heavier and rougher than in the female; B. the male pelvis is narrow and deep; C. the male pubic arch is wider than in the female; D. the cavity of the true pelvis is small in the male; E. none of the above.

Arrange the answers in correct sequence.

___ ___ ___ **12.** According to size of the bones, from largest to smallest:
A. femur
B. ulna
C. humerus

___ ___ ___ **13.** From proximal to distal:
A. phalanges
B. metacarpals
C. carpals

Complete the following:

14. Wrist bones are called ________. There are (5? 7? 8? 14?) of them in each wrist.

15. The bones that constitute the palm of the hand are called ________ bones. There are ________ in each hand. The one on the thumb side is numbered (I? V?).

Joints

STUDENT OBJECTIVES

1. Define a joint and identify the factors that determine the degree of movement at a joint.
2. Contrast the structural and functional classification of joints.
3. Discuss and compare the movements possible at various synovial joints.
4. Describe selected joints of the body with respect to the bones that enter into their formation, structural classification, and anatomical components.
5. Describe the causes and symptoms of common joint disorders, including rheumatism, rheumatoid arthritis (RA), osteoarthritis, gouty arthritis, Lyme disease, bursitis, dislocation, and sprain.
6. Define key medical terms associated with joints.

CHAPTER OUTLINE

- **Classification of Joints**
  - Structural Classification
  - Functional Classification
- **Fibrous Joints**
  - Sutures
  - Syndesmoses
  - Gomphoses
- **Cartilaginous Joints**
  - Synchondroses
  - Symphyses
- **Synovial Joints**
  - Structure
  - Contact and Movement at Synovial Joints
  - Types of Movements at Synovial Joints
    - *Gliding*
    - *Angular*
    - *Rotation*
    - *Circumduction*
    - *Special*
  - Types of Synovial Joints
    - *Gliding Joint*
    - *Hinge Joint*
    - *Pivot Joint*
    - *Ellipsoidal Joint*
    - *Saddle Joint*
    - *Ball-and-Socket Joint*
  - Summary of Joints
- **Selected Joints of the Body**
- **Applications to Health**
- **Key Medical Terms Associated with Joints**

Bones are too rigid to bend without damage. Fortunately, flexible connective tissues form joints that hold bones together while still permitting movement, in most cases. All body movements occur at joints. You can appreciate the importance of joints if you recall how a cast over the knee joint makes walking difficult or how a splint on a finger limits the ability to manipulate small objects.

A ***joint (articulation)*** is a point of contact between bones, between cartilage and bones, or between teeth and bones. The scientific study of joints is called ***arthrology*** (ar-THROL-ō-jē; *arthro* = joint; *logos* = study of). A joint's structure determines how it functions. Some joints permit no movement, others permit slight movement, and still others afford considerable movement.

Generally, the closer the bones fit at the point of contact, the stronger the joint. Tightly fitted joints, however, restrict movement. The looser the fit, the greater the movement. Unfortunately, loosely fitted joints are prone to dislocation (displacement). Several other factors also affect joint movement: the precise manner in which the articulating bones fit together; the flexibility (tension or tautness) of the connective tissue ligaments and joint capsules that bind the bones together; and the position of ligaments, muscles, and tendons (which will be discussed in detail later).

CLASSIFICATION OF JOINTS

Based on anatomic characteristics, joints are categorized into structural classes, and based on the type of movement they permit, joints are categorized into functional classes.

STRUCTURAL CLASSIFICATION

The structural classification of joints is based on the presence or absence of a joint cavity (a space between the articulating bones) and the type of connective tissue that binds the bones together. Structurally, a joint is classified as: (1) ***fibrous,*** if there is no joint cavity and the bones are held together by fibrous connective tissue; (2) ***cartilaginous,*** if there is no joint cavity and the bones are held together by cartilage; or (3) ***synovial,*** if there is a joint cavity and the bones forming the joint are united by a surrounding articular capsule and frequently by accessory ligaments (to be described in detail later).

FUNCTIONAL CLASSIFICATION

The functional classification of joints takes into account the degree of movement they permit. Functionally, a joint is classified as: (1) a ***synarthrosis*** (sin′-ar-THRŌ-sis), which is an immovable joint; (2) an ***amphiarthrosis*** (am′-fē-ar-THRŌ-sis), which is a slightly movable joint; or (3) a ***diarthrosis*** (dī-ar-THRŌ-sis), which is a freely movable joint. The terms *synarthroses, amphiarthroses,* and *diarthroses* are plural.

We will discuss the joints of the body based upon their structural classification, referring to their functional classification as well.

FIBROUS JOINTS

Fibrous joints lack a synovial cavity, and the articulating bones are held very closely together by fibrous connective tissue. They permit little or no movement. The three types of fibrous joints are (1) sutures, (2) syndesmoses, and (3) gomphoses. The terms *suture, syndesmosis,* and *gomphosis* are singular.

SUTURES

Sutures (SOO-cherz) are found between bones of the skull. In a suture, the bones are united by a thin layer of dense fibrous connective tissue. An example is the coronal suture between the frontal and parietal bones (see Figure 6-2b). The irregular (inter-digitated) structure of sutures gives them added strength and decreases their chance of fractures. Since sutures are immovable, they are functionally classified as synarthroses. Some sutures, though present during childhood, are replaced by bone in the adult. They are called ***synostoses*** (sin′-os-TŌ-sēz), or bony joints—joints in which there is a complete fusion of bone across the suture line. An example is the frontal suture between the left and right sides of the frontal bone that begins to fuse during infancy (see Figure 6-3a). Synostoses are also functionally classified as synarthroses.

SYNDESMOSES

Syndesmoses (sin′-dez-MŌ-sēz) are fibrous joints in which there is considerably more fibrous connective tissue than in sutures. The fit between the bones is not quite as tight. The fibrous connective tissue forms an interosseous membrane or ligament. Syndesmoses are slightly movable because the bones are separated more than in a suture and some flexibility is permitted by the interosseous membrane or ligament. Syndesmoses are functionally classified as amphiarthroses and typically permit slight movement. An example of a syndesmosis is the distal articulation of the tibia and fibula (see Figure 7-13).

GOMPHOSES

Gomphoses (gom-FŌ-sēz) are fibrous joints in which a cone-shaped peg fits into a socket. The substance between the bones is the periodontal ligament. Gomphoses are functionally classified as synarthroses. Examples are the articulations of the roots of the teeth with the alveoli (sockets) of the maxillae and mandible (see Figure 23-7).

CARTILAGINOUS JOINTS

Another joint that has no synovial cavity is a ***cartilaginous joint.*** Here the articulating bones are tightly connected by cartilage. Like fibrous joints, they allow little or no movement. The two types of cartilaginous joints are (1) synchondroses and (2) symphyses. The terms *synchondrosis* and *symphysis* are singular.

SYNCHONDROSES

Synchondroses (sin′-kon-DRŌ-sēz) are cartilaginous joints in which the connecting material is hyaline cartilage. The most common type of synchondrosis is the epiphyseal plate (see Figure 5-5). Such a joint connects the epiphysis and diaphysis of a growing bone and is immovable. Thus, these joints are synarthroses. Since the hyaline cartilage is eventually replaced by bone when growth ceases, the joint is temporary. It is replaced by a synostosis. Another example of a synchondrosis is the joint between the first rib and the sternum. The cartilage in this joint undergoes ossification during adult life.

SYMPHYSES

Symphyses (SIM-fi-sēz) are cartilaginous joints in which the connecting material is a broad, flat disc of fibrocartilage. These joints are found between bodies of vertebrae (see Figure 8-7b). A portion of the intervertebral disc is cartilaginous material. The pubic symphysis between the anterior surfaces of the hipbones is another example (see Figure 7-7). These joints are slightly movable, or amphiarthroses.

SYNOVIAL JOINTS

STRUCTURE

Joints in which there is a space between articulating bones are called ***synovial*** (si-NŌ-vē-al) ***joints.*** The space is called a ***synovial (joint) cavity*** (Figure 8-1). Because of this cavity and because of the arrangement of the articular capsule and accessory ligaments, synovial joints are freely movable. Thus, synovial joints are functionally classified as diarthroses.

Synovial joints are also characterized by the presence of ***articular cartilage.*** Articular cartilage, which is hyaline cartilage, covers the surfaces of the articulating bones but does not bind the bones together.

Synovial joints are surrounded by a sleevelike ***articular capsule*** that encloses the synovial cavity and unites the articulating bones. The articular capsule is composed of two layers. The outer layer, the ***fibrous capsule,*** consists of dense, irregular connective tissue. It is attached to the periosteum of the articulating bones at a variable distance from the edge of the articular cartilage. The flexibility of the fibrous capsule permits movement at a joint, whereas its great tensile strength resists dislocation. The fibers of some fibrous capsules are arranged in parallel bundles and are therefore highly adapted to resist recurrent strain. Such fibers are called ***ligaments*** and are given special names. The strength of the ligaments is one of the principal factors in holding bone to bone.

The inner layer of the articular capsule is formed by a ***synovial membrane.*** The synovial membrane is composed of areolar connective tissue with elastic fibers and a variable amount of adipose tissue. It secretes ***synovial fluid,*** which lubricates the joint and provides nourishment for the articular cartilage. Synovial fluid also contains phagocytic cells that remove microbes and debris resulting from wear and tear in the joint. Synovial fluid consists of hyaluronic acid and an interstitial fluid formed from blood plasma and is similar in appearance and consistency to uncooked egg white. When there is no joint movement, the fluid is quite viscous, but as movement increases, the fluid becomes less viscous. The amount of synovial fluid varies in different joints of the body, ranging from a thin, viscous layer to about 3.5 ml (about ⅛ oz) of free fluid in a large joint such as the knee. The amount present in each joint is sufficient only to form a thin film over the surfaces within an articular capsule where it reduces friction and supplies nutrients to and removes metabolic wastes from the cartilage cells of the articular cartilage. (Recall that cartilage is avascular.)

An interesting feature of some synovial joints is the ***cracking sound*** produced when the joint is pulled suddenly. If you pull on a synovial joint and expand its volume, the pressure of the joint fluid decreases (a partial vacuum arises) because the fluid suddenly occupies a larger volume. As the fluid bounces against the articular cartilage in the partial vacuum, the cracking sound is produced. Water sloshed up and down in a sealed, partially filled glass tube makes a similar clinking sound if the air has been pumped out of the tube. Shaking an unopened bottle of catsup (which is vacuum-packed) produces similar sounds.

Many synovial joints also contain ***accessory ligaments*** (*ligare* = to bind), which are called extracapsular ligaments and intracapsular ligaments. ***Extracapsular ligaments*** are outside of the articular capsule. An example is the fibular (lateral) collateral ligament of the knee joint (see Figure 8-12e). ***Intracapsular ligaments*** occur within the articular capsule but are excluded from the synovial cavity by folds of the synovial membrane. Examples are the cruciate ligaments of the knee joint (see Figure 8-12e).

CLINICAL APPLICATION

Artificial Ligaments

Artificial ligaments are used to support or replace severely torn ligaments, especially in the knee. One such device is a carbon fiber implant. The implant consists of carbon fibers coated with a plastic called polylactic acid. The coated fibers are sewn in and around torn ligaments

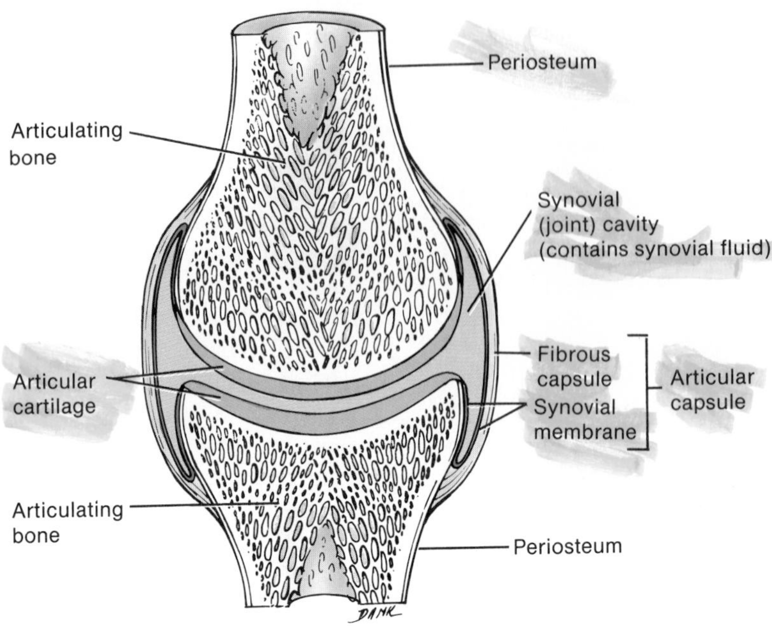

(a) Frontal section of a generalized synovial joint

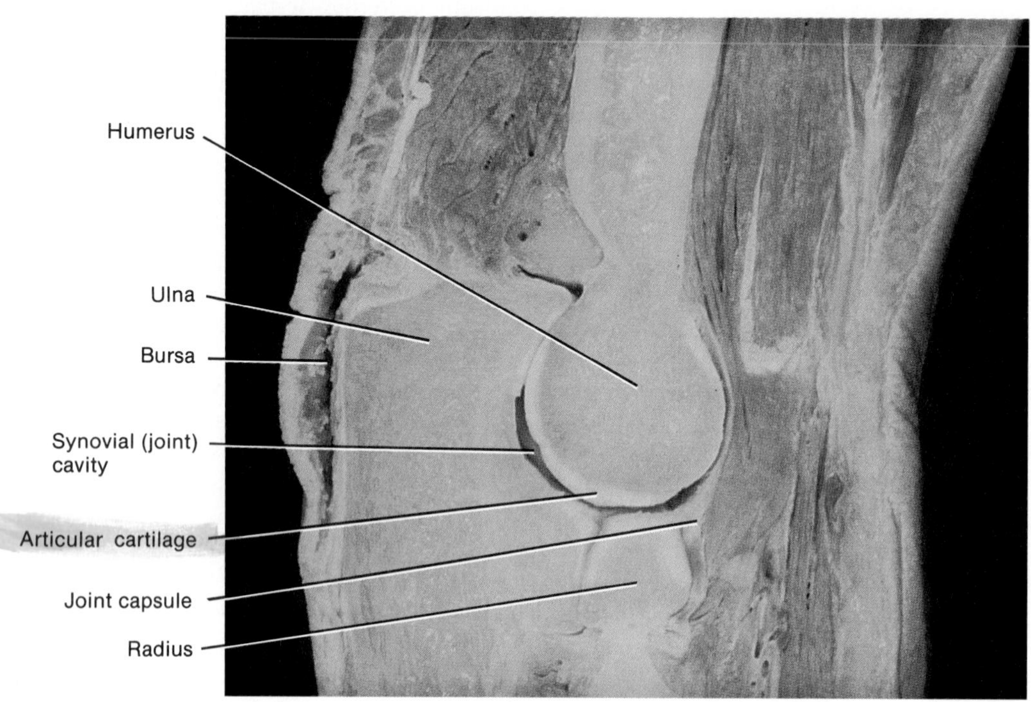

(b) Sagittal section of elbow joint

FIGURE 8-1 Synovial joint. (b) Courtesy of J. A. Gosling, P. F. Harris, et al., *Atlas of Human Anatomy,* Gower Medical Publishing Ltd., 2nd ed., 1991.

and tendons to reinforce them and to provide a scaffolding around which the body's own collagenous fibers grow. Within two weeks, the polylactic acid is absorbed by the body and the carbon fibers eventually fracture. By this time, the fibers are completely clad in collagen produced by fibroblasts.

Inside some synovial joints are pads of fibrocartilage that lie between the articular surfaces of the bones and are attached by their margins to the fibrous capsule. These pads are called ***articular discs (menisci).*** The discs usually subdivide the synovial cavity into two separate spaces (see Figure 8-12e). Articular discs allow two bones of different shapes to fit tightly; they modify the shape of the joint surfaces of the articulating bones. Articular discs also help to maintain the stability of the joint and direct the flow of synovial fluid to areas of greatest friction.

CLINICAL APPLICATION

Torn Cartilage and Arthroscopy

A tearing of articular discs in the knee, commonly called ***torn cartilage,*** occurs frequently among athletes. Such damaged cartilage requires surgical removal (meniscectomy) or it will begin to wear and cause arthritis. At one time, knee joint surgery for torn cartilage necessitated cutting through layers of healthy tissue and removing much, if not all, of the cartilage. This procedure is usually painful and expensive and does not always provide full recovery. These problems largely have been overcome by arthroscopy.

Arthroscopy (ar-THROS-kō-pē; *arthro* = joint; *skopein* = to view) refers to examination of the interior of a joint, usually the knee, using an arthroscope, a lighted instrument the diameter of a pencil. After a local or general anesthetic is given, the arthroscope is inserted into the knee joint through an incision as small as one-quarter inch. A second small incision is made to insert a tube through which a saline (salt) solution is injected into the joint. (Alternately, the saline solution may be introduced through the arthroscope.) If necessary, another small incision is used for insertion of an instrument that shaves off and reshapes the damaged cartilage and then suctions out the shaved cartilage along with the salt solution. Some orthopedic surgeons attach a lightweight television camera to the arthroscope so that the image from inside the knee can be projected onto a screen. Since arthroscopy requires only small incisions, recovery is more rapid than with conventional surgery, although a cast or splint may be worn for several days, depending on the extent of the procedure. Although arthroscopy is used mostly to remove torn cartilage, it also can be used to determine the nature and extent of damage following knee injury; to repair cruciate ligaments in the knee; to perform surgery on other joints and obtain tissue samples for analysis; to monitor the progression of disease and the effects of therapy; and to plan, if necessary, additional surgical procedures.

The various movements of the body create friction between moving parts. To reduce this friction, saclike structures called ***bursae*** are situated in the body tissues. These sacs resemble joint capsules in that their walls consist of connective tissue lined by a synovial membrane. They are also filled with a fluid similar to synovial fluid. Bursae are located between the skin and bone in places where skin rubs over bone. They are also found between tendons and bones, muscles and bones, and ligaments and bones (see Figure 8-12c). Such fluid-filled sacs cushion the movement of one part of the body over another. An inflammation of a bursa is called ***bursitis.***

CONTACT AND MOVEMENT AT SYNOVIAL JOINTS

Several factors contribute to keeping the articular surfaces of synovial joints in contact. How the surfaces contact one another, in turn, determines the type and extent of motion that is possible.

First is the ***structure or shape of the articulating bones,*** which determines how they fit together. An interlocking shape is very obvious at the hip joint, where the head of the femur articulates with the acetabulum of the hipbone. This type of fit allows rotational movement. Other shapes permit a diversity of motions, as will be described shortly.

A second factor is the ***strength and tension (tautness) of the joint ligaments.*** The different components of a fibrous capsule are tense only when the joint is in certain positions. Tense ligaments not only restrict the range of movement but also direct the movement of the articulating bones with respect to each other. In the knee joint, for example, the major ligaments are lax when the knee is bent but tense when the knee is straightened. Also, when the knee is straightened, the surfaces of the articulating bones are in fullest contact with each other.

A third factor that holds joints together but also restricts movement is the ***arrangement and tension of the muscles*** around the joint. Muscle tension reinforces the restraint placed on a joint by ligaments. A good example of the effect of muscle tension on a joint is seen at the hip joint. When the thigh is raised with the knee straight, the movement is restricted by the tension of the hamstring muscles on the posterior surface of the thigh. But if the knee is bent, the tension on the hamstring muscles is lessened, and the thigh can be raised further.

In a few joints, the ***apposition*** (placing together) ***of soft parts*** may limit mobility. For example, if you bend your arm at the elbow, it can bend no further once the muscles of the anterior surface of the forearm press against the muscles of the upper arm.

Joint flexibility may also be affected by ***hormones***. For example, relaxin, a hormone produced by the placenta and ovaries, relaxes the pubic symphysis and ligaments between the sacrum and hipbone and sacrum and coccyx toward the end of pregnancy. This allows expansion of the birth canal, which facilitates delivery.

TYPES OF MOVEMENTS AT SYNOVIAL JOINTS

Gliding

A ***gliding movement*** is the simplest kind that can occur at a joint. One surface moves back and forth and from side to side over another surface without angular or rotary motion. Some joints that glide are those between the carpals and between the tarsals. The heads and tubercles of ribs glide on the bodies and transverse processes of vertebrae. Also, the clavicle glides on the sternum and scapula.

Angular

Angular movements increase or decrease the angle between bones. Among the angular movements are flexion, extension, abduction, and adduction (Figure 8-2). ***Flexion*** involves a decrease in the angle between the surfaces of the articulating bones. Examples of flexion include bending the head forward (the joint is between the occipital bone and the atlas), bending the elbow, and bending the knee.

Extension involves an increase in the angle between the surfaces of the articulating bones. Extension restores a body part to its anatomical position after it has been

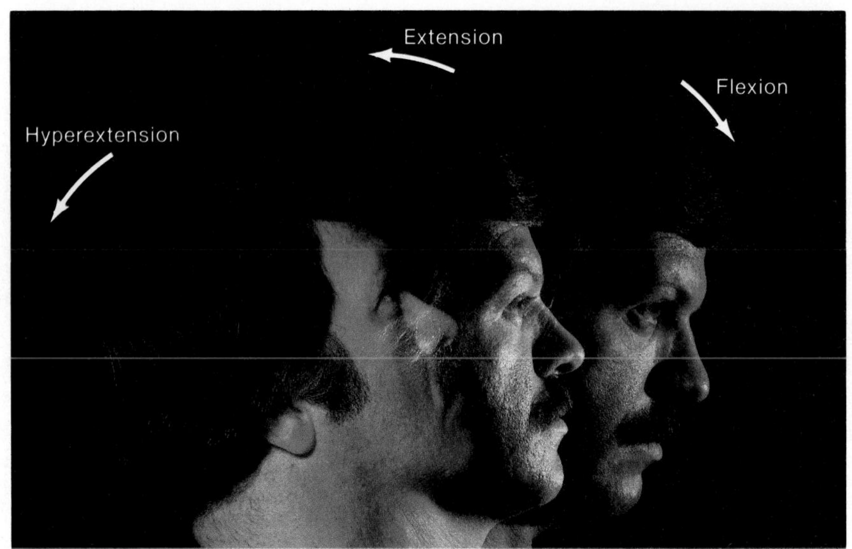

(a) Atlanto-occipital and cervical intervertebral joints

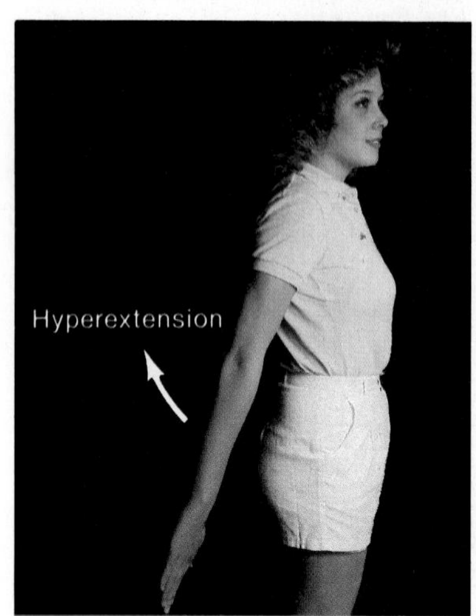

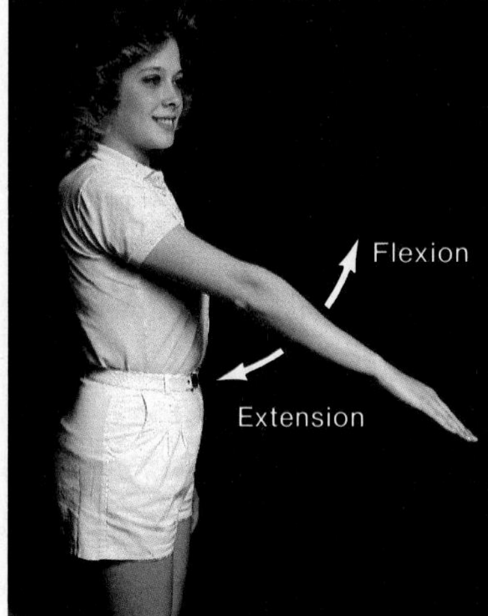

(b) Shoulder joint

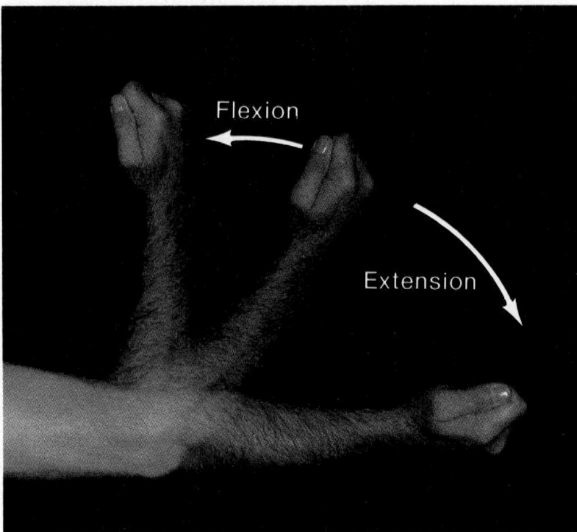

(c) Elbow joint

FIGURE 8-2 Angular movements at synovial joints. (Copyright © 1983 by Gerard J. Tortora. Courtesy of Lynne and James Borghesi.)

flexed. Examples of extension are returning the head to the anatomical position after flexion, straightening the arm after flexion, and straightening the leg after flexion. Continuation of extension beyond the anatomical position, as in bending the head backward, is called ***hyperextension.***

Abduction usually means movement of a bone *away from* the midline of the body. An example of abduction is moving the arm upward and away from the body until it is held straight out at right angles to the chest. With the fingers and toes, however, the midline of the body is not used as the line of reference. Abduction of the fingers (not the thumb) is a movement away from an imaginary line drawn through the middle finger; in other words, it is spreading the fingers. Abduction of the thumb moves the thumb away from the plane of the palm at a right angle to the palm. Abduction of the toes is relative to an imaginary line drawn through the second toe.

Adduction is usually movement of a part *toward* the midline of the body. An example of adduction is returning the arm to the side after abduction. As in abduction, adduction of the fingers (not the thumb) is relative to the middle finger, and adduction of the toes is relative to the second toe. In adduction of the thumb, the thumb moves toward the plane of the palm at a right angle to the palm.

Rotation

Rotation is the movement of a bone around its own longitudinal axis. During rotation, no other motion is permitted. In ***medial rotation,*** the anterior surface of a bone or extremity

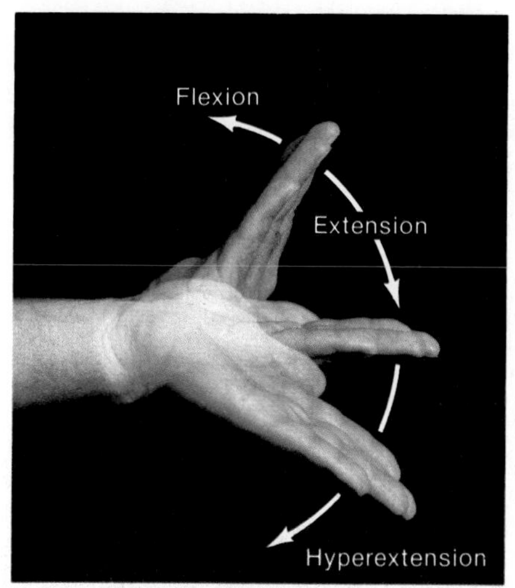

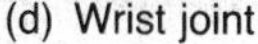
(d) Wrist joint

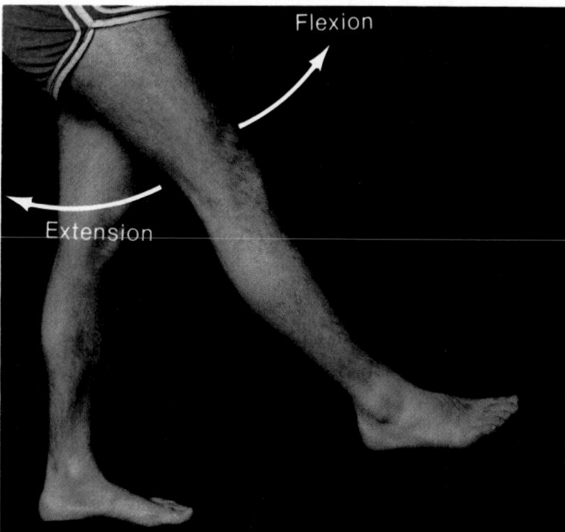

(e) Hip joint

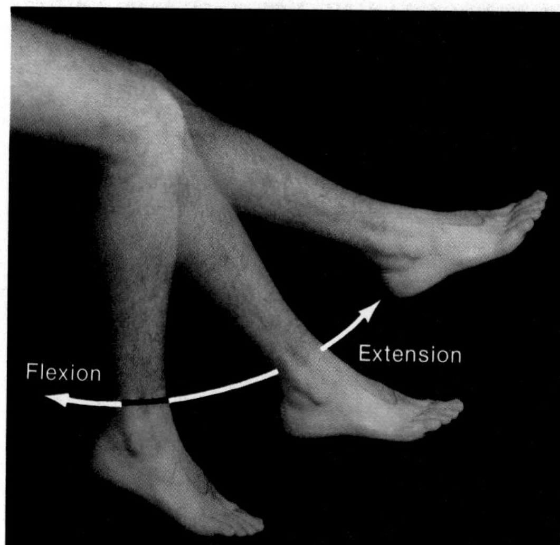

(f) Knee joint

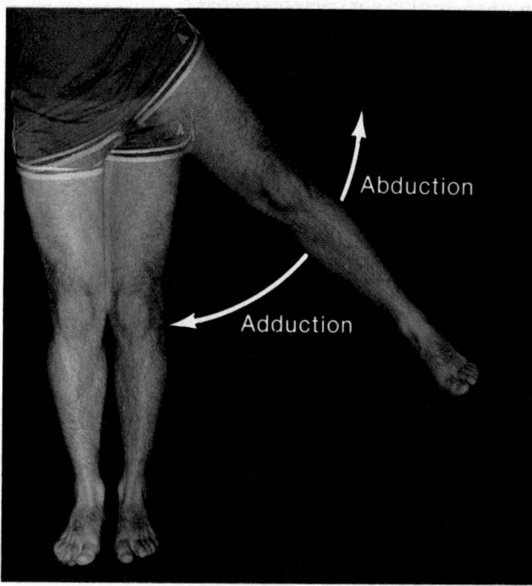

(g) Hip joint

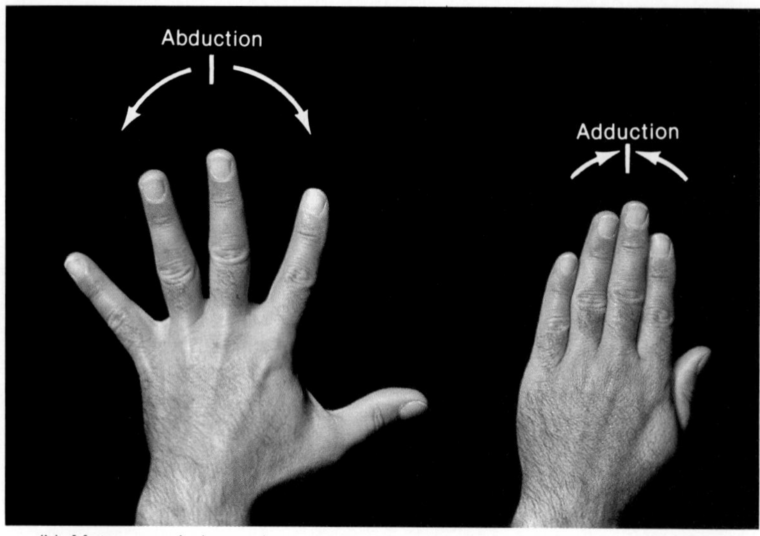

(h) Metacarpophalangeal joints

(i) Metacarpophalangeal joints

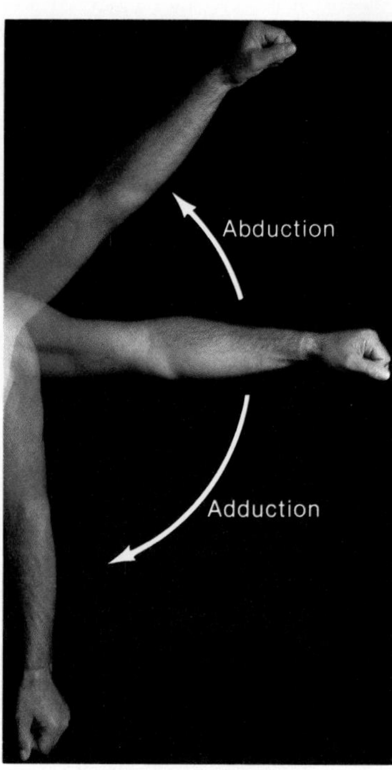

(j) Shoulder joint

moves toward the midline. In ***lateral rotation,*** the anterior surface moves away from the midline. We rotate the atlas around the dens of the axis when we shake the head from side to side indicating ''no'' (Figure 8-3). Rotation of the humerus turns the anterior surface of the humerus to face either medially or laterally.

Circumduction

Circumduction is a movement in which the distal end of a bone moves in a circle while the proximal end remains stable. The bone outlines a cone in the air. Circumduction typically involves flexion, abduction, adduction, extension, and rotation. It involves a 360° rotation. An example is moving the outstretched arm in a circle to wind up to pitch a ball (Figure 8-4).

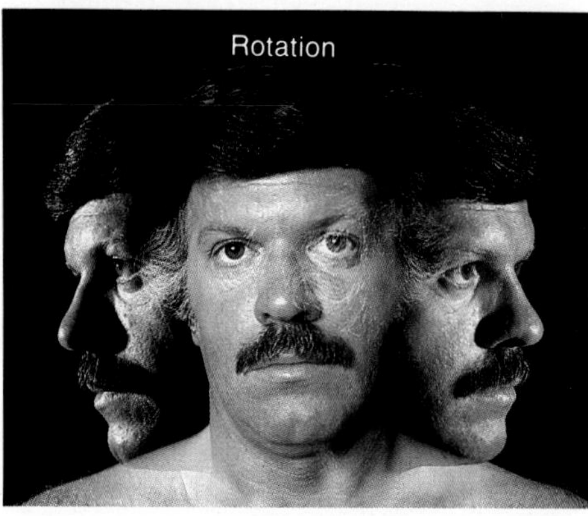

Atlantoaxial joint

FIGURE 8-3 Rotation. (Copyright © 1983 by Gerard J. Tortora.)

Special

Special movements are those found only at the joints indicated in Figure 8-5. ***Inversion*** is the movement of the sole of the foot inward (medially) so that the soles face toward each other. ***Eversion*** is the movement of the sole outward (laterally) so that the soles face away from each other. ***Dorsiflexion*** involves bending of the foot in the direction of the dorsum (upper surface). ***Plantar flexion*** involves bending the foot in the direction of the plantar surface (sole).

Protraction is the movement of the mandible or shoulder girdle forward on a plane parallel to the ground. Thrusting the jaw outward is protraction of the mandible. Bringing your arms forward until the elbows touch requires protraction of the clavicle or shoulder girdle. ***Retraction*** is the movement of a protracted part of the body backward on a plane parallel to the ground. Pulling the lower jaw back in line with the upper jaw is retraction of the mandible.

Supination is a movement of the forearm in which the palm of the hand is turned anterior or superior. To demonstrate supination, flex your forearm at the elbow to prevent rotation of the humerus in the shoulder joint. ***Pronation*** is a movement of the forearm in which the palm is turned posterior or inferior.

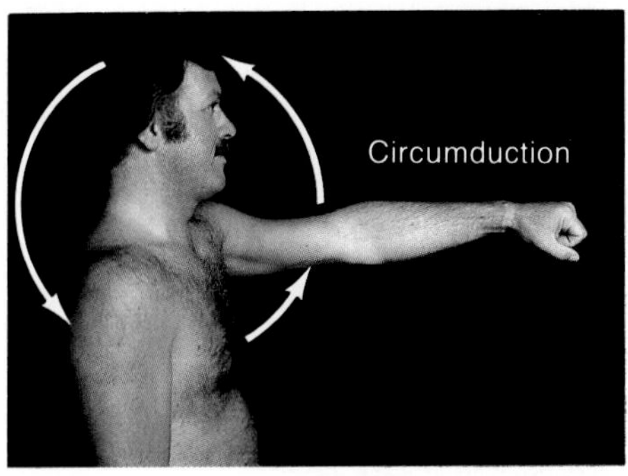

Shoulder joint

FIGURE 8-4 Circumduction. (Copyright © 1983 by Gerard J. Tortora.)

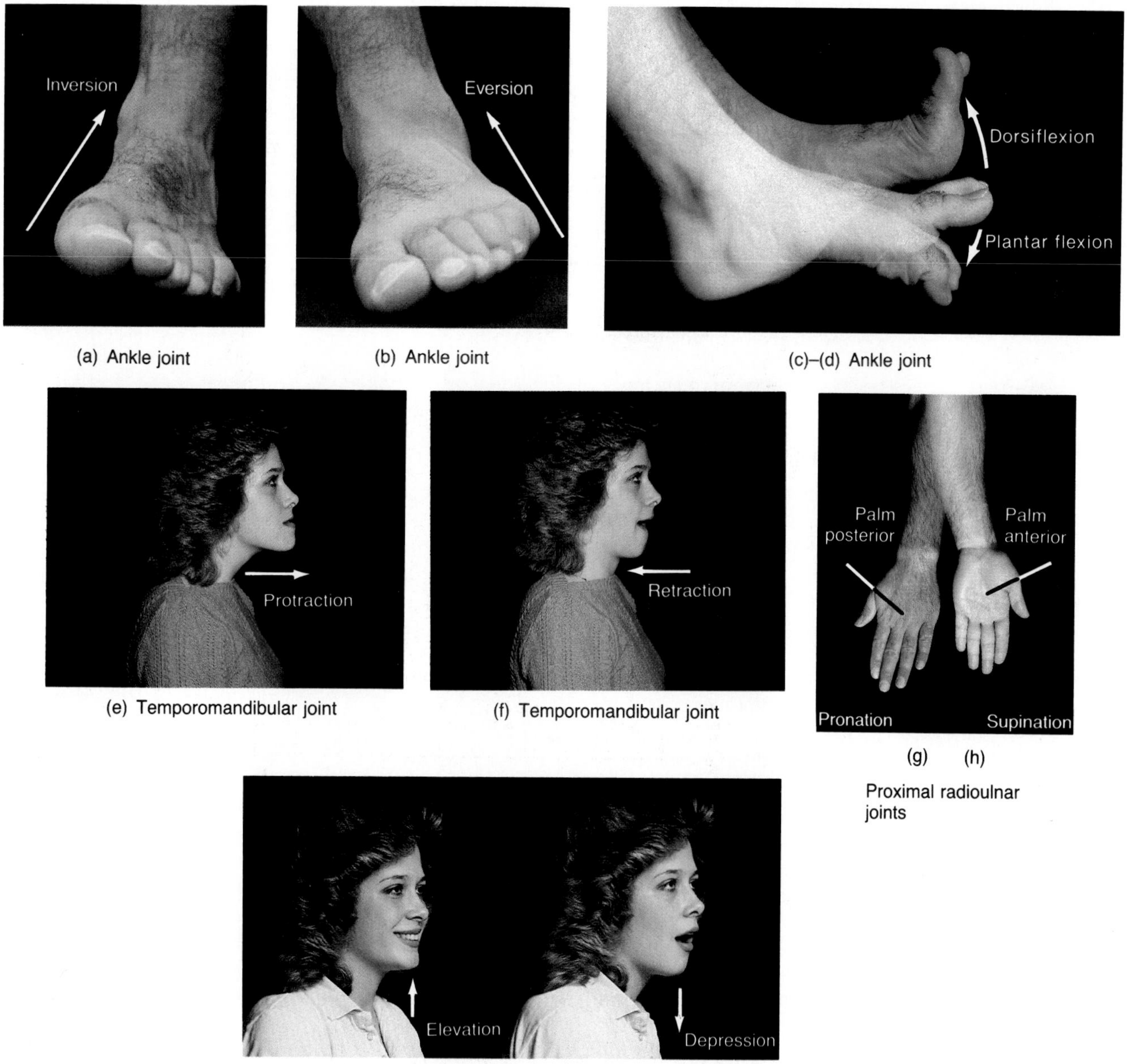

(a) Ankle joint

(b) Ankle joint

(c)–(d) Ankle joint

(e) Temporomandibular joint

(f) Temporomandibular joint

(g) (h)

Proximal radioulnar joints

(i) Temporomandibular joints (j)

FIGURE 8-5 Special movements. (Copyright © 1983 by Gerard J. Tortora. Courtesy of Lynne and James Borghesi.)

Elevation is an upward movement of a part of the body. You elevate your mandible when you close your mouth. ***Depression*** is a downward movement of a part of the body. You depress your mandible when you open your mouth. The shoulders can also be elevated and depressed.

A summary of movements that occur at synovial joints is presented in Exhibit 8-1.

TYPES OF SYNOVIAL JOINTS

Although all synovial joints have a generally similar structure, variations exist in the shape of the articulating surfaces. Accordingly, synovial joints are divided into six subtypes: gliding, hinge, pivot, ellipsoidal, saddle, and ball-and-socket joints.

Gliding Joint

The articulating surfaces of bones in ***gliding joints*** or ***arthrodia*** (ar-THRŌ-dē-a) are usually flat. Only side-to-side and back-and-forth movements are permitted (Figure 8-6a). Twisting and rotation are inhibited at gliding joints, generally because ligaments or adjacent bones restrict the range of movement. Since gliding joints do not move around an axis, they are referred to as ***nonaxial.*** Examples are the joints between carpal bones, tarsal bones, the sternum and clavicle, and the scapula and clavicle.

Hinge Joint

A ***hinge*** or ***ginglymus*** (JIN-gli-mus) ***joint*** is one in which the convex surface of one bone fits into the concave surface of another bone. Movement is primarily in a single plane, and the joint is therefore known as ***monaxial*** or ***uniaxial*** (Figure 8-6b). The motion is similar to that of a hinged door. Movement is flexion and extension. Examples of hinge joints are the elbow, ankle, and interphalangeal joints. The movement allowed by a hinge joint is illustrated by flexion and extension at the elbow (see Figure 8-2c).

EXHIBIT 8-1
Summary of Movements at Synovial Joints

MOVEMENT	DEFINITION
GLIDING	One surface moves back and forth and from side to side over another surface without angular or rotary motion.
ANGULAR	There is an increase or decrease in the angle between bones.
Flexion	Involves a decrease in the angle between the surfaces of articulating bones.
Extension	Involves an increase in the angle between the surfaces of articulating bones.
Hyperextension	Continuation of extension beyond the anatomical position.
Abduction	Movement of a bone away from the midline.
Adduction	Movement of a bone toward the midline.
ROTATION	Movement of a bone around its longitudinal axis; may be medial or lateral.
CIRCUMDUCTION	A movement in which the distal end of a bone moves in a circle while the proximal end remains stable.
SPECIAL	Occurs at specific joints.
Inversion	Movement of the sole of the foot inward so that the soles face toward each other.
Eversion	Movement of the sole of the foot outward so that the soles face away from each other.
Dorsiflexion	Bending the foot in the direction of the dorsum (upper surface).
Plantar Flexion	Bending the foot in the direction of the plantar surface (sole).
Protraction	Movement of the mandible or shoulder girdle forward on a plane parallel to the ground.
Retraction	Movement of a protracted part backward on a plane parallel to the ground.
Supination	Movement of the forearm in which the palm is turned anterior or superior.
Pronation	Movement of the forearm in which the palm is turned posterior or inferior.
Elevation	Movement of a part of the body upward.
Depression	Movement of a part of the body downward.

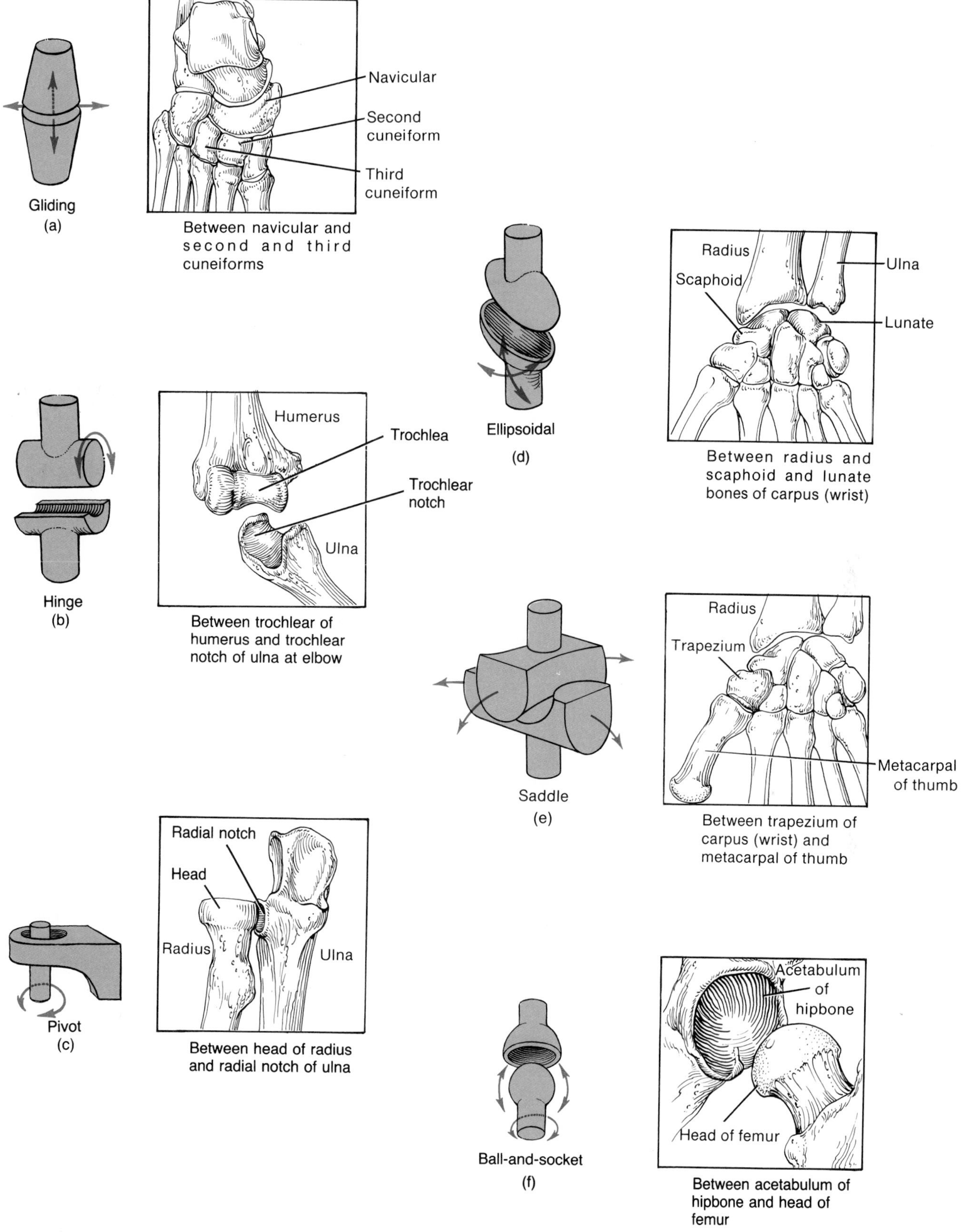

FIGURE 8-6 Subtypes of synovial joints. For each subtype shown, there is a simplified diagram on the left and the actual bones that form the joint on the right.

Pivot Joint

In a ***pivot*** or ***trochoid*** (TRŌ-koyd) ***joint,*** a rounded, pointed, or conical surface of one bone articulates within a ring formed partly by another bone and partly by a ligament. The primary movement permitted is rotation, and the joint is therefore ***monaxial*** (Figure 8-6c). Examples include the joint between the atlas and axis (atlantoaxial) and between the proximal ends of the radius and ulna. Movement at a pivot joint is illustrated by supination and pronation of the palms and rotation of the head from side to side to signify "no" (see Figure 8-3).

Ellipsoidal Joint

In an ***ellipsoidal*** or ***condyloid*** (KON-di-loyd) ***joint,*** an oval-shaped condyle of one bone fits into an elliptical cavity of another bone. Since the joint permits side-to-side and back-and-forth movements, it is ***biaxial*** (Figure 8-6d). The joint at the wrist between the radius and carpals is ellipsoidal. The movement permitted by such a joint is illustrated when you flex and extend (see Figure 8-2d) and abduct and adduct and circumduct the wrist.

Saddle Joint

In a ***saddle*** or ***sellaris*** (sel-A-ris) ***joint,*** the articular surface of one bone is saddle-shaped and the articular surface of the other bone is shaped like a rider sitting in the saddle. Essentially, the saddle joint is a modified ellipsoidal joint in which the movement is somewhat freer. Movements at a saddle joint are from side to side and back and forth. Thus, the joint is ***biaxial*** (Figure 8-6e). The joint between the trapezium of the carpus and metacarpal proximal to the thumb is an example of a saddle joint.

Ball-and-Socket Joint

A ***ball-and-socket*** or ***spheroid*** (SFĒ-royd) ***joint*** consists of a ball-like surface of one bone fitted into a cuplike depression of another bone. Such a joint permits ***triaxial*** movement, or movement in three planes of motion: flexion–extension, abduction–adduction, and rotation–circumduction (Figure 8-6f). Examples of ball-and-socket joints are the shoulder joint and hip joint. The range of movement at a ball-and-socket joint is illustrated by circumduction of the arm (Figure 8-4).

SUMMARY OF JOINTS

The summary of joints presented in Exhibit 8-2 is based on the anatomy of the joints. If we rearrange the types of joints into a classification based on movement, we arrive at the following:

Synarthroses: immovable joints
1. ***Sutures***
2. ***Synchondroses***
3. ***Gomphoses***

Amphiarthroses: slightly movable joints
1. ***Syndesmoses***
2. ***Symphyses***

Diarthroses: freely movable joints
1. ***Gliding***
2. ***Hinge***
3. ***Pivot***
4. ***Ellipsoidal***
5. ***Saddle***
6. ***Ball-and-socket***

SELECTED JOINTS OF THE BODY

We shall now examine in some detail selected joints of the body. In order to simplify your learning efforts, a series of exhibits have been prepared. Each exhibit considers a specific joint and contains (1) a definition—that is, a description of the bones that form the joint; (2) the type of joint (its structural classification); and (3) its anatomical components—a description of the major connecting ligaments, articular disc, articular capsule, and other distinguishing features of the joint. Each exhibit also refers you to an illustration of the joint.

EXHIBIT 8-2

Summary of Joints

TYPE	DESCRIPTION	MOVEMENT	EXAMPLES
FIBROUS	No synovial (joint) cavity; bones held together by a thin layer of fibrous tissue or dense fibrous tissue.		
Sutures	Found only between bones of the skull; articulating bones separated by a thin layer of fibrous tissue.	None (synarthroses).	Lambdoid suture between occipital and parietal bones.
Syndesmoses	Articulating bones united by dense fibrous tissue.	Slight (amphiarthroses).	Distal ends of tibia and fibula.
Gomphoses	Cone-shaped peg fits into a socket; articulating bones separated by periodontal ligament.	None (synarthroses).	Roots of teeth in alveoli (sockets).
CARTILAGINOUS	No synovial cavity; articulating bones united by cartilage.		
Synchondroses	Connecting material is hyaline cartilage.	None (synarthroses).	Temporary joint between the diaphysis and epiphyses of a long bone and permanent joint between true ribs and sternum.
Symphyses	Connecting material is a broad, flat disc of fibrocartilage.	Slight (amphiarthroses).	Intervertebral joints and pubic symphysis.
SYNOVIAL	Synovial cavity and articular cartilage present; articular capsule composed of an outer fibrous capsule and an inner synovial membrane; may contain accessory ligaments, articular discs (menisci), and bursae.	Freely movable (diarthroses).	
Gliding	Articulating surfaces usually flat.	Nonaxial.	Intercarpal and intertarsal joints.
Hinge	Spool-like surface fits into a concave surface.	Monaxial (flexion–extension).	Elbow, ankle, and interphalangeal joints.
Pivot	Rounded, pointed, or concave surface fits into a ring formed partly by bone and partly by a ligament.	Monaxial (rotation).	Atlantoaxial and proximal radioulnar joints.
Ellipsoidal	Oval-shaped condyle fits into an elliptical cavity.	Biaxial (flexion–extension, abduction–adduction).	Radiocarpal joint.
Saddle	Articular surface of one bone is saddle-shaped, and the articular surface of the other bone is shaped like a rider sitting in the saddle.	Biaxial (flexion–extension, abduction–adduction).	Carpometacarpal proximal to the thumb.
Ball-and-Socket	Ball-like surface fits into a cuplike depression.	Triaxial (flexion–extension, abduction–adduction, rotation–circumduction).	Shoulder and hip joints.

EXHIBIT 8-3

Intervertebral Joints (Figure 8-7)

DEFINITION Joints formed between (1) vertebral bodies and (2) vertebral arches.

TYPE OF JOINT Joints between vertebral bodies—cartilaginous (fibrocartilage), symphysis type. Joints between vertebral arches—synovial, gliding (arthrodial) type.

ANATOMICAL COMPONENTS OF JOINTS BETWEEN VERTEBRAL BODIES

1. ***Anterior longitudinal ligament.*** Broad, strong band that extends along the anterior surfaces of the vertebral bodies from the axis to the sacrum. It is firmly attached to the intervertebral discs.
2. ***Posterior longitudinal ligament.*** Extends along the posterior surfaces of the vertebral bodies within the vertebral canal from the axis to the sacrum. The free surface of the ligament is separated from the spinal dura mater by areolar connective tissue.
3. ***Intervertebral discs.*** Unite with adjacent surfaces of the vertebral bodies from the axis to the sacrum. Each disc is composed of a peripheral ***annulus fibrosus*** consisting of fibrous tissue and fibrocartilage and a central ***nucleus pulposus*** composed of a soft, pulpy, highly elastic substance (see Figure 6-12c).

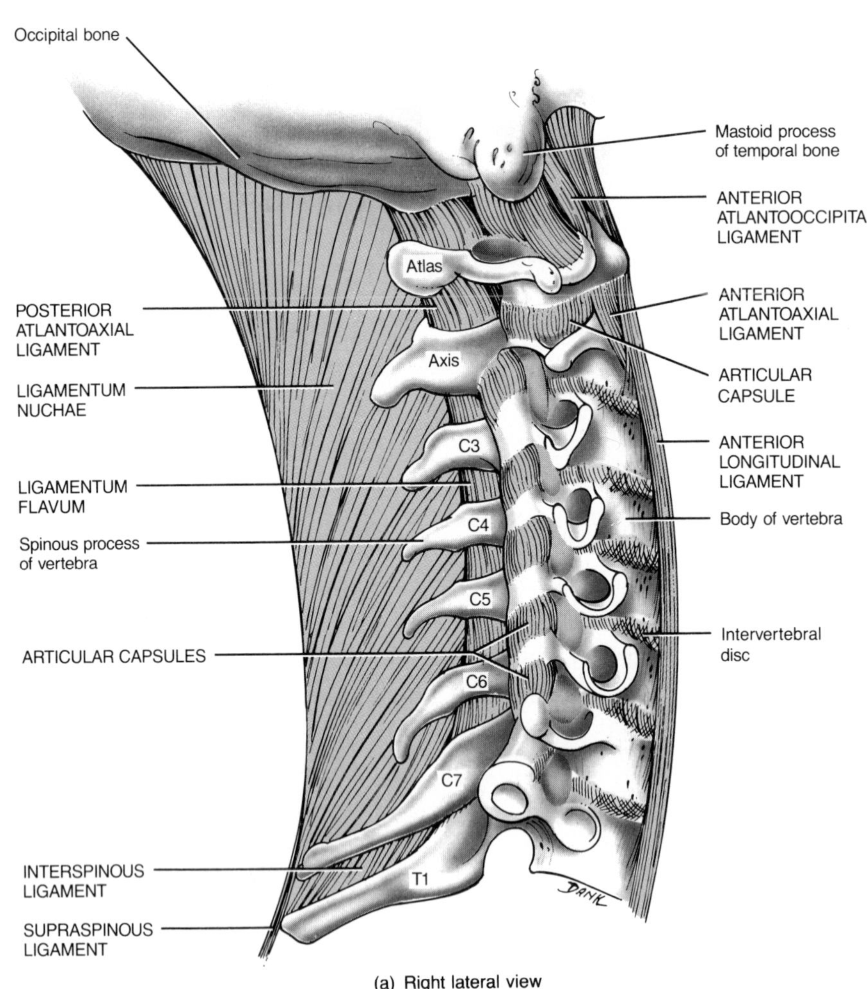

(a) Right lateral view

FIGURE 8-7 Intervertebral joints.

ANATOMICAL COMPONENTS OF JOINTS BETWEEN VERTEBRAL ARCHES

1. ***Articular capsules.*** Thin, loose ligaments attached to the margins of the articular processes of adjacent vertebrae.
2. ***Ligamenta flava.*** Contain elastic tissue and connect the laminae of adjacent vertebrae from the axis to the first segment of the sacrum.
3. ***Supraspinous ligament.*** Strong fibrous cord that connects the spinous processes from the seventh cervical vertebra to the sacrum.
4. ***Ligamentum nuchae.*** Fibrous ligament that represents an enlargement of the supraspinous ligament in the neck. It extends from the external occipital protuberance of the occipital bone to the spinous process of the seventh cervical vertebra.
5. ***Interspinous ligaments.*** Relatively weak bands that run between adjacent spinous processes.
6. ***Intertransverse ligaments.*** Bands between transverse processes that are readily apparent in the lumbar region.

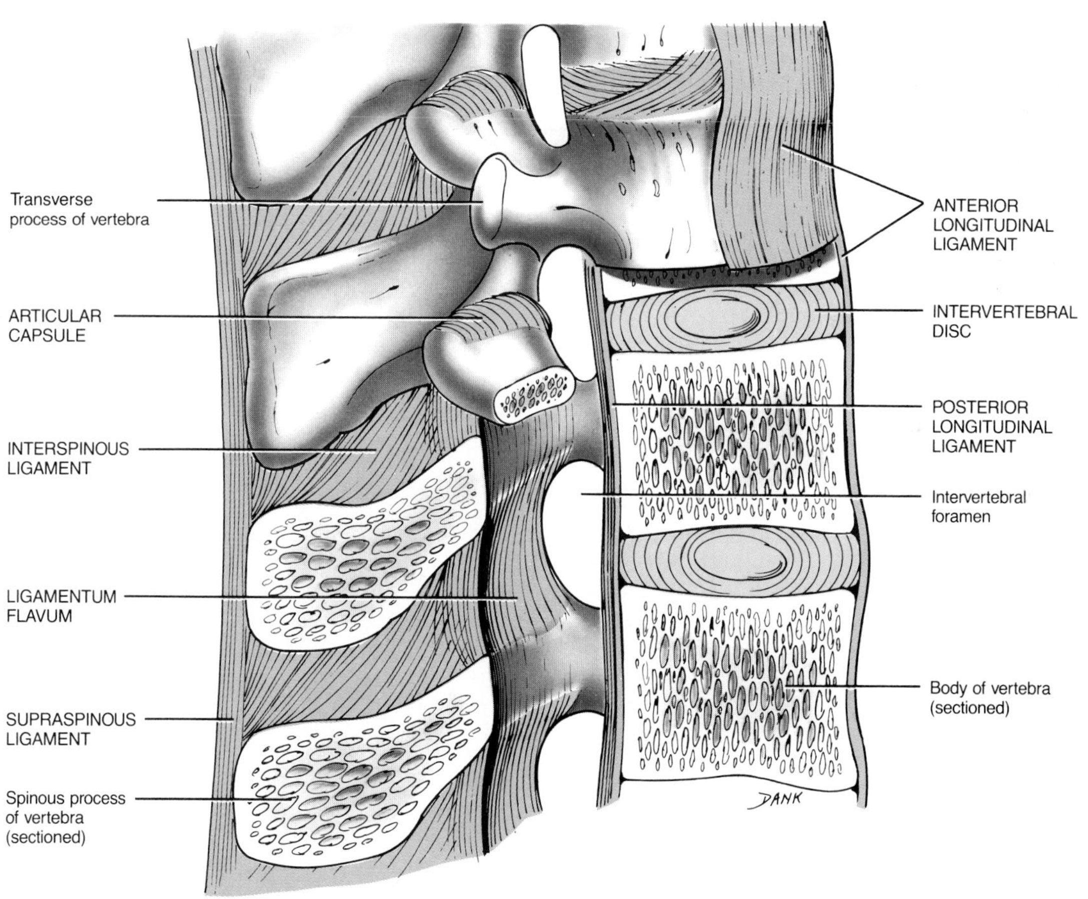

(b) Median view

EXHIBIT 8-4

Shoulder (Humeroscapular or Glenohumeral) Joint (Figure 8-8)

DEFINITION Joint formed by the head of the humerus and the glenoid cavity of the scapula.

TYPE OF JOINT Synovial joint, ball-and-socket (spheroid) type.

ANATOMICAL COMPONENTS

1. ***Articular capsule.*** Loose sac that completely envelopes the joint, extending from the circumference of the glenoid cavity to the anatomical neck of the humerus.
2. ***Coracohumeral ligament.*** Strong, broad ligament that extends from the coracoid process of the scapula to the greater tubercle of the humerus.
3. ***Glenohumeral ligaments.*** Three thickenings of the articular capsule over the anterior surface of the joint.
4. ***Transverse humeral ligament.*** Narrow sheet extending from the greater tubercle to the lesser tubercle of the humerus.
5. ***Glenoid labrum.*** Narrow rim of fibrocartilage around the edge of the glenoid cavity.
6. Among the ***bursae*** associated with the shoulder joint are:
   a. ***Subscapular bursa*** between the tendon of the subscapularis muscle and the underlying joint capsule.
   b. ***Subdeltoid bursa*** between the deltoid muscle and joint capsule.
   c. ***Subacromial bursa*** between the acromion and joint capsule.
   d. ***Subcoracoid bursa,*** either lying between the coracoid process and joint capsule or appearing as an extension from the subacromial bursa.

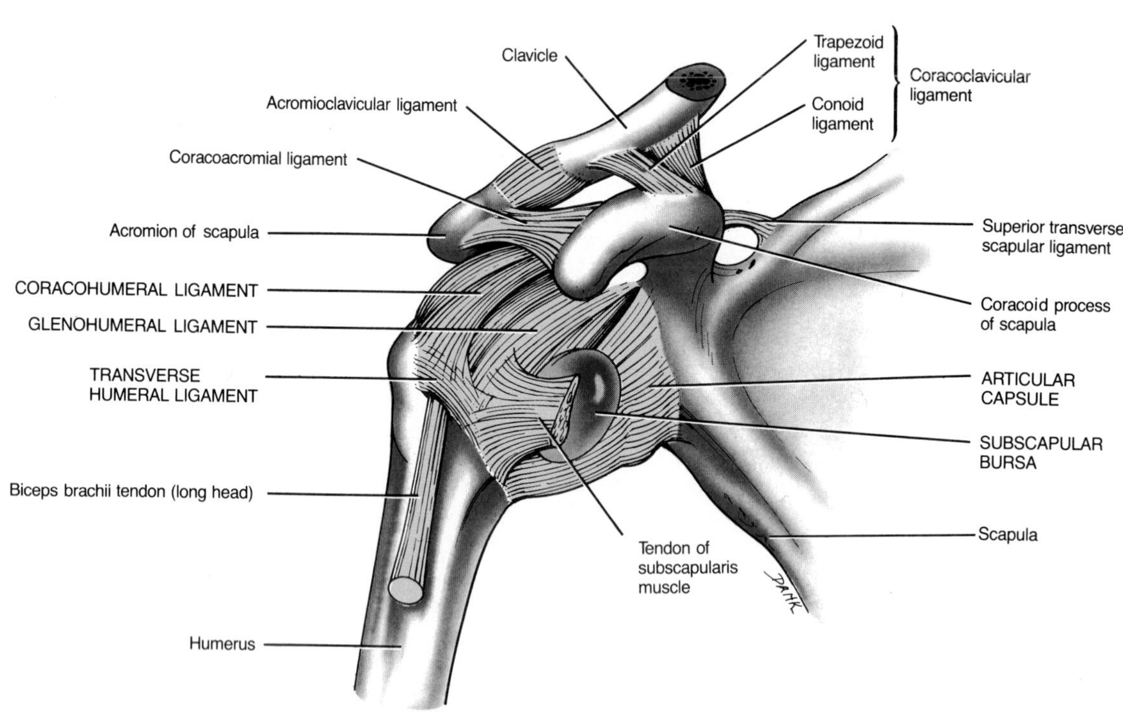

FIGURE 8-8 Shoulder (humeroscapular) joint.

EXHIBIT 8-5
Elbow Joint (Figure 8-9)

DEFINITION Joint formed by the trochlea of the humerus, the trochlear notch of the ulna, and the head of the radius.

TYPE OF JOINT Synovial joint, hinge (ginglymus) type.

ANATOMICAL COMPONENTS

1. ***Articular capsule.*** The anterior part covers the anterior part of the joint from the radial and coronoid fossae of the humerus to the coronoid process of the ulna and the annular ligament of the radius. The posterior part extends from the capitulum, olecranon fossa, and lateral epicondyle of the humerus to the annular ligament of the radius, the olecranon of the ulna, and the ulna posterior to the radial notch.
2. ***Ulnar collateral ligament.*** Thick, triangular ligament that extends from the medial epicondyle of the humerus to the coronoid process and olecranon of the ulna.
3. ***Radial collateral ligament.*** Strong, triangular ligament that extends from the lateral epicondyle of the humerus to the annular ligament of the radius and the radial notch of the ulna.

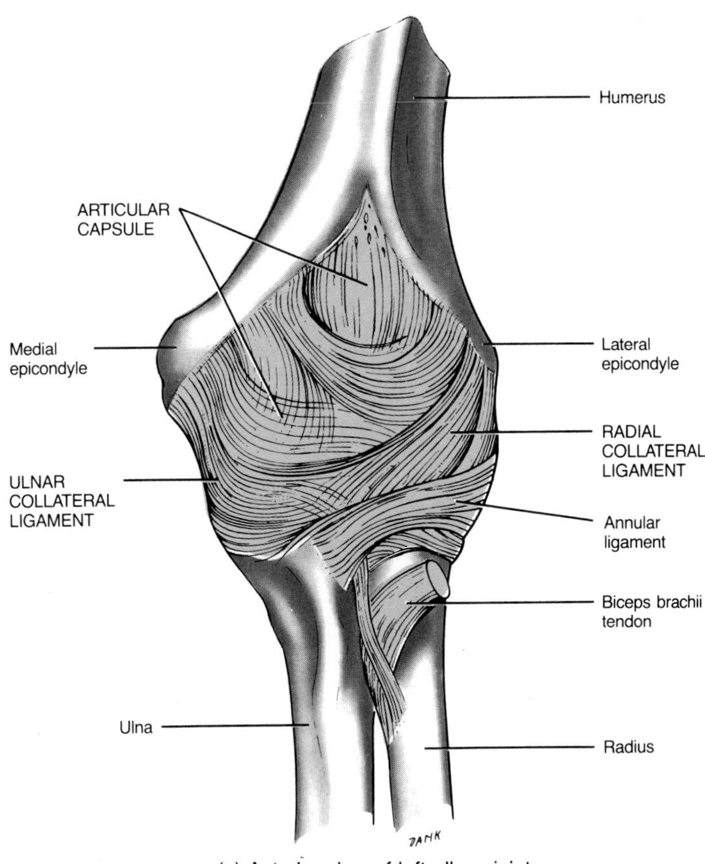

(a) Anterior view of left elbow joint

FIGURE 8-9 Elbow joint. (b) Courtesy of J. A. Gosling, P. F. Harris, et al., *Atlas of Human Anatomy,* Gower Medical Publishing Ltd., 2nd ed., 1991.

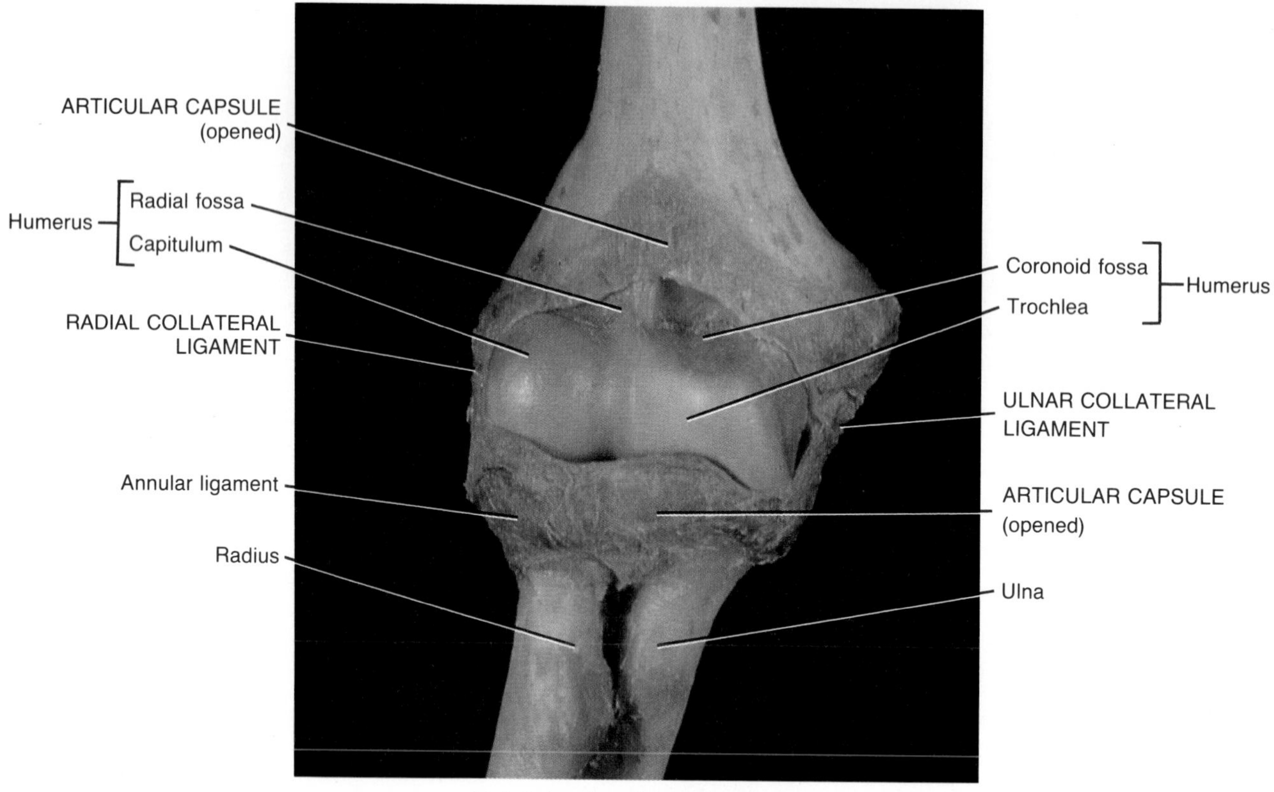

(b) Anterior view of right elbow joint

CLINICAL APPLICATION

Tennis Elbow, Little-League Elbow

Tennis elbow most commonly refers to pain at or near the lateral epicondyle of the humerus, usually caused by an improperly executed backhand. In this maneuver, the extensor muscles, especially the extensor carpi radialis brevis muscle, strain or sprain. Also, the extensors rub and roll over the lateral epicondyle and head of the radius, resulting in pain.

Little-league elbow typically develops as a result of a heavy pitching schedule, especially among youngsters. The disorder occurs mostly on the medial side of the elbow, in which there is some degree of involvement of the medial epicondylar epiphysis that may cause it to enlarge, fragment, or separate. Disorders on the lateral side of the elbow are more severe and chronic and are caused by the repeated jamming of the radial head against the capitulum of the humerus, which results in necrosis of the head of the radius, osteochondritis, and osteoarthritis.

EXHIBIT 8-6

Wrist (Radiocarpal) Joint (Figure 8-10)

DEFINITION Joint formed by the distal end of the radius, the distal surface of the articular disc separating the carpal and distal radioulnar joint, and the scaphoid, lunate, and triquetral carpal bones.

TYPE OF JOINT Synovial joint, ellipsoidal (condyloid) type.

ANATOMICAL COMPONENTS

1. ***Articular capsule.*** Extends from the styloid processes of the ulna and radius to the proximal row of carpal bones.
2. ***Palmar radiocarpal ligament.*** Thick, strong ligament that extends from the anterior border of the distal end of the radius to the proximal row of carpals. Some longer fibers extend to the capitate in the distal row.
3. ***Dorsal radiocarpal ligament.*** Extends from the posterior border of the distal end of the radius to the proximal row of carpals, especially the triquetral bone (not illustrated).
4. ***Ulnar collateral ligament.*** Extends from the styloid process of the ulna to the triquetral and pisiform bones and the transverse carpal ligament.
5. ***Radial collateral ligament.*** Extends from the styloid process of the radius to the scaphoid and pisiform bones.

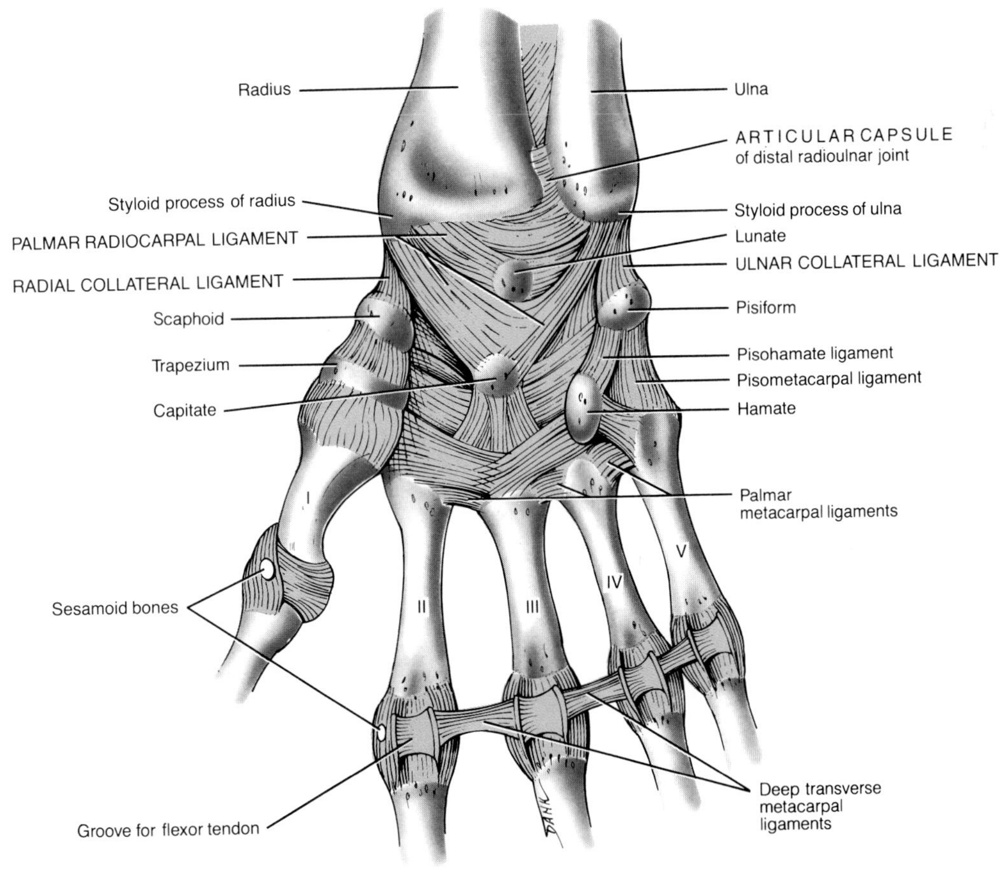

FIGURE 8-10 Wrist (radiocarpal) joint.

EXHIBIT 8-7

Hip (Coxal) Joint (Figure 8-11)

DEFINITION Joint formed by the head of the femur and the acetabulum of the hipbone.

TYPE OF JOINT Synovial, ball-and-socket (spheroid) type.

ANATOMICAL COMPONENTS

1. ***Articular capsule.*** Extends from the rim of the acetabulum to the neck of the femur. One of the strongest ligaments of the body, the capsule consists of circular and longitudinal fibers. The circular fibers, called the ***zona orbicularis,*** form a collar around the neck of the femur. The longitudinal fibers are reinforced by accessory ligaments known as the iliofemoral ligament, pubofemoral ligament, and ischiofemoral ligament.
2. ***Iliofemoral ligament.*** Thickened portion of the articular capsule that extends from the anterior inferior iliac spine of the hipbone to the intertrochanteric line of the femur.

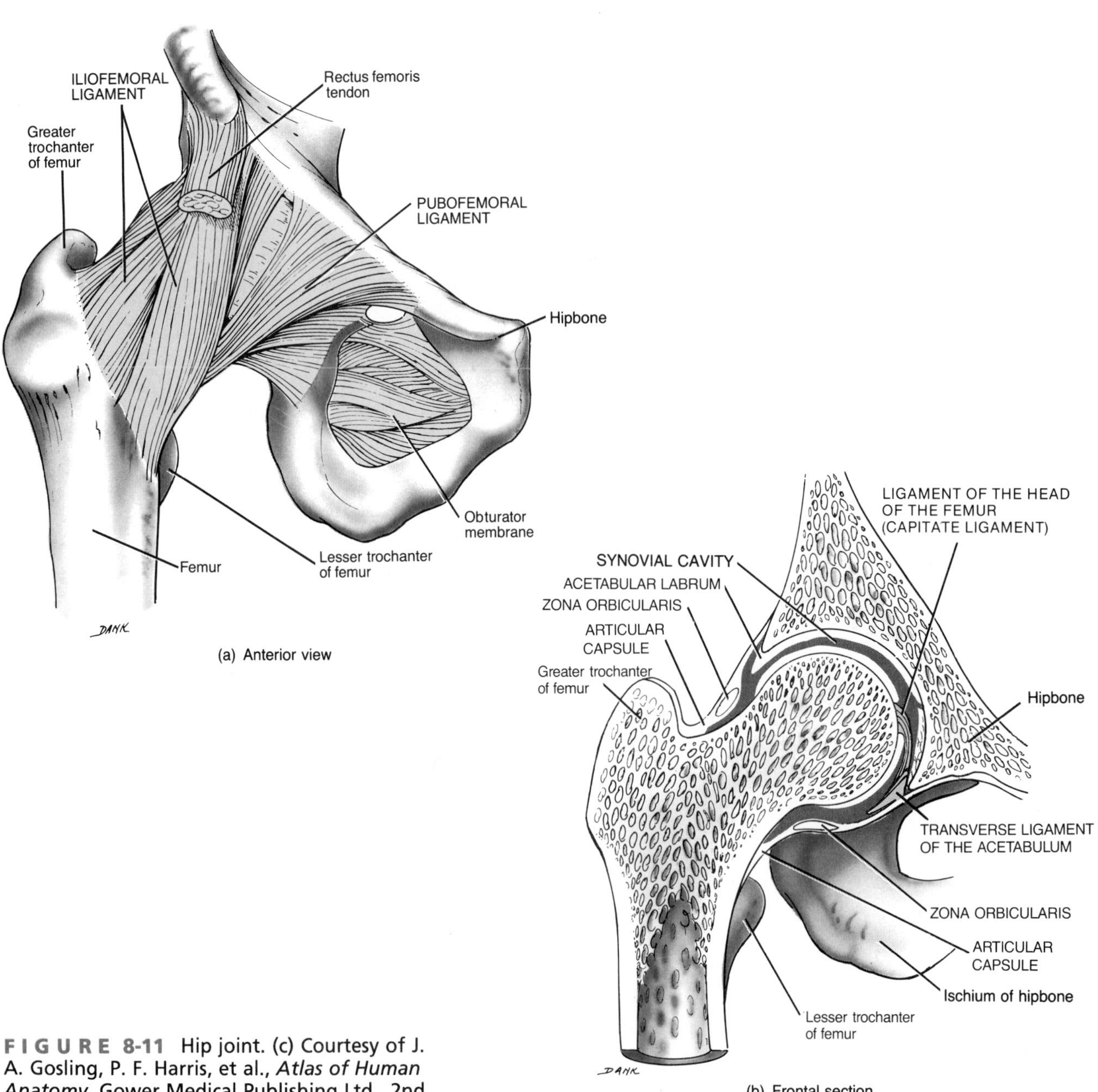

FIGURE 8-11 Hip joint. (c) Courtesy of J. A. Gosling, P. F. Harris, et al., *Atlas of Human Anatomy,* Gower Medical Publishing Ltd., 2nd ed., 1991.

3. ***Pubofemoral ligament.*** Thickened portion of the articular capsule that extends from the pubic part of the rim of the acetabulum to the neck of the femur.
4. ***Ischiofemoral ligament.*** Thickened portion of the articular capsule that extends from the ischial wall of the acetabulum to the neck of the femur (not illustrated).
5. ***Ligament of the head of the femur (capitate ligament).*** Flat, triangular band that extends from the fossa of the acetabulum to the head of the femur.
6. ***Acetabular labrum.*** Fibrocartilage rim attached to the margin of the acetabulum.
7. ***Transverse ligament of the acetabulum.*** Strong ligament that crosses over the acetabular notch, converting it to a foramen. It supports part of the acetabular labrum and is connected with the ligament of the head of the femur and the articular capsule.

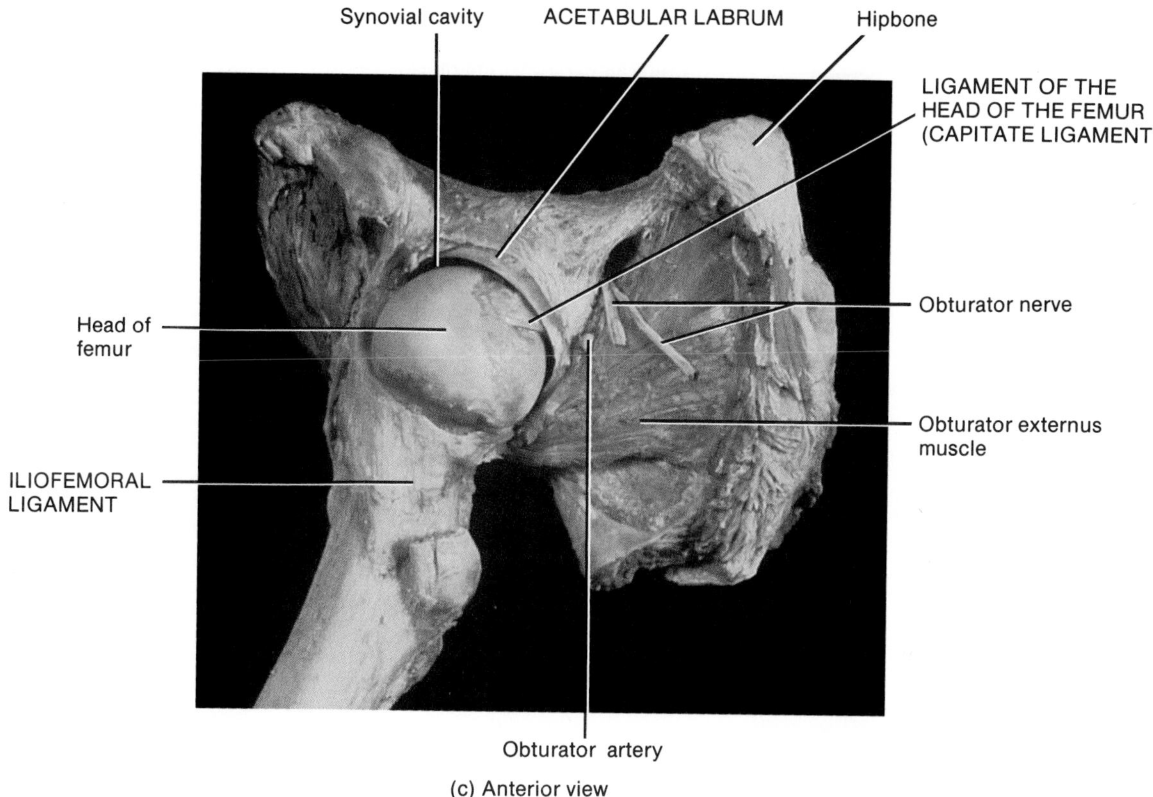

(c) Anterior view

CLINICAL APPLICATION

Arthroplasty

Arthroplasty (AR-thrō-plas′-tē; *arthro* = joint; *plasty* = plastic repair of) refers to surgical replacement of joints. Each year in the United States over 150,000 ***total hip replacements (hip arthroplasty)*** are performed. Total hip replacement means that the procedure involves both the acetabulum and head of the femur. Thousands of partial hip replacements, involving only the femur, are also performed annually. In total hip replacement, the damaged portions of the acetabulum and head of the femur are replaced by prefabricated prostheses (artificial devices). The acetabular component consists of polyethylene, whereas the femoral component is composed of cobalt-chrome. These materials are designed to withstand a high degree of stress. Once the appropriate acetabular and femoral components are selected, they are attached to the healthy portion of bone with acrylic cement, which forms an interlocking mechanical bond with bone.

Total hip replacement is indicated for individuals with osteoarthritis, rheumatoid arthritis, hip fractures, dislocations, metabolic bone diseases (osteoporosis and osteomalacia), or congenital or developmental deformities of the hip.

EXHIBIT 8-8

Knee (Tibiofemoral) Joint (Figure 8-12)

DEFINITION The largest joint of the body, actually consisting of three joints: (1) an intermediate patellofemoral joint between the patella and the patellar surface of the femur; (2) a lateral tibiofemoral joint between the lateral condyle of the femur, lateral meniscus, and lateral condyle of the tibia; and (3) a medial tibiofemoral joint between the medial condyle of the femur, medial meniscus, and medial condyle of the tibia.

TYPE OF JOINT Patellofemoral joint—partly synovial, gliding (arthrodial) type. Lateral and medial tibiofemoral joints—synovial, hinge (ginglymus) type.

ANATOMICAL COMPONENTS

1. ***Articular capsule.*** No complete, independent capsule unites the bones. The ligamentous sheath surrounding the joint consists mostly of muscle tendons or expansions of them. There are, however, some capsular fibers connecting the articulating bones.
2. ***Medial and lateral patellar retinacula.*** Fused tendons of insertion of the quadriceps femoris muscle and the fascia lata (deep fascia of thigh) that strengthen the anterior surface of the joint.
3. ***Patellar ligament.*** Central portion of the common tendon of insertion of the quadriceps femoris muscle that extends from the patella to the tibial tuberosity. This ligament also strengthens the anterior surface of the joint. The posterior surface of the ligament is separated from the synovial membrane of the joint by an ***infrapatellar fat pad.***

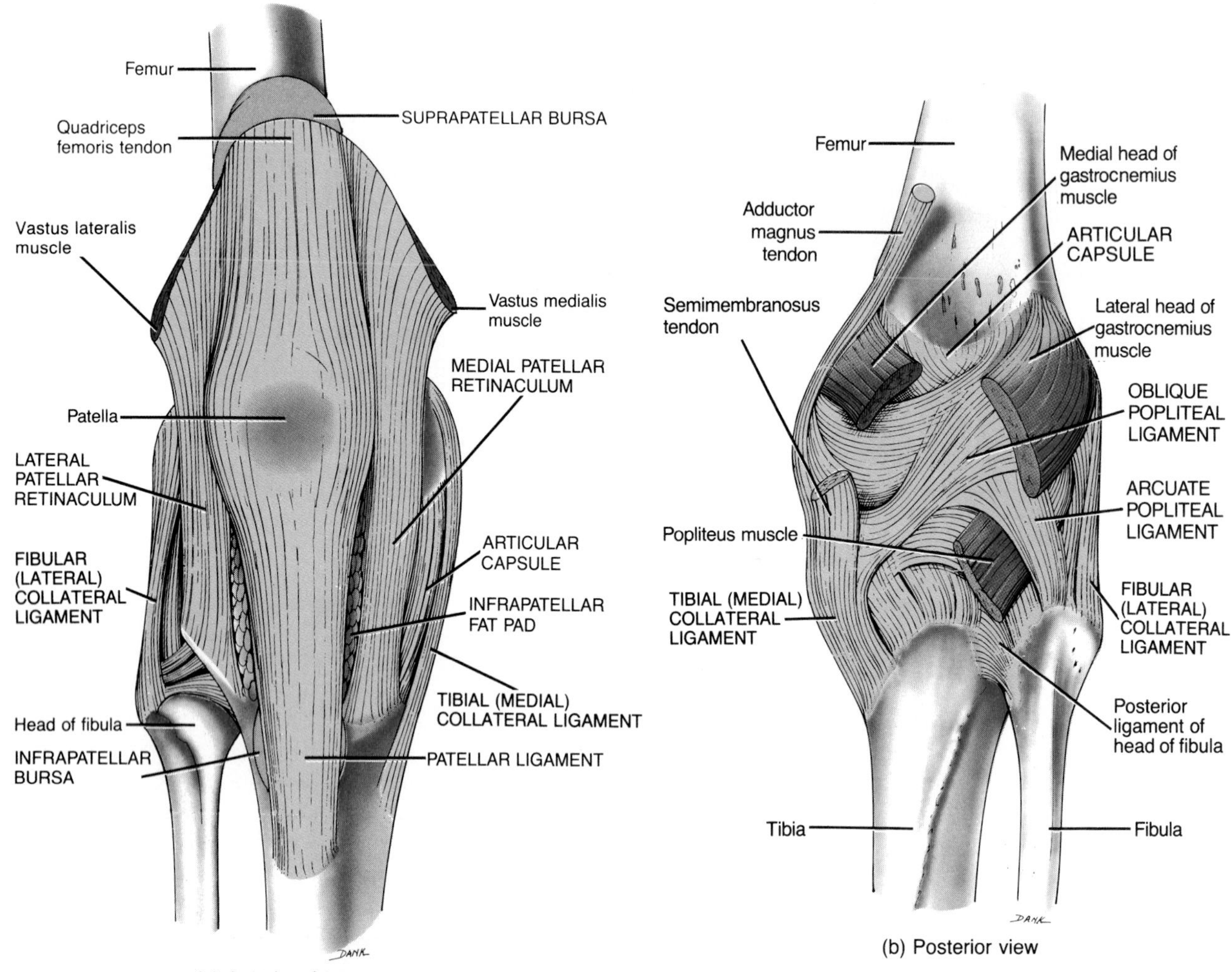

FIGURE 8-12 Knee (tibiofemoral) joint. (d) and (f) Courtesy of J. A. Gosling, P. F. Harris, et al., *Atlas of Human Anatomy,* Gower Medical Publishing Ltd., 2nd ed., 1991.

4. ***Oblique popliteal ligament.*** Broad, flat ligament that extends from the intercondylar fossa of the femur to the head of the tibia. The tendon of the semimembranosus muscle is superficial to the ligament and passes from the medial condyle of the tibia to the lateral condyle of the femur. The ligament and tendon afford strength for the posterior surface of the joint.
5. ***Arcuate popliteal ligament.*** Extends from the lateral condyle of the femur to the styloid process of the head of the fibula. It strengthens the lower lateral part of the posterior surface of the joint.
6. ***Tibial (medial) collateral ligament.*** Broad, flat ligament on the medial surface of the joint that extends from the medial condyle of the femur to the medial condyle of the tibia. The ligament is crossed by tendons of the sartorius, gracilis, and semitendinosus muscles, all of which strengthen the medial aspect of the joint.

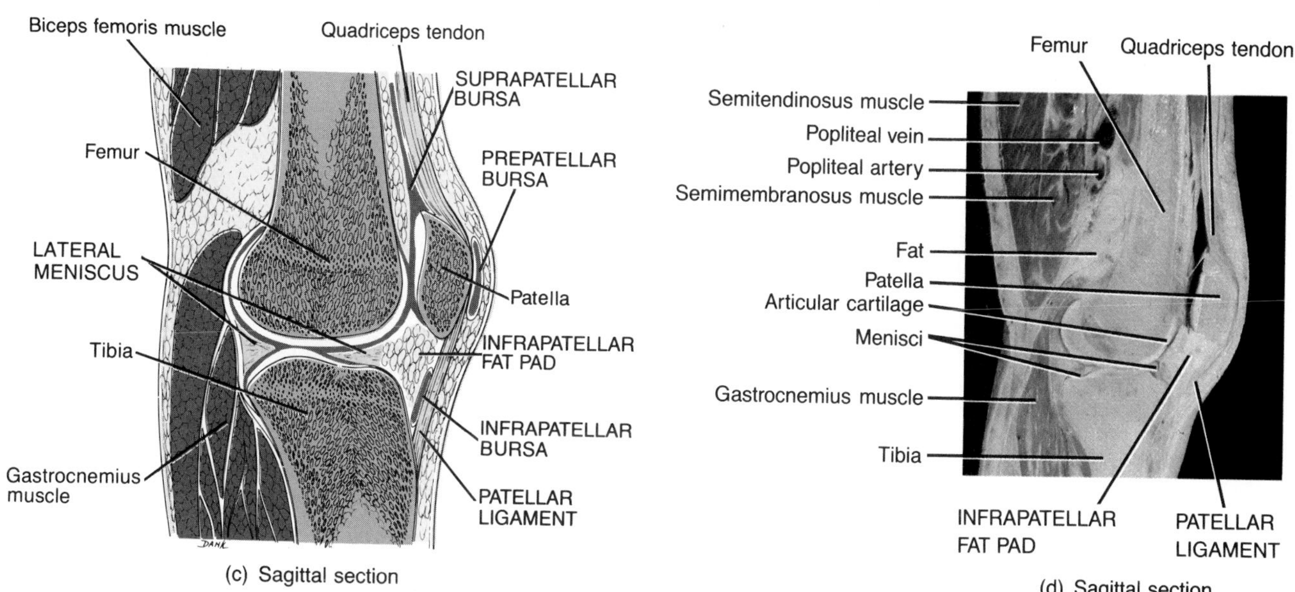

(c) Sagittal section

(d) Sagittal section

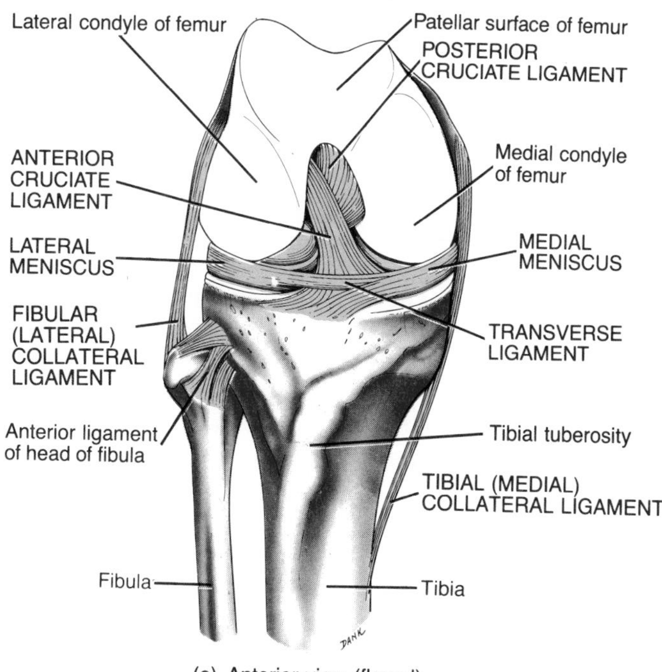

(e) Anterior view (flexed)

EXHIBIT 8-8 (*Continued*)

7. ***Fibular (lateral) collateral ligament.*** Strong, rounded ligament on the lateral surface of the joint that extends from the lateral condyle of the femur to the lateral side of the head of the fibula. The ligament is covered by the tendon of the biceps femoris muscle. The tendon of the popliteal muscle is deep to the tendon.
8. ***Intra-articular ligaments.*** Ligaments within the capsule that connect the tibia and femur.
   a. ***Anterior cruciate ligament.*** Extends posteriorly and laterally from the area anterior to the intercondylar eminence of the tibia to the posterior part of the medial surface of the lateral condyle of the femur. This ligament is stretched or torn in about 70 percent of all serious knee injuries. Both O. J. Simpson and Gale Sayers had their careers in professional football ended by torn anterior cruciate ligaments.
   b. ***Posterior cruciate ligament.*** Extends anteriorly and medially from a depression on the posterior intercondylar area of the tibia and lateral meniscus to the anterior part of the medial surface of the medial condyle of the femur.
9. ***Articular discs.*** Fibrocartilage discs between the tibial and femoral condyles.
   a. ***Medial meniscus.*** Semicircular piece of fibrocartilage (C-shaped). Its anterior end is attached to the anterior intercondylar fossa of the tibia, in front of the anterior cruciate ligament. Its posterior end is attached to the posterior intercondylar fossa of the tibia between the attachments of the posterior cruciate ligament and lateral meniscus.
   b. ***Lateral meniscus.*** Nearly circular piece of fibrocartilage (approaches an incomplete O in shape). Its anterior end is attached anterior to the intercondylar eminence of the tibia and lateral and posterior to the anterior cruciate ligament. Its posterior end is attached posterior to the intercondylar eminence of the tibia and anterior to the posterior end of the medial meniscus. The medial and lateral menisci are connected to each other by the ***transverse ligament*** and to the margins of the head of the tibia by the ***coronary ligaments.***
10. The principal ***bursae*** of the knee include the following:
    a. ***Anterior bursae:*** (1) between the patella and skin (***prepatellar bursa***), (2) between upper part of tibia and patellar ligament (***infrapatellar bursa***), (3) between lower part of tibial tuberosity and skin, and (4) between lower part of femur and deep surface of quadriceps femoris muscle (***suprapatellar bursa***).
    b. ***Medial bursae:*** (1) between medial head of gastrocnemius muscle and the articular capsule, (2) superficial to the tibial collateral ligament between the ligament and tendons of the sartorius, gracilis, and semitendinosus muscles, (3) deep to the tibial collateral ligament between the ligament and the tendon of the semimembranosus muscle, (4) between the tendon of the semimembranosus muscle and the head of the tibia, and (5) between the tendons of the semimembranosus and semitendinosus muscles.
    c. ***Lateral bursae:*** (1) between the lateral head of the gastrocnemius muscle and articular capsule, (2) between the tendon of the biceps femoris muscle and fibular collateral ligament, (3) between the tendon of the popliteal muscle and fibular collateral ligament, and (4) between the lateral condyle of the femur and the popliteal muscle.

CLINICAL APPLICATION

Knee Injuries

The most common type of ***knee injury*** in football is rupture of the tibial (medial) collateral ligament, often associated with tearing of the anterior cruciate ligament and medial meniscus (torn cartilage). It is caused by a blow to the lateral side of the knee. When a knee is examined for such an injury, the three C's are kept in mind: collateral ligament, cruciate ligament, and cartilage.

Patellofemoral stress syndrome ("runner's knee") is the single most common problem in runners. During normal flexion and extension of the knee, the patella tracks (glides) up and down in the groove between the femoral condyles. In patellofemoral stress syndrome, normal tracking does not occur; instead, the patella tracks laterally, and the increased pressure of abnormal tracking causes the associated pain. The pain is usually described as an aching or tenderness around or under the patella. The pain typically occurs after a person has been sitting for a while, especially after exercise. The syndrome may be due to excessive pronation of the foot or tight hamstring muscles. A common cause of runner's knee is constantly walking, running, or jogging on the same side of the road. Since roads are high in the middle and slope down on the sides, the slope stresses the knee that is closer to the center of the road.

A ***swollen knee*** may occur immediately or be delayed. The immediate swelling is due to escape of blood from rupture of the anterior cruciate ligament, torn menisci,

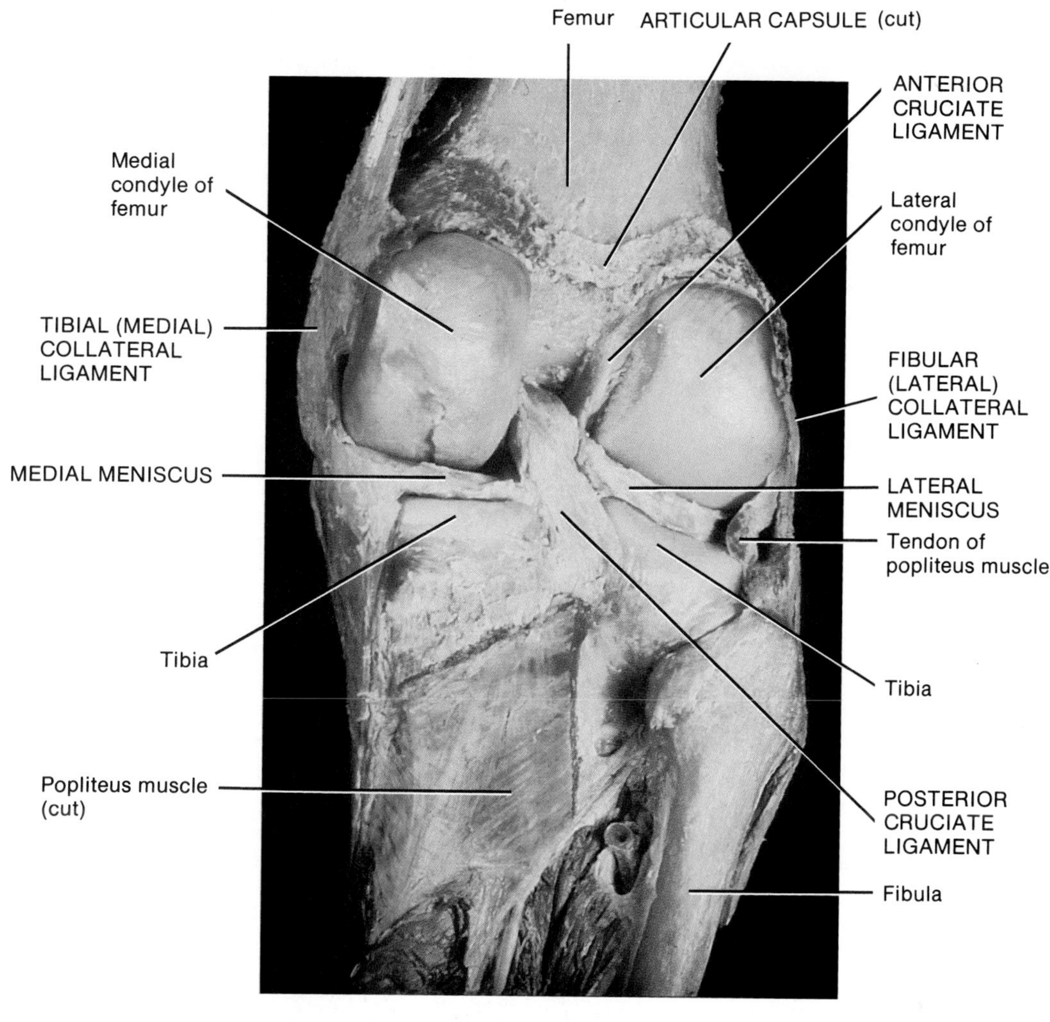

(f) Posterior view

fractures, or collateral ligament sprains. Delayed swelling is due to an excessive production of synovial fluid as a result of conditions that irritate the synovial membrane.

A ***dislocated knee*** refers to the displacement of the tibia relative to the femur. Accordingly, such dislocations are classified as anterior, posterior, medial, lateral, or rotatory. The most common type is anterior dislocation, resulting from hyperextension of the knee. A frequent consequence of a dislocated knee is damage to the popliteal artery.

In November 1987, a group of orthopedic surgeons at the Hospital of the University of Pennsylvania performed the ***first transplant of an entire human knee.*** The surgery was necessitated by a potentially malignant tumor on the knee of a 32-year-old female recipient. The donor was an 18-year-old male. Once the tumor was removed, the patient's knee joint was removed and replaced with the donor's joint. The donor knee joint, about 41 cm (16 in.) long, consisted of the lower portion of a femur that was connected to the recipient's femur by a rod; the upper portion of the tibia and head of the fibula that were connected by a metal plate; the patella; medial and lateral menisci; intracapsular and extracapsular ligaments; and certain tendons. The recipient's own muscles, nerves, and blood vessels were used. Since rejection occurs in only about 3 to 5 percent of bone transplant cases, no immunosuppressive drugs were needed.

EXHIBIT 8-9

Ankle (Talocrucal) Joints (Figure 8–13)

DEFINITION	***Talo*** means talus and ***crural*** means leg. Joints between (1) the distal end of the tibia and its medial malleolus and the talus and (2) the lateral malleolus of the fibula and the talus.
TYPE OF JOINT	Both joints—synovial, hinge (ginglymus) type.
ANATOMICAL COMPONENTS	1. ***Articular capsule.*** Surrounds the joint and extends from the borders of the tibia and malleoli to the talus. 2. ***Deltoid (medial) ligament.*** Strong, triangular liagment that extends from the medial malleolus of the tibia to the navicular, calcaneus, and talus.

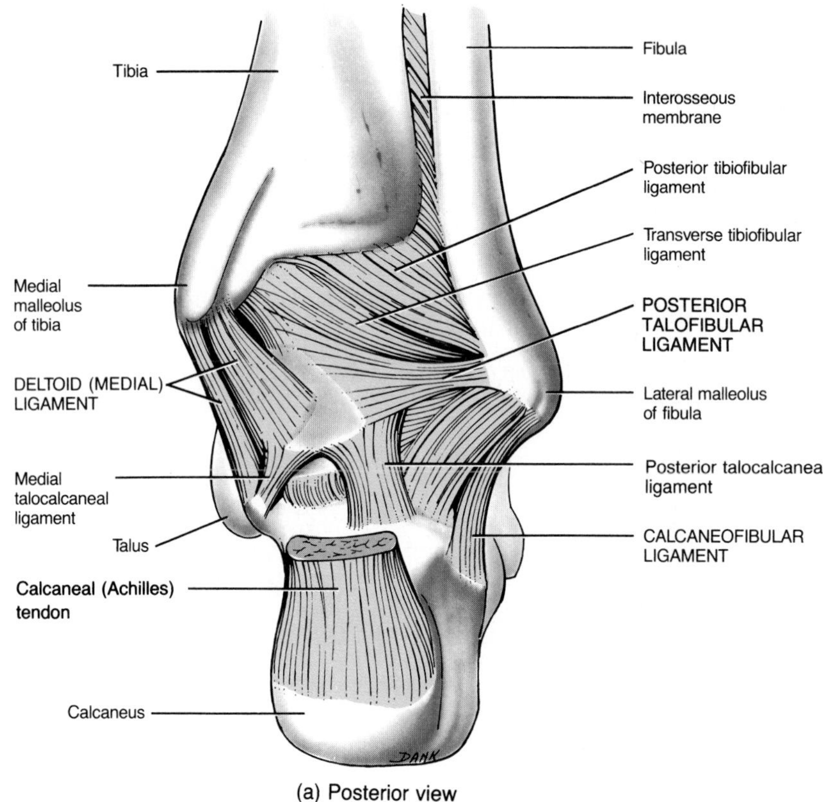

FIGURE 8-13 Ankle (talocrural) joints. (b) Courtesy of J. A. Gosling, P. F. Harris, et al., *Atlas of Human Anatomy,* Gower Medical Publishing Ltd., 2nd ed., 1991.

CLINICAL APPLICATION

Ankle Sprain

A sprain involves twisting of a joint, partial rupture of its ligaments, or other injury to components of a joint such as blood vessels, nerves, muscles, and tendons. The most common ***ankle sprain*** is a lateral sprain in which the foot is turned inward. A person is more likely to turn the foot inward rather than outward because of normal adduction that occurs during running and because the lateral ligament is weaker than the deltoid (medial) ligament. Most lateral ankle sprains involve the anterior talofibular ligament and frequently occur on irregular surfaces. Depending on the severity of the sprain, treatment may include strapping with elastic bandages or tape, elevation, cast immobilization, or surgery.

3. ***Anterior talofibular ligament.*** Extends from the anterior margin of the lateral malleolus of the fibula to the talus.
4. ***Posterior talofibular ligament.*** Extends from the posterior margin of the lateral malleolus of the fibula to the talus.
5. ***Calcaneofibular ligament.*** Extends from the apex of the lateral mallelous of the fibula to the talus. Together, the anterior talofibullar, posterior talofibular, and calcaneofibular ligaments are referred to as the ***lateral ligament.***

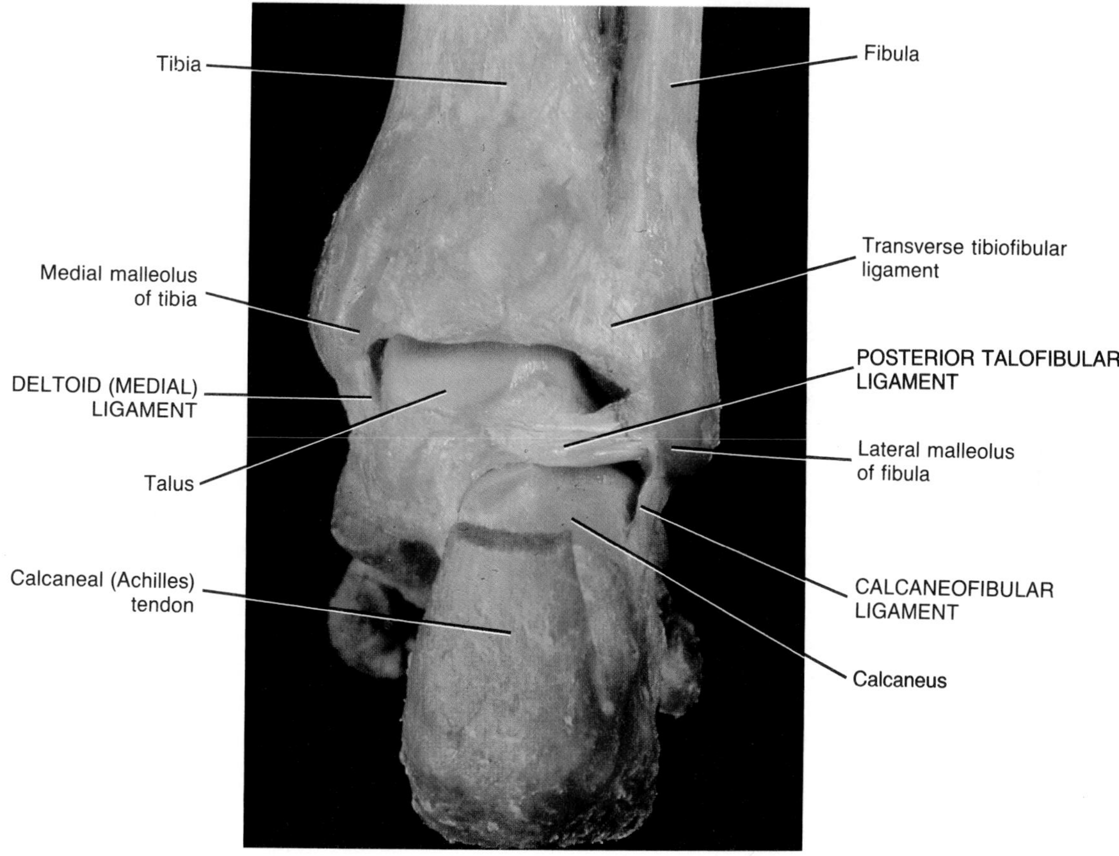

(b) Posterior view

APPLICATIONS TO HEALTH

RHEUMATISM

Rheumatism (*rheumat* = subject to flux) refers to any painful state of the supporting structures of the body—its bones, ligaments, joints, tendons, or muscles. Arthritis is a form of rheumatism in which the joints have become inflamed.

ARTHRITIS

The term ***arthritis*** refers to many different diseases, most of which are characterized by inflammation of one or more joints. Inflammation, pain, and stiffness may also be present in adjacent parts of the body, such as the muscles near the joint.

Rheumatoid Arthritis (RA)

Rheumatoid (ROO-ma-toyd) ***arthritis*** (***RA***) is an autoimmune disease in which the body attacks its own tissues, in this case its own cartilage and joint linings. It is characterized by inflammation of the joint, swelling, pain, and loss of function. Usually, this form occurs bilaterally—if your left wrist is affected, your right wrist is likely to be affected also, though usually not to the same degree.

The primary symptom of rheumatoid arthritis is inflammation of the synovial membrane. If untreated, the membrane thickens and synovial fluid accumulates. The resulting pressure causes pain and tenderness. The membrane then produces an abnormal granulation tissue called a ***pannus,*** which adheres to the surface of the articular cartilage. The pannus formation sometimes erodes the cartilage completely. When the cartilage is destroyed, fibrous tissue joins the exposed bone ends. The tissue ossifies and fuses the joint so that it is immovable—the ultimate crippling effect of rheumatoid arthritis. Most cases do not progress to this stage, but the range of motion of the joint is greatly inhibited by the severe inflammation and swelling. The growth of the pannus is what causes the distortion of the fingers that is so typical of the clinical appearance of hands that have been affected by rheumatoid arthritis.

Treatment is aimed at reducing pain and inflammation and preserving muscle strength and joint function. Therapies include adequate rest; anti-inflammatory drugs, such as aspirin and in selected patients steroids; exercises to maintain full range of joint motion; heat and other forms of physical therapy; and weight loss to relieve pressure on weight-bearing joints. In severe cases, damaged joints may be surgically replaced, either partly or entirely, with artificial joints.

Osteoarthritis

Osteoarthritis (os′-tē-ō-ar-THRĪ-tis) is a degenerative joint disease that apparently results from a combination of aging, irritation of the joints, and wear and abrasion. It is commonly known as "wear-and-tear" arthritis.

Osteoarthritis is a noninflammatory, progressive disorder of movable joints, particularly weight-bearing joints. It is characterized by the deterioration of articular cartilage and by formation of new bone in the subchondral areas and at the margins of the joint. The cartilage slowly degenerates, and as the bone ends become exposed, small bumps, or ***spurs,*** of new osseous tissue are deposited on them. These spurs decrease the space of the joint cavity and restrict joint movement. Unlike rheumatoid arthritis, osteoarthritis usually affects only the articular cartilage. The synovial membrane is rarely destroyed, and other tissues are unaffected. The main distinction between osteoarthritis and rheumatoid arthritis is that the former strikes the big joints (knees, hips) first, whereas the latter strikes the small joints first.

Gouty Athritis

Uric acid (a substance that gives urine its name) is a waste product produced during the metabolism of nucleic acid (DNA and RNA) subunits. The person who suffers from ***gout*** either produces excessive amounts of uric acid or is not able to excrete normal amounts. The result is a buildup of uric acid in the blood. This excess acid then reacts with sodium to form a salt called sodium urate. Crystals of this salt accumulate in soft tissues. Typical sites are the kidneys and the cartilage of the ears and joints.

In ***gouty*** (GOW-tē) ***arthritis,*** sodium urate crystals are deposited in the soft tissues of the joints. The crystals irritate the cartilage, causing inflammation, swelling, and acute pain. Eventually, the crystals destroy all the joint tissues. If the disorder is not treated, the ends of the articulating bones fuse and the joint becomes immovable.

Although other forms of arthritis cannot be treated with complete success, the treatment of gouty arthritis with the use of various drugs has been quite effective. A chemical called colchicine has been utilized periodically since the sixth century to relieve the pain, swelling, and tissue destruction that occur during attacks of gouty arthritis. The drug allopurinal is used for the treatment of gouty arthritis because it prevents the formation of uric acid without interfering with nucleic acid synthesis. Relief from an attack of gouty arthritis may be obtained by rest and the use of antiinflammatory medication.

LYME DISEASE

In 1975, a cluster of disease cases in young people that were first diagnosed as rheumatoid arthritis were reported near the city of Lyme, Connecticut. Today this disease is known as ***Lyme disease*** and it has spread into many states across the United States. It is caused by a bacterium (*Borrelia burgdorferi*) transmitted to humans by deer ticks and other ticks that are so small their bites often go unnoticed.

Within a few weeks of the tick bite a rash may appear at the site, although some people never develop one. The typical rash resembles a bull's eye, but there are many variations. The rash often is accompanied by joint stiffness, fever and chills, headache, stiff neck, nausea, and low back pain. In the next stage some patients develop cardiac abnormalities, indicated by weakness, lightheadedness, chest pain, and an irregular heartbeat. Other patients develop neurologic problems, most often Bell's palsy (sudden paralysis of one side of the face). In the third stage, arthritis is the principal complication. It usually involves the larger joints such as the knee, ankle, hip, elbow, or wrist. Lyme disease responds well to antibiotics, especially if they are given in the early stages of the disease.

BURSITIS

An acute chronic inflammation of a bursa is called ***bursitis.*** The condition may be caused by trauma, by an acute or chronic infection (including syphilis and tuberculosis), or by rheumatoid arthritis. Repeated excessive friction often results in a bursitis with local inflammation and the accumulation of fluid. Bunions are frequently associated with a friction bursitis over the head of the first metatarsal bone. Symptoms include pain, swelling, tenderness, and limited motion. The prepatellar or subcutaneous infrapatellar bursae may become inflammed in individuals who spend a great deal of time kneeling. This bursitis is usually called ***"housemaid's knee"*** (***"carpet layer's knee"***).

DISLOCATION

A ***dislocation*** or ***luxation*** (luks-Ā-shun) is the displacement of a bone from a joint with tearing of ligaments, tendons, and articular capsules. A partial or incomplete dislocation is called a ***subluxation.*** The most common dislocations are those involving the fingers or shoulders. Symptoms include loss of motion, temporary paralysis of the involved joint, pain, swelling, and occasional shock. A dislocation is usually caused by a blow or fall, although unusual physical effort may lead to this condition.

SPRAIN AND STRAIN

A ***sprain*** is the forcible twisting of a joint with partial rupture or other injury to its attachments without dislocation. It occurs when the attachments are stressed beyond their normal capacity. There may be damage to the associated blood vessels, muscles, tendons, ligaments, or nerves. Severe sprains may be so painful that the joint cannot be moved. There is considerable swelling and pain may occur owing to underlying hemorrhage from ruptured blood vessels. The ankle joint is most often sprained; the low back area is another frequent location for sprains. A sprain is more serious than a ***strain,*** which is the overstretching of a muscle.

KEY MEDICAL TERMS ASSOCIATED WITH JOINTS

Ankylosis (ang′-ki-LŌ-sus; *ankyle* = stiff joint; *osis* = condition) Severe or complete loss of movement at a joint.

Arthralgia (ar-THRAL-jē-a; *arth* = joint; *algia* = pain) Pain in a joint.

Arthrosis (ar-THRŌ-sis) Refers to an articulation; also a disease of a joint.

Bursectomy (bur-SEK-tō-mē; *ectomy* = removal of) Removal of a bursa.

Chondritis (kon-DRĪ-tis; *chondro* = cartilage) Inflammation of cartilage.

Rheumatology (roo′-ma-TOL-ō-jē; *rheumat* = subject to flux) The study of joints; the field of medicine devoted to joint diseases and related conditions.

Synovitis (sin′-ō-VĪ-tis; *synov* = joint) Inflammation of a synovial membrane in a joint.

STUDY OUTLINE

Classification of Joints (p. 182)

1. A joint (articulation) is a point of contact between two or more bones.
2. Structural classification is based on the presence of a synovial (joint) cavity and type of connecting tissue. Structurally, joints are classified as fibrous, cartilaginous, or synovial.
3. Functional classification of joints is based on the degree of movement permitted. Joints may be synarthroses, amphiarthroses, or diarthroses.

Fibrous Joints (p. 182)

1. Bones held by fibrous connective tissue, with no synovial cavity, are fibrous joints.
2. These joints include immovable sutures (found in the skull), slightly movable syndesmoses (such as the distal tibiofibular articulation), and immovable gomphoses (roots of teeth in alveoli of mandible and maxilla).

Cartilaginous Joints (p. 183)

1. Bones held together by cartilage, with no synovial cavity, are cartilaginous joints.
2. These joints include immovable synchondroses united by hyaline cartilage (temporary cartilage between diaphysis and epiphyses) and partially movable symphyses united by fibrocartilage (the pubic symphysis).

Synovial Joints (p. 183)

1. Synovial joints contain a synovial cavity, articular cartilage, and a synovial membrane; some also contain ligaments, articular discs, and bursae.
2. All synovial joints are freely movable.
3. Contact and movements at synovial joints are determined by the structure (shape) of articulating bones, strength and tension of ligaments, muscle arrangement and tension, apposition of soft parts, and hormones.
4. Types of movements at synovial joints include gliding movements, angular movements, rotation, circumduction, and special movements.
5. Types of synovial joints include gliding joints (between wrist bones), hinge joints (elbow), pivot joints (between radius and ulna), ellipsoidal joints (between radius and wrist), saddle joints (between trapezium of wrist and metacarpal of thumb), and ball-and-socket joints (shoulder and coxal).
6. A joint may be described according to the number of planes of movement it allows as nonaxial, monaxial, biaxial, or triaxial.

Selected Joints of the Body (p. 192)

1. Intervertebrals are between vertebral bodies and between vertebral arches.
2. Shoulder (humeroscapular) is between the humerus and scapula.

3. Elbow is between the humerus and ulna and between the humerus and radius.
4. Wrist (radiocarpal) is between the radius and scaphoid, lunate, and triquetral carpal bones.
5. Hip (coxal) is between the femur and hipbone.
6. Knee (tibiofemoral) is between the patella and femur and between the femur and tibia.
7. Ankle (talocrucal) is between the tibia and talus and between the fibula and talus.

Applications to Health (p. 207)

1. Rheumatism is a painful state of supporting body structures such as bones, ligaments, tendons, joints, and muscles.
2. Arthritis refers to several disorders characterized by inflammation of joints, often accompanied by stiffness of adjacent structures.
3. Rheumatoid arthritis (RA) refers to inflammation of a joint accompanied by pain, swelling, and loss of function.
4. Osteoarthritis is a degenerative joint disease characterized by deterioration of articular cartilage and spur formation.
5. Gouty arthritis is a condition in which sodium urate crystals are deposited in the soft tissues of joints and eventually destroy the tissues.
6. Lyme disease is caused by a bacterium and transmitted by deer ticks and other ticks. The disease is characterized by a rash, fever and chills, headache, stiff neck, cardiac and neurological disorders, and arthritis.
7. Bursitis is an acute or chronic inflammation of bursae.
8. A dislocation, or luxation, is a displacement of a bone from its joint; a partial dislocation is called a subluxation.
9. A sprain is the forcible wrenching or twisting of a joint with partial rupture to its attachments without dislocation, whereas a strain is the stretching of a muscle.

REVIEW QUESTIONS

1. Define a joint. What factors determine the degree of movement at joints? (p. 182)
2. Distinguish among the three kinds of joints on the basis of structure and function. List the subtypes. Be sure to include degree of movement and specific examples. (p. 182)
3. Explain the components of a synovial joint. Indicate the relation of ligaments and tendons to the strength of the joint and restrictions on movement. (p. 183)
4. Explain how the articulating bones in a synovial joint are held together. (p. 185)
5. What is an accessory ligament? Define the two principal types. (p. 183)
6. What are articular discs? Why are they important? (p. 185)
7. Describe the principle and importance of arthroscopy. (p. 185)
8. What are bursae? What is their function? (p. 185)
9. Define the following principal movements: gliding, angular, rotation, circumduction, and special. Name a joint where each occurs. (p. 186)
10. Have another person assume the anatomical position and execute each of the movements at joints discussed in the text. Reverse roles, and see whether you can execute the same movements. (p. 186)
11. Contrast nonaxial, monaxial, biaxial, and triaxial planes of movement. Give examples of each, and name a joint at which each occurs. (p. 190)
12. For each joint of the body discussed in Exhibits 8-3 through 8-9, be sure that you can name the bones that form the joint, identify the joint by type, and list the anatomical components of the joint. (p. 194)
13. Define the following: rheumatoid arthritis (RA), osteoarthritis, gouty arthritis, Lyme disease, bursitis, dislocation, sprain, and strain. (p. 207)
14. Refer to the glossary of key medical terms associated with joints. Be sure that you can define each term. (p. 209)

SELF QUIZ

1. Match the following:

___ a. decrease in angle between surfaces of bones		A. abduction
___ b. simplest kind of movement that can occur at a joint; no angular or rotary motion involved; for example, ribs moving against vertebrae		B. adduction
___ c. state of entire body when it is in anatomical position		C. circumduction
___ d. movement away from the midline of the body		D. dorsiflexion
___ e. movement of a bone around its own axis		E. extension
___ f. position of foot when heel is on the floor and rest of foot is raised		F. flexion
		G. gliding
		H. inversion
		I. plantar flexion
		J. rotation

2. Indicate whether the following structures are associated with the shoulder, hip, or knee joints.
   a. medial and lateral menisci: ______
   b. anterior and posterior cruciate ligaments: ______
   c. pubofemoral ligaments: ______
   d. subdeltoid bursa: ______
   e. fibular collateral ligament: ______
   f. glenohumeral ligament: ______

Complete the following:

3. Fibrous joints (have? lack?) a synovial cavity.
4. The tibia/fibula joint is a fibrous joint that has (more? less?) flexibility than sutures. The joint is known as a (syndesmosis? gomphosis?).

5. Synchondroses involve (hyaline? fibrous?) cartilage between bone. An example is the ________ cartilage between diaphysis and epiphysis of a growing bone.

6. Match the following:

___	**a.** monaxial joint; only rotation possible	**A.**	ball-and-socket
___	**b.** joint between carpal and metacarpal proximal to the thumb joint	**B.**	ellipsoidal
___	**c.** shoulder and hip joints	**C.**	gliding
___	**d.** spool-like surface articulated with concave surface	**D.**	hinge
___	**e.** monaxial joint; only flexion and extension possible	**E.**	pivot
___	**f.** biaxial joints (two answers)	**F.**	saddle

Choose the one best answer to these questions:

___ **7.** Which structure is extracapsular in location?
A. acetabular labrum; B. posterior cruciate ligament; C. ligament of the head of the femur; D. tibial collateral ligament; E. articular cartilage.

___ **8.** All these structures are associated with the knee joint *except*
A. glenoid labrum; B. patellar ligament; C. infrapatellar bursa; D. medial meniscus; E. tibial collateral ligament.

___ **9.** The joint between the epiphysis and the diaphysis of a long bone is
(1) one example of a synchondrosis.
(2) a synarthrosis.
(3) temporary and becomes a synostosis.
A. (1) only; B. (2) only; C. (3) only; D. all of the above; E. (1) and (3).

___ **10.** A joint
(1) may also be called an articulation.
(2) is formed where one bone joins another bone.
(3) is formed where a cartilage joins a bone.
(4) would be classified as freely movable if it were anywhere but the skull.
A. (1) only; B. (2) only; C. (3) only; D. (4) only; E. (1), (2), and (3).

___ **11.** The pubic symphysis
(1) represents one type of immovable joint (synarthrosis).
(2) is one of the subtypes of movable joints (diarthrosis).
(3) is a joint found in the vertebral column.
(4) is a good example of an amphiarthrosis (slightly movable joint).
A. (1) only; B. (2) only; C. (3) only; D. (4) only; E. (2) and (3).

___ **12.** Diarthrotic joints (diarthroses)
(1) are enclosed in a synovial-lined capsule and have ligaments uniting the bones of the joint.
(2) include relatively few of the joints of the body.
(3) always permit flexion, extension, abduction, and adduction.
A. (1) only; B. (2) only; C. (3) only; D. all of the above; E. none of the above.

___ **13.** The synovial membrane
(1) secretes synovial fluid into a joint cavity.
(2) lines the fibrous capsule of movable joints.
(3) is composed of hyaline cartilage in movable joints.
A. (1) only; B. (2) only; C. (3) only; D. (1) and (2); E. none of the above.

14. Match the following:

___	**a.** contains glenoid labrum	**A.**	intervertebral joints
___	**b.** anterior and posterior longitudinal ligaments, as well as ligamenta flava, add stability to these joints	**B.**	hip joint
___	**c.** ankle joint	**C.**	wrist joint
___	**d.** contains zona orbicularis	**D.**	talocrural
___	**e.** radial and ulnar collateral ligaments located here	**E.**	shoulder joint

___ **15.** Which of the following pairs of terms are correctly (closely) associated or matched?
A. shoulder–acetabulum; B. knee joint–cruciate ligaments; C. symphysis–epiphyseal line; D. ellipsoid joint–interphalangeal; E. shoulder joint–dens.

___ **16.** Which of the following pairs of terms is correctly (closely) associated or matched?
A. knee joint–semilunar cartilages (menisci); B. knee–acetabulum; C. inversion–shoulder joint; D. cartilaginous joint–joint cavity; E. deltoid ligament–radiocarpal joint.

Circle T (true) or F (false) for the following:

T F 17. Joints that are very stable tend to have more mobility than joints that are unstable.

T F 18. All fibrous joints are synarthrotic and all cartilaginous joints are amphiarthrotic.

T F 19. In synovial joints synovial membranes cover the surfaces of articular cartilages.

T F 20. Bursae are saclike structures that reduce friction at joints.

T F 21. Sutures, syndesmoses, and symphyses are kinds of fibrous joints.

Arrange the answers in correct sequence.

___ ___ ___ **22.** From most mobile to least mobile:
A. amphiarthrotic
B. diarthrotic
C. synarthrotic

Muscle Tissue

9

STUDENT OBJECTIVES

1. List the general characteristics and functions of muscle tissue.
2. Compare the location, microscopic appearance, nervous control, and functions of the three kinds of muscle tissue.
3. Explain the relation of blood vessels and nerves to skeletal muscles.
4. Outline the principal events associated with the sliding-filament mechanism of muscle contraction.
5. Describe the structure and importance of a neuromuscular junction and a motor unit.
6. Compare the structure and function of the three types of skeletal muscle fibers.
7. Explain the effects of aging on muscle tissue.
8. Describe the development of the muscular system.
9. Define such common muscular disorders as fibrosis, fibromyalgia, muscular dystrophies, myasthenia gravis (MG), and abnormal contractions.
10. Define key medical terms associated with the muscular system.

CHAPTER OUTLINE

- **Characteristics**
- **Functions**
- **Types**
- **Skeletal Muscle Tissue**

Connective Tissue Components
Nerve and Blood Supply
Histology

- **Contraction**

Sliding-Filament Mechanism
Neuromuscular Junction
Motor Unit
Mechanism
All-or-None Principle
Muscle Length and Force of Contraction
Muscle Tone
Muscular Atrophy and Hypertrophy
Types of Skeletal Muscle Fibers

- **Cardiac Muscle Tissue**
- **Smooth (Visceral) Muscle Tissue**
- **Regeneration of Muscle Tissue**
- **Aging and Muscle Tissue**
- **Developmental Anatomy of the Muscular System**
- **Applications to Health**
- **Key Medical Terms Associated with the Muscular System**

Although bones and joints provide leverage and form the framework of the body, they are not capable of moving the body by themselves. Motion is an essential body function that results from the contraction and relaxation of muscles.

Muscle tissue is highly specialized for actively generating force and constitutes about 40 to 50 percent of total body weight. The scientific study of muscles is known as ***myology*** (mī-OL-ō-jē; *myo* = muscle; *logos* = study of).

The developmental anatomy of the muscular system is considered at the end of the chapter.

CHARACTERISTICS

Muscle tissue has four principal characteristics.

1. ***Excitability,*** a property of both muscle and nerve tissue, is the ability to receive and respond to stimuli. A stimulus is a change in the internal or external environment strong enough to initiate an electric current that results in an action potential.
2. ***Contractility*** is the ability of muscle tissue to contract—that is, to shorten and thicken—and thereby generate forces that enable work to be done when a sufficient stimulus is received.
3. ***Extensibility*** is the ability of muscle tissue to be stretched (extended) without damaging the tissue. Many skeletal muscles are arranged in opposing pairs. While one contracts, the other relaxes and stretches.
4. ***Elasticity*** is the ability of muscle tissue to return to its original shape after contraction or extension.

FUNCTIONS

Through contraction, muscle performs three important functions:

1. ***Motion (both reflex and voluntary).*** Motion is obvious in movements such as walking and running, and in localized movements, such as grasping a pencil, nodding the head, or chest movements involved with breathing. All these movements rely on the integrated functioning of the bones, joints, and skeletal muscles attached to the bones. Less noticeable kinds of motion produced by muscles are the beating of the heart to transport blood throughout the body, churning of food in the stomach, pushing of food through the intestines, contraction of the gallbladder to release bile, and contraction of the urinary bladder to expel urine.
2. ***Maintenance of posture.*** In addition to the movement function, muscle tissue also enables the body to maintain posture. The contraction of skeletal muscles holds the body in stationary positions, such as standing and sitting.
3. ***Heat production.*** The third function of muscle tissue is heat production. Skeletal muscle contractions produce most of the heat generated in the body and are thereby important in maintaining normal body temperature. It has been estimated that as much as 85 percent of all body heat is generated by muscle contractions.

TYPES

Types of muscle tissue are categorized by location, histology (microscopic structure), and nervous and other modes of control.

Skeletal muscle tissue, which is named for its location, is attached primarily to bones and moves the skeleton. (Some skeletal muscles are also attached to skin, other muscles, or deep fascia.) Skeletal muscle tissue is ***striated*** because *striations,* or alternating light and dark bandlike structures, are visible when the tissue is examined under a microscope (see Exhibit 3-3). It is a ***voluntary*** muscle tissue because it can be made to contract and relax by conscious control.

Cardiac muscle tissue forms the bulk of the wall of the heart. It is ***striated*** at the microscopic level and ***involuntary;*** that is, its contraction is usually not under conscious control. Cardiac muscle has its own built-in rhythm (autorhythmicity) that causes the heart to beat about 75 times a minute. Faster or slower rates are possible because heart rate is also influenced by involuntary nerves and certain hormones.

Smooth (visceral) muscle tissue is involved with internal processes related to maintaining the internal environment. It is located in the walls of hollow internal structures, such as blood vessels, the stomach, and the intestines. It is also found in the skin attached to hair follicles. It is referred to as ***nonstriated*** because it lacks striations at the microscopic level. It is usually ***involuntary*** muscle tissue which often has its own built-in rhythm and is also influenced by involuntary nerves and certain hormones.

Thus, all muscle tissues are classified in the following way: (1) skeletal, striated, voluntary muscle tissue; (2) cardiac, striated, involuntary muscle tissue; and (3) smooth, nonstriated, involuntary muscle tissue.

SKELETAL MUSCLE TISSUE

To understand the fundamental mechanisms of muscle movement, you will need some knowledge of the connective tissue components, nerve and blood supply, and histology of skeletal muscle.

CONNECTIVE TISSUE COMPONENTS

The term ***fascia*** (FASH-ē-a) is applied to a sheet or broad band of fibrous connective tissue beneath the skin or around muscles and other organs of the body. ***Superficial fascia (subcutaneous layer)*** is immediately deep to the skin (Figure 9-1b). It covers the entire body and varies in thickness in

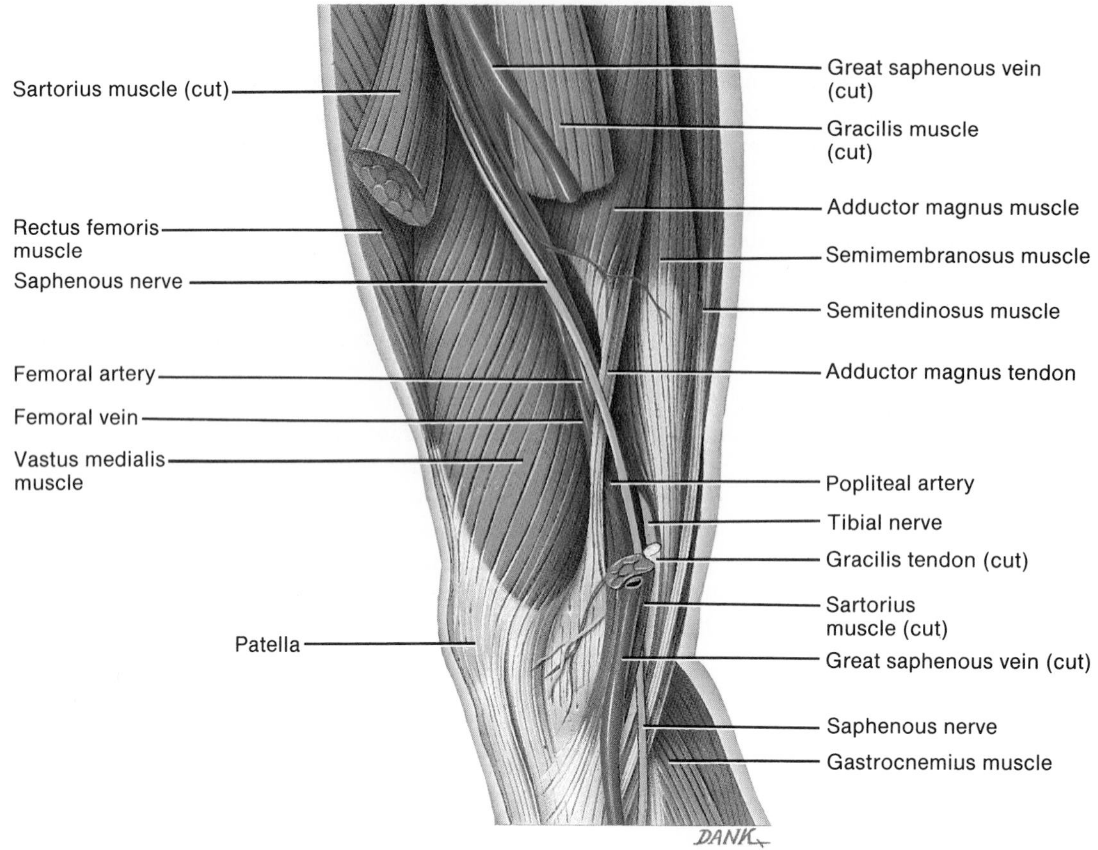

(a) Medial view of right thigh and knee

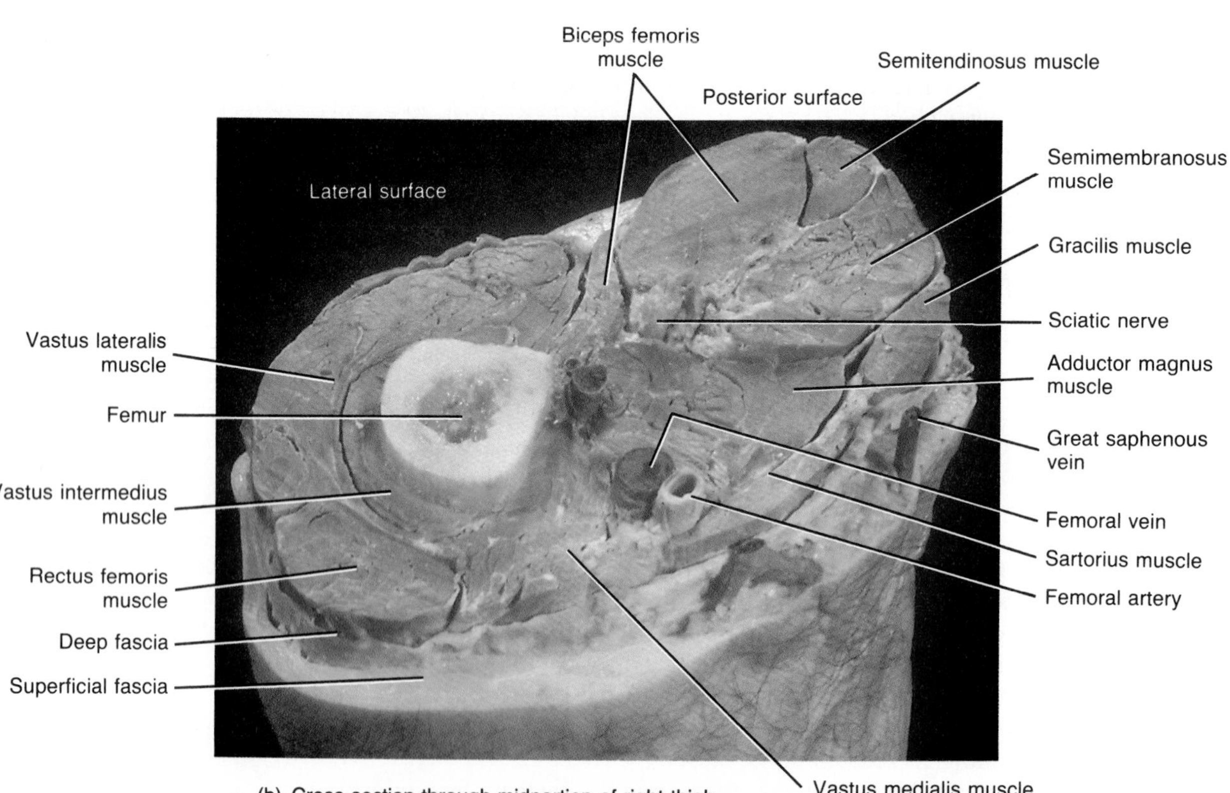

(b) Cross section through midportion of right thigh

FIGURE 9-1 Relation of blood vessels and nerves to skeletal muscles. (b) Courtesy of J. A. Gosling, P. F. Harris, et al., *Atlas of Human Anatomy,* Gower Medical Publishing Ltd., 2nd ed., 1991.

different regions. On the back, or dorsum, of the hand it is quite thin, whereas over the inferior abdominal wall it is thick. The superficial fascia is composed of adipose tissue and areolar connective tissue. The outer layer usually contains fat and varies considerably in thickness, whereas the inner layer is thin and elastic. Between the two layers are found arteries, veins, lymphatics, nerves, the mammary glands, and the facial muscles. Superficial fascia has a number of important functions.

1. It serves as a storehouse for water and particularly for fat. Much of the fat of an overweight person is in the superficial fascia.
2. It forms a layer of insulation protecting the body from loss of heat.
3. It provides mechanical protection from blows.
4. It provides a pathway for nerves and vessels to enter and exit muscles.

Deep fascia is by far the more extensive type of fascia (Figure 9-1b). It is a dense connective tissue that lines the body wall and extremities and holds muscles together, separating them into functioning groups. Functionally, deep fascia allows free movement of muscles, carries nerves and blood and lymphatic vessels, fills spaces between muscles, and sometimes provides the origin (one point of attachment to a bone) for muscles.

CLINICAL APPLICATION

Fasciae and Surgery

Lines of fusion of fascial sheets are essentially avascular and for this reason they are often sites for ***surgical incisions.*** Surgeons also prefer fascial junctional areas for anchoring of sutures because of the strength of fasciae and the strong union that results from wound healing.

Three layers of fibrous connective tissue extend from the deep fascia to further protect and strengthen skeletal muscles (see Figure 9-2b). The entire muscle is wrapped with a substantial quantity of fibrous connective tissue called the ***epimysium*** (ep′-i-MĪZ-ē-um). Bundles of fibers (cells) called ***fasciculi*** (fa-SIK-yoo-lī) or ***fascicles*** (FAS-i-kuls) are covered by a fibrous connective tissue called the ***perimysium*** (per′-i-MĪZ-ē-um). ***Endomysium*** (en′-dō-MĪZ-ē-um) is a fibrous connective tissue that penetrates into the interior of each fascicle and surrounds the muscle fibers.

Epimysium, perimysium, and endomysium are all continuous with the connective tissue that attaches the muscle to another structure, such as bone or other muscle. All three elements may be extended beyond the muscle fibers as a ***tendon*** (*tendere* = to stretch out)—a cord of connective tissue that attaches a muscle to the periosteum of a bone. When the connective tissue elements extend as a broad, flat layer, the tendon is called an ***aponeurosis.*** This structure also attaches to the coverings of a bone or another muscle. An example is the aponeurosis of the external oblique muscle (see Figure 10-12a). When a muscle contracts, the tendon and its corresponding bone or muscle are pulled toward the contracting muscle. In this way skeletal muscles produce movement.

Certain tendons, especially those of the wrist and ankle, are enclosed by tubes of fibrous connective tissue called ***tendon (synovial) sheaths.*** They are similar in structure to bursae. The inner layer of a tendon sheath, the ***visceral layer,*** is attached to the surface of the tendon. The outer layer is known as the ***parietal layer*** (see Figure 10-20a). Between the layers is a cavity that contains a film of synovial fluid. Tendon sheaths permit tendons to slide back and forth more easily.

CLINICAL APPLICATION

Tenosynovitis

Tenosynovitis (ten′-ō-sin-ō-VĪ-tis) refers to inflammation involving the tendons, tendon sheaths, and synovial membranes surrounding certain joints. The wrists, shoulders, elbows (tennis elbow), finger joints (trigger finger), ankles, feet, and associated tendons are most often affected. The affected sheaths may become visibly swollen because of fluid accumulation, or they may remain dry. Local tenderness is variable, and there may be disabling pain with movement of the body part. The condition often follows some form of trauma, strain, or excessive exercise.

NERVE AND BLOOD SUPPLY

Skeletal muscles are well supplied with nerves and blood vessels. The innervation and vascularization are directly related to contraction, the chief characteristic of muscle. For a skeletal muscle fiber (cell) to contract, it must first be stimulated by an action potential (nerve impulse) from a nerve cell. Muscle contraction also requires a good deal of energy and therefore large amounts of nutrients and oxygen. Moreover, the waste products of these energy-producing reactions must be eliminated. Thus, prolonged muscle action depends on a rich blood supply to deliver nutrients and oxygen and remove wastes.

Generally, an artery and one or two veins accompany each nerve that penetrates a skeletal muscle. The larger branches of the blood vessels accompany the nerve branches through the connective tissue of the muscle (see Figure 9-1). Microscopic blood vessels called capillaries are plentiful within the endomysium. Each muscle fiber is thus in close contact with one or more capillaries. Each skeletal muscle fiber also makes contact with a portion of a nerve cell.

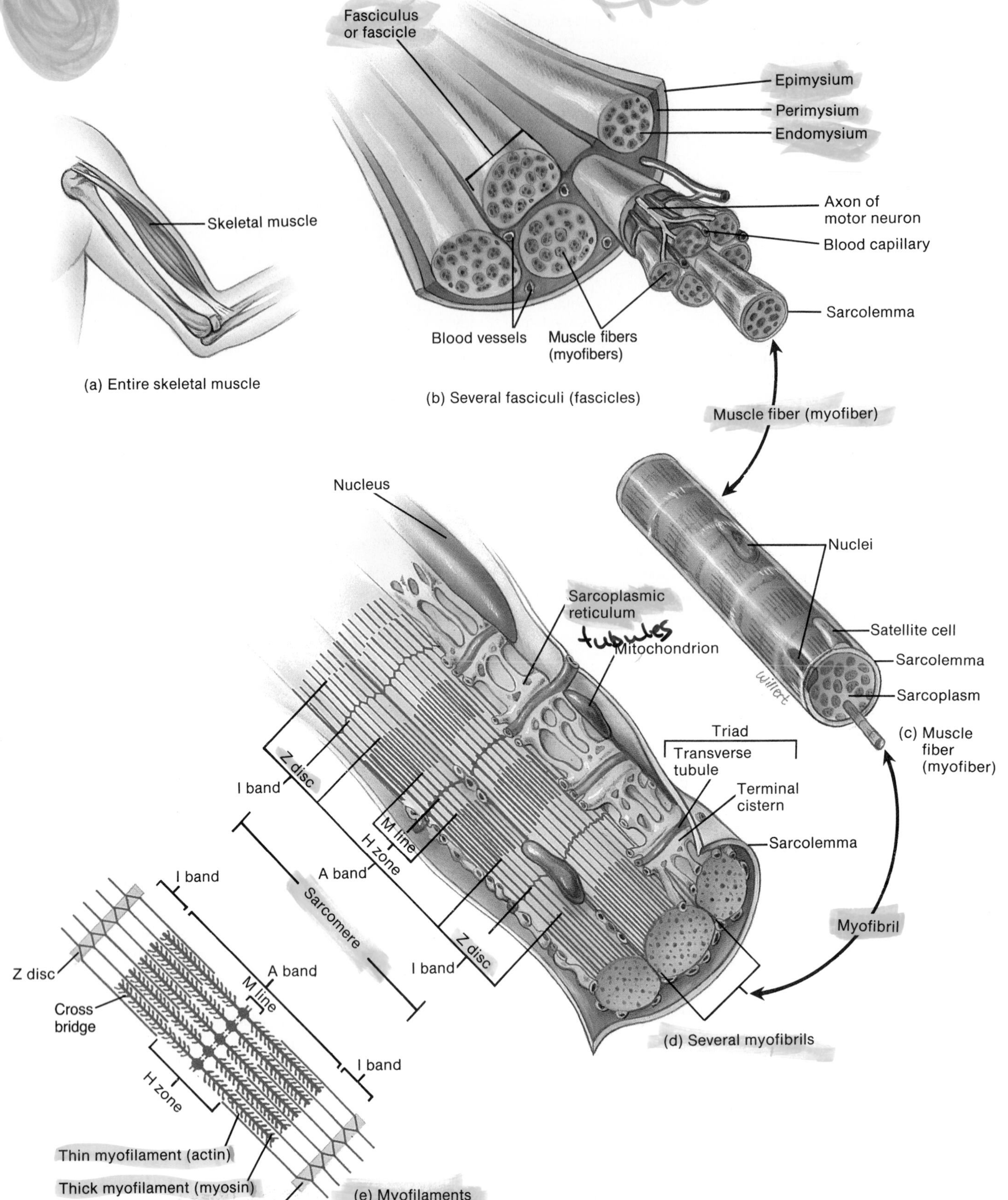

Skeletal muscle

(a) Entire skeletal muscle

Fasciculus or fascicle
Epimysium
Perimysium
Endomysium
Axon of motor neuron
Blood capillary
Sarcolemma
Blood vessels
Muscle fibers (myofibers)

(b) Several fasciculi (fascicles)

Muscle fiber (myofiber)

Nuclei
Satellite cell
Sarcolemma
Sarcoplasm

(c) Muscle fiber (myofiber)

Nucleus
Sarcoplasmic reticulum
Mitochondrion
Triad
Transverse tubule
Terminal cistern
Sarcolemma
Z disc
I band
M line
H zone
A band
Sarcomere
Z disc
I band
Myofibril

(d) Several myofibrils

I band
Z disc
A band
M line
Cross bridge
H zone
I band
Thin myofilament (actin)
Thick myofilament (myosin)
Z disc

(e) Myofilaments

FIGURE 9-2 Organization of a skeletal muscle from gross to molecular levels. (f) Courtesy of D. E. Kelly, from *Introduction to the Musculoskeletal System* by Cornelius Rosse and D. Kay Clawson, Harper & Row, Publishers, Inc., New York, 1970.

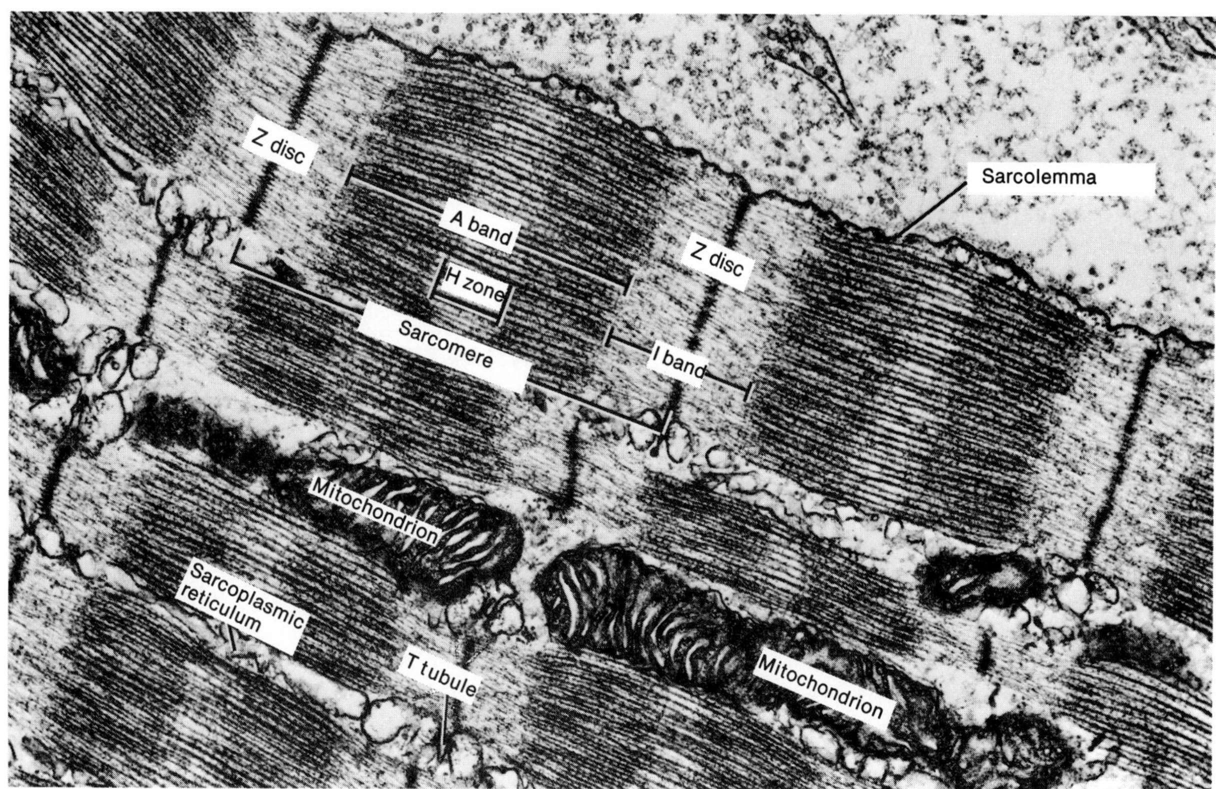

(f) Electron micrograph of several sarcomeres, 35,000×

HISTOLOGY

When a typical skeletal muscle is teased apart and viewed microscopically, it can be seen to consist of thousands of elongated cylindrical cells called ***muscle fibers*** or ***myofibers*** (Figure 9-2c; see also Exhibit 3-3). These fibers lie parallel to one another and range from 10 to 100 μm in diameter. Some fibers may reach lengths of 30 cm (12 in.) or more. Each muscle fiber is enveloped by a plasma membrane called the ***sarcolemma*** (*sarco* = flesh; *lemma* = sheath). The sarcolemma surrounds a quantity of cytoplasm called ***sarcoplasm.*** Within the sarcoplasm of a muscle fiber and lying close to the sarcolemma are many nuclei. Skeletal muscle fibers are thus multinucleate. Since skeletal muscle fibers are unusually long and metabolically very active, it is not surprising that they have many nuclei to direct protein synthesis. Immediately beneath the sarcolemma, around the nuclei, and in the sarcoplasm between myofibrils are found numerous mitochondria. The large numbers of mitochondria are related to the large amounts of energy in the form of adenosine triphosphate (ATP) that have to be generated by muscle tissue in order to contract. Recall that ATP stores energy for cellular use. The sarcoplasm also contains myofibrils (to be described shortly), special high-energy molecules (also to be described shortly), enzymes, and ***sarcoplasmic reticulum*** (sar′-kō-PLAZ-mik re-TIK-yoo-lum), a network of membrane-enclosed tubules comparable to smooth endoplasmic reticulum (Figure 9-2d). Dilated sacs of sarcoplasmic reticulum, called ***terminal cisterns,*** form ringlike channels around myofibrils. Running perpendicularly to the sarcoplasmic reticulum are ***transverse tubules (T tubules),*** which are extensions of the sarcolemma that open to the outside of the fiber. A ***triad*** consists of a transverse tubule and the segments of sarcoplasmic reticulum (terminal cisterns) on either side. Skeletal muscle fibers also contain varying amounts of ***myoglobin,*** a reddish pigment similar to hemoglobin in blood. Myoglobin stores oxygen until needed by mitochondria to generate ATP.

A highly magnified view of skeletal muscle fibers reveals that they are composed of cylindrical structures, about 1 or 2 μm in diameter, called ***myofibrils*** (Figure 9-2d). Ranging in number from several hundred to several thousand, myofibrils run longitudinally through the muscle fiber and consist of two kinds of even smaller structures called ***myofilaments.*** The ***thin (actin) myofilaments*** are about 6 nm in diameter. The ***thick (myosin) myofilaments*** are about 16 nm in diameter.

The myofilaments of a myofibril do not extend the entire length of a muscle fiber—they are arranged in compartments called ***sarcomeres.*** Sarcomeres are separated from one another by narrow zones of dense material called ***Z discs (lines).*** Within a sarcomere, certain areas can be distinguished (Figure 9-2e,f). A dark, dense area, called the ***A (anisotropic) band,*** represents the length of thick myofilaments. The sides of the A band are darkened by the overlapping of thick and thin myofilaments. The length of darkening depends on the extent of overlapping. As you will see later, the greater the degree of contraction, the greater the overlapping of thick and thin myofilaments. A light-colored, less dense area called the ***I (isotropic) band*** is composed of thin myofilaments only. This combination of alternating dark A bands and light I bands gives the muscle fiber its striated (striped) appearance. A narrow ***H zone*** is a region in the center of the A band that contains thick myofilaments only. In the center of the H zone is the ***M line,*** a series of fine threadlike protein molecules that connect the middle parts of adjacent thick myofilaments.

Thin myofilaments are anchored to the Z discs and project in both directions. They are composed mostly of the protein ***actin.*** The actin molecules are arranged in two single strands that entwine helically and give the thin myofilaments their characteristic shape (Figure 9-3a). Each actin molecule contains a ***myosin-binding site*** that interacts with a cross bridge of a myosin molecule (to be described shortly). Besides actin, the thin myofilaments contain two other protein molecules, ***tropomyosin*** and ***troponin,*** that are involved in the regulation of muscle contractions. Tropomyosin is arranged in strands that are loosely attached to the actin helices. Troponin is located at regular intervals on the surface of tropomyosin. Together tropomyosin and troponin are referred to as the ***tropomyosin–troponin complex.***

Thick myofilaments overlap the free ends of the thin myofilaments and occupy the A band region of a sarcomere. They are anchored to Z discs by recently discovered myofilaments called ***elastic myofilaments.*** Thick myofilaments are composed mostly of the protein ***myosin.*** A myosin molecule is shaped like a golf club. The tails (handles of the golf club) are arranged parallel to each other, forming the shaft of the thick myofilament. The heads of the golf clubs project outward from the shaft and are arranged spirally on the surface of the shaft. The projecting heads are referred to as ***cross bridges*** and contain an ***actin-binding site*** and an ***ATP-binding site*** (Figure 9-3b).

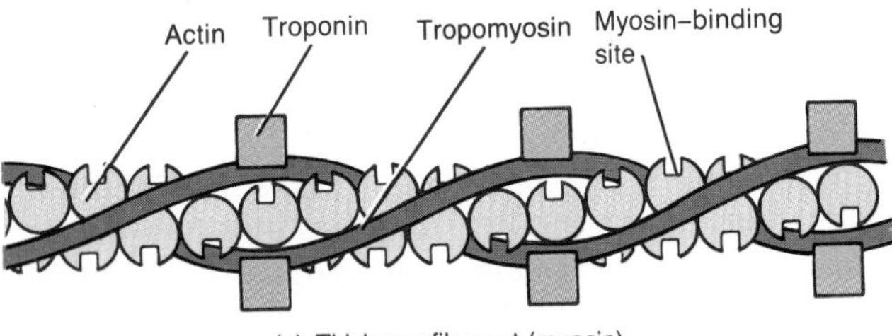

(a) Thick myofilament (myosin)

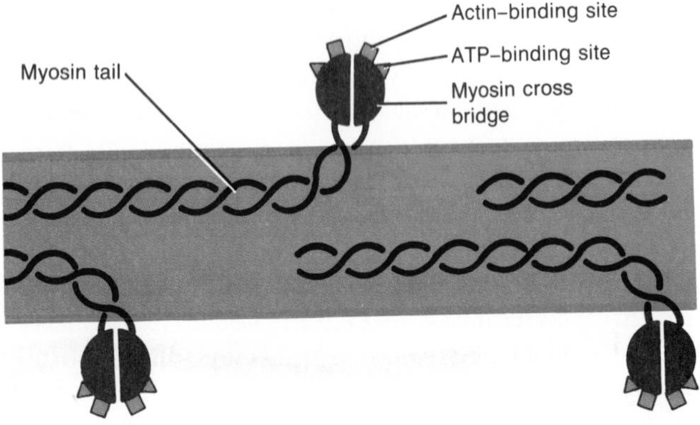

(b) Thin myofilament (actin)

FIGURE 9-3 Detailed structure of myofilaments.

CONTRACTION

SLIDING-FILAMENT MECHANISM

During muscle contraction, myosin cross bridges slide inward toward the H zone. The sarcomere shortens, but the lengths of the thin and thick myofilaments do not change. The myosin cross bridges of the thick myofilaments connect with portions of actin of the thin myofilaments. The myosin cross bridges move like the oars of a boat on the surface of the thin myofilaments, and the thin and thick myofilaments slide past each other as the cross bridges pull on the thin myofilaments. As the thin myofilaments move past the thick myofilaments, the H zone narrows and even disappears when the thin myofilaments meet at the center of the sarcomere (Figure 9-4). In fact, the myosin cross bridges may pull the thin myofilaments of each sarcomere so far inward that their ends overlap. As the thin myofilaments slide inward, the Z discs are drawn toward each other and the sarcomere is shortened. The sliding of myofilaments and shortening of sarcomeres causes the shortening of the muscle fibers. All these events associated with the movement of myofilaments are known as the ***sliding-filament mechanism*** of muscle contraction.

NEUROMUSCULAR JUNCTION

For a skeletal muscle fiber to contract, a stimulus must be applied to it. The stimulus is delivered by a nerve cell, or ***neuron.*** A neuron has a threadlike process called a fiber, or axon, that may run 91 cm (3 ft) or more to a muscle. A bundle of such fibers from many different neurons composes a nerve. A neuron that stimulates muscle tissue is called a ***motor neuron.***

On entering a skeletal muscle, the axon of a motor neuron branches into axon terminals that approach—but do not touch—a portion of the sarcolemma of a muscle fiber. The region of the sarcolemma adjacent to the axon terminal is known as the ***motor end plate.*** The term ***neuromuscular junction*** or ***myoneural junction*** refers to the axon terminal of a motor neuron together with the motor end plate (Figure 9-5). Close examination of a neuromuscular junction reveals that the distal ends of the axon terminals are expanded into bulblike structures called ***synaptic end bulbs.*** The bulbs contain membrane-enclosed sacs, the ***synaptic vesicles,*** that store chemicals called ***neurotransmitters.*** These chemicals

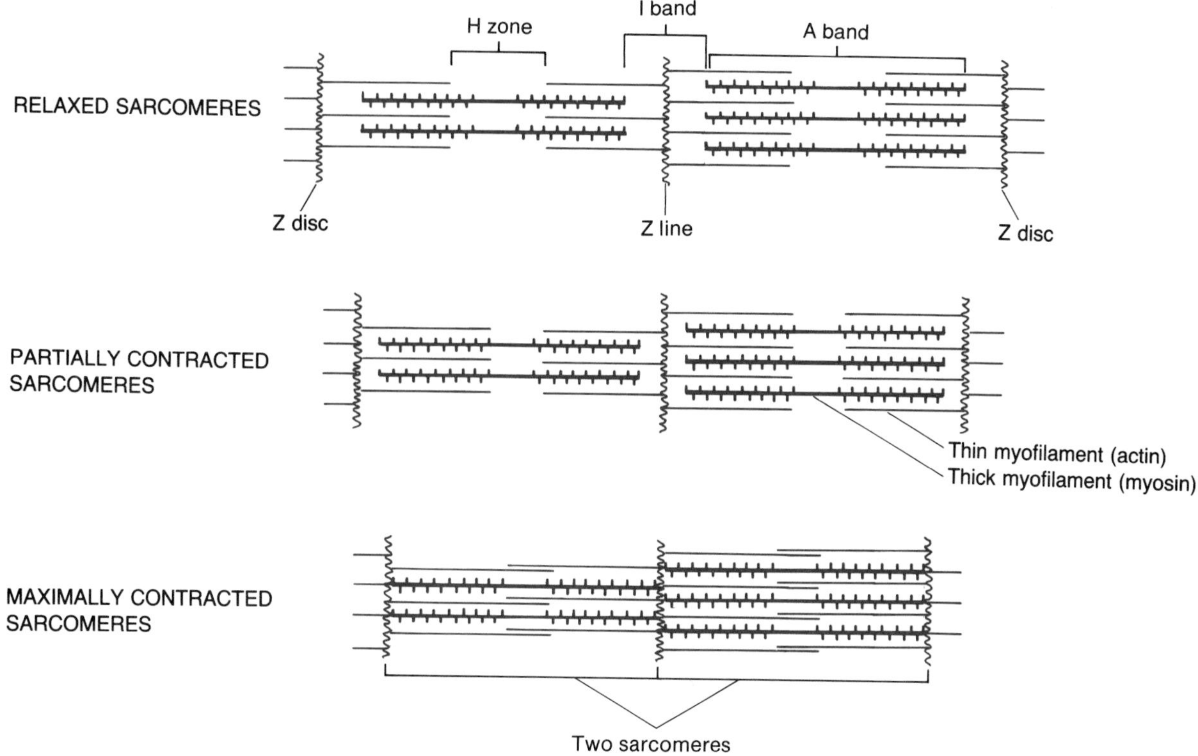

FIGURE 9-4 Sliding-filament mechanism of muscle contraction. Shown are the positions of the parts of two sarcomeres in various states. Note the movement of the thin myofilaments and the relative size of the sarcomeres.

determine whether an impulse is passed on to a muscle (or gland or another nerve cell). The invaginated area of the sarcolemma under the axon terminal is referred to as a ***synaptic gutter (trough)***, and the space between the axon terminal and sarcolemma is known as a ***synaptic cleft***. There are numerous folds of the sarcolemma along the synaptic gutter, called ***subneural clefts***, which greatly increase the surface area of the synaptic gutter. This arrangement allows for increased numbers of receptor site molecules that are able to bond to the neurotransmitter. The neurotransmitter released at neuromuscular junctions is ***acetylcholine (ACh)***.

In the vast majority of skeletal muscle fibers, there is only one neuromuscular junction for each fiber, and the junction is located in the middle of the fibers. Thus, the muscle action potential (impulse) spreads from the center of the sarcolemma of the fiber to the ends and through the transverse tubules as well, thereby permitting nearly simultaneous contraction of all sarcomeres in the fiber as the action potential spreads deeply into the sarcoplasm by way of the transverse tubules.

MOTOR UNIT

A motor neuron (nerve cell), together with all the muscle fibers it stimulates, is referred to as a ***motor unit*** (Figure 9-6). A single motor neuron may innervate about 150 muscle fibers, depending on the region of the body. This means that stimulation of one neuron will cause the simultaneous contraction of about 150 muscle fibers. In addition, all the muscle fibers of a motor unit will contract and relax together. Muscles that control precise movements, such as the extrinsic (external) eye muscles, have fewer than 10 muscle fibers to each motor unit. Muscles of the body that are responsible for gross movements, such as the biceps brachii in the arm and gastrocnemius in the leg, may have as many as 2000 muscle fibers in each motor unit.

Stimulation of a motor neuron produces a contraction in all the muscle fibers in a particular motor unit. Accordingly, the total strength of contraction is varied by adjusting the number of motor units that are activated. The process of increasing the number of active motor units is called ***recruitment (motor unit summation)***, and is determined by the needs of the body at a given time. The various motor neurons to a whole muscle fire asynchronously; that is, while some are excited, others are inhibited. Therefore, all the motor units are not contracting at the same time. This pattern of firing of motor neurons prevents fatigue while maintaining contraction by allowing a brief rest for the inactive units. The alternating motor units relieve one another so smoothly that the contraction can be sustained for long periods. It also helps to maintain a state of partial contraction in a relaxed skeletal muscle, a phenomenon

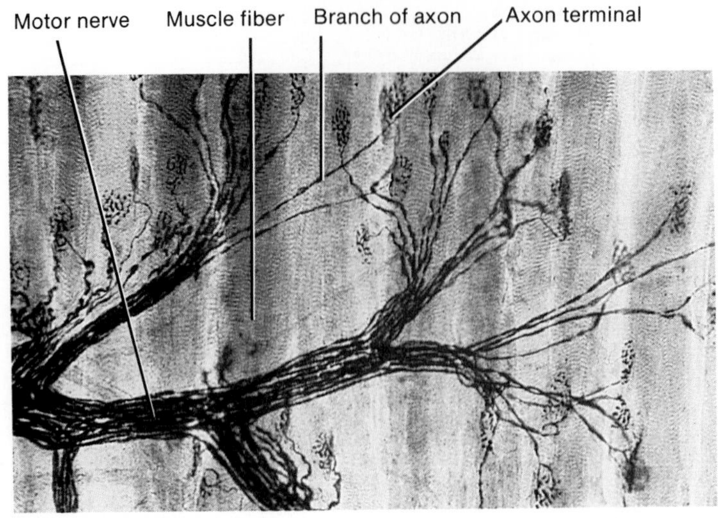

(a) Photomicrograph (450 x)

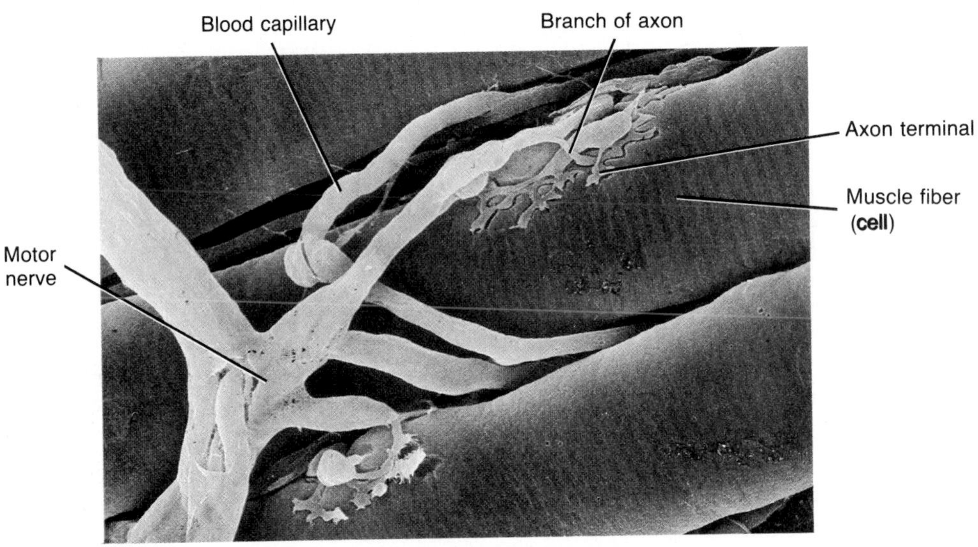

(b) Scanning electron micrograph, 1,650×

FIGURE 9-5 Neuromuscular junction. (a) © Biophoto, SPL, Photo Researchers. (b) (FUJIJA)

called muscle tone (to be described later in the chapter). Also, recruitment is one factor responsible for producing smooth movements during a muscle contraction, rather than a series of jerky movements.

MECHANISM

In lieu of a detailed discussion of the mechanism of muscle contraction, the process is outlined in Exhibit 9-1. By following the step-by-step summary, you should be able to understand the basics.

CLINICAL APPLICATION

Rigor Mortis
Following death, certain chemical changes occur in muscle tissue that affect the status of the muscles. Because of a lack of ATP, myosin cross bridges remain attached to the actin myofilaments, thus preventing relaxation. The resulting condition, in which muscles are in a state of rigidity (cannot contract or stretch), is called ***rigor mortis*** (rigidity of death). The time elapsing between death and the onset

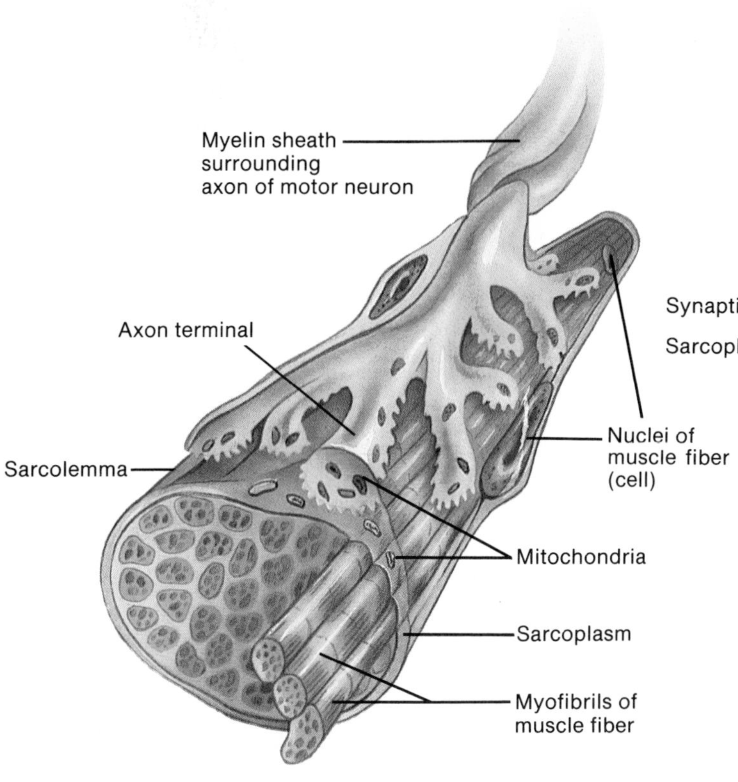

Myelin sheath
surrounding
axon of motor neuron
Axon terminal
Sarcolemma
Nuclei of
muscle fiber
(cell)
Mitochondria
Sarcoplasm
Myofibrils of
muscle fiber

(c) Diagram based on photomicrograph

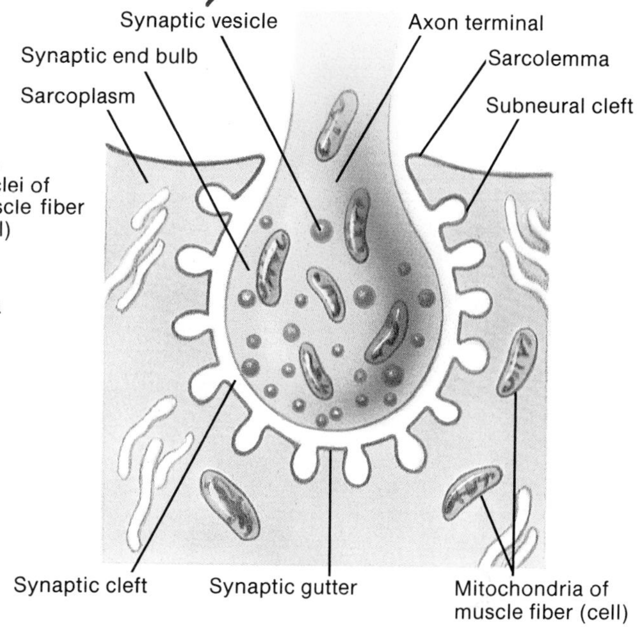

Synaptic vesicle
Axon terminal
Synaptic end bulb
Sarcolemma
Sarcoplasm
Subneural cleft
Synaptic cleft
Synaptic gutter
Mitochondria of
muscle fiber (cell)

(d) Enlargement based on electron micrograph

of rigor mortis varies greatly among individuals. Those who have had long, wasting illnesses undergo rigor mortis more quickly. Rigor mortis is not a permanent state that continues unabated after death. Depending on conditions, it lasts about 24 hours, then begins to abate, and disappears after another 12 hours when tissues begin to disintegrate, although a residual stiffness in the joints will remain.

ALL-OR-NONE PRINCIPLE

The weakest stimulus from a neuron that can still initiate a contraction is called a ***threshold (liminal) stimulus.*** A stimulus of lesser intensity, which cannot initiate contraction, is referred to as a ***subthreshold (subliminal) stimulus.*** According to the ***all-or-none principle,*** once a threshold or greater stimulus is applied, individual muscle fibers of a motor unit will contract to their fullest extent or will not contract at all, provided conditions remain constant. In other words *individual muscle fibers* do not partly contract. The principle does not mean that the entire muscle must be either fully relaxed or fully contracted because some of the many motor units that compose the entire muscle are contracting and others are relaxing. Thus, the muscle as a whole can have graded contractions (varying magnitudes) in order to perform a specific task. For example, the muscles of the arm do not contract to the same extent

Spinal cord
Neuromuscular
junction
Muscle fibers (cells)

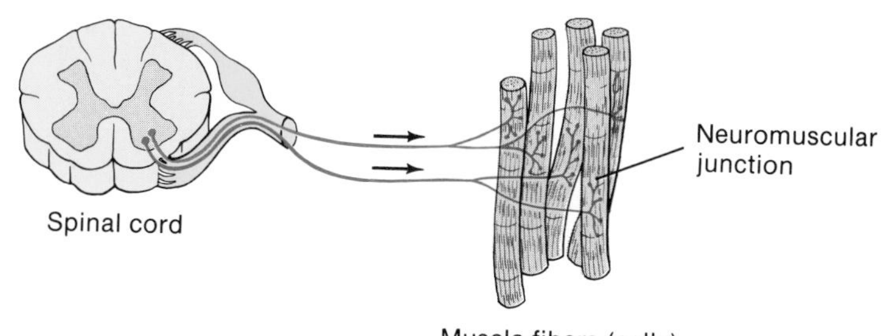

FIGURE 9-6 Motor unit. Shown are two motor neurons, one in red and one in blue, each supplying its respective muscle fibers (cells).

EXHIBIT 9-1

Summary of Events Involved in Contraction and Relaxation of a Skeletal Muscle Fiber

1. A nerve impulse (nerve action potential) causes synaptic vesicles in motor axon synaptic end bulbs to release acetylcholine (ACh).
2. Acetylcholine diffuses across the synaptic cleft within the neuromuscular junction and initiates a muscle action potential that spreads over the surface of the sarcolemma and transverse tubules.
3. The muscle action potential enters the transverse tubules and sarcoplasmic reticulum and stimulates the sarcoplasmic reticulum to release calcium ions from storage into the sarcoplasm.
4. Calcium ions combine with troponin, causing the tropomyosin–troponin complex to move, thus exposing the myosin-binding sites on actin.
5. When a muscle action potential stimulates a muscle fiber, ATPase splits ATP into adenosine diphosphate (ADP) + P and energy is released. The released energy activates (energizes) myosin cross bridges, which combine with the exposed myosin-binding sites on actin, and apply force that causes the myosin cross bridges to move toward the H zone (power stroke). This movement results in the sliding of the thin myofilaments past thick myofilaments.
6. The sliding draws the Z discs toward each other, the sarcomere shortens, the muscle fibers contract, and the muscle contracts.
7. Acetylcholine is inactivated by acetylcholinesterase (AChE). As a result, ACh no longer has any effect at the neuromuscular junction.
8. Once no more muscle action potentials are generated, calcium ions are actively transported back into the sarcoplasmic reticulum by calsequestrin and calcium ATPase, using energy from ATP breakdown.
9. The low calcium concentration in the sarcoplasm permits the tropomyosin–troponin complex to reattach to actin. As a result, myosin-binding sites of actin become covered, myosin cross bridges separate from actin, ADP is resynthesized into ATP (which reattaches to the ATP-binding site of the myosin cross bridge), and the thin myofilaments return to their relaxed position.
10. Sarcomeres return to their resting lengths, muscle fibers relax, and the muscle relaxes.

when you lift a 10-pound weight as compared with a 1-pound weight. But they do contract in a smooth, graded fashion to lift either weight. The strength of contraction may be decreased by fatigue, lack of nutrients, or lack of oxygen.

MUSCLE LENGTH AND FORCE OF CONTRACTION

Having already considered the sliding-filament mechanism of muscle contraction, we can now examine the relationship between muscle length and force of contraction (tension). As you already know, a skeletal muscle fiber contracts when myosin cross bridges of thick myofilaments connect with portions of thin myofilaments within a sarcomere. As it turns out, a muscle fiber develops its greatest tension when there is maximum overlap between thick and thin myofilaments (Figure 9-7). At this length, the optimal length, the maximum number of myosin cross bridges make contact with thin myofilaments to bring about the greatest force of contraction. As a muscle fiber is stretched, fewer and fewer myosin cross bridges make contact with thin myofilaments, and the force of contraction progressively decreases. In fact, if a muscle fiber is stretched to 175 percent its optimal length, no myosin cross bridges attach to thin myofilaments and no contraction occurs. At lengths less than the optimum length, the force of contraction also decreases. This is because extreme shortening of sarcomeres causes thin myofilaments to overlap and thick myofilaments to crumple as they run into Z discs, resulting in fewer myosin cross bridge contacts with thin myofilaments. In general, changes in resting muscle fiber length greater or less than the optimum length rarely exceed 30 percent.

MUSCLE TONE

A muscle may be in a state of partial contraction because at any given time some fibers in a muscle are contracted, whereas others are relaxed. This contraction tightens a muscle, but there may not be enough fibers contracting at the time to produce movement. Recruitment (asynchronous firing) allows the contraction to be sustained for long periods.

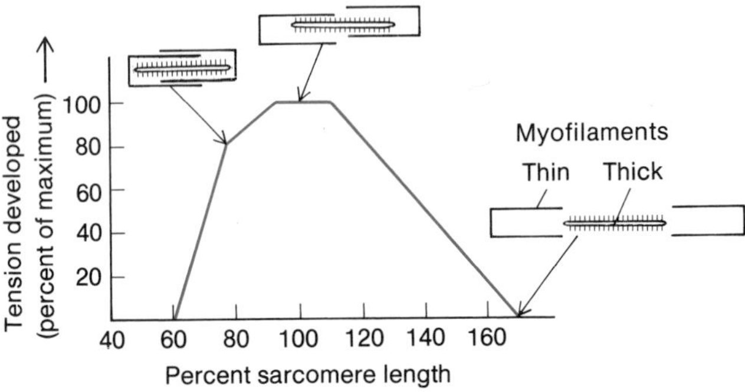

FIGURE 9-7 Length-tension relationship in skeletal muscle fibers (cells). Maximum tension is produced at sarcomere length 100 percent.

A sustained partial contraction of portions of a skeletal muscle results in ***muscle tone*** and occurs even in a relaxed muscle. Tone is essential for maintaining posture.

CLINICAL APPLICATION

Hypotonia and Hypertonia

Abnormalities of muscle tone are expressed as hypotonia or hypertonia. ***Hypotonia*** refers to decreased or lost muscle tone. Such muscles are said to be ***flaccid*** (FLAK-sid or FLAS-sid). Flaccid muscles are loose, their normal rounded contour is replaced by a flattened appearance, and the affected limbs are hyperextended. As you will see later, certain disorders of the nervous system may result in ***flaccid paralysis,*** which is characterized by loss of muscle tone, loss or reduction of tendon reflexes, and atrophy (wasting away) and degeneration of muscles.

Hypertonia refers to increased muscle tone and is expressed in two ways: spasticity or rigidity. ***Spasticity*** is characterized by increased muscle tone (stiffness) associated with an increase in tendon reflexes and pathological reflexes (Babinski sign). Spastic muscles exhibit increased resistance to passive movement, followed by a sudden or gradual release of resistance (clasp-knife reaction). As you will see later, certain disorders of the nervous system may result in ***spastic paralysis,*** partial paralysis in which the muscle exhibits spasticity. ***Rigidity*** refers to increased muscle tone, although reflexes are not affected. If rigidity is uniform throughout a range of movement, it is called lead-pipe rigidity; if rigidity is interrupted by a series of jerks, it is called cogwheel rigidity, as in Parkinson's disease.

MUSCULAR ATROPHY AND HYPERTROPHY

Muscular atrophy (A-trō-fē) is a wasting away of muscles. Individual muscle fibers decrease in size owing to a progressive loss of myofibrils. Muscles may become atrophied if they are not used. The result is termed ***disuse atrophy.*** Bedridden individuals and people with casts may experience atrophy because the flow of impulses to the inactive muscle is greatly reduced. If the nerve supply to a muscle is cut, it will undergo complete atrophy. This is termed ***denervation atrophy.*** In about six months to two years, the muscle will be one-quarter its original size and the muscle fibers will be replaced by fibrous tissue. The transition to fibrous tissue, when complete, cannot be reversed.

Muscular hypertrophy (hī-PER-trō-fē) is the opposite of atrophy. It refers to an increase in the diameters of muscle fibers owing to the production of more myofibrils, mitochondria, sarcoplasmic reticulum, nutrients (glycogen and triglycerides), and energy-supplying molecules (ATP and creatine phosphate). Hypertrophic muscles are capable of more forceful contractions. Weak muscular activity does not produce significant hypertrophy. It results from very forceful muscular activity or repetitive muscular activity at moderate levels. During childhood, the increase in the *size* of muscle fibers appears to be at least partially under the control of human growth hormone (hGH), which is produced by the anterior pituitary gland. In males, a further increase in the size of muscle fibers appears to be due to the hormone testosterone, produced by the testes. The influence of testosterone accounts for the generally larger and better defined muscles in males than females. More forceful muscular contractions, as in weight lifting, also contribute to generally larger muscles in males.

CLINICAL APPLICATION

Muscle-Building Anabolic Steroids

Steroids are chemical substances derived from cholesterol and serve many useful functions in the body (discussed in later chapters). Most steroids are hormones. In recent years, attention has been focused on the use of ***muscle-building anabolic steroids*** by athletes. These steroids, a derivative of the hormone testosterone, are used to build muscle proteins and therefore increase strength and endurance during athletic events. However, anabolic steroids may cause a number of damaging and devastating side effects, including liver cancer, kidney damage, increased risk of heart disease, muscle spasm, increased cholesterol, stunted growth in young people, increased irritability and aggressive behavior, psychotic symptoms (including hallucinations), manic episodes, major depression, and mood swings. In females, additional adverse effects include sterility, the development of facial hair, deepening of the voice, atrophy of the breasts and uterus, enlargement of the clitoris, and irregularities of menstruation. In males, additional side effects include testicular atrophy, baldness, excessive development of breast glands, and diminished hormone secretion and sperm production by the testes. Steroids may also be addictive.

TYPES OF SKELETAL MUSCLE FIBERS

All skeletal muscle fibers are not alike in structure or function. For example, skeletal muscle fibers vary in color depending on their content of ***myoglobin,*** a reddish pigment similar to hemoglobin in blood. Myoglobin stores oxygen until needed by mitochondria, the organelles in which ATP generation occurs. Skeletal muscle fibers that have a high myoglobin content are referred to as ***red muscle fibers.*** Conversely, skeletal muscle fibers that have a low content of myoglobin are called ***white muscle fibers.*** Red muscle fibers are smaller in diameter than white muscle fibers, and red muscle fibers have more mitochondria and more blood capillaries.

Skeletal muscle fibers contract with different velocities, depending on how rapidly they split ATP. Faster-contracting fibers have greater ability to split ATP. In addition, skeletal muscle fibers vary with respect to the metabolic processes they use to generate ATP and how quickly they fatigue.

On the basis of various structural and functional characteristics, skeletal muscle fibers are classified into three types:

1. ***Type I fibers.*** These fibers, also called ***slow twitch, slow oxidative,*** or ***fatigue resistant fibers,*** contain large amounts of myoglobin, many mitochondria, and many blood capillaries and have a high capacity to generate ATP by aerobic metabolic processes. Type I fibers are red. Such fibers also split ATP at a slow rate and, as a result, contraction velocity is slow. The fibers are very resistant to fatigue. Such fibers are found in large numbers, as in the postural muscles of the neck, for example.
2. ***Type II B fibers.*** These fibers, also called ***fast twitch, fast glycolytic,*** or ***fatigable fibers,*** have a low content of myoglobin, relatively few mitochondria, and relatively few blood capillaries. They do, however, contain large amounts of glycogen. Type II B fibers are white and are geared to generate ATP by anaerobic metabolic processes, which are not able to supply skeletal muscle fibers continuously with sufficient ATP. Accordingly, these fibers fatigue easily. They are the largest diameter fiber and they split ATP at a fast rate so that contraction is strong and fast. Muscles of the arm contain many of these fibers.
3. ***Type II A fibers.*** These fibers, also called ***fast twitch, fast oxidative,*** or ***fatigue resistant fibers,*** contain large amounts of myoglobin, many mitochondria, and many blood capillaries. These fibers are red and have a very high capacity for generating ATP by aerobic metabolic processes. Such fibers also split ATP at a very rapid rate, and as a result, contraction velocity is fast. Type II A fibers are resistant to fatigue but not quite as much as type I fibers. In humans, these fibers occur infrequently.

Most skeletal muscles of the body are a mixture of all three types of skeletal muscle fibers, but their proportion varies depending on the usual action of the muscle. For example, postural muscles of the neck, back, and legs have a higher proportion of type I fibers. Muscles of the shoulders and arms are not constantly active but are used intermittently, usually for short periods of time, to produce large amounts of tension such as in lifting and throwing. These muscles have a higher proportion of type II B fibers. Leg muscles not only support the body but are also used for walking and running. Such muscles have higher proportions of type I and type II B fibers.

Even though most skeletal muscles are a mixture of all three types of skeletal muscle fibers, all the skeletal muscle fibers of any one motor unit are the same. In addition, the different skeletal muscle fibers in a muscle may be used in various ways, depending on need. For example, if only a weak contraction is needed to perform a task, only type I fibers are activated by their motor units. If a stronger contraction is needed, the motor units of type II A fibers are activated. And, if a maximal contraction is required, motor units of type II B fibers are activated as well. Activation of various motor units is determined in the brain and spinal cord.

Although the number of the different skeletal muscle fibers does not change, the characteristics of those present can be altered. Various types of exercises can bring about changes in the fibers in a skeletal muscle. Endurance-type exercises, such as running or swimming, cause a gradual transformation of type II B fibers into type II A fibers. The transformed muscle fibers show a slight increase in diameter, mitochondria, blood capillaries, and strength. Endurance exercises result in cardiovascular and respiratory changes that cause skeletal muscles to receive better supplies of oxygen and carbohydrates but do not increase muscle mass. On the other hand, exercises that require great strength for short periods of time, such as weight lifting, produce an increase in the size and strength of type II B fibers. The increase in size is due to increased synthesis of thin and thick myofilaments. The overall result is that the person develops large muscles.

CLINICAL APPLICATION

Electromyography

Whenever an action potential passes over a skeletal muscle, a small portion of the action potential also spreads to the skin. The recording and study of electrical changes that occur in muscle tissue is called ***electromyography*** or ***EMG*** (*electro* = electricity; *myo* = muscle; *graph* = to write). The record of the electrical changes is known as an ***electromyogram.*** To record electromyograms, a flat metal disc electrode is placed on the skin over the muscle to be tested. Then a thin sterile needle attached by wires to a recording machine is inserted through the skin into the muscle. The electrical activity of the muscle is recorded at rest and during contraction and displayed as electrical waves on an oscilloscope at the same time that the activity is reproduced as sounds over a speaker. Electromyograms may be used to help determine the cause of muscular weakness or paralysis, to evaluate involuntary muscle twitching, to determine why abnormal levels of muscle enzymes (e.g., creatine phosphokinase) appear in blood, and to serve as a component of biofeedback studies.

CARDIAC MUSCLE TISSUE

The principal constituent of the heart wall is ***cardiac muscle tissue.*** Although it is striated in appearance like skeletal muscle, it is involuntary. The fibers of cardiac muscle tissue are roughly quadrangular and usually have only a single centrally located nucleus (see Exhibit 3-3). Skeletal muscle fibers contain several nuclei that are peripherally located.

The thin sarcolemma of cardiac muscle fibers is similar to that of skeletal muscle, but the sarcoplasm is more abundant and the mitochondria are larger and more numerous. Cardiac muscle fibers have the same arrangement of actin and myosin and the same bands, zones, and lines as skeletal muscle fibers (Figure 9-8). The transverse tubules of mammalian cardiac muscle are larger than those of skeletal muscle and are located at the Z discs rather than at the A–I band junctions as in skeletal muscle fibers. The sarcoplasmic reticulum of cardiac muscle is less well developed than that in skeletal muscle.

Note in Exhibit 3-3 that cardiac muscle fibers branch and interconnect with each other. Recall that skeletal muscle fibers are arranged in parallel fashion. Cardiac muscle fibers are also physically connected to each other by cell junctions called ***gap junctions,*** which permit a direct conduction of muscle action potentials from one fiber to another (see Figure 3-1). Cardiac muscle fibers form two separate networks. The muscular walls and partition of the upper chambers of the heart (atria) compose one network. The muscular walls and partition of the lower chambers of the heart (ventricles) compose the other network. Each fiber in a network is separated from the next fiber by an irregular transverse thickening of the sarcolemma called an ***intercalated*** (in-TER-ka-lāt-ed) ***disc.*** These discs contain desmosomes, which hold fibers together, and gap junctions, which allow muscle action potentials to spread from one muscle fiber to another. When a single fiber of either network is stimulated, all the fibers in the network become stimulated as well. Thus, each network contracts as a functional unit. As you will see in Chapter 13, when the fibers of the atria contract as a unit, blood moves into the ventricles. Then, when the ventricular fibers contract as a unit, blood is pumped into arteries.

Under normal resting conditions, cardiac muscle tissue contracts and relaxes rapidly, continuously, and rhythmically about 75 times a minute without stopping. This is a major physiological difference between cardiac and skeletal muscle tissue. Accordingly, cardiac muscle tissue requires a constant supply of oxygen. Energy generation occurs in numerous large mitochondria. Another difference is the source of stimulation. Skeletal muscle tissue ordinarily contracts only when stimulated by acetylcholine released by a nerve impulse from a motor neuron. In contrast, cardiac muscle tissue can contract without extrinsic (outside) nerve or hormonal stimulation. Its source of stimulation is a conducting network of specialized intrinsic (internal) cardiac muscle fibers within the heart. Nerve stimulation merely causes the conducting tissue to increase or decrease its rate of discharge. Some types of smooth muscle fibers and nerve cells in the brain and spinal cord also possess spontaneous, rhythmical self-excitation, a phenomenon referred to as ***autorhythmicity.***

Another difference between cardiac and skeletal muscle tissue is that cardiac muscle tissue remains contracted 10 to 15 times longer than skeletal muscle tissue. This is because there is a prolonged delivery of Ca^{2+} ions into the sarcoplasm. In cardiac muscle fibers, Ca^{2+} ions are derived from sarcoplasmic reticulum (as in skeletal muscle fibers) and from extracellular fluids. Although the release of Ca^{2+} ions from the sarcoplasmic reticulum is fast, passage of Ca^{2+} ions from extracellular fluid through the sarcolemma is much slower, thus accounting for the prolonged contraction of cardiac muscle fibers.

Cardiac muscle tissue also has an extra-long refractory period (period of lost excitability), lasting several tenths of a second, that allows time for the heart to relax between beats. The long refractory period permits heart rate to be increased significantly but prevents the heart itself from undergoing *tetanus*—that is, a sustained contraction. Tetanus of heart muscle would stop blood flow through the body and result in death.

SMOOTH (VISCERAL) MUSCLE TISSUE

Like cardiac muscle tissue, ***smooth (visceral) muscle tissue*** is usually involuntary. However, it is nonstriated. Smooth muscle fibers are considerably smaller than skeletal muscle fibers. A single fiber of smooth muscle tissue is about 5 to 10 μm in diameter and 30 to 200 μm long. Each fiber is thickest in the midregion and tapers at each end. Within the fiber is a single, oval, centrally located nucleus (Figure 9-9; see also Exhibit 3-3). The sarcoplasm of smooth muscle fibers contains thick myofilaments that are longer than those of skeletal muscle fibers and thin myofilaments. They are not arranged as orderly sarcomeres, however, as in striated muscle. In smooth muscle fibers, the ratio of thin to thick myofilaments is 16:1; in skeletal muscle fibers, the ratio is 2:1. Smooth muscle fibers also contain ***intermediate filaments.*** Since the various myofilaments have no regular pattern of organization and since there are no A or I bands of sarcomeres, smooth muscle fibers have no characteristic striations, and thus the name *smooth.*

Intermediate filaments are attached to structures called ***dense bodies,*** which have characteristics similar to Z discs in striated muscle fibers. Some dense bodies are dispersed throughout the cytoplasm; others are attached to the sarcolemma. Bundles of intermediate filaments stretch from one dense body to another (Figure 9-9b). The sliding filament mechanism involving thick and thin myofilaments during contraction generates tension that is transmitted to intermediate filaments. These filaments in turn pull on the dense bodies attached to the sarcolemma, causing a lengthwise shortening of the muscle fiber. Note in Figure 9-9b that shortening of the muscle fiber produces a bubblelike expansion of the sarcolemma. Evidence suggests that a smooth muscle fiber contracts in a corkscrewlike manner; the fiber twists in a helix as it shortens and rotates in the opposite direction as it lengthens.

Smooth muscle fibers also possess a less well developed sarcoplasmic reticulum than skeletal muscle fibers. Small

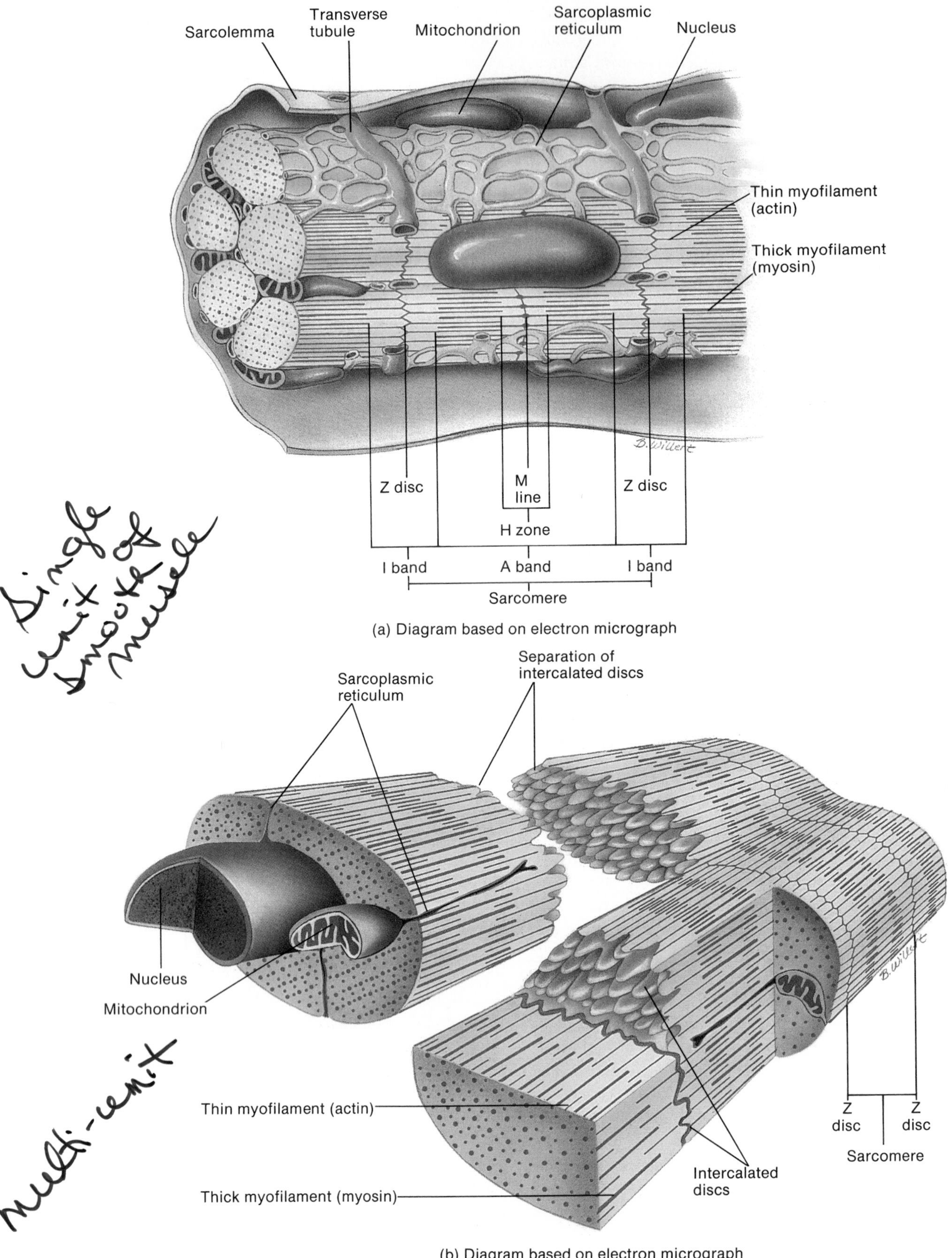

FIGURE 9-8 Histology of cardiac muscle tissue. Note the intercalated discs in (b).

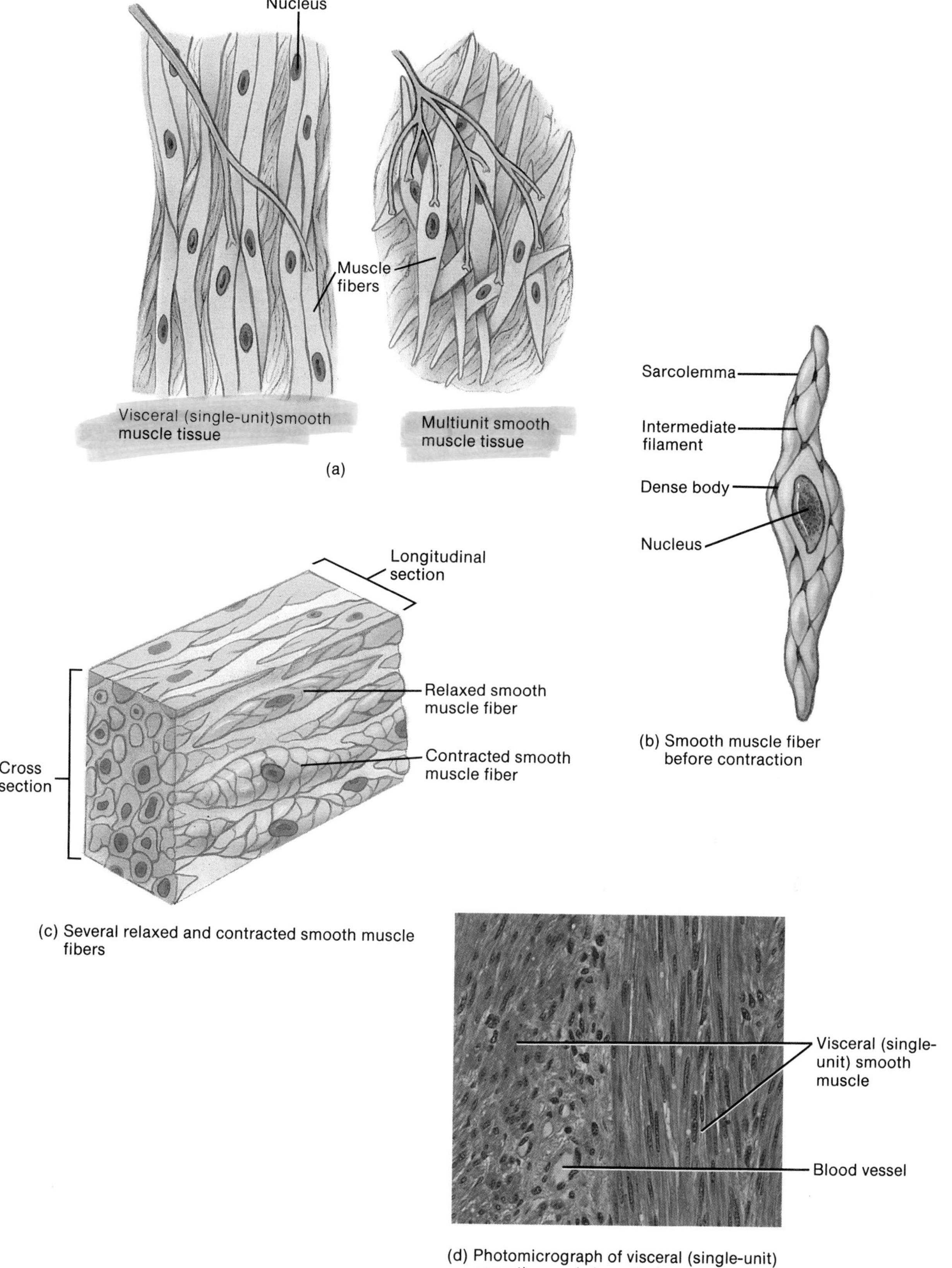

FIGURE 9-9 Histology of smooth muscle tissue. (e) Courtesy of Douglas Merrill, 100×.

subsurface vesicles termed ***caveolae*** open onto the surface of the smooth muscle fiber and are thought to function in a way similar to the transverse tubules of striated fibers—namely, to carry muscle action potentials into the fibers. Tubules of the sarcoplasmic reticulum are associated with longitudinal rows of caveolae.

Two kinds of smooth muscle tissue, visceral and multiunit, are recognized (Figure 9-9a). The more common type is called ***visceral (single-unit) muscle tissue.*** It is found in wraparound sheets that form part of the walls of small arteries and veins and hollow viscera such as the stomach, intestines, uterus, and urinary bladder. The terms *smooth muscle tissue* and *visceral muscle tissue* are sometimes used interchangeably. The fibers in visceral muscle tissue are tightly bound together to form a continuous network. They contain gap junctions that facilitate conduction of muscle action potentials between fibers. When a neurotransmitter, hormone, or autorhythmic signal stimulates one fiber, the muscle action potential spreads to the other fibers so that contraction occurs in a wave over many adjacent fibers. Thus, whereas skeletal muscle fibers contract as individual units, visceral muscle cells contract in sequence as the action potential spreads from one cell to another.

The second kind of smooth muscle tissue, ***multiunit smooth muscle tissue,*** consists of individual fibers, each with its own motor nerve endings. Whereas stimulation of a single visceral muscle fiber causes contraction of many adjacent fibers, stimulation of a single multiunit fiber causes contraction of only that fiber. In this respect, multiunit muscle tissue is like skeletal muscle tissue. Multiunit smooth muscle tissue is found in the walls of large arteries, in large airways to the lungs, in the arrector pili muscles that attach to hair follicles, and in the intrinsic (internal) muscles of the eye, such as the radial and circular muscles of the iris.

Although the principles of contraction are essentially the same in smooth and striated muscle tissue, smooth muscle tissue exhibits several important physiological differences. First of all, the duration of contraction and relaxation of smooth muscle fibers is about 5 to 500 times longer than in skeletal muscle fibers. Smooth muscle fibers derive their Ca^{2+} ions from sarcoplasmic reticulum and extracellular fluid. Since there are no transverse tubules in smooth muscle fibers, it takes longer for Ca^{2+} ions to reach the deep filaments in the center of the fiber to trigger the contractile process, accounting in part for the slow activation and prolonged contraction of smooth muscle. Moreover, smooth muscle fibers do not contain calcium-binding troponin to facilitate myosin cross bridge attachment to thin myofilaments. They use a different mechanism that takes more time and thus contributes to prolonged contraction.

Not only do Ca^{2+} ions move slowly into deep filaments, but they also move slowly out of the muscle fiber following contraction. Therefore, relaxation is delayed and the prolonged stay of Ca^{2+} ions in the fibers provides for tone, the state of continued partial contraction. Smooth muscle tissue can exhibit sustained, long-term tone, which is important in the gastrointestinal tract where the walls of the tract maintain a steady pressure on the contents of the tract, in the walls of blood vessels called arterioles that maintain a steady pressure on blood, and in the wall of the urinary bladder that maintains a steady pressure on urine.

Some smooth muscle fibers contract in response to nerve impulses from the autonomic (involuntary) nervous system. Thus, smooth muscle is normally not under voluntary control. Other smooth muscle fibers contract in response to hormones or local factors such as pH, oxygen and carbon dioxide levels, temperature, and ion concentrations.

Finally, unlike skeletal muscle fibers, smooth muscle fibers can stretch considerably without developing tension. When smooth muscle fibers are stretched, they initially develop increased tension. However, there is an almost immediate decrease in the tension. This phenomenon is referred to as ***stress–relaxation*** and is important because it permits smooth muscle to accommodate great changes in size while still retaining the ability to contract effectively. Thus, the smooth muscle in the wall of hollow organs such as the stomach, intestines, and urinary bladder can stretch as the viscera distend, while the pressure within them remains the same.

A summary of the principal characteristics of the three types of muscle tissue is presented in Exhibit 9-2.

REGENERATION OF MUSCLE TISSUE

Skeletal muscle fibers cannot divide. After the first year of life, all growth of skeletal muscle is due to enlargement of existing cells (hypertrophy), rather than increase in the number of fibers (hyperplasia). Skeletal muscle fibers, however, can be replaced on an individual basis by new fibers derived from ***satellite cells,*** which are dormant stem cells found in association with skeletal muscle fibers. During rapid postnatal growth, satellite cells lengthen existing skeletal muscle fibers by fusing with them. They also persist as a lifelong source of cells that can fuse with each other to form new skeletal muscle fibers. However, the number of new skeletal muscle fibers formed by this mechanism is not sufficient to compensate for any significant skeletal muscle damage. In cases of such damage, skeletal muscle tissue is replaced by fibrous scar tissue. For this reason, skeletal muscle tissue has only limited powers of regeneration.

Cardiac muscle fibers, like those of skeletal muscle tissue, do not appear to have any capacity for division in the human body. (They have been induced experimentally to undergo division under controlled laboratory conditions.) The fibers increase in size by hypertrophy. In addition, cardiac muscle fibers do not associate with cells comparable to satellite cells. Healing of cardiac muscle tissue is by scar formation. Cardiac muscle tissue has no powers of regeneration.

EXHIBIT 9-2

Summary of the Principal Characteristics of Muscle Tissue

CHARACTERISTIC	SKELETAL MUSCLE	CARDIAC MUSCLE	SMOOTH MUSCLE
Location	Attached primarily to bones.	Heart.	Walls of hollow viscera, blood vessels, iris, arrector pili.
Microscopic Appearance	Striated, multinucleated, unbranched fibers.	Striated uninucleated, branched fibers with intercalated discs.	Nonstriated (smooth) uninucleated, spindle-shaped fibers.
Nervous Control	Voluntary.	Involuntary.	Involuntary.
Sarcomeres	Yes.	Yes.	No.
Transverse Tubules	Yes.	Yes.	No.
Gap Junctions	No.	Yes.	Yes, in visceral smooth muscle.
Cell Size (Diameter)	Very large (10–100 μm).	Large (about 14 μm).	Small (3–8 μm).
Source of Calcium	Sarcoplasmic reticulum.	Sarcoplasmic reticulum and extracellular fluid.	Sarcoplasmic reticulum and extracellular fluid.
Speed of Contraction	Fast.	Moderate.	Slow.
Capacity for Division	None.	None.	Limited.
Capacity for Regeneration	Limited.	None.	Considerable compared with other muscle tissues but limited compared with tissues such as epithelium.

Smooth muscle tissue, like skeletal and cardiac muscle tissue, can undergo hypertrophy. In addition, certain smooth muscle fibers, such as those in the uterus, retain their capacity for division and thus can grow by hyperplasia. Also, new smooth muscle fibers can arise from cells called ***pericytes,*** stem cells found in association with the endothelium of blood capillaries and venules. It is also known that smooth muscle fibers can proliferate in certain pathological conditions such as occur in the development of atherosclerosis (Chapter 13). Compared with the other two types of muscle tissue, smooth muscle tissue has a considerably higher, though still limited, power of regeneration when compared with other tissues, such as epithelium.

AGING AND MUSCLE TISSUE

Beginning at about 30 years of age there is a progressive loss of skeletal muscle mass that is largely replaced by fat. Accompanying the loss of muscle mass is a decrease in maximal strength and a diminishing of muscle reflexes.

In some muscles, there may be a selective loss of muscle fibers of a given type. With aging, the relative number of type I fibers appears to increase, either because of atrophy of the other fiber types or their conversion into type I fibers. Whether this result is an effect of aging itself or merely reflects the more limited physical activity of older people is still an unresolved question.

DEVELOPMENTAL ANATOMY OF THE MUSCULAR SYSTEM

In this brief discussion of the development of the muscular system we shall concentrate mostly on skeletal muscles. Except for the muscles of the iris of the eyes and the arrector pili muscles attached to hairs, all muscles of the body are derived from ***mesoderm.*** As the mesoderm develops, a portion of it becomes arranged in dense columns on either side of the developing nervous system. These columns of mesoderm undergo segmentation into a series of blocks of cells called ***somites*** (Figure 9-10a). The first pair of somites appears on the twentieth day. Eventually, 44 pairs of somites are formed by the thirtieth day.

With the exception of the skeletal muscles of the head and extremities, *skeletal muscles* develop from the ***mesoderm of somites.*** Since there are very few somites in the head region of the embryo, most of the skeletal muscles there develop from ***somatic mesoderm*** in the head region. The skeletal muscles of the limbs develop from masses of somatic mesoderm around developing bones in embryonic limb buds (origins of future extremities: see Figure 5-6).

The cells of a somite are differentiated into three regions: (1) ***myotome,*** which forms most of the skeletal muscles; (2) ***dermatome,*** which forms the connective tissues, including the dermis, under the epidermis; and (3) ***sclerotome,*** which gives rise to the vertebrae (Figure 9-10b). *Fasciae,*

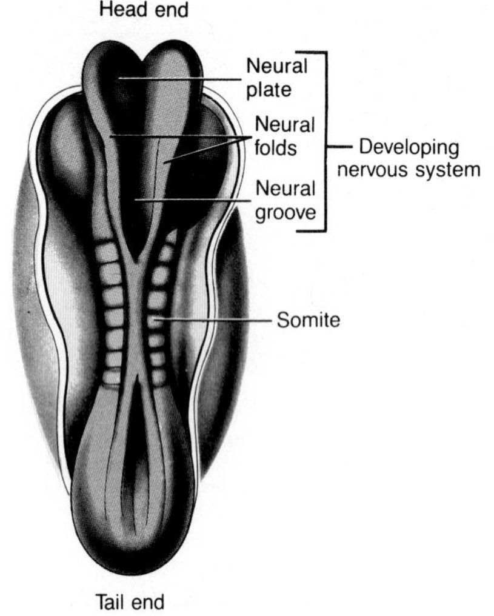

(a) Dorsal aspect of an embryo showing somites

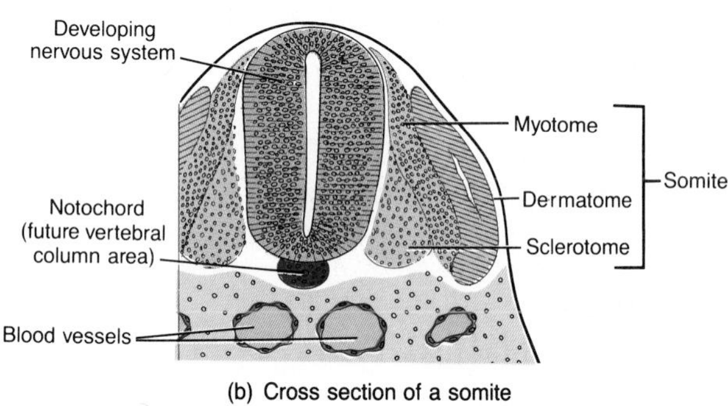

(b) Cross section of a somite

FIGURE 9-10 Development of the muscular system.

ligaments, and *aponeuroses* may form as a result of degeneration of all or parts of myotomes.

Smooth muscle develops from ***mesodermal cells*** that migrate to and envelop the developing gastrointestinal tract and viscera.

Cardiac muscle develops from ***mesodermal cells*** that migrate to and envelop the developing heart while it is still in the form of primitive heart tubes (see Figure 13-8).

APPLICATIONS TO HEALTH

FIBROSIS

The formation of fibrous (containing fibers) connective tissue in locations where it normally does not exist is called ***fibrosis***. Skeletal and cardiac muscle fibers cannot undergo mitosis, and dead muscle fibers are normally replaced with fibrous connective tissue. Fibrosis, then, is often a consequence of muscle injury or degeneration.

FIBROMYALGIA

Fibromyalgia (*algia* = painful condition) refers to a group of common nonarticular rheumatic disorders, occurring primarily among females and characterized by pain, tenderness, and stiffness of muscles, tendons, and surrounding soft tissues. The disorders specifically affect the fibrous connective tissue components of muscles, tendons, and ligaments. Fibromyalgia may be caused or aggravated by physical or mental stress, trauma, exposure to dampness or cold, poor sleep, or a rheumatic condition. Frequent sites at which fibromyalgia occur include the lumbar region ***(lumbago)***, neck, chest, and thighs ***(charley horse)***.

MUSCULAR DYSTROPHIES

Muscular dystrophies (*dystrophy* = degeneration) are hereditary muscle-destroying diseases. The diseases are characterized by degeneration of individual muscle fibers (cells), which leads to a progressive atrophy of the skeletal muscle. Usually, the voluntary skeletal muscles are weakened equally on both sides of the body, whereas the internal muscles, such as the diaphragm, are not affected. Histologically, the changes that occur include variation in muscle fiber size, degeneration of fibers, and deposition of fat. Muscular dystrophies are classified by mode of inheritance, age of onset, and clinical features. The most common form is called ***Duchenne*** (du-SHĀN) ***muscular dystrophy (DMD)***. The gene responsible for DMD has been identified, and its DNA code has been worked out. It is hoped that this information will lead to replacement therapy and a halt to muscle loss.

Muscular dystrophies are due to genetic defects that may result in faulty metabolism of potassium, protein deficiency, or inability of the body to utilize creatine. Just recently, scientists discovered that a protein they named ***dystrophin*** is present in normal muscle tissue but absent in persons with DMD. The function of dystrophin is unknown. According to one hypothesis, lack of dystrophin may result in leakage of calcium ions from the sarcoplasmic reticulum into sarcoplasm, which may activate an enzyme (phospholipase A) that degenerates muscle fibers.

Muscular dystrophies are diagnosed on the basis of muscle enzyme studies, electromyography, and muscle biopsy.

MYASTHENIA GRAVIS (MG)

Myasthenia (mī-as-THĒ-nē-a) ***gravis (MG)*** is a weakness of skeletal muscles. It is caused by an abnormality at the neuromuscular junction that prevents muscle fibers from contracting. Recall that motor neurons stimulate skeletal muscle fibers to contract by releasing acetylcholine (ACh). Myasthenia gravis is an autoimmune disorder caused by antibodies directed against ACh receptors of the muscle fiber sarcolemma. The antibodies bind to the receptors and hinder the attachment of ACh to the receptors. As the disease progresses, more neuromuscular junctions become affected. The muscle becomes increasingly weaker and may eventually cease to function altogether.

Initial symptoms include a weakness of the eye muscles, which may produce double vision, and difficulty in swallowing. Later, the individual has difficulty chewing and talking. Eventually, the muscles of the limbs may become involved. Death may result from paralysis of the respiratory muscles, but usually the disorder does not progress to this stage.

Anticholinesterase drugs (neostigmine and pyridostigmine) have been the primary treatment for the disease. They act as inhibitors of acetylcholinesterase, thus raising the level of ACh to bind with available receptors. More recently, steroid drugs, such as prednisone, have been used with great success to reduce antibody levels. Immunosuppressant drugs are also used to decrease the production of antibodies that interfere with normal muscle contraction. Another treatment involves *plasmapheresis,* a procedure that separates blood cells from the plasma that contains the unwanted antibodies. The blood cells are then mixed with a plasma substitute and pumped back into the individual. In some individuals, surgical removal of the thymus gland is indicated (thymectomy).

ABNORMAL CONTRACTIONS

One kind of abnormal muscular contraction is a ***spasm,*** a sudden involuntary contraction of large groups of muscles. (Cerebral palsy is characterized by generalized spastic contractions.) ***Tremor*** is a rhythmic, involuntary, purposeless contraction of opposing muscle groups. (A resting tremor occurs in Parkinson's disease.) A ***fasciculation*** is an involuntary, brief twitch of a muscle visible under the skin. It occurs irregularly and is not associated with movement of the affected muscle. (Fasciculations may be seen in multiple sclerosis, discussed in Chapter 13, or amyotrophic lateral sclerosis, also called Lou Gehrig's disease.) A ***fibrillation*** is similar to a fasciculation except that it is not visible under the skin. It is recorded by electromyography. A ***tic*** is a spasmodic twitching made involuntarily by muscles that are ordinarily under voluntary control. Twitching of the eyelid and facial muscles are examples. In general, tics are of psychological origin.

KEY MEDICAL TERMS ASSOCIATED WITH THE MUSCULAR SYSTEM

Gangrene (GANG-rēn; *gangraena* = an eating sore) Death of a soft tissue, such as muscle, that results from interruption of its blood supply. One type is caused by various species of *Clostridium,* bacteria that live anaerobically in the soil.

Myalgia (mi-AL-jē-a; *algia* = painful condition) Pain in or associated with muscles.

Myoma (mī-Ō-ma; *oma* = tumor) A tumor consisting of muscle tissue.

Myomalacia (mī′-ō-ma-LĀ-shē-a; *malaco* = soft) Softening of a muscle.

Myopathy (mī-OP-a-thē; *pathos* = disease) Any disease of muscle tissue.

Myosclerosis (mī′-ō-skle-RŌ-sis; *scler* = hard) Hardening of a muscle.

Myositis (mī′-ō-SĪ-tis; *itis* = inflammation of) Inflammation of muscle fibers (cells).

Myospasm (MĪ-o-spazm) Spasm of a muscle.

Myotonia (mī-ō-TŌ-nē-a; *tonia* = tension) Increased muscular excitability and contractility with decreased power of relaxation; tonic spasm of the muscle.

Paralysis (pa-RAL-a-sis; *para* = beyond; *lyein* = to loosen) Loss or impairment of motor (muscular) function resulting from a lesion of nervous or muscular origin.

Trichinosis (trik′-i-NŌ-sis) A myositis caused by the parasitic worm *Trichinella spiralis,* which may be found in the muscles of humans, rats, and pigs. People contract the disease by eating insufficiently cooked infected pork.

Volkmann's contracture (FŌLK-manz kon-TRAK-tur; *contra* = against) Permanent contraction of a muscle due to replacement of destroyed muscle fibers (cells) with fibrous tissue that lacks ability to stretch. Destruction of muscle fibers may occur from interference with circulation caused by a tight bandage, a piece of elastic, or a cast.

Wryneck (RĪ-neck) or **torticollis** (*tortus* = twisted; *collum* = neck) Contracted state of several superficial and deep muscles of the neck that produces twisting of the neck and an unnatural position of the head; one of the common triggering factors is vigorous activity.

STUDY OUTLINE

Characteristics (p. 213)

1. Excitability is the property of receiving and responding to stimuli.
2. Contractility is the ability to shorten and thicken (contract).
3. Extensibility is the ability to be stretched (extended).
4. Elasticity is the ability to return to original shape after contraction or extension.

Functions (p. 213)

1. Through contraction, muscle tissue performs three important functions.
2. These functions are motion, maintenance of posture, and heat production.

Types (p. 213)

1. Skeletal muscle tissue is primarily attached to bones. It is striated and voluntary.
2. Cardiac muscle tissue forms the wall of the heart. It is striated and involuntary.
3. Visceral muscle tissue is located in viscera. It is nonstriated (smooth) and involuntary.

Skeletal Muscle Tissue (p. 213)

Connective Tissue Components (p. 213)

1. The term *fascia* is applied to a sheet or broad band of fibrous connective tissue underneath the skin (superficial) or around muscles (deep).
2. Other connective tissue components are epimysium, covering the entire muscle; perimysium, covering fasciculi; and endomysium, covering fibers; all are extensions of deep fascia.
3. Tendons and aponeuroses are extensions of connective tissue beyond muscle cells that attach the muscle to bone or other muscle.

Nerve and Blood Supply (p. 215)

1. Nerves convey impulses for muscular contraction.
2. Blood provides nutrients and oxygen for contraction.

Histology (p. 217)

1. Skeletal muscle consists of fibers (cells) covered by a sarcolemma. The fibers contain sarcoplasm, mitochondria, nuclei, sarcoplasmic reticulum, and transverse tubules.
2. Each fiber contains myofibrils that consist of thin and thick myofilaments. The myofilaments are compartmentalized into sarcomeres.
3. Thin myofilaments are composed of actin, tropomyosin, and troponin; thick myofilaments consist mostly of myosin.
4. Projecting myosin heads are called cross bridges and contain actin- and ATP-binding sites.

Contraction (p. 218)

Sliding-Filament Mechanism (p. 218)

1. Actual contraction is brought about when the thin myofilaments of a sarcomere slide toward each other as the myosin cross bridges pull on the actin myofilaments (sliding-filament mechanism).
2. This contraction causes sarcomeres to shorten.

Neuromuscular Junction (p. 218)

1. A motor neuron transmits a nerve impulse (nerve action potential) to a skeletal muscle where it serves as a stimulus for contraction.
2. A neuromuscular junction refers to an axon terminal of a motor neuron and the portion of the muscle fiber sarcolemma in close approximation with it (motor end plate).

Motor Unit (p. 219)

1. A motor neuron and the muscle fibers it stimulates form a motor unit.
2. A single motor unit may innervate as few as 10 or as many as 2000 muscle fibers.

Mechanism (p. 221)

1. When a nerve impulse (nerve action potential) reaches an axon terminal, the synaptic vesicles of the terminal release acetylcholine (ACh), which ultimately initiates a muscle action potential in the muscle fiber sarcolemma that then travels into the transverse tubules and causes the sarcoplasmic reticulum to release some of its stored Ca^{2+} into the sarcoplasm.
2. The muscle action potential releases calcium ions that combine with troponin, causing it to pull on tropomyosin to change its orientation, thus exposing myosin-binding sites on actin.
3. ATPase splits ATP into ADP + P and the released energy activates (energizes) myosin cross bridges.
4. Activated cross bridges attach to actin and a change in the orientation of the cross bridge occurs (power stroke); their movement results in the sliding of thin myofilaments.

All-or-None Principle (p. 221)

1. The weakest stimulus capable of causing contraction is a liminal (threshold) stimulus.
2. A stimulus not capable of inducing contraction is a subliminal (subthreshold) stimulus.
3. Muscle fibers of a motor unit contract to their fullest extent or not at all.

Muscle Length and Force of Contraction (p. 222)

1. A muscle fiber develops its greatest tension when there is maximum overlap between thick and thin myofilaments (optimum length).
2. As a fiber is stretched or shortened, force of contraction continuously decreases.

Muscle Tone (p. 222)

1. A sustained partial contraction of portions of a skeletal muscle results in muscle tone.
2. Tone is essential for maintaining posture.
3. Flaccidity is a condition of less-than-normal tone. Atrophy is a wasting away or decrease in size; hypertrophy is an enlargement or overgrowth.

Muscular Atrophy and Hypertrophy (p. 223)

1. Muscular atrophy refers to a state of wasting away of muscles.
2. Muscular hypertrophy refers to an increase in the diameter of muscle fibers.

Types of Skeletal Muscle Fibers (p. 223)

1. On the basis of structure and function, skeletal muscle fibers are classified as type I, type II B, and type II A.
2. Most skeletal muscles contain a mixture of all three fiber types, their proportions varying with the usual action of the muscle.
3. Various exercises can modify the types of skeletal muscle fibers.

Cardiac Muscle Tissue (p. 224)

1. This muscle is found only in the heart. It is striated and involuntary.
2. The fibers are quadrangular and usually contain a single centrally placed nucleus.

3. Compared to skeletal muscle tissue, cardiac muscle tissue has more sarcoplasm, more mitochondria, less well-developed sarcoplasmic reticulum, and larger transverse tubules located at Z lines rather than at A–I band junctions. Myofilaments are not arranged in discrete myofibrils.
4. The fibers branch freely and are connected via gap junctions.
5. Intercalated discs provide strength and aid in conduction of muscle action potentials by way of gap junctions located in the discs.
6. Unlike skeletal muscle tissue, cardiac muscle tissue contracts and relaxes rapidly, continuously, and rhythmically. Energy is supplied by glycogen and fat in large, numerous mitochondria.
7. Cardiac muscle tissue can contract without extrinsic stimulation and can remain contracted longer than skeletal muscle tissue.
8. Cardiac muscle tissue has a long refractory period, which prevents tetanus.

Smooth (Visceral) Muscle Tissue (p. 225)

1. Smooth muscle is nonstriated and involuntary.
2. Smooth muscle fibers contain intermediate filaments, dense bodies (function as Z discs), and caveolae (function as transverse tubules).
3. Visceral (single-unit) smooth muscle is found in the walls of viscera. The fibers are arranged in a network.
4. Multiunit smooth muscle is found in blood vessels and the eye. The fibers operate singly rather than as a unit.
5. The duration of contraction and relaxation of smooth muscle is longer than in skeletal muscle.
6. Smooth muscle fibers contract in response to nerve impulses, hormones, and local factors.
7. Smooth muscle fibers can stretch considerably without developing tension.

Regeneration of Muscle Tissue (p. 228)

1. Skeletal muscle fibers cannot divide and have limited powers of regeneration.
2. Cardiac muscle fibers cannot divide or regenerate.
3. Smooth muscle fibers have limited capacity for division and regeneration.

Aging and Muscle Tissue (p. 229)

1. Beginning at about 30 years of age, there is a progressive loss of skeletal muscle, which is replaced by fat.
2. There is also a decrease in muscle strength and diminished muscle reflexes.

Developmental Anatomy of the Muscular System (p. 229)

1. With few exceptions, muscles develop from mesoderm.
2. Skeletal muscles of the head and extremities develop from somatic mesoderm; the remainder of the skeletal muscles develop from the mesoderm of somites.

Applications to Health (p. 230)

1. Fibromyalgia refers to a group of nonarticular rheumatic disorders characterized by pain, tenderness, and stiffness of muscles, tendons, and ligaments. Frequent sites are the lower back (lumbago) and thigh (charley horse).
2. Muscular dystrophies refer to hereditary diseases of muscles characterized by degeneration of individual muscle fibers.
3. Myasthenia gravis (MG) is a disease characterized by great muscular weakness and fatigability resulting from improper neuromuscular transmission.
4. Abnormal contractions include spasm, tremor, fasciculation, fibrillation, and tic.

REVIEW QUESTIONS

1. How is the skeletal system related to the muscular system? What are the four characteristics of muscle tissue? (p. 213)
2. What are the three basic functions of the muscular system? (p. 213)
3. How can the three types of muscle tissue be distinguished on the basis of location, microscopic appearance, and nervous control? (p. 213)
4. What is fascia? Distinguish between superficial and deep fascia. (p. 213)
5. Define epimysium, perimysium, endomysium, tendon, and aponeurosis. Describe the nerve and blood supply to a skeletal muscle. (p. 215)
6. Describe the microscopic structure of skeletal muscle tissue. (p. 217)
7. In considering the contraction of skeletal muscle tissue, describe the following: neuromuscular junction (p. 218), motor unit (p. 219), and sliding-filament mechanism (p. 218).
8. What is the all-or-none principle? Relate it to a threshold and subthreshold stimulus. (p. 221)
9. What is muscle tone? Why is it important? (p. 222) Distinguish between muscular atrophy and hypertrophy. (p. 223)
10. Compare type I, type II B, and type II A fibers with respect to structure and function. (p. 224)
11. Compare skeletal, cardiac, and smooth muscle with regard to differences in structure, physiology, and capacity for division and regeneration. (p. 229)
12. Describe the effects of aging on muscle tissue. (p. 229)
13. Describe how the muscular system develops. (p. 229)
14. Define fibrosis, fibromyalgia, muscular dystrophies, myasthenia gravis (MG), spasm, tremor, fasciculation, fibrillation, and tic. (p. 230)
15. Define the following: tenosynovitis (p. 215), rigor mortis (p. 221), hypotonia (p. 223), flaccidity (p. 223), hypertonia (p. 223), spasticity (p. 223), rigidity (p. 223), muscle-building anabolic steroids (p. 223), and electromyography (p. 224).
16. Refer to the glossary of key medical terms associated with the muscular system. Be sure that you can define each term. (p. 231)

SELF QUIZ

1. Contrast the kinds of fascia by indicating which of the following are characteristics of superficial fascia (S) or deep fascia (D).
___ **a.** immediately under the skin
___ **b.** composed of dense connective tissue
___ **c.** contains much fat and so serves as insulator of the body
___ **d.** also called subcutaneous layer
___ **e.** forms epimysium, perimysium, and endomysium that holds muscles into functional groups

Complete the following:

2. Each muscle fiber (also known as a ________) is surrounded by a membrane called the ________ and contains cytoplasm called ________.
3. A transverse tubule, along with sarcoplasmic reticulum (terminal cisterns) on either side, is called a ________. The hundreds or thousands of myofibrils in a skeletal muscle fiber each consist of bundles of thick and thin ________.
4. To describe a sarcomere during contraction, complete each statement with one of the following: lengthens, shortens, or stays the same length.
   a. The sarcomere ________.
   b. Each thick myofilament (A band) ________.
   c. Each thin myofilament ________.
   d. The I band ________.
   e. The H zone ________.
5. Smooth muscle fibers are (cylinder? spindle?) shaped with (several nuclei? one nucleus?) per cell. They (do? do not?) contain actin and myosin. However, because of the irregular arrangement of these filaments, smooth muscle tissue appears (striated? nonstriated or ''smooth''?). In general, smooth muscle contracts and relaxes more (rapidly? slowly?) than skeletal muscle does, and smooth muscle holds the contraction for a (shorter? longer?) period of time than skeletal muscle does.
6. Which of the three types of muscle tissue develops from mesoderm? ________. Part of the mesoderm forms columns on either side of the developing nervous system. This tissue segments into blocks of tissue called ________.
7. Match the muscle types listed at the right with descriptions below:

___ **a.** involuntary muscle found in blood vessels and intestine		**A.** cardiac
___ **b.** involuntary striated muscle		**B.** skeletal
___ **c.** striated voluntary muscle attached to bones		**C.** smooth

8. Match the following:

___ **a.** invagination of deep fascia that surrounds muscle fibers and bundles and holds muscles into functional groups	**A.** aponeurosis
___ **b.** cord of dense connective tissue that attaches muscle into periosteum	**B.** endomysium, perimysium, epimysium
___ **c.** similar in function to tendon but consists of a broad, flat layer of dense connective tissue	**C.** tendon
___ **d.** tube of fibrous connective tissue lined with synovial membrane that permits tendons to slide easily, as in wrist and ankle	**D.** tendon sheath

9. Arrange the answers in correct sequence:
___ ___ ___ **a.** According to the amount of muscle tissue they surround, from most to least:
   A. perimysium
   B. endomysium
   C. epimysium
___ ___ ___ **b.** From largest to smallest:
   A. myofibril
   B. myofilament
   C. muscle fiber

Choose the one best answer to these questions:

___ **10.** All of the following molecules are parts of thin myofilaments *except*
A. actin; B. myosin; C. tropomyosin; D. troponin.

___ **11.** Choose the one statement that is false:
A. A band refers to the anisotropic band; B. the A band is darker than the I band; C. thick myofilaments reach the Z disc in relaxed muscle; D. the H zone contains thick myofilaments, but not thin ones; E. thick myofilaments are made of myosin.

___ **12.** The ability of muscle tissue to receive and respond to a stimulus is referred to as
A. contractility; B. excitability; C. elasticity; D. extensibility; E. conductivity.

___ **13.** Which of the following is *not* performed by muscles?
A. motion; B. excretion; C. maintenance of posture; D. heat production.

___ **14.** Choose the false statement about cardiac muscle:
A. cardiac muscle has a long refractory period; B. cardiac muscle has one centrally located nucleus per fiber; C. cardiac muscle cells are called cardiac muscle fibers; D. cardiac muscle fibers are separated by intercalated discs; E. cardiac muscle fibers are spindle shaped with no striations.

___ **15.** A motor unit is defined as a
A. nerve and a muscle; B. single neuron and a single muscle fiber; C. neuron and the muscle fibers it supplies; D. single muscle fiber and the nerves that innervate it; E. muscle and the motor and sensory nerves that innervate it.

___ **16.** The all-or-none principle applied to muscle means
A. individual motor units contract in a graded fashion, that is, from maximum to minimum; B. individual motor units have only one type of contraction; C. muscles contract with minimal stimuli; D. muscles contract by summation of stimuli; E. the whole muscle contracts rather than single motor units.

___ **17.** Skeletal muscle tissue
A. has the ability to contract rhythmically by itself; B. is composed of long, spindlelike cells, each containing a single nucleus; C. has the ability to contract when stimulated; D. is present in the walls of arteries; E. is not described by any of the above.

Circle T (true) or F (false) for the following:

T F 18. A muscle fiber develops its greatest tension when there is minimum overlap between thick and thin myofilaments.

T F 19. Spasticity and rigidity are types of hypertonia.

T F 20. Muscular atrophy and hypertrophy are opposite conditions.

T F 21. Type I skeletal muscle fibers contract forcefully and rapidly.

The Muscular System

10

STUDENT OBJECTIVES

1. Describe the relation between bones and skeletal muscles in producing body movements.
2. Define a lever and fulcrum and compare the three classes of levers on the basis of placement of the fulcrum, effort, and resistance.
3. Identify the various arrangements of muscle fibers in a skeletal muscle and relate the arrangements to the strength of contraction and range of movement.
4. Discuss most body movements as activities of groups of muscles by explaining the roles of the prime mover, antagonist, synergist, and fixator.
5. Define the criteria employed in naming skeletal muscles.
6. Identify the principal skeletal muscles in different regions of the body by name, origin, insertion, action, and innervation.

CHAPTER OUTLINE

- **How Skeletal Muscles Produce Movement**

  Origin and Insertion

  Lever Systems and Leverage

  Arrangement of Fasciculi

  Group Actions
- **Naming Skeletal Muscles**
- **Principal Skeletal Muscles**

The term ***muscle tissue*** refers to all the contractile tissues of the body: skeletal, cardiac, and smooth muscle. The ***muscular system,*** however, refers to the *skeletal* muscle system: the skeletal muscle tissue and connective tissues that make up individual muscle organs, such as the biceps brachii muscle. Cardiac muscle tissue is located in the heart and is therefore considered part of the cardiovascular system. Smooth muscle tissue of the intestine is part of the digestive system, whereas smooth muscle tissue of the urinary bladder is part of the urinary system. In this chapter we discuss only the muscular system. We will see how skeletal muscles produce movement, and we will describe the principal skeletal muscles.

HOW SKELETAL MUSCLES PRODUCE MOVEMENT

ORIGIN AND INSERTION

Skeletal muscles produce movements by exerting force on tendons, which in turn pull on bones or other structures, such as skin. Most muscles cross at least one joint and are attached to the articulating bones that form the joint (Figure 10-1). When such a muscle contracts, it draws one articulating bone toward the other. The two articulating bones usually do not move equally in response to the contraction. One is held nearly in its original position because

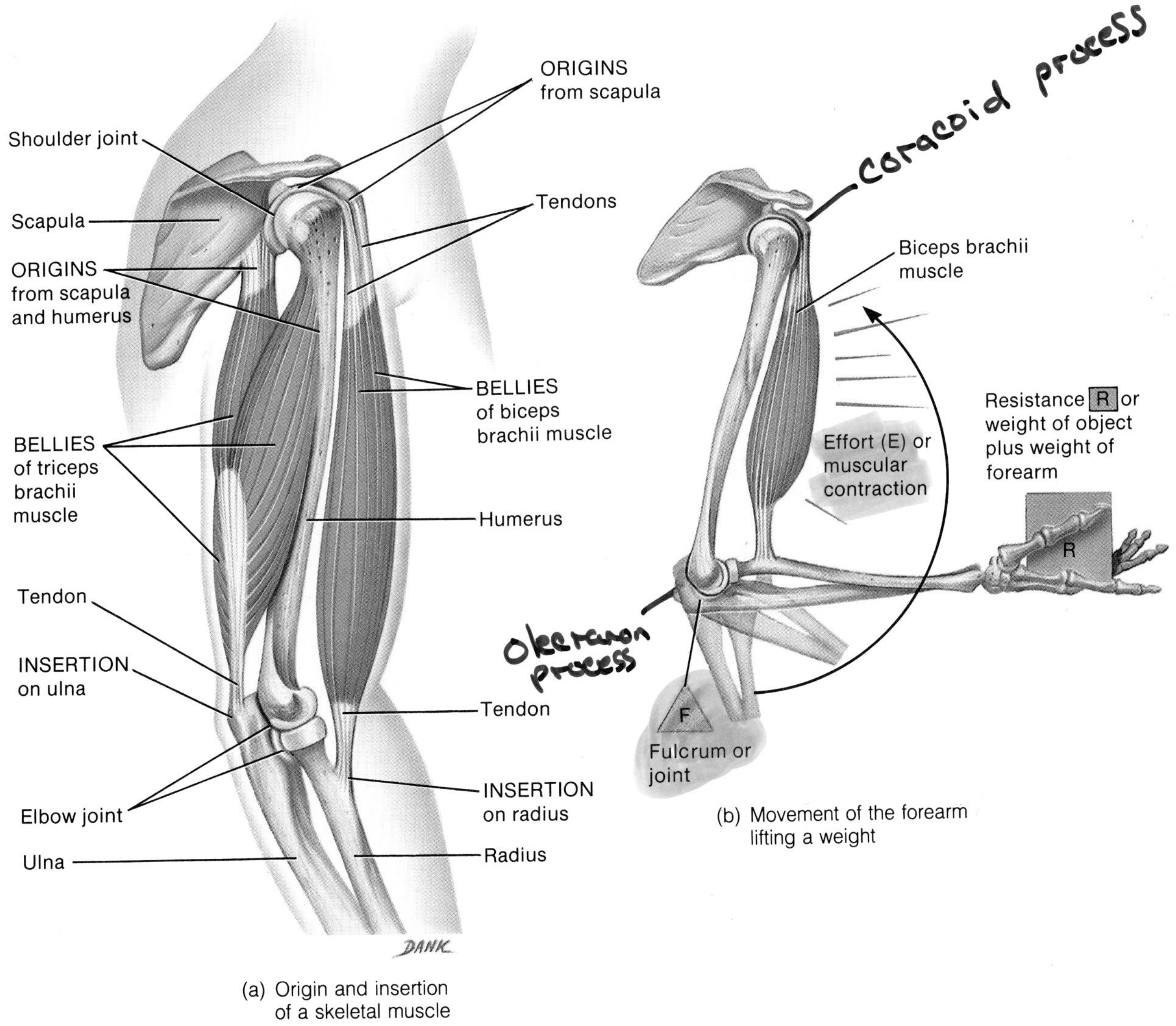

FIGURE 10-1 Relationship of skeletal muscles to bones. (a) Skeletal muscles produce movements by pulling on bones. (b) Bones serve as levers, and joints act as fulcrums for the levers. Here the lever-fulcrum principle is illustrated by the movement of the forearm. Note where the resistance and effort are applied in this example.

other muscles contract to pull it in the opposite direction or because its structure makes it less movable. Ordinarily, the attachment of a muscle tendon to the stationary bone is called the ***origin.*** The attachment of the other muscle tendon to the movable bone is the ***insertion.*** A good analogy is a spring on a door. The part of the spring attached to the door represents the insertion; the part attached to the frame is the origin. The fleshy portion of the muscle between the tendons of the origin and insertion is called the ***belly (gaster).*** The origin is usually proximal and the insertion distal, especially in the extremities. In addition, muscles that move a body part often do not cover the moving part. Figure 10-1a shows that although one of the functions of the biceps brachii muscle is to move the forearm, the belly of the muscle lies over the humerus, not the forearm.

LEVER SYSTEMS AND LEVERAGE

In producing a body movement, bones act as levers and joints function as fulcrums of these levers. A ***lever*** may be defined as a rigid rod that moves about on some fixed point called a ***fulcrum.*** A fulcrum may be symbolized as △F. A lever is acted on at two different points by two different forces: the ***resistance*** [R] and the ***effort*** (E). Whereas resistance is the force that opposes movement, effort is the force exerted to achieve an action. The resistance may be the weight of a part of the body that is to be moved. The effort is the muscular contraction that must be applied to the bone at the insertion to produce motion if the effort exceeds the resistance (load). Consider the biceps brachii flexing the forearm at the elbow as a weight is lifted (Figure 10-1b). When the forearm is raised, the elbow is the fulcrum. The weight of the forearm plus the weight in the hand is the resistance. The shortening due to the force of contraction of the biceps brachii pulling the forearm up is the effort.

Levers are categorized into three types according to the positions of the fulcrum, the effort, and the resistance.

1. In ***first-class levers,*** the fulcrum is between the effort and resistance (Figure 10-2a), symbolized by EFR. An example of a first-class lever is a seesaw. There are not many first-class levers in the body. One example

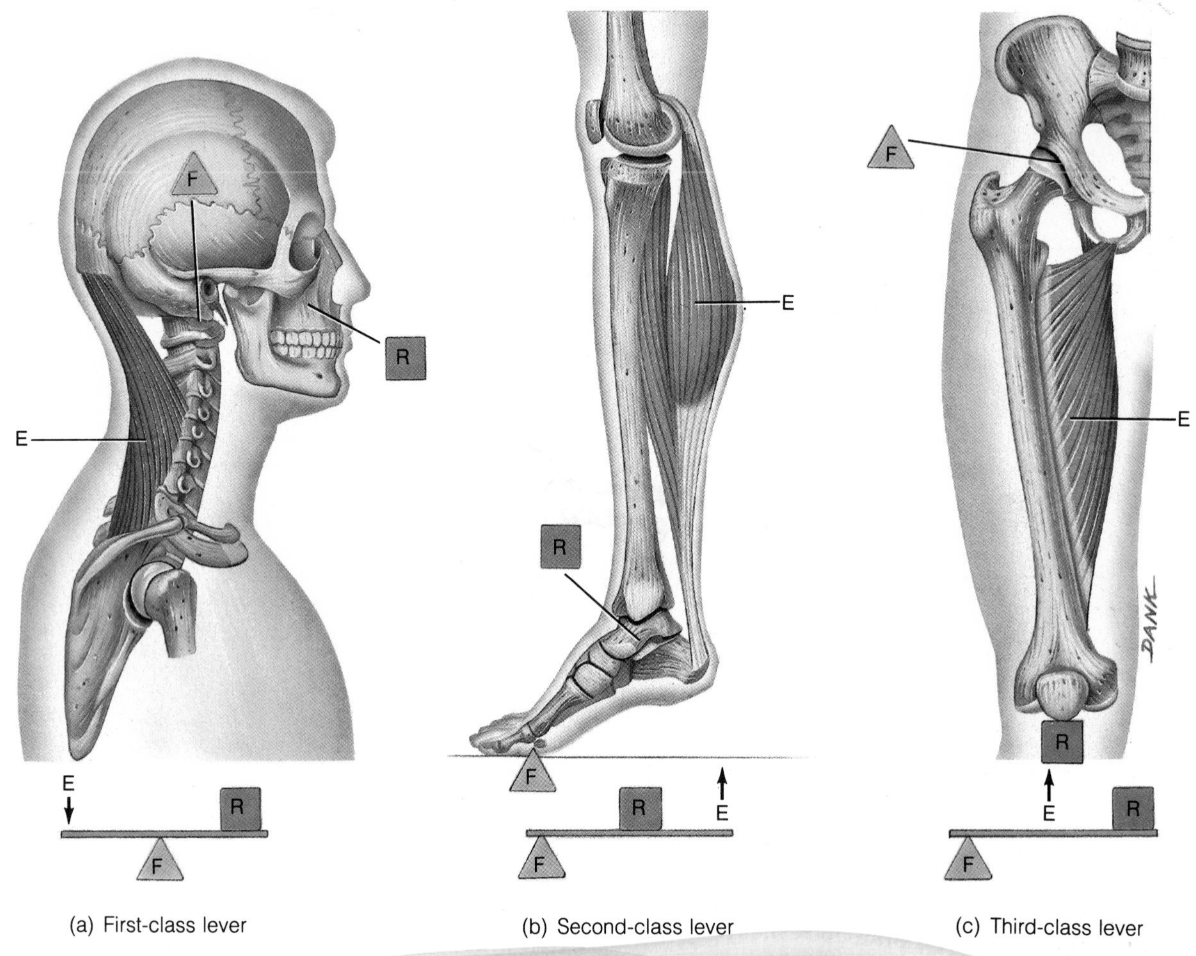

(a) First-class lever (b) Second-class lever (c) Third-class lever

FIGURE 10-2 Classes of levers. Each is defined on the basis of the placement of the fulcrum, effort, and resistance. A third-class lever is also shown in Figure 10-1b.

is the head resting on the vertebral column. When the head is raised, the facial portion of the skull is the resistance. The joint between the atlas and occipital bone (atlanto-occipital joint) is the fulcrum. The contraction of the muscles of the back is the effort.

2. ***Second-class levers*** have the fulcrum at one end, the effort at the opposite end, and the resistance between them (Figure 10-2b), symbolized by FRE. They operate like a wheelbarrow. Most authorities agree that there are very few examples of second-class levers in the body. One example is raising the body on the toes. The body is the resistance, the ball of the foot is the fulcrum, and the contraction of the calf muscles to pull the heel upward is the effort.
3. ***Third-class levers*** consist of the fulcrum at one end, the resistance at the opposite end, and the effort between them (Figure 10-2c), symbolized by FER. They are the most common levers in the body. One example is adduction of the thigh, in which the weight of the thigh is the resistance, the hip joint is the fulcrum, and contraction of the adductor muscles is the effort. Another example is flexing the forearm at the elbow. As we have seen, the weight of the forearm is the resistance, the contraction of the biceps brachii is the effort, and the elbow joint is the fulcrum (see Figure 10-1b).

Leverage, the mechanical advantage gained by a lever, is largely responsible for a muscle's strength and range of movement. Consider strength first. Suppose we have two muscles of the same strength crossing and acting on the same joint. Assume also that one is attached farther from the joint and one is nearer. The muscle attached farther will produce the more powerful movement. Thus, strength of movement depends on the placement of muscle attachments.

In considering range of movement, again assume that we have two muscles of the same strength crossing and acting on the same joint and that one is attached farther from the joint than the other. The muscle inserting closer to the joint will produce the greater range and speed of movement. Thus, range of movement also depends on the placement of muscle attachments. Since strength increases with distance from the joint and range of movement decreases, maximal strength and maximal range are incompatible; strength and range vary inversely.

ARRANGEMENT OF FASCICULI

Recall from Chapter 9 that skeletal muscle fibers (cells) are arranged within the muscle in bundles called fasciculi (fascicles). The muscle fibers are arranged in a parallel fashion within each bundle, but the arrangement of the fasciculi with respect to the tendons may take several characteristic patterns.

One pattern is called ***parallel.*** The fasciculi are parallel with the longitudinal axis and terminate at either end in flat tendons. The muscle is typically quadrilateral in shape. An example is the stylohyoid muscle (see Figure 10-7). In a modification of the parallel arrangement, called ***fusiform,*** the fasciculi are nearly parallel with the longitudinal axis and terminate at either end in flat tendons, but the muscle tapers toward the tendons, where the diameter is less than that of the belly. An example is the digastric muscle (see Figure 10-10).

In another pattern, called ***convergent,*** a broad origin of fasciculi converges to a narrow, restricted insertion. Such a pattern gives the muscle a triangular shape. An example is the pectoralis major muscle (see Figure 10-12a).

In the ***pennate*** pattern, the fasciculi are short in relation to the entire length of the muscle, and the tendon extends nearly the entire length of the muscle. The fasciculi are directed obliquely toward the tendon like the plumes of a feather. If the fasciculi are arranged on only one side of a tendon, as in the extensor digitorum longus muscle, the muscle is referred to as ***unipennate*** (see Figure 10-24c,d). If the fasciculi are arranged on both sides of a centrally positioned tendon, as in the rectus femoris muscle, the muscle is referred to as ***bipennate*** (see Figure 10-22a). In ***multipennate*** muscles, such as the deltoid muscle, there are several tendons in which the fibers attach obliquely from many directions (see Figure 10-17b).

The final distinct pattern is referred to as ***circular.*** The fasciculi are arranged in a circular pattern and enclose an orifice. An example is the orbicularis oris muscle (see Figure 10-4).

Fascicular arrangement is correlated with the power of a muscle and range of movement. When a muscle fiber contracts, it shortens to a length just slightly greater than half of its resting length. Thus, the longer the fibers in a muscle, the greater the range of movement it can produce. By contrast, the strength of a muscle depends on the total number of fibers it contains, since a short fiber can contract as forcefully as a long one. Because a given muscle can contain either a small number of long fibers or a large number of short fibers, fascicular arrangement represents a compromise between power and range of movement. Pennate muscles, for example, have a large number of fasciculi distributed over their tendons, giving them greater power, but a smaller range of movement. Parallel muscles, on the other hand, have comparatively few fasciculi that extend the length of the muscle. Thus, they have a greater range of movement but less power.

GROUP ACTIONS

Most movements require several skeletal muscles acting in groups rather than individually. Also, most skeletal muscles are arranged in opposing (antagonistic) pairs at joints, that is, flexors–extensors, abductors–adductors, and so on. Consider flexing the forearm at the elbow, for example. A muscle that causes a desired action is referred to as the ***prime mover (agonist).*** In this instance, the biceps brachii is the prime mover (see Figure 10-18a). Simultaneous with the contraction of the biceps brachii, another muscle, called the ***antagonist,*** is relaxing. In this movement, the triceps brachii serves as the antagonist (see Figure 10-18b). The antagonist has an effect opposite to that of the prime mover; that is, the antagonist relaxes and yields to the movement of the prime mover. You should not assume, however, that the biceps brachii is always the prime mover and the triceps brachii is always the antagonist. For example, when extending the forearm at the elbow, the triceps brachii serves as the prime mover and the biceps brachii functions as the antagonist; their roles are reversed. Note that if the prime mover and antagonist contracted simultaneously with equal force, there would be no movement.

In addition to prime movers and antagonists, most movements also involve muscles called ***synergists,*** which serve to steady a movement, thus preventing unwanted movements and helping the prime mover function more efficiently. For example, flex your hand at the wrist and then make a fist. Note how difficult this is to do. Now, extend your hand at the wrist and then make a fist. Note how much easier it is to clench your fist. In this case, the extensor muscles of the wrist act as synergists in cooperation with the flexor muscles of the fingers acting as prime movers. The extensor muscles of the fingers serve as antagonists (see Figure 10-19c).

Some synergist muscles in a group also act as ***fixators,*** which stabilize the origin of the prime mover so that the prime mover can act more efficiently. For example, the scapula is a freely movable bone in the pectoral (shoulder) girdle that serves as a firm origin for several muscles that move the arm. However, for the scapula to do this, it must be held steady. This is accomplished by fixator muscles that hold the scapula firmly against the back of the chest. In abduction of the arm, the deltoid muscle serves as the prime mover, whereas fixators (pectoralis minor, rhomboideus major, rhomboideus minor, trapezius, subclavius, and serratus anterior muscles) hold the scapula firmly (see Figure 10-16). These fixators stabilize the scapula that serves as the attachment site for the origin of the deltoid muscle, whereas the insertion of the muscle pulls on the humerus to abduct the arm. Under different conditions and depending on the movement and which point is fixed, many muscles act, at various times, as prime movers, antagonists, synergists, or fixators.

NAMING SKELETAL MUSCLES

The names of most of the nearly 700 skeletal muscles are based on several types of characteristics. Learning the terms used to indicate specific characteristics will help you remember the names of muscles.

FUNCTION

1. Muscle names may indicate the ***direction of the muscle fibers.*** *Rectus* fibers usually run parallel to the midline of the body. *Transverse* fibers run perpendicular to the midline. *Oblique* fibers are diagonal to the midline. Muscles named according to directions of fibers include the rectus abdominis, transversus abdominis, and external oblique.
2. A muscle may be named according to ***location.*** The temporalis is near the temporal bone. The tibialis anterior is near the front of the tibia.
3. ***Size*** is another characteristic. *Maximus* means largest; *minimus,* smallest; *longus,* long; and *brevis,* short. Examples include the gluteus maximus, gluteus minimus, adductor longus, and peroneus brevis.
4. Some muscles are named for their ***number of origins.*** The biceps brachii has two origins; the triceps brachii, three; and the quadriceps femoris, four.
5. Other muscles are named on the basis of ***shape.*** Common examples include the deltoid (meaning triangular), trapezius (meaning trapezoid), serratus anterior (meaning saw-toothed), and rhomboideus major (meaning rhomboid or diamond shaped).
6. Muscles may be named after their ***origin*** and ***insertion.*** The sternocleidomastoid originates on the sternum and clavicle and inserts at the mastoid process of the temporal bone; the stylohyoid originates on the styloid process of the temporal bone and inserts at the hyoid bone.
7. Still another characteristic of muscles used for naming is ***action.*** Exhibit 10-1 lists the principal actions of muscles, their definitions, and examples of muscles that perform the actions. For convenience, the actions are grouped as antagonistic pairs where possible.

PRINCIPAL SKELETAL MUSCLES

Exhibits 10-2 through 10-24 list the principal muscles of the body with their origins, insertions, actions, and innervations. (By no means have all the muscles of the body been included.) An **overview** section in each Exhibit provides a general orientation to the muscles under consideration. Refer to Chapters 6 and 7 to review bone markings, since they serve as points of origin and insertion for muscles. The muscles are divided into groups according to the part of the body on which they act. If you have mastered the naming of the muscles, their actions will have more meaning. Figure 10-3 shows general anterior and posterior views of the muscular system. Do not try to memorize all these muscles yet. As you study groups of muscles in the following exhibits, refer to Figure 10-3 to see how each group is related to all others.

The figures that accompany the exhibits contain superficial and deep, anterior and posterior, or medial and lateral views to show each muscle's position as clearly as possible. An attempt has been made to show the relationship of the muscles under consideration to other muscles in the area you are studying.

EXHIBIT 10-1

Principal Actions of Muscles

ACTION	DEFINITION	EXAMPLE
Flexor	Decreases the angle at a joint.	Flexor carpi radialis (see Exhibit 10-18).
Extensor	Increases the angle at a joint.	Extensor carpi ulnaris (see Exhibit 10-18).
Abductor	Moves a bone away from the midline.	Abductor pollicis brevis (see Exhibit 10-19).
Adductor	Moves a bone closer to the midline.	Adductor longus (see Exhibit 10-21).
Levator	Produces an upward movement.	Levator scapulae (see Exhibit 10-15).
Depressor	Produces a downward movement.	Depressor labii inferioris (see Exhibit 10-2).
Supinator	Turns the palm upward or anteriorly.	Supinator (see Exhibit 10-17).
Pronator	Turns the palm downward or posteriorly.	Pronator teres (see Exhibit 10-17).
Sphincter	Decreases the size of an opening.	External anal sphincter (see Exhibit 10-14).
Tensor	Makes a body part more rigid.	Tensor fasciae latae (see Exhibit 10-21).
Rotator	Moves a bone around its longitudinal axis.	Obturator externus (see Exhibit 10-21).

In order to further assist your learning efforts, the following list is provided to enable you to see the order in which the skeletal muscles will be studied according to region.

Exhibit 10-2. Muscles of Facial Expression (Figure 10-4), pp. 244–247

Exhibit 10-3. Muscles That Move the Lower Jaw (Figure 10-5), pp. 248–250

Exhibit 10-4. Muscles That Move the Eyeballs—Extrinsic Muscles (Figure 10-6), pp. 251–252

Exhibit 10-5. Muscles That Move the Tongue—Extrinsic Muscles (Figure 10-7), p. 253

Exhibit 10-6. Muscles of the Soft Palate (Figure 10-8), pp. 254–255

Exhibit 10-7. Muscles of the Pharynx (Figure 10-9), pp. 256–258

Exhibit 10-8. Muscles of the Floor of the Oral Cavity (Figure 10-10), p. 259

Exhibit 10-9. Muscles of the Larynx (Figure 10-11), pp. 260–261

Exhibit 10-10. Muscles That Move the Head, p. 262

Exhibit 10-11. Muscles That Act on the Abdominal Wall (Figure 10-12), pp. 263–265

Exhibit 10-12. Muscles Used in Breathing (Figure 10-13), pp. 266–267

Exhibit 10-13. Muscles of the Pelvic Floor (Figure 10-14), pp. 268–269

Exhibit 10-14. Muscles of the Perineum (Figure 10-15), pp. 270–271

Exhibit 10-15. Muscles That Move the Shoulder (Pectoral) Girdle (Figure 10-16), pp. 272–274

Exhibit 10-16. Muscles That Move the Arm (Humerus) (Figure 10-17), pp. 275–278

Exhibit 10-17. Muscles That Move the Forearm (Radius and Ulna) (Figure 10-18), pp. 279–281

Exhibit 10-18. Muscles That Move the Wrist, Hand, and Fingers (Figure 10-19), pp. 282–287

Exhibit 10-19. Intrinsic Muscles of the Hand (Figure 10-20), pp. 288–291

Exhibit 10-20. Muscles That Move the Vertebral Column (Figure 10-21), pp. 292–295

Exhibit 10-21. Muscles That Move the Thigh (Femur) (Figure 10-22), pp. 296–302

Exhibit 10-22. Muscles That Act on the Leg (Tibia and Fibula) (Figures 10-22 and 10-23), pp. 303–304

Exhibit 10-23. Muscles That Move the Foot and Toes (Figure 10-24), pp. 305–310

Exhibit 10-24. Intrinsic Muscles of the Foot (Figure 10-25), pp. 311–313

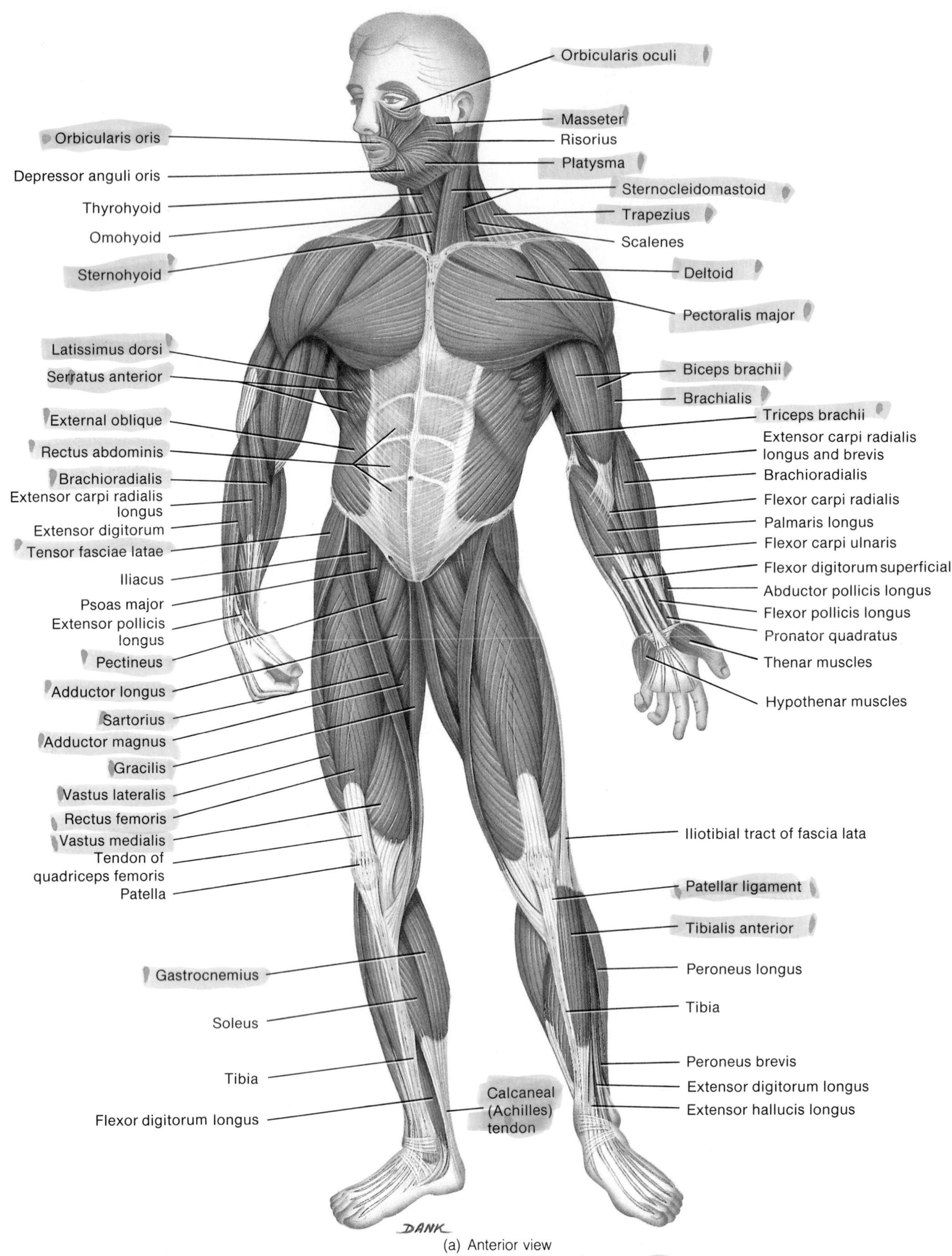

(a) Anterior view

FIGURE 10-3 Principal superficial skeletal muscles.

Will ask for the names on Lab tests by locating w/pin on cat

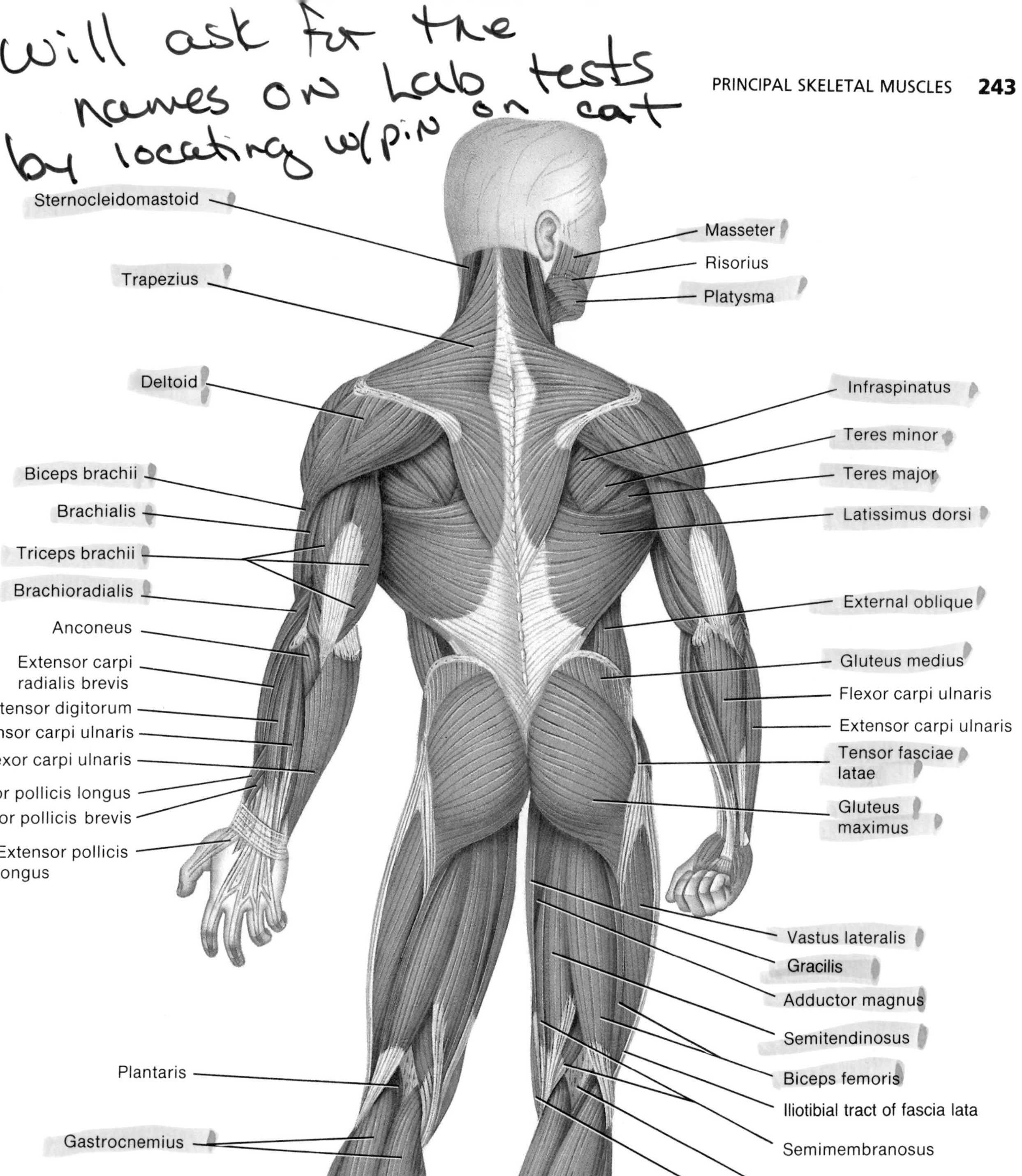

(b) Posterior view

EXHIBIT 10-2
Muscles of Facial Expression (Figure 10-4)

Overview: The muscles in this group provide humans with the ability to express a wide variety of emotions, including frowning, surprise, fear, and happiness. The muscles themselves lie within the layers of superficial fascia. As a rule, they arise from the fascia or bones of the skull and insert into the skin. Because of their insertions, the muscles of facial expression move the skin rather than a joint when they contract.

MUSCLE	ORIGIN	INSERTION	ACTION	INNERVATION
Epicranius (*epi* = over; *crani* = skull)	This muscle is divisible into two portions: the frontalis over the frontal bone and the occipitalis over the occipital bone. The two muscles are united by a strong aponeurosis, the galea aponeurotica, which covers the superior and lateral surfaces of the skull.			
Frontalis (*front* = forehead)	Galea aponeurotica.	Skin superior to supraorbital line.	Draws scalp forward, raises eyebrows, and wrinkles skin of forehead horizontally.	Facial (VII) nerve.
Occipitalis (*occipito* = base of skull)	Occipital bone and mastoid process of temporal bone.	Galea aponeurotica.	Draws scalp backward.	Facial (VII) nerve.
Orbicularis Oris (*orb* = circular; *or* = mouth)	Muscle fibers surrounding opening of mouth.	Skin at corner of mouth.	Closes lips, compresses lips against teeth, protrudes lips, and shapes lips during speech.	Facial (VII) nerve.
Zygomaticus Major (*zygomatic* = cheek bone; *major* = greater)	Zygomatic bone.	Skin at angle of mouth and orbicularis oris.	Draws angle of mouth upward and outward as in smiling or laughing.	Facial (VII) nerve.
Levator Labii Superioris (*levator* = raises or elevates; *labii* = lip; *superioris* = upper)	Superior to infraorbital foramen of maxilla.	Skin at angle of mouth and orbicularis oris.	Elevates (raises) upper lip.	Facial (VII) nerve.
Depressor Labii Inferioris (*depressor* = depresses or lowers; *inferioris* = lower)	Mandible.	Skin of lower lip.	Depresses (lowers) lower lip.	Facial (VII) nerve.
Buccinator (*bucc* = cheek)	Alveolar processes of maxilla and mandible and pterygomandibular raphe (fibrous band extending from the pterygoid hamulus to the mandible).	Orbicularis oris.	Major cheek muscle; compresses cheek as in blowing air out of mouth and causes cheeks to cave in, producing the action of sucking.	Facial (VII) nerve.
Mentalis (*mentum* = chin)	Mandible.	Skin of chin.	Elevates and protrudes lower lip and pulls skin of chin up as in pouting.	Facial (VII) nerve.
Platysma (*platy* = flat, broad)	Fascia over deltoid and pectoralis major muscles.	Mandible, muscles around angle of mouth, and skin of lower face.	Draws outer part of lower lip downward and backward as in pouting; depresses mandible.	Facial (VII) nerve.
Risorius (*risor* = laughter)	Fascia over parotid (salivary) gland.	Skin at angle of mouth.	Draws angle of mouth laterally as in tenseness.	Facial (VII) nerve.
Orbicularis Oculi (*ocul* = eye)	Medial wall of orbit.	Circular path around orbit.	Closes eye.	Facial (VII) nerve.

EXHIBIT 10-2 (*Continued*)

MUSCLE	ORIGIN	INSERTION	ACTION	INNERVATION
Corrugator Supercilli (*corrugo* = wrinkle; *supercilium* = eyebrow)	Medial end of superciliary arch of frontal bone.	Skin of eyebrow.	Draws eyebrow downward as in frowning.	Facial (VII) nerve.
Levator Palpebrae Superioris (*palpebrae* = eyelids) (see also Figure 10-6a)	Roof of orbit (lesser wing of sphenoid bone).	Skin of upper eyelid.	Elevates upper eyelid.	Oculomotor (III) nerve.

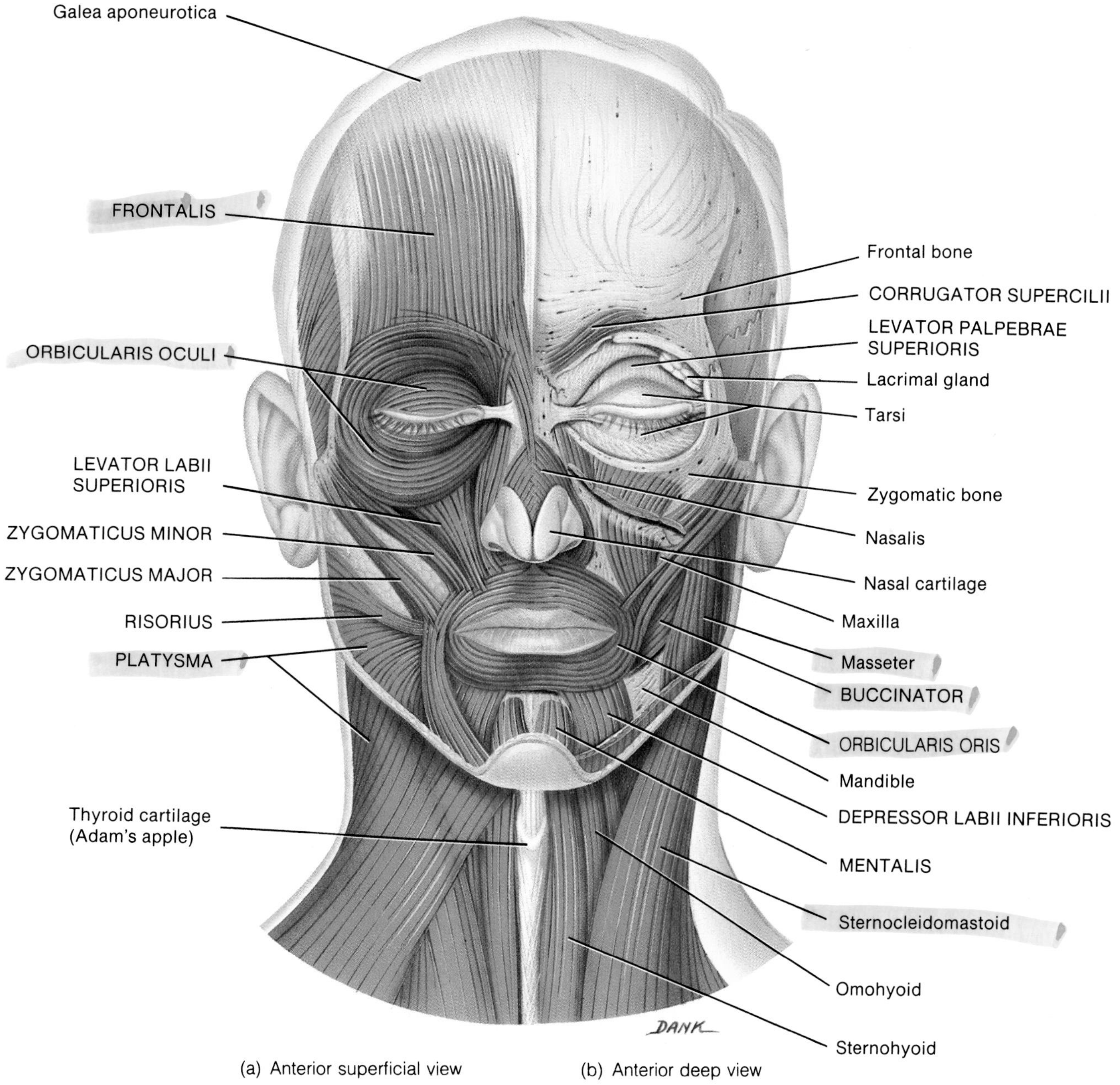

FIGURE 10-4 Muscles of facial expression. (d) Courtesy of J. A. Gosling, P. F. Harris, et al., *Atlas of Human Anatomy,* Gower Medical Publishing Ltd., 2nd ed., 1991.

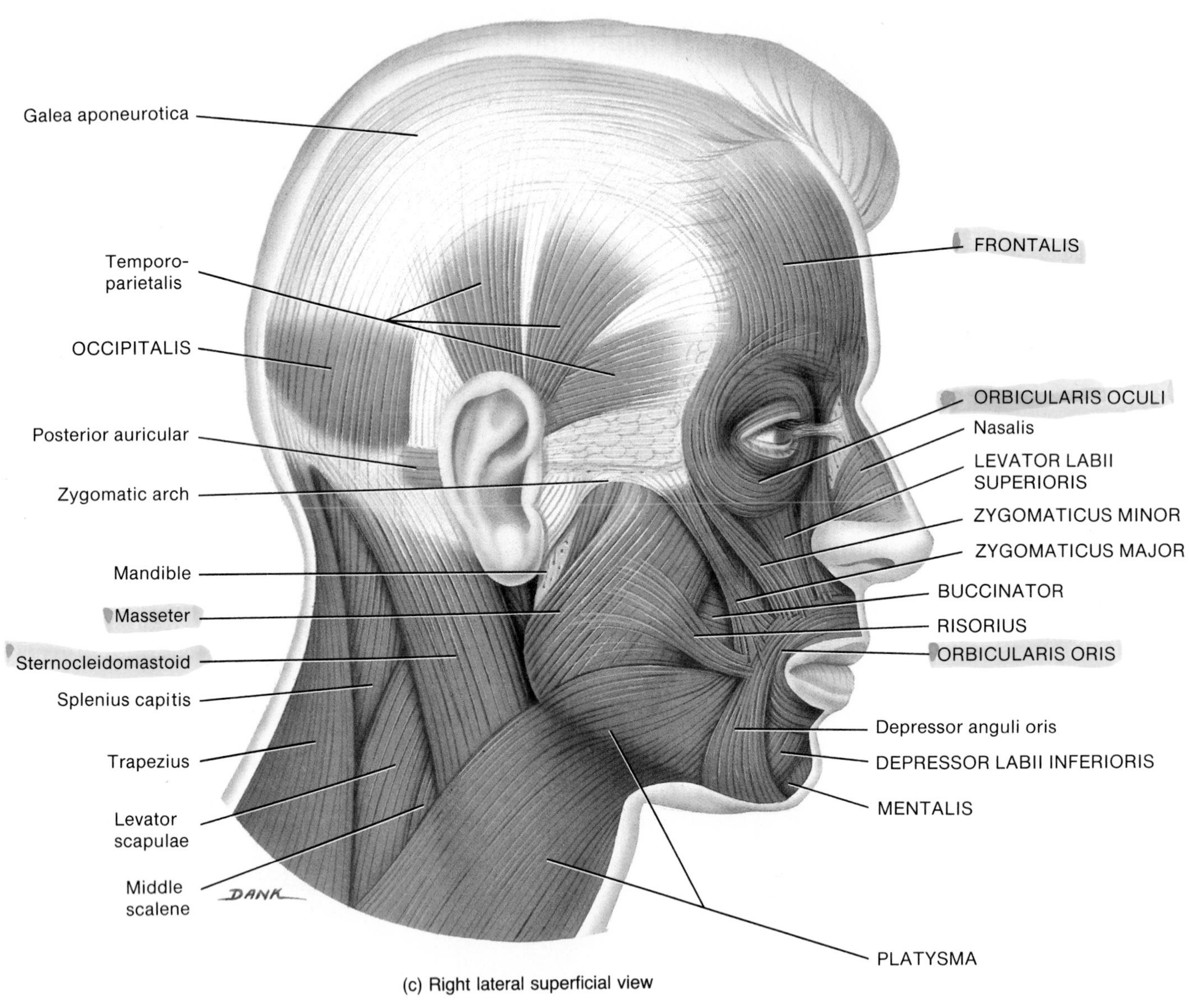

(c) Right lateral superficial view

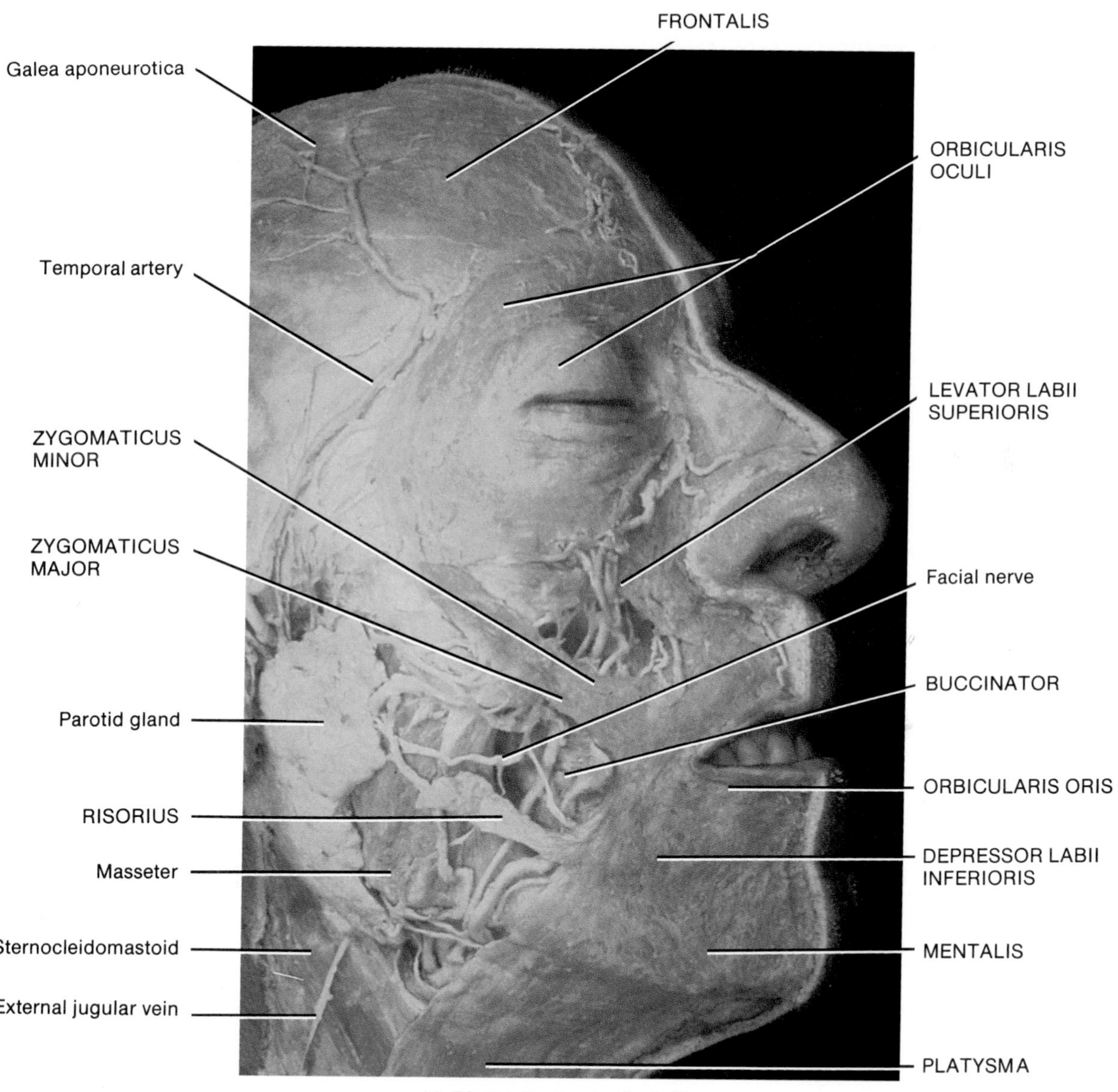

(d) Right lateral superfical view

EXHIBIT 10-3

Muscles That Move the Lower Jaw (Figure 10-5)

Overview: Muscles that move the lower jaw are also known as muscles of mastication because they are involved in biting and chewing. These muscles also assist in speech.

MUSCLE	ORIGIN	INSERTION	ACTION	INNERVATION
Masseter (*maseter* = chewer)	Maxilla and zygomatic arch.	Angle and ramus of mandible.	Elevates mandible as in closing mouth, assists in side to side movement of mandible, and protracts (protrudes) mandible.	Mandibular division of trigeminal (V) nerve.
Temporalis (*tempora* = temples)	Parietal bone.	Coronoid process of mandible.	Elevates and retracts mandible and assists in side to side movement of mandible.	Mandibular division of trigeminal (V) nerve.
Medial Pterygoid (*medial* = closer to midline; *pterygoid* = like a wing; pterygoid plate of sphenoid bone)	Medial surface of lateral pterygoid plate of sphenoid; maxilla.	Angle and ramus of mandible.	Elevates and protracts mandible and moves mandible from side to side.	Mandibular division of trigeminal (V) nerve.
Lateral Pterygoid (*lateral* = farther from midline)	Greater wing and lateral surface of lateral pterygoid plate of sphenoid.	Condyle of mandible; temporomandibular articulation.	Protracts mandible, opens mouth, and moves mandible from side to side.	Mandibular division of trigeminal (V) nerve.

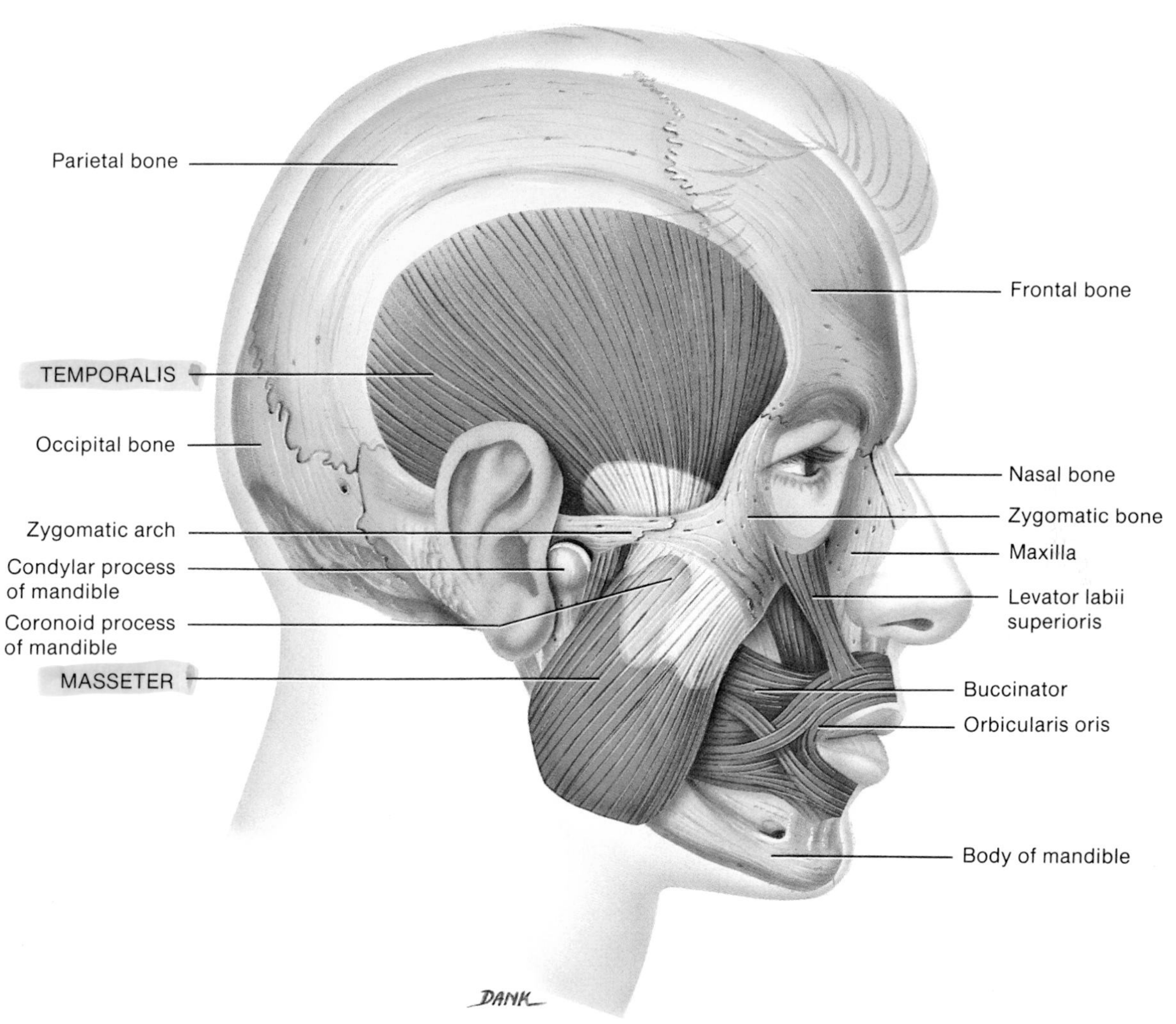

(a) Right lateral superficial view

FIGURE 10-5 Muscles that move the lower jaw.

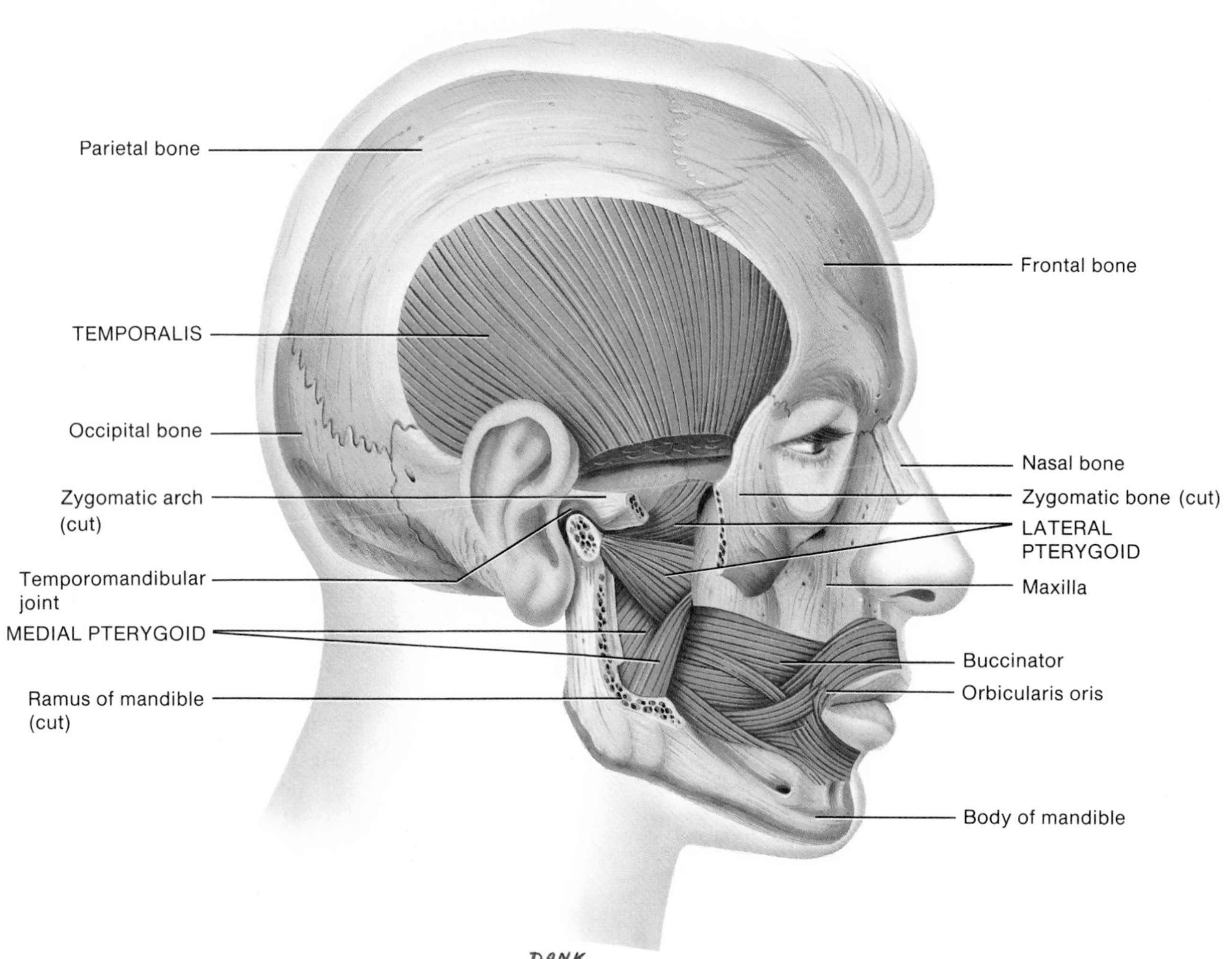

(b) Right lateral deep view

EXHIBIT 10-4

Muscles That Move the Eyeballs—Extrinsic Muscles (Figure 10-6)

Overview: Muscles associated with the eyeball are of two principal types: extrinsic and intrinsic. ***Extrinsic muscles*** originate outside the eyeball and are inserted on its outer surface (sclera). ***Intrinsic muscles*** originate and insert entirely within the eyeball.

Movements of the eyeballs are controlled by three pairs of extrinsic muscles. Two pairs of rectus muscles move the eyeball in the direction indicated by their respective names—superior, inferior, lateral, and medial. One pair of muscles, the oblique muscles—superior and inferior—rotate the eyeball on its axis. The extrinsic muscles of the eyeballs are among the fastest contracting and most precisely controlled skeletal muscles of the body.

MUSCLE	ORIGIN	INSERTION	ACTION	INNERVATION
Superior Rectus (*superior* = above; *rectus* = in this case, muscle fibers running parallel to long axis of eyeball)	Tendinous ring attached to bony orbit around optic foramen.	Superior and central part of eyeball.	Rolls eyeball upward.	Oculomotor (III) nerve.
Inferior Rectus (*inferior* = below)	Same as above.	Inferior and central part of eyeball.	Rolls eyeball downward.	Oculomotor (III) nerve.
Lateral Rectus	Same as above.	Lateral side of eyeball.	Rolls eyeball laterally.	Abducens (VI) nerve.
Medial Rectus	Same as above.	Medial side of eyeball.	Rolls eyeball medially.	Oculomotor (III) nerve.
Superior Oblique (*oblique* = in this case, muscle fibers running diagonally to long axis of eyeball)	Same as above.	Eyeball between superior and lateral recti. The muscle inserts in a round tendon that moves through a ring of fibrocartilaginous tissue called the trochlea (*trochlea* = pulley).	Rotates eyeball on its axis; directs cornea downward and laterally.	Trochlear (IV) nerve.
Inferior Oblique	Maxilla (front of orbital cavity).	Eyeball between inferior and lateral recti.	Rotates eyeball on its axis; directs cornea upward and laterally.	Oculomotor (III) nerve.

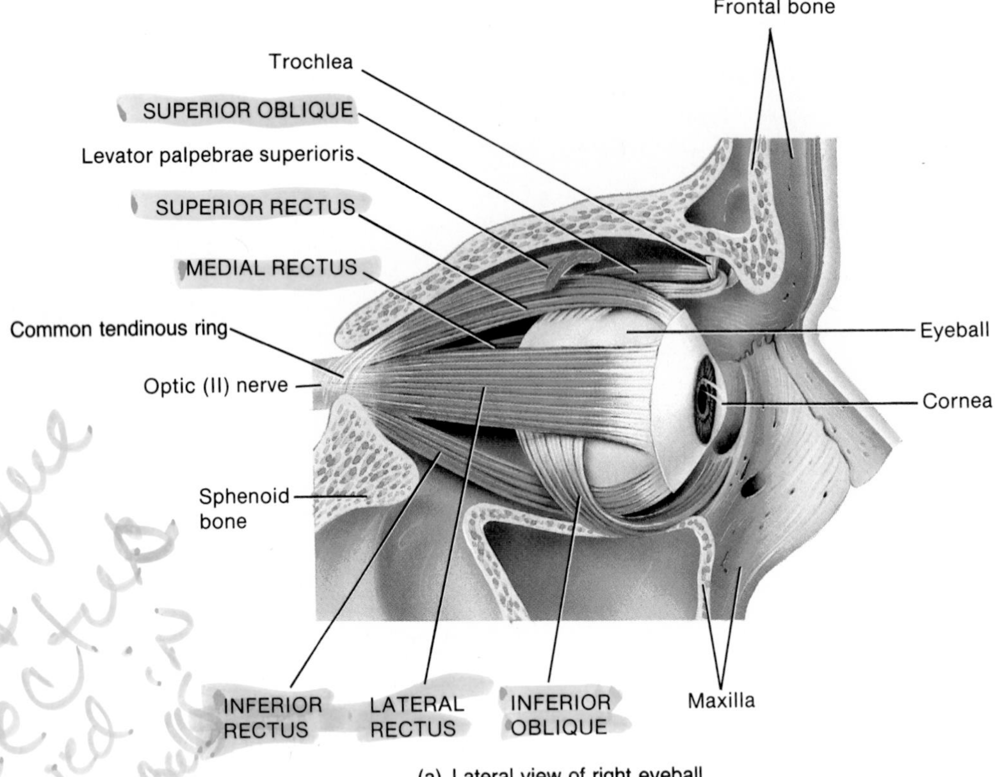

(a) Lateral view of right eyeball

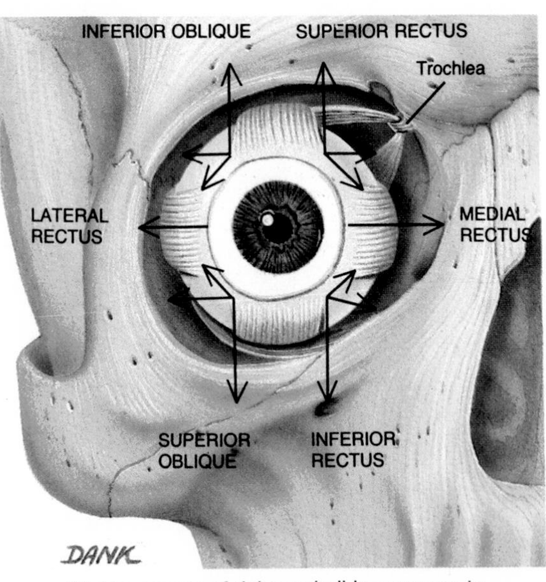

(b) Movements of right eyeball in response to contraction of its extrinsic muscles

FIGURE 10-6 Extrinsic muscles of the eyeball.

EXHIBIT 10-5

Muscles That Move the Tongue—Extrinsic Muscles (Figure 10-7)

Overview: The tongue is divided into lateral halves by a median fibrous septum. The septum extends throughout the length of the tongue and is attached inferiorly to the hyoid bone. Like the muscles of the eyeball, muscles of the tongue are of two principal types—extrinsic and intrinsic. ***Extrinsic muscles*** originate outside the tongue and insert into it. ***Intrinsic muscles*** originate and insert within the tongue. The extrinsic and intrinsic muscles of the tongue are arranged in both lateral halves of the tongue.

MUSCLE	ORIGIN	INSERTION	ACTION	INNERVATION
Genioglossus (*geneion* = chin; *glossus* = tongue)	Mandible.	Undersurface of tongue and hyoid bone.	Depresses tongue and thrusts it forward (protraction).	Hypoglossal (XII) nerve.
Styloglossus (*stylo* = stake or pole; styloid process of temporal bone)	Styloid process of temporal bone.	Side and undersurface of tongue.	Elevates tongue and draws it backward (retraction).	Hypoglossal (XII) nerve.
Palatoglossus (*palato* = palate)	Anterior surface of soft palate.	Side of tongue.	Elevates posterior portion of tongue and draws soft palate down on tongue.	Pharyngeal plexus.
Hyoglossus	Body of hyoid bone.	Side of tongue.	Depresses tongue and draws down its sides.	Hypoglossal (XII) nerve.

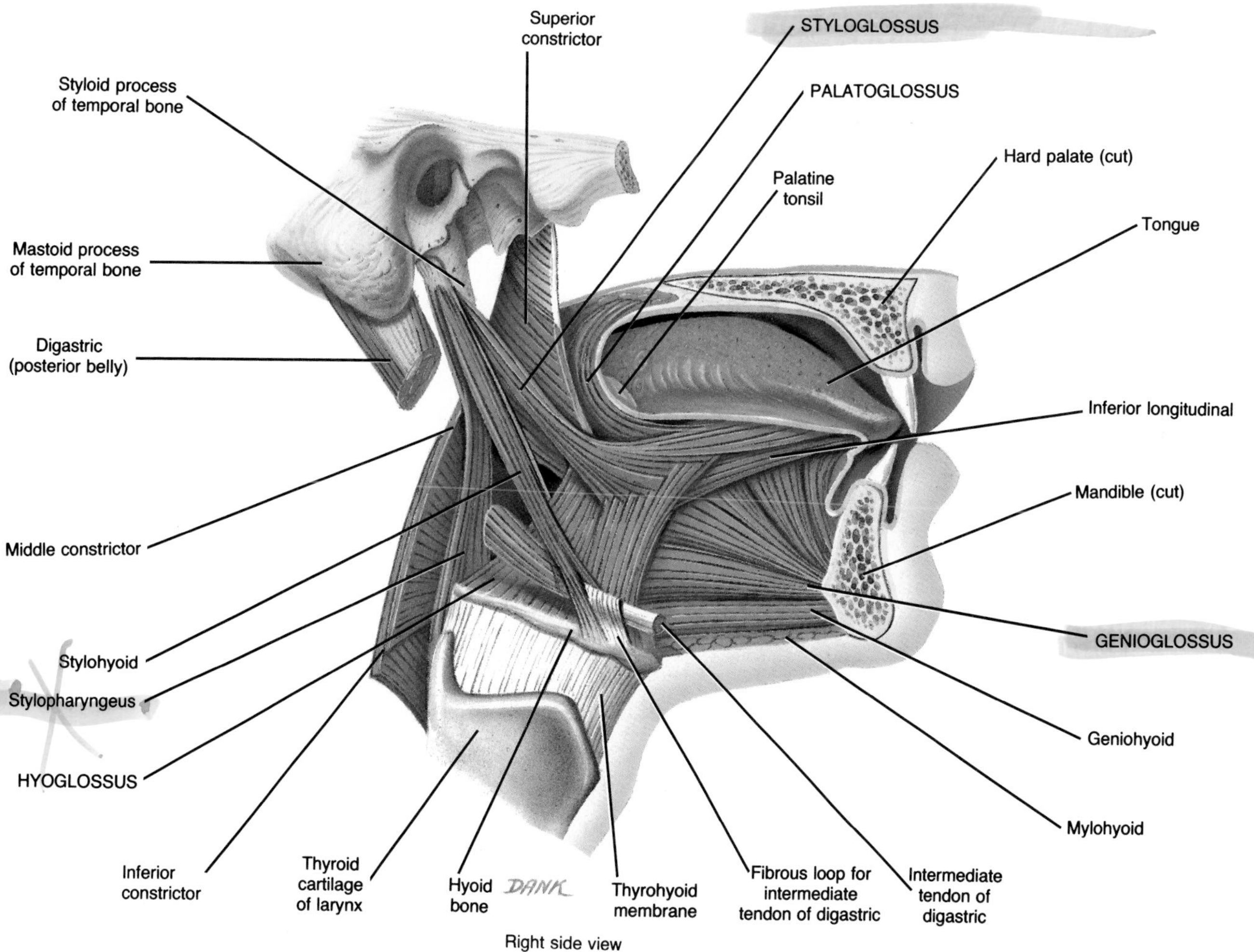

FIGURE 10-7 Muscles that move the tongue.

EXHIBIT 10-6

Muscles of the Soft Palate (Figure 10-8)

Overview: The soft palate is mainly muscular in structure and attaches anteriorly to the hard palate and blends posteriorly with the pharynx. The soft palate forms the posterior aspect of the partition that separates the oral (mouth) cavity from the pharynx. Hanging down from the free edge of the soft palate is a nipplelike structure, the uvula. During swallowing, the muscles of the soft palate tighten and elevate it, thus preventing food from entering the nasal cavities and openings of the auditory (Eustachian) tubes.

MUSCLE	ORIGIN	INSERTION	ACTION	INNERVATION
Levator Veli Palatini (*levator* = raises; *velum* = veil; *palato* = palate) (see also Figure 10-9)	Petrous portion of temporal bone and medial wall of auditory (Eustachian) tube.	Blends with corresponding muscle of opposite side.	Elevates soft palate during swallowing.	Pharyngeal plexus.
Tensor Veli Palatini (*tensor* = makes tense) (see also Figure 10-9)	Medial pterygoid plate of sphenoid bone, spine of sphenoid, lateral wall of auditory (Eustachian) tube.	Palatine aponeurosis and palatine bone.	Tenses (tightens) soft palate during swallowing.	Mandibular branch of trigeminal (V) nerve.
Musculus Uvulae (*uvulae* = uvula)	Posterior border of hard palate and palatine aponeurosis.	Uvula.	Tenses (tightens) and raises uvula.	Pharyngeal plexus.
Palatoglossus (*palato* = palate; *glossus* = tongue) (see also Figure 10-7)	Anterior surface of soft palate.	Side of tongue.	Elevates posterior portion of tongue and draws soft palate down on tongue.	Pharyngeal plexus.
Palatopharyngeus (*pharyngo* = pharynx)	Posterior border of hard palate and palatine aponeurosis.	Posterior border of thyroid cartilage and lateral and posterior wall of pharynx.	Elevates larynx and pharynx and helps close nasopharynx during swallowing.	Pharyngeal plexus.

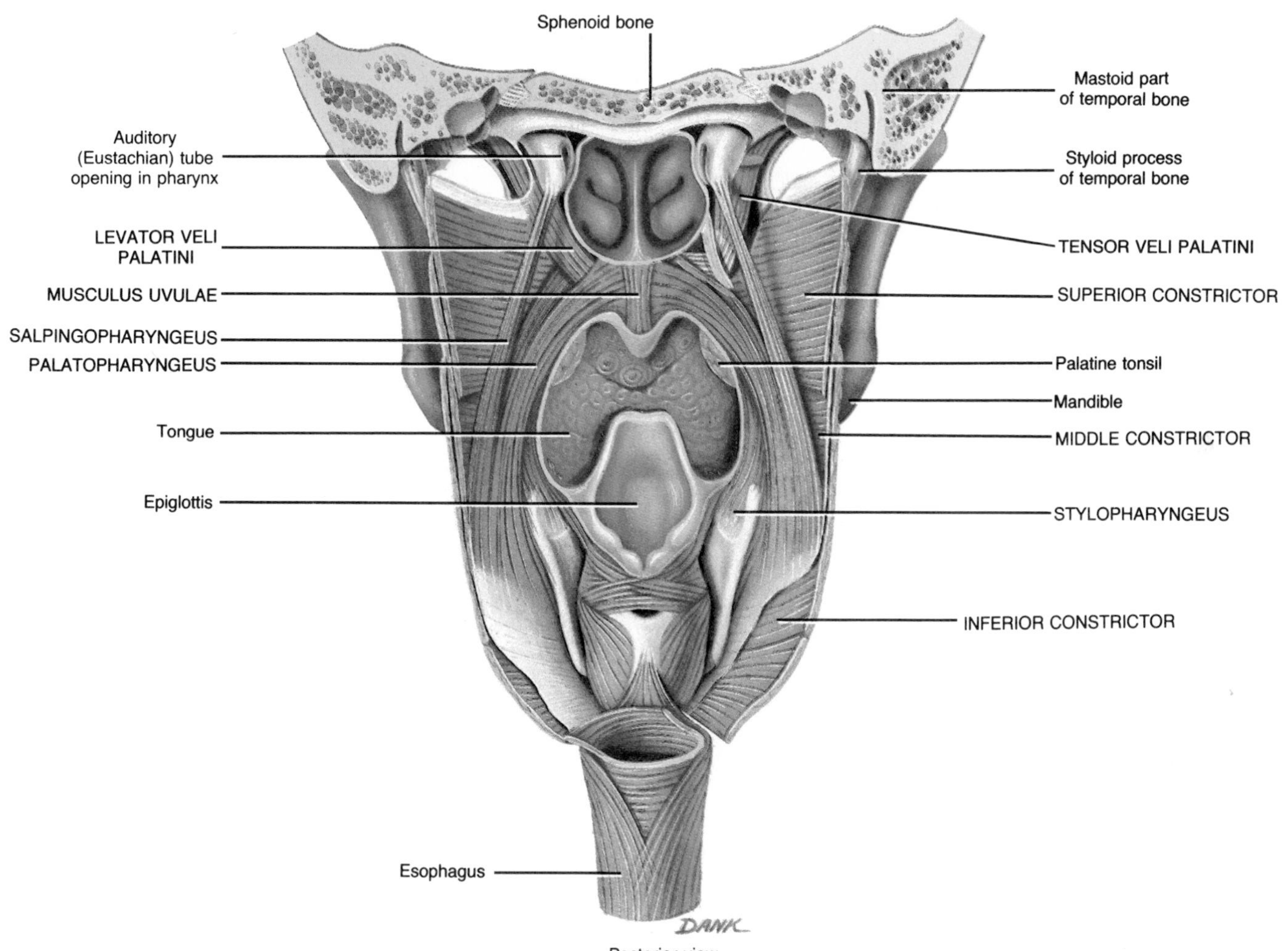

FIGURE 10-8 Muscles of the soft palate.

EXHIBIT 10-7

Muscles of the Pharynx (Figure 10-9)

Overview: The pharynx (throat) is a funnel-shaped, muscular tube located posterior to the nasal cavity, mouth, and larynx (voice box). The muscles are arranged in two layers—an outer circular layer and an inner longitudinal layer. The ***circular layer*** is composed of the three constrictors, each overlapping the one above. The remaining muscles constitute the ***longitudinal layer.***

MUSCLE	ORIGIN	INSERTION	ACTION	INNERVATION
CIRCULAR LAYER				
Inferior Constrictor (*inferior* = below; *constrictor* = decreases diameter of a lumen)	Cricoid and thyroid cartilages of larynx.	Posterior median raphe of pharynx.	Constricts inferior portion of pharynx to propel a bolus into esophagus.	Pharyngeal plexus.
Middle Constrictor	Greater and lesser horns of hyoid bone and stylohyoid ligament.	Posterior median raphe of pharynx.	Constricts middle portion of pharynx to propel a bolus into esophagus.	Pharyngeal plexus.
Superior Constrictor (*superior* = above)	Pterygoid process, pterygomandibular raphe, and mylohyoid line of mandible.	Posterior median raphe of pharynx.	Constricts superior portion of pharynx to propel a bolus into esophagus.	Pharyngeal plexus.
LONGITUDINAL LAYER				
Stylopharyngeus (*stylo* = stake or pole; styloid process of temporal bone; *pharyngo* = pharynx) (see also Figure 10-7)	Medial side of base of styloid process.	Lateral aspects of pharynx and thyroid cartilage.	Elevates larynx and dilates pharynx to help bolus descend.	Glossopharyngeal (IX) nerve.
Salpingopharyngeus (*salping* = pertaining to the auditory or uterine tube) (see also Figure 10-8)	Inferior portion of auditory (Eustachian) tube.	Posterior fibers of palatopharyngeus muscle.	Elevates superior portion of lateral wall of pharynx during swallowing and opens orifice of auditory (Eustachian) tube.	Pharyngeal plexus.
Palatopharyngeus (*palato* = palate)	Soft palate.	Posterior border of thyroid cartilage and lateral and posterior wall of pharynx.	Elevates larynx and pharynx and helps close nasopharynx during swallowing.	Pharyngeal plexus.

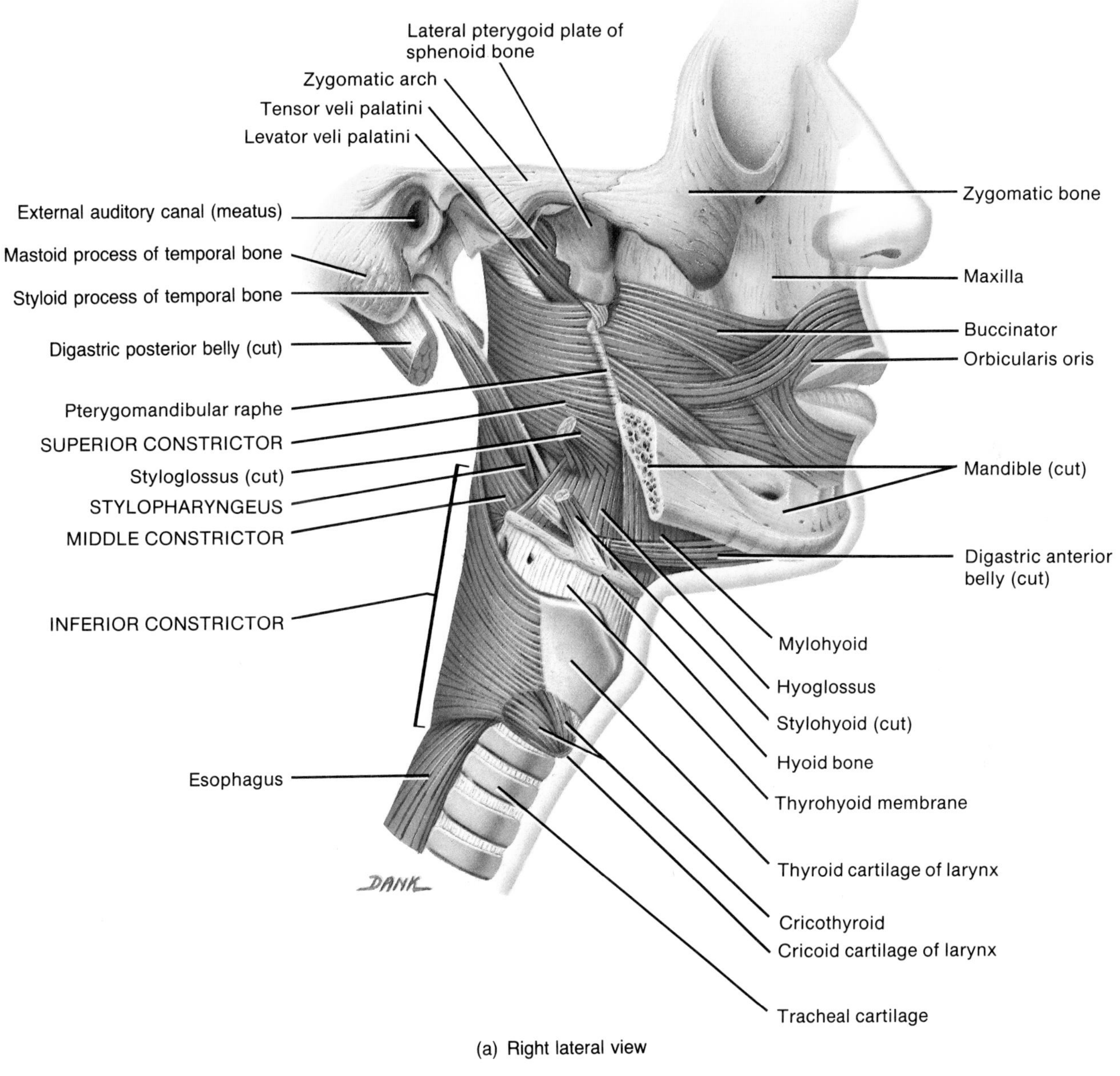

(a) Right lateral view

FIGURE 10-9 Muscles of the pharynx. (b) Courtesy of J. A. Gosling, P. F. Harris, et al., *Atlas of Human Anatomy,* Gower Medical Publishing Ltd., 2nd ed., 1991.

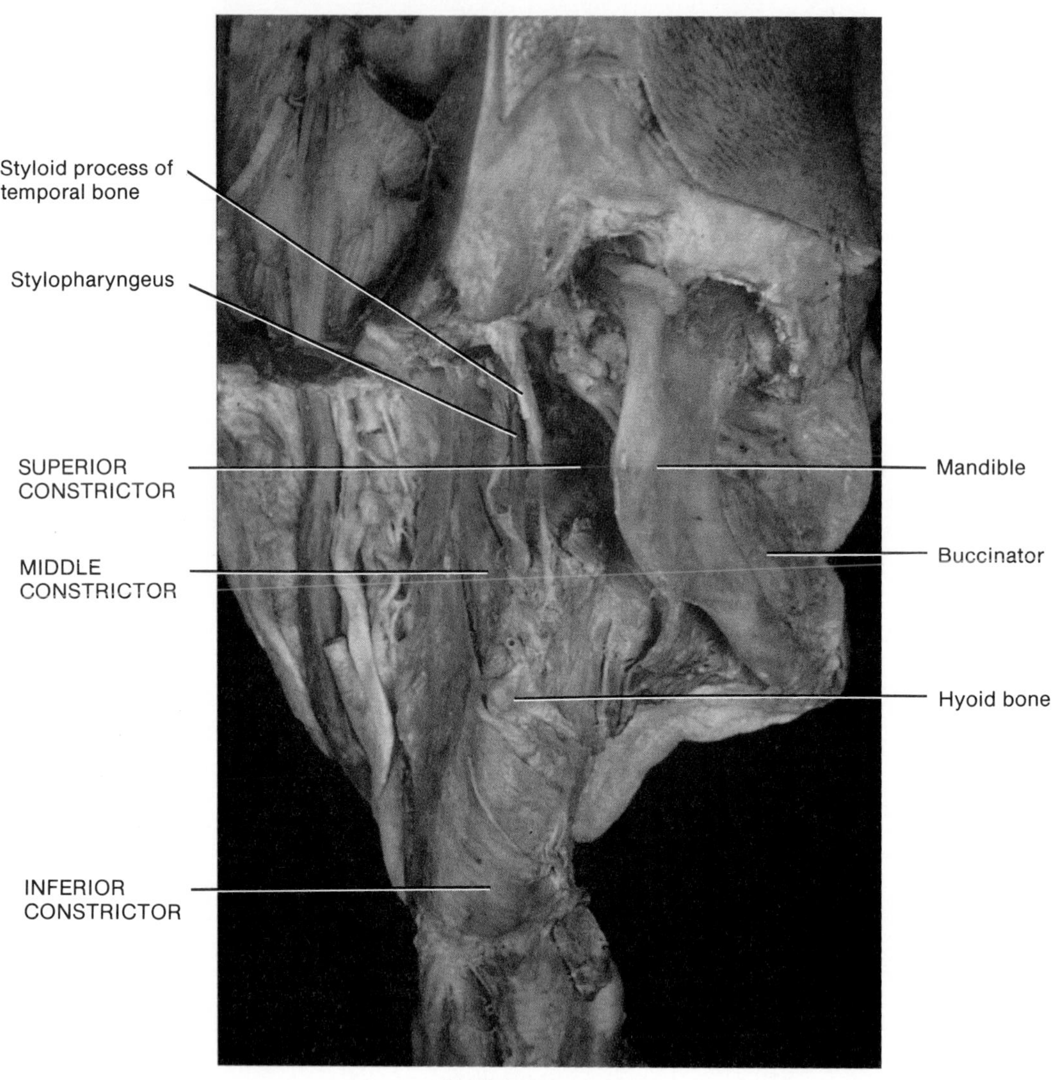

(b) Right lateral view

EXHIBIT 10-8
Muscles of the Floor of the Oral Cavity (Figure 10-10)

Overview: As a group, these muscles are referred to as ***suprahyoid muscles.*** They lie superior to the hyoid bone, and all insert into it. The digastric muscle consists of an anterior belly and a posterior belly united by an intermediate tendon that is held in position by a fibrous loop (see also Figure 10-7).

MUSCLE	ORIGIN	INSERTION	ACTION	INNERVATION
Digastric (*di* = two; *gaster* = belly)	Anterior belly from inner side of lower border of mandible; posterior belly from mastoid process of temporal bone.	Body of hyoid bone via an intermediate tendon.	Elevates hyoid bone and depresses mandible as in opening the mouth.	Anterior belly from mandibular division of trigeminal (V) nerve; posterior belly from facial (VII) nerve.
Stylohyoid (*stylo* = stake or pole, styloid process of temporal bone; *hyoedes* = U-shaped, pertaining to hyoid bone) (see also Figure 10-7)	Styloid process of temporal bone.	Body of hyoid bone.	Elevates hyoid bone and draws it posteriorly.	Facial (VII) nerve.
Mylohyoid	Inner surface of mandible.	Body of hyoid bone.	Elevates hyoid bone and floor of mouth and depresses mandible.	Mandibular division of trigeminal (V) nerve.
Geniohyoid (*geneion* = chin) (see also Figure 10-7)	Inner surface of mandible.	Body of hyoid bone.	Elevates hyoid bone, draws hyoid bone and tongue anteriorly, and depresses mandible.	Cervical nerve C1.

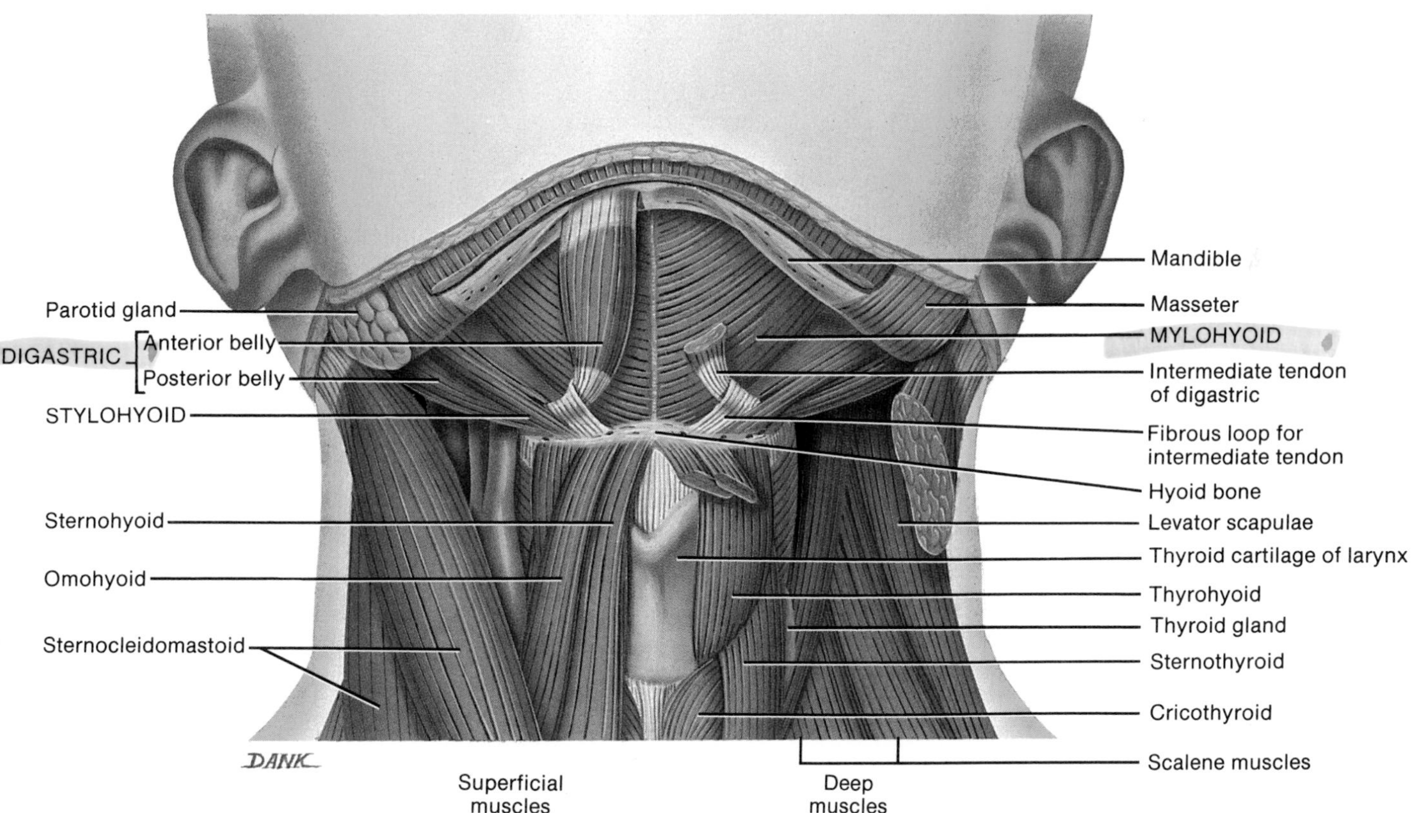

FIGURE 10-10 Muscles of the floor of the oral cavity and front of the neck.

EXHIBIT 10-9

Muscles of the Larynx (Figure 10-11)

Overview: The muscles of the larynx, like those of the eyeballs and tongue, are grouped into extrinsic and intrinsic. The extrinsic muscles of the larynx marked with an asterisk (*) are together referred to as ***infrahyoid (strap) muscles.*** They lie inferior to the hyoid bone. The omohyoid muscle, like the digastric muscle, is composed of two bellies and an intermediate tendon. In this case, however, the two bellies are referred to as superior and inferior, rather than anterior and posterior.

MUSCLE	ORIGIN	INSERTION	ACTION	INNERVATION
EXTRINSIC				
Omohyoid* (*omo* = relationship to the shoulder; *hyoedes* = U-shaped; pertaining to hyoid bone)	Superior border of scapula and superior transverse ligament.	Body of hyoid bone.	Depresses hyoid bone.	Branches of ansa cervicalis (C1–C3).
Sternohyoid* (*sterno* = sternum)	Medial end of clavicle and manubrium of sternum.	Body of hyoid bone.	Depresses hyoid bone.	Branches of ansa cervicalis (C1–C3).
Sternothyroid* (*thyro* = thyroid gland)	Manubrium of sternum.	Thyroid cartilage of larynx.	Depresses thyroid cartilage.	Branches of ansa cervicalis (C1–C3).
Thyrohyoid*	Thyroid cartilage of larynx.	Greater horn of hyoid bone.	Elevates thyroid cartilage and depresses hyoid bone.	Branches of ansa cervicalis (C1–C2) and descending hypoglossal (XII) nerve.
Stylopharyngeus	See Exhibit 10-7.			
Palatopharyngeus	See Exhibit 10-7.			
Inferior Constrictor	See Exhibit 10-7.			
Middle Constrictor	See Exhibit 10-7.			
INTRINSIC				
Cricothyroid (*crico* = cricoid cartilage of larynx)	Anterior and lateral portion of cricoid cartilage of larynx.	Anterior border of inferior cornu of thyroid cartilage of larynx and posterior part of inferior border of lamina of thyroid cartilage.	Produces tension and elongation of vocal folds.	External laryngeal branch of vagus (X) nerve.
Posterior Cricoarytenoid (*arytaina* = shaped like a jug)	Posterior surface of cricoid cartilage.	Posterior surface of muscular process of arytenoid cartilage of larynx.	Opens glottis.	Recurrent laryngeal branch of vagus (X) nerve.
Lateral Cricoarytenoid	Superior border of cricoid cartilage.	Anterior surface of muscular process of arytenoid cartilage.	Closes glottis.	Recurrent laryngeal branch of vagus (X) nerve.
Arytenoid	Posterior surface and lateral border of one arytenoid cartilage.	Corresponding parts of opposite arytenoid cartilage.	Closes glottis.	Recurrent laryngeal branch of vagus (X) nerve.
Thyroarytenoid	Inferior portion of angle of thyroid cartilage and middle of cricothyroid ligament.	Base and anterior surface of arytenoid cartilage.	Shortens and relaxes vocal folds.	Recurrent laryngeal branch of vagus (X) nerve.

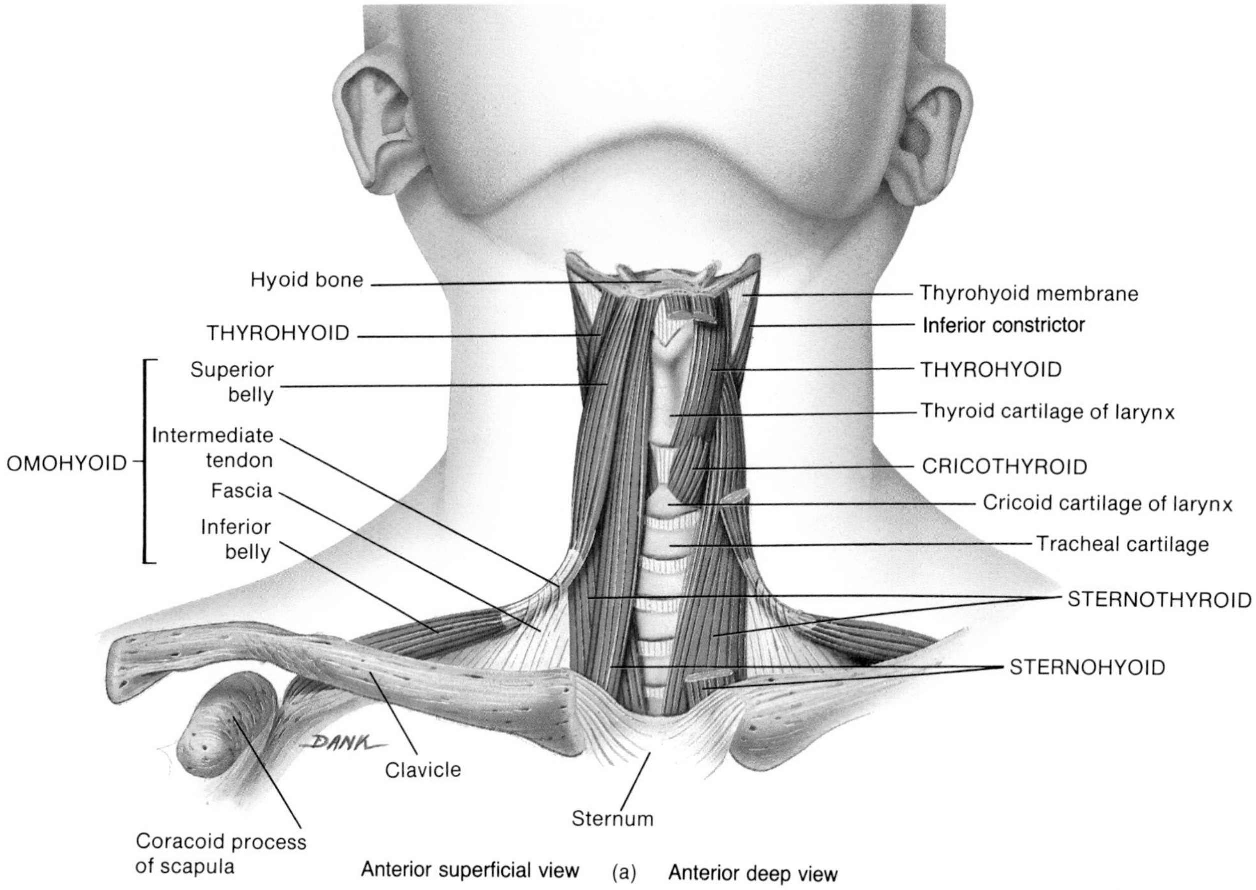

Hyoid bone
THYROHYOID
Superior belly
Intermediate tendon
OMOHYOID
Fascia
Inferior belly
Thyrohyoid membrane
Inferior constrictor
THYROHYOID
Thyroid cartilage of larynx
CRICOTHYROID
Cricoid cartilage of larynx
Tracheal cartilage
STERNOTHYROID
STERNOHYOID
DANK
Clavicle
Sternum
Coracoid process of scapula
Anterior superficial view
(a)
Anterior deep view

Epiglottis
Hyoid bone
Thyrohyoid membrane
Thyroepiglottis
Aryepiglottis
ARYTENOID (oblique and transverse)
THYROARYTENOID
Thyroid cartilage of larynx (cut)
Thyroid cartilage of larynx
POSTERIOR CRICOARYTENOID
LATERAL CRICOARYTENOID
Cricoid cartilage of larynx
CRICOTHYROID
Tracheal cartilage

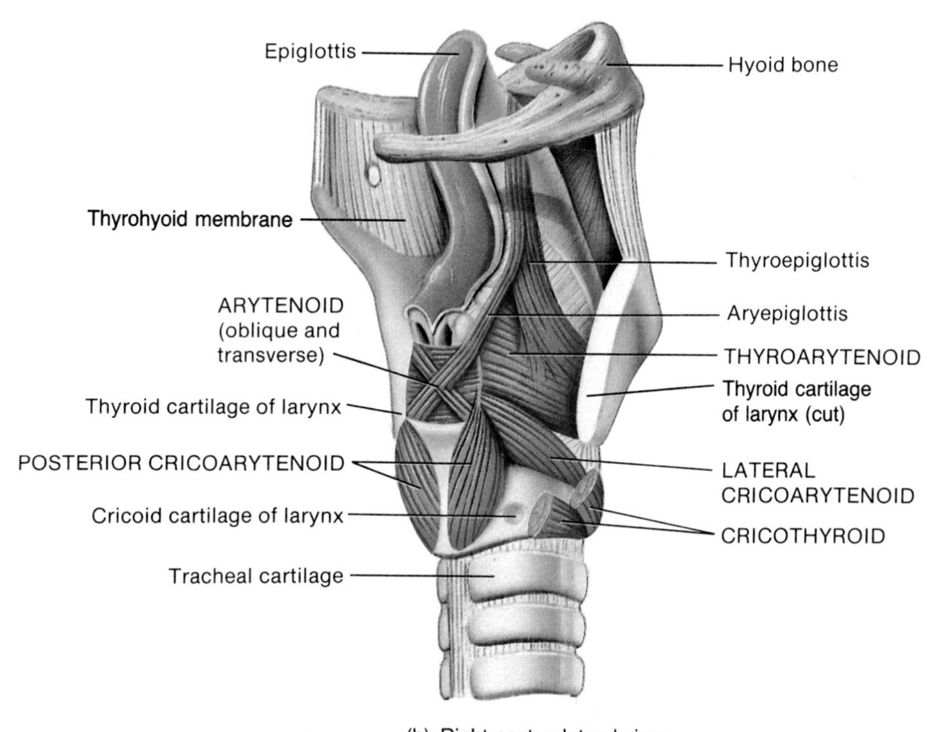

(b) Right posterolateral view

FIGURE 10-11 Muscles of the larynx.

EXHIBIT 10-10
Muscles That Move the Head

Overview: The cervical region is divided by the sternocleidomastoid muscle into two principal triangles—anterior and posterior. The ***anterior triangle*** is bordered superiorly by the mandible, inferiorly by the sternum, medially by the cervical midline, and laterally by the anterior border of the sternocleidomastoid muscle. The ***posterior triangle*** is bordered inferiorly by the clavicle, anteriorly by the posterior border of the sternocleidomastoid muscle, and posteriorly by the anterior border of the trapezius muscle. Subsidiary triangles exist within the two principal triangles.

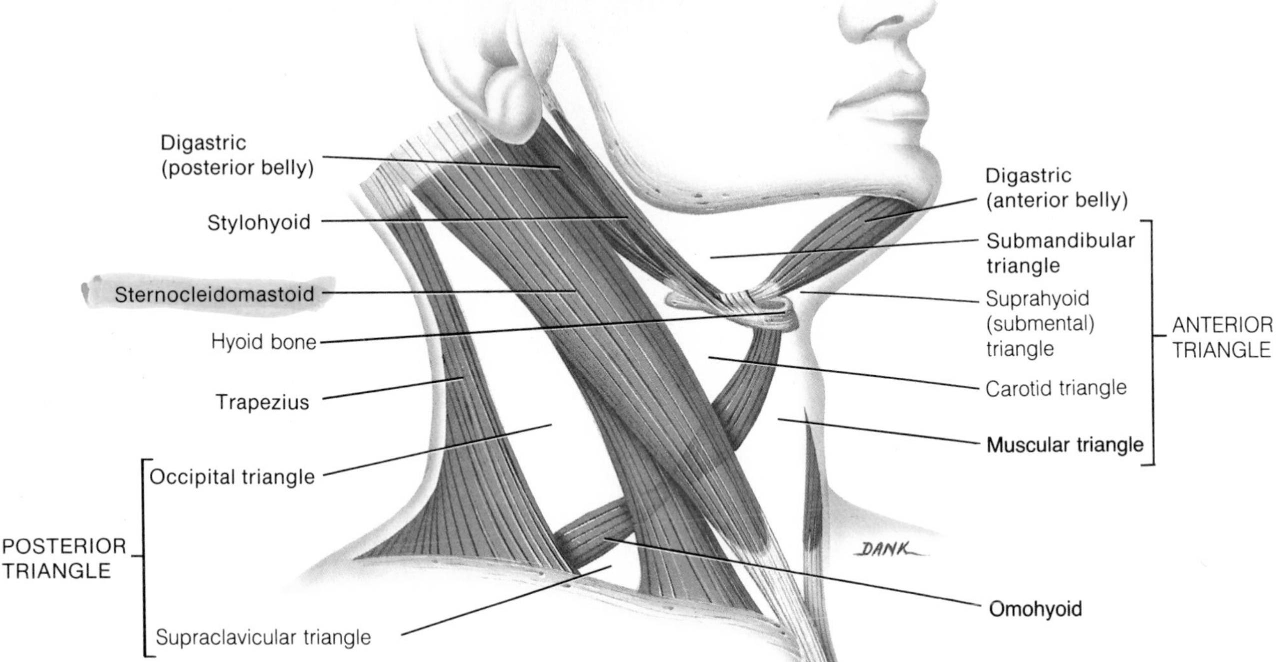

Triangles of the neck viewed from right side

MUSCLE	ORIGIN	INSERTION	ACTION	INNERVATION
Sternocleidomastoid (*sternum* = breastbone; *cleido* = clavicle; *mastoid* = mastoid process of temporal bone) (see Figure 10-16c)	Sternum and clavicle.	Mastoid process of temporal bone.	Contraction of both muscles flexes the cervical part of the vertebral column and draws the head forward; contraction of one muscle rotates face toward side opposite contracting muscle.	Accessory (XI) nerve; cervical nerves C2–C3.
Semispinalis Capitis (*semi* = half; *spine* = spinous process; *caput* = head) (see Figure 10-21)	Articular process of seventh cervical vertebra and transverse processes of first six thoracic vertebrae.	Occipital bone between superior and inferior nuchal lines.	Both muscles extend head; contraction of one muscle rotates face toward same side as contracting muscle.	Dorsal rami of spinal nerves.
Splenius Capitis (*splenion* = bandage) (see Figure 10-21)	Ligamentum nuchae and spinous processes of seventh cervical vertebra and first three or four thoracic vertebrae.	Occipital bone and mastoid process of temporal bone.	Both muscles extend head; contraction of one rotates it to same side as contracting muscle.	Dorsal rami of middle and lower cervical nerves.
Longissimus Capitis (*longissimus* = longest) (see Figure 10-21)	Transverse processes of last four cervical and upper four thoracic vertebrae.	Mastoid process of temporal bone.	Extends head and rotates face toward same side as contracting muscle.	Dorsal rami of middle and lower cervical nerves.

EXHIBIT 10-11

Muscles That Act on the Abdominal Wall (Figure 10-12)

Overview: The anterolateral abdominal wall is composed of skin, fascia, and four pairs of flat, sheetlike muscles: rectus abdominis, external oblique, internal oblique, and transversus abdominis. The anterior surfaces of the rectus abdominis muscles are interrupted by three transverse fibrous bands of tissue called ***tendinous intersections,*** believed to be remnants of septa that separated myotomes during embryological development. The aponeuroses of the external oblique, internal oblique, and transversus abdominis muscles meet at the midline to form the ***linea alba*** (white line), a tough, fibrous band that extends from the xiphoid process of the sternum to the pubic symphysis. The inferior free border of the external oblique aponeurosis, plus some collagenous fibers, forms the ***inguinal ligament,*** which runs from the anterior superior iliac spine to the pubic tubercle (see Figure 10-22). The ligament demarcates the thigh and body wall.

Just superior to the medial end of the inguinal ligament is a triangular slit in the aponeurosis referred to as the ***superficial inguinal ring,*** the outer opening of the ***inguinal canal*** (see Figure 25-8). The canal contains the spermatic cord and ilioinguinal nerve in males and round ligament of the uterus and ilioinguinal nerve in females.

The posterior abdominal wall is formed by the lumbar vertebrae, parts of the ilia of the hipbones, psoas major muscle (described in Exhibit 10-21), quadratus lumborum muscle, and iliacus muscle (also described in Exhibit 10-21). Whereas the anterolateral abdominal wall is contractile and distensible, the posterior abdominal wall is bulky and stable by comparison.

MUSCLE	ORIGIN	INSERTION	ACTION	INNERVATION
Rectus Abdominis (*rectus* = fibers parallel to midline; *abdomino* = abdomen)	Pubic crest and pubic symphysis.	Cartilage of fifth to seventh ribs and xiphoid process.	Compresses abdomen to aid in defecation, urination, forced expiration, and childbirth and flexes vertebral column.	Branches of thoracic nerves T7–T12.
External Oblique (*external* = closer to surface; *oblique* = fibers diagonal to midline)	Lower eight ribs.	Iliac crest and linea alba (midline aponeurosis).	Contraction of both compresses abdomen; contraction of one side alone bends vertebral column laterally; rotates vertebral column.	Branches of thoracic nerves T7–T12 and iliohypogastric nerve.
Internal Oblique (*internal* = farther from surface)	Iliac crest, inguinal ligament, and thoracolumbar fascia.	Cartilage of last three or four ribs and linea alba.	Compresses abdomen; contraction of one side alone bends vertebral column laterally; rotates vertebral column.	Branches of thoracic nerves T8–T12, iliohypogastric, and ilioinguinal nerves.
Transversus Abdominis (*transverse* = fibers perpendicular to midline)	Iliac crest, inguinal ligament, lumbar fascia, and cartilages of last six ribs.	Xiphoid process, linea alba, and pubis.	Compresses abdomen.	Branches of thoracic nerves T8–T12, iliohypogastric, and ilioinguinal nerves.
Quadratus Lumborum (*quad* = four; *lumbo* = lumbar region) (see Figure 10-13)	Iliac crest and iliolumbar ligament.	Lower border of twelfth rib and transverse processes of first four lumbar vertebrae.	Contraction of one side bends vertebral column laterally and pulls thoracic cage toward pelvis.	Branches of thoracic nerve T1 and lumbar nerves L1–L3 or L1–L4.

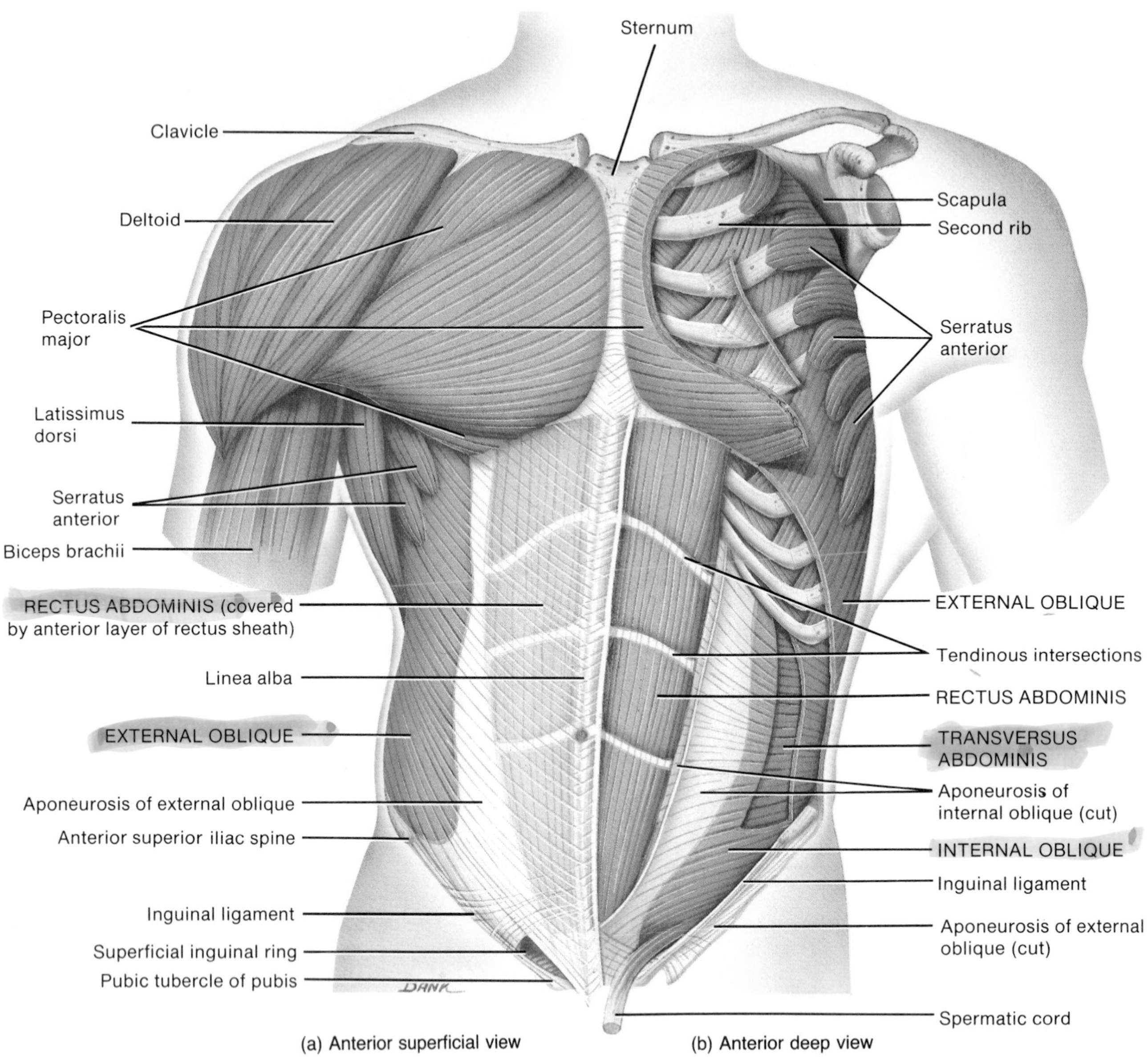

FIGURE 10-12 Muscles of the anterior abdominal wall. (d) Courtesy of J. A. Gosling, P. F. Harris, et al., *Atlas of Human Anatomy,* Gower Medical Publishing Ltd., 2nd ed., 1991.

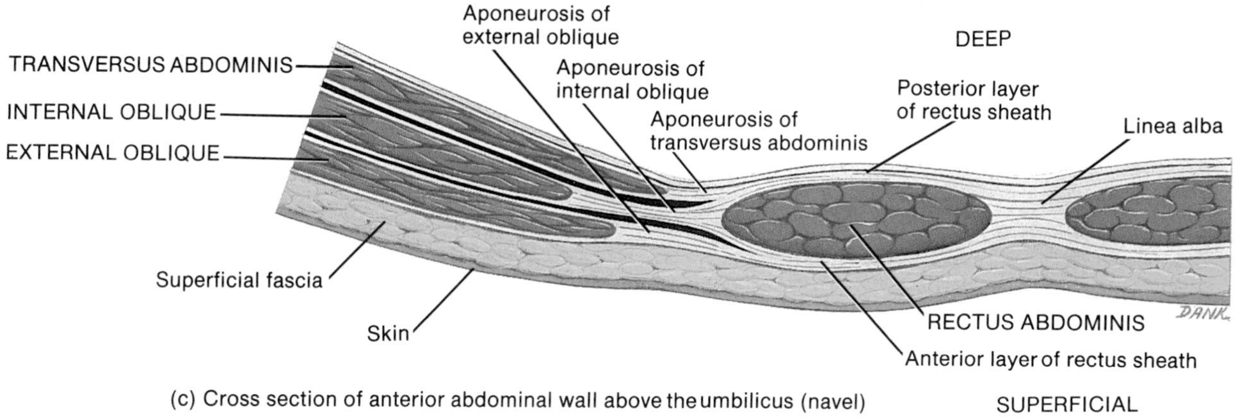

(c) Cross section of anterior abdominal wall above the umbilicus (navel)

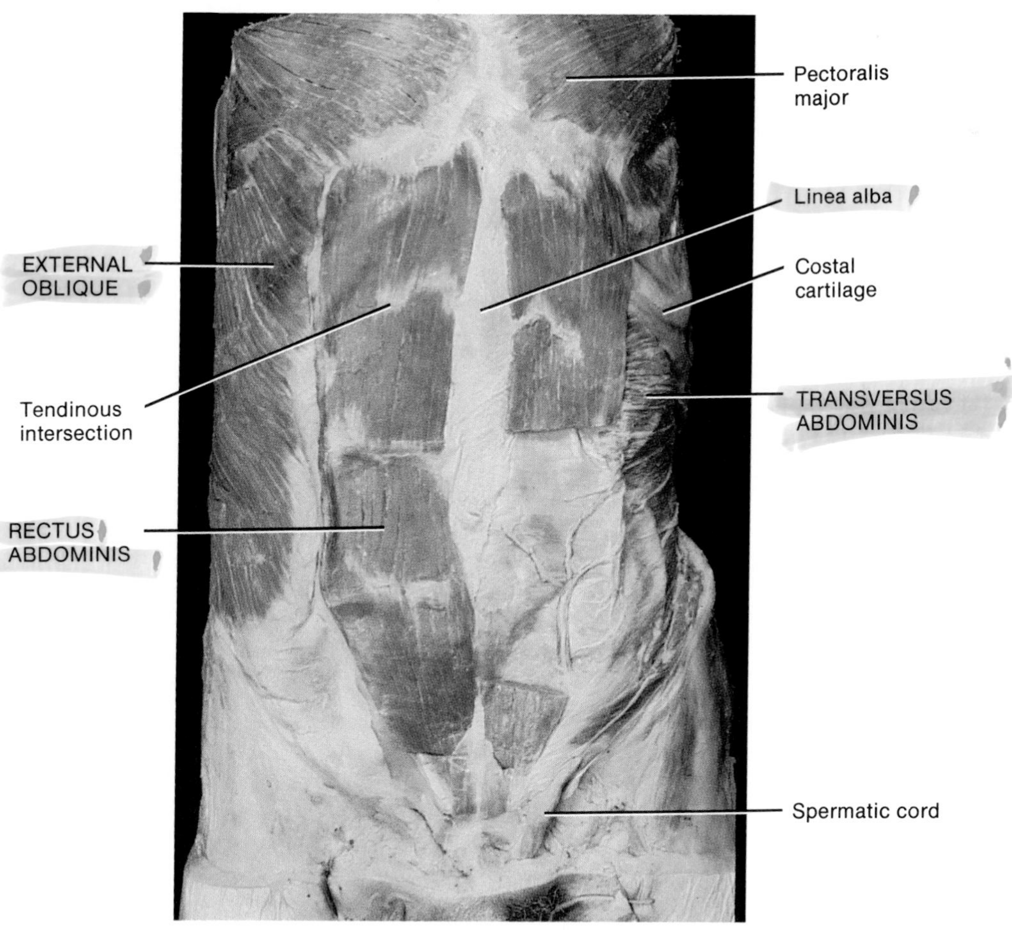

(d) Anterior superficial view

EXHIBIT 10-12

Muscles Used in Breathing (Figure 10-13)

Overview: The muscles described here are attached to the ribs and by their contraction and relaxation alter the size of the thoracic cavity during normal breathing. In forced breathing, other muscles are involved as well. Essentially, inspiration occurs when the thoracic cavity increases in size. Expiration occurs when the thoracic cavity decreases in size.

The diaphragm, one of the muscles used in breathing, is dome-shaped and has several openings through which various structures pass between the thorax and abdomen. These structures include the aorta along with the thoracic duct and azygos vein, the esophagus with accompanying vagus (X) nerves, and the inferior vena cava along with the phrenic nerve.

MUSCLE	ORIGIN	INSERTION	ACTION	INNERVATION
Diaphragm (*dia* = across; *phragma* = wall)	Xiphoid process, costal cartilages of last six ribs, and lumbar vertebrae.	Central tendon (strong aponeurosis that serves as the tendon of insertion for all muscular fibers of the diaphragm).	Forms floor of thoracic cavity; pulls central tendon downward during inspiration and as dome of diaphragm flattens increases vertical length of thorax.	Phrenic nerve.
External Intercostals (*external* = closer to surface; *inter* = between; *costa* = rib)	Inferior border of rib above.	Superior border of rib below.	May elevate ribs during inspiration and thus increase lateral and anteroposterior dimensions of thorax.	Intercostal nerves.
Internal Intercostals (*internal* = farther from surface)	Superior border of rib below.	Inferior border of rib above.	May draw adjacent ribs together during forced expiration and thus decrease lateral and anteroposterior dimensions of thorax.	Intercostal nerves.

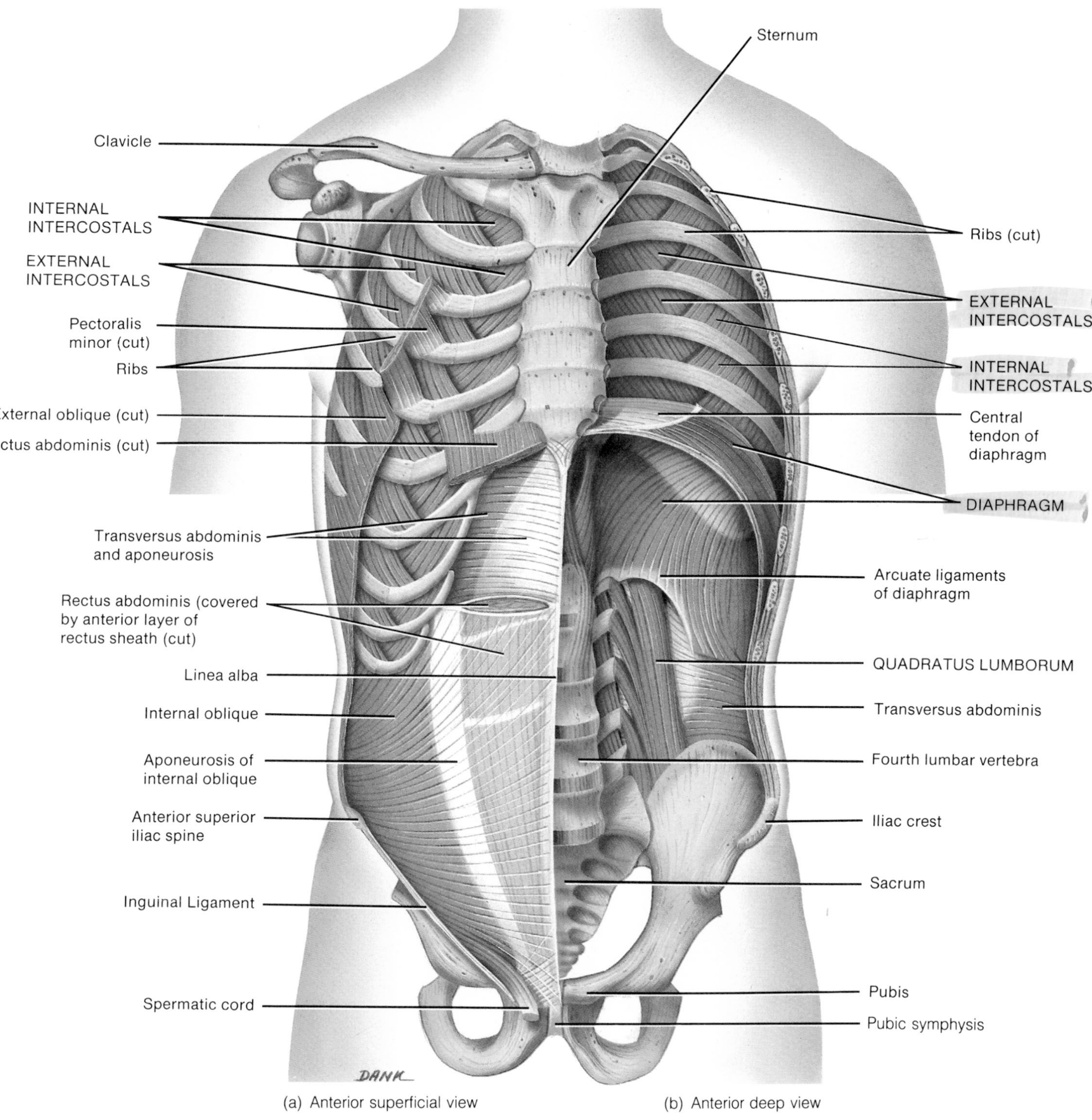

FIGURE 10-13 Muscles used in breathing.

EXHIBIT 10-13

Muscles of the Pelvic Floor (Figure 10-14)

Overview: The muscles of the pelvic floor, together with the fascia covering their external and internal surfaces, are referred to as the ***pelvic diaphragm.*** The diaphragm is funnel-shaped and forms the floor of the abdominopelvic cavity, where it supports the pelvic viscera. It is pierced by the anal canal and urethra in both sexes and also by the vagina in the female.

MUSCLE	ORIGIN	INSERTION	ACTION	INNERVATION
Levator Ani (*levator* = raises; *ani* = anus)	This muscle is divisible into two parts, the pubococcygeus muscle and the iliococcygeus muscle.			
Pubococcygeus (*pubo* = pubis; *coccygeus* = coccyx)	Pubis.	Coccyx, urethra, anal canal, central tendon of perineum, and anococcygeal raphe.	Supports and slightly raises pelvic floor, resists increased intra-abdominal pressure, and draws anus toward pubis and constricts it.	Sacral nerves S3–S4 or S4 and perineal branch of pudendal nerve.
Iliococcygeus (*ilio* = ilium)	Ischial spine.	Coccyx.	Supports and slightly raises pelvic floor, resists increased intra-abdominal pressure, and draws anus toward pubis and constricts it.	Sacral nerves S3–S4 or S4 and perineal branch of pudendal nerve.
Coccygeus	Ischial spine.	Lower sacrum and upper coccyx.	Supports and slightly raises pelvic floor, resists intra-abdominal pressure, and pulls coccyx forward following defecation or parturition.	Sacral nerve S3 or S4.

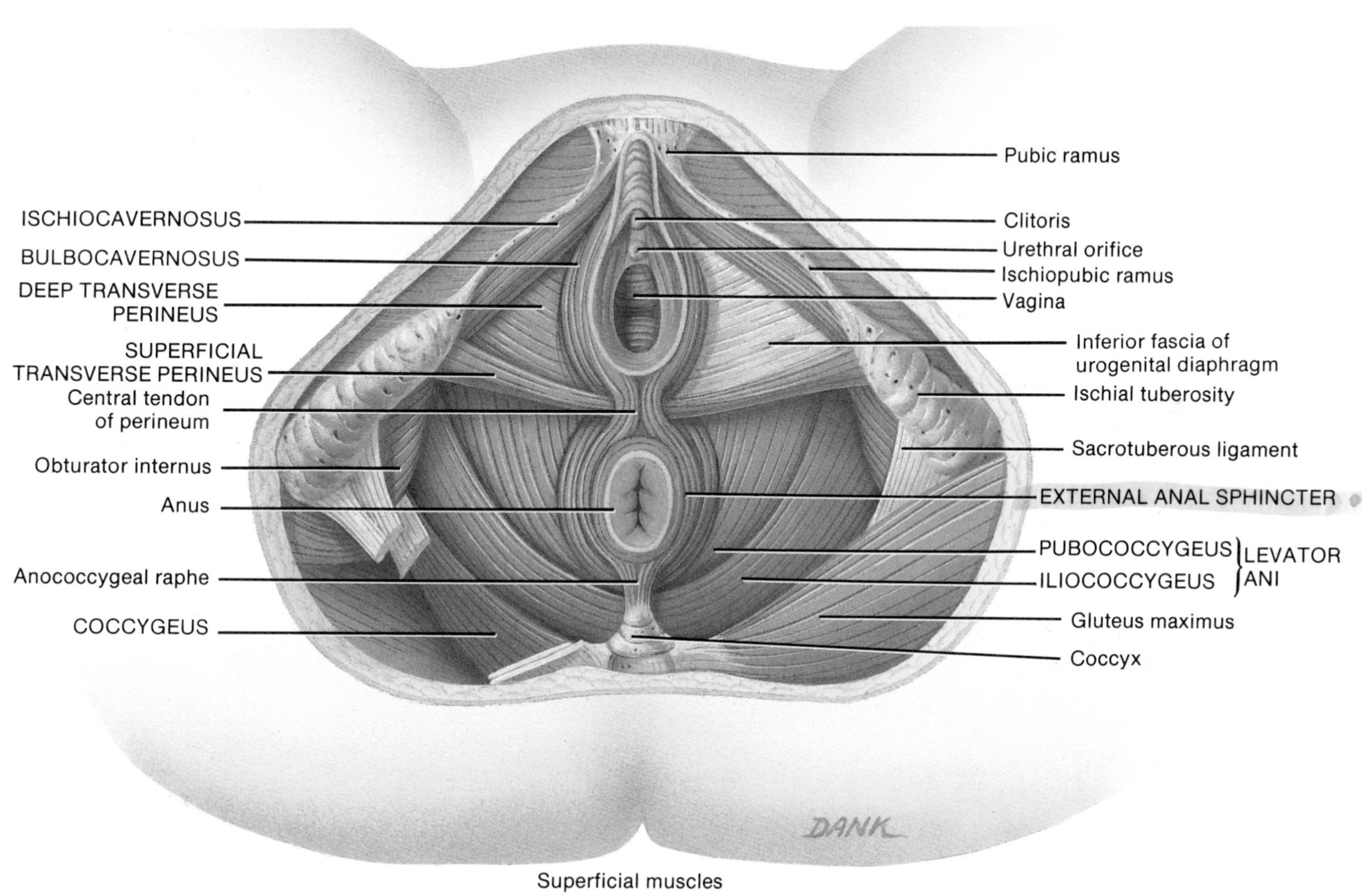

FIGURE 10-14 Muscles of the pelvic floor seen in the female perineum.

EXHIBIT 10-14

Muscles of the Perineum (Figure 10-15; see also Figure 10-14)

Overview: The ***perineum*** is the entire outlet of the pelvis. It is a diamond-shaped area at the lower end of the trunk between the thighs and buttocks. It is bordered anteriorly by the pubic symphysis, laterally by the ischial tuberosities, and posteriorly by the coccyx. A transverse line drawn between the ischial tuberosities divides the perineum into an anterior ***urogenital triangle*** that contains the external genitals and a posterior ***anal triangle*** that contains the anus (see Figure 25-21). The deep transverse perineus, the urethral sphincter, and a fibrous membrane constitute the ***urogenital diaphragm.*** It surrounds the urogenital ducts and helps to strengthen the pelvic floor.

MUSCLE	ORIGIN	INSERTION	ACTION	INNERVATION
Superficial Transverse Perineus (*superficial* = closer to surface; *transverse* = across; *perineus* = perineum)	Ischial tuberosity.	Central tendon of perineum.	Helps to stabilize the central tendon of the perineum.	Perineal branch of pudendal nerve.
Bulbocavernosus (*bulbus* = bulb; *caverna* = hollow place)	Central tendon of perineum.	Inferior fascia of urogenital diaphragm, corpus spongiosum of penis, and deep fascia on dorsum of penis in male; pubic arch and root and dorsum of clitoris in female.	Helps expel last drops of urine during micturition, helps propel semen along urethra, and may assist in erection of the penis in male; decreases vaginal orifice and assists in erection of clitoris in female.	Perineal branch of pudendal nerve.
Ischiocavernosus (*ischion* = hip)	Ischial tuberosity and ischial and pubic rami.	Corpus cavernosum of penis in male and clitoris in female.	May maintain erection of penis in male and clitoris in female.	Perineal branch of pudendal nerve.
Deep Transverse Perineus (*deep* = farther from surface)	Ischial rami.	Central tendon of perineum.	Helps expel last drops of urine and semen in male and urine in female.	Perineal branch of pudendal nerve.
Urethral Sphincter (*sphincter* = circular muscle that decreases size of an opening; *urethrae* = urethra)	Ischial and pubic rami.	Median raphe in male and vaginal wall in female.	Helps expel last drops of urine and semen in male and urine in female.	Perineal branch of pudendal nerve.
External Anal Sphincter	Anococcygeal raphe.	Central tendon of perineum.	Keeps anal canal and orifice closed.	Sacral nerve S4 and inferior rectal branch of pudendal nerve.

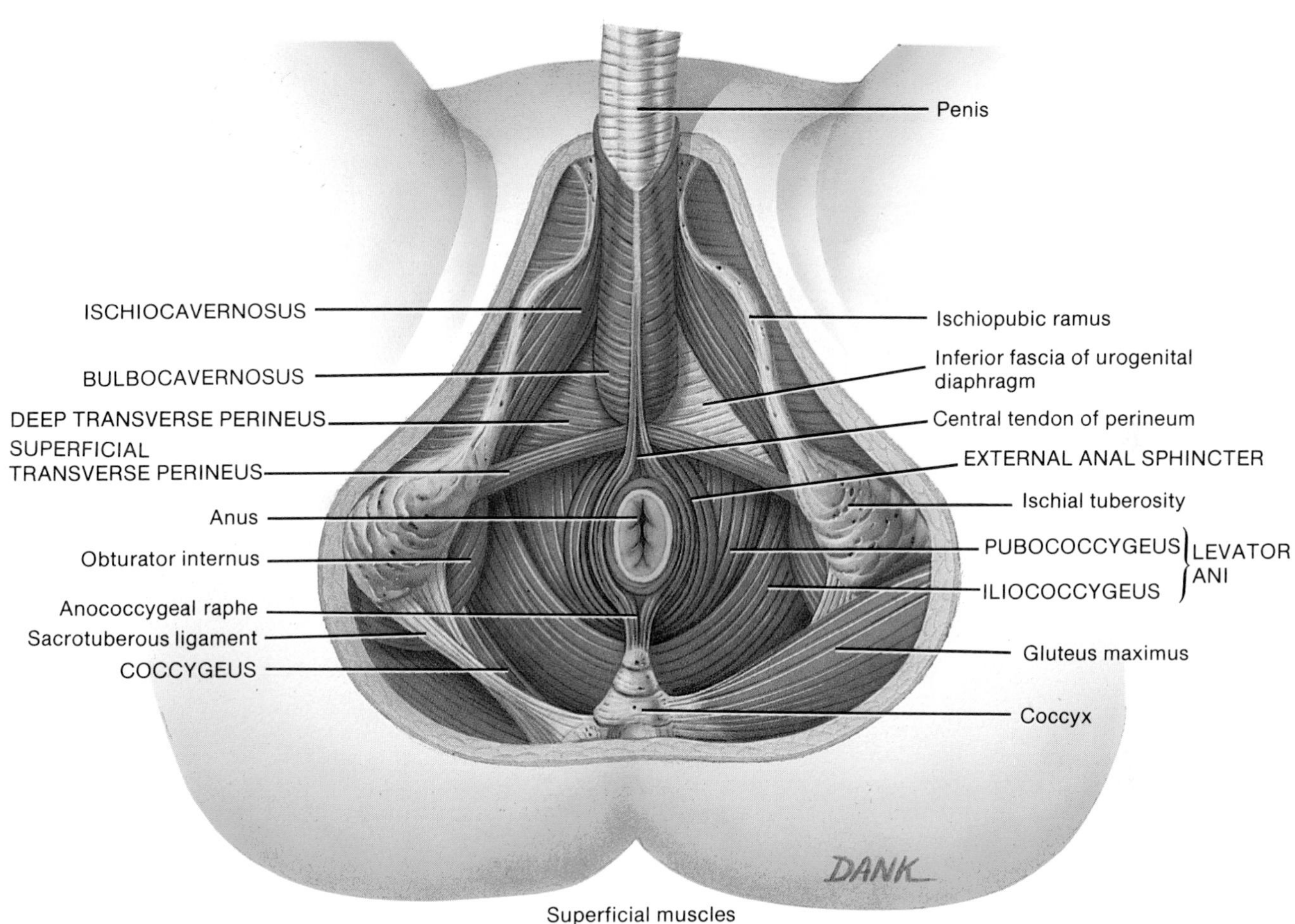

Superficial muscles

FIGURE 10-15 Muscles of the male perineum.

EXHIBIT 10-15

Muscles That Move the Shoulder (Pectoral) Girdle (Figure 10-16)

Overview: Muscles that move the shoulder (pectoral) girdle originate on the axial skeleton and insert on the clavicle or scapula. The muscles can be distinguished into ***anterior*** and ***posterior*** groups. The principal action of the muscles is to stabilize the scapula so that it can function as a stable point of origin for most of the muscles that move the humerus (arm).

MUSCLE	ORIGIN	INSERTION	ACTION	INNERVATION
ANTERIOR				
Subclavius (*sub* = under; *clavius* = clavicle)	First rib.	Clavicle.	Depresses clavicle.	Nerve to subclavius.
Pectoralis Minor (*pectus* = breast, chest, thorax; *minor* = lesser)	Third through fifth ribs.	Coracoid process of scapula.	Depresses and moves scapula anteriorly and elevates third through fifth ribs during forced inspiration when scapula is fixed.	Medial pectoral nerve.
Serratus Anterior (*serratus* = saw-toothed; *anterior* = front)	Upper eight or nine ribs.	Vertebral border and inferior angle of scapula.	Rotates scapula upward and laterally and elevates ribs when scapula is fixed.	Long thoracic nerve.
POSTERIOR				
Trapezius (*trapezoides* = trapezoid-shaped)	Superior nuchal line of occipital bone, ligamentum nuchae, and spines of seventh cervical and all thoracic vertebrae.	Clavicle and acromion and spine of scapula.	Elevates clavicle, adducts scapula, rotates scapula upward, elevates or depresses scapula, and extends head.	Accessory (XI) nerve and cervical nerves C3–C4.
Levator Scapulae (*levator* = raises; *scapulae* = scapula)	Upper four or five cervical vertebrae.	Superior vertebral border of scapula.	Elevates scapula and slightly rotates it downward.	Dorsal scapular nerve and cervical nerves C3–C5.
Rhomboideus Major (*rhomboides* = rhomboid or diamond-shaped)	Spines of second to fifth thoracic vertebrae.	Vertebral border of scapula below spine.	Adducts scapula and slightly rotates it downward.	Dorsal scapular nerve.
Rhomboideus Minor	Spines of seventh cervical and first thoracic vertebrae.	Vertebral border of scapula above spine.	Adducts scapula and slightly rotates it downward.	Dorsal scapular nerve.

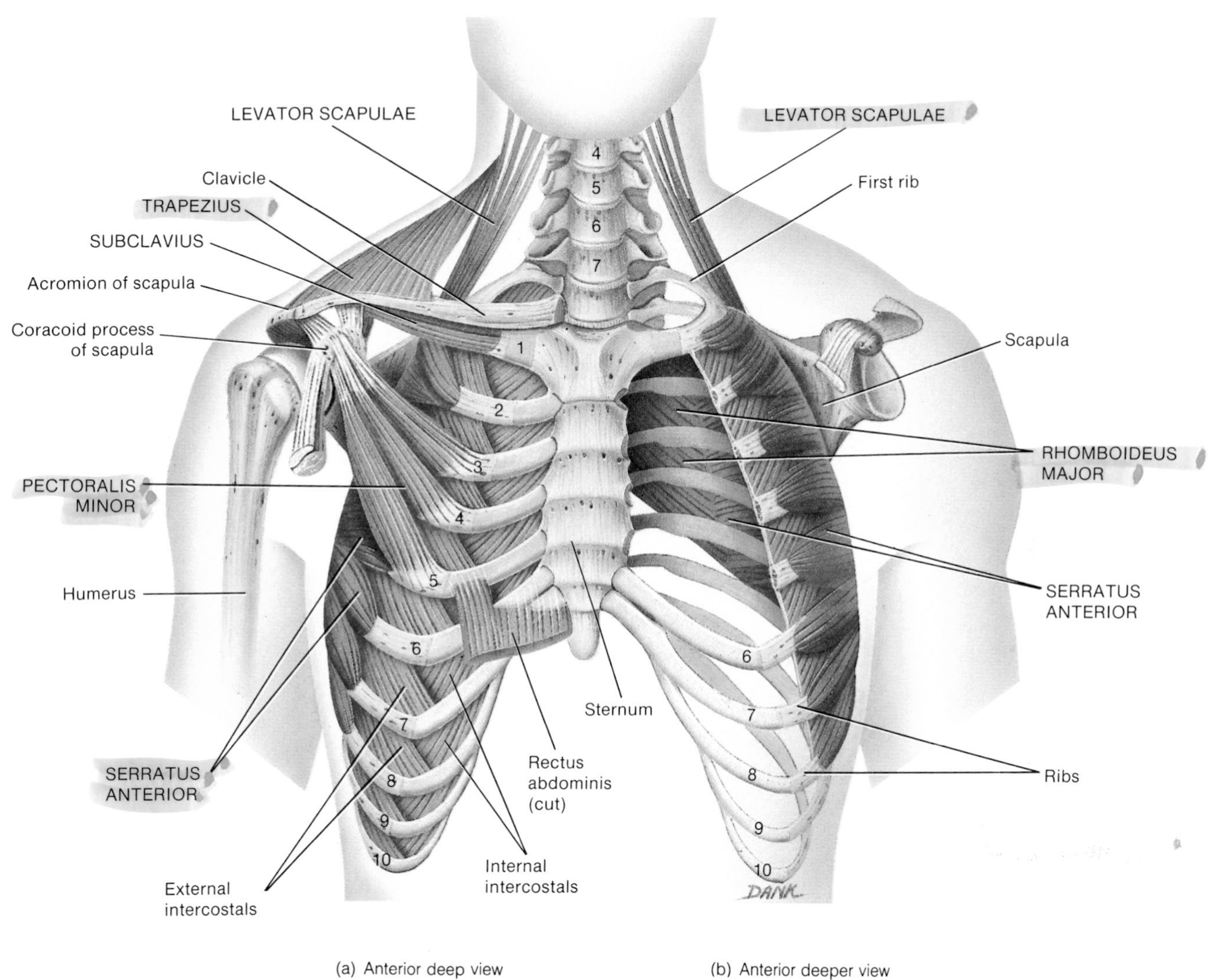

(a) Anterior deep view

(b) Anterior deeper view

FIGURE 10-16 Muscles that move the shoulder (pectoral) girdle.

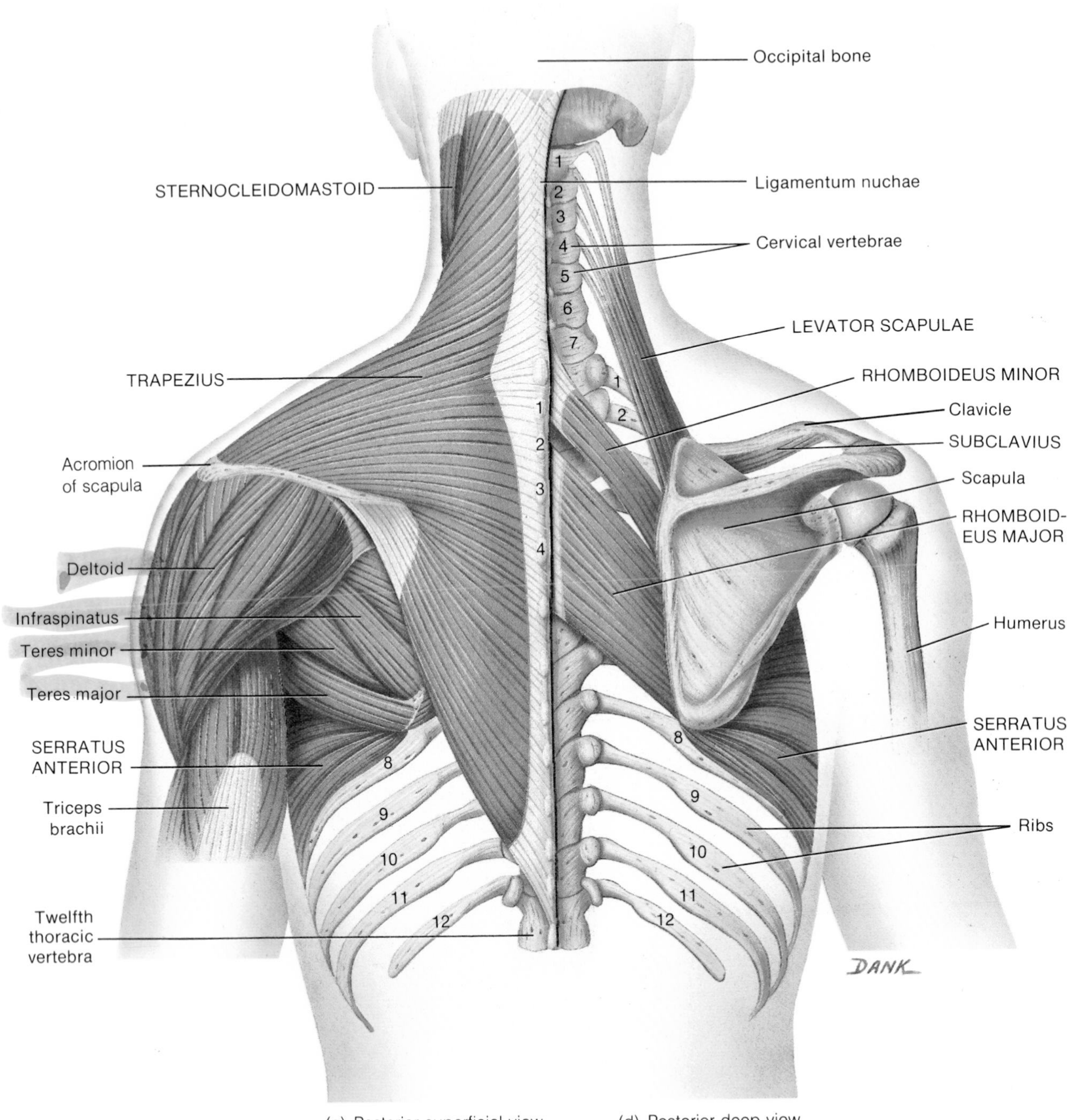

(c) Posterior superficial view (d) Posterior deep view

EXHIBIT 10-16

Muscles That Move the Arm (Humerus) (Figure 10-17)

Overview: Of the nine muscles that cross the shoulder joint, only two of them (pectoralis major and latissimus dorsi) do not originate on the scapula. These two muscles are thus designated as ***axial muscles,*** since they originate on the axial skeleton. The remaining seven muscles, the ***scapular muscles,*** arise from the scapula.

The strength and stability of the shoulder joint are not provided by the shape of the articulating bones or its ligaments. Instead, four deep muscles of the shoulder and their tendons—subscapularis, supraspinatus, infraspinatus, and teres minor—strengthen and stabilize the shoulder joint. The muscles and their tendons are so arranged as to form a nearly complete circle around the joint. This arrangement is referred to as the ***rotator (musculotendinous) cuff*** and is a common site of injury to baseball pitchers, especially tearing of the supraspinatus muscle tendon. This tendon is especially predisposed to wear-and-tear changes because of its location between the head of the humerus and acromion of the scapula, which compresses the tendon during shoulder movements.

After you have studied the muscles in this exhibit, arrange them according to the following actions: flexion, extension, abduction, adduction, medial rotation, and lateral rotation. (The same muscle can be used more than once.)

MUSCLE	ORIGIN	INSERTION	ACTION	INNERVATION
AXIAL				
Pectoralis Major (see also Figure 10-12a)	Clavicle, sternum, cartilages of second to sixth ribs.	Intertubercular sulcus of humerus.	Flexes, adducts, and rotates arm medially.	Medial and lateral pectoral nerve.
Latissimus Dorsi (*latissimus* = widest; *dorsum* = back)	Spines of lower six thoracic vertebrae, lumbar vertebrae, crests of sacrum and ilium, lower four ribs.	Intertubercular sulcus of humerus.	Extends, adducts, and rotates arm medially; draws arm downward and backward.	Thoracodorsal nerve.
SCAPULAR				
Deltoid (*delta* = triangular)	Acromial extremity of clavicle and acromion and spine of scapula.	Deltoid tuberosity of humerus.	Abducts, flexes, extends, and medially and laterally rotates arm.	Axillary nerve.
Subscapularis (*sub* = below; *scapularis* = scapula)	Subscapular fossa of scapula.	Lesser tubercle of humerus.	Rotates arm medially.	Upper and lower subscapular nerves.
Supraspinatus (*supra* = above; *spinatus* = spine of scapula)	Fossa superior to spine of scapula.	Greater tubercle of humerus.	Assists deltoid muscle in abducting arm.	Suprascapular nerve.
Infraspinatus (*infra* = below)	Fossa inferior to spine of scapula.	Greater tubercle of humerus.	Rotates arm laterally; adducts arm.	Suprascapular nerve.
Teres Major (*teres* = long and round)	Inferior angle of scapula.	Intertubercular sulcus of humerus.	Extends arm; assists in adduction and medial rotation of arm.	Lower subscapular nerve.
Teres Minor	Inferior lateral border of scapula.	Greater tubercle of humerus.	Rotates arm laterally; extends and adducts arm.	Axillary nerve.
Coracobrachialis (*coraco* = coracoid process)	Coracoid process of scapula.	Middle of medial surface of shaft of humerus.	Flexes and adducts arm.	Musculocutaneous nerve.

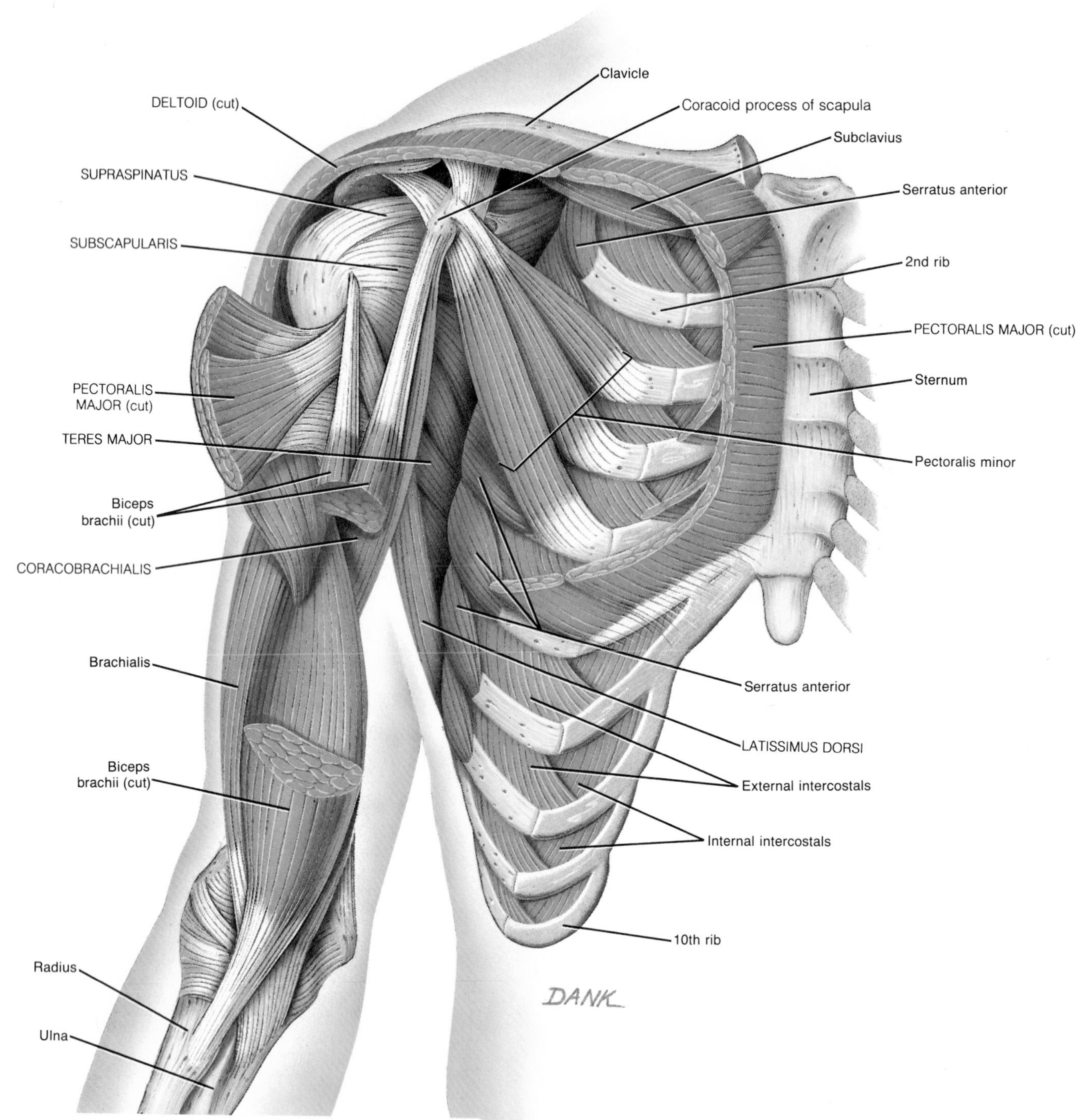

(a) Anterior deep view

FIGURE 10-17 Muscles that move the arm (humerus). (d) Courtesy of J. A. Gosling, P. F. Harris, et al., *Atlas of Human Anatomy,* Gower Medical Publishing Ltd., 2nd ed., 1991.

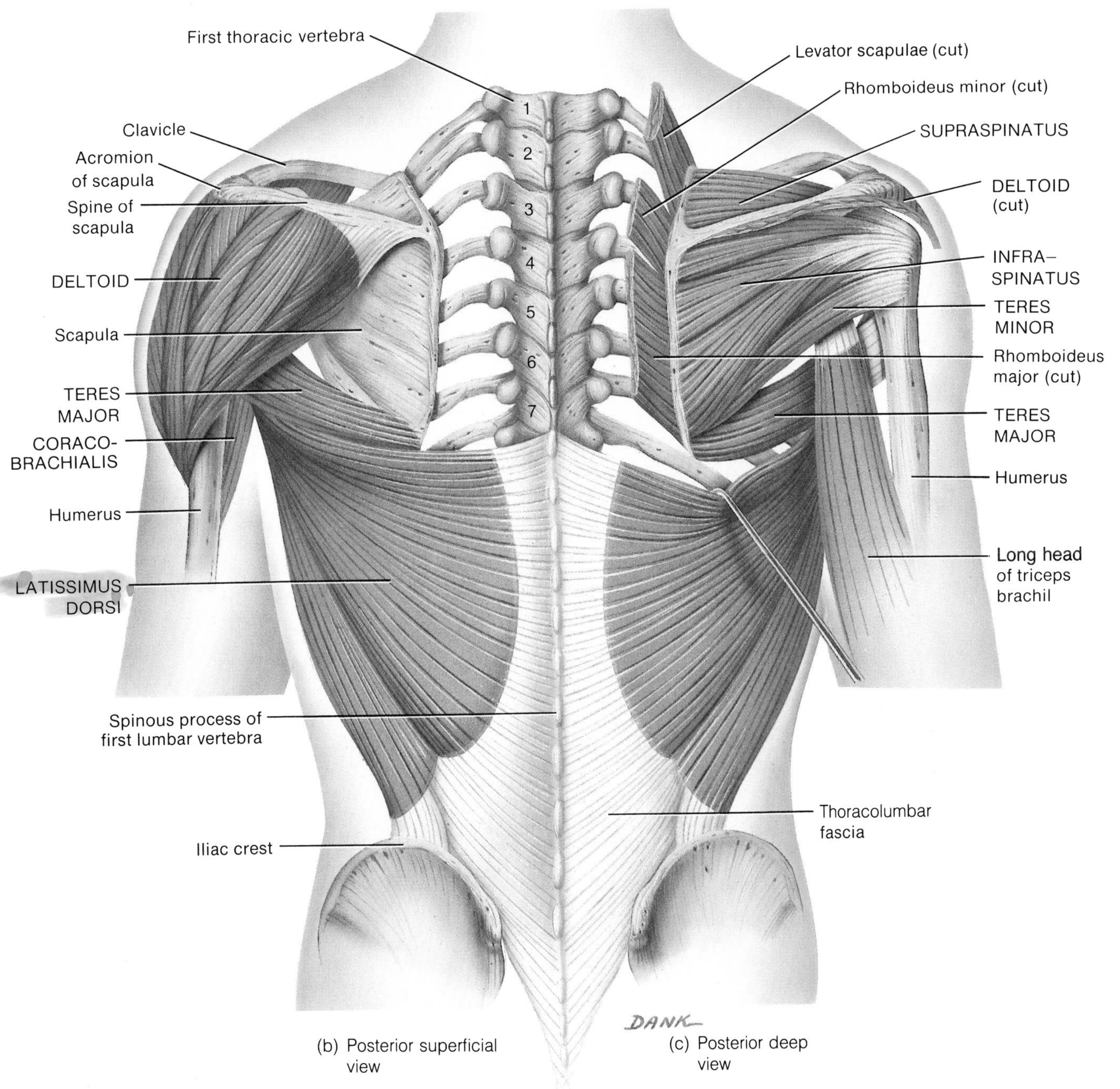

(b) Posterior superficial view

(c) Posterior deep view

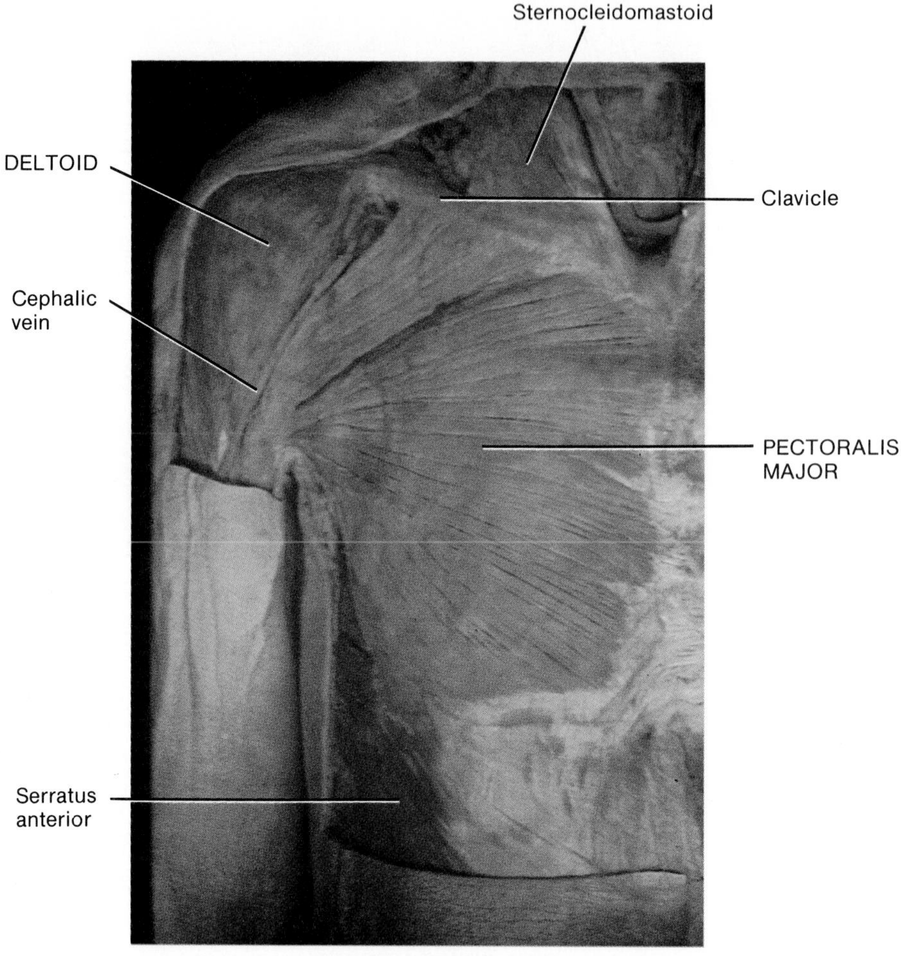

(d) Anterior superfical view

EXHIBIT 10-17

Muscles That Move the Forearm (Radius and Ulna) (Figure 10-18)

Overview: Most of the muscles that move the forearm (radius and ulna) are divided into ***flexors*** and ***extensors.*** Recall that the elbow joint is a hinge joint, capable only of flexion and extension under normal conditions. Whereas the biceps brachii, brachialis, and brachioradialis are flexors of the elbow joint, the triceps brachii and anconeus are extensors. Other muscles that move the forearm are concerned with pronation and supination.

MUSCLE	ORIGIN	INSERTION	ACTION	INNERVATION
FLEXORS				
Biceps Brachii (*biceps* = two heads of origin; *brachion* = arm)	Long head originates from tubercle above glenoid cavity; short head originates from coracoid process of scapula.	Radial tuberosity and bicipital aponeurosis.	Flexes and supinates forearm; flexes arm.	Musculocutaneous nerve.
Brachialis	Distal, anterior surface of humerus.	Tuberosity and coronoid process of ulna.	Flexes forearm.	Musculocutaneous and radial nerves.
Brachioradialis (*radialis* = radius) (see also Figure 10-19a)	Supracondyloid ridge of humerus.	Superior to styloid process of radius.	Flexes forearm; semisupinates and semipronates forearm.	Radial nerve.
EXTENSORS				
Triceps Brachii (*triceps* = three heads of origin)	Long head originates from infraglenoid tuberosity of scapula; lateral head originates from lateral and posterior surface of humerus superior to radial groove; medial head originates from posterior surface of humerus inferior to radial groove.	Olecranon of ulna.	Extends forearm; extends arm.	Radial nerve.
Anconeus (*anconeal* = pertaining to elbow) (see Figure 10-18b)	Lateral epicondyle of humerus.	Olecranon and superior portion of shaft of ulna.	Extends forearm.	Radial nerve.
PRONATORS				
Pronator Teres (*pronation* = turning palm downward or posteriorly) (see Figure 10-19a)	Medial epicondyle of humerus and coronoid process of ulna.	Midlateral surface of radius.	Pronates forearm and hand and flexes forearm.	Median nerve.
Pronator Quadratus (*quadratus* = squared, four-sided) (see Figure 10-19a,b)	Distal portion of shaft of ulna.	Distal portion of shaft of radius.	Pronates forearm and hand.	Median nerve.
SUPINATOR				
Supinator (*supination* = turning palm upward or anteriorly) (see Figure 10-19b)	Lateral epicondyle of humerus and ridge of ulna.	Lateral surface of proximal one-third of radius.	Supinates forearm and hand.	Deep radial nerve.

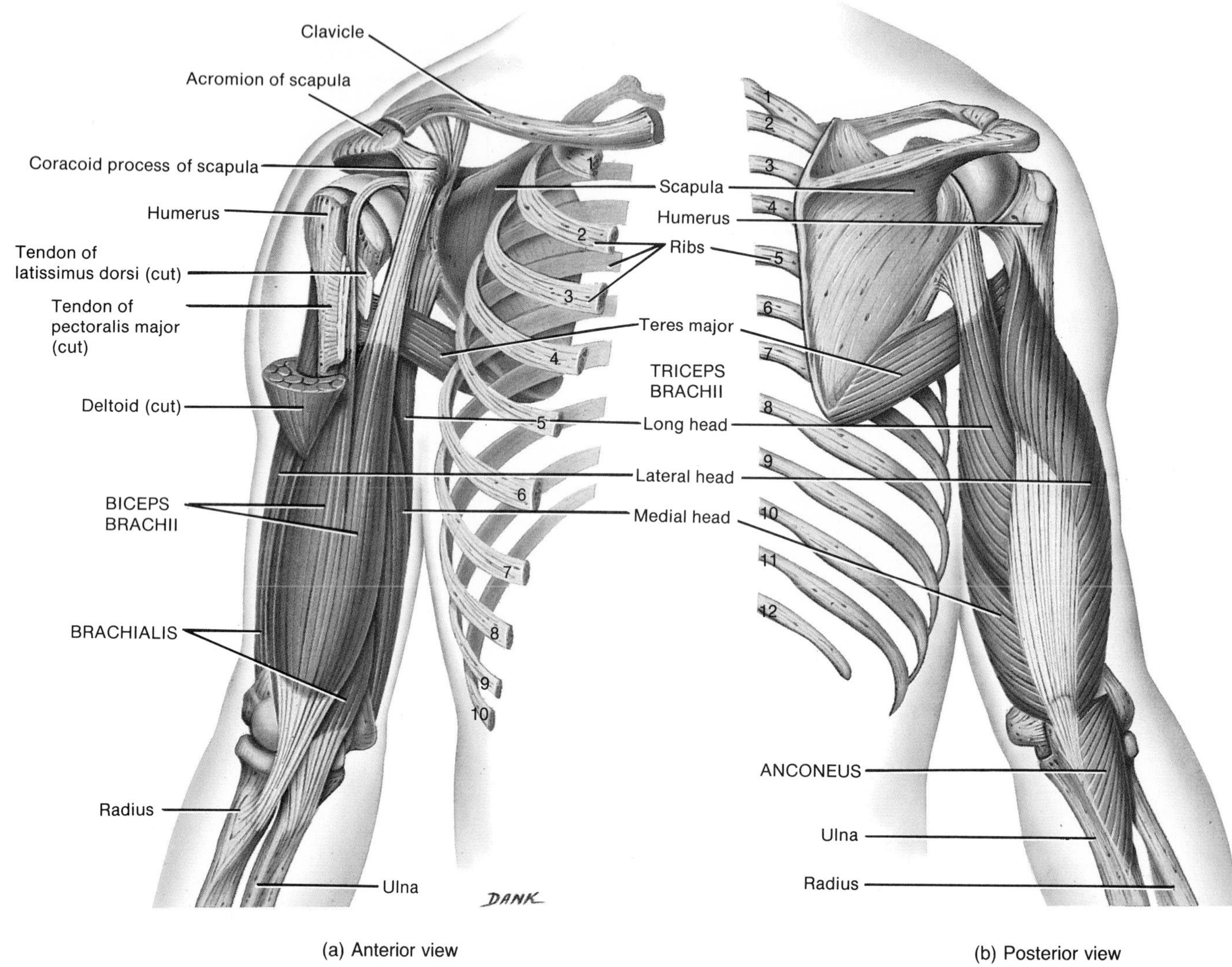

FIGURE 10-18 Muscles that move the forearm (ulna and radius). (c) Courtesy of J. A. Gosling, P. F. Harris, et al., *Atlas of Human Anatomy,* Gower Medical Publishing Ltd., 2nd ed., 1991.

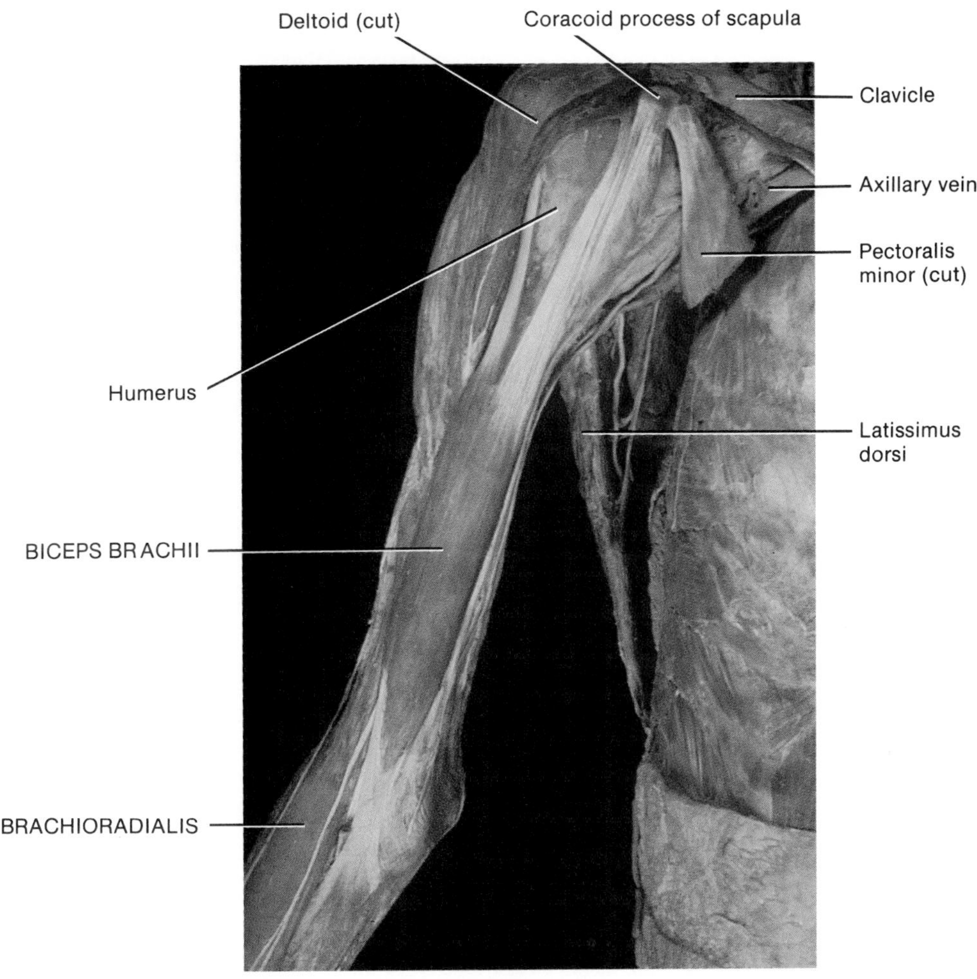

(c) Anterior view

EXHIBIT 10-18

Muscles That Move the Wrist, Hand, and Fingers (Figure 10-19)

Overview: Muscles that move the wrist, hand, and fingers are many and varied. However, as you will see, their names for the most part give some indication of their origin, insertion, or action. On the basis of location and function, the muscles are divided into two groups—anterior and posterior. The ***anterior muscles*** function as flexors. They originate on the humerus and typically insert on the carpals, metacarpals, and phalanges. The bellies of these muscles form the bulk of the proximal forearm. The ***posterior muscles*** function as extensors. These muscles arise on the humerus and insert on the metacarpals and phalanges. Each of the two principal groups is also divided into superficial and deep muscles.

The tendons of the muscles of the forearm that attach to the wrist or continue into the hand, along with blood vessels and nerves, are held close to bones by strong fascial structures. The tendons are also surrounded by tendon sheaths. At the wrist, the deep fascia is thickened into fibrous bands called retinacula (*retinere* = retain). The ***flexor retinaculum* (*transverse carpal ligament*)** is located over the palmar surface of the carpal bones. Through it pass the long flexor tendons of the digits and wrist and the median nerve. The ***extensor retinaculum* (*dorsal carpal ligament*)** is located over the dorsal surface of the carpal bones. Through it pass the extensor tendons of the wrist and digits.

After you have studied the muscles in this exhibit, arrange them according to the following actions: flexion, extension, abduction, adduction, supination, and pronation. (The same muscles can be used more than once.)

MUSCLE	ORIGIN	INSERTION	ACTION	INNERVATION
ANTERIOR GROUP (flexors)				
Superficial				
Flexor Carpi Radialis (*flexor* = decreases angle at joint; *carpus* = wrist; *radialis* = radius)	Medial epicondyle of humerus.	Second and third metacarpals.	Flexes and abducts wrist.	Median nerve.
Palmaris Longus (*palma* = palm; *longus* = long)	Medial epicondyle of humerus.	Flexor retinaculum and palmar aponeurosis.	Flexes wrist.	Median nerve.
Flexor Carpi Ulnaris (*ulnaris* = ulna)	Medial epicondyle of humerus and upper posterior border of ulna.	Pisiform, hamate, and fifth metacarpal.	Flexes and adducts wrist.	Ulnar nerve.
Flexor Digitorum Superficialis (*digit* = finger or toe; *superficialis* = closer to surface)	Medial epicondyle of humerus, coronoid process of ulna, and oblique line of radius.	Middle phalanges.	Flexes middle phalanges of each finger.	Median nerve.
Deep				
Flexor Digitorum Profundus (*profundus* = deep)	Anterior medial surface of body of ulna.	Bases of distal phalanges.	Flexes distal phalanges of each finger.	Median and ulnar nerves.
Flexor Pollicis Longus (*pollex* = thumb)	Anterior surface of radius and interosseous membrane (sheet of fibrous tissue that holds bones together).	Base of distal phalanx of thumb.	Flexes thumb.	Median nerve.

MUSCLE	ORIGIN	INSERTION	ACTION	INNERVATION
POSTERIOR GROUP (extensors)				
Superficial				
Extensor Carpi Radialis Longus (*extensor* = increases angle at joint)	Lateral epicondyle of humerus.	Second metacarpal.	Extends and abducts wrist.	Radial nerve.
Extensor Carpi Radialis Brevis (*brevis* = short)	Lateral epicondyle of humerus.	Third metacarpal.	Extends and abducts wrist.	Radial nerve.
Extensor Digitorum	Lateral epicondyle of humerus.	Second through fifth distal and middle phalanges.	Extends phalanges.	Radial nerve.
Extensor Digiti Minimi (*minimi* = little finger)	Tendon of extensor digitorum.	Tendon of extensor digitorum on fifth phalanx.	Extends little finger.	Deep radial nerve.
Extensor Carpi Ulnaris	Lateral epicondyle of humerus and posterior border of ulna.	Fifth metacarpal.	Extends and adducts wrist.	Deep radial nerve.
Deep				
Abductor Pollicis Longus (*abductor* = moves part away from midline)	Posterior surface of middle of radius and ulna and interosseous membrane.	First metacarpal.	Extends thumb and abducts wrist.	Deep radial nerve.
Extensor Pollicis Brevis	Posterior surface of middle of radius and interosseous membrane.	Base of proximal phalanx of thumb.	Extends thumb and abducts wrist.	Deep radial nerve.
Extensor Pollicis Longus	Posterior surface of middle of ulna and interosseous membrane.	Base of distal phalanx of thumb.	Extends thumb and abducts wrist.	Deep radial nerve.
Extensor Indicis (*indicis* = index)	Posterior surface of ulna.	Tendon of extensor digitorum of index finger.	Extends index finger.	Deep radial nerve.

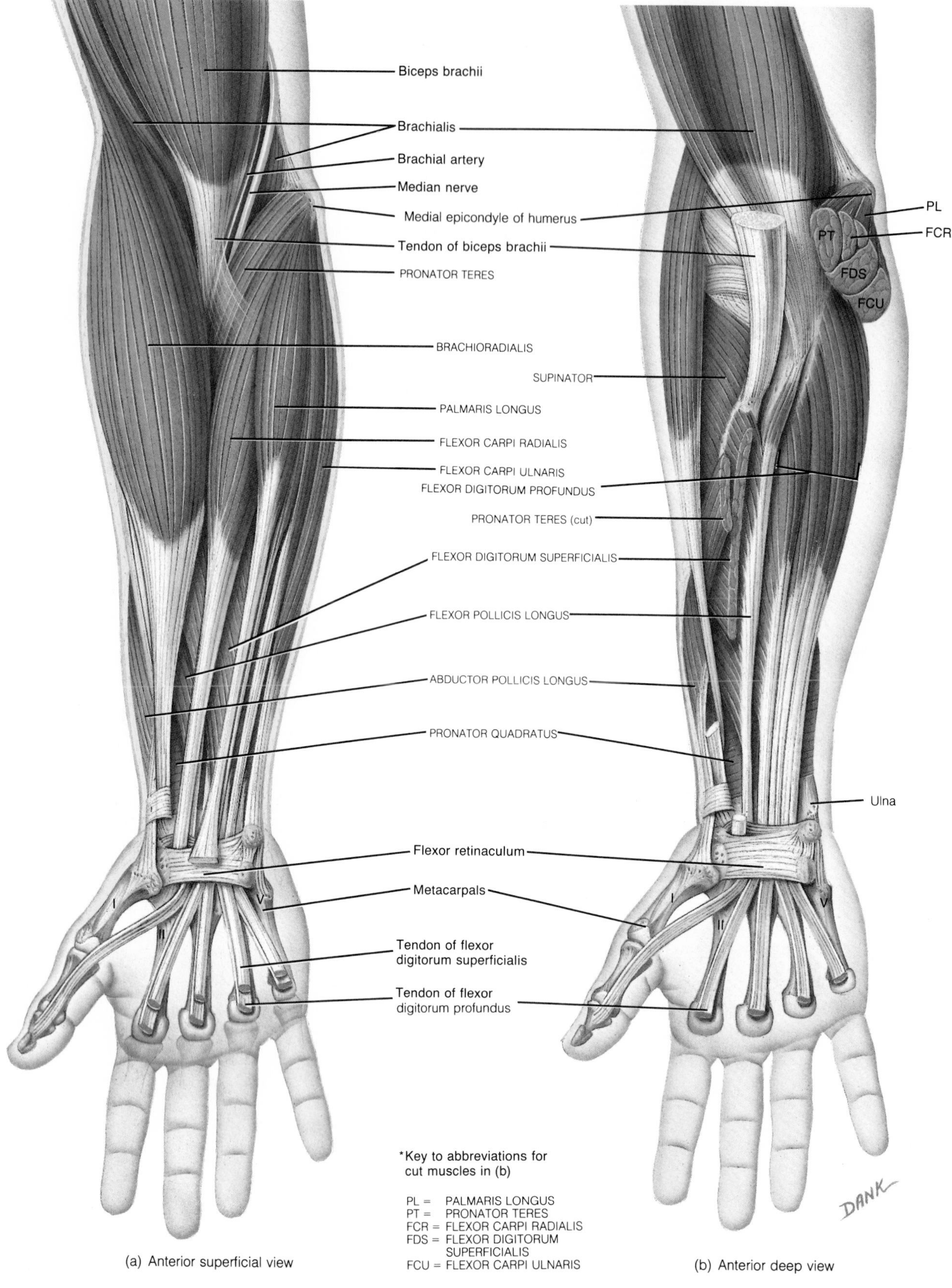

FIGURE 10-19 Muscles that move the wrist, hand, and fingers. (g) and (h) Courtesy of J. A. Gosling, P. F. Harris, et al., *Atlas of Human Anatomy,* Gower Medical Publishing Ltd., 2nd ed., 1991.

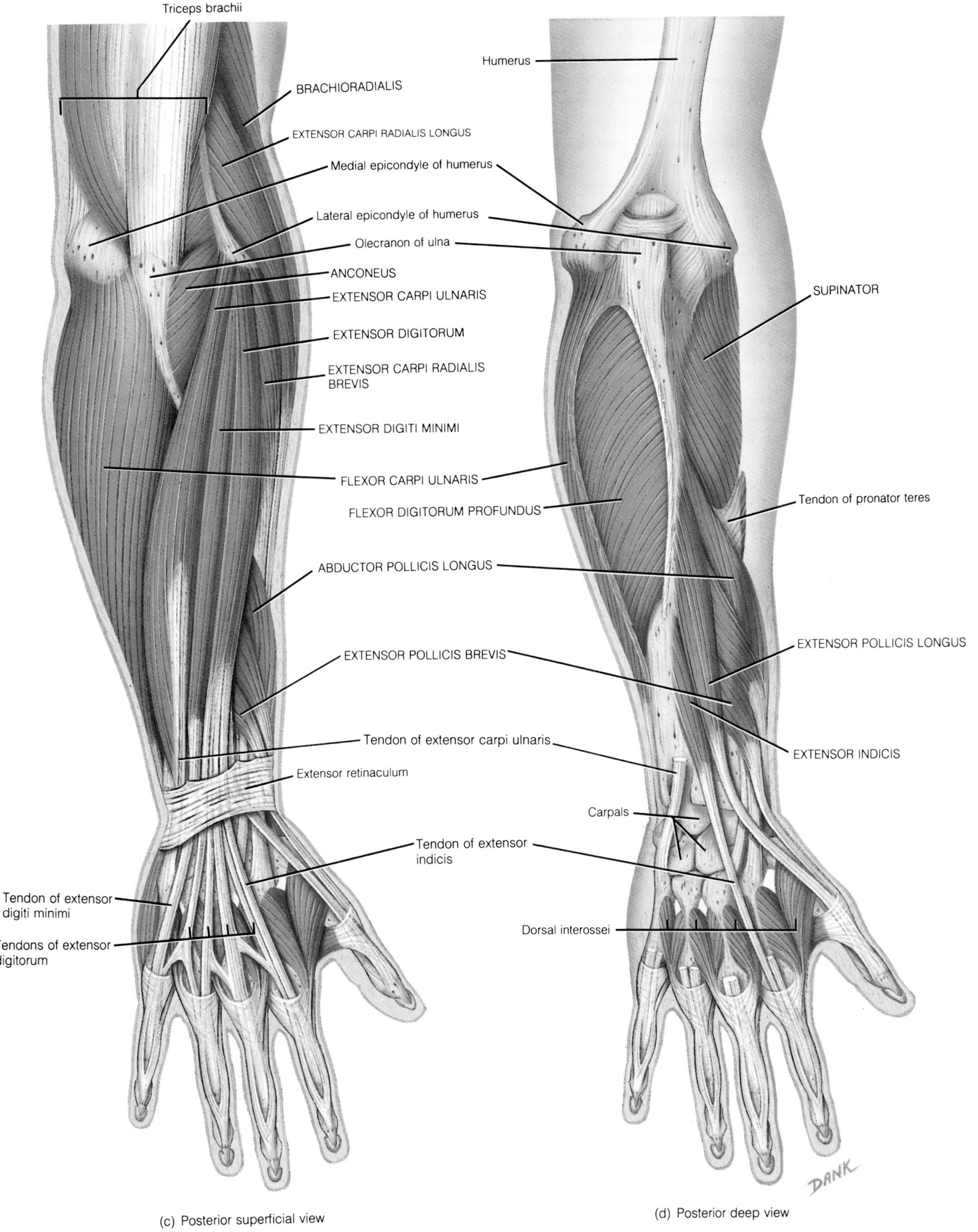

(c) Posterior superficial view

(d) Posterior deep view

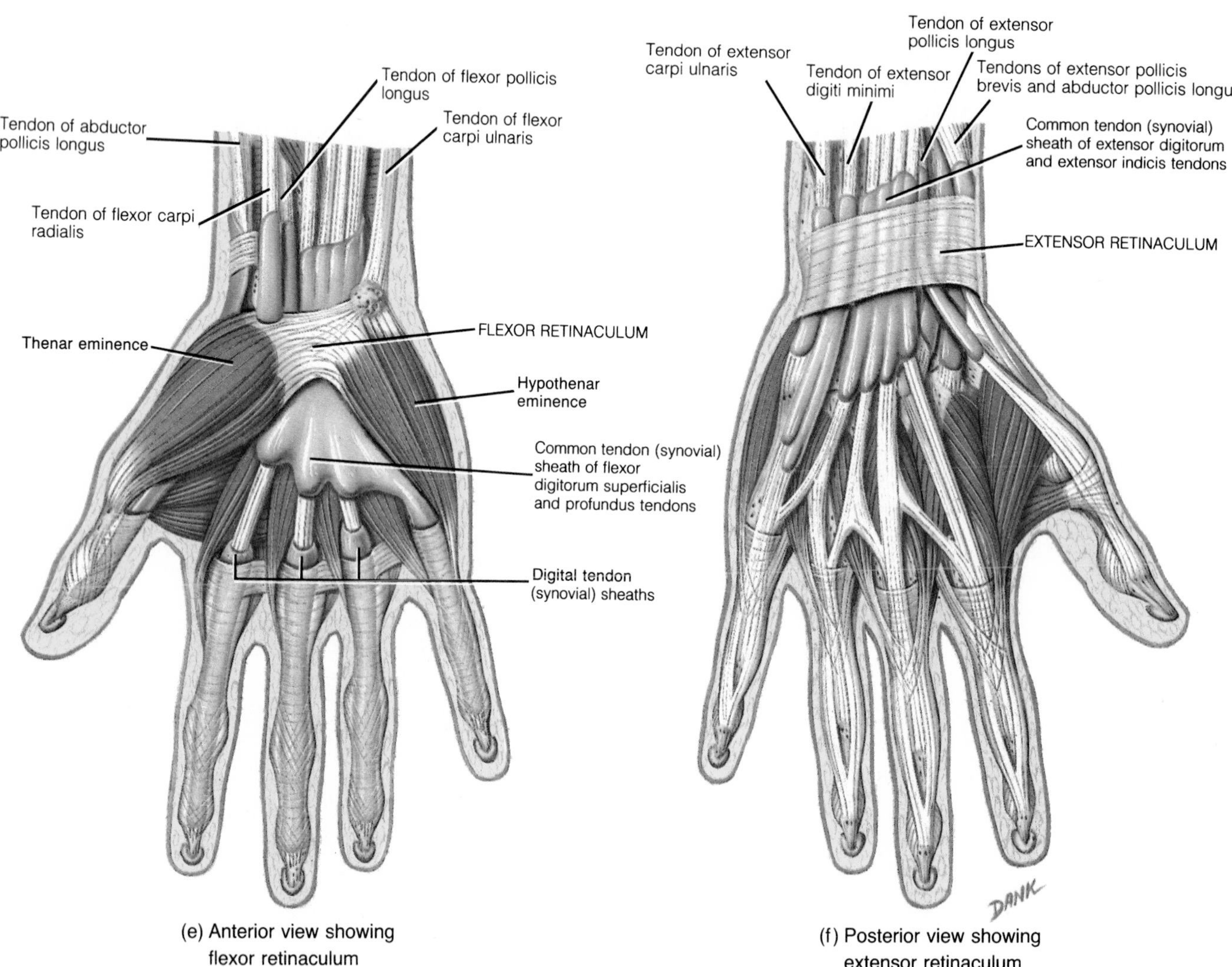

(e) Anterior view showing flexor retinaculum

(f) Posterior view showing extensor retinaculum

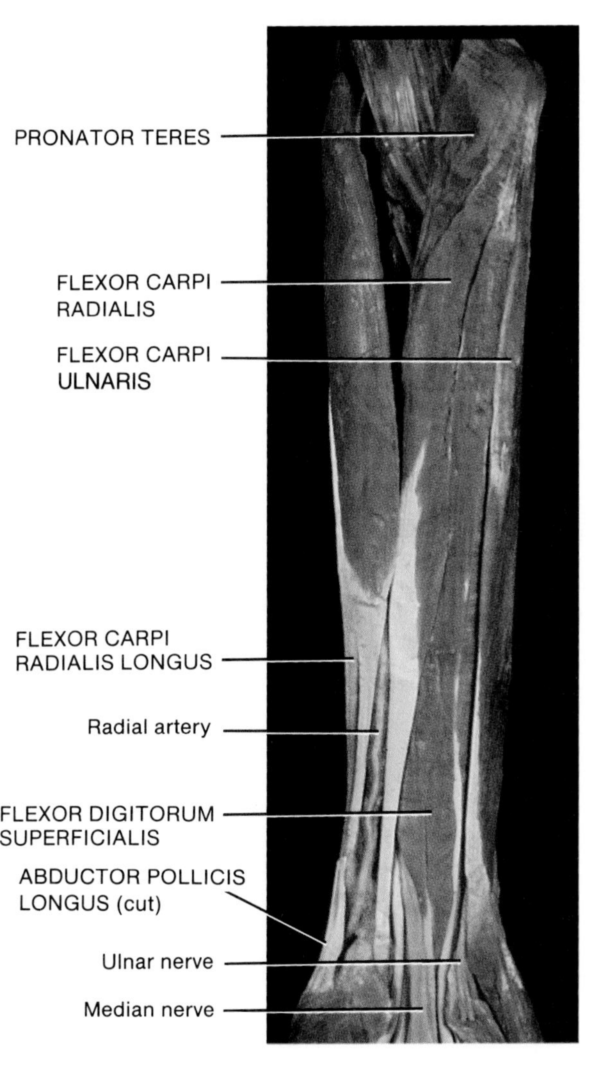

(g) Anterior superfical view

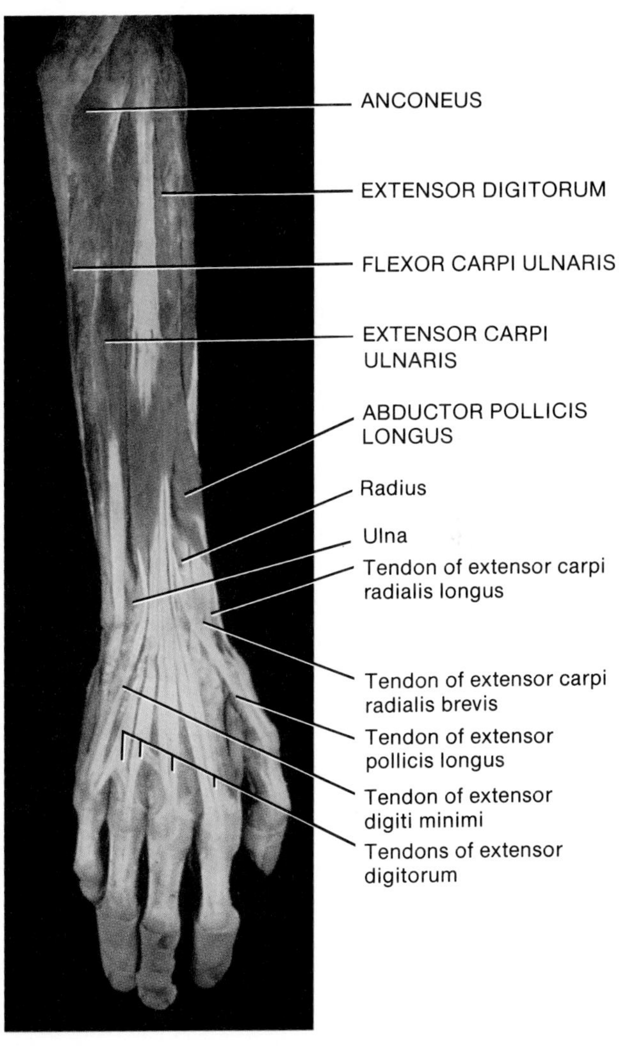

(h) Posterior superfical view

EXHIBIT 10-19

Intrinsic Muscles of the Hand (Figure 10-20)

Overview: Several of the muscles discussed in Exhibit 10-18 help to move the digits in various ways. In addition, there are muscles in the palmar surface of the hand called ***intrinsic muscles*** that also help to move the digits. Such muscles are so named because their origins and insertions are both within the hands. These muscles assist in the intricate and precise movements that are characteristic of the human hand.

The intrinsic muscles of the hand are divided into three principal groups—thenar, hypothenar, and intermediate. The four ***thenar*** (THĒ-nar) ***muscles*** act on the thumb and form the ***thenar eminence.*** The four ***hypothenar*** (HĪ-pō-thē′-nar) ***muscles*** act on the little finger and form the ***hypothenar eminence.*** The 11 ***intermediate (midpalmar) muscles*** act on all the digits, except the thumb.

The functional importance of the hand is readily apparent when one considers that certain hand injuries can result in permanent disability. In fact, most of the efficiency of the hand depends on the movements of the thumb. The general activities of the hand are free motion, power grip (forcible movement of the fingers and thumb against the palm, as in squeezing), precision handling (a change in position of a handled object that requires exact control of finger and thumb positions, as in winding a watch or threading a needle), and pinch (compression between the thumb and index finger or between the thumb and first two fingers).

Movement of the thumb is very important in the precise activities of the hand. The five principal movements of the thumb, illustrated below, are flexion (movement of the thumb at right angles to the fingers), extension (movement in the opposite direction), abduction (movement of the thumb anteriorly and laterally), adduction (movement of the thumb toward the index finger), and opposition (movement of the thumb across the palm so that the tip of the thumb meets the tips of the fingers). Opposition is the single most distinctive digital movement that gives humans their characteristic tool-making and tool-using capacity.

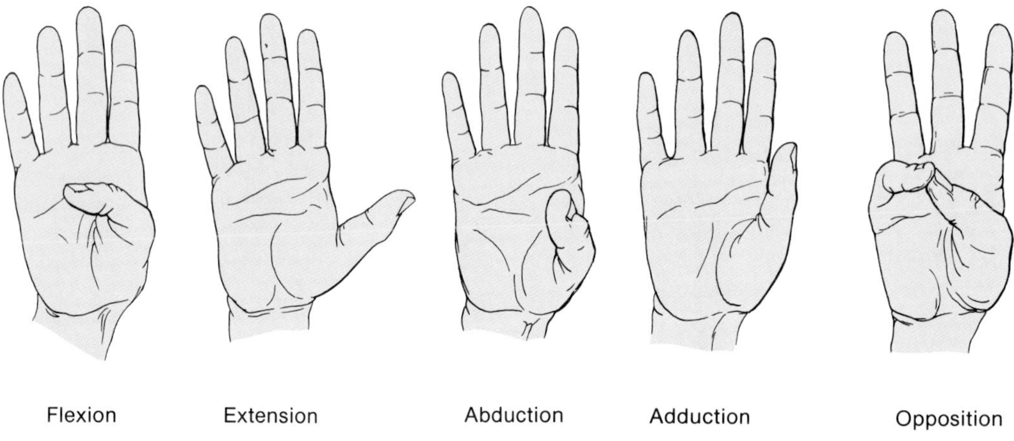

Principal movements of the thumb

MUSCLE	ORIGIN	INSERTION	ACTION	INNERVATION
THENAR				
Abductor Pollicis Brevis (*abductor* = moves part away from middle; *pollex* = thumb; *brevis* = short)	Flexor retinaculum, scaphoid, and trapezium.	Proximal phalanx of thumb.	Abducts thumb.	Median nerve.
Opponens Pollicis (*opponens* = opposes)	Flexor retinaculum and trapezium.	Metacarpal of thumb.	Draws thumb across palm to meet little finger (opposition).	Median nerve.
Flexor Pollicis Brevis (*flexor* = decreases angle at joint)	Flexor retinaculum, trapezium, and first metacarpal.	Proximal phalanx of thumb.	Flexes and adducts thumb.	Median and ulnar nerve.
Adductor Pollicis (*adductor* = moves part toward midline)	Capitate and second and third metacarpals.	Proximal phalanx of thumb.	Adducts thumb.	Ulnar nerve.
HYPOTHENAR				
Palmaris Brevis[a] (*palma* = palm)	Flexor retinaculum and palmar aponeurosis.	Skin on ulnar border of palm of hand.	Draws skin toward middle of palm as in clenching fist.	Ulnar nerve.

MUSCLE	ORIGIN	INSERTION	ACTION	INNERVATION
Abductor Digiti Minimi (*digit* = finger or toe; *minimi* = little finger)	Pisiform and tendon of flexor carpi ulnaris.	Proximal phalanx of little finger.	Abducts little finger.	Ulnar nerve
Flexor Digiti Minimi Brevis	Flexor retinaculum and hamate.	Proximal phalanx of little finger.	Flexes little finger.	Ulnar nerve.
Opponens Digiti Minimi	Flexor retinaculum and hamate.	Metacarpal of little finger.	Draws little finger across palm to meet thumb.	Ulnar nerve.
INTERMEDIATE (Midpalmar)				
Lumbricals (four muscles)	Tendons of flexor digitorum profundus.	Tendons of extensor digitorum.	Extend interphalangeal joints and flex metacarpophalangeal joints.	Median and ulnar nerve.
Dorsal Interossei (four muscles; *dorsal* = back surface; *inter* = between; *ossei* = bones)	Adjacent sides of metacarpals.	Proximal phalanx of second, third, and fourth fingers.	Abduct fingers from middle finger; flex fingers at metacarpophalangeal joints; and extend interphalangeal joints.	Ulnar nerve.
Palmar Interossei (three muscles)	Medial side of second metacarpal and lateral sides of fourth and fifth metacarpals.	Proximal phalanx of same finger.	Adduct fingers toward middle finger; flex fingers at metacarpophalangeal joints; and extend interphalangeal joints.	Ulnar nerve.

[a] Not illustrated.

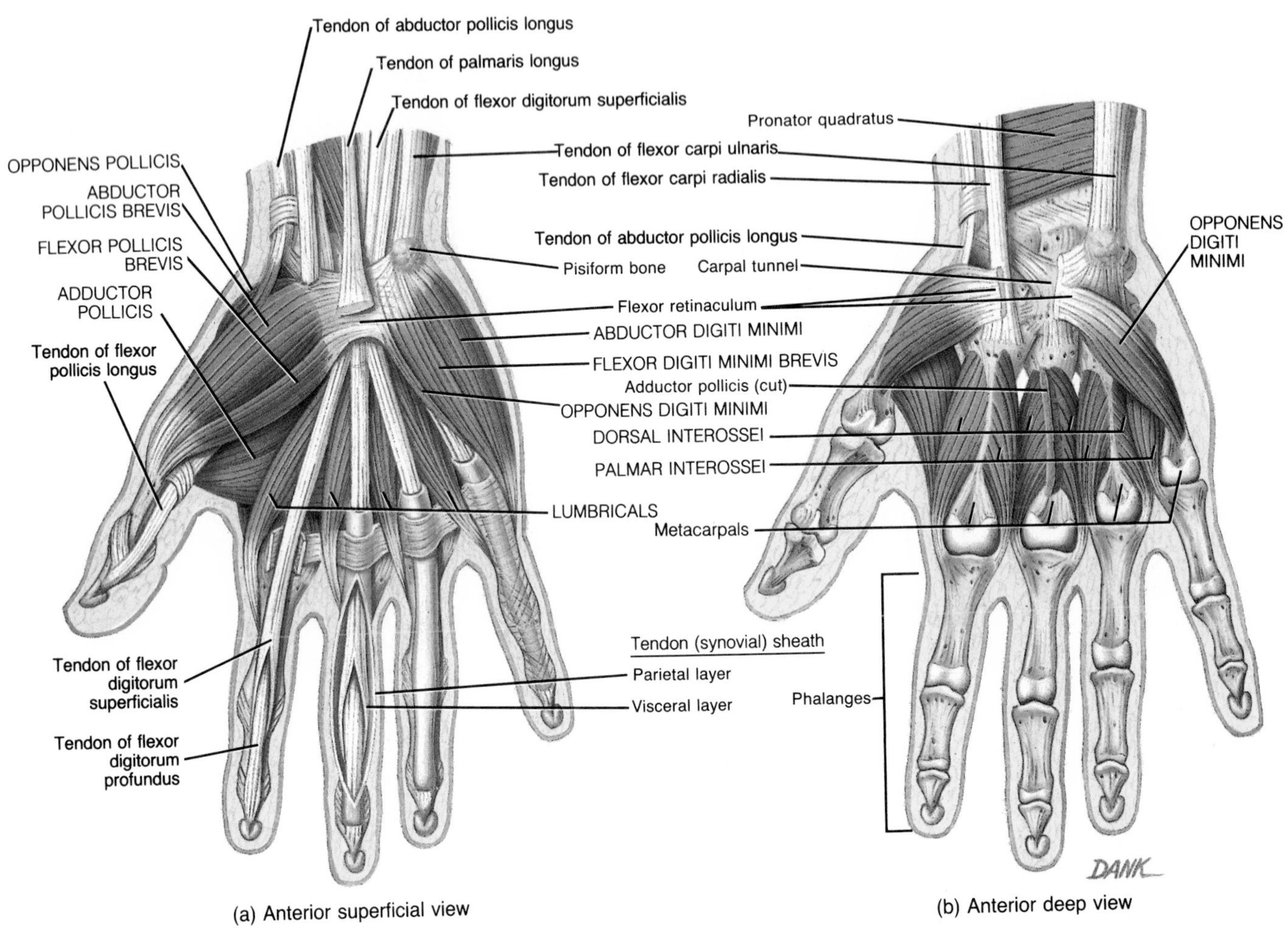

FIGURE 10-20 Palmar view of the intrinsic muscles of the right hand. (c) Courtesy of J. A. Gosling, P. F. Harris, et al., *Atlas of Human Anatomy,* Gower Medical Publishing Ltd., 2nd ed., 1991.

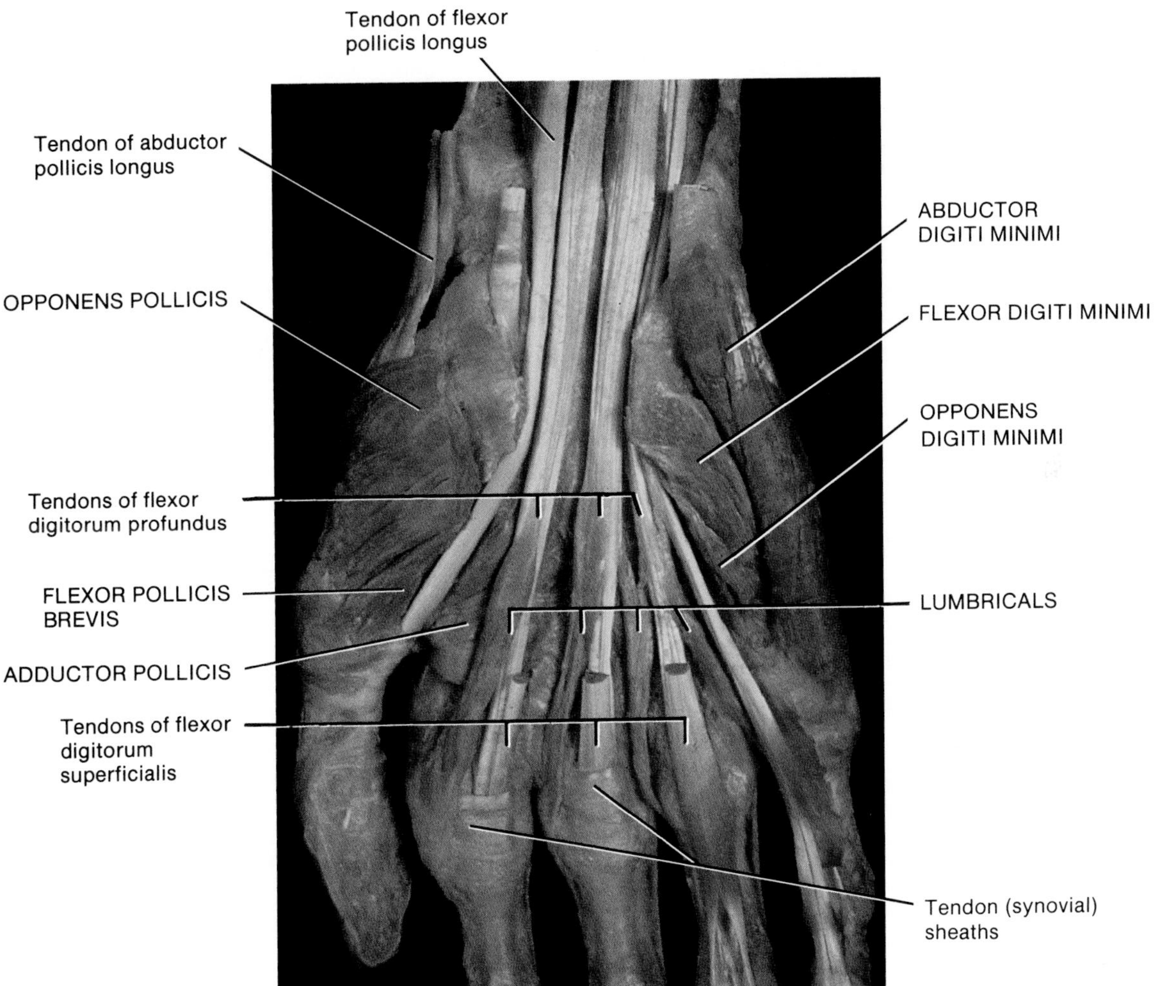

(c) Anterior superficial view

EXHIBIT 10-20

Muscles That Move the Vertebral Column (Figure 10-21)

Overview: The muscles that move the vertebral column are quite complex because they have multiple origins and insertions and there is considerable overlapping among them. One way to group the muscles is on the basis of the general direction of the muscle bundles and their approximate lengths. For example, the ***splenius muscles*** arise from the midline and run laterally and superiorly to their insertions. The ***erector spinae (sacrospinalis) muscle*** arises from either the midline or more laterally but usually runs almost longitudinally, with neither a marked outward nor inward direction as it is traced superiorly. The ***transversospinalis muscles*** arise laterally but run toward the midline as they are traced superiorly. Deep to these three muscle groups are small ***segmental muscles*** that run between spinous processes or transverse processes of vertebrae. Since the scalene muscles also assist in moving the vertebral column, they are included in this exhibit.

Note in Exhibit 10-11 that the rectus abdominis and quadratus lumborum muscles also assume a role in moving the vertebral column.

MUSCLE	ORIGIN	INSERTION	ACTION	INNERVATION
SPLENIUS				
Splenius Capitis (*splenium* = bandage; *caput* = head)	Ligamentum nuchae and spinous processes of seventh cervical vertebra and first three or four thoracic vertebrae.	Occipital bone and mastoid process of temporal bone.	Acting together, they extend the head and neck; acting singly, each laterally flexes and rotates head to same side.	Dorsal rami of middle cervical nerves.
Splenius Cervicis (*cervix* = neck)	Spinous processes of third through sixth thoracic vertebrae.	Transverse processes of first two or four cervical vertebrae.	Acting together, they extend the head and neck; acting singly, each laterally flexes and rotates head to same side.	Dorsal rami of lower cervical nerves.
ERECTOR SPINAE (SACROSPINALIS)	This is the largest muscular mass of the back and consists of three groupings: iliocostalis, longissimus, and spinalis. These groups, in turn, consist of a series of overlapping muscles. The iliocostalis group is laterally placed, the longissimus group is intermediate in placement, and the spinalis group is medially placed.			
Iliocostalis (Lateral) Group				
Iliocostalis Lumborum (*ilium* = flank; *costa* = rib)	Iliac crest.	Lower six ribs.	Extends lumbar region of vertebral column.	Dorsal rami of lumbar nerves.
Iliocostalis Thoracis (*thorax* = chest)	Lower six ribs.	Upper six ribs.	Maintains erect position of spine.	Dorsal rami of thoracic (intercostal) nerves.
Iliocostalis Cervicis	First six ribs.	Transverse processes of fourth to sixth cervical vertebrae.	Extends cervical region of vertebral column.	Dorsal rami of cervical nerves.
Longissimus (Intermediate) Group				
Longissimus Thoracis (*longissimus* = longest)	Transverse processes of lumbar vertebrae.	Transverse processes of all thoracic and upper lumbar vertebrae and ninth and tenth ribs.	Extends thoracic region of vertebral column.	Dorsal rami of spinal nerves.
Longissimus Cervicis	Transverse processes of fourth and fifth thoracic vertebrae.	Transverse processes of second to sixth cervical vertebrae.	Extends cervical region of vertebral column.	Dorsal rami of spinal nerves.
Longissimus Capitis	Transverse processes of last four cervical and upper four thoracic vertebrae.	Mastoid process of temporal bone.	Extends head and rotates it to opposite side.	Dorsal rami of middle and lower cervical nerves.

MUSCLE	ORIGIN	INSERTION	ACTION	INNERVATION
Spinalis (Medial) Group				
Spinalis Thoracis (*spinalis* = vertebral column)	Spinous processes of upper lumbar and lower thoracic vertebrae.	Spinous processes of upper thoracic vertebrae.	Extends vertebral column.	Dorsal rami of spinal nerves.
Spinalis Cervicis	Ligamentum nuchae and spinous process of seventh cervical vertebra.	Spinous process of axis.	Extends vertebral column.	Dorsal rami of spinal nerves.
Spinalis Capitis	Arises with semispinalis capitis.	Inserts with semispinalis capitis.	Extends vertebral column.	Dorsal rami of spinal nerves.
TRANSVERSOSPINALIS				
Semispinalis Thoracis (*semi* = partially or one-half)	Transverse processes of sixth to tenth thoracic vertebrae.	Spinous processes of first four thoracic and last two cervical vertebrae.	Extends vertebral column and rotates it to opposite side.	Dorsal rami of thoracic and cervical spinal nerves.
Semispinalis Cervicis	Transverse processes of first five or six thoracic vertebrae.	Spinous processes of first to fifth cervical vertebrae.	Extends vertebral column and rotates it to opposite side.	Dorsal rami of thoracic and cervical spinal nerves.
Semispinalis Capitis	Transverse processes of first six or seven thoracic vertebrae and seventh cervical vertebra and articular processes of fourth, fifth, and sixth cervical vertebrae.	Occipital bone.	Extends vertebral column and rotates it to opposite side.	Dorsal rami of cervical nerves.
Multifidus (*multi* = many; *findere* = to split)	Sacrum, ilium, transverse processes of lumbar, thoracic, and lower four cervical vertebrae.	Spinous process of a higher vertebra.	Extends vertebral column and rotates it to opposite side.	Dorsal rami of spinal nerves.
Rotatores (*rotate* = turn on an axis)	Transverse processes of all vertebrae.	Spinous process of vertebra above the one of origin.	Extends vertebral column and rotates it to opposite side.	Dorsal rami of spinal nerves.
SEGMENTAL				
Interspinales (*inter* = between)	Superior surface of all spinous processes.	Inferior surface of spinous process of vertebra above the one of origin.	Extends vertebral column.	Dorsal rami of spinal nerves.
Intertransversarii (*inter* = between)	Transverse processes of all vertebrae.	Transverse process of vertebra above the one of origin.	Laterally flexes vertebral column.	Dorsal and ventral rami of spinal nerves.
SCALENE				
Anterior Scalene (*anterior* = front; *skalenos* = uneven)	Transverse processes of third through sixth cervical vertebrae.	First rib.	Flexes and rotates neck and assists in inspiration.	Ventral rami of fifth and sixth cervical nerves.
Middle Scalene	Transverse processes of last six cervical vertebrae.	First rib.	Flexes and rotates neck and assists in inspiration.	Ventral rami of third through eighth cervical nerves.
Posterior Scalene	Transverse processes of fourth through sixth cervical vertebrae.	Second rib.	Flexes and rotates neck and assists in inspiration.	Ventral rami of last three cervical nerves.

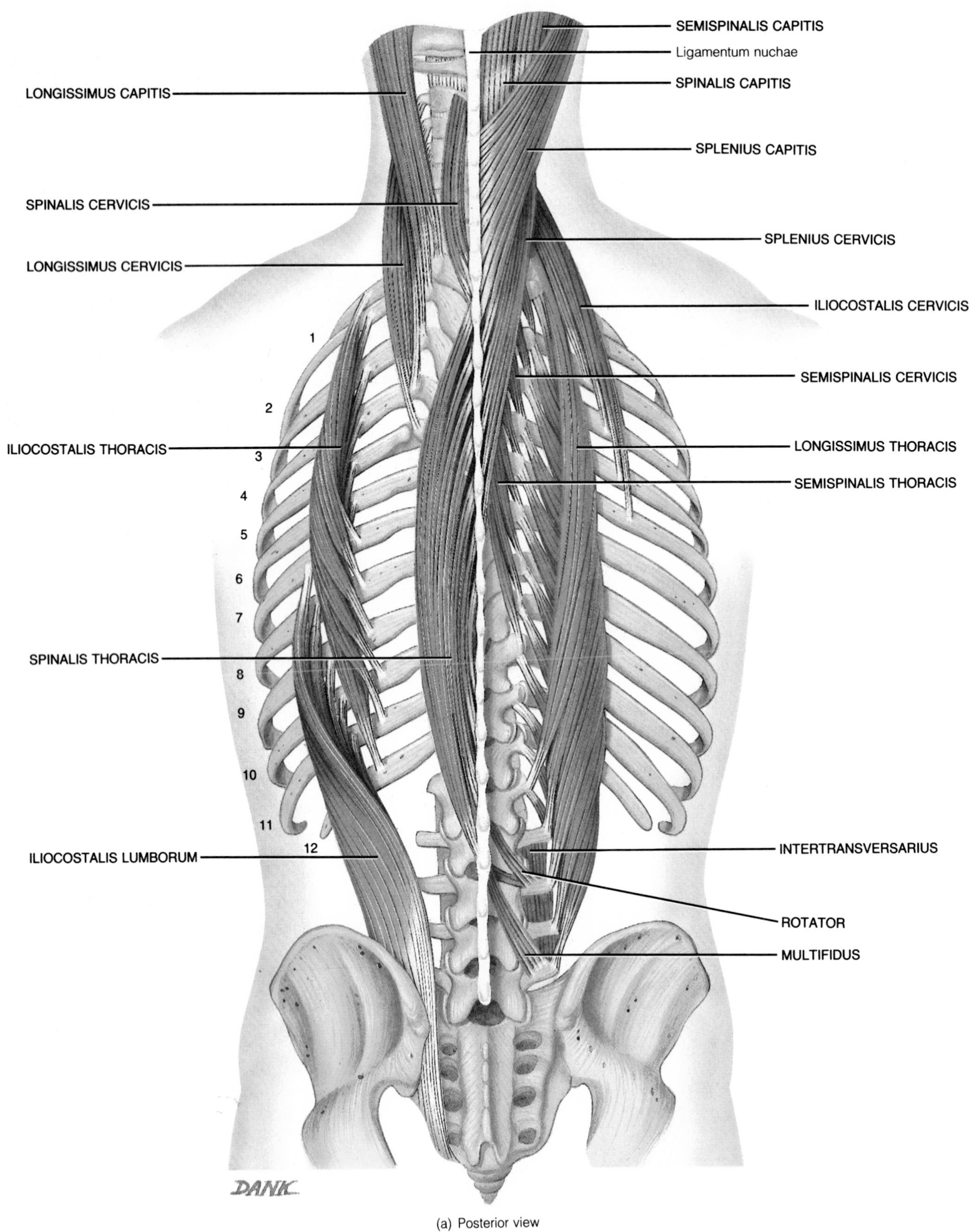

(a) Posterior view

FIGURE 10-21 Muscles that move the vertebral column.

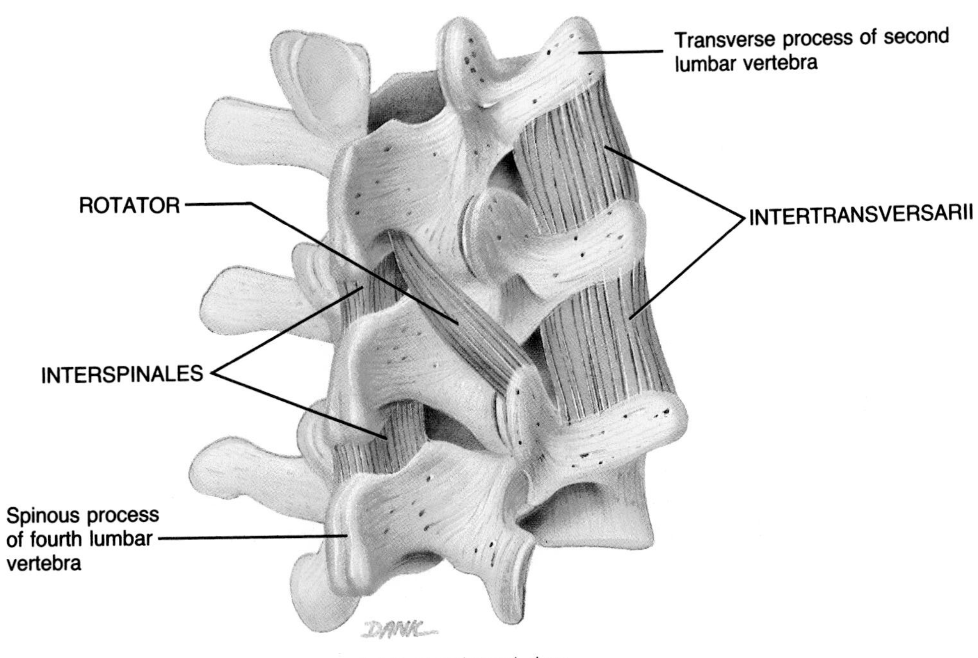

(b) Posterolateral view

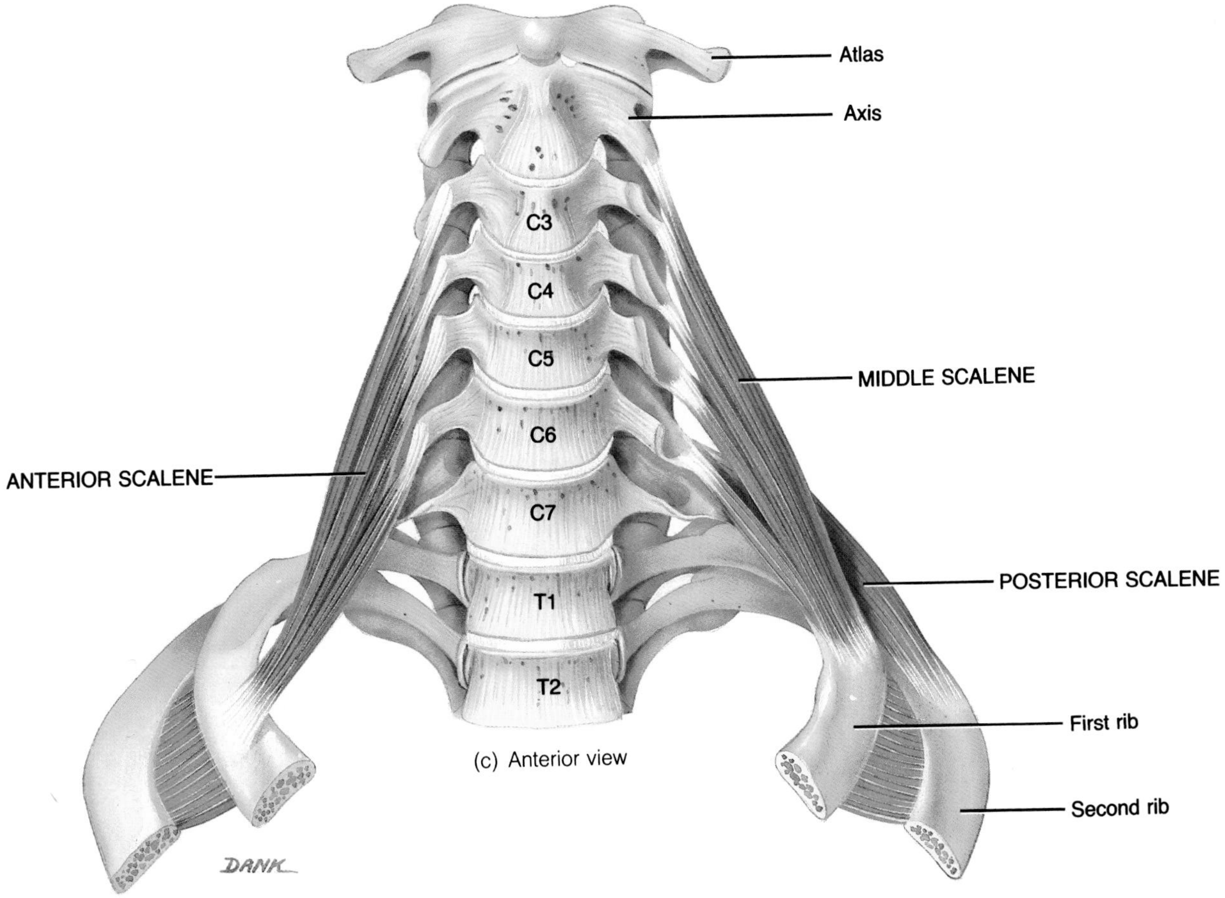

Atlas

Axis

C3

C4

C5

C6

C7

T1

T2

MIDDLE SCALENE

ANTERIOR SCALENE

POSTERIOR SCALENE

First rib

Second rib

DANK

(c) Anterior view

EXHIBIT 10-21

Muscles That Move the Thigh (Femur) (Figure 10-22)

Overview: As you will see, muscles of the lower extremities are larger and more powerful than those of the upper extremities since lower extremity muscles function in stability, locomotion, and maintenance of posture. Upper extremity muscles are characterized by versatility of movement. In addition, muscles of the lower extremities frequently cross two joints and act equally on both.

The majority of muscles that act on the thigh (femur) originate on the pelvic (hip) girdle and insert on the femur. The anterior muscles are the psoas major and iliacus, together referred to as the iliopsoas muscle. The remaining muscles (except for the pectineus, adductors, and tensor fasciae latae) are posterior muscles. Technically, the pectineus and adductors are components of the medial compartment of the thigh, but they are included in this exhibit because they act on the thigh. The tensor fasciae latae muscle is laterally placed. The ***fascia lata*** is a deep fascia of the thigh that encircles the entire thigh. It is well developed laterally, where together with the tendons of the gluteus maximus and tensor fasciae latae muscles it forms a structure called the ***iliotibial tract.*** The tract inserts into the lateral condyle of the tibia.

After you have studied the muscles in this exhibit, arrange them according to the following actions: flexion, extension, abduction, adduction, medial rotation, and lateral rotation. (The same muscles can be used more than once.)

MUSCLE	ORIGIN	INSERTION	ACTION	INNERVATION
Psoas Major (*psoa* = muscle of loin)	Transverse processes and bodies of lumbar vertebrae.	Lesser trochanter of femur.	Flexes and rotates thigh laterally; flexes vertebral column.	Lumbar nerves L2–L3.
Iliacus (*iliac* = ilium)	Iliac fossa.	Tendon of psoas major.	Flexes and rotates thigh laterally and flexes vertebral column.	Femoral nerve.
Gluteus Maximus (*glutos* = buttock; *maximus* = largest; largest single muscle in body)	Iliac crest, sacrum, coccyx, and aponeurosis of sacrospinalis.	Iliotibial tract of fascia lata and gluteal tuberosity of femur.	Extends and rotates thigh laterally.	Inferior gluteal nerve.
Gluteus Medius (*media* = middle)	Ilium.	Greater trochanter of femur.	Abducts and rotates thigh medially.	Superior gluteal nerve.
Gluteus Minimus (*minimus* = smallest)	Ilium.	Greater trochanter of femur.	Abducts and rotates thigh medially.	Superior gluteal nerve.
Tensor Fasciae Latae (*tensor* = makes tense; *fascia* = band; *latus* = wide)	Iliac crest.	Tibia by way of the iliotibial tract.	Flexes and abducts thigh.	Superior gluteal nerve.
Piriformis (*pirum* = pear; *forma* = shape)	Sacrum anteriorly.	Superior border of greater trochanter of femur.	Rotates thigh laterally and abducts it.	Sacral nerves S2 or S1–S2.
Obturator Internus (*obturator* = obturator foramen; *internus* = inside)	Inner surface of obturator foramen, pubis, and ischium.	Greater trochanter of femur.	Rotates thigh laterally and abducts it.	Nerve to obturator internus.
Obturator Externus (*externus* = outside)	Outer surface of obturator membrane.	Trochanteric fossa of femur.	Rotates thigh laterally.	Obturator nerve.
Superior Gemellus (*superior* = above; *gemellus* = twins)	Ischial spine.	Greater trochanter of femur.	Rotates thigh laterally and abducts it.	Nerve to obturator internus.
Inferior Gemellus (*inferior* = below)	Ischial tuberosity.	Greater trochanter of femur.	Rotates thigh laterally and abducts it.	Nerve to quadratus femoris.
Quadratus Femoris (*quad* = four; *femoris* = femur)	Ischial tuberosity.	Small tubercle on posterior femur.	Laterally rotates and adducts thigh.	Nerve to quadratus femoris.

MUSCLE	ORIGIN	INSERTION	ACTION	INNERVATION
Adductor Longus (*adductor* = moves part closer to midline; *longus* = long)	Pubic crest and pubic symphysis.	Linea aspera of femur.	Adducts, laterally rotates, and flexes thigh.	Obturator nerve.
Adductor Brevis (*brevis* = short)	Inferior ramus of pubis.	Upper half of linea aspera of femur.	Adducts, laterally rotates, and flexes thigh.	Obturator nerve.
Adductor Magnus (*magnus* = large)	Inferior ramus of pubis and ischium to ischial tuberosity.	Linea aspera of femur.	Adducts, flexes, laterally rotates, and extends thigh (anterior part flexes; posterior part extends).	Obturator and sciatic nerves.
Pectineus (*pecten* = comb-shaped)	Superior ramus of pubis.	Pectineal line of femur, between lesser trochanter and linea aspera.	Flexes and adducts thigh.	Femoral nerve.

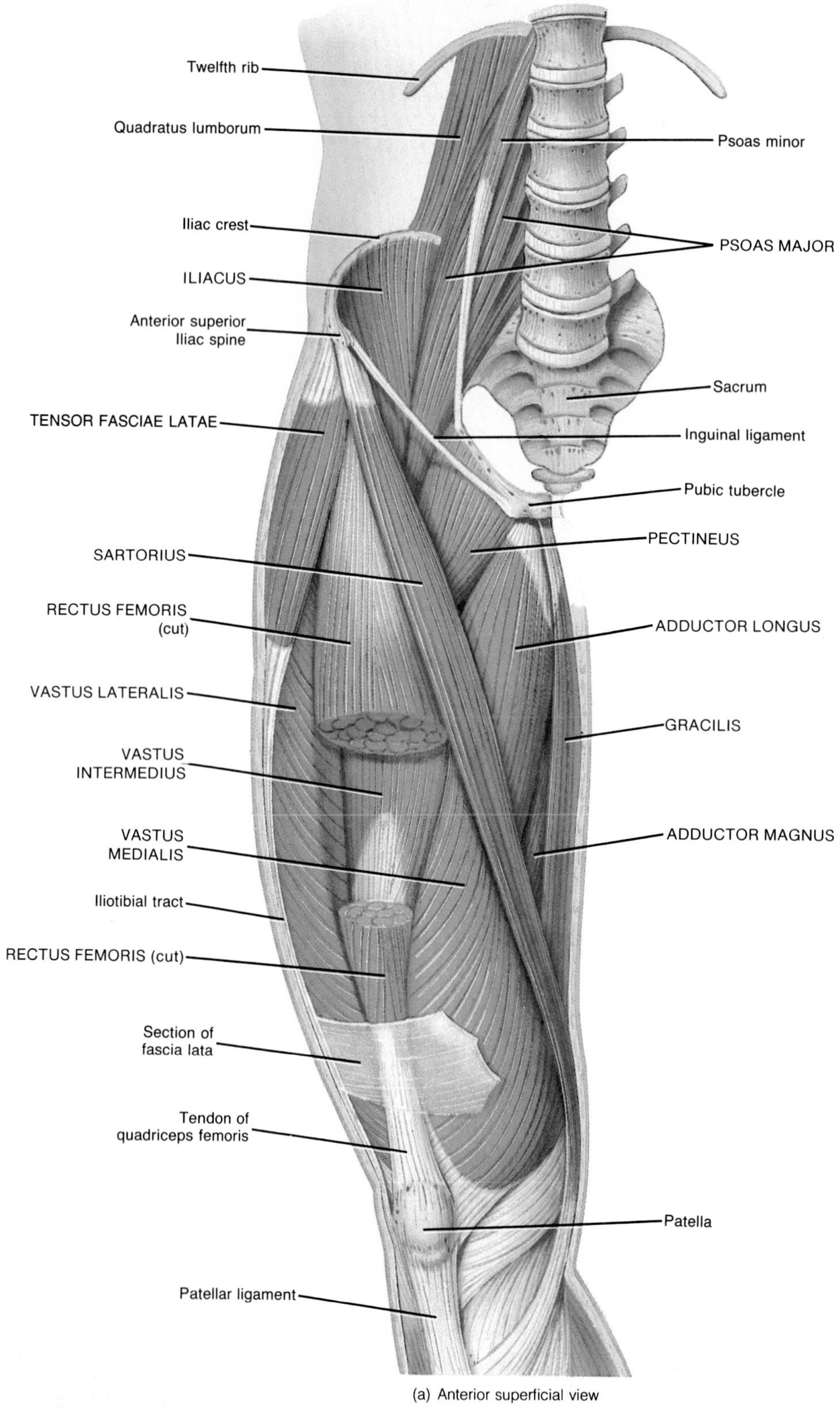

(a) Anterior superficial view

FIGURE 10-22 Muscles that move the thigh (femur). (e) and (f) Courtesy of J. A. Gosling, P. F. Harris, et al., *Atlas of Human Anatomy,* Gower Medical Publishing Ltd., 2nd ed., 1991.

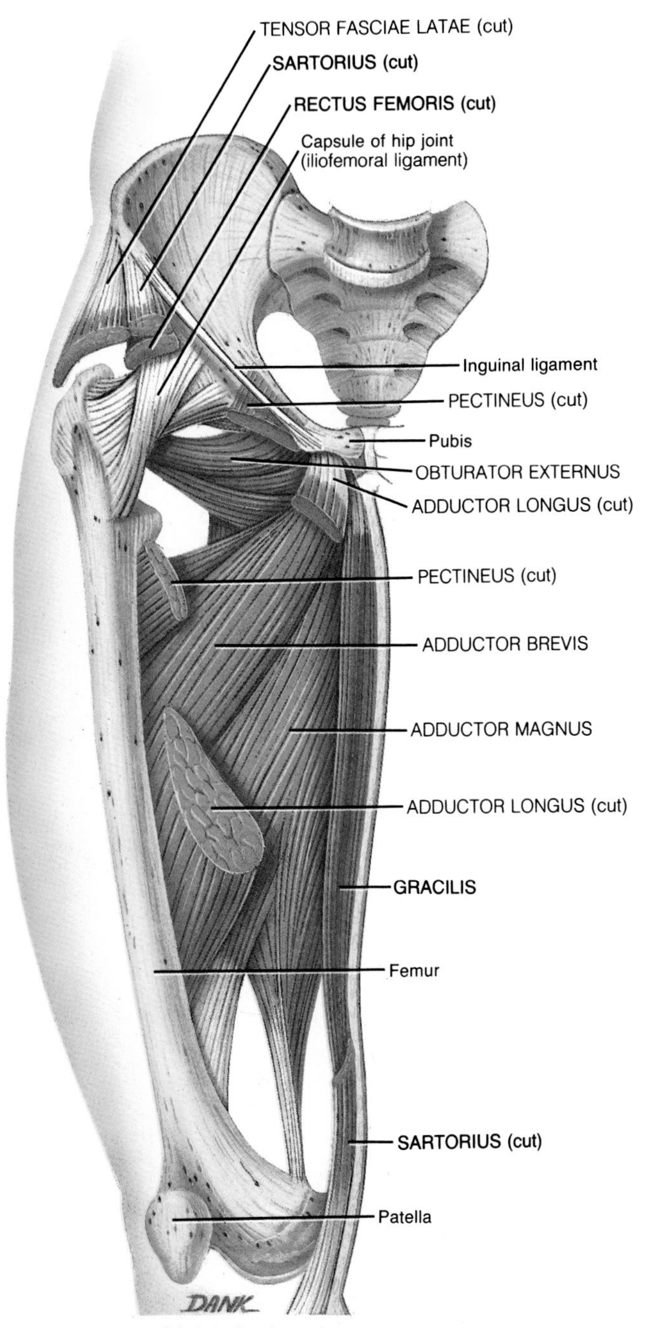

(b) Anterior deep view

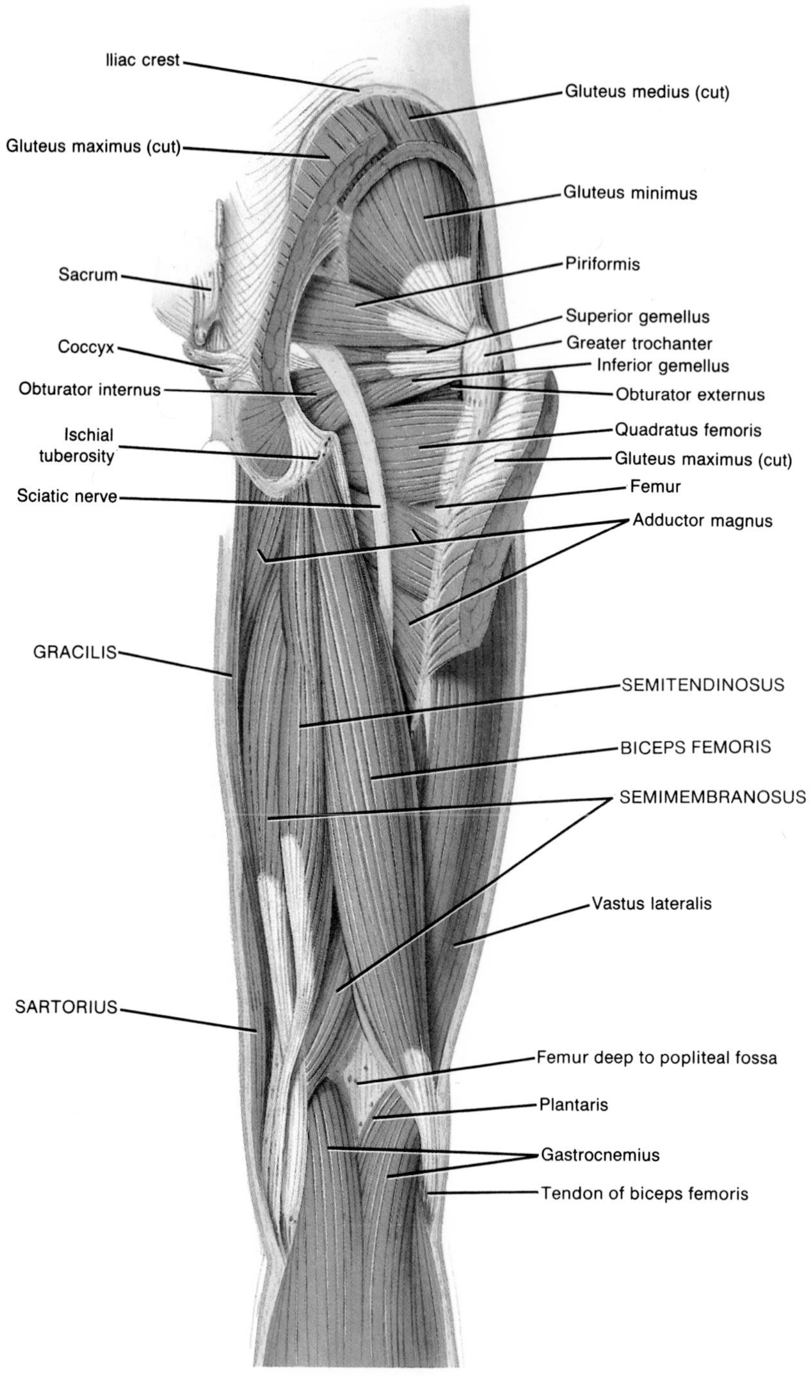

(c) Posterior deep view

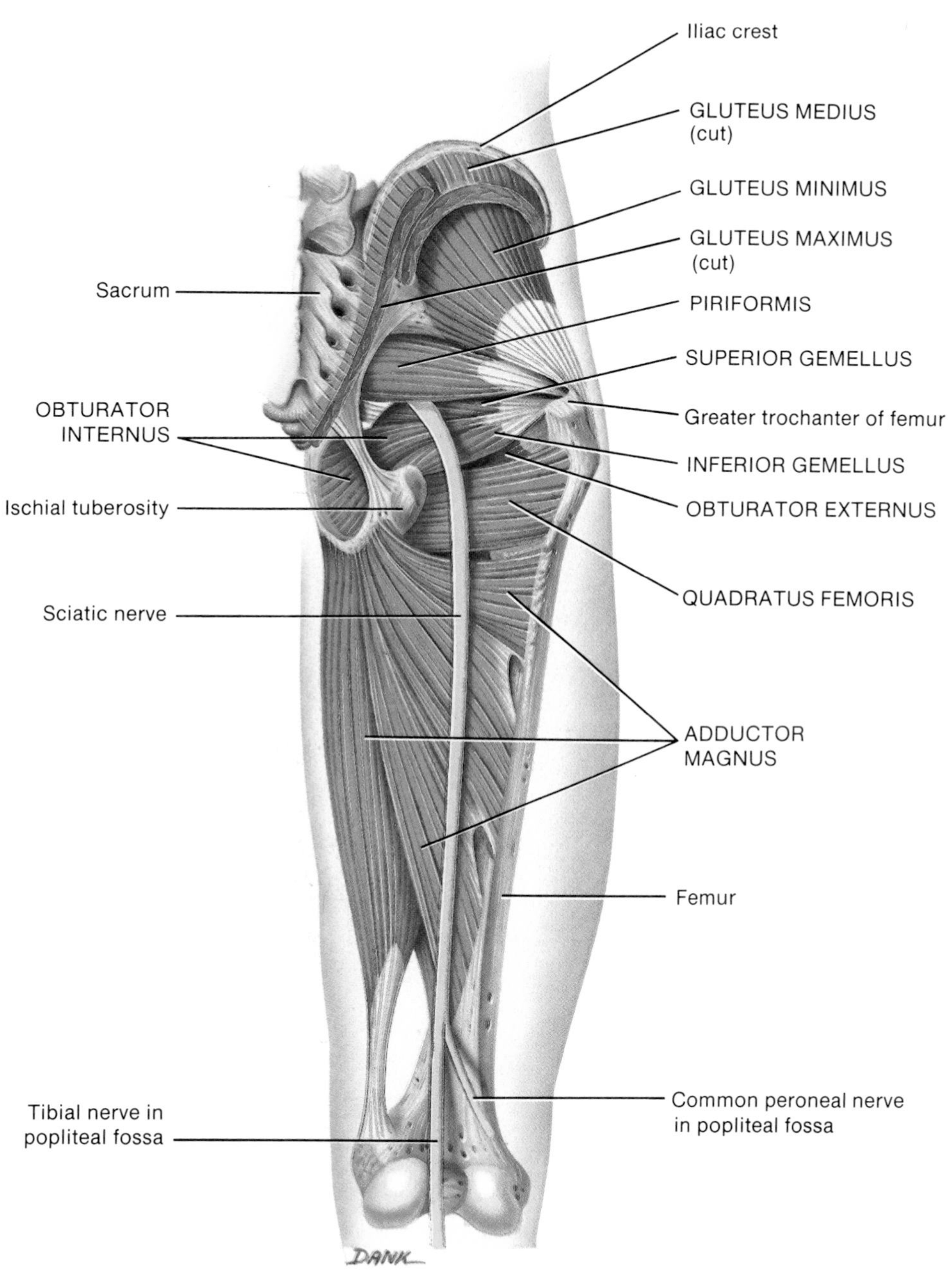

(d) Posterior deeper view

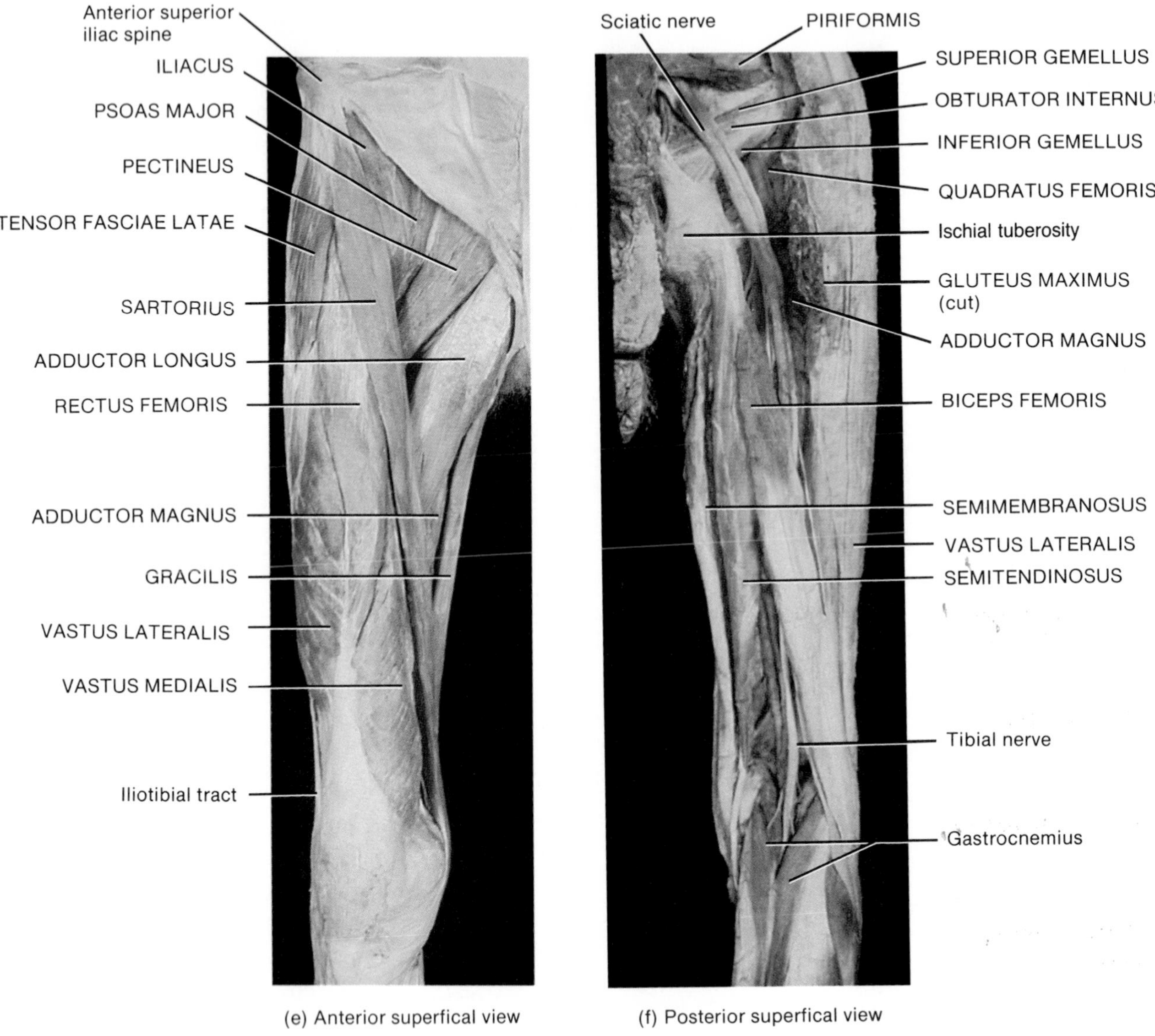

(e) Anterior superfical view

(f) Posterior superfical view

EXHIBIT 10-22

Muscles That Act on the Leg (Tibia and Fibula) (Figures 10-22 and 10-23)

Overview: The muscles that act on the leg (tibia and fibula) originate in the hip and thigh and are separated into compartments by deep fascia. The ***medial (adductor) compartment*** is so named because its muscles adduct the thigh. It is innervated by the obturator nerve. As noted earlier, the adductor magnus, adductor longus, adductor brevis, and pectineus muscles, components of the medial compartment, are included in Exhibit 10-21 because they act on the femur. The gracilis, the other muscle in the medial compartment, not only adducts the thigh but also flexes the leg. For this reason, it is included in this exhibit.

The ***anterior (extensor) compartment*** is so designated because its muscles act to extend the leg, and some also flex the thigh. It is composed of the quadriceps femoris and sartorius muscles and is innervated by the femoral nerve. The quadriceps femoris muscle is a composite muscle that includes four distinct parts, usually described as four separate muscles (rectus femoris, vastus lateralis, vastus intermedius, and vastus medialis). The common tendon for the four muscles is known as the ***patellar ligament*** and attaches to the tibial tuberosity. The rectus femoris and sartorius muscles are also flexors of the thigh.

The ***posterior (flexor) compartment*** is so named because its muscles flex the leg (but also extend the thigh). It is innervated by branches of the sciatic nerve. Included are the hamstrings (biceps femoris, semitendinosus, and semimembranosus). The hamstrings are so named because their tendons are long and stringlike in the popliteal area. The ***popliteal fossa*** is a diamond-shaped space on the posterior aspect of the knee bordered laterally by the tendons of the biceps femoris and medially by the semitendinosus and semimembranosus muscles.

MUSCLE	ORIGIN	INSERTION	ACTION	INNERVATION
MEDIAL (ADDUCTOR) COMPARTMENT				
Adductor Magnus **Adductor Longus** **Adductor Brevis** **Pectineus**	See Exhibit 10-21			
Gracilis (*gracilis* = slender)	Pubic symphysis and pubic arch.	Medial surface of body of tibia.	Adducts thigh and flexes leg.	Obturator nerve.
ANTERIOR (EXTENSOR) COMPARTMENT				
Quadriceps Femoris (*quadriceps* = four heads of origin; *femoris* = femur)				
Rectus Femoris (*rectus* = fibers parallel to midline)	Anterior inferior iliac spine.	Upper border of patella.	All four heads extend leg; rectus portion alone also flexes thigh.	Femoral nerve.
Vastus Lateralis (*vastus* = large; *lateralis* = lateral)	Greater trochanter and linea aspera of femur.	Tibial tuberosity through patellar ligament (tendon of quadriceps).		Femoral nerve.
Vastus Medialis (*medialis* = medial)	Linea aspera of femur.			Femoral nerve.
Vastus Intermedius (*intermedius* = middle)	Anterior and lateral surfaces of body of femur.			Femoral nerve.
Sartorius (*sartor* = tailor; refers to cross-legged position of tailors; longest muscle in body)	Anterior superior spine of ilium.	Medial surface of body of tibia.	Flexes leg; flexes thigh and rotates it laterally, thus crossing leg.	Femoral nerve.

EXHIBIT 10-22 (*Continued*)

MUSCLE	ORIGIN	INSERTION	ACTION	INNERVATION
POSTERIOR (FLEXOR) COMPARTMENT				
Hamstrings	A collective designation for three separate muscles.			
Biceps Femoris (*biceps* = two heads of origin)	Long head arises from ischial tuberosity; short head arises from linea aspera of femur.	Head of fibula and lateral condyle of tibia.	Flexes leg and extends thigh.	Tibial and common peroneal nerves from sciatic nerve.
Semitendinosus (*semi* = half; *tendo* = tendon)	Ischial tuberosity.	Proximal part of medial surface of body of tibia.	Flexes leg and extends thigh.	Tibial nerve from sciatic nerve.
Semimembranosus (*membran* = membrane)	Ischial tuberosity.	Medial condyle of tibia.	Flexes leg and extends thigh.	Tibial nerve from sciatic nerve.

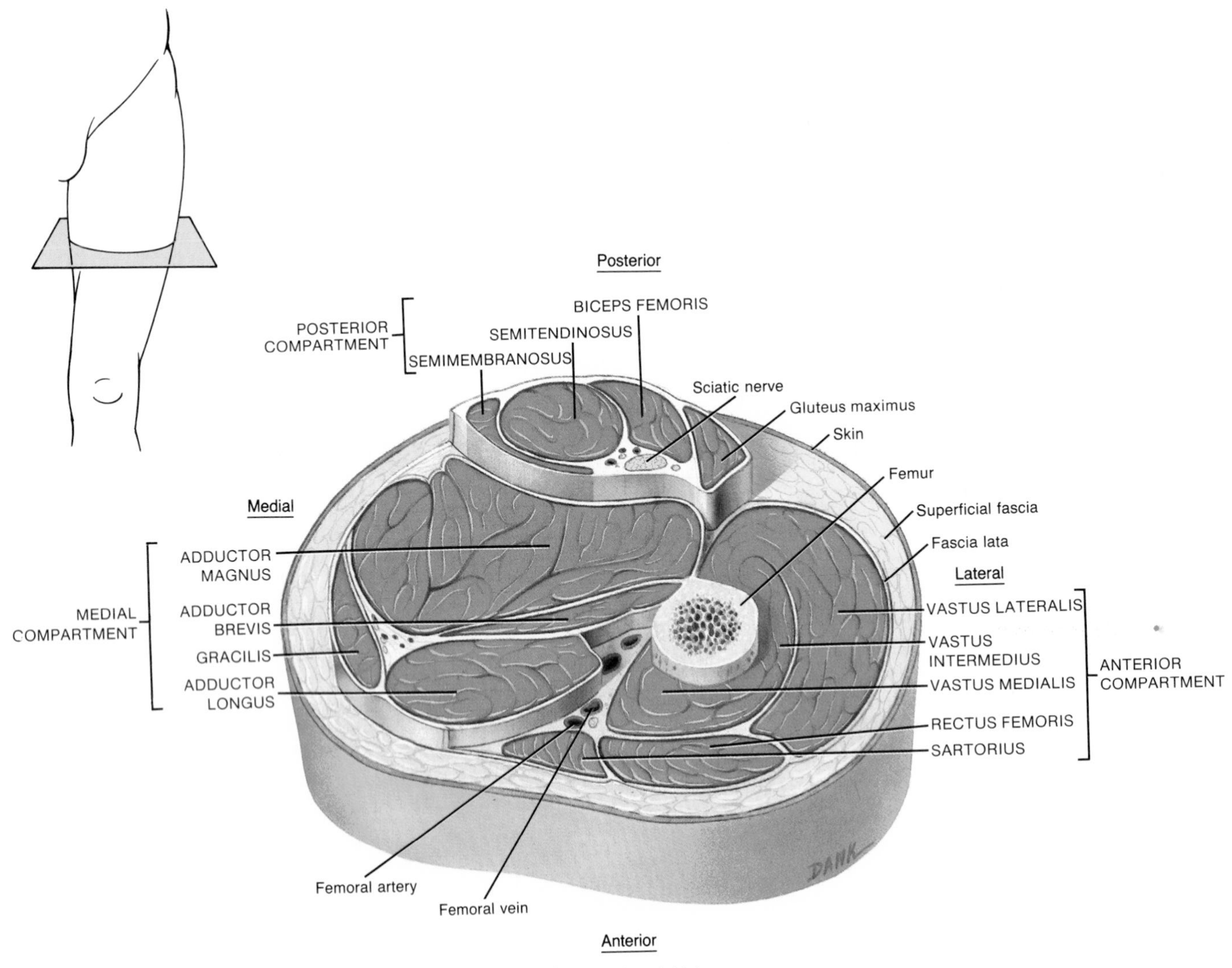

FIGURE 10-23 Muscles that act on the leg.

EXHIBIT 10-23

Muscles That Move the Foot and Toes (Figure 10-24)

Overview: The musculature of the leg, like that of the thigh, can be distinguished into three compartments by deep fascia. In addition, all the muscles in the respective compartments are innervated by the same nerve. The ***anterior compartment*** consists of muscles that dorsiflex the foot and are innervated by the deep peroneal nerve. In a situation analogous to the wrist, the tendons of the muscles of the anterior compartment are held firmly to the ankle by thickenings of deep fascia called the ***superior extensor retinaculum* (*transverse ligament of the ankle*)** and ***inferior extensor retinaculum* (*cruciate ligament of the ankle*).**

The ***lateral* (*peroneal*) *compartment*** contains two muscles that plantar flex and evert the foot. They are supplied by the superficial peroneal nerve.

The ***posterior compartment*** consists of muscles that are divisible into superficial and deep groups. All are innervated by the tibial nerve. All three superficial muscles share a common tendon of insertion, the calcaneal (Achilles) tendon that inserts into the calcaneus bone of the ankle. The superficial muscles are plantar flexors of the foot. Of the four deep muscles, three plantar flex the foot.

MUSCLE	ORIGIN	INSERTION	ACTION	INNERVATION
ANTERIOR COMPARTMENT				
Tibialis Anterior (*tibialis* = tibia; *anterior* = front)	Lateral condyle and body of tibia and interosseous membrane (sheet of fibrous tissue that holds bones together).	First metatarsal and first (medial) cuneiform.	Dorsiflexes and inverts foot.	Deep peroneal nerve.
Extensor Hallucis Longus (*extensor* = increases angle at joint; *hallucis* = hallux or great toe; *longus* = long)	Anterior surface of fibula and interosseous membrane.	Distal phalanx of great toe.	Dorsiflexes and inverts foot and extends great toe.	Deep peroneal nerve.
Extensor Digitorum Longus	Lateral condyle of tibia, anterior surface of fibula, and interosseous membrane.	Middle and distal phalanges of four outer toes.	Dorsiflexes and everts foot and extends toes.	Deep peroneal nerve.
Peroneus Tertius (*perone* = fibula; *tertius* = third)	Distal third of fibula and interosseous membrane.	Fifth metatarsal.	Dorsiflexes and everts foot.	Deep peroneal nerve.
LATERAL (PERONEAL) COMPARTMENT				
Peroneus Longus	Head and body of fibula and lateral condyle of tibia.	First metatarsal and first cuneiform.	Plantar flexes and everts foot.	Superficial peroneal nerve.
Peroneus Brevis (*brevis* = short)	Body of fibula.	Fifth metatarsal.	Plantar flexes and everts foot.	Superficial peroneal nerve.
POSTERIOR COMPARTMENT				
Superficial				
Gastrocnemius (*gaster* = belly; *kneme* = leg)	Lateral and medial condyles of femur and capsule of knee.	Calcaneus by way of calcaneal (Achilles) tendon.[a]	Plantar flexes foot and flexes leg.	Tibial nerve.
Soleus (*soleus* = sole of foot)	Head of fibula and medial border of tibia.	Calcaneus by way of calcaneal (Achilles) tendon.	Plantar flexes foot.	Tibial nerve.
Plantaris (*plantar* = sole of foot)	Femur above lateral condyle.	Calcaneus by way of calcaneal (Achilles) tendon.	Plantar flexes foot.	Tibial nerve.
Deep				
Popliteus (*poples* = posterior surface of knee)	Lateral condyle of femur.	Proximal tibia.	Flexes and medially rotates leg.	Tibial nerve.

EXHIBIT 10-23 (*Continued*)

MUSCLE	ORIGIN	INSERTION	ACTION	INNERVATION
Flexor Hallucis Longus (*flexor* = decreases angle at joint)	Lower two-thirds of fibula.	Distal phalanx of great toe.	Plantar flexes and inverts foot and flexes great toe.	Tibial nerve.
Flexor Digitorum Longus (*digitorum* = finger or toe)	Posterior surface of tibia.	Distal phalanges of four outer toes.	Plantar flexes and inverts foot and flexes toes.	Tibial nerve.
Tibialis Posterior (*posterior* = back)	Tibia, fibula, and interosseous membrane.	Second, third, and fourth metatarsals; navicular; all three cuneiforms; and cuboid.	Plantar flexes and inverts foot.	Tibial nerve.

[a] The calcaneal (Achilles) tendon, the ***strongest tendon of the body,*** is able to withstand a 1000-pound force without tearing. Despite this, however, the calcaneal tendon ruptures more frequently than any other tendon because of the tremendous pressures placed on it during competitive sports.

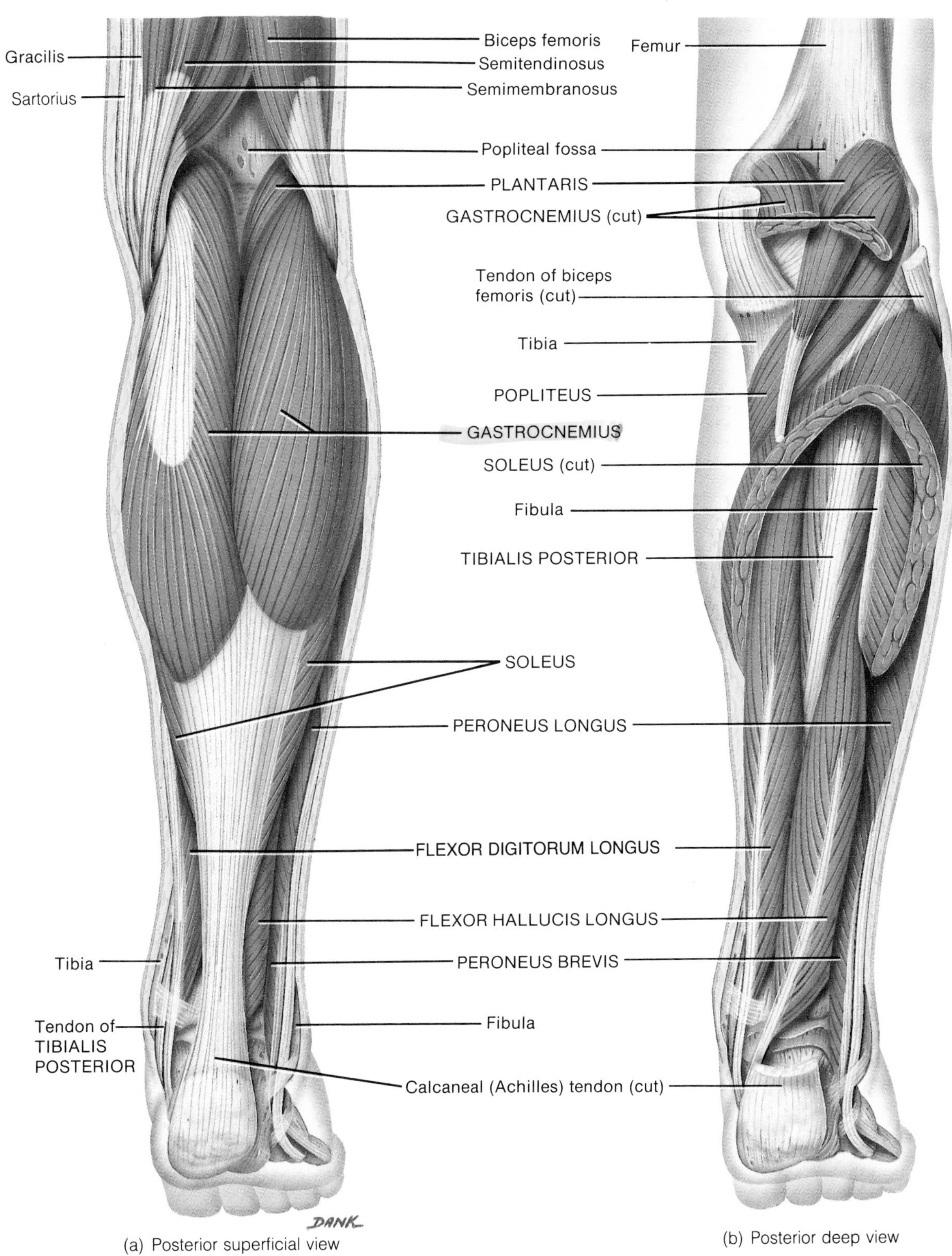

FIGURE 10-24 Muscles that move the foot and toes. (e) and (f) Courtesy of J. A. Gosling, P. F. Harris, et al., *Atlas of Human Anatomy,* Gower Medical Publishing Ltd., 2nd ed., 1991.

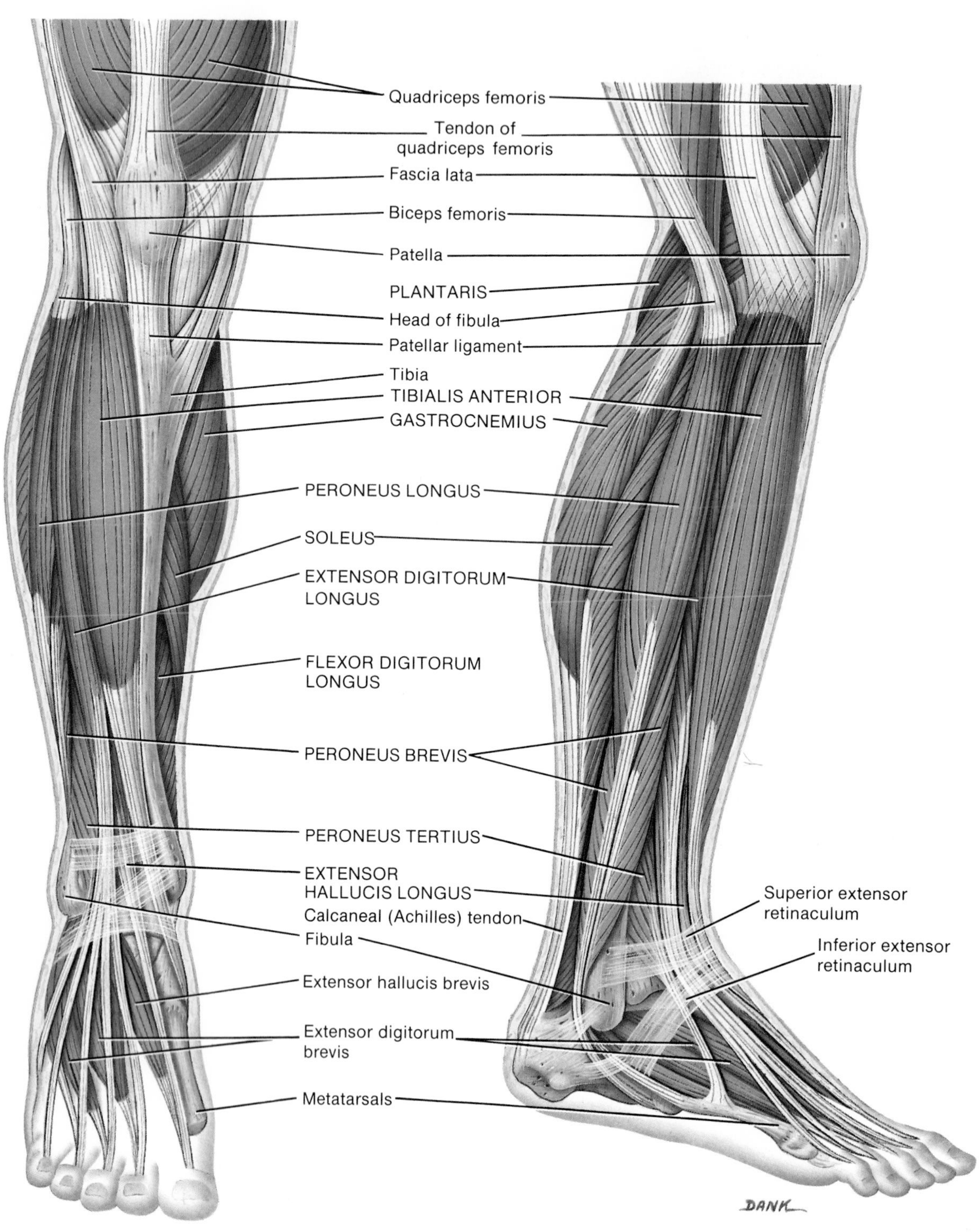

(c) Anterior superficial view

(d) Right lateral superficial view

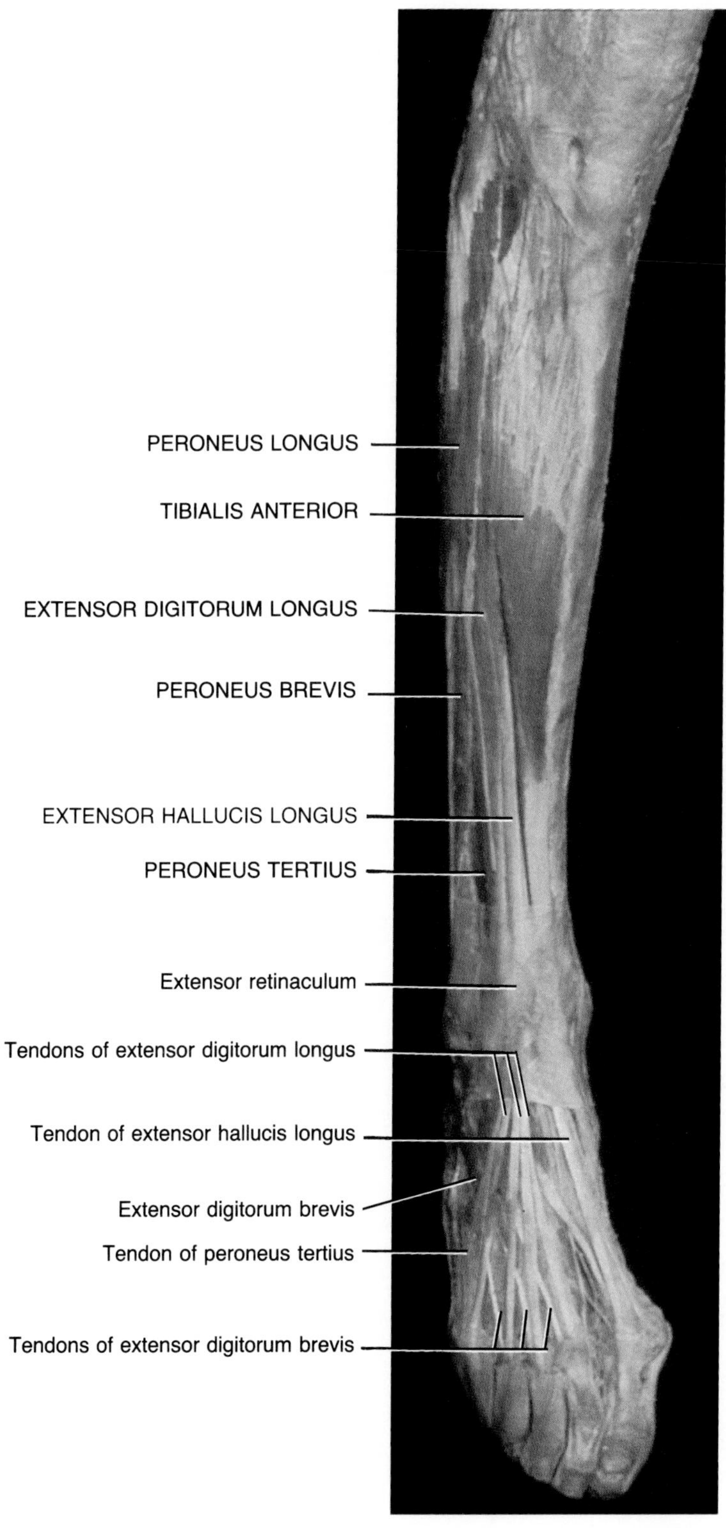

(e) Anterior superficial view

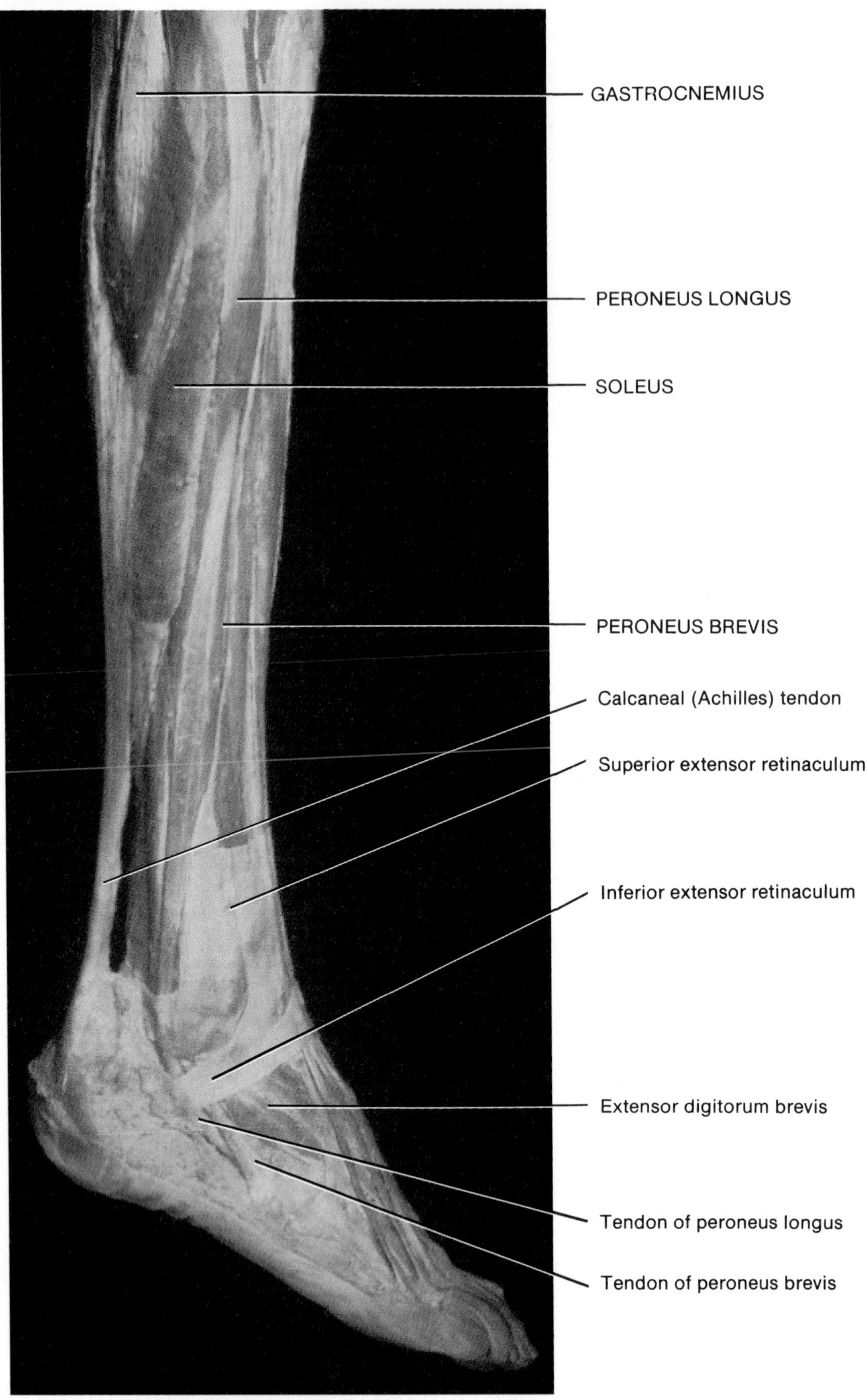

(f) Right lateral superfical view

EXHIBIT 10-24

Intrinsic Muscles of the Foot (Figure 10-25)

Overview: The intrinsic muscles of the foot are, for the most part, comparable to those in the hand. Where the muscles of the hand are specialized for precise and intricate movements, those of the foot are limited to support and locomotion. The deep fascia of the foot forms the ***plantar aponeurosis* (*fascia*)** that extends from the calcaneus to the phalanges. The aponeurosis supports the longitudinal arch of the foot and transmits the flexor tendons of the foot.

The intrinsic musculature of the foot is divided into two groups—dorsal and plantar. There is only one ***dorsal muscle.*** The ***plantar muscles*** are arranged in four layers, the most superficial layer being referred to as the first layer.

MUSCLE	ORIGIN	INSERTION	ACTION	INNERVATION
DORSAL				
Extensor Digitorum Brevis[a] (*extensor* = increases angle at joint; *digit* = finger or toe; *brevis* = short)	Calcaneus.	Tendon of extensor digitorum longus and proximal phalanx of great toe.	Extends first through fourth toes.	Deep peroneal nerve.
PLANTAR				
First (Superficial) Layer				
Abductor Hallucis (*abductor* = moves part away from midline; *hallucis* = hallux or great toe)	Calcaneus and plantar aponeurosis.	Proximal phalanx of great toe.	Abducts great toe and flexes metatarsophalangeal joint.	Medial plantar nerve.
Flexor Digitorum Brevis (*flexor* = decreases angle at joint)	Calcaneus and plantar aponeurosis.	Middle phalanx of second through fifth toes.	Flexes second through fifth toes.	Medial plantar nerve.
Abductor Digiti Minimi (*minimi* = small toe)	Calcaneus and plantar aponeurosis.	Proximal phalanx of small toe.	Abducts and flexes small toe.	Lateral plantar nerve.
Second Layer				
Quadratus Plantae (*quad* = four; *planta* = sole of foot)	Calcaneus.	Tendon of flexor digitorum longus.	Flexes second through fifth toes.	Lateral plantar nerve.
Lumbricals	Tendons of flexor digitorum longus.	Tendons of extensor digitorum longus.	Extend interphalangeal joints and flex metatarsophalangeal joints of second through fifth toes.	Medial and lateral plantar nerves.
Third Layer				
Flexor Hallucis Brevis	Cuboid and third (lateral) cuneiform.	Proximal phalanx of great toe.	Flexes great toe.	Medial plantar nerve.
Adductor Hallucis	Second through fourth metatarsals and ligaments of metatarsophalangeal joints.	Proximal phalanx of great toe.	Adducts and flexes great toe.	Lateral plantar nerve.
Flexor Digiti Minimi Brevis	Fifth metatarsal.	Proximal phalanx of small toe.	Flexes small toe.	Lateral plantar nerve.

[a] Not illustrated.

EXHIBIT 10-24 (Continued)

MUSCLE	ORIGIN	INSERTION	ACTION	INNERVATION
Fourth (Deep) Layer				
Dorsal Interossei	Adjacent side of metatarsals.	Proximal phalanges, both sides of second toe, lateral side of third and fourth toes.	Abduct toes and flex proximal phalanges.	Lateral plantar nerve.
Plantar Interossei	Third, fourth, and fifth metatarsals.	Proximal phalanges of same toes.	Adduct third, fourth, and fifth toes and flex proximal phalanges.	Lateral plantar nerve.

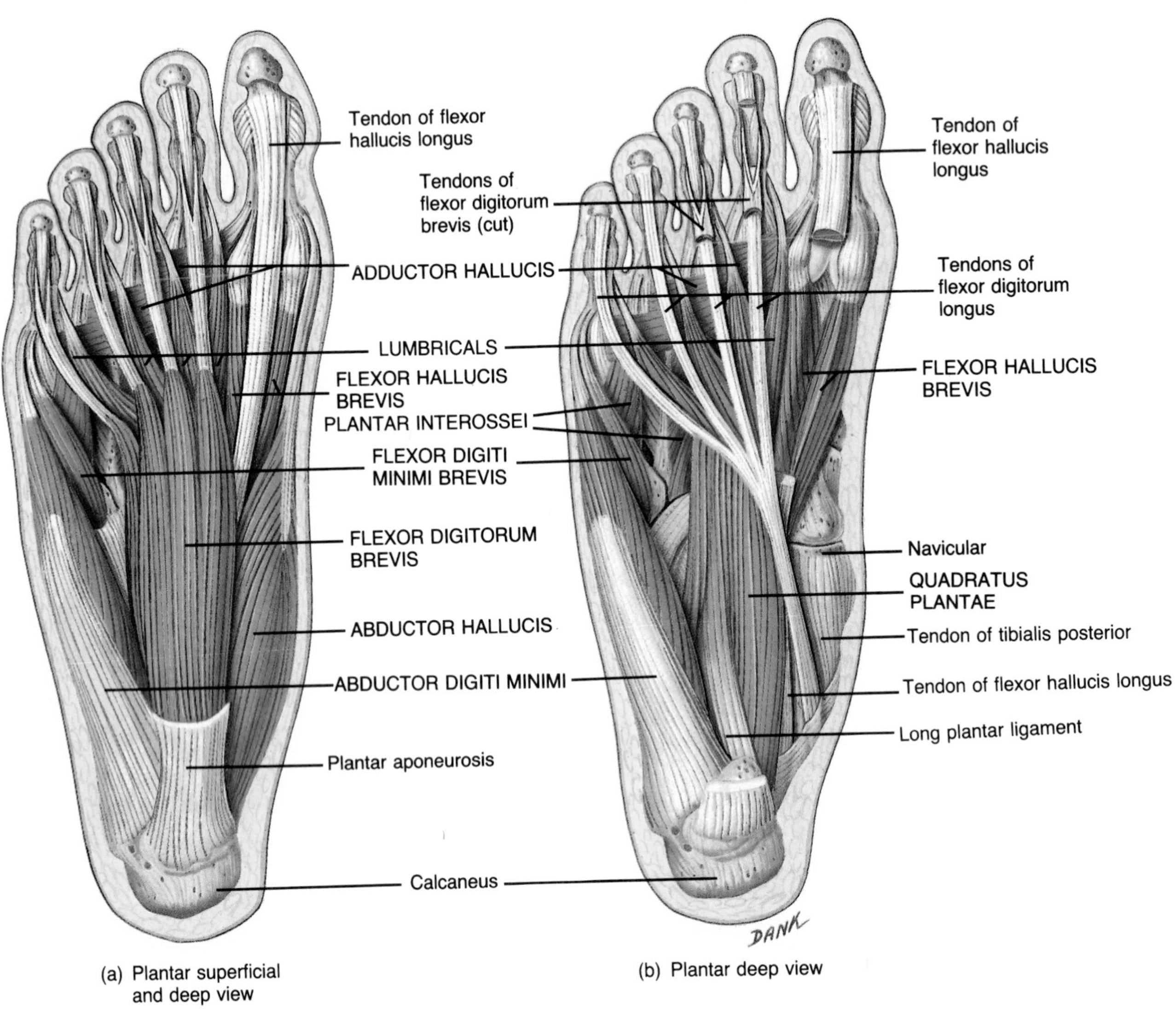

FIGURE 10-25 Intrinsic muscles of the foot. (e) Courtesy of J. A. Gosling, P. F. Harris, et al., *Atlas of Human Anatomy,* Gower Medical Publishing Ltd., 2nd ed., 1991.

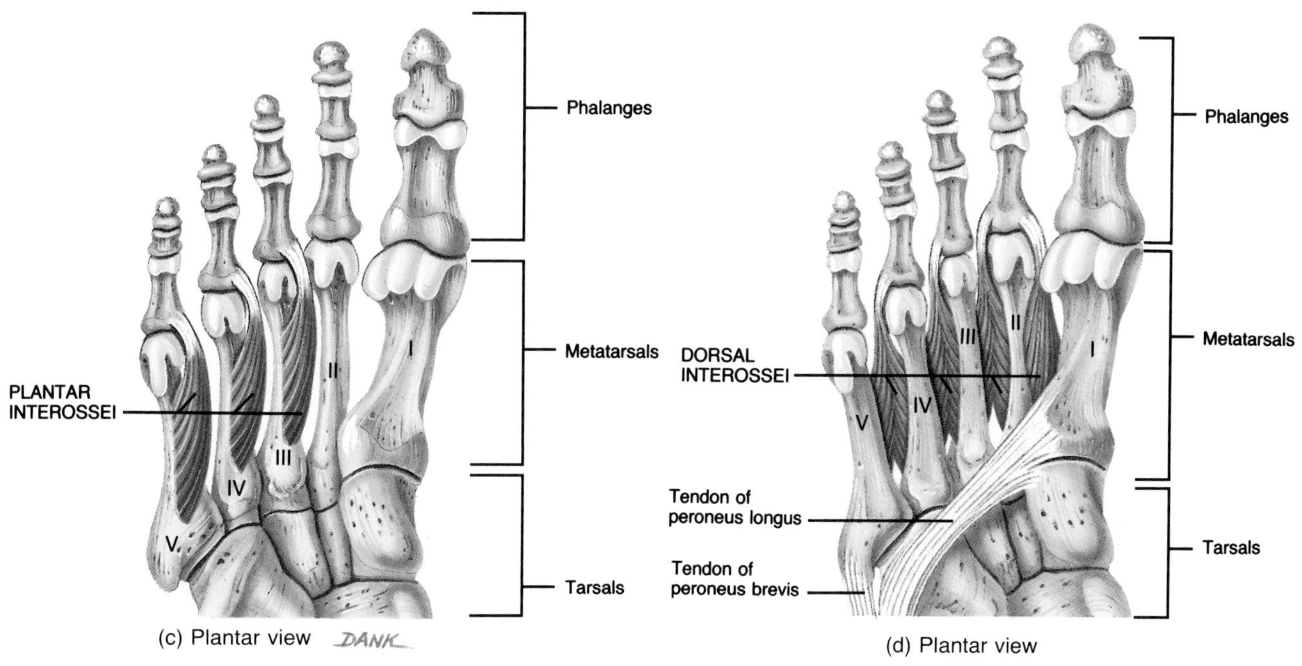

(c) Plantar view

(d) Plantar view

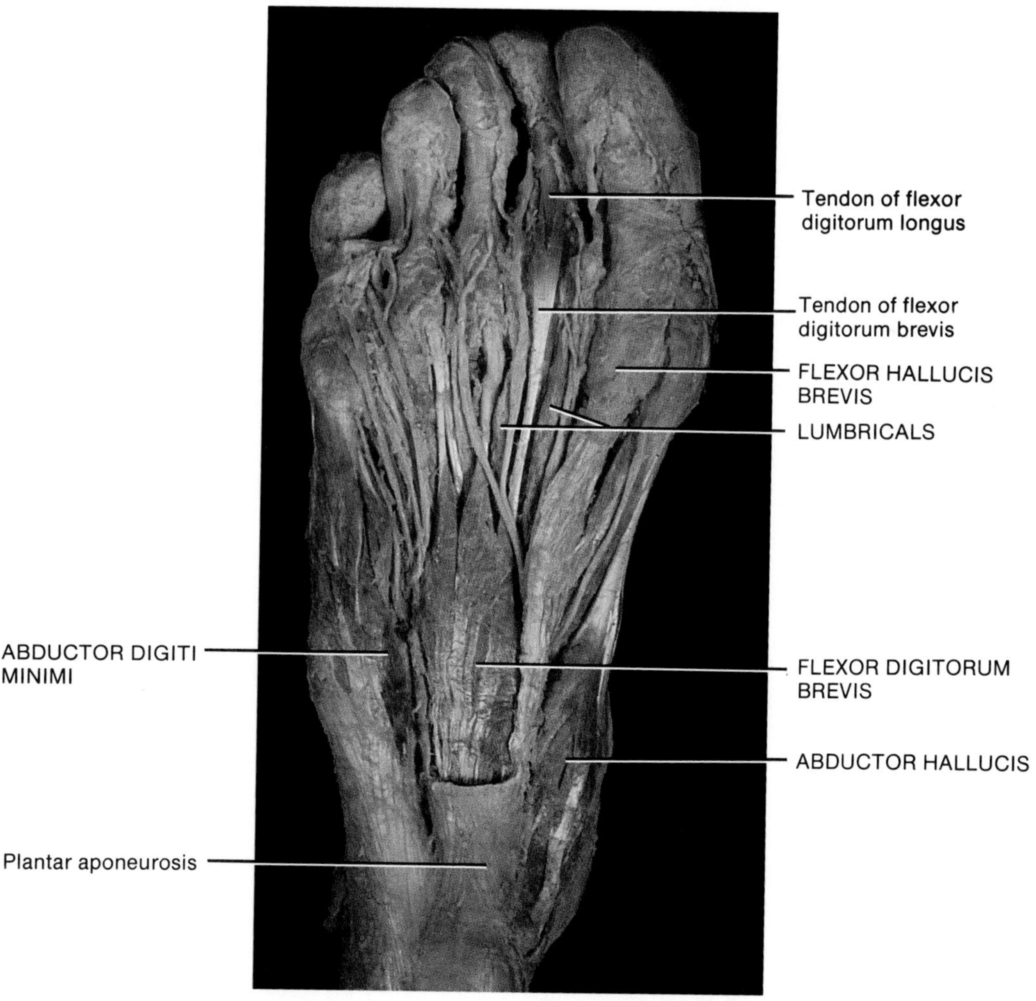

(e) Plantar superficial and deep view

STUDY OUTLINE

How Skeletal Muscles Produce Movement (p. 237)

1. Skeletal muscles produce movement by pulling on bones.
2. The attachment to the stationary bone is the origin. The attachment to the movable bone is the insertion.
3. Bones serve as levers and joints serve as fulcrums. The lever is acted on by two different forces: resistance and effort.
4. Levers are categorized into three types—first class, second class, and third class—according to the position of the fulcrum, effort, and resistance on the lever.
5. Fascicular arrangements include parallel, convergent, pennate, and circular. Fascicular arrangement is correlated with the power of a muscle and the range of movement.
6. The prime mover produces the desired action. The antagonist produces an opposite action. The synergist assists the prime mover by reducing unnecessary movement. The fixator stabilizes the origin of the prime mover so that it can act more efficiently.

Naming Skeletal Muscles (p. 240)

1. Skeletal muscles are named on the basis of distinctive criteria: direction of fibers, location, size, number of origins (or heads), shape, origin and insertion, and action.

Principal Skeletal Muscles (p. 240)

1. The principal skeletal muscles of the body are grouped according to region in Exhibits 10-2 through 10-24.

REVIEW QUESTIONS

1. What is meant by the muscular system? (p. 237)
2. Using the terms *origin, insertion,* and *belly* in your discussion, describe how skeletal muscles produce body movements by pulling on bones. (p. 238)
3. What is a lever? Fulcrum? Apply these terms to the body and indicate the nature of the forces that act on levers. Describe the three classes of levers and provide one example for each in the body. (p. 238)
4. Describe the various arrangements of fasciculi. How is fascicular arrangement correlated with the strength of a muscle and its range of movement? (p. 239)
5. Define the role of the prime mover (agonist), antagonist, synergist, and fixator in producing body movements. (p. 239)
6. Select at random several muscles presented in Exhibits 10-2 through 10-24 and see whether you can determine the criterion or criteria employed for naming each. In addition, refer to the prefixes, suffixes, roots, and definitions in each exhibit as a guide. Select as many muscles as you wish, as long as you feel you understand the concept involved.
7. What muscles would you use to do the following: (a) frown, (b) pout, (c) show surprise, (d) show your upper teeth, (e) pucker your lips, (f) squint, (g) blow up a balloon, (h) smile? (p. 244)
8. What are the principal muscles that move the mandible? Give the function of each. (p. 248)
9. What would happen if you lost tone in the masseter and temporalis muscles? (p. 248)
10. What extrinsic muscles move the eyeball? In which direction does each muscle move the eyeball? (p. 250)
11. Describe the action of each of the muscles acting on the tongue. (p. 252)
12. What tongue, facial, and mandibular muscles would you use when chewing food? (p. 252)
13. What muscles tense and elevate the soft palate? (p. 253)
14. What muscles constrict the pharynx? What muscle dilates the pharynx? Distinguish the circular and longitudinal layers. (p. 254)
15. Describe the muscles involved, and their actions, in moving the hyoid bone. (p. 257)
16. Describe the actions of the extrinsic and intrinsic muscles of the larynx. (p. 258)
17. What muscles are responsible for moving the head? And how do they move the head? (p. 260)
18. What muscles would you use to signify "yes" and "no" by moving your head? (p. 260)
19. Describe the composition of the anterolateral and posterior abdominal wall. (p. 262)
20. What muscles accomplish compression of the abdominal wall? (p. 262)
21. What are the principal muscles involved in breathing? What are their actions? (p. 263)
22. Describe the actions of the muscles of the pelvic floor. What is the pelvic diaphragm? (p. 265)
23. Describe the actions of the muscles of the perineum. What is the urogenital diaphragm? (p. 266)
24. In what directions is the pectoral (shoulder) girdle drawn? What muscles accomplish these movements? (p. 268)
25. What muscles are used to raise your shoulders, lower your shoulders, join your hands behind your back, and join your hands in front of your chest? (p. 268)
26. What movements are possible at the shoulder joint? What muscles accomplish these movements? (p. 270)
27. Distinguish axial and scapular muscles involved in moving the arm. (p. 270)
28. What muscles move the arm? In which directions do these movements occur? (p. 270)
29. Organize the muscles that move the forearm into flexors and extensors. What muscles move the forearm and what actions are used when striking a match? (p. 275)
30. Discuss the various movements possible at the wrist, hand, and fingers. What muscles accomplish these movements? (p. 279)
31. How many muscles and actions of the wrist, hand, and fingers that are used when writing can you list? (p. 279)
32. What is the flexor retinaculum? Extensor retinaculum? (p. 276)
33. What muscles form the thenar eminence? What are their functions? (p. 282)

34. What muscles form the hypothenar eminence? What are their functions? (p. 282)
35. Discuss the various muscles and movements of the vertebral column. How are the muscles grouped? (p. 285)
36. What muscles accomplish movements of the thigh? What actions are produced by these muscles? What is the iliotibial tract? (p. 289)
37. Organize the muscles that act on the leg into medial, anterior, and posterior compartments. What is the popliteal fossa? (p. 294)
38. What muscles act at the knee joint? What kinds of movements do these muscles perform? (p. 296)
39. Determine the muscles and their actions listed in Exhibit 10-22 that you would use to climb a ladder to a diving board, dive into the water, swim the length of a pool, and then sit at pool side. (p. 296)
40. Name the muscles that plantar flex, evert, and dorsiflex the foot. What is the superior extensor retinaculum? Inferior extensor retinaculum? (p. 297)
41. How are the intrinsic muscles of the foot organized? (p. 300)
42. In which directions are the toes moved? What muscles bring about these movements? (p. 300)

SELF QUIZ

Complete the following:

1. In flexion, your forearm serves as a rigid rod, or ________, that moves about a fixed point, called a ________ (your elbow joint, in this case).
2. The attachment of a muscle to the bone that moves is called the ________ end of the muscle.
3. Hyperextend your head as if to look at the sky. The weight of your face and jaw serves as the ((E)? △F? [R]?), while your neck muscles provide the ((E)? △F? [R]?). The fulcrum is the joint between the atlas and ________ bone. This is an example of a ________-class lever.
4. A muscle that contracts to cause the desired action is called the ________. Muscles that assist or cooperate with the muscle that causes the desired action are known as ________.
5. Name the skeletal muscle that:
   a. produces a frowning expression: ________
   b. allows you to blow air out of your mouth and produce a sucking action: ________
6. Essentially, all the muscles controlling facial expression receive nerve impulses via the ________ nerve, which is cranial nerve (III? V? VII?).
7. Two large muscles help you to close your mouth forcefully, as in chewing. Both of these act by (lowering the maxilla? elevating the mandible?). The ________ covers your temple and the ________ covers the ramus of the mandible.
8. All the rectus muscles move the eyeball in the direction (opposite to? the same as?) that given in the muscle name. For example, the superior rectus moves the eyeball (superiorly? inferiorly?).
9. Contraction of the diaphragm flattens the dome, causing the size of the thorax to (increase? decrease?), as occurs during (inspiration? expiration?).
10. The deep transverse perineus, urethral sphincter, and a fibrous membrane constitute the ________ diaphragm.
11. Muscles that move the pectoral (shoulder) girdle, originate on the axial skeleton and insert on the clavicle are ________.
12. The majority of muscles that act on the thigh originate in the pelvic (hip) girdle and insert on the ________.
13. Whereas the ________ muscles act on the thumb, the ________ muscles act on the little finger.
14. The deep fascia organized into fibrous bands through which flexor tendons of the digits and wrist pass is the ________ (transverse carpal ligament).
15. Examine your own forearm, palm, and fingers. There is more muscle mass on the (anterior? posterior?) surface. You therefore have more muscles that can (flex? extend?) your wrist and fingers.
16. Of the muscles that move the vertebral column, the deepest group is the ________ muscles.
17. Quadriceps femoris is the main muscle mass on the (anterior? posterior?) surface of the thigh. The name of this muscle mass indicates it has ________ heads of origin. All converge to insert on the ________ bone by means of the patellar ligament. All therefore cause (flexion? extension?) of the leg. Feel their contraction as you extend your own leg. The hamstrings are (synergists? antagonists?) of the quadriceps. Hamstrings are located on the (anterior? posterior?) surface of the thigh.
18. Of the muscles that move the foot and toes, those of the posterior compartment are innervated by the ________ nerve.
19. All three of the hamstring muscles originate on the ________ and insert just below the knee. The actions they cause are (flexion? extension?) of the thigh and (flexion? extension?) of the leg.
20. The deep fascia of the foot forms the ________ aponeurosis that supports the longitudinal arch and transmits the flexor tendons of the foot.

Choose the one best answer to these questions:

___ 21. Which generalization concerning movement by skeletal muscles is *not* true?
A. muscles produce movements by pulling on bones; B. during contractions the two articulating bones move equally; C. the tendon attachment to the more stationary bone is the origin; D. bones serve as levers and joints as fulcrums of the levers; E. in the extremities, the insertion is usually distal.

___ 22. The muscle
A. that is used to elevate the scapula and extend the head is the trapezius; B. that is considered part of the abdominal muscle group is the latissimus dorsi; C. that is found on the lateral and posterior part of the thigh is the sartorius; D. that is considered one of the hamstring muscles is the rectus femoris; E. that is used to extend the elbow and abduct the arm is the triceps brachii.

___ **23.** The action of the pectoralis major muscle is to A. abduct the arm and rotate the arm laterally; B. flex, adduct, and rotate the arm medially; C. adduct the arm and rotate the arm laterally; D. abduct the arm and rotate the arm medially; E. abduct and raise the arm.

___ **24.** What muscle is located on the posterior surface of the arm?
A. biceps brachii; B. biceps femoris; C. pronator teres; D. vastus lateralis; E. triceps brachii.

___ **25.** Of these muscles, which one adducts the thigh?
A. gluteus medius; B. superior gemellus; C. inferior gemellus; D. pectineus; E. obturator internus.

26. For each of the following, indicate the type of clue that each part of the name gives:

___ **a.** rectus abdominis	**A.** action
___ **b.** gluteus maximus	**B.** direction of fibers
___ **c.** biceps brachii	**C.** location
___ **d.** sternocleidomastoid	**D.** number of heads or origins
___ **e.** adductor longus	**E.** points of attachment of origin and insertion
	F. size or shape

27. Match the following:

___ **a.** located superior to the hyoid (suprahyoid), these form the floor of the oral cavity

___ **b.** "strap muscles" that are infrahyoid, these muscles cover the anterior of the larynx and trachea; they permit you to elevate your thyroid cartilage (Adam's apple") during swallowing to prevent food from entering your larynx (try it)

___ **c.** forming part of the wall of the pharynx, these muscles elevate the larynx and pharynx during swallowing; and they close off the nasopharynx (so that food does not back up into the nasal cavity) during swallowing

___ **d.** three muscles that squeeze the pharynx to propel a bolus of food into the esophagus during swallowing

___ **e.** muscles permitting tongue movements

A. styloglossus, hypoglossus, and genioglossus

B. superior, middle, and inferior constrictors

C. stylopharyngeus, salpingopharyngeus, and palatopharyngeus

D. diagastric, stylohyoid, mylohyoid, and geniohyoid

E. thyrohyoid, sternohyoid, omohyoid, and sternothyroid

Surface Anatomy

11

STUDENT OBJECTIVES

1. Define surface anatomy.
2. Identify the principal regions of the human body.
3. Describe the surface anatomy features of the head.
4. Describe the surface anatomy features of the neck.
5. Describe the surface anatomy features of the trunk.
6. Describe the surface anatomy features of the upper extremities.
7. Describe the surface anatomy features of the lower extremities.

CHAPTER OUTLINE

- **Head**
- **Neck**
- **Trunk**
- **Upper Extremity**
- **Lower Extremity**

In Chapter 1, several branches of anatomy were defined and their importance to an understanding of the structure of the body was noted. Now that you have a fairly good idea how the body is organized, we can take a closer look at surface anatomy. Very simply, ***surface anatomy*** is the study of the form and markings of the surface of the body. A knowledge of surface anatomy will help you identify certain superficial structures through visual inspection or palpation through the skin. ***Palpation*** means to feel with the hand. Knowledge of surface anatomy is important in health-related activities such as taking pulse, listening to internal organs, drawing blood, and inserting needles and tubes.

To introduce you to surface anatomy, we shall consider some key features of each of the principal regions of the body. These regions may be reviewed in Figure 1-2.

HEAD

The ***head*** (***cephalic region,*** or ***caput***) contains the brain and sense organs (eyes, ears, nose, tongue) and is divided into the cranium and face. The ***cranium*** (***skull,*** or ***brain case***) consists of a ***frontal region*** (front of skull, which

EXHIBIT 11-1
Surface Anatomy of the Head

HEAD (Figure 11-1)

1. ***Orbital (ocular) region.*** Includes eyeball, eyelids, and eyebrows.
2. ***Nasal region.*** Nose.
3. ***Buccal region.*** Cheek.
4. ***Oral region.*** Mouth.
5. ***Mental region.*** Anterior portion of mandible.
6. ***Occipital region.*** Base of skull.
7. ***Auricular region.*** Ear.
8. ***Parietal region.*** Crown of skull.
9. ***Temporal region.*** Side of skull.
10. ***Frontal region.*** Front of skull.

EYE (Figure 11-2)

1. ***Pupil.*** Opening of center of iris of eyeball for light transmission.
2. ***Iris.*** Circular pigmented muscular membrane behind cornea.
3. ***Sclera.*** "White" of eye; a coat of fibrous tissue that covers entire eyeball except for cornea.
4. ***Conjunctiva.*** Membrane that covers exposed surface of eyeball and lines eyelids.
5. ***Palpebrae (eyelids).*** Folds of skin and muscle lined on their inner aspect by conjunctiva.
6. ***Palpebral fissure.*** Space between eyelids when they are open; when the eye is closed, it lies just below the level of the pupil.
7. ***Medial commissure.*** Site of union of upper and lower eyelids near nose.
8. ***Lateral commissure.*** Site of union of upper and lower eyelids away from nose.
9. ***Lacrimal caruncle.*** Fleshy, yellowish projection of medial commissure that contains modified sudoriferous (sweat) and sebaceous (oil) glands.

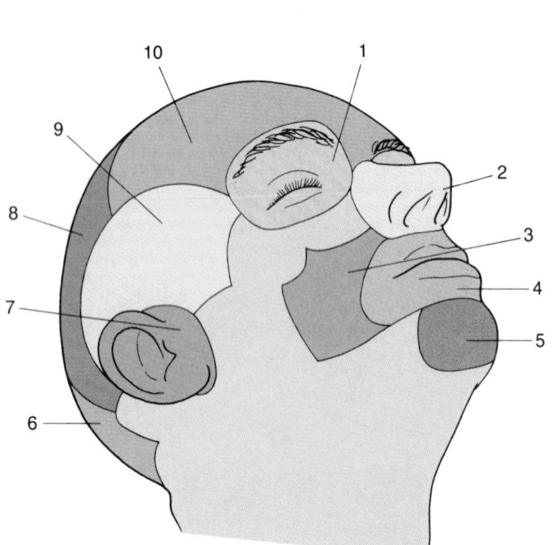

FIGURE 11-1 Regions of the head.

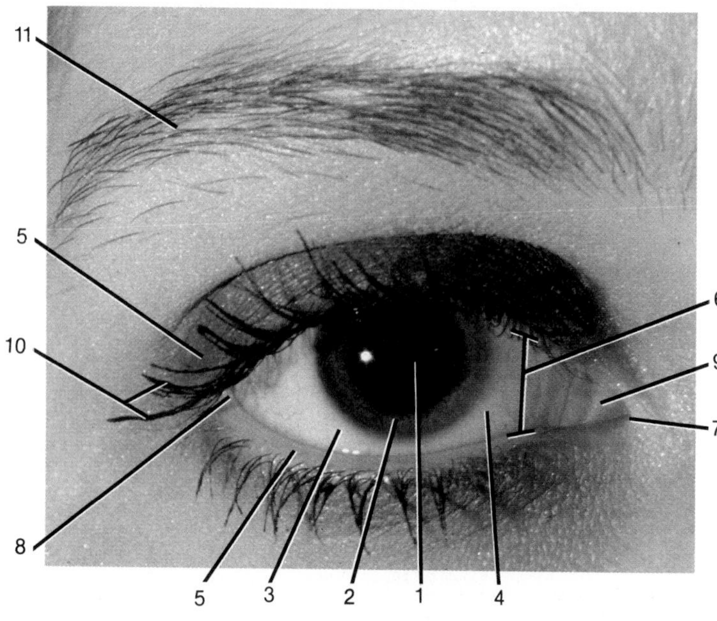

FIGURE 11-2 Surface anatomy of the right eye. (Copyright © 1982 by Gerard J. Tortora. Courtesy of Lynne Borghesi.)

includes the forehead), ***parietal region*** (crown of skull, or vertex), ***temporal region*** (side of skull, or tempora), and ***occipital region*** (base of skull, or occiput).

The ***face*** is subdivided into several regions. Among the prominent ones are the ***orbital*** or ***ocular region,*** which includes the eyeballs (bulbi oculorum), eyebrows (supercilia), and eyelids (palpebrae); ***nasal region*** (nose); ***oral region*** (mouth); ***mental region*** (anterior portion of mandible); ***buccal region*** (cheek); and ***auricular region*** (ear).

Examine Exhibit 11-1. It contains illustrations of various surface features of the head with an accompanying description of each of the features.

NECK

The ***neck*** (***collum***) is divided into an ***anterior cervical region,*** two ***lateral cervical regions,*** and a ***posterior region*** (***nucha***).

The most prominent structure in the midline of the anterior cervical region is the ***thyroid cartilage*** (***Adam's apple***) of the larynx. Just superior to it, the ***hyoid bone*** can be palpated. Inferior to the thyroid cartilage, the ***cricoid cartilage*** of the larynx can be felt. This cartilage is used as a landmark in locating the rings of cartilage in the trachea

10. ***Eyelashes.*** Hairs on margins of eyelids, usually arranged in two or three rows.
11. ***Supercilia (eyebrows).*** Several rows of hairs superior to upper eyelids.

EAR (Figure 11-3)

1. ***Auricle.*** Portion of external ear not contained in head; also called pinna.
2. ***Tragus.*** Cartilaginous projection anterior to external auditory canal.
3. ***Antitragus.*** Cartilaginous projection opposite tragus.
4. ***Concha.*** Hollow of auricle.
5. ***Helix.*** Superior and posterior free margin of auricle.
6. ***Antihelix.*** Semicircular ridge superior and posterior to tragus.
7. ***Triangular fossa.*** Depression in superior portion of antihelix.
8. ***Lobule.*** Inferior portion of auricle devoid of cartilage.
9. ***External auditory canal (meatus).*** Canal about 3 cm (1 in.) long extending from external ear to eardrum (tympanic membrane).

NOSE AND LIPS (Figure 11-4)[a]

1. ***Root.*** Superior attachment of nose at forehead located between eyes.
2. ***Apex.*** Tip of nose.
3. ***Dorsum nasi.*** Rounded anterior border connecting root and apex; in profile, may be straight, convex, concave, or wavy.
4. ***External naris.*** External opening into nose.
5. ***Bridge.*** Superior portion of dorsum nasi, superficial to nasal bones.
6. ***Lips.*** Upper and lower fleshy borders of oral cavity.

[a] Details of the mouth are shown in Figure 23-4.

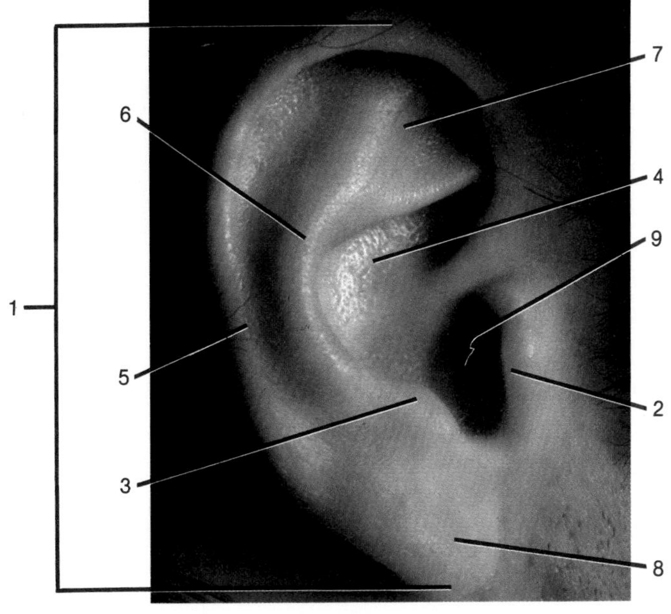

FIGURE 11-3 Surface anatomy of the right ear. (Copyright © 1982 by Gerard J. Tortora. Courtesy of James Borghesi.)

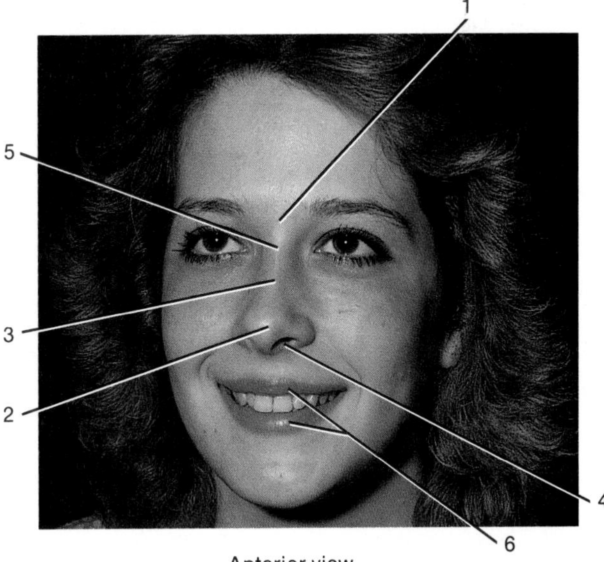

FIGURE 11-4 Surface anatomy of the nose and lips. (Copyright © 1982 by Gerard J. Tortora. Courtesy of Lynne Borghesi.)

(windpipe) when performing a tracheostomy (see Figure 22-4). The incision is made through the second, third, or fourth tracheal rings and a tube is inserted to assist breathing.

A major portion of the lateral cervical regions is formed by the ***sternocleidomastoid muscles.*** Each muscle extends from the mastoid process of the temporal bone, felt as a bump behind the auricle of the ear, to the sternum and clavicle. The carotid (neck) pulse may be detected by palpating the carotid artery along the anterior border of the sternocleidomastoid muscle. Each sternocleidomastoid muscle divides its portion of the neck into an anterior triangle and a posterior (lateral) triangle (see Exhibit 10-10). The ***anterior triangle*** is bordered superiorly by the mandible, inferiorly by the sternum, medially by the cervical midline, and laterally by the anterior border of the sternocleidomastoid muscle. The ***posterior (lateral) triangle*** is bordered inferiorly by the clavicle, anteriorly by the posterior border of the sternocleidomastoid muscle, and posteriorly by the anterior border of the trapezius muscle.

A muscle that extends downward and outward from the base of the skull and occupies a portion of the lateral cervical region is the ***trapezius muscle.*** A "stiff neck" is frequently associated with an inflammation of this muscle.

A very prominent vein that runs along the lateral surface of the neck is the ***external jugular vein.*** It is readily seen if you are angry or if your collar is too tight. The locations of a few surface features of the neck are shown in Exhibit 11-2.

EXHIBIT 11-2

Surface Anatomy of the Neck (Figure 11-5)

1. ***Anterior triangle of neck.*** Bordered superiorly by mandible, inferiorly by sternum, medially by cervical midline, and laterally by anterior border of sternocleidomastoid muscle. (See also Exhibit 10-10.)
2. ***Posterior triangle of neck.*** Bordered inferiorly by clavicle, anteriorly by posterior border of sternocleidomastoid muscle, and posteriorly by anterior border of trapezius muscle. (See also Exhibit 10-10.)
3. ***Trapezius muscle.*** Occupies a portion of lateral surface of neck; helps raise, lower, and draw shoulders backward and extends head.
4. ***Sternocleidomastoid muscle.*** Forms major portion of lateral surface of neck; flexes head and rotates it to opposite side; carotid (neck) pulse is felt along its anterior border.
5. ***Cricoid cartilage.*** Inferior laryngeal cartilage that attaches larynx to trachea; can be palpated by running your fingertip down from your chin over the thyroid cartilage (after you pass the cricoid cartilage, your fingertip sinks in); used as a landmark for a tracheostomy.
6. ***Thyroid cartilage (Adam's apple).*** Triangular laryngeal cartilage in the midline of the anterior cervical region.
7. ***Hyoid bone.*** Lies just superior to the thyroid cartilage opposite the superior border of C4; it is the first resistant structure palpated in the midline below the chin.

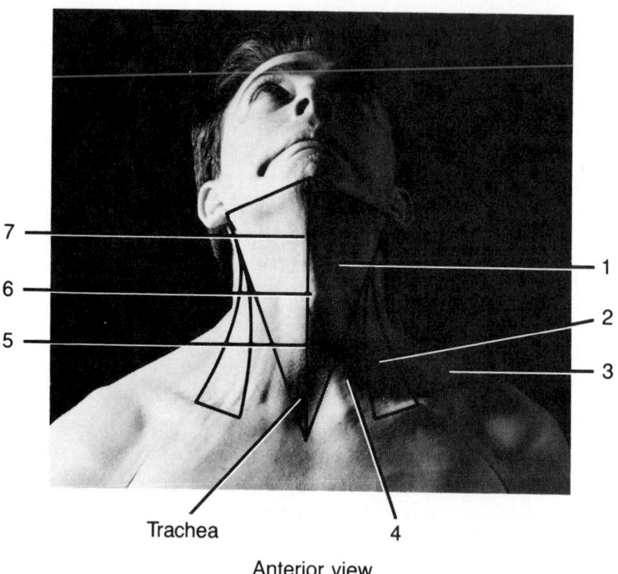

FIGURE 11-5 Surface anatomy of the neck. (Courtesy of Carroll H. Weiss, Camera M. D. Studios, Inc.)

TRUNK

The ***trunk*** is divided into the back, chest, abdomen, and pelvis.

One of the most striking surface features of the ***back (dorsum)*** is the ***vertebral spines,*** the posteriorly pointed projections of the vertebrae. A very prominent vertebral spine is the ***vertebra prominens*** of the seventh cervical vertebra. It is easily seen when the head is flexed. Another easily identifiable surface landmark of the back is the ***scapula.*** In fact, several parts of the scapula (axillary border, vertebral border, inferior angle, spine, and acromion) may also be seen or palpated. In lean individuals the ***ribs*** may also be seen. The vertebral border of the scapula crosses ribs 2 to 7. Among the superficial muscles of the back that can be seen are the ***latissimus dorsi, erector spinae (sacrospinalis), infraspinatus, trapezius,*** and ***teres major.*** The ***triangle of auscultation*** is formed by the latissimus dorsi muscle, trapezius muscle, and vertebral border of the scapula. The space between the muscles permits respiratory sounds to be clearly heard with a stethoscope. These features, as well as the other surface features of the back, are described in Exhibit 11-3.

The ***chest (thorax)*** presents a number of anatomical landmarks. At its superior region are the ***clavicles.*** The ***sternum*** lies in the midline of the chest and is divisible into a superior manubrium, a middle body, and an inferior xiphoid process. Its superior border attaches to the clavicles. Between the medial ends of the clavicles is a depression on the superior surface of the sternum called the ***suprasternal notch.*** The trachea can be palpated posterior to the suprasternal notch. The ***sternal angle*** is formed by a junction line between the manubrium and body of the sternum and is palpable under the skin. It locates the costal cartilage of the second rib and is the most reliable surface landmark of the chest.

EXHIBIT 11-3
Surface Anatomy of the Trunk

BACK (Figure 11-6)

1. ***Vertebral spines.*** Posteriorly pointed projections of vertebrae. The spine of C2 is the first bony prominence encountered when the finger is drawn downward in the midline; the spine of C7 is the upper of the two prominences found at the base of the neck; the spine of T1 is the lower prominence at the base of the neck; the spine of T3 is about at the same level as the spine of the scapula; the spine of T7 is about opposite the inferior angle of the scapula; a line passing through the highest points of the iliac crests, called the supracristal line, passes through the spine of L4.
2. ***Scapula.*** Shoulder blade (label line is on vertebral border). The scapula lies on the back over the second to seventh ribs.
3. ***Latissimus dorsi muscle.*** Broad muscle of back that helps draw shoulders backward and downward.
4. ***Erector spinae (sacrospinalis) muscle.*** Parallel to vertebral column on either side of midline between the twelfth rib and iliac crest; moves vertebral column in various directions.
5. ***Infraspinatus muscle.*** Located inferior to spine of scapula; helps laterally rotate humerus.
6. ***Trapezius muscle.*** Extends from cervical and thoracic vertebrae to spine of scapula and lateral end of clavicle; occupies portion of lateral surface of neck (see Figure 11-5) and forms posterior border of posterior triangle of neck; helps raise, lower, and draw shoulders backward and extend head.
7. ***Teres major muscle.*** Located inferior to infraspinatus; together with tendon of latissimus dorsi muscle, helps form inferior border of posterior wall of axilla (posterior axillary fold); helps extend, adduct, and medially rotate humerus.
8. ***Posterior axillary fold.*** Formed by the latissimus dorsi and teres major muscles; can be palpated between the finger and thumb.
9. ***Triangle of auscultation.*** Triangle formed by latissimus dorsi muscle, trapezius muscle, and vertebral border of scapula. The space between the muscles in the region permits respiratory sounds to be heard clearly with a stethoscope. (Not illustrated here).

CHEST (Figure 11-7)

1. ***Clavicle.*** Collarbone. Visible at the junction of the neck and thorax.
2. ***Suprasternal notch of sternum.*** Depression on superior border of manubrium of sternum between medial ends of clavicles; trachea can be palpated in the notch.
3. ***Manubrium of sternum.*** Superior portion of sternum at the same levels as the bodies of the third and fourth thoracic vertebrae and anterior to the arch of the aorta.
4. ***Sternal angle of sternum.*** Formed by junction between manubrium and body of sternum, located about 4 cm (1½ in.) below suprasternal notch. It locates the costal cartilage of the second rib and is the starting point from which ribs are counted.
5. ***Body of sternum.*** Midportion of sternum anterior to heart and the vertebral bodies of T5–T8.
6. ***Xiphoid process of sternum.*** Inferior portion of sternum, between the seventh costal cartilages.
7. ***Costal margin.*** Inferior edges of costal cartilages of ribs 7 through 10. The first costal cartilage lies below the medial end of the clavicle; the seventh costal cartilage is the lowest to articulate directly with the sternum; the tenth costal cartilage forms the lowest part of the costal margin, when viewed anteriorly.

continued

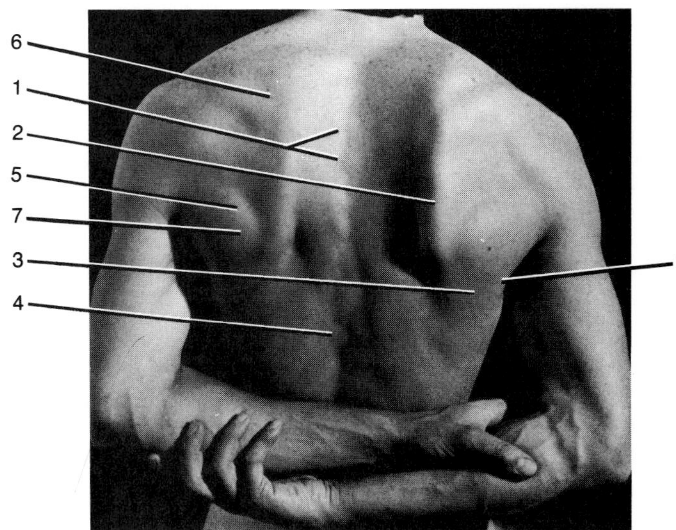

FIGURE 11-6 Surface anatomy of the back. (Courtesy of Victor B. Eichler, copyright © 1980.)

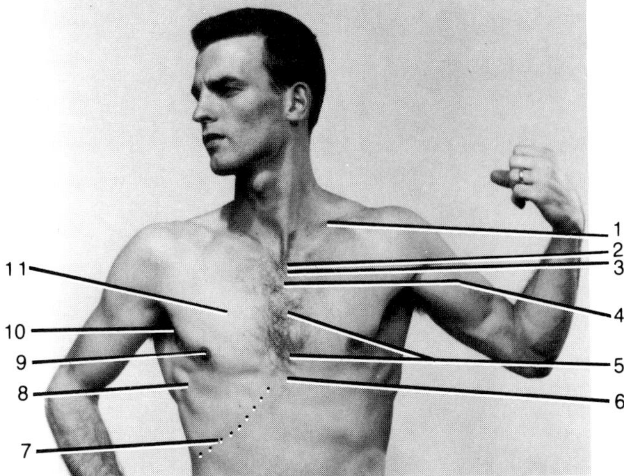

FIGURE 11-7 Surface anatomy of the chest. (Courtesy of Vincent P. Destro, Mayo Foundation.)

EXHIBIT 11-3 (Continued)

8. ***Serratus anterior muscle.*** Inferior and lateral to pectoralis major muscle; helps rotate scapula superiorly and elevate ribs; also illustrated in Figure 11-8.
   Ribs. Form bony cage of thoracic cavity (not illustrated here; see Figure 6-18). The apex beat of the heart in adults is heard in the left fifth intercostal space, just medial to the left midclavicular line.
   Mammary glands. Accessory organs of the female reproductive system located inside the breasts. They overlie the pectoralis major muscle (two-thirds) and serratus anterior muscle (one-third). After puberty, they enlarge to their hemispherical shape, and in young adult females, they extend from the second through sixth ribs and from the lateral margin of the sternum to the midaxillary line (see Figure 25-22).
9. ***Nipples.*** Superficial to fourth intercostal space or fifth rib about 10 cm (4 in.) from the midline in males and most females. The position of the nipples in females is variable depending on the size and pendulousness of the breasts. The right dome of the diaphragm is just inferior to the right nipple, the left dome is about 2–3 cm (1 in.) inferior to the left nipple, and the central tendon is at the level of the junction of the body of the sternum and xiphoid process.
10. ***Anterior axillary fold.*** Formed by the lateral border of the pectoralis major muscle; can be palpated between the fingers and thumb.
11. ***Pectoralis major muscle.*** Principal upper chest muscle; flexes, adducts, and medially rotates humerus. In the male, the inferior border of the muscle forms a curved line leading to the anterior wall of the axilla and serves as a guide to the fifth rib. In the female, the inferior border is mostly covered by the breast.

ABDOMEN AND PELVIS (Figure 11-8)

1. ***Umbilicus.*** Also called navel; previous site of attachment of umbilical cord in fetus. It is level with the intervertebral disc between the bodies of L3 and L4. The abdominal aorta bifurcates into the right and left common iliac arteries anterior to the body of vertebra L4. The inferior vena cava lies to the right of the abdominal aorta and is wider; it arises anterior to the body of vertebra L5.
2. ***External oblique muscle.*** Located inferior to serratus anterior; helps compress abdomen and bend vertebral column laterally. Its lower border is the inguinal ligament, a structure along which hernias frequently occur.
3. ***Rectus abdominis muscle.*** Located just lateral to midline of abdomen; helps compress abdomen and flex vertebral column.
4. ***Linea alba.*** Flat, tendinous raphe forming a furrow along midline between rectus abdominis muscles. The furrow extends from the xiphoid process to the pubic symphysis. It is broad above the umbilicus and narrow below it. The linea alba is a frequently selected site for abdominal surgery since an incision through it severs no muscles and only a few blood vessels and nerves.
5. ***Tendinous intersection.*** Fibrous band that runs transversely or obliquely across the rectus abdominis muscles. One intersection is at the level of the umbilicus, one at the level of the xiphoid process, and one midway between.
6. ***McBurney's point.*** An important landmark at the junction of the lateral and middle thirds of a line joining the umbilicus and anterior superior iliac spine. An oblique incision through McBurney's point is made for appendectomy. Pressure of the finger on McBurney's point produces tenderness in acute appendicitis.
   Pubic symphysis. Anterior joint of hipbones; palpated as a firm resistance in the midline at the inferior portion of the anterior abdominal wall (see Figure 7-7).

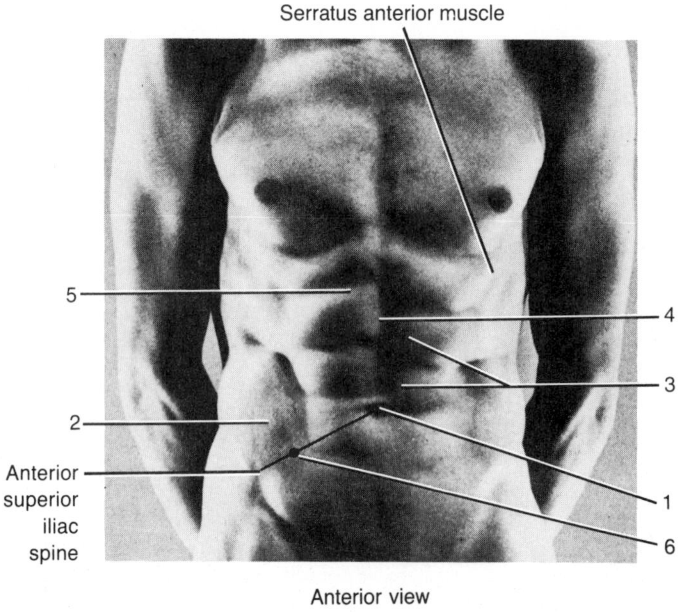

Serratus anterior muscle
5
4
3
2
Anterior superior iliac spine
1
6
Anterior view

FIGURE 11-8 Surface anatomy of the abdomen. (Courtesy of R. D. Lockhard, *Living Anatomy,* Faber & Faber, London, 1974.)

It marks the starting point from which the ribs are counted. At or inferior to the sternal angle and slightly to the right, the trachea *bifurcates* (divides in two) into the left and right primary bronchi. The inferior portion of the sternum, the ***xiphoid process,*** may be palpated. Also visible or palpable are the ***ribs.*** The apex beat of the heart in adults is heard in the left, fifth intercostal span, just medial to the left midclavicular line. The apex of the lung rises as high as the neck of the first rib, superior to the clavicle. The ***costal margins,*** the inferior edges of the costal cartilages of ribs 7 through 10, can also be seen or palpated. The margins are very near the iliac crests, sometimes only 1–2 cm away. Among the more prominent superficial chest muscles are the ***pectoralis major*** and ***serratus anterior.*** These and other surface features of the chest are described in Exhibit 11-3. The surface features of the heart are shown in Figure 13-5.

The ***abdomen*** and ***pelvis*** have already been discussed in terms of their nine regions or four quadrants (see Figures 1-8 and 1-9). External features of the abdomen and pelvis include the ***umbilicus, linea alba, external oblique muscle, rectus abdominis muscle, tendinous intersections,*** and ***pubic symphysis.*** The surface features of the abdomen and pelvis are described in Exhibit 11-3.

UPPER EXTREMITY

The ***upper extremity*** consists of the armpit (axilla), shoulder, arm, elbow, forearm, wrist, and hand.

At the ***shoulder (acromial region),*** moving laterally along the top of the clavicle, it is possible to palpate a slight elevation at the lateral end of the clavicle, the ***acromioclavicular joint.*** Less than 2.5 cm (1 in.) distal to this joint, one can also feel the ***acromion*** of the scapula, which forms the tip of the shoulder. The rounded prominence of the shoulder is formed by the ***deltoid muscle,*** a frequent site for intramuscular injections (Exhibit 11-4).

EXHIBIT 11-4
Surface Anatomy of the Upper Extremity

SHOULDER (Figure 11-9)

1. ***Acromion.*** Expanded end of spine of scapula; forms tip of shoulder; clearly visible in some individuals and can be palpated about 2.5 cm (1 in.) distal to acromioclavicular joint.
2. ***Deltoid muscle.*** Triangular muscle that forms rounded prominence of shoulder; primary action is to abduct arm.

ARM AND ELBOW (Figure 11-10)

1. ***Biceps brachii muscle.*** Forms bulk of anterior surface of arm; helps flex forearm.
2. ***Triceps brachii muscle.*** Forms bulk of posterior surface of arm; helps extend forearm.
3. ***Medial epicondyle.*** Medial projection at distal end of humerus.
4. ***Lateral epicondyle.*** Lateral projection at distal end of humerus.
5. ***Olecranon.*** Projection of proximal end of ulna; forms elbow.
6. ***Cubital fossa.*** Triangular space in anterior region of elbow bounded proximally by an imaginary line between humeral epicondyles, laterally by the medial border of the brachioradialis muscle, and medially by the lateral border of the pronator teres muscle; contains tendon of biceps brachii muscle, brachial artery and its terminal branches (radial and ulnar arteries), and parts of median and radial nerves.
7. ***Median cubital vein*** (not illustrated here; see Figure 14-13a). Crosses cubital fossa obliquely and is useful for veni puncture.
8. ***Brachial artery*** (not illustrated here; see Figure 14-8a). Continuation of axillary artery that passes posterior to coracobrachialis muscle and then medial to biceps brachii muscle. It enters the middle of the cubital fossa and passes

continued

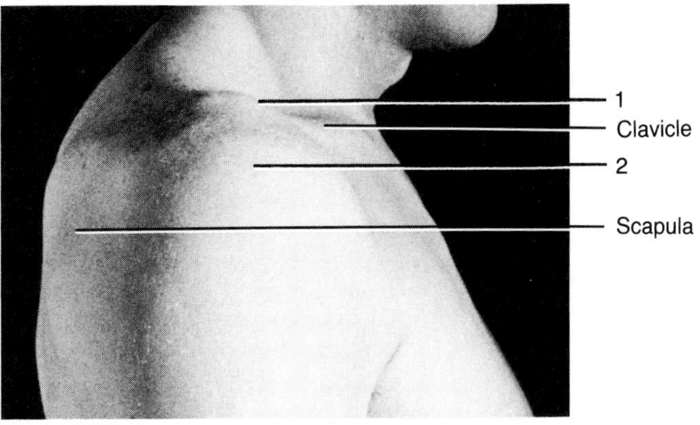

Lateral view

FIGURE 11-9 Surface anatomy of the right shoulder. (Courtesy of Victor B. Eichler, copyright © 1980.)

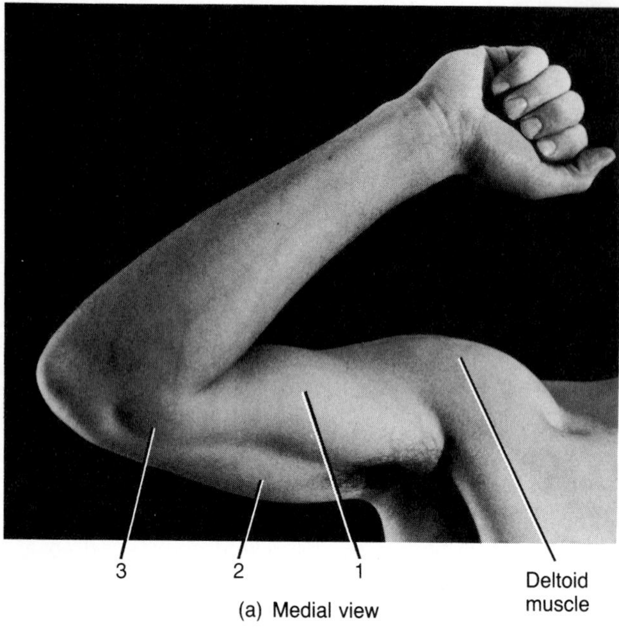

(a) Medial view

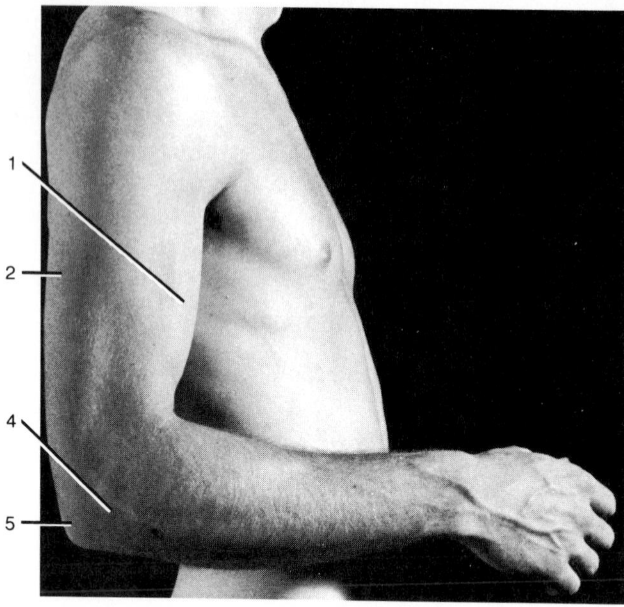

(b) Lateral view

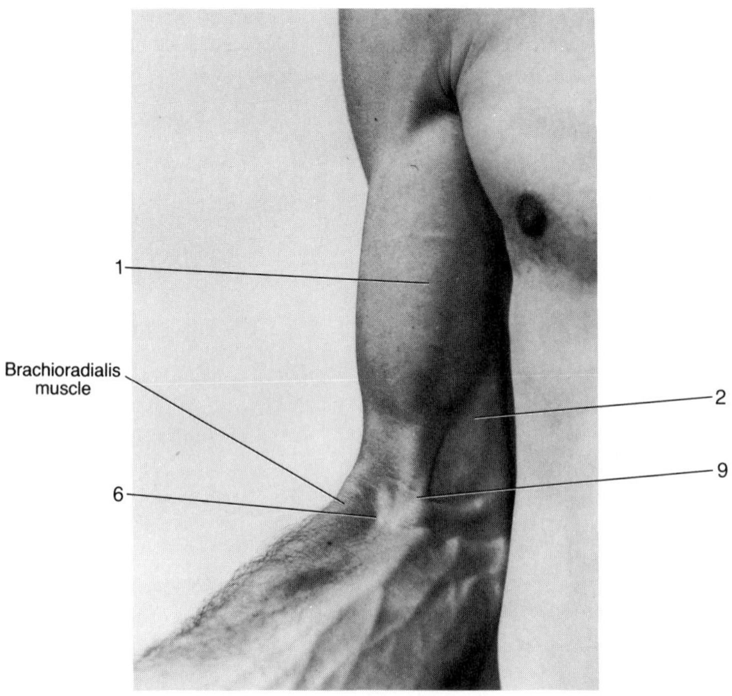

(c) Anterior view

FIGURE 11-10 Surface anatomy of the arm and elbow. (a) and (b) Courtesy of Victor B. Eichler, copyright © 1980. (c) Courtesy of Vincent P. Destro, Mayo Foundation.

EXHIBIT 11-4 (*Continued*)

under the bicipital aponeurosis, which separates it from the median cubital vein. The artery is frequently used to take blood pressure. Pressure may be applied to it in cases of severe hemorrhage in the forearm and hand.

9. ***Bicipital aponeurosis.*** An aponeurotic band that inserts the biceps brachii muscle into the deep fascia in the medial aspect of the forearm. It can be felt when the muscle contracts.

FOREARM (Figure 11-11)

1. ***Styloid process of ulna.*** Projection of distal end of head of ulna at medial side of wrist.
2. ***Flexor carpi radialis muscle.*** Located along midportion of forearm; helps flex wrist.
3. ***Flexor carpi ulnaris muscle.*** Located at medial aspect of forearm; helps flex wrist.
4. ***Brachioradialis muscle.*** Located at superior and lateral aspect of forearm; helps flex forearm. See also Figure 11-10c.

WRIST (Figure 11-12)

1. ***Wrist creases.*** Three more or less constant lines on anterior aspect of wrist (proximal, middle, and distal) where skin is firmly attached to underlying deep fascia.
2. ***Tendon of palmaris longus muscle.*** If you make a fist, the tendon can be seen on anterior surface of wrist nearer ulna; muscle helps to flex wrist (this muscle is not present in about 13 percent of the population).
3. ***Tendon of flexor carpi radialis muscle.*** Tendon on anterior surface of wrist lateral to tendon of palmaris longus.
4. ***Radial artery.*** Can be palpated just medial to styloid process of radius; frequently used to take pulse.
5. ***Pisiform bone.*** Medial bone of proximal carpals; easily palpated as a projection distal and anterior to styloid process of ulna.
6. ***Tendon of extensor pollicis brevis muscle.*** Tendon closer to styloid process of radius along posterior surface of wrist; best seen when thumb is bent backward; muscle extends thumb.
7. ***Tendon of extensor pollicis longus muscle.*** Tendon closer to styloid process of ulna along posterior surface of wrist; best seen when thumb is bent backward; muscle extends thumb.
8. ***"Anatomical snuffbox."*** Depression between tendons of extensor pollicis brevis and extensor pollicis longus muscles; styloid process of the radius, the base of the first metacarpal, trapezium, scaphoid, and radial artery can all be palpated in the depression.

continued

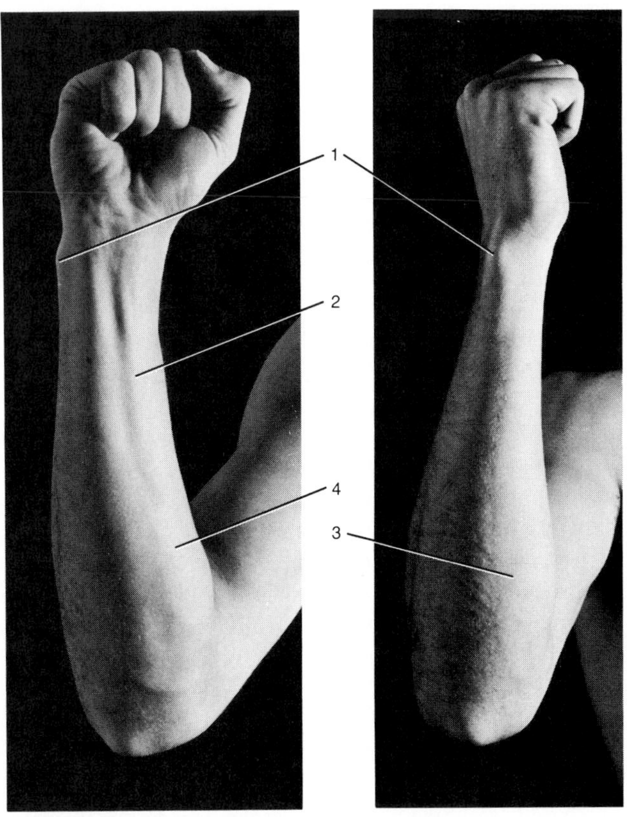

FIGURE 11-11 Surface anatomy of the right forearm. (Courtesy of Victor B. Eichler, copyright © 1980.)

EXHIBIT 11-4 (Continued)

HAND (Figure 11-13)

1. ***"Knuckles."*** Commonly refers to dorsal aspects of distal ends of metacarpals II, III, IV, and V; also includes dorsal aspects of metacarpophalangeal and interphalangeal joints.
2. ***Thenar eminence.*** Lateral rounded contour on palm of hand formed by muscles of thumb.
3. ***Hypothenar eminence.*** Medial rounded contour on palm of hand formed by muscles of little finger.
4. ***Digital flexion creases.*** Skin creases on anterior surface of fingers.
5. ***Palmar flexion creases.*** Skin creases on palm of hand.
6. ***Tendon of extensor digiti minimi muscle.*** Extensor tendon in line with phalanx V (little finger); muscle extends little finger.
7. ***Tendons of extensor digitorum muscle.*** Extensor tendons in line with phalanges II, III, and IV; muscle extends fingers and wrist.
8. ***Dorsal venous arch.*** Superficial veins on dorsum of hand that form cephalic vein.

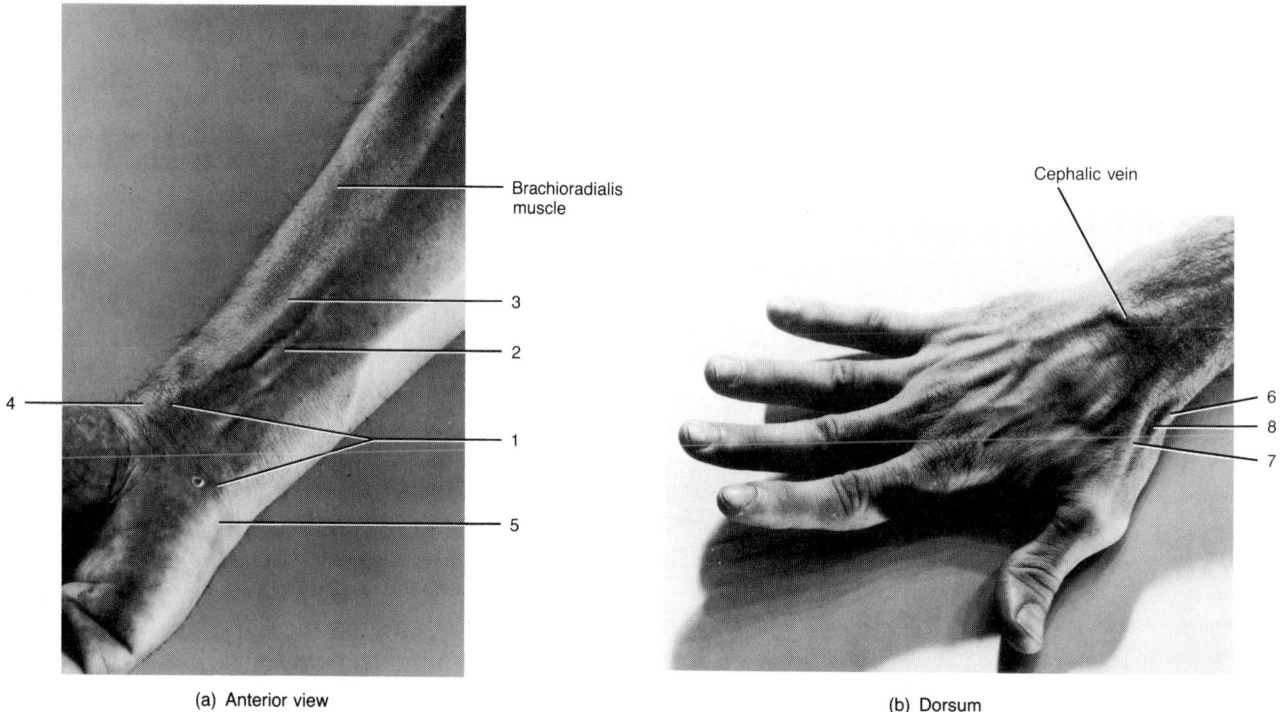

FIGURE 11-12 Surface anatomy of the right forearm. (Courtesy of Vincent P. Destro, Mayo Foundation.)

Most of the anterior surface of the ***arm (brachium)*** is occupied by the ***biceps brachii muscle,*** whereas most of the posterior surface is occupied by the ***triceps brachii muscle.*** The ***brachial artery*** provides the main arterial supply to the arm and is frequently used to take blood pressure and may be used as a site to apply pressure in cases of severe hemorrhage in forearm and hand. It begins at the lower border of the teres major muscle as the continuation of the axillary artery and terminates at the level of the neck of the radius by dividing into the ulnar and radial arteries (Exhibit 11-4).

At the ***elbow (cubitus)*** it is possible to locate three bony protuberances. The ***medial*** and ***lateral epicondyles*** of the humerus form visible eminences on the dorsum of the elbow. The ***olecranon*** of the ulna forms the large eminence in the middle of the dorsum of the elbow and lies between and slightly superior to the epicondyles when the forearm is extended. The ***ulnar nerve*** can be palpated in a groove behind the medial epicondyle. The triangular space of the anterior region of the elbow is the ***cubital fossa.*** The ***median cubital vein*** usually crosses the cubital fossa obliquely. This vein is the one frequently selected for removal of blood for diagnosis, transfusions, and intravenous therapy (Exhibit 11-4).

One of the most prominent landmarks of the ***forearm (antebrachium)*** is the ***styloid process*** of the ulna. It may be seen as a protuberance on the medial side of the wrist. The ulna is the medial bone of the forearm, and the radius is the lateral bone of the forearm. The entire length of the ulna may be palpated. On the lateral side of the upper

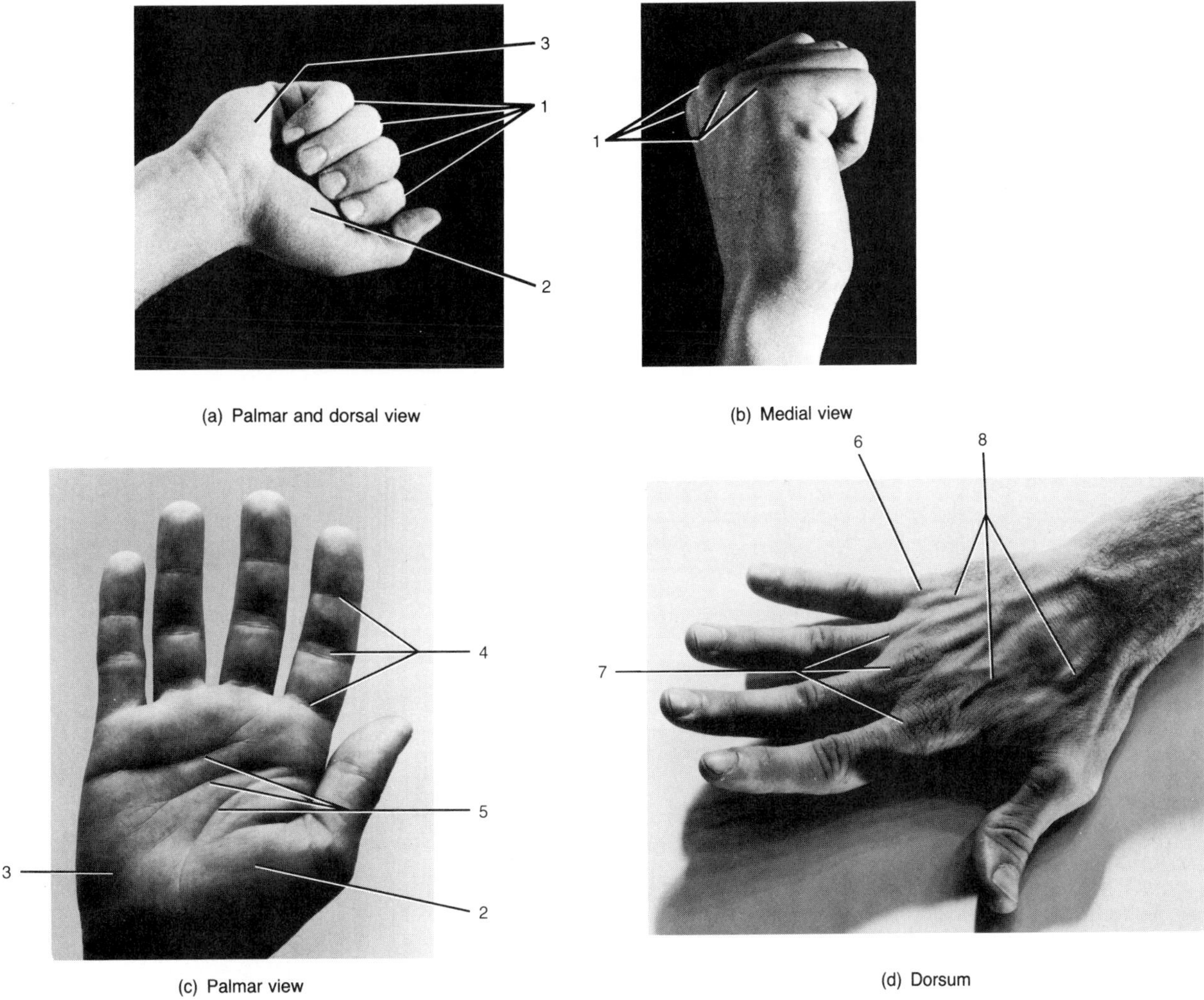

FIGURE 11-13 Surface anatomy of the right hand. (a) and (b) Courtesy of Victor B. Eichler, copyright © 1980. (c) Courtesy of Carroll H. Weiss, Camera M. D. Studios, Inc. (d) Courtesy of Vincent P. Destro, Mayo Foundation.

forearm is the ***brachioradialis muscle.*** Next to it is the ***flexor carpi radialis muscle.*** On the medial side of the upper forearm is the ***flexor carpi ulnaris muscle*** (Exhibit 11-4).

At the ***wrist (carpus),*** several structures may be palpated or seen. On the anterior surface are ***wrist creases,*** designated as proximal, middle, and distal, where the skin is firmly attached to the underlying deep fascia. On the anterior surface of the wrist, it is possible to see the ***tendon of the palmaris longus muscle*** by making a fist. Next to this tendon, as you move toward the thumb, you can feel the ***tendon of the flexor carpi radialis muscle.*** If you continue toward the thumb, you can palpate the ***radial artery*** just medial to the styloid process of the radius. This artery is frequently used to take the pulse. The ***pisiform bone,*** the medial bone of the proximal carpals, can be palpated as a projection distal and anterior to the styloid process of the ulna. If the thumb is bent backward, two prominent tendons may be located along the posterior surface of the wrist. The one closer to the styloid process of the radius is the ***tendon of the extensor pollicis brevis muscle.*** The one closer to the styloid process of the ulna is the ***tendon of the extensor pollicis longus muscle.*** The depression between these two tendons is known as the ***"anatomical snuffbox."*** By palpating the depression, you can feel the radial artery (Exhibit 11-4).

On the dorsum of the ***hand (manus)*** the distal ends of the second through fifth metacarpal bones are commonly referred to as the ***"knuckles."*** The term also includes the joints between metacarpals and phalanges and the joints between the phalanges of the fingers. The ***dorsal venous arch,*** the superficial veins on the dorsum of the hand, can be displayed by compressing the blood vessels at the wrist for a few moments as the hand is opened and closed. The ***tendon of the extensor digiti minimi muscle*** can be seen in line with phalanx V (little finger), and the ***tendons of***

the extensor digitorum muscle can be seen in line with phalanges II, III, and IV. By examining the palm of the hand, it is possible to see a number of ***skin creases,*** as well as the location of the joints of the fingers. The skin creases of the palm are collectively called ***palmar flexion creases.*** The skin creases on the anterior surface of the fingers are called ***digital flexion creases.*** Also on the palm are the ***thenar eminence,*** a rounded contour formed by thumb muscles, and the ***hypothenar eminence,*** a rounded contour formed by muscles of the little finger (Exhibit 11-4).

LOWER EXTREMITY

The ***lower extremity*** consists of the buttocks, thigh, knee, leg, ankle, and foot.

The outline of the superior border of the ***buttock (gluteal region)*** is formed by the ***iliac crest.*** The ***posterior superior iliac spine,*** the posterior termination of the iliac crest, lies deep to a dimple (skin depression) about 4 cm (1.5 in.) lateral to the midline. Most of the prominence of the buttocks is formed by the ***gluteus maximus*** and ***gluteus medius muscles.*** The sciatic nerve lies deep to the gluteus maximus muscle. The depression that separates the buttocks is the ***gluteal (natal) cleft,*** whereas the inferior limit of the buttock formed by the inferior margin of the gluteus maximus muscle is called the ***gluteal fold.*** The bony prominence in each buttock is the ***ischial tuberosity*** of the hipbone. This structure bears the weight of your body when you are seated. About 20 cm (8 in.) below the highest portion of the iliac crest, the ***greater trochanter*** of the femur can be felt on the lateral side of the thigh. The iliac crest and greater trochanter are useful landmarks when giving an intramuscular injection in the gluteus medius muscle (Exhibit 11-5).

Among the prominent superficial muscles on the anterior surface of the ***thigh (femoral region)*** are the ***sartorius*** and

EXHIBIT 11-5
Surface Anatomy of the Lower Extremity

BUTTOCKS AND THIGH (Figure 11-14)

1. ***Iliac crest.*** Superior margin of ilium of hipbone; forms outline of superior border of buttock; when you rest your hands on your hips, they rest on the iliac crests.
2. ***Posterior superior iliac spine.*** Posterior termination of iliac crest; lies deep to a dimple (skin depression) about 4 cm (1.5 in.) lateral to midline; dimple forms because skin and underlying fascia are attached to bone. The spine marks the inferior limit of cerebrospinal fluid in the subarachnoid space around the spinal cord.
3. ***Gluteus maximus muscle.*** Forms major portion of prominence of buttock; extends and laterally rotates thigh.
4. ***Gluteus medius muscle.*** Superolateral to gluteus maximus; abducts and medially rotates thigh; frequent site for intramuscular injections.
5. ***Gluteal (natal) cleft.*** Depression along midline that separates the buttocks; it extends as high as the fourth or third sacral vertebra.
6. ***Gluteal fold.*** Inferior limit of buttock formed by inferior margin of gluteus maximus muscle.
7. ***Ischial tuberosity.*** Bone prominence of ischium of hipbone; bears weight of body when seated.
8. ***Greater trochanter.*** Projection of proximal end of femur on lateral surface of thigh felt and seen in front of hollow on side of hip; can be palpated about 20 cm (8 in.) inferior to iliac crest.
9. ***Hamstrings.*** Superficial posterior thigh muscles that flex leg and extend thigh.
10. ***Sartorius muscle.*** Superficial anterior thigh muscle that flexes leg and flexes and laterally rotates thigh.
11. ***Rectus femoris muscle.*** Component of quadriceps femoris group located at midportion of anterior aspect of thigh; extends leg (in conjunction with other components of quadriceps femoris) and assists in flexing thigh (acting alone).
12. ***Vastus lateralis muscle.*** Component of quadriceps femoris group located at anterolateral aspect of thigh; extends leg.
13. ***Vastus medialis muscle.*** Component of quadriceps femoris group located at anteromedial aspect of thigh; extends leg.
14. ***Adductor magnus muscle.*** Located on medial aspect of thigh; adducts, flexes, and extends thigh.

KNEE (Figure 11-15)

1. ***Patella.*** Kneecap; large sesamoid bone located within quadriceps femoris tendon on anterior surface of knee along midline; margins of condyles (described shortly) can be felt on either side of it.
2. ***Medial condyle of femur.*** Medial projection of distal end of femur.
3. ***Medial condyle of tibia.*** Medial projection of proximal end of tibia.
4. ***Patellar ligament.*** Continuation of quadriceps femoris tendon inferior to patella; infrapatellar fat pads are located on both sides.
5. ***Popliteal fossa.*** Diamond-shaped space on posterior aspect of knee visible when knee is flexed; fossa is bordered superolaterally by the biceps femoris muscle, superomedially by the semimembranosus and semitendinosus muscle, and inferolaterally and inferomedially by the lateral and medial heads of the gastrocnemius muscle, respectively (see also Figures 11-16 and 10-24a).
6. ***Lateral condyle of tibia.*** Lateral projection of proximal end of tibia.
7. ***Lateral condyle of femur.*** Lateral projection of distal end of femur.

LEG AND ANKLE (Figure 11-16)

Tibial tuberosity. Bony prominence of tibia into which patellar ligament inserts (see Figure 11-15a).

Tibialis anterior muscle. Located at anterior surface of leg along midportion; dorsiflexes and inverts foot (see Figure 11-15).

1. ***Gastrocnemius muscle.*** Forms bulk of midportion and upper portion of posterior aspect of leg; plantar flexes foot.

continued

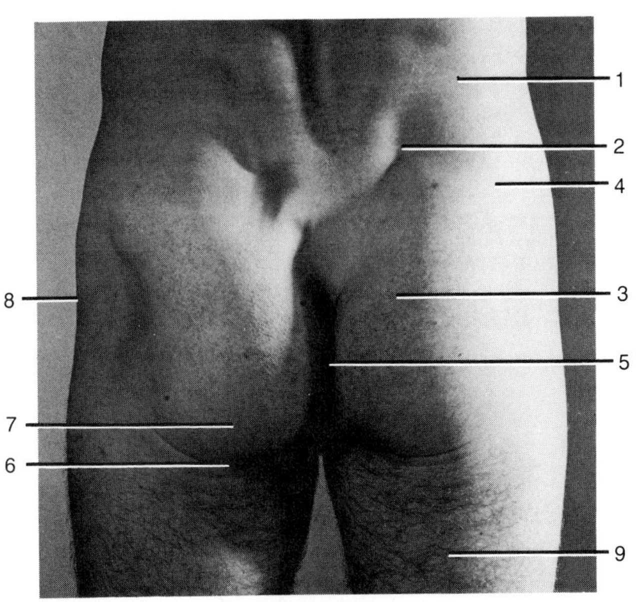

(a) Posterior view

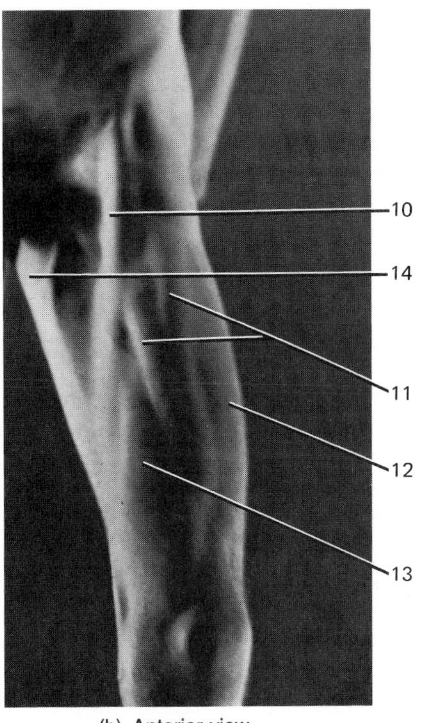

(b) Anterior view

FIGURE 11-14 Surface anatomy of the buttocks and thigh. (a) Courtesy of Lester Bergman & Associates. (b) Courtesy of R. D. Lockhard, *Living Anatomy*, Faber & Faber, London, 1974.

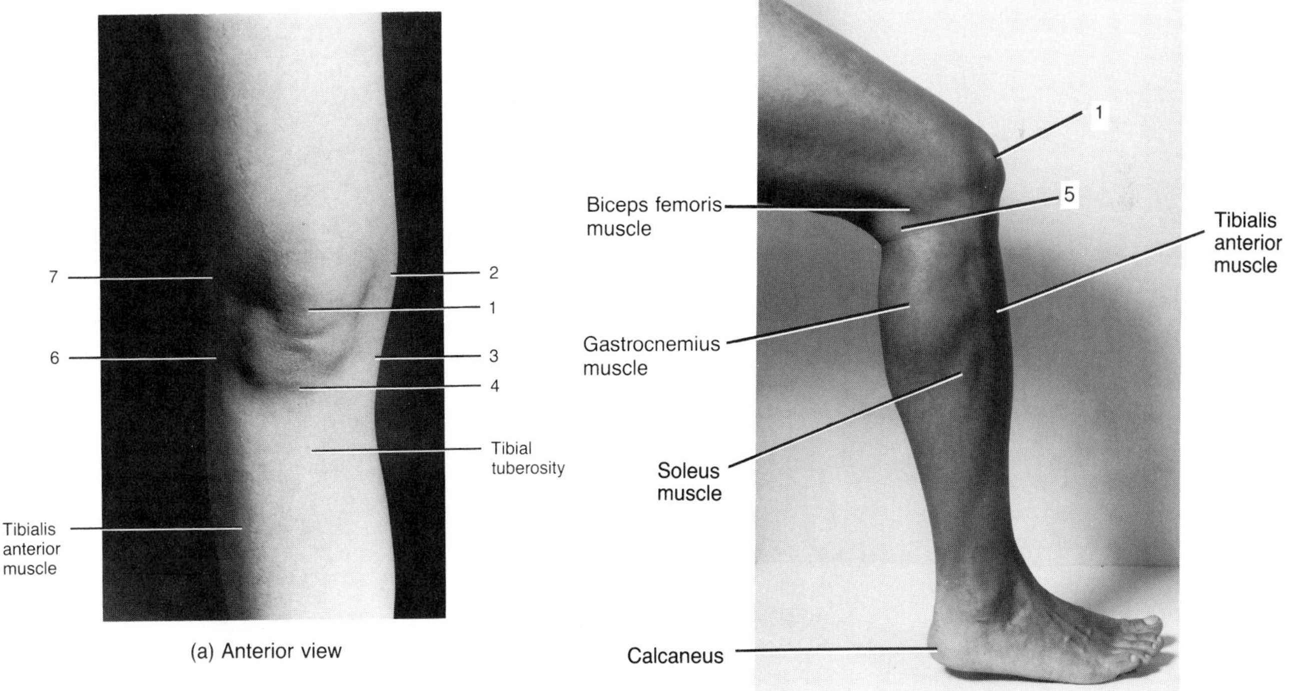

(a) Anterior view

(b) Posterolateral view

FIGURE 11-15 Surface anatomy of the right knee. (a) Courtesy of Carroll H. Weiss, Camera M. D. Studios, Inc. (b) Courtesy of J. J. Martin, Mayo Clinic.

EXHIBIT 11-5 (*Continued*)

2. ***Soleus muscle.*** Located deep to gastrocnemius muscle; plantar flexes foot (see Figure 11-15b).
3. ***Calcaneal (Achilles) tendon.*** Conspicuous tendon of gastrocnemius and soleus muscles that inserts into calcaneus (heel) bone of foot.
4. ***Lateral malleolus of fibula.*** Projection of distal end of fibula that forms lateral prominence of ankle. The head of the fibula, at the proximal end of the bone, lies at the same level as the tibial tuberosity.
5. ***Medial malleolus of tibia.*** Projection of distal end of tibia that forms medial prominence of ankle.

FOOT (Figure 11-17)

1. ***Calcaneus.*** Heel bone (see also Figure 11-15b).
2. ***Dorsal venous arch.*** Superficial veins on dorsum of foot that unite to form small and great saphenous veins.
3. ***Tendons of extensor digitorum longus muscle.*** Visible in line with phalanges II through V; muscle extends toes and dorsiflexes and everts foot.
4. ***Tendon of extensor hallucis longus muscle.*** Visible in line with phalanx I (great toe); muscle extends great toe and dorsiflexes ankle. Pulsations in the dorsalis pedis artery may be felt in most people just lateral to this tendon when the blood vessel passes over the navicular and cuneiform bones.

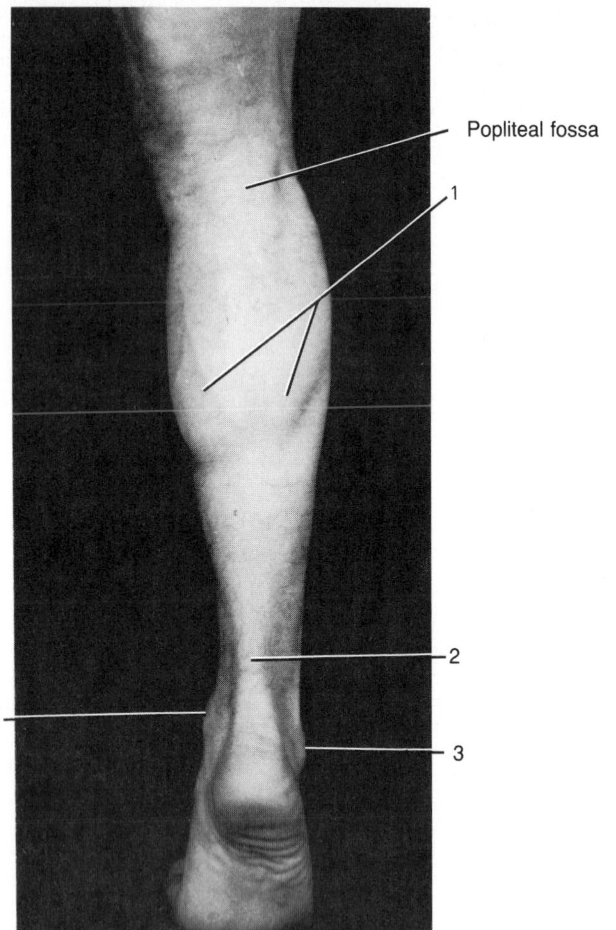

Posterior view

FIGURE 11-16 Surface anatomy of the right leg and ankle. (Courtesy of Donald Castellaro and Richard Sollazzo.)

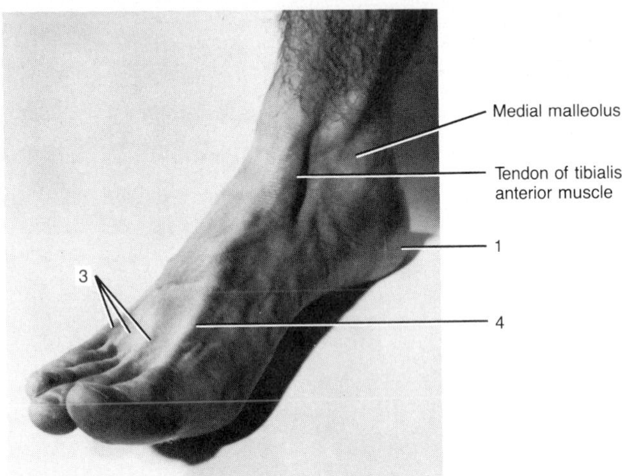

(a) Dorsomedial view

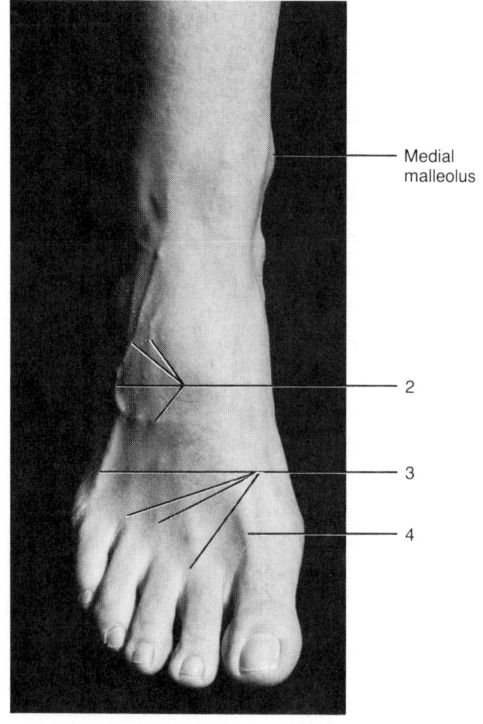

(b) Dorsum

FIGURE 11-17 Surface anatomy of the right foot. (a) Courtesy of Vincent P. Destro, Mayo Foundation. (b) Courtesy of Carroll H. Weiss, Camera M. D. Studios, Inc.

three of the four components of the ***quadriceps femoris*** (***vastus lateralis, vastus medialis,*** and ***rectus femoris***). The vastus lateralis is frequently used as an injection site for diabetics when administering insulin. The superficial medial thigh muscles include the ***adductor magnus, adductor brevis, adductor longus, gracilis, obturator externus,*** and ***pectineus.*** The superficial posterior thigh muscles are the ***hamstrings*** (***semitendinosus, semimembranosus,*** and ***biceps femoris***). The ***femoral triangle*** is a large space formed by the inguinal ligament superiorly, the sartorius muscle laterally, and the adductor longus muscle medially (see Figure 10-22a). The triangle contains the femoral artery, vein, and nerve; lymph nodes; and the terminal portion of the great saphenous vein. The triangle is an important arterial pressure point in cases of severe hemorrhage of the lower extremity (Exhibit 11-5).

On the anterior surface of the ***knee*** (***genu***), the ***patella,*** or kneecap, is observable. Below it is the ***patellar ligament.*** On the posterior surface of the knee is a diamond-shaped space, the ***popliteal fossa.*** Just below the patella, on either side of the patellar ligament, the ***medial*** and ***lateral condyles*** of the femur and tibia can be felt (Exhibit 11-5).

The bony prominence inferior to the patella in the middle of the ***leg*** (***crus***) is the ***tibial tuberosity.*** The tibia is the medial bone of the leg, and the fibula is the lateral bone of the leg. Prominent superficial muscles of the leg include the ***tibialis anterior, gastrocnemius,*** and ***soleus*** (Exhibit 11-5).

At the ***ankle*** (***tarsus***) the ***medial malleolus*** of the tibia and the ***lateral malleolus*** of the fibula can be noted as two prominent eminences (Exhibit 11-5).

On the dorsum of the ***foot*** you can see the ***tendon of the extensor hallucis longus muscle*** in line with phalanx I (great toe). Just lateral to this tendon, the dorsalis pedis arterial pulse may be felt. Also on the dorsum of the foot are the ***tendons of the extensor digitorum longus muscle*** in line with phalanges II through V, and the ***dorsal venous arch,*** superficial veins that unite to form the small and great saphenous veins. The great saphenous vein is the longest vein of the body. Its position about 2.5 cm (1 in.) in front of the medial malleolus of the tibia is constant and knowledge of this point may be lifesaving when an urgent transfusion is needed in obese or collapsed patients when other veins cannot be detected. Arising from the ***heel bone*** (***calcaneus***) is the ***calcaneal*** (***Achilles***) ***tendon*** (Exhibit 11-5).

STUDY OUTLINE

1. Surface anatomy is the study of the form and markings of the surface of the body.
2. Surface anatomy features may be noted by visual inspection or palpation.
3. The principal regions of the body used to study surface anatomy are the head, neck, trunk, upper extremity, and lower extremity.
4. A review of surface anatomy is presented in Exhibits 11-1–11-5.

REVIEW QUESTIONS

1. Define surface anatomy. What is meant by palpation (p. 318)?
2. List the principal regions of the body and their subdivisions. (p. 318)
3. Using Exhibit 11-1 as an outline, locate as many of the surface features of the head as you can on your partner's body, wall charts, models, photographs, and skeletons. (p. 318)
4. Using Exhibit 11-2 as an outline, locate as many of the surface features of the neck as you can on your partner's body, wall charts, models, photographs, and skeletons. (p. 320)
5. Using Exhibit 11-3 as an outline, locate as many of the surface features of the trunk as you can on your partner's body, wall charts, models, photographs, and skeletons. (p. 321)
6. Using Exhibit 11-4 as an outline, locate as many of the surface features of the upper extremity as you can on your partner's body, wall charts, models, photographs, and skeletons. (p. 323)
7. Using Exhibit 11-5 as an outline, locate as many of the surface features of the lower extremity as you can on your partner's body, wall charts, models, photographs, and skeletons. (p. 328)

SELF QUIZ

1. Complete the following table relating common terms to anatomical terms.

COMMON TERM	ANATOMICAL TERM
a.	Cephalic region
b. Skull or brain case	
c.	Temporal region
d. Eyebrows	
e.	Palpebrae
f. Mouth	
g. Cheek	
h.	Auricular region

i.	Tarsus
j. Forearm	
k.	Manus
l. Fingers and toes	
m. Buttocks	
n.	Crus

Complete the following:

2. The head is divided into two regions: these are ________ and ________.

3. Each lateral cervical region can be divided into two triangles by the diagonally positioned ________ muscle. The anterior triangle thus formed is bordered superiorly by the ________ and medially by the ________. The posterior (lateral) triangle is bordered inferiorly by the ________ and posteriorly by the anterior border of the ________ muscle.

4. Circle the answer that best fits each description:

a. frequent site of intramuscular injections in the upper extremity: (flexor carpi ulnaris? trapezius? deltoid?).
b. muscle producing bulge on the superior and lateral aspect of the forearm: (brachioradialis? coracobrachialis? brachialis?).
c. slight elevation at the lateral end of the clavicle: (sternal angle? manubrium? acromioclavicular joint?).
d. groove posterior to the medial epicondyle: (site of the inner ankle? site of the ulnar nerve? site of the radial nerve?).
e. location of a vein often used for intravenous therapy on anterior surface of elbow joint: (cubital fossa? lateral epicondyle? popliteal fossa?).
f. "anatomical snuffbox": (depression between tendons of two muscles that move the thumb? depression between tendons of two muscles that move the forearm?).
g. "knuckles": (proximal ends of proximal phalanges of the hand? distal ends of second through fifth metacarpals?).
h. protuberance at distal end of the ulna: (coronoid process? olecranon process? styloid process?).
i. superior border of the buttock: (iliac crest? greater trochanter? ischial tuberosity?).
j. frequently used as an injection site of insulin in diabetics: (vastus intermedius? vastus lateralis? adductor longus?).
k. bony prominence below the patella: (patellar ligament? calcaneal (Achilles) tendon? tibial tuberosity?).

5. Arrange the answers in correct sequences:

__ __ __ **a.** From anterior to posterior:
A. parietal region
B. nasal region
C. frontal region

__ __ __ **b.** From superior to inferior:
A. orbital region
B. oral region
C. nasal region

__ __ __ **c.** From superior to inferior:
A. xiphoid process
B. umbilicus
C. hyoid bone

__ __ __ **d.** From proximal to distal:
A. antebrachium
B. brachium
C. carpus

__ __ __ **e.** From proximal to distal:
A. phalanges
B. proximal wrist crease
C. distal wrist crease

__ __ __ **f.** From medial to lateral:
A. acromioclavicular joint
B. deltoid muscle
C. sternal angle

__ __ __ **g.** From anterior to posterior of the ear:
A. concha
B. external auditory meatus
C. helix

__ __ __ **h.** From medial to lateral of the eye:
A. medial commissure
B. iris
C. lateral commissure

Circle T (true) or F (false) for the following:

T F 6. The ulnar nerve lies in a groove behind the medial epicondyle.
T F 7. The ischial spine bears most of the weight of a person in the sitting position.
T F 8. All of the following surface features are located on the lower extremities: tibial tuberosity, calcaneal (Achilles) tendon, lateral malleolus.
T F 9. The triceps brachii is located on the posterior surface of the body.
T F 10. The epicondyles are markings on the ulna.
T F 11. The trapezius muscle forms the posterior border of the posterior triangle of the lateral cervical region.
T F 12. The linea alba is located on the anterior surface of the abdomen.
T F 13. A landmark for performing a tracheostomy is the cricoid cartilage.
T F 14. The trapezius muscle is a useful landmark for locating the carotid artery.
T F 15. The triangle of auscultation is formed by the latissimus dorsi, trapezius, and scapula.
T F 16. The apex beat of the heart is heard in the seventh intercostal space.
T F 17. The radial artery is frequently used to take blood pressure.
T F 18. The femoral triangle is bordered by the inguinal ligament, sartorius, and adductor longus.
T F 19. The dorsalis pedis muscle is just lateral to the tendon of the extensor digitorum longus muscle.
T F 20. The longest vein in the body is the femoral vein.

The Cardiovascular System: Blood

12

STUDENT OBJECTIVES

1. Describe the principal physical characteristics of blood and its functions in the body.
2. Compare the origins, histology, and functions of the formed elements in blood.
3. List the components of plasma and explain their importance.
4. Contrast the causes and clinical symptoms of various types of anemia, polycythemia, infectious mononucleosis (IM), chronic fatigue syndrome, and leukemia.
5. Define key medical terms associated with blood.

CHAPTER OUTLINE

- **Functions of Blood**
- **Physical Characteristics of Blood**
- **Components of Blood**
  - Formed Elements
  - Plasma
- **Origin of Blood Cells**
- **Erythrocytes (Red Blood Cells)**
  - Structure
  - Functions
  - Life Span and Number
  - Production
  - Blood Group Systems
- **Leukocytes (White Blood Cells)**
  - Structure and Types
  - Functions
    - *Neutrophils and Monocytes*
    - *Eosinophils*
    - *Basophils*
    - *Lymphocytes*
  - Life Span and Number
  - Production
- **Thrombocytes (Platelets)**
  - Structure
  - Function
  - Life Span and Number
  - Production
- **Applications to Health**
- **Key Medical Terms Associated with Blood**

The blood, heart, and blood vessels together make up the ***cardiovascular system.*** In this chapter we shall consider blood. The heart and blood vessels are discussed in the next two chapters. A closely related system, the ***lymphatic system,*** consists of a fluid called lymph, lymphatic vessels that transport lymph, and structures containing lymphatic tissue (large numbers of white blood cells called lymphocytes). This system will be discussed in detail in Chapter 15.

Functionally, the cardiovascular system is designed to transport blood to cells of the body. Blood delivers substances such as oxygen, nutrients, and hormones and removes carbon dioxide and various waste products from cells. Blood also functions in regulating pH and body temperature, preventing hemorrhaging, and combating disease. In order to perform its functions, blood must circulate throughout the body. This is accomplished by the heart, which serves as a pump, and blood vessels, which bring blood from the heart to body cells and from body cells to the heart.

The substance that bathes cells is called ***interstitial fluid*** (also known as ***intercellular*** or ***tissue fluid***). Interstitial fluid, in turn, is serviced by blood and lymph. Blood picks up oxygen from the lungs, nutrients from the gastrointestinal tract, hormones from endocrine glands, and enzymes from still other parts of the body. It transports these substances to the tissues where they diffuse from capillaries (microscopic blood vessels) into interstitial fluid. In the interstitial fluid, the substances are passed on to the cells and exchanged for wastes.

Since the blood services the tissues of the body, it can also be an important medium for the transport of disease-causing organisms. The lymphatic system protects the body from the spread of disease. Lymph picks up materials, including wastes, from interstitial fluid, cleanses them of bacteria, and returns them to the blood. The blood then carries the wastes to the lungs, kidneys, and sweat glands, where they are eliminated from the body. The blood also takes wastes to the liver, where they are processed.

We shall first study the cardiovascular system by taking a look at the substance known as ***blood.*** The branch of science concerned with the study of blood and blood-forming tissues and the disorders associated with them is called ***hematology*** (hēm-a-TOL-ō-jē; *hem* = blood; *logos* = study of).

The developmental anatomy of blood and blood vessels is considered in Chapter 14.

FUNCTIONS OF BLOOD

Blood is a liquid connective tissue that performs a number of critical functions.

1. ***It transports:*** oxygen from the lungs to the cells of the body; carbon dioxide from the cells to the lungs; nutrients from the gastrointestinal tract to the cells; waste products from cells; hormones from endocrine glands to the cells; heat from various cells.
2. ***It regulates:*** pH through buffers; normal body temperature through the heat-absorbing and coolant properties of its water content; the water content of cells, principally through dissolved sodium ions (Na^+) and proteins.
3. ***It protects against:*** blood loss by means of the clotting mechanism; foreign microbes and toxins through phagocytic white blood cells or specialized plasma proteins such as antibodies, interferon, and complement.

PHYSICAL CHARACTERISTICS OF BLOOD

The red body fluid that flows through all the vessels except the lymph vessels is called ***blood.*** Blood is a viscous fluid—it is thicker than water. Water is considered to have a viscosity of 1.0. The viscosity of blood, by comparison, ranges from 4.5 to 5.5. It flows more slowly than water, at least in part because of its viscosity. The adhesive quality of blood, or its stickiness, may be demonstrated by touching it. Blood is also slightly heavier than water.

Other physical characteristics of blood include a temperature of about 38°C (100.4°F), a pH range of 7.35 to 7.45 (slightly alkaline), and a salt (NaCl) concentration of 0.90 percent.

Blood constitutes about 8 percent of the total body weight. The blood volume of an average-sized male is 5 to 6 liters (5 to 6 qt). The average-sized female has 4 to 5 liters.

The physical characteristics of blood are summarized in Exhibit 12-1.

CLINICAL APPLICATION

Withdrawing Blood

Blood samples for laboratory testing may be obtained in several ways. ***Venipuncture,*** the most frequently used procedure, involves withdrawal of blood from a vein. Veins are used instead of arteries because they are closer to the surface and more readily accessible and they contain blood at a much lower pressure. A commonly used vein is the median cubital vein in front of the elbow (see Figure 14-13a). A tourniquet is wrapped around the arm to stop blood flow through the veins. This makes the veins below the tourniquet stand out. Opening and closing the fist has the same effect.

Another procedure used to withdraw blood is the ***fingerstick.*** Using a sterile needle or lancet, a drop or two of capillary blood is taken from a finger, earlobe, or heel of the foot for evaluation.

Finally, an ***arterial stick*** may be used to withdraw blood. The sample is most frequently taken from the radial artery in the wrist or the femoral artery in the thigh.

EXHIBIT 12-1

Summary of Physical Characteristics of Blood

Viscosity	4.5–5.5 (by comparison to water)
Temperature	38°C (100.4°F)
pH	7.35–7.45
Salinity	0.90 percent
Total body weight	8 percent
Volume	5–6 liters for average male; 4–5 liters for average female

COMPONENTS OF BLOOD

Blood is composed of two portions: formed elements (cells and cell-like structures) and plasma (liquid containing dissolved substances). The formed elements compose about 45 percent of the volume of blood; plasma constitutes about 55 percent (Figure 12-1).

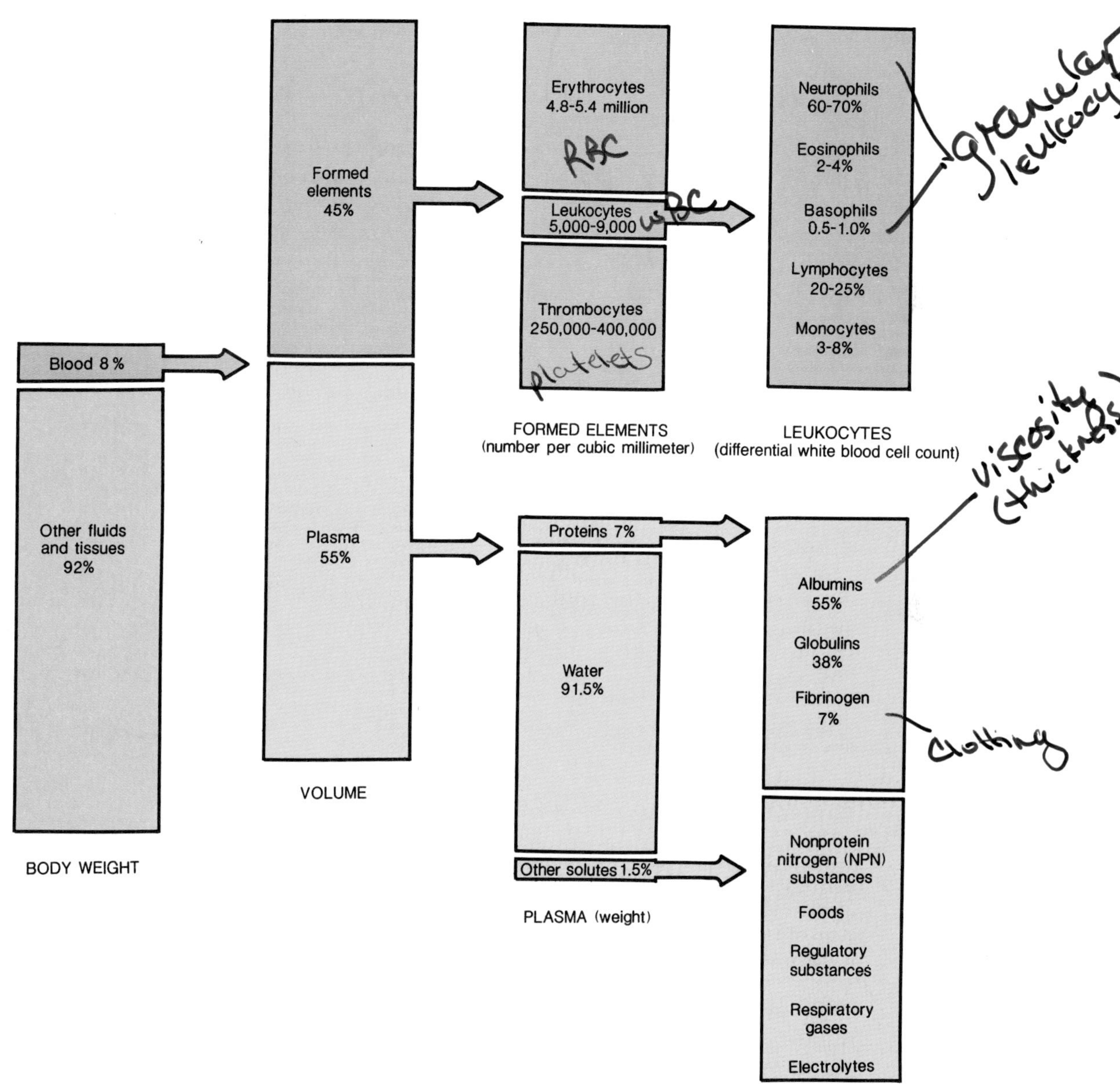

FIGURE 12-1 Components of blood in a normal adult. The sizes of the boxes do not necessarily reflect the proportions of the various substances contained within them. (Modified from Melloni's Illustrated Medical Dictionary.)

FORMED ELEMENTS

In clinical practice, the most common classification of the ***formed elements*** of the blood is the following (Figure 12-2).

Erythrocytes (red blood cells)

Leukocytes (white blood cells)
- **Granular leukocytes** (granulocytes)
  - Neutrophils
  - Eosinophils
  - Basophils
- **Agranular leukocytes** (agranulocytes)
  - Lymphocytes
  - Monocytes

Thrombocytes (platelets)

FIGURE 12-2 Blood cells. (b) Shown in the center is a blood smear at a magnification of 160×. The more numerous cells are red blood cells, the larger cells with darker staining nuclei are white blood cells, and the specklike objects are thrombocytes. The principal types of white blood cells are shown at a magnification of 500×. Blood smear courtesy of Merrill; eosinophil, basophil, monocyte, red blood cells, and platelets, © Biophoto, SPL, Photo Researchers; neutrophil courtesy of Ed Reschke; lymphocyte courtesy of Lester V. Bergman & Associates. (c) From *Tissues and Organs: A Text-Atlas of Scanning Electron Microscopy* by Richard G. Kessel and Randy H. Kardin. Copyright © 1979 W. H. Freeman and Company.

PLASMA

When the formed elements are removed from blood, a straw-colored liquid called ***plasma*** is left. Plasma consists of 91.5 percent water and 8.5 percent solutes. Exhibit 12-2 outlines the chemical composition of plasma. Some of the proteins in plasma are also found elsewhere in the body, but in blood they are called ***plasma proteins. Albumins,*** which are synthesized by the liver, constitute 55 percent of plasma proteins. The concentration of albumins is about four times higher in plasma than in interstitial fluid and they are the principal reason that the osmotic pressure of the blood is greater than that of tissue fluid. Albumins can therefore help regulate blood volume by preventing the water in the blood from moving into the interstitial fluid and can draw water from the tissue fluid. Water moves by osmosis from an area of high water (low solute) concentration to an area of low water (high solute) concentration. ***Globulins,*** also produced by the liver, comprise 38 percent of plasma proteins and include the antibody proteins released by plasma cells. Gamma globulin is especially well known because it is able to form an antigen–antibody complex with the protein of the hepatitis and measles viruses and the tetanus bacterium, among others. ***Fibrinogen*** makes

EXHIBIT 12-2

Chemical Composition and Description of Substances in Plasma

CONSTITUENT	DESCRIPTION
WATER	Liquid portion of blood; constitutes about 91.5 percent of plasma. Ninety percent of water derived from absorption from gastrointestinal (GI) tract; 10 percent from cellular respiration. Acts as solvent and suspending medium for solid components of blood and absorbs, transports, and releases heat.
SOLUTES	Constitute about 8.5 percent of plasma.
Proteins	
Albumins	Smallest plasma proteins. Produced by liver and provide blood with viscosity, a factor related to maintenance and regulation of blood pressure. Also exert considerable osmotic pressure to maintain water balance between blood and tissues and regulate blood volume.
Globulins	Protein group to which antibodies belong. Gamma globulins attack measles, hepatitis, and polio viruses, and tetanus bacterium.
Fibrinogen	Produced by liver. Plays essential role in clotting.
Nonprotein nitrogen (NPN) substances	Contain nitrogen but are not proteins. Represent breakdown products of protein metabolism and are carried by blood to organs of excretion. Include urea, uric acid, creatine, creatinine, and ammonium salts.
Nutrients	Products of digestion passed into blood for distribution to all body cells. Include amino acids (from proteins), glucose (from carbohydrates), and fatty acids and glycerol (from fats).
Regulatory substances	Enzymes, produced by body cells, to catalyze chemical reactions. Hormones, produced by endocrine glands, to regulate growth and development in body.
Respiratory gases	Oxygen (O_2) and carbon dioxide (CO_2). Whereas O_2 is more closely associated with hemoglobin or red blood cells, CO_2 is more closely associated with plasma.
Electrolytes	Inorganic salts of plasma. Cations include Na^+, K^+, Ca^{2+}, Mg^{2+}; anions include Cl^-, $HPO_4{}^{3-}$, $SO_4{}^{2-}$, $HCO_3{}^-$. Help maintain osmotic pressure, normal pH, physiological balance between tissues and blood.

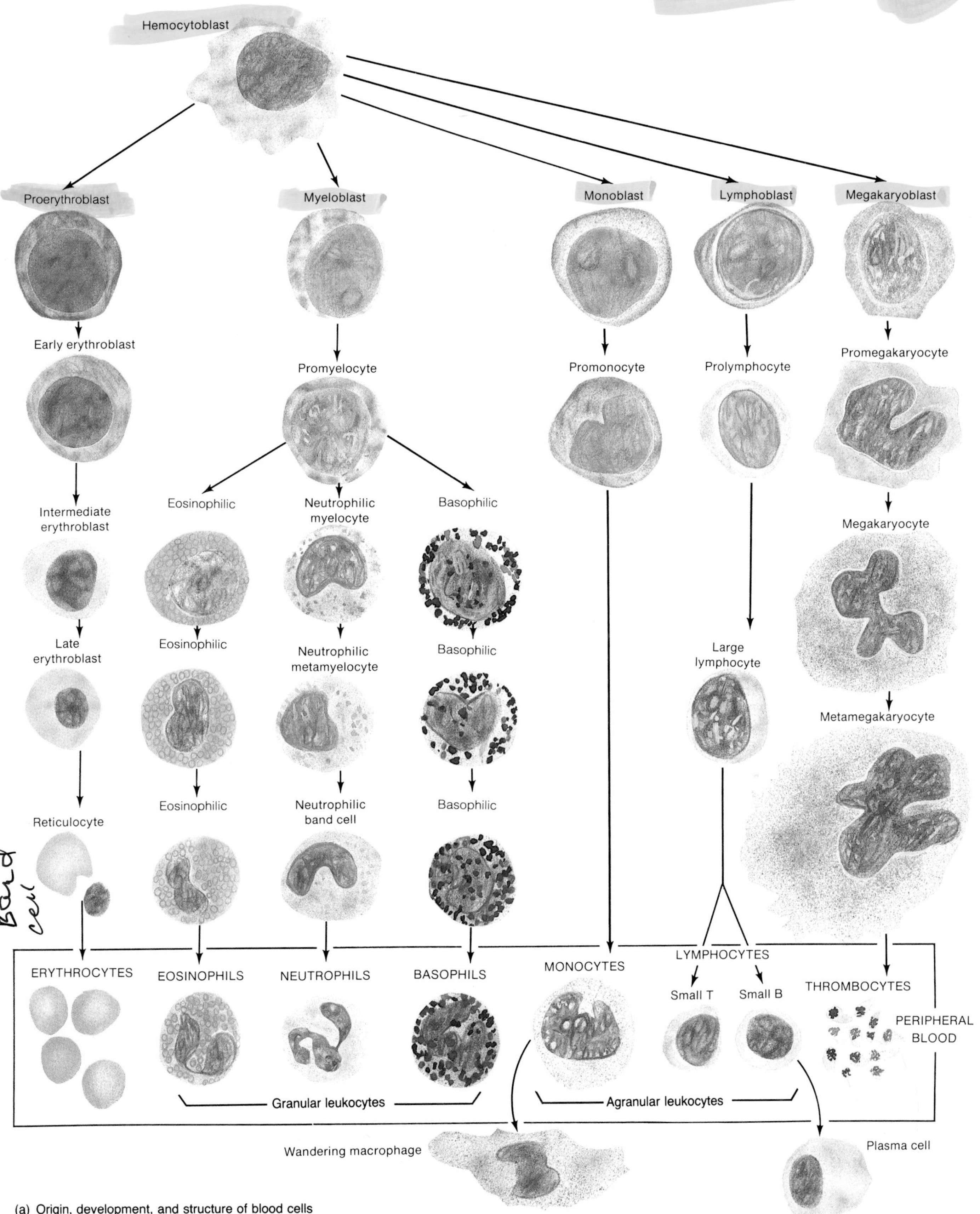

(a) Origin, development, and structure of blood cells

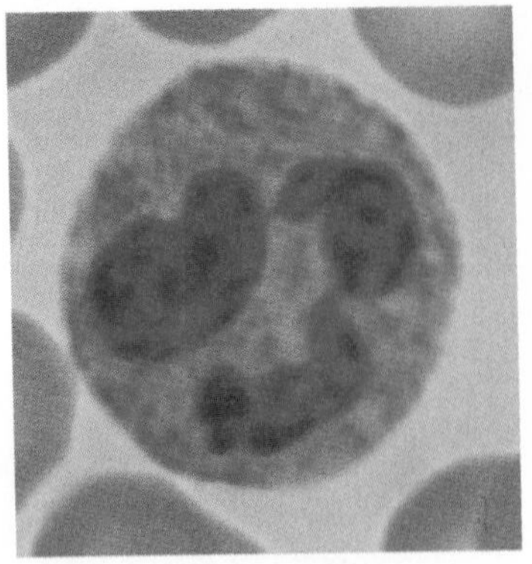

NEUTROPHIL

EOSINOPHIL

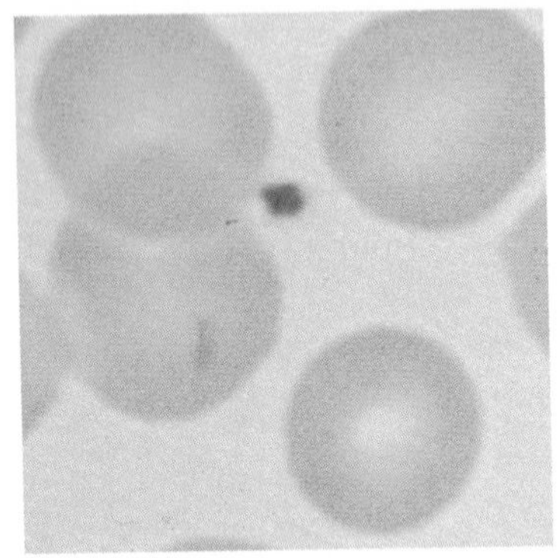

RED BLOOD CELLS AND PLATELET

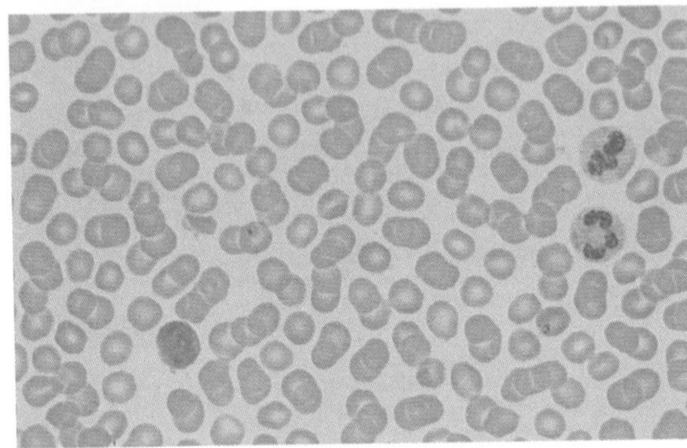

BLOOD SMEAR

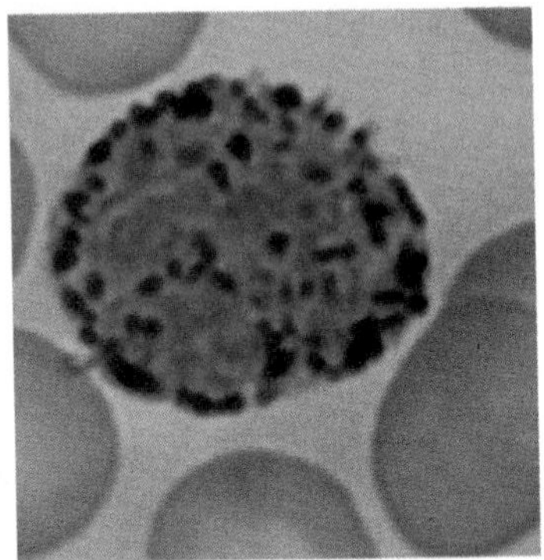

BASOPHIL

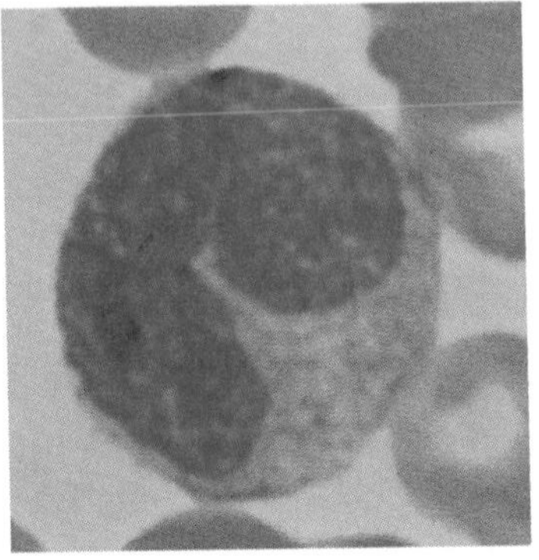

MONOCYTE

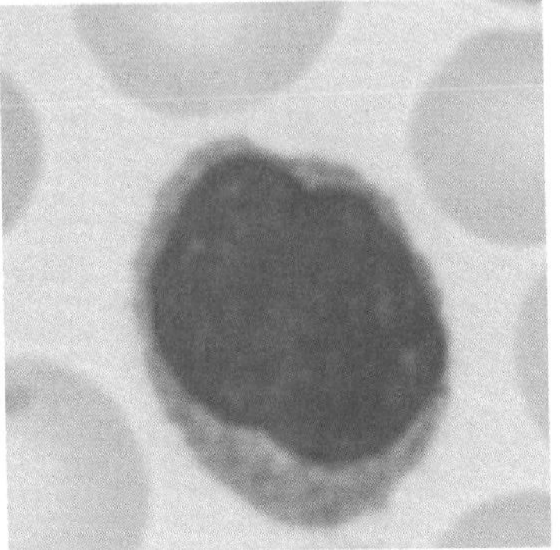

LARGE LYMPHOCYTE

(b) Photomicrographs of blood cells

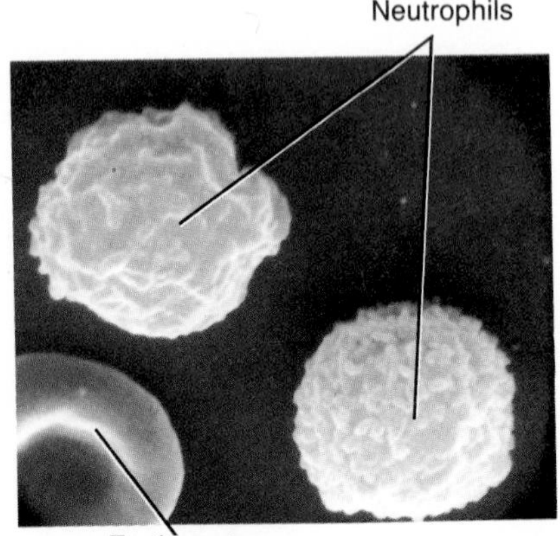

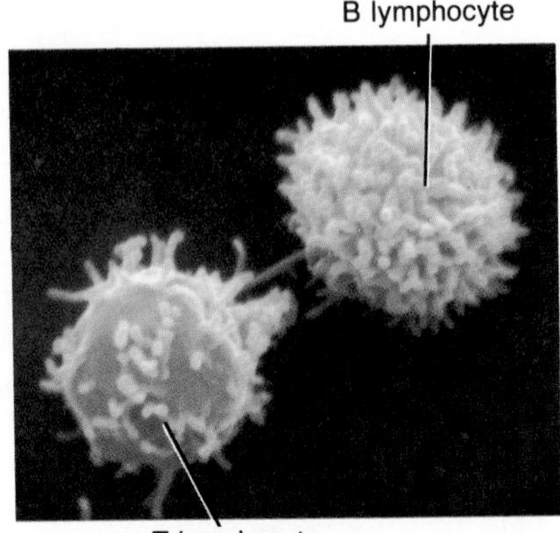

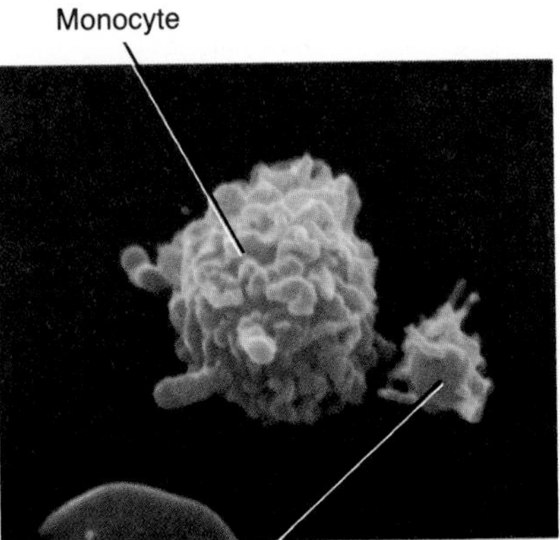

(c) Scanning electron micrographs of blood cells, 7120×

up about 7 percent of plasma proteins and takes part in the blood-clotting mechanism along with the platelets. It is also produced by the liver.

Other solutes in plasma include nonprotein nitrogen (NPN) substances, which are breakdown products of protein metabolism (urea, uric acid, creatinine, ammonia); products of digestion (amino acids, glucose, fatty acids, glycerol); enzymes; hormones; respiratory gases (O_2 and CO_2); and electrolytes (Na^+, K^+, Ca^{2+}, Mg^{2+}, Cl^-, HCO_3^-, SO_4^{2-}, and HPO_4^{3-}).

CLINICAL APPLICATION

Plasmapheresis

Plasmapheresis (plaz′-ma-fe-RĒ-sis; *aphairesis* = removal), also called ***therapeutic plasma exchange (TPE)***, refers to a procedure in which blood is withdrawn from the body, its components are selectively separated, the undesirable component causing disease is removed, and the remainder is returned to the body. Among the substances removed are toxins, metabolic substances, and antibodies.

The ability of plasmapheresis to remove antibodies and other immunologically active substances has made the procedure useful for neurological conditions in which autoimmunity is believed to play a role. About half of plasmapheresis procedures are done on patients with neurological disorders such as myasthenia gravis and Guillain-Barré syndrome. Plasmapheresis may also be used in the treatment of some anemias, multiple myeloma, thrombocytemia, posttransfusion Rh incompatibility, life-threatening sickle-cell crisis, kidney diseases, systemic lupus erythematosus (SLE), rheumatoid arthritis (RA), and certain drug overdoses. Because of some potentially serious complications of plasmapheresis (hypovolemia, shock, congestive heart failure, pulmonary edema, muscle twitching, thrombosis), the procedure is used discriminately and for short periods of time when more conservative therapy is unsuccessful.

ORIGIN OF BLOOD CELLS

The process by which blood cells are formed is called ***hemopoiesis*** (hē-mō-poy-Ē-sis), or ***hematopoiesis.*** During embryonic and fetal life, there are several centers for blood cell production. The yolk sac, liver, spleen, thymus gland, lymph nodes, and bone marrow all participate at various times in producing the formed elements. In the adult, however, we can pinpoint the production process to the red bone marrow (myeloid tissue) in the proximal epiphyses of the humerus and femur; flat bones of the cranium, sternum, ribs, vertebrae, and pelvis; and lymphoid tissue. Red blood cells, granular leukocytes, and platelets are produced in red bone marrow. Agranular leukocytes arise from both myeloid tissue and from lymphoid tissue—spleen, tonsils, thymus gland, lymph nodes. Undifferentiated mesenchymal cells in red blood marrow are transformed into ***hemocytoblasts*** (hē′-mō-SĪ-tō-blasts), immature cells that are eventually capable of developing into mature blood cells (Figure 12-2a). The hemocytoblasts undergo differentiation into five types of cells from which the major types of blood cells develop.

1. ***Proerythroblasts*** (***rubriblasts***) form mature erythrocytes.
2. ***Myeloblasts*** form mature neutrophils, eosinophils, and basophils.
3. ***Megakaryoblasts*** form mature thrombocytes (platelets).
4. ***Lymphoblasts*** form mature lymphocytes.
5. ***Monoblasts*** form mature monocytes.

ERYTHROCYTES (RED BLOOD CELLS)

STRUCTURE

Microscopically, ***erythrocytes*** (e-RITH-rō-sīts), or ***red blood cells*** (***RBCs***), or ***red blood corpuscles,*** appear as biconcave discs averaging about 8 μm in diameter (Figures 12-2b and 12-3a). Mature red blood cells are quite simple in structure. They lack a nucleus and other organelles and can neither reproduce nor carry on extensive metabolic activities. The plasma membrane is selectively permeable and consists of a protein (stromatin) and lipids (lecithin and cholesterol). The membrane encloses cytoplasm and a red pigment called ***hemoglobin.*** Hemoglobin, which constitutes about 33 percent of the cell weight, functions primarily in carrying oxygen to body cells and is responsible for the red color of blood. Normal values for hemoglobin are 14 to 20 g/100 ml of blood in infants, 12 to 15 g/100 ml in adult females, and 14 to 16.5 g/100 ml in adult males. As you will see later, certain proteins (antigens) on the surfaces of red blood cells are responsible for the various blood groups. The ABO and Rh groups are examples.

FUNCTIONS

The hemoglobin in erythrocytes combines with oxygen to form oxyhemoglobin, and with carbon dioxide to form carbaminohemoglobin, and then transports them through blood vessels. (Most carbon dioxide is transported as the bicarbonate ion, HCO_3^-). A hemoglobin molecule consists of a protein called ***globin,*** which is composed of four polypeptide chains and combines reversibly with carbon dioxide (carbaminohemoglobin). It also consists of four nonprotein pigments called ***hemes,*** each of which contains iron (as Fe^{2+}) that can combine reversibly with an oxygen molecule (Figure 12-3b). As erythrocytes pass through the lungs, each of the four iron atoms in the hemoglobin molecules

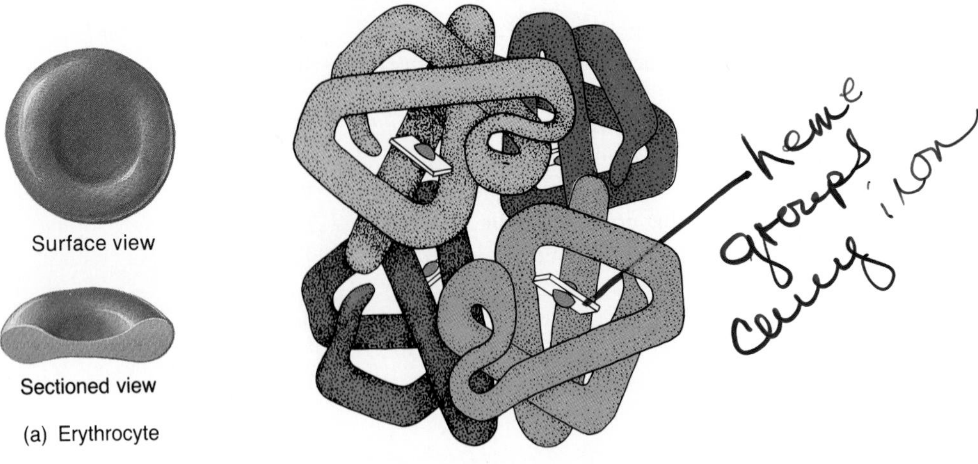

FIGURE 12-3 Erythrocytes. (a) Shape of an erythrocyte. (b) The four globin (protein) portions of the molecule are indicated in green and blue and the four heme (iron-containing) portions are in the center of each globin molecule. (Modified from R. E. Dickerson and I. Geis.)

combines with a molecule of oxygen (oxyhemoglobin). Since erythrocytes lack a nucleus, the additional area greatly increases their capacity for carrying oxygen. Moreover, since they lack mitochondria and generate ATP anaerobically (without oxygen), they do not consume any of the oxygen that they transport.

CLINICAL APPLICATION

Induced Erythrocytemia (Blood Doping)

In recent years, some athletes have become involved in a practice called ***induced erythrocytemia (blood doping).*** It involves removing red blood cells from the body and then reinjecting them a few days before an athletic event. Since red blood cells carry oxygen, it is presumed that the increased oxygen-carrying capacity of the blood increases muscular performance. In one major study, it has been shown that induced erythrocytemia does improve athletic performance in endurance events. Moreover, the procedure is considered dishonest by the International Olympics Committee, and there is still the possibility that it can lead to some negative effects. For example, there can be decreased delivery of oxygen to body cells if too many red blood cells are introduced. This results in decreased blood flow. In addition, there is the possibility of the introduction of hepatitis and influenza viruses into the blood.

Red blood cells are highly specialized for their transport function. They contain a large number of hemoglobin molecules, which increases their oxygen-carrying capacity. One estimate is that each erythrocyte can carry 280 million molecules of hemoglobin. The biconcave shape of a red blood cell has a much greater surface area than, say, a sphere or a cube. The erythrocyte thus presents a large surface area for the diffusion of gas molecules that pass through the membrane to combine with hemoglobin.

LIFE SPAN AND NUMBER

Red blood cells live only about 120 days because of wear and tear on their fragile plasma membranes as they squeeze through blood capillaries. The plasma membranes and other parts of worn-out red blood cells are removed from circulation by fixed phagocytic macrophages in the spleen and liver and recycled. The hemoglobin molecules are split apart, the iron is reused, and the rest of the molecule is converted into other substances for reuse (globin) or elimination (heme).

A healthy male has about 5.4 million red blood cells per cubic millimeter (mm^3) of blood, and a healthy female has about 4.8 million. The higher value in the male is caused by higher levels of testosterone, which stimulates the production of red blood cells. To maintain normal quantities of erythrocytes, the body must produce new mature cells at the astonishing rate of at least 2 million per second. In the adult, production takes place in the red bone marrow in the spongy bone of the cranium, ribs, sternum, bodies of vertebrae, proximal epiphyses of the humerus and femur, and hipbones.

PRODUCTION

The process by which erythrocytes are formed is called ***erythropoiesis*** (e-rith′-rō-poy-Ē-sis). It starts in red bone marrow with the transformation of a hemocytoblast into a proerythroblast (Figure 12-2a). The ***proerythroblast (rubriblast)*** gives rise to an ***early erythroblast (prorubricyte),*** which then develops into an ***intermediate erythroblast (rubricyte),*** the first cell in the sequence that begins to synthesize hemoglobin. The intermediate erythroblast next develops into a ***late erythroblast (metarubricyte),*** in which hemoglobin synthesis is at a maximum. In the next stage, the late erythroblast develops into a ***reticulocyte,*** a cell in the developmental sequence that contains about 34 percent hemoglobin, retains some mitochondria and ribosomes, and loses its nucleus. After the nucleus is lost, the cell is known as a reticulocyte. Reticulocytes pass from bone marrow into the bloodstream by squeezing between the endothelial cells of blood capillaries. Reticulocytes generally become ***erythrocytes,*** or mature red blood cells, within one to two days after their release from bone marrow. (A nucleated red blood cell found in red marrow, but rarely found in blood, is called a ***normoblast.***) The normal proportion of reticulocytes in blood is between 0.5 and 1.5 percent.

Normally erythropoiesis and red blood cell destruction proceed at the same pace. But if the body suddenly needs more erythrocytes or if erythropoiesis is not keeping up with red blood cell destruction, a homeostatic mechanism steps up erythrocyte production. The mechanism is triggered by the reduced supply of oxygen for body cells that results from a reduced number of erythrocytes or other dysfunctions that reduce oxygen delivery to tissues (hypoxia). If certain kidney cells become oxygen-deficient, they release an enzyme called ***renal erythropoietic factor*** that converts a plasma protein (possibly produced by the liver) into the hormone ***erythropoietin*** (*poiem* = to make). This hormone circulates through the blood to the red bone marrow, where it stimulates more hemocytoblasts to develop into red blood cells. Erythropoietin is also produced by other body tissues, particularly the liver.

CLINICAL APPLICATION

Reticulocyte Count and Hematocrit

The rate of erythropoiesis is measured by a ***reticulocyte*** (re-TIK-yoo-lō-sīt) ***count.*** Some reticulocytes are normally released into the bloodstream before they become mature red blood cells. A high reticulocyte count (greater than 1.5 percent) might indicate the response to bleeding, hemolysis (rapid breakdown of erythrocytes), or the response to iron therapy in someone who is iron-deficient. A low reticulocyte count (less than 0.5 percent) in the presence of anemia might indicate inability of the bone marrow to respond, owing to a nutritional deficiency, pernicious anemia, or leukemia.

Hematocrit (he-MAT-ō-krit), or ***Hct,*** is the percentage of red blood cells in blood. A hematocrit of 40 means that 40 percent of the volume of blood is composed of red blood cells. The test is used to diagnose anemia and polycythemia (an increase in the percentage of the red blood cells) and abnormal states of hydration. The normal range of hematocrit for females is 38–46 percent (average 42); for males it is 40–54 percent (average 47). A significant drop in hematocrit constitutes anemia, which may vary from mild (hematocrit of 35) to severe (hematocrit of less than 15). Polycythemic blood may have a hematocrit of 65 percent or higher. Athletes not uncommonly have a higher-than-average hematocrit, and the average hematocrit of persons living in mountainous terrain is greater than that of persons living at sea level.

BLOOD GROUP SYSTEMS

The surfaces of erythrocytes contain genetically determined antigens called ***agglutinogens*** (ag′-loo-TIN-ō-jens). There are at least 300 blood group systems that can be detected on the surface of red blood cells. The two major blood group classifications are the ABO and Rh systems.

ABO blood grouping is based on two agglutinogens, symbolized as *A* and *B*. Individuals whose erythrocytes manufacture only agglutinogen *A* are said to have blood type A. Those who manufacture only agglutionogen *B* are type B. Individuals who manufacture both *A* and *B* are type AB. Those who manufacture neither are type O.

The ***Rh blood grouping system*** is so named because it was first worked out in the blood of the *Rhesus* monkey. Individuals whose erythrocytes have the Rh agglutinogens (D antigens) are designated Rh^+. Those who lack Rh agglutinogens are designated Rh^-.

LEUKOCYTES (WHITE BLOOD CELLS)

STRUCTURE AND TYPES

Unlike red blood cells, ***leukocytes,*** also spelled ***leucocytes*** (LOO-kō-sīts), or ***white blood cells (WBCs),*** have nuclei and do not contain hemoglobin (see Figure 12-2).

Leukocytes fall into two major groups. The first group is the ***granular leukocytes.*** They develop from red bone marrow, have conspicuous granules in the cytoplasm, and possess lobed nuclei. The three kinds of granular leukocytes (***polymorphs, polymorphonuclear leukocytes,*** or ***PMNS***) are ***neutrophils*** (10 to 12 μm in diameter), ***eosinophils*** (10 to 12 μm in diameter), and ***basophils*** (8 to 10 μm in diameter). The nuclei of neutrophils have two to six lobes, connected by very thin strands. As the cells age, the extent of lobulation increases. Fine, evenly distributed pale lilac colored granules are found in the cytoplasm when a commonly used stain is applied to the cells. Eosinophils contain nuclei that are usually bilobed, with the lobes connected by a thin strand or thick isthmus. The cytoplasm is packed with large, uniform-sized granules that do not cover or obscure the nucleus. The granules stain red-orange with a commonly used stain. The nuclei of basophils are bilobed or irregular in shape, often in the form of a letter S. The cytoplasmic granules are round, variable in size, stain blue-black, and commonly obscure the nucleus.

The second principal group of leukocytes is the ***agranular leukocytes.*** They develop from lymphoid and myeloid tissue (red bone marrow), and no cytoplasmic granules can be seen under a light microscope, owing to their small size and poor staining qualities. The two kinds of agranular leukocytes are ***lymphocytes*** (7 to 15 μm in diameter) and ***monocytes*** (14 to 19 μm in diameter). The nuclei of lymphocytes are darkly stained, round, or slightly indented. The cytoplasm stains sky blue, and forms a rim around the nucleus. The nuclei of monocytes are usually indented or kidney-shaped, and the cytoplasm has a foamy appearnace.

Just as red blood cells have surface proteins, so do white blood cells and all other nucleated cells in the body. These proteins, called ***HLA (human leukocyte–associated) antigens,*** are unique for each person (except for identical twins) and can be used to identify a tissue. HLA antigens are determined by a group of genes on a single chromosome (chromosome 6). This group of genes is called the ***major histocompatibility complex (MHC).*** Since HLA antigens are determined by the MHC, they are also referred to as ***histocompatibility antigens.***

FUNCTIONS

The skin and mucous membranes of the body are continuously exposed to microbes and their toxins. Some of these microbes are capable of invading deeper tissues to cause disease, and once they enter the body, the general function of leukocytes is to combat them by phagocytosis or antibody production.

Neutrophils and Monocytes

Neutrophils and monocytes are actively ***phagocytotic***—they can ingest bacteria and dispose of dead matter (see Figure 2-4). Neutrophils (NOO-trō-fils) are the most active leukocytes in responding to tissue destruction by bacteria. In addition to carrying on phagocytosis, they release the enzyme lysozyme, which destroys certain bacteria. Apparently, monocytes (MON-ō-sīts) take longer to reach the site of infection than do neutrophils, but once they arrive, they do so in larger numbers and destroy more microbes. Monocytes that have migrated to infected tissues and differentiated into phagocytes are called ***wandering macrophages.*** They clean up cellular debris and microbes following an infection.

A number of different chemicals in inflamed tissue attract phagocytes toward the tissue. This phenomenon is called ***chemotaxis.*** Among the substances that provide stimuli for chemotaxis are toxins produced by microbes and degenerative products of damaged tissues and specialized products called kinins.

Most keukocytes possess, to some degree, the ability to migrate through the minute spaces between the cells that form the walls of capillaries and into connective and epithelial tissue. This movement through capillary walls is called ***diapedesis*** (di′-a-pe-DĒ-sis). First, part of the cell membrane projects outward. Then the cytoplasm and nucleus flow into the projection. Finally, the rest of the membrane moves into place. Another projection is make, and so on, until the cell has migrated to its destination.

Neutrophils also contain ***defensins,*** peptides that exhibit a broad range of antibiotic activity against bacteria, fungi, and viruses.

Eosinophils

Eosinophils (ē′-ō-SIN-ō-fils) are believed to release substances that combat the effects of histamine and other mediators of inflammation in allergic reactions. Eosinophils leave the capillaries, enter the tissue fluid, and phagocytize antigen–antibody complexes. Eosinophils are also effective against certain parasitic worms. Thus, a high eosinophil count frequently indicates an allergic condition or a parasitic infection.

Basophils

Basophils (BĀ-sō-fils) are also believed to be involved in allergic reactions. Basophils leave the capillaries, enter the tissues, and liberate heparin, histamine, and serotonin. These substances intensify the overall inflmmatory reaction and are involved in hypersensitivity (allergic) reactions (Chapter 15).

Lymphocytes

Lymphocytes (LIM-fō-sīts) are involved in the production of ***antibodies*** (AN-ti-bod′-ēz), which are special proteins that inactivate antigens. ***Antigens*** (AN-ti-jens) are substances that will stimulate the production of antibodies and are capable of reacting specifically with the antibody. Most antigens are proteins, and most are not synthesized by the body. Many of the proteins that make up the cell structures and enzymes of bacteria are antigens. The toxins released by bacteria are also antigens. When antigens enter the body, they react chemically with substances in the lymphocytes and stimulate some lymphocytes, called ***B cells,*** to become ***plasma cells*** (Figure 12-4). The plasma cells then produce antibodies, globulin-type proteins that attach to antigens in a very precise way. A specific antibody will generally attach only to a certain antigen. However, unlike enzymes, which enhance the reactivity of the substrate, antibodies "cover" their antigens so the antigens cannot come in contact with other chemicals in the body. In this way, bacterial poisons can be sealed up and rendered harmless. The bacteria themselves are destroyed by the antibodies. This process

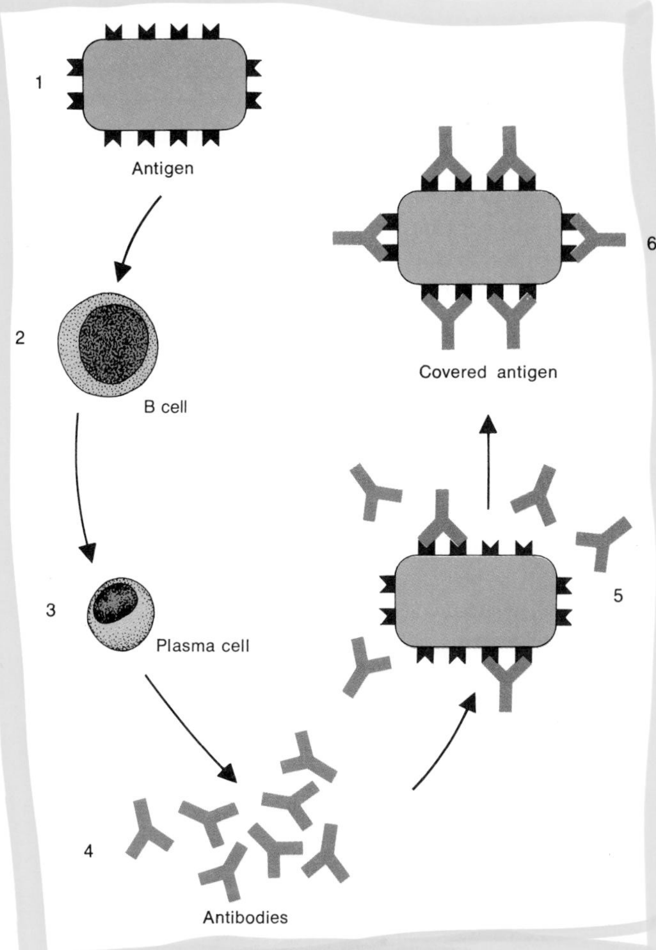

FIGURE 12-4 Antigen–antibody response. An antigen entering the body stimulates a B cell to develop into an antibody-producing plasma cell. The antibodies attach to the antigen, cover it, and render it harmless.

is called the ***antigen–antibody response***. Phagocytes in tissues destroy the antigen–antibody complexes.

Other lymphocytès are called ***T cells***. One group of T cells, the ***cytotoxic (killer) T cells***, are activated by certain antigens and react by destroying them directly or indirectly by recruiting other lymphoctyes and macrophages. T cells are especially effective against bacteria, viruses, fungi, transplanted cells, and cancer cells.

The antigen–antibody response helps us combat infection and gives us immunity to some diseases. It is also responsible for transfusion reactions, allergies, and the body's rejection of organs transplanted from an individual with a different genetic makeup.

An increase in the number of white cells present in the blood typically indicates a state of inflammation or infection. Because each type of white cell plays a different role, determining the *percentage* of each type in the blood assists in diagnosing the condition. This test is called a differential white blood cell count and should not be confused with a total count, which determines the *number* of white blood cells per cubic millimeter (mm^3) of blood.

CLINICAL APPLICATION

Differential White Blood Cell Count

A ***differential white blood cell count*** is a rountine part of the complete blood count (CBC), to be discussed shortly, that may be helpful in evaluating infection or inflammation, determining the effects of possible poisoning by chemicals or drugs, monitoring blood disorders (e.g., leukemia) and effects of chemotherapy, or detecting allergic reactions and parasitic infections. The percentage of each type of white blood cell in a normal differential white blood cell count is as follows:

	Percent
Neutrophils	60–70
Eosinophils	2–4
Basophils	0.5–1
Lymphoctyes	20–25
Monocytes	3–8
	100

An increased neutrophil count might result from infections (bacterial), burns, stress, or inflammation; a decreased count might be caused by radiation, certain drugs, vitamin B_{12} deficiency, or systemic lupus erythematosus (SLE). An increase in the number of eosinophils could indicate allergic reactions, parasitic infections, autoimmune disease, or adrenal insufficiency; a decrease could be caused by certain drugs, stress, or Cushing's syndrome. Basophils could be elevated in some types of allergic responses, leukemias, cancers, and hypothyroidism, decreases could occur during pregnancy, ovulation, stress, and hyperthyroidism. High lymphocyte counts could indicate viral infections, immune diseases, and some leukemias, low counts might occur as a result of prolonged severe illness, high steroid levels, and immunosuppression. Finally, a high monocyte count could result from certain viral or fungal infections, tuberculosis (TB), some leukemias, and chronic diseases; below-normal monocyte levels rarely occur.

LIFE SPAN AND NUMBER

Bacteria exist everywhere in the environment and have continuous access to the body through the mouth, nose, and pores of the skin. Furthermore, many cells, especially those of epithelial tissue, age and die, and their remains must be disposed of daily. Even when the body is healthy, the leukocytes actively ingest bacteria and debris. However, a leukocyte can phagocytize only a certain number of substances before they interfere with the leukocyte's normal metabolic activities and bring on its death. Consequently, the life span of most leukocytes is very short. In a healthy body, some white blood cells can live up to several months, but most live only a few days. During a period of infection, they may live only a few hours. Aged and dead leukocytes are thought to be removed by fixed phagocytic macrophages in the liver and spleen and locally within connective tissue.

Leukocytes are far less numerous than red blood cells, averaging from 5,000 to 10,000 cells per cubic millimeter of blood. Red blood cells, therefore, outnumber white blood cells about 700:1. The term ***leukocytosis*** (loo'-kō-sī-TŌ-sis) refers to an increase in the number of white blood cells. If the increase exceeds 10,000/mm^3, a pathological condition is usually indicated. An abnormally low level of white blood cells (below 5,000/mm^3) is termed ***leukopenia*** (loo-kō-PĒ-nē-a).

PRODUCTION

Granular leukocytes are produced in red bone marrow (myeloid tissue); agranular leukocytes are produced in both myeloid and lymphoid tissue. The developmental sequences for the five types of leukocytes are shown in Figure 12-2a.

Recall that red blood cells develop under the influence of the hormone erythropoietin, one of the hematopoietins. White blood cells also develop under the influence of hematopoietins, but these are called ***colony-stimulating factors (CSFs).*** In addition to stimulating the development of various white blood cells, CSFs also enhance their function. Because of their roles in immunity, CSFs are also being tested to determine their effects on bolstering body defenses. Trials are in progress to evaluate the effectiveness of CSFs in treating AIDS, cancer, and bone marrow suppression.

CLINICAL APPLICATION

Bone Marrow Transplant

Bone marrow, as indicated earlier, contains hemocytoblasts that differentiate into erythrocytes, leukocytes, and platelets. Among the leukocytes are cells that combat infections and cause tissue rejection. Until recently, the use of ***bone marrow transplant,*** the transfer of bone marrow from a donor to a recipient, required that the marrow of the donor be obtained from a relative and closely matched to that of the recipient. In cases where the match is not identical, T cells in donor marrow, which are part of the immune system, recognize the recipient's tissues as foreign and attack them. The principal tissues rejected in this way are in the skin, liver, and gastrointestinal tract. Several recent advances now permit bone marrow transplants in which the tissue match between donor and recipient is not even close.

Donor marrow is aspirated from the hipbones, mixed with heparin (an anticoagulant), and then passed through screens. The suspension of bone marrow cells is treated to remove T cells and given to the recipient just like a blood transfusion. The cells in the suspension pass through the lungs, enter general circulation, and reseed and grow in the marrow cavities of the recipient's bones.

Bone marrow transplants have been used to treat aplastic anemia, certain types of leukemia, and severe combined immunodeficiency disease (SCID), an inherited deficiency of infection-fighting blood cells. The technique is also being expanded to treat other kinds of leukemia, non-Hodgkin's lymphoma, thalassemia, multiple myeloma, sickle-cell anemia (SCA), and hemolytic anemia.

THROMBOCYTES (PLATELETS)

STRUCTURE

In addition to the immature cell types that develop into erythrocytes and leukocytes, hemocytoblasts differentiate into still another kind of cell, called a megakaryoblast (see Figure 12-2a). Megakaryoblasts are ultimately transformed into megakaryocytes, large cells that shed fragments of cytoplasm. Each fragment becomes enclosed by a piece of the cell membrane and is called a ***thrombocyte*** (THROM-bō-sīt), or ***platelet.*** Thrombocytes break off from the megakaryocytes in bone marrow and then enter the blood circulation, whereas the megakaryocytes stay in the marrow. Platelets are disc-shaped structures without a nucleus. They average from 2 to 4 μm in diameter.

FUNCTION

Platelets repair slightly damaged blood vessels and initiate a chain of reactions that results in blood clotting.

LIFE SPAN AND NUMBER

Platelets have a short life span, normally just five to nine days. Aged and dead platelets are removed by fixed phagocytic macrophages in the spleen and liver. Between 250,000 and 400,000 platelets are present in each cubic millimeter of blood.

PRODUCTION

Platelets are produced in red bone marrow according to the development sequence shown in Figure 12-2a.

A summary of the formed elements in blood is presented in Exhibit 12-3.

CLINICAL APPLICATION

Complete Blood Count (CBC)

A ***complete blood count (CBC)*** is a valuable test that screens for anemia and various infections. Usually included

EXHIBIT 12-3

Summary of the Formed Elements in Blood

FORMED ELEMENT	NUMBER	DIAMETER (in μm)	LIFE SPAN	FUNCTION
ERYTHROCYTE (RED BLOOD CELL)	4.8 million/mm³ in females; 5.4 million/mm³ in males.	8	120 days.	Transports oxygen and carbon dioxide.
LEUKOCYTE (WHITE BLOOD CELL)	5,000–10,000/mm³.		Few hours to a few days.[a]	
Granular				
Neutrophil	60–70% of total.	10–12		Phagocytosis.
Eosinophil	2–4% of total.	10–12		Combats the effects of histamine in allergic reactions, phagocytizes antigen–antibody complexes, and destroys certain parasitic worms.
Basophil	0.5–1% of total.	8–10		Liberates heparin, histamine, and serotonin in allergic reactions that intensify the overall inflammatory response.
Agranular				
Lymphocyte	20–25% of total.	7–15		Immunity (antigen–antibody reactions).
Monocyte	3–8% of total.	14–19		Phagocytosis (after transforming into wandering macrophages).
THROMBOCYTE (PLATELET)	250,000–400,000/mm³.	2–4	5–9 days.	Blood clotting.

[a] Some lymphocytes, called T and B memory cells, can live throughout life once they are established. Most white blood cells, however, have life spans ranging from a few hours to a few days.

are a determination of red blood cell count, hemoglobin, hematocrit, white blood cell count, differential white blood cell count, and platelet count.

APPLICATIONS TO HEALTH

ANEMIA

Anemia is a condition in which the oxygen-carrying capacity of the blood is impaired; it is a sign, not a diagnosis. Many kinds of anemia exist, all characterized by insufficient erythrocytes or hemoglobin. These conditions lead to fatigue and intolerance to cold, both of which are related to lack of oxygen needed for energy and heat production, and to paleness, which is due to low hemoglobin content.

Nutritional Anemia

Nutritional anemia arises from an inadequate diet, one without sufficient amounts of iron, the necessary amino acids, or vitamin B_{12}.

Pernicious Anemia

Pernicious anemia is the insufficient production of erythrocytes resulting from an inability of the stomach to produce intrinsic factor, which allows absorption of vitamin B_{12}.

Hemorrhagic Anemia

An excessive loss of erythrocytes through bleeding is called ***hemorrhagic anemia.*** Common causes are large wounds, stomach ulcers, and heavy menstrual bleeding.

Hemolytic Anemia

If erythrocyte cell membranes rupture prematurely, the cells remain as "ghosts," and their hemoglobin pours out into the plasma. A characteristic sign of this condition, called ***hemolytic anemia,*** is distortion in the shape of erythrocytes. It may result from inherent defects, such as hemoglobin defects, abnormal red blood cell enzymes, or defects of the red blood cell membrane. Agents that may cause hemolytic anemia are parasites, toxins, and antibodies from incompatible blood (Rh^- mother and Rh^+ fetus, for instance).

Hemolytic disease of the newborn (erythroblastosis fetalis) is an example of a hemolytic anemia.

The term ***thalassemia*** (thal′-a-SĒ-mē-a) represents a group of hereditary hemolytic anemias, resulting from a defect in the synthesis of hemoglobin, which produces extremely thin and fragile erythrocytes. It occurs primarily in populations from countries bordering the Mediterranean Sea. Treatment generally consists of blood transfusions.

Aplastic Anemia

Destruction or inhibition of the red bone marrow results in ***aplastic anemia.*** Typically, the marrow is replaced by fatty tissue, fibrous tissue, or tumor cells. Toxins, gamma radiation, and certain medications are causes. Many of the medications inhibit the enzymes involved in hemopoiesis.

Sickle-Cell Anemia (SCA)

The erythrocytes of a person with ***sickle-cell anemia (SCA)*** manufacture an abnormal kind of hemoglobin. When such an erythrocyte gives up its oxygen to the interstitial fluid, the abnormal hemoglobin tends to lose its integrity in places of low oxygen tension and forms long, stiff, rodlike structures that bend the erythrocyte into a sickle shape (Figure 12-5). The sickled cells rupture easily. Even though erythropoiesis is stimulated by the loss of the cells, it cannot keep pace with the hemolysis. The individual consequently suffers from a hemolytic anemia that reduces the amount of oxygen that can be supplied to the tissues. Prolonged oxygen reduction may eventually cause extensive tissue damage. Furthermore, because of the shape of the sickled cells, they tend to get stuck in blood vessels and can cut off blood supply to an organ altogether.

Sickle-cell anemia is an inherited disorder and is characterized by several symptoms. In young children, hand–foot syndrome is present, in which there is swelling and pain in the wrists and feet. Older patients experience abdominal pain and pain in the back and extremities without swelling. Complications include neurological disorders (meningitis, seizures, stroke), impaired pulmonary function, orthopedic abnormalities (femoral head necrosis, osteomyelitis), genitourinary tract disorders (involuntary urination, blood in urine, kidney failure), ocular disturbances (hemorrhage, detached retina, blindness), and obstetric complications (convulsions, coma, infection).

There are several blood tests designed to determine sickle-cell anemia. In the most common screening test, a small amount of blood is taken from a finger, oxygen is removed, and the sample is examined for sickled cells.

Treatment consists of administration of analgesics to relieve pain, antibiotics to counter infections, and transfusion therapy.

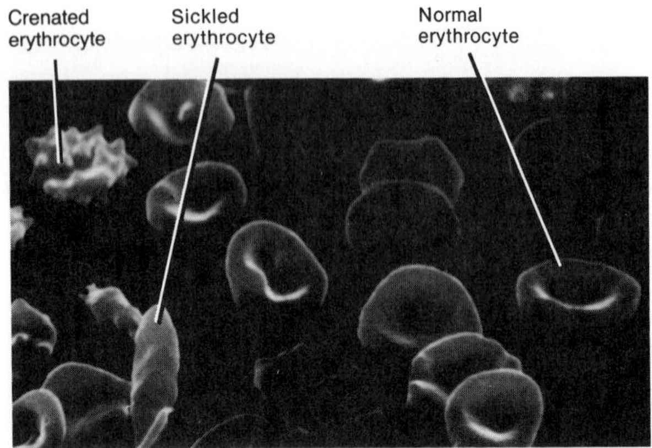

Scanning electron micrograph of erythrocytes in sickle-cell anemia, 2000×

FIGURE 12-5 Sickle-cell anemia. (Courtesy of Fisher Scientific Company and S.T.E.M. Laboratories, Inc., Copyright © 1975.)

POLYCYTHEMIA

The term ***polycythemia*** (pol′-ē-sī-THĒ-mē-a) refers to a disorder characterized by a hematocrit (Hct) that is elevated significantly above the normal upper limit of about 55. There is an increased blood viscosity associated with the elevated hematocrit. The increased viscosity causes a rise in blood pressure and contributes to thrombosis (blood clot formation) and hemorrhage. The thrombosis results from too many red blood cells piling up as they try to enter smaller vessels. The hemorrhage is due to widespread hyperemia (unusually large amount of blood in an organ).

INFECTIOUS MONONUCLEOSIS (IM)

Infectious mononucleosis (IM) is a contagious disease primarily affecting lymphoid tissue throughout the body but also affecting the blood. It is caused by the *Epstein-Barr virus (EBV)*, the same agent that has been linked to Burkitt's lymphoma, nasopharyngeal carcinoma, and Hodgkin's disease. It occurs mainly in children and young adults, with the peak incidence at 15 to 20 years of age. The ratio of females to males affected by IM is 3:1. The virus most commonly enters the body through intimate oral contact, multiplies in lymphoid tissues, and spreads into the blood where it infects and multiplies in B lymphocytes, the primary host cells. As a result of this infection, the B lymphocytes become enlarged and abnormal in appearance and resemble monocytes, the primary reason for which the disease receives its name, mononucleosis. Infectious mononucleosis is characterized by an elevated white blood cell count with an abnormally high percentage of lymphocytes. Symptoms include fatigue, headache, dizziness, sore throat, lymphadenopathy (enlarged and tender lymph nodes), fever, brilliant red throat and soft palate, stiff neck, cough, and malaise. The spleen may also enlarge. Secondary complications may be hematologic (hemolytic anemia, neutropenia), respiratory (pneumonia), neurologic (meningitis, encephalitis, seizures,

Guillain-Barré syndrome), cardiovascular (myocarditis), gastrointestinal (hepatitis, splenic rupture), and renal (nephritis, nephrosis). There is no cure for infectious mononucleosis, and treatment consists of watching for and treating complications. Usually the disease runs its course in a few weeks, and the individual generally suffers no permanent ill effects.

CHRONIC FATIGUE SYNDROME

Chronic fatigue syndrome typically occurs among young adults, primarily females. Diagnosis of the disease must include the following: (1) new, extreme fatigue that impairs normal activities for at least six months; (2) the absence of known diseases (cancer, infections, drug abuse, toxicity, or psychiatric disorders) that might produce similar symptoms; and (3) at least 8 of these 11 indications that persist or recur over six months: mild fever or chills, sore throat, painful lymph nodes, general muscle weakness, muscle pain, fatigue for more than 24 hours after mild exercise, headache that differs in type and severity from past ones, joint pain without swelling, neuropsychological complaints (irritability, confusion, depression), sleep disturbances, and development of the initial symptoms over a few hours to a few days. The diagnosis can also be made if the patient reports 6 of the 11 symptoms just indicated plus observation by a physician of two of these three physical signs: low-grade fever, inflamed throat, and enlarged lymph nodes in the neck or axilla.

Among the possible causes of chronic fatigue syndrome are emotional factors such as depression and excess stress and physical factors such as a viral or bacterial infection or reactivation of an existing virus. There is no effective treatment for the disease.

LEUKEMIA

Clinically, ***leukemia*** is classified on the basis of the duration and character of the disease, that is, acute or chronic. In its simplest terms, acute leukemia refers to a malignant disease of blood-forming tissues characterized by uncontrolled production and accumulation of immature leukocytes. In chronic leukemia, there is an accumulation of mature leukocytes in the bloodstream because they do not die at the end of their normal life span. The *human T cell leukemia-lymphoma virus-1* (*HTLV-1*) is strongly associated with some types of leukemia. Leukemia is also classified according to the identity and site of origins of the predominant cell involved, such as myelocytic (myelogenous, myeloblastic, granulocytic), lymphocytic (lymphogenous, lymphatic), and monocytic.

In acute leukemia, the anemia and bleeding problems commonly seen result from the crowding out of normal bone marrow cells by the overproduction of immature cells, preventing normal production of red blood cells and platelets. One cause of death from acute leukemia is internal hemorrhaging, especially cerebral hemorrhage that destroys the vital centers in the brain. Perhaps the most frequent cause of death is uncontrolled infection due to lack of mature or normal leukocytes. The abnormal accumulation of immature leukocytes may be reduced by using x-rays and antileukemic drugs. Partial or complete remissions may be induced, with some lasting as long as 15 years.

KEY MEDICAL TERMS ASSOCIATED WITH BLOOD

Acute normovolemic (nor′-mō-vō-LĒ-mik) **hemodilution** (hē′-mō-dī-LOO-shun) Removal of blood immediately before surgery and replacing it with a cell-free crystalloid or colloid solution to maintain normal blood volume for adequate circulation. At the conclusion of surgery, when hemostasis (blood clotting) has been controlled, the collected blood is returned to the body.

Autologous (aw-TOL-o-gus; *auto* = self) **preoperative transfusion** (trans-FYOO-zhun) Donating one's own blood for up to six weeks before elective surgery to ensure an abundant supply; reduce transfusion complications such as those that may be associated with AIDS, hepatitis, and other blood-borne diseases; avoid a fatal transfusion reaction; ease the demand on blood banks; and provide an alternative to some individuals who object to receiving banked blood for religious reasons. The ability of patients to predonate greater volumes of blood is enhanced by administration of recombinant erythropoietin. It is not indicated for individuals who cannot donate 500 ml of blood in a short period of time and who have preexisting anemia, bleeding disorders, infection, unstable blood pressure, and malignancy involving bone marrow. Also called **predonation.**

Autologous intraoperative transfusion (AIT) Procedure in which blood lost during surgery is suctioned from the patient, treated with an anticoagulant, filtered of debris, and centrifuged to recover the red blood cells. Then the red blood cells are washed in saline solution and reinfused into the patient.

Blood bank A stored supply of blood for future use by the donor or other individuals. Since blood banks have now assumed additional and diverse functions (immunohematology reference work, continuing medical education, bone and tissue storage, and clinical consultation), they are more appropriately referred to as **centers of transfusion medicine.**

Citrated (SIT-rāt-ed) **whole blood** Whole blood protected from coagulation by CPD (citrate phosphate dextrose) or a similar compound.

Cyanosis (sī′-a-NŌ-sis; *cyano* = blue) A reduced hemoglobin (unoxygenated) concentration of blood of more than 5 g/dl that results in a dark purple discoloration most easily seen in the nail beds and mucous membranes.

Direct (immediate) transfusion (*trans* = through) Transfer of blood directly from one person to another without exposing the blood to a storage container.

Exchange transfusion Removing blood from the recipient while alternately replacing it with donor blood. This method is used for treating hemolytic disease of the newborn (HDN) and poisoning. The treatment may be intrauterine.

Gamma globulin (GLOB-yoo-lin) Solution of globulins from nonhuman blood consisting of antibodies that react with specific pathogens, such as measles, epidemic hepatitis, tetanus, and possibly poliomyelitis viruses. It is prepared by injecting the specific virus into animals, removing blood from the animals after antibodies have accumulated, isolating antibodies, and injecting them into a human for short-term immunity.

Hemochromatosis (hē-mō-krō′-ma-TŌ-sis; *heme* = iron; *chroma* = color) Disorder of iron metabolism characterized by excess deposits of iron in tissues, especially the liver and pancreas, that result in bronze coloration of the skin, cirrhosis, diabetes mellitus, and bone and joint abnormalities.

Hemorrhage (HEM-or-ij; *rrhage* = bursting forth) Bleeding, either internal (from blood vessels into tissues) or external (from blood vessels directly to the surface of the body).

Indirect (mediate) transfusion Transfer of blood from a donor to a container and then to the recipient, permitting blood to be stored for an emergency. The blood may be separated into its components so that a patient will receive only a needed part.

Multiple myeloma (mī′-e-LŌ-ma) Malignant disorder of plasma cells in bone marrow; symptoms (pain, osteoporosis, hypercalcemia, thrombocytopenia, kidney damage) are caused by the growing tumor cell mass or antibodies produced by malignant cells.

Platelet (PLĀT-let) **concentrates** A preparation of platelets obtained from freshly drawn whole blood and used for transfusions in platelet-deficiency disorders such as hemophilia.

Porphyria (por-FĒ-rē-a or por-FI-rē-a) Any of a group of inherited disorders caused by the accumulation in the body of substances called porphyrins (precursor substances formed during the synthesis of hemoglobin and other important molecules). The accumulation is the result of inherited enzyme deficiencies. Symptoms of porphyria include a rash or skin blistering brought on by sunlight, abdominal pain, and nervous system disturbances from certain drugs, such as barbiturates and alcohol.

Reciprocal (re-CIP-rō-cal) **transfusion** Transfer of blood from a person who has recovered from a contagious infection into the vessels of a patient suffering with the same infection. An equal amount of blood is returned from the patient to the well person.

Septicemia (sep′-ti-SĒ-mē-a; *sep* = decay; *emia* = condition of blood) Toxins or disease-causing bacteria in the blood. Also called "blood poisoning."

Thrombocytopenia (throm′-bō-sī′-tō-PĒ-nē-a; *thrombo* = clot; *penia* = poverty) Very low platelet count that results in a tendency to bleed from capillaries.

Transfusion (trans-FYOO-zhun) Transfer of whole blood, blood components (red blood cells only or plasma only), or bone marrow directly into the bloodstream.

Venesection (vēn′-e-SEK-shun; *veno* = vein) Opening of a vein for withdrawal of blood. Although **phlebotomy** (fle-BŌT-ō-mē; *phlebo* = vein; *tome* = to cut) is a synonym for venesection, in clinical practice phlebotomy refers to therapeutic bloodletting, such as might be done to remove a pint of blood to lower the viscosity of blood of a patient with polycythemia.

Whole blood Blood containing all formed elements, plasma, and plasma solutes in natural concentration.

STUDY OUTLINE

Functions of Blood (p. 334)

1. Blood transports oxygen, carbon dioxide, nutrients, wastes, and hormones.
2. It helps to regulate pH, body temperature, and water content of cells.
3. It prevents blood loss through clotting and combats toxins and microbes through special combat-unit cells.

Physical Characteristics of Blood (p. 334)

1. The cardiovascular system consists of blood, the heart, and blood vessels. The lymphatic system consists of lymph, lymph vessels, and lymph glands.
2. Physical characteristics of blood include viscosity, 4.5 to 5.5; temperature, 38°C (100.4°F); pH, 7.35 to 7.45; and a salt (NaCl) concentration of 0.90 percent. Blood constitutes about 8 percent of body weight.
3. Blood samples may be obtained by venipuncture, fingerstick, or arterial stick.

Components of Blood (p. 335)

1. The formed elements in blood include erythrocytes (red blood cells), leukocytes (white blood cells), and thrombocytes (platelets).
2. The liquid portion of blood, called plasma, consists of 91.5 percent water and 8.5 percent solutes.
3. Principal solutes include proteins (albumins, globulins, fibrinogen), nonprotein nitrogen (NPN) substances, foods, hormones, respiratory gases, and electrolytes.
4. Plasmapheresis involves withdrawing blood from the body, selectively separating its components, removing undesirable components, and returning it to the body.

Origin of Blood Cells (p. 339)

1. Blood cells are formed by a process called hemopoiesis.
2. Red bone marrow (myeloid tissue) is responsible for producing red blood cells, granular leukocytes, and platelets; lymphoid tissue and myeloid tissue produce agranular leukocytes.

Erythrocytes (Red Blood Cells) (p. 339)

1. Erythrocytes are biconcave discs without nuclei and containing hemoglobin.
2. The function of red blood cells is to transport oxygen and carbon dioxide.
3. Red blood cells live about 120 days. A healthy male has about 5.4 million/mm^3 of blood; a healthy female, about 4.8 million/mm^3.

4. Erythrocyte formation, called erythropoiesis, occurs in adult red marrow of certain bones.
5. Antigens on the surfaces of erythrocytes provide the basis for the ABO and Rh blood group systems.
6. A reticulocyte count is a diagnostic test that indicates the rate of erythropoiesis.
7. A hematocrit (Hct) measures the percentage of red blood cells in whole blood.

Leukocytes (White Blood Cells) (p. 341)

1. Leukocytes are nucleated cells. Two principal types are granular (neutrophils, eosinophils, basophils) and agranular (lymphocytes and monocytes).
2. The general function of leukocytes is to combat inflammation and infection. Neutrophils and monocytes (wandering macrophages) do so through phagocytosis.
3. Eosinophils combat the effects of histamine in allergic reactions, phagocytize antigen–antibody complexes, and combat parasitic worms; basophils liberate heparin, histamine, and serotonin in allergic reactions that intensify the inflammatory response.
4. Lymphocytes, in response to the presence of foreign substances called antigens, differentiate into tissue plasma cells that produce antibodies. Antibodies attach to the antigens and render them harmless. This antigen–antibody response combats infection and provides immunity.
5. A differential white blood cell count is a diagnostic test in which specific white blood cells are enumerated.
6. White blood cells usually live for only a few hours or a few days. Normal blood contains 5,000 to 10,000/mm^3.

Thrombocytes (Platelets) (p. 344)

1. Thrombocytes are disc-shaped structures without nuclei.
2. They are formed from megakaryocytes and are involved in clotting.
3. Normal blood contains 250,000 to 400,000/mm^3.
4. A complete blood count (CBC) is used to determine blood cell counts, hemoglobin, and hematocrit and comments about blood cell morphology.

Applications to Health (p. 344)

1. Anemia is a decreased erythrocyte count or hemoglobin deficiency. Kinds of anemia include nutritional, pernicious, hemorrhagic, hemolytic, aplastic, and sickle-cell anemia (SCA).
2. Polycythemia is an abnormal increase in the number of erythrocytes.
3. Infectious mononucleosis (IM) is a contagious disease that primarily affects lymphoid tissue. It is characterized by an elevated white blood cell count, with an abnormally high percentage of lymphocytes. The cause is the Epstein-Barr virus (EBV).
4. Chronic fatigue syndrome is characterized by extreme fatigue for at least six months and the absence of known diseases that might produce similar symptoms. Among the symptoms are sore throat, headache, muscular aches, fever and chills, fatigue, joint pain, and neurological defects.
5. Leukemia is a malignant disease of blood-forming tissues characterized by the uncontrolled production of white blood cells that interferes with normal clotting and vital body activities.

REVIEW QUESTIONS

1. How are blood, interstitial fluid, and lymph related to the maintenance of homeostasis? (p. 334)
2. Distinguish between the cardiovascular system and lymphatic system. (p. 334)
3. List the functions of blood and their relationship to other systems of the body. (p. 334)
4. List the principal physical characteristics of blood. (p. 334)
5. Distinguish between plasma and formed elements. (p. 335)
6. What are the major constituents in plasma? What do they do? What is the difference between plasma and serum? (p. 336)
7. Describe the origin of blood cells. (p. 339)
8. Describe the microscopic appearance of erythrocytes. What is the function of erythrocytes? (p. 339)
9. Define erythropoiesis. Relate erythropoiesis to red blood cell count. What factors accelerate and decelerate erythropoiesis? (p. 340)
10. Describe the classification of leukocytes and describe the microscopic appearance. What are their functions? (p. 340)
11. What is the importance of diapedesis, chemotaxis, and phagocytosis in fighting bacterial invasion? (p. 342)
12. Distinguish between leukocytosis and leukopenia. (p. 344)
13. Describe the antigen–antibody response. How is it protective? (p. 342)
14. Describe the structure and function of thrombocytes. (p. 344)
15. Compare erythrocytes, leukocytes, and thrombocytes with respect to size, number per mm^3, and life span. (p. 345)
16. Define anemia. Contrast the causes of nutritional, pernicious, hemorrhagic, hemolytic, aplastic, and sickle-cell anemia (SCA). (p. 345)
17. What is infectious mononucleosis (IM)? (p. 346)
18. Describe the symptoms of chronic fatigue syndrome. (p. 343)
19. What is leukemia, and what are the causes of some of its symptoms? (p. 347)
20. Briefly describe the diagnostic value for the following blood tests: reticulocyte count (p. 340), hematocrit (Hct) (p. 341), differential white blood cell count (p. 343), and complete blood count (CBC) (p. 344).
21. Refer to the glossary of key medical terms associated with blood. Be sure that you can define each term (p. 347).

SELF QUIZ

1. Match the following:

___ a. constitute the largest percentage (about 70 percent) of leukocytes
___ b. make up 20–25 percent of leukocytes
___ c. are involved in immunity; form plasma cells for antibody production
___ d. are involved in allergic response; release serotonin, heparin, and histamine
___ e. are involved in allergic response; release antihistamines
___ f. develop into wandering macrophages that clean up sites of infection
___ g. are important in phagocytosis (two answers)
___ h. are classified as granular leukocytes (three answers)

A. basophils
C. lymphocytes
D. monocytes
E. neutrophils

Complete the following:

2. Blood is a connective tissue that consists of about ________ percent intercellular material and about ________ percent cells or formed elements. The intercellular material is a liquid named ________. The process of blood formation is called ________.

3. Arrange, in order, these stages in erythropoiesis. Write letters on the lines provided. ________ ________ ________ ________ A. late erythroblast formed; B. reticulocyte loses nucleus; C. erythroblast starts to synthesize hemoglobin; D. reticulocyte moves out of marrow into general circulation.

4. During infection, it is likely that a person's leukocyte count will (increase? decrease?) A count of (4,000? 8,000? 15,000?) leukocytes/mm^3 blood is most likely. This condition is known as (leukocytosis? leukopenia?). In order to confirm changes in counts of specific types of leukocytes, a ________ count may be taken.

Choose the one best answer to these questions.

___ 5. Choose the false statement about neutrophils:
A. they are actively phagocytic; B. they are the most abundant type of leukocyte; C. their number decreases during most infections; D. an increase in their number would be a form of leukocytosis; E. they are granular leukocytes.

___ 6. The normal red blood cell count in healthy males is ________ RBCs/mm^3.
A. 5.4 million; B. 2 million; C. 0.5 million; D. 250,000; E. 8,000.

___ 7. All of the following types of formed elements are produced in red bone marrow (myeloid tissue) *except*
A. neutrophils; B. basophils; C. platelets; D. erythrocytes; E. lymphocytes.

___ 8. Megakaryocytes are involved in the formation of
A. red blood cells; B. basophils; C. lymphocytes; D. platelets; E. neutrophils.

___ 9. Choose the false statement about blood:
A. blood is thicker than water. B. It normally has a pH of 6.5–7.0; C. the male human body normally contains about 5–6 liters of it; D. it normally consists of more plasma than cells; E. it constitutes about 8 percent of total body weight.

___ 10. Cells that form mature red blood cells are
A. proerythroblasts; B. lymphoblasts; C. myeloblasts; D. megakaryoblasts; E. monoblasts.

___ 11. Which of the following statements about the function of blood is *not* true?
A. it transports gases, nutrients, and wastes; B. it plays a role in the regulation of temperature; C. it transports tissue cells; D. it regulates the water content of cells; E. it regulates pH.

___ 12. The parent (stem) cell of all blood cells is the
A. myeloblast; B. megakaryoblast; C. hemocytoblast; D. monoblast; E. proerythroblast.

___ 13. The ability of white blood cells to crawl through capillaries and reach an injured or infected body tissue is
A. diapedesis; B. phagocytosis; C. leukocytosis; D. leukopenia; E. pinocytosis.

___ 14. Which of these values is *not* considered normal for blood?
A. viscosity of 4.5–5.5; B. temperature of 98.6°F; C. pH of 7.35–7.45; D. volume of 4–6 liters; E. salt concentration of 0.90 percent.

___ 15. Which of the following numbers of white blood cells per cubic millimeter of blood represents leukopenia?
A. 12,000; B. 16,000; C. 4,000; D. 10,000; E. 8,000.

___ 16. Erythrocytes
A. are nonnucleated, biconcave discs; B. are usually called white blood corpuscles; C. are formed in the lungs; D. contain fibroblasts; E. are phagocytes.

___ 17. The percentage of formed elements in the blood, by volume, is called the
A. red blood cell count; B. white blood cell count; C. hematocrit; D. serum volume; E. Cushny determination.

___ 18. One function of platelets is to aid in the
A. transport of carbon dioxide; B. production of vitamin K; C. utilization of calcium and phosphorus; D. destruction of bacteria; E. clotting of blood.

Circle T (true) or F (false) for the following:

T F 19. The liquid portion of blood that contains the clotting proteins is called plasma.

T F 20. The three kinds of granular leukocytes are neutrophils, monocytes, and eosinophils.

T F 21. The pigment hemoglobin is responsible for the color of blood and the carriage of respiratory gases.

T F 22. The process by which blood cells are formed is called fibrinolysis.

T F 23. A high lymphocyte count may indicate that an allergic reaction is taking place in the body.

T F 24. A count of 150,000 thrombocytes per cubic millimeter of blood would probably interfere with the blood's ability to clot.

T F 25. Blood groups, such as ABO and Rh, are distinguished on the basis of antigens on the surfaces of leukocytes.

The Cardiovascular System: The Heart

13

STUDENT OBJECTIVES

1. Describe the location of the heart and identify its borders.
2. Describe the structure of the pericardium and heart wall.
3. Identify and describe the chambers, great vessels, valves, and cardiac skeleton of the heart.
4. Describe the surface anatomy features of the heart.
5. Discuss the route of blood in coronary (cardiac) circulation.
6. Explain the structural and functional features of the conduction system of the heart and describe an electrocardiogram (ECG) and explain its significance.
7. List the risk factors involved in heart disease.
8. Describe the development of the heart.
9. Describe how atherosclerosis and coronary artery spasm contribute to coronary artery disease (CAD).
10. Define coarctation of the aorta, patent ductus arteriosus, septal defects, valvular stenosis, tetralogy of Fallot, atrioventricular (AV) block, ventricular fibrillation (VF), congestive heart failure (CHF), and cor pulmonale (CP).
11. Define key medical terms associated with the heart.

CHAPTER OUTLINE

- **Location**
- **Pericardium**
- **Heart Wall**
- **Chambers of the Heart**
- **Great Vessels of the Heart**
- **Valves of the Heart**

Atrioventricular (AV) Valves
Semilunar Valves
Skeleton of the Heart
Surface Projection

- **Blood Supply**
- **Conduction System**
- **Electrocardiogram (ECG)**
- **Cardiac Cycle**
- **Autonomic Control**
- **Artificial Heart**
- **Heart–Lung Machine**
- **Risk Factors in Heart Disease**
- **Developmental Anatomy of the Heart**
- **Applications to Health**
- **Key Medical Terms Associated with the Heart**

The ***heart*** is the center of the cardiovascular system. Whereas the term *cardio* refers to the heart, the term *vascular* refers to blood vessels (or an abundant blood supply). The heart is a hollow, muscular organ that weighs about 342 grams (11 oz) and beats over 100,000 times a day to pump 7200 liters (1900 gallons) of blood per day through more than 96,600 kilometers (60,000 miles) of blood vessels. The blood vessels form a network of tubes that carry blood from the heart to the tissues of the body and then return it to the heart. The specific aspects of the heart that we shall consider are its location, covering, wall and chambers, great vessels, valves, surface anatomy, conduction system, cardiac cycle (heartbeat), and autonomic control. Several disorders related to the heart will also be considered.

The study of the normal heart and diseases associated with it is known as ***cardiology*** (kar-dē-OL-ō-jē).

The developmental anatomy of the heart is considered at the end of the chapter.

LOCATION

The heart is situated between the lungs in the mediastinum. Recall that the mediastinum is the space, actually the mass of tissue, between the lungs that extends from the sternum to the vertebral column. Specifically, the heart is in the middle mediastinum (see Figures 1-7 and 22-7e). About two-thirds of its mass lies to the left of the body's midline (Figure 13-1). The heart is shaped like a blunt cone about the size of your closed fist—12 cm (5 in.) long, 9 cm (3.5 in.) wide at its broadest point, and 6 cm (2.5 in.) thick.

Its pointed end, the ***apex,*** is formed by the tip of the left ventricle, projects inferiorly, anteriorly, and to the left, and lies superior to the central tendon of the diaphragm. Anteriorly, the apex is in the fifth intercostal space.

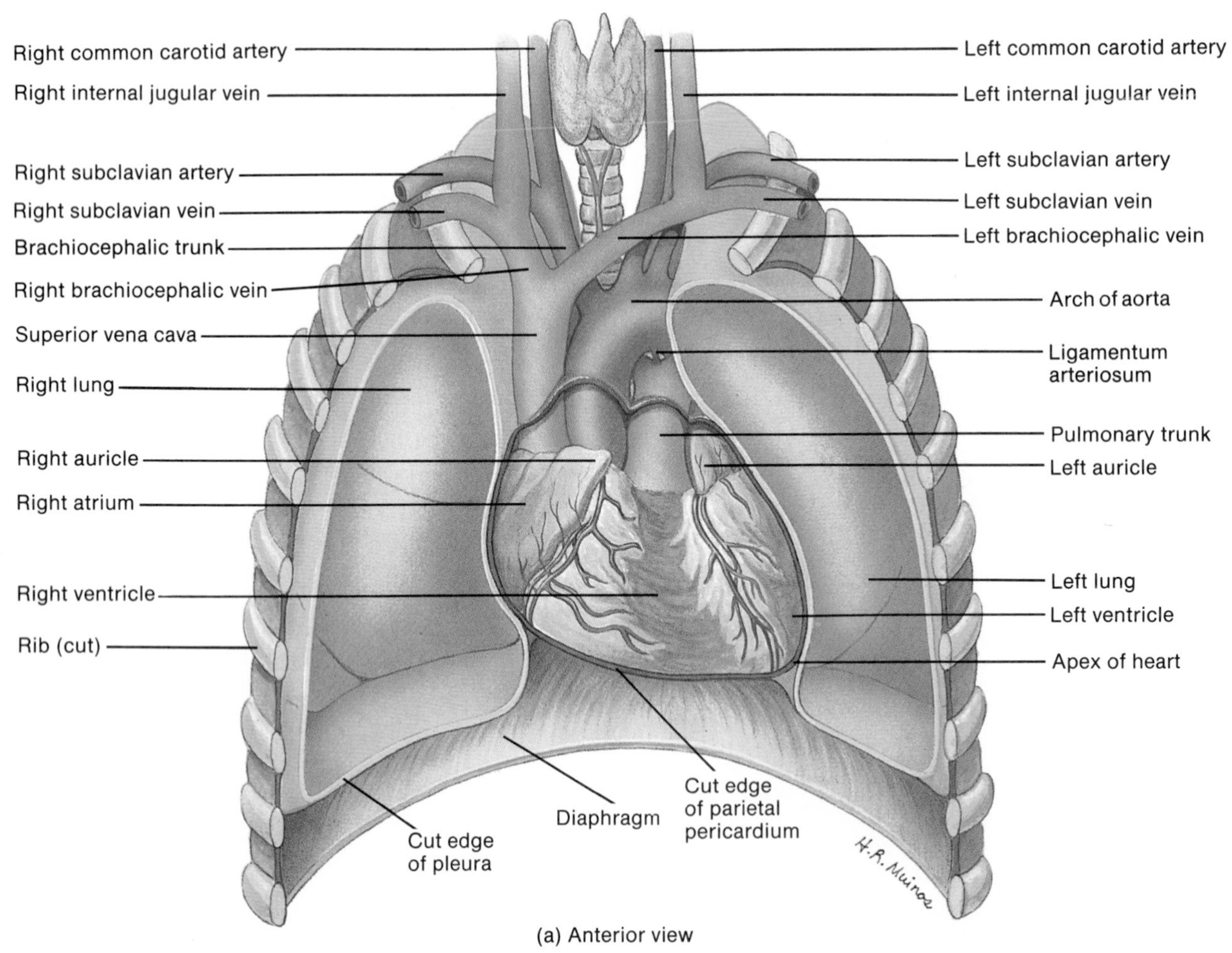

FIGURE 13-1 Position of the heart and associated blood vessels in the thoracic cavity. In this and subsequent illustrations, vessels that carry oxygenated blood are colored red; vessels that carry deoxygenated blood are colored blue. (b) Courtesy of Stephen A. Kieffer and E. Robert Heitzman, *An Atlas of Cross-Sectional Anatomy,* Harper & Row, Publishers, Inc., New York, 1979.

The ***base*** of the heart projects superiorly, posteriorly, and to the right. It is formed by the atria, mostly the left atrium. It lies opposite the fifth to ninth thoracic vertebrae. Anteriorly, it lies just inferior to the second rib.

The ***sternocostal (anterior) surface*** is formed mainly by the right ventricle, right atrium, and left ventricle, whereas the ***diaphragmatic (inferior) surface*** is formed by the left and right ventricles, mostly the left.

PERICARDIUM

The heart is enclosed and held in place by the ***pericardium,*** an ingenious structure designed to confine the heart to its position in the mediastinum, yet allow it sufficient freedom of movement so that it can contract vigorously and rapidly when the need arises.

The pericardium consists of two portions referred to as the fibrous pericardium and the serous pericardium (Figure 13-2a). The outer ***fibrous pericardium*** consists of very heavy fibrous connective tissue. The fibrous pericardium resembles a bag that rests on the diaphragm with its open end fused to the connective tissues of the great vessels entering and leaving the heart. The fibrous pericardium prevents overdistension of the heart, provides a tough protective membrane around the heart, and anchors the heart in the mediastinum. The lateral surfaces of the fibrous pericardium lie against the parietal pleurae, the outer coverings of the lungs. The

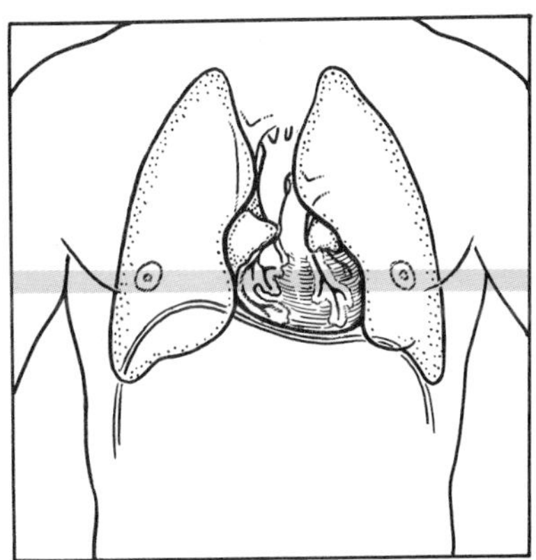

(b) Cross section

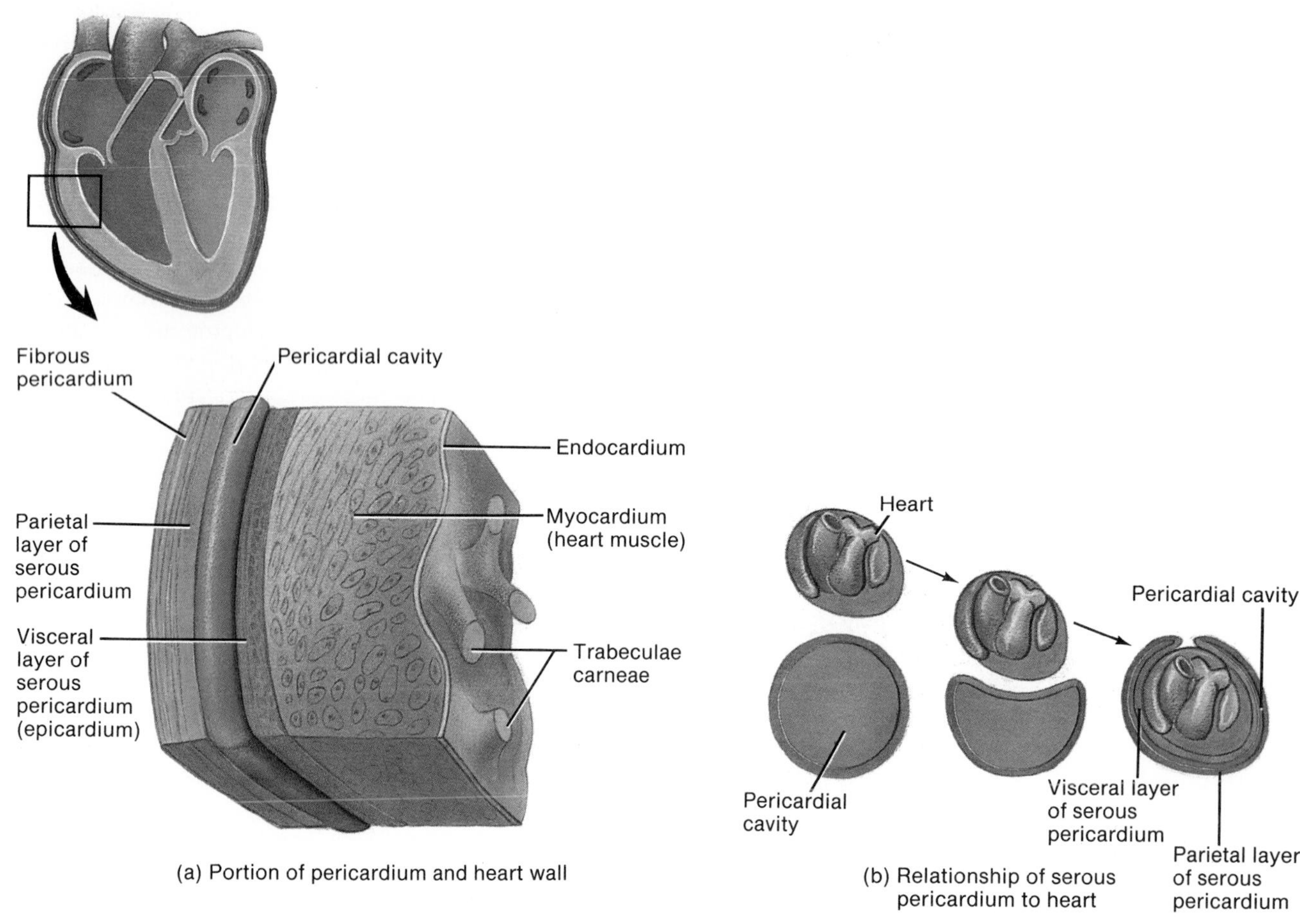

FIGURE 13-2 Pericardium and heart wall.

inner ***serous pericardium*** is a thinner, more delicate membrane that forms a double layer around the heart (Figure 13-2b). The outer ***parietal layer*** of the serous pericardium is directly deep to the fibrous pericardium. The inner ***visceral layer*** of the serous pericardium, also called the ***epicardium,*** is beneath the parietal layer, attached to the myocardium (muscle) of the heart. Between the parietal and visceral layers of the serous pericardium is a thin film of serous fluid that holds the two layers together much like a thin film of water binds two microscope slides. The serous fluid, known as ***pericardial fluid,*** is an ultrafiltrate of plasma and prevents friction between the membranes as the heart moves. There are up to 150 ml of pericardial fluid. The space occupied by the pericardial fluid is a potential space (not an actual space), called the ***pericardial cavity.***

An inflammation of the pericardium is known as ***pericarditis.*** Pericarditis with a buildup of pericardial fluid or extensive bleeding into the pericardium, if untreated, is a life-threatening condition. Since the pericardium cannot stretch to accommodate the excessive fluid or blood buildup, the heart is subjected to compression. This compression is known as ***cardiac tamponade*** (tam′-pon-ĀD) and can result in cardiac failure.

HEART WALL

The wall of the heart (Figure 13-2a) is divided into three layers: the epicardium (external layer), myocardium (middle layer), and endocardium (inner layer). The ***epicardium*** (also called the visceral layer of the serous pericardium) is the thin, transparent outer layer of the wall. It is composed of mesothelium and delicate connective tissue (lamina propria).

The ***myocardium,*** which is cardiac muscle tissue, constitutes the bulk of the heart. Cardiac muscle fibers (cells) are involuntary, striated, and branched, and the tissue is arranged in interlacing bundles of fibers. The myocardium is responsible for the contraction of the heart.

The ***endocardium*** is a thin layer of endothelium overlying a thin layer of connective tissue. It lines the inside of the myocardium and covers the valves of the heart and the tendons that hold them open. It is continuous with the endothelial lining of the large blood vessels of the heart.

Inflammation of the epicardium, myocardium, and endocardium is referred to as ***epicarditis, myocarditis,*** and ***endocarditis,*** respectively.

CHAMBERS OF THE HEART

The interior of the heart is divided into four cavities called ***chambers*** that receive circulating blood (Figure 13-3). The two superior chambers are called the right and left ***atria.*** Each atrium has an appendage called an ***auricle*** (OR-i-kul; *auris* = ear), so named because its shape resembles a dog's ear. The auricle increases the atrium's surface area. The lining of the atria is smooth, except for their anterior walls and the lining of the auricles, which contain projecting muscle bundles that are parallel to one another and resemble the teeth of a comb: the ***pectinate*** (PEK-ti-nāt) ***muscles.*** These bundles give the lining of the auricles a ridged appearance.

The atria are separated by a partition called the ***interatrial septum.*** A prominent feature of this septum is an oval depression, the ***fossa ovalis,*** which corresponds to the site of the foramen ovale, an opening in the interatrial septum of the fetal heart through which blood passes directly from the right atrium to the left atrium, thus bypassing the lungs. The fossa ovalis faces the opening of the inferior vena cava and is located in the septal wall of the atrium (see Figure 14-18a).

The two inferior chambers are the right and left ***ventricles.*** They are separated by an ***interventricular septum.*** The irregular surface of ridges and folds of the myocardium in the ventricles are known as the ***trabeculae carneae*** (tra-BEK-yoo-lē KAR-nē-e).

The muscle tissue of the atria and ventricles is separated by connective tissue that also forms the valves. This "cardiac skeleton" effectively divides the myocardium into two separate muscle masses. Externally, a groove known as the ***coronary sulcus*** (SUL-kus) separates the atria from the ventricles. It encircles the heart and houses the coronary sinus and circumflex branch of the left coronary artery. The ***anterior interventricular sulcus*** and ***posterior interventricular sulcus*** separate the right and left ventricles externally. The sulci contain coronary blood vessels and a variable amount of fat (Figure 13-3a,c).

The thickness of the four chambers varies according to function (Figure 13-3d). The atria are thin-walled because they need only enough cardiac muscle tissue to deliver the blood into the ventricles with the aid of gravity and a reduced pressure created by the expanding ventricles. The right ventricle has a thicker layer of myocardium than the atria, because it must send blood to the lungs and back around to the left atrium. The left ventricle has the thickest

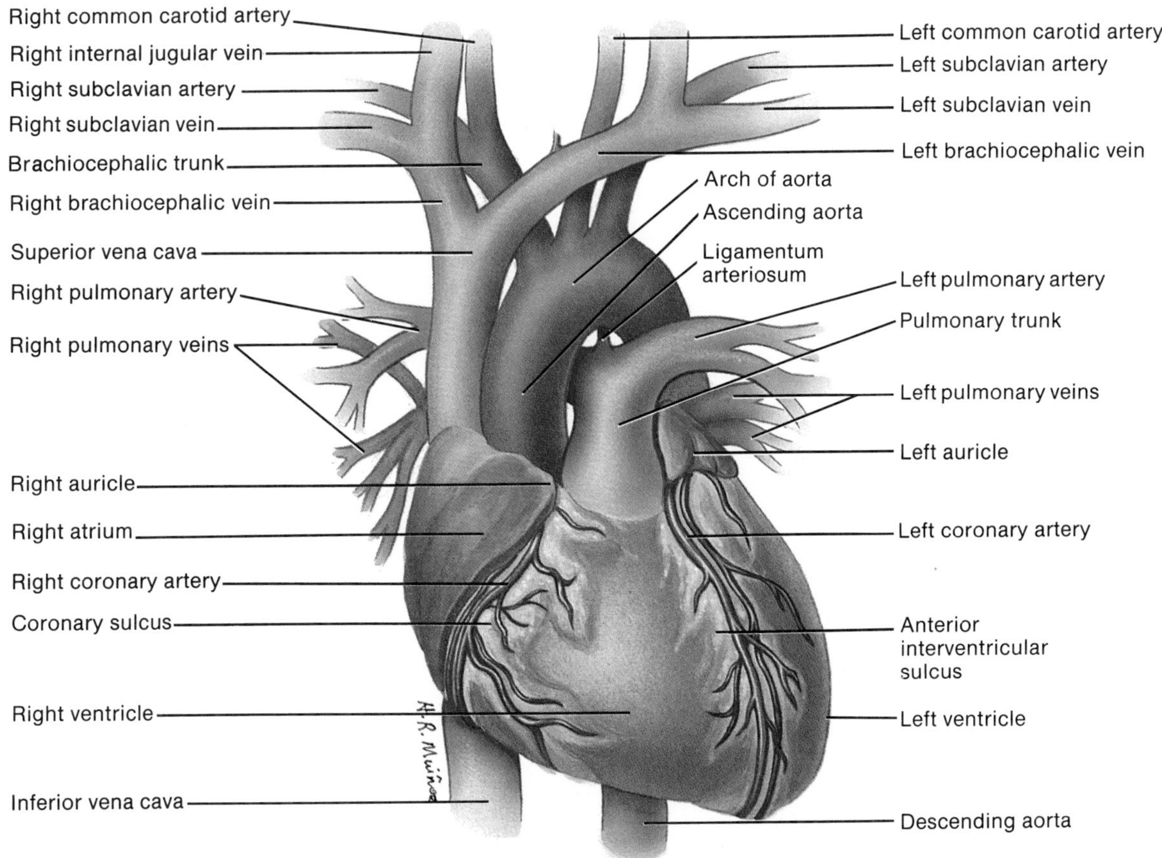

FIGURE 13-3 Structure of the heart. (b) Courtesy of J. A. Gosling, P. F. Harris, et al., *Atlas of Human Anatomy,* Gower Medical Publishing Ltd., 2nd ed., 1991.

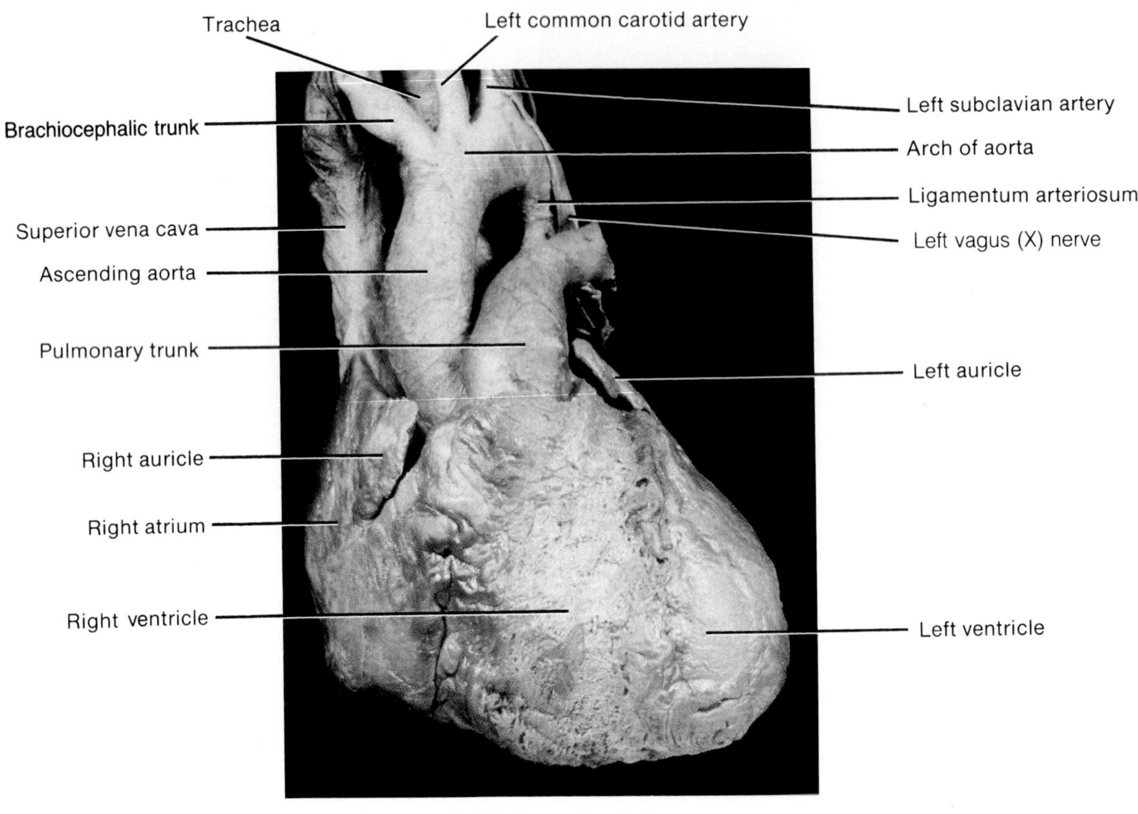

(b) Anterior external view

Left internal jugular vein
Left subclavian vein
Left common carotid artery
Left subclavian artery
Left brachiocephalic vein
Arch of aorta
Descending aorta
Left pulmonary artery
Pulmonary trunk
Left pulmonary veins
Left auricle
Left atrium
Coronary sinus
Left ventricle
Posterior interventricular sulcus
Right common carotid artery
Right internal jugular vein
Right subclavian artery
Right subclavian vein
Brachiocephalic trunk
Right brachiocephalic vein
Superior vena cava
Ligamentum arteriosum
Ascending aorta
Right pulmonary artery
Right pulmonary veins
Right atrium
Right coronary artery
Right ventricle
Inferior vena cava

(c) Posterior external view

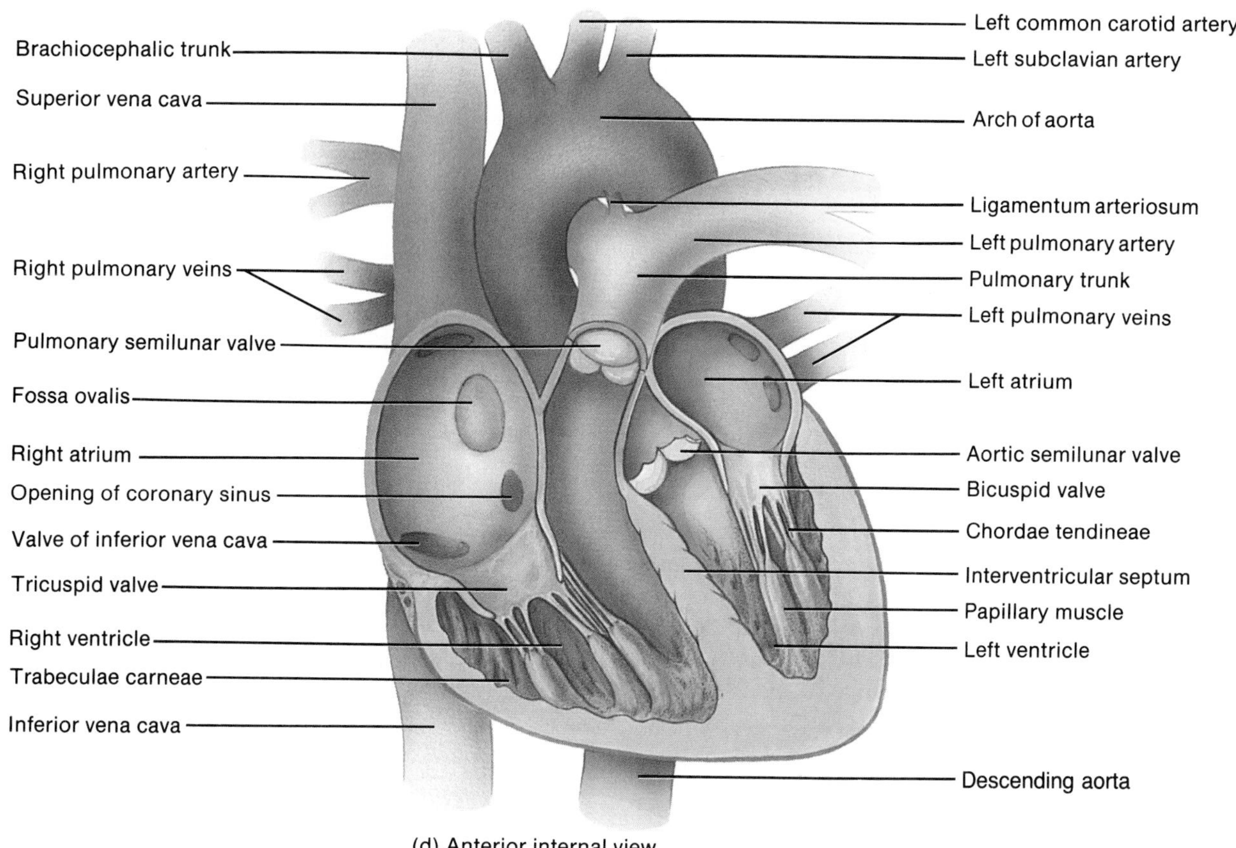

(d) Anterior internal view

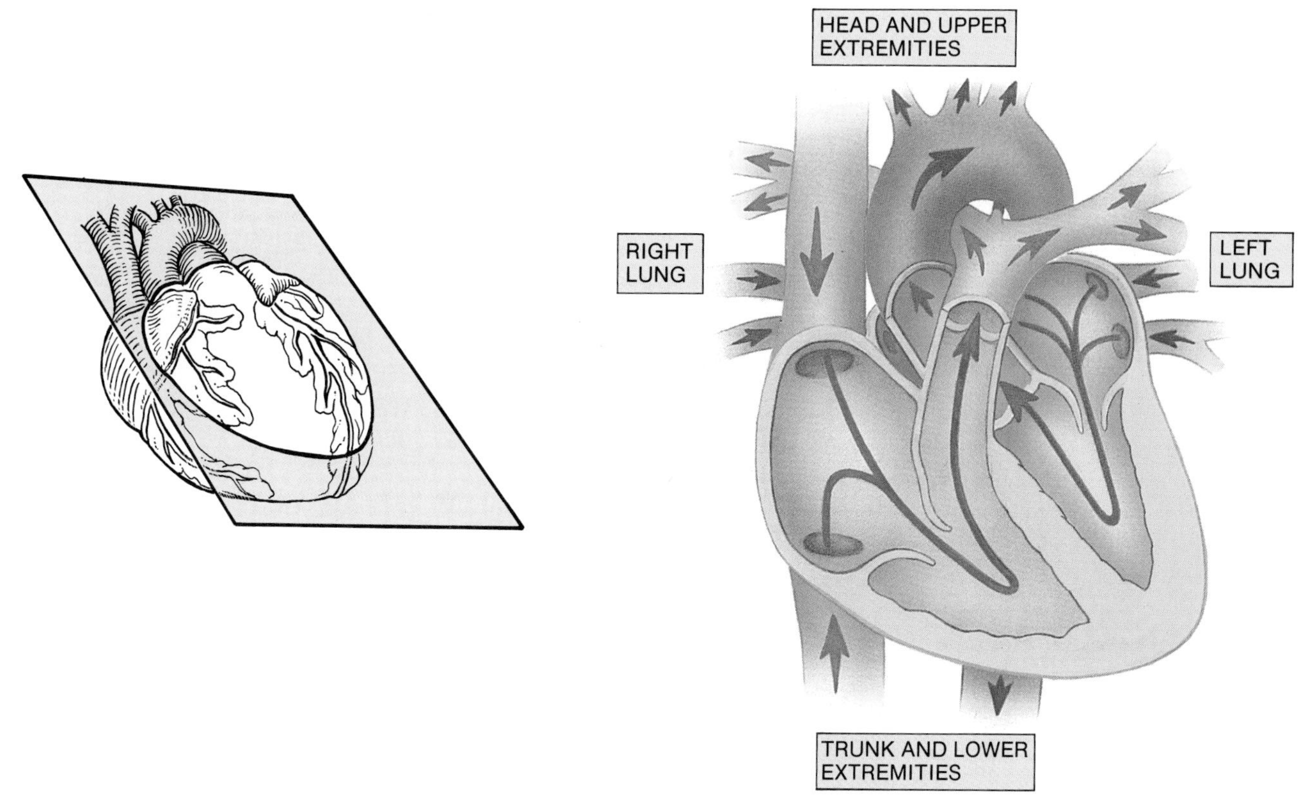

(e) Path of blood through heart

wall, since it must pump blood at high pressure through literally thousands of miles of vessels in the head, trunk, and extremities.

GREAT VESSELS OF THE HEART

The right atrium receives *deoxygenated blood* (blood that has given up some of its oxygen to cells) from all parts of the body except the lungs. It receives the blood through three veins. In general, the ***superior vena cava (SVC)*** brings blood from parts of the body superior to the heart; in general, the ***inferior vena cava (IVC)*** brings blood from parts of the body inferior to the heart; and the ***coronary sinus*** drains blood from most of the vessels supplying the wall of the heart (Figure 13-3c). The right atrium then delivers the blood into the right ventricle, which pumps it into the ***pulmonary trunk.*** The pulmonary trunk divides into a ***right*** and ***left pulmonary artery,*** each of which carries blood to the lungs. In the lungs, the blood releases its carbon dioxide and takes on oxygen so that the blood becomes *oxygenated* (blood that has not given up its oxygen to cells). The blood returns to the heart via four ***pulmonary veins*** that empty into the left atrium. The blood then passes into the left ventricle, which pumps the blood into the ***ascending aorta*** (*aorte* = to suspend, because the aorta was once believed to suspend the heart). From here the blood is passed into the ***coronary arteries, arch of the aorta, thoracic aorta,*** and ***abdominal aorta.*** These blood vessels and their branches transport the blood to the heart and other body parts, except the lungs.

During fetal life there is a temporary blood vessel, called the ductus arteriosus, that connects the pulmonary trunk with the aorta (see Figure 14-18a). Its purpose is to redirect blood so that only a small volume enters the nonfunctioning fetal lungs. The ductus arteriosus normally closes shortly after birth, leaving a remnant known as the ***ligamentum arteriosum.***

VALVES OF THE HEART

As each chamber of the heart contracts, it pushes a portion of blood into a ventricle or out of the heart through an artery. The heart has structures composed of dense irregular connective tissue covered by endothelium called ***valves*** that prevent the backflow of blood.

ATRIOVENTRICULAR (AV) VALVES

Atrioventricular (AV) valves lie between the atria and ventricles (Figure 13-3d). The right atrioventricular valve between the right atrium and right ventricle is also called the ***tricuspid valve*** because it consists of three cusps (flaps). These cusps are fibrous tissues that grow out of the walls of the heart and are covered with endocardium. The pointed ends of the cusps project into the ventricle. Tendonlike fibrous cords called ***chordae tendineae*** (KOR-dē ten-DIN-ē-ē) connect the pointed ends and undersurfaces to small conical projections—the ***papillary muscles*** (muscular columns)—located on the inner surface of the ventricles. The chordae tendineae are like the spokes of an umbrella and keep the valves from pushing up into the atria when the ventricles contract, much the same way that an umbrella resists being blown inside out by a very strong wind. This action prevents backflow of blood to decrease the work load of the heart. The left atrioventricular valve between the left atrium and left ventricle is called the ***bicuspid (mitral) valve.*** It has two cusps that work in the same way as the cusps of the tricuspid valve. Its cusps are also attached by way of the chordae tendineae to papillary muscles.

In order for blood to pass from an atrium to a ventricle, an atrioventricular valve must open. The valves open and close as the result of pressure differences across them. When blood moves from an atrium to a ventricle, the valve is pushed open, the papillary muscles relax, and the chordae tendineae slacken (Figure 13-4a). When a ventricle contracts, the pressure of the blood drives the cusps upward until their edges meet and close the opening (Figure 13-4b). At the same time, contraction of the papillary muscles and tightening of the chordae tendineae help prevent the valve from swinging upward into the atrium.

SEMILUNAR VALVES

Both arteries that leave the heart have valves that prevent blood from flowing back into the heart. These are the ***semilunar valves*** (see Figure 13-3d). The ***pulmonary semilunar valve*** lies in the opening where the pulmonary trunk leaves the right ventricle. The ***aortic semilunar valve*** is situated at the opening between the left ventricle and the aorta.

Both valves consist of three semilunar (half-moon or crescent-shaped) cusps. Each cusp is attached by its convex margin to the artery wall. The free borders of the cusps curve outward and project into the opening inside the blood vessel. Like the atrioventricular valves, the semilunar valves permit blood to flow in one direction only; in this case, the flow is from the ventricles into the arteries.

SKELETON OF THE HEART

In addition to cardiac muscle tissue, the heart wall also consists of dense connective tissue that forms the ***skeleton of the heart.*** The skeleton forms the foundation to which the heart valves attach, serves as an attachment for cardiac muscle fibers, and acts as an electrical insulator between atria and ventricles. Essentially, the skeleton consists of dense connective tissue rings that surround the valves of the heart, fuse with one another, and merge with the interventricular septum.

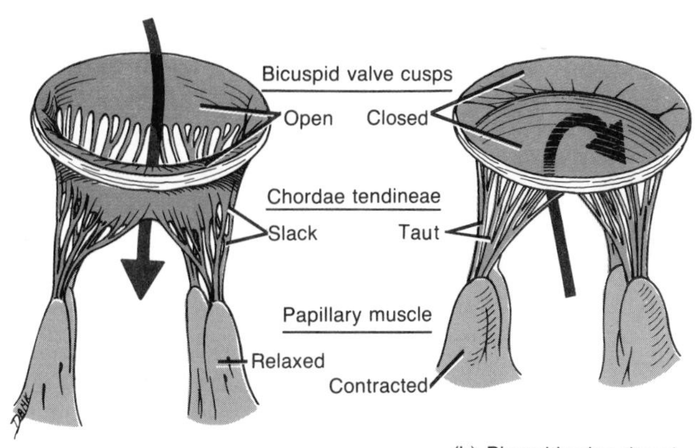

(a) Bicuspid valve open

(b) Bicuspid valve closed

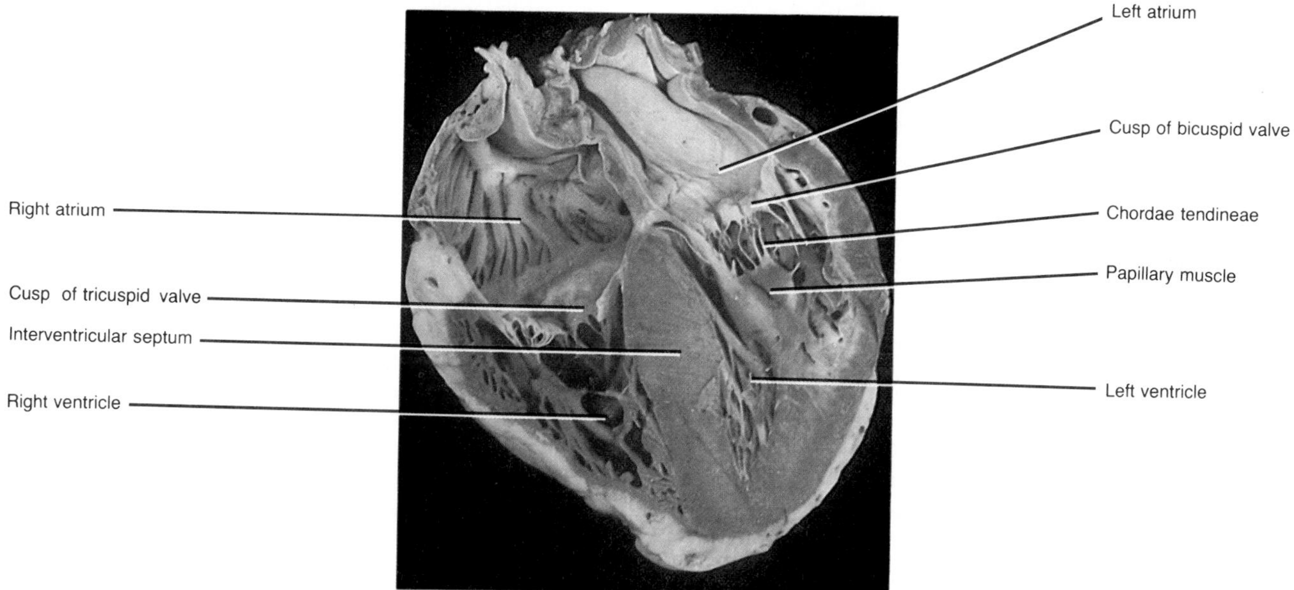

(c) Anterior internal view

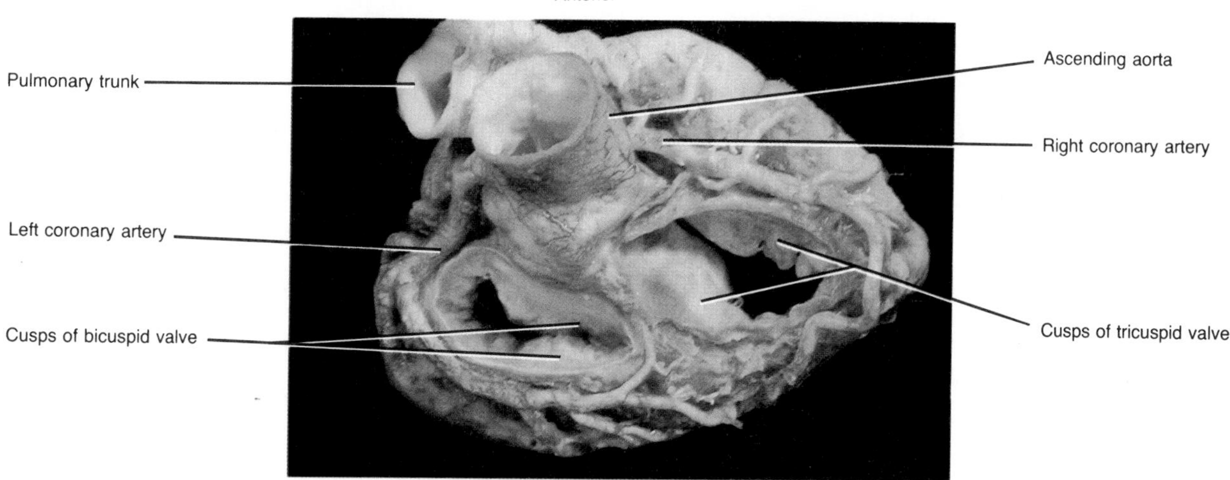

(d) Superior view (atria removed)

FIGURE 13-4 Atrioventricular (AV) valves. The bicuspid and tricuspid valves operate in a similar manner. (c) and (d) Courtesy of J. Willis Hurst et al., *Atlas of the Heart,* Gower Medical Publishing Ltd., 1988.

SURFACE PROJECTION

The location of the valves of the heart may be identified by surface projection (Figure 13-5). The pulmonary and aortic semilunar valves are represented on the surface by a line about 2.5 cm (1 in.) in length. The pulmonary semilunar valve lies horizontally behind the inner end of the left third costal cartilage and the adjoining part of the sternum. The aortic semilunar valve is placed obliquely behind the left side of the sternum, at the level of the third intercostal space. The tricuspid valve lies behind the sternum, extending from the midline at the level of the fourth costal cartilage down toward the right sixth chondrosternal junction. The bicuspid valve lies behind the left side of the sternum obliquely at the level of the fourth costal cartilage. It is represented by a line about 3 cm in length.

Although heart sounds are produced in part by the closure of valves, they are not necessarily heard best over these valves. Each sound tends to be clearest in a slightly different location closest to the surface of the body (Figure 13-5).

CLINICAL APPLICATION

Heart Murmur

Heart sounds provide valuable information about the valves. Besides the familiar sounds, some persons also have a third or fourth heart sound. Because this sound may suggest the cantering of a horse, the term "gallop rhythm" is used to describe these extra heart sounds, which are often associated with heart disease. Other abnormal sounds include "snaps," "knocks," "rubs," and "clicks." A ***heart murmur*** is an abnormal sound that consists of a flow noise that is heard before or after the lubb-dupp, or that may mask the normal heart sounds. Some murmurs are caused by turbulent blood flow around valves due to abnormal anatomy or increased volume of flow. Among the valvular abnormalities that may contribute to murmurs are ***mitral stenosis*** (narrowing of the mitral valve by scar formation or a congenital defect), ***mitral insufficiency*** (backflow of blood from the left ventricle into the left atrium due to a damaged mitral valve or ruptured chordae tendineae), ***aortic stenosis*** (narrowing of the aortic semilunar valve), and ***aortic insufficiency*** (back-flow of blood from the aorta into the left ventricle). Another cause of a heart murmur is due to ***mitral valve prolapse (MVP),*** an inherited disorder in which a portion of a mitral valve is pushed back too far (prolapsed) during contraction owing to expansion of the cusps and elongation of the chordae tendineae. This condition is usually asymptomatic and may be found in up to 10 percent of otherwise healthy young men and women. Nevertheless, in some patients the condition is associated with various problems, and the individual who has it must be examined periodically by a cardiologist. Not all murmurs are abnormal.

BLOOD SUPPLY

The wall of the heart, like any other tissue, has its own blood vessels. Nutrients could not possibly diffuse through all the layers of cells that make up the heart tissue. The flow of blood through the numerous vessels that pierce the myocardium is called ***coronary (cardiac) circulation*** (Figure 13-6). The term ***coronary*** means a crown or circle and refers to the fact that the coronary blood vessels circle the heart as a crown circles the head.

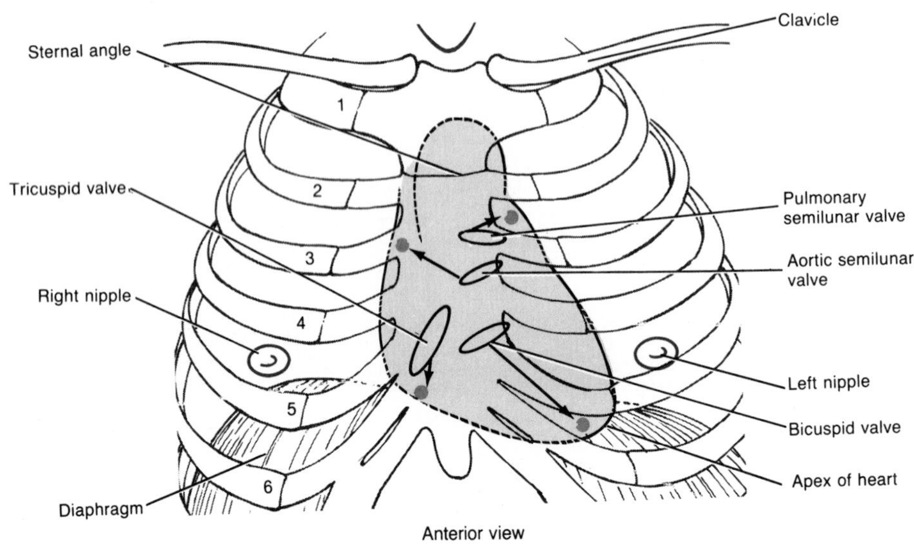

FIGURE 13-5 Surface projection of the heart. The red circles indicate where heart sounds caused by the respective valves are best heard.

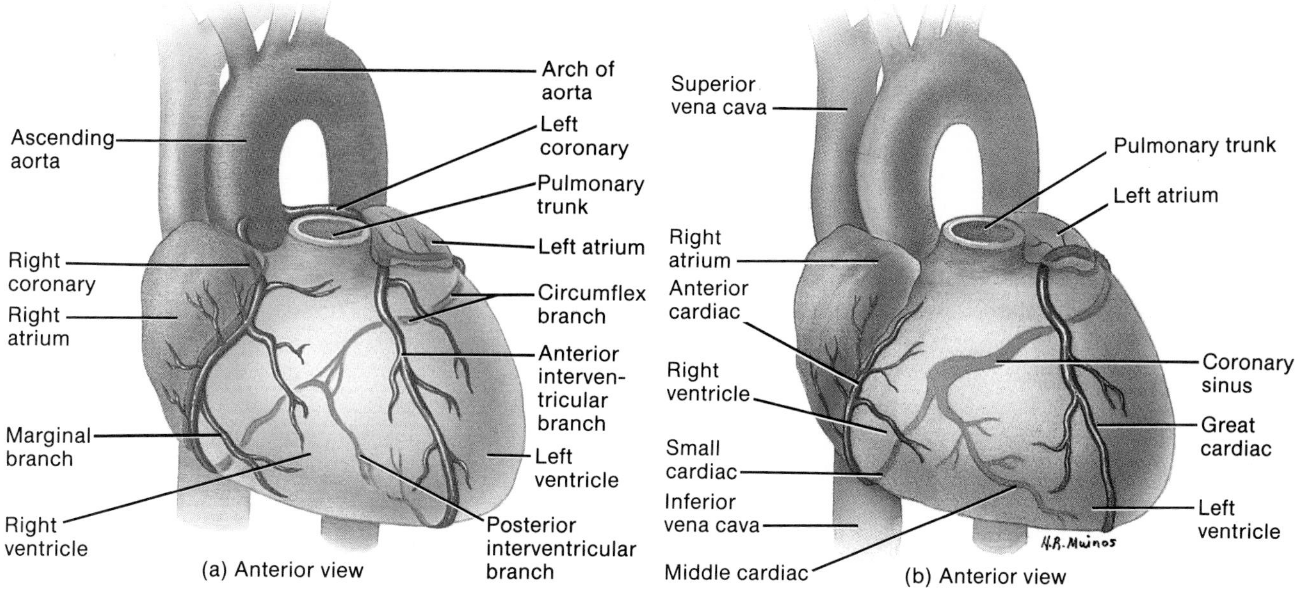

FIGURE 13-6 Coronary (cardiac) circulation. (a) Arterial distribution. (b) Venous drainage.

The vessels that serve the myocardium include the ***left coronary artery,*** which originates as a branch of the ascending aorta. The artery runs under the left auricle and divides into the anterior interventricular and circumflex branches. The ***anterior interventricular branch*** follows the anterior interventricular sulcus and supplies oxygenated blood to the walls of both ventricles. The ***circumflex branch*** distributes oxygenated blood to the walls of the left ventricle and left atrium.

In January 1988, Pete Maravich, the leading scorer in National Collegiate Athletic Association history and a member of the Basketball Hall of Fame, collapsed during a pickup basketball game and died a short time later. He had a very rare defect, a missing left coronary artery. As a result, his heart was enlarged and weakened owing to a continuous decreased oxygen supply. People who have this condition usually die before age 20.

The ***right coronary artery*** also originates as a branch of the ascending aorta. In its course, it supplies small branches to the right atrium. It runs under the right auricle and divides into the posterior interventricular and marginal branches. The ***posterior interventricular branch*** follows the posterior interventricular sulcus and supplies the walls of the two ventricles with oxygenated blood. The ***marginal branch*** transports oxygenated blood to the myocardium of the right ventricle. The left ventricle receives the most abundant blood supply because of the enormous work it must do.

As blood passes through the coronary system of the heart, it delivers oxygen and nutrients and collects carbon dioxide and wastes. Most of the deoxygenated blood, which carries the carbon dioxide and wastes, is collected by a large vein on the back of the heart, the ***coronary sinus,*** which empties into the right atrium. A vascular sinus is a vein with a thin wall that has no smooth muscle to alter its diameter. The principal tributaries of the coronary sinus are the ***great cardiac vein,*** which drains the anterior aspect of the heart, and the ***middle cardiac vein,*** which drains the posterior aspect of the heart.

Most parts of the body receive branches from more than one artery, and where two or more arteries supply the same region, they usually connect with each other. This arrangement is referred to as *anastomosis* (a-nas-tō-MŌ-sis). Anastomoses between arteries provide collateral circulation (alternate routes) for blood to reach a particular organ or tissue. The myocardium contains numerous anastomoses either connecting branches of one coronary artery or between branches of different coronary arteries. When a major coronary blood vessel is about 90 percent obstructed, blood will flow through the collateral vessels. Although most collaterals in the heart are quite small, heart muscle can remain alive as long as it receives as little as 10 to 15 percent of its normal supply.

CLINICAL APPLICATION

Angina Pectoris and Myocardial Infarction

Most heart problems result from faulty coronary circulation. If a reduced oxygen supply weakens cells but does not actually kill them, the condition is called ***ischemia*** (is-KĒ-mē-a). ***Angina pectoris*** (an-JĪ-na, or AN-ji-na,

PEK-to-ris), meaning "chest pain," results from ischemia of the myocardium. Angina is typically described as a tightness or choking sensation or a squeezing, pressure-type discomfort and is usually of short duration. It occurs during exertion but is relieved by rest. (Pain impulses originating from most visceral muscles are referred to an area on the surface of the body. The pain associated with angina pectoris is referred to the neck, chin, or down the left arm to the elbow.) Common causes of angina include stress, strenuous exertion after a heavy meal, atherosclerosis, coronary artery spasm, high blood pressure (hypertension), fever, anemia, hyperthyroidism, and aortic stenosis. The symptoms of angina pectoris include chest pain, accompanied by tightness or pressure, labored breathing, and a sensation of foreboding. Sometimes weakness, dizziness, and perspiration occur. ***Silent myocardial ischemia*** refers to the development of electrocardiographic evidence of decreased blood flow to the heart in the absence of any clinical symptoms, such as chest pain or tightness. It is quite common in patients who have recovered from a heart attack.

A much more serious problem is ***myocardial infarction*** (in-FARK-shun), or ***MI,*** commonly called a heart attack. ***Infarction*** means the death of an area of tissue because of an interrupted blood supply. Myocardial infarction may result from a thrombus (stationary blood clot) or embolus (blood clot transported by blood) in one of the coronary arteries. The tissue distal to the obstruction dies and is replaced by noncontractile scar tissue. Thus, the heart muscle loses at least some of its strength. The aftereffects depend partly on the size and location of the infarcted, or dead, area. In addition to killing normal heart tissue, the infarction may disturb the conducting system of the heart (described shortly). Therefore, heart attacks may cause sudden death due to ventricular fibrillation. The most modern treatment for a myocardial infarction is to perform coronary angiography and inject a clot-dissolving chemical such as t-PA or to do coronary angioplasty (to be discussed at the end of the chapter).

There has been considerable publicity concerning the use of low doses of aspirin to reduce the rate of acute myocardial infarction. However, the benefits of aspirin may not be the same for everyone, and aspirin may not be indicated for people at low risk for MI because of the risk of cerebral bleeding (aspirin interferes with blood clotting). Moreover, some people develop gastrointestinal problems associated with taking aspirin. Most experts agree that routine aspirin therapy should be discouraged. The drug should be considered on a case-by-case basis.

CONDUCTION SYSTEM

Recall from Chapter 9 that cardiac muscle fibers form two separate networks—one atrial and one ventricular. Each fiber is in physical contact with other fibers in the networks by transverse thickenings of the sarcolemma called ***intercalated discs*** (see Figure 9-8b). Within the discs are connections called gap junctions that aid in the conduction of muscle action potentials between cardiac muscle fibers. The gap junctions provide low-resistance bridges for the spread of excitation (muscle action potentials) from one fiber to another. Thus, the spread of excitation between fibers is rapid, and the atria contract as one unit and the ventricles as another. The intercalated discs also function in adhesion of cardiac muscle fibers so that they do not pull apart. Each network contracts as a functional unit.

The heart is innervated by the autonomic nervous system (Chapter 19), but the autonomic neurons only increase or decrease the time it takes to complete a cardiac cycle (heartbeat); that is, they do not initiate contraction. The chamber walls can go on contracting and relaxing, contracting and relaxing, without any direct stimulus from the nervous system. This action is possible because the heart has an intrinsic regulating system called the ***conduction system.*** The conduction system is composed of specialized muscle tissue that generates and distributes the electrical impulses that stimulate the cardiac muscle fibers (cells) to contract. These tissues are the sinoatrial (sinuatrial) or SA node, the atrioventricular (AV) node, the atrioventricular (AV) bundle (bundle of His), the bundle branches, and the conduction myofibers (Purkinje fibers). The cells of the conduction system develop during embryological life from certain cardiac muscle fibers and provide autorhythmicity (self-rhythm) for the heart.

The ***sinoatrial (sinuatrial) node,*** also known as the ***SA node,*** or ***pacemaker,*** is a compact mass of cells located in the right atrial wall inferior to the opening of the superior vena cava (Figure 13-7a). The SA node initiates each cardiac cycle and thereby sets the basic pace for the heart rate. The SA node spontaneously depolarizes and generates action potentials faster than other components of the conduction system and myocardium. As a result, action potentials from the SA node spread to other areas of the conduction system and myocardium and stimulate them so frequently that they are not able to generate action potentials at their own inherent rates. Thus, the faster rate of discharge of the SA node sets the rhythm for the rest of the heart—hence, its common name, pacemaker. The rate set by the SA node may be altered by nerve impulses from the autonomic nervous system or by certain blood-borne chemicals such as thyroid hormones and epinephrine.

Once a muscle action potential is initiated by the SA node, the impulse spreads out over both atria through gap junctions between cardiac muscle fibers (cells), causing them to contract, and at the same time depolarizes the

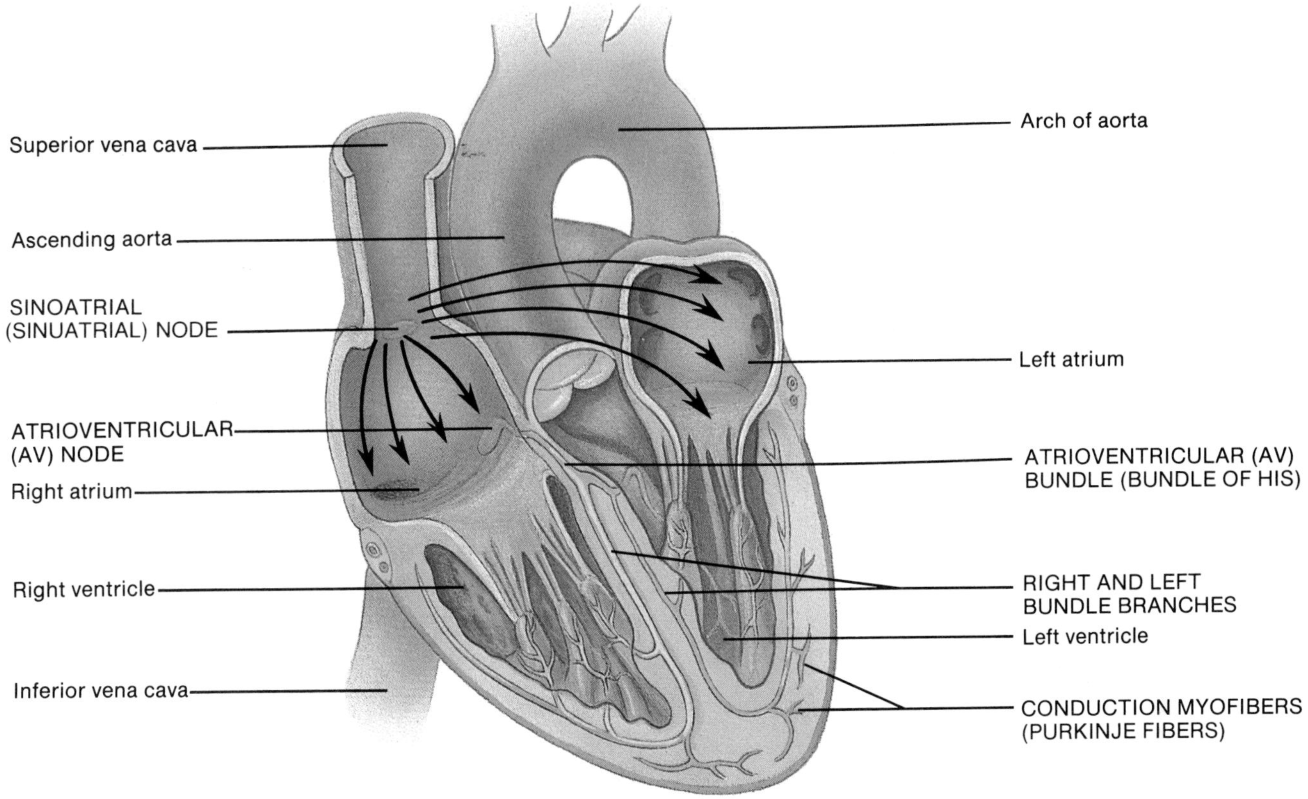

Superior vena cava
Ascending aorta
SINOATRIAL (SINUATRIAL) NODE
ATRIOVENTRICULAR (AV) NODE
Right atrium
Right ventricle
Inferior vena cava
Arch of aorta
Left atrium
ATRIOVENTRICULAR (AV) BUNDLE (BUNDLE OF HIS)
RIGHT AND LEFT BUNDLE BRANCHES
Left ventricle
CONDUCTION MYOFIBERS (PURKINJE FIBERS)

(a) Anterior view

R
P
T
Q
S
1.0
0.5
0
—0.5
Millivolts (mV)
0
0.2
0.4
0.6
0.8
Seconds

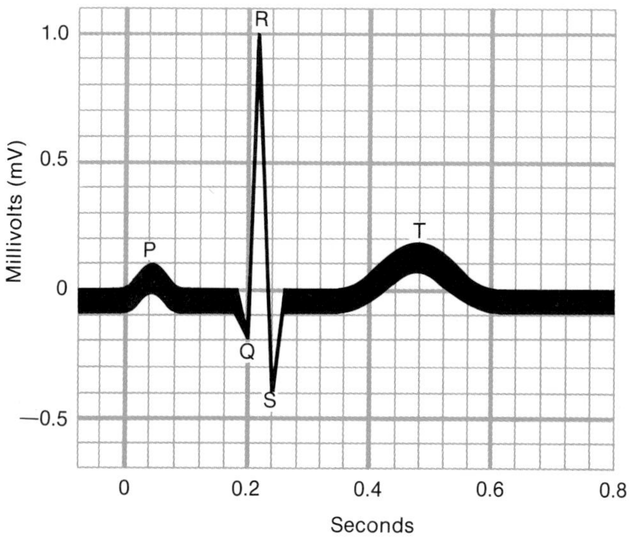

(b) Normal electrocardiogram (Lead II)

FIGURE 13-7 Conduction system of the heart. (a) Location of the nodes and bundles of the conduction system. The arrows indicate the flow of electricity (impulses) through the atria. (b) Normal electrocardiogram of a single heartbeat, enlarged for emphasis.

atrioventricular (AV) node. Owing to its location near the inferior portion of the interatrial septum, the AV node is one of the last portions of the atria to be depolarized. Because of size and electrical properties, muscle action potentials pass slowly through the AV node. This allows time for the atria to empty their blood into the ventricles before the ventricles begin their contraction; the atria finish their contraction before the ventricles begin theirs.

From the AV node, a tract of conducting fibers called the ***atrioventricular (AV) bundle (bundle of His)*** runs to the top of the interventricular septum. It then continues down both sides of the septum as the ***right*** and ***left bundle branches.*** The AV bundle distributes the action potential over the medial surfaces of the ventricles. Actual contraction of the ventricles is stimulated by ***conduction myofibers (Purkinje fibers)*** that emerge from the bundle branches and pass into the fibers of the myocardium of the ventricles.

ELECTROCARDIOGRAM (ECG OR EKG)

In a normal heartbeat, the two atria contract simultaneously while the two ventricles relax. Then, when the two ventricles contract, the two atria relax. The term ***systole*** (SIS-tō-lē; *systellein* = draw together) refers to the phase of contraction; ***diastole*** (dī-AS-tō-lē = pause) is the phase of relaxation. A ***cardiac cycle,*** or complete heartbeat, consists of the systole and diastole of both atria plus the systole and diastole of both ventricles.

Impulse transmission through the atria and ventricles generates electrical currents that can be detected on the body's surface. A recording of the electrical changes that accompany the cardiac cycle is called an ***electrocardiogram*** (e-lek′-trō-KAR-dē-ō-gram) (***ECG*** or ***EKG***)**.** The instrument used to record the changes is an ***electrocardiograph.*** An ECG is performed to evaluate possible symptoms of heart disease (chest pain, palpitations, dizziness, faintness), to detect abnormal cardiac rhythms and conduction patterns, to follow the course of recovery after a heart attack, to monitor the effectiveness and side effects of certain drugs that might affect the heart, to check the functioning of artificial pacemakers, and to evaluate the status of the heart prior to surgery. Some physicians include an ECG as part of a routine physical examination for individuals over 40. In addition to its screening value, this baseline ECG might be compared with a subsequent ECG done after onset of symptoms. Although an ECG monitors cardiac status, it does not detect all instances of heart disease, and it cannot predict what cardiac event might occur in the future.

Each portion of the cardiac cycle produces a different electrical impulse. These impulses are transmitted from the electrodes to a recording pen that graphs the impulses as a series of up-and-down waves called ***deflection waves.*** In a typical record (Figure 13-7b), three clearly recognizable waves accompany each cardiac cycle. The first, called the ***P wave,*** is a small upward wave. It indicates atrial depolarization—the spread of an impulse from the SA node through the muscle of the two atria. A fraction of a second after the P wave begins, the atria contract. The second wave, called the ***QRS wave (complex),*** begins as a downward deflection, continues as a large, upright, triangular wave, and ends as a downward wave at its base. This deflection represents ventricular depolarization—that is, the spread of the electrical impulse through the ventricles. The third recognizable deflection is a dome-shaped ***T wave,*** which indicates ventricular repolarization. There is no deflection to show atrial repolarization because the stronger QRS wave masks this event.

For some people, an ***ambulatory ECG*** is indicated. In this procedure a ***Holter monitor*** is used. It is a lightweight portable recorder that is connected to several electrodes attached to the person's chest. These detect electrical impulses from the heart that are recorded on magnetic tape in the monitor and later analyzed by computer. While wearing a Holter monitor for 24 hours or longer, an individual goes about normal daily activities, thus, the term *ambulatory* (walking) ECG. An ambulatory ECG is especially important (1) in diagnosing rhythm disorders in the conduction system that might not show up during the brief period of time used for recording a resting ECG and (2) in evaluating the effectiveness of heart surgery, drugs, or an artificial pacemaker.

CLINICAL APPLICATION

Artificial Pacemaker

When major elements of the conduction system are disrupted, an irregular heart rhythm may occur. In one type of rhythm disturbance, the ventricles fail to receive atrial impulses, causing the ventricles and atria to beat independently of each other. In patients with such a condition, normal heart rhythm can be restored and maintained with an ***artificial pacemaker,*** a device that sends out small electrical charges that stimulate the heart. It consists of three basic parts: a *pulse generator,* which contains the battery cells and produces the impulse; a *lead,* which is a flexible wire connected to the pulse generator that delivers the impulse to the electrode; and an *electrode,* which makes contact with a portion of the heart and delivers the charge to the heart. Many of the newer pacemakers, called activity-adjusted pacemakers, automatically speed up the heartbeat during exercise.

CARDIAC CYCLE

Two phenomena control the movement of blood through the heart as part of its cardiac cycle (heartbeat): the opening and closing of the valves and the contraction and relaxation of the myocardium. Both these activities occur without direct stimulation from the nervous system. The valves are controlled by pressure changes in each heart chamber. The contraction of the cardiac muscle is stimulated by its conduction system.

If we assume that the average heart rate (HR) is about 75 times per minute, then each cardiac cycle requires about 0.8 sec. During the first 0.1 sec, the atria contract and the ventricles relax. The atrioventricular (AV) valves are open, and the semilunar valves are closed. For the next 0.3 sec, the atria are relaxing and the ventricles are contracting. During the first part of this period, all valves are closed; during the second part, the semilunar valves are open. The last 0.4 sec of the cycle is the ***relaxation,*** or ***quiescent*** (kwī-ES-ent), ***period*** and all chambers are in diastole. In a complete cycle, then, the atria are in systole 0.1 sec and in diastole 0.7 sec; the ventricles are in systole 0.3 sec and in diastole 0.5 sec. For the first part of the quiescent period, all valves are closed; during the latter part, the AV valves open and blood starts draining into the ventricles. When the heart beats faster than normal, the quiescent period is shortened accordingly.

The act of listening to sounds within the body is called ***auscultation*** (aws-kul-TĀ-shun; *auscultare* = to listen) and it is usually done with a stethoscope. The sound of the heartbeat comes primarily from turbulence in blood flow created by the closure of the valves, not from the contraction of the heart muscle. The first sound, which can be described as a ***lubb*** sound, is a long, booming sound. The lubb is the sound created by the closure of the AV valves soon after ventricular systole begins. The second sound, which is heard as a short, sharp sound, can be described as a ***dupp*** sound. Dupp is the sound created as the semilunar valves close toward the end of ventricular systole. A pause between the second sound and the first sound of the next cycle is about two times longer than the pause between the first and second sound of each cycle. Thus, the cardiac cycle can be heard as a lubb, dupp, pause; lubb, dupp, pause; lubb, dupp, pause.

AUTONOMIC CONTROL

The sinoatrial (SA) node initiates contraction and, if left to itself, would set an unvarying heart rate. However, the body's need for blood supply varies under different conditions. There are several regulatory mechanisms stimulated by factors such as chemicals present in the body, temperature, emotional state, and age. The most important control of heart rate and strength of contraction is the effect of the autonomic nervous system.

Within the medulla of the brain is a group of neurons called the ***cardioacceleratory center (CAC).*** Arising from this center are sympathetic fibers that travel down a tract in the spinal cord and then pass outward in the ***cardiac (accelerator) nerves*** and innervate the SA node, AV node, and portions of the myocardium. When the cardioacceleratory center is stimulated, nerve impulses travel along the sympathetic fibers and cause them to release norepinephrine (NE), which increases the rate of heartbeat and the strength of contraction.

The medulla also contains a group of neurons that form the ***cardioinhibitory center (CIC).*** Arising from this center are parasympathetic fibers that reach the heart via the ***vagus (X) nerve,*** a cranial nerve. These fibers innervate the SA node and AV node. When this center is stimulated, nerve impulses transmitted along the parasympathetic fibers cause the release of acetylcholine (ACh), which decreases the rate of heartbeat.

The autonomic control of the heart is therefore the result of opposing sympathetic (stimulatory) and parasympathetic (inhibitory) influences. Sensory impulses from receptors in different parts of the cardiovascular system act on the centers so that a balance between stimulation and inhibition is maintained. Nerve cells capable of responding to changes in blood pressure are called ***baroreceptors (pressoreceptors).*** These receptors affect the rate of heartbeat.

Heart rate may also be influenced by certain chemicals (epinephrine, sodium, potassium), temperature, emotions, gender, and age.

ARTIFICIAL HEART

On December 2, 1982, Dr. Barney B. Clark, a 61-year-old retired dentist, made medical history by becoming the first human to receive an ***artificial heart.*** The 7½-hour operation was performed by Dr. William DeVries, who was a surgeon at the University of Utah Medical Center at the time.

The artificial heart, known as Jarvik-7 after its inventor, Dr. Robert Jarvik, consists of an aluminum base and a pair of rigid plastic chambers that serve as ventricles. Each chamber contains a flexible diaphragm driven by compressed air that forces blood past mechanical valves into the major blood vessels. The surgery consisted of removing the ventricles from Dr. Clark's heart but leaving the atria intact. Then Dacron connectors were sutured onto the atria, aorta, and pulmonary trunk. The artificial heart was then snapped into position via the connectors. Essentially, the artificial heart is designed to pump sufficient blood to sustain only moderate activity.

The artificial heart came under some criticism because of complications experienced by recipients. Complications include acute renal failure, postoperative bleeding, hemolytic anemia, thrombi, infections, seizures, and strokes. In fact, in January 1990, the Food and Drug Administration banned use of the Jarvik-7. Research is under way to develop an artificial heart that uses an electrically powered implant to replace the current model, which uses a bulky external power source to drive a mechanical pump inside the body by compressed air. Also under development is a ***left ventricular assist device (LVAD),*** which boosts the action of a failing heart (left ventricle) rather than replacing both ventricles. The LVAD takes blood from a patient's weakened left ventricle and pumps it into the aorta. It is powered by a battery pack strapped outside the body. Both artificial hearts and LVADs are used primarily as ***bridges to transplants;*** that is, they are used as temporary mechanical support devices until a donor heart transplantation can be performed. A more recent development is the ***Hemopump.*** It consists of a propellerlike pump, about ¼ in. wide and ½ in. long, that is threaded through an artery in the groin into the left ventricle. There the blades of the pump whirl at about 25,000 revolutions per minute, pulling blood out of the left ventricle into the aorta.

CLINICAL APPLICATION

Three-Way Transplant

Teams of surgeons at two hospitals in Baltimore performed a ***three-way transplant*** in May 1987. At the University of Maryland Hospital, physicians removed the heart and lungs from an accident victim who had been declared brain dead. The organs were then transported to Johns Hopkins Medical Center and transplanted into a person whose lungs were destroyed by cystic fibrosis (CF) but whose heart was healthy. The healthy heart was removed and transplanted into another person who was in need of a heart transplant. The procedure, referred to as a "domino donor" organ exchange, marked the first time in U.S. medical history that a healthy human heart was taken from a living person and transplanted into another human.

HEART–LUNG MACHINE

Some surgical procedures, such as those that involve heart transplantation, open-heart surgery, and coronary artery bypass surgery, make it necessary to use a ***heart–lung machine.*** This device performs two complex roles simultaneously: (1) it pumps blood (functioning as a heart); and (2) it removes carbon dioxide (CO_2) from blood and adds oxygen (O_2) to it (functioning as the lungs). The principle of the heart–lung machine is fairly simple. Deoxygenated blood from the venae cavae is removed from the body and passed through an oxygenator, where carbon dioxide is removed and oxygen is added. Then the oxygenated blood is passed through a temperature controller, where it may be rewarmed to body temperature or cooled. Next, the blood passes through a filter to remove emboli and is finally returned to the body into the arterial system at the proper pressure. If necessary, drugs, anesthetics, and transfusions may be added to the circuit.

To surgically repair certain heart abnormalities, it is necessary to slow down a patient's heart. One such method is called ***hypothermia*** (hī-pō-THER-mē-a), which refers to a low body temperature. In various surgical procedures, it refers to a deliberate cooling of the body to slow metabolism and reduce the oxygen needs of the tissues. Thus, the heart and brain can withstand short periods of interrupted or reduced blood flow. Lost blood is then replaced by transfusion during and after the operation.

RISK FACTORS IN HEART DISEASE

It is estimated that one in every five persons who reaches age 60 will have a myocardial infarction (heart attack). One in every four persons between 30 and 60 has the potential to be stricken. Heart disease is epidemic in the United States, despite the fact that some of the causes can be foreseen and prevented. The results of research indicate that people who develop combinations of certain risk factors are likely to have heart attacks. ***Risk factors*** are characteristic symptoms or signs present in a person free of disease that are statistically associated with an excessive rate of development of a disease. Among the major risk factors in heart disease are:

1. High blood cholesterol level.
2. High blood pressure.
3. Cigarette smoking.
4. Obesity.
5. Lack of regular exercise.
6. Diabetes mellitus.
7. Genetic predisposition (family history of heart disease at an early age).
8. Gender (male).
9. Age.
10. Fibrinogen level.
11. Left ventricular hypertrophy (enlarged left ventricle).

The first five risk factors can be modified and all contribute to increasing the heart's work load. High blood cholesterol is discussed shortly, and hypertension is discussed in the next chapter. Cigarette smoking, through the effects of nicotine, stimulates the adrenal gland to oversecrete aldosterone, epinephrine, and norepinephrine (NE)—the latter two being powerful vasoconstrictors. Overweight people develop miles of extra capillaries to nourish fat tissue. The heart has to work harder to pump the blood through more vessels. Without exercise, venous return gets less help from contracting skeletal muscles. In addition, regular exercise strengthens the smooth muscle of blood vessels and enables them to assist general circulation. Exercise also increases cardiac efficiency and output. In diabetes mellitus, fat metabolism dominates glucose metabolism. As a result, cholesterol levels get progressively higher and result in plaque formation, a situation that may lead to high blood pressure. Blood pressure is the force of blood exerted on the wall of an artery. It is determined by rate and force of heartbeat, blood volume, and resistance (opposition) to blood flow as a result of friction between the blood and walls of blood vessels and blood viscosity. High blood pressure drives fat into the vessel wall, encouraging atherosclerosis. Up to age 50, there is a 10- to 15-year lag in the extent of heart disease in females compared with males; after that, the rates of disease in both sexes are similar. Regardless of sex, the incidence of heart disease increases with age. Regarding fibrinogen, the higher the level, the higher the risk of heart disease (more so in males than females). Fibrinogen enhances blood clot formation. Both hypertension and obesity contribute to left ventricular hypertrophy.

DEVELOPMENTAL ANATOMY OF THE HEART

The *heart,* a derivative of ***mesoderm,*** begins to develop before the end of the third week of gestation. It begins its development in the ventral region of the embryo beneath the foregut (see Figure 23-20). The first step is the formation of a pair of tubes, the ***endothelial (endocardial) tubes,*** from mesodermal cells (Figure 13-8). These tubes then unite to form a common tube, the ***primitive heart tube.*** Next, the primitive heart tube develops into five regions: ***ventricle, bulbus cordis, atrium, sinus venosus,*** and ***truncus arteriosus.*** Since the bulbus cordis and ventricle subsequently grow more rapidly than the others, and the heart grows more rapidly than its superior and inferior attachments, the heart assumes a U-shape and later an S-shape. The flexures of the heart reorient the regions so that the atrium and sinus venosus eventually come to lie superior to the bulbus cordis, ventricle, and truncus arteriosus.

At about the seventh week, a partition, the ***interatrial septum,*** forms in the atrial region, dividing it into a *right* and *left atrium.* The opening in the partition is the ***foramen ovale,*** which normally closes at birth and later forms a depression called the *fossa ovalis.* An ***interventricular septum*** also develops and partitions the ventricular region into a *right* and *left ventricle.* The bulbus cordis and truncus arteriosus divide into two vessels, the *aorta* (arising from the left ventricle) and the *pulmonary trunk* (arising from the right ventricle). Recall that the ductus arteriosus is a temporary vessel between the aorta and pulmonary trunk until birth. The great veins of the heart, *superior vena cava* and *inferior vena cava,* develop from the venous end of the primitive heart tube.

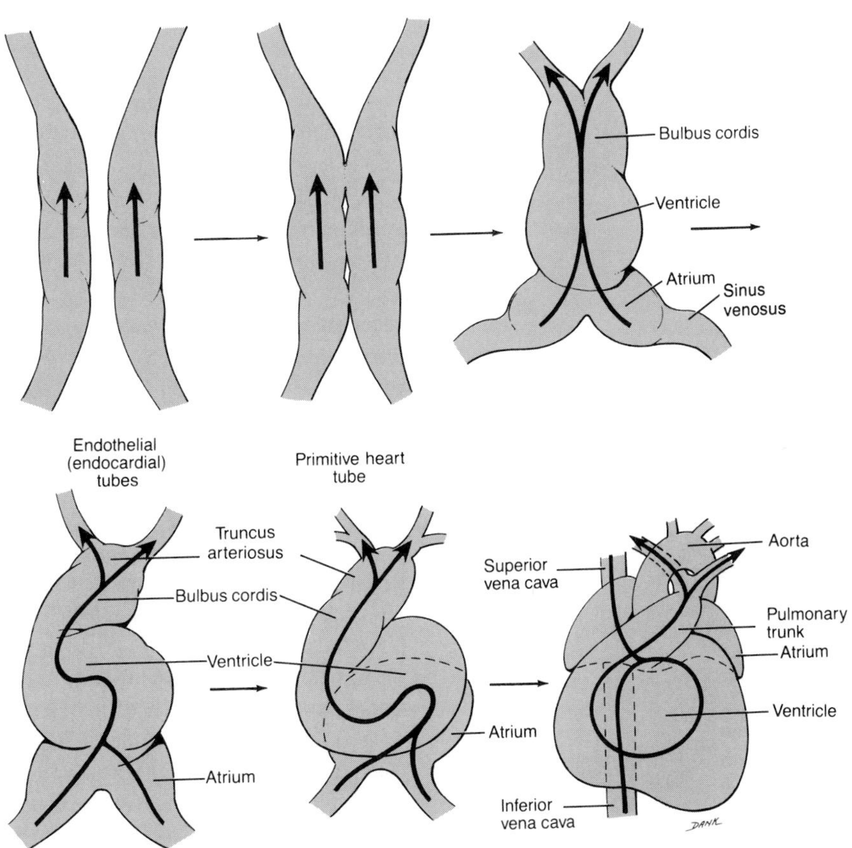

FIGURE 13-8 Development of the heart. The arrows show the direction of the flow of blood from the venous to the arterial end.

APPLICATIONS TO HEALTH

CORONARY ARTERY DISEASE (CAD)

Coronary artery disease (CAD) is a condition in which the heart muscle receives an inadequate amount of blood because of an interruption of its blood supply. Two of the principal causes are atherosclerosis and coronary artery spasm. Another is a thrombus or an embolus in a coronary artery.

Atherosclerosis

Atherosclerosis (ath′-er-ō-skle-RŌ-sis) is a process in which fatty substances, especially cholesterol and triglycerides (ingested fats), are deposited in the walls of medium-sized and large arteries in response to certain stimuli. In time, a lesion called an ***atherosclerotic plaque*** develops (Figure 13-9). As it grows, it may obstruct blood flow in the affected artery and damage the tissues that the artery supplies. An additional danger is that the plaque provides a roughened surface that causes a blood clot to form. If the clot breaks off and forms an embolus (blood clot transported by blood), it may obstruct small arteries and capillaries quite a distance from the site of formation.

Cholesterol is a major building block of all plasma membranes and a key compound for the synthesis of steroid hormones and bile salts. Some cholesterol is present in foods (eggs, dairy products, organ meats, beef, pork, and processed luncheon meats); most is synthesized by the liver. As it turns out, foods containing cholesterol are not the main source of cholesterol in the blood. It is the type of fat eaten. When saturated fats are broken down in the body, the liver uses some of the breakdown products to produce cholesterol. Saturated fats can increase blood cholesterol level as much as 25 percent.

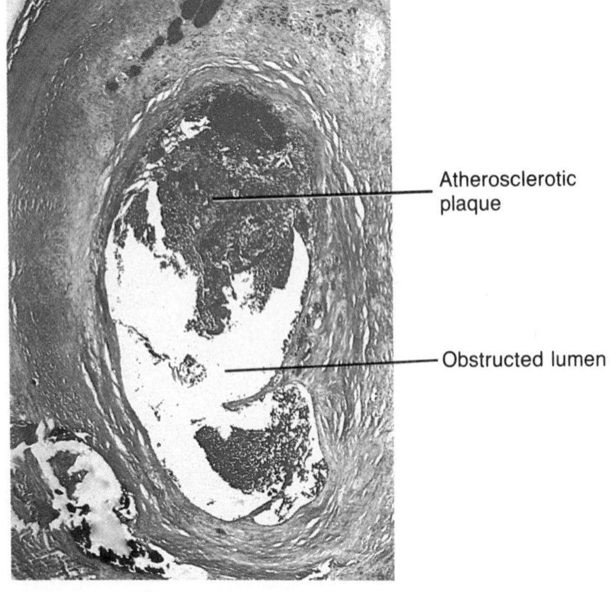

FIGURE 13-9 Photomicrograph of an artery partially obstructed by an atherosclerotic plaque. (Courtesy of Sklar/Photo Researchers.)

Cholesterol and triglycerides cannot dissolve in water and thus cannot travel in the blood in their unaltered forms. They are made water-soluble by combining with proteins produced by the liver and intestine (apoproteins), and the complexes thus formed are called ***lipoproteins,*** which vary in size, weight, and density. Two of the major classes are called ***low-density lipoproteins (LDLs)*** and ***high-density lipoproteins (HDLs).*** There is an important difference between the two: LDL contains 60 to 70 percent of total serum cholesterol and seems to pick up cholesterol and deposit it in body cells including, under abnormal conditions, smooth muscle fibers (cells) in arteries. As blood levels of LDL *increase,* the risk of coronary artery disease *increases.* LDL may be calculated by using the following formula: LDL = total cholesterol − HDL − (triglycerides/5). HDL, which contains 20 to 30 percent of total serum cholesterol, seems to gather cholesterol from body cells and transport it to the liver for elimination. As blood levels of HDL *increase,* the risk of coronary artery disease *decreases.* Most tissues of the body contain LDL receptors. The liver, suprarenal (adrenal) glands, and ovaries, because of their large requirement for cholesterol, have the most LDL receptors. Once LDL attaches to its receptor, it is taken into the cell by receptor-mediated endocytosis (Chapter 2). Within the cell, the LDL is broken down, and the cholesterol is released to serve the cell's needs. In this way, LDL is removed from the blood. Once a cell has sufficient cholesterol for its activities, a negative feedback system inhibits the cell from synthesizing new LDL receptors. It is believed that some people have too few LDL receptors, owing to various environmental and genetic factors, and thus are more susceptible to atherosclerosis. A third type of lipoprotein, known as a ***very low-density lipoprotein (VLDL),*** contains about 10 to 15 percent of total serum cholesterol and a large amount of triglyceride (fat). It also plays a role in the development of atherosclerotic plaques, but its role is not clear.

The amount of total cholesterol (HDL, LDL, and VLDL) in blood is one commonly used indicator of risk for CAD. In general, as the total cholesterol level increases above 150 mg/dl (a deciliter is 1⁄10 of a liter), the risk of CAD slowly begins to rise. Above 200 mg/dl, the risk increases even more. The chance of a heart attack has been found to double with every 50 mg/dl increase in total cholesterol once the level goes over 200 mg/dl. The risk of developing CAD may be predicted by determining the ratio of HDL to total cholesterol. The risk ratio is determined by dividing total cholesterol by HDL. For example, a person with a total cholesterol of 200 and an HDL of 65 has a risk ratio of about 3. Ratios above 4 are considered undesirable; the higher the ratio, the greater the risk of developing CAD.

Several ways to improve the HDL:LDL ratio are regular sustained exercise, diet (reducing animal fats, especially), and giving up cigarettes.

Unsaturated oils from fish, seal, and whale contain fatty acids referred to as ***omega-3 fatty acids.*** These fatty acids appear to have a role in protecting against heart disease. Specifically, they are believed to decrease serum cholesterol and low-density lipoprotein (LDL) levels; decrease the tendency of blood platelets to clump together, thus decreasing the likelihood of blood clot formation; decrease blood viscosity; suppress production of clotting factors; enhance the production of an anticlotting agent; and discourage renarrowing of arteries previously opened by insertion of a tiny balloon. Oily fish, such as salmon, bluefish, herring, sardines, anchovies, and mackerel, are good sources of omega-3 fatty acids. Shellfish also contain omega-3 fatty acids. Among the potential side effects of taking high concentrations of fish oil supplements are weight gain, vitamin A and D toxicity, vitamin E deficiency, and possibly increased LDL levels. At the present time, most researchers recommend increasing the intake of seafood rather than taking supplements.

Among the therapies used to reduce cholesterol levels are exercise, diet, and drugs. Regular physical activity at aerobic and nearly aerobic levels tends to raise HDL. Dietary changes are aimed at reducing the intake of total fat, saturated fats, and cholesterol. This means restricting intake of foods containing butterfat (butter, cheese, ice cream, cream, and whole milk), fatty meats, egg yolks, organ meats, and certain oils (coconut and palm). Oat bran and rice bran are also believed to have a role in reducing total cholesterol. Among the drugs used to treat high blood cholesterol levels are cholestyramine (Questran) and colestipol (Colestid), both bile acid sequestrants; nicotinic acid (Liponicin); lovastatin (Mevacor), an HMG CoA reductase inhibitor; gemifbrozil (Lopid); and probucol (Lorelco).

Treatment of CAD varies with the nature and urgency of symptoms. Among the treatment options are drug therapy (nitroglycerine, beta blockers, and thrombolytic agents) and various surgical and nonsurgical procedures.

Coronary artery bypass grafting (CABG) is one way of increasing the blood supply to the heart. It is a surgical procedure in which a portion of a blood vessel is removed from another part of the body. A segment of the grafted blood vessel is then sutured between the aorta and the unblocked portion of the coronary artery, distal to the obstruction (Figure 13-10a). If more than one artery is clogged, additional bypasses may be made.

An innovative procedure called ***cardiomyoplasty*** (*cardio* = heart; *myo* = muscle; *plasty* = to mold or shape) is providing hope for patients with end-stage heart failure who cannot withstand the rigors of coronary artery bypass grafting or a transplant. In cardiomyoplasty, surgeons remove a rib, free one of the latissimus dorsi muscles from its connective tissue tendons, and wrap it over the heart. The remainder of the muscle, which is not wrapped around the heart, is left attached to the posterior body wall along with its blood and nerve supply. A pacemaker is placed between the heart and latissimus dorsi muscle to stimulate the muscle's motor neurons, which in turn, cause the skeletal muscle to contract. The pacemaker coordinates the type II (fast-twitch) fibers of the latissimus dorsi muscle with the slow-twitch fibers of the heart muscle.

A nonsurgical procedure used to treat CAD is referred to as ***percutaneous transluminal coronary angioplasty (PTCA)*** (*percutaneous* = through the skin; *trans* = across; *lumen* = channel in a tube; *angio* = blood vessel; *plasty* = to mold or shape). Like coronary artery bypass grafting, it is an attempt to increase the blood supply to the heart muscle. In this procedure, a balloon catheter is inserted into an artery of an arm or leg and gently guided through the arterial system under x-ray observation (Figure 13-10b). Once in place in an obstructed coronary artery, the balloon is inflated for a few seconds, thus compressing the obstructing material against the vessel wall and permitting increased blood flow. PTCA is most frequently used to relieve angina pectoris. Since about 30 percent of balloon-opened arteries restenose (narrow from reformed obstructions), special devices called ***stents*** are being used to keep the arteries patent (open). A stent is a stainless steel device, resembling a spring coil, that is permanently placed in an artery to maintain patency, permitting blood to circulate (Figure 13-10c). A stent is inserted in a blood vessel via a balloon catheter. In addition, it has been found that megadoses of omega-3 fatty acids discourage restenosis. A modification of PTCA, called ***percutaneous balloon valvuloplasty,*** is used to treat faulty heart valves.

Another nonsurgical technique for opening clogged arteries is a procedure called ***laser angioplasty.*** In one variation of this procedure, a laser vaporizes the atherosclerotic plaque and makes a channel through the blood vessel obstruction. Then a balloon catheter is inserted, and the balloon is inflated to widen the vessel.

Two of the latest nonsurgical techniques for clearing arteries are balloon–laser welding and catheter artherectomy. In ***balloon–laser welding,*** an artery is first widened by PTCA. On the last balloon inflation, a laser heats the surrounding tissue sufficiently to stretch and weld the arterial wall into a smooth surface. In ***catheter artherectomy,*** a rotating drill shaves off plaque. Shavings are trapped for removal.

Coronary Artery Spasm

Atherosclerosis results in a fixed obstruction to blood flow. Obstruction can also be caused by ***coronary artery spasm,*** a condition in which the smooth muscle of a coronary artery undergoes a sudden contraction, resulting in vasoconstriction. Coronary artery spasm typically occurs in individuals with atherosclerosis and may result in chest pain during rest (variant angina), chest pain during exertion (typical angina), heart attacks, and sudden death. Although the

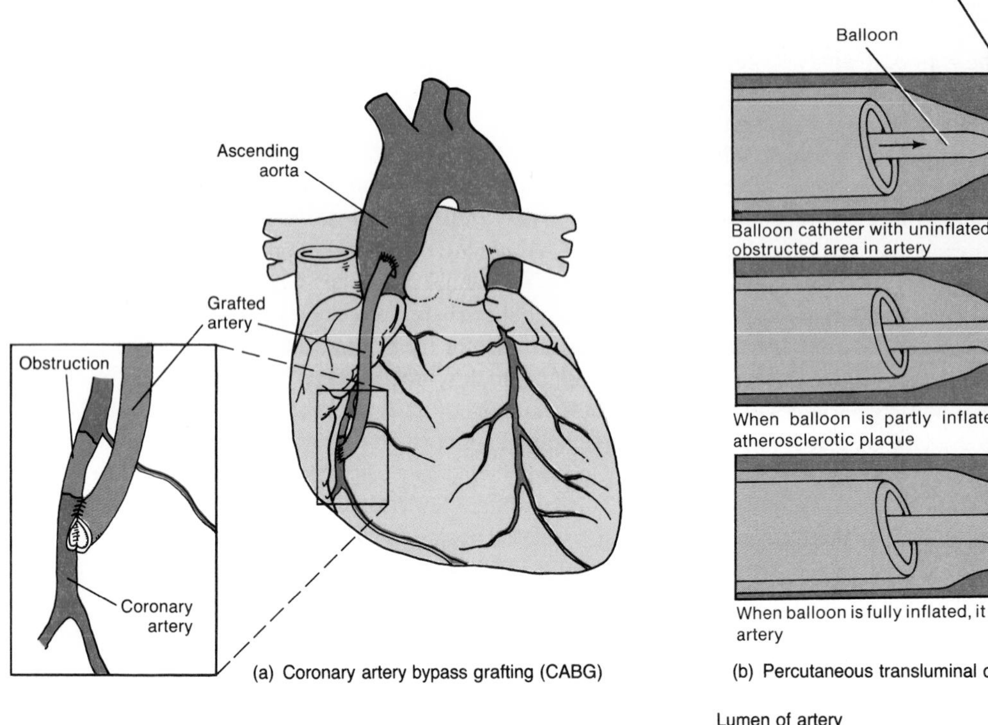

(a) Coronary artery bypass grafting (CABG)

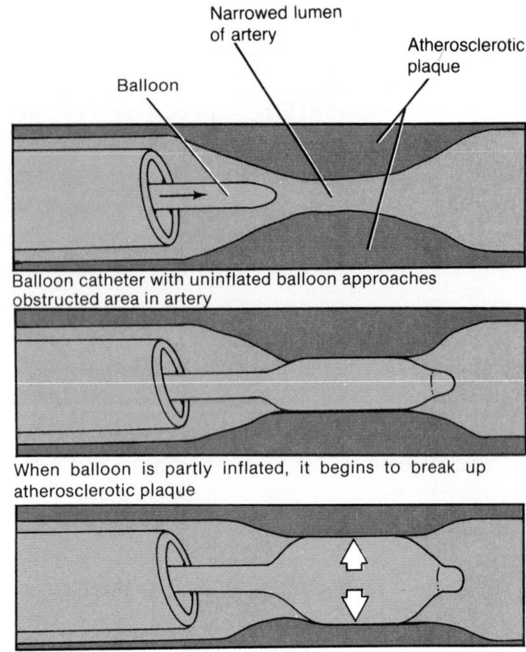

(b) Percutaneous transluminal coronary angioplasty (PTCA)

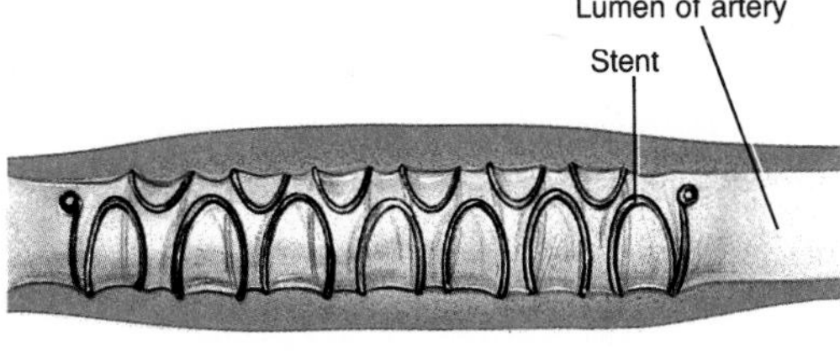

(c) Stent in artery

FIGURE 13-10 Several procedures for reestablishing blood flow in occluded coronary arteries. In coronary artery bypass grafting (CABG), shown in (a), the internal mammary artery is removed from the patient's chest. At a point distal to the obstruction, the grafted artery is sutured to the coronary artery.

causes of coronary artery spasm are unknown, several factors are receiving attention. These include smoking, stress, and a vasoconstrictor chemical released by platelets. There is considerable interest in aspirin and other drugs that might inhibit the vasoconstrictor chemical. Coronary artery spasm is treated with long-acting nitrates (nitroglycerin) and calcium-blocking agents.

CONGENITAL DEFECTS

A defect that exists at birth, and usually before, is called a ***congenital defect.***

One congenital defect that can occur is ***coarctation*** (kō′-ark-TĀ-shun) ***of the aorta*** (Figure 13-11a). In this condition, a segment of the aorta is too narrow. As a result, the flow of oxygenated blood to the body is reduced, the left ventricle is forced to pump harder, and high blood pressure develops. The condition may be corrected by insertion of a synthetic graft.

Another common congenital defect is ***patent ductus arteriosus*** (Figure 13-11b). The ductus arteriosus, a temporary blood vessel between the aorta and the pulmonary trunk, normally closes shortly after birth in response to increased oxygenation of blood flowing through the ductus arteriosus. In some babies, the ductus arteriosus remains open. As a result, aortic blood flows into the lower-pressure pulmonary trunk, thus increasing the pulmonary trunk blood pressure and overworking both ventricles and the heart. Treatment consists of a surgical procedure in which the ductus arteriosus is ligated or cut and the cut ends are tied off.

A ***septal defect*** is an opening in the septum that separates the interior of the heart into a left and right side. In one type of ***interatrial septal defect*** there is a failure of the fetal foramen ovale between the two atria to close after

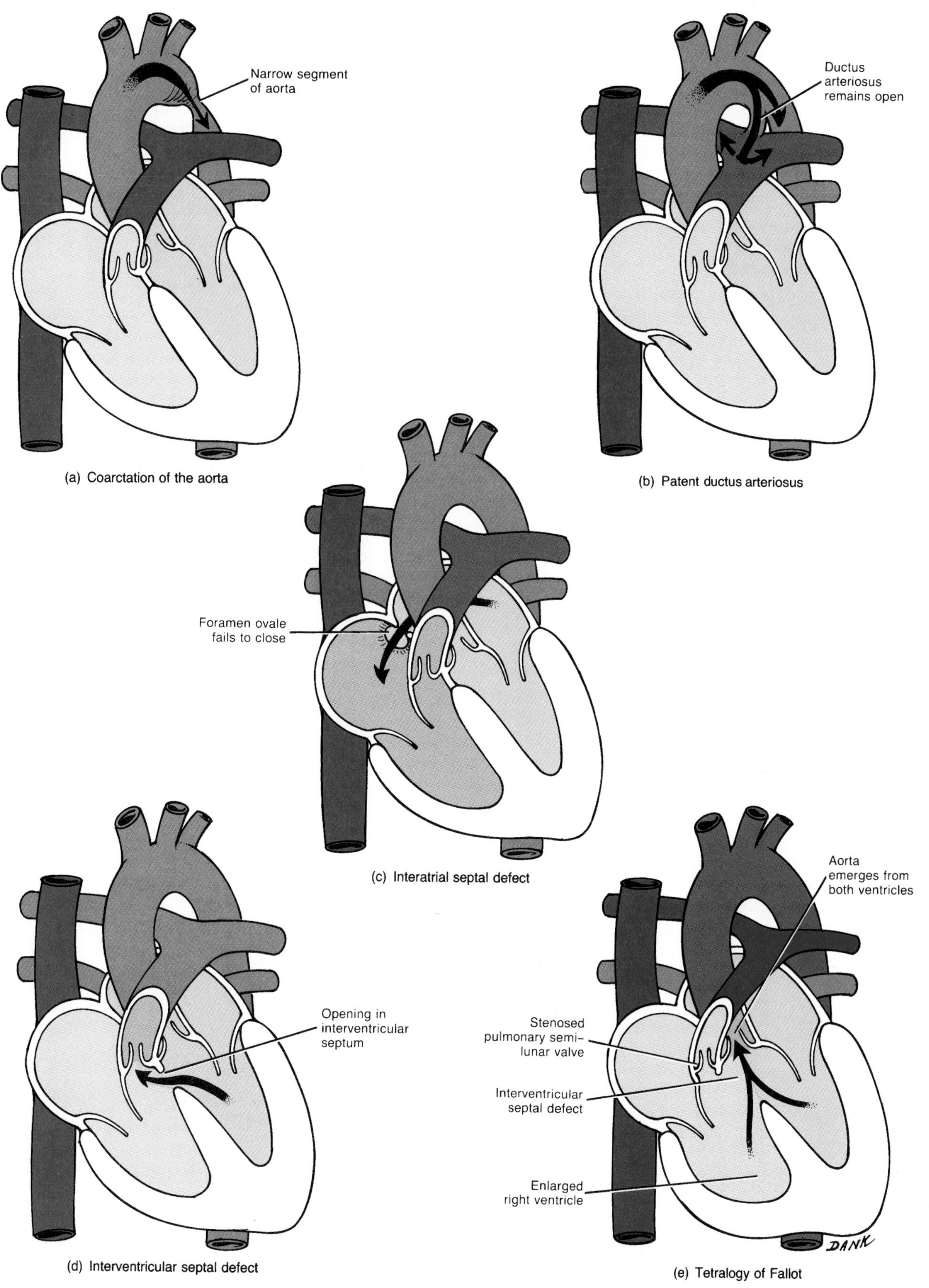

FIGURE 13-11 Some common congenital heart defects.

birth (Figure 13-11c). It allows a good deal of blood to flow from the left atrium to the right without going through systemic circulation. This defect overloads the pulmonary circulation, produces fatigue, and increases respiratory infections. ***Interventricular septal defect*** is caused by an incomplete closure of the interventricular septum (Figure 13-11d). It permits oxygenated blood to flow directly from the left ventricle into the right ventricle where it mixes with deoxygenated blood. Septal openings can now be sewn shut or covered with synthetic patches inserted via a catheter.

Valvular stenosis is a narrowing of one of the valves regulating blood flow in the heart. All stenoses are serious because they place a severe work load on the heart by making it work harder to push the blood through the abnormally narrow valve openings. As a result of mitral valve stenosis, blood pressure is increased. Most stenosed valves are totally replaced with artificial valves.

Tetralogy of Fallot (tet-RAL-ō-jē of fal-Ō) is a combination of four defects: an interventricular septal defect, an aorta that emerges from both ventricles instead of from the left ventricle only, a stenosed pulmonary semilunar valve, and an enlarged right ventricle (Figure 13-11e). The condition is an example of a right-to-left shunt, in which blood is shunted from the right ventricle to the left ventricle without going through pulmonary circulation. Because there is stenosis of the pulmonary semilunar valve, the increased right ventricular pressure forces deoxygenated blood from the right ventricle to enter the left ventricle through the interventricular septum. As a result, deoxygenated blood gets mixed with the oxygenated blood that is pumped into systemic circulation. Also, because the aorta emerges from the right ventricle and the pulmonary artery is stenosed, very little blood gets to the lungs, and pulmonary circulation is bypassed almost completely. ***Cyanosis*** (sī-a-NŌ-sis) refers to a reduced hemoglobin (unoxygenated) concentration in blood of more than 5 g/dl. It causes a blue or dark purple discoloration that is most easily seen in nail beds and mucous membranes. For this reason, tetralogy of Fallot is one of the conditions that causes a "blue baby." Chronic lung disorders and suffocation also result in cyanosis.

It is possible to correct cases of tetralogy of Fallot when the patient is of proper age and condition. Open-heart operations are performed in which the narrowed pulmonary valve is cut open and the interventricular septal defect is sealed with a Dacron patch.

ARRHYTHMIAS

Arrhythmia (a-RITH-mē-a) is a general term that refers to an abnormality or irregularity in the heart rhythm. More physicians are using the term ***dysrhythmia*** since it implies an abnormal rhythm, whereas arrhythmia implies no rhythm. An arrhythmia results when there is a disturbance in the conduction system of the heart, either due to faulty production of electrical impulses or faulty conduction of impulses as they pass through the system.

There are many different types of arrhythmias that can occur, some normal and some quite serious. Arrhythmias may be caused by factors such as caffeine, nicotine, alcohol, anxiety, certain drugs, hyperthyroidism, potassium deficiency, and certain heart diseases. Serious arrhythmias can result in cardiac arrest if the heart cannot supply its own oxygen demands, as well as those of the rest of the body. They can be controlled, and the normal heart rhythm can be reestablished, if they are detected and treated early enough.

One very serious arrhythmia is called ***ventricular fibrillation (VF)***. It almost always indicates imminent cardiac arrest and death unless corrected quickly. It is characterized by asynchronous, haphazard, ventricular muscle contractions. The rate may be rapid or slow. The impulse travels to the different parts of the ventricles at different rates. Thus, part of the ventricle may be contracting while other parts are still unstimulated. Ventricular contraction becomes ineffective, and circulatory failure and death occur. Ventricular fibrillation may be caused by coronary artery occlusion. It sometimes occurs during surgical procedures on the heart or pericardium. It may be the cause of death in electrocution. It is possible to pass a very strong electric current through the ventricles for a short interval of time and thereby stop ventricular fibrillation. This technique, called ***defibrillation***, involves placing electrodes on two sides of the heart and applying a controlled current.

Another form of arrhythmia arises when a small region of the heart outside the pacemaker becomes more excitable than normal, causing an occasional abnormal impulse to be generated between normal impulses. The region from which the abnormal impulse is generated is called an ***ectopic focus***. As a wave of depolarization spreads outward from the ectopic focus, it causes a ***ventricular premature contraction (VPC)***, also called a ***premature ventricular contraction (PVC)***. The contraction occurs early in diastole before the SA node is normally scheduled to discharge its impulses. A person with such a ventricular premature contraction might feel a thump in the chest. The contractions may be relatively benign and may be caused by emotional stress, excessive intake of stimulants such as caffeine or nicotine, and lack of sleep. In other cases, the contractions may indicate an underlying pathology.

CONGESTIVE HEART FAILURE (CHF)

Congestive heart failure (CHF) may be defined as a chronic or acute state that results when the heart is not capable of supplying the oxygen demands of the body. Symptoms and signs of CHF include fatigue, peripheral and pulmonary edema, and visceral congestion. These symptoms are produced by diminished blood flow to the various tissues of the body and by accumulation of excess blood in the various organs because the heart is unable to pump out the blood returned to it by the great veins. Both diminished blood flow and accumulation of excess blood in the organs occur together, but certain symptoms result from congestion, whereas others are produced by poor tissue nutrition. Among the causes of CHF are long-standing hypertension and myocardial infarction.

COR PULMONALE (CP)

Cor pulmonale (kor pul-mōn-ALE; *cor* = heart; *pulmon* = lung), or ***CP,*** refers to right ventricular hypertrophy from disorders that bring about hypertension (high blood pressure) in the pulmonary circulation. Certain disorders such as those involving the respiratory center in the brain, lung diseases, diseases involving airways in the lungs, deformities of the thoracic cage, and neuromuscular disorders can result in hypoxia in the lungs. As a result, vasoconstriction of pulmonary arteries and arterioles occurs, and blood flow to the lungs is decreased. Also, any conditions such as tumors, aneurysms, and diseased blood vessels in the lungs can result in decreased blood flow. The reduction in blood flow and increased resistance in the lungs bring about pulmonary hypertension. Ultimately, this condition causes increased pressure in the pulmonary arteries and pulmonary trunk, resulting in right ventricular hypertrophy.

Individuals with CP may exhibit fatigue, weakness, dyspnea, right upper quadrant (RUQ) pain caused by an enlarged liver, warm and cyanotic extremities, cyanosis of the face, distended neck veins, chest pain, and edema.

KEY MEDICAL TERMS ASSOCIATED WITH THE HEART

Angiocardiography (an′-jē-ō-kar′-dē-OG-ra-fē; *angio* = vessel; *cardio* = heart; *graph* = writing) X-ray examination of the heart and great blood vessels after injection of a radiopaque dye into the bloodstream.

Cardiac arrest (KAR-dē-ak a-REST) A clinical term meaning cessation of an effective heartbeat. The heart may be completely stopped (cardiac standstill) or quivering ineffectively (ventricular fibrillation).

Cardiomegaly (kar′-dē-ō-MEG-a-lē; *mega* = large) Heart enlargement. Long-distance runners and weight lifters frequently develop heart enlargement as a natural adaptation to increased work load produced by regular exercise (physiologic cardiomegaly). The enlargement may also be due to disease (pathologic cardiomegaly).

Commissurotomy (kom′-i-shur-OT-ō-mē) An operation that is performed to widen the opening in a heart valve that has become narrowed by scar tissue.

Incompetent valve (in-KOM-pe-tent VALV) Any valve that does not close properly, thus permitting a backflow of blood; also called **valvular insufficiency.**

Palpitation (pal′-pi-TĀ-shun) A fluttering of the heart or abnormal rate or rhythm of the heart.

Pancarditis (pan′-kar-DĪ-tis; *pan* = all) Inflammation of the whole heart including the inner layer (endocardium), heart muscle (myocardium), and outer sac (pericardium).

Paroxysmal tachycardia (par′-ok-SIZ-mal tak′-e-KAR-dē-ā) A period of rapid heartbeats that begins and ends suddenly.

Stokes–Adams syndrome Sudden attacks of unconsciousness, sometimes with convulsions, that may accompany heart block.

Sudden cardiac death The unexpected cessation of circulation and breathing due to an underlying heart disease such as ischemia, myocardial infarction, or a disturbance in cardiac rhythm.

STUDY OUTLINE

Location (p. 352)

1. The heart is situated between the lungs in the mediastinum.
2. About two-thirds of its mass is to the left of the midline.

Pericardium (p. 353)

1. The pericardium consists of an outer fibrous pericardium and an inner serous pericardium.
2. The serous pericardium is composed of a parietal and visceral layer.
3. Between the parietal and visceral layers of the serous pericardium is the pericardial cavity, a potential space filled with pericardial fluid that prevents friction between the two membranes.

Wall; Chambers; Vessels; and Valves (pp. 354–358)

1. The wall of the heart has three layers; epicardium, myocardium, and endocardium.
2. The chambers include two upper atria and two lower ventricles.
3. The blood flows through the heart from the superior and inferior venae cavae and the coronary sinus to the right atrium, through the tricuspid valve to the right ventricle, through the pulmonary trunk to the lungs, through the pulmonary veins into the left atrium, through the bicuspid valve to the left ventricle, and out through the aorta.
4. Valves prevent backflow of blood in the heart.
5. Atrioventricular (AV) valves, between the atria and their ventricles, are the tricuspid valve on the right side of the heart and the bicuspid (mitral) valve on the left.
6. The chordae tendineae and their muscles stop blood from backing into the atria.
7. The two arteries that leave the heart both have a semilunar valve. They are the aortic semilunar valve in the aorta and pulmonary semilunar valve in the pulmonary trunk.
8. The heart valves may be identified by surface projection.
9. The skeleton of the heart consists of fibrous connective tissue around the valves of the heart and fused with each other; the skeleton anchors the valves and cardiac fibers and forms an electrical insulator between atria and ventricles.

Blood Supply (p. 360)

1. The coronary (cardiac) circulation delivers oxygenated blood to the myocardium and removes carbon dioxide from it.
2. Deoxygenated blood returns to the right atrium via the coronary sinus.
3. Complications of this system are angina pectoris and myocardial infarction (MI).

Conduction System (p. 362)

1. The conduction system consists of tissue specialized for generation and conduction of action potentials.
2. Components of this system are the sinoatrial (SA) node (pacemaker), atrioventricular (AV) node, atrioventricular (AV) bundle (bundle of His), bundle branches, and conduction myofibers (Purkinje fibers).

Electrocardiogram (ECG) (p. 364)

1. The record of electrical changes during each cardiac cycle is referred to as an electrocardiogram (ECG).
2. A normal ECG consists of a P wave (spread of impulse from SA node over atria), QRS wave (spread of impulse through ventricles), and T wave (ventricular repolarization).
3. The ECG is invaluable in diagnosing abnormal cardiac rhythms and conduction patterns, detecting the presence of fetal life, determining the presence of several fetuses, and following the course of recovery from a heart attack.
4. An artificial pacemaker may be used to restore an abnormal cardiac rhythm.

Cardiac Cycle (p. 364)

1. With an average heartbeat of 75/min, a complete cardiac cycle requires 0.8 sec.
2. The first heart sound (lubb) represents the closing of the atrioventricular valves. The second sound (dupp) represents the closing of semilunar valves.

Autonomic Control (p. 365)

1. Heart rate and strength of contraction may be increased by sympathetic stimulation from the cardioacceleratory center (CAC) in the medulla and decreased by parasympathetic stimulation from the cardioinhibitory center (CIC) in the medulla.
2. Baroreceptors are nerve cells that respond to changes in blood pressure. They act on the cardiac centers in the medulla.
3. Other influences on heart rate include chemicals (epinephrine, sodium, potassium), temperature, emotion, sex, and age.

Artificial Heart (p. 365)

1. The Jarvik artificial heart consists of an aluminum base and rigid plastic chambers that serve as ventricles.
2. When it is implanted, the recipient's atria are left intact. (It is now banned by the FDA.)
3. Two recently developed bridges to transplant devices are the left ventricular assist device (LVAD) and the Hemopump.

Heart–Lung Machine (p. 366)

1. Hypothermia (deliberate body cooling) and the heart–lung bypass permit open-heart surgery.
2. The heart–lung machine pumps blood (functioning as a heart) and oxygenates blood and removes carbon dioxide (functioning as lungs).

Risk Factors in Heart Disease (p. 366)

1. Risk factors in heart disease include high blood cholesterol, high blood pressure, cigarette smoking, obesity, lack of regular exercise, diabetes mellitus, genetic disposition, gender (male), age, fibrinogen level, and left ventricular hypertrophy.

Developmental Anatomy of the Heart (p. 367)

1. The heart develops from mesoderm.
2. The endothelial tubes develop into the four-chambered heart and great vessels of the heart.

Applications to Health (p. 368)

1. Coronary artery disease (CAD) refers to a condition in which the myocardium receives inadequate blood due to atherosclerosis, coronary artery spasm, or thrombi or emboli; it may be treated with drugs or surgically.
2. Congenital heart defects include coarctation of the aorta, patent ductus arteriosus, septal defects, valvular stenosis, and tetralogy of Fallot.
3. Arrhythmias include heart block, flutter, and fibrillation.
4. Congestive heart failure (CHF) results when the heart cannot supply the oxygen demands of the body.
5. Cor pulmonale (CP) refers to right ventricular hypertrophy secondary to pulmonary hypertension.

REVIEW QUESTIONS

1. Describe the location of the heart in the mediastinum and identify the various borders and surfaces of the heart. (p. 352)
2. Distinguish the subdivisions of the pericardium. What is the purpose of this structure? (p. 353)
3. Compare the three layers of the heart wall according to composition, location, and function. (p. 354)
4. Define atria and ventricles. What vessels enter or exit the atria and ventricles? (p. 355)
5. Describe the principal valves in the heart and how they operate. (p. 358)
6. Describe the components of the skeleton of the heart. (p. 358)
7. Describe how the valves of the heart may be identified by surface projection. (p. 360)
8. Describe the route of blood in coronary (cardiac) circulation. Distinguish between angina pectoris and myocardial infarction (MI). (p. 360)
9. Describe the structure and function of the heart's conducting system. (p. 362)

10. Define and label the deflection waves of a normal electrocardiogram (ECG). Explain why the ECG is an important diagnostic tool. (p. 364)
11. Define a cardiac cycle and describe its timing. What causes heart sounds? (p. 364)
12. Distinguish between the cardioacceleratory center (CAC) and cardioinhibitory center (CIC) with respect to the regulation of heart rate. (p. 365)
13. Describe the Jarvik artificial heart. What is the left ventricular assist device (LVAD)? How does the Hemopump operate? (p. 365)
14. Describe the operation of the heart–lung machine. (p. 366)
15. Describe the risk factors involved in heart disease. (p. 366)
16. Describe how the heart develops. (p. 367)
17. What is coronary artery disease (CAD)? What factors contribute to it? Explain. (p. 368)
18. Define each of the following congenital heart defects: coarctation of the aorta, patent ductus arteriosus, septal defects, valvular stenosis, and tetralogy of Fallot. (p. 370)
19. Define an arrhythmia. Give an example. (p. 372)
20. Describe the symptoms of congestive heart failure (CHF) and cor pulmonale (CP). (p. 372)
21. Refer to the glossary of key medical terms associated with the heart. Be sure that you can define each term. (p. 373)

SELF QUIZ

1. Match the following:

___ **a.** also called the mitral valve		**A.**	aortic semilunar
___ **b.** prevents backflow of blood from right ventricle to right atrium		**B.**	bicuspid
___ **c.** prevents backflow of blood from pulmonary trunk to right ventricle		**C.**	pulmonary semilunar
___ **d.** prevents backflow of blood into left atrium		**D.**	tricuspid
___ **e.** have half-moon-shaped leaflets or cusps (two answers)			
___ **f.** also called atrioventricular (AV) valve (two answers)			

Complete the following:

2. The heart is derived from ________-derm. The heart begins to develop during the (third? fifth? seventh?) week. Its initial formation consists of two endothelial tubes that unite to form the ________ tube.
3. The first heart sound (lubb? dupp?) is created by turbulence of blood at the (opening? closing?) of the ________ valves.
4. Oxygenated blood returns from the lungs to the heart by means of the pulmonary ________. They empty into the ________ of the heart.
5. The pointed part of the heart, called the ________, lies in the ________ intercostal space. The base of the heart is located just inferior to your ________.
6. The heart is divided into superior chambers called ________ and inferior chambers called ________.
7. Blood from all parts of the body except the lungs flows into a chamber of the heart named the ________. Blood from superior body parts enters the heart via the vein called the ________. Blood from vessels supplying heart tissue returns to the right atrium via the vessel named the ________.

Choose the one best answer to these questions.

___ **8.** The structure that encloses the heart and holds it in place is the ________
A. ectocardium; B. myocardium; C. diaphragm; D. endocardium; E. pericardium.

___ **9.** The right atrium of the heart
(1) contains the pacemaker.
(2) receives blood from the superior and inferior vena cavae.
(3) receives blood directly from the lungs.
(4) empties into the aorta.
A. (1) only; B. (2) only; C. (3) only; D. (4) only; E. (1) and (2).

___ **10.** Stimulation of the vagus (X) nerve
A. increases conductivity of the heart tissue; B. increases heart rate; C. decreases excitability of the heart; D. decreases the inhibitory mechanisms of the heart; E. has no effect on the heart.

___ **11.** Which of these vessels carries oxygenated blood?
A. superior vena cava; B. coronary sinus; C. inferior vena cava; D. pulmonary vein; E. pulmonary artery.

___ **12.** Which vessel is *not* associated with the right side of the heart?
A. superior vena cava; B. pulmonary trunk; C. inferior vena cava; D. aorta; E. coronary sinus.

___ **13.** Which sequences correctly represent the conduction of an impulse through the heart?
A. SA node, AV node, AV bundle, bundle branches;
B. SA node, AV bundle, AV node, bundle branches;
C. AV node, SA node, AV bundle, bundle branches;
D. SA node, bundle branches, AV node, AV bundle.

___ **14.** Heart sounds are produced by the
A. contraction of the myocardium; B. closure of the AV and semilunar valves; C. the flow of blood in the atria; D. the flow of blood in the ventricles; E. expansion and recoil of the aorta.

___ **15.** The groove that separates the atria from the ventricles is the
A. coronary sulcus; B. interatrial septum; C. interventricular sulcus; D. interventricular septum; E. fossa ovalis.

___ **16.** Choose the false statement about heart structure:
A. the heart chamber with the thickest wall is the left ventricle; B. the apex of the heart is more superior in location than the base; C. the heart has four chambers; D. the left ventricle forms the apex and most of the left border of the heart; E. pectinate muscles are found in atrial walls.

___ **17.** Which of the following structures are located in ventricles?
A. trabeculae carneae; B. fossa ovalis; C. ligamentum arteriosum; D. pectinate muscles; E. orifice of coronary sinus.

Arrange the answers in correct sequence.

__ __ __ __ __ **18.** From most superficial to deepest:
A. epicardium (visceral pericardium)
B. myocardium
C. parietal pericardium, fibrous layer
D. pericardial space containing pericardial fluid
E. parietal pericardium, serous layer

__ __ __ __ __ **19.** Route of a red blood cell now in the right atrium:
A. left atrium
B. left ventricle
C. right ventricle
D. pulmonary artery
E. pulmonary vein

The Cardiovascular System: Blood Vessels

14

STUDENT OBJECTIVES

1. Contrast the structure and function of arteries, arterioles, capillaries, venules, and veins.
2. Identify the principal arteries and veins of systemic circulation.
3. Identify the major blood vessels of pulmonary circulation.
4. Trace the route of blood involved in hepatic portal circulation and explain its importance.
5. Contrast fetal and adult circulation.
6. Describe the effects of aging on the cardiovascular system.
7. Describe the development of blood vessels and blood.
8. List the causes and symptoms of hypertension, aneurysms, and deep-venous thrombosis (DVT).
9. Define key medical terms associated with blood vessels.

CHAPTER OUTLINE

- **Arteries**
  - Elastic (Conducting) Arteries
  - Muscular (Distributing) Arteries
  - Anastomoses
- **Arterioles**
- **Capillaries**
- **Venules**
- **Veins**
- **Blood Reservoirs**
- **Circulatory Routes**
  - Systemic Circulation
  - Hepatic Portal Circulation
  - Pulmonary Circulation
  - Fetal Circulation
- **Aging and the Cardiovascular System**
- **Developmental Anatomy of Blood Vessels and Blood**
- **Applications to Health**
- **Key Medical Terms Associated with Blood Vessels**

Blood vessels form a network of tubes that carry blood away from the heart, transport it to the tissues of the body, and then return it to the heart. ***Arteries*** are vessels that carry blood from the heart to the tissues. Large, elastic arteries leave the heart and divide into medium-sized, muscular arteries that branch out into the various regions of the body. Medium-sized arteries then divide into small arteries, which in turn divide into still smaller arteries called ***arterioles.*** As the arterioles enter a tissue, they branch into countless microscopic vessels called ***capillaries.*** Through the walls of capillaries, substances are exchanged between the blood and body tissues. Before leaving the tissue, groups of capillaries reunite to form small veins called ***venules,*** which in turn merge to form progressively larger tubes called veins. ***Veins,*** then, are blood vessels that convey blood from the tissues back to the heart. Since blood vessels require oxygen and nutrients just like other tissues of the body, they also have blood vessels in their own walls called ***vasa vasorum.***

The developmental anatomy of blood vessels and blood will be considered later in the chapter.

ARTERIES

Arteries (*aer* = air; *tereo* = to carry; arteries found empty at death were once thought to contain only air) have walls constructed of three coats or tunics and a hollow center, called a ***lumen,*** through which blood flows (Figure 14-1). The inner coat of an arterial wall, the ***tunica interna (intima),*** is composed of a lining of *endothelium* (simple squamous epithelium) that is in contact with the blood, a *basement membrane,* and a layer of elastic tissue called the *internal elastic lamina.* The middle coat, or ***tunica media,*** is usually the thickest layer. It consists of elastic fibers and smooth muscle. The outer coat, the ***tunica externa (adventitia),*** is composed principally of elastic and collagenous fibers. An *external elastic lamina* may separate the tunica externa from the tunica media.

As a result of the structure of the middle coat especially, arteries have two major properties: elasticity and contractility. When the ventricles of the heart contract and eject blood into the large arteries, they expand to accommodate the extra blood. Then, as the ventricles relax, the elastic recoil of the arteries forces the blood onward. The contractility of an artery comes from its smooth muscle, which is arranged longitudinally and in rings around the lumen somewhat like a doughnut and is innervated by sympathetic branches of the autonomic nervous system. When there is sympathetic stimulation, the smooth muscle contracts, squeezes the wall around the lumen, and narrows the vessel. Such a decrease in the size of the lumen is called ***vasoconstriction.*** Conversely, when sympathetic stimulation decreases, the smooth muscle fibers relax and the size of the arterial lumen increases. This increase is called ***vasodilation*** and is usually due to the inhibition of vasoconstriction.

The contractility of arteries also serves a function in stopping bleeding—called *vascular spasm.* The blood flowing through an artery is under a great deal of pressure. Thus, great quantities of blood can be quickly lost from a broken artery. When an artery is cut, its wall constricts due to contraction of the smooth muscle so that blood does not escape quite so rapidly. However, there is a limit as to how much vasoconstriction can help.

ELASTIC (CONDUCTING) ARTERIES

Large arteries are referred to as ***elastic (conducting) arteries.*** They include the aorta and brachiocephalic, common carotid, subclavian, vertebral, and common iliac arteries. The walls of elastic arteries are relatively thin in proportion to their diameters, and their tunica media contains more elastic fibers and less smooth muscle. As the heart alternately contracts and relaxes, the rate of blood flow tends to be intermittent. When the heart contracts and forces blood into the aorta, the walls of the elastic arteries stretch to accommodate the surge of blood and store the pressure energy. During relaxation of the heart, the walls of the elastic arteries recoil to create pressure, moving the blood forward in a more continuous flow. Elastic arteries are called conducting arteries because they *conduct* blood from the heart to medium-sized muscular arteries.

MUSCULAR (DISTRIBUTING) ARTERIES

Medium-sized arteries are called ***muscular (distributing) arteries.*** They include the axillary, brachial, radial, intercostal, splenic, mesenteric, femoral, popliteal, and tibial arteries. Their tunica media contains more smooth muscle than elastic fibers, and they are capable of greater vasoconstriction and vasodilation to adjust the volume of blood to suit the needs of the structure supplied. The walls of muscular arteries are relatively thick, mainly due to the large amounts of smooth muscle. Muscular arteries are called distributing arteries because they *distribute* blood to various parts of the body.

ANASTOMOSES

Most tissues of the body receive blood from more than one artery. In such areas the distal ends of the vessels unite. The junction of two or more vessels supplying the same body region is called an ***anastomosis*** (a-nas-tō-MŌ-sis). Anastomoses may also occur between the origins of veins and between arterioles and venules. Anastomoses between arteries provide alternate routes by which blood can reach a tissue or organ. Thus, if a vessel is occluded by disease, injury, or surgery, circulation to a part of the body is not necessarily stopped. The alternate route of blood to a body part through an anastomosis is known as ***collateral circulation.*** An alternate blood route may also be from

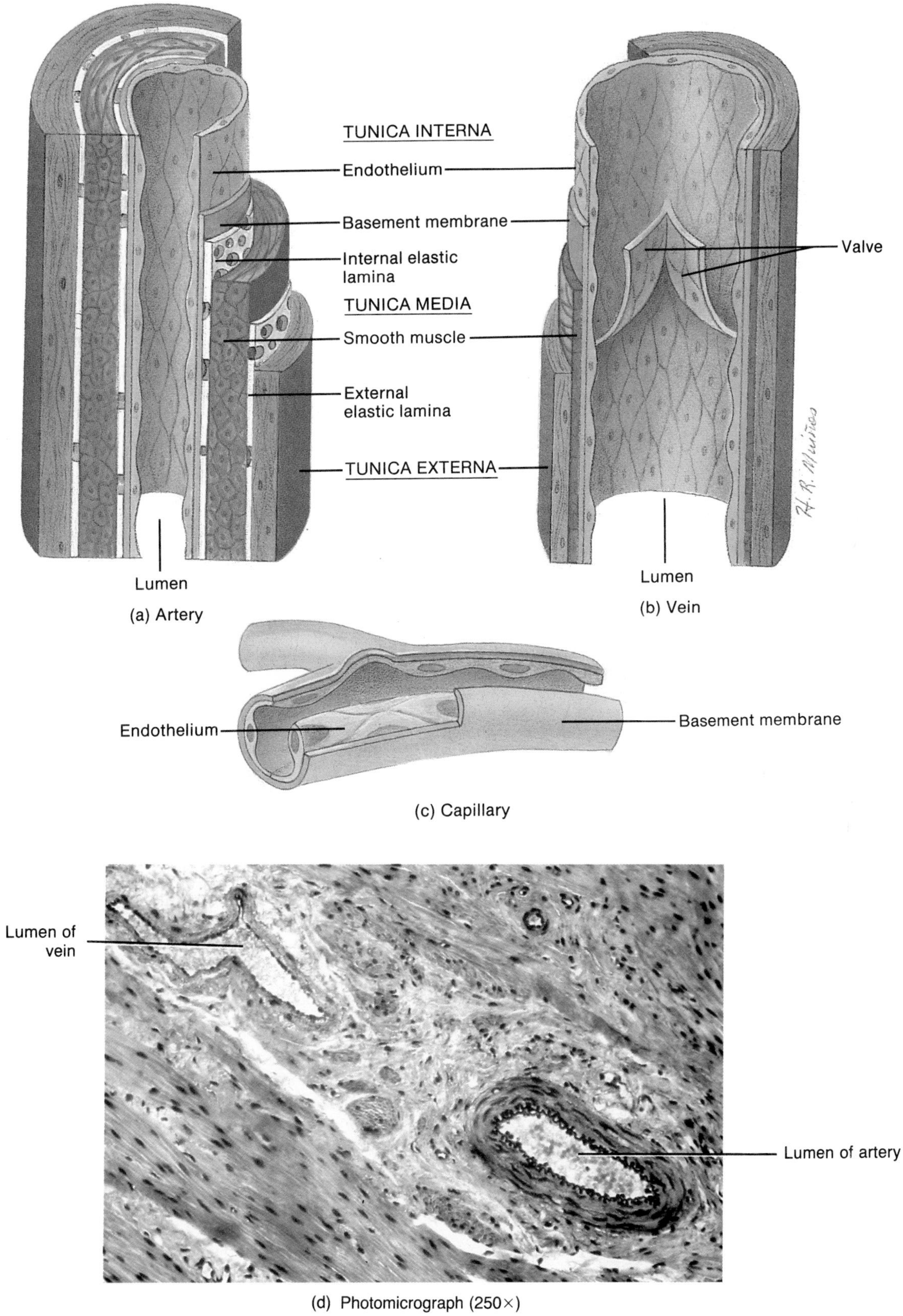

TUNICA INTERNA
Endothelium
Basement membrane
Internal elastic lamina
TUNICA MEDIA
Smooth muscle
External elastic lamina
TUNICA EXTERNA
Valve
Lumen
(a) Artery
Lumen
(b) Vein

Endothelium
Basement membrane
(c) Capillary

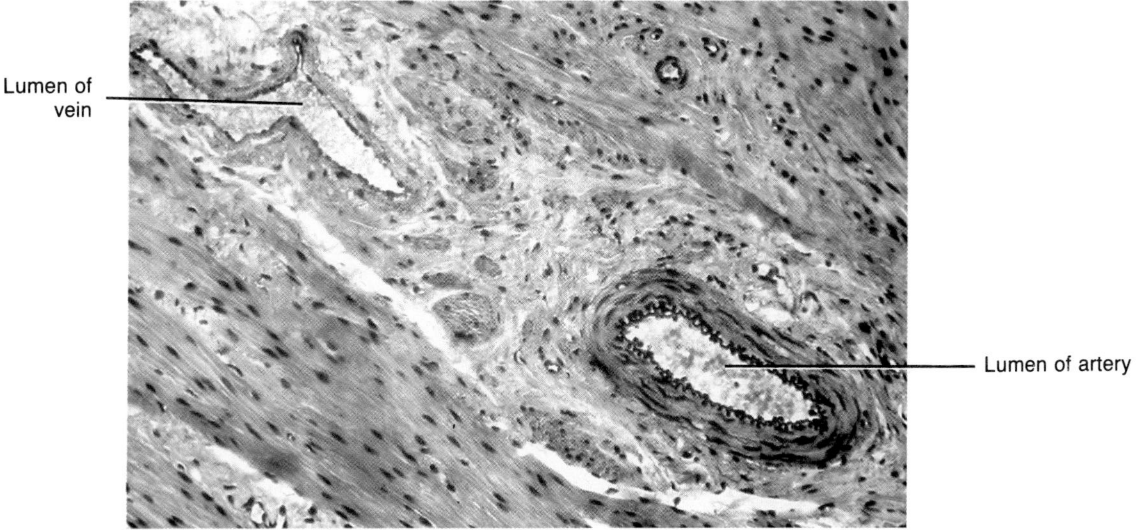

(d) Photomicrograph (250×)

FIGURE 14-1 Comparative structure of blood vessels. The relative size of the capillary in (c) is enlarged. Note in (d) that the lumen of a vein is larger than that of an artery, but the wall of the vein is thinner and the vein frequently appears collapsed (flattened). (Courtesy of Andrew J. Kuntzman.)

nonanastomosing vessels that supply the same region of the body.

Arteries that do not anastomose are known as ***end arteries.*** Occlusion of an end artery interrupts the blood supply to a whole segment of an organ, producing necrosis (death) of that segment.

ARTERIOLES

An ***arteriole*** is a very small, almost microscopic artery that delivers blood to capillaries. Arterioles closer to the arteries from which they branch have a tunica interna like that of arteries, a tunica media composed of smooth muscle and very few elastic fibers, and a tunica externa composed mostly of elastic and collagenous fibers. As arterioles get smaller in size, the tunics change character so that arterioles closest to capillaries consist of little more than a layer of endothelium surrounded by a few scattered smooth muscle fibers (cells) (Figure 14-2).

Arterioles play a key role in regulating blood flow from arteries into capillaries. The smooth muscle of arterioles, like that of arteries, is subject to vasoconstriction and vasodilation. During vasoconstriction, blood flow into capillaries is restricted; during vasodilation, the flow is significantly increased. A change in diameter of arterioles can also significantly affect blood pressure.

CAPILLARIES

Capillaries are microscopic vessels that usually connect arterioles and venules. They are found near almost every cell in the body. The distribution of capillaries in the body varies with the activity of the tissue. For example, in places where activity is higher, such as muscles, the liver, kidneys, lungs, and nervous system, there are rich capillary supplies. In areas where activity is lower, such as tendons and ligaments, the capillary supply is not as extensive. The epidermis, cornea and lens of the eye, and cartilage are devoid of capillaries.

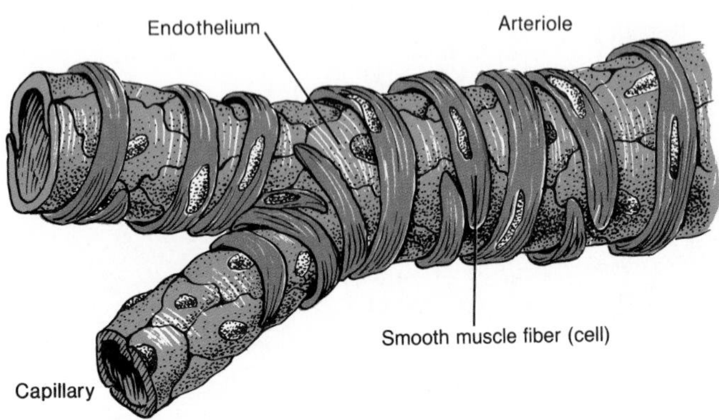

FIGURE 14-2 Structure of an arteriole.

The primary function of capillaries is to permit the exchange of nutrients and wastes between the blood and tissue cells. The structure of the capillaries is admirably suited to this purpose. Capillary walls are composed of only a single layer of cells (endothelium) and a basement membrane (see Figure 14-1c). They have no tunica media or tunica externa. Thus, a substance in the blood must pass through the plasma membrane of just one cell to reach tissue cells. This vital exchange of materials occurs only through capillary walls; the thick walls of arteries and veins present too great a barrier.

Although capillaries pass directly from arterioles to venules in some places in the body, in other places they form extensive branching networks. These networks increase the surface area for diffusion and filtration and thereby allow a rapid exchange of large quantities of materials. In most tissues, blood normally flows through only a small portion of the capillary network when metabolic needs are low. But when a tissue becomes active, the entire capillary network fills with blood.

The flow of blood through capillaries is regulated by vessels with smooth muscle in their walls. A ***metarteriole*** (*met* = beyond) is a vessel that emerges from an arteriole, passes through the capillary network, and empties into a venule (Figure 14-3). The proximal portions of metarterioles are surrounded by scattered smooth muscle fibers whose contraction and relaxation help to regulate the amount and force of blood. The distal portion of a metarteriole has no smooth muscle fibers and is called a ***thoroughfare channel.*** It serves as a low-resistance channel that increases blood flow. ***True capillaries*** emerge from arterioles or metarterioles and are not on the direct flow route from arteriole to venule. At their sites of origin, there is a ring of smooth muscle fibers called a ***precapillary sphincter*** that controls the flow of blood entering a true capillary.

Some capillaries of the body, such as those found in muscle tissue and other locations, are referred to as ***continuous capillaries,*** so named because the cytoplasm of the endothelial cells is continuous when viewed in cross section through a microscope; the cytoplasm appears as an uninterrupted ring, except for the endothelial junction. Other capillaries of the body are referred to as ***fenestrated capillaries.*** They differ from continuous capillaries in that their endothelial cells have numerous fenestrae (pores) where the cytoplasm is very thin or absent. The fenestrae range from 70 to 100 nm in diameter and are closed by a thin diaphragm, except in the capillaries in the kidneys where they are assumed to be open. Fenestrated capillaries are also found in the villi of the small intestine, choroid plexuses of the ventricles in the brain, ciliary processes of the eyes, and endocrine glands.

Microscopic blood vessels in certain parts of the body, such as the liver, are termed ***sinusoids.*** They are wider

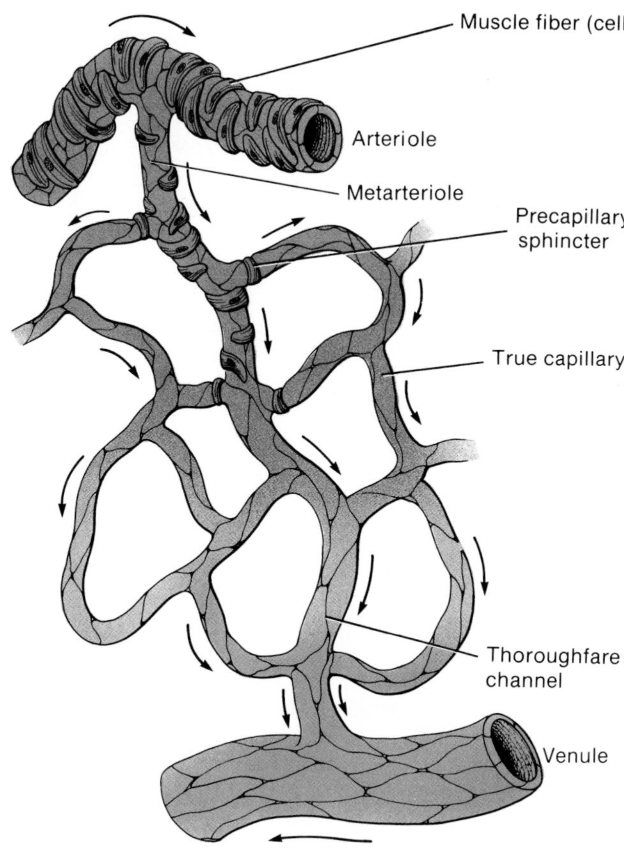

(a) Details of a capillary network

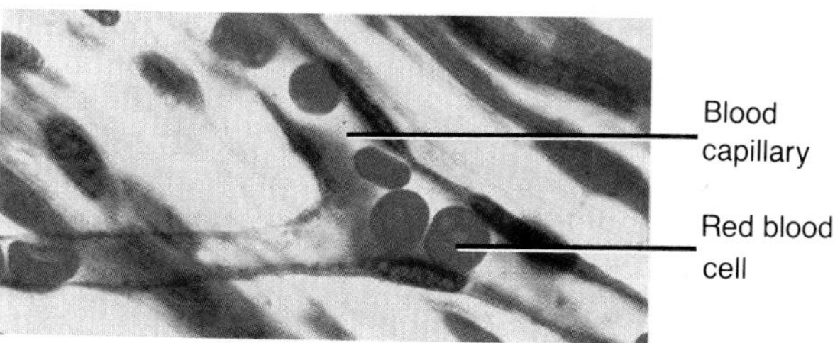

(b) Photomicrograph

FIGURE 14-3 Blood capillaries. Note in (b) that some of the red blood cells pass through sideways. (Courtesy of Lennart Nilsson, *Behold Man,* Delacourt Press, Dell Publishing.)

than capillaries and more tortuous. Also, instead of the usual endothelial lining, sinusoids contain spaces between endothelial cells, and the basal lamina is incomplete or absent. In addition, sinusoids contain specialized lining cells that are adapted to the function of the tissue. For example, in the liver, sinusoids contain phagocytic cells called ***stellate reticuloendothelial (Kupffer's) cells.*** Like capillaries, sinusoids convey blood from arterioles to venules. Other regions containing sinusoids include the spleen, anterior pituitary gland, parathyroid glands, adrenal cortex, and bone marrow.

In performing their function of permitting the exchange of substances between blood and body cells, substances enter and leave capillaries through four basic routes: through junctions that anchor endothelial cells, via pinocytic vesicles, directly across capillary membranes, and through fenestrations.

VENULES

When several capillaries unite, they form small veins called ***venules,*** which collect blood from capillaries and drain it into veins. The venules closest to the capillaries consist of a tunica interna of endothelium and a tunica externa of connective tissue. As the venules approach the veins, they also contain the tunica media characteristic of veins.

VEINS

Veins are composed of essentially the same three coats as arteries, but there are variations in their relative thickness. The tunica interna of veins is extremely thin compared with that of their accompanying arteries. In addition, the tunica media of veins is much thinner than that of accompanying arteries, and the tunica externa is thicker in veins (see Figure 14-1). Despite these differences, veins are still distensible enough to adapt to variations in the volume and pressure of blood passing through them.

By the time the blood leaves the capillaries and moves into the veins, it has lost a great deal of pressure. The difference in pressure can be observed in the blood flow from a cut vessel; blood leaves a cut vein in an even, slow flow rather than in the rapid spurts characteristic of a cut artery. Most of the structural differences between arteries and veins reflect this pressure difference. For example, the walls of veins are not as strong as those of arteries. The low pressure in veins, however, has its disadvantages. When you stand, the pressure pushing blood up the veins in your lower extremities is barely enough to balance the force of gravity pushing it back down. For this reason, many veins, especially those in the limbs, contain valves that prevent backflow (Figure 14-4). Normal valves ensure the flow of blood toward the heart.

CLINICAL APPLICATION

Varicose Veins

In people with weak venous valves, gravity forces large quantities of blood back down into distal parts of the vein. This pressure overloads the vein and pushes its wall outward. After repeated overloading, the walls lose their elasticity and become stretched and flabby. Such dilated and tortuous veins caused by incompetent valves are called ***varicose veins (VVs).*** They may be due to heredity, mechanical factors (prolonged standing and pregnancy), or aging. Because a varicosed wall is not able to exert a

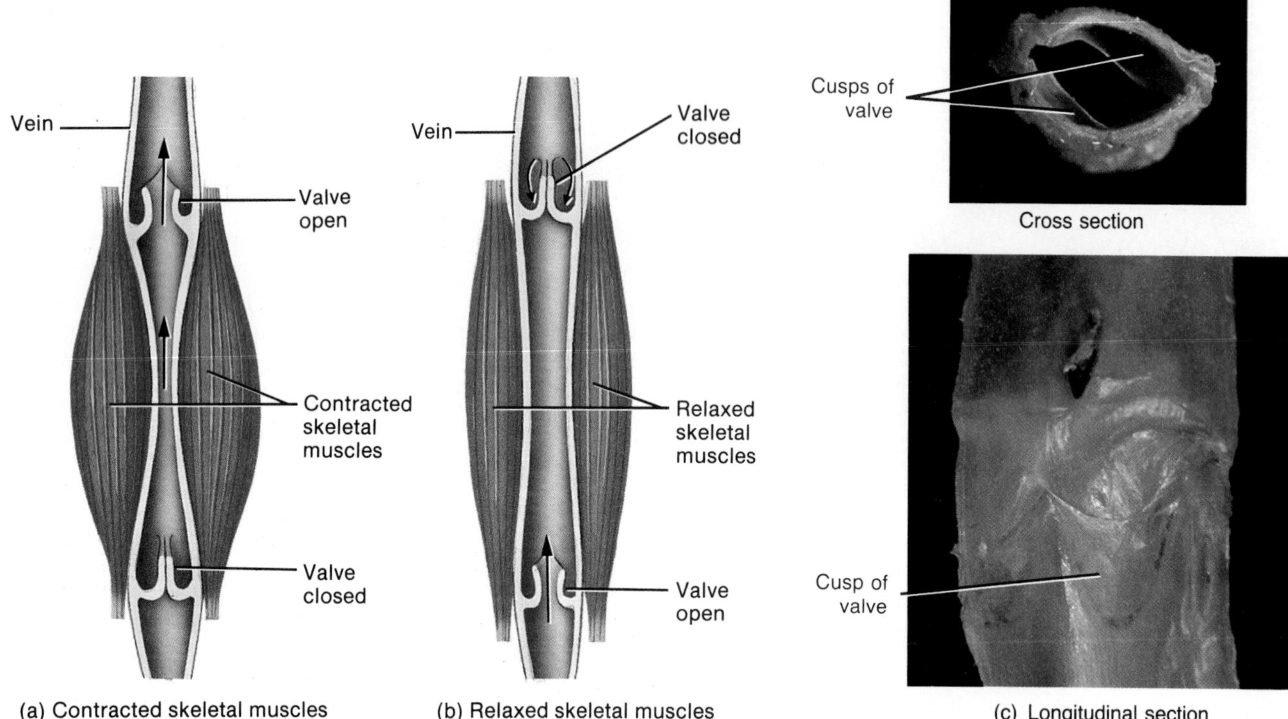

FIGURE 14-4 Role of skeletal muscle contractions and venous valves in returning blood to the heart. (a) When skeletal muscles contract, the valves open, and blood is forced toward the heart. (b) When skeletal muscles relax, the valves close to prevent the backflowing of blood from the heart. (c) Photographs of a one-way valve in a vein. (Courtesy of J. A. Gosling, P. F. Harris, et al., *Atlas of Human Anatomy,* Gower Medical Publishing Ltd., 2nd ed., 1991.)

firm resistance against the blood, blood tends to accumulate in the pouched-out area of the vein, causing it to swell and forcing fluid into the surrounding tissue. Veins close to the surface of the legs are highly susceptible to varicosities. Veins that lie deeper are not as vulnerable because surrounding skeletal muscles prevent their walls from overstretching.

Varicose veins may be treated by several methods, depending on the severity of the condition. These include: (1) frequent periods of rest with elevation of the lower extremities; (2) external pressure with elastic stockings or bandages; (3) a procedure known as sclerotherapy, an intravenous injection of sclerosing chemicals that collapses the veins and prevents blood from flowing into them, thus eliminating the purple-blue discoloration; and (4) surgery, in which the saphenous veins and their incompetent valves are ligated and removed ("stripping").

A mild form of varicose veins in the legs is referred to as ***spider burst veins,*** a condition that is most commonly treated by sclerotherapy.

A ***vascular (venous) sinus*** is a vein with a thin endothelial wall that has no smooth muscle to alter its diameter. Surrounding dense connective tissue replaces the tunica media and tunica externa to provide support. Intracranial vascular sinuses, which are supported by the dura mater, return cerebrospinal fluid and deoxygenated blood from the brain to the heart. Another example of a vascular sinus is the coronary sinus of the heart.

BLOOD RESERVOIRS

The volume of blood in various parts of the cardiovascular system varies considerably. Veins, venules, and venous sinuses contain about 59 percent of the blood in the system; arteries, about 13 percent; pulmonary vessels, about 12 percent; the heart, about 9 percent; and arterioles and capillaries, about 7 percent. Since systemic veins contain so much of the blood, they are referred to as ***blood reservoirs.*** They serve as storage depots for blood, which can be moved quickly to other parts of the body if the need arises. When there is increased muscular activity, the vasomotor center sends increasing sympathetic impulses to veins that serve as blood reservoirs. The result is vasoconstriction, which permits the distribution of blood from venous reservoirs to skeletal muscles, where it is needed most. A similar mechanism operates in cases of hemorrhage, when blood volume and pressure decrease. Vasoconstriction of veins in venous reservoirs helps to compensate for the blood loss. Among the principal blood reservoirs are the veins of the abdominal organs (especially the liver and spleen) and the veins of the skin.

CIRCULATORY ROUTES

Arteries, arterioles, capillaries, venules, and veins are organized into definite routes that circulate blood throughout the body. We can now look at the basic routes the blood takes as it is transported through its vessels.

Figure 14-5 shows a number of basic ***circulatory routes*** through which the blood travels. The two basic postnatal (after birth) routes are systemic and pulmonary. ***Systemic circulation*** includes all the blood vessels that carry oxygenated blood that leaves the left ventricle through the aorta and reaches all systemic capillaries. It also includes the blood vessels that carry deoxygenated blood that returns to the right atrium after traveling to all the organs including the nutrient arteries to the lungs. Three of the many subdivisions of the systemic circulation are ***coronary*** (***cardiac***)

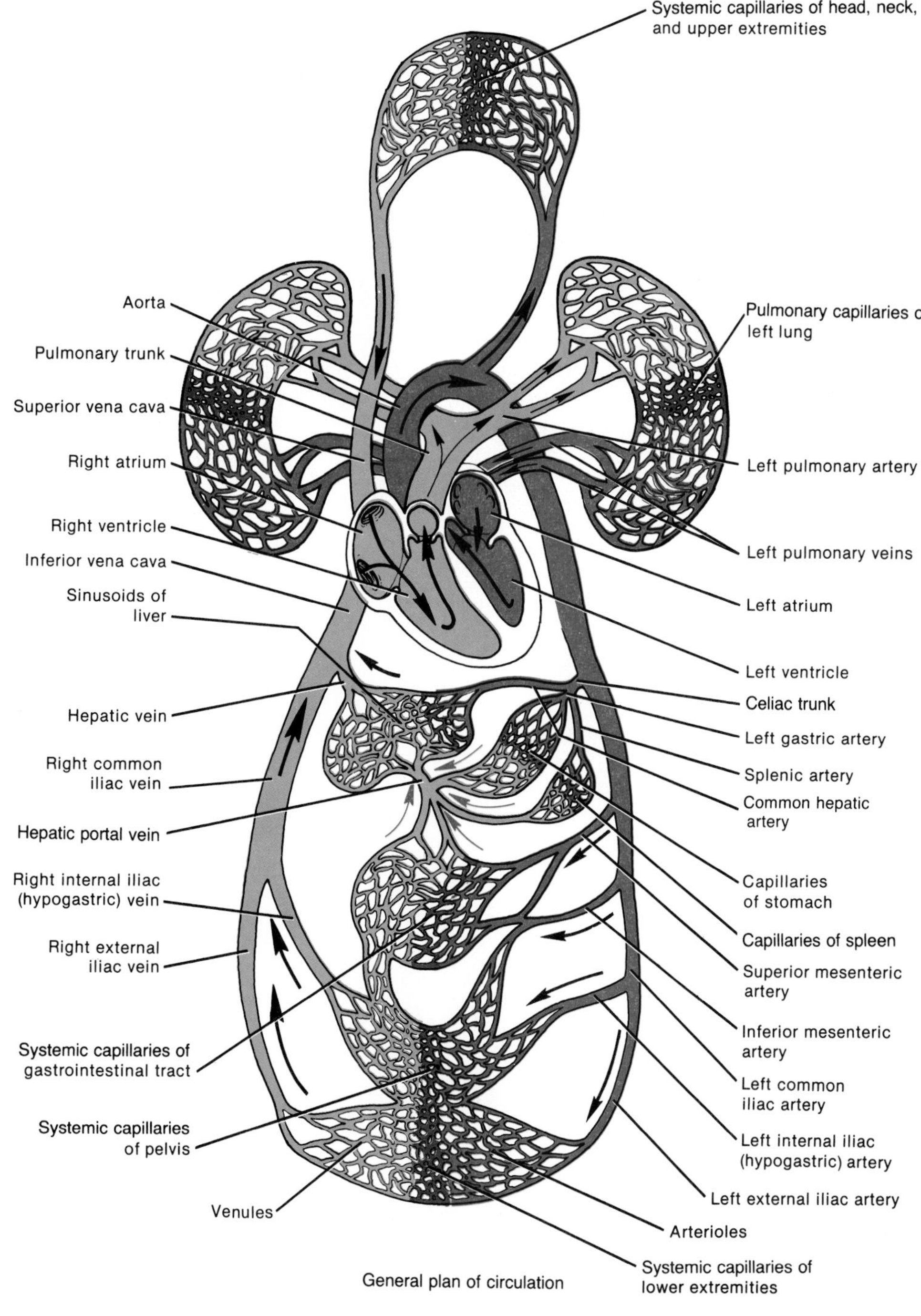

FIGURE 14-5 Circulatory routes. The overall pattern of systemic circulation is indicated by heavy black arrows; pulmonary circulation, by thin black arrows in the pulmonary blood vessels. Hepatic portal circulation, a subdivision of systemic circulation, is indicated by thin red arrows. Refer to Figure 13-6 for the details of coronary (cardiac) circulation and to Figure 14-18 for the details of fetal circulation.

circulation (see Figure 13-6), which supplies the myocardium of the heart; ***cerebral circulation,*** which supplies the brain; and the ***hepatic portal circulation,*** which runs from the gastrointestinal tract to the liver (see Figure 14-16). Blood leaving the aorta and traveling through the systemic arteries is a bright red color. As it moves through capillaries, it loses some of its oxygen and takes on carbon dioxide, so that blood in systemic veins is a dark red color.

When blood returns to the heart from the systemic route, it is eventually pumped out of the right ventricle through the ***pulmonary circulation*** to the lungs (see Figure 14-17). In pulmonary capillaries of the air sacs of the lungs, it loses some of its carbon dioxide and takes on oxygen. It is now bright red again. It returns to the left atrium of the heart and reenters the systemic circulation as it is pumped out by the left ventricle.

Another major route—***fetal circulation***—exists only in the fetus and contains special structures that allow the developing fetus to exchange materials with its mother (see Figure 14-18).

SYSTEMIC CIRCULATION

The flow of blood from the left ventricle to all parts of the body (except the air sacs of the lungs) and back to the right atrium is called the ***systemic circulation.*** The purpose of systemic circulation is to carry oxygen and nutrients to body tissues and to remove carbon dioxide and other wastes and heat from the tissues. All systemic arteries branch from the ***aorta,*** which arises from the left ventricle of the heart.

As the aorta emerges from the left ventricle, it passes upward and posterior to the pulmonary trunk. At this point, it is called the ***ascending aorta.*** The ascending aorta gives off two coronary branches to the heart muscle. Then it turns to the left, forming the ***arch of the aorta,*** before descending to the level of the fourth thoracic vertebra as the ***descending aorta.*** The descending aorta lies close to the vertebral bodies, passes through the diaphragm, and divides at the level of the fourth lumbar vertebra into two ***common iliac arteries,*** which carry blood to the lower extremities. The section of the descending aorta between the arch of the aorta and the diaphragm is referred to as the ***thoracic aorta.*** The section between the diaphragm and the common iliac arteries is termed the ***abdominal aorta.*** Each section of the aorta gives off arteries that continue to branch into distributing arteries leading to organs and finally into the arterioles and capillaries that service the systemic tissues (except the air sacs of lungs).

Blood is returned to the heart through the systemic veins. All the veins of the systemic circulation flow into either the ***superior vena cava, inferior vena cava,*** or the ***coronary sinus.*** They in turn empty into the right atrium. The principal arteries and veins of systemic circulation are described and illustrated in Exhibits 14-1 through 14-12 and Figures 14-6 through 14-15.

EXHIBIT 14-1

Aorta and Its Branches (Figure 14-6)

Overview: The ***aorta*** (ā-OR-ta) is the largest artery of the body, averaging about 3 cm (1 in.) in diameter. It begins at the left ventricle and contains a valve at its origin, called the aortic semilunar valve (see Figure 13-3d), which prevents backflow of blood into the left ventricle during its diastole (relaxation). As indicated previously, the principal divisions of the aorta are the ascending aorta, arch of the aorta, thoracic aorta, and abdominal aorta.

DIVISION OF AORTA	ARTERIAL BRANCH	REGION SUPPLIED
Ascending Aorta	Right and left coronary	Heart.
Arch of Aorta	Brachiocephalic trunk → Right common carotid	Right side of head and neck.
	Brachiocephalic trunk → Right subclavian	Right upper extremity.
	Left common carotid	Left side of head and neck.
	Left subclavian	Left upper extremity.
Thoracic Aorta	Intercostals	Intercostal and chest muscles and pleurae.
	Superior phrenics	Posterior and superior surfaces of diaphragm.
	Bronchials	Bronchi of lungs.
	Esophageals	Esophagus.
Abdominal Aorta	Inferior phrenics	Inferior surface of diaphragm.
	Celiac → Common hepatic	Liver.
	Celiac → Left gastric	Stomach and esophagus.
	Celiac → Splenic	Spleen, pancreas, and stomach.
	Superior mesenteric	Small intestine, cecum, ascending and transverse colons, and pancreas.
	Suprarenals	Adrenal (suprarenal) glands.
	Renals	Kidneys.
	Gonadals → Testiculars	Testes.
	or	
	Gonadals → Ovarians	Ovaries.
	Inferior mesenteric	Transverse, descending, and sigmoid colons and rectum.
	Common iliacs → External iliacs	Lower extremities.
	Common iliacs → Internal iliacs (hypogastrics)	Uterus, prostate gland, muscles of buttocks, and urinary bladder.

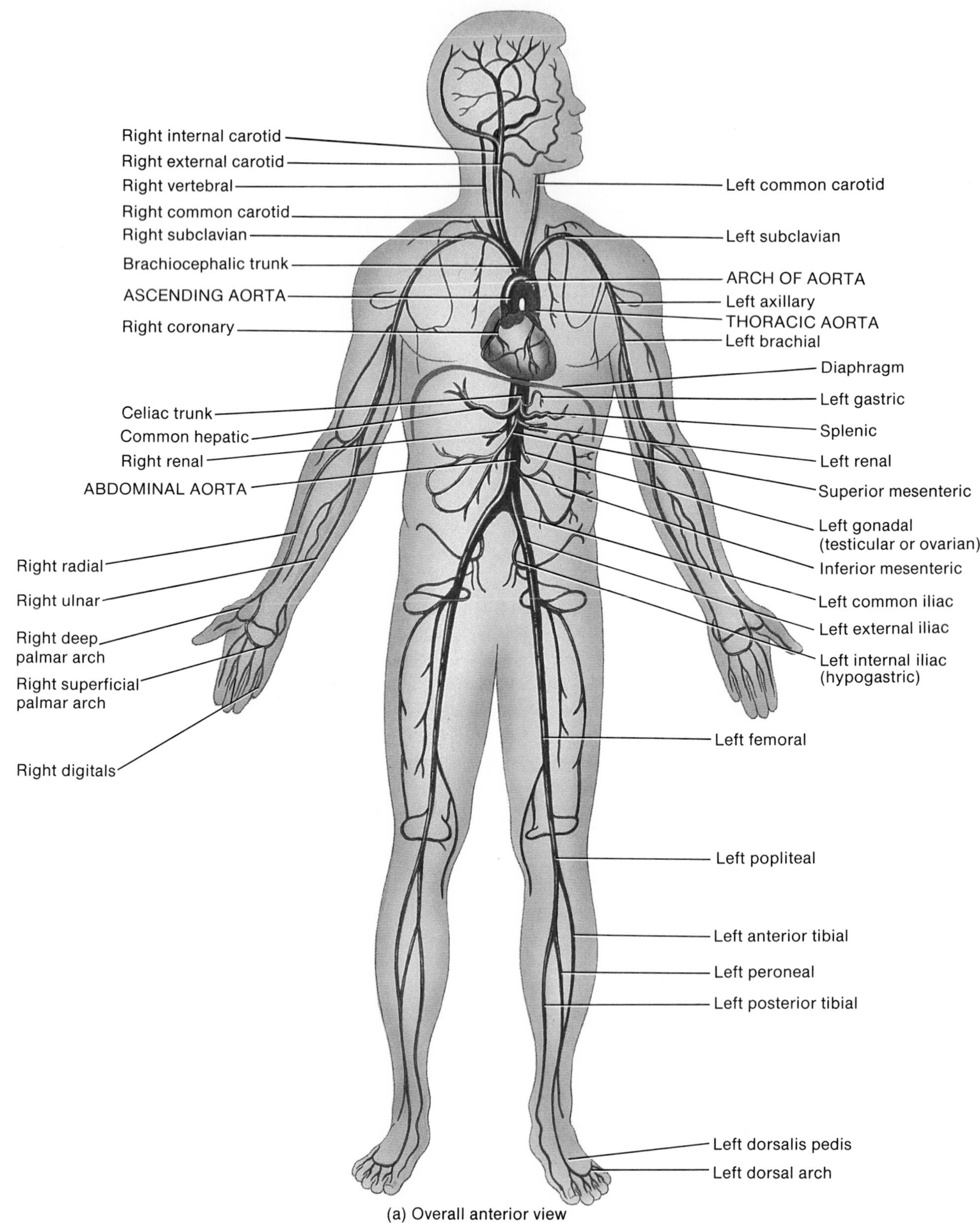

FIGURE 14-6 Aorta and its principal branches.

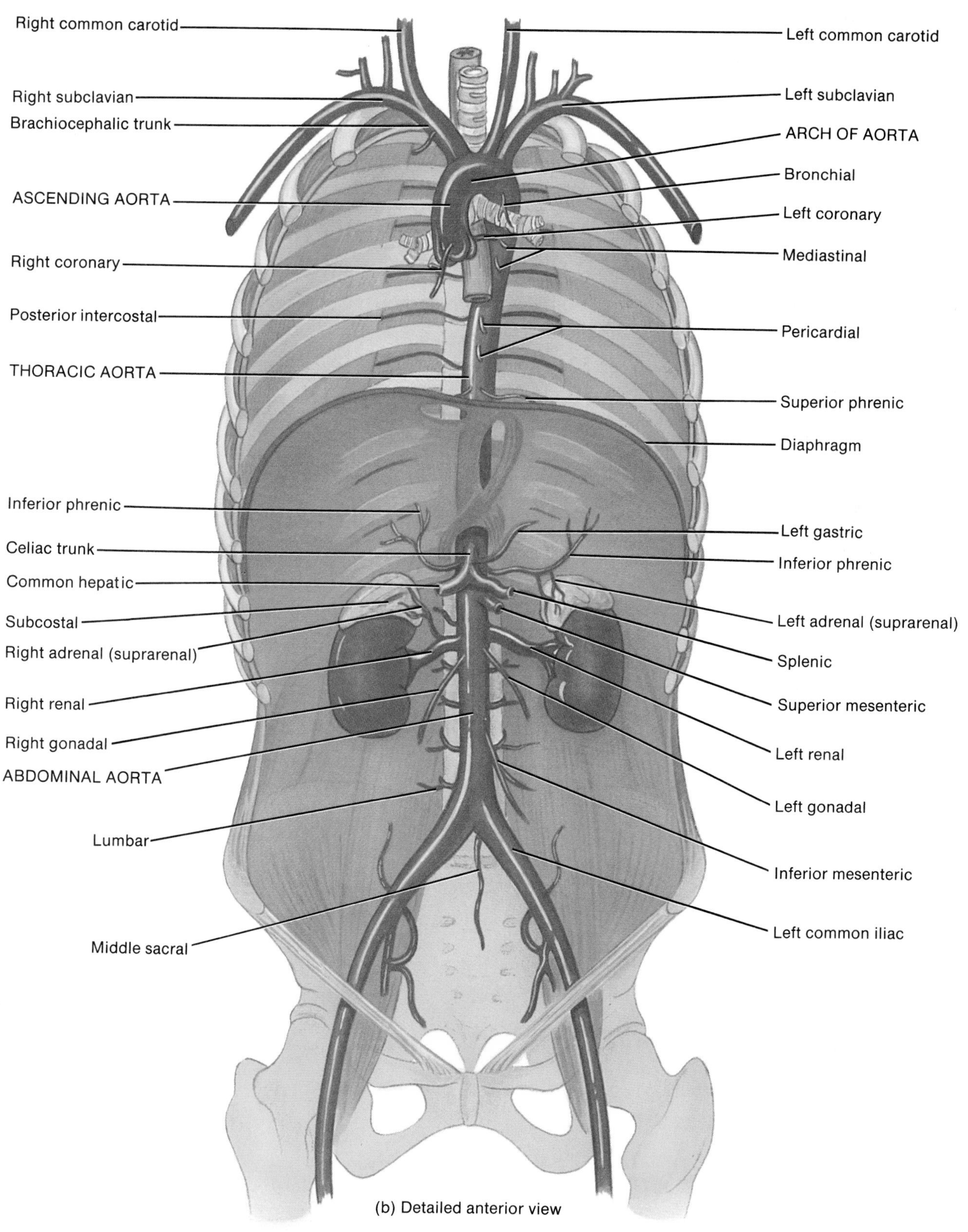

(b) Detailed anterior view

EXHIBIT 14-2

Ascending Aorta (Figure 14-7)

Overview: The ***ascending aorta*** is the first division of the aorta, about 5 cm (2 in.) in length. It is directed upward, forward, and to the right and ends at the level of the sternal angle where it becomes the arch of the aorta. The beginning of the ascending aorta is covered by the pulmonary trunk and right auricle; the right pulmonary artery is behind it. At its origin the ascending aorta contains three dilations, called aortic sinuses. Two of these, the right and left sinuses, give rise to the right and left coronary arteries, respectively.

BRANCH	DESCRIPTION AND REGION SUPPLIED
Coronary Arteries	Right and left branches arise from ascending aorta just superior to aortic semilunar valve. They form crown around heart, giving off branches to atrial and ventricular myocardium.

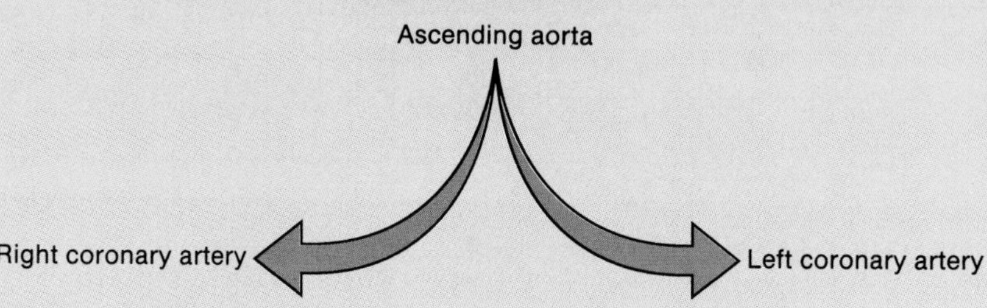

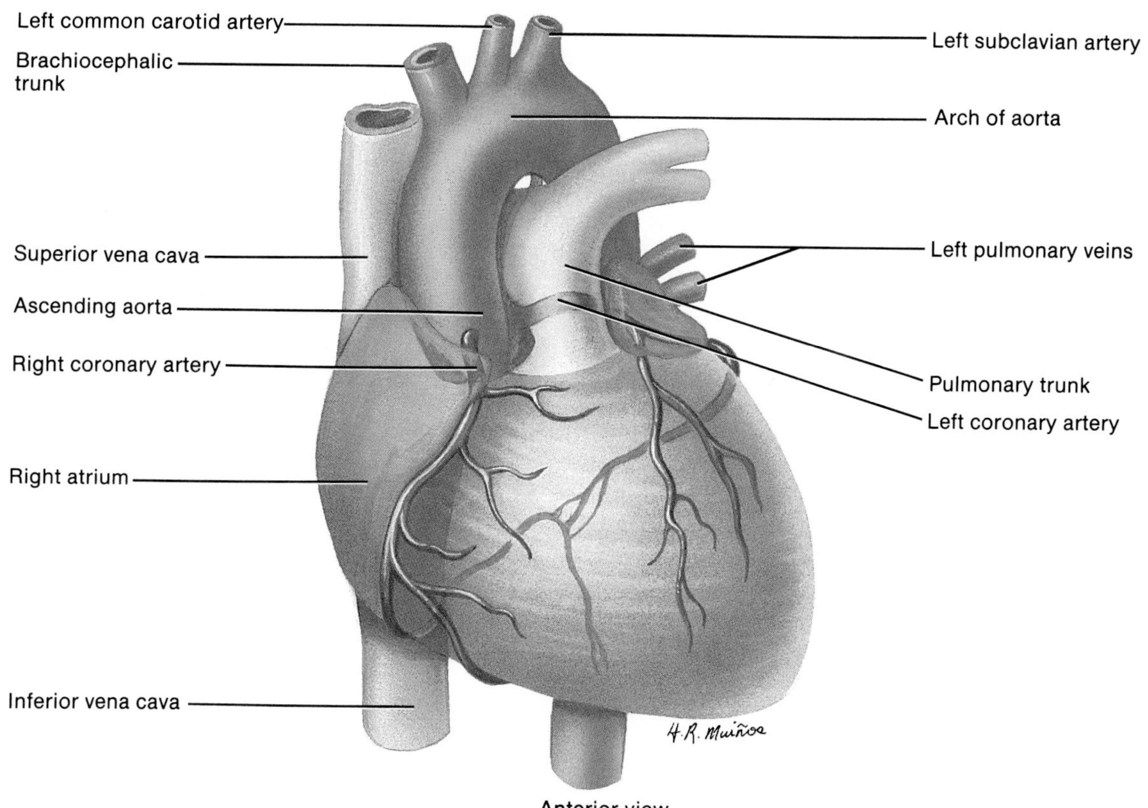

FIGURE 14-7 Ascending aorta and its branches.

EXHIBIT 14-3

Arch of Aorta (Figure 14-8)

Overview: The ***arch of the aorta*** is about 4.5 cm (2 in.) in length and is the continuation of the ascending aorta that emerges from the pericardium behind the sternum at the level of the sternal angle. Initially, the arch is directed upward, backward and to the left and then downward on the left side of the body of the fourth thoracic vertebra. Actually, the arch is directed not only from right to left, but from anterior to posterior as well. The arch of the aorta terminates at the level of the disc between the fourth and fifth thoracic vertebrae where it becomes the thoracic aorta. The thymus gland lies in front of the arch of the aorta, whereas the trachea lies behind it.

Three major arteries branch from the arch of the aorta. In order of their origination, they are the brachiocephalic trunk, left common carotid artery, and left subclavian artery.

BRANCH	DESCRIPTION AND REGION SUPPLIED
Brachiocephalic	***Brachiocephalic trunk*** is the first and largest branch off arch of aorta. It divides to form a right subclavian artery and right common carotid artery. ***Right subclavian artery*** extends from brachiocephalic trunk to first rib and then passes into armpit (axilla) and supplies arm, forearm, and hand. Continuation of right subclavian into axilla is called ***axillary artery.***[a] From here, it continues into arm as ***brachial artery.*** At bend of elbow, brachial artery divides into medial ***ulnar*** and lateral ***radial arteries.*** These vessels pass down to palm, one on each side of forearm. In palm, branches of two arteries anastomose to form two palmar arches—***superficial palmar arch*** and ***deep palmar arch.*** From these arches arise ***digital arteries,*** which supply fingers and thumb. Before passing into axilla, right subclavian gives off major branch to brain called ***vertebral artery.*** Right vertebral artery passes through foramina of transverse processes of cervical vertebrae and enters skull through foramen magnum to reach undersurface of brain. Here it unites with left vertebral artery to form ***basilar artery.*** ***Right common carotid artery*** passes upward in neck. At upper level of larynx, it divides into ***right external*** and ***right internal carotid arteries.*** External carotid supplies right side of thyroid gland, tongue, throat, face, ear, scalp, and dura mater. Internal carotid supplies brain, right eye, and right sides of forehead and nose. Inside the cranium, anastomoses of left and right internal carotids along with basilar artery form a somewhat hexagonal arrangement of blood vessels at base of brain near sella turcica called ***cerebral arterial circle (circle of Willis).*** From this circle arise arteries supplying most of the brain. Essentially, cerebral arterial circle is formed by union of ***anterior cerebral arteries*** (branches of internal carotids) and ***posterior cerebral arteries*** (branches of basilar artery). Posterior cerebral arteries are connected with internal carotids by ***posterior communicating arteries.*** Anterior cerebral arteries are connected by ***anterior communicating arteries.*** The ***internal carotid arteries*** are also considered part of cerebral arterial circle. The function of the cerebral arterial circle is to equalize blood pressure to brain and provide alternate routes for blood to brain, should arteries become damaged.
Left Common Carotid	***Left common carotid*** is second branch off arch of aorta (see Figure 14-7). Corresponding to right common carotid, it divides into basically same branches with same names, except that arteries are now labeled "left" instead of "right."
Left Subclavian	***Left subclavian artery*** is third branch off arch of aorta (see Figure 14-7). It distributes blood to left vertebral artery and vessels of left upper extremity. Arteries branching from left subclavian are named like those of right subclavian.

[a] The right subclavian artery is a good example of the practice of giving the same vessel different names as it passes through different regions.

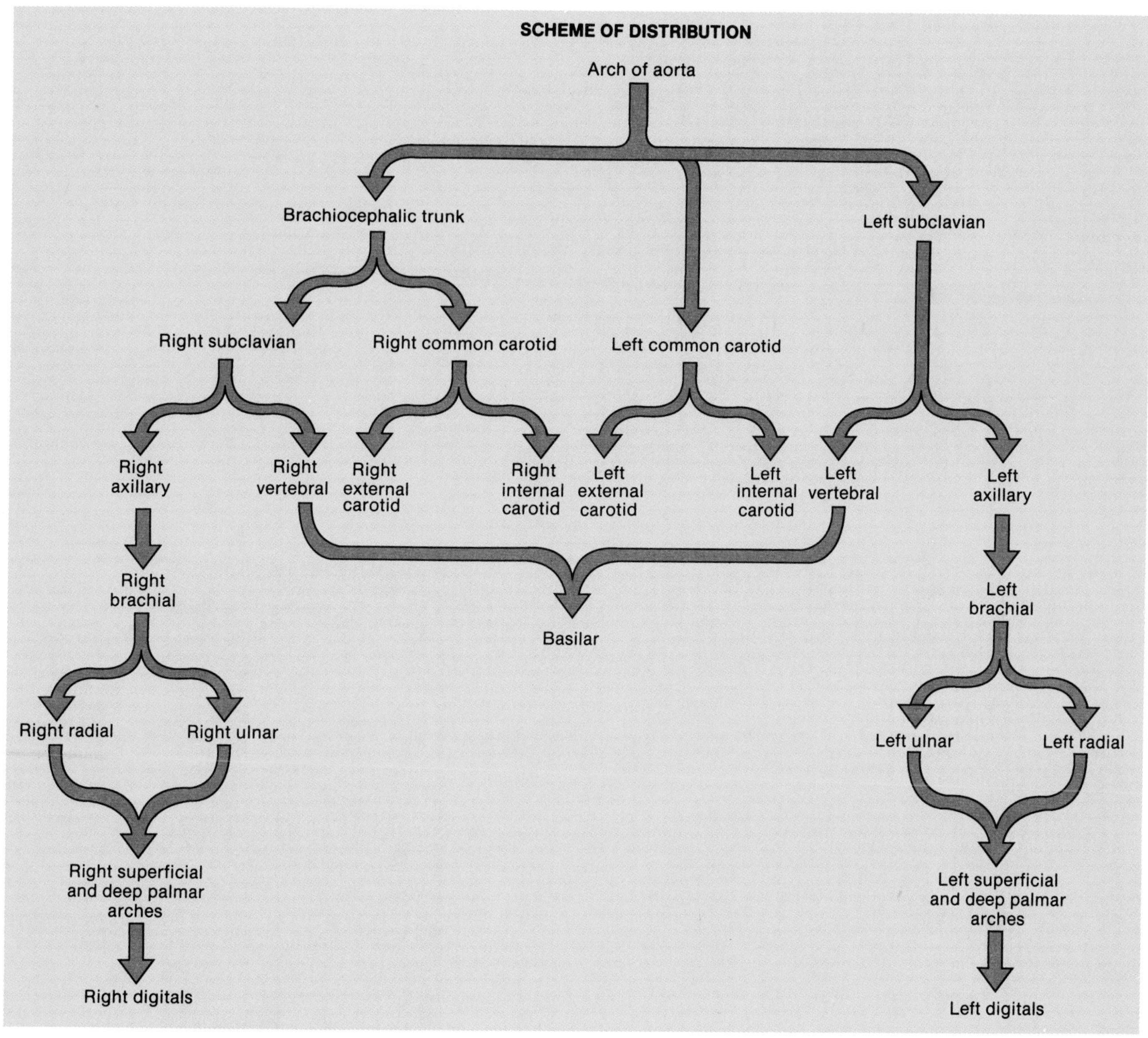
SCHEME OF DISTRIBUTION
Arch of aorta
Brachiocephalic trunk
Left subclavian
Right subclavian
Right common carotid
Left common carotid
Right axillary
Right vertebral
Right external carotid
Right internal carotid
Left external carotid
Left internal carotid
Left vertebral
Left axillary
Right brachial
Basilar
Left brachial
Right radial
Right ulnar
Left ulnar
Left radial
Right superficial and deep palmar arches
Left superficial and deep palmar arches
Right digitals
Left digitals

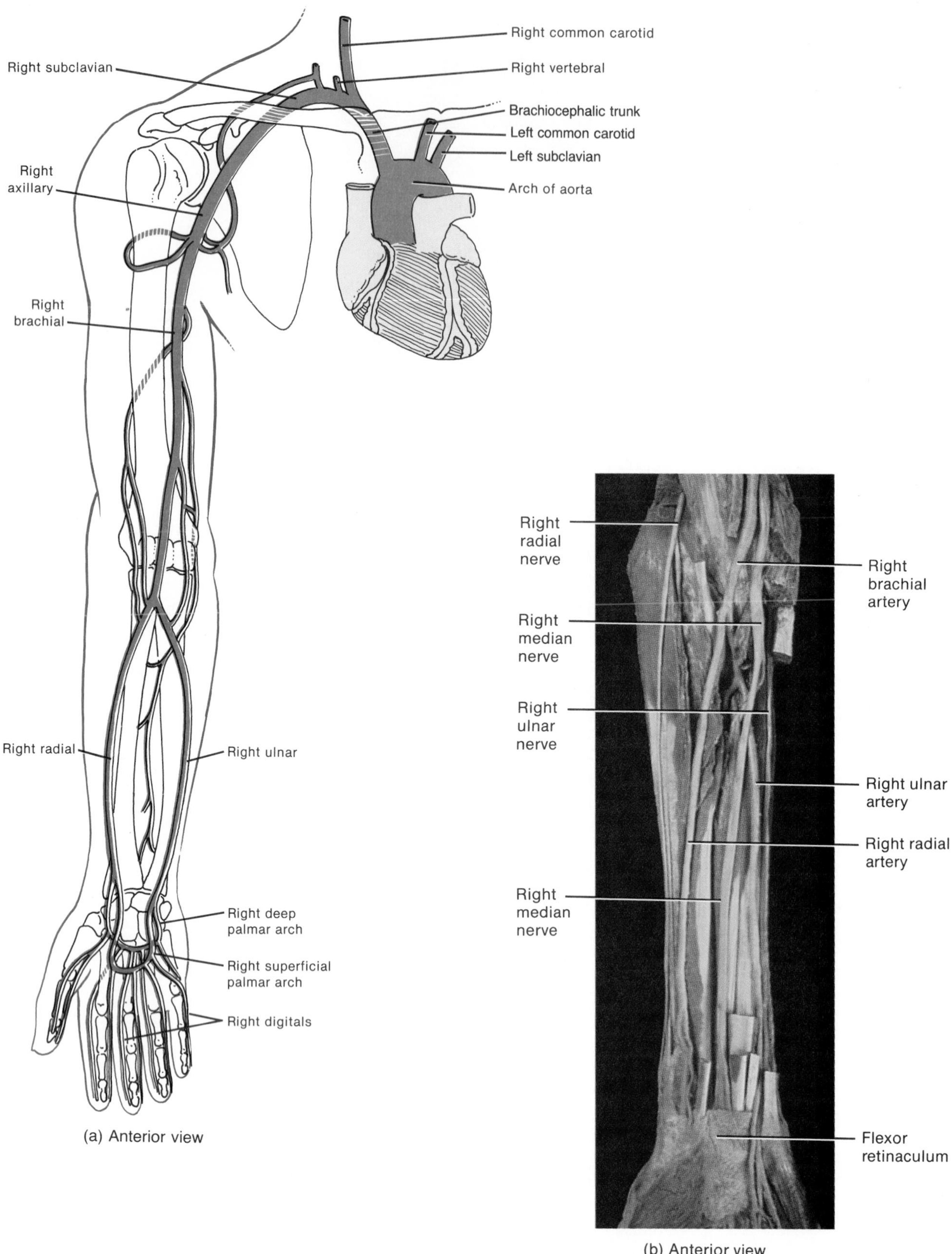

FIGURE 14-8 Arch of the aorta and its branches. (b) and (c) Courtesy of J. A. Gosling, P. F. Harris, et al., *Atlas of Human Anatomy,* Gower Medical Publishing Ltd., 2nd ed., 1991. In (e), note the arteries that comprise the cerebral arterial circle (circle of Willis).

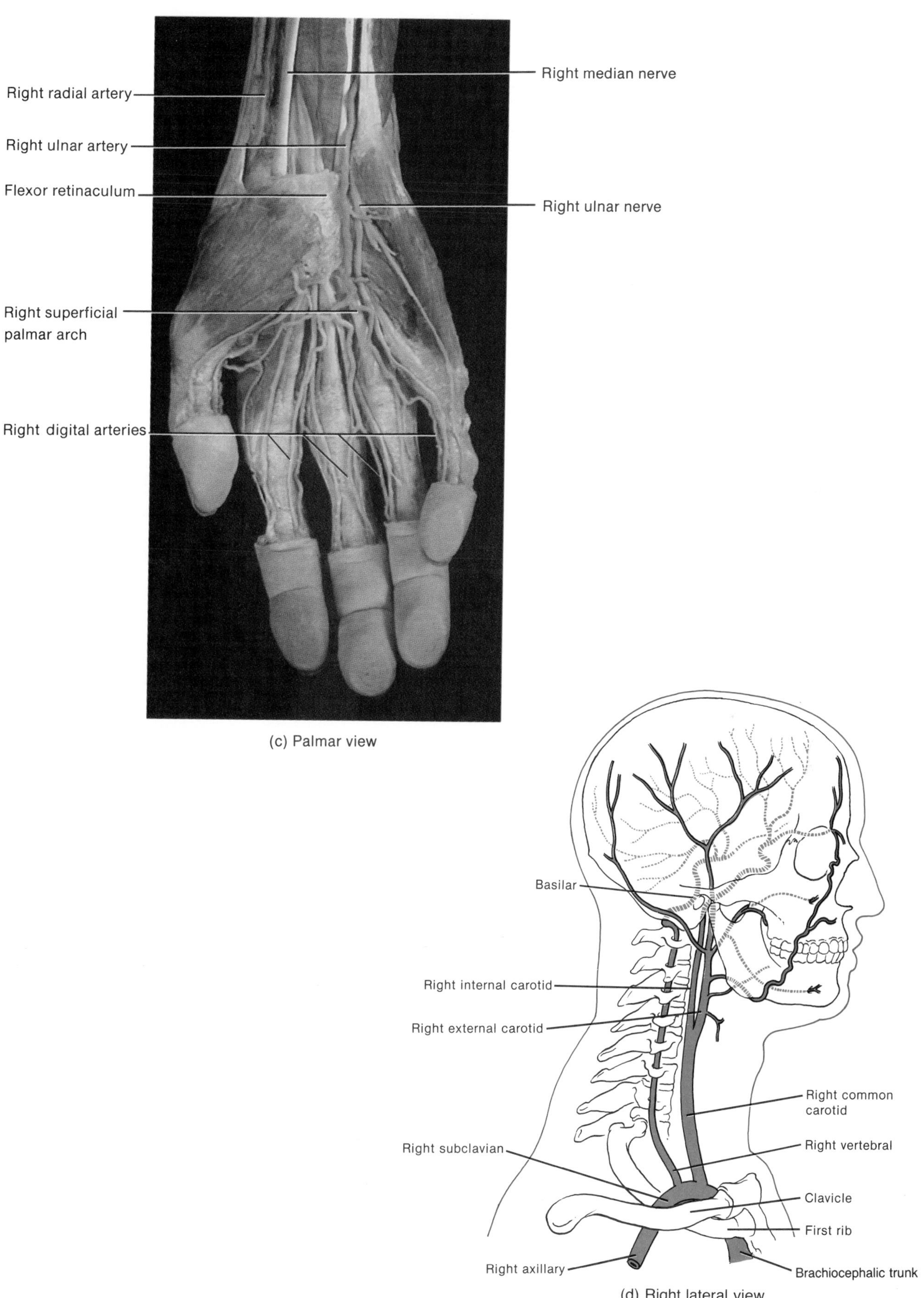

(c) Palmar view

(d) Right lateral view

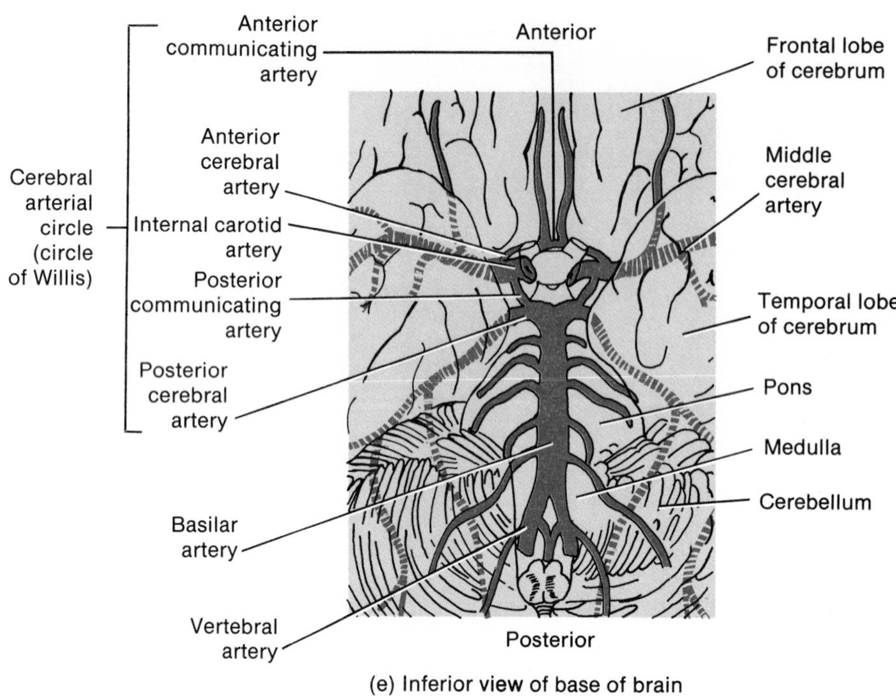

(e) Inferior view of base of brain

EXHIBIT 14-4

Thoracic Aorta (Figure 14-9)

Overview: The ***thoracic aorta*** is about 20 cm (8 in.) long and is a continuation of the arch of the aorta. It begins at the level of the disc between the fourth and fifth thoracic vertebrae where it lies to the left of the vertebral column. As it descends, it moves closer to the midline and terminates at an opening of the diaphragm (aortic hiatus) in front of the vertebral column at the level of the intervertebral disc between the twelfth thoracic and first lumbar vertebrae.

Along its course the thoracic aorta sends off numerous small arteries to viscera (*visceral branches*) and body wall structures (*parietal branches*).

BRANCH	DESCRIPTION AND REGION SUPPLIED
VISCERAL	
Pericardial	Several minute ***pericardial arteries*** supply blood to posterior aspect of pericardium.
Bronchial	One right and two left ***bronchial arteries*** supply the bronchial tubes, visceral pleurae, bronchial lymph nodes, and esophagus. (Whereas the right bronchial artery arises from the third posterior intercostal artery or upper left bronchial artery, the two left bronchial arteries arise from the thoracic aorta.)
Esophageal	Four or five ***esophageal arteries*** supply the esophagus.
Mediastinal	Numerous small ***mediastinal arteries*** supply blood to structures in the posterior mediastinum.
PARIETAL	
Posterior Intercostal	Nine pairs of ***posterior intercostal arteries*** supply the intercostal, pectoral, and abdominal muscles; overlying subcutaneous tissue and skin; mammary glands; and vertebral canal and its contents.
Subcostal	The left and right ***subcostal arteries*** have a distribution similar to that of the posterior intercostals.
Superior Phrenic	Small ***superior phrenic arteries*** supply the posterior and superior surfaces of the diaphragm.

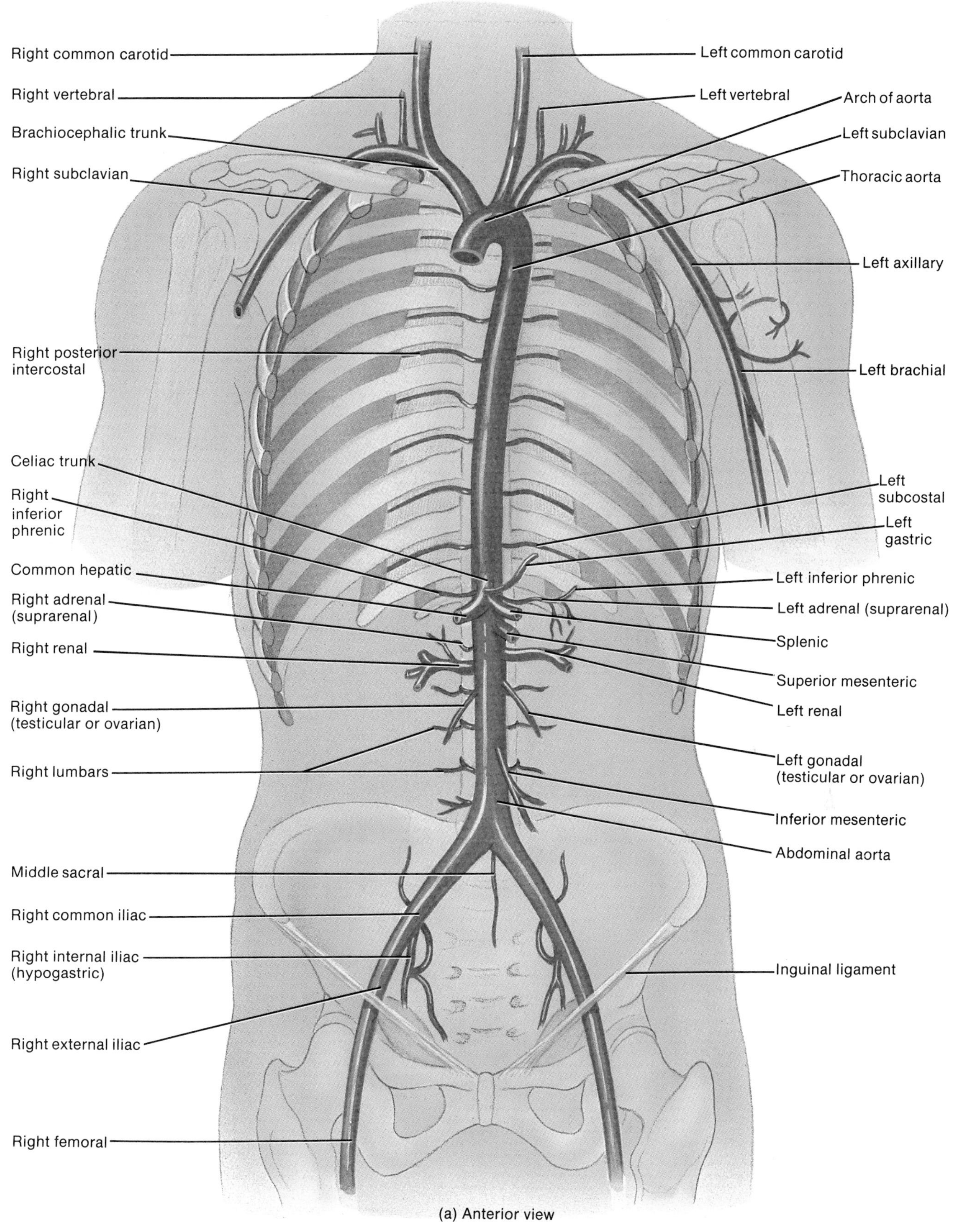

(a) Anterior view

FIGURE 14-9 Thoracic and abdominal aorta and their principal branches. (b) Courtesy of J. A. Gosling, P. F. Harris, et al., *Atlas of Human Anatomy,* Gower Medical Publishing Ltd., 2nd ed., 1991.

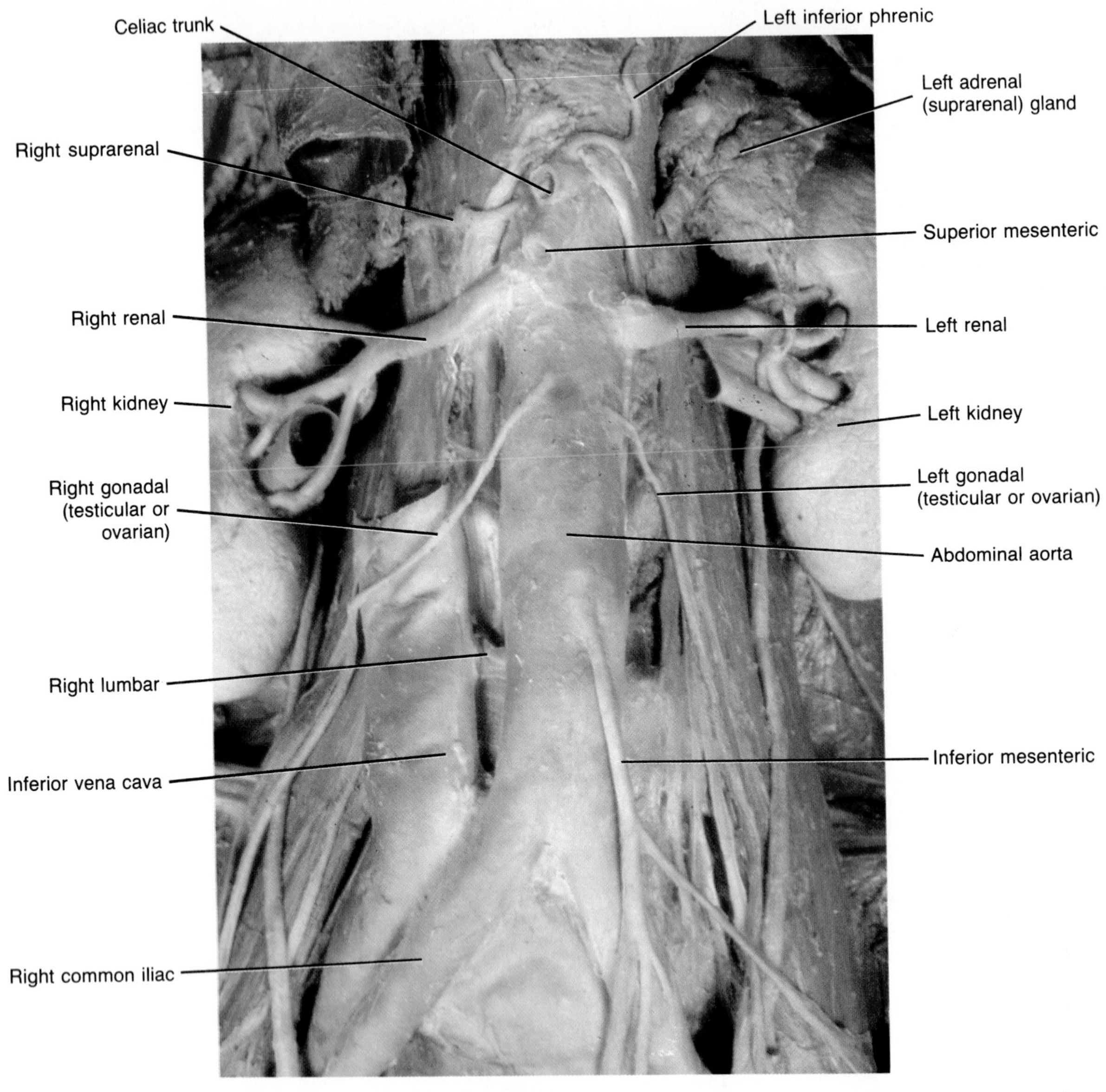

(b) Anterior view

Stomach (turned upward)
Esophageal branch of left gastric
Left gastroepiploic
Right gastroepiploic
Short gastrics
Left gastric
Right gastric
Hepatic artery proper
Common hepatic
Splenic
Spleen
Gastroduodenal
Pancreatic
Superior pancreaticoduodenal
Jejunum of small intestine
Pancreas
Superior mesenteric
Abdominal aorta
Duodenum of small intestine
DANK

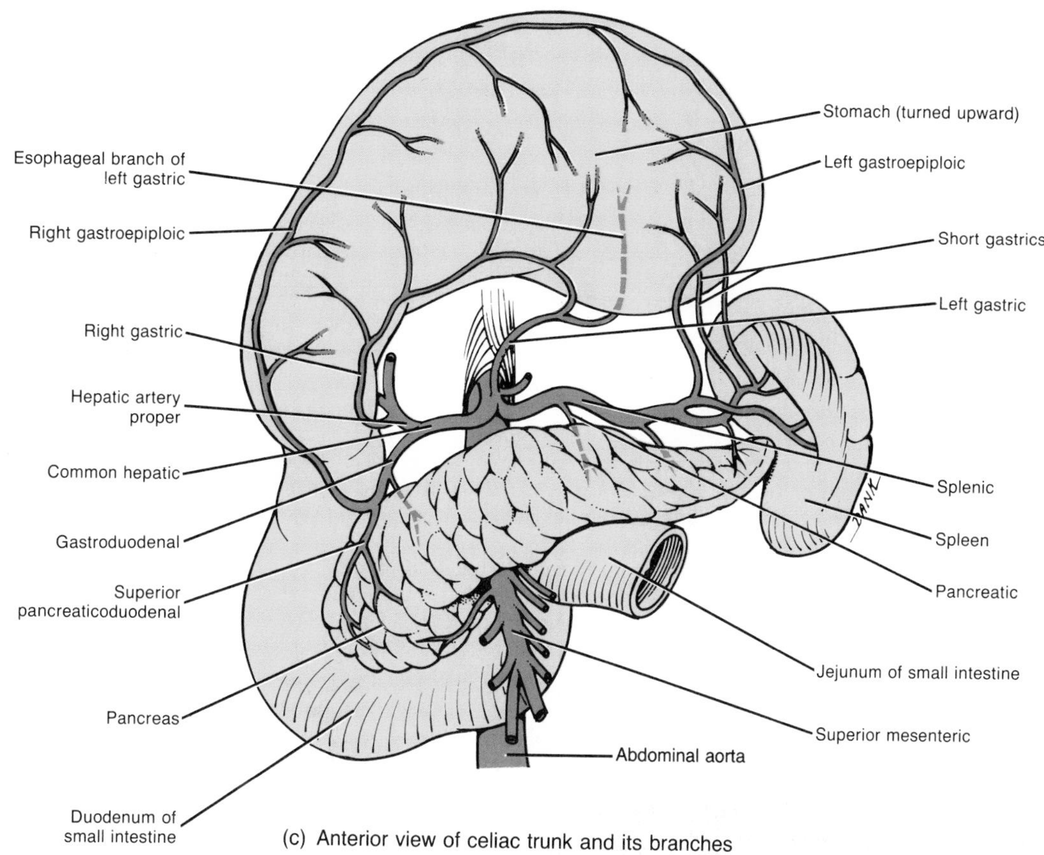

(c) Anterior view of celiac trunk and its branches

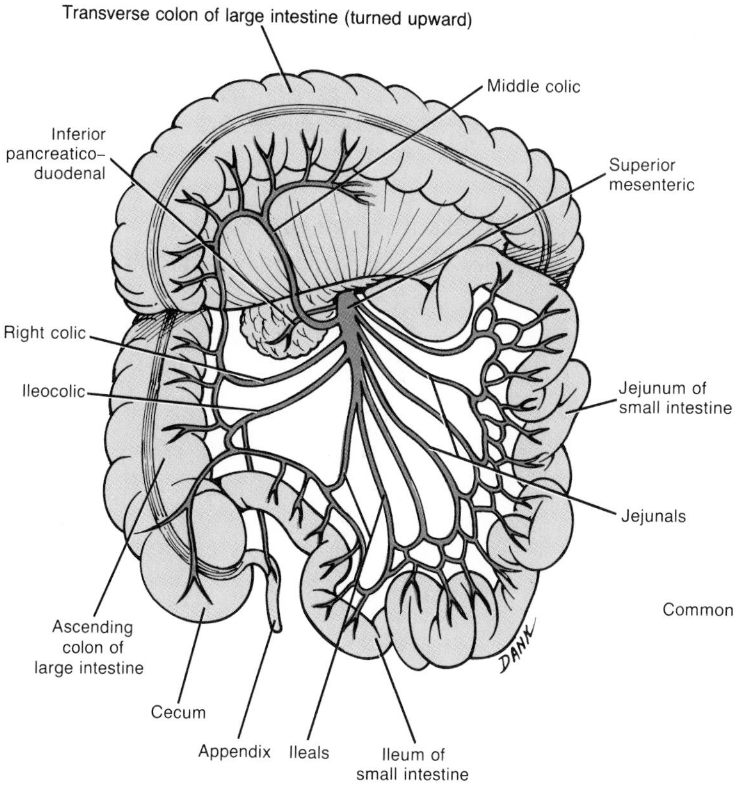

Transverse colon of large intestine (turned upward)
Middle colic
Inferior pancreatico-duodenal
Superior mesenteric
Right colic
Ileocolic
Jejunum of small intestine
Jejunals
Ascending colon of large intestine
Cecum
Appendix
Ileals
Ileum of small intestine
DANK

(d) Anterior view of superior mesenteric artery and its branches

Transverse colon of large intestine (turned upward)
Superior mesenteric
Left colic
Inferior mesenteric
Abdominal aorta
Descending colon of large intestine
Common iliac
Sigmoids
Sigmoid colon of large intestine
Superior rectal
Rectum of large intestine
DANK

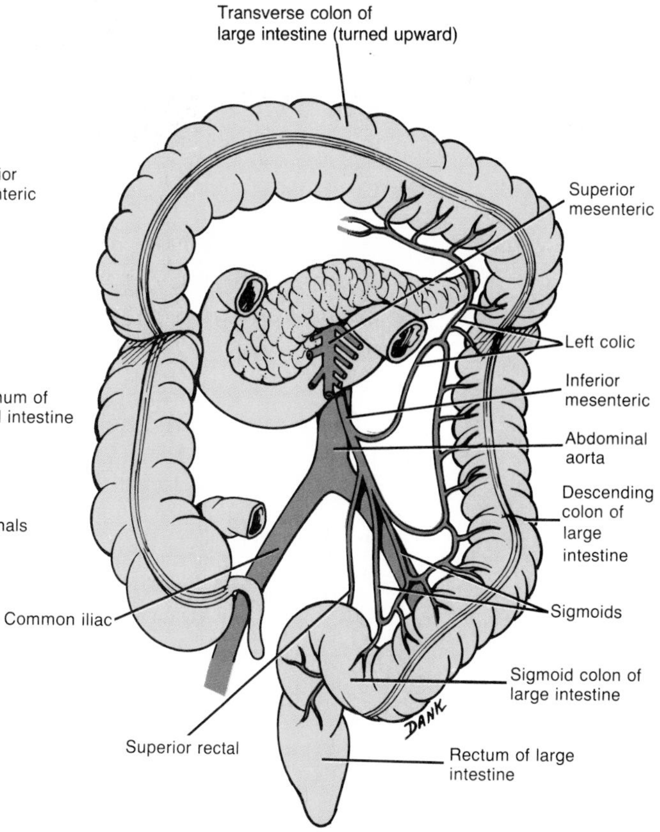

(e) Anterior view of inferior mesenteric artery and its branches

EXHIBIT 14-5

Abdominal Aorta (Figure 14-9)

Overview: The ***abdominal aorta*** is the continuation of the thoracic aorta. It begins at the aortic hiatus in the diaphragm and ends at about the level of the fourth lumbar vertebra, where it divides into right and left common iliac arteries. The abdominal aorta lies in front of the vertebral column.

As with the thoracic aorta, the abdominal aorta gives off visceral and parietal branches. The unpaired visceral branches arise from the anterior surface of the aorta and include the celiac trunk and the superior and inferior mesenteric arteries. The paired visceral branches arise from the lateral surfaces of the aorta and include the suprarenal, renal, and gonadal arteries. The paired parietal branches arise from the posterolateral surfaces of the aorta and include the inferior phrenic and lumbar arteries. The unpaired parietal artery is the middle sacral.

BRANCH	DESCRIPTION AND REGION SUPPLIED
VISCERAL	
Celiac Trunk	***Celiac trunk*** is first visceral aortic branch below diaphragm. It has three branches: (1) ***common hepatic artery,*** (2) ***left gastric artery,*** and (3) ***splenic artery.*** The common hepatic artery has three main branches: (1) ***hepatic artery proper,*** a continuation of the common hepatic artery, which supplies the liver and gallbladder; (2) ***right gastric artery,*** which supplies the stomach and duodenum; and (3) ***gastroduodenal artery,*** which supplies the stomach, duodenum, and pancreas. The left gastric artery supplies the stomach, and its ***esophageal branch*** supplies the esophagus. The splenic artery supplies the spleen and has three main branches: (1) ***pancreatic arteries,*** which supply the pancreas; (2) ***left gastroepiploic artery,*** which supplies the stomach and greater omentum; and (3) ***short gastric arteries,*** which supply the stomach.
Superior Mesenteric	The ***superior mesenteric artery*** has several principal branches: (1) ***inferior pancreaticoduodenal artery,*** which supplies the pancreas and duodenum; (2) **jejunal** and ***ileal arteries,*** which supply the jejunum and ileum, respectively; (3) ***ileocolic artery,*** which supplies the ileum and ascending colon; (4) ***right colic artery,*** which supplies the ascending colon; and (5) ***middle colic artery,*** which supplies the transverse colon.
Suprarenals	Right and left ***suprarenal arteries*** supply blood to adrenal (suprarenal) glands. The glands are also supplied by branches of the renal and inferior phrenic arteries.
Renals	Right and left ***renal arteries*** carry blood to kidneys and adrenal (suprarenal) glands.
Gonadals (Testiculars or Ovarians)	Right and left ***testicular arteries*** extend into scrotum and terminate in testes; right and left ***ovarian arteries*** are distributed to ovaries.
Inferior Mesenteric	The principal branches of the ***inferior mesenteric artery*** are the (1) ***left colic artery,*** which supplies the transverse and descending colons; (2) ***sigmoid arteries,*** which supply the descending and sigmoid colons; and (3) ***superior rectal artery,*** which supplies the rectum.
PARIETAL	
Inferior Phrenics	***Inferior phrenic arteries*** are distributed to undersurface of diaphragm and adrenal (suprarenal) glands.
Lumbars	***Lumbar arteries*** supply spinal cord and its meninges and muscles and skin of lumbar region of back.
Middle Sacral	***Middle sacral artery*** supplies sacrum, coccyx, and rectum.

EXHIBIT 14-6

Arteries of Pelvis and Lower Extremities (Figure 14-10)

Overview: The ***internal iliac arteries*** enter the pelvic cavity in front of the sacroiliac joint and supply most of the blood to the pelvic viscera and wall. The ***external iliacs*** travel along the brim of the lesser (true) pelvis. Behind the midportion of the inguinal ligament each external iliac artery enters the thigh, where its name changes to the femoral artery.

BRANCH	DESCRIPTION AND REGION SUPPLIED
Common Iliacs	At about level of fourth lumbar vertebra, abdominal aorta divides into right and left ***common iliac arteries.*** Each passes downward about 5 cm (2 in.) and gives rise to two branches: internal iliac and external iliac.
Internal Iliacs	***Internal iliac (hypogastric) arteries*** form branches that supply psoas major, gluteal muscles, quadratus lumborum, and medial side of each thigh, urinary bladder, rectum, prostate gland, ductus (vas) deferens, uterus, and vagina.
External Iliacs	***External iliac arteries*** diverge through greater (false) pelvis and enter thighs to become right and left ***femoral arteries.*** Both femorals send branches back up to genitals and wall of abdomen. Other branches run to muscles of thigh. Femoral continues down medial and posterior side of thigh at back of knee joint, where it becomes ***popliteal artery.*** Between knee and ankle, popliteal runs down back of leg and is called ***posterior tibial artery.*** Below knee, ***peroneal artery*** branches off posterior tibial to supply structures on medial side of fibula and calcaneus. In calf, ***anterior tibial artery*** branches off popliteal and runs along front of leg. At ankle, it becomes ***dorsalis pedis artery.*** At ankle, posterior tibial divides into ***medial*** and ***lateral plantar arteries.***

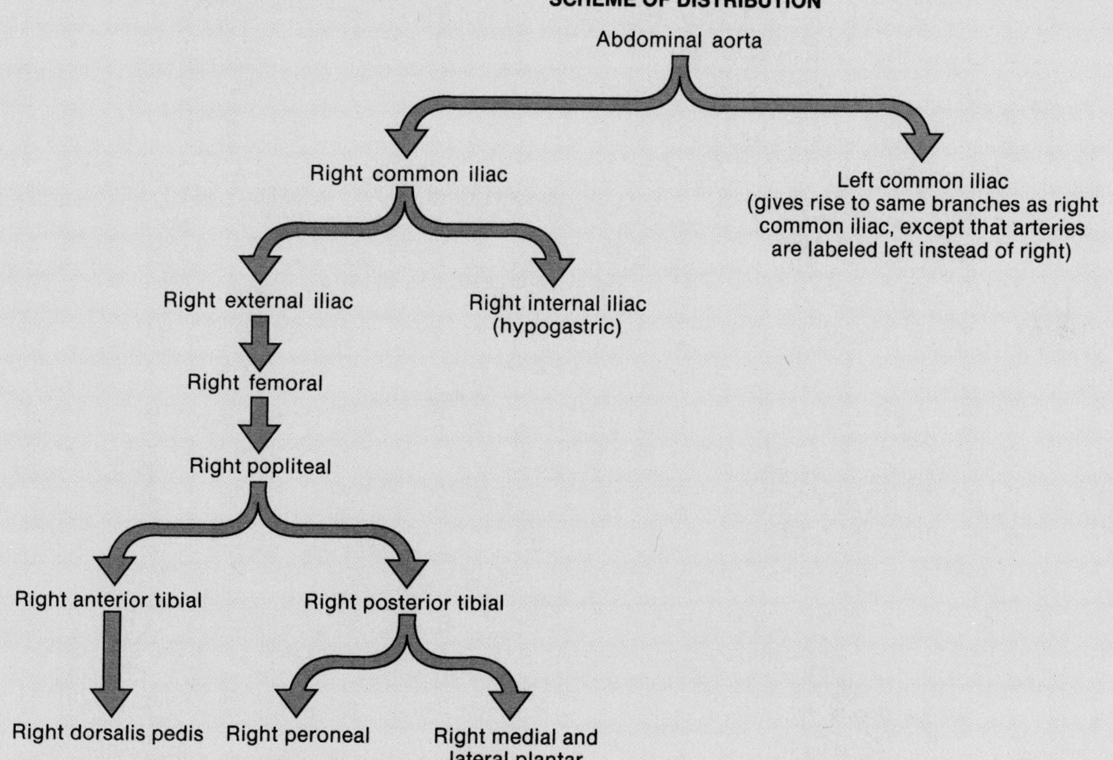

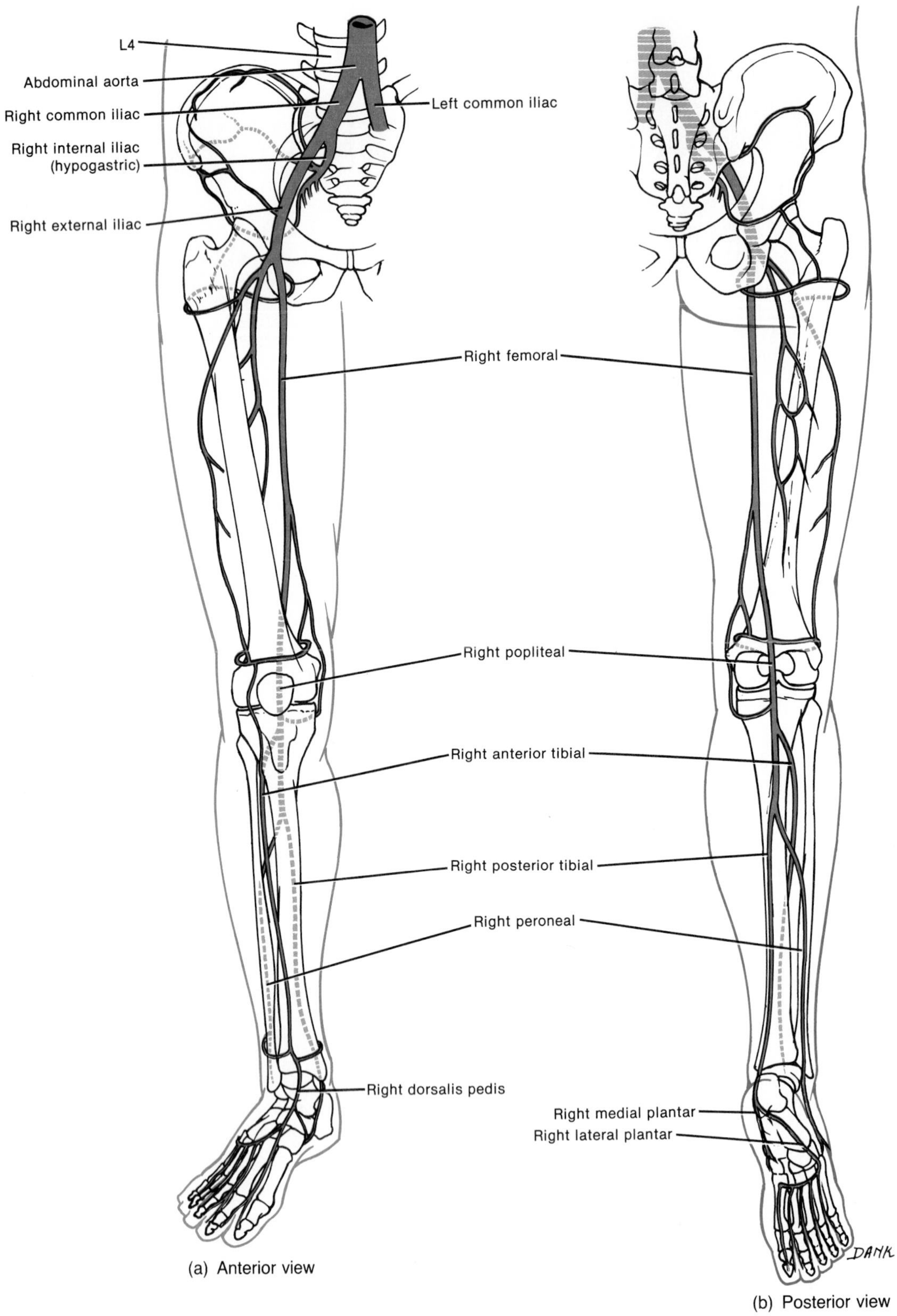

FIGURE 14-10 Arteries of the pelvis and right lower extremity. (c) and (d) Courtesy of J. A. Gosling, P. F. Harris, et al., *Atlas of Human Anatomy,* Gower Medical Publishing Ltd., 2nd ed., 1991.

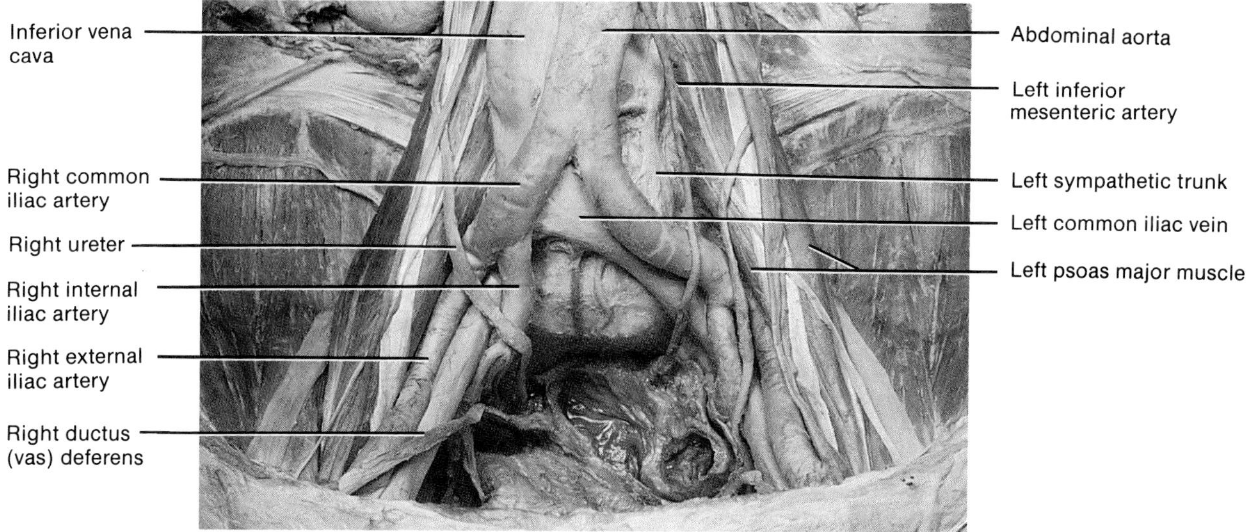

(c) Anterior view

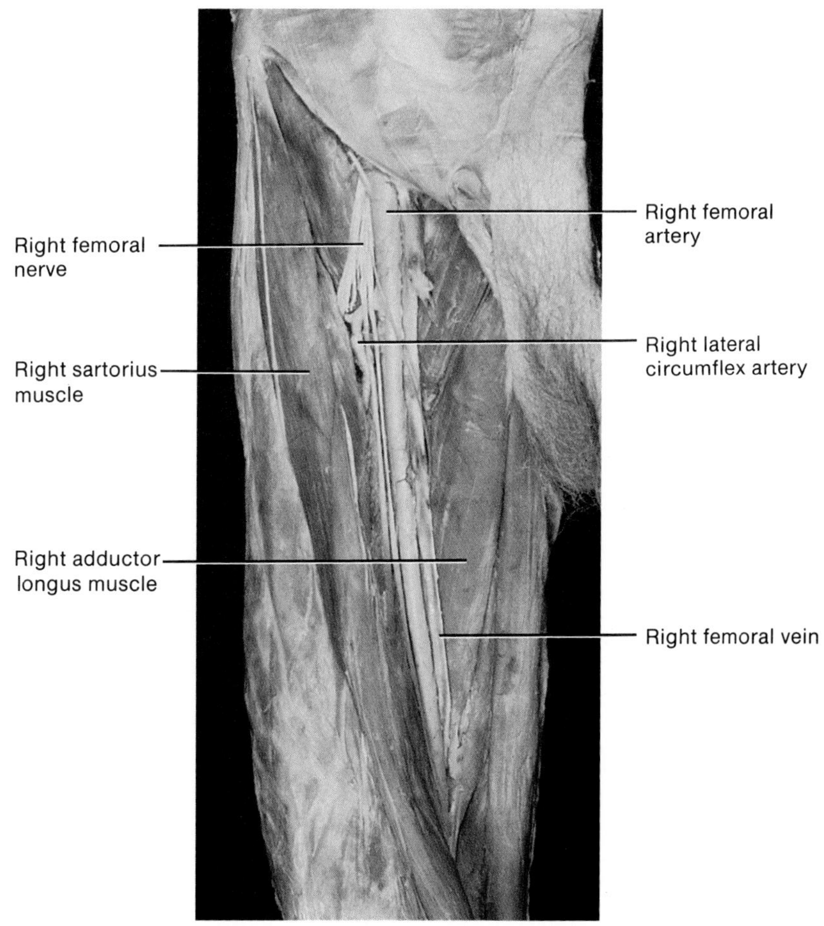

(d) Anterior view

EXHIBIT 14-7

Veins of Systemic Circulation (Figure 14-11)

Overview: Deoxygenated blood returns to the right atrium from three veins: ***coronary sinus, superior vena cava* (*SVC*),** and ***inferior vena cava* (*IVC*).** The coronary sinus receives blood from the cardiac veins; the superior vena cava receives blood from veins superior to the diaphragm, except the lungs, that is, veins from the head, neck, upper extremities, and thoracic wall. The inferior vena cava receives blood from veins inferior to the diaphragm, that is, veins from the lower extremities, most of the abdominal wall, and abdominal viscera.

VEIN	DESCRIPTION AND REGION DRAINED
Coronary Sinus	***Coronary sinus*** receives almost all venous blood from myocardium. It is located in coronary sulcus (see Figure 13-3c) and opens into right atrium between orifice of inferior vena cava and tricuspid valve.
Superior Vena Cava (SVC)	***Superior vena cava*** is about 7.5 cm (3 in.) long and empties its blood into the upper part of right atrium. It begins posterior to right first costal cartilage by the union of the right and left brachiocephalic veins and ends at level of right third costal cartilage, where it enters right atrium.
Inferior Vena Cava (IVC)	***Inferior vena cava*** is the largest diameter vein in the body, about 3.5 cm (1½ in.) in diameter. It begins anterior to fifth lumbar vertebra by union of common iliac veins, ascends behind the peritoneum to the right of the midline, pierces the costal tendon of the diaphragm at the level of the eighth thoracic vertebra, and enters the lower part of the right atrium.

CLINICAL APPLICATION

Compression of Inferior Vena Cava

The inferior vena cava is commonly compressed during the later stages of pregnancy owing to the enlargement of the uterus. This compression often produces edema of the ankles and feet and may result in temporary varicose veins.

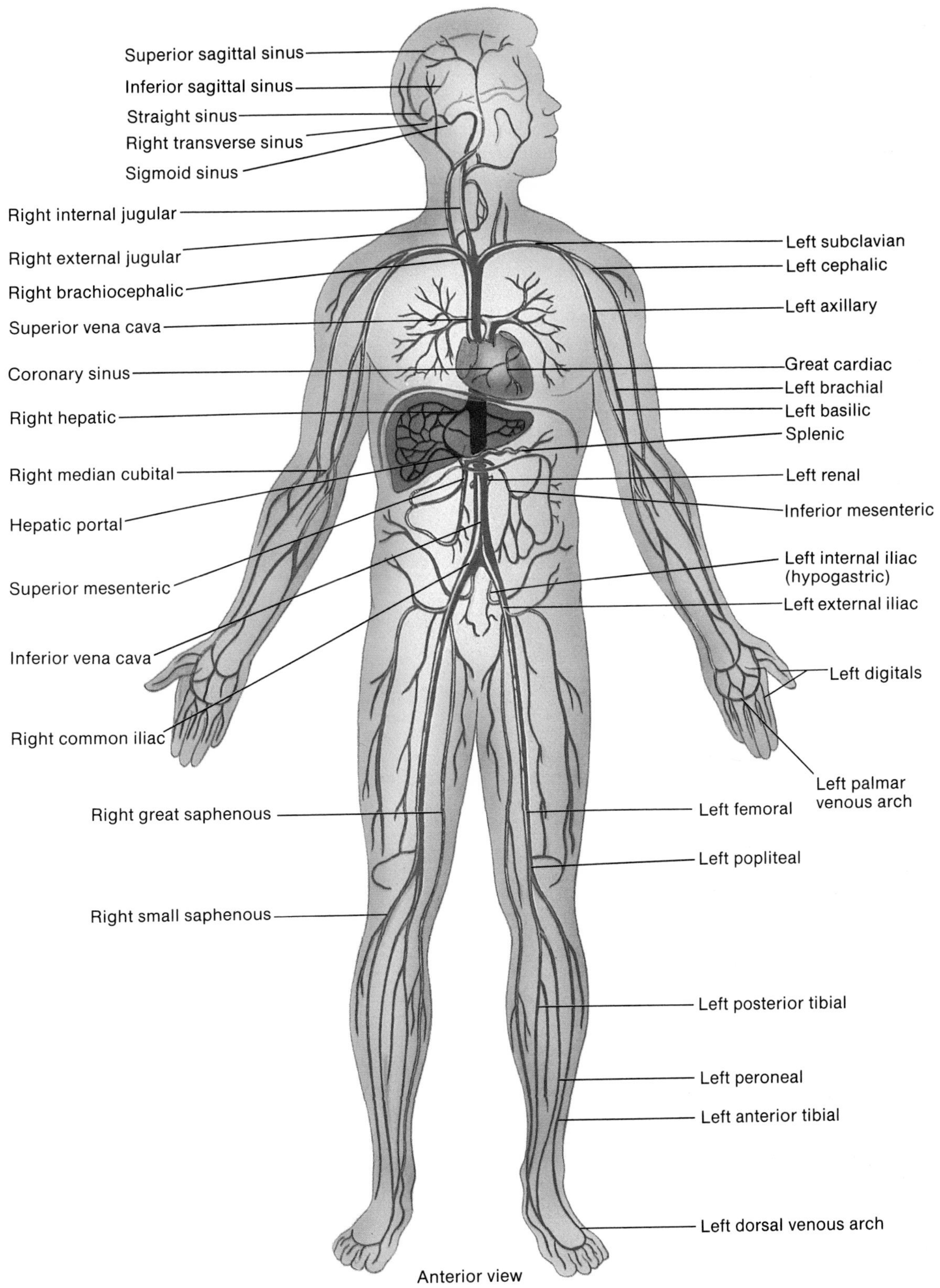

FIGURE 14-11 Principal veins.

EXHIBIT 14-8

Veins of Head and Neck (Figure 14-12)

Overview: The majority of blood draining from the head is pushed into three pairs of veins: ***internal jugular, external jugular,*** and ***vertebral.*** Within the cranium, all veins lead to the internal jugular veins.

VEIN	DESCRIPTION AND REGION DRAINED
Internal Jugulars	Right and left ***internal jugular veins*** receive blood from face and neck. They arise as continuation of ***sigmoid sinuses*** at base of skull. Intracranial vascular sinuses are located between layers of dura mater and receive blood from brain. Other sinuses that drain into internal jugular include ***superior sagittal sinus, inferior sagittal sinus, straight sinus,*** and ***transverse (lateral) sinuses.*** Internal jugulars descend on either side of neck and pass behind clavicles, where they join with right and left subclavian veins. Unions of internal jugulars and subclavians form right and left brachiocephalic veins. From here blood flows into superior vena cava.
External Jugulars	Right and left ***external jugular veins*** run down neck along outside of internal jugulars. They drain blood from parotid (salivary) glands, facial muscles, scalp, and other superficial structures into subclavian veins.
Vertebrals	Right and left ***vertebral veins*** descend through transverse foramina of cervical vertebrae and enter brachiocephalic veins. They drain deep structures of neck such as vertebrae and muscles.

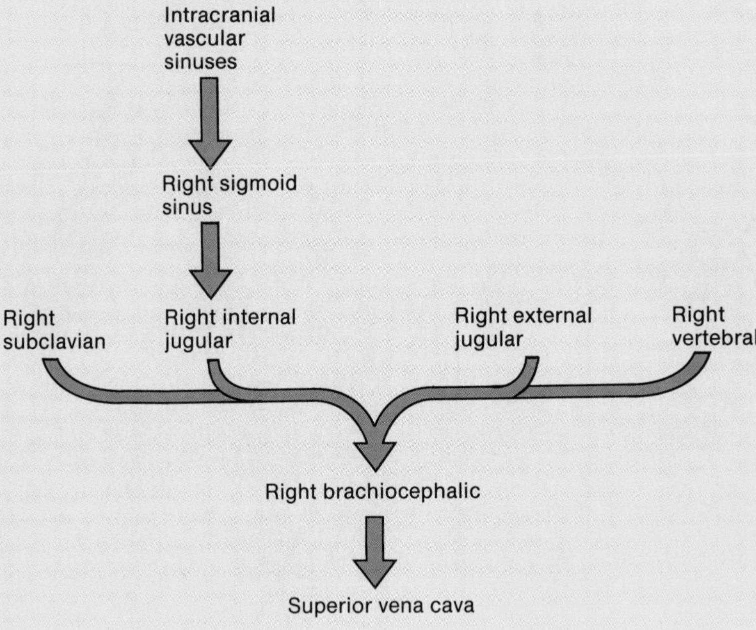

CLINICAL APPLICATION

Distension of Jugular Vein

In cases of heart failure, the venous pressure in the right atrium may rise. In such patients the pressure in the column of blood in the external jugular vein rises so that, even with the patient at rest and sitting in a chair, the external jugular vein will be visibly distended. Temporary distension of the vein is often seen in healthy adults when the intrathoracic pressure is raised as a result of coughing or physical exertion.

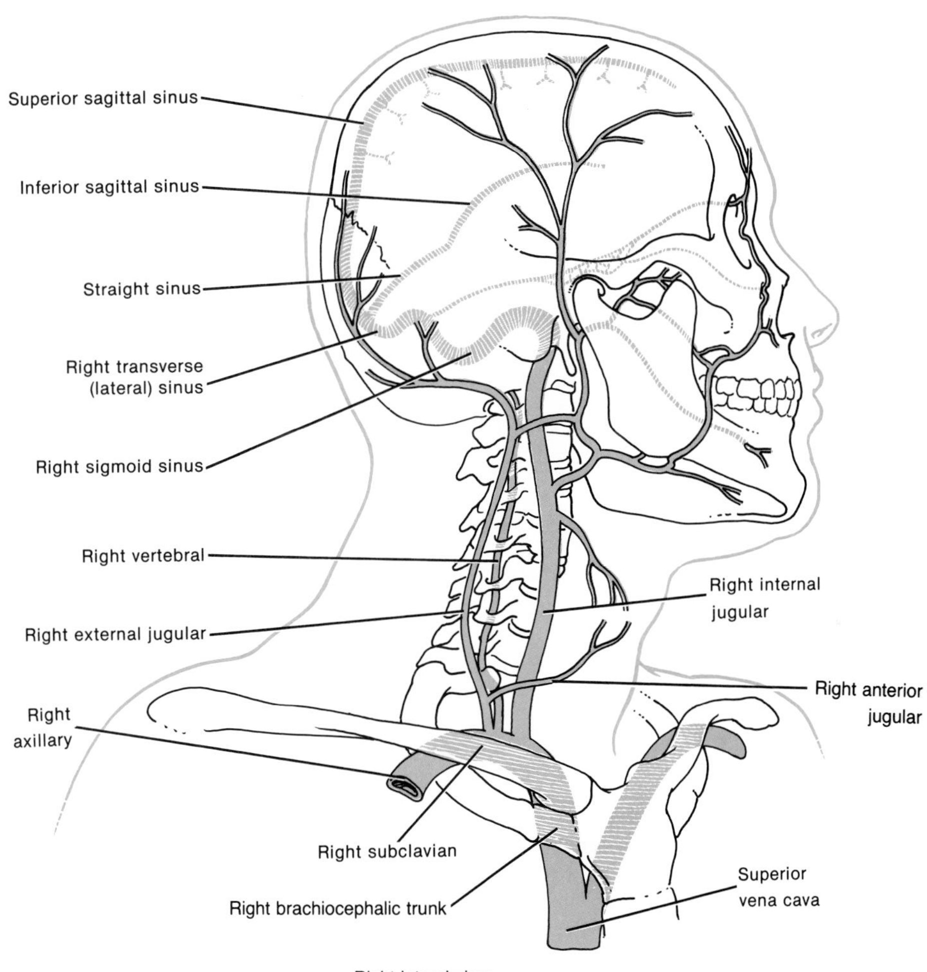

FIGURE 14-12 Principal veins of the head and neck.

EXHIBIT 14-9

Veins of Upper Extremities (Figure 14-13)

Overview: Blood from each upper extremity is returned to the heart by superficial and deep veins. Both sets of veins contain valves. ***Superficial veins*** are located just below the skin, are larger than deep veins, and are often visible. They anastomose extensively with each other and deep veins. ***Deep veins*** are located deep in the body. They usually accompany arteries, and many have the same names as corresponding arteries. Most deep veins are paired vessels.

VEIN	DESCRIPTION AND REGION DRAINED
SUPERFICIAL	
Cephalics	***Cephalic vein*** of each upper extremity begins in the medial part of ***dorsal venous arch*** and winds upward around radial border of forearm. In front of elbow, it is connected to basilic vein by the ***median cubital vein.*** Just below elbow, cephalic vein unites with ***accessory cephalic vein*** to form cephalic vein of upper extremity. Ultimately, cephalic vein empties into axillary vein.
Basilics	***Basilic vein*** of each upper extremity originates in the ulnar part of ***dorsal venous arch.*** It extends along posterior surface of ulna to point below elbow where it receives ***median cubital vein.*** If a vein must be punctured for an injection, transfusion, or removal of a blood sample, median cubitals are preferred. After receiving the median cubital vein, the basilic continues ascending on medial side until it reaches middle of the upper arm. There it penetrates the tissues deeply and runs alongside the brachial artery until it joins the brachial vein. As the basilic and brachial veins merge in the axillary area, they form the axillary vein.
Median Antebrachials	***Median antebrachial veins*** drain ***palmar venous arch,*** ascend on ulnar side of anterior forearm, and end in median cubital veins.
DEEP	
Radials	***Radial veins*** receive ***dorsal metacarpal veins.***
Ulnars	***Ulnar veins*** receive tributaries from ***palmar venous arch.*** Radial and ulnar veins unite in bend of elbow to form brachial veins.
Brachials	Located on either side of brachial artery, ***brachial veins*** join into axillary veins.
Axillaries	***Axillary veins*** are a continuation of brachials and basilics. Axillaries end at first rib, where they become subclavians.
Subclavians	Right and left ***subclavian veins*** unite with internal jugulars to form brachiocephalic veins. Thoracic duct of lymphatic system delivers lymph into left subclavian veins at junction with internal jugular. Right lymphatic duct delivers lymph into right subclavian vein at corresponding junction.

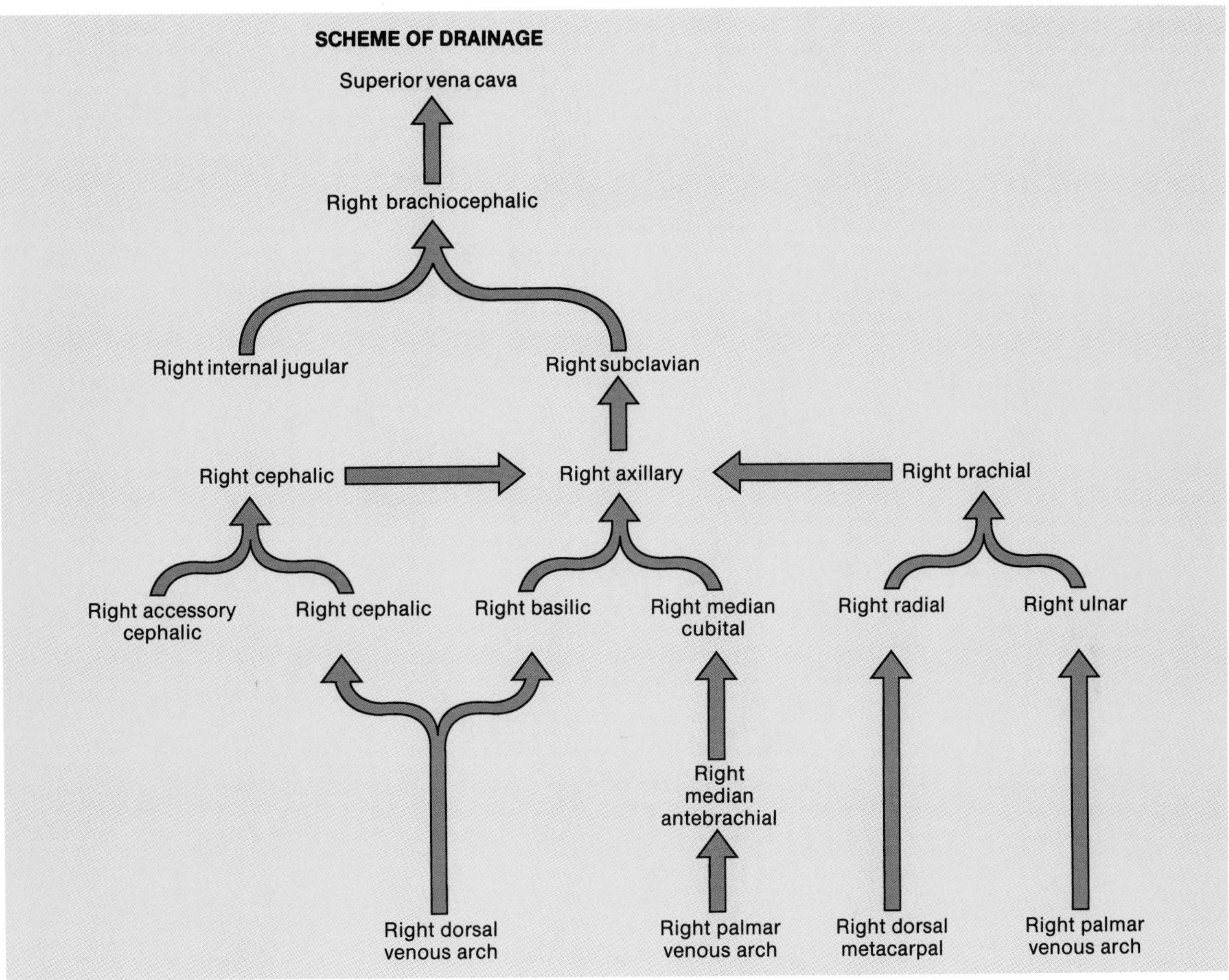
SCHEME OF DRAINAGE
Superior vena cava
Right brachiocephalic
Right internal jugular
Right subclavian
Right cephalic
Right axillary
Right brachial
Right accessory cephalic
Right cephalic
Right basilic
Right median cubital
Right radial
Right ulnar
Right median antebrachial
Right dorsal venous arch
Right palmar venous arch
Right dorsal metacarpal
Right palmar venous arch

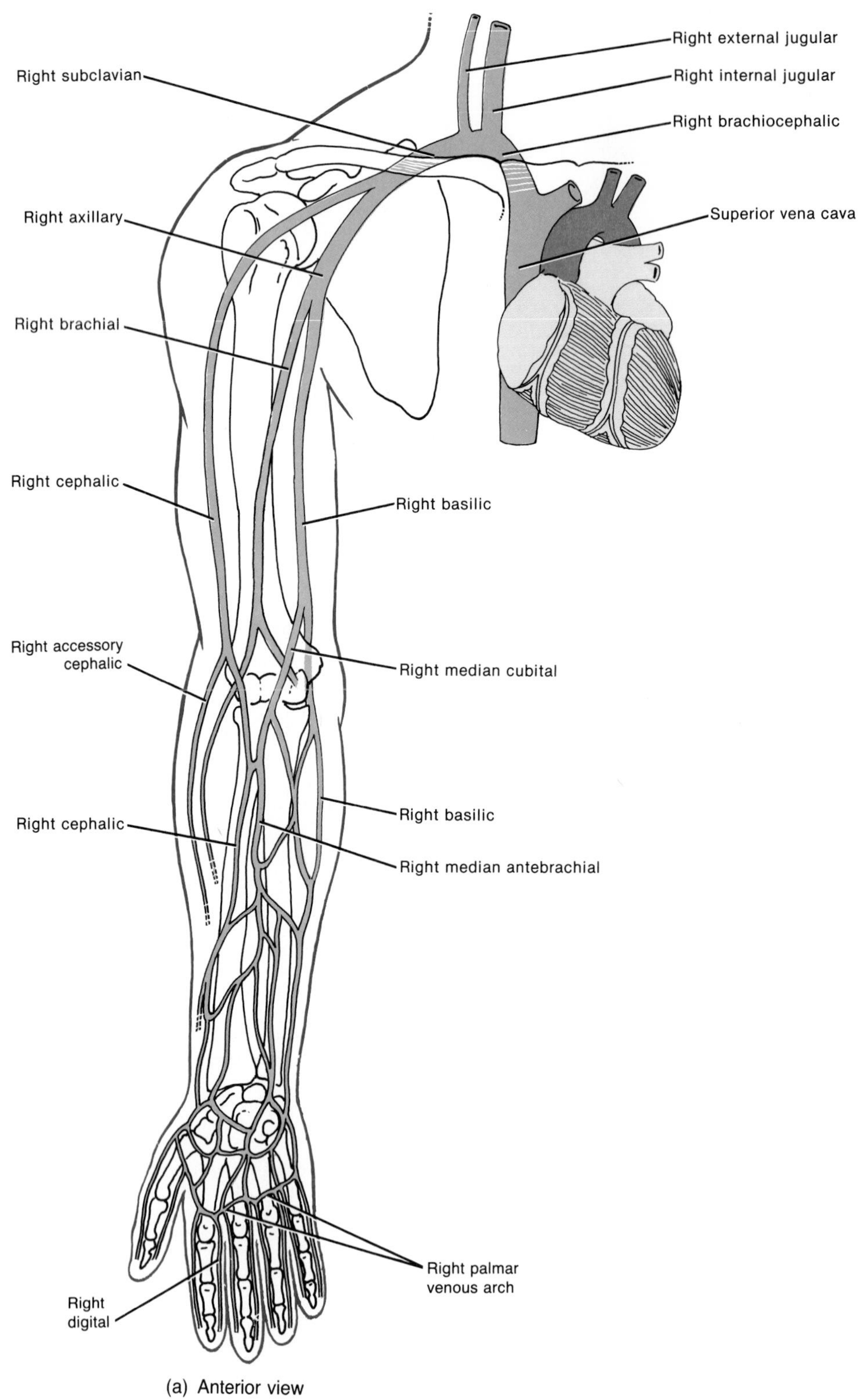

(a) Anterior view

FIGURE 14-13 Principal veins of the right upper extremity (b) Courtesy of J. A. Gosling, P. F. Harris, et al., *Atlas of Human Anatomy,* Gower Medical Publishing Ltd., 2nd ed., 1991.

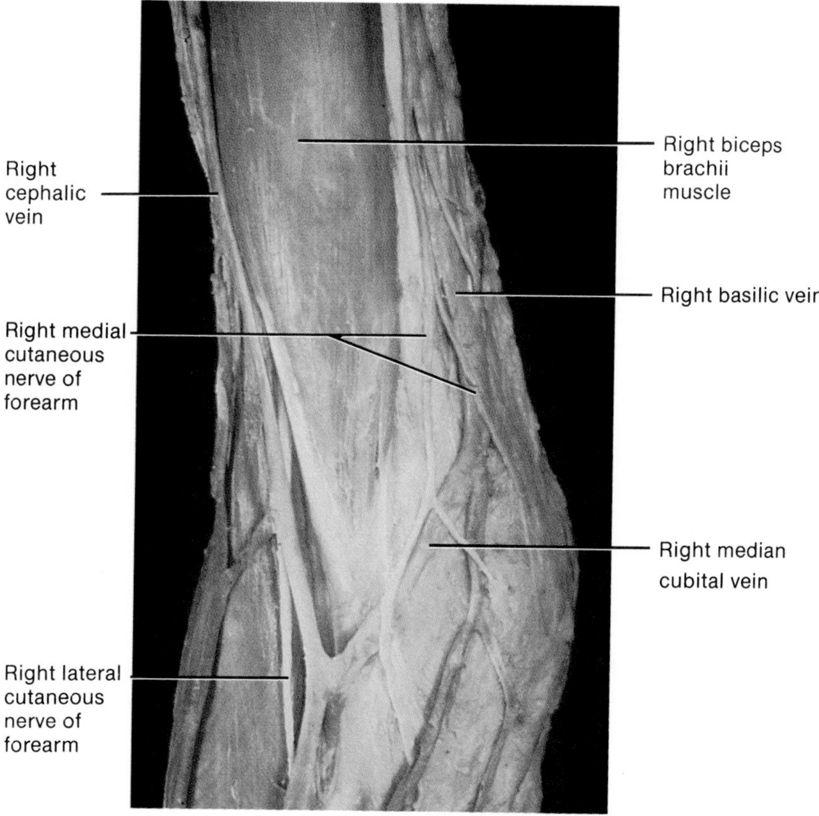

(b) Anterior view

EXHIBIT 14-10

Veins of Thorax (Figure 14-14)

Overview: Although brachiocephalic veins drain some portions of thorax, most thoracic structures are drained by a network of veins called the ***azygos system.*** The system consists of a network of veins on each side of vertebral column: azygos, hemiazygos, and accessory hemiazygos. They show considerable variation in origin, course, tributaries, anastomoses, and termination. Ultimately, they empty into the superior vena cava.

VEIN	DESCRIPTION AND REGION DRAINED
Brachiocephalic	Right and left ***brachiocephalic veins,*** formed by union of subclavians and internal jugulars, drain blood from head, neck, upper extremities, mammary glands, and upper thorax. Brachiocephalics unite to form superior vena cava.
Azygos Veins	***Azygos veins,*** besides collecting blood from thorax, may serve as bypass for inferior vena cava that drains blood from lower body. Several small veins directly link azygos veins with inferior vena cava. Large veins that drain lower extremities and abdomen dump blood into azygos. If inferior vena cava or hepatic portal vein becomes obstructed, azygos veins can return blood from lower body to superior vena cava.
Azygos	***Azygos vein*** lies in front of vertebral column, slightly right of midline. It begins as continuation of right ascending lumbar vein. It connects with inferior vena cava, right common iliac, and lumbar veins. Azygos receives blood from ***right intercostal veins*** that drain chest muscles; from hemiazygos and accessory hemiazygos veins; from several ***esophageal, mediastinal,*** and ***pericardial veins;*** and from right ***bronchial vein.*** Vein ascends to fourth thoracic vertebra, arches over right lung, and empties into superior vena cava.
Hemiazygos	***Hemiazygos vein*** is in front of vertebral column and slightly left of midline. It begins as continuation of left ascending lumbar vein. It receives blood from lower four or five ***intercostal veins*** and some ***esophageal*** and ***mediastinal veins.*** At level of ninth thoracic vertebra, it joins azygos vein.
Accessory Hemiazygos	***Accessory hemiazygos vein*** is also in front and to left of vertebral column. It receives blood from three to four ***intercostal veins*** and left ***bronchial vein.*** It joins azygos at level of eighth thoracic vertebra.

EXHIBIT 14-11

Veins of Abdomen and Pelvis (Figure 14-14)

Overview: Blood from the abdominopelvic viscera and abdominal wall is returned to the heart via the ***inferior vena cava.*** Numerous small veins enter the inferior vena cava and most carry return flow from parietal branches of the abdominal aorta; their names correspond to the names of the arteries. The inferior vena cava does not directly receive veins from the gastrointestinal tract, spleen, pancreas, and gallbladder. These organs pass their blood into a common vein, the hepatic portal vein, which delivers the blood to the liver. From here, the blood is drained into the hepatic veins, which then enter the inferior vena cava. This special flow of venous blood is called ***hepatic portal circulation,*** which will be described shortly.

VEIN	DESCRIPTION AND REGION DRAINED
Inferior Vena Cava	***Inferior vena cava*** is formed by union of two common iliac veins that drain lower extremities and abdomen. Inferior vena cava extends upward through abdomen and thorax to right atrium.
Common Iliacs	***Common iliac veins*** are formed by union of internal (hypogastric) and external iliac veins and represent distal continuation of inferior vena cava at its bifurcation.
Internal Iliacs	Tributaries of ***internal iliac (hypogastric) veins*** basically correspond to branches of internal iliac arteries. Internal iliacs drain gluteal muscles, medial side of thigh, urinary bladder, rectum, prostate gland, ductus (vas) deferens, uterus, and vagina.
External Iliacs	***External iliac veins*** are continuation of femoral veins and receive blood from lower extremities and inferior part of anterior abdominal wall.
Renals	***Renal veins*** drain kidneys.
Gonadals (Testicular or Ovarian)	***Testicular veins*** drain testes (left testicular vein empties into left renal vein and right testicular vein drains into inferior vena cava); ***ovarian veins*** drain ovaries (left ovarian vein empties into left renal vein and right ovarian vein drains into inferior vena cava).
Suprarenals	***Suprarenal veins*** drain adrenal (suprarenal) glands (left suprarenal vein empties into left renal vein).
Inferior Phrenics	***Inferior phrenic veins*** drain diaphragm (left inferior phrenic vein sends tributary to left renal vein).
Hepatics	***Hepatic veins*** drain liver.
Lumbars	A series of parallel ***lumbar veins*** drain blood from both sides of posterior abdominal wall. Lumbars connect at right angles with right and left ***ascending lumbar veins,*** which form origin of corresponding azygos or hemiazygos vein. Lumbars drain blood into ascending lumbars and then run to inferior vena cava, where they release remainder of flow.

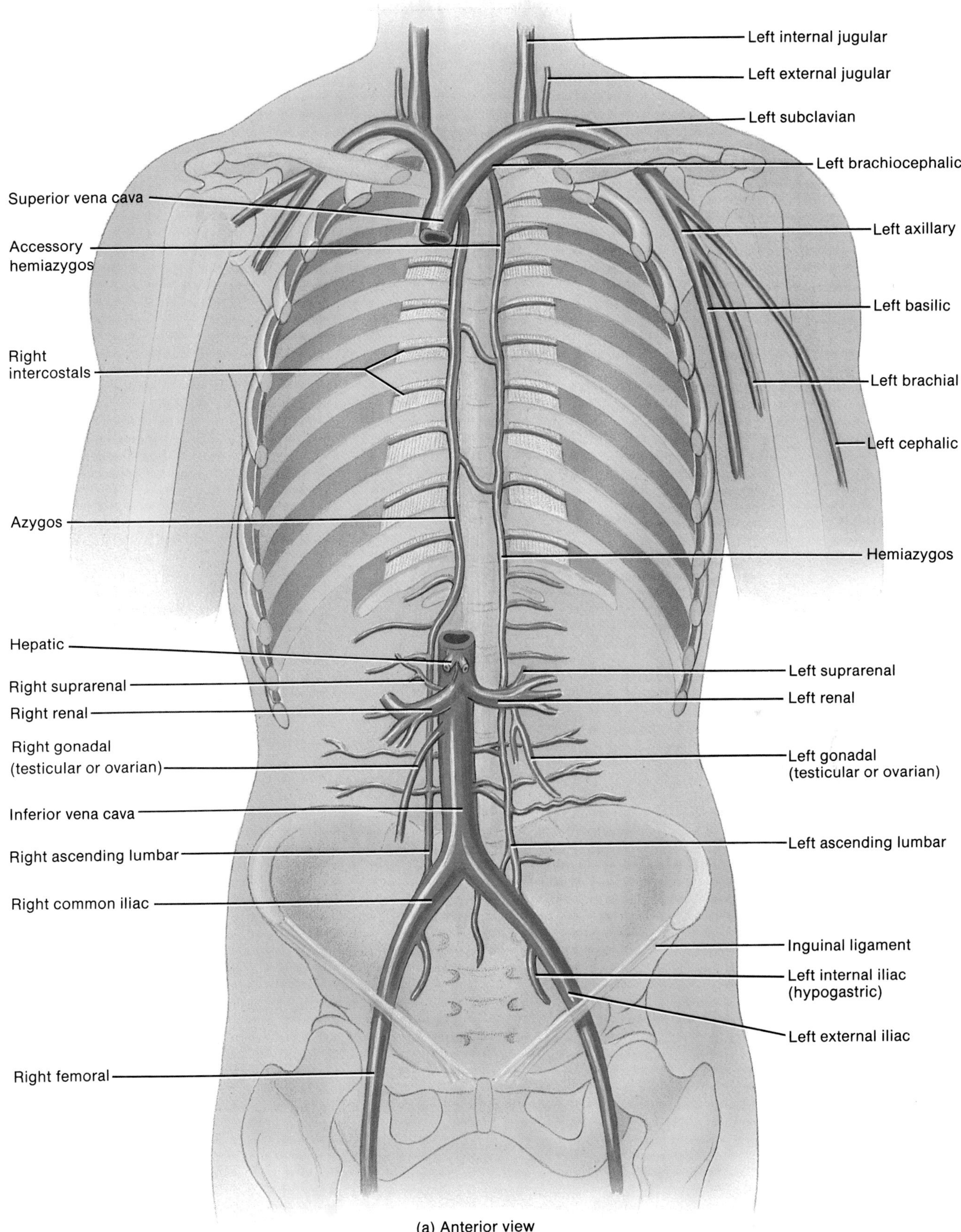

(a) Anterior view

FIGURE 14-14 Principal veins of the thorax, abdomen, and pelvis. (b) and (c) Courtesy of J. A. Gosling, P. F. Harris, et al., *Atlas of Human Anatomy,* Gower Medical Publishing Ltd., 2nd ed., 1991.

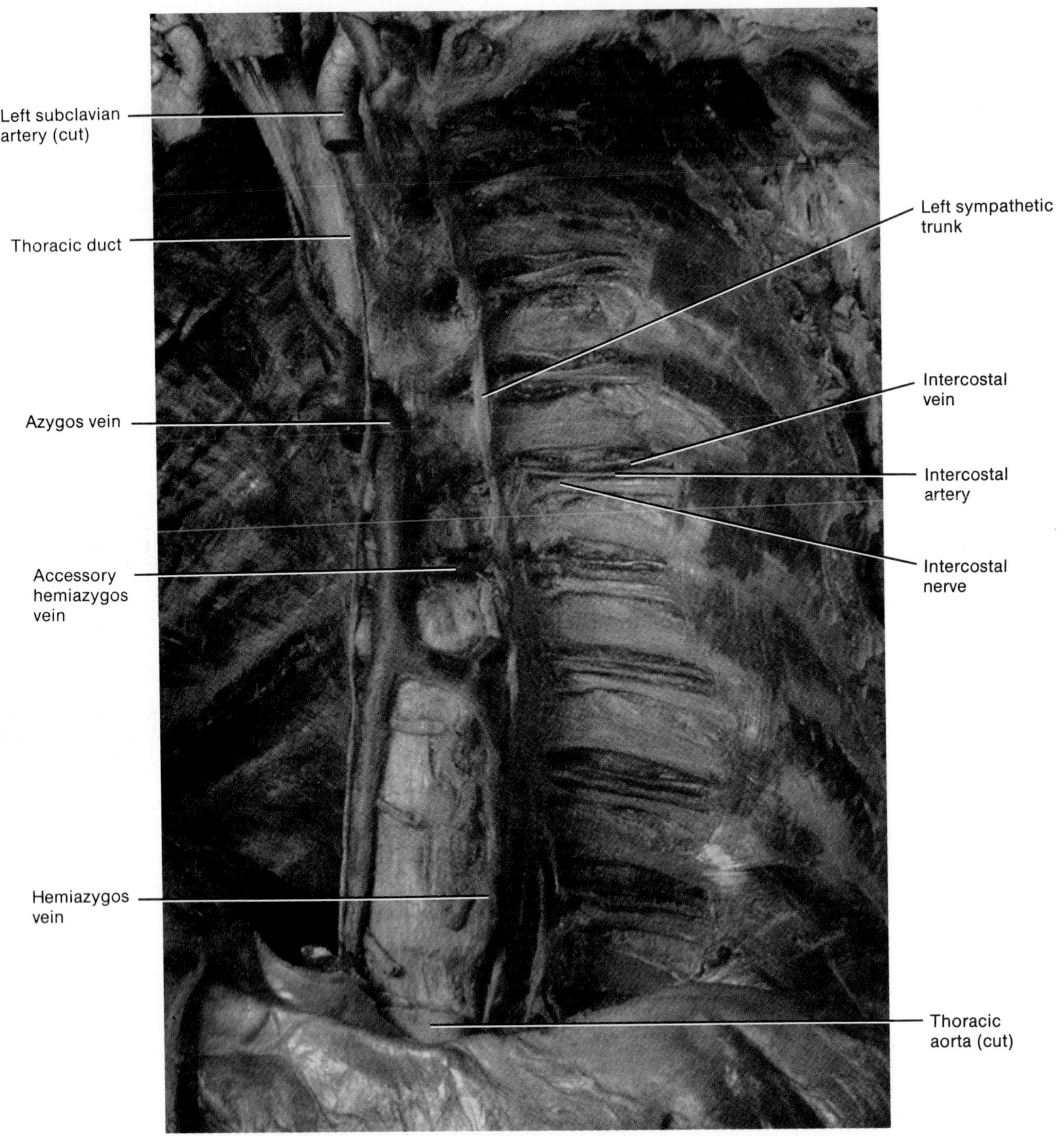

(b) Anterior view

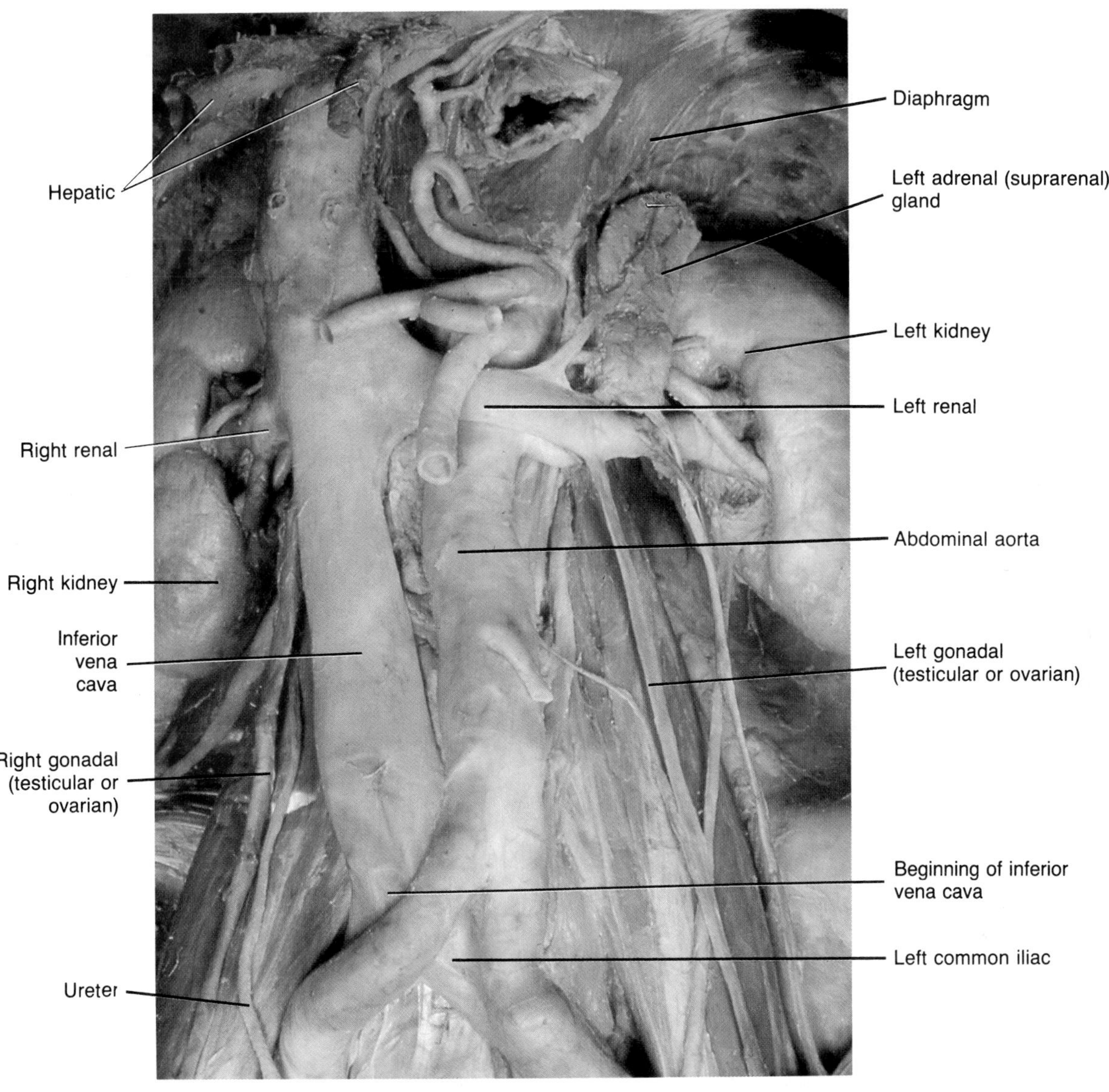

(c) Anterior view

EXHIBIT 14-12

Veins of Lower Extremities (Figure 14-15)

Overview: Blood from each lower extremity is drained by ***superficial*** and ***deep veins.*** The superficial veins frequently anastomose with each other and with deep veins along their length. Deep veins, for the most part, have the same names as their accompanying arteries.

VEIN	DESCRIPTION AND REGION DRAINED
SUPERFICIAL VEINS	
Great Saphenous	***Great saphenous vein,*** longest vein in body, begins at medial end of ***dorsal venous arch*** of foot. It passes in front of medial malleolus and then upward along medial aspect of leg and thigh just deep to the skin. It receives tributaries from superficial tissues and connects with deep veins as well. It empties into femoral vein in groin.
Small Saphenous	***Small saphenous vein*** begins at lateral end of dorsal venous arch of foot. It passes behind lateral malleolus and ascends under skin of back of leg. It receives blood from foot and posterior portion of leg. It empties into popliteal vein behind knee.

EXHIBIT 14-12 (*Continued*)

VEIN	DESCRIPTION AND REGION DRAINED
DEEP VEINS	
Posterior Tibial	***Posterior tibial vein*** is formed by union of ***medial*** and ***lateral plantar veins*** behind medial malleolus. It ascends deep in muscle at back of leg, receives blood from ***peroneal vein,*** and unites with anterior tibial vein just below knee.
Anterior Tibial	***Anterior tibial vein*** is upward continuation of ***dorsalis pedis veins*** in foot. It runs between tibia and fibula and unites with posterior tibial to form popliteal vein.
Popliteal	***Popliteal vein,*** just behind knee, receives blood from anterior and posterior tibials and small saphenous vein.
Femoral	***Femoral vein*** is upward continuation of popliteal just above knee. Femorals run up posterior surface of thighs and drain deep structures of thighs. After receiving great saphenous veins in groin, they continue as right and left external iliac veins.

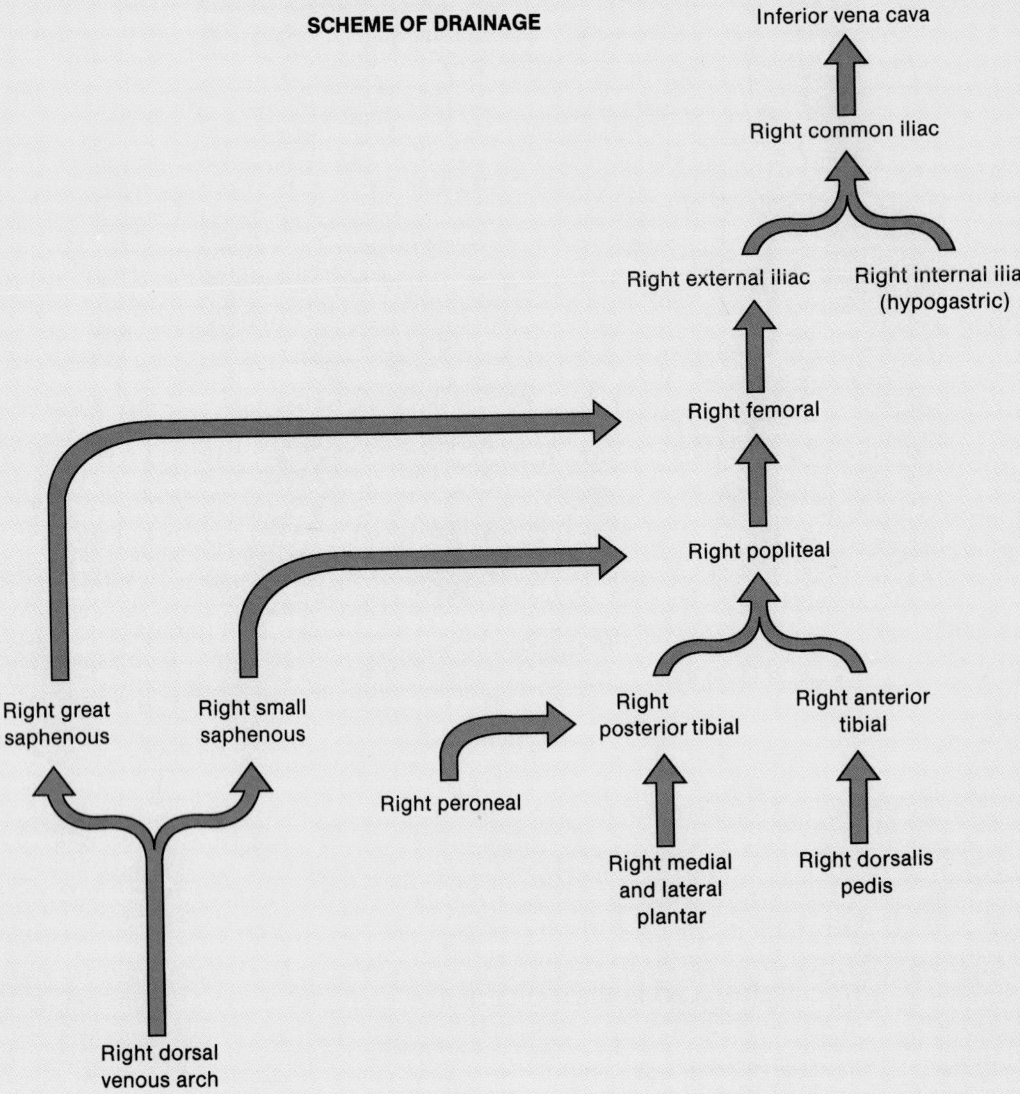

SCHEME OF DRAINAGE

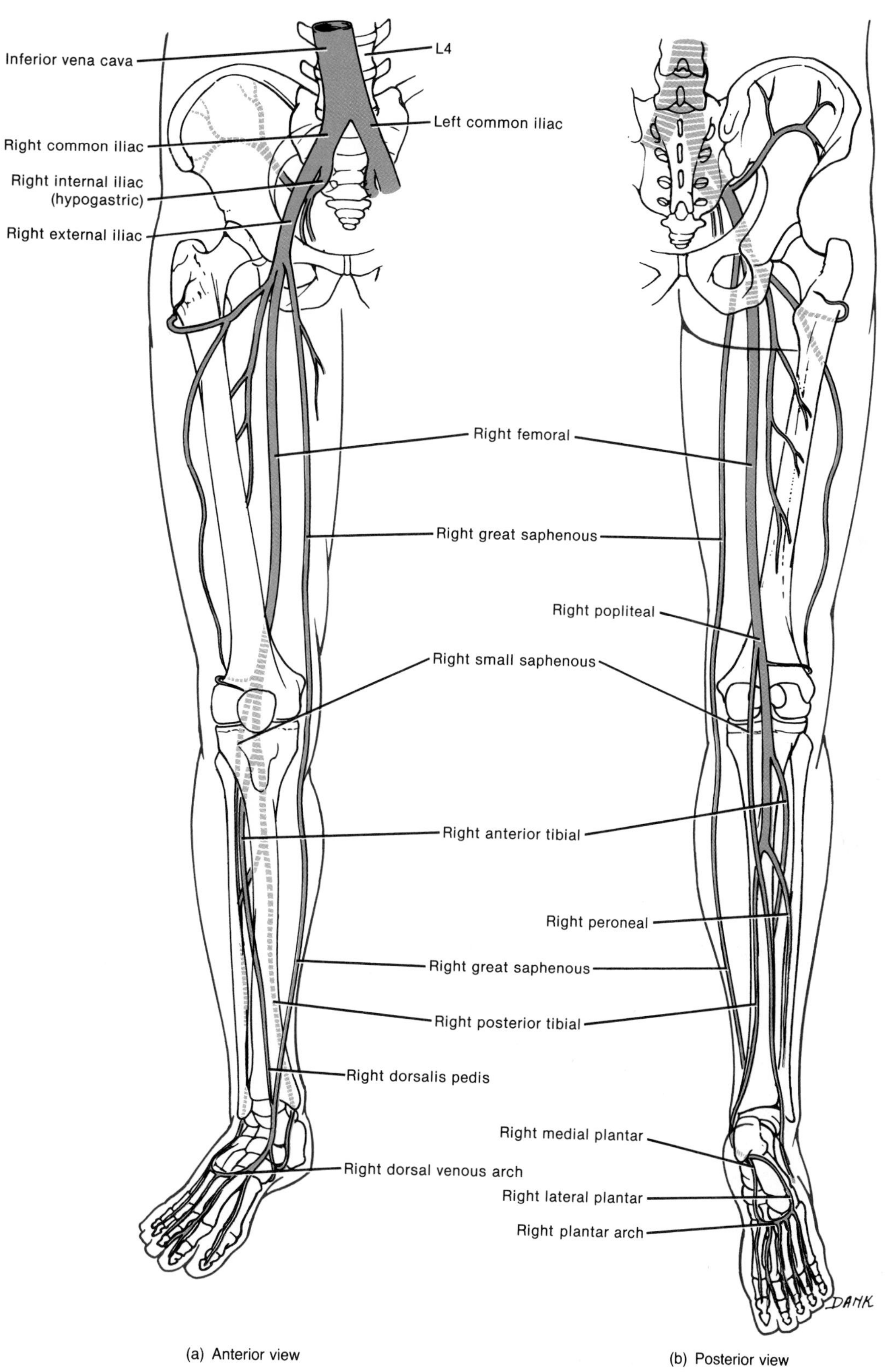

FIGURE 14-15 Principal veins of the pelvis and right lower extremity.

CLINICAL APPLICATION

Administration of Intravenous Fluids

The great saphenous vein is very constant in its position anterior to the medial malleolus. It is frequently used for prolonged administration of intravenous fluids. This feature is particularly important in very young babies and in patients of any age who are in shock and whose veins are collapsed. It and the small saphenous vein are subject to varicosity.

HEPATIC PORTAL CIRCULATION

Blood enters the liver from two sources. The hepatic artery delivers oxygenated blood from the systemic circulation; the hepatic portal vein delivers deoxygenated blood from the gastrointestinal tract, spleen, pancreas, and gallbladder. The term ***hepatic portal circulation*** refers to the flow of venous blood from the gastrointestinal organs and spleen to the liver before returning to the heart (Figure 14-16). During the absorptive state, hepatic portal blood is rich with substances absorbed from the gastrointestinal tract. The liver monitors these substances before they pass into the general circulation. For example, the liver stores nutrients such as glucose. It also modifies other digested substances so they may be used by cells, detoxifies harmful substances that have been absorbed by the gastrointestinal tract, and destroys bacteria by phagocytosis.

The hepatic portal system includes veins that drain blood from the pancreas, spleen, stomach, intestines, and gallbladder and transport it to the hepatic portal vein of the liver.

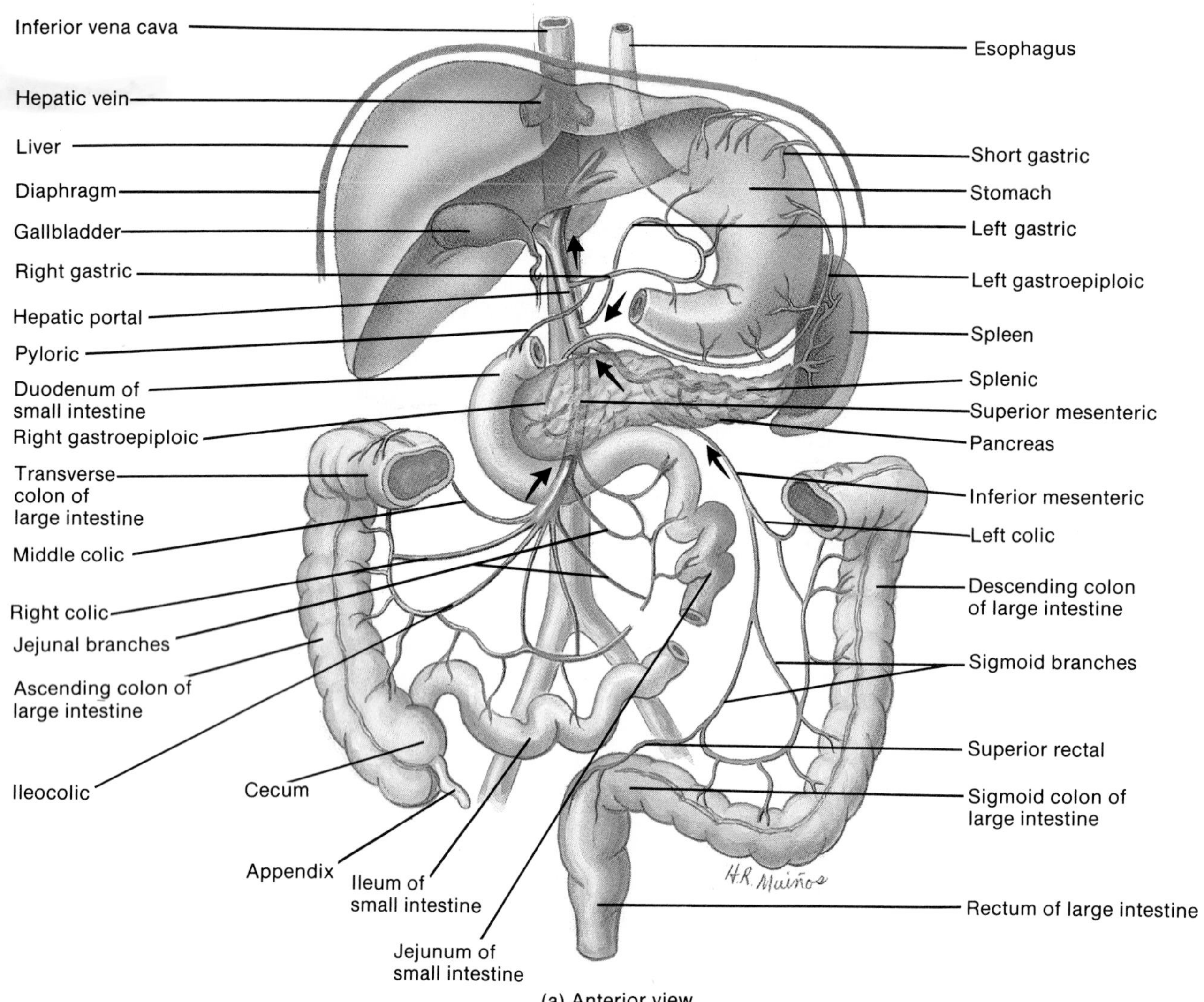

(a) Anterior view

FIGURE 14-16 Hepatic portal circulation. The scheme of blood flow through the liver, including arterial circulation, is shown in (b). Deoxygenated blood is indicated in blue; oxygenated blood in red.

The ***hepatic portal vein*** is formed by the union of the superior mesenteric and splenic veins. The ***superior mesenteric vein*** drains blood from the small intestine and portions of the large intestine and stomach. The ***splenic vein*** drains the spleen and receives tributaries from the stomach, pancreas, and portions of the colon. The tributaries from the stomach are the ***gastric, pyloric,*** and ***gastroepiploic veins.*** The ***pancreatic veins*** come from the pancreas, and the ***inferior mesenteric veins*** come from the portions of the colon. Before the hepatic portal vein enters the liver, it receives the ***cystic vein*** from the gallbladder and other veins. Ultimately, deoxygenated blood leaves the liver through the ***hepatic veins,*** which enter the inferior vena cava.

PULMONARY CIRCULATION

The flow of deoxygenated blood from the right ventricle to the air sacs of the lungs and the return of oxygenated blood from the air sacs of the lungs to the left atrium is called ***pulmonary circulation*** (Figure 14-17). The ***pulmonary trunk*** emerges from the right ventricle and passes upward, backward, and to the left. It then divides into two branches: the ***right pulmonary artery*** runs to the right lung; the ***left pulmonary artery*** goes to the left lung. The pulmonary arteries are the only postnatal arteries that carry deoxygenated blood. On entering the lungs, the branches of the pulmonary arteries divide and subdivide until ultimately they form capillaries around the alveoli (air sacs) in the lungs. Carbon dioxide is passed from the blood into the alveoli to be breathed out of the lungs. Oxygen breathed in by the lungs is passed from the alveoli into the blood. The capillaries unite, venules and veins are formed, and eventually two ***pulmonary veins*** exit from each lung and transport the oxygenated blood to the left atrium. The pulmonary veins are the only postnatal veins that carry oxygenated blood. Contractions of the left ventricle then send the blood into the systemic circulation.

FETAL CIRCULATION

The circulatory system of a fetus, called ***fetal circulation,*** differs from an adult's because the lungs, kidneys, and gastrointestinal tract of a fetus are not functioning. The fetus derives its oxygen and nutrients from the maternal blood and eliminates its carbon dioxide and wastes into the maternal blood (Figure 14-18).

The exchange of materials between fetal and maternal circulation occurs through a structure called the ***placenta*** (pla-SEN-ta). It is attached to the umbilicus (navel) of the fetus by the umbilical (um-BIL-i-kal) cord, and it communicates with the mother through countless small blood vessels that emerge from the uterine wall. The umbilical cord contains blood vessels that branch into capillaries in the placenta. Wastes from the fetal blood diffuse out of the capillaries, into spaces containing maternal blood (intervillous spaces) in the placenta, and finally into the mother's uterine blood vessels (see Figure 26-9). Nutrients travel the opposite route—from the maternal blood vessels to the intervillous spaces to the fetal capillaries. Normally, there is no mixing of maternal and fetal blood since all exchanges occur through capillaries.

Blood passes from the fetus to the placenta via two ***umbilical arteries.*** These branches of the internal iliac (hypogastric) arteries are included in the umbilical cord. At the placenta, the blood picks up oxygen and nutrients and eliminates carbon dioxide and wastes. The oxygenated blood

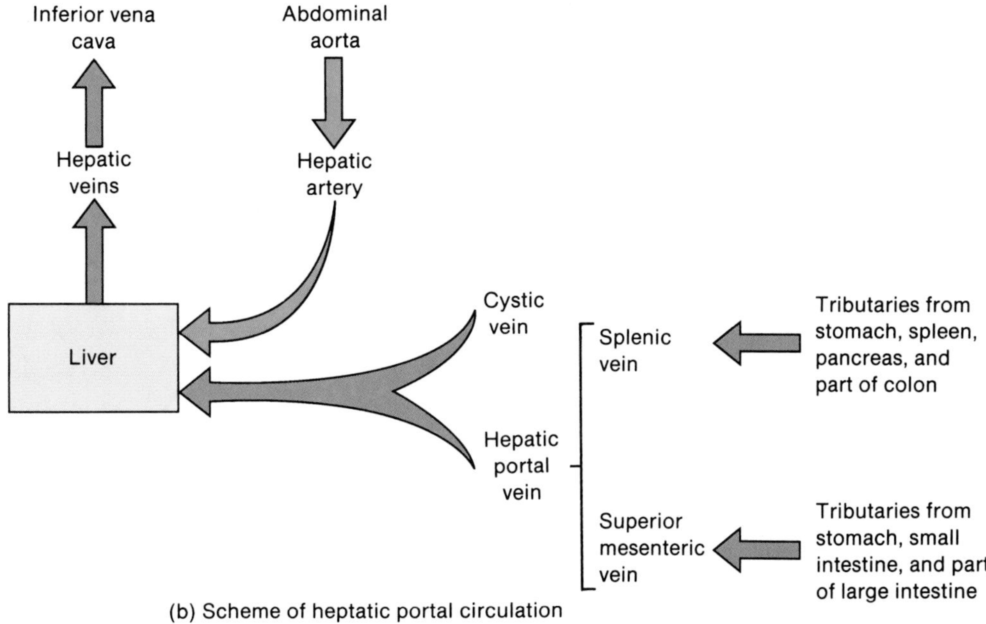

(b) Scheme of heptatic portal circulation

returns from the placenta via a single ***umbilical vein.*** This vein ascends to the liver of the fetus, where it divides into two branches. Some blood flows through the branch that joins the hepatic portal vein and enters the liver. Although the fetal liver manufactures red blood cells, it does not function in digestion. Therefore, most of the blood flows into the second branch, the ***ductus venosus*** (DUK-tus ve-NŌ-sus). The ductus venosus passes its blood to the inferior vena cava, bypassing the liver.

In general, circulation through other portions of the fetus is not unlike postnatal circulation. Deoxygenated blood returning from the lower regions is mingled with oxygenated

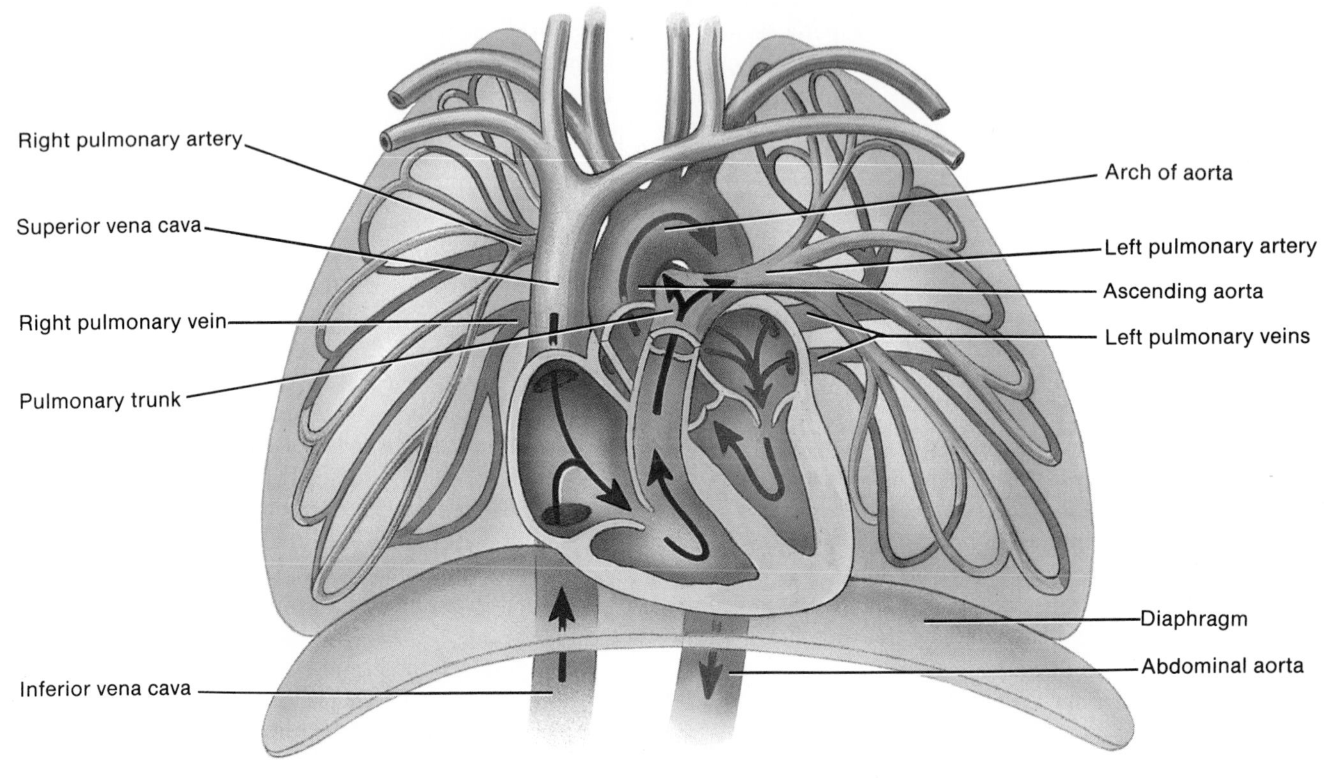

(a) Anterior view

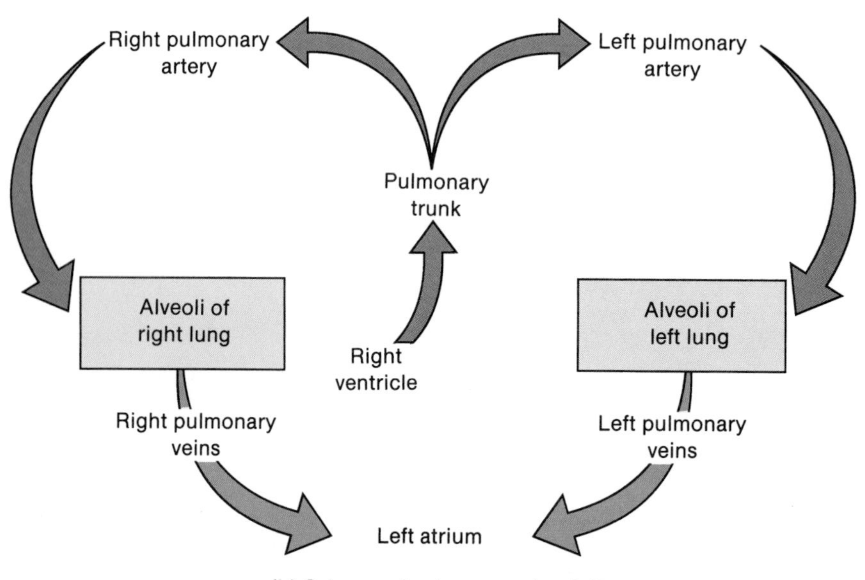

(b) Scheme of pulmonary circulation

FIGURE 14-17 Pulmonary circulation.

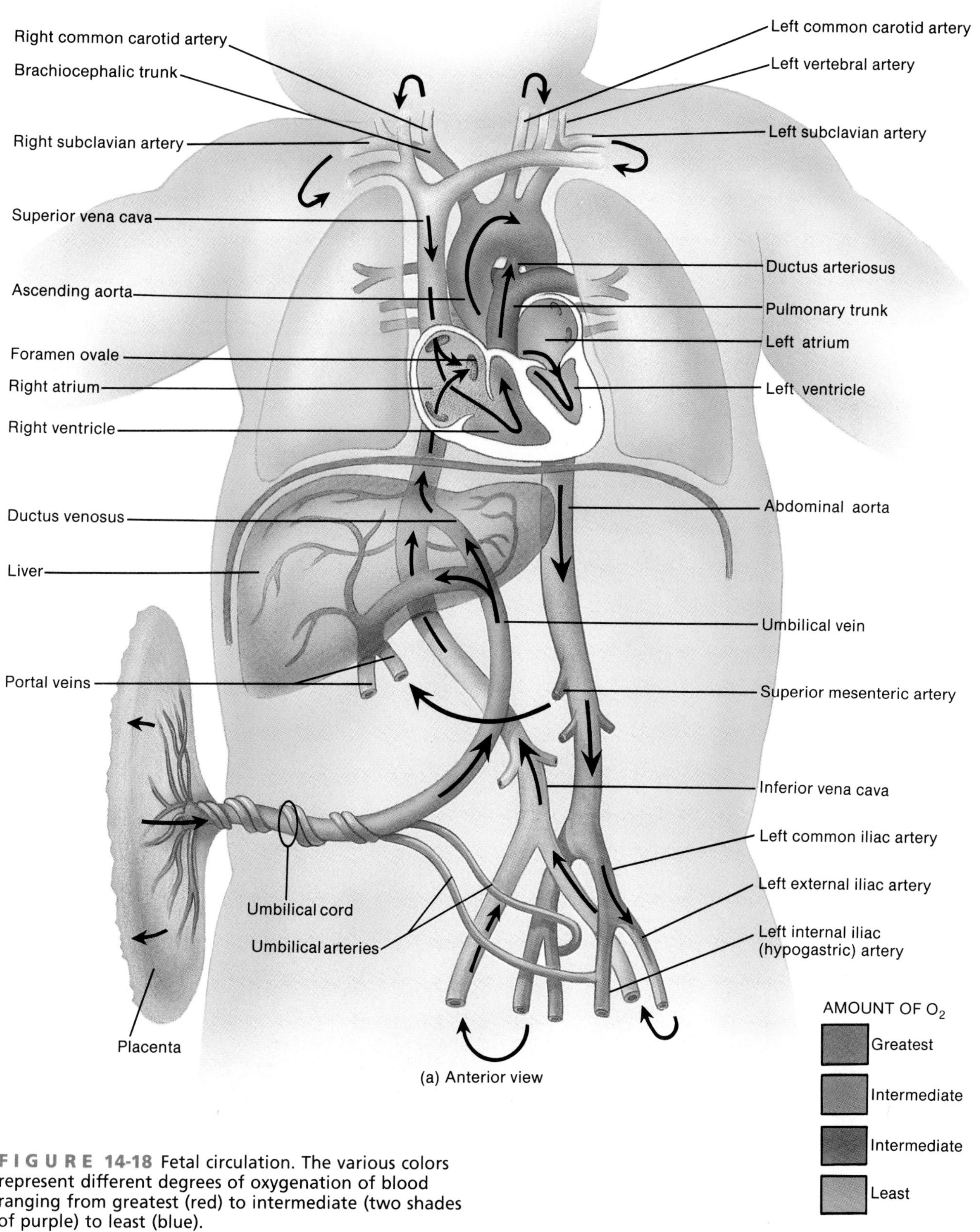

(a) Anterior view

FIGURE 14-18 Fetal circulation. The various colors represent different degrees of oxygenation of blood ranging from greatest (red) to intermediate (two shades of purple) to least (blue).

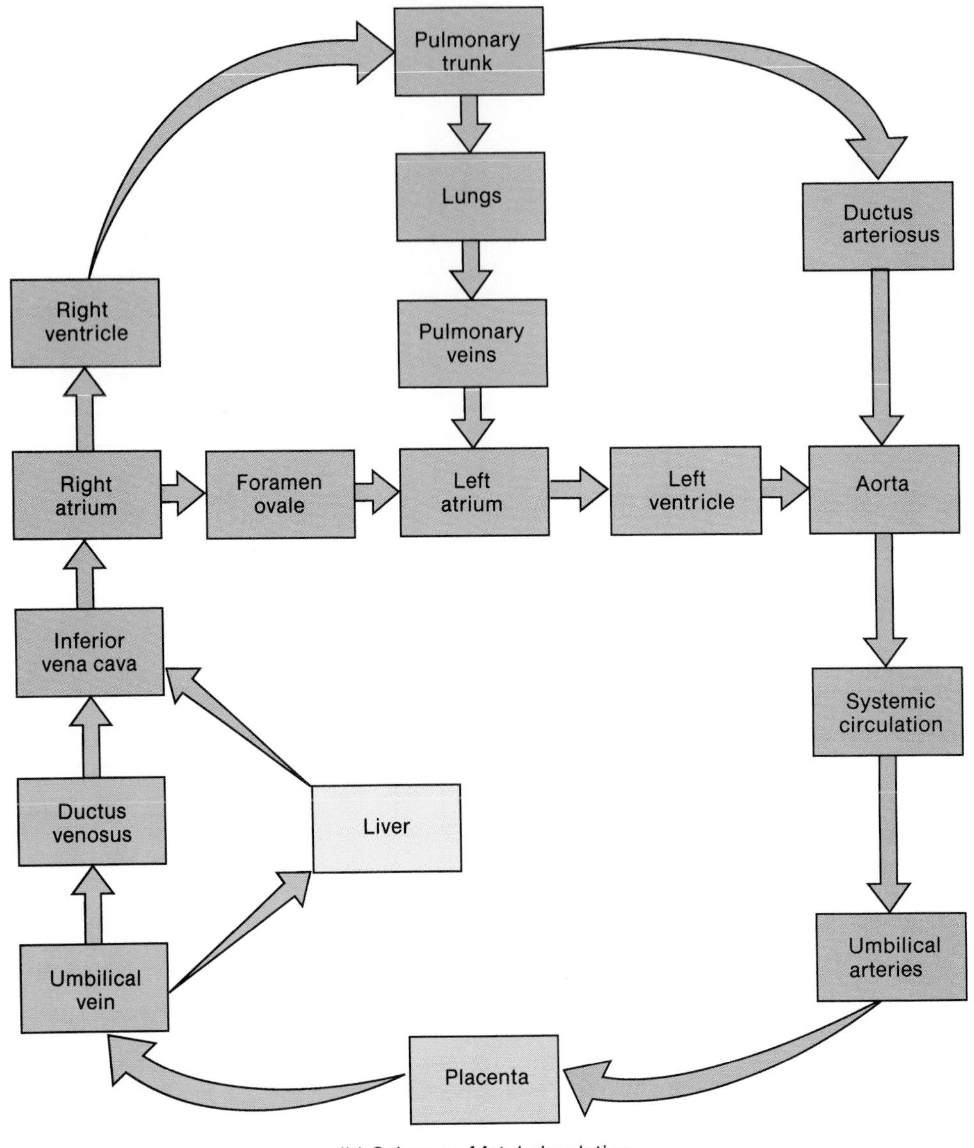

(b) Scheme of fetal circulation

blood from the ductus venosus in the inferior vena cava. This mixed blood then enters the right atrium. The circulation of blood through the upper portion of the fetus is also similar to postnatal flow. Deoxygenated blood returning from the upper regions of the fetus is collected by the superior vena cava, and it also passes into the right atrium.

Most of the blood does not pass through the right ventricle to the lungs, as it does in postnatal circulation, since the fetal lungs do not operate. In the fetus, an opening called the ***foramen ovale*** (fō-RĀ-men ō-VAL-ē) exists in the septum between the right and left atria. A valve in the inferior vena cava directs about a third of the blood through the foramen ovale so that it may be sent directly into the systemic circulation. The blood that does descend into the right ventricle is pumped into the pulmonary trunk, but little of this blood actually reaches the lungs. Most blood in the pulmonary trunk is sent through the ***ductus arteriosus*** (ar-tē-rē-Ō-sus). This vessel connecting the pulmonary trunk with the aorta enables most blood to bypass the fetal lungs. The blood in the aorta is carried to all parts of the fetus through its systemic branches. When the common iliac arteries branch into the external and internal iliacs, part of the blood flows into the internal iliacs. It then goes to the umbilical arteries and back to the placenta for another exchange of materials. The only vessel that carries fully oxygenated blood is the umbilical vein.

At birth, when pulmonary (lung), renal, digestive, and liver functions are established, the special structures of fetal circulation are no longer needed and the following changes occur.

1. The umbilical arteries vasoconstrict shut and atrophy to become the ***lateral umbilical ligaments.***

2. The umbilical vein vasoconstricts shut and becomes the ***ligamentum teres* (*round ligament*)** of the liver.
3. The placenta is delivered by the mother as the ***"afterbirth."***
4. The ductus venosus vasoconstricts shut and becomes the ***ligamentum venosum,*** a fibrous cord in the liver.
5. The foramen ovale normally closes shortly after birth to become the ***fossa ovalis,*** a depression in the interatrial septum.
6. The ductus arteriosus closes by vasoconstriction and atrophies and becomes the ***ligamentum arteriosum.***

Anatomical defects resulting from failure of these changes to occur are described in Chapter 13.

AGING AND THE CARDIOVASCULAR SYSTEM

General changes associated with aging and the cardiovascular system include loss of extensibility of the aorta, reduction in cardiac muscle fiber (cell) size, progressive loss of cardiac muscular strength, a reduced output of blood by the heart, a decline in maximum heart rate, and an increase in systolic blood pressure. Total blood cholesterol tends to increase with age, as does low-density lipoprotein (LDL); high-density lipoprotein (HDL) tends to decrease. There is an increase in the incidence of coronary artery disease (CAD), the major cause of heart disease and death in older Americans. Congestive heart failure (CHF), a set of symptoms associated with impaired pumping performance of the heart, also occurs. Changes in blood vessels, such as hardening of the arteries and cholesterol deposits in arteries that serve brain tissue, reduce nourishment to the brain and result in the malfunction or death of brain cells. By age 80, cerebral blood flow is 20 percent less and renal blood flow is 50 percent less than in the same person at age 30.

DEVELOPMENTAL ANATOMY OF BLOOD VESSELS AND BLOOD

Since the human egg and yolk sac have little yolk to nourish the developing embryo, blood and blood vessel formation starts as early as 15 to 16 days. The development begins in the ***mesoderm*** of the yolk sac, chorion, and body stalk (see Figure 26-6c).

Blood vessels develop from isolated masses and cords of mesenchyme in the mesoderm called ***blood islands*** (Figure 14-19). Spaces soon appear in the islands and become the lumens of the blood vessels. Some of the mesenchymal cells immediately around the spaces give rise to the *endothelial lining of the blood vessels.* Mesenchyme around the endothelium forms the *tunics* (intima, media, externa) of the larger blood vessels. Growth and fusion of blood islands form an extensive network of blood vessels throughout the embryo.

Blood plasma and blood cells are produced by the endothelial cells and appear in the blood vessels of the yolk sac and allantois quite early. Blood formation in the embryo itself begins at about the second month in the liver and spleen, a little later in bone marrow, and much later in lymph nodes.

APPLICATIONS TO HEALTH

HYPERTENSION

Hypertension, or high blood pressure, is the most common disease affecting the heart and blood vessels. Statistics indicate that hypertension afflicts one out of every five American adults. Although there is some disagreement as to what defines hypertension, a strong consensus has emerged suggesting that a blood pressure of 120/80 is normal and desirable in a healthy adult. Borderline high blood pressure is defined as diastolic pressure between 85 and 89. Mild high blood pressure is diastolic pressure between 90 and 104. Moderate high blood pressure is diastolic pressure between 105 and 114. Severe high blood pressure is diastolic pressure of 115 or higher. Isolated systolic hypertension is systolic pressure greater than 160 in those whose diastolic pressure is less than 90. As noted in Chapter 13, the lower the

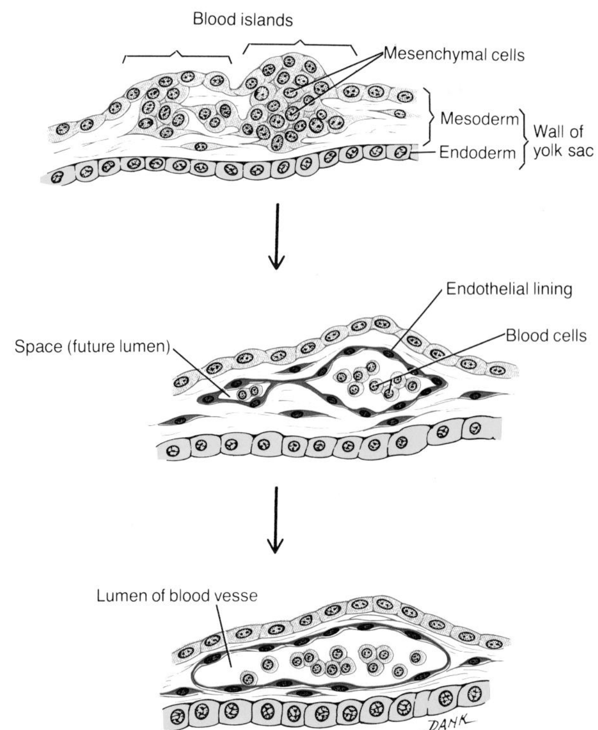

FIGURE 14-19 Development of blood vessels and blood cells from blood islands.

blood pressure, the less the risk of coronary artery disease (CAD). The lowest functioning pressure for an individual is the optimum pressure for that person.

Primary hypertension (essential) is a persistently elevated blood pressure that cannot be attributed to any particular organic cause. Approximately 90 to 95 percent of all hypertension cases fit this definition. The remaining percent are ***secondary hypertension.*** Secondary hypertension has an identifiable underlying cause such as kidney disease or adrenal hypersecretion.

High blood pressure is of considerable concern because of the harm it can do to the heart, brain, and kidneys if it remains uncontrolled. The heart is most commonly affected by high blood pressure. When pressure is high, the heart uses more energy in pumping against the increased resistance caused by the elevated arterial blood pressure. Because of the increased effort, the heart muscle thickens and the heart becomes enlarged. The heart also needs more oxygen. If it cannot meet the demands put on it, angina pectoris or even myocardial infarction may develop. Hypertension is also a factor in the development of atherosclerosis. Continued high blood pressure may produce a cerebral vascular accident (CVA), or stroke. In this case, severe strain has been imposed on the cerebral arteries that supply the brain. These arteries are usually less protected by surrounding tissues than are the major arteries in other parts of the body. These weakened cerebral arteries may finally rupture, and a brain hemorrhage follows.

The kidneys are also prime targets of hypertension. The principal site of damage is in the arterioles that supply them. The continual high blood pressure pushing against the walls of the arterioles causes them to thicken, thus narrowing the lumen. The blood supply to the kidneys is thereby gradually reduced. In response, the kidneys may secrete renin, which raises the blood pressure even higher and complicates the problem. The reduced blood flow to the kidney cells may eventually lead to the death of the cells.

Although medical science cannot cure primary hypertension, certain forms of secondary hypertension, if diagnosed early, can be cured by removing the underlying cause. And even in some cases of primary hypertension, measures such as weight loss in an obese person or salt reduction in one with a history of high salt (NaCl) intake may result in a dramatic reduction in blood pressure. Evidence suggests that less fat and more potassium and calcium may also lead to a reduction in blood pressure, whereas magnesium deficiency may cause hypertension. Since nicotine is a vasoconstrictor, it elevates blood pressure. Stopping smoking may help to decrease blood pressure. As indicated earlier, regular exercise can help to reduce hypertension. Since alcohol and caffeine raise blood pressure, lowering intake can help to reduce it. In recent years, some physicians have advocated relaxation techniques (yoga, meditation, and biofeedback) to treat hypertension. For those individuals who require medication, a number of drugs are available. Many people can be treated with *diuretics,* which eliminate large amounts of water and sodium, thus decreasing blood volume and reducing blood pressure. *Vasodilators* are often used in combination with diuretics. They relax the smooth muscle in arterial walls, causing vasodilation and thus lowering blood pressure by lowering peripheral resistance. *Beta blockers* reduce blood pressure by inhibiting secretion of renin and decreasing rate and force of heartbeat. Beta blockers are often used in combination with diuretics. *Calcium channel blockers* prevent movement of Ca^{2+} ions into blood vessels and heart cells, causing relaxation and lowering of blood pressure.

ANEURYSM

An ***aneurysm*** (AN-yoo-rizm) is a thin, weakened section of the wall of an artery or a vein that bulges outward, forming a balloonlike sac of the blood vessel. Common causes of aneurysms include atherosclerosis, syphilis, congenital blood vessel defects, and trauma. If an aneurysm goes untreated, it grows larger and larger until the blood vessel wall becomes so thin that it may burst, causing massive hemorrhage with shock, severe pain, stroke, or death, depending on which vessel is involved. Even an unruptured aneurysm can lead to damage by interrupting blood flow or putting pressure on adjacent blood vessels, organs, or bones.

Surgical repair of an aneurysm consists of temporarily clamping the damaged artery above and below the aneurysm and then excising (surgically removing) the aneurysm. A graft, usually of Dacron, is then sutured to healthy segments of the artery to reestablish normal blood flow.

CORONARY ARTERY DISEASE (CAD)

In Chapter 13 it was indicated that the common causes of heart disease are related to inadequate coronary blood supply, anatomical disorders, and arrhythmias. ***Coronary artery disease (CAD)*** is a condition in which the heart muscle receives inadequate blood because of an interruption of blood supply. Depending on the degree of interruption, symptoms can range from a mild chest pain to a full-scale heart attack. The underlying causes of CAD are many and varied. Two of the principal ones are atherosclerosis and coronary artery spasm, both of which were discussed in detail in Chapter 13.

DEEP-VENOUS THROMBOSIS (DVT)

Venous thrombosis, the presence of a thrombus (blood clot) in a vein, typically occurs in deep veins of the lower extremities. In such cases, the condition is referred to as ***deep-venous thrombosis (DVT).*** The two most serious complications of DVT are *pulmonary embolism,* in which the thrombus dislodges and finds its way into the pulmonary arterial blood flow, and *postphlebitic syndrome,* which consists of edema, pain, and skin changes due to destruction of venous valves. Diagnosis is by ultrasound and treatment consists of anticoagulant therapy, elevation of the extremity, fibrinolytic therapy (streptokinase or urokinase), and, in rare cases, thrombectomy.

KEY MEDICAL TERMS ASSOCIATED WITH BLOOD VESSELS

Aortography (ā′-or-TOG-ra-fē) X-ray examination of the aorta and its main branches after injection of radiopaque dye.

Arteritis (ar′-te-RĪ-tis; *itis* = inflammation of) Inflammation of an artery, probably due to an autoimmune response.

Carotid endarterectomy (ka-ROT-id end′-ar-ter-EK-tō-mē) The removal of atherosclerotic plaque from the carotid artery to restore greater blood flow to the brain.

Claudication (klaw′-di-KĀ-shun) Pain and lameness or limping caused by defective circulation of the blood in the vessels of the limbs.

Hypotension (hī′-pō-TEN-shun; *hypo* = below; *tension* = pressure) Low blood pressure; most commonly used to describe an acute drop in blood pressure, as occurs during excessive blood loss.

Normotensive (nor′-mō-TEN-siv) Characterized by normal blood pressure.

Occlusion (o-KLOO-shun) The closure or obstruction of the lumen of a structure such as a blood vessel.

Orthostatic (or′-tho-STAT-ik) **hypotension** (*ortho* = straight; *statikos* = causing to stand) An excessive lowering of systemic blood pressure with the assumption of an erect or semi-erect posture; it is usually a sign of a disease. May be caused by excessive fluid loss, certain drugs (antihypertensives), and cardiovascular or neurogenic factors. Also called **postural hypotension.**

Phlebitis (fle-BĪ-tis; *phleb* = vein) Inflammation of a vein, often in a leg.

Reynaud's (rā-NOZ) **disease** A vascular disorder, primarily of females, characterized by bilateral attacks of ischemia, usually of the fingers and toes, in which the skin becomes pale and exhibits burning and pain; it is brought on by cold or emotional stimuli.

Shunt A passage between two blood vessels or between the two sides of the heart.

Thrombectomy (throm-BEK-tō-mē; *thrombo* = clot) An operation to remove a blood clot from a blood vessel.

Thrombophlebitis (throm′-bō-fle-BĪ-tis) Inflammation of a vein with clot formation. Superficial thrombophlebitis occurs in veins under the skin, especially the calf.

White coat (office) hypertension A syndrome found in patients who have elevated blood pressures while being examined by health-care personnel, but are otherwise normotensive.

STUDY OUTLINE

Arteries (p. 378)

1. Arteries carry blood away from the heart. The wall of an artery consists of a tunica interna, tunica media (which maintains elasticity and contractility), and tunica externa.
2. Large arteries are referred to as elastic (conducting) arteries, and medium-sized arteries are called muscular (distributing) arteries.
3. Many arteries anastomose; that is, the distal ends of two or more vessels unite. An alternate blood route from an anastomosis is called collateral circulation. Arteries that do not anastomose are called end arteries.

Arterioles (p. 380)

1. Arterioles are small arteries that deliver blood to capillaries.
2. Through constriction and dilation, they assume a key role in regulating blood flow from arteries into capillaries and in altering peripheral resistance and thus arterial blood pressure.

Capillaries (p. 380)

1. Capillaries are microscopic blood vessels through which materials are exchanged between blood and tissue cells; some capillaries are continuous, whereas others are fenestrated.
2. Capillaries branch to form an extensive capillary network throughout the tissue. This network increases the surface area, allowing a rapid exchange of large quantities of materials.
3. Precapillary sphincters regulate blood flow through capillaries.
4. Microscopic blood vessels in the liver are called sinusoids.

Venules (p. 381)

1. Venules are small vessels that continue from capillaries and merge to form veins.
2. They drain blood from capillaries into veins.

Veins (p. 381)

1. Veins consist of the same three tunics as arteries but have less elastic tissue and smooth muscle.
2. They contain valves to prevent backflow of blood.
3. Weak valves can lead to varicose veins or hemorrhoids.
4. Vascular (venous) sinuses are veins with very thin walls.

Blood Reservoirs (p. 382)

1. Systemic veins are collectively called blood reservoirs.
2. They store blood, which through vasoconstriction can move to other parts of the body if the need arises.
3. The principal reservoirs are the veins of the abdominal organs (liver and spleen) and skin.

Circulatory Routes (p. 383)

1. The two principal postnatal circulatory routes are systemic and pulmonary.
2. Three of the several subdivisions of the systemic circulation are coronary (cardiac) circulation, hepatic portal circulation, and cerebral.
3. Fetal circulation occurs in the fetus.

Systemic Circulation (p. 384)

1. The systemic circulation takes oxygenated blood from the left ventricle through the aorta to all parts of the body, including lung tissue (but does *not* supply the air sacs of the lungs) and returns the deoxygenated blood to the right atrium.
2. The aorta is divided into the ascending aorta, the arch of the aorta, and the descending aorta. Each section gives off arteries that branch to supply the whole body.
3. Blood is returned to the heart through the systemic veins. All the veins of the systemic circulation flow into either the superior or inferior venae cavae or the coronary sinus. They in turn empty into the right atrium.

Hepatic Portal Circulation (p. 416)

1. The hepatic portal circulation collects blood from the veins of the pancreas, spleen, stomach, intestines, and gallbladder and directs it into the hepatic portal vein of the liver.
2. This circulation enables the liver to utilize nutrients and detoxify harmful substances in the blood.

Pulmonary Circulation (p. 417)

1. The pulmonary circulation takes deoxygenated blood from the right ventricle to the air sacs of the lungs and returns oxygenated blood from the lungs to the left atrium.
2. It allows blood to be oxygenated for systemic circulation.

Fetal Circulation (p. 417)

1. The fetal circulation involves the exchange of materials between fetus and mother.
2. The fetus derives its oxygen and nutrients and eliminates its carbon dioxide and wastes through the maternal blood supply by means of a structure called the placenta.
3. At birth, when pulmonary (lung), digestive, and liver functions are established, the special structures of fetal circulation are no longer needed.

Aging and the Cardiovascular System (p. 421)

1. General changes include loss of elasticity of blood vessels, reduction in cardiac muscle size, and reduced cardiac output.
2. The incidence of coronary artery disease (CAD), congestive heart failure (CHF), and atherosclerosis increases with age.

Developmental Anatomy of Blood and Blood Vessels (p. 421)

1. Blood vessels develop from isolated masses of mesenchyme in mesoderm called blood islands.
2. Blood is produced by the endothelium of blood vessels.

Applications to Health (p. 421)

1. Hypertension, or high blood pressure, is classified as primary and secondary.
2. An aneurysm is a thin, weakened section of the wall of an artery or vein that bulges outward, forming a balloonlike sac.
3. Deep-venous thrombosis (DVT) refers to a blood clot in a deep vein, especially in lower extremities.

REVIEW QUESTIONS

1. Describe the structural and functional differences among arteries, arterioles, capillaries, venules, and veins. (p. 378)
2. Discuss the importance of the elasticity and contractility of arteries. (p. 378)
3. Distinguish between elastic (conducting) and muscular (distributing) arteries in terms of location, histology, and function. What is an anastomosis? What is collateral circulation? (p. 378)
4. Describe how capillaries are structurally adapted for exchanging materials between blood and body cells. (p. 380)
5. Define varicose veins and describe how they are treated. (p. 381)
6. What are blood reservoirs? Why are they important? (p. 382)
7. What is meant by a circulatory route? Define systemic circulation. (p. 383)
8. Diagram the major divisions of the aorta, their principal arterial branches, and the regions supplied. (p. 385)
9. Trace a drop of blood from the arch of the aorta through its systemic circulatory route to the tip of the big toe on your left foot and back to the heart again. Remember that the major branches of the arch are the brachiocephalic artery, left common carotid artery, and left subclavian artery. Be sure to also indicate which veins return the blood to the heart. (p. 386)
10. What is the cerebral arterial circle (circle of Willis)? Why is it important? (p. 390)
11. What are visceral branches of an artery? Parietal branches? What major organs are supplied by branches of the thoracic aorta? How is blood returned from these organs to the heart? (p. 392)
12. What organs are supplied by the celiac, superior mesenteric, renal, inferior mesenteric, inferior phrenic, and middle sacral arteries? How is blood returned to the heart? (p. 396)
13. Trace a drop of blood from the brachiocephalic artery into the digits of the right upper extremity and back again to the right atrium. (p. 394)
14. What are the three major groups of systemic veins? (p. 402)
15. What is hepatic portal circulation? Describe the route by means of a diagram. Why is this route significant? (p. 416)
16. Define pulmonary circuation. Prepare a diagram to indicate the route. What is the purpose of the route? (p. 417)
17. Discuss in detail the anatomy and physiology of fetal circulation. Be sure to indicate the function of the umbilical arteries, umbilical vein, ductus venosus, foramen ovale, and ductus arteriosus. (p. 417)
18. Describe the effects of aging on the cardiovascular system. (p. 421)
19. Describe the development of blood vessels and blood. (p. 421)
20. How does hypertension affect the body? How is hypertension treated? (p. 421)
21. What is an aneurysm? Why is an aneurysm a serious problem? (p. 422)
22. What is deep-venous thrombosis (DVT)? (p. 422)
23. Refer to the glossary of key medical terms associated with blood vessels. Be sure that you can define each term. (p. 423)

SELF QUIZ

Complete the following:

1. For each of the following pairs of fetal vessels, circle the vessel with higher oxygen content. A. umbilical artery/umbilical vein; B. ductus arteriosus/ductus venosus; C. femoral artery/femoral vein; D. pulmonary artery/pulmonary vein; E. aorta/thoracic portion of inferior vena cava.
2. Virtually all blood from gastrointestinal organs, as well as blood from the spleen, empties into veins that lead to the single ________ vein. This vessel enters the undersurface of the liver.
3. The aorta ends at about the level of the ________ vertebra by dividing into right and left ________.
4. Besides supplying the arms, the subclavian arteries each send a branch that ascends the neck through foramina in cervical vertebrae. These are the ________ arteries. They join at the base of the brain to form the ________ artery.
5. The major artery from which all systemic arteries branch is the ________. It exits from the chamber of the heart known as the (right? left?) ventricle. The first arteries to branch off this artery are the ________ arteries.
6. The three branches of the celiac trunk are the ________, ________, and ________.
7. Name the three main vessels that empty venous blood into the right atrium of the heart:
A. ________ B. ________ C. ________
8. Blood from all the cranial vascular sinuses eventually drains into the ________ veins, which descend in the neck.
9. In order for blood to flow from the left brachial veins to the left brachial artery, which of these must it pass through? (a brachial capillary? the heart and a lung?)
10. In order for blood to flow from the left brachial vein to the right arm, must it pass through the heart? (Yes? No?) Both sides of the heart, that is, right and left? (Yes? No?) One lung? (Yes? No?)

Choose *all* correct answers to questions 11 and 12.

___ **11.** In the most direct route from the left leg to the left arm of an adult, blood must pass through all these structures: A. inferior vena cava; B. brachiocephalic artery; C. capillaries in lung; D. hepatic portal vein; E. left subclavian artery; F. right ventricle of heart; G. left external iliac vein.

___ **12.** In the most direct route from the fetal right ventricle to the fetal left leg, blood must pass through all these structures:
A. aorta; B. umbilical artery; C. lung; D. ductus arteriosus; E. ductus venosus; F. left ventricle; G. left common iliac artery.

Choose the one best answer to these questions.

___ **13.** Choose the false statement:
A. in order for blood to pass from a vein to an artery, it must pass through chambers of the heart; B. in its passage from an artery to a vein a red blood cell must ordinarily travel through a capillary; C. the wall of the femoral artery is thicker than the wall of the femoral vein; D. most of the smooth muscle in arteries is in the tunica interna.

___ **14.** Choose the false statement:
A. arteries contain valves, but veins do not; B. decrease in the size of the lumen of a blood vessel by contraction of a smooth muscle is called vasoconstriction; C. end arteries are vessels that do not anastomose; D. sinusoids are wider and more tortuous (winding) than capillaries.

___ **15.** A small vessel connecting the pulmonary trunk with the aorta and bypassing the fetal lungs is the
A. foramen ovale; B. ductus venosus; C. ductus arteriosus; D. fossa ovalis; E. vasa vasorum.

___ **16.** Which statement best describes arteries?
A. all carry oxygenated blood to the heart; B. all contain valves to prevent the backflow of blood; C. all carry blood away from the heart; D. only large arteries are lined with endothelium; E. all branch from the descending aorta.

___ **17.** Which statement is *not* true of veins?
A. their tunica interna is thinner than in arteries; B. their tunica externa is thicker than in arteries; C. most veins in the extremities have valves; D. they always carry deoxygenated blood; E. all ultimately empty into the inferior vena cava.

___ **18.** All these vessels are in the leg or foot *except* the
A. saphenous vein; B. azygos vein; C. peroneal artery; D. dorsalis pedis artery; E. popliteal artery.

___ **19.** Which vessel returns blood to the heart from systemic circulation?
A. pulmonary artery; B. pulmonary vein; C. superior vena cava; D. aorta; E. subclavian artery.

___ **20.** In fetal circulation, blood passes from the right atrium to the left atrium through the
A. ductus venosus; B. ductus arteriosus; C. umbilical vein; D. foramen ovale; E. umbilical artery.

___ **21.** Which of the following are involved in pulmonary circulation?
A. superior vena cava, right atrium, and left ventricle; B. inferior vena cava, right atrium, left ventricle; C. right ventricle, pulmonary trunk, and left atrium; D. left ventricle, aorta, and inferior vena cava; E. superior vena cava, right atrium, right ventricle.

___ **22.** In fetal circulation, the blood containing the highest amount of oxygen is found in the
A. umbilical arteries; B. ductus venosus; C. aorta; D. umbilical vein; E. ductus arteriosus.

___ **23.** Which of the following is/are true?
(1) The left common carotid artery branches off the brachiocephalic artery.
(2) The right subclavian artery branches off the brachiocephalic artery.
(3) The right common carotid artery branches off the brachiocephalic artery.
A. (1) only; B. (2) only; C. (3) only; D. (2) and (3); E. none of the above.

___ **24.** Which of the following arteries do(es) *not* arise from the arch of the aorta?
A. brachiocephalic; B. left common carotid; C. right common carotid; D. left subclavian; E. all of the above *do* arise from the arch of the aorta.

___ **25.** Blood is supplied to the pelvic viscera by way of the A. inferior vena cava; B. anterior tibial artery; C. external iliac artery; D. internal iliac artery; E. femoral artery.

___ **26.** The superior vena cava
(1) is one of the three veins carrying blood to the right atrium.
(2) is formed by joining of the right and left brachiocephalic veins.
(3) is joined by the azygos vein in the upper posterior part of the thoracic cavity.
A. (1) only; B. (2) only; C. (3) only; D. all of the above; E. (1) and (2).

___ **27.** The vessel that brings blood from the liver and other abdominal organs to the heart is
A. the portal vein; B. the hepatic vein; C. the abdominal aorta; D. the inferior vena cava; E. none of the above.

___ **28.** From which parts of the body does blood come, draining into the right brachiocephalic vein?
A. head, right arm and shoulder, upper right thoracic region; B. head and neck, both arms and shoulders; C. right side of head and neck; D. right side of the body; E. right side of head and neck, right arm and shoulder, upper right thoracic region.

Arrange the answers in correct sequence.

___ ___ ___ ___ **29.** Route of a drop of blood from the right side of the heart to the left side of the heart:
A. pulmonary artery
B. arterioles
C. capillaries
D. venules and veins

___ ___ ___ ___ ___ **30.** Route of a drop of blood from small intestine to heart:
A. superior mesenteric vein
B. hepatic portal vein
C. small vessels within the liver
D. hepatic vein
E. inferior vena cava

The Lymphatic System

15

STUDENT OBJECTIVES

1. Describe the components of the lymphatic system and list their functions.
2. Describe the structure and origin of lymphatic vessels and contrast them with veins.
3. Describe the histological aspects of lymph nodes and explain their functions.
4. Trace the general plan of lymph circulation from lymphatic vessels into the thoracic duct or right lymphatic duct.
5. Describe the principal lymph nodes of the head and neck, extremities, and trunk, their location and the areas they drain.
6. Describe the effects of aging on the lymphatic system.
7. Describe the development of the lymphatic system.
8. Describe the clinical symptoms of the following disorders: acquired immune deficiency syndrome (AIDS), autoimmune diseases, severe combined immunodeficiency (SCID), hypersensitivity (allergy), and Hodgkin's disease (HD).
9. Define key medical terms associated with the lymphatic system.

CHAPTER OUTLINE

- **Lymphatic Vessels**
- **Lymphatic Tissue**

Lymph Nodes
Tonsils
Spleen
Thymus Gland

- **Lymph Circulation**

Route
Thoracic (Left Lymphatic) Duct
Right Lymphatic Duct
Maintenance

- **Principal Groups of Lymph Nodes**
- **Aging and the Immune System**
- **Developmental Anatomy of the Lymphatic System**
- **Applications to Health**
- **Key Medical Terms Associated with the Lymphatic System**

The ***lymphatic*** (lim-FAT-ik) ***system*** consists of a fluid called lymph, vessels that transport lymph called lymphatic vessels (lymphatics), and a number of structures and organs that contain lymphatic (lymphoid) tissue (see Figure 15-2b). Essentially, lymphatic tissue is a specialized form of reticular connective tissue that contains large numbers of lymphocytes. The stroma (framework) of most lymphatic tissue is a meshwork of reticular fibers and reticular cells (fibroblasts and fixed macrophages).

Lymphatic tissue occurs in the body in various ways. Lymphatic tissue not enclosed by a capsule is referred to as ***diffuse lymphatic tissue.*** It is the simplest form of lymphatic tissue and is found in the lamina propria (connective tissue) of mucous membranes of the gastrointestinal tract, respiratory passageways, urinary tract, and reproductive tract. It is also normally found in small amounts in the stroma of almost every organ of the body.

Lymphatic nodules also do not have capsules and are oval-shaped concentrations of lymphatic tissue that usually consist of a central, lighter-staining region consisting of large lymphocytes (***germinal center***) and a peripheral, darker-staining region of small lymphocytes (***cortex***). Most lymphatic nodules are solitary, small, and discrete. Such nodules are found randomly in the lamina propria of mucous membranes of the gastrointestinal tract, respiratory passageways, urinary tract, and reproductive tract. Some lymphatic nodules occur in multiple, large aggregations in specific parts of the body. Among these are the tonsils in the pharyngeal region and aggregated lymphatic follicles (Peyer's patches) in the ileum of the small intestine (Chapter 23). Aggregations of lymphatic nodules also occur in the appendix.

Lymphatic organs of the body—the lymph nodes, spleen, and thymus gland—all contain lymphatic tissue enclosed by a connective tissue capsule. Since bone marrow produces lymphocytes, it may also be considered a component of the lymphatic system.

The lymphatic system has several functions:

1. Lymphatic vessels drain tissue spaces that have protein-containing fluid (interstitial fluid) that escapes from blood capillaries. The proteins, which cannot be directly reabsorbed by blood vessels, are returned to the cardiovascular system by lymphatic vessels.
2. Lymphatic vessels transport fats from the gastrointestinal tract to the blood.
3. Lymphatic tissue functions in surveillance and defense; that is, lymphocytes, with the aid of macrophages, protect the body from foreign cells, microbes, and cancer cells. Lymphocytes recognize foreign cells and substances, microbes, and cancer cells and respond to them in two general ways. Some lymphocytes (T cells) destroy them directly or indirectly by releasing various substances. Other lymphocytes (B cells) differentiate into plasma cells that secrete antibodies against foreign substances to help eliminate them.

Overall, the lymphatic system concentrates foreign substances in certain lymphatic organs, circulates lymphocytes through the organs to make contact with the foreign substances, and destroys the foreign substances and eliminates them from the body.

The developmental anatomy of the lymphatic system is considered later in the chapter.

LYMPHATIC VESSELS

Lymphatic vessels originate as ***lymph capillaries,*** microscopic vessels in spaces between cells (Figure 15-1a). Lymph capillaries may occur singly or in extensive plexuses. They are found throughout the body, with the exception of avascular tissue, the central nervous system, splenic pulp, and bone marrow. They are slightly larger and more permeable in one direction only than blood capillaries.

Lymph capillaries also differ from blood capillaries in that they end blindly; blood capillaries have an arterial and a venous end. In addition, lymph capillaries are structurally adapted to ensure the return of proteins to the cardiovascular system when they leak out of blood capillaries. Close examination of lymph capillaries reveals that there are minute openings between endothelial cells making up the capillary wall that permit fluid to flow easily into the capillary but prevent the flow of fluid out of the capillary, much like a one-way valve would operate (Figure 15-1b). Note also that the outer surfaces of the endothelial cells of the capillary wall are attached to the surrounding tissue by structures called ***anchoring filaments.*** During edema, there is an excessive accumulation of fluid in the tissue, causing tissue swelling. This swelling produces a pull on the anchoring filaments, making the openings between cells even larger so that more fluid can flow into the lymph capillary.

Just as blood capillaries converge to form venules and veins, lymph capillaries unite to form larger and larger lymph vessels called ***lymphatic vessels*** (Figure 15-2a,b). Lymphatic vessels resemble veins in structure but have thinner walls and more valves and contain lymph nodes at various intervals along their length. Lymphatic vessels of the skin travel in subcutaneous tissue and generally follow veins. Lymphatic vessels of the viscera generally follow arteries, forming plexuses around them. Ultimately, lymphatic vessels deliver their lymph into two main channels—the thoracic duct and the right lymphatic duct. These ducts will be described shortly.

CLINICAL APPLICATION

Lymphangiography

Lymphangiography (lim-fan'-jē-OG-ra-fē) is the x-ray examination of lymphatic vessels and lymph organs after they are filled with a radiopaque substance. Such an x-ray is called a ***lymphangiogram*** (lim-FAN-jē-ō-gram). Lymphangiograms are used to detect and stage

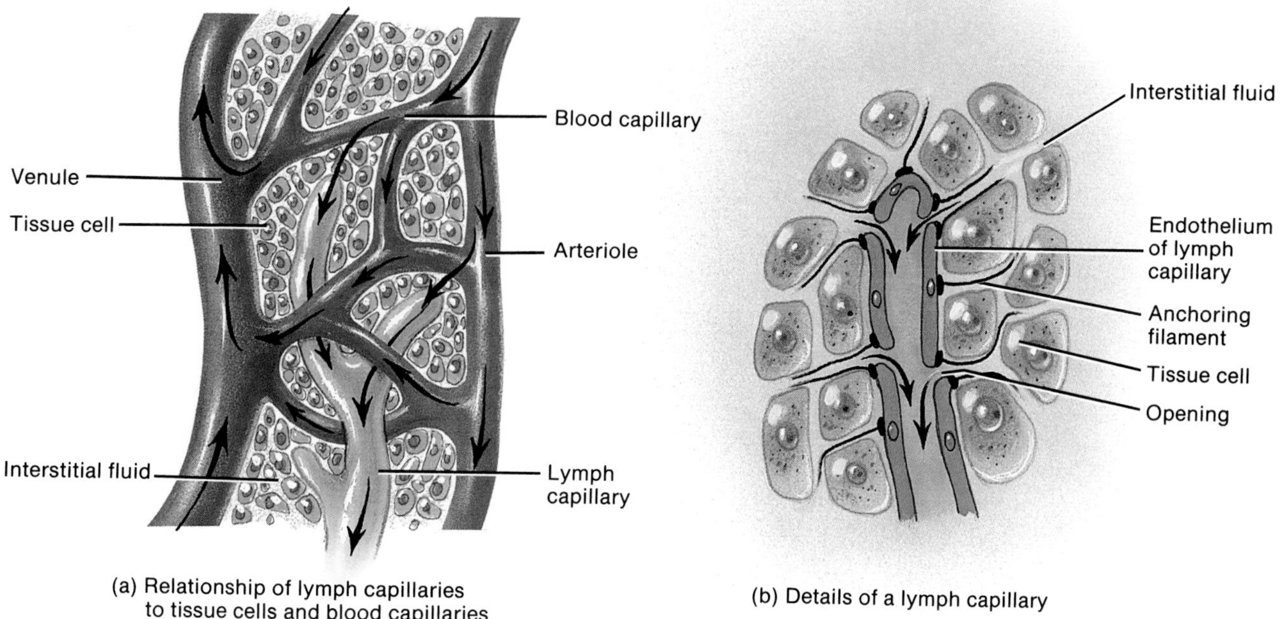

(a) Relationship of lymph capillaries to tissue cells and blood capillaries

(b) Details of a lymph capillary

FIGURE 15-1 Lymph capillaries.

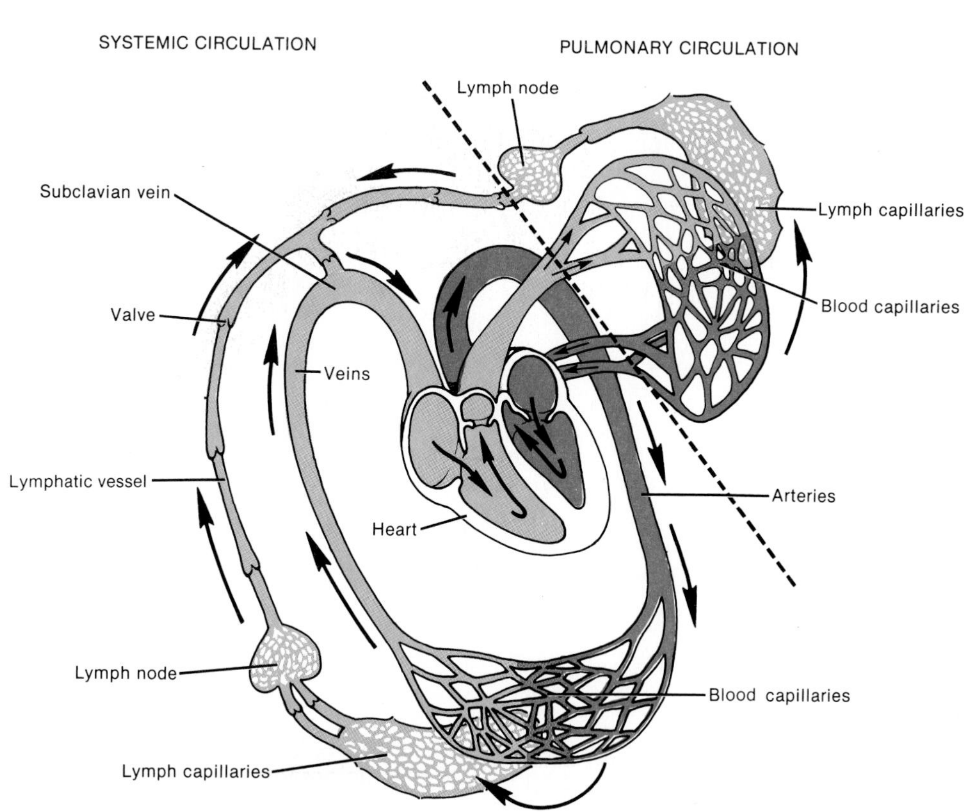

(a) Relationship of lymphatic system to cardiovascular system

FIGURE 15-2 Lymphatic system. The gold area in (c) indicates those portions of the body drained by the right lymphatic duct. The light green area of the body is drained by the thoracic duct.

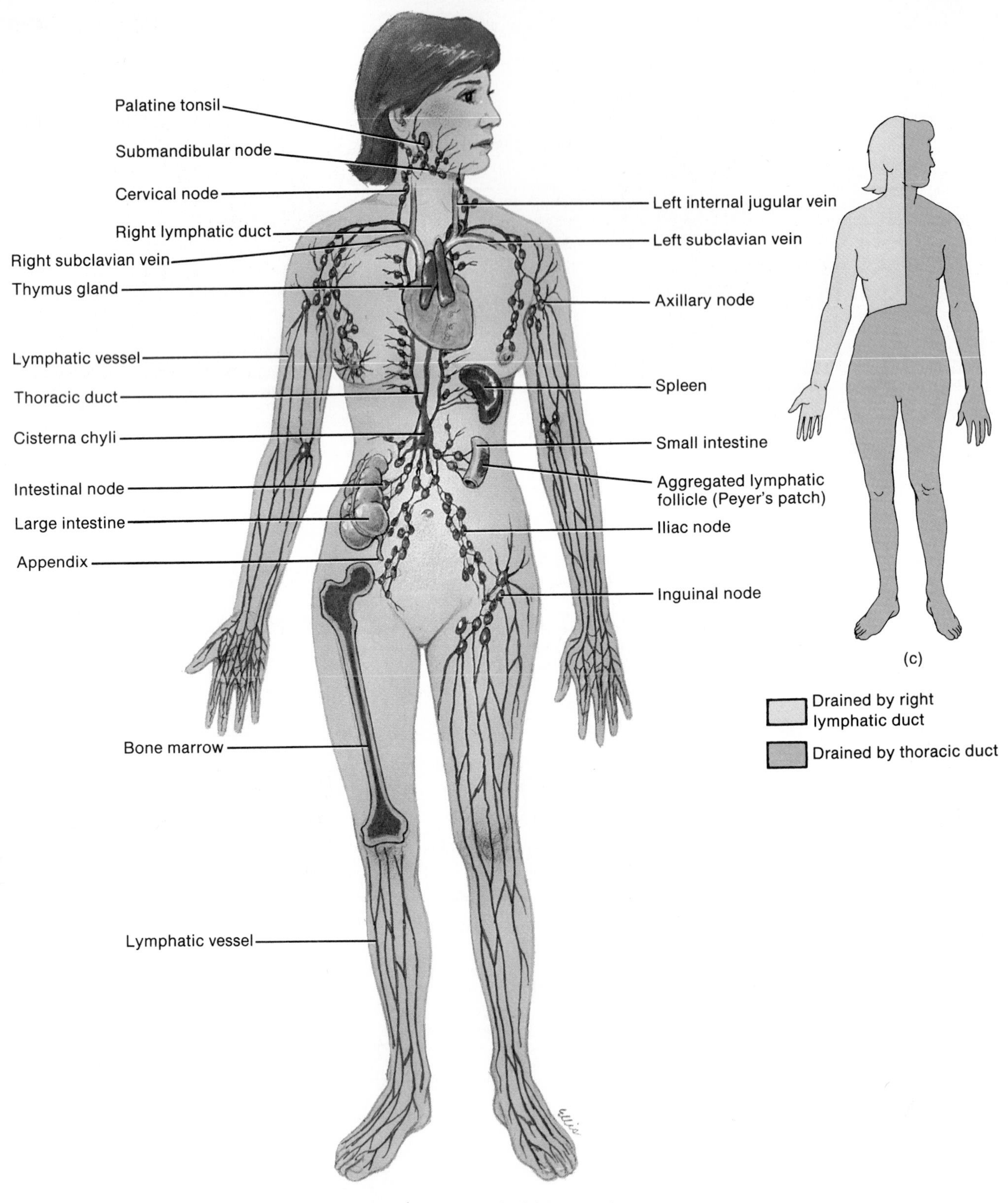

(b) Anterior view of principal components of lymphatic system

lymphomas (lymphatic tissue tumors), to identify cancerous involvement of lymph nodes, to help determine the cause of certain forms of edema, to locate enlarged lymph nodes for surgical or radiotherapeutic treatment, and to evaluate the effectiveness of chemotherapy and radiation therapy in treating cancer.

LYMPHATIC TISSUE

LYMPH NODES

The oval or bean-shaped structures located along the length of lymphatic vessels are called ***lymph nodes***. They range from 1 to 25 mm (0.04 to 1 in.) in length. A lymph node contains a slight depression on one side called a ***hilus*** (HĪ-lus), or ***hilum,*** where blood vessels and efferent lymphatic vessels leave the node (Figure 15-3a). Each node is covered by a ***capsule*** of dense connective tissue that extends into the node. The capsular extensions are called ***trabeculae*** (tra-BEK-yoo-lē). Internal to the capsule is a supporting network of reticular fibers and reticular cells (fibroblasts and macrophages). The capsule, trabeculae, and reticular fibers and cells constitute the stroma (framework) of a lymph node. The interior of a lymph node is specialized into two regions: cortex and medulla. The outer ***cortex*** contains densely packed lymphocytes arranged in masses called ***lymphatic nodules.*** The nodules often contain lighter-staining central areas, the ***germinal centers,*** where lymphocytes are produced. The inner region of a lymph node is called the ***medulla.*** In the medulla, the lymphocytes are arranged in strands called ***medullary cords.*** These cords also contain macrophages and plasma cells.

The circulation of lymph through a node involves afferent lymphatic vessels, sinuses in the node, and efferent lymphatic vessels. ***Afferent lymphatic vessels*** enter the convex surface of the node at several points. They contain valves that open toward the node so that the lymph is directed *inward.* Once inside the node, the lymph enters the sinuses, which are a series of irregular channels. Lymph from the afferent lymphatic vessels enters the ***cortical sinuses*** just inside the capsule. From here it circulates to the ***medullary***

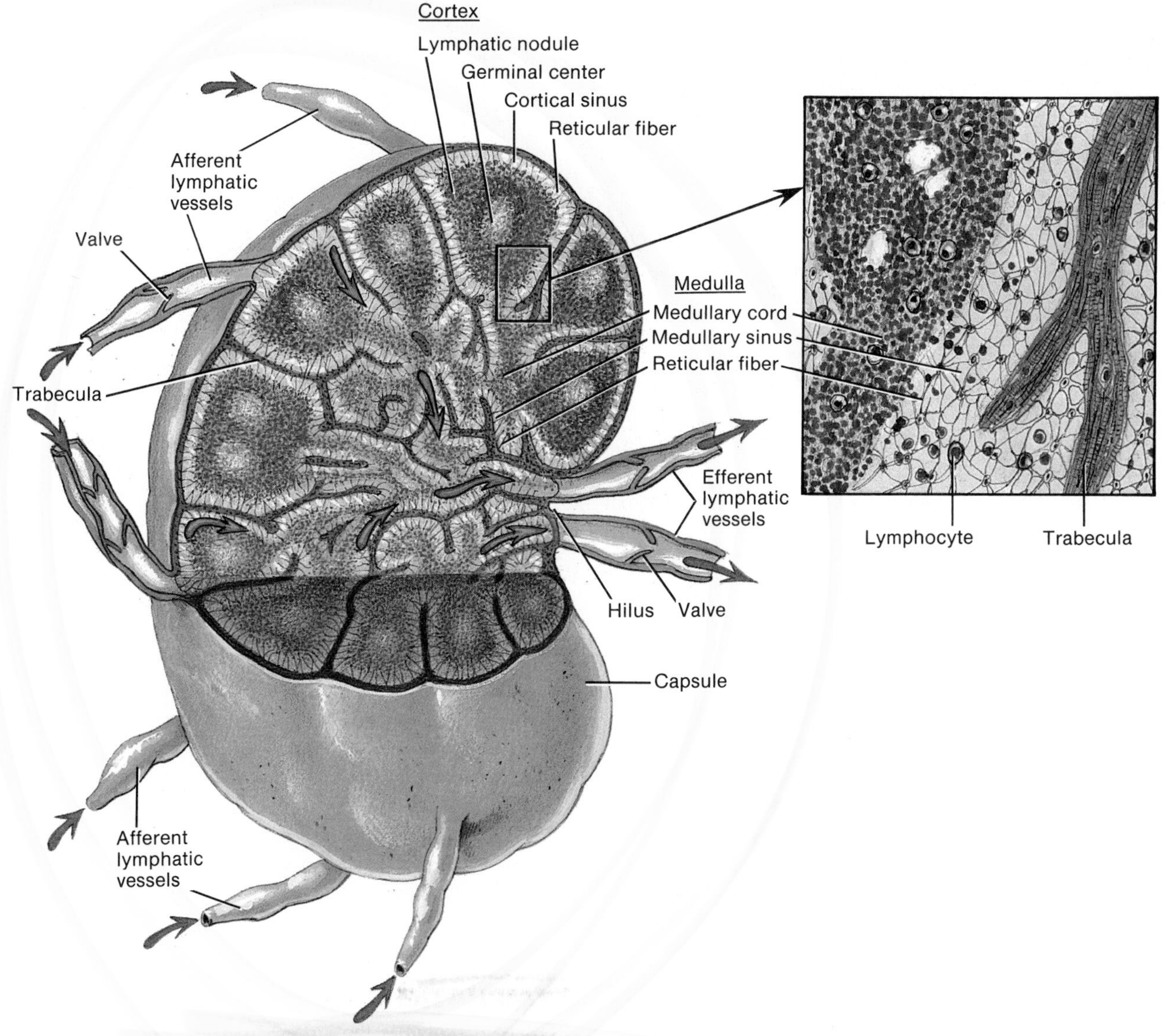

(a) Sectioned lymph node showing structure and path of lymph

FIGURE 15-3 Structure of a lymph node. (b) Courtesy of Lester Bergman and Associates. (c) Courtesy of Leroy, Biocosmos, Photo Researchers.

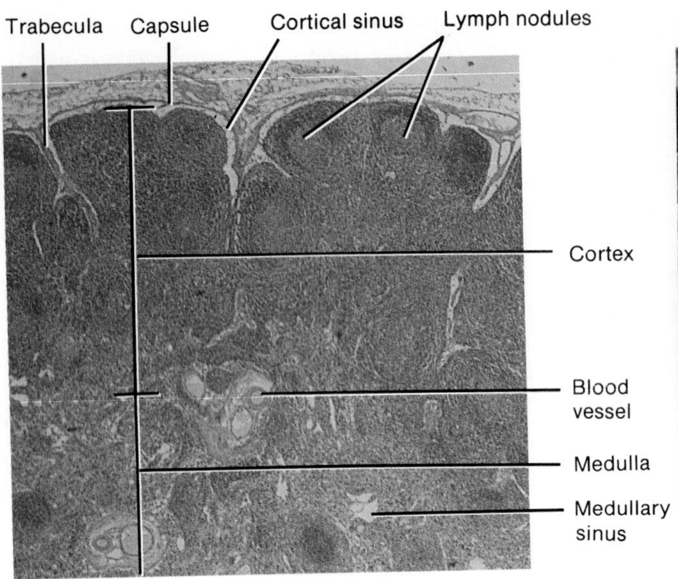

(b) Photomicrograph of a portion of a lymph node, (17 x)

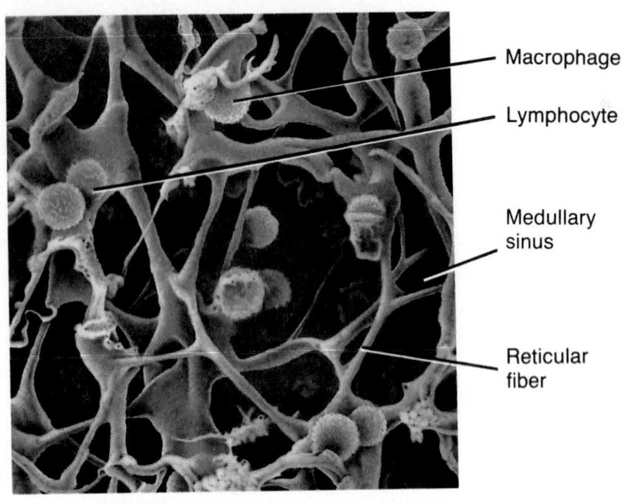

(c) Illustration of a scanning electron micrograph of a portion of a lymph node, 1,000×

sinuses between the medullary cords. From these sinuses the lymph usually circulates into one or two ***efferent lymphatic vessels.*** Efferent lymphatic vessels are wider than the afferent vessels and contain valves that open away from the node to convey lymph *out* of the node.

Lymph nodes are scattered throughout the body, usually in groups (see Figure 15-2b). Typically, these groups are arranged in two sets: ***superficial*** and ***deep.***

Lymph passing from tissue spaces through lymphatic vessels on its way back to the cardiovascular system is filtered through lymph nodes. As lymph passes through the nodes, it is filtered of foreign substances. These substances are trapped by the reticular fibers within the node. Then macrophages destroy the foreign substances by phagocytosis. T cells may destroy them by releasing various products, and/or B cells develop into plasma cells that produce antibodies that destroy them. Lymph nodes also produce lymphocytes, some of which can circulate to other parts of the body.

Histological features of a lymph node are shown in Figure 15-3b.

CLINICAL APPLICATION

Metastasis Through Lymphatic System

Knowledge of the location of the lymph nodes and the direction of lymph flow is important in the diagnosis and prognosis of the spread of cancer by ***metastasis.*** Cancer cells may be spread by way of the lymphatic system and produce aggregates of tumor cells where they lodge. Such secondary tumor sites are predictable by the direction of lymph flow from the organ primarily involved. (Cancer may also spread via the cardiovascular system and local extension.) The hallmark of a lymph node that is cancerous is that it feels enlarged, firm, and nontender. Most lymph nodes that undergo infectious enlargement, by contrast, are not firm and are very tender.

TONSILS

Tonsils are multiple aggregations of large lymphatic nodules embedded in a mucous membrane. The tonsils are arranged in a ring at the junction of the oral cavity and pharynx. The ***pharyngeal*** (fa-RIN-jē-al) ***tonsil*** or ***adenoid*** is a group of lymphatic nodules embedded in the posterior wall of the nasopharynx (see Figure 22-2a). The paired ***palatine*** (PAL-a-tīn) ***tonsils*** are situated in the tonsillar fossae between the pharyngopalatine and palatoglossal arches (see Figure 23-4). They are the ones commonly removed by a tonsillectomy. The ***lingual*** (LIN-gwal) ***tonsil*** is an aggregation of lymphatic nodules located at the base of the tongue and may also have to be removed by a tonsillectomy (see Figure 20-11b).

The tonsils are situated strategically to protect against invasion of foreign substances that are inhaled or ingested. Functionally, the tonsils produce lymphocytes and antibodies.

SPLEEN

The oval ***spleen*** is the largest mass of lymphatic tissue in the body, measuring about 12 cm (5 in.) in length. It is situated in the left hypochondriac region between the fundus of the stomach and diaphragm (Figure 15-4a). See also Figure 1-8d. Its ***visceral surface*** (Figure 15-4b) contains

Esophagus (cut)
Splenic artery
Splenic vein
Left renal vein
Inferior mesenteric vein
Spleen
Diaphragm
Tail of pancreas
Left kidney
Colon

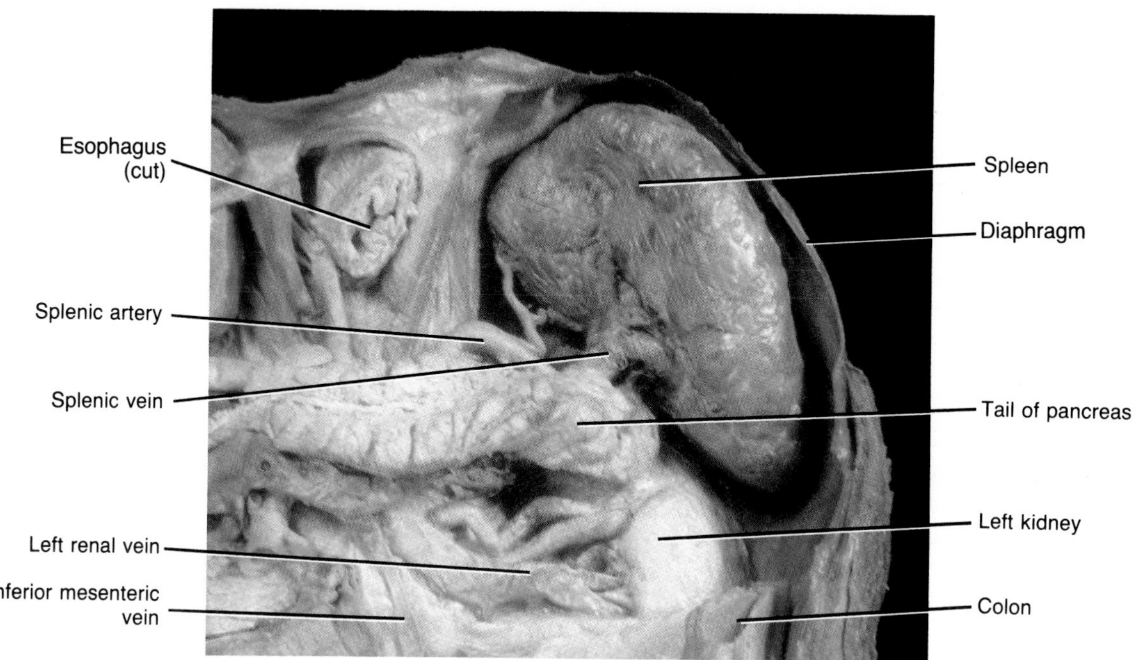

(a) Superior view

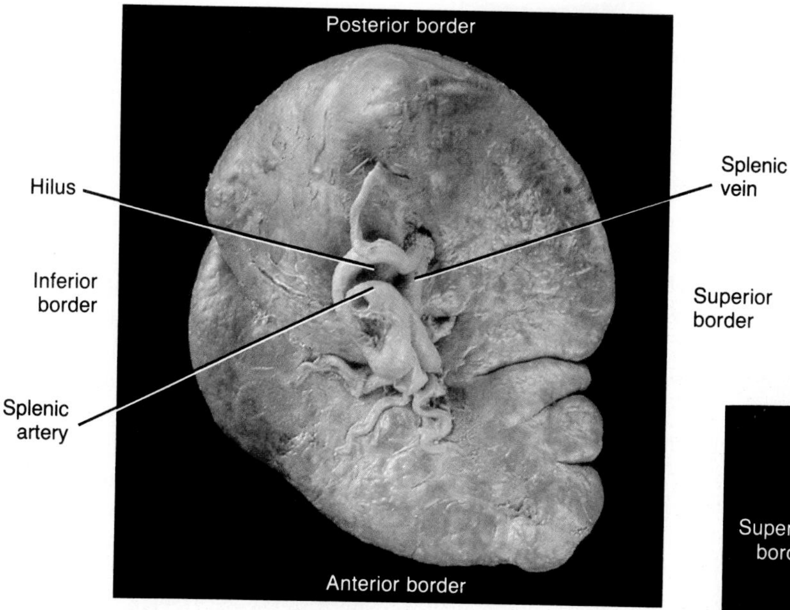

Posterior border
Hilus
Inferior border
Splenic artery
Splenic vein
Superior border
Anterior border

(b) Visceral surface

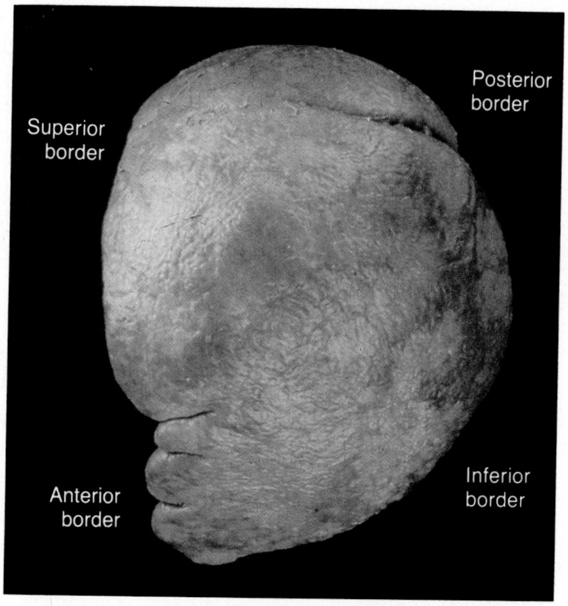

Superior border
Posterior border
Anterior border
Inferior border

(c) Diaphragmatic surface

FIGURE 15-4 Gross structure of the spleen. (a) Photograph of the spleen in relation to abdominal viscera. The stomach has been removed. (b) Photograph of visceral surface. (c) Photograph of diaphragmatic surface. (Photographs courtesy of J. A. Gosling, P. F. Harris, et al., *Atlas of Human Anatomy,* Gower Medical Publishing Ltd., 2nd ed., 1991.)

the contours of the organs adjacent to it—the gastric impression (stomach), renal impression (left kidney), and colic impression (left flexure of colon). The ***diaphragmatic*** (dī-a-fra-MAT-ik) ***surface*** is smooth and convex and conforms to the concave surface of the diaphragm (Figure 15-4c).

The spleen is surrounded by a capsule of dense connective tissue and scattered smooth muscle fibers (cells). The capsule, in turn, is covered by a serous membrane, the peritoneum. Like lymph nodes, the spleen contains a hilus, trabeculae, reticular fibers, and reticular cells. The capsule, trabeculae, reticular fibers, and reticular cells constitute the stroma of the spleen.

The parenchyma of the spleen consists of two different kinds of tissue, called white pulp and red pulp (Figure 15-5). ***White pulp*** is essentially lymphatic tissue, mostly lymphocytes, arranged around arteries called central arteries. In various areas the lymphocytes are thickened into lymphatic nodules, referred to as ***splenic nodules (Malpighian corpuscles).*** The ***red pulp*** consists of ***venous sinuses*** filled with blood and cords of splenic tissue called ***splenic (Billroth's) cords.*** Veins are closely associated with the red pulp. Splenic cords consist of erythrocytes, macrophages, lymphocytes, plasma cells, and granulocytes.

The splenic artery and vein and the efferent lymphatic vessels pass through the hilus. Since the spleen has no afferent lymphatic vessels or lymph sinuses, it does not filter lymph. One key splenic function related to immunity is the production of B cells, which develop into antibody-producing plasma cells. The spleen also phagocytizes bacteria and worn-out and damaged red blood cells and platelets. In addition, the spleen stores and releases blood in case of demand, such as during hemorrhage. Sympathetic impulses cause the smooth muscle of the capsule of the spleen to contract. During early fetal development, the spleen participates in blood cell formation.

About 10 percent of the population has ***accessory spleens.*** They are most commonly found near the hilus of the primary spleen or embedded in the tail of the pancreas. In general, accessory spleens are about 1 cm (0.5 in.) or less in diameter.

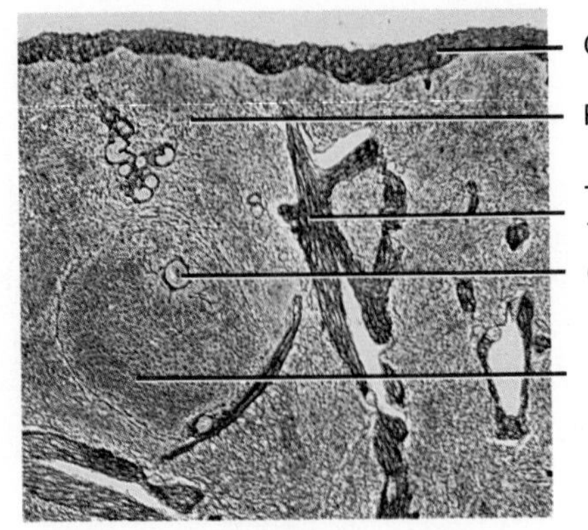

Photomicrograph of portion of the spleen (approx. 23 x)

FIGURE 15-5 Histology of the spleen. © Biophoto, SPL, Photo Researchers.

CLINICAL APPLICATION

Ruptured Spleen and Splenectomy

The spleen is the most frequently damaged organ in cases of abdominal trauma, particularly those involving severe blows over the lower left chest or upper abdomen that fracture the protecting ribs. Such a crushing injury may ***rupture the spleen,*** which causes severe intraperitoneal hemorrhage and shock. Prompt removal of the spleen, called a ***splenectomy,*** is needed to prevent the patient from bleeding to death. The functions of the spleen are then assumed by other structures, particularly bone marrow and the liver.

THYMUS GLAND

Usually a bilobed lymphatic organ, the ***thymus gland*** is located in the superior mediastinum, posterior to the sternum and between the lungs (Figure 15-6a). The two ***thymic lobes*** are held in close proximity by an enveloping layer of connective tissue. Each lobe is enclosed by a connective tissue ***capsule.*** The capsule gives off extensions into the lobes called ***trabeculae,*** which divide the lobes into ***lobules*** (Figure 15-6b). Each lobule consists of a deeply staining peripheral ***cortex*** and a lighter-staining central ***medulla.*** The cortex is composed almost entirely of small, medium, and large tightly packed lymphocytes held in place by reticular tissue fibers. Since the reticular tissue of the thymus gland differs in origin and structure from that usually found in other lymphatic organs, it is referred to as ***epithelioreticular*** supporting tissue. The medulla consists mostly of epithelial cells and more widely scattered lymphocytes, and its reticulum is more cellular than fibrous. In addition, the medulla contains characteristic ***thymic (Hassall's) corpuscles,*** concentric layers of epithelial cells (Figure 15-6c). Their significance is unknown.

The thymus gland is conspicuous in the infant, and it reaches its maximum size of about 40 grams during puberty. After puberty, much of the thymic tissue is replaced by fat and connective tissue. By the time the person reaches maturity, the gland has atrophied but still continues to be functional.

Its role in immunity is to synthesize hormones that help produce and distribute to other lymphoid organs T cells that destroy invading microbes directly or indirectly by producing various substances. Because of its synthesis of hormones, the thymus gland is an endocrine gland.

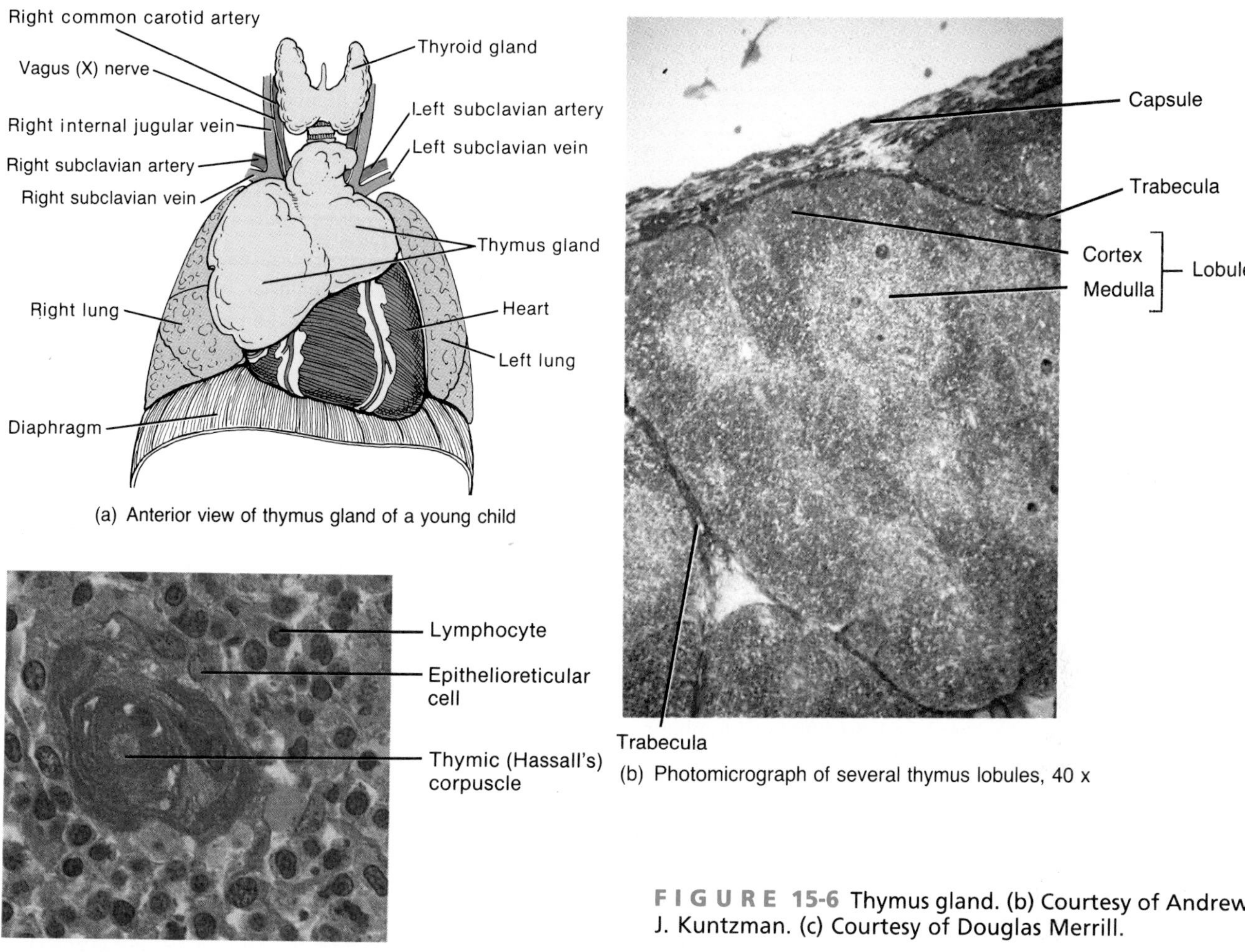

(a) Anterior view of thymus gland of a young child

(b) Photomicrograph of several thymus lobules, 40 x

(c) Photomicrograph of thymic corpuscle (620 x)

FIGURE 15-6 Thymus gland. (b) Courtesy of Andrew J. Kuntzman. (c) Courtesy of Douglas Merrill.

LYMPH CIRCULATION

ROUTE

When plasma is filtered by blood capillaries, it passes into the interstitial spaces; it is then known as interstitial fluid. When this fluid passes from interstitial spaces into lymph capillaries, it is called ***lymph*** (*lympha* = clear water). Lymph from lymph capillaries is then passed to lymphatic vessels that run toward lymph nodes. At the nodes, afferent vessels penetrate the capsules at numerous points, and the lymph passes through the sinuses of the nodes. Efferent vessels from the nodes either run with afferent vessels into another node of the same group or pass on to another group of nodes. From the most proximal group of each chain of nodes, the efferent vessels unite to form ***lymph trunks.*** The principal trunks are the ***lumbar, intestinal, bronchomediastinal, subclavian,*** and ***jugular trunks*** (Figure 15-7a).

Thoracic (Left Lymphatic) Duct

The principal trunks pass their lymph into two main channels, the thoracic duct and the right lymphatic duct. The ***thoracic (left lymphatic) duct*** is about 38 to 45 cm (15 to 18 in.) in length and begins as a dilation in front of the second lumbar vertebra called the ***cisterna chyli*** (sis-TER-na KĪ-lē) (Figure 15-7a, b). The thoracic duct is the main collecting duct of the lymphatic system and receives lymph from the left side of the head, neck, and chest, the left upper extremity, and the entire body below the ribs (see Figure 15-2c).

The cisterna chyli receives lymph from the right and left lumbar trunks and from the intestinal trunk. The lumbar trunks drain lymph from the lower extremities, wall and viscera of the pelvis, kidneys, adrenals (suprarenals), and the deep lymphatics from most of the abdominal wall. The intestinal trunk drains lymph from the stomach, intestines, pancreas, spleen, and visceral surface of the liver. In the

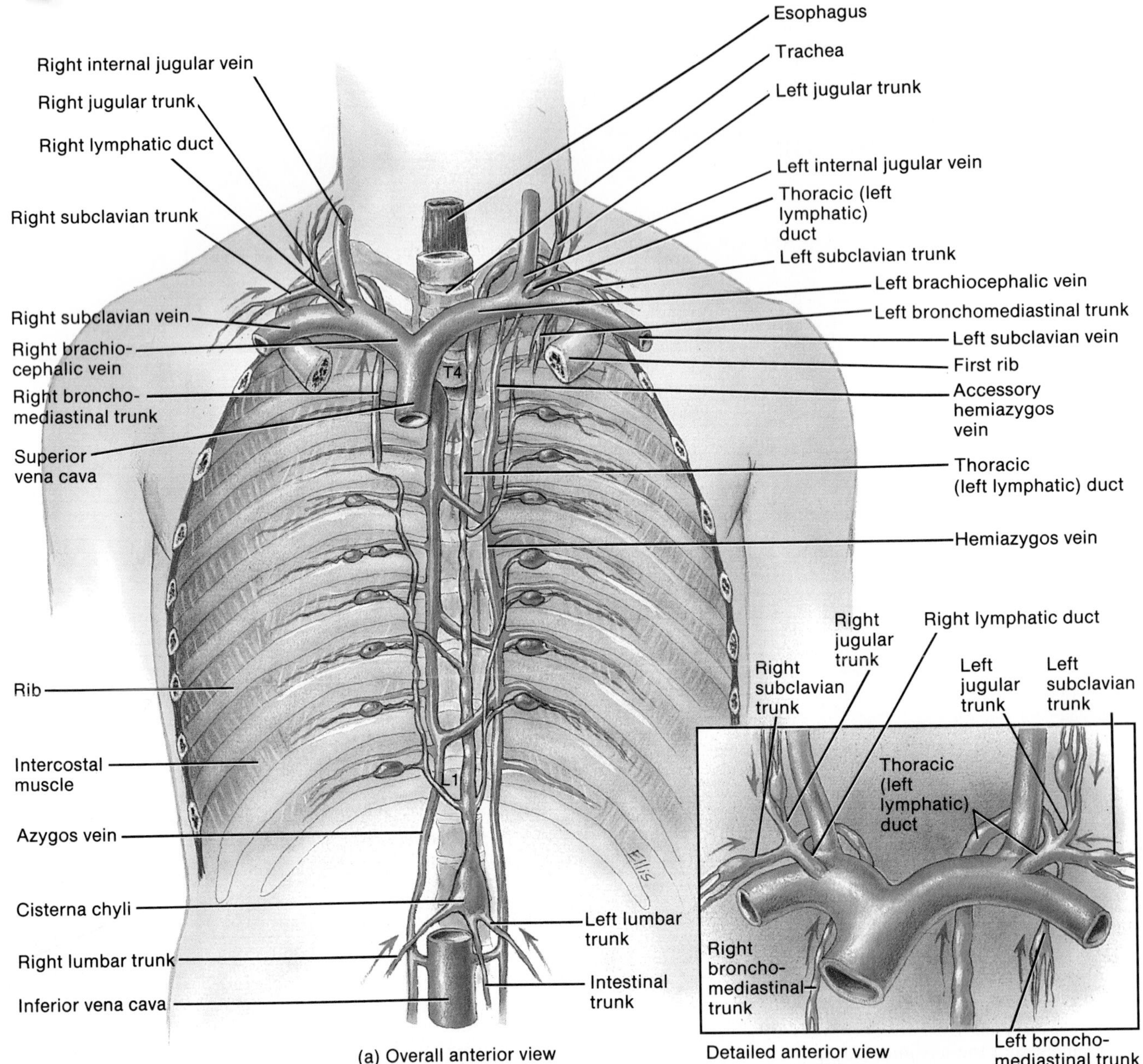

FIGURE 15-7 Scheme of lymphatic circulation. (a) Relation of lymph trunks to the thoracic duct and right lymphatic duct. See also Figure 15-2b. (b) Photograph of the origin of the cisterna chyli. Note the duplication of the thoracic duct after it emerges from the cisterna chyli. The single duct is apparent at the top of the photograph. (Courtesy of J. A. Gosling, P. F. Harris, et al., *Atlas of Human Anatomy*, Gower Medical Publishing Ltd., 2nd ed., 1991.)

neck, the thoracic duct also receives lymph from the left jugular, left subclavian, and left bronchomediastinal trunks. The left jugular trunk drains lymph from the left side of the head and neck; the left subclavian trunk drains lymph from the left upper extremity; and the left bronchomediastinal trunk drains lymph from the left side of the deeper parts of the anterior thoracic wall, upper part of the anterior abdominal wall, anterior part of the diaphragm, left lung, and left side of the heart.

Right Lymphatic Duct

The ***right lymphatic duct*** is about 1.25 cm (0.5 in.) long and drains lymph from the upper right side of the body

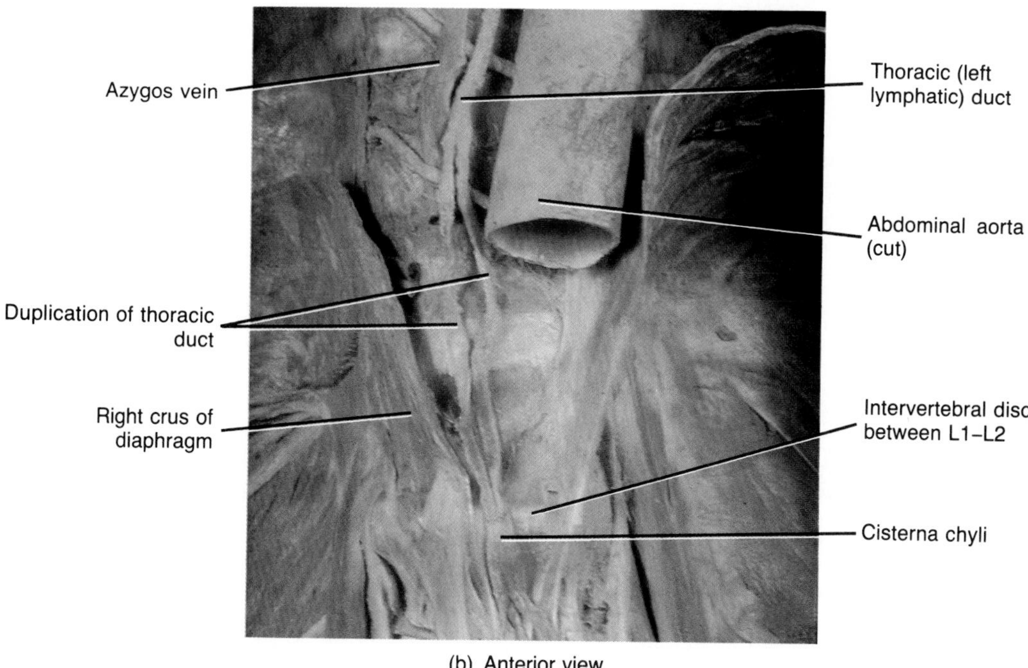

(b) Anterior view

(see Figure 15-2c). The right lymphatic duct collects lymph from its trunks as follows (Figure 15-7a). It receives lymph from the right jugular trunk, which drains the right side of the head and neck, from the right subclavian trunk, which drains the right upper extremity, and from the right bronchomediastinal trunk, which drains the right side of the thorax, right lung, right side of the heart, and part of the convex surface of the liver.

Ultimately, the thoracic duct empties all its lymph into the junction of the left internal jugular vein and left subclavian vein, and the right lymphatic duct empties all its lymph into the junction of the right internal jugular vein and right subclavian vein. Thus, lymph is drained back into the blood, and the cycle repeats itself continuously.

MAINTENANCE

The flow of lymph from tissue spaces to the large lymphatic ducts to the subclavian veins is maintained primarily by the milking action of skeletal muscles. Skeletal muscle contractions compress lymphatic vessels and force lymph toward the subclavian veins. Lymphatic vessels, like veins, contain valves, and the valves ensure the movement of lymph toward the subclavian veins.

Another factor that maintains lymph flow is respiratory movements. These movements create a pressure gradient between the two ends of the lymphatic system. Lymph flows from the abdominal region, where the pressure is higher, toward the thoracic region, where it is lower as each inhalation occurs.

CLINICAL APPLICATION

Edema

Edema, an excessive accumulation of interstitial fluid in tissue spaces, may be caused by an obstruction, such as an infected node or a blockage of vessels, in the pathway between the lymph capillaries and the subclavian veins. Another cause is excessive lymph formation and increased permeability of blood capillary walls. A rise in capillary blood pressure, in which interstitial fluid is formed faster than it is passed into lymphatics, also may result in edema.

PRINCIPAL GROUPS OF LYMPH NODES

Lymph nodes usually appear in groups and typically are arranged in two sets: ***superficial*** and ***deep.***

Exhibits 15-1 through 15-5 list the principal groups of lymph nodes of the body by region and the general areas of the body they drain.

EXHIBIT 15-1

Principal Lymph Nodes of the Head and Neck (Figure 15-8)

LYMPH NODES	LOCATION AND AREAS DRAINED
LYMPH NODES OF THE HEAD	
Occipital	Near trapezius and semispinalis capitis muscles. They drain the occipital portion of scalp and upper neck.
Retroauricular	Behind ear. They drain skin of ear and posterior parietal region of scalp.
Preauricular	In front of ear. They drain auricle of ear and temporal region of the scalp.
Parotid	Embedded in and below parotid gland. They drain root of nose, eyelids, anterior temporal region, external auditory meatus, tympanic cavity, nasopharynx, and posterior portions of nasal cavity.
Facial	Consist of three groups: infraorbital, buccal, and mandibular.
Infraorbital	Below the orbit. They drain eyelids and conjunctiva.
Buccal	At angle of mouth. They drain the skin and mucous membrane of nose and cheek.
Mandibular	Over mandible. They drain the skin and mucous membrane of nose and cheek.
LYMPH NODES OF THE NECK	
Submandibular	Along inferior border of mandible. They drain chin, lips, nose, nasal cavity, cheeks, gums, lower surface of palate, and anterior portion of tongue.
Submental	Between digastric muscles. They drain chin, lower lip, cheeks, tip of tongue, and floor of mouth.
Superficial Cervical	Along external jugular vein. They drain lower part of ear and parotid region.
Deep Cervical	Largest group of nodes in neck, consisting of numerous large nodes forming a chain extending from base of skull to root of neck. They are arbitrarily divided into superior deep cervical nodes and inferior deep cervical nodes.
Superior Deep Cervical	Under sternocleidomastoid muscle. They drain posterior head and neck, auricle, tongue, larynx, esophagus, thyroid gland, nasopharynx, nasal cavity, palate, and tonsils.
Inferior Deep Cervical	Near subclavian vein. They drain posterior scalp and neck, superficial pectoral region, and part of arm.

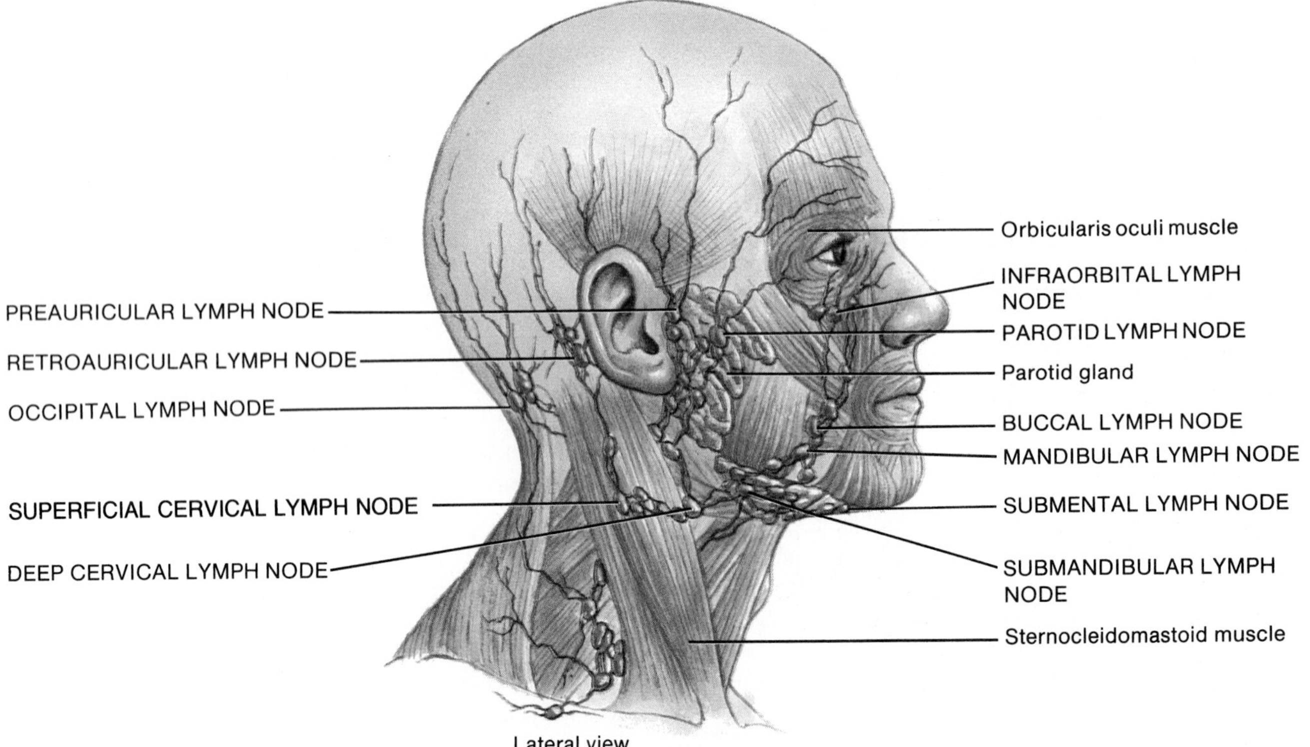

FIGURE 15-8 Principal lymph nodes of the head and neck.

EXHIBIT 15-2

Principal Lymph Nodes of the Upper Extremities (Figure 15-9)

LYMPH NODES	LOCATION AND AREAS DRAINED
Supratrochlear	Above medial epicondyle of humerus. They drain medial fingers, palm, and forearm.
Deltopectoral	Below clavicle. They drain lymphatic vessels on radial side of upper extremity.
Axillary	Most deep lymph nodes of the upper extremities are in the axilla and are called the axillary nodes. They are large in size and may be grouped as follows.
	Medial and posterior aspects of axillary artery. They drain most of whole upper extremity. Since infection or malignancy of upper extremity may cause tenderness and swelling in axilla, the axillary nodes, especially the lateral group, are clinically important since they filter lymph from much of upper extremity.
Pectoral (Anterior)	Along inferior border of the pectoralis minor muscle. They drain skin and muscles of anterior and lateral thoracic walls and central and lateral portions of mammary gland.
Subscapular (Posterior)	Along subscapular artery. They drain skin and muscles of posterior part of neck and thoracic wall.
Central (Intermediate)	Base of axilla embedded in adipose tissue. They drain lateral, pectoral (anterior), and subscapular (posterior) nodes.
Subclavicular (Apical)	Posterior and superior to pectoralis minor muscle. They drain deltopectoral nodes.

CLINICAL APPLICATION

Breast Cancer and Lymphatic Draining

Most of the lymph drainage of the breast (more than 75 percent) is to the pectoral (anterior) group of axillary lymph nodes. In breast cancer, it is possible that cancer cells may leave the breast and lodge in the pectoral nodes. From here, metastasis may develop in other axillary nodes. Most of the remaining lymphatic drainage is to the sternal lymph nodes (see also Figure 15-12).

EXHIBIT 15-3

Principal Lymph Nodes of the Lower Extremities (Figure 15-10)

LYMPH NODES	LOCATION AND AREAS DRAINED
Popliteal	In adipose tissue in popliteal fossa. They drain knee and portions of leg and foot, especially heel.
Superficial Inguinal	Parallel to saphenous vein. They drain anterior and lateral abdominal wall to level of umbilicus, gluteal region, external genitals, perineal region, and entire superficial lymphatics of lower extremity.
Deep Inguinal	Medial to femoral vein. They drain deep lymphatics of lower extremity, penis, and clitoris.

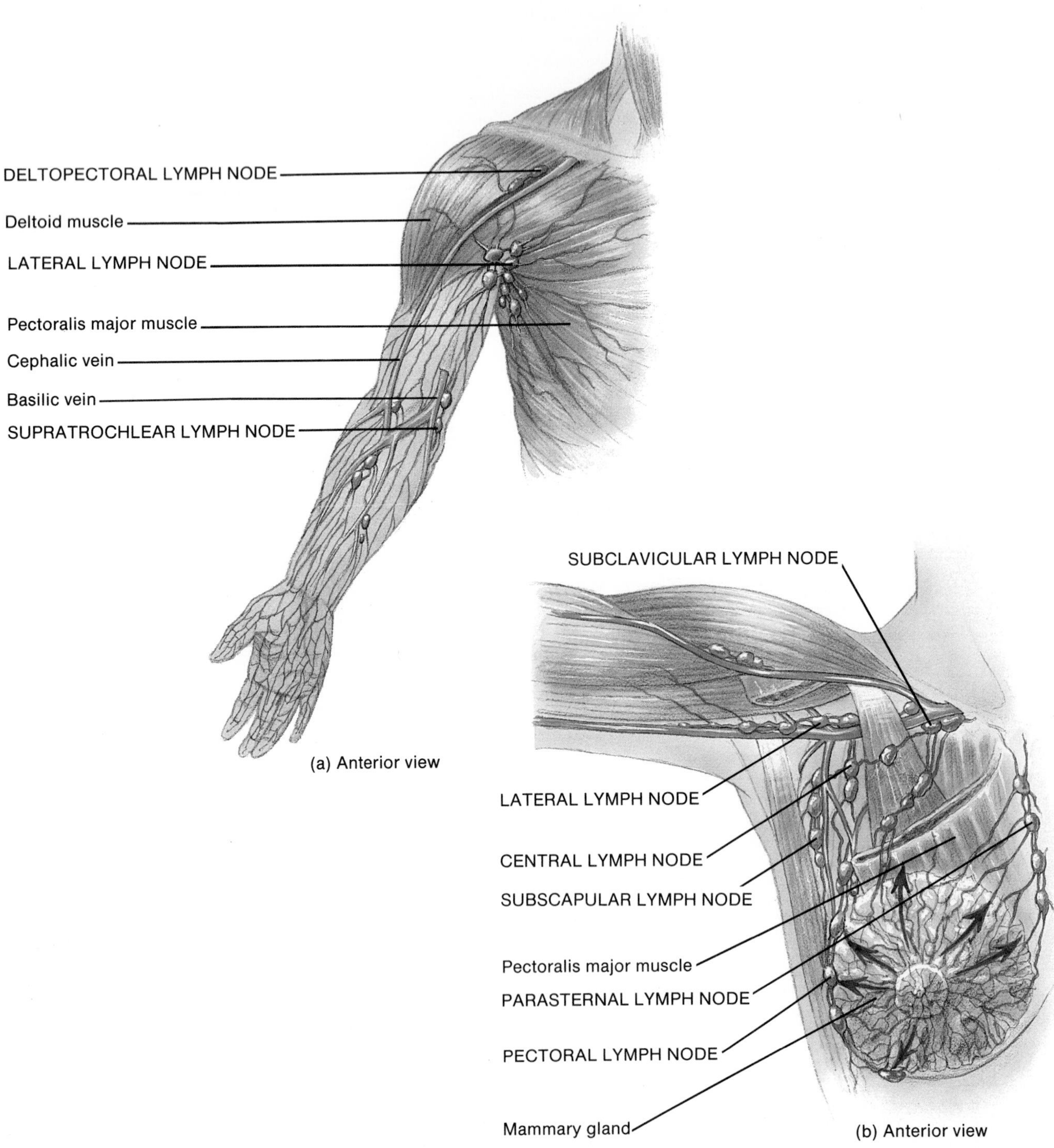

FIGURE 15-9 Principal lymph nodes of the upper extremities. Shown in (b) are the axillary lymph nodes. The direction of drainage is indicated by arrows.

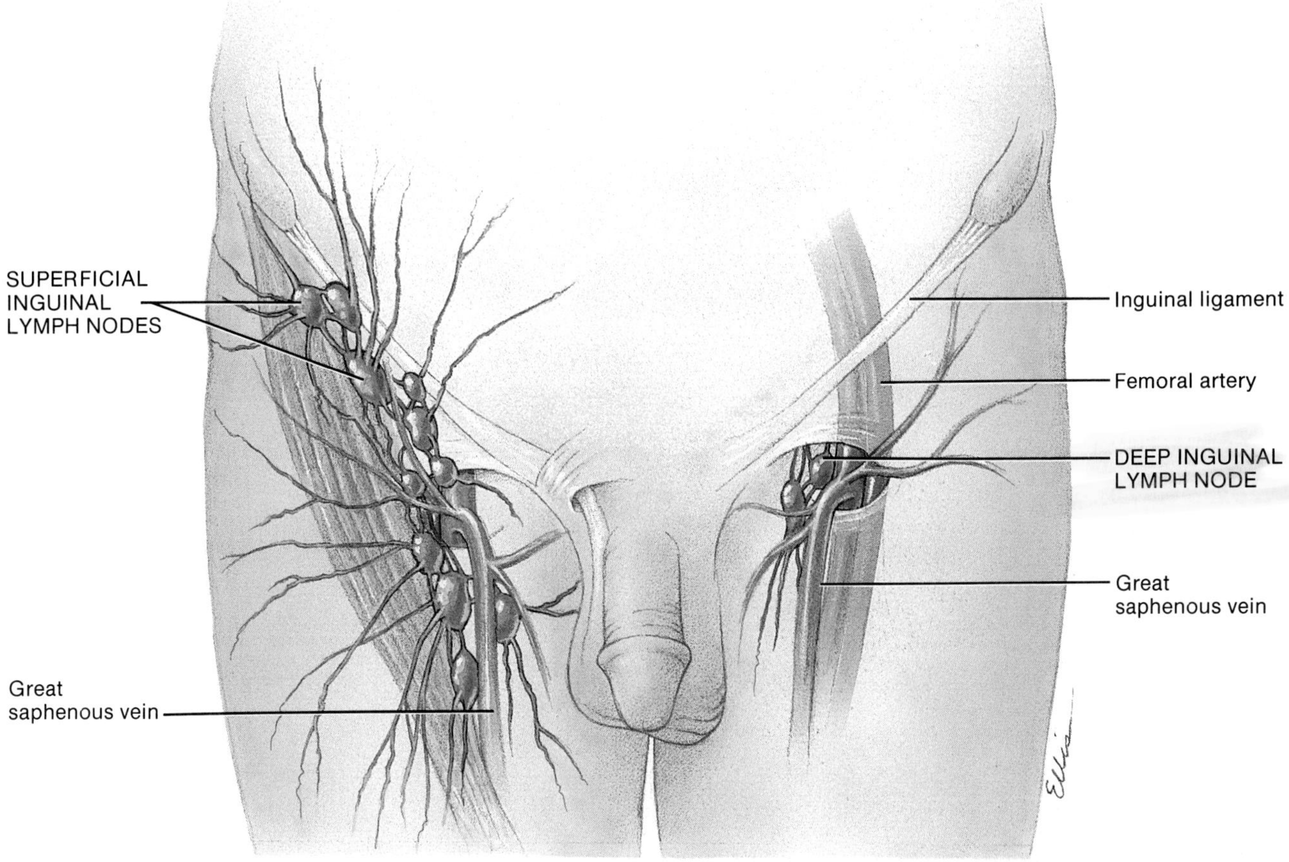

SUPERFICIAL INGUINAL LYMPH NODES
Inguinal ligament
Femoral artery
DEEP INGUINAL LYMPH NODE
Great saphenous vein
Great saphenous vein

(a) Anterior view

Popliteal vein
POPLITEAL LYMPH NODES
Small saphenous vein

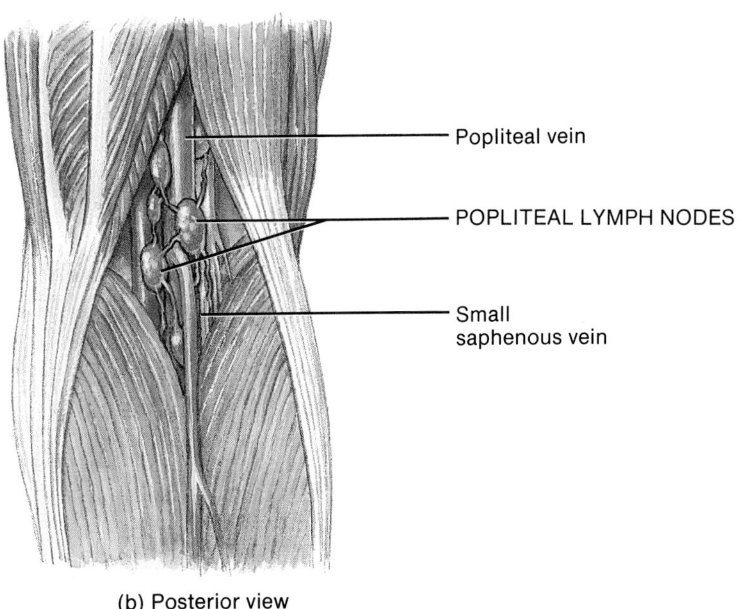

(b) Posterior view

FIGURE 15-10 Principal lymph nodes of the lower extremities.

EXHIBIT 15-4

Principal Lymph Nodes of the Abdomen and Pelvis (Figure 15-11)

Lymph nodes of the abdomen and pelvis are divided into ***parietal lymph nodes*** that are retroperitoneal (behind the parietal peritoneum) and in close association with larger blood vessels and ***visceral lymph nodes*** found in association with visceral arteries.

LYMPH NODES	LOCATION AND AREAS DRAINED
PARIETAL	
External Iliac	Arranged about external iliac vessels. They drain the deep lymphatics of abdominal wall below umbilicus, adductor region of thigh, urinary bladder, prostate gland, ductus (vas) deferens, seminal vesicles, prostatic and membranous urethra, uterine (Fallopian) tubes, uterus, and vagina.
Common Iliac	Arranged along course of common iliac vessels. They drain pelvic viscera.
Internal Iliac	Near internal iliac artery. They drain pelvic viscera, perineum, gluteal region, and posterior surface of thigh.

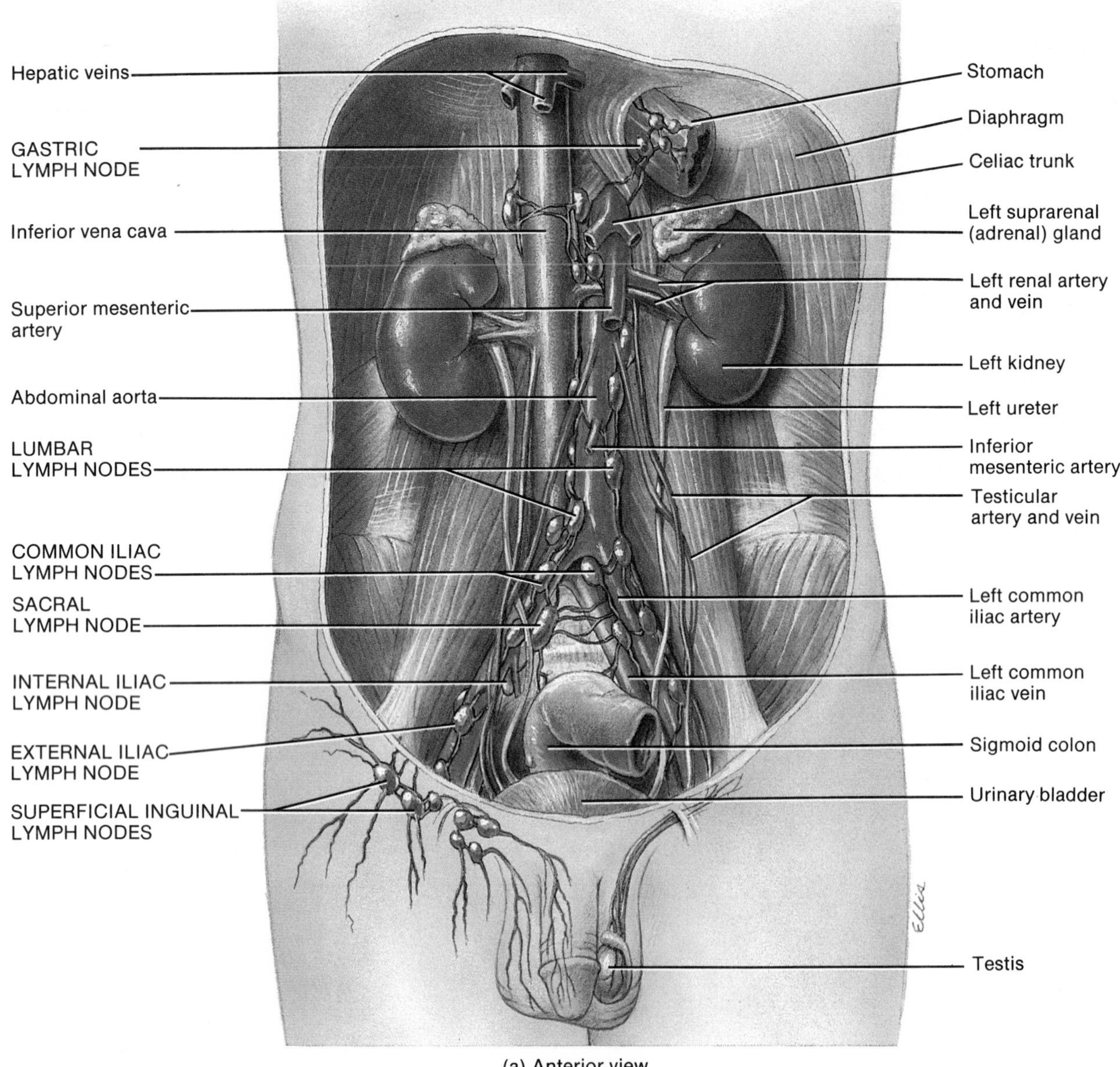

(a) Anterior view

FIGURE 15-11 Principal lymph nodes of the abdomen and pelvis. See also Figure 15-2b.

LYMPH NODES	LOCATION AND AREAS DRAINED
Sacral	In hollow of sacrum. They drain rectum, prostate gland, and posterior pelvic wall.
Lumbar	From aortic bifurcation to diaphragm; arranged around aorta and designated as ***right lateral aortic nodes, left lateral aortic nodes, preaortic nodes,*** and ***retroaortic nodes.*** They drain the efferents from testes, ovaries, uterine (Fallopian) tubes, uterus, kidneys, adrenal (suprarenal) glands, abdominal surface of diaphragm, and lateral abdominal wall.
VISCERAL	
Celiac	Consist of three groups of nodes: gastric, hepatic, and pancreaticosplenic.
Gastric	Lie along lesser and greater curvatures of stomach. They drain lesser curvature of stomach, inferior, anterior, and posterior aspects of stomach, and esophagus.
Hepatic	Along the hepatic artery. They drain stomach, duodenum, liver, gallbladder, and pancreas.
Pancreaticosplenic	Along splenic artery. They drain stomach, spleen, and pancreas.
Superior Mesenteric	These nodes are divided into mesenteric, ileocolic, and mesocolic groups:
Mesenteric	Along superior mesenteric artery. They drain jejunum and all parts of ileum, except for terminal portion.
Ileocolic	Along ileocolic artery. They drain terminal portion of ileum, appendix, cecum, and ascending colon.
Transverse Mesocolic	Between layers of transverse mesocolon. They drain descending iliac and sigmoid parts of colon.
Inferior Mesenteric	Near left colic, sigmoid, and superior rectal arteries. They drain descending, iliac, and sigmoid parts of colon, superior part of rectum, and superior anal canal.

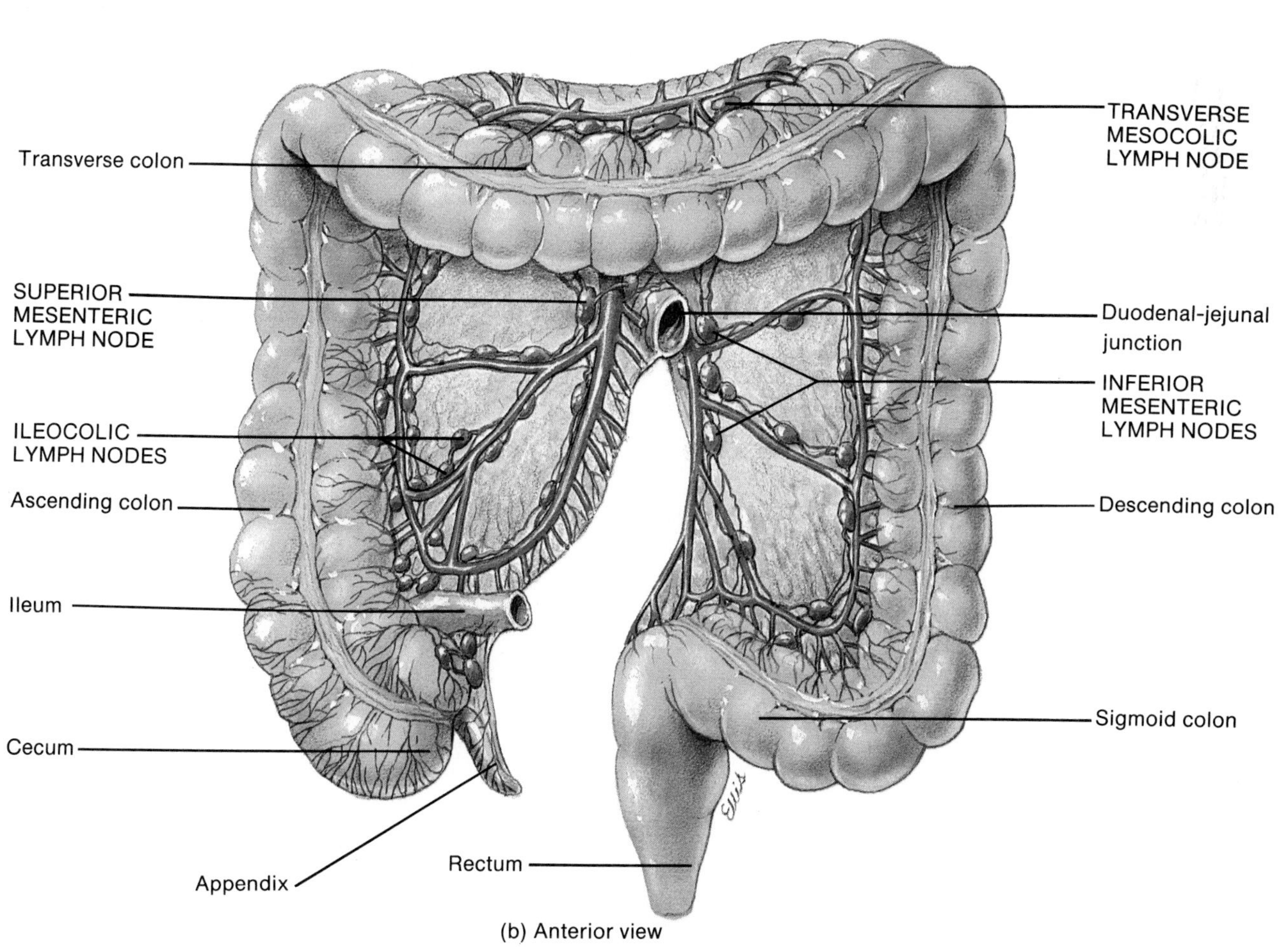

(b) Anterior view

EXHIBIT 15-5

Principal Lymph Nodes of the Thorax (Figure 15-12)

Lymph nodes of the thorax are divided into ***parietal lymph nodes,*** which drain the wall of the thorax, and ***visceral lymph nodes,*** which drain the viscera.

LYMPH NODES	LOCATION AND AREAS DRAINED
PARIETAL	
Sternal (Parasternal)	Alongside internal thoracic artery. They drain central and lateral parts of mammary gland, deeper structures of anterior abdominal wall above umbilicus, diaphragmatic surface of liver, and deeper parts of anterior portion of thoracic wall.
Intercostal	Near heads of ribs at posterior parts of intercostal spaces. They drain posterolateral aspect of thoracic wall.
Phrenic (Diaphragmatic)	Located on thoracic aspect of the diaphragm and divisible into three sets called anterior phrenic, middle phrenic, and posterior phrenic.
Anterior Phrenic	Behind base of xiphoid process. They drain convex surface of liver, diaphragm, and anterior abdominal wall.
Middle Phrenic	Close to phrenic nerves where they pierce diaphragm. They drain middle part of diaphragm and convex surface of liver.
Posterior Phrenic	Back of diaphragm near aorta. They drain posterior part of diaphragm.
VISCERAL	
Anterior Mediastinal	Anterior part of superior mediastinum anterior to arch of aorta. They drain thymus gland and pericardium.
Posterior Mediastinal	Posterior to pericardium. They drain esophagus, posterior aspect of the pericardium, diaphragm, and convex surface of liver.

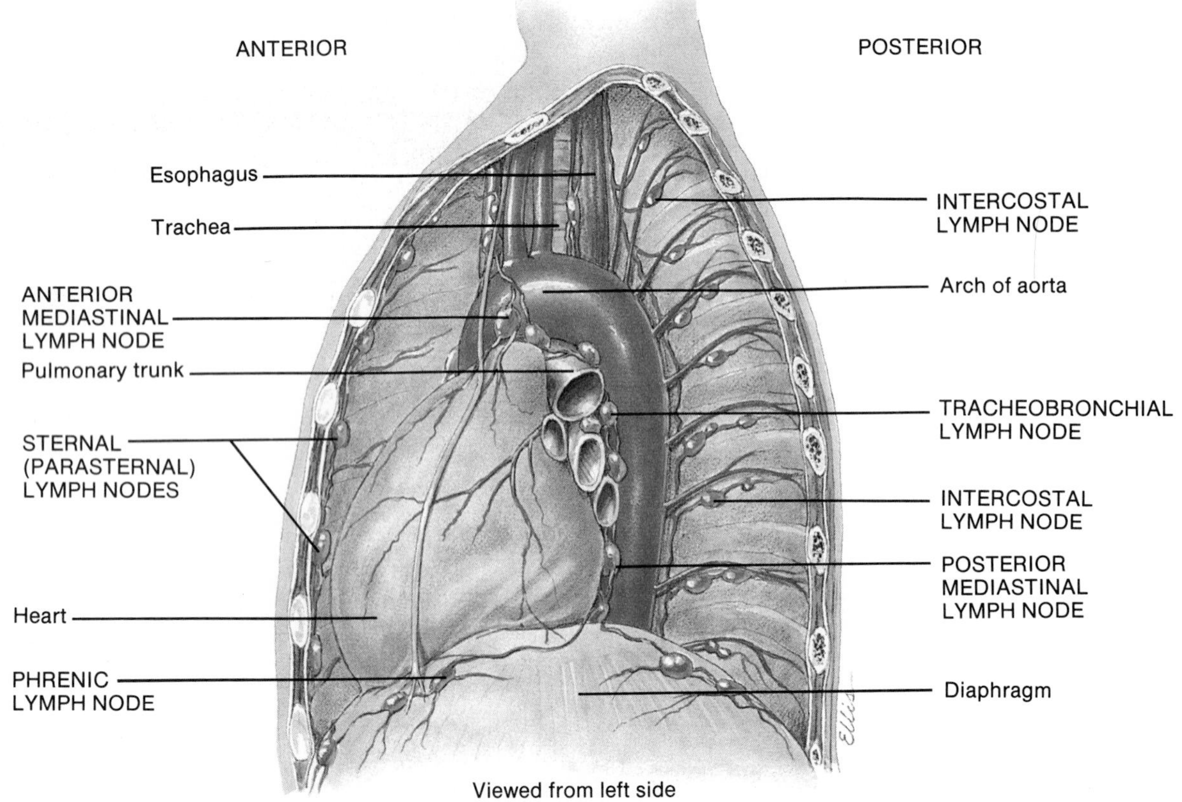

FIGURE 15-12 Principal lymph nodes of the thorax.

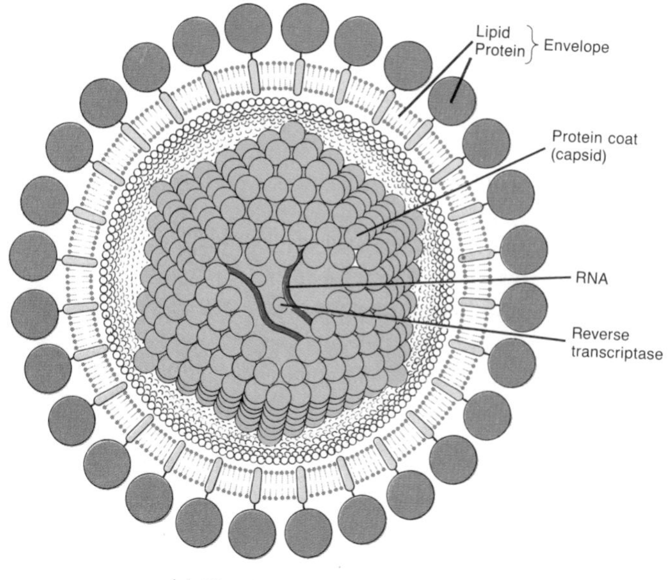

(a) Structure of HIV

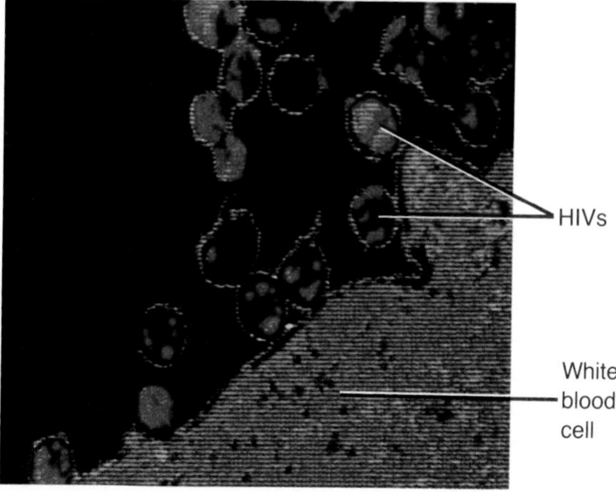

(b) Electron micrograph of several HIVs targeting a white blood cell at a magnification of 93,500×

FIGURE 15-14 Human immunodeficiency virus (HIV), the causative agent of AIDS. (b) Courtesy of Wagner, Herbert Stock/Phototake.

B; human T cell lymphotropic virus type I (HTLV I), which is associated with adult T cell leukemia; and human T cell lymphotropic virus type II (HTLV II), which is associated with hairy cell leukemia.

Infection with HIV begins when the virus binds to a protein receptor, called the CD4 receptor, on the surface of a subpopulation of T cells called helper T (T4) cells. Through very complicated mechanisms, helper T cells are killed and, as the process progresses, there is a decline in immune functioning. Helper T cells cooperate with B cells to amplify antibody production. The reduction in the number of T cells inhibits antibody production by descendants of B cells (plasma cells) against HIV. Helper T cells also stimulate the destruction of infected cells by cytotoxic (killer) T cells and natural killer (NK) cells. Helper T cells also influence the activity of monocytes and macrophages, which engulf infected cells and foreign particles. Since helper T cells orchestrate a large portion of our immune defense, their destruction leads to collapse of the immune system and susceptibility to opportunistic infections such as Kaposi's sarcoma (KS) and *Pneumocystis carinii* pneumonia (PCP). AIDS victims are also subject to tuberculosis, thrush, oral hairy leukoplakia, cytomegalovirus, a form of herpes that attacks the central nervous system, cryptococcal meningitis, central nervous system toxoplasmosis, oral and esophageal candidiasis, cryptosporidiosis, a severe form of arthritis called Reiter's syndrome, salmonellosis, and psoriasis. The AIDS virus also attacks macrophages; brain cells (where it causes AIDS dementia complex); endothelial cells that line various organs, body cavities, and blood vessels; bone marrow cells; and possibly colon cells. Heart and liver cells are also possible target cells. The virus might kill these cells or multiply in them, thus spreading the virus.

At present, macrophages are receiving a great deal of attention as pivotal cells in the development of AIDS. Research suggests that macrophages may actually be the first and possibly the only cells invaded by the AIDS virus. It seems that once the virus invades macrophages, it then spreads to helper T cells. As a result, helper T cells are killed and macrophages do not function properly in the immune response. Macrophages serve as reservoirs for the virus, and even though the viruses multiply within the macrophages, they are not themselves killed. A very important consequence of this is that commonly used screening tests that detect antibodies to the AIDS virus in the blood may be negative even though the person has the AIDS virus hidden in macrophages. While stored in macrophages, the virus does not trigger the production of antibodies that would appear in blood. A new test, called the Cetus test, is used to detect the presence of AIDS viruses in macrophages, rather than AIDS antibodies in blood.

Soon after infection with the AIDS virus, the host develops antibodies against several proteins in the virus. The presence of these antibodies in blood is used as a basis for diagnosing AIDS. Normally, antibodies so developed are protective, but are not necessarily so in the case of the AIDS virus.

Symptoms

The various stages through which an individual passes are correlated with decreasing helper T cell counts. Following exposure to HIV, there are usually no symptoms at first, and it may take from six weeks to one year before HIV is detected by standard antibody tests. Although most people have no symptoms when AIDS is first diagnosed, some develop mononucleosis-like symptoms such as fatigue, fever, swollen glands, headache, and even encephalitis. Within a few weeks, these symptoms disappear. In the next stage of the disease, which lasts from three to five years, chronically swollen lymph nodes develop in the neck, armpits, and groin. Following this, helper T cell counts decline even more, and the patients fail to respond to most skin tests that measure delayed hypersensitivity, a measure of the individual's ability to produce a cellular immune response against specific proteins injected under the skin. Progression to the next stage is evidenced by a total absence of delayed hypersensitivity. It is during this stage that opportunistic infections develop (thrush, hairy leukoplakia, Kaposi's sarcoma, *Pneumocystis carinii* pneumonia, toxoplasmosis, candidiasis, and so on). In the terminal stages of HIV infection, many patients suffer from AIDS dementia complex, which is characterized by a progressive loss of function in motor activities, cognition, and behavior.

HIV Outside the Body

Outside the body, the AIDS virus is fragile and can be eliminated in a number of ways. Dishwashing and clothes-washing by exposing the virus to 135°F (56°C) for 10 minutes will kill HIV. Chemicals such as hydrogen peroxide (H_2O_2), rubbing alcohol, Lysol, household bleach, and germicidal skin cleaner (such as Betadine and and Hibiclens) are very effective against the virus. Standard chlorination is also sufficient to kill HIV in swimming pools and hot tubs.

Transmission

Although HIV has been isolated from a number of body fluids, such as blood, semen, vaginal fluid, cerebrospinal fluid, tears, and saliva, it appears that the fluids that provide sufficient virus for transmission seem to be limited to blood, semen, and vaginal secretions. The virus is found in macrophages in these fluids. The macrophages also serve as a reservoir of the AIDS virus. It is possible that macrophages pass the virus to helper T cells, resulting in their destruction. The sites best suited for establishment of infection after exposure appear to be the cardiovascular system, open wounds of the skin, the penis, vagina, and rectum.

HIV is effectively transmitted by sexual contact between males, from males to females, and from females to males through vaginal or anal intercourse with infected persons. HIV is also effectively transmitted through exchanges of blood, such as by contaminated hypodermic needles and neddlestick, open wound, or mucous membrane exposure in health-care workers. The virus is also transmitted from infected mothers to their infants before or during birth or through breast-feeding. It does not appear that individuals become infected as a result of routine, nonintimate contacts. Thousands of family members, coworkers, and friends of AIDS patients do not have AIDS.

No evidence exists that AIDS can be spread through *normal* kissing (some experts feel that prolonged, vigorous, wet deep kissing that results in breaks or tears in the lining of the mouth or if preexisting sores are present could theoretically transmit the virus). Also, although mosquitoes can carry the AIDS virus for several days, there is no evidence that the virus can multiply inside mosquitoes or that they are capable of transmitting the disease. It also appears that health-care personnel who take proper routine barrier precautions (gloves, masks, safety glasses) are not at risk.

Drugs and Vaccines Against HIV

Medical scientists are mounting what is probably the greatest concentrated effort ever to find a cure for a single virus disease. One of the problems with finding a drug to treat AIDS is that HIV can lie dormant in body cells. In addition, HIV can infect a variety of cells, including those in the central nervous system that are protected by the blood–brain barrier. Added to this is the problem of opportunistic infections, which may be very difficult to treat. Antiviral therapy is aimed at disrupting various points in the viral life cycle. Some research centers on preventing the attachment of the virus to the host cell plasma membrane. Other research is designed to counterattack the virus once it gets inside a host cell and takes over the machinery of the host cell. Still other research is aimed at inhibiting assembly of viruses within the host cell and their subsequent release. Most drugs now used against the AIDS virus are directed against reverse transcriptase.

The drug currently most widely tried against the AIDS virus and given approval by the Food and Drug Administration (FDA) is zidovudine (Retrovir), popularly still known as azidothymidine (AZT). The action of this drug is to inhibit the action of reverse transcriptase, thus preventing the virus from making DNA from RNA. It has been shown to improve survival in people with advanced AIDS and to delay progression to AIDS in outwardly healthy people infected with HIV. Clinical improvements among patients taking zidovudine are weight gain, increased energy, and neurological improvements (reversal of loss of mental function and dementia). Side effects include severe bone marrow damage, anemia, and suppression of the immune system. In addition, the beneficial effects of zidovudine appear to diminish after prolonged usage, and strains of HIV resistant to the drug have appeared. A drug, similar in action to zidovudine, called 2-3-dideoxyinosine or ddI, is now undergoing clinical trials. Evidence thus far suggests that ddI elevates levels of helper T (T4) cells and does not cause anemia. Side effects include pancreatitis and pains in the legs and feet. Like zidovudine, ddI brings about weight gain and increased energy levels. Still another drug receiving considerable attention and undergoing clinical trials is soluble CD4. The drug mimics the CD4 receptor on the surface of helper T (T4) cells. By combining with HIV in the blood, soluble CD4 hampers viral attachment to host cell receptors. The drug appears to slow progression of AIDS and does not cause any significant side effects. Alpha interferon is believed to attach at the final stage of virus production. It is being tried in trials both alone and in combination with other drugs.

In addition to drug development and testing, considerable emphasis is also being given to development of a vaccine against AIDS. The purpose of a vaccine is to stimulate the production of antibodies against the virus that will kill the virus directly and to bolster immune defenses after the virus has invaded the body. The principal experimental vaccines under study make use of various subunits of the AIDS virus. In a process that is largely trial and error, researchers have selected different elements of the virus (proteins from the envelope or coat) that are believed to be most likely to produce the broadest range of antibodies. Although some scientists propose using the entire killed AIDS virus in a vaccine to stimulate antibody production, there is some concern that some of the viruses might remain alive and thus cause disease.

The development of an effective AIDS vaccine has been impeded because of lack of understanding of protective immunity against HIV, the ability of HIV to mutate so quickly, and inability to use most animals for experimentation since the virus will not grow within them. In addition, since the AIDS virus can remain protected in certain cells, antibodies cannot attack it. Moreover, clinical trials have raised safety concerns, and there may be a shortage of volunteers in the United States.

The objective of developing an AIDS vaccine is to determine which part of HIV elicits the most powerful immune response to inhibit the virus. At present, most studies are concentrated on proteins in the envelope of the HIV virus as the component most likely to elicit the most powerful immune response. Others focus on internal components of HIV rather than envelope antigens. Dr. Jonas Salk is using the whole virus minus its outer envelope in an attempt to trigger a protective immune response. This method is similar to the one he used in the 1950s to develop the polio vaccine. In one experiment using whole viruses, researchers inoculated nine rhesus monkeys with a vaccine for the monkey version of AIDS and then exposed them to the active virus. Nine months later, none of the monkeys developed the disease and only one showed any signs of infection.

Prevention of Transmission

At the present time, there are no drugs or vaccines to prevent AIDS. The only means of prevention is to stop transmission of the virus. Sexual transmission of the AIDS virus can be prevented if infected persons do not have vaginal, oral, or anal intercourse with susceptible persons, or, if during intercourse, effective barrier techniques (condoms with spermicides such as nonoxynol 9) are used. Infection from donated blood and blood products can be prevented by testing blood for evidence of the AIDS virus. AIDS transmitted by needles and syringes could be avoided if the use of intravenous drugs is stopped or if unsterilized infection paraphernalia are not used. Mother-to-infant infections could be avoided if infected females did not become pregnant. If these measures are to be effective, they must be part of an overall program involving education, counseling, screening individuals at high risk, tracing contacts, and modifying behavior.

Until there is effective drug therapy or an effective vaccine, preventing the spread of AIDS must rely on education and safer sexual practices.

AUTOIMMUNE DISEASES

Under normal conditions, the body's immune mechanism is able to recognize its own tissues and chemicals. It normally does not produce T cells or B cells against its own substances. Such recognition of self is called ***immunologic tolerance.*** Although the mechanism of tolerance is not completely understood, it is believed that a subpopulation of T cells called suppressor T cells may inhibit the differentiation of B cells into antibody-producing plasma cells or inhibit helper T cells that cooperate with B cells to amplify antibody production.

At times, however, immunologic tolerance breaks down, and this leads to an ***autoimmune disease (autoimmunity).*** For reasons still not understood, certain tissues undergo changes causing the immune system to recognize them as foreign antigens and produce an immune attack against them. Autoimmune diseases are immunologic responses mediated by antibodies against a person's own tissue antigens. Among human autoimmune diseases are rheumatoid arthritis (RA), systemic lupus erythematosus (SLE), thyroiditis, rheumatic fever, glomerulonephritis, encephalomyelitis, hemolytic and pernicious anemias, Addison's disease, Graves' disease, type I diabetes mellitus, myasthenia gravis, and multiple sclerosis (MS).

SEVERE COMBINED IMMUNODEFICIENCY (SCID)

Severe combined immunodeficiency (SCID) is a rare immunodeficiency disease in which both B cells and T cells are missing or inactive in providing immunity. Perhaps the most famous patient with SCID was David, the "bubble boy," who lived in a sterile plastic chamber for all but 15 days of his life; he died on February 22, 1984, at age 12. He was the oldest untreated survivor of SCID.

David was placed in the sterile chamber shortly after birth to protect him from microbes that his body could not fight. Many years later, in an effort to correct his disorder, David underwent a bone marrow transplant from his older sister, who was the most closely, but still not perfectly, matched donor available. Eighty days after the transplant and still in a germ-free environment, David developed some of the symptoms of infectious mononucleosis (IM), a condition caused by the Epstein-Barr virus (EBV). David was brought out of isolation for easier treatment, with the hope that the transplant would provide the same protection as his sterile plastic chamber. Unfortunately, the transplant was not successful, and David died about four months after it was performed. It was discovered that what killed David was not the immediate failure of the transplant, but cancer, as a result of the EBV. This case is a very clear demonstration of a virus causing a cancer in humans.

HYPERSENSITIVITY (ALLERGY)

A person who is overly reactive to an antigen is said to be ***hypersensitive (allergic).*** Whenever an allergic reaction occurs, there is tissue injury. The antigens that induce an allergic reaction are called ***allergens.*** Almost any substance can be an allergen for some individual. Common allergens include certain foods (milk, peanuts, shellfish, eggs), antibiotics (penicillin, tetracycline), vitamins (thiamine, folic acid), protein drugs (insulin, ACTH, estradiol), vaccines (pertussis, typhoid), venoms (honeybee, wasp, snake), cosmetics, chemicals in plants such as poison ivy, pollens, dust, molds, iodine-containing dyes used in certain x-ray procedures, and even microbes.

There are four basic types of hypersensitivity reactions: type I (anaphylaxis), type II (cytotoxic), type III (immune complex), and type IV (cell-mediated). The first three involve antibodies; the last involves T cells.

Type I (anaphylaxis) reactions are the most common and occur within a few minutes after a person sensitized to an allergen is reexposed to it. ***Anaphylaxis*** (an′-a-fi-LAK-sis) literally means "against protection" and results from the interaction of one class of antibodies (IgE) with mast cells and basophils. In response to certain allergens, some people produce IgE antibodies that bind to the surfaces of mast cells and basophils. This binding is what causes a person to be allergic to the allergen. Whereas basophils circulate in blood, mast cells are especially numerous in connective tissue of the skin, respiratory system, and endothelium of blood vessels. In response to the attachment of IgE antibodies to basophils and mast cells, the cells release chemicals called ***mediators of anaphylaxis,*** among which are histamine, prostaglandins (PGs), leukotrines, and kinin. Collectively, the mediators increase blood capillary permeability, increase smooth muscle contraction in the lungs, and increase mucus secretion. As a result, a person may experience inflammatory responses, difficulty in breathing from constricted bronchial tubes, and a "runny" nose from excess mucus secretion.

Some anaphylactic reactions, such as hives, eczema, swelling of the lips or tongue, abdominal cramps, and diarrhea are referred to as ***localized*** (affecting one part or a limited area). Other anaphylactic reactions are considered ***systemic*** (affecting several parts or the entire body). An example is acute anaphylaxis (anaphylactic shock), which may occur in a susceptible individual who has just received a triggering drug or been stung by a wasp or (more rarely) ingested a certain causative food such as seafood in persons with a marked sensitivity to iodine. The person develops respiratory symptoms (wheezing and shortness of breath) as bronchioles constrict, usually accompanied by cardiovascular failure and collapse due to vasodilation and fluid loss from blood. The treatment includes epinephrine and usually injectable corticosteroids which reverse the respiratory and cardiovascular effects, as well as supportive measures. It also includes the prevention of future such events by education and by having the person purchase and wear a Medic-Alert bracelet or necklace.

Type II (cytotoxic) reactions are caused by classes of antibodies (IgG or IgM) directed against antigens on a person's blood cells (red blood cells, lymphocytes, or platelets) or tissue cells. Type II reactions, which may occur in incompatible transfusion reactions, damage cells by lysis.

Type III (immune complex) reactions involve antigens (not part of a host tissue cell), certain classes of antibodies (IgA or IgM), and complement. When certain ratios of antigen to antibody occur, the complexes are small and escape phagocytosis. The complexes become trapped in the basement membrane under the endothelium of blood vessels, activate complement, and cause an inflammation. Conditions that so arise include glomerulonephritis, systemic lupus erythematosus (SLE), and rheumatoid arthritis (RA).

Type IV (cell-mediated) reactions involve T cells and often are not apparent for a day or more. Type IV reactions occur when allergens that bind to tissue cells are phagocytized by macrophages and presented to receptors on the surfaces of T cells. As a result, there is a proliferation of T cells, which respond by destroying the allergens. An example of a type IV reaction that involves the skin is the familiar skin test for tuberculosis. Tissue rejection is another example.

TISSUE REJECTION

Transplantation is the replacement of an injured or diseased tissue or organ. Usually, the body recognizes the proteins in the transplanted tissue or organ as foreign and produces antibodies against them. This phenomenon is known as ***tissue rejection.*** Rejection can be somewhat reduced by matching donor and recipient HLA antigens and by administering drugs that inhibit the body's ability to form antibodies. White blood cells and other nucleated cells have surface antigens called human leukocyte associated (HLA) antigens. These antigens are unique for each person, except identical twins. The more closely matched the HLA antigens between donor and recipient, the less the likelihood of tissue rejection.

Types of Transplants

The most successful transplants are ***autografts,*** transplants in which one's own tissue is grafted to another part of the body (such as skin grafts for burn treatment or plastic surgery), and ***isografts,*** transplants in which the donor and recipient have identical genetic backgrounds.

An ***allograft*** is a transplant between individuals of the same species but with different genetic backgrounds. The success of this type of transplant has been moderate. Frequently, it is used as a temporary measure until the damaged or diseased tissue is able to repair itself. Skin transplants from other individuals and blood transfusions might properly be considered allografts. The one organ allograft that has been quite successful is the thymus. Children born without a thymus can now receive the gland from an aborted fetus. The thymus-deficient child cannot produce antibodies and thus cannot reject the transplant. Rejection later on indicates that the child is manufacturing antibodies and no longer needs the organ.

A ***xenograft*** is a transplant between animals of different species. This type of transplantation is used primarily as a physiological dressing over severe burns.

Until recently, ***immunosuppressive drugs*** suppressed not only the recipient's immune rejection of the donor organ but also the immune response to all antigens as well, thereby causing patients to become very susceptible to infectious diseases. A drug called *cyclosporine,* derived from a fungus, has largely overcome this problem with regard to kidney, heart, and liver transplants. A selective immunosuppressive drug, it inhibits T cells that are responsible for tissue rejection but has only a minimal effect on B cells. Thus, rejection is avoided and resistance against disease is still maintained.

As an alternative to immunosuppressive drugs, some scientists are experimenting with total irradiation of lymph nodes, a procedure called ***total lymphoid irradiation (TLI).*** The radiation kills cytotoxic (killer) T cells, helper T cells, and suppressor T cells, and thus prevents rejection. It is believed that the cells reappear from 1 to 12 months after irradiation and that the suppressor T cells become sensitive to the transplant and inhibit cytotoxic and helper T cells that would attack the transplant.

It is recommended that testing for AIDS antibodies or virus should be done for all donors of tissues or organs intended for transplantation, as well as donors of blood, semen, or ova.

HODGKIN'S DISEASE (HD)

Hodgkin's disease (HD) is a form of cancer, usually arising in lymph nodes, the cause of which is unknown. It may, however, arise from a combination of genetic predisposition, disturbance of the immune system, and an infectious agent (Epstein-Barr virus). The histological diagnosis of the disease is made by the presence of large, malignant, multinucleate cells in the affected lymph node called Reed-Sternberg cells. The disease is initially characterized by a painless, nontender enlargement of one or more lymph nodes, most commonly in the neck but occasionally in the axilla, inguinal, or femoral region. About one-quarter to one-third of patients also have an unexplained and persistent fever and/or night sweats. Fatigue and weight loss are also associated complaints, as is pruritus (itching). Treatment consists of radiation therapy, chemotherapy, and combinations of the two. Hodgkin's disease is considered to be a curable malignancy.

KEY MEDICAL TERMS ASSOCIATED WITH THE LYMPHATIC SYSTEM

Adenitis (ad′-e-NĪ-tis; *adeno* = gland; *itis* = inflammation of) Enlarged, tender, and inflamed lymph nodes resulting from an infection.

Elephantiasis (el′-e-fan-TĪ-a-sis) Long-standing edema of one or both lower extremities, and sometimes of the arms or other body parts, that is due to lymphatic obstruction. The involved part is tremendously swollen and hardened, and the skin surface folds and produces fissures, causing it to resemble the leg of an elephant. Elephantiasis may be due to filariasis, an infestation by a parasitic worm. Other causes include congestive heart failure and chronic obstruction of the lymphatic vessels.

Hypersplenism (hī′-per-SPLĒN-izm; *hyper* = over) Abnormal splenic activity due to splenic enlargement and associated with an increased rate of destruction of normal blood cells.

Lymphadenectomy (lim-fad′-e-NEK-tō-mē; *ectomy* = removal) Removal of a lymph node.

Lymphadenopathy (lim-fad′-e-NOP-a-thē; *patho* = disease) Enlarged, sometimes tender lymph glands.

Lymphangioma (lim-fan′-jē-Ō-ma; *angio* = vessel; *oma* = tumor). A benign tumor of the lymphatic vessels.

Lymphangitis (lim′-fan-JĪ-tis) Inflammation of the lymphatic vessels.

Lymphedema (lim′-fe-DĒ-ma; *edema* = swelling) Accumulation of lymph fluid producing subcutaneous tissue swelling.

Lymphoma (lim′-FŌ-ma) Any tumor composed of lymphatic tissue.

Lymphostasis (lim-FŌS-tā-sis; *stasis* = halt) A lymph flow stoppage.

Splenomegaly (splē′-nō-MEG-a-lē; *mega* = large) Enlarged spleen.

STUDY OUTLINE

Lymphatic Vessels (p. 428)

1. The lymphatic system consists of lymph, lymphatic vessels, and structures and organs that contain lymphatic tissue (specialized reticular tissue containing large numbers of lymphocytes).
2. Among the lymphatic tissue–containing components of the lymphatic system are diffuse lymphatic tissue, lymphatic nodules, and lymphatic organs (lymph nodes, spleen, and thymus gland).
3. Lymphatic vessels begin as blind-ended lymph capillaries in tissue spaces between cells.
4. Lymph capillaries merge to form larger vessels, called lymphatic vessels, that ultimately converge into the thoracic duct or right lymphatic duct.
5. Lymphatic vessels have thinner walls and more valves than veins.

Lymphatic Tissue (p. 430)

1. Lymph nodes are oval structures located along lymphatic vessels.
2. Lymph enters nodes through afferent lymphatic vessels and exits through efferent lymphatic vessels.
3. Lymph passing through the nodes is filtered. Lymph nodes also produce lymphocytes.
4. Tonsils are multiple aggregations of large lymphatic nodules embedded in mucous membranes. They include the pharyngeal, palatine, and lingual tonsils.
5. The spleen is the largest mass of lymphatic tissue in the body and functions in production of lymphocytes and antibodies, phagocytosis of bacteria and worn-out red blood cells, and storage of blood.
6. The thymus gland functions in immunity by producing T cells.

Lymph Circulation (p. 435)

1. The passage of lymph is from interstitial fluid to lymph capillaries to lymphatic vessels to lymph trunks to the thoracic duct or right lymphatic duct to the subclavian veins.
2. Lymph flows as a result of skeletal muscle contractions and respiratory movements. It is also aided by valves in the lymphatic vessels.

Principal Groups of Lymph Nodes (p. 437)

1. Lymph nodes are scattered throughout the body as superficial and deep groups.
2. The principal groups of lymph nodes are found in the head and neck, upper extremities, lower extremities, abdomen and pelvis, and thorax.

Aging and the Immune System (p. 445)

1. With advancing age, individuals become more susceptible to infections and malignancies, response to vaccines is decreased, and more autoantibodies are produced.
2. Cellular and humoral responses also diminish.

Developmental Anatomy of the Lymphatic System (p. 445)

1. Lymphatic vessels develop from lymph sacs, which develop from veins. Thus, they are derived from mesoderm.
2. Lymph nodes develop from lymph sacs that become invaded by mesenchymal cells.

Applications to Health (p. 446)

1. Acquired immune deficiency syndrome (AIDS) lowers the body's immunity by decreasing the number of helper T cells and reversing the ratio of helper T cells to suppressor T cells. AIDS victims frequently develop Kaposi's sarcoma (KS) and *Pneumocystis carinii* pneumonia (PCP).
2. Autoimmune diseases result when the body does not recognize "self" antigens and produces antibodies against them. Several human autoimmune diseases are rheumatoid arthritis (RA), systemic lupus erythematosus (SLE), rheumatic fever, hemolytic and pernicious anemias, myasthenia gravis, and multiple sclerosis (MS).
3. Severe combined immunodeficiency (SCID) is an immunodeficiency disease in which both B cells and T cells are missing or inactive in providing immunity.
4. Hypersensitivity (allergy) is overreactivity to an antigen. Localized anaphylactic reactions include hay fever, asthma, eczema, and hives; acute anaphylaxis is a severe reaction with systemic effects.
5. Tissue rejection of a transplanted tissue or organ involves antibody production against the proteins (antigens) in the transplant. It may be overcome with immunosuppressive drugs.
6. Hodgkin's disease (HD) is a malignant disorder, usually arising in lymph nodes.

REVIEW QUESTIONS

1. Identify the components and functions of the lymphatic system. (p. 428)
2. How do lymphatic vessels originate? Compare veins and lymphatic vessels with regard to structure. (p. 428)
3. What is a lymphangiogram? What is its diagnostic value? (p. 428)
4. Describe the structure of a lymph node. What functions do lymph nodes serve? (p. 430)
5. Identify the tonsils by location. (p. 432)
6. Describe the location, gross anatomy, histology, and functions of the spleen. What is a splenectomy? (p. 432)
7. Describe the role of the thymus gland in immunity. (p. 434)
8. Construct a diagram to indicate the route of lymph circulation. (p. 435)
9. List and explain the various factors involved in the maintenance of lymph circulation. (p. 437)
10. Define edema. What are some of its causes? (p. 437)
11. For each of the following regions of the body, list the major

lymph nodes and the areas of the body they drain: head, neck, upper extremities, lower extremities, abdomen and pelvis, and thorax. (p. 438)

12. Describe how the lymphatic system develops. (p. 445)
13. Describe the symptoms of acquired immune deficiency syndrome (AIDS). What are the complications of AIDS? (p. 446)
14. Define an autoimmune disease. Give several examples. (p. 449)
15. Describe the symptoms of severe combined immunodeficiency (SCID). (p. 449)
16. Define hypersensitivity (allergy) and distinguish among the four types of hypersensitivity reactions. (p. 450)
17. Why does tissue rejection occur? How is this problem overcome? (p. 450)
18. What is Hodgkin's disease (HD)? (p. 451)

SELF QUIZ

Complete the following:

1. Lymph is conveyed into a node at several points by (afferent? efferent?) lymphatic vessels.
2. The thoracic duct starts in the lumbar region as a dilation known as the ________.
3. The lymphatic system derives from ________-derm, beginning about the (fifth? seventh? tenth?) week of gestation.

Circle T (true) or F (false) for the following.

T F 4. Most of the lymph nodes that drain viscera bear the same names as arteries that supply these viscera.

T F 5. The palatine tonsils are the ones most often removed in a tonsillectomy.

T F 6. Blood capillaries are more permeable than lymph capillaries.

T F 7. Lymph nodes are distributed evenly throughout the body, with equal numbers in all tissues.

T F 8. Lymph may pass through several lymph nodes in a number of regions before returning to blood.

Arrange the answers in correct sequence.

__ __ __ __ __ 9. Pathway of lymph from the stomach to blood plasma:
- **A.** thoracic duct
- **B.** lymphatic vessels and lymph nodes
- **C.** lymph capillaries
- **D.** interstitial fluid
- **E.** left subclavian vein

__ __ __ __ 10. Flow of lymph through a lymph node:
- **A.** afferent lymphatic vessel
- **B.** medullary sinus
- **C.** cortical sinus
- **D.** efferent lymphatic vessel

11. Match the following:

__ **a.** these nodes drain the knee

__ **b.** these nodes would be most likely to enlarge as defense against a serious infection following a cut on the little finger

__ **c.** infection of the skin of the leg or external genitals would cause these nodes to become swollen

__ **d.** these nodes drain the deep lymphatic vessels of the lower extremity

__ **e.** during infections of the nasal cavity or nasopharynx, these nodes are likely to swell; they are the largest group of nodes in the neck

__ **f.** lymph from infected urinary bladder or uterus (pelvic viscera) would pass first into these nodes

__ **g.** located adjacent to the abdominal aorta, they receive efferents from common iliacs and also drain kidneys

__ **h.** these visceral nodes drain the stomach, liver, spleen, and pancreas

__ **i.** these nodes lie along the superior mesenteric artery and drain structures supplied by this artery

__ **j.** these nodes drain and provide first defense against infections of the lower respiratory passageways and the lungs

A. celiac
B. deep cervical
C. deep inguinal
D. external, internal, and common iliacs
E. lumbar
F. popliteal
G. superior mesenteric
H. superficial inguinal
I. supratrochlear
J. tracheal, bronchial, and pulmonary

Choose the one best answer to these questions.

__ 12. The lymphatic system has the important function(s) of A. controlling body temperature by evaporation of sweat; B. manufacturing all white blood cells; C. transporting fluids out to, and back from, the body tissues; D. returning fluid and proteins to the cardiovascular system; E. all of the above.

__ 13. The spleen
A. serves as a site for production of B cells; B. is an organ in which phagocytosis of aged erythrocytes takes place; C. is a blood reservoir; D. is a site of blood formation in the fetus; E. is described by all of the above.

__ 14. The flow of lymph is
(1) aided by a pressure gradient difference—toward the thoracic region.
(2) maintained primarily by the milking action of muscles.
(3) possible because of valves in lymphatic vessels.
A. (1) only; B. (2) only; C. (3) only; D. all of the above; E. (2) and (3).

__ 15. Which of the following lymph nodes drains the leg?
A. axillary; B. anterior mediastinal; C. deltopectoral; D. deep inguinal; E. superior mesenteric.

__ 16. Which of the following lymph organs is located in the abdomen?
A. adenoids; B. lingual tonsil; C. thymus; D. spleen; E. deep cervical lymph nodes.

___ **17.** Both the thoracic duct and the right lymphatic duct empty directly into
A. axillary lymph nodes; B. superior vena cava; C. cisterna chyli; D. subclavian arteries; E. junction of internal jugular and subclavian veins.

___ **18.** All of these are major sites of lymphatic tissue *except*
A. tonsils; B. thymus; C. kidneys; D. spleen; E. lymph nodes.

___ **19.** Choose the false statement:
A. lymph capillaries are more permeable than blood capillaries; B. lymphatic vessels have thinner walls than veins; C. like arteries, lymphatic vessels contain no valves; D. lymph capillaries are blind-ended; E. most lymph drains into the thoracic duct.

Nervous Tissue

16

STUDENT OBJECTIVES

1. Classify the organs of the nervous system into central and peripheral divisions.
2. Contrast the histological characteristics and functions of neuroglia and neurons.
3. Classify neurons by structure and function.
4. Describe the capacity of nervous tissue for regeneration.
5. Explain the organization of neurons in the nervous system.

CHAPTER OUTLINE

- **Organization**
- **Histology**
  - Neuroglia
  - Neurons
    - *Structure*
    - *Structural Variation*
    - *Classification*
- **Nerve Impulse**
  - Speed
  - Synapses
  - Neurotransmitters
- **Regeneration**
  - Chromatolysis
  - Wallerian Degeneration
  - Retrograde Degeneration
  - Repair
- **Organization of Neurons**

The ***nervous system*** is one of the body's principal control and integrating centers. The other is the endocrine system. In humans, the nervous system serves three broad functions: sensory, integrative, and motor. First, it senses certain changes within the body and in the outside environment; this is its sensory function. Second, it interprets the changes; this is its integrative function. Third, it responds to the interpretation by initiating action in the form of muscular contractions or glandular secretions; this is its motor function.

Through sensation, integration, and response, the nervous system represents the body's most rapid means of maintaining homeostasis. Its split-second reactions, carried out by nerve impulses, can normally make the adjustments necessary to keep the body functioning efficiently. As you will see later, the nervous system shares the maintenance of homeostasis with the endocrine system. Although the adjustments made by hormones secreted by endocrine glands are usually slower than those made by nerve impulses, they are no less effective.

The branch of medical science that deals with the normal functioning and disorders of the nervous system is called ***neurology*** (noo-ROL-ō-jē; *neuro* = nerve or nervous system; *logos* = study of).

The developmental anatomy of the nervous system is considered in Chapter 18.

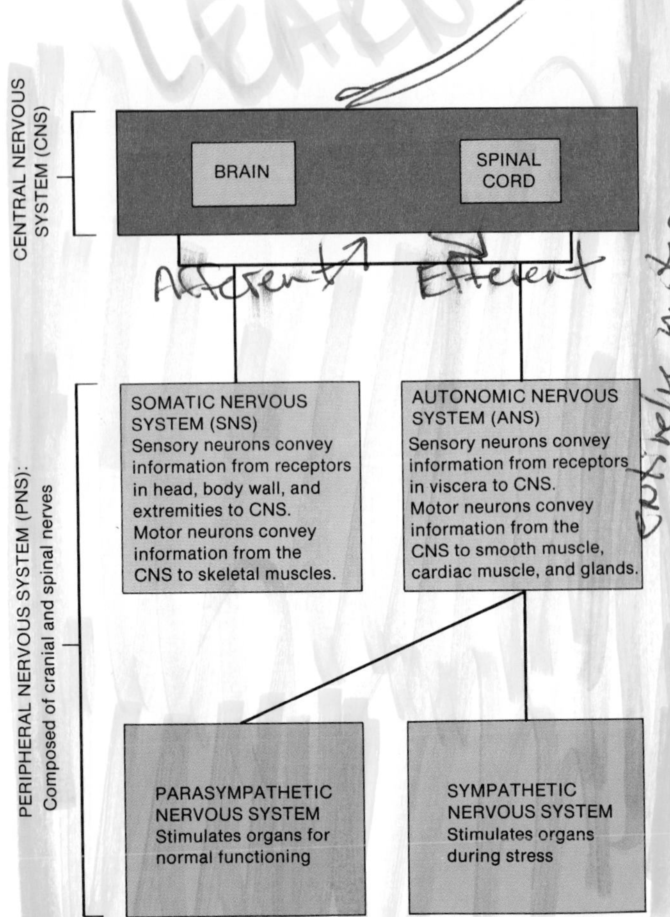

FIGURE 16-1 Organization of the nervous system.

ORGANIZATION

The nervous system has two principal divisions: the ***central nervous system (CNS)*** and ***peripheral*** (pe-RIF-er-al) ***nervous system (PNS)*** (Figure 16-1).

The CNS consists of the brain and spinal cord and is the control center for the entire nervous system. Within the CNS, incoming sensory information is interpreted and correlated, thoughts and emotions are generated, muscles may be stimulated to contract, and glands may be directed to secrete via outgoing nerve impulses delivered by nerves.

The PNS consists of nerves emerging from the brain (cranial nerves) and nerves emerging from the spinal cord (spinal nerves). Portions of these nerves carry impulses inward from receptors to the CNS while other portions carry impulses outward from the CNS to muscles and glands. The sensory (afferent) component of the PNS consists of nerve cells, called sensory (afferent) neurons, that conduct impulses from receptors in various parts of the body to the CNS. The motor (efferent) component consists of nerve cells, called motor (efferent) neurons, that conduct impulses from the CNS to muscles and glands.

Based primarily on the part of the body that responds, the PNS may be subdivided into a ***somatic*** (*soma* = body) ***nervous system (SNS)*** and ***autonomic*** (*auto* = self; *nomos* = law) ***nervous system (ANS).*** The SNS consists of sensory neurons that convey information from cutaneous and special sense receptors primarily in the head, body wall, and extremities to the CNS and motor neurons from the CNS that conduct impulses to skeletal muscle only. Since its motor responses are under conscious control, the SNS is voluntary.

The ANS consists of sensory neurons that convey homeostatic information from receptors primarily in the viscera to the CNS and motor neurons from the CNS that conduct impulses to smooth muscle, cardiac muscle, and glands. Since the motor responses of the ANS are under subconscious control, it is involuntary.

HISTOLOGY

Despite the complexity of the nervous system, it consists of only two principal kinds of cells; neurons (nerve cells) and neuroglia. Neurons are highly specialized for nerve impulse conduction and for all special functions attributed to the nervous system: thinking, controlling muscle activity, regulating glands. Neuroglia serve as a special supporting and protective component of the nervous system.

NEUROGLIA

The cells of the nervous system that perform the functions of support and protection are called ***neuroglia*** (noo-ROGlē-a; *neuro* = nerve; *glia* = glue) or ***glial cells*** (Figure 16-2). Neuroglia are derived from ectoderm. They are generally smaller than neurons and outnumber them by 5 to 10 times. Many neuroglia form a supporting network by twining around neurons or lining certain structures in the brain and spinal cord. Others bind nervous tissue to supporting structures and attach the neurons to their blood vessels. A few types of neuroglia also serve specialized functions. For example, some produce a phospholipid covering, called a myelin sheath, around nerve fibers, which increases the speed of nerve impulse conduction and insulates the fibers. Certain small neuroglia are phagocytic; they protect the central nervous system from disease by engulfing invading microbes and clearing away debris. Neuroglia are of clinical interest because they are a common source of tumors (gliomas) of the nervous system. It is estimated that gliomas account for 40 to 45 percent of brain tumors. Unfortunately, gliomas are very invasive.

Exhibit 16-1 lists the neuroglial cells and summarizes their functions.

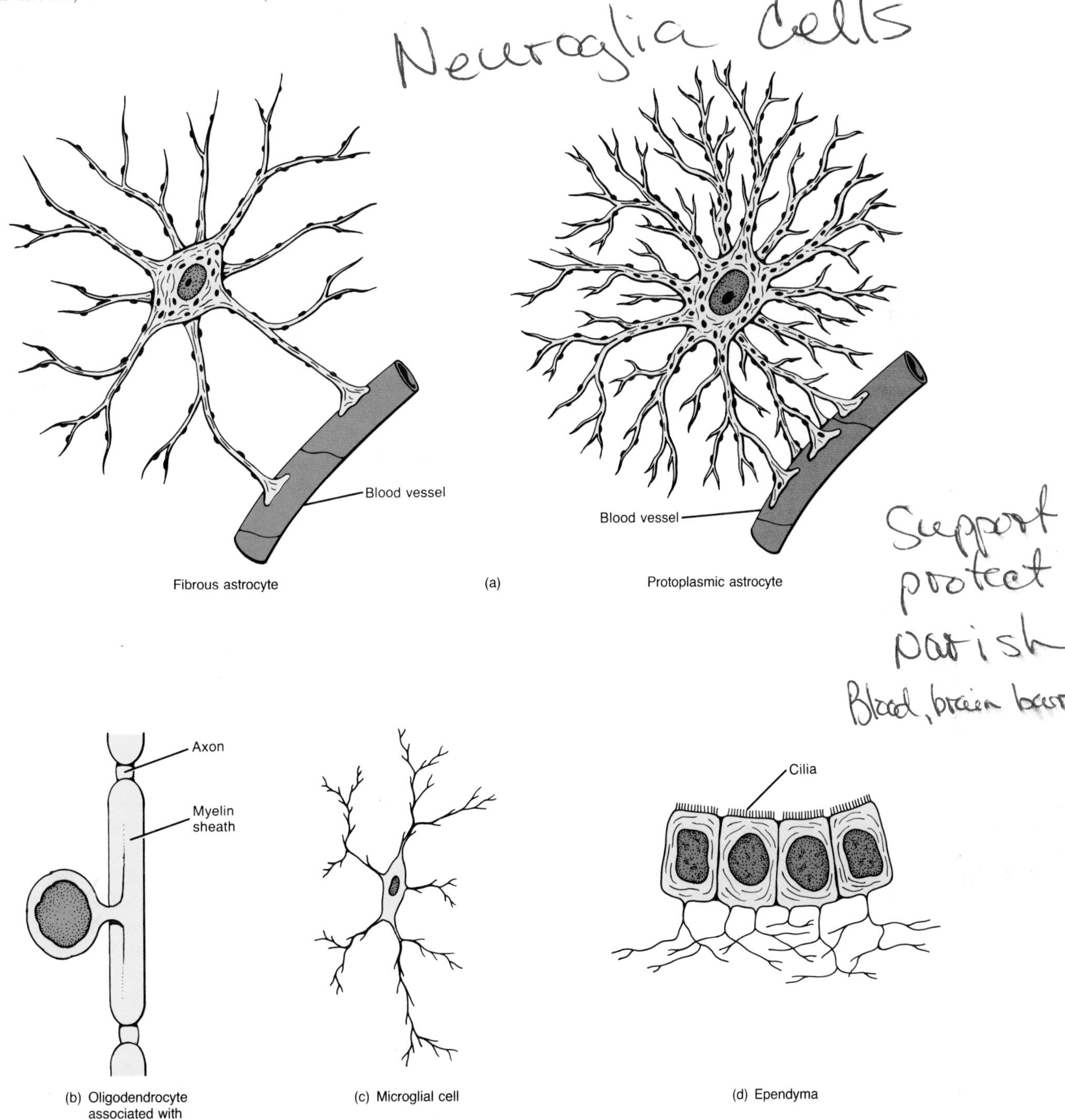

FIGURE 16-2 Histology of neuroglia. A neurolemmocyte (Schwann cell) is illustrated in Figure 16-3.

EXHIBIT 16-1

Neuroglia of Nervous System

TYPE	DESCRIPTION	FUNCTION
Astrocytes (*astro* = star; *cyte* = cell)	Star-shaped cells with numerous processes. ***Protoplasmic astrocytes*** are found in the gray matter of the CNS, and ***fibrous astrocytes*** are found in the white matter of the CNS.	Neuroglia of the CNS that assist in normal brain physiology by participating in the metabolism of neurotransmitters (glutamate and gamma-aminobutyric acid) and maintain the proper balance of potassium ions for generation of nerve action potentials by brain cells; participate in brain development by assisting nerve cells to migrate from points of origin to their final destinations and help to form the blood–brain barrier, which regulates the passage of substances into the brain (Chapter 18); twine around nerve cells to form a supporting network; provide a link between nerve cells and blood vessels.
Oligodendrocytes (*oligo* = few; *dendro* = tree)	Resemble astrocytes in some ways, but processes are fewer and shorter.	Give support by forming semirigid connective tissue rows between neurons in CNS; produce a phospholipid myelin sheath around axons and dendrites of nerve cells of CNS.
Microglia (*micro* = small; *glia* = glue)	Small cells with few processes; derived from monocytes; normally stationary but may migrate to site of injury; also called ***brain macrophages.***	Engulf and destroy microbes and cellular debris in the CNS; may migrate to area of injured nervous tissue and function as small macrophage in the CNS.
Ependyma (Ependymocytes) (*ependyma* = upper garment)	Epithelial cells arranged in a single layer and ranging in shape from squamous to columnar; many are ciliated.	Form a continuous epithelial lining for the ventricles of the brain (spaces that form and circulate cerebrospinal fluid) and the central canal of the spinal cord; probably assist in the circulation of cerebrospinal fluid (CSF) in these areas.
Neurolemmocytes (Schwann Cells)	Flattened cells arranged around axons.	Produce a phospholipid myelin sheath around axons and dendrites of nerve cells of PNS.
Satellite Cells	Flattened cells arranged around the cell bodies of ganglia (collections of nerve cell bodies outside the CNS).	Support ganglia of the PNS.

NEURONS

Neurons are responsible for conducting nerve impulses from one part of the body to another. They are the basic information processing units of the nervous system.

Structure

Most neurons consist of three distinct portions: (1) cell body, (2) dendrites, and (3) axon (Figure 16-3a). The ***cell body*** contains a well-defined nucleus and nucleolus surrounded by a granular cytoplasm. Within the cytoplasm are typical organelles such as lysosomes, mitochondria, and Golgi complexes. Many neurons also contain cytoplasmic inclusions such as ***lipofuscin*** pigment that occurs as clumps of yellowish brown granules. Lipofuscin may be a by-product of lysosomal activity. Although its significance is unknown, lipofuscin is related to aging; the amount of pigment increases with age. Also located in the cytoplasm are structures characteristic of neurons: chromatophilic substance and neurofibrils. The ***chromatophilic substance*** (***Nissl bodies***) is an orderly arrangement of granular (rough) endoplasmic reticulum whose function is protein synthesis. Newly synthesized proteins pass from the cell body into the neuronal processes, mainly the axon, at the rate of about 1 mm (0.04 in.) per day. These proteins replace those lost during metabolism and are used for growth of neurons and regeneration of peripheral nerve fibers. ***Neurofibrils*** are long, thin fibrils composed of intermediate filaments. They may assume a function in support and the transportation of nutrients.

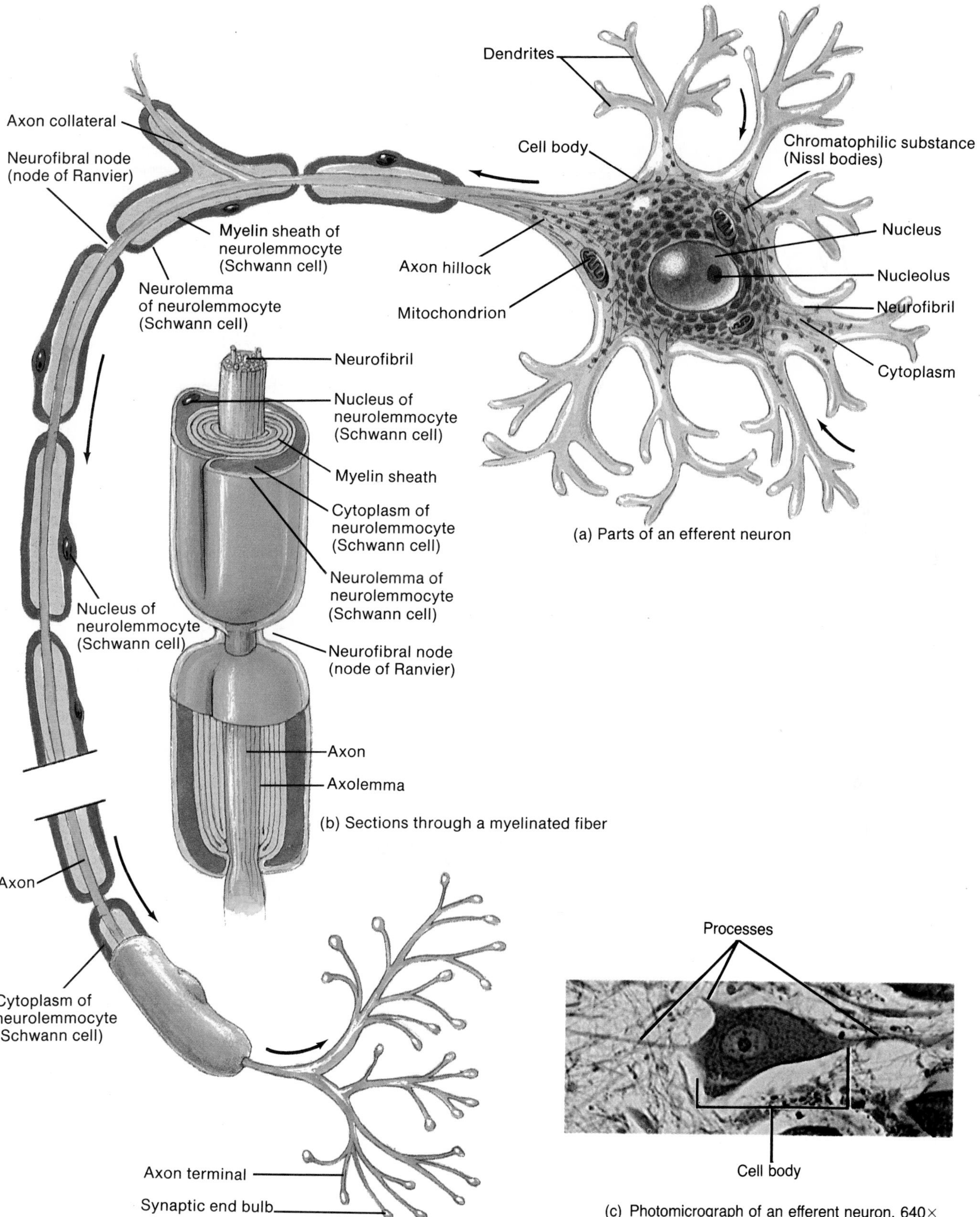

FIGURE 16-3 Structure of a typical neuron as exemplified by an efferent (motor) neuron. The arrows in (a) indicate the direction in which nerve impulses travel. The break indicates that the process is actually longer than shown. (c) Courtesy of Biophoto Associates/Photo Researchers.

Neurons have two kinds of cytoplasmic processes: dendrites and axons. ***Dendrites*** (*dendro* = tree) are usually short, thick, highly branched extensions of the cytoplasm of the cell body. They typically contain chromatophilic substance, mitochondria, and other cytoplasmic organelles. A neuron usually has several main dendrites. Their function is to conduct nerve impulses toward the cell body.

The second type of cytoplasmic process, called an ***axon,*** is usually a single long, thin projection that is highly specialized and conducts nerve impulses away from the cell body to another neuron or muscular or glandular tissue. It usually originates from the cell body as a small conical elevation called the ***axon hillock.*** An axon contains mitochondria and neurofibrils but no chromatophilic substance; thus, it does not carry on protein synthesis. Its cytoplasm, called ***axoplasm,*** is surrounded by a plasma membrane known as the ***axolemma*** (*lemma* = sheath or husk). Axons vary in length from a few millimeters (1 mm = 0.04 in.) in the brain to a meter (3.28 ft) or more between the spinal cord and toes. Along the length of an axon, there may be side branches called ***axon collaterals.*** The axon and its collaterals terminate by branching into many fine filaments called ***axon terminals.*** The distal ends of axon terminals are expanded into bulblike structures called ***synaptic end bulbs.*** They contain membrane-enclosed sacs called ***synaptic vesicles*** that store chemicals called neurotransmitters that determine whether or not impulses pass from one neuron to another or from a neuron to another tissue (muscle or gland).

The term ***nerve fiber*** is a general term for any nerve cell process projecting from the cell body. (More commonly, it refers to an axon and its sheaths.) As you will see later, a nerve is a group of nerve fibers outside the CNS. Figure 16-3b shows two sectional planes of a nerve fiber of the peripheral nervous system. Many nerve fibers, especially ones outside the CNS, are surrounded by a multilayered, white, phospholipid, segmented covering called the ***myelin sheath.*** Fibers containing such a covering are ***myelinated,*** whereas those without it are ***unmyelinated*** (see Figure 16-4). The function of the myelin sheath is to increase the speed of nerve impulse conduction and to insulate and maintain the fiber. Myelin is responsible for the color of the white matter in the nerves, brain, and spinal cord. As you will see later, certain diseases such as multiple sclerosis and Tay-Sachs disease are related to destruction of myelin sheaths.

The myelin sheath of nerve fibers of the peripheral nervous system is produced by flattened neuroglia called ***neurolemmocytes (Schwann cells),*** located along the axons. In the formation of a sheath, a developing neurolemmocyte encircles the fiber until its ends meet and overlap (Figure 16-4). The cell then winds around the fiber many times; as it does so, the cytoplasm and nucleus are pushed to the outside layer. The inner portion, consisting of up to 20–30 layers of neurolemmocyte membrane, is the myelin sheath. The peripheral nucleated cytoplasmic layer of the neurolemmocyte (the outer layer that encloses the sheath) is called the ***neurolemma (sheath of Schwann).***

The neurolemma is found only around fibers of the peripheral nervous system. Its function is to assist in the regeneration of injured axons and dendrites by forming a tube in which a regenerating axon or dendrite grows (see Figure 16-9d). Between the segments of the myelin sheath are unmyelinated gaps, called ***neurofibral nodes (nodes of Ran-***

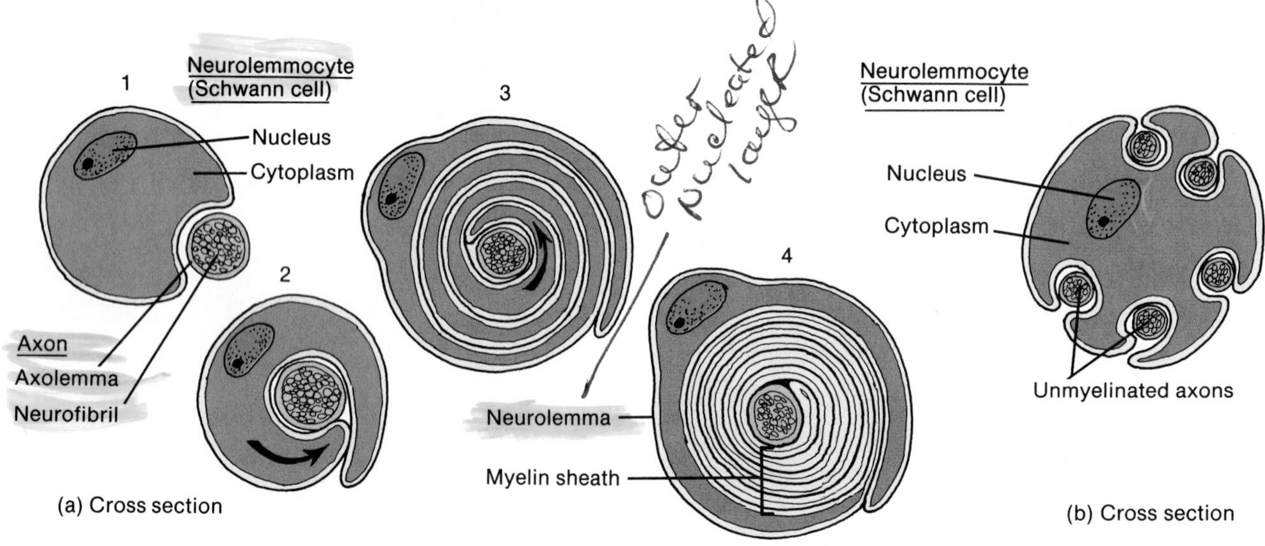

FIGURE 16-4 Comparison between myelinated and unmyelinated axons. (a) Stages in the formation of a myelin sheath by a neurolemmocyte (Schwann cell). (b) Unmyelinated axon. (c) Copyright © 1980, Biology Media, Schultz, Photo Researchers.

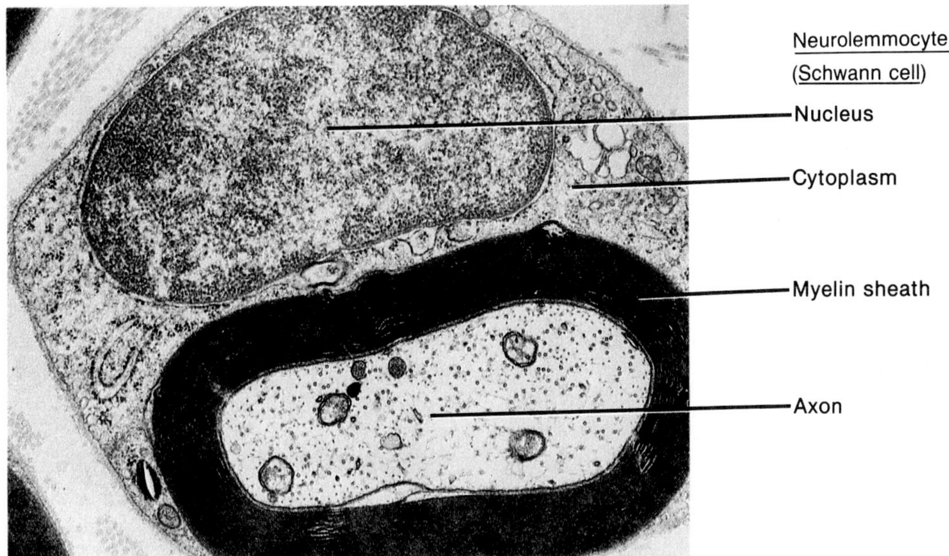

Electron micrograph of a cross section of a myelinated axon

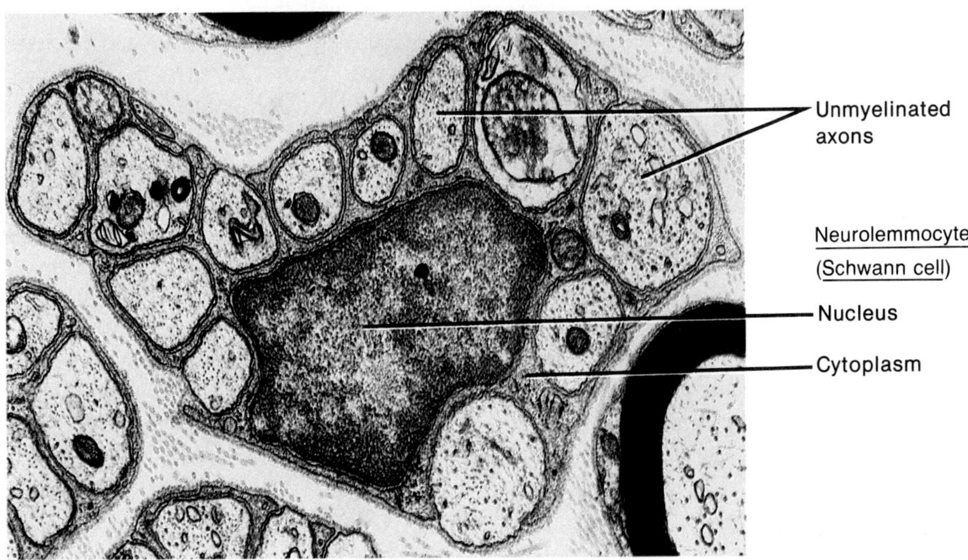

(c) Electron micrograph of a cross section of several unmyelinated axons

vier) (RON-vē-ā). Unmyelinated fibers are also enclosed by neurolemmocytes. However, they do not have multiple wrappings; they contain only a neurolemma.

Nerve fibers of the central nervous system may also be myelinated or unmyelinated. Myelination of CNS fibers is accomplished by oligodendrocytes in somewhat the same manner that neurolemmocytes myelinate PNS fibers (see Figure 16-2b), with one important difference. Oligodendrocytes merely deposit the sheath. There is no neurolemma, so the fibers of the central nervous system cannot regenerate. Myelinated fibers of the CNS also contain neurofibral nodes, but they are not so numerous.

Myelin sheaths are first laid down during the later part of fetal development and during the first year of life. The amount of myelin increases from birth to maturity, and its presence greatly increases the rate of nerve impulse conduction. Since myelination is still in progress during infancy, an infant's responses to stimuli are not as rapid or coordinated as those of an older child or an adult.

Structural Variation

Although all neurons conform to the general plan described, there are considerable differences in structure. For example, cell bodies range in diameter from 5 μm for the smallest cells to 135 μm for large motor neurons. The pattern of dendritic branching is varied and distinctive for neurons in different parts of the body. The axons of very small neurons are only a fraction of a millimeter in length and lack a myelin sheath, whereas axons of large neurons are more than a meter long and are usually enclosed in a myelin sheath.

Classification

The different neurons in the body may be classified by structure and function.

The structural classification is based on the number of processes extending from the cell body (Figure 16-5). ***Multipolar neurons*** usually have several dendrites and one axon. Most neurons in the brain and spinal cord are of this type. ***Bipolar neurons*** have one dendrite and one axon and are found in the retina of the eye, inner ear, and olfactory area. ***Unipolar (pseudounipolar) neurons*** have only one process extending from the cell body. The single process divides into a central branch, which functions as an axon, and a peripheral branch, which functions as a dendrite. Unipolar neurons originate in the embryo as bipolar neurons, and during development the axon and dendrite fuse into a single process. Unipolar neurons are found in posterior (sensory) root ganglia of spinal nerves and the ganglia of cranial nerves that carry general somatic sensory impulses. Ganglia are groups of neuron cell bodies outside the CNS.

The functional classification of neurons is based on the direction in which they transmit impulses. ***Sensory (afferent) neurons*** transmit impulses from receptors in the skin, sense organs, muscles, joints, and viscera to the brain and spinal cord. They are usually unipolar (Figure 16-6). ***Motor (efferent) neurons*** convey impulses from the brain and spinal cord to effectors, which may be either muscles or glands (see Figure 16-3a), and from higher to lower centers of the CNS. Other neurons, called ***association (connecting*** or ***interneuron) neurons,*** usually carry impulses from sensory neurons to motor neurons and are located in the brain and spinal cord only. Examples of association neurons are the ***granule cell*** and ***Purkinje*** (pur-KIN-jē) ***cell*** in the cortex (outer zone) of the cerebellum (Figure 16-7). Most neurons in the body, perhaps 90 percent, are association neurons.

The processes of afferent and efferent neurons are arranged into bundles called ***nerves*** if outside the CNS or ***fiber tracts*** if inside the CNS. (Axons of association neurons can also form fiber tracts.) Since nerves lie outside the central nervous system, they belong to the peripheral nervous system. The functional components of nerves are the nerve fibers, which may be grouped according to the following scheme.

1. ***General somatic afferent fibers*** conduct nerve impulses from the skin, skeletal muscles, and joints to the central nervous system.
2. ***General somatic efferent fibers*** conduct nerve impulses from the central nervous system to skeletal muscles. Impulses over these fibers cause the contraction of skeletal muscles.
3. ***General visceral afferent fibers*** convey nerve impulses from the viscera and blood vessels to the CNS.
4. ***General visceral efferent fibers*** belong to the autonomic nervous system and are also called ***autonomic fibers.*** They convey nerve impulses from the CNS to help control contractions of smooth and cardiac muscle and rate of secretion by glands. entirely motor

In addition to being grouped as nerves, neural tissue is also organized into other structures such as ganglia, tracts, nuclei, and horns. These structures are described in Chapter 17.

Two striking features of neurons are (1) their highly developed ability to generate and conduct electrical

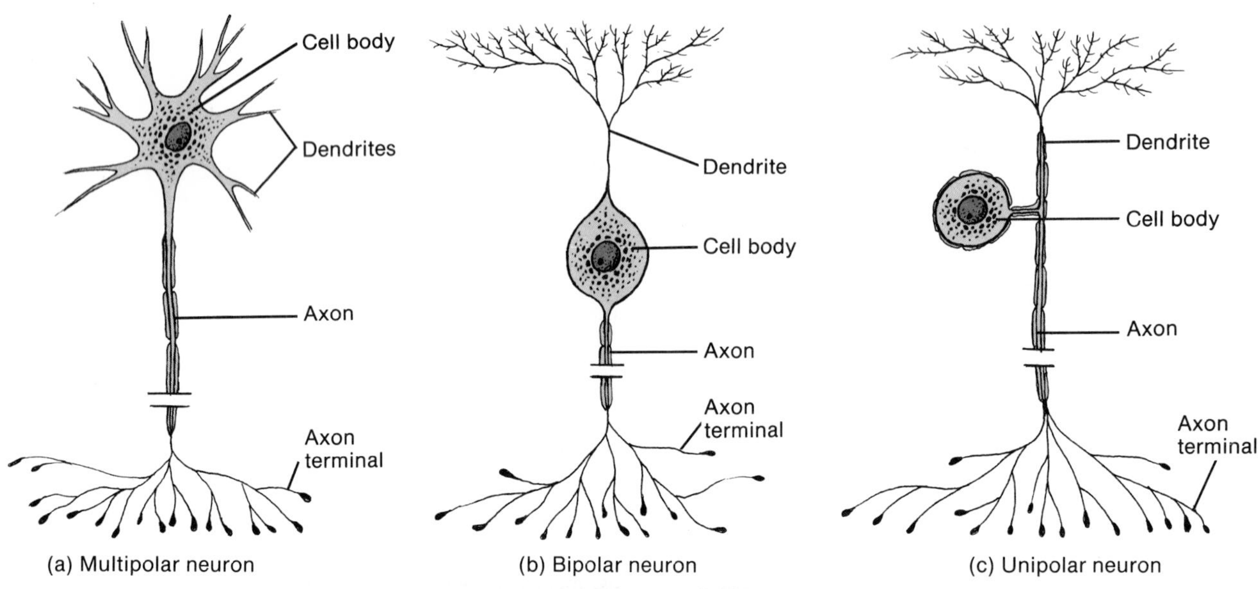

FIGURE 16-5 Structural classification of neurons.

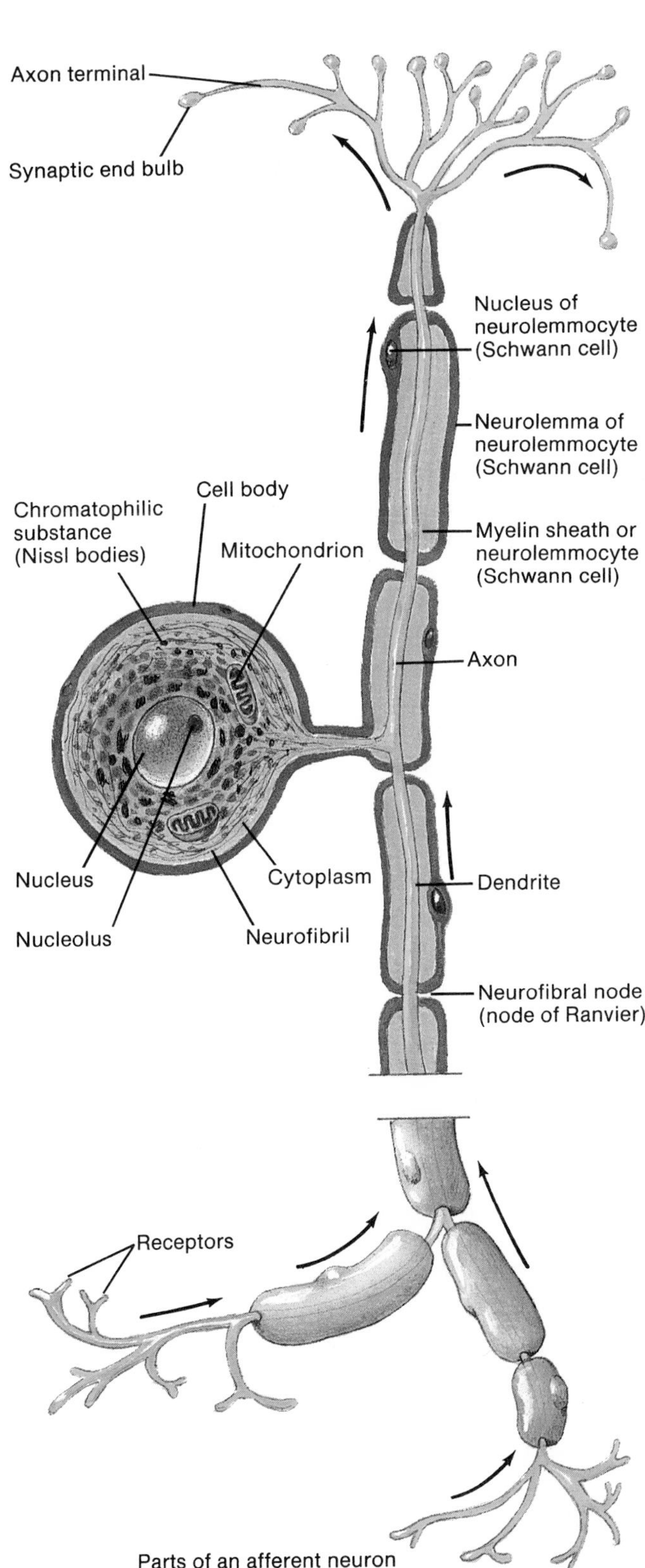

FIGURE 16-6 Structure of a typical afferent (sensory) neuron. Arrows indicate the direction in which the nerve impulses travel. The break indicates that the process is actually longer than shown.

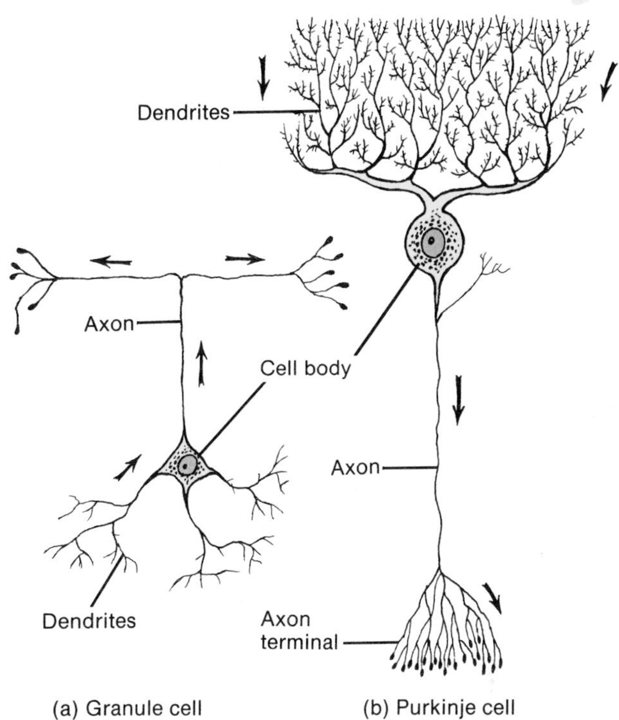

FIGURE 16-7 Association (connecting) neurons. Arrows indicate direction in which nerve impulses travel.

messages called nerve impulses and (2) their limited ability to regenerate.

NERVE IMPULSE

Although it is beyond the scope of this text to describe the details of a nerve impulse, certain concepts must be understood to know how the nervous system works. Very simply, a ***nerve impulse (nerve action potential)*** is a tiny electrochemical current of negativity that self-propagates along the surface of the membrane of a neuron. Among other things, a nerve impulse depends on the movement of sodium, potassium, and other ions between interstitial fluid and the inside of a neuron. For a nerve impulse to begin, a stimulus of adequate strength must be applied to the neuron. A ***stimulus*** is a change in the environment that may be of sufficient strength to initiate a nerve impulse. The ability of a neuron to respond to a stimulus and convert it into a nerve impulse is known as ***excitability.***

SPEED

The nerve impulse is the most rapid way that the body can respond to environmental changes. It provides the quickest means for achieving homeostasis. The speed of a nerve impulse is determined by the size, type, and physiological condition of the nerve fiber. For example, myelinated fibers with the largest diameters can transmit impulses at speeds up to about 100 m (328 ft)/sec. Unmyelinated fibers with the smallest diameters can conduct impulses at the rate of about 0.5 m (1.5 ft)/sec.

SYNAPSES

Impulses are not only conducted along the length of a neuron, but also transmitted from one neuron to another or from a neuron to another cell such as a muscle fiber or glandular cell across a ***synapse***—a junction between the cells. Within a synapse the involved cells approach one another but do not quite touch. The synapse is essential for homeostasis because of its ability to transmit certain impulses and inhibit others. Most diseases of the brain and many psychiatric disorders result from a disruption of synaptic communication. Synapses are also the sites of action for most drugs that affect the brain—both therapeutic and addictive substances.

Impulses from a neuron to a muscle fiber (cell) are transmitted across a ***neuromuscular junction,*** which was discussed in Chapter 9 (see Figure 9-5). Between a neuron and glandular cell, the impulse is transmitted across a ***neuroglandular junction.*** Together, these junctions are referred to as ***neuroeffector junctions.***

There are two types of synapses: electrical and chemical. At ***electrical synapses*** nerve impulses pass from one neuron to another through small, tubular, protein structures called ***gap junctions.*** Recall that gap junctions are also found between visceral (smooth) and cardiac muscle fibers for conducting muscle action potentials between fibers.

In a ***chemical synapse*** a neuron secretes a chemical substance called a ***neurotransmitter*** that acts on receptors of integral proteins of the next neuron, a muscle fiber at a neuromuscular junction, or a glandular cell at a neuroglandular junction. Many synapses in the central nervous system are chemical synapses. Neurotransmitters are made by the neuron, usually from amino acids, and are stored in the synaptic end bulbs in synaptic vesicles, small membrane-enclosed sacs. Perhaps the best studied neurotransmitter is ***acetylcholine (ACh),*** which is released by many neurons outside the CNS, some neurons inside the CNS, and at neuromuscular junctions.

At this point we will consider the synapse between neurons. Figure 16-8 shows the three parts of a synapse between

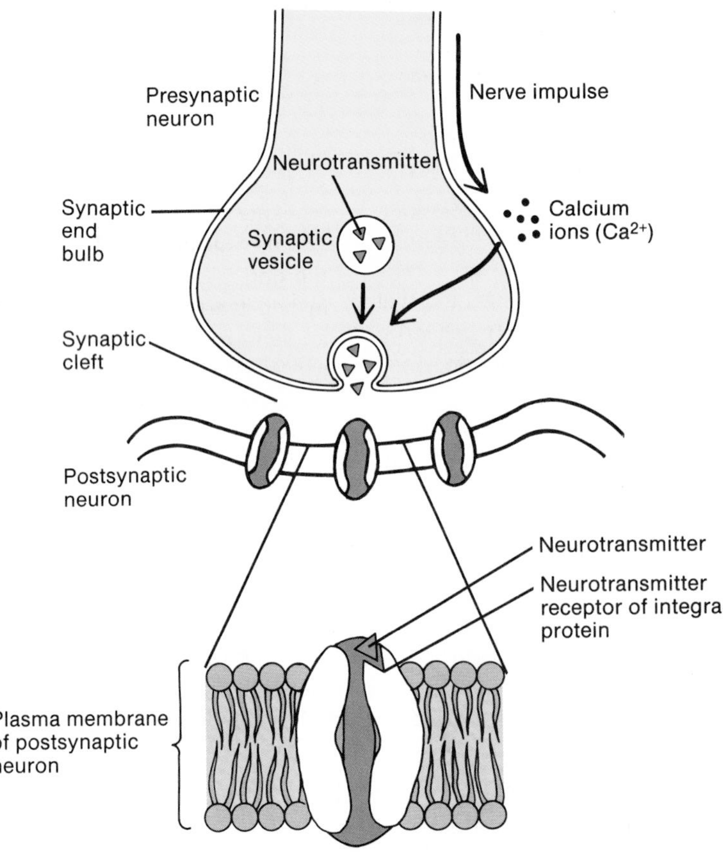

FIGURE 16-8 Synapse between neurons. Depending on the nature of the neurotransmitter and how it interacts with receptors of the postsynaptic neuron, a nerve impulse may be generated or inhibited across the synapse.

neurons. The ***presynaptic neuron*** is a neuron located before a synapse. The ***postsynaptic neuron*** is located after a synapse. The space between is a minute gap filled with extracellular fluid, about 20 nm across, called the ***synaptic cleft.*** The synaptic end bulbs of a presynaptic neuron may synapse with the dendrites, cell body, axon hillock, or axon terminal of a postsynaptic neuron. Accordingly, synapses may be classified as ***axodendritic, axosomatic,*** and ***axoaxonic.*** The synaptic end-bulbs from a single presynaptic neuron may synapse with several postsynaptic neurons. Such an arrangement, called ***divergence,*** permits a single presynaptic neuron to influence several postsynaptic neurons or several muscle fibers or gland cells at the same time (see Figure 16-10a). In another arrangement, called ***convergence,*** the synaptic end-bulbs of several presynaptic neurons synapse with a single postsynaptic neuron (see Figure 16-10b). This arrangement permits stimulation or inhibition of the postsynaptic neuron.

NEUROTRANSMITTERS

How are neurotransmitters released? When a nerve impulse arrives at the synaptic end bulb of a presynaptic neuron, calcium channels in the bulb open, calcium (Ca^{2+}) ions flood in from interstitial fluid, attract synaptic vesicles to the plasma membrane, and help liberate the neurotransmitter molecules from the vesicles into the synaptic cleft. In general, a neurotransmitter will either produce an excitatory transmission (helps create a new nerve impulse) or an inhibitory transmission (prevents further nerve impulses).

At a chemical synapse there is only ***one-way impulse conduction***—from a presynaptic axon to a postsynaptic cell—because only synaptic end bulbs of presynaptic neurons can release neurotransmitters. As a result, nerve impulses must move forward over their pathways. They cannot back up into another presynaptic neuron, a situation that would seriously disrupt homeostasis. Such a mechanism is crucial in preventing nerve impulse conduction along improper pathways.

A single postsynaptic neuron receives synapses from many presynaptic neurons. Some neurotransmitters lead to excitation and some produce inhibition. The sum of all the effects, excitatory and inhibitory, determines the final effect on the postsynaptic neuron. Thus, the postsynaptic neuron is an ***integrator.*** It receives signals, integrates them, and then responds accordingly. The postsynaptic neuron may respond in the following ways:

1. If the excitatory effect is greater than the inhibitory effect but less than the threshold (minimal) level of stimulation, the result is ***facilitation***—that is, near excitation so that subsequent stimuli can generate a nerve impulse.
2. If the excitatory effect is greater than the inhibitory effect but equal to or higher than the threshold level of stimulation, the result is ***generation of one or many nerve impulses.***
3. If the inhibitory effect is greater than the excitatory effect, the result is ***inability to generate a nerve impulse.***

CLINICAL APPLICATION

Alteration of Synaptic Conduction

There are many ways that ***synaptic conduction can be altered*** by disease, drugs, and pressure. In Chapter 9 it was noted that *myasthenia gravis* results from antibodies directed against acetylcholine receptors on skeletal muscle fiber membranes at neuromuscular junctions, causing dysfunctions in skeletal muscular contractions. *Alkalosis,* an increase in pH above 7.45, results in increased excitability of neurons that can cause lightheadedness, numbness around the mouth, tingling in the fingertips, nervousness, and muscle spasms and convulsions. *Acidosis,* a decrease in pH below 7.35, results in a progressive depression of neuronal activity that can produce apathy, weakness, and coma.

The plant derivative *curare* competes for acetylcholine receptor sites and can thus prevent muscular contractions. Curare-like drugs are often used in surgery to increase muscle relaxation. *Neostigmine* is an anticholinesterase agent that combines with acetylcholinesterase to inactivate it for several hours. Neostigmine is the antidote to curare used in surgery or in case of accidental curare poisoning and can be used to treat myasthenia gravis.

Diisopropyl fluorophosphate is a very powerful nerve gas found in many insecticides. It inactivates acetylcholinesterase for up to several weeks, making it a particularly lethal drug. It may cause nausea, diarrhea, sweating, bronchial constriction, excess respiratory mucus, generalized weakness, and fasciculation of skeletal muscles. The *botulinum toxin* inhibits the release of acetylcholine, thus inhibiting muscle contraction. It is the substance involved in one type of food poisoning. *Hypnotics, tranquilizers,* and *anesthetics* depress synaptic conduction by increasing the threshold for excitation of neurons, whereas *caffeine, benzedrine,* and *nicotine* reduce the threshold for excitation of neurons and result in facilitation.

Crack, a potent form of cocaine that is smoked rather than sniffed, interferes with the normal functioning of neurotransmitters—dopamine, norepinephrine, and serotonin—that are involved in the regulation of mood and motor functions. Once a neurotransmitter has accomplished its function, it is inactivated and returned to the presynaptic neuron for resynthesis. Initially, crack inhibits the inactivation of the neurotransmitters. The resultant buildup of dopamine, in particular, has been linked to feelings of euphoria. Inhibition or inactivation of neurotransmitters also can cause convulsions, accelerated and abnormal heart rate, vasoconstriction and high blood pressure, weight loss, insomnia, and susceptibility to disease. Repeated use of crack may produce temporary shortage of neurotransmitters, resulting in depression, anxiety, and craving for more crack. Heavy, prolonged use of crack may eventually deplete neurotransmitters to the point where euphoria no longer occurs and depression is persistent.

Pressure has an effect on nerve impulse transmission also. If excessive or prolonged pressure is applied to a nerve, as when crossing one's legs, impulse transmission is interrupted, and part of the body may "go to sleep," producing a tingling sensation. This sensation is caused by an accumulation of waste products and a depressed circulation of blood.

REGENERATION

Unlike the cells of epithelial tissue, neurons have only limited powers of ***regeneration***—that is, a natural ability to renew themselves. Around 6 months of age, virtually all developing nerve cells lose their mitotic spindle and their ability to reproduce. Thus, when a neuron is damaged or destroyed, it cannot be replaced by the daughter cells of other neurons. A neuron destroyed is permanently lost, and only some types of damage may be repaired.

In the peripheral nervous system, damage to some types of myelinated axons and dendrites can be repaired if the cell body remains intact, if the cell that performs the myelination remains active, and if a completely severed nerve is surgically reattached. If the myelinating cell is a neurolemmocyte (Schwann cell), it helps regeneration. These cells proliferate following axonal damage, and their neurolemmas form a tube that assists in regeneration. Axons in the central nervous system are myelinated by oligodendrocytes, which do not form neurolemmas to assist in regeneration and do not survive following axonal damage. An added complication in the CNS is that following axonal damage, astrocytes appear to stop axons from regenerating by activating what is called a *physiological stop pathway*. In addition, following axonal damage, astrocyte proliferation causes rapid scar tissue formation, and scar tissue is an actual physical barrier to regeneration. Thus, an injury to the brain or spinal cord is permanent. An injury to a nerve in the PNS may repair itself before scar tissue forms, and so some nerve function may be restored.

When there is damage to an axon (or to dendrites of somatic afferent neurons), there are usually changes that occur in the cell body of the affected neuron, called chromatolysis. In addition, there are always changes that occur in the portion of the axon distal to the site of injury, called Wallerian degeneration, and in the portion of the axon proximal to the site of injury, called retrograde degeneration. Chromatolysis occurs in essentially the same way, whether the damaged fiber is in the central or peripheral nervous system. The Wallerian degeneration reaction, however, depends on whether the fiber is central or peripheral.

CHROMATOLYSIS

About 24 to 48 hours after injury to a process of a central or peripheral neuron, the chromatophilic substance (Nissl bodies), normally arranged in an orderly fashion in an uninjured cell body, breaks down into finely granular masses. This alteration is called ***chromatolysis*** (krō′-ma-TOL-i-sis; *chromo* = color; *lysis* = dissolution). It begins between the axon hillock and nucleus but spreads throughout the cell body. As a result of chromatolysis, the cell body swells, and the swelling reaches its maximum between 10 and 20 days after injury (Figure 16-9b). Chromatolysis results in a loss of ribosomes by the rough endoplasmic reticulum

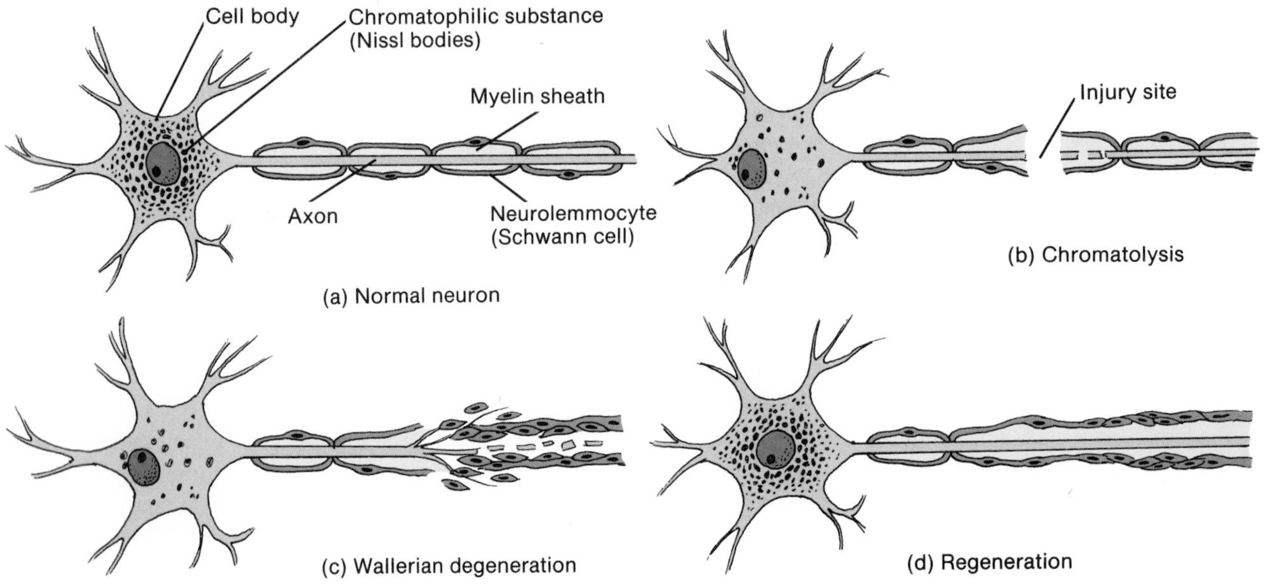

FIGURE 16-9 Peripheral nerve damage and repair.

and an increase in the number of free ribosomes. Another sign of the chromatolysis is the off-center position of the nucleus in the cell body. This change makes it possible to identify the cell bodies of damaged fibers through a microscope.

WALLERIAN DEGENERATION

The part of the process distal to the damage becomes slightly swollen and then breaks up into fragments by the third to fifth day. The myelin sheath around the axon or dendrite also undergoes degeneration (Figure 16-9c). Degeneration of the distal portion of the axon or dendrite and myelin sheath is called ***Wallerian degeneration.*** Following degeneration, there is phagocytosis of the remains by macrophages.

Even though there is degeneration of the axon or dendrite and myelin sheath, the neurolemma of the neurolemmocytes remains. The neurolemmocytes on either side of the site of injury multiply by mitosis and grow toward each other and attempt to form a tube across the injured area. The tube provides a means for new axons or dendrites to grow from the proximal area across the injured area into the distal area previously occupied by the original nerve fiber (Figure 16-9d). The growth of new axons or dendrites will not occur if the gap at the site of injury is too large or if the gap becomes filled with dense collagenous fibers.

RETROGRADE DEGENERATION

The changes in the proximal portion of the axon, called ***retrograde degeneration,*** are similar to those that occur during Wallerian degeneration. The main difference in retrograde degeneration is that the changes occur only as far as the first neurofibral node (node of Ranvier).

REPAIR

Following chromatolysis, there are signs of recovery in the cell body. There is an acceleration of RNA and protein synthesis, which favors regeneration of the axon. Recovery often takes several months and involves the restoration of normal levels of RNA, proteins, and the chromatophilic substance to their usual, uninjured patterns.

Accelerated protein synthesis is required for repair of the damaged axon. The proteins synthesized in the cell body pass into the empty lumen of the tube formed by neurolemmocytes, by axoplasmic flow at the rate of about 1 mm (0.04 in.) per day. The proteins assist in regenerating the damaged axon. During the first few days following damage, buds of regenerating axons or dendrites begin to invade the tube formed by the neurolemmocytes. Axons or dendrites from the proximal area grow at the rate of about 1.5 mm (0.06 in.) per day across the area of damage, find their way into the distal neurolemmal tubes, and grow toward the distally located receptors and effectors. Thus, sensory and motor connections are reestablished. In time, a new myelin sheath is also produced by the neurolemmocytes. However, function is never completely restored after a nerve is severed.

ORGANIZATION OF NEURONS

The central nervous system contains billions of neurons. Their arrangement is not haphazard. They are organized into definite patterns called ***neuronal pools.*** Each pool differs from all others and has its own role in regulating homeostasis.

Neuronal pools in the CNS are arranged in patterns over which the impulses are conducted. These are termed ***circuits. Simple series circuits*** are arranged so that a presynaptic neuron stimulates a single neuron in a pool. The single neuron then stimulates another, and so on. In other words, the impulse is relayed from one neuron to another in succession as new impulses are generated at each synapse.

Most circuits, however, are more complex. In a ***diverging circuit,*** the nerve impulse from a single presynaptic neuron causes the stimulation of increasing numbers of cells along the circuit (Figure 16-10a). An example of such a circuit is a single motor neuron in the brain stimulating numerous other motor neurons in the spinal cord that, in turn, leave the spinal cord where each stimulates many skeletal muscle fibers. Thus, a single impulse may result in the contraction of several skeletal muscle fibers. In another kind of diverging circuit, impulses from one pathway are relayed to other pathways, so the same information travels in various directions at the same time. This circuit is common along sensory pathways of the nervous system.

Another kind of circuit is called a ***converging circuit*** (Figure 16-10b). In one pattern of convergence, the postsynaptic neuron receives impulses from several fibers of the same source. Here, there is the possibility of strong excitation or inhibition. In a second pattern, the postsynaptic neuron receives impulses from several different sources. Here, there is a possibility of reacting the same way to different stimuli. Suppose your reaction to vomit is distinctly unpleasant. The smell of vomit (one kind of stimulus), the sight of vomit (another kind), or just reading about vomit (still another kind) might all have the same effect on you—an unpleasant one.

Some circuits in your body are constructed so that once the synaptic cell is stimulated, it will cause the postsynaptic cell to transmit a series of nerve impulses. One such circuit is called a ***reverberating (oscillatory) circuit*** (Figure 16-10c). In this pattern, the incoming impulse stimulates the first neuron, which stimulates the second, which stimulates the third, and so on. Branches from the second and third neurons synapse with the first, however, sending the impulse back through the circuit again and again. A central feature of the reverberating circuit is that once fired, the output signal may last from a few seconds to many hours. The

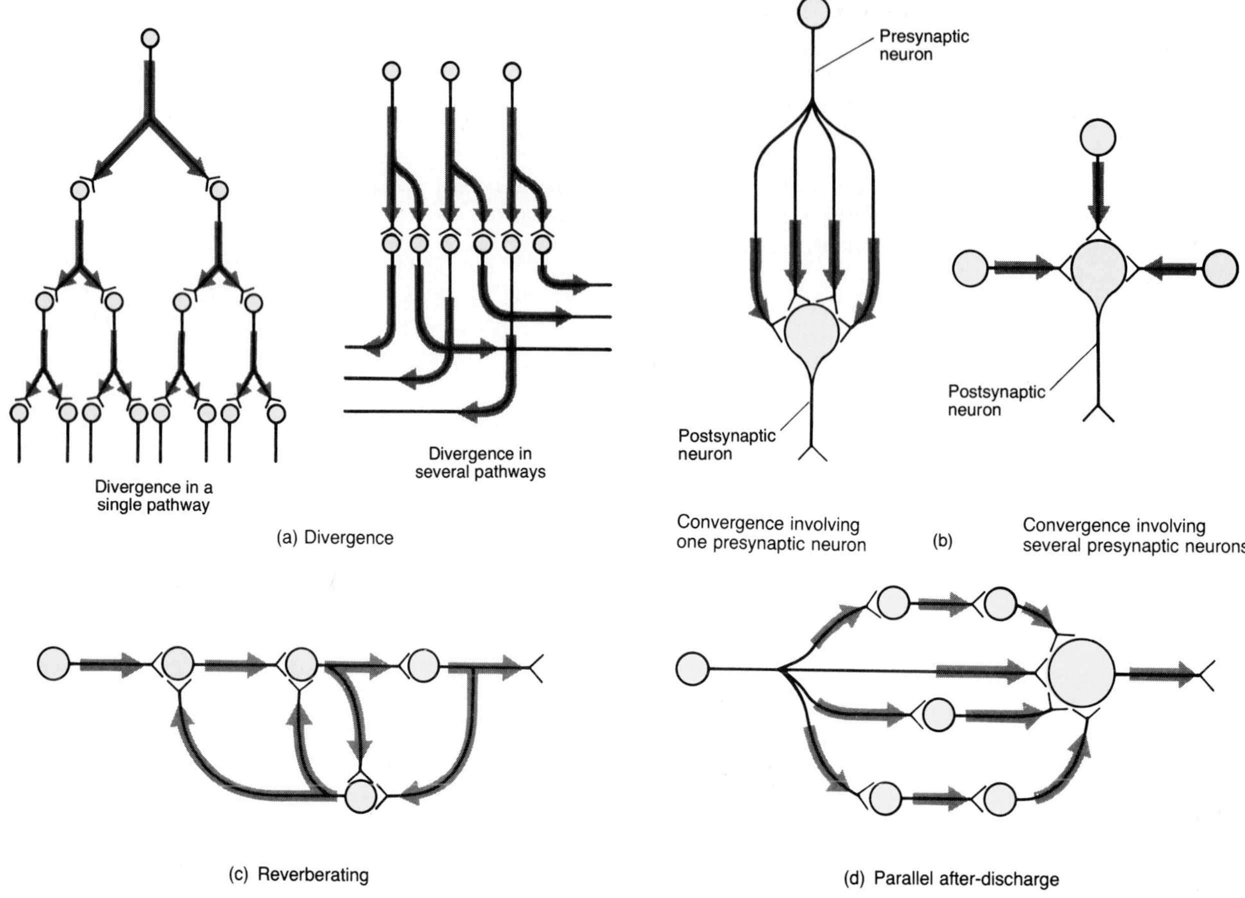

FIGURE 16-10 Circuits of neuronal pools.

duration depends on the number and arrangement of neurons in the circuit. Among the body responses thought to be the result of output signals from reverberating circuits are the rate of breathing, coordinated muscular activities, waking up, and sleeping (when reverberation stops). Some scientists think reverberating circuits are related to short-term memory. One form of epilepsy (grand mal) is probably caused by abnormal reverberating circuits.

A final circuit worth consideration is the ***parallel after-discharge circuit*** (Figure 16-10d). Like a reverberating circuit, a parallel after-discharge circuit is constructed so the postsynaptic cell transmits a series of nerve impulses. In a parallel after-discharge circuit, a single presynaptic cell stimulates a group of neurons, each of which synapses with a common postsynaptic cell. The advantage of this circuit is that the postsynaptic neuron can send out a stream of impulses in succession as they are received. The impulses leave the postsynaptic neuron once every 1/2000 sec. This circuit has no feedback system. Once all the neurons in the circuit have transmitted their impulses to the postsynaptic neuron, the circuit is broken. It is thought that the parallel after-discharge circuit is employed for precise activities such as mathematical calculations.

STUDY OUTLINE

Organization (p. 456)

1. The nervous system helps control and integrate all body activities by sensing changes (sensory), interpreting them (integrative), and reacting to them (motor).
2. The central nervous system (CNS) consists of the brain and spinal cord.
3. The peripheral nervous system (PNS) consists of nerves emerging from the brain (cranial nerves) and nerves emerging from the spinal cord (spinal nerves).
4. The afferent component of the PNS consists of sensory neurons that conduct impulses from receptors to the CNS; the efferent component consists of motor neurons that conduct impulses from the CNS to muscles and glands.

5. The PNS is divided into a somatic nervous system (SNS) and an autonomic nervous system (ANS).
6. The SNS contains sensory neurons that convey impulses from cutaneous and special sense receptors in the head, body wall, and extremities and motor neurons that conduct impulses from the CNS to skeletal muscle tissue.
7. The ANS consists of sensory neurons that convey impulses from receptors in viscera and motor neurons that convey impulses from the CNS to smooth muscle tissue, cardiac muscle tissue, and glands.

Histology (p. 456)

Neuroglia (p. 456)

1. Neuroglia are specialized tissue cells that support neurons, attach neurons to blood vessels, produce the myelin sheath around axons, and carry out phagocytosis.
2. Neuroglia include astrocytes, oligodendrocytes, microglia, ependyma, neurolemmocytes (Schwann cells), and satellite cells.

Neurons (p. 458)

1. Most neurons, or nerve cells, consist of a cell body, dendrites that pick up stimuli and convey nerve impulses to the cell body, and usually a single axon. The axon conducts nerve impulses from the neuron to the dendrites or cell body of another neuron or to an effector organ of the body (muscle or gland).
2. On the basis of structure, neurons are multipolar, bipolar, and unipolar.
3. On the basis of function, sensory (afferent) neurons conduct nerve impulses to the central nervous system; association (connecting) neurons conduct nerve impulses to other neurons, including motor neurons; and motor (efferent) neurons conduct nerve impulses to effectors.
4. A nerve is a bundle of processes of afferent or efferent neurons outside the central nervous system.

Nerve Impulse (p. 463)

1. A nerve impulse (nerve action potential) is a tiny electrochemical current that travels along the surface of the membrane of a neuron.
2. The ability of a neuron to respond to a strong enough stimulus and convert it into a nerve impulse is called excitability.
3. The speed of a nerve impulse is determined by the size, type, and physiological condition of the nerve fiber.
4. The junction between neurons and muscle fibers or glandular cells is called a synapse. Synapses may be electrical or chemical.
5. Impulse conduction across a chemical synapse requires a neurotransmitter.
6. Impulse conduction at a chemical synapse is one-way conduction.
7. A postsynaptic neuron is an integrator. It may respond by facilitation, generation of an impulse, or inhibition of an impulse.
8. Synaptic conduction may be altered by disease, drugs, and pressure.

Regeneration (p. 466)

1. At about 6 months of age, the neuron loses its mitotic apparatus and is no longer able to divide.
2. Nerve fibers in the PNS have a neurolemma and are thus capable of regeneration.
3. Axons and dendrites in the CNS do not have a neurolemma, and therefore injury to the brain or spinal cord is permanent.
4. Following peripheral nerve damage, regeneration is accompanied by chromatolysis, Wallerian degeneration, retrograde degeneration, and repair.

Organization of Neurons (p. 467)

1. Neurons in the central nervous system are organized into definite patterns called neuronal pools.
2. Neuronal pools are organized into circuits. These include simple series, diverging, converging, reverberating, and parallel after-discharge circuits.

REVIEW QUESTIONS

1. Describe the three basic functions of the nervous system. (p. 456)
2. Distinguish between the central and peripheral nervous systems, and describe the functions of each subdivision. (p. 456)
3. Relate the terms *voluntary* and *involuntary* to the nervous system. (p. 456)
4. What are neuroglia? List the principal types and their functions. Why are they important clinically? (p. 456)
5. Define a neuron. Diagram and label a neuron. Next to each part list its function. (p. 458)
6. What is a myelin sheath? How is it formed? (p. 460)
7. Define the neurolemma (sheath of Schwann). Why is it important? (p. 460)
8. Discuss the structural classification of neurons. Give an example of each. (p. 462)
9. What are the structural differences between a typical afferent and efferent neuron? Give several examples of association neurons. (p. 462)
10. Describe the functional classification of neurons. (p. 462)
11. Distinguish among the following kinds of fibers: general somatic afferent, general somatic efferent, general visceral afferent, and general visceral efferent. (p. 462)
12. Define a nerve impulse. What is excitability? (p. 463)
13. What factors determine the speed of nerve impulse conduction? (p. 463)
14. What is a synapse? Distinguish between an electrical and chemical synapse. (p. 464)
15. How are nerve impulses transmitted across a chemical synapse? (p. 465)
16. Why does one-way impulse conduction occur at a chemical synapse? (p. 465)

17. Why is the postsynaptic neuron called an integrator? (p. 465)
18. How do disease, drugs, crack, and pressure affect synaptic transmission? (p. 465)
19. What determines neuron regeneration? (p. 466)
20. Outline the principal events that occur as part of chromatolysis, Wallerian degeneration, and retrograde degeneration, and repair following peripheral nerve damage. (p. 466)
21. What is a neuron circuit? Distinguish among simple series, diverging, converging, reverberating, and parallel after-discharge circuits. (p. 467)

SELF QUIZ

Choose the one best answer to these questions.

___ **1.** Damage to the retina of the eye might involve which kind of neuron?
A. unipolar; B. multipolar; C. bipolar; D. tripolar; E. pseudounipolar.

___ **2.** The myelin sheath of central nervous system (CNS) neurons is produced by
A. neurolemmocytes (Schwann cells); B. astrocytes; C. oligodendrocytes; D. microglia; E. ependyma.

___ **3.** During an infection of the nervous system, you would expect to find an increase in the number of
A. microglia; B. astrocytes; C. oligodendrocytes; D. association neurons; E. ependyma.

___ **4.** Which fibers would conduct a nerve impulse fastest?
A. myelinated fibers with small diameter; B. unmyelinated fibers with small diameter; C. myelinated fibers with large diameter; D. unmyelinated fibers with large diameter; E. myelinated fibers with intermediate diameter.

___ **5.** The point of contact between a nerve fiber and a muscle or a gland is called the
A. axon hillock; B. exteroceptor; C. neuroeffector junction; D. association neuron; E. ganglion.

___ **6.** A neuron cannot undergo regeneration if
(1) its cell body has been destroyed.
(2) its axon has no myelin sheath.
(3) its axon has no neurolemma.
A. (1) only; B. (2) only; C. (3) only; D. any of the above; E. either (1) or (3).

___ **7.** Which of the following is true?
A. bipolar neurons are those with two axons; B. a typical motor neuron is a unipolar neuron; C. sensory neurons are usually bipolar; D. most unipolar neurons are found in the CNS; E. multipolar neurons possess numerous dendrites and only one axon.

___ **8.** Which of the following, regarding neurons, is/are true?
A. the axon of a motor neuron extends from the cell body to an effector; B. a neuron usually has many axons, connected to other neurons; C. sensory and motor neurons have dendrites, whereas association neurons do not; D. the dendrite of one neuron connects with a dendrite of the next neuron; E. all of the above.

___ **9.** There are several types of specialized cells called neuroglia. Which of the following pertain(s) correctly to one or the other of these types?
(1) attach neurons to blood vessels
(2) conduct nerve impulses
(3) produce a myelin sheath around CNS neurons
(4) engulf and destroy microbes
A. (1) only; B. (2) only; C. (3) only; D. (4) only; E. (1), (3), and (4).

___ **10.** Which of the following is/are false?
A. the CNS consists of the brain and cranial nerves; B. the PNS consists of cranial and spinal nerves; C. motor responses of the somatic nervous system are under conscious control; D. motor responses of the autonomic nervous system are under subconscious control; E. all of the above.

11. Match the following:

___ **a.** contains nucleus; cannot regenerate since it lacks mitotic apparatus
___ **b.** yellowish pigment that increases with age; appears to be by-product of lysosomes
___ **c.** provide energy for neurons
___ **d.** long, thin fibrils composed of intermediate filaments; may function in transport
___ **e.** orderly arrangement of rough ER; site of protein synthesis
___ **f.** conducts nerve impulses toward cell body
___ **g.** conducts nerve impulses away from cell body; has synaptic end bulbs that contain neurotransmitters
___ **h.** fine filaments that are branching ends of axon collaterals

A. axon
B. axon terminal
C. cell body
D. chromatophilic substance (Nissl bodies)
E. dendrite
F. lipofuscin
G. mitochondria
H. neurofibrils

Complete the following.

12. The junction between cells is called a ________.
13. The part of a neuron that conducts nerve impulses toward the cell body is the ________.
14. Sacs in the axon terminals that store neurotransmitters are called ________.
15. In a ________ circuit, a nerve impulse from a single presynaptic neuron causes stimulation of several postsynaptic neurons.

Circle T (true) or F (false) for the following.

T F 16. Neurotransmitters are released at chemical synapses.
T F 17. A neuron with several dendrites and one axon is classified as a sensory neuron.
T F 18. The ability of a neuron to respond to a stimulus and convert it to a nerve impulse is called excitability.
T F 19. Neuroglia are a common source of tumors of the nervous system.
T F 20. General somatic afferent neurons conduct nerve impulses from viscera to the CNS.

The Spinal Cord and the Spinal Nerves

17

STUDENT OBJECTIVES

1. Describe how neural tissue is grouped.
2. Explain how the spinal cord is protected.
3. Describe the gross anatomical features of the spinal cord.
4. Describe the structure and location of the spinal meninges.
5. Explain the functions of the spinal cord as a conduction pathway and a reflex center.
6. List the location, origin, termination, and function of the principal ascending and descending tracts of the spinal cord.
7. Describe the composition, coverings, and branches of a spinal nerve.
8. Explain the composition and distribution of the cervical, brachial, lumbar, and sacral plexuses.
9. Define a dermatome and state its clinical importance.
10. Describe spinal cord injury and list the immediate and long-range effects.
11. Explain the causes and symptoms of neuritis, sciatica, and shingles.

CHAPTER OUTLINE

■ **Grouping of Neural Tissue**
■ **Spinal Cord**
Protection and Coverings
Vertebral Canal
Meninges
General Features
Structure in Cross Section
Functions
Impulse Conduction
Reflex Center
■ **Spinal Nerves**
Names
Composition and Coverings
Distribution
Branches
Plexuses
Intercostal (Thoracic) Nerves
Dermatomes
■ **Applications to Health**

In this chapter, our main concern will be to study the structure and function of the spinal cord and the nerves that originate from it. Keep in mind, however, that the spinal cord is continuous with the brain and that together they constitute the central nervous system.

GROUPING OF NEURAL TISSUE

The term ***white matter*** refers to aggregations of myelinated processes from many neurons supported by oligodendrocytes or neurolemmocytes (Schwann cells). Myelin is a phospholipid and has a whitish color that gives white matter its name. The ***gray matter*** of the nervous system contains either nerve cell bodies and dendrites or bundles of unmyelinated axons and neuroglia. The absence of myelin in these areas accounts for their gray color.

A ***nerve*** is a bundle of fibers (axons and/or dendrites) *outside* the central nervous system. Since the dendrites of somatic afferent neurons and axons of somatic efferent neurons of the peripheral nervous system are myelinated, most nerves are white matter. Nerve cell bodies outside the central nervous system are generally grouped with other nerve cell bodies to form clumps termed ***ganglia*** (GANG-lē-a; *ganglion* = knot). Ganglia, since they are made up principally of nerve cell bodies, are masses of gray matter.

A ***tract*** is a bundle of myelinated fibers inside the central nervous system. Tracts may run long distances up or down the spinal cord. Tracts also exist in the brain and connect parts of the brain with each other and with the spinal cord. Spinal tracts that conduct impulses up the cord to tracts in the brain are concerned with sensory impulses and are called ***ascending tracts.*** Spinal tracts that carry impulses down the cord are motor tracts called ***descending tracts*** and are continuous with motor tracts of the brain. The major tracts consist of mostly myelinated fibers and are therefore white matter.

A ***nucleus*** is a mass of unmyelinated nerve cell bodies and dendrites in the central nervous system. Nuclei form gray matter. ***Horns*** are the chief areas of gray matter in the spinal cord. ***Columns*** are regions of white matter in the spinal cord that surround horns (see Figure 17-3).

SPINAL CORD

PROTECTION AND COVERINGS

Vertebral Canal

The spinal cord is located in the vertebral (spinal) canal of the vertebral column. The canal is formed by the vertebral foramina of all the vertebrae arranged continuously (see Figure 6-13a). Since the wall of the vertebral canal is essentially a ring of bone surrounding the spinal cord, the cord is well protected. Additional protection is provided by the meninges, cerebrospinal fluid, and the vertebral ligaments.

Meninges

The ***meninges*** (me-NIN-jēz) are connective tissue coverings that run continuously around the spinal cord and brain (***meninx*** is the singular form). Those associated specifically with the cord are known as ***spinal meninges*** (Figure 17-1). Those surrounding the brain are called ***cranial meninges.*** The outer spinal meninx is called the ***dura mater*** (DYOO-ra MĀ-ter), meaning "tough mother," since it is composed of dense connective tissue. It forms a tube from the level of the second sacral vertebra to the foramen magnum, where it is continuous with the dura mater of the brain. Between the dura mater and the wall of the vertebral canal is the ***epidural space,*** which is filled with fat, connective tissue, and blood vessels. It serves as padding around the cord. The epidural space inferior to the second lumbar vertebra is the site for the injection of anesthetics, such as a saddleblock for childbirth.

The middle spinal meninx is called the ***arachnoid*** (a-RAK-noyd; *arachne* = spiderlike) because of its delicate weblike appearance. The arachnoid consists of delicate collagenous fibers and some elastic fibers. It forms a tube

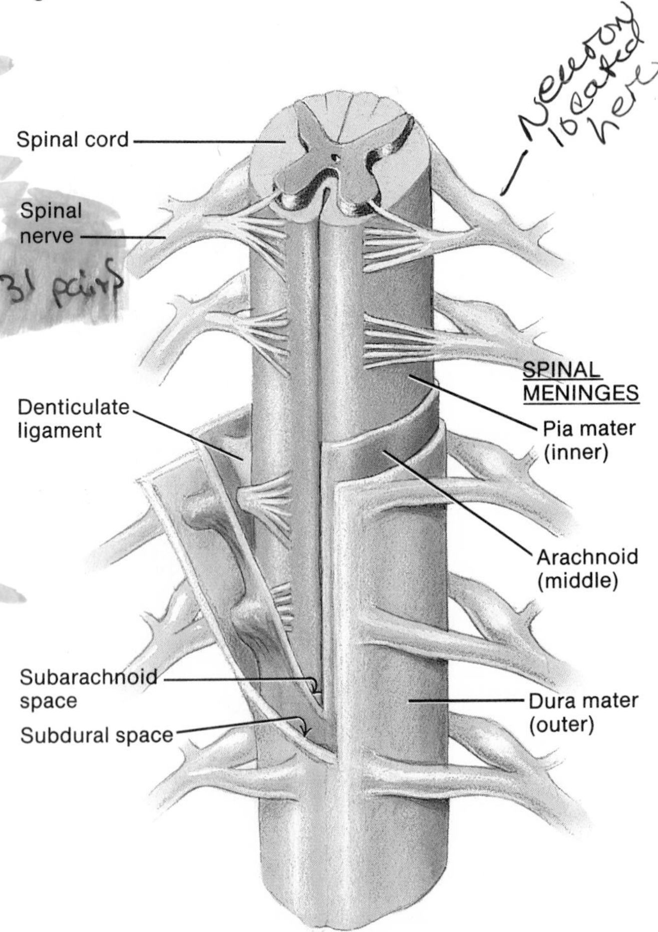

(a) Sections through spinal cord

FIGURE 17-1 Spinal meninges. (b) and (c) Courtesy of N. Gluhbegovic and T. H. Williams, *The Human Brain: A Photographic Guide,* Harper & Row, Publishers, Inc., New York, 1980.

inside the dura mater and is also continuous with the arachnoid of the brain. Between the dura mater and the arachnoid is the ***subdural space,*** which contains lymphatic fluid.

The inner meninx, known as the ***pia mater*** (PĪ-a MĀ-ter), or delicate mother, is a thin transparent connective tissue layer that adheres to the surface of the spinal cord and brain. It consists of interlacing bundles of collagenous fibers and some fine elastic fibers and contains numerous blood vessels. Between the arachnoid and the pia mater is the ***subarachnoid space,*** where cerebrospinal fluid circulates.

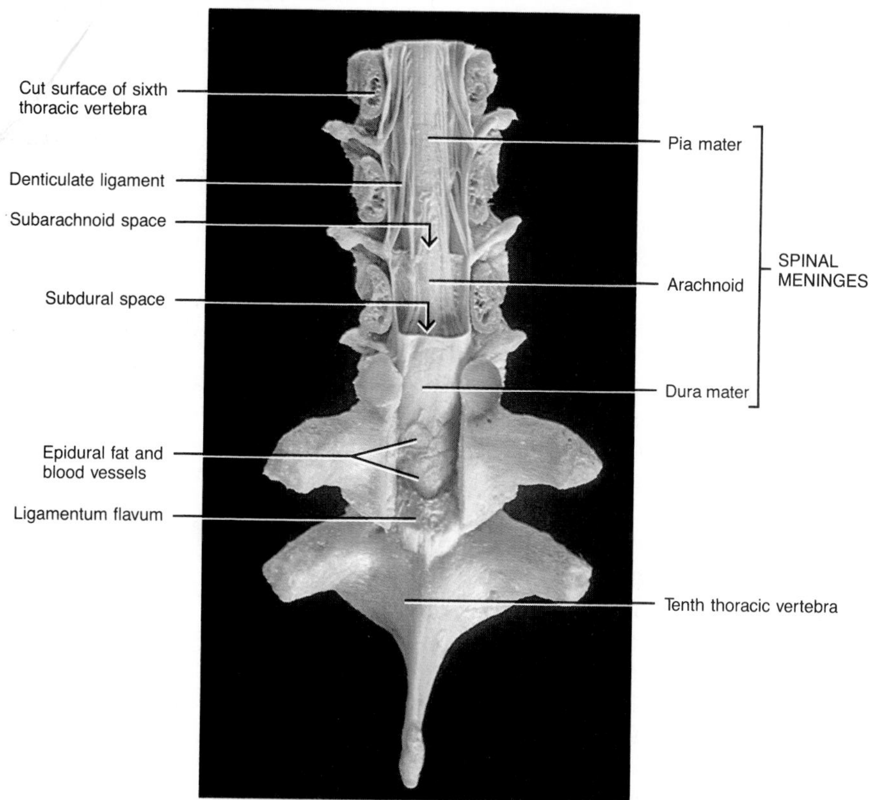

(b) Photograph of the posterior aspect of the spinal cord

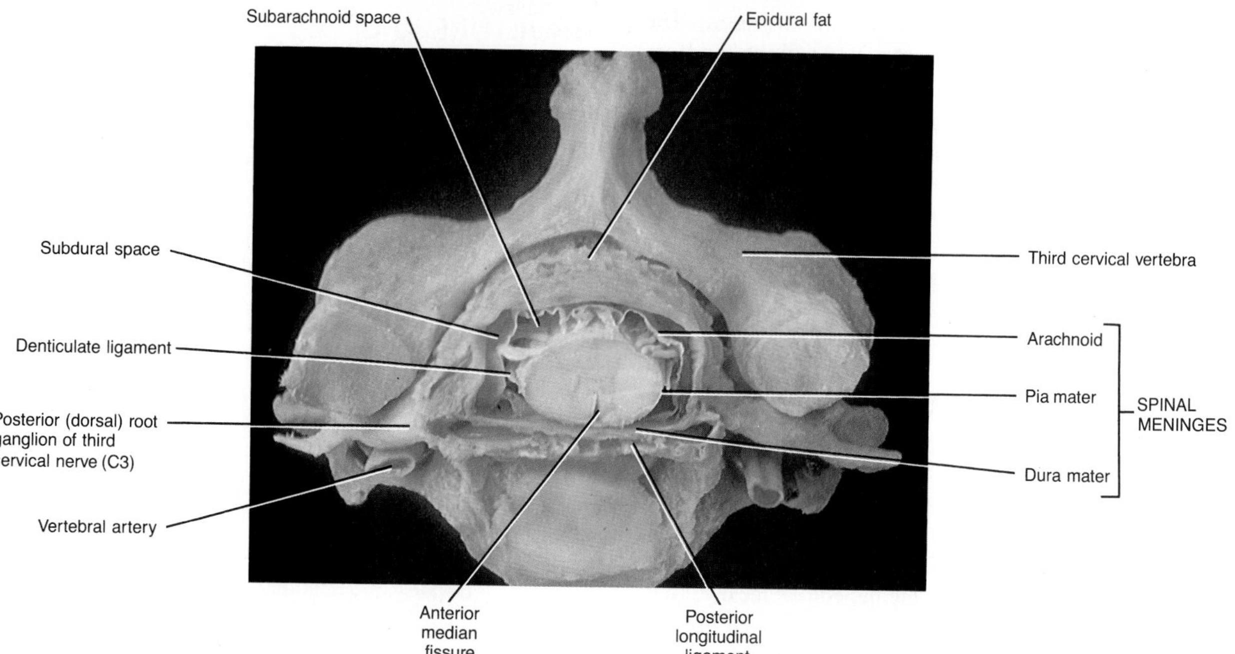

(c) Photograph of a cross section of the spinal cord

All three spinal meninges cover the spinal nerves up to the point of exit from the spinal column through the intervertebral foramina. The spinal cord is suspended in the middle of its dural sheath by membranous extensions of the pia mater. These extensions, the ***denticulate*** (den-TIK-yoo-lāt) ***ligaments,*** are attached laterally to the dura mater along the length of the cord between the ventral and dorsal nerve roots on either side (Figure 17-1). The ligaments protect the spinal cord against shock and sudden displacement.

Inflammation of the meninges in general is known as ***meningitis.*** Inflammation of the arachnoid and pia mater, specifically, is called ***leptomeningitis,*** the most common form of meningitis.

CLINICAL APPLICATION

Spinal Puncture

A ***spinal (lumbar) puncture (tap)*** is normally performed between the third and fourth or fourth and fifth lumbar vertebrae. (A line drawn across the highest points of the iliac crests passes through the spinous process of the fourth lumbar vertebra.) Because the spinal cord ends at the upper border of the second lumbar vertebra, a spinal puncture poses no danger to the cord. In one commonly used position, the patient lies on one side, drawing the knees and chest together to separate the vertebrae slightly. A local anesthetic is given, and a long needle is inserted into the subarachnoid space. Once the needle is in place, a pressure-measuring device (manometer) is attached to measure the pressure of the CSF. Next, a sample of CSF, usually between 5 to 10 ml in three or more separate tubes, is carefully withdrawn for analysis. Before the needle is withdrawn, the pressure of the CSF is again recorded.

Spinal punctures are used to withdraw cerebrospinal fluid (CSF) for diagnostic purposes; to administer chemotherapy; to measure CSF pressure; to evaluate the effects of treatment; and to introduce antibiotics (as in the case of meningitis), contrast media for other procedures, and anesthetics. In a radiologic technique called ***myelography*** (mī-e-LOG-ra-fē), a spinal puncture is used to introduce a contrast medium into the spinal subarachnoid space to determine or rule out the presence of lesions (tumors and herniated discs, for example) within and around the spinal cord.

GENERAL FEATURES

The ***spinal cord*** is a cylindrical structure that is slightly flattened anteriorly and posteriorly. It begins as a continuation of the medulla oblongata, the inferior part of the brain stem, and extends from the foramen magnum of the occipital bone to the level of the upper border of the second lumbar vertebra (Figure 17-2). The length of the adult spinal cord ranges from 42 to 45 cm (16 to 18 in.). The diameter of the cord is about 2.54 cm (1 in.) in the midthoracic region but is somewhat larger in the lower cervical and midlumbar regions.

When the cord is viewed externally, two conspicuous enlargements can be seen. The superior enlargement, the ***cervical enlargement,*** extends from the fourth cervical to the first thoracic vertebra. Nerves that supply the upper extremities arise from the cervical enlargement. The inferior enlargement, called the ***lumbar enlargement,*** extends from the ninth to the twelfth thoracic vertebra. Nerves that supply the lower extremities arise from the lumbar enlargement.

Below the lumbar enlargement, the spinal cord tapers to a conical portion known as the ***conus medullaris*** (KŌ-nus med-yoo-LAR-is). It ends at the level of the intervertebral disc between the first and second lumbar vertebrae. Arising from the conus medullaris is the ***filum terminale*** (FĪ-lum ter-mi-NAL-ē), a nonnervous fibrous tissue of the spinal cord that extends inferiorly to attach to the coccyx. The filum terminale consists mostly of pia mater, the innermost of three meninges that cover and protect the spinal cord and brain. Some nerves that arise from the lower portion of the cord do not leave the vertebral column immediately. They angle inferiorly in the vertebral canal like wisps of flowing hair from the end of the cord. They are appropriately named the ***cauda equina*** (KAW-da ē-KWĪ-na), meaning "horse's tail."

The spinal cord is a series of 31 sections called ***spinal segments,*** each giving rise to a pair of spinal nerves. The cord is divided into right and left sides by two grooves (see Figure 17-3). The ***anterior median fissure*** is a deep, wide groove on the anterior (ventral) surface, and the ***posterior median sulcus*** is a shallower, narrow groove on the posterior (dorsal) surface.

STRUCTURE IN CROSS SECTION

The spinal cord consists of both gray and white matter. Figure 17-3 shows that the gray matter forms an H-shaped area within the white matter. The gray matter consists primarily of nerve cell bodies and unmyelinated axons and dendrites of association and motor neurons. The white matter consists of bundles of myelinated axons of motor and sensory neurons and forms the sensory and motor tracts of the spinal cord.

The cross bar of the H is formed by the ***gray commissure*** (KOM-mi-shur). In the center of the gray commissure is a small space called the ***central canal.*** This canal, which contains cerebrospiral fluid, runs the length of the spinal cord and is continuous with the fourth ventricle (a space that contains cerebrospinal fluid) of the medulla of the brain. Anterior to the gray commissure is the ***anterior (ventral) white commissure,*** which connects the white matter of the right and left sides of the spinal cord.

The sides of the H are further subdivided into regions called ***horns.*** The regions closer to the front of the cord are called ***anterior (ventral) gray horns,*** those closer to

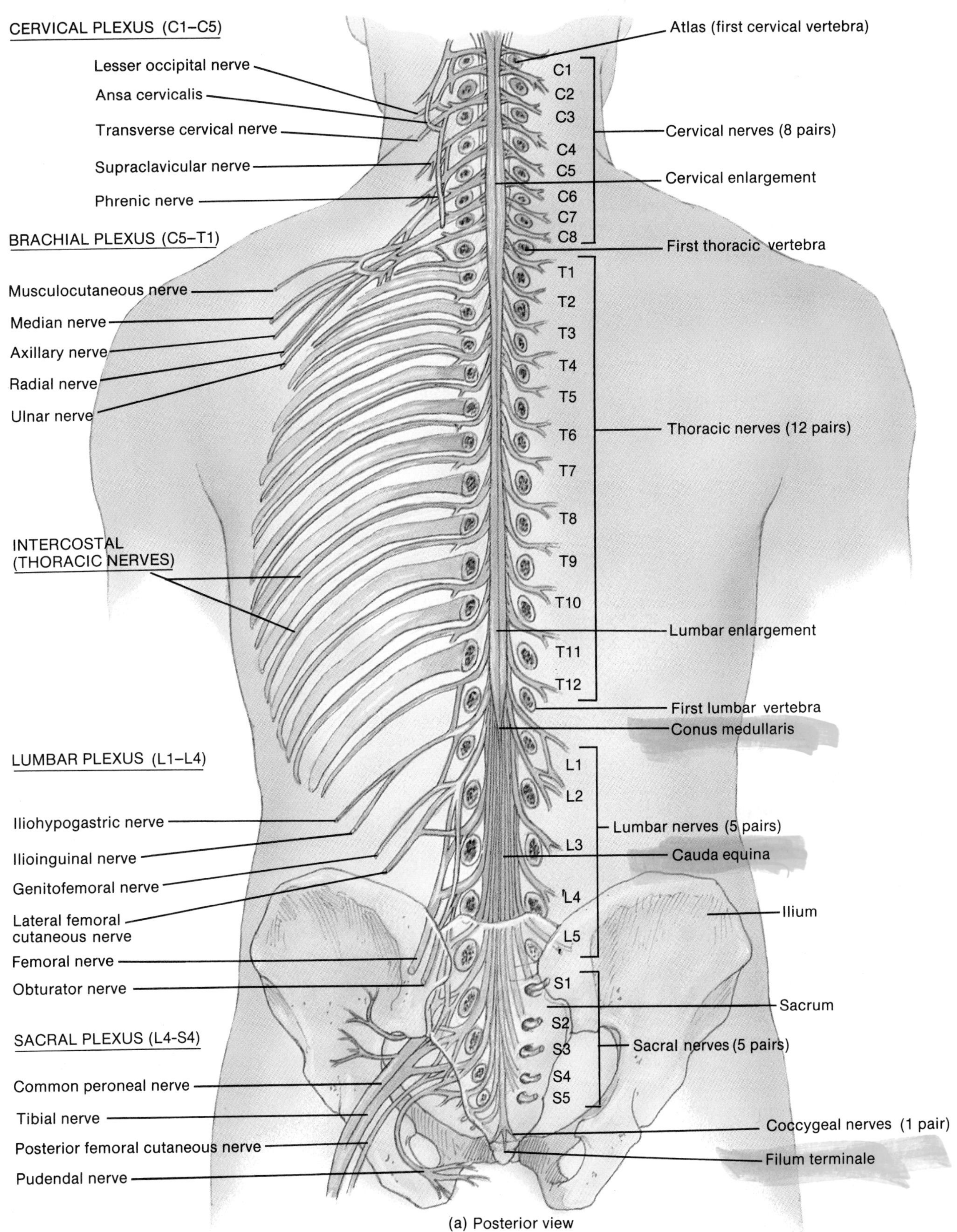

(a) Posterior view

FIGURE 17-2 Spinal cord and spinal nerves. (a) Overview. (b) Photograph of the inferior portion of the medulla and the superior six segments of the cervical portion of the spinal cord. (c) Photograph of the conus medullaris and cauda equina. (Photographs courtesy of N. Gluhbegovic and T. H. Williams, *The Human Brain: A Photographic Guide,* Harper & Row, Publishers, Inc., New York, 1980.)

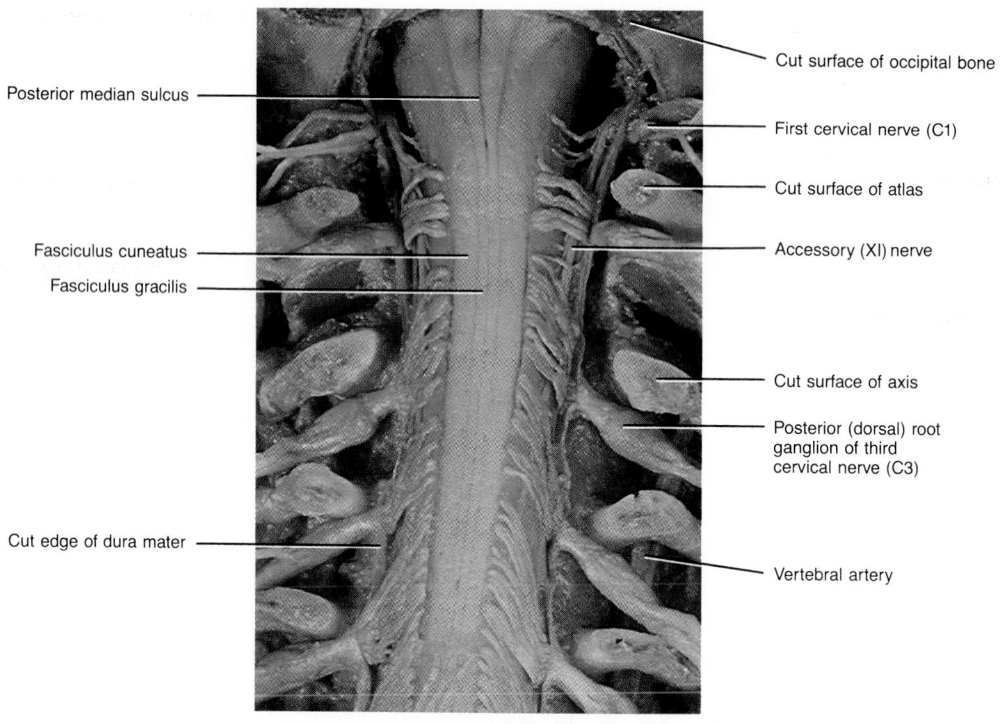

(b) Posterior view

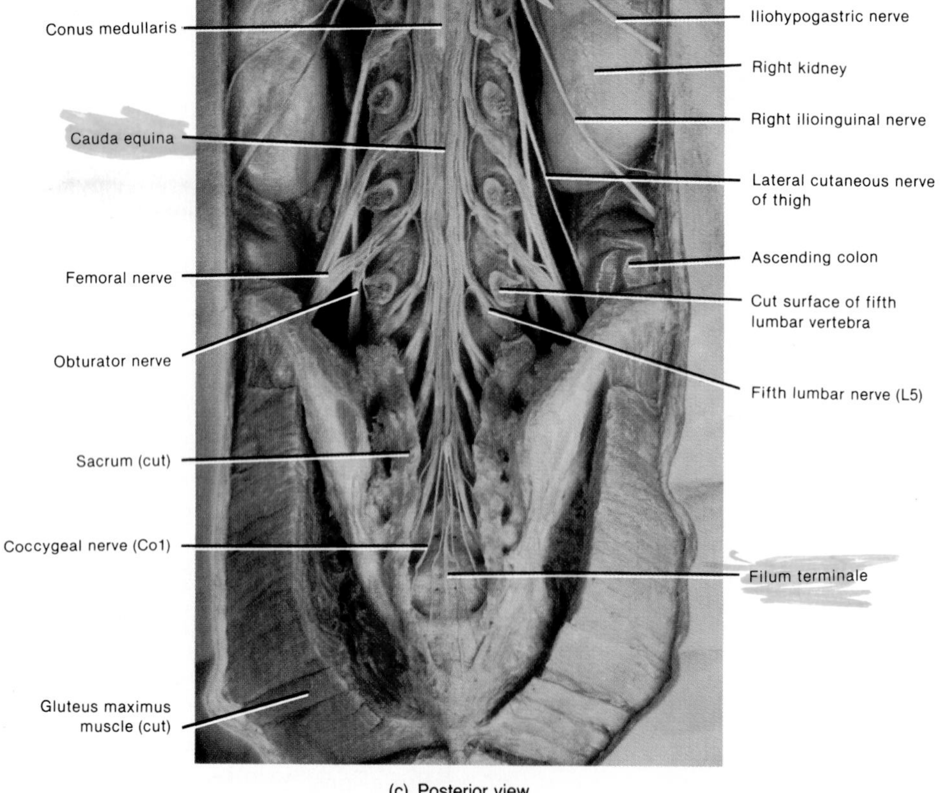

(c) Posterior view

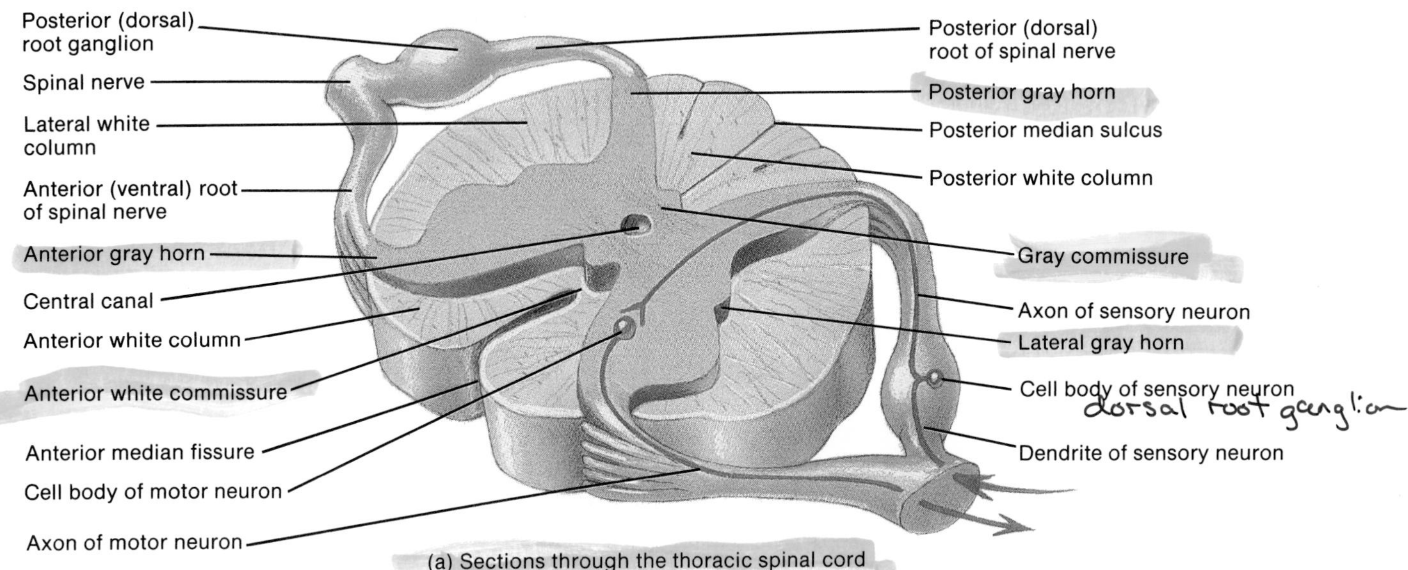

(a) Sections through the thoracic spinal cord

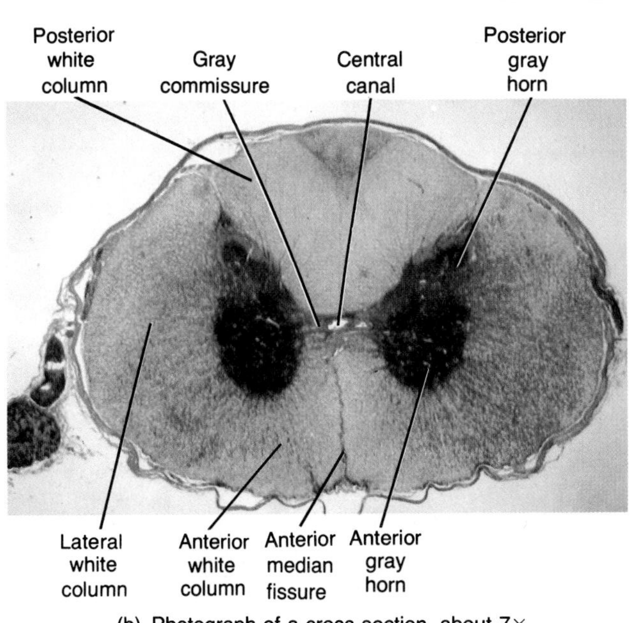

(b) Photograph of a cross section, about 7×

FIGURE 17-3 Spinal cord. (a) The organization of gray and white matter in the spinal cord as seen in cross section. Note what is inside the posterior root ganglion, posterior root of the spinal nerve, anterior root of the spinal nerve, and the spinal nerve. In this and other illustrations of cross sections of the spinal cord, dendrites are not shown in relation to cell bodies of motor or association neurons for purposes of simplicity. (b) Photograph of the spinal cord at the seventh cervical segment. (Courtesy of Victor B. Eichler, Ph.D., Wichita, Kansas.)

the back of the cord are referred to as ***posterior* (*dorsal*) *gray horns,*** and those between the anterior and posterior gray horns are intermediate ***lateral gray horns.*** The lateral gray horns are present only in the thoracic, upper lumbar, and sacral segments of the cord.

The gray matter of the cord also contains several nuclei that serve as relay stations for nerve impulses and origins for certain nerves. Nuclei are clusters of nerve cell bodies and dendrites in the spinal cord and brain.

The white matter, like the gray matter, is also organized into regions. The anterior and posterior gray horns divide the white matter on each side into three broad areas called ***columns*: *anterior* (*ventral*) *white columns, posterior* (*dorsal*) *white columns,*** and ***lateral white columns.*** Each column in turn consists of distinct bundles of myelinated fibers that run within the cord. These bundles are called ***tracts.*** The long ***ascending tracts*** consist of sensory axons that conduct impulses that enter the spinal cord upward to the brain. The long ***descending tracts*** consist of motor axons that conduct impulses from the brain downward into the spinal cord, where they synapse with other neurons whose axons pass out to muscles and glands. Thus, the ascending tracts are sensory tracts and the descending tracts are motor tracts.

FUNCTIONS

A major function of the spinal tracts in the spinal cord is to convey sensory impulses from the periphery to the brain and to conduct motor impulses from the brain to the periphery. A second principal function is to provide a means of integrating reflexes. Both functions are essential to maintaining homeostasis.

Impulse Conduction

The vital function of conveying sensory and motor information to and from the brain is carried out by the ascending and descending tracts of the cord. The names of the tracts typically indicate the white column in which the tract travels, where the cell bodies of the tract originate, and where the axons of the tract terminate. Since the origin and termination are specified, the direction of impulse conduction is also indicated by the name. For example, the anterior spinothalamic tract is located in the *anterior* white column, originates in the *spinal cord,* and terminates in the *thalamus* (a region of the brain). It is therefore an ascending (sensory) tract and conveys nerve impulses from the cord upward to the brain.

The principal ascending and descending tracts are listed in Exhibit 17-1 and shown in Figure 17-4.

As sensory information is conveyed from receptors to ascending (sensory) tracts, it becomes part of a large pool

EXHIBIT 17-1

Selected Ascending and Descending Tracts of Spinal Cord (see Figure 17-4)

TRACT	LOCATION (WHITE COLUMN)	ORIGIN	TERMINATION	FUNCTION
ASCENDING				
Anterior (Ventral) Spinothalamic (see also Figure 20-7)	Anterior (ventral) column.	Posterior (dorsal) gray horn on one side of cord but crosses to opposite side of brain.	Thalamus; impulses eventually conveyed to cerebral cortex.	Conveys sensations for touch and pressure from one side of body to opposite side of thalamus. Eventually, sensations reach cerebral cortex.
Lateral Spinothalamic (see also Figure 20-6)	Lateral column.	Posterior (dorsal) gray horn on one side of cord but crosses to opposite side of brain.	Thalamus; impulses eventually conveyed to cerebral cortex.	Conveys sensations for pain and temperature from one side of body to opposite side of thalamus. Eventually, sensations reach cerebral cortex.
Fasciculus Gracilis and Fasciculus Cuneatus (see also Figure 20-5)	Posterior (dorsal) column.	Axons of afferent neurons from periphery that enter posterior (dorsal) column on one side of cord and rise to same side of brain.	Nucleus gracilis and nucleus cuneatus of medulla; impulses eventually conveyed to opposite side of cerebral cortex.	Convey sensations from one side of body to opposite side of medulla for touch; two-point discrimination (ability to distinguish that two points on skin are touched even though close together); conscious proprioception (conscious awareness of precise position of body parts and their direction of movement); stereognosis (ability to recognize size, shape, and texture of object); weight discrimination (ability to assess weight of an object); and vibration. Eventually, impulses over these tracts reach cerebral cortex, thus allowing the sensations to be consciously perceived.
Posterior (Dorsal) Spinocerebellar	Posterior (dorsal) portion of lateral column.	Posterior (dorsal) gray horn on one side of cord and rises to same side of brain.	Cerebellum.	Conveys sensations from one side of body to same side of cerebellum for subconscious proprioception.
Anterior (Ventral) Spinocerebellar	Anterior (ventral) portion of lateral column.	Posterior (dorsal) gray horn on one side of cord; contains both crossed and uncrossed fibers.	Cerebellum.	Conveys sensations from both sides of body to cerebellum for subconscious proprioception.

of sensory input. We do not respond actively to every bit of input the central nervous system receives. Rather, each piece of incoming information is integrated with all the other information arriving from activated sensory receptors. The integration process occurs not just once but at many stations along the tracts of the central nervous system with the help of association neurons. It occurs within the spinal cord and the brain stem, cerebellum, and cerebrum of the brain. As a result, a motor response to make a muscle contract or a gland secrete can be initiated at any of these levels. The cerebral motor cortex assumes the major role for controlling precise, discrete, muscular movements.

When the input reaches the highest center, sensory–motor integration occurs. This involves not only using information contained within that center but also information coming to it from other centers in the central nervous system. After integration occurs, the output of the center is sent down the spinal cord in descending (motor) tracts that result in muscular contraction or glandular secretion.

Reflex Center

The second principal function of the spinal cord is to serve as a center for some ***reflexes,*** which are fast responses to changes (stimuli) in the internal or external environment that allow the body to maintain homeostasis. Spinal nerves are the paths of communication between the spinal cord tracts and the periphery. Figure 17-3 reveals that each spinal

TRACT	LOCATION (WHITE COLUMN)	ORIGIN	TERMINATION	FUNCTION
DESCENDING				
Lateral Corticospinal (see also Figure 20-9)	Lateral column.	Cerebral cortex on one side of brain but crosses in base of medulla to opposite side of cord.	Anterior (ventral) gray horn.	Conveys motor impulses from one side of cortex to anterior gray horn of opposite side. Eventually, impulses reach skeletal muscles on opposite side of body that coordinate precise, discrete movements.
Anterior (Ventral) Corticospinal (see also Figure 20-9)	Anterior (ventral) column.	Cerebral cortex on one side of brain, uncrossed in medulla, but crosses to opposite side of cord.	Anterior (ventral) gray horn.	Conveys motor impulses from one side of cortex to anterior gray horn of same side. Impulses cross to opposite side in spinal cord and reach skeletal muscles that coordinate movements of the axial skeleton.
Rubrospinal	Lateral column.	Midbrain (red nucleus) on one side of brain but crosses to opposite side of cord.	Anterior (ventral) gray horn.	Conveys motor impulses from one side of midbrain to skeletal muscles on opposite side of body that are concerned with precise, discrete movements.
Tectospinal	Anterior (ventral) column.	Midbrain on one side of brain but crosses to opposite side of cord.	Anterior (ventral) gray horn.	Conveys motor impulses from one side of midbrain to skeletal muscles on opposite side of body that control movements of head in response to auditory, visual, and cutaneous stimuli.
Vestibulospinal	Anterior (ventral) column.	Medulla on one side of brain and descends to same side of cord.	Anterior (ventral) gray horn.	Conveys motor impulses from one side of medulla to skeletal muscles on same side of body that regulate body tone in response to movements of head (equilibrium).
Lateral Reticulospinal	Lateral column.	Medulla on one side of brain and descends mainly to same side of cord.	Anterior (ventral) gray horn.	Conveys motor impulses from one side of medulla to axial skeleton muscles and proximal limb muscles that inhibit extensor reflexes and muscle tone.

EXHIBIT 17-1

Selected Ascending and Descending Tracts of Spinal Cord (*Continued*)

TRACT	LOCATION (WHITE COLUMN)	ORIGIN	TERMINATION	FUNCTION
Anterior (Ventral) or Medial Reticulospinal	Anterior (ventral) column.	Pons on one side of brain and descends mainly to same side of cord.	Anterior (ventral) gray horn.	Conveys motor impulses from one side of pons to axial skeleton muscles and proximal limb muscles that facilitate extensor reflexes and muscle tone.

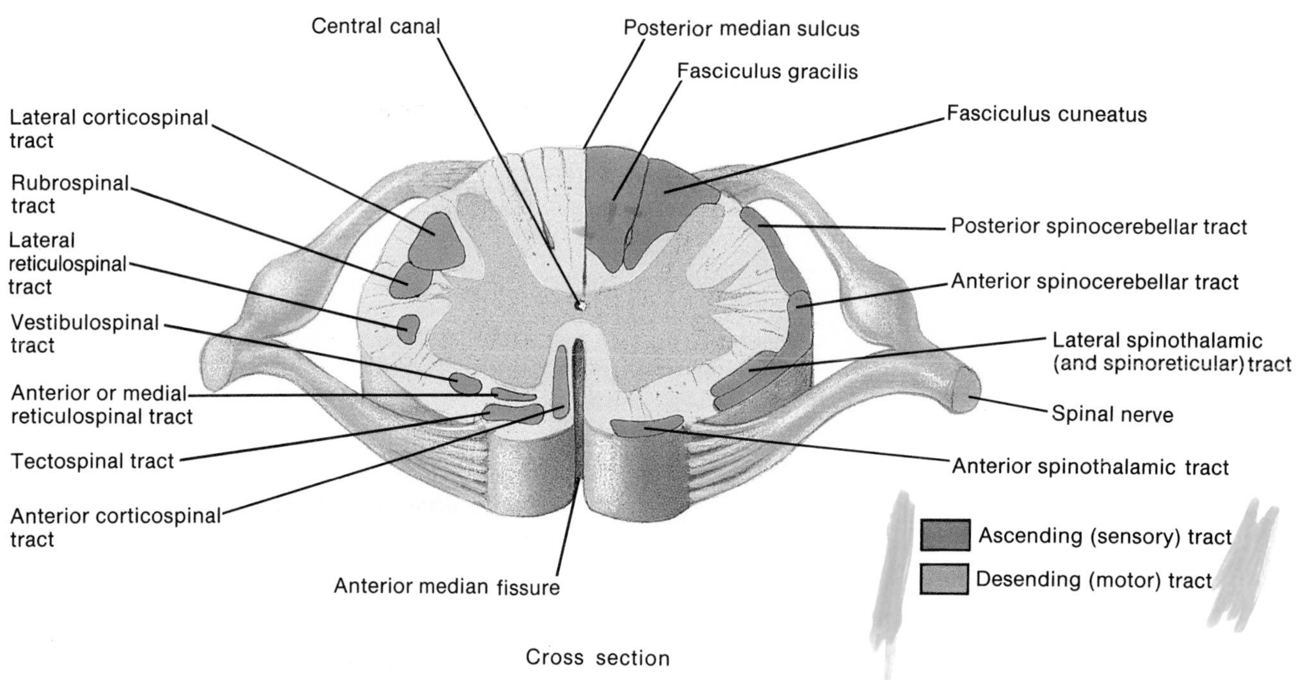

FIGURE 17-4 Selected tracts of the spinal cord.

nerve is connected to a segment of the cord by two points of attachment called roots. The ***posterior*** or ***dorsal*** (***sensory***) ***root*** contains sensory nerve fibers only and conducts nerve impulses from the periphery to the spinal cord. These fibers extend into the posterior (dorsal) gray horn. Each dorsal root also has a swelling, the ***posterior*** or ***dorsal*** (***sensory***) ***root ganglion,*** which contains the cell bodies of the sensory neurons from the periphery. The other point of attachment of a spinal nerve to the cord is the ***anterior*** or ***ventral*** (***motor***) ***root.*** It contains motor neuron axons only and conducts impulses from the spinal cord to the periphery.

The cell bodies of the motor neurons are located in the gray matter of the cord. If the motor nerve impulse supplies a skeletal muscle, the cell bodies are located in the anterior (ventral) gray horn. If, however, the motor nerve impulse supplies smooth muscle, cardiac muscle, or a gland through the autonomic nervous system, the cell bodies are located in the lateral gray horn.

The path a nerve impulse follows from its origin in the dendrites or cell body of a neuron in one part of the body to its termination elsewhere in the body is called a ***conduction pathway.*** All conduction pathways consist of circuits of neurons. The simplest kind of pathway is known as a ***reflex arc,*** the functional unit of the nervous system. A reflex arc contains two or more types of neurons over which nerve impulses are conducted from a receptor to the brain or spinal cord by way of sensory neurons and then to an effector by way of motor neurons. The basic components of a reflex arc are as follows (Figure 17-5).

1. ***Receptor.*** The distal end of a dendrite or a sensory structure associated with the distal end of a dendrite.

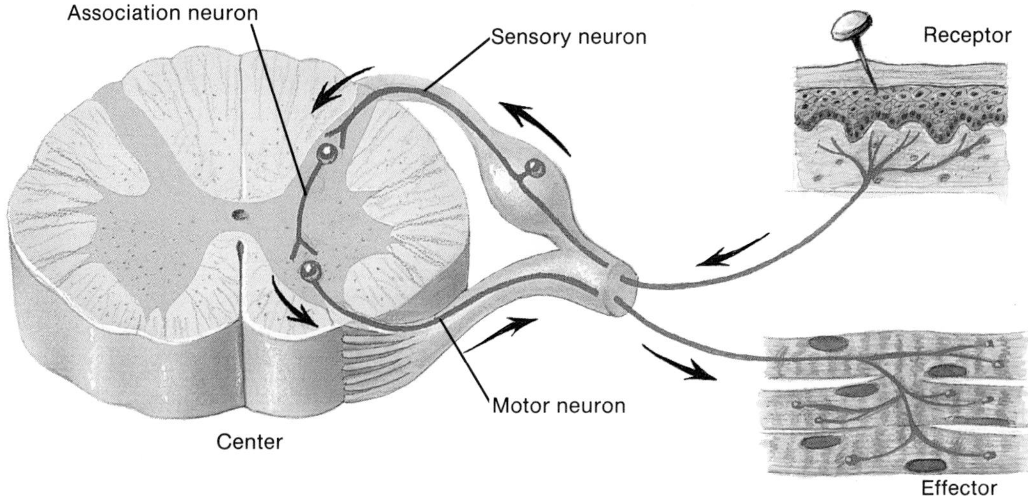

FIGURE 17-5 Components of a generalized reflex arc. The arrows show the direction of nerve impulse conduction.

Its role in the reflex arc is to respond to a specific change in the internal or external environment (stimulus) by initiating a nerve impulse in a sensory neuron by way of a generator potential (a local depolarization of the receptor cell membrane).

2. ***Sensory neuron.*** A neuron that passes the nerve impulse from the receptor to its axonal termination in the central nervous system (CNS).
3. ***Center.*** A region in the CNS where an incoming sensory impulse may be transmitted or inhibited. In the center of simple reflex arcs, the sensory neuron directly generates the impulse in the motor neuron. In more complex reflex arcs, the sensory impulse is relayed to an association neuron, which may relay the impulse to other association neurons as well as to a motor neuron.
4. ***Motor neuron.*** A neuron that transmits the impulse generated by the sensory or association neuron in the center to the part of the body that will respond, such as a muscle or gland.
5. ***Effector.*** The part of the body that responds to the motor nerve impulse. This response is called a reflex and involves either an increase or decrease in muscle contraction or an increase or decrease in secretion by glands.

Reflexes allow the body to maintain homeostasis. They are associated not only with skeletal muscle contraction but also with body functions such as heart rate, respiration, digestion, urination, and defecation, which involve cardiac and smooth muscle and glands. Reflexes carried out by the spinal cord alone are called ***spinal reflexes.*** Reflexes that result in the contraction of skeletal muscles are known as ***somatic reflexes.*** Reflexes that involve brain centers and cranial nerves are called ***cranial reflexes.*** Those that cause the contraction of smooth or cardiac muscle or secretion by glands are ***visceral (autonomic) reflexes.***

SPINAL NERVES

NAMES

Spinal nerves connect the central nervous system to sensory receptors, muscles, and glands and are components of the somatic portion of the peripheral nervous system. The 31 pairs of spinal nerves are named and numbered according to the region and level of the spinal cord from which they emerge (see Figure 17-2). The first cervical pair emerges between the atlas (first cervical vertebra) and the occipital bone. All other spinal nerves leave the vertebral column from the intervertebral foramina between adjoining vertebrae. There are 8 pairs of cervical, 12 pairs of thoracic, 5 pairs of lumbar, 5 pairs of sacral, and 1 pair of coccygeal nerves.

During fetal life, the spinal cord and vertebral column grow at different rates, the cord growing more slowly. Thus, not all the spinal cord segments are in line with their corresponding vertebrae. Remember that the spinal cord terminates near the level of the upper border of the second lumbar vertebra. Thus, the lower lumbar, sacral, and coccygeal nerves must descend more and more to reach their foramina before emerging from the vertebral column. This arrangement constitutes the cauda equina.

COMPOSITION AND COVERINGS

A ***spinal nerve*** has two points of attachment to the cord: a posterior root and an anterior root (see Figure 17-3a). The posterior and anterior roots unite to form a spinal nerve at the intervertebral foramen. Since the posterior root contains sensory fibers and the anterior root contains motor fibers, a spinal nerve is a ***mixed nerve,*** at least at its origin.

The posterior (dorsal) root ganglion contains cell bodies of sensory neurons.

In Figure 17-6, you can see that spinal nerves (as well as cranial nerves) contain many fibers surrounded by different coverings. The individual fibers, whether myelinated or unmyelinated, are wrapped in a connective tissue called the ***endoneurium*** (en′-dō-NYOO-rē-um). Groups of fibers with their endoneurium are arranged in bundles called ***fascicles,*** and each bundle is wrapped in connective tissue called the ***perineurium*** (per′-i-NYOO-rē-um). The outermost covering around the entire nerve is the ***epineurium*** (ep′-i-NYOO-rē-um). The spinal meninges fuse with the epineu-

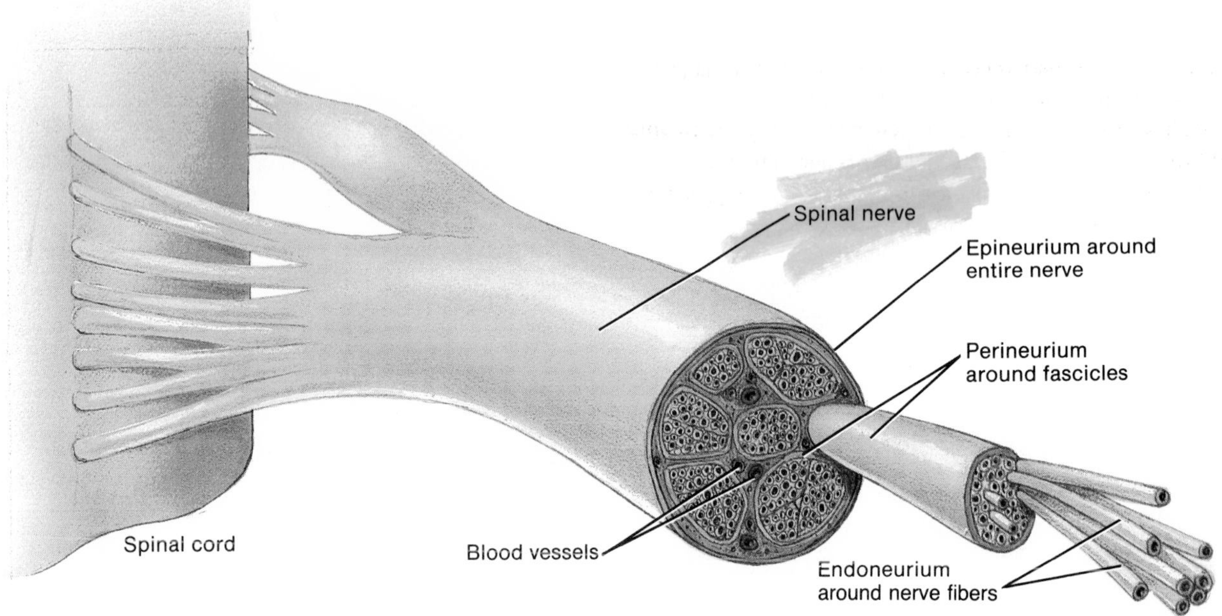

(a) Cross section

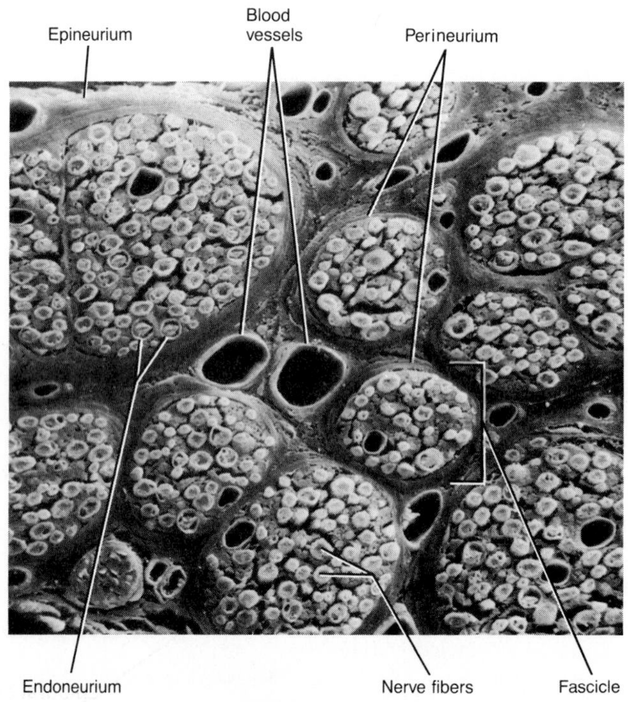

(b) Scanning electron micrograph of a cross section of several spinal nerves, 900×

FIGURE 17-6 Coverings of a spinal nerve. (b) Courtesy of Richard G. Kessel and Randy H. Kardon, from *Tissues and Organs: A Text-Atlas of Scanning Electron Microscopy.* Copyright © 1979 W. H. Freeman & Co.

rium as the nerve exits through the intervertebral foramen. Blood vessels, which nourish nerves, are located under the epineurium and perineurium and sometimes under the endoneurium.

DISTRIBUTION

Branches

Shortly after a spinal nerve leaves its intervertebral foramen, it divides into several branches (Figure 17-7). These branches are known as ***rami*** (RĀ-mī). The ***dorsal ramus*** (RĀ-mus) innervates the deep muscles and skin of the dorsal surface of the trunk. The ***ventral ramus*** of a spinal nerve innervates the muscles and structures of the extremities and the lateral and ventral trunk. In addition to dorsal and ventral rami, spinal nerves also give off a ***meningeal branch.*** This branch reenters the spinal canal through the intervertebral foramen and supplies the vertebrae, vertebral ligaments, blood vessels of the spinal cord, and the meninges. Other branches of a spinal nerve are the ***rami communicantes*** (kō-myoo-ni-KAN-tēz), components of the autonomic nervous system whose structure and function are discussed in Chapter 19. Singular is ramus communicans.

Plexuses

The ventral rami of spinal nerves, except for thoracic nerves T2–T11, do not go directly to the body structures they supply. Instead, they form networks on both right and left sides of the body by joining with adjacent nerves. Such a network is called a ***plexus*** (*plexus* = braid). The principal plexuses are the cervical plexus, brachial plexus, lumbar plexus, and sacral plexus (see Figure 17-2a). Emerging from the plexuses are nerves bearing names that are often descriptive of the general regions they supply or the course they take. Each of the nerves, in turn, may have several branches named for the specific structures they innervate.

■ ***Cervical Plexus*** The ***cervical plexus*** is formed by the ventral rami of the first four cervical nerves (C1–C4) with contributions from C5. There is one on each side of the neck alongside the first four cervical vertebrae (Figure 17-8). The ***roots*** of the plexus indicated in the diagram are the ventral rami. The cervical plexus supplies the skin and muscles of the head, neck, and upper part of the shoulders. Branches of the cervical plexus also connect with the accessory (XI) and hypoglossal (XII) cranial nerves. The phrenic nerves are a major pair of nerves arising from the cervical plexuses that supply the motor fibers to the diaphragm. Damage to the spinal cord above the origin of the phrenic nerves results in paralysis of the diaphragm, since the phrenic nerves no longer send nerve impulses to the diaphragm. Contractions of the diaphragm are essential for normal breathing.

Exhibit 17-2 summarizes the nerves and distributions of the cervical plexus. The relationship of the cervical plexus to the other plexuses is shown in Figure 17-2a.

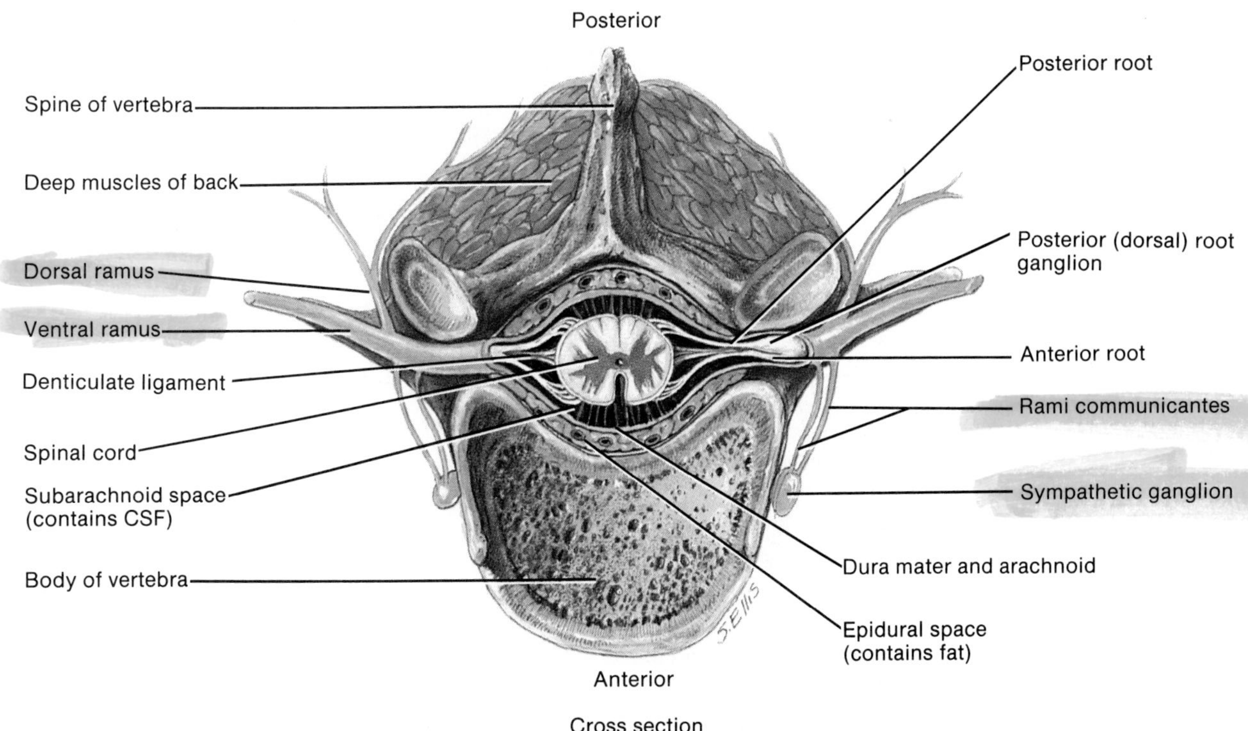

FIGURE 17-7 Branches of a typical spinal nerve.

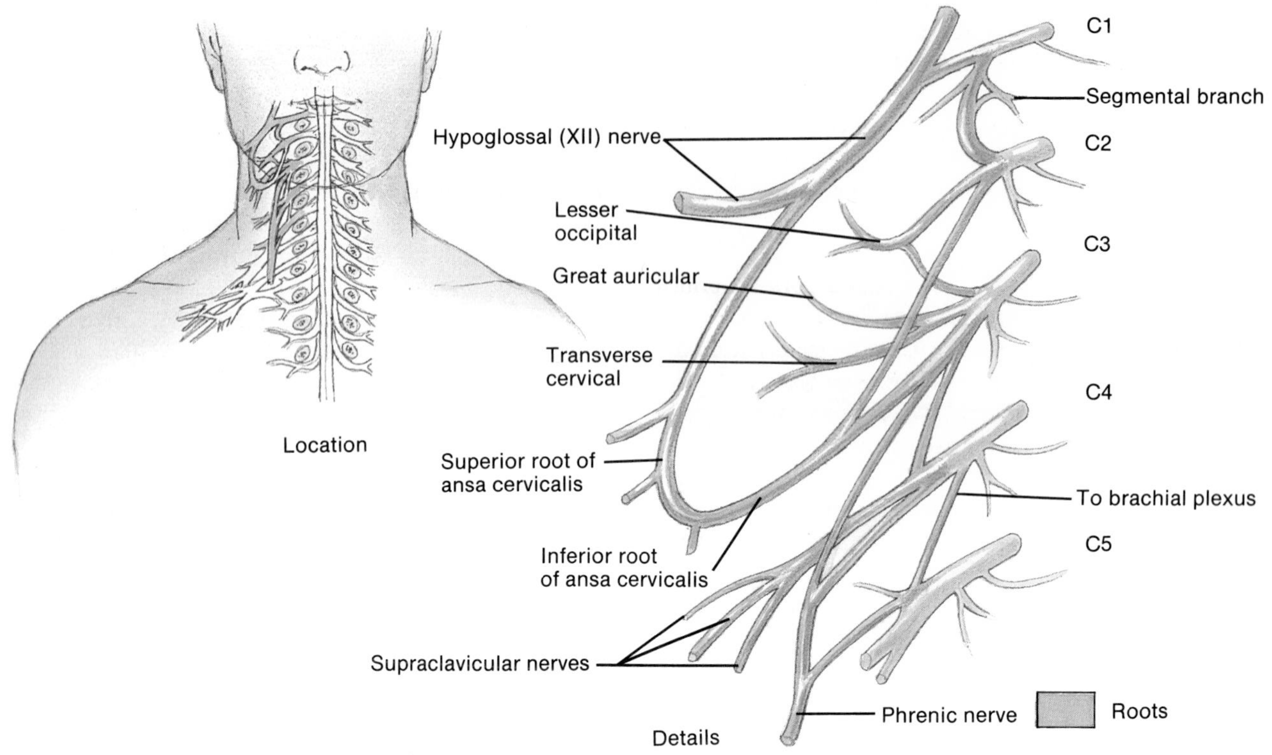

FIGURE 17-8 Cervical plexus. Consult Exhibit 17-2 so that you can determine the distribution of each of the nerves of the cervical plexus.

EXHIBIT 17-2

Cervical Plexus

NERVE	ORIGIN	DISTRIBUTION
SUPERFICIAL OR SENSORY BRANCHES		
Lesser Occipital	C2.	Skin of scalp behind and above ear.
Greater Auricular	C2–C3.	Skin in front of, below, and over ear and over parotid glands.
Transverse Cervical	C2–C3.	Skin over anterior aspect of neck.
Supraclavicular	C3–C4.	Skin over upper portion of chest and shoulder.
DEEP OR LARGELY MOTOR BRANCHES		
Ansa Cervicalis	This nerve is divided into a superior root and an inferior root.	
Superior Root	C1.	Infrahyoid (omohyoid, sternohyoid, sternothyroid, and thyrohyoid) and geniohyoid muscles of neck.
Inferior Root	C2–C3.	Omohyoid, sternohyoid, and sternothyroid muscles of neck.
Phrenic	C3–C5.	Diaphragm between thorax and abdomen.
Segmental Branches	C1–C5.	Prevertebral (deep) muscles of neck, levator scapulae, and middle scalene muscles.

■ ***Brachial Plexus*** The ***brachial plexus*** is formed by the ventral rami of spinal nerves C5–C8 and T1. On either side of the last four cervical and first thoracic vertebrae, the brachial plexus extends downward and laterally, passes over the first rib behind the clavicle, and then enters the axilla (Figure 17-9). The brachial plexus constitutes the entire nerve supply for the upper extremities and shoulder region.

The ***roots*** of the brachial plexus, like those of the cervical plexus, are the ventral rami of the spinal nerves. The roots of C5 and C6 unite to form the ***superior trunk,*** C7 becomes the ***middle trunk,*** and C8 and T1 form the ***inferior trunk.*** Each trunk, in turn, divides into an ***anterior division*** and a ***posterior division.*** The divisions then unite to form cords. The ***posterior cord*** is formed by the union of the posterior divisions of the superior, middle, and inferior trunks. The ***medial cord*** is formed as a continuation of the anterior division of the inferior trunk. The ***lateral cord*** is formed by the union of the anterior divisions of the superior and middle trunk. The peripheral nerves arise mainly from the cords. Thus, the brachial plexus begins as roots that unite to form trunks, the trunks branch into divisions, the divisions form cords, and the cords give rise to the peripheral nerves.

Five important nerves arising from the brachial plexus are the axillary, musculocutaneous, radial, median, and ulnar. The ***axillary nerve*** supplies the deltoid and teres minor muscles. The ***musculocutaneous nerve*** supplies the flexors of the arm and forearm. The ***radial nerve*** supplies the muscles on the posterior aspect of the arm and forearm. The ***median nerve*** supplies most of the muscles of the anterior forearm and some of the muscles in the palm. The ***ulnar nerve*** supplies the anteromedial muscles of the forearm and most of the muscles of the palm.

CLINICAL APPLICATION

Injuries to Brachial Plexus

Prolonged use of a crutch that presses into the axilla may result in injury to a portion of the brachial plexus. The usual ***crutch palsy*** involves the posterior cord of the brachial plexus or, more frequently, just the radial nerve, which, in general, supplies extensors.

Radial nerve damage is indicated by wrist drop: inability to extend the hand at the wrist. Care must be taken not to injure the radial and axillary nerves when deltoid intramuscular injections are given. The radial nerve may also be injured when a cast is applied too tightly around the midhumerus. ***Median nerve damage*** is indicated by numbness, tingling, and pain in the palm and fingers; weak thumb movements; and inability to pronate the forearm and difficulty in flexing the wrist properly. Compression of the median nerve inside the carpal tunnel, formed anteriorly by the flexor retinaculum (transverse

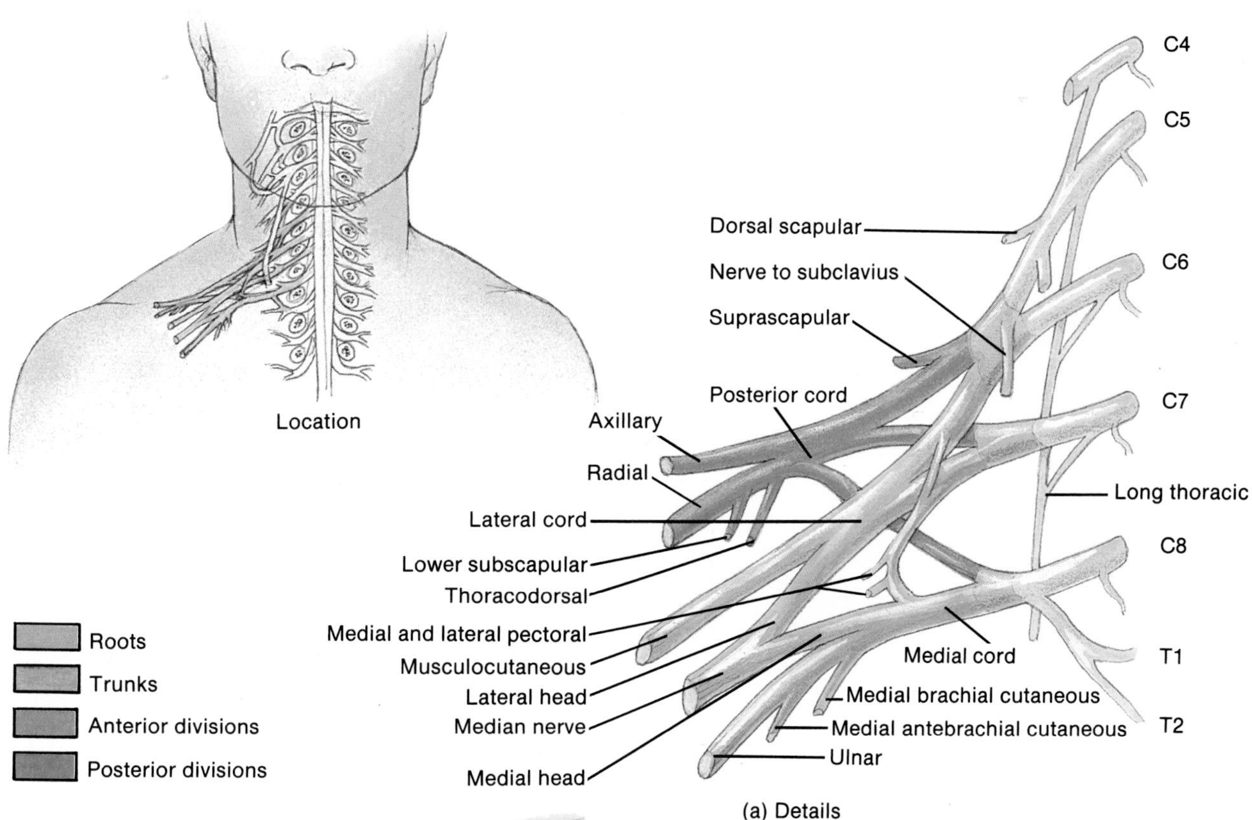

FIGURE 17-9 Brachial plexus. Consult Exhibit 17-3 so that you can determine the distribution of each of the nerves of the plexus. (c) Courtesy of J. A. Gosling, P. F. Harris, et al., *Atlas of Human Anatomy,* Gower Medical Publishing Ltd., 2nd ed., 1991.

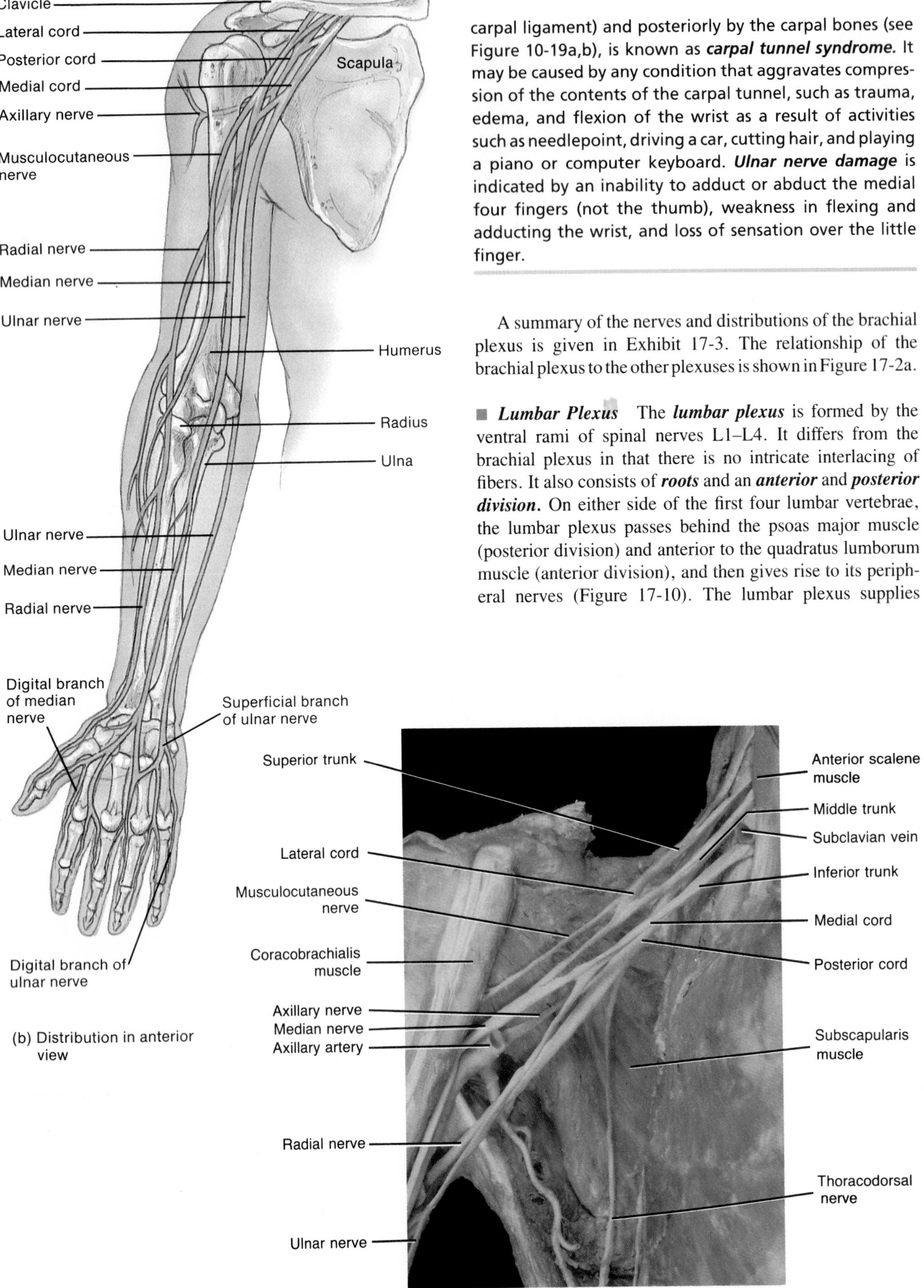

(b) Distribution in anterior view

(c) Photograph of anterior view

carpal ligament) and posteriorly by the carpal bones (see Figure 10-19a,b), is known as ***carpal tunnel syndrome.*** It may be caused by any condition that aggravates compression of the contents of the carpal tunnel, such as trauma, edema, and flexion of the wrist as a result of activities such as needlepoint, driving a car, cutting hair, and playing a piano or computer keyboard. ***Ulnar nerve damage*** is indicated by an inability to adduct or abduct the medial four fingers (not the thumb), weakness in flexing and adducting the wrist, and loss of sensation over the little finger.

A summary of the nerves and distributions of the brachial plexus is given in Exhibit 17-3. The relationship of the brachial plexus to the other plexuses is shown in Figure 17-2a.

■ ***Lumbar Plexus*** The ***lumbar plexus*** is formed by the ventral rami of spinal nerves L1–L4. It differs from the brachial plexus in that there is no intricate interlacing of fibers. It also consists of ***roots*** and an ***anterior*** and ***posterior division.*** On either side of the first four lumbar vertebrae, the lumbar plexus passes behind the psoas major muscle (posterior division) and anterior to the quadratus lumborum muscle (anterior division), and then gives rise to its peripheral nerves (Figure 17-10). The lumbar plexus supplies

EXHIBIT 17-3

Brachial Plexus

NERVE	ORIGIN	DISTRIBUTION
ROOT		
Dorsal Scapular	C5.	Levator scapulae, rhomboideus major, and rhomboideus minor muscles.
Long Thoracic	C5–C7.	Serratus anterior muscle.
TRUNK		
Nerve to Subclavius	C5–C6.	Subclavius muscle.
Suprascapular	C5–C6.	Supraspinatus and infraspinatus muscles.
LATERAL CORD		
Musculocutaneous	C5–C7.	Coracobrachialis, biceps brachii, and brachialis muscles.
Median (Lateral Head)	C5–C7.	See distribution for ***Median (medial head)*** in this exhibit.
Lateral Pectoral	C5–C7.	Pectoralis major muscle.
POSTERIOR CORD		
Upper Subscapular	C5–C6.	Subscapularis muscle.
Thoracodorsal	C6–C8.	Latissimus dorsi muscle.
Lower Subscapular	C5–C6.	Subscapularis and teres major muscles.
Axillary (Circumflex)	C5–C6.	Deltoid and teres minor muscles; skin over deltoid and upper posterior aspect of arm.
Radial	C5–C8 and T1.	Extensor muscles of arm and forearm (triceps brachii, brachioradialis, extensor carpi radialis longus, extensor digitorum, extensor carpi ulnaris, extensor carpi radialis brevis, extensor indicis); skin of posterior arm and forearm, lateral two-thirds of dorsum of hand, and fingers over proximal and middle phalanges.
MEDIAL CORD		
Medial Pectoral	C8–T1.	Pectoralis major and pectoralis minor muscles.
Medial Brachial Cutaneous	C8–T1.	Skin of medial and posterior aspects of lower third of arm.
Medial Antebrachial Cutaneous	C8–T1.	Skin of medial and posterior aspects of forearm.
Median (Medial Head)	C5–C8 and T1.	Medial and lateral heads form median nerve. Distributed to flexors of forearm (pronator teres, flexor carpi radialis, flexor digitorum superficialis, lateral half of flexor digitorum profundus) except flexor carpi ulnaris; skin of lateral two-thirds of palm of hand and fingers.
Ulnar	C8–T1.	Flexor carpi ulnaris and flexor digitorum profundus muscles; skin of medial side of hand, little finger, and medial half of ring finger.
OTHER CUTANEOUS DISTRIBUTIONS		
Intercostobrachial	Second intercostal nerve.	Skin over medial side of arm.
Superior Lateral Brachial Cutaneous	Axillary.	Skin over lower half of deltoid and triceps brachii muscles.
Posterior Brachial Cutaneous	Radial.	Skin over posterior aspect of arm.
Inferior Lateral Brachial Cutaneous	Radial.	Skin over lower half of lateral and anterior aspect of arm.
Lateral Antebrachial Cutaneous	Musculocutaneous.	Skin over lateral aspect of forearm.
Posterior Antebrachial Cutaneous	Radial.	Skin over posterior aspect of forearm.

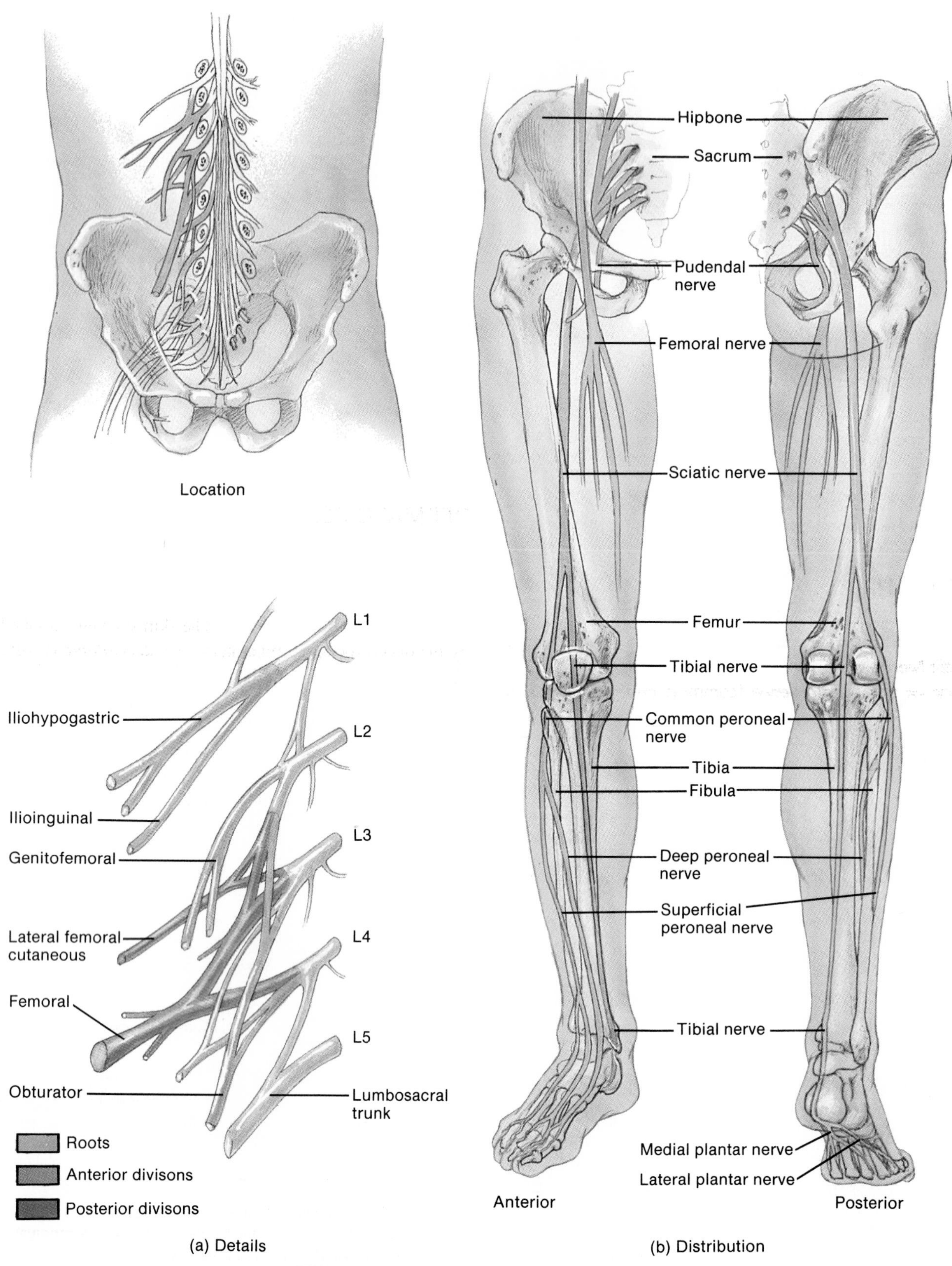

FIGURE 17-10 Lumbar plexus. Consult Exhibit 17-4 so that you can determine the distribution of the nerves of the lumbar plexus.

the anterolateral abdominal wall, external genitals, and part of the lower extremity. The largest nerve arising from the lumbar plexus is the femoral nerve.

CLINICAL APPLICATION

Femoral Nerve Injury

Injury to the femoral nerve is indicated by an inability to extend the leg and by loss of sensation in the skin over the anteromedial aspect of the thigh and leg.

A summary of the nerves and distributions of the lumbar plexus is presented in Exhibit 17-4. The relationship of the lumbar plexus to the other plexuses is shown in Figure 17-2a.

■ *Sacral Plexus* The ***sacral plexus*** is formed by the ventral rami of spinal nerves L4–L5 and S1–S4. It is situated largely in front of the sacrum (Figure 17-11). Like the lumbar plexus, it contains ***roots*** and an ***anterior*** and ***posterior division.*** The sacral plexus supplies the buttocks, perineum, and lower extremities. The largest nerve arising from the sacral plexus—and in fact the largest nerve in the body—is the sciatic nerve. The sciatic nerve supplies the entire musculature of the leg and foot.

CLINICAL APPLICATION

Sciatic Nerve Injury

Injury to the sciatic nerve (common peroneal portion) and its branches results in pain that may extend from the buttock down the back of the leg, foot drop, an inability to dorsiflex the foot, and loss of sensation over the leg and foot. This nerve may be injured because of a slipped disc, dislocated hip, osteoarthritis of the lumbosacral spine, pressure from the uterus during pregnancy, or an improperly administered gluteal intramuscular injection.

A summary of the nerves and distributions of the sacral plexus is given in Exhibit 17-5. The relationship of the sacral plexus to the other plexuses is shown in Figure 17-2a.

Intercostal (Thoracic) Nerves

The ventral rami of spinal nerves T2–T11 do not enter into the formation of plexuses and are known as ***intercostal (thoracic) nerves.*** These nerves are distributed directly to the structures they supply in intercostal spaces (see Figure 17-2a). After leaving its intervertebral foramen, the ventral ramus of nerve T2 supplies the intercostal muscles of the second intercostal space and the skin of the axilla and posteromedial aspect of the arm. Nerves T3–T6 pass in the costal grooves of the ribs and are distributed to the intercostal muscles and skin of the anterior and lateral chest wall. Nerves T7–T11 supply the intercostal muscles and the abdominal muscles and overlying skin. The dorsal rami of the intercostal nerves supply the deep back muscles and skin of the dorsal aspect of the thorax.

DERMATOMES

The skin over the entire body is supplied segmentally by spinal nerves; that is, the spinal nerves innervate specific, constant segments of the skin. All spinal nerves except C1 supply branches to the skin. The skin segment supplied by the dorsal root of a spinal nerve is a ***dermatome*** (Figure 17-12).

In the neck and trunk, the dermatomes form consecutive bands of skin. In the trunk, there is an overlap of adjacent dermatome nerve supply. Thus there is little loss of sensation if only a single nerve supply to a dermatome is interrupted. Most of the skin of the face and scalp is supplied by the trigeminal (V) cranial nerve.

Since physicians know which spinal nerves are associated with each dermatome, it is possible to determine which

EXHIBIT 17-4

Lumbar Plexus

NERVE	ORIGIN	DISTRIBUTION
Iliohypogastric	L1.	Muscles of anterolateral abdominal wall (external oblique, internal oblique, transversus abdominis); skin of lower abdomen and buttock.
Ilioinguinal	L1.	Muscles of anterolateral abdominal wall as indicated above; skin of upper medial aspect of thigh, root of penis and scrotum in male, and labia majora and mons pubis in female.
Genitofemoral	L1–L2.	Cremaster muscle; skin over middle anterior surface of thigh, scrotum in male, and labia majora in female.
Lateral Femoral Cutaneous	L2–L3.	Skin over lateral, anterior, and posterior aspects of thigh.
Femoral	L2–L4.	Flexor muscles of thigh (iliacus, psoas major, pectineus, rectus femoris, sartorius); extensor muscles of leg (rectus femoris, vastus lateralis, vastus medialis, vastus intermedius); skin on front and over medial aspect of thigh and medial side of leg and foot.
Obturator	L2–L4.	Adductor muscles of leg (obturator externus, pectineus, adductor longus, adductor brevis, adductor magnus, gracilis); skin over medial aspect of thigh.

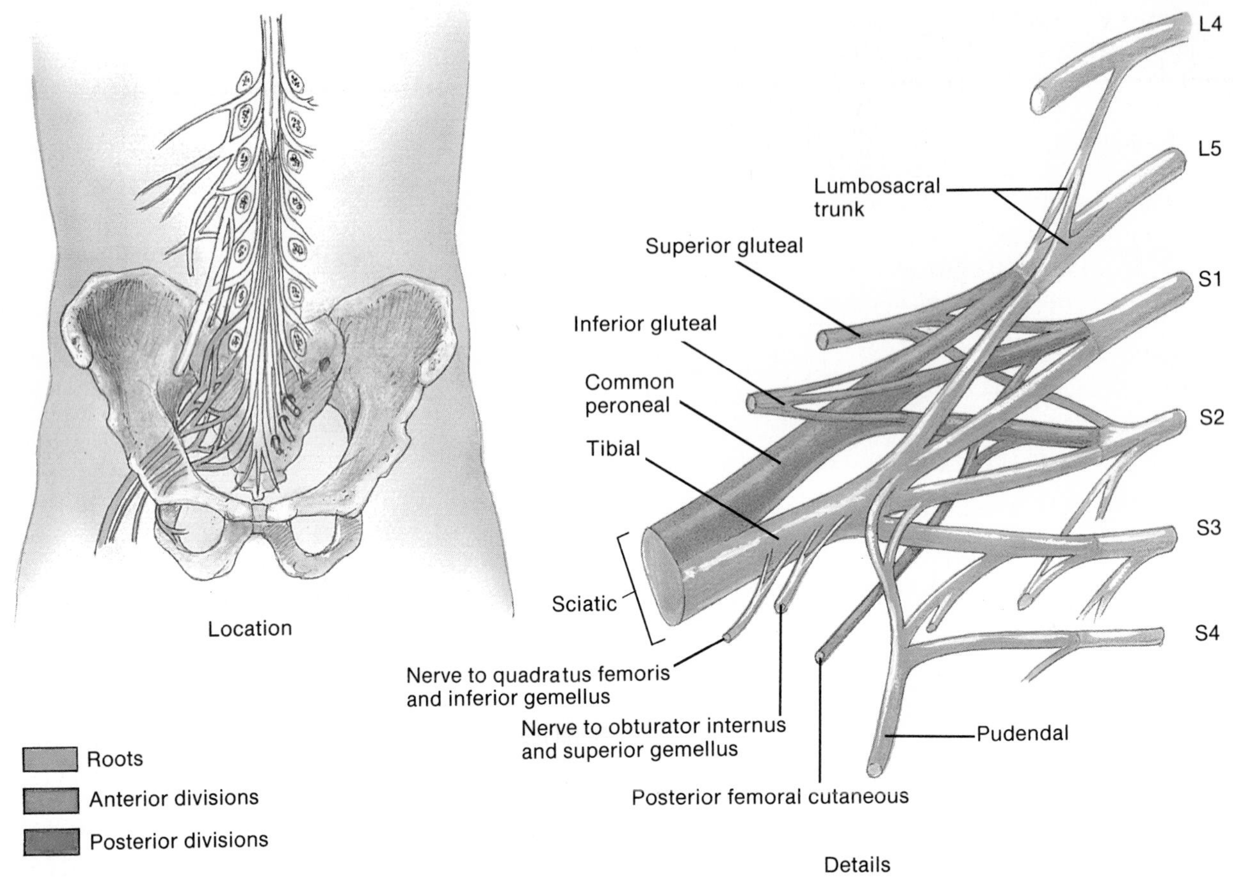

FIGURE 17-11 Sacral plexus. Refer to Figure 17-10b and Exhibit 17-5 so that you can determine the distribution of each of the nerves of the sacral plexus.

segment of the spinal cord or spinal nerve is malfunctioning. If a dermatome is stimulated and the sensation is not perceived, it can be assumed that the nerves supplying the dermatome are involved.

APPLICATIONS TO HEALTH

SPINAL CORD INJURY

The spinal cord may be damaged by compression from a tumor within or adjacent to the spinal cord, herniated intervertebral discs, blood clots, degenerative and demyelinating disorders, fracture or dislocation of the vertebrae enclosing it, penetrating wounds caused by projectile metal fragments, or other traumatic events such as automobile accidents. Depending on the location and extent of the injury, paralysis may occur. ***Paralysis*** refers to the total loss of motor function resulting from damage to nervous or muscle tissue. The various types of paralysis may be classified as follows: ***monoplegia*** (*mono* = one; *plege* = stroke), paralysis of one extremity only; ***diplegia*** (*di* = two), paralysis of both upper extremities or both lower extremities; ***paraplegia*** (*para* = beyond), paralysis of both lower extremities; ***hemiplegia*** (*hemi* = half), paralysis of the upper extremity, trunk, and lower extremity on one side of the body; and ***quadriplegia*** (*quad* = four), paralysis of the two upper and two lower extremities.

Complete transection of the spinal cord means that the cord is cut transversely and severed from one side to the other, thus cutting all ascending and descending tracts. It results in a loss of all sensations and voluntary movement below the level of the transection. If the upper cervical cord is transected, quadriplegia results; if the transection is between the cervical and lumbar enlargements, paraplegia results. ***Hemisection*** of the spinal cord is partial transection. It is characterized, below the hemisection, by a loss of proprioception, tactile discrimination, and feeling of vibration on the same side as the injury; paralysis on the same side; and loss of feelings of pain and temperature on the opposite side. If the hemisection is of the upper cervical cord, hemiplegia results; if the hemisection is of the thoracic cord, paralysis of one lower extremity results (monoplegia).

Following transection, there is an initial period of ***spinal shock*** that lasts from a few days to several weeks. During this period, all reflex activity is abolished, a condition called ***areflexia*** (a'-rē-FLEK-sē-a). In time, however, there is a return of reflex activity. The first reflex to return is a stretch

EXHIBIT 17-5

Sacral Plexus

NERVE	ORIGIN	DISTRIBUTION
Superior Gluteal	L4–L5 and S1.	Gluteus minimus and gluteus medius muscles and tensor fasciae latae.
Inferior Gluteal	L5–S2.	Gluteus maximus muscle.
Nerve to Piriformis	S1–S2.	Piriformis muscle.
Nerve to Quadratus Femoris	L4–L5 and S1.	Quadratus femoris and inferior gemellus muscles.
Nerve to Obturator Internus	L5–S2.	Obturator internus and superior gemellus muscles.
Perforating Cutaneous	S2–S3.	Skin over lower medial aspect of buttock.
Posterior Femoral Cutaneous	S1–S3.	Skin over anal region, lower lateral aspect of buttock, upper posterior aspect of thigh, upper part of calf, scrotum in male, and labia majora in female.
Sciatic	L4–S3.	Actually two nerves—tibial and common peroneal—bound together by common sheath of connective tissue. It splits into its two divisions, usually at knee. (See below for distributions.) As sciatic nerve descends through thigh, it sends branches to hamstring muscles (biceps femoris, semitendinosus, semimembranosus) and adductor magnus.
Tibial	L4–S3.	Gastrocnemius, plantaris, soleus, popliteus, tibialis posterior, flexor digitorum longus, and flexor hallucis longus muscles. Branches of tibial nerve in foot are medial plantar nerve and lateral plantar nerve.
Medial Plantar		Abductor hallucis, flexor digitorum brevis, and flexor hallucis brevis muscles; skin over medial two-thirds of plantar surface of foot.
Lateral Plantar		Remaining muscles of foot not supplied by medial plantar nerve; skin over lateral third of plantar surface of foot.
Common Peroneal	L4–S2.	Divides into a superficial peroneal and a deep peroneal branch.
Superficial Peroneal		Peroneus longus and peroneus brevis muscles; skin over distal third of anterior aspect of leg and dorsum of foot.
Deep Peroneal		Tibialis anterior, extensor hallucis longus, peroneus tertius, and extensor digitorum longus and brevis muscles; skin on adjacent sides of great and second toes.
Pudendal	S2–S4.	Muscles of perineum; skin of penis and scrotum in male and clitoris, labia majora, labia minora, and lower vagina in female.

reflex (knee jerk). Its reappearance may take several days. Next, the flexion reflexes return, over a period of up to several months. Then the crossed extensor reflexes return. In some cases, males may not be able to attain an erection or ejaculation. Moreover, urinary bladder and bowel functions are no longer under voluntary control. An experimental procedure called ***electroejaculation*** has been used with some success in helping males with spinal cord injuries to ejaculate. In the procedure, a probe is inserted into the rectum and attached to a device that delivers an electric current in gradually increasing increments until ejaculation occurs. The man's sperm is then used to inseminate the woman.

Recent studies show that patients with spinal cord injury have an improved outcome when given an antiinflammatory drug called methylprednisolone, a corticosteroid drug, within eight hours of the injury.

NEURITIS

Neuritis is inflammation of a single nerve, which could lead to sciatica; two or more nerves in separate areas, as a result of systemic lupus erythematosus; or many nerves simultaneously, as typified by Guillain-Barré syndrome. It may result from irritation to the nerve produced by direct blows, bone fractures, contusions, or penetrating injuries. Additional causes include vitamin deficiency (usually thiamine) or poisons such as carbon monoxide, carbon tetrachloride, heavy metals, and some drugs.

SCIATICA

Sciatica (sī-AT-i-ka) is a type of neuritis characterized by severe pain along the path of the sciatic nerve or its branches.

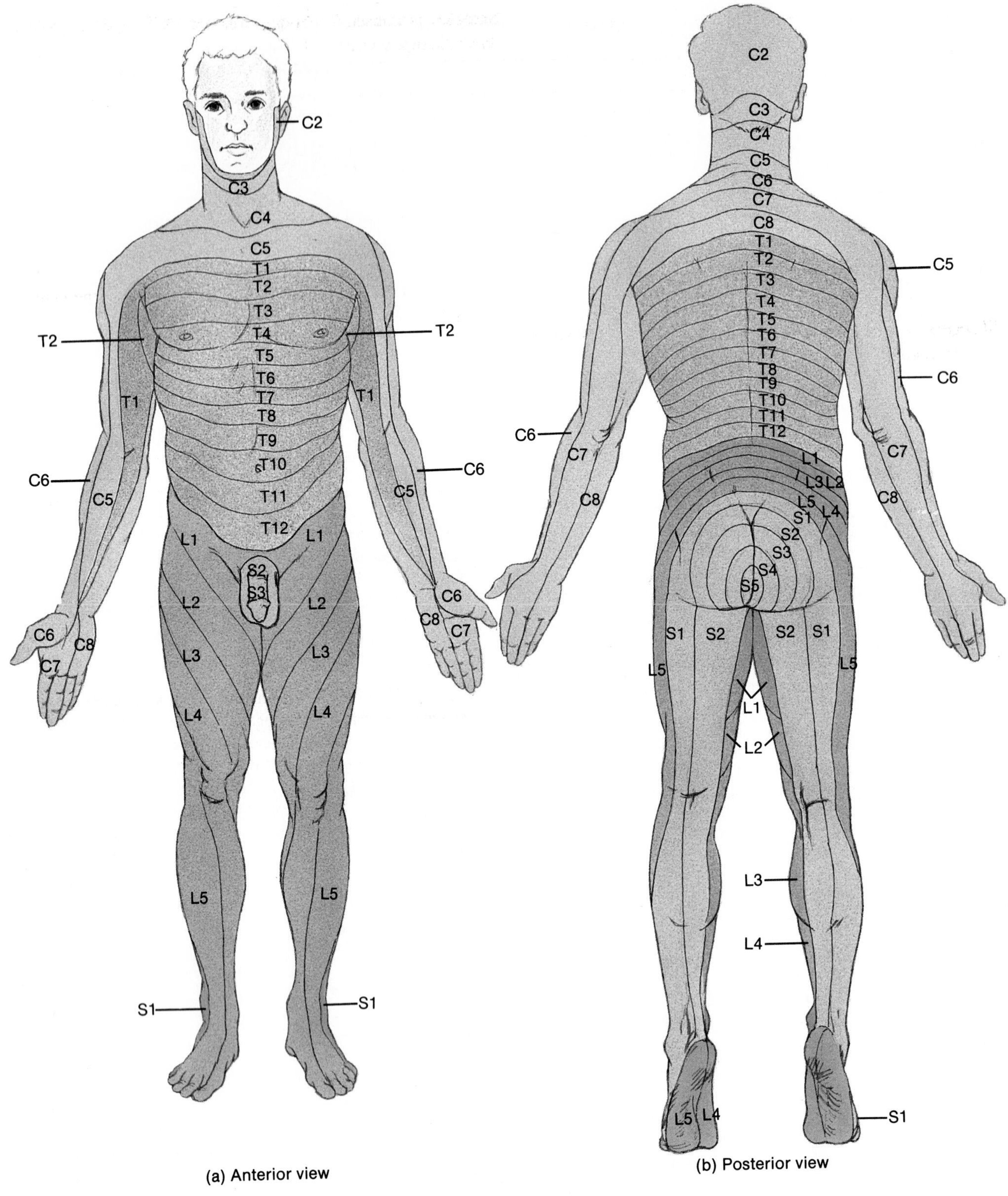

FIGURE 17-12 Distribution of spinal nerves to dermatomes. Illustrations of this nature can only approximate and cannot clearly indicate the degree of overlap of the cutaneous nerve innervations.

Because of its length and size, the sciatic nerve is exposed to many kinds of injury. Inflammation of or injury to the nerve causes pain that passes from the back or thigh down its length into the leg, foot, and toes.

Probably the most common cause of sciatica is a herniated (slipped) intervertebral disc. Other causes include irritation from osteoarthritis, back injuries, or pressure on the nerve from certain types of exertion. Sciatica may be associated with diabetes mellitus, gout, or vitamin deficiencies. Other cases are idiopathic (unknown).

SHINGLES

Shingles is an acute infection of the peripheral nervous system and is frequently a relapse of childhood chickenpox. Shingles is caused by herpes zoster (HER-pēz ZOS-ter), the chickenpox virus. Following recovery from chickenpox, the virus retreats to posterior (dorsal) root ganglia. If activated, the immune system usually prevents it from spreading. However, from time to time, the activated virus overcomes a weakened immune system, leaves the ganglion and travels down sensory neurons, causing the pain, and invades the skin where the neurons end, causing a characteristic line of skin blisters and discoloration of the skin. The line of blisters has a shape corresponding to the distribution of a particular nerve. The intercostal nerves in the waist area are most commonly affected. Almost always, the skin rash is limited to one side of the body.

STUDY OUTLINE

Grouping of Neural Tissue (p. 472)

1. White matter is an aggregation of myelinated axons and associated neuroglia.
2. Gray matter is a collection of unmyelinated neuron cell bodies and dendrites or unmyelinated axons along with associated neuroglia.
3. A nerve is a bundle of axons and/or dendrites outside the central nervous system.
4. A ganglion is a collection of unmyelinated cell bodies outside the CNS.
5. A tract is a bundle of myelinated fibers of similar function in the CNS.
6. A nucleus is a mass of unmyelinated neuron cell bodies and dendrites present as gray matter in the brain and spinal cord.
7. A horn is an area of gray matter in the spinal cord.
8. A column is a region of white matter in the spinal cord that surrounds a horn.

Spinal Cord (p. 472)

Protection and Coverings (p. 472)

1. The spinal cord is protected by the vertebral canal, meninges, epidural fat and connective tissue, cerebrospinal fluid, and vertebral ligaments.
2. The meninges are three connective tissue coverings that run continuously around the spinal cord and brain: dura mater, arachnoid, and pia mater.
3. Removal of cerebrospinal fluid from the subarachnoid space is called a spinal (lumbar) tap (puncture). The procedure is used to diagnose pathologies and to introduce antibiotics or contrast media.

General Features (p. 474)

1. The spinal cord begins as a continuation of the medulla oblongata and terminates at about the second lumbar vertebra.
2. It contains cervical and lumbar enlargements that serve as points of origin for nerves to the extremities.
3. The tapered portion of the spinal cord is the conus medullaris, from which arise the filum terminale and cauda equina.
4. The spinal cord is partially divided into right and left sides by the anterior median fissure and posterior median sulcus.
5. The gray matter in the spinal cord is divided into horns and the white matter into columns.
6. In the center of the spinal cord is the central canal, which runs the length of the spinal cord and contains cerebrospinal fluid.
7. There are ascending (sensory) tracts and descending (motor) tracts.

Structure in Cross Section (p. 474)

1. Parts of the spinal cord observed in cross section are the gray commissure; central canal; anterior, posterior, and lateral gray horns; anterior, posterior, and lateral white columns; and ascending and descending tracts.
2. The spinal cord conveys sensory and motor information by way of the ascending and descending tracts, respectively.

Functions (p. 478)

1. A major function of the spinal cord is to convey sensory nerve impulses from the periphery to the brain and to conduct motor impulses from the brain to the periphery.
2. Another function is to serve as a reflex center. The posterior root, posterior root ganglion, and anterior root are involved in conveying an impulse.
3. A reflex arc is the shortest route that can be taken by an impulse from a receptor to an effector. Its basic components are a receptor, a sensory neuron, a center, a motor neuron, and an effector.
4. A reflex is a quick, involuntary response to a stimulus that passes along a reflex arc. Reflexes represent the body's principal mechanisms for responding to certain changes (stimuli) in the internal and external environment.

Spinal Nerves (p. 481)

Names (p. 481)

1. The 31 pairs of spinal nerves are named and numbered according to the region and level of the spinal cord from which they emerge.
2. There are 8 pairs of cervical, 12 pairs of thoracic, 5 pairs of lumbar, 5 pairs of sacral, and 1 pair of coccygeal nerves.

Composition and Coverings (p. 481)

1. Spinal nerves are attached to the spinal cord by means of a posterior root and an anterior root. All spinal nerves are mixed.
2. Spinal nerves are covered by endoneurium, perineurium, and epineurium.

Distribution (p. 483)

1. Branches of a spinal nerve include the dorsal ramus, ventral ramus, meningeal branch, and rami communicantes.
2. The ventral rami of spinal nerves, except for T2–T11, form networks of nerves called plexuses.
3. Emerging from the plexuses are nerves bearing names that are often descriptive of the general regions they supply or the course they take.
4. The cervical plexus supplies the skin and muscles of the head, neck, and upper part of the shoulders; connects with some cranial nerves; and supplies the diaphragm.
5. The brachial plexus constitutes the nerve supply for the upper extremities and a number of neck and shoulder muscles.
6. The lumbar plexus supplies the anterolateral abdominal wall, external genitals, and part of the lower extremities.
7. The sacral plexus supplies the buttocks, perineum, and lower extremities.
8. Ventral rami of nerves T2–T11 do not form plexuses and are called intercostal (thoracic) nerves. They are distributed directly to the structures they supply in intercostal spaces.

Dermatomes (p. 489)

1. All spinal nerves except C1 innervate specific, constant segments of the skin. The skin segments are called dermatomes.
2. Knowledge of dermatomes helps a physician to determine which segment of the spinal cord or a spinal nerve is malfunctioning.

Applications to Health (p. 490)

1. Spinal cord injury may result in paralysis, which may be classified as monoplegia, diplegia, paraplegia, hemiplegia, or quadriplegia.
2. Transection is followed by a period of loss of reflex activity called areflexia.
3. Inflammation of nerves is known as neuritis.
4. Neuritis of the sciatic nerve and its branches is called sciatica.
5. Shingles is an acute infection of peripheral nerves.

REVIEW QUESTIONS

1. Define the following groupings of neural tissue: white matter, gray matter, nerve, ganglion, tract, nucleus, horn, and column. (p. 472)
2. Describe the bony covering of the spinal cord. (p. 472)
3. Explain the location and composition of the spinal meninges. Describe the location of the epidural, subdural, and subarachnoid spaces. Define meningitis. (p. 472)
4. What is a spinal (lumbar) puncture? Why is it performed? (p. 472)
5. Describe the location of the spinal cord. What are the cervical and lumbar enlargements? (p. 474)
6. Define conus medullaris, filum terminale, and cauda equina. What is a spinal segment? How is the spinal cord partially divided into a right and left side? (p. 474)
7. Based upon your knowledge of the structure of the spinal cord in cross section, define the following: gray commissure, central canal, anterior gray horn, lateral gray horn, posterior gray horn, anterior white column, lateral white column, posterior white column, ascending tract, and descending tract. (p. 474)
8. Describe the function of the spinal cord as a conduction pathway. Using Exhibit 17-1 as a guide, be sure that you can list the location, origin, termination, and function of the principal ascending and descending tracts. (p. 478)
9. Describe how the spinal cord serves as a reflex center. (p. 479)
10. What is a reflex arc? List and define the components of a reflex arc. (p. 480)
11. Define a spinal nerve. Why are all spinal nerves classified as mixed nerves? (p. 481)
12. Describe how a spinal nerve is attached to the spinal cord. (p. 481)
13. Explain how a spinal nerve is enveloped by its connective tissue coverings. (p. 482)
14. How are spinal nerves named and numbered? (p. 481)
15. Describe the branches and innervations of a typical spinal nerve. (p. 483)
16. What is a plexus? Describe the principal plexuses and the regions they supply. (p. 483)
17. What are intercostal (thoracic) nerves? (p. 489)
18. Define a dermatome. Why is a knowledge of dermatomes important? (p. 489)
19. Distinguish the following types of paralysis: monoplegia, diplegia, paraplegia, hemiplegia, and quadriplegia. (p. 490)
20. Define complete transection and hemisection. What are the consequences of each? What is spinal shock? (p. 490)
21. Distinguish between sciatica and neuritis. (p. 491)
22. What is shingles? (p. 493)

SELF QUIZ

1. Match the following:

___ a. tapering inferior end of spinal cord

___ b. any region of spinal cord from which one pair of spinal nerves arises

A. cauda equina
B. conus medullaris
C. filum terminale
D. spinal segment

___ **c.** nonnervous extension of pia mater; anchors spinal cord in place
___ **d.** "horse's tail"; extension of spinal nerve roots in lumbar and sacral regions within subarachnoid space

Complete the following.

2. Individual nerve fibers are wrapped in a connective tissue covering known as (endo-? epi-? peri-?)neurium. Groups of nerve fibers are held in bundles (fascicles) by ________-neurium. The entire nerve is wrapped with ________-neurium.

3. Cell bodies of (sensory? motor?) neurons are located in the posterior (dorsal) root ganglia. Cell bodies of motor neurons leading to skeletal muscles are located in the ________ gray horns of the cord, while visceral efferent cell bodies lie in the ________ gray horns.

4. Tracts are located in (gray horns? white columns?). They appear white since they consist of bundles of (myelinated? unmyelinated?) nerve fibers. Ascending tracts are all (sensory? motor?), conveying impulses between the spinal cord and the ________.

5. Match the following:

___ **a.** provides the entire nerve supply for the arm
___ **b.** contains origin of phrenic nerve (nerve that supplies diaphragm)
___ **c.** forms median, radial, and axillary nerves
___ **d.** not a plexus at all, but rather segmentally arranged nerves
___ **e.** supplies nerves to scalp, neck, and part of shoulder and chest
___ **f.** supplies fibers to the femoral nerve, which innervates the quadriceps, so injury to this plexus would interfere with actions such as touching the toes
___ **g.** forms the largest nerve in the body (sciatic) that supplies posterior of thigh and the leg

A. brachial
B. cervical
C. intercostal
D. lumbar
E. sacral

Choose the one best answer to these questions.

6. The branch of a spinal nerve that innervates the vertebrae, vertebral ligaments, and blood vessels of the spinal cord is known as the
A. ramus communicans; B. dorsal ramus; C. meningeal branch; D. ventral ramus; E. white ramus.

7. Which of the following statements does *not* describe the spinal cord?
A. it begins as a continuation of the medulla oblongata; B. it extends from the foramen magnum of the occipital bone to the level of the second lumbar vertebra; C. it has three enlargements where the majority of spinal nerves originate; D. it contains 31 spinal segments; E. it contains ascending and descending tracts.

8. A reflex arc
(1) always includes at least a sensory and a motor neuron.
(2) always has its center in the brain or the spinal cord.
(3) always terminates in muscle or gland.
A. (1) only; B. (2) only; C. (3) only; D. all of the above; E. none of the above.

9. A ganglion is an aggregation of nerve cell bodies
A. inside the brain or cord; B. outside the brain or cord; C. only in the cord; D. only in the brain; E. inside receptors.

10. The posterior (dorsal) root ganglion contains
A. cell bodies of motor neurons; B. cell bodies of sensory neurons; C. cranial nerve axons; D. synapses; E. all of the above.

11. A region of the spinal cord containing cell bodies of motor neurons is
A. the anterior horn of gray matter; B. the posterior horn of gray matter; C. the lateral column of white matter; D. the anterior column of white matter; E. any part of gray matter.

12. The order of the meningeal layers from outside in is
A. arachnoid, pia, dura; B. pia, arachnoid, dura; C. dura, arachnoid, pia; D. dura, pia, arachnoid; E. arachnoid, dura, pia.

13. All these tracts are sensory *except*
A. anterior spinothalamic; B. lateral spinothalamic; C. fasciculus cuneatus; D. lateral corticospinal; E. posterior spinocerebellar.

14. Choose the false statement about the spinal cord:
A. it has enlargements in the cervical and lumbar areas; B. it lies in the vertebral foramen; C. it extends from the medulla to the sacrum; D. it is surrounded by meninges; E. in cross section an H-shaped area of gray matter can be found.

15. Which of the following is the correct sequence of parts along which a nerve impulse normally travels?
A. receptor–axon–dendrite; B. dendrite–cell body–axon; C. axon–cell body–effector; D. dendrite–receptor–effector; E. cell body–axon–dendrite.

16. Which sequence best represents the course of a nerve impulse over a reflex arc?
A. receptor, center, sensory neuron, motor neuron, effector;
B. effector, sensory neuron, center, motor neuron, receptor;
C. receptor, sensory neuron, center, motor neuron, effector;
D. receptor, motor neuron, center, sensory neuron, effector;
E. effector, center, sensory neuron, motor neuron, receptor.

17. Match the following:

___ **a.** conveys nerve impulses that tell you that you touched a hot object
___ **b.** allow you to be conscious of the position of your body parts; help you assess weight and shape of objects
___ **c.** located in anterior white column; starts in spinal cord and ends in thalamus
___ **d.** sends nerve impulses from your brain to enable you to move your skeletal muscles voluntarily
___ **e.** conveys nerve impulses from one side of the medulla to the same side of your head to regulate head movements, for example, when you are trying to maintain your balance
___ **f.** controls head movements on opposite side, so the right tract causes you to turn to your left when you hear or see something toward your left

A. anterior spinothalmic
B. anterior and lateral corticospinal
C. fasciculus cuneatus and fasciculus gracilis
D. lateral spinothalamic
E. tectospinal
F. vestibulospinal

Circle T (true) or F (false) for the following.

T F 18. The two main functions of the spinal cord are that it serves as a reflex center and it is the site where sensations are felt.

T F 19. Posterior (dorsal) roots of spinal nerves are sensory, anterior roots are motor, and spinal nerves are mixed.

T F 20. A tract is a bundle of nerve fibers inside the central nervous system.

T F 21. The anterior (ventral) root of a spinal nerve contains axons and dendrites of both motor and sensory neurons.

T F 22. The networks formed by the ventral rami of spinal nerves, except for T2–T11, are called plexuses.

T F 23. The deeper, wider groove along the outer surface of the spinal cord is the posterior median sulcus.

24. Arrange the answers in correct sequence.

___ ___ ___ **a.** From superficial to deep:
- **A.** subarachnoid space
- **B.** epidural space
- **C.** dura mater

___ ___ ___ **b.** From anterior to posterior in the spinal cord:
- **A.** fasciculus gracilis and cuneatus
- **B.** anterior spinothalamic tract
- **C.** central canal of the spinal cord

___ ___ ___ ___ **c.** The plexuses, from superior to inferior:
- **A.** lumbar
- **B.** brachial
- **C.** cervical
- **D.** sacral

The Brain and the Cranial Nerves

18

STUDENT OBJECTIVES

1. Identify the principal parts of the brain and describe how the brain is protected.
2. Explain the formation and circulation of cerebrospinal fluid (CSF).
3. Describe the blood supply to the brain and the concept of the blood–brain barrier (BBB).
4. Compare the components of the brain stem, diencephalon, cerebrum, and cerebellum with regard to structure and function.
5. Describe the structure and functions of the limbic system.
6. Compare the sensory, motor, and association areas of the cerebrum.
7. Explain the concept of brain lateralization.
8. Define a cranial nerve and identify the 12 pairs of cranial nerves by name, number, type, location, and function.
9. Describe the effects of aging on the nervous system.
10. Describe the development of the nervous system.
11. List the clinical symptoms of these disorders of the nervous system: cerebrovascular accidents (CVAs), epilepsy, transient ischemic attacks (TIAs), brain tumors, poliomyelitis, cerebral palsy (CP), Parkinson's disease (PD), multiple sclerosis (MS), dyslexia, Tay-Sachs disease, headache, trigeminal neuralgia, Reye's syndrome (RS), Alzheimer's disease (AD), and delirium.
12. Define key medical terms associated with the central nervous system.

CHAPTER OUTLINE

- **Brain**

Principal Parts
Protection and Coverings
Cerebrospinal Fluid (CSF)
Blood Supply
Brain Stem
    Medulla Oblongata
    Pons
    Midbrain
Diencephalon
    Thalamus
    Hypothalamus
    Epithalamus
    Subthalamus
Cerebrum
    Lobes
    White Matter
    Basal Ganglia (Cerebral Nuclei)
    Limbic System
    Functional Areas of Cerebral Cortex
    Electroencephalogram (EEG)
Brain Lateralization (Split-Brain Concept)
Cerebellum
    Structure
    Functions

- **Cranial Nerves**
- **Aging and the Nervous System**
- **Developmental Anatomy of the Nervous System**
- **Applications to Health**
- **Key Medical Terms Associated with the Central Nervous System**

Now we shall consider the principal parts of the brain, how the brain is protected, and how it is related to the spinal cord and to the 12 pairs of cranial nerves.

The developmental anatomy of the brain will be explained in detail at the end of the chapter.

BRAIN

PRINCIPAL PARTS

The ***brain*** of an average adult is made up of about 1000 billion neurons and is one of the largest organs of the body, weighing about 1300 g (3 lb). Figure 18-1 shows that the brain is mushroom-shaped and divided into four principal parts: brain stem, diencephalon, cerebrum, and cerebellum. In some cases, embryological names are retained when distinguishing the various parts of the brain. The ***brain stem,*** the stalk of the mushroom, consists of the medulla oblongata, pons, and midbrain or mesencephalon (mes-en-SEF-a-lon). The lower end of the brain stem is a continuation of the spinal cord. Above the brain stem is the ***diencephalon*** (dī-en-SEF-a-lon), consisting of the thalamus, hypothalamus, epithalamus, and subthalamus. The ***cerebrum*** spreads over the diencephalon. The cerebrum constitutes about seven-eighths of the total weight of the brain and occupies most of the cranium. Inferior to the cerebrum and posterior to the brain stem is the ***cerebellum.***

The brain develops very rapidly during the first few years of life. Growth is due mainly to an increase in the size of cells already present, proliferation and growth of neuroglia, development of synaptic contacts and dendritic branching, and myelination of the various fiber tracts.

PROTECTION AND COVERINGS

The brain is protected by the cranial bones (see Figure 6-2). Like the spinal cord, the brain is also protected by

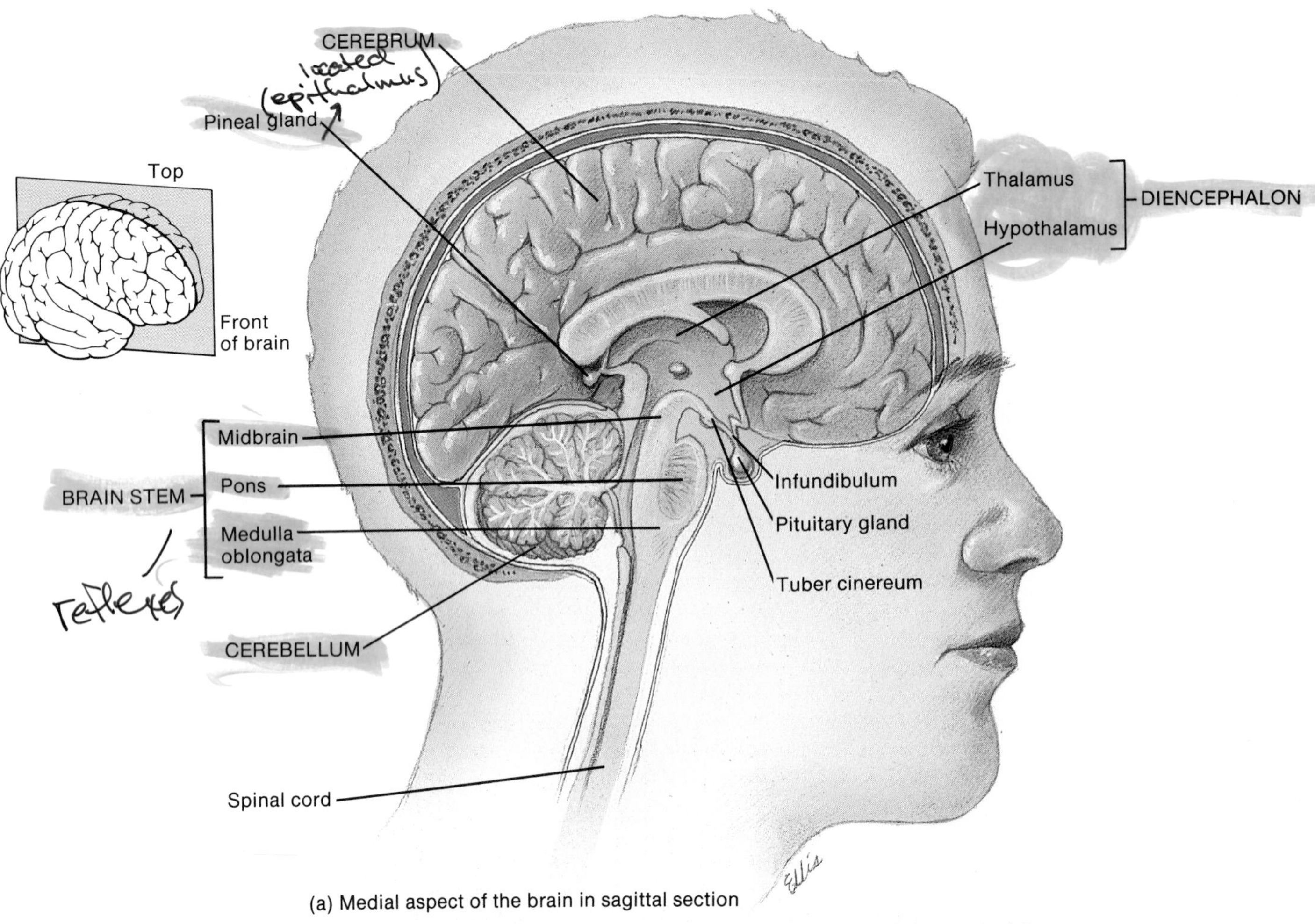

(a) Medial aspect of the brain in sagittal section

FIGURE 18-1 Brain. The infundibulum and pituitary gland shown in (a) are discussed in conjunction with the endocrine system in Chapter 21. (b) © 1984, Rotker, SPL, Photo Researchers.

meninges. The ***cranial meninges*** surround the brain, are continuous with the spinal meninges, and have the same basic structure and bear the same names as the spinal meninges: the outermost ***dura mater,*** middle ***arachnoid,*** and innermost ***pia mater*** (Figure 18-2).

The cranial dura mater consists of two layers. The thicker, outer layer (endosteal layer) tightly adheres to the cranial bones and serves as periosteum. The thinner, inner layer (meningeal layer) includes a mesothelial layer on its smooth surface. The spinal dura mater corresponds to the meningeal layer of the cranial dura mater.

CEREBROSPINAL FLUID (CSF)

The brain, as well as the rest of the central nervous system, is further protected against injury by ***cerebrospinal fluid (CSF).*** This fluid continuously circulates through the subarachnoid space around the brain and spinal cord and through cavities within the brain. The subarachnoid space is the area between the arachnoid and pia mater. CSF is an excellent shock absorber and thus helps protect the delicate brain and spinal cord from trauma.

The ***ventricles*** (VEN-tri-kuls) are cavities in the brain that communicate with each other, with the central canal of the spinal cord, and with the subarachnoid space (Figure 18-2a, b). Each of the two ***lateral ventricles*** is located in a hemisphere (side) of the cerebrum. The ***third ventricle*** is a vertical slit between and inferior to the right and left halves of the thalamus and between the lateral ventricles. Each lateral ventricle communicates with the third ventricle by a narrow, oval opening, the ***interventricular foramen (foramen of Monro).*** The ***fourth ventricle*** lies between the inferior portion of the brain stem and the cerebellum. It communicates with the third ventricle via the ***cerebral aqueduct,*** which passes through the midbrain. The roof of the fourth ventricle has three openings: a ***median aperture (of Magendie)*** and two ***lateral apertures (of Luschka).*** Through these openings, fluid in the fourth ventricle may flow into the subarachnoid space of the brain and cord.

The entire central nervous system contains between 80 and 150 ml (3 to 5 oz) of cerebrospinal fluid. It is a clear, colorless liquid of watery consistency. Chemically, it contains proteins, glucose, urea, and salts. It also contains some lymphocytes. Cerebrospinal fluid has two principal functions: protection and circulation. The fluid serves as a shock-absorbing medium that protects the brain and spinal cord from jolts that would otherwise cause them to crash against the bony walls of the cranial and vertebral cavities. The fluid also buoys the brain so that it "floats" in the cranial cavity. With regard to its circulatory function, CSF delivers nutritive substances filtered from blood to the brain and spinal cord and removes wastes and toxic substances produced by brain and spinal cord cells.

Cerebrospinal fluid is formed primarily by filtration and secretion from networks of capillaries in the pia mater and ependymal cells in the ventricles called ***choroid*** (KŌ-royd; *chorion* = delicate) ***plexuses*** (Figure 18-2a). Various components of the choroid plexuses form a ***blood–cerebrospinal fluid barrier*** that permits certain substances to enter the fluid but prohibits others. Such a barrier protects the brain

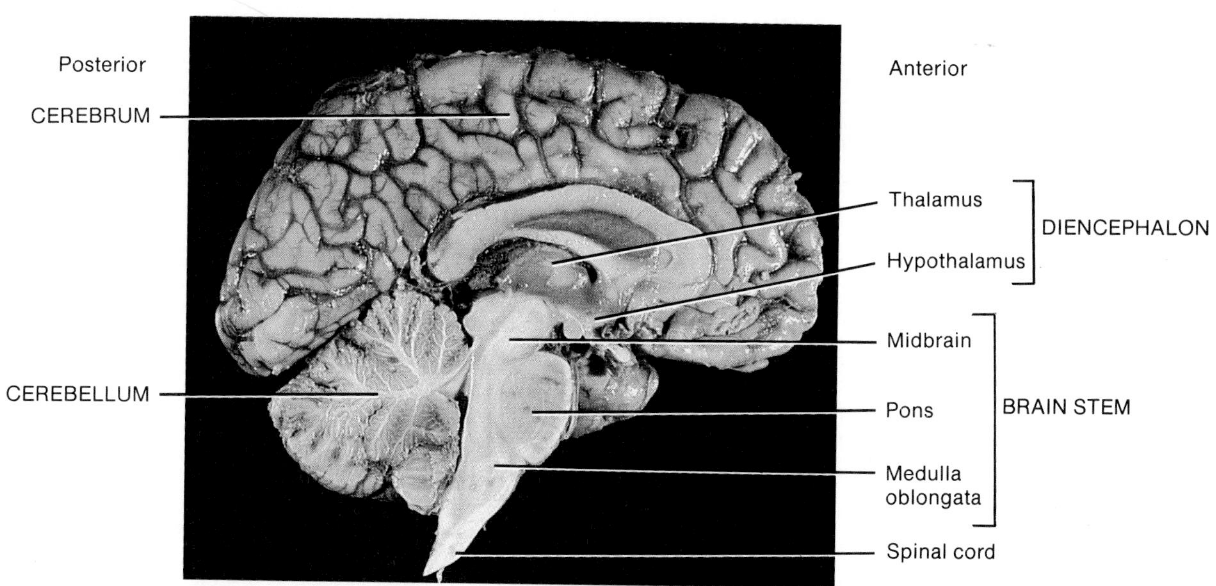

(b) Medial aspect of brain in sagittal section

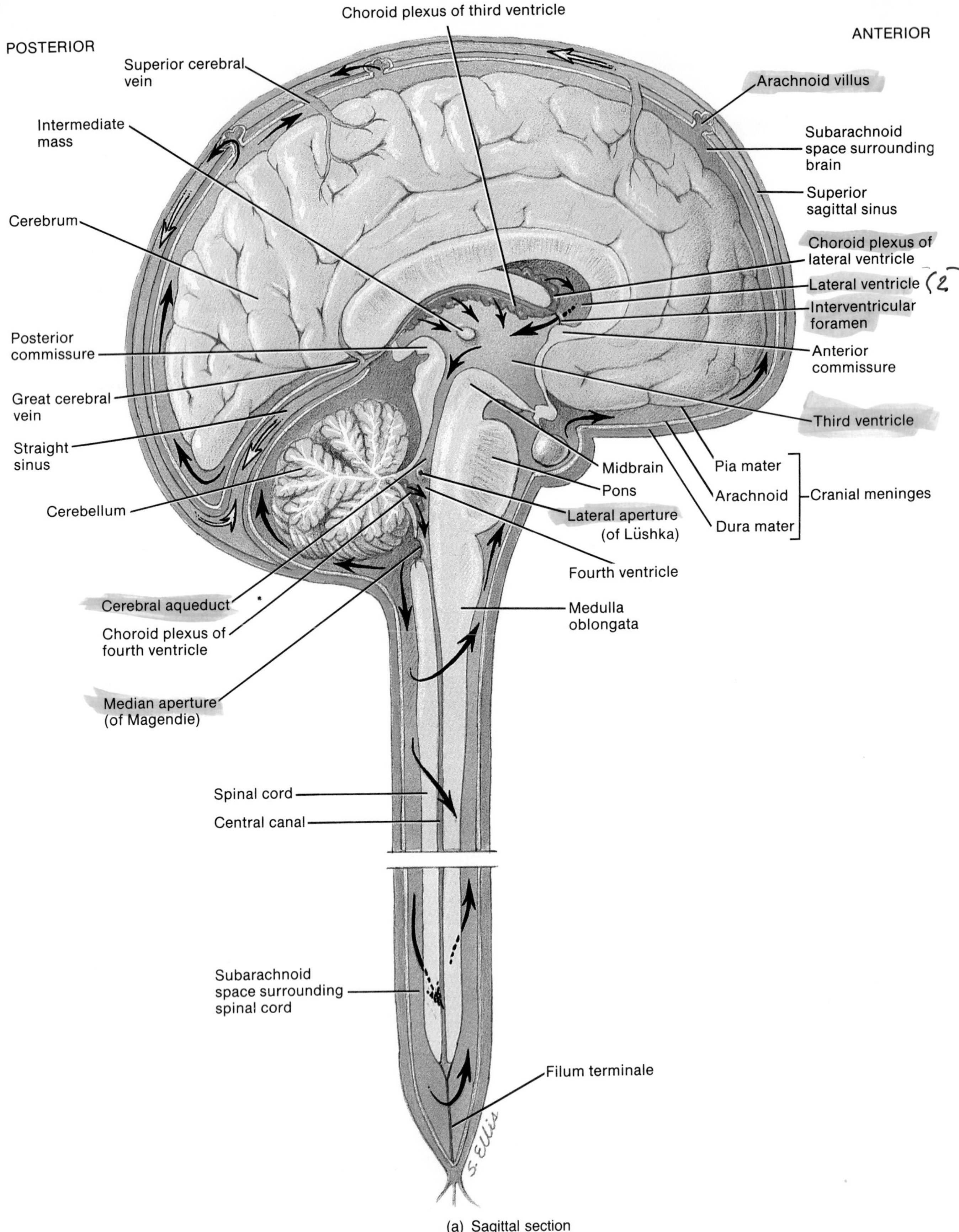

(a) Sagittal section

FIGURE 18-2 Meninges and ventricles of the brain. (a) Brain, ventricles, spinal cord, and meninges. Arrows indicate the direction of flow of cerebrospinal fluid. (b) Ventricles projected to surface. (c) Superior portion of the brain, showing the relationship of the superior sagittal sinus to the arachnoid villi.

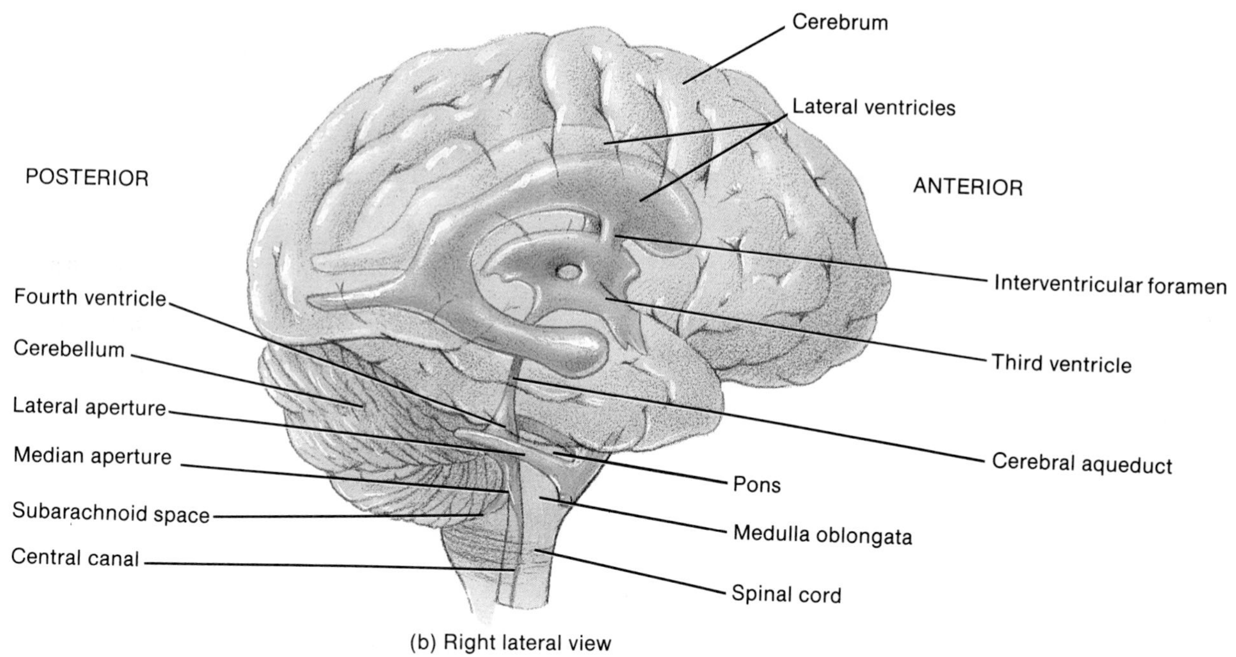

(b) Right lateral view

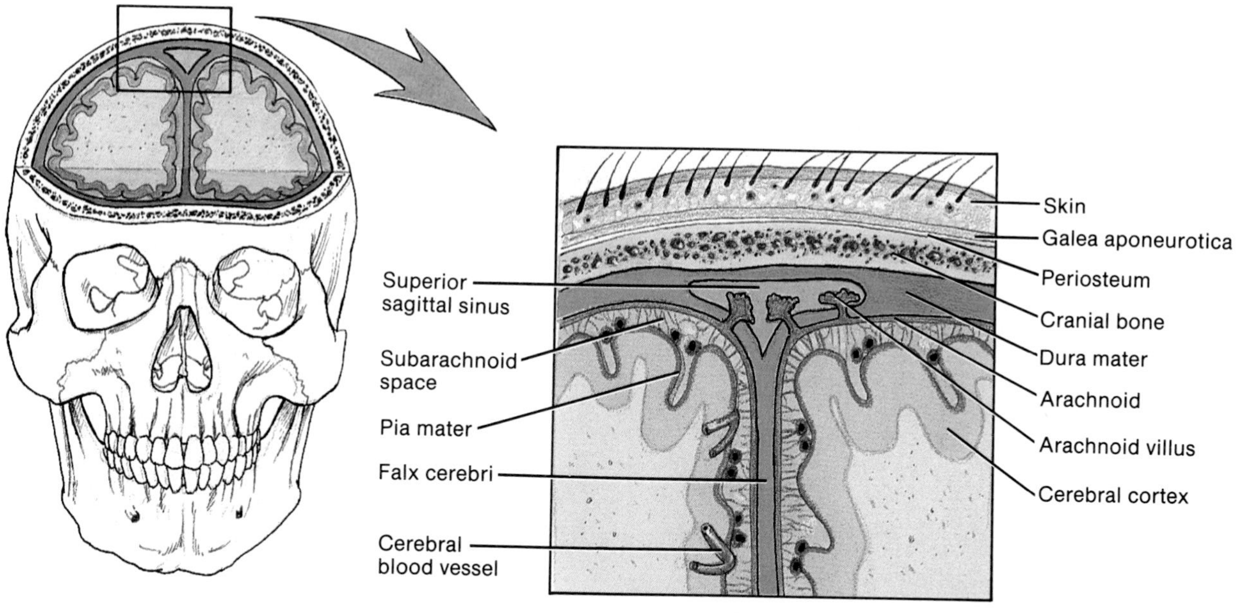

(c) Frontal section

and spinal cord from harmful substances. The fluid formed in the choroid plexuses of the lateral ventricles circulates through the interventricular foramina to the third ventricle, where more fluid is added by the choroid plexus of the third ventricle. It then flows through the cerebral aqueduct into the fourth ventricle. Here, there are contributions from the choroid plexus of the fourth ventricle. The fluid then circulates through the apertures of the fourth ventricle into the subarachnoid space around the posterior surface of the brain. It also passes downward to the subarachnoid space around the posterior surface of the spinal cord, up the anterior surface of the spinal cord, and around the anterior part of the brain. From there it is gradually reabsorbed into a blood vascular sinus (vein) called the superior sagittal sinus. The actual reabsorption occurs through ***arachnoid villi***—finger-like projections of the arachnoid that push into the dural venous sinuses, especially the superior sagittal sinus (Figure 18-2c). Normally, cerebrospinal fluid is reabsorbed as rapidly as it is formed.

The formation, circulation, and absorption of cerebrospinal fluid are summarized in Figure 18-3.

CLINICAL APPLICATION

Hydrocephalus

If an obstruction, such as a tumor or a congenital blockage, or an inflammation arises in the brain and interferes with the drainage of cerebrospinal fluid from the ventricles into the subarachnoid space, large amounts of fluid accumulate in the ventricles. Fluid pressure inside the brain increases, and if the fontanels have not yet closed, the head bulges to relieve the pressure. This condition is called ***internal (noncommunicating) hydrocephalus*** (*hydro* = water; *enkephalos* = brain). If an obstruction interferes with drainage somewhere in the subarachnoid space and cerebrospinal fluid accumulates inside the space, the condition is termed ***external (communicating) hydrocephalus.*** Hydrocephalus responds dramatically to ventricular drainage and diversion of CSF.

BLOOD SUPPLY

The brain is well supplied with oxygen and nutrients mainly by blood vessels that form the cerebral arterial circle (circle of Willis). Cerebral circulation is outlined in Exhibit 14-3 and Figure 14-8e. Blood vessels that enter brain tissue pass along the surface of the brain and, as they penetrate inward, they are surrounded by a loose-fitting layer of pia mater. The space between the penetrating blood vessel and pia mater is called a ***perivascular space.***

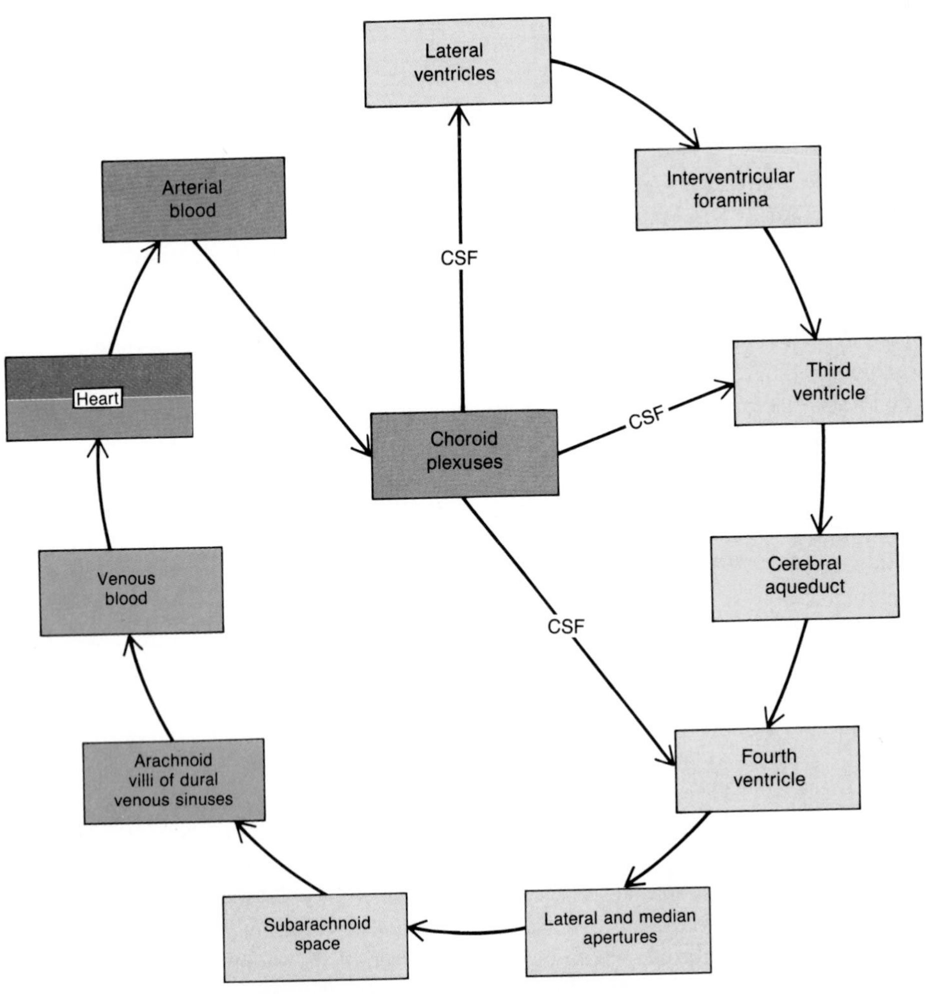

FIGURE 18-3 Summary of the formation, circulation, and absorption of cerebrospinal fluid (CSF).

Although the brain composes only about 2 percent of total body weight, it utilizes about 20 percent of the oxygen used by the entire body at rest. The brain is one of the most metabolically active organs of the body, and the amount of oxygen it uses varies with the degree of mental activity. If the blood flow to the brain is interrupted even briefly, unconsciousness may result. A one- or two-minute interruption may impair the brain cells by starving them of oxygen, and if the cells are totally deprived of oxygen for about four minutes, many are permanently injured. Lysosomes of brain cells are sensitive to decreased oxygen concentration. If the condition persists long enough, lysosomes break open and release enzymes that bring about self-destruction of brain cells. Occasionally, during childbirth the oxygen supply from the mother's blood is interrupted before the baby leaves the birth canal and can breathe. Often such babies are stillborn or suffer permanent brain damage that may result in mental retardation, epilepsy, or paralysis.

Blood supplying the brain also contains glucose, the principal source of energy for brain cells. Because carbohydrate storage in the brain is limited, the supply of glucose must be continuous. If blood entering the brain has a low glucose level, mental confusion, dizziness, convulsions, and loss of consciousness may occur.

Glucose, oxygen, carbon dioxide, water, and most lipid-soluble substances, such as alcohol and anesthetics, pass rapidly from the circulating blood into brain cells. Other substances, such as creatinine, urea, and sodium, potassium, and chloride ions, enter quite slowly. Still other substances—proteins and most antibiotics—do not pass at all from the blood into brain cells. The differential rates of passage of certain materials from the blood into most parts of the brain are based upon a concept called the ***blood–brain barrier (BBB).*** The barrier is absent or less selective in the choroid plexuses, area postrema (portion of medulla where vomiting is triggered), portions of the hypothalamus (median eminence), and pituitary and pineal glands. Electron micrograph studies of the capillaries of the brain reveal that they differ structurally from other capillaries. Brain capillaries are constructed of more densely packed cells and are surrounded by terminations of processes of large numbers of astrocytes (one of the types of neuroglia) and a continuous basement membrane.

Current evidence indicates that astrocytes produce a substance that influences the capillaries and confers on them the ability to selectively pass various substances but inhibit others. The blood–brain barrier functions as a selective barrier to protect brain cells from harmful substances. An injury to the brain due to trauma, inflammation, or toxins causes a breakdown of the blood–brain barrier, permitting the passage of normally restricted substances into brain tissue. Recent evidence suggests that the AIDS virus may penetrate the blood–brain barrier, producing dementia (irreversible deterioration of mental state), acute meningitis, back spasm, and other neurologic disorders before other symptoms of AIDS become apparent.

BRAIN STEM

Medulla Oblongata

The ***medulla oblongata*** (me-DULL-la ob′-long-GA-ta), or simply ***medulla,*** is a continuation of the upper portion of the spinal cord and forms the inferior part of the brain stem (Figure 18-4). Its position in relation to the other parts of the brain may be noted in Figure 18-1. It lies just superior to the level of the foramen magnum and extends upward to the inferior portion of the pons. The medulla measures 3 cm (about 1 in.) in length.

The medulla contains all ascending and descending tracts that communicate between the spinal cord and various parts of the brain. These tracts constitute the white matter of the medulla. Some tracts cross as they pass through the medulla. Let us see how this crossing occurs and what it means.

On the ventral side of the medulla are two structures called ***pyramids*** (Figures 18-4 and 18-5). The pyramids are composed of the largest motor tracts that pass from the outer region of the cerebrum (cerebral cortex) to the spinal cord. Just above the junction of the medulla with the spinal cord, most of the fibers in the left pyramid cross to the right side, and most of the fibers in the right pyramid cross to the left. This crossing is called the ***decussation*** (dē′-ku-SĀ-shun) ***of pyramids.*** Decussation explains why motor areas of one side of the cerebral cortex control muscular movements on the opposite side of the body. The principal motor fibers that undergo decussation descend in the lateral corticospinal tracts (see Figure 20-9). These fibers originate in the cerebral cortex and pass inferiorly to the medulla. The fibers cross in the pyramids and descend in the lateral columns of the spinal cord, terminating in the anterior gray horns. Here synapses occur with motor neurons that terminate in skeletal muscles. As a result of the crossing, fibers that originate in the left cerebral cortex activate muscles on the right side of the body, and fibers that originate in the right cerebral cortex activate muscles on the left side.

The dorsal side of the medulla contains two pairs of prominent nuclei: the right and left ***nucleus gracilis*** (gras-I-lis; *gracilis* = slender) and ***nucleus cuneatus*** (kyoo-nē-Ā-tus; *cuneus* = wedge). These nuclei receive sensory fibers from ascending tracts (right and left fasciculus gracilis and fasciculus cuneatus) of the spinal cord and relay the sensory information to the opposite side of the medulla (see Figure 20-5). The information is conveyed to the thalamus and then to the sensory areas of the cerebral cortex. Nearly all sensory impulses initiated on one side of the body cross in the medulla or spinal cord and are perceived in the opposite side of the cerebral cortex.

In addition to its function as a conduction pathway for motor and sensory impulses between the brain and spinal cord, the medulla also contains an area of dispersed gray matter containing some white fibers. This region is called the ***reticular formation*** (Figure 18-6). Actually, portions of the reticular formation are also located in the spinal cord, pons, midbrain, and diencephalon. A part of the reticular formation called the ***reticular activating system*** **(*RAS*)** functions in consciousness and awakening from sleep.

CLINICAL APPLICATION

Unconsciousness
The most common knockout blow is one that makes contact with the mandible. Such a blow twists and distorts the brain stem and overwhelms the reticular activating system of the reticular formation by sending a sudden volley of nerve impulses to the brain, resulting in ***unconsciousness.***

Within the medulla are also three vital reflex centers. The ***cardiac center*** regulates the rate of heartbeat and force of contraction, the ***medullary rhythmicity area*** of the ***respiratory center*** adjusts the basic rhythm of breathing, and the ***vasomotor*** **(*vasoconstrictor*)** ***center*** regulates the diameter of blood vessels. Other centers in the medulla are considered nonvital and coordinate swallowing, vomiting, coughing, sneezing, and hiccuping.

The medulla also contains the nuclei of origin for several pairs of cranial nerves (see Figures 18-4 and 18-5). These are the cochlear and vestibular branches of the vestibulocochlear (VIII) nerves, which are concerned with hearing and equilibrium (there is also a nucleus for the vestibular branches in the pons); the glossopharyngeal (IX) nerves,

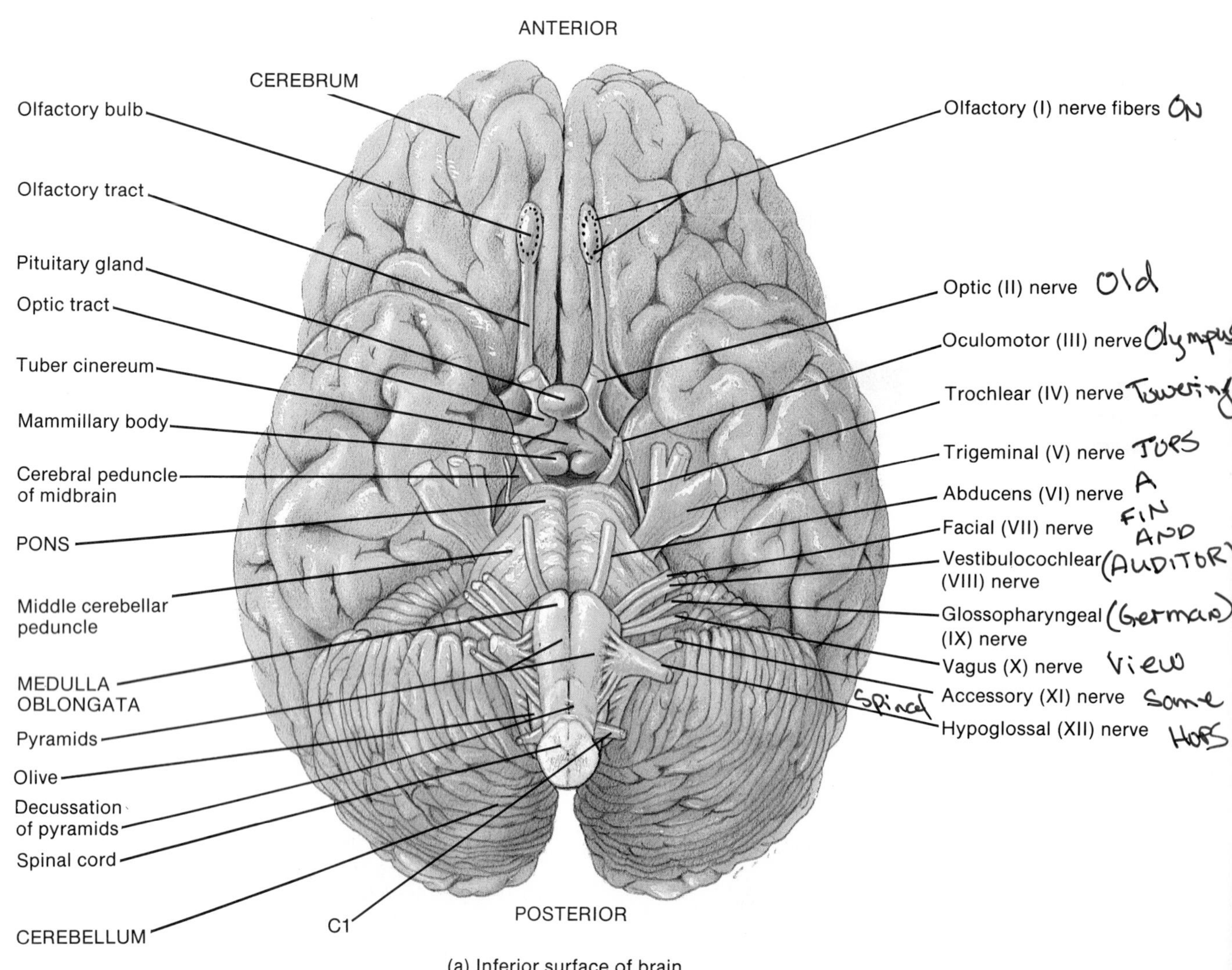

(a) Inferior surface of brain

FIGURE 18-4 Brain stem. (a) Diagram of the brain stem in relation to the cranial nerves and associated structures. (b) Courtesy of Martin Rotker, Phototake. (c) through (e) Courtesy of N. Gluhbegovic and T. H. Williams, *The Human Brain: A Photographic Guide,* Harper & Row, Publishers, Inc., Hagerstown, MD, 1980.

Anterior

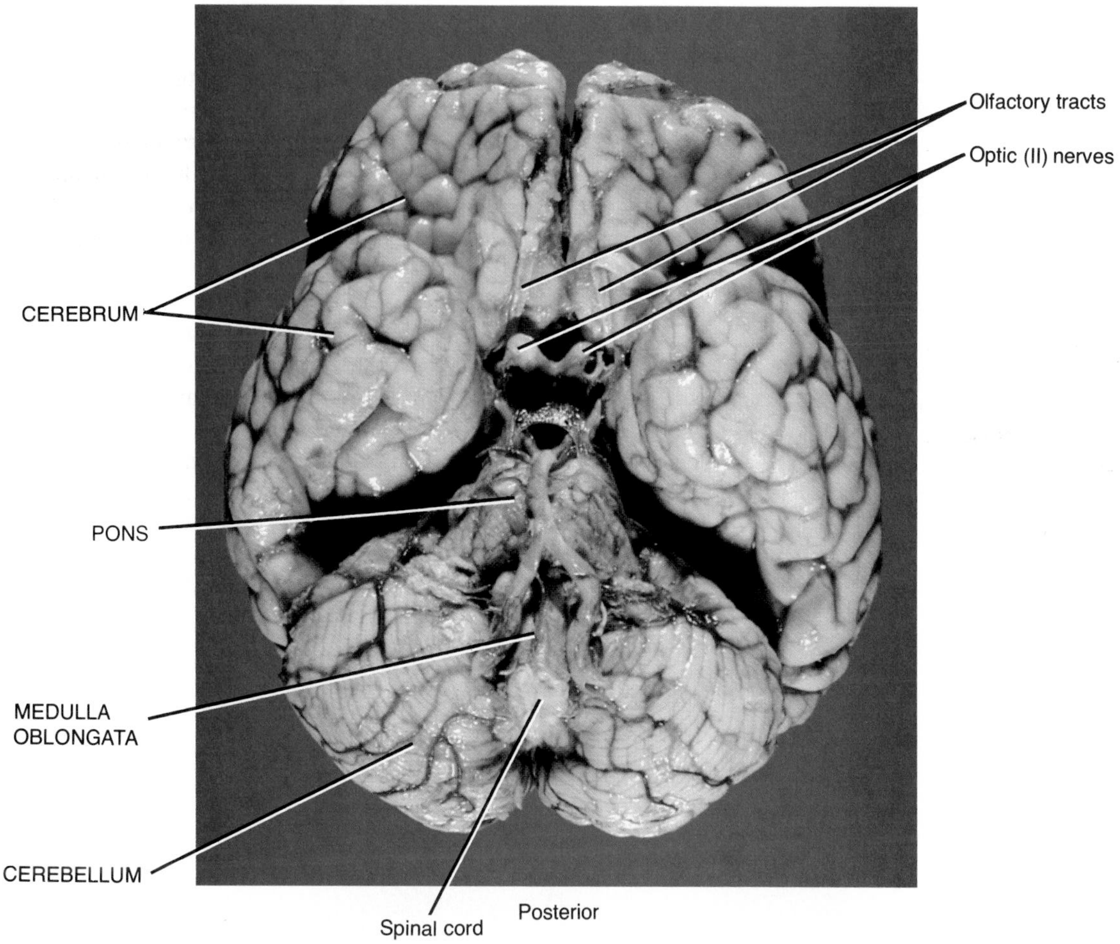

Olfactory tracts
Optic (II) nerves
CEREBRUM
PONS
MEDULLA
OBLONGATA
CEREBELLUM
Spinal cord

Posterior

(b) Inferior surface of brain

Trigeminal (V) nerve
PONS
Facial (VII) nerve
Vestibulocochlear
(VIII) nerve
Olive of medulla
Pyramids
of medulla
Decussation
of pyramids
Anterior median
fissure of
spinal cord

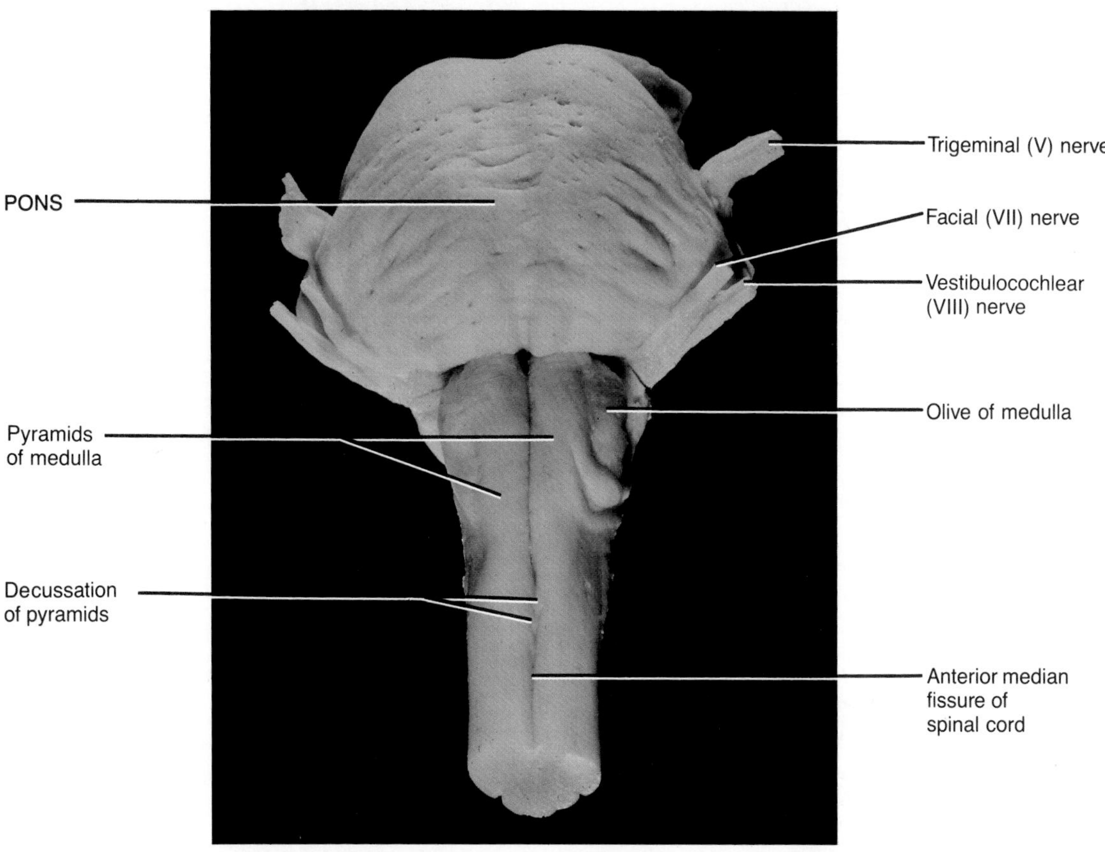

(c) Ventral surface of brain stem

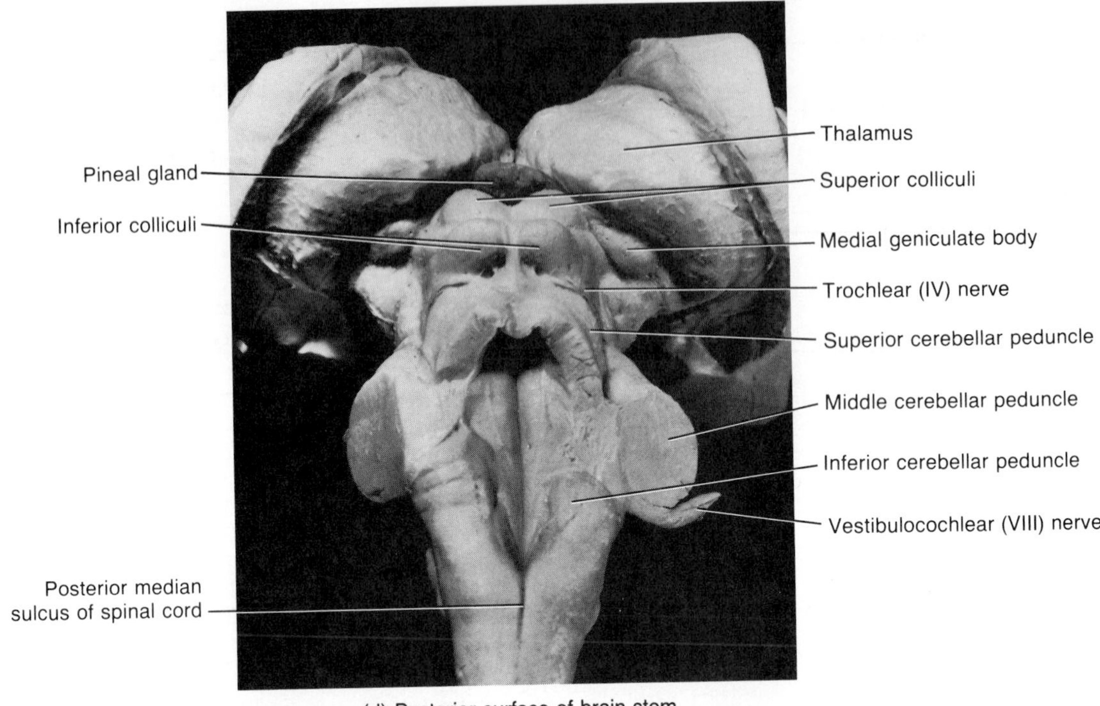

(d) Posterior surface of brain stem

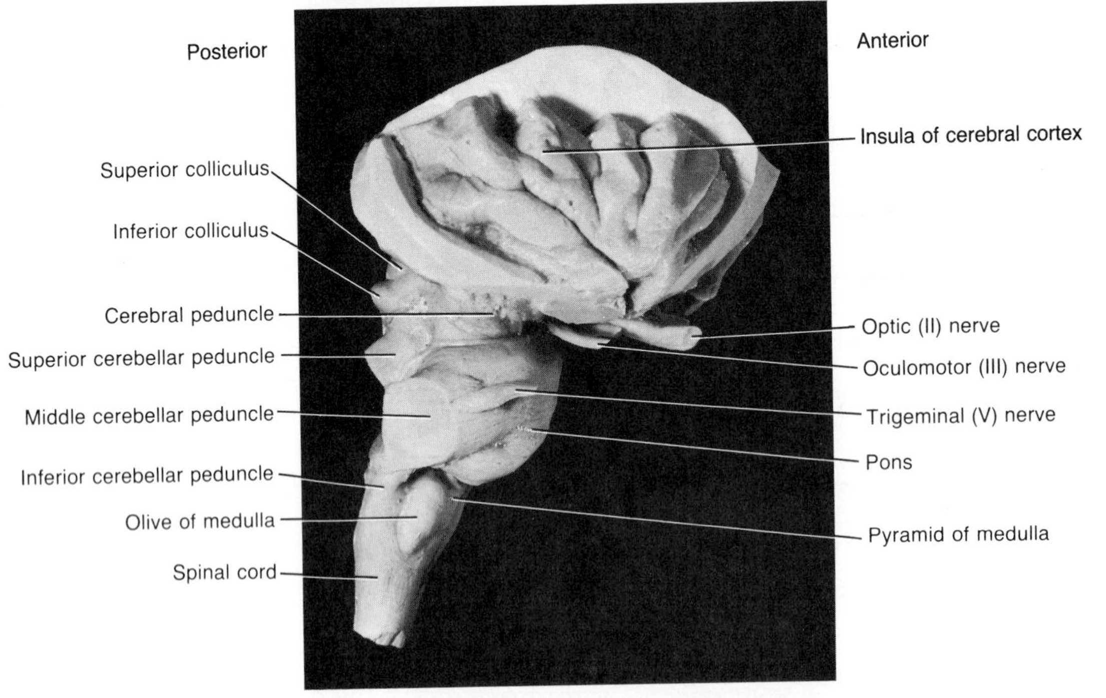

(e) Right lateral view of brain stem

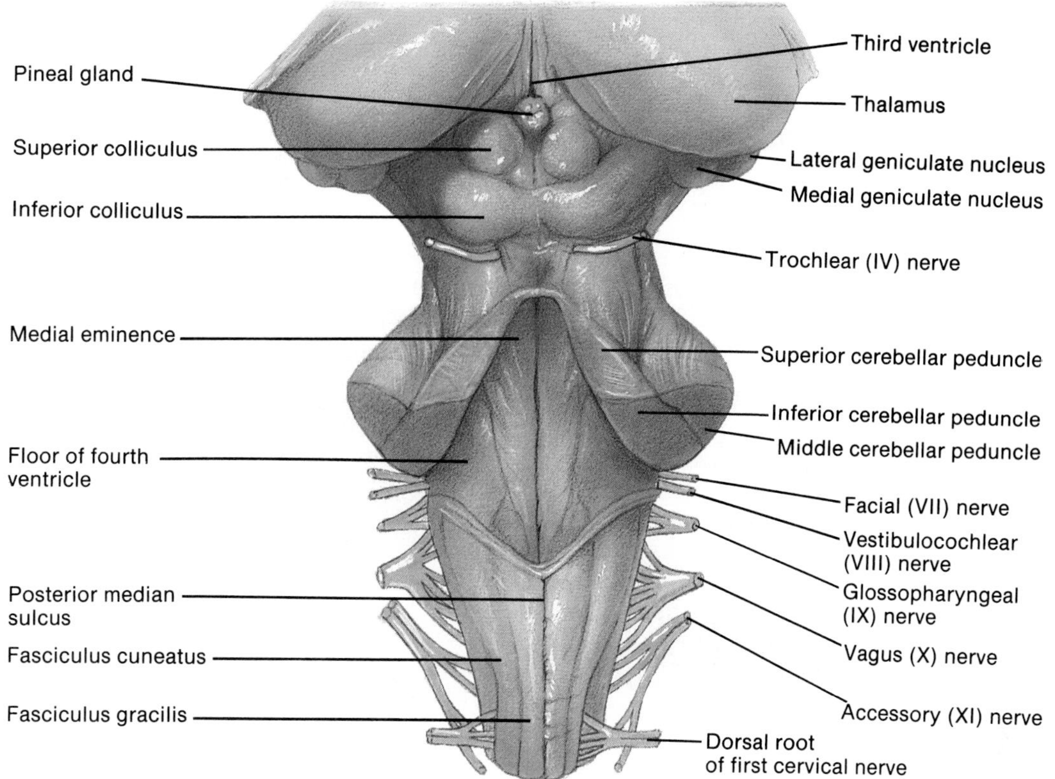

(a) Ventral surface of brain stem

Fourth ventricle
Vagus nucleus (dorsal motor)
Hypoglossal nucleus
Vagus (X) nerve
Vagus (X) nerve
Olive
Inferior olivary nucleus
Hypoglossal (XII) nerve
Hypoglossal (XII) nerve
Pyramid
Decussation of pyramids
Lateral corticospinal tract fibers
C1
Ventral corticospinal tract fibers
Spinal cord

(b) Cross section and ventral surface of medulla

FIGURE 18-5 Medulla. (a) Relationship of the medulla to the other components of the brain stem following removal of the cerebellum. (b) Details of the medulla showing the decussation of pyramids.

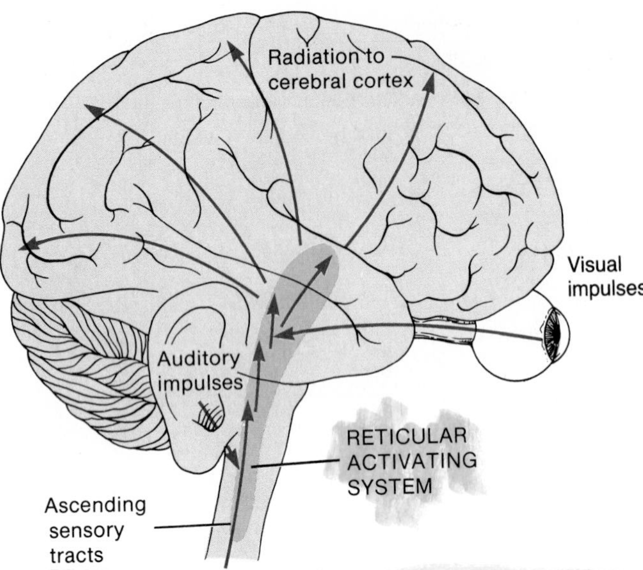

FIGURE 18-6 The reticular activating system (RAS) of the reticular formation.

which relay nerve impulses related to swallowing, salivation, and taste; the vagus (X) nerves, which relay nerve impulses to and from many thoracic and abdominal viscera; the cranial portion of the accessory (XI) nerves (a part of this nerve, the spinal portion, originates in the upper five cervical segments of the spinal cord), which conveys nerve impulses related to head and shoulder movements; and the hypoglossal (XII) nerves, which convey nerve impulses that involve tongue movements.

On each lateral surface of the medulla is an oval projection called the ***olive*** (see Figure 18-4), which contains an ***inferior olivary nucleus*** and two ***accessory olivary nuclei.*** The nuclei are connected to the cerebellum by fibers. The inferior olivary nucleus projects fibers to the part of the cerebellum that ensures the efficiency of voluntary movements, especially precision ones. The accessory olivary nuclei project nerve fibers to regions of the cerebellum concerned with the maintenance of equilibrium, postural changes, and locomotion.

Also associated with the medulla is the greater part of the ***vestibular nuclear complex.*** This nuclear group consists of the ***lateral, medial,*** and ***inferior vestibular nuclei*** in the medulla and the ***superior vestibular nucleus*** in the pons. As you will see later (Chapter 20), the vestibular nuclei assume an important role in helping the body maintain its sense of equilibrium.

In view of the many vital activities controlled by the medulla, it is not surprising that a hard blow to the base of the skull can be fatal. Nonfatal medullary injury may be indicated by cranial nerve malfunctions on the same side of the body as the area of medullary injury, paralysis and loss of sensation on the opposite side of the body, and irregularities in respiratory control.

Pons

The relationship of the ***pons*** to other parts of the brain can be seen in Figures 18-1 and 18-4. The pons, which means "bridge," lies directly above the medulla and anterior to the cerebellum. It measures about 2.5 cm (1 in.) in length. Like the medulla, the pons consists of nuclei and white fibers present as tracts that are scattered throughout. As the name implies, the pons is a bridge connecting the spinal cord with the brain and parts of the brain with each other. These connections are provided by fibers that run in two principal directions. The transverse fibers connect the right and left sides of the cerebellum through the ***middle cerebellar peduncles*** (pe-DUNG-kulz). The longitudinal fibers of the pons belong to the motor and sensory tracts that connect the spinal cord or medulla with the upper parts of the brain stem.

The nuclei for certain paired cranial nerves are also contained in the pons (see Figures 18-4a and 18-16). These include the trigeminal (V) nerves, which relay nerve impulses for chewing and for sensations of the head and face; the abducens (VI) nerves, which regulate certain eyeball movements; the facial (VII) nerves, which conduct impulses related to taste, salivation, and facial expression; and the vestibular branches of the vestibulocochlear (VIII) nerves, which are concerned with equilibrium.

Other important nuclei in the reticular formation of the pons are the ***pneumotaxic*** (noo-mō-TAK-sik) ***area*** and the ***apneustic*** (ap-NOO-stik) ***area*** of the ***respiratory center.*** Together with the medullary rhythmicity area in the medulla, they help control respiration (breathing movements).

Midbrain

The ***midbrain,*** or ***mesencephalon*** (*meso* = middle; *enkephalos* = brain), extends from the pons to the lower portion of the diencephalon (see Figures 18-1, 18-4, and 18-5). It is about 2.5 cm (1 in.) in length. The cerebral aqueduct passes through the midbrain and connects the third ventricle above with the fourth ventricle below.

The ventral portion of the midbrain contains a pair of fiber bundles referred to as ***cerebral peduncles.*** The cerebral peduncles contain some motor fibers that convey nerve impulses from the cerebral cortex to the pons, medulla, and spinal cord. They also contain sensory fibers that pass from the spinal cord to the medulla and then from the pons to the thalamus. The cerebral peduncles constitute the main connection for tracts between upper parts of the brain and lower parts of the brain and the spinal cord.

The dorsal portion of the midbrain is called the ***tectum*** (*tectum*=roof) and contains four rounded elevations: the ***corpora quadrigemina*** (KOR-po-ra kwad-ri-JEM-in-a). Two of the elevations are known as the ***superior colliculi*** (ko-LIK-yoo-lī). These serve as reflex centers for movements of the eyeballs and head and neck in response to visual and other stimuli. The other two elevations, the ***infe-***

rior colliculi, serve as reflex centers for movements of the head and trunk in response to auditory stimuli. The midbrain also contains the ***substantia nigra*** (sub-STAN-shē-a NĪ-gra), a large, heavily pigmented nucleus near the cerebral peduncles.

A major nucleus in the reticular formation of the midbrain is the ***red nucleus.*** Fibers from the cerebellum and cerebral cortex terminate in the red nucleus. The red nucleus is also the origin of cell bodies of the descending rubrospinal tract. Other nuclei in the midbrain are associated with cranial nerves (see Figures 18-4 and 18-16). These include the oculomotor (III) nerves, which mediate some movements of the eyeballs and changes in pupil size and lens shape, and the trochlear (IV) nerves, which conduct impulses that move the eyeballs.

A structure called the ***medial lemniscus*** (*lemniskos* = ribbon or band) extends through the medulla, pons, and midbrain. The medial lemniscus is a band of white fibers containing axons that convey impulses for fine touch, proprioception, pressure, and vibrations from the medulla to the thalamus.

DIENCEPHALON

The ***diencephalon*** (*dia* = through; *enkephalos* = brain) consists of the thalamus, hypothalamus, epithalamus, and subthalamus. The relationship of these structures to the rest of the brain is shown in Figure 18-1.

Thalamus

The ***thalamus*** (THAL-a-mus; *thalamos* = inner chamber) is an oval structure above the midbrain that measures about 3 cm (1 in.) in length and constitutes four-fifths of the diencephalon. It consists of paired oval masses of mostly gray matter organized into nuclei that form the lateral walls of the third ventricle (Figure 18-7). The masses are frequently joined by a bridge of gray matter that crosses the third ventricle called the ***intermediate mass.*** Each mass is deeply embedded in a cerebral hemisphere and is bounded laterally by the ***internal capsule.***

Although the thalamic masses are primarily gray matter, some portions are white matter. Among the white matter

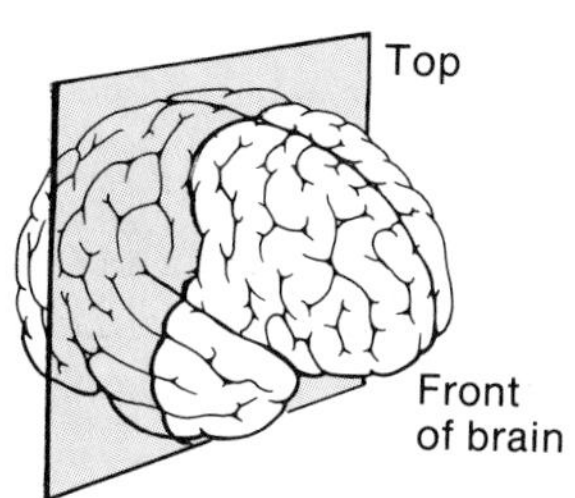

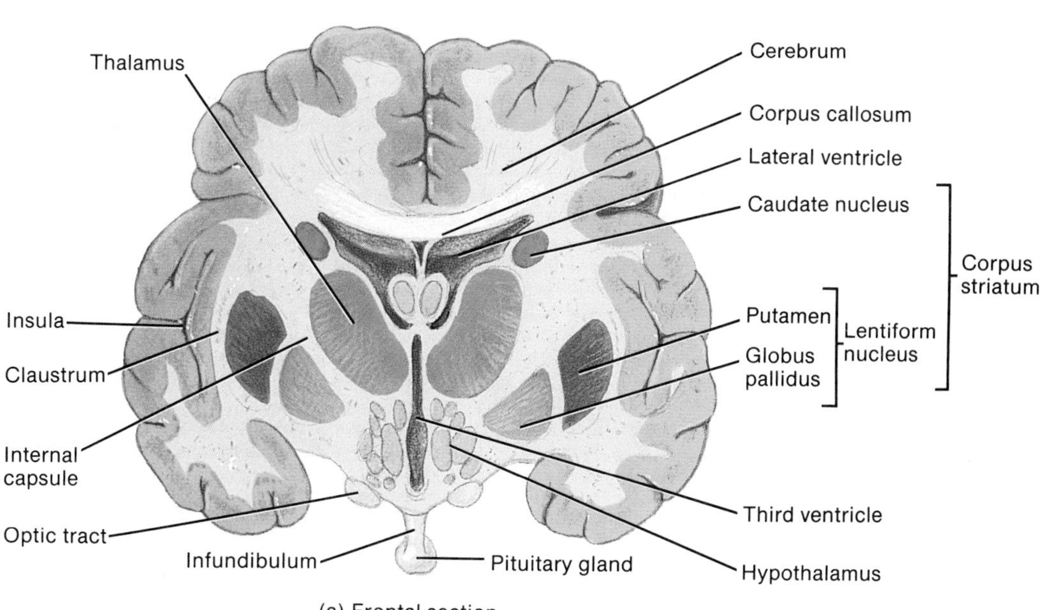

(a) Frontal section

FIGURE 18-7 Thalamus. (a) Thalamus and associated structures. (b) Courtesy of Stephen A. Kieffer and E. Robert Heitzman, *An Atlas of Cross-Sectional Anatomy,* Harper & Row, Publishers, Inc., Hagerstown, MD, 1979. (c) Thalamic nuclei. The arrows indicate some of the connections between the thalamus and cerebral cortex.

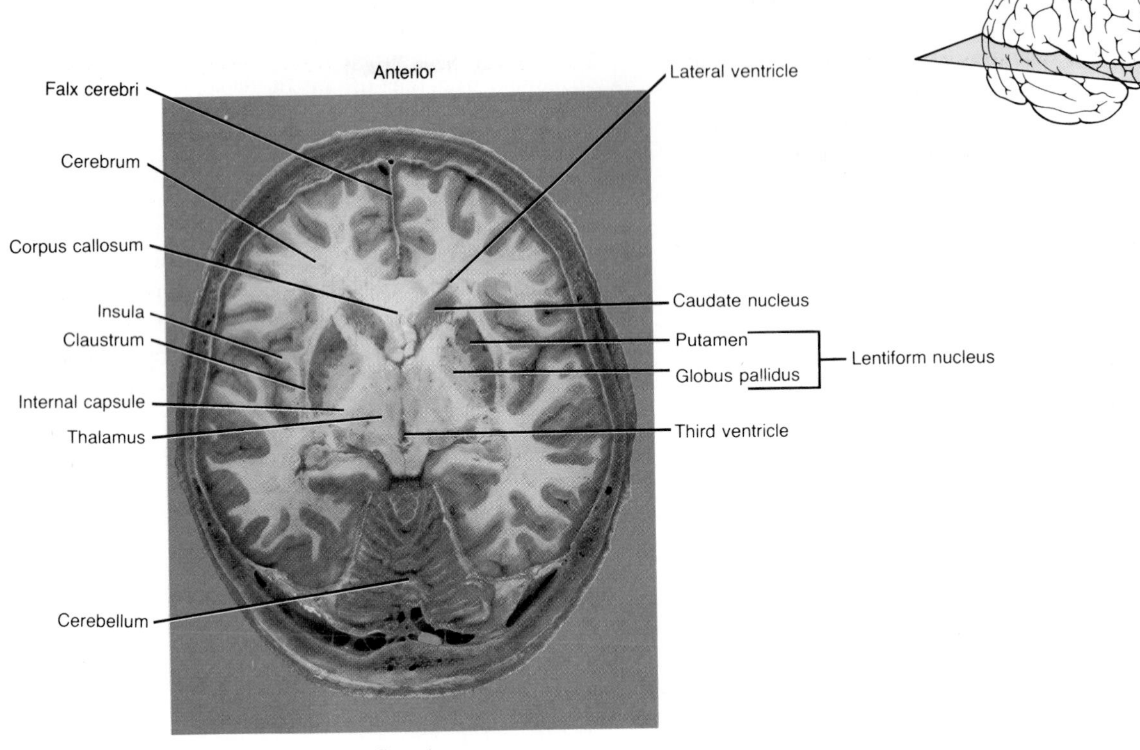

(b) Cross section of brain

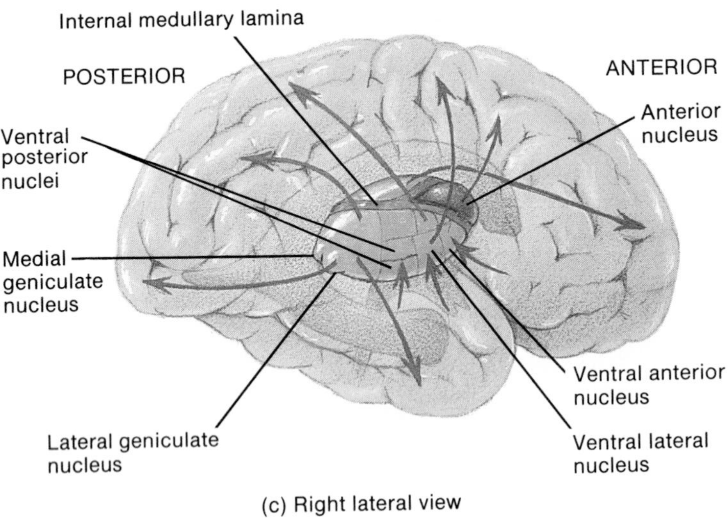

(c) Right lateral view

portions are the ***stratum zonale,*** which covers the dorsal surface; the ***external medullary lamina,*** covering the lateral surface; and the ***internal medullary lamina,*** which divides the gray matter masses into an anterior nuclear group, a medial nuclear group, and a lateral nuclear group.

Within each group are nuclei that assume various roles. Some nuclei in the thalamus serve as relay stations for all sensory impulses, except smell, to the cerebral cortex. These include the ***medial geniculate*** (je-NIK-yoo-lāt) ***nuclei*** (hearing), the ***lateral geniculate nuclei*** (vision), and the ***ventral posterior nuclei*** (taste and general sensations such as touch, pressure, vibration, heat, cold, and pain). Other nuclei are centers for synapses in the somatic motor system. These include the ***ventral lateral nuclei*** (voluntary motor actions) and ***ventral anterior nuclei*** (voluntary motor actions and arousal). (See Figure 18-7c.) The thalamus is the principal relay station for sensory impulses that reach the cerebral cortex from the spinal cord, brain stem, cerebellum, and parts of the cerebrum.

The thalamus also functions as an interpretation center for some sensory impulses, such as pain, temperature, light touch, and pressure. The thalamus also contains a ***reticular nucleus*** in its reticular formation, which in some way seems to modify neuronal activity in the thalamus, and an ***anterior nucleus*** in the floor of the lateral ventricle, which is concerned with certain emotions and memory.

Hypothalamus

The ***hypothalamus*** (*hypo* = under) is a small portion of the diencephalon located below the thalamus. Its relationship to other parts of the brain is shown in Figures 18-1 and 18-7a. The hypothalamus forms the floor and part of the lateral walls of the third ventricle. It is partially protected by the sella turcica of the sphenoid bone.

Information from the external environment ultimately comes to the hypothalamus via afferent pathways originating in the peripheral sense organs. Impulses from sound, taste, smell, and somatic receptors all come to the hypothalamus. Afferent impulses, monitoring the internal environment, arise from the internal viscera and reach the hypothalamus. Other receptors in parts of the hypothalamus itself continually monitor water concentration, certain hormone concentrations, and the temperature of blood. And, as you will see shortly, the hypothalamus has several very important connections with the pituitary gland and is itself able to produce a variety of hormones.

The hypothalamus is divided into several nuclei in four major regions (Figure 18-8):

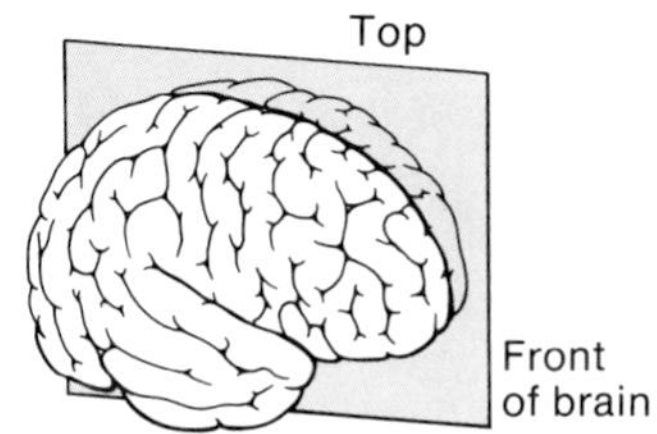

FIGURE 18-8 Hypothalamus. Selected areas of the hypothalamus and a three-dimensional representation of hypothalamic nuclei (after Netter).

1. ***Supraoptic region.*** This anterior region lies above the optic chiasma (point of crossing of optic nerves) and contains the paraventricular nucleus, supraoptic nucleus, anterior hypothalamic nucleus, and suprachiasmatic nucleus.
2. ***Tuberal region.*** This middle region is the widest portion of the hypothalamus. On its ventral surface are the ***tuber cinereum*** (si-NE-rē-um), ***infundibulum,*** and ***median eminence.*** The tuber cinereum is an elevated mass of gray matter that contains neurons that transport regulating hormones (or factors) from the hypothalamus to the infundibulum, a stalklike structure that attaches the pituitary gland to the hypothalamus. The median eminence is a slightly raised region which encircles the site at which the infundibulum becomes the stalk of the pituitary gland. Within the infundibulum, axons of neurons in the tuber cinereum form the tubero-hypophyseal tract that transports regulating hormones (or factors) into blood vessels (see also Figure 21-2). From here, the regulating hormones (or factors) are transported via blood vessels to the anterior lobe of the pituitary gland. Also found in the infundibulum are nerve fibers from some hypothalamic nuclei that form the hypothalamic hypophyseal tract (see Figure 21-5). This tract transports hormones (oxytocin and antidiuretic hormone) made in the supraoptic and paraventricular nuclei to the posterior pituitary gland where they are stored and released. Nuclei associated with the tuberal region include the ventromedial, dorsomedial, and arcuate nuclei.
3. ***Mammillary region.*** This region is posterior to the tuberal region and contains the mammillary bodies and posterior hypothalamic nucleus. The ***mammillary bodies*** are two, small, rounded bodies placed side by side that are posterior to the tuber cinereum. The bodies serve as relay stations for olfactory neurons that are involved in reflexes related to the sense of smell.
4. ***Preoptic region.*** This region is in front of the supraoptic region and is usually considered part of the hypothalamus because it functions in regulating certain autonomic activities in conjunction with the hypothalamus. The preoptic region contains the preoptic periventricular nucleus, medial preoptic nucleus, and lateral preoptic nucleus.

Despite its small size, nuclei in the hypothalamus control many body activities, most of them related to homeostasis. Although differentiation of the hypothalamic nuclei is far from precise, it is possible to identify certain nuclei. Some of these nuclei are more readily identified in lower animals and are more distinct in fetuses than adults. Also, within a given nucleus there may be several kinds of cells that can be differentiated histologically. The localization of function, with a few exceptions, is not specific to the individual nuclei; certain functions tend to overlap nuclear boundaries.

The chief functions of the hypothalamus are as follows:

1. It controls and integrates activities of the autonomic nervous system, which regulates contraction of smooth muscle and cardiac muscle and secretions of many glands. This function is accomplished by axons of neurons whose dendrites and cell bodies are in hypothalamic nuclei. The axons form tracts from the hypothalamus to sympathetic and parasympathetic nuclei in the brain stem and spinal cord. Through the autonomic nervous system, the hypothalamus is the main regulator of visceral activities. It regulates heart rate, movement of food through the gastrointestinal tract, and contraction of the urinary bladder.
2. It is involved in the reception and integration of sensory impulses from the viscera.
3. It is the principal connection between the nervous system and the endocrine system—the two major control systems of the body. The hypothalamus lies just above the pituitary. When the hypothalamus detects certain changes in the body, it releases chemicals called regulating hormones (or factors) that stimulate or inhibit specific cells in the anterior pituitary gland. The anterior pituitary then releases or holds back hormones that regulate various physiological activities of the body. The hypothalamus also produces two hormones, antidiuretic hormone (ADH) and oxytocin (OT), which are transported to and stored in the posterior pituitary gland. ADH decreases urine volume and OT brings about uterine contractions during labor and assists in milk ejection by the mammary glands. The hormones are released from storage when needed by the body.
4. It is the center for mind-over-body phenomena. When the cerebral cortex interprets strong emotions, it often sends nerve impulses along the tracts that connect the cortex with the hypothalamus. The hypothalamus then directs impulses via the autonomic nervous system and also releases chemicals that stimulate the anterior pituitary gland. The result can be a wide range of changes in body activities. For instance, when you are under stress, impulses leave the hypothalamus to stimulate your heart to beat faster. Likewise, continued psychological stress can produce long-term abnormalities in body function that result in serious illness. These so-called psychosomatic disorders are definitely real.
5. It is associated with feelings of rage and aggression.
6. It controls normal body temperature. Certain cells of the hypothalamus serve as a thermostat. If blood flowing through the hypothalamus is above normal body temperature, the hypothalamus directs nerve impulses along the autonomic nervous system to stimulate activities that promote heat loss. Heat can be lost through relaxation of the smooth muscle in the blood vessels, causing vasodilation of cutaneous vessels and increased heat loss from the skin. Heat loss also is enhanced by sweating. Conversely, if the temperature of the blood is

below normal, the hypothalamus generates impulses that promote heat retention. Heat can be retained through the constriction of cutaneous blood vessels, cessation of sweating, and by shivering.

7. It regulates food intake through two centers. The ***feeding (hunger) center*** is responsible for hunger sensations. When sufficient food has been ingested, the ***satiety*** (sa-TĪ-e-tē) ***center*** is stimulated and sends out nerve impulses that inhibit the feeding center.
8. It contains a ***thirst center.*** Certain cells in the hypothalamus are stimulated when the extracellular fluid volume is reduced. The stimulated cells produce the sensation of thirst.
9. It is one of the centers that maintains the waking state and sleep patterns.
10. It exhibits properties of a self-sustained oscillator and, as such, acts as a pacemaker to drive many biological rhythms, such as daily cycles that involve the release of certain hormones.

Epithalamus

The ***epithalamus*** (*epi* = above) helps to form the superior margin (roof) of the diencephalon. Among the structures that comprise the epithalamus is the pineal gland (see Figure 18-1a), whose structure and functions will be discussed in Chapter 21.

Subthalamus

The ***subthalamus*** (*sub* = under or below) lies inferior to the thalamus. Two nuclei, the substantia nigra and red nucleus, extend from the midbrain into the subthalamus.

CEREBRUM

Supported on the diencephalon and brain stem and forming the bulk of the brain is the ***cerebrum*** (see Figure 18-1). The surface of the cerebrum is composed of gray matter 2 to 4 mm (0.08 to 0.16 in.) thick and is referred to as the ***cerebral cortex*** (*cortex* = kind of bark). The cortex, containing billions of cells, consists of six layers of nerve cell bodies in most areas. Beneath the cortex lies the cerebral white matter.

During embryonic development, when there is a rapid increase in brain size, the gray matter of the cortex enlarges out of proportion to the underlying white matter. As a result, the cortical region rolls and folds upon itself. The folds are called ***gyri*** (JĪ-rī) or ***convolutions*** (Figure 18-9). The deep grooves between folds are referred to as ***fissures;*** the shallow grooves between folds are ***sulci*** (SUL-sī). The most prominent fissure, the ***longitudinal fissure,*** nearly separates the cerebrum into right and left halves, or ***hemispheres.*** The hemispheres, however, are connected internally by a large bundle of transverse fibers composed of

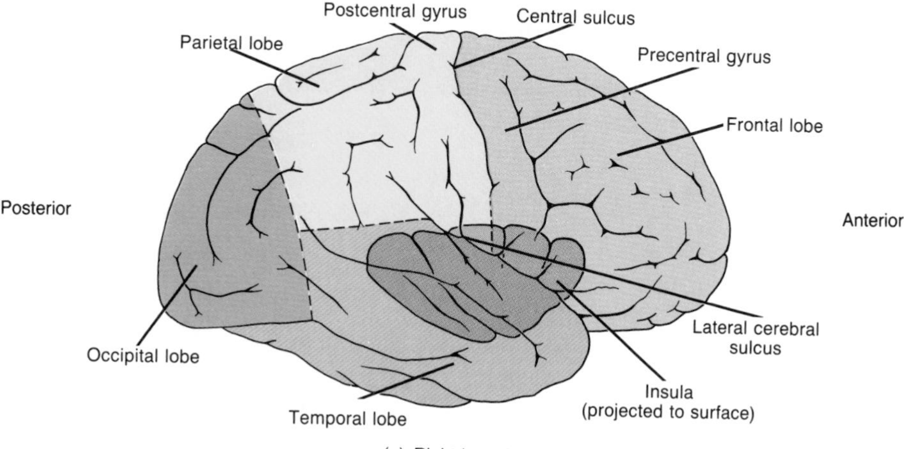

FIGURE 18-9 Lobes and fissures of the cerebrum. Since the insula in (a) cannot be seen externally, it has been projected to the surface. It can be seen in Figures 18-7a, b and 18-9c. (b) Courtesy of Martin Rotker, Phototake. In (c), a portion of the cerebrum has been removed. (Courtesy of N. Gluhbegovic and T. H. Williams, *The Human Brain: A Photographic Guide,* Harper & Row, Publishers, Inc., Hagerstown, MD, 1980.) The insert to the left in (d) indicates the relative differences among a gyrus, sulcus, and fissure. (e) Courtesy of Martin Rotker, Phototake. (f) Courtesy of Stephen A. Kieffer and E. Robert Heitzman, *An Atlas of Cross-Sectional Anatomy,* Harper & Row, Publishers, Inc., Hagerstown, MD, 1979.

Central sulcus
Postcentral gyrus
Precentral gyrus
Parietal lobe
Frontal lobe
Occipital lobe
Lateral cerebral sulcus
Anterior
Posterior
Cerebellum
Temporal lobe

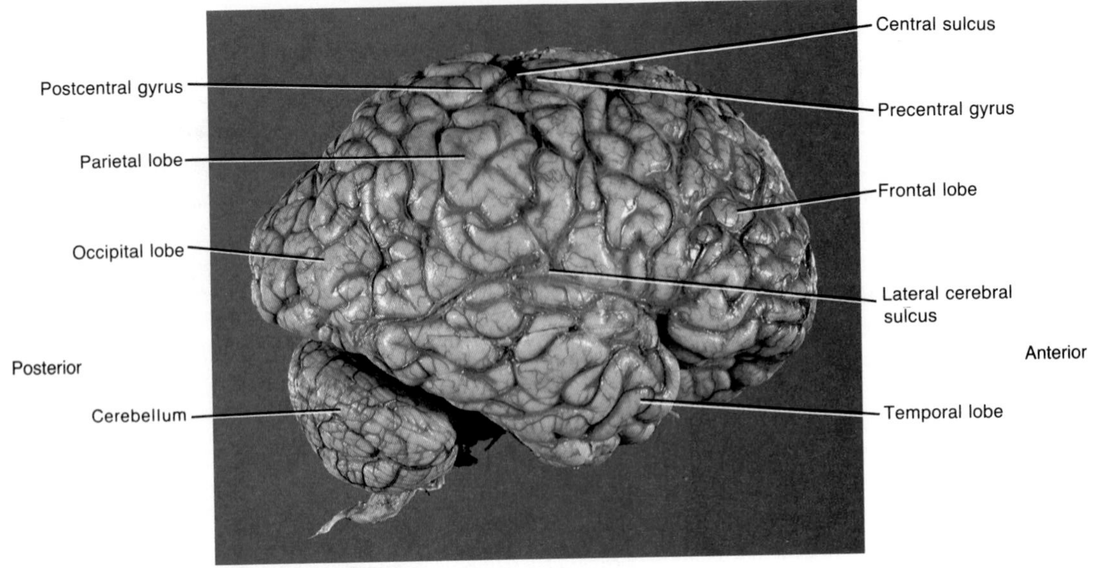

(b) Right lateral view

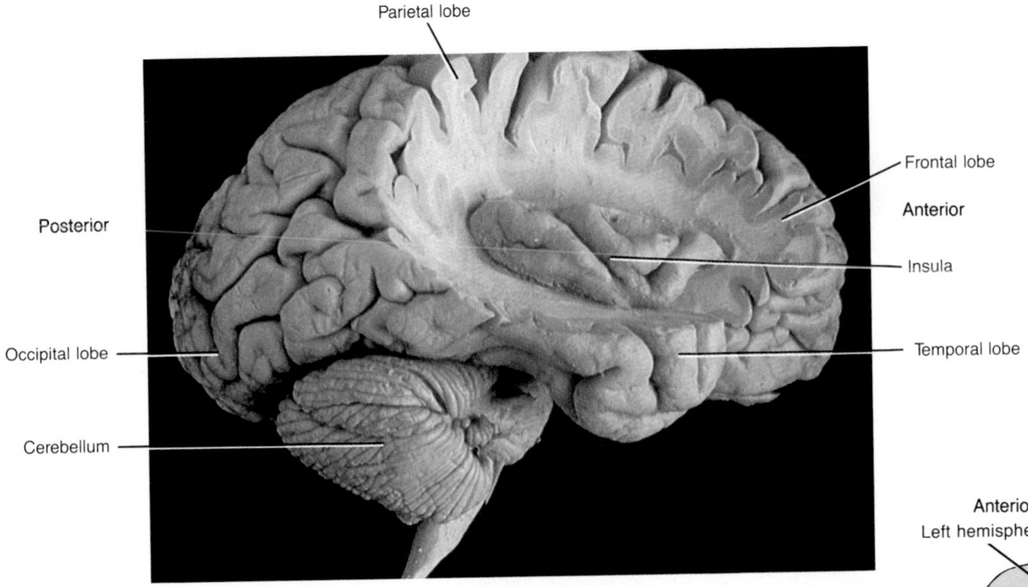

Parietal lobe
Frontal lobe
Anterior
Posterior
Insula
Temporal lobe
Occipital lobe
Cerebellum

(c) Right lateral view

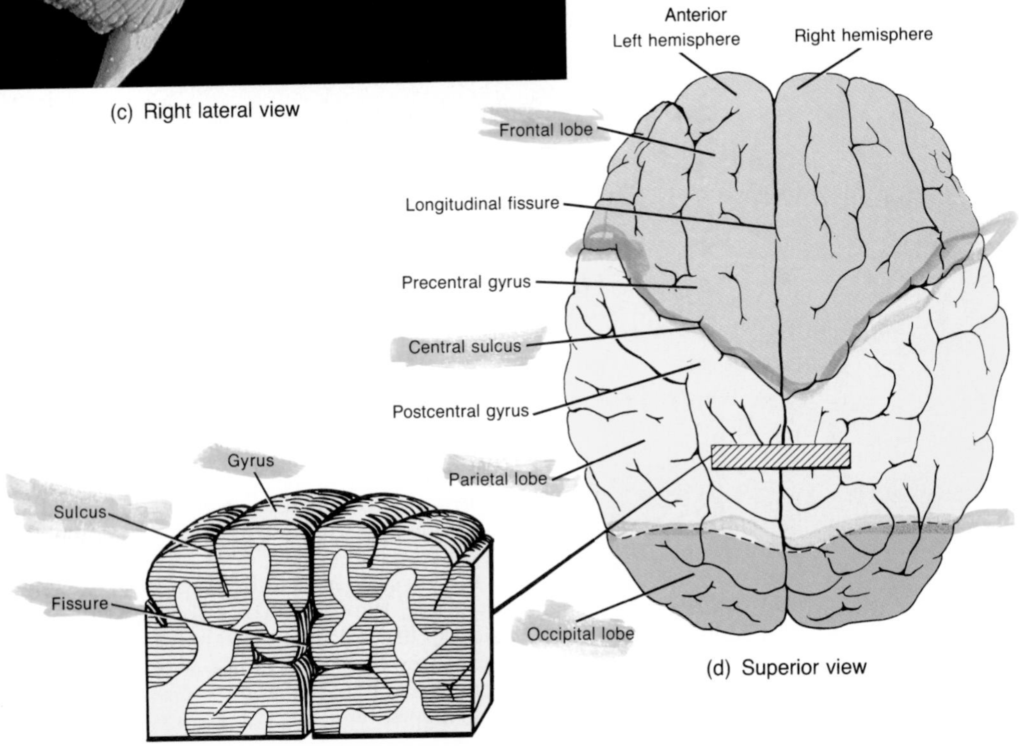

Anterior
Left hemisphere
Right hemisphere
Frontal lobe
Longitudinal fissure
Precentral gyrus
Central sulcus
Postcentral gyrus
Parietal lobe
Gyrus
Sulcus
Fissure
Occipital lobe

(d) Superior view

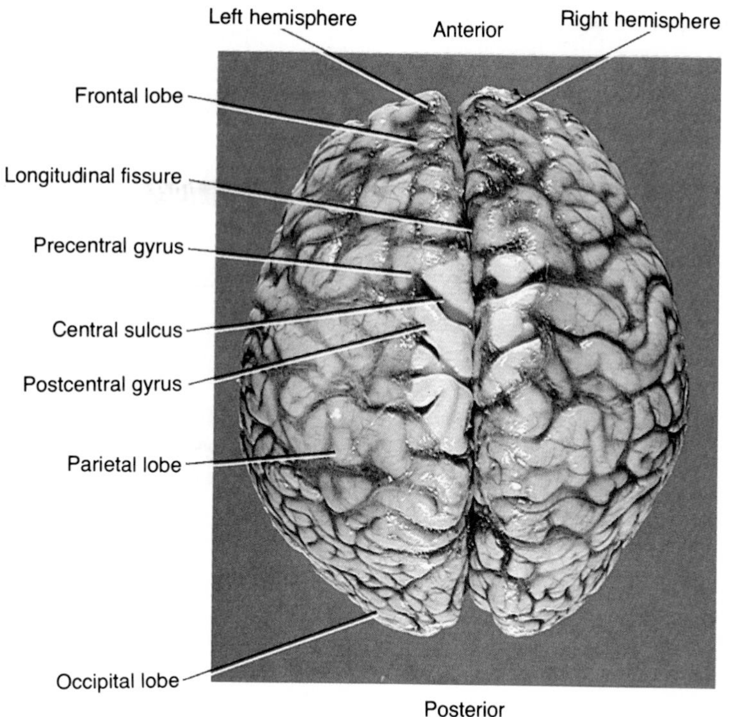

(e) Superior view

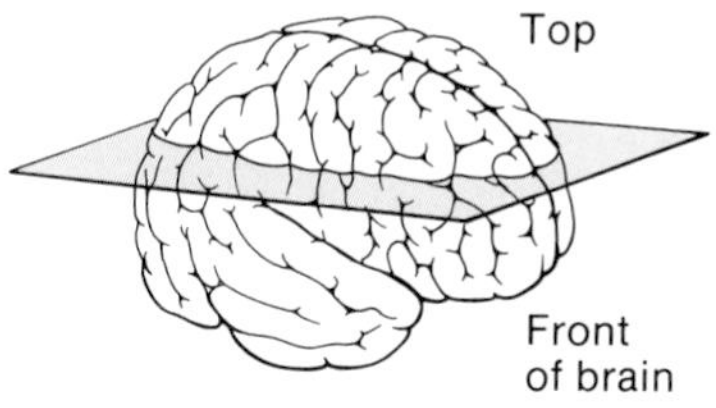

Top
Front
of brain

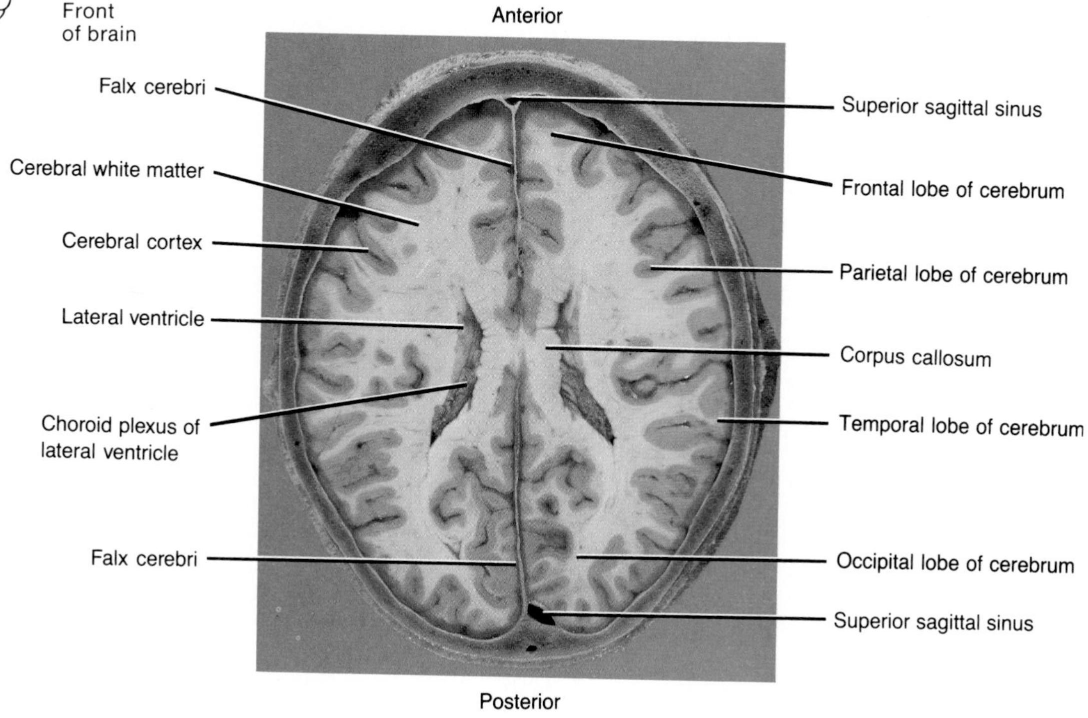

Anterior

Falx cerebri
Cerebral white matter
Cerebral cortex
Lateral ventricle
Choroid plexus of
lateral ventricle
Falx cerebri
Superior sagittal sinus
Frontal lobe of cerebrum
Parietal lobe of cerebrum
Corpus callosum
Temporal lobe of cerebrum
Occipital lobe of cerebrum
Superior sagittal sinus

Posterior

(f) Cross section

white matter called the ***corpus callosum*** (kal-LŌ-sum; *corpus* = body; *callous* = hard). Between the hemispheres is an extension of the cranial dura mater called the ***falx*** (FALKS) ***cerebri*** (***cerebral fold***). It encloses the superior and inferior sagittal sinuses.

Lobes

Each cerebral hemisphere is further subdivided into four lobes by sulci or fissures. The ***central sulcus*** separates the ***frontal lobe*** from the ***parietal lobe.*** A major gyrus, the ***precentral gyrus,*** is located immediately anterior to the central sulcus. The gyrus is a landmark for the primary motor area of the cerebral cortex. Another major gyrus, the ***postcentral gyrus,*** is located immediately posterior to the central sulcus. This gyrus is a landmark for the general sensory area of the cerebral cortex. The ***lateral cerebral sulcus*** separates the ***frontal lobe*** from the ***temporal lobe.*** The ***parietooccipital sulcus*** separates the ***parietal lobe*** from the ***occipital lobe.*** Another prominent fissure, the ***transverse fissure,*** separates the cerebrum from the cerebellum. The frontal lobe, parietal lobe, temporal lobe, and occipital lobe are named after the bones that cover them. A fifth part of the cerebrum, the ***insula,*** lies deep within the lateral cerebral fissure, under the parietal, frontal, and temporal lobes. It cannot be seen in an external view of the brain (Figure 18-9a, b).

As you will see later, the olfactory (I) and optic (II) nerves are associated with specific lobes of the cerebrum.

White Matter

The white matter underlying the cortex consists of myelinated axons running in three principal directions (Figure 18-10).

1. ***Association fibers*** connect and transmit nerve impulses between gyri in the same hemisphere.
2. ***Commissural fibers*** transmit impulses from the gyri in one cerebral hemisphere to the corresponding gyri in the opposite cerebral hemisphere. Three important groups of commissural fibers are the ***corpus callosum, anterior commissure,*** and ***posterior commissure.***
3. ***Projection fibers*** form ascending and descending tracts that transmit impulses from the cerebrum to other parts of the brain and spinal cord. The internal capsule is an example.

Basal Ganglia (Cerebral Nuclei)

The ***basal ganglia*** (***cerebral nuclei***) are paired masses of gray matter in each cerebral hemisphere (Figures 18-7 and 18-11). The largest of the basal ganglia of each hemisphere is the ***corpus striatum*** (strī-Ā-tum; *corpus* = body; *striatum* = striped). It consists of the ***caudate*** (*cauda* = tail) ***nucleus*** and the ***lentiform*** (*lenticula* = shaped like a lentil or lens) ***nucleus.*** The lentiform nucleus, in turn, is subdivided into a lateral portion called the ***putamen*** (pu-TĀ-men; *putamen* = shell) and a medial portion called the ***globus pallidus*** (*globus* = ball; *pallid* = pale).

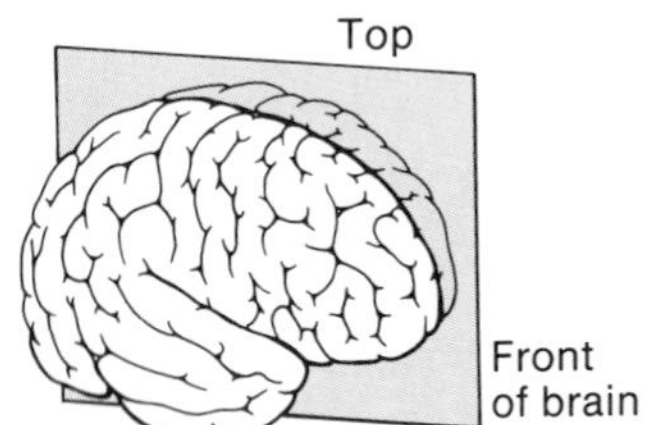

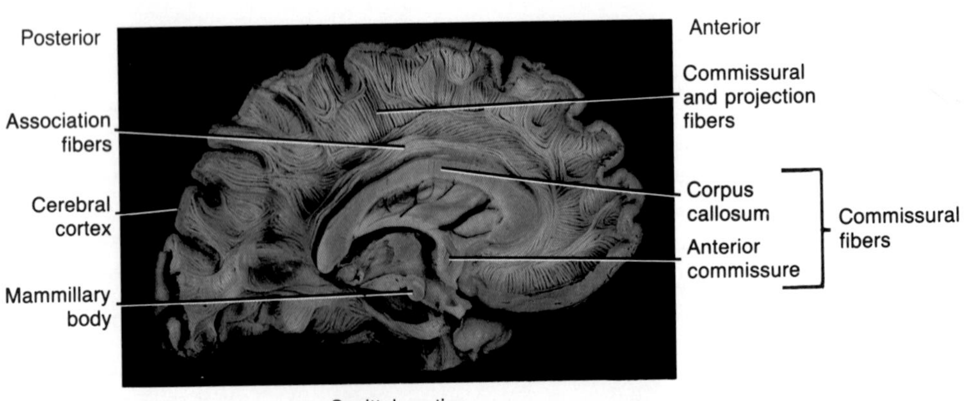

FIGURE 18-10 White matter tracts of the left cerebral hemisphere. (Courtesy of N. Gluhbegovic and T. H. Williams, *The Human Brain: A Photographic Guide,* Harper & Row, Publishers, Inc., Hagerstown, MD, 1980.)

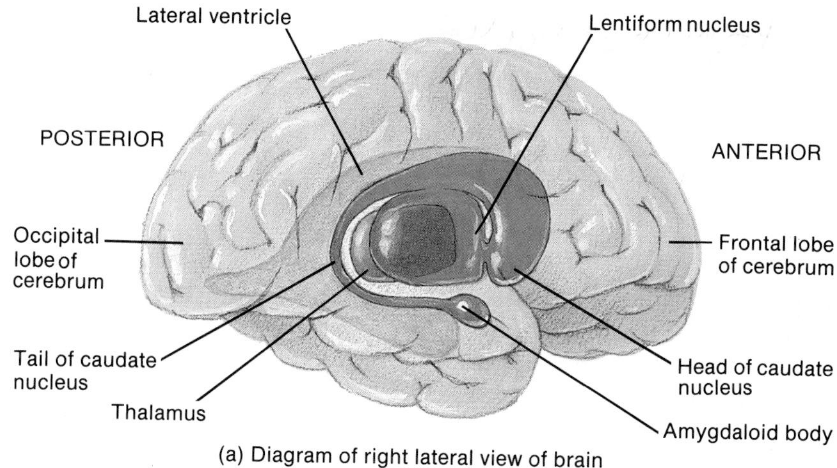

(a) Diagram of right lateral view of brain

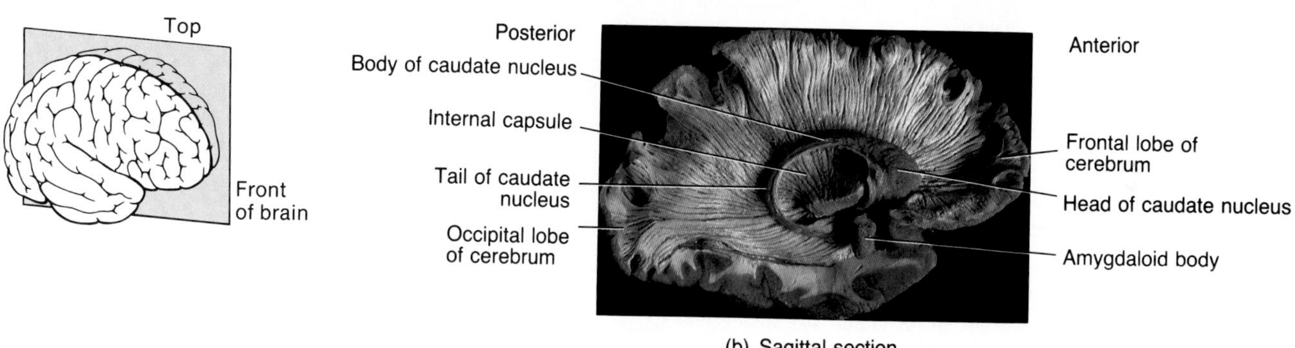

(b) Sagittal section

FIGURE 18-11 Basal ganglia. (a) In this diagram of the cerebrum, the basal ganglia have been projected to the surface. Refer to Figure 18-7a for the positions of the basal ganglia in the frontal section of the cerebrum. (b) Courtesy of N. Gluhbegovic and T. H. Williams, *The Human Brain: A Photographic Guide,* Harper & Row, Publishers, Inc., Hagerstown, MD, 1980.

The portion of the ***internal capsule*** passing between the lentiform nucleus and the caudate nucleus and between the lentiform nucleus and thalamus is sometimes considered part of the corpus striatum. The internal capsule is made up of a group of sensory and motor white matter tracts that connect the cerebral cortex with the brain stem and spinal cord.

Other structures frequently considered part of the basal ganglia are the ***substantia nigra, subthalamic nucleus,*** and ***red nucleus.*** The substantia nigra is a large nucleus in the midbrain whose axons terminate in the caudate nucleus and putamen. The subthalamic nucleus lies against the internal capsule. Its major connection is with the globus pallidus.

The basal ganglia are interconnected by many fibers. They are also connected to the cerebral cortex, thalamus, and hypothalamus. The caudate nucleus and the putamen control large automatic movements of skeletal muscles, such as swinging the arms while walking. Such gross movements are also consciously controlled by the cerebral cortex. The globus pallidus is concerned with the regulation of muscle tone required for specific body movements.

CLINICAL APPLICATION

Damage to Basal Ganglia

Damage to the basal ganglia results in abnormal body movements, such as uncontrollable shaking, called ***tremor,*** and ***involuntary movements of skeletal muscles.*** Moreover, destruction of a substantial portion of the caudate nucleus results in almost total ***paralysis*** of the side of the body opposite to the damage. The caudate nucleus is an area often affected by a stroke.

A lesion in the subthalamic nucleus results in a motor disturbance on the opposite side of the body called ***hemiballismus*** (*hemi* = half; *ballismos* = jumping), which is characterized by involuntary movements occurring suddenly with great force and rapidity. The movements are purposeless and generally of the withdrawal type, although they may be jerky. The spontaneous movements affect the proximal portions of the extremities most severely, especially the arms.

Limbic System

Primitive feelings

Certain components of the cerebral hemispheres and diencephalon constitute the ***limbic*** (*limbus* = border) ***system.*** Among its components are the following regions of gray matter (Figure 18-12).

1. ***Limbic lobe (cortex).*** Largest components are the ***parahippocampal*** and ***cingulate gyri,*** both gyri of the cerebral hemispheres, and ***hippocampus,*** an extension of the parahippocampal gyrus that extends into the floor of the lateral ventricle.
2. ***Dentate gyrus.*** A cerebral gyrus between the hippocampus and parahippocampal gyrus.
3. ***Amygdaloid body (amygdala).*** Several groups of neurons located at the tail end of the caudate nucleus.
4. ***Septal nuclei.*** Nuclei within the septal area, a region formed by the regions under the corpus callosum and a cerebral gyrus (paraterminal).
5. ***Mammillary bodies of the hypothalamus.*** Two round masses close to the midline near the cerebral peduncles.
6. ***Anterior nucleus of the thalamus.*** Located in the floor of the lateral ventricle.
7. ***Olfactory bulbs.*** Flattened bodies of the olfactory (I) nerves that rest on the cribriform plate.
8. ***Bundles of interconnecting myelinated axons.*** Bundles that interconnect various components of the limbic system and include the fornix, stria terminalis, stria medullaris, medial forebrain bundle, and mammillothalamic tract.

The limbic system is a wishbone-shaped group of structures that encircles the brain stem and functions in the emotional aspects of behavior related to survival. The hippocampus, together with portions of the cerebrum, also functions in memory. Memory impairment results from lesions in the limbic system. People with such damage forget recent events and cannot commit anything to memory. How the limbic system functions in memory is not clear. Although behavior is a function of the entire nervous system, the limbic system controls most of its involuntary aspects. Experiments on the limbic system of monkeys and other animals indicate that the amygdaloid nucleus assumes a major role in controlling the overall pattern of behavior.

Other experiments have shown that the limbic system is associated with pleasure and pain. When certain areas

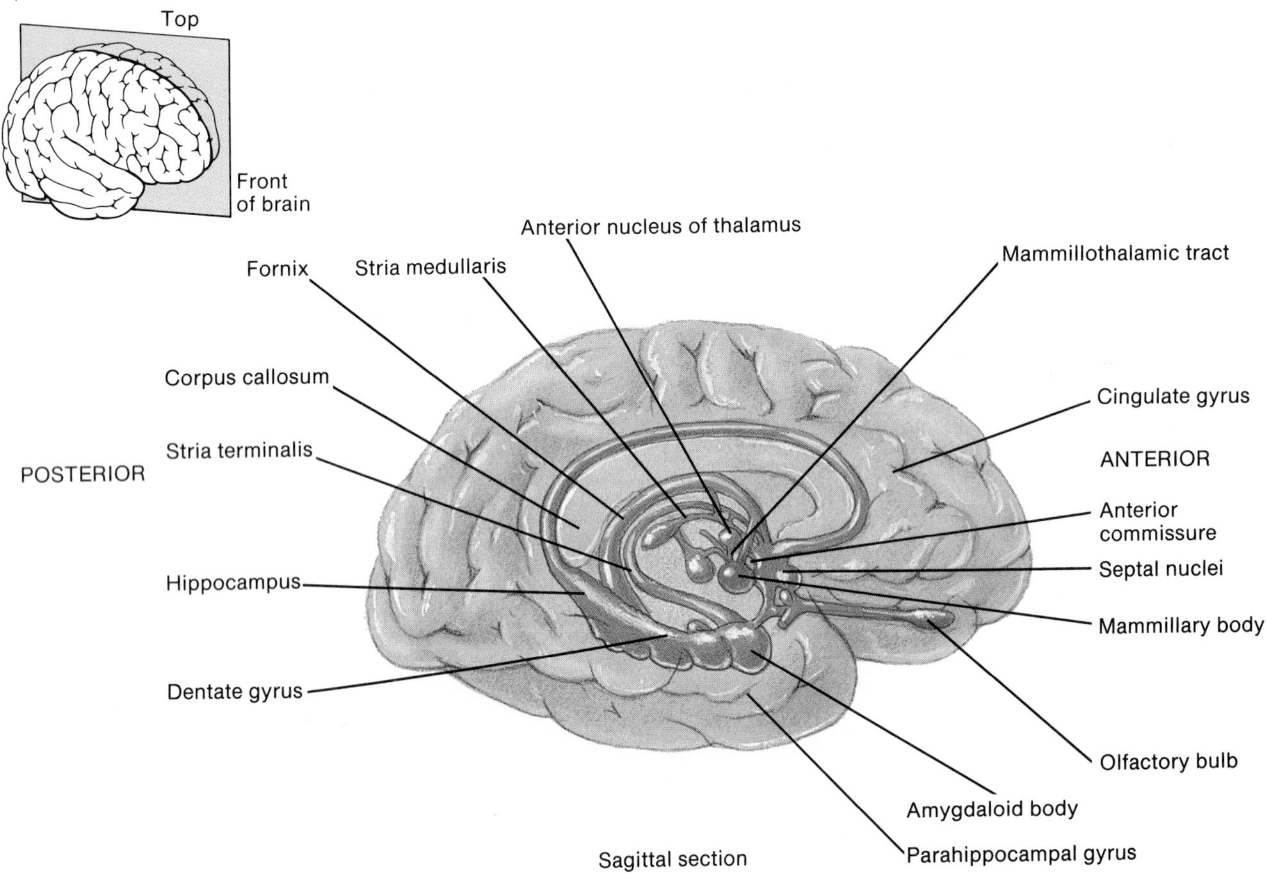

FIGURE 18-12 Selected components of the limbic system and surrounding structures.

of the limbic system of the hypothalamus, thalamus, and midbrain are stimulated in animals, their reactions indicate they are experiencing intense punishment. When other areas are stimulated, the animals' reactions indicate they are experiencing extreme pleasure. In still other studies, stimulation of the amygdaloid body or certain nuclei of the hypothalamus results in a behavioral pattern called rage. The animal assumes a defensive posture—extending its claws, raising its tail, hissing, spitting, growling, and opening its eyes wide. Stimulating other areas of the limbic system results in an opposite behavioral pattern: docility, tameness, and affection. Because the limbic system assumes a primary function in emotions such as pain, pleasure, anger, rage, fear, sorrow, sexual feelings, docility, and affection, it is sometimes called the "visceral" or "emotional" brain.

CLINICAL APPLICATION

Brain Injuries

Brain injuries are commonly associated with head injuries and result from displacement and distortion of neuronal tissue at the moment of impact. The various degrees of brain injury are described by the following terms.

1. ***Concussion.*** An abrupt but temporary loss of consciousness following a blow to the head or a sudden stopping of a moving head. A concussion produces no visible bruising of the brain, but posttraumatic amnesia may occur.
2. ***Contusion.*** A visible bruising of the brain due to trauma and blood leaking from microscopic vessels. The pia mater is stripped from the brain over the injured area and may be torn, allowing blood to enter the subarachnoid space. A contusion usually results in an extended loss of consciousness, ranging from several minutes to many hours.
3. ***Laceration.*** Tearing of the brain, usually from a skull fracture or gunshot wound. A laceration results in rupture of large blood vessels with bleeding into the brain and subarachnoid space. Consequences include cerebral hematoma, edema, and increased intracranial pressure.

Although trauma to the head can lead to brain injuries, not all of the damage is due to the impact alone. Most of the damage is probably due to the release of large numbers of free radicals, that is, charged oxygen molecules from damaged cells. (Brain cells recovering from the effects of a stroke or cardiac arrest also release excess free radicals.) Free radicals cause damage by disrupting cellular DNA and enzymes and altering plasma membrane permeability. Investigations are under way to develop compounds to counter the effects of free radicals.

Functional Areas of Cerebral Cortex

The functions of the cerebrum are numerous and complex. In a general way, the cerebral cortex is divided into sensory, motor, and association areas. The ***sensory areas*** interpret sensory impulses, the ***motor areas*** control muscular movement, and the ***association areas*** are concerned with emotional and intellectual processes.

■ ***Sensory Areas*** The ***primary somesthetic*** (sō-mes-THET-ik; *soma* = body; *aisthesis* = sensation) ***area*** or ***general sensory area*** is located directly posterior to the central sulcus of the cerebrum in the postcentral gyrus of the parietal lobe. It extends from the longitudinal fissure on the top of the cerebrum to the lateral cerebral sulcus. In Figure 18-13 the general sensory area is designated by the areas numbered 1, 2, and 3.*

The primary somesthetic area receives sensations from cutaneous, muscular, and visceral receptors in various parts of the body. Each point of the area receives sensations from specific parts of the body, and essentially the entire body is spatially represented in it. The size of the portion of the sensory area receiving stimuli from body parts is not dependent on the size of the part but on the number of receptors the part contains. For example, a larger portion of the sensory area receives impulses from the lips than from the thorax (see Figure 20-4). The major function of the primary somesthetic area is to localize exactly the points of the body where the sensations originate. The thalamus is capable of localizing sensations in a general way; that is, it receives sensations from large areas of the body but cannot distinguish precisely between specific areas of stimulation. This ability is reserved for the primary somesthetic area of the cortex.

The ***secondary somesthetic area*** is a small region in the posterior wall of the lateral sulcus in line with the postcentral gyrus. It is involved mainly in less discriminative aspects of sensation.

Posterior to the primary somesthetic area is the ***somesthetic association area.*** It corresponds to the areas numbered 5 and 7 in Figure 18-13. The somesthetic association area receives input from the thalamus, other lower portions of the brain, and the primary somesthetic area. Its role is to integrate and interpret sensations. This area permits you to determine the exact shape and texture of an object without looking at it, to determine the orientation of one object to another as they are felt, and to sense the relationship of one body part to another. Another role of the somesthetic association area is the storage of memories of past sensory experiences. Thus, you can compare sensations with previous experiences.

* These numbers, as well as most of the others shown, are based on K. Brodmann's cytoarchitectural map of the cerebral cortex. His map, first published in 1909, attempts to correlate structure and function.

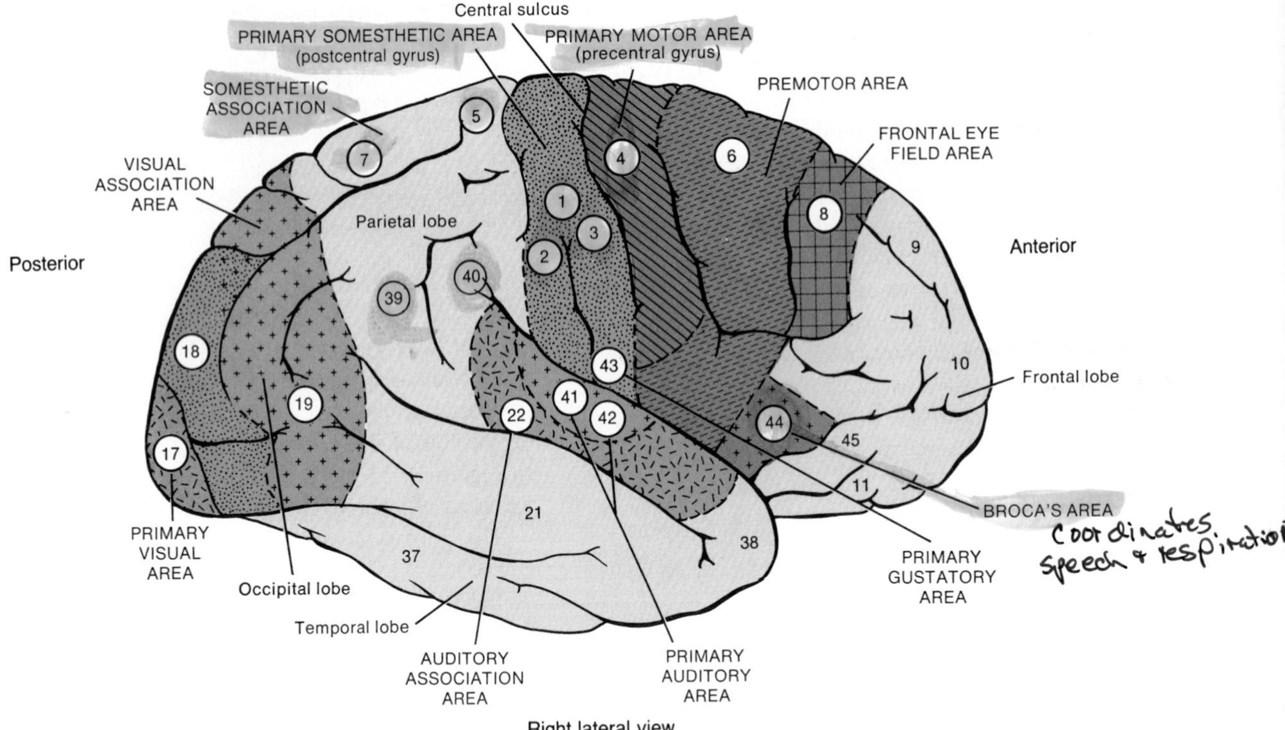

FIGURE 18-13 Sensory and motor areas of the right cerebral hemisphere. Although Broca's area is in the left hemisphere of most people, it is shown here to indicate its location.

Other sensory areas of the cerebral cortex include:

1. ***Primary visual area*** (area 17). Located in the medial surface of the occipital lobe and occasionally extends around to the lateral surface. It receives sensory impulses from the eyes and interprets shape, color, and movement.
2. ***Visual association area*** (areas 18 and 19). Located in the occipital lobe. It receives sensory impulses from the primary visual area and the thalamus. It relates present to past visual experiences with recognition and evaluation of what is seen.
3. ***Primary auditory area*** (areas 41 and 42). Located in the superior part of the temporal lobe near the lateral cerebral sulcus. It interprets the basic characteristics of sound such as pitch and rhythm. Whereas the anterolateral portion of the auditory area responds to low pitches, the posterolateral portion responds to high pitches.
4. ***Auditory association (Wernicke's) area*** (area 22). Located inferior to the primary auditory area in the temporal lobe. It determines whether a sound is speech, music, or noise. It also interprets the meaning of speech by translating words into thoughts.
5. ***Primary gustatory area*** (area 43). Located at the base of the postcentral gyrus above the lateral cerebral sulcus in the parietal cortex. It interprets sensations related to taste.
6. ***Primary olfactory area.*** Located in the temporal lobe of the medial aspect. It interprets sensations related to smell.
7. ***Gnostic*** (NOS-tik; *gnosis* = knowledge) ***area*** (areas 5, 7, 39, and 40). A ***common integrative area*** located among the somesthetic, visual, and auditory association areas. The gnostic area receives nerve impulses from these areas, as well as from the taste and smell areas, the thalamus, and lower portions of the brain stem. It integrates sensory interpretations from the association areas and impulses from other areas so that a common thought can be formed from the various sensory inputs. It then transmits signals to other parts of the brain to cause the appropriate response to the sensory signal.

■ ***Motor Areas*** The ***primary motor area*** (area 4) is located in the precentral gyrus of the frontal lobe (Figure 18-13). Like the primary somesthetic area, the primary motor area consists of regions that control specific muscles or groups of muscles (see Figure 20-8). Stimulation of a specific point of the primary motor area results in a muscular contraction, usually on the opposite side of the body.

The ***premotor area*** (area 6) is anterior to the primary motor area. It is concerned with learned motor activities of a complex and sequential nature. It generates nerve impulses that cause a specific group of muscles to contract in a specific sequence, for example, writing. Thus, the premotor area controls skilled movements.

The ***frontal eye field area*** (area 8) in the frontal cortex is sometimes included in the premotor area. This area controls voluntary scanning movements of the eyes—searching for a word in a dictionary, for instance.

The ***language areas*** are also significant parts of the motor cortex. The translation of speech or written words into thought involves sensory areas—primary auditory, auditory association, primary visual, visual association, and gnostic—as just described. The translation of thoughts into speech involves the ***motor speech area*** (area 44) or ***Broca's*** (BRŌ-kaz) ***area,*** located in the frontal lobe just superior to the lateral cerebral sulcus. From this area, a sequence of nerve impulses is sent to the premotor regions that control the muscles of the larynx, pharynx, and mouth. The impulses from the premotor area to the muscles result in specific, coordinated contractions that enable you to speak. Simultaneously, impulses are sent from Broca's area to the primary motor area. From here, impulses reach your breathing muscles to regulate the proper flow of air past the vocal cords. The coordinated contractions of your speech and breathing muscles enable you to translate your thoughts into speech.

CLINICAL APPLICATION

Speech Area Injuries

Broca's area and other language areas are located in the left cerebral hemisphere of most individuals, regardless of whether they are left-handed or right-handed. Injury to the sensory or motor speech areas results in ***aphasia*** (a-FĀ-zē-a; *a* = without; *phasis* = speech), an inability to speak; ***agraphia*** (*a* = without; *graph* = write), an inability to write; ***word deafness,*** an inability to understand spoken words; or ***word blindness,*** an inability to understand written words.

■ ***Association Areas*** The ***association areas*** of the cerebrum are made up of association tracts that connect motor and sensory areas (see Figure 18-10). The association region of the cortex occupies the greater portion of the lateral surfaces of the occipital, parietal, and temporal lobes, and the frontal lobes anterior to the motor areas. The association areas are concerned with memory, emotions, reasoning, will, judgment, personality traits, and intelligence.

Electroencephalogram (EEG)

Brain cells can generate electrical activity as a result of literally millions of nerve impulses (nerve action potentials) and other changes in membrane potentials of individual neurons. These electrical potentials added together are called ***brain waves*** and indicate electrical activity of the cerebral cortex. Brain waves pass easily through the skull and can be detected by sensors called electrodes. A record of such waves is called an ***electroencephalogram (EEG).***

An EEG is used to diagnose epilepsy and other seizure disorders, infectious diseases, tumors, trauma, hematomas, metabolic abnormalities, degenerative diseases, and periods of unconsciousness and confusion. In some cases an EEG can be used to furnish information regarding sleep and wakefulness. An EEG may also be used as one criterion to confirm brain death in which two flat EEGs (complete absence of brain waves) are taken 24 hours apart. Other criteria include unconsciousness, lack of spontaneous breathing, absence of a light response and dilation of the pupil (black hole in center of colored portion of eyeball), absence of reflexes (although local spinal reflexes may be present), unresponsiveness, and lack of normal muscle tone or strength.

BRAIN LATERALIZATION (SPLIT-BRAIN CONCEPT)

Gross examination of the brain would suggest that it is the same on both sides. However, detailed examination by the use of CT scans reveals certain anatomical differences between the two hemispheres. For example, in left-handed people the parietal and occipital lobes of the right hemisphere are usually narrower than the corresponding lobes of the left hemisphere. In addition, the frontal lobe of the left hemisphere of such individuals is typically narrower than that of the right hemisphere.

In addition to the structural differences between both sides of the brain, there are also several important functional differences (Figure 18-14). It has been shown that the left hemisphere in most people is more important for spoken and written language, numerical and scientific skills, ability to use and understand sign language, and reasoning. Conversely, it has been shown that the right hemisphere is more important for left-hand control; musical and artistic awareness; space and pattern perception; insight; imagination; and generating mental images of sight, sound, touch, taste, and smell in order to compare relationships.

CEREBELLUM

The ***cerebellum*** is the second-largest portion of the brain (almost one-eighth of the brain's mass) and occupies the inferior and posterior aspects of the cranial cavity. Specifically, it is posterior to the medulla and pons and inferior to the occipital lobes of the cerebrum (see Figure 18-1). It is separated from the cerebrum by the ***transverse fissure*** and by an extension of the cranial dura mater called the ***tentorium*** (*tentorium* = tent) ***cerebelli.*** The tentorium cerebelli partially encloses the transverse sinuses and supports the occipital lobes of the cerebral hemispheres.

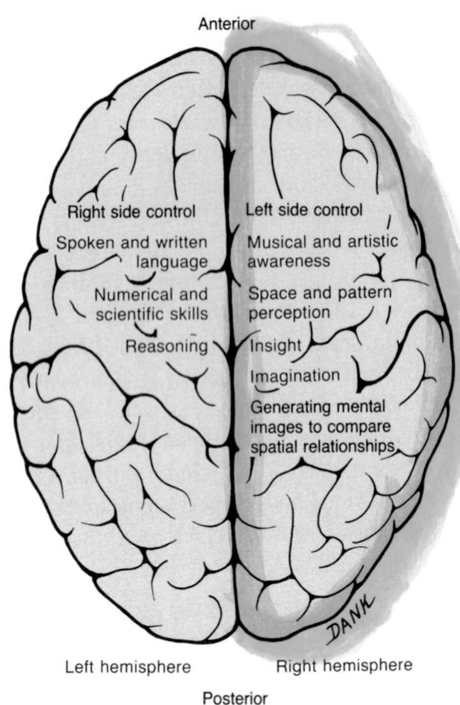

FIGURE 18-14 Summary of the principal functional differences between the left and right cerebral hemispheres.

Structure

The cerebellum is shaped somewhat like a butterfly. The central constricted area is the ***vermis,*** which means "worm-shaped," and the lateral "wings" or lobes are referred to as ***hemispheres*** (Figure 18-15). Each hemisphere consists of lobes that are separated by deep and distinct fissures. The ***anterior lobe*** and ***posterior lobe*** are concerned with subconscious movements of skeletal muscles. The ***flocculonodular lobe*** is concerned with the sense of equilibrium (see Chapter 20). Between the hemispheres is an extension of the cranial dura mater: the ***falx*** (= sickle-shaped) ***cerebelli.*** It passes only a short distance between the cerebellar hemispheres and contains the occipital sinus (vein).

The surface of the cerebellum, called the ***cortex,*** consists of gray matter in a series of slender, parallel ridges called ***folia.*** They are less prominent than the convolutions of the cerebral cortex. Beneath the gray matter are ***white matter tracts*** (***arbor vitae***) that resemble branches of a tree. Deep within the white matter are masses of gray matter, the ***cerebellar nuclei.*** The nuclei give rise to nerve fibers that convey information out of the cerebellum to other parts of the nervous system (to both other brain centers and the spinal cord).

The cerebellum is attached to the brain stem by three paired bundles of fibers (tracts) called ***cerebellar peduncles*** (see Figures 18-4 and 18-5). ***Inferior cerebellar peduncles*** connect the cerebellum with the medulla at the base of the brain stem and with the spinal cord. These peduncles contain both afferent and efferent fibers and thus bring information into and out of the cerebellum. ***Middle cerebellar peduncles*** connect the cerebellum with the pons. These peduncles contain only afferent fibers and thus bring information into the cerebellum. ***Superior cerebellar peduncles*** connect the cerebellum with the midbrain. These peduncles contain mostly efferent fibers and thus mostly bring information out of the cerebellum.

Functions

Functionally, the cerebellum is an area of the brain concerned with coordinating subconscious contractions of skeletal muscles. The cerebellar peduncles are the fiber tracts that channel information into and out of the cerebellum. The cerebellum constantly receives input signals from proprioceptors in muscles, tendons, and joints, receptors for equilibrium, and visual receptors of the eyes. Such input permits the cerebellum to collect information on the physical status of the body with regard to posture, equilibrium, and all movements at joints. In addition, when other motor areas of the brain, such as the motor cortex of the cerebrum and basal ganglia, send signals to skeletal muscles, they also send a duplicate set of signals to the cerebellum. The cerebellum compares this input information regarding the actual status of the body with the intended movement determined by motor areas of the brain (cerebrum and basal ganglia). If the intent of these motor areas is not being attained by the skeletal muscles, the cerebellum detects the variation and sends feedback signals to the motor areas to either stimulate or inhibit the activity of skeletal muscles. This interaction produces smooth, coordinated movements by way of our body's skeletal muscles.

The cerebellum also functions in maintaining equilibrium and controlling posture. For example, receptors for equilibrium send nerve impulses to the cerebellum, informing it of body position. When the direction of movement changes, the cerebellum sends corrective signals to the motor cortex of the cerebrum. The motor cortex then sends signals over motor tracts to somatic motor neurons to skeletal muscles to reposition the body.

Another function of the cerebellum is related to predicting the future position of a body part during a particular movement. Just before a moving part of the body reaches its intended position, the cerebellum sends signals over motor tracts to somatic motor neurons to skeletal muscles to slow the moving part and stop it at a specific point. This function of the cerebellum is used in actions such as walking.

There is some evidence that the cerebellum may play a role in a person's emotional development, modulating sensations of anger and pleasure, allowing normal emotional expression and interpretation.

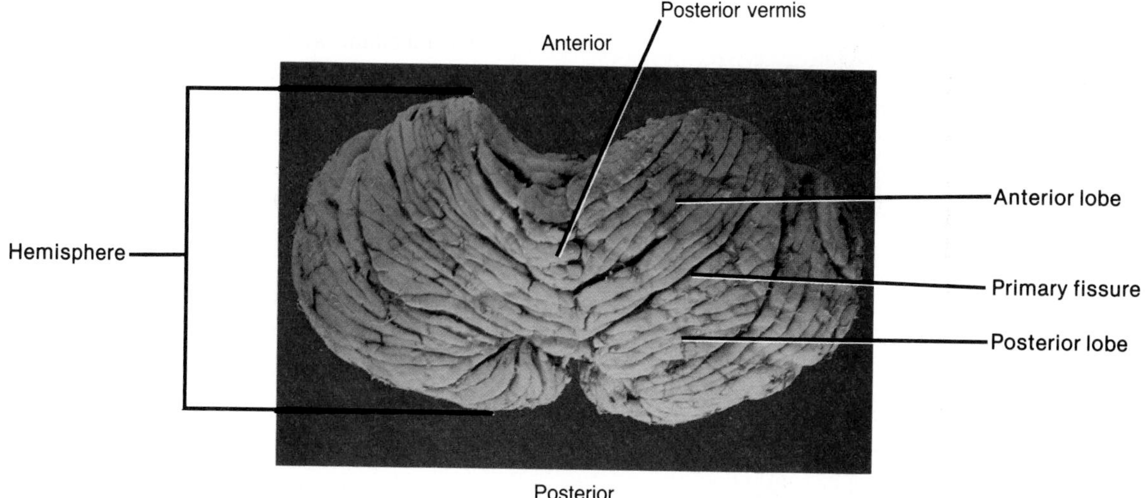

(a) Superior view

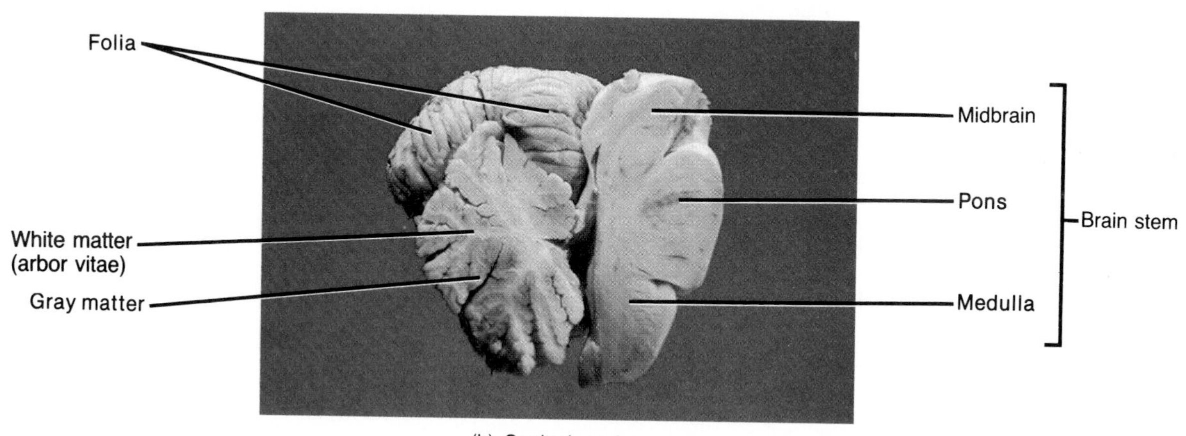

(b) Sagittal section

FIGURE 18-15 Cerebellum. Courtesy of Lester V. Bergman & Associates, Inc.

CLINICAL APPLICATION

Damage to Cerebellum

Damage to the cerebellum through trauma or disease is characterized by certain symptoms involving skeletal muscles on the same side of the body as the damage. The effects are on the same side of the body as the damaged side of the cerebellum because of a double crossing of tracts within the cerebellum. There may be lack of muscle coordination, called ***ataxia*** (*a* = without; *taxis* = order). Blindfolded people with ataxia cannot touch the tip of their nose with a finger because they cannot coordinate movement with their sense of where a body part is located. Another sign of ataxia is a change in the speech pattern due to a lack of coordination of speech muscles. Cerebellar damage may also result in ***disturbances of gait,*** in which the subject staggers or cannot coordinate normal walking movements, and ***severe dizziness.***

A summary of the functions of the various parts of the brain is presented in Exhibit 18-1.

CRANIAL NERVES

Cranial nerves, like spinal nerves, are part of the peripheral nervous system (PNS). Of the 12 pairs of cranial nerves, 10 originate from the brain stem, but all leave the skull through foramina of the skull (see Exhibit 6-4). The cranial nerves are designated with Roman numerals and with names. The Roman numerals indicate the order in which the nerves arise from the brain (front to back). The names indicate the distribution or function.

Some cranial nerves contain only sensory fibers and thus are called ***sensory nerves.*** The remainder contain both sensory and motor fibers and are referred to as ***mixed nerves.*** At one time it was believed that some cranial nerves (oculomotor, trochlear, abducens, accessory, and hypoglossal)

EXHIBIT 18-1

Summary of Functions of Principal Parts of Brain

PART	FUNCTION
BRAIN STEM	
Medulla	Relays motor and sensory impulses between other parts of the brain and the spinal cord. Reticular formation (also in pons, midbrain, and diencephalon) functions in consciousness and arousal. Vital reflex centers regulate heartbeat, breathing (together with pons), and blood vessel diameter. Nonvital reflex centers coordinate swallowing, vomiting, coughing, sneezing, and hiccuping. Contains nuclei of origin for cranial nerves VIII, IX, X, XI, and XII. Vestibular nuclear complex helps maintain equilibrium.
Pons	Relays impulses within the brain and between parts of the brain and spinal cord. Contains nuclei of origin for cranial nerves V, VI, VII, and VIII. Pneumotaxic area and apneustic area, together with the medulla, help control breathing.
Midbrain	Relays motor impulses from the cerebral cortex to the pons and spinal cord and relays sensory impulses from the spinal cord to the thalamus. Superior colliculi coordinate movements of the eyeballs in response to visual and other stimuli, and the inferior colliculi coordinate movements of the head and trunk in response to auditory stimuli. Contains nuclei of origin for cranial nerves III and IV.
DIENCEPHALON	
Thalamus	Several nuclei serve as relay stations for all sensory impulses, except smell, to the cerebral cortex. Relays motor impulses from the cerebral cortex to the spinal cord. Interprets pain, temperature, light touch, and pressure sensations. Anterior nucleus functions in emotions and memory.
Hypothalamus	Controls and integrates the autonomic nervous system. Receives hormonal signals and sensory impulses from viscera. Regulates the pituitary gland. Center for mind-over-body phenomena. Secretes regulating hormones (or factors). Functions in rage and aggression. Controls normal body temperature, food intake, and thirst. Helps maintain the waking state and sleep. Functions as a self-sustained oscillator that drives many biological rhythms.
Epithalamus	Contains pineal gland (see Chapter 21).
Subthalamus	Contains substantia nigra and red nucleus, also considered components of basal ganglia, which modulate subconscious skeletal muscle movements and muscle tone.
CEREBRUM	Sensory areas interpret sensory impulses, motor areas control muscular movement, and association areas function in emotional and intellectual processes. Basal ganglia control gross muscle movements and regulate muscle tone. Limbic system functions in emotional aspects of behavior related to survival.
CEREBELLUM	Controls subconscious skeletal muscle contractions required for coordination, posture, and balance. Assumes a role in emotional development, modulating sensations of anger and pleasure.

were entirely motor. However, it is now known that these cranial nerves also contain some sensory fibers from proprioceptors (receptors that detect muscle contraction) in muscles they innervate. Although these nerves are mixed, they are primarily motor in function, serving to stimulate skeletal muscle contraction. The cell bodies of sensory fibers are found outside the brain, whereas the cell bodies of motor fibers lie in nuclei within the brain (Figure 18-16).

Although the cranial nerves are mentioned singly in the following description of their type, location, and function, remember that they are paired structures.

OLFACTORY (I)

The ***olfactory (I) nerve*** is entirely sensory and conveys nerve impulses related to smell. It arises as bipolar neurons from the olfactory mucosa of the nasal cavity (see Figure 20-10). The dendrites and cell bodies of these neurons are generally limited to the mucosa covering the superior nasal conchae and the adjacent nasal septum. Axons from the neurons pass through the cribriform plate of the ethmoid bone and synapse with other olfactory neurons in the ***olfactory bulb,*** an extension of the brain lying above the cribri-

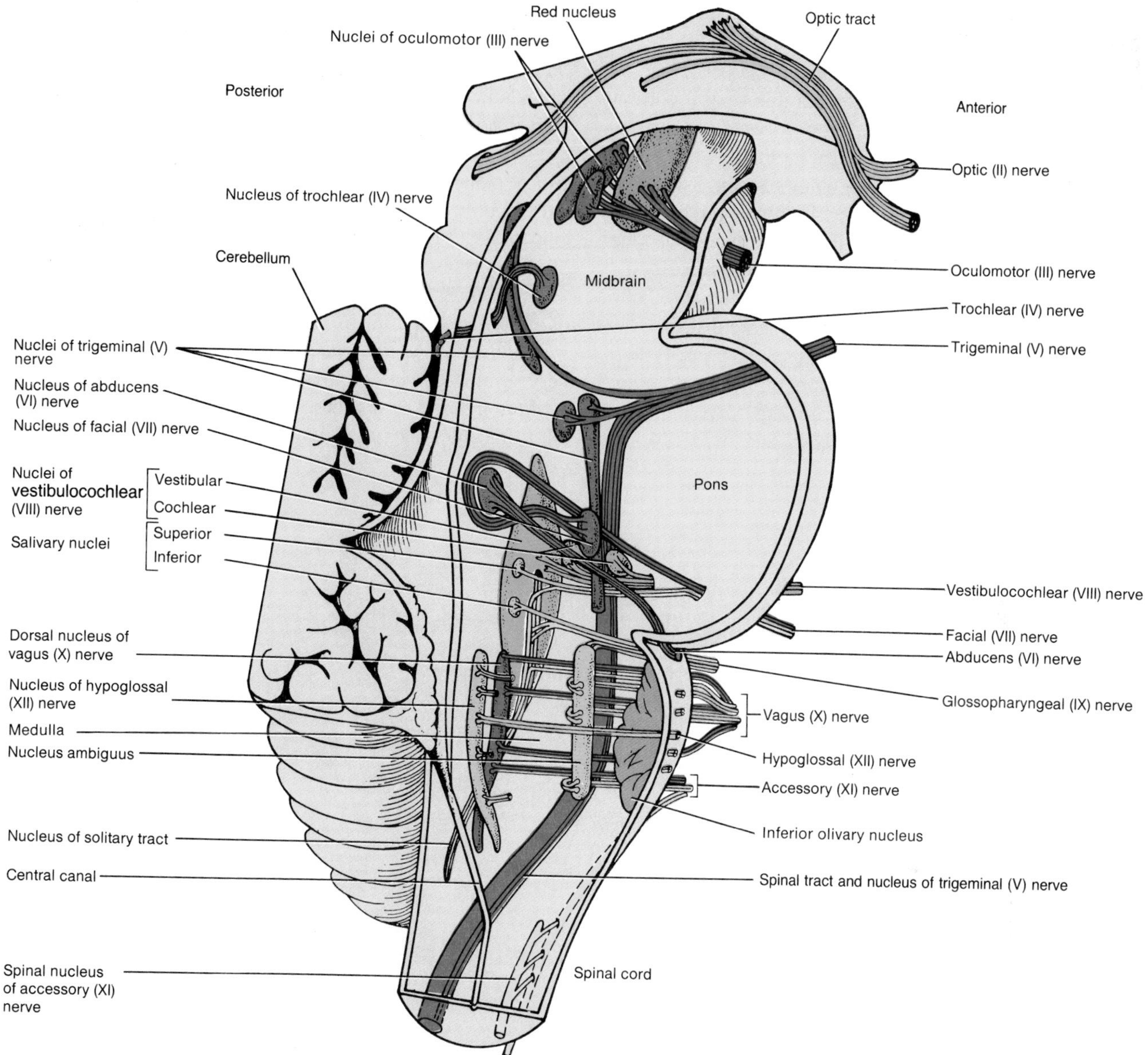

FIGURE 18-16 Nuclei of cranial nerves.

form plate. The axons of these neurons make up the ***olfactory tract.*** The fibers from the tract terminate in the primary olfactory area in the cerebral cortex.

OPTIC (II)

The ***optic (II) nerve*** is entirely sensory and conveys nerve impulses related to vision. Signals initiated by rods and cones of the retina are relayed by bipolar neurons to ganglion cells (see Figure 20-13). Axons of the ganglion cells, the optic nerve fibers, exit the optic foramen, after which the two optic nerves unite to form the ***optic chiasma*** (kī-AZ-ma). Within the chiasma, fibers from the medial half of each retina cross to the opposite side; those from the lateral half remain on the same side. From the chiasma the regrouped fibers pass posteriorly to the ***optic tracts.*** From the optic tracts the majority of fibers terminate in a nucleus (lateral geniculate) in the thalamus. They then synapse with neurons that pass to the visual areas of the cerebral cortex (see Figure 20-14). Some fibers from the optic chiasma terminate in the superior colliculi of the midbrain. They synapse with neurons whose fibers terminate in the nuclei that convey impulses to the oculomotor (III), trochlear (IV), and abducens (VI) nerves—nerves that control the extrinsic (external) and intrinsic (internal) eye muscles. Through this relay there are widespread motor responses to light stimuli.

OCULOMOTOR (III)

The ***oculomotor (III) nerve*** is a mixed cranial nerve. It originates from neurons in a nucleus in the ventral portion of the midbrain (Figure 18-17a). It runs forward and divides into superior and inferior divisions, both of which pass through the superior orbital fissure into the orbit. The superior branch is distributed to the superior rectus (an extrinsic eyeball muscle) and the levator palpebrae superioris (the muscle of the upper eyelid). The inferior branch is distributed to the medial rectus, inferior rectus, and inferior oblique muscles—all extrinsic eyeball muscles. These distributions to the levator palpebrae superioris and extrinsic eyeball muscles constitute the motor portion of the oculomotor nerve. Through these distributions are sent nerve impulses that control movements of the eyeball and upper eyelid.

The inferior branch of the oculomotor nerve also provides parasympathetic innervation to the ***ciliary ganglion,*** a relay center of the autonomic nervous system that connects a nucleus in the midbrain with the intrinsic eyeball muscles. These intrinsic muscles include the ciliary muscle of the eyeball and the sphincter muscle of the iris. Through the ciliary ganglion the oculomotor nerve controls the smooth muscle (ciliary muscle) responsible for adjustment of the lens for near vision and the smooth muscle (sphincter muscle of iris) responsible for constriction of the pupil.

The sensory portion of the oculomotor nerve consists of afferent fibers from proprioceptors in the eyeball muscles supplied by the nerve to the midbrain. These fibers convey nerve impulses related to muscle sense (proprioception).

Although the oculomotor (III), trochlear (IV), accessory (XI), and hypoglossal (XII) cranial nerves are referred to

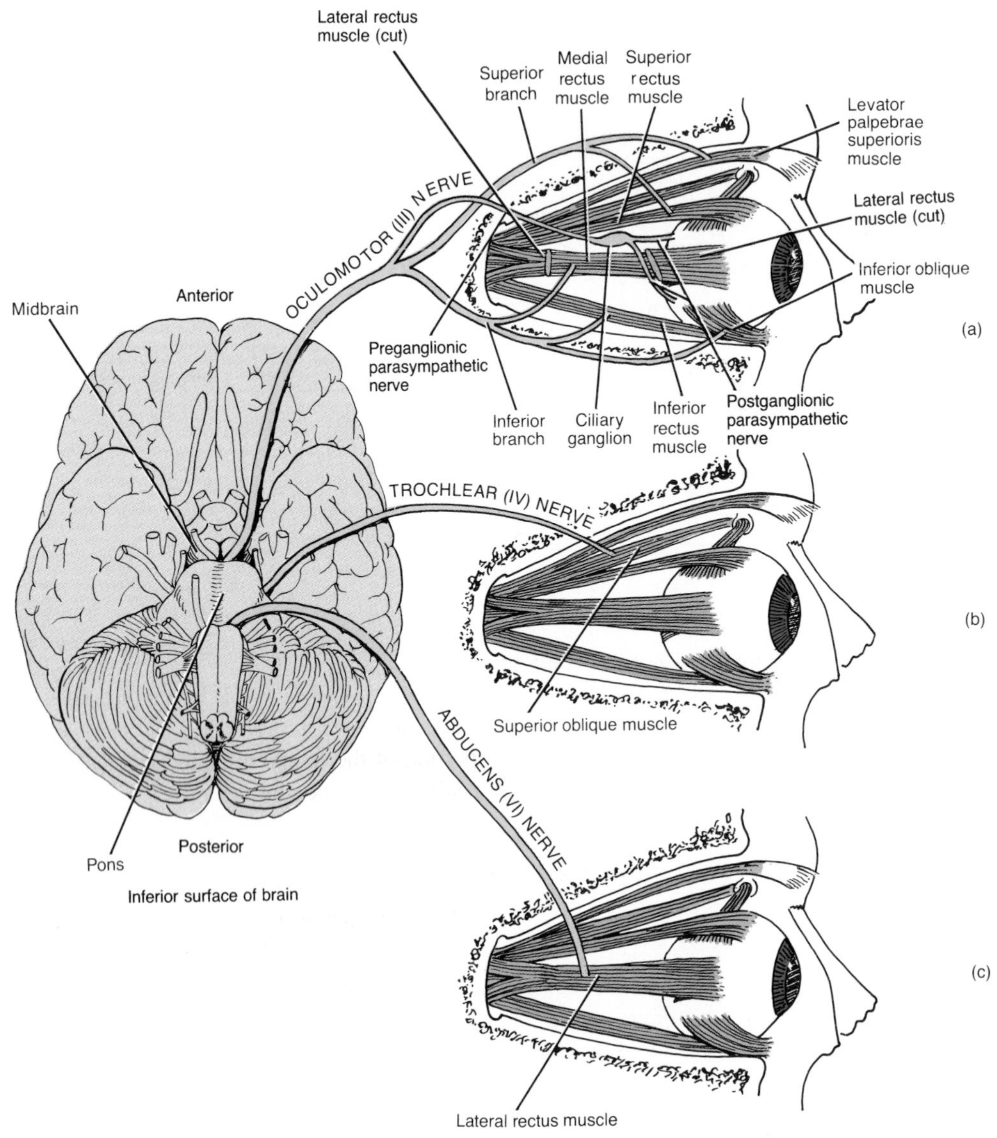

FIGURE 18-17 (a) Oculomotor (III) nerve. (b) Trochlear (IV) nerve. (c) Abducens (VI) nerve.

as mixed nerves because they contain fibers from proprioceptors, they are primarily motor, serving to stimulate skeletal muscle contractions.

TROCHLEAR (IV)

The ***trochlear*** (TROK-lē-ar) ***(IV) nerve*** is a mixed cranial nerve. It is the smallest of the 12 cranial nerves and is also unique because it is the only cranial nerve to arise from the posterior aspect of the brain stem. The motor portion originates in a nucleus in the midbrain, and axons from the nucleus pass through the superior orbital fissure of the orbit (Figure 18-17b). The motor fibers innervate the superior oblique muscle of the eyeball, another extrinsic eyeball muscle. The trochlear nerve controls movement of the eyeball.

The sensory portion of the trochlear nerve consists of afferent fibers that run from proprioceptors in the superior oblique muscle to a nucleus of the nerve in the midbrain. The sensory portion is responsible for muscle sense, the awareness of the activity of muscles.

TRIGEMINAL (V)

The ***trigeminal (V) nerve*** is a mixed cranial nerve and the largest of the cranial nerves. As indicated by its name, the trigeminal nerve has three branches: ophthalmic (of-THAL-mik), maxillary, and mandibular (Figure 18-18). The trigeminal nerve contains two roots on the ventrolateral surface of the pons. The large sensory root has a swelling called the ***semilunar (Gasserian) ganglion*** located in a fossa on the inner surface of the petrous portion of the temporal bone. The ganglion contains cell bodies of most of the primary sensory neurons. From this ganglion, the ***ophthalmic branch,*** the smallest branch, enters the orbit via the superior orbital fissure; the ***maxillary branch,*** intermediate in size between the ophthalmic and mandibular branches, enters the foramen rotundum; and the ***mandibular branch,*** the largest branch, exits through the foramen ovale. The smaller motor root originates in a nucleus in the pons. The motor fibers join the mandibular branch and supply the muscles of mastication. These motor fibers, which control chewing movements, constitute the motor portion of the trigeminal nerve and nerve to the mylohyoid and anterior belly of the digastric muscles.

The sensory portion of the trigeminal nerve delivers nerve impulses related to touch, pain, and temperature and consists of the ophthalmic, maxillary, and mandibular branches. The ophthalmic branch receives sensory fibers from the skin over the upper eyelid, eyeball, lacrimal glands, upper part of the nasal cavity, side of the nose, forehead, and anterior half of the scalp. The maxillary branch receives sensory fibers from the mucosa of the nose, palate, parts of the pharynx, upper teeth, upper lip, and lower eyelid. The mandibular branch transmits sensory fibers from the anterior two-thirds of the tongue (not taste), cheek and mucosa deep to it, lower teeth, skin over the mandible and side of the head in front of the ear, and mucosa of the floor of the mouth. Sensory fibers from the three branches of the trigeminal nerve enter the semilunar ganglion and terminate in nuclei in the pons. There are also sensory fibers from proprioceptors in the muscles of mastication.

CLINICAL APPLICATION

Dental Anesthesia

The inferior alveolar nerve, a branch of the mandibular nerve, supplies all the teeth in one-half of the mandible and is frequently ***anesthetized in dental procedures.*** The same procedure will anesthetize the lower lip because the mental nerve is a branch of the inferior alveolar nerve. Since the lingual nerve runs very close to the inferior alveolar near the mental foramen, it too is often anesthetized at the same time. For anesthesia to the upper teeth, the superior alveolar nerve endings, which are branches of the maxillary branch, are blocked by inserting the needle beneath the mucous membrane; the anesthetic solution is then infiltrated slowly throughout the area of the roots of the teeth to be treated.

ABDUCENS (VI)

The ***abducens*** (ab-DOO-sens) ***(VI) nerve*** is a mixed cranial nerve that originates from a nucleus in the pons (see Figure 18-17c). The motor fibers extend from the nucleus to the lateral rectus muscle of the eyeball, an extrinsic eyeball muscle. Nerve impulses over the fibers bring about abduction of the eyeball, the reason for the name of the nerve. The sensory fibers run from proprioceptors in the lateral rectus muscle to the pons and mediate muscle sense. The abducens nerve reaches the lateral rectus muscle through the superior orbital fissure of the orbit.

FACIAL (VII)

The ***facial (VII) nerve*** is a mixed cranial nerve. Its motor fibers originate from a nucleus in the pons, enter the petrous portion of the temporal bone, and are distributed to facial, scalp, and neck muscles (Figure 18-19). Nerve impulses along these fibers cause contraction of the muscles of facial expression and posterior belly of the digastric and stylohyoid muscles. Some motor fibers are also distributed via parasympathetics to the lacrimal, nasal, and palatine glands, and saliva-producing sublingual and submandibular glands.

The sensory fibers extend from the taste buds of the anterior two-thirds of the tongue to the ***geniculate ganglion,*** a swelling of the facial nerve. From here, the fibers pass to a nucleus in the pons, which sends fibers to the thalamus for relay to the gustatory area of the cerebral cortex. The sensory portion of the facial nerve also conveys deep general sensations from the face. There are also sensory fibers from proprioceptors in the muscles of the face and scalp.

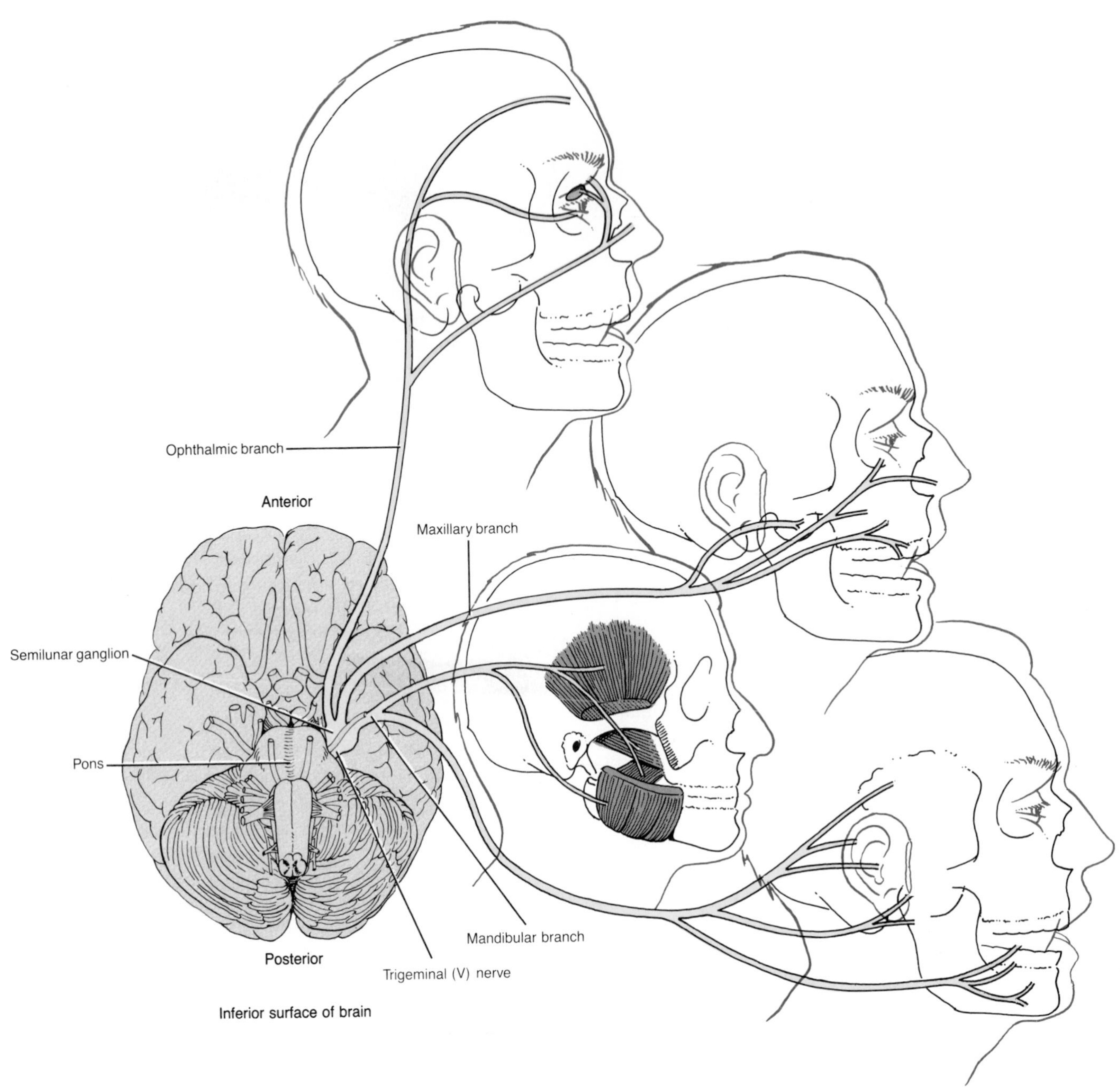

FIGURE 18-18 Trigeminal (V) nerve.

VESTIBULOCOCHLEAR (VIII)

The ***vestibulocochlear*** (ves-tib′-yoo-lō-KŌK-lē-ar) ***(VIII) nerve,*** formerly known as the ***acoustic nerve,*** is another sensory cranial nerve. It consists of two branches: the cochlear (auditory) branch and the vestibular branch (see Figure 20-17c). The ***cochlear branch,*** which conveys nerve impulses associated with hearing, arises in the spiral organ (organ of Corti) in the cochlea of the internal ear (Chapter 20). The cell bodies of the cochlear branch are located in the ***spiral ganglion*** of the cochlea. From here the axons synapse in nuclei in the medulla. The next neurons synapse in the inferior colliculus and pass on to the medial geniculate nucleus in the thalamus. Ultimately, the fibers synapse with neurons that relay the nerve impulses to the auditory areas of the cerebral cortex.

The ***vestibular branch*** arises in the semicircular canals, the saccule, and the utricle of the inner ear (Chapter 20). Fibers from the semicircular canals, saccule, and utricle extend to the ***vestibular ganglion,*** where the cell bodies are contained. The fibers of the neurons with cell bodies in the vestibular ganglion terminate in nuclei in the thalamus.

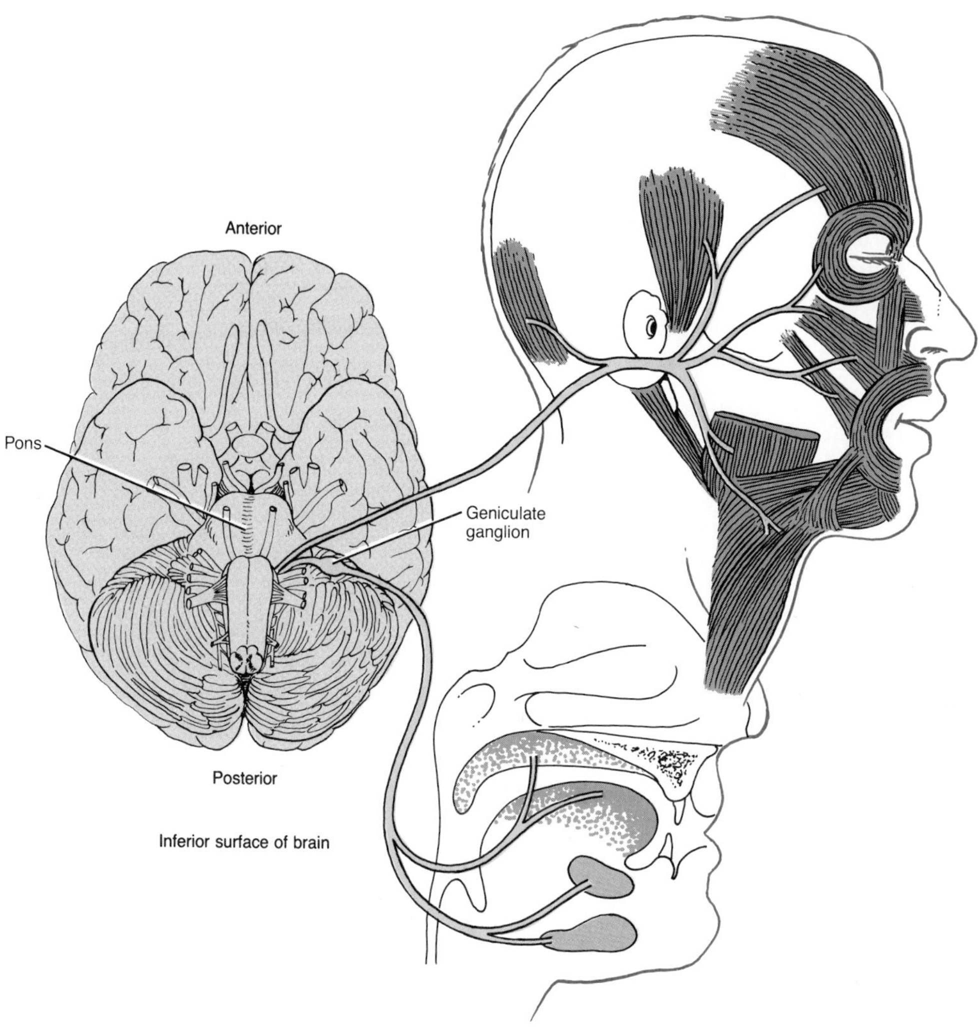

FIGURE 18-19 Facial (VII) nerve.

Some fibers also enter the cerebellum. The vestibular branch transmits impulses related to equilibrium.

GLOSSOPHARYNGEAL (IX)

The ***glossopharyngeal*** (glos′-ō-fa-RIN-jē-al) ***(IX) nerve*** is a mixed cranial nerve. Its motor fibers originate in nuclei in the medulla (Figure 18-20). The nerve exits the skull through the jugular foramen. The motor fibers are distributed to the stylopharyngeus muscle and parotid gland (parasympathetic) to mediate the secretion of saliva.

The sensory fibers of the glossopharyngeal nerve supply the pharynx and taste buds of the posterior third of the tongue. Some sensory fibers also originate from receptors in the carotid sinus, which assumes a major role in blood pressure regulation. The sensory fibers terminate in a nucleus in the thalamus. There are also sensory fibers from proprioceptors in the muscles innervated by this nerve.

VAGUS (X)

The ***vagus (X) nerve*** is a mixed cranial nerve that is widely distributed from the head and neck into the thorax and abdomen (Figure 18-21). The nerve derives its name from its wide distribution (*vagrant* or *wandering*). In the neck, it is positioned between and behind the internal jugular vein and common carotid artery. Motor fibers of the vagus nerve originate in nuclei of the medulla and terminate in the muscles of the respiratory passageways, lungs, heart, esophagus, stomach, small intestine, most of the large intestine, and gallbladder. Nerve impulses along the motor fibers generate visceral, cardiac, and skeletal muscle movement. Parasympathetic fibers innervate involuntary muscles and glands of the gastrointestinal tract.

Sensory fibers of the vagus nerve supply essentially the same structures as the motor fibers. They convey nerve impulses for various sensations from the larynx, the viscera,

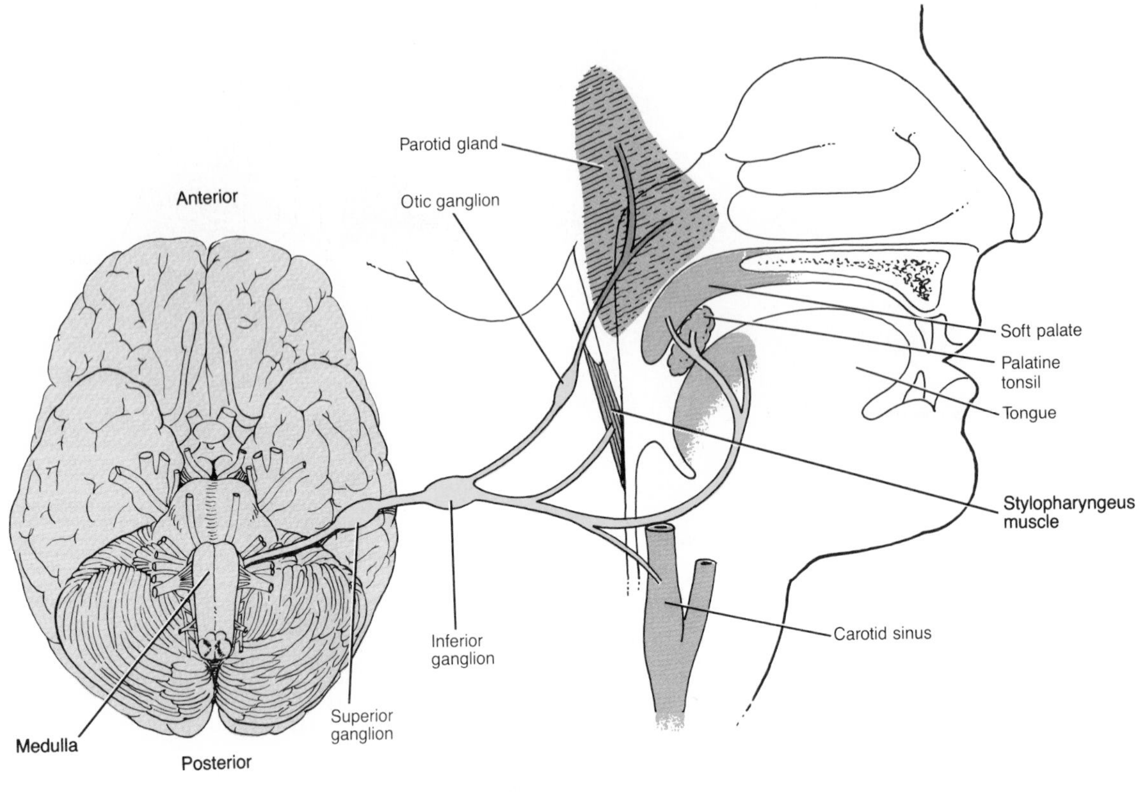

FIGURE 18-20 Glossopharyngeal (IX) nerve.

and the ear. The fibers terminate in the medulla and pons. There are also sensory fibers from proprioceptors in the muscles supplied by this nerve.

ACCESSORY (XI)

The ***accessory (XI) nerve*** (formerly the ***spinal accessory nerve***) is a mixed cranial nerve. It differs from all other cranial nerves in that it originates from both the brain stem and the spinal cord (Figure 18-22). The ***cranial portion*** originates from nuclei in the medulla, passes through the jugular foramen, and supplies the voluntary muscles of the pharynx, larynx, and soft palate that are used in swallowing. The ***spinal portion*** originates in the anterior gray horn of the first five segments of the cervical portion of the spinal cord. The fibers from the segments join, enter the foramen magnum, and exit through the jugular foramen along with the cranial portion.

The spinal portion conveys motor impulses to the sternocleidomastoid and trapezius muscles to coordinate head movements. The sensory fibers originate from proprioceptors in the muscles supplied by its motor neurons and terminate in upper cervical posterior root ganglia.

HYPOGLOSSAL (XII)

The ***hypoglossal (XII) nerve*** is a mixed cranial nerve. The motor fibers originate in a nucleus in the medulla, pass through the hypoglossal canal, and supply the muscles of the tongue (Figure 18-23). These fibers conduct nerve impulses related to speech and swallowing.

The sensory portion of the hypoglossal nerve consists of fibers originating from proprioceptors in the tongue muscles and terminating in the medulla. The sensory fibers conduct nerve impulses for muscle sense.

A summary of cranial nerves and clinical applications related to dysfunction is presented in Exhibit 18-2.

AGING AND THE NERVOUS SYSTEM

One of the effects of aging on the nervous system is that neurons are lost. Associated with this decline is a decreased capacity for sending nerve impulses to and from the brain. Conduction velocity decreases, voluntary motor movements slow down, and the reflex time for skeletal muscles increases. Deep reflexes may diminish and superficial reflexes

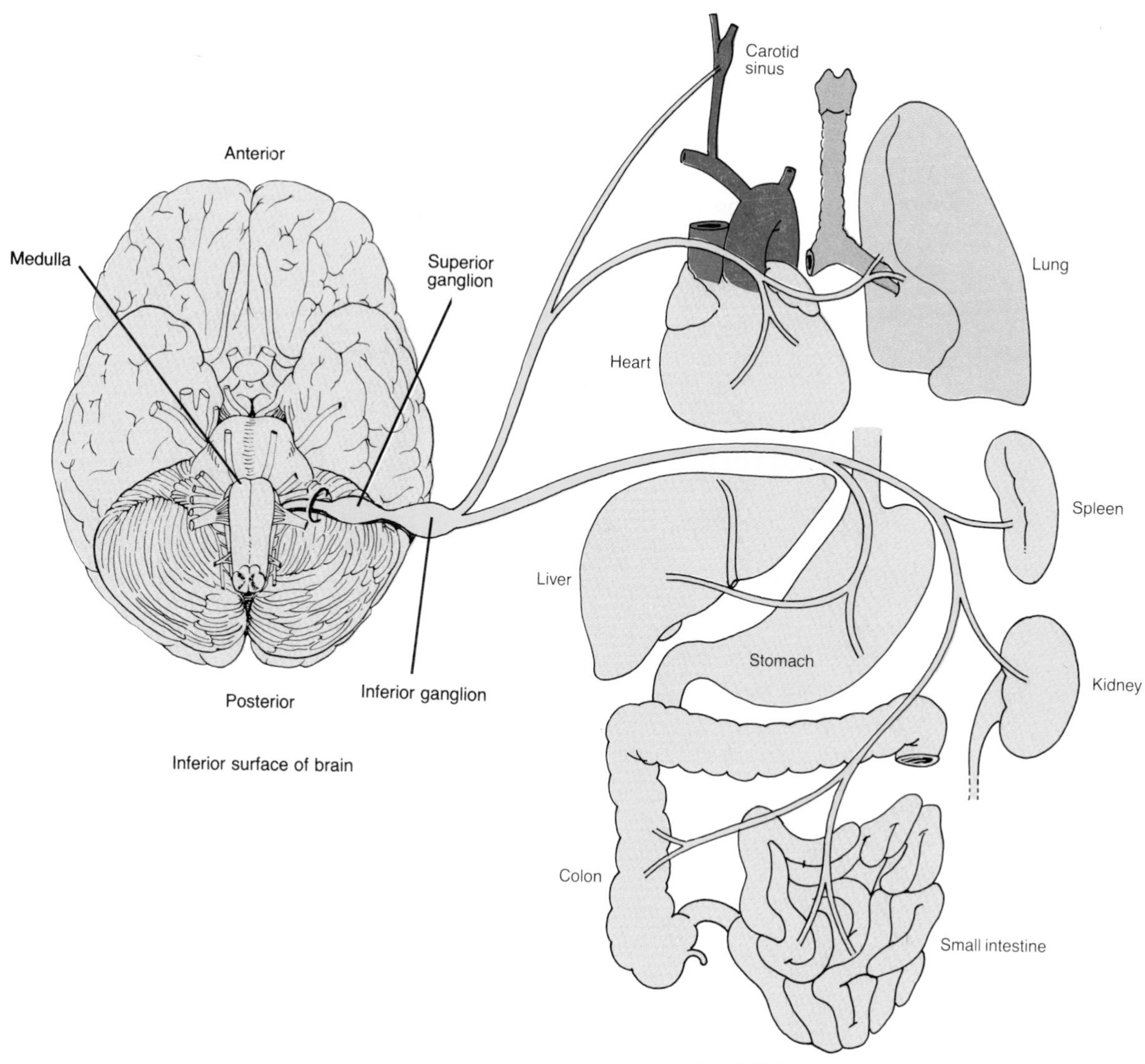

FIGURE 18-21 Vagus (X) nerve.

may be lost. Parkinson's disease is the most common movement disorder involving the central nervous system.

DEVELOPMENTAL ANATOMY OF THE NERVOUS SYSTEM

The development of the nervous system begins early in the third week of development with a thickening of the ***ectoderm*** called the ***neural plate*** (Figure 18-24). The plate folds inward and forms a longitudinal groove, the ***neural groove.*** The raised edges of the neural plate are called ***neural folds.*** As development continues, the neural folds increase in height, meet, and form a tube, the ***neural tube.***

The cells of the wall that encloses the neural tube differentiate into three kinds. The outer or ***marginal layer*** develops into the *white matter* of the nervous system; the middle or ***mantle layer*** develops into the *gray matter* of the system; and the inner or ***ependymal layer*** eventually forms the *lining of the ventricles* of the central nervous system.

The ***neural crest*** is a mass of tissue between the neural tube and the ectoderm (Figure 18-24c). It becomes differentiated and eventually forms the *posterior (dorsal) root ganglia of spinal nerves, spinal nerves, ganglia of cranial nerves, ganglia of the autonomic nervous system,* and the *adrenal medulla.*

When the neural tube is formed from the neural plate, the anterior portion of the neural tube develops into three enlarged areas called vesicles: (1) ***forebrain vesicle (prosencephalon),*** (2) ***midbrain vesicle (mesencephalon),*** and (3) ***hindbrain vesicle (rhombencephalon)*** (Figure 18-25). The vesicles are fluid-filled enlargements that develop by the fourth week of gestation. Since they are the first vesicles to form, they are called ***primary vesicles.*** As development

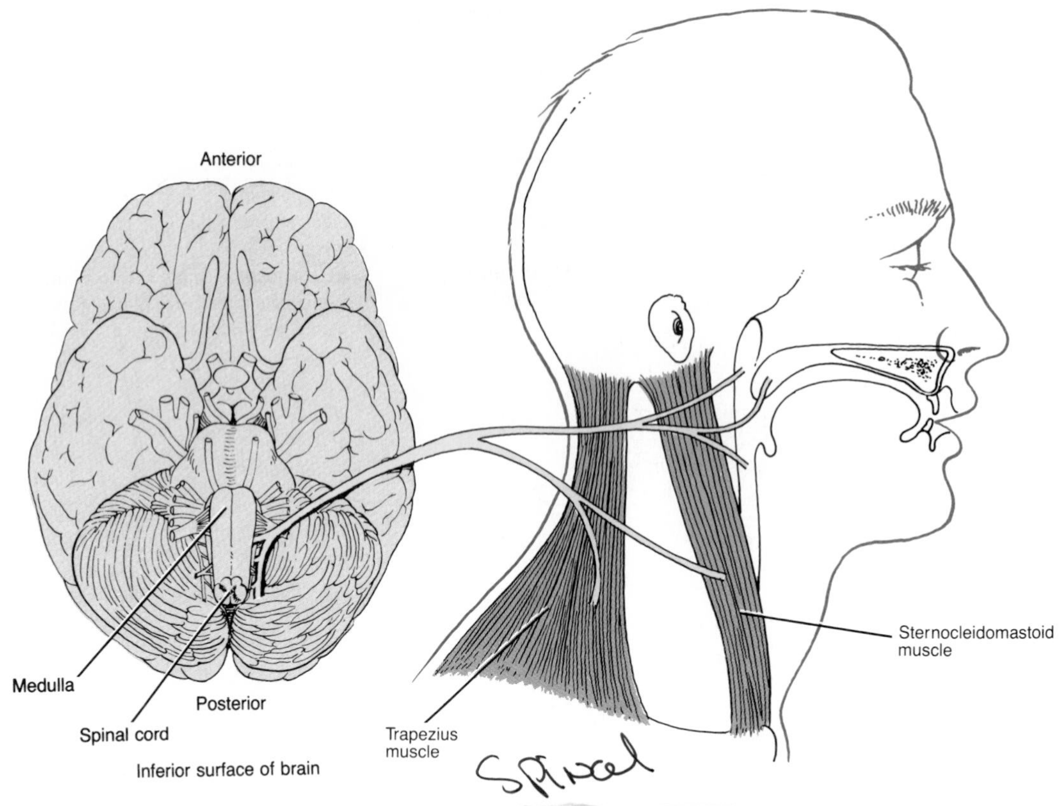

FIGURE 18-22 Accessory (XI) nerve.

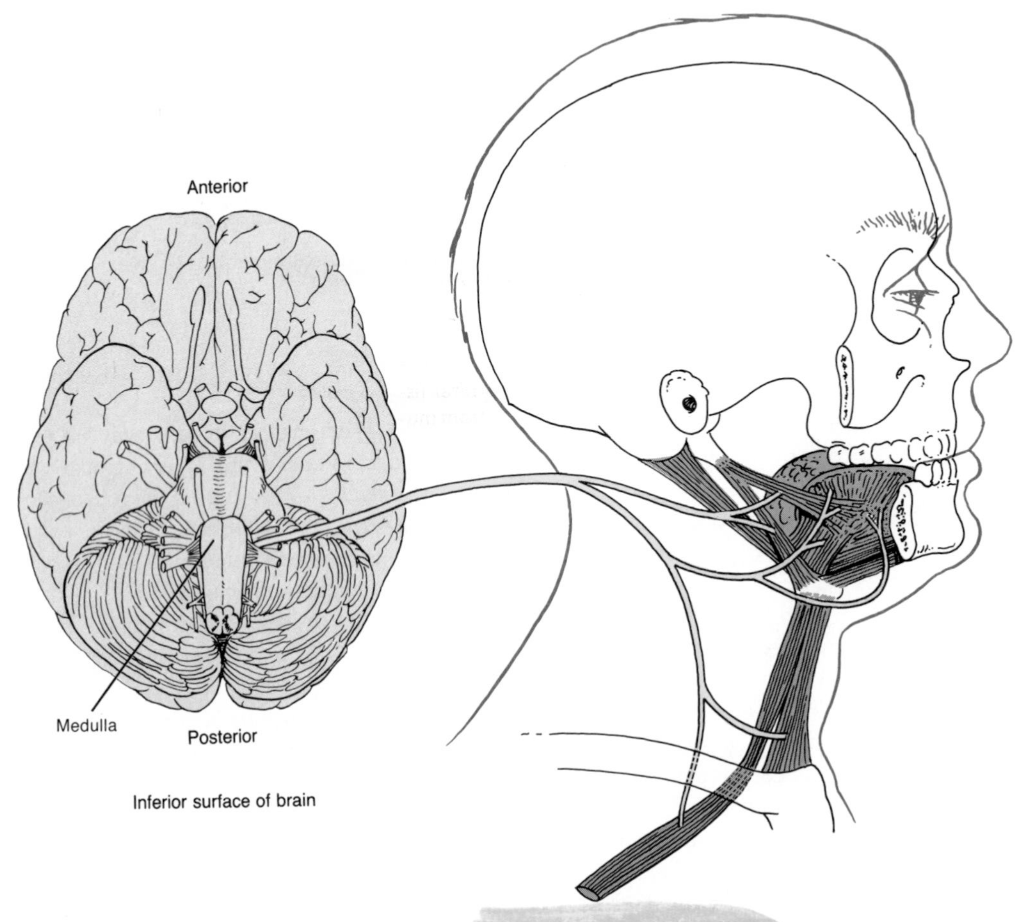

FIGURE 18-23 Hypoglossal (XII) nerve.

EXHIBIT 18-2

Summary of Cranial Nerves[a]

NERVE (TYPE)	LOCATION	FUNCTION AND CLINICAL APPLICATION
Olfactory (I) (sensory)	Arises in olfactory mucosa, passes through cribriform plate of ethmoid bone, through olfactory bulb and olfactory tract, and terminates in primary olfactory areas of cerebral cortex.	Function: smell. Clinical application: Loss of the sense of smell, called *anosmia,* may result from head injuries in which the cribriform plate of the ethmoid bone is fractured and from lesions along the olfactory pathway.
Optic (II) (sensory)	Arises in retina of the eye, passes through optic foramen, forms optic chiasma, passes through optic tracts, lateral geniculate nucleus in thalamus, and terminates in visual areas of cerebral cortex.	Function: vision. Clinical application: Fractures in the orbit, lesions along the visual pathway, and diseases of the nervous system may result in visual field defects and loss of visual acuity. A defect of vision is called *anopsia.*
Oculomotor (III) (mixed, primarily motor)	Motor portion: originates in midbrain, passes through superior orbital fissure, and is distributed to levator palpebrae superioris of upper eyelid and four extrinsic eyeball muscles (superior rectus, medial rectus, inferior rectus, and inferior oblique); parasympathetic innervation to ciliary muscle of eyeball and sphincter muscle of iris. Sensory portion: consists of afferent fibers from proprioceptors in eyeball muscles that passes through superior orbital fissure and terminates in midbrain.	Motor function: movement of eyelid and eyeball, accommodation of lens for near vision, and constriction of pupil. Sensory function: muscle sense (proprioception). Clinical application: a lesion in the nerve causes *strabismus* (squinting), *ptosis* (drooping) of the upper eyelid, pupil dilation, the movement of the eyeball downward and outward on the damaged side, a loss of accommodation for near vision, and double vision (*diplopia*).
Trochlear (IV) (mixed, primarily motor)	Motor portion: originates in midbrain, passes through superior orbital fissure, and is distributed to superior oblique muscle, an extrinsic eyeball muscle. Sensory portion: consists of afferent fibers from proprioceptors in superior oblique muscles that pass through superior orbital fissure and terminates in midbrain.	Motor function: movement of eyeball. Sensory function: muscle sense (proprioception). Clinical application: In trochlear nerve paralysis, diplopia and strabismus occur.
Trigeminal (V) (mixed)	Motor portion: part of the mandibular branch, originates in pons, passes through foramen ovale, and terminates in muscles of mastication, anterior belly of digastric and mylohyoid muscles. Sensory portion: consists of three branches: *ophthalmic*—contains sensory fibers from skin over upper eyelid, eyeball, lacrimal glands, nasal cavity, side of nose, forehead, and anterior half of scalp and passes through superior orbital fissure; *maxillary*—contains sensory fibers from mucosa of nose, palate, parts of pharynx, upper teeth, upper lip, and lower eyelid and passes through foramen rotundum; *mandibular*—contains general sensory fibers (but not special sense of taste) from anterior two-thirds of tongue, lower teeth, skin over mandible, cheek and mucosa deep to it, and side of head in front of ear and passes through foramen ovale. The three branches terminate in pons. Sensory portion also consists of afferent fibers from proprioceptors in muscles of mastication.	Motor function: chewing. Sensory function: conveys sensations for touch, pain, and temperature from structures supplied; muscle sense (proprioception). Clinical application: Injury results in paralysis of the muscles of mastication and a loss of sensation of touch and temperature. *Neuralgia* (pain) of one or more branches of trigeminal nerve is called *trigeminal neuralgia* (*tic douloureux*).
Abducens (VI) (mixed, primarily motor)	Motor portion: originates in pons, passes through superior orbital fissure, and is distributed to lateral rectus muscle, an extrinsic eyeball muscle. Sensory portion: consists of afferent fibers from proprioceptors in lateral rectus muscle that pass through superior orbital fissure and terminates in pons.	Motor function: movement of eyeball. Sensory function: muscle sense (proprioception). Clinical application: With damage to this nerve, the affected eyeball cannot move laterally beyond the midpoint and the eye is usually directed medially.

(Continued)

EXHIBIT 18-2

Summary of Cranial Nerves (*Continued*)

NERVE (TYPE)	LOCATION	FUNCTION AND CLINICAL APPLICATION
Facial (VII) (mixed)	Motor portion: originates in pons, passes through stylomastoid foramen, and is distributed to facial, scalp, and neck muscles; parasympathetic distribution to lacrimal, sublingual, submandibular, nasal, and palatine glands. Sensory portion: arises from taste buds on anterior two-thirds of tongue, passes through stylomastoid foramen, passes through geniculate ganglion, a nucleus in pons that sends fibers to thalamus for relay to gustatory areas of cerebral cortex. Also consists of afferent fibers from proprioceptors in muscles of face and scalp.	Motor function: facial expression and secretion of saliva and tears. Sensory function: muscle sense (proprioception) and taste. Clinical application: Injury produces paralysis of the facial muscles, called *Bell's palsy;* loss of taste; and loss of ability to close the eyes, even during sleep.
Vestibulocochlear (VIII) (sensory)	Cochlear branch: arises in spiral organ (organ of Corti), forms spiral ganglion, passes through nuclei in the medulla, and terminates in thalamus. Fibers synapse with neurons that relay impulses to auditory areas of cerebral cortex. Vestibular branch: arises in semicircular canals, saccule, and utricle and forms vestibular ganglion; fibers terminate in nuclei in thalamus.	Cochlear branch function: conveys impulses associated with hearing. Vestibular branch function: conveys impulses associated with equilibrium. Clinical application: Injury to the cochlear branch may cause *tinnitus* (ringing) or deafness. Injury to the vestibular branch may cause *vertigo* (a subjective feeling of rotation), *ataxia,* and *nystagmus* (involuntary rapid movement of the eyeball).
Glossopharyngeal (IX) (mixed)	Motor portion: originates in medulla, passes through jugular foramen, and is distributed to stylopharyngeus muscle; parasympathetic distribution to parotid gland. Sensory portion: arises from taste buds on posterior one-third of tongue and from carotid sinus, passes through jugular foramen, and terminates in thalamus. Also consists of afferent fibers from proprioceptors in swallowing muscles supplied.	Motor function: secretion of saliva. Sensory function: taste and regulation of blood pressure; muscle sense (proprioception). Clinical application: Injury results in difficulty during swallowing, reduced secretion of saliva, loss of sensation in the throat, and loss of taste.
Vagus (X) (mixed)	Motor portion: originates in medulla, passes through jugular foramen, and terminates in muscles of respiratory passageways, lungs, esophagus, heart, stomach, small intestine, most of large intestine, and gallbladder; parasympathetic fibers innervate involuntary muscles and glands of the gastrointestinal (GI) tract. Sensory portion: arises from essentially same structures supplied by motor fibers, passes through jugular foramen, and terminates in medulla and pons. Also consists of afferent fibers from proprioceptors in muscles supplied.	Motor function: visceral muscle movement. Sensory function: sensations from organs supplied; muscle sense (proprioception). Clinical application: Severing of both nerves in the upper body interferes with swallowing, paralyzes vocal cords, and interrupts sensations from many organs. Injury to both nerves in the abdominal area has little effect, since the abdominal organs are also supplied by autonomic fibers from the spinal cord.
Accessory (XI) (mixed, primarily motor)	Motor portion: consists of a cranial portion and a spinal portion. Cranial portion originates from medulla, passes through jugular foramen, and supplies voluntary muscles of pharynx, larynx, and soft palate. Spinal portion originates from anterior gray horn of first five cervical segments of spinal cord, passes through jugular foramen, and supplies sternocleidomastoid and trapezius muscles. Sensory portion: consists of afferent fibers from proprioceptors in muscles supplied and passes through jugular foramen.	Motor function: cranial portion mediates swallowing movements; spinal portion mediates movement of head. Sensory function: muscle sense (proprioception). Clinical application: If damaged, the sternocleidomastoid and trapezius muscles become paralyzed, with resulting inability to raise the shoulders and difficulty in turning the head.

NERVE (TYPE)	LOCATION	FUNCTION AND CLINICAL APPLICATION
Hypoglossal (XII) (mixed, primarily motor)	Motor portion: originates in medulla, passes through hypoglossal canal, and supplies muscles of tongue. Sensory portion: consists of fibers from proprioceptors in tongue muscles that pass through hypoglossal canal and terminate in medulla.	Motor function: movement of tongue during speech and swallowing. Sensory function: muscle sense (proprioception). Clinical application: Injury results in difficulty in chewing, speaking, and swallowing. The tongue, when protruded, curls toward the affected side and the affected side becomes atrophied, shrunken, and deeply furrowed.

[a] A mnemonic device used to remember the names of the nerves is: "Oh, oh, oh, to touch and feel very green vegetables—AH!" The initial letter of each word corresponds to the initial letter of each pair of cranial nerves.

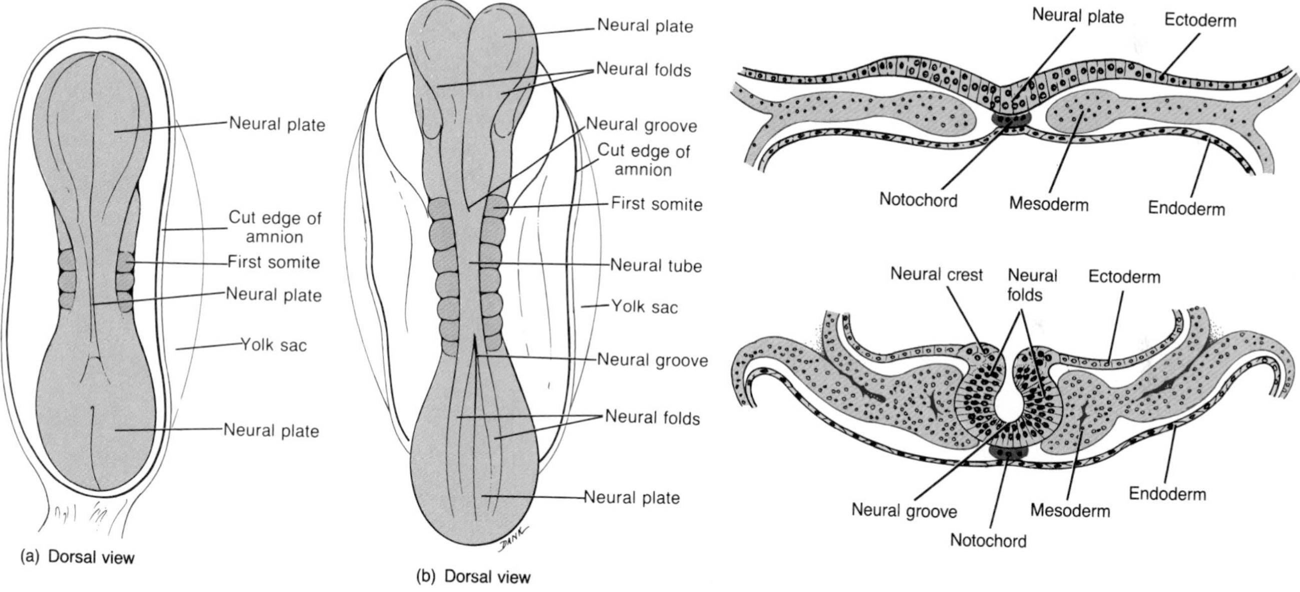

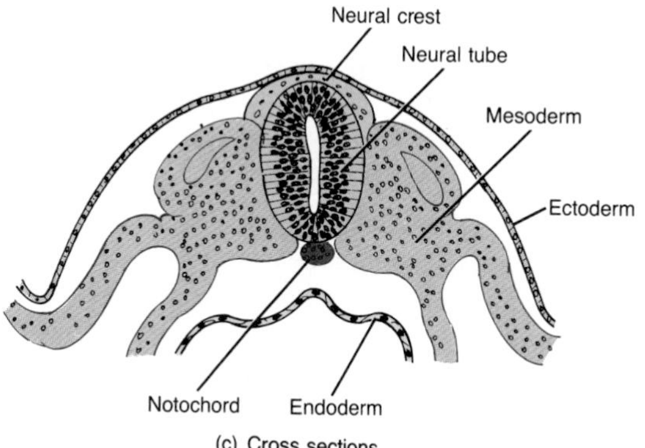

FIGURE 18-24 Origin of the nervous system. (a) Embryo with three pairs of somites showing the neural plate. (b) Embryo with seven pairs of somites in which the neural folds have united just medial to the somites, forming the early neural tube. (c) Embryo showing the formation of the neural tube.

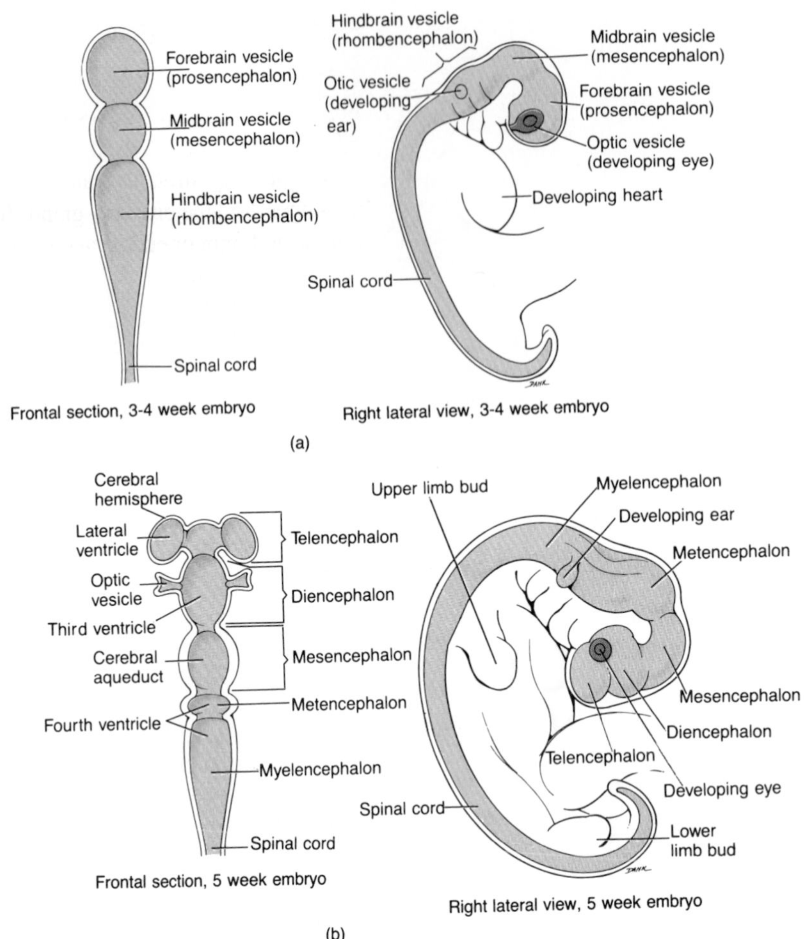

FIGURE 18-25 Development of the brain and spinal cord. (a) Primary vesicles of the neural tube. (b) Secondary vesicles.

progresses, the vesicular region undergoes several flexures (bends), resulting in subdivision of the three primary vesicles; by the fifth week of development the embryonic brain consists of five ***secondary vesicles.*** The forebrain vesicle (prosencephalon) divides into an anterior ***telencephalon*** and a posterior ***diencephalon;*** the midbrain vesicle (mesencephalon) remains unchanged; the hindbrain vesicle (rhombencephalon) divides into an anterior ***metencephalon*** and a posterior ***myelencephalon.***

Ultimately, the telencephalon develops into the *cerebral hemispheres* and *basal ganglia;* the diencephalon develops into the *thalamus, hypothalamus,* and *pineal gland;* the midbrain vesicle (mesencephalon) develops into the *midbrain;* the metencephalon develops into the *pons* and *cerebellum;* and the myelencephalon develops into the *medulla oblongata.* The cavities within the vesicles develop into the *ventricles* of the brain, and the fluid within them is *cerebrospinal fluid.* The area of the neural tube posterior to the myelencephalon gives rise to the *spinal cord.*

APPLICATIONS TO HEALTH

CEREBROVASCULAR ACCIDENT (CVA)

The most common brain disorder is a ***cerebrovascular accident (CVA),*** also called a ***stroke.*** A CVA is characterized by a relatively abrupt onset of persisting neurological symptoms attributable to the destruction of brain tissue (infarction) resulting from disorders in the blood vessels that supply the brain. CVAs may be classified into two principal types: (1) *ischemic,* the most common type, due to a decreased blood supply; and (2) *hemorrhagic,* due to a blood vessel in the brain that bursts. Common causes of CVAs are intracerebral hemorrhage (rupture of a blood vessel in the pia mater or brain), emboli (blood clots), and atherosclerosis (formation of plaques) of the cerebral arteries.

Among the risk factors implicated in CVAs are high blood pressure, heart disease, narrowed carotid arteries, transient ischemic attacks (TIAs), diabetes, smoking, obesity, high plasma fibrinogen level, maternal history of CVA (for males), and excessive alcohol intake.

EPILEPSY

Epilepsy is the second most common neurological disorder after stroke. It is characterized by short, recurrent, periodic attacks of motor, sensory, or psychological malfunction. The attacks, called ***epileptic seizures,*** are initiated by abnormal and irregular discharges of electricity from millions of neurons in the brain. The discharges stimulate many of the neurons to send nerve impulses over their conduction pathways. As a result, a person undergoing an attack may contract skeletal muscles involuntarily. Lights, noise, or smells may be sensed when the eyes, ears, and nose actually have not been stimulated. The electrical discharges may also inhibit certain brain centers. For instance, the waking center in the brain may be so depressed that the person loses consciousness.

Symptomatic causes of epilepsy include brain damage at birth, the most common cause; metabolic disturbances (hypoglycemia, hypocalcemia, uremia, hypoxia); infections (encephalitis or meningitis); toxins (alcohol, tranquilizers, hallucinogens); vascular disturbances (hemorrhage, hypotension); head injuries; and tumors and abscesses of the brain. Most epileptic seizures, however, are idiopathic; that is, they have no demonstrable cause. Epilepsy almost never affects intelligence.

Epileptic seizures can be eliminated or alleviated by drugs that make neurons more difficult to stimulate.

TRANSIENT ISCHEMIC ATTACK (TIA)

A ***transient ischemic attack*** **(*TIA*)** is an episode of temporary, focal, nonconvulsive, cerebral dysfunction caused by an interference of the blood supply to the brain. Symptoms include dizziness, weakness, numbness, or paralysis in a limb or in one-half of the body; drooping of one side of the face; headache; slurred speech or difficulty understanding speech; or a partial loss of vision or double vision. Sometimes nausea or vomiting also occur. The onset of symptoms is sudden and reaches maximum intensity almost immediately. A TIA usually persists for 2 to 15 minutes and only rarely as long as 24 hours. Each TIA leaves no persistent neurologic deficits. The causes of the ischemia that lead to TIAs are emboli, atherosclerosis, impaired blood flow due to hemodynamic disruptions, and hematologic disorders (polycythemia, thrombocytosis, and sickle-cell anemia).

It is estimated that about one-third of patients with a TIA will have a CVA within five years. Therapy for TIAs includes antiplatelet-aggregating agents such as aspirin, anticoagulants, cerebral artery bypass grafting, and carotid endarterectomy (excision of the atheromatous tunica intima of an artery).

BRAIN TUMORS

A ***brain tumor*** refers to any benign or malignant growth within the cranium. Tumors may arise from neuroglia in the cerebrum, brain stem, and cerebellum or from supporting or neighboring structures such as cranial nerve coverings, meninges, and the pituitary gland. Intracranial pressure from a growing tumor or edema associated with the tumor produce the characteristic signs and symptoms of a brain tumor. Among these are headache, altered consciousness, and vomiting. Brain tumors can also result in seizures, visual problems, cranial nerve abnormalities, hormonal syndromes, personality changes, dementia, and sensory or motor deficits. Treatment of brain tumors involves surgery, radiation therapy, and chemotherapy.

POLIOMYELITIS

Poliomyelitis **(*infantile paralysis*),** or simply ***polio,*** is most common during childhood and is caused by a virus called poliovirus. The onset of the disease is marked by fever, severe headache, a stiff neck and back, deep muscle pain and weakness, and loss of certain somatic reflexes. In its most serious form, called ***bulbar polio,*** the virus spreads via blood to the central nervous system, where it destroys the motor nerve cell bodies, specifically those in the anterior horns of the spinal cord and in the nuclei of the cranial nerves. Injury to the spinal gray matter is the basis for the name of this disease (*polio* = gray matter; *myel* = spinal cord). Destruction of the anterior horns produces paralysis. The first sign of bulbar polio is difficulty in swallowing, breathing, and speaking. Poliomyelitis can cause death from respiratory or heart failure if the virus invades the brain cells of the vital medullary centers. The incidence of polio in the United States has decreased markedly since the availability of polio vaccines (Salk vaccine and, more recently, Sabin vaccine).

CEREBRAL PALSY (CP)

The term ***cerebral palsy*** **(*CP*)** refers to a group of motor disorders resulting in muscular incoordination and loss of muscle control. It is caused by damage to the motor areas of the brain during fetal life, birth, or infancy. One cause is infection of the mother with German measles during the first three months of pregnancy when certain cells in the fetus are dividing and differentiating in order to lay down the basic structures of the brain. These cells can be abnormally changed by toxin from the measles virus. Radiation during fetal life, temporary oxygen starvation during birth, and hydrocephalus during infancy may also damage brain cells. About 70 percent of cerebral palsy victims appear to be mentally retarded. The apparent mental slowness, however, is often due to the person's inability to speak or hear well. Such individuals are often more mentally acute than they appear. Cerebral palsy is not a progressive disease; it does not worsen as time elapses. Once the damage is done, however, it is irreversible.

PARKINSON'S DISEASE (PD)

Parkinson's disease (PD) or ***parkinsonism*** is a progressive disorder of the central nervous system that begins inconspicuously and typically affects its victims around age 60. The cause is unknown, but there are indications that an environmental agent might be involved. The disease is related to pathological changes in the substantia nigra and basal ganglia. The substantia nigra contains cell bodies of neurons that produce the neurotransmitter dopamine (DA) in their axon terminals. The axon terminals release DA in the basal ganglia of the cerebrum. Recall that basal ganglia regulate subconscious contractions of skeletal muscles that aid activities also consciously controlled by the motor areas of the cerebral cortex—swinging the arms when walking, for example. In Parkinson's disease there is a degeneration of DA-producing neurons in the substantia nigra, and the severe reduction of DA in the basal ganglia brings about most of the symptoms of Parkinson's disease.

Diminished levels of DA cause unnecessary skeletal muscle movements that often interfere with voluntary movement. For instance, the muscles of the upper extremity may alternately contract and relax, causing the hand to shake. This shaking is called ***tremor,*** the most common symptom of Parkinson's disease. The tremor may spread to the lower extremity of the same side and then to the extremity of the opposite side. Motor performance is also impaired by ***bradykinesia*** (*brady* = slow; *kinesis* = motion), in which activities such as shaving, cutting food, and buttoning a shirt take longer and become increasingly more difficult. Muscular movements are performed not only slowly but with decreasing range of motion (***hypokinesia***). For example, as handwriting continues, letters get smaller, become poorly formed, and eventually become illegible. Some muscles may contract continuously, causing ***rigidity*** of the involved body part. Rigidity of the facial muscles gives the face a masklike appearance. The expression is characterized by a wide-eyed, unblinking stare and a slightly open mouth with uncontrolled drooling. Decreased DA production also results in impaired walking, in which steps become shorter and shuffling and arm swing diminishes. There are also changes that lead to stooped posture and loss of postural reflexes, autonomic dysfunction (constipation, retention), sensory complaints (pain, numbness, tingling), and sustained muscle spasms. Vision, hearing, and intelligence are unaffected by the disorder, indicating that Parkinson's disease does not attack the cerebral cortex.

Treatment of the symptoms of Parkinson's disease is directed toward increasing levels of DA. Although people with Parkinson's disease do not manufacture enough DA, injections of it are useless; the blood–brain barrier stops it. However, symptoms are somewhat relieved by a drug developed in the 1960s called levodopa, a precursor of DA. Administered by itself, levodopa may elevate brain levels of DA, causing undesirable side effects such as low blood pressure, nausea, mental changes, and liver dysfunction. Levodopa has been combined with carbidopa, which inhibits the formation of DA outside the brain. The combined drugs diminish the undesirable side effects. Levodopa, however, does not slow the disease process and as more affected brain cells die, the drug becomes useless. Recently, patients with parkinsonism have been given a drug called deprenyl. Early results indicate that it can prevent symptoms for nearly a year in newly diagnosed patients and has minimal side effects.

In 1989, physicians in Sweden transplanted fetal nerve tissue (dopamine-rich mesencephalic tissue) into the brain (putamen) of a patient with severe Parkinson's disease. Following the surgery, the patient exhibited diminished symptoms and less rigidity and sluggishness. Results are still being evaluated.

MULTIPLE SCLEROSIS (MS)

Multiple sclerosis (MS) is the progressive destruction of the myelin sheaths of neurons in the central nervous system accompanied by disappearance of oligodendrocytes and the proliferation of astrocytes. The sheaths deteriorate to ***scleroses,*** which are hardened scars or plaques, in multiple regions. The destruction of myelin sheaths interferes with the transmission of nerve impulses from one neuron to another, literally short-circuiting conduction pathways. Usually, the first symptoms occur in early adult life. The average age of onset is 33. The frequency of flare-ups is greatest during the first three to four years of the disease, but a first attack, which may have been so mild as to escape medical attention, may not be followed by another attack for 10 to 20 years.

Among the first symptoms of MS are muscular weakness of one or more extremities; abnormal sensations, such as burning or pins and needles; visual impairment that includes blurring, double vision, and problems with color and light perception; uncoordination; vertigo; and sphincter impairment that results in urinary problems, such as urinary urgency. Following a period of remission during which the symptoms temporarily disappear, a new series of plaques develop, and the victim suffers a second attack. One attack follows another over the years, usually every year or two. Each time the plaques form, some neurons are damaged by the hardening of their sheaths, whereas others are uninjured by their plaques. The result is a progressive loss of function interspersed with remission periods during which the undamaged neurons regain their ability to conduct nerve impulses.

The symptoms of MS depend on the areas of the central nervous system most heavily laden with plaques. Sclerosis of the white matter of the spinal cord is common. As the sheaths of the neurons in the corticospinal tracts deteriorate, the patient loses the ability to contract skeletal muscles. Damage to the ascending tracts produces numbness and short-circuits impulses related to position of body parts and flexion of joints. Damage to either set of tracts also destroys spinal cord reflexes.

The clinical course of MS is unpredictable. During typical episodes, symptoms worsen over a period of a few days to two to three weeks and then remit. Relapses occur at an average rate of 0.5 per year during the initial five years, although this rate is highly variable. Some patients experience complete remission following relapses, whereas others gradually accumulate neurologic problems. Many patients suffer multiple attacks but are never disabled. About 20 percent of MS patients have symptoms and signs that appear slowly and steadily, without a clear relapsing–remitting pattern. Such a course often occurs in late-onset patients (over 40) and is frequently associated with severe disability. About two-thirds of MS patients are ambulatory 25 years after the onset of their disease, one-half will be working 10 years from the onset, and one-third will have unrestricted functions. MS does not predictably shorten life except in a minority of patients who are bedridden or succumb to a urinary tract infection or pneumonia.

Although the etiology of MS is unclear, there is some evidence that it might result from a viral infection that precipitates an autoimmune response. Viruses may trigger the destruction of myelin-producing oligodendrocytes by the antibodies and cytotoxic (killer) cells of the body's immune system. Like other demyelinating diseases, MS is incurable. However, in view of the evidence that it might be an autoimmune disease, immunosuppressive therapy is widely used (glucocorticoids such as prednisone). A recent treatment consists of administering cyclophosphamide (a powerful anticancer drug that suppresses the immune system) in combination with adrenocorticotropic hormone or ACTH (a pituitary gland hormone that stimulates secretion of hormones containing cortisone) to patients with active MS. An experimental therapy for MS is the use of a synthetic protein called copolymer (Cop 1), a decoy that spares destruction of a similar protein in oligodendrocytes. Another treatment is the use of colchicine, a drug used to treat gout. Electrical stimulation of the spinal cord can also improve function in certain patients. Improvement has also been shown in patients who are administered pure oxygen while in a pressure chamber (hyperbaric oxygen). This result supports the idea that the destruction of myelin occurs preferentially in parts of the brain that are relatively low in oxygen. Treatment is also directed at management of complications such as spasticity, facial neuralgia and twitching, urinary bladder problems, and constipation.

DYSLEXIA

Dyslexia (dis-LEK-sē-a; *dys* = difficulty; *lexis* = words) is an impairment of the brain's ability to translate images received from the eyes or ears into understandable language. The condition is unrelated to basic intellectual capacity, but it causes a mysterious difficulty in handling words and symbols. Apparently, some peculiarity in the brain's organizational pattern distorts the ability to read, write, and count. Letters in words seem transposed, reversed, or upside down—*dog* becomes *god; b* changes identity with *d;* a sign saying "OIL" inverts into "710." Frequently, dyslectics reread only portions of paragraphs, skip words, and change the order of letters in a word. Many dyslectics cannot orient themselves in the three dimensions of space and may show bodily awkwardness.

The exact cause of dyslexia is unknown, since it is unaccompanied by outward scars of detectable neurological damage and its symptoms vary from victim to victim. It occurs almost three times as often among boys as among girls. It has been variously attributed to defective vision, brain damage, abnormal brain development, lead in the air, physical trauma, or oxygen deprivation during birth. A recent theory holds that it might be related to a complex language deficiency involving inability to represent and access the sound of a word in order to help remember it, inability to break down words into components, poor vocabulary development, and difficulty discriminating grammatical differences among words and phrases.

A technique used primarily to diagnose dyslexia is called ***brain electrical activity mapping (BEAM).*** It is a noninvasive procedure that measures and displays the electrical activity of the brain on a color television screen and compares the image produced with a normal image. BEAM may have other diagnostic applications for learning disabilities, schizophrenia, depression, dementia, epilepsy, and early tumor occurrence and recurrence.

Positron emission tomography (PET) scans (see Exhibit 1-6) are also providing information about dyslexia. For example, PET scans have demonstrated that in persons with dyslexia the left side of the brain is more active than the right, the language region is less active than normal, and the visual discrimination region is more active.

TAY-SACHS DISEASE

Tay-Sachs disease is a central nervous system affliction that brings death before age 5. The Tay-Sachs gene is carried mostly by individuals descended from the Ashkenazi Jews of Eastern Europe. Approximately 1 in 30 of their offspring will carry the trait (if both parents carry the gene, 1 in 4 children will inherit the disease). The disease involves the neuronal degeneration of the central nervous system because of excessive amounts of a lipid called ganglioside in the neurons of the brain. The substance accumulates because of a deficient lysosomal enzyme. The afflicted child develops normally until the age of 4 to 8 months. Then the symptoms follow a course of progressive degeneration: paralysis, blindness, inability to eat, decubitus ulcers, and death from infection. Amniocentesis can detect the trait prenatally and a blood test can be done to detect the trait postnatally. There is no known cure.

HEADACHE

One of the most common human afflictions is ***headache,*** or ***cephalgia*** (*enkephalos* = brain; *algia* = painful condition). Based on origin, two general types are distinguished: intracranial and extracranial. Serious headaches of intracranial origin are caused by brain tumors, blood vessel abnormalities, inflammation of the brain or meninges, decrease in oxygen supply to the brain, and damage to brain cells. Extracranial headaches are related to infections of the eyes, ears, nose, and sinuses and are commonly felt as headaches because of the location of these structures.

Most headaches require no special treatment. Analgesic and tranquilizing compounds are generally effective for tension headaches but not for migraine headaches. Drugs that constrict the blood vessels can be helpful for migraine. Biofeedback training and dietary changes may also be of some help. Tension headaches are the most common form of extracranial headache; they are associated with stress, fatigue, and anxiety; and classically they occur in the occipital and temporal muscles.

TRIGEMINAL NEURALGIA (TIC DOULOUREUX)

As noted earlier, pain arising from irritation of the trigeminal (V) nerve is known as ***trigeminal neuralgia,*** or ***tic douloureux*** (doo-loo-ROO). The disorder is characterized by brief but extreme pain in the face and forehead on the affected side. The characteristic pain consists of red-hot, needlelike jabs lasting a few seconds and building to a searing pain like a hot poker being dragged or stuck into the face, lasting 10 to 15 seconds. Many patients describe sensitive regions around the mouth and nose that can cause an attack when touched. Eating, drinking, washing the face, and exposure to cold may also bring on an attack.

Treatment may be palliative (relieving symptoms without curing the disease) or surgical. One technique involves alcohol injections directly into the semilunar (Gasserian) ganglion, which controls the trigeminal (V) nerve. This method is superior to open surgery because it is safer. Moreover, it can bring lasting pain relief with preservation of touch sensation in the face and release of the patient 24 hours after treatment. Acupuncture can also provide relief in some patients.

REYE'S SYNDROME (RS)

Reye's syndrome (RS), first described in 1963 by the Australian pathologist R. Douglas Reye, seems to occur following a viral infection, particularly chickenpox or influenza. Aspirin at normal doses is believed to be a risk factor in the development of RS. The majority of persons affected are children or teenagers. The disease is characterized by vomiting and brain dysfunction (disorientation, lethargy, and personality changes) and may progress to coma. Also, the liver becomes infiltrated with small lipid droplets and loses some of its ability to detoxify ammonia.

Brain dysfunction and death are typically caused by swelling of brain cells. The pressure not only kills the cells directly but also results in hypoxia that kills them indirectly. The survival rate is about 70 percent. Swelling may result in irreversible brain damage, including mental retardation, in children who survive. Therapy is directed at controlling the swelling.

ALZHEIMER'S DISEASE (AD)

Alzheimer's (ALTZ-hī-merz) ***disease,*** or ***AD,*** is a disabling neurological disorder that afflicts about 11 percent of the U.S. population over age 65. Its causes are unknown, its effects are irreversible, and it has no cure. The disease claims more than 120,000 lives a year, making it the fourth leading cause of death among the elderly following heart disease, cancer, and stroke.

Victims of AD initially have trouble remembering recent events. Next they become more confused and forgetful, often repeating questions or getting lost while traveling to previously familiar places. Disorientation grows, memories of past events disappear, and there may be episodes of paranoia, hallucination, or violent changes in mood. As their minds continue to deteriorate, they lose their ability to read, write, talk, eat, walk, or take care of themselves. Finally, the disease culminates in dementia, the loss of reason. A person with AD usually dies of some complication that affects bedridden patients, such as pneumonia.

At present, there is no diagnostic test for AD. Autopsy findings, however, clearly show several characteristic pathologies in the brains of Alzheimer's victims. Although there is normally gradual loss of neurons in the brain associated with aging, the rate of loss in persons with AD is greater than normal, especially in regions of the brain that are important for memory and other intellectual processes, such as the cerebral cortex and hippocampus. Another pathological finding is the presence of ***neurofibrillary tangles,*** bundles of fibrous proteins, in the cell bodies of neurons in the cerebral cortex, hippocampus, and brain stem. Also present is ***amyloid,*** pathological protein-rich accumulations. In some cases, amyloid surrounds and invades cerebral blood vessels. This form is called ***cerebrovascular amyloid,*** which collects in the middle muscular layer of the blood vessels and eventually results in hemorrhage. Another form of amyloid is a component of ***neuritic (senile) plaques,*** which consist of abnormal axons and axon terminals that surround amyloid. Neuritic plaques are abundant in the cerebral cortex, hippocampus, and amygdala.

Amyloid forms a protein called A68, which is found in neurofibrillary tangles. Both neurofibrillary tangles and A68 are found in only two groups of adults: persons with

AD and persons with Down syndrome (DS). DS is a genetic defect that is the leading cause of mental retardation (see Chapter 26). A68 is emerging as a key factor in the onset of AD. In normal infants, A68 is believed to cause programmed death of surplus neurons, a process that is completed by age 2, after the A68 disappears from the brain. However, it reappears in the brains of AD patients. It is suspected that A68 may cause AD by triggering the growth of neurofibrillary tangles or mediate programmed cell death as it does in normal infants.

It has recently been learned that a gene on chromosome 21 codes for amyloid, and the extra gene, as found in DS, could lead to excessive amyloid production. Such information suggests that a common genetic defect might be responsible for both AD and DS.

One hypothesis concerning the possible cause of AD centers on the observation that the enzyme choline acetyltransferase is found in very low concentrations in the brains of AD victims. The enzyme is needed for the synthesis of the neurotransmitter acetylcholine (ACh) in axon terminals. Levels of the enzyme are quite low in the cerebral cortex and hippocampus of AD patients owing to the loss of axon terminals in these regions, and thus acetylcholine levels are also low. Decreased levels of acetylcholine are correlated with reduced transmission of nerve impulses, impulses required for memory.

Other hypotheses suggest that AD may be caused by the loss of neurons due to the inheritance of faulty genes, an abnormal accumulation of proteins in the brain, a slow-acting virus, environmental toxins such as aluminum, and decreased blood flow to the brain resulting in inadequate amounts of oxygen and glucose.

DELIRIUM

Delirium (de-LIR-ē-um; *deliria* = off the tract), also called ***acute confusional state (ACS),*** is a transient disorder of abnormal cognition (perception, thinking, and memory) and disordered attention that is accompanied by disturbances of the sleep–wake cycle and psychomotor behavior (hyperactivity or hypoactivity of movements and speech). Delirium is also accompanied by emotions that range from apathy or depression to fear or rage and signs of sympathetic nervous system arousal (tachycardia, sweating, dilated pupils, elevated blood pressure, and pallor or flushing of the face). Among the conditions that cause delirium are brain disorders (infections, tumors, trauma, epilepsy, and stroke); systemic diseases that affect the brain (infections, metabolic disturbances, and cardiovascular diseases); intoxication with medical and recreational drugs and environmental poisons; and withdrawal from substances of abuse such as alcohol and sedative-hypnotic drugs.

Treatment consists of reduction or elimination of drugs, especially anticholinergic drugs; psychiatric evaluation; maintaining proper fluid and electrolyte balance and nutrition; and sedation.

KEY MEDICAL TERMS ASSOCIATED WITH THE CENTRAL NERVOUS SYSTEM

Agnosia (ag-NŌ-zē-a; *a* = without; *gnosis* = knowledge) Inability to recognize the significance of sensory stimuli such as auditory, visual, olfactory, gustatory, and tactile.

Analgesia (an-al-JĒ-zē-a; *an* = without; *algia* = painful condition) Pain relief.

Anesthesia (an′-es-THĒ-zē-a; *esthesia* = feeling) Loss of feeling.

Apraxia (a-PRAK-sē-a; *pratto* = to do) Inability to carry out purposeful movements in the absence of paralysis.

Coma (KŌ-ma) Abnormally deep unconsciousness with an absence of voluntary response to stimuli and with varying degrees of reflex activity. It may be due to illness or to an injury.

Dementia (de-MEN-shē-a; *de* = away from; *mens* = mind) An organic mental disorder that results in permanent or progressive general loss of intellectual abilities such as impairment of memory, judgment, and abstract thinking and changes in personality. The most common cause is Alzheimer's disease. Others are cerebrovascular disease, central nervous system infection, brain tumors or trauma, pernicious anemia, external hydrocephalus, and neurological diseases (Parkinson's disease, multiple sclerosis, and Huntington's chorea).

Electroconvulsive therapy (ECT) (e-lek′-trō-con-VUL-siv THER-a-pē) A form of shock therapy in which convulsions are induced by the passage of a brief electric current through the brain. A patient undergoing ECT is properly anesthetized and given a muscle relaxant to minimize the convulsions. ECT is regarded as an important therapeutic option in the treatment of severe depression and acute mania. Side effects include acute confusional states and memory deficits.

Huntington's chorea (HUNT-ing-tunz kō-RĒ-a; *choreia* = dance) A rare hereditary disease characterized by involuntary jerky movements and mental deterioration that terminates in dementia.

Lethargy (LETH-ar-jē) A condition of functional torpor or sluggishness.

Nerve block Loss of sensation in a region, such as in local dental anesthesia, due to injection of a local anesthetic.

Neuralgia (noo-RAL-jē-a; *neur* = nerve) Attacks of pain along the entire course or branch of a peripheral sensory nerve.

Paralysis (pa-RAL-a-sis) Diminished or total loss of motor function resulting from damage to nervous tissue or a muscle.

Spastic (SPAS-tik; *spas* = draw or pull) An increase in muscle tone (stiffness) associated with an increase in tendon reflexes and abnormal reflexes (Babinski sign).

Stupor (STOO-por) Unresponsiveness from which a patient can be aroused only briefly and by vigorous and repeated stimulation.

Torpor (TOR-por) State of lethargy and sluggishness that precedes stupor, which precedes semicoma, which precedes coma.

Viral encephalitis (VĪ-ral en′-sef-a-LĪ-tis) An acute inflammation of the brain caused by a direct attack by various viruses or by an allergic reaction to any of the many viruses that are normally harmless to the central nervous system. If the virus affects the spinal cord as well, it is called **encephalomyelitis.**

STUDY OUTLINE

Brain (p. 498)

Principal Parts (p. 498)

1. During embryological development, brain vesicles are formed and serve as forerunners of various parts of the brain.
2. The diencephalon develops into the thalamus and hypothalamus, the telencephalon forms the cerebrum, the mesencephalon develops into the midbrain, the myelencephalon forms the medulla, and the metencephalon develops into the pons and cerebellum.
3. The principal parts of the brain are the brain stem, diencephalon, cerebrum, and cerebellum.

Protection and Coverings (p. 498)

1. The brain is protected by cranial bones, meninges, and cerebrospinal fluid.
2. The cranial meninges are continuous with the spinal meninges and are named dura mater, arachnoid, and pia mater.

Cerebrospinal Fluid (CSF) (p. 499)

1. Cerebrospinal fluid is formed in the choroid plexuses and circulates through the subarachnoid space, ventricles, and central canal. Most of the fluid is absorbed by the arachnoid villi of the superior sagittal sinus.
2. Cerebrospinal fluid protects by serving as a shock absorber. It also delivers nutritive substances from the blood and removes wastes.
3. If cerebrospinal fluid accumulates in the ventricles, it is called internal hydrocephalus. If it accumulates in the subarachnoid space, it is called external hydrocephalus.

Blood Supply (p. 502)

1. The blood supply to the brain is via the cerebral arterial circle (circle of Willis).
2. Any interruption of the oxygen supply to the brain can result in weakening, permanent damage, or death of brain cells. Interruption of the mother's blood supply to a child during childbirth before it can breathe may result in paralysis, mental retardation, epilepsy, or death.
3. Glucose deficiency may produce dizziness, convulsions, and unconsciousness.
4. The blood–brain barrier (BBB) is a concept that explains the differential rates of passage of certain material from the blood into the brain.

Brain Stem (p. 503)

1. The medulla oblongata is continuous with the upper part of the spinal cord and contains portions of both motor and sensory tracts. It contains nuclei that are reflex centers for regulation of heart rate, respiratory rate, vasoconstriction, swallowing, coughing, vomiting, sneezing, and hiccuping. It also contains the nuclei of origin for cranial nerves VIII (cochlear and vestibular branches) through XII.
2. The pons is superior to the medulla. It connects the spinal cord with the brain and links parts of the brain with one another by way of tracts. It relays nerve impulses related to voluntary skeletal movements from the cerebral cortex to the cerebellum. It contains the nuclei for cranial nerves V through VII and the vestibular branch of VIII. The reticular formation of the pons contains the pneumotaxic and apneustic centers, which help control respiration.
3. The midbrain connects the pons and diencephalon. It conveys motor impulses from the cerebrum to the cerebellum and spinal cord, sends sensory impulses from cord to thalamus, and regulates auditory and visual reflexes. It also contains the nuclei of origin for cranial nerves III and IV.

Diencephalon (p. 509)

1. The diencephalon consists primarily of the thalamus and hypothalamus.
2. The thalamus is superior to the midbrain and contains nuclei that serve as relay stations for all sensory impulses, except smell, to the cerebral cortex. It also registers conscious recognition of pain and temperature and some awareness of light touch and pressure.
3. The hypothalamus is inferior to the thalamus. It controls and integrates the autonomic nervous system, receives sensory impulses from viscera, connects the nervous and endocrine systems, secretes a variety of regulating hormones (or factors), coordinates mind-over-body phenomena, functions in rage and aggression, controls body temperature, regulates food and fluid intake, maintains the waking state and sleep patterns, and acts as a self-sustained oscillator that drives biological rhythms.

Cerebrum (p. 513)

1. The cerebrum is the largest part of the brain. Its cortex contains convolutions, fissures, and sulci.
2. The cerebral lobes are named the frontal, parietal, temporal, and occipital.
3. The white matter is under the cortex and consists of myelinated axons running in three principal directions.
4. The basal ganglia (cerebral nuclei) are paired masses of gray matter in the cerebral hemispheres. They help to control muscular movements.
5. The limbic system is found in the cerebral hemispheres and diencephalon. It functions in emotional aspects of behavior and memory.
6. The sensory areas of the cerebral cortex are concerned with the interpretation of sensory impulses. The motor areas are the regions that govern muscular movement. The association areas are concerned with emotional and intellectual processes.
7. Positron emission tomography (PET) helps identify which parts of the brain are involved in specific sensory and motor activities.
8. Brain waves generated by the cerebral cortex are recorded as an electroencephalogram (EEG). It may be used to diagnose epilepsy, infections, and tumors.

Brain Lateralization (Split-Brain Concept) (p. 521)

1. Recent research indicates that the two hemispheres of the brain are not bilaterally symmetrical, either anatomically or functionally.
2. The left hemisphere is more important for right-handed control, spoken and written language, numerical and scientific skills, and reasoning.

___ **f.** feelings of hunger, fullness, and thirst stimulate centers here so that you can respond accordingly
___ **g.** all sensations except smell are relayed through here
___ **h.** regulation of heart, blood pressure, and respiration occurs by centers located here
___ **i.** it constitutes four-fifths of the diencephalon
___ **j.** it lies under the third ventricle, forming its floor
___ **k.** it forms most of side walls of the third ventricle
___ **l.** tumor in this region would compress cerebral aqueduct and cause internal hydrocephalus
___ **m.** it releases chemicals called regulating hormones (or factors) that control hormones

Choose the one best answer to these questions.

___ **13.** The meninx that adheres to the surface of the brain and spinal cord and contains blood vessels is the
A. arachnoid; B. dura mater; C. pia mater; D. mia mater; E. perineurium.

___ **14.** The reason that the motor areas of the right cerebral cortex control voluntary movements on the left side of the body is because
A. the cerebrum contains projection fibers; B. the cerebellum controls voluntary movements; C. the medulla contains decussating pyramids; D. the pons connects the spinal cord with the brain; E. the midbrain reroutes all motor impulses.

___ **15.** A patient exhibits the following signs: irregular fluctuations in body temperature, loss of appetite, lack of sensation of thirst, and psychosomatic disorders. What portion of the brain may be malfunctioning?
A. cerebrum; B. midbrain; C. hypothalamus; D. medulla; E. pons.

___ **16.** The vital reflex centers for the control of heart beat, respiration, and blood vessel diameter are located in the
A. pons; B. medulla; C. cerebrum; D. cerebellum; E. midbrain.

___ **17.** An obstruction in the interventricular foramen would interfere with the flow of cerebrospinal fluid into the
A. lateral ventricles; B. third ventricle; C. fourth ventricle; D. subarachnoid space of the spinal cord; E. subdural space of the brain.

___ **18.** The corpus callosum is a bridge of white fibers that connects the
A. midbrain and medulla; B. cerebral hemispheres; C. cerebrum and cerebellum; D. pons and cerebellum; E. cerebellar hemispheres.

___ **19.** The primary motor area is located in the
A. precentral gyrus; B. basal ganglia; C. corpus callosum; D. hypothalamus; E. postcentral gyrus.

___ **20.** Which of the following would you expect to observe in a patient with a tumor of the cerebellum?
(1) loss of general sensation;
(2) inability to execute any voluntary movements;
(3) inability to execute smooth, steady movement.
A. (1) only; B. (2) only; C. (3) only; D. none of the above; E. all of the above.

___ **21.** Which of the following indicates the correct order in which cranial nerves originate from the base of the brain?
A. optic, olfactory, trigeminal, trochlear, ophthalmic; B. olfactory, ophthalmic, oculomotor, trochlear, trigeminal; C. olfactory, optic, trochlear, trigeminal, facial; D. olfactory, optic, oculomotor, trochlear, trigeminal; E. oculomotor, optic, trigeminal, trochlear, facial.

___ **22.** Damage to the occipital lobe of the cerebrum would most likely cause
A. loss of hearing; B. loss of vision; C. loss of ability to smell; D. paralysis; E. loss of muscle sense (proprioception).

23. Arrange the answers in correct sequence.

___ ___ ___ **a.** From anterior to posterior:
A. fourth ventricle
B. pons and medulla
C. cerebellum

___ ___ ___ **b.** From superior to inferior:
A. thalamus
B. hypothalamus
C. corpus callosum

Circle T (true) or F (false) for the following.

T F 24. The thalamus, hypothalamus, and cerebrum are all developed from the forebrain.
T F 25. The language areas are located in the cerebellar cortex.
T F 26. The limbic system functions in the control of behavior.
T F 27. The reticular formation controls arousal and consciousness.
T F 28. The fourth ventricle communicates with the subarachnoid space of the brain and spinal cord through the cerebral aqueduct.
T F 29. The dural extension between cerebral hemispheres is known as the falx cerebri.
T F 30. The cerebral nuclei that control large subconscious movements, such as swinging the arms while walking, are the caudate and putamen.

The Autonomic Nervous System

19

STUDENT OBJECTIVES

1. Compare the structural and functional differences between the somatic efferent and autonomic portions of the nervous system.
2. Identify the principal structural features of the autonomic nervous sytem.
3. Compare the sympathetic and parasympathetic divisions of the autonomic nervous system in terms of structure, physiology, and neurotransmitters released.
4. Describe the various postsynaptic receptors involved in autonomic responses.
5. Explain the role of the hypothalamus and its relation to the sympathetic and parasympathetic divisions.
6. Explain the relation between biofeedback and meditation and the autonomic nervous system.

CHAPTER OUTLINE

■ **Somatic Efferent and Autonomic Nervous Systems**
■ **Structure of the Autonomic Nervous System**
Visceral Efferent Pathways
Preganglionic Neurons
Autonomic Ganglia
Postganglionic Neurons
Sympathetic Division
Parasympathetic Division
■ **Physiology of the Autonomic Nervous System**
Neurotransmitters
Receptors
Activities
■ **Visceral Autonomic Reflexes**
■ **Control by Higher Centers**
Biofeedback
Meditation

The portion of the nervous system that regulates the activities of smooth muscle, cardiac muscle, and certain glands is the ***autonomic nervous system (ANS).*** Structurally, the system consists of visceral efferent neurons organized into nerves, ganglia, and plexuses. (There are also visceral receptors and afferent neurons involved in the ANS.) Functionally, it usually operates without conscious control. The system was originally named *autonomic* because physiologists thought it functioned with no control from the central nervous system, that it was autonomous or self-governing. It is now known that the ANS is neither structurally nor functionally independent of the central nervous system (CNS). It is regulated by centers in the brain, in particular by the cerebral cortex, hypothalamus, and medulla oblongata. However, the old terminology has been retained, and since the ANS does differ from the somatic nervous system in some ways, the two are separated for convenience of study.

SOMATIC EFFERENT AND AUTONOMIC NERVOUS SYSTEMS

Whereas the somatic efferent nervous system produces conscious movement in skeletal muscles, the autonomic nervous system (visceral efferent nervous system) regulates visceral activities, and it generally does so involuntarily and automatically. Examples of visceral activities regulated by the ANS are changes in the size of the pupil, accommodation for near vision, dilation of blood vessels, adjustment of the rate and force of the heartbeat, movements of the gastrointestinal tract, and secretion by most glands. These activities usually lie beyond conscious control. They are automatic.

The ANS is generally considered to be entirely motor. All its axons are efferent fibers, which transmit nerve impulses from the CNS to visceral effectors. Autonomic fibers are called ***visceral efferent fibers. Visceral effectors*** include cardiac muscle, smooth muscle, and glandular epithelium. This does not mean that there are no afferent (sensory) impulses from visceral effectors, however. Impulses that give rise to visceral sensations pass over visceral afferent neurons that have cell bodies located in the posterior (dorsal) root ganglia of spinal nerves and ganglia of some cranial nerves. Some functions of these afferent neurons were described in discussing the cranial and spinal nerves. The hypothalamus, which largely controls the autonomic nervous system, also receives impulses along multisynaptic pathways from the visceral sensory fibers, as well as from some somatic sensory fibers.

In the motor portion of the neural pathway of the somatic efferent system, the axon of an efferent neuron extends from the CNS and synapses directly on skeletal muscle fibers. In the neural pathway of the ANS are two types of efferent neurons and a ganglion between them. The axon of the first neuron extends from the central nervous system to a ganglion, where it synapses with the cell body of the second efferent neuron. It is the axon of this second neuron that ultimately synapses on a visceral effector. Also, whereas fibers of somatic efferent neurons release acetylcholine (ACh) as their neurotransmitter, fibers of autonomic efferent neurons release either ACh or norepinephrine (NE).

The ANS consists of two principal divisions: the ***sympathetic*** and the ***parasympathetic.*** Many organs innervated by the ANS receive visceral efferent neurons from both components of it—one set from the sympathetic division and another from the parasympathetic division. In general, impulses transmitted by the fibers of one division stimulate the organ to start or increase activity, whereas impulses from the other division decrease the organ's activity. Organs that receive impulses from both sympathetic and parasympathetic fibers are said to have ***dual innervation.*** Thus, autonomic innervation may be excitatory or inhibitory. In the somatic efferent nervous system, only one kind of motor neuron innervates a skeletal muscle and, moreover, innervation is always excitatory. When a somatic neuron stimulates a skeletal muscle, the muscle becomes active. When the neuron ceases to stimulate the muscle, contraction stops altogether.

A summary of the principal differences between the somatic efferent and autonomic nervous system is presented in Exhibit 19-1.

STRUCTURE OF THE AUTONOMIC NERVOUS SYSTEM

VISCERAL EFFERENT PATHWAYS

Autonomic visceral efferent pathways almost always consist of two efferent (motor) neurons. One extends from the CNS to a ganglion. The other extends directly from the ganglion to the effector (muscle or gland).

The first of the visceral efferent neurons in an autonomic pathway is called a ***preganglionic neuron*** (Figure 19-1). Its cell body is in the brain or spinal cord. Its myelinated axon, called a ***preganglionic fiber,*** passes out of the central nervous system as part of a cranial or spinal nerve. At some point, the fiber separates from the nerve and travels to an autonomic ganglion, where it synapses with the dendrites or cell body of the postganglionic neuron, the second neuron in the visceral efferent pathway.

The ***postganglionic neuron*** lies entirely outside the CNS. Its cell body and dendrites (if it has dendrites) are located in the autonomic ganglion, where the synapse with one or more preganglionic fibers occurs. The axon of a postganglionic neuron, called a ***postganglionic fiber,*** is unmyelinated and terminates in a visceral effector.

Thus, preganglionic neurons convey efferent impulses from the central nervous system to autonomic ganglia. Postganglionic neurons relay the impulses from autonomic ganglia to visceral effectors along a ***two motor neuron pathway.***

EXHIBIT 19-1

Comparison of Somatic Efferent and Autonomic Nervous Systems

	SOMATIC EFFERENT	AUTONOMIC
Effectors	Skeletal muscles.	Cardiac muscle, smooth muscle, glandular epithelium.
Type of Control	Voluntary.	Involuntary.
Neural Pathway	The axon of one efferent neuron extends from CNS and synapses directly on skeletal muscle fibers.	The axon of one efferent neuron extends from the CNS and synapses with another efferent neuron in a ganglion, the axon of the second neuron synapses on a visceral effector.
Action on Effector	Always excitatory.	May be excitatory or inhibitory, depending on whether stimulation is sympathetic or parasympathetic and the effector innervated.
Neurotransmitters	Acetylcholine (ACh).	Acetylcholine (ACh) or norepinephrine (NE).

Preganglionic Neurons

In the sympathetic division the preganglionic neurons have their cell bodies in the lateral gray horns of the 12 thoracic segments and first 2 or 3 lumbar segments of the spinal cord (Figure 19-2). It is for this reason that the sympathetic division is also called the ***thoracolumbar*** (thō′-ra-kō-LUM-bar) ***division,*** and the fibers of the sympathetic preganglionic neurons are known as the ***thoracolumbar outflow.***

The cell bodies of the preganglionic neurons of the parasympathetic division are located in the nuclei of cranial nerves III, VII, IX, and X in the brain stem and in the lateral gray horns of the second through fourth sacral segments of the spinal cord. Hence, the parasympathetic division is also known as the ***craniosacral division,*** and the fibers of the parasympathetic preganglionic neurons are referred to as the ***craniosacral outflow.***

Autonomic Ganglia

Autonomic pathways almost always include ***autonomic ganglia,*** where synapses between preganglionic and postganglionic visceral efferent neurons occur. Autonomic ganglia differ from posterior root ganglia. The latter contain cell bodies of sensory neurons; no synapses occur in them.

The autonomic ganglia may be divided into three general groups. The ***sympathetic trunk (vertebral chain) ganglia*** are a series of ganglia that lie in a vertical row on either side of the vertebral column, extending from the base of the skull to the coccyx (Figure 19-3). They are also known as ***paravertebral (lateral) ganglia.*** They receive preganglionic fibers only from the sympathetic division (see Figure 19-2). Because of this, sympathetic preganglionic fibers tend to be short.

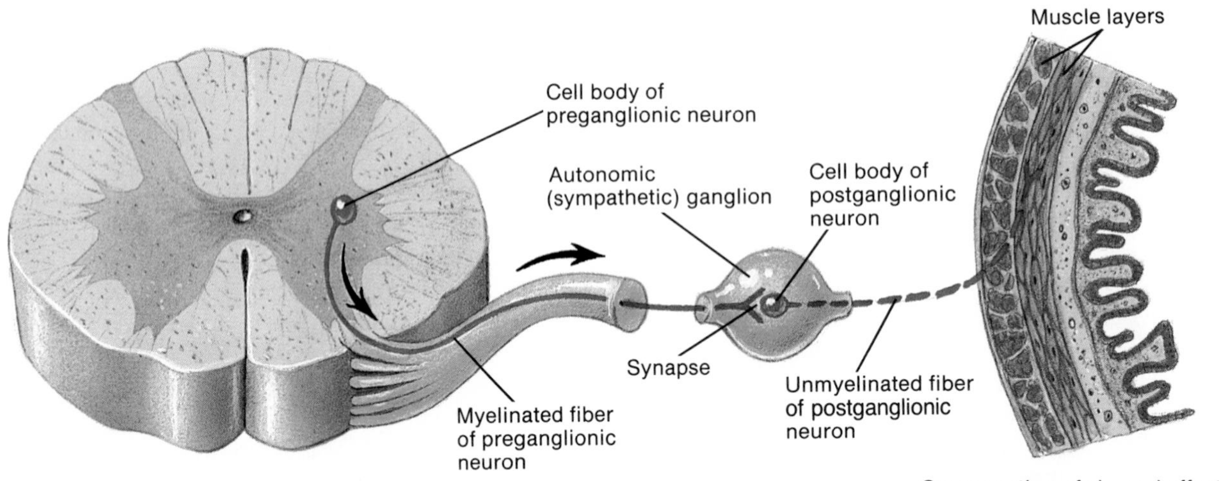

FIGURE 19-1 Relationship between preganglionic and postganglionic (sympathetic) neurons.

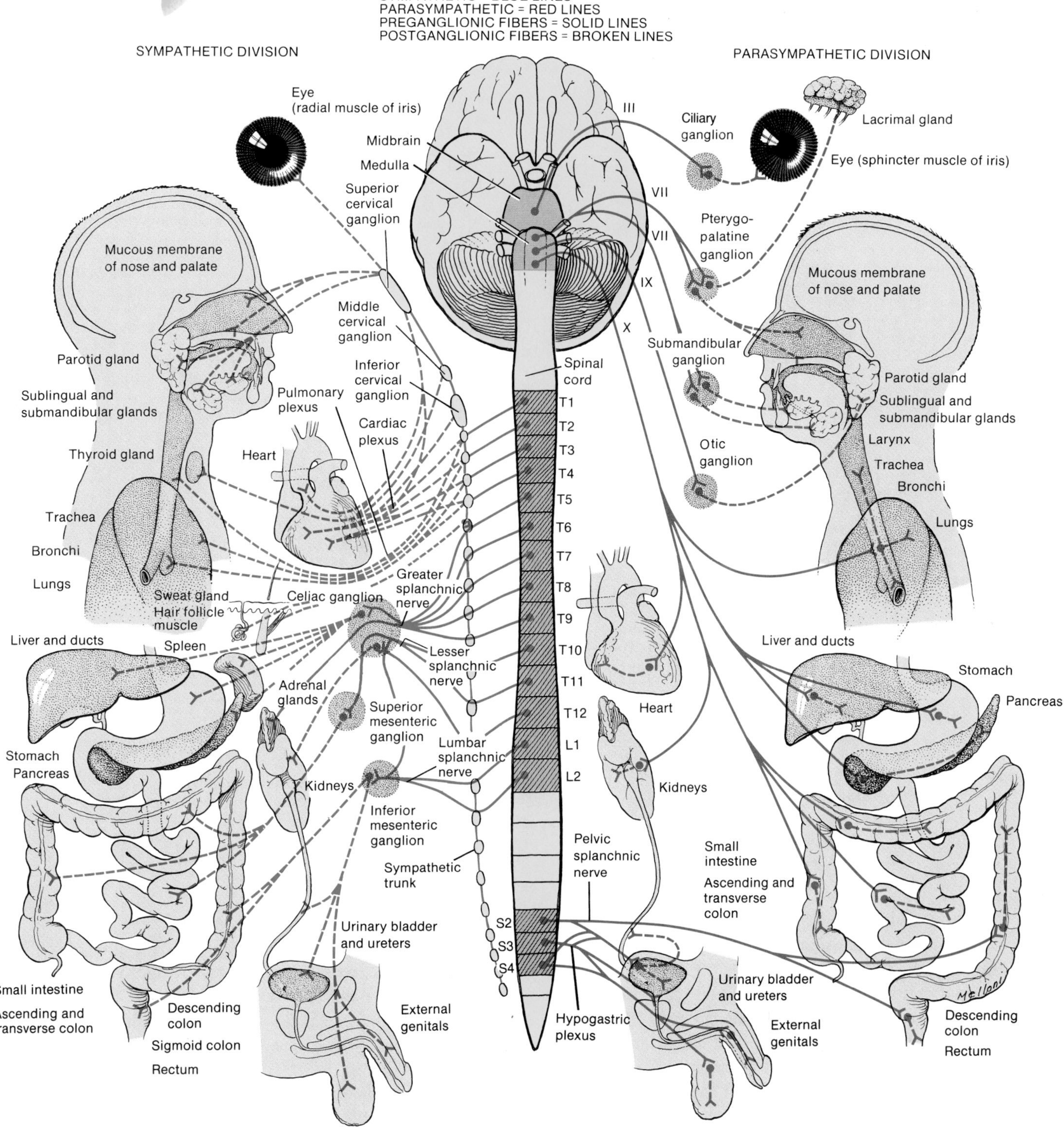

FIGURE 19-2 Structure of the autonomic nervous system. Although the parasympathetic division is shown only on the right side of the figure and the sympathetic division is shown only on the left side, keep in mind that each division is actually on both sides of the body (bilateral symmetry).

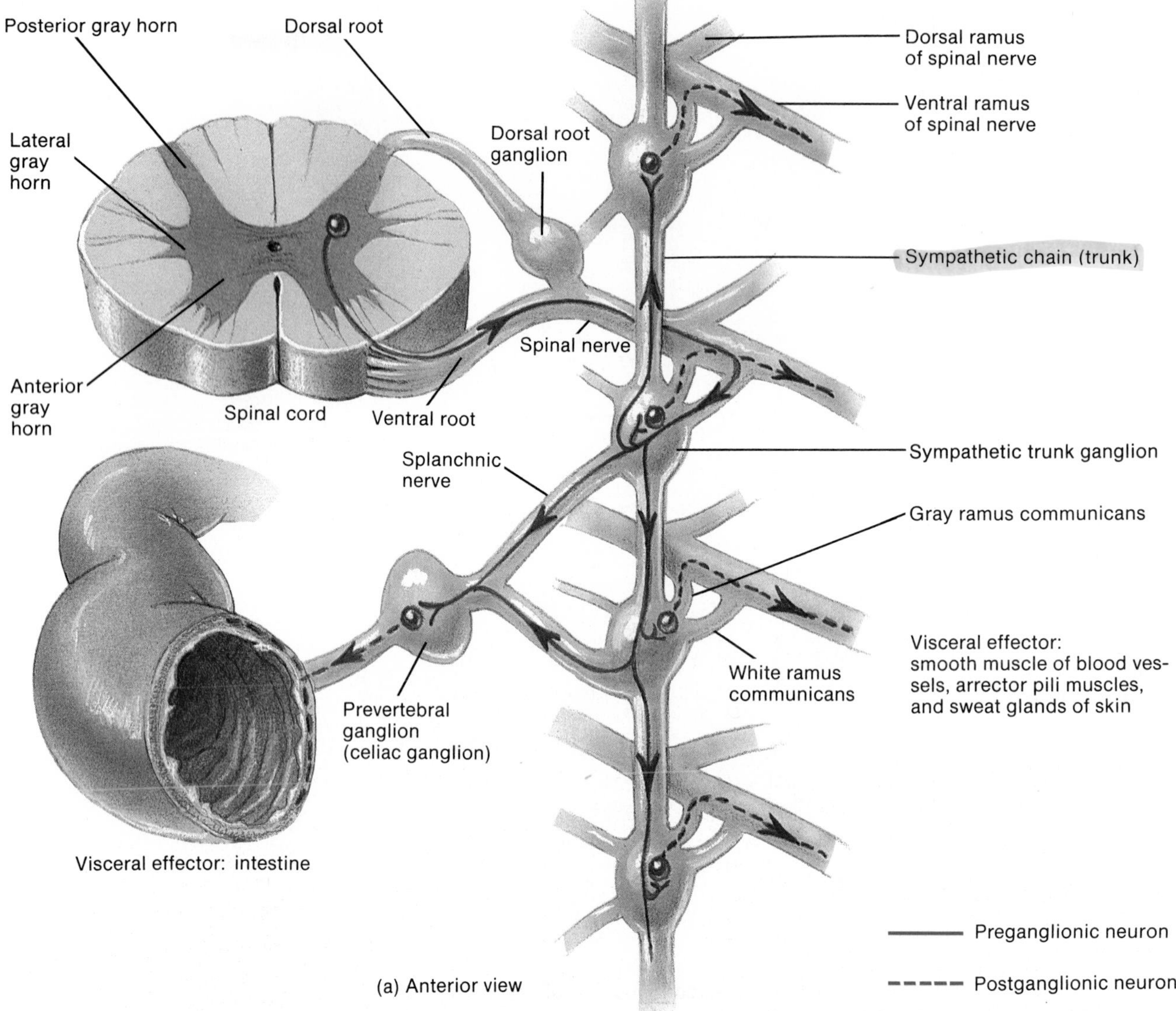

FIGURE 19-3 Ganglia and rami communicantes of the sympathetic division of the autonomic nervous system. (b) Courtesy of N. Gluhbegovic and T. H. Williams, *The Human Brain: A Photographic Guide,* Harper & Row, Publishers, Inc., Hagerstown, MD, 1980.

The second kind of autonomic ganglion also belongs to the sympathetic division. It is called a ***prevertebral (collateral) ganglion*** (Figure 19-3). The ganglia of this group lie anterior to the spinal column and close to the large abdominal arteries from which their names are derived. Examples of prevertebral ganglia so named are the celiac ganglion, on either side of the celiac artery just below the diaphragm; the superior mesenteric ganglion, near the beginning of the superior mesenteric artery in the upper abdomen; and the inferior mesenteric ganglion, located near the beginning of the inferior mesenteric artery in the middle of the abdomen (see Figure 19-2). Prevertebral ganglia receive preganglionic fibers from the sympathetic division.

The third kind of autonomic ganglion belongs to the parasympathetic division and is called a ***terminal (intramural) ganglion.*** The ganglia of this group are located at the end of a visceral efferent pathway very close to visceral effectors or actually within the walls of visceral effectors. Terminal ganglia receive preganglionic fibers from the parasympathetic division. Since preganglionic fibers travel from the CNS to terminal ganglia close to or within the organ innervated, parasympathetic preganglionic fibers tend to be long (see Figure 19-2).

In addition to autonomic ganglia, the ANS contains ***autonomic plexuses.*** Slender nerve fibers from ganglia containing postganglionic nerve cell bodies arranged in a branching network constitute an autonomic plexus.

Postganglionic Neurons

Axons from preganglionic neurons of the sympathetic division pass to ganglia of the sympathetic trunk (see Figure

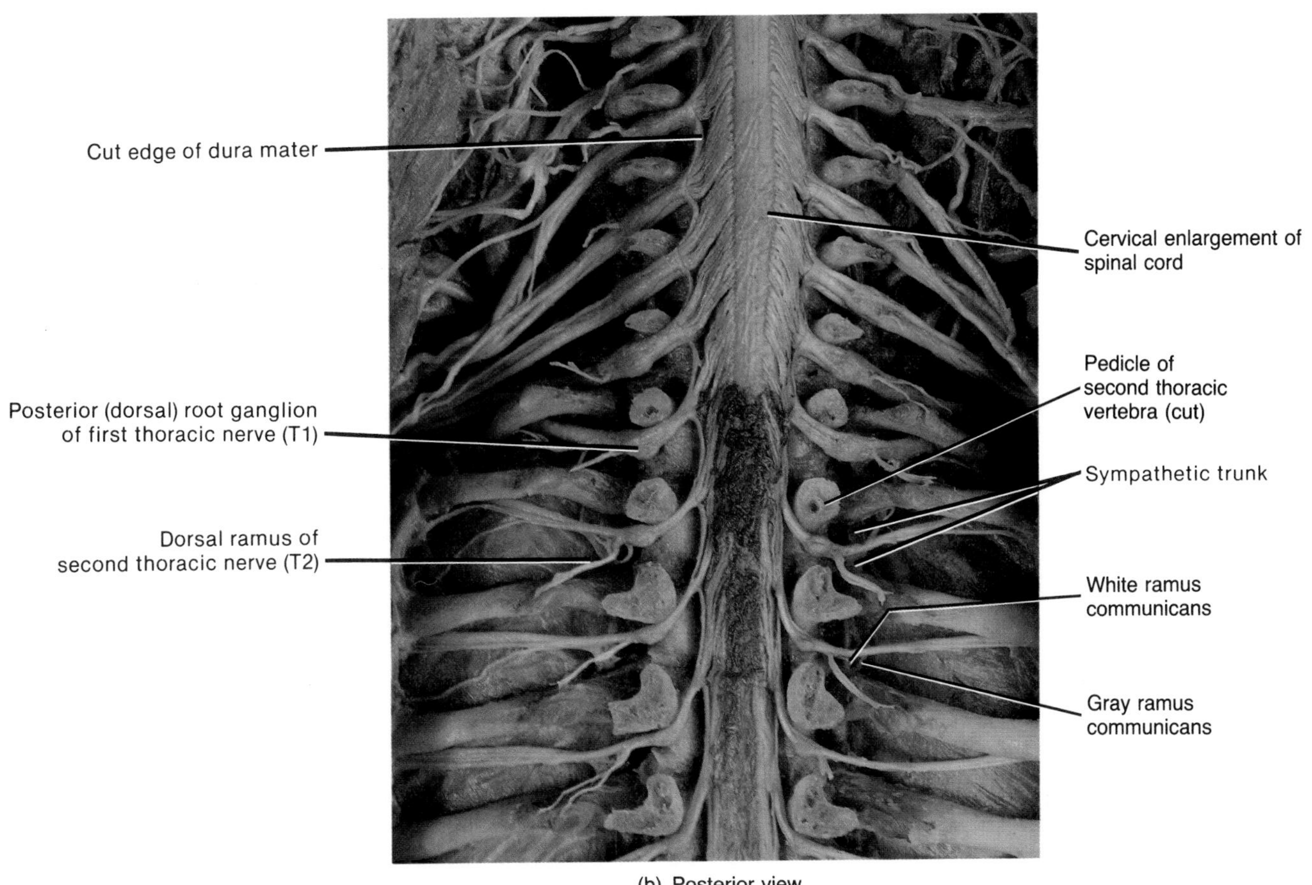

(b) Posterior view

19-1). They can either synapse in the sympathetic chain ganglia with postganglionic neurons, or they can continue, without synapsing, through the chain ganglia to end at a prevertebral ganglion where synapses with the postganglionic neurons can take place. Each sympathetic preganglionic fiber synapses with 20 or more postganglionic fibers in the ganglion, and the postganglionic fibers pass to several visceral effectors. After exiting their ganglia, the postganglionic fibers innervate their visceral effectors.

Axons from preganglionic neurons of the parasympathetic division pass to terminal ganglia near or within a visceral effector. In the ganglion, the presynaptic neuron usually synapses with only four or five postsynaptic neurons to a single visceral effector. After exiting their ganglia, the postsynaptic fibers supply their visceral effectors.

With this background in mind, we can now examine some specific structural features of the sympathetic and parasympathetic divisions of the autonomic nervous system.

SYMPATHETIC DIVISION

The preganglionic fibers of the sympathetic division have their cell bodies located in the lateral gray horns of all the thoracic segments and first two or three lumbar segments of the spinal cord (see Figure 19-2). The preganglionic fibers are myelinated and leave the spinal cord through the ventral root of a spinal nerve along with the somatic efferent fibers at the same segmental levels. After exiting through the intervertebral foramina, the preganglionic sympathetic fibers enter a white ramus to pass to the nearest sympathetic trunk ganglion on the same side. Collectively, the white rami are called the ***white rami communicantes*** (kō-myoo-ni-KAN-tēz). Their name indicates that they contain myelinated fibers. Only thoracic and upper lumbar nerves have white rami communicantes. The white rami communicantes connect the ventral ramus of the spinal nerve with the ganglia of the sympathetic trunk.

The paired sympathetic trunks are situated anterolaterally to the spinal cord, one on either side. Each consists of a series of ganglia arranged more or less segmentally. The divisions of the sympathetic trunk are named on the basis of location. Typically, there are 22 ganglia in each chain: 3 cervical, 11 thoracic, 4 lumbar, and 4 sacral. Although the trunk extends downward from the neck, thorax, and abdomen to the coccyx, it receives preganglionic fibers only from the thoracic and lumbar segments of the spinal cord (see Figure 19-2).

The cervical portion of each sympathetic trunk is located in the neck anterior to the prevertebral muscles. It is subdivided into a superior, middle, and inferior ganglion (see Figure 19-2). The ***superior cervical ganglion*** is posterior to the internal carotid artery and anterior to the transverse processes of the second cervical vertebra. Postganglionic fibers leaving the ganglion serve the head, where they are distributed to the sweat glands, the smooth muscle of the eye and blood vessels of the face, the nasal mucosa, and the submandibular, sublingual, and parotid salivary glands. Gray rami communicantes (described shortly) from the ganglion also pass to the upper two to four cervical spinal nerves. The ***middle cervical ganglion*** is situated near the sixth cervical vertebra at the level of the cricoid cartilage. Postganglionic fibers from it innervate the heart. The ***inferior cervical ganglion*** is located near the first rib, anterior to the transverse processes of the seventh cervical vertebra. Its postganglionic fibers also supply the heart.

The thoracic portion of each sympathetic trunk usually consists of 11 segmentally arranged ganglia, lying ventral to the necks of the corresponding ribs. This portion of the sympathetic trunk receives most of the sympathetic preganglionic fibers. Postganglionic fibers from the thoracic sympathetic trunk innervate the heart, lungs, bronchi, and other thoracic viscera.

The lumbar portion of each sympathetic trunk is found on either side of the corresponding lumbar vertebrae. The sacral portion of the sympathetic trunk lies in the pelvic cavity on the medial side of the sacral foramina. Postganglionic fibers from the lumbar and sacral sympathetic chain ganglia are distributed with the respective spinal nerves via gray rami, or they may join the hypogastric plexus via direct visceral branches.

When a preganglionic fiber of a white ramus communicans enters the sympathetic trunk, it may terminate (synapse) in several ways. Some fibers synapse in the first ganglion at the level of entry. Others pass up or down the sympathetic trunk for a variable distance to form the fibers on which the ganglia are strung. These fibers, known as ***sympathetic chains*** (see Figure 19-3), may not synapse until they reach a ganglion in the cervical or sacral area. Most rejoin the spinal nerves before supplying peripheral visceral effectors such as sweat glands and the smooth muscle in blood vessels and around hair follicles in the extremities. The ***gray ramus communicans*** (kō-MYOO-ni-kanz) is the structure containing the postganglionic fibers that connect the ganglion of the sympathetic trunk to the spinal nerve (see Figure 19-3). The fibers are unmyelinated. All spinal nerves have gray rami communicantes. Gray rami communicantes outnumber the white rami, since there is a gray ramus leading to each of the 31 pairs of spinal nerves.

In most cases, a sympathetic preganglionic fiber terminates by synapsing with a large number, usually 20 or more, of postganglionic cell bodies in a ganglion. Often the postganglionic fibers then terminate in widely separated organs of the body. Thus, an impulse that starts in a single preganglionic neuron may reach several visceral effectors. For this reason, most sympathetic responses have widespread effects on the body.

Some preganglionic fibers pass through the sympathetic trunk without terminating in the trunk. Beyond the trunk, they form nerves known as ***splanchnic*** (SPLANK-nik) ***nerves*** (see Figure 19-2). After passing through the trunk of ganglia, the splanchnic nerves from the thoracic area terminate in the ***celiac*** (SĒ-lē-ak) ***ganglion*** or ***solar plexus.*** In the plexus, the preganglionic fibers synapse in ganglia with postganglionic cell bodies. These ganglia are prevertebral ganglia. The greater splanchnic nerve passes to the celiac ganglion of the celiac plexus. From here, postganglionic fibers are distributed to the stomach, spleen, liver, kidney, and small intestine. The lesser splanchnic nerve passes through the celiac plexus to the superior mesenteric ganglion of the superior mesenteric plexus. Postganglionic fibers from this ganglion innervate the small intestine and colon. The lowest splanchnic nerve, not always present, enters the renal plexus. Postganglionics supply the renal arterioles and ureter. The lumbar splanchnic nerve enters the inferior mesenteric plexus. In the plexus, the preganglionic fibers synapse with postganglionic fibers in the inferior mesenteric ganglion. These fibers pass through the hypogastric plexus and supply the distal colon and rectum, urinary bladder, and genital organs. As noted earlier, the postganglionic fibers leaving the prevertebral ganglia follow the course of various arteries to abdominal and pelvic visceral effectors.

Sympathetic preganglionic fibers that innvervate the medulla of each of the adrenal glands travel in splanchnic nerves but do not synapse before terminating in the gland. This means that there are no sympathetic postganglionic fibers to the adrenal medulla. This is the only exception to the usual pattern of two efferent neurons in a chain in an autonomic efferent pathway.

PARASYMPATHETIC DIVISION

The preganglionic cell bodies of the parasympathetic division are found in nuclei in the brain stem and the lateral gray horn of the second through fourth sacral segments of the spinal cord (see Figure 19-2). Their fibers emerge as part of a cranial nerve or as part of the ventral root of a spinal nerve. The ***cranial parasympathetic outflow*** consists of preganglionic fibers that leave the brain stem by way of the oculomotor (III) nerves, facial (VII) nerves, glossopharyngeal (IX) nerves, and vagus (X) nerves. The ***sacral parasympathetic outflow*** consists of preganglionic fibers that leave the ventral roots of the second through fourth sacral nerves. The preganglionic fibers of both the cranial and sacral outflows end in terminal ganglia, where they synapse with postganglionic neurons. We will first look at the cranial outflow.

The cranial outflow has five components: four pairs of ganglia and the plexuses associated with the vagus (X) nerve. The four pairs of cranial parasympathetic ganglia innervate structures in the head and are located close to the organs they innervate. The ***ciliary ganglion*** is near the back of an orbit lateral to each optic (II) nerve. Preganglionic fibers pass with the oculomotor (III) nerve to the ciliary ganglion. Postganglionic fibers from the ganglion innervate smooth muscle cells in the eyeball. Each ***pterygopalatine*** (ter′-i-gō-PAL-a-tin) ***ganglion*** is situated lateral to a sphenopalatine foramen. It receives preganglionic fibers from the facial (VII) nerve and transmits postganglionic fibers to the nasal mucosa, palate, pharynx, and lacrimal gland. Each ***submandibular ganglion*** is found near the duct of a submandibular salivary gland. It receives preganglionic fibers from the facial (VII) nerve and transmits postganglionic fibers that innervate the submandibular and sublingual salivary glands. The ***otic ganglia*** are situated just below each foramen ovale. The otic ganglion receives preganglionic fibers from the glossopharyngeal (IX) nerve and transmits postganglionic fibers that innervate the parotid salivary gland. Ganglia associated with the cranial outflow are classified as terminal ganglia. Since the terminal ganglia are close to their visceral effectors, postganglionic parasympathetic fibers are short. Postganglionic sympathetic fibers, by contrast, are relatively long.

The last component of the cranial outflow consists of the preganglionic fibers that leave the brain via the vagus (X) nerves. This component has the most extensive distribution of the parasympathetic fibers, providing about 80 percent of the craniosacral outflow. Each vagus (X) nerve enters into the formation of several plexuses in the thorax and abdomen. As it passes through the thorax, it sends fibers to the ***superficial cardiac plexus*** in the arch of the aorta and the ***deep cardiac plexus*** anterior to the branching of the trachea. These plexuses contain terminal ganglia, and the postganglionic parasympathetic fibers emerging from them supply the heart. Also in the thorax is the ***pulmonary plexus,*** anterior and posterior to roots of the lungs and within the lungs themselves. It receives preganglionic fibers from the vagus and transmits postganglionic parasympathetic fibers to the lungs and bronchi. Other plexuses associated with the vagus (X) nerve will be described in later chapters in conjunction with the appropriate thoracic, abdominal, and pelvic viscera. Postganglionic fibers from these plexuses innervate viscera such as the liver, pancreas, stomach, kidneys, small intestine, and part of the colon.

The sacral parasympathetic outflow consists of preganglionic fibers from the ventral roots of the second through fourth sacral nerves. Collectively, they form the ***pelvic splanchnic nerve.*** These nerves synapse with parasympathetic postganglionic neurons located in terminal ganglia in the walls of the innervated viscera. From the ganglia, parasympathetic postganglionic fibers innervate the colon, ureters, urinary bladder, and reproductive organs.

The salient structural features of the sympathetic and parasympathetic divisions are compared in Exhibit 19-2.

EXHIBIT 19-2

Structural Features of Sympathetic and Parasympathetic Divisions

SYMPATHETIC	PARASYMPATHETIC
Forms thoracolumbar outflow.	Forms craniosacral outflow.
Contains sympathetic trunk and prevertebral ganglia.	Contains terminal ganglia.
Ganglia are close to the CNS and distant from visceral effectors.	Ganglia are near or within visceral effectors.
Each preganglionic fiber synapses with many postganglionic neurons that pass to many visceral effectors.	Each preganglionic fiber usually synapses with four or five postganglionic neurons that pass to a single visceral effector.
Distributed throughout the body, including the skin.	Distribution limited primarily to head and viscera of thorax, abdomen, and pelvis.

PHYSIOLOGY OF THE AUTONOMIC NERVOUS SYSTEM

NEUROTRANSMITTERS

Autonomic fibers, like other axons of the nervous system, release neurotransmitters at synapses as well as at points of contact with visceral effectors (smooth and cardiac muscle and glands). These latter points are called ***neuroeffector junctions.*** Neuroeffector junctions may be either neuromuscular or neuroglandular junctions. On the basis of the neurotransmitter produced, autonomic fibers may be classified as either cholinergic or adrenergic (Figure 19-4).

Cholinergic (kō′-lin-ER-jik) ***fibers*** release ***acetylcholine (ACh)*** and include the following: (1) all sympathetic and parasympathetic preganglionic axons, (2) all parasympathetic postganglionic axons, and (3) a few sympathetic postganglionic axons. The cholinergic sympathetic postganglionic axons include those to sweat glands and a few blood vessels in skeletal muscles. Since acetylcholine is quickly inactivated by the enzyme ***acetylcholinesterase (AChE),*** the effects of cholinergic fibers are short-lived and local.

Adrenergic (ad′-ren-ER-jik) ***fibers*** release ***norepinephrine (NE).*** Most sympathetic postganglionic axons are adrenergic. Since norepinephrine is inactivated much more slowly by ***catechol-O-methyltransferase (COMT)*** or ***monoamine oxidase (MAO)*** than acetylcholine is inactivated by acetylcholinesterase, and since norepinephrine may enter

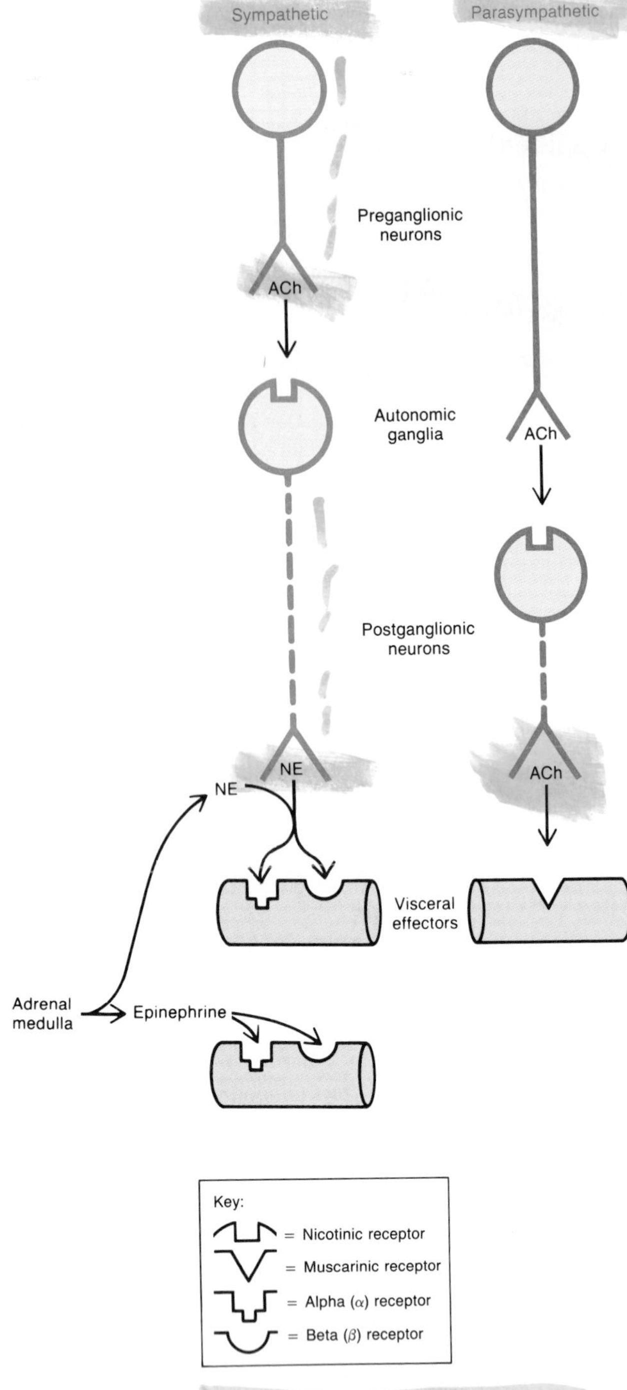

FIGURE 19-4 Neurotransmitters and receptors associated with the autonomic nervous system.

the bloodstream, the effects of sympathetic stimulation are longer lasting and more widespread than parasympathetic stimulation. The action of NE produced by sympathetic postganglionic axons is augmented by both NE and epinephrine secreted by the adrenal medulla of each of the two adrenal glands. Since both substances are secreted into the blood, their effects are more sustained than NE released by axons. NE and epinephrine released by the adrenal medullae are ultimately destroyed by enzymes in the liver once they have acted on their visceral effectors.

RECEPTORS

ACh is synthesized and stored in an inactive form in synaptic vesicles in the axon terminals of the cholinergic fibers. The release of ACh from the axon terminals requires calcium ions (Ca^{2+}) from the interstitial fluid and comes in bursts of thousands of molecules. ACh then diffuses the short distance across the synaptic cleft to bind with and activate receptors on the postsynaptic membrane. Then, the ACh is rapidly inactivated by AChE, an enzyme that can be inhibited by a variety of drugs such as physostigmine and neostigmine.

The actual effect produced when ACh activates its receptors is determined by the type of receptor and the type of cell (Figure 19-4). The two types of ACh postsynaptic receptors are known as nicotinic receptors and muscarinic receptors. ***Nicotinic receptors*** are found on both sympathetic and parasympathetic postganglionic neurons. These receptors are so named because the actions of ACh on such receptors are similar to those produced by nicotine. ***Muscarinic receptors*** are found on all effectors innervated by parasympathetic postganglionic axons and some effectors innervated by sympathetic postganglionic axons. These postsynaptic receptors are so named because the actions of ACh on such receptors are similar to those produced by muscarine, a toxin produced by a mushroom.

NE is released from most postganglionic sympathetic fibers. It is synthesized and stored in synaptic vesicles located in the axon terminals of adrenergic fibers. When an action potential reaches the axon terminal, NE is rapidly released into the synaptic cleft. The molecules diffuse across the cleft and combine with specific receptors on the postsynaptic membrane to elicit a specific effector response.

The effects of NE and epinephrine, like those of ACh, are also determined by the type of postsynaptic receptor with which they interact. Such receptors are found on visceral effectors innervated by most sympathetic postganglionic axons and are referred to as ***alpha* (α) *receptors*** and ***beta* (β) *receptors*** (Figure 19-4). Both alpha and beta receptors are distinguished by the specific responses they elicit and by their selective combination with drugs that activate or block them.

Although cells of most effectors contain either alpha or beta receptors, some visceral effector cells contain both. NE, in general, stimulates alpha receptors to a greater extent than beta receptors, and epinephrine, in general, stimulates both alpha and beta receptors.

Exhibit 19-3 shows the types of receptors present on the cells of visceral effectors and their response to autonomic stimulation. Note that there are some very important (and

EXHIBIT 19-3

Activities of Autonomic Nervous System

VISCERAL EFFECTOR	SYMPATHETIC RECEPTOR	EFFECT OF SYMPATHETIC STIMULATION	EFFECT OF PARASYMPATHETIC STIMULATION
GLANDS			
Sweat	α	Stimulates local secretion.	Stimulates generalized secretion.
Lacrimal (Tear)	—	No known functional innervation.	Stimulates secretion.
Adrenal Medulla	—	Promotes epinephrine and norepinephrine secretion.	No known functional innervation.
Liver	β	Promotes glycogenolysis and gluconeogenesis: decreases bile secretion.	Promotes glycogen synthesis; increases bile secretion.
Kidney	β	Secretion of renin.	No effect.
Pancreas	α, β	Inhibits secretion of enzymes and insulin; promotes secretion of glucagon.	Promotes secretion of enzymes and insulin.
SMOOTH MUSCLE			
Radial Muscle of Iris	α	Contraction that results in dilation of pupil.	No known functional innervation.
Sphincter Muscle of Iris	α	No known functional innervation.	Contraction that results in constriction of pupil.
Ciliary Muscle of Eye	β	Relaxation for far vision.	Contraction for near vision.
Salivary Glands	α, β	Vasoconstriction, which decreases secretion.	Stimulates vasodilation.
Gastric Glands	—	Vasoconstriction, which inhibits secretion (?).	Stimulates secretion.
Intestinal Glands	α	Vasoconstriction, which inhibits secretion.	Stimulates secretion.
Lungs (Muscle of Bronchi)	β	Dilation.	Constriction.
Heart (Muscle of Coronary Vessels)	β	Dilation.	Constriction.
Skin and Mucosa Arterioles	α	Constriction.	No known functional innervation for most.
Skeletal Muscle Arterioles	α, β	Constriction or dilation.	No known functional innervation.
Abdominal Viscera Arterioles	α, β	Constriction.	No known functional innervation for most.
Cerebral Arterioles	α	Slight constriction.	No known functional innervation.
Systemic Veins	α, β	Constriction and dilation.	No known functional innervation.
Gallbladder and Ducts	β	Relaxation.	Contraction.
Stomach	α, β	Decreases motility and tone; contracts sphincters.	Increases motility and tone; relaxes sphincters.
Intestines	α, β	Decreases motility and tone; contracts sphincters.	Increases motility and tone; relaxes sphincters.
Kidney	α, β	Constriction of blood vessels that results in decreased urine volume and decreased renin secretion.	No known functional innvervation.
Ureter	α	Increases motility.	Decreases motility.
Spleen	α, β	Contraction and discharge of stored blood into general circulation.	No known functional innervation.
Urinary Bladder	α, β	Relaxation of muscular wall; contraction of internal sphincter.	Contraction of muscular wall; relaxation of internal sphincter.
Arrector Pili of Hair Follicles	α	Contraction that results in erection of hairs.	No known functional innervation.
Uterus	α, β	Stimulates contraction if pregnant; inhibits contraction if nonpregnant.	Minimal effect.
Sex Organs	α	In male, contraction of smooth muscle of ductus (vas) deferens, seminal vesicle, prostate; results in ejaculation. In female, reverse uterine peristalsis.	Vasodilation and erection in both sexes.
CARDIAC MUSCLE			
Heart	β	Increases rate and force of contraction.	Decreases rate and force of contraction (especially the atria).

perplexing) exceptions to the general rule describing alpha and beta receptors. For example, heart muscle has beta receptors that operate as excitatory receptors and cause the heart muscle to contract more forcibly. Other visceral effector cells possess both types of receptors. In these situations, the majority rules. The beta receptors greatly outnumber the alpha receptors in the smooth muscle fibers (cells) of the blood vessels that flow through skeletal muscles. Therefore, an injection of NE, which binds only to alpha receptors, attaches onto the surface of such smooth muscle fibers and causes a vasoconstriction. An injection of epinephrine, however, would result in a vasodilation because epinephrine acts equally on both types of receptors. But since there is a preponderance of beta receptors, the effect is that of the beta receptor—that is, vasodilation caused by relaxation of smooth muscle fibers. Adding to the complexity of this situation is the fact that the effects on skeletal muscle blood vessels are primarily due to responses to local metabolites.

ACTIVITIES

Most visceral effectors have ***dual innervation;*** that is, they receive fibers from both the sympathetic and the parasympathetic divisions. In these cases, impulses from one division stimulate the organ's activities, whereas impulses from the other division inhibit the organ's activities. The stimulating division may be either the sympathetic or the parasympathetic, depending on the organ. For example, sympathetic impulses increase heart activity, whereas parasympathetic impulses decrease it. On the other hand, parasympathetic impulses increase digestive activities, whereas sympathetic impulses inhibit them. The actions of the two systems are carefully integrated to help maintain homeostasis. A summary of the activities of the autonomic nervous system is presented in Exhibit 19-3.

The parasympathetic division is primarily concerned with activities that conserve and restore body energy during times of rest or recovery of the body. It is an ***energy conservation-restorative system.*** Under normal body conditions, for instance, parasympathetic impulses to the digestive glands and the smooth muscle of the gastrointestinal tract dominate over sympathetic impulses. Thus, energy-supplying food can be digested and absorbed by the body.

The sympathetic division, in contrast, is primarily concerned with processes involving the expenditure of energy as when one is in a condition of stress resulting from either physical or emotional activities. When the body is in homeostasis, the main function of the sympathetic division is to counteract the parasympathetic effects just enough to carry out normal processes requiring energy. During extreme stress, however, the sympathetic dominates the parasympathetic. When people are confronted with a stress condition, for example, their bodies become alert and they sometimes perform feats of unusual strength. Fear stimulates the sympathetic division as do a variety of other emotions and physical activities.

Activation of the sympathetic division sets into operation a series of physiological responses collectively called the ***fight-or-flight response.*** It produces the following effects.

1. The pupils of the eyes dilate.
2. The heart rate and force of contraction increase and blood pressure increases.
3. The blood vessels of the skin and viscera constrict.
4. The remainder of the blood vessels dilate. This causes a faster flow of blood into the dilated blood vessels of skeletal muscles, cardiac muscle, and lungs—organs involved in fighting off danger.
5. Rapid and deeper breathing occurs and the bronchioles dilate to allow faster movement of air in and out of the lungs.
6. Blood sugar level rises as liver glycogen is converted to glucose to supply the body's additional energy needs.
7. The medullae of the adrenal glands are stimulated to produce epinephrine and norepinephrine, hormones that intensify and prolong the sympathetic effects noted previously.
8. Processes that are not essential for meeting the stress situation are inhibited. For example, muscular movements of the gastrointestinal tract and digestive secretions are slowed down or even stopped.

CLINICAL APPLICATION

Horner's Syndrome

If the cervical sympathetic trunk is cut or damaged on one side, the sympathetic supply to that side of the head is removed, resulting in ***Horner's syndrome,*** in which the patient exhibits (on the affected side): ptosis (drooping of the upper eyelid), slight elevation of the lower eyelid, narrowing of the space between eyelids, enophthalmos (the eye appears sunken), miosis (constricted pupil), and anhidrosis (lack of sweating).

VISCERAL AUTONOMIC REFLEXES

A ***visceral autonomic reflex*** adjusts the activity of a visceral effector. In other words, it results in the contraction or relaxation of smooth or cardiac muscle or a change in the rate of secretion by a gland. Such reflexes assume a key role in activities involved in homeostasis such as regulating heart action, blood pressure, respiration, digestion, defecation, and urinary bladder functions.

A visceral autonomic reflex arc consists of the following components.

1. ***Receptor.*** The receptor is the distal end of an afferent neuron.
2. ***Afferent neuron.*** This neuron, either a somatic afferent or visceral afferent neuron, conducts the sensory impulse to the spinal cord or brain.
3. ***Association neurons.*** These neurons are found in the central nervous system.
4. ***Visceral efferent preganglionic neuron.*** In the thoracic and abdominal regions, this neuron is in the lateral gray horn of the spinal cord. The axon passes through the ventral root of the spinal nerve, the spinal nerve itself, and the white ramus communicans. It then enters a sympathetic trunk or prevertebral ganglion, where it synapses with a postganglionic neuron. In the cranial and sacral regions, the visceral efferent preganglionic axon leaves the central nervous system and passes to a terminal ganglion, where it synapses with a postganglionic neuron. The role of the visceral efferent preganglionic neuron is to convey a motor impulse from the brain or spinal cord to an autonomic ganglion.
5. ***Visceral efferent postganglionic neuron.*** This neuron conducts a motor impulse from a visceral efferent preganglionic neuron to the visceral effector.
6. ***Visceral effector.*** A visceral effector is smooth muscle, cardiac muscle, or a gland. Alteration of the rate of activity in the effector is the response.

The basic difference between a somatic reflex arc and a visceral autonomic reflex arc is that in a somatic reflex arc only one efferent neuron is involved. In a visceral autonomic reflex arc, two efferent neurons are involved.

Visceral sensations do not always reach the cerebral cortex. Most remain at subconscious levels. Under normal conditions, you are not aware of muscular contractions of the digestive organs, heartbeat, changes in the diameter of blood vessels, and pupil dilation and constriction. Your body adjusts such visceral activities by visceral reflex arcs whose centers are in the spinal cord or lower regions of the brain. Among such centers are the cardiac, respiratory, vasomotor, swallowing, and vomiting centers in the medulla and the temperature control center in the hypothalamus. Stimuli delivered by somatic or visceral afferent neurons synapse in these centers, and the returning motor impulses conducted by visceral efferent neurons bring about an adjustment in the visceral effector usually without conscious recognition. The impulses are interpreted and acted on subconsciously. More intense visceral sensations do give rise to conscious recognition: hunger, nausea, fullness of the urinary bladder and rectum, and pain from damaged viscera.

CONTROL BY HIGHER CENTERS

The autonomic nervous system is not a separate nervous system. Axons from many parts of the central nervous system are connected to both the sympathetic and the parasympathetic divisions of the ANS and thus exert considerable control over it. The thalamus sorts incoming impulses before they reach the cerebral cortex. The cerebral cortex then turns over control and integration of visceral activities to the hypothalamus. It is at the level of the hypothalamus that the major control and integration of the ANS is exerted and much of this control is by way of hypothalamic influence on centers in the medulla and spinal cord.

The hypothalamus receives input from areas of the nervous system concerned with emotions, visceral functions, olfaction (smell), gustation (taste), as well as changes in temperature, osmolarity, and levels of various substances in blood. Anatomically, the hypothalamus is connected to both the sympathetic and the parasympathetic divisions of the ANS by axons of neurons whose dendrites and cell bodies are in various hypothalamic nuclei. The axons form tracts from the hypothalamus to sympathetic and parasympathetic nuclei in the brain stem and spinal cord through relays in the reticular formation. The posterior and lateral portions of the hypothalamus appear to control the sympathetic division. When these areas are stimulated, there is an increase in visceral activities—an increase in heart rate and force of beat, a rise in blood pressure due to vasoconstriction of blood vessels, an increase in the rate and depth of respiration, dilation of the pupils, and inhibition of the gastrointestinal tract. On the other hand, the anterior and medial portions of the hypothalamus seem to control the parasympathetic division. Stimulation of these areas results in a decrease in heart rate, lowering of blood pressure, constriction of the pupils, and increased secretion and motility of the gastrointestinal tract.

Control of the ANS by the cerebral cortex occurs primarily during emotional stress. In extreme anxiety, which can result from either conscious or subconscious stimulation in the cerebral cortex, the cortex can stimulate the hypothalamus as part of the limbic system. In turn, the cardiac and vasomotor centers of the medulla are stimulated, thereby increasing heart rate and force of beat, and blood pressure. If the cortex is stimulated by hearing bad news or experiencing an extremely unpleasant sight, the stimulation causes vasodilation of blood vessels, a lowering of blood pressure, and fainting.

Evidence of even more direct control of visceral responses is provided by data gathered from studies of biofeedback and meditation.

BIOFEEDBACK

In the simplest terms, ***biofeedback*** is a process in which people get constant signals, or feedback, about various visceral biological functions such as blood pressure, heart rate, and muscle tension. By using special monitoring devices, they can control these visceral functions consciously to a limited degree.

In a study conducted at the Menninger Foundation,* subjects suffering from migraine headaches received instructions in the use of a monitor that registers the skin temperature of the right index finger. Subjects were also given a typewritten sheet containing two sets of phrases. The first set was designed to help them relax the entire body. The second set was designed to bring about an increased flow of blood in the hands. When skin temperature increased, the monitor emitted a high-pitched sound. In time, the monitor was abandoned.

Once the subjects learned how to vasodilate blood vessels of their extremities, the migraine headaches lessened. Since migraine headaches are believed to involve a distension of blood vessels in the head, the shunting of blood from head to hands relieved the distension and thus the pain.

Other experiments have shown that biofeedback can be applied to childbirth. The results of studies indicate that the women in a state of reduced nervousness need less medication during labor, and labor time is shortened.

There is no way to determine where biofeedback will lead. Perhaps the outstanding contribution of biofeedback research has been to demonstrate that the autonomic nervous system is not completely autonomous. Visceral responses can be controlled consciously to some degree. Current therapeutic applications of biofeedback include treatment of asthma, Raynaud's disease, hypertension, gastrointestinal disorders, fecal incontinence, anxiety, pain, and neuromuscular rehabilitation following cerebrovascular accidents (CVAs).

* Much of the following discussion of the use of biofeedback for the treatment of migraine headaches is based on information provided by Dr. Joseph D. Sargent of the Menninger Foundation, Topeka, Kansas.

MEDITATION

Yoga, which literally means "union," is defined as a higher state of consciousness achieved through a fully rested and relaxed body and a fully awake and relaxed mind. One widely practiced technique for achieving higher consciousness is called ***transcendental meditation (TM).*** One sits in a comfortable position with the eyes closed and concentrates on a suitable sound or thought.

Research indicates that transcendental meditation can alter physiological responses. Oxygen consumption decreases drastically along with carbon dioxide elimination. Subjects have experienced a reduction in metabolic rate and blood pressure. Researchers have also observed a decrease in heart rate, an increase in the intensity of alpha brain waves, a sharp decrease in the amount of lactic acid in the blood, and an increase in the skin's electrical resistance. These last four responses are characteristic of a highly relaxed state of mind. Alpha waves are brain waves that are found in the EEGs of almost all normal individuals in a resting, but awake, state; they disappear during sleep.

These responses have been called an ***integrated response***—essentially, a hypometabolic state resulting from inactivation of the sympathetic division of the autonomic nervous system. The response is the exact opposite of the fight-or-flight response, which is a hyperactive state of the sympathetic division. The existence of the integrated response suggests that the central nervous system does exert some control over the autonomic nervous system.

STUDY OUTLINE

Somatic Efferent and Autonomic Nervous Systems (p. 547)

1. The somatic efferent nervous system allows conscious control of movement in skeletal muscles.
2. The ANS, or visceral efferent nervous system, regulates visceral activities—that is, activities of smooth muscle, cardiac muscle, and certain glands—and it usually operates without conscious control.
3. It is regulated by centers in the brain and spinal cord, in particular by the cerebral cortex, the hypothalamus, and the medulla oblongata.
4. In the somatic nervous system, there is only a single neuron along the route to a skeletal muscle fiber; in the ANS, there are two efferent neurons—one from the CNS to a ganglion and one from a ganglion to a visceral effector.
5. Somatic efferent neurons release acetylcholine (ACh), and autonomic efferent neurons release either acetylcholine (ACh) or norepinephrine (NE).

Structure of the Autonomic Nervous System (p. 547)

1. The autonomic nervous system consists of visceral efferent neurons organized into nerves, ganglia, and plexuses.
2. It is considered entirely motor but also involves visceral receptors and visceral afferent neurons. All autonomic axons are efferent fibers.
3. Efferent neurons are preganglionic (with myelinated axons) and postganglionic (with unmyelinated axons).
4. The ANS consists of two principal divisions: sympathetic (thoracolumbar) and parasympathetic (craniosacral).
5. Autonomic ganglia are classified as sympathetic trunk ganglia (on both sides of spinal column), prevertebral ganglia (anterior to spinal column), and terminal ganglia (near or inside visceral effectors).

Physiology of the Autonomic Nervous System (p. 553)

1. Autonomic fibers release neurotransmitters at synapses. On the basis of the neurotransmitter produced, these fibers may be classified as cholinergic or adrenergic.
2. Cholinergic fibers release acetylcholine (ACh). Adrenergic fibers produce norepinephrine (NE).
3. Acetylcholine (ACh) interacts with nicotinic receptors on postganglionic neurons and muscarinic receptors on certain visceral effectors.

4. Norepinephrine (NE) generally interacts with alpha receptors on visceral effectors, and epinephrine generally interacts with alpha and beta receptors on visceral effectors.
5. Sympathetic responses are widespread and, in general, are concerned with energy expenditure. Parasympathetic responses are restricted and are typically concerned with energy restoration and conservation.

Visceral Autonomic Reflexes (p. 556)

1. A visceral autonomic reflex adjusts the activity of a visceral effector.
2. A visceral autonomic reflex arc consists of a receptor, afferent neuron, association neuron, visceral efferent preganglionic neuron, visceral efferent postganglionic neuron, and visceral effector.

Control by Higher Centers (p. 557)

1. The hypothalamus controls and integrates the ANS. It is connected to both the sympathetic and the parasympathetic divisions.
2. Biofeedback is a process in which people learn to monitor visceral functions and to control them consciously. It has been used to control heart rate, alleviate migraine headaches, and make childbirth easier.
3. Yoga is a higher consciousness achieved through a fully rested and relaxed body and a fully awake and relaxed mind.
4. Transcendental meditation (TM) produces the following physiological responses: decreased oxygen consumption and carbon dioxide elimination, reduced metabolic rate, decreased intensity of alpha brain waves, a sharp decreased amount of lactic acid in the blood, and an increased electrical resistance of the skin.

REVIEW QUESTIONS

1. What are the principal components of the autonomic nervous system? What is its general function? Why is it called involuntary? (p. 547)
2. What are the principal differences between the voluntary nervous system and the ANS? (p. 547)
3. Relate the role of visceral efferent fibers and visceral effectors to the ANS. (p. 547)
4. Distinguish between preganglionic neurons and postganglionic neurons with respect to location and function. (p. 548)
5. What is an autonomic ganglion? Describe the location and function of the three types of autonomic ganglia. Define white and gray rami communicantes. (p. 548)
6. On what basis are the sympathetic and parasympathetic divisions of the autonomic nervous system differentiated anatomically and functionally? (p. 551)
7. Discuss the distinction between cholinergic and adrenergic fibers of the ANS. (p. 553)
8. How is acetylcholine (ACh) related to nicotinic and muscarinic receptors? (p. 554)
9. How are alpha and beta receptors related to norepinephrine (NE) and epinephrine? (p. 554)
10. Give examples of the antagonistic effects of the sympathetic and parasympathetic divisions of the ANS. (p. 555)
11. Summarize the principal functional differences between the voluntary nervous system and the autonomic nervous system. (p. 547)
12. Give the *sympathetic response* in a fear situation for each of the following body parts: hair follicles, iris of eye, lungs, spleen, adrenal medulla, kidneys, urinary bladder, stomach, intestines, gallbladder, liver, heart, arterioles of the abdominal viscera, skeletal muscles, and skin and mucosa. (p. 555)
13. Define a visceral autonomic reflex and give three examples. (p. 556)
14. Describe how the hypothalamus controls and integrates the ANS. (p. 557)
15. Define biofeedback. Explain how it could be useful. (p. 557)
16. What is transcendental meditation (TM)? How is the integrated response related to the ANS? (p. 558)
17. What is Horner's syndrome? (p. 556)

SELF QUIZ

1. Write S if the description applies to the sympathetic division of the autonomic nervous system, P if it applies to the parasympathetic division, and P,S if it applies to both:

___ **a.** also called thoracolumbar outflow
___ **b.** has long preganglionic fibers leading to terminal ganglia and very short postganglionic fibers
___ **c.** has relatively short preganglionic fibers and long postganglionic fibers
___ **d.** sends some preganglionic fibers through cranial nerves
___ **e.** has some preganglionic fibers synapsing in sympathetic trunk ganglia
___ **f.** has more widespread effect in the body, affecting more organs
___ **g.** has some fibers running in gray rami to supply sweat glands, hair muscles, and blood vessels
___ **h.** has fibers in white rami (connecting spinal nerve with sympathetic trunk ganglia)
___ **i.** contains fibers that supply viscera with motor impulses
___ **j.** celiac and superior mesenteric ganglia are sites of postganglionic neuron cell bodies

Choose all correct answers to the following.

___ **2.** Which activities are characteristic of the stress response, or fight-or-flight reaction?
A. the liver breaks down glycogen to glucose; B. the heart rate decreases; C. kidneys increase urine production since blood is shunted to kidneys; D. there is increased blood flow to genitalia, causing erect state; E. hairs stand on end (''goose pimples'') due to contraction of arrector pili muscles; F. in general, the sympathetic system is active.

___ 3. Choose all true statements about gray rami:
A. they contain only sympathetic nerve fibers; B. they contain only postganglionic nerve fibers; C. they carry nerve impulses from sympathetic trunk ganglia to spinal nerves; D. they are located at all levels of the vertebral column (from C1 to coccyx); E. they carry nerve impulses between paravertebral ganglia and prevertebral ganglia; F. they carry preganglionic neurons from anterior ramus of spinal nerve to trunk ganglion.

___ 4. Which are structural features of the parasympathetic system?
A. ganglia are close to the central nervous system and distant from the effector; B. forms the craniosacral outflow; C. distributed throughout the body, including extremities; D. supplies nerves to blood vessels, sweat glands, and adrenal (suprarenal) gland; E. has some of its nerve fibers passing through paravertebral ganglia.

5. Use arrows to show whether parasympathetic (P) or sympathetic (S) fibers stimulate (↑) or inhibit (↓) each of the following activities. Use a dash (—) to indicate that there is no parasympathetic innervation. The first one is done for you.

a. P ↓ S ↑ dilation of pupil
b. P ___ S ___ heart rate
c. P ___ S ___ constriction of skin blood vessels
d. P ___ S ___ salivation and digestive organ contractions
e. P ___ S ___ dilation of bronchioles for easier breathing
f. P ___ S ___ contraction of urinary bladder
g. P ___ S ___ contraction of pili of hair follicles causing "goose bumps"
h. P ___ S ___ contraction of spleen that transfers some of its blood to general circulation, causing increase in blood pressure
i. P ___ S ___ release of epinephrine and norepinephrine (NE) from adrenal medulla
j. P ___ S ___ coping with stress, "fight-or-flight" response
k. P ___ S ___ conservation of energy, "energy conservation-restorative system"
l. P ___ S ___ erection of genitalia

Circle T (true) or F (false) for the following.

T F 6. Synapsing occurs in both sympathetic and parasympathetic ganglia.

T F 7. In a visceral autonomic reflex arc only one efferent neuron is involved.

T F 8. The sympathetic system has a more widespread effect in the body than the parasympathetic does.

T F 9. In general, the parasympathetic division of the ANS has long preganglionic fibers and short postganglionic fibers.

Choose the one best answer to these questions.

___ 10. Which statement about postganglionic neurons is false?
A. they all lie entirely outside the CNS; B. their axons are nonmyelinated; C. they terminate in visceral effectors; D. their cell bodies lie in the lateral gray matter of the cord; E. they are very short in the parasympathetic system.

___ 11. All of the following axons are cholinergic *except:*
A. parasympathetic preganglionic; B. parasympathetic postganglionic; C. sympathetic preganglionic; D. sympathetic postganglionic to sweat glands; E. sympathetic postganglionic to heart muscle.

___ 12. The ANS provides the chief nervous control in which of these activities?
A. following a moving object with the eyes; B. moving a hand reflexly from a hot object; C. typing; D. digesting food; E. writing an essay.

___ 13. Which of the following is not a visceral effector?
A. smooth muscle of the iris; B. pancreas; C. heart; D. skeletal muscle; E. salivary gland.

___ 14. The ANS
A. has two parts—the parasympathetic, which controls all normal functions, and the sympathetic, which controls the same functions but to a greater degree; B. is the part of our nervous system controlling all reflexes; C. does not function when the body is subjected to stress situations; D. has two divisions that act antagonistically—one counteracts the effects of the other; E. is that part of our anatomy that controls the contraction of skeletal, smooth, and cardiac muscle tissue.

___ 15. The preganglionic autonomic nerve fibers that arise from the thoracic and lumbar parts of the spinal cord
(1) are also called preganglionic sympathetic fibers.
(2) synapse with postganglionic fibers of the parasympathetic nervous system.
(3) form the dorsal roots of the thoracic and lumbar spinal nerves.
(4) are also called preganglionic parasympathetic fibers.
A. (1) only; B. (2) only; C. (3) only; D. (4) only; E. (2) and (4).

___ 16. Which of the following statements is true regarding the parasympathetic ganglia of the ANS?
A. they contain afferent neuron cell bodies only; B. they are located in or on the walls of the viscera to which their nerve fibers are going; C. they lie within the spinal cord; D. they consist of a double chain of structures along the spinal column; E. none of the above.

___ 17. Cholinergic fibers are thought to include
A. all preganglionic axons; B. almost all postganglionic parasympathetic axons; C. a few postganglionic sympathetic axons; D. all axons of somatic motor neurons; E. all of the above.

___ 18. Which of the following would indicate increased parasympathetic activity?
A. "cotton mouth" from reduced salivation; B. increased gastric secretion; C. rise of blood pressure; D. decreased flow of blood through the skin; E. increased blood glucose level.

___ 19. The axons of neurons lying within the central nervous system that connect with the autonomic nervous system are
A. myelinated fibers; B. postganglionic fibers; C. preganglionic fibers; D. cranial nerve fibers; E. none of the above.

Sensory and Motor Systems

20

STUDENT OBJECTIVES

1. Define a sensation and list the characteristics of sensations.
2. Classify receptors on the basis of location, stimulus detected, and simplicity or complexity.
3. List the location and function of the receptors for tactile sensations (touch, pressure, vibration), thermoreceptive sensations (heat and cold), and pain.
4. Describe how sensory impulses are conveyed from receptors to the brain.
5. Compare the course of the pyramidal and extrapyramidal motor pathways.
6. Locate the receptors for olfaction and describe the neural pathway for smell.
7. Identify the gustatory receptors and describe the neural pathway for taste.
8. List and describe the structural divisions of the eye and identify the afferent pathway of light impulses to the brain.
9. Describe the anatomical subdivisions of the ear and the auditory pathway.
10. Identify the receptor organs for static and dynamic equilibrium and describe their neural pathway.
11. Describe the effects of aging on the special senses.
12. Contrast the causes and symptoms of cataracts, glaucoma, conjunctivitis, trachoma, deafness, labyrinthine disease, Ménière's syndrome, vertigo, otitis media, and motion sickness.
13. Define key medical terms associated with sensory structures.

CHAPTER OUTLINE

■ **Sensations**
Definition
Characteristics
Classification of Receptors
Location
Stimulus Detected
Simplicity or Complexity
■ **General Senses**
Cutaneous Sensations
Tactile Sensations
Thermoreceptive Sensations
Pain Sensations
Proprioceptive Sensations
Receptors
Levels of Sensation
■ **Sensory Pathways**
Somatosensory Cortex
Proprioception, Discriminative Touch, Two-Point Discrimination, and Vibration
Pain, Temperature, Light Touch, and Pressure
Cerebellar Tracts
■ **Motor Pathways**
Linkage of Sensory Input and Motor Responses
Motor Cortex
Pyramidal Pathways
Extrapyramidal Pathways
■ **Special Senses**
Olfactory Sensations
Structure of Receptors
Olfactory Pathway
Gustatory Sensations
Structure of Receptors
Gustatory Pathway
Visual Sensations
Accessory Structures of Eye
Structure of Eyeball
Visual Pathway
Auditory Sensations and Equilibrium
External (Outer) Ear
Middle Ear
Internal (Inner) Ear
Auditory Pathway
Mechanism of Equilibrium
■ **Aging and the Special Senses**
■ **Applications to Health**
■ **Key Medical Terms Associated with Sensory Structures**

SENSATIONS

The central nervous system (CNS) requires a continual flow of information to regulate homeostasis and initiate appropriate responses to changes in the internal and external environments. At any given time, our brains receive and respond to many varieties of information. However, we are aware only of the information on which we consciously focus. The CNS selects only those bits of information that are important for the moment, and it is only those bits of information that are brought to our conscious level. There is no question that we would collapse into nervous wrecks if our consciousness were forced to deal with all the information arriving at once. The conscious mind is turned off to protect itself from overstimulation.

Your ability to sense stimuli is vital to your survival. If pain could not be sensed, burns would be common. An inflamed appendix or stomach ulcer would progress unnoticed. A lack of sight would increase the risk of injury from unseen obstacles, a loss of smell would allow harmful gas to be inhaled, a loss of hearing would prevent recognition of automobile horns, and a lack of taste would allow toxic substances to be ingested. In short, if you could not "sense" your environment and make the necessary homeostatic adjustments, you could not survive very well on your own.

DEFINITION

In its broadest context, ***sensation*** refers to a state of awareness of external or internal conditions of the body. ***Perception*** refers to the conscious registration of a sensory stimulus. For a sensation to occur, four prerequisites must be satisfied.

1. A ***stimulus,*** or change in the environment, capable of initiating a response by the nervous system must be present.
2. A ***receptor*** or ***sense organ*** must pick up the stimulus and transduce (convert) it to a nerve signal in the form of a generator or receptor potential to be described shortly. A receptor or sense organ may be viewed as specialized nervous tissue that is extremely sensitive to internal or external stimuli.
3. The nerve impulse must be ***conducted*** along a neural pathway from the receptor or sense organ to the brain.
4. A region of the brain must ***translate*** the nerve impulse into a sensation.

Most receptors are capable of converting a specific stimulus into a nerve impulse. The stimulus may be light, heat, pressure, mechanical energy, or chemical energy. Usually, the stimulus causes the membrane of the receptor to depolarize (become more positive). This depolarization is called a ***generator potential.***

The generator potential is a graded response within limits; the magnitude increases with stimulus strength and frequency. When the generator potential reaches the threshold level, it initiates a nerve impulse (nerve action potential). Once initiated, the nerve impulse is propagated along the nerve fiber. Whereas a generator potential is a local graded response, a nerve impulse obeys the all-or-none principle. The function of a generator potential is to transduce a stimulus into a nerve impulse.

A receptor may be quite simple. It may consist of the dendrites of a single neuron in the skin that are sensitive to pain stimuli; or it may be contained in a complex organ such as the eye. Regardless of complexity, all sense receptors contain the dendrites of sensory neurons. The dendrites occur either alone or in close association with specialized cells of other tissues.

Once a stimulus is received by a receptor and converted into a nerve impulse, the impulse is conducted along an afferent pathway that enters either the spinal cord or the brain. Many sensory impulses are conducted to the sensory areas of the cerebral cortex. It is in this region that stimuli produce conscious sensations. Sensory impulses that terminate in the spinal cord or brain stem can initiate motor activities but typically do not produce conscious sensation. The thalamus detects pain sensations but cannot distinguish the intensity or location from which they arise. This ability to distinguish is a function of the cerebrum.

CHARACTERISTICS

Most conscious sensations or perceptions occur in the cortical regions of the brain. In other words, you see, hear, and feel in the brain. You seem to see with your eyes, hear with your ears, and feel pain in an injured part of your body only because the cortex interprets the sensation as coming from the stimulated sense receptor. The term ***projection*** describes this process by which the brain refers sensations to their point of stimulation.

A second characteristic of many sensations is ***adaptation***—that is, a decrease in sensitivity to continued stimuli. In fact, the perception of a sensation may actually disappear even though the stimulus is still being applied. For example, when you first get into a tub of hot water, you probably feel a burning sensation, but soon the sensation decreases to one of comfortable warmth even though the stimulus (hot water) is still present. In time, the sensation of warmth disappears completely. Other examples of adaptation include placing a ring on your finger, putting on your shoes or hat, sitting on a chair, and pushing your glasses up onto the top of your head. Adaptation results from a change in a receptor, a change in a structure associated with a receptor, or inhibitory feedback from the brain. Receptors vary in their ability to adapt. ***Rapidly adapting (phasic) receptors,*** such as those associated with pressure, touch, and smell, adapt very quickly. Such receptors play a major role in signaling changes in a particular sensation. ***Slowly adapting (tonic) receptors,*** such as those associated with pain, body position, and detecting chemicals in blood, adapt slowly.

These receptors are important in signaling information regarding steady states of the body.

Sensations may also be characterized by ***afterimages;*** that is, some sensations persist even though the stimulus has been removed. This phenomenon is the reverse of adaptation. One common example of afterimage occurs when you look at a bright light and then look away or close your eyes. You still see the light for several seconds or minutes afterward.

Another characteristic of sensations is ***modality:*** the specific type of sensation felt. The sensation may be one of temperature change, pain, pressure, touch, body position, equilibrium, hearing, vision, smell, or taste. In other words, the distinct property by which one sensation may be distinguished from another is its modality.

CLASSIFICATION OF RECEPTORS

Location

One convenient method of classifying receptors is by their location. ***Exteroceptors*** (eks′-ter-ō-SEP-tors) provide information about the external environment. They are sensitive to stimuli outside the body and transmit sensations of hearing, sight, smell, taste, touch, pressure, temperature, and pain. Exteroceptors are located at or near the surface of the body.

Visceroceptors (vis′-er-ō-SEP-tors), or ***enteroceptors,*** provide information about the internal environment. These sensations arise from within the body and may be felt as pain, pressure, fatigue, hunger, thirst, and nausea. Visceroceptors are located in blood vessels and viscera.

Proprioceptors (prō′-prē-ō-SEP-tors) provide information about body position and movement. Such sensations give us information about muscle tension, the position and activity of our joints, and equilibrium. These receptors are located in muscles, tendons, joints, and the internal ear.

Stimulus Detected

Another method of classifying receptors is by the type of stimuli they detect. ***Mechanoreceptors*** detect mechanical deformation of the receptor itself or in adjacent cells. Stimuli so detected include those related to touch, pressure, vibration, proprioception, hearing, equilibrium, and blood pressure. ***Thermoreceptors*** detect changes in temperature. ***Nociceptors*** detect pain, usually as a result of physical or chemical damage to tissues. ***Photoreceptors*** detect light on the retina of the eye. ***Chemoreceptors*** detect taste in the mouth, smell in the nose, and chemicals in body fluids, such as oxygen, carbon dioxide, water, and glucose.

Simplicity or Complexity

As will be described shortly, receptors may also be classified according to the simplicity or complexity of their structure and the neural pathway involved. ***Simple receptors*** and neural pathways are associated with ***general senses.*** The receptors for general sensations are numerous and widespread. Examples include cutaneous sensations such as touch, pressure, vibration, heat, cold, and pain. ***Complex receptors*** and neural pathways are associated with ***special senses.*** The receptors for each special sense are found in only one or two specific areas of the body. Among the special senses are smell, taste, sight, equilibrium, and hearing.

GENERAL SENSES

CUTANEOUS SENSATIONS

Cutaneous (*cuta* = skin) ***sensations*** include tactile sensations (touch, pressure, vibration), thermoreceptive sensations (cold and heat), and pain. The receptors for these sensations are in the skin, connective tissue under the skin, mucous membranes, and the beginning and end of the gastrointestinal tract.

The cutaneous receptors are distributed over the body surface in such a way that certain parts of the body are densely populated with receptors and other parts contain only a few. Such an unequal distribution of receptors is called ***punctate distribution.*** Areas of the body that have few cutaneous receptors are insensitive; those containing many are very sensitive.

Cutaneous receptors have simple structures. They consist of the dendrites of sensory neurons that may or may not be enclosed in a capsule of epithelial or connective tissue. Nerve impulses generated by cutaneous receptors pass along somatic afferent neurons in spinal and cranial nerves, through the thalamus, to the general sensory area of the parietal lobe of the cortex (see Figures 20-5 through 20-7).

Tactile Sensations

Even though the ***tactile*** (*tact* = touch) ***sensations*** are divided into separate sensations (touch, pressure, and vibration), they are all detected by the same types of receptors, called ***mechanoreceptors,*** which are receptors that transduce mechanical deformation into generator potentials.

■ ***Touch*** ***Touch sensations*** generally result from stimulation of tactile receptors in the skin or tissues immediately beneath the skin. ***Light touch*** refers to the ability to perceive that something has touched the skin, although its exact location, shape, size, or texture cannot be determined. ***Discriminative touch*** refers to the ability to recognize exactly what point of the body is touched.

Tactile receptors for touch include hair root plexuses, free nerve endings, tactile discs, corpuscles of touch, and type II cutaneous mechanoreceptors (Figure 20-1). ***Hair root plexuses*** are dendrites arranged in networks around hair follicles. They are not surrounded by supportive or

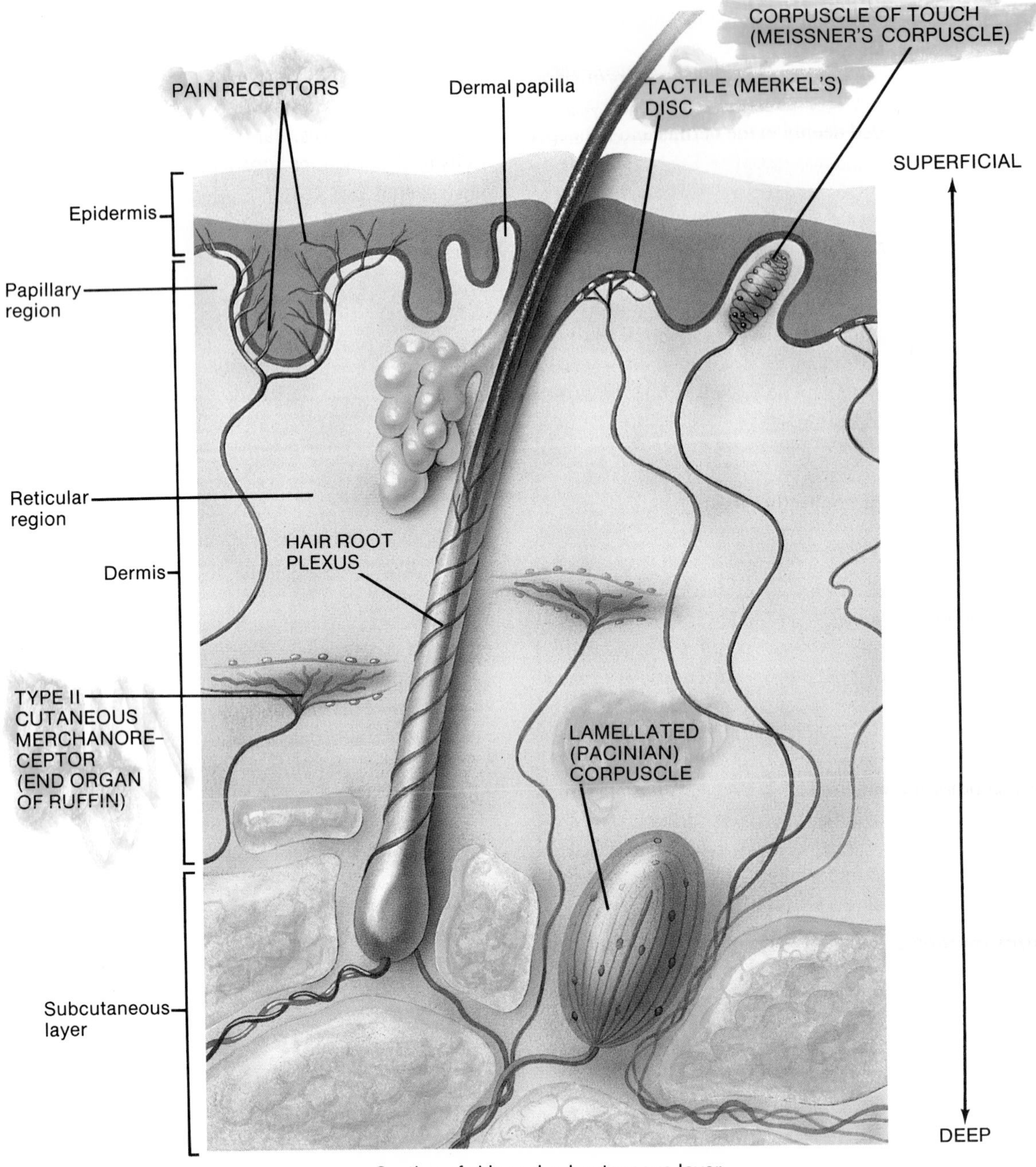

FIGURE 20-1 Structure and location of cutaneous receptors.

protective structures. If a hair shaft is moved, the dendrites are stimulated. Hair root plexuses detect movements mainly on the surface of the body when hairs are disturbed, the hair functioning as a lever.

Other receptors that are not surrounded by supportive or protective structures are called ***free (naked) nerve endings.*** Free nerve endings are found everywhere in the skin and many other tissues. Although they are very important pain receptors, free nerve endings also respond to objects that are in continuous contact with the skin, such as clothing.

Tactile, or ***Merkel's*** (MER-kelz) ***discs,*** are modified epidermal cells in the stratum basale of hairless skin. Their basal ends are in contact with dendrites of sensory neurons. Tactile discs are distributed in many of the same locations as corpuscles of touch and also function in discriminative touch.

Corpuscles of touch, or ***Meissner's*** (MĪS-nerz) ***corpuscles,*** are egg-shaped receptors for discriminative touch containing a mass of dendrites enclosed by connective tissue. They are located in the dermal papillae of the skin and are most numerous in the fingertips, palms of the hands, and soles of the feet. They are also abundant in the eyelids, tip of the tongue, lips, nipples, clitoris, and tip of penis.

Type II cutaneous mechanoreceptors, or ***end organs of Ruffini,*** are embedded deeply in the dermis and in deeper tissues of the body. They detect heavy and continuous touch sensations.

■ ***Pressure*** ***Pressure sensations*** generally result from stimulation of tactile receptors in deeper tissues and are longer lasting and have less variation in intensity than touch sensations. Pressure is really sustained touch. Moreover, pressure is felt over a larger area than touch.

Pressure receptors are free nerve endings, type II cutaneous mechanoreceptors, and lamellated corpuscles. ***Lamellated,*** or ***Pacinian*** (pa-SIN-ē-an), ***corpuscles*** (Figure 20-1) are oval structures composed of a capsule resembling an onion that consist of connective tissue layers enclosing dendrites. Lamellated corpuscles are located in the subcutaneous tissue under the skin, the deep submucosal tissues that lie under mucous membranes, in serous membranes, around joints and tendons, in the perimysium of muscles, in the mammary glands, in the external genitalia of both sexes, and in certain viscera, such as the pancreas.

■ ***Vibration*** ***Vibration sensations*** result from rapidly repetitive sensory signals from tactile receptors.

The receptors for vibration sensations are corpuscles of touch and lamellated corpuscles. Whereas corpuscles of touch detect low-frequency vibration, lamellated corpuscles detect higher-frequency vibration.

Thermoreceptive Sensations

The ***thermoreceptive*** (*therm* = heat) ***sensations*** are hot and cold. The exact nature of ***thermoreceptive receptors*** is not known, but they might be free (naked) nerve endings.

Pain Sensations

Pain is indispensable for a normal life. It provides us with information about tissue-damaging (noxious) stimuli and thus often enables us to protect ourselves from greater damage. It is pain that initiates our search for medical assistance, and it is our subjective description and indication of the location of the pain that helps to pinpoint the underlying cause of disease.

The receptors for ***pain,*** called ***nociceptors*** (nō′-sē-SEP-tors; *noci* = harmful), are simply free (naked) nerve endings, the branching ends of the dendrites of certain sensory neurons (Figure 20-1). Pain receptors are found in practically every tissue of the body. They may respond to any type of stimulus if it is strong enough to cause tissue damage. When stimuli for other sensations, such as touch, pressure, heat, and cold, reach a certain threshold, they stimulate the sensation of pain as well. Excessive stimulation of a sense organ causes pain. Additional stimuli for pain receptors include excessive distension or dilation of a structure, prolonged muscular contractions, muscle spasms, inadequate blood flow to an organ, or the presence of certain chemical substances. Pain receptors, because of their sensitivity to all stimuli, perform a protective function by identifying changes that may endanger the body. Pain receptors adapt only slightly or not at all. Adaptation is the decrease or disappearance of the perception of a sensation even though the stimulus is still present. If there were adaptation to pain, it would cease to be sensed and irreparable damage could result.

Sensory impulses for pain are conducted to the central nervous system along spinal and cranial nerves (see Figure 20-6). The lateral spinothalamic tracts of the spinal cord relay impulses to the thalamus. From here the impulses may be relayed to the postcentral gyrus of the parietal lobe. Recognition of the kind and intensity of most pain is ultimately localized in the cerebral cortex. Some awareness of pain occurs at subcortical levels such as the thalamus.

Pain may be classified on the basis of speed of onset and duration: acute and chronic. ***Acute pain*** occurs very rapidly, usually within 0.1 second after a stimulus is applied, and is not felt in deeper tissues of the body. This type of pain is also known as sharp, fast, and pricking pain. The pain felt from a needle puncture or knife cut to the skin are examples of acute pain. ***Chronic pain,*** by contrast, begins after a second or more and then gradually increases over a period of several seconds or minutes. This type of pain may be excruciating and is also referred to as burning, aching, throbbing, and slow pain. Chronic pain can occur both in the skin and deeper tissues or internal organs. An example is the pain associated with a toothache.

Pain may be divided into two types on the basis of the location of stimulated receptors: somatic and visceral. ***Somatic pain*** arises from stimulation of receptors in the skin, in which it is called ***superficial somatic pain,*** or from stimulation of receptors in skeletal muscles, joints, tendons, and fascia, then called ***deep somatic pain. Visceral pain*** results from stimulation of receptors in the viscera.

Although receptors for somatic and visceral pain are similar, viscera do not evoke the same pain response as somatic tissue. For example, highly *localized* damage to certain viscera, such as cutting the intestine in two in a patient who is awake, causes very little, if any, pain. But, if stimulation is *diffuse,* involving large areas, visceral pain can be severe. Such stimulation might result from distension, spasms, or ischemia. There are even some viscera that are almost entirely insensitive to pain of any type. Examples are parts of the liver and lungs.

The ability of the cerebral cortex to locate the origin of pain is related to past experience. In most instances of somatic pain and in some instances of visceral pain, the cortex accurately projects the pain back to the stimulated area. If you burn your finger, you feel the pain in your finger. If the pleural membranes are inflamed, you experience pain in the chest. In most instances of visceral pain, however, the sensation is not projected back to the point of stimulation. Rather, the pain may be felt in or just under the skin that overlies the stimulated organ. The pain may also be felt in a surface area far from the stimulated organ. This phenomenon is called ***referred pain.*** In general, the area to which the pain is referred and the visceral organ involved are served by the same segment of the spinal cord. Consider the following example. Afferent fibers from the heart as well as from the skin over the heart and along the medial aspect of the left upper extremity enter spinal cord segments T1–T4. Thus, the pain of a heart attack is typically felt in the skin over the heart and along the left arm. Figure 20-2 illustrates cutaneous regions to which visceral pain may be referred.

CLINICAL APPLICATION

Phantom Pain

A kind of pain frequently experienced by patients who have had a limb amputated is called ***phantom pain (phantom limb sensation).*** They still experience sensations such as itching, pressure, tingling, or pain in the extremity as if the limb were still there. This probably occurs because the remaining proximal portions of the sensory nerves that previously received nerve impulses from the limb are being stimulated by the trauma of the amputation. Stimuli from these nerves are interpreted by the brain as coming from the nonexistent (phantom) limb.

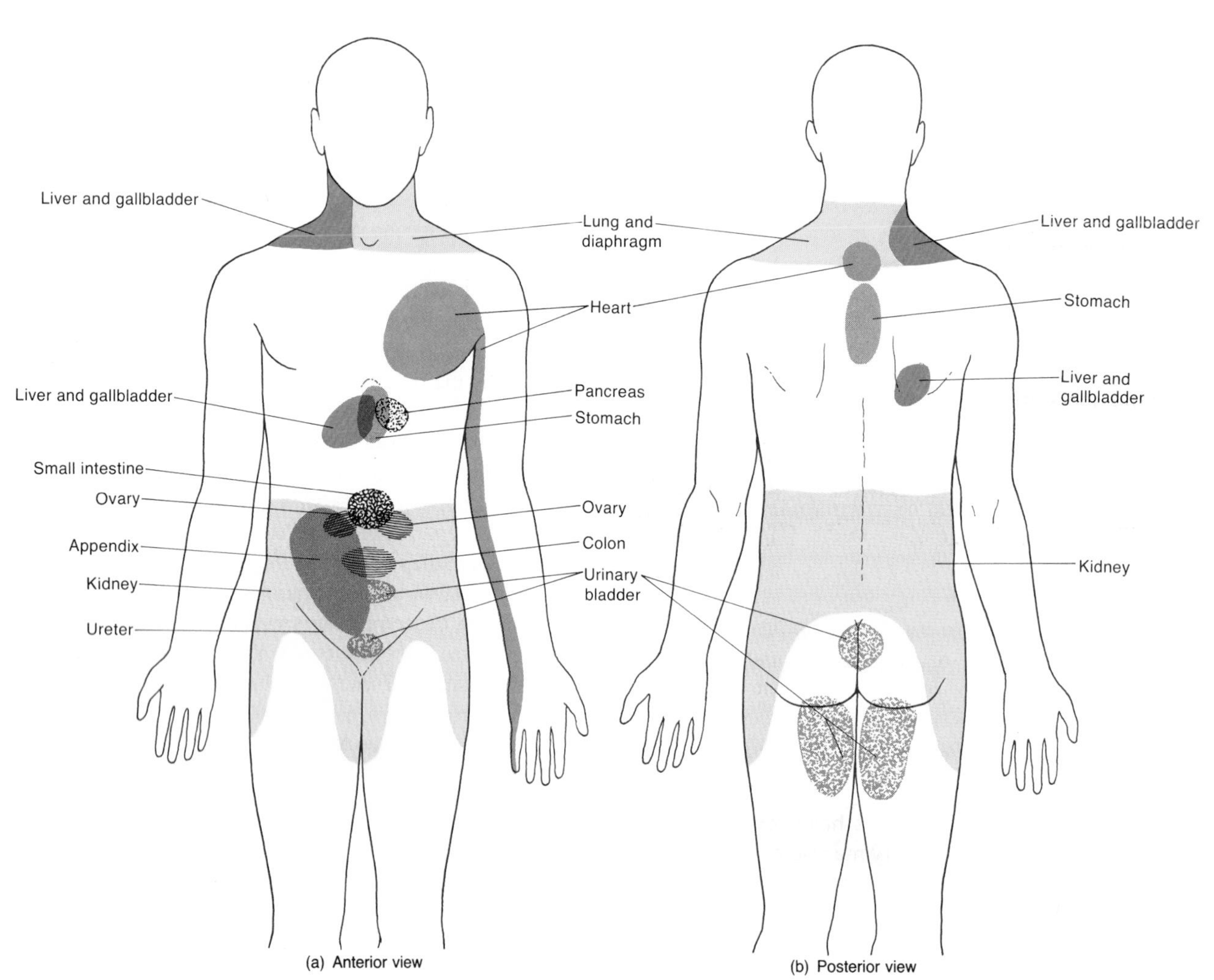

FIGURE 20-2 Referred pain. The colored parts of the diagrams indicate cutaneous areas to which visceral pain is referred.

Pain sensations may be controlled by interrupting the pain impulse between the receptors and the interpretation centers of the brain—chemically, surgically, or by other means. Most pain sensations respond to pain-reducing drugs, which, in general, act to inhibit nerve impulse conduction at synapses.

CLINICAL APPLICATION

Anesthesia

During certain surgical or diagnostic procedures, ***anesthesia*** (an′-es-THĒ-zē-a; *an* = without; *algia* = painful condition) is required to remove the sensation of pain while still maintaining the stability of the patient's organ systems. Two commonly used forms of anesthesia are general and spinal. ***General anesthesia*** not only removes the sensation of pain but also produces unconsciousness and sometimes muscular relaxation. Usually, general anesthesia involves use of more than one drug. First, an injection of a drug is given to induce unconsciousness. Then a drug to alleviate pain is given, either by injection or by inhalation. For certain surgical procedures, a third drug is given to relax muscles in order to prevent spasms during surgery. In this case, a respirator is used to take the place of the patient's own breathing. Continuous monitoring of a patient's blood pressure, heart rate, respirations, and eye reflexes during surgery are used as indicators of the level of unconsciousness and muscular relaxation.

Spinal anesthesia, a form of local anesthesia, involves injection of a drug via a spinal puncture into the subarachnoid space in order to render a person insensitive to pain. The procedure is widely used for surgery performed below the diaphragm such as hernia repair, procedures on the hips and lower extremities, and operations involving the rectum, urinary bladder, prostate gland, and other pelvic structures. During the procedure, a person may be wide awake or, more commonly, sedated. Among the advantages of spinal anesthesia are the following: (1) there is almost no risk of toxic reactions that sometimes occur with general anesthesia, (2) patients with diseased kidneys or livers who could have a problem disposing of general anesthetic drugs are at less risk with spinal anesthesia, (3) no artificial breathing devices are required, (4) normal functioning of the lungs is unaffected so that people with respiratory problems have fewer complications, and (5) emergency surgery may be performed even if a person has not been fasting. (Fasting is required prior to general anesthesia because of the danger of vomiting and aspiration of the vomited material into the respiratory system.) Some disadvantages of spinal anesthesia are brief periods of headache, back pain, difficulty in urinating, and lowered blood pressure.

PROPRIOCEPTIVE SENSATIONS

An awareness of the activities of muscles, tendons, and joints and of equilibrium is provided by the ***proprioceptive*** (*proprio* = one's own), or ***kinesthetic*** (kin′-es-THET-ik), ***sense.*** It informs us of the degree to which muscles are contracted, the amount of tension created in the tendons, the change of position of a joint, and the orientation of the head relative to the ground and in response to movements (equilibrium). The proprioceptive sense enables us to recognize the location and rate of movement of one body part in relation to others. It also allows us to estimate weight and determine the muscular work necessary to perform a task. With the proprioceptive sense, we can judge the position and movements of our limbs, without using our eyes, when we walk, type, or dress in the dark.

Receptors

Proprioceptive receptors are located in skeletal muscles, tendons in and around synovial joints, and the internal ear.

■ *Muscle Spindles* ***Muscle spindles*** are delicate proprioceptive receptors interspersed among skeletal muscle fibers (cells) and oriented parallel to the fibers (Figure 20-3a). The ends of the spindles are anchored to the endomysium and perimysium. Muscle spindles consist of 3 to 10 specialized muscle fibers called ***intrafusal muscle fibers,*** which are partially enclosed in a connective tissue capsule. The spindles are surrounded by skeletal muscle fibers of the muscle, called ***extrafusal muscle fibers.*** The central region of each intrafusal muscle fiber has few or no actin and myosin myofilaments and an accumulation of nuclei. In some intrafusal muscle fibers, the nuclei bunch at the center (***nuclear bag fibers***); in others, the nuclei form a chain at the center (***nuclear chain fibers***). The central region of the intrafusal muscle fibers cannot contract and represents the sensory receptor area for a spindle.

Although the central receptor area cannot contract because it lacks myofilaments, it does contain two types of afferent (sensory) fibers. A large sensory fiber, called a ***type Ia fiber,*** innervates the central area of the intrafusal muscle fibers. The branches of the Ia fiber, called ***primary*** (***annulospiral***) ***endings,*** wrap around the center of the intrafusal muscle fibers. When the central part of the spindle is stretched, the primary endings are stimulated and send nerve impulses to the spinal cord at exceedingly great velocities. The central receptor area of some muscle spindles is also innervated by two sensory fibers called ***type II fibers.*** Their branches, known as ***secondary*** (***flower spray***) ***endings,*** are located on either side of the primary ending. Secondary endings are also stimulated when the central part of the spindle is stretched, and they, too, send impulses to the spinal cord.

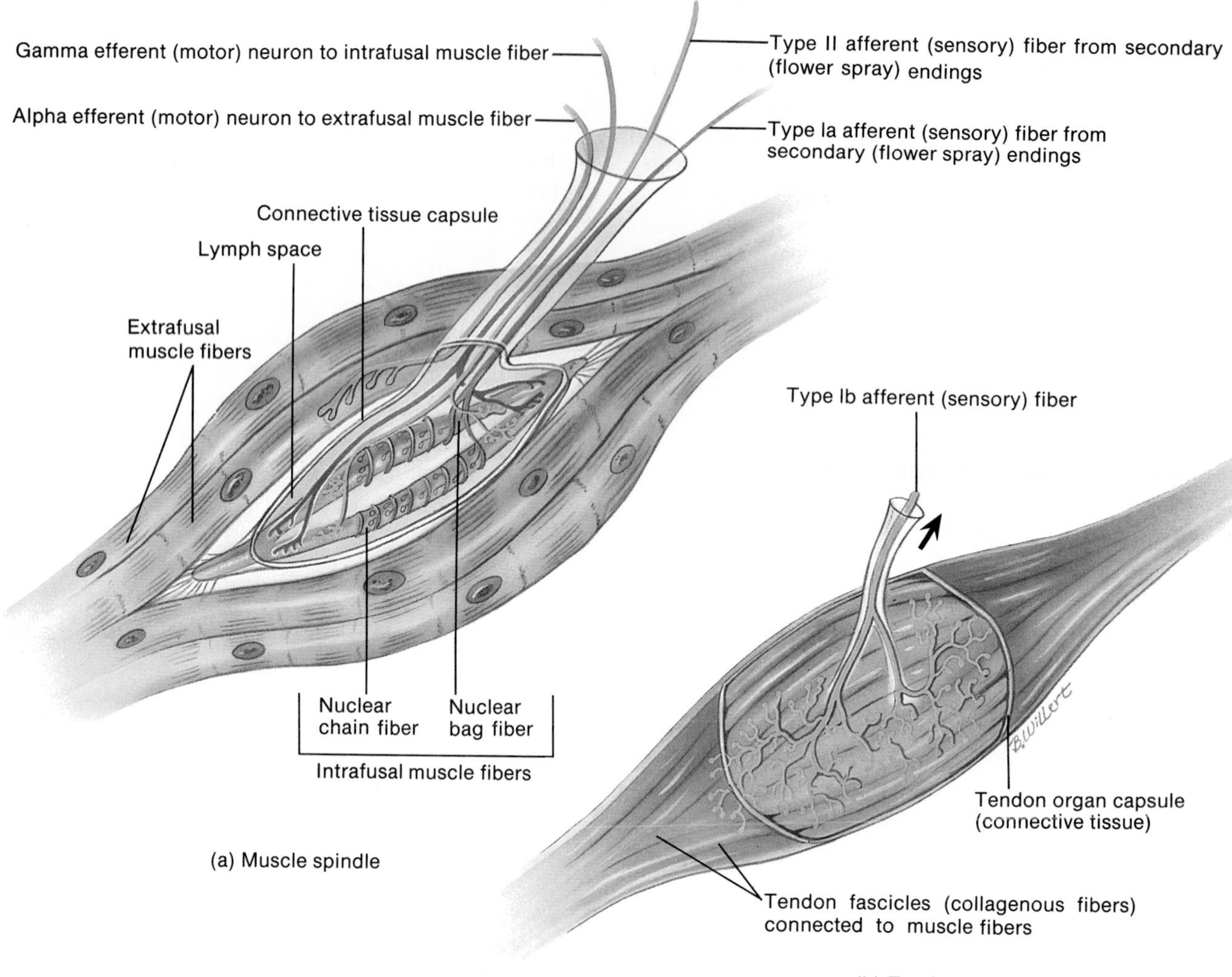

FIGURE 20-3 Proprioceptive receptors.

The ends of the intrafusal muscle fibers contain actin and myosin myofilaments and represent the contractile portions of the fibers. The ends of the fibers contract when stimulated by ***gamma efferent (motor) neurons.*** These neurons are small motor neurons located in the anterior gray horn of the spinal cord. The fibers of some gamma efferent neurons end as axon terminals on the ends of the intrafusal fibers. Extrafusal muscle fibers are innervated by large motor neurons called ***alpha efferent (motor) neurons.*** These neurons are also located in the anterior gray horn of the spinal cord near gamma efferent neurons.

Muscle spindles are stimulated in response to both sudden and prolonged stretch on the central areas of the intrafusal muscle fibers. The muscle spindles monitor changes in the length of a skeletal muscle by responding to the rate and degree of change in length. This information is relayed to the central nervous system to assist in the coordination and efficiency of muscle contraction (see Figure 20-5).

■ ***Tendon Organs*** ***Tendon organs*** **(*Golgi tendon organs*)** are proprioceptive receptors found at the junction of a tendon with a muscle. They help protect tendons and their associated muscles from damage resulting from excessive tension and also function as contraction receptors; that is, they monitor the force of contraction of each muscle. Each consists of a thin capsule of connective tissue, called a ***tendon organ capsule,*** that encloses a few ***tendon fascicles*** (collagenous fibers) (Figure 20-3b). The capsule is penetrated by one or more ***type Ib*** afferent (sensory) fibers whose terminal branches entwine among and around the collagenous fibers. When tension is applied to a tendon, tendon organs are stimulated, and the information is relayed to the central nervous system (see Figure 20-5).

■ ***Joint Kinesthetic Receptors*** There are several types of ***joint kinesthetic receptors*** within and around the articular capsules of synovial joints. Encapsulated receptors, similar to type II cutaneous mechanoreceptors (end organs of Ruffini), are present in the capsules of joints and respond to pressure. Small lamellated (Pacinian) corpuscles in the connective tissue outside articular capsules are receptors that respond to acceleration and deceleration. Articular ligaments contain receptors similar to tendon organs that mediate reflex inhibition of the adjacent muscles when excessive strain is placed on the joint.

■ ***Maculae and Cristae*** The proprioceptors in the internal ear are the ***macula*** of the saccule and the utricle and ***cristae*** in the semicircular ducts. Their function in equilibrium is discussed later in the chapter.

Proprioceptors adapt only slightly. This feature is advantageous since the brain must be apprised of the status of different parts of the body at all times so that adjustments can be made to ensure coordination.

The afferent pathway for muscle sense consists of impulses generated by proprioceptors via cranial and spinal nerves to the central nervous system (see Figure 20-5). Impulses for conscious proprioception pass along ascending tracts in the cord, where they are related to the thalamus and cerebral cortex. The sensation is registered in the general sensory area in the parietal lobe of the cerebral cortex posterior to the central sulcus. Proprioceptive impulses that result in reflex action pass to the cerebellum along spinocerebellar tracts and contribute to subconscious proprioception.

LEVELS OF SENSATION

As we have said, a receptor converts a stimulus into a nerve impulse by way of a generator or receptor potential, and only after that impulse has been conducted to a region of the spinal cord or brain can it be translated into a sensation. The nature of the sensation and the type of reaction generated vary with the level of the central nervous system at which the sensation is translated.

Sensory fibers terminating in the spinal cord can generate spinal reflexes without immediate action by the brain. Sensory fibers terminating in the lower brain stem bring about far more complex motor reactions than simple spinal reflexes. When sensory impulses reach the lower brain stem, they cause subconscious motor reactions. Sensory impulses that reach the thalamus can be localized crudely in the body. At the thalamic levels, sensations are sorted by modality—that is, identified as the *specific* sensation of touch, pressure, pain, position, hearing, or taste. When sensory information reaches the cerebral cortex, we experience precise localization. It is at this level that memories of previous sensory information are stored and the perception of sensation occurs on the basis of past experience.

SENSORY PATHWAYS

SOMATOSENSORY CORTEX

Most sensory information from receptors on one side of the body crosses over to the opposite side in the spinal cord or brain stem and then ascends to the ***somatosensory cortex*** (***primary somesthetic*** or ***general sensory area***) of the cerebral cortex where conscious sensations result (see Figure 18-13). Areas of the somatosensory cortex have been mapped out that represent the termination of sensory information from all parts of the body. Figure 20-4 shows the location and areas of representation of the somatosensory cortex of the right cerebral hemisphere. The left cerebral hemisphere has a duplicate somatosensory cortex.

Note that some parts of the body are represented by large areas in the somatosensory cortex, including the lips, face, tongue, and thumb. Other parts of the body, such as the trunk and lower extremities, are represented by relatively small areas. The relative sizes of the areas in the somatosensory cortex are directly proportional to the number of specialized sensory receptors in each respective part of

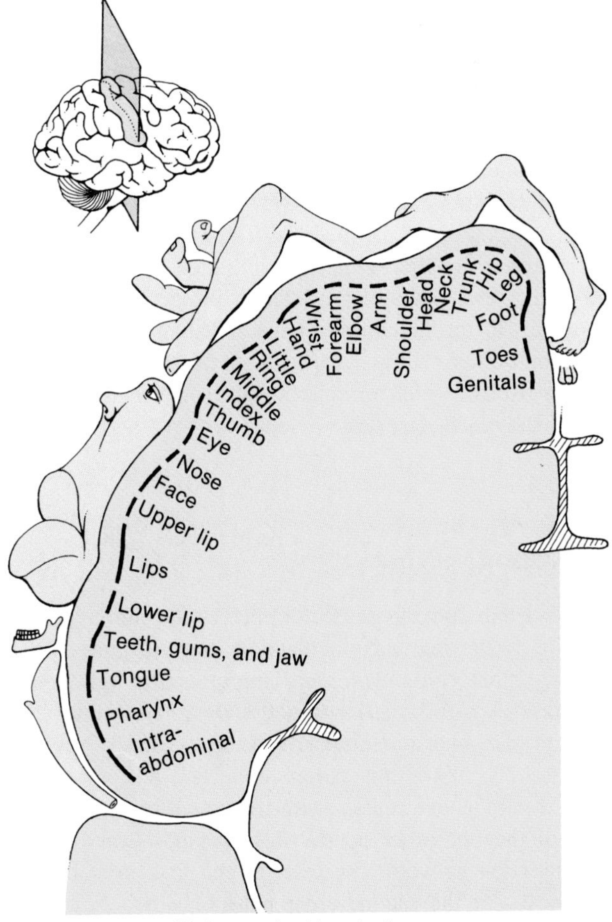

FIGURE 20-4 Somatosensory cortex. (After Penfield and Rasmussen.)

the body. Thus, there are numerous receptors in the skin of the lips but relatively few in the skin of the trunk. Essentially, the size of the area for a particular part of the body is determined by the functional importance of the part, and increased sensitivity is related to the size of the area.

Let us now examine how sensory information is transmitted from cutaneous receptors to the central nervous system. You will find it helpful to review the principal ascending and descending tracts of the spinal cord (see Exhibit 17-1 and Figure 17-4).

PROPRIOCEPTION, DISCRIMINATIVE TOUCH, TWO-POINT DISCRIMINATION, AND VIBRATION

Conscious proprioception, discriminative touch, two-point discrimination, and vibration involve the fasciculus gracilis or fasciculus cuneatus tracts. Three separate sensory neurons (Figure 20-5) are involved in transmitting sensory information to the cerebral cortex. The ***first-order neuron*** extends from the receptor into the spinal cord and up to the medulla on the same side of the body. The cell body of the first-order neuron is in the posterior (dorsal) root ganglion of a spinal nerve, and the axon ascends through the cord to the medulla. The axon of the first-order neuron synapses with the ***second-order neuron,*** the axon of which passes from the medulla upward to the thalamus. The cell body of the second-order neuron is located in the nucleus cuneatus or nucleus gracilis of the medulla. The axon of the second-order neuron crosses to the opposite side of the medulla and enters the medial lemniscus, a projection tract that terminates at the thalamus. In the thalamus, the axon of the second-order neuron synapses with the ***third-order neuron,*** the axon of which terminates in the somesthetic sensory area of the cerebral cortex.

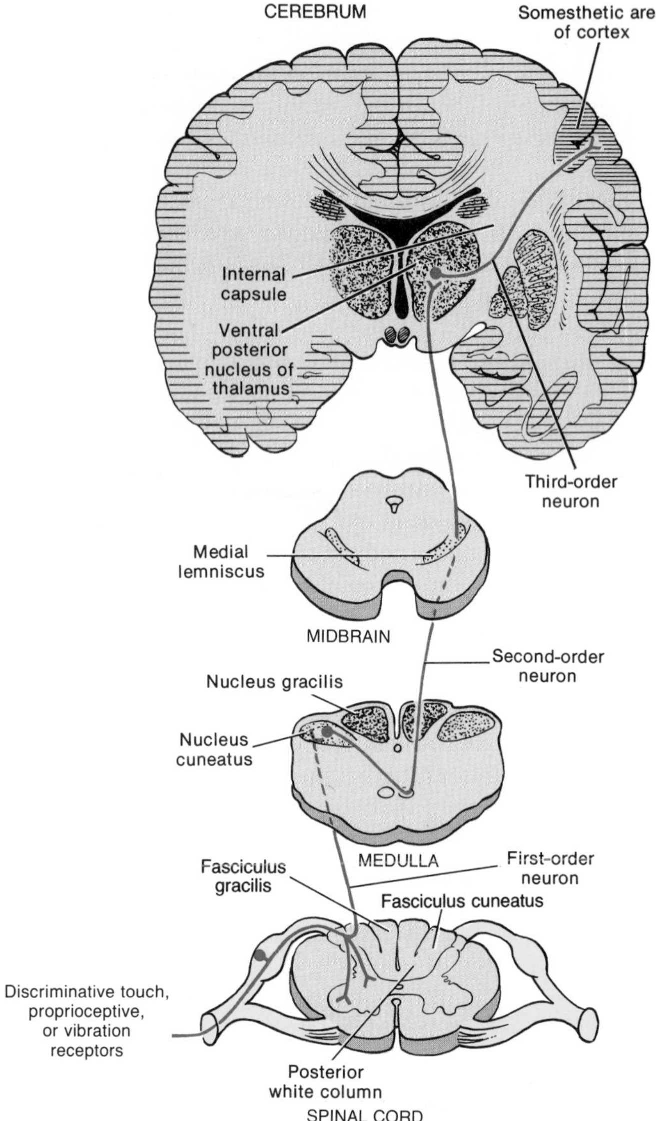

FIGURE 20-5 Sensory pathway for conscious proprioception, discriminative touch, two-point discrimination, and vibration.

PAIN, TEMPERATURE, LIGHT TOUCH, AND PRESSURE

The tract that conveys sensory impulses for pain and temperature is the lateral spinothalamic tract (Figure 20-6). The tract that conveys sensory impulses for light touch and pressure is the anterior spinothalamic tract (Figure 20-7). Both pathways are composed of three orders of sensory neurons. The first-order neuron connects a receptor of the neck, trunk, or extremities with the spinal cord. The cell body of the first-order neuron is in the posterior (dorsal) root ganglion as well. The axon of the first-order neuron synapses with the second-order neuron, which is located in the posterior gray horn of the spinal cord. The axon of the second-order neuron crosses to the opposite side of the spinal cord and passes upward to the brain stem in the lateral spinothalamic tract or anterior spinothalamic tract. The axon from the second-order neuron terminates in the thalamus. There, the axon of the second-order neuron synapses with the third-order neuron. The axon of the third-order neuron terminates in the somesthetic sensory area of the cerebral cortex.

CEREBELLAR TRACTS

Both the ***posterior spinocerebellar tract*** and ***anterior spinocerebellar tract*** participate in conveying impulses concerned with subconscious proprioception and thus assume a role in posture and muscle tone (see Exhibit 17-1).

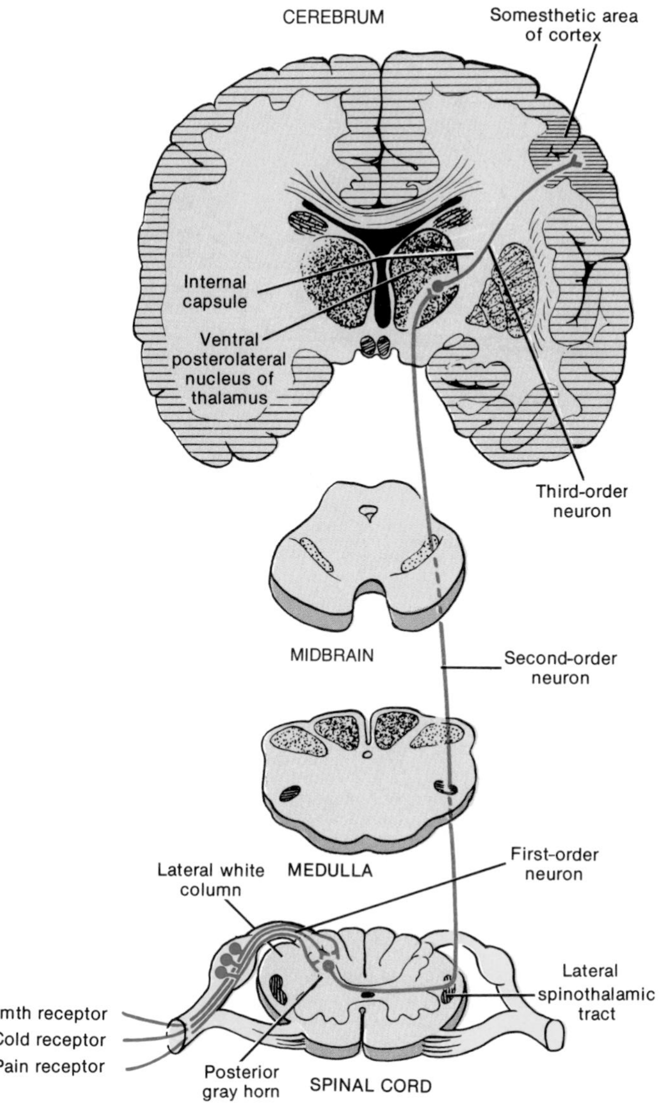

FIGURE 20-6 Sensory pathway for pain and temperature.

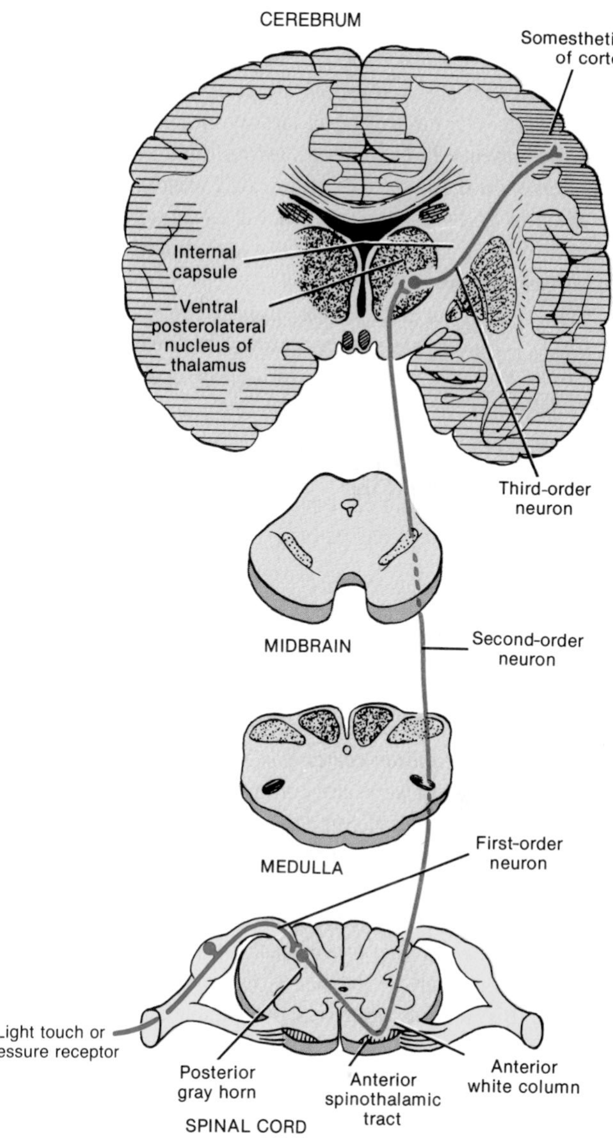

FIGURE 20-7 Sensory pathway for light touch and pressure.

MOTOR PATHWAYS

After receiving and interpreting sensations, the central nervous system generates nerve impulses to direct responses to that sensory input. Now we will look at the transmission of motor impulses that result in the movement of skeletal muscles.

LINKAGE OF SENSORY INPUT AND MOTOR RESPONSES

Sensory systems function to keep the CNS informed of certain changes in the external and internal environment. Responses to this information are brought about by the motor systems, which enable us to move about and alter the secretory rate of certain glands and change our relationship to the world around us. As sensory information is conveyed into the CNS, it becomes part of a large pool of sensory input. We do not actively respond to every bit of input the CNS receives. Rather, the incoming information is integrated with other information arriving from all other operating sensory receptors. The integration process occurs not just once but at many stations along the pathways of the CNS and at both a conscious and subconscious level. It occurs within the spinal cord, brain stem, cerebellum, and cerebral motor cortex. As a result, a motor response to make a muscle contract or a gland secrete can be initiated at any of these levels. The cerebral motor cortex assumes the major role for controlling precise, discrete muscular movements. The basal ganglia largely integrate semivoluntary movements like walking, swimming, and laughing. The cerebellum assists the motor cortex and basal ganglia by making body movements smooth and coordinated and by contributing significantly to the maintenance of normal posture and balance.

When the input reaches the highest center, the process of sensory–motor integration occurs. This process involves not only the utilization of information contained within that center but also the information impinging on that center that is delivered from other centers within the central nervous system. After integration occurs, the output of the center is sent down the spinal cord in two major descending motor pathways: the pyramidal pathways and the extrapyramidal pathways.

MOTOR CORTEX

Just as the somatosensory cortex has been mapped to indicate the termination of sensory information from many parts of the body, the ***motor cortex*** has been mapped to indicate which groups of muscles are controlled by its specific areas (Figure 20-8). The right cerebral hemisphere has a duplicate of the motor cortex in the left cerebral hemisphere. Note that the different muscle groups are not represented equally in the motor cortex. In general, the degree of representation is proportional to the preciseness of movement required of a particular part of the body. For example, the thumb, fingers, lips, tongue, and vocal cords have large representations. The trunk has a relatively small representation. By comparing Figures 20-4 and 20-8, you will see that somatosensory and motor representations are similar but not identical for the same part of the body.

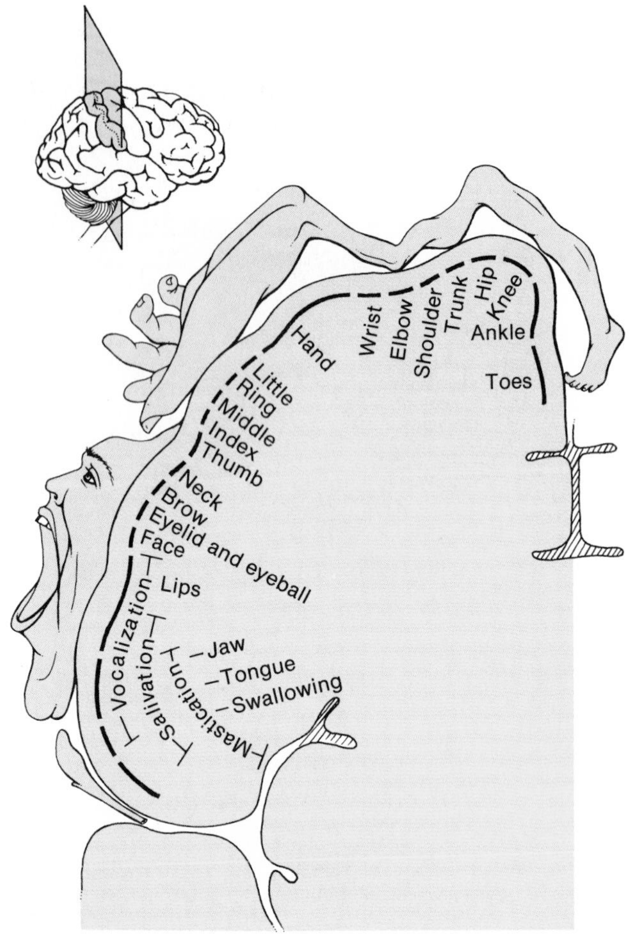

FIGURE 20-8 Motor cortex. (After Penfield and Rasmussen.)

PYRAMIDAL PATHWAYS

Voluntary motor impulses are conveyed from the motor areas of the brain to somatic efferent neurons (voluntary motor neurons) leading to skeletal muscles via the ***pyramidal*** (pi-RAM-i-dal) ***pathways*** (Figure 20-9). Most pyramidal axons originate from cell bodies in the precentral gyrus. They descend through the internal capsule of the cerebrum, and most cross to the opposite side of the brain. They terminate in nuclei of cranial nerves that innervate voluntary muscles or in the anterior gray horn of the spinal cord. A short association neuron probably completes the connection of the pyramidal axons with the cell bodies of the somatic (voluntary) motor neurons that activate voluntary muscles.

The pathways over which the nerve impulses travel from the motor cortex to skeletal muscles have two components: ***upper motor neurons*** (***pyramidal fibers***) in the brain and spinal cord and ***lower motor neurons*** (***peripheral fibers***) which extend from the spinal cord to the muscle fibers. (Lower motor neurons also originate in motor nuclei of cranial nerves within the brain and extend out to muscle fibers in the head.)

The various tracts of the pyramidal pathways convey nerve impulses from the cerebral cortex that result in precise muscular movements. These tracts include the ***lateral corticospinal, anterior corticospinal,*** and ***corticobulbar*** (see Exhibit 17-1 and Figure 17-4).

EXTRAPYRAMIDAL PATHWAYS

The ***extrapyramidal pathways*** include all descending tracts other than the pyramidal tracts. Generally, these include tracts that begin in the basal ganglia and reticular formation. The main extrapyramidal tracts are the rubrospinal, tectospinal, vestibulospinal, and reticulospinal (see Exhibit 17-1 and Figure 17-4).

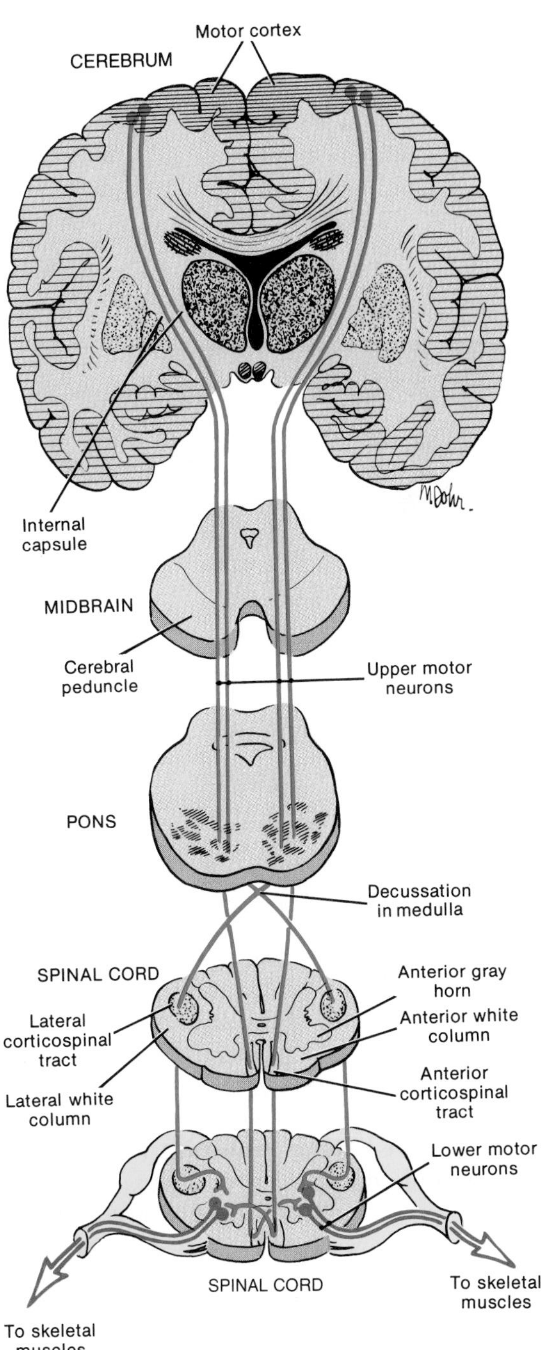

FIGURE 20-9 Pyramidal pathways.

CLINICAL APPLICATION

Flaccid and Spastic Paralysis

The lower motor and upper motor neurons are important clinically. If the lower motor neuron is damaged or diseased, there is neither voluntary nor reflex action of the muscle it innervates, and the muscle remains flaccid (decreased or lost muscle tone), a condition called ***flaccid paralysis.*** Injury or disease of upper motor neurons in a motor pathway is characterized by varying degrees of spasticity (increased muscle tone), exaggerated reflexes, and pathological reflexes, such as the Babinski sign. This condition is called ***spastic paralysis.*** The Babinski sign is dorsiflexion of the great toe accompanied by fanning of the lateral toes in response to stroking the plantar surface along the outer border of the foot. As we have seen, this response is normal only in infants; the normal adult response is plantar flexion of the toes.

SPECIAL SENSES

The special senses—smell, taste, sight, hearing, and equilibrium—have receptor organs that are structurally more complex than receptors for general sensations. The sense of smell is the least specialized, as opposed to the sense of sight, which is the most specialized. Like the general senses, however, the special senses allow us to detect changes in our environment.

OLFACTORY SENSATIONS

Structure of Receptors

The receptors for the ***olfactory*** (ol-FAK-tō-rē; *olfact* = smell) ***sense,*** or sense of smell, are located in the nasal epithelium in the superior portion of the nasal cavity on either side of the nasal septum (Figure 20-10). The nasal epithelium consists of three principal kinds of cells: supporting, olfactory, and basal. The ***supporting (sustentacular) cells*** are columnar epithelial cells of the mucous membrane lining the nose. The ***olfactory cells*** are bipolar neurons whose cell bodies lie between the supporting cells. The distal (free) end of each olfactory cell contains a dendrite that terminates in a swelling (***olfactory vesicle***) from which six to eight cilia, called ***olfactory hairs,*** radiate. The hairs are believed to react to odors and then to stimulate the olfactory cells, thus initiating the olfactory response. The proximal (basal) part of each olfactory cell contains a single process that represents its axon. ***Basal cells*** lie between the bases of the supporting cells and produce new olfactory cells. Within the connective tissue beneath the olfactory epithelium are ***olfactory (Bowman's) glands*** that produce mucus, which is carried to the surface of the epithelium by ducts. The secretion moistens the surface of the olfactory epithelium and serves as a solvent for odoriferous substances. The continuous secretion of mucus also serves to freshen the surface film of fluid and prevents continuous stimulation of olfactory hairs by the same odor. Continuous stimulation is also prevented by rapid adaptation of olfactory receptors.

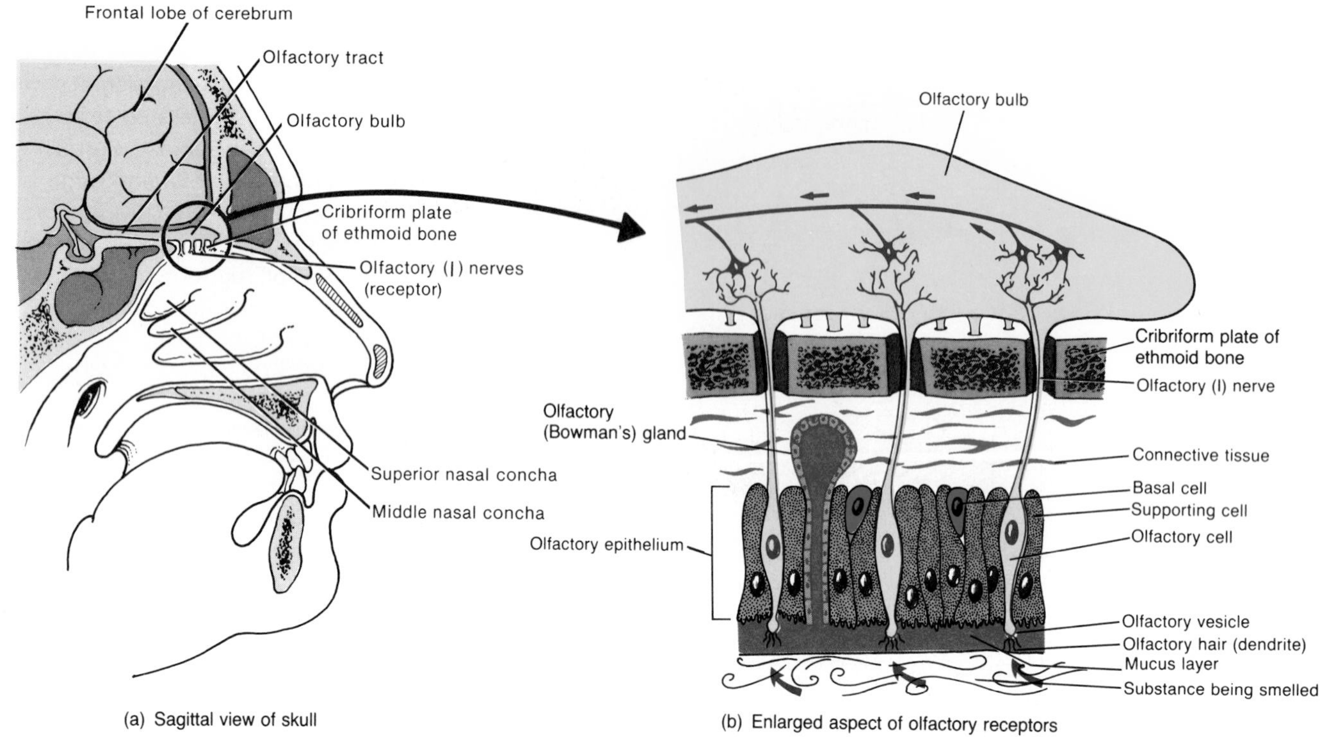

FIGURE 20-10 Olfactory receptors. (a) Location in nasal cavity. (b) Details.

Olfactory Pathway

The unmyelinated axons of the olfactory receptor cells unite to form the ***olfactory (I) nerves,*** which pass through foramina in the cribriform plate of the ethmoid bone (Figure 20-10a,b). The olfactory (I) nerves terminate in paired masses of gray matter in the brain called the ***olfactory bulbs.*** The olfactory bulbs lie beneath the frontal lobes of the cerebrum on either side of the crista galli of the ethmoid bone. The first synapse of the olfactory neural pathway occurs in the olfactory bulbs between the axons of the olfactory (I) nerves and the dendrites of neurons inside the olfactory bulbs. Axons of these neurons run posteriorly to form the ***olfactory tract.*** From here, nerve impulses are conveyed to the primary olfactory area of the cerebral cortex. In the cerebral cortex, the impulses are interpreted as odor and give rise to the sensation of smell. Impulses related to olfaction do not pass through the thalamus.

Both the supporting cells of the nasal epithelium and olfactory glands are innervated by branches of the facial (VII) nerve. The trigeminal (V) nerve receives stimuli of pain, cold, heat, tickling, and pressure.

GUSTATORY SENSATIONS

Structure of Receptors

The receptors for ***gustatory*** (GUS-ta-tō′-rē; *gust* = taste) ***sensations,*** or sensations of taste, are located in the taste buds (Figure 20-11). The nearly 2000 taste buds are most numerous on the tongue, but they are also found on the soft palate and in the throat. The ***taste buds*** are oval bodies consisting of three kinds of cells: supporting, gustatory, and basal. The ***supporting (sustentacular) cells*** are a specialized epithelium that forms a capsule, inside each of which are 4 to 20 ***gustatory cells.*** Each gustatory cell contains a hairlike process ***(gustatory hair)*** that projects to the external surface through an opening in the taste bud called the ***taste pore.*** Gustatory cells make contact with taste stimuli through the taste pore. ***Basal cells*** are found at the periphery of the taste bud near the basal lamina. These cells produce supporting and gustatory cells whose average life span is about 10 days.

Taste buds are found in some elevations on the tongue called ***papillae*** (pa-PIL-ē). The papillae give the upper surface of the tongue its rough appearance. ***Circumvallate*** (ser-kum-VAL-āt) or ***vallate papillae,*** the largest type, are circular and form an inverted V-shaped row at the posterior portion of the tongue. ***Fungiform*** (FUN-ji-form; meaning mushroom-shaped) ***papillae*** are knob-like elevations found primarily on the tip and sides of the tongue. All circumvallate and most fungiform papillae contain taste buds. ***Filiform*** (FIL-i-form) ***papillae*** are pointed threadlike structures that cover the anterior two-thirds of the tongue. They rarely contain taste buds.

Despite the many substances we seem to taste, there are basically only four primary taste sensations: sour, salt,

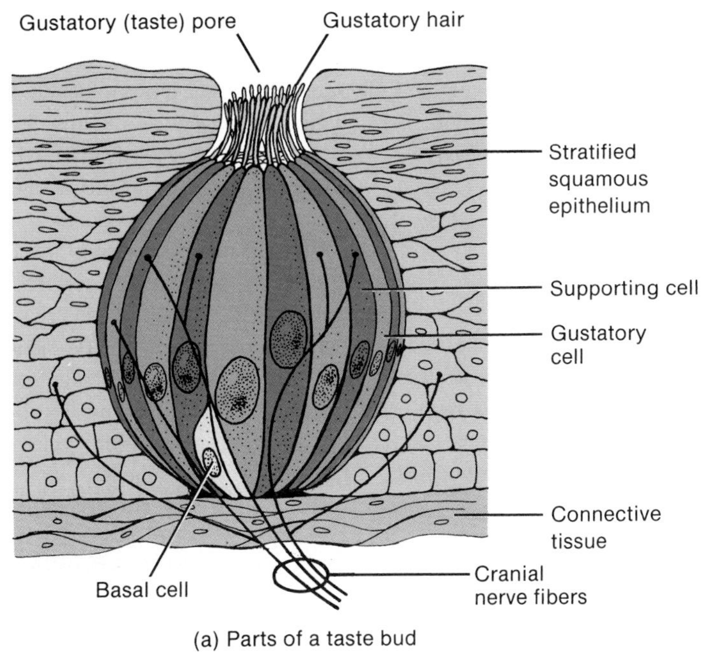

(a) Parts of a taste bud

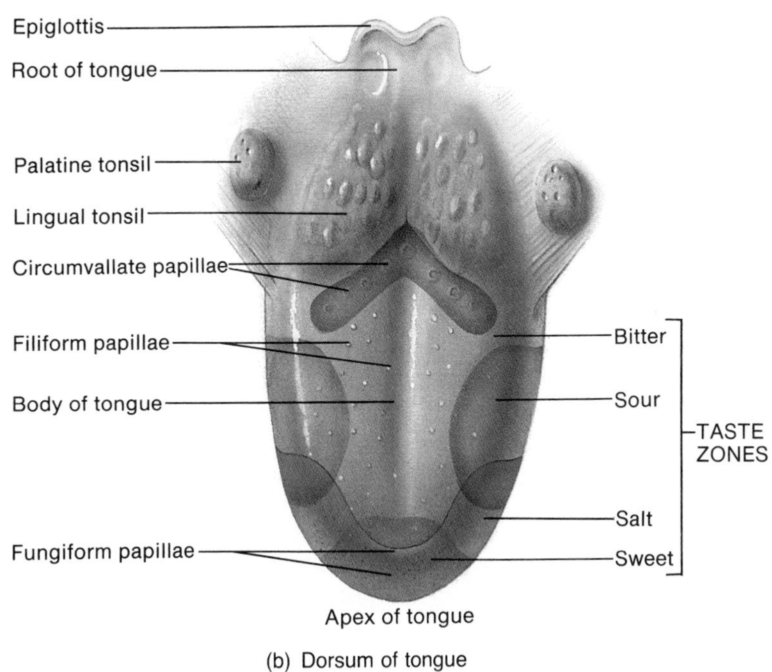

(b) Dorsum of tongue

FIGURE 20-11 Gustatory receptors. Locations of papillae and four taste zones are shown in (b).

bitter, and sweet. All other "tastes," such as chocolate, pepper, and coffee, are combinations of these four that are modified by accompanying olfactory sensations.

Persons with colds or allergies sometimes complain that they cannot taste their food. Although their taste sensations may be operating normally, their olfactory sensations are not. This illustrates that much of what we think of as taste is actually smell. Odors from foods pass upward into the nasopharynx (portion of the throat behind the nose) and stimulate the olfactory system. In fact, a given concentration of a substance will stimulate the olfactory system thousands of times more than it stimulates the gustatory system.

Each of the four primary tastes is caused by a different response to different chemicals. Certain regions of the tongue react more strongly than others to certain taste sensations. Although the tip of the tongue reacts to all four primary taste sensations, it is highly sensitive to sweet and salty substances. The posterior portion of the tongue is highly sensitive to bitter substances. The lateral edges of the tongue are more sensitive to sour substances (Figure 20-11b).

Gustatory Pathway

The cranial nerves that supply afferent fibers to taste buds are the facial (VII), which supplies the anterior two-thirds of the tongue; the glossopharyngeal (IX), which supplies the posterior one-third of the tongue; and the vagus (X), which supplies the throat and epiglottis. Taste impulses are conveyed from the gustatory cells in taste buds along the nerves to the medulla and then to the thalamus. They terminate in the primary gustatory area in the parietal lobe of the cerebral cortex.

VISUAL SENSATIONS

The study of the structure, function, and diseases of the eye is known as ***ophthalmology*** (of′-thal-MOL-ō-jē; *ophthalmo* = eye; *logos* = study of). A physician who specializes in the diagnosis and treatment of eye disorders with drugs, surgery, and corrective lenses is known as an ***ophthalmologist,*** whereas an ***optometrist*** is a specialist with a doctorate in optometry who is licensed to examine and test the eyes and treat visual defects by prescribing corrective lenses. An ***optician*** is a technician who fits, adjusts, and dispenses corrective lens prescribed by an ophthalmologist or optometrist.

The structures related to vision are the eyeball, the optic (II) nerve, the brain, and a number of accessory structures.

Accessory Structures of Eye

Among the ***accessory structures*** of the eye are the eyebrows, eyelids, eyelashes, and the lacrimal (tearing) apparatus (Figure 20-12). The ***eyebrows*** form a transverse arch at the junction of the upper eyelid and forehead. Structurally, they resemble the hairy scalp. The skin of the eyebrows is richly supplied with sebaceous (oil) glands. The hairs are generally coarse and directed laterally. Deep to the skin of the eyebrows are the fibers of the orbicularis oculi muscles. The eyebrows help to protect the eyeballs from foreign objects, perspiration, and the direct rays of the sun.

The upper and lower ***eyelids,*** or ***palpebrae*** (PAL-pe-brē), have several important roles. They shade the eyes during sleep, protect the eyes from excessive light and foreign objects, and spread lubricating secretions over the eyeballs. The upper eyelid is more movable than the lower and contains in its superior region a special levator muscle known as the ***levator palpebrae superioris.*** The space between the upper and lower eyelids that exposes the eyeball is called the ***palpebral fissure.*** Its angles are known as the ***lateral commissure*** (KOM-i-shūr), which is narrower and closer to the temporal bone, and the ***medial commissure,*** which is broader and nearer the nasal bone. In the medial commissure, there is a small, reddish elevation, the ***lacrimal caruncle*** (KAR-ung-kul), containing sebaceous (oil) and sudoriferous (sweat) glands. A whitish material secreted by the caruncle collects in the medial commissure.

From superficial to deep, each eyelid consists of epidermis, dermis, subcutaneous tissue, fibers of the orbicularis oculi muscle, a tarsal plate, tarsal glands, and a conjunctiva. The ***tarsal plate*** is a thick fold of connective tissue that forms much of the inner wall of each eyelid and gives form and support to the eyelids. Embedded in grooves on the deep surface of each tarsal plate is a row of elongated glands known as ***tarsal*** or ***Meibomian*** (mī-BŌ-mē-an) ***glands.*** These are modified sebaceous glands, and their oily secretion helps keep the eyelids from adhering to each other. Infection of the tarsal glands produces a tumor or cyst on the eyelid called a ***chalazion*** (ka-LĀ-zē-on). The ***conjunctiva*** (kon′-junk-TĪ-va) is a thin mucous membrane. It is called the ***palpebral conjunctiva*** when it lines the inner aspect of the eyelids. It is called the ***bulbar (ocular) conjunctiva*** when it is reflected from the eyelids onto the anterior surface of the eyeball. When the blood vessels of the bulbar conjunctiva are dilated and congested due to local irritation or infection, the person has bloodshot eyes.

Projecting from the border of each eyelid, anterior to the tarsal glands, is a row of short, thick hairs, the ***eyelashes.*** In the upper lid, they are long and turn upward; in the lower lid, they are short and usually turn downward. Sebaceous glands at the base of the hair follicles of the eyelashes, called ***sebaceous ciliary glands*** or ***glands of Zeis*** (ZĪS), pour a lubricating fluid into the follicles. Infection of these glands is called a ***sty.***

The ***lacrimal*** (*lacrima* = tear) ***apparatus*** is a term used for a group of structures that manufactures and drains tears. These structures are the lacrimal glands, excretory lacrimal ducts, lacrimal canals, lacrimal sacs, and nasolacrimal ducts. A ***lacrimal gland*** is a compound tubuloacinar gland located at the superior anterolateral portion of each orbit. Each is about the size and shape of an almond. Leading from the lacrimal glands are 6 to 12 ***excretory lacrimal ducts*** that empty lacrimal fluid, or tears, onto the surface of the conjunctiva of the upper lid. From here the lacrimal fluid passes medially and enters two small openings called ***lacrimal puncta*** that appear as two small pores, one in each eyelid, at the medial commissure of the eye. The lacrimal secretion then passes into two ducts, the ***lacrimal canals,*** and is next conveyed into the lacrimal sac. The lacrimal canals are located in the lacrimal grooves of the lacrimal bones. The ***lacrimal sac*** is the superior expanded portion of the ***nasolacrimal duct,*** a canal that transports the lacrimal secretion into the inferior meatus of the nose.

The ***lacrimal secretion*** is a watery solution containing salts, some mucus, and a bactericidal enzyme called ***lyso-***

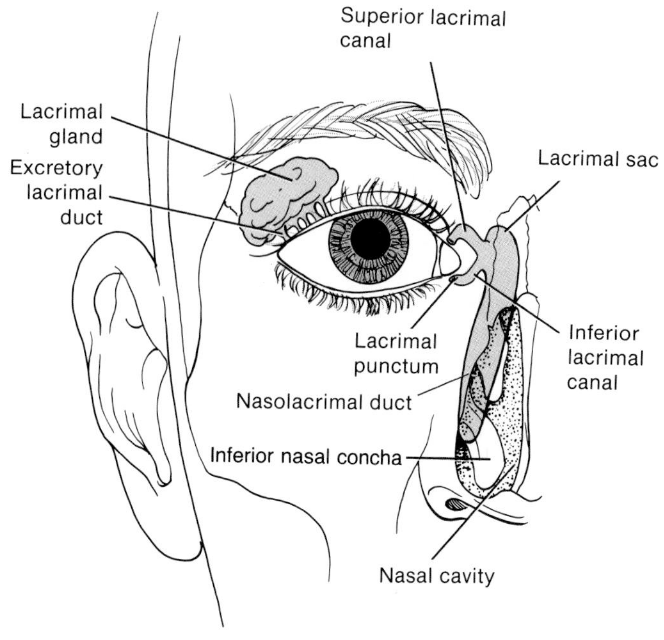

(a) Anterior view

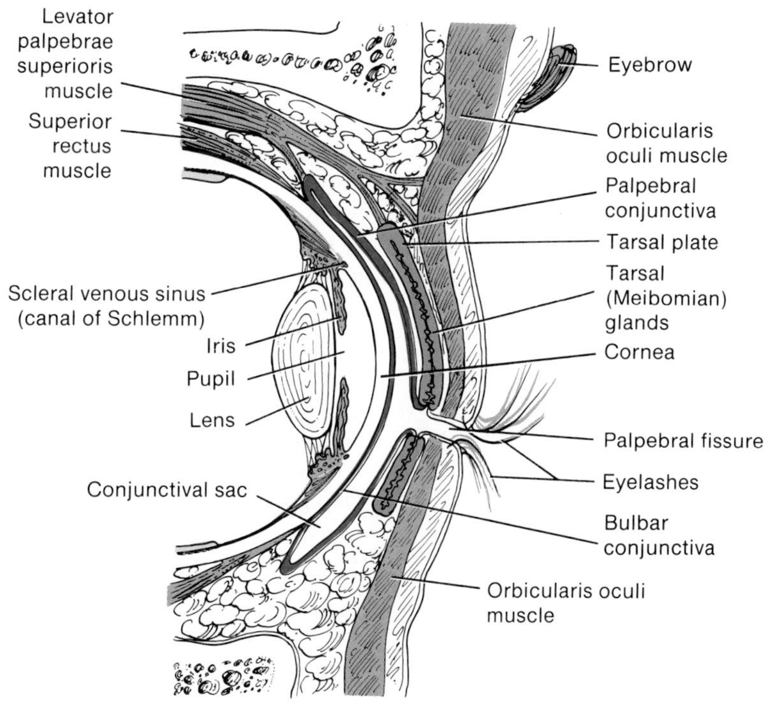

(b) Sagittal section

FIGURE 20-12 Accessory structures of the eye.

zyme. It cleans, lubricates, and moistens the eyeball. After being secreted by the lacrimal glands, it is spread medially over the surface of the eyeball by the blinking of the eyelids. Usually, 1 ml per day is produced by each gland.

The surface anatomy of the accessory structures of the eye and the eyeball is shown in Figure 11-2.

CLINICAL APPLICATION

"Watery" Eyes

Normally, the lacrimal secretion is carried away by evaporation or by passing into the lacrimal canals and then into the nasal cavity as fast as it is produced. If, however, an irritating substance makes contact with the conjunctiva, the lacrimal glands are stimulated to oversecrete. Tears then accumulate more rapidly than they can be carried away, a condition commonly referred to as ***"watery" eyes.*** This is a protective mechanism, since the tears dilute and wash away the irritating substance. "Watery" eyes also occur when an inflammation of the nasal mucosa, such as a cold, obstructs the nasolacrimal ducts so that drainage of tears is blocked. Humans are unique in that they have the ability to cry to express certain emotions. In response to parasympathetic stimulation, the lacrimal glands produce excessive tears that may spill over the edges of the eyelids and even fill the nasal cavity with fluid. Crying may indicate happiness or sadness.

Structure of Eyeball

The adult ***eyeball*** measures about 2.5 cm (1 in.) in diameter. Of its total surface area, only the anterior one-sixth is exposed. The remainder is recessed and protected by the orbit into which it fits. Anatomically, the eyeball can be divided into three layers: fibrous tunic, vascular tunic, and retina or nervous tunic (Figure 20-13a).

■ ***Fibrous Tunic*** The ***fibrous tunic*** is the outer coat of the eyeball. It can be divided into two regions: the posterior portion is the sclera, and the anterior portion is the cornea. The ***sclera*** (SKLE-ra; *skleros* = hard), the "white of the eye," is a white coat of dense fibrous tissue that covers all the eyeball except the most anterior portion, the cornea. The sclera gives shape to the eyeball and protects its inner parts. Its posterior surface is pierced by the optic (II) nerve. The ***cornea*** (KOR-nē-a) is a nonvascular, transparent, fibrous coat through which the iris can be seen. The cornea's outer surface is covered by an epithelial layer continuous with the epithelium of the bulbar conjunctiva. At the junction of the sclera and cornea is a venous sinus known as the ***scleral venous sinus,*** or ***canal of Schlemm.***

CLINICAL APPLICATION

Corneal Surgery

Corneal transplants are the most common organ transplant operation, and they are considered to be the most successful type of transplant since they are rarely rejected. The reason is that the cornea is avascular, and antibodies that might cause rejection do not circulate there. (The cornea receives nourishment from tears and a watery fluid in the eyeball called aqueous humor.) The surgical procedure is performed under an operating microscope at magnifications of between 6× and 40×. A general or local anesthesia is used. The defective cornea, usually 7 to 8 mm (about 0.3 in.) in diameter, is excised and replaced with a donor cornea of similar diameter. The transplanted cornea is sewn into place with nylon sutures, which stay in place for about a year because of slow corneal healing. Patients are later fitted with hard or soft contact lenses or eyeglasses. (The shortage of donated corneas has been partially overcome by the development of artificial corneas made of plastic.)

In the process of normal vision, light rays pass through the cornea, as well as other structures, to form a clear image on the retina (nervous tunic). In myopia (nearsightedness), the eyeball is elongated or the lens is thickened, and images of distant objects fall short of the retina, causing blurred vision. Many people with myopia may be helped by a surgical procedure called ***radial keratotomy (RK).*** In the procedure, several microscopic incisions, like the spokes of a wheel, are made in the cornea. This flattens the cornea and improves vision without using corrective lenses. Because of various problems associated with the procedure (undercorrected vision, overcorrected vision, and extra sensitivity to glare at night or changes in visual acuity during the day), it appears that it is best suited to people with severe myopia.

An alternative procedure to radial keratotomy for treating myopia is known as ***epikeratoplasty*** (ep′-ē-KER-a-tō-plas′-tē). In one modification of this procedure, a small circular area of defective cornea is removed and a commercially prepared piece of donor cornea, shaped to provide proper curvature, is sewn into its place.

■ ***Vascular Tunic*** The ***vascular tunic*** or ***uvea*** (YOO-vē-a) is the middle layer of the eyeball and is composed of three portions: choroid, ciliary body, and iris (Figure 20-13a). The ***choroid*** (KŌ-royd), the posterior portion of the vascular tunic, is a thin, dark brown membrane that lines most of the internal surface of the sclera. It contains numerous blood vessels and a large amount of pigment. The choroid absorbs light rays so that they are not reflected within the eyeball. Through its blood supply, it nourishes the retina. Like the sclera, it is pierced by the optic (II) nerve at the back of the eyeball.

In the anterior portion of the vascular tunic, the choroid becomes the ***ciliary*** (SIL-ē-ar′-ē) ***body.*** It is the thickest portion of the vascular tunic. It extends from the ***ora serrata*** (Ō-ra ser-RĀ-ta) of the retina (nervous tunic) to a point just behind the sclerocorneal junction. (The ora serrata is simply the jagged margin of the retina.) The ciliary body consists of the ciliary processes and ciliary muscle. The ***ciliary processes*** consist of protrusions or folds on the internal surface of the ciliary body that secrete aqueous humor. The ***ciliary muscle*** is a smooth muscle that alters the shape of the lens for near or far vision.

The ***iris*** (*irid* = colored circle) is the third portion of the vascular tunic. It is the colored portion seen through the cornea and consists of circular iris and radial iris smooth muscle fibers (cells) arranged to form a doughnut-shaped structure. The black hole in the center of the iris is the ***pupil,*** the area through which light enters the eyeball. The iris is suspended between the cornea and the lens and is attached at its outer margin to the ciliary process. A principal function of the iris is to regulate the amount of light entering the posterior cavity of the eyeball. When the eye is stimulated by bright light, the circular iris (constrictor pupillae) muscles contract and decrease the size of the pupil (constriction). When the eye must adjust to dim light, the radial iris (dilator pupillae) muscles contract and increase the pupil's size (dilation).

■ ***Retina (Nervous Tunic)*** The third and inner coat of the eye, the ***retina (nervous tunic),*** lies only in the posterior portion of the eye. Images formed by the lens are focused on the retina. The retina consists of an inner nervous tissue layer (visual portion) and an outer pigmented layer (nonvisual portion). The retina lines the choroid. At the edge of the ciliary body, it terminates in a scalloped border called the ora serrata, at which the nervous layer or visual portion of the retina ends. The pigmented layer extends anteriorly

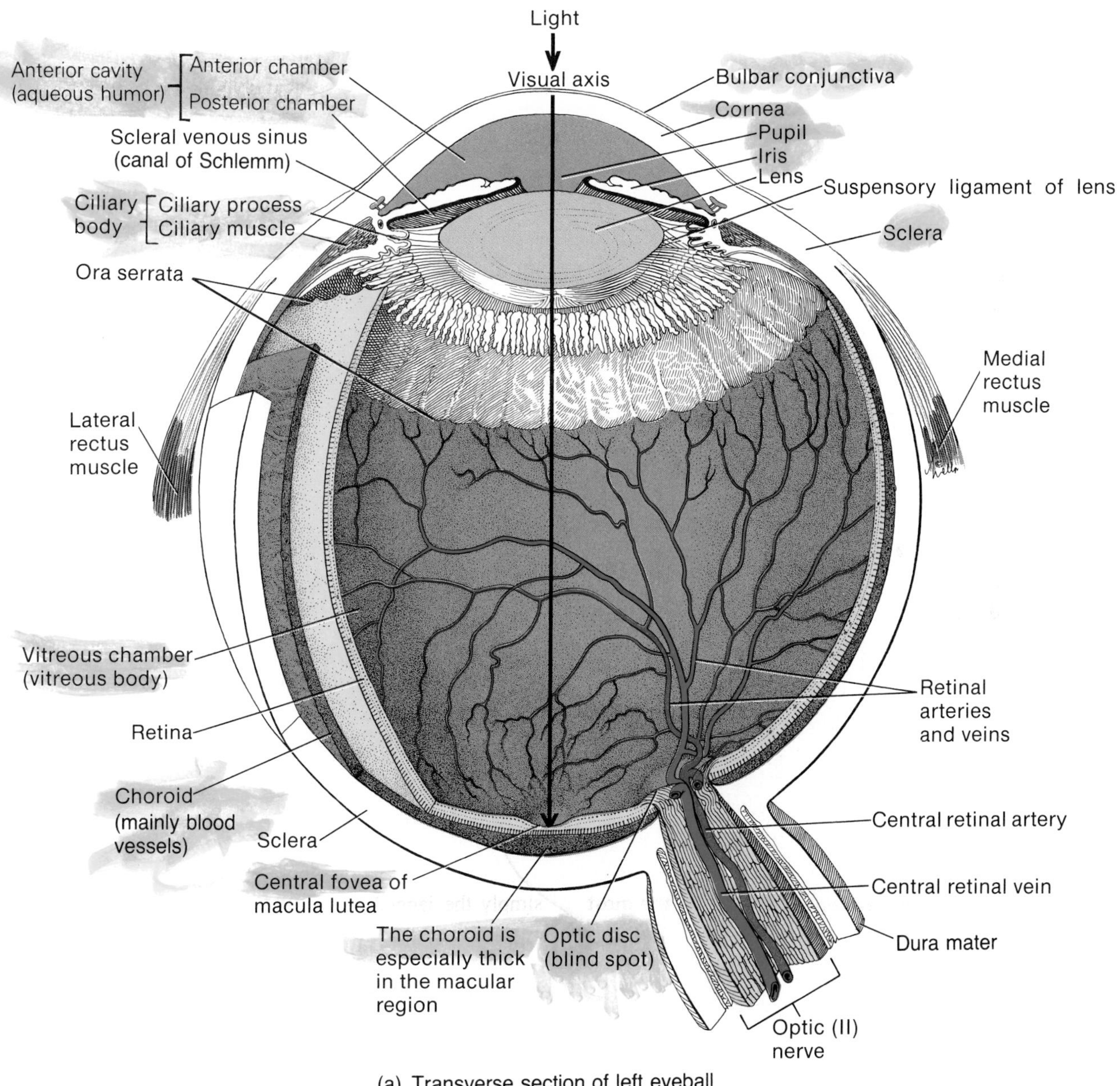

(a) Transverse section of left eyeball

FIGURE 20-13 Structure of the eyeball. The diagram in (c) is exaggerated for emphasis. The arrows pointing upward indicate the direction of the signal passing through the nervous layer of the retina, which ultimately results in a nerve impulse that passes into the optic (II) nerve.

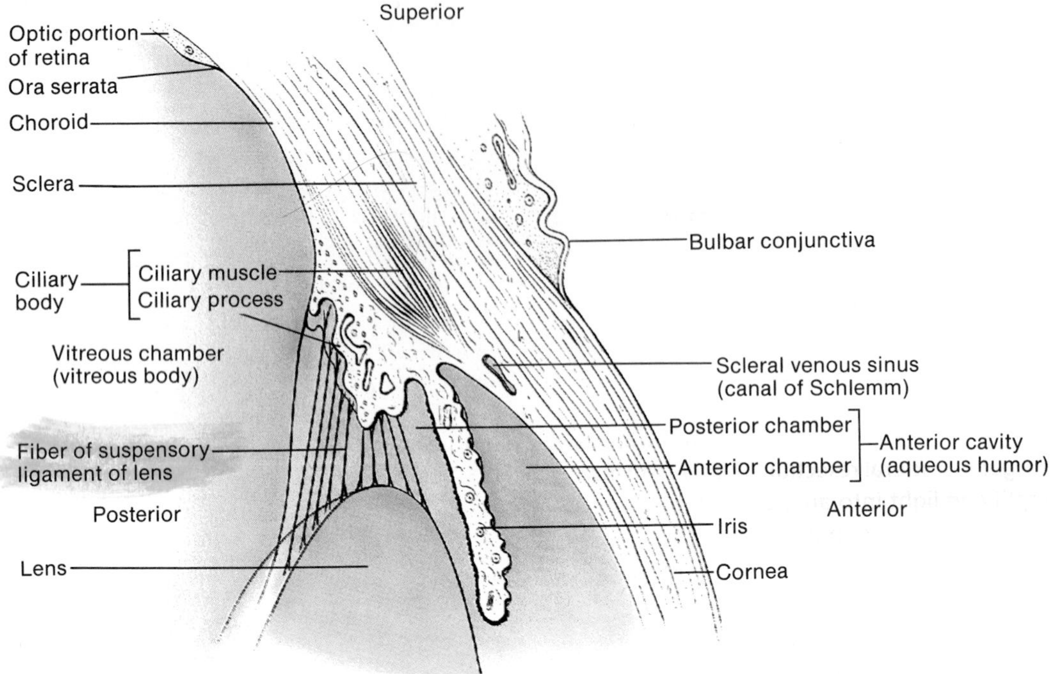

(b) Section through anterior portion of eyeball at sclerocorneal junction

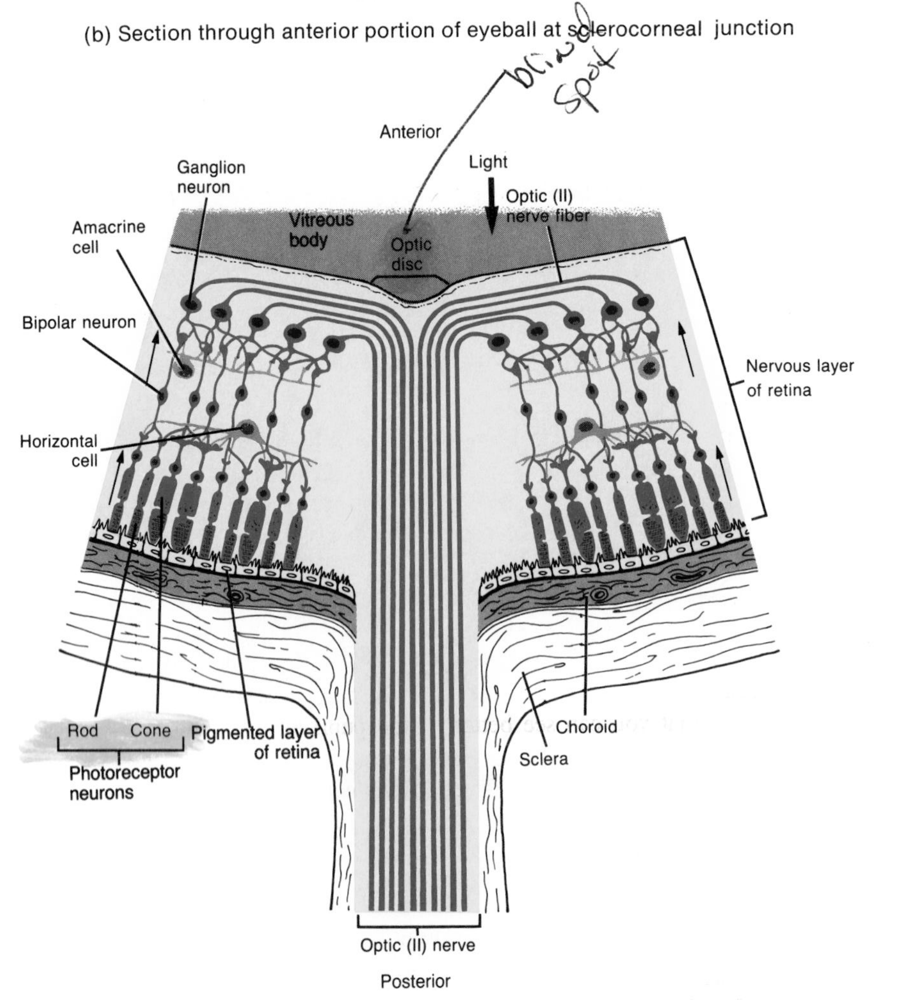

(c) Diagram of microscopic structure of retina

over the back of the ciliary body and the iris as the nonvisual portion of the retina.

CLINICAL APPLICATION

Detached Retina

Detachment of the retina may occur in trauma, such as a blow to the head, or may be secondary to various intraocular disorders. The actual detachment occurs between the inner nervous tissue layer and outer pigmented layer. Fluid accumulates between these layers, forcing the thin, pliable retina to billow out, resulting in distorted vision and blindness in the corresponding field of vision. The retina may be reattached by photocoagulation by laser beam, cryosurgery, or scleral resection. (A laser is an instrument for amplifying light into an intense, highly concentrated beam that is strong enough to burn tissue.)

CLINICAL APPLICATION

Senile Macular Degeneration

In a disease called ***senile macular degeneration* (*SMD*),** new blood vessels grow over the macula lutea. The effect ranges from distorted vision to blindness. SMD accounts for nearly all new cases of blindness in people over 65. Its cause is unknown. Laser beam treatment has been used effectively in arresting blood vessel proliferation and restoring normal vision in some cases. In order to determine whether SMD exists, a simple test can be performed without any assistance or special instruments. Stare, one eye at a time (covering the other with your hand), at any long straight line, such as a door frame. If either eye perceives the line as bent or twisted, or if a black spot appears, a physician should be informed immediately.

The nervous layer of the retina contains three zones of neurons (Figure 20-13c). These three zones, named in the order in which they conduct nerve signals, are ***photoreceptor neurons, bipolar neurons,*** and ***ganglion neurons.*** The dendrites of the photoreceptor neurons are called rods or cones because of the shape of their outer segments. They are visual receptors highly specialized for stimulation by light rays. Functionally, rods and cones develop generator potentials. ***Rods*** are specialized for black-and-white vision in dim light. They also allow us to discriminate between different shades of dark and light and permit us to see shapes and movement. ***Cones*** are specialized for color vision and sharpness of vision (***visual acuity***) in bright light. This is why we cannot see color by moonlight. It is estimated that there are 3 million cones and 100 million rods. Cones are most densely concentrated in the ***central fovea,*** a small depression in the center of the macula lutea. The ***macula lutea*** (MAK-yoo-la LOO-tē-a; *macula* = spot; *lutea* = yellow) is in the exact center of the posterior portion of the retina, corresponding to the visual axis of the eye. The fovea is the area of sharpest vision because of the high concentration of cones. Rods are absent from the fovea and macula and increase in density toward the periphery of the retina. It is for this reason that you can see better at night while not looking directly at an object.

The principal blood supply of the retina is from the ***central retinal artery,*** a branch of the ophthalmic artery. The central retinal artery enters the retina at about the middle of the optic disc, an area where the optic (II) nerve exits the eyeball (Figure 20-13a). After emerging through the disc, the central retinal artery divides into superior and inferior branches, each of which subdivides into nasal and temporal branches. The ***central retinal vein*** drains blood from the retina through the optic disc.

When information has passed through the photoreceptor neurons, it is conducted across synapses to the bipolar neurons in the intermediate zone of the nervous layer of the retina. From here it is passed to the ganglion neurons. These cells transmit their signals through optic (II) nerve fibers to the brain in the form of nerve impulses. Many rods connect with one bipolar neuron, and many of these bipolar neurons transmit impulses to one ganglion cell. This arrangement greatly lowers visual acuity, but it permits summation effects to occur so that low levels of light can stimulate a ganglion cell that would not respond had it been connected directly to a cone. The synaptic connections thus contribute much to the difference in visual acuity and light sensitivity.

The axons of the ganglion neurons extend posteriorly to a small area of the retina called the ***optic disc*** (***blind spot***). This region contains openings through which the axons of the ganglion neurons exit as the optic (II) nerve. Since it contains no rods or cones, an image striking it cannot be seen. Thus, it is called the blind spot.

■ ***Lens*** In addition to the fibrous tunic, vascular tunic, and retina, the eyeball itself contains the lens, just behind the pupil and iris. The ***lens*** is constructed of three basic parts: lens capsule, lens epithelium, and lens fibers. The ***lens capsule*** surrounds the entire lens. It is formed by the epithelium of the lens and consists of reticular fibers embedded in a matrix of glycoproteins and sulfated glycosaminoglycan. The ***lens epithelium*** is composed of simple cuboidal epithelium found only on the anterior surface of the lens. At the equator (peripheral edge) of the lens, the lens epithelial cells elongate into columnar cells called ***lens fibers.*** These cells compose the bulk of the lens and are formed by the division and differentiation of the lens epithelial cells. Lens fibers lose their nuclei and consist of a few microtubules and occasional mitochondria in their cytoplasm. As new lens fibers are added, the lens enlarges. Normally, the lens is perfectly transparent. It is enclosed by a clear connective tissue capsule and held in position by ***suspensory ligaments.*** The lens helps focus light rays for clear vision. A loss of transparency of the lens is known as a ***cataract.***

■ ***Interior*** The interior of the eyeball is a large space divided into two cavities by the lens: anterior cavity and vitreous chamber. The ***anterior cavity,*** the division anterior to the lens, is further divided into the ***anterior chamber,*** which lies behind the cornea and in front of the iris, and the ***posterior chamber,*** which lies behind the iris and in front of the suspensory ligaments and lens. The anterior cavity is filled with a watery fluid, similar to cerebrospinal fluid, called the ***aqueous*** (*aqua* = water) ***humor.*** The fluid is believed to be secreted into the posterior chamber by choroid plexuses of the ciliary processes of the ciliary bodies behind the iris. Once the fluid is formed, it passes into the posterior chamber and then passes forward between the iris and the lens, through the pupil, into the anterior chamber. From the anterior chamber, the aqueous humor, which is continually produced, is drained off into the scleral venous sinus (canal of Schlemm) and then into the blood. The anterior chamber thus serves a function similar to the subarachnoid space around the brain and spinal cord. The scleral venous sinus is analogous to a venous sinus of the dura mater. The pressure in the eye, called ***intraocular pressure (IOP),*** is produced mainly by the aqueous humor. The intraocular pressure, along with the vitreous body, maintains the shape of the eyeball and keeps the retina smoothly applied to the choroid so that clear images will result. Normal intraocular pressure (about 16 mm Hg) is maintained by drainage of the aqueous humor through the scleral venous sinus. Excessive intraocular pressure, called ***glaucoma*** (glaw-KŌ-ma), results in degeneration of the retina and blindness. Besides maintaining intraocular pressure, the aqueous humor is also the principal link between the cardiovascular system and the lens and cornea. Neither the lens nor the cornea has blood vessels.

The second, and larger, cavity of the eyeball is the ***vitreous chamber*** (***posterior cavity***). It lies between the lens and the retina and contains a jellylike substance called the ***vitreous body.*** This substance contributes to intraocular pressure, helps to prevent the eyeball from collapsing, and holds the retina flush against the internal portions of the eyeball. The vitreous body, unlike the aqueous humor, does not undergo constant replacement. It is formed during embryonic life and is not replaced thereafter.

A summary of structures associated with the eyeball is presented in Exhibit 20-1.

EXHIBIT 20-1

Summary of Structures Associated with the Eyeball

STRUCTURE	FUNCTION
Fibrous Tunic	
Sclera	Provides shape and protects inner parts.
Cornea	Admits and refracts light. (Refraction is the bending of light rays at the surface of two media with different densities.)
Vascular Tunic	
Choroid	Provides blood supply and absorbs light.
Ciliary Body	Secretes aqueous humor and alters shape of lens for near or far vision (accommodation).
Iris	Regulates amount of light that enters eyeball.
Retina (Nervous Tunic)	Receives light, converts light into generator potentials and nerve impulses, and transmits impulses to the optic (II) nerve.
Lens	Refracts light.
Anterior Cavity	Contains aqueous humor that helps maintain shape of eyeball and refracts light.
Vitreous Chamber (Posterior Cavity)	Contains vitreous body that helps maintain shape of eyeball, keeps retina applied to choroid, and refracts light.

CLINICAL APPLICATION

Ophthalmoscopy

Ophthalmoscopy (of′-thal-MOS-kō-pē; *ophthalmo* = eye; *skopein* = to view) is a procedure used to evaluate the interior (fundus) of the eye and especially to detect retinal changes associated with hypertension, diabetes mellitus, atherosclerosis, and increased intracranial pressure resulting from head injury, brain tumor, meningitis, or encephalitis.

An ophthalmoscope is used to examine the interior of the eyeball, with or without the use of dilating drops. The instrument contains a light source and a set of mirrors or prism to reflect light so that the interior of the eyeball can be examined. The room lights are turned down or off and the patient is instructed to look at a specific object straight ahead while the examiner directs light to each eye. Such a test permits direct examination of the retina (nervous tunic), optic disc, optic nerve (II), macula lutea, and retinal blood vessels.

Visual Pathway

Before light can reach the rods and cones of the retina to result in image formation, it must pass through the cornea, aqueous humor, pupil, lens, and vitreous body. For vision to occur, light reaching the rods and cones must form an image on the retina. The resulting nerve impulses must then be conducted to the visual area of the cerebral cortex.

For light to stimulate rods and cones to produce generator potentials, light must pass through ganglion and bipolar cells before reaching rods and cones. Note also that there are two special types of cells present in the retina (nervous tunic) called ***horizontal cells*** and ***amacrine cells.***

Once generator potentials are developed by rods and cones, the potentials induce signals in both bipolar neurons and horizontal cells, probably by means of neurotransmitters. In response to the neurotransmitters, bipolar neurons become excited and horizontal cells become inhibited. Functionally, bipolar neurons transmit the excitatory visual signal from rods and cones to ganglion cells. Horizontal cells transmit inhibitory signals to bipolar neurons in the areas lateral to excited rods and cones. This lateral inhibition enhances contrasts in the visual scene between areas of the retina (nervous tunic) that are strongly stimulated and adjacent areas that are weakly stimulated. Horizontal cells also assist in the differentiation of various colors. Amacrine cells, which are also excited by bipolar neurons, synapse with ganglion cells and transmit information to them that signals a change in the level of illumination of the retina.

When bipolar neurons transmit excitatory visual signals to ganglion cells, the ganglion cells become depolarized and initiate nerve impulses. The cell bodies of the ganglion cells lie in the retina, and their axons leave the eyeball via the ***optic (II) nerve*** (Figure 20-13c). The axons pass through the ***optic chiasma*** (kī-AZ-ma), a crossing point of the optic (II) nerves. Some fibers cross to the opposite side. Others remain uncrossed. After passing through the optic chiasma, the fibers, now part of the ***optic tract,*** enter the brain and terminate in the lateral geniculate nucleus of the thalamus. Here the fibers synapse with third-order neurons whose axons pass to the visual areas located in the occipital lobes of the cerebral cortex.

Analysis of the afferent pathway to the brain reveals that the visual field of each eye is divided into two regions: the ***nasal (medial) half*** and the ***temporal (lateral) half.*** For each eye, light rays from an object in the nasal half of the visual field fall on the temporal half of the retina. Light rays from an object in the temporal half of the visual field fall on the nasal half of the retina (Figure 20-14). Also, light rays from objects at the top of the visual field of each eye fall on the inferior portion of the retina, and light rays from objects at the bottom of the visual field fall on the superior portion of the retina. In the optic chiasma, nerve fibers from the nasal halves of the retinas cross and continue on to the lateral geniculate nuclei of the thalamus; nerve fibers from the temporal halves of the retinas do not cross, but continue directly on to the lateral geniculate nuclei. As a result, the primary visual area of the cerebral cortex of the right occipital lobe interprets visual sensations from the left side of an object via nerve impulses from the temporal half of the retina of the right eye and the nasal half of the retina of the left eye. The primary visual area of the cerebral cortex of the left occipital lobe interprets visual sensations from the right side of an object via impulses from the nasal half of the right eye and the temporal half of the left eye.

CLINICAL APPLICATION

Scotoma

A ***scotoma*** (skō-TŌ-ma) or ***blind spot*** in the field of vision, other than the normal blind spot (optic disc), may indicate a brain tumor along one of the afferent pathways. For instance, a symptom of a tumor in the right optic tract might be an inability to see the left side of a normal field of vision without moving the eyeball.

AUDITORY SENSATIONS AND EQUILIBRIUM

In addition to containing receptors for sound waves, the ear also contains receptors for equilibrium. Anatomically, the ear is divided into three principal regions: the external (outer) ear, middle ear, and internal (inner) ear.

External (Outer) Ear

The ***external (outer) ear*** is structurally designed to collect sound waves and direct them inward (Figure 20-15a,b). It consists of the auricle, external auditory canal, and tympanic membrane, also called the eardrum.

The ***auricle (pinna)*** is a flap of elastic cartilage shaped like the flared end of a trumpet and covered by thick skin. The rim of the auricle is called the ***helix;*** the inferior portion is the ***lobule.*** The auricle is attached to the head by ligaments and muscles. Additional surface anatomy features of the external ear are shown in Figure 11-3.

The ***external auditory*** (*audire* = hearing) ***canal (meatus)*** is a curved tube about 2.5 cm (1 in.) in length that lies in the temporal bone. It leads from the auricle to the eardrum. The wall of the canal consists of bone lined with cartilage that is continuous with the cartilage of the pinna. The cartilage in the external auditory canal is covered with thin, highly sensitive skin. Near the exterior opening, the canal contains a few hairs and specialized sebaceous glands called ***ceruminous*** (se-ROO-mi-nus) ***glands*** that secrete ***cerumen*** (earwax). The combination of hairs and cerumen (se-ROO-min) helps to prevent dust and foreign objects from entering

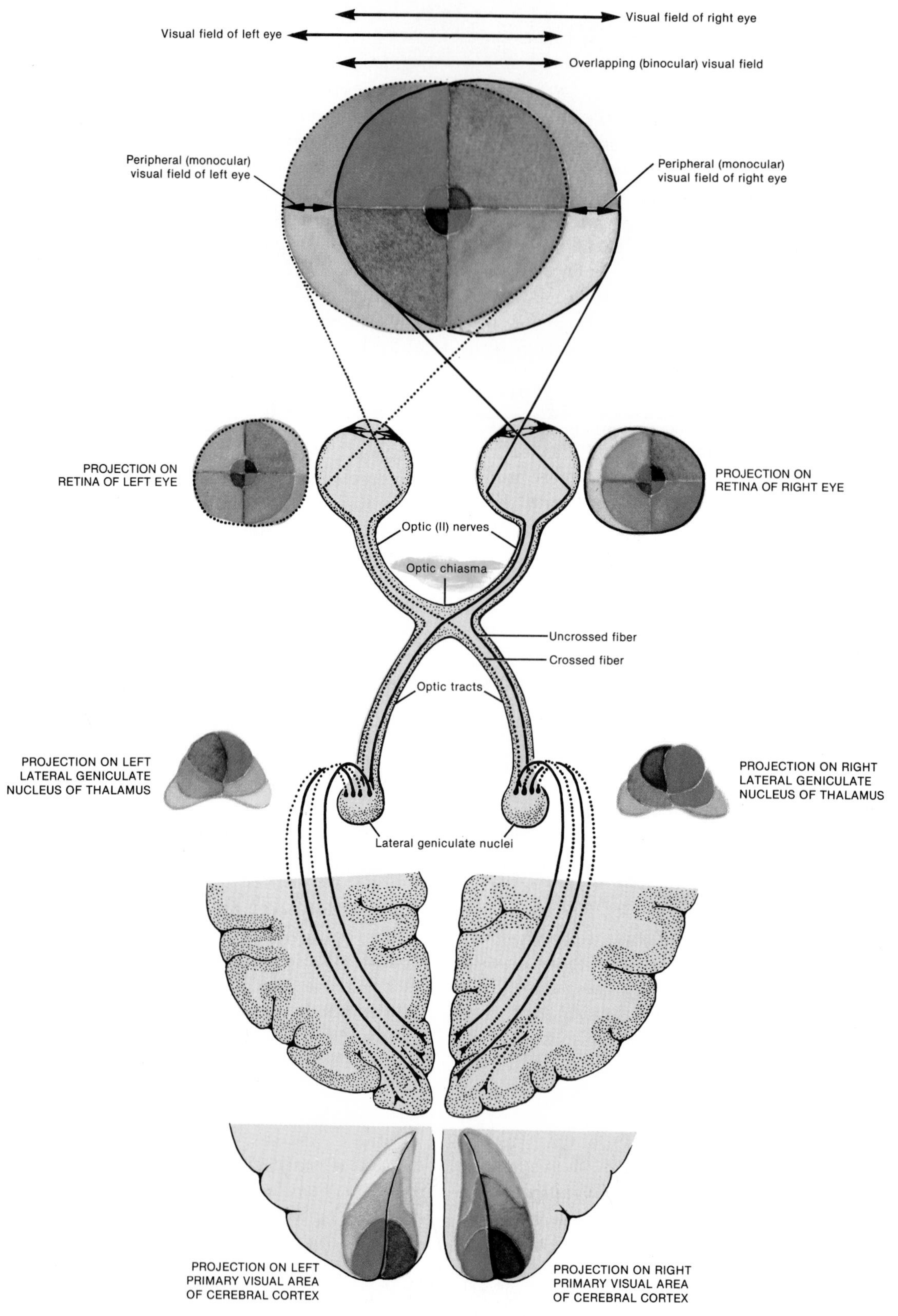

(a) Diagram of olfactory pathway

FIGURE 20-14 Afferent pathway for visual impulses. In (a), the dark circle in the center of the visual fields is the macula lutea. The center of the macula lutea is the central fovea, the area of sharpest vision. (b) Courtesy of N. Gluhbegovic and T. H. Williams, *The Human Brain: A Photographic Guide,* Harper & Row, Publishers, Inc., Hagerstown, MD, 1980.

Frontal lobe of cerebrum
Olfactory tract
Infundibulum
Mammillary body
Globus pallidus
Putamen
Insula
Superior cerebellar peduncle
Cerebral aqueduct
Corpus callosum
Occipital lobe of cerebrum
Optic (II) nerve
Optic chiasma
Optic tract
Lateral geniculate nucleus of thalamus
Fibers to visual area of cerebral cortex in occipital lobe

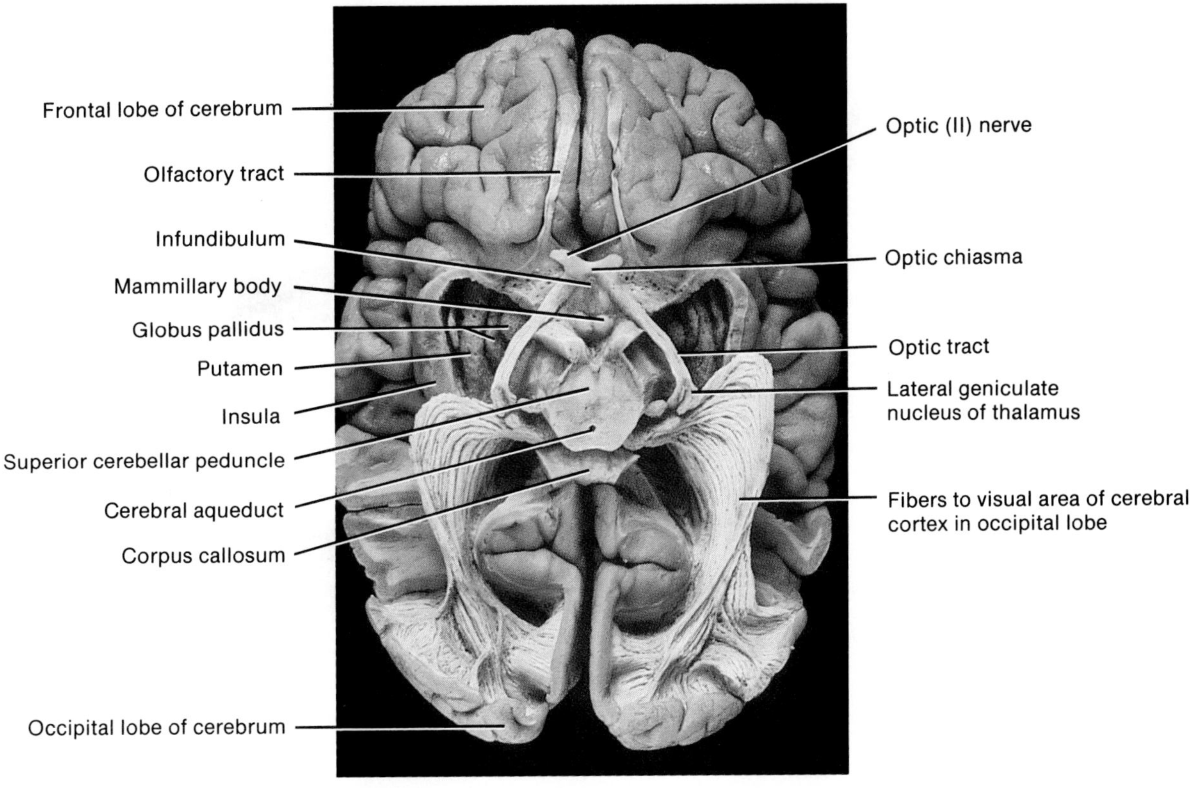

(b) Photograph of inferior aspect of brain

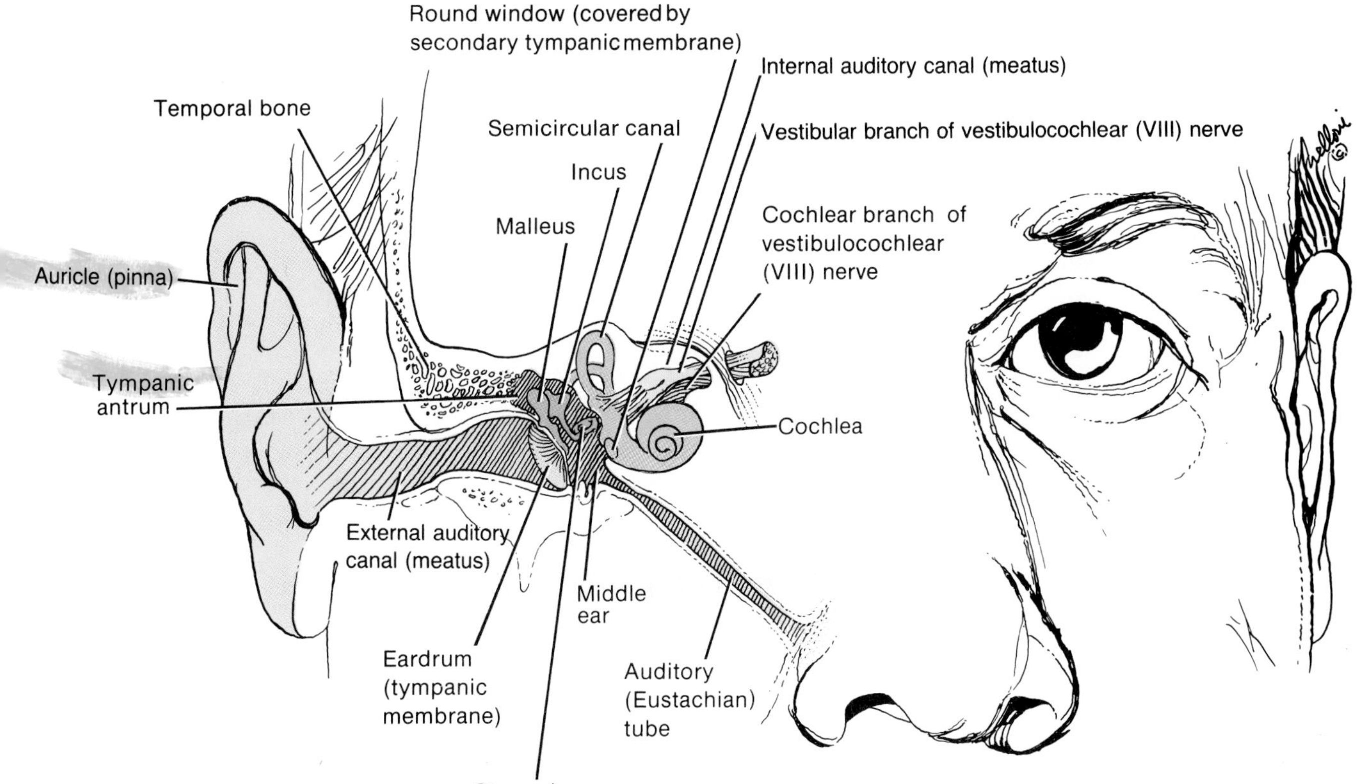

Round window (covered by secondary tympanic membrane)
Internal auditory canal (meatus)
Temporal bone
Semicircular canal
Vestibular branch of vestibulocochlear (VIII) nerve
Incus
Malleus
Cochlear branch of vestibulocochlear (VIII) nerve
Auricle (pinna)
Tympanic antrum
Cochlea
External auditory canal (meatus)
Middle ear
Eardrum (tympanic membrane)
Auditory (Eustachian) tube
Stapes in oval window

(a) Frontal section through the right side of the skull

FIGURE 20-15 Structure of the auditory apparatus. In (a), note the divisions of the ear into external, middle, and internal portions.

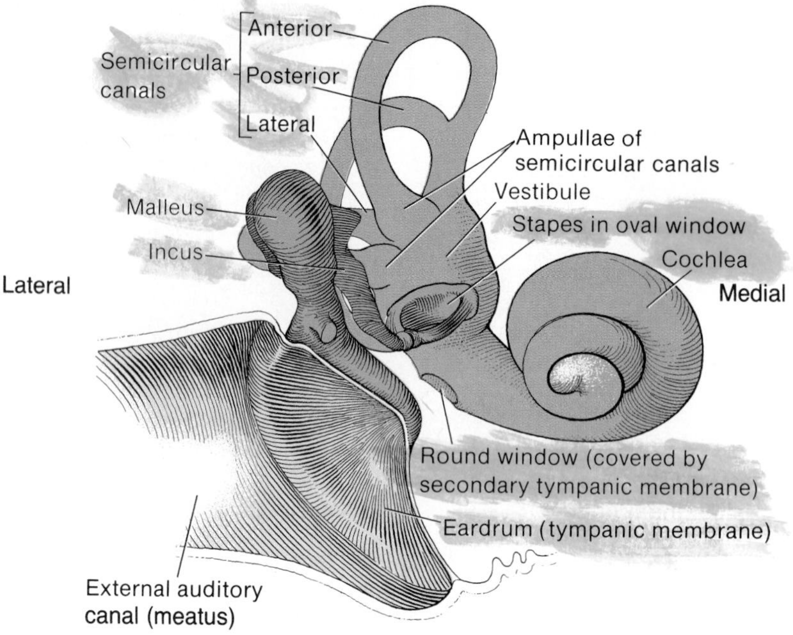

(b) Details of the middle ear and bony labyrinth of the internal ear

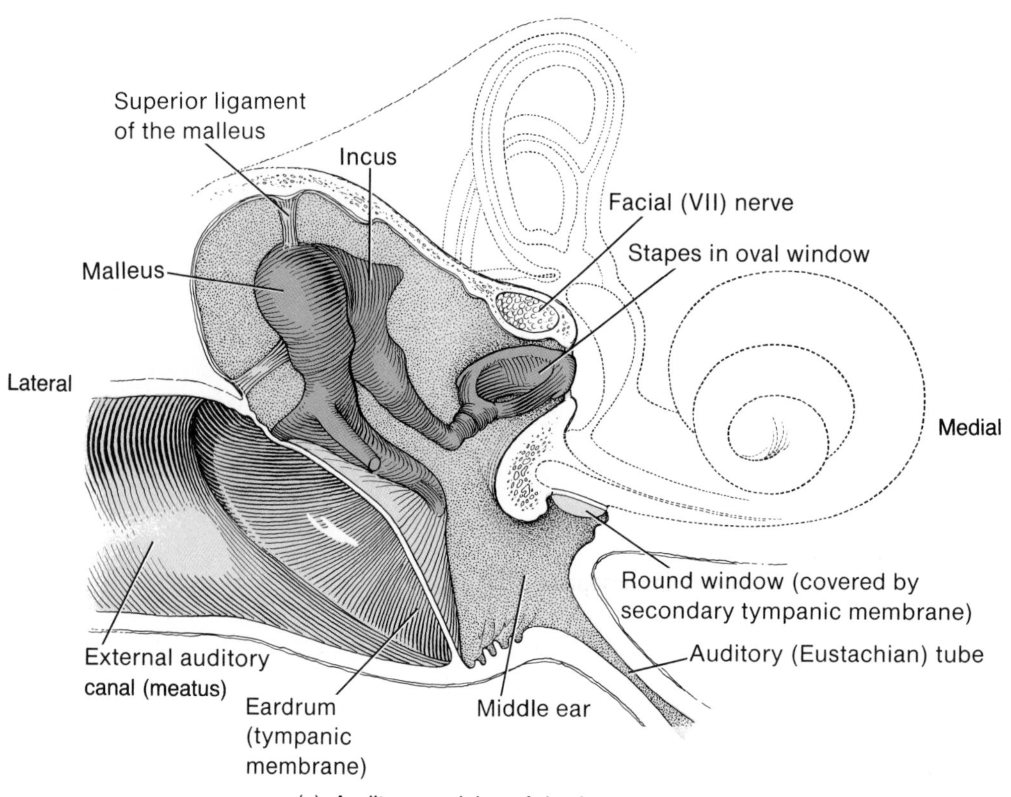

(c) Auditory ossicles of the inner ear

the ear. Usually, cerumen dries up and falls out of the ear canal. Sometimes, however, it builds up in the canal and impairs hearing (impacted cerumen).

The ***tympanic*** (tim-PAN-ik; *tympano* = drum) ***membrane (eardrum)*** is a thin, semitransparent partition of fibrous connective tissue between the external auditory canal and middle ear. Its external surface is concave and covered with skin. Its internal surface is convex and covered with a mucous membrane.

CLINICAL APPLICATION

Perforated Eardrum

A ***perforated eardrum*** is a hole in the tympanic membrane that reduces sound transmission. The condition is characterized by acute pain initially, ringing or roaring in the affected ear, hearing impairment, and sometimes dizziness. Causes of perforated eardrums include compressed air (explosion), scuba diving, trauma (skull fracture or from objects such as ear swabs), or acute middle ear infections. If the condition does not resolve by itself, a surgical procedure called myringoplasty can be performed, in which a graft of skin, a vein, or fascia is used to repair the perforation. Or, the wound edges may be cauterized and a piece of blood-soaked Gelfoam (absorbable gelatin sponge) is placed over the defect. New tissues will grow over the Gelfoam.

Middle Ear

The ***middle ear (tympanic cavity)*** is a small, epithelial-lined, air-filled cavity hollowed out of the temporal bone (Figure 20-15b,c). It is separated from the external ear by the eardrum and from the internal ear by a thin bony partition that contains two small openings: the oval window and the round window.

The posterior wall of the middle ear communicates with the mastoid air cells of the temporal bone through a chamber called the ***tympanic antrum.*** This anatomical fact explains why a middle ear infection may spread to the temporal bone, causing mastoiditis, or even to the brain.

The anterior wall of the middle ear contains an opening that leads directly into the ***auditory (Eustachian) tube.*** The auditory tube connects the middle ear with the nasopharynx of the throat. Through this passageway, infections may travel from the throat and nose to the ear. The function of the tube is to equalize air pressure on both sides of the tympanic membrane. Abrupt changes in external or internal air pressure might otherwise cause the eardrum to rupture. During swallowing and yawning, the tube opens to allow atmospheric air to enter or leave the middle ear until the internal pressure equals the external pressure. If the pressure is not relieved, intense pain, hearing impairment, tinnitus, and vertigo could develop. Any sudden pressure changes against the eardrum may be equalized by deliberately swallowing or pinching the nose closed, closing the mouth, and gently forcing air from the lungs into the nasopharynx.

Extending across the middle ear are three exceedingly small bones called ***auditory ossicles*** (OS-si-kuls). The bones, named for their shape, are the malleus, incus, and stapes, commonly called the hammer, anvil, and stirrup, respectively. They are connected by synovial joints. The "handle" of the ***malleus*** is attached to the internal surface of the tympanic membrane. Its head articulates with the body of the incus. The ***incus*** is the intermediate bone in the series and articulates with the head of the stapes. The base or footplate of the ***stapes*** fits into a small membranous structure in the thin bony partition between the middle and inner ear called the ***oval window*** or ***fenestra vestibuli*** (fe-NES-tra ves-TIB-yoo-lī). Directly below the oval window is another opening, the ***round window*** or ***fenestra cochleae*** (fe-NES-tra KŌK-lē-ē). This opening is enclosed by a membrane called the ***secondary tympanic membrane.*** Auditory ossicles are attached to the middle ear by ligaments.

Three ligaments are associated with the malleus (Figure 20-16). The ***anterior ligament of the malleus*** is attached by one end to the anterior process of the malleus and by the other to the anterior wall of the middle ear. The ***superior ligament of the malleus*** extends from the head of the malleus to the roof of the middle ear. The ***lateral ligament of the malleus*** extends from the neck of the malleus to the lateral wall of the middle ear. Two ligaments are associated with the incus. The ***posterior ligament of the incus*** extends from the short crus (a projection) of the incus to the posterior wall of the middle ear, and the ***superior ligament of the***

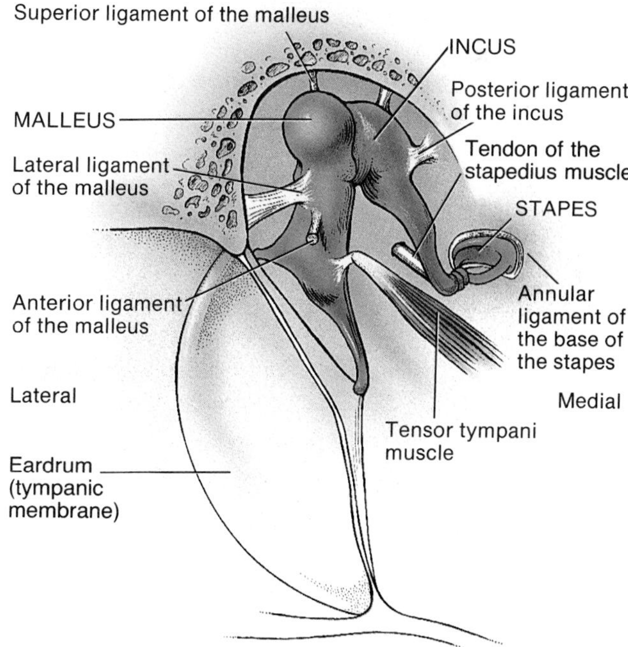

FIGURE 20-16 Ligaments and muscles of the auditory ossicles.

incus extends from the body of the incus to the roof of the middle ear. A single ligament, the ***annular ligament of the base of the stapes,*** is associated with the stapes. It extends from the base of the stapes to the fenestra vestibuli.

In addition to the ligaments, there are also two skeletal muscles attached to the ossicles. The ***tensor tympani muscle*** draws the malleus medially, thus limiting movement and increasing tension on the tympanic membrane to prevent damage to the inner ear when exposed to loud sounds. Since this response is slow, it only protects the inner ear from prolonged loud noises, not brief ones, such as a gunshot.

The ***stapedius muscle*** is the smallest of all skeletal muscles. Its action is to draw the stapes posteriorly, thus preventing it from moving too much. Like the tensor tympani muscle, the stapedius muscle has a protective function in that it dampens (checks) large vibrations that result from loud noises. Paralysis of the stapedius muscle is therefore associated with ***hyperacusia*** (acuteness of hearing).

Internal (Inner) Ear

The ***internal (inner) ear*** is also called the ***labyrinth*** (LAB-i-rinth) because of its complicated series of canals. Structurally, it consists of two main divisions: an outer bony labyrinth and an inner membranous labyrinth that fits in the bony labyrinth (Figure 20–17a). The ***bony labyrinth*** is a series of cavities in the petrous portion of the temporal bone. It can be divided into three areas named on the basis of shape: the vestibule, cochlea, and semicircular canals. The bony labyrinth is lined with periosteum and contains a fluid called ***perilymph.*** This fluid, which is chemically similar to cerebrospinal fluid, surrounds the ***membranous labyrinth,*** a series of sacs and tubes lying inside and having the same general form as the bony labyrinth. The membranous labyrinth is lined with epithelium and contains a fluid called ***endolymph,*** which is chemically similar to intracellular fluid.

The ***vestibule*** constitutes the oval central portion of the bony labyrinth. The membranous labyrinth in the vestibule consists of two sacs called the ***utricle*** (YOO-tri-kul; *utricle* = little bag) and ***saccule*** (SAK-yool). These sacs are connected to each other by a small duct.

Projecting upward and posteriorly from the vestibule are the three bony ***semicircular canals.*** Each is arranged at approximately right angles to the other two. On the basis of their positions, they are called the anterior, posterior, and lateral canals. The anterior and posterior semicircular canals are oriented vertically; the lateral one is oriented horizontally. One end of each canal enlarges into a swelling called the ***ampulla*** (am-POOL-la; *ampulla* = little jar). The portions of the membranous labyrinth that lie inside the bony semicircular canals are called the ***semicircular ducts (membranous semicircular canals).*** These structures are almost identical in shape to the semicircular canals and communicate with the utricle of the vestibule.

Lying in front of the vestibule is the ***cochlea*** (KŌK-lē-a), so designated because of its resemblance to a snail's shell (Figure 20-17a-c). The cochlea consists of a bony spiral canal that makes about 2¾ turns around a central bony core called the ***modiolus.*** A cross section through the cochlea shows the canal is divided into three separate channels by partitions that together have the shape of the letter Y (Figure 20-17b-d). The stem of the Y is a bony shelf that protrudes into the canal; the wings of the Y are composed mainly of membranous labyrinth. The channel above the bony partition is the ***scala vestibuli,*** which ends at the oval window; the channel below is the ***scala tympani,*** which terminates at the round window. The scala vestibuli and scala tympani both contain perilymph and are completely separated except at an opening at the apex of the cochlea called the ***helicotrema.*** The cochlea adjoins the wall of the vestibule, into which the scala vestibuli opens. The perilymph of the vestibule is continuous with that of the scala vestibuli. The third channel (between the wings of the Y) is the membranous labyrinth: the ***cochlear duct (scala media).*** The cochlear duct is separated from the scala vestibuli by the ***vestibular membrane.*** It is separated from the scala tympani by the ***basilar membrane.***

Resting on the basilar membrane is the ***spiral organ (organ of Corti),*** the organ of hearing (Figure 20-17d, e). The spiral organ is a series of epithelial cells on the inner surface of the basilar membrane. It consists of a number of supporting cells and hair cells, which are the receptors for auditory sensations. The inner hair cells are medially placed in a single row and extend the entire length of the cochlea. The outer hair cells are arranged in several rows throughout the cochlea. The hair cells have long hairlike processes at their free ends that extend into the endolymph of the cochlear duct. The basal ends of the hair cells are in contact with fibers of the cochlear branch of the vestibulocochlear (VIII) nerve. Projecting over and in contact with the hair cells of the spiral organ is the ***tectorial*** (*tectum* = cover) ***membrane,*** a delicate and flexible gelatinous membrane.

Hair cells of the spiral organ are easily damaged by exposure to high-intensity noises such as those produced by engines of jet planes, revved-up motorcycles, and loud music. As a result of the noises, hair cells become arranged in disorganized patterns, or they and their supporting cells may degenerate. High-intensity noises may produce permanent hearing damage that typically begins in the high-frequency tone range and therefore is not noticed until, in some cases, it is fairly advanced. Only a part of the hearing loss is reversible; this is why ear protection should be worn by those exposed to high-intensity noises.

Auditory Pathway

Sound waves result from the alternate compression and decompression of air molecules. They originate from a vibrating object, much the same way that waves travel over

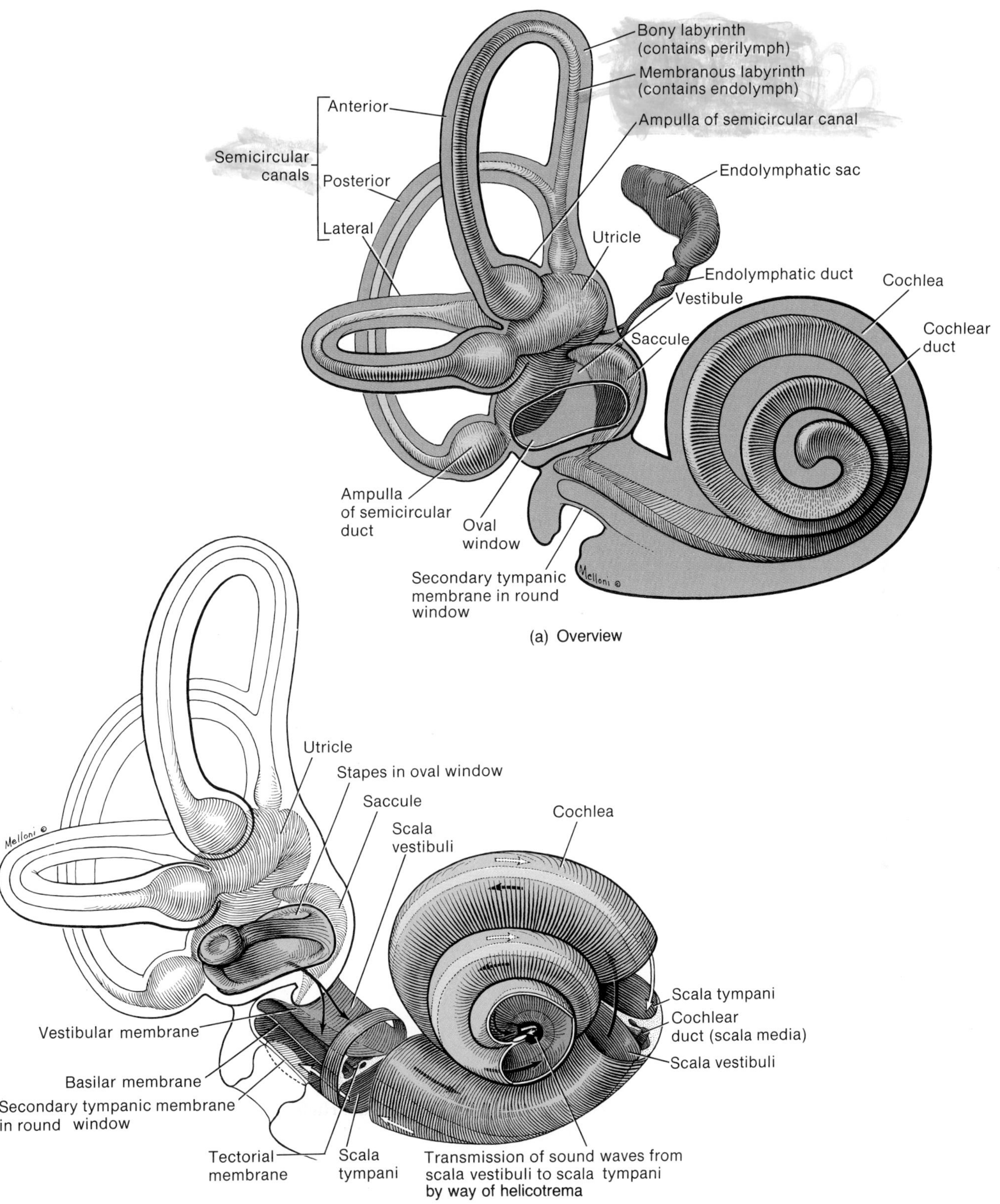

FIGURE 20-17 Details of the internal ear. (a) The outer, blue-colored area belongs to the bony labyrinth; the inner, pink-colored area belongs to the membranous labyrinth. (b) Relationship of the scala tympani, cochlear duct, and scala vestibuli. The arrows indicate the transmission of sound waves, which are discussed shortly. (c) Origin of the vestibular and cochlear branches of the vestibulocochlear (VIII) nerve. (d) Details of portion of cochlea. (e) Details of spiral organ (organ of Corti).

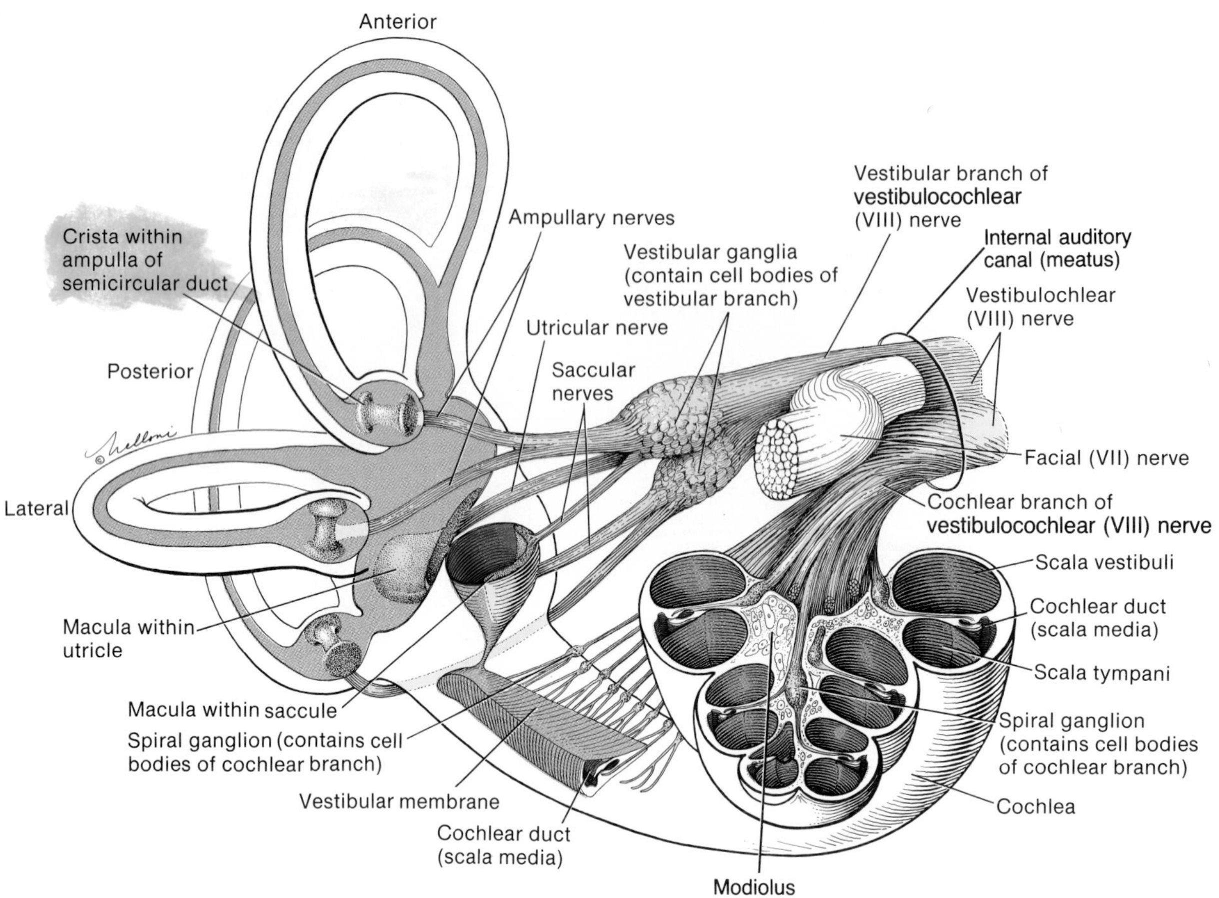

(c) Components of the vestibulocochlear (VIII) nerve

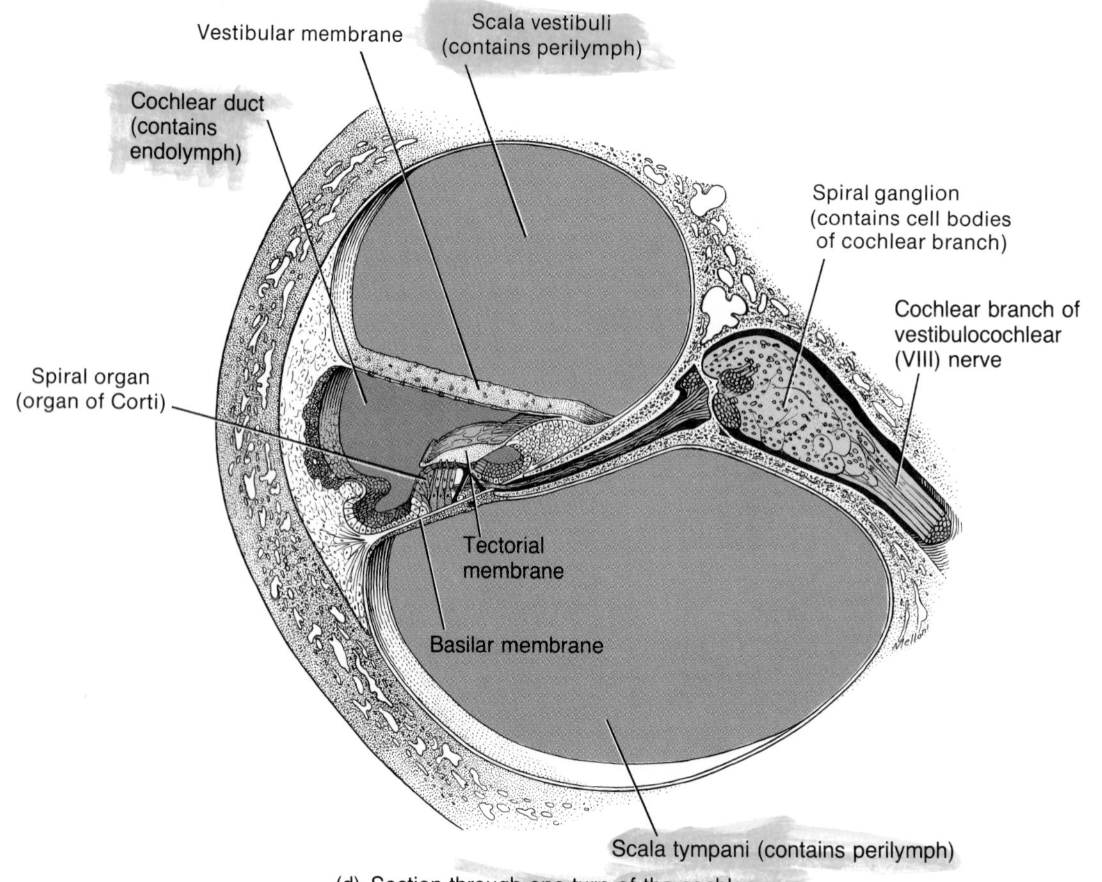

(d) Section through one turn of the cochlea

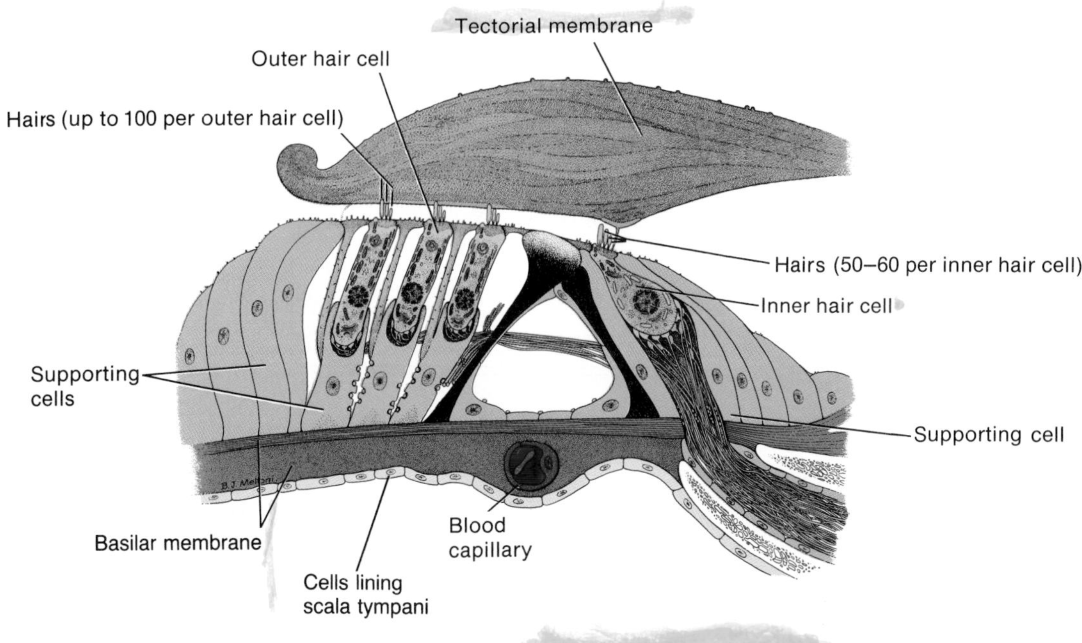

(e) Enlargement of the spiral organ (organ of Corti)

the surface of water. The events involved in the physiology of hearing and the conduction of auditory impulses to the brain are as follows (Figure 20-18):

1. Sound waves that reach the ear are directed by the auricle into the external auditory canal.
2. When the waves strike the tympanic membrane, the alternate compression and decompression of the air causes the membrane to vibrate (move forward and backward). The distance the membrane moves is always very small and is relative to the force and velocity of the sound waves. It vibrates slowly in response to low-frequency sounds and rapidly in response to high-frequency sounds.
3. The central area of the tympanic membrane is connected to the malleus, which also starts to vibrate. The vibration is then picked up by the incus, which transmits the vibration to the stapes.
4. As the stapes moves back and forth, it pushes the membrane of the oval window in and out. If sound waves passed directly to the oval window without passing through the tympanic membrane and auditory ossicles, hearing would be inadequate. A minimal amount of sound energy is required to transmit sound waves through the perilymph of the cochlea. Since the tympanic membrane has a surface area about 22 times larger than that of the oval window, it can collect about 22 times more sound energy. This energy is sufficient to transmit sound waves through the perilymph.
5. The movement of the oval window sets up waves in the perilymph.
6. As the oval window bulges inward, it pushes the perilymph of the scala vestibuli, and pressure waves are propagated through the scala vestibuli. If the inward movement of the stapes at the oval window is slow, the pressure in the perilymph pushes the perilymph from the scala vestibuli to the scala tympani and eventually to the round window, causing it to bulge outward into the middle ear. See number 9 in the illustration.
7. As the pressure moves through the perilymph of the scala vestibuli, it pushes the vestibular membrane inward and increases the pressure of the endolymph inside the cochlear duct.
8. As a result, the basilar membrane moves slightly and bulges into the scala tympani. Slow movement of the oval window has only a *very slight* effect on the basilar membrane.
9. As the pressure moves through the scala tympani, perilymph moves toward the round window, causing it to bulge outward into the middle ear. If the stapes vibrates rapidly, pressure in the perilymph does not have time to pass from the scala vestibuli to the scala tympani to the round window. Instead, pressure in the perilymph in the scala vestibuli is transmitted through the basilar membrane and eventually to the round window. As a result, the area of the basilar membrane near the oval and round windows vibrates.
10. When the basilar membrane vibrates, the hair cells of the spiral organ move against the tectorial membrane. The movement of the hairs develops receptor potentials that ultimately lead to the generation of nerve impulses.

The function of hair cells is to convert a mechanical force (stimulus) into an electrical signal (nerve impulse). It is believed to occur as follows. When the hairs at the top of the cell are moved, the hair cell membrane depolar-

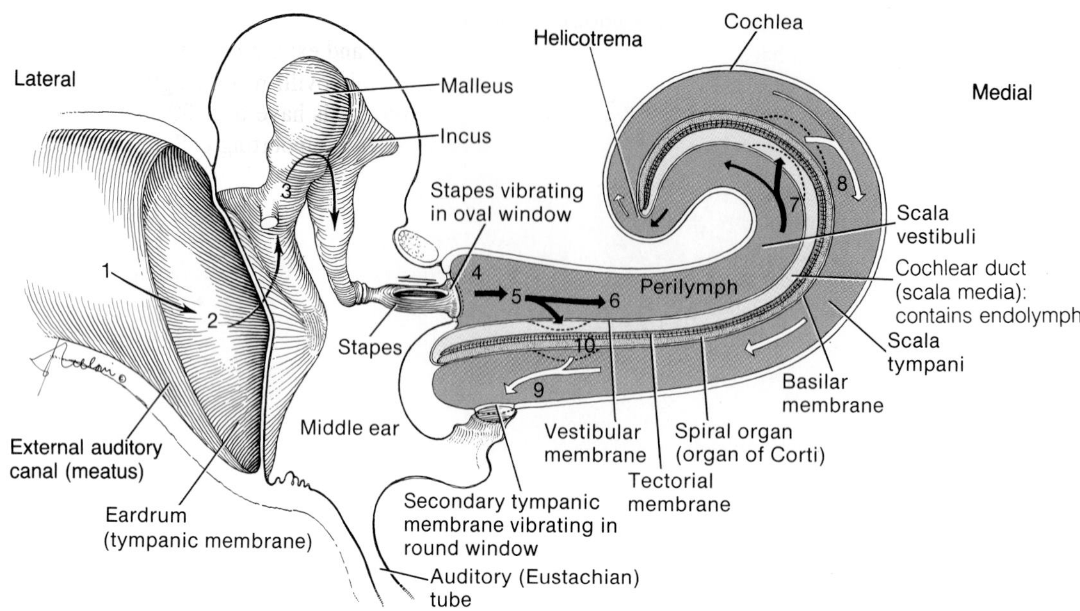

FIGURE 20-18 Mechanism of hearing. The numbers correspond to the events listed in the text. The cochlea has been uncoiled in order to schematize the transmission of sound waves and their subsequent distortion of the vestibular and basilar membranes of the cochlear duct.

izes, producing the receptor potential. Depolarization spreads through the cell and causes the release of neurotransmitter from the hair cell, which excites a sensory nerve fiber at the base of the hair cell. There is some evidence that the neurotransmitter might be glutamate or gamma aminobutyric acid (GABA).

The nerve impulses are then passed on to the cochlear branch of the vestibulocochlear (VIII) nerve (see Figure 20-17c) and cochlear nuclei in the medulla. Here most impulses cross to the opposite side and then travel to the midbrain, thalamus, and finally to the auditory area of the temporal lobe of the cerebral cortex.

Differences in pitch are related to differences in width of the basilar membrane and sound waves of various frequencies that cause specific regions of the basilar membrane to vibrate more intensely than others. The membrane is less flexible at the base of the cochlea (portion closer to the oval window) and more flexible near the apex of the cochlea. High-frequency or high-pitched sounds cause the basilar membrane to vibrate near the base of the cochlea, and low-frequency or low-pitched sounds cause the basilar membrane to vibrate near the apex of the cochlea. Loudness is determined by the intensity of sound waves. High-intensity sound waves cause greater vibration of the basilar membrane. Thus, more hair cells are stimulated and more impulses reach the brain.

CLINICAL APPLICATION

Artificial Ears

Artificial ears (cochlear implants) are devices that translate sounds into electronic signals that can be interpreted by the brain. They take the place of hair cells of the spiral organ (organ of Corti), which normally convert sound waves into electrical signals carried to the brain. The implants are used for individuals with sensorineural deafness—deafness due to disease or injury that has destroyed hair cells of the spiral organ. Such people make up the majority of those who have no hearing in either ear.

The device consists of electrodes implanted in the cochlea that are connected to a plug attached to the outside of the head. A microprocessor pack, about the size of a small transistor radio, is worn on the person's belt or in a shirt pocket and connected by a cord to the plug. Sound waves enter a tiny microphone in the ear, similar to that of a usual hearing aid, and travel down the cord into the pack where they are converted into electrical signals. The signals then travel to the electrodes in the cochlea, where they stimulate nerve endings and transmit the signals to the brain via the cochlear branch of the vestibulocochlear (VIII) nerve. The sounds heard are crude compared to normal hearing. However, the sounds provide a sense of rhythm and loudness and information about noises such as telephones, automobiles, and the pitch and cadence of speech.

Mechanism of Equilibrium

There are two kinds of ***equilibrium.*** One, called ***static equilibrium,*** refers to the orientation of the body (mainly the head) relative to the ground (gravity). The second kind, ***dynamic equilibrium,*** is the maintenance of body position (mainly the head) in response to sudden movements such as rotation, acceleration, and deceleration. The receptor organs for equilibrium are the maculae of the saccule and the utricle and cristae in the semicircular ducts.

■ ***Static Equilibrium*** The walls of both the utricle and saccule contain a small, thickened region called a ***macula*** (Figure 20-19). The maculae are the receptors that are concerned with static equilibrium. They provide sensory information regarding the orientation of the head in space and are essential for maintaining posture as we stand or sit.

Microscopically, the two maculae resemble the spiral organ (organ of Corti). They consist of differentiated neuroepithelial cells that are innervated by the vestibular branch of the vestibulocochlear (VIII) nerve. The maculae are anatomically located in planes perpendicular to one another and possess two kinds of cells: ***hair (receptor) cells*** and ***supporting cells.*** Two shapes of hair cells have been identified: one is more flask-shaped and the other is more cylindrical. Both show long extensions of the cell membrane consisting of many ***stereocilia*** (they are actually microvilli) and one ***kinocilium*** (a conventional cilium) anchored firmly to its basal body and extending beyond the longest microvilli. Some of the microvilli reach lengths of over 100 μm, and some receptor cells have over 80 such projections.

The columnar supporting cells of the maculae are scattered between the hair cells. Floating directly over the hair cells is a thick, gelatinous, glycoprotein layer, probably secreted by the supporting cells, called the ***otolithic membrane.*** A layer of calcium carbonate crystals, called ***otoliths*** (*oto* = ear; *lithos* = stone), extends over the entire surface of the otolithic membrane. The specific gravity of these otoliths is about 3, which makes them much denser than the endolymph fluid that fills the rest of the utricle.

The otolithic membrane sits on top of the macula like a discus on a greased cookie sheet. If you tilt your head forward, the otolithic membrane (the discus in our analogy) slides downhill over the hair cells in the direction determined by the tilt of your head. As the otoliths move, they pull on the gelatinous layer, which pulls on the stereocilia and makes them bend. The movement of the stereocilia initiates nerve impulses by way of receptor potentials that are then transmitted to the vestibular branch of the vestibulocochlear (VIII) nerve (see Figure 20-17c).

Most of the vestibular branch fibers enter the brain stem and terminate in the vestibular nuclear complex in the medulla. The remaining fibers enter the flocculonodular lobe of the cerebellum through the inferior cerebellar peduncle. Bidirectional pathways connect the vestibular nuclei and cerebellum. Fibers from all the vestibular nuclei form the medial longitudinal tracts that extend from the brain stem into the cervical portion of the spinal cord. The tracts send

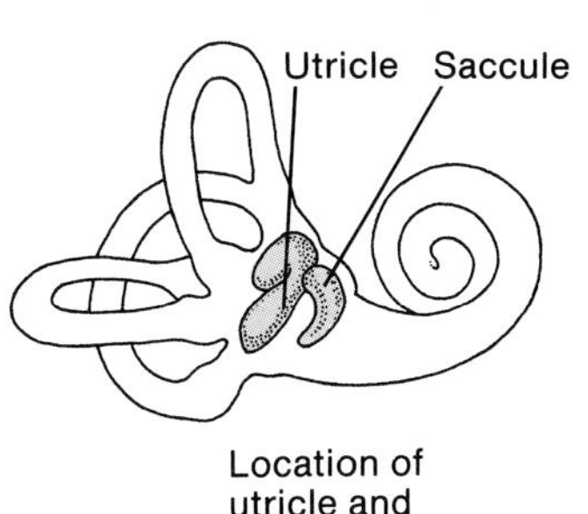

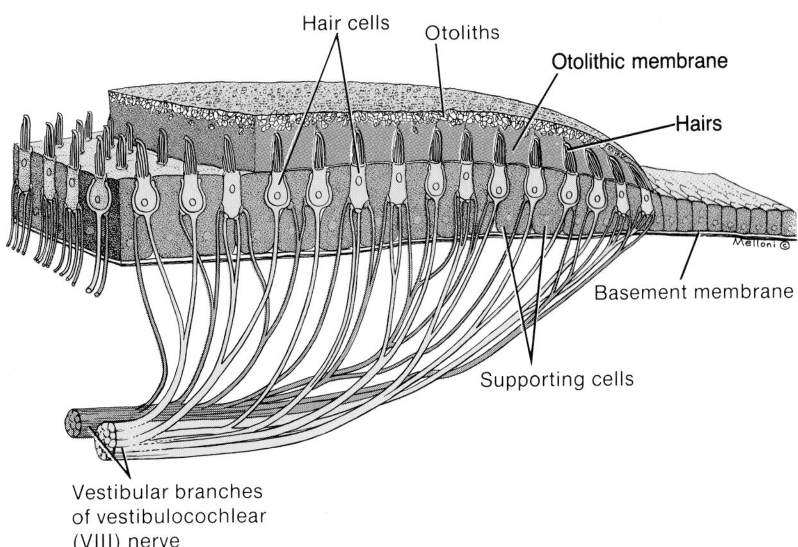

(a) Overall structure of a section of the macula

FIGURE 20-19 Structure of the macula.

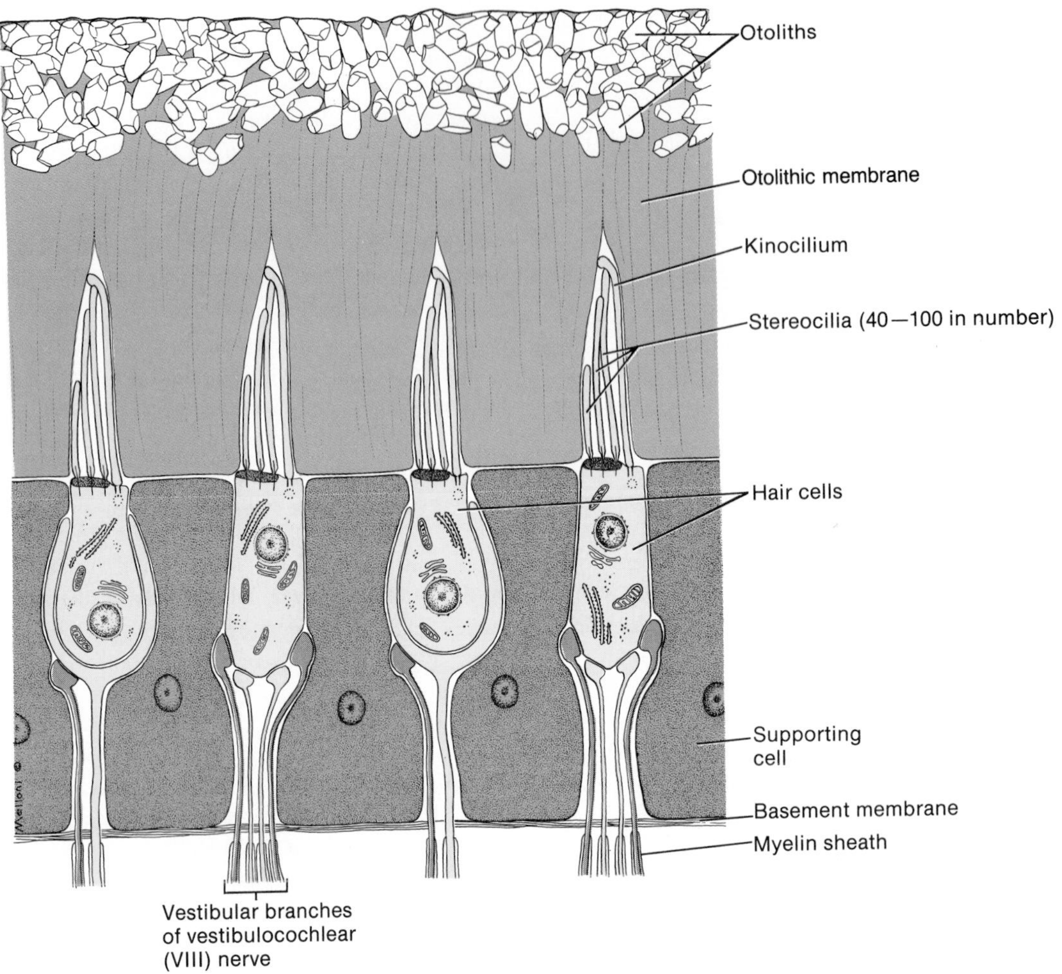

(b) Details of several hair cells

nerve impulses to the nuclei of cranial nerves that control eye movements [oculomotor (III), trochlear (IV), and abducens (VI)] and to the accessory (XI) nerve nucleus that helps control movements of the head and neck. In addition, fibers from the lateral vestibular nucleus form the vestibulospinal tract, which conveys impulses to skeletal muscles that regulate body tone in response to head movements. Various pathways between the vestibular nuclei, cerebellum, and cerebrum enable the cerebellum to assume a key role in helping the body maintain static equilibrium. The cerebellum continuously receives updated sensory information from the utricle and saccule concerning static equilibrium. Using this information, the cerebellum monitors and makes corrective adjustments in the motor activities that originate in the cerebral cortex. Essentially, the cerebellum sends continuous nerve impulses to the motor areas of the cerebrum, in response to input from the utricle and saccule, causing the motor system to increase or decrease its impulses to specific skeletal muscles in order to maintain static equilibrium.

■ ***Dynamic Equilibrium*** The cristae in the three semicircular ducts maintain dynamic equilibrium (Figure 20-20). The ducts are positioned at right angles to one another in three planes: the anterior and posterior semicircular ducts are oriented vertically, and the lateral semicircular duct is oriented horizontally. This positioning permits detection of an imbalance in three planes. In the ampulla, the dilated portion of each duct, there is a small elevation called the ***crista.*** Each crista is composed of a group of ***hair (receptor) cells*** and ***supporting cells*** covered by a mass of gelatinous material called the ***cupula.*** When the head moves due to rotation of the body, the endolymph in the semicircular ducts flows over the hairs and bends them. The movement of the hairs stimulates sensory neurons, and the nerve impulses, produced by the resulting receptor potentials, pass over the vestibular branch of the vestibulocochlear (VIII) nerve. The impulses follow the same pathways as those involved in static equilibrium and are eventually sent to the muscles that must contract to maintain body balance in the new position.

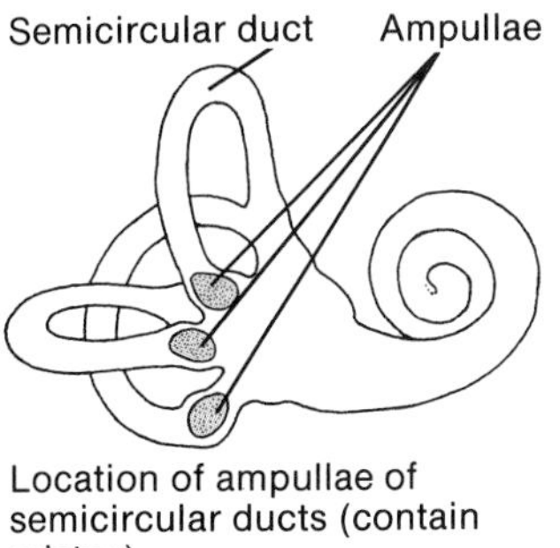

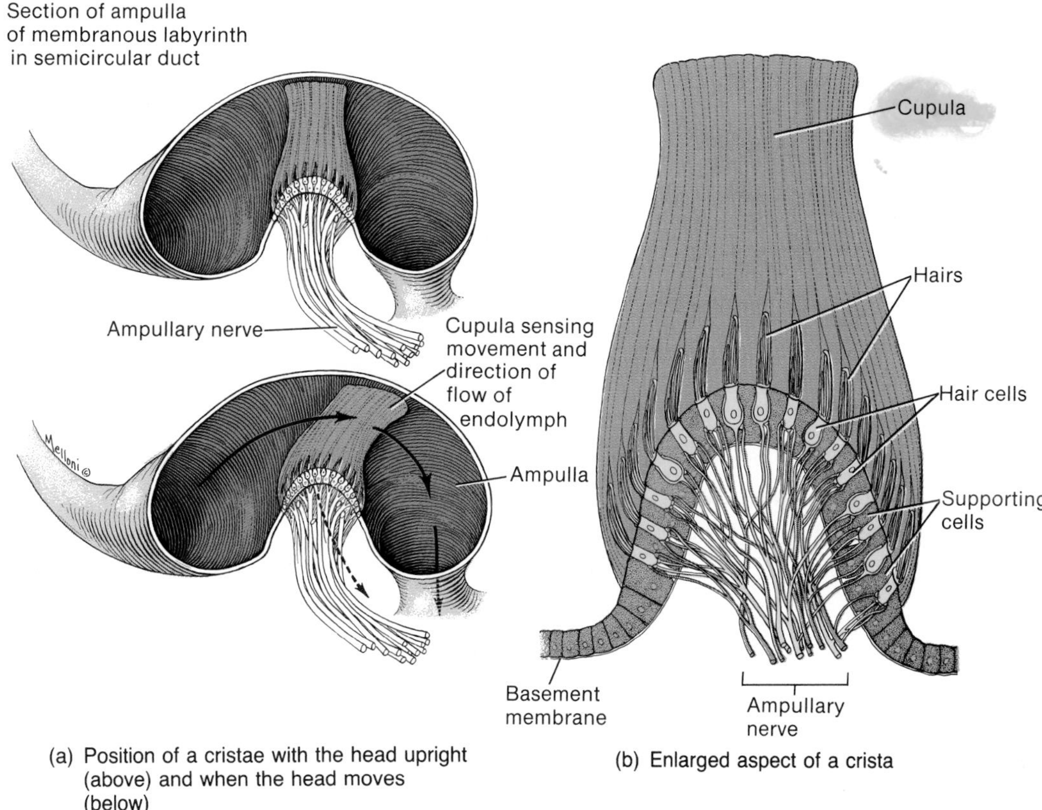

FIGURE 20-20 Semicircular ducts and dynamic equilibrium. The ampullary nerves in (b) are branches of the vestibular division of the vestibulocochlear (VIII) nerve.

A summary of the structures of the ear related to hearing and equilibrium is presented in Exhibit 20-2.

AGING AND THE SPECIAL SENSES

Degenerative changes and disease states involving the sense organs can alter vision, hearing, taste, and smell. The disorders that represent the most common visual problems and may be responsible for serious loss of vision are presbyopia (inability to focus on nearby objects), cataracts (cloudiness of the lens), and glaucoma (excessive fluid pressure in the eyeball). Impaired hearing associated with aging, known as presbycusis, is usually the result of changes in important structures of the inner ear.

APPLICATIONS TO HEALTH

CATARACT

The most prevalent disorder resulting in blindness is cataract formation. A ***cataract,*** meaning ''waterfall,'' is a clouding of the lens or its capsule so that it becomes opaque or

EXHIBIT 20-2

Summary of Structures of the Ear Related to Hearing and Equilibrium

STRUCTURE	FUNCTION
External (Outer) Ear	
Auricle (Pinna)	Collects sound waves.
External Auditory Canal (Meatus)	Directs sound waves to tympanic membrane.
Tympanic Membrane	Sound waves cause it to vibrate, which in turn causes the malleus to vibrate.
Middle Ear	
Auditory (Eustachian) Tube	Equalizes pressure on both sides of the tympanic membrane.
Auditory Ossicles	Transmit sound waves from tympanic membrane to oval window.
Internal (Inner) Ear	
Utricle	Contains macula, receptor for static equilibrium.
Saccule	Contains macula, receptor for static equilibrium.
Semicircular Ducts	Contain cristae, receptors for dynamic equilibrium.
Cochlea	Contains a series of fluids, channels, and membranes that transmit sound waves to the spiral organ (organ of Corti), the organ of hearing; the spiral organ generates nerve impulses and transmits them to the cochlear branch of the vestibulocochlear (VIII) nerve.

milk white. Two basic processes involved in cataract formation are the breakdown of the normal lens protein and an influx of water into the lens. As a result, the light from an object, which normally passes directly through the lens to produce a sharp image, produces only a degraded image. If the cataract is severe enough, no image at all is produced. Cataracts are associated with aging but may also be caused by injury, exposure to radiation (ultraviolet B), certain medications (long-term use of steroids), or complications of other diseases. It is possible to surgically remove a defective lens and replace it with an artificial one.

GLAUCOMA

Glaucoma is a group of disorders characterized by an abnormally high intraocular pressure (IOP) owing to a buildup of aqueous humor inside the eyeball. The aqueous humor does not return into the bloodstream through the scleral venous sinus (canal of Schlemm) as quickly as it is formed. The fluid accumulates and, by compressing the lens into the vitreous body, puts pressure on the neurons of the retina. If the pressure continues over a long period of time, glaucoma can progress from mild visual impairment to a point where neurons of the retina are destroyed, resulting in degeneration of the optic disc, visual field defects, and blindness. Treatment is through drugs such as timolol (Timoptic) and acetazolamide (Diamox), or laser surgery.

CONJUNCTIVITIS (PINKEYE)

Many different eye inflammations exist, but the most common type is ***conjunctivitis (pinkeye)***, an inflammation of the conjunctiva. Conjunctivitis can be caused by bacteria such as pneumococci, staphylococci, or *Hemophilis influenzae*. In such cases, the inflammation is very contagious. This epidemic type is common in children. Conjunctivitis may also be caused by a number of irritants, in which case the inflammation is not contagious. Irritants include dust, smoke, wind, pollutants in the air, and excessive glare.

TRACHOMA

A serious form of chronic contagious conjunctivitis is known as ***trachoma*** (tra-KŌ-ma), which is caused by a bacterium called *Chlamydia trachomatis*. Trachoma is characterized by many granulations or fleshy projections on the eyelids. If untreated, these projections can irritate and inflame the cornea and reduce vision. The disease produces an excessive growth of subconjunctival tissue and the invasion of blood vessels into the upper half of the front of the cornea. The disease progresses until it covers the entire cornea, bringing about a loss of vision because of corneal opacity.

Antibiotics, such as tetracycline and the sulfa drugs, kill the organisms that cause trachoma and have reduced the seriousness of this infection.

DEAFNESS

Deafness is the lack of the sense of hearing or significant hearing loss. ***Sensorineural deafness*** is caused by impairment of the cochlea or cochlear branch of the vestibulocochlear (VIII) nerve. ***Conduction deafness*** is caused by impairment of the external and middle ear mechanisms for transmitting sounds into the cochlea. Among the factors that contribute to deafness are genetic factors; otosclerosis, the deposition of new bone around the oval window; repeated exposure to loud noise, which destroys hair cells of the spiral organ (organ of Corti); certain drugs such as streptomycin; swimmer's ear and other infections; impacted cerumen; injury to the tympanic membrane; and aging, which results in thickening of the tympanic membrane, stiffening of the joints of the auditory ossicles, and decreased numbers of hair cells owing to diminished cell division.

LABYRINTHINE DISEASE

Labyrinthine (lab′-i-RIN-thēn) ***disease*** refers to a malfunction of the inner ear that is characterized by attacks of hearing loss, tinnitus (ringing in the ears), vertigo (sensation of spinning or movement), nausea, and vomiting. There may also be a blurring of vision, nystagmus (rapid, involuntary movement of the eyeballs), and a tendency to fall in a certain direction.

Among the causes of labyrinthine disease are: (1) infection of the inner ear, (2) trauma from brain concussion producing hemorrhage or splitting of the labyrinth, (3) cardiovascular diseases such as atherosclerosis and blood vessel disturbances, (4) congenital malformation of the labyrinth, (5) excessive formation of endolymph, (6) allergy, (7) blood abnormalities, and (8) aging.

MÉNIÈRE'S SYNDROME

Ménière's (men-YAIRZ) ***syndrome*** is characterized by an increased amount of endolymph that enlarges the labyrinth. Among the symptoms are fluctuating hearing loss, attacks of vertigo, and roaring tinnitus. Etiology of Ménière's syndrome is unknown. It is now thought that there is either an overproduction or underabsorption of endolymph in the cochlear duct. The hearing loss is caused by distortions in the basilar membrane of the cochlea. The classic type of Ménière's syndrome involves both the semicircular canals and the cochlea. The natural course of Ménière's syndrome may stretch out over a period of years, with the result being almost total destruction of hearing. Treatment consists of drugs to relieve the vertigo or surgery that creates an opening in the labyrinth, severing the vestibular branch of the vestibulocochlear (VIII) nerve, or removal of the labyrinth.

VERTIGO

Vertigo (*vertex* = whorl) is a sensation of spinning or movement in which the world is revolving or the person is revolving in space. Depending on its cause, vertigo may be classified as (1) ***peripheral,*** which originates in the ear, has a sudden onset, and lasts from minutes to hours; (2) ***central,*** which is caused by an abnormality of the central nervous system and may persist for more than three weeks; and (3) ***psychogenic,*** in which the cause is of psychological origin.

OTITIS MEDIA

Otitis media is an acute infection of the middle ear, caused primarily by bacteria such as *Streptococcus pneumoniae* and *Hemophilus influenzae*. It is characterized by pain, malaise, fever, and a reddening and outward bulging of the eardrum, which may rupture unless prompt treatment is given (such as draining pus from the middle ear). Abnormal function of the auditory (Eustachian) tube appears to be the most important mechanism in the pathogenesis of middle ear disease. In this case, bacteria from the nasopharynx, which are a primary cause of middle ear infection, are allowed to enter the middle ear.

MOTION SICKNESS

Motion sickness is a functional disorder brought on by repetitive angular, linear, or vertical motion and characterized by various symptoms, primarily nausea and vomiting. Warning symptoms include yawning, salivation, pallor, hyperventilation, profuse cold sweating, and prolonged drowsiness. Specific kinds of motion sickness include seasickness, airsickness, carsickness, and swing sickness.

Etiology is excessive stimulation of the vestibular apparatus by motion. Nerve impulses pass from the internal ear to the vomiting center in the medulla. Visual stimuli and emotional factors like fear and anxiety can also contribute to motion sickness. Ideally, treatment with drugs of susceptible individuals should be instituted prior to their entering into conditions that would produce motion sickness, since prevention is more successful than treatment of symptoms once they have developed. Among the drugs used to prevent motion sickness are dimenhydrinate (Dramamine), meclizine (Antivert), and promethazine (Phenegran).

KEY MEDICAL TERMS ASSOCIATED WITH SENSORY STRUCTURES

Achromatopsia (a-krō′-ma-TOP-sē-a; *a* = without; *chrom* = color) Complete color blindness.

Ametropia (am′-e-TRŌ-pē-a; *ametro* = disproportionate; *ops* = eye) Refractive defect of the eye resulting in an inability to focus images properly on the retina.

Anopsia (an-OP-sē-a; *opsia* = vision) A defect of vision.

Astereognosis (a-ster-ē-og-NŌ-sis; *stereos* = solid; *gnosis* = knowledge) Loss of ability to recognize objects or to appreciate their form by touching them.

Blepharitis (blef-a-RĪ-tis; *blepharo* = eyelid; *itis* = inflammation of) An inflammation of the eyelid.

Dyskinesia (dis′-ki-NĒ-zē-a; *dys* = difficult; *kinesis* = movement) Abnormality of motor function characterized by involuntary, purposeless movements.

Epiphora (e-PIF-ō-ra; *epi* = above) Abnormal overflow of tears.

Eustachitis (yoo′-stā-KĪ-tis) An inflammation or infection of the auditory (Eustachian) tube.

Exotropia (ek′-sō-TRŌ-pē-a; *ex* = out; *tropia* = turning) Turning outward of the eyes.

Keratitis (ker′-a-TĪ-tis; *kerato* = cornea) An inflammation or infection of the cornea.

Kinesthesis (kin′-es-THĒ-sis; *aisthesis* = sensation) The sense of perception of movement.

Labyrinthitis (lab′-i-rin-THĪ-tis) An inflammation of the labyrinth (inner ear).

Mydriasis (mi-DRĒ-a-sis) Dilated pupil.

Myringitis (mir′-in-JĪ-tis; *myringa* = eardrum) An inflammation of the eardrum; also called **tympanitis.**

Nystagmus (nis-TAG-mus; *nystazein* = to nod) A rapid involuntary movement of the eyeballs, possibly caused by a disease of the central nervous system. It is associated with conditions that cause vertigo.

Otalgia (o-TAL-jē-a; *oto* = ear; *algia* = pain) Earache.

Otosclerosis (ō-tō-skle-RŌ-sis; *oto* = ear; *sclerosis* = hardening) Pathological process that may be hereditary in which new bone is deposited around the oval window. The result may be immobilization of the stapes, leading to deafness.

Photophobia (fō′-tō-FŌ-bē-a; *photo* = light; *phobia* = fear) Abnormal visual intolerance to light.

Presbyopia (pres′-bē-Ō-pē-a; *presby* = old) Inability to focus on nearby objects owing to loss of elasticity of the crystalline lens. The loss is usually caused by aging.

Ptosis (TŌ-sis; *ptosis* = fall) Falling or drooping of the eyelid. (This term is also used for the slipping of any organ below its normal position.)

Retinoblastoma (ret′-i-nō-blas-TŌ-ma; *blast* = bud; *oma* = tumor) A tumor arising from immature retinal cells and accounting for 2 percent of childhood malignancies.

Strabismus (stra-BIZ-mus) An imbalance in the extrinsic eye muscles that produces a squint (formerly referred to as "cross-eyes"). **Amblyopia** is the term used to describe the loss of vision in an otherwise normal eye that, because of muscle imbalance, cannot focus in sync with the other eye.

Tinnitus (ti-NĪ-tus) A ringing, roaring, or clicking in the ears.

STUDY OUTLINE

Sensations (p. 562)

Definition (p. 562)

1. Sensation is a state of awareness of external and internal conditions of the body.
2. The prerequisites for a sensation to occur are reception of a stimulus, conversion of the stimulus into a nerve impulse through the generation of a receptor or generator potential by a receptor, conduction of the impulse to the brain, and translation of the impulse into a sensation by a region of the brain.
3. Each stimulus is capable of causing the membrane of a receptor to depolarize, resulting in a generator potential or receptor potential.

Characteristics (p. 562)

1. Projection occurs when the brain refers a sensation to the point of stimulation.
2. Adaptation is the loss of sensation even though the stimulus is still applied.
3. An afterimage is the persistence of a sensation even though the stimulus is removed.
4. Modality is the property by which one sensation is distinguished from another.

Classification of Receptors (p. 563)

1. According to location, receptors are classified as exteroceptors, visceroceptors, and proprioceptors.
2. On the basis of type of stimulus detected, receptors are classified as mechanoreceptors, thermoreceptors, nociceptors, photoreceptors, and chemoreceptors.
3. In terms of simplicity or complexity, simple receptors are associated with general senses, and complex receptors are associated with special senses.

General Senses (p. 563)

Cutaneous Sensations (p. 563)

1. Cutaneous sensations include tactile sensations (touch, pressure, vibration), thermoreceptive sensations (heat and cold), and pain. Receptors for these sensations are located in the skin, connective tissue under the skin, mucous membranes, and the ends of the gastrointestinal tract.
2. Receptors for touch are hair root plexuses, free nerve endings, tactile (Merkel's) discs, corpuscles of touch (Meissner's corpuscles), and type II cutaneous mechanoreceptors (end organs of Ruffini). Receptors for pressure are free nerve endings, type II cutaneous mechanoreceptors, and lamellated (Pacinian) corpuscles. Receptors for vibration are corpuscles of touch and lamellated corpuscles.
3. Pain receptors (nociceptors) are located in nearly every body tissue. Pain may be acute or chronic.
4. Two kinds of pain recognized in the parietal lobe of the cortex are somatic and visceral.
5. Referred pain is felt in the skin near or away from the organ sending pain impulses.
6. Phantom pain is the sensation of pain in a limb that has been amputated.
7. Pain impulses may be inhibited by drugs or surgery.

Proprioceptive Sensations (p. 567)

1. Receptors located in skeletal muscles, tendons in and around joints, and the internal ear convey nerve impulses related to muscle tone, movement of body parts, and body position.
2. The receptors include muscle spindles, tendon organs (Golgi tendon organs), joint kinesthetic receptors, and the maculae and cristae.

Levels of Sensation (p. 569)

1. Sensory fibers terminating in the lower brain stem bring about far more complex motor reactions than simple spinal reflexes.
2. When sensory impulses reach the lower brain stem, they cause subconscious motor reactions.
3. Sensory impulses that reach the thalamus can be localized crudely in the body.
4. When sensory impulses reach the cerebral cortex, we experience precise localization.

Sensory Pathways (p. 569)

1. Sensory information from all parts of the body terminates in a specific area of the somatosensory cortex.
2. The neural pathway for proprioception, discriminative touch, two-point discrimination, and vibration involves the fasciculus gracilis and fasciculus cuneatus.
3. The neural pathway for pain and temperature is the lateral spinothalamic tract.
4. The neural pathway for light touch and pressure is in the anterior spinothalamic tract.
5. The pathways to the cerebellum are the anterior and posterior spinocerebellar tracts.

Motor Pathways (p. 571)

1. Incoming sensory information is added to, subtracted from, or integrated with other information arriving from all other operating sensory receptors.
2. The integration process occurs at many stations along the pathways of the central nervous system such as within the spinal cord, brain stem, cerebellum, and cerebral cortex.
3. A motor response to make a muscle contract or a gland secrete can be initiated at any of these stations or levels.
4. The muscles of all parts of the body are controlled by a specific area of the motor cortex.
5. Voluntary motor impulses are conveyed from the brain through the spinal cord along the pyramidal pathways and the extrapyramidal pathways.
6. Pyramidal pathways include the lateral corticospinal, anterior corticospinal, and corticobulbar tracts.
7. Major extrapyramidal tracts are the rubrospinal, tectospinal, vestibulospinal, and reticulospinal tracts.

Special Senses (p. 573)

Olfactory Sensations (p. 573)

1. The receptors for olfaction, the olfactory cells, are in the nasal epithelium.
2. Olfactory cells convey impulses to olfactory (I) nerves, olfactory bulbs, olfactory tracts, and cerebral cortex.

Gustatory Sensations (p. 574)

1. The receptors for gustation, the gustatory cells, are located in taste buds.
2. The four primary tastes are salty, sweet, sour, and bitter.
3. Gustatory cells convey impulses to cranial nerves V, VII, IX, and X, the medulla, thalamus, and cerebral cortex.

Visual Sensations (p. 576)

1. Accessory structures of the eyes include the eyebrows, eyelids, eyelashes, and the lacrimal apparatus.
2. The eye is constructed of three coats: (a) fibrous tunic (sclera and cornea); (b) vascular tunic (choroid, ciliary body, and iris); and (c) retina (nervous tunic), which contains rods and cones.
3. The anterior cavity contains aqueous humor; the vitreous chamber contains vitreous body.
4. Rods and cones develop generator potentials and ganglion cells initiate nerve impulses.
5. Impulses from ganglion cells are conveyed through the retina to the optic (II) nerve, the optic chiasma, the optic tract, the thalamus, and the cortex.

Auditory Sensations and Equilibrium (p. 583)

1. The ear consists of three anatomical subdivisions: (a) the external or outer ear (auricle, external auditory canal, and tympanic membrane); (b) the middle ear (auditory or Eustachian tube, ossicles, oval window, and round window); and (c) the internal or inner ear (bony labyrinth and membranous labyrinth). The internal ear contains the spiral organ (organ of Corti), the organ of hearing.
2. Sound waves enter the external auditory canal, strike the tympanic membrane, pass through the ossicles, strike the oval window, set up waves in the perilymph, strike the vestibular membrane and scala tympani, increase pressure in the endolymph, strike the basilar membrane, and stimulate hairs on the spiral organ (organ of Corti). A nerve impulse is then initiated.
3. Static equilibrium is the orientation of the body relative to the pull of gravity. The maculae of the utricle and saccule are the sense organs of static equilibrium.
4. Dynamic equilibrium is the maintenance of body position in response to movement. The cristae in the semicircular ducts are the sense organs of dynamic equilibrium.

Aging and the Special Senses (p. 595)

1. Degenerative changes and disease can alter vision, hearing, taste, and smell.
2. Common age-related disorders include presbyopia, cataracts, glaucoma, and presbycusis.

Applications to Health (p. 595)

1. Cataract is the loss of transparency of the lens or capsule.
2. Glaucoma is abnormally high intraocular pressure (IOP), which destroys neurons of the retina.
3. Conjunctivitis is an inflammation of the conjunctiva.
4. Trachoma is a chronic, contagious inflammation of the conjunctiva.
5. Deafness is the lack of the sense of hearing or significant hearing loss. It is classified as sensorineural or conduction.
6. Labyrinthine disease is basically a malfunction of the inner ear that has a variety of causes.
7. Ménière's syndrome is the malfunction of the inner ear that may cause deafness and loss of equilibrium.
8. Vertigo is a sensation of motion that is classified as peripheral, central, or psychogenic.
9. Otitis media is an acute infection of the middle ear.
10. Motion sickness is a functional disorder precipitated by repetitive angular, linear, or vertical motion.

REVIEW QUESTIONS

1. Define a sensation and a sense receptor. What prerequisites are necessary for the perception of a sensation? (p. 562)
2. Describe the following characteristics of a sensation: projection, adaptation, afterimage, modality. (p. 562)
3. Classify receptors on the basis of location, stimulus detected, and simplicity or complexity. (p. 563)
4. Distinguish between a general sense and a special sense. (p. 563)
5. What is a cutaneous sensation? Distinguish tactile, thermoreceptive, and pain sensations. (p. 563)
6. For each of the following cutaneous sensations, describe the receptor involved in terms of structure, function, and location: touch, pressure, vibration, and pain. (p. 563)
7. Why are pain receptors important? Differentiate somatic pain, visceral pain, referred pain, and phantom pain. (p. 566)
8. What is the proprioceptive sense? Where are the receptors for this sense located? (p. 567)
9. Describe the structure of muscle spindles, tendon organs (Golgi tendon organs), and joint kinesthetic receptors. (p. 567)
10. Describe the various levels of sensation in the central nervous system. (p. 569)
11. Describe how various parts of the body are represented in the somatosensory cortex. (p. 569)
12. Describe the sensory pathways involved in proprioception, pain, and pressure. (p. 570)
13. Describe how various parts of the body are represented in the motor cortex. (p. 572)
14. Which pathways control voluntary motor impulses from the brain through the spinal cord? How are they distinguished? (p. 572)
15. Identify the receptors for olfaction. Describe the neural pathway for olfaction. (p. 573)
16. Identify the receptors for gustation. Describe the neural pathway for gustation. (p. 574)
17. How are papillae related to taste buds? Describe the structure and location of the papillae. (p. 574)
18. Describe the structure and importance of the following accessory structures of the eye: eyelids, eyelashes, and eyebrows. (p. 576)
19. What is the function of the lacrimal apparatus? Explain how it operates. (p. 576)
20. By means of a labeled diagram, indicate the principal anatomical structures of the eye. (p. 579)
21. Describe the location and contents of the chambers of the eye. What is intraocular pressure (IOP)? How is the scleral venous sinus (canal of Schlemm) related to this pressure? (p. 582)
22. Describe the path of a visual impulse from the optic (II) nerve to the brain. (p. 583)
23. Diagram the principal parts of the outer, middle, and inner ear. Describe the function of each part labeled. (p. 585)
24. Explain the events involved in the transmission of sound from the auricle to the spiral organ (organ of Corti). (p. 591)
25. What is the afferent pathway for sound impulses from the cochlear branch of the vestibulocochlear (VIII) nerve to the brain? (p. 592)
26. Compare the function of the maculae in the saccule and utricle in maintaining static equilibrium with the role of the cristae in the semicircular ducts in maintaining dynamic equilibrium. (p. 593)
27. Describe the path of a nerve impulse that results in static and dynamic equilibrium. (p. 593)
28. Describe the effects of aging on the special senses. (p. 595)
29. Define each of the following: cataract, glaucoma, conjunctivitis, trachoma, deafness, labyrinthine disease, Ménière's syndrome, vertigo, otitis media, and motion sickness. (p. 595)
30. Refer to the glossary of key medical terms associated with the sensory structures. Be sure that you can define each term. (p. 597)

SELF QUIZ

Complete the following:

1. The final common pathway from the central nervous system to skeletal muscles consists of ________ neurons.
2. The neuron that crosses to the opposite side in sensory pathways is usually the ________ neuron.
3. In general, the extrapyramidal tracts begin in the ________ and end in the ________.
4. Taste buds are located on elevated projections of the tongue called ________.
5. In the macula, the gelatinous membrane is embedded with calcium carbonate crystals called ________. These respond to gravity in such a way that the macula is the receptor for (static? dynamic?) equilibrium.
6. Show the route of nerve impulses along the pyramidal pathway by listing in correct sequence the structures that comprise the pathway.

   __ __ __ __ __ __ __ __

   A. anterior gray horn (lower motor neuron)
   B. midbrain and pons
   C. effector (skeletal muscle)
   D. internal capsule
   E. lateral corticospinal tract
   F. medulla, decussation site
   G. precentral gyrus (upper motor neuron)
   H. anterior root of spinal nerve

7. Match the following:

___ **a.** sensitive to movement of hair shaft	**A.** corpuscle of touch (Meissner's corpuscles)	
___ **b.** egg-shaped masses located in dermal papillae, especially in fingertips, palms, and soles that are receptors for discriminative touch	**B.** tactile (Merkel's) discs	
___ **c.** onion-shaped structures sensitive to pressure	**C.** lamellated (Pacinian) corpuscles	
___ **d.** modified epidermal cells in the stratum basale of hairless skin	**D.** hair root plexuses	

Arrange the answers in correct sequence.

___ ___ ___ ___ **8.** Levels of sensation, from those causing simplest, least precise reflexes to those causing most complex and precise responses:
A. thalamus
B. brain stem
C. cerebral cortex
D. spinal cord

___ ___ ___ ___ ___ **9.** Pathway for conduction of most nerve impulses for voluntary movement of muscles:
A. anterior gray horn of the spinal cord
B. precentral gyrus
C. internal capsule
D. location where decussation occurs
E. lateral corticospinal tract

___ ___ ___ ___ ___ **10.** Order of nerve impulses along conduction pathway for smell:
A. olfactory bulb
B. olfactory hairs
C. olfactory (I) nerves
D. olfactory tract
E. primary olfactory area of cortex

___ ___ ___ ___ ___ **11.** From anterior to posterior:
A. anterior chamber
B. iris
C. lens
D. vitreous chamber (posterior cavity)
E. posterior chamber

___ ___ ___ **12.** Layers of the eye, from superficial to deep:
A. sclera
B. retina
C. choroid

___ ___ ___ ___ **13.** From anterior to posterior:
A. vitreous body
B. optic (II) nerve
C. cornea
D. lens

___ ___ ___ ___ **14.** Pathway of aqueous humor, from site of formation to destination:
A. anterior chamber
B. scleral venous sinus (canal of Schlemm)
C. ciliary body
D. posterior chamber

___ ___ ___ ___ ___ **15.** Pathway of sound waves:
A. external auditory canal
B. stapes
C. malleus and incus
D. oval window
E. tympanic membrane

___ ___ ___ ___ ___ **16.** Pathway of tears, from site of formation to entrance to nose:
A. lacrimal gland and ducts
B. lacrimal sac
C. nasolacrimal duct
D. surface of conjunctiva
E. lacrimal puncta and lacrimal canals

17. Match the following:

___ **a.** "white of the eye"	**A.** central fovea
___ **b.** a clear structure, composed of protein layers arranged like an onion	**B.** choroid
___ **c.** blind spot; area in which there are no cones or rods	**C.** cornea
___ **d.** area of sharpest vision; area of densest concentration of cones	**D.** ciliary muscle
___ **e.** nonvascular, transparent, fibrous coat; most anterior eye structure	**E.** iris
___ **f.** layer containing neurons; if detached, causes blindness	**F.** lens
___ **g.** dark brown layer; prevents reflection of light rays; also nourishes eyeball since it is vascular	**G.** optic disc
___ **h.** a hole; appears black, like a circular doorway leading into a dark room	**H.** ora serrata
___ **i.** regulates the amount of light entering the eye; colored part of the eye	**I.** pupil
___ **j.** attaches to the lens by means of radially arranged fibers called the suspensory ligaments	**J.** retina
___ **k.** serrated margin of the retina	**K.** sclera
___ **l.** located at the junction of sclera and cornea; drains aqueous humor	**L.** scleral venous sinus (canal of Schlemm)

Choose the one best answer to these questions.

___ **18.** All of these sensations are conveyed by the fasciculus gracilis or fasciculus cuneatus *except*
A. pain and temperature; B. proprioception; C. discriminative touch; D. vibration; E. two-point discrimination.

___ **19.** Receptors for hearing are located in the
A. middle ear; B. cochlea; C. semicircular canals; D. tympanic membrane; E. vestibule.

___ **20.** The incorrect interpretation of pain as having come from regions far from the actual site of pain is known as
A. visceral pain; B. phantom pain; C. referred pain; D. somatosensory pain; E. somatic pain.

___ **21.** Upper motor neurons
A. carry impulses from the cerebellum to the spinal cord; B. carry impulses from the frontal lobe to the cerebellum; C. carry impulses from the cerebral cortex to the spinal cord; D. coordinate reflexes; E. synapse with lower motor neurons in the spinal ganglia.

___ **22.** A layer of tissue that is continuous over the inner surfaces of the eyelids and is reflected over the edge of the outer surface of the cornea is the
A. choroid; B. conjunctiva; C. sclera; D. suspensory ligament; E. palpebra.

___ **23.** Which of the following is/are not part(s) of the vascular tunic of the eyeball?
A. cornea; B. choroid; C. ciliary body; D. iris; E. all of the above.

___ **24.** Rods function in
(1) color vision.
(2) dim light vision.
(3) bright light vision.
A. (1) only; B. (2) only; C. (3) only; D. all of the above; E. none of the above.

___ **25.** Which of these receptors does not belong with the others?
A. muscle spindles; B. tactile (Merkel's) discs; C. tendon (Golgi) organs; D. joint kinesthetic receptors.

26. Match the following. Not all answers will be used.

___ **a.** tube used to equalize pressure on either side of tympanic membrane	**A.** auditory (Eustachian) tube
___ **b.** chamber posterior to middle ear; permits middle ear infection to spread to cause mastoiditis	**B.** incus
___ **c.** eardrum	**C.** malleus
___ **d.** structure on which stapes exerts pistonlike action	**D.** oval window
___ **e.** flared portion of the outer ear	**E.** auricle
___ **f.** ossicle adjacent to eardrum	**F.** round window
___ **g.** anvil-shaped ear bone	**G.** stapes
	H. tympanic antrum
	I. tympanic membrane

The Endocrine System

21

STUDENT OBJECTIVES

1. Define an endocrine gland and an exocrine gland, and list the endocrine glands of the body.
2. Describe the structural and functional divisions of the pituitary gland into adenohypophysis and neurohypophysis.
3. Describe the location, histology, and blood and nerve supply of the pituitary gland.
4. Discuss the symptoms of pituitary dwarfism, giantism, acromegaly, and diabetes insipidus as pituitary gland disorders.
5. Describe the location, histology, and blood and nerve supply of the thyroid gland.
6. Discuss the symptoms of cretinism, myxedema, exophthalmic goiter, and simple goiter as thyroid gland disorders.
7. Describe the location, histology, and blood and nerve supply of the parathyroid glands.
8. Discuss the symptoms of tetany and osteitis fibrosa cystica as parathyroid gland disorders.
9. Describe the location, histology, and blood and nerve supply of the adrenal (suprarenal) glands.
10. Explain the subdivisions of the adrenal (suprarenal) glands into cortical and medullary portions.
11. Discuss the symptoms of aldosteronism, Addison's disease, Cushing's syndrome, and adrenogenital syndrome as adrenal cortical disorders.
12. Discuss the symptoms of pheochromocytoma as an adrenal medullary disorder.
13. Describe the location, histology, and blood and nerve supply of the pancreas.
14. Discuss the symptoms of diabetes mellitus and hyperinsulinism as endocrine disorders of the pancreas.
15. Describe the location, histology, and blood and nerve supply of the pineal gland.
16. Describe the location, histology, and blood and nerve supply of the thymus gland.
17. Describe the effects of aging on the endocrine system.
18. Describe the development of the endocrine system.
19. Define key medical terms associated with the endocrine system.

CHAPTER OUTLINE

- **Endocrine Glands**
- **Pituitary (Hypophysis)**
  Adenohypophysis
  Neurohypophysis
- **Thyroid**
- **Parathyroids**
- **Adrenals (Suprarenals)**
  Adrenal Cortex
  Adrenal Medulla
- **Pancreas**
- **Ovaries and Testes**
- **Pineal Gland (Epiphysis Cerebri)**
- **Thymus**
- **Aging and the Endocrine System**
- **Developmental Anatomy of the Endocrine System**
- **Other Endocrine Tissues**
- **Key Medical Terms Associated with the Endocrine System**

Two regulatory systems are involved in transmitting messages and correlating various body functions: the nervous system and endocrine system. The nervous system controls homeostasis through impulses delivered via neurons. The endocrine system affects bodily activities by releasing chemical messengers, called hormones, into the bloodstream. Whereas the nervous system sends messages to a specific set of cells (muscle fibers, gland cells, or other neurons), the endocrine system as a whole sends messages to cells in virtually any part of the body. The nervous system causes muscles to contract or relax and glands to secrete more or less of their product; the endocrine system brings about changes in the metabolic activities of almost all body tissues. Thus the endocrine system not only helps to regulate activity of smooth and cardiac muscle and some glands; it significantly affects virtually all other tissues as well. Neurons tend to act within a few milliseconds; hormones can take up to several hours or more to bring about their responses. Also, the effects of nervous system stimulation are generally brief compared with the effects of endocrine stimulation.

Obviously, the body could not function if the two great control systems were to pull in opposite directions. The nervous and endocrine systems coordinate their activities like an interlocking supersystem. Certain parts of the nervous system stimulate or inhibit the release of hormones, and certain hormones, in turn, are quite capable of stimulating or inhibiting the flow of nerve impulses.

Although the effects of hormones are many and varied, their actions can be categorized into seven broad areas:

1. They help to control the internal environment by regulating its chemical composition and volume.
2. They respond to marked changes in environmental conditions to help the body cope with emergency demands such as infection, trauma, emotional stress, dehydration, starvation, hemorrhage, and temperature extremes.
3. They assume a role in the smooth, sequential integration of growth and development.
4. They contribute to the basic processes of reproduction, including gamete (egg and sperm) production, fertilization, nourishment of the embryo and fetus, delivery, and nourishment of the newborn.
5. They help regulate organic metabolism and energy balance.
6. They help regulate certain components of the immune system.
7. Some hormones cause muscles to contract, for example, motilin causes contraction of intestinal smooth muscle.

The science concerned with the structure and functions of the endocrine glands and the diagnosis and treatment of disorders of the endocrine system is called ***endocrinology*** (en′-dō-kri-NOL-ō-jē; *endo* = within; *crin* = to secrete; *logos* = study of).

The developmental anatomy of the endocrine system will be considered later in the chapter.

ENDOCRINE GLANDS

The endocrine glands make up the ***endocrine system.*** The body contains two kinds of glands: exocrine and endocrine. ***Exocrine glands*** secrete their products into ducts, and the ducts carry the secretions into body cavities, into the lumens of various organs, or to the body's surface. Exocrine glands include sudoriferous (sweat), sebaceous (oil), mucous, and digestive glands. ***Endocrine glands,*** by contrast, secrete their products (hormones) into the extracellular space around the secretory cells rather than into ducts. The secretion then passes into capillaries to be transported in the blood. The endocrine glands of the body include the pituitary (hypophysis), thyroid, parathyroids, adrenals (suprarenals), pineal (epiphysis cerebri), and thymus gland. In addition, there are several organs of the body that contain endocrine tissue but are not exclusively endocrine glands. These organs include the pancreas, ovaries, testes, kidneys, stomach, small intestine, skin, heart, and placenta. The location of many organs of the endocrine system and endocrine-containing organs is illustrated in Figure 21-1.

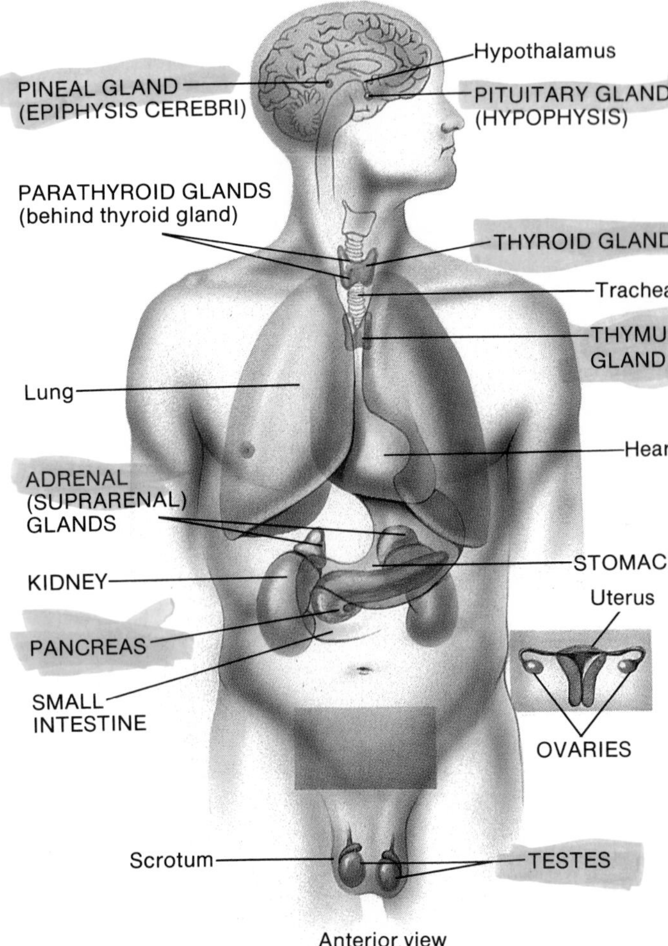

FIGURE 21-1 Location of many endocrine glands, organs containing endocrine tissue, and associated structures.

The secretions of endocrine glands are called ***hormones*** (*hormone* = set in motion). The one thing all hormones have in common is the function of maintaining homeostasis by changing the rate of the physiological activities of cells. The amount of hormone released by an endocrine gland or tissue is determined by the body's *need* for the hormone at any given time. This need is the basis on which the endocrine system operates. Hormone-producing cells have available to them information from sensing and signaling systems that permit these cells to regulate the amount and duration of hormone release. Secretion is normally regulated so that there is no overproduction or underproduction of a particular hormone.

Once a hormone is released by secretory cells, it is carried by the blood to ***target cells,*** which respond to the hormone. All cells are target cells for one or more hormones, but not all cells respond to a particular hormone. Target cells contain receptors that bind the hormone so that it can produce the effect. The combination of hormone and receptor activates a chain of events within the target cell in which the physiological effects of the hormone are expressed. Once the target cells respond to the hormone, the response must be recognized by the secretory cell by some type of feedback signal. Hormones that have accomplished their goals are degraded by target cells, the liver, and kidneys and excreted by the liver or kidneys.

PITUITARY (HYPOPHYSIS)

The hormones of the ***pituitary gland,*** also called the ***hypophysis*** (hī-POF-i-sis), regulate so many body activities that the pituitary has been nicknamed the "master gland." It is a round structure and surprisingly small, measuring about 1.3 cm (0.5 in.) in diameter. The pituitary gland lies in the sella turcica of the sphenoid bone. Posterior to the optic chiasma is a grayish protuberance, the ***tuber cinereum,*** which is part of the hypothalamus (see Figure 18-4a). The ***median eminence*** is a raised portion of the tuber cinereum (see Figure 18-5a) to which is attached the ***infundibulum,*** a stalklike structure that attaches the pituitary gland to the hypothalamus (see Figure 21-2).

The pituitary gland is divided structurally and functionally into an anterior lobe and a posterior lobe. Both are closely associated with the hypothalamus, but only the posterior lobe is neurally connected to the hypothalamus. The ***anterior lobe*** constitutes about 75 percent of the total weight of the gland. It is derived from an outgrowth of ectoderm in the embryonic pharyngeal region called the hypophyseal (Rathke's) pouch in the mouth area (see Figure 21-17b). Accordingly, the anterior lobe contains many glandular epithelial cells and forms the glandular part of the pituitary.

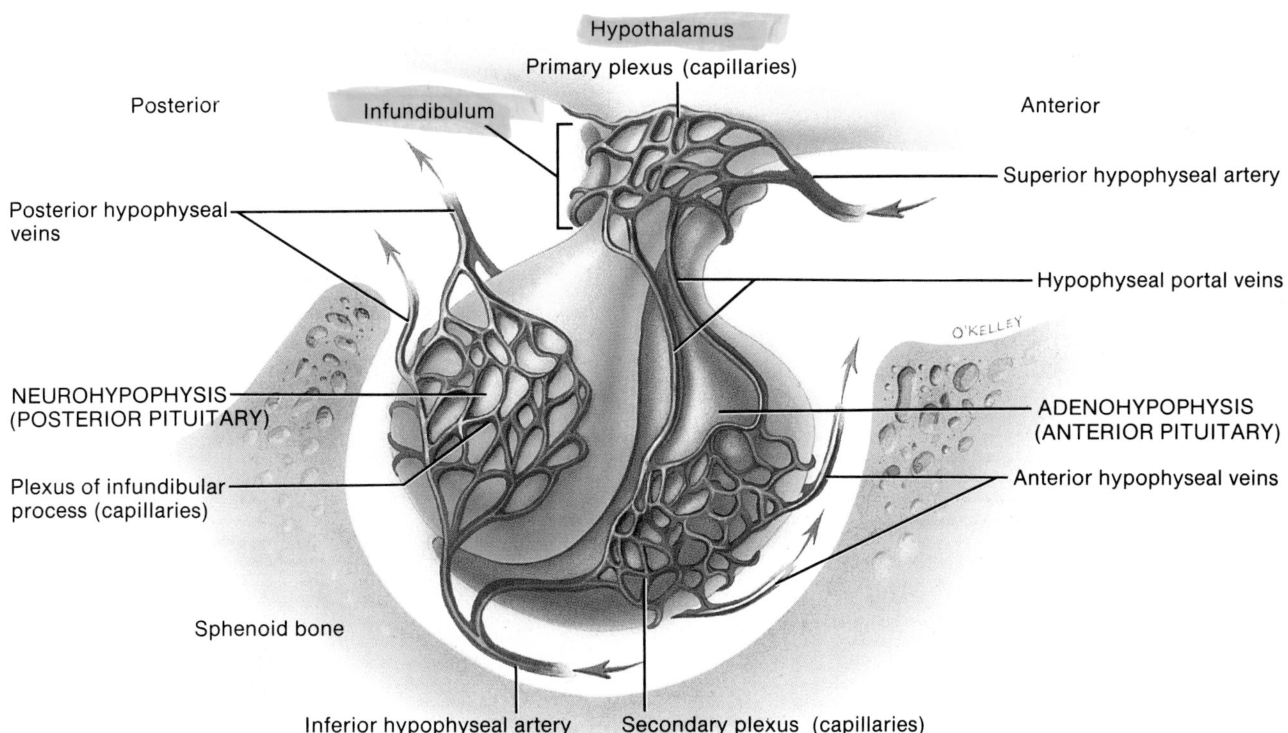

FIGURE 21-2 Blood supply of the pituitary gland (hypophysis).

A system of blood vessels connects the anterior lobe with the hypothalamus.

The ***posterior lobe*** is also derived from the ectoderm, but from an outgrowth called the neurohypophyseal bud (see Figure 21-17b). Accordingly, the posterior lobe contains axon terminations of neurons whose cell bodies are located in the hypothalamus. The nerve fibers that terminate in the posterior lobe are associated with supporting cells called pituicytes.

Between the lobes is a small, relatively avascular zone, the ***pars intermedia.*** Although it is much larger and more clearly defined in structure and function in some lower vertebrates, its role in humans is obscure.

ADENOHYPOPHYSIS

The anterior lobe of the pituitary is also called the ***adenohypophysis*** (ad′-e-nō-hī-POF-i-sis; *adeno* = glandular). It releases hormones that regulate a whole range of bodily activities from growth to reproduction. The release of these hormones is either stimulated or inhibited by chemical secretions from the hypothalamus called ***regulating hormones*** (or ***factors***), which constitute an important link between the nervous and endocrine systems.

The hypothalamic regulating hormones (or factors) are delivered to the adenohypophysis through a series of blood vessels. The blood supply to the adenohypophysis and infundibulum is derived principally from several ***superior hypophyseal*** (hī′-po-FIZ-ē-al) ***arteries.*** These arteries are branches of the internal carotid and posterior communicating arteries (Figure 21-2). The superior hypophyseal arteries form a network or plexus of capillaries, the ***primary plexus,*** in the infundibulum near the inferior portion of the hypothalamus. Regulating hormones (or factors) from the hypothalamus diffuse into this plexus. This plexus drains into the ***hypophyseal portal veins*** that pass down the infundibulum. At the inferior portion of the infundibulum, the veins form a ***secondary plexus*** of capillaries in the adenohypophysis. From this plexus, hormones of the adenohypophysis pass into the anterior hypophyseal veins for distribution to tissue cells. Such a delivery system permits regulating hormones (or factors) to act quickly on the adenohypophysis without first circulating through the heart. The short route prevents dilution or destruction of the regulating hormones (or factors).

When the adenohypophysis receives proper stimulation from the hypothalamus via regulating hormones (or factors), its glandular cells secrete any one of seven hormones.* Special staining techniques have established the division of glandular cells into five principal types (Figure 21-3):

* Human chorionic gonadotropin (hCG) is a hormone that was believed to be produced only by the placenta to help maintain pregnancy. Recent evidence suggests that it is also produced by the pituitary gland and that its production is stimulated by a regulating hormone produced by the hypothalamus called gonadotropin-releasing hormone (GnRH).

1. ***Somatotroph cells*** produce ***human growth hormone (hGH),*** which controls general body growth.
2. ***Lactotroph cells*** synthesize ***prolactin (PRL),*** which initiates milk production by the mammary glands.
3. ***Corticolipotroph cells*** synthesize ***adrenocorticotropic hormone (ACTH),*** which stimulates the adrenal cortex to secrete its hormones, and ***melanocyte-stimulating hormone (MSH),*** which is believed to be related to skin pigmentation.
4. ***Thyrotroph cells*** manufacture ***thyroid-stimulating hormone (TSH),*** which controls the thyroid gland.
5. ***Gonadotroph cells*** produce ***follicle-stimulating hormone (FSH),*** which stimulates the production of eggs and sperm in the ovaries and testes, respectively, and ***luteinizing hormone (LH),*** which stimulates other sexual and reproductive activities.

Except for human growth hormone (hGH), melanocyte-stimulating hormone (MSH), and prolactin (PRL), all the secretions of the adenohypophosis are referred to as ***tropic hormones*** (*trop* = turn on), which means that they stimulate other endocrine glands. Follicle-stimulating hormone (FSH) and luteinizing hormone (LH) are also called ***gonadotropic*** (gō-nad-ō-TRŌ-pik) ***hormones*** because they regulate the functions of the gonads (ovaries and testes). The gonads are the endocrine glands that produce sex hormones, or steroids.

CLINICAL APPLICATION

Pituitary Dwarfism, Giantism, and Acromegaly

Disorders of the endocrine system, in general, involve ***hyposecretion*** (underproduction) ***hypersecretion*** (overproduction) of hormones.

Among the clinically interesting disorders related to the adenohypophysis are those involving hGH. If hGH is hyposecreted during the growth years, bone growth is slow, and the epiphyseal plates close before normal height is reached. This condition is called ***pituitary dwarfism.*** Other organs of the body also fail to grow, and the pituitary dwarf is childlike in many physical respects. Treatment requires administration of hGH during childhood before the epiphyseal plates close. Dwarfism is also caused by other conditions, in which administration of hGH is not corrective.

Hypersecretion of hGH during childhood results in ***giantism (gigantism),*** an abnormal increase in the length of long bones. As a result, the person grows to be very large, but body proportions are about normal. Hypersecretion during adulthood is called ***acromegaly*** (ak′-rō-MEG-a-lē), which is shown in Figure 21-4; hGH cannot produce further lengthening of the long bones because the epiphyseal plates are already closed. Instead, the bones of the hands, feet, cheeks, and jaws thicken, and other tissues also grow. The eyelids, lips, tongue, and nose enlarge,

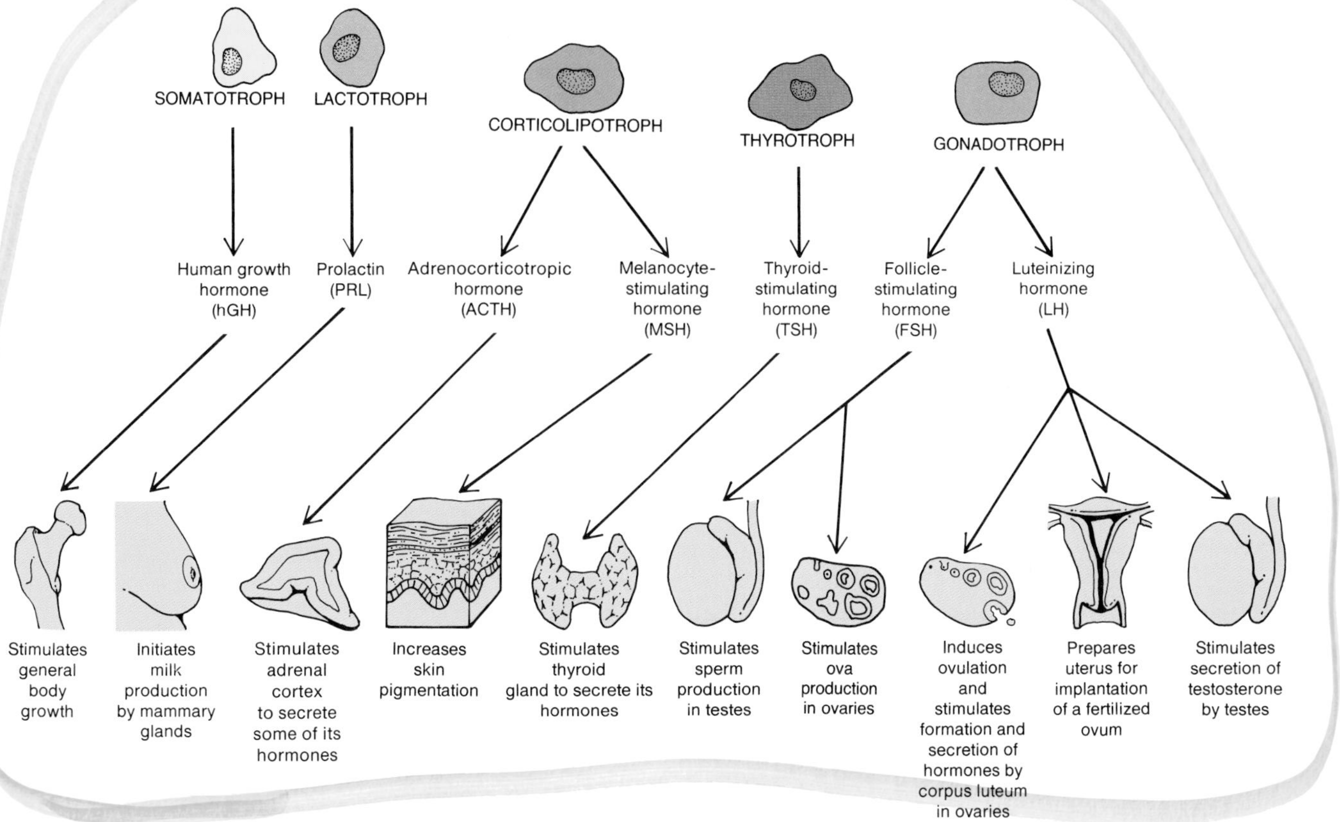

SOMATOTROPH
LACTOTROPH
CORTICOLIPOTROPH
THYROTROPH
GONADOTROPH

Human growth hormone (hGH)
Prolactin (PRL)
Adrenocorticotropic hormone (ACTH)
Melanocyte-stimulating hormone (MSH)
Thyroid-stimulating hormone (TSH)
Follicle-stimulating hormone (FSH)
Luteinizing hormone (LH)

Stimulates general body growth
Initiates milk production by mammary glands
Stimulates adrenal cortex to secrete some of its hormones
Increases skin pigmentation
Stimulates thyroid gland to secrete its hormones
Stimulates sperm production in testes
Stimulates ova production in ovaries
Induces ovulation and stimulates formation and secretion of hormones by corpus luteum in ovaries
Prepares uterus for implantation of a fertilized ovum
Stimulates secretion of testosterone by testes

FIGURE 21-3 Cells of the adenohypophysis as revealed by special stains. Most cells that produce human growth hormone (somatotrophs) and prolactin (lactotrophs) are separate cells. However, some normal and tumor cells are single cells that produce both human growth hormone and prolactin. The corticolipotroph cell produces adrenocorticotropic hormone and melanocyte-stimulating hormone. Thyrotroph cells synthesize thyroid-stimulating hormone. Most gonadotroph cells produce both follicle-stimulating hormone and luteinizing hormone. However, a few separate cells may exist, some producing follicle-stimulating hormone and some producing luteinizing hormone. (Adapted from a slide provided by Calvin Ezrin, M.D., Clinical Professor of Medicine, U.C.L.A., and Adjunct Professor of Pathology, University of Toronto.)

and the skin thickens and furrows, especially on the forehead and soles of the feet. Diabetes mellitus is also a complication.

NEUROHYPOPHYSIS

In a strict sense, the posterior lobe, or ***neurohypophysis,*** is not an endocrine gland, since it does not synthesize hormones. Instead, it stores and releases two hormones. The posterior lobe consists of cells called ***pituicytes*** (pi-TOO-i-sītz), which are similar in appearance to the neuroglia of the nervous system. It also contains axon terminals of secretory neurons of the hypothalamus (Figure 21-5). Such neurons are called ***neurosecretory cells.*** The cell bodies of the neurons originate in nuclei (paraventricular and supraoptic) in the hypothalamus. The fibers project from the hypothalamus, form the ***hypothalamic-hypophyseal tract,*** and terminate on blood capillaries in the neurohypophysis.

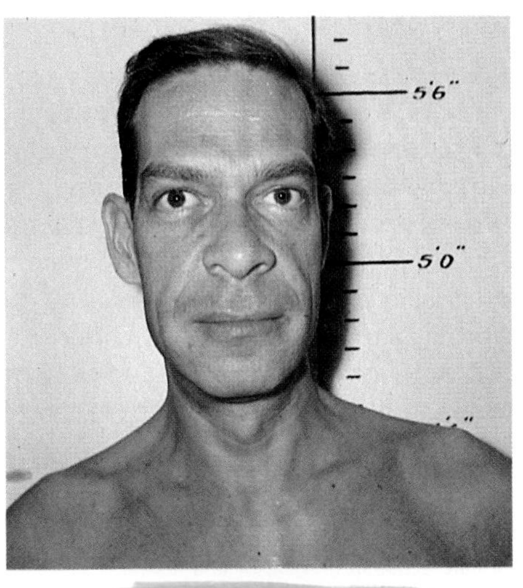

FIGURE 21-4 An individual with acromegaly. (Courtesy of Lester Bergman & Associates, Inc.)

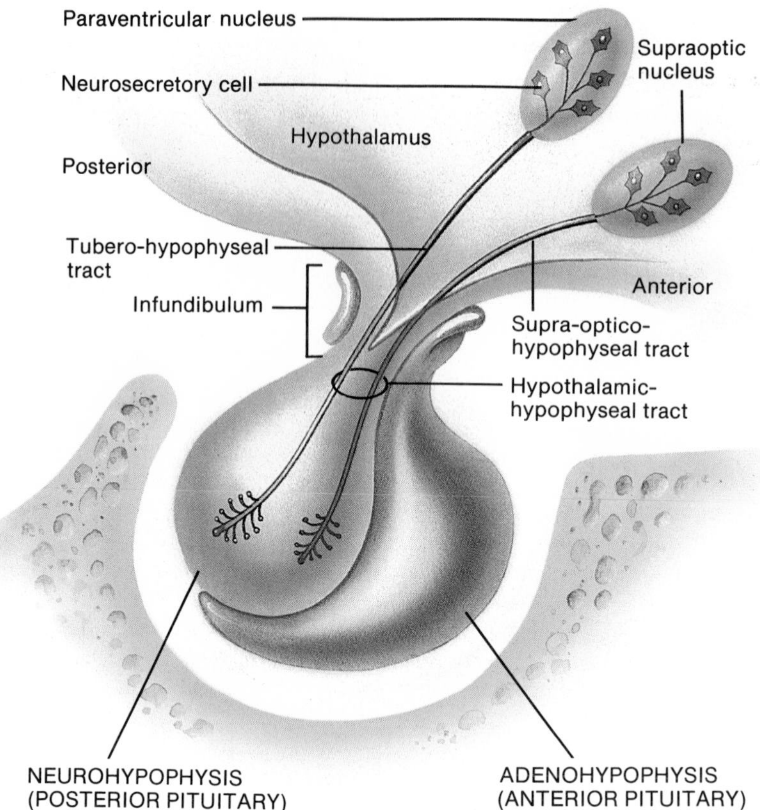

FIGURE 21-5 Hypothalamic-hypophyseal tract.

The cell bodies of the neurosecretory cells produce two hormones: ***oxytocin (OT)*** and ***antidiuretic hormone (ADH).*** OT stimulates contraction of the smooth muscle of the pregnant uterus during labor and stimulates contractile cells around the ducts of the mammary glands to eject milk. ADH prevents excessive urine production by bringing about water reabsorption and secondarily causes blood pressure to rise by bringing about constriction of arterioles.

The blood supply to the neurohypophysis is from the ***inferior hypophyseal arteries,*** derived from the internal carotid arteries (see Figure 21-2). In the neurohypophysis, the inferior hypophyseal arteries form a plexus of capillaries called the ***plexus of the infundibular process.*** From this plexus, hormones stored in the neurohypophysis pass into the ***posterior hypophyseal veins*** for distribution to tissue cells.

CLINICAL APPLICATION

Diabetes Insipidus
The principal abnormality associated with dysfunction of the neurohypophysis is ***diabetes insipidus*** (in-SIP-i-dus). *Diabetes* means "overflow" and *insipidus* means "tasteless." This disorder should not be confused with diabetes mellitus (*meli* = honey), a disorder of the pancreas characterized by glucose in the urine. Diabetes insipidus is the result of a hyposecretion of ADH, usually caused by damage to the neurohypophysis or the hypothalamic paraventricular and supraoptic nuclei. Symptoms include excretion of large amounts of urine and subsequent dehydration and thirst. A major problem associated with this condition is inability to concentrate urine. Because of this, a person with severe diabetes insipidus may die of dehydration if deprived of water for only a day or so. Diabetes insipidus is treated by administering ADH.

A summary of pituitary gland hormones, actions, and disorders is presented in Exhibit 21-1.

THYROID

The ***thyroid gland*** is located just below the larynx. The right and left ***lateral lobes*** lie one on either side of the trachea. The lobes are connected by a mass of tissue called an ***isthmus*** (IS-mus) that lies in front of the trachea, just below the cricoid cartilage (Figure 21-6). The ***pyramidal lobe,*** when present, extends upward from the isthmus and may be attached to the hyoid bone by a fibrous or fibromuscular band. The gland has a rich blood supply, receiving about 80 to 120 ml of blood per minute. Thus, the thyroid

EXHIBIT 21-1

Summary of Pituitary Gland Hormones, Principal Actions, and Selected Disorders

HORMONE	PRINCIPAL ACTIONS	SELECTED DISORDERS
ADENOHYPOPHYSEAL HORMONES		
Human Growth Hormone (hGH)	Growth of body cells; protein anabolism; elevation of blood glucose concentration.	Hyposecretion of hGH during the growth years results in pituitary dwarfism; hypersecretion of hGH during the growth years results in giantism; hypersecretion of hGH during adulthood results in acromegaly.
Thyroid-Stimulating Hormone (TSH)	Controls secretion of thyroid hormones by thyroid gland.	Hypersecretion of thyroid hormones through the action of TSH causes exophthalmic goiter (to be discussed later).
Adrenocorticotropic Hormone (ACTH)	Controls secretion of some hormones by adrenal cortex (mainly cortisol).	Hyposecretion of glucocorticoids through the action of ACTH results in Addison's disease (to be discussed later).
Follicle-Stimulating Hormone (FSH)	In female, initiates development of ova and induces ovarian secretion of estrogens. In male, stimulates testes to produce sperm.	
Luteinizing Hormone (LH) [also called interstitial cell-stimulating hormone (ICSH) in male]	In female, together with FSH, stimulates ovulation and formation of progesterone-producing corpus luteum. In male, stimulates interstitial cells in testes to develop and produce testosterone.	
Prolactin (PRL)	In females, together with other hormones, initiates and maintains effect of luteinizing hormone in promoting milk secretion by the mammary glands; in males, it enhances the production of testosterone.	
Melanocyte-Stimulating Hormone (MSH)	Stimulates dispersion of melanin granules in melanocytes.	
NEUROHYPOPHYSEAL HORMONES		
Oxytocin (OT)	Stimulates contraction of smooth muscle cells of pregnant uterus during labor and stimulates contraction of contractile cells of mammary glands for milk ejection.	
Antidiuretic Hormone (ADH)	Principal effect is to decrease urine volume; also raises blood pressure by constricting arterioles during severe hemorrhage.	Hyposecretion of ADH results in diabetes insipidus.

gland can deliver high levels of hormones in a short period of time, if necessary.

Histologically, the thyroid gland is composed of spherical sacs called ***thyroid follicles*** (Figure 21-7). The wall of each follicle consists of two types of cells. Those cells that reach the surface of the lumen of the follicle are called ***follicular cells,*** and those that do not reach the lumen are called ***parafollicular cells,*** or ***C cells.*** When the cells are inactive, they tend to be low cuboidal to squamous, but when actively secreting hormones, they become more columnar. The follicular cells manufacture ***thyroxine*** (thī-ROK-sēn), or T_4, since it contains four atoms of iodine, and ***triiodothyronine*** (trī-ī′-ōd-ō-THĪ-rō-nēn), or T_3, since it contains three atoms of iodine. Together these hormones are referred to as the ***thyroid hormones.*** Thyroxine is normally secreted in greater quantity than triiodothyronine, but triiodothyronine is three to four times more potent. Moreover, in peripheral tissues, especially the liver and lungs, much of the thyroxine is converted into triiodothyronine. Both hormones are functionally similar. They control metabolism, regulate growth and development, and increase reactivity of the nervous system. The parafollicular cells produce ***calcitonin*** (kal-si-TŌ-nin), or ***CT.*** This hormone decreases blood levels of calcium and phosphate by inhibiting bone breakdown (osteoclastic activity) and accelerating calcium absorption by bones.

One of the thyroid gland's unique features is its ability to store hormones and release them in a steady flow over a long period of time. In the follicle cells, iodide is oxidized to iodine. Through a series of enzymatically controlled

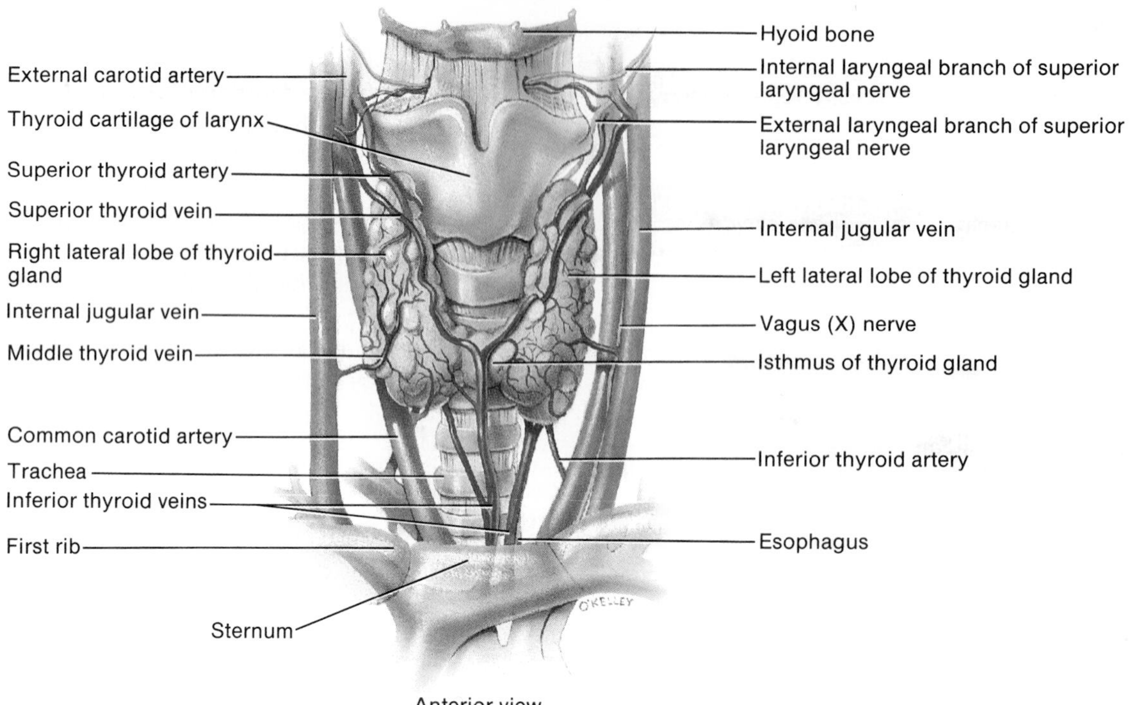

FIGURE 21-6 Location and blood supply of the thyroid gland.

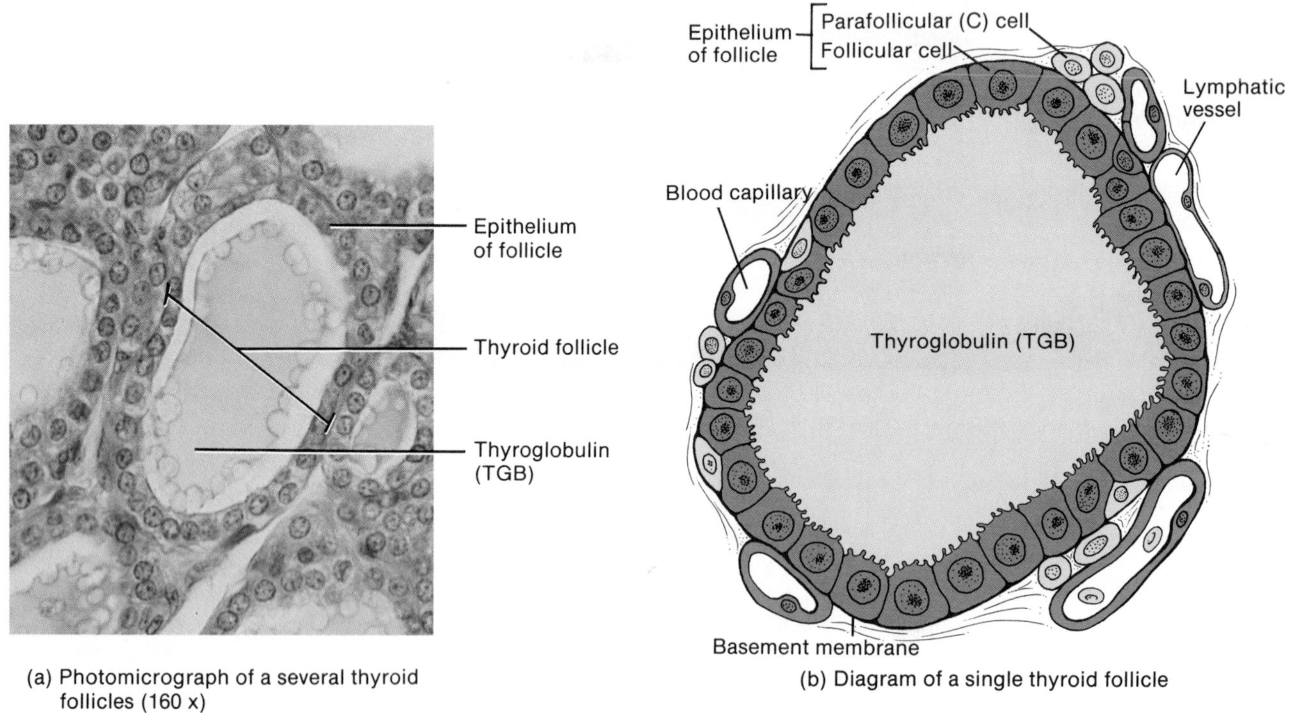

FIGURE 21-7 Histology of the thyroid gland. (a) Courtesy of Lester V. Bergman & Associates, Inc.

reactions, the iodine combines with the amino acid tyrosine to form the thyroid hormones. This combination occurs within a large glycoprotein molecule, called ***thyroglobulin (TGB)***, which is secreted by the follicle cells into the follicle.

The main blood supply of the thyroid gland is from the superior thyroid artery, a branch of the external carotid artery, and the inferior thyroid artery, a branch of the thyrocervical trunk from the subclavian artery. The thyroid is drained by the superior and middle thyroid veins, which pass into the internal jugular veins, and the inferior thyroid veins, which join the brachiocephalic veins or internal jugular veins.

The nerve supply of the thyroid consists of postganglionic fibers from the superior and middle cervical sympathetic ganglia. Preganglionic fibers from the ganglia are derived from the second through seventh thoracic segments of the spinal cord.

CLINICAL APPLICATION

Cretinism, Myxedema, and Goiter

Hyposecretion of thyroid hormones during fetal life or infancy results in ***cretinism*** (KRĒ-tin-izm), which is shown in Figure 21-8a. Two outstanding clinical symptoms of the cretin are dwarfism and mental retardation. The first is caused by failure of the skeleton to grow and mature; the second is caused by failure of the brain to develop fully. Recall that one function of thyroid hormones is to control tissue growth and development. Cretins also exhibit retarded sexual development and a yellowish skin color. Flat pads of fat develop, giving the cretin a characteristic round face and thick nose; a large, thick, protruding tongue; and protruding abdomen. Because carbohydrates are stored rather than utilized, energy-producing metabolic reactions are slow; thus the cretin has a low body temperature and a slow heart rate, and suffers from general lethargy. If the condition is diagnosed early, the symptoms can be eliminated by administering thyroid hormones.

Hypothyroidism during the adult years produces ***myxedema*** (mix-e-DĒ-ma). A hallmark of this disorder is an edema that causes the facial tissues to swell and look puffy. Like the cretin, the person with myxedema suffers

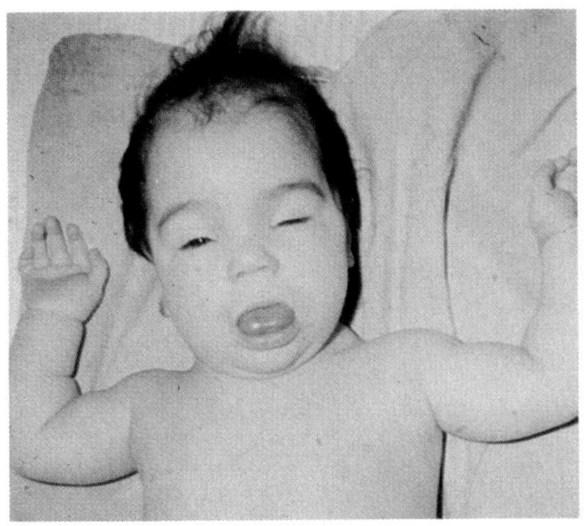

(a) Cretinism

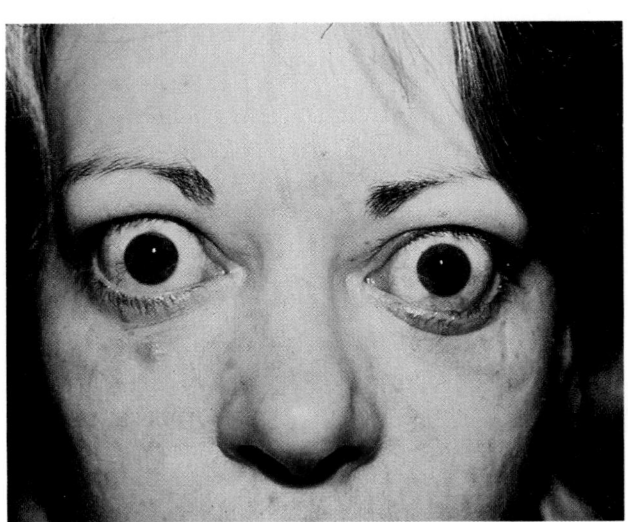

(b) Exophthalmos

(c) Simple goiter

FIGURE 21-8 Abnormalities related to the thyroid gland. (a) Courtesy of Lester Bergman & Associates, Inc. (b) Courtesy of Lester Bergman & Associates, Inc. (c) Copyright © Kay, Peter Arnold.

from slow heart rate, low body temperature, sensitivity to cold, hypersensitivity to certain drugs (narcotics, barbiturates, and anesthetics), dry hair and skin, muscular weakness, general lethargy, and a tendency to gain weight easily. The long-term effect of a slow heart rate may overwork the heart muscle, causing the heart to enlarge. Because the brain has already reached maturity, the person with myxedema does not experience mental retardation. However, in moderately severe cases, nerve reactivity may be so dulled that the person lacks mental alertness. Myxedema occurs about five times more frequently in females than in males. Its symptoms are alleviated by the administration of thyroid hormones.

Hypersecretion of thyroid hormones may be due to an autoimmune disease called ***exophthalmic*** (ek′-sof-THAL-mik) ***goiter*** (GOY-ter), also called ***Grave's disease.*** This disease, like myxedema, is also more frequent in females. One of its primary symptoms is an enlarged thyroid, called a ***goiter,*** which may be two to three times its original size. Two other symptoms are an edema behind the eye, which causes the eye to protrude **(*exophthalmos*),** which is shown in Figure 21-8b, and an abnormally high metabolic rate. The high metabolic rate produces a range of effects that are generally opposite to those of myxedema—increased pulse, high body temperature, heat intolerance, and moist, flushed skin. The person loses weight and is usually full of "nervous" energy. The thyroid hormones also increase the responsiveness of the nervous system, causing the person to become irritable and exhibit tremors of the extended fingers. Hyperthyroidism is usually treated by administration of antithyroid drugs that suppress thyroid hormone synthesis, by administration of radioactive iodine that selectively destroys thyroid cells, or by surgical removal of part of the gland.

Goiter is a symptom of many thyroid disorders. In North America, it is typically caused by ingestion of excess iodine (owing to its presence in bread, iodized salt, and certain vegetables, such as turnips and rutabagas). It may also be genetically caused. This condition is called ***simple goiter*** (Figure 21-8c). Simple goiter may also be caused by a lower-than-average amount of iodine in the diet. This cause is more prevalent in Third World countries. It may also develop if iodine intake is not increased during certain conditions that put a high demand on the body for thyroxine, such as pregnancy, frequent exposure to cold, and high-fat and high-protein diets.

A summary of thyroid gland hormones, actions, and disorders is presented in Exhibit 21-2.

PARATHYROIDS

Typically embedded on the posterior surfaces of the lateral lobes of the thyroid gland are small, round masses of tissue called the ***parathyroid glands.*** Usually, two parathyroids, superior and inferior, are attached to each lateral thyroid lobe (Figure 21-9).

Histologically, the parathyroids contain two kinds of epithelial cells (Figure 21-10). The more numerous cells, called ***principal (chief) cells,*** are believed to be the major synthesizer of ***parathyroid hormone (PTH).*** Some researchers believe that the other kind of cell, called an ***oxyphil cell,*** synthesizes a reserve capacity of the hormone. Functionally, PTH increases blood calcium and magnesium levels and decreases blood phosphate level by increasing the rate of calcium absorption from the gastrointestinal tract into the blood; increases the number and activity of osteoclasts; increases calcium absorption by the kidneys; increases phosphate excretion by the kidneys; and activates vitamin D.

The parathyroids are abundantly supplied with blood from branches of the superior and inferior thyroid arteries. Blood is drained by the superior, middle, and inferior thyroid veins. The nerve supply of the parathyroids is derived from

EXHIBIT 21-2

Summary of Thyroid Gland Hormones, Principal Actions, and Selected Disorders

HORMONE	PRINCIPAL ACTIONS	SELECTED DISORDERS
Thyroid Hormones		
Thyroxine (T_4)	Regulates organic metabolism, growth and development, and activity of nervous system.	Hyposecretion of thyroid hormones during the growth years results in cretinism; hypothyroidism during adult years results in myxedema; hypersecretion of thyroid hormones results in exophthalmic goiter; iodine excess or deficiency produces simple goiter.
Triiodothyronine (T_3)	Same as above.	
Calcitonin (CT)	Lowers blood levels of calcium by accelerating calcium absorption by bones.	

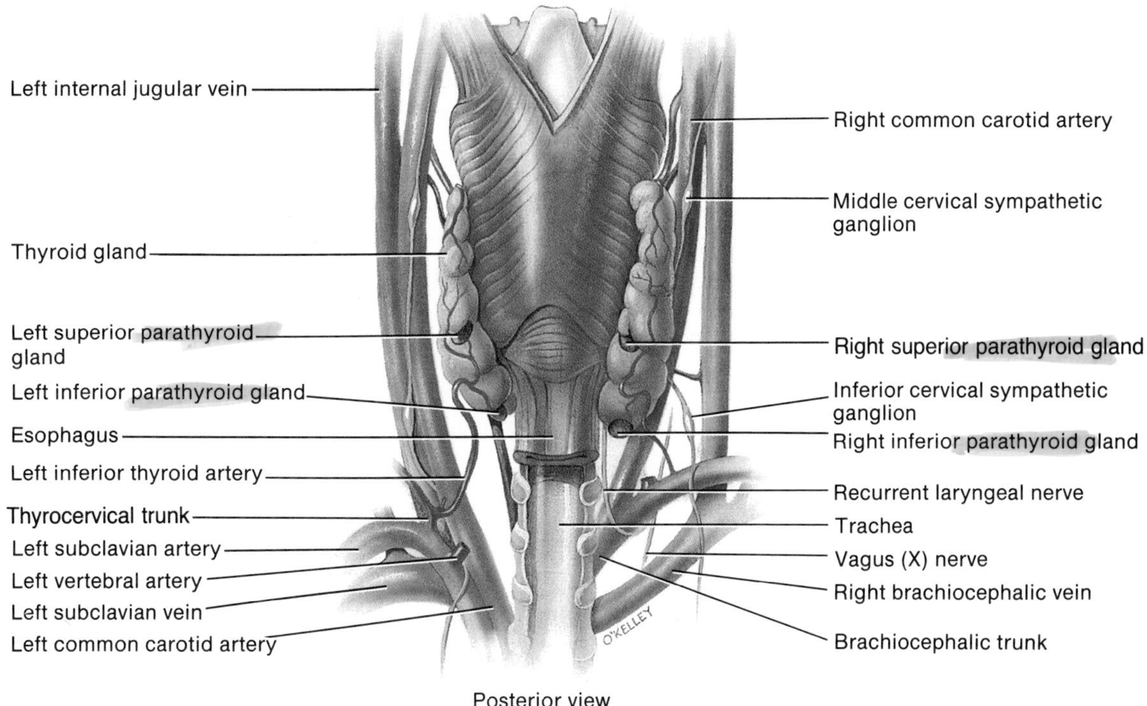

FIGURE 21-9 Location and blood supply of the parathyroid glands.

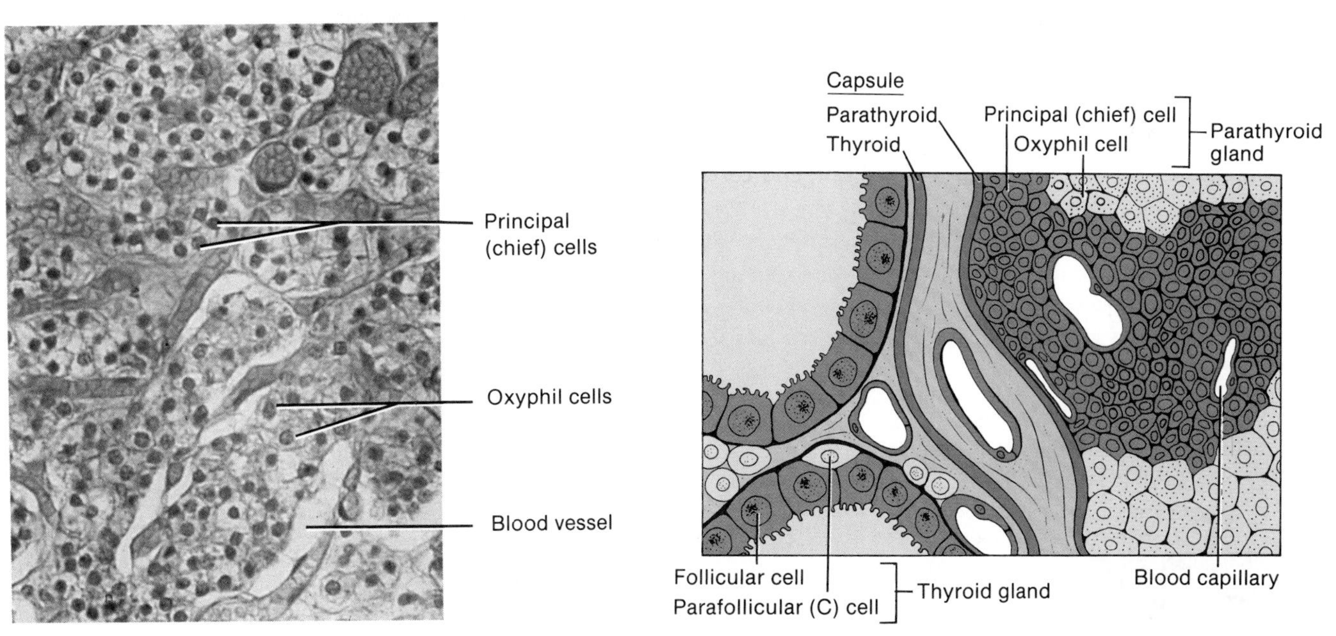

(a) Photomicrograph of parathyroid gland (340 x)

(b) Diagram of a portion of the thyroid gland (left) and parathyroid gland (right)

FIGURE 21-10 Histology of the parathyroid glands. (a) © Biophoto, SPL, Photo Researchers.

the thyroid branches of cervical sympathetic ganglia and appear to be vasomotor in function.

CLINICAL APPLICATION

Tetany and Osteitis Fibrosa Cystica

A normal amount of calcium in the extracellular fluid is necessary to maintain the resting state of neurons. A deficiency of calcium caused by ***hypoparathyroidism*** causes neurons to depolarize without the usual stimulus. As a result, nerve impulses increase and result in muscle twitches, spasms, and convulsions. This condition is called ***tetany.*** Hypoparathyroidism results from surgical removal of the parathyroids or from damage caused by parathyroid disease, infection, hemorrhage, or mechanical injury.

Hyperparathyroidism causes demineralization of bone. If uncorrected, this condition may lead to ***osteitis fibrosa cystica,*** so named because the areas of destroyed bone tissue are replaced by cavities that fill with fibrous tissue. The bones thus become deformed and are highly susceptible to fracture. Hyperparathyroidism is usually caused by a tumor in the parathyroids.

A summary of parathyroid hormones, actions, and disorders is presented in Exhibit 21-3.

ADRENALS (SUPRARENALS)

The body has two ***adrenal (suprarenal) glands,*** one of which is located superior to each kidney (Figure 21-11). Each adrenal gland is structurally and functionally differentiated into two regions: the outer ***adrenal cortex,*** which makes up the bulk of the gland, and the inner ***adrenal medulla*** (Figure 21-12). Whereas the adrenal cortex is derived from mesoderm of a developing embryo, the adrenal medulla is derived from the ectoderm. Since their origins are different, they also produce different hormones. Covering the gland is a connective tissue capsule. The adrenals, like the thyroid, are among the more vascular organs of the body.

ADRENAL CORTEX

Histologically, the adrenal cortex is subdivided into three zones (Figure 21-12). Each zone has a different cellular arrangement and secretes different groups of steroid hormones. The outer zone, directly underneath the connective tissue capsule, is referred to as the ***zona glomerulosa.*** It composes about 15 percent of the total cortical volume. Its cells are arranged in arched loops or round balls. Its primary secretions are a group of hormones called mineralocorticoids (min′-er-al-ō-KOR-ti-koyds). The principal mineralocorticoid is ***aldosterone*** (al-dō-STER-ōn), which causes the kidneys to absorb sodium and water and to increase potassium excretion.

CLINICAL APPLICATION

Aldosteronism

Hypersecretion of the mineralocorticoid aldosterone results in ***aldosteronism,*** characterized by an increase in sodium and a decrease in potassium concentration in the blood and a usually correctable form of hypertension (high blood pressure). If potassium depletion is great, neurons cannot depolarize, and muscular paralysis results. Hypersecretion also brings about excessive retention of sodium and water. The water increases the volume of the blood and causes the hypertension.

The middle zone, or ***zona fasciculata,*** is the widest of the three zones and consists of cells arranged in long, straight cords. The zona fasciculata secretes mainly ***glucocorticoids*** (gloo′-kō-KŌR-ti-koyds). These include ***cortisol (hydrocortisone), corticosterone,*** and ***cortisone.*** Together the hormones work with other hormones to promote normal metabolism, help resist stress, and decrease edema caused by the inflammatory response.

CLINICAL APPLICATION

Addison's Disease and Cushing's Syndrome

Hyposecretion of glucocorticoids (and aldosterone) results in the condition called ***Addison's disease*** **(primary**

EXHIBIT 21-3

Summary of Parathyroid Gland Hormone, Principal Actions, and Selected Disorders

HORMONE	PRINCIPAL ACTIONS	SELECTED DISORDERS
Parathyroid Hormone (PTH)	Increases blood calcium and magnesium levels and decreases blood phosphate level by increasing rate of calcium and magnesium absorption from gastrointestinal tract into blood; increases number and activity of osteoclasts; increases calcium absorption by kidneys; increases phosphate excretion by kidneys; and activates vitamin D.	Hypoparathyroidism results in tetany. Hyperparathyroidism produces osteitis fibrosa cystica.

Inferior phrenic arteries
Right superior suprarenal arteries
Left superior suprarenal arteries
Right adrenal (suparenal) gland
Left adrenal (suprarenal) gland
Celiac trunk
Right middle suprarenal artery
Left middle suprarenal artery
Left inferior suprarenal artery
Left inferior suprarenal artery
Left renal artery
Right renal artery
Left renal vein
Right renal vein
O'KELLEY
Superior mesenteric artery
Inferior vena cava
Abdominal aorta
Anterior view

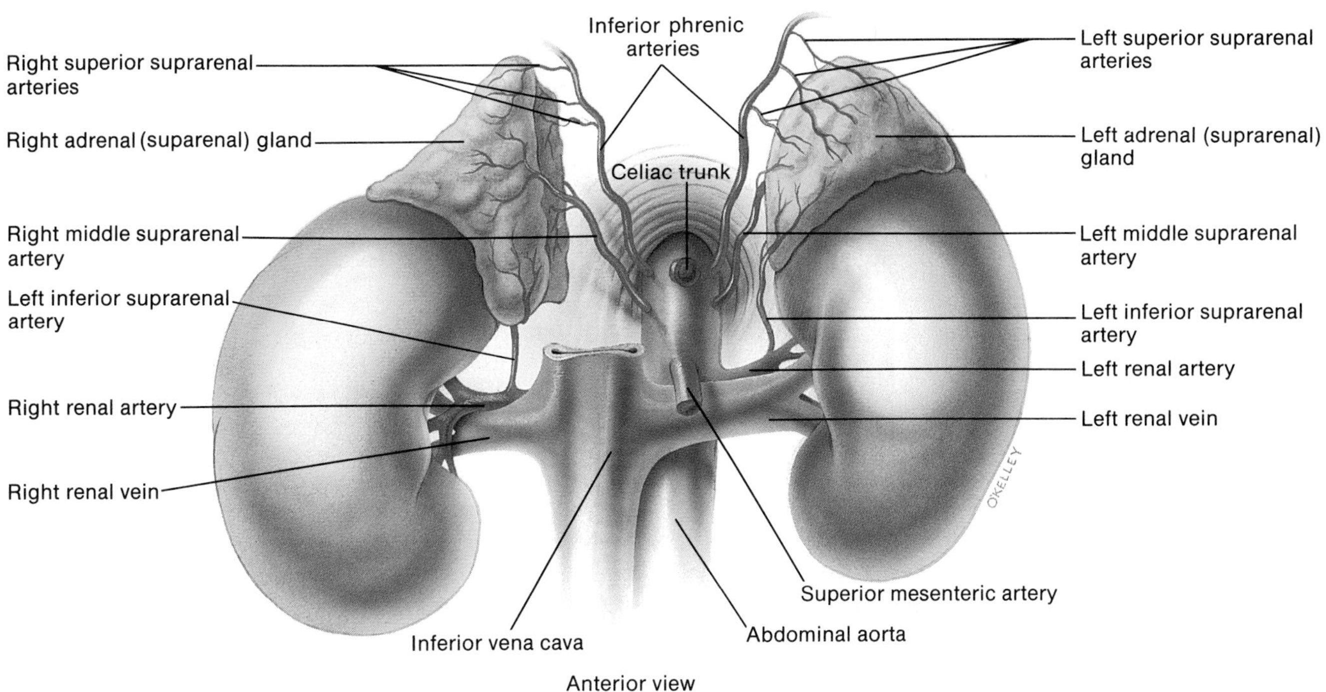

FIGURE 21-11 Location and blood supply of the adrenal (suprarenal) glands.

Superficial
Capsule
Zona glomerulosa
Zona fasciculata
Adrenal cortex
Zona reticularis
Adrenal medulla
Deep
(a) Photomicrograph of the subdivisions of the adrenal gland, 45×
Capsule
Zona glomerulosa
Mineralocorticoids (mainly aldosterone)
water balance
Zona fasciculata
Glucocorticoids (mainly cortisol)
Zona reticularis
Gonadocorticoids (estrogens and androgens)
Chromaffin cells of medulla
Epinephrine and norepinephrine (NE)
(b) Origin of hormones of adrenal gland

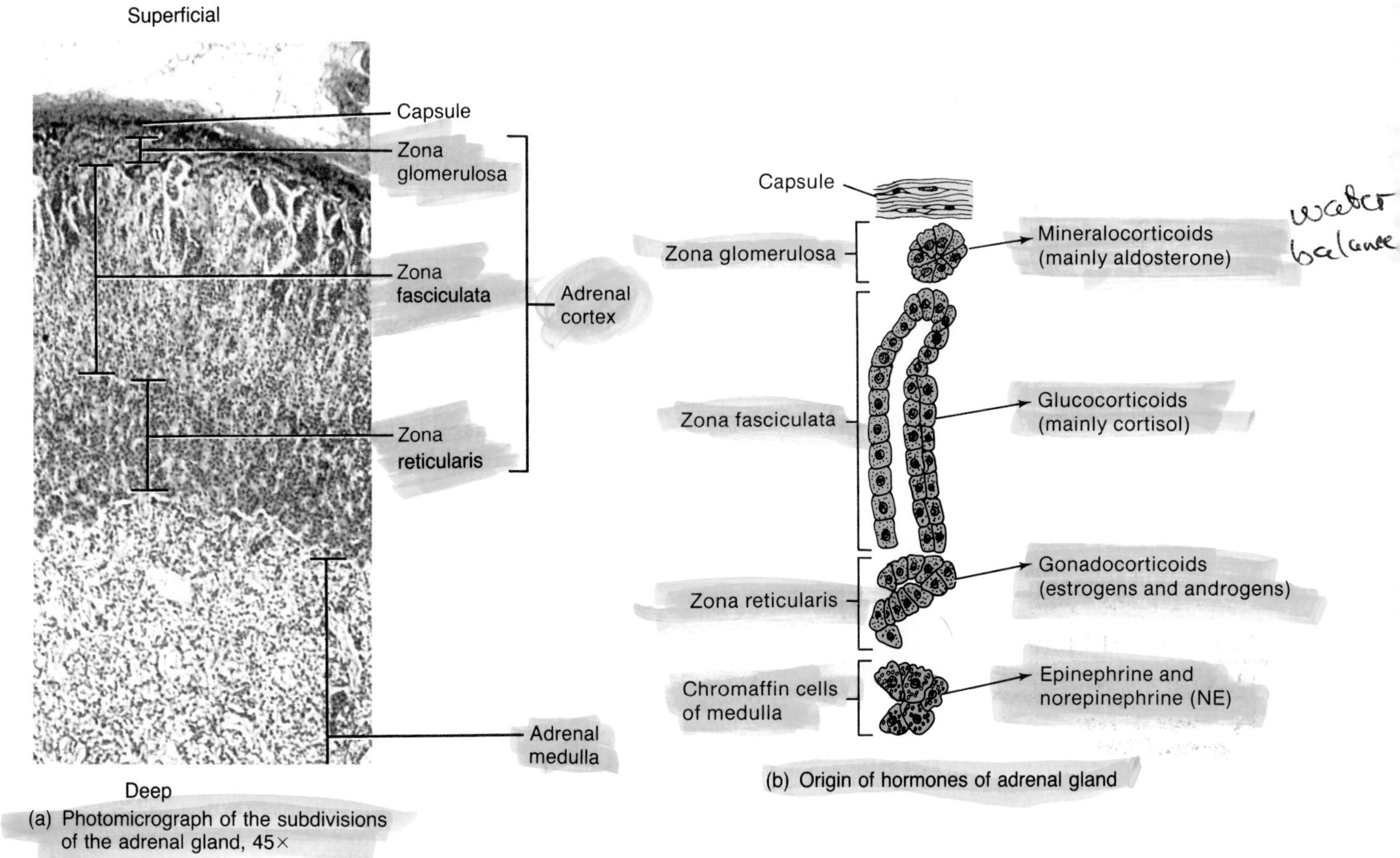

FIGURE 21-12 Histology of the adrenal (suprarenal) glands. (a) Courtesy of Andrew J. Kuntzman.

***adrenocortical insufficiency*).** Clinical symptoms include mental lethargy, anorexia, nausea and vomiting, weight loss, and hypoglycemia, which leads to muscular weakness. Increased potassium and decreased sodium lead to low blood pressure, dehydration, decreased cardiac output, arrhythmias, and potential cardiac arrest. Excessive skin pigmentation, especially in sun-exposed areas and in mucous membranes, also occurs.

Cushing's syndrome is a hypersecretion of glucocorticoids, especially cortisol and cortisone (Figure 21-13). The condition is characterized by the redistribution of fat. The result is spindly arms and legs, increased fat in the chest and abdomen, and rounded face. There is also muscle wasting and weakness. Facial skin is flushed, and the skin covering the abdomen develops stretch marks (striae). The individual also bruises easily, and wound healing is poor. Other symptoms include hyperglycemia, osteoporosis, hypertension, increased susceptibility to infection, decreased resistance to stress, and mood swings. The most common cause of a cushinoid appearance is the administration of a steroid hormone such as prednisone to a transplant recipient or in the treatment of asthma, arthritis, or a blood disorder.

The inner zone, the ***zona reticularis,*** contains cords of cells that branch freely. This zone synthesizes minute amounts of hormones, mainly ***gonadocorticoids*** (gō-na-dō-KŌR-ti-koyds), or ***sex hormones.*** These are estrogens and androgens. Estrogens are several closely related female sex hormones that are also produced by the ovaries and placenta. Androgens are male sex hormones. An important androgen, called testosterone, is produced by the testes. The concentration of sex hormones secreted by normal adult adrenals is usually so low that their effects are insignificant. In females, androgens contribute to sex drive (libido) and may produce small amounts of estrogens after menopause. Androgens produced by the adrenal cortex also assist in the prepubertal growth spurt and early development of axillary and pubic hair in boys and girls.

CLINICAL APPLICATION

Adrenogenital Syndrome and Gynecomastia

The ***adrenogenital syndrome*** usually refers to a group of enzyme deficiencies that block the synthesis of glucocorticoids. In an attempt to compensate, the anterior pituitary secretes more ACTH. As a result, excess androgens are produced, causing ***virilism,*** or masculinization. For instance, the female develops extremely virile characteristics such as growth of a beard (Figure 21-14), development of a much deeper voice, occasionally the development of baldness, development of a masculine distribution of hair on the body and on the pubis, growth of the clitoris that resembles a penis, atrophy of the breasts, infrequent or absent menstruation, and deposition of proteins in the skin and muscles, producing typical masculine characteristics. Such virilism may also result from tumors of the adrenal gland called ***virilizing adenomas*** (*aden* = gland; *oma* = tumor).

In the prepubertal male, the syndrome causes the same characteristics as in the female, plus rapid development of the male sexual organs and creation of male sexual desires. In the adult male, the virilizing characteristics of the adrenogenital syndrome are usually completely obscured by the normal virilizing characteristics of the testosterone secreted by the testes. As a result, it is often difficult

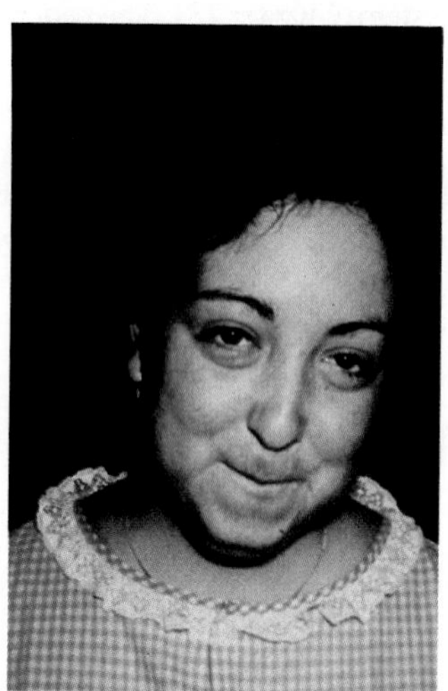

Cushing's syndrome

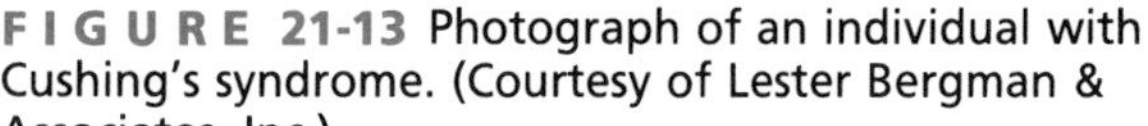

FIGURE 21-13 Photograph of an individual with Cushing's syndrome. (Courtesy of Lester Bergman & Associates, Inc.)

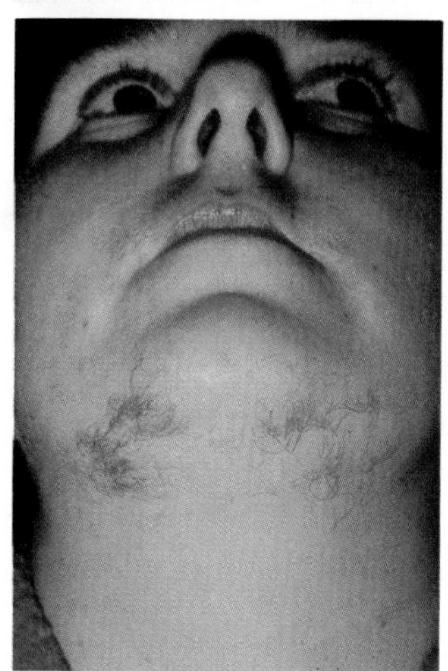

Adrenogenital syndrome

FIGURE 21-14 Photograph of an individual with adrenogenital syndrome. (Courtesy of Lester Bergman & Associates, Inc.)

to make a diagnosis of adrenogenital syndrome in the male adult. However, an occasional adrenal tumor secretes sufficient quantities of feminizing hormones (estrogens) that the male patient develops ***gynecomastia*** (*gyneca* = woman; *mast* = breast), which means excessive growth (benign) of the male mammary glands. Such a tumor is called a ***feminizing adenoma.*** Gynecomastia is also associated with androgen-deficiency states, pulmonary diseases, chest wall trauma, psychological stress, and certain drugs (such as alcohol, cimetidine, and digitalis derivatives). As a rule, specific treatment of gynecomastia is not indicated; however, medical treatment (antiestrogen or synthetic androgen) may be given or surgery may be undertaken for cosmetic reasons or a chronic condition.

ADRENAL MEDULLA

The adrenal medulla consists of hormone-producing cells, called ***chromaffin*** (krō-MAF-in) ***cells,*** which surround large blood-containing sinuses (see Figure 21-12). Chromaffin cells develop from the same source as the postganglionic cells of the sympathetic division of the autonomic nervous system. They are directly innervated by preganglionic cells of the sympathetic division of the autonomic nervous system and may be regarded as postganglionic cells that are specialized to secrete. In all other visceral effectors, preganglionic sympathetic fibers first synapse with postganglionic neurons before innervating the effector. In the adrenal medulla, however, the preganglionic fibers pass directly into the chromaffin cells of the gland. The secretion of hormones from the chromaffin cells is directly controlled by the autonomic nervous system, and innervation by the preganglionic fibers allows the gland to respond rapidly to a stimulus.

The two principal hormones synthesized by the adrenal medulla are ***epinephrine*** and ***norepinephrine (NE),*** also called adrenaline and noradrenaline, respectively. Epinephrine constitutes about 80 percent of the total secretion of the gland and is more potent in its action than norepinephrine. Both hormones are ***sympathomimetic*** (sim′-pa-thō-mi-MET-ik); that is, they mimic effects of the sympathetic division of the autonomic nervous system. To a large extent, they are responsible for the fight-or-flight response. Like the glucocorticoids of the adrenal cortices, these hormones help the body resist stress. Unlike the cortical hormones, the medullary hormones are not essential for life.

The main arteries that supply the adrenal glands are the several superior suprarenal arteries arising from the inferior phrenic artery, the middle suprarenal artery from the aorta, and the inferior suprarenal arteries from the renal arteries. The suprarenal vein of the right adrenal gland drains into the inferior vena cava, whereas the suprarenal vein of the left adrenal gland empties into the left renal vein.

The principal nerve supply to the adrenal glands is from preganglionic fibers from the thoracic splanchnic nerves, which pass through the celiac and associated sympathetic plexuses. These myelinated fibers end on the secretory cells of the gland found in a region of the medulla.

CLINICAL APPLICATION

Pheochromocytomas

Tumors of the chromaffin cells of the adrenal medulla, called ***pheochromocytomas*** (fē-ō-krō′-mō-sī-TŌ-mas), cause hypersecretion of the medullary hormones. Such tumors are usually benign. The oversecretion causes rapid heart rate, headache, high blood pressure, high levels of sugar in the blood and urine, an elevated basal metabolic rate (BMR), flushing of the face, nervousness, sweating, decreased gastrointestinal motility, and vertigo. Since the medullary hormones create the same effects as sympathetic nervous stimulation, hypersecretion puts the individual into a prolonged version of the fight-or-flight response. This condition ultimately wears out the body, and the individual eventually suffers from general weakness. The only definitive treatment of pheochromocytomas is surgical removal of the tumor(s). Pheochromocytoma is one of the curable causes of hypertension that is tested for when a person has episodes of hypertension.

A summary of adrenal gland hormones, actions, and disorders is presented in Exhibit 21-4.

PANCREAS

The ***pancreas*** can be classified as both an endocrine and an exocrine gland. Thus, it is referred to as a ***heterocrine gland.*** We shall treat its endocrine functions at this point; its exocrine functions will be discussed in the chapter on the digestive system (Chapter 23). The pancreas is a flattened organ located posterior and slightly inferior to the stomach (Figure 21-15). The adult pancreas consists of a head, body, and tail.

The endocrine portion of the pancreas consists of clusters of cells called ***pancreatic islets,*** or ***islets of Langerhans*** (LAHNG-er-hanz) (Figure 21-16). Four kinds of cells are found in these clusters: (1) ***alpha cells*** that secrete the hormone ***glucagon,*** which increases blood sugar level; (2) ***beta cells*** that secrete the hormone ***insulin,*** which decreases blood sugar level, increases lipid synthesis, and stimulates protein synthesis; (3) ***delta cells*** that secrete ***growth hormone-inhibiting factor (GHIF),*** or ***somatostatin,*** a hormone that inhibits the secretion of insulin and glucagon; and (4) ***F cells*** that secrete ***pancreatic polypeptide,*** a hormone that regulates the release of pancreatic digestive enzymes. The islets are infiltrated by blood capillaries and surrounded by cells (acini) that form the exocrine part of the gland.

The arterial supply of the pancreas is from the superior and inferior pancreaticoduodenal arteries and from the

EXHIBIT 21-4

Summary of Hormones Produced by the Adrenal Glands, Principal Actions, and Selected Disorders

HORMONE	PRINCIPAL ACTIONS	SELECTED DISORDERS
Adrenal Cortical Hormones		
Mineralocorticoids (Mainly Aldosterone)	Increase blood levels of sodium and water and decrease blood levels of potassium.	Hypersecretion of aldosterone results in aldosteronism.
Glucocorticoids (Mainly Cortisol)	Help promote normal organic metabolism, provide resistance to stress, and counter inflammatory response.	Hyposecretion of glucocorticoids produces Addison's disease; hypersecretion results in Cushing's syndrome.
Gonadocorticoids	Concentrations secreted by adults are so low that their effects are usually insignificant.	Adrenogenital syndrome inhibits synthesis of glucocorticoids that results in excess production of ACTH and androgens, causing virilism. The release of sufficient feminizing hormones in males causes gynecomastia.
Adrenal Medullary Hormones		
Epinephrine	Sympathomimetic—that is, produces effects that mimic those of the sympathetic division of the autonomic nervous system (ANS) during stress.	Hypersecretion of medullary hormones results in a prolonged fight-or-flight response.
Norepinephrine (NE)	Same as above.	

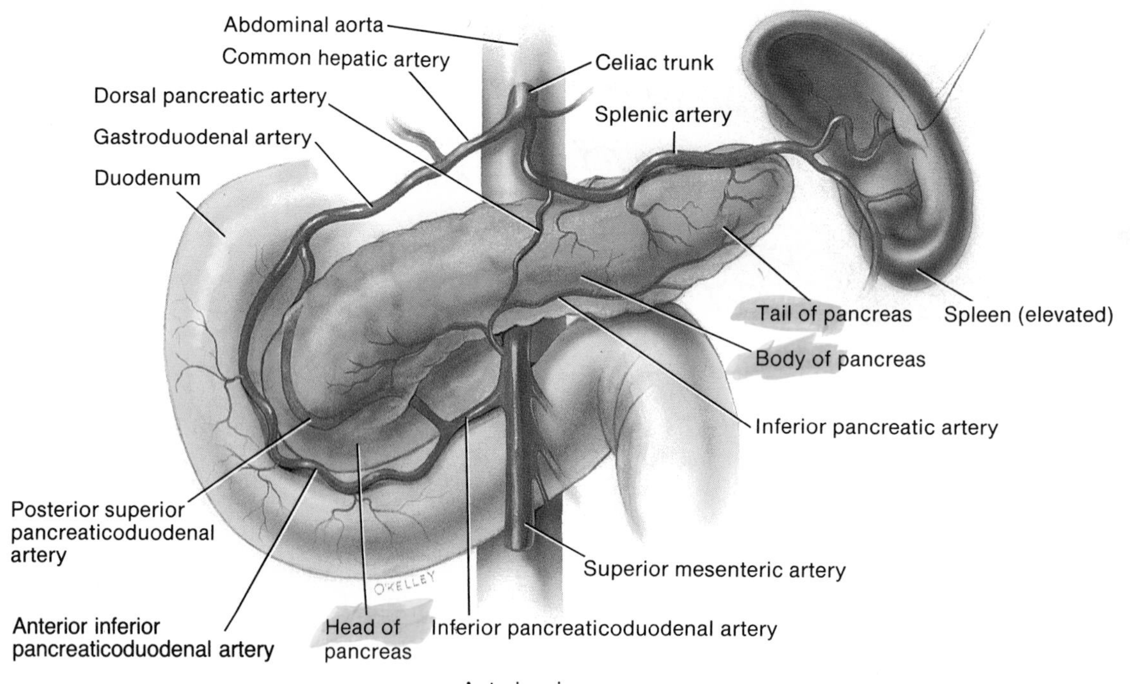

FIGURE 21-15 Location and blood supply of the pancreas.

Endocrine as well as exocrine

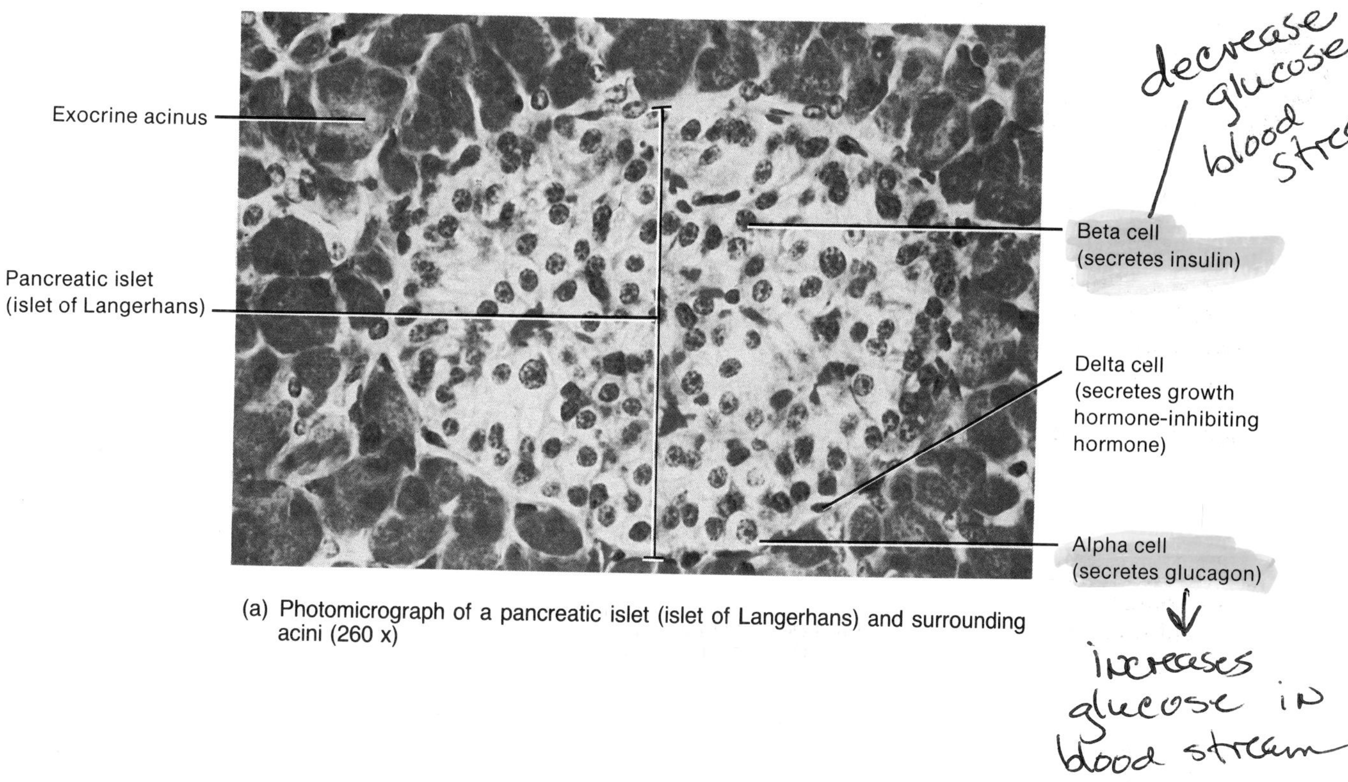

(a) Photomicrograph of a pancreatic islet (islet of Langerhans) and surrounding acini (260 x)

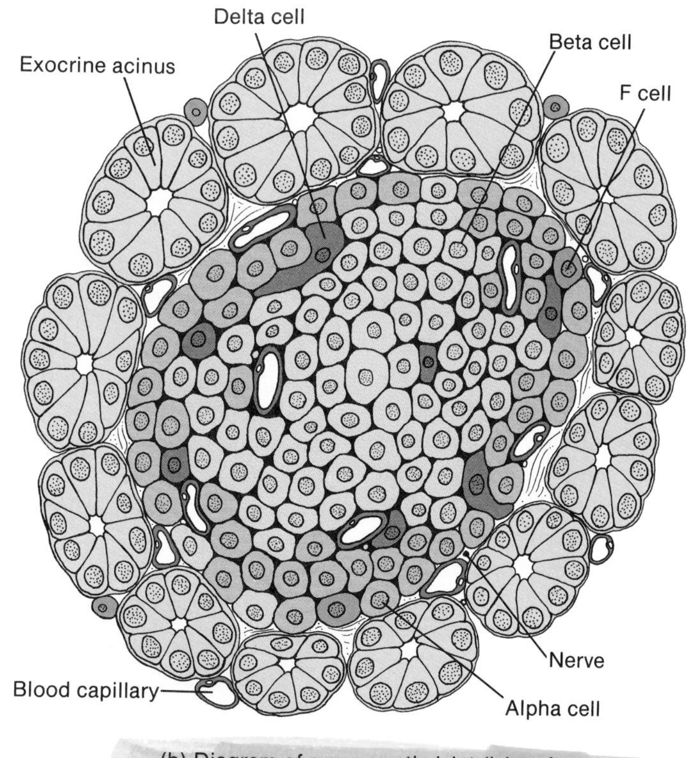

(b) Diagram of a pancreatic islet (islet of Langerhans) and surrounding acini

FIGURE 21-16 Histology of the pancreas. (a) © Biophoto, SPL, Photo Researchers. Of the nearly almost one million islets in the pancreas, most are located in the tail of the pancreas.

splenic and superior mesenteric arteries. The veins, in general, correspond to the arteries. Venous blood reaches the hepatic portal vein by means of the splenic and superior mesenteric veins.

The nerves to the pancreas are autonomic nerves derived from the celiac and superior mesenteric plexuses. Included are preganglionic vagal, postganglionic sympathetic, and afferent fibers. Parasympathetic vagal fibers are said to terminate at both acinar (exocrine) and islet (endocrine) cells. Although the innervation is presumed to influence enzyme formation, pancreatic secretion is controlled largely by the hormones secretin and cholecystokinin (CCK). The sympathetic fibers enter the islets and also end on blood vessels and are vasomotor and accompanied by afferent fibers, especially for pain.

CLINICAL APPLICATION

Diabetes Mellitus and Hyperinsulinism (hyperglycemia)

Diabetes mellitus (MEL-i-tus) is not a single hereditary disease but a heterogeneous group of diseases, all of which ultimately lead to an elevation of glucose in the blood (hyperglycemia) and loss of glucose in the urine as hyperglycemia increases. Diabetes mellitus is also characterized by the three "polys": an inability to reabsorb water, resulting in increased urine production ***(polyuria)***; excessive thirst ***(polydipsia)***; and excessive eating ***(polyphagia)***.

Two major types of diabetes mellitus are distinguished: type I and type II. ***Type I diabetes,*** which occurs abruptly, is characterized by an absolute deficiency of insulin due to a marked decline in the number of insulin-producing beta cells (caused by the autoimmune destruction of beta cells), even though target cells contain insulin receptors. Type I diabetes is called ***insulin-dependent diabetes*** because periodic administration of insulin is required to treat it. It is also known as ***juvenile-onset diabetes*** because it most commonly develops in people younger than age 20, though it persists throughout life. Although people who develop type I diabetes appear to have certain genes that make them more susceptible, some triggering factor is required. Viral infection seems to be such a factor. The deficiency of insulin accelerates the breakdown of the body's reserve of fat, resulting in the production of organic acids called ketones. The result is a form of acidosis called ***ketosis,*** which lowers the pH of the blood and can cause death. The catabolism of stored fats and proteins also causes weight loss. As lipids are transported by the blood from storage depots to cells, lipid particles are deposited on the walls of blood vessels. The deposition leads to atherosclerosis and a multitude of cardiovascular problems including cerebrovascular insufficiency, ischemic heart disease, peripheral vascular disease, and gangrene. One of the major complications of diabetes is loss of vision due to cataracts (excessive blood sugar chemically attaches to lens proteins, causing cloudiness) or damage to blood vessels of the retina. Severe kidney problems also may result from damage to renal blood vessels.

Based on studies with the drug cyclosporine, which is used to counter the rejection of an organ following a transplant, plus other data, it is now clear that type I diabetes is an autoimmune disease. Several studies have shown that cyclosporine can interrupt the destruction of beta cells.

Research is now under way to prevent or improve type I diabetes by using transplants of islet cells. In some cases, patients are given clusters of islet cells that have been treated to render them incapable of inducing rejection by the recipient. In other cases, clusters of islet cells are encapsulated so that the insulin can get out, but elements responsible for rejection cannot get in. In still another procedure, patients have been injected with fetal islet cells. Transplantation of the entire pancreas in conjunction with immunosuppressive drugs is also being tried. Before any of these procedures can be used in larger numbers of patients, considerably more research is needed.

Type II diabetes is much more common than type I, representing more than 90 percent of all cases. Type II diabetes most often occurs in people who are over 40 and overweight. Since type II diabetes usually occurs later in life, it is called ***maturity-onset diabetes.*** Clinical symptoms are mild, and the high glucose levels in the blood can usually be controlled by diet, exercise, and/or with antidiabetic drugs such as *glyburide* (DiaBeta). Many type II diabetics have a sufficient amount or even a surplus of insulin in the blood. For these individuals, diabetes arises not from a shortage of insulin but probably from defects in the molecular machinery that mediates the action of insulin on its target cells. Cells in many parts of the body, especially those in skeletal muscles and the liver, become less sensitive to insulin because they have fewer insulin receptors. Type II diabetes is therefore called ***non–insulin-dependent diabetes.*** However, some elderly persons with this type of diabetes are required to administer insulin to themselves daily since the cells may be slightly sensitive to insulin.

Hyperinsulinism (hypoglycemia) is much rarer than hyposecretion and is generally the result of a malignant tumor in an islet. It can also be caused by hyperplasia of islets. The principal symptom is a decreased blood glucose level, which stimulates the secretion of epinephrine, glucagon, and hGH. As a consequence, anxiety, sweating, tremor, increased heart rate, hunger, and weakness occur. Moreover, brain cells do not have enough glucose to function efficiently. This condition leads to mental disorientation, convulsions, unconsciousness, and shock. Surgery may be required to remove the tumor or hyperplastic pancreatic tissue.

EXHIBIT 21-5

Summary of Hormones Produced by the Pancreas, Principal Actions, and Selected Disorders

HORMONE	PRINCIPAL ACTIONS	SELECTED DISORDERS
Glucagon	Raises blood sugar level by accelerating breakdown of glycogen into glucose in liver (glycogenolysis) and conversion of other nutrients into glucose in liver (gluconeogenesis) and releasing glucose into blood.	
Insulin	Lowers blood sugar level by accelerating transport of glucose into cells, converting glucose into glycogen (glycogenesis), and decreasing glycogenolysis and gluconeogenesis; also increases lipid synthesis and stimulates protein synthesis.	An absolute deficiency of insulin or a defect in the molecular machinery that mediates the action of insulin on target cells produces diabetes mellitus. Hypersecretion of insulin results in hyperinsulinism.
Growth Hormone-Inhibiting Hormone (GHIH) or Somatostatin	Inhibits secretion of insulin and glucagon.	
Pancreatic Polypeptide	Regulates release of pancreatic digestive enzymes.	

A summary of pancreatic hormones, actions, and disorders is presented in Exhibit 21-5.

OVARIES AND TESTES

The female gonads, called the ***ovaries***, are paired oval bodies located in the pelvic cavity. The ovaries produce female sex hormones called ***estrogens*** and ***progesterone***. These hormones are responsible for the development and maintenance of female sexual characteristics. Along with the gonadotropic hormones of the pituitary gland, the sex hormones also regulate the menstrual cycle, maintain pregnancy, and prepare the mammary glands for lactation. The ovaries (and placenta) also produce a hormone called ***relaxin***, which relaxes the pubic symphysis and helps dilate the uterine cervix toward the end of pregnancy. The ovaries also produce ***inhibin***, a hormone that inhibits secretion of FSH (and to a lesser extent, LH) and that might be important in decreasing secretion of FSH and LH toward the end of the menstrual cycle.

The male has two oval glands, called ***testes***, that lie in a pouchlike structure called the scrotum. The testes produce ***testosterone***, the primary male sex hormone, which stimulates the development and maintenance of male sexual characteristics, regulates sperm production, and stimulates descent of testes before birth. The testes also produce the hormone inhibin that, like in the female, inhibits secretion of FSH. The detailed structure of the ovaries and testes and the specific roles of gonadotropic hormones and sex hormones will be discussed in Chapter 25.

A summary of hormones produced by the ovaries and testes and their actions is presented in Exhibit 21-6.

EXHIBIT 21-6

Summary of Hormones of the Ovaries and Testes and Their Principal Actions

HORMONE	PRINCIPAL ACTIONS
Ovarian Hormones	
Estrogens and Progesterone	Development and maintenance of female sexual characteristics. Together with gonadotropic hormones of the adenohypophysis, they also regulate the menstrual cycle, maintain pregnancy, prepare the mammary glands for lactation, and regulate oogenesis.
Relaxin	Relaxes pubic symphysis and helps dilate uterine cervix near the end of pregnancy.
Inhibin	Inhibits secretion of FSH toward the end of the menstrual cycle.
Testicular Hormones	
Testosterone	Development and maintenance of male sexual characteristics, regulation of sperm production, and stimulation of descent of testes before birth.
Inhibin	Inhibits secretion of FSH to control sperm production.

PINEAL GLAND (EPIPHYSIS CEREBRI)

The endocrine gland attached to the roof of the third ventricle is known as the ***pineal*** (PĪN-ē-al) ***gland*** (because of its resemblance to a pine cone), or ***epiphysis cerebri*** (see Figure 21-1). The gland is covered by a capsule formed by the pia mater and consists of masses of ***neuroglial cells*** and parenchymal secretory cells called ***pinealocytes*** (pin-ē-AL-ō-sīts). Around the cells are scattered postganglionic sympathetic fibers. The pineal gland starts to accumulate calcium at about the time of puberty. Such calcium deposits are referred to as ***brain sand.*** Contrary to a once widely held belief, there is no evidence that the pineal gland atrophies with age and that the presence of brain sand is an indication of atrophy. In fact, the presence of brain sand may even indicate increased secretory activity.

Although many anatomical facts concerning the pineal gland have been known for years, its physiology is still somewhat obscure. One hormone secreted by the pineal gland is ***melatonin,*** which appears to inhibit reproductive activities by inhibiting gonadotropic hormones. Melatonin may also be related to regulation of core body temperature and sleep. Some evidence also exists that the pineal secretes a second hormone called ***adrenoglomerulotropin*** (a-drē′-nō-glō-mer′-yoo-lō-TRŌ-pin). This hormone may stimulate the adrenal cortex to secrete aldosterone. Other substances found in the pineal gland include norepinephrine (NE), serotonin, histamine, gonadotropin-releasing hormone (GnRH), and gamma aminobutyric acid (GABA).

The posterior cerebral artery supplies the pineal gland with blood, and the great cerebral vein drains it.

A summary of hormones produced by the pineal gland and their actions is presented in Exhibit 21-7.

THYMUS

Because of its role in immunity, the details of the structure and functions of the ***thymus gland*** are discussed in Chapter 15, which deals with the lymphatic system and immunity. At this point, only its hormonal role in immunity will be discussed.

One of the types of white blood cells is referred to as lymphocytes. These lymphocytes in turn are divided into two types, called T cells and B cells, on the basis of their specific roles in immunity. Hormones produced by the thymus gland, called ***thymosin, thymic humoral factor (THF),*** ***thymic factor (TF),*** and ***thymopoietin,*** promote the proliferation and maturation of T cells, which destroy foreign microbes and substances. There is also some evidence that thymic hormones may retard the aging process.

The arterial supply of the thymus is derived mainly from the internal thoracic and inferior thyroid vessels. The veins that drain the thymus are the internal thoracic, brachiocephalic, and thyroid veins. Postganglionic sympathetic and parasympathetic fibers supply the gland.

A summary of hormones produced by the thymus gland and their actions is presented in Exhibit 21-8.

AGING AND THE ENDOCRINE SYSTEM

The endocrine system exhibits a variety of changes, and many researchers look to this system with the hope of finding the key to the aging process. Disorders of the endocrine system are not frequent, but when they do occur, most often they are related to pathologic changes rather than age. Diabetes mellitus and thyroid disorders, especially hypothyroidism, are relatively common endocrine problems that have a significant effect on health and function. Male sex hormone levels decrease with age, but elderly males can still produce active sperm in normal numbers.

DEVELOPMENTAL ANATOMY OF THE ENDOCRINE SYSTEM

The development of the endocrine system is not as localized as the development of other systems. The endocrine organs develop in widely separated parts of the embryo.

The *pituitary gland* (*hypophysis*) originates from two different regions of the ***ectoderm.*** The *neurohypophysis* (posterior lobe) is derived from an outgrowth of ectoderm called the ***neurohypophyseal bud,*** located on the floor of the hypothalamus (Figure 21-17a,b). The *infundibulum,* also an outgrowth of the neurohypophyseal bud, connects the neurohypophysis to the hypothalamus. The *adenohypophysis* (anterior lobe) is derived from an outgrowth of ***ectoderm***

EXHIBIT 21-7

Summary of Hormones of the Pineal Gland and Their Principal Actions

HORMONE	PRINCIPAL ACTIONS
Melatonin	May inhibit reproductive activities by inhibiting gonadotropic hormones.
Adrenoglomerulotropin	May stimulate the adrenal cortex to secrete aldosterone.

EXHIBIT 21-8

Summary of Hormones Produced by the Thymus Gland and Their Principal Actions

HORMONES	PRINCIPAL ACTIONS
Thymosin, Thymic Humoral Factor (THF), Thymic Factor (TF), and Thymopoietin	Promote proliferation and maturation of T cells.

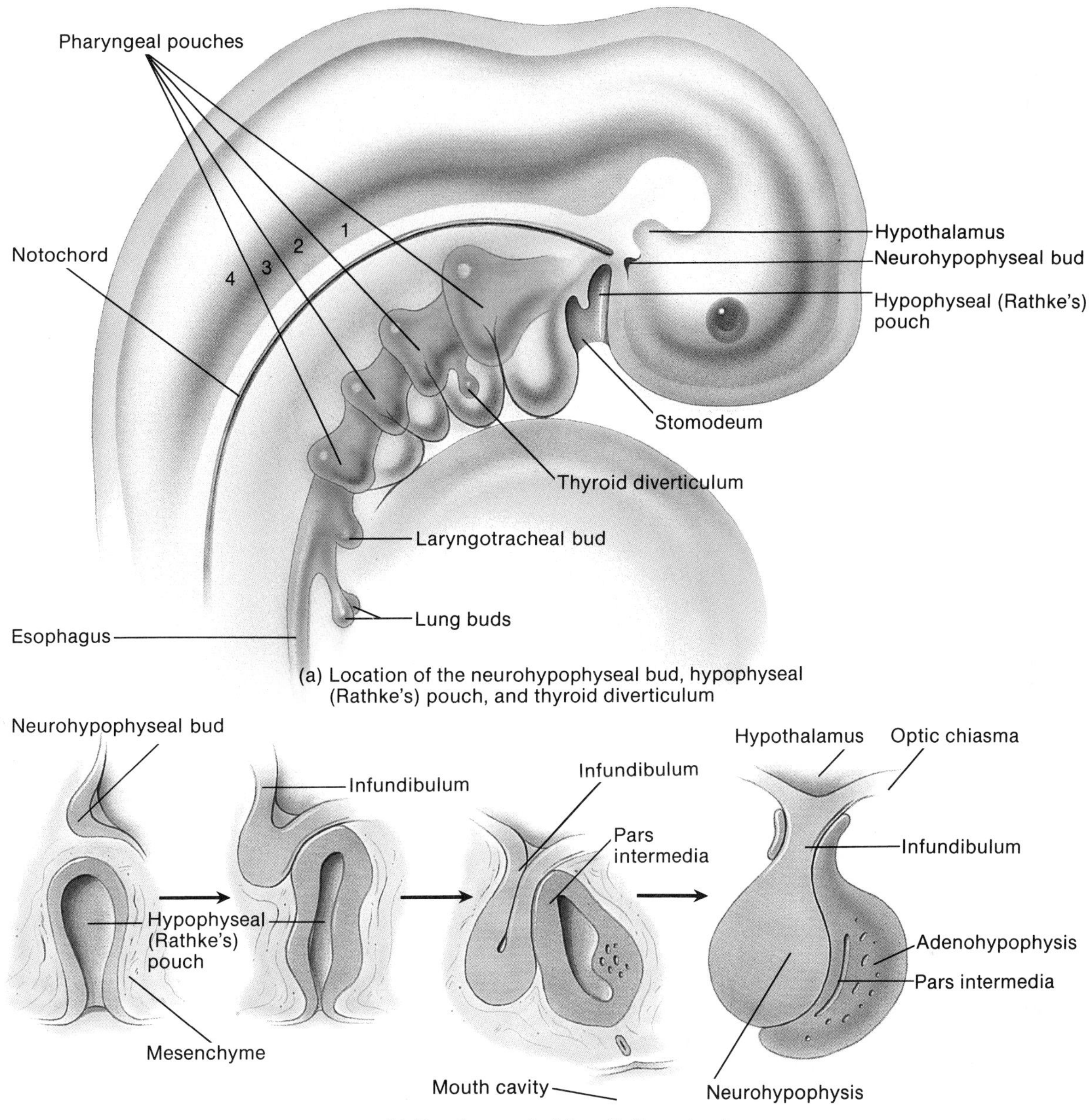

FIGURE 21-17 Development of the endocrine system.

from the roof of the stomodeum (mouth) called the ***hypophyseal (Rathke's) pouch.*** The pouch grows toward the neurohypophyseal bud, and the pouch loses its connection with the roof of the mouth.

The *thyroid gland* develops as a midventral outgrowth of ***endoderm,*** called the ***thyroid diverticulum,*** from the floor of the pharynx at the level of the second pair of pharyngeal pouches (Figure 21-17b). The outgrowth projects inferiorly and differentiates into the right and left lateral lobes and the isthmus of the gland.

The *parathyroid glands* develop from ***endoderm*** as outgrowths from the third and fourth ***pharyngeal pouches*** (Figure 21-17a).

The adrenal cortex and adrenal medulla have completely different embryological origins. The *adrenal cortex* is derived from intermediate ***mesoderm*** from the same region that produces the gonads (see Figure 25-25). The *adrenal medulla* is ***ectodermal*** in origin and is derived from the ***neural crest,*** which also produces sympathetic ganglia and other structures of the nervous system (see Figure 18-24c).

The *pancreas* develops from ***endoderm*** from dorsal and ventral outgrowths of the part of the ***foregut*** that later becomes the duodenum (see Figure 23-20). The two outgrowths eventually fuse to form the pancreas. The origin of the ovaries and testes will be discussed in Chapter 25.

The *pineal gland* arises from ***ectoderm*** of the ***diencephalon*** (see Figure 18-25b), as an outgrowth between the thalamus and colliculi.

The *thymus gland* arises from ***endoderm*** from the third ***pharyngeal pouches*** (Figure 21-17b).

OTHER ENDOCRINE TISSUES

Before leaving our discussion of hormones we should note that body tissues other than those usually classified as endocrine glands also contain endocrine tissue and thus secrete hormones. The gastrointestinal tract synthesizes several hormones that regulate digestion in the stomach and small intestine. Among these hormones are ***stomach gastrin, enteric gastrin, secretin, cholecystokinin (CCK), enterocrinin, gastric inhibitory peptide (GIP),*** and ***motilin.***

The placenta produces ***human chorionic gonadotropin (hCG), estrogens, progesterone, relaxin,*** and ***human chorionic somatomammotropin (hCS),*** all of which are related to pregnancy (see Chapter 26).

When the kidneys (and liver, to a lesser extent) become hypoxic (subject to below-normal levels of oxygen), it is believed that they release an enzyme called ***renal erythropoietic factor.*** It is secreted into the blood, where it acts on a plasma protein to convert it to a hormone called ***erythropoietin*** (ē-rith′-rō-POY-ē-tin), which stimulates red blood cell production. The kidneys also help bring about the activation of the hormone ***vitamin D.*** The skin produces vitamin D from inactive precursor molecules in the presence of sunlight.

Cardiac muscle fibers (cells) of the atria (upper chambers) of the heart produce a peptide hormone, called ***atrial natriuretic factor (ANF),*** that is secreted when they are stretched, as might occur in response to increased blood volume, a factor that increases blood pressure. The general effect of ANF is an increase in sodium and water excretion and blood vessel dilation; the result is a decrease in blood pressure. ANF also acts directly on the hypothalamus to inhibit the secretion of ADH. The overall function of ANF is therefore to lower blood pressure.

KEY MEDICAL TERMS ASSOCIATED WITH THE ENDOCRINE SYSTEM

Hyperplasia (hī′-per-PLĀ-zē-a; *hyper* = over, *plas* = grow) Increase in the number of cells due to an increase in the frequency of cell division.

Hypoplasia (hi′-pō-PLĀ-zē-a; *hypo* = under) Defective development of tissue.

Neuroblastoma (noo′-rō-blas-TŌ-ma; *neuro* = nerve) Malignant tumor arising from the adrenal medulla associated with metastases to bone.

Thyroid (THĪ-roid) **storm** An aggravation of all symptoms of hyperthyroidism characterized by unregulated hypermetabolism with fever and rapid heart rate; results from trauma, surgery, and unusual emotional stress or labor.

STUDY OUTLINE

Endocrine Glands (p. 604)

1. Both the endocrine and nervous systems assume a role in maintaining homeostasis.
2. Hormones help regulate the internal environment, respond to stress, help regulate growth and development, and contribute to reproductive processes.
3. Exocrine glands (sweat, sebaceous, digestive) secrete their products through ducts into body cavities or onto body surfaces.
4. Endocrine glands secrete hormones into the blood.
5. The amount of hormone released is determined by the body's need for the hormone.
6. Cells that respond to the effects of hormones are called target cells.
7. The combination of hormone and receptor activates a chain of events in a target cell in which the physiological effects of the hormone are expressed.

Pituitary (Hypophysis) (p. 605)

1. The pituitary gland is located in the sella turcica and is differentiated into the adenohypophysis (the anterior lobe and glandular portion), neurohypophysis (the posterior lobe and nervous portion), and pars intermedia (relatively small and poorly defined avascular zone between lobes).
2. Hormones of the adenohypophysis are released or inhibited by regulating hormones (or factors) produced by the hypothalamus.
3. The blood supply to the adenohypophysis is from the superior hypophyseal arteries. It transports hypothalamic regulating hormones (or factors).
4. Histologically, the adenohypophysis consists of somatotroph cells that produce human growth hormone (hGH), which regulates growth; lactotroph cells that produce prolactin (PRL), which helps initiate milk secretion; corticolipotroph cells that

secrete adrenocorticotropic hormone (ACTH), which regulates the activities of the adrenal cortex, and melanocyte-stimulating hormone (MSH), which increases skin pigmentation; thyrotroph cells that secrete thyroid-stimulating hormone (TSH), which regulates thyroid gland activities; and gonadotroph cells that synthesize follicle-stimulating hormone (FSH), which regulates the activities of the ovaries and testes; and luteinizing hormone (LH), which regulates female and male reproductive activities (ovulation, stimulation of secretion of progesterone and testosterone), and prepares uterus for implantation.
5. Disorders associated with improper levels of hGH are pituitary dwarfism, giantism, and acromegaly.
6. The neural connection between the hypothalamus and neurohypophysis is via the hypothalamic-hypophyseal tract.
7. Hormones made by the hypothalamus and stored in the neurohypophysis are oxytocin, or OT (stimulates contraction of uterus and ejection of milk), and antidiuretic hormone, or ADH (stimulates water reabsorption by the kidneys and arteriole constriction).
8. A disorder associated with dysfunction of the neurohypophysis is diabetes insipidus.

Thyroid (p. 608)
1. The thyroid gland is located below the larynx.
2. Histologically, the thyroid consists of thyroid follicles composed of follicular cells, which secrete the thyroid hormones thyroxine (T_4) and triiodothyronine (T_3), and parafollicular cells, which secrete calcitonin (CT).
3. Thyroid hormones are synthesized from iodine and tyrosine within thyroglobulin (TGB).
4. Thyroid hormones regulate the rate of metabolism, growth and development, and the reactivity of the nervous system.
5. Cretinism, myxedema, exophthalmic goiter, and simple goiter are disorders associated with dysfunction of the thyroid gland.
6. Calcitonin (CT) lowers the blood level of calcium.

Parathyroids (p. 612)
1. The parathyroids are embedded on the posterior surfaces of the lateral lobes of the thyroid.
2. Histologically, the parathyroids consist of principal and oxyphil cells.
3. Parathyroid hormone (PTH) regulates the homeostasis of calcium and phosphate by increasing blood calcium level and decreasing blood phosphate level.
4. Tetany and osteitis fibrosa cystica are disorders associated with the parathyroid glands.

Adrenals (Suprarenals) (p. 614)
1. The adrenal glands are located superior to the kidneys. They consist of an outer cortex and inner medulla.
2. Histologically, the cortex is divided into a zona glomerulosa, zona fasciculata, and zona reticularis; the medulla consists of chromaffin cells.
3. Cortical secretions are mineralocorticoids, glucocorticoids, and gonadocorticoids.
4. Mineralocorticoids (e.g., aldosterone) increase sodium and water reabsorption and decrease potassium reabsorption.
5. A dysfunction related to aldosterone secretion is aldosteronism.
6. Glucocorticoids (e.g., cortisol) promote normal metabolism, help resist stress, and serve as anti-inflammatories.
7. Disorders associated with glucocorticoid secretion are Addison's disease and Cushing's syndrome.
8. Gonadocorticoids secreted by the adrenal cortex usually have minimal effects. Excessive production results in adrenogenital syndrome.
9. Medullary secretions are epinephrine and norepinephrine (NE), which produce effects similar to sympathetic responses. They are released under stress.
10. Tumors of medullary chromaffin cells are called pheochromocytomas.

Pancreas (p. 617)
1. The pancreas is posterior and slightly inferior to the stomach.
2. Histologically, it consists of pancreatic islets, or islets of Langerhans (endocrine cells), and acini (enzyme-producing cells). Four types of cells in the endocrine portion are alpha cells, beta cells, delta cells, and F cells.
3. Alpha cells secrete glucagon, beta cells secrete insulin, delta cells secrete growth hormone-inhibiting factor (GHIF), and F cells secrete pancreatic polypeptide.
4. Glucagon increases blood sugar level, whereas insulin decreases blood sugar level.
5. Disorders associated with insulin production are diabetes mellitus and hyperinsulinism.

Ovaries and Testes (p. 621)
1. Ovaries are located in the pelvic cavity and produce sex hormones related to development and maintenance of female sexual characteristics, menstrual cycle, pregnancy, lactation, and normal reproductive functions.
2. Testes lie inside the scrotum and produce sex hormones related to the development and maintenance of male sexual characteristics and to normal reproductive functions.

Pineal Gland (Epiphysis Cerebri) (p. 621)
1. The pineal gland is attached to the roof of the third ventricle.
2. Histologically, it consists of secretory parenchymal cells called pinealocytes, neuroglial cells, and scattered postganglionic sympathetic fibers. Calcium-containing deposits are referred to as brain sand.
3. It secretes melatonin (possibly regulates reproductive activities by inhibiting gonadotropic hormones) and adrenoglomerulotropin (may stimulate adrenal cortex to secrete aldosterone).

Thymus (p. 622)
1. The thymus is a bilobed lymphatic gland located in the superior mediastinum posterior to the sternum and between the lungs.
2. The thymus gland secretes several hormones related to immunity.
3. Thymosin, thymic humoral factor (THF), thymic factor (TF), and thymopoietin promote the maturation of T cells.

Aging and the Endocrine System (p. 622)
1. Most endocrine disorders are related to pathologies rather than age.
2. Diabetes mellitus and thyroid disorders are among the more important endocrine disorders.

Developmental Anatomy of the Endocrine System (p. 622)

1. The development of the endocrine system is not as localized as other systems.
2. The adenohypophysis arises from the hypophyseal (Rathke's) pouch; the neurohypophysis develops from the hypophyseal bud.
3. The thyroid gland develops from the thyroid diverticulum.
4. The parathyroid glands and thymus gland develop from pharyngeal pouches.
5. The adrenal cortex arises from mesoderm; the adrenal medulla develops from ectoderm (neural crest).
6. The pancreas develops from the foregut; the pineal gland develops from the diencephalon.

Other Endocrine Tissues (p. 624)

1. The gastrointestinal tract synthesizes stomach and enteric gastrin, secretin, cholecystokinin (CCK), enterocrinin, and gastric inhibitory peptide (GIP).
2. The placenta produces human chorionic gonadotropin (hCG), estrogens, progesterone, relaxin, and human chorionic somatomammotropin (hCS).
3. The kidneys release an enzyme that produces erythropoietin.
4. The skin synthesizes vitamin D.
5. The atria of the heart produce atrial natriuretic factor (ANF).

REVIEW QUESTIONS

1. Distinguish between an endocrine gland and an exocrine gland. What are the seven principal actions of hormones? (p. 604)
2. What is a hormone? Distinguish between tropic and gonadotropic hormones. (p. 604)
3. Explain how receptors are related to hormones. (p. 605)
4. In what respect is the pituitary gland actually two glands? (p. 605)
5. Describe the histology of the adenohypophysis. Why does the anterior lobe of the gland have such an abundant blood supply? (p. 606)
6. What hormones are produced by the adenohypophysis? What are their functions? (p. 606)
7. Describe the clinical symptoms of pituitary dwarfism, gigantism, and acromegaly. (p. 606)
8. Discuss the histology of the neurohypophysis and the function and regulation of its hormones. Describe the structure and importance of the hypothalamic-hypophyseal tract. (p. 607)
9. What are the clinical symptoms of diabetes insipidus? (p. 608)
10. Describe the location and histology of the thyroid gland. (p. 608)
11. Discuss the physiological effects of the thyroid hormones. (p. 608)
12. Discuss the clinical symptoms of cretinism, myxedema, exophthalmic goiter, and simple goiter. (p. 611)
13. Describe the function of calcitonin (CT). (p. 612)
14. Where are the parathyroids located? What is their histology? (p. 612)
15. What are the functions of parathyroid hormone (PTH)? (p. 614)
16. Discuss the clinical symptoms of tetany and osteitis fibrosa cystica. (p. 614)
17. Compare the adrenal cortex and adrenal medulla with regard to location and histology. (p. 614)
18. Describe the hormones produced by the adrenal cortex in terms of type and normal function. (p. 616)
19. Describe the clinical symptoms of aldosteronism, Addison's disease, Cushing's syndrome, and adrenogenital syndrome. (p. 614)
20. What relationship does the adrenal medulla have to the autonomic nervous system? What is the action of adrenal medullary hormones? (p. 617)
21. What is a pheochromocytoma? (p. 617)
22. Describe the location of the pancreas and the histology of the pancreatic islets (islets of Langerhans). (p. 617)
23. What are the actions of glucagon and insulin? (p. 621)
24. Describe the clinical symptoms of diabetes mellitus and hyperinsulinism. Distinguish the two types of diabetes mellitus. (p. 620)
25. Why are the ovaries and testes considered to be endocrine glands? (p. 621)
26. Where is the pineal gland located? What are its assumed functions? (p. 621)
27. How are hormones of the thymus gland related to immunity? (p. 622)
28. Describe the effects of aging on the endocrine system. (p. 622)
29. Describe the development of the endocrine system. (p. 622)
30. List the hormones secreted by the gastrointestinal tract, placenta, kidneys, skin, and heart. (p. 624)
31. Refer to the glossary of key medical terms associated with the endocrine system. Be sure that you can define each term. (p. 624)

SELF QUIZ

Complete the following:

1. Glucagon, produced by (alpha? beta? delta?) cells, (increases? decreases?) blood sugar in two ways.
2. Arrange in order the names of vessels that supply blood to the adenohypophysis. Use lines provided. __ __ __ __

A. hypophyseal portal veins
B. primary plexus
C. superior hypophyseal arteries
D. secondary plexus

Choose the one best answer to these questions.

___ **3.** One of the endocrine glands develops from the sympathetic nervous system during embryological development. Considering your knowledge of function, this endocrine gland is the
A. pancreas; B. adrenal medulla; C. anterior pituitary; D. thymus; E. posterior pituitary.

___ **4.** A generalized anti-inflammatory reaction is most closely associated with
A. glucocorticoids; B. mineralocorticoids; C. parathyroid hormone (PTH); D. anterior pituitary hormones; E. insulin.

___ **5.** A chemical produced by the hypothalamus that causes an endocrine gland to secrete a hormone is called a
A. gonadotropic hormone; B. tropic hormone; C. regulating hormone (or factor); D. target hormone; E. neurotransmitter.

___ **6.** If a person is diagnosed as having a high metabolic rate, which endocrine gland is probably malfunctioning?
A. parathyroids; B. thymus; C. posterior pituitary; D. thyroid; E. pancreas.

___ **7.** A tumor of the beta cells of the pancreatic islets (islets of Langerhans) would probably affect the body's ability to
A. lower blood sugar level; B. raise blood sugar level; C. lower blood calcium level; D. raise blood calcium level; E. regulate metabolism.

___ **8.** Which hormone is involved in milk production?
A. melanocyte-stimulating hormone (MSH); B. follicle-stimulating hormone (FSH); C. prolactin (PRL); D. glucagon.

___ **9.** A suspected role of the pineal gland is
A. regulation of electrolyte balance; B. water retention; C. regulation of the reproductive activities; D. retention of glucose; E. regulation of calcium metabolism.

___ **10.** The sella turcica and infundibulum are associated with the
A. pineal; B. thyroid; C. thymus; D. pituitary; E. parathyroids.

___ **11.** The stomach, pancreas, testes, and ovaries have something in common. These organs
A. are all influenced by hormones from the anterior pituitary; B. have tissues that are derived from embryological ectoderm; C. form hormones that influence secondary sex characteristics; D. receive their blood supply from the superior mesenteric artery; E. are considered to be both exocrine and endocrine.

___ **12.** A hormone
(1) is a chemical substance secreted into the body fluids.
(2) is secreted by one cell or a group of cells.
(3) exerts a physiological effect on other cells of the body than those that produced it.
A. (1) only; B. (2) only; C. (3) only; D. all of the above; E. (1) and (2).

___ **13.** Which of the following terms is used when referring to hormones that control the activity of other endocrine structures?
A. cryptic; B. trophic; C. troponic; D. tonic; E. tachyptic.

___ **14.** Which endocrine gland consists of lobes made up of follicles that contain the hormone secreted in a colloidal form?
A. pituitary; B. parathyroid; C. thyroid; D. adrenal medulla; E. pancreatic islets.

___ **15.** Which of the endocrine glands are "double" glands; that is, each consists of two endocrine tissues having different embryological origins?
A. testis and ovary; B. pituitary and adrenal (suprarenal); C. adrenal (suprarenal) and thyroid; D. thyroid and parathyroid; E. pancreas and thymus.

___ **16.** Which one of these glands is called the "emergency gland" and helps the body meet sudden stress?
A. pituitary; B. pancreas; C. thyroid; D. thymus; E. adrenal (suprarenal).

___ **17.** Whereas an exocrine gland secretes into a duct, an endocrine gland secretes into
A. open cavities; B. closed cavities; C. blood; D. muscle tissue; E. neural tissue.

Circle T (true) or F (false) for the following.

T F 18. Whereas the anterior pituitary contains axons of neurons, the posterior pituitary contains glandular cells.

T F 19. The middle region of hormone-secreting cells of the adrenal cortex is called the zona reticularis.

T F 20. The amount of a hormone released by an endocrine gland or tissue is determined by the body's need for the hormone at any given time.

T F 21. The gland that assumes a role in the maturation of T cells is the pineal.

T F 22. Chromaffin cells are the principal secreting cells of the pancreas.

T F 23. Another term for the anterior pituitary gland is the neurohypophysis.

T F 24. The outer region of the adrenal (suprarenal) gland is called the cortex.

25. Match the following:

___ **a.** develops from the foregut area that later becomes part of small intestine
___ **b.** derived from the roof of the stomodeum (mouth) called the hypophyseal (Rathke's) pouch
___ **c.** originates from the neural crest, which also produces sympathetic ganglia
___ **d.** derived from tissue from the same region that forms gonads
___ **e.** arise from the third and fourth pharyngeal pouches (two answers)

A. adrenal cortex
B. adrenal medulla
C. anterior pituitary
D. pancreas
E. parathyroid
F. thymus

The Respiratory System

22

STUDENT OBJECTIVES

1. Identify the organs of the respiratory system.
2. Compare the structure and function of the external and internal nose.
3. Differentiate the three anatomical regions of the pharynx and describe their roles in respiration.
4. Describe the structure of the larynx and explain its function in respiration and voice production.
5. Describe the location and structure of the tubes that form the bronchial tree.
6. Describe the coverings of the lungs, the division of the lungs into lobes, and the composition of a lobule of the lung.
7. Explain the structure of the alveolar–capillary (respiratory) membrane and its function in the diffusion of respiratory gases.
8. Describe the effects of aging on the respiratory system.
9. Describe the development of the respiratory system.
10. Define bronchogenic carcinoma (lung cancer), bronchial asthma, bronchitis, emphysema, pneumonia, tuberculosis (TB), respiratory distress syndrome (RDS) of the newborn, respiratory failure, sudden infant death syndrome (SIDS), coryza (common cold), influenza (flu), pulmonary embolism, pulmonary edema, carbon monoxide (CO) poisoning, and smoke inhalation injury as disorders of the respiratory system.
11. Define key medical terms associated with the respiratory system.

CHAPTER OUTLINE

- **Organs**

Nose
Pharynx
Larynx
Trachea
Bronchi
Lungs
Gross Anatomy
Lobes and Fissures
Lobules
Alveolar–Capillary (Respiratory) Membrane
Blood and Nerve Supply

- **Aging and the Respiratory System**
- **Developmental Anatomy of the Respiratory System**
- **Applications to Health**
- **Key Medical Terms Associated with the Respiratory System**

Cells need a continuous supply of oxygen (O_2) for various metabolic reactions that release energy from nutrient molecules, some of which is stored in ATP for cellular use. As a result of these reactions, cells also release quantities of carbon dioxide (CO_2). Since an excessive amount of carbon dioxide produces acid conditions that are poisonous to cells, the excess gas must be eliminated quickly and efficiently. The two systems that supply oxygen and eliminate carbon dioxide are the cardiovascular system and the respiratory system. The respiratory system also contains receptors for the sense of smell, filters inspired air, produces sounds, and helps eliminate wastes. The ***respiratory system*** consists of the nose, pharynx, larynx, trachea, bronchi, and lungs (Figure 22-1). The cardiovascular system transports the gases in the blood between the lungs and the cells. Structurally, the term ***upper respiratory system*** refers to the nose, pharynx, and associated structures; the ***lower respiratory system*** refers to the larynx, trachea, bronchi, and lungs. Functionally, the respiratory system consists of two portions. The ***conducting portion*** consists of a system of interconnecting cavities and tubes that conduct air into the lungs—nose, pharynx, larynx, trachea, and bronchi. The ***respiratory portion*** consists of those portions of the respiratory system where the exchange of respiratory gases occurs—respiratory bronchioles, alveolar ducts, and alveoli.

The overall exchange of gases among the atmosphere, blood, and cells is called ***respiration.*** Three basic processes are involved. The first process, ***pulmonary*** (*pulmo* = lung) ***ventilation,*** or breathing, is the *inspiration* (inflow) and *expiration* (outflow) of air between the atmosphere and the lungs. The second and third processes involve the exchange of gases within the body. ***External (pulmonary)***

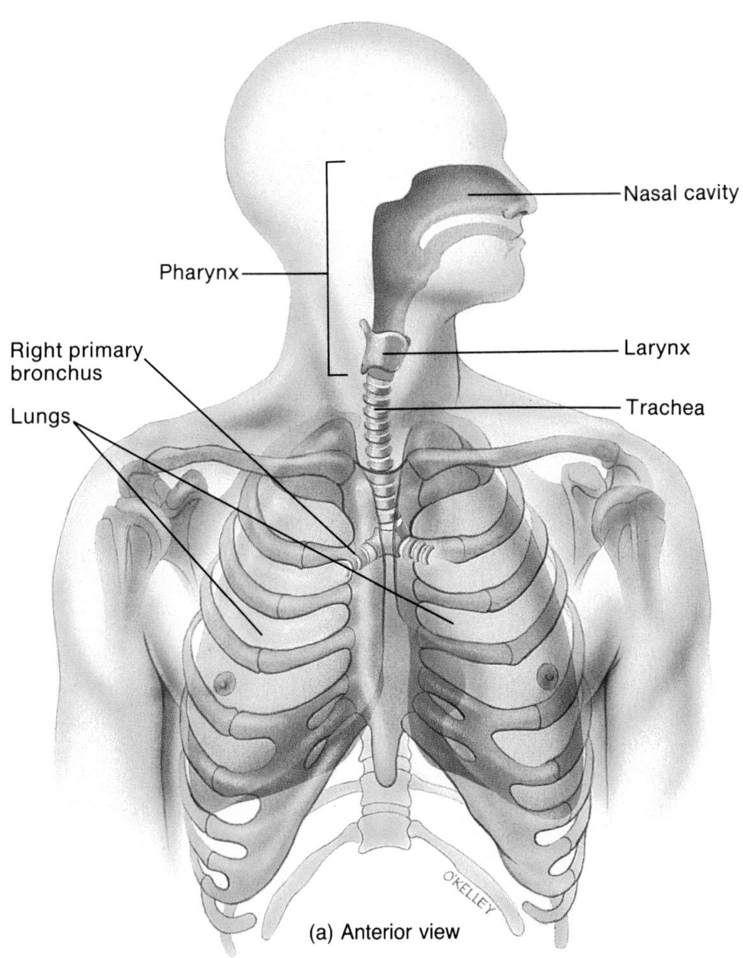

FIGURE 22-1 Organs of the respiratory system in relation to surrounding structures. The photograph of lungs in (b) was taken after removal of anterolateral thoracic wall and parietal pleura. In this specimen, the lungs occupy more of the mediastinum than normal. (Courtesy of J. A. Gosling, P. F. Harris, et al., *Atlas of Human Anatomy,* Gower Medical Publishing Ltd., 2nd ed., 1991.)

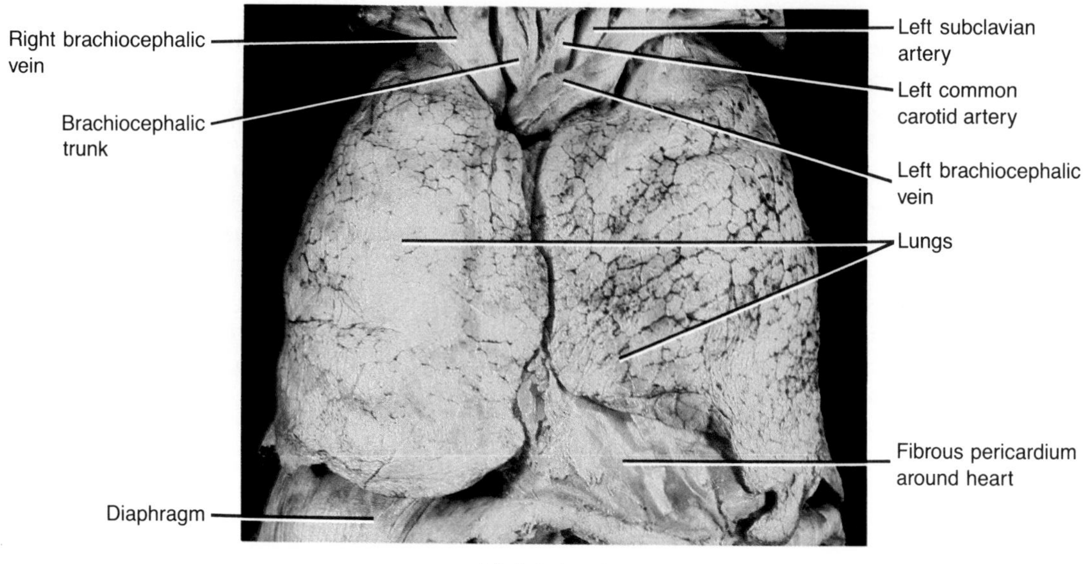

(b) Anterior view

respiration is the exchange of gases between the lungs and blood. ***Internal (tissue) respiration*** is the exchange of gases between the blood and the cells.

The developmental anatomy of the respiratory system will be considered later in the chapter.

ORGANS

NOSE

The ***nose*** has an external portion and an internal portion inside the skull (Figure 22-2). The external portion consists of a supporting framework of bone and cartilage covered with skin and lined with mucous membrane. The bridge of the nose is formed by the nasal bones, which hold it in a fixed position. Because it has a framework of pliable cartilage, the rest of the external nose is somewhat flexible. On the undersurface of the external nose are two openings called the ***external nares*** (NA-rēz; *sing.*, naris), or ***nostrils.*** The surface anatomy of the nose is shown in Figure 11-4.

The internal portion of the nose is a large cavity in the skull that lies inferior to the cranium and superior to the mouth. Anteriorly, the internal nose merges with the external nose, and posteriorly it communicates with the throat (pharynx) through two openings called the ***internal nares (choanae).*** Four paranasal sinuses (frontal, sphenoidal, maxillary, and ethmoidal) and the nasolacrimal ducts also open into the internal nose. The lateral walls of the internal nose are formed by the ethmoid, maxillae, lacrimals, palatines, inferior nasal conchae (turbinates), and pterygoid processes of the sphenoid bone. The ethmoid also forms the roof. The floor is formed by the soft palate and palatine bones and palatine process of the maxilla, which together comprise the hard palate.

The inside of both the external and internal nose consists of a ***nasal cavity,*** divided into right and left sides by a vertical partition called the ***nasal septum.*** The anterior portion of the septum is made primarily of cartilage. The remainder is formed by the vomer, perpendicular plate of the ethmoid, maxillae, and palatine bones (see Figure 6-7a).

CLINICAL APPLICATION

Rhinoplasty

Rhinoplasty (RĪ-nō-plas'-tē; *rhino* = nose; *plassein* = to form), also called a "nose job," is a surgical procedure in which the structure of the external nose is altered. Although it is frequently done for cosmetic reasons, it is sometimes performed to repair a fractured nose or deviated nasal septum. In the procedure, a local or general anesthetic is given, and with instruments inserted through the nostrils, the nasal bones are fractured and repositioned to achieve the desired shape. Any extra bone fragments or cartilage are shaved down and removed through the nostrils. Both an internal packing and an external splint keep the nose in the desired position while it heals. In some cases, only the cartilage has to be reshaped to achieve the desired appearance.

The nasal cavity is divided into three distinct regions: (1) vestibule, (2) respiratory area, and (3) olfactory area. The ***vestibule*** is the anterior portion of the nasal cavity just inside of the nostrils that is surrounded by cartilage. (The upper nasal cavity is surrounded by bone.) The vestibule is lined by skin (stratified squamous epithelium) containing coarse hairs that filter large dust particles present in inspired air. Except for the olfactory area to be described shortly, the remainder of the nasal cavity is composed of the ***respiratory area.*** This area is lined by pseudostratified columnar epithelium with many goblet cells. The respiratory area contains three shelves formed by projections of the superior, middle, and inferior nasal conchae (turbinates) that extend out of the lateral walls of the nasal cavity. The conchae, almost reaching the septum, subdivide each side of the nasal cavity into a series of groovelike passageways—the ***superior, middle,*** and ***inferior nasal meatuses*** (the singular form is ***meatus***). As inspired air whirls around the conchae and meatuses, it is warmed by the capillaries. Mucus secreted by the goblet cells moistens the air and traps dust particles. Drainage from the nasolacrimal ducts and perhaps secretions from the paranasal sinuses also help moisten the air. The cilia move the mucus–dust packages to the pharynx so that they can be eliminated from the body. The ***olfactory area*** is found in the roof of the nasal cavity above the superior nasal conchae. It contains pseudostratified columnar epithelium with reduced cilia and no goblet cells. It does, however, have olfactory cells that function in the sense of smell (see Figure 20-10).

The interior structures of the nose are specialized for three functions: (1) incoming air is warmed by blood capillaries, moistened by mucus secreted by goblet cells and filtered by mucus, which traps dust particles, and cilia, which move the particles to the throat for elimination; (2) olfactory stimuli are received; and (3) large, hollow resonating chambers modify speech sounds.

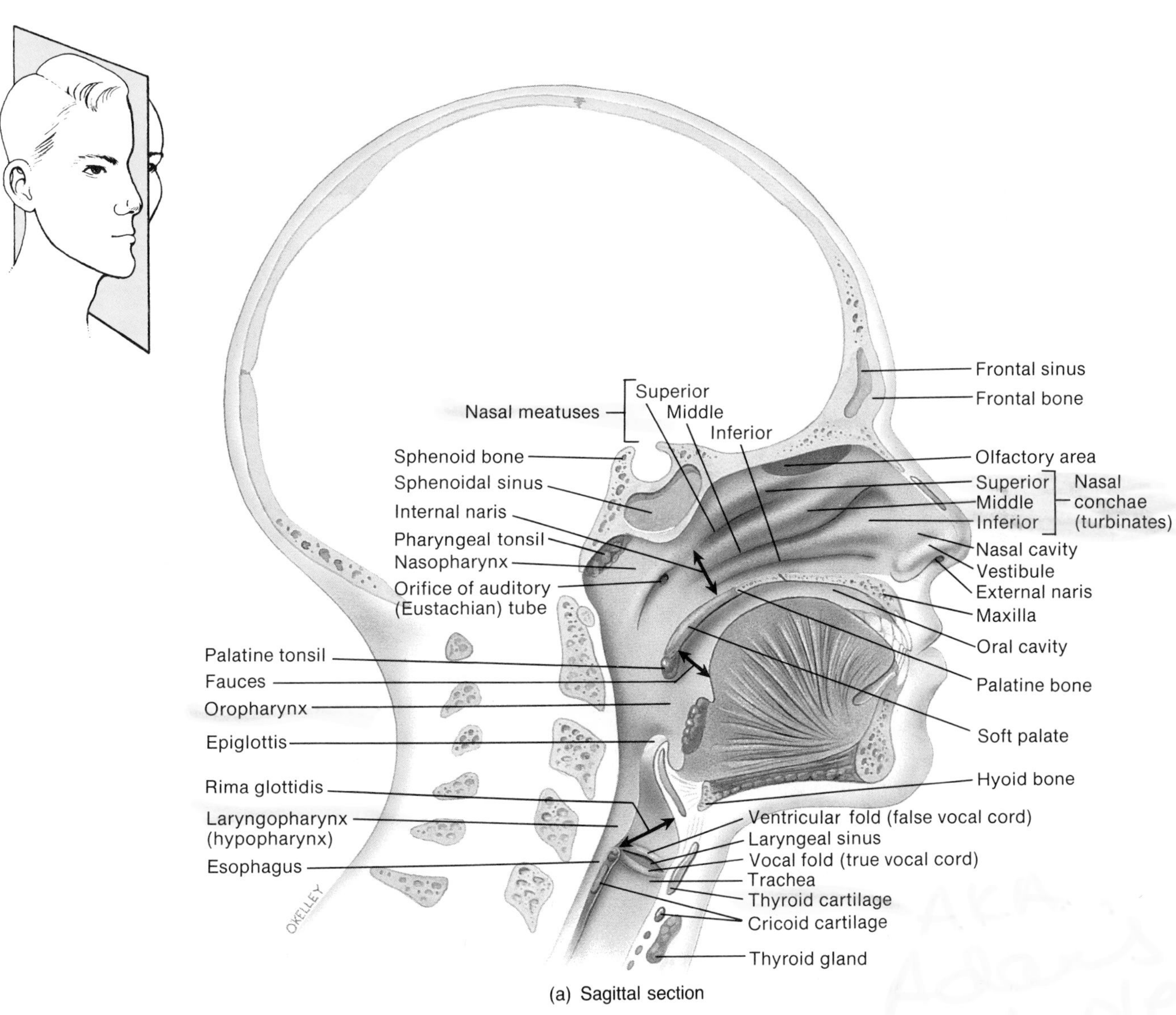

FIGURE 22-2 Respiratory organs in the head and neck. (a) Diagram of the left side of the head and neck with the nasal septum removed. (b) Photograph of the left side of the nasal cavity with the nasal septum removed. (Courtesy of J. A. Gosling, P. F. Harris, et al., *Atlas of Human Anatomy,* Gower Medical Publishing Ltd., 2nd ed., 1991.)

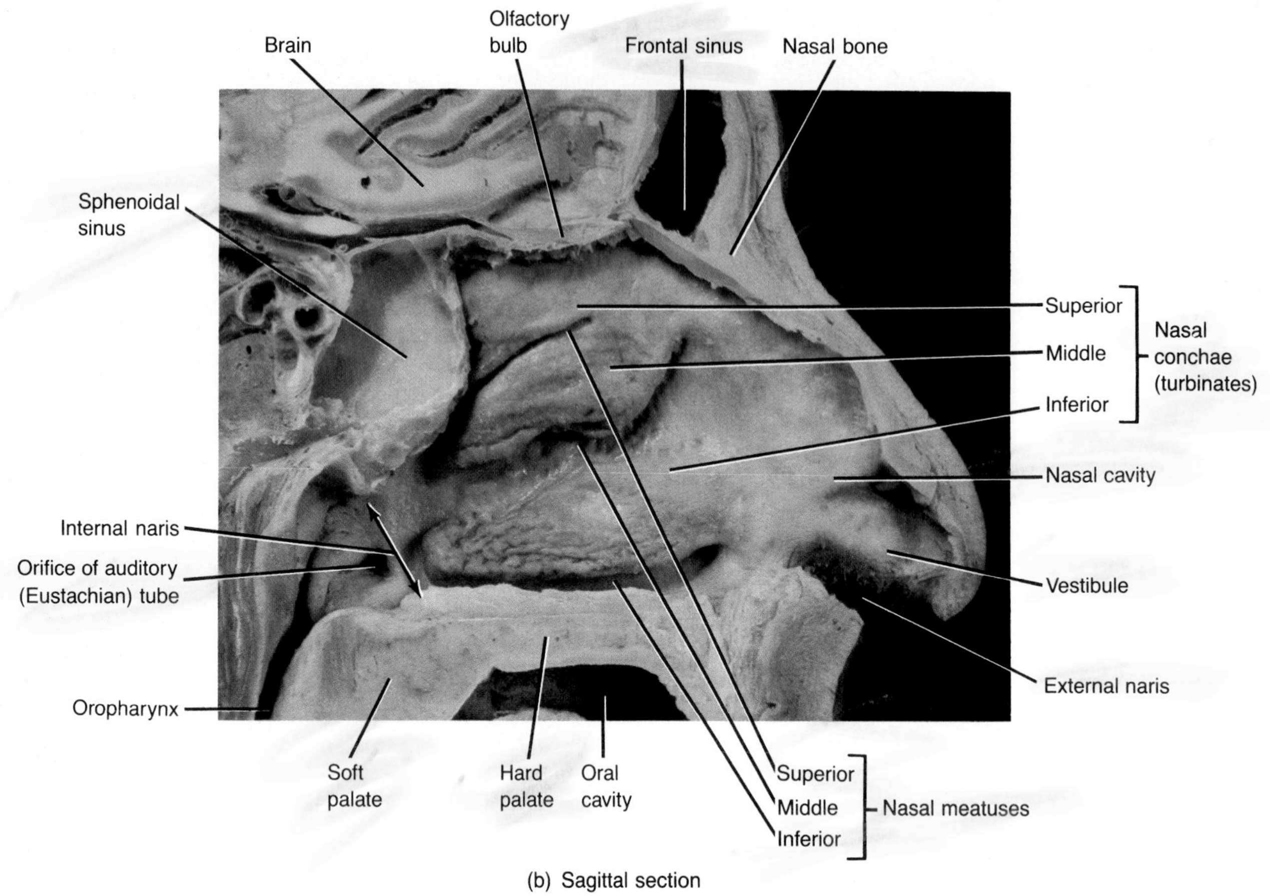

(b) Sagittal section

CLINICAL APPLICATION

Nosebleed (Epistaxis)

Nosebleed, or ***epistaxis*** (ep′-i-STAK-sis), is common because of the exposure of the nose to trauma and the extensive blood supply of the nose. In addition to trauma, other causes of nosebleed are intranasal infection, allergy, bleeding disorders, and neoplasms. Bleeding, either arterial or venous, usually occurs on the anterior part of the septum and can be arrested by cautery with silver nitrate, electrocautery, and firm packing of the external nares. If the point of bleeding is in the posterior region, plugging of both the external and internal nares may be necessary. In extreme emergency, the external carotid artery may have to be ligated (tied) in order to control the hemorrhage.

The arterial supply to the nasal cavity is principally from the sphenopalatine branch of the maxillary artery. The remainder is supplied by the ophthalmic artery. The veins of the nasal cavity drain into the sphenopalatine vein, the facial vein, and the ophthalmic vein.

The nerve supply of the nasal cavity consists of olfactory cells in the olfactory epithelium associated with the olfactory nerve (see Figure 20-10) and the nerves of general sensation. These nerves are branches of the ophthalmic and maxillary divisions of the trigeminal (V) nerve.

PHARYNX

The ***pharynx*** (FAR-inks), or throat, is a somewhat funnel-shaped tube about 13 cm (5 in.) long that starts at the internal nares and extends to the level of the cricoid cartilage (Figure 22-3). It lies just posterior to the nasal cavity, oral cavity, and larynx and just anterior to the cervical vertebrae. Its wall is composed of skeletal muscles and lined with mucous membrane. The functions of the pharynx are to serve as a passageway for air and food and to provide a resonating chamber for speech sounds.

The branch of medicine that deals with the diagnosis and treatment of diseases of the ears, nose, and throat is called ***otorhinolaryngology*** (ō′-tō-rī′-nō-lar′-in-GOL-ō-jē; *oto* = ear; *rhino* = nose).

The uppermost portion of the pharynx, called the ***naso-pharynx,*** lies posterior to the internal nasal cavity and ex-

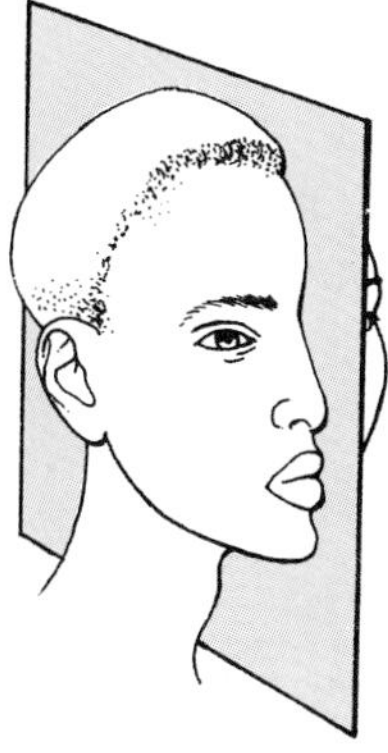

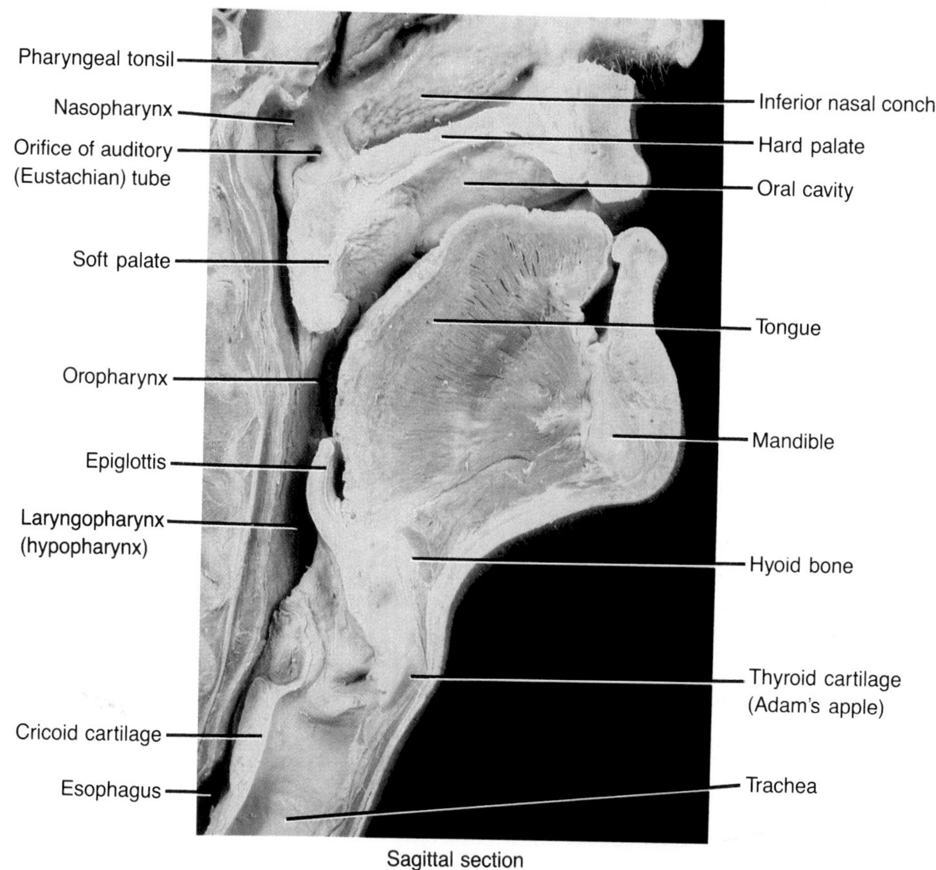

FIGURE 22-3 Photograph of a portion of the left side of the head and neck with the nasal septum removed. (Courtesy of J. A. Gosling, P. F. Harris, et al., *Atlas of Human Anatomy*. Gower Medical Publishing Ltd., 2nd ed., 1991.)

tends to the plane of the soft palate. There are four openings in its wall: two internal nares and two openings that lead into the auditory (Eustachian) tubes. The posterior wall also contains the pharyngeal tonsil, or adenoid. Through the internal nares, the nasopharynx receives air from the nasal cavities and receives the packages of dust-laden mucus. It is lined with pseudostratified ciliated epithelium, and the cilia move the mucus down toward the pharynx. The nasopharynx also exchanges small amounts of air with the auditory (Eustachian) tubes so that the air pressure inside the middle ear equals the pressure of the atmospheric air flowing through the nose and pharynx.

The middle portion of the pharynx, the ***oropharynx,*** lies posterior to the oral cavity and extends from the soft palate inferiorly to the level of the hyoid bone. It has only one opening, the ***fauces*** (FAW-sēz), the opening from the mouth. It is lined by stratified squamous epithelium. This portion of the pharynx is both respiratory and digestive in function, since it is a common passageway for air, food, and drink. Two pairs of tonsils, the palatine and lingual tonsils, are found in the oropharynx (see also Figure 20-11b).

The lowest portion of the pharynx, the ***laryngopharynx*** (la-rin′-gō-FAR-inks), or ***hypopharynx,*** extends downward from the hyoid bone and becomes continuous with the esophagus (food tube) posteriorly and the larynx (voice box) anteriorly. Like the oropharynx, the laryngopharynx is a respiratory and a digestive pathway and is lined by stratified squamous epithelium.

The arteries of the pharynx are the ascending pharyngeal, the ascending palatine branch of the facial, the descending palatine and pharyngeal branches of the maxillary, and the muscular branches of the superior thyroid artery. The veins of the pharynx drain into the pterygoid plexus and the internal jugular vein.

Most of the muscles of the pharynx (see Figure 10-9) are innervated by the pharyngeal plexus. This plexus is formed by the pharyngeal branches of the glossopharyngeal

(IX), vagal (X), and cranial portion of the accessory (XI) nerves and the superior cervical sympathetic ganglion.

LARYNX

The ***larynx,*** or voice box, is a short passageway that connects the pharynx with the trachea. It lies in the midline of the neck anterior to the fourth through sixth cervical vertebrae (C4–C6).

The wall of the larynx is composed of nine pieces of cartilage (Figure 22-4). Three are single and three are paired. The three single pieces are the thyroid cartilage, epiglottis (epiglottic cartilage), and cricoid cartilage. Of the paired cartilages, the arytenoid cartilages are the most important.

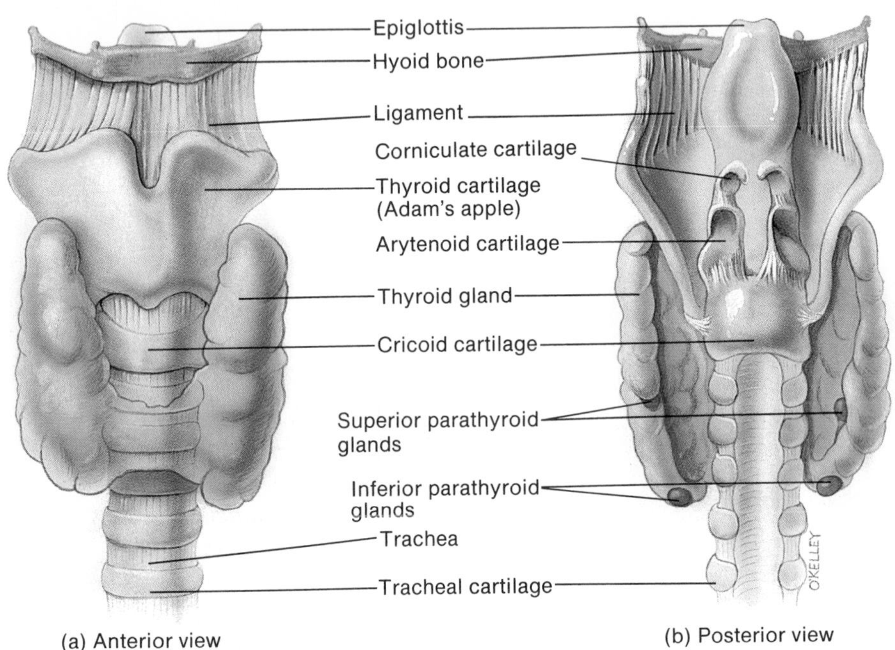

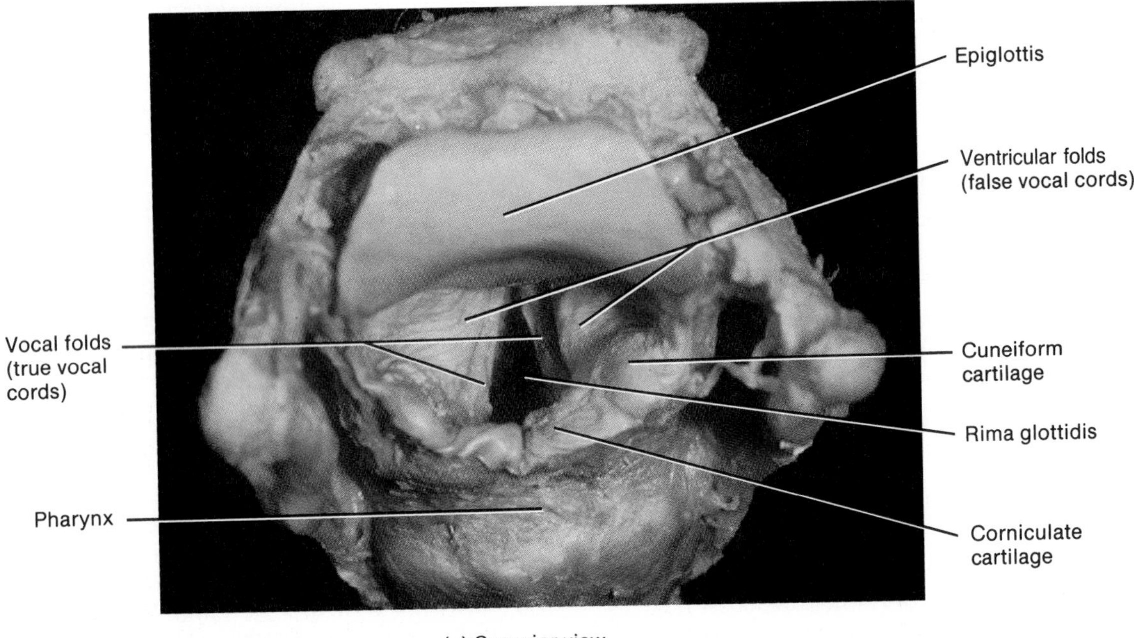

FIGURE 22-4 Larynx. (c) Courtesy of J. A. Gosling, P. F. Harris, et al., *Atlas of Human Anatomy,* Gower Medical Publishing Ltd., 2nd ed., 1991.

The paired corniculate and cuneiform cartilages are of lesser significance.

The ***thyroid cartilage*** (***Adam's apple***) consists of two fused plates that form the anterior wall of the larynx and give it its triangular shape. It is larger in males than in females.

The ***epiglottis*** (*epi* = above; *glotta* = tongue) is a large, leaf-shaped piece of cartilage (see also Figure 22-3). The "stem" of the epiglottis is attached to the anterior rim of the thyroid cartilage, but the "leaf" portion is unattached and free to move up and down like a trapdoor. During swallowing, there is elevation of the larynx, which causes the free edge of the epiglottis to form a lid over the glottis, closing it off. The ***glottis*** consists of the vocal folds (true vocal cords) in the larynx and the space between them (***rima glottidis***). In this way, the larynx is closed off, and liquids and foods are routed into the esophagus and kept out of the larynx and air passageways below it. When anything but air passes into the larynx, a cough reflex attempts to expel the material.

The ***cricoid*** (KRĪ-koyd) ***cartilage*** is a ring of cartilage forming the inferior wall of the larynx. It is attached to the first ring of cartilage of the trachea. It is a clinically important landmark for making an emergency airway into the trachea.

The paired ***arytenoid*** (ar′-i-TĒ-noyd) ***cartilages*** are triangular in shape and located at the posterior, superior border of the cricoid cartilage. They attach to the vocal folds and intrinsic pharyngeal muscles and by their action can move the vocal folds.

The paired ***corniculate*** (kor-NIK-yoo-lāt) ***cartilages*** are horn-shaped. One is located at the apex of each arytenoid cartilage. The paired ***cuneiform*** (kyoo-NĒ-i-form) ***cartilages*** are club-shaped cartilages anterior to the corniculate cartilages.

The epithelium lining the larynx below the vocal folds is pseudostratified columnar. It consists of ciliated columnar cells, goblet cells, and basal cells, and it helps trap dust not removed in the upper passages.

The mucous membrane of the larynx is arranged into two pairs of folds—an upper pair called the ***ventricular folds*** (***false vocal cords***) and a lower pair called simply the ***vocal folds*** (***true vocal cords***). The space between the ventricular folds is known as the ***rima vestibuli***. The ***laryngeal sinus*** is a lateral expansion of the middle portion of the laryngeal cavity between the vestibular folds above and the vocal folds below. When the ventricular folds are brought together, they function in holding the breath against pressure in the thoracic cavity, such as might occur when a person exerts a strain while lifting a heavy object. The mucous membrane of the vocal folds is lined by nonkeratinized stratified squamous epithelium. Under the membrane lie bands of elastic ligaments stretched between pieces of rigid cartilage like the strings on a guitar. Skeletal muscles of the larynx, called intrinsic muscles, are attached internally to the pieces of rigid cartilage and to the vocal folds themselves. When the muscles contract, they pull the strings of elastic ligaments tight and stretch the vocal folds out into the air passageways so that the glottis is narrowed. If air is directed against the vocal folds, they vibrate and set up sound waves in the column of air in the pharynx, nose, and mouth. The greater the pressure of air, the louder the sound.

Pitch is controlled by the tension on the vocal folds. If they are pulled taut by the muscles, they vibrate more rapidly, and a higher pitch results. Lower sounds are produced by decreasing the muscular tension on the vocal folds. Vocal folds are usually thicker and longer in males than in females, and therefore they vibrate more slowly. Thus, men generally have a lower range of pitch than women.

Sound originates from the vibration of the vocal folds, but other structures are necessary for converting the sound into recognizable speech. The pharynx, mouth, nasal cavity, and paranasal sinuses all act as resonating chambers that give the voice its human and individual quality. By constricting and relaxing the muscles in the wall of the pharynx, we produce the vowel sounds. Muscles of the face, tongue, and lips help us enunciate words.

CLINICAL APPLICATION

Laryngitis and Cancer of Larynx

Laryngitis is an inflammation of the larynx that is most often caused by a respiratory infection or irritants such as cigarette smoke. Inflammation of the vocal folds causes hoarseness or loss of voice by interfering with the contraction of the folds or by causing them to swell to the point where they cannot vibrate freely. Many long-term smokers acquire a permanent hoarseness from the damage done by chronic inflammation.

Cancer of the larynx is found almost exclusively in individuals who smoke. The condition is characterized by hoarseness, pain on swallowing, or pain radiating to an ear. Treatment consists of radiation therapy or surgery.

The arteries of the larynx are the superior laryngeal and inferior laryngeal. The superior and inferior laryngeal veins accompany the arteries. The superior laryngeal vein empties into the superior thyroid vein, and the inferior vein empties into the inferior thyroid vein.

The nerves of the larynx are the superior and recurrent (inferior) laryngeal branches of the vagus (X) nerve.

TRACHEA

The ***trachea*** (TRĀ-kē-a), or windpipe, is a tubular passageway for air about 12 cm (4.5 in.) in length and 2.5 cm (1 in.) in diameter. It is located anterior to the esophagus and extends from the larynx to the fifth thoracic vertebra (T5), where it divides into right and left primary bronchi (see Figure 22-6).

The wall of the trachea (Figure 22-5a) consists of a mucosa, submucosa, cartilaginous layer, and adventitia (outer layer of areolar connective tissue). The tracheal epithelium of the mucosa is pseudostratified columnar. It consists of ciliated columnar cells that reach the luminal surface, goblet cells, and basal cells that do not reach the luminal surface (Figure 22-5b). The epithelium provides the same protection against dust as the membrane lining the larynx. Seromucous glands and their ducts are present in the submucosa. The cartilaginous layer consists of 16 to 20 horizontal incomplete rings of hyaline cartilage that look like a series of letter C's stacked one on top of another. The open parts of the C's face the esophagus and permit it to expand slightly into the trachea during swallowing (Figure 22-5c). Transverse smooth muscle fibers, called the ***trachealis muscle,*** and elastic connective tissue, attach the open ends of the cartilage rings. The solid parts of the C's provide a rigid support so the tracheal wall does not collapse inward and obstruct the air passageway. In some situations, however, such as crushing injuries to the chest, the rings of cartilage may not be strong enough to overcome collapse and obstruction of the trachea.

At the point where the trachea divides into right and left primary bronchi, there is an internal ridge called the ***carina*** (ka-RĪ-na). It is formed by a posterior and somewhat inferior projection of the last tracheal cartilage. The mucous membrane of the carina is one of the most sensitive areas of the respiratory system and is associated with the cough reflex. Widening and distortion of the carina, which can be seen in an examination by bronchoscopy, is a serious prognostic sign, since it usually indicates a carcinoma of the lymph nodes around the region where the trachea divides. ***Bronchoscopy*** is the visual examination of the bronchi through a ***bronchoscope,*** an illuminated tubular instrument that can be passed through the trachea into the bronchi.

CLINICAL APPLICATION

Tracheostomy and Intubation

Occasionally, the respiratory passageways are unable to protect themselves from obstruction. The rings of cartilage may accidentally be crushed; the mucous membrane may become inflamed and swell so much that it closes off

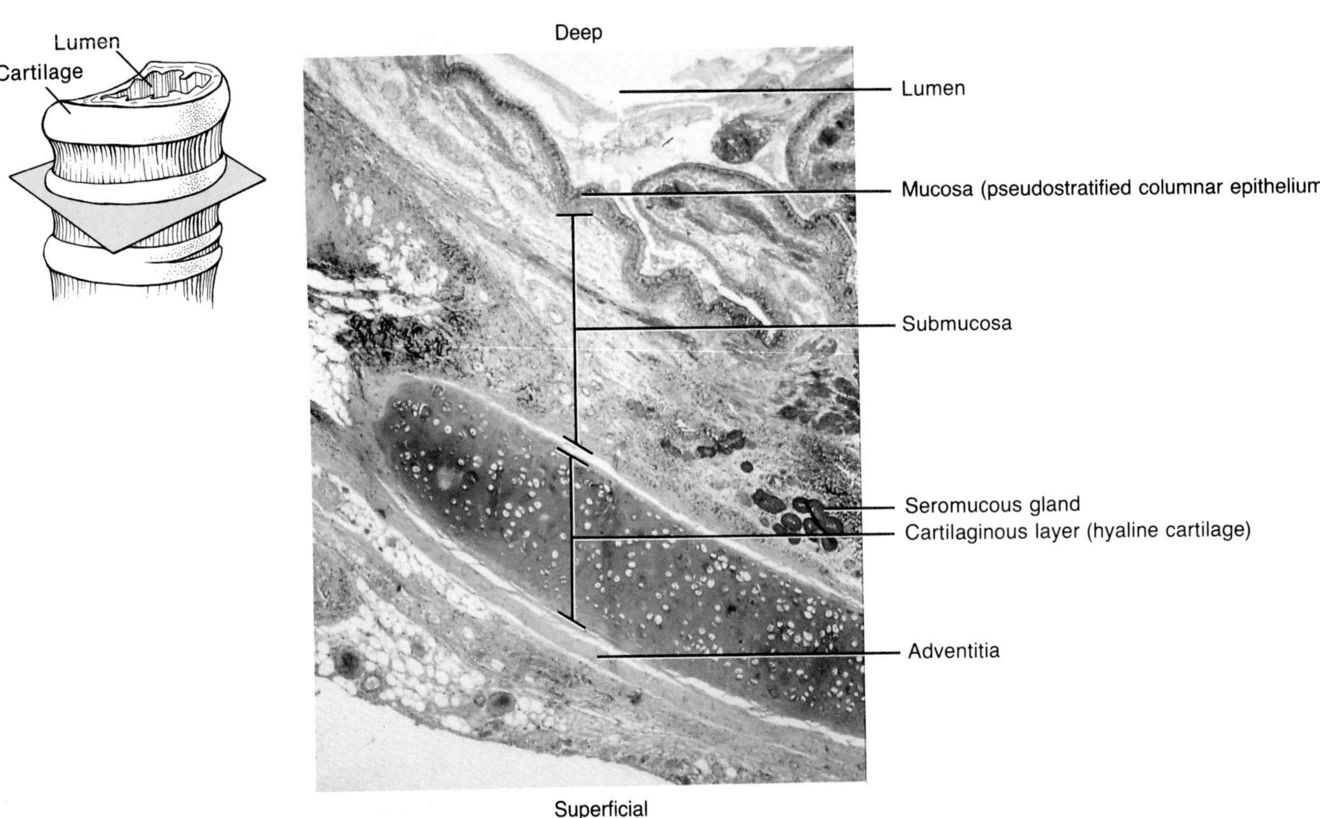

(a) Photomicrograph of a portion of the tracheal wall, 80×

FIGURE 22-5 Histology of the trachea. (a) and (b) Courtesy of Andrew J. Kuntzman. (c) © Cunningham, Visuals Unlimited.

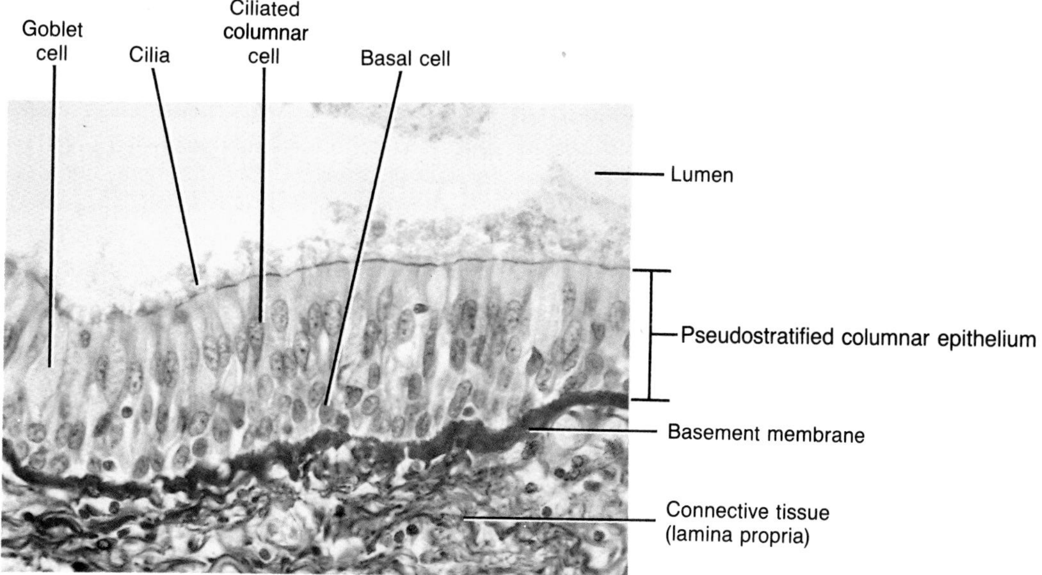

(b) Photomicrograph of enlarged aspect of tracheal epithelium, 600×

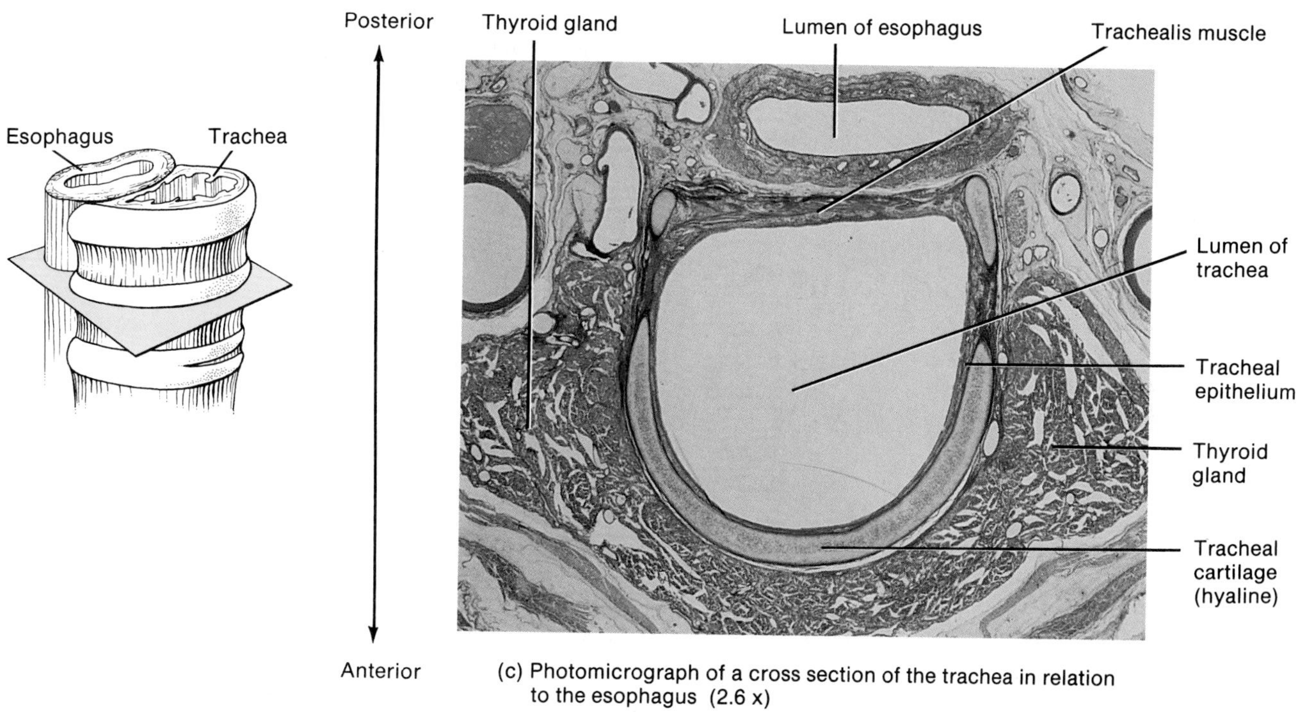

(c) Photomicrograph of a cross section of the trachea in relation to the esophagus (2.6 x)

the air passageways; inflamed membranes secrete a great deal of mucus that may clog the lower respiratory passageways; a large object may be breathed in (aspirated) while the rima glottidis is open; or an aspirated foreign object may cause spasm of the laryngeal muscles. The passageways must be cleared quickly. If the obstruction is above the level of the larynx, a ***tracheostomy*** (trā-kē-OS-tō-mē) may be performed. A skin incision is made, followed by a short longitudinal incision into the trachea inferior to the cricoid cartilage. The patient breathes through a metal or plastic tracheal tube inserted through the incision. Another method is ***intubation.*** A tube is inserted into the mouth or nose and passed down through the larynx and trachea. The firm wall of the tube pushes back any flexible obstruction, and the inside of the tube provides a passageway for air. If mucus is clogging the trachea, it can be suctioned out through the tube.

The arteries of the trachea are branches of the inferior thyroid, internal thoracic, and bronchial arteries. The veins of the trachea terminate in the inferior thyroid veins.

The smooth muscle and glands of the trachea are innervated parasympathetically via the vagus (X) nerve directly and by its recurrent laryngeal branches. Sympathetic innervation is through branches from the sympathetic trunk and its ganglia.

BRONCHI

The trachea terminates in the chest by dividing at the sternal angle into a ***right primary bronchus*** (BRON-kus), which goes to the right lung, and a ***left primary bronchus,*** which goes to the left lung (Figure 22-6). The right primary bronchus is more vertical, shorter, and wider than the left. As a result, foreign objects in the air passageways are more likely to enter the right primary bronchus than the left and frequently lodge in it. Like the trachea, the primary bronchi (BRON-kē) contain incomplete rings of cartilage and are lined by pseudostratified columnar epithelium.

On entering the lungs, the primary bronchi divide to form smaller bronchi—the ***secondary (lobar) bronchi,*** one for each lobe of the lung (the right lung has three lobes; the left lung has two). The secondary bronchi continue to branch, forming still smaller bronchi, called ***tertiary (segmental) bronchi,*** that divide into ***bronchioles.*** Bronchioles, in turn, branch into even smaller tubes called ***terminal bronchioles.*** This continuous branching from the trachea resembles a tree trunk with its branches and is commonly referred to as the ***bronchial tree.***

Bronchography (bron-KOG-ra-fē) is a technique for examining the bronchial tree. An intratracheal catheter is passed into the mouth or nose, through the rima glottidis, and into the trachea. Then an opaque contrast medium, usually containing iodine, is introduced by means of gravity

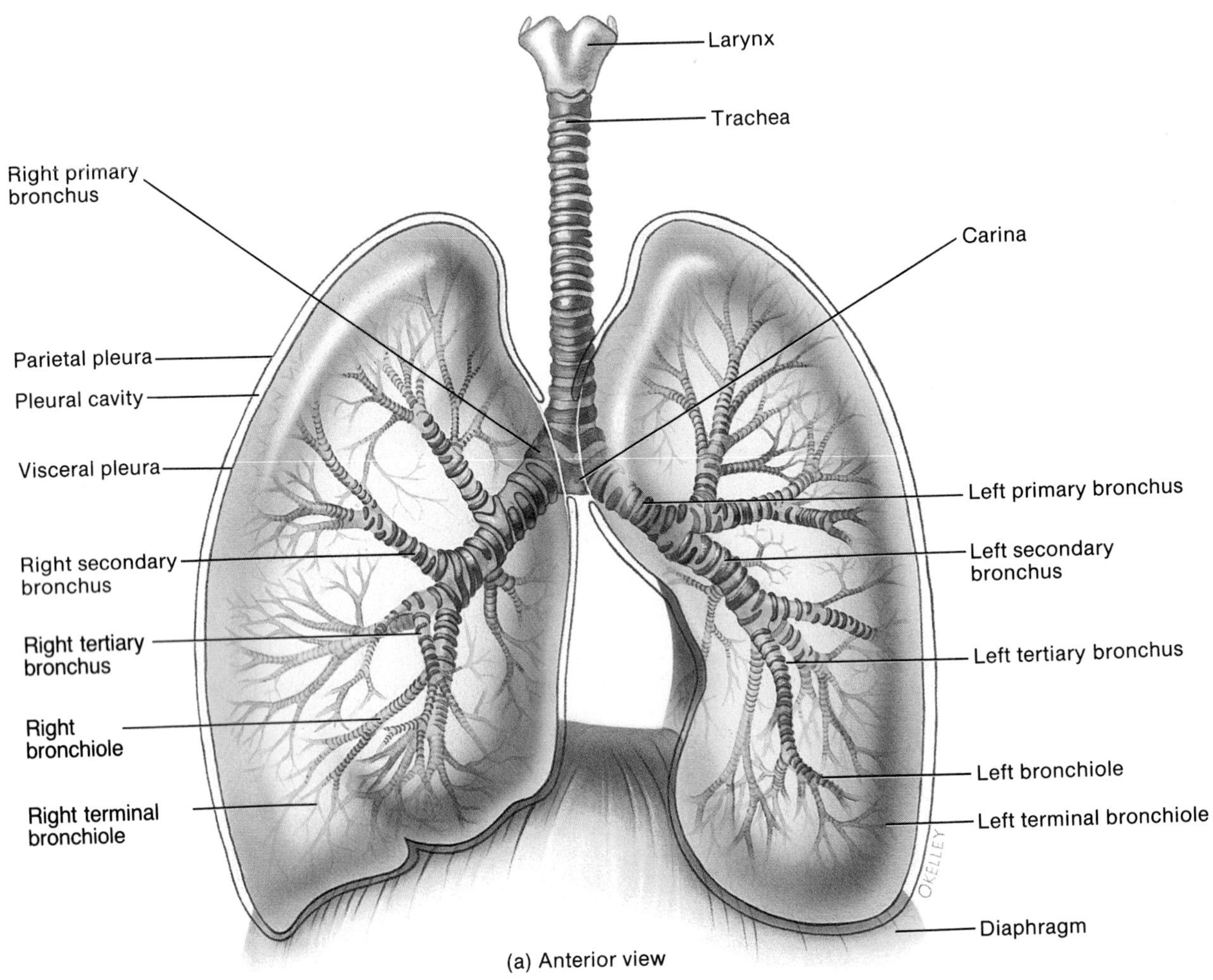

FIGURE 22-6 Bronchial tree. (b) Courtesy of John H. Juhl. From Lester W. Paul and John H. Juhl, *The Essentials of Roentgen Interpretation,* 4th ed., J. B. Lippincott, Philadelphia, 1987.

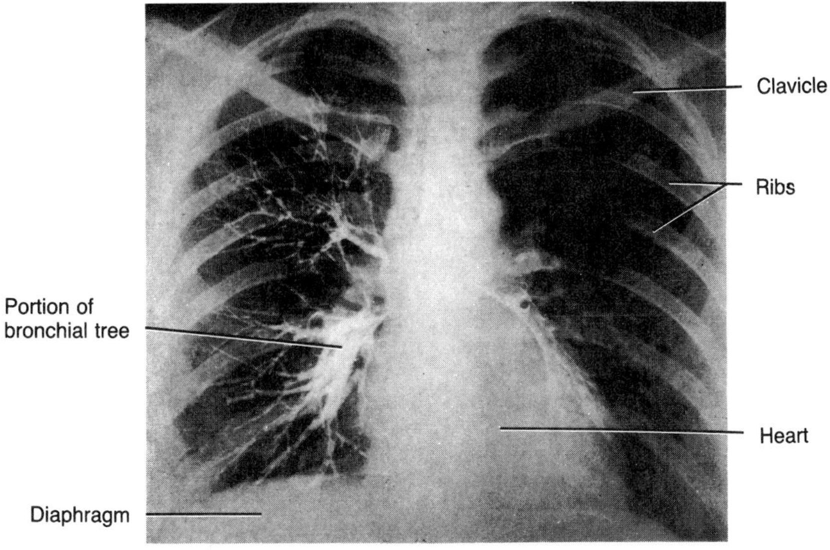

(b) Anteroposterior bronchogram

into the trachea and distributed through the bronchial branches. X-rays of the chest in various positions are taken, and the developed film, a ***bronchogram*** (BRON-kō-gram), provides a picture of the tree (Figure 22-6b).

As the branching becomes more extensive in the bronchial tree, several structural changes occur. First, rings of cartilage are replaced by plates of cartilage that finally disappear in the bronchioles. Second, the mucous membrane changes from pseudostratified columnar to simple cuboidal in the terminal bronchioles. Therefore, inspired debris that reaches the terminal bronchioles must be removed by macrophages. Third, as the cartilage decreases, the amount of smooth muscle increases. Smooth muscle encircles the lumen in spiral bands and its contraction is affected by both the autonomic nervous system and various chemicals. The parasympathetic division of the autonomic nervous system and mediators of allergic reactions, such as histamine, cause bronchiole constriction, whereas the sympathetic division and epinephrine cause bronchiole dilation. The fact that bronchiole walls contain smooth muscle and no cartilage is clinically significant. During an asthma attack the muscles go into spasm. Because there is no supporting cartilage, the spasms can close off the air passageways. Movement of air through constricted bronchial tubes causes breathing to sound louder.

The blood supply to the bronchi is via the left bronchial and right bronchial arteries. The veins that drain the bronchi are the right bronchial vein, which enters the azygos vein, and the left bronchial vein, which empties into the hemiazygos vein or the left superior intercostal vein.

LUNGS

The ***lungs*** (*lunge* = light, since the lungs float) are paired cone-shaped organs lying in the thoracic cavity. They are separated from each other by the heart and other structures in the mediastinum (see Figure 13-1). Two layers of serous membrane, collectively called the ***pleural membrane,*** enclose and protect each lung. The outer layer is attached to the wall of the thoracic cavity and is called the ***parietal pleura.*** The inner layer, the ***visceral pleura,*** covers the lungs themselves. Between the visceral and parietal pleura is a small potential space, the ***pleural cavity,*** which contains a lubricating fluid (pleural fluid) secreted by the membranes (see Figure 1-7d). This fluid prevents friction between the membranes and allows them to move easily on one another during breathing.

CLINICAL APPLICATION

Pneumothorax, Hemothorax, and Pleurisy

In certain conditions, the pleural cavity may fill with air (***pneumothorax;*** *pneumo* = air or breath), blood (***hemothorax***), or pus. Air in the pleural cavity, most commonly introduced in a surgical opening of the chest or as a result of a stab or gunshot wound, may cause the lung to collapse (atelectasis). Fluid can be drained from the pleural cavity by inserting a needle, usually posteriorly through the seventh intercostal space. The needle is passed along the superior border of the lower rib to avoid damage to the intercostal nerves and blood vessels. Below the seventh intercostal space there is danger of penetrating the diaphragm.

Inflammation of the pleural membrane, or ***pleurisy,*** may in its early stages cause pain due to friction between the parietal and visceral layers of the pleura. If the infection persists, fluid accumulates in the pleural space, a condition known as ***pleural effusion.*** One cause of pleural effusion is cancer that may have arisen in the lung or reached the pleura from some other cancer site.

Gross Anatomy

The lungs extend from the diaphragm to a point about 1.5 to 2.5 cm (0.75–1 in.) superior to the clavicles and lie against the ribs anteriorly and posteriorly. The broad inferior portion of the lung, the ***base,*** is concave and fits over the convex area of the diaphragm (Figure 22-7). The narrow superior portion of the lung is termed the ***apex (cupula).*** The surface of the lung lying against the ribs, the ***costal surface,*** is rounded to match the curvature of the ribs. The ***mediastinal (medial) surface*** of each lung contains a region, the ***hilus,*** through which bronchi, pulmonary blood vessels, lymphatic vessels, and nerves enter and exit. These structures are held together by the pleura and connective tissue and constitute the ***root*** of the lung. Medially, the left lung also contains a concavity, the ***cardiac notch,*** in which the heart lies.

The right lung is thicker and broader than the left. It is also somewhat shorter than the left because the diaphragm is higher on the right side to accommodate the liver that lies below it.

Lobes and Fissures

Each lung is divided into lobes by one or more fissures (Figure 22-7). Both lungs have an ***oblique fissure,*** which extends downward and forward. The right lung also has a ***horizontal fissure.*** The oblique fissure in the left lung separates the ***superior lobe*** from the ***inferior lobe.*** The upper part of the oblique fissure of the right lung separates the superior lobe from the inferior lobe, whereas the lower part of the oblique fissure separates the inferior lobe from the ***middle lobe.*** The horizontal fissure of the right lung subdivides the superior lobe, thus forming a middle lobe.

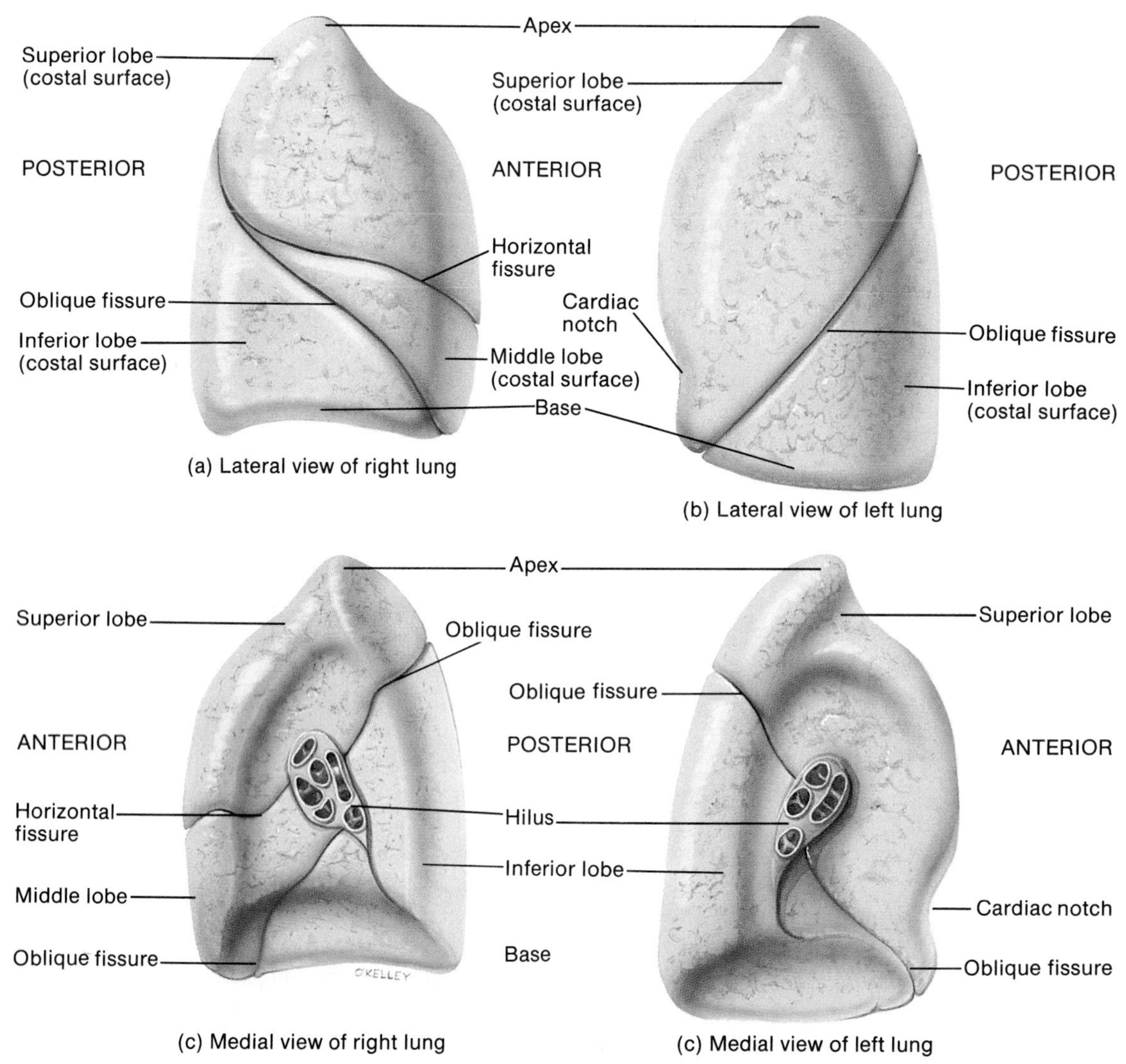

FIGURE 22-7 Lungs. (e) Courtesy of J. A. Gosling, P. F. Harris, et al., *Atlas of Human Anatomy,* Gower Medical Publishing Ltd., 2nd ed., 1991.

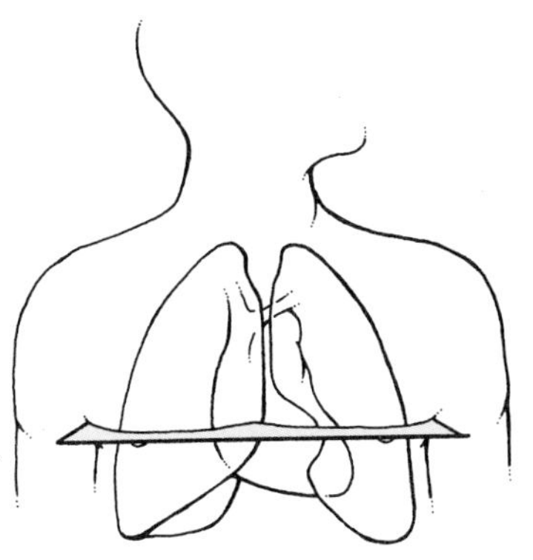

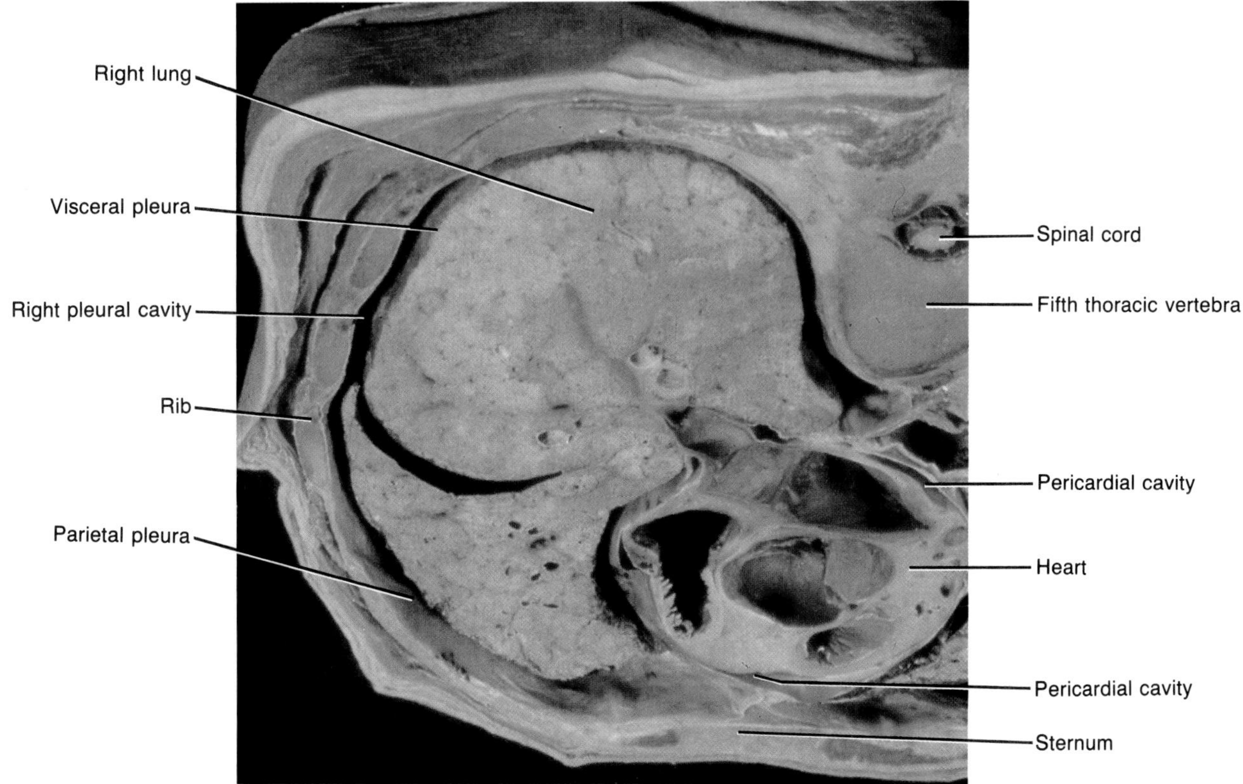

(e) Photograph of a cross section through the thorax

Each lobe receives its own secondary (lobar) bronchus. Thus, the right primary bronchus gives rise to three secondary (lobar) bronchi called the ***superior, middle,*** and ***inferior secondary (lobar) bronchi.*** The left primary bronchus gives rise to a ***superior*** and an ***inferior secondary (lobar) bronchus.*** Within the substance of the lung the secondary bronchi give rise to the ***tertiary (segmental) bronchi,*** which are constant in both origin and distribution. There are 10 tertiary bronchi in each lung. The segment of lung tissue that each supplies is called a ***bronchopulmonary segment*** (Figure 22-8). Accordingly, there are also 10 bronchopulmonary segments in each lung. Bronchial and pulmonary disorders, such as tumors or abscesses, may be localized in a bronchopulmonary segment and may be surgically removed without seriously disrupting surrounding lung tissue.

Lobules

Each bronchopulmonary segment of the lungs is broken up into many small compartments called ***lobules*** (Figure 22-9a). Each lobule is wrapped in elastic connective tissue and contains a lymphatic vessel, an arteriole, a venule, and a branch from a terminal bronchiole. Terminal bronchioles subdivide into microscopic branches called ***respiratory bronchioles.*** As the respiratory bronchioles penetrate more deeply into the lungs, the epithelial lining changes from cuboidal to squamous. Respiratory bronchioles in turn subdivide into several (2–11) ***alveolar ducts (atria).***

Around the circumference of the alveolar ducts are numerous alveoli and alveolar sacs. An ***alveolus*** (al-VĒ-ō-lus) is a cup-shaped outpouching lined by epithelium and

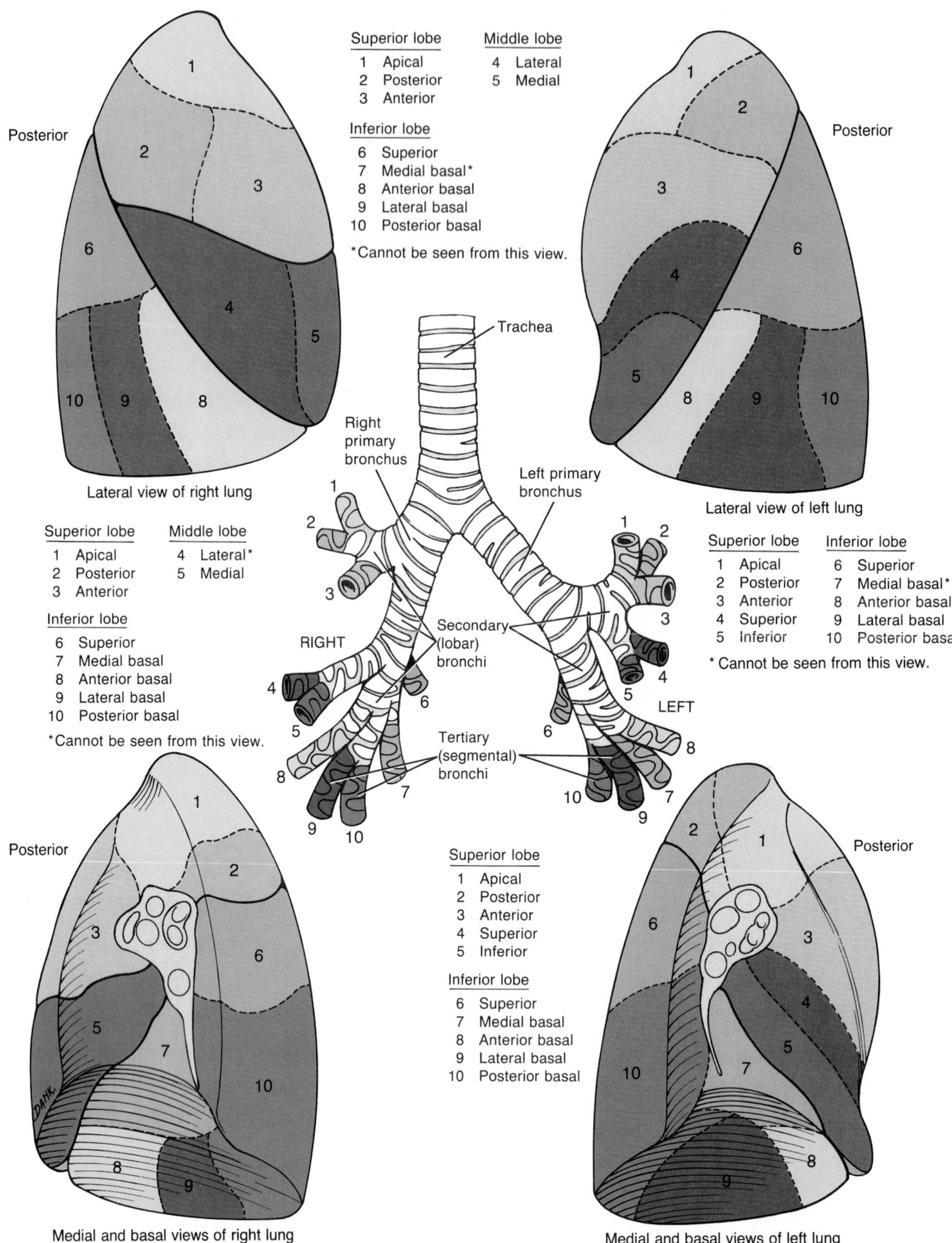

FIGURE 22-8 Bronchopulmonary segments of the lungs. The bronchial branches are shown in the center of the figure. The bronchopulmonary segments are numbered and named for convenience.

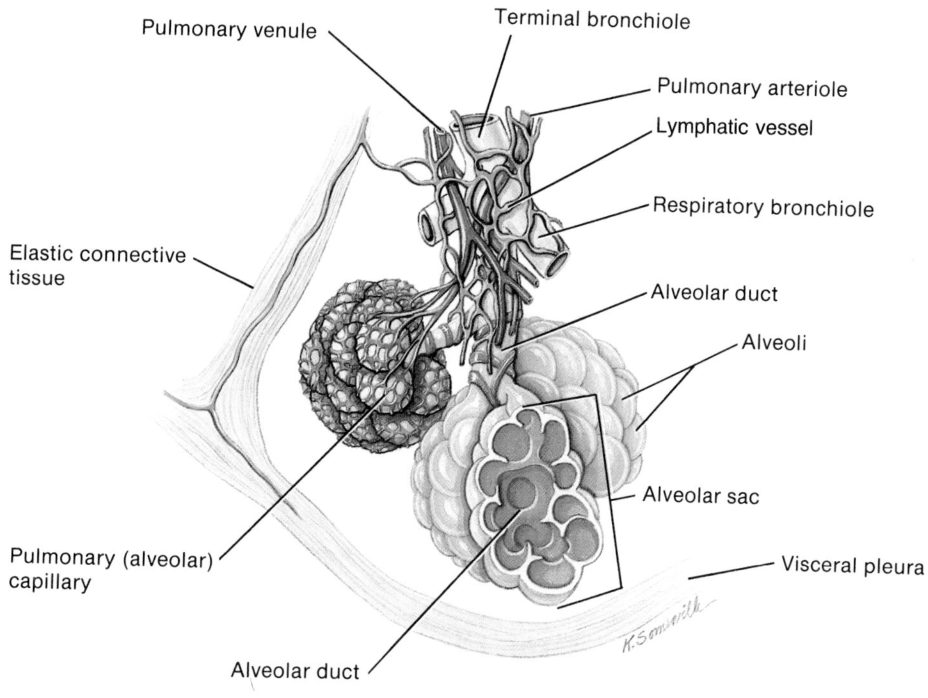

(a) Portion of a lobule of the lung

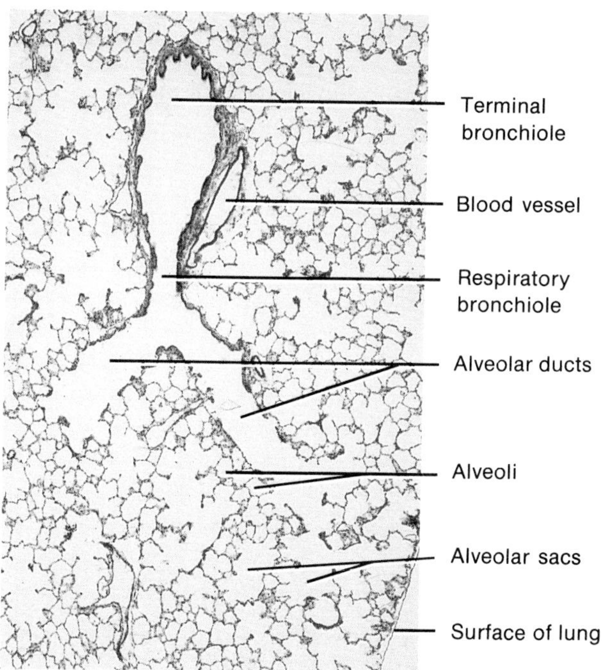

(b) Photomicrograph of the lung (approx. 30 x)

FIGURE 22-9 Histology of the lungs. (b) © Biophoto, SPL, Photo Researchers.

supported by a thin elastic basement membrane. ***Alveolar sacs*** are two or more alveoli that share a common opening (Figure 22-9). The alveolar walls consist of several types of epithelial cells (Figure 22-10a,b). ***Type I alveolar (squamous pulmonary epithelial) cells*** form a continuous lining of the alveolar wall, except for occasional type II cells. ***Type II alveolar (septal) cells*** are dispersed among the squamous pulmonary epithelial cells. Type II alveolar cells produce a phospholipid substance called ***surfactant*** (sur-FAK-tant), which lowers surface tension. Free ***alveolar macrophages (dust cells)*** are highly phagocytic cells that remove dust particles or other debris from alveolar spaces. Also found are monocytes, white blood cells that become transformed into alveolar macrophages, and fibroblasts. Also present between lining cells of the alveolar wall are reticular and elastic fibers. Deep to the layer of type I alveolar cells is an elastic basement membrane. Over the alveoli, the arteriole and venule disperse into a capillary network. The walls of the blood capillaries consist of a single layer of endothelial cells and basement membrane.

CLINICAL APPLICATION

Nebulization

Many respiratory disorders are treated by means of **nebulization** (neb-yoo-li-ZĀ-shun). This procedure consists of administering medication in the form of droplets that are suspended in air to select areas of the respiratory tract. The patient inhales the medication as a fine mist. Nebulization therapy can be used with many different types of drugs, such as chemicals that relax the smooth

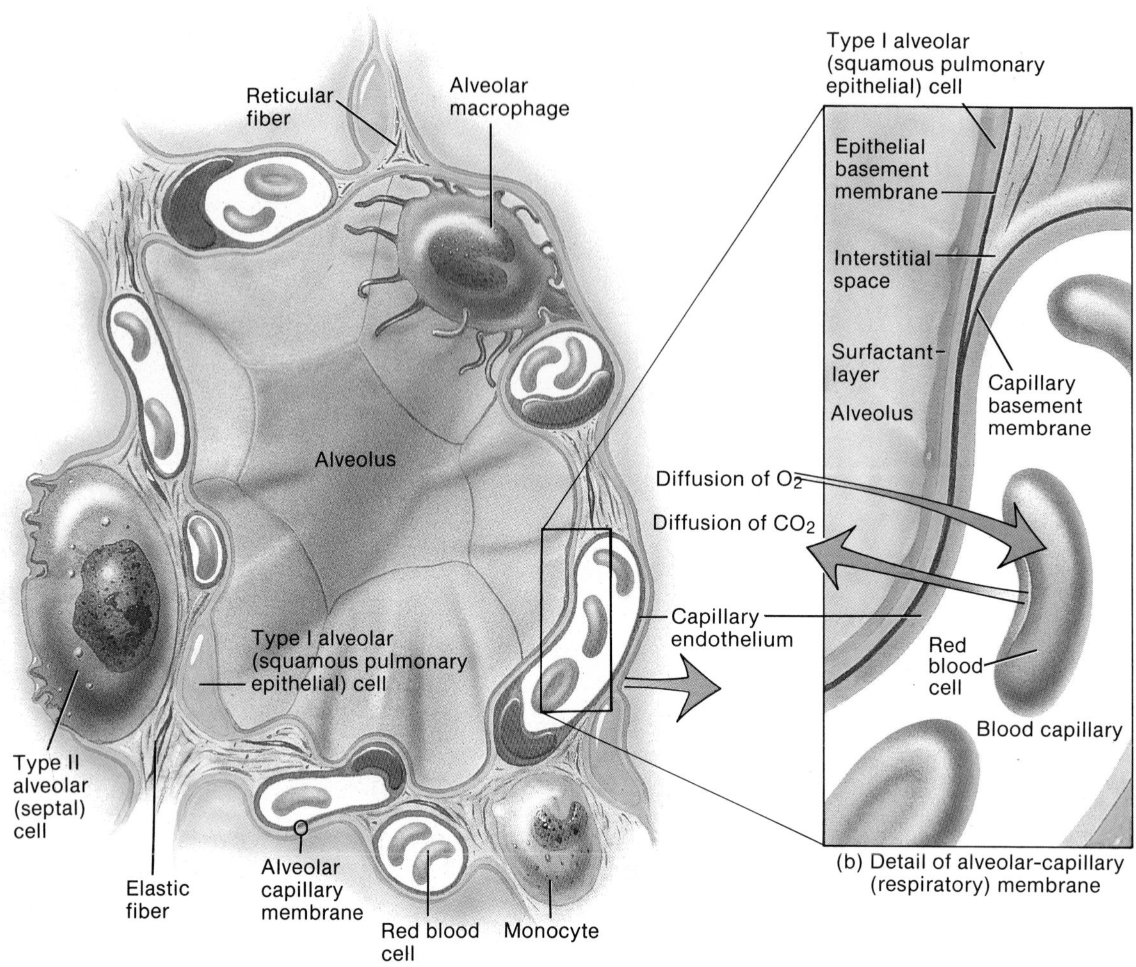

FIGURE 22-10 Alveolar–capillary membrane. (Courtesy of Richard K. Kessel and Randy H. Kardon, *Tissues and Organs: A Text-Atlas of Scanning Electron Microscopy.* Copyright © 1979 by Scientific American, Inc.)

muscle of the respiratory passageways, chemicals that reduce the thickness of mucus, and antibiotics.

Alveolar–Capillary (Respiratory) Membrane

The exchange of respiratory gases between the lungs and blood takes place by diffusion across alveolar and capillary walls. This membrane, through which the respiratory gases move, is collectively known as the ***alveolar–capillary (respiratory) membrane*** (Figure 22-10b). It consists of:

1. A layer of type I alveolar cells, type II alveolar cells, and free alveolar macrophages that constitute the ***alveolar (epithelial) wall.***
2. An ***epithelial basement membrane*** underneath the alveolar wall.
3. A ***capillary basement membrane*** that is often fused to the epithelial basement membrane.
4. The ***endothelial cells*** of the capillary.

Despite the several layers, the alveolar–capillary membrane averages only 0.5 μm in thickness, about 1⁄16 the diameter of a red blood cell. This is of considerable importance to the rapid diffusion of respiratory gases. Moreover, it has been estimated that the lungs contain 30 million alveoli, providing an immense surface area of 70 m^2 (753 ft^2) for the exchange of gases.

Blood and Nerve Supply

There is a double blood supply to the lungs. Deoxygenated blood passes through the pulmonary trunk, which divides

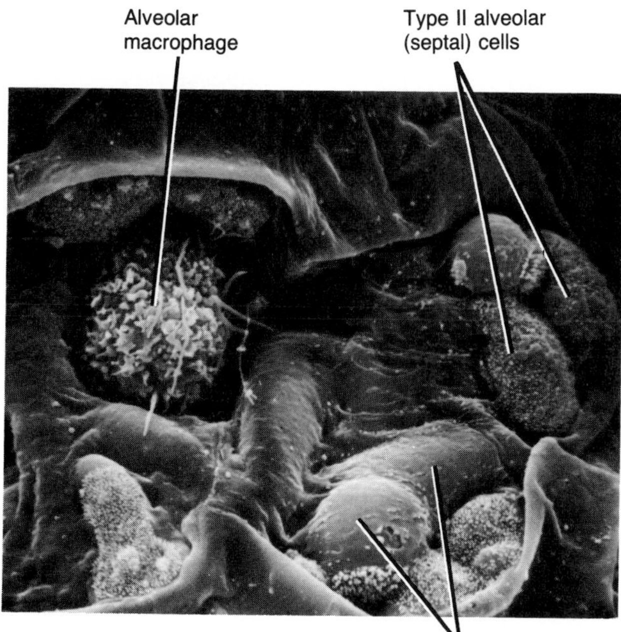

(c) Scanning electron micrograph of an alveolus, 3,430×

into a left pulmonary artery that enters the left lung and a right pulmonary artery that enters the right lung. The venous return of the oxygenated blood is by way of the pulmonary veins, typically two in number on each side—the right and left superior and inferior pulmonary veins. All four veins drain into the left atrium (see Figure 14-17).

Oxygenated blood is delivered through bronchial arteries, direct branches of the aorta. There are communications between the two systems, and most blood returns via pulmonary veins. Some blood, however, drains into bronchial veins, branches of the azygos system.

The nerve supply of the lungs is derived from the pulmonary plexus, located anterior and posterior to the roots of the lungs. The pulmonary plexus is formed by branches of the vagus (X) nerves and sympathetic trunks. Efferent parasympathetic fibers arise from the dorsal nucleus of the vagus (X) nerve, whereas efferent sympathetic fibers are postganglionic fibers of the second to fifth thoracic paravertebral ganglia of the sympathetic trunk.

AGING AND THE RESPIRATORY SYSTEM

The cardiovascular and respiratory systems operate as a unit, and damage or disease in one of these organ systems is often secondarily reflected in the other. In pulmonary heart disease, the right side of the heart enlarges in response to certain lung diseases. This condition is known as *cor pulmonale* or *right ventricular hypertrophy*. With advancing age, the airways and tissues of the respiratory tract, including the air sacs, become less elastic and more rigid. In addition to the lungs becoming less elastic, the chest wall also becomes more rigid. As a result, there is a decrease in pulmonary lung capacity. In fact, vital capacity (the amount of air moved by maximal inspiration followed by maximal expiration) can decrease as much as 35 percent by age 70. Also, there is a decrease in blood levels of oxygen, decreased activity of alveolar macrophages, and diminished ciliary action of the epithelium lining the respiratory tract. Owing to all these age-related factors, elderly people are more susceptible to pneumonia, bronchitis, emphysema, and other pulmonary disorders.

DEVELOPMENTAL ANATOMY OF THE RESPIRATORY SYSTEM

The development of the mouth and pharynx is considered in Chapter 23, in which the digestive system is discussed. Here we will consider the remainder of the respiratory system.

At about 4 weeks of fetal development, the respiratory system begins as an outgrowth of the ***endoderm*** of the foregut (precursor of some digestive organs) just behind the pharynx. This outgrowth is called the ***laryngotracheal bud*** (see Figure 21-17a). As the bud grows, it elongates and differentiates into the future *larynx* and other structures as well. Its proximal end maintains a slitlike opening into the pharynx called the *glottis*. The middle portion of the bud gives rise to the *trachea*. The distal portion divides into two ***lung buds,*** which grow into the *bronchi* and *lungs* (Figure 22-11).

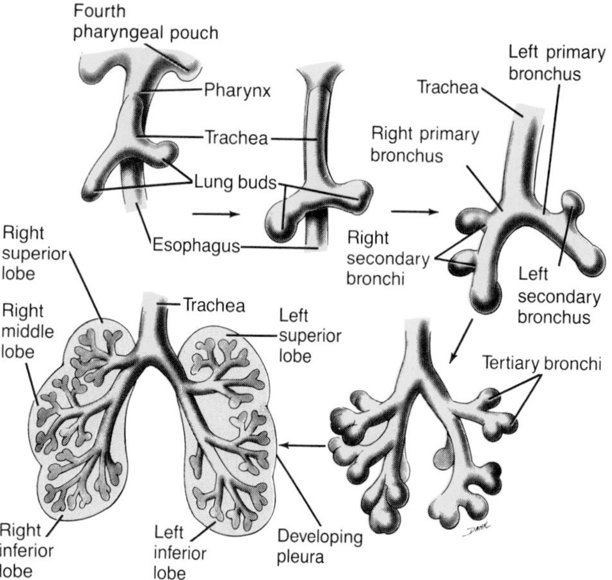

FIGURE 22-11 Development of the bronchial tubes and lungs.

As the lung buds develop, they branch and rebranch and give rise to all the *bronchial tubes*. After the sixth month, the closed terminal portions of the tubes dilate and become the *alveoli* of the lungs. The smooth muscle, cartilage, and connective tissues of the bronchial tubes and the pleural sacs of the lungs are contributed by ***mesenchymal (mesodermal) cells.***

APPLICATIONS TO HEALTH

BRONCHOGENIC CARCINOMA (LUNG CANCER)

A common lung cancer, ***bronchogenic carcinoma,*** starts in the walls of the bronchi. The constant irritation by inhaled smoke and pollutants causes the goblet cells of the bronchial epithelium to enlarge. They respond by secreting excessive mucus. The basal cells also respond to the stress by undergoing cell division so fast that they push into the area occupied by the goblet and columnar cells. Many researchers believe that if the stress is removed at this point, the epithelium can return to normal.

If, however, the stress persists, more and more mucus is secreted and the cilia become less effective. As a result, mucus is not carried toward the throat but remains trapped in the bronchial tubes. The individual then develops a "smoker's cough." Moreover, the constant irritation from the pollutant slowly destroys the alveoli, which are replaced with thick, inelastic connective tissue. Alveoli are destroyed by protein-digesting enzymes produced by leukocytes and macrophages in response to the stress. Mucus that has accumulated becomes trapped in the air sacs. Millions of sacs rupture, reducing the diffusion surface for the exchange of oxygen and carbon dioxide. The individual has now developed emphysema. If the stress is removed at this point, there is little chance for improvement. Alveolar tissue that has been destroyed cannot be repaired. But removal of the stress can stop further destruction of lung tissue.

If the stress still continues, the emphysema gets progressively worse, and the basal cells of the bronchial tubes continue to divide and break through the basement membrane. At this point the stage is set for bronchogenic carcinoma. Columnar and goblet cells disappear and may be replaced with squamous cancer cells. If this happens, the malignant growth spreads through the lung and may block a bronchial tube. If the obstruction occurs in a large bronchial tube, very little oxygen enters the lung, and disease-producing bacteria thrive on the mucoid secretions, bringing a host of infectious diseases. Treatment involves surgical removal of the diseased lung. However, metastasis (spreading) of the growth through the lymphatic or cardiovascular system may result in new growths in other parts of the body, particularly the brain and liver.

Other factors may be associated with lung cancer. For instance, breast, stomach, and prostate malignancies can metastasize to the lungs. People who apparently have not been exposed to pollutants do occasionally develop bronchogenic carcinoma. However, the occurrence of bronchogenic carcinoma is probably more than 20 times as high in cigarette smokers as it is in nonsmokers.

BRONCHIAL ASTHMA

Bronchial asthma is a reaction, usually allergic, characterized by attacks of wheezing and difficult breathing. Attacks are brought on by spasms of the smooth muscles in the walls of the smaller bronchi and bronchioles, causing the passageways to close partially. The patient has trouble exhaling, and the alveoli may remain inflated during expiration. Usually the mucous membranes that line the respiratory passageways become irritated and secrete excessive amounts of mucus that clog the bronchi and bronchioles and worsen the attack.

BRONCHITIS

Bronchitis is inflammation of the bronchi characterized by enlargement of glands and goblet cells lining the bronchial airways. The typical symptom is a productive cough in which a thick greenish-yellow sputum is raised. This secretion signifies the presence of the underlying infection that is causing excessive secretion of mucus. Cigarette smoking remains the leading cause of chronic bronchitis—that is, bronchitis that lasts for at least three months of the year for two successive years.

EMPHYSEMA

In ***emphysema*** (em′-fi-SĒ-ma), the alveolar walls lose their elasticity and remain filled with air during expiration. Reduced forced expiratory volume is the first symptom. As increasing numbers of alveoli are damaged, the lungs become permanently inflated because they have lost elasticity. To adjust to the increased lung size, the size of the chest cage increases, resulting in a "barrel chest." The patient has to work voluntarily to exhale. Oxygen diffusion across the damaged alveolar–capillary membrane is reduced, blood O_2 is somewhat lowered, and any mild exercise that raises the oxygen requirements of cells leaves the patient breathless.

As the disease progresses, the alveoli are replaced with thick fibrous connective tissue. Even carbon dioxide does not diffuse easily through this fibrous tissue. High carbon dioxide levels produce acid conditions that are toxic to brain cells. Consequently, the inspiratory area becomes less active and the respiration rate slows down, further aggravating the problem. The compressed and damaged capillaries around the deteriorating alveoli may no longer be able to receive blood. Resistance to blood flow increases in the pulmonary trunk and the right ventricle overworks as it attempts to force blood through the remaining capillaries.

Emphysema is generally caused by a long-term irritation. Cigarette smoke, air pollution, and occupational exposure to industrial dust are the most common irritants.

Diseases such as bronchial asthma, bronchitis, and emphysema have in common some degree of obstruction of the air passageways. The general term ***chronic obstructive pulmonary disease (COPD)*** is used to refer to these disorders. Among the symptoms that might indicate significant airflow obstruction are cough, wheezing, and labored breathing.

PNEUMONIA

Pneumonia refers to an acute infection or inflammation of the alveoli. The alveolar sacs fill up with fluid and dead white blood cells, reducing the amount of air space in the lungs. (One of the cardinal signs of inflammation is edema.) Oxygen has difficulty diffusing through the inflamed alveoli, and the blood O_2 may be drastically reduced. Blood CO_2 usually remains normal because carbon dioxide diffuses through the alveoli more easily than oxygen does.

TUBERCULOSIS (TB)

The bacterium *Mycobacterium tuberculosis* produces an infectious, communicable disease called ***tuberculosis (TB).*** TB most often affects the lungs and the pleurae. The bacteria destroy parts of the lung tissue, and the tissue is replaced by fibrous connective tissue. Because the connective tissue is inelastic and thick, the affected areas of the lungs do not snap back during expiration, and air is retained. Gases no longer diffuse easily through the fibrous tissue. Patients with AIDS have a higher-than-normal incidence of tuberculosis.

RESPIRATORY DISTRESS SYNDROME (RDS) OF THE NEWBORN

Respiratory distress syndrome (RDS) of the newborn is also called ***glassy lung disease*** or ***hyaline membrane disease (HMD).*** Before birth, the respiratory passages are filled with fluid. Part of this fluid is amniotic fluid inhaled during respiratory movements in the uterus. The remainder is produced by the glands and goblet cells of the respiratory epithelium.

At birth, this fluid-filled passageway must become an air-filled passageway, and the collapsed primitive alveoli must expand and function in gaseous exchange. The success of this transition depends largely on surfactant, the phospholipid produced by type II alveolar cells that lowers surface tension. The presence of surfactant can be detected by amniocentesis.

In the newborn whose lungs are deficient in surfactant, the effort required for the first breath is essentially the same as that required in normal newborns. However, during expiration after the first inspiration, the surface tension of the alveoli increases as the alveoli deflate and the alveoli collapse almost to their original uninflated state.

Idiopathic RDS of the newborn usually appears within a few hours after birth. (An *idiopathic* condition is one that occurs spontaneously in an individual from an unknown or obscure cause.) Affected infants show difficult and labored breathing. Death may occur soon after the onset of respiratory difficulty or may be delayed for a few days, although many infants survive. RDS of the newborn occurs frequently in premature infants and also in infants of diabetic mothers, particularly if the diabetes is untreated or poorly controlled.

RESPIRATORY FAILURE

Respiratory failure refers to a condition in which the respiratory system cannot supply sufficient oxygen to maintain metabolism or cannot eliminate enough carbon dioxide to prevent respiratory acidosis (excess acid). Respiratory failure always causes dysfunction in other organs as well.

Among the causes of respiratory failure are lung disorders; mechanical disorders that disturb the chest wall or neuromusculature; depression of the respiratory center by drugs, strokes, or trauma; and carbon monoxide poisoning.

Symptoms include disorientation, malaise, headache, muscular weakness, difficulty with sleep, breathlessness, coughing, rapid pulse, irregular heartbeat, cyanosis, hypertension, edema, and coma. Treatment consists of providing adequate oxygenation and reversing the respiratory acidosis.

SUDDEN INFANT DEATH SYNDROME (SIDS)

Sudden infant death syndrome (SIDS), also called crib death, kills more infants between the ages of 1 week and 12 months than any other disease. SIDS occurs without warning. Although about half of its victims had an upper respiratory infection within two weeks of death, the babies tended to be otherwise remarkably healthy. Two other findings are important: the apparently *silent* nature of death and the *disarray* frequently found at the scene.

Given this death scene, it is not surprising that for centuries "crib death" was mistakenly thought to be due to suffocation. Most medical people, however, believe that a healthy child *cannot* smother in its bedclothes. The diaper of a SIDS victim is usually full of stool and urine, and the urinary bladder is almost always empty. External examination reveals only a blood-tinged froth that often exudes from the nostrils. These signs all point to a sudden lethal episode associated with motor activity, not to a gradually deepening coma leading to death.

CORYZA (COMMON COLD) AND INFLUENZA (FLU)

Several viruses are responsible for ***coryza (common cold).*** A group of viruses, called rhinoviruses, is responsible for about 40 percent of all colds in adults. Typical symptoms include sneezing, excessive nasal secretion, dry cough, and congestion. The uncomplicated common cold is not usually accompanied by a fever. Complications include sinusitis, asthma, bronchitis, ear infections, and laryngitis. Direct transfer of nasal secretions on hands accounts for nearly all transmission of viruses that cause coryza, not sneezing. Since common colds are caused by viruses, antibiotics are of no use in treatment. Although over-the-counter drugs (aspirin, antihistamines, and decongestants) may lessen the severity of certain symptoms of the common cold, they do not lessen recovery time.

Influenza (flu) is also caused by a virus. Its symptoms include chills, fever (usually higher than 101°F), headache, and muscular aches. Coldlike symptoms appear as the fever subsides.

PULMONARY EMBOLISM

Pulmonary embolism refers to the presence of a blood clot or other foreign substance in a pulmonary arterial vessel that obstructs circulation to lung tissue. The immediate effect is complete or partial obstruction of blood flow, resulting in dysfunction of the affected lung tissue. A large embolus can produce death in a few minutes.

PULMONARY EDEMA

Pulmonary edema refers to an abnormal accumulation of interstitial fluid in the interstitial spaces and alveoli of the lungs. The edema may arise from increased pulmonary capillary permeability (pulmonary origin) or increased pulmonary capillary pressure (cardiac origin). This latter cause may coincide with congestive heart failure. The most common symptom is dyspnea. Others include wheezing, tachypnea (rapid respirations), restlessness, feeling of suffocation, cyanosis, pallor (paleness), and diaphoresis (excessive respiration).

CARBON MONOXIDE (CO) POISONING

Carbon monoxide (CO) is a colorless and odorless gas found in exhaust fumes from automobiles and in tobacco smoke. It is a by-product of burning carbon-containing material such as coal and wood. It combines with hemoglobin very much as oxygen does, except that the combination of carbon monoxide and hemoglobin is more than 200 times as tenacious as the combination of oxygen and hemoglobin. In addition, in concentrations as small as 0.1 percent, carbon monoxide will combine with half the hemoglobin molecules. Thus, the oxygen-carrying capacity of the blood is reduced by one-half. Increased levels of carbon monoxide lead to hypoxia, and the result is ***carbon monoxide poisoning.*** The condition may be treated by administering pure oxygen, which slowly replaces the carbon monoxide combined with the hemoglobin.

SMOKE INHALATION INJURY

When smoke is inhaled, the lungs are injured directly by heat from flames and substances in fumes. ***Smoke inhalation injury*** has three components that occur in sequence: (1) inhibition of oxygen delivery and utilization, (2) upper airway injury from heat, and (3) lung damage from acids and aldehydes in smoke.

KEY MEDICAL TERMS ASSOCIATED WITH THE RESPIRATORY SYSTEM

Apnea (AP-nē-a; *a* = without; *pnoia* = breath) Absence of ventilatory movements.

Asphyxia (as-FIK-sē-a; *sphyxis* = pulse) Oxygen starvation due to low atmospheric oxygen or interference with ventilation, external respiration, or internal respiration.

Aspiration (as′-pi-RĀ-shun; *spirare* = breathe) Inhalation of a foreign substance such as water, food, or foreign body into the bronchial tree; drawing of a substance in or out by suction.

Atelectasis (at′-e-LEK-ta-sis; *ateles* = incomplete; *ektasis* = dilation) A collapsed lung or portion of a lung.

Bronchiectasis (bron′-kē-EK-ta-sis; *ektasis* = dilation) A chronic dilation of the bronchi or bronchioles.

Cardiopulmonary resuscitation (kar′-dē-ō-PUL-mo-ner-ē re-sus′-i-TĀ-shun) **(CPR)** The artificial establishment of normal or near-normal respiration and circulation. The **ABC's** of cardiopulmonary resuscitation are **Airway, Breathing,** and **Circulation,** meaning the rescuer must establish an airway, provide artificial ventilation if breathing has stopped, and reestablish circulation if there is inadequate cardiac action. The procedure should be performed in that order (A, B, C).

Cheyne-Stokes respiration (CHĀN STŌKS res′-pi-RĀ-shun) A repeated cycle of irregular breathing beginning with shallow breaths that increase in depth and rapidity, then decrease and cease altogether for 15 to 20 seconds. Cheyne-Stokes is normal in infants. It is also often seen just before death from pulmonary, cerebral, cardiac, and kidney disease.

Diphtheria (dif-THĒ-rē-a; *diphthera* = membrane) An acute bacterial infection that causes the mucous membranes of the oropharynx, nasopharynx, and larynx to enlarge and become leathery. Enlarged membranes may obstruct airways and cause death from asphyxiation.

Dyspnea (DISP-nē-a; *dys* = painful, difficult; *pnoia* = breath) Painful or labored breathing.

Heimlich (abdominal thrust) maneuver (HĪM-lik ma-NOO-ver) First-aid procedure designed to clear the air passageways of obstructing objects. It is performed by applying a quick upward thrust that causes sudden elevation of the diaphragm and forceful, rapid expulsion of air in the lungs; this action forces air out of the trachea to eject the obstructing object.

The Heimlich maneuver is also used to expel water from the lungs of near-drowning victims before resuscitation is begun.

Hemoptysis (hē-MOP-ti-sis; *hemo* = blood; *ptein* = to spit) Spitting of blood from the respiratory tract.

Hypoxia (hī-POK-sē-a; *hypo* = below, under) Reduction in oxygen supply to cells.

Orthopnea (or′-thop-NĒ-a; *ortho* = straight) Dyspnea that occurs in the horizontal position. It is an abnormal finding because a normal individual can tolerate the reduction in vital capacity that accompanies the recumbent position.

Pneumonectomy (noo′-mō-NEK-tō-mē; *pneumo* = lung; *tome* = cutting) Surgical removal of a lung.

Rales (RĀLS) Sounds sometimes heard in the lungs that resemble bubbling or rattling. Rales are to the lungs what murmurs are to the heart. There are many types of rales, but each is due to the presence of an abnormal amount or type of fluid or mucus inside the bronchi or alveoli, or to bronchoconstriction so that air cannot enter or leave the lungs normally.

Respirator (RES-pi-rā′-tor) An apparatus fitted to a mask over the nose and mouth, or hooked directly to an endotracheal or tracheotomy tube, that is used to assist or support ventilation or to provide nebulized medication to the air passages under pressure.

Rhinitis (rī-NĪ-tis; *rhino* = nose) Chronic or acute inflammation of the mucous membrane of the nose.

Tachypnea (tak′-ip-NĒ-a; *tachy* = rapid; *pnoia* = breath) Rapid breathing.

STUDY OUTLINE

Organs (p. 630)

1. Respiratory organs include the nose, pharynx, larynx, trachea, bronchi, and lungs.
2. They act with the cardiovascular system to supply oxygen and remove carbon dioxide from the blood.

Nose (p. 630)

1. The external portion is made of cartilage and skin and is lined with mucous membrane. Openings to the exterior are the external nares.
2. The internal portion communicates with the nasopharynx through the internal nares and with the paranasal sinuses.
3. The nasal cavity is divided by a septum and consists of three regions: vestibule, respiratory area, and olfactory area.
4. The nose is adapted for warming, moistening, and filtering air, for olfaction, and for speech.

Pharynx (p. 632)

1. The pharynx (throat) is a muscular tube lined by a mucous membrane.
2. The anatomic regions are nasopharynx, oropharynx, and laryngopharynx.
3. The nasopharynx functions in respiration. The oropharynx and laryngopharynx function both in digestion and in respiration.

Larynx (p. 634)

1. The larynx (voice box) is a passageway that connects the pharynx with the trachea.
2. It contains the thyroid cartilage (Adam's apple); the epiglottis, which prevents food from entering the larynx; the cricoid cartilage, which connects the larynx and trachea; and the paired arytenoid, corniculate, and cuneiform cartilages.
3. The larynx contains vocal folds, which produce sound. Taut folds produce high pitches, and relaxed ones produce low pitches.

Trachea (p. 635)

1. The trachea (windpipe) extends from the larynx to the primary bronchi.
2. It is composed of smooth muscle and C-shaped rings of cartilage and is lined with pseudostratified columnar epithelium.
3. Two methods of bypassing obstructions from the respiratory passageways are tracheostomy and intubation.

Bronchi (p. 638)

1. The bronchial tree consists of the trachea, primary bronchi, secondary bronchi, tertiary bronchi, bronchioles, and terminal bronchioles. Walls of bronchi contain incomplete rings of cartilage; walls of bronchioles contain smooth muscle.
2. A bronchogram is an x-ray of the tree after introduction of an opaque contrast medium usually containing iodine.

Lungs (p. 639)

1. Lungs are paired organs in the thoracic cavity. They are enclosed by the pleural membrane. The parietal pleura is the outer layer; the visceral pleura is the inner layer.
2. The right lung has three lobes separated by two fissures; the left lung has two lobes separated by one fissure and a depression, the cardiac notch.
3. The secondary bronchi give rise to branches called segmental bronchi, which supply segments of lung tissue called bronchopulmonary segments.
4. Each bronchopulmonary segment consists of lobules, which contain lymphatic vessels, arterioles, venules, terminal bronchioles, respiratory bronchioles, alveolar ducts, alveolar sacs, and alveoli.
5. Gas exchange occurs across the alveolar–capillary (respiratory) membranes.

Aging and the Respiratory System (p. 645)

1. Aging results in decreased vital capacity, decreased blood level of oxygen, and diminished alveolar macrophage activity.
2. Elderly people are more susceptible to pneumonia, emphysema, bronchitis, and other pulmonary disorders.

Developmental Anatomy of the Respiratory System (p. 645)

1. The respiratory system begins as an outgrowth of endoderm called the laryngotracheal bud.

2. Smooth muscle, cartilage, and connective tissue of the bronchial tubes and pleural sacs develop from mesoderm.

Applications to Health (p. 646)

1. In bronchogenic carcinoma, bronchial epithelial cells are replaced by cancer cells after constant irritation has disrupted the normal growth, division, and function of the epithelial cells.
2. Bronchial asthma is characterized by spasms of smooth muscle in bronchial tubes that result in partial closure of air passageways; inflammation; inflated alveoli; and excess mucus production.
3. Bronchitis is an inflammation of the bronchial tubes.
4. Emphysema is characterized by deterioration of alveoli leading to loss of their elasticity. Symptoms are reduced expiratory volume, inflated lungs at the end of expiration, and enlarged chest.
5. Pneumonia is an acute inflammation or infection of alveoli.
6. Tuberculosis is an inflammation of pleura and lungs produced by the organism *Mycobacterium tuberculosis*.
7. Respiratory distress syndrome (RDS) of the newborn is an infant disorder in which surfactant is lacking and alveolar ducts and alveoli have a glassy appearance.
8. In respiratory failure the respiratory system cannot supply sufficient oxygen or eliminate sufficient carbon dioxide.
9. Sudden infant death syndrome (SIDS) appears to be associated with a sudden lethal episode of motor activity.
10. Coryza (common cold) is caused by viruses and is usually not accompanied by a fever, whereas influenza (flu) is usually accompanied by a fever greater than 101°F.
11. Pulmonary embolism refers to a blood clot or other foreign substance in a pulmonary arterial vessel that obstructs blood flow to lung tissue.
12. Pulmonary edema is an accumulation of interstitial fluid in interstitial spaces and alveoli of the lungs.
13. Carbon monoxide poisoning occurs when carbon monoxide combines with hemoglobin instead of oxygen. The result is hypoxia.
14. Smoke inhalation injury has three sequential components: (a) inhibition of oxygen delivery and utilization, (b) upper airway injury from heat, and (c) lung damage from acids and aldehydes in smoke.

REVIEW QUESTIONS

1. What organs make up the respiratory system? What functions do the respiratory and cardiovascular systems have in common? (p. 629)
2. Describe the structure of the external and internal nose, and describe their functions in filtering, warming, and moistening air. (p. 630)
3. What is the pharynx? Differentiate the three anatomical regions of the pharynx and indicate their roles in respiration. (p. 632)
4. Describe the structure of the larynx and explain how it functions in respiration and voice production. (p. 634)
5. Describe the location and structure of the trachea. (p. 635)
6. What is the bronchial tree? Describe its structure. What is a bronchogram? (p. 638)
7. Where are the lungs located? Distinguish the parietal pleura from the visceral pleura. (p. 639)
8. Define each of the following parts of a lung: base, apex, costal surface, medial surface, hilus, root, cardiac notch, and lobe. (p. 640)
9. What is a lobule of the lung? Describe its composition and function in respiration. (p. 641)
10. What is a bronchopulmonary segment? (p. 641)
11. Describe the histology and function of the alveolar–capillary (respiratory) membrane. (p. 644)
12. Describe the effects of aging on the respiratory system. (p. 645)
13. Describe the development of the respiratory system. (p. 645)
14. For each of the following disorders, list the principal clinical symptoms: bronchogenic carcinoma (lung cancer), bronchial asthma, bronchitis, emphysema, pneumonia, tuberculosis (TB), respiratory distress syndrome (RDS) of the newborn, respiratory failure, sudden infant death syndrome (SIDS), coryza (common cold), influenza (flu), pulmonary embolism, pulmonary edema, carbon monoxide (CO) poisoning, and smoke inhalation injury. (p. 646)
15. Refer to the key medical terms associated with the respiratory system. Be sure that you can define each term. (p. 648)

SELF QUIZ

Complete the following:

1. The broad, inferior portion of the lung that sits on the diaphragm is called the ________. The upper narrow apex of each lung extends just superior to the ________. The costal surfaces lie against the ________.
2. Arrange in order the structures through which air passes as it enters a lobule en route to alveoli. ________ ________ ________

   A. alveolar ducts; B. respiratory bronchiole; C. terminal bronchiole.
3. Answer these questions with right or left. Which lung is thicker and broader? ________. In which lung is the cardiac notch located? ________. Which lung has just two lobes and hence only two lobar bronchi? ________. Which lung has a horizontal fissure? ________.
4. The laryngotracheal bud is derived from ________-derm.
5. Match the following:

___	**a.** nasal conchae and meatuses are located here	**A.** larynx
		B. laryngopharynx

___ **b.** enlarged adenoids here may block openings into auditory (Eustachian) tubes, causing middle ear infections
___ **c.** palatine tonsils are located here
___ **d.** this structure leads directly into the esophagus
___ **e.** cricoid, epiglottis, and thyroid cartilages are here
___ **f.** vocal cords here enable voice production
___ **g.** internal nares (choanae) are located between these two structures (two answers)
___ **h.** fauces are located between these two structures (two answers)

C. mouth
D. nose
E. nasopharynx
F. oropharynx

6. Arrange the answers in correct sequence.

___ ___ ___ **a.** From most superficial to deepest:
A. parietal pleura
B. visceral pleura
C. pleural cavity

___ ___ ___ ___ ___ **b.** From superior to inferior:
A. bronchioles
B. bronchi
C. larynx
D. pharynx
E. trachea

___ ___ ___ ___ ___ **c.** Pathway of inspired air:
A. external nares
B. internal nares
C. meatuses
D. nasopharynx
E. vestibule

___ ___ ___ ___ ___ ___ **d.** Pathway of expired air:
A. alveolar ducts
B. bronchioles
C. secondary bronchi
D. primary bronchi
E. tertiary bronchi
F. alveoli

Choose the one best answer to these questions.

___ **7.** The mucous membranes of the nose
A. moisten, warm, and filter the inhaled air; B. contain only alveoli; C. contain only capillaries; D. are not involved in the sense of smell; E. clean, cool, and dry the inhaled air.

___ **8.** The nasal septum
(1) has as one of its parts the vomer.
(2) includes part of the ethmoid bone, the perpendicular palate.
(3) is partly cartilage—the septal cartilage.
(4) includes the nasal conchae.
A. (1) only; B. (2) only; C. (3) only; D. (4) only; E. (1), (2), and (3).

___ **9.** Which of the following is/are true regarding the pharynx?
(1) It serves as a common passage for both the respiratory and digestive systems.
(2) Adenoids, or pharyngeal tonsils, are located in the lower part, the laryngopharynx.
(3) The nasal cavity connects with it through two internal nares, into the nasopharynx.
A. (1) only: B. (2) only; C. (3) only; D. (1) and (3); E. all of the above.

___ **10.** The auditory (Eustachian) tube opens into the lateral wall of the
A. inner ear; B. nasopharynx; C. oropharynx; D. laryngopharynx; E. nasal cavity.

___ **11.** The flaplike structure that reflexly prevents aspiration of solid and liquid substances into the trachea
(1) contains a cartilage within it.
(2) is called the epiglottis.
(3) is called the larynx.
(4) is called the thyroid cartilage.
A. (1) only; B. (2) only; C. (3) only; D. (4) only; E. (1) and (2).

___ **12.** The cartilage ''rings'' in tracheal walls
(1) are made of tough fibrocartilage.
(2) are complete rings of cartilage just like those of the bronchi.
(3) keep the lumen open.
A. (1) only; B. (2) only; C. (3) only; D. all of the above; E. none of the above.

___ **13.** The pleurae
A. consist of a single layer of tissue; B. line the inside of the lungs; C. are separated by a layer of cells that secrete mucus; D. are separated by a thin film of pleural fluid; E. fill the pleural cavity.

___ **14.** Which is not a component of the conduction portion of the respiratory system?
A. pharynx; B. larynx; C. bronchi; D. trachea; E. alveoli

___ **15.** Of the structures listed, the smallest in diameter would be the
A. left primary bronchus; B. respiratory bronchioles; C. secondary bronchi; D. alveolar ducts; E. right primary bronchus.

___ **16.** Each section of lung tissue supplied by a tertiary bronchus is called a
A. lobe; B. lobule; C. bronchopulmonary segment; D. hilus; E. fissure.

Circle T (true) or F (false) for the following.

T F 17. Phagocytic cells in the alveoli are known as alveolar macrophages.

T F 18. The opening from the oral cavity into the oropharynx is called the fauces.

T F 19. The medial concavity of the left lung that receives the heart is the pulmonary depression.

T F 20. The cricoid cartilages attach the epiglottis to the glottis.

T F 21. Voice production is a function of the ventricular folds.

The Digestive System

23

STUDENT OBJECTIVES

1. Define digestion and identify the organs of the gastrointestinal (GI) tract and the accessory organs of digestion.
2. Describe the structure of the wall of the gastrointestinal tract.
3. Define the mesentery, mesocolon, falciform ligament, lesser omentum, and greater omentum as extensions of the peritoneum.
4. Describe the structure of the tongue.
5. Identify the location and histology of the salivary glands.
6. Identify the parts of a typical tooth and compare deciduous and permanent dentitions.
7. Describe the anatomy, histology, and functions of the esophagus, stomach, pancreas, liver, gallbladder (GB), small intestine, and large intestine.
8. Describe the effects of aging on the digestive system.
9. Describe the development of the digestive system.
10. Describe the clinical symptoms of the following disorders: dental caries, periodontal disease, peritonitis, peptic ulcers, appendicitis, gastrointestinal tumors, diverticulitis, cystic fibrosis (CF), cirrhosis, hepatitis, gallstones, obesity, anorexia nervosa, and bulimia.
11. Define key medical terms associated with the digestive system.

CHAPTER OUTLINE

- **Digestive Processes**
- **Organization**

General Histology
Mucosa
Submucosa
Muscularis
Serosa
Peritoneum

- **Mouth (Oral Cavity)**

Tongue
Salivary Glands
Teeth
Dental Terminology
Dentitions
Blood and Nerve Supply

- **Pharynx**
- **Esophagus**

Histology
Activities
Blood and Nerve Supply

- **Stomach**

Anatomy
Histology
Activities
Blood and Nerve Supply

- **Pancreas**

Anatomy
Histology
Activities
Blood and Nerve Supply

- **Liver**

Anatomy
Histology
Activities
Blood and Nerve Supply

- **Gallbladder (GB)**

Histology
Activities
Blood and Nerve Supply

- **Small Intestine**

Anatomy
Histology
Activities
Blood and Nerve Supply

- **Large Intestine**

Anatomy
Histology
Activities
Blood and Nerve Supply

- **Aging and the Digestive System**
- **Developmental Anatomy of the Digestive System**
- **Applications to Health**
- **Key Medical Terms Associated with the Digestive System**

Food is vital for life because it is the source of energy that drives the chemical reactions occurring in every cell and provides matter that is used to form new tissue or to repair damaged tissue. Energy is needed for muscle contraction, conduction of nerve impulses, and secretory and absorptive activities of many cells. Food as it is consumed, however, is not in a state suitable for use as an energy source by any cell. The food must be broken down into molecules small enough to be transported through the plasma (cell) membranes. The breaking down of larger food molecules into molecules small enough for use by body cells is called ***digestion,*** and the organs that collectively perform this function compose the ***digestive system.***

The medical specialty that deals with the structure, function, diagnosis, and treatment of diseases of the stomach and intestines is called ***gastroenterology*** (gas′-trō-en′-ter-OL-ō-jē; *gastro* = stomach; *enteron* = intestines).

The developmental anatomy of the digestive system is considered later in the chapter.

DIGESTIVE PROCESSES

The digestive system prepares food for consumption by the cells through five basic activities.

1. ***Ingestion.*** Taking food into the body (eating).
2. ***Movement of food.*** Passage of food along the gastrointestinal tract.
3. ***Digestion.*** The breakdown of food by both chemical and mechanical processes.
4. ***Absorption.*** The passage of digested food from the gastrointestinal tract into the cardiovascular and lymphatic systems for distribution to cells.
5. ***Defecation.*** The elimination of indigestible substances from the gastrointestinal tract.

Chemical digestion is a series of catabolic (hydrolysis) reactions that use enzymes to break down the large carbohydrate, lipid, and protein molecules that we eat into smaller molecules that are absorbable and usable by body cells. (Enzymes are substances that speed up chemical reactions without themselves being changed.) The products of digestion are small enough to pass through the epithelial cells of the wall of the gastrointestinal tract, into the blood and lymph capillaries, and eventually into the body's cells. ***Mechanical digestion*** consists of various movements of the gastrointestinal tract that aid chemical digestion. Food is prepared by the teeth before it can be swallowed. Then the smooth muscles of the stomach and small intestine churn the food so it is thoroughly mixed with enzymes that digest foods.

ORGANIZATION

The organs of digestion are traditionally divided into two main groups. First is the ***gastrointestinal (GI) tract,*** or ***alimentary canal,*** a continuous tube running through the ventral body cavity and extending from the mouth to the anus (Figure 23-1). The relationship of the digestive organs to the nine regions of the abdominopelvic cavity may be reviewed in Figure 1-8b. The length of a tract taken from a cadaver is about 9 m (30 ft). In a living person it is somewhat shorter because the muscles in its wall are in a state of tone. Organs composing the gastrointestinal tract include the mouth, pharynx, esophagus, stomach, small intestine, and large intestine. The GI tract contains the food from the time it is eaten until it is digested and prepared for elimination. Muscular contractions in the wall of the GI tract break down the food physically by churning it. Secretions produced by cells along the tract break down the food chemically.

The second group of organs composing the digestive system consists of the ***accessory structures***—the teeth, tongue, salivary glands, liver, gallbladder, and pancreas. Teeth protrude into the GI tract and aid in the physical breakdown of food. The tongue assists in mastication (chewing) and deglutition (swallowing). The other accessory structures, except for the tongue, lie totally outside the tract and produce or store secretions that aid in the chemical breakdown of food. These secretions are released into the tract through ducts.

GENERAL HISTOLOGY

The wall of the GI tract, especially from the esophagus to the anal canal, has the same basic arrangement of tissues. The four layers or tunics of the tract from the inside out are the mucosa, submucosa, muscularis, and serosa or adventitia (Figure 23-2).

Mucosa

The ***mucosa,*** or inner lining of the tract, is a mucous membrane. Three layers compose the membrane in the GI tract: (1) a lining layer of ***epithelium*** in direct contact with the contents of the GI tract, (2) an underlying layer of areolar connective tissue called the ***lamina propria,*** and (3) a thin layer of smooth muscle called the ***muscularis mucosae.***

The epithelial layer is composed of nonkeratinized cells that are stratified in the mouth, oropharynx, laryngopharynx, and esophagus but are simple throughout the rest of the tract. The functions of the stratified epithelium are protection and secretion. The functions of the simple epithelium are secretion and absorption.

The lamina propria is made of areolar connective tissue containing many blood and lymph vessels and scattered lymphatic nodules, masses of lymphatic tissue that are not encapsulated. This layer supports the epithelium, binds it to the muscularis mucosae, and provides it with a blood and lymph supply. The blood and lymph vessels are the avenues by which nutrients in the tract reach the other tissues of the body. The lymphatic tissue also protects against

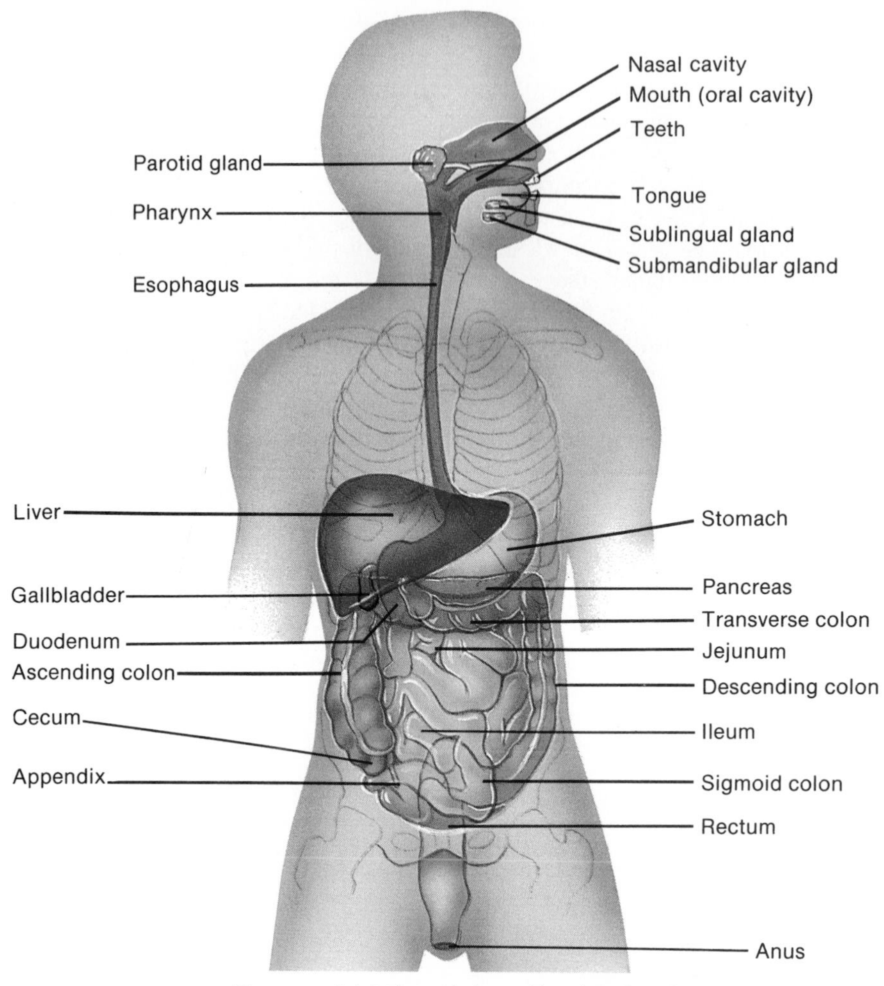

FIGURE 23-1 Organs of the digestive system and related structures.

disease. Remember that the GI tract is in contact with the outside environment and contains food that often carries harmful bacteria. Unlike the skin, the mucous membrane of the tract is not protected from bacterial entry by keratin.

The muscularis mucosae contains smooth muscle fibers (cells) arranged in two layers (longitudinal and circular) that throw the mucous membrane of the intestine into small folds that increase the digestive and absorptive area. With one exception, which will be described later, the other three coats of the intestine contain no glandular epithelium.

Submucosa

The ***submucosa*** consists of dense connective tissue that binds the mucosa to the third tunic, the muscularis. It is highly vascular and contains a portion of the ***submucosal plexus* (*plexus of Meissner*)**, which is part of the autonomic nerve supply to the muscularis mucosae. This plexus is also important in controlling secretions by the GI tract.

Muscularis

The ***muscularis*** of the mouth, pharynx, and upper esophagus consists in part of skeletal muscle that produces voluntary swallowing. Throughout the rest of the tract, the muscularis consists of smooth muscle that is generally found in two sheets: an inner sheet of circular fibers and an outer sheet of longitudinal fibers. Involuntary contractions of the smooth muscles help break down food physically, mix it with digestive secretions, and propel it through the tract. The muscularis also contains the major nerve supply to the gastrointestinal tract—the ***myenteric plexus* (*plexus of Auerbach*)**, which consists of fibers from both autonomic divisions. This plexus mostly controls GI motility.

Serosa

The ***serosa*** is the outermost layer of most portions of the GI tract. It is a serous membrane composed of connective

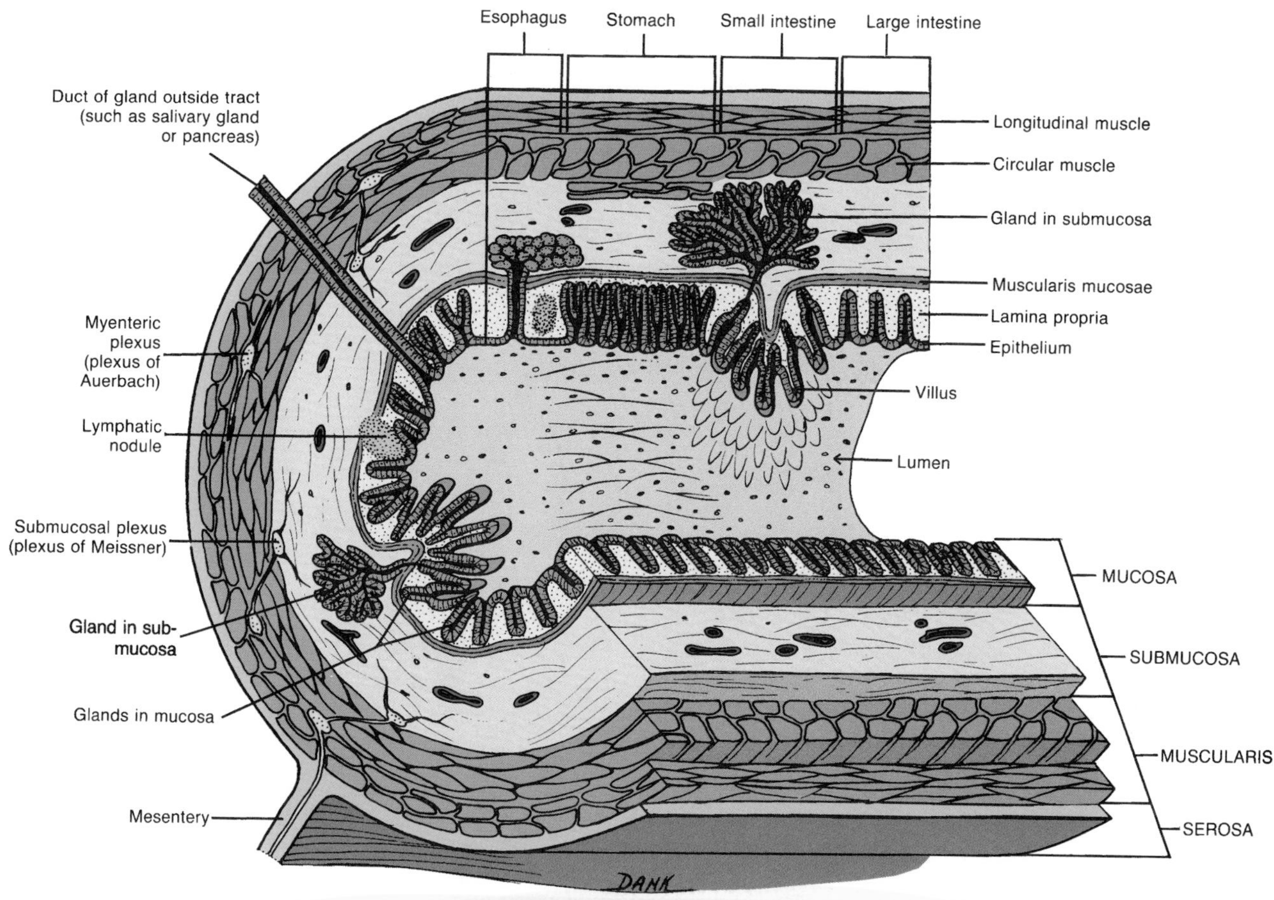

FIGURE 23-2 Composite of various sections of the gastrointestinal tract seen in a three-dimensional drawing depicting the various layers and related structures.

tissue and epithelium. This layer, below the diaphragm, is also called the ***visceral peritoneum*** and forms a portion of the peritoneum, which we shall now describe in detail.

PERITONEUM

The ***peritoneum*** (per′-i-tō-NĒ-um; *peri* = around; *tonos* = tension) is the largest serous membrane of the body. Serous membranes are also associated with the heart (pericardium) and lungs (pleurae). Serous membranes consist of a layer of simple squamous epithelium (called mesothelium) and an underlying supporting layer of connective tissue. The ***parietal peritoneum*** lines the wall of the abdominal cavity. The ***visceral peritoneum*** covers some of the organs and constitutes their serosa. The potential space between the parietal and visceral portions of the peritoneum is called the ***peritoneal cavity*** and contains serous fluid. In certain diseases, the peritoneal cavity may become distended by several liters of fluid so that it forms an actual space. Such an accumulation of serous fluid is called ***ascites*** (a-SĪ-tēz). As you will see later, some organs lie on the posterior abdominal wall and are covered by peritoneum on their anterior surfaces only. Such organs, including the pancreas, duodenum, ascending colon, descending colon, and kidneys, are said to be ***retroperitoneal.***

Unlike the pericardium and pleurae, the peritoneum contains large folds that weave between the viscera. The folds bind the organs to each other and to the walls of the cavity and contain the blood and lymphatic vessels and the nerves that supply the abdominal organs. One extension of the peritoneum is called the ***mesentery*** (MEZ-en-ter′-ē; *meso* = middle; *enteron* = intestine). It is an outward fold of the serous coat of the small intestine (Figure 23-3). The tip of the fold is attached to the posterior abdominal wall. The mesentery binds the small intestine to the wall. A similar fold of parietal peritoneum, called the ***mesocolon*** (mez′-ō-KŌ-lon), binds the large intestine to the posterior body wall. It also carries blood and lymphatic vessels to the intestines.

protects

Other important peritoneal folds are the falciform ligament, lesser omentum, and greater omentum. The ***falciform*** (FAL-si-form) ***ligament*** attaches the liver to the anterior

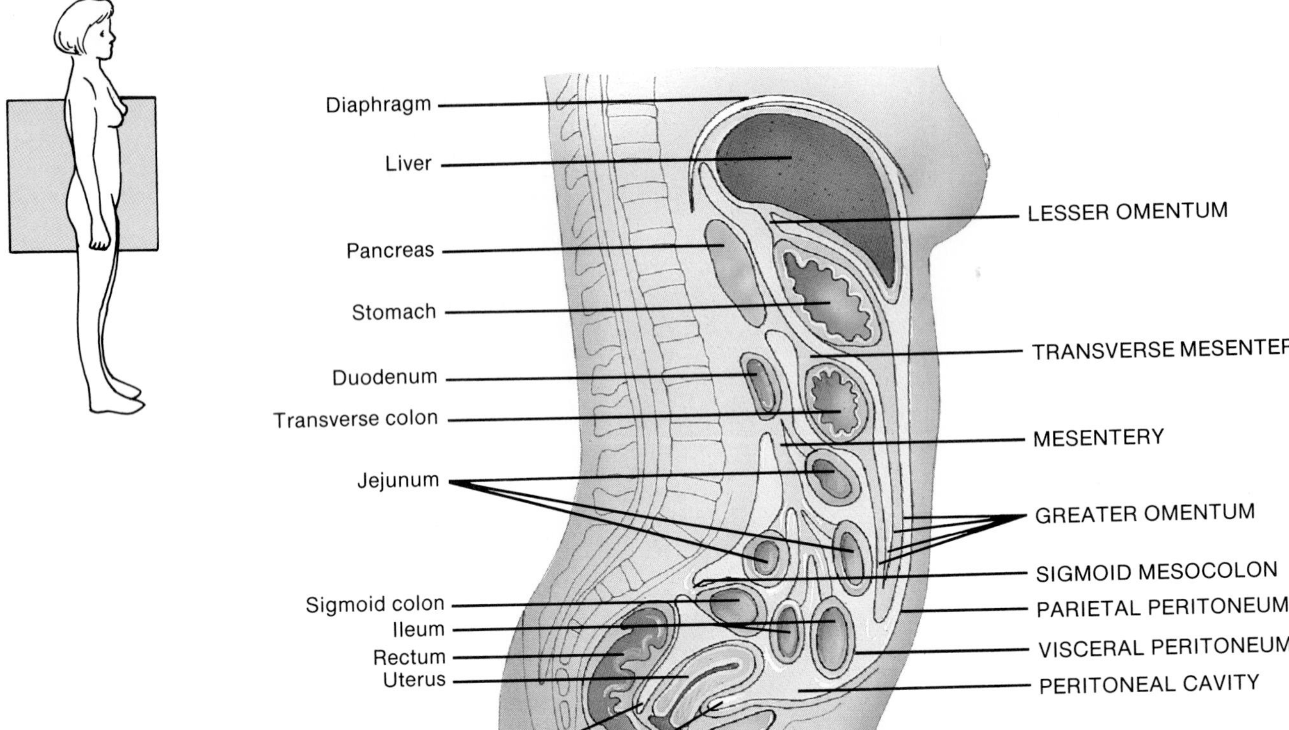

FIGURE 23-3 Relationship of the peritoneal extensions to each other.

abdominal wall and diaphragm. The ***lesser omentum*** (ō-MENT-um) arises as two folds in the serosa of the stomach and duodenum suspending the stomach and duodenum from the liver. The ***greater omentum*** is the largest peritoneal fold that drapes over the transverse colon and coils of the small intestine. It is a double sheet that folds upon itself and thus is a four-layered structure. It is attached along the stomach and duodenum, passes downward over the small intestine for a variable distance, and then turns upward to the transverse colon where it is attached to it. Because the greater omentum contains large quantities of adipose tissue, it commonly is called the "fatty apron." The greater omentum contains numerous lymph nodes. If an infection occurs in the intestine, plasma cells formed in the lymph nodes combat the infection and help prevent it from spreading to the peritoneum.

MOUTH (ORAL CAVITY)

The ***mouth,*** also referred to as the ***oral*** or ***buccal*** (BUK-al) ***cavity,*** is formed by the cheeks, hard and soft palates, and tongue (Figure 23-4). Forming the lateral walls of the oral cavity are the ***cheeks***—muscular structures covered on the outside by skin and lined by nonkeratinized stratified squamous epithelium. The anterior portions of the cheeks terminate in the superior and inferior lips.

The ***lips (labia)*** are fleshy folds surrounding the orifice of the mouth. They are covered on the outside by skin and on the inside by a mucous membrane. The transition zone where the two kinds of covering tissue meet is called the ***vermilion*** (ver-MIL-yon). This portion of the lips is nonkeratinized, and the color of the blood in the underlying

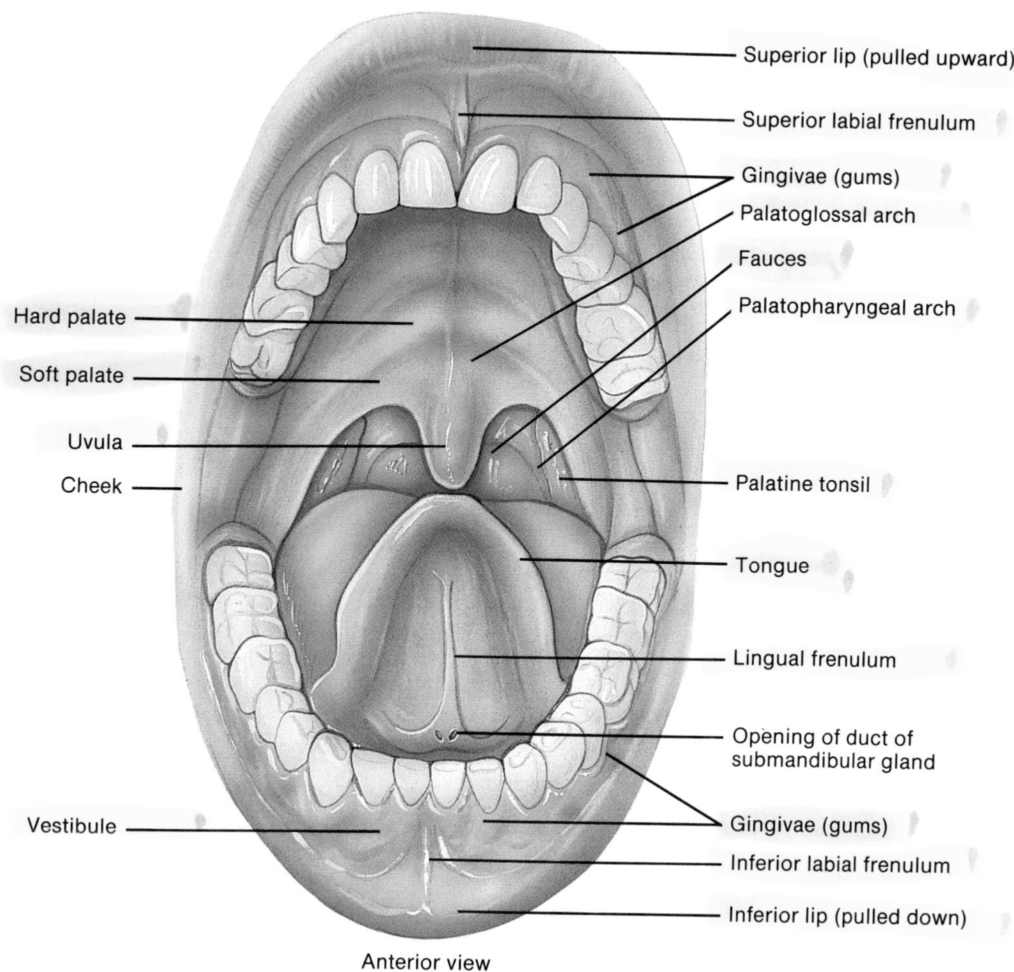

FIGURE 23-4 Structures of the mouth (oral cavity).

blood vessels is visible through the transparent surface layer of the vermilion. The inner surface of each lip is attached to its corresponding gum by a midline fold of mucous membrane called the ***labial frenulum*** (LĀ-bē-al FREN-yoo-lum).

The orbicularis oris muscle and connective tissue lie between the external integumentary covering and the internal mucosal lining. During chewing, the cheeks and lips help keep food between the upper and lower teeth. They also assist in speech.

The ***vestibule*** of the oral cavity is bounded externally by the cheeks and lips and internally by the gums and teeth. The ***oral cavity proper*** extends from the vestibule to the ***fauces*** (FAW-sēs; *fauces* = passages), the opening between the oral cavity and the pharynx or throat.

The ***hard palate,*** the anterior portion of the roof of the mouth, is formed by the maxillae and palatine bones, is covered by mucous membrane, and forms a bony partition between the oral and nasal cavities. The ***soft palate*** forms the posterior portion of the roof of the mouth. It is an arch-shaped muscular partition between the oropharynx and nasopharynx and is lined by mucous membrane.

Hanging from the free border of the soft palate is a conical muscular process called the ***uvula*** (YOU-vyoo-la). On either side of the base of the uvula are two muscular folds that run down the lateral side of the soft palate. Anteriorly, the ***palatoglossal arch (anterior pillar)*** extends inferiorly, laterally, and anteriorly to the side of the base of the tongue. Posteriorly, the ***palatopharyngeal*** (PAL-a-tō-fa-rin′-jē-al) ***arch (posterior pillar)*** projects inferiorly, laterally, and posteriorly to the side of the pharynx. The palatine tonsils are situated between the arches, and the lingual tonsil is situated at the base of the tongue. At the posterior border of the soft palate, the mouth opens into the oropharynx through the fauces (Figure 23-4).

TONGUE

The ***tongue,*** together with its associated muscles, forms the floor of the oral cavity. It is an accessory structure of the digestive system composed of skeletal muscle covered with mucous membrane (see Figure 20-11b). The tongue is divided into symmetrical lateral halves by a median septum that extends throughout its entire length and is attached inferiorly to the hyoid bone. Each half of the tongue consists of an identical complement of extrinsic and intrinsic muscles.

The ***extrinsic muscles*** of the tongue originate outside the tongue and insert into it. They include the hyoglossus, genioglossus, and styloglossus (see Figure 10-7). The extrinsic muscles move the tongue from side to side and in and out. These movements maneuver food for chewing, shape the food into a rounded mass, and force the food to the back of the mouth for swallowing. They also form the floor of the mouth and hold the tongue in position. The ***intrinsic muscles*** originate and insert within the tongue and alter the shape and size of the tongue for speech and swallowing. The intrinsic muscles include the longitudinalis superior, longitudinalis inferior, transversus linguae, and verticalis linguae (Figure 23-5a). The ***lingual frenulum,*** a fold of mucous membrane in the midline of the undersurface of the tongue, aids in limiting the movement of the tongue posteriorly.

CLINICAL APPLICATION

Ankyloglossia

If the lingual frenulum is too short, tongue movements are restricted, speech is faulty, and the person is said to be "tongue-tied." This congenital problem is referred to as ***ankyloglossia*** (ang'-ki-lō-GLOSS-ē-a). It can be corrected by cutting the lingual frenulum.

The upper surface and sides of the tongue are covered with ***papillae*** (pa-PIL-ē), projections of the lamina propria covered with epithelium (Figure 23-5b,c). ***Filiform papillae*** are conical projections distributed in parallel rows over the anterior two-thirds of the tongue. They are whitish and contain no taste buds. ***Fungiform papillae*** are mushroomlike elevations distributed among the filiform papillae and are more numerous near the tip of the tongue. They appear as red dots on the surface of the tongue, and most of them contain taste buds. ***Circumvallate papillae,*** 10 to 12 in number, are arranged in the form of an inverted V on the posterior surface of the tongue, and all of them contain taste buds. Note the taste zones of the tongue in Figure 20-11b.

SALIVARY GLANDS

Saliva is a fluid that is continuously secreted by glands associated with the mouth. Ordinarily, just enough saliva is secreted to keep the mucous membranes of the mouth and pharynx moist, but when food enters the mouth, secretion increases so the saliva can lubricate, dissolve, and begin the chemical breakdown of the food. The mucous membrane lining the cheeks contains many small glands, the ***buccal glands,*** that secrete small amounts of saliva. There are also minor salivary glands in the lips, tongue, and hard and soft palates that secrete small amounts of saliva. However, most saliva is secreted by the ***major salivary glands*** (hereafter referred to as simply the salivary glands), accessory structures that lie outside the mouth and pour their contents into ducts that empty into the oral cavity. There are three pairs of salivary glands: parotid, submandibular (submaxillary), and sublingual glands (Figure 23-6a).

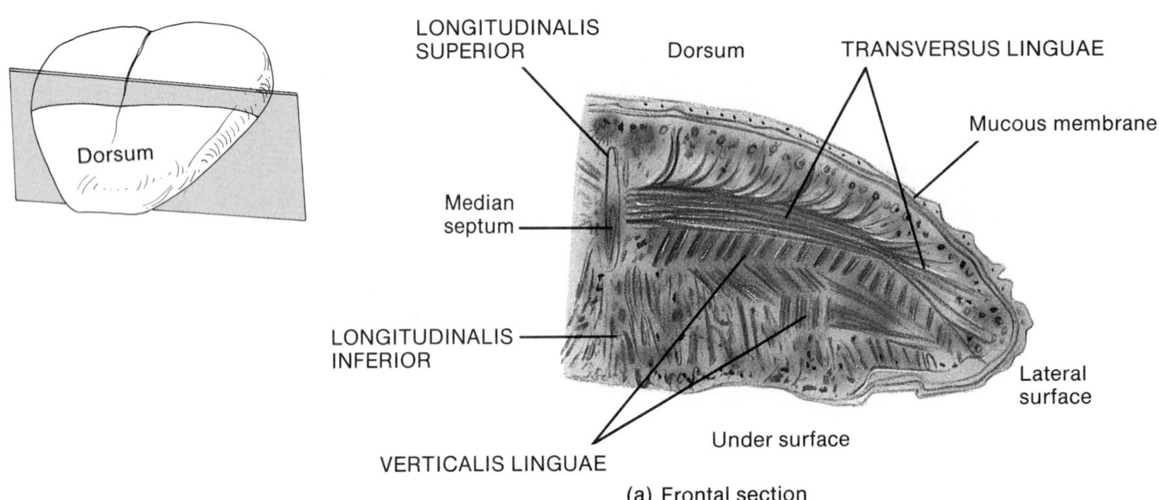

FIGURE 23-5 Tongue. (a) Left side of the tongue showing its intrinsic muscles. (b) and (c) Courtesy of Richard K. Kessel and Randy H. Kardon, *Tissues and Organs: A Text-Atlas of Scanning Electron Microscopy.* Copyright © 1979 by Scientific American, Inc.

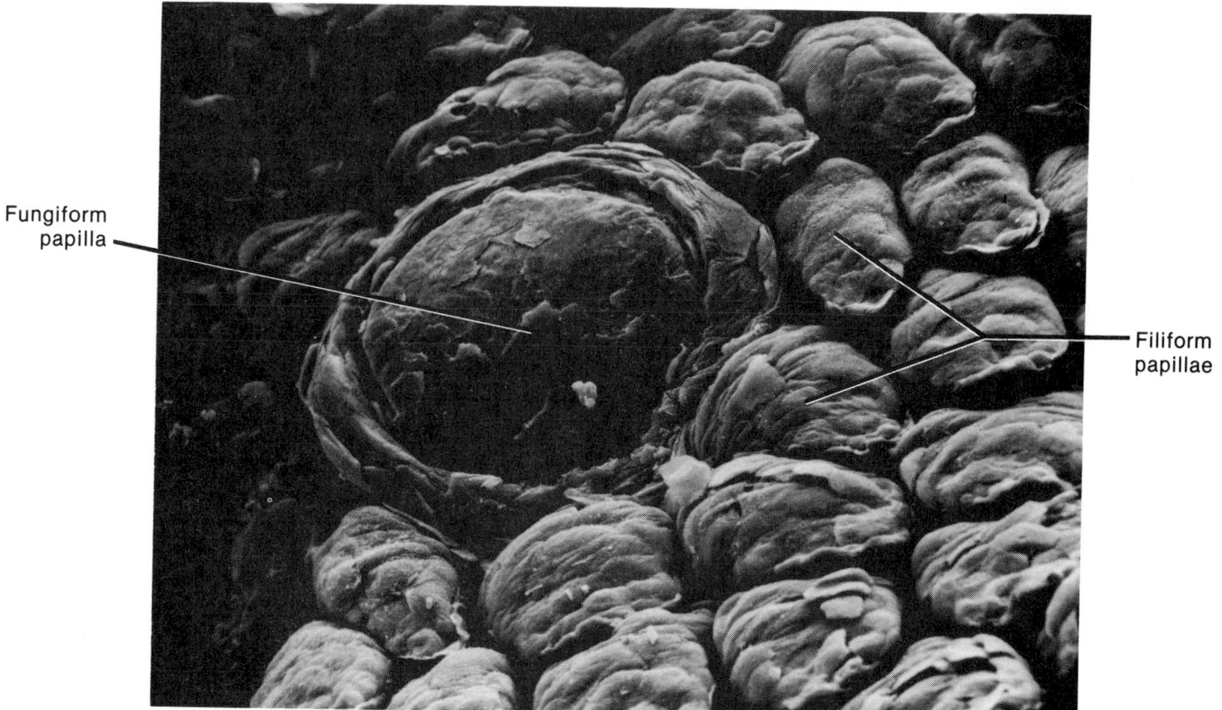

(b) Fungiform papilla surrounded by filiform papillae, 500×

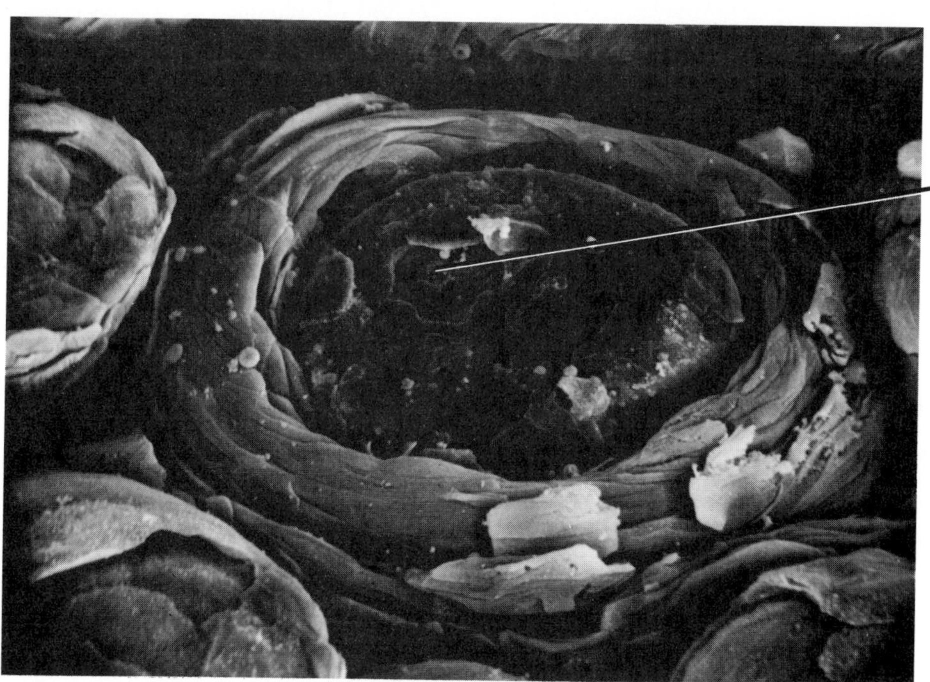

(c) Circumvallate papilla, 500×

The ***parotid glands*** are located inferior and anterior to the ears between the skin and the masseter muscle. They are compound tubuloacinar glands. Each secretes into the oral cavity vestibule via a duct, called the ***parotid (Stensen's) duct,*** that pierces the buccinator muscle to open into the vestibule opposite the upper second molar tooth. The ***submandibular glands,*** which are compound acinar glands, are found beneath the base of the tongue in the posterior part of the floor of the mouth (Figure 23-6b). Their ducts, the ***submandibular (Wharton's) ducts,*** run superficially under the mucosa on either side of the midline of the floor of the mouth and enter the oral cavity proper on either side of the lingual frenulum. The ***sublingual glands,*** also compound acinar glands, are superior to the submandibular

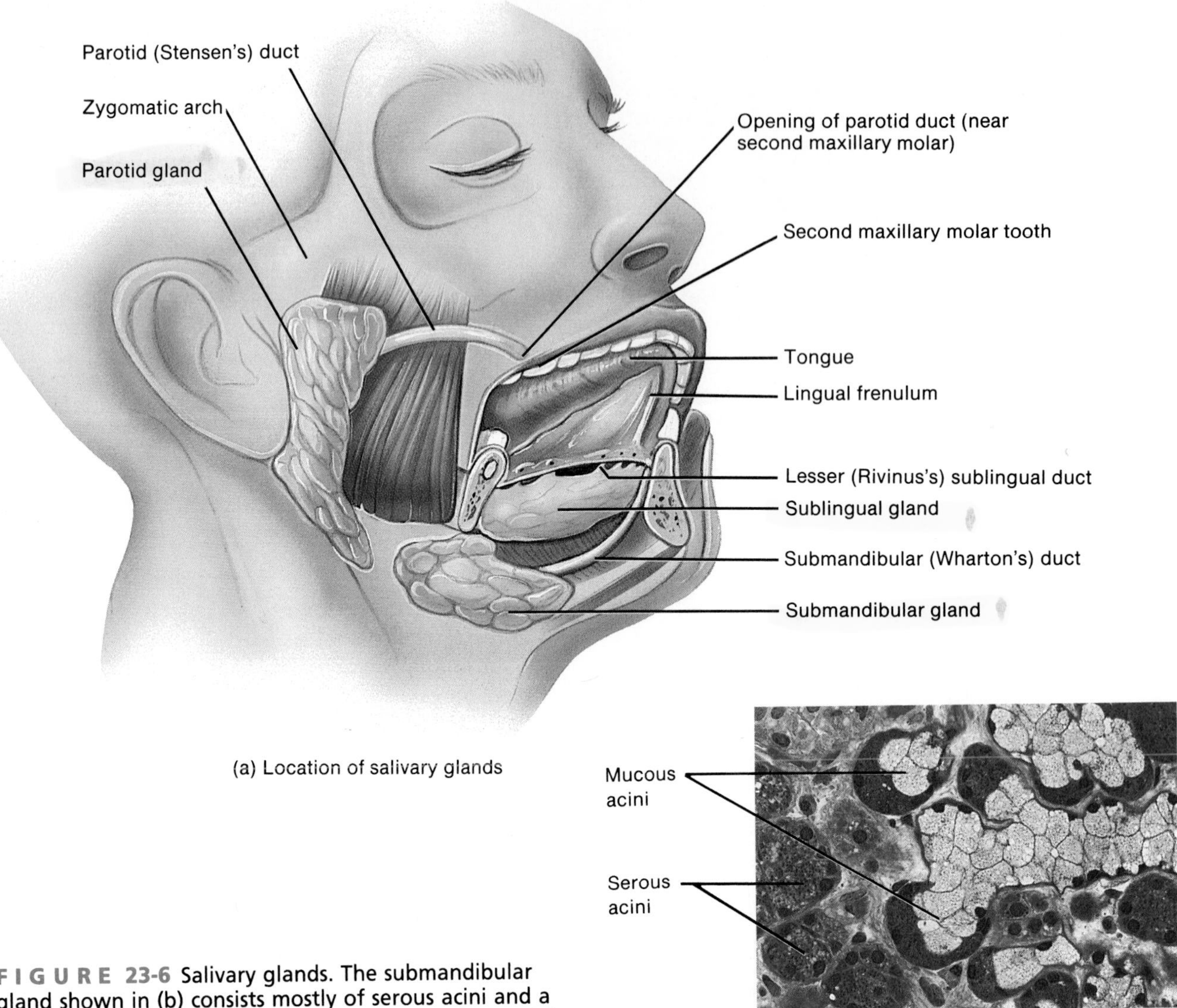

FIGURE 23-6 Salivary glands. The submandibular gland shown in (b) consists mostly of serous acini and a few mucous acini. The parotids consist of all serous acini, and the sublinguals consist of mostly mucous acini and a few serous acini. (Courtesy of Douglas Merrill.)

glands, and their ducts, the ***lesser sublingual (Rivinus's) ducts,*** open into the floor of the mouth in the oral cavity proper.

CLINICAL APPLICATION

Mumps

Although any of the salivary glands may become infected as a result of a nasopharyngeal infection, the parotids are typically the target of the mumps virus (myxovirus). ***Mumps*** is an inflammation and enlargement of the parotid glands accompanied by moderate fever, malaise, and extreme pain in the throat, especially when swallowing sour foods or acid juices. Swelling occurs on one or both sides of the face, just anterior to the ramus of the mandible. In about 20 to 35 percent of males past puberty, the testes may also become inflamed, and although it rarely occurs, sterility is a possible consequence. In some people, aseptic meningitis, pancreatitis, and hearing loss may also occur as complications.

The parotid gland receives its blood supply from branches of the external carotid artery and is drained by vessels that are tributaries of the external jugular vein. The submandibular gland is supplied by branches of the facial artery and drained by tributaries of the facial vein. The sublingual gland is supplied by the sublingual branch of the lingual artery and the submental branch of the facial artery and is drained by tributaries of the sublingual and submental veins.

The salivary glands receive both sympathetic and parasympathetic innervation. The sympathetic fibers form plexuses on the blood vessels that supply the glands and serve as vasoconstrictors. The parotid gland receives sympathetic fibers from the plexus on the external carotid artery, whereas the submandibular and sublingual glands receive sympathetic fibers that contribute to the sympathetic plexus and accompany the facial artery to the glands. The parasympathetic fibers of the glands consist of secretomotor fibers to the glands.

The fluids secreted by the buccal glands, minor salivary glands, and the three pairs of major salivary glands constitute ***saliva.*** Amounts of saliva secreted daily vary considerably but range from 1000 to 1500 ml. Chemically, saliva is 99.5 percent water and 0.5 percent solutes. Among the solutes are salts, dissolved gases, and various organic substances, including urea and uric acid, serum albumin and globulin, mucin, the bacteriolytic enzyme lysozyme, and the digestive enzyme salivary amylase, are also present. ***Salivary amylase*** initiates the breakdown of starch. Saliva continues to be secreted heavily some time after food is swallowed. This flow of saliva washes out the mouth and dilutes and buffers the chemical remnants of irritating substances.

TEETH

The ***teeth (dentes)*** are accessory structures of the digestive system located in sockets of the alveolar processes of the mandible and maxillae. The alveolar processes are covered by the ***gingivae*** (jin-JI-vē), or gums, which extend slightly into each socket forming the gingival sulcus (Figure 23-7). The sockets are lined by the ***periodontal ligament,*** which consists of dense fibrous connective tissue and is attached to the socket walls and the cemental surface of the roots. Thus, it anchors the teeth in position and also acts as a shock absorber to dissipate the forces of chewing.

A typical tooth consists of three principal portions. The ***crown*** is the exposed portion above the level of the gums. The ***root*** consists of one to three projections embedded in

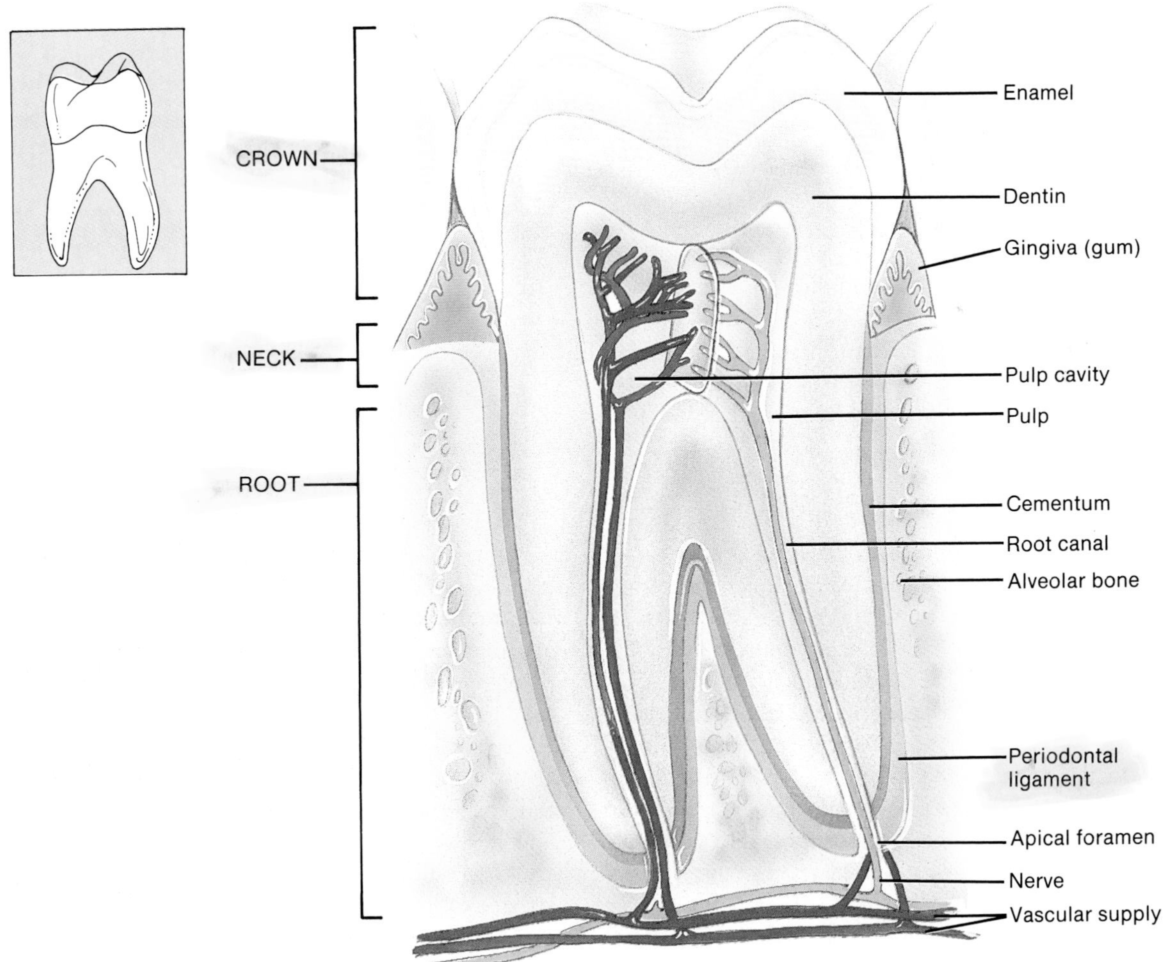

FIGURE 23-7 Parts of a typical tooth.

the socket. The ***neck*** is the constricted junction line of the crown and the root near the gumline.

Teeth are composed primarily of ***dentin,*** a calcified connective tissue that gives the tooth its basic shape and rigidity. The dentin encloses a cavity. The enlarged part of the cavity, the ***pulp cavity,*** lies in the crown and is filled with ***pulp,*** a connective tissue containing blood vessels, nerves, and lymphatic vessels. Narrow extensions of the pulp cavity run through the root of the tooth and are called ***root canals.*** Each root canal has an opening at its base, the ***apical foramen.*** Through the foramen enter blood vessels bearing nourishment, lymphatic vessels affording protection, and nerves providing sensation. The dentin of the crown is covered by ***enamel*** that consists primarily of calcium phosphate and calcium carbonate. Enamel is the hardest substance in the body and protects the tooth from the wear of chewing. It is also a barrier against acids that easily dissolve the dentin. The dentin of the root is covered by ***cementum,*** another bonelike substance, which attaches the root to the periodontal ligament.

The branch of dentistry that is concerned with the prevention, diagnosis, and treatment of diseases that affect the pulp, root, periodontal ligament, and alveolar bone is known as ***endodontics*** (en′-dō-DON-tiks; *endo* = within; *odous* = tooth).

CLINICAL APPLICATION

Root Canal Therapy

Root canal therapy refers to a procedure, accomplished in several phases, in which all traces of pulp tissue are removed from the pulp cavity and root canal of a badly diseased tooth. After a hole is made in the tooth, the root canal is filed out and irrigated to remove bacteria. Then the canal is treated with medication and sealed tightly. The damaged crown is then repaired.

Dental Terminology

Because of the arrangement of the teeth in the mouth, it is necessary to use terms other than anterior, posterior, medial, and lateral in describing the surfaces of the teeth. Accordingly, the following directional terms are used. ***Labial*** refers to the surface of a tooth in contact with or directed toward the lips. ***Buccal*** refers to the surface in contact with or directed toward the cheeks. ***Lingual*** is restricted to the teeth of the lower jaw and refers to the surface directed toward the tongue. ***Palatal,*** on the other hand, is restricted to the teeth of the upper jaw and refers to the surface directed toward the palate. The term ***mesial*** designates the anterior or medial side of the tooth relative to its position in the dental arch. ***Distal*** refers to the posterior or lateral side of the tooth relative to its position in the dental arch. Essentially, mesial and distal refer to the sides of the adjacent teeth that are in contact with each other. Finally, ***occlusal*** refers to the biting surface of a tooth.

Dentitions

Humans have two ***dentitions,*** or sets of teeth, a phenomenon called ***diphyodonty*** (dif-i-ō-DONT-ē; *phyein* = to produce; *odous* = tooth). The first of these—the ***deciduous (primary) teeth, milk teeth,*** or ***baby teeth***—begin to erupt at about 6 months of age, and one pair appears at about each month thereafter until all 20 are present. Figure 23-8a illustrates the deciduous teeth. The incisors, which are closest to the midline, are chisel-shaped and adapted for cutting into food. They are referred to as either ***central*** or ***lateral incisors*** on the basis of their position. Next to the incisors, moving posteriorly, are the ***cuspids (canines),*** which have a pointed surface called a cusp. Cuspids are used to tear and shred food. The incisors and cuspids have only one root apiece. Behind them lie the ***first*** and ***second molars,*** which have four cusps. Upper molars have three roots; lower molars have two roots. The molars crush and grind food.

All the deciduous teeth are lost—generally between 6 and 12 years of age—and are replaced by the ***permanent (secondary) dentition*** (Figure 23-8b). The permanent dentition contains 32 teeth that appear between age 6 and adulthood. It resembles the deciduous dentition with the following exceptions. The deciduous molars are replaced with the ***first*** and ***second premolars (bicuspids),*** which have two cusps and one root (upper first bicuspids have two roots) and are used for crushing and grinding. The permanent molars erupt into the mouth behind the bicuspids. They do not replace any deciduous teeth and erupt as the jaw grows to accommodate them—the ***first molars*** at age 6, the ***second molars*** at age 12, the ***third molars (wisdom teeth)*** after age 18. The human jaw has become smaller through time and often does not afford enough room behind the second molars for the eruption of the third molars. In this case, the third molars remain embedded in the alveolar bone and are said to be "impacted." Most often they cause pressure and pain and must be surgically removed. In some individuals, third molars may be dwarfed in size or may not develop at all.

Blood and Nerve Supply

The arteries that supply blood to the teeth are distributed to the pulp cavity and surrounding periodontal ligament. The upper incisors and cuspids are supplied by anterior superior alveolar branches of the maxillary artery; the upper premolars and molars are supplied by the posterior superior alveolar branches of the maxillary artery; the lower incisors and cuspids are supplied by the incisive branches of the inferior alveolar artery; and the lower premolars and molars are supplied by the dental branches of the inferior alveolar artery.

The teeth receive sensory fibers from branches of the maxillary and mandibular divisions of the trigeminal (V) nerve—the upper teeth from branches of the maxillary division and the lower teeth from branches of the mandibular division.

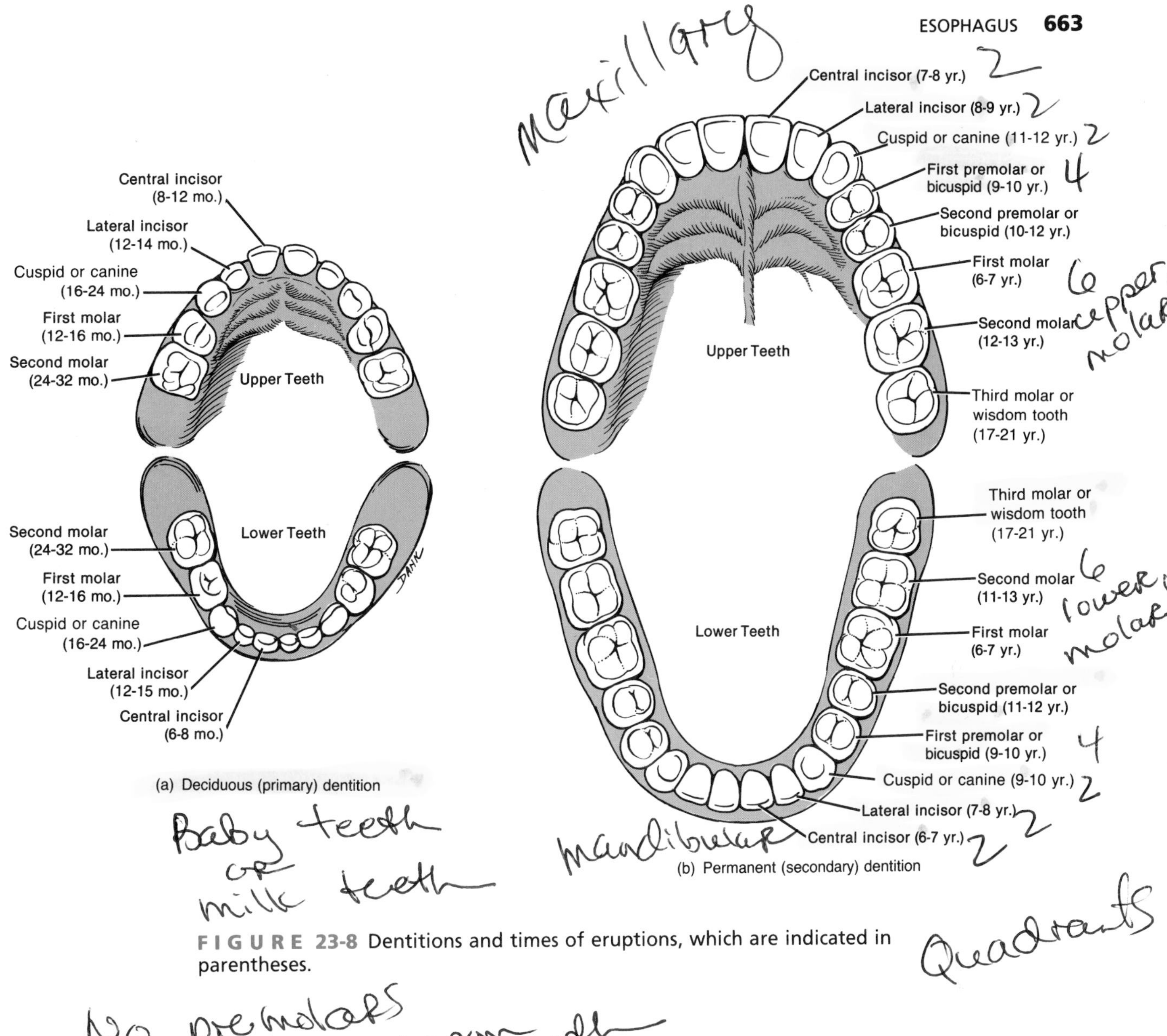

FIGURE 23-8 Dentitions and times of eruptions, which are indicated in parentheses.

PHARYNX

Through chewing, or ***mastication,*** the tongue manipulates food, the teeth grind it, and the food is mixed with saliva. As a result, the food is reduced to a soft, flexible mass called a ***bolus*** that is easily swallowed. When food is first swallowed, it passes from the mouth into the pharynx.

The ***pharynx*** is a funnel-shaped tube that extends from the internal nares to the esophagus posteriorly and the larynx anteriorly (see Figure 22-1a). The pharynx is composed of skeletal muscle and lined by mucous membrane. Whereas the nasopharynx functions only in respiration, both the oropharynx and laryngopharynx have digestive as well as respiratory functions. Food that is swallowed passes from the mouth into the oropharynx and laryngopharynx before passing into the esophagus. Muscular contractions of the oropharynx and laryngopharynx help propel food into the esophagus.

Swallowing, or ***deglutition*** (dē-gloo-TISH-un), is a mechanism that moves food from the mouth to the stomach. It is helped by saliva and mucus and involves the mouth, pharynx, and esophagus. Swallowing is divided into three stages: (1) the voluntary stage, in which the bolus is moved into the oropharynx; (2) the pharyngeal stage, the involuntary passage of the bolus through the pharynx into the esophagus; and (3) the esophageal stage, the involuntary passage of the bolus through the esophagus into the stomach.

ESOPHAGUS

The ***esophagus*** (e-SOF-a-gus) is a muscular, collapsible tube that lies behind the trachea. It is about 23 to 25 cm (10 in.) long and begins at the end of the laryngopharynx, passes through the mediastinum anterior to the vertebral column, pierces the diaphragm through an opening called the ***esophageal hiatus,*** and terminates in the superior portion of the stomach (see Figure 23-1).

HISTOLOGY

The ***mucosa*** of the esophagus consists of nonkeratinized stratified squamous epithelium, lamina propria, and a muscularis mucosae (Figure 23-9c). Near the stomach, the mucosa of the esophagus also contains mucous glands. The ***submucosa*** contains connective tissue, blood vessels, and mucous glands. The ***muscularis*** of the upper third is striated, the middle third is striated and smooth, and the lower third is smooth. The outer layer is known as the ***adventitia*** (ad-ven-TISH-ya) rather than the serosa because the areolar connective tissue of the layer is not covered by epithelium (mesothelium) and because the connective tissue merges with the connective tissue of surrounding structures.

ACTIVITIES

The esophagus does not produce digestive enzymes and does not carry on absorption. It secretes mucus and transports food to the stomach. The passage of food from the laryngopharynx into the esophagus is regulated by a sphincter (thick

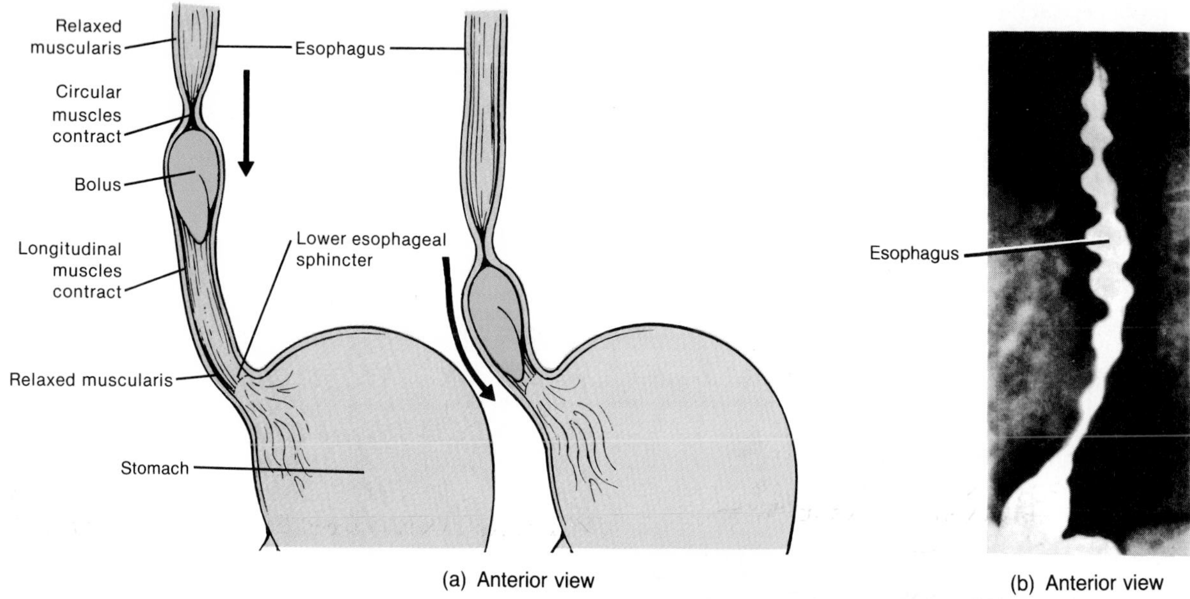

(a) Anterior view

(b) Anterior view

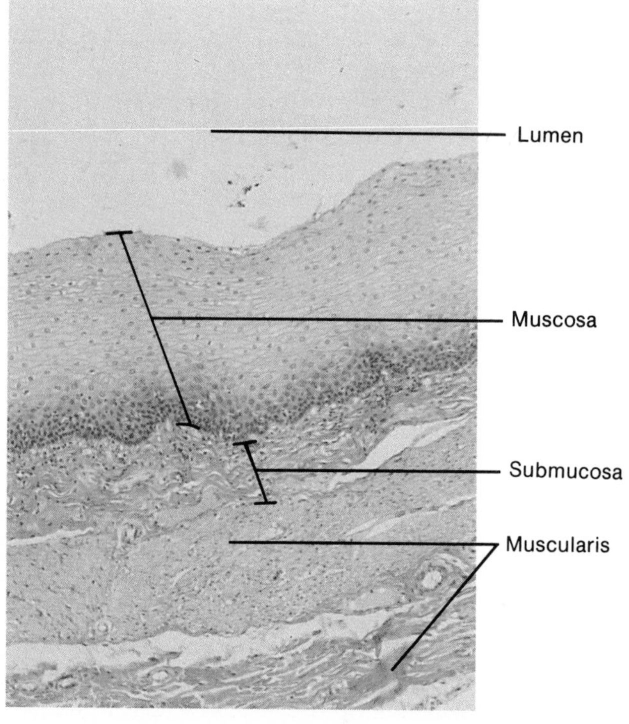

(c) Photomicrograph of a portion of the wall of the human esophagus (400 x)

FIGURE 23-9 Esophagus. (a) Diagram of peristalsis. (b) X-ray of peristalsis made during fluoroscopic examination while a patient was swallowing barium. (Courtesy of Lester W. Paul and John H. Juhl, *The Essentials of Roentgen Interpretation,* 3rd ed., Harper & Row, Publishers, Inc., New York, 1972.) (c) Histology of the esophagus. (© Bruce Iverson) An enlarged aspect of the mucosa of the esophagus is shown in Exhibit 3-1, Stratified squamous.

circle of muscle around an opening) at the entrance to the esophagus called the ***upper esophageal*** (e-sof′-a-JĒ-al) ***sphincter.*** It consists of the cricopharyngeus muscle attached to the cricoid cartilage. The elevation of the larynx during the pharyngeal stage of swallowing causes the sphincter to relax, and the bolus enters the esophagus. The sphincter also relaxes during expiration.

During the esophageal stage of swallowing, food is pushed through the esophagus by involuntary muscular movements called ***peristalsis*** (per′-is-STAL-sis) (Figure 23-9a,b). Peristalsis, which also occurs in other portions of the gastrointestinal tract, is a function of the muscularis and is controlled by the medulla. In the section of the esophagus lying just above and around the top of the bolus, the circular muscle fibers (cells) contract. The contraction constricts the esophageal wall and squeezes the bolus downward. Meanwhile, longitudinal fibers lying around the bottom of and just below the bolus also contract. Contraction of the longitudinal fibers shortens this lower section, pushing its walls outward so it can receive the bolus. The contractions are repeated in a wave that moves down the esophagus, pushing the food toward the stomach. Passage of the bolus is further facilitated by glands that secrete mucus.

Just above the level of the diaphragm, the esophagus is slightly narrowed. This narrowing has been attributed to a physiological sphincter in the inferior part of the esophagus known as the ***lower esophageal (gastroesophageal) sphincter.*** The lower esophageal sphincter relaxes during swallowing and thus aids the passage of the bolus from the esophagus into the stomach.

CLINICAL APPLICATION

Achalasia and Heartburn

If the lower esophageal sphincter fails to relax normally as food approaches, the condition is called ***achalasia*** (ak′-a-LĀ-zē-a; *a* = without; *chalasis* = relaxation). As a result, food passage from the esophagus into the stomach is greatly impeded. A whole meal may become lodged in the esophagus, entering the stomach very slowly. Distension of the esophagus results in chest pain that is often confused with pain originating from the heart. The condition is caused by malfunction of the myenteric plexus (plexus of Auerbach).

If, on the other hand, the lower esophageal sphincter fails to close adequately after food has entered the stomach, the stomach contents can enter the lower esophagus. Hydrochloric acid (HCl) from the stomach contents can irritate the esophageal wall, resulting in a burning sensation. The sensation is known as ***heartburn*** because it is experienced in the region very near the heart, although it is not related to any cardiac problem. Heartburn can be treated by taking antacids (Tums, Gelusil, Rolaids, Maalox) that neutralize the hydrochloric acid and lessen the severity of the burning sensation and discomfort. In addition to the use of antacids for heartburn, other measures are important. Symptoms are less likely to occur if food is eaten in smaller amounts at a meal (no gorging), and heartburn is less of a problem if the person does not lie down right after a meal. Some people with persistent heartburn have a hiatal hernia (to be defined later).

BLOOD AND NERVE SUPPLY

The arteries of the esophagus are derived from the arteries along its length: inferior thyroid, thoracic aorta, intercostal arteries, phrenic, and left gastric arteries. It is drained by the adjacent veins. Innervation of the esophagus is by recurrent laryngeal nerves, the cervical sympathetic chain, and vagi.

STOMACH

The ***stomach*** is a J-shaped enlargement of the GI tract directly under the diaphragm in the epigastric, umbilical, and left hypochondriac regions of the abdomen (see Figure 1-8b). The superior portion of the stomach is a continuation of the esophagus. The inferior portion empties into the duodenum, the first part of the small intestine. Within each individual, the position and size of the stomach vary continually. For instance, the diaphragm pushes the stomach downward with each inspiration and pulls it upward with each expiration. Empty, it is about the size of a large sausage, but it can stretch to accommodate large amounts of food.

ANATOMY

The stomach is divided by gross anatomists into four areas: cardia, fundus, body, and pylorus (Figure 23-10). The ***cardia*** surrounds the lower esophageal sphincter. The rounded portion above and to the left of the cardia is the ***fundus*** (*fundus* = bottom). Below the fundus is the large central portion of the stomach, called the ***body.*** The narrow, inferior region is the ***pylorus.*** The concave medial border of the stomach is called the ***lesser curvature,*** and the convex lateral border is the ***greater curvature.*** The pylorus communicates with the duodenum of the small intestine via a sphincter called the ***pyloric*** (*pyle* = gate; *ouros* = guard) ***sphincter (valve).***

CLINICAL APPLICATION

Pylorospasm and Pyloric Stenosis

Two abnormalities of the pyloric sphincter can occur in infants.

Pylorospasm is characterized by failure of the muscle fibers encircling the opening to relax normally. It can be caused by hypertrophy or continuous spasm of the sphincter and usually occurs between the second and twelfth weeks after birth. Ingested food does not pass

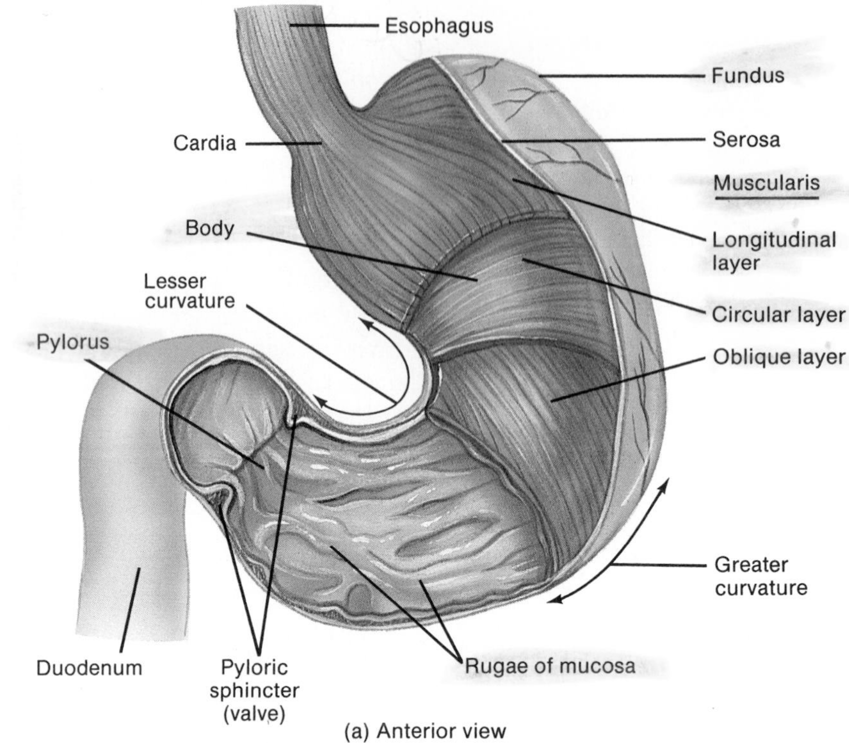

(a) Anterior view

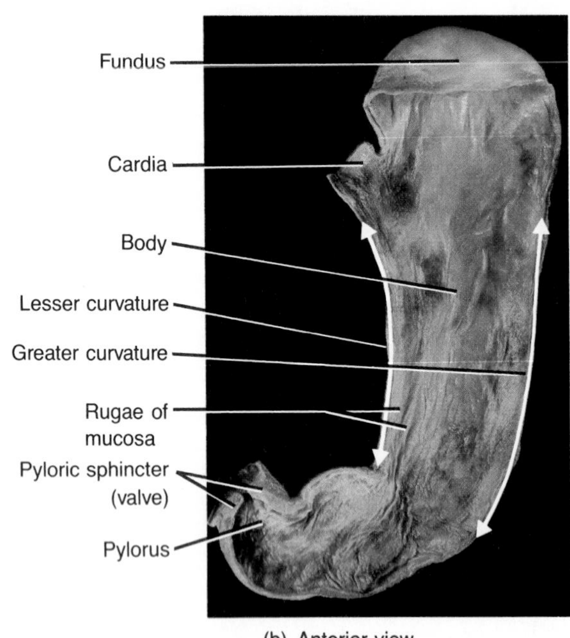

(b) Anterior view

FIGURE 23-10 External and internal anatomy of the stomach. (a) Diagram. (b) Photograph of the internal surface showing rugae. (Courtesy of J. A. Gosling, P. F. Harris, et al., *Atlas of Human Anatomy,* Gower Medical Publishing Ltd., 2nd ed., 1991.)

easily from the stomach to the small intestine, the stomach becomes overly full, and the infant vomits frequently to relieve the pressure. Pylorospasm is treated by drugs that relax the muscle fibers of the sphincter.

Pyloric stenosis is a narrowing of the pyloric sphincter caused by a tumorlike mass that apparently is formed by enlargement of the circular muscle fibers. The hallmark symptom is projectile vomiting—the spraying of liquid vomitus some feet from the infant. Pyloric stenosis must be surgically corrected.

HISTOLOGY

The stomach wall is composed of the same four basic layers as the rest of the GI tract, with certain modifications. When the stomach is empty, the ***mucosa*** lies in large folds, called ***rugae*** (ROO-jē), that can be seen with the naked eye (Figure 23-10b). Microscopic inspection of the mucosa reveals a layer of simple columnar epithelium (surface mucous cells) containing many narrow openings that extend down into the lamina propria called ***gastric pits*** (Figure 23-11). At the bottoms of the pits are the orifices of ***gastric glands.*** Each gland consists of four types of secreting cells: zymogenic, parietal, mucous, and enteroendocrine. The ***zymogenic (peptic)*** or ***chief cells*** secrete the principal gastric enzyme precursor, pepsinogen. Hydrochloric acid, involved in the conversion of pepsinogen to the active enzyme pepsin, and intrinsic factor, involved in the absorption of vitamin B_{12} for red blood cell production, are produced by the ***parietal (oxyntic) cells.*** You may recall from Chapter 12 that inability to produce intrinsic factor can result in pernicious anemia. The ***mucous cells*** secrete mucus. Secretions of the zymogenic, parietal, and mucous cells are collectively called ***gastric juice.*** The ***enteroendocrine cells*** secrete stomach gastrin, a hormone that stimulates secretion of hydro-

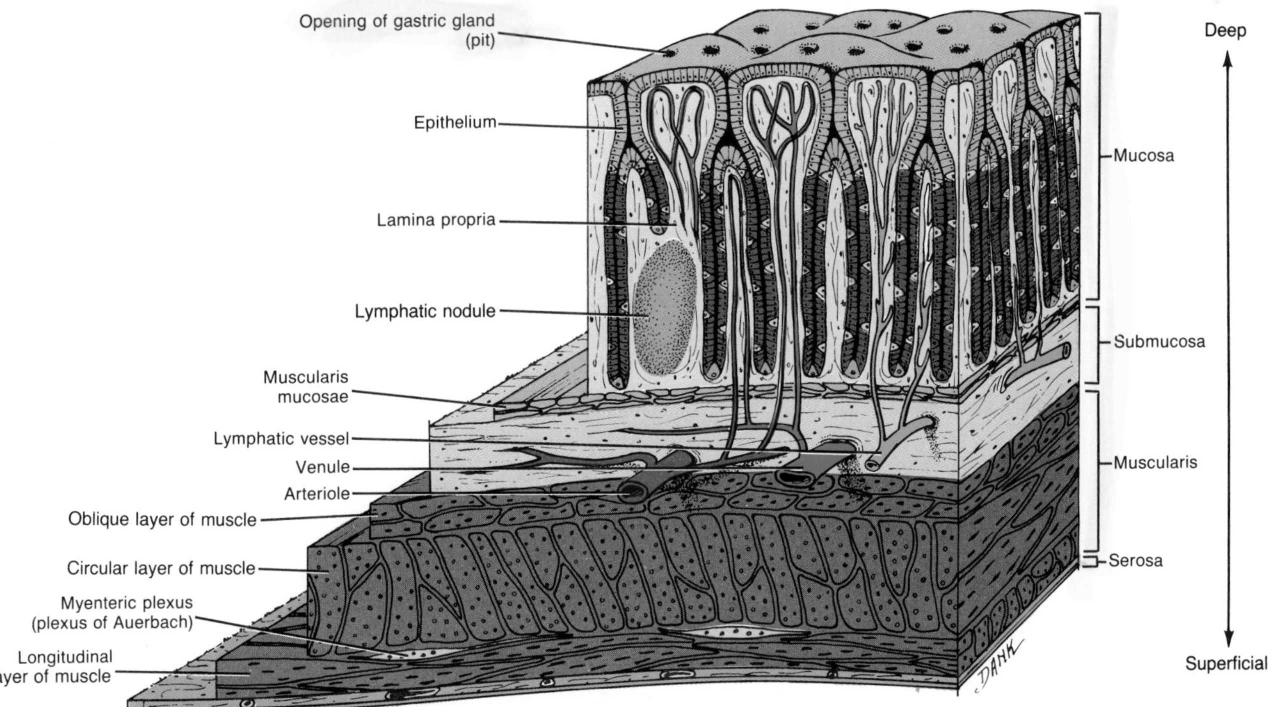

(a) Sectional views of layers of the stomach

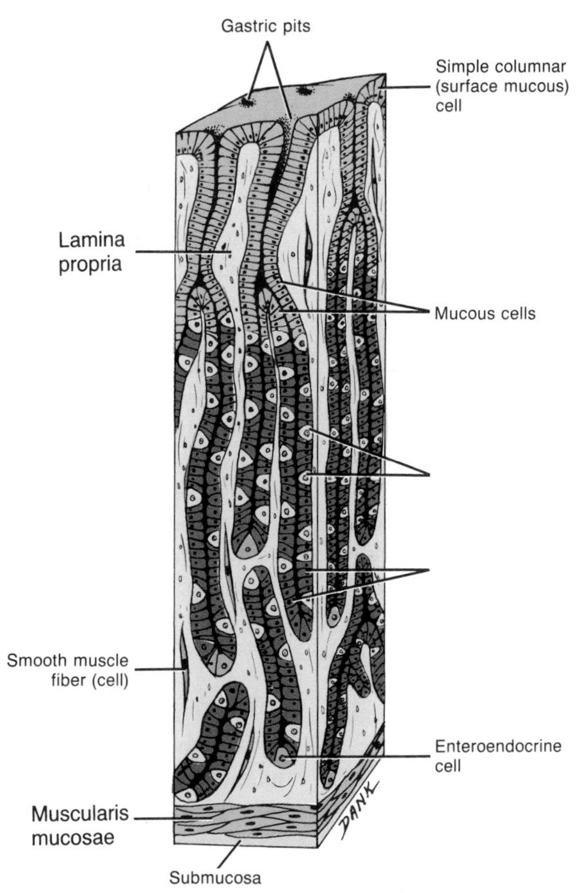

(b) Sectional views of the stomach mucosa

FIGURE 23-11 Histology of the stomach. (c) © 1988, Ed Reschke. (d) Courtesy of Lester V. Bergman & Associates, Inc.

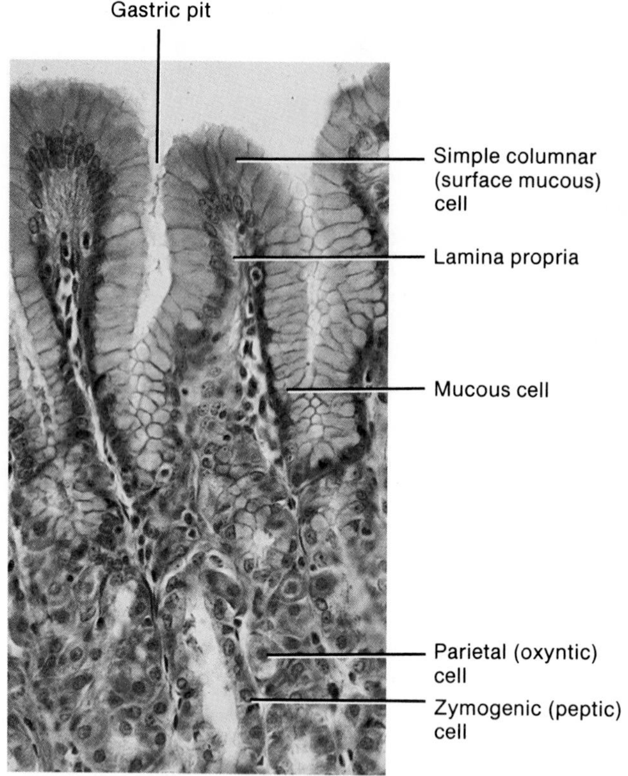

(c) Photomicrograph of fundic mucosa (1120 x)

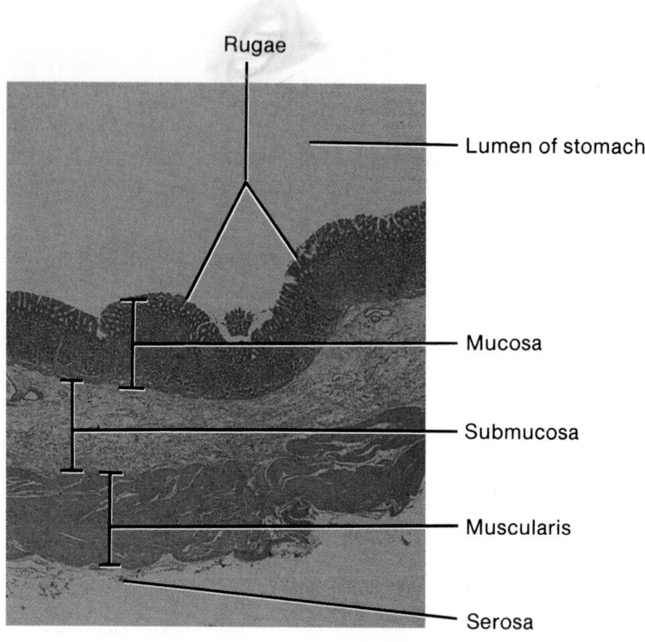

(d) Photomicrograph of fundic wall, (8 x)

chloric acid and pepsinogen, contracts the lower esophageal sphincter, mildly increases motility of the GI tract, and relaxes the pyloric sphincter.

CLINICAL APPLICATION

Gastroscopy

The general term ***endoscopy*** refers to visual inspection of any cavity of the body using an endoscope, an illuminated tube with lenses. Endoscopes can be used to visualize the entire gastrointestinal tract. In fact, endoscopes can be fitted with special devices that also remove foreign objects from the esophagus and stomach, dilate strictures in the esophagus, remove small gallstones, temporarily stop bleeding, biopsy lesions, and remove polyps from the colon.

Endoscopic examination of the stomach is called ***gastroscopy*** (gas-TROS-kō-pē; *gastro* = stomach; *skopein* = to examine). During the procedure, the patient is sedated, and a local anesthetic is applied to the back of the throat. The gastroscope is guided through the mouth, throat, and esophagus into the stomach.

The ***submucosa*** of the stomach is composed of areolar connective tissue, which connects the mucosa to the muscularis.

The ***muscularis,*** unlike that in other areas of the gastrointestinal tract, has three layers of smooth muscle: an outer longitudinal layer, a middle circular layer, and an inner oblique layer. The oblique layer is limited mostly to the body of the stomach. This arrangement of fibers allows the stomach to contract in a variety of ways to churn food, break it into small particles, mix it with gastric juice, and pass it to the duodenum.

The ***serosa*** covering the stomach is part of the visceral peritoneum. At the lesser curvature, the two layers of the visceral peritoneum come together and extend upward to the liver as the lesser omentum. At the greater curvature, the visceral peritoneum continues downward as the greater omentum hanging over the intestines.

ACTIVITIES

Several minutes after food enters the stomach, gentle, rippling, peristaltic movements called ***mixing waves*** pass over the stomach every 15 to 25 seconds. These waves macerate food, mix it with the secretions of the gastric glands, and reduce it to a thin liquid called ***chyme*** (kīm). Few mixing waves are observed in the fundus, which is primarily a storage area. Foods may remain in the fundus for an hour or more without becoming mixed with gastric juice. During this time, salivary digestion continues.

As digestion proceeds in the stomach, more vigorous mixing waves begin at the body of the stomach and intensify as they reach the pylorus. The pyloric sphincter normally remains almost, but not completely, closed. As food reaches the pylorus, each mixing wave forces a small amount of the gastric contents into the duodenum through the pyloric sphincter. Most of the food is forced back into the body of the stomach, where it is subjected to further mixing. The next wave pushes it forward again and forces a little more into the duodenum. The forward and backward movement of the gastric contents are responsible for almost all the mixing in the stomach.

The principal chemical activity of the stomach is to begin the digestion of proteins. In the adult, digestion is achieved primarily through the enzyme ***pepsin.*** Another enzyme of the stomach, ***gastric lipase,*** splits the butterfat molecules found in milk. This enzyme has a limited role in the adult stomach. Adults rely almost exclusively on an enzyme found in the small intestine (pancreatic lipase) to digest fats. The infant stomach also secretes ***rennin,*** which is important in the digestion of milk (not to be confused with renin, the enzyme produced by the kidneys). Rennin and calcium act on the casein of milk to produce a curd. The coagulation prevents too rapid a passage of milk from the stomach into the duodenum (first portion of the small intestine). Rennin is absent in the gastric secretions of adults.

The stomach empties all its contents into the duodenum two to six hours after ingestion. Food rich in carbohydrate leaves the stomach in a few hours. Protein foods are somewhat slower, and emptying is slowest after a meal containing large amounts of fat.

CLINICAL APPLICATION

Vomiting

Excessive gastric emptying in the wrong direction sometimes occurs. ***Vomiting*** is the forcible expulsion of the contents of the upper GI tract (stomach and sometimes duodenum) through the mouth. The strongest stimuli for vomiting are irritation and distension of the stomach. Other stimuli include unpleasant sights, dizziness, and certain drugs such as morphine and derivatives of digitalis. Nerve impulses are transmitted to the vomiting center in the medulla, and returning impulses to the upper GI tract organs, diaphragm, and abdominal muscles bring about the vomiting act. Basically, vomiting involves squeezing the stomach between the diaphragm and abdominal muscles and expelling of the contents through the open esophageal sphincters. Prolonged vomiting, especially in infants and elderly people, can be serious because the loss of gastric juice and fluids can lead to disturbances in fluid and acid–base balance.

The stomach wall is impermeable to the passage of most materials into the blood, so most substances are not absorbed until they reach the small intestine. However, the stomach does participate in the absorption of some water, electrolytes, certain drugs (especially aspirin), and alcohol. The absorption of alcohol by the stomach of females is faster than that in males. The difference is attributed to smaller amounts of the enzyme alcohol dehydrogenase in the stomachs of females. The enzyme breaks down alcohol in the stomach, reducing the amount of alcohol that enters the blood.

BLOOD AND NERVE SUPPLY

The arterial supply of the stomach is derived from the celiac artery. The right and left gastric arteries form an anastomosing arch along the lesser curvature, and the right and left gastroepiploic arteries form a similar arch on the greater curvature. Short gastric arteries supply the fundus. The veins of the same name accompany the arteries and drain, directly or indirectly, into the hepatic portal vein.

The vagi convey parasympathetic fibers to the stomach. These fibers form synapses within the submucosal plexus in the submucosa and the myenteric plexus (plexus of Auerbach) in the muscularis. The sympathetic nerves arise from the celiac ganglia and the nerves reach the stomach along the branches of the celiac artery.

PANCREAS

The next organ of the GI tract involved in the breakdown of food is the small intestine. Chemical digestion in the small intestine depends not only on its own secretions but also on activities of three accessory structures of digestion outside the gastrointestinal tract: the pancreas, liver, and gallbladder.

ANATOMY

The ***pancreas*** is an oblong tubuloacinar gland about 12.5 cm (5 in.) long and 2.5 cm (1 in.) thick. It lies posterior to the greater curvature of the stomach and is connected, usually by two ducts, to the duodenum. The pancreas is divided into a head, body, and tail. The ***head*** is the expanded portion near the C-shaped curve of the duodenum. Moving superiorly and to the left of the head are the centrally located ***body*** and the terminal tapering ***tail*** (Figure 23-12).

Pancreatic secretions pass from the secreting cells in the pancreas to small ducts that unite to form the two ducts that convey the secretions into the small intestine. The larger of the two ducts is called the ***pancreatic duct (duct of Wirsung).*** In most people, the pancreatic duct unites with the common bile duct from the liver and gallbladder and enters the duodenum in a common duct called the ***hepatopancreatic ampulla (ampulla of Vater).*** The ampulla opens on an elevation of the duodenal mucosa known as the ***major duodenal papilla,*** about 10 cm (4 in.) below

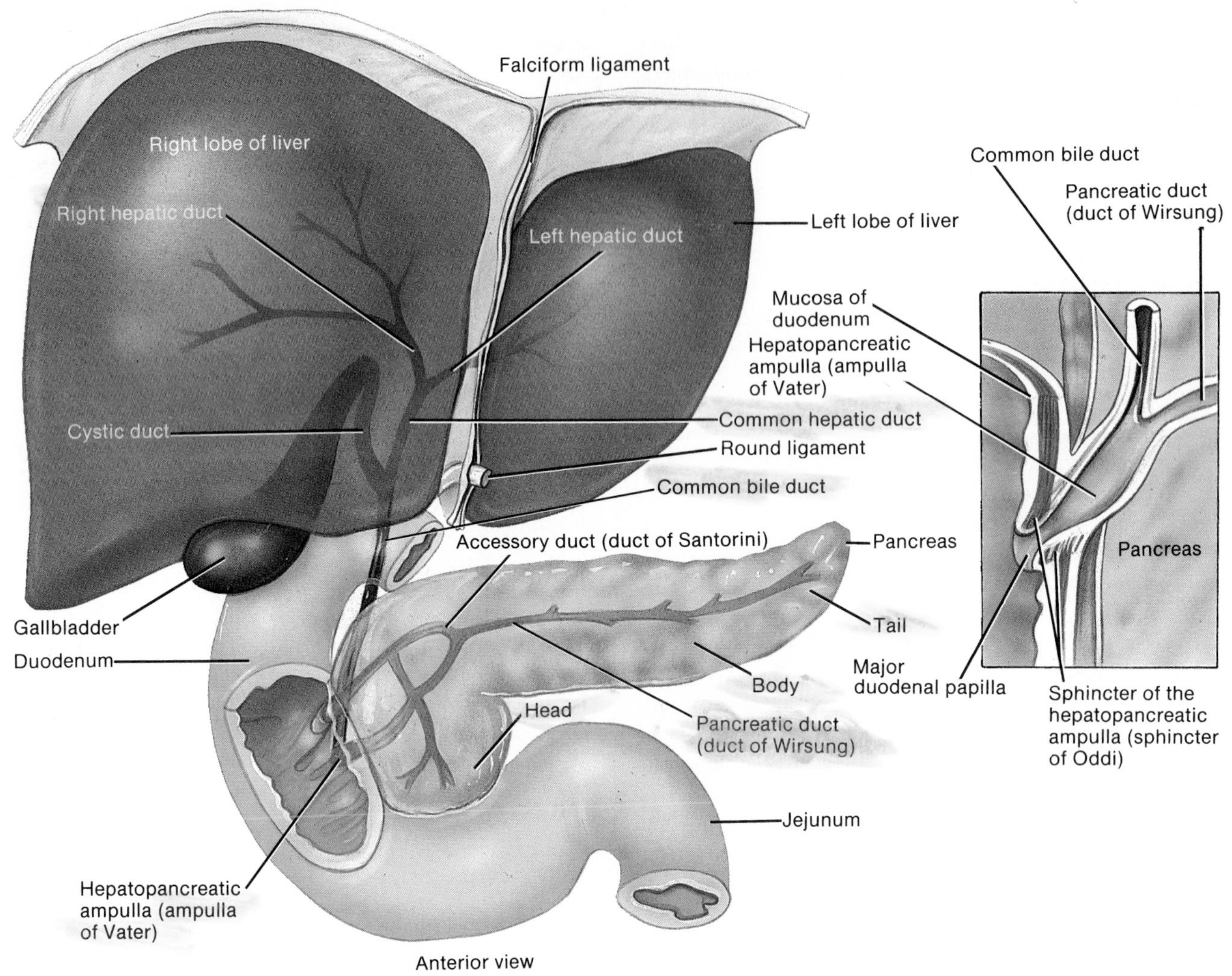

FIGURE 23-12 Pancreas. Shown is the relation of pancreas to liver, gallbladder, and duodenum. The insert shows details of the common bile duct and pancreatic duct forming the hepatopancreatic ampulla (of Vater) emptying into the duodenum.

the pylorus of the stomach. The smaller of the two ducts is the ***accessory duct (duct of Santorini)***, which leads from the pancreas and empties into the duodenum about 2.5 cm (1 in.) above the hepatopancreatic ampulla.

HISTOLOGY

The pancreas is made up of small clusters of glandular epithelial cells. About 1 percent of the clusters of cells, the ***pancreatic islets (islets of Langerhans)***, form the endocrine portion of the pancreas and consist of alpha, beta, delta, and F cells that secrete hormones (glucagon, insulin, growth hormone-inhibiting factor, and pancreatic polypeptide respectively). The functions of these hormones may be reviewed in Chapter 21. The remaining 99 percent of the clusters of cells, called ***acini*** (AS-i-nī), constitute the exocrine portions of the organ (see Figure 21-16). Secreting cells of the acini release a mixture of digestive enzymes called ***pancreatic juice.***

ACTIVITIES

Each day the pancreas produces 1200 to 1500 ml (about 1.2 to 1.5 qt) of pancreatic juice, a clear, colorless liquid. It consists mostly of water, some salts, sodium bicarbonate, and enzymes. The sodium bicarbonate gives pancreatic juice a slightly alkaline pH (7.1 to 8.2) that stops the action of pepsin from the stomach and creates the proper environment for the enzymes in the small intestine. The enzymes in pancreatic juice include a carbohydrate-digesting enzyme called ***pancreatic amylase;*** several protein-digesting enzymes called ***trypsin*** (TRIP-sin), ***chymotrypsin*** (kī′-mō-

TRIP-sin), and ***carboxypeptidase*** (kar-bok′-sē-PEP-ti-dās); the principal fat-digesting enzyme in the adult body called ***pancreatic lipase;*** and nucleic acid–digesting enzymes called ***ribonuclease*** and ***deoxyribonuclease.***

BLOOD AND NERVE SUPPLY

The vascular and nerve supply of the pancreas may be reviewed in Chapter 21.

LIVER

The ***liver*** weighs about 1.4 kg (about 3 lb) in the average adult. It is located under the diaphragm and occupies most of the right hypochondrium and part of the epigastrium of the abdomen (see Figure 1-8c).

ANATOMY

The liver is almost completely covered by peritoneum and completely covered by a dense connective tissue layer that lies beneath the peritoneum. It is divided into two principal lobes—a large ***right lobe*** and a smaller ***left lobe***—separated by the ***falciform ligament*** (Figure 23-13). The right lobe is considered by many anatomists to consist of an inferior ***quadrate lobe*** and a posterior ***caudate lobe.*** However, on the basis of internal morphology, primarily the distribution of blood, the quadrate and caudate lobes more appropriately belong to the left lobe. The falciform ligament is a reflection of the parietal peritoneum, which extends from the undersurface of the diaphragm to the superior surface of the liver, between the two principal lobes of the liver. In the free border of the falciform ligament is the ***ligamentum teres (round ligament).*** It extends from the liver to the umbilicus. The ligamentum teres is a fibrous cord that is the remnant of the umbilical vein of the fetus.

CLINICAL APPLICATION

Living-Donor Liver Transplant

In November 1989, a team of surgeons at the University of Chicago Medical Center performed the first ***living-donor liver transplant*** in the United States. In the 14-hour procedure, a 21-month-old child received part of the left lobe of her mother's liver. The child suffered from biliary atresia, the closure or absence of some or all the major bile ducts. Since the liver is capable of regeneration, the mother's liver returned to its normal size in about two months and the child's liver is expected to grow as she does. Living-donor transplants have also been done for the pancreas, kidneys, and bone marrow.

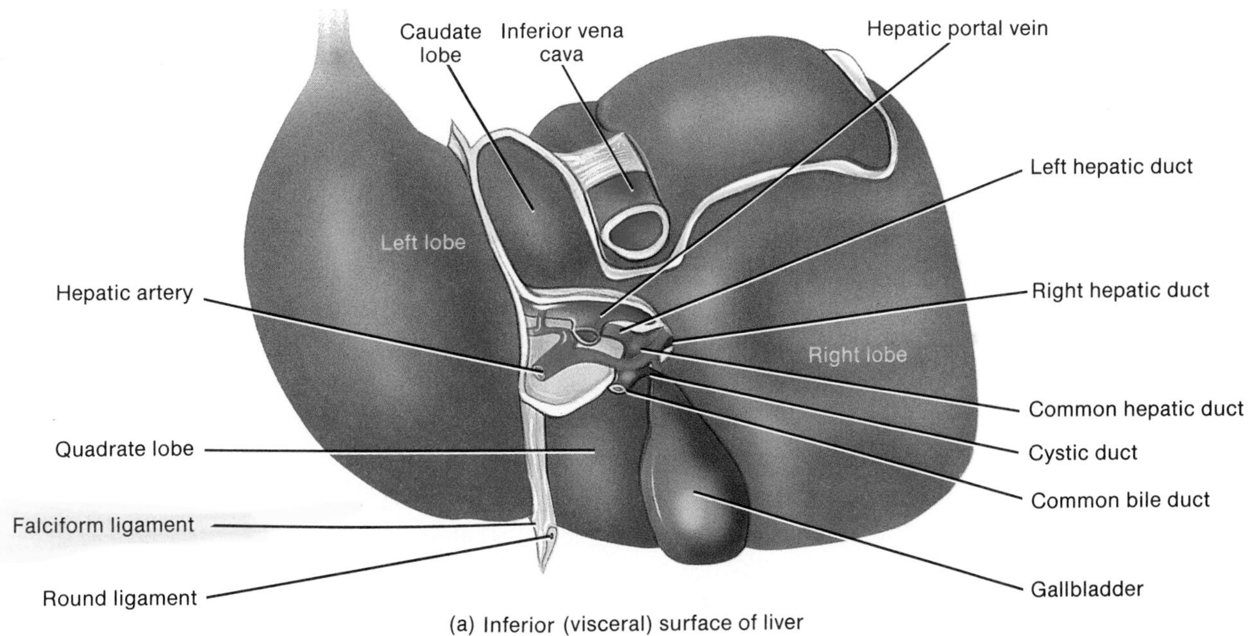

FIGURE 23-13 External anatomy of the liver. The anterior view is illustrated in Figure 23-12. (b) Courtesy of J. A. Gosling, P. F. Harris, et al., *Atlas of Human Anatomy,* Gower Medical Publishing Ltd., 2nd ed., 1991. (c) Courtesy of Stephen A. Kieffer and E. Robert Heitzman, *An Atlas of Cross-Sectional Anatomy,* Harper & Row, Publishers, Inc., New York, 1979.

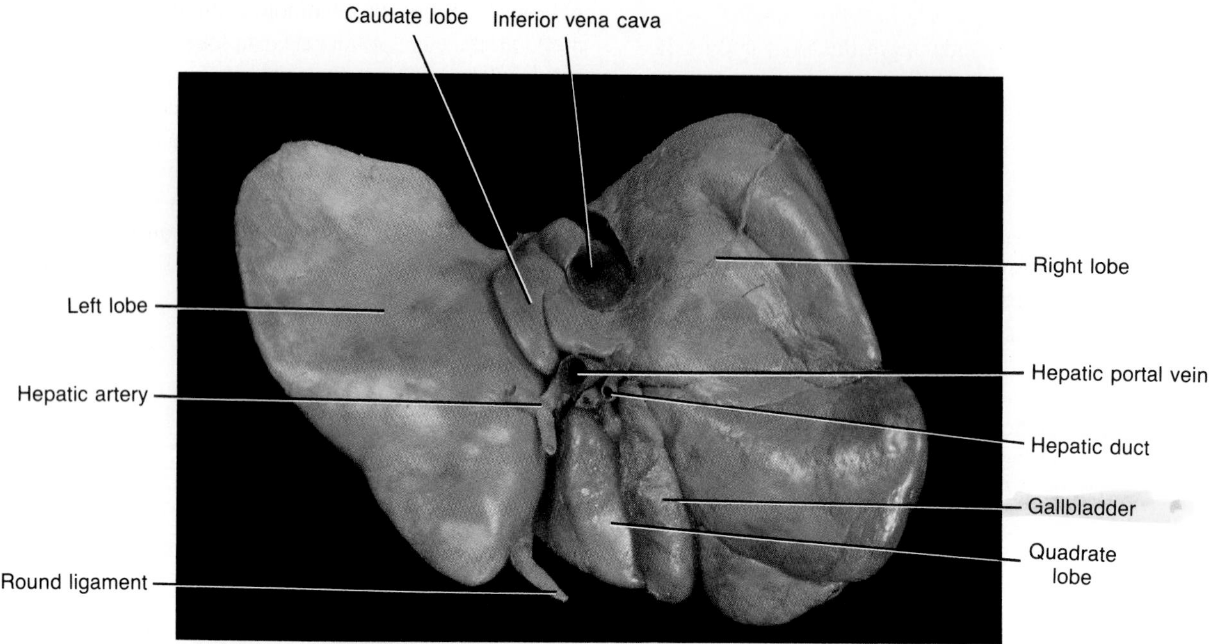

(b) Photograph of inferior (visceral) surface of liver

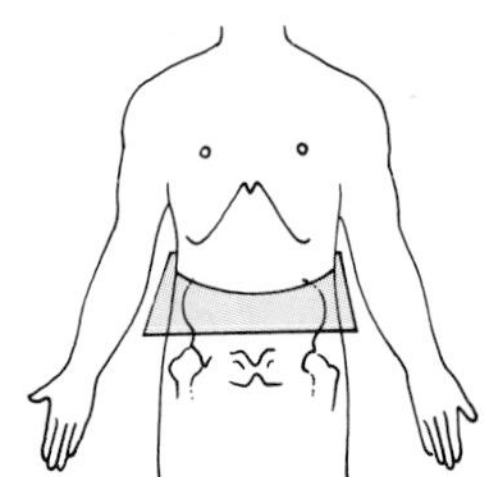

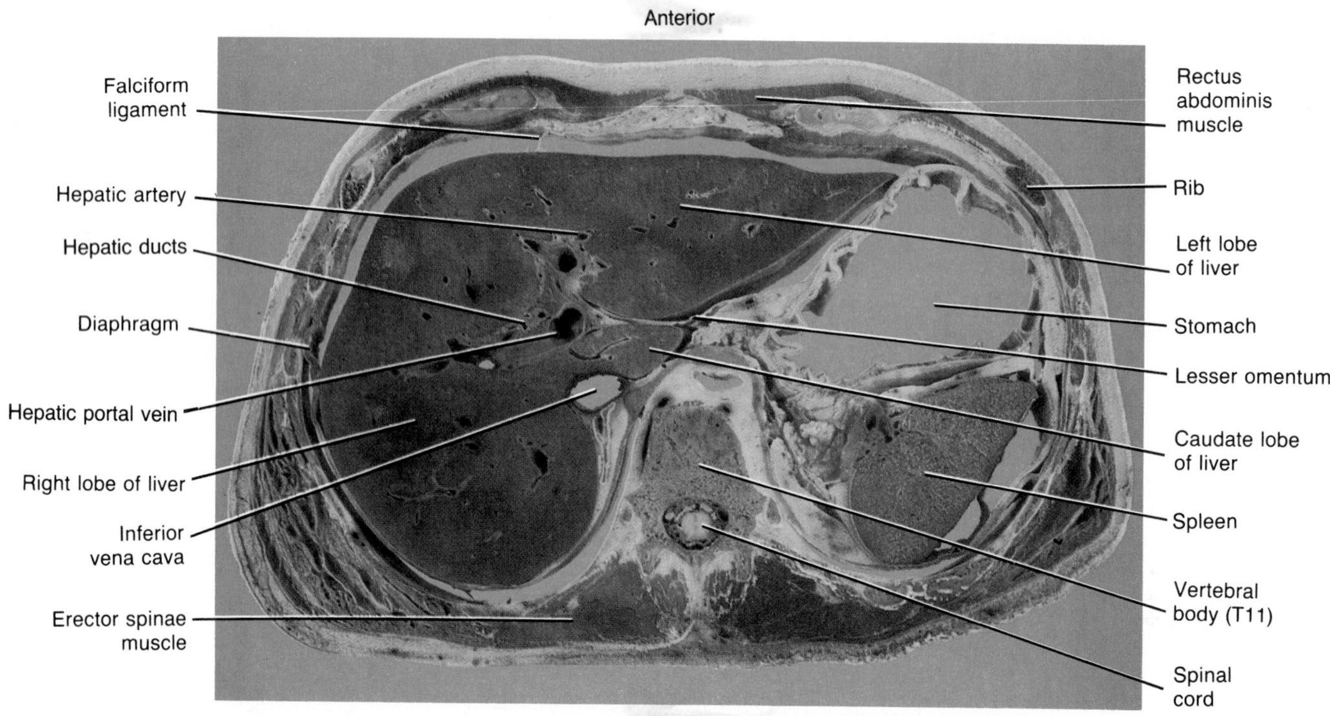

(c) Photograph of a cross section of the abdomen

HISTOLOGY

The lobes of the liver are made up of numerous functional units called ***lobules,*** which may be seen under a microscope (Figure 23-14). A lobule consists of epithelial cells, called ***hepatic (liver) cells*** or ***hepatocytes,*** arranged in irregular, branching, interconnected plates around a ***central vein.*** These cells secrete bile. Between the plates of cells are endothelial-lined spaces called ***sinusoids,*** through which blood passes. The sinusoids are also partly lined with phagocytic cells, termed ***stellate reticuloendothelial (Kupffer's) cells,*** that destroy worn-out white and red blood cells, bacteria, and toxic substances. The liver contains sinusoids instead of typical capillaries.

Each day, the hepatic cells secrete 800 to 1000 ml (about 1 qt) of ***bile,*** a yellow, brownish, or olive-green liquid. It has a pH of 7.6 to 8.6. Bile consists mostly of water and bile salts, cholesterol, a phospholipid called lecithin, bile pigments, and several ions. Bile is partially an excretory product and partially a digestive secretion. Bile salts assume a role in ***emulsification,*** the breakdown of large fat globules into a suspension of fat droplets about 1 μm in diameter, and absorption of fats following their digestion. The tiny fat droplets present a very large surface area for the action of the pancreatic lipase necessary for rapid fat digestion. Cholesterol is made soluble in bile by bile salts and lecithin. The principal bile pigment is ***bilirubin.*** When worn-out red blood cells are broken down, iron, globin, and bilirubin (derived from heme) are released. The iron and globin are recycled, but some of the bilirubin is excreted into the bile ducts. Bilirubin eventually is broken down in the intestine, and one of its breakdown products (urobilinogen) gives feces their color.

Bile enters ***bile capillaries*** or ***canaliculi*** (kan′-a-LIK-yoo-lī) that empty into small ducts. These small ducts eventually merge to form the larger ***right*** and ***left hepatic ducts,*** which unite to leave the liver as the ***common hepatic duct*** (see Figures 23-12 and 23-13a). Further on, the common hepatic duct joins the ***cystic duct*** from the gallbladder. The two tubes become the ***common bile duct.*** The common bile duct and pancreatic duct enter the duodenum in a common duct called the ***hepatopancreatic ampulla (ampulla of Vater).***

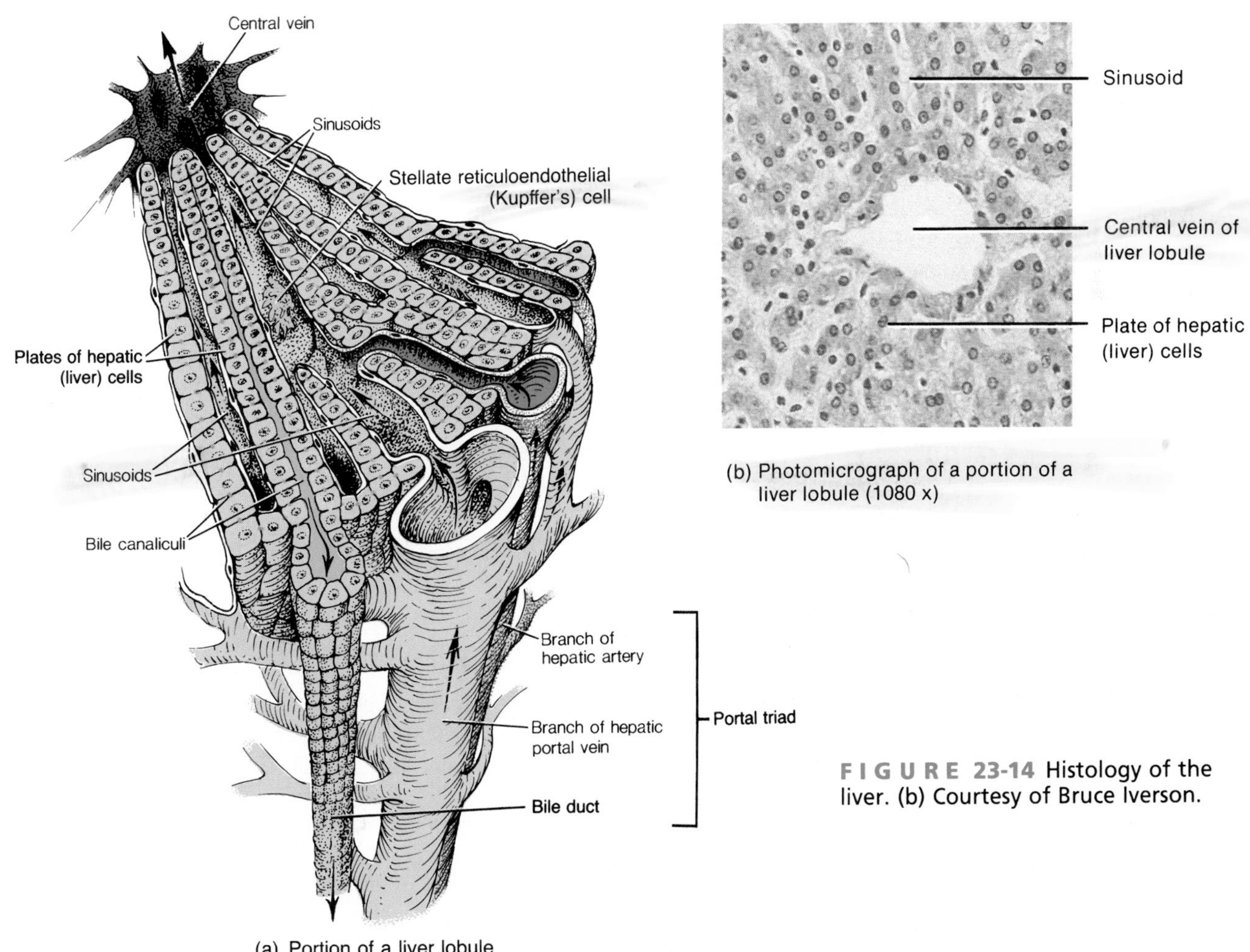

(a) Portion of a liver lobule

(b) Photomicrograph of a portion of a liver lobule (1080 x)

FIGURE 23-14 Histology of the liver. (b) Courtesy of Bruce Iverson.

CLINICAL APPLICATION

Jaundice

Jaundice is a yellowish coloration of the sclera of the eyes, skin, and mucous membranes due to a buildup of bilirubin in the body. After bilirubin is formed from the breakdown of the heme pigment in worn-out red blood cells, it is transported to the liver where it is processed and eventually excreted into bile. The three main categories of jaundice are:

1. ***Prehepatic (hemolytic) jaundice,*** which is due to excess production of bilirubin. In hemolytic anemia, for example, bilirubin may be produced so rapidly and in such quantities that the liver cannot excrete it fast enough to avoid the occurrence of jaundice. Since the liver of a newborn functions poorly for the first week or so, many babies experience a mild form of jaundice called ***neonatal (physiological) jaundice*** that disappears as the liver matures. Neonatal jaundice is usually treated by exposing the infant to blue light, which converts bilirubin into substances the liver can excrete. A new treatment consists of administering a new drug (Sn-protoporphyrin) that combines with the enzyme that catalyzes heme breakdown, thus blocking production of bilirubin.
2. ***Hepatic (medical) jaundice,*** which is due to dysfunction of liver cells. Certain congenital liver diseases result in jaundice because of a derangement of hepatic enzyme systems. Jaundice accompanying cirrhosis of the liver results from a loss of functional hepatic lobules. Hepatitis, by producing swelling within the liver lobules, blocks the bile canaliculi and produces jaundice.
3. ***Extrahepatic (surgical) jaundice,*** which is due to interference with the removal of bile from the hepatobiliary system. It is also known as ***obstructive jaundice.*** Causes include gallstones and cancer of the bowel or the head of the pancreas. Surgery must usually be performed to relieve the obstruction.

ACTIVITIES

The liver performs many vital functions, including the following:

1. ***Carbohydrate metabolism.*** In carbohydrate metabolism, the liver is especially important in maintaining a normal blood glucose level. For example, the liver can convert glucose to glycogen (glycogenesis) when blood sugar level is high and convert glycogen to glucose (glycogenolysis) when blood sugar level is low. The liver can also convert amino acids to glucose (gluconeogenesis) when blood sugar level is low; convert other sugars, such as fructose and galactose into glucose; and convert glucose to fats.
2. ***Fat metabolism.*** With respect to fat metabolism, the liver breaks down fatty acids into acetyl coenzyme A, a process called beta oxidation; converts excess acetyl coenzyme A into ketone bodies (ketogenesis); synthesizes lipoproteins that transport fatty acids, fats, and cholesterol to and from body cells; synthesizes cholesterol and phospholipids and breaks down cholesterol to bile salts; and stores fats.
3. ***Protein metabolism.*** Without the role of the liver in protein metabolism, death would occur in a few days. Among the functions of the liver related to protein metabolism are: deamination (removal of the amino group, NH_2) of amino acids so that they can be used for energy or converted to carbohydrates or fats; conversion of ammonia (NH_3), a toxic substance, into the much less toxic urea for excretion in urine (ammonia is produced from deamination and by bacteria in the gastrointestinal tract), synthesis of most plasma proteins, such as angiotensinogen, alpha and beta globulins, albumin, prothrombin, and fibrinogen (together with mast cells, the liver also produces an anticoagulant, heparin); and transamination, the transfer of an amino group from an amino acid to another substance (α-keto acid) in order to convert one amino acid into another.
4. ***Removal of drugs and hormones.*** The liver can detoxify or excrete into bile drugs such as penicillin, ampicillin, erythromycin, and sulfonamides. It can also chemically alter or excrete steroid hormones, such as estrogens and aldosterone, and thyroxine.
5. ***Excretion of bile.*** Bilirubin, derived from the heme of worn-out red blood cells, is absorbed by the liver from the blood and secreted into bile. Most of the bilirubin in bile is metabolized in the intestine by bacteria and eliminated in feces.
6. ***Synthesis of bile salts.*** Bile salts are used in the small intestine for the emulsification and absorption of fats, cholesterol, phospholipids, and lipoproteins.
7. ***Storage.*** In addition to glycogen, the liver stores vitamins (A, B_{12}, C, D, E, and K) and minerals (iron and copper).
8. ***Phagocytosis.*** The stellate reticuloendothelial (Kupffer's) cells of the liver phagocytize worn-out red and white blood cells and some bacteria.
9. ***Activation of vitamin D.*** The liver and kidneys participate in the activation of vitamin D.

BLOOD AND NERVE SUPPLY

The liver receives a double supply of blood. From the hepatic artery it obtains oxygenated blood, and from the hepatic portal vein it receives deoxygenated blood containing newly absorbed nutrients (see Figures 14-16 and 23-14a). Branches of both the hepatic artery and the hepatic portal vein carry the blood into the sinusoids of the lobules, where oxygen, most of the nutrients, and certain poisons are extracted by the hepatic cells. The phagocytic reticuloendothelial (Kupffer's) cells lining the sinusoids remove microbes and bits of foreign matter from the blood. Nutrients are stored or used to make new materials. The poisons are stored or detoxified. Products manufactured by the hepatic cells and nutrients needed by other cells are secreted back into the blood. The blood then drains into the central vein and eventually passes into a hepatic vein, inferior vena cava, and back to the heart. Unlike the other products of the liver, bile normally is not secreted into the bloodstream.

Branches of the hepatic portal vein, hepatic artery, and bile duct typically accompany each other in their distribution through the liver. Collectively, these three structures are referred to as a ***portal triad*** (Figure 23-14a).

The nerve supply to the liver consists of parasympathetic innervation from the vagus nerves and sympathetic innervation from the greater splanchnic nerves through the celiac ganglia.

GALLBLADDER (GB)

The ***gallbladder (GB)*** is a pear-shaped sac about 7 to 10 cm (3 to 4 in.) long. It is located in a fossa of the visceral surface of the liver (see Figures 23-12a and 23-13).

HISTOLOGY

The mucosa of the gallbladder consists of simple columnar epithelium arranged in rugae resembling those of the stomach (Figure 23-15). The gallbladder lacks a submucosa. The middle, muscular coat of the wall consists of smooth muscle fibers (cells). Contraction of these fibers by hormonal stimulation and parasympathetic stimulation ejects the contents of the gallbladder into the ***cystic duct.*** The outer coat is the visceral peritoneum.

ACTIVITIES

The functions of the gallbladder are to store and concentrate bile (up to 10-fold) until it is needed in the small intestine. In the concentration process, water and many ions are absorbed by the mucosa of the gallbladder. Bile from the liver enters the small intestine through the common bile duct. When the small intestine is empty, a valve around the hepatopancreatic ampulla (ampulla of Vater) called the ***sphincter of the hepatopancreatic ampulla (sphincter of Oddi)*** closes, and the backed-up bile flows into the cystic duct to the gallbladder for storage (see Figure 23-12a).

BLOOD AND NERVE SUPPLY

The gallbladder is supplied by the cystic artery, which usually arises from the right hepatic artery. The cystic veins drain the gallbladder. The nerves to the gallbladder include branches from the celiac plexus and the vagus (X) nerve.

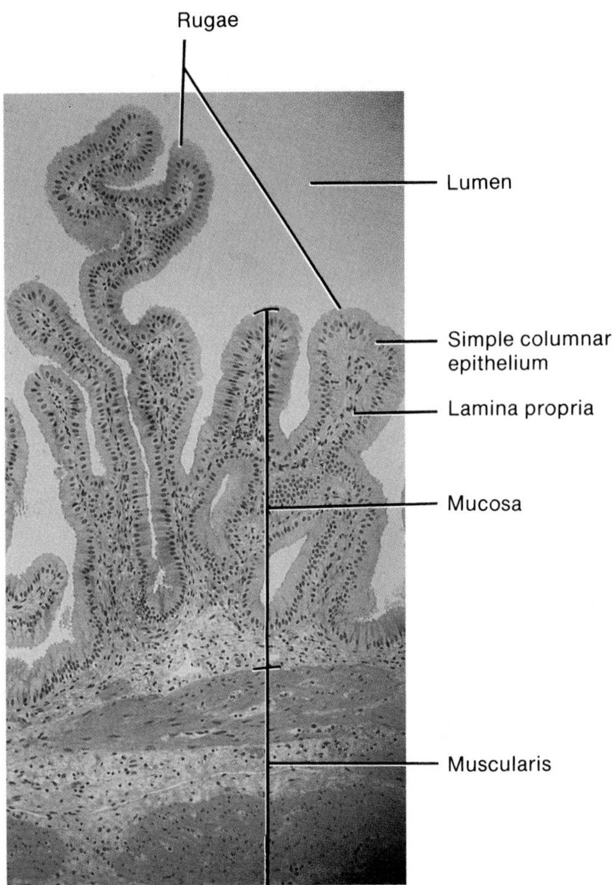

FIGURE 23-15 Histology of the gallbladder. © 1988, Ed Reschke.

SMALL INTESTINE

The major portions of digestion and absorption occur in a long tube called the ***small intestine.*** The small intestine begins at the pyloric sphincter of the stomach, coils through the central and lower part of the abdominal cavity, and eventually opens into the large intestine. It averages 2.5 cm (1 in.) in diameter and about 6.35 m (21 ft) in length in a cadaver.

ANATOMY

The small intestine is divided into three segments (see Figure 23-1). The ***duodenum*** (doo′-ō-DĒ-num), the shortest part, originates at the pyloric sphincter of the stomach and extends about 25 cm (10 in.) until it merges with the jejunum. *Duodenum* means "12"; the structure is 12 fingers' breadth in length. The ***jejunum*** (jē-JOO-num) is about 2.5 m (8 ft) long and extends to the ileum. *Jejunum* means "empty," since at death it is found empty. The final portion of the small intestine, the ***ileum*** (IL-ē-um; *eileos* = twisted) measures about 3.6 m (12 ft) and joins the large intestine at the ***ileocecal*** (il′-ē-ō-SĒ-kal) ***sphincter*** (***valve***).

HISTOLOGY

The wall of the small intestine is composed of the same four coats that make up most of the GI tract. However, both the mucosa and the submucosa are modified to allow the small intestine to complete the processes of digestion and absorption (Figure 23-16).

The ***mucosa*** contains many pits lined with glandular epithelium. These pits—the ***intestinal glands*** or ***crypts of Lieberkühn*** (LĒ-ber-kēn)—secrete intestinal juice. The submucosa of the duodenum contains ***duodenal*** (***Brunner's***) ***glands,*** which secrete an alkaline mucus that protects the wall of the small intestine from the action of the enzymes and aids in neutralizing acid in the chyme. Some of the epithelial cells in the mucosa and submucosa have been transformed to goblet cells, which secrete additional mucus.

Since almost all the digestion and absorption of nutrients occurs in the small intestine, its structure is specially adapted for this function. Its length alone provides a large surface area for digestion and absorption, and that area is further increased by modifications in the structure of its wall. The epithelium of the mucosa consists of simple columnar epithelium and contains goblet cells and absorptive cells. The absorptive cells possess ***microvilli,*** fingerlike projections of the plasma membrane. Larger amounts of digested nutrients diffuse into the absorptive cells of the intestinal wall because the microvilli increase the surface area of the plasma membrane. They also increase the surface area for digestion.

The mucosa lies in a series of ***villi,*** projections 0.5 to 1 mm high, giving the intestinal mucosa its velvety appearance. The enormous number of villi (10 to 40 per square millimeter) vastly increases the surface area of the epithelium available for absorption and digestion. Each villus has a core of lamina propria, the connective tissue layer of the mucosa. Embedded in this connective tissue are an arteriole, a venule, a capillary network, and a ***lacteal*** (LAK-tē-al), which is a lymphatic vessel. Nutrients that diffuse through the epithelial cells that cover the villus are able to pass through the capillary walls and the lacteal and enter the cardiovascular and lymphatic systems, respectively.

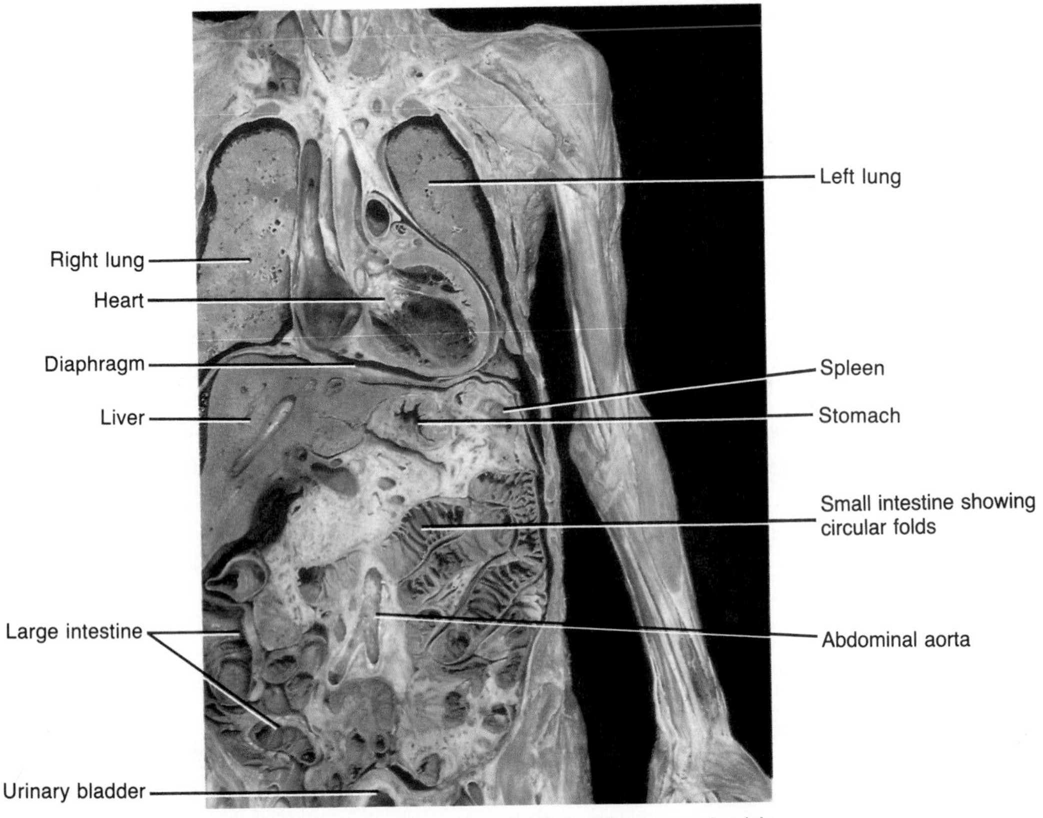

(a) Photograph of frontal section of chest, abdomen, and pelvis

FIGURE 23-16 Small intestine. Shown are various structures that adapt the small intestine for digestion and absorption. Note the circular folds in (a). Courtesy of J. A. Gosling, P. F. Harris, et al., *Atlas of Human Anatomy,* Gower Medical Publishing Ltd., 1985.

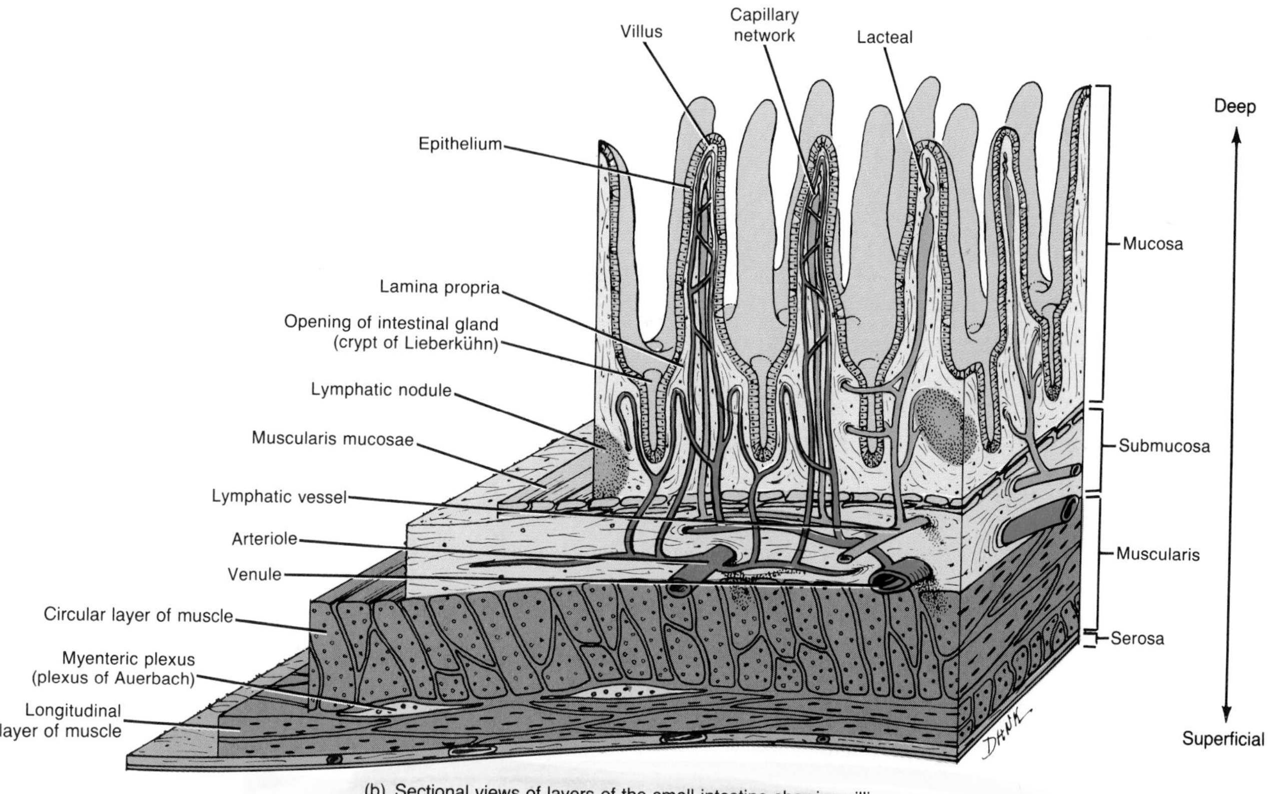

(b) Sectional views of layers of the small intestine showing villi

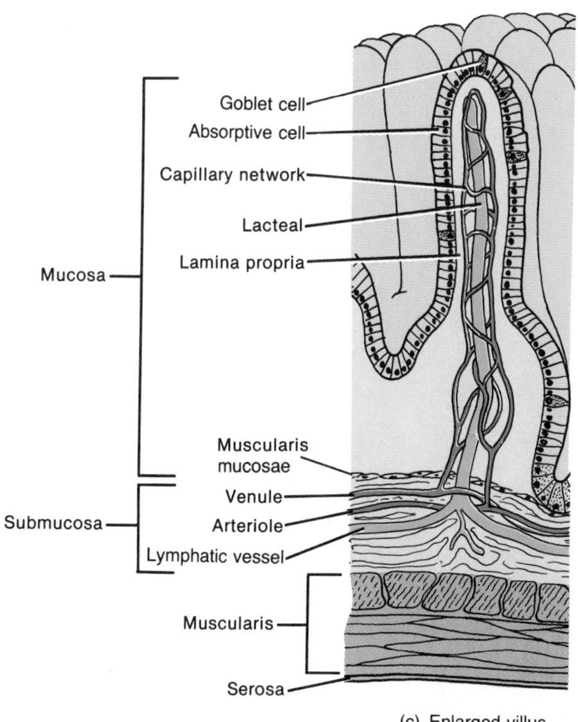

(c) Enlarged villus

In addition to the microvilli and villi, a third set of projections called ***circular folds,*** or ***plicae circulares*** (PLĪ-kē SER-kyoo-lar-es), further increases the surface area for absorption and digestion. The folds are permanent ridges, about 10 mm (0.4 in.) high, in the mucosa. Some of the folds extend all the way around the intestine, and others extend only partway around. The folds begin near the proximal portion of the duodenum and terminate at about the midportion of the ileum. The circular folds enhance absorption by causing the chyme to spiral, rather than to move in a straight line, as it passes through the small intestine. Since the folds and villi decrease in size in the distal ileum, most absorption occurs in the duodenum and jejunum.

The ***muscularis*** of the small intestine consists of two layers of smooth muscle. The outer, thinner layer contains longitudinally arranged fibers (cells). The inner, thicker layer contains circularly arranged fibers. Except for a major portion of the duodenum, the serosa (or visceral peritoneum) completely covers the small intestine. Additional histological aspects of the small intestine are shown in Figure 23-17.

There is an abundance of lymphatic tissue in the form of lymphatic nodules, masses of lymphatic tissue not covered by a capsule wall. ***Solitary lymphatic nodules*** are most numerous in the lower part of the ileum. Groups of lymphatic nodules, referred to as ***aggregated lymphatic follicles (Peyer's patches),*** are numerous in the ileum. Their purpose is to prevent bacteria from entering the bloodstream.

ACTIVITIES

Chyme entering the small intestine contains partially digested carbohydrates, partially digested proteins, and essentially undigested lipids. The completion of the digestion

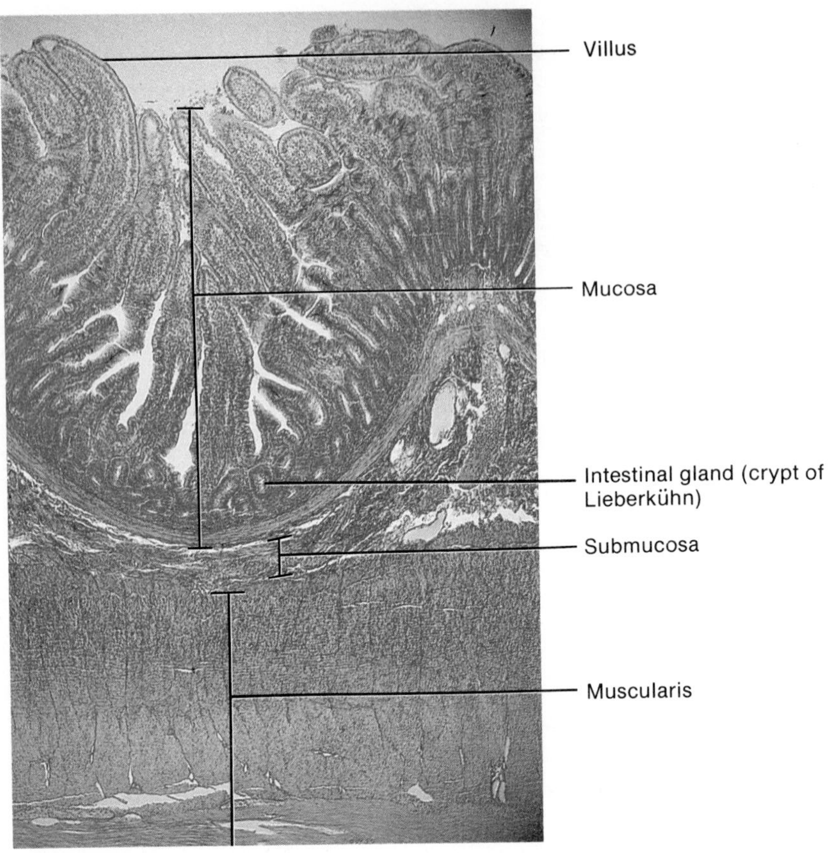

(a) Photomicrograph of portion of wall of duodenum, (425 x)

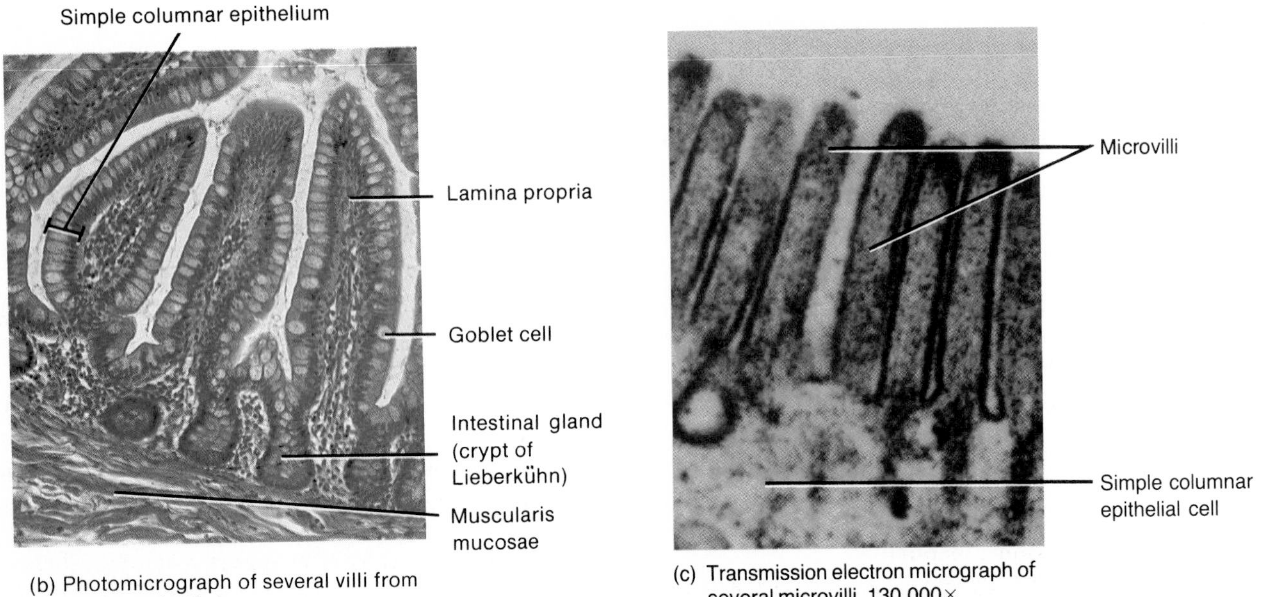

(b) Photomicrograph of several villi from the ileum (125 x)

(c) Transmission electron micrograph of several microvilli, 130,000×

FIGURE 23-17 Histology of the small intestine. (a) © Birke, Peter Arnold. (b) Courtesy of Ed Reschke. (c) Bhatnagar, Visuals Unlimited.

of carbohydrates into monosaccharides, proteins into amino acids, and lipids into fatty acids, glycerol, and glycerides is a collective effort of pancreatic juice, bile, and intestinal juice in the small intestine.

All the chemical and mechanical phases of digestion from the mouth down through the small intestine are directed toward changing food into forms that can pass through the epithelial cells lining the mucosa into the underlying blood and lymphatic vessels. Passage of these digested nutrients from the GI tract into the blood or lymph is called ***absorption.***

Intestinal juice is a clear yellow fluid secreted in amounts of about 2 to 3 liters (about 2 to 3 qt) a day. It has a pH of 7.6, which is slightly alkaline, and contains water and mucus. The juice is rapidly reabsorbed by the villi and provides a vehicle for the absorption of substances from chyme as they come in contact with the villi. The intestinal enzymes are formed in the epithelial cells that line the villi. Some digestion by enzymes of the small intestine occurs within the cells on the surfaces of their microvilli. As small intestinal cells containing enzymes slough off into the lumen of the intestine, they break apart and release small quantities of enzymes that digest food in the chyme. Thus, some digestion by enzymes of the small intestine occurs in or on the epithelial cells that line the villi, rather than in the lumen exclusively, as in other parts of the gastrointestinal tract. Among the enzymes produced by small intestinal cells are three carbohydrate-digesting enzymes called ***maltase, sucrase,*** and ***lactase;*** protein-digesting enzymes called ***peptidases*** (***aminopeptidase*** and ***dipeptidase***); and two nucleic acid–digesting enzymes, ***ribonuclease*** and ***deoxyribonuclease.***

The movements of the small intestine are divided into two types: segmentation and peristalsis. ***Segmentation*** is the major movement of the small intestine. It is strictly a localized contraction in areas containing food. It mixes chyme with the digestive juices and brings the particles of food into contact with the mucosa for absorption. It does not push the intestinal contents along the tract. Segmentation starts with the contractions of circular muscle fibers in several neighboring portions of the small intestine, an action that constricts the intestine into segments. Next, muscle fibers that encircle the middle of each segment also contract, dividing each segment again. Finally, the fibers that contracted first relax, and each small segment unites with an adjoining small segment so that large segments are formed. This sequence of events is repeated 12 to 16 times a minute, sloshing the chyme back and forth. Segmentation depends mainly on intestinal distension, which initiates nerve impulses to the central nervous system. Returning parasympathetic impulses increase motility. Sympathetic impulses decrease intestinal motility.

Peristalsis propels the chyme onward through the intestinal tract. Peristaltic contractions in the small intestine are normally very weak compared with those in the esophagus or stomach. Chyme moves through the small intestine at a rate of about 1 cm/min. Thus, chyme remains in the small intestine for three to five hours. Peristalsis, like segmentation, is initiated by distension and controlled by the autonomic nervous system.

About 90 percent of all absorption of nutrients takes place throughout the length of the small intestine. The other 10 percent occurs in the stomach and large intestine. Any undigested or unabsorbed material left in the small intestine is passed on to the large intestine. Absorption of materials in the small intestine occurs specifically through the villi and depends on diffusion, facilitated diffusion, osmosis, and active transport.

BLOOD AND NERVE SUPPLY

The arterial blood supply of the small intestine is from the superior mesenteric artery and the gastroduodenal artery, coming from the hepatic artery of the celiac trunk. Blood is returned by way of the superior mesenteric vein, which, with the splenic vein, forms the hepatic portal vein.

The nerves to the small intestine are supplied by the superior mesenteric plexus. The branches of the plexus contain postganglionic sympathetic fibers, preganglionic parasympathetic fibers, and afferent fibers. The afferent fibers are both vagal and of spinal nerves. In the wall of the small intestine are two autonomic plexuses: the myenteric plexus between the muscular layers and the submucosal plexus in the submucosa. The nerve fibers are derived chiefly from the sympathetic division of the autonomic nervous system and partly from the vagus (X) nerve.

LARGE INTESTINE

The overall functions of the large intestine are the completion of absorption, the manufacture of certain vitamins, the formation of feces, and the expulsion of feces from the body.

ANATOMY

The ***large intestine*** is about 1.5 m (5 ft) in length and averages 6.5 cm (2.5 in.) in diameter. It extends from the ileum to the anus and is attached to the posterior abdominal wall by a fold of parietal peritoneum called ***mesocolon.*** Structurally, the large intestine is divided into three principal regions: cecum, colon, and rectum (Figure 23-18a).

The opening from the ileum into the large intestine is guarded by a fold of mucous membrane called the ***ileocecal sphincter*** (***valve***)**.** This structure allows materials from the small intestine to pass into the large intestine. Hanging below the ileocecal valve is the ***cecum,*** a blind pouch about 6 cm (2.5 in.) long. Attached to the cecum is a twisted, coiled tube, measuring about 8 cm (3 in.) in length, called the ***vermiform appendix*** (*vermis* = worm; *appendix* = appendage). The mesentery of the appendix, called the ***meso-***

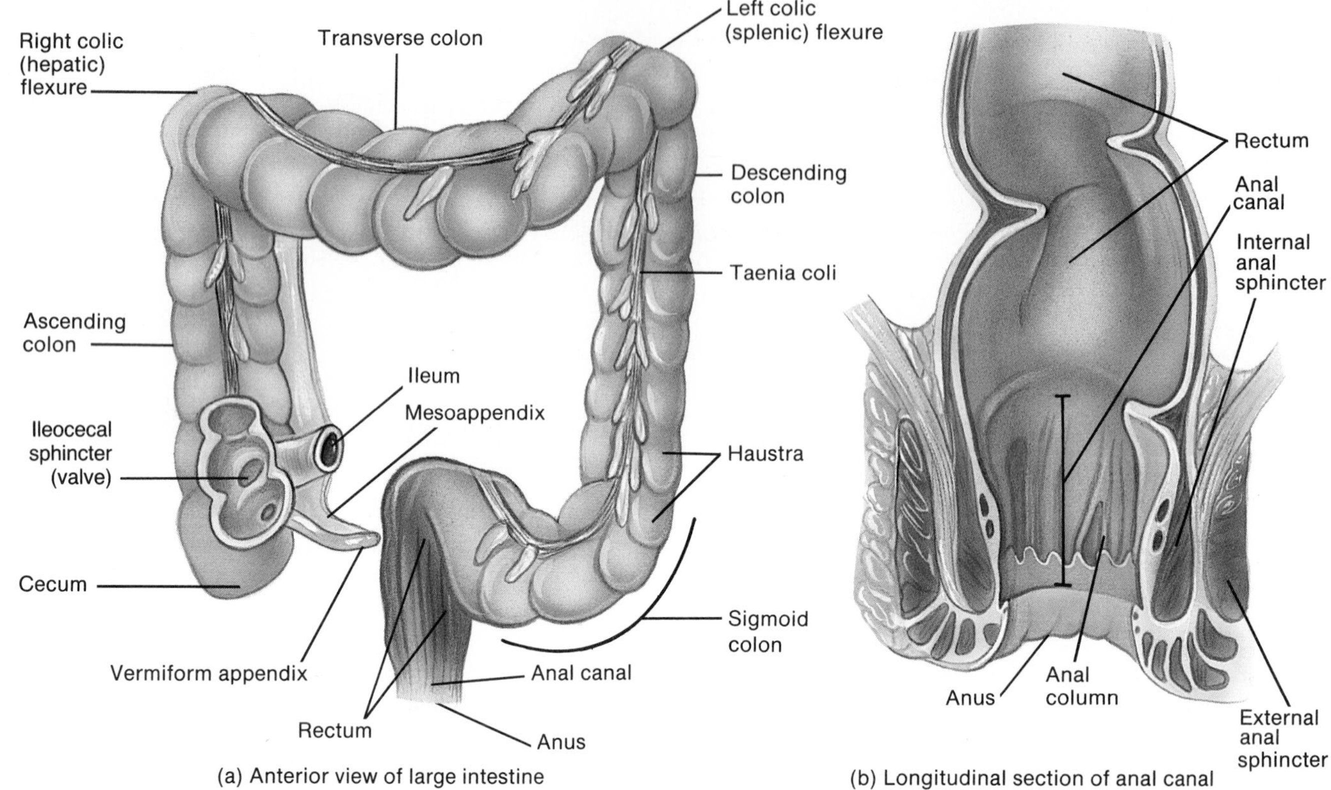

FIGURE 23-18 Large intestine.

appendix, attaches the appendix to the inferior part of the ileum and adjacent part of the posterior abdominal wall.

The open end of the cecum merges with a long tube called the ***colon*** (*kolon* = food passage). The colon is divided into ascending, transverse, descending, and sigmoid portions. The ***ascending colon,*** which is retroperitoneal (external to the peritoneum), ascends on the right side of the abdomen, reaches the undersurface of the liver, and turns abruptly to the left. Here it forms the ***right colic (hepatic) flexure.*** The colon continues across the abdomen to the left side as the ***transverse colon.*** It curves beneath the lower end of the spleen on the left side as the ***left colic (splenic) flexure*** and passes downward to the level of the iliac crest as the ***descending colon,*** also a retroperitoneal structure. The ***sigmoid colon*** begins near the left iliac crest, projects inward to the midline, and terminates as the rectum at about the level of the third sacral vertebra.

The ***rectum,*** the last 20 cm (8 in.) of the GI tract, lies anterior to the sacrum and coccyx. The terminal 2 to 3 cm (1 in.) of the rectum is called the ***anal canal*** (Figure 23-18b). The mucous membrane of the anal canal is arranged in longitudinal folds called ***anal columns*** that contain a network of arteries and veins. The opening of the anal canal to the exterior is called the ***anus.*** It is guarded by an internal sphincter of smooth muscle (involuntary) and an external sphincter of skeletal muscle (voluntary). Normally the anus is closed except during the elimination of the wastes of digestion.

The medical specialty that deals with the diagnosis and treatment of disorders of the rectum and anus is called ***proctology*** (prok-TOL-ō-jē; *proct* = rectum; *logos* = study of).

CLINICAL APPLICATION

Hemorrhoids

As noted in Chapter 14, varicosities in any veins involve inflammation and enlargement. Varicosities of the rectal veins are known as ***hemorrhoids (piles).*** Hemorrhoids develop when the veins are put under pressure and become engorged with blood. If the pressure continues, the wall of the vein stretches. Such a distended vessel oozes blood, and bleeding or itching are usually the first signs that a hemorrhoid has developed. Stretching of a vein also favors clot formation, further aggravating swelling and pain. Initially contained within the anus (first degree), they gradually enlarge until they prolapse or extend outward on defecation (second degree) and finally remain prolapsed through the anal orifice (third degree). Hemorrhoids may be caused by constipation. Repeated straining during defecation forces blood down into the superior hemorrhoidal plexus, increasing pressure in these veins. Constipation is related to low-fiber diets, especially in North America. One hypothesis links increased intra-abdominal pressure, caused by straining during evacuation of firm feces, directly to hemorrhoids or varicose veins. The most widely used treatment for hemorrhoids that bleed consistently and cause severe pain and discomfort is ***rubber band ligation,*** in which a rubber band is tied around the hemorrhoid, cutting off its blood supply. In a few days, the hemorrhoid dries up and falls off. A newer treatment is called ***infrared photocoagulation,*** in which a very high-energy light beam coagulates the hemorrhoid. (A laser can be used for the same purpose.) Other treatments include sclerotherapy (injection of a substance that causes atrophy of the hemorrhoid) and surgery.

HISTOLOGY

The wall of the large intestine differs from that of the small intestine in several respects. No villi or permanent circular folds are found in the mucosa, which does, however, contain simple columnar epithelium with numerous goblet cells (Figure 23-19). The columnar cells function primarily in water absorption. The goblet cells secrete mucus that lubricates the colonic contents as they pass through the colon. Both columnar and mucous cells are located in long, straight, tubular intestinal glands that extend the full thickness of the mucosa. Solitary lymphatic nodules are also found in the mucosa. The submucosa of the large intestine is similar to that found in the rest of the gastrointestinal tract. The muscularis consists of an external layer of longitudinal muscles and an internal layer of circular muscles. Unlike other parts of the gastrointestinal tract, portions of the longitudinal muscles are thickened, forming three conspicuous longitudinal bands referred to as ***taeniae coli*** (TĒ-nē-a KŌ-lī). Each band runs the length of most of the large intestine (see Figure 23-18a). Tonic contractions of the bands gather the colon into a series of pouches called ***haustra*** (HAWS-tra; *haustrum* = shaped like a pouch), which give the colon its puckered appearance. The serosa of the large intestine is part of the visceral peritoneum. Small pouches of visceral peritoneum filled with fat are attached to taeniae coli and are called ***epiploic appendages.***

ACTIVITIES

The passage of chyme from the ileum into the cecum is regulated by the action of the ileocecal sphincter. The valve normally remains mildly contracted so that the passage of chyme into the cecum is usually a slow process. Immediately following a meal, there is a ***gastroileal reflex*** in which ileal peristalsis is intensified and any chyme in the ileum is forced into the cecum. The hormone stomach gastrin also relaxes the sphincter. Whenever the cecum is distended, the degree of contraction of the ileocecal sphincter is intensified.

Movements of the colon begin when substances enter through the ileocecal sphincter. Since chyme moves through the small intestine at a fairly constant rate, the time required for a meal to pass into the colon is determined by gastric evacuation time. As food passes through the ileocecal sphincter, it fills the cecum and accumulates in the ascending colon.

One movement characteristic of the large intestine is ***haustral churning.*** In this process, the haustra remain relaxed and distended while they fill up. When the distension reaches a certain point, the walls contract and squeeze the contents into the next haustrum. ***Peristalsis*** also occurs, although at a slower rate than in other portions of the tract (3 to 12 contractions per minute). A final type of movement is ***mass peristalsis,*** a strong peristaltic wave that begins at about the middle of the transverse colon and drives the colonic contents into the rectum. Food in the stomach initiates this reflex action in the colon. Thus, mass peristalsis usually takes place three or four times a day, during a meal or immediately after.

The last stage of digestion occurs through bacterial action, not through enzymes secreted by the colon. Mucus is secreted by the glands of the large intestine, but no enzymes are secreted. Chyme is prepared for elimination by the action of bacteria. These bacteria ferment any remaining carbohydrates and release hydrogen, carbon dioxide, and methane gas. These gases contribute to flatus (gas) in the colon. They also convert remaining proteins to amino acids and break down the amino acids into simpler substances: indole, skatole, hydrogen sulfide, and fatty acids. Some of the indole and skatole are carried off in the feces and contribute to their odor. The rest are absorbed and transported to the liver, where they are converted to less toxic compounds and excreted in the urine. Bacteria also decompose bilirubin to simpler pigments (urobilinogen), which give feces their brown color. Several vitamins needed for normal metabolism, including some B vitamins and vitamin K, are synthesized by bacterial action and absorbed.

By the time the chyme has remained in the large intestine for 3 to 10 hours, it has become solid or semisolid as a result of absorption, principally of water, and is now known as ***feces.*** Chemically, feces consist of water, inorganic salts, sloughed-off epithelial cells from the mucosa of the gastrointestinal tract, bacteria, products of bacterial decomposition, and undigested parts of food.

Although most water absorption occurs in the small intestine, the large intestine absorbs enough to make it an important organ in maintaining the body's water balance. Of the 0.5 to 1.0 liter that enters the large intestine, all but about 100 ml is absorbed. The absorption is greatest in the cecum and ascending colon. The large intestine also absorbs electrolytes, including sodium and chloride, and some vitamins.

CLINICAL APPLICATION

Occult Blood

The term ***occult blood*** refers to blood that is hidden; it is not detectable by the human eye. The main diagnostic value of occult blood testing is to screen for colorectal cancer. Two substances frequently examined for occult blood are feces and urine. There are several types of products available for at-home testing for hidden blood in feces. The tests are based on color changes when reagents are added to feces. The presence of occult blood in urine may be detected at home by using dip-and-read reagent strips.

Mass peristaltic movements push fecal material from the sigmoid colon into the rectum. The resulting distension

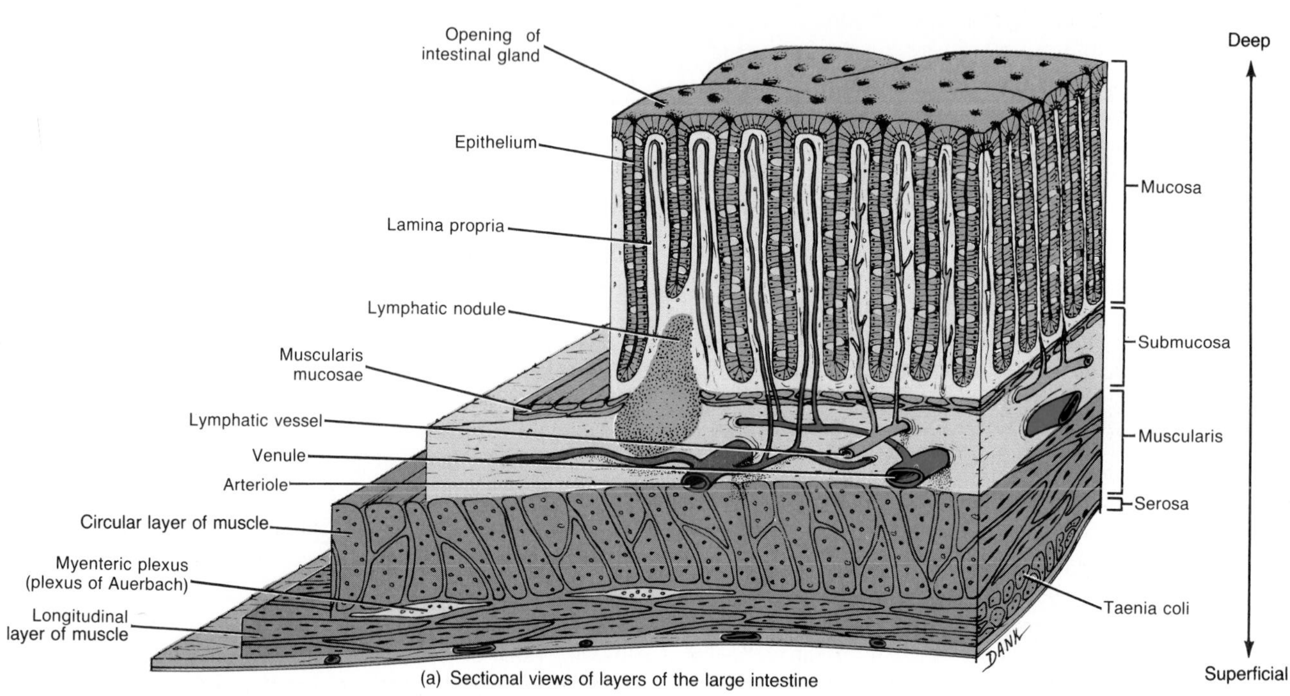

(a) Sectional views of layers of the large intestine

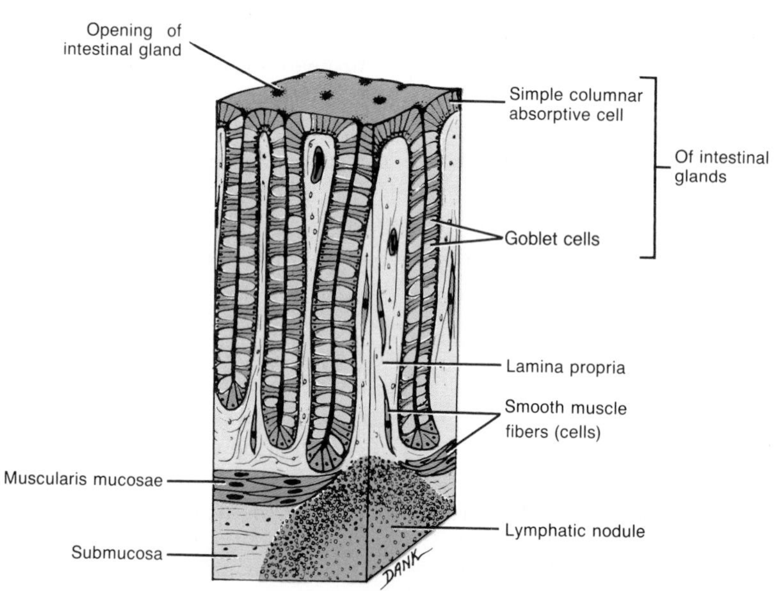

(b) Sectional views of the large intestinal mucosa

FIGURE 23-19 Histology of the large intestine. (c) Courtesy of Bruce Iverson. (d) Courtesy of CNRI, Photo Researchers.

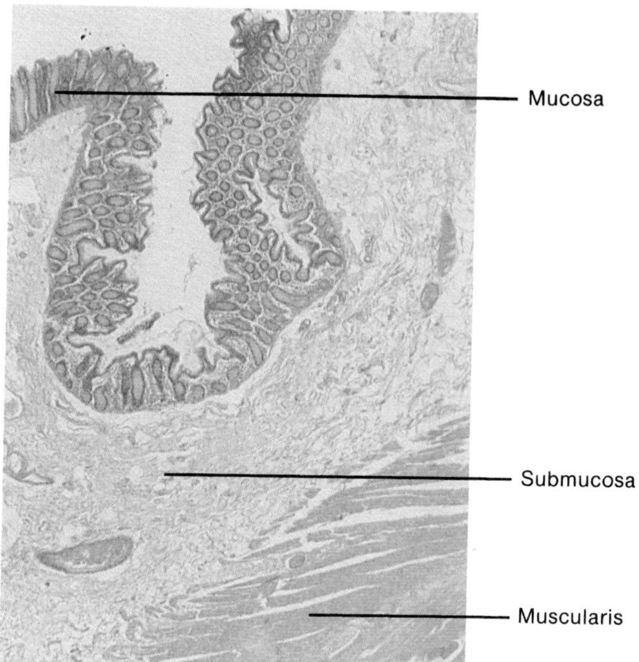

(c) Photomicrograph of a portion of the wall of the large intestine (100 x)

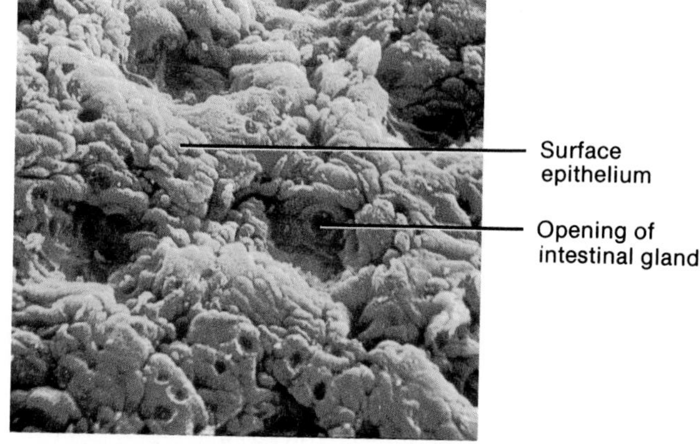

(d) Scanning electron micrograph of large intestinal mucosa, (600 x)

of the rectal wall stimulates pressure-sensitive receptors, initiating a reflex for ***defecation,*** the emptying of the rectum. The defecation reflex occurs as follows. In response to distension of the rectal wall, the receptors send nerve impulses to the sacral spinal cord. Motor impulses from the cord travel along parasympathetic nerves back to the descending colon, sigmoid colon, rectum, and anus. Contraction of the longitudinal rectal muscles shortens the rectum, thereby increasing the pressure inside it. The pressure along with voluntary contractions of the diaphragm and abdominal muscles forces the internal sphincter open, and the feces are expelled through the anus. The external sphincter is voluntarily controlled. If it is voluntarily relaxed, defecation occurs; if it is voluntarily constricted, defecation can be postponed. Voluntary contractions of the diaphragm and abdominal muscles aid defecation by increasing the pressure inside the abdomen, which pushes the walls of the sigmoid colon and rectum inward. If defecation does not occur, the feces back into the sigmoid colon until the next wave of mass peristalsis again stimulates the pressure-sensitive receptors, creating the desire to defecate.

In infants, the defecation reflex causes automatic emptying of the rectum without the voluntary control of the external anal sphincter. In certain instances of spinal cord injury, the reflex is abolished and defecation requires supportive measures, such as cathartics (laxatives).

Diarrhea refers to frequent defecation of liquid feces caused by increased motility of the intestines. Since chyme passes too quickly through the small intestine and feces pass too quickly through the large intestine, there is not enough time for absorption. Like vomiting, diarrhea can result in dehydration and electrolyte imbalances. Diarrhea may be caused by stress and microbes that irritate the gastrointestinal mucosa.

Constipation refers to infrequent or difficult defecation. It is caused by decreased motility of the intestines in which feces remain in the colon for prolonged periods of time. As it does so, there is considerable water absorption, and feces become dry and hard. Constipation may be caused by improper bowel habits, spasms of the colon, insufficient bulk in the diet, inadequate fluid intake, lack of exercise, and emotions. Usual treatment for constipation is a mild cathartic, such as milk of magnesia, that induces defecation. However, many physicians maintain that laxatives are habit-forming, and that adding bulk to the diet, increasing one's amount of exercise, and improving fluid intake are safer ways of controlling this common problem.

BLOOD AND NERVE SUPPLY

The arterial supply of the cecum and colon is derived from branches of the superior mesenteric and inferior mesenteric arteries. The venous return is by way of the superior and inferior mesenteric veins ultimately to the hepatic portal vein and into the liver. The arterial supply of the rectum and anal canal is derived from the superior, middle, and inferior rectal arteries. The rectal veins correspond to the rectal arteries.

The nerves to the large intestine consist of sympathetic, parasympathetic, and afferent components. The sympathetic innervation is derived from the celiac, superior, and inferior mesenteric ganglia and superior and inferior mesenteric plexuses. The fibers reach the viscera by way of the thoracic and lumbar splanchnic nerves. The parasympathetic innervation is derived from the vagus (X) and pelvic splanchnic nerves.

AGING AND THE DIGESTIVE SYSTEM

Overall general changes associated with aging of the digestive system include decreasing secretory mechanisms, decreasing motility (muscular movement) of the digestive organs, loss of strength and tone of the muscular tissue and its supporting structures, changes in neurosensory feedback regarding enzyme and hormone release, and diminished response to pain and internal sensations. Specific changes include reduced sensitivity to mouth irritation and sores, loss of taste, pyorrhea, difficulty in swallowing, hiatal hernia, cancer of the esophagus, gastritis, peptic ulcer, and gastric cancer. Changes in the small intestine include duodenal ulcers, appendicitis, malabsorption, and maldigestion. Other pathologies that increase in incidence are gallbladder problems, jaundice, cirrhosis, and acute pancreatitis. Changes associated with the large intestine, such as constipation, cancer of the colon or rectum, hemorrhoids, and diverticular disease of the colon, also occur.

DEVELOPMENTAL ANATOMY OF THE DIGESTIVE SYSTEM

About the fourteenth day after fertilization, the cells of the endoderm form a cavity referred to as the ***primitive gut*** (Figure 23-20). Soon after the mesoderm forms and splits into two layers (somatic and splanchnic), the splanchnic mesoderm associates with the endoderm of the primitive gut. Thus the primitive gut has a double-layered wall. The ***endodermal layer*** gives rise to the *epithelial lining* and *glands* of most of the gastrointestinal tract, and the ***mesodermal layer*** produces its *smooth muscle* and *connective tissue*.

The primitive gut elongates, and about the latter part of the third week it differentiates into an anterior ***foregut,*** a central ***midgut,*** and a posterior ***hindgut.*** Until the fifth week of development, the midgut opens into the yolk sac. After that time, the yolk sac constricts, detaches from the midgut, and the midgut seals. In the region of the foregut, a depression consisting of ***ectoderm,*** the ***stomodeum,*** appears. This depression develops into the *oral cavity*. The ***oral membrane*** that separates the foregut from the stomodeum ruptures during the fourth week of development and thus the foregut becomes continuous with the outside of the embryo through the oral cavity. Another depression consisting of ectoderm, the ***proctodeum,*** forms in the hindgut and goes on to develop into the *anus*. The ***cloacal membrane,*** which separates the hindgut from the proctodeum, ruptures, and the hindgut becomes continuous with the outside of the embryo through the anus. Thus, the gastrointestinal tract forms a continuous tube from the mouth to the anus.

The foregut develops into the *pharynx, esophagus, stomach,* and a *portion of the duodenum*. The midgut is transformed into the *remainder of the duodenum,* the *jejunum,* the *ileum,* and *portions of the large intestine* (cecum, appendix, ascending colon, and most of the transverse colon). The hindgut develops into the *remainder of the large intestine,* except for a portion of the anal canal that is derived from the proctodeum.

As development progresses, the endoderm at various places along the foregut develops into hollow buds that grow into the mesoderm. These buds will develop into the *salivary glands, liver, gallbladder,* and *pancreas*. Each of the glands retains a connection with the gastrointestinal tract through ducts.

APPLICATIONS TO HEALTH

DENTAL CARIES

Dental caries, or tooth decay, involve a gradual demineralization (softening) of the enamel and dentin. If untreated, various microorganisms may invade the pulp, causing inflammation and infection, with subsequent death (necrosis) of the pulp and abscess of the alveolar bone surrounding the root's apex. Such teeth are treated by root canal therapy.

The process of dental caries is initiated when bacteria act on sugars, giving off acids that demineralize the enamel. Microbes that digest sugar into lactic acid are common in the mouth cavity. One that seems to be cariogenic (caries-causing) is the bacterium *Streptococcus mutans*. ***Dextran,*** a sticky polysaccharide produced from sucrose, forms a capsule around the bacteria, causing them to stick to the teeth. Masses of bacterial cells, dextran, and other debris adhering to teeth are collectively called ***dental plaque.*** Saliva cannot reach the tooth surface to buffer the acid because the plaque covers the teeth. Brushing the teeth immediately after eating removes the plaque from flat surfaces before the bacteria have a chance to go to work. Dentists also suggest that the plaque between the teeth be removed every 24 hours with dental floss or by flushing with a water irrigation device or by using an antiplaque dental rinse.

Preventive measures other than brushing, flossing, and irrigation include prenatal diet supplements (chiefly vitamin D, calcium, and phosphorus), fluoride treatments to protect against acids during the period when teeth are being calcified, and dental sealing. In this last procedure, pits and fissures that serve as reservoirs for dental plaque are sealed by the application of a permanent, durable plastic sealant. The sealant is applied to the prepared biting surfaces of

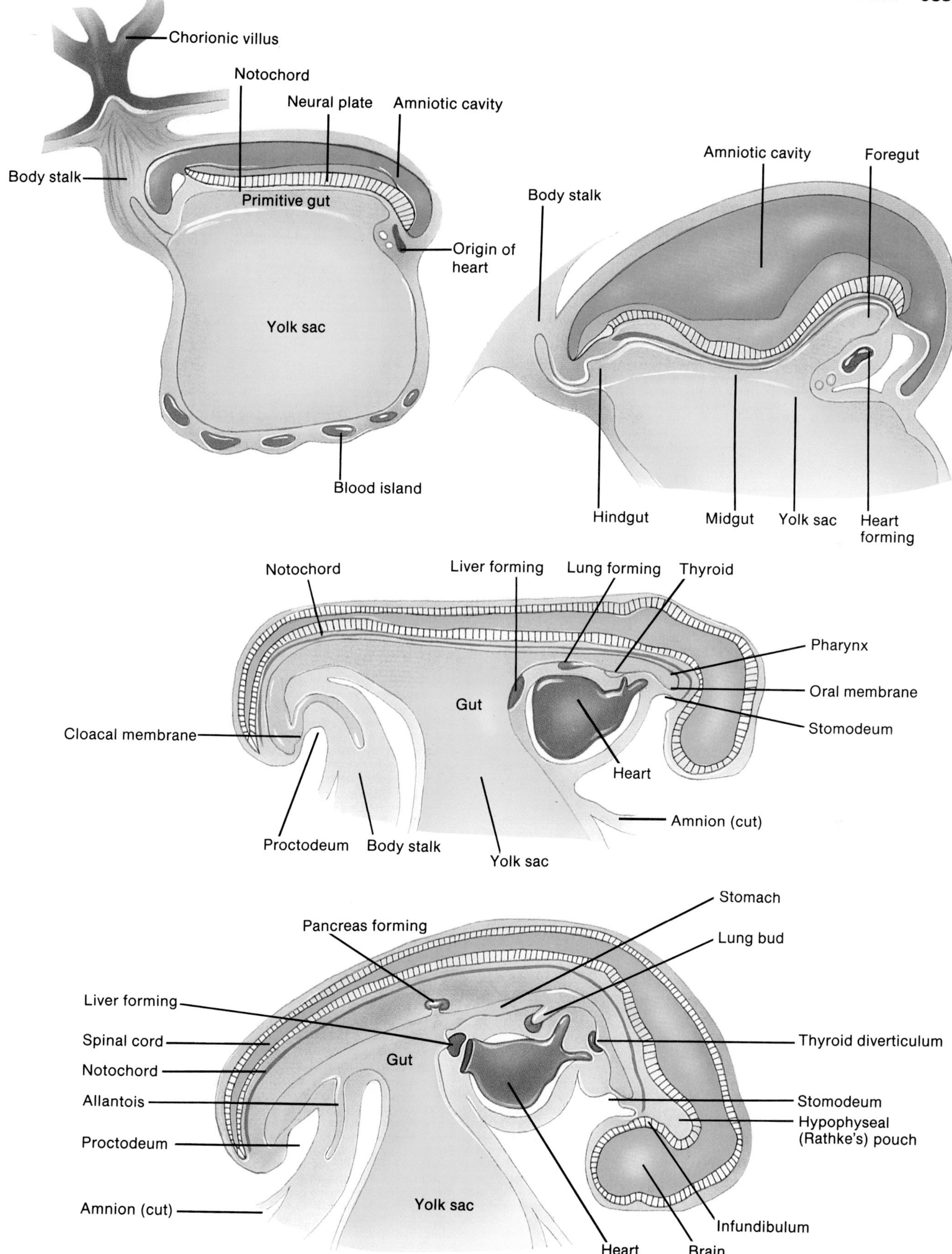

FIGURE 23-20 Development of the digestive system.

the molar teeth. Dental sealants are used primarily for children and adolescents, although some adults could benefit from them.

PERIODONTAL DISEASE

Periodontal disease is a collective term for a variety of conditions characterized by inflammation and degeneration of the gingivae, alveolar bone, periodontal ligament, and cementum. One such condition is called ***pyorrhea.*** The initial symptoms are enlargement and inflammation of the soft tissue and bleeding gums. Without treatment, the soft tissue may deteriorate and the alveolar bone may be resorbed, causing loosening of the teeth and recession of the gums. Periodontal diseases are frequently caused by poor oral hygiene; by local irritants, such as bacteria, impacted food, and cigarette smoke; or by a poor "bite."

PERITONITIS

Peritonitis is an acute inflammation of the serous membrane lining the abdominal cavity and covering the abdominal viscera (peritoneum). One possible cause is contamination of the peritoneum by pathogenic bacteria from the external environment. This contamination could result from accidental or surgical wounds in the abdominal wall or from perforation or rupture of organs with consequent exposure to the outside environment. Another possible cause is perforation of the walls of organs that contain bacteria or chemicals beneficial to the organ but toxic to the peritoneum.

PEPTIC ULCERS

An ***ulcer*** is a craterlike lesion in a membrane. Ulcers that develop in areas of the gastrointestinal tract exposed to acid gastric juice are called ***peptic ulcers.*** Most occur on the lesser curvature of the stomach, where they are called ***gastric ulcers,*** or in the first part of the duodenum, where they are called ***duodenal ulcers.***

Hypersecretion of acid gastric juice seems to be the immediate cause of duodenal ulcers. In gastric ulcer patients, because the stomach wall is highly adapted to resist gastric juice through the secretion of mucus, the cause may be hyposecretion of mucus. Hypersecretion of pepsin also may contribute to ulcer formation.

Among the factors believed to stimulate an increase in acid secretion are emotions, cigarette smoking, certain foods or medications (alcohol, coffee, aspirin), and overstimulation of the vagus (X) nerve. Normally, the mucous membrane lining the stomach and duodenal walls resists the secretions of hydrochloric acid and pepsin. In some people, however, this resistance breaks down and an ulcer develops. Some evidence suggests that peptic ulcers may be caused by the bacterium *Campylobacter pyloridis.*

APPENDICITIS

Appendicitis is an inflammation of the vermiform appendix. It is preceded by obstruction of the lumen of the appendix by fecal material, inflammation, a foreign body, carcinoma of the cecum, stenosis, or kinking of the organ. The infection that follows may result in edema, ischemia, gangrene, and perforation. Rupture of the appendix develops into peritonitis.

Typically, appendicitis begins with referred pain in the umbilical region of the abdomen, followed by anorexia (lack or loss of appetite for food), nausea, and vomiting. After several hours, the pain localizes in the right lower quadrant (RLQ) and is continuous, dull or severe, and intensified by coughing, sneezing, or body movements.

Early appendectomy (surgical removal of the appendix) is recommended in all suspected cases because it is safer to operate than to risk gangrene, rupture, and peritonitis.

TUMORS

Both benign and malignant ***tumors*** can occur in all parts of the gastrointestinal tract. Although benign growths are much more common than malignant ones, cancers of the gastrointestinal tract are responsible for 30 percent of all deaths from cancer in the United States.

Colorectal cancer is one of the most common malignant diseases, ranking second to cancer of the lungs in males and third to cancer of the lungs and breasts in females. The overall mortality rate is nearly 60 percent. Over 50 percent of colorectal cancers occur in the sigmoid colon and rectum. Both environmental and genetic factors affect the development of colorectal cancer. Dietary fiber, retinoids, calcium, and selenium may be protective, whereas intake of animal fat and protein may cause an increase in the disease. Genetics plays a very important role in that an inherited predisposition contributes to more than half of all cases of colorectal cancer. There is considerable evidence that most colorectal cancers develop from benign polyps. Signs and symptoms of colorectal cancer include changes in the normal pattern of bowel habits (diarrhea, constipation), cramping, abdominal pain, and rectal bleeding, either visible or occult. Screening for colorectal cancer includes fecal occult blood testing, digital rectal examination, sigmoidoscopy, colonoscopy, and barium enema. The only definitive treatment for gastrointestinal tumors, if they cannot be removed endoscopically, is surgery.

DIVERTICULITIS

Diverticula are saclike outpouchings of the wall of the colon in places where the muscularis has become weak. The development of diverticula is called ***diverticulosis.*** Many people who develop diverticulosis are asymptomatic and experience no complications. About 15 percent of people with diverticulosis will eventually develop an inflammation within diverticula, a condition known as ***diverticulitis.*** Research indicates that diverticula form because of lack of sufficient bulk in the colon during segmentation.

CYSTIC FIBROSIS (CF)

Cystic fibrosis (CF) is an inherited disease of the exocrine glands that affects the pancreas, respiratory passageways, and salivary and sweat glands. It is the most common lethal genetic disease of Caucasians—5 percent of the population are thought to be genetic carriers. The cause of cystic fibrosis is linked to an inability of chloride ions (Cl^-) to cross epithelial cells in affected regions of the body. It is characterized by the production of thick exocrine secretions that do not drain easily from the respiratory passageways. The buildup of the secretions leads to inflammation and replacement of injured cells with connective tissue that blocks these passageways. One of the prominent features is blockage of the pancreatic ducts so that the digestive enzymes cannot reach the intestine. Since pancreatic juice contains the only fat-digesting enzyme, the person fails to absorb fats or fat-soluble vitamins and thus suffers from vitamin A, D, and K deficiency diseases. Calcium needs fat to be absorbed, so tetany also may result.

CIRRHOSIS

Cirrhosis refers to a distorted or scarred liver as a result of chronic inflammation. The parenchymal (functional) liver cells are replaced by fibrous or adipose connective tissue. The symptoms of cirrhosis include jaundice, edema in the legs, uncontrolled bleeding, and increased sensitivity to drugs. Cirrhosis may be caused by hepatitis (inflammation of the liver), certain chemicals that destroy liver cells, parasites that infect the liver, and alcoholism.

HEPATITIS

Hepatitis refers to inflammation of the liver and can be caused by viruses, drugs, and chemicals, including alcohol. Clinically, several types are recognized.

Hepatitis A (infectious hepatitis) is caused by hepatitis A virus and is spread by fecal contamination of food, clothing, toys, eating utensils, and so forth (fecal–oral route). It is generally a mild disease of children and young adults characterized by anorexia, malaise, nausea, diarrhea, fever, and chills. Eventually, jaundice appears. It does not cause lasting liver damage. Most people recover in four to six weeks.

Hepatitis B (serum hepatitis) is caused by hepatitis B virus and is spread primarily by sexual contact and contaminated syringes and transfusion equipment. It can also be spread by saliva and tears. Hepatitis B virus can be present for years or even a lifetime and can produce cirrhosis and possibly cancer of the liver. Persons who harbor the active hepatitis B virus are at risk for cirrhosis and also become carriers. A vaccine produced through recombinant DNA technology (Recombivax HB) is available for hepatitis B.

Hepatitis C (non-A, non-B hepatitis) is a form of hepatitis that cannot be traced to either hepatitis A or hepatitis B viruses. It is clinically similar to hepatitis B and is often spread by blood transfusions. It accounts for considerably more posttransfusion hepatitis than that related to hepatitis B. The hepatitis C virus can cause cirrhosis and possibly liver cancer.

GALLSTONES

The fusion of crystals of cholesterol in bile is the beginning of 95 percent of all ***gallstones (biliary calculi).*** Following their formation, gallstones gradually grow in size and number and may cause minimal, intermittent, or complete obstruction to the flow of bile from the gallbladder into the duct system. If obstruction of the outlet occurs and the gallbladder cannot empty as it normally does after eating, the pressure within it increases, and the individual may have intense pain or discomfort (***biliary colic***). Jaundice, due to the inability to secrete bilirubin into the intestine, will accompany complete biliary obstruction.

OBESITY

Obesity is defined as a body weight 10 to 20 percent above a desirable standard as the result of an excessive accumulation of fat. There is little doubt that even moderate obesity is hazardous to health. It is implicated as a risk factor in cardiovascular disease, hypertension, pulmonary disease, diabetes mellitus (type II), arthritis, certain cancers (uterus and colon), varicose veins, and gallbladder disease. Also, loss of body fat in obese individuals has been shown to elevate HDL cholesterol, the type associated with prevention of cardiovascular disease.

Classification

In ***hypertrophic (adult-onset) obesity,*** there is an increase in the amount of fat in adipocytes but no increase in the number of fat cells. Such individuals tend to be thin or average in weight until about age 20 to 40, at which time weight gain begins. The gain may be associated with an imbalance between caloric intake and utilization. People with hypertrophic obesity tend to have fat distributed in central locations (''middle-age spread''), and their problem is easier to treat. The second category of obesity is ***hyperplastic (lifelong) obesity,*** in which there is an increase in the number of adipocytes, as well as an increase in the amount of fat within them. These individuals tend to be obese as children and have a large spurt in weight gain during adolescence. After adolescence, the number of adipocytes remains about the same throughout life. Such people are generally grossly obese, and the distribution of fat is both central and peripheral. Treatment is considerably more difficult.

Causes

In a relatively few cases, obesity may result from trauma or tumors in the food-regulating centers in the hypothalamus. In most cases of obesity, no specific cause can be identified. Contributing factors include overeating habits taught early in life, overeating to relieve tension, and social customs. Recently, obesity has been strongly linked to genetic factors. Studies indicate that some people inherit a low metabolic rate and that they become obese not because they eat too much but because they burn calories too slowly. In addition, it has been learned that abnormally low levels of a protein, called *adipsin,* produced by adipocytes, may be linked to obesity. A possible explanation is that adequate amounts of adipsin may signal the satiety center of the hypothalamus to diminish appetite.

Treatment

Reduction of body weight involves keeping caloric intake well below energy expenditure. The goals during weight decrease are:

1. Loss of body fat with a minimal accompanying breakdown of lean tissue.
2. Maintenance of physical and emotional fitness during the reducing period.
3. Establishment of eating and exercise habits to maintain weight at the recommended level.

Morbid obesity refers to obese individuals whose body weight exceeds their ideal weight by more than 100 percent. The condition is so named because it is associated with serious and life-threatening conditions such as hypertension, diabetes mellitus, and atherosclerosis. For individuals with morbid obesity, surgery may be an alternative to control weight. The suitable patients for surgical procedures are individuals whose weight is more than 45.5 kg (100 lb) over the ideal body weight for height or 100 percent over the ideal and who have tried and failed to permanently lose substantial weight on medically supervised reduction regimens.

ANOREXIA NERVOSA

Anorexia nervosa is a chronic disorder characterized by self-induced weight loss, body image and other perceptual disturbances, and physiological changes that result from nutritional depletion. The subconsciously self-imposed starvation appears to be a response to emotional conflicts about self-identification and acceptance of a normal adult sex role. Patients with anorexia nervosa have a fixation on weight control and, often, the insistence of having a bowel movement every day despite a lack of adequate food intake. They abuse cathartics, which worsens the fluid/electrolyte/nutrient deficiencies. The disorder is found predominantly in young, single females and may be inherited. Abnormal patterns of menstruation, amenorrhea (absence of menstruation), and a lowered basal metabolic rate reflect the depressant effects of the starvation. Individuals may become emaciated and may ultimately die of starvation or one of its complications. Also associated with the disorder are osteoporosis, depression, and brain abnormalities coupled with impaired mental performance.

BULIMIA

A disorder that typically affects single, middle-class, young, white females is known as ***bulimia*** (*bous* = ox; *limos* = hunger), or ***binge–purge syndrome.*** It is characterized by overeating at least twice a week followed by purging by self-induced vomiting, strict dieting or fasting, vigorous exercise, or use of laxatives or diuretics. This binge–purge cycle occurs in response to fears of being overweight, stress, depression, and physiological disorders such as hypothalamic tumors.

Bulimia can upset the body's electrolyte balance and increase susceptibility to flu, salivary gland infections that result in bilateral parotid gland enlargement, pharyngeal scratches from self-induced gagging, dry skin, acne, muscle spasms, loss of hair, kidney and liver diseases, erosion of dental enamel from stomach acids, ulcers, hernias, constipation, and hormone imbalances. Although the cause of bulimia is unknown, some evidence suggests that it may be related to impaired release of cholecystokinin (CCK), a hormone that induces satiety, the sensation of being full to satisfaction.

DIETARY FIBER (ROUGHAGE) AND GI DISORDERS

A deficiency in our diet that has received recent attention is lack of dietary fiber or roughage. Dietary fiber consists of indigestible plant substances, such as cellulose, lignin, and pectin, found in fruits, vegetables, grains, and beans. Fiber may be classified as ***insoluble,*** which does not dissolve in water, and ***soluble,*** which does dissolve in water. Insoluble fiber includes the woody or structural parts of plants such as fruit and vegetable skins and the bran coating around wheat and corn kernels. Insoluble fiber passes through the GI tract largely unchanged and speeds up the passage of material through the tract. Soluble fiber is found in abundance in beans, oats, barley, broccoli, prunes, apples, and citrus fruits. It has the consistency of a gel and tends to slow the passage of material through the tract. Both insoluble and soluble fiber are important for digestive functioning. People who choose a fiber-rich, unrefined diet will greatly reduce their chances of developing diseases of overnutrition (such as obesity, diabetes, gallstones, and coronary heart disease), diseases of the underworked mouth (caries and periodontal disease), and diseases of an underfed large bowel (constipation, varicose veins, hemorrhoids, colon spasm, diverticulitis, appendicitis, and large intestinal cancer). Each of these conditions is directly related to the digestion and metabolism of food and the operation of the digestive system. There is also evidence that insoluble fiber may help protect against colon cancer and that soluble fiber may help lower cholesterol (LDL).

KEY MEDICAL TERMS ASSOCIATED WITH THE DIGESTIVE SYSTEM

Borborygmus (bor′-bō-RIG-mus) A rumbling noise caused by the propulsion of gas through the intestines.

Botulism (BOCH-yoo-lism; *botulus* = sausage) A type of food poisoning caused by a toxin produced by *Clostridium botulinum.* The bacterium is ingested when improperly cooked or preserved foods are eaten. The toxin inhibits nerve impulse conduction at synapses by inhibiting the release of acetylcholine. Symptoms include paralysis, nausea, vomiting, blurred or double vision, difficulty in speech, difficulty in swallowing, dryness of mouth, and general weakness.

Canker (KANG-ker) **sore** Painful ulcer on the mucous membrane of the mouth that affects females more frequently than males and usually occurs between ages 10 to 40; may be an autoimmune reaction.

Cholecystitis (kō′-lē-sis-TĪ-tis; *chole* = bile; *kystis* = bladder; *itis* = inflammation of) Inflammation of the gallbladder that often leads to infection. Some cases are caused by obstruction of the cystic duct with bile stones. Stagnating bile salts irritate the mucosa. Dead mucosal cells provide a medium for the growth of bacteria.

Cholelithiasis (kō′-lē-li-THĪ-a-sis; *lithos* = stone) The presence of gallstones.

Colitis (ko-LĪ-tis) Inflammation of the mucosa of the colon and rectum in which absorption of water and salts is reduced, producing watery, bloody feces and, in severe cases, dehydration and salt depletion. Spasms of the irritated muscularis produce cramps.

Colostomy (ko-LOS-tō-mē; *stomoun* = provide an opening) The diversion of the fecal stream through an opening in the colon, creating a surgical "stoma" (artificial opening) that is affixed to the exterior of the abdominal wall. This opening serves as a substitute anus through which feces are eliminated. A temporary colostomy may be done to allow a badly inflamed colon to rest and heal. If the rectum is removed for malignancy, the colostomy provides a permanent outlet for feces.

Dysphagia (dis-FĀ-jē-a; *dys* = abnormal; *phagein* = to eat) Difficulty in swallowing that may be caused by inflammation, paralysis, obstruction, or trauma.

Enteritis (en′-ter-Ī-tis; *enteron* = intestine) An inflammation of the intestine, particularly the small intestine.

Flatus (FLĀ-tus) Air (gas) in the stomach or intestine, usually expelled through the anus. If the gas is expelled through the mouth, it is called **eructation** or **belching** (burping). Flatus may result from gas released during the breakdown of foods in the stomach or large intestine or from swallowing air or gas-containing substances such as carbonated drinks.

Gastrectomy (gas-TREK-tō-mē; *gastro* = stomach; *tome* = excision) Removal of a portion of or the entire stomach.

Hernia (HER-nē-a) Protrusion of an organ or part of an organ through a membrane or cavity wall, usually the abdominal cavity. Diaphragmatic (hiatal) hernia is the protrusion of the lower esophagus, stomach, or intestine into the thoracic cavity through the opening in the diaphragm (esophageal hiatus) that allows passage of the esophagus. Umbilical hernia is usually a mild defect that contains the protrusion of a portion of peritoneum through the navel area of the abdominal wall. Inguinal hernia is the protrusion of the hernial sac into the inguinal opening. It may contain a portion of the bowel in an advanced stage and may extend into the scrotal compartment in males, causing strangulation of the herniated part.

Inflammatory bowel (in-FLAM-a-tō′-rē BOW-el) **disease** Disorder that exists in two forms: (1) Crohn's disease (inflammation of the gastrointestinal tract, especially the distal ileum and proximal colon, in which the inflammation may extend from the mucosa through the serosa) and (2) ulcerative colitis (inflammation of the mucosa of the gastrointestinal tract, usually limited to the large intestine and usually accompanied by rectal bleeding).

Irritable bowel (IR-i-ta-bul BOW-el) **syndrome (IBS)** Disease of the entire gastrointestinal tract in which persons with this condition may react to stress by developing symptoms such as cramping and abdominal pain associated with alternating patterns of diarrhea and constipation. Excessive amounts of mucus may appear in the stools, and other symptoms include flatulence, nausea, and loss of appetite. The condition is also known as **irritable colon** or **spastic colitis.**

Malocclusion (mal′-ō-KLOO-zhun; *mal* = disease; *occlusio* = to fit together) Condition in which the upper and lower teeth do not close together.

Nausea (NAW-sē-a; *nausia* = seasickness) Discomfort characterized by a loss of appetite and the sensation of impending vomiting. Its causes include local irritation of the gastrointestinal tract, a systemic disease, brain disease or injury, overexertion, or the effects of medication or drug overdosage.

Pancreatitis (pan′-krē-a-TĪ-tis) Inflammation of the pancreas, as may occur in association with mumps. In a more severe condition, known as **acute pancreatitis,** which is associated with heavy alcohol intake or biliary tract obstruction, the pancreatic cells may release trypsin instead of trypsinogen, and the trypsin begins to digest the pancreatic cells. The patient with acute pancreatitis usually responds to treatment, but recurrent attacks are the rule.

Traveler's diarrhea Infectious disease of the gastrointestinal tract that results in loose, urgent bowel movements, cramping, abdominal pain, malaise, nausea, and occasionally fever and dehydration. It is acquired through ingestion of food or water that has become contaminated with fecal material containing mostly bacteria (especially *Escherichia coli*). Viruses or protozoan parasites are less frequently involved. Commonly referred to as **Montezuma's revenge, turista,** and **Tut's tummy.**

STUDY OUTLINE

Digestive Processes (p. 653)

1. Food is prepared for use by cells by five basic activities: ingestion, movement, mechanical and chemical digestion, absorption, and defecation.
2. Chemical digestion is a series of catabolic (hydrolysis) reactions that break down large carbohydrate, lipid, and protein food molecules into smaller molecules that are usable by body cells.
3. Mechanical digestion consists of movements that aid chemical digestion.
4. Absorption is the passage of end products of digestion from the gastrointestinal tract into blood or lymph for distribution to cells.
5. Defecation is emptying of the rectum.

Organization (p. 653)

1. The organs of digestion are usually divided into two main groups: those composing the gastrointestinal (GI) tract, or alimentary canal, and accessory structures.
2. The GI tract is a continuous tube running through the ventral body cavity from the mouth to the anus.
3. The accessory structures include the teeth, tongue, salivary glands, liver, gallbladder, and pancreas.
4. The basic arrangement of layers in the alimentary canal from the inside outward is the mucosa, submucosa, muscularis, and serosa (visceral peritoneum).
5. Extensions of the peritoneum include the mesentery, mesocolon, falciform ligament, lesser omentum, and greater omentum.

Mouth (Oral Cavity) (p. 656)

1. The mouth is formed by the cheeks, hard and soft palates, lips, and tongue, which aid mechanical digestion.
2. The vestibule is the space between the cheeks and lips and teeth and gums.
3. The oral cavity proper extends from the vestibule to the fauces.

Tongue (p. 657)

1. The tongue, together with its associated muscles, forms the floor of the oral cavity. It is composed of skeletal muscle covered with mucous membrane.
2. The upper surface and sides of the tongue are covered with papillae. Some papillae contain taste buds.

Salivary Glands (p. 658)

1. The major portion of saliva is secreted by the salivary glands, which lie outside the mouth and pour their contents into ducts that empty into the oral cavity.
2. There are three pairs of salivary glands: parotid, submandibular (submaxillary), and sublingual glands.
3. Saliva lubricates food and starts the chemical digestion of carbohydrates.

Teeth (p. 661)

1. The teeth, or dentes, project into the mouth and are adapted for mechanical digestion.
2. A typical tooth consists of three principal portions: crown, root, and neck.
3. Teeth are composed primarily of dentin and are covered by enamel, the hardest substance in the body.
4. There are two dentitions—deciduous and permanent.
5. Through mastication, food is mixed with saliva and shaped into a bolus.
6. Salivary amylase initiates the breakdown of starch.

Pharynx (p. 663)

1. Deglutition, or swallowing, moves a bolus from the mouth to the stomach.
2. It consists of a voluntary stage, pharyngeal stage (involuntary), and esophageal stage (involuntary).

Esophagus (p. 663)

1. The esophagus is a collapsible, muscular tube that connects the pharynx to the stomach.
2. It passes a bolus into the stomach by peristalsis.
3. It contains an upper and lower esophageal sphincter.

Stomach (p. 665)

1. The stomach begins at the bottom of the esophagus and ends at the pyloric sphincter.
2. The gross anatomic subdivisions of the stomach include the cardia, fundus, body, and pylorus.
3. Adaptations of the stomach for digestion include rugae; glands that produce mucus, hydrochloric acid, a protein-digesting enzyme, intrinsic factor, and stomach gastrin; and a three-layered muscularis for efficient mechanical movement.
4. Mechanical digestion consists of mixing waves.
5. Chemical digestion consists of the breakdown of proteins by pepsin.
6. Among the substances absorbed are some water, certain electrolytes and drugs, and alcohol.

Pancreas (p. 669)

1. The pancreas is divisible into a head, body, and tail and is connected to the duodenum via the pancreatic duct (duct of Wirsung) and accessory duct (duct of Santorini).
2. Pancreatic islets (islets of Langerhans) secrete hormones, and acini secrete pancreatic juice.
3. Pancreatic juice contains enzymes that digest starch (pancreatic amylase), proteins (trypsin, chymotrypsin, and carboxypeptidase), fats (pancreatic lipase), and nucleotides (nucleases).

Liver (p. 671)

1. The liver is divisible into left and right lobes; associated with the right lobe are the caudate and quadrate lobes.
2. The lobes of the liver are made up of lobules that contain hepatic cells, sinusoids, stellate reticuloendothelial (Kupffer's) cells, and a central vein.
3. Hepatic cells produce bile that is transported by a duct system to the gallbladder for concentration and temporary storage.
4. Bile's contribution to digestion is the emulsification of neutral fats.
5. The liver also functions in carbohydrate, fat, and protein metabolism; excretion of bile; synthesis of bile salts; removal of

drugs and hormones; storage of vitamins and minerals; phagocytosis; and activation of vitamin D.
6. Bile secretion is regulated by nervous and hormonal mechanisms.

Gallbladder (GB) (p. 675)

1. The gallbladder (GB) is a sac located in a fossa on the visceral surface of the liver
2. The gallbladder stores and concentrates bile.

Small Intestine (p. 675)

1. The small intestine extends from the pyloric sphincter to the ileocecal sphincter.
2. It is divided into duodenum, jejunum, and ileum.
3. It is highly adapted for digestion and absorption. Its glands produce enzymes and mucus, and the microvilli, villi, and circular folds of its wall provide a large surface area for digestion and absorption.
4. Some intestinal enzymes break down foods inside epithelial cells of the mucosa.
5. Intestinal enzymes, pancreatic enzymes, and bile contribute to the complete digestion of carbohydrates into monosaccharides; proteins into amino acids; fats into fatty acids, glycerol, and glycerides; and nucleic acids into pentoses and nitrogenous bases in the small intestine.
6. Mechanical digestion in the small intestine involves segmentation and peristalsis.

Large Intestine (p. 679)

1. The large intestine extends from the ileocecal sphincter to the anus.
2. Its subdivisions include the cecum, colon, and rectum.
3. The mucosa contains numerous goblet cells, and the muscularis consists of taeniae coli.
4. Mechanical movements of the large intestine include haustral churning, peristalsis, and mass peristalsis.
5. The last stages of chemical digestion occur in the large intestine through bacterial, rather than enzymatic, action. Substances are further broken down and some vitamins are synthesized.
6. The large intestine absorbs water, electrolytes, and vitamins.
7. Feces consists of water, inorganic salts, epithelial cells, bacteria, and undigested foods.
8. The elimination of feces from the rectum is called defecation.
9. Defecation is a reflex action aided by voluntary contractions of the diaphragm and abdominal muscles.

Aging and the Digestive System (p. 684)

1. General changes include decreased secretory mechanisms, decreased motility, and loss of tone.
2. Specific changes include loss of taste, pyorrhea, hernias, ulcers, constipation, hemorrhoids, and diverticular diseases.

Developmental Anatomy of the Digestive System (p. 684)

1. The endoderm of the primitive gut forms the epithelium and glands of most of the gastrointestinal tract.
2. The mesoderm of the primitive gut forms the smooth muscle and connective tissue of the gastrointestinal tract.

Applications to Health (p. 684)

1. Dental caries are started by acid-producing bacteria that reside in dental plaque.
2. Periodontal diseases are characterized by inflammation and degeneration of gingivae, alveolar bone, periodontal membrane, and cementum.
3. Peritonitis is inflammation of the peritoneum.
4. Peptic ulcers are craterlike lesions that develop in the mucous membrane of the gastrointestinal tract in areas exposed to gastric juice.
5. Appendicitis is an inflammation of the vermiform appendix resulting from obstruction of the lumen of the appendix by inflammation, a foreign body, carcinoma of the cecum, stenosis, or kinking of the organ.
6. Tumors may occur in any portion of the gastrointestinal tract. One of the most common malignancies is colorectal cancer.
7. Diverticulitis is the inflammation of diverticula in the colon.
8. Cystic fibrosis (CF) is characterized by the production of thick exocrine secretions, especially in the pancreas, respiratory passageways, and salivary and sweat glands.
9. Cirrhosis is a condition in which cells of the liver damaged by chronic inflammation are replaced by fibrous or adipose connective tissue.
10. Hepatitis is an inflammation of the liver. Types include hepatitis A, hepatitis B, and hepatitis C.
11. The fusion of individual crystals of cholesterol is the beginning of 95 percent of all gallstones. Gallstones can cause obstruction to the outflow of bile in any portion of the duct system.
12. Obesity is a body weight 10 to 20 percent above a desirable standard as a result of an excessive accumulation of fat.
13. Anorexia nervosa is a disorder characterized by bizarre eating patterns.
14. Bulimia is a binge–purge syndrome of behavior in which uncontrollable overeating is followed by forced vomiting or overdoses of laxatives.

REVIEW QUESTIONS

1. Define digestion and describe the five phases involved. Distinguish between chemical and mechanical digestion. (p. 653)
2. Identify, in sequence, the organs of the gastrointestinal (GI) tract. How does the gastrointestinal tract differ from the accessory structures of digestion? (p. 653)
3. Describe the structure of each of the four layers of the gastrointestinal tract. (p. 653)
4. What is the peritoneum? Describe the location and function of the mesentery, mesocolon, falciform ligament, lesser omentum, and greater omentum. (p. 655)

5. What structures form the mouth or oral cavity? (p. 656)
6. Make a simple diagram of the tongue. Indicate the location of the papillae and the four taste zones. (p. 561)
7. Describe the location of the salivary glands and their ducts. (p. 658)
8. How are salivary glands distinguished histologically? (p. 659)
9. What are the principal parts of a typical tooth? What are the functions of each part? (p. 661)
10. Contrast the functions of incisors, cuspids, premolars, and molars. (p. 662)
11. Define deglutition. List the sequence of events involved in passing a bolus from the mouth to the stomach. (p. 663)
12. Describe the location and histology of the esophagus. What is its role in digestion? (p. 663)
13. Describe the location of the stomach. List and briefly explain the anatomical features of the stomach. (p. 665)
14. What is the importance of rugae, zymogenic cells, parietal cells, mucous cells, and enteroendocrine cells in the stomach? (p. 666)
15. Describe mechanical digestion in the stomach. What is the role of pepsin? (p. 668)
16. Where is the pancreas located? Describe the duct system connecting the pancreas to the duodenum. (p. 669)
17. What are pancreatic acini? Contrast their functions with those of the pancreatic islets (islets of Langerhans). (p. 670)
18. Where is the liver located? What are its principal functions? (p. 671)
19. Describe the anatomy of the liver. Draw a labeled diagram of a liver lobule. (p. 671)
20. Once bile has been formed by the liver, how is it collected and transported to the gallbladder for storage? What is the function of bile? (p. 673)
21. Where is the gallbladder (GB) located? How is it connected to the duodenum? (p. 675)
22. What are the subdivisions of the small intestine? How are the mucosa and submucosa of the small intestine adapted for digestion and absorption? (p. 675)
23. Describe the movements in the small intestine. (p. 679)
24. Define absorption. (p. 679)
25. What are the principal subdivisions of the large intestine? How does the muscularis of the large intestine differ from that of the rest of the gastrointestinal tract? What are haustra? (p. 679)
26. Describe the mechanical movements that occur in the large intestine. (p. 681)
27. Explain the activities of the large intestine that change its contents into feces. Define defecation. How does it occur? (p. 681)
28. Describe the effect of aging on the digestive system. (p. 684)
29. Describe the development of the digestive system. (p. 684)
30. Describe the causes (where known) and clinical symptoms for: dental caries, periodontal disease, peritonitis, peptic ulcers, appendicitis, tumors, diverticulitis, cystic fibrosis, hepatitis, gallstones, obesity, anorexia nervosa, and bulimia. (p. 684)
31. Refer to the glossary of key medical terms associated with the digestive system. Be sure that you can define each term. (p. 689)

SELF QUIZ

Complete the following:

1. The most important enzyme released by the stomach is ________.
2. Indicate which germ layer, endoderm (E) or mesoderm (M), gives rise to each of these structures:
   ___ a. epithelial lining and digestive glands of the gastrointestinal tract
   ___ b. liver, gallbladder, and pancreas
   ___ c. muscularis layer and connective tissue of submucosa
3. How many teeth are in a complete permanent dentition? ________ How many teeth would be in a complete child's dentition, as in a 4-year-old? ________ The four centrally located teeth are named ________. Lateral to these are ________. Posterior to these teeth are ________ and finally ________.
4. Emptying of feces from the rectum is called ________.
5. The duct that conveys bile from the gallbladder is the ________ duct.
6. The digestion of starch begins in the ________.
7. Permanent folds in the mucosa and submucosa of the small intestine are called ________.
8. The terminal 2–3 cm (1 in.) of the rectum is called the ________.
9. The ________ border of the stomach is the greater curvature.
10. Arrange the answers in correct sequence.
   ___ ___ ___ ___ a. Gastrointestinal tract wall, from deepest to most superficial:
   A. mucosa
   B. muscularis
   C. serosa
   D. submucosa
   ___ ___ ___ ___ ___ b. Pathway of chyme:
   A. ileum
   B. jejunum
   C. cecum
   D. duodenum
   E. pylorus
   ___ ___ ___ ___ ___ c. Pathway of bile:
   A. bile canaliculi
   B. common bile duct
   C. common hepatic duct
   D. right and left hepatic ducts
   E. hepatopancreatic ampulla (ampulla of Vater) and duodenum
   ___ ___ ___ ___ ___ d. Pathway of wastes:
   A. ascending colon
   B. transverse colon
   C. sigmoid colon

D. descending colon
E. rectum

11. Match the following:

___ **a.** attaches liver to anterior abdominal wall	**A.** falciform ligament
___ **b.** binds small intestine to posterior abdominal wall	**B.** greater omentum
___ **c.** binds part of large intestine to posterior abdominal wall	**C.** lesser omentum
___ **d.** "fatty apron"; covers and helps prevent infection in small intestine	**D.** mesentery
___ **e.** suspends stomach and duodenum from liver	**E.** mesocolon

Choose the one best answer to these questions.

___ **12.** Pyloric stenosis will interfere most directly with the passage of materials from
A. the esophagus into the stomach; B. the pharynx into the esophagus; C. the ileum into the cecum; D. the stomach into the duodenum; E. the ileum into the jejunum.

___ **13.** The cells of the gastric glands that produce secretions directly involved in chemical digestion are the
A. mucous cells; B. parietal cells; C. zymogenic cells; D. pancreatic islets (islets of Langerhans); E. goblet cells.

___ **14.** Which anatomical subdivision of the stomach is closest to the duodenum?
A. body; B. pyloris; C. fundus; D. cardia; E. ileocecal sphincter.

___ **15.** An organ of the digestive system that assumes a role in phagocytosis, manufacturing plasma proteins, detoxification, and interconversions of nutrients is the
A. stomach; B. liver; C. small intestine; D. pancreas; E. gallbladder.

___ **16.** Which of the following is/are characteristic of the large intestine?
(1) It is divided into ascending, transverse, and descending portions.
(2) It contains bacteria that synthesize certain nutritional factors such as vitamins.
(3) It serves as the main absorptive surface for digesting foods.
(4) It absorbs much of the water remaining in the waste materials.
A. (1) only; B. (2) only; C. (3) only; D. (4) only; E. (1), (2), and (4).

___ **17.** A human whose gallbladder is removed may at first have difficulty with
A. carbohydrate digestion; B. fat emulsification; C. protein oxidation; D. starch assimilation; E. carbohydrate assimilation.

___ **18.** One of the following digestive juices contains enzymes that digest carbohydrates, fats, and proteins. That digestive juice is
A. pancreatic juice; B. bile; C. saliva; D. gastric juice; E. none of the above since no one digestive juice contains enzymes for digesting all three classes of foods.

___ **19.** Which of the following is *not* considered to be an accessory organ of digestion?
A. teeth; B. tongue; C. esophagus; D. liver; E. pancreas.

___ **20.** Enzymes produced in the small intestine function to
A. complete digestion of carbohydrates, fats, and proteins; B. initiate digestion of fats and complete digestion of carbohydrates and proteins; C. initiate digestion of proteins and fats and complete digestion of carbohydrates; D. initiate digestion of proteins and complete digestion of fats and carbohydrates; E. stimulate the flow of both gastric and pancreatic juices.

___ **21.** The esophagus
(1) extends from the pharynx to the stomach.
(2) is approximately 23–25 cm (9–10 in.) long in the adult.
(3) is posterior to the trachea, anterior to the vertebral column, and passes through the diaphragm.
A. (1) only; B. (2) only; C. (3) only; D. all of the above; E. (1) and (2).

___ **22.** Villi
A. are present in the lining of the stomach; B. cover the surface of the circular folds of the small intestine; C. line the appendix; D. are present in the cecum; E. are present above the esophagus.

___ **23.** Enamel, the hardest substance in the body, covers the crown of a tooth. The roots of a tooth are covered by a hardened bonelike tissue that is harder and denser than bone and is called
A. the gingivae; B. cementum; C. dentin; D. the periodontal ligament; E. pulp.

___ **24.** The process of swallowing is also known as
A. eructation; B. regurgitation; C. deglutition; D. mastication; E. peristalsis.

___ **25.** The parotid glands
(1) are the largest of the salivary glands.
(2) are located just anterior to the ear and deep to and behind the mandibular ramus.
(3) empty into the mouth by a duct that passes through a cheek muscle.
A. (1) only; B. (2) only; C. (3) only; D. all of the above; E. (1) and (3).

___ **26.** Which of the following is/are true for *every* part of the gastrointestinal tract from the lower third of the esophagus to the anus?
A. enzyme-secreting cells in the mucosa; B. muscular wall of smooth muscle; C. lymphoid follicles (Peyer's patches) in the submucosa; D. epiploic appendages on the outer serosa; E. all of the above.

___ **27.** The terminal portion of the small intestine is the
A. ileum; B. cecum; C. duodenum; D. jejunum; E. appendix.

The Urinary System

24

STUDENT OBJECTIVES

1. Identify the external and internal gross anatomical features of the kidneys.
2. Define the structural features of a nephron.
3. Describe the blood and nerve supply to the kidneys.
4. Describe the location, structure, and function of the ureters.
5. Describe the location, structure, histology, and function of the urinary bladder.
6. Describe the location, structure, and function of the urethra.
7. Describe the effects of aging on the urinary system.
8. Describe the development of the urinary system.
9. Discuss the causes of renal calculi, glomerulonephritis, pyelitis, pyelonephritis, cystitis, nephrotic syndrome, polycystic disease, renal failure, and urinary tract infections (UTIs).
10. Define key medical terms associated with the urinary system.

CHAPTER OUTLINE

- **Kidneys**
  - External Anatomy
  - Internal Anatomy
  - Nephron
  - Blood and Nerve Supply
  - Juxtaglomerular Apparatus (JGA)
  - Physiology
- **Hemodialysis Therapy**
- **Ureters**
  - Structure
  - Histology
  - Physiology
  - Blood and Nerve Supply
- **Urinary Bladder**
  - Structure
  - Histology
  - Physiology
  - Blood and Nerve Supply
- **Urethra**
  - Histology
  - Physiology
- **Aging and the Urinary System**
- **Developmental Anatomy of the Urinary System**
- **Applications to Health**
- **Key Medical Terms Associated with the Urinary System**

The metabolism of nutrients results in the production of wastes by body cells, including carbon dioxide and excess water and heat. Protein catabolism produces toxic nitrogenous wastes such as ammonia and much less toxic urea. In addition, many of the essential ions such as sodium, chloride, sulfate, phosphate, and hydrogen tend to accumulate in excess of the body's needs. All the toxic materials and the excess essential materials must be eliminated.

The primary function of the ***urinary system*** is to help keep the body in homeostasis by controlling the composition and volume of blood. It does so by removing and restoring selected amounts of water and solutes. Two kidneys, two ureters, one urinary bladder, and a single urethra make up the system (Figure 24-1). The kidneys regulate the composition and volume of the blood and remove wastes from the blood in the form of urine. They excrete selected amounts of various wastes, assume a role in erythropoiesis by forming renal erythropoietic factor, help control blood pH, help regulate blood pressure by secreting renin (which activates the renin–angiotensin pathway), and participate in the activation of vitamin D. Urine is excreted from each kidney through its ureter and is stored in the urinary bladder until it is expelled from the body through the urethra. Other systems that aid in waste elimination are the respiratory, integumentary, and digestive systems (Exhibit 24-1).

The specialized branch of medicine that deals with structure, function, and diseases of the male and female urinary systems and the male reproductive system is known as ***nephrology*** (nef-ROL-ō-jē; *neph* = kidney; *logos* = study of). The branch of surgery related to male and female urinary systems and male reproductive system is referred to as ***urology*** (yoo-ROL-ō-jē; *uro* = urine or urinary tract).

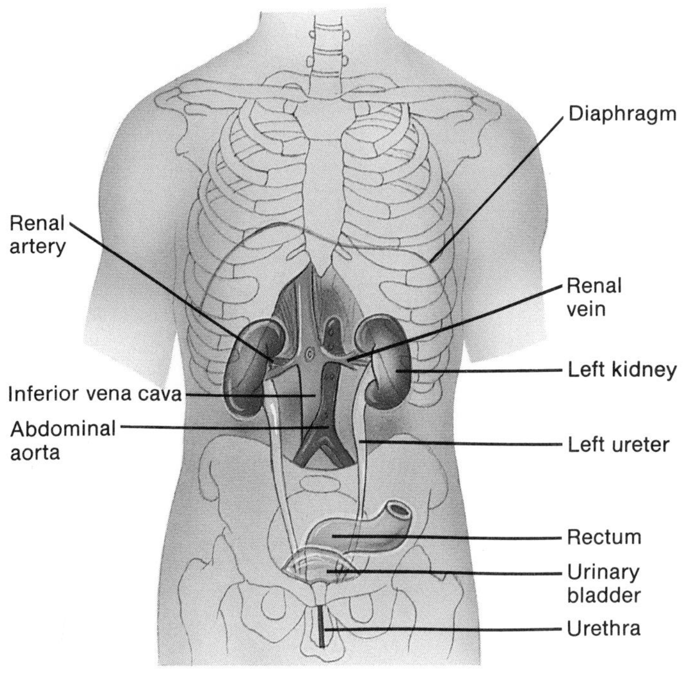

FIGURE 24-1 Organs of the male urinary system in relation to surrounding structures.

EXHIBIT 24-1

Excretory Organs and Products Eliminated

EXCRETORY ORGANS	PRODUCTS ELIMINATED: PRIMARY	PRODUCTS ELIMINATED: SECONDARY
Kidneys	Water, nitrogenous wastes from protein catabolism, and inorganic salts.	Heat and carbon dioxide.
Lungs	Carbon dioxide.	Heat and water.
Skin (Sudoriferous Glands)	Heat.	Carbon dioxide, water, salts, and urea.
Gastrointestinal (GI) Tract	Solid wastes and secretions.	Carbon dioxide, water, salts, and heat.

The developmental anatomy of the urinary system will be considered at the end of the chapter.

KIDNEYS

The paired ***kidneys*** are reddish organs that resemble kidney beans in shape. They are found just above the waist between the parietal peritoneum and the posterior wall of the abdomen. Since they are external to the peritoneal lining of the abdominal cavity, their placement is described as ***retroperitoneal*** (re′-trō-per-i-tō-NĒ-al). Other retroperitoneal structures include the ureters and adrenal (suprarenal) glands. Relative to the vertebral column, the kidneys are located between the levels of the last thoracic and third lumbar vertebrae and are partially protected by the eleventh and twelfth pairs of ribs. The right kidney is slightly lower than the left because of the large area occupied by the liver.

EXTERNAL ANATOMY

The average adult kidney measures about 10 to 12 cm (4 to 5 in.) long, 5.0 to 7.5 cm (2 to 3 in.) wide, and 2.5 cm (1 in.) thick. Its concave medial border faces the vertebral column. Near the center of the concave border is a notch called the ***hilus,*** through which the ureter leaves the kidney. Blood and lymphatic vessels and nerves also enter and exit the kidney through the hilus (Figure 24-2). The hilus is the entrance to a cavity in the kidney called the ***renal sinus.***

Three layers of tissue surround each kidney. The innermost layer, the ***renal capsule,*** is a smooth, transparent, fibrous membrane that can easily be stripped off the kidney

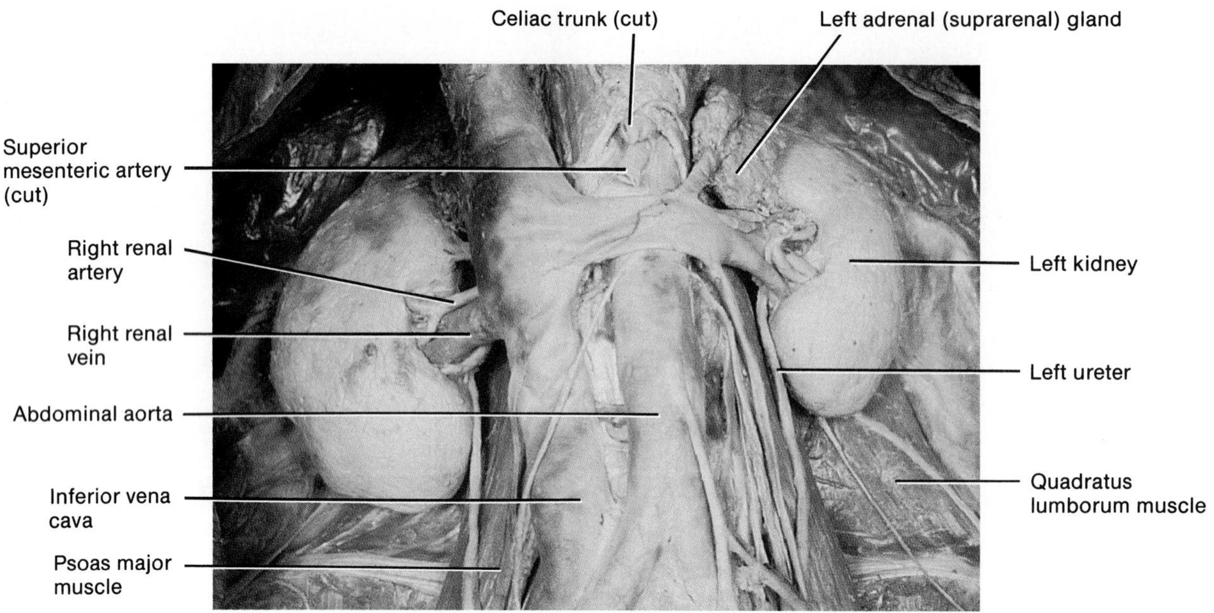

(a) Anterior external view and associated structures

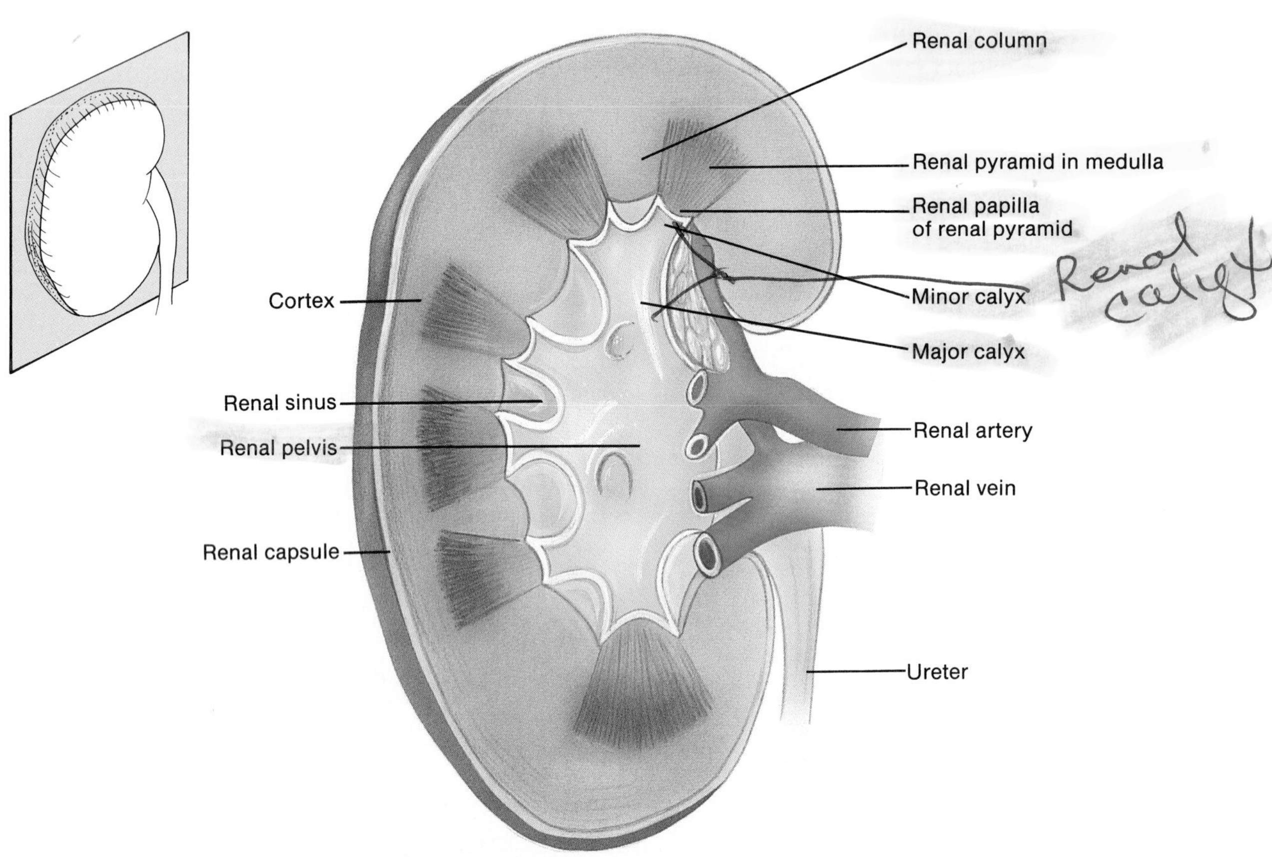

(b) Coronal section of right kidney

FIGURE 24-2 Kidney. (a) and (c) Photographs courtesy of J. A. Gosling, P. F. Harris, et al., *Atlas of Human Anatomy,* Gower Medical Publishing Ltd., 2nd ed., 1991. (d) Courtesy of Stephen A. Kieffer and E. Robert Heitzman, *An Atlas of Cross-Sectional Anatomy,* Harper & Row, Publishers, Inc., New York, 1979.

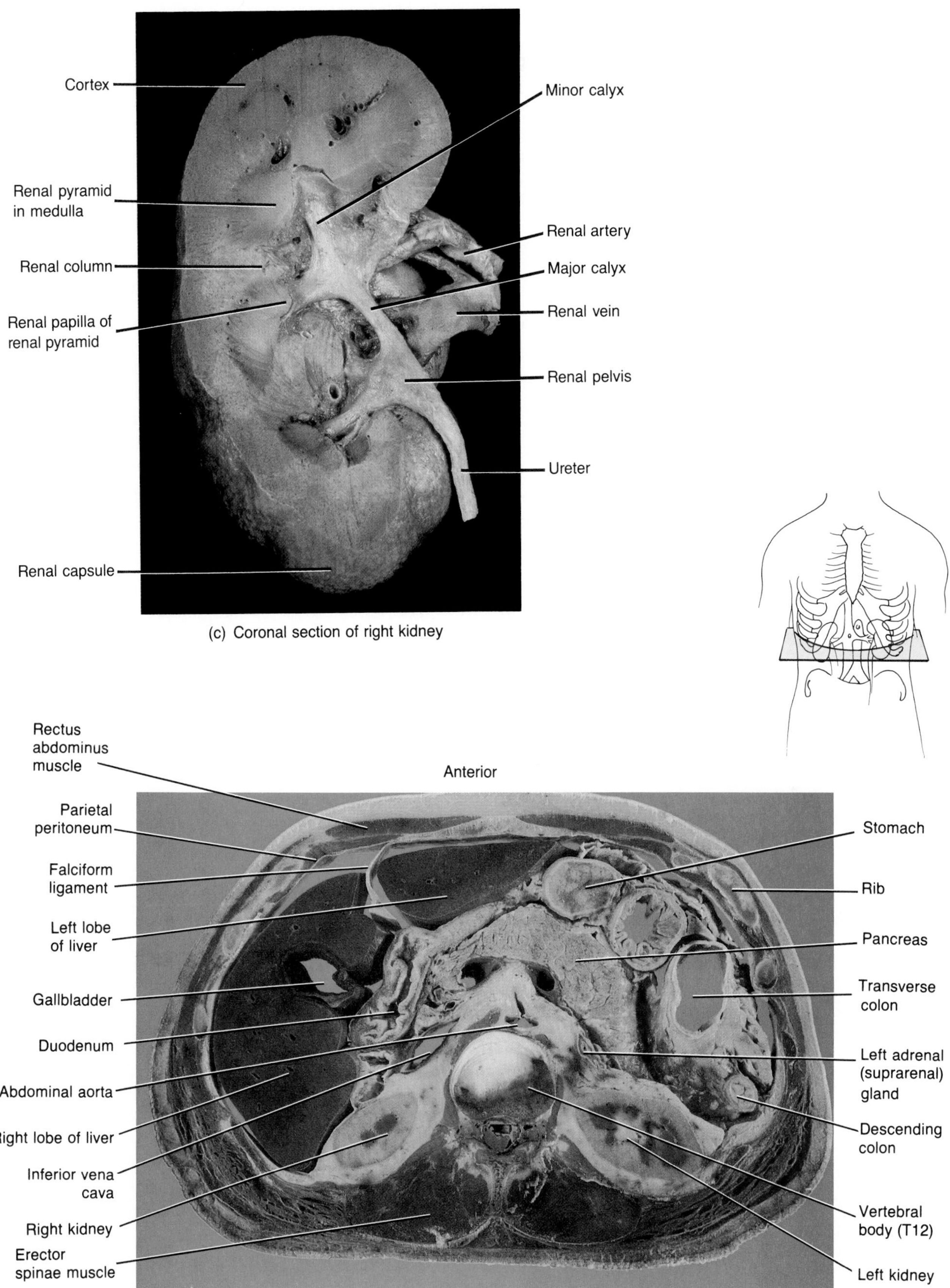

(c) Coronal section of right kidney

(d) Cross section of abdomen

and is continuous with the outer coat of the ureter at the hilus. It serves as a barrier against trauma and the spread of infection to the kidney. The second layer, the ***adipose capsule,*** is a mass of fatty tissue surrounding the renal capsule. It also protects the kidney from trauma and holds it firmly in place within the abdominal cavity. The outermost layer, the ***renal fascia,*** is a thin layer of fibrous connective tissue that anchors the kidney to its surrounding structures and to the abdominal wall.

CLINICAL APPLICATION

Floating Kidney (Nephrotosis)

Floating kidney, or ***nephrotosis*** (nef′-ro-TŌ-sis; *ptosis* = falling), is a downward displacement or dropping of the kidney. It occurs when the kidney is no longer held in place securely by the adjacent organs or its covering of fat and slips from its normal position. Individuals, especially thin people, in whom either the adipose capsule or renal fascia is deficient, may develop nephrotosis. It is dangerous because it may cause kinking of the ureter with reflux of urine and retrograde pressure. Pain occurs if the ureter is twisted. Also, if the kidneys drop below the rib cage, they become susceptible to blows and penetrating injuries.

INTERNAL ANATOMY

A coronal (frontal) section through a kidney reveals an outer reddish area called the ***cortex*** and an inner reddish-brown region called the ***medulla*** (Figure 24-2). Within the medulla are 8 to 18 striated triangular structures termed ***renal (medullary) pyramids.*** The striated (striped) appearance is due to the presence of straight tubules and blood vessels. The bases of the pyramids face the cortical area, and their apices, called ***renal papillae,*** are directed toward the center of the kidney. The cortex is the smooth-textured area extending from the renal capsule to the bases of the pyramids and into the spaces between them. The cortex is divided into an outer cortical zone and an inner juxtamedullary zone. Portions of the cortex extend between the renal pyramids to form the ***renal columns.***

Together the cortex and renal pyramids constitute the ***parenchyma*** (functional portion) of the kidney. Structurally, the parenchyma of each kidney consists of approximately 1 million microscopic units called nephrons, the functional units of the kidney. They help regulate blood composition and form urine. Associated with nephrons are collecting tubules and a vascular supply.

In the renal sinus of the kidney is a large cavity called the ***renal pelvis.*** The edge of the pelvis contains cuplike extensions called ***major*** and ***minor calyces*** (KĀ-li-sēz; *calyx* = cup). There are 2 or 3 major calyces and 8 to 18 minor calyces. Each minor calyx collects urine from collecting tubules of the pyramids. From the major calyces, the urine drains into the pelvis and out into the ureter.

NEPHRON

The functional unit of the kidney is the ***nephron*** (NEF-ron) (Figure 24-3). Nephrons have several functions related to homeostasis. They filter blood; that is, they permit some substances to pass into the kidneys, while keeping others out. As the filtered liquid (filtrate) moves through the nephrons, it is further processed by nephrons by the addition of some substances (wastes and excess substances) and the removal of others (useful materials). As a result of the activities of nephrons, urine is formed.

Essentially, a nephron consists of two portions: a ***renal tubule*** and a tuft (knot) of capillaries, called the glomerulus. The renal tubule begins as a double-walled epithelial cup, called the ***glomerular (Bowman's) capsule,*** lying in the cortex of the kidney. The outer wall, or ***parietal layer,*** is composed of simple squamous epithelium (Figure 24-4). It is separated from the inner wall, known as the ***visceral layer,*** by the ***capsular space.*** The visceral layer consists of specialized epithelial cells called podocytes. The capsule surrounds a capillary network called the ***glomerulus*** (glō-MER-yoo-lus; *glomus* = ball; *ulus* = small). Collectively, the glomerular capsule and its enclosed glomerulus constitute a ***renal corpuscle*** (KŌR-pus-sul; *corpus* = body; *cle* = tiny).

The visceral layer of the glomerular capsule and the endothelium of the glomerulus form an ***endothelial–capsular membrane.*** This membrane consists of the following parts, listed here in the order in which substances filtered by the kidney must pass through them.

1. ***Endothelium of the glomerulus.*** This single layer of endothelial cells has completely opened pores (fenestrated) averaging 50–100 nm in diameter. It restricts the passage of blood cells.
2. ***Basement membrane of the glomerulus.*** This extracellular membrane lies beneath the endothelium and contains no pores. It consists of fibrils in a glycoprotein matrix. It restricts the passage of large-sized proteins.
3. ***Epithelium of visceral layer of the glomerular (Bowman's) capsule.*** These epithelial cells, because of their peculiar shape, are called ***podocytes.*** The podocytes contain footlike structures called ***pedicels*** (PED-i-sels). The pedicels are arranged parallel to the circumference of the glomerulus and cover the basement membrane, except for spaces between them called ***filtration slits (slit pores).*** Pedicels contain numerous thin contractile filaments that are believed to regulate the passage of substances through the filtration slits. In addition, another factor that helps regulate the passage of substances through the filtration slits is a thin membrane, the ***slit membrane,*** that extends between filtration slits. This membrane restricts the passage of intermediate-sized proteins.

The endothelial–capsular membrane filters water and small solutes from blood plasma. Large molecules, such as proteins, and the formed elements in blood do not nor-

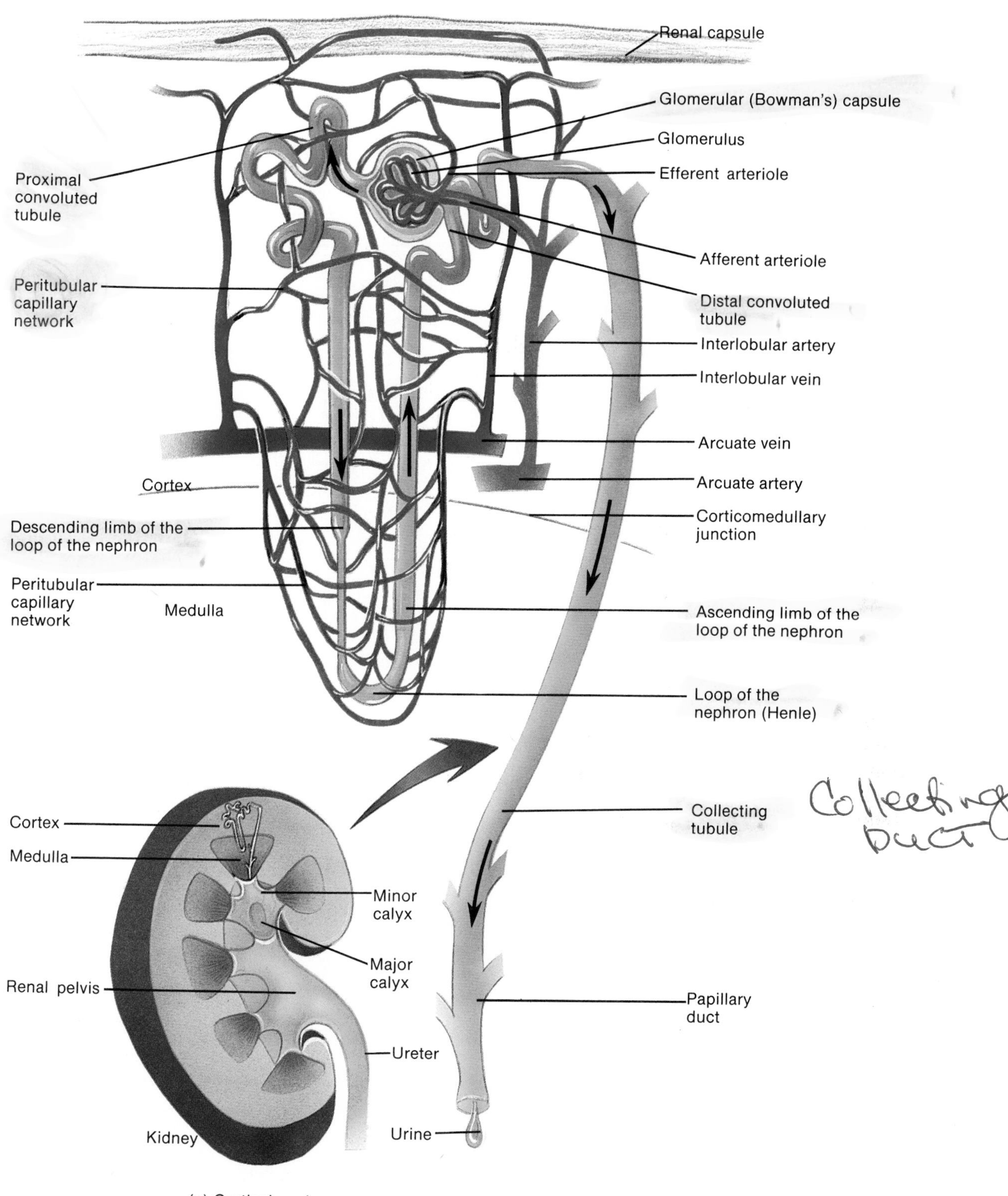

(a) Cortical nephron

FIGURE 24-3 Nephrons

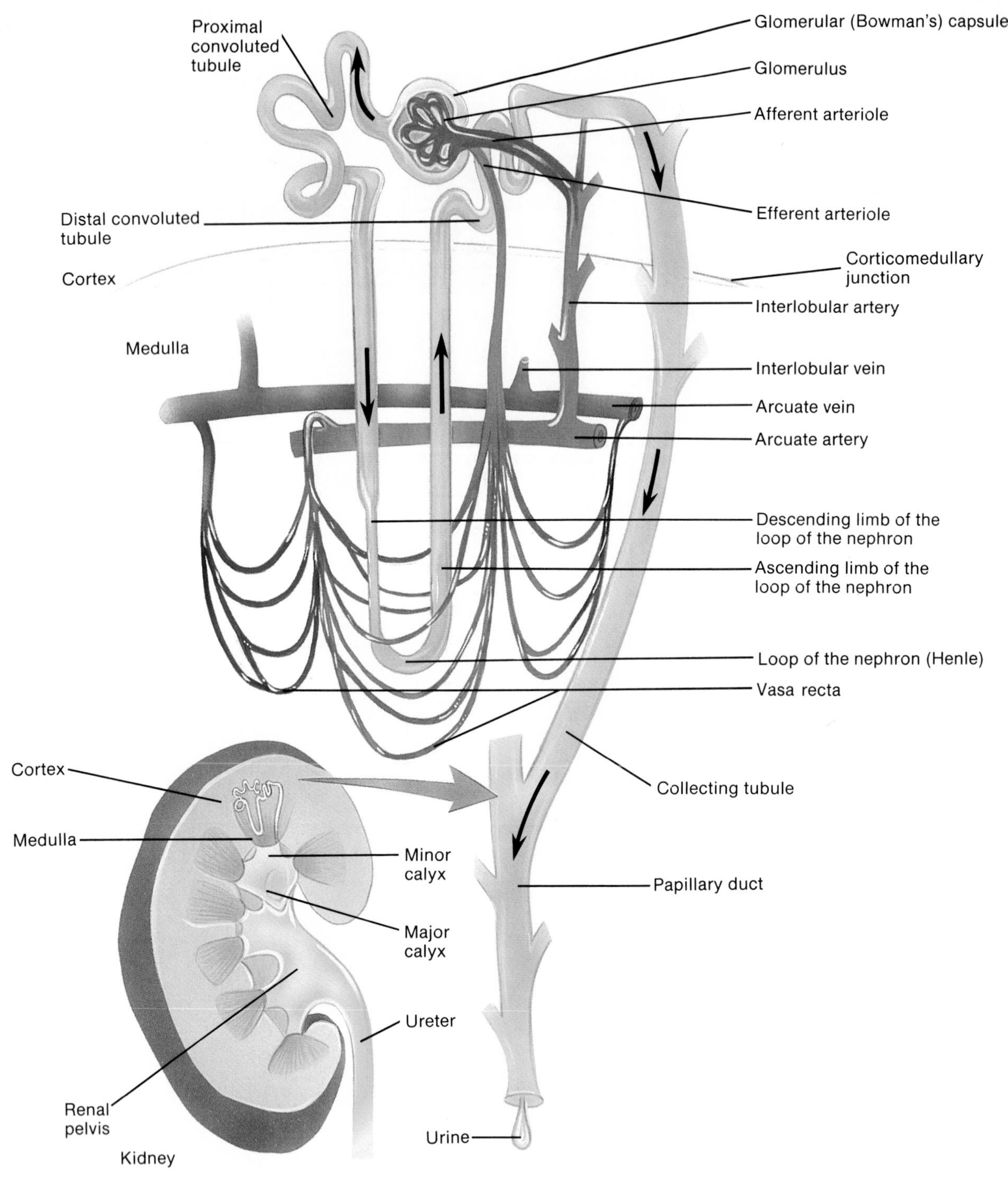

(b) Juxtamedullary nephron

mally pass through it. The water and solutes that are filtered out of the blood pass into the capsular space between the visceral and parietal layers of the glomerular (Bowman's) capsule and then into the renal tubule.

The glomerular capsule opens into the next portion of the renal tubule, called the ***proximal convoluted tubule,*** which also lies in the cortex. *Convoluted* means the tubule is coiled rather than straight; *proximal* signifies that the glomerular (Bowman's) capsule is the origin of the tubule. The wall of the proximal convoluted tubule consists of cuboidal epithelium with microvilli. These surface specializations, like those of the small intestine, increase the surface area for reabsorption and secretion.

Nephrons are frequently classified into two kinds. A ***cortical nephron*** usually has its glomerulus in the outer cortical zone, and the remainder of the nephron usually penetrates only into the outer region of the medulla (see Figure 24-3a). A ***juxtamedullary nephron*** usually has its glomerulus close to the corticomedullary junction, and other parts of the nephron penetrate deeply into the medulla (see Figure 24-3b).

In a juxtamedullary nephron, the proximal convoluted tubule straightens, becomes thinner, and dips into the medulla, where it is called the ***descending (thin) limb of the loop of the nephron.*** This section consists of squamous epithelium. The tubule then increases in diameter after it bends into a U-shaped structure called the ***loop of the nephron (loop of Henle).*** It then ascends toward the cortex as the ***ascending (thick) limb of the loop of the nephron,*** which consists of cuboidal and low columnar epithelium.

In the cortex the tubule again becomes convoluted. Because of its distance from the point of origin at the glomerular (Bowman's) capsule, this section is referred to as the ***distal convoluted tubule.*** The cells of the distal tubule, like those of the proximal tubule, are cuboidal. Unlike the cells of the proximal tubule, however, the cells of the distal tubule have few microvilli. In a cortical nephron the proximal section runs into the distal tubule through shorter limbs

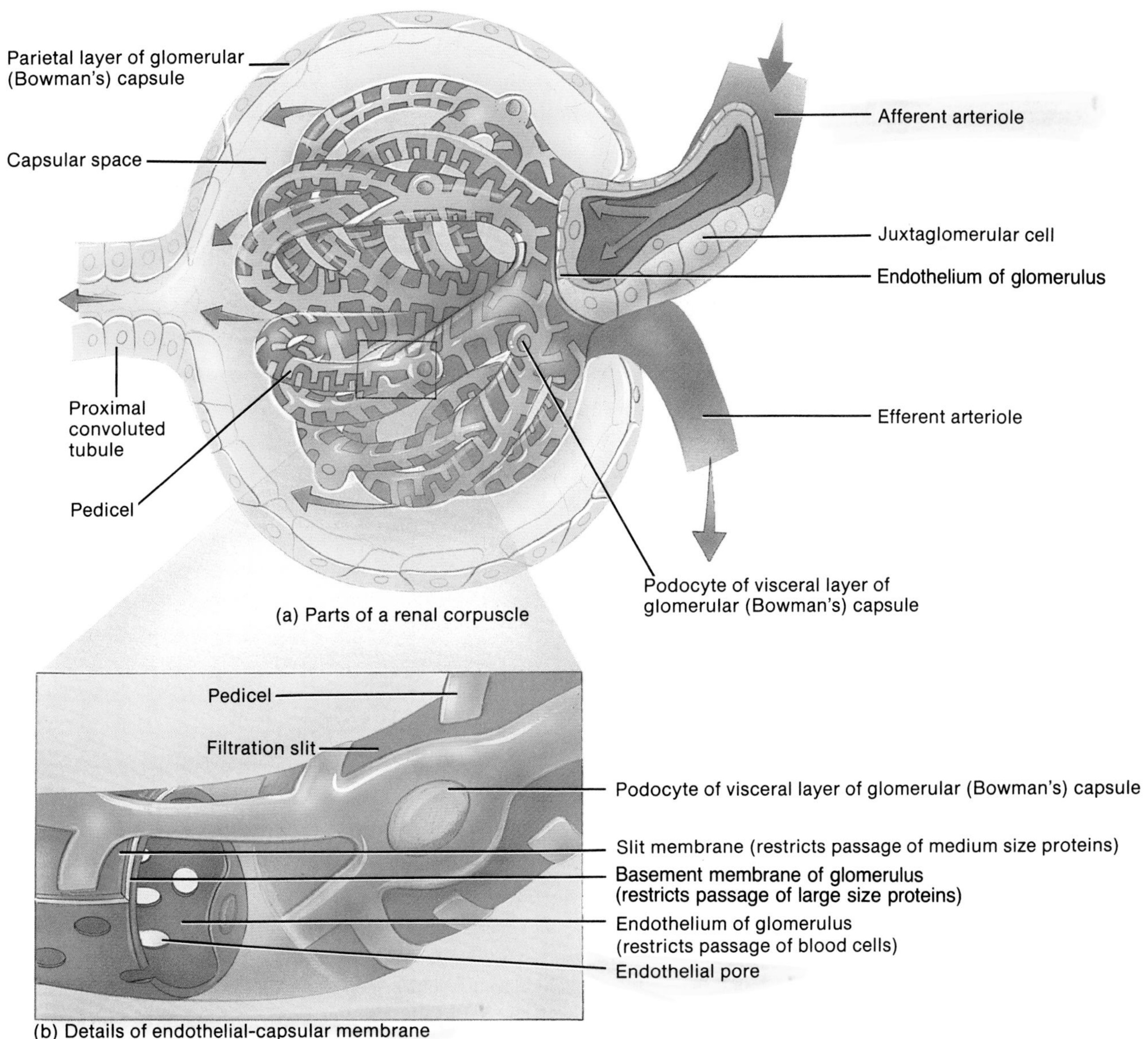

FIGURE 24-4 Endothelial–capsular membrane. In (b), the size of the filtration slits has been exaggerated for emphasis. (c) Courtesy of CNRI, Photo Researchers. (d) and (e) Courtesy of Richard K. Kessel and Randy H. Kardon, *Tissues and Organs: A Text-Atlas of Scanning Electron Microscopy.* Copyright © 1979 by Scientific American, Inc.

(c) Scanning electron micrograph of a renal corpuscle, 2000×

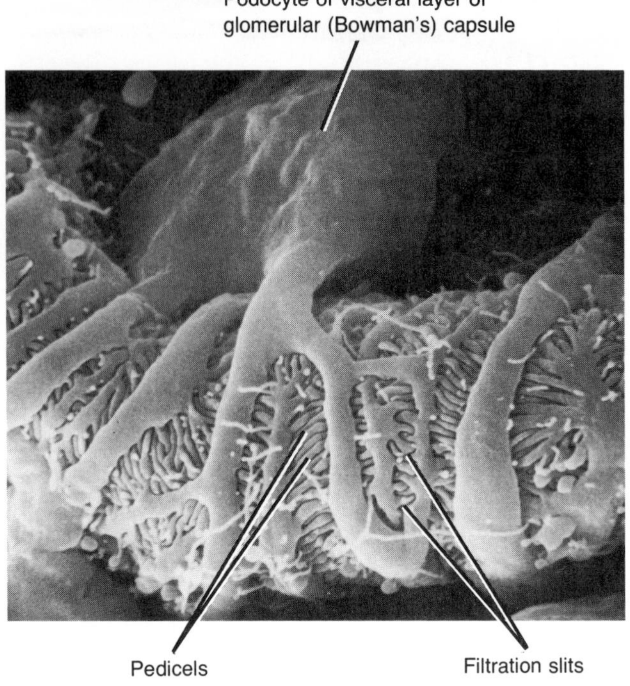

(d) Scanning electron micrograph of a podocyte, 4500×

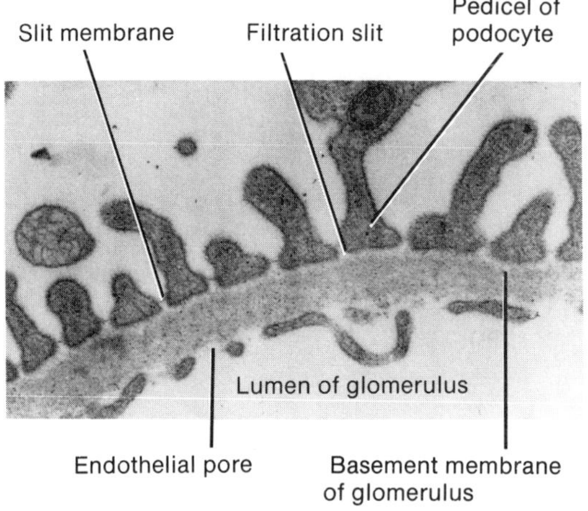

(e) Transmission electron micrograph of the endothelial-capsular membrane, (42,700 x)

and loop of the nephron. In the cortex the distal tubules of both types of nephrons empty into ***collecting tubules.***

In the medulla the collecting tubules combine to form ***papillary ducts,*** which open at the renal papillae into the minor calyces. On the average, there are 30 papillary ducts per renal papilla. Cells of the collecting tubules are cuboidal; those of the papillary ducts are columnar.

The histology of parts of a nephron and glomerulus is shown in Figure 24-5.

BLOOD AND NERVE SUPPLY

Nephrons are largely responsible for removing wastes from the blood and regulating its fluid and electrolyte content. Thus, they are abundantly supplied with blood vessels. The right and left ***renal arteries*** transport about one-fourth the total cardiac output to the kidneys (Figure 24-6). Approximately 1200 ml passes through the kidneys every minute—about 20–25 percent of the total cardiac output.

Capsular space
Glomerulus
Renal corpuscle
Glomerular (Bowman's) capsule
Proximal convoluted tubule
Distal convoluted tubule

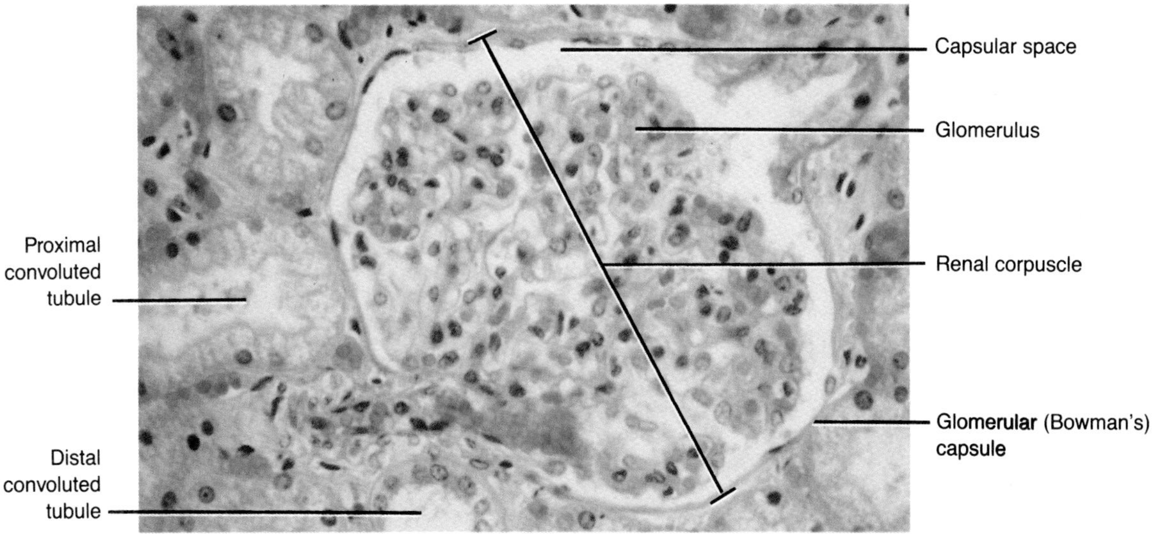

Photomicrograph of a renal corpuscle and surrounding renal tubules, 400×

FIGURE 24-5 Histology of a nephron. (Courtesy of Andrew J. Kuntzman.)

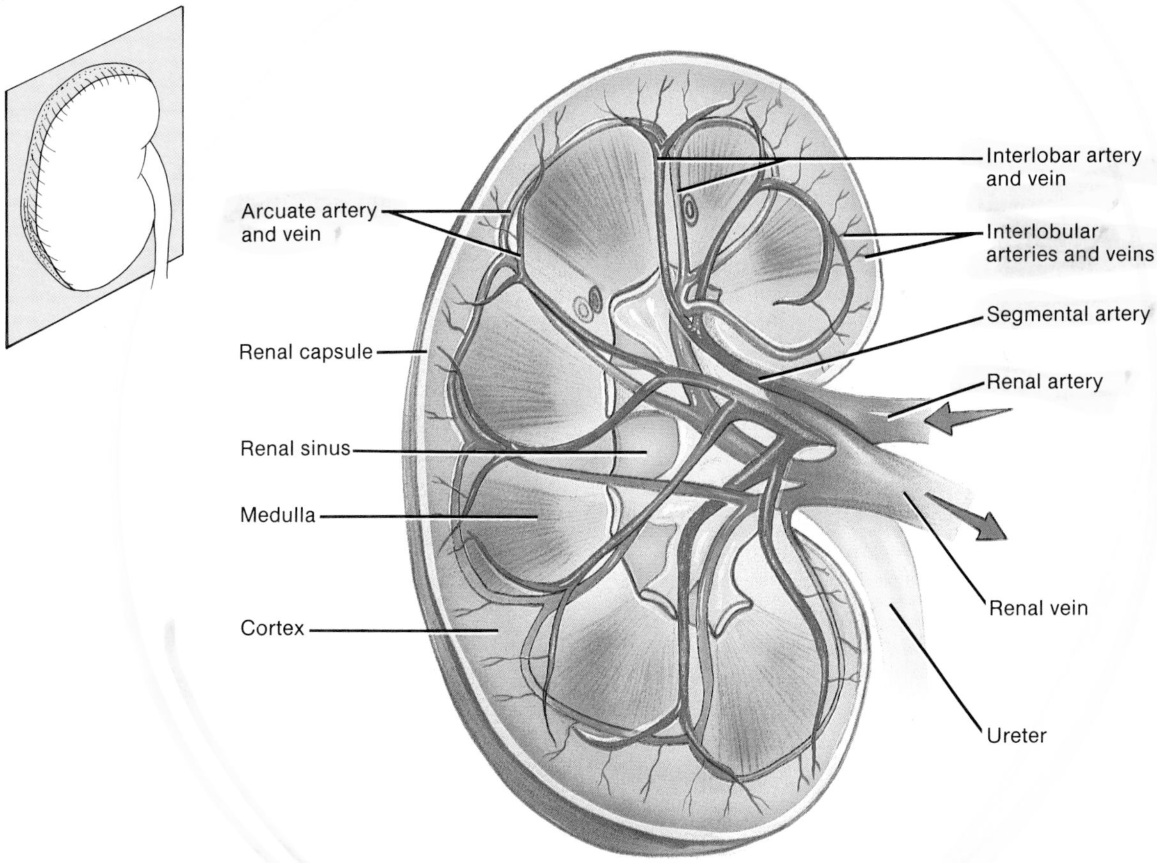

Interlobar artery and vein
Arcuate artery and vein
Interlobular arteries and veins
Segmental artery
Renal capsule
Renal artery
Renal sinus
Medulla
Renal vein
Cortex
Ureter

(a) Coronal section

FIGURE 24-6 Blood supply of the right kidney. (b) is designed to show the sequence of blood flow, not the anatomical location of blood vessels, which is shown in (a), or the direction of flow.

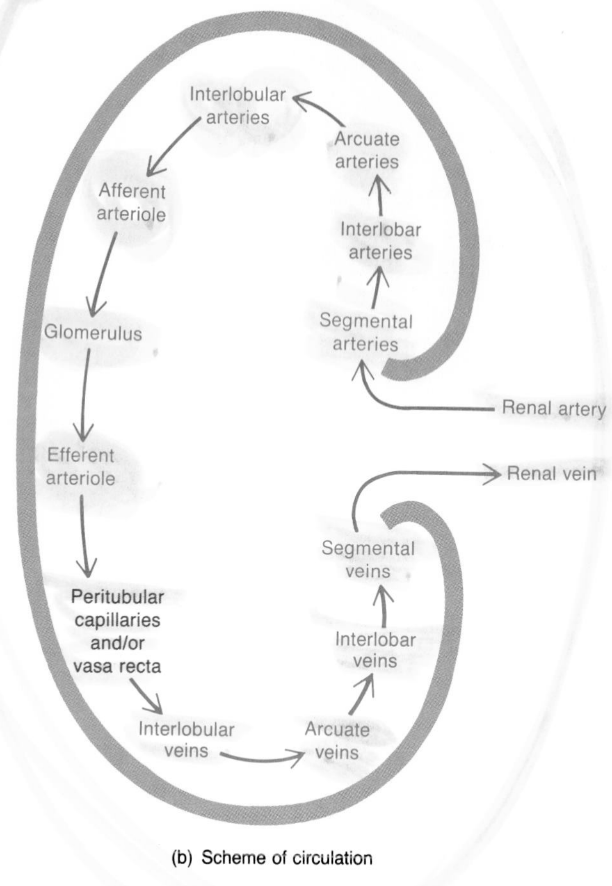

(b) Scheme of circulation

Before or immediately after entering the hilus, the renal artery typically divides into a larger anterior branch and a smaller posterior branch. From these branches, five ***segmental arteries*** originate, each supplying a particular segment of the kidneys. Each segmental artery gives off several branches that enter the parenchyma and pass as the ***interlobar arteries*** between the renal pyramids in the renal columns. At the bases of the pyramids the interlobar arteries arch between the medulla and cortex; here they are known as the ***arcuate arteries.*** Divisions of the arcuate arteries produce a series of ***interlobular arteries,*** which enter the cortex and divide into ***afferent arterioles*** (see Figure 24-3).

One afferent arteriole is distributed to each glomerular (Bowman's) capsule, where the arteriole divides into the tangled capillary network called the ***glomerulus.*** The glomerular capillaries then reunite to form an ***efferent arteriole,*** which leads away from the capsule and is smaller in diameter than the afferent arteriole. This variation in diameter helps raise the glomerular blood pressure to a level significantly higher than the pressure in capillaries elsewhere in the body. The afferent–efferent arteriole situation is unique because blood usually flows out of capillaries into venules and not into other arterioles.

Each efferent arteriole of a cortical nephron divides to form a network of capillaries, called the ***peritubular capillaries,*** around the convoluted tubules. The efferent arteriole of a juxtamedullary nephron also forms peritubular capillaries. In addition, it forms long loops of thin-walled vessels called ***vasa recta*** that dip down alongside the loop of the nephron into the medullary region of the papilla.

The peritubular capillaries eventually reunite to form ***peritubular venules*** and then ***interlobular veins.*** The blood then drains through the ***arcuate veins*** to the ***interlobar veins*** running between the pyramids, then the ***segmental veins,*** and leaves the kidney through a single ***renal vein*** that exists at the hilus. The vasa recta pass blood into the interlobular veins. From here it goes to the arcuate veins, to the interlobar veins, and then into the renal vein.

The nerve supply to the kidneys is derived from the ***renal plexus*** of the sympathetic division of the autonomic nervous system. Nerves from the plexus accompany the renal arteries and their branches and are distributed to the vessels. Because the nerves are vasomotor, they regulate the circulation of blood in the kidney by regulating the diameters of the arterioles.

JUXTAGLOMERULAR APPARATUS (JGA)

The smooth muscle fibers (cells) of the tunica media adjacent to the afferent arteriole (and sometimes efferent arteriole) are modified in several ways. Their nuclei are round (instead of long), and their cytoplasm contains granules (instead of myofibrils). Such modified muscle fibers are called ***juxtaglomerular cells.*** The cells of the distal convoluted tubule adjacent to the afferent and efferent arterioles are considerably narrower and taller than the other cells. Collectively, these cells are known as the ***macula densa.*** Together with the modified cells of the afferent arteriole they constitute the ***juxtaglomerular apparatus,*** or ***JGA*** (Figure 24-7), which helps regulate renal blood pressure. In response to low blood pressure, juxtaglomerular cells secrete the enzyme renin. It initiates a sequence of events in which the adrenal cortex secretes aldosterone, which helps elevate blood pressure back to normal.

PHYSIOLOGY

The major work of the urinary system is done by the nephrons. The other parts of the system are primarily passageways and storage areas. Nephrons carry out three important functions: (1) they control blood concentration and volume by removing selected amounts of water and solutes; (2) they help regulate blood pH; and (3) they also remove toxic wastes from the blood. As the nephrons go about these activities, they remove many materials from the blood, return the ones that the body requires, and eliminate the remainder. The eliminated materials are collectively called ***urine.*** The entire volume of blood in the body is filtered by the kidneys approximately 60 times a day.

Urine formation requires three principal processes: glomerular filtration, tubular reabsorption, and tubular secretion (Figure 24-8).

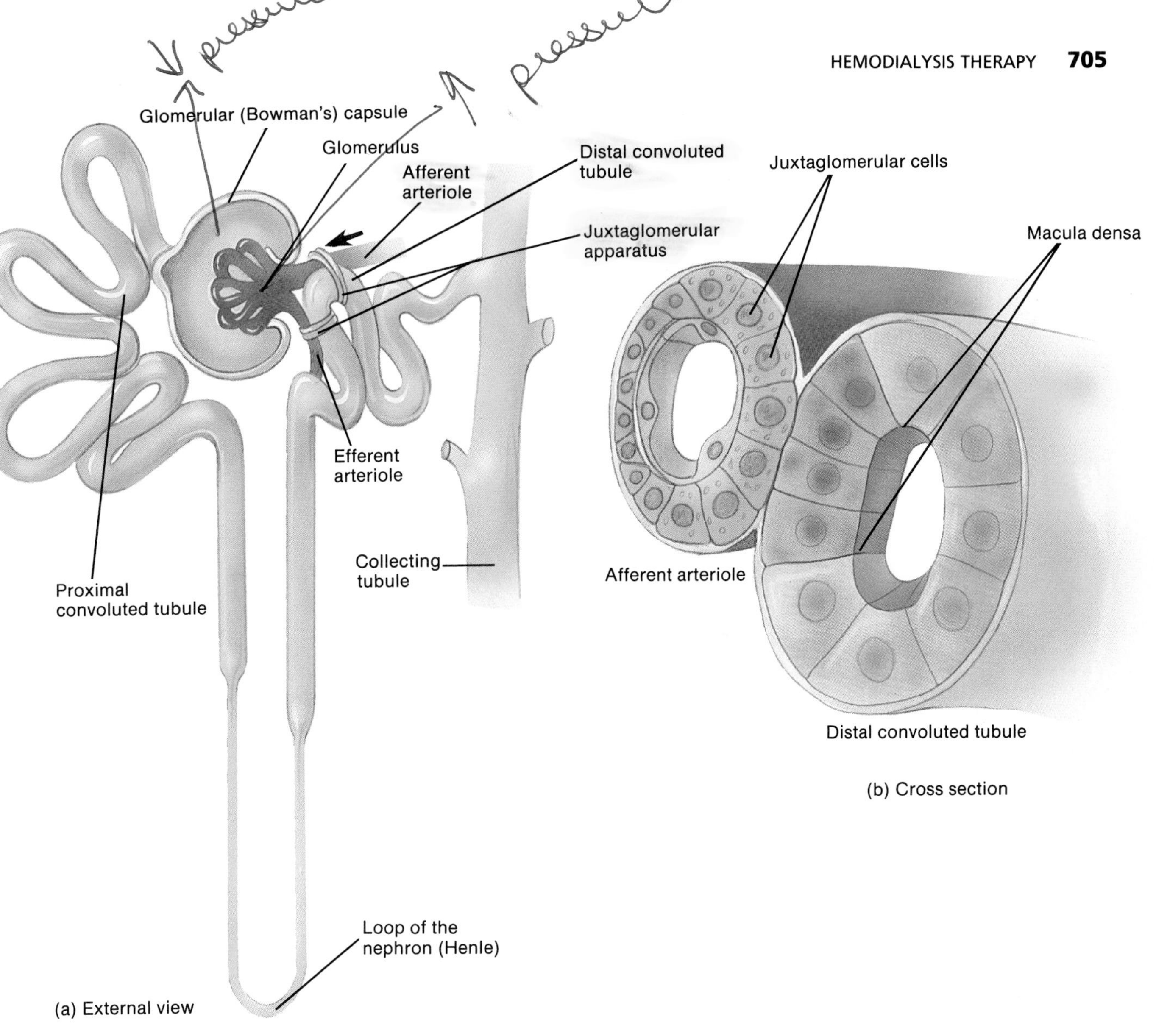

FIGURE 24-7 Juxtaglomerular apparatus. In (b), the macula densa adjacent to the efferent arteriole is not illustrated.

In ***glomerular filtration,*** blood entering the glomerulus is filtered by the endothelial–capsular membrane. The filtered fluid (filtrate) consists of all the substances in blood except for the formed elements and most proteins. The blood pressure in the glomerulus is the main force involved in filtering the blood. As the filtrate flows through the renal tubules, about 99 percent of it moves from the renal tubules into blood in the peritubular capillaries or vasa recta, a process called ***tubular reabsorption.*** Among the substances reabsorbed are water, glucose, amino acids, and ions such as sodium, potassium, calcium, chloride, and phosphate. Tubular reabsorption permits the body to retain most of its nutrients. In ***tubular secretion,*** substances pass from the blood into filtrate. Among these are hydrogen ions, ammonia, creatine, and certain drugs. Tubular secretion rids the body of certain materials and helps control normal pH.

HEMODIALYSIS THERAPY

If the kidneys are impaired by disease or injury, the blood must be filtered by an artificial device. Such filtering of the blood is called ***hemodialysis.*** *Dialysis* means the separation of large nondiffusible particles from smaller diffusible ones through a selectively permeable membrane. One of the best-known devices for accomplishing dialysis is the artificial kidney machine (Figure 24-9). A tube connects it with the patient's radial artery. The blood is pumped

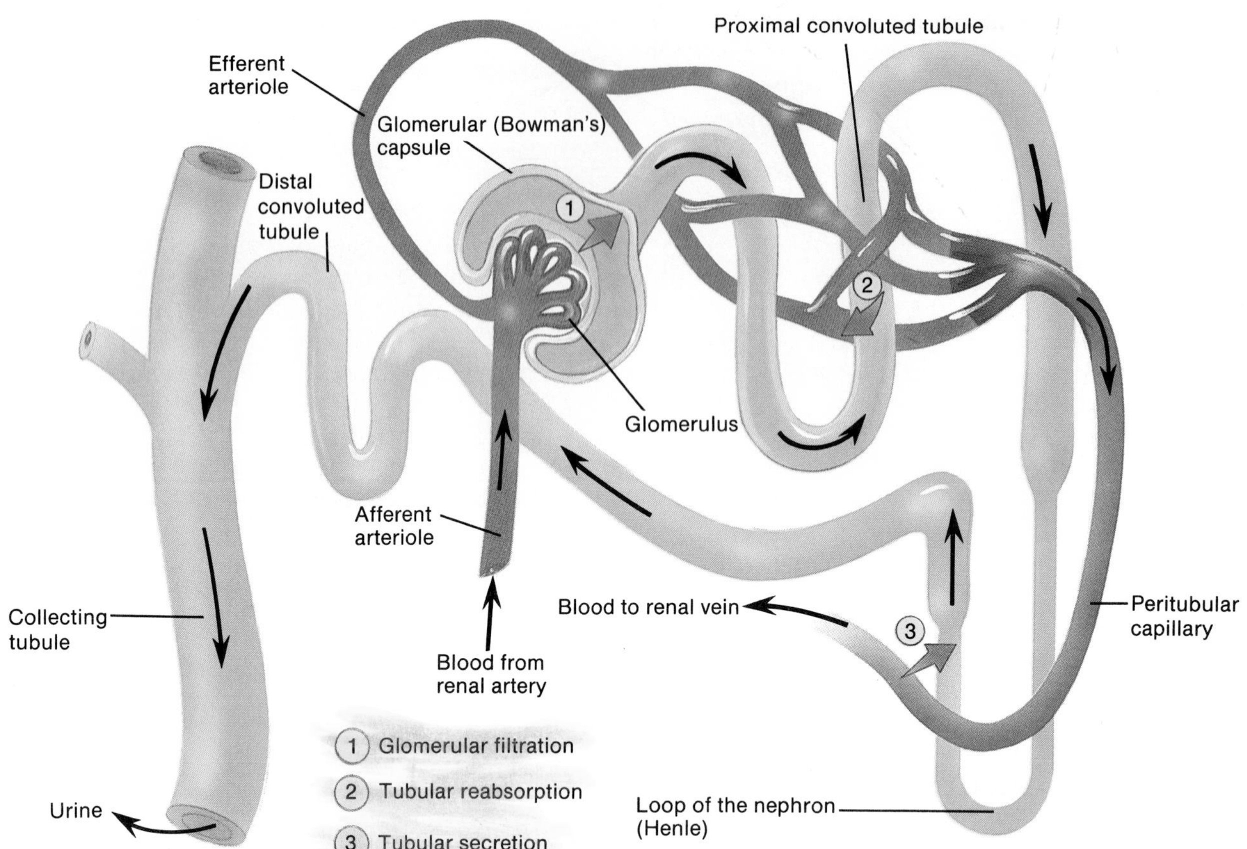

FIGURE 24-8 Summary of functions of a nephron. ① As blood flows from the glomerulus into the glomerular (Bowman's) capsule, it is filtered by the endothelial–capsular membrane. As the filtrate passes through the nephron ②, certain components are selectively reabsorbed into blood and ③ other substances are secreted into the filtrate for elimination in urine.

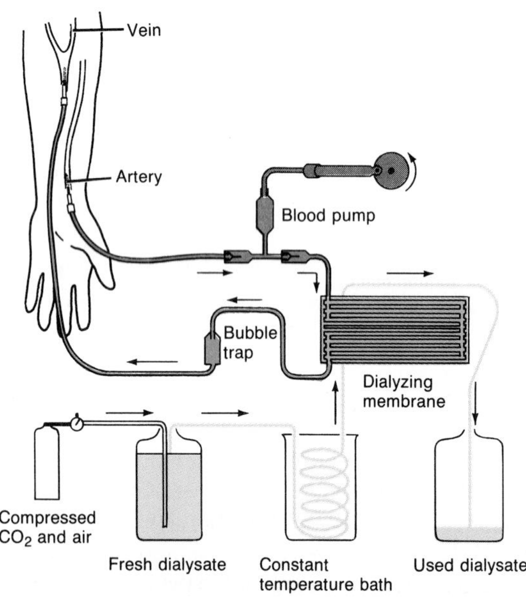

FIGURE 24-9 Operation of an artificial kidney. The blood route is indicated in red and blue. The route of the dialysis is indicated in gold.

from the artery through the tubes to one side of a selectively permeable dialyzing membrane made of cellulose acetate. The other side of the membrane is continually washed with an artificial solution called the dialysate. The blood that passes through the artificial kidney is treated with anticoagulant (heparin). Only about 500 ml of the patient's blood is in the machine at a time.

All substances (including wastes) in the blood except protein molecules and blood cells can diffuse back and forth across the selectively permeable membrane. The electrolyte level of the plasma is controlled by keeping the dialysate electrolytes at the same concentration found in normal plasma. Any excess plasma electrolytes move down the concentration gradient and into the dialysate. If the plasma electrolyte level is normal, it is in equilibrium with the dialysate, and there is no net gain or loss of electrolytes. Since the dialysate contains no wastes, substances such as urea move down the concentration gradient and into the dialysate. Thus, wastes are removed and normal electrolyte balance is maintained. Hemodialysis typically is performed three times a week, each session lasting four to six hours.

A great advantage of the kidney machine is that nutrition can be bolstered by placing large quantities of glucose in the dialysate. While the blood gives up its wastes, the glucose diffuses into the blood. Thus, the kidney machine beautifully accomplishes the principal function of the fundamental unit of the kidney—the nephron.

There are obvious drawbacks to the artificial kidney, however. Anticoagulants must be added to the blood during dialysis. A large amount of the patient's blood must flow through this apparatus to make the treatment effective, and so the slow rate at which the blood can be processed makes the treatment time-consuming, and blood cells can be damaged in the process. To date, no artificial kidney has been implanted permanently.

Continuous ambulatory peritoneal dialysis (CAPD) is more convenient and less time-consuming for many patients. CAPD uses the peritoneum instead of cellulose acetate as the dialyzing membrane. Since the peritoneum is a selectively permeable membrane, it permits rapid bidirectional transfer of substances. A catheter is placed in the patient's peritoneal cavity and connected to a supply of dialysate. Gravity feeds the solution into the abdominal cavity from its plastic container. When the process is complete, the dialysate is returned from the abdominal cavity to the plastic container and then discarded. Danger of infection is always present when this process is used.

URETERS

Once urine is formed by the nephrons and passed into collecting tubules, it drains through papillary ducts into the calyces surrounding the renal papillae. The minor calyces join to become the major calyces that unite to become the renal pelvis. From the pelvis, the urine drains into the ureters and is carried by peristalsis to the urinary bladder. From the urinary bladder, the urine is discharged from the body through the single urethra. From the minor calyces on, the urine is in no way modified in either volume or composition.

STRUCTURE

The body has two ***ureters*** (YOO-re-ters)—one for each kidney. Each ureter is an extension of the pelvis of the kidney and extends 25 to 30 cm (10 to 12 in.) to the urinary bladder (see Figure 24-1). As the ureters descend, their thick walls increase in diameter, but at their widest point they measure less than 1.7 cm (0.7 in.) in diameter. Like the kidneys, the ureters are retroperitoneal in placement. The ureters enter the urinary bladder at the superior lateral angle of its base.

Although there are no anatomical valves at the openings of the ureters into the urinary bladder, there is a functional one that is quite effective. Since the ureters pass obliquely through the wall of the urinary bladder, pressure in the urinary bladder compresses the ureters and prevents backflow of urine when pressure builds up in the urinary bladder as it fills during urination. When this physiological valve is not operating, it is possible for cystitis (urinary bladder inflammation) to develop into kidney infection.

HISTOLOGY

Three coats of tissue form the wall of the ureters (Figure 24-10). The inner coat, or mucosa, is a mucous membrane with transitional epithelium (see Exhibit 3-1, stratified epithelium). The solute concentration and pH of urine differs drastically from the internal environment of cells that form the wall of the ureters. Mucus secreted by the mucosa prevents the cells from coming in contact with urine. Throughout most of the length of the ureters, the second or middle coat, the muscularis, is composed of inner longitudinal and outer circular layers of smooth muscle. The muscularis of the distal third of the ureters also contains a layer of outer longitudinal muscle. Peristalsis is the major function of the muscularis. The third, or external, coat of the ureters is a fibrous coat (fibrosa). Extensions of the fibrous coat anchor the ureters in place.

PHYSIOLOGY

The principal function of the ureters is to transport urine from the renal pelvis into the urinary bladder. Urine is carried through the ureters primarily by peristaltic contractions of the muscular walls of the ureters, but hydrostatic pressure and gravity also contribute. Peristaltic waves pass from the kidney to the urinary bladder, varying in rate from one to five per minute, depending on the amount of urine formation.

BLOOD AND NERVE SUPPLY

The arterial supply of the ureters is from the renal, testicular or ovarian, common iliac, and inferior vesical arteries (arising from the internal iliac artery, a trunk with the internal pudendal and superior gluteal arteries, or a branch of the internal pudendal artery). The veins terminate in the corresponding trunks.

The ureters are innervated by the renal plexuses, which are supplied by sympathetic and parasympathetic fibers from the lesser and lowest splanchnic nerves.

URINARY BLADDER

The ***urinary bladder*** is a hollow muscular organ situated in the pelvic cavity posterior to the pubic symphysis (see Figure 24-1). In the male, it is directly anterior to the rectum. In the female, it is anterior to the vagina and inferior to the uterus. It is a freely movable organ held in position by folds of the peritoneum. The shape of the urinary bladder depends on how much urine it contains. When empty, the

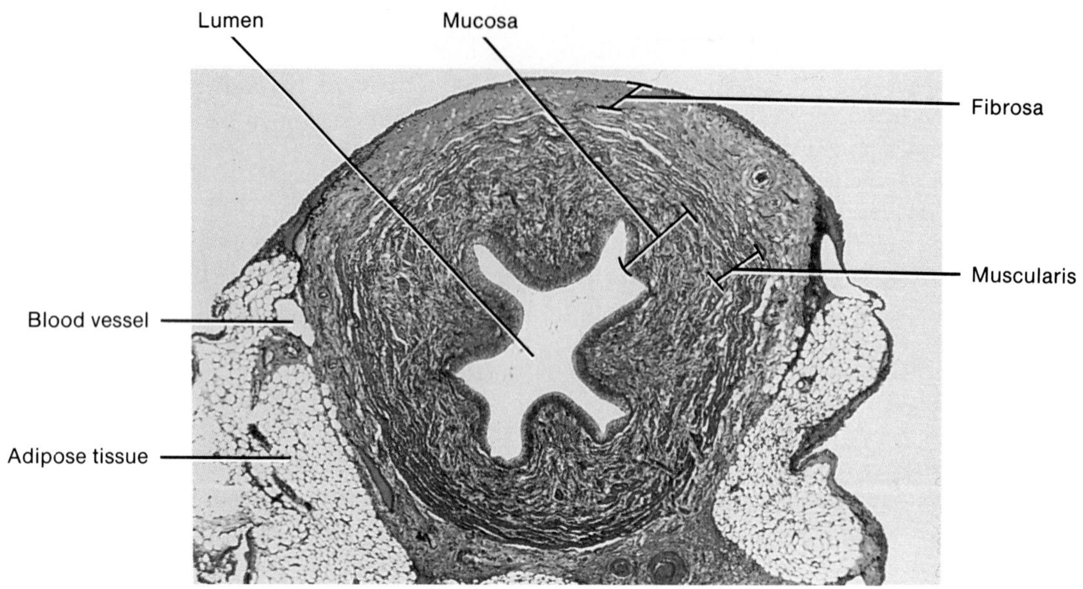

Photomicrograh of cross section of ureter (15x)

FIGURE 24-10 Histology of the ureter. The details of the mucosa are shown in Exhibit 3-1, Transitional epithelium. © Biophoto, SPL, Photo Researchers.

size of the lumen is decreased, and the wall appears thicker. It becomes spherical when slightly distended. As urine volume increases, it becomes pear-shaped and rises into the abdominal cavity.

STRUCTURE

At the base of the urinary bladder is a small triangular area, the ***trigone*** (TRĪ-gōn), that points anteriorly (Figure 24-11). The opening into the urethra, the ***internal urethral orifice,*** is found in the apex of this triangle. At the two points of the base, the ureters drain into the urinary bladder. It is easily identified because the mucosa is firmly bound to the muscularis so that the trigone is typically smooth.

HISTOLOGY

Four coats make up the wall of the urinary bladder (Figure 24-12). The mucosa, the innermost coat, is a mucous membrane containing transitional epithelium, which is able to stretch—a marked advantage for an organ that must continually inflate and deflate. Rugae (folds in the mucosa) are also present. The second coat, the submucosa, is a layer of connective tissue that connects the mucosa and muscular coats. The third coat—a muscular one called the ***detrusor*** (de-TROO-ser) ***muscle***—consists of three layers of smooth muscle: inner longitudinal, middle circular, and outer longitudinal muscles. In the area around the opening to the urethra the circular fibers form an ***internal urethral sphincter*** muscle. Below the internal sphincter is the ***external urethral sphincter,*** which is composed of skeletal muscle and is a modification of the urogenital diaphragm muscle. The outermost coat, the serous coat, is formed by the peritoneum and covers only the superior surface of the organ.

PHYSIOLOGY

Urine is expelled from the urinary bladder by an act called ***micturition*** (mik′-too-RISH-un), commonly known as urination or voiding. This response is brought about by a combination of involuntary and voluntary nerve impulses. The average capacity of the urinary bladder is 700 to 800 ml. When the amount of urine in the urinary bladder exceeds 200 to 400 ml, stretch receptors in the urinary bladder wall transmit impulses to the lower portion of the spinal cord. These impulses, by way of sensory tracts to the cortex, initiate a conscious desire to expel urine and, by way of a center in the sacral cord, a subconscious reflex referred to as the ***micturition reflex.*** Parasympathetic impulses transmitted from the micturition reflex center of the sacral area of the spinal cord reach the urinary bladder wall and internal urethral sphincter, bringing about contraction of the detrusor muscle of the urinary bladder and relaxation of the internal urethral sphincter. Then the conscious portion of the brain sends impulses to the external urethral sphincter, the sphincter relaxes, and urination takes place. Although emptying the urinary bladder is controlled by reflex, it may be initiated voluntarily and stopped at will because of cerebral control of the external urethral sphincter and certain muscles of the urogenital (pelvic) diaphragm.

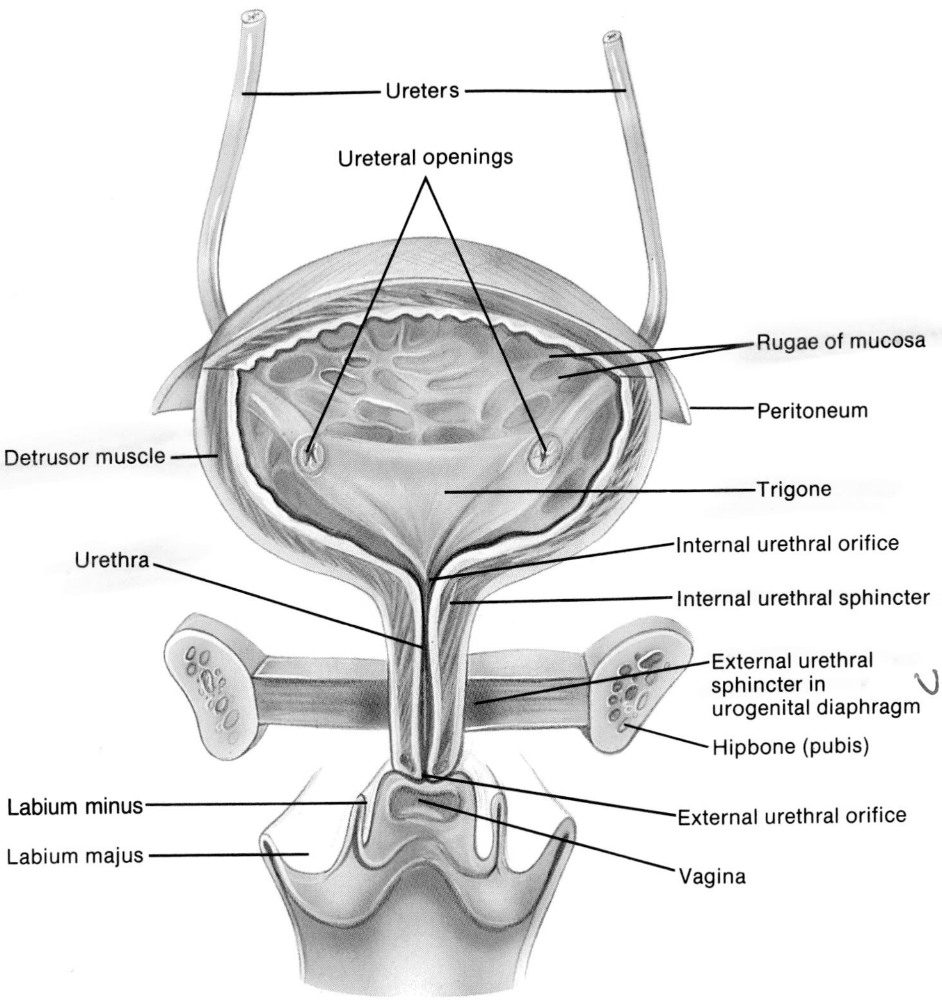

Ureters
Ureteral openings
Rugae of mucosa
Peritoneum
Detrusor muscle
Trigone
Urethra
Internal urethral orifice
Internal urethral sphincter
External urethral sphincter in urogenital diaphragm
Hipbone (pubis)
Labium minus
Labium majus
External urethral orifice
Vagina

(a) Coronal section, viewed from anterior

Sigmoid colon
Uterus
Urinary bladder
Vesicouterine pouch
Pubic symphysis
Vagina
Urethra
Posterior
Anterior

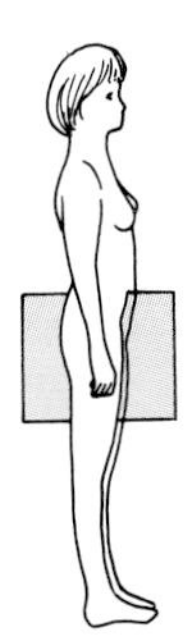

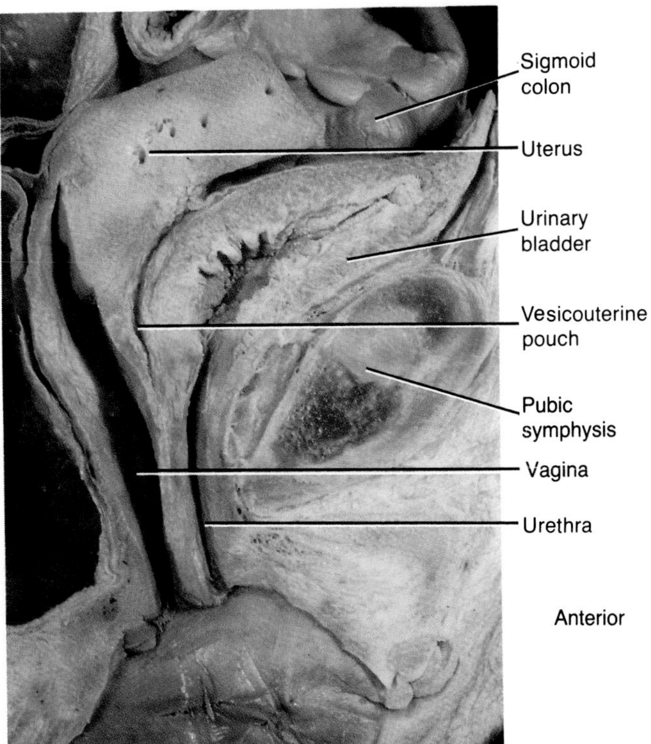

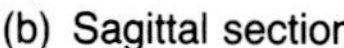

(b) Sagittal section

FIGURE 24-11 Urinary bladder and female urethra. (b) Courtesy of J. A. Gosling, P. F. Harris, et al., *Atlas of Human Anatomy,* Gower Medical Publishing Ltd., 2nd ed., 1991. See also Figure 25-12c.

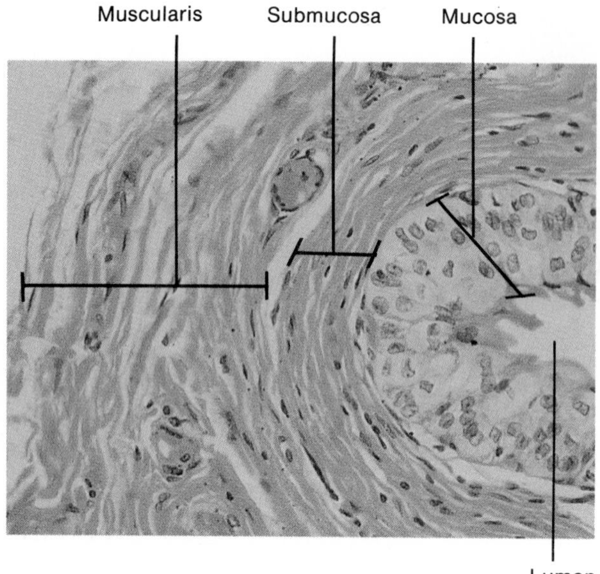

Photomicrograph of a portion of the wall of the urinary bladder, (400 x)

FIGURE 24-12 Histology of the urinary bladder. The details of the mucosa of the urinary bladder are shown in Exhibit 3-1, Stratified transitional. © Iverson, Visuals Unlimited.

CLINICAL APPLICATION

Incontinence and Retention

A lack of voluntary control over micturition is referred to as ***incontinence.*** In infants about 2 years old and under, incontinence is normal because neurons to the external urethral sphincter muscle are not completely developed. Infants void whenever the urinary bladder is sufficiently distended to arouse a reflex stimulus. Proper training overcomes incontinence if the latter is not caused by emotional stress or irritation of the urinary bladder.

Involuntary micturition in the adult may occur as a result of unconsciousness, injury to the spinal nerves controlling the urinary bladder, irritation due to abnormal constituents in urine, disease of the urinary bladder, damage to the external urethral sphincter, and inability of the detrusor muscle to relax due to emotional stress.

Retention, a failure to completely or normally void urine, may be due to an obstruction in the urethra or neck of the urinary bladder, nervous contraction of the urethra, or lack of sensation to urinate.

BLOOD AND NERVE SUPPLY

The arteries of the urinary bladder are the superior vesical (arises from the umbilical artery), the middle vesical (arises from the umbilical artery or a branch of the superior vesical), and the inferior vesical (arises from the internal iliac artery, a trunk with the internal pudendal and superior gluteal arteries, or a branch of the internal pudendal artery). The veins from the urinary bladder pass to the internal iliac trunk.

The nerves are derived partly from the hypogastric sympathetic plexus and partly from the second and third sacral nerves. The fibers from the sacral nerves constitute the nervi erigentes.

URETHRA

The ***urethra*** is a small tube leading from the floor of the urinary bladder to the exterior of the body (see Figure 24-11). In females, it lies directly posterior to the pubic symphysis and is in front of the anterior wall of the vagina. Its undilated diameter is about 6 mm (0.25 in.), and its length is approximately 3.8 cm (1.5 in.). The female urethra is directed obliquely, inferiorly, and anteriorly. The opening of the urethra to the exterior, the ***external urethral orifice,*** is located between the clitoris and vaginal opening.

In males the urethra is about 20 cm (8 in.) long. Immediately below the urinary bladder it passes vertically through the prostate gland (prostatic urethra), then pierces the urogenital diaphragm (membranous urethra), and finally pierces the penis (spongy urethra) and takes a curved course through its body (see Figures 25-1 and 25-10).

HISTOLOGY

The wall of the female urethra consists of three coats: an inner mucous coat, an intermediate thin layer of spongy tissue containing a plexus of veins, and an outer muscular coat that is continuous with that of the urinary bladder and consists of circularly arranged smooth muscle fibers (cells). The mucosa is usually lined with transitional epithelium near the urinary bladder. The remainder consists of stratified squamous epithelium with areas of stratified columnar or pseudostratified columnar epithelium.

The male urethra is composed of two coats: an inner mucous membrane and an outer submucous tissue that connects the urethra with the structures through which it passes. The mucosa varies in different regions. The mucosa of the ***prostatic urethra*** is continuous with that of the urinary bladder and is lined by transitional epithelium. The mucosa of the ***membranous urethra*** is lined by pseudostratified columnar epithelium. The ***spongy urethra*** is lined mostly by pseudostratified columnar epithelium. Near its opening to the exterior it is lined by stratified squamous epithelium. In the spongy urethra, especially, there are glands, called ***urethral (Littré) glands,*** that produce mucus for lubrication during sexual intercourse.

PHYSIOLOGY

The urethra is the terminal portion of the urinary system. It serves as the passageway for discharging urine from the body. The male urethra also serves as the duct through which reproductive fluid (semen) is discharged from the body.

AGING AND THE URINARY SYSTEM

The effectiveness of kidney function decreases with aging, and by age 70 the filtering mechanism is only about one-half what it was at age 40. Urinary incontinence and urinary tract infections are two common problems associated with aging of the urinary system. Other pathologies include polyuria (excessive urine production), nocturia (excessive urination at night), increased frequency of urination, dysuria (painful urination), retention (failure to release urine from the urinary bladder), and hematuria (blood in the urine). Changes and diseases in the kidney include acute and chronic kidney inflammations and renal calculi (kidney stones). Since water balance and thirst are altered, elderly persons are susceptible to dehydration. The prostate gland is often implicated in various disorders of the urinary tract, and cancer of the prostate is the most frequent malignancy in elderly males.

DEVELOPMENTAL ANATOMY OF THE URINARY SYSTEM

As early as the third week of fetal development, a portion of the mesoderm along the posterior half of the dorsal side of the embryo, the ***intermediate mesoderm,*** differentiates into the kidneys. Three pairs of kidneys form within the intermediate mesoderm in successive time periods; pronephros, mesonephros, and metanephros (Figure 24-13). Only the last one remains as the functional kidneys of the adult.

The first kidney to form, the ***pronephros,*** is the superior of the three. Associated with its formation is a tube, the ***pronephric duct.*** This duct empties into the ***cloaca,*** which is the dilated caudal end of the gut derived from ***endoderm.*** The pronephros begins to degenerate during the fourth week and is completely gone by the sixth week. The pronephric ducts, however, remain.

The pronephros is replaced by the second kidney, the ***mesonephros.*** With its appearance, the retained portion of the pronephric duct, which connects to the mesonephros, becomes known as the ***mesonephric duct.*** The mesonephros begins to degenerate by the sixth week and is just about completely gone by the eighth week.

At about the fifth week, an outgrowth, called a ***ureteric bud,*** develops from the distal end of the mesonephric duct near the cloaca. This bud is the developing ***metanephros.*** As it grows toward the head of the embryo, its end widens to form the *pelvis* of the kidney with its *calyces* and associated *collecting tubules.* The unexpanded portion of the bud, the ***metanephric duct,*** becomes the *ureter.* The *nephrons,* the functional units of the kidney, arise from the intermediate mesoderm around each ureteric bud.

During the development, the cloaca divides into a ***urogenital sinus,*** into which urinary and genital ducts empty, and a *rectum* that discharges into the anal canal. The *urinary bladder* develops from the urogenital sinus. In the female, the *urethra* develops from lengthening of the short duct that extends from the urinary bladder to the urogenital sinus. The *vestibule,* into which the urinary and genital ducts empty, is also derived from the urogenital sinus. In the male, the urethra is considerably longer and more complicated but is also derived from the urogenital sinus.

APPLICATIONS TO HEALTH

RENAL CALCULI (KIDNEY STONES)

Occasionally, the crystals of salts found in urine may solidify into insoluble stones called ***renal calculi (kidney stones).*** Conditions leading to stone formation include the ingestion of excessive mineral salts, a decrease in the amount of water intake, abnormally alkaline or acidic urine, and overactivity of the parathyroid glands. Kidney stones usually form in the pelvis of the kidney, where they cause pain, hematuria (red blood cells in urine), and pyuria (pus in the urine).

For kidney stones that become painful or obstructive, surgical removal is the typical alternative. However, there are now several new methods of treatment that do not involve conventional surgery. One procedure, called ***extracorporeal shock wave lithotripsy (ESWL),*** involves the use of ultrasound waves generated by an instrument called a lithotripter. Once the stones are shattered, the fragments are eliminated via the urine. In another procedure, a laser fiber is inserted into the urinary tract and the laser device uses short blasts to shatter the stones. This procedure is used for larger kidney stones that become lodged in the ureters. One other procedure is called ***percutaneous ultrasonic lithotripsy (PUL).*** A tube resembling a cystoscope is inserted into the kidney through a small opening in the back. The surgeon can then pick out small stones whole, or shatter large stones with ultrasound waves, and remove the fragments by suction.

GLOMERULONEPHRITIS (BRIGHT'S DISEASE)

Glomerulonephritis (Bright's disease) is an inflammation of the kidney that involves the glomeruli. One of the most common causes is an allergic reaction to the toxins given off by streptococci bacteria that have recently infected another part of the body, especially the throat. The glomeruli become so inflamed, swollen, and engorged with blood

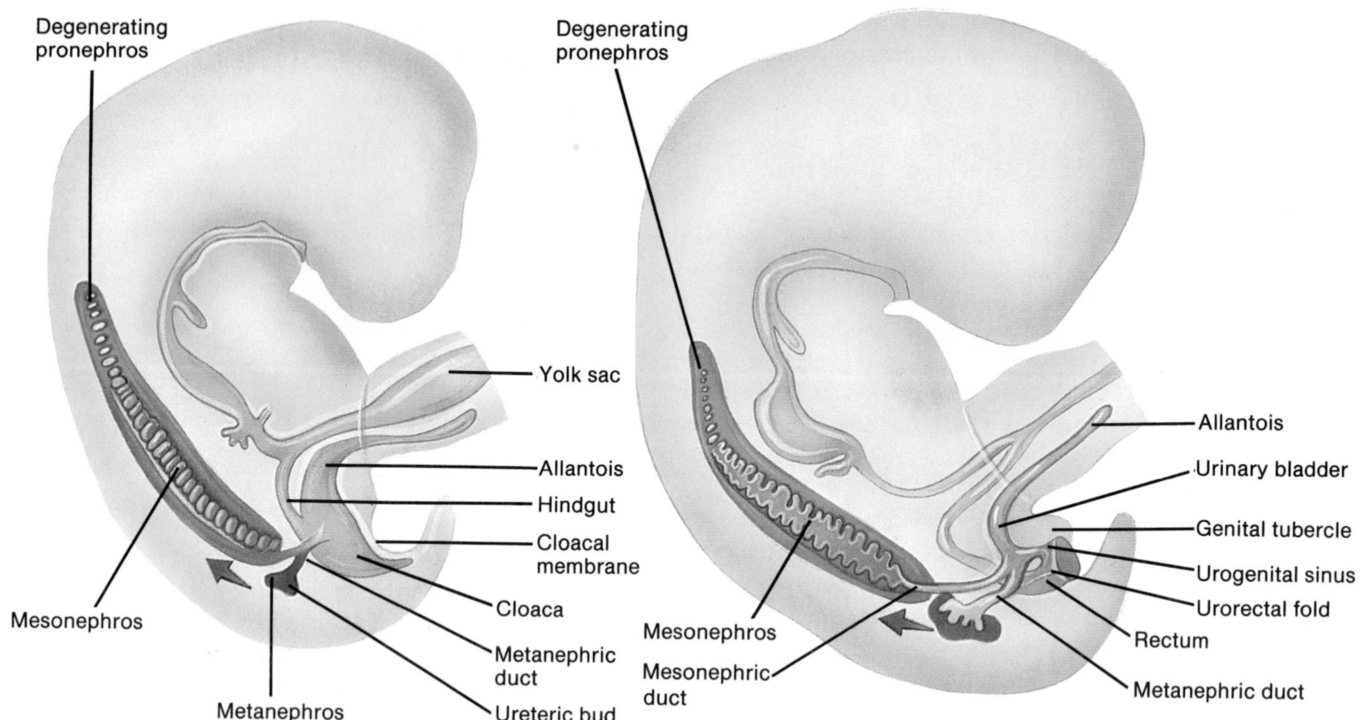

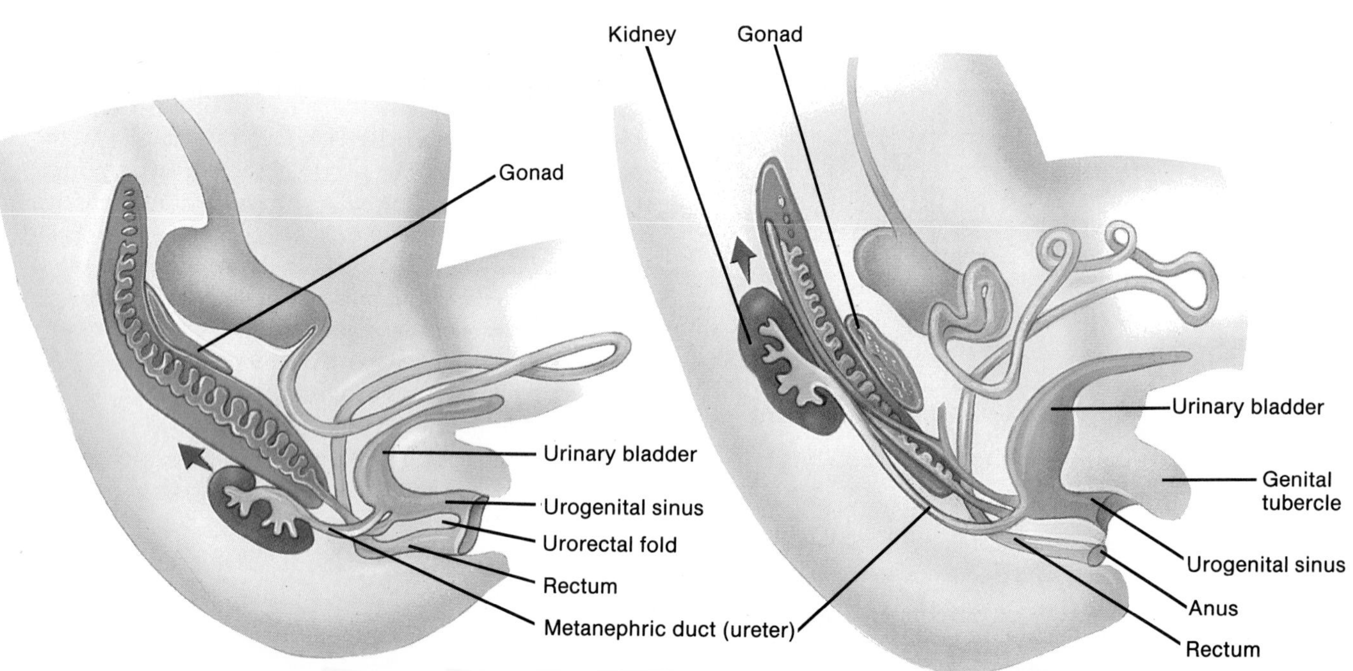

FIGURE 24-13 Development of the urinary system from about the fifth through twelfth weeks.

that the endothelial–capsular membranes become highly permeable and allow blood cells and proteins to enter the filtrate. Thus, the urine contains many erythrocytes and much protein.

PYELITIS AND PYELONEPHRITIS

Pyelitis is an inflammation of the renal pelvis and its calyces. ***Pyelonephritis,*** an inflammation of one or both kidneys, involves the nephrons and the renal pelvis. The disease is generally a complication of infection elsewhere in the body. In females, it is often a complication of lower urinary tract infections. In about 75 percent of the cases the cause is the bacterium *Escherichia coli.* Should pyelonephritis become chronic, scar tissue forms in the kidneys and severely impairs their function.

CYSTITIS

Cystitis is an inflammation of the urinary bladder. It may be caused by bacterial infection, chemicals, or mechanical injury. Symptoms include burning on urination or painful urination, urgency and frequent urination, and low back pain. Bed-wetting may also occur.

NEPHROTIC SYNDROME

Nephrotic syndrome refers to protein in the urine, primarily albumin, that results in low blood level of albumin, edema, and high blood levels of cholesterol, phospholipids, and triglycerides. The protein in urine is due to an increased permeability of the endothelial–capsular membrane, permitting proteins to escape from the blood. Among the causes are diabetes mellitus, rheumatoid arthritis, systemic lupus erythematosus, lymphoma, leukemia, bacterial and viral infections, certain drugs, hypertension, and sickle-cell anemia.

POLYCYSTIC DISEASE

Polycystic disease may be caused by a defect in the renal tubular system that deforms nephrons and results in cystlike dilations along their course. It is the most common inherited disorder of the kidneys. The kidney tissue is riddled with cysts, small holes, and fluid-filled bubbles ranging in size from a pinhead to the diameter of an egg. These cysts gradually increase until they squeeze out the normal tissue, interfering with kidney function and causing uremia (toxic levels of urea in blood). The chief symptom is weight gain. The kidneys themselves may enlarge from their normal 0.25 kg (0.5 lb) to as much as 14 kg (30 lb). Although the disease is progressive, its advance can be slowed by diet, drugs, and regulation of fluid intake.

RENAL FAILURE

Renal failure is a decrease or cessation of glomerular filtration. In ***acute renal failure (ARF)*** the kidneys abruptly stop working entirely or almost entirely. The main feature of ARF is suppression of urine flow, usually characterized by *oliguria* (*olig* = scanty), daily urine output less than 250 ml, or *anuria,* daily urine output less than 50 ml. Causes include low blood volume, decreased cardiac output, damaged renal tubules, and kidney stones.

Chronic renal failure (CRF) refers to the progressive and generally irreversible decline in glomerular filtration rate (GFR). It may result from chronic glomerulonephritis, pyelonephritis, or traumatic loss of kidney tissue. Individuals with CRF are candidates for hemodialysis therapy and kidney transplantation.

Renal failure causes edema from salt and water retention; acidosis due to inability of the kidneys to excrete acidic substances; increased levels of urea due to impaired renal excretion of metabolic waste products; elevated potassium levels that can lead to cardiac arrest; anemia, since the kidneys no longer produce renal erythropoietic factor required for red blood cell production; and osteomalacia, since the kidneys are no longer able to convert vitamin D to its active form for calcium absorption from the small intestine.

URINARY TRACT INFECTIONS (UTIs)

The term ***urinary tract infection (UTI)*** is used to describe either an infection of a part of the urinary system or the presence of large numbers of microbes in urine. Included are ***significant bacteriuria*** (the presence of bacteria in urine in sufficient numbers to indicate active infection), ***asymptomatic bacteriuria*** (the multiplication of large numbers of bacteria in urine without producing symptoms), ***urethritis*** (inflammation of the urethra), ***cystitis*** (inflammation of the urinary bladder), and ***pyelonephritis*** (inflammation of the kidneys).

Symptoms associated with UTI include burning on urination or painful urination, urinary urgency and frequency, pubic and back pain, passage of cloudy or blood-tinged urine, chills, fever, nausea, vomiting, and urethral discharge, usually in males.

KEY MEDICAL TERMS ASSOCIATED WITH THE URINARY SYSTEM

Azotemia (az-ō-TĒ-mē-a; *azo* = nitrogen-containing; *emia* = condition of blood) Presence of urea or other nitrogenous elements in the blood.

Cystocele (SIS-tō-sēl; *cyst* = bladder; *cele* = cyst) Hernia of the urinary bladder.

Dysuria (dis-YOO-rē-a; *dys* = painful; *uria* = urine) Painful urination.

Enuresis (en′-yoo-RĒ-sis; *enourein* = to void urine) Bed-wetting; may be due to faulty toilet training, to some psychological or emotional disturbance, or rarely to some physical disorder. Also referred to as **nocturia.**

Intravenous pyelogram (in′-tra-VĒ-nus PĪ-e-lō-gram′), or **IVP** (*intra* = within; *veno* = vein; *pyelo* = pelvis of kidney; *gram* = written or recorded) X-ray film of the kidneys after venous injections of a dye.

Nephroblastoma (nef′-rō-blas-TŌ-ma; *neph* = kidney; *blastos* = germ or forming; *oma* = tumor) Embryonal carcinosarcoma; a malignant tumor of the kidneys arising from epithelial and connective tissue.

Polyuria (pol′-ē-YOO-rē-a; *poly* = much) Excessive urine formation.

Stricture (STRIK-chur) Narrowing of the lumen of a canal or hollow organ, as may occur in the ureter, urethra, or any other tubular structure in the body.

Uremia (yoo-RĒ-mē-a; *emia* = condition of blood) Toxic levels of urea in the blood resulting from severe malfunction of the kidneys.

Urethritis (yoo′-rē-THRĪ-tis) Inflammation of the urethra, caused by highly acidic urine, the presence of bacteria, or constriction of the urethral passage.

STUDY OUTLINE

1. The primary function of the urinary system is to regulate the concentration and volume of blood by removing and restoring selected amounts of water and solutes. It also excretes wastes.
2. The organs of the urinary system are the kidneys, ureters, urinary bladder, and urethra.

Kidneys (p. 695)

1. The kidneys are retroperitoneal organs attached to the posterior abdominal wall.
2. Three layers of tissue surround the kidneys: renal capsule, adipose capsule, and renal fascia.
3. Internally, the kidneys consist of a cortex, medulla, pyramids, papillae, columns, calyces, and a pelvis.
4. The nephron is the functional unit of the kidneys.
5. Each juxtamedullary nephron consists of a glomerular (Bowman's) capsule, glomerulus, proximal convoluted tubule, descending limb of the loop of the nephron, loop of the nephron (loop of Henle), ascending limb of the loop of the nephron, distal convoluted tubule, and collecting tubule.
6. The filtering unit of a nephron is the endothelial–capsular membrane. It consists of the glomerular endothelium, glomerular basement membrane, and epithelium (podocytes) of the visceral layer of the glomerular (Bowman's) capsule.
7. The extensive flow of blood through the kidney begins in the renal artery and terminates in the renal vein.
8. The nerve supply to the kidney is derived from the renal plexus.
9. The juxtaglomerular apparatus (JGA) consists of the juxtaglomerular cells and the macula densa of the distal convoluted tubule.

Hemodialysis Therapy (p. 705)

1. Filtering blood through an artificial device is called hemodialysis.
2. The kidney machine filters the blood of wastes and adds nutrients; a recent variation is called continuous ambulatory peritoneal dialysis (CAPD).

Ureters (p. 707)

1. The ureters are retroperitoneal and consist of a mucosa, muscularis, and fibrous coat.
2. The ureters transport urine from the renal pelvis to the urinary bladder, primarily by peristalsis.

Urinary Bladder (p. 707)

1. The urinary bladder is posterior to the pubic symphysis. Its function is to store urine prior to micturition.
2. Histologically, the urinary bladder consists of a mucosa (with rugae), a muscularis (detrusor muscle), and a serous coat.
3. A lack of control over micturition is called incontinence; failure to void urine completely or normally is referred to as retention.

Urethra (p. 710)

1. The urethra is a tube leading from the floor of the urinary bladder to the exterior.
2. Its function is to discharge urine from the body.

Aging and the Urinary System (p. 711)

1. After age 40, kidney function decreases.
2. Common problems related to aging include incontinence, urinary tract infections, prostate disorders, and renal calculi.

Developmental Anatomy of the Urinary System (p. 711)

1. The kidneys develop from intermediate mesoderm.
2. They develop in the following sequence: pronephros, mesonephros, metanephros.

Applications to Health (p. 711)

1. Crystals of salts in urine may solidify into renal calculi (kidney stones).
2. Glomerulonephritis is an inflammation of the glomeruli of the kidney.
3. Pyelitis is an inflammation of the kidney pelvis and calyces; pyelonephritis is an inflammation of the nephrons and renal pelvis of one or both kidneys.
4. Cystitis is an inflammation of the urinary bladder.
5. Nephrotic syndrome is characterized by protein in the urine due to increased endothelial–capsular membrane permeability.
6. Polycystic disease is an inherited kidney disease in which nephrons are deformed.
7. Renal failure is classified as acute and chronic.
8. The term urinary tract infection (UTI) refers to either an infection of a part of the urinary system or the presence of large numbers of microbes in urine.

REVIEW QUESTIONS

1. What are the functions of the urinary system? What organs compose the system? (p. 695)
2. Describe the location of the kidneys. Why are they said to be retroperitoneal? (p. 695)
3. Prepare a labeled diagram that illustrates the principal external and internal features of the kidney. What is a floating kidney? (p. 696)
4. What is a nephron? List and describe the parts of a nephron from the glomerular (Bowman's) capsule to the collecting tubule. (p. 698)
5. Describe the structure of the endothelial–capsular membrane. How is the membrane adapted for filtration? (p. 698)
6. Distinguish between cortical and juxtaglomerular nephrons. How are nephrons supplied with blood? (p. 700)
7. Describe the structure and importance of the juxtaglomerular apparatus (JGA). (p. 704)
8. What is hemodialysis? Briefly describe the operation of an artificial kidney. What is continuous ambulatory peritoneal dialysis (CAPD)? (p. 705)
9. Describe the structure, histology, and function of the ureters. (p. 707)
10. How is the urinary bladder adapted to its storage function? (p. 000). Describe the micturition reflex. (p. 707)
11. Contrast the causes of incontinence and retention. (p. 710)
12. Compare the location of the urethra in the male and female. (p. 710)
13. Describe the effects of aging on the urinary system. (p. 711)
14. Describe the development of the urinary system. (p. 711)
15. Define each of the following: renal calculi, glomerulonephritis, pyelitis, pyelonephritis, cystitis, nephrotic syndrome, polycystic disease, renal failure, and urinary tract infections (UTIs). (p. 711)
16. Refer to the glossary of key medical terms associated with the urinary system. Be sure that you can define each term. (p. 713)

SELF QUIZ

Complete the following:

1. The cells of the endothelial–capsular membrane that form filtration slits are called ________.
2. The musculature of the urinary bladder is called the ________ muscle.
3. Artificial filtering of blood as a result of kidney damage or disease is referred to as ________.
4. The glomerular (Bowman's) capsule and its enclosed glomerular capillaries constitute a ________.
5. A ________ consists of a renal corpuscle and a renal tubule.
6. The triangular-shaped structures inside the medulla of a kidney are called the ________.
7. The ________ carries urine from the kidney to the urinary bladder.
8. Arrange the answers in correct sequence.

   __ __ __ **a.** From superior to inferior:
   A. ureter
   B. urinary bladder
   C. urethra

   __ __ __ __ **b.** From most superficial to deepest:
   A. renal capsule
   B. renal medulla
   C. renal cortex
   D. renal pelvis

   __ __ __ __ __ **c.** Pathway of blood:
   A. afferent arteriole
   B. peritubular capillaries and vasa recta
   C. glomerular capillaries
   D. venules and veins
   E. efferent arteriole

   __ __ __ __ __ **d.** Pathway of glomerular filtrate:
   A. ascending limb of loop of nephron
   B. descending limb of loop of nephron
   C. collecting tubule
   D. distal convoluted tubule
   E. proximal convoluted tubule

   __ __ __ __ __ **e.** Pathway of blood:
   A. arcuate arteries
   B. interlobular arteries
   C. renal arteries
   D. afferent arteriole
   E. interlobar arteries

Choose the one best answer to these questions.

__ **9.** Choose the one false statement:
A. the efferent arteriole normally has a larger diameter than the afferent arteriole: B. glomerular capillaries have higher blood pressure than other capillaries of the body; C. blood in glomerular capillaries flows into arterioles, not into venules; D. vasa recta pass blood from peritubular capillaries toward veins.

__ **10.** All of the following structures are parts of the urinary bladder *except*
A. trigone; B. transitional epithelium; C. major calyces; D. detrusor muscle; E. rugae.

__ **11.** Which is a normal function of the urinary bladder?
A. oliguria; B. nephrosis; C. calculi; D. micturition; E. peristalsis.

__ **12.** Which of the following statements is/are true?
(1) The kidneys are located posterior to the peritoneum—that is, retroperitoneally.
(2) Normal kidneys are enclosed in capsules of fatty tissue.
(3) The left kidney is usually located lower than the right kidney.

(4) The hilus is on the concave medial side of the kidney.

A. (1) only; B. (2) only; C. (3) only; D. (4) only; E. (1), (2), and (4).

___ **13.** Which of the following is/are characteristic or descriptive of the kidney?

(1) The cavity of the hilus of the kidney is called the pelvis and represents the upper expanded portion of the ureter.

(2) The kidney exhibits an inner darkened area, the medulla, and an outer pale area, the cortex.

(3) Each kidney is supplied by a renal artery and a renal vein.

A. (1) only; B. (2) only; C. (3) only; D. all of the above; E. (1) and (2).

___ **14.** Urine that leaves the distal convoluted tubule passes through these structures in which sequence?

A. collecting tubule, hilus, calyx, ureter; B. collecting tubule, calyx, pelvis, ureter; C. calyx, collecting tubule, pelvis, ureter; D. calyx, hilus, pelvis, ureter; E. collecting tubule, hilus, ureter, calyx.

___ **15.** An obstruction in the glomerulus would first affect the flow of blood into the

A. renal artery; B. efferent arteriole; C. afferent arteriole; D. intralobular artery; E. renal vein.

___ **16.** The trigone, a helpful landmark in the urinary bladder, is a triangular area bounded by

A. the orifices of the ejaculatory ducts and the urethra; B. the internal urethral orifice and the lower border of the detrusor muscle; C. the ureteral and the internal urethral orifices; D. the top of the fundus and the ureteral orifices; E. the major and minor calyces.

___ **17.** The notch on the medial surface of the kidney through which blood vessels enter and exit is called the

A. medulla; B. major calyx; C. hilus; D. renal column; E. minor calyx.

___ **18.** The epithelium of the urinary bladder that permits distention is

A. stratified squamous; B. transitional; C. simple squamous; D. pseudostratified columnar; E. simple cuboidal.

___ **19.** The *primary* function of the urinary system is to

A. remove waste products from the body; B. regulate metabolism; C. help keep the body in homeostasis by controlling the composition and volume of blood; D. regulate blood pH; E. regulate body temperature.

___ **20.** The functional unit of the kidney is the

A. nephron; B. minor calyx; C. major calyx; D. hilus; E. renal pelvis.

The Reproductive Systems

25

STUDENT OBJECTIVES

1. Define reproduction and classify the organs of reproduction by function.
2. Explain the structure, histology, and functions of the testes.
3. Define meiosis and explain the principal events of spermatogenesis.
4. Describe the seminiferous tubules, straight tubules, and rete testis as components of the duct system of the testes.
5. Describe the location, structure, histology, and functions of the ductus epididymis, ductus (vas) deferens, and ejaculatory duct.
6. Explain the location and functions of the seminal vesicles, prostate gland, and bulbourethral (Cowper's) glands, the accessory sex glands.
7. Explain the structure and functions of the penis.
8. Describe the location, histology, and functions of the ovaries.
9. Describe the principal events of oogenesis.
10. Explain the location, structure, histology, and functions of the uterine (Fallopian) tubes.
11. Describe the events and importance of the menstrual cycle.
12. Describe the location, structure, and functions of the uterus and vagina.
13. Describe the components of the vulva and explain their functions.
14. Explain the structure and histology of the mammary glands.
15. Describe the various types of birth control.
16. Describe the effects of aging on the reproductive systems.
17. Describe the development of the reproductive systems.
18. Explain the symptoms and causes of selected sexually transmitted diseases (STDs).
19. Describe the symptoms and causes of male disorders (testicular cancer, prostate dysfunctions, impotence, and infertility) and female disorders (amenorrhea, dysmenorrhea, premenstrual syndrome [PMS], toxic shock syndrome [TSS], ovarian cysts, endometriosis, infertility, breast tumors, cervical cancer, pelvic inflammatory disease [PID], and vulvovaginal candidiasis).
20. Define key medical terms associated with the reproductive systems.

CHAPTER OUTLINE

- **Male Reproductive System**
Scrotum
Testes
  Spermatogenesis
  Spermatozoa
Ducts
  Ducts of the Testis
  Epididymis
  Ductus (Vas) Deferens
  Ejaculatory Ducts
  Urethra
Accessory Sex Glands
Semen (Seminal Fluid)
Penis
- **Female Reproductive System**
Ovaries
  Oogenesis
Uterine (Fallopian) Tubes
Uterus
Menstrual Cycle
Vagina
Vulva
Perineum
Mammary Glands
  Structure
  Breast Cancer
- **Birth Control (BC)**
Sterilization
Hormonal
Intrauterine Devices (IUDs)
Barrier
Chemical
Physiologic
Coitus Interruptus (Withdrawal)
Induced Abortion
Male Contraception
- **Aging and the Reproductive Systems**
- **Developmental Anatomy of the Reproductive Systems**
- **Applications to Health**
- **Key Medical Terms Associated with the Reproductive Systems**

Reproduction is the mechanism by which the thread of life is sustained. In one sense, reproduction is the process by which a single cell duplicates its genetic material, allowing an organism to grow and repair itself; thus, reproduction is also the process by which genetic material is passed from generation to generation. In this regard, reproduction maintains the continuation of the species.

The organs of the male and female reproductive systems may be grouped by function. The testes and ovaries, also called ***gonads*** (*gonos* = seed), function in the production of gametes—sperm cells and ova, respectively. The gonads also secrete hormones. The production of gametes and their discharge into ducts classifies the gonads as exocrine glands, whereas their production of hormones classifies them as endocrine glands. The ***ducts*** of the reproductive system transport, receive, and store gametes. Still other reproductive organs, called ***accessory sex glands,*** produce materials that support gametes.

The developmental anatomy of the reproductive systems is considered later in the chapter.

MALE REPRODUCTIVE SYSTEM

The organs of the male reproductive system are the testes, or male gonads, that produce sperm and hormones; a number of ducts that either store or transport sperm to the exterior; accessory sex glands that add secretions constituting part of the semen; and several supporting structures, including the penis (Figure 25-1).

SCROTUM

The ***scrotum*** (SKRŌ-tum) is a cutaneous outpouching of the abdomen consisting of loose skin and superficial fascia (Figure 25-1). It is the supporting structure for the testes. Externally, it looks like a single pouch of skin separated into lateral portions by a median ridge called the ***raphe*** (RĀ-fē; *rafe* = seam). Internally, it is divided by a septum into two sacs, each containing a single testis. The septum consists of superficial fascia and muscle tissue called the ***dartos*** (DAR-tōs), which consists of bundles of smooth muscle fibers (cells). The dartos muscle is also found in the subcutaneous tissue of the scrotum and is directly continuous with the subcutaneous tissue of the abdominal wall. The dartos muscle causes wrinkling of the skin of the scrotum.

The location of the scrotum and the contraction of its muscle fibers regulate the temperature of the testes. The production and survival of sperm require a temperature that is lower than normal core body temperature. Because the scrotum is outside the body cavities, it provides an environment about 3°C below normal body temperature. The ***cremaster*** (krē-MAS-ter; *kremaster* = suspender) ***muscle*** (see Figure 25-8) is a small band of skeletal muscle that arises from the middle of the inguinal ligament as a continuation of the internal oblique muscle and inserts into the pubis of the hipbone and sheath of the rectus abdominis muscle. The cremaster muscle elevates the testes during sexual arousal, on exposure to cold, and during times of fear, moving them closer to the pelvic cavity where they can absorb body heat. Exposure to warmth reverses the process. The dartos muscle also is reflexly controlled to help assure that testis temperature is maintained about 3°C below core body temperature.

The blood supply of the scrotum is derived from the internal pudendal branch of the internal iliac, the cremasteric branch of the inferior epigastric artery, and the external pudendal artery from the femoral artery. The scrotal veins follow the arteries.

The scrotal nerves are derived from the pudendal, posterior cutaneous of the thigh, and ilioinguinal nerves.

TESTES

The ***testes,*** or ***testicles,*** are paired oval glands measuring about 5 cm (2 in.) in length and 2.5 cm (1 in.) in diameter (Figure 25-2). Each weighs about 10 to 15 g. The testes develop high on the embryo's posterior abdominal wall and usually begin their descent into the scrotum through the inguinal canals during the latter half of the seventh month of fetal development (see Figure 25-8).

CLINICAL APPLICATION

Cryptorchidism

When the testes do not descend, the condition is referred to as ***cryptorchidism*** (krip-TOR-ki-dizm). The condition occurs in about 3 percent of full-term infants and about 30 percent of premature infants. Bilateral cryptorchidism results in sterility because the cells involved in the initial development of sperm cells are destroyed by the higher body temperature of the pelvic cavity. However, the testes will still secrete testosterone. Also, the probability of testicular cancer is 30 to 50 times greater in cryptorchid testes. The testes of about four-fifths of boys with cryptorchidism will descend spontaneously during the first year of life. When the testes remain undescended, injections of human chorionic gonadotropin (hCG) given at 2 to 5 years of age may stimulate descent. Surgical descent of the testes, known as *orchidopexy,* may become necessary and should be done at about age 5.

The testes are partially covered by a serous membrane called the ***tunica vaginalis,*** an outpocketing of the peritoneum formed during the descent of the testes. Internal to the tunica vaginalis is a dense white fibrous capsule, the ***tunica albuginea*** (al′-byoo-JIN-ē-a), that extends inward and divides each testis into a series of internal compartments called ***lobules.*** Each of the 200 to 300 lobules contains one to three tightly coiled tubules, the convoluted ***seminifer-***

ous tubules, that produce sperm by a process called ***spermatogenesis.*** This process is considered shortly.

A cross section through a seminiferous tubule reveals that it is lined with spermatogenic cells in various stages of development (Figure 25-3). Spermatogenic cells represent successive stages in a continuous process of differentiation of male germ cells. The most immature spermatogenic cells, the ***spermatogonia,*** are located against the basement membrane. Toward the lumen of the tube, one can see layers of progressively more mature cells. In order of advancing maturity, these cells are primary spermatocytes, secondary spermatocytes, and spermatids. By the time a ***sperm cell,*** or ***spermatozoon*** (sper′-ma-tōn-ZŌ-on), has nearly reached maturity, it is in the lumen of the tubule and begins to be moved through a series of ducts.

Embedded between the developing sperm cells in the tubules are ***sustentacular*** (sus′-ten-TAK-yoo-lar), or ***Sertoli, cells.*** Just internal to the basement membrane, sustentacular cells are joined to one another by junctional points that form a ***blood–testis barrier.*** The barrier is important because spermatozoa and developing cells produce surface antigens that are recognized as foreign by the immune system. The barrier prevents an immune response against the antigens by isolating the cells from the blood. Such an immune response is seen following vasectomy (to be described shortly) in which sperm-specific antibodies,

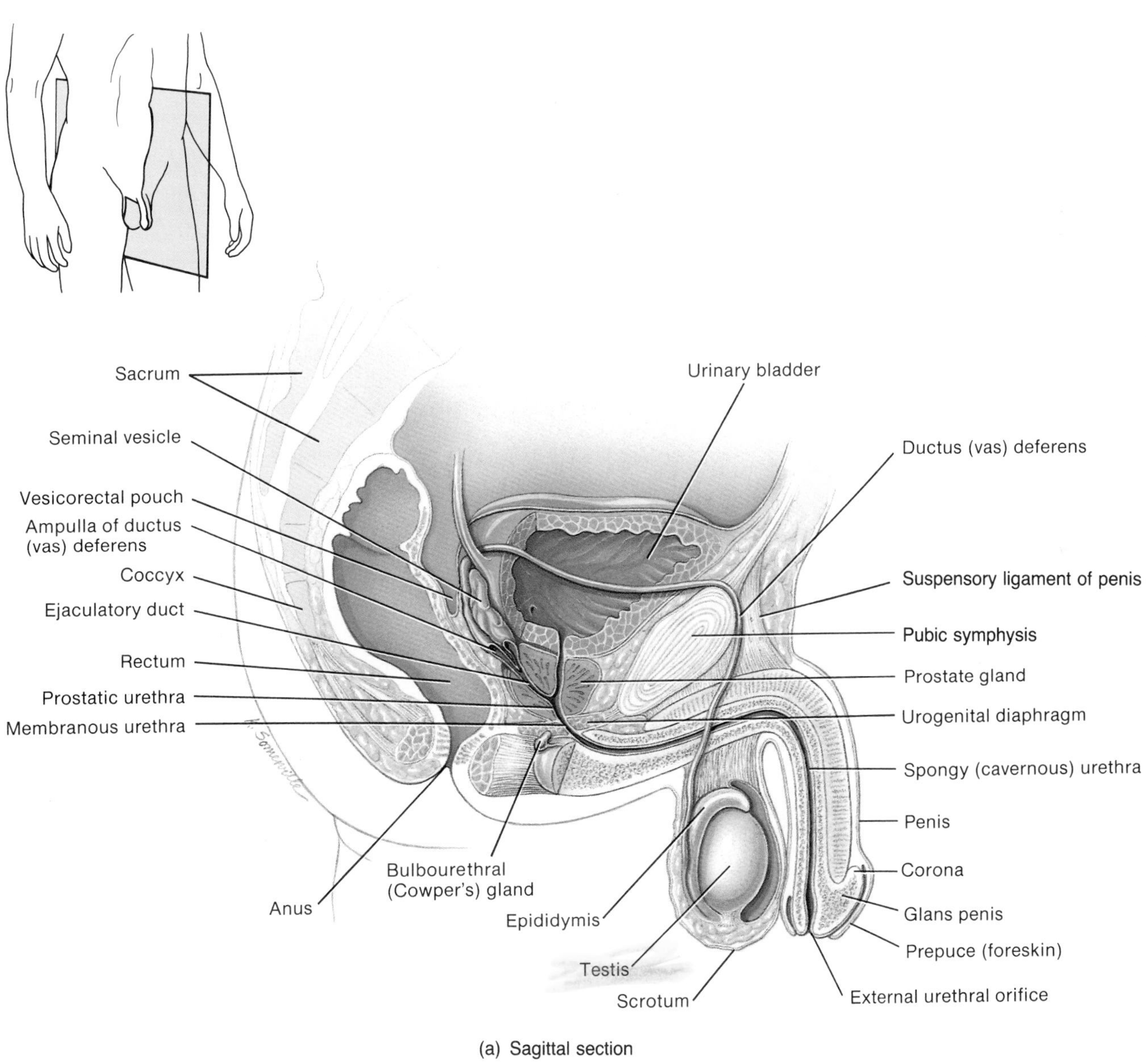

(a) Sagittal section

FIGURE 25-1 Male organs of reproduction and surrounding structures seen in sagittal section. (b) Courtesy of J. A. Gosling, P. F. Harris, et al., *Atlas of Human Anatomy,* Gower Medical Publishing Ltd., 1985.

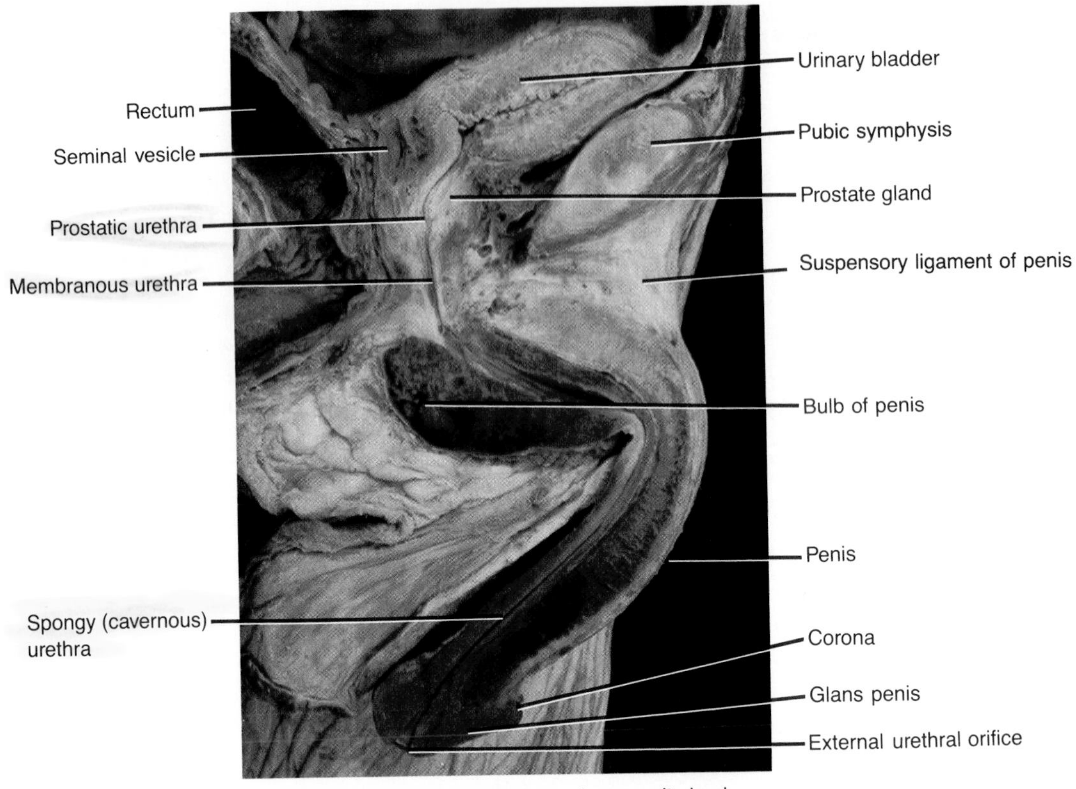

(b) Sagittal section with the cadaver on its back

FIGURE 25-1 ***(continued)***

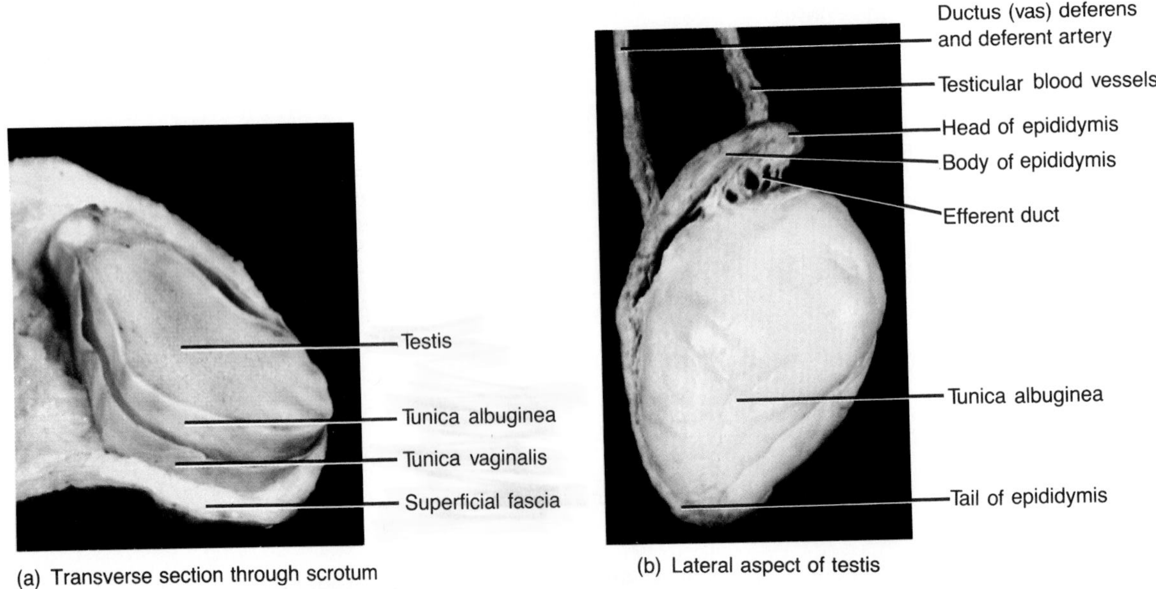

(a) Transverse section through scrotum

(b) Lateral aspect of testis

FIGURE 25-2 Anatomy of the testes. (a) and (b) Courtesy of J. A. Gosling, P. F. Harris, et al., *Atlas of Human Anatomy,* Gower Medical Publishing Ltd., 2nd ed., 1991.

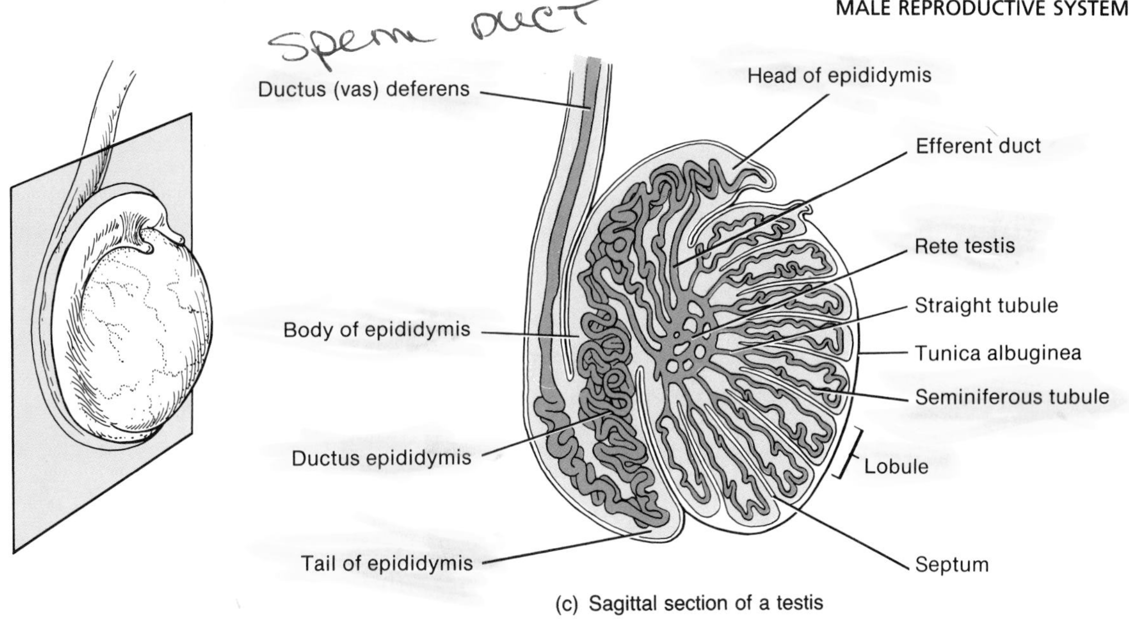

(c) Sagittal section of a testis

FIGURE 25-2 *(continued)*

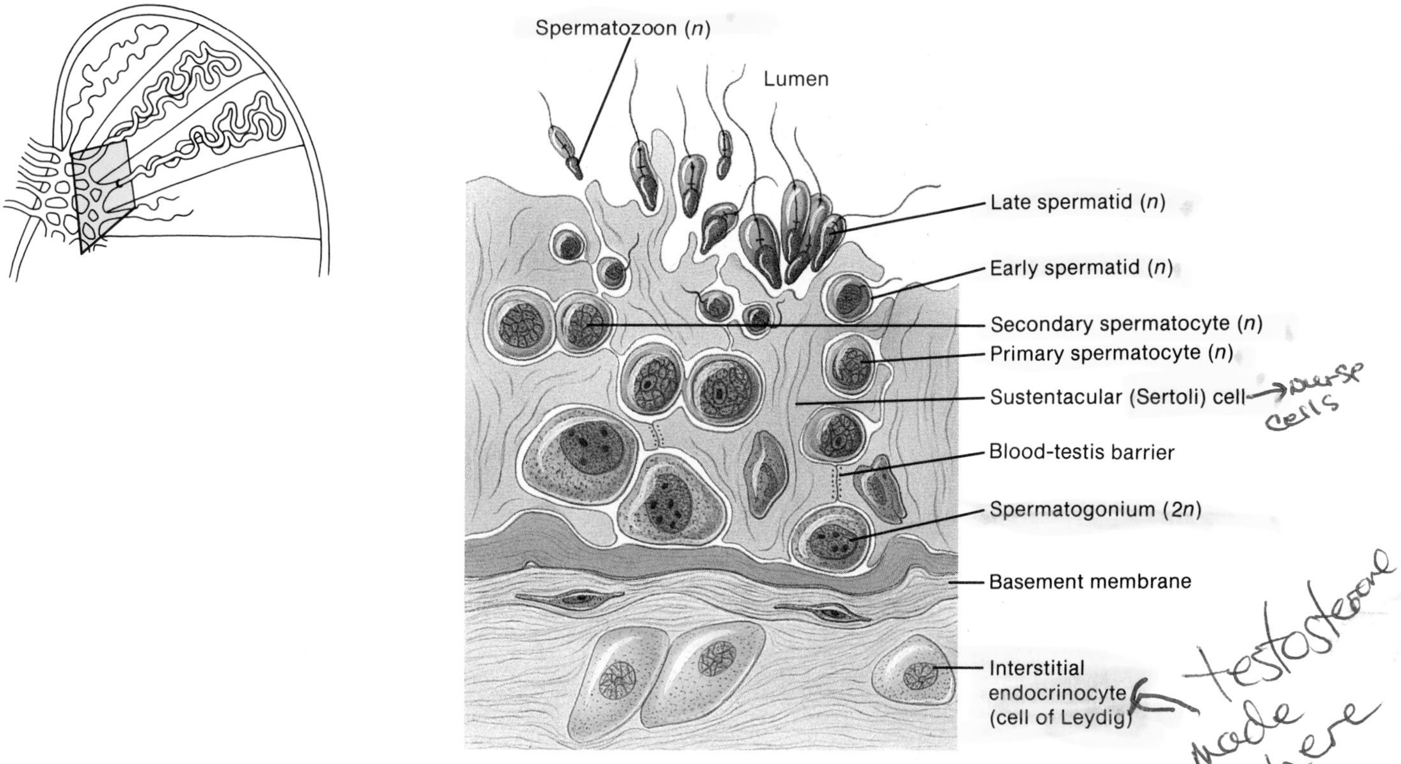

Spermatozoon (n)
Lumen
Late spermatid (n)
Early spermatid (n)
Secondary spermatocyte (n)
Primary spermatocyte (n)
Sustentacular (Sertoli) cell
Blood-testis barrier
Spermatogonium ($2n$)
Basement membrane
Interstitial endocrinocyte (cell of Leydig)

(a) Cross section of a portion of a seminiferous tubule

FIGURE 25-3 Histology of the testes. The stages of spermatogenesis are shown in (a). (b) Courtesy of Ed Reschke.(c) Courtesy of Richard K. Kessel and Randy H. Kardon, *Tissues and Organs: A Text-Atlas of Scanning Electron Microscopy.* Copyright © 1979 by Scientific American, Inc.

Spermatozoa
Basement membrane
Interstitial endocrinocyte
(cell of Leydig)
Lumen of seminiferous tubule
Spermatogonium
Spermatocyte
Spermatid

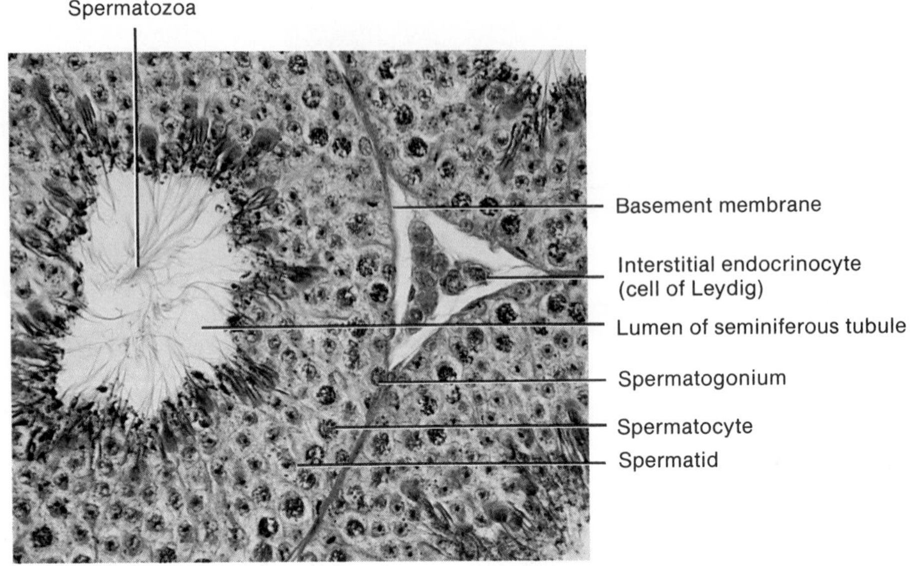

(b) Photomicrograph of a cross section of several seminiferous tubules (1280 x)

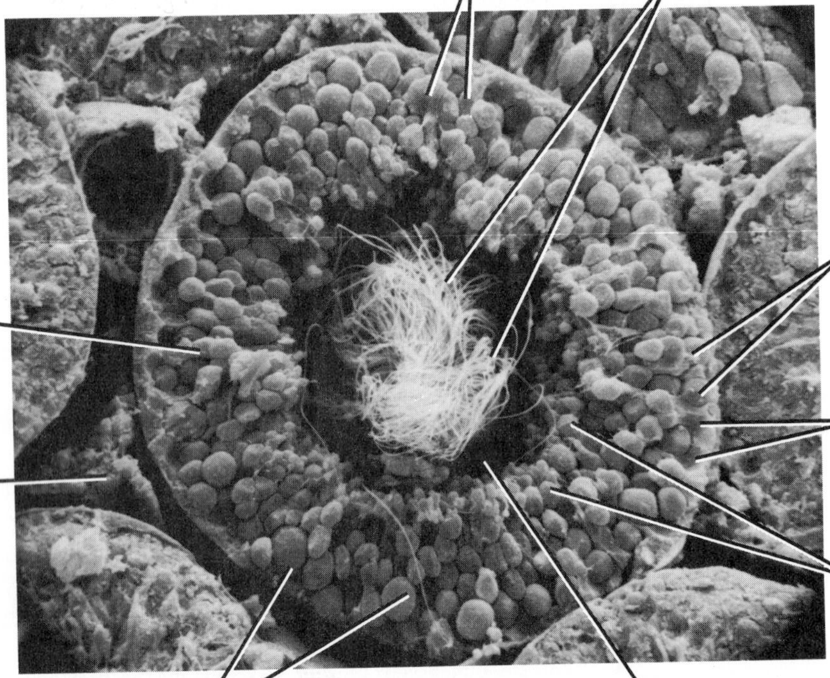

Sustentacular
(Sertoli) cells
Tails of
spermatozoa
Spermatogonia
Sustentacular
(Sertoli) cell
Sustentacular
(Sertoli) cells
Interstitial endocrinocytes
(cells of Leydig)
Spermatids
Primary
spermatocytes
Lumen

(c) Scanning electron micrograph of a cross section of a seminiferous tubule, 400×

produced in cells of the immune system, are exposed to spermatozoa that no longer remain isolated in the reproductive tract. Sustentacular cells support and protect developing spermatogenic cells; nourish spermatocytes, spermatids, and spermatozoa; phagocytize degenerating spermatogenic cells; control movements of spermatogenic cells and the release of spermatozoa into the lumen of the seminiferous tubule; and secrete the hormone inhibin that inhibits secretion of follicle-stimulating hormone (FSH) and thus helps regulate sperm production and androgen-binding protein, a substance required for sperm production that concentrates testosterone in the seminiferous tubule. Between the seminiferous tubules are clusters of ***interstitial endocrinocytes (interstitial cells of Leydig).*** These cells secrete the male hormone testosterone, the most important androgen (Exhibit 25-1).

Spermatogenesis

The process by which the seminiferous tubules of the testes produce haploid (n) spermatozoa involves several phases, including meiosis and mitosis, and is called ***spermatogenesis*** (sper'-ma-tō-JEN-e-sis). Before reading the following discussion of spermatogenesis, you should review the details of meiosis in Chapter 2 (see Figure 2-17). At this point, a few key concepts should be kept in mind.

1. In sexual reproduction, a new organism is produced by the union and fusion of sex cells called ***gametes*** (*gameto* = to marry). Male gametes, produced in the testes, are called sperm cells, and female gametes, produced in the ovaries, are called ova.
2. The cell resulting from the union and fusion of gametes, called a ***zygote*** (*zygo* = joined), contains a mixture of chromosomes (DNA) from the two parents. Through repeated mitotic cell divisions, a zygote develops into a new organism.
3. Gametes differ from all other body cells (somatic cells) in that they contain the ***haploid*** (one-half) ***chromosome number,*** symbolized as n. In humans, this number is 23, which composes a single set of chromosomes. Uninucleated somatic cells contain the ***diploid chromosome number,*** symbolized $2n$. In humans, this number is 46, which composes two sets of chromosomes.
4. In a diploid cell, two chromosomes that belong to a pair are called ***homologous*** (*homo* = same) ***chromosomes (homologues).*** In human diploid cells, 22 of the 23 pairs of chromosomes are morphologically similar and are called ***autosomes.*** The other pair comprises the ***sex chromosomes,*** designated as X and Y. In the female, the homologous pair of sex chromosomes consists of two X chromosomes; in the male, the pair consists of an X and a Y.
5. If gametes were diploid ($2n$), like somatic cells, the zygote would contain twice the diploid number ($4n$), and with every succeeding generation the chromosome number would continue to double and normal development could not occur.
6. This continual doubling of the chromosome number does not occur because of meiosis, a process of cell division by which gametes produced in the testes and ovaries receive the haploid chromosome number. Thus, when haploid (n) gametes fuse, the zygote contains the diploid chromosome number ($2n$) and can undergo normal development.

In humans, spermatogenesis takes about 74 days. The seminiferous tubules are lined with immature cells called ***spermatogonia*** (sper'-ma-tō-GŌ-nē-a; *sperm* = seed; *gonium* = generation or offspring), or sperm mother cells (Figures 25-3a and 25-4). These cells develop from ***primordial*** (*primordialis* = primitive or early form) ***germ cells*** that arise from yolk sac endoderm and enter the testes early in development. In the embryonic testes, the primordial germ cells differentiate into spermatogonia but remain dormant until they begin to undergo mitotic proliferation at puberty. Spermatogonia contain the diploid ($2n$) chromosome number and represent a heterogenous group of cells

EXHIBIT 25-1

Summary of Hormones Secreted by Testes and Ovaries

HORMONE	FUNCTIONS
Testosterone	Secreted by testes and controls development, growth, and maintenance of male sex organs; stimulates bone growth, protein anabolism, sexual behavior, final maturation of sperm, and development of male secondary sex characteristics; stimulates descent of testes.
Inhibin	Secreted by testes and ovaries and inhibits secretion of FSH.
Estrogens	Secreted by ovaries and control development and maintenance of female reproductive structures, especially the endometrium, secondary sex characteristics, and breasts; control fluid and electrolyte balance; increase protein anabolism.
Progesterone	Secreted by ovaries and works with estrogens to prepare endometrium for implantation of a fertilized ovum and mammary glands for milk secretion.
Relaxin	Secreted by ovaries and relaxes pubic symphysis plus associated ligaments and helps dilate uterine cervix to facilitate delivery.

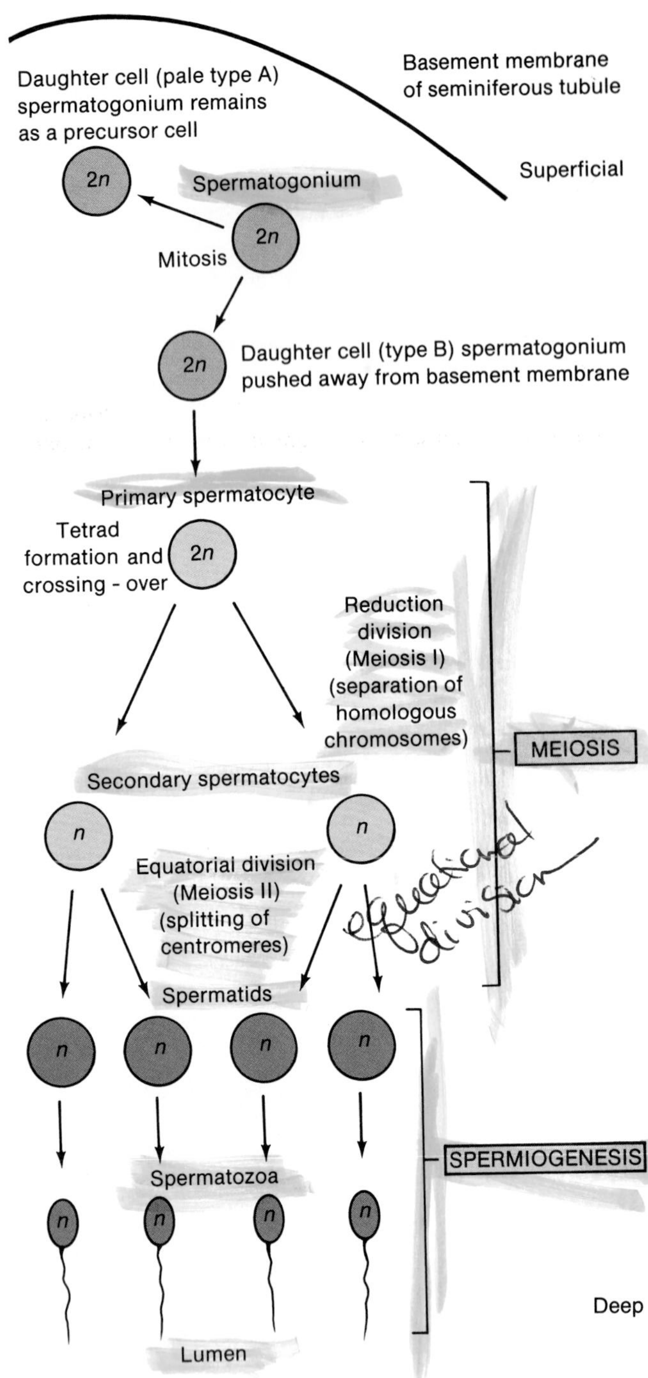

FIGURE 25-4 Spermatogenesis.

in which three subtypes can be distinguished. These are referred to as *pale type A, dark type A,* and *type B* and are distinguished by the appearance of their nuclear chromatin. Pale type A spermatogonia remain relatively undifferentiated and capable of extensive mitotic division. Following division, some of the daughter cells remain undifferentiated and serve as a reservoir of precursor cells to prevent depletion of the stem cell population. Such cells remain near the basement membrane. The remainder of the daughter cells differentiate into type B spermatogonia. These cells lose contact with the basement membrane of the seminiferous tubule, undergo certain developmental changes, and become known as ***primary spermatocytes*** (SPER-ma-tō-sītz′). Primary spermatocytes, like spermatogonia, are diploid ($2n$); that is, they have 46 chromosomes. Dark type A spermatogonia are believed to represent reserve stem cells, only becoming activated if pale type A cells become critically depleted.

■ ***Reduction Division* (*Meiosis I*)** Each primary spermatocyte enlarges before dividing. Then two nuclear divisions take place as part of meiosis. In the first, DNA is replicated and 46 chromosomes (each made up of two identical chromatids) form and move toward the equatorial plane of the nucleus. There they line up by homologous pairs so that there are 23 pairs of duplicated chromosomes in the center of the nucleus. This pairing of homologous chromosomes is called ***synapsis.*** The four chromatids of each homologous pair then become associated with each other to form a ***tetrad.*** In a tetrad, portions of one chromatid may be exchanged with portions of another. This process, called ***crossing-over,*** permits an exchange of genes among chromatids (see Figure 2-18) that results in the recombination of genes. Thus, the spermatozoa eventually produced are genetically unlike each other and unlike the cell that produced them—one reason for the great variation among humans. Next, the meiotic spindle forms and the kinetochore microtubules organized by the centromere extend toward the poles of the cell. As the pairs separate, one member of each pair migrates to opposite poles of the dividing nucleus. The random arrangement of chromosome pairs on the spindle is another reason for variation among humans. The cells formed by the first nuclear division (reduction division) are called ***secondary spermatocytes.*** Each cell has 23 chromosomes—the haploid number. Each chromosome of the secondary spermatocytes, however, is made up of two identical chromatids. Moreover, the genes of the chromosomes of secondary spermatocytes may be rearranged as a result of crossing-over.

■ ***Equatorial Division* (*Meiosis II*)** The second nuclear division of meiosis is ***equatorial division.*** There is no replication of DNA. The chromosomes (each composed of two identical chromatids) line up in single file around the equatorial plane, and the chromatids of each chromosome separate from each other. The cells formed from the equatorial division are called ***spermatids.*** Each contains half the original chromosome number, or 23 chromosomes, and is haploid. Each primary spermatocyte therefore produces four spermatids by meiosis (reduction division and equatorial division). Spermatids lie close to the lumen of the seminiferous tubule.

During spermatogenesis, a very interesting and unique process occurs. As the sperm cells proliferate, they fail to complete cytoplasmic separation (cytokinesis) so that all the daughter cells, except for the least-differentiated spermatogonia, remain continuous via cytoplasmic bridges. These cytoplasmic bridges persist until development of the spermatozoa is complete, at which point they float out individually into the lumen of the seminiferous tubule. Thus, the offspring of an original spermatogonium remain in cytoplasmic communication through their entire development. This pattern of development undoubtedly accounts for the synchronized production of spermatozoa in any given area of a seminiferous tubule. This pattern may have survival value in that half the spermatozoa contain an X chromosome and half a Y chromosome. The X chromosome probably carries many essential genes that are lacking on the Y chromosome, and if it were not for the cytoplasmic bridges between the developing sperm, the Y-bearing spermatozoon perhaps could not survive, with the result that no males could be produced in the next generation.

■ ***Spermiogenesis*** The final stage of spermatogenesis, called ***spermiogenesis*** (sper′-mē-ō-JEN-e-sis), involves the maturation of spermatids into spermatozoa. Each spermatid embeds in a sustentacular (Sertoli) cell and develops a head with an acrosome (described shortly) and a flagellum (tail). Sustentacular cells extend from the basement membrane to the lumen of the seminiferous tubule, where they nourish the developing spermatids. Since there is no cell division in spermiogenesis, each spermatid develops into a single ***spermatozoon (sperm cell).*** The release of a spermatozoon from a sustentacular cell is known as ***spermiation.***

Spermatozoa enter the lumen of the seminiferous tubule and migrate to the ductus epididymis, where in 10 to 14 days they complete their maturation and become capable of fertilizing an ovum. Spermatozoa are also stored in the ductus (vas) deferens. Here, they can retain their fertility for up to several months.

Spermatozoa

Spermatozoa are produced or matured at the rate of about 300 million per day, and, once ejaculated, have a life expectancy of about 48 hours within the female reproductive tract. A spermatozoon is highly adapted for reaching and penetrating a female ovum. It is composed of a head, a midpiece, and a tail (Figure 25-5). Within the ***head*** are the nuclear material and a dense granule called the ***acrosome*** (*acro* = atop), which develops from the Golgi complex and contains enzymes (hyaluronidase and proteinases) that facilitate penetration of the sperm cell into a secondary oocyte. The acrosome is basically a specialized lysosome. Numerous mitochondria in the ***midpiece*** carry on the metabolism that provides energy for locomotion. The ***tail,*** a typical flagellum, propels the sperm along its way.

DUCTS

Ducts of the Testis

Following their production, spermatozoa are moved through the convoluted seminiferous tubules to the ***straight tubules*** (see Figure 25-2c). The straight tubules lead to a network

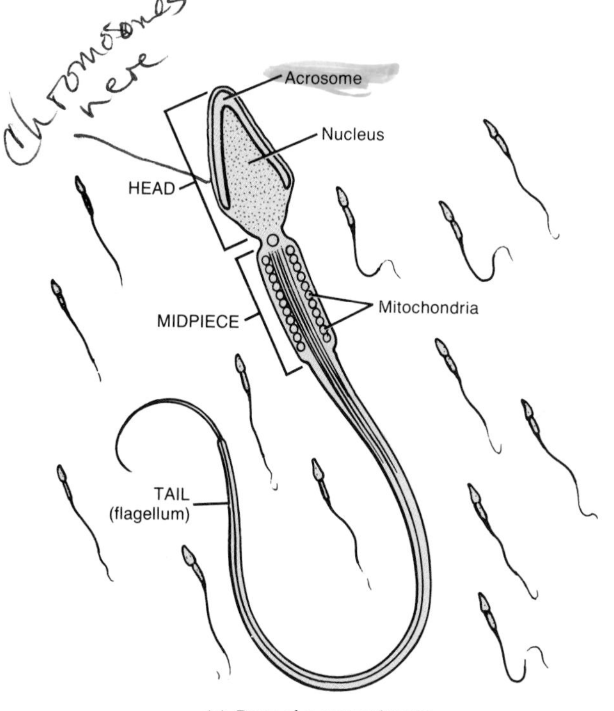

(a) Parts of a spermatozoon

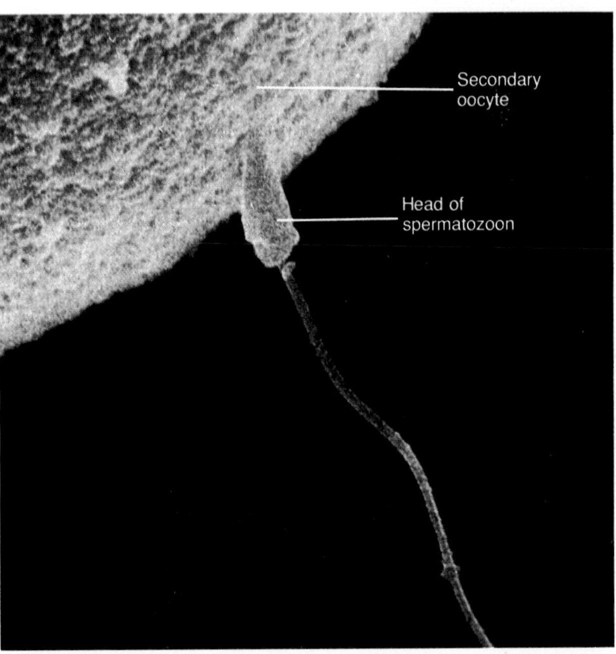

(b) Scanning electron micrograph of a spermatozoon in contact with a secondary oocyte, 1100×

FIGURE 25-5 Spermatozoa. (b) Courtesy of Fawcett-Phillips, Science Photo Library, Photo Researchers.

of ducts in the testis called the ***rete*** (RĒ-tē) ***testis.*** Some of the cells lining the rete testis possess cilia that probably help move the sperm along. The sperm are next transported out of the testis into an adjacent organ, the epididymis.

Epididymis

The sperm are transported out of the testis through a series of coiled ***efferent ducts*** in the epididymis that empty into a single tube called the ductus epididymis. Morphological changes occur in the spermatozoa during their passage through the epididymis.

The ***epididymis*** (ep′-i-DID-i-mis; *epi* = above; *didymos* = testis) is a comma-shaped organ that lies along the posterior border of the testis (see Figures 25-1 and 25-2) and consists mostly of a tightly coiled tube, the ***ductus epididymis.*** The larger, superior portion of the epididymis is known as the ***head.*** In the head, the efferent ducts join the ductus epididymis. The ***body*** is the narrow midportion of the epididymis. The ***tail*** is the smaller, inferior portion. At its distal end, the tail of the epididymis continues as the ductus (vas) deferens.

The ductus epididymis is a tightly coiled structure that would measure about 6 m (20 ft) in length and 1 mm in diameter if it were straightened out. The epididymis measures only about 3.8 cm (1½ in.). The ductus epididymis is lined with pseudostratified columnar epithelium and encircled by layers of smooth muscle. The free surfaces of the columnar cells contain long, branching microvilli called ***stereocilia*** (Figure 25-6).

Functionally, the ductus epididymis is the site of sperm maturation (increased motility and fertility potential). They require between 10 and 14 days to complete their maturation—that is, to become capable of fertilizing an ovum. The ductus epididymis also stores spermatozoa and propels them toward the urethra during emission by peristaltic contraction of its smooth muscle. Spermatozoa may remain in storage in the ductus epididymis up to several months. After that, they are expelled from the epididymis or reabsorbed in the epididymis.

Ductus (Vas) Deferens

Within the tail of the epididymis, the ductus epididymis becomes less convoluted, its diameter increases, and at this point it is referred to as the ***ductus (vas) deferens*** or ***seminal duct*** (see Figure 25-1). The ductus (vas) deferens, about 45 cm (18 in.) long, ascends along the posterior border of the testis, penetrates the inguinal canal, and enters the pelvic cavity, where it loops over the side and down the posterior surface of the urinary bladder (see Figure 25-1a). The dilated terminal portion of the ductus (vas)

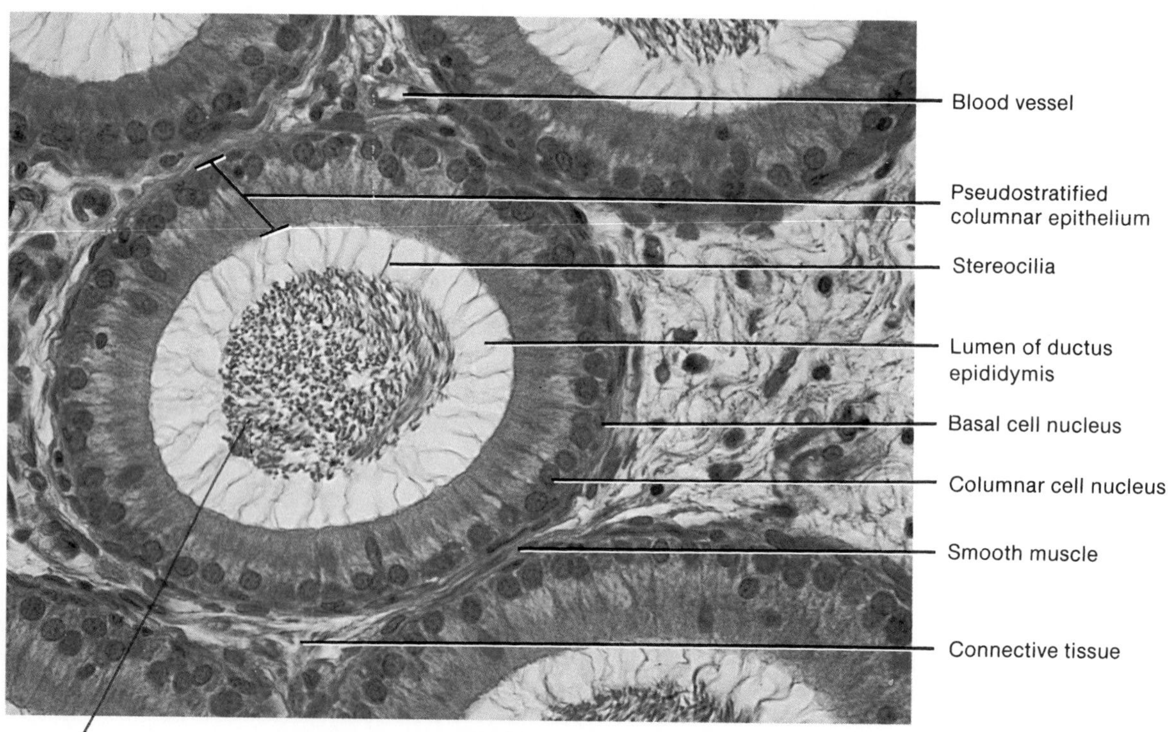

Photomicrograph of a cross section of the ductus epididymis (460 x)

FIGURE 25-6 Histology of the ductus epididymis. Courtesy of Ed Reschke.

deferens is known as the ***ampulla*** (am-POOL-la). The ductus (vas) deferens is lined with pseudostratified columnar epithelium and contains a heavy coat of three layers of muscle (Figure 25-7). Functionally, a portion of the ductus (vas) deferens stores sperm and it conveys sperm from the epididymis toward the urethra during emission by peristaltic contractions of the muscular coat.

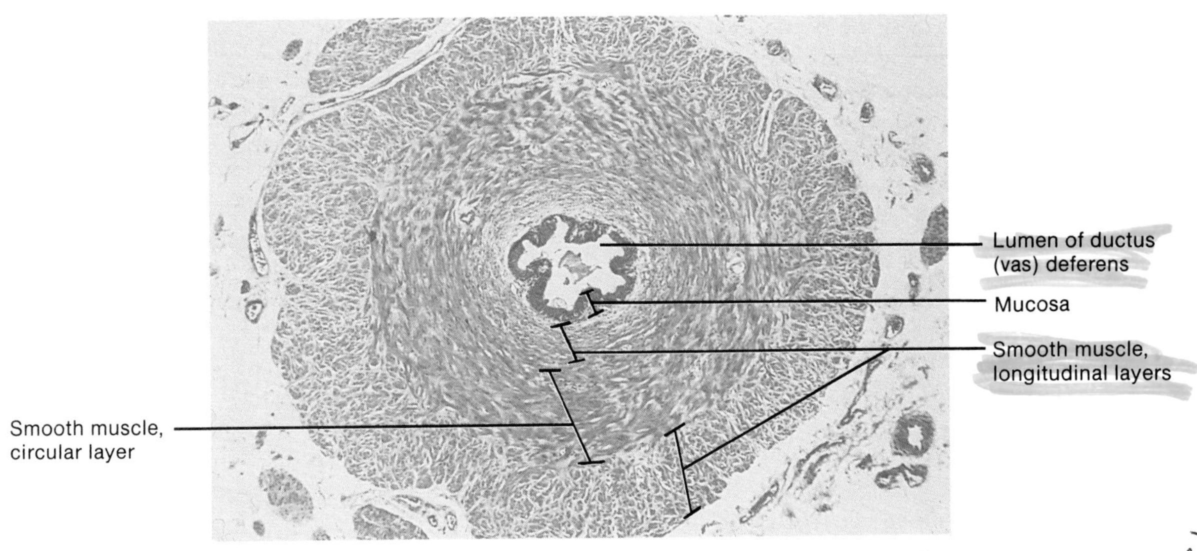

(a) Photomicrograh of ductus (vas) deferens in cross section (36x) (Sperm duct)

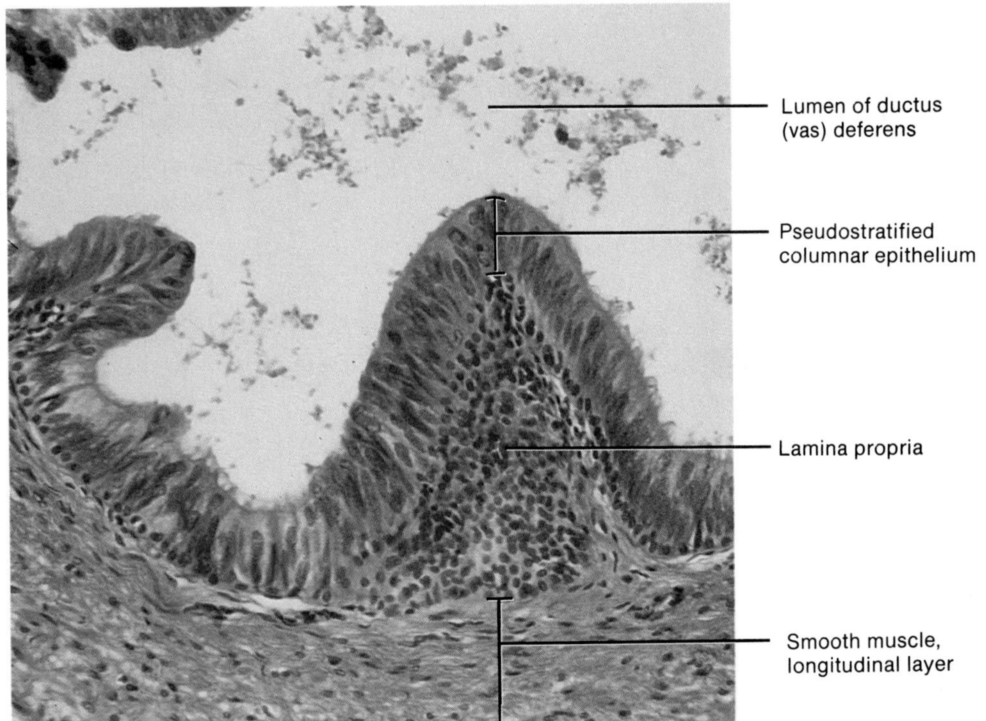

(b) Photomicrograh of mucosa of ductus (vas) deferens (200 x)

FIGURE 25-7 Histology of the ductus (vas) deferens. (a) © Biophoto, SPL, Photo Researchers. (b) Courtesy of Michler, SPL, Photo Researchers.

CLINICAL APPLICATION

Vasectomy

One method of sterilization of males is called ***vasectomy,*** a relatively uncomplicated procedure, typically performed under local anesthesia, in which a portion of each ductus (vas) deferens is removed. In the procedure, an incision is made in the scrotum, the ducts are located, each is tied in two places, and the portion between the ties is removed. Although sperm production continues in the testes, the sperm cannot reach the exterior because the ducts are cut, and the sperm degenerate and are destroyed by phagocytosis. Vasectomy has no effect on sexual desire and performance, and if performed correctly, it is virtually 100 percent effective. Although the procedure may be reversed, only about 45–60 percent of males subsequently achieve fertility.

Traveling with the ductus (vas) deferens as it ascends in the scrotum are the testicular artery, autonomic nerves, veins that drain the testes (pampiniform plexus), lymphatic vessels, and the cremaster muscle. These structures constitute the ***spermatic cord,*** a supporting structure of the male reproductive system (Figure 25-8). The cremaster muscle, which also surrounds the testes, elevates the testes during sexual stimulation and exposure to cold. The spermatic cord and ilioinguinal nerve pass through the ***inguinal*** (IN-gwin-al) ***canal*** in the male. The canal is an oblique passageway in the anterior abdominal wall just superior and parallel to the medial half of the inguinal ligament. The canal is about 4 to 5 cm (1.6 to 2.0 in.) long. It originates at the ***deep (abdominal) inguinal ring,*** a slitlike opening in the aponeurosis of the transversus abdominis muscle. The canal terminates at the ***superficial (subcutaneous) inguinal ring,*** a somewhat triangular opening in the aponeurosis of the external oblique muscle. In the female, the round ligament

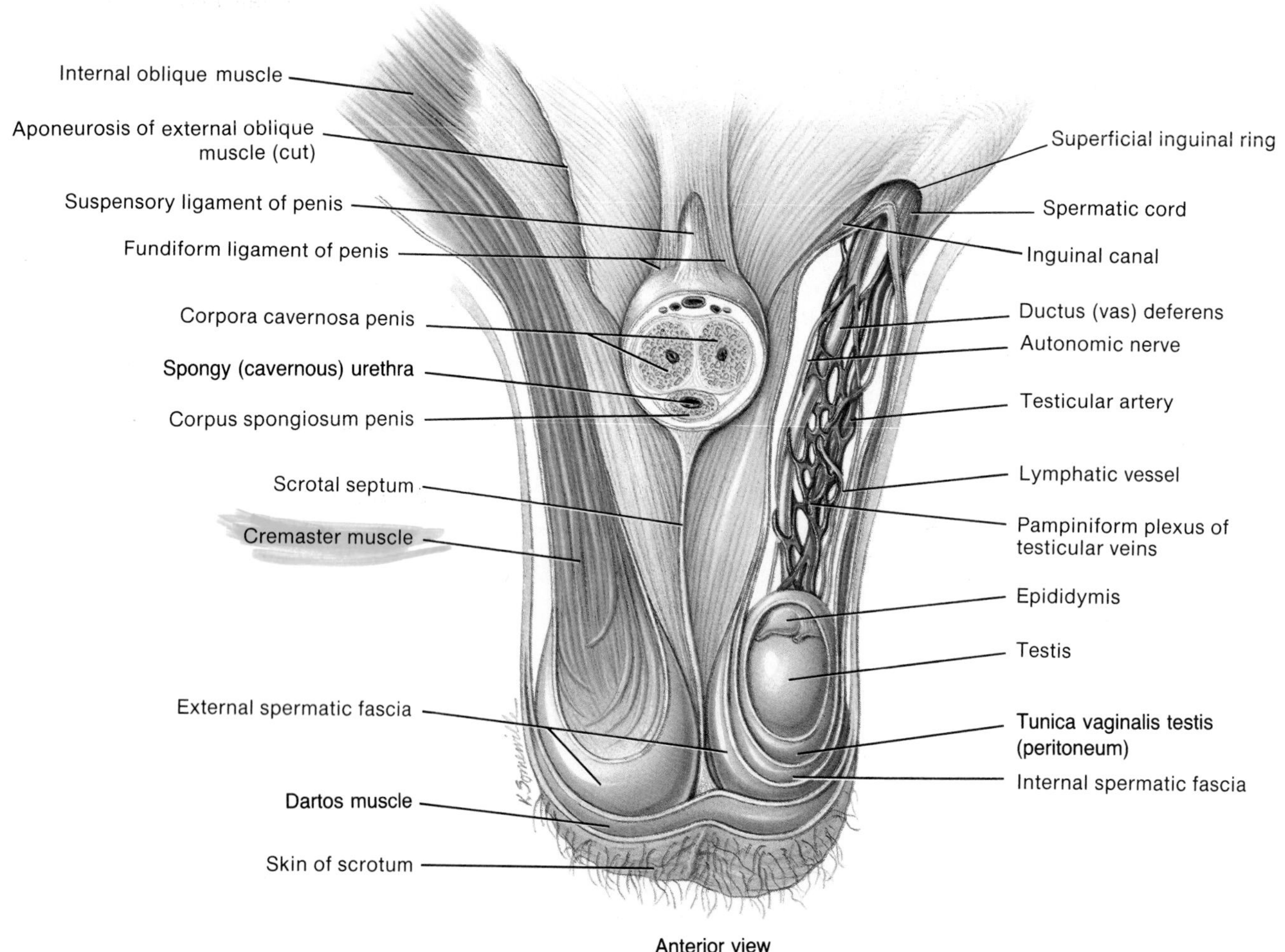

FIGURE 25-8 Spermatic cord and inguinal canal. The left spermatic cord has been opened to expose its contents.

of the uterus and ilioinguinal nerve pass through the inguinal canal.

CLINICAL APPLICATION

Inguinal Hernias

The inguinal region represents a weak area in the abdominal wall. It is frequently the site of an ***inguinal hernia***—a rupture or separation of a portion of the abdominal wall of the inguinal triangle resulting in the protrusion of a part of an organ. In an *indirect inguinal hernia* the herniation protrudes through the inguinal ring and follows the round ligament or spermatic cord. A *direct inguinal hernia* goes through the posterior inguinal wall. Inguinal hernias occur much less frequently in females.

Ejaculatory Ducts

Posterior to the urinary bladder are the ***ejaculatory*** (e-JAK-yoo-la-tō′-rē) ***ducts*** (Figure 25-9). Each duct is about 2 cm (1 in.) long and is formed by the union of the duct from the seminal vesicle and ductus (vas) deferens. The ejaculatory ducts eject spermatozoa into the prostatic urethra just prior to ejaculation.

Urethra

The ***urethra*** is the terminal duct of the system, serving as a passageway for spermatozoa or urine. In the male, the urethra passes through the prostate gland, the urogenital diaphragm, and the penis. It measures about 20 cm (8 in.) in length and is subdivided into three parts (see Figures

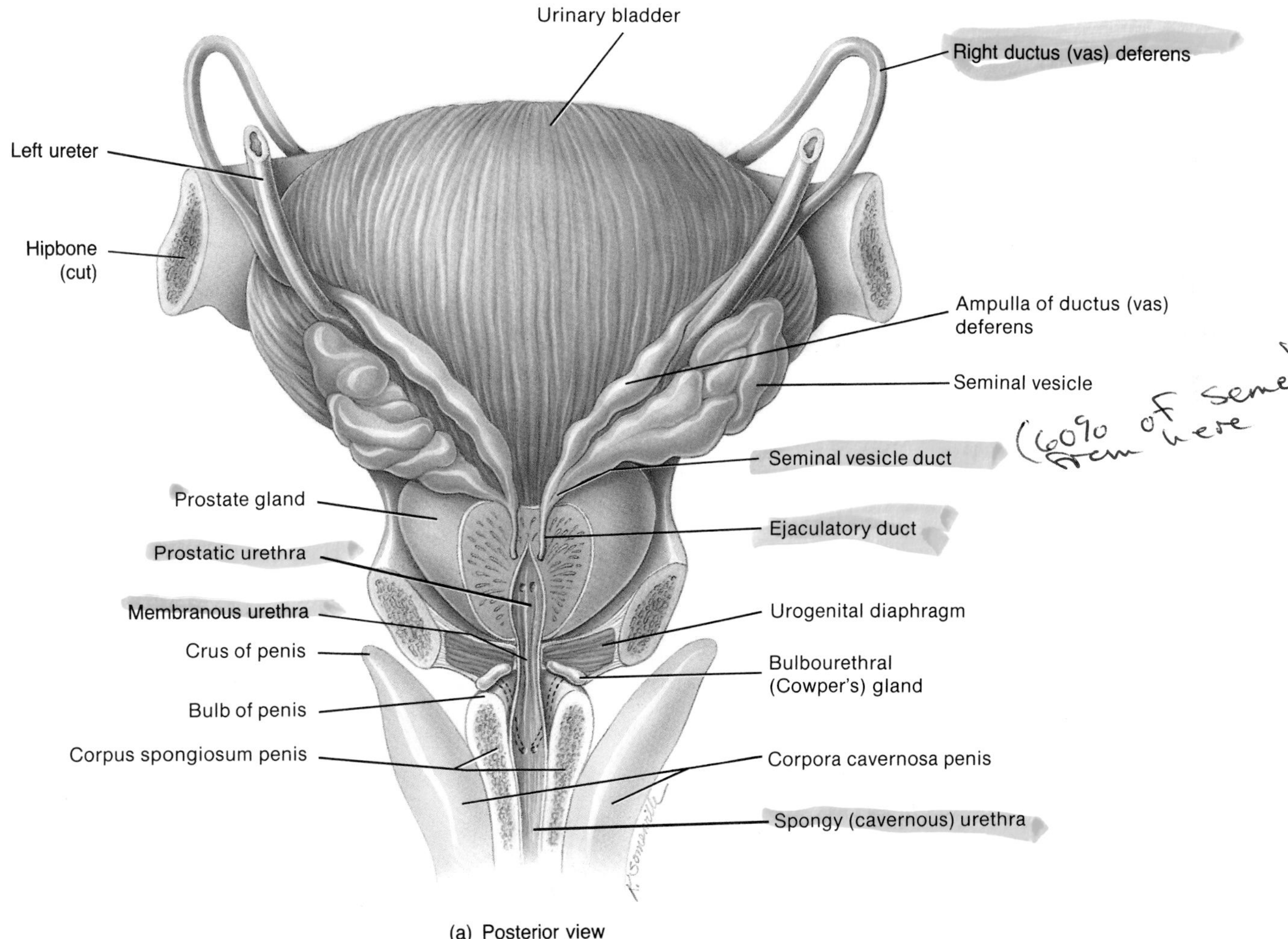

FIGURE 25-9 Male reproductive organs in relation to surrounding structures. (b) Courtesy of J. A. Gosling, P. F. Harris, et al., *Atlas of Human Anatomy,* Gower Medical Publishing Ltd., 2nd ed., 1991.

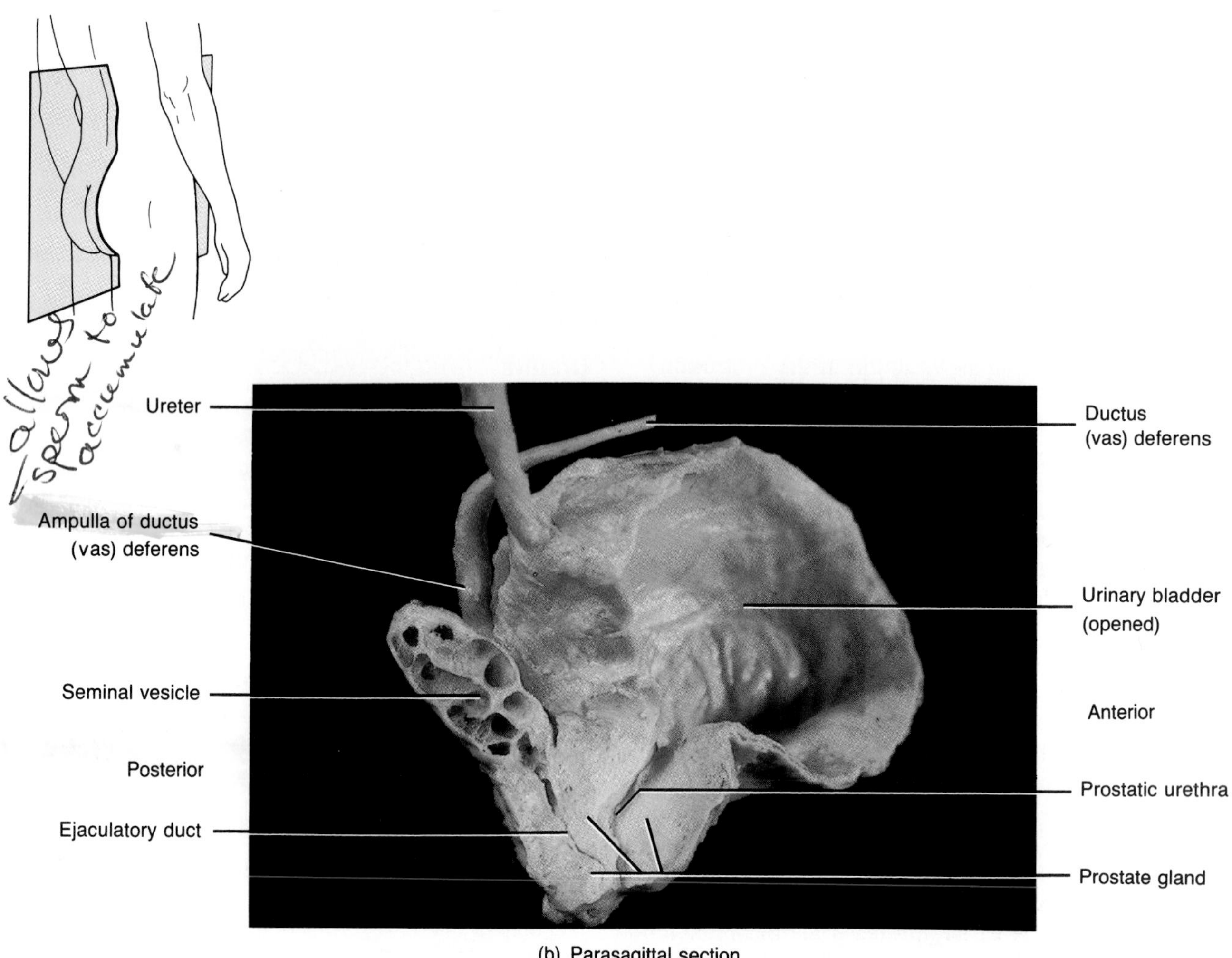

(b) Parasagittal section

25-1 and 25-9). The ***prostatic urethra*** is 2 to 3 cm (1 in.) long and passes through the prostate gland. It continues inferiorly, and as it passes through the urogenital diaphragm, a muscular partition between the two ischiopubic rami, it is known as the ***membranous urethra.*** The membranous portion is about 1 cm (0.5 in.) long. As it passes through the corpus spongiosum of the penis, it is known as the ***spongy (cavernous) urethra.*** This portion is about 15 cm (6 in.) long. The spongy urethra enters the bulb of the penis and terminates at the ***external urethral orifice.*** The histology of the male urethra may be reviewed in Chapter 24.

ACCESSORY SEX GLANDS

Whereas the ducts of the male reproductive system store and transport sperm cells, the ***accessory sex glands*** secrete most of the liquid portion of semen. The paired ***seminal vesicles*** (VES-i-kuls) are convoluted pouchlike structures, about 5 cm (2 in.) long, lying posterior to and at the base of the urinary bladder in front of the rectum (Figure 25-9). They secrete an alkaline, viscous fluid, rich in the sugar fructose, and pass it into the ejaculatory duct. This secretion provides a carbohydrate (fructose) that is used as an energy source by sperm. The seminal vesicles also produce prostaglandins, which contribute to sperm motility and viability. It constitutes about 60 percent of the volume of semen. The alkaline nature of the fluid helps to neutralize acid in the female tract. This acid would inactivate and kill sperm if not neutralized.

The ***prostate*** (PROS-tāt) ***gland*** is a single gland about the size of a chestnut (Figure 25-9). It is inferior to the urinary bladder and surrounds the superior portion of the urethra. The prostate secretes an alkaline fluid rich in citric acid, and prostatic acid phosphatase into the prostatic urethra through numerous prostatic ducts. The prostatic secretion constitutes 13 to 33 percent of the volume of semen and contributes to sperm motility and viability. The prostate gland slowly increases in size from birth to puberty, and then a rapid growth spurt occurs. The size attained by the third decade remains stable until about age 45, when enlargement may occur.

CLINICAL APPLICATION

Cancer of the Prostate Gland

In cases of ***cancer of the prostate gland,*** *prostatic acid phosphatase* is released by the prostate gland into the blood. Elevated levels of prostatic acid phosphatase typically indicate prostate cancer and that the tumor has spread to other parts of the body, especially bone. Another test measures the level of *prostate-specific antigen* in blood. This substance is a protease produced only by prostate epithelial cells, normal and malignant. The amount of prostate-specific antigen is also elevated in cancer of the prostate gland.

The paired ***bulbourethral*** (bul'-bō-yoo-RĒ-thral), or ***Cowper's, glands*** are about the size of peas. They are located beneath the prostate on either side of the membranous urethra within the urogenital diaphragm (Figure 25-9). The bulbourethral glands secrete an alkaline substance that protects sperm by neutralizing the acid environment of the urethra and mucus that lubricates the end of the penis during sexual intercourse. Their ducts open into the spongy urethra.

SEMEN (SEMINAL FLUID)

Semen (***seminal fluid***) is a mixture of sperm and the secretion of the seminal vesicles, prostate gland, and bulbourethral glands. The average volume of semen for each ejaculation is 2.5 to 5 ml, and the average range of spermatozoa ejaculated is 50 to 150 million/ml. When the number of spermatozoa falls below 20 million/ml, the male is likely to be infertile. The very large number is required because only a small percentage eventually reach the ovum. And although only a single spermatozoon fertilizes an ovum, fertilization seems to require the combined action at the ovum of a larger number of them. The intercellular material of the cells covering the ovum presents a barrier to the sperm. This barrier is digested by the hyaluronidase and proteinases secreted by the acrosomes of sperm, resulting in the dispersion of the cells surrounding the ovum. A single sperm does not produce enough of these enzymes to dissolve the barrier. A passageway through which one sperm may enter can be created only by the action of many sperm cells.

Semen has a slightly alkaline pH of 7.20 to 7.60. The prostatic secretion gives semen a milky appearance, and fluids from the seminal vesicles and bulbourethral glands give it a mucoid consistency. Semen provides spermatozoa with a transportation medium and nutrients. It neutralizes the acid environment of the male urethra and the female vagina. It also contains enzymes that activate sperm after ejaculation.

Semen contains an antibiotic, ***seminalplasmin,*** that has the ability to destroy a number of bacteria. Since both semen and the lower female reproductive tract contain bacteria, the antibiotic activity of seminalplasmin may keep these bacteria under control to help ensure fertilization.

Once ejaculated into the vagina, liquid semen coagulates rapidly because of a clotting enzyme produced by the prostate gland that acts on a substance produced by the seminal vesicle. This clot liquefies in about 5 to 20 minutes because of another enzyme produced by the prostate gland. Abnormal or delayed liquefaction of coagulated semen may cause complete or partial immobilization of spermatozoa, thus inhibiting their movement through the cervix of the uterus.

CLINICAL APPLICATION

Semen Analysis

Semen analysis is the most valuable test in evaluating male sterility. Among the criteria analyzed are the following:

1. ***Volume.*** A low volume might suggest an anatomical or functional defect or inflammation.
2. ***Motility.*** This refers to the percentage of motile spermatozoa (40–60 percent) and quality of movement (forward and progressive).
3. ***Count.*** Sperm counts below 20 million/ml could indicate sterility.
4. ***Liquefaction.*** Delayed liquefaction of more than two hours suggest inflammation of accessory sex glands or enzyme defects in the secretory products of the glands.
5. ***Morphology.*** No more than about 35 percent of spermatozoa should have abnormal morphology.
6. ***Autoagglutination.*** Agglutination does not occur normally.
7. ***pH.*** A rise in pH could indicate prostatitis.
8. ***Fructose.*** This sugar is present in a normal ejaculate. Its absence indicates obstruction or congenital absence of the ejaculatory ducts or seminal vesicles.

A normal semen analysis does not guarantee fertility; the absence of spermatozoa or zero motility are the only definitive signs of sterility.

PENIS

The ***penis*** is used to introduce spermatozoa into the vagina (Figure 25-10). The penis is cylindrical in shape and consists of a body, root, and glans penis. The ***body*** of the penis is composed of three cylindrical masses of tissue, each bound by fibrous tissue (***tunica albuginea***). The two dorsolateral masses are called the ***corpora cavernosa penis.*** The smaller midventral mass, the ***corpus spongiosum penis,*** contains the spongy urethra. All three masses are enclosed by fascia and skin and consist of erectile tissue permeated by blood sinuses. Under the influence of sexual stimulation (visual, tactile, auditory, olfactory, and imaginative), the arteries supplying the penis dilate, and large quantities of blood enter the blood sinuses. Expansion of these spaces compresses the veins draining the penis, so most entering blood is retained. These vascular changes result in an ***erection,*** a parasympathetic reflex. The penis returns to its flaccid state when the arteries constrict and pressure on the veins is relieved. During ***ejaculation,*** the propulsion of semen from the urethra to the exterior, which is a sympathetic reflex, the smooth muscle sphincter at the base of the urinary bladder is closed. Thus, urine is not expelled during ejaculation and semen does not enter the urinary bladder.

The ***root*** of the penis is the attached portion and consists of the ***bulb of the penis,*** the expanded portion of the base of the corpus spongiosum penis, and the ***crura*** (sing., ***crus***) ***of the penis,*** the separated and tapered portion of the corpora cavernosa penis. The bulb of the penis is attached to the

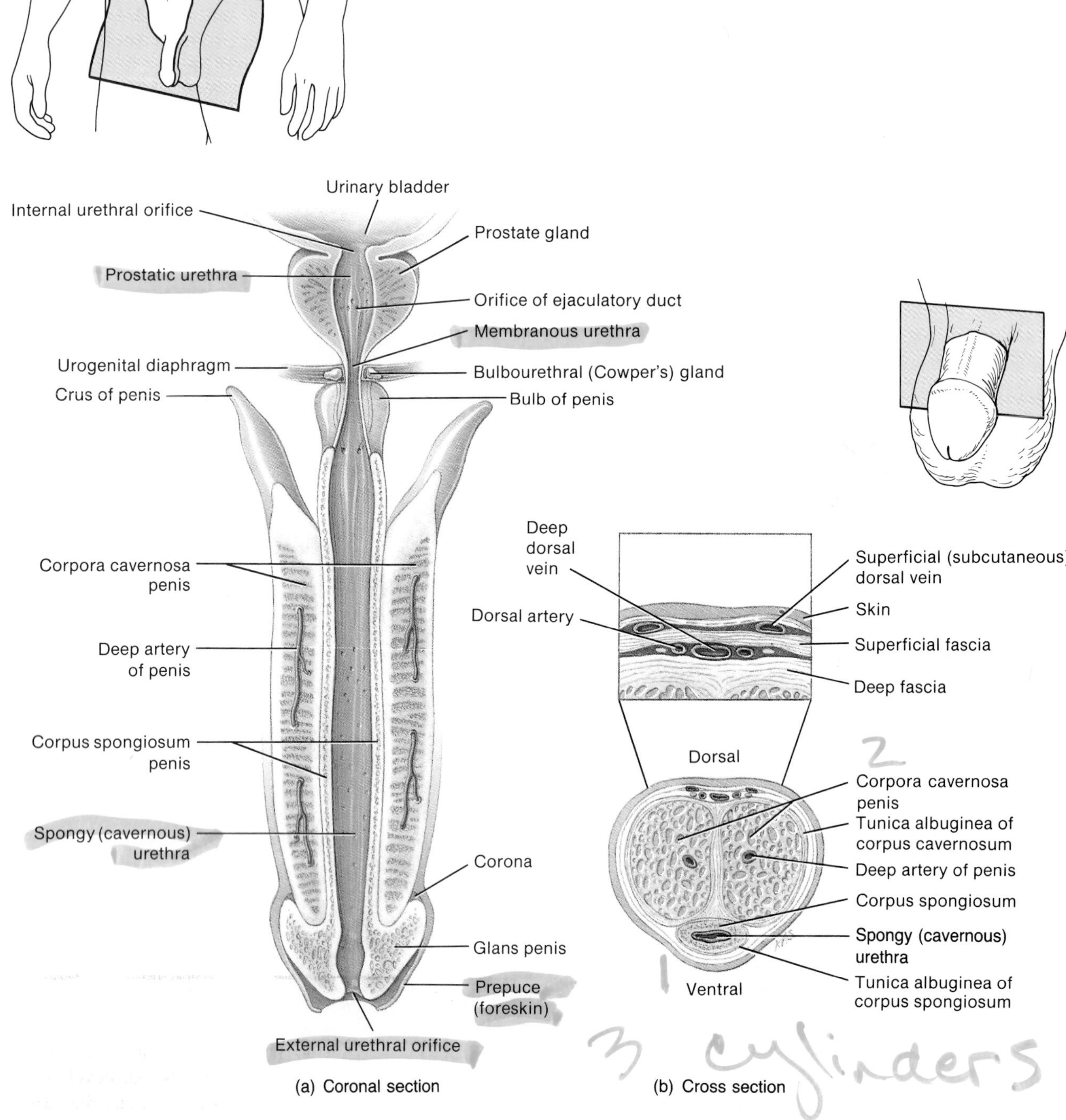

FIGURE 25-10 Internal structure of the penis. (c) Courtesy of J. A. Gosling, P. F. Harris, et al., *Atlas of Human Anatomy,* Gower Medical Publishing Ltd., 2nd ed., 1991.

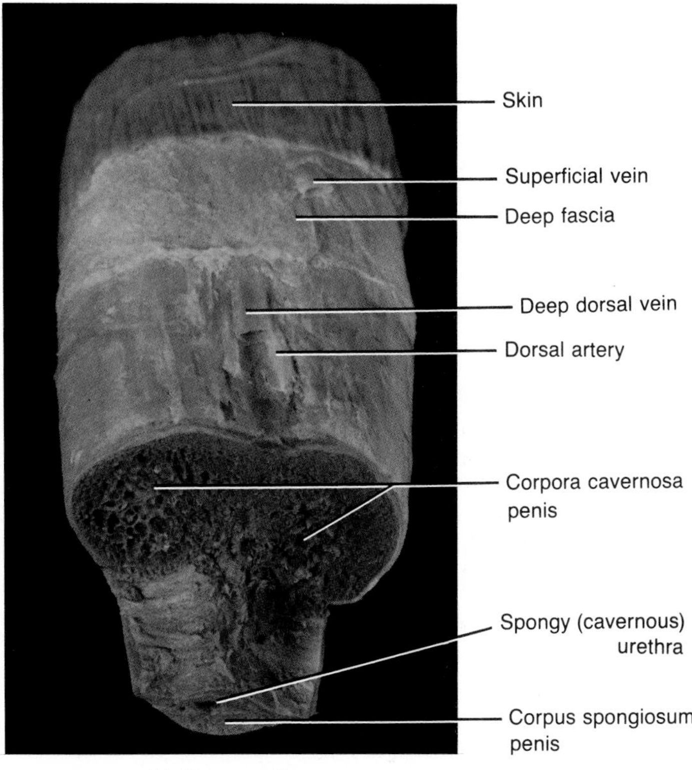

(c) Cross section

inferior surface of the urogenital diaphragm and enclosed by the bulbocavernosus muscle. Each crus of the penis is attached to the ischial and pubic rami and surrounded by the ischiocavernosus muscle.

The distal end of the corpus spongiosum penis is a slightly enlarged region called the ***glans penis,*** which means "shaped like an acorn." The margin of the glans penis is referred to as the ***corona.*** The distal urethra enlarges within the glans penis and forms a terminal slitlike opening, the ***external urethral orifice.*** Covering the glans is the loosely fitting ***prepuce*** (PRE-pyoos), or ***foreskin.***

CLINICAL APPLICATION

Circumcision

Circumcision (*circumcido* = to cut around) is a surgical procedure in which part or all of the prepuce is removed. It is usually performed in the delivery room or by the third or fourth day after birth (or on the eighth day as part of a Jewish religious rite). There is no consensus among physicians regarding circumcision. Some physicians are opposed to circumcision, except for religious reasons, and feel that there is no medical justification for performing it. Other physicians feel that circumcised boys have a far lower risk of urinary tract infections, protection against penile cancer, and possibly a lower risk for many sexually transmitted diseases.

The penis has a very rich blood supply from the internal pudendal artery and the femoral artery. The veins drain into corresponding vessels.

The sensory nerves to the penis are branches from the pudendal and ilioinguinal nerves. The corpora have a sympathetic and parasympathetic supply. As a result of parasympathetic stimulation, the blood vessels dilate, increasing the flow of blood into the erectile tissue. The result is that blood is trapped within the penis and erection is maintained. At ejaculation, sympathetic stimulation causes the smooth muscle located in the walls of the ducts and accessory glands of the reproductive tract to contract and propel the sperm and secretions along their course. The musculature of the penis, which is supplied by the pudendal nerve, also contracts at ejaculation. The muscles include the bulbocavernosus muscle, which overlies the bulb of the penis, the ischiocavernosus muscles on either side of the penis, and the superficial transverse perineus muscles on either side of the bulb of the penis (see Figure 10–15).

FEMALE REPRODUCTIVE SYSTEM

The female organs of reproduction include the ovaries, which produce secondary oocytes (cells that develop into mature ova or eggs following fertilization) and the female sex hormones progesterone, estrogens, relaxin, and inhibin; the uterine (Fallopian) tubes, which transport ova to the uterus (womb); the vagina; and external organs that constitute the vulva, or pudendum (Figure 25-11). The mammary glands also are considered part of the female reproductive system.

The specialized branch of medicine that deals with the diagnosis and treatment of diseases of the female reproductive system is called ***gynecology*** (gī′-ne-KOL-ō-jē; *gyneco* = woman).

OVARIES

The ***ovaries*** (*ovarium* = egg receptacle), or female gonads, are paired glands resembling unshelled almonds in size and shape. They are homologous to the testes. (***Homologous*** means that two organs correspond in structure, position, and origin, but not necessarily in function.) The ovaries descend to the brim of the pelvis during the third month of development. They are positioned in the upper pelvic cavity, one on each side of the uterus. The ovaries are maintained in position by a series of ligaments (Figure 25-12). They are attached to the broad ligament of the uterus, which is itself part of the parietal peritoneum, by a double-layered fold of peritoneum called the ***mesovarium.*** The ovaries are anchored to the uterus by the ***ovarian ligament*** and are attached to the pelvic wall by the ***suspensory ligament.*** Each ovary also contains a ***hilus,*** the point of entrance for blood vessels and nerves and along which the mesovarium is attached.

The microscope reveals that each ovary consists of the following parts (Figure 25-13).

1. ***Germinal epithelium.*** A layer of simple epithelium (low cuboidal or squamous) that covers the free surface of the ovary and is continuous with the mesothelium that covers the mesovarium. The term *germinal epithelium* is a misnomer since it does not give rise to ova, although at one time it was believed that it did. It is now known that the cells that give rise to ova arise from the endoderm of the yolk sac and migrate to the ovaries.
2. ***Tunica albuginea.*** A capsule of collagenous connective tissue immediately deep to the germinal epithelium.
3. ***Stroma.*** A region of connective tissue deep to the tunica albuginea and composed of an outer, dense layer called

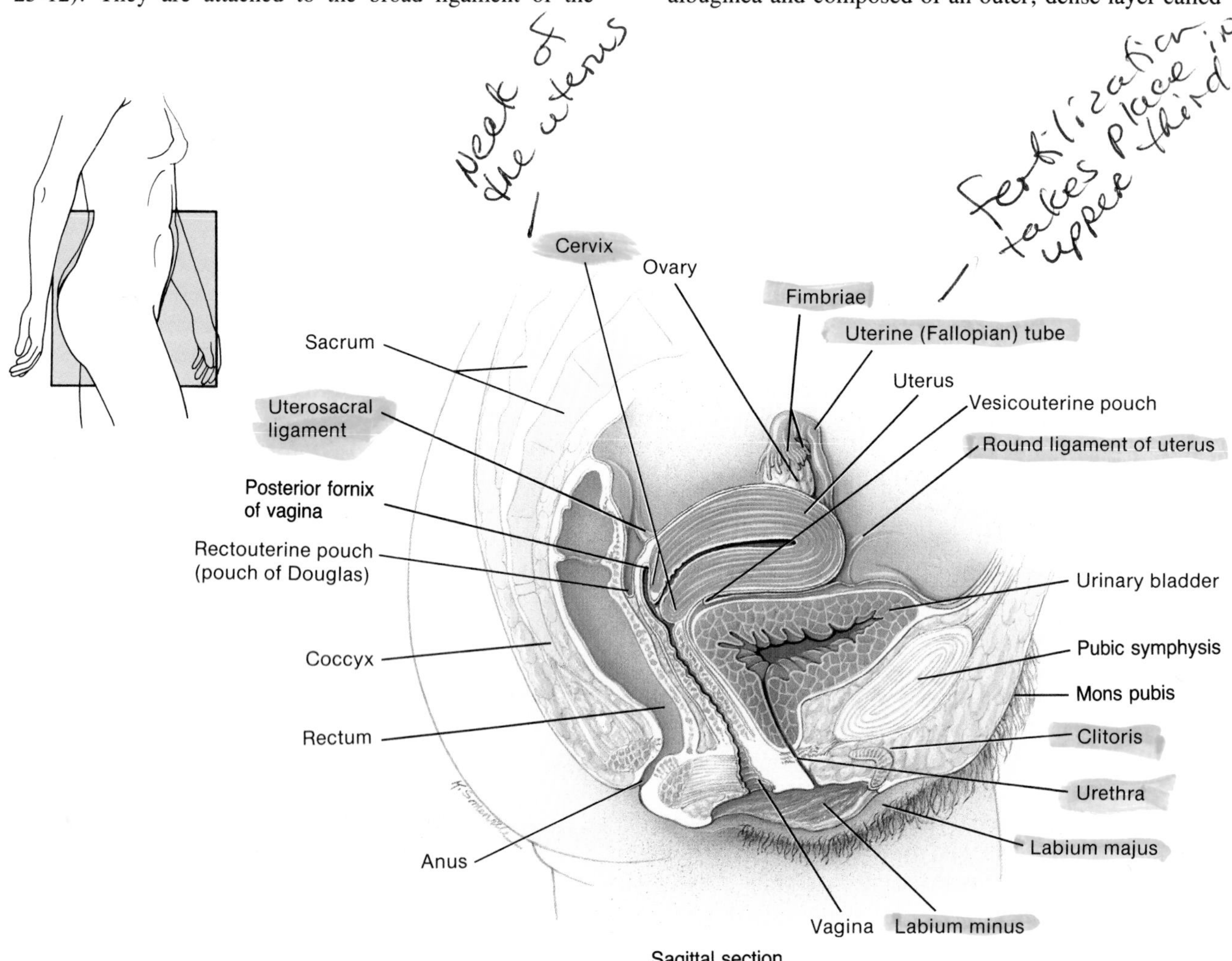

FIGURE 25-11 Female organs of reproduction and surrounding structures.

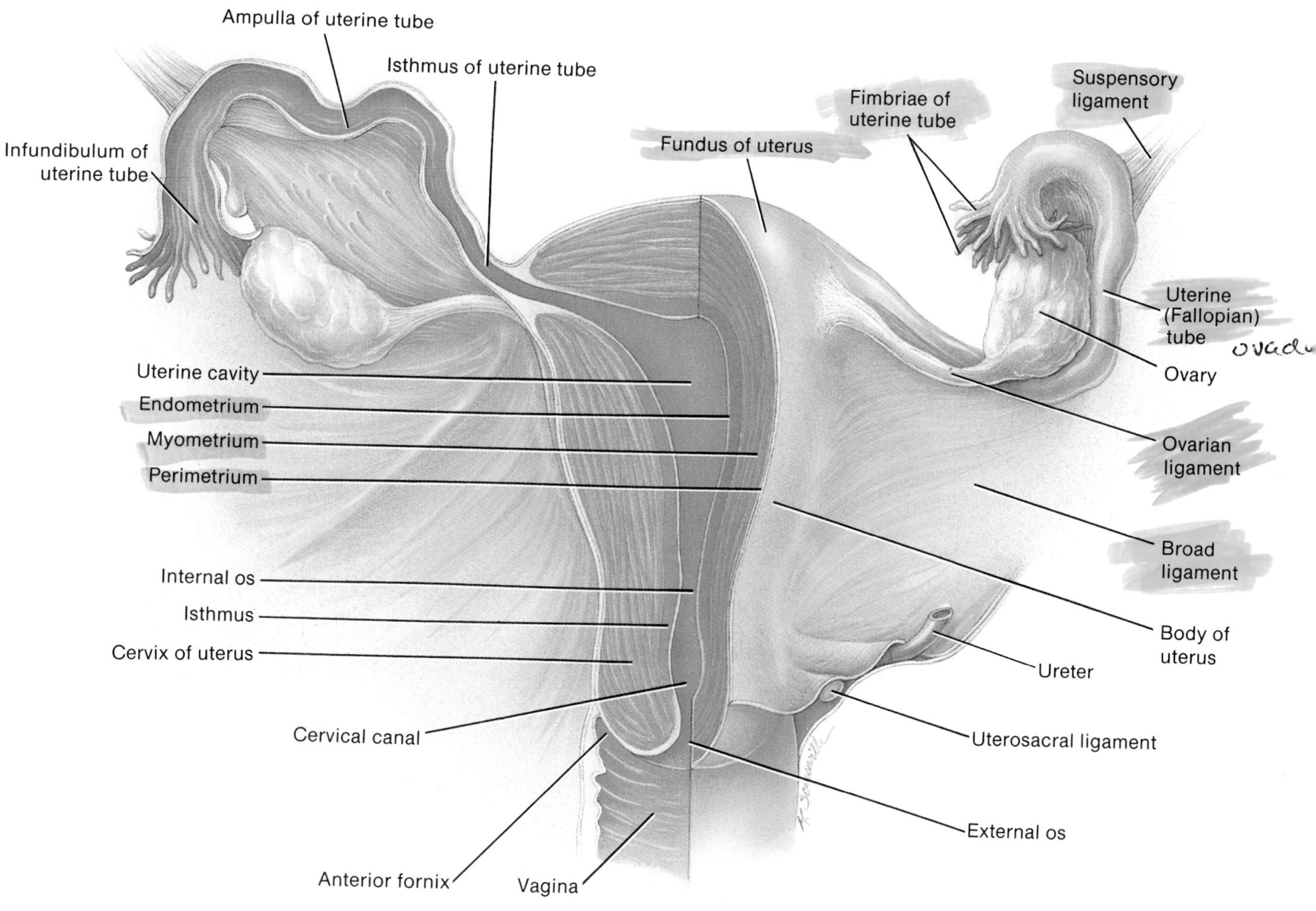

Ampulla of uterine tube
Isthmus of uterine tube
Infundibulum of uterine tube
Fundus of uterus
Fimbriae of uterine tube
Suspensory ligament
Uterine (Fallopian) tube
ovadu
Ovary
Uterine cavity
Endometrium
Myometrium
Perimetrium
Ovarian ligament
Broad ligament
Internal os
Isthmus
Cervix of uterus
Body of uterus
Ureter
Cervical canal
Uterosacral ligament
External os
Anterior fornix
Vagina

(a) Posterior view

Isthmus of uterine tube
Uterine (Fallopian) tube
Infundibulum
Fundus of uterus
Ovary
Ovarian ligament
Uterine cavity
Broad ligament
Body of uterus
Cervical canal
Vagina

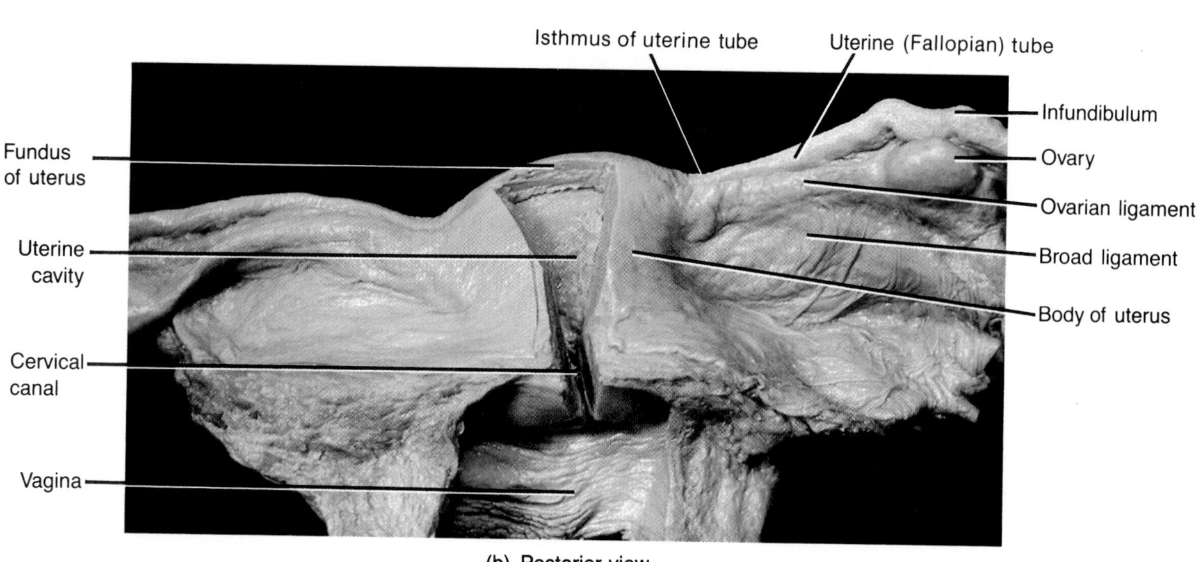

(b) Posterior view

FIGURE 25-12 Uterus and associated structures. In (a), the left side of the figure has been sectioned to show internal structures. In (b), part of the posterior wall of the uterus has been removed. (b) and (c) Courtesy of J. A. Gosling et al, Atlas of Human Anatomy, Gower Medical Publishing Ltd., 2nd. ed., 1991.

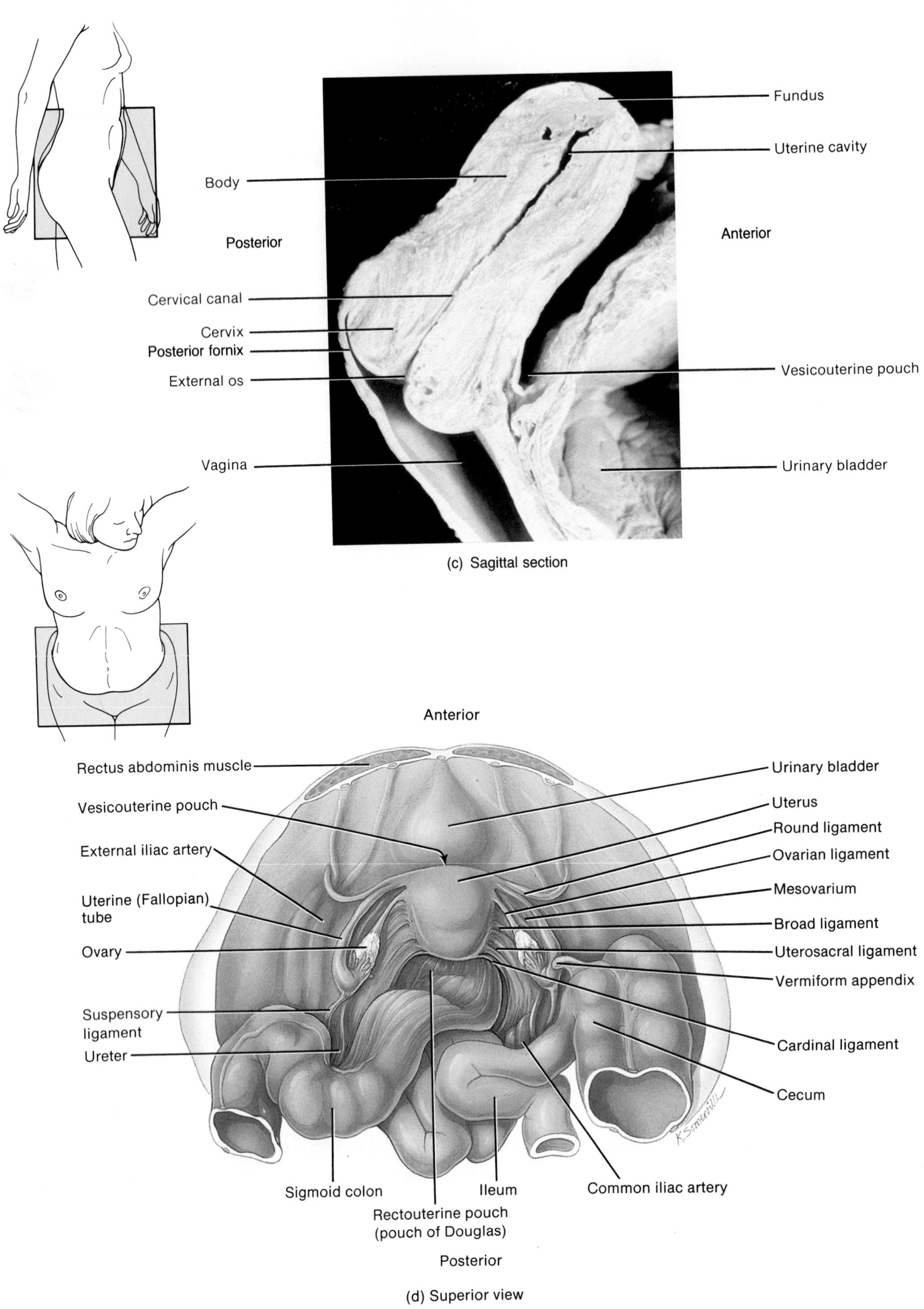

(c) Sagittal section

(d) Superior view

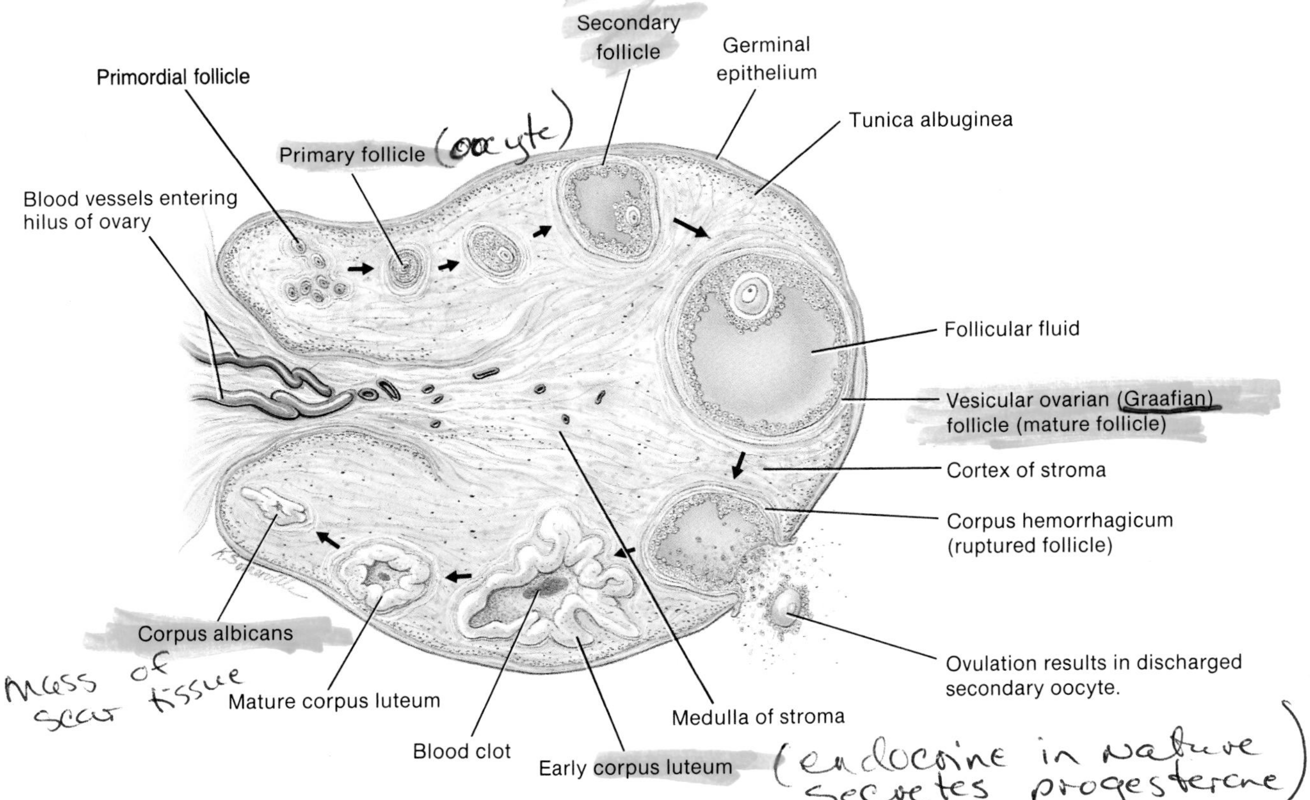

(secondary oocyte)
Secondary follicle
Germinal epithelium
Primordial follicle
Tunica albuginea
Primary follicle (oocyte)
Blood vessels entering hilus of ovary
Follicular fluid
Vesicular ovarian (Graafian) follicle (mature follicle)
Cortex of stroma
Corpus hemorrhagicum (ruptured follicle)
Corpus albicans
mass of scar tissue
Ovulation results in discharged secondary oocyte.
Mature corpus luteum
Medulla of stroma
Blood clot
Early corpus luteum
(endocrine in nature secretes progesterone also secretes estrogen)

(a) Sectional view

Germinal epithelium
Primordial follicles
Tunica albuginea
Cortex
Primary follicle

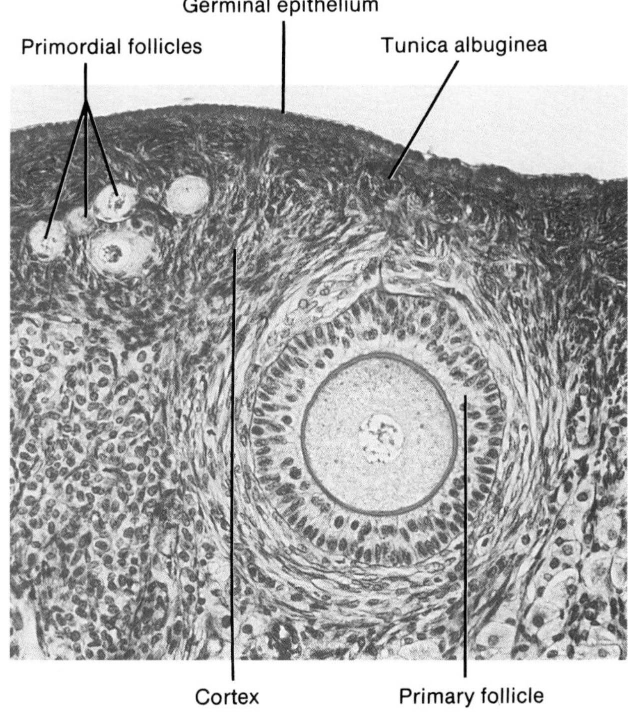

(b) Photomicrograph of the cortex of the ovary (approx. 200 x)

FIGURE 25-13 Histology of the ovary. In (a), the arrows indicate the sequence of developmental stages that occur as part of the ovarian cycle. (b) and (c) © Biophoto, SPL, Photo Researchers.

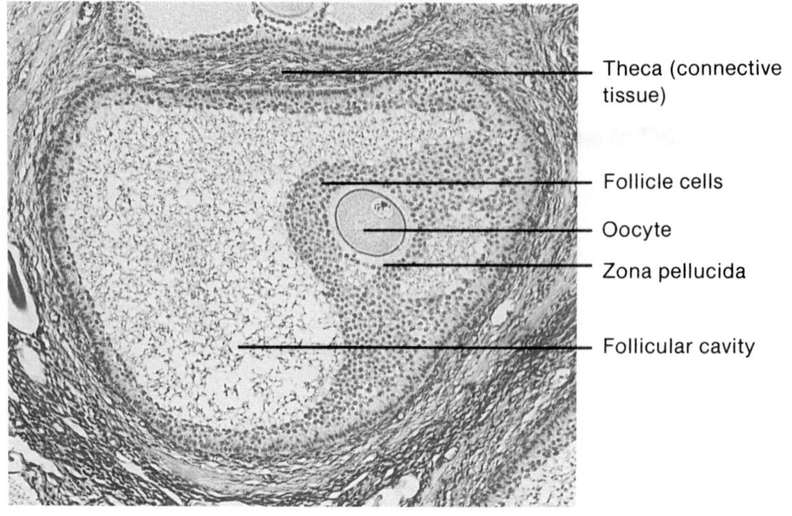

(c) Photomicrograph of a vesicular ovarian (Graafian) follicle (60 x)

the *cortex* and an inner, loose layer known as the *medulla.* The cortex contains ovarian follicles.

4. ***Ovarian follicles.*** Oocytes (immature ova) and their surrounding protective glandular tissues called follicles in various stages of development.
5. ***Vesicular ovarian (Graafian) follicle.*** A relatively large, fluid-filled follicle containing an immature ovum and its surrounding tissues. The follicle secretes hormones called estrogens.
6. ***Corpus luteum.*** A mature vesicular ovarian follicle that has ruptured to expel a secondary oocyte (potential mature ovum), a process known as ovulation. The corpus luteum continues to produce the hormones progesterone, estrogens, relaxin, and inhibin until it degenerates and turns into fibrous tissue (corpus albicans).

The ovarian blood supply is furnished by the ovarian arteries, which anastomose with branches of the uterine arteries. The ovaries are drained by the ovarian veins. On the right side they drain into the inferior vena cava, and on the left side they drain into the renal vein.

Sympathetic and parasympathetic nerve fibers to the ovaries terminate on the blood vessels and enter the substance of the ovaries.

Oogenesis

The formation of haploid (n) ova in the ovary involves several phases, including meiosis, and is referred to as ***oogenesis*** (ō′-ō-JEN-e-sis). With some exceptions, oogenesis occurs in essentially the same manner as spermatogenesis.

■ ***Reduction Division (Meiosis I)*** During early fetal development, primordial germ cells migrate from the endoderm of the yolk sac to the ovaries. There, germ cells differentiate into ***oogonia*** (ō′-o-GŌ-nē-a; *oo* = egg), cells that can give rise to other cells that develop into ova (Figure 25-14). Oogonia are diploid ($2n$) cells that divide mitotically to produce a large population of cells. At about the third month of prenatal development, oogonia divide and develop into larger diploid ($2n$) cells called ***primary oocytes*** (Ō-o-sītz). These cells enter prophase of reduction division (meiosis I) but do not complete it until after the female reaches puberty. Each primary oocyte is surrounded by a single layer of flattened epithelial cells (follicular), an d the entire structure is called a ***primordial follicle.*** Primordial follicles do not begin further development until they are stimulated by follicle-stimulating hormone (FSH) from the anterior pituitary gland, which has responded to gonadotropin-releasing hormone (GnRH) from the hypothalamus. Many primordial follicles degenerate (atresia) before birth and are known as ***atretic follicles.***

Starting with puberty, several primordial follicles respond each month to the rising level of FSH and form ***primary follicles,*** which are surrounded by a single layer of cuboidal-shaped follicular cells. As the preovulatory phase of the menstrual cycle proceeds and luteinizing hormone (LH) is secreted from the anterior pituitary, one of the primary follicles reaches a stage in which meiosis resumes and the diploid primary oocyte completes reduction division (meiosis I). Synapsis, tetrad formation, and crossing-over occur, and two cells of unequal size, both with 23 chromosomes (n) of two chromatids each, are produced. The smaller cell, called the ***first polar body,*** is essentially a packet of discarded nuclear material. The larger cell, known as the ***secondary oocyte,*** receives most of the cytoplasm. Each secondary oocyte is surrounded by several layers of first cuboidal then columnar epithelial cells, and the entire struc-

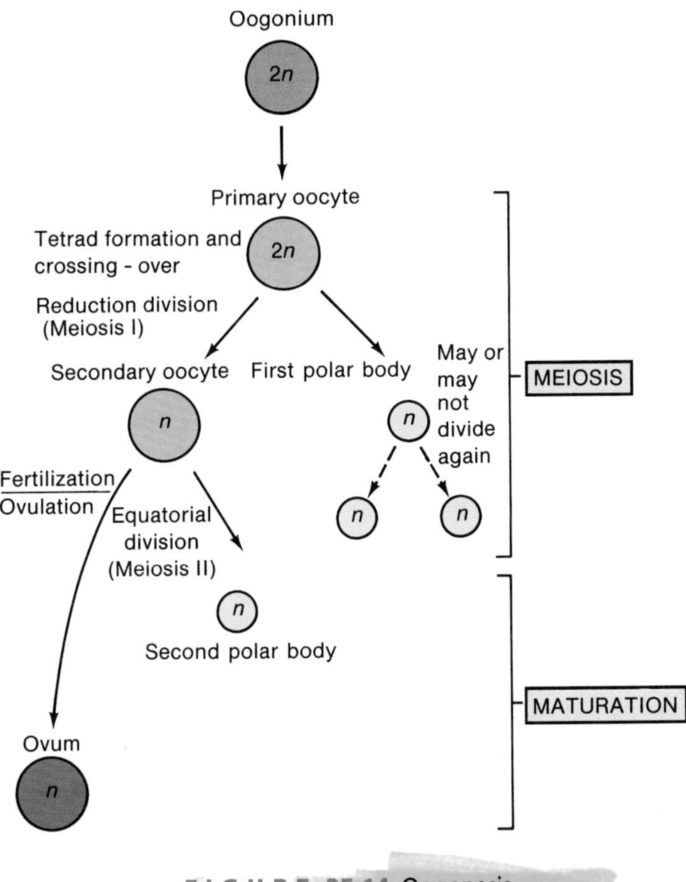

FIGURE 25-14 Oogenesis.

ture is called a ***secondary (growing) follicle.*** Once a secondary oocyte is formed, it proceeds to the metaphase of equatorial division (meiosis II) and then stops at this stage. The equatorial division (meiosis II) is completed following ovulation and fertilization.

■ ***Equatorial Division (Meiosis II)*** At ovulation, the secondary oocyte with its polar body and some surrounding supporting cells is discharged. The discharged secondary oocyte enters the uterine (Fallopian) tube and, if spermatozoa are present and fertilization occurs, the second division, the equatorial division (meiosis II), is completed.

■ ***Maturation*** The secondary oocyte produces two cells of unequal size, both of them haploid (n). The larger cell eventually develops into an ***ovum,*** or mature egg; the smaller is the ***second polar body.***

The first polar body may undergo another division to produce two polar bodies. If it does, meiosis of the primary oocyte results in a single haploid (n), secondary oocyte and three haploid (n) polar bodies. In any event, all polar bodies disintegrate. Thus, each oogonium produces a single secondary oocyte, whereas each spermatocyte produces four spermatozoa. Spermatogenesis and oogenesis differ in other ways as well. Spermatogenesis is a continuous process that begins in puberty and continues throughout life; oogenesis begins at menstruation and ends at menopause. Also, spermatozoa are quite small, have flagella for locomotion, and contain few nutrients; a secondary oocyte is larger, lacks flagella, and contains more nutrients for nourishment until implantation occurs in the uterus.

UTERINE (FALLOPIAN) TUBES

The female body contains two ***uterine (Fallopian) tubes,*** also called ***oviducts,*** that extend laterally from the uterus and transport the ova from the ovaries to the uterus (see Figure 25-12). Measuring about 10 cm (4 in.) long, the tubes are positioned between the folds of the broad ligaments of the uterus. The funnel-shaped open distal end of each tube, called the ***infundibulum,*** lies close to the ovary and is surrounded by a fringe of fingerlike projections called ***fimbriae*** (FIM-brē-ē). One fimbria is attached to the lateral end of the ovary. As you will see later, fimbriae help to carry a secondary oocyte into the uterine tube following ovulation. From the infundibulum, the uterine tube extends medially and inferiorly and attaches to the superior lateral angle of the uterus. The ***ampulla*** of the uterine tube is the widest, longest portion, making up about two-thirds of its length. The ***isthmus*** of the uterine tube is the short, narrow, thick-walled portion that joins the uterus.

Histologically, the uterine tubes are composed of three layers. The internal ***mucosa*** contains ciliated columnar cells and secretory cells, which contain microvilli and are believed to aid the movement and nutrition of the ovum (Figure 25-15). The middle layer, the ***muscularis,*** is composed of a thick, circular region of smooth muscle and an outer, thin, longitudinal region of smooth muscle. Peristaltic contractions of the muscularis and the ciliated action of the mucosa help move the ovum down into the uterus. The outer layer of the uterine tubes is a serous membrane, the ***serosa.***

About once a month a vesicular ovarian (Graafian) follicle (developed from a secondary follicle) ruptures, releasing a secondary oocyte, a process called ***ovulation.*** The oocyte is swept into the uterine tube by the ciliary action of the epithelium of the infundibulum, which becomes associated with the surface of the most mature vesicular ovarian follicle just before ovulation occurs. The oocyte is then moved along the tube by ciliary action that is supplemented by the peristaltic contractions of the muscularis. If the oocyte is fertilized by a sperm cell, it usually occurs in the ampulla of the uterine tube. Fertilization may occur at any time up to about 24 hours following ovulation. With fertilization the secondary oocyte completes meiosis II, in which the oocyte produces a larger cell that develops into an ovum (mature egg) and a smaller second polar body. The fertilized ovum is now referred to as a zygote. Following a number of cell divisions, it descends into the uterus within seven days and at this point is called a blastocyst. An unfertilized secondary oocyte disintegrates.

The uterine tubes are supplied by branches of the uterine and ovarian arteries (see Figure 25-17). Venous return is via the uterine veins.

The uterine tubes are supplied with sympathetic and parasympathetic nerve fibers from the hypogastric plexus and the pelvic splanchnic nerves. The fibers are distributed to the muscular coat of the tubes and their blood vessels.

UTERUS

Part of the pathway for sperm to reach the uterine (Fallopian) tubes is the ***uterus.*** It is also the site of menstruation, implantation of a fertilized ovum, development of the fetus during pregnancy, and labor. Situated between the urinary bladder and the rectum, the uterus is shaped like an inverted pear (see Figures 25-11 and 25-12). Before the first pregnancy, the adult uterus measures approximately 7.5 cm (3 in.) long, 5 cm (2 in.) wide, and 2.5 cm (1 in.) thick.

Anatomical subdivisions of the uterus include the dome-shaped portion above the uterine tubes called the ***fundus,*** the major tapering central portion called the ***body,*** and the inferior narrow portion opening into the vagina called the ***cervix.***

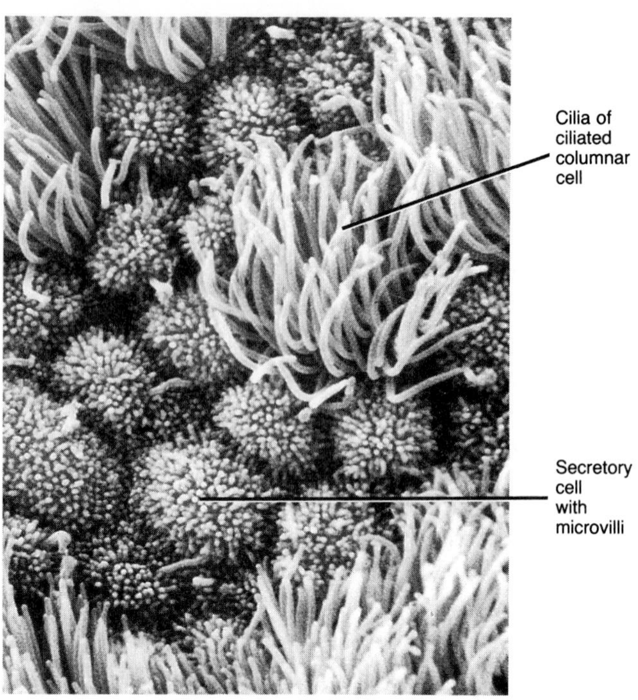

FIGURE 25-15 Histology of the uterine (Fallopian) tube. (Courtesy of Photo Researchers, Inc., © D. W. Fawcett/Gaddum-Rosse.)

The secretory cells of the mucosa of the cervix produce a secretion called ***cervical mucus,*** a mixture of water, glycoprotein, serum-type proteins, lipids, enzymes, and inorganic salts. Females of reproductive age secrete 20 to 60 ml of mucus per day. The uterine cervix and its secretions are important in reproduction. Cervical mucus is less viscous and thus more receptive to spermatozoa at or near the time of ovulation. At other times, the mucus is more viscous and forms a plug (***cervical plug***) that impedes sperm penetration. The mucus also supplements the energy requirements of spermatozoa. Both the cervix and mucus serve as a sperm reservoir, protect spermatozoa from the hostile environment of the vagina, protect spermatozoa from phagocytes, and may assume a role in capacitation—a functional change that spermatozoa undergo in the female reproductive tract in order to fertilize a secondary oocyte.

Between the body and the cervix is the ***isthmus*** (IS-mus), a constricted region about 1 cm (0.5 in.) long. The interior of the body of the uterus is called the ***uterine cavity,*** and the interior of the narrow cervix is called the ***cervical canal.*** The junction of the isthmus with the cervical canal is the ***internal os.*** The ***external os*** is the place where the cervix opens into the vagina.

CLINICAL APPLICATION

Pap Smear

Early diagnosis of ***cancer of the uterus*** is accomplished by the ***Papanicolaou*** (pap′-a-NIK-ō-la-oo) ***test,*** or ***Pap***

smear. In this generally painless procedure, a few cells from the part of the vagina surrounding the cervix and the cervix itself are removed with a swab and examined microscopically. Malignant cells have a characteristic appearance and indicate an early stage of cancer, even before symptoms occur. Estimates indicate that the Pap smear is more than 90 percent reliable in detecting cancer of the cervix. Females at risk of developing cancer of the uterus should undergo annual gynecological examinations that include a Pap smear. Females with a history of normal Pap smears and no risk factors probably should have a gynecological examination and Pap smear every two or three years until age 50 and annually thereafter. In all cases, the decision as to how frequently the test should be performed should be made by a physician in consultation with the patient.

To rule out invasive carcinoma, a ***cone biopsy*** of the cervix is performed. A cone biopsy is a hospital procedure in which an inverted cone of tissue is excised. It requires an anesthetic and is usually done only when abnormal cells have been detected. In another procedure, ***punch biopsy*** is combined with an ***endocervical curettage*** (ku-re-TAZH), or ***ECC;*** this combination has a high degree of diagnostic accuracy. In a punch biopsy a disc or segment of tissue is excised. Curettage is a procedure in which the cervix is dilated and the endometrium (lining) of the uterus is scraped with a spoon-shaped instrument called a curette. This procedure is commonly called a ***D and C.*** If the carcinoma has spread beyond the lining, treatment may involve complete or partial removal of the uterus, called a ***hysterectomy,*** or radiation treatment.

Normally, the uterus is flexed between the uterine body and the cervix. In this position, called ***anteflexion,*** the body of the uterus projects anteriorly and slightly superiorly over the urinary bladder, and the cervix projects inferiorly and posteriorly and enters the anterior wall of the vagina at nearly a right angle. Several structures that are either extensions of the parietal peritoneum or fibromuscular cords, referred to as ligaments, maintain the position of the uterus (see Figure 25-12d). The paired ***broad ligaments*** are double folds of parietal peritoneum attaching the uterus to either side of the pelvic cavity. Uterine blood vessels and nerves pass through the broad ligaments. The paired ***uterosacral ligaments,*** also peritoneal extensions, lie on either side of the rectum and connect the uterus to the sacrum. The ***cardinal (lateral cervical) ligaments*** extend below the bases of the broad ligaments between the pelvic wall and the cervix and vagina. These ligaments contain smooth muscle, uterine blood vessels, and nerves and are the chief ligaments that maintain the position of the uterus and help keep it from dropping down into the vagina. The ***round ligaments*** are bands of fibrous connective tissue between the layers of the broad ligament. They extend from a point on the uterus just below the uterine (Fallopian) tubes to a portion of the labia majora of the external genitalia. Although the ligaments normally maintain the anteflexed position of the uterus, they also afford the uterine body some movement. As a result, the uterus may become malpositioned. A posterior tilting of the uterus is called ***retroflexion.***

Histologically, the uterus consists of three layers of tissue (Figure 25-16). The outer layer, the ***perimetrium (serosa)*** is part of the visceral peritoneum. Laterally, it becomes the broad ligament. Anteriorly, it is reflected over the urinary bladder and forms a shallow pouch, the ***vesicouterine*** (ves′-i-kō-YOO-ter-in) ***pouch*** (see Figures 25-11 and 25-12c). Posteriorly, it is reflected onto the rectum and forms a deep pouch, the ***rectouterine*** (rek-tō-YOO-ter-in) ***pouch (pouch of Douglas)***—the lowest point in the pelvic cavity.

The middle layer of the uterus, the ***myometrium,*** forms the bulk of the uterine wall. This layer consists of three layers of smooth muscle fibers and is thickest in the fundus and thinnest in the cervix. During childbirth, coordinated contractions of the muscles help expel the fetus from the body of the uterus.

The inner layer of the uterus, the ***endometrium,*** is very vascular and is composed of (1) a surface layer of simple columnar epithelium (ciliated and secretory cells), (2) uterine (endometrial) glands that develop as invaginations of

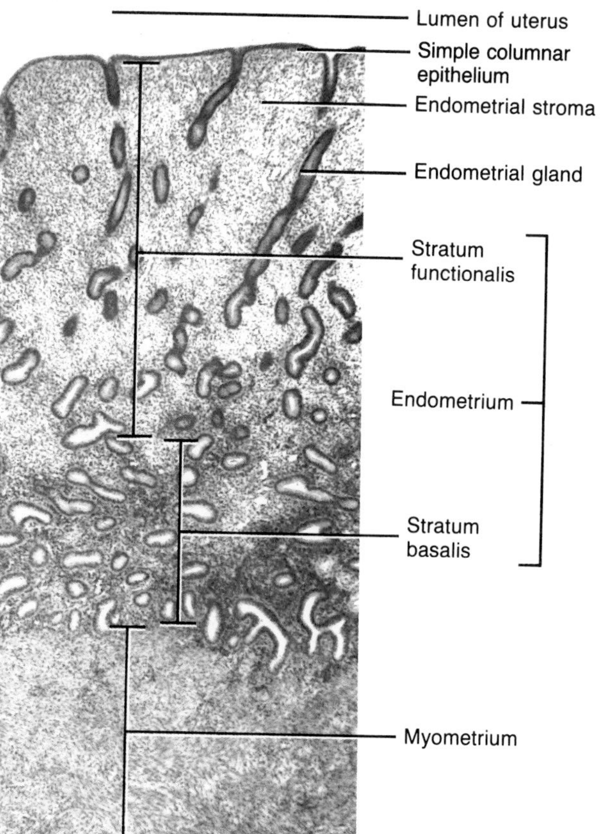

FIGURE 25-16 Histology of the uterus. (Courtesy of Andrew J. Kuntzman.)

the surface epithelium, and (3) endometrial stroma, a very thick region of lamina propria (connective tissue). The endometrium is divided into two layers. The ***stratum functionalis,*** the layer closer to the uterine cavity, is shed during menstruation. The second layer, the ***stratum basalis*** (bā-SAL-is), is permanent. Its function is to produce a new functionalis following menstruation.

CLINICAL APPLICATION

Colposcopy

Colposcopy (kol-POS-ko-pē) is a procedure used to evaluate the status of the mucosa of the vagina and cervix. Colposcopy is the direct examination of the vaginal and cervical mucosa with a magnifying device (a colposcope) similar to a low-power binocular microscope. Various instruments that magnify the mucous membrane from about 6 to 40 times its actual size are commercially available. The application of a 3 percent solution of acetic acid removes mucus and enhances the appearance of mucosal columnar epithelium.

Blood is supplied to the uterus by branches of the internal iliac artery called ***uterine arteries*** (Figure 25-17). Branches called ***arcuate arteries*** are arranged in a circular fashion in the myometrium and give off ***radial arteries*** that penetrate deeply into the myometrium. Just before the branches enter the endometrium, they divide into two kinds of arterioles. The ***straight arterioles*** terminate in the basalis and supply it with the materials necessary to regenerate the functionalis. The ***spiral arterioles*** penetrate the functionalis and change markedly during the menstrual cycle. The uterus is drained by the ***uterine veins.***

Sympathetic and parasympathetic fibers are supplied to the uterus via the hypogastric and pelvic plexuses. Both sets of nerves terminate on the uterine vessels. The myometrium is believed to be innervated by the sympathetic fibers alone.

MENSTRUAL CYCLE

The principal events of the menstrual cycle can be correlated with those of the ovarian cycle and changes in the endometrium. All are hormonally controlled events.

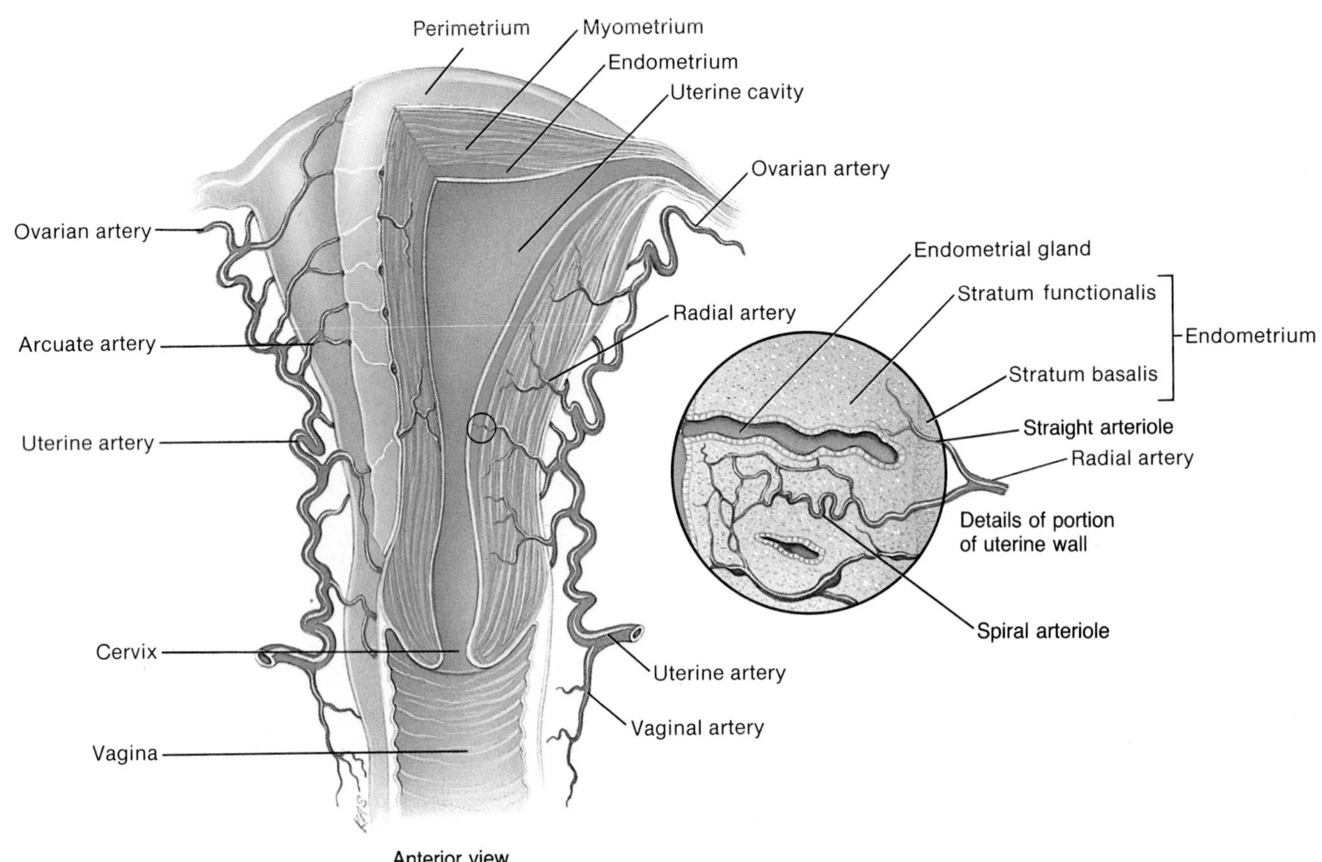

FIGURE 25-17 Blood supply of the uterus. The insert shows details of the blood vessels of the endometrium.

The ***menstrual cycle*** is a series of changes in the endometrium of a nonpregnant female. Each month, the endometrium is prepared to receive an already fertilized ovum that eventually normally develops into an embryo and then into a fetus until delivery. If no fertilization occurs, the stratum functionalis portion of the endometrium is shed. The ***ovarian cycle*** is a monthly series of events associated with the maturation of an ovum.

Hormonal Control

The menstrual cycle, ovarian cycle, and other changes associated with puberty in the female are controlled by a regulating factor from the hypothalamus called gonadotropin-releasing hormone (GnRH). Its influence is shown in Figure 25-18. GnRH stimulates the release of follicle-stimulating hormone (FSH) from the anterior pituitary. FSH stimulates the initial development of the ovarian follicles and the secretion of estrogens by the follicles. GnRH also stimulates the release of another anterior pituitary hormone—the luteinizing hormone (LH), which stimulates the further development of ovarian follicles, brings about ovulation, and stimulates the production of estrogens, progesterone, inhibin, and relaxin by ovarian cells of the corpus luteum.

At least six different estrogens have been isolated from the plasma of human females. Of these, β-estradiol exerts the major effect. It is synthesized from cholesterol or acetyl

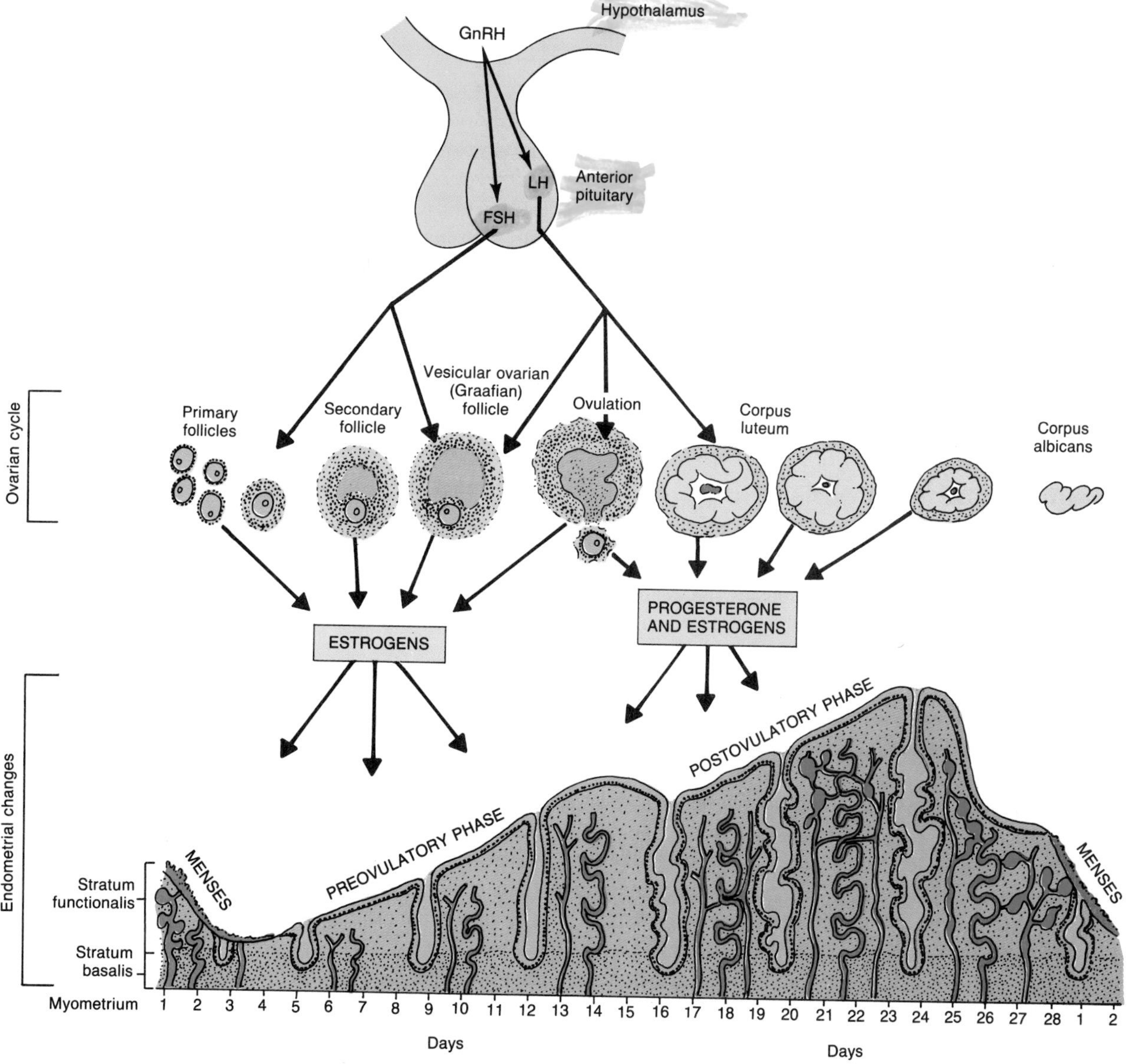

FIGURE 25-18 Correlation of menstrual and ovarian cycles with the hypothalamic and anterior pituitary gland hormones. In the cycle shown, fertilization and implantation have not occurred.

coenzyme A in the ovaries. As reference is made to estrogens in subsequent discussions, keep in mind that β-estradiol is the principal estrogen.

Estrogens, the hormones of growth, have three main functions. First is the development and maintenance of female reproductive structures, especially the endometrial lining of the uterus, secondary sex characteristics, and the breasts. The secondary sex characteristics include fat distribution to the breasts, abdomen, mons pubis, and hips; voice pitch; broad pelvis; and hair pattern. Second, they help control fluid and electrolyte balance. Third, they increase protein anabolism. In this regard, estrogens are synergistic with human growth hormone (hGH). High levels of estrogens in the blood inhibit the release of GnRH by the hypothalamus, which in turn inhibits the secretion of FSH by the anterior pituitary gland. This inhibition provides the basis for the action of one kind of contraceptive pill.

Progesterone, the hormone of maturation, works with estrogens to prepare the endometrium for implantation of a fertilized ovum and the mammary glands for milk secretion. High levels of progesterone also inhibit GnRH and prolactin.

Inhibin is secreted by the corpus luteum and sustentacular (Sertoli) cells of the testis. Its function is to inhibit secretion of FSH and, to a lesser extent, LH. Inhibin might be important in decreasing secretion of FSH and LH toward the end of the menstrual cycle.

Relaxin is produced by the corpus luteum (and placenta) during pregnancy and goes into operation near the end of pregnancy. It relaxes the pubic symphysis and helps dilate the uterine cervix to facilitate delivery.

Menstrual Phase (Menstruation)

The duration of the menstrual cycle ranges from 24 to 35 days. For this discussion, we shall assume an average duration of 28 days. Events occurring during the menstrual cycle may be divided into three phases: the menstrual phase, preovulatory phase, and postovulatory phase (Figure 25-18).

The ***menstrual phase,*** also called ***menstruation*** or the ***menses,*** is the periodic discharge of 25 to 65 ml of blood, tissue fluid, mucus, and epithelial cells. It is caused by a sudden reduction in estrogens and progesterone and lasts for approximately the first 5 days of the cycle. The first day of the ovarian cycle is designated as the first day of menstruation. The discharge is associated with endometrial changes in which the stratum functionalis layer degenerates and patchy areas of bleeding develop. Small areas of the stratum functionalis detach one at a time (total detachment would result in hemorrhage), the uterine glands discharge their contents and collapse, and tissue fluid is discharged. The menstrual flow passes from the uterine cavity to the cervix and through the vagina to the exterior. Generally, the flow terminates by the fifth day of the cycle. At this time the entire stratum functionalis has been shed, and the endometrium is very thin because only the stratum basalis remains.

During the menstrual phase, the ovarian cycle is also in operation. Ovarian follicles, called ***primary follicles,*** begin their development. At birth, each ovary contains about 200,000 such follicles, each consisting of a primary oocyte (potential ovum) surrounded by a single flattened layer of epithelial (follicular) cells. During the early part of each menstrual phase, 20 to 25 primary follicles start to produce very low levels of estrogens. Toward the end of the menstrual phase (days 4 to 5), about 20 of the primary follicles develop into ***secondary (growing) follicles.*** A secondary follicle consists of a secondary oocyte and several layers of cells formed by division of the single layer of epithelial cells around a primary follicle. The epithelial cells of the secondary follicle are cuboidal and later columnar and are called *granular (granulosa) cells.* As a secondary follicle continues to grow, it forms a clear glycoprotein layer between the secondary oocyte and granular cells called the ***zona pellucida*** (pe-LOO-si-da). Also, the granular cells secrete follicular (fō-LIK-yoo-lar) fluid that forces the secondary oocyte to the edge of the secondary follicle and fills the follicular cavity or antrum (see Figure 25-13a). The production of estrogens by the secondary follicles elevates the level of estrogens in the blood slightly. Ovarian follicle development is the result of GnRH secretion by the hypothalamus, which in turn stimulates FSH production by the anterior pituitary. During this part of the cycle, FSH secretion is relatively high. Although about 20 follicles begin development each cycle, usually only 1 attains maturity. The others undergo atresia (death).

Preovulatory Phase

The ***preovulatory phase,*** the second phase of the menstrual cycle, is the time between menstruation and ovulation. This phase of the menstrual cycle is more variable in length than the other phases. It lasts from days 6 to 13 in a 28-day cycle.

FSH and LH stimulate the ovarian follicles to produce more estrogens, and this increase in estrogens stimulates the repair of the endometrium. Cells of the stratum basalis undergo mitosis and produce a new stratum functionalis. As the endometrium thickens, the short, straight endometrial glands develop and the arterioles coil and lengthen as they penetrate the stratum functionalis. The thickness of the endometrium approximately doubles to about 4 to 6 mm. Because of the proliferation of endometrial cells, the preovulatory phase is also termed the ***proliferative phase.*** Still another name is the ***follicular phase*** because of increasing secretion of estrogens by the developing follicle. Functionally, estrogens are the dominant ovarian hormones during this phase of the menstrual cycle.

During the preovulatory phase, one of the secondary follicles in the ovary matures into a ***vesicular ovarian (Graafian) follicle*** or ***mature follicle,*** a follicle ready for ovulation.

This follicle produces a bulge on the surface of the ovary. During the maturation process, the follicle increases its estrogen production. Early in the preovulatory phase, FSH is the dominant hormone of the anterior pituitary, but close to the time of ovulation, LH is secreted in increasing quantities. Moreover, small amounts of progesterone may be produced by the vesicular ovarian (Graafian) follicle a day or two before ovulation.

Ovulation

Ovulation, the rupture of the vesicular ovarian (Graafian) follicle with release of the secondary oocyte into the pelvic cavity, usually occurs on day 14 in a 28-day cycle. During ovulation, the secondary oocyte remains surrounded by its zona pellucida and a covering of follicle cells directly around it. These cells are referred to as the ***corona radiata.*** It generally takes 10 to 14 days for a primary follicle to develop into a vesicular ovarian (Graafian) follicle, and it is during this time that the developing ovum completes reduction division (meiosis I) and reaches metaphase of equatorial division (meiosis II). The developing ovum is in this stage when it is discharged during ovulation. The fimbriae of the uterine tubes drape over the ovaries and become active near the time of ovulation. Movements of the fimbriae and ciliary action create currents in the peritoneal serous fluid that carry the secondary oocyte into the uterine tube.

Just prior to ovulation, the high level of estrogens that developed during the preovulatory phase exerts a positive feedback directly on both LH and GnRH. LH release increases sharply because of the direct effect of estrogens on the anterior pituitary and also because of the increase in secretion of GnRH by the hypothalamus. This causes the anterior pituitary to release a surge of LH. Without this surge of LH, ovulation will not occur. (An over-the-counter home test that detects the LH surge associated with ovulation is now available. The test predicts ovulation a day in advance.) FSH also increases at this time, but not as dramatically as LH because FSH is stimulated only by the increase in GnRH. Following ovulation, the vesicular ovarian (Graafian) follicle collapses, and blood within it forms a clot called the ***corpus hemorrhagicum.*** The clot is eventually absorbed by the remaining follicular cells. In time, the follicular cells enlarge, change character, and form the ***corpus luteum,*** or yellow body, under the influence of LH, which also stimulates the corpus luteum to secrete estrogens and progesterone.

CLINICAL APPLICATION

Signs of Ovulation

One ***sign of ovulation*** involves ***basal temperature*** (body temperature at rest). When menstruation ceases, the temperature is taken immediately upon awakening each morning and marked on a chart. An increase in temperature, usually between 0.4 to 0.6°F, typically occurs about 14 days after the start of the last menstrual cycle and is due to increasing levels of progesterone. The 24 hours following this rise in temperature is the period immediately following ovulation and is generally considered the best time to become pregnant. The accuracy of the determination depends on many factors including individual variations, the accuracy of the temperature readings, and any factor other than the ovarian cycle that might affect body temperature.

Another sign of ovulation is the amount and consistency of ***cervical mucus.*** Secretion of cervical mucus is regulated by estrogens and progesterone. At midcycle, near the time of ovulation, increasing levels of estrogens cause secretory cells of the cervix to produce large amounts of cervical mucus. About a day or two before ovulation, the quantity of mucus frequently begins to decrease and usually disappears a few days after ovulation. More important, as ovulation approaches, the mucus becomes clear, stretchy (it may stretch from 2.54 to 15.24 cm, that is, 1 to 6 inches), and slippery and causes feelings of lubrication, slipperiness, or wetness on the outer lips (labia majora) of the external genitals. This is the more fertile type of mucus and indicates the time of greatest fertility. Around the day of ovulation, cervical mucus becomes nonstretchy, tacky, thicker, and more opaque and then disappears by a few days after ovulation. This less fertile mucus is produced in response to the influence of progesterone.

The cervix also exhibits signs of ovulation. The external os opens, the cervix rises, and the cervix becomes softer. There is also abundant cervical mucus.

Some females also experience a pain in the area of one or both ovaries around the time of ovulation. Such pain is called ***mittelschmerz*** (MIT-el-shmarts), meaning "pain in the middle," and may last for several hours to a day or two.

Postovulatory Phase

The ***postovulatory phase*** of the menstrual cycle is the most constant in duration and lasts from days 15 to 28 in a 28-day cycle. It represents the time between ovulation and the onset of the next menses. Following ovulation, LH secretion stimulates the development of the corpus luteum. The corpus luteum then secretes increasing quantities of estrogens and progesterone. Progesterone is responsible for preparing the endometrium to receive a fertilized ovum. Preparatory activities include secretory activity of the endometrial glands that causes them to appear tortuously coiled, vascularization of the superficial endometrium, thickening of the endometrium, glycogen storage, and an increase in the amount of tissue fluid. These preparatory changes are maximal about one week after ovulation, and they correspond to the anticipated arrival of the fertilized ovum. During the latter postovulatory phase, FSH secretion again gradually increases and LH secretion decreases. The functionally dominant ovarian hormone during this phase is progesterone. The relation of progesterone to prostaglandins in causing painful menstruation will be considered at the end of the chapter.

If fertilization and implantation do not occur, the rising levels of progesterone and estrogens from the corpus luteum inhibit GnRH and LH secretion. As a result, the corpus luteum degenerates and becomes the ***corpus albicans,*** or white body. The decreased secretion of progesterone and estrogens by the degenerating corpus luteum then initiates another menstrual period. In addition, the decreased levels of progesterone and estrogens in the blood bring about a new output of the anterior pituitary hormones—especially FSH in response to an increased output of GnRH by the hypothalamus. Thus, a new ovarian cycle is initiated.

If, however, fertilization and implantation do occur, the corpus luteum is maintained until the placenta takes over its hormone-producing functions. During this time, the corpus luteum secretes estrogens and progesterone. The corpus luteum is maintained by ***human chorionic*** (kō-rē-ON-ik) ***gonadotropin (hCG),*** a hormone produced by the developing placenta. As you will see later, the presence of hCG is an indication that a female is pregnant. The placenta itself secretes estrogens to support pregnancy and progesterone to support pregnancy and breast development for lactation. Once the placenta begins its secretion, the role of the corpus luteum becomes minor.

Menarche and Menopause

The menstrual cycle normally occurs once each month from ***menarche*** (me-NAR-kē), the first menses, to ***menopause*** (*mens* = monthly; *pausa* = to stop), the last menses. The advent of menopause is signaled by the ***climacteric*** (klī-MAK-ter-ik)—menstrual cycles become less frequent. The climacteric, which typically begins between ages 40 and 50, results from the failure of the ovaries to respond to the stimulation of gonadotropic hormones from the anterior pituitary. Some women experience hot flashes, copious sweating, headache, hair loss, muscular pains, vaginal dryness, insomnia, depression, weight gain, and emotional instability. In the postmenopausal woman there will be some atrophy of the ovaries, uterine (Fallopian) tubes, uterus, vagina, external genitalia, and breasts. Osteoporosis is also a possible occurrence.

The cause of menopause is related to a decreasing ability of aging ovaries to respond to FSH and LH. As a result, there is a decrease in the production of estrogens, progesterone, and ova by the ovaries. Throughout a woman's sexual life, some of the primary ovarian follicles grow into vesicular ovarian follicles with each sexual cycle, and eventually most of them degenerate. As the number of primary follicles diminishes, the production of estrogens by the ovary decreases. Alterations in GnRH release patterns and decreased responsiveness to it by cells of the anterior pituitary gland that secrete LH also contribute to the onset of menopause.

VAGINA

The ***vagina*** serves as a passageway for spermatozoa and the menstrual flow. It is also the receptacle for the penis during coitus, or sexual intercourse, and the lower portion of the birth canal. It is a tubular, fibromuscular organ lined with mucous membrane and measures about 10 cm (4 in.) in length, extending from the cervix to the vestibule (see Figures 25-11 and 25-12). Situated between the urinary bladder and the rectum, it is directed superiorly and posteriorly, where it attaches to the uterus. A recess, called the ***fornix*** (*fornix* = arch or vault), surrounds the vaginal attachment to the cervix. The fornix makes it possible for a female to use contraceptive diaphragms.

Histologically, the mucosa of the vagina is continuous with that of the uterus and consists of stratified squamous epithelium and connective tissue that lies in a series of transverse folds, the ***rugae*** (Figure 25-19). The muscularis is composed of smooth muscle that can stretch considerably. This distension is important because the vagina receives the penis during sexual intercourse and serves as the lower portion of the birth canal. At the lower end of the vaginal opening, the ***vaginal orifice,*** there may be a thin fold of vascularized mucous membrane called the ***hymen*** (*hymen* = membrane), which forms a border around the orifice, partially closing it (see Figure 25-20).

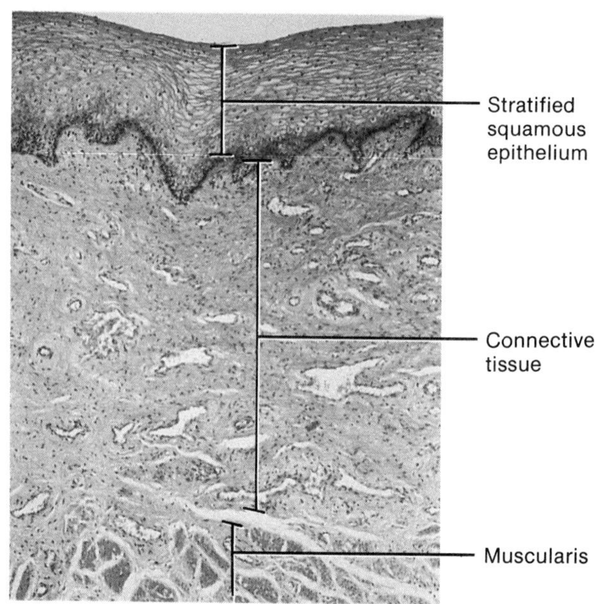

Photomicrograph of portion of wall of vagina (50x)

FIGURE 25-19 Histology of the vagina. © Biophoto, SPL, Photo Researchers.

CLINICAL APPLICATION

Imperforate Hymen
Sometimes the hymen completely covers the orifice, a condition called ***imperforate*** (im-PER-fō-rāt) ***hymen.*** Surgery is required to open the orifice to permit the discharge of the menstrual flow.

The mucosa of the vagina contains large amounts of glycogen, which upon decomposition produces organic acids. These acids create a low pH environment that retards microbial growth; however, the acidity is also injurious to sperm cells. Semen neutralizes the acidity of the vagina to ensure survival of the sperm.

VULVA

The term ***vulva*** (VUL-va; *volvere* = to wrap around), or ***pudendum*** (pyoo-DEN-dum), is a collective designation for the external genitalia of the female (Figure 25-20). Its components are as follows.

The ***mons pubis,*** an elevation of adipose tissue covered by skin and coarse pubic hair, is situated over the pubic symphysis. It lies anterior to the vaginal and urethral openings. From the mons pubis, two longitudinal folds of skin, the ***labia majora*** (LĀ-bē-a ma-JŌ-ra), extend inferiorly and posteriorly. The labia majora are homologous to the scrotum. The labia majora contain an abundance of adipose tissue and sebaceous (oil) and sudoriferous (sweat) glands; they are covered by pubic hair. Medial to the labia majora are two folds of skin called the ***labia minora*** (mī-NŌ-ra). Unlike the labia majora, the labia minora are devoid of pubic hair and fat and have few sudoriferous (sweat) glands. They do, however, contain numerous sebaceous (oil) glands.

The ***clitoris*** (KLI-to-ris) is a richly innervated structure that contains a small, cylindrical mass of erectile tissue and nerves. It is located at the anterior junction of the labia minora. A layer of skin called the ***prepuce*** (foreskin) is formed at the point where the labia minora unite and covers the body of the clitoris. The exposed portion of the clitoris is the ***glans.*** The clitoris is homologous to the penis of the male. Like the penis, the clitoris is capable of enlargement upon tactile stimulation and assumes a role in sexual excitement of the female.

The cleft between the labia minora is called the ***vestibule.*** Within the vestibule are the hymen (if present), vaginal orifice, urethral orifice, and the openings of several ducts. The ***vaginal orifice,*** the opening of the vagina to the exterior, occupies the greater portion of the vestibule and is bordered by the hymen. The ***bulb of the vestibule*** consists of two elongated masses of erectile tissue just deep to the labia on either side of the vaginal orifice. The bulb becomes engorged with blood during sexual arousal, narrowing the vaginal orifice and placing pressure on the penis during

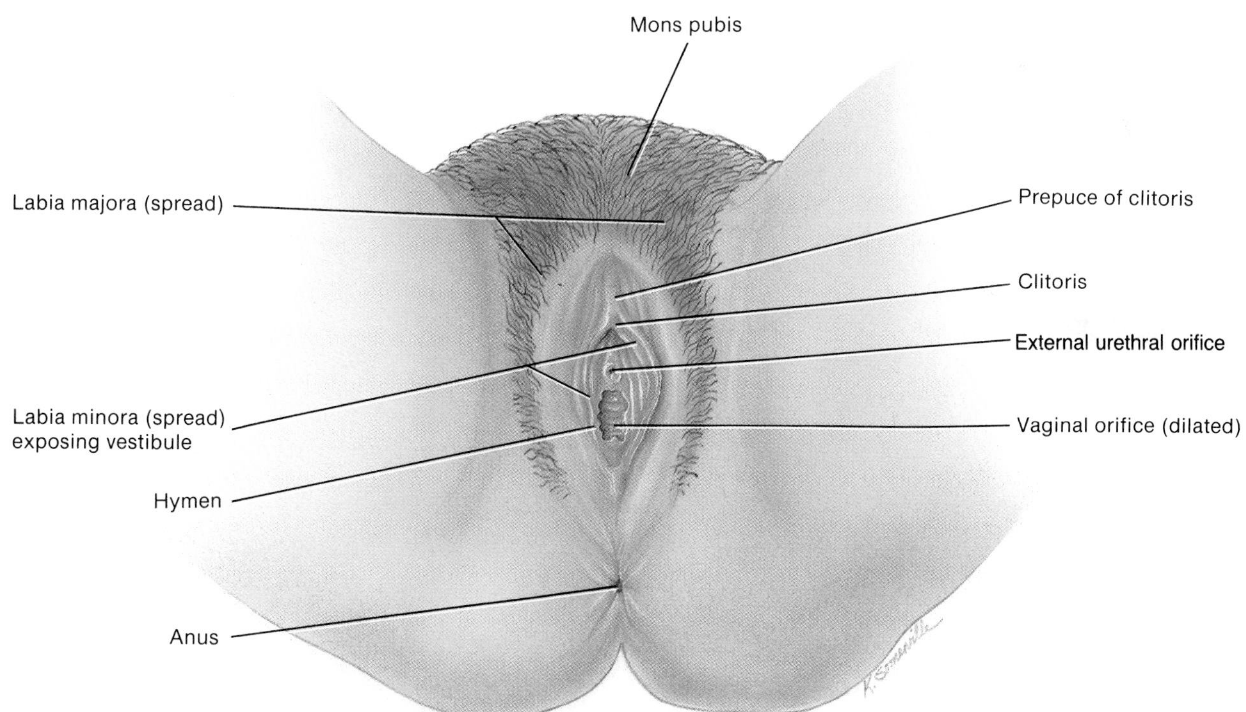

FIGURE 25-20 Components of the vulva.

intercourse. The bulb is homologous to the corpus spongiosum penis and bulb of the penis. Anterior to the vaginal orifice and posterior to the clitoris is the ***external urethral orifice,*** the opening of the urethra to the exterior. On either side of the external urethral orifice are the openings of the ducts of the ***paraurethral (Skene's) glands,*** which are embedded in the wall of the urethra and secrete mucus. The paraurethral glands are homologous to the male prostate. On either side of the vaginal orifice itself are the ***greater vestibular (Bartholin's) glands.*** These glands open by ducts into a groove between the hymen and labia minora and produce a mucoid secretion that supplements lubrication during sexual intercourse. The greater vestibular glands are homologous to the male bulbourethral (Cowper's) glands. A number of ***lesser vestibular glands,*** whose orifices are microscopic, open into the vestibule.

CLINICAL APPLICATION

Diagnosis of Pregnancy

One important sign in the ***diagnosis of pregnancy*** is a bluish discoloration of the vulva and vagina due to venous congestion. The discoloration appears at about the eighth to twelfth week and increases in intensity as the pregnancy progresses.

PERINEUM

The ***perineum*** (per'-i-NĒ-um) is the diamond-shaped area between the thighs and buttocks of both males and females. It is bounded anteriorly by the pubic symphysis, laterally by the ischial tuberosities, and posteriorly by the coccyx. A transverse line drawn between the ischial tuberosities divides the perineum into an anterior ***urogenital*** (yoo'-rō-JEN-i-tal) ***triangle*** that contains the external genitalia and a posterior ***anal triangle*** that contains the anus (Figure 25-21).

CLINICAL APPLICATION

Episiotomy

The perineal region is stretched during childbirth as the fetal head stretches the vaginal epithelium, subcutaneous fat, and superficial transverse perineal muscle. ***Episiotomy*** (e-piz'-ē-OT-ō-mē) is an option that the mother may choose in order to prevent undue stretching and possibly tearing of this region. The purpose of episiotomy—a cut made with surgical scissors—is to enlarge the perineal opening to make room for the fetal head. In effect, a controlled cut is substituted for a jagged, uncontrolled laceration. The incision is closed in layers with a continuous suture that is resorbed by the body within a few weeks

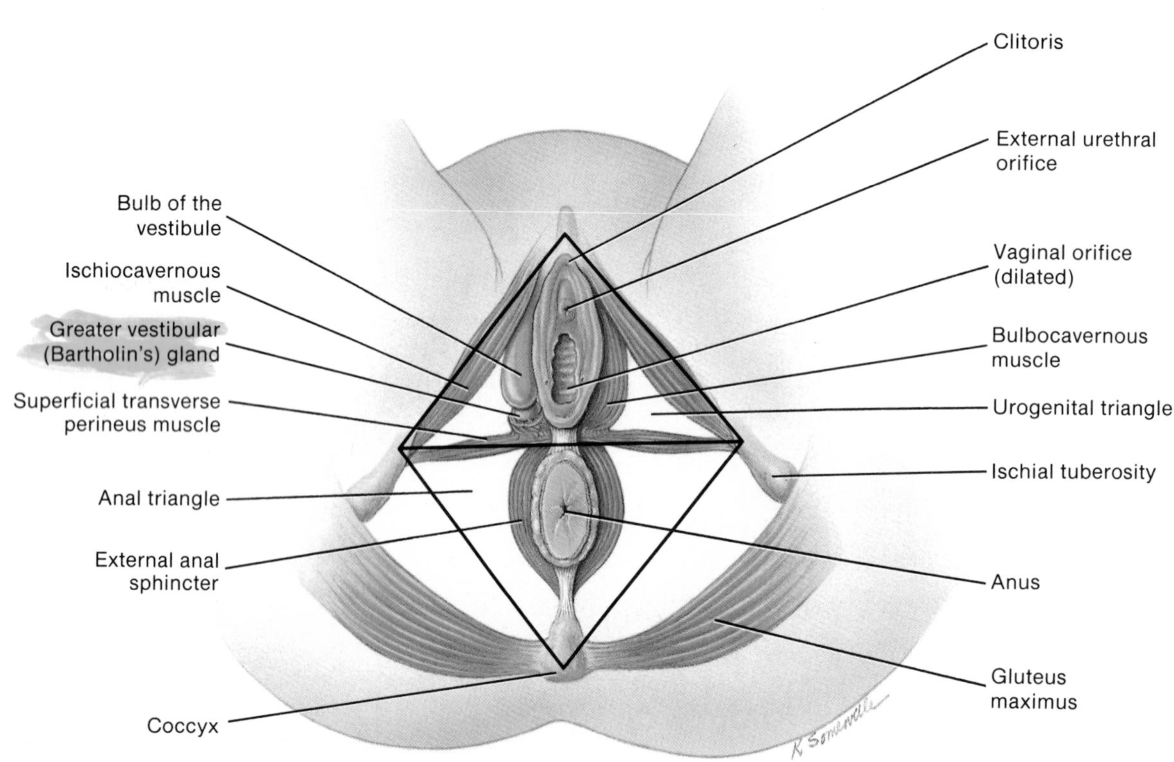

FIGURE 25-21 Perineum. Borders seen in the female. The male perineum is shown in Figure 10-15.

so that stitches do not have to be removed. However, not all women undergoing childbirth need or elect to have an episiotomy.

MAMMARY GLANDS

Structure

The ***mammary glands*** are modified sudoriferous (sweat) glands (branched tubuloalveolar) that lie over the pectoralis major and serratus anterior muscles and are attached to them by a layer of connective tissue (Figure 25-22). Internally, each mammary gland consists of 15 to 20 ***lobes,*** or compartments, separated by adipose tissue. The amount of adipose tissue determines the size of the breasts. However, breast size has nothing to do with the amount of milk produced. In each lobe are several smaller compartments called ***lobules,*** composed of connective tissue in which clusters of milk-secreting cells referred to as ***alveoli*** are embedded (Figure 25-23). Alveoli are arranged in grapelike clusters. Alveoli convey the milk into a series of ***secondary tubules.*** From here the milk passes into the ***mammary ducts.*** As the mammary ducts approach the nipple, they expand to form sinuses called ***lactiferous sinuses,*** where milk may be stored. The sinuses continue as ***lactiferous ducts*** that terminate in the ***nipple.*** Each lactiferous duct conveys milk from one of the lobes to the exterior, although some may join before reaching the surface. The circular pigmented area of skin surrounding the nipple is called the ***areola*** (a-RĒ-ō-la). It appears rough because it contains modified sebaceous (oil) glands. Strands of connective tissue called the ***suspensory ligaments of the breast (Cooper's ligaments)*** run between the skin and deep fascia and support the breasts.

The essential function of the mammary glands is milk secretion and ejection, together called ***lactation.*** The secretion of milk is due largely to the hormone prolactin (PRL), with contributions from progesterone and estrogens. The ejection of milk occurs in the presence of oxytocin (OT), which is released from the posterior pituitary gland in response to sucking.

Breast Cancer

■ ***Introduction*** Early detection—especially by breast self-examination (BSE) and mammography—is still the most promising method to increase the survival rate for ***breast cancer.*** It is estimated that 95 percent of breast cancer is first detected by women themselves. Each month

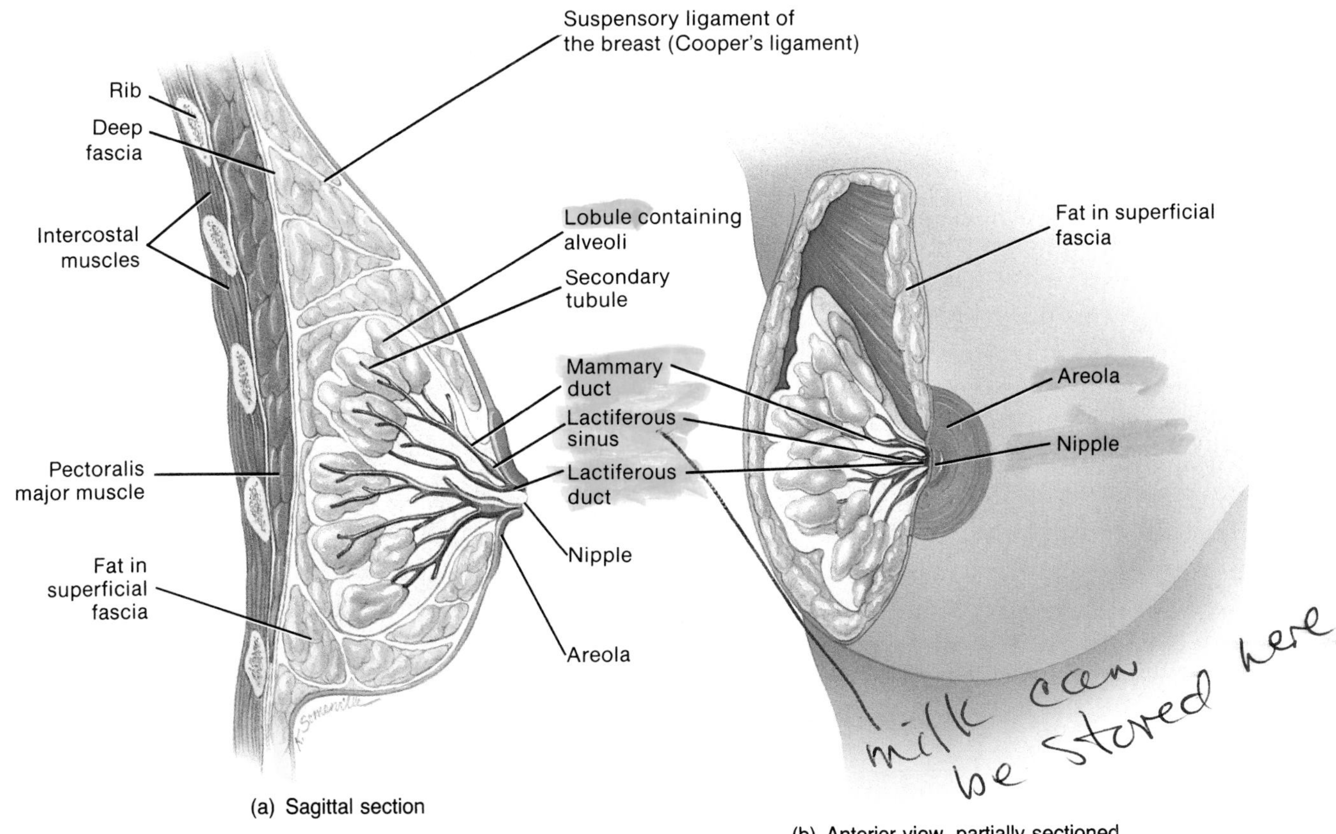

FIGURE 25-22 Mammary glands. (c) Courtesy of J. A. Gosling, P. F. Harris, et al., *Atlas of Human Anatomy,* Gower Medical Publishing Ltd., 2nd ed., 1991.

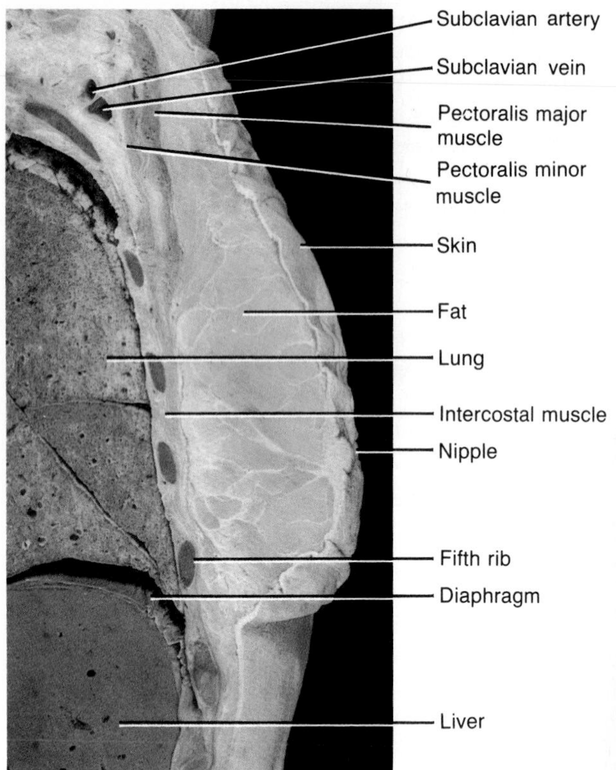

(c) Sagittal section

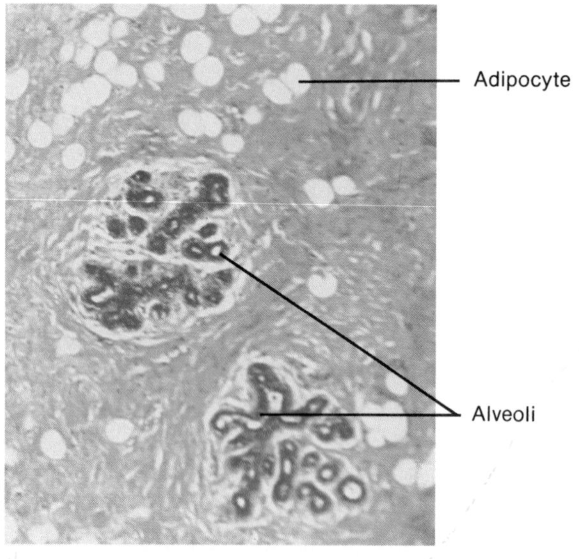

Photomicrograph of a nonlactating mammary gland (approx. 100 x)

FIGURE 25-23 Histology of the mammary glands. © Biophoto, SPL, Photo Researchers.

after the menstrual period the breasts should be thoroughly examined for lumps, puckering of the skin, or nipple retraction or discharge.

■ ***Risk Factors*** Among the factors that clearly increase the risk of breast cancer development are (1) a family history of breast cancer, especially in a mother or sister; (2) never having a child or having a first child after age 34; (3) previous cancer in one breast; (4) exposure to ionizing radiation; and (5) excessive fat and alcohol intake. Females who take birth control pills do not have a higher risk of developing breast cancer than females who do not. But cigarette smoking may increase the incidence of breast cancer, especially in postmenopausal females.

The American Cancer Society recommends the following steps in order to help diagnose breast cancer as early as possible:

1. A mammogram should be taken between the ages of 35 and 39, to be used later for comparison (baseline mammogram).
2. A physician should examine the breasts every three years when a female is between the ages of 20 and 40, and every year after 40.
3. Females with no symptoms should have a mammogram every year or two between ages 40 and 49, and every year after 50.
4. Females of any age with a history of breast cancer, a strong family history of the disease, or other risk factors should consult a physician to determine a schedule for mammography.
5. All females over 20 should develop the habit of monthly breast self-examination (BSE).

■ ***Detection*** The most effective technique for routinely detecting tumors less than 1.27 cm (0.5 in.) in diameter is ***mammography*** (mam-OG-ra-fē; *mammae* = breast; *graphein* = to record). One reason that mammography is so useful is that it can detect small calcium deposits, called microcalcifications, in breast tissue. Such calcifications frequently indicate the presence of a tumor.

Mammography is used to evaluate symptoms of breast disease such as lumps, nipple discharges, retraction of nipples, dimpling of skin, or persistent breast pain. The test is also used to screen asymptomatic women as well as those having an increased risk for breast cancer. The test may be performed between ages 35 to 39 to provide a baseline mammogram, an image to be used later for comparison.

The mammographic image, called a ***mammogram*** (Figure 25-24), is obtained by placing the breasts, one at a time, on a flat surface and using a compressor to smooth the breast for better imaging. Usually, two images are taken of each breast, one from the top and one from the side. There are two basic types of mammography: xeromammography and film-screen mammography. In *xeromammogra-*

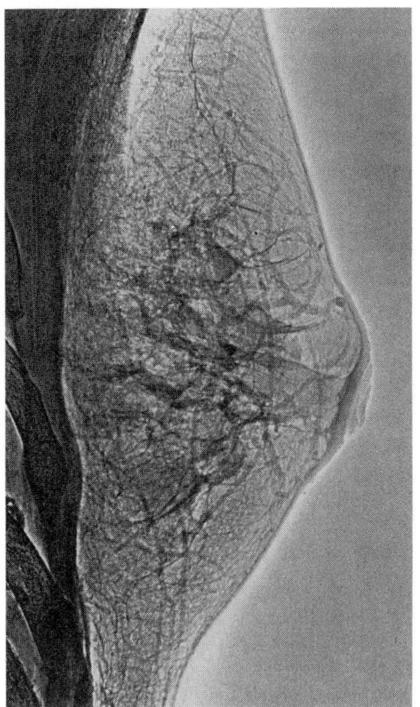

(a) Normal breast

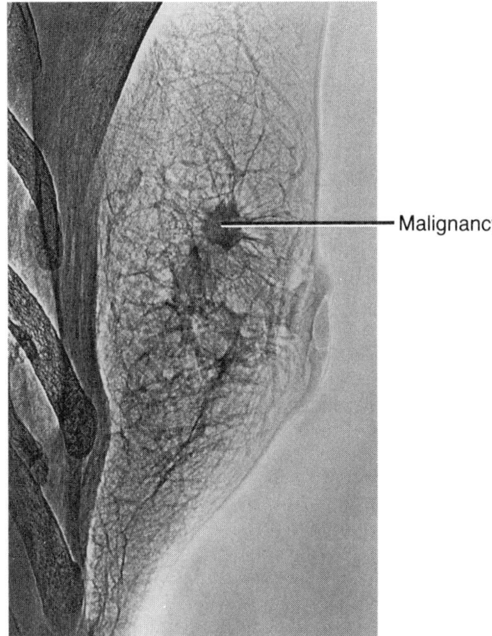

(b) Malignant breast

FIGURE 25-24 Photographs of xeroradiographic mammograms. (Courtesy of Xerox Medical Systems, Pasadena, CA.)

phy, x-rays are beamed onto a specially coated metal plate, and the blue-on-white image produced provides the physician with detail of thicker as well as thinner portions of the breast. In *film-screen mammography,* x-rays are beamed onto a fluorescent screen and the image is produced on x-ray film. Breast compression is the key to effective mammography. It increases image detail, decreases blurring and overlap of tissue, and reduces radiation dosage.

One of the most recent breast cancer–detecting procedures is ***ultrasound (US)*****.** The procedure is performed while the patient lies on her stomach in a specially designed hospital bed with her breasts immersed in a tank of water. Ultrasound produces images using a device that first emits a pulse of high-frequency sound and then records the echo on a monitor. Although ultrasound can detect neither microcalcifications nor tumors less than 1 cm in diameter, it can be used to determine whether a lump is a benign cyst or a malignant tumor.

Another technique combines ***computed tomography with mammography (CT/M)*** and appears to overcome some of the limitations of mammography. The procedure is based on the fact that breast carcinoma has an abnormal affinity for iodide. CT scans are made before and after the rapid intravenous infusion of an iodide contrast material. Comparison of the initial density of a suspected lesion with the density following infusion of the iodide gives an indication of the status of the tumor. CT/M affords definitive diagnostic help in instances where the mammographic and physical examinations are inconclusive and appears to be a significantly improved method of breast cancer diagnosis.

Long-term studies are now under way to determine the effectiveness of an experimental home device for breast cancer detection. It is called a ***breast cancer screening indicator*****.** It is composed of two plastic discs that contain heat-sensitive chemicals. The discs are worn inside a woman's brassiere for 15 minutes each month so that she can observe color changes that could indicate a breast abnormality. The screening device is said to detect tumors as small as 2 mm in diameter, their size about 2 to 10 years before they become palpable or can be seen by mammography.

■ ***Treatment*** Treatment for breast cancer may involve hormone therapy, chemotherapy, ***lumpectomy*** (removal of just the tumor and immediate surrounding tissue), a modified or radical mastectomy, or a combination of these. A ***radical mastectomy*** involves removal of the affected breast along with the underlying pectoral muscles and the axillary lymph nodes. Metastasis of cancerous cells is usually through the lymphatic vessels (see Figure 15-9) or blood vessels. Radiation treatment and chemotherapy may follow the surgery to ensure the destruction of any stray cancer cells.

By using silicone implants and skin, fat, and muscles from other parts of the body, breast reconstruction following a radical mastectomy can be accomplished. Using these techniques, it is possible to reconstruct a natural-looking breast.

BIRTH CONTROL (BC)

Although there is no single, ideal method of ***birth control (BC)***, several types of contraceptive methods are available, each with its own advantages and disadvantages. The methods discussed here are sterilization, hormonal, intrauterine, barrier, chemical, physiologic, coitus interruptus (withdrawal), and induced abortion.

STERILIZATION

One means of ***sterilization*** of males is ***vasectomy*** (discussed earlier in the chapter). Sterilization in females generally is achieved by performing a ***tubal ligation*** (lī-GĀ-shun). In one procedure, an incision is made into the abdominal cavity, the uterine (Fallopian) tubes are squeezed, and a small loop called a knuckle is made. A suture is tied tightly at the base of the knuckle, and the knuckle is then cut. After four or five days the suture is digested by body fluids, and the two severed ends of the tubes separate. The ovum is thus prevented from passing to the uterus, and the sperm cannot reach the ovum.

Tubal ligation may also be performed by ***laparoscopy*** (lap′-a-ROS-kō-pē; *lapara* = flank; *skopein* = to examine). Following local or general anesthesia, a small incision is made in the abdominal wall in or just below the umbilicus (navel). A needle is inserted through the incision, and gas is slowly injected to inflate the abdomen. (This gas creates a space inside the abdomen for better visualization and easier manipulation of instruments.) After the needle is withdrawn, a ***laparoscope*** (lighted tube) is inserted through the incision so that the physician can view the abdominal and pelvic viscera. Other instruments, such as forceps, probes, and small scissors, may be inserted through a second incision, usually at the pubic hairline. Thus, the technique may also be used to remove fluids and tissues for biopsy, drain ovarian cysts, and cut adhesions. Although tubal ligation, like vasectomy, may be reversed, about 50–80 percent of females subsequently achieve fertility.

HORMONAL

The ***hormonal method,*** also called ***oral contraception*** or ***the pill,*** has found rapid and widespread use. Although several pills are available, the one most commonly used contains a high concentration of progesterone and a low concentration of estrogens (combination pill). These two hormones act on the anterior pituitary to decrease the secretion of FSH and LH by inhibiting GnRH release by the hypothalamus. The low levels of FSH and LH usually prevent ovulation, and thus pregnancy cannot occur. Even if ovulation does occur, as it does in some cases, oral contraceptives also alter cervical mucus so that it is more hostile to spermatozoa and may make the endometrium less receptive to implantation.

Women for whom all oral contraceptives are contraindicated include those with a history of thromboembolic disorders (predisposition to blood clotting), cerebral blood vessel damage, hypertension, liver malfunction, heart disease, or cancer of the breast or reproductive system. Recent reports link pill users with an increased risk of infertility. About 40 percent of all pill users experience side effects—generally minor problems such as nausea, weight gain, headache, irregular menses, spotting between periods, and amenorrhea. The statistics on the life-threatening conditions associated with the pill such as blood clots, heart attacks, liver tumors, and gallbladder disease are somewhat more reassuring. For all the problems combined, fewer than 3 deaths occur per 100,000 users under age 30, 4 among those 30 to 35, 10 among those 35 to 39, and 18 among women over 40. The major exception is that women who take the pill and smoke face far higher odds of developing heart attack and stroke than do nonsmoking pill users.

Among the noncontraceptive benefits of oral contraceptives are menstrual regulation, decreased menstrual flow, and prevention of functional ovarian cysts. Some evidence also suggests a protective effect of the pill against endometrial and ovarian cancer.

In December 1990, the FDA approved the use of ***Norplant*** in the United States. This method uses six capsules surgically implanted under the skin of the arm that slowly release hormones that inhibit ovulation. Norplant is 99 percent effective and exerts its effectiveness for five years. Removal of the capsules, which may be done at any time, does not affect fertility. A hormonal method not yet available in the United States is ***Depo-Provera,*** an injectable hormone given every three months, which is released slowly from muscle and inhibits ovulation.

Modified human chorionic gonadotropin (hCG), a hormone produced by the placenta, has shown promise in laboratory animals as a chemical method of birth control. In its modified form, hCG prevents implantation of a fertilized egg or terminates an already established pregnancy. Modified versions of gonadotropin-releasing hormone (GnRH) are also being tested as contraceptives that inhibit ovulation.

INTRAUTERINE DEVICES (IUDs)

An ***intrauterine device (IUD)*** is a small object made of plastic, copper, or stainless steel that is inserted into the cavity of the uterus. It is not clear how IUDs operate. Some investigators believe they cause changes in the uterine lining, which, in turn, produce a substance that destroys either the sperm or fertilized ovum. The dangers associated with the use of IUDs in some females include the device falling out, pelvic inflammatory disease (PID), infertility, and excess menstrual bleeding and pain. (Females in monogamous relationships are at a lesser risk of developing PID.) Because of these risks, IUDs are rapidly declining in popularity. Lawsuits resulting from damage claims have caused most manufacturers of copper IUDs in the United

States to stop production and sales. An IUD that is similar to the famous copper T is now available in the United States. It is called Para Gard model T380A. A female who uses it must sign or initial a consent form to indicate that she is well aware of the potential risks involved.

BARRIER

Barrier methods are designed to prevent spermatozoa from gaining access to the uterine cavity and uterine tubes. Among the barrier methods are condoms, diaphragms, and cervical caps.

The ***condom*** is a nonporous, elastic (latex or similar material) covering placed over the penis that prevents deposition of sperm in the female reproductive tract. Proper use of condoms, especially when used with sperm-killing chemicals, with each act of sexual intercourse can reduce, but not eliminate, risk of sexually transmitted diseases (STDs). Individuals likely to become infected or known to be infected with human immunodeficiency virus (HIV) should be aware that condom use cannot completely eliminate the risk of transmission to themselves or to others since condoms have a failure rate of up to 10 percent.

The ***diaphragm*** is a dome-shaped structure that fits over the cervix and is generally used in conjunction with a sperm-killing chemical. The diaphragm stops the sperm from passing into the cervix. The chemical kills the sperm cells. Toxic shock syndrome (TSS) and recurrent urinary tract infections are associated with diaphragm use in some females.

The ***cervical cap*** (Prentif cavity-Rim), is a thimble-shaped contraceptive device made of latex or plastic that measures about 3.9 cm (1.5 in.) in diameter. It fits snugly over the cervix of the uterus and is held in position by suction. Like the diaphragm, the cervical cap is used with a spermicide. Also, like the diaphragm, it must be fitted initially by a physician or other trained personnel. Among the advantages of the cervical cap over the diaphragm are the following: the cap can be worn up to 48 hours versus 24 hours for the diaphragm and, since the cap fits tightly and rarely leaks, it is not necessary to reintroduce spermicide before intercourse. The cervical cap is prescribed only to females with normal Pap smears, and users should undergo a follow-up Pap smear after three months. The cervical cap is contraindicated for females with toxic shock syndrome (TSS), known or suspected cervical or uterine malignancies, and current vaginal or cervical infections.

CHEMICAL

Chemical methods of contraception include spermicidal agents. Various foams, creams, jellies, suppositories, and douches that contain spermicidal agents make the vagina and cervix unfavorable for sperm survival. The most widely used spermicides are nonoxynol-9 and octoxynol-9. The action is to disrupt plasma membranes of spermatozoa, thus killing them. They are most effective when used in conjunction with a diaphragm. Several studies have found no increase in the overall frequency of birth defects in association with the use of spermicides. A recent spermicidal development is a ***contraceptive sponge.*** It is a nonprescription polyurethane sponge that contains nonoxynol-9. This spermicide decreases the incidence of chlamydia and gonorrhea but slightly increases the risk of developing vaginal infections caused by *Candida,* a fungus. The sponge is placed in the vagina where it releases spermicide for up to 24 hours and also acts as a physical barrier to sperm. The sponge is equal to the diaphragm in effectiveness. Some cases of TSS have been reported among users of the contraceptive sponge.

PHYSIOLOGIC

Physiologic methods are based upon knowledge of certain physiologic events that occur during the menstrual cycle. In females with normal menstrual cycles, especially, the physiologic events help to predict during which days of the cycle ovulation is likely to occur.

The first physiologic method, developed in the 1930s, is known as the ***rhythm method.*** It takes advantage of the fact that a secondary oocyte is fertilizable for only 24 hours and is available only during a period of three to five days in each menstrual cycle. During this time, the couple refrains from intercourse (three days before ovulation, the day of ovulation, and three days after ovulation). Its effectiveness is limited by the fact that few women have absolutely regular cycles.

Another natural family planning system, developed during the 1950s and 1960s, is the ***sympto-thermal method.*** According to this method, couples are instructed to know and understand certain signs of fertility and infertility. The signs of ovulation include increased basal body temperature; the production of clear, stretchy cervical mucus; opening of the external os; elevation of the cervix; softening of the cervix; abundant cervical mucus; and pain associated with ovulation (mittelschmerz). If the couple refrains from sexual intercourse when the signs of ovulation are present, the likelihood of pregnancy is significantly decreased.

COITUS INTERRUPTUS (WITHDRAWAL)

Coitus (KŌ-i-tus; *coitio* = coming together) ***interruptus*** refers to withdrawal of the penis from the vagina just prior to ejaculation. Failures with this method are related to either failure to withdraw before ejaculation or pre-ejaculatory escape of sperm-containing fluid from the urethra.

INDUCED ABORTION

Abortion refers to the premature expulsion from the uterus of the products of conception. An abortion may be spontaneous (naturally occurring) or induced (intentionally performed). When birth control methods are not practiced or are not successful, ***induced abortion*** may be performed. Induced abortions may involve vacuum aspiration (suction), a saline solution, surgical evacuation (scraping), or use of drugs such as mifepristone (RU486) and epostane.

In late 1986, a team of French physicians reported a new approach to birth control. It involves using a substance called ***mifepristone (RU486)*** that blocks the action of progesterone. Progesterone prepares the uterine endometrium for implantation and then maintains the uterine lining after implantation. If progesterone levels fall during pregnancy or if the hormone is inhibited from acting, menstruation occurs, and the embryo is sloughed off along with the uterine lining. Mifepristone occupies the endometrial receptor sites for progesterone and, in effect, blocks the access of progesterone to the endometrium. As a result, it initiates breakdown of the uterine lining, detachment of the embryo, contractility of the myometrium, and softening and dilation of the cervix. All of these responses contribute to expulsion of the embryo and uterine lining (miscarriage). Mifepristone can be taken up to five weeks after conception. One side effect of the drug is uterine bleeding, which averages 11 days and ranges from 5 to 21 days. The drug is not yet available in the United States but is available in France and China. In 1988, Dutch scientists reported another drug used to terminate early pregnancy (during the first eight weeks) called ***epostane.*** It inhibits progesterone synthesis. The most frequent side effect following epostane use is nausea. Epostane is currently unavailable in the United States.

A summary of methods of birth control is presented in Exhibit 25-2.

MALE CONTRACEPTION

The quest for an efficient male oral contraceptive has generally been disappointing. However, research is continuing and the search is aimed at inhibiting hypothalamic–pituitary function, directly inhibiting spermatogenesis, or inhibiting epididymal function.

Injections of long-acting testosterone suppresses secretion of gonadotropin-releasing hormone (GnRH), in turn inhibiting the release of follicle-stimulating hormone (FSH), which initiates spermatogenesis, and luteinizing hormone (LH), which helps develop mature spermatozoa. Unfortunately, this approach has proved successful in only half the subjects tested.

In males given long-acting analogues (modified compounds with similar structure) of luteinizing hormone-releasing hormone (LHRH), gonadotropin levels decrease with a resultant decrease in testosterone levels and associated decreases in sperm count and motility. Although combining testosterone injections with LHRH antagonists results in returning testosterone levels to normal, success has been inconsistent in totally inhibiting spermatogenesis.

Recently, a drug called *pyrimethamine,* which is used clinically to control malaria, has been given to mice as a possible male contraceptive. In studies to date it has been shown to cause arrest of spermatogenesis and result in male infertility, which is reversible.

Finally, alteration of epididymal function as an approach toward male contraception has become an important area of research. Certain drugs, such as *sulfasalazine,* probably alter epididymal function by interfering with normal carbohydrate metabolism in the epididymal epithelium, thereby impairing sperm maturation. An advantage to selective inhibition of epididymal function is that there is no risk of disrupting testicular function, and hence there is a greater possibility of finding a reversible procedure and there are minimal adverse genetic effects.

AGING AND THE REPRODUCTIVE SYSTEMS

Although there are major age-specific physical changes in structure and function, few age-specific disorders are associated with the reproductive system. In the male, the decreasing production of testosterone produces less muscle strength, fewer viable sperm, and decreased sexual desire. However, abundant spermatozoa may be found even in old age. Most of the age-dependent pathologies do not affect general health, except for prostate problems that could become serious and fatal.

The female reproductive system has a time-limited span of fertility between the first menstrual cycle and menopause. The system demonstrates an age-dependent decline in fertility, possibly as a result of less frequent ovulation and the declining ability of the uterine (Fallopian) tubes and uterus to support the young embryo. There is a decrease in the production of progesterone and estrogens. Menopause is only one of a series of phases that leads to reduced fertility, irregular or absent menstruation, and a variety of physical changes. Uterine cancer peaks at about 65 years of age, but cervical cancer is more common in younger women, and breast cancer is the leading cause of death among women between the ages of 40 and 60. Prolapse (falling down or sinking) of the uterus is possibly the most common complaint among female geriatric patients.

DEVELOPMENTAL ANATOMY OF THE REPRODUCTIVE SYSTEMS

The *gonads* develop from the ***intermediate mesoderm.*** By the sixth week, they appear as bulges that protrude into the ventral body cavity (Figure 25-25). The gonads develop near the mesonephric ducts. A second pair of ducts, the ***paramesonephric (Müller's) ducts,*** develop lateral to the mesonephric ducts. Both sets of ducts empty into the urogenital sinus. An embryo contains primitive gonads for both sexes. Differentiation into a male depends on the presence of a gene on the Y chromosome called *testis-determining factor (TDF)* and the release of testosterone, whereas

EXHIBIT 25-2

Summary of Birth Control (BC) Methods

METHOD	COMMENTS
Sterilization	Procedure involving severing ductus (vas) deferens in males (vasectomy) and uterine (Fallopian) tubes in females (tubal ligation and laparoscopy). Failure rate is less than 1 percent.[a]
Hormonal	Except for total abstinence or surgical sterilization, oral contraception is the most effective contraceptive known. Side effects include nausea, occasional light bleeding between periods, breast tenderness or enlargement, fluid retention, and weight gain. Should not be used by women who have cardiovascular conditions (thromboembolic disorders, cerebrovascular disease, heart disease, hypertension), liver malfunction, cancer or neoplasia of breast or reproductive organs, or by women who smoke. Pill users may have an increased risk of infertility. Failure rate is about 2 percent.
Intrauterine Device (IUD)	Small object (loop, coil, T, or 7) made of plastic, copper, or stainless steel and inserted into uterus by physician. May be left in place for long periods of time (some must be changed every two to three years). Does not require continued attention by user. Some women cannot use them because of expulsion, bleeding, or discomfort. Not recommended for women who have not had children because uterus is too small and cervical canal too narrow. Use of IUDs has diminished because of problems experienced by some females such as pelvic inflammatory disease (PID) and infertility. Para Gard model T380A is now available in the United States, but a consent form indicating the potential risks must be initialed or signed. Failure rate is about 5 percent.
Barrier	A condom is a thin, strong sheath of rubber or similar material worn by male to prevent sperm from entering vagina. Failures caused by sheath tearing or slipping off after climax or not putting the sheath on soon enough. If used correctly and consistently, especially with a spermicide, their effectiveness is similar to that of diaphragm. Failure rate is about 10 percent.
	A diaphragm is a flexible rubber dome inserted into vagina to cover cervix, providing barrier to sperm. Usually used with spermicidal cream or jelly. Must be left in place at least 6 hours after intercourse and may be left in place as long as 24 hours. Must be fitted by physician or other trained personnel and refitted every two years and after each pregnancy. Offers high level of protection if used with spermicide. Occasional failures caused by improper insertion or displacement during sexual intercourse. Failure rate is about 20 percent.
	A cervical cap is a thimble-shaped latex device that fits snugly over the cervix of the uterus. Used with a spermicide, it must be fitted by a physician or other trained personnel. May be left in place for up to 48 hours, and it is not necessary to reintroduce spermicide before sexual intercourse. Prescribed for females with normal Pap smears, and users should undergo a follow-up after three months. Failure rate is about 10 to 20 percent.
Chemical	Sperm-killing chemicals are inserted into vagina to coat vaginal surfaces and cervical opening. Provides protection for about one hour. Effective when used alone but significantly more effective when used with diaphragm or condom. The contraceptive sponge is made of polyurethane and releases a spermicide for up to 24 hours. A few cases of toxic shock syndrome (TSS) have been reported among users of the contraceptive sponge. Failure rate is about 20 percent.
Physiologic	In the rhythm method, sexual intercourse is avoided just before and just after ovulation (about seven days). Failure rate is about 20 percent even in females with regular menses.
	In the sympto-thermal method, signs of ovulation are noted (increased basal body temperature, clear and stretchy cervical mucus, opening of the external os, elevation and softening of the cervix, abundant cervical mucus, and pain associated with ovulation), and sexual intercourse is avoided. Failure rate is about 20 percent.
Coitus Interruptus	Withdrawal of penis from vagina before ejaculation occurs. Failure rate is 20 to 25 percent.
Induced Abortion	Involves vacuum aspiration (suction), saline solution, surgical evacuation (scraping) to remove prematurely products of conception, or drugs such as mifepristone (RU486) and epostane.

[a] Failure rates are based on estimated pregnancy rates (percentage) in the first year of use, assuming variations in consistency of use.

differentiation into a female depends on the absence of TDF and the absence of testosterone. At about the seventh week, the gonads are clearly differentiated into ovaries or testes.

In the male embryo, the *testes* connect to the mesonephric duct through a series of tubules that become the *seminiferous tubules*. Continued development of the mesonephric ducts produces the *efferent ducts, ductus epididymis, ductus (vas) deferens, ejaculatory ducts,* and *seminal vesicle*. The *prostate* and *bulbourethral (Cowper's) glands* are ***endodermal*** outgrowths of the urethra. Shortly after the gonads differentiate into testes, the paramesonephric ducts degenerate without

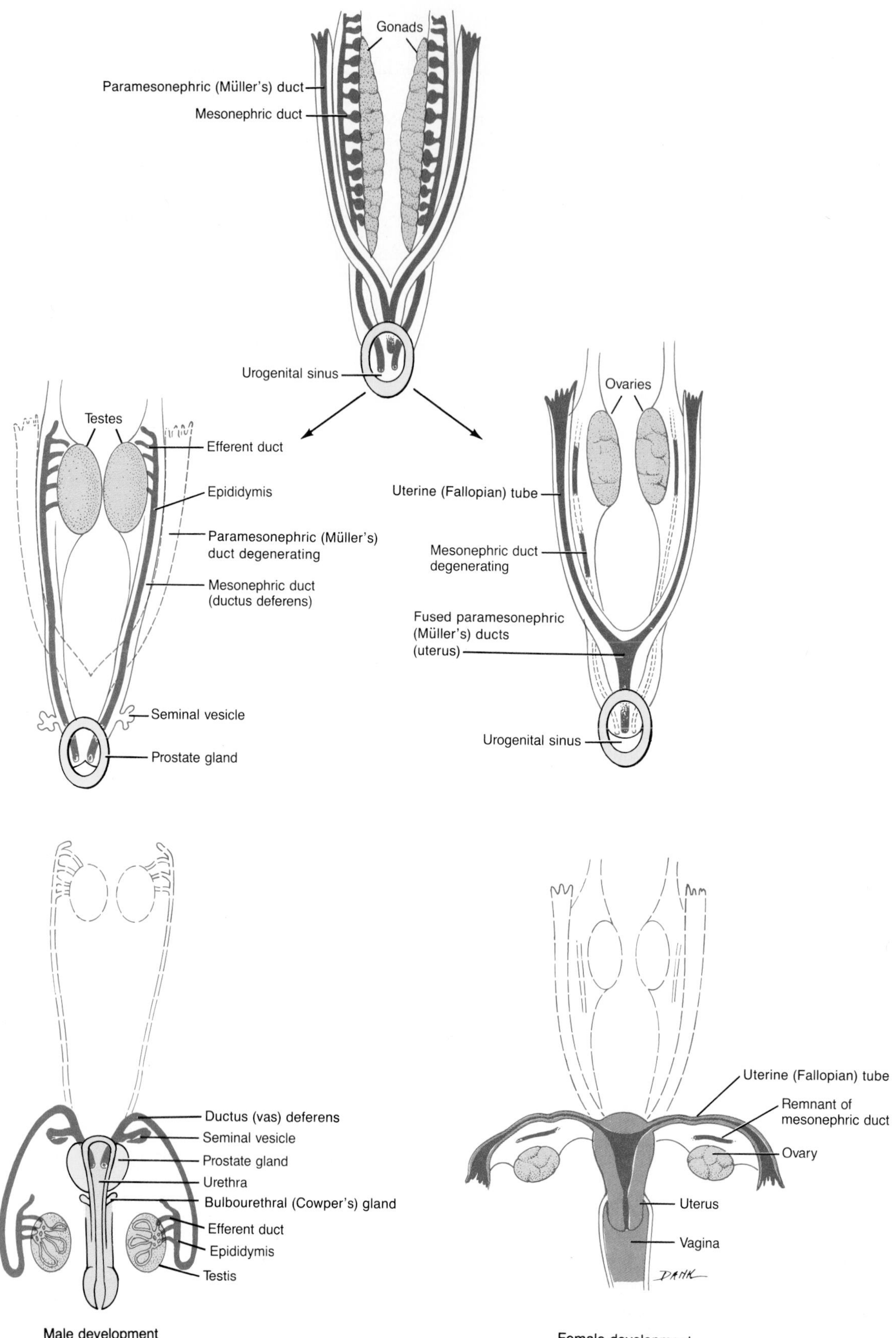

FIGURE 25-25 Development of the internal reproductive systems.

contributing any functional structures to the male reproductive system.

In the female embryo, the gonads develop into *ovaries*. At about the same time, the distal ends of the paramesonephric ducts fuse to form the *uterus* and *vagina*. The unfused portions become the *uterine (Fallopian) tubes*. The *greater (Bartholin's)* and *lesser vestibular glands* develop from ***endodermal*** outgrowths of the vestibule. The mesonephric ducts in the female degenerate without contributing any functional structures to the female reproductive system.

The *external genitals* of both male and female embryos also remain undifferentiated until about the eighth week. Before differentiation, all embryos have an elevated region, the ***genital tubercle,*** a point between the tail (future coccyx) and the umbilical cord, where the mesonephric and paramesonephric ducts open to the exterior (Figure 25-26). The tubercle consists of a ***urethral groove*** (opening into the urogenital sinus), paired ***urethral folds,*** and paired ***labioscrotal swellings.***

In the male embryo, the genital tubercle elongates and develops into a *penis*. Fusion of the urethral folds forms the *spongy (cavernous) urethra* and leaves an opening only at the distal end of the penis, the *external urethral orifice*. The labioscrotal swellings develop into the *scrotum*. In the female, the genital tubercle gives rise to the *clitoris*. The urethral folds remain open as the *labia minora,* and the labioscrotal swellings become the *labia majora*. The urethral groove becomes the *vestibule*.

APPLICATIONS TO HEALTH

SEXUALLY TRANSMITTED DISEASES (STDs)

The general term ***sexually transmitted disease (STD)*** is applied to any of the large group of diseases that can be spread by sexual contact. The group includes conditions traditionally specified as ***venereal diseases (VD),*** from Venus, goddess of love, such as gonorrhea, syphilis, and genital herpes, and several other conditions that are contracted sexually, or may be contracted otherwise, but are then transmitted to a sexual partner. AIDS, which is an STD but may also be contracted in other ways, has already been discussed in Chapter 15.

Gonorrhea

Gonorrhea (or ***"clap"***) is an infectious sexually transmitted disease that affects primarily the mucous membrane of the urogenital tract, the rectum, and occasionally the eyes. The disease is caused by the bacterium *Neisseria gonorrhoeae*. In the United States, nearly 1 million new cases of gonorrhea are reported annually. However, the actual number is probably between 3 to 4 million. Most cases are in the 15- to 24-year-old age group. Discharges from the involved mucous membranes are the source of infection, and the bacteria are transmitted by direct contact, usually sexual or during passage of a newborn through the birth canal.

Males usually suffer inflammation of the urethra with pus and painful urination. Fibrosis sometimes occurs in an advanced stage, causing stricture of the urethra. There also may be involvement of the epididymis and prostate gland. In females, infection may occur in the urethra, vagina, and cervix, and there may be a discharge of pus. However, infected females often harbor the disease without any symptoms until it has progressed to a more advanced stage. If the uterine (Fallopian) tubes become involved, pelvic inflammation may follow. Peritonitis, or inflammation of the peritoneum, is a very serious disorder. The infection should be treated and controlled immediately because, if neglected, sterility or death may result. Although antibiotics have greatly reduced the mortality rate of acute peritonitis, it is estimated that between 50,000 and 80,000 women are made sterile by gonorrhea every year as a result of scar tissue formation that closes the uterine tubes. If the bacteria are transmitted to the eyes of a newborn in the birth canal, blindness can result.

Administration of a 1 percent silver nitrate solution or penicillin in the infant's eyes prevents infection. Penicillin or tetracycline are the drugs of choice for the treatment of gonorrhea in adults. However, the incidence of antibiotic-resistant gonorrhea has increased substantially since 1984.

Syphilis

Syphilis is a sexually transmitted disease caused by the bacterium *Treponema pallidum*. In the United States, it affects about 85,000 persons per year. Although the incidence among male homosexuals is 25 to 30 percent higher than in the general population, significant increases in reported cases among heterosexuals have been observed since 1985. The highest incidence is in the 20- to 39-year-old age group. It is acquired through sexual contact or transmitted through the placenta to a fetus. The disease progresses through several stages: primary, secondary, latent, and sometimes tertiary. During the ***primary stage,*** the chief symptom is an open sore, called a ***chancre*** (pronounced SHANKG-ker), at the point of contact. The chancre heals within one to five weeks. From 6 to 24 weeks later, symptoms such as a skin rash, fever, and aches in the joints and muscles usher in the ***secondary stage.*** These symptoms also eventually disappear (in about 4 to 12 weeks), and the disease ceases to be infectious, but a blood test for the presence of the bacteria generally remains positive. During this "symptomless" period, called the ***latent stage,*** the bacteria may invade body organs. When signs of organ degeneration appear, the disease is said to be in the ***tertiary stage.***

If the syphilis bacteria attack the organs of the nervous system, the tertiary stage is called ***neurosyphilis.*** Neurosyphilis may take different forms, depending on the tissue

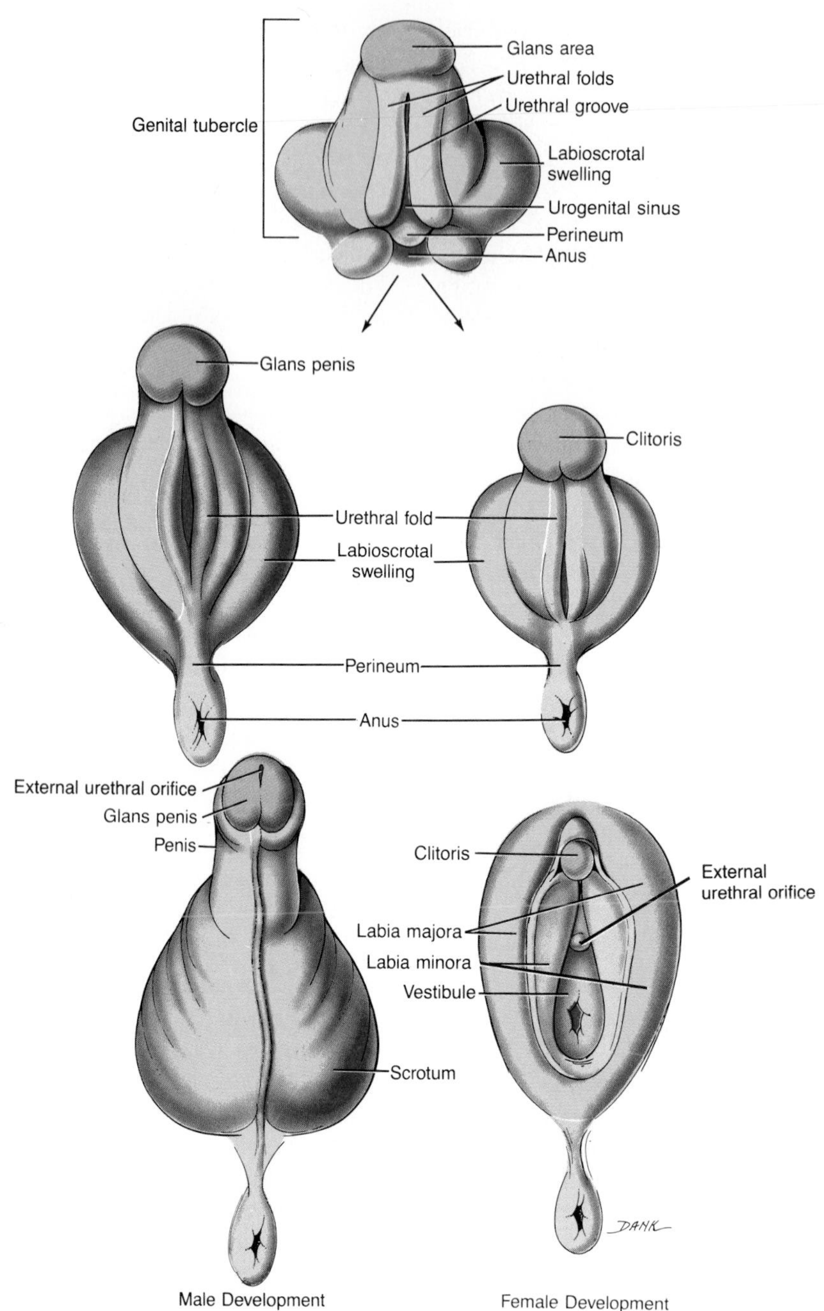

FIGURE 25-26 Development of the external genitals.

involved. For instance, about two years after the onset of the disease, the bacteria may attack the meninges, producing meningitis. The blood vessels that supply the brain may also become infected. In this case, symptoms depend on the parts of the brain destroyed by oxygen and glucose starvation. Cerebellar damage is manifested by uncoordinated movements in such activities as writing. As the motor areas become extensively damaged, victims may be unable to control urine and bowel movements. Eventually, they may become bedridden, unable even to feed themselves. Damage to the cerebral cortex produces memory loss and personality changes that range from irritability to hallucinations.

Infection of the fetus with syphilis can occur after the fifth month. Infection of the mother is not necessarily followed by fetal infection, provided that the placenta remains intact. But once the bacteria gain access to fetal circulation, there is nothing to impede their growth and multiplication. As many as 80 percent of children born to untreated syphilitic mothers will be infected in the uterus if the fetus is exposed at the onset or in the early stages of the disease. About 25 percent of the fetuses will die within the uterus. Most of the survivors will arrive prematurely, but 30 percent will die shortly after birth. Of the infected and untreated children surviving infancy, about 40 percent will develop symptomatic syphilis during their lifetimes.

Syphilis can be treated with antibiotics (penicillin) during the primary, secondary, and latent periods. Certain forms of neurosyphilis may also be successfully treated, but the prognosis for others is very poor. Noticeable symptoms do not always appear during the first two stages of the disease. Syphilis, however, is usually diagnosed through a blood test whether noticeable symptoms appear or not. The importance of these blood tests and follow-up treatments cannot be overemphasized.

Some evidence suggests that AIDS may alter the course of neurosyphilis by accelerating its progression, possibly by impairing macrophages and antibody production.

Genital Herpes

Another sexually transmitted disease, ***genital herpes,*** is common in the United States. Each year, between 400,000 and 600,000 new cases are reported. The sexual transmission of the herpes simplex virus is well established. Unlike syphilis and gonorrhea, genital herpes is incurable. Type I herpes simplex virus is the virus that causes the majority of infections above the waist, such as cold sores. Type II herpes simplex virus causes most infections below the waist, such as painful genital blisters on the prepuce, glans penis, and penile shaft in males and on the vulva or sometimes high up in the vagina in females. The blisters disappear and reappear in most patients, but the virus itself remains in the body.

Genital herpes virus infection causes considerable discomfort, and there is an extraordinarily high rate of recurrence of the symptoms. The infection is usually characterized by fever, chills, flulike symptoms, lymphadenopathy, and numerous clusters of genital blisters. For pregnant women with genital herpes symptoms at the time of delivery, a cesarean section will usually prevent complications in the child. Complications range from a mild asymptomatic infection to central nervous system damage to death.

Treatment of the symptoms involves pain medication, saline compresses, sexual abstinence for the duration of the eruption, and use of an oral drug called acyclovir (Zovirax). This drug interferes with viral DNA replication but not with host cell DNA replication. Acyclovir speeds the healing and sometimes reduces the pain of initial genital herpes infections and shortens the duration of lesions in patients with recurrent genital herpes. A topically applied ointment that contains Inter Vir-A (Immuvir), an antiviral substance, is another drug used to treat genital herpes. Inter Vir-A provides rapid relief for the pain, itching, and burning associated with genital herpes. An experimental genital herpes vaccine will involve human testing shortly.

Chlamydia

Chlamydia (kla-MID-ē-a) is a sexually transmitted disease caused by the bacterium *Chlamydia trachomatis* (*chlamys* = cloak; the bacterium cannot grow outside the body; it cloaks itself inside cells to divide). At present, chlamydia is the most prevalent and one of the most damaging of the sexually transmitted diseases. It affects between 3 and 5 million persons annually. In males, urethritis is the principal result. It is characterized by burning on urination, frequency of urination, painful urination, and low back pain. In females, urethritis may spread through the reproductive tract and develop into inflammation of the uterine (Fallopian) tubes, which increases the risk of ectopic pregnancy and sterility. As in gonorrhea, the organism may be passed from mother to infant during childbirth, infecting the eyes. Treatment consists of the administration of tetracycline or doxycycline.

Trichomoniasis

The microorganism *Trichomonas vaginalis,* a flagellated protozoan (one-celled animal), causes ***trichomoniasis,*** an inflammation of the mucous membrane of the vagina in females and the urethra in males. *T. vaginalis* is a common inhabitant of the vagina of females and urethra of males. If the normal acidity of the vagina is disrupted, the protozoan may overgrow the normal microbial population and cause trichomoniasis. Symptoms include a yellow vaginal discharge with a particularly offensive odor and severe vaginal itch in women. Men can have it without symptoms but can transmit it to women nonetheless. Sexual partners must be treated simultaneously. The drug of choice is metronidazole.

Genital Warts

Warts are an infectious disease caused by viruses. Sexual transmission of ***genital warts*** is common and is caused by the human papilloma virus (HPV). It is estimated that nearly 1 million persons a year develop genital warts. Patients with a history of genital warts may be at increased risk for certain types of cancer (cervical, vaginal, anal, vulval, and penile). There is no cure for genital warts. Treatment consists of cryotherapy with liquid nitrogen, electrocautery, excision, laser surgery, and topical application of podophyllin in tincture of benzoin. Alpha interferon is also used to treat genital warts.

MALE DISORDERS

Testicular Cancer

Testicular cancer occurs most often between the ages of 15 and 34 and is one of the most common cancers seen in young males. Although the cause is unknown, the condition is associated with males with a history of undescended or late-descended testes. Most testicular cancers arise from the sperm-producing cells. An early sign of testicular cancer is a mass in the testis, often associated with pain or discomfort. Treatment involves removal of the diseased testis.

Prostate Disorders

The prostate gland is susceptible to infection, enlargement, and benign and malignant tumors. Because the prostate surrounds the urethra, any of these disorders can obstruct the flow of urine. Prolonged obstruction may result in serious changes in the urinary bladder, ureters, and kidneys and may perpetuate urinary tract infections. An experimental treatment that consists of widening a narrowed urethra with a balloon catheter (balloon urethroplasty) is undergoing clinical trials. If the obstruction cannot be relieved by other means, surgical removal of part of or the entire gland is indicated. The surgical procedure is called ***prostatectomy*** (pros′-ta-TEK-tō-mē).

Acute and chronic infections of the prostate gland are common in postpubescent males, often in association with inflammation of the urethra. In ***acute prostatitis,*** the prostate gland becomes swollen and tender. Appropriate antibiotic therapy, bed rest, and above-normal fluid intake are effective treatment.

Chronic prostatitis is one of the most common chronic infections in men of the middle and later years. On examination, the prostate gland feels enlarged, soft, and extremely tender, and its surface outline is irregular. This disease frequently produces no symptoms, but the prostate is believed to harbor infectious microorganisms responsible for some allergic conditions, arthritis, and inflammation of nerves (neuritis), muscles (myositis), and the iris (iritis).

An enlarged prostate gland, increasing to two to four times larger than normal, occurs in approximately one-third of all males over age 60. The condition is called ***benign prostatic hyperplasia*** (***BPH***) and is characterized by nocturia (bed-wetting), hesitancy in urination, decreased force of urinary stream, postvoiding dribbling, and a sensation of incomplete emptying. The condition may be corrected surgically by a procedure called ***transurethral resection of the prostate*** (***TURP***), in which pieces of the gland are removed using a special cystoscope inserted into the urethra. The enlarged condition usually can be detected by rectal examination.

Prostate cancer is the second leading cause of death from cancer in men in the United States, and it is responsible for approximately 26,000 deaths annually. Its incidence is related to age, race, occupation, geography, and ethnic origin. Both benign and malignant growths are common in elderly men. Both types of tumors put pressure on the urethra, making urination painful and difficult. At times, the excessive back pressure destroys kidney tissue and gives rise to an increased susceptibility to infection. Therefore, even when the tumor is benign, surgery is indicated. Prostate cancer may be detected by digital rectal examination and fine-needle aspiration. In addition, there is a new procedure called ***transrectal ultrasonography,*** wherein a rectal probe is used to bounce sound waves off the prostate gland. The returning echoes are converted into an image that can be viewed on a monitor and printed on paper. The procedure can detect tumors as small as a grain of rice.

Sexual Functional Abnormalities

Impotence (*impotenia* = lack of strength) is the inability of an adult male to attain or hold an erection long enough for sexual intercourse. Impotence could be the result of diabetes mellitus, physical abnormalities of the penis, systemic disorders such as syphilis, vascular disturbances (arterial or venous obstructions), neurological disorders, testosterone deficiency, drugs (alcohol, antidepressants, antihistamines, antihypertensives, narcotics, nicotine, and tranquilizers), or psychic factors such as fear of causing pregnancy, fear of sexually transmitted diseases, religious inhibitions, or emotional immaturity. In selected individuals, penile implants may be indicated. Penile injections of papaverine (Pavabid), a vasodilator, and phentolamine mesylate (Regitine), an alpha-adrenergic blocker, can produce excellent effects in overcoming both physical and psychological impotence.

Male infertility (***sterility***) is an inability to fertilize the ovum. It does not imply impotence. Male fertility requires production of adequate amounts of viable, normal spermatozoa by the testes, unobstructed transportation of sperm through the seminal tract, and satisfactory deposition in the vagina. The tubules of the testes are sensitive to many factors—x-rays, infections, toxins, malnutrition, and significantly higher-than-normal scrotal temperatures—that may cause degenerative changes and produce male sterility. If inadequate spermatozoa production is suspected, a sperm analysis should be performed. At least one type of infertility may be improved by administration of vitamin C.

FEMALE DISORDERS

Menstrual Abnormalities

Because menstruation reflects not only the health of the uterus but also the health of the endocrine glands that control it, the ovaries and the pituitary gland, disorders of the female reproductive system frequently involve menstrual disorders.

Amenorrhea (ā-men′-ō-RĒ-a; *a* = without; *men* = month; *rhein* = to flow) is the absence of menstruation. If a woman has never menstruated, the condition is called ***primary amenorrhea.*** Primary amenorrhea can be caused by endocrine disorders, most often in the pituitary gland and hypothalamus, or by a genetically caused abnormal development of the ovaries or uterus. ***Secondary amenorrhea,*** the skipping of one or more periods, is commonly experienced by women at some time during their lives. Changes in body weight, either gains or losses, often cause amenorrhea. Obesity may disturb ovarian function, and similarly, the extreme weight loss that characterizes anorexia nervosa often leads to a suspension of menstrual flow. When amenorrhea is unrelated to weight, analysis of levels of estrogens often reveals deficiencies of pituitary and ovarian hormones. Amenorrhea may also be caused by continuous involvement in rigorous athletic training (exercise-associated amenorrhea). This condition may be related to increased levels of glucocorticoids and decreased levels of estrogens, progesterone, and prolactin.

Dysmenorrhea (dis′-men-ō-RĒ-a; *dys* = difficult) refers to pain associated with menstruation and is usually reserved to describe an individual with menstrual symptoms that are severe enough to prevent her from functioning normally for one or more days each month. ***Primary dysmenorrhea*** is painful menstruation with no detectable organic disease. The pain of primary dysmenorrhea is thought to result from uterine contractions, probably associated with uterine muscle ischemia and prostaglandins produced by the uterus. Prostaglandins are known to stimulate uterine contractions, but they cannot do so in the presence of high levels of progesterone. As we have noted earlier, progesterone levels are high during the last half of the menstrual cycle. During this time, prostaglandins are apparently inhibited by progesterone from producing uterine contractions. However, if pregnancy does not occur, progesterone levels drop rapidly and prostaglandin production increases. This increase causes the uterus to contract and slough off its lining and may result in dysmenorrhea. In addition to pain, other signs and symptoms may include headache, nausea, diarrhea or constipation, and urinary frequency. Primary dysmenorrhea is less of a problem after pregnancy and vaginal delivery, perhaps because of enlargement of the endocervical canal. Drugs that inhibit prostaglandin synthesis (naproxen and ibuprofen) are used to treat primary dysmenorrhea.

Secondary dysmenorrhea is painful menstruation that is frequently associated with a pelvic pathology. Some cases are caused by uterine tumors, ovarian cysts, pelvic inflammatory disease (PID), endometriosis, and intrauterine devices (IUDs). Treatment is aimed at correction of the underlying cause.

Abnormal uterine bleeding includes menstruation of excessive duration or excessive amount, diminished menstrual flow, too frequent menstruation, intermenstrual bleeding, and postmenopausal bleeding. These abnormalities may be caused by disordered hormonal regulation, emotional factors, fibroid tumors of the uterus, or systemic diseases.

Premenstrual syndrome (PMS) is a term usually reserved for severe physical and emotional distress occurring late in the postovulatory phase of the menstrual cycle and sometimes overlapping with menstruation. Signs and symptoms usually increase in severity until the onset of menstruation and then dramatically disappear. Among the signs and symptoms are edema, weight gain, breast swelling and tenderness, abdominal distension, backache, joint pain, constipation, skin eruptions, fatigue and lethargy, greater need for sleep, depression or anxiety, irritability, mood swings, headache, poor coordination and clumsiness, and cravings for sweet or salty foods. The basic cause of PMS is unknown. Although PMS is related to the cyclic production of ovarian hormones, the symptoms are not directly due to changes in the levels of these hormones. Treatment is individualized, depending on the type and severity of symptoms, and may include dietary changes, exercise, over-the-counter drugs (aspirin or acetaminophen), psychoactive drugs (sedatives, tranquilizers, and antidepressants), prostaglandins (Ponstel), diuretics, hormone therapy (progesterone), and vitamin B_6. One approach to management of PMS provides medical, psychological, and social support that may include education of the patient and her family; elimination of fears or inappropriate beliefs regarding the menstrual cycle; alteration in coping style; change in life-style, occupation, or family relationships; and use of appropriate medications.

Toxic Shock Syndrome (TSS)

Toxic shock syndrome (TSS), first described in 1978, is primarily a disease of previously healthy, young, menstruating females who use tampons. It is also recognized in males, children, and nonmenstruating females. Clinically, TSS is characterized by high fever up to 40.6°C (105°F), sore throat or very tender mouth, headache, fatigue, irritability, muscle soreness and tenderness, conjunctivitis, diarrhea and vomiting, abdominal pain, vaginal irritation, and erythematous rash. Other symptoms include lethargy, unresponsiveness, memory loss, hypotension, peripheral vasoconstriction, respiratory distress syndrome, intravascular coagulation, decreased platelet count, renal failure, circulatory shock, and liver involvement.

Toxin-producing strains of the bacterium *Staphylococcus aureus* are necessary for development of the disease. Actually, it appears that a virus has become incorporated into *S. aureus,* causing the bacterium to produce the toxins. Although all tampon users are at some risk for developing TSS, the risk is increased considerably by females who use highly absorbent tampons. Apparently, high-absorbency tampons provide a substrate on which the bacteria grow and produce toxins. There is some evidence that the absorption of magnesium by the fibers of the tampon is a factor in toxic shock syndrome. As the metal is absorbed, its absence slows microbial growth in the reproductive tract but stimulates the staphylococci to produce toxin. TSS can also occur as a complication of influenza and influenzalike illness and use of contraceptive sponges. Initial therapy is directed at correcting all homeostatic imbalances as quickly as possible. Antistaphylococcal antibiotics, such as penicillin or clindamycin, are also administered. In severe cases, high doses of corticosteroids are also administered.

Read

Ovarian Cysts

Ovarian cysts are fluid-containing tumors of the ovary. Follicular cysts may occur in the ovaries of elderly women, in ovaries that have inflammatory diseases, and in menstruating females. They have thin walls and contain a serous albuminous material. Cysts may also arise from the corpus luteum or the endometrium.

Endometriosis

Endometriosis (en′-dō-mē-trē-Ō-sis; *endo* = within; *metri* = uterus; *osis* = condition) is a benign condition characterized by the growth of endometrial tissue outside the uterus. The tissue enters the pelvic cavity via the open uterine (Fallopian) tubes and may be found in any of several sites—on the ovaries, rectouterine pouch, surface of the uterus, sigmoid colon, pelvic and abdominal lymph nodes, cervix, abdominal wall, kidneys, and urinary bladder. One theory for the development of endometriosis is that there is regurgitation of menstrual flow through the uterine tubes. Another theory is that migrational events during embryonic development are somehow altered. Endometriosis is common in women 25 to 40 years of age who have not had children. Symptoms include premenstrual pain or unusual menstrual pain. The unusual pain is caused by the displaced tissue sloughing off at the same time the normal uterine endometrium is being shed during menstruation. Infertility can be a consequence. Treatment usually consists of hormone therapy, modified GnRH (nafarelin), videolaseroscopy (laparoscope with camera and laser), or conventional surgery. Endometriosis disappears at menopause or when the ovaries are removed.

Female Infertility

Female infertility, or the inability to conceive, occurs in about 10 percent of married females in the United States. Once it is established that ovulation occurs regularly, the reproductive tract is examined for functional and anatomical disorders to determine the possibility of union of the sperm and the ovum in the uterine tube. Female infertility may be caused by tubal obstruction, ovarian disease, or certain conditions of the uterus. An upset in hormone balance, so that the endometrium is not adequately prepared to receive the fertilized ovum, may also be the problem. Some research suggests that an autoimmune disease might underlie many cases of infertility. Infertility treatment may involve the use of fertility drugs, donor (artificial) insemination, or surgery. Gynecologists are now using a procedure called ***transcervical balloon tuboplasty*** to clear obstructions in the uterine tubes. The technique, borrowed from the cardiology procedure to unclog coronary arteries, consists of inserting a catheter through the cervix of the uterus and into the uterine tube. Then a balloon is inflated, compressing the obstruction.

Disorders Involving the Breasts

The breasts of females are highly susceptible to cysts and tumors. Men are also susceptible to breast tumors, but certain breast cancers are 100 times more common in women.

In the female, the benign ***fibroadenoma*** is a common tumor of the breast. It occurs most frequently in young women. Fibroadenomas have a firm rubbery consistency and are easily moved about within the mammary tissue. The usual treatment is excision of the growth. The breast itself is not removed.

Breast cancer has one of the highest fatality rates of all cancers affecting women, but it is rare in men. In the female, breast cancer is rarely seen before age 30, and its occurrence rises rapidly after menopause. Breast cancer is generally not painful until it becomes quite advanced, so often it is not discovered early or, if noted, is ignored. Any lump, no matter how small, should be reported to a doctor at once. New evidence links some breast cancers to loss of protective anti-oncogenes.

Cervical Cancer

Another common disorder of the female reproductive tract is ***cervical cancer,*** carcinoma of the cervix of the uterus. The condition starts with ***cervical dysplasia*** (dis-PLĀ-sē-a), a change in the shape, growth, and number of the cervical cells. If the condition is minimal, the cells may regress to normal. If it is severe, it may progress to cancer. Cervical cancer may be detected in most cases in its earliest stages by a Pap smear. There is some evidence linking cervical cancer to penile virus (papillomavirus) infections of male sexual partners. Depending on the progress of the disease, treatment may consist of excision of lesions, radiotherapy, chemotherapy, and hysterectomy.

Pelvic Inflammatory Disease (PID)

Pelvic inflammatory disease (PID) is a collective term for any extensive bacterial infection (primarily involving *Chlamydia trachomatis, Neisseria gonorrhoeae, Bacteroides, Peptostreptococcus,* and *Gardnerella vaginalis*) of the pelvic organs, especially the uterus, uterine (Fallopian) tubes, or ovaries. A vaginal or uterine infection may spread into the uterine tube (***salpingitis***) or even farther into the abdominal cavity, where it infects the peritoneum (***peritonitis***). Diagnosis of PID depends on three findings: abdominal tenderness; cervical tenderness; and ovarian, uterine tube, and uterine ligament tenderness. In addition, diagnosis is based on at least one of the following: fever, leukocytosis, pelvic abscess or inflammation, purulent cervical discharge, and the presence of certain bacteria in smears. Early treatment with bed rest and antibiotics (cefoxitin, penicillin, tetracycline, doxycycline) can stop the spread of PID.

Vulvovaginal Candidiasis

Candida albicans is a yeastlike fungus that commonly grows on mucous membranes of the gastrointestinal and genitourinary tracts. The organism is responsible for ***vulvovaginal candidiasis,*** the most common form of vaginitis. It is characterized by severe itching; a thick, yellow, cheesy discharge; a yeasty odor; and pain. The disorder, experienced at least once by about 75 percent of females, is usually a result of proliferation of the fungus following antibiotic therapy for another condition. Predisposing conditions include use of oral contraceptives, cortisonelike medications, pregnancy, and diabetes. Treatment is by topical (clostrinazole) or oral (ketoconazole) antibiotics.

KEY MEDICAL TERMS ASSOCIATED WITH THE REPRODUCTIVE SYSTEMS

Castration (kas-TRĀ-shun; *castrare* = to prune) Removal, inactivation, or destruction of the testes.

Colposcopy (kol-POS-ko-pē) Direct examination of the cervical and vaginal mucosa with a magnifying device (colposcope).

Colpotomy (kol-POT-ō-mē; *colp* = vagina; *tome* = cutting) Incision of the vagina.

Culdoscopy (kul-DOS-kō-pē; *skopein* = to examine) A procedure in which a culdoscope (endoscope) is used to view the pelvic cavity. The approach is through the vagina.

Culdotomy (kul-DOT-ō-mē; *tome* = cutting) An incision into (or needle aspiration of) the cul-de-sac to test for the presence of pelvic bleeding.

Hermaphroditism (her-MAF-rō-di-tizm') Presence of both male and female sex organs in one individual.

Hypospadias (hī-pō-SPĀ-dē-as; *hypo* = below; *span* = to draw) A displaced urethral opening. In the male, the opening may be on the underside of the penis, at the penoscrotal junction, between the scrotal folds, or in the perineum. In the female, the urethra opens into the vagina.

Leukorrhea (loo'-kō-RĒ-a; *leuco* = white; *rrhea* = discharge) A nonbloody vaginal discharge that may occur at any age and affects most women at some time.

Oophorectomy (ō-of-ō-REK-tō-mē; *oophoro* = bearing eggs) Removal of the ovaries.

Salpingectomy (sal'-pin-JEK-tō-mē; *salpingo* = tube) Excision of a uterine (Fallopian) tube.

Smegma (SMEG-ma; *smegma* = soap) The secretion, consisting principally of desquamated epithelial cells, found chiefly about the external genitalia and especially under the foreskin of the male.

Vaginitis (vaj'-i'-NĪ-tis) Inflammation of the vagina.

STUDY OUTLINE

Male Reproductive System (p. 718)

1. Reproduction is the process by which genetic material is passed on from one generation to the next.
2. The organs of reproduction are grouped as gonads (produce gametes), ducts (transport and store gametes), and accessory sex glands (produce materials that support gametes).
3. The male structures of reproduction include the testes, ductus epididymis, ductus (vas) deferens, ejaculatory duct, urethra, seminal vesicles, prostate gland, bulbourethral (Cowper's) glands, and penis.

Scrotum (p. 718)

1. The scrotum is a cutaneous outpouching of the abdomen that supports the testes.
2. It regulates the temperature of the testes by contraction of the cremaster muscle, which elevates them and brings them closer to the pelvic cavity or relaxes causing testes to move farther from the pelvic cavity.

Testes (p. 718)

1. The testes are oval-shaped glands (gonads) in the scrotum containing seminiferous tubules, in which sperm cells are made; sustentacular (Sertoli) cells, which nourish sperm cells and secrete inhibin; and interstitial endocrinocytes (cells of Leydig), which produce the male sex hormone testosterone.
2. Failure of the testes to descend is called cryptorchidism.
3. Ova and sperm are collectively called gametes, or sex cells, and are produced in gonads.
4. Uninucleated somatic cells divide by mitosis, the process in which each daughter cell receives the full complement of 23 chromosome pairs (46 chromosomes). Somatic cells are said to be diploid ($2n$).
5. Immature gametes divide by meiosis, in which the pairs of chromosomes are split so that the mature gamete has only 23 chromosomes. It is said to be haploid (n).
6. Spermatogenesis occurs in the testes. It results in the formation of four haploid spermatozoa from each primary spermatocyte.
7. Spermatogenesis is a process in which immature spermatogonia develop into mature spermatozoa. The spermatogenesis sequence includes reduction division (meiosis I), equatorial division (meiosis II), and spermiogenesis.
8. Mature spermatozoa consist of a head, midpiece, and tail. Their function is to fertilize an ovum.

Ducts (p. 725)

1. The duct system of the testes includes the seminiferous tubules, straight tubules, and rete testis.
2. Sperm are transported out of the testes through the efferent ducts.
3. The ductus epididymis is lined by stereocilia and is the site of sperm maturation and storage.

4. The ductus (vas) deferens stores sperm and propels them toward the urethra during ejaculation.
5. Alteration of the ductus (vas) deferens to prevent fertilization is called vasectomy.
6. The ejaculatory ducts are formed by the union of the ducts from the seminal vesicles and ductus (vas) deferens and eject spermatozoa into the prostatic urethra.
7. The male urethra is subdivided into three portions: prostatic, membranous, and spongy (cavernous).

Accessory Sex Glands (p. 730)

1. The seminal vesicles secrete an alkaline, viscous fluid that constitutes about 60 percent of the volume of semen and contributes to sperm viability.
2. The prostate gland secretes a slightly alkaline fluid that constitutes about 13 to 33 percent of the volume of semen and contributes to sperm motility.
3. The bulbourethral (Cowper's) gland secretes mucus for lubrication and a substance that neutralizes acid.

Semen (p. 731)

1. Semen (seminal fluid) is a mixture of spermatozoa and accessory sex gland secretions that provides the fluid in which spermatozoa are transported, provides nutrients, and neutralizes the acidity of the male urethra and female vagina.

Penis (p. 731)

1. The penis is the male organ of copulation that consists of a root, body, and glans penis.
2. Expansion of its blood sinuses under the influence of sexual excitation is called erection.

Female Reproductive System (p. 733)

1. The female organs of reproduction include the ovaries (gonads), uterine (Fallopian) tubes, uterus, vagina, and vulva.
2. The mammary glands are considered part of the reproductive system.

Ovaries (p. 734)

1. The ovaries are female gonads located in the upper pelvic cavity, on either side of the uterus.
2. They produce secondary oocytes, discharge secondary oocytes (ovulation), and secrete estrogens, progesterone, inhibin, and relaxin.
3. Oogenesis occurs in the ovaries. It results in the formation of a single haploid secondary oocyte.
4. The oogenesis sequence includes reduction division (meiosis I), equatorial division (meiosis II), and maturation.

Uterine (Fallopian) Tubes (p. 739)

1. The uterine (Fallopian) tubes transport ova from the ovaries to the uterus and are the normal sites of fertilization.
2. Ciliated cells and peristaltic contractions help move a secondary oocyte toward the uterus.

Uterus (p. 740)

1. The uterus is an organ shaped like an inverted pear that functions in transporting spermatozoa, menstruation, implantation of a fertilized ovum, development of a fetus during pregnancy, and labor.
2. The uterus is normally held in position by a series of ligaments.
3. Histologically, the uterus consists of an outer perimetrium, middle myometrium, and inner endometrium.

Endocrine Cycle (p. 742)

1. The function of the menstrual cycle is to prepare the endometrium each month for the reception of a fertilized egg.
2. The menstrual and ovarian cycles are controlled by GnRH, which stimulates the release of FSH and LH.
3. FSH stimulates the initial development of ovarian follicles and secretion of estrogens by the ovaries. LH stimulates further development of ovarian follicles, ovulation, and the secretion of estrogens and progesterone by the ovaries.
4. Estrogens stimulate the growth, development, and maintenance of female reproductive structures; stimulate the development of secondary sex characteristics; regulate fluid and electrolyte balance; and stimulate protein anabolism.
5. Progesterone works with estrogens to prepare the endometrium for implantation and the mammary glands for milk secretion.
6. Relaxin relaxes the pubic symphysis and helps dilate the uterine cervix to facilitate delivery, and increases sperm motility.
7. During the menstrual phase, the stratum functionalis layer of the endometrium is shed with a discharge of blood, tissue fluid, mucus, and epithelial cells. Primary follicles develop into secondary follicles.
8. During the preovulatory phase, endometrial repair occurs. A secondary follicle develops into a vesicular (Graafian) follicle. Estrogens are the dominant ovarian hormones.
9. Ovulation is the rupture of a vesicular (Graafian) follicle and the release of a secondary oocyte into the pelvic cavity brought about by inhibition of FSH and a surge of LH. Signs of ovulation include increased basal body temperature; clear, stretchy cervical mucus; changes in the uterine cervix, and ovarian pain.
10. During the postovulatory phase, the endometrium thickens in anticipation of implantation. Progesterone is the dominant ovarian hormone.
11. If fertilization and implantation do not occur, the corpus luteum degenerates, and low levels of estrogens and progesterone initiate another menstrual and ovarian cycle.
12. If fertilization and implantation do occur, the corpus luteum is maintained by placental hCG, and the corpus luteum and placenta secrete estrogens and progesterone to support pregnancy and breast development for lactation.
13. The female climacteric is the time immediately before menopause, the cessation of the sexual cycles.

Vagina (p. 746)

1. The vagina is a passageway for spermatozoa and the menstrual flow, the receptacle of the penis during sexual intercourse, and the lower portion of the birth canal.
2. It is capable of considerable distension to accomplish its functions.

Vulva (p. 747)

1. The vulva is a collective term for the external genitals of the female.
2. It consists of the mons pubis, labia majora, labia minora, clitoris, vestibule, vaginal and urethral orifices, hymen, bulb of the vestibule, paraurethral (Skene's), greater vestibular (Bartholin's), and lesser vestibular glands.

Perineum (p. 748)

1. The perineum is a diamond-shaped area at the inferior end of the trunk between the thighs and buttocks.
2. An incision in the perineal skin prior to delivery is called an episiotomy.

Mammary Glands (p. 749)

1. The mammary glands are modified sweat glands (branched tubuloalveolar) over the pectoralis major muscles. Their function is to secrete and eject milk (lactation).
2. Mammary gland development is dependent on estrogens and progesterone.
3. Milk secretion is mainly due to prolactin (PRL), and milk ejection is stimulated by oxytocin (OT).
4. Screening for breast cancer may involve mammography, ultrasound (US), and computed tomography combined with mammography (CT/M).

Birth Control (BC) (p. 752)

1. Methods include sterilization (vasectomy, tubal ligation), hormonal, intrauterine devices, barriers (condom, diaphragm, cervical cap), chemical (spermicides), physiologic (rhythm, sympto-thermal method), coitus interruptus, and induced abortion.
2. Contraceptive pills of the combination type contain estrogens and progesterone in concentrations that decrease the secretion of FSH and LH and thereby inhibit ovulation.
3. Development of a male contraceptive is aimed at inhibiting hypothalamic–pituitary function, directly inhibiting spermatogenesis, or inhibiting epididymal function.

Aging and the Reproductive Systems (p. 754)

1. In the male, decreased levels of testosterone decrease muscle strength, sexual desire, and viable sperm; prostate disorders are common.
2. In the female, levels of progesterone and estrogens decrease, resulting in changes in menstruation; uterine and breast cancer increase in incidence.

Developmental Anatomy of the Reproductive Systems (p. 754)

1. The gonads develop from intermediate mesoderm and are differentiated into ovaries or testes by about the seventh week of fetal development.
2. The external genitals develop from the genital tubercle.

Applications to Health (p. 757)

1. Sexually transmitted diseases (STDs) are diseases spread by sexual contact and include gonorrhea, syphilis, genital herpes, chlamydia, trichomoniasis, and genital warts.
2. Testicular cancer originates in sperm-producing cells.
3. Conditions that affect the prostate gland are prostatitis, benign prostatic hyperplasia (BPH), and cancer.
4. Impotence is the inability of the male to attain or hold an erection long enough for intercourse.
5. Male infertility is the inability of a male's sperm to fertilize an ovum.
6. Menstrual disorders include amenorrhea, dysmenorrhea, abnormal bleeding, and premenstrual syndrome (PMS).
7. Toxic shock syndrome (TSS) includes widespread homeostatic imbalances and is a reaction to toxins produced by *Staphylococcus aureus*.
8. Ovarian cysts are tumors that contain fluid.
9. Endometriosis refers to the growth of uterine tissue outside the uterus.
10. Female infertility is the inability of the female to conceive.
11. The mammary glands are susceptible to benign fibroadenomas and malignant tumors. The removal of a malignant breast, pectoral muscles, and lymph nodes is called a radical mastectomy.
12. Cervical cancer starts with dysplasia and can be diagnosed by a Pap test.
13. Pelvic inflammatory disease (PID) refers to bacterial infection of pelvic organs.
14. Vulvovaginal candidiasis is a form of vaginitis caused by the yeastlike fungus *Candida albicans*.

REVIEW QUESTIONS

1. Define reproduction. Describe how the reproductive organs are classified and list the male and female organs of reproduction. (p. 718)
2. Describe the function of the scrotum in protecting the testes from temperature fluctuations. (p. 718)
3. Describe the internal structure of a testis. Where are the sperm cells made? (p. 719)
4. Describe the principal events of spermatogenesis. Why is meiosis important? Distinguish between haploid (n) and diploid ($2n$) cells. (p. 723)
5. Identify the principal parts of a spermatozoon. List the functions of each. (p. 725)
6. Which ducts are involved in transporting sperm *within* the testes? (p. 725)
7. Describe the location, structure, and functions of the ductus epididymis, ductus (vas) deferens, and ejaculatory duct. (p. 726)
8. What is the spermatic cord? (p. 728)
9. Give the location of the three subdivisions of the male urethra. (p. 729)
10. Trace the course of spermatozoa through the male system of ducts from the seminiferous tubules through the urethra. (p. 725)
11. Briefly explain the locations and functions of the seminal vesicles, prostate gland, and bulbourethral (Cowper's) glands. How is cancer of the prostate gland detected by a blood test? (p. 730)
12. What is semen? What is its function? (p. 731)
13. How is the penis structurally adapted as an organ of copulation? How does an erection occur? (p. 731)
14. How are the ovaries held in position in the pelvic cavity? (p. 734)
15. Describe the microscopic structure of an ovary. What are the functions of the ovaries? (p. 734)
16. Describe the principal events of oogenesis. (p. 738)

17. Where are the uterine (Fallopian) tubes located? What is their function? (p. 739)
18. Diagram the principal parts of the uterus. (p. 740)
19. Describe the arrangement of ligaments that hold the uterus in its normal position. What is retroflexion? (p. 741)
20. Describe the histology of the uterus. (p. 741)
21. Discuss the blood supply to the uterus. Why is an abundant blood supply important? (p. 742)
22. What is the function of each of the following in the menstrual and ovarian cycles: GnRH, FSH, LH, estrogens, progesterone, and inhibin? (p. 742)
23. Briefly outline the major events of each phase of the menstrual cycle and correlate them with the events of the ovarian cycle. (p. 742)
24. What is the function of the vagina? Describe its histology. (p. 746)
25. List the parts of the vulva and the functions of each part. (p. 747)
26. Describe the structure of the mammary glands. How are they supported? (p. 749)
27. Describe the passage of milk from the clusters of alveolar cells of the mammary gland to the nipple. (p. 749)
28. How is breast cancer detected? How is it treated? (p. 750)
29. Briefly describe the various methods of birth control (BC) and the effectiveness of each. (p. 752)
30. Explain the effects of aging on the reproductive systems. (p. 754)
31. Describe the development of the reproductive systems. (p. 754)
32. Define a sexually transmitted disease (STD). Describe the cause, clinical symptoms, and treatment of gonorrhea, syphilis, genital herpes, chlamydia, trichomoniasis, and genital warts. (p. 757)
33. What is testicular cancer? (p. 759)
34. Describe several disorders that affect the prostate gland. (p. 760)
35. What are some of the causes of amenorrhea, dysmenorrhea, and abnormal uterine bleeding? (p. 760)
36. Describe the clinical symptoms of premenstrual syndrome (PMS) and toxic shock syndrome (TSS). (p. 761)
37. What are ovarian cysts? Define endometriosis. (p. 762)
38. What is cervical cancer? Relate the condition to cervical dysplasia. (p. 762)
39. Define pelvic inflammatory disease (PID) and vulvovaginal candidiasis. (p. 762)
40. Refer to the glossary of key medical terms at the end of the chapter. Be sure you can define each term. (p. 763)

SELF QUIZ

Complete the following:

1. Identify male homologues for each of the following:
   a. labia majora: ________
   b. clitoris: ________
   c. paraurethral (Skene's) glands: ________
   d. greater vestibular (Bartholin's) glands: ________
2. Gametes are (haploid? diploid?); human gametes contain ________ chromosomes. Fusion of gametes produces a cell called a ________. This cell, and all cells of the organism derived from it, contain the (haploid? diploid?) number of chromosomes, written as (n? $2n$?). These chromosomes, received from both sperm and ovum, are said to exist in ________ pairs.
3. Write E (endoderm) or M (mesoderm) next to the structures below to indicate their embryonic origins.
   __ **a.** ovaries and testes
   __ **b.** prostate and bulbourethral (Cowper's) glands
   __ **c.** greater (Bartholin's) vestibular and lesser vestibular glands
4. The process by which the testes produce spermatozoa involves meiosis and is referred to as ________.
5. Which of the following cells is *least* mature? (primary spermatocyte, spermatid, spermatogonium, secondary spermatocyte)
6. The most important ligaments in prevention of drooping (prolapse) of the uterus are the ________ ligaments, which are attached to the base of the uterus. Most of the uterus consists of ________-metrium. This layer is (smooth muscle? epithelium?). The innermost layer of the uterus is the ________-metrium. Which portion of it is shed during menstruation? Stratum (basalis? functionalis?). Which arteries supply the stratum functionalis? (spiral? straight?)
7. The two dorsolaterally located masses of tissue within the penis are called the ________ penis.
8. The ovary is attached to the uterus by the ________ ligament.
9. The rupture of the vesicular ovarian (Graafian) follicle with release of a secondary oocyte into the pelvic cavity is called ________.
10. If fertilization and implantation do not occur, the corpus luteum degenerates and becomes the ________.
11. The organ that serves as a passageway for the menstrual flow, the receptacle for the penis during copulation, and the lower portion of the birth canal is the ________.
12. Arrange the answers in correct sequence.
    __ __ __ **a.** From most external to deepest:
    A. endometrium
    B. serosa
    C. myometrium
    __ __ __ **b.** Portions of the urethra, from proximal (closest to urinary bladder) to distal (closest to outside of the body):
    A. prostatic
    B. membranous
    C. spongy
    __ __ __ **c.** From anterior to posterior:
    A. uterus and vagina
    B. urinary bladder and urethra
    C. rectum and anus

____ ____ ____ ____ **d.** From anterior to posterior:
A. anus
B. vaginal orifice
C. clitoris
D. urethral orifice

____ ____ ____ ____ ____ **e.** Pathway of sperm:
A. ejaculatory duct
B. testis
C. urethra
D. ductus (vas) deferens
E. epididymis

____ ____ ____ ____ ____ **f.** Pathway of milk in breasts:
A. lactiferous sinuses
B. alveoli
C. secondary tubules
D. mammary ducts
E. lactiferous ducts

____ ____ ____ ____ ____ ____ **g.** Pathway of sperm entering female reproductive system:
A. uterine cavity
B. cervical canal
C. external os
D. internal os
E. uterine (Fallopian) tube
F. vagina

____ ____ ____ **h.** Sequence involved in spermatogenesis:
A. spermiogenesis
B. reduction division
C. equatorial division

13. Match the following:

____ **a.** uses high concentrations of progesterone and low concentrations of estrogens
____ **b.** elastic covering placed over penis
____ **c.** signs of fertility and infertility are noted
____ **d.** penis is withdrawn prior to ejaculation
____ **e.** tubal ligation and vasectomy
____ **f.** small object is inserted into uterus
____ **g.** may involve vacuum aspiration and surgical evacuation

A. sterilization
B. hormonal contraception
C. IUD
D. condom
E. sympto-thermal method
F. coitus interruptus
G. induced abortion

Choose the one best answer to these questions.

____ **14.** The ovaries
(1) are female gonads.
(2) produce secondary oocytes and female hormones.
(3) are homologous to the testes.
(4) are attached to the pelvic wall by the suspensory ligament.
A. (1), (3), (4); B. (2), (3), (4); C. (1), (2), (4); D. (1) and (4) only; E. all of the above.

____ **15.** Which is *not* a component of the vulva?
A. labia majora; B. clitoris; C. greater vestibular (Bartholin's) glands; D. ovaries; E. labia minora.

____ **16.** Which of these structures is the male gonad?
A. seminal vesicle; B. scrotum; C. testis; D. prostate gland; E. penis.

____ **17.** Which of the following produce(s) a secretion that helps maintain the motility and viability of spermatozoa?
A. prostate; B. penis; C. bulbourethral (Cowper's) glands; D. ejaculatory duct; E. all of the above.

____ **18.** In the human male, the highly coiled duct in which sperm are stored for maturation before being released is called the
A. epididymis; B. urethra; C. seminal vesicle; D. ductus (vas) deferens; E. ejaculatory duct.

____ **19.** The site where fertilization normally occurs is
A. about one-third the way down the uterine (Fallopian) tube; B. the uterine wall, somewhere in the fundus; C. the cervix of the uterus; D. the abdominal cavity; E. the vagina.

____ **20.** The vagina
(1) is the anterior part of the uterus, sometimes called the cervix.
(2) is a thin-walled organ whose external opening lies between two folds of skin, the labia minora.
(3) is the site of the growth of the fetus; it is also called the womb.
(4) has its inner mucosa slough off each month as the menstrual flow.
A. (1) only; B. (2) only; C. (3) only; D. (4) only; E. (3) and (4).

____ **21.** As part of oogenesis,
(1) oogonia develop into primary oocytes.
(2) four haploid (n) secondary oocytes are produced from each oogonium.
(3) further development of primary follicles requires stimulation by FSH.
(4) equatorial division of a secondary oocyte is completed only if ovulation and fertilization occur.
A. (1), (2), (3); B. (2), (3), (4); C. (1), (3), (4); D. (1), (2), (4); E. all of the above.

22. Match the following:

____ **a.** 200–300 of these per testis
____ **b.** tightly coiled tubes (1–3 per lobule) composed of cells that develop into sperm
____ **c.** cells between developing sperm cells that form the blood–testis barrier and provide nourishment
____ **d.** cells located between seminiferous tubules that secrete testosterone

A. interstitial endocrinocytes (cells of Leydig)
B. lobule
C. sustentacular (Sertoli) cells
D. seminiferous tubules

Developmental Anatomy

26

STUDENT OBJECTIVES

1. Explain the activities associated with fertilization, morula formation, blastocyst development, and implantation.
2. Describe the various types of in vitro fertilization (IVF).
3. Discuss the formation of the primary germ layers, embryonic membranes, placenta, and umbilical cord as the principal events of the embryonic period.
4. List representative body structures produced by the primary germ layers.
5. Discuss the principal body changes associated with fetal growth.
6. Describe some of the anatomical and physiological changes associated with gestation.
7. Describe several prenatal diagnostic tests.
8. Explain the events associated with the three stages of labor.
9. Define key medical terms associated with developmental anatomy.

CHAPTER OUTLINE

■ **Development During Pregnancy**
Fertilization and Implantation
Fertilization
Formation of the Morula
Development of the Blastocyst
Implantation
In Vitro Fertilization (IVF)
■ **Embryonic Development**
Beginnings of Organ Systems
Embryonic Membranes
Placenta and Umbilical Cord
■ **Fetal Growth**
■ **Gestation**
Structural and Functional Changes
Exercise and Pregnancy
■ **Prenatal Diagnostic Tests**
Amniocentesis
Chorionic Villus Sampling (CVS)
Fetal Ultrasonography
Electronic Fetal Monitoring (EFM)
Alphafetoprotein (AFP) Test
■ **Parturition and Labor**
■ **Key Medical Terms Associated with Developmental Anatomy**

Developmental anatomy is the study of the sequence of events from the fertilization of a secondary oocyte to the formation of an adult organism. As we look at the sequence from fertilization to birth, we will consider fertilization, implantation, placental development, embryonic development, fetal growth, gestation, parturition, and labor. The sequential development of an embryo and fetus is an exceedingly complex series of events that are precisely coordinated and controlled.

DEVELOPMENT DURING PREGNANCY

Once spermatozoa and ova are developed through meiosis and maturation and the spermatozoa are deposited in the vagina, pregnancy can occur. ***Pregnancy*** is a sequence of events that normally includes fertilization, implantation, embryonic growth, and fetal growth that terminates in birth.

FERTILIZATION AND IMPLANTATION

Fertilization

Once a spermatozoon makes contact with a secondary oocyte, a process called ***syngamy,*** the spermatozoon penetrates the secondary oocyte, and their nuclei meet and fuse, a process called ***fertilization.*** Of the 300 to 500 million sperm cells introduced into the vagina, very few, perhaps only several hundred to several thousand, arrive in the vicinity of the oocyte. Fertilization normally occurs in the uterine (Fallopian) tube when the oocyte is about one-third of the way down the tube, usually within 24 hours after ovulation. Peristaltic contractions and the action of cilia transport the oocyte through the uterine tube. The mechanism by which sperm reach the uterine tube is apparently related to several factors. Sperm probably swim up the female tract by means of whiplike movements of their flagella. In addition, the acrosome of sperm produces an enzyme called ***acrosin*** that stimulates sperm motility and migration within the female reproductive tract.

In addition to assisting in the transport of sperm, the female reproductive tract also confers on sperm the capacity to fertilize a secondary oocyte. Although sperm undergo maturation in the epididymis, they are still not able to fertilize an oocyte until they have remained in the female reproductive tract for about 10 hours. The functional changes that sperm undergo in the female reproductive tract that allow them to fertilize a secondary oocyte are referred to as ***capacitation*** (ka′-pas′-i′-TĀ-shun). During this process, substances in secretions in the female reproductive tract cause the membrane around the acrosome to become fragile so that several enzymes—hyaluronidase, acrosin, and neuraminidase—are released by the acrosomes. The enzymes help sperm penetrate the ***corona radiata,*** several layers of follicular cells around the oocyte, and a gelatinous glycoprotein layer internal to the corona radiata called the ***zona pellucida*** (pe-LOO-si-da) (Figure 26-1a). Spermatozoa bind

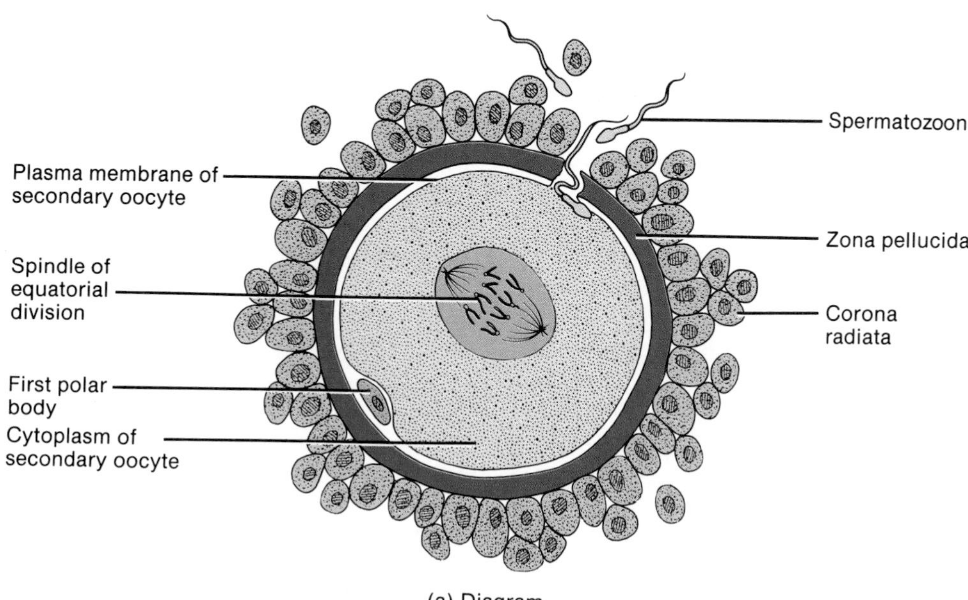

FIGURE 26-1 Fertilization and implantation. (a) Spermatozoon moving through the corona radiata and zona pellucida on its way to reach the nucleus of a secondary oocyte. (b) Spermatozoon moving through the zona pellucida on its way to reach the nucleus of a secondary oocyte. (Courtesy of Fawcett, Photo Researchers, Inc.). (c) Female and male pronuclei. (Courtesy of Carolina Biological Supply Company.) Figure 25-5b shows a spermatozoon in contact with the surface of a secondary oocyte.

(b) Scanning electron micrograph

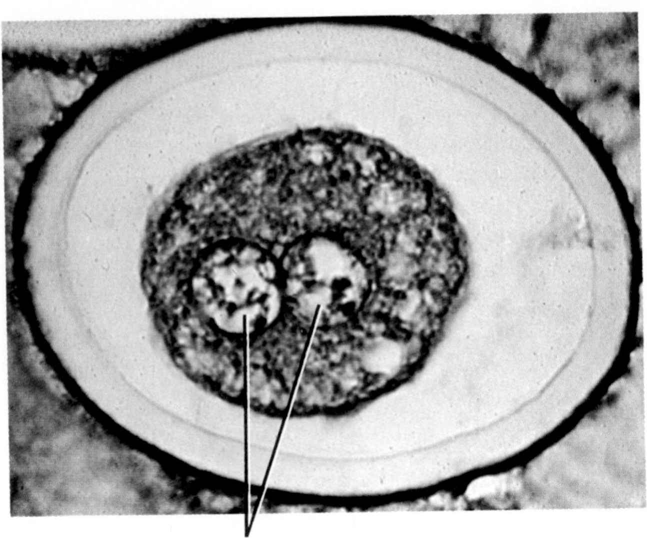

(c) Photomicrograph

to receptors in the zona pellucida. Once this is accomplished, normally only one spermatozoon enters and fertilizes a secondary oocyte because once union is achieved, the ionic changes in the plasma membrane of the oocyte block the entry of other sperm, and enzymes produced by the fertilized ovum (egg) alter receptor sites so that sperm already bound are detached and others are prevented from binding. In this way, polyspermy, fertilization by more than one spermatozoon, is prevented. Once a spermatozoon has entered a secondary oocyte, the oocyte completes equatorial division (meiosis II). The secondary oocyte divides into a larger ovum (mature egg) and a smaller second polar body that fragments and disintegrates.

When a spermatozoon has entered a secondary oocyte, the tail is shed and the nucleus in the head develops into a structure called the ***male pronucleus.*** The nucleus of the oocyte develops into a ***female pronucleus*** (Figure 26-1c). After the pronuclei are formed, they fuse to produce a ***segmentation nucleus.*** The segmentation nucleus contains 23 chromosomes (n) from the male pronucleus and 23 chromosomes (n) from the female pronucleus. Thus, the fusion of the haploid (n) pronuclei restores the diploid number ($2n$). The fertilized ovum, consisting of a segmentation nucleus, cytoplasm, and zona pellucida, is called a ***zygote*** (*zygotos* = yoked together).

Dizygotic (fraternal) twins are produced from the independent release of two ova and the subsequent fertilization of each by different spermatozoa. They are the same age and are in the uterus at the same time, but they are genetically as dissimilar as any other siblings. They may or may not be the same sex. ***Monozygotic (identical) twins*** are derived from a single fertilized ovum that splits at an early stage in development. They contain the same genetic material and are always the same sex.

On rare occasions, monozygotic twins may be joined in varying degrees, from slight skin fusion to sharing of limbs, trunks, and viscera. Such twins are called ***conjoined (Siamese) twins.*** In September 1987, a surgical team at Johns Hopkins University performed a unique operation in which 7-month-old conjoined twins were successfully separated after being joined at the head since birth. The 22-hour procedure, performed by a 70-member team, was exceedingly complex and involved a total stoppage of circulation and hypothermia for about an hour. This technique was necessary to prevent hemorrhage while separating a shared sagittal sinus between the infants and to reduce brain activity to near zero to reduce brain swelling. In all known previous attempts to separate conjoined twins joined at their heads, one or both infants died or suffered serious neurological impairment. In this case the twins suffered only slight brain damage.

CLINICAL APPLICATION

Ectopic Pregnancy

Ectopic (*ektos* = outside; *topos* = place) ***pregnancy (EP)*** refers to the development of an embryo or fetus outside the uterine cavity. The majority occur in the uterine (Fallopian) tube, usually in the ampullar and infundibular portions. Some occur in the ovaries, abdomen, uterine cervix, and broad ligaments. The basic cause of a tubal pregnancy is impaired passage of the fertilized ovum through the uterine tube related to factors such as pelvic inflammatory disease (PID), previous uterine tube surgery, previous ectopic pregnancy, repeated elective abortions, pelvic tumors, and developmental abnormalities. Ectopic pregnancy may be characterized by one or two missed periods, followed by bleeding and acute abdominal and pelvic pain. Strong risk factors for EP include current use of an intrauterine device (IUD), a history of pelvic inflammatory disease (PID) or infertility, and prior surgery involving the uterine tubes.

Unless removed or discharged from a uterine tube, the developing embryo can rupture the tube, often resulting in death of the mother. Standard practice for terminating an ectopic pregnancy in the uterine tube is to remove the tube, occasionally with its associated ovary. An alternative procedure involves injecting prostaglandin F into the uterine tube directly above the improperly implanted embryo. Prostaglandin F induces contractions that expel the embryo.

Formation of the Morula

Immediately after fertilization, rapid mitotic cell division of the zygote takes place. This early division of the zygote is called ***cleavage.*** During this time, the dividing cells are contained by the zona pellucida. Although cleavage increases the number of cells, it does not result in an increase in the size of the developing organism.

The first cleavage is completed after about 36 hours, and each succeeding division takes slightly less time (Figure 26-2). By the second day after conception, the second cleavage is completed. By the end of the third day, there are 16 cells. The progressively smaller cells produced by cleavage are called ***blastomeres*** (BLAS-tō-mērz; *blast* = germ, sprout). A few days after fertilization, the successive cleavages have produced a solid mass of cells, still surrounded by zona pellucida, the ***morula*** (MOR-yoo-la) or mulberry, which is about the same size as the original zygote.

Development of the Blastocyst

As the number of cells in the morula increases, it moves from the original site of fertilization down through the ciliated uterine (Fallopian) tube toward the uterus and enters the uterine cavity. By this time, the dense cluster of cells is altered to form a hollow ball of cells. The mass is now referred to as a ***blastocyst*** (Figure 26-3).

The blastocyst is differentiated into an outer covering of cells called the ***trophoblast*** (TRŌF-ō-blast; *troph* = nourish), an ***inner cell mass (embryoblast),*** and an internal fluid-filled cavity called the ***blastocele*** (BLAS-tō-sēl). The trophoblast ultimately forms part of the membranes composing the fetal portion of the placenta, a structure through which components of maternal and fetal blood are exchanged; the inner cell mass develops into the embryo.

Implantation

The blastocyst remains free within the cavity of the uterus from two to four days before it actually attaches to the uterine wall. During this time, the zona pellucida degenerates and nourishment is provided by secretions of glands of the endometrium, sometimes called uterine milk. The attachment of the blastocyst to the endometrium occurs seven to eight days after fertilization and is called ***implantation*** (Figure 26-4). At this time, the endometrium is in its postovulatory phase. As the blastocyst becomes implanted, the trophoblast separates into two layers in the region of contact between the blastocyst and endometrium. These layers are an outer ***syncytiotrophoblast*** (sin-sīt′-ē-ō-TRŌF-ō-blast; *synctium* = multinucleate mass) that contains no cell boundaries and an inner ***cytotrophoblast*** (sī-tō-TRŌF-ō-blast) that is composed of distinct cells. During implantation, the syncytiotrophoblast secretes enzymes that enable the blastocyst to penetrate the uterine lining. The enzymes digest and liquefy the endometrial cells. The fluid and nutrients further nourish the burrowing blastocyst for about a week after implantation. Eventually the blastocyst becomes buried in the endometrium, usually on the posterior wall of the fundus or body of the uterus and nutrients are delivered through the placenta for the subsequent growth and development of the embryo and fetus.

Since a developing embryo, and later a fetus, contains genes from the father as well as the mother, it is essentially

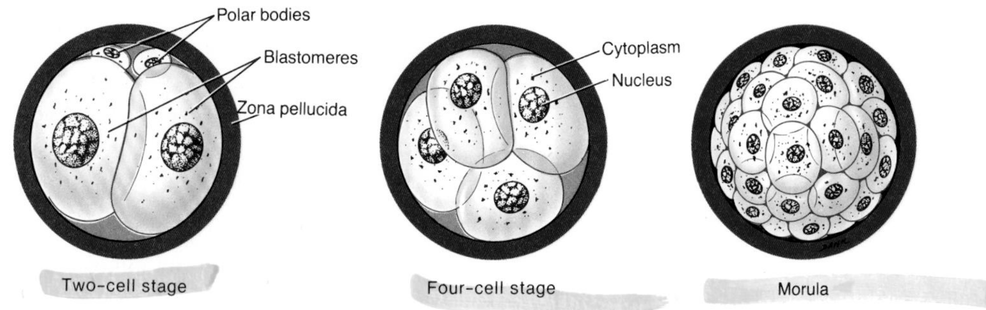

FIGURE 26-2 Formation of the morula. About 72 hours after fertilization, the morula enters the uterus at about the 8- to 12-cell stage. For simplicity, the corona radiata is not included (see Figure 26-1a).

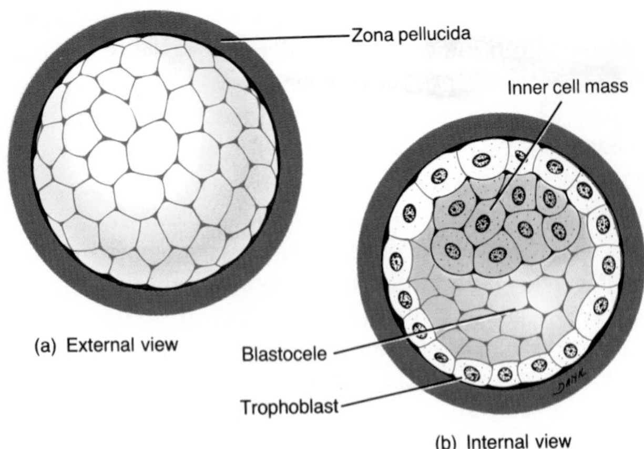

FIGURE 26-3 Blastocyst.

a foreign graft and should be rejected as such by the mother. The trophoblast is the only tissue of the developing organism that makes contact with the mother's uterus. Even though trophoblast cells have paternal antigens that could provoke a rejection response, some mechanism in the uterus prevents this in order to permit development of the fetus to term. One suggestion for the mechanism that prevents rejection is that the mother makes antibodies that mask paternal antigens so that other components of her immune system cannot recognize and attack the antigens.

A summary of the principal events associated with fertilization and implantation is shown in Figure 26-5.

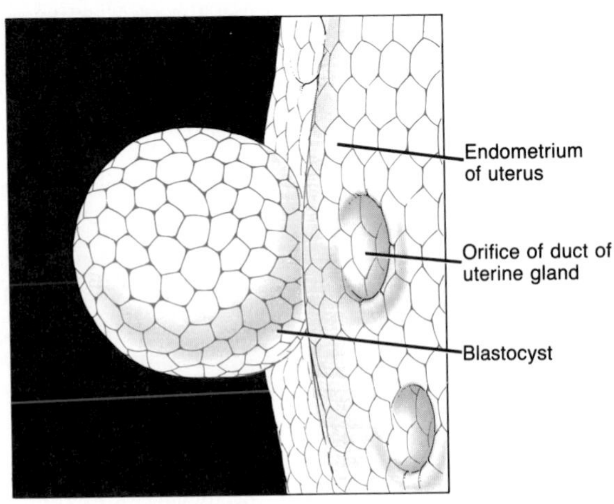

(a) External view, about 5 days after fertilization

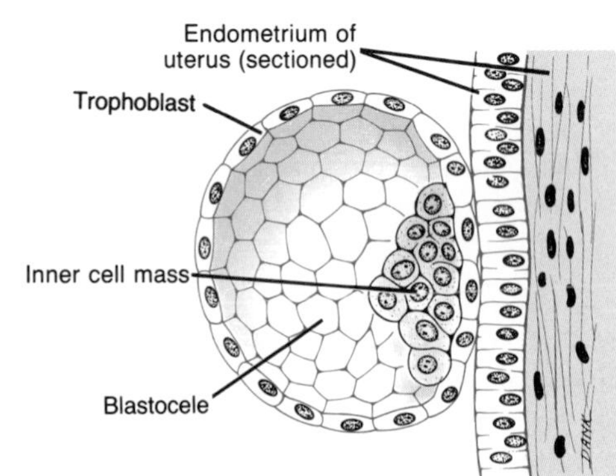

(b) Internal view, about 6 days after fertilization

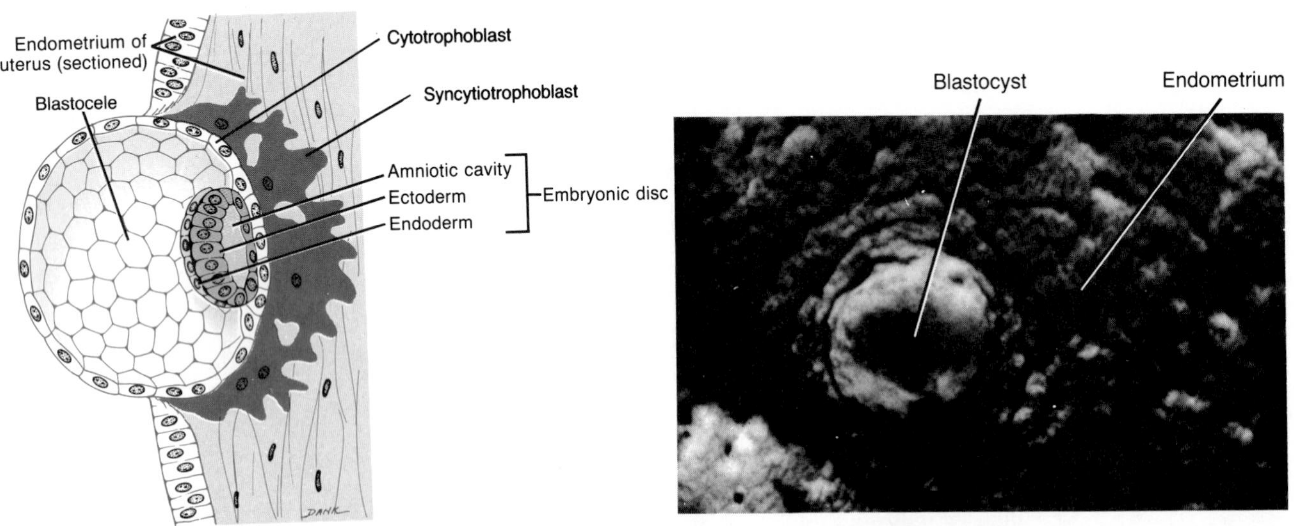

(c) Internal view, about 7 days after fertilization

(d) Photomicrograph

FIGURE 26-4 Implantation. Shown is the blastocyst in relation to the endometrium of the uterus at various time intervals after fertilization. (d) From *From Conception to Birth: The Drama of Life's Beginnings* by Roberts Rugh, Landrum B. Shettles with Richard Einhorn. Copyright © 1971 by Roberts Rugh and Landrum B. Shettles. By permission of Harper & Row, Publishers, Inc.

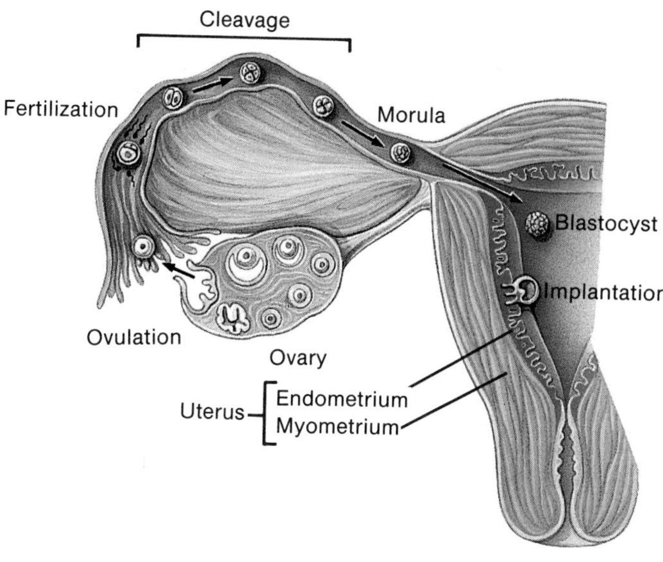

FIGURE 26-5 Summary of events associated with fertilization and implantation.

CLINICAL APPLICATION

Morning Sickness

In the early months of pregnancy, ***emesis gravidarum (morning sickness)*** may occur, characterized by episodes of nausea and possibly vomiting that are most likely to occur in the morning. The cause is unknown, but the high levels of human chorionic gonadotropin (hCG) secreted by the placenta have been implicated. In some individuals the severity of these symptoms requires hospitalization for intravenous feeding, and the condition is then known as ***hyperemesis gravidarum.***

IN VITRO FERTILIZATION (IVF)

On July 12, 1978, Louise Joy Brown was born near Manchester, England. Her birth was the first recorded case of ***in vitro fertilization (IVF)***—fertilization in a glass dish. In this procedure for IVF, the female is given follicle-stimulating hormone (FSH) soon after menstruation, so that several secondary oocytes, rather than the typical single one, will be produced (superovulation). Administration of luteinizing hormone (LH) may also ensure the maturation of the secondary oocytes. Next, a small incision is made near the umbilicus, and the secondary oocytes are aspirated from the follicles and placed in a medium that simulates the fluids in the female reproductive tract. The secondary oocytes are then transferred to a solution of the male's sperm. Once fertilization has taken place, the fertilized ovum is put in another medium and observed for cleavage. When the fertilized ovum reaches the 8-cell or 16-cell stage, it is introduced into the uterus for implantation and subsequent growth. The growth and developmental sequences that occur are similar to those in internal fertilization. It is also possible to freeze unused embryos (cryopreservation) to permit parents a successive pregnancy several years later or allow a second attempt at implantation if the first attempt is unsuccessful. The first instance in the United States of two successful pregnancies from a single IVF procedure occurred on November 19, 1988, when a woman gave birth to two daughters at the same time two years after the birth of her son.

Embryo transfer is a type of IVF in which a husband's seminal fluid is used to artificially inseminate a fertile secondary oocyte donor, and following fertilization, the morula or blastocyst is transferred from the donor to the infertile wife who carries it to term. Embryo transfer is indicated for females who are infertile; who have surgically untreatable blocked uterine (Fallopian) tubes; or who are afraid of passing on their own genes because they are carriers of a serious genetic disorder. In the procedure, the donor is monitored to ascertain the time of ovulation by checking her blood levels of luteinizing hormone (LH) and by ultrasound. The woman is also monitored to make sure that her ovarian cycle is synchronized with that of the donor. Once ovulation occurs in the donor, she is donor (artificially) inseminated with the husband's (or another male's) seminal fluid. Four days later, a morula or blastocyst is flushed from the donor's uterus through a soft plastic catheter and transferred to the uterus of the wife, where it grows and develops until the time of birth. Embryo transfer is an office procedure that requires no anesthesia.

Gamete intrafallopian transfer (GIFT) is a type of IVF that is essentially an attempt to mimic the normal process of conception by uniting sperm and secondary oocyte in the prospective mother's uterine (Fallopian) tubes. In the procedure, the female is given FSH and LH to stimulate the production of several secondary oocytes. The secondary oocytes are aspirated with a laparoscope fitted with a suction device, mixed with a solution of the male's sperm outside the body, and then immediately inserted into the uterine (Fallopian) tubes.

In ***transvaginal oocyte retrieval,*** another type of IVF, the female is given hormones to produce several secondary oocytes, a needle is placed through the vaginal wall and guided to the ovaries by ultrasound, suction is applied to the needle, and the secondary oocytes are removed and placed in a solution outside the body. The male's sperm are added to the solution, and the fertilized ova are then implanted in the uterus.

EMBRYONIC DEVELOPMENT

The first two months of development are generally considered the ***embryonic period.*** During this period the developing human is called an ***embryo*** (*bryein* = grow). The study of development from the fertilized egg through the eighth week in utero is referred to as ***embryology*** (em-brē-OL-ō-jē). The months of development after the second month are considered the ***fetal period,*** and during this time the developing human is called a ***fetus*** (*feo* = to bring forth). By the end of the embryonic period, the rudiments of all the principal adult organs are present, the embryonic membranes are developed, and the placenta is functioning.

BEGINNINGS OF ORGAN SYSTEMS

Following implantation, the first major event of the embryonic period occurs. The inner cell mass of the blastocyst begins to differentiate into the three ***primary germ layers:*** ectoderm, endoderm, and mesoderm. They are the embryonic tissues from which all tissues and organs of the body will develop. The various movements of cell groups leading to the establishment of the primary germ layers are referred to as ***gastrulation.***

In the human the germ layers form so quickly that it is difficult to determine the exact sequence of events. Within eight days after fertilization, the top layer of cells of the inner cell mass proliferates and forms the amnion (a fetal membrane) and a space, the ***amniotic (amnionic) cavity,*** over the inner cell mass. The upper layer of cells of the inner cell mass that is closer to the amniotic cavity develops into the ***ectoderm.*** The bottom layer of the inner cell mass that borders the blastocele develops into the ***endoderm.***

About the twelfth day after fertilization, striking changes appear (Figure 26-6a). The cells below the amniotic cavity are called the ***embryonic disc.*** They will form the embryo. At this stage, the embryonic disc contains ectodermal and endodermal cells; the mesodermal cells are scattered external to the disc. The cells of the endodermal layer have been dividing rapidly, so that groups of them now extend downward in a circle, forming the yolk sac, another fetal membrane. The cells of the ***mesoderm,*** which develop between the ectodermal and endodermal layers, also have been dividing, and many have left the area of the embryonic disc and can be seen around the structures that are becoming fetal membranes.

About the fourteenth day, the cells of the embryonic disc differentiate into three distinct layers: the upper ectoderm, the middle mesoderm, and the lower endoderm (Figure 26-6b). The mesoderm in the disc soon splits into two layers, and the space between the layers becomes the ***extraembryonic coelom.***

As the embryo develops (Figure 26-6c), the endoderm becomes the epithelial lining of the gastrointestinal tract, respiratory tract, and a number of other organs. The mesoderm forms muscle, bone, other connective tissue, and the peritoneum. The ectoderm develops into the skin and nervous system.

Exhibit 26-1 provides more details about the fates of these primary germ layers.

EMBRYONIC MEMBRANES

The ***embryonic membranes*** form as part of the second major event of the embryonic period (Figure 26-7). These membranes lie outside the embryo and protect and nourish the embryo and, later, the fetus. The membranes are the yolk sac, amnion, chorion, and allantois.

The ***yolk sac*** is an endoderm-lined membrane that, in many species, provides the primary or exclusive nutrient for the embryo (Figure 26-8; see also Figures 26-6c and 26-7). However, the human embryo receives its nourishment from the endometrium, and the yolk sac remains small. In humans, the yolk sac provides an early site of blood formation. It also contains cells that differentiate into the primitive germ cells (spermatogonia and oogonia). During

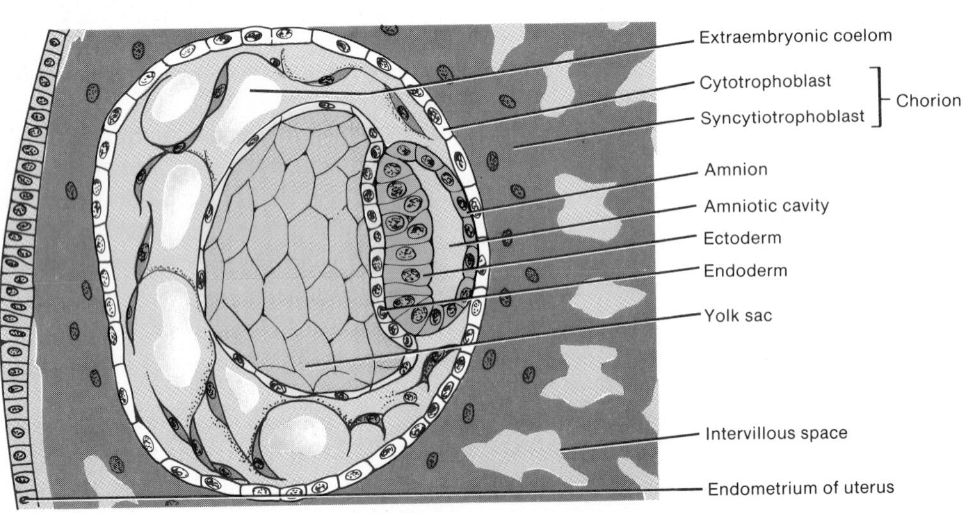

(a) Internal view, about 12 days after fertilization

FIGURE 26-6 Formation of the primary germ layers and associated structures.

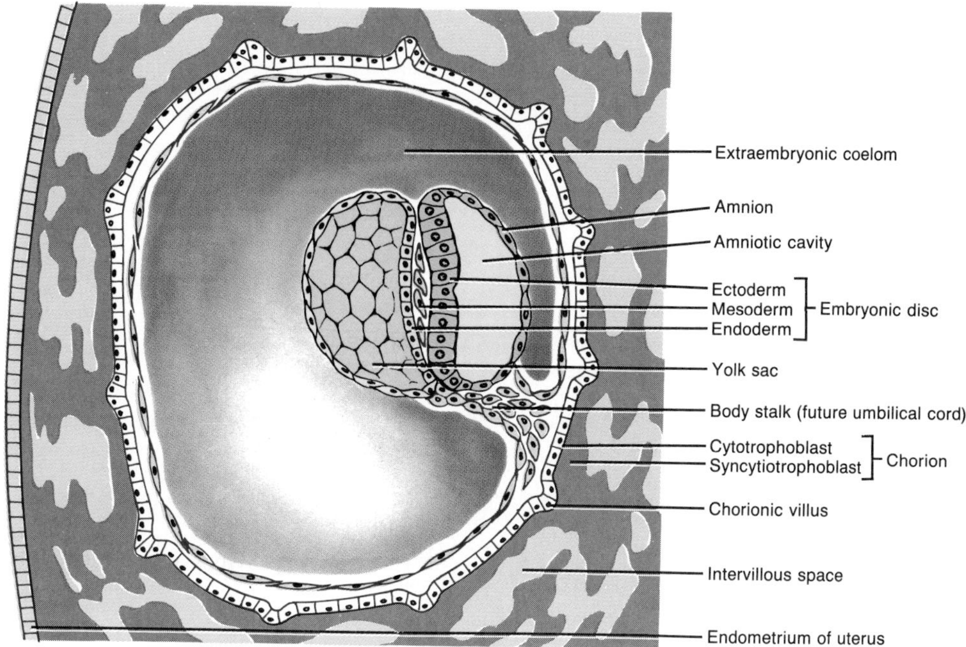

(b) Internal view, about 14 days after fertilization

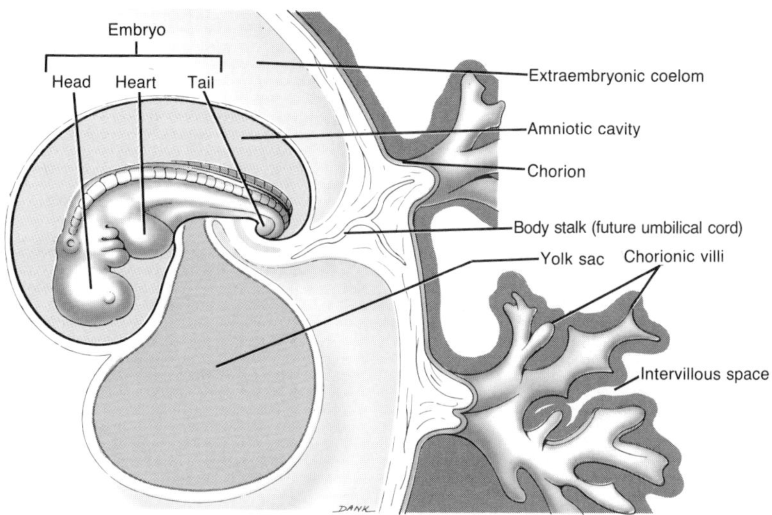

(c) External view, about 25 days after fertilization

an early stage of development, the yolk sac becomes a nonfunctional part of the umbilical cord.

The ***amnion*** is a thin, protective membrane that initially overlies the embryonic disc and is formed by the eighth day following fertilization. As the embryo grows, the amnion entirely surrounds the embryo, creating a cavity that becomes filled with ***amniotic fluid,*** or ***AF*** (Figure 26-8). Most amniotic fluid is initially derived from a filtrate of maternal blood. Later, the fetus makes daily contributions to the fluid by excreting urine into the amniotic cavity. Amniotic fluid serves as a shock absorber for the fetus and assists in the regulation of fetal body temperature and prevents adhesions between the skin of the fetus and surrounding tissues. Embryonic cells are sloughed off into amniotic fluid; they can be examined in the procedure called ***amniocentesis*** (am′-nē-ō-sen-TĒ-sis), which is discussed later. The amnion usually ruptures just before birth and with its fluid constitutes the ''bag of waters'' (BOW).

The ***chorion*** (KŌ-rē-on) is derived from the trophectoderm of the blastocyst and the mesoderm that lines the trophoblast. It surrounds the embryo and, later, the fetus. Eventually, the chorion becomes the principal embryonic part of the placenta, the structure through which materials are exchanged between mother and fetus. The amnion also

EXHIBIT 26-1

Structures Produced by the Three Primary Germ Layers

Endoderm	Mesoderm	Ectoderm
Epithelium of gastrointestinal tract (except the oral cavity and anal canal) and the epithelium of its glands.	All skeletal, most smooth, and all cardiac muscle.	All nervous tissue.
Epithelium of urinary bladder, gallbladder, and liver.	Cartilage, bone, and other connective tissues.	Epidermis of skin.
Epithelium of pharynx, auditory (Eustachian) tube, tonsils, larynx, trachea, bronchi, and lungs.	Blood, bone marrow, and lymphoid tissue.	Hair follicles, arrector pili muscles, nails, and epithelium of skin glands (sebaceous and sudoriferous).
Epithelium of thyroid, parathyroid, pancreas, and thymus glands.	Endothelium of blood vessels and lymphatic vessels.	Lens, cornea, and internal eye muscles.
Epithelium of prostate and bulbourethral (Cowper's) glands, vagina, vestibule, urethra, and associated glands such as the greater (Bartholin's) vestibular and lesser vestibular glands.	Dermis of skin.	Internal and external ear.
	Fibrous tunic and vascular tunic of eye.	Neuroepithelium of sense organs.
	Middle ear.	Epithelium of oral cavity, nasal cavity, paranasal sinuses, salivary glands, and anal canal.
	Mesothelium of ventral body and joint cavities.	Epithelium of pineal gland, pituitary gland (hypophysis), and adrenal medulla.
	Epithelium of kidneys and ureters.	
	Epithelium of adrenal cortex.	
	Epithelium of gonads and genital ducts.	

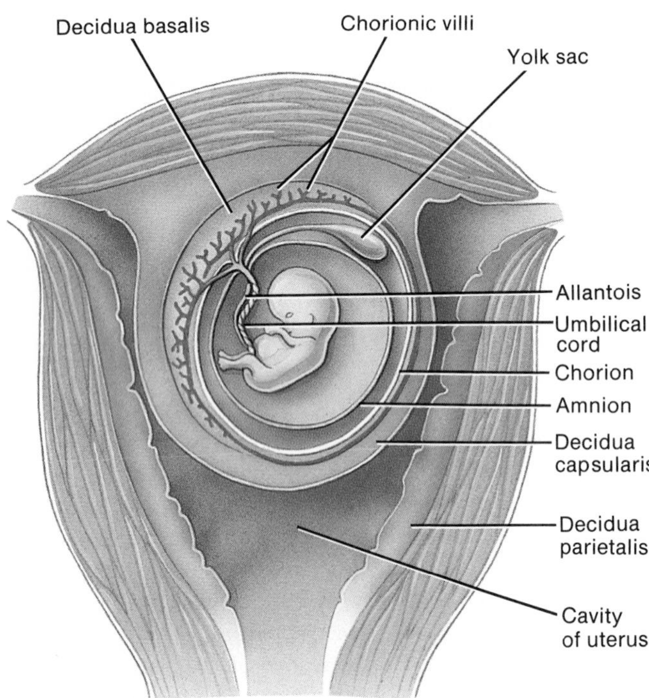

FIGURE 26-7 Embryonic membranes.

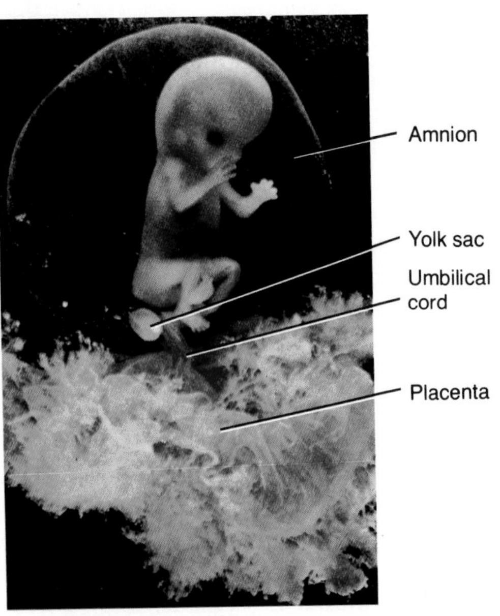

FIGURE 26-8 Amnion, yolk sac, umbilical cord, and placenta. (From *From Conception to Birth: The Drama of Life's Beginnings* by Roberts Rugh, Landrum B. Shettles with Richard Einhorn. Copyright © 1971 by Roberts Rugh and Landrum B. Shettles. By permission of Harper & Row, Publishers, Inc.)

surrounds the fetus and eventually fuses to the inner layer of chorion.

The ***allantois*** (a-LAN-tō-is; *allas* = sausage) is a small, vascularized, fingerlike outpouching of the yolk sac. It serves as an early site of blood formation. Later its blood vessels serve as connections in the placenta between mother and fetus. This connection is the umbilical cord.

PLACENTA AND UMBILICAL CORD

Development of the placenta, the third major event of the embryonic period, is accomplished by the third month of pregnancy. The ***placenta*** (pla-SEN-ta) has the shape of a flat cake when fully developed and is formed by the chorion of the embryo and a portion of the endometrium (decidua basalis) of the mother (Figure 26-9). Functionally, the placenta allows oxygen and nutrients to diffuse into fetal blood from maternal blood and allows carbon dioxide and wastes to diffuse from fetal blood into maternal blood. (Almost all drugs, including alcohol, pass the placenta freely.) In addition, the placenta provides some degree of protection since most microorganisms cannot cross it. (Viruses associated with diseases such as AIDS, German measles, chickenpox, measles, encephalitis, and poliomyelitis may pass through the placenta.) The placenta also stores nutrients such as carbohydrates, proteins, calcium, and iron, which are released into fetal circulation as required. Finally, the placenta produces several hormones that are necessary to maintain pregnancy.

If implantation occurs, a portion of the endometrium becomes modified and is known as the ***decidua*** (dē-SID-yoo-a). The decidua includes all but the deepest layer of the endometrium and is shed when the fetus is delivered. Different regions of the decidua, all areas of the stratum functionalis, are named on the basis of their positions relative to the site of the implanted, fertilized ovum (Figure 26-10). The ***decidua basalis*** is the portion of the endometrium that underlies the embryo, between the chorion and the stratum basalis of the uterus. The decidua basalis becomes the maternal part of the placenta. The ***decidua capsularis*** is the portion of the endometrium that overlies the embryo, between the embryo and the uterine cavity. The ***decidua parietalis*** (par-rī-ē-TAL-us) is the remaining modified endometrium that lines the uterus during pregnancy.

During embryonic life, fingerlike projections of the chorion, called ***chorionic villi*** (kō′-rē-ON-ik VIL-ī), grow into the decidua basalis of the endometrium (see Figures 26-6 and 26-7). These will contain fetal blood vessels of the allantois. They continue growing until they are bathed in maternal blood sinuses called ***intervillous*** (in-ter-VIL-us) ***spaces***. Thus, maternal and fetal blood vessels are brought into proximity. It should be noted, however, that maternal and fetal blood do not normally mix. Oxygen and nutrients from the mother's blood diffuse into the capillaries of the villi where all exchange occurs between fetal and maternal blood. From the capillaries the nutrients circulate into the umbilical vein. Wastes leave the fetus through the umbilical arteries, pass into the capillaries of the villi, and diffuse into the maternal blood. The ***umbilical*** (um-BIL-i-kul) ***cord*** is a vascular connection between mother and fetus that consists of an outer layer of amnion containing the umbilical arteries and umbilical vein, supported internally by mucous connective tissue (Wharton's jelly) from the allantois.

At delivery, the placenta detaches from the uterus and is referred to as the ***afterbirth***. At this time, the umbilical cord is severed, leaving the baby on its own. The scar that marks the site of the entry of the fetal umbilical cord into the abdomen is the ***umbilicus (navel)***.

Pharmaceutical houses use human placenta as a source of hormones, drugs, and blood. Portions of placentas are also used for burn coverage. The placental and umbilical cord veins are used in blood vessel grafts.

CLINICAL APPLICATION

Placenta Previa, Fetomaternal Hemorrhage, and Umbilical Cord Accidents

In some cases, part or all of the placenta becomes implanted in the lower portion of the uterus, near or over the internal os of the cervix. This condition is called ***placenta previa*** (PRĒ-vē-a; *previa* = before or in front of) and occurs in approximately 1 in 250 live births. The condition occurs 10 to 20 times more frequently in association with spontaneous abortion. It is also associated with fetal abnormalities, twin gestation, and multiple uterine curettages. The most important symptom is sudden, painless, bright red vaginal bleeding in the third trimester. Cesarean section is the preferred method of delivery in placenta previa.

Fetomaternal hemorrhage refers to the entrance of blood into fetal circulation brought on by a dysfunction in placental circulation. Although the condition occurs in at least 50 percent of all pregnancies, in most instances blood loss is so small that the pregnancy is not adversely affected. In some situations, however, the hemorrhage can compromise the fetus and lead to serious complications later in the same pregnancy or a future pregnancy. Among the causes of fetomaternal hemorrhage are trauma, rapid deceleration, placental and umbilical cord abnormalities, amniocentesis, intrauterine fetal surgery, umbilical vein thrombosis, and operative delivery (e.g., cesarean section). As a result of fetomaternal hemorrhage, certain complications may result. Examples include intrauterine fetal death, hypovolemic shock and anemia in the newborn, edema of the newborn, fetal cardiac arrhythmia, and anaphylactic shock in the mother.

Two of the most frequently encountered ***umbilical cord accidents*** are prolapse and entrapment. In ***prolapse,*** the umbilical cord descends in advance of the fetus during delivery. In ***entrapment,*** circulation through the cord is compromised because of pressure between the fetus and uterine wall. These, plus other conditions, such as knots

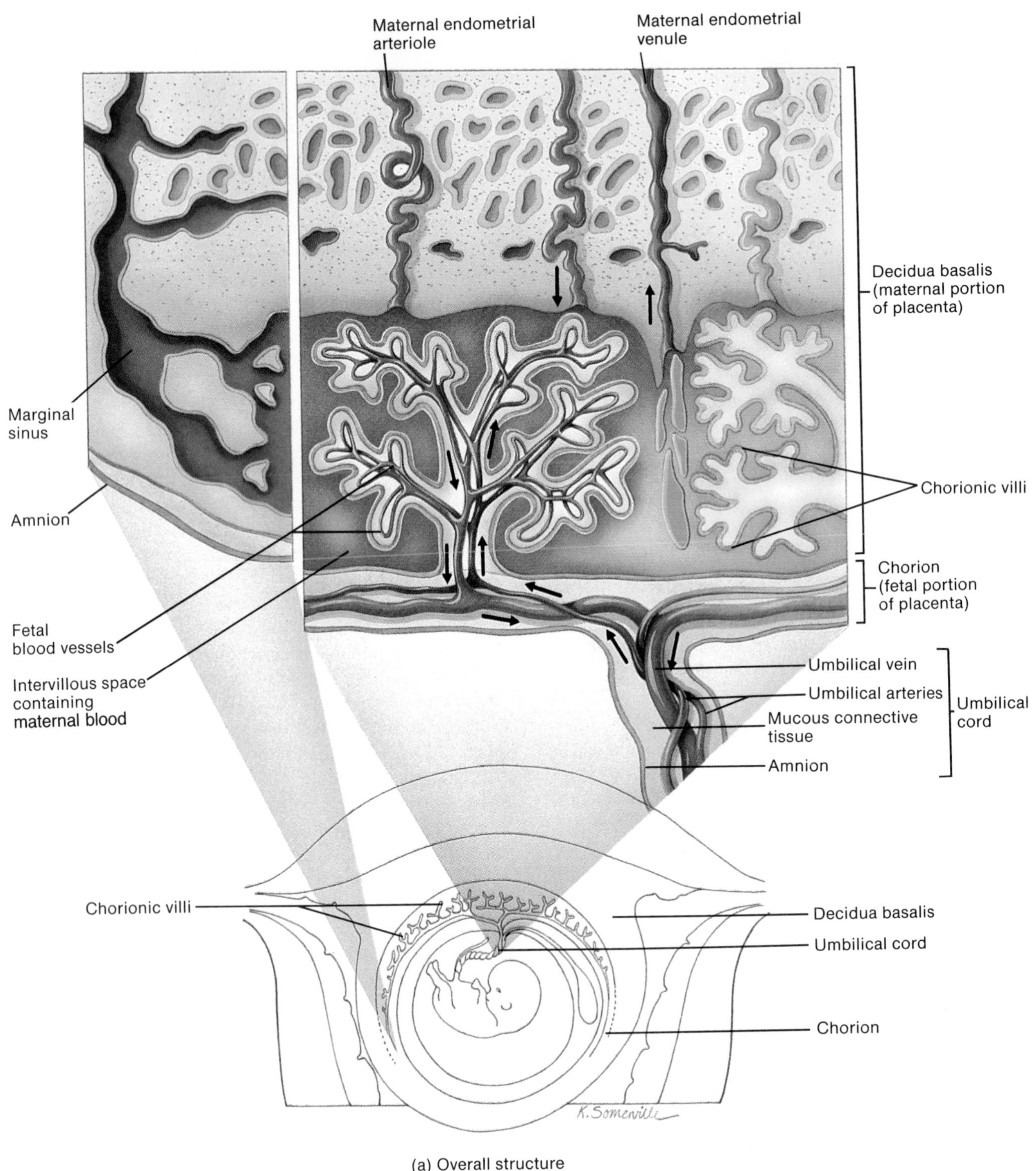

(a) Overall structure

FIGURE 26-9 Placenta and umbilical cord. Whereas the umbilical arteries carry deoxygenated blood, the umbilical vein carries oxygenated blood. (b) and (c) Courtesy of M. A. Colin England, *A Colour Atlas of Life Before Birth*, Year Book Medical Publishers, Chicago.

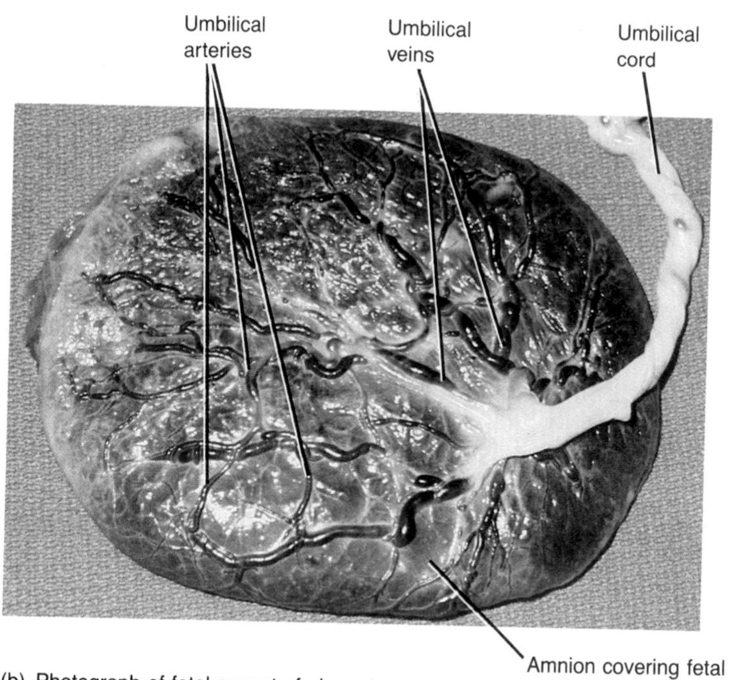

(b) Photograph of fetal aspect of placenta

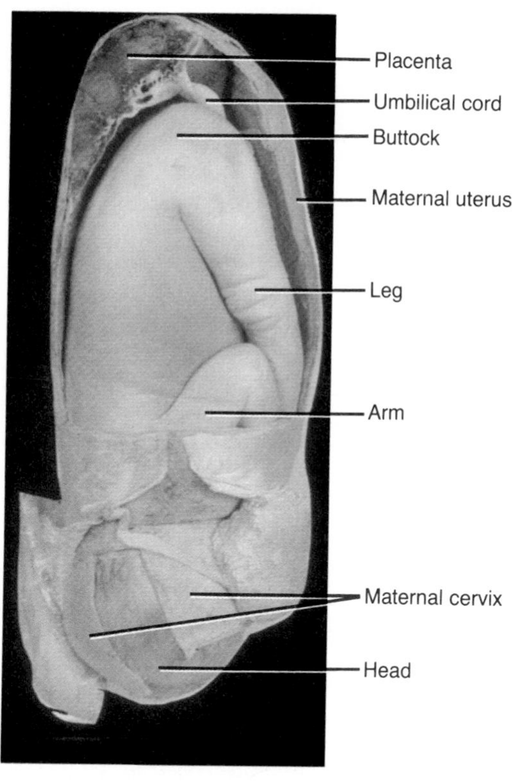

(c) Photograph of placenta during second stage of labor

in the cord, strictures, and thrombosis, could compromise the oxygen supply to the fetus, resulting in fetal damage or death.

FETAL GROWTH

During the ***fetal period,*** organs established by the primary germ layers grow rapidly. The organism takes on a human appearance. A summary of changes associated with the embryonic and fetal periods is presented in Exhibit 26-2.

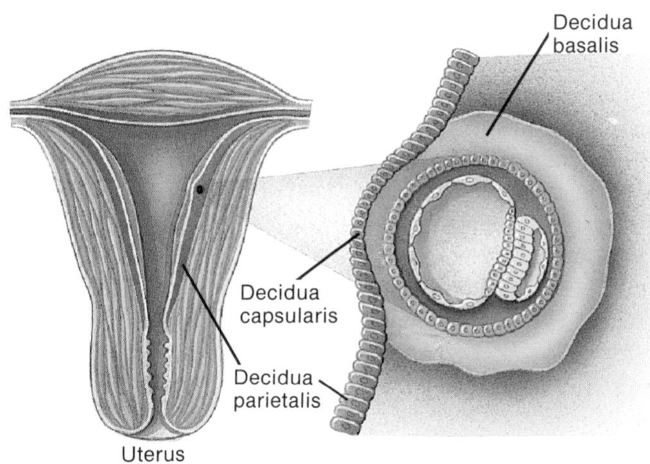

FIGURE 26-10 Regions of the decidua.

CLINICAL APPLICATION

Fetal Surgery and Fetal-Tissue Implantation

Fetal surgery is a new medical field that had its beginnings in 1985. In a pioneering operation, a team of surgeons removed a 23-week-old fetus from its mother's uterus, operated to correct a blocked urinary tract, and then returned the fetus to the uterus. Nine weeks later, a healthy baby was delivered. Since then, surgeons have performed procedures to repair diaphragmatic hernias and are experimenting on research animals with methods to correct spina bifida and hydrocephalus. It has been observed that surgery on fetuses does not leave any scars, although the reason is not known. It is hoped that surgeons can perform craniofacial surgery before birth to correct conditions such as cleft lip without leaving scars.

Another relatively new therapy for treating certain diseases is known as ***fetal-tissue implantation.*** In the procedure, tissue is used from aborted fetuses in order to correct certain defects. For example, surgeons in China have been transplanting fetal pancreatic islet (islet of Langerhans) cells to treat type I diabetes since 1982. In 1988, surgeons in Mexico City and the United States transplanted fetal brain cells into the brains of patients with Parkinson's disease. It is hoped that fetal liver tissue may be transplanted to cure hereditary blood disorders such as thalassemia.

Fetal cells have the advantage of being immunologically naive; that is, they have not yet developed all the

EXHIBIT 26-2

Changes Associated with Embryonic and Fetal Growth

End of Month	Approximate Size and Weight	Representative Changes
1	0.6 cm (3/16 in.)	Eyes, nose, and ears not yet visible. Backbone and vertebral canal form. Small buds that will develop into arms and legs form. Heart forms and starts beating. Body systems begin to form. Central nervous system appears at the start of the third week.
2	3 cm (1¼ in.) 1 g (1/30 oz)	Eyes far apart, eyelids fused, nose flat. Ossification begins. Limbs become distinct as arms and legs. Digits are well formed. Major blood vessels form. Many internal organs continue to develop.
3	7.5 cm (3 in.) 28 g (1 oz)	Eyes almost fully developed but eyelids still fused, nose develops bridge, and external ears are present. Ossification continues. Extremities are fully formed and nails develop. Heartbeat can be detected. Urine starts to form. Fetus begins to move, but the movement cannot be detected by mother. Body systems continue to develop.
4	18 cm (6½–7 in.) 113 g (4 oz)	Head large in proportion to rest of body. Face takes on human features and hair appears on head. Skin bright pink. Many bones ossified, and joints begin to form. Rapid development of body systems.
5	25–30 cm (10–12 in.) 227–454 g (½–1 lb)	Head less disproportionate to rest of body. Fine hair (lanugo) covers body. Skin still bright pink. Brown fat forms and is the site of heat production. Fetal movements commonly felt by mother. Rapid development of body systems.
6	27–35 cm (11–14 in.) 567–781 g (1¼–1½ lb)	Head becomes even less disproportionate to rest of body. Eyelids separate and eyelashes form. Substantial weight gain. Skin wrinkled and pink. Type II alveolar cells begin to produce surfactant.
7	32–42 cm (13–17 in.) 1135–1362 g (2½–3 lb)	Head and body more proportionate. Skin wrinkled and pink. Seven-month fetus (premature baby) is capable of survival since lungs are capable of breathing air and central nervous system can control respiration and body temperature. Fetus assumes an upside-down position.
8	41–45 cm (16½–18 in.) 2043–2270 g (4½–5 lb)	Subcutaneous fat deposited. Skin less wrinkled. Testes descend into scrotum. Bones of head are soft. Chances of survival much greater at end of eighth month.
9	50 cm (20 in.) 3178–3405 g (7–7½ lb)	Additional subcutaneous fat accumulates. Lanugo shed. Nails extend to tips of fingers and maybe even beyond.

antigens that allow a recipient's immune system to reject the cells. In addition, fetal cells are usually not mature enough to cause graft-versus-host disease, in which the tissues of a transplant recipient are attacked by implanted adult cells. Moreover, fetal nerve cells have the ability to regenerate and thus have the potential to repair damaged brain or spinal cord tissue.

GESTATION

The time a zygote, embryo, or fetus is carried in the female reproductive tract is called ***gestation*** (jes-TĀ-shun). The total human gestation period is about 280 days from the beginning of the last menstrual period. The specialized branch of medicine that deals with pregnancy, labor, and the period of time immediately following delivery is called ***obstetrics*** (ob-STET-riks; *obstetrix* = midwife).

STRUCTURAL AND FUNCTIONAL CHANGES

By about the end of the third month of gestation, the uterus occupies most of the pelvic cavity, and as the fetus continues to grow, the uterus extends higher and higher into the abdominal cavity. In fact, toward the end of a full-term pregnancy, the uterus occupies practically all the abdominal cavity, reaching above the costal margin nearly to the xiphoid process of the sternum (Figure 26-11), causing displacement of the maternal intestines, liver, and stomach upward, elevation of the diaphragm, and widening of the thoracic cavity. In the pelvic cavity, there is compression of the ureters and urinary bladder.

In addition to the anatomical changes associated with pregnancy, there are also certain pregnancy-induced physiological changes. General changes include weight gain due to the fetus, amniotic fluid, placenta, uterine enlargement, and increased total body water; increased proteins, fat, and mineral storage; marked breast enlargement in anticipation of lactation; and lower back pain caused by lordosis.

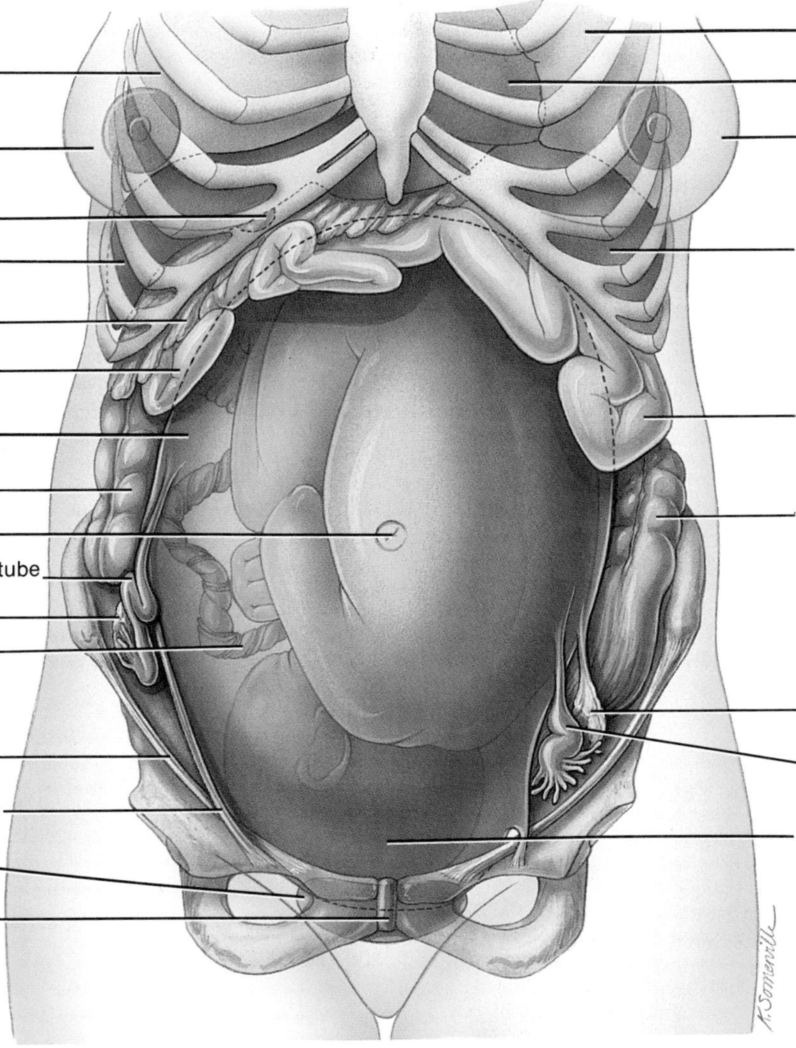

Right lung
Right mammary gland
Gallbladder
Liver
Greater omentum
Small intestine
Uterine wall
Ascending colon
Maternal umbilicus
Right uterine (Fallopian) tube
Right ovary
Umbilical cord
Inguinal ligament
Round ligament of uterus
Urinary bladder
Pubic symphysis
Left lung
Pericardium of heart
Left mammary gland
Stomach
Small intestine
Descending colon
Left ovary
Left uterine (Fallopian) tube
Head of fetus

(a) Anterior view

Maternal uterus
Umbilical cord
Maternal ovary

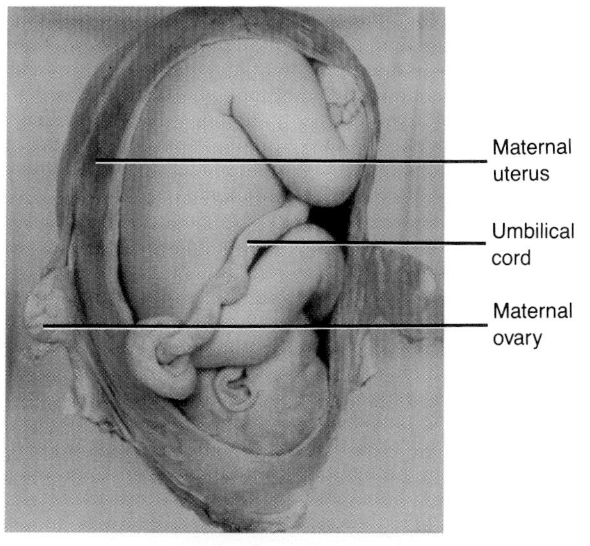

(b) Anterior view

FIGURE 26-11 Normal fetal position during a full-term pregnancy. (a) Diagram. (b) Photograph. (Courtesy of M. A. Colin England, *A Colour Atlas of Life Before Birth,* Year Book Medical Publishers, Chicago.)

With respect to the cardiovascular system, there is an increase in the output of blood by the heart by about 20 to 30 percent by the twenty-seventh week because of increased maternal blood flow to the placenta and increased metabolism; an increase in heart rate by about 10 to 15 percent; and an increase in blood volume up to 30 to 50 percent, mostly during the latter half of pregnancy. When a pregnant female is lying on her back, the enlarged uterus may compress the aorta, resulting in diminished blood flow to the uterus. Hormonal changes associated with pregnancy and compression of the inferior vena cava also can produce varicose veins.

Pulmonary function is also altered during pregnancy in that the amount of air entering and exiting the lungs during normal breathing can increase 30 to 40 percent, and airway resistance in the bronchial tree can decrease up to 36 percent. There is also an increase in total body oxygen consumption by about 10 to 20 percent. Dyspnea (labored breathing) also occurs.

With regard to the gastrointestinal tract, there is an increase in appetite and a general decrease in motility that can result in constipation and a delay in gastric emptying time. Nausea, vomiting, and heartburn also occur.

Pressure on the urinary bladder by the enlarging uterus can produce urinary symptoms, such as frequency, urgency, and stress incontinence. Other conditions related to the urinary system include an increase in renal plasma flow up to 35 percent, an increase in glomerular filtration rate up to 40 percent, and a decrease in ureteral muscle tone.

Changes in the skin during pregnancy are more apparent in some patients than others. Included are increased pigmentation around the eyes and cheekbones in a masklike pattern (chloasma), in the areolae of the breasts, and in the linea alba of the lower abdomen (linea nigra). Striae (stretch marks) over the abdomen occur as the uterus enlarges, and hair loss also increases.

Changes in the reproductive system include edema and increased vascularity of the vulva and increased pliability and vascularity of the vagina. The uterus increases in weight from its nonpregnant state of 60 to 80 g to 900 to 1200 g at term. This increase is due to hyperplasia of muscle fibers (cells) in the myometrium in early pregnancy and hypertrophy of muscle fibers during the second and third trimesters.

EXERCISE AND PREGNANCY

Since pregnancy results in so many major body changes, it has an impact on a female's ability to exercise. In early pregnancy, there are only a few changes that affect exercise. Accordingly, the mother may tire much earlier than usual, or she may become lethargic. Morning sickness may also curtail regular exercise. As the pregnancy develops, weight is gained and posture changes. As a result, more energy is needed to perform activities, and certain maneuvers (sudden stopping, changes in direction, rapid movements) are difficult to execute. In even later stages of pregnancy, certain joints, especially the pubic symphysis, become less stable in response to elevated levels of progesterone. As compensation, many females walk with widely spread legs and a shuffling motion.

Although during exercise blood shifts from viscera (including the uterus) to the muscles and skin, there is no evidence of placental insufficiency. The heat generated during exercise may cause dehydration and further increase body temperature. During early pregnancy, especially, excessive exercise and heat buildup should be avoided since elevated body temperature has been implicated in neural tube defects. Exercise has no known effect on lactation, provided the female remains hydrated and wears a bra with good support. Moderate physical activity does not endanger the fetuses of healthy females who have a normal pregnancy.

Among the benefits of exercise during pregnancy are improvement in oxygen capacity, greater sense of well-being, and fewer minor complaints.

PRENATAL DIAGNOSTIC TESTS

Several tests are available to detect genetic disorders and assess fetal well-being. Here we will describe amniocentesis, chorionic villi sampling (CVS), fetal ultrasonography, electronic fetal monitoring (EFM), and the alphafetoprotein (AFP) test.

AMNIOCENTESIS

Amniocentesis (am′-nē-ō-sen-TĒ-sis; *amnio* = amnion; *kentesis* = puncture) is a procedure that is used to test for the presence of certain chromosomal abnormalities and genetic disorders, such as Down syndrome (DS), spina bifida, hemophilia, Tay-Sachs disease, sickle-cell anemia, and certain muscular dystrophies or to determine fetal maturity and well-being near the time of the delivery. About 300 chromosomal disorders and over 50 biochemical defects can be detected through amniocentesis.

Down syndrome (DS) is a disorder that results from an error in cell division called ***nondisjunction.*** In this situation, sister chromatids fail to separate properly during anaphase of mitosis (or equatorial division of meiosis), or homologous chromosomes fail to separate properly during reduction division of meiosis. As a result, the chromatids or chromosomes pass to the same daughter cell. Down syndrome is characterized by mental retardation, retarded physical development (short stature and stubby fingers), distinctive facial structures (large tongue, flat profile, broad skull, slanting eyes, and round head), and malformation of the heart, ears, hands, and feet (Figure 26-12). Sexual maturity is rarely attained.

FIGURE 26-12 Photograph of an individual with Down syndrome. (Courtesy of Michael Mohan and Ohio School Pictures.)

Individuals with the disorder usually have 47 chromosomes instead of the normal 46 (an extra chromosome in the twenty-first pair). A cell that has one or more chromosomes of a set added or deleted is called an ***aneuploid*** (an′-yoo-PLOID). A monosomic cell ($2n - 1$) has a missing chromosome; a trisomic cell ($2n + 1$) has an added chromosome. Since the most common form of Down syndrome is characterized by an extra chromosome in the twenty-first pair, it is also known as ***trisomy 21.***

When both parents are known or suspected to be genetic carriers of any one of these disorders, or when maternal age approaches 35, amniocentesis is advised. The procedure is also advised when there is concern of a preterm delivery, a medical condition that necessitates an early delivery, and for patients who are Rh^- sensitized. Amniocentesis can also determine gender.

Using ultrasound, and palpation, the position of the fetus and placenta are first determined. After the skin is prepared with an antiseptic, a local anesthetic is given, a hypodermic needle is inserted through the mother's abdominal wall and uterus into the amniotic cavity, and about 10 to 20 ml of fluid is removed (Figure 26-13a). The test is usually done at 14–16 weeks of gestation. Cells and fluid are subjected to microscopic examination, biochemical testing, and chromosome studies.

CHORIONIC VILLUS SAMPLING (CVS)

Chorionic (ko-rē-ON-ik) ***villus*** (VIL-us) ***sampling (CVS)*** determines the same defects as amniocentesis, but it offers several advantages. It can be performed earlier, usually at 8 to 10 weeks of gestation. Moreover, the procedure does not require penetration of the abdomen, uterus, or amniotic cavity. The safety of the procedure is believed to be comparable to that for amniocentesis, although some feel that the risk to the fetus is slightly greater than with amniocentesis. A catheter is placed through the vagina (or a spinal needle is inserted through the abdominal cavity) into the uterus and then to the chorionic villi under ultrasound guidance (Figure 26-13b). About 30 mg of tissue is suctioned out and prepared for chromosomal analysis. Chorion cells and fetal cells contain identical genetic information.

FETAL ULTRASONOGRAPHY

Fetal ultrasonography (ul′-tra-son-OG-ra-fē) is used only when there is some clinical question about the normal progress of the pregnancy. By far the most common use of diagnostic ultrasound is to determine true fetal age when the date of conception is uncertain. It is also used to evaluate fetal viability and growth, determine fetal position, ascertain multiple pregnancies, identify fetal–maternal abnormalities, and serve as an adjunct to special procedures such as amniocentesis. Ultrasound is not used routinely to determine the sex of a fetus; it is performed only for a specific medical indication.

An instrument (transducer) that emits high-frequency sound waves is passed back and forth over the abdomen. The reflected sound waves from the developing fetus are picked up by the transducer and converted to an image on a screen (see Exhibit 1-6, Ultrasound). This image is called a ***sonogram.*** Since the urinary bladder serves as a landmark

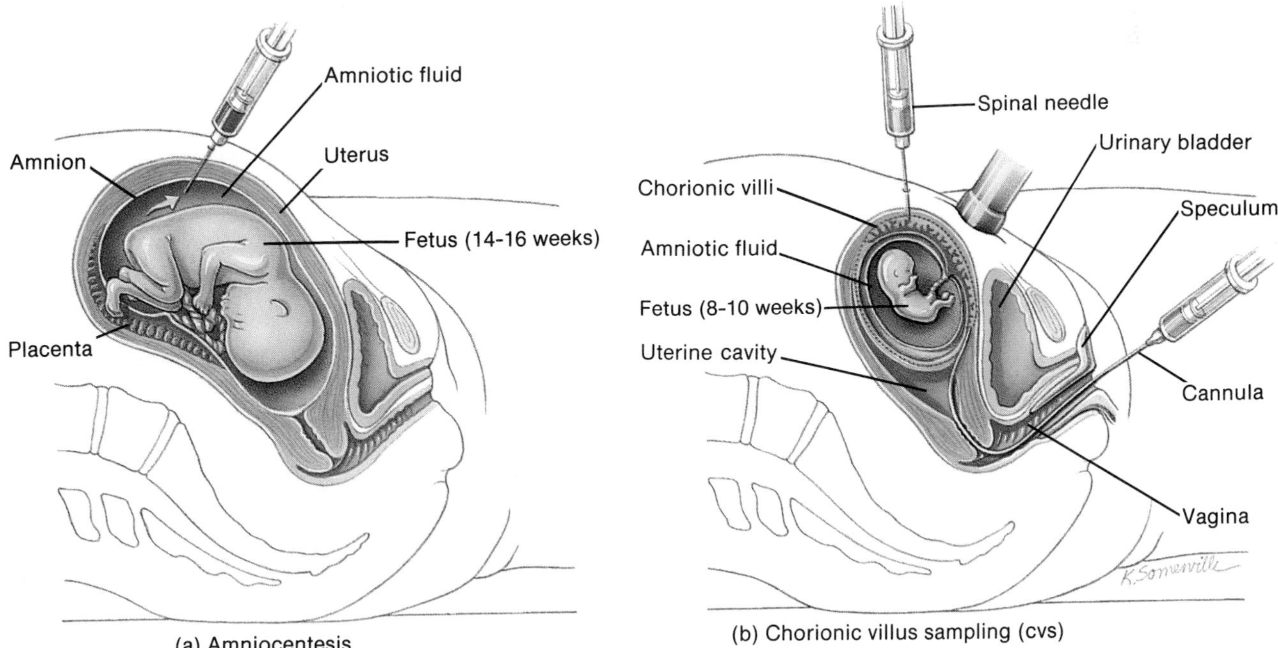

FIGURE 26-13 Prenatal diagnostic techniques.

during the procedure, the patient needs to drink liquids and not void in order to maintain a full bladder.

ELECTRONIC FETAL MONITORING (EFM)

***Electronic fetal monitoring* (*EFM*)** records fetal heart rate and maternal uterine contractions and is used to evaluate fetal well-being during labor and to detect any early signs of potential problems. EFM is also used to monitor the fetus during special tests that may be given prior to labor (nonstress test and oxytocin challenge test). Generally, EFM is used to monitor high-risk pregnancies.

In ***external fetal monitoring,*** two rubber straps are placed around the abdomen. Attached to the straps are sensors that detect fetal heart rate using ultrasound and uterine contractions or fetal movements. Measurements are recorded by a small machine that traces them on moving graph paper. External fetal monitoring can be done at any time, including early labor before the cervix dilates and the amniotic sac ruptures. In ***internal fetal monitoring,*** fetal heart rate is measured through an electrode placed through the mother's vagina and attached to the fetal scalp. A sensor attached to a catheter is also inserted through the vagina and placed in the uterus. Measurements are also made on graph paper. Internal fetal monitoring can be used only if cervical dilation has occurred and the amniotic sac has ruptured.

The ***nonstress test*** uses EFM to check a fetus's well-being before labor begins. After an external monitoring belt is applied to the abdomen, fetal movements are noted on the recording of the fetal heart rate. In this way, fetal movements and heart rate are timed simultaneously. A normal test shows that the fetus moves at least two to three times during a 20-minute period and the heart rate increases with each movement.

The ***oxytocin challenge*** or ***stress test*** uses EFM to check fetal well-being also before labor begins. After two external monitoring belts are applied to the abdomen, fetal heart rate and maternal uterine contractions are measured. A small amount of oxytocin is given intravenously to bring about uterine contractions. Fetal response to the contractions is observed. A normal fetus can adjust to the decreased amount of oxygen that accompanies a contraction, as evidenced by a heart rate that remains the same or increases. This suggests fetal well-being when natural contractions occur during labor. If, instead, the test shows a decreased heart rate during contractions, then fetal distress may possibly occur during delivery and a cesarean section may be indicated.

ALPHAFETOPROTEIN (AFP) TEST

Alphafetoprotein (al'-fa-fē'-tō-PRŌ-tēn), or ***AFP*** is a substance produced by liver cells of fetuses and adults and by the ovaries and testes. In adults, the level of AFP may be elevated with cancer of the liver, testicles, and ovaries. AFP can also be used to monitor response to therapy for these malignancies. When done on pregnant women, the test can help detect a neural tube defect, a twin pregnancy, or the need for further studies such as fetal ultrasonography.

AFP levels may be determined by a blood sample or amniotic fluid sample, which should be obtained at 15 to 16 weeks of gestation.

PARTURITION AND LABOR

The term ***parturition*** (par'-too-RISH-un) refers to birth. Parturition is accompanied by a sequence of events commonly called ***labor.*** The onset of labor is apparently related to a complex interaction of many factors. Just prior to birth, the muscles of the uterus contract rhythmically and forcefully. Both placental and ovarian hormones seem to play a role in these contractions. Since progesterone inhibits uterine contractions, labor cannot take place until its effects are diminished. At the end of gestation, the level of estrogens in the mother's blood is sufficient to overcome the inhibiting effects of progesterone, since the progesterone level falls, and labor commences. It has been suggested that some factor released by the placenta, fetus, or mother rather suddenly overcomes the inhibiting effects of progesterone so that estrogens can exert their effect. Prostaglandins may also play a role in labor. Oxytocin from the posterior pituitary gland also stimulates uterine contractions, and relaxin assists by relaxing the pubic symphysis and helping to dilate the uterine cervix.

Uterine contractions occur in waves, quite similar to peristaltic waves, that start at the top of the uterus and move downward. These waves expel the fetus. ***True labor*** begins when pains occur at regular intervals. The pains correspond to uterine contractions. As the interval between contractions shortens, the contractions intensify. Another sign of true labor in some females is localization of pain in the back, which is intensified by walking. A reliable indication of true labor is the "show" and dilation of the cervix. The "show" is a discharge of a blood-containing mucus that accumulates in the cervical canal during labor. In ***false labor,*** pain is felt in the abdomen at irregular intervals. The pain does not intensify and is not altered significantly by walking. There is no "show" and no cervical dilation.

Labor can be divided into three stages (Figure 26-14):

1. The ***stage of dilation*** is the time from the onset of labor to the complete dilation of the cervix. During this stage, there are regular contractions of the uterus, usually a rupturing of the amniotic sac, and complete dilation (10 cm) of the cervix. If the amniotic sac does not rupture spontaneously, it is done artificially.
2. The ***stage of expulsion*** is the time from complete cervical dilation to delivery.
3. The ***placental stage*** is the time after delivery until the placenta or "afterbirth" is expelled by powerful uterine contractions. These contractions also constrict blood ves-

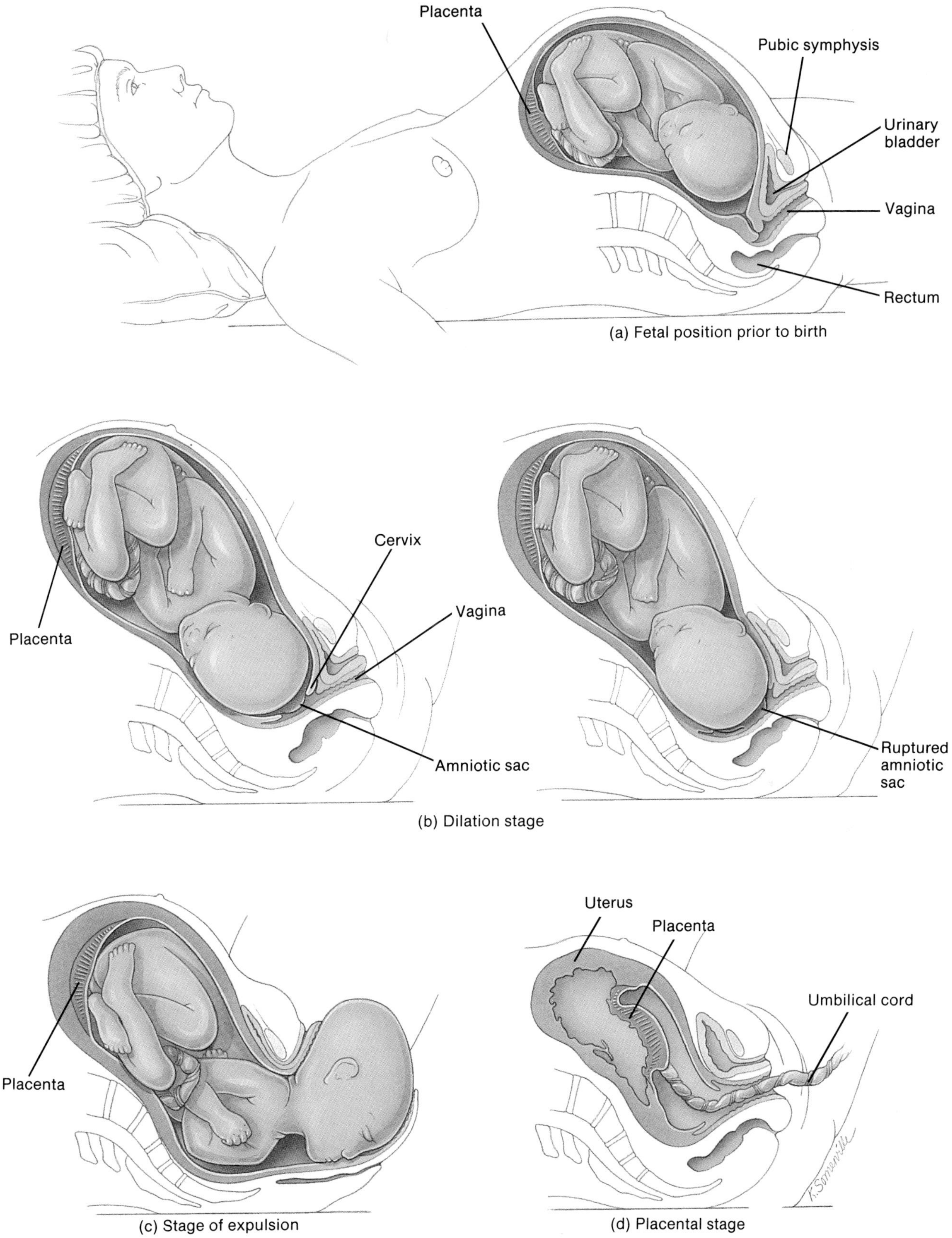

Placenta
Pubic symphysis
Urinary bladder
Vagina
Rectum
(a) Fetal position prior to birth
Cervix
Vagina
Placenta
Amniotic sac
Ruptured amniotic sac
(b) Dilation stage
Placenta
(c) Stage of expulsion
Uterus
Placenta
Umbilical cord
(d) Placental stage

FIGURE 26-14 Parturition. In (b), note protrusion of the amnionic sac through the partly dilated cervix (left) and rupture of the amnionic sac and complete dilation of cervix (right).

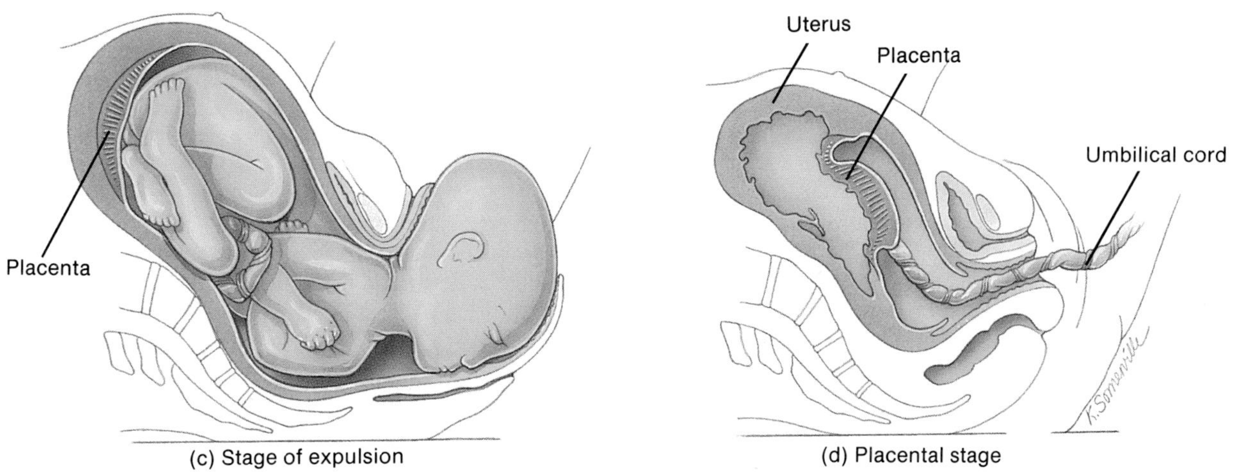

sels that were torn during delivery. In this way, the possibility of hemorrhage is reduced.

During labor, the fetus may be squeezed through the birth canal for up to several hours. As a result, the fetal head is compressed, and there is some degree of intermittent hypoxia due to compression of the umbilical cord and placenta during uterine contractions. In response to this compression, the adrenal medulla of a fetus secretes very high levels of epinephrine and norepinephrine (NE), the "fight-or-flight" hormones. Much of the protection afforded against the stresses of the birth process and preparation of the infant to survive extrauterine life are provided by the adrenal medullary hormones. Among other functions, the hormones clear the lungs and alter their physiology for breathing outside the uterus, mobilize readily usable nutrients for cellular metabolism, and promote a rich vascular supply to the brain and heart.

CLINICAL APPLICATION

Dystocia and Cesarean Section

Dystocia (dis-TŌ-sē-a), or difficult labor, may result from impaired uterine forces, an abnormal position (presentation) of the fetus, or a birth canal of inadequate size to permit vaginal birth. In these instances, and in certain conditions of fetal or maternal distress occurring during labor, it may be necessary to deliver the baby via a ***cesarean*** (*caedere* = to cut) ***section (C-section).*** In this procedure, a low, horizontal incision is made through the abdominal wall near the pubic hairline and lower portion of the uterus, through which the baby and placenta are removed. Even a history of multiple C-sections need not exclude a pregnant woman from attempting a vaginal delivery.

Following delivery of the baby and placenta, there is a period of time that lasts about six weeks during which the reproductive organs and maternal physiology return to the prepregnancy state. This is called the ***puerperium*** (pyoo′-er-PE-rē-um). Through a process of tissue catabolism, the uterus undergoes a remarkable reduction in size, called ***involution*** (in′-vō-LOO-shun). It is related primarily to a decrease in cytoplasm and cell size of myometrial cells. The cervix loses its elasticity and assumes its prepregnancy firmness. For up to about 10 days or more after delivery, there is a uterine discharge called ***lochia*** (LŌ-kē-a), which consists initially of blood and later serous fluid derived from the former placental site.

KEY MEDICAL TERMS ASSOCIATED WITH DEVELOPMENTAL ANATOMY

Abortion (a-BOR-shun) Premature expulsion from the uterus of the products of conception—embryo or nonviable fetus. May be caused by abnormal development of embryo, placental abnormalities, endocrine disturbances, certain diseases, trauma, and stress. **Induced (nonspontaneous) abortions** are brought on intentionally by methods such as vacuum aspiration (suction curettage), up to the twelfth week of pregnancy; dilation and evacuation (D & E), commonly performed between the thirteenth and fifteenth weeks of pregnancy and sometimes up to twenty weeks; and use of saline or prostaglandin preparations to induce labor and delivery, usually after the fifteenth week of pregnancy. **Spontaneous abortions (miscarriages)** occur without apparent cause. Recent evidence suggests that as many as one-third of all successful fertilizations end in spontaneous abortions, most of which occur even before a female or her physician is aware that she is pregnant.

Breech presentation A malpresentation in which the fetal buttocks or lower extremities present into the maternal pelvis; most common cause is prematurity.

Karyotype (KAR-ē-ō-tīp; *karyon* = nucleus) The chromosomal elements typical of a cell, drawn in their true proportions, based on the average of measurements determined in a number of cells. Useful in judging whether or not chromosomes are normal in number and structure.

Lethal gene (LĒ-thal jēn; *lethum* = death) A gene that, when expressed, results in death either in the embryonic state or shortly after birth.

Mutation (myoo-TĀ-shun; *mutare* = change) A permanent heritable change in a gene that causes it to have a different effect than it had previously.

Preeclampsia (prē′-e-KLAMP-sē-a) A syndrome characterized by sudden hypertension, large amounts of protein in urine, and generalized edema; might be related to autoimmune or allergic reaction due to the presence of a fetus; when the condition is also associated with convulsions and coma, it is referred to as **eclampsia.**

Puerperal (pyoo-ER-per-al; *puer* = child; *parere* = to bring forth) **fever** Infectious disease of childbirth, also called puerperal sepsis and childbed fever. The disease results from an infection originating in the birth canal and affects the endometrium. It may spread to other pelvic structures and lead to septicemia.

STUDY OUTLINE

Development During Pregnancy (p. 769)

1. Pregnancy is a sequence of events that includes fertilization.
2. Its various events are hormonally controlled.

Fertilization and Implantation (p. 769)

1. Fertilization refers to the penetration of a secondary oocyte by a sperm cell and the subsequent union of the sperm and oocyte nuclei to form a zygote.

2. Penetration is facilitated by enzymes produced by sperm acrosomes.
3. Normally, only one sperm fertilizes a secondary oocyte.
4. Early rapid cell division of a zygote is called cleavage, and the cells produced by cleavage are called blastomeres.
5. The solid mass of cells produced by cleavage is a morula.
6. The morula develops into a blastocyst, a hollow ball of cells differentiated into a trophoblast (future embryonic membranes) and inner cell mass (future embryo).
7. The attachment of a blastocyst to the endometrium is called implantation; it occurs by enzymatic degradation of the endometrium.
8. In vitro fertilization (IVF) refers to the fertilization of a secondary oocyte outside the body and the subsequent implantation of the zygote. Variations include embryo transfer, gamete intrafallopian transfer (GIFT), and transvaginal oocyte retrieval.

Embryonic Development (p. 773)
1. During embryonic growth, the primary germ layers and embryonic membranes are formed and the placenta is functioning.
2. The primary germ layers—ectoderm, mesoderm, and endoderm—form all tissues of the developing organism. These are summarized in Exhibit 26-1.
3. Embryonic membranes include the yolk sac, amnion, chorion, and allantois.
4. Fetal and maternal materials are exchanged through the placenta.

Fetal Growth (p. 779)
1. During the fetal period, organs established by the primary germ layers grow rapidly.
2. The principal changes associated with fetal growth are summarized in Exhibit 26-2.

Gestation (p. 780)
1. The time an embryo or fetus is carried in the uterus is called gestation.
2. Human gestation lasts about 280 days from the beginning of the last menstrual period.
3. During gestation, several anatomical and physiological changes occur.

Prenatal Diagnostic Tests (p. 782)
1. Amniocentesis is the withdrawal of amniotic fluid. It can be used to diagnose inherited biochemical defects and chromosomal disorders, such as hemophilia, Tay-Sachs disease, sickle-cell anemia, and Down syndrome.
2. Chorionic villus sampling (CVS) involves withdrawal of chorionic villi for chromosomal analysis. CVS can be done sooner than amniocentesis, and the results are available sooner.
3. In fetal ultrasonography, an image of a fetus is displayed on a screen.
4. Electronic fetal monitoring (EFM) records fetal heart rate and maternal uterine contractions.
5. The presence of abnormal levels of alphafetoprotein (AFP) can help detect neural tube defects.

Parturition and Labor (p. 784)
1. Parturition refers to birth and is accompanied by a sequence of events called labor.
2. The birth of a baby involves dilation of the cervix, expulsion of the fetus, and delivery of the placenta.

REVIEW QUESTIONS

1. Define developmental anatomy. (p. 769)
2. Define fertilization. Where does it normally occur? How is a morula formed? (p. 769)
3. Explain how dizygotic (fraternal) and monozygotic (identical) twins are produced. (p. 770)
4. Describe the components of a blastocyst. (p. 771)
5. What is implantation? How does the fertilized ovum implant itself? Why is an implanted ovum usually not rejected by the mother? (p. 771)
6. Describe the various types of in vitro fertilization (IVF). (p. 773)
7. Define the embryonic period and the fetal period. (p. 773)
8. List several body structures formed by the endoderm, mesoderm, and ectoderm. (p. 776)
9. What is an embryonic membrane? Describe the functions of the four embryonic membranes. (p. 774)
10. Explain the importance of the placenta and umbilical cord to fetal growth. (p. 777)
11. Outline some of the major developmental changes during fetal growth. (p. 780)
12. Define gestation and parturition. (p. 780)
13. Describe several anatomical and physiological changes that occur during gestation. (p. 780)
14. Explain the effects of pregnancy on exercise and exercise on pregnancy. (p. 782)
15. Explain the procedure and diagnostic value of the following: amniocentesis, chorionic villus sampling (CVS) , fetal ultrasonography, and electronic fetal monitoring (EFM). (p. 782)
16. Distinguish between false and true labor. Describe what happens during the stage of dilation, the stage of expulsion, and the placental stage of delivery. (p. 784)
17. Refer to the glossary of key medical terms associated with development. Be sure that you can define each term. (p. 786)

SELF QUIZ

Choose the one best answer to these questions.

___ 1. Which of these statements is true?
(1) Capacitation occurs in the male reproductive tract.
(2) The acrosome produces enzymes that help a sperm cell penetrate a secondary oocyte.
(3) A segmentation nucleus contains the diploid ($2n$) number of chromosomes.
(4) The zona pellucida is a gelatinous covering around the acrosome of a sperm cell.
A. (1), (2), (3); B. (2), (3), (4); C. (2), (3); D. (1), (3); E. all are true.

___ 2. Implantation of a developing individual (blastocyst stage) usually occurs about ______ after fertilization.
A. 3 weeks; B. 1 week; C. 1 day; D. 7 hours; E. 7 minutes.

___ 3. During pregnancy
(1) the uterus extends upward into the abdominal cavity.
(2) there is an increased pulse rate.
(3) there is a decrease in air entering the lungs.
(4) frequency and urgency of urination may occur.
A. (1), (2), (4); B. (1), (2), (3); C. (2), (3), (4); D. (2) and (3) only; E. all are true.

___ 4. Which procedures involve use of the mother's *own* secondary oocytes for fertilization?
(1) Embryo transfer.
(2) Gamete intrafallopian transfer (GIFT).
(3) Transvaginal oocyte retrieval.
A. (2), (3); B. (1), (3); C. (1), (2); D. none is correct; E. all are correct.

___ 5. Which of the following is *not* an embryonic membrane?
A. amnion; B. membranous labyrinth; C. chorion; D. allantois; E. yolk sac.

___ 6. The placenta is formed by the union of the decidua basalis of the endometrium with the
A. yolk sac; B. amnion; C. chorion; D. umbilicus; E. allantois.

___ 7. The "bag of waters" that ruptures just before birth is the
A. placenta; B. allantois; C. chorion; D. amnion; E. umbilical cord.

Complete the following:

8. Which portion of the decidua is located between the embryo and uterine cavity? (basalis, capsularis, parietalis)

9. The embryonic tissues from which all tissues and organs develop are called ______ layers.

10. In a developing embryo, the upper layer of cells of the inner cell mass near the amniotic cavity develop into ______, whereas the bottom layer of the inner cell mass that borders the blastocoel develops into ______.

11. Identify descriptions of the three phases of labor (first, second, or third).
___ a. stage of expulsion: from complete cervical dilation through delivery of the baby
___ b. time after the delivery of the baby until the placenta ("afterbirth") is expelled; the placental stage
___ c. time from onset of labor to complete dilation of the cervix; the stage of dilation

12. Amniocentesis involves withdrawal of ______ fluid, usually at about (2–4? 8–10? 16–20?) weeks after conception. CVS (meaning ______) is a procedure most often performed at about (2–4? 8–10? 16–20?) weeks. It (does? does not?) involve penetration of the uterine cavity.

13. Write the name of each fetal membrane next to its description.
a. Originally formed from ectoderm, this membrane encloses fluid that acts as a shock absorber for the developing baby: ______.
b. Derived from mesoderm and trophoblast, it becomes the principal embryonic part of the placenta: ______.
c. Endoderm-lined membrane serving as exclusive nutrient supply for embryos of some species: ______.
d. A small membrane that forms umbilical blood vessels: ______.

14. Identify what structures will ultimately form from these two parts of the implanted blastocyst. Trophoblast: ______; inner cell mass: ______.

15. The early rapid division of a zygote that results in the production of smaller cells called blastomeres is known as ______.

16. The normal gestation period is about ______ days from the beginning of the last menstrual period.

17. Successive divisions of a zygote produce a solid mass of cells called the ______.

18. Match the following:

___ a. epithelial lining of all gastrointestinal, respiratory, and genitourinary tracts except near openings to the exterior of the body		A. ectoderm B. endoderm C. mesoderm
___ b. epidermis of skin, epithelial lining of entrances to the body (such as mouth, nose, and anus), hair, nails		
___ c. all of the skeletal system (bone, cartilage, joint cavities)		
___ d. muscle (skeletal, smooth, and cardiac)		
___ e. blood and all blood and lymphatic vessels		
___ f. entire nervous system		
___ g. thyroid, parathyroid, thymus, and pancreas		

19. Arrange the answers in correct sequence.

___ ___ ___ ___ a. From most superficial to deepest (closest to embryo):
A. amnion
B. amniotic cavity
C. chorion
D. decidua

___ ___ ___ ___ ___ b. Stages in development:
A. morula
B. blastocyst
C. zygote
D. fetus
E. embryo

Appendix Answers to Self Quizzes

Chapter 1

1. a. armpit b. brachium c. head d. thorax e. cervical **2.** C **3.** D **4.** B **5.** B **6.** E **7.** C **8.** E **9.** A **10.** B **11.** E **12.** C **13.** E **14.** B **15.** E **16.** C **17.** right iliac **18.** anterior **19.** ipsi **20.** midsagittal

Chapter 2

1. C **2.** C **3.** A **4.** C **5.** A **6.** A **7.** D **8.** E **9.** a. J b. L c. C d. E e. F f. K g. I h. G i. H j. A k. D l. B **10.** eating **11.** phospholipids **12.** meiosis **13.** S **14.** 23, *n* **15.** a. A b. D c. E d. C e. B

Chapter 3

1. C **2.** A **3.** C **4.** B **5.** D **6.** A **7.** E **8.** D **9.** C **10.** E **11.** E **12.** B **13.** C **14.** a. F b. B c. H d. J e. L f. D g. K h. G i. C j. M k. A l. N m. E n. I **15.** osseous, lamellae, osteocytes, lacunae **16.** visceral, parietal **17.** mesenchyme **18.** basement membrane **19.** vascular, plasma **20.** exocrine

Chapter 4

1. C **2.** A **3.** C **4.** B **5.** A **6.** a. A b. E c. C d. D e. B **7.** melanocyte **8.** ecto, fourth, meso **9.** matrix **10.** papillae, touch **11.** epidermis, epithelium **12.** dermis, connective tissue **13.** A, C, B **14.** B, C, A **15.** A. sudoriferous B. sebaceous C. ceruminous **16.** T **17.** T **18.** F **19.** F **20.** F

Chapter 5

1. a. A b. D c. C d. E e. B **2.** C **3.** D **4.** B **5.** E **6.** B **7.** E **8.** C **9.** E **10.** matrix **11.** meso-, mesenchyme **12.** hyaline cartilage and fibrous membranes **13.** a. lacunae b. osteocytes c. lamellae d. osteons (Haversian canals) e. medullary f. canaliculi g. perforating (Volkmann's) canals **14.** a. B, C, A b. A, C, B **15.** T **16.** F **17.** T **18.** T **19.** T **20.** F **21.** T **22.** F

Chapter 6

1. a. E b. N c. D d. G e. L f. H g. J h. F i. I j. B k. A l. K m. M n. C **2.** D **3.** C **4.** B **5.** A **6.** D **7.** C **8.** E **9.** E **10.** B **11.** D **12.** C **13.** A **14.** A **15.** C **16.** E **17.** B **18.** C **19.** a. A, B, C b. C, B, A c. A, C, B **20.** a. A b. A c. C d. E e. C f. E g. B h. D

Chapter 7

1. D **2.** B **3.** E **4.** B **5.** C **6.** B **7.** D **8.** D **9.** B **10.** E **11.** C **12.** A, C, B **13.** C, B, A **14.** carpals, 8 **15.** metacarpals, 5, I

Chapter 8

1. a. F b. G c. E d. A e. J f. D **2.** a. knee b. knee c. hip d. shoulder e. knee f. shoulder **3.** lack **4.** more, syndesmoses **5.** hyaline, epiphyseal **6.** a. E b. F c. A d. D e. D f. B, F **7.** D **8.** A **9.** D **10.** E **11.** D **12.** A **13.** D **14 a.** E **b.** A **c.** D **d.** B **e.** C **15.** B **16.** A **17.** F **18.** F **19.** F **20.** T **21.** F **22.** B, A, C

Chapter 9

1. a. S b. D c. S d. S e. D **2.** myofiber, sarcolemma, sarcoplasm **3.** triad, myofilaments **4.** shortens, stays the same length, stays the same length, shortens, shortens or disappears **5.** spindle, one nucleus, do, nonstriated or smooth, slowly, longer **6.** skeletal, somites **7.** a. C b. A c. B **8.** a. B b. C c. A d. D **9.** a. C, A, B b. C, A, B **10.** B **11.** C **12.** B **13.** B **14.** E **15.** C **16.** B **17.** C **18.** F **19.** T **20.** T **21.** F

Chapter 10

1. lever, fulcrum **2.** insertion **3.** R, E, occipital, first **4.** prime mover (agonist), antagonists **5.** a. frontalis b. buccinator **6.** facial, VII **7.** elevating the mandible, temporalis, masseter **8.** the same as, superiorly **9.** increase, inspiration **10.** urogenital **11.** scapula **12.** femur **13.** thenar, hypothenar **14.** flexor retinaculum **15.** anterior, flex **16.** segmental **17.** anterior, 4, tibia, extension, antagonists, posterior **18.** tibial **19.** ischium, extension, flexion **20.** plantar **21.** B **22.** A **23.** B **24.** E **25.** D **26.** a. B, C b. C, F c. D, C d. E e. A, F **27.** a. D b. E c. C d. B e. A

Chapter 11

1. a. head b. cranium c. side of skull d. supercilia e. eyelids f. oral region g. buccal h. ear i. ankle j. antebrachium k. hand l. phalanges (digits) m. gluteal region n. leg **2.** cranium, face **3.** sternocleidomastoid, mandible, cervical midline, clavicle, trapezius **4.** a. deltoid b. brachioradialis c. acromioclavicular joint d. site of the ulnar nerve e. cubital fossa f. depression between tendons of two muscles that move the thumb g. distal ends of second through fifth metacarpals h. styloid process i. iliac crest j. vastus lateralis k. tibial tuberosity **5.** a. B, C, A b. A, C, B c. C, A, B d. B, A, C e. B, C, A f. C, A, B g. B, A, C h. A, B, C **6.** T **7.** F **8.** T **9.** T **10.** F **11.** T **12.** T **13.** T **14.** F **15.** T **16.** F **17.** F **18.** T **19.** F **20.** F

Chapter 12

1. a. E b. C c. C d. A e. B f. D g. D, E h. A, B, E **2.** 55, 45, plasma, hemopoiesis **3.** C, A, B, D **4.** in, 15,000, leukocytosis, differential white blood cell **5.** C **6.** A **7.** E **8.** D **9.** B **10.** A **11.** C **12.** C **13.** A **14.** B **15.** C **16.** A **17.** C **18.** E **19.** T **20.** F **21.** T **22.** F **23.** F **24.** T **25.** F

Chapter 13

1. a. B b. D c. C d. B e. A, C f. B, D **2.** meso, third, primitive heart **3.** lubb, closing, atrioventricular **4.** veins, left atrium **5.** apex, fifth, second rib **6.** atria, ventricles **7.** right atrium, superior vena cava, coronary sinus **8.** E **9.** E **10.** C **11.** D **12.** D **13.** A **14.** B **15.** A **16.** B **17.** A **18.** C, E, D, A, B **19.** C, D, E, A, B

Chapter 14

1. A. umbilical vein; B. ductus venosus; C. femoral artery; D. pulmonary artery; E. thoracic portion of inferior vena cava **2.** hepatic portal **3.** fourth lumbar, common iliac arteries **4.** vertebral, basilar **5.** aorta, coronary **6.** common hepatic, left gastric, splenic **7.** A. superior vena cava; B. inferior vena cava; C. coronary sinus (answers can be in any sequence) **8.** internal jugular **9.** the heart and a lung **10.** Yes, Yes, Yes **11.** A, C, E, F, G **12.** A, D, G **13.** D **14.** A **15.** C **16.** C **17.** D **18.** B **19.** C **20.** D **21.** C **22.** D **23.** D **24.** C **25.** D **26.** D **27.** D **28.** E **29.** A, B, C, D **30.** A, B, C, D, E

Chapter 15

1. afferent **2.** cisterna chyli **3.** meso, fifth **4.** T **5.** T **6.** F **7.** F **8.** T **9.** D, C, B, A, E **10.** A, C, B, D **11.** a. F b. I c. H d. C e. B f. D g. E h. A i. G j. J **12.** D **13.** E **14.** D **15.** D **16.** D **17.** E **18.** C **19.** C

Chapter 16

1. C **2.** C **3.** A **4.** C **5.** C **6.** E **7.** E **8.** A **9.** E **10.** A **11.** a. C b. F c. G d. H e. D f. E g. A h. B **12.** synapse **13.** dendrite **14.** synaptic vesicles **15.** parallel after-discharge **16.** T **17.** F **18.** T **19.** T **20.** F

Chapter 17

1. a. B b. D c. C d. A **2.** endo, peri, epi **3.** sensory, anterior, lateral **4.** white columns, myelinated, sensory, brain **5.** a. A b. B c. A d. C e. B f. D g. E **6.** C **7.** C **8.** D **9.** B **10.** B **11.** A **12.** C **13.** D **14.** C **15.** B **16.** C **17.** a. D b. C c. A d. B e. F f. E **18.** F **19.** T **20.** T **21.** F **22.** T **23.** F **24.** a. B, C, A, b. B, C, A, c. C, B, A, D

Chapter 18

1. cuneatus **2.** inferior **3.** tentorium cerebelli **4.** association **5.** chiasma, tracts **6.** cortex, gray, cell bodies, gyri, fissures, sulci **7.** basal ganglia, gray **8.** ecto, neural **9.** a. accessory; b. vestibulocochlear; c. vagus; d. trigeminal; e. trigeminal; f. oculomotor; g. facial; h. facial, glossopharyngeal; i. olfactory, optic, vestibulocochlear (answers can be in any sequence) **10.** A, C, D **11.** B, D, E **12.** a. A b. C c. C d. D e. B f. A g. E h. B i. E j. A k. E l. C m. A **13.** C **14.** C **15.** C **16.** B **17.** B **18.** B **19.** A **20.** C **21.** D **22.** B **23.** a. B, A, C b. C, A, B **24.** T **25.** F **26.** T **27.** T **28.** F **29.** T **30.** T

Chapter 19

1. a. S b. P c. S d. P e. S f. S g. S h. S i. P, S j. S **2.** A, E, F **3.** A, C, D **4.** B **5.** a. P ↓ S ↑ b. P ↓ S ↑ c. P = S ↑ d. P ↑ S ↓ e. P ↓ S ↑ f. P ↑ S ↓ g. P = S ↑ h. P = S ↑ i. P ↑ S = j. P ↓ S ↑ k. P ↑ S ↓ l. P ↑ S ↓ **6.** T **7.** F **8.** T **9.** T **10.** D **11.** E **12.** D **13.** D **14.** D **15.** A **16.** B **17.** E **18.** B **19.** C

Chapter 20

1. upper motor **2.** second-order **3.** basal ganglia, reticular formation **4.** papillae **5.** otoliths, static **6.** G, D, B, F, E, A, H, C **7.** a. D b. A c. C d. B **8.** D, B, A, C **9.** B, C, D, E, A **10.** B, C, A, D, E **11.** A, B, E, C, D **12.** A, C, B **13.** C, D, A, B **14.** C, D, A, B **15.** A, E, C, B, D **16.** A, D, E, B, C **17.** a. K b. F c. G d. A e. C f. J g. B h. I i. E j. D k. H l. L **18.** A **19.** B **20.** C **21.** C **22.** B **23.** A **24.** B **25.** B **26.** a. A b. H c. I d. D e. E f. C g. B

Chapter 21

1. alpha, increases **2.** SHA, PP, HPV, SP **3.** B **4.** A **5.** C **6.** D **7.** A **8.** C **9.** C **10.** D **11.** E **12.** D **13.** B **14.** C **15.** B **16.** E **17.** C **18.** F **19.** F **20.** T **21.** F **22.** F **23.** F **24.** T **25.** a. D b. C c. B d. A e. E, F

Chapter 22

1. base, clavicle, ribs **2.** C, B, A **3.** right, left, left, right **4.** endo **5.** a. D b. E c. F d. B e. A f. A g. D, E h. C. F **6.** a. A, C, B b. D, C, E, B, A c. A, E, C, B, D d. D, C, E, B, A, F **7.** A **8.** E **9.** D **10.** B **11.** E **12.** C **13.** D **14.** E **15.** D **16.** C **17.** T **18.** T **19.** F **20.** F **21.** F

Chapter 23

1. pepsin **2.** a. E b. E c. M **3.** 32, 20, incisors, cuspids, premolars, molars **4.** defecation **5.** cystic **6.** mouth **7.** circular folds **8.** anal canal **9.** lateral **10** a. A, D, B, C b. E, D, B, A, C c. A, D, C, B, E d. A, B, D, C, E **11.** a. A b. D c. E d. B e. C **12.** D **13.** C **14.** B **15.** B **16.** E **17.** B **18.** A **19.** C **20.** A **21.** D **22.** B **23.** B **24.** C **25.** D **26.** B **27.** A

1. podocytes **2.** detrusor **3.** hemodialysis **4.** renal corpuscle **5.** nephron **6.** renal pyramids **7.** ureter **8.** a. A, B, C b. A, C, B, D c. A, C, E, B. D d. E, B, A, D, C e. C, E, A, B, D **9.** A **10.** C **11.** D **12.** E **13.** D **14.** B **15.** B **16.** C **17.** C **18.** B **19.** C **20.** A

Chapter 25

1. a. scrotum; b. penis; c. prostate gland; d. bulbourethral (Cowper's) glands **2.** haploid, 23, zygote, diploid, 2*n*, homologous **3.** a. M; b. E; c. E **4.** spermatogenesis **5.** spermatogonium **6.** cardinal, myo, smooth muscle, endo, functionalis, spiral **7.** corpora cavernosa **8.** ovarian **9.** ovulation **10.** corpus albicans **11.** vagina **12.** a. B, C, A b. A, B, C c. B, A, C d. C, D, B, A e. B, E, D, A, C f. B, C, D, A, E g. F, C, B, D, A, E h. B, C, A **13.** a. B b. D c. E d. F e. A f. C g. G **14.** E **15.** E **16.** C **17.** A **18.** A **19.** A **20.** B **21.** C **22.** a. B b. D c. C d. A

Chapter 26

1. C **2.** B **3.** A **4.** A **5.** B **6.** C **7.** D **8.** capsularis **9.** primary germ **10.** ectoderm, endoderm **11.** a. second; b. third; c. first **12.** amniotic, 16–20; chorionic villus sampling, 8–10, does not **13.** a. amnion, b. chorion, c. yolk sac, d. allantois **14.** fetal portion of placenta, embryo **15.** cleavage **16.** 280 **17.** morula **18.** a. B b. A c. C d. C e. C f. A g. B **19.** a. D, C, A, B b. C, A, B, E, D

Glossary of Combining Forms, Word Roots, Prefixes, and Suffixes

Many medical terms are "compound" words; that is, they are made up of one or more word roots or combining forms of word roots with prefixes or suffixes. For example, *leukocyte* (white blood cell) is a combination of *leuko*, the combining form for the word root meaning "white," and *cyt*, the word root meaning "cell." Learning the medical meanings of the fundamental word parts will enable you to analyze many long, complicated terms.

The following list includes some of the most commonly used combining forms, word roots, prefixes, and suffixes used in making medical terms and an example for each.

COMBINING FORMS AND WORD ROOTS

Acou-, Acu- **hearing** Acoustics (a-KOO-stiks), the science of sounds or hearing.

Acr-, Acro- **extremity** Acromegaly (ak'-rō-MEG-a-lē), hyperplasia of the nose, jaws, fingers, and toes.

Aden-, Adeno- **gland** Adenoma (ad-en'-Ō-ma), a tumor with a gland-like structure.

Alg-, Algia- **pain** Neuralgia (nyoo-RAL-ja), pain along the course of a nerve.

Angi- **vessel** Angiocardiography (an'-jē-ō-kard-ē-OG-ra-fē), roentgenography of the great blood vessels and heart after intravenous injection of radiopaque fluid.

Arthr-, Arthro- **joint** Arthropathy (ar-THROP-a-thē), disease of a joint.

Aut-, Auto- **self** Autolysis (aw-TOL-i-sis), destruction of cells of the body by their own enzymes, even after death.

Bio- **life, living** Biopsy (BĪ-op-sē), examination of tissue removed from a living body.

Blast- **germ, bud** Blastocyte (BLAS-tō-sīt), an embryonic or undifferentiated cell.

Blephar- **eyelid** Blepharitis (blef-a-RĪT-is), inflammation of the eyelids.

Brachi- **arm** Brachialis (brā-kē-AL-is), muscle that flexes the forearm.

Bronch- **trachea, windpipe** Bronchoscopy (bron-KOS-kō-pē), direct visual examination of the bronchi.

Bucc- **cheek** Buccocervical (bū-kō-SER-vi-kal), pertaining to the cheek and neck.

Capit- **head** Decapitate (dē-KAP-i-tāt), to remove the head.

Carcin- **cancer** Carcinogenic (kar-sin-ō-JEN-ik), causing cancer.

Cardi-, Cardia-, Cardio- **heart** Cardiogram (KARD-ē-o-gram), a recording of the force and form of the heart's movements.

Cephal- **head** Hydrocephalus (hī-drō-SEF-a-lus), enlargement of the head due to an abnormal accumulation of fluid.

Cerebro- **brain** Cerebrospinal (se-rē'-brō-SPĪN-al) fluid, fluid contained within the cranium and spinal canal.

Cheil- **lip** Cheilosis (kī-LŌ-sis), dry scaling of the lips.

Chole- **bile, gall** Cholecystogram (kō-lē-SIS-tō-gram), roentgenogram of the gallbladder.

Chondr-, Chondri-, Chondrio- **cartilage** Chondrocyte (KON-drō-sīt), a cartilage cell.

Chrom-, Chromat-, Chromato- **color** Hyperchromic (hī-per-KRŌ-mik), highly colored.

Cili- **eyelash** Supercilia (soo'-per-SIL-ē-a), eyebrow (hairs above eyelash).

Colpo- **vagina** Colpotomy (kol-POT-ō-mē), incision into the wall of the vagina.

Cor-, coron- **heart** Coronary (KOR-ō-na-rē), arteries supplying blood to the heart muscle.

Cost- **rib** Costal (KOS-tal), pertaining to a rib.

Crani- **skull** Craniotomy (krā-nē-OT-ō-mē), surgical opening of the skull.

Cry-, Cryo- **cold** Cryosurgery (krī-ō-SERJ-e-rē), surgical procedure using a very cold liquid nitrogen probe.

Cut- **skin** Subcutaneous (sub-kyoo-TĀ-nē-us), under the skin.

Cysti-, Cysto- **sac, bladder** Cystoscope (SIS-tō-skōp), instrument for interior examination of the urinary bladder.

Cyt-, Cyto-, Cyte- **cell** Cytology (sī-TOL-ō-jē), the study of cells.

Dactyl-, Dactylo- **digits** (usually fingers, but sometimes toes) Polydactylism (pol-ē-DAK-til-ism), above normal number of fingers or toes.

Derma-, Dermato- **skin** Dermatosis (der-ma-TŌ-sis), any skin disease.

Dura- **hard** Dura mater (DYOO-ra MĀ-ter), outer membrane covering brain and spinal cord.

Entero- **intestine** Enteritis (ent-e-RĪT-is), inflammation of the intestine.

Erythro- **red** Erythrocyte (e-RITH-rō-sīt), red blood cell.

Galacto- **milk** Galactose (ga-LAK-tōse), a milk sugar.

Gastr- **stomach** Gastrointestinal (gas'-trō-in-TES-tin-al), pertaining to the stomach and intestine.

Gloss-, Glosso- **tongue** Hypoglossal (hī'-pō-GLOS-al), located under the tongue.

Glyco- **sugar** Glycosuria (glī'-kō-SUR-ē-a), sugar in the urine.

Gravid- **pregnant** Gravidity (gra-VID-i-tē), condition of being pregnant.

Gyn-, Gyne-, Gynec- **female, women** Gynecology (gīn'-e-KOL-ō-jē), the medical specialty dealing with disorders of the female reproductive system.

Hem-, Hemat- **blood** Hematoma (hē'-ma-TŌ-ma), a tumor or swelling filled with blood.

Hepar-, Hepato- **liver** Hepatitis (hep-a-TĪT-is), inflammation of the liver.

Hist-, Histio- **tissue** Histology (his-TOL-ō-jē), the study of tissues.

Hydr- **water** Hydrocele (HĪ-drō-sēl), accumulation of fluid in a saclike cavity.

Hyster- **uterus** Hysterectomy (his'-te-REK-tō-mē), surgical removal of the uterus.

Ileo- **ielum** Ileocecal (il'-ē-ō-SĒ-kal) valve, folds at the opening between ileum and cecum.

Ilio- **ilium** Iliosacral (il'-ē-ō-SĀ-kral), pertaining to ilium and sacrum.

Kines- **motion** Kinesiology (ki-nē-sē-OL-ō-jē), study of movement of body parts.

Labi- **lip** Labial (LĀ-bē-al), pertaining to a lip.
Lachry-, Lacri- **tears** Nasolacrimal (nā-zō-LAK-rim-al), pertaining to the nose and lacrimal apparatus.
Laparo- **loin, flank, abdomen** Laparoscopy (lap'-a-ROS-kō-pē), examination of the interior of the abdomen by means of a laparoscope.
Leuco-, Leuko- **white** Leucocyte (LYOO-kō-sīt), white blood cell.
Lingua- **tongue** Lingual (LIN-gwal), pertaining to the tongue.
Lip-, Lipo **fat** Lipoma (lī-PŌ-ma), a fatty tumor.
Lith- **stone** Lithiasis (li-THĒ-a-sis), the formation of stones.
Lumbo- **lower back, loin** Lumbar (LUM-bar), pertaining to the loin.

Macul- **spot, blotch** Macula (MAK-yoo-la), spot or blotch.
Malign- **bad, harmful** Malignant (ma-LIG-nant), condition that gets worse and results in death.
Mamm- **breast** Mammography (ma-MOG-ra-fē), roentgenography of the mammary gland.
Mast- **breast** Mastitis (ma-STĪT-is), inflammation of the mammary gland.
Meningo- **membrane** Meningitis (men-in-JĪT-is), inflammation of the membranes of spinal cord and brain.
Metro- **uterus** Endometrium (en'-dō-MĒ-trē-um), lining of the uterus.
Morpho- **form, shape** Morphology (mor-FOL-o-jē), the study of form and structure of the things.
Myelo- **marrow, spinal cord** Poliomyelitis (pō'-lē-ō-mī'-a-LĪT-is), inflammation of the gray matter of the spinal cord.
Myo- **muscle** Myocardium (mī-ō-KARD-ē-um), heart muscle.

Necro- **corpse, dead** Necrosis (ne-KRŌ-sis), death of areas of tissue surrounded by healthy tissue.
Nephro- **kidney** Nephrosis (ne-FRŌ-sis), degeneration of kidney tissue.
Neuro- **nerve** Neuroblastoma (nyoor'-ō-blas-TŌ-ma), malignant tumor of the nervous system composed of embryonic nerve cells.

Oculo- **eye** Binocular (bī-NOK-yoo-lar), pertaining to the two eyes.
Odont- **tooth** Orthodontic (or-thō-DONT-ik), pertaining to the proper positioning and relationship of the teeth.
Onco- **mass, tumor** Oncology (ong-KOL-ō-jē), study of tumors.
Oo- **egg** Oocyte (Ō-ō-sīt), original egg cell.
Oophor- **ovary, egg carrier** Oophorectomy (ō'-ōf-o-REK-tō-mē), surgical removal of ovaries.
Ophthalm- **eye** Ophthalmology (of '-thal-MOL-ō-jē), the study of the eye and its diseases.
Or- **mouth** Oral (Ō-ral), pertaining to the mouth.
Orchido- **testicle** Orchidectomy (or'-ki-DEK-tō-me), surgical removal of a testicle.
Osmo- **odor, sense of smell** Anosmia (an-OZ-mē-a), absence of sense of smell.
Oss-, Osseo-, Osteo- **bone** Osteoma (os-tē-Ō-ma), bone tumor.
Oto- **ear** Otosclerosis (ō'-tō-skle-RŌ-sis), formation of bone in the labyrinth of the ear.

Palpebr- **eyelid** Palpebra (PAL-pe-bra), eyelid.
Part- **birth, delivery, labor** Parturition (par'-too-RISH-un), act of giving birth.
Patho- **disease** Pathogenic (path'-ō-JEN-ik), causing disease.
Ped- **children** Pediatrician (pēd-ē-a-TRISH-an), medical specialist in the treatment of children.
Peps- **digest** Peptic (PEP-tik), pertaining to digestion.
Phag-, Phago- **to eat** Phagocytosis (fag'-ō-sī-TŌ-sis), the process by which cells ingest particulate matter.
Philic-, Philo- **to like, have an affinity for** Hydrophilic (hī-drō-FIL-ik), having an affinity for water.
Phleb- **vein** Phlebitis (fle-BĪT-is), inflammation of the veins.
Phon- **voice, sound** Phonogram (FŌ-nō-gram), record made of sound.
Phren- **diaphragm** Phrenic (FREN-ik), pertaining to the diaphragm.
Pilo- **hair** Depilatory (de-PIL-a-tō-re), hair remover.
Pneumo- **lung, air** Pneumothorax (nyoo-mō-THŌR-aks), air in the thoracic cavity.
Pod- **foot** Podiatry (po-DĪ-a-trē), the diagnosis and treatment of foot disorders.
Procto- **anus, rectum** Proctoscopy (prok-TOS-kō-pē), instrumental examination of the rectum.
Psycho- **soul, mind** Psychiatry (sī-KĪ-a-trē), treatment of mental disorders.
Pulmon- **lung** Pulmonary (PUL-mō-ner'-ē), pertaining to the lungs.
Pyle-, Pyloro **opening, passage** Pyloric (pī-LOR-ik), pertaining to the pylorus of the stomach.
Pyo- **pus** Pyuria (pī-YOOR-ē-a), pus in the urine.

Ren- **kidneys** Renal (RĒ-nal), pertaining to the kidney.
Rhin- **nose** Rhinitis (ri-NĪT-is), inflammation of nasal mucosa.

Salpingo- **uterine (Fallopian) tube** Salpingitis (sal'-pin-JĪ-tis), inflammation of the uterine (Fallopian) tubes.
Scler-, Sclero- **hard** Atherosclerosis (ath'-er-ō-skle-RŌ-sis), hardening of the arteries.
Sep-, Septic- **toxic condition due to microorganisms** Septicemia (sep'-ti-SĒ-mē-a), presence of bacterial toxins in the blood (blood poisoning).
Soma-, Somato- **body** Somatotropic (sō-mat-ō-TRŌ-pik), having a stimulating effect on body growth.
Somni- **sleep** Insomnia (in-SOM-nē-a), inability to sleep.
Sten- **narrow** Stenosis (ste-NŌ-sis), narrowing of a duct or canal.
Stasis-, Stat- **stand still** Homeostasis (hō'-mē-ō-STĀ-sis), achievement of a steady state.

Tegument- **skin, covering** Integumentary (in-teg-yoo-MEN-ta-rē), pertaining to the skin.
Therm- **heat** Thermometer (ther-MOM-et-er), instrument used to measure and record heat.
Thromb- **clot, lump** Thrombus (THROM-bus), clot in a blood vessel or heart.
Tox-, Toxic- **poison** Toxemia (tok-SĒ-mē-a), poisonous substances in the blood.
Trich- **hair** Trichosis (trik-Ō-sis), disease of the hair.
Tympan- **eardrum** Tympanic (tim-PAN-ik) membrane, eardrum.

Vas- **vessel, duct** Cerebrovascular (se-rē-brō-VAS-kyoo-lar), pertaining to the blood vessels of the cerebrum of the brain.
Viscer- **organ** Visceral (VIS-e-ral), pertaining to the abdominal organs.

Zoo- **animal** Zoology (zō-OL-o-jē), the study of animals.

PREFIXES

A-, An- **without, lack of, deficient** Anesthesia (an'-es-THĒ-zha), without sensation.
Ab- **away from, from** Abnormal (ab-NOR-mal), away from normal.
Ad- **to, near, toward** Adduction (a-DUK-shun), movement of an extremity toward the axis of the body.
Alb- **white** Albino (al-BĪ-no), person whose skin, hair, and eyes lack the pigment melanin.
Alveol- **cavity, socket** Alveolus (al-VĒ-ō-lus), air sac in the lung.
Ambi- **both sides** Ambidextrous (am'-bi-DEK-strus), able to use either hand.

Ambly- **dull** Amblyaphia (am-blē-A-fē-a), dull sense of touch.

Andro- **male, masculine** Androgen (AN-drō-jen), male sex hormone.

Ankyl(o)- **bent, fusion** Ankylosed (ANG-ki-lōsd), fused joint.

Ante- **before** Antepartum (ant-ē-PAR-tum), before delivery of a baby.

Anti- **against** Anticoagulant (an-tī-kō-AG-yoo-lant), a substance that prevents coagulation of blood.

Basi- **base, foundation** Basal (BĀ-sal), located near the base.

Bi- **two, double, both** Biceps (BĪ-seps), a muscle with two heads of origin.

Bili- **bile, gall** Biliary (BIL-ē-er-ē), pertaining to bile, bile ducts, or gallbladder.

Brachy- **short** Brachyesophagus (brā-kē-e-SOF-a-gus), short esophagus.

Brady- **slow** Bradycardia (brād'-ē-KARD-ē-a), abnormal slowness of the heartbeat.

Cata- **down, lower, under, against** Catabolism (ka-TAB-a-lizm), metabolic breakdown into simpler substances.

Circum- **around** Circumrenal (ser-kum-RĒN-al), around the kidney.

Cirrh- **yellow** Cirrhosis (si-RŌ-sis), liver disorder that causes yellowing of skin.

Co-, Con-, Com- **with, together** Congenital (kon-JEN-i-tal), existing at birth.

Contra- **against, opposite** Contraception (kon-tra-SEP-shun), the prevention of conception.

Crypt- **hidden, concealed** Cryptorchidism (krip-TOR-ka-dizm'), undescended or hidden testes.

Cyano- **blue** Cyanosis (sī-a-NŌ-sis), bluish discoloration due to inadequate oxygen.

De- **down, from** Decay (de-KĀ), waste away from normal.

Demi-, hemi- **half** Hemiplegia (hem'-ē-PLĒ-jē-a), paralysis on one side of the body.

Di-, Diplo- **two** Diploid (DIP-loyd), having double the haploid number of chromosomes.

Dis- **separation, apart, away from** Disarticulate (dis'-ar-TIK-yoo-lāt'), to separate at a joint.

Dys- **painful, difficult** Dyspnea (disp-NĒ-a), difficult breathing.

E-, Ec-, Ex- **out from, out of** Eccentric (ek-SEN-trik), not located at the center.

Ecto-, Exo- **outside** Ectopic (ek-TOP-ik) pregnancy, gestation outside the uterine cavity.

Em-, En- **in, on** Empyema (em'-pī-Ē-ma), pus in a body cavity.

End-, Endo- **inside** Endocardium (en'-dō-KARD-ē-um), membrane lining the inner surface of the heart.

Epi- **upon, on, above** Epidermis (ep'-i-DER-mis), outermost layer of skin.

Eu- **well** Eupnea (YOOP-nē-a), normal breathing.

Ex-, Exo- **out, away from** Exocrine (EK-sō-krin), excreting outwardly or away from.

Extra- **outside, beyond, in addition to** Extracellular (ek'-stra-SEL-yoo-lar), outside of the cell.

Fore- **before, in front of** Forehead (FOR-hed), anterior part of head.

Gen- **originate, produce, form** Pathogen (PATH-ō-jen), disease producer.

Gingiv- **gum** Gingivitis (jin'-je-VĪ-tus), inflammation of the gums.

Hemi- **half** Hemiplegia (hem-ē-PLĒ-jē-a), paralysis of only half of the body.

Heter-, Hetero- **other, different** Heterogeneous (het'-e-rō-JEN-ē-us), composed of different substances.

Homeo-, Homo- **unchanging, the same, steady** Homeostasis (hō'-mē-ō-STA-sis), achievement of a steady state.

Hyper- **beyond, excessive** Hyperglycemia (hī-per-glī-SĒ-mē-a), excessive amount of sugar in the blood.

Hypo- **under, below, deficient** Hypodermic (hī-pō-DER-mik), below the skin or dermis.

Idio- **self, one's own, separate** Idiopathic (id'-ē-ō-PATH-ik), a disease without recognizable cause.

In-, Im- **in, inside, not** Incontinent (in-KON-ti-nent), not able to retain urine or feces.

Infra- **beneath** Infraorbital (in'-fra-OR-bi-tal), beneath the orbit.

Inter- **among, between** Intercostal (int'-er-KOS-tal), between the ribs.

Intra **within, inside** Intracellular (in'-tra-SEL-yoo-lar), inside the cell.

Iso- **equal, like** Isogenic (ī-sō-JEN-ik), alike in morphological development.

Later- **side** Lateral (LAT-er-al), pertaining to a side or farther from the midline.

Lepto- **small, slender, thin** Leptodermic (lep'-tō-DER-mik), having thin skin.

Macro- **large, great** Macrophage (MAK-rō-fāj), large phagocytic cell.

Mal- **bad, abnormal** Malnutrition (mal'-noo-TRISH-un), lack of necessary food substances.

Medi-, Meso- **middle** Medial (MĒD-ē-al), nearer to midline.

Mega-, Megalo- **great, large** Megakaryocyte (meg'-a-KAR-ē-ō-sīt), giant cell of bone marrow.

Melan- **black** Melanin (MEL-a-nin), black or dark brown pigment found in skin and hair.

Meta- **after, beyond** Metacarpus (met'-a-KAR-pus), the part of the hand between the wrist and fingers.

Micro- **small** Microtome (MĪ-krō-tōm), instrument for preparing very thin slices of tissue for microscopic examination.

Mono- **one** Monorchid (mon-OR-kid), having one testicle.

Neo- **new** Neonatal (nē-ō-NĀT-al), pertaining to the first 4 weeks after birth.

Noct(i)- **night** Nocturia (nok-TOO-rē-a), urination occuring at night.

Null(i)- **none** Nullipara (nu-LIP-a-ra), woman with no children.

Nyct- **night** Nyctalopia (nik'-ta-LŌ-pē-a), night blindness.

Oligo- **small, deficient** Oliguria (ol-ig-YOO-rē-a), abnormally small amount of urine.

Ortho- **straight, normal** Orthopnea (or-thop-NĒ-a), inability to breathe in any position except when straight or erect.

Pan- **all** Pancarditis (pan-kar-DĪ-tis), inflammation of the entire heart.

Para- **near, beyond, apart from, beside** Paranasal (par-a-NĀ-zal), near the nose.

Per- **through** Percutaneous (per'-kyoo-TĀ-nē-us), through the skin.

Peri- **around** Pericardium (per'-i-KARD-ē-um), membrane or sac around the heart.

Poly- **much, many** Polycythemia (pol'-i-sī-THĒ-mē-a), an excess of red blood cells.

Post- **after, beyond** Postnatal (pōst-NĀT-al), after birth.

Pre-, Pro **before, in front of** Prenatal (prē-NĀT-al), before birth.

Prim- **first** Primary (PRĪ-me-rē), first in time or order.

Proto- **first** Protocol (PRŌ-tō-kol), clinical report made from first notes taken.

Pseud-, Pseudo- **false** Pseudoangina (soo'-dō-an-JĪ-na), false angina.

Retro- **backward, located behind** Retroperitoneal (re'-trō-per'-it-on-Ē-al), located behind the peritoneum.

Schizo- **split, divide** Schizophrenia (skiz'-ō-FRE-nē-a), split personality mental disorder.

Semi- **half** Semicircular (sem'-i-SER-kyoo-lar) canals, canals in the shape of a half circle.

Sub- **under, beneath, below** Submucosa (sub'-myoo-KŌ-sa), tissue layer under a mucous membrane.

Super- **above, beyond** Superficial (soo-per-FISH-al), confined to the surface.

Supra- **above, over** Suprarenal (soo-pra-RĒN-al), adrenal gland above the kidney.

Sym-, Syn- **with, together, joined** Syndrome (SIN-drōm), all the symptoms of a disease considered as a whole.

Tachy- **rapid** Tachycardia (tak'-i-KARD-ē-a), rapid heart action.

Terat(o)- **malformed fetus** Teratogen (TER-a-tō-jen), an agent that caused development of a malformed fetus.

Tetra-, quadra- **four** Tetrad (TET-rad), group of four with something in common.

Trans- **across, through, beyond** Transudation (trans-yoo-DĀ-shun), oozing of a fluid through pores.

Tri- **three** Trigone (TRĪ-gon), a triangular space, as at the base of the bladder.

SUFFIXES

-able **capable of, having ability to** Viable (VĪ-a-bal), capable of living.

-ac, -al **pertaining to** Cardiac (KARD-ē-ak), pertaining to the heart.

-agra **severe pain** Myagra (mī-AG-ra), severe muscle pain.

-an, -ian **pertaining to** Circadian (ser-KĀ-dē-an), pertaining to a cycle of active and inactive periods.

-ant **having the characteristic of** Malignant (ma-LIG-nant), having the characteristic of badness.

-ary **connected with** Ciliary (SIL-ē-ar-ē), resembling any hairlike structure.

-asis, -asia, -esis, -osis **condition or state of** Hemostasis (hē-mō-STĀ-sis), stopping of bleeding or circulation.

-asthenia **weakness** Myasthenia (mi-as-THĒ-nē-a), weakness of skeletal muscles.

-ation **process, action, condition** Inspiration (in-spi-RĀ-shun), process of drawing air into lungs.

-cel, -cele **swelling, an enlarged space or cavity** Meningocele (men-IN-gō-sēl), enlargement of the meninges.

-centesis **puncture, usually for drainage** Amniocentesis (am'-nē-ō-sen-TĒ-sis), withdrawal of amniotic fluid.

-cid, -cide, -cis **cut, kill, destroy** Germicide (jer-mi-SĪD), a substance that kills germs.

-ectasia, -ectasis **stretching, dilation** Bronchiectasis (bron-kē-EK-ta-sis), dilation of a bronchus or bronchi.

-ectomize, ectomy **excision of, removal of** Thyroidectomy (thī-royd-EK-tō-mē), surgical removal of a thyroid gland.

-ema **swelling, distension** Emphysema (em'-fi-SĒ-ma), swelling of air sacs in lungs.

-emia **condition of blood** Lipemia (lip-Ē-mē-a), abnormally high concentration of fat in the blood.

-esis **condition, process** Enuresis (en'-yoo-RĒ-sis), condition of involuntary urination.

-esthesia **sensation, feeling** Anesthesia (an'-es-THĒ-zē-a), total or partial loss of feeling.

-ferent **carry** Efferent (EF-e-rent), carrying away from a center.

-form **shape** Fusiform (FYOO-zi-form), spindle-shaped.

-gen **agent that produces or originates** Pathogen (PATH-ō-jen), microorganism or substance capable of producing a disease.

-genic **produced from, producing** Pyogenic (pī-ō-JEN-ik), producing pus.

-gram **record, that which is recorded** Electrocardiogram (e-lek'-trō-KARD-ē-ō-gram), record of heart action.

-graph **instrument for recording** Electroencephalograph (e-lek'-trō-en-SEF-a-lō-graf), instrument for recording electrical activity of the brain.

-ia **state, condition** Hypermetropia (hī'-per-me-TRŌ-pē-a), condition of farsightedness.

-iatrics, iatry **medical practice specialities** Pediatrics (pēd-ē-A-triks), medical science relating to care of children and treatment of their diseases.

-ician **person associated with** Technician (tek-NISH-an), person skilled in a technical field.

-ics **art or science of** Optics (OP-tiks), science of light and vision.

-ion **action, condition resulting from action** Incision (in-SIZH-un), act or result of cutting into flesh.

-ism, **condition, state** Rheumatism (ROO-ma-tizm), inflammation, especially of muscles and joints.

-ist **one who practices** Internist (in-TER-nist), one who practices internal medicine.

-itis **inflammation** Neuritis (nyoo-RĪT-is), inflammation of a nerve or nerves.

-ive **relating to** Sedative (SED-a-tive), relating to a pain or tension reliever.

-logy, -ology **the study or science of** Physiology (fiz-ē-OL-ō-jē), the study of function of body parts.

-lyso, -lysis **solution, dissolve, loosening** Hemolysis (hē-MOL-i-sis), dissolution of red blood cells.

-malacia **softening** Osteomalacia (os'-tē-ō-ma-LĀ-shē-a), softening of bone.

-megaly **enlarged** Cardiomegaly (kar'-dē-ō-MEG-a-lē), enlarged heart.

-oid **resembling** Lipoid (li-POYD), resembling fat.

-ologist **specialist** Dermatologist (der-ma-TOL-ō-gist), specialist in the study of the skin.

-oma **tumor** Fibroma (fi-BRŌ-ma), tumor composed mostly of fibrous tissue.

-ory **pertaining to** Sensory (SENS-o-rē), pertaining to sensation.

-ose **full of** Adipose (AD-i-pōz), characterized by presence of fat.

-osis **condition, disease** Necrosis (ne-KRŌ-sis), condition of death of cells.

-ostomy **create an opening** Colostomy (kō-LOS-tō-me), surgical creation of an opening between the colon and body surface.

-otomy **surgical incision** Tracheotomy (trā-kē-OT-ō-me), surgical incision of the trachea.

-pathy **disease** Neuropathy (nyoo-ROP-a-thē), disease of the peripheral nervous system.

-penia **deficiency** Thrombocytopenia (throm'-bō-sīt'-o-PĒ-nē-a), deficiency of thrombocytes in the blood.

-phobe, -phobia **fear of, aversion to** Hydrophobia (hī-drō-FŌ-bē-a), fear of water.

-plasia, -plasty **development, formation** Rhinoplasty (RĪ-nō-plas-tē), surgical reconstruction of the nose.

-plegia, -plexy **stroke, paralysis** Apoplexy (AP-ō-plek-sē), sudden loss of consciousness and paralysis.

-pnea **to breathe** Apnea (AP-nē-a), temporary absence of respiration, following a period of overbreathing.

-poiesis **production** Hematopoiesis (he-mat'-a-poy-Ē-sis), formation and development of red blood cells.

-ptosis **falling, sagging** Blepharoptosis (blef'-a-rō-TŌ-sis), dropping of upper eyelid.

-rrhage **bursting forth, abnormal discharge** Hemorrhage (HEM-or-rij), bursting forth of blood.

-rrhea **flow, discharge** Diarrhea (dī-a-RĒ-a), abnormal frequency of bowel evacuation, the stools with a more or less fluid consistency.

-scope **instrument for viewing** Bronchoscope (BRON-kō-skōp), instrument used to examine the interior of a bronchus.

-stomy **creation of a mouth or artificial opening** Tracheostomy (trā-kē-OST-ō-mē), creation of an opening in the trachea.

-tic, -ulnar **pertaining to** Diagnostic (dī'-ag-NOS-tik), pertaining to diagnosis.

-tomy **cutting into, incision into** Laparatomy (lap-a-ROT-ō-mē), an abdominal incision to gain access to the peritoneal cavity.

-tripsy **crushing** Lithotripsy (LITH-ō-trip'-sē), crushing of a calculus (stone).

-trophy **state relating to nutrition or growth** Hypertrophy (hī-PER-trō-fē), excessive growth of an organ or part.

-tropic **turning toward, influencing, changing** Gonadotropic (gō-nad-a-TRŌ-pic), influencing the gonads.

-uria **urine** Polyuria (pol-ē-YOOR-ē-a), excessive secretion of urine.

Abatement (a-BĀT-ment) A decrease in the seriousness of a disorder or in the severity of pain or other symptoms.

Abdomen (ab-DŌ-men or AB-dō-men) The area between the diaphragm and pelvis.

Abdominal (ab-DŌM-i-nal) ***cavity*** Superior portion of the abdominopelvic cavity that contains the stomach, spleen, liver, gallbladder, pancreas, small intestine, and most of the large intestine.

Abdominal thrust maneuver A first-aid procedure for choking. Employs a quick, upward thrust against the diaphragm that forces air out of the lungs with sufficient force to eject any lodged material. Also called the ***Heimlich*** (HĪM-lik) ***maneuver.***

Abdominopelvic (ab-dom'-i-nō-PEL-vic) ***cavity*** Inferior component of the ventral body cavity that is subdivided into an upper abdominal cavity and a lower pelvic cavity.

Abduction (ab-DUK-shun) Movement away from the axis or midline of the body or one of its parts.

Abortion (a-BOR-shun) The premature loss (spontaneous) or removal (induced) of the embryo or nonviable fetus; any failure in the normal process of developing or maturing.

Abrasion (a-BRĀ-shun) A portion of skin that has been scraped away.

Abscess (AB-ses) A localized collection of pus and liquefied tissue in a cavity.

Absorption (ab-SORP-shun) The taking up of liquids by solids or of gases by solids or liquids; intake of fluids or other substances by cells of the skin or mucous membranes; the passage of digested foods from the gastrointestinal tract into blood or lymph.

Accessory duct A duct of the pancreas that empties into the duodenum about 2.5 cm (1 in.) superior to the ampulla of Vater (hepatopancreatic ampulla). Also called the ***duct of Santorini*** (san'-tō-RE-ne).

Accretion (a-KRĒ-shun) A mass of material that has accumulated in a space or cavity; the adhesion of parts.

Acetabulum (as'-e-TAB-yoo-lum) The rounded cavity on the external surface of the hipbone that receives the head of the femur.

Acetylcholine (as'-ē-til-KŌ-lēn) ***(ACh)*** A neurotransmitter, liberated at synapses in the central nervous system, that stimulates skeletal muscle contraction.

Achilles tendon *See* ***Calcaneal tendon.***

Acid (AS-id) A proton donor, or substance that dissociates into hydrogen ions (H^+) and anions, characterized by an excess of hydrogen ions and a pH less than 7.

Acinar (AS-i-nar) Flasklike.

Acini (AS-i-nē) Masses of cells in the pancreas that secrete digestive enzymes.

Acne (AK-nē) Inflammation of sebaceous (oil) glands that usually begins at puberty; the basic acne lesions in order of increasing severity are comedones, papules, pustules, and cysts.

Acoustic (a-KOOS-tik) Pertaining to sound or the sense of hearing.

Acquired immune deficiency syndrome (AIDS) A disorder characterized by a positive HIV-antibody test and certain indicator diseases (Kaposi's sarcoma, *Pneumocystis carinii* pneumonia, tuberculosis, fungus diseases, etc.). A deficiency of helper T cells and a reversed ratio of helper T cells to suppressor T cells that results in fever or night sweats, coughing, sore throat, fatigue, body aches, weight loss, and enlarged lymph nodes. Caused by a virus called human immunodeficiency virus (HIV).

Acromegaly (ak'-rō-MEG-a-lē) Condition caused by hypersecretion of human growth hormone (hGH) during adulthood characterized by thickened bones and enlargement of other tissues.

Acrosome (AK-rō-sōm) A dense granule in the head of a spermatozoon that contains enzymes that facilitate the penetration of a spermatozoon into a secondary oocyte.

Actin (AK-tin) The contractile protein that makes up thin myofilaments in muscle fiber (cell).

Action potential A wave of negativity that self-propagates along the outside surface of the membrane of a neuron or muscle fiber (cell); a rapid change in membrane potential that involves a depolarization following a repolarization. Also called a ***nerve action potential (nerve impulse)*** as it relates to a neuron and a ***muscle action potential*** as it relates to a muscle fiber (cell).

Active transport The movement of substances, usually ions, across cell membranes, against a concentration gradient, requiring the expenditure of energy (ATP).

Acuity (a-KYOO-i-tē) Clearness or sharpness, usually of vision.

Acupuncture (AK-ū-punk'-chur) The insertion of a needle into a tissue for the purpose of drawing fluid or relieving pain. It is also an ancient Chinese practice employed to cure illnesses by inserting needles into specific locations of the skin.

Acute (a-KYOOT) Having rapid onset, severe symptoms, and a short course; not chronic.

Adam's apple *See* ***Thyroid cartilage.***

Adaptation (ad'-ap-TĀ-shun) The adjustment of the pupil of the eye to light variations. The property by which a neuron relays a decreased frequency of action potentials from a receptor even though the strength of the stimulus remains constant. The decrease in perception of a sensation over time while the stimulus is still present.

Addison's (AD-i-sonz) ***disease*** Disorder caused by hyposecretion of glucocorticoids (and aldosterone) characterized by muscular weakness, hypoglycemia, mental lethargy, anorexia, nausea and vomiting, weight loss, low blood pressure, dehydration, and excessive skin and mucous membrane pigmentation.

Adduction (ad-DUK-shun) Movement toward the axis or midline of the body or one of its parts.

Adenohypophysis (ad'-e-nō-hī-POF-i-sis) The anterior portion of the pituitary gland.

Adenoids (AD-e-noyds) The pharyngeal tonsils.

Adenosine triphosphate (a-DEN-ō-sēn trī-FOS-fāt) ***(ATP)*** The universal energy-carrying molecule manufactured in all living cells as a means of capturing and storing energy. It consists of the purine base *adenine* and the five-carbon sugar *ribose,* to which are added, in linear array, three *phosphate* molecules.

Adhesion (ad-HĒ-zhun) Abnormal joining of parts to each other.

Adipocyte (AD-i-pō-sīt) Fat cell, derived from a fibroblast.

Adrenal cortex (a-DRĒ-nal KOR-teks) The outer portion of an adrenal gland, divided into three zones, each of which has a different cellular arrangement and secretes different hormones.

Adrenal (a-DRĒ-nal) ***glands*** Two glands located superior to each kidney. Also called the ***suprarenal*** (soo'-pra-RĒ-nal) ***glands.***

Adrenal medulla (me-DUL-a) The inner portion of an adrenal gland, consisting of cells that secrete epinephrine and norepinephrine (NE) in response to the stimulation of preganglionic sympathetic neurons.

Adrenergic (ad'-ren-ER-jik) ***fiber*** A nerve fiber that when stimulated releases norepinephrine (noradrenaline) at a synapse.

Adrenocorticotropic (ad-rē'-nō-kor-ti-kō-TRŌP-ik) ***hormone (ACTH)*** A hormone produced by the adenohypophysis (anterior lobe) of the pituitary gland that influences the production and secretion of certain hormones of the adrenal cortex.

Adrenoglomerulotropin (a-drē'-nō-glō-mer'-yoo-lō-TRŌ-pin) A hormone secreted by the pineal gland that may stimulate aldosterone secretion.

Adventitia (ad-ven-TISH-ya) The outermost covering of a structure or organ.

Aerobic (air-Ō-bik) Requiring molecular oxygen.

Afferent arteriole (AF-er-ent ar-TĒ-rē-ōl) A blood vessel of a kidney that breaks up into the capillary network called a glomerulus; there is one afferent arteriole for each glomerulus.

Afferent neuron (NOO-ron) A neuron that carries a nerve impulse toward the central nervous system. Also called a ***sensory neuron.***

Afterimage Persistence of a sensation even though the stimulus has been removed.

Agglutination (a-gloo'-ti-NĀ-shun) Clumping of microorganisms or blood corpuscles; typically an antigen-antibody reaction.

Agglutinin (a-GLOO-ti-nin) A specific principle or antibody in blood serum capable of causing the clumping of bacteria, blood corpuscles, or particles. Also called an ***isoantibody.***

Agglutinogen (ag'-loo-TIN-ō-gen) A genetically determined antigen located on the surface of erythrocytes; basis for the ABO grouping and Rh system of blood classsification. Also called an ***isoantigen.***

Aggregated lymphatic follicles Aggregated lymph nodules that are most numerous in the ileum. Also called ***Peyer's*** (PĪ-erz) ***patches.***

Aging Normal process accompanied by a progressive alteration of the body's homeostatic adaptive responses.

Agnosia (ag-NŌ-zē-a) A loss of the ability to recognize the meaning of stimuli from the various senses (visual, auditory, touch).

Agraphia (a-GRAF-ē-a) An inability to write.

Albinism (AL-bin-izm) Abnormal, nonpathological, partial or total absence of pigment in skin, hair, and eyes.

Albumin (al-BYOO-min) The most abundant (60 percent) and smallest of the plasma proteins, which functions primarily to regulate osmotic pressure of plasma.

Albuminuria (al-byoo'-min-UR-ēa) Presence of albumin in the urine.

Aldosterone (al-do-STĒR-ōn) A mineralocorticoid produced by the adrenal cortex that brings about sodium and water reabsorption and potassium excretion.

Aldosteronism (al'-do-STER-ōn-izm') Condition caused by hypersecretion of aldosterone that results in increased sodium concentration and decreased potassium concentration in blood and characterized by muscular paralysis, high blood pressure, and edema.

Alimentary (al-i-MEN-ta-rē) Pertaining to nutrition.

Alkaline (AL-ka-līn) Containing more hydroxyl ions (OH^-) than hydrogen ions (H^+) to produce a pH of more than 7.

Alkalosis (al-ka-LŌ-sis) A condition in which blood pH ranges from 7.45 to 8.00 or higher.

Allantois (a-LAN-tō-is) A small, vascularized membrane between the chorion and amnion of the fetus that serves as an early site for blood formation.

Allele (a-LĒL) Genes that control the same inherited trait (such as height or eye color) that are located on the same position (locus) on homologous chromosomes.

Allergen (AL-er-jen) An antigen that evokes a hypersensitivity reaction.

Allergic (a-LER-jik) Pertaining to or sensitive to an allergen.

All-or-none principle In muscle physiology, muscle fibers (cells) of a motor unit contract to their fullest extent or not at all. In neuron physiology, if a stimulus is strong enough to initiate an action potential, a nerve impulse is transmitted along the entire neuron at a constant and minimum strength.

Alpha (AL-fa) ***cell*** A cell in the pancreatic islets (islets of Langerhans) in the pancreas that secretes glucagon.

Alpha receptor Receptor found on visceral effectors innervated by most sympathetic postganglionic axons; in general, stimulation of alpha receptors leads to excitation.

Alveolar-capillary (al-VĒ-ō-lar) ***membrane*** Structure in the lungs consisting of the alveolar wall and basement membrane and a capillary endothelium and basement membrane through which the diffusion of respiratory gases occurs. Also called the ***respiratory membrane.***

Alveolar duct Branch of a respiratory bronchiole around which alveoli and alveolar sacs are arranged.

Alveolar macrophage (MAK-rō-fāj) Cell found in the alveolar walls of the lungs that is highly phagocytic. Also called a ***dust cell.***

Alveolar sac A collection or cluster of alveoli that share a common opening.

Alveolus (al-VĒ-ō-lus) A small hollow or cavity; an air sac in the lungs; milk-secreting portion of a mammary gland.

Alzheimer's (ALTZ-hī-merz) ***disease (AD)*** Disabling neurological disorder characterized by dysfunction and death of specific cerebral neurons resulting in widespread intellectual impairment, personality changes, and fluctuations in alertness.

Ambulatory (AM-byoo-la-tō'-rē) Capable of walking.

Amenorrhea (ā-men-ō-RĒ-a) Absence of menstruation.

Amnesia (am-NĒ-zē-a) A lack or loss of memory.

Amniocentesis (am'-nē-ō-sen-TĒ-sis) Removal of amniotic fluid by inserting a needle transabdominally into the amniotic cavity.

Amnion (AM-nē-on) The innermost fetal membrane; a thin transparent sac that holds the fetus suspended in amniotic fluid. Also, called the ***"bag of waters."***

Amniotic (am'-nē-OT-ik) ***fluid*** Fluid in the amniotic cavity, the space between the developing embryo (or fetus) and amnion; the fluid is initially produced as a filtrate from maternal blood and later from fetal urine.

Amorphous (a-MOR-fus) Without definite shape or differentiation in structure; pertains to solids without crystalline structure.

Amphiarthrosis (am'-fē-ar-THRŌ-sis) Articulation midway between diarthrosis and synarthrosis, in which the articulating bony surfaces are separated by an elastic substance to which both are attached, so that the mobility is slight.

Ampulla (am-POOL-la) A saclike dilation of a canal.

Ampulla of Vater *See* ***Hepatopancreatic ampulla.***

Amyotrophic (a-mē-ō-TROF-ik) ***lateral sclerosis (ALS)*** Progressive neuromuscular disease characterized by degeneration of motor neurons in the spinal cord that leads to muscular weakness. Also called ***Lou Gehrig's disease.***

Anabolism (a-NAB-ō-lizm) Synthetic energy-requiring reactions whereby small molecules are built up into larger ones.

Anaerobic (an-AIR-ō-bik) Not requiring molecular oxygen.

Anal (Ā-nal) ***canal*** The terminal 2 or 3 cm (1 in.) of the rectum; opens to the exterior of the anus.

Anal column A longitudinal fold in the mucous membrane of the anal canal that contains a network of arteries and veins.

Analgesia (an-al-JĒ-zē-a) Pain relief.

Anal triangle The subdivision of the female or male perineum that contains the anus.

Anaphase (AN-a-fāz) The third stage of mitosis in which the chromatids that have separated at the centromeres move to opposite poles of the cell.

Anaphylaxis (an'-a-fi-LAK-sis) Against protection; a hypersensitivity (allergic) reaction in which IgE antibodies attach to mast cells and basophils, causing them to produce mediators of anaphylaxis (histamine, leukotrines, kinins, and prostaglandins) that bring about increased blood permeability, increased smooth muscle contraction, and increased mucus production. Examples are hayfever, hives, and anaphylactic shock.

Anastomosis (a-nas-tō-MŌ-sis) An end-to-end union or joining together of blood vessels, lymphatic vessels, or nerves.

Anatomic dead space The volume of air that is inhaled but remains in spaces in the upper respiratory system and does not reach the alveoli to participate in gas exchange; about 150 ml.

Anatomical (an'-a-TOM-i-kal) ***position*** A position of the body universally used in anatomical descriptions in which the body is erect, facing the observer, the upper extremities are at the sides, the palms of the hands are facing forward, and the feet are on the floor.

Anatomy (a-NAT-ō-mē) The structure or study of structure of the body and the relation of its parts to each other.

Androgen (AN-drō-jen) Substance producing or stimulating male characteristics, such as the male hormone testosterone.

Anemia (a-NĒ-mē-a) Condition of the blood in which the number of functional red blood cells or their hemoglobin content is below normal.

Anesthesia (an'-es-THĒ-zē-a) A total or partial loss of feeling or sensation, usually defined with respect to loss of pain sensation; may be general or local.

Aneurysm (AN-yoo-rizm) A saclike enlargement of a blood vessel caused by a weakening of its wall.

Angina pectoris (an-JĪ-na *or* AN-ji-na PEK-tō-ris) A pain in the chest related to reduced coronary circulation that may or may not involve heart or artery disease.

Angiography (an-jē-OG-ra-fē) X-ray examination of blood vessels after injection of a radiopaque substance.

Ankyloglossia (ang'-ki-lō-GLOSS-ē-a) "Tongue-tied"; restriction of tongue movements by a short lingual frenulum.

Ankylosis (ang'-ki-LŌ-sus) Severe or complete loss of movement at a joint.

Anomaly (a-NOM-a-lē) An abnormality that may be a developmental (congenital) defect; a variant from the usual standard.

Anopsia (an-OP-sē-a) A defect of vision.

Anorexia nervosa (an-ō-REK-sē-a ner-VŌ-sa) A chronic disorder characterized by self-induced weight loss, body-image and other perceptual disturbances, and physiologic changes that result from nutritional depletion.

Anosmia (an-OZ-mē-a) Loss of the sense of smell.

Anoxia (an-OK-sē-a) Deficiency of oxygen.

Antagonist (an-TAG-ō-nist) A muscle that has an action opposite that of the prime mover (agonist) and yields to the movement of the prime mover.

Antepartum (an-tē-PAR-tum) Before delivery of the child; occurring (to the mother) before childbirth.

Anterior (an-TĒR-ē-or) Nearer to or at the front of the body. Also called ***ventral.***

Anterior root The structure composed of axons of motor or efferent fibers that emerges from the anterior aspect of the spinal cord and extends laterally to join a posterior root, forming a spinal nerve. Also called a ***ventral root.***

Antibiotic (an'-ti-bī-OT-ik) Literally, "antilife"; a chemical produced by a microorganism that is able to inhibit the growth of or kill other microorganisms.

Antibody (AN-ti-bod'-ē) A protein produced by certain cells in the body in the presence of a specific antigen; the antibody combines with that antigen to neutralize, inhibit, or destroy it.

Anticoagulant (an-tī-cō-AG-yoo-lant) A substance that is able to delay, suppress, or prevent the clotting of blood.

Antidiuretic (an'-ti-dī-yoo-RET-ik) Substance that inhibits urine formation.

Antidiuretic hormone (ADH) Hormone produced by neurosecretory cells in the paraventricular and supraoptic nuclei of the hypothalamus that stimulates water reabsorption from kidney cells into the blood and vasoconstriction of arterioles.

Antigen (AN-ti-jen) Any substance that when introduced into the tissues or blood induces the formation of antibodies and reacts only with its specific antibodies.

Anti-oncogene (ONG-kō-jēn) A gene that can cause cancer when inappropriately inactivated.

Antrum (AN-trum) Any nearly closed cavity or chamber, especially one within a bone, such as a sinus.

Anulus fibrosus (AN-yoo-lus fī-BRŌ-sus) A ring of fibrous tissue and fibrocartilage that encircles the pulpy substance (nucleus pulposus) of an intervertebral disc.

Anuria (a-NOO-rē-a) A daily urine output of less than 50 ml.

Anus (Ā-nus) The distal end and outlet of the rectum.

Aorta (ā-OR-ta) The main systemic trunk of the arterial system of the body; emerges from the left ventricle.

Aortic (ā-OR-tik) ***body*** Receptor on or near the arch of the aorta that responds to alterations in blood levels of oxygen, carbon dioxide, and hydrogen ions.

Aperture (AP-er-chur) An opening or orifice.

Apex (Ā-peks) The pointed end of a conical structure, such as the apex of the heart.

Aphasia (a-FĀ-zē-a) Loss of ability to express oneself properly through speech or loss of verbal comprehension.

Apnea (AP-nē-a) Temporary cessation of breathing.

Apneustic (ap-NOO-stik) ***area*** Portion of the respiratory center in the pons that sends stimulatory nerve impulses to the inspiratory area that activate and prolong inspiration and inhibit expiration.

Apocrine (AP-ō-krin) ***gland*** A type of gland in which the secretory products gather at the free end of the secreting cell and are pinched off, along with some of the cytoplasm, to become the secretion, as in mammary glands.

Aponeurosis (ap'-ō-noo-RŌ-sis) A sheetlike tendon joining one muscle with another or with bone.

Appendage (a-PEN-dij) A structure attached to the body.

Appendicitis (a-pen-di-SĪ-tis) Inflammation of the vermiform appendix.

Aqueduct (AK-we-duct) A canal or passage, especially for the conduction of a liquid.

Aqueous humor (AK-wē-us HYOO-mor) The watery fluid, similar in composition to cerebrospinal fluid, that fills the anterior cavity of the eye.

Arachnoid (a-RAK-noyd) The middle of the three coverings (meninges) of the brain or spinal cord.

Arachnoid villus (VIL-us) Berrylike tuft of arachnoid that protrudes into the superior sagittal sinus and through which cerebrospinal fluid is reabsorbed into the bloodstream.

Arbor vitae (AR-bōr VĒ-tē) The treelike appearance of the white matter tracts of the cerebellum when seen in midsagittal section. A series of branching ridges within the cervix of the uterus.

Arch of the aorta (ā-OR-ta) The most superior portion of the aorta, lying between the ascending and descending segments of the aorta.

Areflexia (a'-rē-FLEK-sē-a) Absence of reflexes.

Areola (a-RĒ-ō-la) Any tiny space in a tissue. The pigmented ring around the nipple of the breast.

Arm The portion of the upper extremity from the shoulder to the elbow.

Arrector pili (a-REK-tor PI-lē) Smooth muscles attached to hairs; contraction pulls the hairs into a more vertical position, resulting in "goose bumps."

Arrhythmia (a-RITH-mē-a) Irregular heart rhythm. Also called a ***dysrhythmia.***

Arteriogram (ar-TĒR-ē-ō-gram) A roentgenogram of an artery after injection of a radiopaque substance into the blood.

Arteriole (ar-TĒ-rē-ōl) A small, almost microscopic, artery that delivers blood to a capillary.

Artery (AR-ter-ē) A blood vessel that carries blood away from the heart.

Arthritis (ar-THRĪ-tis) Inflammation of a joint.

Arthrology (ar-THROL-ō-jē) The study or description of joints.

Arthroplasty (AR-thrō-plas'-tē) Surgical replacement of joints.

Arthroscopy (ar-THROS-co-pē) A procedure for examining the interior of a joint, usually the knee, by inserting an arthroscope into a small incision; used to determine extent of damage, remove torn cartilage, repair cruciate ligaments, and obtain samples for analysis.

Articular (ar-TIK-yoo-lar) ***capsule*** Sleevelike structure around a synovial joint composed of a fibrous capsule and a synovial membrane.

Articular cartilage (KAR-ti-lij) Hyaline cartilage attached to articular bone surfaces.

Articular disc Fibrocartilage pad between articular surfaces of bones of some synovial joints. Also called a ***meniscus*** (men-IS-cus).

Articulate (ar-TIK-yoo-lāt) To join together as a joint to permit motion between parts.

Articulation (ar-tik'-yoo-LĀ-shun) A joint; a point of contact between bones, cartilage and bones, or teeth and bones.

Artificial pacemaker A device that generates and delivers electrical signals to the heart to maintain a regular heart rhythm.

Arytenoid (ar'-i-TĒ-noyd) Ladle-shaped.

Arytenoid (ar'-i-TĒ-noyd) ***cartilages*** A pair of small, pyramidal cartilages of the larynx that attach to the vocal folds and intrinsic pharyngeal muscles and can move the vocal folds.

Ascending colon (KŌ-lon) The portion of the large intestine that passes upward from the cecum to the lower edge of the liver where it bends at the right colic (hepatic) flexure to become the transverse colon.

Ascites (as-SĪ-tēz) Serous fluid in the peritoneal cavity.

Aseptic (ā-SEP-tik) Free from any infectious or septic material.

Asphyxia (as-FIX-ē-a) Unconsciousness due to interference with the oxygen supply of the blood.

Aspiration (as'-pi-RĀ-shun) Inhalation of a foreign substance (water, food, or foreign body) into the bronchial tree; drainage of a substance in or out by suction.

Association area A portion of the cerebral cortex connected by many motor and sensory fibers to other parts of the cortex. The association areas are concerned with motor patterns, memory, concepts of word-hearing and word-seeing, reasoning, will, judgment, and personality traits.

Association neuron (NOO-ron) A nerve cell lying completely within the central nervous system that carries nerve impulses from sensory neurons to motor neurons. Also called a ***connecting neuron.***

Astereognosis (as-ter'-ē-ōg-NŌ-sis) Inability to recognize objects or forms by touch.

Asthenia (as-THĒ-nē-a) Lack or loss of strength; debility.

Astigmatism (a-STIG-ma-tizm) An irregularity of the lens or cornea of the eye causing the image to be out of focus and producing faulty vision.

Astrocyte (AS-trō-sīt) A neuroglial cell having a star shape that supports neurons in the brain and spinal cord and attaches the neurons to blood vessels.

Ataxia (a-TAK-sē-a) A lack of muscular coordination, lack of precision.

Atelectasis (at'-ē-LEK-ta-sis) A collapsed or airless state of all or part of the lung, which may be acute or chronic.

Atherosclerosis (ath'-er-ō-skle-RŌ-sis) A process in which fatty substances (cholesterol and triglycerides) are deposited in the walls of medium and large arteries in response to certain stimuli (hypertension, carbon monoxide, dietary cholesterol). Following endothelial damage, monocytes stick to the tunica interna, develop into macrophages, and take up cholesterol and low-density lipoproteins. Smooth muscle fibers (cells) in the tunica media ingest cholesterol. This results in the formation of an atherosclerotic plaque that decreases the size of the arterial lumen.

Atherosclerotic (ath'-er-ō-skle-RO-tic) ***plaque*** (PLAK) A lesion that results from accumulated cholesterol and smooth muscle fibers (cells) of the tunica media of an artery; may become obstructive.

Atresia (a-TRĒ-zē-a) Abnormal closure of a passage, or absence of a normal body opening.

Atrial fibrillation (Ā-trē-al fib-ri-LĀ-shun) Asynchronous contraction of the atria that results in the cessation of atrial pumping.

Atrial natriuretic (na'-trē-yoo-RET-ik) ***factor (ANF)*** Peptide hormone produced by the atria of the heart in response to their stretching that inhibits aldosterone production and thus lowers blood pressure.

Atrioventricular (AV) (ā'-trē-ō-ven-TRIK-yoo-lar) ***bundle*** The portion of the conduction system of the heart that begins at the atrioventricular (AV) node, passes through the cardiac skeleton separating the atria and the ventricles, then runs a short distance down the interventricular septum before splitting into right and left bundle branches. Also called the ***bundle of His (HISS).***

Atrioventricular (AV) node The portion of the conduction system of the heart made up of a compact mass of conducting cells located near the orifice of the coronary sinus in the right atrial wall.

Atrioventricular (AV) valve A structure made up of membranous flaps or cusps that allows blood to flow in one direction only, from an atrium into a ventricle.

Atrium (Ā-trē-um) A superior chamber of the heart.

Atrophy (AT-rō-fē) Wasting away or decrease in size of a part, due to a failure, abnormality of nutrition, or lack of use.

Auditory ossicle (AW-di-tō-rē OS-si-kul) One of the three small bones of the middle ear called the malleus, incus, and stapes.

Auditory tube The tube that connects the middle ear with the nose and nasopharynx region of the throat. Also called the ***Eustachian*** (yoo-STĀ-kē-an) ***tube.***

Auricle (OR-i-kul) The projecting part of the external ear composed of elastic cartilage and covered by skin and shaped like the flared end of a trumpet. Also called the ***pinna*** (PIN-na).

Auscultation (aws-kul-TĀ-shun) Examination by listening to sounds in the body.

Autoimmunity An immunologic response against a person's own tissue antigen.

Autologous preoperative transfusion (aw-TOL-ō-gus prē-OP-er-a-tiv trans-FYOO-zhun) Donating one's own blood up to six weeks before elective surgery to ensure an abundant supply and reduce transfusion complications such as those that may be associated with diseases such as AIDS and hepatitis. Also called ***predonation.***

Autolysis (aw-TOL-i-sis) Spontaneous self-destruction of cells by their own digestive enzymes upon death or a pathological process or during normal embryological development.

Autonomic ganglion (aw'-tō-NOM-ik GANG-lē-on) A cluster of sympathetic or parasympathetic cell bodies located outside the central nervous system.

Autonomic nervous system (ANS) Visceral efferent neurons, both sympathetic and parasympthetic, that transmit nerve impulses from the central nervous system to smooth muscle, cardiac muscle, and glands; so named because this portion of the nervous system was thought to be self-governing or spontaneous.

Autonomic plexus (PLEK-sus) An extensive network of sympathetic and parasympathetic fibers; the cardiac, celiac, and pelvic plexuses are located in the thorax, abdomen, and pelvis, respectively.

Autophagy (aw-TOF-a-jē) Process by which worn-out organelles are digested within lysosomes.

Autopsy (AW-top-sē) The examination of the body after death.

Autosome (AW-tō-sōm) Any chromosome other than the pair of sex chromosomes.

Axilla (ak-SIL-a) The small hollow beneath the arm where it joins the body at the shoulders. Also called the ***armpit.***

Axon (AK-son) The usually single, long process of a nerve cell that carries a nerve impulse away from the cell body.

Axon terminal Terminal branch of an axon and its collateral. Also called a ***telodendrium*** (tel-ō-DEN-drē-um).

Azygos (AZ-ī-gos) An anatomical structure that is not paired; occurring singly.

Babinski (ba-BIN-skē) ***sign*** Extension of the great toe, with or without fanning of the other toes, in response to stimulation of the outer margin of the sole of the foot; normal up to 1½ years of age.

Back The posterior part of the body; the dorsum.

Ball-and-socket joint A synovial joint in which the rounded surface of one bone moves within a cup-shaped depression or fossa of another bone, as in the shoulder or hip joint. Also called a ***spheroid*** (SFĒ-roid) ***joint.***

Bartholin's glands *See* ***Greater vestibular glands.***

Basal ganglia (GANG-glē-a) Paired clusters of cell bodies that make up the central gray matter in each cerebral hemisphere, including the caudate nucleus, lentiform nucleus, claustrum, and amygdaloid body. Also called ***cerebral nuclei*** (SER-e-bral NOO-klē-ī).

Basal metabolic (BĀ-sal met'-a-BOL-ik) ***rate (BMR)*** The rate of metabolism measured under standard or basal conditions.

Base The broadest part of a pyramidal structure. A nonacid or a proton acceptor, characterized by excess of hydroxide ions (OH^-) and a pH greater than 7. A ring-shaped, nitrogen-containing organic molecule that is one of the components of a nucleotide, for example, adenine, guanine, cytosine, thymine, and uracil.

Basement membrane Thin, extracellular layer consisting of basal lamina secreted by epithelial cells and reticular lamina secreted by connective tissue cells.

Basilar (BAS-i-lar) ***membrane*** A membrane in the cochlea of the inner ear that separates the cochlear duct from the scala tympani and on which the spiral organ (organ of Corti) rests.

Basophil (BĀ-sō-fil) A type of white blood cell characterized by a pale nucleus and large granules that stain readily with basic dyes.

B cell A lymphocyte that develops into a plasma cell that produces antibodies or a memory cell.

Belly The abdomen. The gaster or prominent, fleshy part of a skeletal muscle.

Benign (be-NĪN) Not malignant; favorable for recovery; a mild disease.

Beta (BĀ-ta) ***cell*** A cell in the pancreatic islets (islets of Langerhans) in the pancreas that secretes insulin.

Beta receptor Receptor found on visceral effectors innervated by most sympathetic postganglionic axons; in general, stimulation of beta receptors leads to inhibition.

Bicuspid (bī-KUS-pid) ***valve*** Atrioventricular (AV) valve on the left side of the heart. Also called the ***mitral valve.***

Bifurcate (bī-FUR-kāt) Having two branches or divisions; forked.

Bilateral (bī-LAT-er-al) Pertaining to two sides of the body.

Bile (BĪL) A secretion of the liver consisting of water, bile salts, bile pigments, cholesterol, lecithin, and several ions; it assumes a role in emulsification of fats prior to their digestion.

Biliary (BIL-ē-er-ē) Relating to bile, the gallbladder, or the bile ducts.

Biliary calculi (BIL-ē-er-ē CAL-kyoo-lē) Gallstones formed by the crystallization of cholesterol in bile.

Bilirubin (bil-ē-ROO-bin) A red pigment that is one of the end products of hemoglobin breakdown in the liver cells and is excreted as a waste material in the bile.

Bilirubinuria (bil-ē-roo-bi-NOO-rē-a) The presence of above-normal levels of bilirubin in urine.

Biliverdin ((bil-ē-VER-din) A green pigment that is one of the first products of hemoglobin breakdown in the liver cells and is converted to bilirubin or excreted as a waste material in bile.

Biofeedback Process by which an individual gets constant signals (feedback) about varous visceral biological functions.

Biopsy (BĪ-op-sē) Removal of tissue or other material from the living body for examination, usually microscopic.

Blastocele (BLAS-tō-sēl) The fluid-filled cavity within the blastocyst.

Blastocyst (BLAS-tō-sist) In the development of an embryo, a hollow ball of cells that consists of blastocele (the internal cavity), trophoblast (outer cells), and inner cell mass.

Blastomere (BLAS-tō-mēr) One of the cells resulting from the cleavage of a fertilized ovum.

Blastula (BLAS-tyoo-la) An early stage in the development of a zygote.

Blepharism (BLEF-a-rizm) Spasm of the eyelids; continuous blinking.

Blind spot Area in the retina at the end of the optic (II) nerve in which there are no light receptor cells.

Blood The fluid that circulates through the heart, arteries, capillaries, and veins and that constitutes the chief means of transport within the body.

Blood-brain barrier (BBB) A special mechanism that prevents the passage of certain substances the blood to the cerebrospinal fluid and brain.

Blood island Isolated mass and cord of mesenchyme in the mesoderm from which blood vessels develop.

Blood pressure (BP) Pressure exerted by blood as it presses against and attempts to stretch blood vessels, especially arteries; the force is generated by the rate and force of heartbeat; clinically, a measure of the pressure in arteries during ventricular systole and ventricular diastole.

Blood reservoir (REZ-er-vwar) Systemic veins that contain large amounts of blood that can be moved quickly to parts of the body requiring the blood.

Blood-testis barrier A barrier formed by sustentacular (Sertoli) cells that prevents an immune response against antigens produced by spermatozoa and developing cells by isolating the cells from the blood.

Body cavity A space within the body that contains various internal organs.

Bolus (BŌ-lus) A soft, rounded mass, usually food, that is swallowed.

Bony labyrinth (LAB-i-rinth) A series of cavities within the petrous portion of the temporal bone forming the vestibule, cochlea, and semicircular canals of the inner ear.

Bowman's capsule *See* ***Glomerular capsule.***

Brachial plexus (BRĀ-kē-al PLEK-sus) A network of nerve fibers of the anterior rami of spinal nerves C5, C6, C7, C8, and T1. The nerves that emerge from the brachial plexus supply the upper extremity.

Brain A mass of nervous tissue located in the cranial cavity.

Brain electrical activity mapping (BEAM) Noninvasive procedure that measures and displays the electrical activity of the brain; used primarily to diagnose epilepsy.

Brain sand Calcium deposits in the pineal gland that are laid down starting at puberty.

Brain stem The portion of the brain immediately superior to the spinal cord, made up of the medulla oblongata, pons, and midbrain.

Brain wave Electrical activity produced as a result of action potentials of brain cells.

Bright's disease *See* ***Glomerulonephritis.***

Broad ligament A double fold of parietal peritoneum attaching the uterus to the side of the pelvic cavity.

Broca's (BRŌ-kaz) ***area*** Motor area of the brain in the frontal lobe that translates thoughts into speech. Also called the ***motor speech area.***

Bronchi (BRONG-kē) Branches of the respiratory passageway including primary bronchi (the two divisions of the trachea), secondary or lobar bronchi (divisions of the primary that are distributed to the lobes of the lung), and tertiary or segmental bronchi (divisions of the secondary that are distributed to bronchopulmonary segments of the lung).

Bronchial asthma (BRONG-kē-al AZ-ma) Usually allergic reaction characterized by smooth muscle spasms in bronchi resulting in wheezing and difficult breathing.

Bronchial tree The trachea, bronchi, and their branching structures.

Bronchiectasis (brong'-kē-EK-ta-sis) A chronic disorder in which there is a loss of the normal tissue and expansion of lung air passages; characterized by difficult breathing, coughing, expectoration of pus, and foul breath.

Bronchiole (BRONG-kē-ol) Branch of a tertiary bronchus further dividing into terminal bronchioles (distributed to lobules of the lung), which divide into respiratory bronchioles (distributed to alveolar sacs).

Bronchitis (brong-KĪ-tis) Inflammation of the bronchi characterized by hypertrophy and hyperplasia of seromucous glands and goblet cells that line the bronchi and resulting in a productive cough.

Bronchogenic carcinoma (brong'-kō-JEN-ik kar'-si-NŌ-ma) Cancer originating in the bronchi.

Bronchogram (BRONG-kō-gram) A roentgenogram of the bronchial tree.

Bronchography (bron-KOG-ra-fē) Technique for examining the bronchial tree in which an opaque contrast medium is introduced into the trachea for distribution to the bronchial branches. The roentgenogram produced is called a bronchogram.

Bronchopulmonary (brong'-kō-PUL-mō-ner-ē) ***segment*** One of the smaller divisions of a lobe of a lung supplied by its own branches of a bronchus.

Bronchoscope (BRONG-kō-skōp) An instrument used to examine the interior of the bronchi of the lungs.

Bronchoscopy (brong-KOS-kō-pē) Visual examination of the interior of the trachea and bronchi with a bronchoscope to biopsy a tumor, clear an obstruction, take cultures, stop bleeding, or deliver drugs.

Bronchus (BRONG-kus) One of the two large branches of the trachea. *Plural,* ***bronchi*** (BRONG-kē).

Brunner's gland *See* ***Duodenal gland.***

Buccal (BUK-al) Pertaining to the cheek or mouth.

Bulb of penis Expanded portion of the base of the corpus spongiosum penis.

Bulbourethral (bul'-bō-yoo-RĒ-thral) ***gland*** One of a pair of glands located inferior to the prostate gland on either side of the urethra that secretes an alkaline fluid into the cavernous urethra. Also called a ***Cowper's*** (KOW-perz) ***gland.***

Bulimia (boo-LIM-ē-a) A disorder characterized by overeating, at least twice a week, followed by purging by self-induced vomiting, strict dieting or fasting, vigorous exercise, or use of laxatives.

Bulk flow The movement of large numbers of ions, molecules, or particles in the same direction as a result of forces that push them (osmotic or hydrostatic pressure).

Bullae (BYOOL-ē) Blisters beneath or within the epidermis.

Bundle branch One of the two branches of the atrioventricular (AV) bundle made up of specialized muscle fibers (cells) that transmit electrical impulses to the ventricles.

Bundle of His *See* ***Atrioventricular (AV) bundle.***

Bunion (BUN-yun) Lateral deviation of the great toe that produces inflammation and thickening of the bursa, bone spurs, and calluses.

Burn An injury in which proteins are destroyed (denatured) as a result of heat (fire, steam), chemicals, electricity, or the ultraviolet rays of the sun.

Bursa (BUR-sa) A sac or pouch of synovial fluid located at friction points, especially about joints.

Bursitis (bur-SĪ-tis) Inflammation of a bursa

Buttocks (BUT-oks) The two fleshy masses on the posterior aspect of the lower trunk, formed by the gluteal muscles.

Cachexia (kah-KEK-sē-ah) A state of ill health, malnutrition, and wasting.

Calcaneal tendon The tendon of the soleus, gastrocnemius, and plantaris muscles at the back of the heel. Also called the ***Achilles*** (a-KIL-ēz) ***tendon.***

Calcify (KAL-si-fī) To harden by deposits of calcium salts.

Calcitonin (kal-si-TŌ-nin) ***(CT)*** A hormone produced by the thyroid gland that lowers the calcium and phosphate levels of the blood by inhibiting bone breakdown and accelerating calcium absorption by bones.

Calculus (KAL-kyoo-lus) A stone, or insoluble mass of crystallized salts or other material, formed within the body, as in the gallbladder, kidney, or urinary bladder.

Callus (KAL-lus) A growth of new bone tissue in and around a fractured area, ultimately replaced by mature bone. An acquired, localized thickening.

Calorie (KAL-ō-rē) A unit of heat. A calorie (cal) is the standard unit and is the amount of heat necessary to raise 1 g of water 1°C from 14° to 15°C. The kilocalorie (kcal), used in metabolic and nutrition studies, is the amount of heat necessary to raise 1,000 g of water 1°C and is equal to 1,000 cal.

Calyx (KĀL-iks) Any cuplike division of the kidney pelvis. *Plural,* ***calyces*** (KĀ-li-sēz).

Canal (ka-NAL) A narrow tube, channel, or passageway.

Canaliculus (kan'-a-LIK-yoo-lus) A small channel or canal, as in bones, where they connect lacunae. *Plural,* ***canaliculi*** (kan'-a-LIK-yoo-lī).

Canal of Schlemm *See* ***Scleral venous sinus.***

Cancellous (KAN-sel-us) Having a reticular or latticework structure, as in spongy tissue of bone.

Cancer (KAN-ser) A malignant tumor of epithelial origin tending to infiltrate and give rise to new growths or metastases. Also called ***carcinoma*** (kar'-si-NŌ-ma).

Canker (KANG-ker) ***sore*** Painful ulcer on the mucous membrane of the mouth that may result from an autoimmune response.

Capacitation (ka'-pas-i-TĀ-shun) The functional changes that sperm undergo in the female reproductive tract that allow them to fertilize a secondary oocyte.

Capillary (KAP-i-lar'-ē) A microscopic blood vessel located between an arteriole and venule through which materials are exchanged between blood and body cells.

Carbohydrate (kar'-bō-HĪ-drāt) An organic compound containing carbon, hydrogen, and oxygen in a particular amount and arrangement and comprised of sugar subunits; usually has the formula $(CH_2O)_n$.

Carbon monoxide (CO) poisoning Hypoxia due to increased levels of carbon monoxide as a result of its preferential and tenacious combination with hemoglobin rather than with oxygen.

Carcinogen (kar-SIN-ō-jen) Any substance that causes cancer.

Carcinoma (kar'-si-NŌ-ma) A malignant tumor consisting of epithelial cells.

Cardiac (KAR-dē-ak) ***arrest*** Cessation of an effective heartbeat in which the heart is completely stopped or in ventricular fibrillation.

Cardiac catheterization (KAR-dē-ak kath'-e-ter-i-ZĀ-shun) Introduction of a catheter into the heart and/or its blood vessels to measure pressure; assess left ventricular function and cardiac output; measure blood flow, oxygen content of blood, and the status of valves and conduction system; and identify valvular and septal defects.

Cardiac (KAR-dē-ak) ***cycle*** A complete heartbeat consisting of systole (contraction) and diastole (relaxation) of both atria plus systole and diastole of both ventricles.

Cardiac muscle An organ specialized for contraction, composed of striated muscle fibers (cells), forming the wall of the heart, and stimulated by an intrinsic conduction system and visceral efferent neurons.

Cardiac notch An angular notch in the anterior border of the left lung.

Cardiac tamponade (tam'-pon-ĀD) Compression of the heart due to excessive fluid or blood in the pericardial sac that could result in cardiac failure.

Cardinal ligament A ligament of the uterus, extending laterally from the cervix and vagina as a continuation of the broad ligament.

Cardioacceleratory (kar-dē-ō-ak-SEL-er-a-tō-rē) ***center (CAC)*** A group of neurons in the medulla from which cardiac nerves (sympathetic) arise; nerve impulses along the nerves release epinephrine that increases the rate and force of heartbeat.

Cardioinhibitory (kar-dē-ō-in-HIB-i-tō-rē) ***center (CIC)*** A group of neurons in the medulla from which parasympathetic fibers that reach the heart via the vagus (X) nerve arise; nerve impulses along the nerves release acetylcholine that decreases the rate and force of heartbeat.

Cardiology (kar-dē-OL-ō-jē) The study of the heart and diseases associated with it.

Cardiopulmonary resuscitation (rē-sus-i-TĀ-shun) ***(CPR)*** A technique employed to restore life or consciousness to a person apparently dead or dying; includes external respiration (exhaled air respiration) and external cardiac massage.

Carina (ka-RĪ-na) A ridge on the inside of the division of the right and left primary bronchi.

Carotid (ka-ROT-id) ***body*** Receptor on or near the carotid sinus that responds to alterations in blood levels of oxygen, carbon dioxide, and hydrogen ions.

Carotid sinus A dilated region of the internal carotid artery immediately above the bifurcation of the common carotid artery that contains receptors that monitor blood pressure.

Carpus (KAR-pus) A collective term for the eight bones of the wrist.

Cartilage (KAR-ti-lij) A type of connective tissue consisting of chondrocytes in lacunae embedded in a dense network of collagenous and elastic fibers and a matrix of chondroitin sulfate.

Cartilaginous (kar'-ti-LAJ-i-nus) ***joint*** A joint without a synovial (joint) cavity where the articulating bones are held tightly together by cartilage, allowing little or no movement.

Caruncle (KAR-ung-kul) A small fleshy eminence, often abnormal.

Cast A small mass of hardened material formed within a cavity in the body and then discharged from the body; can originate in different areas and be composed of various materials.

Castration (kas-TRĀ-shun) The removal of the testes.

Catabolism (ka-TAB-ō-lizm) Chemical reactions that break down complex organic compounds into simple ones with the release of energy.

Cataract (KAT-a-rakt) Loss of transparency of the lens of the eye or its capsule or both.

Catheter (KATH-i-ter) A tube that can be inserted into a body cavity through a canal or into a blood vessel; used to remove fluids, such as urine and blood, and to introduce diagnostic materials or medication.

Cauda equina (KAW-da ē-KWĪ-na) A taillike collection of roots of spinal nerves at the inferior end of the spinal canal.

Caudal (KAW-dal) Pertaining to any taillike structure; inferior in position.

Cecum (SĒ-kum) A blind pouch at the proximal end of the large intestine to which the ileum is attached.

Celiac (SĒ-lē-ak) Pertaining to the abdomen.

Celiac plexus (PLEK-sus) A large mass of ganglia and nerve fibers located at the level of the upper part of the first lumbar vertebra. Also called the ***solar plexus.***

Cell The basic structural and functional unit of all organisms; the smallest structure capable of performing all the activities vital to life.

Cell division Process by which a cell reproduces itself that consists of a nuclear division (mitosis) and a cytoplasmic division (cytokinesis); types include somatic and reproductive cell division.

Cell inclusion A lifeless, often temporary constituent in the cytoplasm of a cell as opposed to an organelle.

Cementum (se-MEN-tum) Calcified tissue covering the root of a tooth.

Center An area in the brain where a particular function is localized.

Center of ossification (os'-i-fi-KĀ-shun) An area in the cartilage model of a future bone where the cartilage cells hypertrophy and then secrete enzymes that result in the calcification of their matrix, resulting in the death of the cartilage cells, followed by the invasion of the area by osteoblasts that then lay down bone.

Central canal A circular channel running longitudinally in the center of an osteon (Haversian system) of mature compact bone, containing blood and lymphatic vessels and nerves. Also called a ***Haversian*** (ha-VĒR-shun) ***canal.*** A microscopic tube running the length of the spinal cord in the gray commissure.

Central fovea (FŌ-vē-a) A cuplike depression in the center of the macula lutea of the retina, containing cones only; the area of clearest vision.

Central nervous system (CNS) That portion of the nervous system that consists of the brain and spinal cord.

Centrioles (SEN-trē-ōlz) Paired, cylindrical structures within a centrosome, each consisting of a ring of microtubules and arranged at right angles to each other.

Centromere (SEN-trō-mēr) The clear, constricted portion of a chromosome where the two chromatids are joined; serves as the point of attachment for the chromosomal microtubules.

Centrosome (SEN-trō-sōm) A rather dense area of cytoplasm, near the nucleus of a cell, containing a pair of centrioles; serves as a center about which microtubules involved in chromosome movement are organized.

Cephalic (se-FAL-ik) Pertaining to the head; superior in position.

Cerebellar peduncle (ser-e-BEL-ar pe-DUNG-kul) A bundle of nerve fibers connecting the cerebellum with the brain stem.

Cerebellum (ser-e-BEL-um) The portion of the brain lying posterior to the medulla and pons, concerned with coordination of movements.

Cerebral aqueduct (SER-ē-bral AK-we-dukt) A channel through the midbrain connecting the third and fourth ventricles and containing cerebrospinal fluid.

Cerebral arterial circle A ring of arteries forming an anastomosis at the base of the brain between the internal carotid and basilar arteries and arteries supplying the brain. Also called the ***circle of Willis.***

Cerebral cortex The surface of the cerebral hemispheres, 2–4 mm thick, consisting of six layers of nerve cell bodies (gray matter) in most areas.

Cerebral palsy (PAL-zē) A group of motor disorders resulting in muscular uncoordination and loss of muscle control and caused by damage to motor areas of the brain (cerebral cortex, basal ganglia, and cerebellum) during fetal life, birth, or infancy.

Cerebral peduncle (pe-DUNG-kul) One of a pair of nerve fiber bundles located on the ventral surface of the midbrain, conducting nerve impulses between the pons and the cerebral hemispheres.

Cerebrospinal (se-rē'-brō-SPĪ-nal) ***fluid (CSF)*** A fluid produced in the choroid plexuses and ependymal cells of the ventricles of the brain that circulates in the ventricles and the subarachnoid space around the brain and spinal cord.

Cerebrovascular (se-rē'-brō-VAS-kyoo-lar) ***accident (CVA)*** Destruction of brain tissue (infarction) resulting from disorders of blood vessels that supply the brain. Also called a ***stroke.***

Cerebrum (SER-ē-brum) The two hemispheres of the forebrain, making up the largest part of the brain.

Ceruminous (se-ROO-mi-nus) ***gland*** A modified sudoriferous (sweat) gland in the external auditory meatus that secretes cerumen (ear wax).

Cervical dysplasia (dis-PLĀ-sē-a) A change in the shape, growth, and number of cervical cells of the uterus that, if severe, may progress to cancer.

Cervical ganglion (SER-vi-kul GANG-glē-on) A cluster of nerve cell bodies of postganglionic sympathetic neurons located in the neck, near the vertebral column.

Cervical mucus A mixture of water, glycoprotein, serum-type proteins, lipids, enzymes, and inorganic salts produced by secreting cells of the mucosa of the cervix.

Cervical plexus (PLEK-sus) A network of neuron fibers formed by the anterior rami of the first four cervical nerves.

Cervix (SER-viks) Neck; any constricted portion of an organ, such as the lower cylindrical part of the uterus.

Cesarean (se-SA-rē-an) ***section*** Procedure in which a low, horizontal incision is made through the abdominal wall and uterus for removal of the baby and placenta. Also called a ***C-section.***

Chalazion (ka-LĀ-zē-on) A small tumor of the eyelid.

Chemoreceptor (kē'-mō-rē-SEP-tor) Receptor outside the central nervous system on or near the carotid and aortic bodies that detects the presence of chemicals.

Chemotaxis (kē-mō-TAK-sis) Attraction of phagocytes to microbes by a chemical stimulus.

Chemotherapy (kē-mō-THER-a-pē) The treatment of illness or disease by chemicals.

Chiasma (kī-AZ-ma) A crossing; especially the crossing of the optic (II) nerve fibers.

Chiropractic (kī'-rō-PRAK-tik) A system of treating disease by using one's hands to manipulate body parts, mostly the vertebral column.

Chlamydia (kla-MID-ē-a) A sexually transmitted disease characterized by burning on urination, frequent and painful urination, and low back pain; may spread to uterine (Fallopian) tubes in females.

Choana (KŌ-a-na) A funnel-shaped structure; the posterior opening of the nasal fossa, or internal naris.

Cholecystectomy (kō'-lē-sis-TEK-tō-mē) Surgical removal of the gallbladder.

Cholesterol (kō-LES-te-rol) Classified as a lipid, the most abundant steroid in animal tissues; located in cell membranes and used for the synthesis of steroid hormones and bile salts.

Cholinergic (kō'-lin-ER-jik) ***fiber*** A nerve ending that liberates acetylcholine at a synapse.

Chondrocyte (KON-drō-sīt) Cell of mature cartilage.

Chondroitin (kon-DROY-tin) ***sulfate*** An amorphous matrix material found outside the cell.

Chordae tendineae (KOR-dē TEN-di-nē-ē) Tendonlike, fibrous cords that connect the heart valves with the papillary muscles.

Chorion (KŌ-rē-on) The outermost fetal membrane that becomes the principal embryonic portion of the placenta; serves a protective and nutritive function.

Chorionic villus (kō'-rē-ON-ik VIL-lus) Fingerlike projection of the chorion that grows into the decidua basalis of the endometrium and contains fetal blood vessels.

Chorionic villus sampling (CVS) The removal of a sample of chorionic villus tissue by means of a catheter to analyze the tissue for prenatal genetic defects.

Choroid (KŌ-royd) One of the vascular coats of the eyeball.

Choroid plexus (PLEK-sus) A vascular structure located in the roof of each of the four ventricles of the brain; produces cerebrospinal fluid.

Chromaffin (krō-MAF-in) ***cell*** Cell that has an affinity for chrome salts, due in part to the presence of the precursors of the neurotransmitter epinephrine; found, among other places, in the adrenal medulla.

Chromatid (KRŌ-ma-tid) One of a pair of identical connected nucleoprotein strands that are joined at the centromere and separate during cell division, each becoming a chromosome of one of the two daughter cells.

Chromatin (KRŌ-ma-tin) The threadlike mass of the genetic material consisting principally of DNA, which is present in the nucleus of a nondividing or interphase cell.

Chromatolysis (krō'-ma-TOL-i-sis) The breakdown of chromatophilic substance (Nissl bodies) into finely granular masses in the cell body of a central or peripheral neuron whose process (axon or dendrite) has been damaged.

Chromatophilic substance Rough endoplasmic reticulum in the cell bodies of neurons that functions in protein synthesis. Also called ***Nissl bodies.***

Chromosome (KRŌ-mō-sōm) One of the 46 small, dark-staining bodies that appear in the nucleus of a human diploid ($2n$) cell during cell division.

Chronic (KRON-ik) Long-term or frequently recurring; applied to a disease that is not acute.

Chyle (kīl) The milky fluid found in the lacteals of the small intestine after digestion.

Chyme (kīm) The semifluid mixture of partly digested food and digestive secretions found in the stomach and small intestine during digestion of a meal.

Cicatris (SIK-a-triks) A scar left by a healed wound.

Ciliary (SIL-ē-ar'-ē) ***body*** One of the three portions of the vascular tunic of the eyeball, the others being the choroid and the iris; includes the ciliary muscle and the ciliary processes.

Ciliary ganglion (GANG-glē-on) A very small parasympathetic ganglion whose preganglionic fibers come from the oculomotor (III) nerve and whose postganglionic fibers carry nerve impulses to the ciliary muscle and the sphincter muscle of the iris.

Cilium (SIL-ē-um) A hair or hairlike process projecting from a cell that may be used to move the entire cell or to move substances along the surface of the cell.

Circadian (ser-KĀ-dē-an) ***rhythm*** A cycle of active and non-active periods in organisms determined by internal mechanisms and repeating about every 24 hours.

Circle of Willis *See* ***Cerebral arterial circle.***

Circular folds Permanent, deep, transverse folds in the mucosa and submucosa of the small intestine that increase the surface area for absorption. Also called ***plicae circulares*** (PLĪ-kē SER-kyoo-lar-ēs).

Circumcision (ser'-kum-SIZH-un) Surgical removal of the foreskin (prepuce), the fold of skin over the glans penis.

Circumduction (ser'-kum-DUK-shun) A movement at a synovial joint in which the distal end of a bone moves in a circle while the proximal end remains relatively stable.

Circumvallate papilla (ser'-kum-VAL-āt pa-PIL-a) One of the circular projections that is arranged in an inverted V-shaped row at the posterior portion of the tongue; the largest of the elevations on the upper surface of the tongue containing taste buds.

Cirrhosis (si-RŌ-sis) A liver disorder in which the parenchymal cells are destroyed and replaced by connective tissue.

Cisterna chyli (sis-TER-na KĪ-lē) The origin of the thoracic duct.

Cleavage The rapid mitotic divisions following the fertilization of a secondary oocyte, resulting in an increased number of progressively smaller cells, called blastomeres, so that the overall size of the zygote remains the same.

Cleft palate Condition in which the palatine processes of the maxillae do not unite before birth; cleft lip, a split in the upper lip, is often associated with cleft palate.

Climacteric (klī-mak-TER-ik) Cessation of the reproductive function in the female or diminution of testicular activity in the male.

Climax The peak period or moments of greatest intensity during sexual excitement.

Clitoris (KLI-to-ris) An erectile organ of the female located at the anterior junction of the labia minora that is homologous to the male penis.

Clot The end result of a series of biochemical reactions that changes liquid plasma into a gelatinous mass; specifically, the conversion of fibrinogen into a tangle of polymerized fibrin molecules.

Coagulation (cō-ag-yoo-LĀ-shun) Process by which a blood clot is formed.

Coarctation (kō'-ark-TĀ-shun) ***of the aorta*** Congenital condition in which the aorta is too narrow and results in reduced blood supply, increased ventricular pumping, and high blood pressure.

Coccyx (KOK-six) The fused bones at the end of the vertebral column.

Cochlea (kŌK-lē-a) A winding, cone-shaped tube forming a portion of the inner ear and containing the spiral organ (organ of Corti).

Cochlear duct The membranous cochlea consisting of a spirally arranged tube enclosed in the bony cochlea and lying along its outer wall. Also called the ***scala media*** (SCA-la MĒ-dē-a).

Coitus (KŌ-i-tus) Sexual intercourse. Also called ***copulation*** (cop-yoo-LĀ-shun).

Colitis (ko-LĪ-tis) Inflammation of the mucosa of the colon and rectum in which absorption of water and salts is reduced, producing watery, bloody feces, and, in severe cases, dehydration and salt depletion. Spasms of the irritated muscularis produce cramps.

Collagen ((KOL-a-jen) A protein that is the main organic constituent of connective tissue.

Collateral circulation The alternate route taken by blood through an anastomosis.

Colliculus (ko-LIK-yoo-lus) A small elevation.

Colon The division of the large intestine consisting of ascending, transverse, descending, and sigmoid portions.

Color blindness Any deviation in the normal perception of colors, resulting from the lack of usually one of the photo-pigments of the cones.

Colostomy (kō-LOS-tō-mē) The diversion of feces through an opening in the colon, created by a surgical opening at the exterior of the abdominal wall.

Colostrum (kō-LOS-trum) A thin, cloudy fluid secreted by the mammary glands a few days prior to or after delivery before true milk is secreted.

Colposcopy (kol-POS-kō-pē) Direct examination of the vaginal and cervical mucosa using a magnifying device; frequently the first procedure performed following an abnormal Pap smear.

Coma (KŌ-ma) Final stage of brain failure that is characterized by total unresponsiveness to all external stimuli.

Commissure (KOM-i-shūr) The angular junction of the eyelids at either corner of the eyes.

Common bile duct A tube formed by the union of the common hepatic duct and the cystic duct that empties bile into the duodenum at the hepatopancreatic ampulla (ampulla of Vater).

Compact (dense) bone Bone tissue with no apparent spaces in which the layers of lamellae are fitted tightly together. Compact bone is found immediately deep to the periosteum and external to spongy bone.

Complete blood count (CBC) Hematology test that usually includes hemoglobin determination, hematocrit, red and white blood cell count, differential white blood cell count, and platelet count.

Computed tomography (tō-MOG-ra-fē) ***(CT)*** X-ray technique that provides a cross-sectional image of any area of the body. Also called ***computed axial tomography (CAT).***

Concha (KONG-ka) A scroll-like bone found in the skull. *Plural,* ***conchae*** (KONG-kē). Also called a ***turbinate*** (TUR-bi-nāt).

Concussion (kon-KUSH-un) Traumatic injury to the brain that produces no visible bruising but may result in abrupt, temporary loss of consciousness.

Conduction myofiber Muscle fiber (cell) in the subendocardial tissue of the heart specialized for conducting an action potential to the myocardium; part of the conduction system of the heart. Also called a ***Purkinje*** (pur-KIN-jē) ***fiber.***

Conduction system An intrinsic regulating system composed of specialized muscle tissue that generates and distributes electrical impulses that stimulate cardiac muscle fibers (cells) to contract.

Conductivity (kon'-duk-TIV-i-tē) The ability to carry the effect of a stimulus from one part of a cell to another; highly developed in nerve and muscle fibers (cells).

Cone The light-sensitive receptor in the retina concerned with color vision.

Cone biopsy (BĪ-op-sē) Removal of a sample of tissue from the cervical os to evaluate for cancer of the uterus.

Congenital (kon-JEN-i-tal) Present at the time of birth.

Congestive heart failure (CHF) Chronic or acute state that results when the heart is not capable of supplying the oxygen demands of the body.

Conjunctiva (kon'-junk-TĪ-va) The delicate membrane covering the eyeball and lining the eyes.

Conjunctivitis (kon-junk'-ti-VĪ-tis) Inflammation of the conjunctiva, the delicate membrane covering the eyeball and lining the eyelids.

Connective tissue The most abundant of the four basic tissue types in the body, performing the functions of binding and supporting; consists of relatively few cells in a great deal of intercellular substance.

Constipation (con-sti-PĀ-shun) Infrequent or difficult defecation caused by decreased motility of the intestines.

Contact inhibition Phenomenon by which migration of a growing cell is stopped when it makes contact with another cell of its own kind.

Contraception (kon'-tra-SEP-shun) The prevention of conception or impregnation without destroying fertility.

Contractility (kon'-trak-TIL-i-tē) The ability of cells or parts of cells to actively generate force to undergo shortening and change form for purposeful movements. Muscle fibers (cells) exhibit a high degree of contractility.

Contralateral (kon'-tra-LAT-er-al) On the opposite side; affecting the opposite side of the body.

Contusion (kon-TOO-shun) Condition in which tissue below the skin is damaged, but the skin is not broken.

Conus medullaris (KŌ-nus med-yoo-LAR-is) The tapered portion of the spinal cord below the lumbar enlargement.

Convergence (con-VER-jens) An anatomical arrangement in which the synaptic end bulbs of several presynaptic neurons terminate on one postsynaptic neuron. The medial movement of the two eyeballs so that both are directed toward a near object being viewed in order to produce a single image.

Convulsion (con-VUL-shun) Violent, involuntary, tetanic contractions of an entire group of muscles.

Cornea ((KOR-nē-a) The nonvascular, transparent fibrous coat through which the iris can be seen.

Corona (kō-RŌ-na) Margin of the glans penis.

Corona radiata Several layers of follicle cells surrounding a secondary oocyte.

Coronal (kō-RŌ-nal) ***plane*** A plane that runs vertical to the ground and divides the body into anterior and posterior portions. Also called ***frontal plane.***

Coronary angiography (KOR-ō-na-rē an'-jē-OG-ra-fē) Procedure in which the severity and location of blocked coronary arteries are visualized by injection of contrast dyes or in which clot-dissolving drugs may be injected into coronary arteries.

Coronary (KOR-ō-na-rē) ***artery bypass grafting (CABG)*** Surgical procedure in which a portion of a blood vessel is removed from another part of the artery and grafted onto a coronary artery so as to bypass an obstruction in the coronary artery.

Coronary (KOR-ō-na-rē) ***artery disease (CAD)*** A condition in which the heart muscle receives inadequate blood due to an interruption of its blood supply.

Coronary artery spasm A condition in which the smooth muscle of a coronary artery undergoes a sudden contraction, resulting in vasoconstriction.

Coronary circulation The pathway followed by the blood from the ascending aorta through the blood vessels supplying the heart and returning to the right atrium. Also called ***cardiac circulation.***

Coronary sinus (SĪ-nus) A wide venous channel on the posterior surface of the heart that collects the blood from the coronary circulation and returns it to the right atrium.

Corpora quadrigemina (KOR-por-a kwad-ri-JEM-in-a) Four small elevations (superior and inferior colliculi) on the dorsal region of the midbrain concerned with visual and auditory functions.

Cor pulmonale (kor pul-mōn-ALE) ***(CP)*** Right ventricular hypertrophy from disorders that bring about hypertension in pulmonary circulation.

Corpus (KOR-pus) The principal part of any organ; any mass or body.

Corpus albicans (KOR-pus-AL-bi-kanz) A white fibrous patch in the ovary that forms after the corpus luteum regresses.

Corpus callosum (kal-LŌ-sum) The great commisure of the brain between the cerebral hemispheres.

Corpuscle of touch The sensory receptor for the sensation of touch; found in the dermal papillae, especially in palms and soles. Also called a ***Meissner's*** (MĪS-nerz) ***corpuscle.***

Corpus luteum (LOO-tē-um) A yellow endocrine gland in the ovary formed when a follicle has discharged its secondary oocyte; secretes estrogens, progesterone, and relaxin.

Corpus striatum (strī-Ā-tum) An area in the interior of each cerebral hemisphere composed of the caudate and lentiform nuclei of the basal ganglia and white matter of the internal capsule, arranged in a striated manner.

Cortex (KOR-teks) An outer layer of an organ. The convoluted layer of gray matter covering each cerebral hemisphere.

Costal (KOS-tal) Pertaining to a rib.

Costal cartilage (KOS-tal KAR-ti-lij) Hyaline cartilage that attaches a rib to the sternum.

Cowper's gland *See* ***Bulbourethral gland.***

Cramp A spasmodic, especially a tonic, contraction of one or many muscles, usually painful.

Cranial (KRĀ-nē-al) ***cavity*** A subdivision of the dorsal body cavity formed by the cranial bones and containing the brain.

Cranial nerve One of 12 pairs of nerves that leave the brain, pass through foramina in the skull, and supply the head, neck, and part of the trunk; each is designated by a Roman numeral and a name.

Craniosacral (krā-nē-ō-SĀ-kral) ***outflow*** The fibers of parasympathetic preganglionic neurons, which have their cell bodies located in nuclei in the brain stem and in the lateral gray matter of the sacral portion of the spinal cord.

Craniotomy (krā'-nē-OT-ō-mē) Any operation on the skull, as for surgery on the brain or decompression of the fetal head in difficult labor.

Cranium (KRĀ-nē-um) The skeleton of the skull that protects the brain and the organs of sight, hearing, and balance; includes the frontal, parietal, temporal, occipital, sphenoid, and ethmoid bones.

Cretinism (KRĒ-tin-izm) Severe congenital thyroid deficiency during childhood leading to physical and mental retardation.

Crista (KRIS-ta) A crest or ridged structure. A small elevation in the ampulla of each semicircular duct that serves as a receptor for dynamic equilibrium.

Crossing-over The exchange of a portion of one chromatid with another in a tetrad during meiosis. It permits an exchange of genes among chromatids and is one factor that results in genetic variation.

Crus (krus) ***of penis*** Separated, tapered portion of the corpora cavernosa penis. *Plural,* ***crura*** (KROO-ra).

Cryosurgery ((KRĪ-ō-ser-jer-ē) The destruction of tissue by application of extreme cold.

Crypt of Lieberkühn *See* ***Intestinal gland.***

Cryptorchidism (krip-TOR-ki-dizm) The condition of undescended testes.

Cupula (KUP-yoo-la) A mass of gelatinous material covering the hair cells of a crista, a receptor in the ampulla of a semicircular canal stimulated when the head moves.

Curvature (KUR-va-tūr) A nonangular deviation of a straight line, as in the greater and lesser curvatures of the stomach. Abnormal curvatures of the vertebral column include kyphosis, lordosis, and scoliosis.

Cushing's syndrome Condition caused by a hypersecretion of glucocorticoids characterized by spindly legs, ''moon face,'' ''buffalo hump,'' pendulous abdomen, flushed facial skin, poor wound healing, hyperglycemia, osteoporosis, weakness, hypertension, and increased susceptibility to disease.

Cutaneous (kyoo-TĀ-nē-us) Pertaining to the skin.

Cyanosis (sī-a-NŌ-sis) Reduced hemoglobin (unoxygenated) concentration of blood of more than 5 g/dl that results in a blue or dark purple discoloration that is most easily seen in nail beds and mucous membranes.

Cyst (SIST) A sac with a distinct connective tissue wall, containing a fluid or other material.

Cystic (SIS-tik) ***duct*** The duct that transports bile from the gallbladder to the common bile duct.

Cystitis (sis-TĪ-tis) Inflammation of the urinary bladder.

Cystoscope (SIS-to-skōp) An instrument used to examine the inside of the urinary bladder.

Cystoscopy (sis-TOS-kō-pē) Direct visual examination of the urinary tract (and prostate gland in males as well) using a cystoscope to evaluate urinary tract disorders and remove tissue for biopsy, kidney stones, urinary bladder tumors, and urine samples.

Cytokinesis (sī'-tō-ki-NĒ-sis) Division of the cytoplasm.

Cytology (sī-TOL-ō-jē) The study of cells.

Cytoplasm (SĪ-tō-plazm) All the cellular contents between the plasma membrane and nucleus. Also called ***protoplasm.***

Cytoskeleton Complex internal structure of cytoplasm consisting of microfilaments, microtubules, and intermediate filaments.

Cytosol (SĪ-tō-sol) The semifluid portion of cytoplasm in which organelles and inclusions are suspended and solutes are dissolved.

Dartos (DAR-tōs) The contractile tissue under the skin of the scrotum.

Deafness Lack of the sense of hearing or a significant hearing loss.

Debility (dē-BIL-i-tē) Weakness of tonicity in functions or organs of the body.

Decibel (DES-i-bel) ***(db)*** A unit that measures relative sound intensity (loudness).

Decidua (dē-SID-yoo-a) That portion of the endometrium of the uterus (all but the deepest layer) that is modified for pregnancy and shed after childbirth.

Deciduous (dē-SID-yoo-us) Falling off or being shed seasonally or at a particular stage of development. In the body, referring to the first set of teeth.

Decubitus (dē-KYOO-bi-tus) ***ulcer*** Tissue destruction due to a constant deficiency of blood to tissues overlying a bony projection that has been subjected to prolonged pressure against an object such as a bed, cast, or splint. Also called ***bedsore, pressure sore,*** or ***trophic ulcer.***

Decussation (dē'-ku-SĀ-shun) A crossing-over; usually refers to the crossing of most of the fibers in the large motor tracts to opposite sides in the medullary pyramids.

Deep Away from the surface of the body.

Deep fascia (FASH-ē-a) A sheet of connective tissue wrapped around a muscle to hold it in place.

Deep inguinal (IN-gwi-nal) ***ring*** A slitlike opening in the aponeurosis of the transversus abdominis muscle that represents the origin of the inguinal canal.

Deep-venous thrombosis ***(DVT)*** The presence of a thrombus in a vein, usually a deep vein of the lower extremities.

Defecation (def-e-KĀ-shun) The discharge of feces from the rectum.

Defibrillation (dē-fib-ri-LĀ-shun) Delivery of a very strong electrical current to the heart in an attempt to stop ventricular fibrillation.

Degeneration (dē-jen-er-Ā-shun) A change from a higher to a lower state; a breakdown in structure.

Deglutition (dē-gloo-TISH-un) The act of swallowing.

Dehydration (dē-hī-DRĀ-shun) Excessive loss of water from the body or its parts.

Delirium (de-LIR-ē-um) A transient disorder of abnormal cognition (perception, thinking, and memory) and disordered attention that is accompanied by disturbances of the sleep-wake cycle and psychomotor behavior (hyperactivity or hypoactivity of movements and speech). Also called ***acute confusional state (ACS).***

Delta cell A cell in the pancreatic islets (islets of Langerhans) in the pancreas that secretes somatostatin.

Dementia (de-MEN-shē-a) An organic mental disorder that results in permanent or progressive general loss of intellectual abilities such as impairment of memory, judgment, and abstract thinking and changes in personality; most common cause is Alzheimer's disease.

Demineralization (de-min'-er-al-i-ZĀ-shun) Loss of calcium and phosphorus from bones.

Dendrite (DEN-drīt) A nerve cell process that carries a nerve impulse toward the cell body.

Dens (denz) Tooth

Dental caries (KA-rēz) Gradual demineralization of the enamel and dentin of a tooth that may invade the pulp and alveolar bone. Also called ***tooth decay.***

Denticulate (den-TIK-yoo-lāt) Finely toothed or serrated; characterized by a series of small, pointed projections.

Dentin (DEN-tin) The osseous tissues of a tooth enclosing the pulp cavity.

Dentition (den-TI-shun) The eruption of teeth. The number, shape, and arrangement of teeth.

Deoxyribonucleic (dē-ok'-sē-ri'-bō-nyoo-KLĒ-ik) ***acid (DNA)*** A nucleic acid in the shape of a double helix constructed of nucleotides consisting of one of four nitrogenous bases (adenine, cytosine, guanine, or thymine), deoxyribose, and a phosphate group; encoded in the nucleotides is genetic information.

Depression (dē-PRESS-shun) Movement in which a part of the body moves downward.

Dermal papilla (pa-PILL-a) Fingerlike projection of the papillary region of the dermis that may contain blood capillaries or corpuscles of touch (Meissner's corpuscles).

Dermatology (der-ma-TOL-ō-jē) The medical specialty dealing with diseases of the skin.

Dermatome (DER-ma-tōm) An instrument for incising the skin or cutting thin transplants of skin. The cutaneous area developed from one embryonic spinal cord segment and receiving most of its innervation from one spinal nerve.

Dermis (DER-mis) A layer of dense connective tissue lying deep to the epidermis; the true skin or corium.

Descending colon (KŌ-lon) The part of the large intestine descending from the left colic (splenic) flexure to the level of the left iliac crest.

Detritus (de-TRI-tus) Particulate matter produced by or remaining after the wearing away or disintegration of a substance or tissue; scales, crusts, or loosened skin.

Detrusor (de-TROO-ser) ***muscle*** Muscle in the wall of the urinary bladder.

Developmental anatomy The study of development from the fertilized egg to the adult form. The branch of anatomy called embryology is generally restricted to the study of development from the fertilized egg through the eighth week in utero.

Diabetes insipidus (dī-a-BĒ-tēz in-SIP-i-dus) Condition caused by hyposecretion of antidiuretic hormone (ADH) and characterized by excretion of large amounts of urine and thirst.

Diabetes mellitus (MEL-i-tus) Hereditary condition caused by hyposecretion of insulin and characterized by hyperglycemia, increased urine production, excessive thirst, and excessive eating.

Diagnosis (dī-ag-NŌ-sis) Distinguishing one disease from another or determining the nature of a disease from signs and symptoms by inspection, palpation, laboratory tests, and other means.

Diapedesis (dī-a-pe-DĒ-sis) The passage of white blood cells through intact blood vessel walls.

Diaphragm (DĪ-a-fram) Any partition that separates one area from another, especially the dome-shaped skeletal muscle between the thoracic and abdominal cavities. Also a dome-shaped structure that fits over the cervix, usually with a spermicide, to prevent conception.

Diaphysis (dī-AF-i-sis) The shaft of a long bone.

Diarrhea (dī-a-RĒ-a) Frequent defecation of liquid feces caused by increased motility of the intestines.

Diarthrosis (dī'-ar-THRŌ-sis) Articulation in which opposing bones move freely, as in a hinge joint.

Diastole (dī-AS-tō-lē) In the cardiac cycle, the phase of relaxation or dilation of the heart muscle, especially of the ventricles.

Diencephalon (dī'-en-SEF-a-lon) A part of the brain consisting primarily of the thalamus and the hypothalamus.

Differential (dif-fer-EN-shal) ***white blood cell count*** Determination of the number of each kind of white blood cell in a sample of 100 cells for diagnostic purposes.

Differentiation (dif'-e-ren'-shē-Ā-shun) Acquisition of specific functions different from those of the original general type.

Diffusion (dif-YOO-zhun) A passive process in which there is a net or greater movement of molecules or ions from a region of high concentration to a region of low concentration until equilibrium is reached.

Digestion (dī-JES-chun) The mechanical and chemical breakdown of food to simple molecules that can be absorbed and used by body cells.

Digital subtraction angiography (an-jē-OG-ra-fē) ***(DSA)*** A medical imaging technique that compares an x-ray image of the same artery of the body before and after a contrast substance containing iodine has been introduced intravenously.

Dilate (DĪ-lāte) To expand or swell.

Dilation (dī-LĀ-shun) ***and curettage*** (ku-re-TAZH) Following dilation of the uterine cervix, the uterine endometrium is scraped with a curette (spoon-shaped instrument). Also called a ***D and C.***

Diploid (DIP-loyd) Having the number of chromosomes characteristically found in the somatic cells of an organism. Symbolized $2n$.

Diplopia (di-PLŌ-pē-a) Double vision.

Disease Any change from a state of health.

Dislocation (dis-lō-KĀ-shun) Displacement of a bone from a joint with tearing of ligaments, tendons, and articular capsules. Also called ***luxation*** (luks-Ā-shun).

Dissect (DIS-sekt) To separate tissues and parts of a cadaver (corpse) or an organ for anatomical study.

Distal (DIS-tal) Farther from the attachment of an extremity to the trunk or a structure; farther from the point of origin.

Diuretic (dī-yoo-RET-ik) A chemical that inhibits sodium reabsorption, reduces antidiuretic hormone (ADH) concentration, and increases urine volume by inhibiting facultative reabsorption of water.

Diurnal (dī-UR-nal) Daily.

Divergence (di-VER-jens) An anatomical arrangement in which the synaptic end bulbs of one presynaptic neuron terminate on several postsynaptic neurons.

Diverticulitis (dī-ver-tik-yoo-LĪ-tis) Inflammation of diverticula, saclike outpouchings of the colonic wall, when the muscularis becomes weak.

Diverticulum (dī-ver-TIK-yoo-lum) A sac or pouch in the wall of a canal or organ, especially in the colon.

Donor insemination (in-sem'-i-NĀ-shun) The deposition of seminal fluid within the vagina or cervix at a time during the menstrual cycle when pregnancy is most likely to occur. It may be homologous (using the husband's semen) or heterologous (using a donor's semen). Also called ***artificial insemination.***

Dorsal body cavity Cavity near the dorsal surface of the body that consists of a cranial cavity and vertebral canal.

Dorsal ramus (RĀ-mus) A branch of a spinal nerve containing motor and sensory fibers supplying the muscles, skin, and bones of the posterior part of the head, neck, and trunk.

Dorsiflexion (dor'-si-FLEK-shun) Bending the foot in the direction of the dorsum (upper surface).

Down syndrome (DS) An inherited defect due to an extra copy of chromosome 21. Symptoms include mental retardation; a small skull, flattened from front to back; a short, flat nose; short fingers; and a widened space between the first two digits of the hand and foot. Also called ***trisomy 21.***

Dropsy (DROP-sē) A condition in which there is abnormal accumulation of water in the tissues and cavities.

Duct of Santorini *See* ***Accessory duct.***

Duct of Wirsung *See* ***Pancreatic duct.***

Ductus arteriosus (DUK-tus ar-tē-rē-Ō-sus) A small vessel connecting the pulmonary trunk with the aorta; found only in the fetus.

Ductus (vas) deferens (DEF-er-ens) The duct that conducts spermatozoa from the epididymis to the ejaculatory duct. Also called the ***seminal duct.***

Ductus epididymis (ep'-i-DID-i-mis) A tightly coiled tube inside the epididymis, distinguished into a head, body, and tail, in which spermatozoa undergo maturation.

Ductus venosus (ve-NŌ-sus) A small vessel in the fetus that helps the circulation bypass the liver.

Duodenal (doo-ō-DĒ-nal) ***gland*** Gland in the submucosa of the duodenum that secretes an alkaline mucus to protect the lining of the small intestine from the action of enzymes and to help neutralize the acid in chyme. Also called ***Brunner's*** (BRUN-erz) ***gland.***

Duodenal papilla (pa-PILL-a) An elevation on the duodenal mucosa that receives the hepatopancreatic ampulla (ampulla of Vater).

Duodenum (doo'-ō-DĒ-num) The first 25 cm (10 in.) of the small intestine.

Dura mater (DYOO-ra MĀ-ter) The outer membrane (meninx) covering the brain and spinal cord.

Dynamic equilibrium (ē-kwi-LIB-rē-um) The maintenance of body position, mainly the head, in response to sudden movements such as rotation.

Dynamic spatial reconstruction (DSR) A technique that has the ability to construct moving, three-dimensional, life-size images of all or part of an internal organ from any view desired.

Dysfunction (dis-FUNK-shun) Absence of complete normal function.

Dyslexia (dis-LEX-sē-a) Impairment of the brain's ability to translate images received from the eyes or ears into understandable language.

Dysmenorrhea (dis'-men-ō-RĒ-a) Painful menstruation.

Dysphagia (dis-FĀ-jē-a) Difficulty in swallowing.

Dysplasia (dis-PLĀ-zē-a) Change in the size, shape, and organization of cells due to chronic irritation or inflammation; may revert to normal if stress is removed or progress to neoplasia.

Dyspnea (DISP-nē-a) Shortness of breath.

Dystocia (dis-TŌ-sē-a) Difficult labor due to factors such as pelvic deformities, malpositioned fetus, and premature rupture of fetal membranes.

Dystrophia (dis-TRŌ-fē-a) Progressive weakening of a muscle.

Dysuria (dis-SOO-rē-a) Painful urination.

Echocardiogram (ek-ō-KAR-dē-ō-gram) A procedure in which high-frequency sound waves directed at the heart are bounced back and the echoes are picked up by a transducer and converted to an image.

Ectoderm The outermost of the three primary germ layers that gives rise to the nervous system and the epidermis of skin and its derivatives.

Ectopic (ek-TOP-ik) Out of the normal location, as in ectopic pregnancy.

Eczema (EK-ze-ma) A skin rash characterized by itching, swelling, blistering, oozing, and scaling of the skin.

Edema (e-DĒ-ma) An abnormal accumulation of interstitial fluid.

Effector (e-FEK-tor) The organ of the body, either a muscle or a gland, that responds to a motor neuron impulse.

Efferent arteriole (EF-er-ent ar-TĒ-rē-ōl) A vessel of the renal vascular system that transports blood from the glomerulus to the peritubular capillary.

Efferent (EF-er-ent) ***ducts*** A series of coiled tubes that transport spermatozoa from the rete testis to the epididymis.

Efferent neuron (NOO-ron) A neuron that conveys nerve impulses from the brain and spinal cord to effectors that may be either muscles or glands. Also called a ***motor neuron.***

Ejaculation (e-jak-yoo-LĀ-shun) The reflex ejection or expulsion of semen from the penis.

Ejaculatory (e-JAK-yoo-la-tō'-rē) ***duct*** A tube that transports spermatozoa from the ductus (vas) deferens to the prostatic urethra.

Elasticity (e-las-TIS-i-tē) The ability of tissue to return to its original shape after contraction or extension.

Electrocardiogram (e-lek'-trō-KAR-dē-ō-gram) ***(ECG*** or *EKG)* A recording of the electrical changes that accompany the cardiac cycle and can be recorded on the surface of the body; may be resting, stress, or ambulatory.

Electroencephalogram (e-lek'-trō-en-SEF-a-lō-gram) ***(EEG)*** A recording of the electrical impulses of the brain to diagnose certain diseases (such as epilepsy), furnish information regarding sleep and wakefulness, and confirm brain death.

Electromyography (e-lek'-trō-mī-OG-ra-fē) Evaluation of the electrical activity of resting and contracting muscle to ascertain causes of muscular weakness, paralysis, involuntary twitching, and abnormal levels of muscle enzymes; also used as part of biofeedback studies.

Elevation (el-e-VĀ-shun) Movement in which a part of the body moves upward.

Ellipsoidal (e-lip-SOY-dal) ***joint*** A synovial joint structured so that an oval-shaped condyle of one bone fits into an elliptical cavity of another bone, permitting side-to-side and back-and-forth movements, as at the joint at the wrist between the radius and carpals. Also called a ***condyloid*** (KON-di-loid) ***joint.***

Embolism (EM-bō-lizm) Obstruction or closure of a vessel by an embolus.

Embolus (EM-bō-lus) A blood clot, bubble of air, fat from broken bones, mass of bacteria, or other debris or foreign material transported by the blood.

Embryo (EM-brē-ō) The young of any organism in an early stage of development; in humans, the developing organism from fertilization to the end of the eighth week in utero.

Embryology (em'-brē-OL-ō-jē) The study of development from the fertilized egg to the end of the eighth week in utero.

Embryo transfer A type of in vitro fertilization in which semen is used to artificially inseminate a fertile secondary oocyte donor and the morula or blastocyst is then transferred from the donor to the infertile woman, who then carries it to term.

Emesis (EM-e-sis) Vomiting.

Emmetropia (em'-e-TRŌ-pē-a) The ideal optical condition of the eyes.

Emphysema (em'-fi-SĒ-ma) A swelling or inflation of air passages due to loss of elasticity in the alveoli.

Emulsification (ē-mul'-si-fi-KĀ-shun) The dispersion of large fat globules to smaller uniformly distributed particles in the presence of bile.

Enamel (e-NAM-el) The hard, white substance covering the crown of a tooth.

Endocardium (en-dō-KAR-dē-um) The layer of the heart wall, composed of endothelium and smooth muscle, that lines the inside of the heart and covers the valves and tendons that hold the valves open.

Endochondral ossification (en'-dō-KON-dral os'-i-fi-KĀ-shun) The replacement of cartilage by bone. Also called ***intracartilaginous*** (in'-tra-kar'-ti-LAJ-i-nus) ***ossification.***

Endocrine (EN-dō-krin) ***gland*** A gland that secretes hormones into the blood; a ductless gland.

Endocrinology (en'-dō-kri-NOL-ō-jē) The science concerned with the structure and functions of endocrine glands and the diagnosis and treatment of disorders of the endocrine system.

Endocytosis (en'-dō-sī-TŌ-sis) The uptake into a cell of large molecules and particles in which a segment of plasma membrane surrounds the substance, encloses it, and brings it in; includes phagocytosis, pinocytosis, and receptor-mediated endocytosis.

Endoderm The innermost of the three primary germ layers of the developing embryo that gives rise to the gastrointestinal tract, urinary bladder and urethra, and respiratory tract.

Endodontics (en'-dō-DON-tiks) The branch of dentistry concerned with the prevention, diagnosis, and treatment of diseases that affect the pulp, root, periodontal ligament, and alveolar bone.

Endolymph (EN-dō-lymf') The fluid within the membranous labyrinth of the inner ear.

Endometriosis (en'-dō-MĒ-trē-ō'-sis) The growth of endometrial tissue outside the uterus.

Endometrium (en'-dō-MĒ-trē-um) The mucous membrane lining the uterus.

Endomysium (en'-dō-MĪZ-ē-um) Invagination of the perimysium separating each individual muscle fiber (cell).

Endoneurium (en'-dō-NYOO-rē-um) Connective tissue wrapping around individual nerve fibers (cells).

Endoplasmic reticulum (en'-do-PLAZ-mik re-TIK-yoo-lum) ***(ER)*** A network of channels running through the cytoplasm of a cell that serves in intracellular transportation, support, storage, synthesis, and packaging of molecules. Portions of ER where ribosomes are attached to the outer surface are called ***granular*** or ***rough reticulum;*** portions that have no ribosomes are called ***agranular*** or ***smooth reticulum.***

End organ of Ruffini *See **Type II cutaneous mechanoreceptor.***

Endorphin (en-DOR-fin) A neuropeptide in the central nervous system that acts as a painkiller.

Endoscope (EN-dō-skōp') An illuminated tube with lenses used to look inside hollow organs such as the stomach (gastroscope) or urinary bladder (cystoscope).

Endoscopy (en-DOS-kō-pē) The visual examination of any cavity of the body using an endoscope, an illuminated tube with lenses.

Endosteum (en-DOS-tē-um) The membrane that lines the medullary cavity of bones, consisting of osteoprogenitor cells and scattered osteoclasts.

Endothelial-capsular (en-dō-THĒ-lē-al) ***membrane*** A filtration membrane in a nephron of a kidney consisting of the endothelium and basement membrane of the glomerulus and the epithelium of the visceral layer of the glomerular (Bowman's) capsule.

Endothelium (en'-dō-THĒ-lē-um) The layer of simple squamous epithelium that lines the cavities of the heart, blood vessels, and lymphatic vessels.

Energy The capacity to do work.

Enkephalin (en-KEF-a-lin) A peptide found in the central nervous system that acts as a painkiller.

Enteroendocrine (en-ter-ō-EN-dō-krin) ***cell*** A stomach cell that secretes the hormone stomach gastrin.

Enuresis (en'-yoo-RĒ-sis) Involuntary discharge of urine, complete or partial, after age 3.

Enzyme (EN-zīm) A substance that affects the speed of chemical changes; an organic catalyst, usually a protein.

Eosinophil (ē'-ō-SIN-ō-fil) A type of white blood cell characterized by granular cytoplasm readily stained by eosin.

Ependyma (e-PEN-de-ma) Neuroglial cells that line ventricles of the brain and probably assist in the circulation of cerebrospinal fluid (CSF). Also called ***ependymocytes*** (e-PEN-di-mō-sītz).

Epicardium (ep'-i-KAR-dē-um) The thin outer layer of the heart wall, composed of serous tissue and mesothelium. Also called the ***visceral pericardium.***

Epidemic (ep'-i-DEM-ik) A disease that occurs above the expected level among individuals in a population.

Epidemiology (ep'-i-dē-mē-OL-ō-jē) Medical science concerned with the occurrence and distribution of disease in human populations.

Epidermis (ep-i-DERM-is) The outermost, thinner layer of skin, composed of stratified squamous epithelium.

Epididymis (ep'-i-DID-i-mis) A comma-shaped organ that lies along the posterior border of the testis and contains the ductus epididymis, in which sperm undergo maturation. *Plural,* ***epididymides*** (ep'-i-DID-i-mi-dēz).

Epidural (ep'-i-DOO-ral) ***space*** A space between the spinal dura mater and the vertebral canal, containing areolar connective tissue and a plexus of veins.

Epiglottis (ep'-i-GLOT-is) A large, leaf-shaped piece of cartilage lying on top of the larynx, with its "stem" attached to the thyroid cartilage and its "leaf" portion unattached and free to move up and down to cover the glottis (vocal folds and rima glottidis).

Epilepsy (EP-i-lep'-sē) Neurological disorder characterized by short, periodic attacks of motor, sensory, or psychological malfunction.

Epimysium (ep'-i-MĪZ-ē-um) Fibrous connective tissue around muscles.

Epinephrine (ep-ē-NEF-rin) Hormone secreted by the adrenal medulla that produces actions similar to those that result from sympathetic stimulation. Also called ***adrenaline*** (a-DREN-a-lin).

Epineurium (ep'-i-NYOO-rē-um) The outermost covering around the entire nerve.

Epiphyseal (ep'-i-FIZ-ē-al) ***line*** The remnant of the epiphyseal plate in a long bone.

Epiphyseal (ep'-i-FIZ-ē-al) ***plate*** The cartilaginous plate between the epiphysis and diaphysis that is responsible for the lengthwise growth of long bones.

Epiphysis (ē-PIF-i-sis) The end of a long bone, usually larger in diameter than the shaft (diaphysis).

Epiphysis cerebri (se-RĒ-brē) Pineal gland.

Episiotomy (e-piz'-ē-OT-ō-mē) A cut made with surgical scissors to avoid tearing of the perineum at the end of the second stage of labor.

Epistaxis (ep'-i-STAK-sis) Loss of blood from the nose due to trauma, infection, allergy, neoplasm, and bleeding disorders. Also called ***nosebleed.***

Epithalamus (ep'-i-THAL-a-mus) Part of the diencephalon superior and posterior to the thalamus, comprising the pineal gland and associated structures.

Epithelial (ep'-i-THĒ-lē-al) ***tissue*** The tissue that forms glands or the outer part of the skin and lines blood vessels, hollow organs, and passages that lead externally from the body.

Eponychium (ep'-ō-NIK-ē-um) Narrow band of stratum corneum at the proximal border of a nail that extends from the margin of the nail wall. Also called the ***cuticle.***

Erection (ē-REK-shun) The enlarged and stiff state of the penis (or clitoris) resulting from the engorgement of the spongy erectile tissue with blood.

Eructation (e-ruk'-TĀ-shun) The forceful expulsion of gas from the stomach. Also called ***belching.***

Erythema (er'-e-THĒ-ma) Skin redness usually caused by engorgement of the capillaries in the lower layers of the skin.

Erythematosus (er-i'-them-a-TŌ-sus) Pertaining to redness.

Erthyrocyte (e-RITH-rō-sīt) Red blood cell.

Erythropoiesis (e-rith'-rō-poy-Ē-sis) The process by which erythrocytes (red blood cells) are formed.

Erythropoietin (e-rith'-rō-POY-ē-tin) A hormone formed from a plasma protein that stimulates erythrocyte (red blood cell) production.

Esophagus (e-SOF-a-gus) A hollow muscular tube connecting the pharynx and the stomach.

Estrogens (ES-tro-jens) Female sex hormones produced by the ovaries concerned with the development and maintenance of female reproductive structures and secondary sex characteristics, fluid and electrolyte balance, and protein anabolism. Examples are β-estradiol, estrone, and estriol.

Etiology (ē'-tē-OL-ō-jē) The study of the causes of disease, including theories of origin and the organisms, if any, involved.

Euphoria (yoo-FŌR-ē-a) A subjectively pleasant feeling of well-being marked by confidence and assurance.

Eupnea (yoop-NĒ-a) Normal quiet breathing.

Eustachian tube *See* ***Auditory tube.***

Euthanasia (yoo'-tha-NĀ-zē-a) The practice of ending a life in case of incurable disease.

Eversion (ē-VER-zhun) The movement of the sole outward at the ankle joint.

Exacerbation (eg-zas'-er-BĀ-shun) An increase in the severity of symptoms or of disease.

Excitability (ek-sīt'-a-BIL-i-tē) The ability of muscle tissue to receive and respond to stimuli; the ability of nerve cells to respond to stimuli and convert them into nerve impulses.

Excrement (EKS-kre-ment) Material cast out from the body as waste, especially fecal matter.

Excretion (eks-KRĒ-shun) The process of eliminating waste products from a cell, tissue, or the entire body; or the products excreted.

Exocrine (EK-sō-krin) ***gland*** A gland that secretes substances into ducts that empty at covering or lining epithelium or directly onto a free surface.

Exocytosis (ex'-ō-sī-TŌ-sis) A process of discharging cellular products too big to go through the membrane. Particles for export are enclosed by Golgi membranes when they are synthesized. Vesicles pinch off from the Golgi complex and carry the enclosed particles to the interior surface of the cell membrane, where the vesicle membrane and plasma membrane fuse and the contents of the vesicle are discharged.

Exophthalmic goiter (ek'-sof-THAL-mik GOY-ter) An autoimmune disease that may result in hypersecretion of thyroid hormones characterized by protrusion of the eyeballs (exophthalmos) and an enlarged thyroid (goiter). Also called ***Graves disease.***

Exophthalmos (ek'-sof-THAL-mus) An abnormal protrusion or bulging of the eyeball.

Expiration (ek-spi-RĀ-shun) Breathing out; expelling air from the lungs into the atmosphere. Also called ***exhalation.***

Extensibility (ek-sten'-si-BIL-i-tē) The ability of muscle tissue to be stretched when pulled.

Extension (ek-STEN-shun) An increase in the angle between two bones; restoring a body part to its anatomical position after flexion.

External Located on or near the surface.

External auditory (AW-di-tōr-ē) ***canal*** or ***meatus*** (mē-Ā-tus) A curved tube in the temporal bone that leads to the middle ear.

External ear The outer ear, consisting of the auricle, external auditory canal, and tympanic membrane or eardrum.

External nares (NA-rēz) The external nostrils, or the openings into the nasal cavity on the exterior of the body.

External respiration The exchange of respiratory gases between the lungs and blood.

Exteroceptor (eks'-ter-ō-SEP-tor) A receptor adapted for the reception of stimuli from outside the body.

Extracellular fluid (ECF) Fluid outside body cells, such as interstitial fluid and plasma.

Extracorporeal (eks'-tra-kor-PŌ-rē-al) The circulation of blood outside the body.

Extrinsic (ek-STRIN-sik) Of external origin.

Exudate (EKS-yoo-dāt) Escaping fluid or semifluid material that oozes from a space that may contain serum, pus, and cellular debris.

Eyebrow The hairy ridge above the eye.

Face The anterior aspect of the head.

Facilitated diffusion (fa-SIL-i-tā-ted dif-YOO-zhun) Diffusion in which a substance not soluble by itself in lipids is transported across a selectively permeable membrane by combining with a carrier substance.

Falciform ligament (FAL-si-form LIG-a-ment) A sheet of parietal peritoneum between the two principal lobes of the liver. The ligamentum teres, or remnant of the umbilical vein, lies within its fold.

Fallopian tube *See* ***Uterine tube.***

Falx cerebelli (FALKS ser'-e-BEL-lē) A small triangular process of the dura mater attached to the occipital bone in the posterior cranial fossa and projecting inward between the two cerebellar hemispheres.

Falx cerebri (SER-e-brē) A fold of the dura mater extending down into the longitudinal fissure between the two cerebral hemispheres.

Fascia (FASH-ē-a) A fibrous membrane covering, supporting, and separating muscles.

Fascicle (FAS-i-kul) A small bundle or cluster, especially of nerve or muscle fibers (cells). Also called a ***fasciculus*** (fa-SIK-yoo-lus). *Plural,* ***fasciculi*** (fa-SIK-yoo-lī).

Fasciculation (fa-sik'-yoo-LĀ-shun) Involuntary brief twitch of a muscle that is visible under the skin and is not associated with the movement of the affected muscle.

Fat A lipid compound formed from one molecule of glycerol and three molecules of fatty acids; the body's most highly concentrated source of energy. Adipose tissue, composed of adipocytes specialized for fat storage and present in the form of soft pads between various organs for support, protection, and insulation.

Fauces (FAW-sēz) The opening from the mouth into the pharynx.

Febrile (FĒ-bril) Feverish; pertaining to a fever.

Feces (FĒ-sēz) Material discharged from the rectum and made up of bacteria, excretions, and food residue. Also called ***stool.***

Feeding (hunger) center A cluster of neurons in the lateral nuclei of the hypothalamus that, when stimulated, brings about feeding.

Fenestration (fen-e-STRĀ-shun) Surgical opening made into the labyrinth of the ear for some conditions of deafness.

Fertilization (fer'-ti-li-ZĀ-shun) Penetration of a secondary oocyte by a spermatozoon and subsequent union of the nuclei of the cells.

Fetal (FĒ-tal) ***alcohol syndrome (FAS)*** Term applied to the effects of intrauterine exposure to alcohol, such as slow growth, defective organs, and mental retardation.

Fetal circulation The cardiovascular system of the fetus, including the placenta and special blood vessels involved in the exchange of materials between fetus and mother.

Fetus (FĒ-tus) The latter stages of the developing young of an animal; in humans, the developing organism in utero from the beginning of the third month to birth.

Fever An elevation in body temperature above its normal temperature of 37°C (98.6°F).

Fibrillation (fi-bri-LĀ-shun) Involuntary brief twitch of a muscle that is not visible under the skin and is not associated with movement of the affected muscle.

Fibrin (FĪ-brin) An insoluble protein that is essential to blood clotting; formed from fibrinogen by the action of thrombin.

Fibrinogen (fī-BRIN-ō-jen) A high-molecular-weight protein in the blood plasma that by the action of thrombin is converted to fibrin.

Fibroblast (FĪ-brō-blast) A large, flat cell that forms collagenous and elastic fibers and intercellular substance of areolar connective tissue.

Fibrocyte (FĪ-brō-sīt) A mature fibroblast that no longer produces fibers or intercellular substance in connective tissue.

Fibromyalgia (fī-bro-mī-AL-jē-a) Groups of common non-articular rheumatic disorders characterized by pain, tenderness, and stiffness of muscles, tendons, and surrounding tissues. Examples of fibromyalgia are lumbago and charleyhorse.

Fibroplasia (fī-brō-PLĀ-zē-a) Period of scar tissue formation.

Fibrosis (fī-BRŌ-sis) Abnormal formation of fibrous tissue.

Fibrous (FĪ-brus) ***joint*** A joint that allows little or no movement, such as a suture and syndesmosis.

Fibrous tunic (TOO-nik) The outer coat of the eyeball, made up of the posterior sclera and the anterior cornea.

Fight-or-flight response The effect of the stimulation of the sympathetic division of the autonomic nervous system.

Filiform papilla (FIL-i-form pa-PIL-a) One of the conical projections that are distributed in parallel rows over the anterior two-thirds of the tongue and contain no taste buds.

Filtrate (fil-TRĀT) The fluid produced when blood is filtered by the endothelial-capsular membrane.

Filum terminale (FĪ-lum ter-mi-NAL-ē) Nonnervous fibrous tissue of the spinal cord that extends inferiorly from the conus medullaris to the coccyx.

Fimbriae (FIM-brē-ē) Fingerlike structures, especially the lateral ends of the uterine (Fallopian) tubes.

Fissure (FISH-ur) A groove, fold, or slit that may be normal or abnormal.

Fistula (FIS-choo-la) An abnormal passage between two organs or between an organ cavity and the outside.

Fixator A muscle that stabilizes the origin of the prime mover so that the prime mover can act more efficiently.

Fixed macrophage (MAK-rō-fāj) Stationary phagocytic cell found in the liver, lungs, brain, spleen, lymph nodes, subcutaneous tissue, and bone marrow. Also called a ***histiocyte*** (HIS-tē-ō-sīt).

Flaccid (FLAS-sid) Relaxed, flabby, or soft; lacking muscle tone.

Flagellum (fla-JEL-um) A hairlike, motile process on the extremity of a bacterium or protozoan. *Plural,* ***flagella*** (fla-JEL-a).

Flatfoot A condition in which the ligaments and tendons of the arches of the foot are weakened and the height of the longitudinal arch decreases.

Flatus (FLĀ-tus) Air (gas) in the stomach or intestines, commonly used to denote passage of gas rectally.

Flexion (FLEK-shun) A folding movement in which there is a decrease in the angle between two bones.

Fluid mosaic (mō-ZĀ-ik) ***model*** Model of plasma membrane structure that depicts the membrane as a mosaic of proteins floating like icebergs in a sea of lipids.

Fluoroscope (FLOOR-ō-skōp) An instrument for visual observation of the body by means of x-ray.

Follicle (FOL-i-kul) A small secretory sac or cavity.

Follicle-stimulating (FOL-i-kul) ***hormone (FSH)*** Hormone secreted by the adenohypophysis (anterior lobe) of the pituitary gland that initiates development of ova and stimulates the ovaries to secrete estrogens in females and initiates sperm production in males.

Fontanel (fon'-ta-NEL) A membrane-covered spot where bone formation is not yet complete, especially between the cranial bones of an infant's skull.

Foot The terminal part of the lower extremity.

Foramen (fo-RĀ-men) A passage or opening; a communication between two cavities of an organ or a hole in a bone for passage of vessels or nerves.

Foramen ovale (ō-VAL-ē) An opening in the fetal heart in the septum between the right and left atria. A hole in the greater wing of the sphenoid bone that transmits the mandibular branch of the trigeminal (V) nerve.

Forearm (FOR-arm) The part of the upper extremity between the elbow and the wrist.

Fornix (FOR-niks) An arch or fold; a tract in the brain made up of association fibers, connecting the hippocampus with the mamillary bodies; a recess around the cervix of the uterus where it protrudes into the vagina.

Fossa (FOS-a) A furrow or shallow depression.

Fourth ventricle (VEN-tri-kul) A cavity within the brain lying between the cerebellum and the medulla and pons.

Fracture (FRAK-chur) Any break in a bone.

Fragile X syndrome Inherited disorder characterized by learning difficulties, mental retardation, and physical abnormalities; due to a defective gene on the X chromosome.

Frenulum (FREN-yoo-lum) A small fold of mucous membrane that connects two parts and limits movement.

Frontal plane A plane at a right angle to a midsagittal plane that divides the body or organs into anterior and posterior portions. Also called a ***coronal*** (kō-RŌ-nal) ***plane.***

Fulminate (FUL-mi-nāt') To occur suddenly with great intensity.

Fundus (FUN-dus) The part of a hollow organ farthest from the opening.

Fungiform papilla (FUN-ji-form pa-PIL-a) A mushroomlike elevation on the upper surface of the tongue appearing as a red dot; papillae contain taste buds.

Furuncle (FYOOR-ung-kul) A boil; painful nodule caused by bacterial infection and inflammation of a hair follicle or sebaceous (oil) gland.

Gallbladder A small pouch that stores bile, located under the liver, which is filled with bile and emptied via the cystic duct.

Gallstone A concretion, usually consisting of cholesterol, formed anywhere between bile canaliculi in the liver and the hepatopancreatic ampulla (ampulla of Vater), where bile enters the duodenum. Also called a ***biliary calculus.***

Gamete (GAM-ēt) A male or female reproductive cell; the spermatozoon or ovum.

Gamete intrafallopian transfer (GIFT) A type of in vitro fertilization in which aspirated secondary oocytes are combined with a solution containing sperm outside the body and then the secondary oocytes are inserted into the uterine (Fallopian) tubes.

Ganglion (GANG-glē-on) A group of nerve cell bodies that lie outside the central nervous system. *Plural,* ***ganglia*** (GANG-glē-a).

Gangrene (GANG-rēn) Death and rotting of a considerable mass of tissue that usually is caused by interruption of blood supply followed by bacterial (*Clostridium*) invasion.

Gastroenterology (gas'-trō-en'-ter-OL-ō-jē) The medical specialty that deals with the structure, function, diagnosis, and treatment of diseases of the stomach and intestines.

Gastrointestinal (gas-trō-in-TES-ti-nal) ***(GI) tract*** A continuous tube running through the ventral body cavity extending from the mouth to the anus. Also called the ***alimentary*** (al'-i-MEN-tar-ē) ***canal.***

Gastroscopy (gas-TROS-kō-pē) Diagnostic procedure in which the interior of the stomach is examined with a gastroscope to detect lesions, biopsy lesions, stop bleeding, and remove foreign objects.

Gastrulation (gas'-troo-LĀ-shun) The various movements of groups of cells that lead to the establishment of the primary germ layers.

Gavage (ga-VAZH) Feeding through a tube passed through the esophagus and into the stomach.

Gene (JĒN) Biological unit of heredity; an ultramicroscopic, self-reproducing DNA particle located in a definite position on a particular chromosome.

Genetic engineering The manufacture and manipulation of genetic material.

Genetics The study of heredity.

Genital herpes (JEN-i-tal HER-pēz) A sexually transmitted disease caused by type II herpes simplex virus.

Genitalia (jen'-i-TĀL-ya) Reproductive organs.

Genome (JĒ-nōm) The complete gene complement of an organism.

Genotype (JĒ-nō-tīp) The total hereditary information carried by an individual; the genetic makeup of an organism.

Geriatrics (jer'-ē-AT-riks) The branch of medicine devoted to the medical problems and care of elderly persons.

Germinal (JER-mi-nal) ***epithelium*** A layer of epithelial cells that covers the ovaries and lines the seminiferous tubules of the testes.

Germinativum (jer'-mi-na-TĒ-vum) Skin layers where new cells are germinated.

Gestation (jes-TĀ-shun) The period of intrauterine fetal development.

Giantism (GĪ-an-tizm) Condition caused by hypersecretion of human growth hormone (hGH) during childhood characterized by excessive bone growth and body size. Also called ***gigantism.***

Gingivae (jin-JI-vē) Gums. They cover the alveolar processes of the mandible and maxilla and extend slightly into each socket.

Gingivitis (jin'-je-VĪ-tis) Inflammation of the gums.

Gland Single or group of specialized epithelial cells that secrete substances.

Glans penis (glanz PĒ-nis) The slightly enlarged region at the distal end of the penis.

Glaucoma (glaw-KŌ-ma) An eye disorder in which there is increased intraocular pressure due to an excess of aqueous humor.

Gliding joint A synovial joint having articulating surfaces that are usually flat, permitting only side-to-side and back-and-forth movements, as between carpal bones, tarsal bones, and the scapula and clavicle. Also called an ***arthrodial*** (ar-THRŌ-dē-al) ***joint.***

Glomerular (glō-MER-yoo-lar) ***capsule*** A double-walled globe at the proximal end of a nephron that encloses the glomerulus. Also called ***Bowman's*** (BŌ-manz) ***capsule.***

Glomerular filtration The first step in urine formation in which substances in blood are filtered at the endothelial-capsular membrane and the filtrate enters the proximal convoluted tubule of a nephron.

Glomerulonephritis (glō-mer-yoo-lō-nef-RĪ-tis) Inflammation of the glomeruli of the kidney that increases the permeability of the endothelial-capsular membrane and permits blood cells and proteins to enter the filtrate. Also called ***Bright's disease.***

Glomerulus (glō-MER-yoo-lus) A rounded mass of nerves or blood vessels, especially the microscopic tuft of capillaries that is surrounded by the glomerular (Bowman's) capsule of each kidney tubule.

Glottis (GLOT-is) The vocal folds (true vocal cords) in the larynx and the space between them (rima glottidis).

Glucagon (GLOO-ka-gon) A hormone produced by the alpha cells of the pancreas that increases the blood glucose level.

Glucocorticoids (gloo-kō-KOR-ti-koyds) A group of hormones of the adrenal cortex.

Glucose (GLOO-kōs) A six-carbon sugar, $C_6H_{12}O_6$; the major energy source for every cell type in the body. Its metabolism is possible by every known living cell for the production of ATP.

Glycogen (GLĪ-kō-jen) A highly branched polymer of glucose containing thousands of subunits; functions as a compact store of glucose molecules in liver and muscle fibers (cells).

Glycosuria (glī'-kō-SOO-rē-a) The presence of glucose in the urine; may be temporary or pathological. Also called ***glucosuria.***

Gnostic (NOS-tik) Pertaining to the faculties of perceiving and recognizing.

Gnostic area Sensory area of the cerebral cortex that receives and integrates sensory input from various parts of the brain so that a common thought can be formed.

Goblet cell A goblet-shaped unicellular gland that secretes mucus. Also called a ***mucous cell.***

Goiter (GOY-ter) An enlargement of the thyroid gland.

Golgi (GOL-jē) ***complex*** An organelle in the cytoplasm of cells consisting of four to eight flattened channels, stacked upon one another, with expanded areas at their ends; functions in packaging secreted proteins, lipid secretion, and carbohydrate synthesis.

Golgi tendon organ *See* ***Tendon organ.***

Gomphosis (gom-FŌ-sis) A fibrous joint in which a cone-shaped peg fits into a socket.

Gonad (GŌ-nad) A gland that produces gametes and hormones; the ovary in the female and the testis in the male.

Gonadocorticoids (gō-na-dō-KOR-ti-koydz) Sex hormones secreted by the adrenal cortex.

Gonadotropic (gō'-nad-ō-TRŌ-pik) ***hormone*** A hormone that regulates the functions of the gonads.

Gonorrhea (gon'-ō-RĒ-a) Infectious, sexually transmitted disease caused by the bacterium *Neisseria gonorrhoeae* and characterized by inflammation of the urogenital mucosa, discharge of pus, and painful urination.

Gout (GOWT) Hereditary condition associated with excessive uric acid in the blood; the acid crystallizes and deposits in joints, kidneys, and soft tissue.

Graafian follicle *See* ***Vesicular ovarian follicle.***

Gray commissure (KOM-i-shur) A narrow strip of gray matter connecting the two lateral gray masses within the spinal cord.

Gray matter Area in the central nervous system and ganglia consisting of nonmyelinated nerve tissue.

Gray ramus communicans (RĀ-mus kō-MYOO-ni-kans) A short nerve containing postganglionic sympathetic fibers; the cell bodies of the fibers are in a sympathetic chain ganglion, and the nonmyelinated axons run by way of the gray ramus to a spinal nerve and then to the periphery to supply smooth muscle in blood vessels, arrector pili muscles, and sweat glands. *Plural,* ***rami communicantes*** (RĀ-mē kō-myoo-ni-KAN-tēz).

Greater omentum (ō-MEN-tum) A large fold in the serosa of the stomach that hangs down like an apron over the front of the intestines.

Greater vestibular (ves-TIB-yoo-lar) ***glands*** A pair of glands on either side of the vaginal orifice that open by a duct into the space between the hymen and the labia minora. Also called ***Bartholin's*** (BAR-to-linz) ***glands.***

Groin (GROYN) The depression between the thigh and the trunk; the inguinal region.

Gross anatomy The branch of anatomy that deals with structures that can be studied without using a microscope. Also called ***macroscopic anatomy.***

Ground substance Amorphous fluid to gel to solid substance produced by connective tissue cells and deposited in the spaces between them.

Growth An increase in size due to an increase in the number of cells or an increase in the size of existing cells as internal components increase in size or an increase in the size of intercellular substances.

Gustatory (GUS-ta-tō'-rē) Pertaining to taste.

Gynecology (gī'-ne-KOL-ō-jē) The branch of medicine dealing with the study and treatment of disorders of the female reproductive system.

Gynecomastia (gīn'-e-kō-MAS-tē-a) Excessive growth (benign) of the male mammary glands due to secretion of sufficient estrogens by an adrenal gland tumor (feminizing adenoma).

Gyrus (JĪ-rus) One of the folds of the cerebral cortex of the brain. *Plural*, ***gyri*** (JĪ-rī). Also called a ***convolution.***

Hair A threadlike structure produced by hair follicles that develops in the dermis. Also called ***pilus*** (PI-lus).

Hair follicle (FOL-li-kul) Structure composed of epithelium surrounding the root of a hair from which hair develops.

Hair root plexus (PLEK-sus) A network of dendrites arranged around the root of a hair as free or naked nerve endings that are stimulated when a hair shaft is moved.

Hallucination (ha-loo'-si-NĀ-shun) A sensory perception of something that does not really exist in the world, that is, a sensory experience created from within the brain.

Hand The terminal portion of an upper extremity, including the carpus, metacarpus, and phalanges.

Haploid (HAP-loyd) Having half the number of chromosomes characteristically found in the somatic cells of an organism; characteristic of mature gametes. Symbolized n.

Hard palate (PAL-at) The anterior portion of the roof of the mouth, formed by the maxillae and palatine bones and lined by mucous membrane.

Haustra (HAWS-tra) The sacculated elevations of the colon.

Haversian canal *See* ***Central canal.***

Haversian system *See* ***Osteon.***

Head The superior part of a human, cephalic to the neck. The superior or proximal part of a structure.

Heart A hollow muscular organ lying slightly to the left of the midline of the chest that pumps the blood through the cardiovascular system.

Heart block An arrhythmia (dysrhythmia) of the heart in which the atria and ventricles contract independently because of a blocking of electrical impulses through the heart at a critical point in the conduction system.

Heartburn Burning sensation in the esophagus due to reflux of hydrochloric acid (HCl) from the stomach.

Heart-lung machine A device that pumps blood, functioning as a heart, and removes carbon dioxide from blood and oxygenates it, functioning as lungs; used during heart transplantation, open-heart surgery, and coronary artery bypass grafting.

Heart murmur (MER-mer) An abnormal sound that consists of a flow noise that is heard before the normal lubb-dupp or that may mask normal heart sounds.

Heimlich maneuver *See* ***Abdominal thrust maneuver.***

Hematocrit (hē-MAT-ō-krit) ***(Hct)*** The percentage of blood made up of red blood cells. Usually calculated by centrifuging a blood sample in a graduated tube and then reading off the volume of red blood cells and total blood.

Hematology (hē'-ma-TOL-ō-jē) The study of blood.

Hematoma (hē'-ma-TŌ-ma) A tumor or swelling filled with blood.

Hematopoiesis (hem'-a-tō-poy-Ē-sis) Blood cell production occurring in the red marrow of bones. Also called ***hemopoiesis*** (hē-mō-poy-Ē-sis).

Hematuria (hē'-ma-TOOR-ē-a) Blood in the urine.

Hemiballismus (hem'-i-ba-LIZ-mus) Violent muscular restlessness of half of the body, especially of the upper extremity.

Hemiplegia (hem-i-PLĒ-jē-a) Paralysis of the upper extremity, trunk, and lower extremity on one side of the body.

Hemocytoblast (hē'-mō-SĪ-tō-blast) Immature stem cell in bone marrow that develops along different lines into all the different mature blood cells.

Hemodialysis (hē'-mō-dī-AL-i-sis) Filtering of the blood by means of an artificial device so that certain substances are removed from the blood as a result of the difference in rates of their diffusion through a selectively permeable membrane while the blood is being circulated outside the body.

Hemoglobin (hē'-mō-GLŌ-bin) ***(Hb)*** A substance in erythrocytes (red blood cells) consisting of the protein globin and the iron-containing red pigment heme and constituting about 33 percent of the cell volume; involved in the transport of oxygen and carbon dioxide.

Hemolytic disease of the newborn A hemolytic anemia of a newborn child that results from the destruction of the infant's red blood cells by antibodies produced by the mother; usually the antibodies are due to an Rh blood type incompatibility. Also called ***erythroblastosis fetalis*** (e-rith'-rō-blas-TŌ-sis fe-TAL-is).

Hemophilia (hē'-mō-FĒL-ē-a) A hereditary blood disorder where there is a deficient production of certain factors involved in blood clotting, resulting in excessive bleeding into joints, deep tissues, and elsewhere.

Hemoptysis (hē-MOP-ti-sis) Spitting of blood from the respiratory tract.

Hemorrhage (HEM-or-rij) Bleeding; the escape of blood from blood vessels, especially when it is profuse.

Hemorrhoids (HEM-ō-royds) Dilated or varicosed blood vessels (usually veins) in the anal region. Also called ***piles.***

Hemostasis (hē-MŌS-tā-sis) The stoppage of bleeding.

Hemostat (HĒ-mō-stat) An agent or instrument used to prevent the flow or escape of blood.

Hepatic (he-PAT-ik) Refers to the liver.

Hepatic duct A duct that receives bile from the bile capillaries. Small hepatic ducts merge to form the larger right and left hepatic ducts that unite to leave the liver as the common hepatic duct.

Hepatic portal circulation The flow of blood from the gastrointestinal organs to the liver before returning to the heart.

Hepatitis (hep-a-TĪ-tis) Inflammation of the liver due to a virus, drugs, or chemicals.

Hepatopancreatic (hep'-a-tō-pan'-krē-A-tik) ***ampulla*** A small, raised area in the duodenum where the combined common bile duct and main pancreatic duct empty into the duodenum. Also called the ***ampulla of Vater*** (VA-ter).

Hering–Breuer reflex *See* ***Inflation reflex.***

Hernia (HER-nē-a) The protrusion or projection of an organ or part of an organ through a membrane or cavity wall, usually the abdominal cavity.

Herniated (her'-nē-Ā-ted) ***disc*** A rupture of an intervertebral disc so that the nucleus pulposus protrudes into the vertebral cavity. Also called a ***slipped disc.***

Heterocrine (HET-er-ō-krin) ***gland*** A gland, such as the pancreas, that is both an exocrine and an endocrine gland.

Hiatus (hī-Ā-tus) An opening; a foramen.

Hilus (HĪ-lus) An area, depression, or pit where blood vessels and nerves enter or leave an organ. Also called a ***hilum.***

Hinge joint A synovial joint in which a convex surface of one bone fits into a concave surface of another bone, such as the elbow, knee, ankle, and interphalangeal joints. Also called a ***ginglymus*** (JIN-gli-mus) ***joint.***

Hirsutism (HER-soot-izm) An excessive growth of hair in females and children, with a distribution similar to that in adult males, due to the conversion of vellus hairs into large terminal hairs in response to higher-than-normal levels of androgens.

Histamine (HISS-ta-mēn) Substance found in many cells, especially mast cells, basophils, and platelets, released when the cells are injured; results in vasodilation, increased permeability of blood vessels, and bronchiole constriction.

Histology (hiss-TOL-ō-jē) Microscopic study of the structure of tissues.

Hives (HĪVZ) Condition of the skin marked by reddened elevated patches that are often itchy; may be caused by infections, trauma, medications, emotional stress, food additives, and certain foods.

Hodgkin's disease (HD) A malignant disorder, usually arising in lymph nodes.

Holocrine (HŌL-ō-krin) ***gland*** A type of gland in which the entire secreting cell, along with its accumulated secretions, makes up the secretory product of the gland, as in the sebaceous (oil) glands.

Holter monitor Electrocardiograph worn by a person while going about everyday routines.

Homeostasis (hō'-mē-ō-STĀ-sis) The condition in which the body's internal environment remains relatively constant, within physiological limits.

Homologous (hō-MOL-ō-gus) Correspondence of two organs in structure, position, and origin.

Homologous chromosomes Two chromosomes that belong to a pair. Also called ***homologues.***

Horizontal plane A plane that runs parallel to the ground and divides the body or organs into superior and inferior portions. Also called a ***transverse plane.***

Hormone (HOR-mōn) A secretion of endocrine tissue that alters the physiological activity of target cells of the body.

Horn Principal area of gray matter in the spinal cord.

Human chorionic gonadotropin (kō-rē-ON-ik gō-nad-ō-TRŌ-pin) ***(hCG)*** A hormone produced by the developing placenta that maintains the corpus luteum.

Human chorionic somatomammotropin (sō-mat-ō-mam-ō-TRŌ-pin) ***(hCS)*** A hormone produced by the chorion of the placenta that may stimulate breast tissue for lactation, enhance body growth, and regulate metabolism.

Human growth hormone (hGH) Hormone secreted by the adenohypophysis (anterior lobe) of the pituitary that brings about growth of body tissues, especially skeletal and muscular. Also known as ***somatotropin*** and ***somatotropic hormone (STH).***

Human leucocyte associated (HLA) antigens Surface proteins on white blood cells and other nucleated cells that are unique for each person (except for identical twins) and are used to type tissues and help prevent rejection.

Hunger center A cluster of neurons in the lateral nuclei of the hypothalamus that, when stimulated, brings about feeding.

Hyaluronic (hī'-a-loo-RON-ik) ***acid*** A viscous, amorphous extracellular material that binds cells together, lubricates joints, and maintains the shape of the eyeballs.

Hydrocele (HĪ-drō-sēl) A fluid-containing sac or tumor. Specifically, a collection of fluid formed in the space along the spermatic cord and in the scrotum.

Hydrocephalus (hī-drō-SEF-a-lus) Abnormal accumulation of cerebrospinal fluid on the brain.

Hydrophobia (hī'-drō-FŌ-bē-a) Rabies; a condition characterized by severe muscle spasms when attempting to drink water. Also, an abnormal fear of water.

Hymen (HĪ-men) A thin fold of vascularized mucous membrane at the vaginal orifice.

Hyperemia (hī'-per-Ē-mē-a) An excess of blood in an area or part of the body.

Hyperextension (hī'-per-ek-STEN-shun) Continuation of extension beyond the anatomical position, as in bending the head backward.

Hyperglycemia (hī'-per-glī-SĒ-mē-a) An elevated blood sugar level.

Hypermetropia (hī'-per-mē-TRŌ-pē-a) A condition in which visual images are focused behind the retina with resulting defective vision of near objects; farsightedness.

Hyperplasia (hī'-per-PLĀ-zē-a) An abnormal increase in the number of normal cells in a tissue or organ, increasing its size.

Hypersecretion (hī'-per-se-KRĒ-shun) Overactivity of glands resulting in excessive secretion.

Hypersensitivity (hī'-per-sen-si-TI-vi-tē) Overreaction to an allergen that results in pathological changes in tissues. Also called ***allergy.***

Hypertension (hī'-per-TEN-shun) High blood pressure.

Hyperthermia (hī'-per-THERM-ē-a) An elevated body temperature.

Hypertonia (hī-per-TŌ-nē-a) Increased muscle tone that is expressed as spasticity or rigidity.

Hypertrophy (hī-PER-trō-fē) An excessive enlargement or overgrowth of tissue without cell division.

Hyperventilation (hī'-per-ven-ti-LĀ-shun) A rate of respiration higher than that required to maintain a normal level of plasma PCO_2.

Hypoglycemia (hī'-pō-glī-SĒ-mē-a) An abnormally low concentration of glucose in the blood; can result from excess insulin (injected or secreted).

Hyponychium (hī'-pō-NIK-ē-um) Free edge of the fingernail.

Hypophyseal (hī-pō-FIZ-ē-al) ***pouch*** An outgrowth of ectoderm from the roof of the stomodeum (mouth) from which the adenohypophysis (anterior lobe) of the pituitary gland develops.

Hypophysis (hī-POF-i-sis) Pituitary gland.

Hypoplasia (hī-pō-PLĀ-zē-a) Defective development of tissue.

Hyposecretion (hī'-pō-se-KRĒ-shun) Underactivity of glands resulting in diminished secretion.

Hypospadias (hī'-pō-SPĀ-dē-as) A displaced urethral opening. In the male, the opening may be on the underside of the penis, at the penoscrotal junction, between the scrotal folds, or in the perineum. In the female, the urethra opens into the vagina.

Hypothalamic-hypophyseal (hī'-pō-thal-AM-ik hī'-po-FIZ-ē-al) ***tract*** A bundle of nerve processes made up of fibers that have their cell bodies in the hypothalamus but release their neurosecretions in the posterior pituitary gland or neurohypophysis.

Hypothalamus (hī'-pō-THAL-a-mus) A portion of the diencephalon, lying beneath the thalamus and forming the floor and part of the wall of the third ventricle.

Hypothermia (hī-pō-THER-mē-a) Lowering of body temperature below 35°C (95°F); in surgical procedures, it refers to deliberate cooling of the body to slow down metabolism and reduce oxygen needs of tissues.

Hypotonia (hī'-pō-TŌ-nē-a) Decreased or lost muscle tone in which muscles appear flaccid.

Hypoxia (hī-POKS-ē-a) Lack of adequate oxygen at the tissue level.

Hysterectomy (his-te-REK-tō-mē) The surgical removal of the uterus.

Ileocecal (il'-ē-ō-SĒ-kal) ***sphincter*** A fold of mucous membrane that guards the opening from the ileum into the large intestine. Also called the ***ileocecal valve.***

Ileum (IL-ē-um) The terminal portion of the small intestine.

Immunity (i-MYOON-i-tē) The state of being resistant to injury, particularly by poisons, foreign proteins, and invading parasites, due to the presence of antibodies.

Immunoglobulin (im-yoo-nō-GLOB-yoo-lin) ***(Ig)*** An antibody synthesized by plasma cells derived from B lymphocytes in response to the introduction of antigen. Immunoglobulins are divided into five kinds (IgG, IgM, IgA, IgD, IgE) based primarily on the larger protein component present in the immunoglobulin.

Immunology (im'-yoo-NOL-ō-jē) The branch of science that deals with the responses of the body when challenged by antigens.

Immunosuppression (im'-yoo-nō-su-PRESH-un) Inhibition of the immune response.

Immunotherapy (im-yoo-nō-THER-a-pē) Attempt to induce the immune system to mount an attack against cancer cells.

Imperforate (im-PER-fō-rāt) Abnormally closed.

Impetigo (im'-pe-TĪ-go) A contagious skin disorder characterized by pustular eruptions.

Implantation (im-plan-TĀ-shun) The insertion of a tissue or a part into the body. The attachment of the blastocyst to the lining of the uterus 7–8 days after fertilization.

Impotence (IM-pō-tens) Weakness; inability to copulate; failure to maintain an erection long enough for sexual intercourse.

Inclusion (in-KLOO-zhun) Temporary structure that contains secretions and storage products of cells.

Incontinence (in-KON-ti-nens) Inability to retain urine, semen, or feces, through loss of sphincter control.

Infant respiratory distress syndrome (RDS) A disease of newborn infants, especially premature ones, in which insufficient amounts of surfactant are produced and breathing is labored. Also called ***hyaline*** (HĪ-a-lin) ***membrane disease (HMD).***

Infarction (in-FARK-shun) The presence of a localized area of necrotic tissue, produced by inadequate oxygenation of the tissue.

Infection (in-FEK-shun) Invasion and multiplication of microorganisms in body tissues, which may be inapparent or characterized by cellular injury.

Infectious mononucleosis (mon-ō-nook'-lē-Ō-sis) ***(IM)*** Contagious disease caused by the Epstein-Barr virus (EBV) and characterized by an elevated mononucleocyte and lymphocyte count, fever, sore throat, stiff neck, cough, and malaise.

Inferior (in-FĒR-ē-or) Away from the head or toward the lower part of a structure. Also called ***caudad*** (KAW-dad).

Inferior vena cava (VĒ-na CĀ-va) ***(IVC)*** Large vein that collects blood from parts of the body inferior to the heart and returns it to the right atrium.

Infertility Inability to conceive or to cause conception. Also called ***sterility.***

Inflammation (in'-fla-MĀ-shun) Localized, protective response to tissue injury designed to destroy, dilute, or wall off the infecting agent or injured tissue; characterized by redness, pain, heat, swelling, and sometimes loss of function.

Inflammatory bowel (in-FLAM-a-tō'-rē BOW-el) ***disease*** Disorder that exists in two forms: (1) Crohn's disease (inflammation of the gastrointestinal tract, especially the distal ileum and proximal colon, in which the inflammation may extend from the mucosa through the serosa); and (2) ulcerative colitis (inflammation of the mucosa of the gastrointestinal tract, usually limited to the large intestine and usually accompanied by rectal bleeding).

Infundibulum (in'-fun-DIB-yoo-lum) The stalklike structure that attaches the pituitary gland (hypophysis) to the hypothalamus of the brain. The funnel-shaped, open, distal end of the uterine (Fallopian) tube.

Ingestion (in-JES-chun) The taking in of food, liquids, or drugs, by mouth.

Inguinal (IN-gwi-nal) Pertaining to the groin.

Inguinal canal An oblique passageway in the anterior abdominal wall just superior and parallel to the medial half of the inguinal ligament that transmits the spermatic cord and ilioinguinal nerve in the male and round ligament of the uterus and ilioinguinal nerve in the female.

Inheritance The acquisition of body characteristics and qualities by transmission of genetic information from parents to offspring.

Inhibin A male sex hormone secreted by sustentacular (Sertoli) cells that inhibits FSH release by the adenohypophysis (anterior pituitary) and thus spermatogenesis.

Inner cell mass A region of cells of a blastocyst that differentiates into the three primary germ layers—ectoderm, mesoderm, and endoderm—from which all tissues and organs develop; also called an ***embryoblast.***

Insertion (in-SER-shun) The manner or place of attachment of a muscle to the bone that it moves.

Insomnia (in-SOM-nē-a) Difficulty in falling asleep and, usually, frequent awakening.

Inspiration (in-spi-RĀ-shun) The act of drawing air into the lungs.

Insula (IN-su-la) A triangular area of cerebral cortex that lies deep within the lateral cerebral fissure, under the parietal, frontal, and temporal lobes, and cannot be seen in an external view of the brain. Also called the ***island*** or ***isle of Reil*** (RĪL).

Insulin (IN-su-lin) A hormone produced by the beta cells of the pancreas that decreases the blood glucose level.

Integumentary (in-teg'-yoo-MEN-tar-ē) Relating to the skin.

Intercalated (in-TER-ka-lāt-ed) ***disc*** An irregular transverse thickening of sarcolemma that contains desmosomes that hold cardiac muscle fibers (cells) together and gap junctions that aid in conduction of muscle action potentials.

Intercostal (in'-ter-KOS-tal) ***nerve*** A nerve supplying a muscle located between the ribs.

Interferon (in'-ter-FĒR-on) ***(IFN)*** Three principal types of protein (alpha, beta, gamma) naturally produced by virus-infected host cells that induce uninfected cells to synthesize antiviral proteins (AVPs) that inhibit intracellular viral replication in uninfected host cells; artifically synthesized through recombinant DNA techniques.

Intermediate Between two structures, one of which is medial and one of which is lateral.

Intermediate filament Cytoplasmic structure, ranging from 8 to 12 nm in diameter, that may provide structural reinforcement and assist in contraction.

Internal Away from the surface of the body.

Internal capsule A tract of projection fibers connecting various parts of the cerebral cortex and lying between the thalamus and the caudate and lentiform nuclei of the basal ganglia.

Internal ear The inner ear or labyrinth, lying inside the temporal bone, containing the organs of hearing and balance.

Internal nares (NA-rēz) The two openings posterior to the nasal cavities opening into the nasopharynx. Also called the ***choanae*** (kō-A-nē).

Internal respiration The exchange of respiratory gases between blood and body cells.

Interphase (IN-ter-fāz) The period during its life cycle when a cell is carrying on every life process except division; the stage between two mitotic divisions. Also called ***metabolic phase.***

Interstitial cell of Leydig *See* ***Interstitial endocrinocyte.***

Interstitial (in'-ter-STISH-al) ***endocrinocyte*** A cell located in the connective tissue between seminiferous tubules in a mature testis that secretes testosterone. Also called an ***interstitial cell of Leydig*** (LĪ-dig).

Interstitial (in'-ter-STISH-al) ***fluid*** The portion of extracellular fluid that fills the microscopic spaces between the cells of tissues; the internal environment of the body. Also called ***intercellular*** or ***tissue fluid.***

Interventricular (in'-ter-ven-TRIK-yoo-lar) ***foramen*** A narrow, oval opening through which the lateral ventricles of the brain communicate with the third ventricle. Also called the ***foramen of Monro.***

Intervertebral (in'-ter-VER-te-bral) ***disc*** A pad of fibrocartilage located between the bodies of two vertebrae.

Intestinal gland Simple tubular gland that opens onto the surface of the intestinal mucosa and secretes digestive enzymes. Also called a ***crypt of Lieberkühn*** (LĒ-ber-kyoon).

Intracellular (in'-tra-SEL-yoo-lar) ***fluid (ICF)*** Fluid located within cells.

Intrafusal (in'-tra-FYOO-zal) ***fibers*** Three to ten specialized muscle fibers (cells), partially enclosed in a connective tissue capsule that is filled with lymph; the fibers compose muscle spindles.

Intramembranous ossification (in'-tra-MEM-bra-nus os'-i'-fi-KĀ-shun) The method of bone formation in which the bone is formed directly in membranous tissue.

Intraocular (in-tra-OC-yoo-lar) ***pressure (IOP)*** Pressure in the eyeball, produced mainly by aqueous humor.

Intrauterine device (IUD) A small metal or plastic object inserted into the uterus for the purpose of preventing pregnancy.

Intrinsic (in-TRIN-sik) Of internal origin; for example, the intrinsic factor, a mucoprotein formed by the gastric mucosa that is necessary for the absorption of vitamin B_{12}.

Intrinsic factor (IF) A glycoprotein synthesized and secreted by the parietal cells of the gastric mucosa that facilitates vitamin B_{12} absorption.

Intubation (in'-too-BĀ-shun) Insertion of a tube through the nose or mouth into the larynx and trachea for entrance of air or to dilate a stricture.

Intussusception (in'-ta-sa-SEP-shun) The infolding (invagination) of one part of the intestine within another segment.

In utero (YOO-ter-ō) Within the uterus.

Invagination (in-vaj'-i-NĀ-shun) The pushing of the wall of a cavity into the cavity itself.

Inversion (in-VER-zhun) The movement of the sole inward at the ankle joint.

In vitro (VĒ-trō) Literally, in glass; outside the living body and in artificial environment such as a laboratory test tube.

In vivo (VĒ-vō) In the living body.

Ipsilateral (ip'-si-LAT-er-al) On the same side, affecting the same side of the body.

Iris The colored portion of the eyeball seen through the cornea that consists of circular and radial smooth muscle; the black hole in the center of the iris is the pupil.

Irritable bowel (IR-i-ta-bul BOW-el) ***syndrome (IBS)*** Disease of the entire gastrointestinal tract in which persons with the condition may react to stress by developing symptoms such as cramping and abdominal pain associated with alternating patterns of diarrhea and constipation. Excessive amounts of mucus may appear in the stools, and other symptoms include flatulence, nausea, and loss of appetite. The condition is also known as ***irritable colon*** or ***spastic colitis.***

Ischemia (is-KĒ-mē-a) A lack of sufficient blood to a part due to obstruction of circulation.

Island of Reil *See* ***Insula.***

Islet of Langerhans *See* ***Pancretic islet.***

Isthmus (IS-mus) A narrow strip of tissue or narrow passage connecting two larger parts.

Jaundice (JAWN-dis) A condition characterized by yellowness of skin, white of eyes, mucous membranes, and body fluids because of a buildup of bilirubin.

Jejunum (jē-JOO-num) The middle portion of the small intestine.

Joint kinesthetic (kin'-es-THET-ik) ***receptor*** A proprioceptive receptor located in a joint, stimulated by joint movement.

Juxtaglomerular (juks-ta-glō-MER-yoo-lar) ***apparatus (JGA)*** Consists of the macula densa (cells of the distal convoluted tubule adjacent to the afferent and efferent arteriole) and juxtaglomerular cells (modified cells of the afferent and sometimes efferent arteriole); secretes renin when blood pressure starts to fall.

Karyotype (KAR-ē-ō-tīp) An arrangement of chromosomes based on shape, size, and position of centromeres.

Keratin (KER-a-tin) An insoluble protein found in the hair, nails, and other keratinized tissues of the epidermis.

Keratinocyte (ker-A-tin'-ō-sīt) The most numerous of the epidermal cells that function in the production of keratin.

Keratosis (ker'-a-TŌ-sis) Formation of a hardened growth of tissue.

Kidney (KID-nē) One of the paired reddish organs located in the lumbar region that regulates the composition and volume of blood and produces urine.

Kidney stone A concretion, usually consisting of calcium oxalate, uric acid, and calcium phosphate crystals, that may form in any portion of the urinary tract. Also called a ***renal calculus.***

Kinesiology (ki-nē'-sē-OL-ō-jē) The study of the movement of body parts.

Kinesthesia (kin-is-THĒ-szē-a) Ability to perceive extent, direction, or weight of movement; muscle sense.

Kinetochore (ki-NĔT-ō-kor) Protein complex attached to the outside of a centromere to which kinetochore microtubules attach.

Korotkoff (kō-ROT-kof) ***sounds*** The various sounds that are heard while taking blood pressure.

Kupffer's cell *See* ***Stellate reticuloendothelial cell.***

Kyphosis (kī-FŌ-sis) An exaggeration of the thoracic curve of the vertebral column, resulting in a "round-shouldered" or hunchback appearance.

Labial frenulum (LĀ-bē-al FREN-yoo-lum) A medial fold of mucous membrane between the inner surface of the lip and the gums.

Labia majora (LĀ-bē-a ma-JO-ra) Two longitudinal folds of skin extending downward and backward from the mons pubis of the female.

Labia minora (min-OR-a) Two small folds of mucous membrane lying medial to the labia majora of the female.

Labium (LĀ-bē-um) A lip. A liplike structure. *Plural,* ***labia*** (LĀ-bē-a).

Labor The process by which the product of conception is expelled from the uterus through the vagina.

Labyrinth (LAB-i-rinth) Intricate communicating passageway, especially in the internal ear.

Labyrinthine (lab-i-RIN-thēn) ***disease*** Malfunction of the internal ear characterized by deafness, tinnitus, vertigo, nausea, and vomiting.

Laceration (las'-er-Ā-shun) Wound or irregular area of the skin.

Lacrimal (LAK-ri-mal) Pertaining to tears.

Lacrimal (LAK-ri-mal) ***canal*** A duct, one on each eyelid, commencing at the punctum at the medial margin of an eyelid and conveying tears medially into the nasolacrimal sac.

Lacrimal gland Secretory cells located at the superior anterolateral portion of each orbit that secrete tears into excretory ducts that open onto the surface of the conjunctiva.

Lacrimal sac The superior expanded portion of the naso-lacrimal duct that receives the tears from a lacrimal canal.

Lactation (lak-TĀ-shun) The secretion and ejection of milk by the mammary glands.

Lacteal (LAK-tē-al) One of many intestinal lymphatic vessels in villi that absorb fat from digested food.

Lacuna (la-KOO-na) A small, hollow space, such as that found in bones in which the osteoblasts lie. *Plural,* ***lacunae*** (la-KOO-nē).

Lambdoidal (lam-DOY-dal) ***suture*** The line of union in the skull between the parietal bones and the occipital bone; sometimes contains sutural bones.

Lamellae (la-MEL-ē) Concentric rings found in compact bone.

Lamellated corpuscle Oval pressure receptor located in subcutaneous tissue and consisting of concentric layers of connective tissue wrapped around an afferent nerve fiber. Also called a ***Pacinian*** (pa-SIN-ē-an) ***corpuscle.***

Lamina (LAM-i-na) A thin, flat layer or membrane, as the flattened part of either side of the arch of a vertebra. *Plural,* ***laminae*** (LAM-i-nē).

Lamina propria (PRO-prē-a) The connnective tissue layer of a mucous membrane.

Lanugo (lan-YOO-gō) Fine downy hairs that cover the fetus.

Laparoscopy (lap'-a-ROS-kō-pē) A procedure in which a laparoscope is inserted through an incision in the abdominal wall to view abdominal and pelvic viscera, remove fluids and tissues for biopsy, drain ovarian cysts, cut adhesions, stop bleeding, and perform tubal ligation.

Large intestine The portion of the gastrointestinal tract extending from the ileum of the small intestine to the anus, divided structurally into the cecum, colon, rectum, and anal canal.

Laryngitis (la-rin-JĪ-tis) Inflammation of the mucous membrane lining the larynx.

Laryngopharynx (la-rin'-gō-FAR-inks) The inferior portion of the pharynx, extending downward from the level of the hyoid bone to divide posteriorly into the esophagus and anteriorly into the larynx. Also called the ***hypopharynx.***

Laryngoscope (la-RIN-gō-skōp) An instrument for examining the larynx.

Laryngotracheal (la-rin'-gō-TRA-kē-al) ***bud*** An outgrowth of endoderm of the foregut from which the respiratory system develops.

Larynx (LAR-inks) The voice box, a short passageway that connects the pharynx with the trachea.

Lateral (LAT-er-al) Farther from the midline of the body or a structure.

Lateral ventricle (VEN-tri-kul) A cavity within a cerebral hemisphere that communicates with the lateral ventricle in the other cerebral hemisphere and with the third ventricle by way of the interventricular foramen.

Leg The part of the lower extremity between the knee and the ankle.

Lens A transparent organ constructed of proteins (crystallins) lying posterior to the pupil and iris of the eyeball and anterior to the vitreous body.

Lesion (LĒ-zhun) Any localized, abnormal change in tissue formation.

Lesser omentum (ō-MEN-tum) A fold of the peritoneum that extends from the liver to the lesser curvature of the stomach and the commencement of the duodenum.

Lesser vestibular (ves-TIB-yoo-lar) ***gland*** One of the paired mucus-secreting glands that have ducts that open on either side of the urethral orifice in the vestibule of the female.

Lethargy (LETH-ar-jē) A condition of drowsiness or indifference.

Leukemia (loo-KĒ-mē-a) A malignant disease of the blood-forming tissues characterized by either uncontrolled production and accumulation of immature leukocytes in which many cells fail to reach maturity (acute) or an accumulation of mature leukocytes in the blood because they do not die at the end of their normal life span (chronic).

Leukocyte (LOO-kō-sīt) A white blood cell.

Leukocytosis (loo'-kō-sī-TŌ-sis) An increase in the number of white blood cells, characteristic of many infections and other disorders.

Leukopenia (loo-kō-PĒ-nē-a) A decrease of the number of white blood cells below 5,000/mm^3.

Leukoplakia (loo-kō-PLĀ-kē-a) A disorder in which there are white patches in the mucous membranes of the tongue, gums, and cheeks.

Libido (li-BĒ-dō) The sexual drive, conscious or unconscious.

Ligament (LIG-a-ment) Dense, regularly arranged connective tissue that attaches bone to bone.

Ligand (LĪ-gand) Chemical in interstitial fluid, usually in a concentration lower than in cells.

Limbic system A portion of the forebrain, sometimes termed the visceral brain, concerned with various aspects of emotion and behavior, that includes the limbic lobe, dentate gyrus, amygdaloid body, septal nuclei, mammillary bodies, anterior thalamic nucleus, olfactory bulbs, and bundles of myelinated axons.

Lingual frenulum (LIN-gwal FREN-yoo-lum) A fold of mucous membrane that connects the tongue to the floor of the mouth.

Lipase (LĪ-pās) A fat-splitting enzyme.

Lipid An organic compound composed of carbon, hydrogen, and oxygen that is usually insoluble in water, but soluble in alcohol, ether, and chloroform; examples include fats, phospholipids, steroids, and prostaglandins.

Lipid profile Blood test that measures total cholesterol, high-density lipoprotein, low-density lipoprotein, and triglycerides, to assess risk for cardiovascular disease.

Lipoma (li-PŌ-ma) A fatty tissue tumor, usually benign.

Lipoprotein (lip'-ō-PRŌ-tēn) Protein containing lipid that is produced by the liver and combines with cholesterol and triglycerides to make them water-soluble for transportation by the cardiovascular system; high levels of low-density lipoproteins (LDL) are associated with increased risk of atherosclerosis, while high levels of high-density lipoproteins (HDL) are associated with decreased risk of atherosclerosis.

Lithotripsy (LITH-ō-trip'-sē) A noninvasive procedure in which shock waves generated by a lithotriptor are used to pulverize kidney stones or gallstones.

Liver Large gland under the diaphragm that occupies most of the right hypochondriac region and part of the epigastric region; functionally, it produces bile salts, heparin, and plasma proteins; converts one nutrient into another; detoxifies substances; stores glycogen, minerals, and vitamins; carries on phagocytosis of blood cells and bacteria; and helps activate vitamin D.

Lobe (lōb) A curved or rounded projection.

Locus coeruleus (LŌ-kus sē-ROO-lē-us) A group of neurons in the brain stem where norepinephrine (NE) is concentrated.

Lordosis (lor-DŌ-sis) An exaggeration of the lumbar curve of the vertebral column.

Lou Gehrig's disease *See* ***Amyotrophic lateral sclerosis.***

Lower extremity The appendage attached at the pelvic (hip) girdle, consisting of the thigh, knee, leg, ankle, foot, and toes.

Lumbar (LUM-bar) Region of the back and side between the ribs and pelvis; loin.

Lumbar plexus (PLEK-sus) A network formed by the anterior branches of spinal nerves L1 through L4.

Lumen (LOO-men) The space within an artery, vein, intestine, or a tube.

Lung One of the two main organs of respiration, lying on either side of the heart in the thoracic cavity.

Lung scan A diagnostic test in which a radioactive substance is detected in the lungs by a scanning camera; used to evaluate for pulmonary embolism, pneumonia, or cancer.

Lunula (LOO-nyoo-la) The moon-shaped white area at the base of a nail.

Luteinizing (LOO-tē-in'-īz-ing) ***hormone (LH)*** A hormone secreted by the adenohypophysis (anterior lobe) of the pituitary gland that stimulates ovulation, progesterone secretion by the corpus luteum, and readies the mammary glands for milk secretion in females and stimulates testosterone secretion by the testes in males.

Lymph (limf) Fluid confined in lymphatic vessels and flowing through the lymphatic system to be returned to the blood.

Lymphangiography (lim-fan'-jē-OG-ra-fē) A procedure by which lymphatic vessels and lymph organs are filled with a radiopaque substance in order to be x-rayed.

Lymphatic (lim-FAT-ik) ***vessel*** A large vessel that collects lymph from lymph capillaries and converges with other lymphatic vessels to form the thoracic and right lymphatic ducts.

Lymphatic tissue A specialized form of reticular tissue that contains large numbers of lymphocytes.

Lymph capillary Blind-ended microscopic lymph vessel that begins in spaces between cells and converges with other lymph capillaries to form lymphatic vessels.

Lymph node An oval or bean-shaped structure located along lymphatic vessels.

Lymphocyte (LIM-fō-sīt) A type of white blood cell, found in lymph nodes, associated with the immune system.

Lysosome (LĪ-sō-sōm) An organelle in the cytoplasm of a cell, enclosed by a single membrane and containing powerful digestive enzymes.

Lysozyme (LĪ-sō-zīm) A bactericidal enzyme found in tears, saliva, and perspiration.

Macrophage (MAK-rō-fāj) Phagocytic cell derived from a monocyte. May be fixed or wandering.

Macula (MAK-yoo-la) A discolored spot or a colored area. A small, thickened region on the wall of the utricle and saccule that serves as a receptor for static equilibrium.

Macula lutea (LOO-tē-a) The yellow spot in the center of the retina.

Magnetic resonance imaging (MRI) A diagnostic procedure that focuses on the nuclei of atoms of a single-element in a tissue, usually hydrogen, to determine if they behave normally in the presence of an external magnetic force; used to indicate the biochemical activity of a tissue. Formerly called ***nuclear magnetic resonance (NMR).***

Malaise (ma-LĀYZ) Discomfort, uneasiness, and indisposition, often indicative of infection.

Malignant (ma-LIG-nant) Referring to diseases that tend to become worse and cause death; especially the invasion and spreading of cancer.

Malignant melanoma (mel'-a-NŌ-ma) A usually dark, malignant tumor of the skin containing melanin.

Malnutrition (mal'-nu-TRISH-un) State of bad or poor nutrition that may be due to inadequate food intake, imbalance of nutrients, malabsorption of nutrients, improper distribution of nutrients, increased nutrient requirements, increased nutrient losses, or overnutrition.

Mammary (MAM-ar-ē) ***gland*** Modified sudoriferous (sweat) gland of the female that secretes milk for the nourishment of the young.

Mammillary (MAM-i-ler-ē) ***bodies*** Two small rounded bodies posterior to the tuber cinereum that are involved in reflexes related to the sense of smell.

Mammography (mam-OG-ra-fē) Procedure for imaging the breasts (xeromammography or film-screen mammography) to evaluate for breast disease or screen for breast cancer.

Marfan (MAR-fan) ***syndrome*** Inherited disorder that results in abnormalities of connective tissue, especially in the skeleton, eyes, and cardiovascular system.

Marrow (MAR-ō) Soft, spongelike material in the cavities of bone. Red marrow produces blood cells; yellow marrow, formed mainly of fatty tissue, has no blood-producing function.

Mast cell A cell found in areolar connective tissue along blood vessels that produces heparin, an anticoagulant. The name given to a basophil after it has left the bloodstream and entered the tissues.

Mastectomy (mas-TEK-tō-mē) Surgical removal of breast tissue.

Mastication (mas'-ti-KĀ-shun) Chewing.

Matrix (MĀ-trix) Ground substance and fibers external to cells of a connective tissue.

Meatus (mē-Ā-tus) A passage or opening, especially the external portion of a canal.

Mechanoreceptor (me-KAN-ō-rē'-sep-tor) Receptor that detects mechanical deformation of the receptor itself or adjacent cells; stimuli so detected include those related to touch, pressure, vibration, proprioception, hearing, equilibrium, and blood pressure.

Medial (MĒ-dē-al) Nearer the midline of the body or a structure.

Medial lemniscus (lem-NIS-kus) A flat band of myelinated nerve fibers extending through the medulla, pons, and midbrain and terminating in the thalamus on the same side. Sensory neurons in this tract transmit impulses for proprioception, fine touch, pressure, and vibration sensations.

Median aperture (AP-er-choor) One of the three openings in the roof of the fourth ventricle through which cerebrospinal fluid enters the subarachnoid space of the brain and cord. Also called the ***foramen of Magendie.***

Mediastinum (mē'-dē-as-TĪ-num) A broad, median partition, actually a mass of tissue found between the pleurae of the lungs that extends from the sternum to the vertebral column.

Medulla (me-DULL-la) An inner layer of an organ, such as the medulla of the kidneys.

Medulla oblongata (ob'-long-GA-ta) The most inferior part of the brain stem.

Medullary (MED-yoo-lar'-ē) ***cavity*** The space within the diaphysis of a bone that contains yellow marrow. Also called the ***marrow cavity.***

Medullary rhythmicity (rith-MIS-i-tē) ***area*** Portion of the respiratory center in the medulla that controls the basic rhythm of respiration.

Meibomian gland *See **Tarsal gland.***

Meiosis (mē-Ō-sis) A type of cell division restricted to sex-cell production involving two successive nuclear divisions that result in daughter cells with the haploid (*n*) number of chromosomes.

Meissner's corpuscle *See **Corpuscle of touch.***

Melanin (MEL-a-nin) A dark black, brown, or yellow pigment found in some parts of the body such as the skin.

Melanocyte (MEL-a-nō-sīt') A pigmented cell located between or beneath cells of the deepest layer of the epidermis that synthesizes melanin.

Melanocyte-stimulating hormone (MSH) A hormone secreted by the adenohypophysis (anterior lobe) of the pituitary gland that stimulates the dispersion of melanin granules in melanocytes in amphibians; continued administration produces darkening of skin in humans.

Melatonin (mel-a-TŌN-in) A hormone secreted by the pineal gland that may inhibit reproductive activities.

Membrane A thin, flexible sheet of tissue composed of an epithelial layer and an underlying connective tissue layer, as in an epithelial membrane, or of areolar connective tissue only, as in a synovial membrane.

Membranous labyrinth (mem-BRA-nus LAB-i-rinth) The portion of the labyrinth of the inner ear that is located inside the bony labyrinth and separated from it by the perilymph; made up of the membranous semicircular canals, the saccule and utricle, and the cochlear duct.

Memory The ability to recall thoughts; commonly classified as short-term (activated) and long-term.

Menarche (me-NAR-kē) Beginning of the menstrual function.

Ménière's (men-YAIRZ) ***syndrome*** A type of labyrinthine disease characterized by fluctuating loss of hearing, vertigo, and tinnitus due to an increased amount of endolymph that enlarges the labyrinth.

Meninges (me-NIN-jēz) Three membranes covering the brain and spinal cord, called the dura mater, arachnoid, and pia mater. *Singular, **meninx*** (MEN-inks).

Meningitis (men-in-JĪ-tis) Inflammation of the meninges, most commonly the pia mater and arachnoid.

Menopause (MEN-ō-pawz) The termination of the menstrual cycles.

Menstrual (MEN-stroo-al) ***cycle*** A series of changes in the endometrium of a nonpregnant female that prepares the lining of the uterus to receive a fertilized ovum.

Menstruation (men'-stroo-Ā-shun) Periodic discharge of blood, tissue fluid, mucus, and epithelial cells that usually lasts for 5 days; caused by a sudden reduction in estrogens and progesterone. Also called the ***menstrual phase*** or ***menses.***

Merocrine (MER-ō-krin) ***gland*** A secretory cell that remains intact throughout the process of formation and discharge of the secretory product, as in the salivary and pancreatic glands.

Mesenchyme (MEZ-en-kīm) An embryonic connective tissue from which all other connective tissues arise.

Mesentery (MEZ-en-ter'-ē) A fold of peritoneum attaching the small intestine to the posterior abdominal wall.

Mesocolon (mez'-ō-KŌ-lon) A fold of peritoneum attaching the colon to the posterior abdominal wall.

Mesoderm The middle of the three primary germ layers that gives rise to connective tissues, blood and blood vessels, and muscles.

Mesothelium (mez'-ō-THĒ-lē-um) The layer of simple squamous epithelium that lines serous cavities.

Mesovarium (mez'-ō-VAR-ē-um) A short fold of peritoneum that attaches an ovary to the broad ligament of the uterus.

Metabolism (me-TAB-ō-lizm) The sum of all the biochemical reactions that occur within an organism, including the synthetic (anabolic) reactions and decomposition (catabolic) reactions.

Metacarpus (met'-a-KAR-pus) A collective term for the five bones that make up the palm of the hand.

Metaphase (MET-a-phāz) The second stage of mitosis in which chromatid pairs line up on the equatorial plane of the cell.

Metaphase plate Midpoint region of the mitotic spindle.

Metaphysis (me-TAF-i-sis) Growing portion of a bone.

Metaplasia (met'-a-PLĀ-zē-a) The transformation of one cell into another.

Metarteriole (met'-ar-TĒ-rē-ōl) A blood vessel that emerges from an arteriole, traverses a capillary network, and empties into a venule.

Metastasis (me-TAS-ta-sis) The spread of cancer to surrounding tissues (local) or to other body sites (distant).

Metatarsus (met'-a-TAR-sus) A collective term for the five bones located in the foot between the tarsals and the phalanges.

Microcephalus (mi-krō-SEF-a-lus) An abnormally small head; premature closing of the anterior fontanel so that the brain has insufficient room for growth, resulting in mental retardation.

Microfilament (mī-krō-FIL-a-ment) Rodlike cytoplasmic structure about 6 nm in diameter; comprises contractile units in muscle fibers (cells) and provides support, shape, and movement in nonmuscle cells.

Microglia (mī-krō-GLĒ-a) Neuroglial cells that carry on phagocytosis. Also called ***brain macrophages*** (MAK-rō-fāj-ez).

Microphage (MĪK-rō-fāj) Granular leukocyte that carries on phagocytosis, especially neutrophils and eosinophils.

Microtrabeculae (mī-krō-tra-BEK-yoo-lē) Three-dimensional meshwork of fine filaments, about 10–15 nm in diameter, that hold together microfilaments, microtubules, and intermediate filaments and together constitute the microtrabecular lattice.

Microtrabecular (mī-krō-tra-BEK-yoo-lar) ***lattice*** (LAT-is) Collective term for microfilaments, microtubules, and intermediate filaments held together by microtrabeculae in cytoplasm.

Microtubule (mī-krō-TOOB-yool') Cylindrical cytoplasmic structure, ranging in diameter from 18 to 30 nm, consisting of the protein tubulin; provides support, structure, and transportation.

Microvilli (mī'-krō-VIL-ē) Microscopic, fingerlike projections of the cell membranes of small intestinal cells that increase surface area for absorption.

Micturition (mik'-too-RISH-un) The act of expelling urine from the urinary bladder. Also called ***urination*** (yoo-ri-NĀ-shun).

Midbrain The part of the brain between the pons and the diencephalon. Also called the ***mesencephalon*** (mes'-en-SEF-a-lon).

Middle ear A small, epithelial-lined cavity hollowed out of the temporal bone, separated from the external ear by the eardrum and from the internal ear by a thin bony partition containing the oval and round windows; extending across the middle ear are the three auditory ossicles. Also called the ***tympanic*** (tim-PAN-ik) ***cavity.***

Midline An imaginary vertical line that divides the body into equal left and right sides.

Midsagittal plane A vertical plane through the midline of the body that divides the body or organs into *equal* right and left sides. Also called a ***median plane.***

Mineral Inorganic, homogeneous solid substance that may perform a function vital to life; examples include calcium, sodium, potassium, iron, phosphorus, and chlorine.

Mineralocorticoids (min'-er-al-ō-KOR-ti-koyds) A group of hormones of the adrenal cortex.

Mitochondrion (mī'-tō-KON-drē-on) A double-membraned organelle that plays a central role in the production of ATP; known as the "powerhouse" of the cell.

Mitosis (mī-TŌ-sis) The orderly division of the nucleus of a cell that ensures that each new daughter nucleus has the same number and kind of chromosomes as the original parent nucleus. The process includes the replication of chromosomes and the distribution of the two sets of chromosomes into two separate and equal nuclei.

Mitotic spindle Football-shaped assembly of microtubules, involved in chromosomal movement during mitosis.

Mitral stenosis (MĪ-tral ste-NŌ-sis) Narrowing of the mitral valve by scar formation or a congenital defect.

Mitral (MĪ-tral) ***valve prolapse*** (PRŌ-laps) or ***MVP*** An inherited disorder in which a portion of a mitral valve is pushed back too far (prolapsed) during contraction due to expansion of the cusps and elongation of the chordae tendineae.

Mittelschmerz ((MIT-el-shmerz) Abdominopelvic pain that supposedly indicates the release of a secondary oocyte from the ovary.

Modality (mō-DAL-i-tē) Any of the specific sensory entities, such as vision, smell, or taste.

Modiolus (mō-DĪ-ō'-lus) The central pillar or column of the cochlea.

Monoclonal antibody (MAb) Antibody produced by in vitro clones of B cells hybridized with cancerous cells.

Monocyte (MON-ō-sīt') A type of white blood cell characterized by agranular cytoplasm; the largest of the leukocytes.

Monounsaturated fat A fat that contains one double covalent bond between its carbon atoms; it is not completely saturated with hydrogen atoms. Examples are olive and peanut oil.

Mons pubis (monz PYOO-bis) The rounded, fatty prominence over the pubic symphysis, covered by coarse pubic hair.

Morbid (MOR-bid) Diseased; pertaining to disease.

Morula (MOR-yoo-la) A solid mass of cells produced by successive cleavages of a fertilized ovum a few days after fertilization.

Motor area The region of the cerebral cortex that governs muscular movement, particularly the precentral gyrus of the frontal lobe.

Motor end plate Portion of the sarcolemma of a muscle fiber (cell) in close approximation with an axon terminal.

Motor unit A motor neuron together with the muscle fibers (cells) it stimulates.

Movement Motion of the whole body, individual organs, single cells, or organelles inside cells.

Mucin (MYOO-sin) A protein found in mucus.

Mucous (MYOO-kus) ***cell*** A unicellular gland that secretes mucus. Also called a ***goblet cell.***

Mucous membrane A membrane that lines a body cavity that opens to the exterior. Also called the ***mucosa*** (myoo-KŌ-sa).

Mucus The thick fluid secretion of mucous glands and mucous membranes.

Multiple sclerosis (skler-Ō-sis) Progressive destruction of myelin sheaths of neurons in the central nervous system, short-circuiting conduction pathways.

Mumps Inflammation and enlargement of the parotid glands accompanied by fever and extreme pain during swallowing.

Muscarinic (mus'-ka-RIN-ik) ***receptor*** Receptor found on all effectors innervated by parasympathetic postganglionic axons and some effectors innervated by sympathetic postganglionic axons; so named because the actions of acetylcholine (ACh) on such receptors are similar to those produced by muscarine.

Muscle An organ composed of one of three types of muscle tissue (skeletal, cardiac, or visceral), specialized for contraction to produce voluntary or involuntary movement of parts of the body.

Muscle spindle An encapsulated receptor in a skeletal muscle, consisting of specialized muscle fiber (cell) and nerve endings, stimulated by changes in length or tension of muscle fibers; a proprioceptor. Also called a ***neuromuscular*** (noo-rō-MUS-kyoo-lar) ***spindle.***

Muscle tissue A tissue specialized to produce motion in response to muscle action potentials by its qualities of contractility, extensibility, elasticity, and excitability.

Muscle tone A sustained, partial contraction of portions of a skeletal muscle in response to activation of stretch receptors.

Muscular dystrophies (DIS-trō-fēz') Inherited muscle-destroying diseases, characterized by degeneration of the individual muscle fibers (cells), which leads to progressive atrophy of the skeletal muscle.

Muscularis (MUS-kyoo-la'-ris) A muscular layer (coat or tunic) of an organ.

Muscularis mucosae (myoo-KŌ-sē) A thin layer of smooth muscle fibers (cells) located in the outermost layer of the mucosa of the gastrointestinal tract, underlying the lamina propria of the mucosa.

Myasthenia (mī-as-THĒ-nē-a) ***gravis*** Weakness of skeletal muscles caused by antibodies directed against acetylcholine receptors that inhibit muscle contraction.

Myelin (MĪ-e-lin) ***sheath*** A white, phospholipid, segmented covering, formed by neurolemmocytes (Schwann cells), around the axons and dendrites of many peripheral neurons.

Myelography (mī-e-LOG-ra-fē) Introduction of a contrast medium into the subarachnoid space of the spinal cord to demonstrate tumors or herniated (slipped) discs within or near the spinal cord.

Myenteric plexus A network of nerve fibers from both autonomic divisions located in the muscularis coat of the small intestine. Also called the ***plexus of Auerbach*** (OW-er-bak).

Myocardial infarction (mī'-ō-KAR-dē-al in-FARK-shun) ***(MI)*** Gross necrosis of myocardial tissue due to interrupted blood supply. Also called a ***heart attack.***

Myocardium (mī'-ō-KAR-dē-um) The middle layer of the heart wall, made up of cardiac muscle, comprising the bulk of the heart, and lying between the epicardium and the endocardium.

Myofibril (mī'-ō-FĪ-bril) A threadlike structure, running longitudinally through a muscle fiber (cell) consisting mainly of thick myofilaments (myosin) and thin myofilaments (actin).

Myoglobin (mī-ō-GLŌ-bin) The oxygen-binding, iron-containing conjugated protein complex present in the sarcoplasm of muscle fibers (cells); contributes the red color to muscle.

Myology (mī-OL-ō-jē) The study of muscles.

Myometrium (mī'-ō-MĒ-trē-um) The smooth muscle layer of the uterus.

Myopia (mī-Ō-pē-a) Defect in vision so that objects can be seen distinctly only when very close to the eyes; nearsightedness.

Myosin (MĪ-ō-sin) The contractile protein that makes up the thick myofilaments of muscle fibers (cells).

Myotonia (mī-ō-TŌ-nē-a) A continuous spasm of muscle; increased muscular irritability and tendency to contract, and less ability to relax.

Myxedema (mix-e-DĒ-ma) Condition caused by hypothyroidism during the adult years characterized by swelling of facial tissues.

Nail A hard plate, composed largely of keratin, that develops from the epidermis of the skin to form a protective covering on the dorsal surface of the distal phalanges of the fingers and toes.

Nail matrix (MĀ-triks) The part of the nail beneath the body and root from which the nail is produced.

Narcosis (nar-KŌ-sis) Unconscious state due to narcotics.

Nasal (NĀ-zal) ***cavity*** A mucosa-lined cavity on either side of the nasal septum that opens onto the face at an external naris and into the nasopharynx at an internal naris.

Nasal septum (SEP-tum) A vertical partition composed of bone (perpendicular plate of ethmoid and vomer) and cartilage, covered with a mucous membrane, separating the nasal cavity into left and right sides.

Nasolacrimal (nā'-zō-LAK-ri-mal) ***duct*** A canal that transports the lacrimal secretion (tears) from the nasolacrimal sac into the nose.

Nasopharynx (nā'-zō-FAR-inks The uppermost portion of the pharynx, lying posterior to the nose and extending down to the soft palate.

Nausea (NAW-sē-a) Discomfort characterized by loss of appetite and sensation of impending vomiting.

Nebulization (neb'-yoo-li-ZĀ-shun) Administration of medication to selected portions of the respiratory tract by droplets suspended in air.

Neck The part of the body connecting the head and the trunk. A constricted portion of an organ such as the neck of the femur or uterus.

Necrosis (ne-KRŌ-sis) Death of a cell or group of cells as a result of disease or injury.

Neonatal (nē'-ō-NĀ-tal) Pertaining to the first 4 weeks after birth.

Neoplasm (NĒ-ō-plazm) A new growth that may be benign or malignant.

Nephritis (ne-FRĪT-is) Inflammation of the kidney.

Nephron (NEF-ron) The functional unit of the kidney.

Nephrotic (ne-FROT-ik) ***syndrome*** A condition in which the endothelial-capsular membrane leaks, allowing large amounts of protein to escape into urine.

Nerve A cordlike bundle of nerve fibers (axons and/or dendrites) and their associated connective tissue coursing together outside the central nervous system.

Nerve impulse A wave of negativity (depolarization) that self-propagates along the outside surface of the plasma membrane of a neuron; also called a ***nerve action potential.***

Nervous tissue Tissue that initiates and transmits nerve impulses to coordinate homeostasis.

Neuralgia (noo-RAL-jē-a) Attacks of pain along the entire course or branch of a peripheral sensory nerve.

Neural plate A thickening of ectoderm that forms early in the third week of development and represents the beginning of the development of the nervous system.

Neuritis (noo-RĪ-tis) Inflammation of a single nerve, two or more nerves in separate areas, or many nerves simultaneously.

Neuroeffector (noo-rō-e-FEK-tor) ***junction*** Collective term for neuromuscular and neuroglandular junctions.

Neurofibral (noo-rō-FĪ-bral) ***node*** A space, along a myelinated nerve fiber, between the individual neurolemmocytes (Schwann cells) that form the myelin sheath and the neurolemma. Also called ***node of Ranvier*** (ron-VĒ-ā).

Neurofibril (noo-rō-FĪ-bril) One of the delicate threads that forms a complicated network in the cytoplasm of the cell body and processes of a neuron.

Neuroglandular (noo-rō-GLAND-yoo-lar) ***junction*** Area of contact between a motor neuron and a gland.

Neuroglia (noo-RŌG-lē-a) Cells of the nervous system that are specialized to perform the functions of connective tissue. The neuroglia of the central nervous system are the astrocytes, oligodendrocytes, microglia, and ependyma; neuroglia of the peripheral nervous system include the neurolemmocytes (Schwann cells) and the ganglion satellite cells. Also called ***glial*** (GLĒ-al) ***cells.***

Neurohypophyseal (noo'-rō-hī'-po-FIZ-ē-al) ***bud*** An outgrowth of ectoderm located on the floor of the hypothalamus that gives rise to the neurohypophysis (posterior lobe) of the pituitary gland.

Neurohypophysis (noo-rō-hī-POF-i-sis) The posterior lobe of the pituitary gland.

Neurolemma (noo-rō-LEM-ma) The peripheral, nucleated cytoplasmic layer of the neurolemmocyte (Schwann cell). Also called ***sheath of Schwann*** (SCHVON).

Neurolemmocyte A neuroglial cell of the peripheral nervous system that forms the myelin sheath and neurolemma of a nerve fiber by wrapping around a nerve fiber in a jelly-roll fashion. Also called a ***Schwann*** (SCHVON) ***cell.***

Neurology (noo-ROL-ō-jē) The branch of science that deals with the normal functioning and disorders of the nervous system.

Neuromuscular (noo-rō-MUS-kyoo-lar) ***junction*** The area of contact between the axon terminal of a motor neuron and a portion of the sarcolemma of a muscle fiber (cell). Also called a ***myoneural*** (mi-o-NOO-ral) ***junction.***

Neuron (NOO-ron) A nerve cell, consisting of a cell body, dendrites, and an axon.

Neurosecretory (noo-rō-SĒC-re-tō-rē) ***cell*** A cell in a nucleus (paraventricular and supraoptic) in the hypothalamus that produces oxytocin (OT) or antidiuretic hormone (ADH), hormones stored in the neurohypophysis of the pituitary gland.

Neurosyphilis (noo-rō-SIF-i-lis) A form of the tertiary stage of syphilis in which various types of nervous tissue are attacked by bacteria and degenerate.

Neurotransmitter One of a variety of molecules synthesized within the nerve axon terminals, released into the synaptic cleft in response to a nerve impulse, and affecting the membrane potential of the postsynaptic neuron. Also called a ***transmitter substance.***

Neutrophil (NOO-trō-fil) A type of white blood cell characterized by granular cytopiasm that stains as readily with acid or basic dyes.

Nicotinic (nik'-ō-TIN-ik) ***receptor*** Receptor found on both sympathetic and parasympathetic postganglionic neurons so named because the actions of acetylcholine (ACh) in such receptors are similar to those produced by nicotine.

Night blindness Poor or no vision in dim light or at night, although good vision is present during bright illumination; frequently caused by a deficiency of vitamin A. Also referred to as ***nyctalopia*** (nik'-ta-LŌ-pē-a).

Nipple A pigmented, wrinkled projection on the surface of the mammary gland that is the location of the openings of the lactiferous ducts for milk release.

Nissl bodies *See* ***Chromatophilic substance.***

Nociceptor (nō'-sē-SEP-tor) A free (naked) nerve ending that detects pain.

Node of Ranvier *See* ***Neurofibral node.***

Nonpigmented granular dendrocytes Two distinct cell types found in the epidermis, also known as ***Langerhans cells*** and ***Granstein cells,*** that differ in their sensitivity to damage by ultraviolet (UV) radiation and their functions in immunity.

Norepinephrine (nor'-ep-ē-NEF-rin) ***(NE)*** A hormone secreted by the adrenal medulla that produces actions similar to those that result form sympathetic stimulation. Also called ***noradrenaline*** (nor-a-DREN-a-lin).

Notochord (NŌ-tō-cord) A flexible rod of embryonic tissue that lies where the future vertebral column will develop.

Nuclear medicine The branch of medicine concerned with the use of radioisotopes in the diagnosis of disease and therapy.

Nuclease (NOO-klē-ās) An enzyme that breaks nucleotides into pentoses and nitrogenous bases; examples are ribonuclease and deoxyribonuclease.

Nucleic (noo-KLĒ-ic) ***acid*** An organic compound that is a long polymer of nucleotides, with each nucleotide containing a pentose sugar, a phosphate group, and one of four possible nitrogenous bases (adenine, cytosine, guanine, and thymine or uracil).

Nucleolus (noo-KLĒ-ō-lus) Nonmembranous spherical body within the nucleus composed of protein, DNA, and RNA that functions in the synthesis and storage of ribosomal RNA.

Nucleus (NOO-klē-us) A spherical or oval organelle of a cell that contains the hereditary factors of the cell, called genes. A cluster of unmyelinated nerve cell bodies in the central nervous system. The central portion of an atom made up of protons and neutrons.

Nucleus cuneatus (kyoo-nē-Ā-tus) A group of nerve cells in the inferior portion of the medulla in which fibers of the fasciculus cuneatus terminate.

Nucleus gracilis (gras-I-lis) A group of nerve cells in the inferior portion of the medulla in which fibers of the fasciculus gracilis terminate.

Nucleus pulposus (pul-PŌ-sus) A soft, pulpy, highly elastic substance in the center of an intervertebral disc, a remnant of the notochord.

Nutrient A chemical substance in food that provides energy, forms new body components, or assists in the functioning of various body processes.

Nystagmus (nis-TAG-mus) Rapid, involuntary, rhythmic movement of the eyeballs; horizontal, rotary, or vertical.

Obesity (ō-BĒS-i-tē) Body weight 10–20 percent over a desirable standard as a result of excessive accumulation of fat. Types of obesity are hypertrophic (adult-onset) and hyperplastic (lifelong).

Obstetrics (ob-STET-riks) The specialized branch of medicine that deals with pregnancy, labor, and the period of time immediately following delivery.

Obturator (OB-tyoo-rā'-ter) Anything that obstructs or closes a cavity or opening.

Occlusion (ō-KLOO-zhun) The act of closure or state of being closed.

Occult (o-KULT) Obscure or hidden from view, as for example, occult blood in stools or urine.

Olfactory (ōl-FAK-tō-rē) Pertaining to smell.

Olfactory bulb A mass of gray matter at the termination of an olfactory (I) nerve, lying beneath the frontal lobe of the cerebrum on either side of the crista galli of the ethmoid bone.

Olfactory cell A bipolar neuron with its cell body lying between supporting cells located in the mucous membrane lining the upper portion of each nasal cavity.

Olfactory tract A bundle of axons that extends from the olfactory bulb posteriorly to the olfactory portion of the cortex.

Oligodendrocyte (o-lig-ō-DEN-drō-sīt) A neuroglial cell that supports neurons and produces a phospholipid myelin sheath around axons of neurons of the central nervous system.

Oligospermia (ol'-i-gō-SPER-mē-a) A deficiency of spermatozoa in the semen.

Oliguria (ol'-i-GYOO-rē-a) Daily urinary output usually less than 250 ml.

Olive A prominent oval mass on each lateral surface of the superior part of the medulla.

Oncogene (ONG-kō-jēn) Gene that has the ability to transform a normal cell into a cancerous cell.

Oncology (ong-KOL-ō-jē) The study of tumors.

Oogenesis (ō'-ō-JEN-e-sis) Formation and development of the ovum.

Oophorectomy (ō'-of-ō-REK-tō-mē) The surgical removal of the ovaries.

Ophthalmic (of-THAL-mik) Pertaining to the eye.

Ophthalmologist (of'-thal-MOL-ō-jist) A physician who specializes in the diagnosis and treatment of eye disorders with drugs, surgery, and corrective lenses.

Ophthalmology (of'-thal-MOL-ō-jē) The study of the structure, function, and diseases of the eye.

Ophthalmoscopy (of'-thal-MOS-kō-pē) Examination of the interior fundus of the eyeball to detect retinal changes associated with hypertension, diabetes mellitus, atherosclerosis, and increased intracranial pressure.

Optic (OP-tik) Refers to the eye, vision, or properties of light.

Optic chiasma (kī-AZ-ma) A crossing point of the optic (II) nerves, anterior to the pituitary gland.

Optic disc A small area of the retina containing openings through which the fibers of the ganglion neurons emerge as the optic (II) nerve. Also called the ***blind spot.***

Optician (op-TISH-an) A technician who fits, adjusts, and dispenses corrective lenses on prescription of an ophthalmologist or optometrist.

Optic tract A bundle of axons that transmits nerve impulses from the retina of the eye between the optic chiasma and the thalamus.

Optometrist (op-TOM-e-trist) Specialist with a doctorate degree in optometry who is licensed to examine and test the eyes and treat visual defects by prescribing corrective lenses.

Oral cholecystogram (kō-lē-SIS-to-gram) X-ray examination of the gallbladder to evaluate for the presence of gallstones, inflammations, and tumors.

Oral contraceptive (OC) A hormone compound, usually a high concentration of progesterone and a low concentration of estrogens, that is swallowed and prevents ovulation, and thus pregnancy. Also called ***"the pill."***

Ora serrata (Ō-ra ser-RĀ-ta) The irregular margin of the retina lying internal and slightly posterior to the junction of the choroid and ciliary body.

Orbit (OR-bit) The bony, pyramid-shaped cavity of the skull that holds the eyeball.

Organ A structure composed of two or more different kinds of tissues with a specific function and usually a recognizable shape.

Organelle (or-gan-EL) A permanent highly organized structure within a cell with characteristic morphology that is specialized to serve a specific function in cellular activities.

Organic (or-GAN-ik) ***compound*** Compound that always contains carbon and hydrogen and the atoms are held together by covalent bonds. Examples include carbohydrates, lipids, protein, and nucleic acids (DNA and RNA).

Organism (OR-ga-nizm) A total living form; one individual.

Orgasm (OR-gazm) Sensory and motor events involved in ejaculation for the male and involuntary contraction of the perineal muscles in the female at the climax of sexual intercourse.

Orifice (OR-i-fis) Any aperture or opening.

Origin (OR-i-jin) The place of attachment of a muscle to the more stationary bone, or the end opposite the insertion.

Oropharynx (or'-ō-FAR-inks) The second portion of the pharynx, lying posterior to the mouth and extending from the soft palate down to the hyoid bone.

Orthopedics (or'-thō-PĒ-diks) The branch of medicine that deals with the preservation and restoration of the skeletal system, articulations, and associated structures.

Orthopnea (or'-thop-NĒ-a) Dyspnea that occurs in the horizontal position.

Osmosis (os-MŌ-sis) The net movement of water molecules through a selectively permeable membrane from an area of high water concentration to an area of lower water concentration until an equilibrium is reached.

Osmotic pressure The pressure required to prevent the movement of pure water into a solution containing solutes when the solutions are separated by a selectively permeable membrane.

Osseous (OS-ē-us) Bony.

Ossicle (OS-si-kul) Small bone, as in the middle ear (malleus, incus, stapes).

Ossification (os'-i-fi-KĀ-shun) Formation of bone. Also called ***osteogenesis.***

Osteoblast (OS-tē-ō-blast') Cell formed from an osteoprogenitor cell that participates in bone formation by secreting some organic components and inorganic salts.

Osteoclast (OS-tē-ō-clast') A large multinuclear cell that develops from a monocyte and destroys or resorbs bone tissue.

Osteocyte (OS-tē-ō-sīt') A mature bone cell that maintains the daily activities of bone tissue.

Osteogenic (os'-tē-ō-JEN-ik) ***layer*** The inner layer of the periosteum that contains cells responsible for forming new bone during growth and repair.

Osteoid (OS-tē-oid) Organic matrix of bone tissue secreted by osteoblasts.

Osteology (os'-tē-OL-ō-jē) The study of bones.

Osteomalacia (os'-tē-ō-ma-LĀ-shē-a) A deficiency of vitamin D in adults causing demineralization and softening of bone.

Osteomyelitis (os'-tē-ō-mī-i-LĪ-tis) Inflammation of bone marrow or of the bone and marrow.

Osteon The basic unit of structure in adult compact bone, consisting of a central (Haversian) canal with its concentrically arranged lamellae, lacunae, osteocytes, and canaliculi. Also called a ***Haversian*** (ha-VĒR-shun) ***system.***

Osteopenia (os'-tē-ō-PĒ-nē-a) Reduced bone mass due to a decrease in the rate of bone synthesis that is insufficient to compensate for normal bone breakdown, such as occurs in osteoporosis and osteomalacia.

Osteoporosis (os'-tē-ō-pō-RŌ-sis) Age-related disorder characterized by decreased bone mass and increased susceptibility to fractures as a result of decreased levels of estrogens.

Osteoprogenitor (os'-tē-ō-prō-JEN-i-tor) ***cell*** Stem cell derived from mesenchyme that has mitotic potential and the ability to differentiate into an osteoblast.

Otalgia (ō-TAL-jē-a) Pain in the ear; earache.

Otic (Ō-tik) Pertaining to the ear.

Otitis media (ō-TĪ-tus MĒ-dē-a) Acute infection of the middle ear characterized by pain, malaise, fever, and an inflamed tympanic membrane, subject to rupture.

Otolith (Ō-tō-lith) A particle of calcium carbonate embedded in the otolithic membrane that functions in maintaining static equilibrium.

Otolithic (ō-tō-LITH-ik) ***membrane*** Thick, gelatinous, glycoprotein layer located directly over hair cells of the macula in the saccule and utricle of the inner ear.

Otorhinolaryngology (ō'-tō-rī-nō-lar'-in-GOL-ō-jē) The branch of medicine that deals with the diagnosis and treatment of diseases of the ears, nose, and throat.

Oval window A small opening between the middle ear and inner ear into which the footplate of the stapes fits. Also called the ***fenestra vestibuli*** (fe-NES-tra ves-TIB-yoo-lē).

Ovarian (ō-VAR-ē-an) ***cycle*** A monthly series of events in the ovary associated with the maturation of an ovum.

Ovarian follicle (FOL-i-kul) A general name for oocytes (immature ova) in any stage of development, along with their surrounding epithelial cells.

Ovarian ligament (LIG-a-ment) A rounded cord of connective tissue that attaches the ovary to the uterus.

Ovary (Ō-var-ē) Female gonad that produces ova and the hormones estrogens, progesterone, and relaxin.

Ovulation (ō-vyoo-LĀ-shun) The rupture of a vesicular ovarian (Graafian) follicle with discharge of a secondary oocyte into the pelvic cavity.

Ovum (Ō-vum) The female reproductive or germ cell; an egg cell.

Oxyhemoglobin (ok'-sē-HĒ-mō-glō-bin) (***HbO_2***) Hemoglobin combined with oxygen.

Oxyphil cell A cell found in the parathyroid gland that secretes parathyroid hormone (PTH).

Oxytocin (ok'-sē-TŌ-sin) (***OT***) A hormone secreted by neurosecretory cells in the paraventricular and supraoptic nuclei of the hypothalamus that stimulates contraction of the smooth muscle fibers (cells) in the pregnant uterus and contractile cells around the ducts of mammary glands.

Pacinian corpuscle *See* ***Lamellated corpuscle.***

Paget's (PAJ-ets) ***disease*** A disorder characterized by a greatly accelerated remodeling process in which osteoclastic resorption is massive and new bone formation by osteoblasts is extensive. As a result, there is an irregular thickening and softening of the bones.

Palate (PAL-at) The horizontal structure separating the oral and the nasal cavities; the roof of the mouth.

Palliative (PAL-ē-a-tiv) Serving to relieve or alleviate without curing.

Palpate (PAL-pāt) To examine by touch; to feel.

Palpitation (pal'-pi-TĀ-shun) A fluttering of the heart or abnormal rate or rhythm of the heart.

Pancreas (PAN-krē-as) A soft, oblong organ lying along the greater curvature of the stomach and connected by a duct to the duodenum. It is both exocrine (secreting pancreatic juice) and endocrine (secreting insulin, glucagon, and somatostatin).

Pancreatic (pan'-krē-AT-ik) ***duct*** A single, large tube that unites with the common bile duct from the liver and gallbladder and drains pancreatic juice into the duodenum at the hepatopancreatic ampulla (ampulla of Vater). Also called the ***duct of Wirsung.***

Pancreatic islet A cluster of endocrine gland cells in the pancreas that secretes insulin, glucagon, and somatostatin. Also called an ***islet of Langerhans*** (LANG-er-hanz).

Papanicolaou (pap'-a-NIK-ō-la-oo) ***test*** A cytological staining test for the detection and diagnosis of premalignant and malignant conditions of the female genital tract. Cells scraped from the genital epithelium are smeared, fixed, stained, and examined microscopically. Also called a ***Pap smear.***

Papilla (pa-PIL-a) A small nipple-shaped projection or elevation.

Paralysis (pa-RAL-a-sis) Loss or impairment of motor function due to a lesion of nervous or muscular origin.

Paranasal sinus (par'-a-NĀ-zal SĪ-nus) A mucus-lined air cavity in a skull bone that communicates with the nasal cavity. Paranasal sinuses are located in the frontal, maxillary, ethmoid, and sphenoid bones.

Paraplegia (par-a-PLĒ-jē-a) Paralysis of both lower extremities.

Parasagittal plane A vertical plane that does not pass through the midline and that divides the body or organs into *unequal* left and right portions.

Parasympathetic (par'-a-sim-pa-THET-ik) ***division*** One of the two subdivisions of the autonomic nervous system, having cell bodies of preganglionic neurons in nuclei in the brain stem and in the lateral gray matter of the sacral portion of the spinal cord; primarily concerned with activities that conserve and restore body energy. Also called the ***craniosacral*** (krā-nē-ō-SĀ-kral) ***division.***

Parathyroid (par'-a-THĪ-royd) ***gland*** One of four small endocrine glands embedded on the posterior surfaces of the lateral lobes of the thyroid gland.

Parathyroid hormone (PTH) A hormone secreted by the parathyroid glands that decreases blood phosphate level and increases blood calcium level.

Paraurethral (par'-a-yoo-RĒ-thral) ***gland*** Gland embedded in the wall of the urethra whose duct opens on either side of the urethral orifice and secretes mucus. Also called ***Skene's*** (SKĒNZ) ***gland.***

Parenchyma (par-EN-ki-ma) The functional parts of any organ, as opposed to tissue that forms its stroma or framework.

Parenteral (par-EN-ter-al) Situated or occurring outside the intestines; referring to introduction of substances into the body other than by way of the intestines such as intradermal, subcutaneous, intramuscular, intravenous, or intraspinal.

Parietal (pa-RĪ-e-tal) Pertaining to or forming the outer wall of a body cavity.

Parietal cell The secreting cell of a gastric gland that produces hydrochloric acid and intrinsic factor. Also called an ***oxyntic cell.***

Parietal pleura (PLOO-ra) The outer layer of the serous pleural membrane that encloses and protects the lungs; the layer that is attached to the wall of the pleural cavity.

Parkinson's disease Progressive degeneration of the basal ganglia and substantia nigra of the cerebrum resulting in decreased production of dopamine (DA) that leads to tremor, slowing of voluntary movements, and muscle weakness. Also called ***Parkinsonism.***

Parotid (pa-ROT-id) ***gland*** One of the paired salivary glands located inferior and anterior to the ears connected to the oral cavity via a duct (Stensen's) that opens into the inside of the cheek opposite the upper second molar tooth.

Paroxysm (PAR-ok-sizm) A sudden periodic attack or recurrence of symptoms of a disease.

Pars intermedia A small avascular zone between the adenohypophysis and neurohypophysis of the pituitary gland.

Parturition (par'-too-RISH-un) Act of giving birth to young; childbirth, delivery.

Patent ductus arteriosus Congenital anatomical heart defect in which the fetal connection (ductus arteriosus) between the aorta and pulmonary trunk remains open instead of closing completely after birth.

Pathogen (PATH-ō-jen) A disease-producing organism.

Pathogenesis (path'-ō-JEN-e-sis) The development of disease or a morbid or pathological state.

Pathological (path'-ō-LOJ-i-kal) Pertaining to or caused by disease.

Pathological (path'-ō-LOJ-i-kal) ***anatomy*** The study of structural changes caused by disease.

Pectinate (PEK-ti-nāt) ***muscles*** Projecting muscle bundles of the anterior atrial walls and the lining of the auricles.

Pectoral (PEK-tō-ral) Pertaining to the chest or breast.

Pediatrician (pē'-dē-a-TRISH-un) A physician who specializes in the care and treatment of children and their illnesses.

Pedicel (PED-i-sel) Footlike structure, as on podocytes of a glomerulus.

Pedicle (PED-i-kul) A short, thick process found on vertebrae.

Pelvic (PEL-vik) ***cavity*** Inferior portion of the abdominopelvic cavity that contains the urinary bladder, sigmoid colon, rectum, and internal female and male reproductive structures.

Pelvic inflammatory disease (PID) Collective term for any extensive bacterial infection of the pelvic organs, especially the uterus, uterine (Fallopian) tubes, and ovaries.

Pelvic splanchnic (PEL-vik SPLANGK-nik) ***nerves*** Preganglionic parasympathetic fibers from the levels of S2, S3, and S4 that supply the urinary bladder, reproductive organs, and the descending and sigmoid colon and rectum.

Pelvimetry (pel-VIM-e-trē) Measurement of the size of the inlet and outlet of the birth canal.

Pelvis The basinlike structure formed by the two hipbones, the sacrum, and the coccyx. The expanded, proximal portion of the ureter, lying within the kidney and into which the major calyces open.

Penis (PĒ-nis) The male copulatory organ, used to introduce spermatozoa into the female vagina.

Pepsin Protein-digesting enzyme secreted by zymogenic (chief) cells of the stomach as the inactive form pepsinogen, which is converted to active pepsin by hydrochloric acid.

Peptic ulcer An ulcer that develops in areas of the gastrointestinal tract exposed to hydrochloric acid; classified as a gastric ulcer if in the lesser curvature of the stomach and as a duodenal ulcer if in the first part of the duodenum.

Percussion (per-KUSH-un) The act of striking (percussing) an underlying part of the body with short, sharp blows as an aid in diagnosing the part by the quality of the sound produced.

Perforating canal A minute passageway by means of which blood vessels and nerves from the periosteum penetrate into compact bone. Also called ***Volkmann's*** (FŌLK-manz) ***canal.***

Pericardial (per'-i-KAR-dē-al) ***cavity*** Small potential space between the visceral and parietal layers of the serous pericardium that contains pericardial fluid.

Pericardium (per'-i-KAR-dē-um) A loose-fitting membrane that encloses the heart, consisting of an outer fibrous layer and an inner serous layer.

Perichondrium (per'-i-KON-drē-um) The membrane that covers cartilage.

Perikaryon (per'-i-KAR-ē-on) The nerve cell body that contains the nucleus and other organelles.

Perilymph (PER-i-lymf) The fluid contained between the bony and membranous labyrinths of the inner ear.

Perimetrium (per-i-MĒ-trē-um) The serosa of the uterus.

Perimysium (per'-i-MĪZ-ē-um) Invagination of the epimysium that divides muscles into bundles.

Perineum (per'-i-NĒ-um) The pelvic floor; the space between the anus and the scrotum in the male and between the anus and the vulva in the female.

Perineurium (per'-i-NYOO-rē-um) Connective tissue wrapping around fascicles in a nerve.

Periodontal (per-ē-ō-DON-tal) ***disease*** A collective term for conditions characterized by degeneration of gingivae, alveolar bone, periodontal ligament, and cementum.

Periodontal ligament Dense fibrous connective tissue that lines the socket walls and covers the cemental surface of the roots of the teeth; it anchors the teeth in position and helps absorb the shock from the forces of chewing.

Periosteal (per'-ē-OS-te-al) ***bud*** Blood vessels and associated osteoblasts, osteoclasts, and red marrow cells that grow into disintegrating calcified cartilage as part of endochondral ossification.

Periosteum (per'-ē-OS-tē-um) The membrane that covers bone and consists of connective tissue, osteoprogenitor cells, and osteoblasts and is essential for bone growth, repair, and nutrition.

Peripheral (pe-RIF-er-al) Located on the outer part or a surface of the body.

Peripheral nervous system (PNS) The part of the nervous system that lies outside the central nervous system—nerves and ganglia.

Periphery (pe-RIF-er-ē) Outer part or a surface of the body; part away from the center.

Peristalsis (per'-i-STAL-sis) Successive muscular contractions along the wall of a hollow muscular structure.

Peritoneum (per'-i-tō-NĒ-um) The largest serous membrane of the body that lines the abdominal cavity and covers the viscera.

Peritonitis (per'-i-tō-NĪ-tis) Inflammation of the peritoneum.

Pernicious (per-NISH-us) Fatal.

Peroxisome (pe-ROKS-ī-sōm) Organelle similar in structure to a lysosome that contains enzymes related to hydrogen peroxide metabolism; abundant in liver cells.

Perspiration Substance produced by sudoriferous (sweat) glands containing water, salts, urea, uric acid, amino acids, ammonia, sugar, lactic acid, and ascorbic acid; helps maintain body temperature and eliminate wastes.

Peyer's patches *See* ***Aggregated lymphatic follicles.***

pH A symbol of the measure of the concentration of hydrogen ions in a solution. The pH scale exends from 0 to 14, with a value of 7 expressing neutrality, values lower than 7 expressing increasing acidity, and values higher than 7 expressing increasing alkalinity.

Phagocytosis (fag'-ō-sī-TŌ-sis) The process by which cells (phagocytes) ingest particulate matter; especially the ingestion and destruction of microbes, cell debris, and other foreign matter.

Phalanx (FĀ-lanks) The bone of a finger or toe. *Plural,* ***phalanges*** (fa-LAN-jēz).

Phantom pain A sensation of pain as originating in a limb that has been amputated.

Pharmacology (far'-ma-KOL-ō-jē) The science that deals with the effects and uses of drugs in the treatment of disease.

Pharynx (FAR-inks) The throat; a tube that starts at the internal nares and runs partway down the neck where it opens into the esophagus posteriorly and the larynx anteriorly.

Phenotype (FĒ-nō-tīp) The observable expression of genotype; physical characteristics of an organism determined by genetic makeup and influenced by interaction between genes and internal and external environmental factors.

Phenylketonuria (fen'-il-kē'-tō-NOO-rē-a) ***(PKU)*** A disorder characterized by an elevation of the amino acid phenylalanine in the blood.

Pheochromocytoma (fē-ō-krō'-mō-sī-TŌ-ma) Tumor of the chromaffin cells of the adrenal medulla that results in hypersecretion of medullary hormones.

Phlebitis (fle-BĪ-tis) Inflammation of a vein, usually in the lower extremities.

Phlebotomy (fle-BOT-ō-me) The cutting of a vein to allow the escape of blood.

Phospholipid (fos'-fō-LIP-id) ***bilayer*** Arrangement of phospholipid molecules in two parallel rows in which the hydrophilic "heads" face outward and the hydrophobic "tails" face inward.

Photoreceptor Receptor that detects light on the retina of the eye.

Physiology (fiz'-ē-OL-ō-jē) Science that deals with the functions of an organism or its parts.

Pia mater (PĪ-a MĀ-ter) The inner membrane (meninx) covering the brain and spinal cord.

Piezoelectric (pē-e-zō-e-LEK-trik) ***effect*** Response of bone, mainly collagen, to stress in which very minute currents of electricity are produced; believed to stimulate osteoblasts to make new bone cells.

Pilonidal (pī-lō-NĪ-dal) Containing hairs resembling a tuft inside a cyst or sinus.

Pineal (PĪN-ē-al) ***gland*** The cone-shaped gland located in the roof of the third ventricle. Also called the ***epiphysis cerebri*** (ē-PIF-i-sis se-RĒ-brē).

Pinealocyte (pin-ē-AL-ō-sīt) Secretory cell of the pineal gland that produces hormones.

Pinocytosis (pi'-nō-sī-TŌ-sis) The process by which cells ingest liquid.

Pituicyte (pi-TOO-i-sīt) Supporting cell of the posterior lobe of the pituitary gland.

Pituitary (pi-TOO-i-tar'-ē) ***dwarfism*** Condition caused by hyposecretion of human growth hormone (hGH) during the growth years and characterized by childlike physical traits in an adult.

Pituitary gland A small endocrine gland lying in the sella turcica of the sphenoid bone and attached to the hypothalamus by the infundibulum; nicknamed the "master gland." Also called the ***hypophysis*** (hī-POF-i-sis).

Pivot joint A synovial joint in which a rounded, pointed, or conical surface of one bone articulates with a ring formed partly by another bone and partly by a ligament, as in the joint between the atlas and axis and between the proximal ends of the radius and ulna. Also called a ***trochoid*** (TRŌ-koid) ***joint.***

Placenta (pla-SEN-ta) The special structure through which the exchange of materials between fetal and maternal circulations occurs. Also called the ***afterbirth.***

Plantar flexion (PLAN-tar FLEK-shun) Bending the foot in the direction of the plantar surface (sole).

Plaque (plak) A cholesterol-containing mass in the tunica media of arteries. A mass of bacterial cells, dextran (polysaccharide), and other debris that adheres to teeth.

Plasma (PLAZ-ma) The extracellular fluid found in blood vessels; blood minus the formed elements.

Plasma cell Cell that produces antibodies and develops from a B cell (lymphocyte).

Plasma (cell) membrane Outer, limiting membrane that separates the cell's internal parts from extracellular fluid and the external environment.

Plasmapheresis (plaz'-ma-fe-RĒ-sis) A procedure in which blood is withdrawn from the body, its components are selectively separated, the undesirable component causing disease is removed, and the remainder is returned to the body. Among the substances removed are toxins, metabolic substances, and antibodies. Also called ***therapeutic plasma exchange (TPE).***

Pleura (PLOOR-a) The serous membrane that covers the lungs and lines the walls of the chest.

Pleural cavity Small potential space between the visceral and parietal pleurae.

Plexus (PLEK-sus) A network of nerves, veins, or lymphatic vessels.

Plexus of Auerbach *See* ***Myenteric plexus.***

Plexus of Meissner *See* ***Submucosal plexus.***

Pneumonia (noo-MŌ-nē-a) Acute infection or inflammation of the alveoli of the lungs.

Pneumotaxic (noo-mō-TAK-sik) ***area*** Portion of the respiratory center in the pons that continually sends inhibitory nerve impulses to the inspiratory area that limit inspiration and facilitate expiration.

Podiatry (pō-DĪ-a-trē) The diagnosis and treatment of foot disorders.

Polar body The smaller cell resulting from the unequal division of cytoplasm during the meiotic divisions of an oocyte. The polar body has no function and is resorbed.

Poliomyelitis (pō'-lē-ō-mī-e-LĪ-tis) Viral infection marked by fever, headache, stiff neck and back, deep muscle pain and weakness, and loss of certain somatic reflexes; a serious form of the disease, ***bulbar polio,*** results in destruction of motor neurons in anterior horns of spinal nerves that leads to paralysis.

Polycythemia (pol'-ē-sī-THĒ-mē-a) Disorder characterized by a hematocrit above the normal level of 55 in which hypertension, thrombosis, and hemorrhage occur.

Polyp (POL-ip) A tumor on a stem found especially on a mucous membrane.

Polysaccharide (pol'-ē-SAK-a-rīd) A carbohydrate in which three or more monosaccharides are joined chemically.

Polyunsaturated fat A fat that contains more than one double covalent bond between its carbon atoms; examples are corn oil, safflower oil, and cottonseed oil.

Polyuria (pol'-ē-YOO-rē-a) An excessive production of urine.

Pons (ponz) The portion of the brain stem that forms a "bridge" between the medulla and the midbrain, anterior to the cerebellum.

Positron emission tomography (PET) A type of radioactive scanning based on the release of gamma rays when positrons collide with negatively charged electrons in body tissues; it indicates where radioisotopes are used in the body.

Postcentral gyrus A gyrus immediately posterior to the central sulcus that contains the general sensory area of the cerebral cortex.

Posterior (pos-TĒR-ē-or) Nearer to or at the back of the body. Also called ***dorsal.***

Posterior root The structure composed of afferent (sensory) fibers lying between a spinal nerve and the dorsolateral aspect of the spinal cord. Also called the ***dorsal (sensory) root.***

Posterior root ganglion A group of cell bodies of sensory (afferent) neurons and their supporting cells located along the posterior root of a spinal nerve. Also called a ***dorsal (sensory) root ganglion*** (GANG-glē-on).

Postganglionic neuron (pōst'-gang-lē-ON-ik NOO-ron) The second visceral efferent neuron in an autonomic pathway, having its cell body and dendrites located in an autonomic ganglion and its unmyelinated axon ending at cardiac muscle, smooth muscle, or a gland.

Postpartum (pōst-PAR-tum) After parturition; occurring after the delivery of a baby.

Postsynaptic (pōst-sin-AP-tik) ***neuron*** The nerve cell that is activated by the release of a neurotransmitter substance from another neuron and carries nerve impulses away from the synapse.

Pouch of Douglas *See* ***Rectouterine pouch.***

Precapillary sphincter (SFINGK-ter) A ring of smooth muscle fibers (cells) at the site of origin of true capillaries that regulate blood flow into true capillaries.

Precentral gyrus A gyrus immediately anterior to the central sulcus that contains the primary motor area of the cerebral cortex.

Preeclampsia (pre'-e'-KLAMP-sē-a) A syndrome characterized by sudden hypertension, large amounts of protein in urine, and generalized edema; it might be related to an autoimmune or allergic reaction due to the presence of a fetus.

Preganglionic (prē'-gang-lē-ON-ik) ***neuron*** The first visceral efferent neuron in an autonomic pathway, with its cell body and dendrites in the brain or spinal cord and its myelinated axon ending at an autonomic ganglion, where it synapses with a postganglionic neuron.

Pregnancy Sequence of events that normally includes fertilization, implantation, embryonic growth, and fetal growth that terminates in birth.

Premenstrual syndrome (PMS) Severe physical and emotional stress occurring late in the postovulatory phase of the menstrual cycle and sometimes overlapping with menstruation.

Premonitory (prē-MON-i-tō-rē) Giving previous warning; as premonitory symptoms.

Prepuce (PRĒ-pyoos) The loose-fitting skin covering the glans of the penis and clitoris. Also called the ***foreskin.***

Presbyopia (prez-bē-Ō-pē-a) A loss of elasticity of the lens of the eye due to advancing age with resulting inability to focus clearly on near objects.

Presynaptic (prē-sin-AP-tik) ***inhibition*** Inhibition of a nerve impulse before it reaches a synapse in which neurotransmitter released by an inhibitory neuron depresses the release of excitatory transmitter at an excitatory neuron.

Presynaptic (prē-sin-AP-tik) ***neuron*** A nerve cell that carries nerve impulses toward a synapse.

Prevertebral ganglion (prē-VERT-e-bral GANG-lē-on) A cluster of cell bodies of postganglionic sympathetic neurons anterior to the spinal column and close to large abdominal arteries. Also called a ***collateral ganglion.***

Primary germ layer One of three layers of embryonic tissue, called ectoderm, mesoderm, and endoderm, that give rise to all tissues and organs of the organism.

Primary motor area A region of the cerebral cortex in the precentral gyrus of the frontal lobe of the cerebrum that controls specific muscles or groups of muscles.

Primary somesthetic (sō-mes-THET-ik) ***area*** A region of the cerebral cortex posterior to the central sulcus in the postcentral gyrus of the parietal lobe of the cerebrum that localizes exactly the points of the body where sensations originate.

Prime mover The muscle directly responsible for producing the desired motion. Also called an ***agonist*** (AG-ō-nist).

Primigravida (prī-mi-GRAV-i-da) A woman pregnant for the first time.

Primitive gut Embryonic structure composed of endoderm and mesoderm that gives rise to most of the gastrointestinal tract.

Primordial (prī-MŌR-dē-al) Existing first; especially primordial egg cells in the ovary.

Primordial (prī-MŌR-dē-al) ***germ cell*** Primitive cell that arises from yolk sac endoderm and enters the ovaries and testes early in their development; gives rise to oogonia and spermatogonia.

Principal cell Cell found in the parathyroid glands that secretes parathyroid hormone (PTH). Also called a ***chief cell.***

Proctology (prok-TOL-ō-jē) The branch of medicine that treats the rectum and its disorders.

Progeny (PROJ-e-nē) Refers to offspring or descendants.

Progesterone (prō-JES-te-rōn) A female sex hormone produced by the ovaries that helps prepare the endometrium for implantation of a fertilized ovum and the mammary glands for milk secretion.

Prognosis (prog-NŌ-sis) A forecast of the probable results of a disorder; the outlook for recovery.

Projection (prō-JEK-shun) The process by which the brain refers sensations to their point of stimulation.

Prolactin (prō-LAK-tin) ***(PRL)*** A hormone secreted by the adenohypophysis (anterior lobe) of the pituitary gland that initiates and maintains milk secretion by the mammary glands.

Prolapse (PRŌ-laps) A dropping or falling down of an organ, especially the uterus or rectum.

Proliferation (pro-lif'-er-Ā-shun) Rapid and repeated reproduction of new parts, especially cells.

Pronation (prō-NĀ-shun) A movement of the forearm in which the palm of the hand is turned posteriorly or inferiorly.

Prophase (PRŌ-fāz) The first stage of mitosis during which chromatid pairs are formed and aggregate around the equatorial plane region of the cell.

Proprioception (prō-prē-ō-SEP-shun) The receipt of information from muscles, tendons, and the labyrinth that enables the brain to determine movements and position of the body and its parts. Also called ***kinesthesia*** (kin'-es-THĒ-zē-a).

Proprioceptor (prō'-prē-ō-SEP-tor) A receptor located in muscles, tendons, or joints that provides information about body position and movements.

Prostaglandin (pros'-ta-GLAN-din) ***(PG)*** A membrane-associated lipid composed of 20-carbon fatty acids with 5 carbon atoms joined to form a cyclopentane ring; synthesized in small quantities and basically mimics the activities of hormones.

Prostatectomy (pros'-ta-TEK-tō-mē) The surgical removal of part of or the entire prostate gland.

Prostate (PROS-tāt) ***gland*** A doughnut-shaped gland inferior to the urinary bladder that surrounds the superior portion of the male urethra and secretes a slightly acid solution that contributes to sperm motility and viability.

Prosthesis (pros-THĒ-sis) An artificial device to replace a missing body part.

Protein An organic compound consisting of carbon, hydrogen, oxygen, nitrogen, and sometimes sulfur and phosphorus, and made up of amino acids linked by peptide bonds.

Proto-oncogene (prō'-tō-ONG-kō-jēn) Gene responsible for some aspect of normal growth and development; it may transform into an oncogene, a gene capable of causing cancer.

Protraction (prō-TRAK-shun) The movement of the mandible or shoulder girdle forward on a plane parallel with the ground.

Proximal (PROK-si-mal) Nearer the attachment of an extremity to the trunk or a structure; nearer to the point of origin.

Pruritus (proo'-RĪ-tus) Itching.

Pseudopodia (soo'-dō-PŌ-dē-a) Temporary, protruding projections of cytoplasm.

Psoriasis (sō-RĪ-a-sis) Chronic skin disease characterized by reddish plaques or papules covered with scales.

Psychosomatic (sī'-kō-sō-MAT-ik) Pertaining to the relation between mind and body. Commonly used to refer to those physiological disorders thought to be caused entirely or partly by emotional disturbances.

Pterygopalatine ganglion (ter'-i-gō-PAL-a-tīn GANG-glē-on) A cluster of cell bodies of parasympathetic postganglionic neurons ending at the lacrimal and nasal glands.

Ptosis (TŌ-sis) Drooping, as of the eyelid or the kidney (nephrotosis).

Puberty (PYOO-ber-tē) The time of life during which the secondary sex characteristics begin to appear and the capability for sexual reproduction is possible; usually between the ages of 10 and 17.

Pubic (PYOO-bik) ***symphysis*** A slightly movable cartilaginous joint between the anterior surfaces of the hipbones.

Pudendum (pyoo-DEN-dum) A collective designation for the external genitalia of the female.

Puerperium (pyoo'-er-PER-ē-um) The state immediately after childbirth, usually 4–6 weeks.

Pulmonary (PUL-mo-ner'-ē) Concerning or affected by the lungs.

Pulmonary circulation The flow of deoxygenated blood from the right ventricle to the lungs and the return of oxygenated blood from the lungs to the left atrium.

Pulmonary edema (e-DĒ-ma) An abnormal accumulation of interstitial fluid in the tissue spaces and alveoli of the lungs due to increased pulmonary capillary permeability or increased pulmonary capillary pressure.

Pulmonary embolism (EM-bō-lizm) ***(PE)*** The presence of a blood clot or other foreign substance in a pulmonary arterial blood vessel that obstructs circulation to lung tissue.

Pulmonary (PUL-mo-ner-ē) ***function tests*** Any number of tests (forced vital capacity, forced expiratory volume in one second, maximum midrespiratory flow, maximum voluntary ventilation) designed to evaluate lung disease and measure pulmonary impairment.

Pulmonary ventilation The inflow (inspiration) and outflow (expiration) of air between the atmosphere and the lungs. Also called ***breathing.***

Pulp cavity A cavity within the crown and neck of a tooth, filled with pulp, a connective tissue containing blood vessels, nerves, and lymphatic vessels.

Pulsating electromagnetic fields (PEMFs) A procedure that uses electrotherapy to treat improperly healing fractures.

Pupil The hole in the center of the iris, the area through which light enters the posterior cavity of the eyeball.

Purkinje fiber *See* ***Conduction myofiber.***

Pus The liquid product of inflammation containing leucocytes or their remains and debris of dead cells.

Pyelitis (pī'-e-LĪ-tis) Inflammation of the kidney pelvis and its calyces.

Pyemia (pī-Ē-mēa) Infection of the blood, with multiple abscesses, caused by pus-forming microorganisms.

Pyloric (pī-LOR-ik) ***sphincter*** A thickened ring of smooth muscle through which the pylorus of the stomach communicates with the duodenum. Also called the ***pyloric valve.***

Pyogenesis (pi'-ō-JEN-e-sis) Formation of pus.

Pyorrhea (pī-ō-RĒ-a) A discharge or flow of pus, especially in the alveoli (sockets) and the tissues of the gums.

Pyramid (PIR-a-mid) A pointed or cone-shaped structure; one of two roughly triangular structures on the ventral side of the medulla composed of the largest motor tracts that run from the cerebral cortex to the spinal cord; a triangular-shaped structure in the renal medulla composed of the straight segments of renal tubules.

Pyramidal (pi-RAM-i-dal) ***pathways*** Collections of motor nerve fibers arising in the brain and passing down through the spinal cord to motor cells in the anterior horns.

Pyrexia (pī-REK-sē-a) A condition in which the temperature is above normal.

Pyuria (pī-YOO-rē-a) The presence of leukocytes and other components of pus in urine.

Quadrant (KWOD-rant) One of four parts.

Quadriplegia (kwod'-ri-PLĒ-jē-a) Paralysis of the two upper and two lower extremities.

Radiographic (rā'-dē-ō-GRAF-ic) ***anatomy*** Diagnostic branch of anatomy that includes the use of x-rays.

Rami communicantes (RĀ-mē ko-myoo-ni-KAN-tēz) Branches of a spinal nerve, *Singular,* ***ramus communicans*** (RĀ-mus ko-MYOO-ni-kans).

Rathke's pouch *See* ***Hypophyseal pouch.***

Raynaud's (rā-NOZ) ***disease*** A vascular disorder, primarily of females, characterized by bilateral attacks of ischemia, usually of the fingers and toes, in which the skin becomes pale and exhibits burning and pain; it is brought on by cold or emotional stimuli.

Receptor A specialized cell or a nerve cell terminal modified to respond to some specific sensory modality, such as touch, pressure, cold, light, or sound, and convert it to a nerve impulse by way of a generator or receptor potential. A specific molecule or arrangement of molecules organized to accept only molecules with a complementary shape.

Receptor-mediated endocytosis A highly selective process in which cells take up large molecules or particles (ligands). In the process, successive compartments called vesicles, endosomes, and CURLs form. Ligands are eventually broken down by enzymes in lysosomes.

Recombinant DNA Synthetic DNA, formed by joining a fragment of DNA from one source to a portion of DNA from another.

Rectouterine pouch A pocket formed by the parietal peritoneum as it moves posteriorly from the surface of the uterus and is reflected onto the rectum; the lowest point in the pelvic cavity. Also called the ***pouch*** or ***cul de sac of Douglas.***

Rectum (REK-tum) The last 20 cm (7 in.) of the gastrointestinal tract, from the sigmoid colon to the anus.

Recumbent (re-KUM-bent) Lying down.

Red nucleus A cluster of cell bodies in the midbrain, occupying a large portion of the tectum and sending fibers into the rubroreticular and rubrospinal tracts.

Red pulp That portion of the spleen that consists of venous sinuses filled with blood and cords of splenic tissue called splenic (Billroth's) cords.

Referred pain Pain that is felt at a site remote from the place of origin.

Reflex Fast response to a change (stimulus) in the internal or external environment that attempts to restore homeostasis; passes over a reflex arc.

Reflex arc The most basic conduction pathway through the nervous system, connecting a receptor and an effector and consisting of a receptor, a sensory neuron, a center in the central nervous system for a synapse, a motor neuron, and an effector.

Refraction (rē-FRAK-shun) The bending of light as it passes from one medium to another.

Regeneration (rē-jen'-er-Ā-shun) The natural renewal of a structure.

Regimen (REJ-i-men) A strictly regulated scheme of diet, exercise, or activity designed to achieve certain ends.

Regional anatomy The division of anatomy dealing with a specific region of the body, such as the head, neck, chest, or abdomen.

Regulating factor Chemical secretion of the hypothalamus whose structure is unknown that can either stimulate or inhibit secretion of hormones of the adenohypophysis (anterior pituitary).

Regulating hormone Chemical secretion of the hypothalamus whose structure is known that can either stimulate or inhibit secretion of hormones of the adenohypophysis (anterior pituitary).

Regurgitation (rē-gur'-ji-TĀ-shun) Return of solids or fluids to the mouth from the stomach; flowing backward of blood through incompletely closed heart valves.

Relapse (RĒ-laps) The return of a disease weeks or months after its apparent cessation.

Relaxin A female hormone produced by the ovaries that relaxes the pubic symphysis and helps dilate the uterine cervix to facilitate delivery.

Remodeling Replacement of old bone by new bone tissue.

Renal (RĒ-nal) Pertaining to the kidney.

Renal corpuscle (KOR-pus'-l) A glomerular (Bowman's) capsule and its enclosed glomerulus.

Renal erythropoietic (ē-rith'-rō-poy-Ē-tik) ***factor*** An enzyme released by the kidneys and liver in response to hypoxia that acts on a plasma protein to bring about the production of erythropoietin, a hormone that stimulates red blood cell production.

Renal failure Inability of the kidneys to function properly, due to abrupt failure (acute) or progressive failure (chronic).

Renal pelvis A cavity in the center of the kidney formed by the expanded, proximal portion of the ureter, lying within the kidney, and into which the major calyces open.

Renal pyramid A triangular structure in the renal medulla composed of the straight segments of renal tubules.

Reproduction (rē'-prō-DUK-shun) Either the formation of new cells for growth, repair, or replacement, or the production of a new individual.

Reproductive cell division Type of cell division in which sperm and egg cells are produced; consists of meiosis and cytokinesis.

Resistance Ability to ward off disease. The hindrance encountered by an electrical charge as it moves through a substance from one point to another. The hindrance encountered by blood as it flows through the vascular system or by air through respiratory passageways.

Respiration (res-pi-RĀ-shun) Overall exchange of gases between the atmosphere, blood, and body cells consisting of pulmonary ventilation, external respiration, and internal respiration.

Respirator (RES-pi-rā'-tor) An apparatus fitted to a mask over the nose and mouth, or hooked directly to an endotracheal or tracheotomy tube, that is used to assist or support ventilation or to provide nebulized medication to the air passages under positive pressure.

Respiratory center Neurons in the reticular formation of the brain stem that regulate the rate of respiration.

Respiratory distress syndrome (RDS) of the newborn A disease of newborn infants, especially premature ones, in which insufficient amounts of surfactant are produced and breathing is labored. Also called ***hyaline*** (HĪ-a-lin) ***membrane disease (HMD).***

Respiratory failure Condition in which the respiratory system cannot supply sufficient oxygen to maintain metabolism or eliminate enough carbon dioxide to prevent respiratory acidosis.

Responsiveness The ability to detect and react to changes in the internal or external environment.

Resuscitation (rē-sus'-i-TĀ-shun) Act of bringing a person back to full consciousness.

Retention (rē-TEN-shun) A failure to void urine due to obstruction, nervous contraction of the urethra, or absence of sensation of desire to urinate.

Rete (RĒ-tē) ***testis*** The network of ducts in the testes.

Reticular (re-TIK-yoo-lar) ***activating system (RAS)*** An extensive network of branched nerve cells running through the core of the brain stem. When these cells are activated, a generalized alert or arousal behavior results.

Reticular formation A network of small groups of nerve cells scattered among bundles of fibers beginning in the medulla as a continuation of the spinal cord and extending upward through the central part of the brain stem.

Reticulocyte (re-TIK-yoo-lō-sīt) An immature red blood cell.

Reticulum (re-TIK-yoo-lum) A network.

Retina (RET-i-na) The inner coat of the eyeball, lying only in the posterior portion of the eye and consisting of nervous tissue and a pigmented layer comprised of epithelial cells lying in contact with the choroid. Also called the ***nervous tunic*** (TOO-nik).

Retraction (rē-TRAK-shun) The movement of a protracted part of the body backward on a plane parallel to the ground, as in pulling the lower jaw back in line with the upper jaw.

Retroflexion (re-trō-FLEK-shun) A malposition of the uterus in which it is tilted posteriorly.

Retrograde degeneration (RE-trō-grād dē-jen-er-Ā-shun) Changes that occur in the proximal portion of a damaged axon only as far as the first neurofibral node (node of Ranvier); similar to changes that occur during Wallerian degeneration.

Retroperitoneal (re'-trō-per-i-tō-NĒ-al) External to the peritoneal lining of the abdominal cavity.

Rheumatism (ROO-ma-tizm') Any painful state of the supporting structures of the body—bones, ligaments, joints, tendons, or muscles.

Rh factor An inherited agglutinogen (antigen) on the surface of red blood cells.

Rhinology (rī-NOL-ō-jē) The study of the nose and its disorders.

Rhinoplasty (RĪ-nō-plas'-tē) Surgical procedure in which the structure of the external nose is altered.

Ribonucleic (rī'-bō-nyoo-KLĒ-ik) ***acid (RNA)*** A single-stranded nucleic acid constructed of nucleotides consisting of one of four possible nitrogenous bases (adenine, cytosine, guanine, or uracil), ribose, and a phosphate group; three types are messenger RNA (mRNA), transfer RNA (tRNA), and ribosomal RNA (rRNA), each of which cooperates with DNA for protein synthesis.

Ribosome (RĪ-bō-sōm) An organelle in the cytoplasm of cells, composed of ribosomal RNA and ribosomal proteins, that synthesizes proteins; nicknamed the "protein factory."

Rickets (RIK-ets) Condition affecting children characterized by soft and deformed bones resulting from inadequate calcium metabolism due to a vitamin D deficiency.

Right lymphatic (lim-FAT-ik) ***duct*** A vessel of the lymphatic system that drains lymph from the upper right side of the body and empties it into the right subclavian vein.

Rigidity (ri-JID-i-tē) Hypertonia characterized by increased muscle tone, but reflexes are not affected.

Rigor mortis State of partial contraction of muscles following death due to lack of ATP that causes cross bridges of thick myofilaments to remain attached to thin myofilaments, thus preventing relaxation.

Rima glottidis (RĪ-ma GLOT-ti-dis) Space between the vocal folds (true vocal cords).

Rod A visual receptor in the retina of the eye that is specialized for vision in dim light.

Roentgen (RENT-gen) The international unit of radiation; a standard quantity of x or gamma radiation.

Roentgenogram (RENT-gen-ō-gram) A photographic image produced by x-rays.

Root canal A narrow extension of the pulp cavity lying within the root of a tooth.

Root of penis Attached portion of penis that consists of the bulb and crura.

Rotation (rō-TĀ-shun) Moving a bone around its own axis, with no other movement.

Round ligament (LIG-a-ment) A band of fibrous connective tissue enclosed between the folds of the broad ligament of the uterus, emerging from a point on the uterus just below the uterine (Fallopian) tube, extending laterally along the pelvic wall, and penetrating the abdominal wall through the deep inguinal ring to end in the labia majora.

Round window A small opening between the middle and inner ear, directly below the oval window, covered by the secondary tympanic membrane. Also called the ***fenestra cochlea*** (fe-NES-tra KŌK-lē-a).

Rugae (ROO-jē) Large folds in the mucosa of an empty hollow organ, such as the stomach and vagina.

Saccule (SAK-yool) The lower and smaller of the two chambers in the membranous labyrinth inside the vestibule of the inner ear containing a receptor organ for static equilibrium.

Sacral hiatus (hi-Ā-tus) Inferior entrance to the vertebral canal formed when the laminae of the fifth sacral vertebra (and sometimes fourth) fail to meet.

Sacral plexus (PLEK-sus) A network formed by the anterior branches of spinal nerves L4 through S3.

Sacral promontory (PROM-on-tor'-ē) The superior surface of the body of the first sacral vertebra that projects anteriorly into the pelvic cavity; a line from the sacral promontory to the superior border of the pubic symphysis divides the abdominal and pelvic cavities.

Saddle joint A synovial joint in which the articular surface of one bone is saddle-shaped and the articular surface of the other bone is shaped like a rider sitting in the saddle, as in the joint between the trapezium and the metacarpal of the thumb. Also called a ***sellaris*** (sel-LA-ris) ***joint.***

Sagittal (SAJ-i-tal) ***plane*** A vertical plane that divides the body or organs into left and right portions. Such a plane may be ***midsagittal (median),*** in which the divisions are equal, or ***parasagittal,*** in which the divisions are unequal.

Saliva (sa-LĪ-va) A clear, alkaline, somewhat viscous secretion produced by the three pairs of salivary glands; contains various salts, mucin, lysozyme, and salivary amylase.

Salivary amylase (SAL-i-ver-ē AM-i-lās) An enzyme in saliva that initiates the chemical breakdown of starch, mostly in the mouth.

Salivary gland One of three pairs of glands that lie outside the mouth and pour their secretory product (called saliva) into ducts that empty into the oral cavity; the parotid, submandibular, and sublingual glands.

Salpingitis (sal'-pin-JĪ-tis) Inflammation of the uterine (Fallopian) or auditory (Eustachian) tube.

Sarcolemma (sar'-kō-LEM-ma) The cell membrane of a muscle fiber (cell), especially of a skeletal muscle fiber.

Sarcoma (sar-KŌ-ma) A connective tissue tumor, often highly malignant.

Sarcomere (SAR-kō-mēr) A contractile unit in a striated muscle fiber (cell) extending from one Z disc to the next Z disc.

Sarcoplasm (SAR-kō-plazm) The cytoplasm of a muscle fiber (cell).

Sarcoplasmic reticulum (sar'-kō-PLAZ-mik re-TIK-yoo-lum) A network of saccules and tubes surrounding myofibrils of a muscle fiber (cell), comparable to endoplasmic reticulum; functions to reabsorb calcium ions during relaxation and to release them to cause contraction.

Satiety (sa-TĪ-e-tē) Fullness or gratification, as of hunger or thirst.

Satiety center A collection of nerve cells located in the ventromedial nuclei of the hypothalamus that, when stimulated, brings about the cessation of eating.

Saturated fat A fat that contains no double bonds between any of its carbon atoms; all are single bonds and all carbon atoms are bonded to the maximum number of hydrogen atoms; found naturally in animal foods such as meat, milk, milk products, and eggs.

Scala tympani (SKA-la TIM-pan-ē) The lower spiral-shaped channel of the bony cochlea, filled with perilymph.

Scala vestibuli (ves-TIB-yoo-lē) The upper spiral-shaped channel of the bony cochlea, filled with perilymph.

Schwann cell *See* ***Neurolemmocyte.***

Sciatica (sī-AT-i-ka) Inflammation and pain along the sciatic nerve; felt at the back of the thigh running down the inside of the leg.

Sclera (SKLE-ra) The white coat of fibrous tissue that forms the outer protective covering over the eyeball except in the most anterior portion; the posterior portion of the fibrous tunic.

Scleral venous sinus A circular venous sinus located at the junction of the sclera and the cornea through which aqueous humor drains from the anterior chamber of the eyball into the blood. Also called the ***canal of Schlemm*** (SHLEM).

Sclerosis (skle-RŌ-sis) A hardening with loss of elasticity of tissues.

Scoliosis (skō'-lē-Ō-sis) An abnormal lateral curvature from the normal vertical line of the backbone.

Scotoma (skō-TŌ-ma) An area of depressed or lost vision within the visual field.

Scrotum (SKRŌ-tum) A skin-covered pouch that contains the testes and their accessory structures.

Sebaceous (se-BĀ-shus) Secreting oil.

Sebaceous (se-BĀ-shus) ***gland*** An exocrine gland in the dermis of the skin, almost always associated with a hair follicle, that secretes sebum. Also called an ***oil gland.***

Sebum (SĒ-bum) Secretion of sebaceous (oil) glands.

Secondary sex characteristic A feature characteristic of the male or female body that develops at puberty under the stimulation of sex hormones but is not directly involved in sexual reproduction, such as distribution of body hair, voice pitch, body shape, and muscle development.

Secretion (se-KRĒ-shun) Production and release from a gland cell of a fluid, especially a functionally useful product as opposed to a waste product.

Selectively permeable membrane A membrane that permits the passage of certain substances, but restricts the passage of others.

Sella turcica (SEL-a TUR-si-ka) A depression on the superior surface of the sphenoid bone that houses the pituitary gland.

Semen (SĒ-men) A fluid discharged at ejaculation by a male that consists of a mixture of spermatozoa and the secretions of the seminal vesicles, prostate gland, and bulbourethral (Cowper's) glands. Also called ***seminal*** (SEM-i-nal) ***fluid.***

Semicircular canals Three bony channels (anterior, posterior, lateral), filled with perilymph, in which lie the membranous semicircular canals filled with endolymph. They contain receptors for equilibrium.

Semicircular ducts The membranous semicircular canals filled with endolymph and floating in the perilymph of the bony semicircular canals. They contain cristae that are concerned with dynamic equilibrium.

Semilunar (sem'-ē-LOO-nar) ***valve*** A valve guarding the entrance into the aorta or the pulmonary trunk from a ventricle of the heart.

Seminal vesicle (SEM-i-nal VES-i-kul) One of a pair of convoluted, pouchlike structures, lying posterior and inferior to the urinary bladder and anterior to the rectum, that secrete a component of semen into the ejaculatory ducts.

Seminiferous tubule (sem'-i-NI-fer-us TOO-byool) A tightly coiled duct, located in a lobule of the testis, where spermatozoa are produced.

Senescence (se-NES-ens) The process of growing old; the period of old age.

Senile macular (MAK-yoo-lar) ***degeneration (SMD)*** A disease in which blood vessels grow over the macula lutea.

Senility (se-NIL-i-tē) A loss of mental or physical ability due to old age.

Sensation A state of awareness of external or internal conditions of the body.

Sensory area A region of the cerebral cortex concerned with the interpretation of sensory impulses.

Sepsis (SEP-sis) A morbid condition that results from the presence in the blood or other body tissues of pathogenic bacteria and their products.

Septal defect An opening in the septum (interatrial or interventricular) between the left and right sides of the heart.

Septicemia (sep'-ti-SĒ-mē-a) Toxins or disease-causing bacteria in blood. Also called ***"blood poisoning."***

Septum (SEP-tum) A wall dividing two cavities.

Serosa (ser-Ō-sa) Any serous membrane. The outermost layer of an organ formed by a serous membrane. The membrane that lines the pleural, pericardial, and peritoneal cavities.

Serous (SIR-us) ***membrane*** A membrane that lines a body cavity that does not open to the exterior. Also called the ***serosa*** (se-RŌ-sa).

Serum Plasma minus its clotting proteins.

Sesamoid (SES-a-moyd) ***bones*** Small bones usually found in tendons.

Sex chromosomes The twenty-third pair of chromosomes, designated X and Y, which determine the genetic sex of an individual; in males, the pair is XY; in females, XX.

Sexual intercourse The insertion of the erect penis of a male into the vagina of a female. Also called ***coitus*** (KŌ-i-tus) or ***copulation.***

Sexually transmitted disease (STD) General term for any of a large number of diseases spread by sexual contact. Also called a ***venereal disease (VD).***

Sheath of Schwann *See* ***Neurolemma.***

Shingles Acute infection of the peripheral nervous system caused by a virus.

Shinsplints Soreness or pain along the tibia probably caused by inflammation of the periosteum brought on by repeated tugging of the muscles and tendons attached to the periosteum. Also called ***tibia stress syndrome.***

Shivering Involuntary contraction of a muscle that generates heat.

Shock Failure of the cardiovascular system to deliver adequate amounts of oxygen and nutrients to meet the metabolic needs of the body due to inadequate cardiac output. It is characterized by hypotension; clammy, cool, and pale skin; sweating; reduced urine formation; altered mental state; acidosis; tachycardia; weak, rapid pulse; and thirst. Types include hypovolemic, cardiogenic, obstructive, neurogenic, and septic.

Shoulder A synovial or diarthrotic joint where the humerus joins the scapula.

Sigmoid colon (SIG-moyd KŌ-lon) The S-shaped portion of the large intestine that begins at the level of the left iliac crest, projects inward to the midline, and terminates at the rectum at about the level of the third sacral vertebra.

Sigmoidoscopy (sig'-moy-DOS-kō-pē) Visualization of the anal canal, rectum, and colon to screen for colorectal cancer, collect biopsy samples, remove polyps, gather specimens for culture, and photograph the intestinal mucosa.

Sign Any objective evidence of disease that can be observed or measured such as a lesion, swelling, or fever.

Sinoatrial (si-nō-Ā-trē-al) ***(SA) node*** A compact mass of cardiac muscle fibers (cells) specialized for conduction, located in the right atrium beneath the opening of the superior vena cava. Also called the ***sinuatrial node*** or ***pacemaker.***

Sinus (SĪ-nus) A hollow in a bone (paranasal sinus) or other tissue; a channel for blood (vascular sinus); any cavity having a narrow opening.

Sinusitis (sīn-yoo-SĪT-is) Inflammation of the mucous membrane of a paranasal sinus.

Sinusoid (SĪN-yoo-soyd) A microscopic space or passage for blood in certain organs such as the liver or spleen.

Skeletal muscle An organ specialized for contraction, composed of striated muscle fibers (cells), supported by connective tissue, attached to a bone by a tendon or an aponeurosis, and stimulated by somatic efferent neurons.

Skene's gland *See* ***Paraurethral gland.***

Skull The skeleton of the head consisting of the cranial and facial bones.

Sliding-filament mechanism The most commonly accepted explanation for muscle contraction in which actin and myosin myofilaments move into interdigitation with each other, decreasing the length of the sarcomeres.

Small intestine A long tube of the gastrointestinal tract that begins at the pyloric sphincter of the stomach, coils through the central and lower part of the abdominal cavity, and ends at the large intestine; divided into three segments: duodenum, jejunum, and ileum.

Smooth muscle An organ specialized for contraction, composed of smooth muscle fibers (cells), located in the walls of hollow internal structures, and innervated by a visceral efferent neuron.

Soft palate (PAL-at) The posterior portion of the roof of the mouth, extending posteriorly from the palatine bones and ending at the uvula. It is a muscular partition lined with mucous membrane.

Somatic cell division Type of cell division in which a single starting cell (parent cell) duplicates itself to produce two identical cells (daughter cells); consists of mitosis and cytokinesis.

Somatic (sō-MAT-ik) ***nervous system (SNS)*** The portion of the peripheral nervous system made up of the somatic efferent fibers that run between the central nervous system and the skeletal muscles and skin.

Somesthetic (sō'-mes-THET-ik) Pertaining to sensations and sensory structures of the body.

Somite (SŌ-mīt) Block of mesodermal cells in a developing embryo that is distinguished into a myotome (which forms most of the skeletal muscles), dermatome (which forms connective tissues), and sclerotome (which forms the vertebrae).

Spasm (spazm) A sudden, involuntary contraction of large groups of muscles.

Spastic (SPAS-tik) An increase in muscle tone (stiffness) associated with an increase in tendon reflexes and abnormal reflexes (Babinski sign).

Spasticity (spas-TIS-i-tē) Hypertonia characterized by increased muscle tone, increased tendon reflexes, and pathological reflexes (Babinski sign).

S period Period of interphase during which chromosomes are replicated preceded by a G_1 period and followed by a G_2 period, when cells grow, metabolize, and produce substances required for division.

Spermatic (sper-MAT-ik) ***cord*** A supporting structure of the male reproductive system, extending from a testis to the deep inguinal ring, that includes the ductus (vas) deferens, arteries, veins, lymphatic vessels, nerves, cremaster muscle, and connective tissue.

Spermatogenesis (sper'-ma-tō-JEN-e-sis) The formation and development of spermatozoa in the seminiferous tubules of the testes.

Spermatozoon (sper'-ma-tō-ZŌ-on) A mature sperm cell.

Spermicide (SPER-mi-sīd') An agent that kills spermatozoa.

Spermiogenesis (sper'-mē-ō-JEN-e-sis) The maturation of spermatids into spermatozoa.

Sphincter (SFINGK-ter) A circular muscle constricting an orifice.

Sphincter of Oddi *See* ***Sphincter of the hepatopancreatic ampulla.***

Sphincter of the hepatopancreatic ampulla A circular muscle at the opening of the common bile and main pancreatic ducts in the duodenum. Also called the ***sphincter of Oddi*** (OD-ē).

Sphygmomanometer (sfig'-mō-ma-NOM-e-ter) An instrument for measuring arterial blood pressure.

Spina bifida (SPĪ-na BIF-i-da) A congenital defect of the vertebral column in which the halves of the neural arch of a vertebra fail to fuse in the midline.

Spinal (SPĪ-nal) ***cord*** A mass of nerve tissue located in the vertebral canal from which 31 pairs of spinal nerves originate.

Spinal nerve One of the 31 pairs of nerves that originate on the spinal cord from posterior and anterior roots.

Spinal shock A period of time, from several days to several weeks, following transection of the spinal cord and characterized by the abolition of all reflex activity.

Spinal (lumbar) tap (puncture) Withdrawal of some of the cerebrospinal fluid from the subarachnoid space in the lumbar region for diagnostic purposes, introduction of various substances, and evaluation of the effects of treatment.

Spinous (SPĪ-nus) ***process*** A sharp or thornlike process or projection. Also called a ***spine.*** A sharp ridge running diagonally across the posterior surface of the scapula.

Spiral organ The organ of hearing, consisting of supporting cells and hair cells that rest on the basilar membrane and extend into the endolymph of the cochlear duct. Also called the ***organ of Corti*** (KOR-tē).

Spirometer (spī-ROM-e-ter) An apparatus used to measure air capacity of the lungs.

Splanchnic (SPLANK-nik) Pertaining to the viscera.

Spleen (SPLĒN) Large mass of lymphatic tissue between the fundus of the stomach and the diaphragm that functions in phagocytosis, production of lymphocytes, and blood storage.

Sprain Forcible wrenching or twisting of a joint with partial rupture or other injury to its attachments without dislocation.

Sputum (SPYOO-tum) Substance ejected from the mouth containing saliva and mucus.

Squamous (SKWĀ-mus) Scalelike.

Starvation (star-VĀ-shun) The loss of energy stores in the form of glycogen, fats, and proteins due to inadequate intake of nutrients or inability to digest, absorb, or metabolize ingested nutrients.

Stasis (STĀ-sis) Stagnation or halt of normal flow of fluids, as blood, urine, or of the intestinal mechanism.

Static equilibrium (ē-kwi-LIB-rē-um) The maintenance of posture in response to changes in the orientation of the body, mainly the head, relative to the ground.

Stellate reticuloendothelial (STEL-āte re-tik'-yoo-lō-en'-dō-THĒ-lē-al) ***cell*** Phagocytic cell that lines a sinusoid of the liver. Also called a ***Kupffer's*** (KOOP-ferz) ***cell.***

Stenosis (sten-Ō-sis) An abnormal narrowing or constriction of a duct or opening.

Stereocilia (ste'-rē-ō-SIL-ē-a) Groups of extremely long, slender, nonmotile microvilli projecting from epithelial cells lining the epididymis.

Sterile (STE-ril) Free from any living microorganisms. Unable to conceive or produce offspring.

Sterilization (ster'-i-li-ZĀ-shun) Elimination of all living microorganisms. The rendering of an individual incapable of reproduction (e.g., castration, vasectomy, hysterectomy).

Sternal puncture Introduction of a wide-bore needle into the marrow cavity of the sternum for aspiration of a sample of red bone marrow.

Stimulus Any change in the environment capable of altering the membrane potential.

Stomach The J-shaped enlargement of the gastrointestinal tract directly under the diaphragm in the epigastric, umbilical, and left hypochondriac regions of the abdomen, between the esophagus and small intestine.

Strabismus (stra-BIZ-mus) A condition in which the visual axes of the two eyes differ, so that they do not fix on the same object.

Straight tubule (TOO-byool) A duct in a testis leading from a convoluted seminiferous tubule to the rete testis.

Stratum (STRĀ-tum) A layer.

Stratum basalis (STRĀ-tum ba-SAL-is) The outer layer of the endometrium, next to the myometrium, that is maintained during menstruation and gestation and produces a new functionalis following menstruation or parturition.

Stratum functionalis (funk'-shun-AL-is) The inner layer of the endometrium, the layer next to the uterine cavity, that is shed during menstruation and that forms the maternal portion of the placenta during gestation.

Stricture (STRIK-cher) A local constriction of a tubular structure.

Stroma (STRŌ-ma) The tissue that forms the ground substance, foundation, or framework of an organ, as opposed to its functional parts.

Stupor (STOO-por) Unresponsiveness from which a patient can be aroused only briefly and by vigorous and repeated stimulation.

Subarachnoid (sub'-a-RAK-noyd) ***space*** A space between the arachnoid and the pia mater that surrounds the brain and spinal cord and through which cerebrospinal fluid circulates.

Subcutaneous (sub'-kyoo-TĀ-nē-us) Beneath the skin. Also called ***hypodermic*** (hī-pō-DER-mik).

Subcutaneous layer A continuous sheet of areolar connective tissue and adipose tissue between the dermis of the skin and the deep fascia of the muscles. Also called the ***superficial fascia*** (FASH-ē-a).

Subdural (sub-DOO-ral) ***space*** A space between the dura mater and the arachnoid of the brain and spinal cord that contains a small amount of fluid.

Sublingual (sub-LING-gwal) ***gland*** One of a pair of salivary glands situated in the floor of the mouth under the mucous membrane and to the side of the lingual frenulum, with a duct (Rivinus's) that opens into the floor of the mouth.

Submandibular (sub'-man-DIB-yoo-lar) ***gland*** One of a pair of salivary glands found beneath the base of the tongue under the mucous membrane in the posterior part of the floor of the mouth, posterior to the sublingual glands, with a duct (Wharton's) situated to the side of the lingual frenulum. Also called the ***submaxillary*** (sub'-MAK-si-ler-ē) ***gland.***

Submucosa (sub-myoo-KŌ-sa) A layer of connective tissue located beneath a mucous membrane, as in the gastrointestinal tract or the urinary bladder; the submucosa connects the mucosa to the muscularis layer.

Submucosal plexus A network of autonomic nerve fibers located in the outer portion of the submucous layer of the small intestine. Also called the ***plexus of Meissner*** (MĪS-ner).

Subserous fascia (sub-SE-rus FASH-ē-a) A layer of connective tissue internal to the deep fascia, lying between the deep fascia and the serous membrane that lines the body cavities.

Subthalamus (sub-THAL-a-mus) Part of the diencephalon inferior to the thalamus; the substantia nigra and red nucleus extend from the midbrain into the subthalamus.

Sudoriferous (soo'-dor-IF-er-us) ***gland*** An apocrine or eccrine exocrine gland in the dermis or subcutaneous layer that produces perspiration. Also called a ***sweat gland.***

Sulcus (SUL-kus) A groove or depression between parts, especially between the convolutions of the brain. *Plural,* ***sulci*** (SUL-sē).

Superficial (soo'-per-FISH-al) Located on or near the surface of the body.

Superficial fascia (FASH-ē-a) A continuous sheet of fibrous connective tissue between the dermis of the skin and the deep fascia of the muscles. Also called ***subcutaneous*** (sub'-kyoo-TĀ-nē-us) ***layer.***

Superficial inguinal (IN-gwi-nal) ***ring*** A triangular opening in the aponeurosis of the external oblique muscle that represents the termination of the inguinal canal.

Superior (soo-PĒR-ē-or) Toward the head or upper part of a structure. Also called ***cephalad*** (SEF-a-lad) or ***craniad.***

Superior vena cava (VĒ-na CĀ-va) ***(SVC)*** Large vein that collects blood from parts of the body superior to the heart and returns it to the right atrium.

Supination (soo-pī-NĀ-shun) A movement of the forearm in which the palm of the hand is turned anteriorly or superiorly.

Suppuration (sup'-yoo-RĀ-shun) Pus formation and discharge.

Surface anatomy The study of the structures that can be identified from the outside of the body.

Surfactant (sur-FAK-tant) A phospholipid substance produced by the lungs that decreases surface tension.

Susceptibility (sus-sep'-ti-BIL-i-tē) Lack of resistance of a body to the deleterious or other effects of an agent such as pathogenic microorganisms.

Suspensory ligament (sus-PEN-so-rē LIG-a-ment) A fold of peritoneum extending laterally from the surface of the ovary to the pelvic wall.

Sustentacular (sus'-ten-TAK-yoo-lar) ***cell*** A supporting cell of seminiferous tubules that produces secretions for supplying nutrients to spermatozoa and the hormone inhibin. Also called a ***Sertoli*** (ser-TŌ-lē) ***cell.***

Sutural (SOO-cher-al) ***bone*** A small bone located within a suture between certain cranial bones.

Suture (SOO-cher) An immovable fibrous joint in the skull where bone surfaces are closely united.

Sympathetic (sim'-pa-THET-ik) ***division*** One of the two subdivisions of the autonomic nervous system, having cell bodies of preganglionic neurons in the lateral gray columns of the thoracic segment and first two or three lumbar segments of the spinal cord; primarily concerned with processes involving the expenditure of energy. Also called the ***thoracolumbar*** (thō'-ra-kō-LUM-bar) ***division.***

Sympathetic trunk ganglion (GANG-glē-on) A cluster of cell bodies of postganglionic sympathetic neurons lateral to the vertebral column, close to the body of a vertebra. These ganglia extend downward through the neck, thorax, and abdomen to the coccyx on both sides of the vertebral column and are connected to one another to form a chain on each side of the vertebral column. Also called ***lateral,*** or ***sympathetic, chain*** or ***vertebral chain ganglia.***

Symphysis (SIM-fi-sis) A line of union. A slightly movable cartilaginous joint such as the pubic symphysis between the anterior surfaces of the hipbones.

Symptom (SIMP-tum) A subjective change in body function not apparent to an observer, such as fever or nausea, that indicates the presence of a disease or disorder of the body.

Synapse (SIN-aps) The junction between the process of two adjacent neurons; the place where the activity of one neuron affects the activity of another; may be electrical or chemical.

Synapsis (sin-AP-sis) The pairing of homologous chromosomes during prophase I of meiosis.

Synaptic (sin-AP-tik) ***cleft*** The narrow gap that separates the axon terminal of one nerve cell from another nerve cell or muscle fiber (cell) and across which a neurotransmitter diffuses to affect the postsynaptic cell.

Synaptic end bulb Expanded distal end of an axon terminal that contains synaptic vesicles. Also called a ***synaptic knob*** or ***end foot.***

Synaptic gutter Invaginated portion of a sarcolemma under an axon terminal. Also called a ***synaptic trough*** (TROF).

Synaptic vesicle Membrane-enclosed sac in a synaptic end bulb that stores neurotransmitters.

Synarthrosis (sin'-ar-THRŌ-sis) An immovable joint.

Synchondrosis (sin'-kon-DRŌ-sis) A cartilaginous joint in which the connecting material is hyaline cartilage.

Syncope (SIN-kō-pē) Faint; a sudden temporary loss of consciousness associated with loss of postural tone and followed by spontaneous recovery; most commonly caused by cerebral ischemia.

Syndesmosis (sin'-dez-MŌ-sis) A fibrous joint in which articulating bones are united by dense fibrous tissue.

Syndrome (SIN-drōm) A group of signs and symptoms that occur together in a pattern that is characteristic of a particular disease or abnormal condition.

Syneresis (si-NER-e-sis) The process of clot retraction.

Synergist (SIN-er-jist) A muscle that assists the prime mover by reducing undesired action or unnecessary movement.

Synostosis (sin'-os-TŌ-sis) A joint in which the dense fibrous connective tissue that unites bones at a suture has been replaced by bone, resulting in a complete fusion across the suture line.

Synovial (si-NŌ-vē-al) ***cavity*** The space between the articulating bones of a synovial (diarthrotic) joint, filled with synovial fluid. Also called a ***joint cavity.***

Synovial fluid Secretion of synovial membranes that lubricates joints and nourishes articular cartilage.

Synovial joint A fully movable or diarthrotic joint in which a synovial (joint) cavity is present between the two articulating bones.

Synovial membrane The inner of the two layers of the articular capsule of a synovial joint, composed of areolar connective tissue that secretes synovial fluid into the synovial (joint) cavity.

Syphilis (SIF-i-lis) A sexually transmitted disease caused by the bacterium *Treponema pallidum.*

System An association of organs that have a common function.

Systemic (sis-TEM-ik) Affecting the whole body; generalized.

Systemic anatomy The study of particular systems of the body, such as the skeletal, muscular, nervous, cardiovascular, or urinary systems.

Systemic circulation The routes through which oxygenated blood flows from the left ventricle through the aorta to all the organs of the body and deoxygenated blood returns to the right atrium.

Systemic lupus erythematosus (er-i-them-a-TŌ-sus) ***(SLE)*** An autoimmune, inflammatory disease that may affect every tissue of the body.

Systole (SIS-tō-lē) In the cardiac cycle, the phase of contraction of the heart muscle, especially of the ventricles.

Systolic (sis-TO-lik) ***blood pressure*** The force exerted by blood on arterial walls during ventricular contraction; the highest pressure measured in the large arteries, about 120 mm Hg under normal conditions for a young, adult male.

Tachycardia (tak'-i-KAR-dē-a) A rapid heartbeat or pulse rate.

Tactile (TAK-tīl) Pertaining to the sense of touch.

Tactile disc Modified epidermal cell in the stratum basale of hairless skin that functions as a cutaneous receptor for discriminative touch. Also called a ***Merkel's*** (MER-kelz) ***disc.***

Taenia coli (TĒ-nē-a KŌ-lī) One of three flat bands of thickened, longitudinal muscles running the length of the large intestine.

Target cell A cell whose activity is affected by a particular hormone.

Tarsal gland Sebaceous (oil) gland that opens on the edge of each eyelid. Also called a ***Meibomian*** (mī-BŌ-mē-an) ***gland.***

Tarsal plate A thin, elongated sheet of connective tissue, one in each eyelid, giving the eyelid form and support. The aponeurosis of the levator palpebrae superioris is attached to the tarsal plate of the superior eyelid.

Tarsus (TAR-sus) A collective term for the seven bones of the ankle.

Tay-Sachs (TĀ SAKS) ***disease*** Inherited, progressive neuronal degeneration of the central nervous system due to a deficient lysosomal enzyme that causes excessive accumulations of a lipid called ganglioside.

T cell A lymphocyte that can differentiate into one of six kinds of cells—killer, helper, suppressor, memory, amplifier, or delayed hypersensitivity—all of which function in cellular immunity.

Tectorial (tek-TŌ-rē-al) ***membrane*** A gelatinous membrane projecting over and in contact with the hair cells of the spiral organ (organ of Corti) in the cochlear duct.

Telophase (TEL-ō-fāz) The final stage of mitosis in which the daughter nuclei become established.

Temporomandibular joint (TMJ) syndrome A disorder of the temporomandibular joint (TMJ) characterized by dull pain around the ear, tenderness of jaw muscles, a clicking or popping noise when opening or closing the mouth, limited or abnormal opening of the mouth, headache, tooth sensitivity, and abnormal wearing of the teeth.

Tendon (TEN-don) A white fibrous cord of dense, regularly arranged connective tissue that attaches muscle to bone.

Tendon organ A proprioceptive receptor, sensitive to changes in muscle tension and force of contraction, found chiefly near the junction of tendons and muscles. Also called a ***Golgi*** (GOL-jē) ***tendon organ.***

Tenosynovitis (ten'-ō-sin-ō-VĪ-tis) Inflammation of a tendon sheath and synovial membrane at a joint.

Tentorium cerebelli (ten-TŌ-rē-um ser'-e-BEL-ē) A transverse shelf of dura mater that forms a partition between the occipital lobe of the cerebral hemispheres and the cerebellum and that covers the cerebellum.

Teratogen (TER-a-tō-jen) Any agent or factor that causes physical defects in a developing embryo.

Terminal ganglion (TER-min-al GANG-lē-on) A cluster of cell bodies of postganglionic parasympathetic neurons either lying very close to the visceral effectors or located within the walls of the visceral effectors supplied by the postganglionic fibers.

Testis (TES-tis) Male gonad that produces sperm and the hormones testosterone and inhibin. Also called a ***testicle.***

Testosterone (tes-TOS-te-rōn) A male sex hormone (androgen) secreted by interstitial endocrinocytes (cells of Leydig) of a mature testis; controls the growth and development of male sex organs, secondary sex characteristics, spermatozoa, and body growth.

Tetanus (TET-a-nus) An infectious disease caused by the toxin of *Clostridium tetani,* characterized by tonic muscle spasms and exaggerated reflexes, lockjaw, and arching of the back. A smooth, sustained contraction produced by a series of very rapid stimuli to a muscle.

Tetany (TET-a-nē) A nervous condition caused by hypoparathyroidism and characterized by intermittent or continuous tonic muscular contractions of the extremities.

Tetralogy of Fallot (tet-RAL-ō-jē of fal-Ō) A combination of four congenital heart defects: (1) constricted pulmonary semilunar valve, (2) interventricular septal opening, (3) emergence of aorta from both ventricles instead of from the left only, and (4) enlarged right ventricle.

Thalamus (THAL-a-mus) A large, oval structure located above the midbrain, consisting of two masses of gray matter covered by a thin layer of white matter.

Thalassemia (thal'-a-SĒ-mē-a) A group of hereditary hemolytic anemias.

Thallium (THAL-ē-um) ***imaging*** Diagnostic procedure used to evaluate blood flow through coronary arteries, cardiac disorders, and effectiveness of drug therapy; thallium concentrates in healthy myocardial tissue.

Therapy (THER-a-pē) The treatment of a disease or disorder.

Thermoreceptor (THER-mō-rē-sep-tor) Receptor that detects changes in temperature.

Thigh The portion of the lower extremity between the hip and the knee.

Third ventricle (VEN-tri-kul) A slitlike cavity between the right and left halves of the thalamus and between the lateral ventricles.

Thoracic (thō-RAS-ik) ***cavity*** Superior component of the ventral body cavity that contains two pleural cavities, the mediastinum, and the pericardial cavity.

Thoracic duct A lymphatic vessel that begins as a dilation called the cisterna chyli, receives lymph from the left side of the head, neck, and chest, the left arm, and the entire body below the ribs, and empties into the left subclavian vein. Also called the ***left lymphatic*** (lim-FAT-ik) ***duct.***

Thoracolumbar (thō'-ra-kō-LUM-bar) ***outflow*** The fibers of the sympathetic preganglionic neurons, which have their cell bodies in the lateral gray columns of the thoracic segment and first two or three lumbar segments of the spinal cord.

Thorax (THŌ-raks) The chest.

Thrombocyte (THROM-bō-sīt) A fragment of cytoplasm enclosed in a cell membrane and lacking a nucleus; found in the circulating blood; plays a role in blood clotting. Also called a ***platelet*** (PLĀT-let).

Thrombophlebitis (throm'-bo-fle-BĪ-tis) A disorder in which inflammation of the wall of a vein is followed by the formation of a blood clot (thrombus).

Thrombosis (throm-BŌ-sis) The formation of a clot in an unbroken blood vessel, usually a vein.

Thrombus A clot formed in an unbroken blood vessel, usually a vein.

Thymectomy (thī-MEK-tō-mē) Surgical removal of the thymus gland.

Thymus (THĪ-mus) ***gland*** A bilobed organ, located in the upper mediastinum posterior to the sternum and between the lungs, that plays a role in the immune mechanism of the body.

Thyroid cartilage (THI-royd KAR-ti-lij) The largest single cartilage of the larynx, consisting of two fused plates that form the anterior wall of the larynx. Also called the ***Adam's apple.***

Thyroid follicle (FOL-i-kul) Spherical sac that forms the parenchyma of the thyroid gland and consists of follicular cells that produce thyroxine (T_4) and triiodothyronine (T_3) and parafollicular cells that produce calcitonin (CT).

Thyroid function tests Tests used to evaluate a swelling or lump in the thyroid gland, ascertain symptoms of abnormal thyroxine levels, monitor responses of thyroid diseases to therapy, and screen newborns for cretinism. Examples are the radioiodine uptake (RAIU), serum T_4 concentration, and serum T_3 concentration tests.

Thyroid gland An endocrine gland with right and left lateral lobes on either side of the trachea connected by an isthmus located in front of the trachea just below the cricoid cartilage.

Thyroid-stimulating hormone (TSH) A hormone secreted by the adenohypophysis (anterior lobe) of the pituitary gland that stimulates the synthesis and secretion of hormones produced by the thyroid gland.

Thyroxine (thī-ROK-sēn) ***(T_4)*** A hormone secreted by the thyroid gland that regulates organic metabolism, growth and development, and the activity of the nervous system.

Tic Spasmodic twitching made involuntarily by muscles that are ordinarily under voluntary control.

Tinnitus (ti-NĪ-tus) A ringing, roaring, or clicking in the ears.

Tissue A group of similar cells and their intercellular substance joined together to perform a specific function.

Tissue factor (TF) A factor, or collection of factors, whose appearance initiates the blood clotting process. Also called ***thromboplastin*** (throm-bō-PLAS-tin).

Tissue plasminogen activator (t-PA) An enzyme that dissolves small blood clots by initiating a process that converts plasminogen to plasmin, which degrades the fibrin of a clot.

Tissue rejection Phenomenon by which the body recognizes the protein (HLA antigens) in transplanted tissues or organs as foreign and produces antibodies against them.

Tongue A large skeletal muscle covered by a mucous membrane located on the floor of the oral cavity.

Tonsil (TON-sil) A multiple aggregation of large lymphatic nodules embedded in mucous membrane.

Topical (TOP-i-kal) Applied to the surface rather than ingested or injected.

Torn cartilage A tearing of an articular disk in the knee.

Torpor (TOR-por) State of lethargy and sluggishness that precedes stupor, which precedes semicoma, which precedes coma.

Toxic (TOK-sik) Pertaining to poison; poisonous.

Toxic shock syndrome (TSS) A disease caused by the bacterium *Staphylococcus aureus,* occurring among menstruating females who use tampons and characterized by high fever, sore throat, headache, fatigue, irritability, and abdominal pain.

Trabecula (tra-BEK-yoo-la) Irregular latticework of thin plate of spongy bone. Fibrous cord of connective tissue serving as supporting fiber by forming a septum extending into an organ from its wall or capsule. *Plural,* ***trabeculae*** (tra-BEK-yoo-lē).

Trabeculae carneae (tra-BEK-yoo-lē KAR-nē-ē) Ridges and folds of the myocardium in the ventricles.

Trachea (TRĀ-kē-a) Tubular air passageway extending from the larynx to the fifth thoracic vertebra. Also called the ***windpipe.***

Tracheostomy (trā-kē-OS-tō-mē) Creation of an opening into the trachea through the neck (below the cricoid cartilage), with insertion of a tube to facilitate passage of air or evacuation of secretions.

Trachoma (tra-KŌ-ma) A chronic infectious disease of the conjunctiva and cornea of the eye caused by *Chlamydia trachomatis.*

Tract A bundle of nerve fibers in the central nervous system.

Transfusion (trans-FYOO-shun) Transfer of whole blood, blood components, or bone marrow directly into the bloodstream.

Transient ischemic (is-KĒ-mik) ***attack (TIA)*** Episode of temporary focal, nonconvulsive cerebral dysfunction caused by interference of the blood supply to the brain.

Transplantation (trans-plan-TĀ-shun) The replacement of injured or diseased tissues or organs with natural ones.

Transvaginal oocyte retrieval In vitro fertilization procedure in which aspirated secondary oocytes are combined with a solution containing sperm outside the body and then the fertilized ova are implanted in the uterus.

Transverse colon (trans-VERS KŌ-lon) The portion of the large intestine extending across the abdomen from right colic (hepatic) flexure to the left colic (splenic) flexure.

Transverse fissure (FISH-er) The deep cleft that separates the cerebrum from the cerebellum.

Transverse tubules (TOO-byools) ***(T tubules)*** Minute, cylindrical invaginations of the muscle fiber (cell) membrane that carry the muscle action potentials deep into the muscle fiber.

Trauma (TRAW-ma) An injury, either a physical wound or psychic disorder, caused by an external agent or force, such as a physical blow or emotional shock; the agent or force that causes the injury.

Traveler's diarrhea Infectious disease of the gastrointestinal tract that results in loose, urgent bowel movements, cramping, abdominal pain, malaise, nausea, and occasionally fever and dehydration. It is acquired through ingestion of food or water that has become contaminated with fecal material containing mostly bacteria (especially *Escherichia coli*). Also called ***Montezuma's revenge, turista,*** and ***Tut's tummy.***

Tremor (TREM-or) Rhythmic, involuntary, purposeless contraction of opposing muscle groups.

Triad (TRĪ-ad) A complex of three units in a muscle fiber (cell) composed of a transverse tubule and the segments of sarcoplasmic reticulum on both sides of it.

Tricuspid (trī-KUS-pid) ***valve*** Atrioventricular (AV) valve on the right side of the heart.

Trigeminal neuralgia (trī-JEM-i-nal noo-RAL-jē-a) Pain in one or more of the branches of the trigeminal (V) nerve. Also called ***tic douloureux*** (doo-loo-ROO).

Trigone (TRĪ-gon) A triangular area at the base of the urinary bladder.

Triiodothyronine (trī-ī-od-ō-THĪ-rō-nēn) ***(T_3)*** A hormone produced by the thyroid gland that regulates organic metabolism, growth and development, and the activity of the nervous system.

Trochlea (TROK-lē-a) A pulleylike surface.

Trophoblast (TRŌF-ō-blast) The outer covering of cells of the blastocyst.

Tropic (TRŌ-pik) ***hormone*** A hormone whose target is another endocrine gland.

Trunk The part of the body to which the upper and lower extremities are attached.

Tubal ligation (lī-GĀ-shun) A sterilization procedure in which the uterine (Fallopian) tubes are tied and cut.

Tuberculosis (too-berk-yoo-LŌ-sis) An infection of the lungs and pleurae caused by *Mycobacterium tuberculosis* resulting in destruction of lung tissue and its replacement by fibrous connective tissue.

Tubular reabsorption The movement of filtrate from renal tubules back into blood in response to the body's specific needs.

Tubular secretion The movement of substances in blood back into filtrate in response to the body's specific needs.

Tumor (TOO-mor) A growth of excess tissue due to an unusually rapid division of cells.

Tumorigenesis (too'-mor-i-JEN-e-sis) Multistep process by which tumors develop.

Tunica albuginea (TOO-ni-ka al'-byoo-JIN-ē-a) A dense layer of white fibrous tissue covering a testis or deep to the surface of an ovary.

Tunica externa (eks-TER-na) The outer coat of an artery or vein, composed mostly of elastic and collagenous fibers. Also called the ***adventitia.***

Tunica interna (in-TER-na) The inner coat of an artery or vein, consisting of a lining of endothelium, basement membrane, and internal elastic lamina. Also called the ***tunica intima*** (IN-ti-ma).

Tunica media (MĒ-dē-a) The middle coat of an artery or vein, composed of smooth muscle and elastic fibers.

Twitch Rapid, jerky contraction of a muscle in response to a single stimulus.

Tympanic antrum (tim-PAN-ik AN-trum) An air space in the posterior wall of the middle ear that leads into the mastoid air cells or sinus.

Tympanic (tim-PAN-ik) ***membrane*** A thin, semitransparent partition of fibrous connective tissue between the external auditory meatus and the middle ear. Also called the ***eardrum.***

Type II cutaneous mechanoreceptor A receptor embedded deeply in the dermis and deeper tissues that detects heavy and continuous touch sensations. Also called an ***end organ of Ruffini.***

Ulcer (UL-ser) An open lesion of the skin or a mucous membrane of the body with loss of substance and necrosis of the tissue.

Ultrasound (US) Medical imaging technique that utilizes high-frequency sound waves to produce an image called a ***sonogram.***

Umbilical (um-BIL-i-kal) Pertaining to the umbilicus or navel.

Umbilical (um-BIL-i-kal) ***cord*** The long, ropelike structure, containing the umbilical arteries and vein that connect the fetus to the placenta.

Umbilicus (um-BIL-i-kus or um-bil-Ī-kus) A small scar on the abdomen that marks the former attachment of the umbilical cord to the fetus. Also called the ***navel.***

Upper extremity The appendage attached at the shoulder girdle, consisting of the arm, forearm, wrist, hand, and fingers.

Uremia (yoo-RĒ-mē-a) Accumulation of toxic levels of urea and other nitrogenous waste products in the blood, usually resulting from severe kidney malfunction.

Ureter (YOO-re-ter) One of two tubes that connect the kidney with the urinary bladder.

Urethra (yoo-RĒ-thra) The duct from the urinary bladder to the exterior of the body that conveys urine in females and urine and semen in males.

Urinalysis The physical, chemical, and microscopic analysis or examination of urine.

Urinary (YOO-ri-ner-ē) ***bladder*** A hollow, muscular organ situated in the pelvic cavity posterior to the pubic symphysis.

Urinary tract infection (UTI) An infection of a part of the urinary tract or the presence of large numbers of microbes in urine.

Urine The fluid produced by the kidneys that contains wastes or excess materials and is excreted from the body through the urethra.

Urobilinogenuria The presence of urobilinogen in urine.

Urogenital (yoo'-rō-JEN-i-tal) ***triangle*** The region of the pelvic floor below the pubic symphysis, bounded by the pubic symphysis and the ischial tuberosities and containing the external genitalia.

Urology (yoo-ROL-ō-jē) The specialized branch of medicine that deals with the structure, function, and diseases of the male and female urinary systems and the male reproductive system.

Urticaria (ur'-ti-KĀ-rē-a) A skin reaction to certain foods, drugs, or other substances to which a person may be allergic; hives.

Uterine (YOO-ter-in) ***tube*** Duct that transports ova from the ovary to the uterus. Also called the ***Fallopian*** (fal-LŌ-pē-an) ***tube*** or ***oviduct.***

Uterosacral ligament (yoo'-ter-ō-SĀ-kral LIG-a-ment) A fibrous band of tissue extending from the cervix of the uterus laterally to attach to the sacrum.

Uterovesical (yoo'-ter-ō-VES-ī-kal) ***pouch*** A shallow pouch formed by the reflection of the peritoneum from the anterior surface of the uterus, at the junction of the cervix and the body, to the posterior surface of the urinary bladder.

Uterus (YOO-te-rus) The hollow, muscular organ in females that is the site of menstruation, implantation, development of the fetus, and labor. Also called the ***womb.***

Utricle (YOO-tri-kul) The larger of the two divisions of the membranous labyrinth located inside the vestibule of the inner ear, containing a receptor organ for static equilibrium.

Uvea (YOO-vē-a) The three structures that together make up the vascular tunic of the eye.

Uvula (YOO-vyoo-la) A soft, fleshy mass, especially the V-shaped pendant part, descending from the soft palate.

Vacuole (VAK-yoo-ōl) Membrane-bound organelle that, in animal cells, frequently functions in temporary storage or transportation.

Vagina (va-JĪ-na) A muscular, tubular organ that leads from the uterus to the vestibule, situated between the urinary bladder and the rectum of the female.

Valvular stenosis (VAL-vyoo-lar STEN-Ō-sis) A narrowing of a heart valve, usually the bicuspid (mitral) valve.

Varicocele (VAR-i-kō-sēl) A twisted vein; especially, the accumulation of blood in the veins of the spermatic cord.

Varicose (VAR-i-kōs) Pertaining to an unnatural swelling, as in the case of a varicose vein.

Vas A vessel or duct.

Vasa recta (REK-ta) Extensions of the efferent arteriole of a juxtaglomerular nephron that run alongside the loop of the nephron (Henle) in the medullary region.

Vasa vasorum (VĀ-sa va-SŌ-rum) Blood vessels that supply nutrients to the larger arteries and veins.

Vascular (VAS-kyoo-lar) Pertaining to or containing many blood vessels.

Vascular spasm Contraction of the smooth muscle in the wall of a damaged blood vessel to prevent blood loss.

Vascular tunic (TOO-nik) The middle layer of the eyeball, composed of the choroid, ciliary body, and iris. Also called the ***uvea*** (YOO-vē-a).

Vascular (venous) sinus A vein with a thin endothelial wall that lacks a tunica media and externa and is supported by surrounding tissue.

Vasectomy (va-SEK-tō-mē) A means of sterilization of males in which a portion of each ductus (vas) deferens is removed.

Vasoconstriction (vāz-ō-kon-STRIK-shun) A decrease in the size of the lumen of a blood vessel caused by contraction of the smooth muscle in the wall of the vessel.

Vasodilation (vās'-ō-DĪ-lā-shun) An increase in the size of the lumen of a blood vessel caused by relaxation of the smooth muscle in the wall of the vessel.

Vasomotion (vāz-ō-MŌ-shun) Intermittent contraction and relaxation of the smooth muscle of the metarterioles and precapillary sphincters that result in an intermittent blood flow.

Vasomotor (vā-sō-MŌ-tor) ***center*** A cluster of neurons in the medulla that controls the diameter of blood vessels, especially arteries.

Vein A blood vessel that conveys blood from tissues back to the heart.

Vena cava (VĒ-na KĀ-va) One of two large veins that open into the right atrium, returning to the heart all of the deoxygenated blood from the systemic circulation except from the coronary circulation.

Venesection (vēn'-e-SEK-shun) Opening of a vein for withdrawal of blood.

Ventral (VEN-tral) Pertaining to the anterior or front side of the body; opposite of dorsal.

Ventral body cavity Cavity near the ventral aspect of the body that contains viscera and consists of a superior thoracic cavity and an inferior abdominopelvic cavity.

Ventral ramus (RĀ-mus) The anterior branch of a spinal nerve, containing sensory and motor fibers to the muscles and skin of the anterior surface of the head, neck, trunk, and the extremities.

Ventricle (VEN-tri-kul) A cavity in the brain or an inferior chamber of the heart.

Ventricular fibrillation (ven-TRIK-yoo-lar fib-ri-LĀ-shun) Asynchronous ventricular contractions that result in cardiovascular failure.

Venule (VEN-yool) A small vein that collects blood from capillaries and delivers it to a vein.

Vermiform appendix (VER-mi-form a-PEN-diks) A twisted, coiled tube attached to the cecum.

Vermilion (ver-MIL-yon) The area of the mouth where the skin on the outside meets the mucous membrane on the inside.

Vermis (VER-mis) The central constricted area of the cerebellum that separates the two cerebellar hemispheres.

Vertebral (VER-te-bral) ***canal*** A cavity within the vertebral column formed by the vertebral foramina of all the vertebrae and containing the spinal cord. Also called the ***spinal canal.***

Vertebral column The 26 vertebrae; encloses and protects the spinal cord and serves as a point of attachment for the ribs and back muscles. Also called the ***spine, spinal column,*** or ***backbone.***

Vertigo (VER-ti-gō) Sensation of spinning or movement.

Vesicle (VES-i-kul) A small bladder or sac containing liquid.

Vesicular ovarian follicle A relatively large, fluid-filled follicle containing an immature ovum and its surrounding tissues that secretes estrogens. Also called a ***Graafian*** (GRAF-ē-an) ***follicle.***

Vestibular (ves-TIB-yoo-lar) ***membrane*** The membrane that separates the cochlear duct from the scala vestibuli.

Vestibule (VES-ti-byool) A small space or cavity at the beginning of a canal, especially the inner ear, larynx, mouth, nose, and vagina.

Villus (VIL-lus) A projection of the intestinal mucosal cells containing connective tissue, blood vessels, and a lymphatic vessel; functions in the absorption of the end products of digestion. *Plural,* ***villi*** (VIL-ī).

Viscera (VIS-er-a) The organs inside the ventral body cavity. *Singular,* ***viscus*** (VIS-kus).

Visceral (VIS-er-al) Pertaining to the organs or to the covering of an organ.

Visceral effector (e-FEK-tor) Cardiac muscle, smooth muscle, and glandular epithelium.

Visceral muscle An organ specialized for contraction, composed of smooth muscle fibers (cells), located in the walls of hollow internal structures, and stimulated by visceral efferent neurons.

Visceral pleura (PLOO-ra) The inner layer of the serous membrane that covers the lungs.

Visceroceptor (vis'-er-ō-SEP-tor) Receptor that provides information about the body's internal environment.

Vital signs Signs necessary to life that include temperature (T), pulse (P), respiratory rate (RR), and blood pressure (BP).

Vitamin An organic molecule necessary in trace amounts that acts as a catalyst in normal metabolic processes in the body.

Vitiligo (vit-i-LĪ-go) Patchy, white spots on the skin due to partial or complete loss of melanocytes.

Vitreous (VIT-rē-us) ***body*** A soft, jellylike substance that fills the vitreous chambers of the eyeball, lying between the lens and the retina.

Vocal folds Pair of mucous membrane folds below the ventricular folds that function in voice production. Also called ***true vocal cords.***

Volkmann's canal *See* ***Perforating canal.***

Vomiting Forcible expulsion of the contents of the upper gastrointestinal tract through the mouth.

Vulva (VUL-va) Collective designation for the external genitalia of the female. Also called the ***pudendum*** (poo-DEN-dum).

Wallerian (wal-LE-rē-an) ***degeneration*** Degeneration of the portion of the axon and myelin sheath of a neuron distal to the site of injury.

Wandering macrophage (MAK-rō-fāj) Phagocytic cell that develops from a monocyte, leaves the blood, and migrates to infected tissues.

Wart Generally benign tumor of epithelial skin cells caused by a virus.

Wheal (hwēl) Elevated lesion of the skin.

White matter Aggregations or bundles of myelinated axons located in the brain and spinal cord.

White matter tract The treelike appearance of the white matter of the cerebellum when seen in midsagittal section. Also called ***arbor vitae*** (AR-bōr VĒ-te). A series of branching ridges within the cervix of the uterus.

White pulp The portion of the spleen composed of lymphatic tissue, mostly lymphocytes, arranged around central arteries; in some areas of white pulp, lymphocytes are thickened into lymphatic nodules called splenic nodules (Malpighian corpuscles).

White ramus communicans (RĀ-mus ko-MYOO-ni-kans) The portion of a preganglionic sympathetic nerve fiber that branches away from the anterior ramus of a spinal nerve to enter the nearest sympathetic trunk ganglion.

Xiphoid (ZĪ-foyd) Sword-shaped. The lowest portion of the sternum.

Yolk sac An extraembryonic membrane that connects with the midgut during early embryonic development, but is nonfunctional in humans.

Zona fasciculata (ZŌ-na fa-sik'-yoo-LA-ta) The middle zone of the adrenal cortex that consists of cells arranged in long, straight cords and that secretes glucocorticoid hormones.

Zona glomerulosa (glo-mer'-yoo-LŌ-sa) The outer zone of the adrenal cortex, directly under the connective tissue covering, that consists of cells arranged in arched loops or round balls and that secretes mineralocorticoid hormones.

Zona pellucida (pe-LOO-si-da) Gelatinous glycoprotein layer internal to the corona radiata that surrounds a secondary oocyte.

Zona reticularis (ret-ik'-yoo-LAR-is) The inner zone of the adrenal cortex, consisting of cords of branching cells that secrete sex hormones, chiefly androgens.

Zygote (ZĪ-gōt) The single cell resulting from the union of a male and female gamete; the fertilized ovum.

Zymogenic (zī'-mō-JEN-ik) ***cell*** One of the cells of a gastric gland that secretes the principal gastric enzyme precursor, pepsinogen. Also called a ***peptic cell.***

Bibliography

Chapter 1

Clemente, C. D. *Anatomy: A Regional Atlas of the Human Body*, 3rd ed. Baltimore: Urban & Schwarzenberg, 1987.
Geller, S. A. ''Autopsy,'' *Scientific American*, March 1983.
Gosling, J. A., P. F. Harris, J. R. Humpherson, I. Whitmore, and P. L. T. Willan. *Atlas of Human Anatomy*. London: Gower Medical, 1985.
Keiffer, S. A., and E. R. Heitzman. *An Atlas of Cross-Sectional Anatomy*. New York: Harper & Row, 1979.
Netter, F. H. *Atlas of Human Anatomy*. Summit, NJ: CIBA, 1989.
Rohen, J. W., and C. Yokochi, *Color Atlas of Anatomy*. Tokyo, New York: Igaku-Shoin, Ltd., 1983.
Sochurek, H. ''Medicine's New Vision,'' *National Geographic*, January 1987.
Yokochi, C., and J. W. Rohen. *Photographic Anatomy of the Human Body*, 2nd ed. Tokyo, New York: Igaku-Shoin, Ltd., 1978.

Chapter 2

Becker, W. M. *The World of the Cell*. Menlo Park, CA: Benjamin/Cummings, 1986.
Begley, S., and M. Hager. ''Brave New Gene Therapy,'' *Newsweek*, 13 February 1989.
Berridge, M. J. ''The Molecular Basis of Communication within the Cell,'' *Scientific American*, October 1985.
Bretscher, M. S. ''The Molecules of the Cell Membrane,'' *Scientific American*, October 1985.
Cairns, J. ''The Treatment of Diseases and the War against Cancer,'' *Scientific American*, November 1985.
Dautry-Varsat, A., and H. F. Lodish. ''How Receptors Bring Proteins and Particles into Cells,'' *Scientific American*, May 1984.
de Duve, C. *A Guided Tour of the Living Cell*. New York: Scientific American Books, 1984.
Feldman, M., and L. Eisenbach. ''What Makes a Tumor Cell Metastatic?'' *Scientific American*, November 1988.
Kartner, N. and V. Ling. ''Multidrug Resistance in Cancer,'' *Scientific American*, March 1989.
Koshland, D. E. ''The Cell Cycle,'' *Science*, 3 November 1989.
Marx, J. L. ''How DNA Viruses May Cause Cancer,'' *Science*, 24 February 1989.
Mazia, D. ''The Cell Cycle,'' *Scientific American*, January 1974.
McIntosh, R. and K. L. McDonald, ''The Mitotic Spindle,'' *Scientific American*, October 1989.
Porter, K. R., and J. B. Tucker. ''The Ground Substance of the Living Cell,'' *Scientific American*, March 1981.
Ptashne, M. ''How Gene Activators Work,'' *Scientific American*, January 1989.
Radman, M., and R. Wagner. ''The High Fidelity of DNA Duplication,'' *Scientific American*, August 1988.
Rothman, J. E. ''The Compartmental Organization of the Golgi Apparatus,'' *Scientific American*, September 1985.
Unwin, N., and R. Henderson. ''The Structure of Proteins in Biological Membranes,'' *Scientific American*, February 1984.
Weinberg, R. A. ''Finding the Anti-Oncogene,'' *Scientific American*, September 1988.

Chapter 3

Cormack, D. H. *Ham's Histology*, 9th ed. Philadelphia: Lippincott, 1987.
Fawcett, D. W. *Bloom and Fawcett: A Textbook of Histology*, 11th ed. Philadelphia: Saunders, 1986.
Kelly, D. E., R. L. Wood, and A. C. Enders. *Bailey's Textbook of Microscopic Anatomy*, 18th ed. Baltimore: Williams & Wilkins, 1984.
Miller, J. ''The Connective Tissue Perspective,'' *Science News*, 20 February 1982.
Synder, S. H. ''The Molecular Basis of Communication Between Cells,'' *Scientific American*, October 1985.
Telford, I. R. and C. F. Bridgman, *Introduction to Functional Histology*. New York: Harper & Row, 1990.
Woerner, E. M., and K. Royalty, ''Marfan Syndrome,'' *Postgraduate Medicine*, April 1990.

Chapter 4

Dahl, M. V. ''Acne: How It Happens and How It's Treated,'' *Modern Medicine*, September 1982.
Edelson, L. E., and J. M. Fink. ''The Immunologic Function of Skin,'' *Scientific American*, June 1985.
Flanagan, B. P. ''Skin Cancer: An Illustrated Guide to Early Diagnosis,'' *Modern Medicine*, October 1985.
Goldberg, L. H., and H. A. Rubin. ''Management of Basal Cell Carcinoma,'' *Postgraduate Medicine*, January 1989.
Levine, N. ''Sunburn: How to Stop the Pain, How to Prevent the Damage,'' *Modern Medicine*, July 1984.
Martin, L. M. ''Nursing Implications of Today's Burn Care Techniques,'' *RN*, May 1989.
Montagna, W. ''The Skin,'' *Scientific American*, June 1989.
Silverberg, N. and L. Silverberg. ''Aging and the Skin,'' *Postgraduate Medicine*, July 1989.
Sober, A. ''Malignant Melanoma: A Guide to Early, Accurate Diagnosis,'' *Modern Medicine*, July 1988.
Wachtel, T. L., ''Major Burns,'' *Postgraduate Medicine*, January 1989.

Chapter 5

Avioli, L. V. ''Osteoporosis: A Guide to Detection,'' *Modern Medicine*, February 1986.

Baran, D. "Diagnosis and Management of Osteoporosis," *Modern Medicine,* March 1989.

Bassett, C. A., S. N. Mitchell, and J. Gaston. "Pulsing Electromagnetic Field Treatment in Ununited Fractures and Failed Arthrodeses," *Journal of the American Medical Association,* February 5, 1982.

Bock, H. (ed.) "Tricky Bone Healing Technique From the Soviet Union," *Medical Update,* March 1989.

Gamble, J. G. *The Musculoskeletal System: Physiological Basis.* New York: Raven Press, 1988.

Gosling, J. A., P. F. Harris, J. R. Humpherson, I. Whitmore, and P. L. T. Willan. *Atlas of Human Anatomy,* Philadelphia: Lippincott, 1985.

Hogan, M. J. (ed.) "TMJ Disorder," *Mayo Clinic Health Letter,* October 1988.

Langone, J. "Back Surgery Without Stitches," *Time,* 5 September 1988.

Netter, F. H. *Musculoskeletal System. Part I, Anatomy, Physiology, and Metabolic Disorders,* Summit, NJ: CIBA, 1987.

Raisz, L. G. "Local and Systemic Factors in the Pathogenesis of Osteoporosis," *New England Journal of Medicine,* 31 March 1988.

Ravnikav, V. "Clinical Considerations in the Diagnosis of Osteoporosis," *Modern Medicine,* May 1988.

Rohen, J. W., and C. Yokochi. *Color Atlas of Anatomy.* Tokyo, New York: Igaku-Shoin, Ltd., 1983.

Rudy, D. R. "Osteoporosis," *Postgraduate Medicine,* August 1989.

Shipman, P., A. Walker, and D. Bichell. *The Human Skeleton.* Cambridge: Harvard University Press, 1985.

Steele, D. G., and C. A. Bramblett. *The Anatomy and Biology of the Human Skeleton.* College Station: Texas A&M University Press, 1988.

Chapter 6

Cristensen, J. B., and I. R. Telford. *Synopsis of Gross Anatomy,* 5th ed. Philadelphia: Lippincott, 1988.

Gosling, J. A., P. F. Harris, J. R. Humpherson, I. Whitmore, and P. L. T. Willan. *Atlas of Human Anatomy.* Philadelphia: Lippincott, 1985.

Hogan, M. J. (ed). "TMJ Disorder," *Mayo Clinic Health Letter,* October 1988.

Netter, F. H. *Musculoskeletal System, Part I, Anatomy, Physiology, and Metabolic Disorders.* Summit, NJ: CIBA, 1987.

Rohen, J. W., and C. Yokochi. *Color Atlas of Anatomy.* Tokyo, New York: Igaku-Shoin, Ltd., 1983.

Shipman, P., A. Walker, and D. Bichell. *The Human Skeleton.* Cambridge, Harvard University Press, 1985.

Steele, D. G., and C. A. Bramblett. *The Anatomy and Biology of the Human Skeleton.* College Station: Texas A&M University Press, 1988.

Chapter 7

Cristensen, J. B., and I. R. Telford, *Synopsis of Gross Anatomy,* 5th ed. Philadelphia: Lippincott, 1988.

Gosling, J. A., P. F. Harris, J. R. Humpherson, I. Whitmore, and P. L. T. Willan. *Atlas of Human Anatomy.* Philadelphia: Lippincott, 1985.

Netter, F. H. *Musculoskeletal System, Part I, Anatomy, Physiology, and Metabolic Disorders.* Summit, NJ: CIBA, 1987.

Rohen, J. W., and C. Yokochi. *Color Atlas of Anatomy.* Tokyo, New York: Igaku-Shoin, Ltd., 1983.

Shipman, P., A. Walker, and D. Bichell. *The Human Skeleton.* Cambridge: Harvard University Press, 1985.

Steele, D. G., and C. A. Bramblett. *The Anatomy and Biology of the Human Skeleton.* College Station: Texas A&M University Press, 1988.

Chapter 8

Allman, W. F. "The Knee," *Science 83,* November 1983.

Beck, M., M. Hager, and V. E. Smith. "Living with Arthritis," *Newsweek,* 20 March 1989.

Bertram, Z., and M. Adams. "Knee Injuries in Sports," *New England Journal of Medicine,* 14 April 1988.

Bienenstock, H. "Diagnosis: Arthritis," *Hospital Medicine,* October 1984.

Edwards, D. D. "Severe Arthritis Under Attack," *Science News,* 19 October 1985.

Epstein, F. H. "The Biology of Osteoarthritis," *New England Journal of Medicine,* 18 May 1989.

Hogan, M. J. (ed.). "Arthroscopy," *Mayo Clinic Health Letter,* March 1989.

______. "Artificial Joints," *Mayo Clinic Health Letter,* Three parts. November and December 1988 and January 1989.

Kiley, J. M. (ed.). "Rheumatoid Arthritis," *Mayo Clinic Health Letter,* December 1986.

Sheon, R. P. "Aches and Pains From Arthritis," *Postgraduate Medicine,* 3 February 1990.

Williams, D. N., and E. S. Schned. "Lyme Disease," *Postgraduate Medicine,* 1 May 1990.

Wilson, J. D. "Tiny Tick, Big Worry," *Newsweek,* 22 May 1989.

Chapter 9

Allman, W. F. "Weight-Lifting: Inside the Pumphouse," *Science,* September 1984.

Bohigian, G. M. "Drug Abuse in Athletes," *Journal of the American Medical Association,* 18 March 1988.

Gamble, J. G. *The Musculoskeletal System: Physiological Basics.* New York: Raven Press, 1988.

Gosling, J. A., P. F. Harris, J. R. Humpherson, I. Whitmore, and P. L. T. Willan. *Atlas of Human Anatomy.* Philadelphia: Lippincott, 1985.

Hinson, M. M. *Kinesiology.* Dubuque: Wm. C. Brown, 1981.

Huxley, H. E. "The Contraction of Muscle," *Scientific American,* November 1968.

Marshall, E. "The Drug of Champions," *Science,* 14 October 1980.

Netter, F. H. *Musculoskeletal System: Anatomy, Physiology, and Metabolic Disorders.* Summit, NJ: CIBA, 1987.

Ravits, J. "Myasthenia Gravis," *Postgraduate Medicine,* January 1988.

Wilson, F. C. *The Musculoskeletal System: Basic Processes and Disorders,* 2nd ed. Philadelphia: Lippincott, 1983.

Windsor, R. E., and D. Dumitru. "Anabolic Steroid Use by Athletes," *Postgraduate Medicine,* September 1988.

Chapter 10

Borg, E., and S. A. Counter, "The Middle-Ear Muscles," *Scientific American,* August 1989.

Clemente, C. D. *Anatomy: A Regional Atlas of the Human Body,* 3rd ed. Baltimore: Urban & Schwarzenberg, 1987.
Ferner, H., and J. Staubesand (eds). *Sobotta Atlas of Human Anatomy.* Vols 1–2. Baltimore: Urban & Schwarzenberg, 1983.
Gosling, J. A., P. F. Harris, J. R. Humpherson, I. Whitmore, and P. L. T. Willan. *Atlas of Human Anatomy.* Philadelphia: Lippincott, 1985.
Hinson, M. M. *Kinesiology.* Dubuque: Wm. C. Brown, 1977.
Lavin, R. J. "The High Pressure Demands of Compartment Syndrome," *RN,* February 1989.
Netter, F. H. *Musculoskeletal System: Anatomy, Physiology, and Metabolic Disorders.* Summit, NJ: CIBA, 1987.
Rohen, J. W., and C. Yokochi. *Color Atlas of Anatomy.* Tokyo, New York: Igaku-Shoin, Ltd., 1983.

Chapter 11

Basmajian, J. V. *Surface Anatomy: An Instructional Manual.* Baltimore: Williams & Wilkins Company, 1977.
Hamilton, W. H., G. Simon, and S. G. I. Hamilton. *Surface and Radiological Anatomy,* 5th ed. Baltimore: Williams & Wilkins Company, 1971.
Keogh, B., and S. Ebbs. *Normal Surface Anatomy With Practical Applications.* Philadelphia: Lippincott, 1984.
Lockhart, R. D. *Living Anatomy,* 7th ed. London: Faber and Faber, 1974.
Royce, J. *Surface Anatomy.* Philadelphia: F. A. Davis Company, 1965.

Chapter 12

Aledort, I. M. "Current Concepts in Diagnosis and Management of Hemophilia," *Hospital Practice,* October 1982.
Bloom, M. (ed). "New Jobs for Man-Made Blood," *Physician's Weekly,* 17 February 1986.
Bock, H. (ed). "New Method For Saving Your Own Blood," *Medical Update,* October 1988.
Butler, S. "Current Trends in Autologous Transfusion," *RN,* November 1989.
Dixon, B. "Of Different Bloods," *Science,* November 1984.
Doolittle, R. F. "Fibrinogen and Fibrin," *Scientific American,* December 1981.
England, J. A. "The Many Faces of Epstein-Barr Virus," *Postgraduate Medicine,* February 1988.
Froberg, J. H. "The Anemias: Causes and Course of Action," *RN,* January, March, and May 1989.
Golde, D. W., and J. C. Ganon. "Hormones that Stimulate the Growth of Blood Cells," *Scientific American,* July 1988.
Goldfinger, S. E. (ed). "Chronic Fatigue Syndrome," *Harvard Medical School Health Letter,* July 1988.
Hogan, M. J. (ed). "Blood Transfusions," *Mayo Clinic Health Letter,* July 1989.
Kiley, J. M. (ed). "Chronic Mononucleosis," *Mayo Clinic Health Letter,* June 1988.
Langone, J. "New Methods for Saving Blood," *Time,* 5 December 1988.
Lefant, C. "Dissolving Blood Clots," *Medical Update,* March 1988.
Sabetta, J. R. "Diagnosis: Infectious Mononucleosis," *Hospital Medicine,* March 1984.
Silberner, J. "Clot Dissolver May Save Heart Tissue," *Science News,* 26 November 1983.
Spivak, J. *Fundamentals of Clinical Hematology,* 2nd ed. New York: Harper & Row, 1984.
Throup, O. A., et al. *Leavell and Thorup's Fundamentals of Clinical Hematology,* 5th ed. Philadelphia: Saunders, 1987.
Vichinsky, E. P., D. Hurst, and B. Lubin. "Sickle Cell Disease: Basic Concepts," *Hospital Medicine,* September 1983.
Zucker, M. D. "The Functioning of Blood Platelets," *Scientific American,* June 1980.

Chapter 13

Alpert, J. S. "Management of Acute and Chronic Myocardial Ischemia," *Modern Medicine,* April 1989.
Becker, B. L. "Cholesterol," *Heartbeat,* June 1988.
Brown, M. S., and J. L. Goldstein. "How LDL Receptors Influence Cholesterol and Atherosclerosis," *Scientific American,* November 1984.
Cantin, M., and J. Genest. "The Heart as an Endocrine Gland," *Scientific American,* February 1986.
Comroe, J. H. "Doctor, You Have Six Minutes," *Science,* January/February 1984.
DeVries, W. C. "The Permanent Artificial Heart," *Journal of the American Medical Association,* 12 February 1988.
Dolan, B., and J. M. Nash. "Searching For Life's Elixir," *Newsweek,* 12 December 1988.
Dusheck, J. "Fish, Fatty Acids, and Physiology," *Science News,* 19 October 1985.
Eisenberg, M. S., L. Bergner, A. P. Hallstrom, and R. O. Cummins. "Sudden Cardiac Death," *Scientific American,* May 1986.
Escher, D. J. W. "Use of Cardiac Pacemakers," *Hospital Practice,* September 1981.
Falk, J. S., B. Kaufman, and M. H. Weil. "Cardiopulmonary Resuscitation: An Update," *Hospital Medicine,* January 1984.
Fowler, R. E. "Acute Myocardial Infarction," *Postgraduate Medicine,* November 1988.
Goldfinger, S. E. (ed). "Balloons: Expanding," *Harvard Medical School Health Letter,* May 1987.
______. "Keeping Up with Cholesterol," *Harvard Medical School Health Letter,* June 1985.
Gwynn, J. T., and M. K. Lawrence. "Current Concepts in the Evaluation and Treatment of Hypercholesterolemia," *Modern Medicine,* March 1989.
Hogan, M. J. (ed). "Cholesterol," *Mayo Clinic Health Letter,* March 1988.
______. "Coronary Atherectomy," *Mayo Clinic Health Letter,* April 1989.
Jarvik, R. K. "The Total Artificial Heart," *Scientific American,* January 1981.
Johnson, R. "A Consumer's Guide to Coronary Angioplasty," *Cardiac Alert,* February 1989.
Johnson, T. R. "The Cholesterol Controversy," *Harvard Medical School Health Letter,* December 1989.
Kiley, J. M. (ed). "Echocardiography," *Mayo Clinic Health Letter,* May 1985.
Leaf, D. A. "Omega-3 Fatty Acids and Coronary Artery Disease," *Postgraduate Medicine,* June 1989.
McIntyre, K. "Cardiac Arrest," *Hospital Medicine,* November 1984.
Netter, F. *The Heart,* CIBA Collection of Medical Illustrations. Summit, NJ: CIBA, 1971.
O'Keefe, J. H., C. J. Lavie, and J. O. O'Keefe. "Dietary Preven-

tion of Coronary Artery Disease,'' *Postgraduate Medicine,* May 1989.

O'Toole, M. T., and A. R. Waldman. ''Chest Pain,'' *RN,* April 1989.

Rios, J. C. (ed). ''Breakthrough Discoveries on How to Avoid a Fatal Heart Attack,'' *Cardiac Alert,* 1988.

Rogers, W. J. ''Diagnosis: Coronary Artery Disease,'' *Hospital Medicine,* August 1984.

Shabeti, R. ''Answers to Questions on Cardiac Catheterization,'' *Hospital Medicine,* August 1982.

Silberner, J. ''Lasers Powering through Coronary Arteries,'' *Science News,* 23 November 1985.

______. ''Anatomy of Atherosclerosis,'' *Science News,* 16 March 1985.

Skluth, H., K. Grauer, and J. Gums. ''Ventricular Arrhythmias,'' *Postgraduate Medicine,* May 1989.

Toufexis, A. ''The Latest Word on What to Eat,'' *Time,* 13 March 1989.

Wehrmacher, W. H. ''Acute Myocardial Infarction,'' *Postgraduate Medicine,* February 1989.

Chapter 14

Blake, P. ''Precision Moves That Counter Cardiogenic Shock,'' *RN,* May 1989.

Bock, H. (ed). ''New Ways to Control Blood Pressure,'' *Medical Update,* July 1988.

Clemente, C. D. *Anatomy: A Regional Atlas of the Human Body,* 3rd ed. Baltimore: Urban & Schwarzenberg, 1987.

Garvas, I., M. Bursztyn, and H. Garvas. ''Changing Trends in Hypertension Therapy,'' *Modern Medicine,* October 1988.

Goldfinger, S. E. (ed). ''Exercise and Well-Being,'' *Harvard Medical School Health Letter,* February 1985.

______. ''High Blood Pressure: Newer Treatments,'' *Harvard Medical School Health Letter,* January 1989.

______. ''High Blood Pressure: A New Look,'' *Harvard Medical School Health Letter,* December 1988.

Gosling, J. A., P. F. Harris, J. R. Humpherson, I. Whitmore, and P. L. T. Willan. *Atlas of Human Anatomy With Integrated Text.* London: Gower Medical, 1985.

Johnson, G. T. (ed). ''The Ups and Downs of Blood Pressure Numbers,'' *Harvard Medical School Health Letter,* November 1987.

Melloni, J. L., I. Dox, H. P. Melloni, and B. J. Melloni. *Melloni's Illustrated Review of Human Anatomy.* Philadelphia: Lippincott, 1988.

Rippe, J. M., A. Ward, J. P. Pocari, and P. S. Freedson. ''Walking for Health and Fitness,'' *Journal of the American Medical Association,* 13 May 1988.

Yokochi, C., J. W. Rohen, and E. L. Weinreb. *Photographic Anatomy of the Human Body,* 3rd ed. Tokyo, New York: Igaku-Shoin, Ltd., 1989.

Chapter 15

Barnes, D. M. ''Obstacles to an AIDS Vaccine,'' *Science,* 6 May 1988.

Bolognesi, D. P. ''Prospects for Prevention of and Early Intervention Against HIV,'' *Journal of the American Medical Association,* 26 May 1989.

Cohen, I. R. ''The Self, the World and Immunity,'' *Scientific American,* April 1988.

Cowley, G., M. Hager, and R. Marshall. ''AIDS: The Next Ten Years'' *Newsweek,* 25 June 1990.

Frolich, E. D. ''Research into AIDS: Status, Prospects, By-products,'' *Hospital Medicine,* August 1988.

Gallo, R. C., and L. Montagnier. ''AIDS in 1988,'' *Scientific American,* October 1988.

Haseltine, W. A., and F. Wong-Staal. ''The Molecular Biology of the AIDS Virus,'' *Scientific American,* October 1988.

Hayward, W. L., and J. W. Curran. ''The Epidemiology of AIDS in the US,'' *Scientific American,* October 1988.

Henry, K., J. Thurn, and D. Anderson. ''Testing for Human Immunodeficiency Virus,'' *Postgraduate Medicine,* January 1989.

Jaroff, L. ''Stop That Germ,'' *Time,* 23 May 1988.

Johnson, R. B. ''Current Concepts: Immunology,'' *New England Journal of Medicine,* 24 March 1988.

Koop, C. E. ''Surgeon General's Report on Acquired Immune Deficiency Syndrome,'' *U.S. Public Health Service Office,* 22 October 1986.

Marx, J. L. ''AIDS Drugs—Coming but Not Here,'' *Science,* 21 April 1989.

Matthews, T. J., and D. P. Bolognese. ''AIDS Vaccines,'' *Scientific American,* October 1988.

Redfield, R. R., and D. S. Burke. ''HIV Infections: The Clinical Picture,'' *Scientific American,* October 1988.

Robertson, S. ''Drugs That Keep AIDS Patients Alive,'' *RN,* February 1989.

Scherrer, P. ''How AIDS Attacks the Brain,'' *American Journal of Nursing,* January 1990.

Scutchfield, F. D., and A. S. Benenson. ''AIDS Update,'' *Postgraduate Medicine,* March 1989.

Silberner, J. ''AIDS: Disease, Research Efforts Advance,'' *Science News,* 27 April 1985.

Tonegawa, S. ''The Molecules of the Immune System,'' *Scientific American,* October 1985.

Weber, J. N., and R. A. Weiss. ''HIV Infection: The Cellular Picture,'' *Scientific American,* October 1988.

Weiss, R. ''HIV Can Linger Years with No Antibodies,'' *Science News,* 3 June 1989.

______. ''AIDS Vaccine: Preliminary but Promising,'' *Science News,* 17 June 1989.

Yarchoan, R., and S. Broder, ''AIDS Therapies,'' *Scientific American,* October 1988.

Chapter 16

Barr, M. L., and J. A. Kiernan. *The Human Nervous System,* 5th ed. Philadelphia: Lippincott, 1988.

Bloom, F. E. ''Neuropeptides,'' *Scientific American,* October 1981.

Byrne, J. H., and S. G. Schultz. *An Introduction to Membrane Transport and Bioelectricity.* New York: Raven Press, 1988.

Dunant, Y., and M. Israël, ''The Release of Acetylcholine,'' *Scientific American,* March 1985.

Gottlieb, D. I. ''GABAergic Neurons,'' *Scientific American,* February 1988.

Kalil, R. E. ''Synapse Formation in the Developing Brain,'' *Scientific American,* December 1989.

Kimelberg, H. K., and M. D. Norenberg, ''Astrocytes,'' *Scientific American,* April 1989.

Llinas, R. R. "Calcium in Synaptic Transmission," *Scientific American,* October 1982.
Miller, J. A. "Cell Communication Equipment: Do-It-Yourself," *Science News,* 14 April 1984.
Morell, P., and W. T. Norton. "Myelin," *Scientific American,* May 1980.
Netter, F. H. *Nervous System: Anatomy and Physiology.* Summit, NJ: CIBA, 1983.
Patton, H. D., A. F. Fuchs, B. Hille, A. M. Scher, and R. Steiner. *Textbook of Physiology: Excitable Cells and Neurophysiology,* 21st ed. Philadelphia: Saunders, 1989.
Stevens, C. F. "The Neuron," *Scientific American,* September 1979.
Weiss, R. "Regenerated Nerves Send First Messengers," *Science News,* 14 October 1989.
_____. "Antibodies Enhance Spinal Nerve Regrowth," *Science News,* 20 January 1990.

Chapter 17

Barinaga, M. "Neuroscientists Track Nerve Development," *Science,* 10 November 1989.
Bruno, M., and S. Katz. "New Hope for Hurt Nerves," *Newsweek,* 7 October 1985.
Goldfinger, S. E. (ed)., "Shingles," *Harvard Medical School Health Letter,* June 1984.
Greenspan, J. "Carpal Tunnel Syndrome," *Postgraduate Medicine,* November 1988.
Johnson, G. T. (ed). "Is Spinal Anesthesia Best for You? *Mayo Clinic Health Letter,* October 1984.
Miller, J. A. "Grow Nerves Grow," *Science News,* 29 March 1986.
Rappaport, S. "Common Peripheral Nerve Injuries," *Hospital Medicine,* June 1984.
Romeo, J. H. "Spinal Cord Injury," *RN,* May 1988.
Weiss, R. "Regenerated Nerves Send First Messages," *Science News,* 14 October 1989.

Chapter 18

Barr, M. L., and J. A. Kiernan. *The Human Nervous System,* 5th ed. Philadelphia: Lippincott, 1988.
Begley, S., J. Carey, and R. Sawhill. "How the Brain Works," *Newsweek,* 7 February 1983.
"The Brain," *Scientific American,* September 1979. (Entire issue devoted to the brain and the human nervous system.)
Bruno, M., and S. Katz. "New Hope for Hurt Nerves," *Newsweek,* 7 October 1985.
Burden, N. "Regional Anesthesia: What Patients and Nurses Need to Know," *RN,* May 1988.
Carpenter, M. B. *Core Text of Neuroanatomy,* 3rd ed. Baltimore: Williams & Wilkins, 1985.
Edwards, D. D. "A Common Medical Denominator," *Science News,* 25 January 1986.
Gelman, D. "How the Brain Recovers," *Newsweek,* 9 April 1990.
Gluhbegovic, N., and T. H. Williams. *The Human Brain: A Photographic Guide.* New York: Harper & Row, 1980.
Goldfinger, S. E. (ed). "Alzheimer's Disease," *Harvard Medical School Health Letter,* April 1988.
_____. "Shingles," *Harvard Medical School Health Letter,* June 1984.
Gorelick, P. B. "Clues to the Mystery of Multiple Sclerosis," *Postgraduate Medicine,* March 1989.
Guyton, A. C. *Basic Neuroscience: Anatomy and Physiology.* Philadelphia: Saunders, 1987.
Kalil, R. E. "Synapse Formation in the Developing Brain," *Scientific American,* December 1989.
Kiley, M. J. (ed). "A Long Goodbye," *Mayo Clinic Health Letter,* January 1988.
_____. "Parkinson's Disease," *Mayo Clinic Health Letter,* May 1989.
Kimelberg, H. K., and M. D. Norenberg. "Astrocytes," *Scientific American,* April 1989.
Koller, W. C. "Diagnosis and Treatment of Parkinson's Disease," *Modern Medicine,* May 1989.
Mattewson, J. "Alzheimer's Disease: Source Searching," *Science News,* 13 July 1985.
Nauta, W. J. H., and M. Feirtag. *Fundamental Neuroanatomy.* New York: W. H. Freeman, 1985.
Spector, R., and C. E. Johanson. "The Mamallian Choroid Plexus," *Scientific American,* November 1989.
Thompson, R. F. *The Brain: An Introduction to Neuroscience.* New York: W. H. Freeman, 1985.
Wurtman, R. J. "Alzheimer's Disease," *Scientific American,* January 1985.

Chapter 19

Agras, W. S. "Relaxation Therapy in Hypertension," *Hospital Practice,* May 1983.
Carney, R. M. "Clinical Applications of Relaxation Training," *Hospital Practice,* July 1983.
Carpenter, M. B. *Core Text of Neuroanatomy,* 3rd ed. Baltimore: Williams & Wilkins, 1985.
Nobach, C. R., and R. J. Demarest. *The Nervous System: Introduction and Review,* 3rd ed. New York: McGraw-Hill, 1986.

Chapter 20

Borg, E., and S. A. Counter. "The Middle-Ear Muscles," *Scientific American,* August 1989.
Franklin, D. "Crafting Sound from Silence," *Science News,* 20 October 1984.
Goldfinger, S. E. (ed). "Hearing Loss and Hearing Aids," *Harvard Medical School Health Letter,* April 1989.
Hogan, M. J. (ed). "Color Blindness," *Mayo Clinic Health Letter,* March 1989.
Hudspeth, A. J. "The Hair Cells of the Inner Ear," *Scientific American,* January 1983.
Kiely, J. M. (ed). "Cataracts," *Mayo Clinic Health Letter,* February 1989.
_____. "Retinal Detachment," *Mayo Clinic Health Letter,* January 1988.
_____. "Your Eyes Can Be Windows to Your Health," *Mayo Clinic Health Letter,* March 1985.
Koretz, J. F., and G. H. Handelman. "How the Human Eye Focuses," *Scientific American,* July 1988.
Loeb, G. E. "The Functional Replacement of the Ear," *Scientific American,* February 1985.
Morrison, A. R. "A Window on the Sleeping Brain," *Scientific American,* April 1983.
Von Bekesy, G. "The Ear," *Scientific American,* August 1975.

Wiet, R. J. "Help for the Hearing-Impaired," *Postgraduate Medicine,* November 1988.

Chapter 21

Atkinson, M. A., and N. K. MacLaren. "What Causes Diabetes?" *Scientific American,* July 1990.
Clark, M. "The Power of Hormones," *Newsweek on Health,* Spring 1987.
Crapo, L. *Hormones: Messengers of Life.* New York: W. H. Freeman, 1985.
de los Santos, E. T., and E. L. Mazzaferri. "Thyroid Function Tests," *Postgraduate Medicine,* April 1989.
Fellman, B. "A Clockwork Gland," *Science,* May 1985.
Fry, W. F. "Benefits of Stress," *Healthline,* January 1984.
Goldfinger, S. E. (ed). "Diabetes: The Long Run," *Harvard Medical School Health Letter,* April 1985.
______. "Insulin and the Diabetic," *Harvard Medical School Health Letter,* March 1985.
______. "The Diseases Called Sugar Diabetes," (Part 1), *Harvard Medical School Health Letter,* February 1985.
Goodman, H. M. *Basic Medical Endocrinology.* New York: Raven Press, 1980.
Griffin, J. E., and S. R. Ojeda. *Textbook of Endocrine Physiology.* New York: Oxford University Press, 1988.
Miller, J. "Eye to (Third) Eye," *Science News,* 9 November 1985.
Molitch, M. E. "Diabetes Mellitus," *Postgraduate Medicine,* March 1989.
Muldoon, T. G., and A. C. Evans. "Hormones and Their Receptors," *Archives of Internal Medicine,* April 1988.
Oldstone, M. B. A. "Viral Alteration of Cell Function," *Scientific American,* August 1989.
Orci, L., J. D. Vassalli, and A. Perrelet. "The Insulin Factory," *Scientific American,* September 1988.
Seyle, H. *Stress in Health and Disease.* London: Butterworths, 1976.
Williams, R. *Textbook of Endocrinology.* 7th ed. Philadelphia: Saunders, 1985.

Chapter 22

Amin, N. M. "Evaluation and Treatment of Pneumonia in Adult Patients," *Modern Medicine,* January 1989.
Carr, D. T. "Malignant Lung Disease," *Hospital Practice,* January 1981.
Eberhart, J. "Common Cold Preventive," *Science News,* 11 January 1986.
Edwards, D. D. "Nicotine: A Drug of Choice?" *Science News,* 18 January 1986.
Heimlich, H. J., and E. A. Patrick. "The Heimlich Maneuver," *Postgraduate Medicine,* May 1990.
Kiley, J. (ed). "Sudden Infant Death Syndrome," *Mayo Clinic Health Letter,* April 1989.
Mines, A. H. *Respiratory Physiology,* 2nd ed. New York: Raven Press, 1986.
Niewuehner, D. E. "Diagnosis Management, and Monitoring of Chronic Pulmonary Obstructive Disease," *Modern Medicine,* February 1989.
Petty, T. L. "Prevention of Emphysema," *Postgraduate Medicine,* November 1989.
Weiss, R. "TB Troubles," *Science News,* 6 February 1988.
Wenger, N. K. "Pulmonary Embolism," *Postgraduate Medicine,* August 1988.

Chapter 23

Bruckstein, A. H. "Chronic Hepatitis," *Postgraduate Medicine,* May 1989.
Council on Scientific Affairs. "Dietary Fiber and Health," *Journal of the American Medical Association,* 28 July 1989.
Cunha, B. A. "The Yellow Patient," *Postgraduate Medicine,* October 1988.
Dolan, B. "A Mother's Gift of Life," *Time,* 11 December 1989.
Farley, D. "Today's Dentistry: A Mouthful of Marvels," *Healthline,* October 1985.
Farley, P. C., and K. H. McFaden, "Colorectal Cancer," *Postgraduate Medicine,* November 1989.
Friedman, G. "Peptic Ulcer Disease," *Clinical Symposia,* 40(5), 1988.
Goldfinger, S. E. (ed). "Shock Waves for Gallstones," *Harvard Medical School Health Letter,* July 1988.
______. "Screening for Bowel Cancer," *Harvard Medical School Health Letter,* April 1986.
Hogan, M. J. (ed). "Gallstones," *Mayo Clinic Health Letter,* May 1989.
______. "Plaque and Tartar," *Mayo Clinic Health Letter,* January 1989.
Jenkins, E. "Large Bowel Exam: When to Use What Test," *Modern Medicine,* March 1984.
Johnson, D. A. "Fecal Occult Blood Testing," *Postgraduate Medicine,* April 1989.
Johnson, G. T. (ed). "The Tragedy of Anorexia Nervosa," *Harvard Medical School Health Letter.* December 1981.
Leff, E. "Hemorrhoids," *Postgraduate Medicine,* November 1987.
Levine, M. P. "Eating Disorders," *Postgraduate Medicine,* November 1988.
Rios, J. E. (ed). "Which Kind of Fiber Will Help You?" *Cardiac Alert,* December 1989.
Steinberg, S. "Cancer and Cuisine," *Science News,* 1 October 1983.
Uvnäs-Moberg, K. "The Gastrointestinal Tract in Growth and Reproduction," *Scientific American,* July 1989.

Chapter 24

Cochran, J. S., S. N. Robinson, V. S. Crane, and D. G. Jones. "Extracorporeal Shock Wave Lithotripsy," *Postgraduate Medicine,* May 1988.
Dretler, S. P. "Stone Crushers," *Harvard Medical School Health Letter,* December 1985.
Nolph, K. D., A. S. Lindblad, and J. W. Novak. "Continuous Ambulatory Peritoneal Dialysis," *New England Journal of Medicine,* 16 June 1988.
Reilly, N. J., and L. C. Torosian. "The New Wave in Lithotripsy," *RN,* March 1988.
Smith, D. A. "Incontinence," *RN,* March 1989.
Spiegel, D. M., M. Burnier, and R. W. Schrier. "Acute Renal Failure," *Postgraduate Medicine,* September 1987.
Twardowski, Z. J. "Peritoneal Dialysis," *Postgraduate Medicine,* April 1989.

Chapter 25

Cole, H. M. "Intrauterine Devices," *Journal of the American Medical Association,* 14 April 1989.

Copeland, L. J. "Screening for Cervical Cancer: The Role of the Pap Test," *Modern Medicine,* January 1987.

Danforth, D. N., and J. R. Scott (eds). *Obstetrics and Gynecology,* 5th ed. Philadelphia: Lippincott, 1986.

Goldfinger, S. E. (ed). "Breast Cancer: Early Decisions," *Harvard Medical School Health Letter,* March 1988.

______. "Circumcision and Urinary Tract Infections," *Harvard Medical School Health Letter,* April 1989.

______. "Prostate Cancer," *Harvard Medical School Health Letter,* September 1988.

Johnson, G. T. (ed). "Cesarean Section," *Harvard Medical School Health Letter,* June 1981.

______. "Sexually Transmitted Diseases," *Harvard Medical School Health Letter,* April 1981.

Kiely, J. M. (ed). "Endometriosis," *Mayo Clinic Health Letter,* March 1987.

______. "Premenstrual Syndrome," *Mayo Clinic Health Letter,* February 1987.

McElhose, P. "The Other STDs: As Dangerous As Ever," *RN,* June 1988.

McKeon, V. A. "Cruel Myths and Clinical Facts About Menopause," *RN,* June 1989.

Mishell, D. R. "Contraception," *New England Journal of Medicine,* 23 March 1989.

Nero, F. A. "When Couples Ask About Infertility," *RN,* November 1988.

Orshan, S. A. "The Pill, the Patient, and You," *RN,* July 1988.

Rivers, R. R. "Breast Cancer," *Postgraduate Medicine,* November 1988.

Sweet, R. L. "Pelvic Inflammatory Disease," *Modern Medicine,* April 1989.

Townsend, C. M. "Management of Breast Cancer," *Clinical Symposia,* 39(4), 1987.

Chapter 26

Begley, S., and J. Carey, "How Human Life Begins," *Newsweek,* 11 January 1982.

Carlson, B. M. *Patten's Foundations of Human Embryology,* 4th ed. New York: McGraw-Hill, 1981.

England, M. A. *Color Atlas of Life Before Birth.* Chicago: Year Book Medical Publishers, 1983.

Fuchs, F. "Genetic Amniocentesis," *Scientific American,* June 1980.

Gehring, W. J. "The Molecular Basis of Development," *Scientific American,* October 1985.

Gold, M. "The Baby Makers," *Science,* April 1985.

Goldfinger, S. E. (ed). "Pregnancy: Age and Outcome," *Harvard Medical School Health Letter,* October 1985.

Grimes, D. A. "Reversible Contraception for the 1980's," *Journal of the American Medical Association,* 3 January 1986.

Hacker, N. F., and J. G. Moore. *Essentials of Obstetrics and Gynecology.* Philadelphia: Saunders, 1986.

Henahan, J. F. "Fertilization, Embryo Transfer Procedures Raise Many Questions," *Journal of the American Medical Association,* 17 August 1984.

Hogan, M. J. "Cesarean Birth," *Mayo Clinic Health Letter,* February 1988.

Holliday, R. "A Different Kind of Inheritance," *Scientific American,* June 1989.

Kantrowitz, B., P. Wingert, and M. Hager. "Preemies," *Newsweek,* 16 May 1988.

Leaf, D. A. "Exercise During Pregnancy," *Postgraduate Medicine,* January 1989.

Miller, J. A. "Window on the Womb," *Science News,* 2 February 1985.

Moore, K. L. *Essentials of Human Embryology.* Toronto: B. C. Decker, 1988.

Seibel, M. M. "A New Era in Reproductive Technology," *New England Journal of Medicine,* 31 March 1988.

Silberner, J. "Babymaking: Expanding Horizons," *Science News,* 14 December 1985.

Singer, S. *Human Genetics.* New York: W. H. Freeman, 1985.

Wassarman, P. M. "Fertilization in Mammals," *Scientific American,* December 1988.

Zuckerman, B. "Effects of Maternal Marijuana and Cocaine Use on Fetal Growth," *New England Journal of Medicine,* 23 March 1989.

Art Credits

Charles Bridgeman 20.14

Leonard Dank 1.4, 4.4a/b, 4.5, 5.1, 5.3a/c/d, 5.4, 5.6, 5.7a-f, 6.1–6.7, 6.8a/b, 6.9–6.12, 6.13a/c, 6.14e, 6.15c, 6.16c, 6.19, 6.20, 7.1, 7.3, 7.4, 7.5a/b, 6.7, 7.7, 7.8c, 7.10, 7.11, 7.13, 7.15, 7.16, 8.1a, 8.7, 8.8, 8.9a, 8.10, 8.11a/b, 8.12a/b/c/e, 8.13a, 9.1a, 9.10, 10.1–10.3, 10.4a-c, 10.5–10.8, 10.9a, 10–10, 10.11, Exhib. 10–10, 10.12a-c, 10.13–10.16, 10.17a-c, 10.18a/b, 10.19a-f, 10.20a/b, 10.21, 10.22a-d, 10.23, 10.24a-d, 10.25a-d, 11.1, 13.4a/b, 13.5, 13.8, 13.10a, 13.11, 14.2, 14.4a/b, 14.5, 14.8a/d/e, 14.9c-e, 14.10a/b, 14.12, 14.13a, 14.15, 14.19, 15.2a, 15.6a, 16.2, 18.6, 18.9a/d, 18.13, 18.14, 18.16–18.25, 20.10, 20.13c, 22.8, 22.11, 23.3, 23.8, 23.9a, 23.11, 23.14a, 23.16b/c, 23.19a/b, 25.5, 25.18, 25.25, 25.26, 26.2, 26.3a-c, 26.6

Marsha Dohrman 20.5, 20.6, 20.7, 20.9

Sharon Ellis 15.1, 15.2b, 15.3, 15.7a, 15.8–15.13, 16.3a/b, 16.4a/b, 16.5–16.7, 16.9, 17.1a, 17.2a, 17.3a, 17.4, 17.5, 17.6a, 17.7–17.9a/b, 17.10–17.12, 18.1a, 18.2, 18.4a, 18.5a/b, 18.7a/c, 18.8, 18.11a, 18.12, 19.1, 19.3a

Lauren Keswick 2.1, 2.2a inset, 2/2b, 2.3, 2.5c, 2.6c, 2.7c, 2.8, 2.9a inset, 2.10c, 2.11, 2.12a/b, 2.15, 2.17, 2.18, 2.19, 3.1, Exhib. 3.2 (p. 76), 4.1a, 4.3, 4.4 insets

Biagio John Melloni 19.2, 20.13a, 20.15, 20.17, 20.18, 20.19, 20.20

Hilda Muinoz 13.1a, b (inset), 13.2, 13.3a/c/d/e, 13.6, 13.7a, 14.1a-c, 14.6, 14.7, 14.9a, 14.11, 14.14a, 14.16a, 14.17a, 14.18a

Lynn O'Kelley 20.1, 20.4, 20.8, 20.11, 20.13b, 20.16, 20.19 inset, 20.20 inset, 21.1, 22.2, 21.5, 21.6, 21.7a, 21.9, 21.10b, 21.11, 21.15, 21.16a, 21.17, 22.1, 22.2, 22.3 inset, 22.4a/b, 22.5 inset, 22.6, inset, 22.6, 22.7a-d, 22.7 inset, 22.10

Nadine Sokol Exhib. 3.1 insets, 12.2a, 12.3a, 13.10b/c, 23.1, 23.3, 23.4, 23.5a, 23.6a, 23.7, 23.10a, 23.12, 23.13a/inset, 23.18, 23.20, 24.1, 24.2b/d (inset), 24.3, 24.4, 24.6, 24.7, 24.8, 24.11 a/inset, 24.13

Kevin Somerville 1.1, Exhib. 1.2, 1.2, 1.3, 1.5a–c, 1.6a-b, 1.7a,c,insets, 1.8a-f, 1.9, 4.2b, 8.6 b/w art, 9.6, Exhib. 10.19, 14.3a, 18.1 inset, 18.5 inset, 18.7 inset, 18.8 inset, 18.9 inset, 18.10 inset, 18.11 inset, 18.12 inset, 22.9a, 25.1, 25.2c, 25.3a, 25.8, 25.9a/inset, 25.10a/b, 25.11, 25.12a/d/insets, 25.13a, 25.17, 25.20, 25.21, 25.22a/b, 26.1, 26.5, 26.7, 26.9a, 26.10, 26.11a, 26.13, 26.14

Beth Willert 7.9, 9.2, 9.5c/d, 9.8, 9.9a-c, 20.3

Index

Page numbers followed by the letter *E* indicate terms to be found in exhibits.

A band, 217
Abdomen
  lymph nodes of, 442E–443E
  muscles that act on walls of, 263E
  surface anatomy of, 322E, 323
  veins of, 410E
Abdominal aorta, 358, 384, 385E, 398E
Abdominopelvic cavity, 15, 15E, 19–21
Abdominopelvic quadrants, 15
Abdominopelvic regions, 15
  representative structures found in, 20E
Abduction, 187, 190E
Abductor muscles, 241
Abnormal cell division, 53–56
  definition, 53
  possible causes of, 55–56
  spread of, 54–55
  treatment, 56
  types of, 53–54
Abnormal contraction of a muscle, 231
Abnormal uterine bleeding, 761
ABO blood grouping system, 341
Abortion, 786
Abrasion, 106
Absorption, 653, 679
Accessory ligaments, 183
Accessory sex glands, 718, 730–731
Accessory spleen, 434
Acetylcholine (ACh), 219, 464, 553
Acetylcholinesterase (AChE), 553
Achalasia, 665
Achondroplasia, 125
Achromatopsia, 597
Acidosis, 465
Acinar gland, 73, 74
Acne, 105
Acquired immune deficiency syndrome (AIDS), 446–449
  drugs and vaccines against HIV, 448–449
  HIV and, 55, 446–447
  prevention of transmission, 449
  symptoms, 447–448
  transmission of, 448
Acromegaly, 606–607
Acromion, 162
Acrosome, 725
Actin, 218
Active processes, 32, 33–34, 36
Active transport, 33–34
Acute normovolemic hemodilution, 347
Acute pain, 565
Acute prostatitis, 760
Acute renal failure (ARF), 713
Adam's apple, 319, 635
Adaptation, 562
Addison's disease, 614–616
Adduction, 187, 190E
Adductor muscle, 241
Adenine nucleotide, 48
Adenitis, 451
Adenohypophysis, 606
Adenoid. *See* Pharyngeal tonsil
Adenosarcoma, 54
Adipocyte, 76, 84
Adipose tissue, 79E, 84–85
Adrenal cortex, 614–616
Adrenal gland, 604, 614–617, 618E
  adrenal cortex, 614–616
  adrenal medulla, 555E, 604, 614, 617
  development of, 622
  disorders of, 614–617, 618E
  hormones of, 614, 616, 618E
Adrenal medulla, 555E, 604, 614, 617
Adrenergic fiber, 553
Adrenocorticotropic hormone (ACTH), 606, 609E
Adrenogenital syndrome, 616
Adrenoglomerulotropin, 622, 622E
Afferent neuron, 555
"Afterbirth," 421, 777
Afterimage, 563
Age spot. *See* Liver spot
Aging
  blood vessels and, 420
  cardiovascular system and, 420
  cells and, 56–57
  digestive system and, 684
  endocrine system and, 622
  free radical theory of, 57
  integumentary system and, 104
  lymphatic system and, 445
  muscle tissue and, 229
  nervous system and, 530
  reproductive system and, 754
  respiratory system and, 645
  skeletal system and, 120
  urinary system and, 711
Agnosia, 541

Agonist muscle, 239–240
Agranular ER, 39
Agranular leukocytes (agranulocytes), 336, 342, 345E
Agraphia, 521
AIDS. *See* Acquired immune deficiency syndrome
Albinism, 98
Albino, 99
Albumin, 336
Aldosterone, 614, 618E
Aldosteronism, 614
Alimentary canal. *See* Gastrointestinal (GI) tract
Alkalosis, 465
Allantois, 777
Allergen, 450
Allergy, 450
Allograft, 451
All-or-none principle, 221–222
Alphafetoprotein, 784
Alpha receptor, 554
Alveolar-capillary membrane, 644
Alveolar duct, 641
Alveolar macrophage, 643
Alzheimer's disease (AD), 540–541
Amacrine cell, 583
Amenorrhea, 760–761
Ametropia, 597
Amniocentesis, 775, 782
Amnion, 775
Amniotic fluid (AF), 775
Amphiarthroses, 182, 192
Amygdaloid nucleus, 518
Amyloid, 548
Anabolic steroid, 223
Anabolism, 5
Anal canal, 680
Analgesia, 541
Anaphase, 50, 51E
Anaphylaxis, 450
Anastomosis, 361, 378, 380
Anatomical position, 9, 10E
Anatomy. *See also* Developmental anatomy; Surface anatomy
  branches of, 2, 318
  characteristics, 5–7
  definition, 2
Anchoring filament, 428
Anemia, 345–340
Anesthesia
  general, 567
  spinal, 567
Aneuploid, 783
Aneurysm, 422
Angina pectoris, 361–362
Angiocardiography, 373
Angular movement, 186–187, 190E
Ankle, 206, 206E–207E
  surface anatomy of, 328E, 331
Ankle joint, 206, 206E, 207E
Ankle sprain, 206
Ankyloglossia, 658
Ankylosis, 209
Annulus fibrosus, 146
Anopsia, 597
Anorexia nervosa, 688
Antagonist muscle, 240
Anterior corticospinal tract, 479E
Anterior reticulospinal tract, 480E
Anterior spinocerebellar tract, 478E, 570
Anterior spinothalamic tract, 478E
Antibody, 342
Anticholinesterase drugs, 231
Antidiuretic hormone (ADH), 608, 609E
Antigen, 342
Anti-oncogene, 55
Anus, 680
Aorta, 384, 385E
  abdominal, 358, 384, 385E, 398E
  arch of, 358, 384, 385E, 390E
  ascending, 358, 384, 385E, 388E
  coarctation of, 370
  descending, 384
  thoracic, 385, 384, 385E, 394E
Aortic insufficiency, 360
Aortography, 423
Aphasia, 521
Apnea, 648
Apocrine gland, 74, 103
Apocrine sweat gland, 103
Aponeurosis, 215
Appendicitis, 686
Appendicular skeleton, 129, 162–179
  divisions of, 131E
Apraxia, 541
Arachnoid, 472, 499
Arches of the foot, 176–178
Arch of the aorta, 358, 384, 385E, 390E
Areflexia, 490
Areola, 749
Areolar connective tissue, 84
Arm
  muscles that move, 275E
  surface anatomy of, 323E, 326
Arrector pili, 102
Arrhythmias, 372
Arterial stick, 334
Arteries, 378–380
  anastomosis, 378–380
  arcuate, 704, 742
  arterial branches of the aorta, 385E
  arterioles, 378, 380
  articular, 120
  basilar, 390E
  brachiocephalic, 390E
  bronchial, 394E
  celiac, 398E
  central retinal, 581
  common hepatic, 398E
  common iliac, 384, 399E
  coronary, 358, 360–361, 388E
  definition, 378
  elastic, 378
  end, 380
  esophageal, 394E

external iliac, 399E
gonadal, 398E
inferior mesenteric, 398E
inferior phrenic, 398E
interlobar, 704
interlobular, 704
internal iliac, 399E
left common carotid, 390E
left gastric, 385E
left pulmonary, 417
left subclavian, 390E
lumbar, 398E
mediastinal, 394E
middle sacral, 398E
muscular, 378
nutrient, 120
pericardial, 394E
periosteal, 120
posterior intercostal, 394E
radial, 325E, 327, 742
renal, 398E
right pulmonary, 417
splenic, 398E
subcostal, 394E
superior mesenteric, 398E
superior phrenic, 394E
suprarenal, 398E
umbilical, 417, 420
uterine, 742

Arteriole, 378, 380
Arteritis, 423
Arthralgia, 209
Arthritis, 207
Arthrodia. *See* Gliding joint
Arthrology, 182
Arthroplasty, 201
Arthroscopy, 185
Arthrosis, 209
Articular cartilage, 86, 112, 116, 183
Articular disc, 185
Articular tubercle, 138
Articulations. *See* Joints
Artificial ear, 592–593
Artificial heart, 365–366
Artificial kidney, 705–707
Artificial pacemaker, 364
Arytenoid cartilage, 635
Ascending aorta, 358, 384, 385E, 388E
Ascending colon, 680
Ascending tract of the spinal cord, 472, 477, 478E
Asphyxia, 648
Aspiration, 648
Association neuron, 557
Astereognosis, 597
Aster microtubule, 50
Astrocyte, 458E
Ataxia, 523
Atelectasis, 648
Atherosclerosis, 368–369
Atherosclerotic plaque, 368
Atretic follicle, 738
Atria, 355
Atrial natriuretic factor (ANF), 624
Atrioventricular (AV) node, 364
Atrioventricular (AV) valve, 358
Atrophy, 57
Auditory ossicle, 587
Auditory sensation, 583–595
auditory pathways, 588–592
external (outer) ear, 583–587, 596E
internal (inner) ear, 588, 596E
mechanism of equilibrium, 593–595
middle ear, 587–588, 596E
Auditory tube, 587
Auricle, 355
Auscultation, 365
Autocrine motility factor (AMF), 54
Autograft, 451
Autoimmune disease, 106, 449
Autologous intraoperative transfusion (AIT), 347
Autologous preoperative transfusion, 347
Autolysis, 42
Autonomic ganglia, 548–550
Autonomic nervous system (ANS), 456, 546–560
activities of, 555E, 556
biofeedback and, 557–558
control of, 557–558
definition, 547
meditation and, 558
neurotransmitters of, 553–554
parasympathetic division, 552–553, 555E
receptors of, 544–556
somatic efferent nervous system compared to, 547–548E
structure of, 547–551
sympathetic division, 551–552, 553E
visceral autonomic reflexes of, 556–557
visceral efferent neurons of, 547
Autophagy, 42
Autopsy, 21
Autorhythmicity, 225
Autosome, 51, 723
Avascular blood vessel, 64
Axial skeleton, 129–157
divisions of, 131E
Axon, 460
Axon collateral, 460
Axon hillock, 460
Axon terminal, 460
Axoplasm, 460
Azotemia, 713
AZT (azidothymidine), 448

Babinski sign, 573
Baby teeth, 662
Back, 320
Backbone. *See* Vertebral column
Baldness, 102
Ball-and-socket joint, 192, 193E
Baroreceptor, 365
Basal cell, 573–574
Basal cell carcinoma (BCC), 105
Basal ganglia, 516–517

Basal lamina, 64
Basement membrane, 64
Basophil, 341–342, 345E
Belly (gaster), 238
Benign prostatic hyperplasia (BPH), 760
Benign tumor, 53
Beta blocker, 369
Beta receptor, 554
Biaxial joint, 192
Bilaterally symmetrical, 8
Bile capillary, 673
Bile colic, 687
Bilirubin, 673
Binge-purge syndrome, 688
Biofeedback, 557–558
Biopsy, 57, 63
Bipennate muscle, 239
Birth control (BC), 752–754, 755E
  barrier, 753, 755E
  chemical, 753, 755E
  coitus interruptus, 753, 755E
  hormonal, 752, 755E
  induced abortion, 753–754, 755E
  intrauterine device, 752–753, 755E
  physiologic, 753, 755E
  sterilization, 752, 755E
Black eye, 134
Blackhead. *See* Comedo
Blastocele, 771
Blastocyst, 771
Blastomere, 771
Blepharitis, 597
Blind spot. *See* Scotoma
Blood, 334–350
  components, 335
  definition, 334
  disorders of, 345–347
  formed elements in, 336, 345E
  functions, 334
  key medical terms, 347
  physical characteristics of, 334
  plasma, 336–339
Blood–brain barrier (BBB), 503
Blood–cerebrospinal fluid barrier, 499
Blood doping. *See* Induced erythrocytemia
Blood reservoir, 382–383
Blood samples, 334
Blood supply
  of bone, 119–120
  of the brain, 502–503
  of the esophagus, 665
  of the heart, 360–361
  of the kidneys, 702–704
  of the large intestine, 683–684
  of the liver, 674–675
  of the lungs, 644–645
  of muscle tissue, 215
  of the skin, 100
  of the small intestine, 679
  of the stomach, 669
  of the teeth, 662
  of the ureter, 707
  of the urinary bladder, 710
Blood–testis barrier, 719–723
Blood tissue, 83E, 86
Blood transfusion. *See* Transfusion
Blood vessels, 377–426
  aging and, 421
  arteries, 378, 380
  arterioles, 378, 380
  capillaries, 378, 380–381
  definition, 378
  development of, 421
  disorders of, 421–423
  of fetal circulation, 417–418, 419
  of hepatic portal circulation, 416–417
  key medical terms, 423
  of pulmonary circulation, 417
  of systemic circulation, 384, 385E, 388E, 390E, 394E, 398E, 399E, 402E, 404E, 406E, 407E, 410E, 413E–414E
  veins, 378, 381–382
  venules, 378
Body cavity, 12–15
Boils, 103
Bolus, 663
Bone
  arch of the foot, 176–178
  arm, 162–166
  blood supply of, 119–120
  carpals, 166
  cervical vertebrae, 148, 150, 151
  clavicle, 162
  coccyx, 152, 154
  compact, 113–114
  cranial, 133–139
  disorders of, 122–123
  ethmoid, 139
  facial, 131, 139–144
  of the female skeleton, 178
  femur, 171, 173
  fibula, 175
  of the finger, 166–167
  flat, 129
  of the foot, 175–176
  of the forearm, 162–166
  formation of, 120, 122
  frontal, 133–134
  growth of, 118–119
  head, 130E
  heel, 331
  hip, 167–170
  histology, 111–114
  humerus, 162–163
  hyoid, 144–145, 320E
  inferior nasal conchae, 143
  irregular, 129
  kneecap, 174
  lacrimal, 143
  long, 129
  of lower extremity, 171–178
  lumbar vertebrae, 152
  of the male skeleton, 178

mandible, 142–143
maxillary, 140–141
metacarpals, 166
metatarsals, 175
nasal, 139
nerve supply of, 120
occipital, 136–137
osseous tissue, 83E, 86
palatine, 143
parietal, 134
patella, 174
pectoral (shoulder) girdle, 162
pelvic girdle, 167–170
phalanges of the finger, 167
phalanges of the foot, 176
pisiform, 166, 325E
radius, 166
replacement of, 119
ribs, 155–156
sacral vertebrae, 152
sacrum, 152
scapula, 162
sesamoid, 129
shinbone, 174–175
short, 129
of the skull, 131–144
sphenoid, 137, 139
spongy, 114
sternum, 154–155
surface markings, 129, 130E
sutural, 129, 132
tarsals, 175
temporal, 134, 136
thighbone, 171, 173
thoracic vertebrae, 151–152
thorax, 154–155
tibia, 174–175
types of, 129
ulna, 164, 166
of upper extremity, 162
vertebral column, 145–154
vomer, 143
zygomatic, 141–142
Bone marrow, red, 111
Bone marrow, yellow, 112
Bone marrow transplantation, 344
Bone tissue, 83E, 86, 111–124
disorders of, 122–123
functions of, 111
histology, 111–114
key medical terms, 125
ossification, 115–118
Borborygmus, 689
Botulism, 689
Botulinum toxin, 465
Bowman's gland. *See* Olfactory gland
Brachial plexus, 484–485, 487E
nerves of, 487E
Bradykinesia, 538
Brain, 498–545
blood supply of, 502–503
brain stem, 498, 503–508, 524E
cerebellum, 498, 521–523, 524E
cerebrospinal fluid, 499–502
cerebrum, 498, 513–521, 524E
coverings, 498
development from brain vesicles, 531–536
diencephalon, 498, 509–513, 524E
injuries to, 579
principal part of, 498, 524E
protection of, 498–499
split-brain concept, 521
Brain electrical activity mapping (BEAM), 539
Brain lateralization. *See* Split-brain concept
Brain sand, 622
Brain stem, 498, 503–508, 524E
medulla oblongata, 503–508, 524E
mesencephalon (midbrain), 508–509, 524E
pons, 508–509, 524E
Brain tumor, 537
Brain vesicle, 531–536
Brain wave, 521
Breast cancer, 749–751, 762
Breast cancer screening indicator, 751
Breathing, muscles used in, 266E
Breech presentation, 786
Bright's disease. *See* Glomerulonephritis
Bronchi, 629, 638–639
Bronchial asthma, 646
Bronchial tree, 638
Bronchiectasis, 648
Bronchitis, 646
Bronchogenic carcinoma, 646
Bronchogram, 639
Bronchography, 638–639
Bronchoscope, 636
Bronchoscopy, 636
Buccal gland, 658
Bulbourethral gland, 731
Bulimia, 688
Bulk flow, 33
Bunion, 178
Burkitt's lymphoma, 55
Burns, 105
Bursa, 185
Bursectomy, 209
Bursitis, 185, 208
Buttocks, surface anatomy of, 328, 328E

Calcaneal tendon, 331
Calcification, 111
Calcified cartilage, 116
Calcitonin, 609, 612E
Canaliculi, 86, 113
Cancer (CA)
breast, 749–751, 762
cell division and, 53–56
cervical, 762
definition, 53
of the larynx, 635
of the lung, 646
possible causes of, 55–56

Cancer (*Continued*)
skin, 105, 106
spread of, 54–55
testicular, 759–760
treatment, 56
types of, 53–54
of the uterus, 740–741
Capillary, 378, 380–381
definition, 378
Carbon monoxide poisoning, 648
Carboxypolypeptidase, 671
Capacitation, 769
Carcinogen, 55
Carcinoma, 53
basal cell, 105
bronchogenic, 646
nasopharyngeal, 55
squamous cell, 105
Cardiac arrest, 373
Cardiac center, 504
Cardiac cycle, 364–365
Cardiac muscle tissue, 87, 88E, 213, 224–225, 229E
Cardioacceleratory center (CAC), 366
Cardioinhibitory center (CIC), 366
Cardiology, 352
Cardiomegaly, 373
Cardiopulmonary resuscitation (CPR), 648
Cardiovascular system
aging and, 420
blood, 334–350
blood vessels, 378–426
the heart, 351–376
Carotene, 99
Carotid foramen, 138
Carpals, 166
Carpal tunnel, 166
Carpal tunnel syndrome, 486
Cartilage, 85
articular, 86, 112, 116, 183
arytenoid, 635
corniculate, 635
costal, 86, 155
cricoid, 635
cuneiform, 635
elastic, 82E, 86
hyaline, 81E, 85–86
Cartilaginous joints, 182, 183, 193E
Castration, 763
Catabolism, 5
Cataract, 582, 595–596
Cauda equina, 474
Caudal anesthesia, 152
Caveolae, 228
Cavities of the body, 12–15
Cell body, 89
Cell division
abnormal, 53–56
cancer, 53–56
definition, 47
events associated with, 51E
reproductive division, 51–53
somatic division, 47–50
Cell junction, 64–66
Cells
aging and, 2, 29, 56–58
definition, 2, 30
division of, 47–53
functions of cell parts, 46E
inclusions, 46E
key medical terms, 57
organelles, 2, 30, 34–36
parts of, 30, 46E
plasma membrane of, 30–34
Cell shapes, 64
Cellular level of structural organization, 2
Cementum, 662
Center, 481
Central canal, 86, 113
Central nervous system (CNS), 7E, 455, 467
aging and, 530
development of, 530–536
disorders of, 536–541
neuroglia of, 458E
neuronal pools in, 467
Centriole, 45
Centromere, 50
Centrosome, 45
Cerebellar peduncle, 522
Cerebellar tract, 570
Cerebellum, 498, 521–523, 524E
damage to, 523
functions of, 522
structure, 522
Cerebral apoplexy, 536–537
Cerebral aqueduct, 499
Cerebral cortex, 513
functional areas of, 519–521
Cerebral palsy (CP), 537–538
Cerebral peduncle, 508
Cerebrospinal fluid (CSF), 499–502
Cerebrovascular accident (CVA), 536–537
Cerebrum, 498, 513–521, 524E
basal ganglia (cerebral nuclei), 516–517
functional areas of cerebral cortex, 519–521
limbic system, 518–519
lobes, 516
white matter, 472, 474–477, 516
Cerumen, 103
Ceruminous gland, 103
Cervical canal, 740
Cervical cancer, 762
Cervical cap, 753
Cervical enlargement, 474
Cervical mucus, 740, 745
Cervical plexus, 483
nerves of, 484E
Cervical plug, 740
Cervical vertebrae, 148, 150, 151
Cesarean section (C-section), 786
Chancre, 757
Chemical contraception, 753–755E

Chemical digestion, 653
Chemical level of structural organization, 2
Chemical synapse, 464
Chemoreceptor, 563
Chemotaxis, 342
Chest, surface anatomy of, 320–323, 321E
Cheyne-Stokes respiration, 648
Chickenpox, 108
Chlamydia, 759
Cholecystitis, 689
Cholecystokinin (CCK), 624
Cholelithiasis, 689
Cholinergic fiber, 553
Chondritis, 209
Chondroblast, 116
Chondrocyte, 85
Chondroitin sulfate, 76
Chondrosarcoma, 54
Chordae tendineae, 358
Chorion, 775
Chorionic villi, 777
Chorionic villus sampling (CVS), 783
Choroid plexus, 499
Chromaffin, 617
Chromatid, 50
Chromatin, 36, 48
Chromatolysis, 466–467
Chromosome, 36, 47
Chronic fatigue syndrome, 347
Chronic obstructive pulmonary disease (COPD), 647
Chronic pain, 565
Chronic prostatitis, 760
Chronic renal failure (CRF), 713
Chyme, 668, 681
Chymotrypsin, 670
Cilia, 46, 72
  function of, 46E
Ciliary body, 578
Ciliary ganglion, 530, 553
Circular pattern of fasciculi, 239
Circumcision, 733
Circumduction, 188, 190E
Cirrhosis, 687
Citrated whole blood, 347
Claudication, 423
Clavicle, 162, 321E
Clawfoot, 178
Cleavage, 771
  lines of, 98
Cleft lip, 141
Cleft palate, 141
Clitoris, 747
Cloaca, 711
Cloacal membrane, 684
Closed fracture, 123
Closed reduction, 123
Coarctation of the aorta, 370
Coccygeal vertebrae, 146, 154
Coccyx, 146, 152, 154
Cochlea, 588
Cochlear duct, 588
Cochlear implant, 592–593
Coitus. *See* Sexual intercourse
Coitus interruptus, 753–755E
Cold sore, 108
Colitis, 689
Collagen implant, 99
Collagenous fiber, 76
Collarbones. *See* Clavicle
Collecting tubule, 702
Colles' fracture, 123
Colon, 680
Colonoscopy, 686
Colostomy, 689
Colposcopy, 742, 743
Column, 472
Columnar cell, 64
Coma, 541
Comedo (blackhead), 103
Comminuted fracture, 123
Commissurotomy, 373
Common cold, 647–648
Complete blood count (CBC), 344–345
Complete fracture, 123
Compound gland, 73, 74
Computed tomography (CT) scanning, 22E
Computed tomography with mammography (CT/M), 751
Concentric lamella, 113
Concussion, 519
Condom, 753, 755E
Conduction deafness, 596
Conduction myofiber, 364
Conduction pathway, 480
Condylar process, 142
Condyle, 130E
Cone, 581
Cone biopsy, 741
Congenital defect, 370
Congestive heart failure (CHF), 372–373
Conjoined twins, 770
Conjunctivitis, 596
Connective tissue, 63, 75–86
  adult, 75–86
  basic elements, 75–76
  classification of, 77
  embryonic, 76E, 84
  general features, 75
  skeletal muscle tissue and, 213–215
Connexon, 66
Constipation, 683
Contact inhibition, 54
Continuous ambulatory peritoneal dialysis (CAPD), 707
Continuous capillary, 380
Contraception, 752–754, 755E
Contraceptive sponge, 753, 755E
Contusion, 106, 519
Conus medullaris, 474
Convergence, 465
Converging circuit, 467
Coracoid process, 162
Corn, 108
Cornea, 578

Corneal transplant, 578
Corniculate cartilage, 635
Coronal suture, 131
Corona radiata, 769
Coronary artery bypass grafting (CABG), 369
Coronary artery disease (CAD), 368–370, 422
Coronary artery spasm, 369–370
Coronary sinus, 358, 361, 384
Coronary sulcus, 355
Coronoid process, 144
Corpora quadrigemina, 508
Cor pulmonale (CP), 373
Corpus albicans, 746
Corpus callosum, 516
Corpuscle of touch, 98, 564–565
Corpus hemorrhagicum, 745
Corpus luteum, 745
Corpus striatum, 516
Cortical nephron, 700
Corticolipotroph cell, 606
Corticosterone, 614, 618E
Cortisol, 614, 618E
Cortisone, 614, 618E
Coryza. *See* Common cold
Costal cartilage, 86, 155
Covergent pattern of fasciculi, 239
Covering and lining epithelium, 64–73
  classification of, 66
Coxal joint. *See* Hip joint
Crack, 465
Cracking sound of synovial joint, 183
Cranial bones, 131, 133–139
  ethmoid bone, 139
  frontal bone, 133–134
  occipital bone, 136
  parietal bone, 134
  sphenoid bone, 137, 139
  temporal bone, 134, 136
Cranial cavity, 15
Cranial fossa, 139
Cranial meninges, 472, 499
Cranial nerves, 523–530, 533E–535E
  abducens, 527, 533E
  accessory, 530, 534E
  clinical application of, 524E, 534E
  facial, 527, 534E
  functions of, 524E, 534E
  glossopharyngeal, 529, 534E
  hypoglossal, 530, 535E
  oculomotor, 526–527, 533E
  olfactory, 525, 533E
  optic, 527, 533E
  trigeminal, 527, 533E
  trochlear, 529–530, 533E
  vagus, 529–530, 534E
  vestibulocochlear, 528–529, 534E
Cranial parasympathetic outflow, 552
Craniosacral outflow, 540
Craniotomy, 125
Cremaster muscle, 718
Crest, 130E
Cretinism, 611
Cricoid cartilage, 635
Cristae, 42, 569
Crista galli, 139
Crossing-over, 52, 724
Cross section, 11–12
Crutch palsy, 485
Cryptorchidism, 718
CT scan. *See* Computed tomography (CT) scanning
Cuboidal cell, 64
Culdoscopy, 763
Culdotomy, 763
Cuneiform cartilage, 635
Curare, 465
CURL (compartment of uncoupling of receptor and ligand), 34
Cushing's syndrome, 614–616
Cuspids, 662
Cutaneous membrane, 87
Cutaneous plexus, 100
Cutaneous sensations, 563–567
Cuticle, 103
Cuticle of the hair, 100
Cyanosis, 347, 372
Cyst, 108
Cystic fibrosis (CF), 687
Cystitis, 713
Cystocele, 713
Cytokinesis, 47, 50, 51E
Cytology, 29
Cytoplasm, 34
  composition of, 34
  definition, 34
  function of, 34, 46E
Cytoskeleton, 44–45
Cytosol, 34, 46
Cytotrophoblast, 771

D and C (dilation and curettage), 741
Dartos, 718
Daughter cell, 47
Deafness, 596
Decidua, 777
Deciduous teeth, 662
Decubitus ulcer, 106
Decussation of pyramids, 503
Deep fascia, 215
Deep-venous thrombosis (DVT), 422
Defecation, 653, 682, 683
Defensin, 342
Defibrillation, 372
Deflection wave, 364
Deglutition, 663
Delirium, 541
Dementia, 541
Demineralization, 122
Dendrite, 89, 460
Dense body, 225
Dense connective tissue, 85
Dense irregular connective tissue, 80E, 85
Dense regular connective tissue, 80E, 85
Dental caries, 684–688

Dental plaque, 684
Dental terminology, 662
Dentes. *See* Teeth
Denticulate ligament, 474
Dentin, 662
Dentition, 662
Deoxyribonuclease, 671, 679
Deoxyribose, 48
Depilatory, 102
Depression, 190, 190E
Depressor muscle, 241
Dermal papilla, 98
Dermatan sulfate, 76
Dermatome, 229, 489–490
Dermis, 97–98
  development of, 104–105
Descending aorta, 384
Descending colon, 680
Descending tract of the spinal cord, 472, 477, 479E–480E
Desmosome, 65
Deterioration, 57
Detrusor muscle, 708
Developmental anatomy, 768–788
  of blood vessels, 421
  of the digestive system, 684
  embryonic development, 773–779, 780E
  of the endocrine system, 622
  of the epidermis, 104–105
  fetal growth, 779–780
  gestation, 780–782
  of the heart, 367
  of the integumentary system, 104–105
  key medical terms, 786
  of the lymphatic system, 445–446
  of the muscular system, 229–230
  of the nervous system, 531–536
  parturition and labor, 784
  during pregnancy, 769–773
  prenatal diagnostic techniques, 782–784
  of the reproductive system, 754–757
  of the respiratory system, 645–646
  of the skeletal system, 120, 122
  of the urinary system, 711
Deviated nasal septum (DNS), 143–144
Dextran, 684
Diabetes insipidus, 608
Diabetes mellitus, 620
Diapedesis, 342
Diaphragm, 753, 755E
Diaphysis, 111
Diarrhea, 683
Diarthroses, 182, 192
Diastole, 364
Diencephalon, 498, 509–513, 524E, 536
  epithalamus, 513, 524E
  hypothalamus, 501–513, 524E
  subthalamus, 513, 534E
  thalamus, 509–511, 524E
Differential white blood cell count, 343
Differentiation, 7
Diffuse lymphatic tissue, 428
Diffusion, 33
Digestion, 653
Digestive system, 7E, 652–693
  aging and, 684
  basic activities of, 653
  definition, 653
  development of, 684
  disorders of, 684–689
  general histology, 653–655
  key medical terms, 689
  organization of the GI tract, 653
  organs of, 656
  peritoneum of, 87, 655–656
Digital subtraction angiography (DSA), 24E
Diphtheria, 648
Diphyodonty, 662
Diplegia, 490
Diploid cell, 51
Diploid chromosome number, 723
Direct (immediate) transfusion, 347
Directional term, 9, 11E
Dislocation, 209
  of the jaw, 143
Displaced fracture, 123
Disturbance of gait, 523
Divergence, 465
Diverging circuit, 467
Diverticulitis, 686–687
Diverticulosis, 686
Dizygotic twins, 770
Dizziness, 523
DNA molecule, 47–48, 50
Dorsal body cavity, 15
Dorsal ramus, 483
Dorsiflexion, 188, 190E
Dorsum sellae, 137
Double helix, 48
Down syndrome (DS), 782
Drugs, passage through blood–brain barrier of, 503. *See also* names of drugs
Dual inversion, 547
Duchenne muscular dystrophy (DMD), 230
Duct, 718
Ductus arteriosus, 420
Ductus (vas) deferens, 726–727
Ductus venosus, 418
Duodenal gland, 676
Duodenal ulcer, 686
Duodenum, 675
Dura mater, 472, 499
Dynamic equilibrium, 594–595
Dynamic spatial reconstructor (DSR), 23E
Dyskinesia, 597
Dyslexia, 539
Dysmenorrhea, 761
Dysphagia, 689
Dysplasia, 57
Dyspnea, 648
Dysrhythmia, 372
Dystocia, 786
Dysuria, 714

Ear
  artificial, 592–593
  disorders of, 595–597
  external (outer), 583–587, 596E
  internal (inner), 588, 596E
  middle (tympanic cavity), 587–588, 596E
  surface anatomy of, 319E
Eardrum, 587
  perforated, 587
Eccrine sweat gland, 103
Ectoderm, 63, 104, 122, 774
  structures produced by, 776E
Ectopic pregnancy (EP), 770–771
Eczema, 108
Edema, 437
Effector, 481
Efferent neuron, 456
Effort, 238
Ejaculation, 732
Ejaculatory duct, 729
Elastic artery, 378
Elastic cartilage, 82E, 86
Elastic connective tissue, 81E, 85
Elastic fiber, 77
Elbow, surface anatomy of, 323E, 326
Elbow joint, 196E
Electrical synapse, 464
Electrocardiogram (ECG), 364
Electrocardiograph, 364
Electroconvulsive therapy (ECT), 541
Electroejaculation, 491
Electroencephalogram (EEG), 521
Electrolysis, 102
Electromagnetic photoreceptor, 563
Electromyogram, 224
Electromyography (EMG), 224
Electronic fetal monitoring (EFM), 784
Electron micrograph (EM), 29
Eleidin, 97
Elephantiasis, 451
Elevation, 190, 190E
Ellipsoidal joint, 192, 193E
Embryo, 77, 84
  definition, 773
  development of, 773–779, 780E
Embryology, 2E, 774
Embryonic connective tissue, 76E, 84
Embryonic development, 773–779, 780E
Embryonic membrane, 774–777
Embryo transfer, 773
Emphysema, 646–647
Enamel, 662
Encephalomyelitis, 541
End artery, 380
Endocarditis, 344
Endocardium, 354
Endocervical curettage (ECC), 741
Endochondrial ossification, 115, 116, 118, 120
Endocrine gland, 71E, 73, 604–622
Endocrine system, 7E, 603–627
  aging and, 622
  development and, 622
  disorders of, 606, 608, 611–612, 614, 616–617, 620
  glands of, 604–622
  hormones of, 604, 622
  key medical terms, 624
  other endocrine tissues, 624
Endocytosis, 34
Endoderm, 63, 104, 623, 645, 774
  structures produced by, 776E
Endolymph, 588
Endometriosis, 762
Endomysium, 215
Endoneurium, 482
Endoplasmic reticulum (ER), 39
  function of, 46E
Endoscopy, 668
Endosome, 34
Endosteum, 112
Endothelial-capsular membrane, 698
Endothelial tube, 367
Energy conservation-restorative system, 556
Enlarged prostate. *See* Benign prostatic hyperplasia (BPH)
Enteric gastrin, 624
Enteritis, 689
Enteroceptor, 563
Enterocrinin, 624
Enteroendocrine cell, 666–668
Enuresis, 714
Eosinophils, 341–342, 345E
Ependymocyte, 458E
Epicarditis, 354
Epicardium, 354
Epicondyle, 130E
Epidemiology, 57
Epidermal groove, 99
Epidermal ridge, 99
Epidermis, 95–97
  development of, 104–105
  epidermal derivatives, 100–102
Epididymis, 726
Epidural space, 472
Epigastric region, 20E
Epiglottis, 635
Epikeratoplasty, 578
Epilepsy, 537
Epileptic seizure, 537
Epimysium, 215
Epinephrine, 617, 618E
Epineurium, 482–483
Epiphora, 597
Epiphyseal line, 119
Epiphyseal plate, 118
Epiphysis, 111
Episiotomy, 748–749
Epithalamus, 513, 524E
Epithelial tissue, 63–64, 653
  covering and lining, 64–73
  general features, 63–64
  glandular, 63, 66E–71E, 73–74

Epithelium. *See* Epithelial tissue
Eponychium. *See* Cuticle
Epstein-Barr virus (EBV), 55
Equilibrium, 593–595
  dynamic, 594–595
  static, 593–594
  structures of the ear related to, 596E
Erection, 731–732
Eructation, 689
Erythema, 108
Erythrocytes (red blood cells), 86, 336, 339–341, 345E
  functions of, 339–340
  life span of, 340
  number, 340
  production of, 340–341
  structure, 339
Erythropoiesis, 340
Esophagus, 663–664
  blood supply of, 665
  nerve supply of, 665
Estrogen replacement therapy (ERT), 122
Estrogens, 621, 621E, 624, 723E, 744
Ethmoidal sinuses, 139
Euchromatin, 36
Eustachian tube. *See* Auditory tube
Eustachitis, 597
Eversion, 188, 190E
Exchange transfusion, 348
Excitability, 463
Exocrine gland, 71E, 604
  definition, 73
  functional classification of, 73–74
  structural classification of, 74
Exocytosis, 34
Exophthalmic goiter, 612
Exotropia, 597
Extension, 186–187, 190E
Extensor muscle, 241
Extensor retinaculum, 282
External hydrocephalus, 502
External occipital protuberance, 137
External respiration, 629–630
External urethral orifice, 748
External urethral sphincter, 708
Exteroceptor, 563
Extracapsular ligament, 183
Extracellular fluid (ECF), 32
Extracellular material, 63
Extracorporeal shock wave lithotripsy (ESWL), 711
Extraembryonic coelom, 774
Extrafusal muscle fiber, 567
Extrapyramidal pathway, 572
Extrinsic muscles, 251E, 253E
Eye
  accessory structures of, 576–578
  disorders of, 577–578, 581, 583
  effect of sympathetic and parasympathetic stimulation on, 555E
  surface anatomy of, 318E–319E
Eyeball
  fibrous tunic of, 578
  interior of, 582
  lens, 581–582
  muscles that move, 251E
  retina, 579–581
  structure, 578
  structures associated with, 582E
  vascular tunic of, 578–579
Eyebrows, 318E, 576
Eye examination, 582
Eyelashes, 319E, 576
Eyelids, 576

Face, surface anatomy of, 319, 318E–319E
Facet, 130E, 147
Facial bones, 131, 133–139
Facial expression muscles, 244E, 245E
Facilitated diffusion, 33
Falciform ligament, 655–656
Fallopian tubes. *See* Uterine tubes
False labor, 784
False rib, 155
Falx cerebri, 516
Fascia, 213
Fascia lata, 291E
Fascicles, 215, 482
Fasciculation, 231
Fasciculi, 215
  arrangement of, 239
Fasciculus cuneatus tract, 478E
Fasciculus gracilis tract, 478E
Fat
  brown, 84
  white, 84–85
Fecal occult-blood testing, 686
Feces, 681
Feeding center, 513
Female gonads. *See* Ovaries
Female infertility, 762
Female reproductive system, 733–751
Female skeleton, 178
Feminizing adenoma, 617
Femur, 171, 173
Fenestrated capillary, 380
Fertilization, 51, 769–771
  external human, 773
Fetal circulation, 417–418, 419
Fetal growth, 779, 780E
Fetal period, 774, 779, 780E
Fetal surgery, 779–780
Fetal-tissue transplantation, 779–780
Fetal ultrasonography, 783–784
Fetomaternal hemorrhage, 777
Fetus, 774
  growth of, 779, 780E
Fiber, dietary, 681
Fibrillation, 231
Fibrinogen, 339
Fibroadenoma, 762
Fibroblast, 75
Fibrocartilage, 82E, 86
Fibromyalgia, 230

Fibromyositis, 230
Fibrosis, 230
Fibrous astrocyte, 458E
Fibrous joint, 182, 193E
Fibrous layer, 112
Fibrous pericardium, 353–354
Fibrous ring, 351–353
Fibrous trigone, 351–353
Fibula, 175
  muscles that act on, 303E, 304E
Fight-or-flight response, 556
Film-screen mammography, 750
Filum terminale, 474
Fingers, muscles that move, 282E, 283E
Fingerstick, 334
First-class lever, 238
Fissure, 130E
Fixator, 240
Flaccid paralysis, 223, 573
Flagella, 46
  function of, 46E
Flatfoot, 178
Flatus, 689
Flexion, 186, 190E
Flexor muscle, 241
Flexor retinaculum, 282
Floating kidney. *See* Nephrotosis
Floating rib, 155
Fluid mosaic model, 30
Follicle-stimulating hormone (FSH), 606, 609E, 743
Fontanel, 130E, 131–132
Foot
  arch of, 176–178
  intrinsic muscles of, 311E, 312E
  muscles that move, 305E, 306E
  surface anatomy of, 328E, 331
Foot plate, 122
Foramen, 130E
Foramen ovale, 367, 420
Foramina, 144, 145E
Forearm
  muscles that move, 279E
  surface anatomy of, 325E, 326–327
Foreskin. *See* Prepuce
Fornix, 746
Fossa, 130E
Fossa ovalis, 355, 421
Fracture (Fx), 123–124
  of clavicle, 162
  Colles', 123
  Pott's, 123, 175
  repair of, 124
  types of, 123–124
  of the vertebral column, 157
Fraternal twins. *See* Dizygotic twins
Freckles, 99
Free nerve ending, 564
Free radical theory of aging, 57
Free ribosomes, 39
Frontal plane, 11
Frontal section, 12
Frontal sinus, 133–134
Frontal squama, 133
Fulcrum, 238
Functional classification of joints, 182
Fusiform, 239

G_1 period, 48
  activity of, 51E
G_2 period, 48
  activity of, 51E
Gallbladder (GB), 675
  effect of sympathetic and parasympathetic stimulation on, 555E
Gallstone, 687
Gamete intrafallopian transfer (GIFT), 773
Gametes, 51, 718, 723
Gamma globulin, 348
Ganglia
  basal, 516–517
  definition, 472
Gangrene, 231
Gap junction, 65–66, 225, 464
Gastrectomy, 689
Gastric gland, 666
Gastric inhibitory peptide (GIP), 624
Gastric lipase, 669
Gastric pit, 666
Gastroenterology, 653
Gastroileal reflex, 681
Gastrointestinal (GI) tract, 653
  products eliminated from, 695E
Gastroscopy, 668
Gastrulation, 774
Generalized animal cell, 29–30
General senses, 563–569
Generator potential, 562
Genetic material, 36
Geniculate ganglion, 527
Genital herpes, 759
Genital wart, 759
Geriatrics, 56
German measles, 106
Gestation, 780–782
Giantism (gigantism), 606
Gingivae, 661
Girdle, 130
Glabella, 133
Glands
  acinar, 74
  adrenal, 604, 614–617
  apocrine, 74, 103
  buccal, 658
  bulbourethral, 731
  ceruminous, 103, 583
  duodenal, 676
  effect of sympathetic and parasympathetic stimulation on, 555E
  endocrine, 71E, 73, 604–622
  exocrine, 71E, 73–74, 604
  gastric, 555E, 666
  holocrine, 74
  intestinal, 555E, 676
  lacrimal, 555E

mammary, 322E, 749–752
mandibular, 659
merocrine, 74
multicellular, 74
parathyroid, 604, 612–614, 614E
parotid, 659
pineal, 621–622
pituitary, 604, 605–608, 609E
prostate, 730
salivary, 658–661
sebaceous, 102–103
sebaceous ciliary, 576
of the skin, 103
sudoriferous, 102
sweat, 555E
tarsal, 576
thymus, 428, 434, 622, 622E
thyroid, 608–612
tubular, 73, 74
tubuloacinar, 73, 74
Glandular epithelium, 63, 71E, 73–74
Glans penis, 733
Glaucoma, 582, 596
Glenoid cavity, 162
Glial cell. *See* Neuroglia cell
Gliding joint, 190, 193E
Gliding movement, 186, 190E
Globulin, 336, 339
Glomerulonephritis, 711–713
Glomerulus, 698
Glottis, 635
Glucagon, 617, 621E
Glucocorticoids, 614, 618E
Glycogen, 46
Goblet cell, 72
Goiter, 611–612
Golgi complex, 39
function of, 46E
Gomphosis, 182, 193E
Gonadocorticoid, 616, 618E
Gonadotroph cell, 606
Gonadotropic hormone, 606
Gonadotropin releasing hormone (GnRH), 743–744
Gonads, 718
removal of, 752
Gonorrhea, 757
Gouty arthritis, 208
Grading malignant tumors, 54
Granstein cell, 95
Granular ER, 39
Granular leukocyte (granulocyte), 336, 341–342, 345E
Gray commisure, 474
Gray matter, 472, 474
Gray ramus communicans, 552
Greenstick fracture, 123
Groove, 130E
Gross anatomy, 2
Ground substance, 76
Growth, 7
Growth hormone-inhibiting factor (GHIF), 617
Gustatory sensations, 574–576
gustatory pathway, 576
receptors, 574–576
Gynecology, 734
Gynecomastia, 616–617
Gyrus (convolution), 513

Hair, 100–102
color, 102
definition, 100
development of distribution, 100, 102
structure, 100–102
Hair cell, 588–589
Hair follicle, 100
Hair root plexus, 102, 564
Hand
intrinsic muscles of, 288E, 289E
muscles that move, 282E, 283E
surface anatomy of, 326E, 327–328
Hand plate, 122
Haploid cell, 51
Haploid chromosome number, 723
Hard palate, 657
Haustral churning, 681
Head
lymph nodes of, 437E
muscles of, 262E
surface anatomy of, 318, 318E
veins of, 404E
Headache, 540
Heart, 351–376
artificial, 365–366
autonomic control of, 365
blood supply of, 360–361
cardiac cycle, 364–365
chambers of, 355, 358
conduction system, 362–364
definition, 352
development of, 367
disorders of, 368–373
effect of sympathetic and parasympathetic stimulation on, 555E
electrocardiogram (ECG), 364
great vessels of, 358
heart wall, 354
key medical terms, 373
location, 352–353
pericardium, 353–354
skeleton of, 358
surface projection of, 360
valves of, 358–362
Heart disease, risk factors in, 366–367
Heart murmur, 360
Heart sound, 364
Heart transplant, three-way, 366
Heimlich maneuver, 648
Hematocrit (Hct), 341
Hematology, 334
Hematopoiesis, 111, 339
Hemiballismus, 517
Hemidesmosome, 65
Hemiplegia, 490

Hemochromatosis, 348
Hemocytoblast, 339
Hemodialysis therapy, 705–707
Hemoglobin, 99, 339
Hemolytic anemia, 345–346
Hemolytic jaundice, 674
Hemopoiesis, 111, 339
Hemoptysis, 648
Hemopump, 366
Hemorrhage, 348
Hemorrhagic anemia, 345
Hemorrhoids, 680–681
Hemothorax, 639
Hepatic cell, 673
Hepatic portal circulation, 416–417
Hepatitis, 687
Hepatitis A, 687
Hepatitis B, 687
Hepatitis C, 687
Hepatitis B virus (HBV), 55
Hermaphroditism, 763
Hernia, 689
Herniated disc, 156
Herpes simplex virus, 55
Heterochromatin, 36
High-density lipoprotein (HDL), 368
Hinge joint, 190, 193E
Hip, 200E–201E
Hip joint, 200E–201E
Hippocampus, 518
Hirsutism, 107
Histology, 2E, 63
Hives, 107
Hodgkin's disease (HD), 55, 451
Holocrine gland, 74
Holter monitor, 364
Homologous chromosomes, 51, 723
Homologous organs, 734
Horizontal plane, 11
Hormones, 605
  regulating, 606
  tropic, 606
Horn, 472
Horner's syndrome, 556
Human chorionic gonadotropin (hCG), 624, 746
Human chorionic somatomammotropin (hCS), 624
Human growth hormone (hGH), 609E
Human immunodeficiency virus (HIV), 55, 446
  drugs and vaccines against, 448–449
  outside the body, 448
  structure and pathogenesis, 446–447
  transmission of, 448
Human leukocyte associated (HLA) antigen, 342
Humeroscapular joint. *See* Shoulder joint
Humerus, 162–163
Huntington's chorea, 541
Hyaline cartilage, 81E, 85–86
Hyaluronic acid, 76
Hyaluronidase, 84
Hymen, 746
Hyperacusia, 588
Hyperemesis gravidarum, 773
Hyperextension, 187, 190E
Hyperinsulinism, 620, 621
Hyperparathyroidism, 614, 614E
Hyperplasia, 57, 624
Hypersecretion, 606
Hypersensitivity, 450
Hypersplenism, 451
Hypertension, 421–422
Hypertonia, 223
Hypertrophy, 57
Hypochondriac region, 20E
Hypodermis, 107
Hypogastric region, 20E
Hypoparathyroidism, 614, 614E
Hypoplasia, 624
Hyposecretion, 606
Hypospadias, 763
Hypotension, 423
Hypothalamus, 511–513, 534E
  mammillary bodies of, 512, 518
Hypotonia, 223
Hypoxia, 648
Hysterectomy, 741
H zone, 218

I band, 218
Identical twins. *See* Monozygotic twins
Idiopathic seizure, 537
Ileum, 676
Iliotibal tract, 296E
Ilium, 168
Immovable joints. *See* Synarthroses
Immunologic tolerance, 449
Immunosuppressive drug, 451
Impacted cerumen, 103
Impacted fracture, 123
Imperforate hymen, 747
Impetigo, 107
Implantation, 771–772
Impotence, 760
Incisor, 663
Inclusion, 30
Incompetent valve, 373
Incontinence, 710
Indirect (mediate) transfusion, 384
Induced abortion, 753, 754, 755E
Induced erythrocytemia, 340
Infarction, 361–362
Infectious hepatitis. *See* Hepatitis A
Infectious mononucleosis (IM), 346–347
Inferior cervical ganglion, 552
Inferior vena cava, 358, 402
Infertility
  female, 762
  male, 760
Inflammatory bowel disease (IBD), 689
Influenza (flu), 647–648
Infrahyoid muscle, 260
Infrared photocoagulation, 681
Infundibulum, 512, 605, 739

Ingestion, 653
Inguinal canal, 728
Inguinal hernia, 728
Inguinal ligament, 263
Inhibin, 621, 621E, 723E
Insertion, 238
Insidious, 57
Insulin, 617, 621E
Insulin-dependent diabetes, 620
Integral protein, 30
Integrated response, 558
Integrator, 465
Integumentary system, 2E, 94–109
  aging and, 104
  definition, 95
  development of, 104–105
  disorders of, 105–106
  key medical terms, 106–107
  skin, 95–106
Interatrial septum, 355, 367
Intercalated disc, 87, 225, 362
Intermediate filament, 44, 225
  function of, 46E
Intermediate junction, 64
Internal hydrocephalus, 502
Internal respiration, 630
Internal urethral orifice, 708
Internal urethral sphincter, 708
Interphase, 51E
Interstitial fluid, 32, 334
Interstitial lamella, 114
Interventricular foramen, 499
Interventricular septum, 355, 367
Intervertebral disc, 146
Intervertebral joints, 194E, 195E
Intestinal gland, 555E, 676
Intestinal juice, 679
Intestines
  effect of sympathetic and parasympathetic stimulation on, 555E
  large, 679–684
  small, 675–679
Intracapsular ligament, 183
Intracellular fluid (ICF), 32
Intracerebral hemorrhage, 536
Intradermal, 107
Intrafusal muscle fiber, 567
Intramembranous ossification, 115, 116, 120
Intraorbital foramen, 141
Intrauterine device (IUD), 752–753, 755E
Intravenous pyelogram, 714
Intrinsic muscle, 288E, 289E
Intubation, 636–637
Inversion, 188, 190E
In vitro fertilization (IVF), 773
Involuntary muscle tissue, 213
Involution, 786
Iris, 578
Irritable bowel syndrome (IBS), 689
Ischemia, 361
Ischium, 168
Islet of Langerhans, 617, 670

Jarvik-7 (artificial heart), 365–366
Jaundice, 374
Jaw
  dislocation of, 143
  lower jaw muscles, 248E
Jejunum, 675
Joint kinesthetic receptor, 568–569
Joints, 182–211
  ankle, 206, 206E–207E
  ball-and-socket, 192, 193E
  cartilaginous, 183, 193E
  cracking sound and, 183
  definition, 182
  disorders of, 207–209
  elbow, 196E, 198
  ellipsoidal, 192, 193E
  fibrous, 182, 193E
  functional classification of, 182
  gliding, 190, 193E
  gomphosis, 182, 193E
  hinge, 190, 193E
  hip, 200E–201E
  intervertebral, 194E–195E
  key medical terms, 209
  knee, 202E–204E, 204–205
  pivot, 192, 193E
  saddle, 192, 193E
  shoulder, 195E
  structural classification of, 182
  symphysis, 183, 193E
  synchondrosis, 183, 193E
  syndesmosis, 182, 193E
  synovial, 182–192, 193E
  temporomandibular, 143
  wrist, 199E, 208–209
Jugular foramen, 138
Juxtaglomerular apparatus (JGA), 704
Juxtamedullary nephron, 700–701

Kaposi's sarcoma (KS), 55, 446, 447
Karyoplasm, 38
Karyotype, 786
Keratan sulfate, 76
Keratin, 73, 95
Keratinization, 97
Keratinocyte, 95
Keratitis, 597
Keratohyalin, 97
Ketosis, 620
Kidneys, 695–707
  anatomy of, 695–702
  artificial, 705–707
  blood supply of, 702
  effect of sympathetic and parasympathetic stimulation on, 555E
  hemodialysis therapy, 705–707
  juxtaglomerular apparatus (JGA), 704
  nephrons, 698–702

Kidneys (*Continued*)
  nerve supply of, 702–702
  products eliminated from, 695E
Kidney stone, 711
Killer T cell, 343
Kinesthesis, 598
Kinetochore, 50
Kinetochore microtubule, 50
Knee, 202E–204E, 204–205
  dislocated, 205
  injury to, 204–205
  runner's knee, 204
  surface anatomy of, 328E, 331
  swollen, 204–205
  transplant, 205
Knee joint, 202E–205E
Kyphosis, 157

Labia. *See* Lips
Labia major, 747
Labia minor, 747
Labile cell, 64
Labor, 784–786
  three stages of, 784–786
Labyrinthine disease, 597
Labyrinthitis, 598
Laceration, 107, 519
Lacotroph cell, 606
Lacrimal apparatus, 576
Lacrimal caruncle, 576
Lacrimal gland, 576
Lacrimal secretion, 576–577
Lactase, 679
Lactation, 749
Lactiferous sinus, 749
Lacuna, 85, 86, 113
Lambdoidal suture, 131
Lamella, 86
Lamellated corpuscle, 98, 565
Laminae, 146
Lamina propria, 86, 653
Langerhans cell, 95
Lanugo, 105
Laparascopy, 752
Large intestine, 679–684
  blood supply of, 683–684
  nerve supply of, 683–684
Laryngeal sinus, 635
Laryngitis, 635
Laryngopharynx, 633
Larynx, 629, 634–635
  cancer of, 635
  muscles of, 260E
Lateral corticospinal tract, 479E
Lateral reticulospinal tract, 479E
Lateral spinothalamic tract, 478E
Lateral ventricle, 499–502
Left iliac region, 20E
Left lumbar region, 20E
Left lymphatic duct, 435–436
Left ventricular assist device (LVAD), 365–366
Leg
  muscles that act on, 303E, 304E
  surface anatomy of, 328E, 331
Lens, 581–582
Lens capsule, 581
Lens epithelium, 581
Lens fiber, 581
Leptomeningitis, 474
Lesser pelvis, 170
Lethal gene, 786
Lethargy, 541
Leukemia, 55, 347
Leukocytes (white blood cells), 76, 86, 336, 341–344, 345E
  functions of, 342–343
  life span of, 343–344
  number, 343–344
  production of, 344
  structure, 341–344
  types, 341–344
Leukorrhea, 763
Levator muscle, 241
Lever, 238
Leverage, 239
Life processes, 5–7
Ligaments, 85, 183
  denticulate, 474
  extracapsular, 183
  falciform, 655–656
  intracapsular, 183
  periodontal, 661
  reconstruction of, 183
Ligand, 34
Limb bud, 120
Limbic system, 518–519
Lines of cleavage, 98
Lingual frenulum, 658
Lingual tonsils, 432
Lipid, 47
Lipoprotein, 368
Lips, 365
  surface anatomy, 319E
Little-league elbow, 198
Liver, 671–675
  blood supply of, 674–675
  effect of sympathetic and parasympathetic stimulation on, 555E
  nerve supply of, 674–675
Liver spot, 99
Living-donor liver transplant, 671
Lobes
  of the cerebrum, 516
  of the lungs, 640–641
Lobules, 641–643
Local disease, 57
Lochia, 786
Long bone, structure of, 129
Longitudinal arch of the foot, 176
Longitudinal axis of the human body, 129
Loose connective tissue, 84–85
Lordosis, 157
Low-density lipoprotein (LDL), 368

Lower extremities, 130, 171–178
  arteries of, 399E
  lymph nodes of, 438E
  surface anatomy of, 328E, 328, 331
  veins of, 413E
Lower jaw, muscles of, 248E
Lower respiratory system, 629
Lumbago, 230
Lumbar enlargement, 474
Lumbar plexus, 486, 489
  nerves of, 489E
Lumbar vertebrae, 152
Lumen, 378
Lumpectomy, 751
Lung cancer, 646
Lungs, 629, 639–645
  alveolar-capillary membrane, 644
  blood supply of, 644–645
  effect of sympathetic and parasympathetic stimulation on, 555E
  fissures, 640–641
  gross anatomy of, 640
  lobes, 640–641
  lobules, 641–643
  nerve supply of, 644–645
  products eliminated from, 695E
Luteinizing hormone (LH), 606, 609E
Luxation, 209
Lyme disease, 208
Lymphadenectomy, 451
Lymphadenopathy, 451
Lymphangiography, 428, 430
Lymphangioma, 451
Lymphangitis, 451
Lymphatic nodule, 428, 431
Lymphatic organs, 428, 432–435
Lymphatic system, 6E, 334, 427–454
  definition, 428
  development of, 445–446
  disorders of, 446–451
  key medical terms, 451
  lymphatic organs, 428, 431–432
  lymphatic tissue, 428, 431–432
  lymphatic vessels, 428
  lymph circulation, 435–437
  lymph nodes, 428, 430–434, 437–446
  relationship of cardiovascular system to, 429
Lymphatic tissue, 428, 430–435
  lymph nodes, 428, 430–434, 437–446
  spleen, 428, 432, 434
  thymus gland, 428, 434
  tonsils, 428, 432
Lymphatic vessel, 428
Lymph capillary, 428
Lymph circulation, 435–537
Lymphedema, 451
Lymph nodes, 428, 430–434, 436
  anterior mediastinal, 444E
  anterior phrenic, 444E
  axillary, 438E
  bronchial, 445E
  bronchopulmonary, 445E
  buccal, 437E
  celiac, 443E
  central, 438E
  common iliac, 442E
  deep cervical, 437E
  deep inguinal, 438E
  deltopectoral, 438E
  external iliac, 442E
  facial, 437E
  gastric, 443E
  hepatic, 443E
  ileocolic, 443E
  inferior deep cervical, 437E
  inferior mesenteric, 443E
  infraorbital, 437E
  intercostal, 444E
  internal iliac, 442E
  lateral, 438E
  lumbar, 443E
  mandibular, 437E
  mesenteric, 443E
  mesocolic, 443E
  middle phrenic, 444E
  occipital, 437E
  pancreaticosplenic, 443E
  parotid, 437E
  pectoral, 438E
  phrenic, 444E
  popliteal, 438E
  posterior mediastinal, 444E
  posterior phrenic, 444E
  preauricular, 437E
  pulmonary, 445E
  retroauricular, 437E
  sacral, 443E
  sternal, 444E
  subclavicular, 438E
  submandibular, 437E
  submental, 437E
  subscapular, 438E
  superficial cervical, 437E
  superficial inguinal, 438E
  superior deep cervical, 437E
  superior mesenteric, 443E
  supratrochlear, 438E
  tracheal, 445E
  tracheobronchial, 445E
Lymphoblast, 339
Lymphocyte, 75, 342, 345E
Lymphoma, 55, 451
Lymphostasis, 451
Lymph sac, 445
Lysosome, 39, 42
  function of, 46E

Macrophage, 75
  alveolar, 643
  wandering, 342
Macroscopic anatomy, 2E
Macula, 569
Macula lutea, 581

Magnetic resonance imaging (MRI), 23E
Male contraception, 754
Male infertility, 760
Male-pattern baldness, 102
Male reproductive system, 718–733
Male skeleton, 178
Malignancy, 53
Malignant melanoma, 99, 105
Malignant tumor, 53–54
Malocclusion, 689
Maltase, 679
Mammary duct, 749
Mammary gland, 322E, 749–752
Mammillary body, 512, 518
Mammillary region, 512
Mammogram, 750
Mammography, 750
Mandible, 142–143
Mandibular fossa, 138
Mandibular ganglion, 553
Mandibular gland, 659
Manubrium, 154
Marfan syndrome, 84
Marrow. *See* Bone marrow
Marrow biopsy, 155
Marrow cavity, 112
Mast cell, 76
Mastectomy, 751
Mastication, 663
Mastoiditis, 138
Mastoid process, 138
Matrix
  of connective tissue, 75
  of mitochondrion, 42
Matrix granule, 42
Mature connective tissue, 78E, 84
Maturity-onset diabetes, 620
Maxillae, 140–141
Measles, 107
Measurement of the human body, 24
Meatus, 130E
Mechanical digestion, 653
Mechanical methods of contraception, 752–753, 755E
Mechanoreceptor, 563
Medial lemniscus, 509
Medial malleolus, 175
Median nerve damage, 485–486
Mediastinum, 15
Medical imaging techniques, 20–24
Meditation, 558
Medulla, 100
Medulla oblongata, 503–508, 524E
Medullary cavity, 112
Megakaryoblast, 339
Meiosis, 52–53
  comparison between mitosis and, 54
Meiosis I, 52
Meiosis II, 52
Meissner's corpuscle. *See* Corpuscle of touch
Melanin, 46, 98
Melanoblast, 98
Melanocyte, 95
Melanocyte-stimulating hormone (MSH), 606, 609E
Melanoma, malignant, 99, 105
Melatonin, 622, 622E
Membrane, 87
Membrane capsule, 85
Menarche, 746
Ménière's syndrome, 597
Meninges, 472
Meningitis, 474
Menopause, 746
Menstrual abnormalities, 760–761
Menstrual cycle, 742–746
Menstruation, 744
Mental foramen, 143
Merocrine gland, 74
Mesencephalon, 508–509, 524E, 531
Mesenchyme, 76E, 84
Mesentery, 655
Mesoappendix, 679–680
Mesocolon, 655, 679
Mesoderm, 63, 104, 105, 122, 229, 367, 445
  structures produced by, 776E
Mesodermal cells, 230
Mesonephros, 711
Metabolism, 5
Metacarpals, 166
Metaphase, 50, 51E
Metaphysis, 111–112, 118
Metaplasia, 57
Metarteriole, 380
Metastasis, 54, 57, 420
Metatarsal, 175
Metencephalon, 536
Metopic suture, 133–134
Microfilament, 44, 65
  function of, 46E
Microglia, 458E
Microtrabeculae, 44
Microtrabecular lattice, 44–45
Microtubule, 44
  function of, 46E
Microvillus, 72, 676
Micturition, 708
Micturition reflex, 708
Midbrain. *See* Mesencephalon
Middle cervical ganglion, 552
Midline, 11
Midsagittal plane, 11
Midsagittal section, 11
Mifepristone (RU486), 754
Milk teeth, 662
Mineralocorticoids, 614, 618E
Mitochondria, 42, 44
  function of, 46E
Mitosis, 48–50
  comparison between meiosis and, 54
  time required for, 50
Mitral stenosis, 360
Mitral valve prolapse (MVP), 360
M line, 218

Molar, 662
Mole, 99
Monaxial joint, 192
Monoamine oxidase (MAO), 553
Monoblast, 339
Monocyte, 75, 342, 345E
Monoplegia, 490
Monozygotic twins, 770
Mons pubis, 747
Morbid obesity, 688
Morning sickness, 773
Morula, 771
Motion sickness, 597
Motor area of the cerebral cortex, 520–521
Motor cortex, 572
Motor end plate, 218
Motor neuron, 218
Motor pathway, 571–573
  extrapyramidal pathway, 572
  linkage of sensory input and motor responses, 571–572
  motor cortex, 572
  pyramidal pathway, 572
Motor unit, 219–220
Mouth, 656–663
  muscle of the floor, 259E
Movable joints. *See* Amphiarthroses; Diarthroses
Movement, 6
Mucosa. *See* Mucous membrane
Mucous connective tissue, 77E, 84
Mucous membrane, 86, 653–654
Multicellular gland, 73, 74
Multiple sclerosis (MS), 538–539
Multiunit smooth muscle tissue, 228
Mumps, 660
Muscarinic receptor, 554
Muscle-building anabolic steroid, 223
Muscle fiber, 217
Muscle spindle, 567–568
Muscle tissue, 87–89, 213–235, 237
  aging and, 229
  blood supply of, 215
  cardiac muscle tissue, 87, 88E, 224–225, 229E
  characteristics of, 213
  classification of, 213
  connective tissue components, 213–215
  contraction of, 218–224
  development of, 229–230
  disorders of, 230–231
  functions, 213
  histology, 217–218
  key medical terms, 231
  nerve supply of, 215
  skeletal muscle tissue, 87, 88E, 213–218, 229E
  smooth muscle tissue, 87, 89E, 225, 228, 229E
  types of, 213, 223–224
Muscle tone, 222
Muscular artery, 378
Muscular atrophy, 223
Muscular dystrophies, 230
Muscular hypertrophy, 223
Muscularis, 654
Muscularis mucosae, 86
Muscular system. *See* Skeletal muscles
Mutation, 786
Myalgia, 231
Myasthenia gravis (MG), 230–231
Mydriasis, 598
Myelencephalon, 536
Myelin sheath, 460
Myeloblast, 339
Myelography, 474
Myeloma, 55
  multiple, 348
Myocardial infarction (MI), 361–362
Myocarditis, 354
Myocardium, 354
Myofiber, 217
Myofibril, 217
Myofilament, 87, 217
Myoglobin, 217, 223
Myology, 213
Myoma, 231
Myomalacia, 231
Myoneural junction, 218
Myopathy, 231
Myosclerosis, 231
Myosin-binding site, 218
Myositis, 231
Myospasm, 231
Myotome, 229
Myotonia, 231
Myringitis, 598
Myxedema, 611–612

Nail matrix, 104
Nails, 103–104
  development of, 104
Nasal cavity, 630
Nasal septum, 139, 630
Nasopharyngeal carcinoma, 55
Nasopharynx, 632–633
Nausea, 689
Navel, 777
Nebulization, 643–644
Neck
  lymph nodes of, 438E
  surface anatomy of, 319–320, 320E
  veins of, 404E
Necrosis, 57
Neonatal jaundice, 674
Neoplasm, 53, 57
Nephroblastoma, 714
Nephron, 698–702
Nephrotic syndrome, 713
Nephrotosis, 698
Nerve, 462, 472
  abducens, 527, 533E
  accessory, 530, 534E
  ansa cervicalis, 484E
  axiliary, 487E
  cardiac, 365
  common peroneal, 491E

Nerve (*Continued*)
cranial, 523–530
deep peroneal, 491E
definition, 472
dorsal scapular, 487E
facial, 527, 534E
femoral, 489, 489E
genitofemoral, 489E
glossopharyngeal, 529, 534E
greater auricular, 484E
hypoglossal, 530, 535E
iliohypogastric, 489E
ilioinguinal, 489E
inferior gluteal, 491E
inferior lateral brachial cutaneous, 487E
inferior root, 484E
intercostal, 489
intercostobrachial, 487E
lateral antebrachial cutaneous, 487E
lateral cord, 487E
lateral femoral cutaneous, 489E
lateral pectoral, 487E
lateral plantar, 491E
lesser occipital, 484E
long thoracic, 487E
lower motor, 572
lower subcapsular, 487E
medial antebrachial cutaneous, 487E
medial brachial cutaneous, 487E
medial cord, 487E
medial pectoral, 487E
medial plantar, 491E
median, 487E
mixed, 481–482, 523
motor branch, 484E
musculocutaneous, 487E
obturator, 489E
obturator internus, 491E
oculomotor, 526–527, 533E
olfactory, 524–525, 533E
optic, 525, 533E, 583
pelvic splanchnic, 553
perforating cutaneous, 491E
phrenic, 484E
piriformis, 491E
posterior antebrachial cutaneous, 487E
posterior brachial cutaneous, 487E
posterior cord, 487E
posterior femoral cutaneous, 491E
pudendal, 491E
quadratus femoris, 491E
radial, 487E
root, 487E
sciatic, 491E
segmental branches, 484E
sensory, 484E, 523
splanchnic, 552
subclavius, 487E
superficial branch, 484E
superficial peroneal, 491E
superior gluteal, 491E
superior lateral brachial cutaneous, 487E
superior root, 484E
supraclavicular, 484E
suprascapular, 487E
thoracodorsal, 487E
tibial, 491E
transverse cervical, 484E
trigeminal, 527, 533E
trochlear, 529–530, 533E
ulnar, 487E
upper subscapular, 487E
vagus, 365, 529–530, 534E
vestibulocochlear, 528–529, 534E
Nerve block, 541
Nerve fiber, 460, 462
Nerve impulse, 463–466
Nerve supply
of bone, 120
of the esophagus, 665
of the kidneys, 702–704
of the large intestine, 683–684
of the liver, 674–675
of the lungs, 644–645
of muscle tissue, 215
of the small intestine, 679
of the stomach, 669
of the teeth, 662
of the ureter, 707
of the urinary bladder, 710
Nervous tissue, 89, 89E, 455–470
classification of, 462
conductivity of, 456–458
definition, 456
histology, 456–458, 460–463
nerve impulse, 463–466
organization of, 456
organization of neurons, 467–468
regeneration of neurons, 466–467
Neural deafness, 596
Neuralgia, 541
Neural tissue, 472
Neuritic plaque, 540
Neuritis, 491
Neuroblastoma, 624
Neuroeffector junction, 464
Neurofibral node, 460–461
Neurofibril, 458
Neurofibrillary tangle, 540
Neuroglandular junction, 464
Neuroglia, 89, 456–457
Neuroglia cell, 456–457
Neurohypophysis, 607–608
Neurolemmocyte, 458E, 460
Neurology, 456
Neuromuscular junction, 218–219, 464
Neuronal pool, 467
Neurons, 89, 218, 458–462
afferent, 456, 462, 587
alpha efferent, 568
association, 462, 557
bipolar, 462, 581

efferent, 456, 462
first-order, 570
gamma efferent, 568
ganglion, 581
motor, 462, 481
multipolar, 462
organization of, 467–468
photoreceptor, 581
postganglionic, 547, 550–551
postsynaptic, 465
preganglionic, 547, 548
presynaptic, 465
regeneration of, 466–467
second-order, 570
sensory, 462, 481
structural variation of, 461–462
structure of, 458–461
third-order, 570
unipolar, 462
upper motor, 572
visceral efferent, 547, 557
visceral efferent postganglionic, 557
visceral efferent preganglionic, 557
Neurosyphilis, 757–759
Neurotransmitter, 218–219, 464, 465, 533–554
Neutrophil, 341–342, 345E
Nevus, 107
Nicotinic receptor, 554
Nipple, 322E, 749
Nitrogen bases, 47
Nitroglycerin, 369
Nociceptor, 563, 565
Node of Ranvier. *See* Neurofibral node
Non-A, non-B (NANB) hepatitis, 687
Nonaxial joint, 190
Nondisplaced fracture, 123
Non-insulin-dependent diabetes, 620
Nonkinetochore microtubule, 50
Nonpigmented granular dendrocyte, 95
Nonstriated muscle tissue, 87, 213
Norepinephrine (NE), 553, 617, 618E
Normal curves of the vertebral column, 146
Normotensive, 423
Norplant, 752
Nose, 629, 630–632
surface anatomy of, 319E
Nosebleed, 632
Nostrils, 630
Notochord, 122
Nuclear envelope, 36
Nuclear magnetic resonance (NMR), 23E
Nuclear medicine, 24E
Nucleotide, 47
Nucleus
function of, 46E
of organelles, 36–39
Nucleus pulposus, 146
Nutrient canal, 120
Nutrient foramen, 120
Nutritional anemia, 345
Nystagmus, 598
Obesity, 687
causes, 688
classification, 687–688
definition, 687
treatment, 688
Obstetrics, 780
Obstructive jaundice, 674
Occlusion, 423
Occult blood, 682
Oil gland. *See* Sebaceous gland
Olfactory area, 631
Olfactory bulb, 518, 524–525, 574
Olfactory foramina, 139
Olfactory gland, 573
Olfactory sense, 573–574
olfactory pathway, 574
receptors, 573
Olfactory tract, 525, 574
Oligodendrocyte, 458E
Oncogene, 55
Oogenesis, 52, 738–739
Oogonia, 738
Oophorectomy, 763
Open fracture, 123
Open reduction, 123
Ophthalmologist, 576
Ophthalmology, 576
Ophthalmoscopy, 582
Optic chiasma, 525, 583
Optic disc, 581
Optician, 576
Optic tracts, 525
Optometrist, 576
Oral cavity. *See* Mouth
Oral contraception (OC), 752, 755E
Ora serrata, 578
Orbits, 144
Organ, 95
Organelles, 2, 30, 34–46
centrioles, 45
cilia, 46
cytoskeleton, 44–45
definition, 2, 36
endoplasmic reticulum, 39
flagella, 46
functions of, 46E
Golgi complex, 39
lysosomes, 39, 42
mitochondria, 42, 44
nucleus, 36–39
peroxisomes, 42
ribosomes, 39
Organism, 4
Organismic level of structural organization, 4
Organ level of structural organization, 2
Origin, 238
Oropharynx, 633
Orthopedics, 113
Orthopnea, 648
Orthostatic hypotension, 423
Osmosis, 33

Osseous tissue. *See* Bone tissue
Ossification, 115–118
  endochondral, 116, 118, 120
  intramembranous, 116, 120
Osteitis, 125
Osteitis fibrosa cystica, 614
Osteoarthritis, 125, 208
Osteoblast, 111, 120
Osteochondroma, 125
Osteoclast, 42, 111
Osteocyte, 86, 111
Osteogenesis, 115
Osteogenic layer, 112
Osteogenic sarcoma, 54
Osteoid, 116
Osteomalacia, 122–123
Osteomyelitis, 123
Osteon, 86, 114
Osteopenia, 125
Osteoporosis, 122
Osteoprogenitor cell, 111
Osteosarcoma, 125
Otalgia, 598
Otic ganglion, 553
Otitis media, 597
Otolithic membrane, 593
Otorhinolaryngology, 632
Otosclerosis, 598
Ovarian cycle, 743
Ovarian cyst, 762
Ovaries, 621, 621E, 734–739
Ovulation, 740, 745
  signs of, 745
Oxytocin (OT), 608, 609E

Pacemaker
  artificial, 364
  SA node, 362
Pacinian corpuscle. *See* Lamellated corpuscle
Paget's disease, 123
Pain sensation, 565–567
Palatine tonsils, 432
Palpation, 318
Palpebral fissure, 576
Palpitation, 373
Pancarditis, 373
Pancreas, 617–621, 621E, 669
  disorders of, 620, 621E
  effect of sympathetic and parasympathetic stimulation on, 555E
  hormones of, 617, 621E
Pancreatic amylase, 670
Pancreatic duct, 669
Pancreatic islet, 617, 670
Pancreatic juice, 670
Pancreatitis, 689
Pannus, 208
Papanicolaou (Pap) test, 740–741
Papillae, 574
Papilla of the hair, 102
Papillary plexus, 100
Papillary region, 98
Parallel after-discharge circuit, 468
Parallel pattern of fasciculi, 239
Paralysis, 231, 491, 541
Paranasal sinus, 130E, 144
Paraplegia, 490
Parasagittal plane, 11
Parasympathetic nervous system, 552–553
  structural features of, 553E
Parathyroid glands, 604, 612–614, 614E
  disorders of, 614, 614E
  hormone of, 612, 614E
Parathyroid hormone (PTH), 612, 614E
Paravertebral ganglia, 548
Parent cell, 47
Parietal layer, 215
Parietal peritoneum, 655
Parkinson's disease (PD), 538
Parotid glands, 659
Paroxysmal tachycardia, 373
Partial fracture, 123
Parturition, 784–786
Passive processes, 32, 33, 36
Patella, 174, 328E
Patellar ligament, 303
Patent ductus arteriosus, 370
Pathological anatomy, 2E
Pathologic fracture, 124
Pectinate muscle, 355
Pectoral girdle, 162
  muscles that move, 272E
Pedicle, 146
Pelvic axis, 170
Pelvic brim, 170
Pelvic cavity, 15, 15E
Pelvic diaphragm, 268E
Pelvic girdle, 167–170
Pelvic inflammatory disease (PID), 762
Pelvic outlet, 170
Pelvimetry, 170–171
Pelvis
  arteries of, 399E
  bones of, 167–170
  comparison of female and male, 178
  lymph nodes of, 442E–443E
  muscles of pelvic floor, 270E
  surface anatomy of, 322E, 323
  veins of, 410E
Penis, 731–733
Pennate, 239
Pepsin, 669
Peptic ulcer, 686
Peptidase, 679
Perception, 562
Percutaneous balloon valvuloplasty, 369
Percutaneous transluminal coronary angioplasty (PTCA), 369
Percutaneous ultrasonic lithotripsy (PUL), 711
Perforating canal, 113
Pericardial cavity, 15, 15E, 354
Pericardial fluid, 354
Pericarditis, 354
Pericardium, 15, 87, 353, 354

Perichondrium, 85, 116
Periderm, 104
Perimysium, 215
Perineum, 748–749
  muscles of, 270E
Perineurium, 482
Perinuclear cisterna, 36
Periodontal disease, 686
Periodontal ligament, 661
Periosteal bud, 116
Periosteum, 112, 116
Peripheral nerve repair, 480
Peripheral nervous system (PNS), 456
Peripheral protein, 30
Peristalsis, 665, 679, 681
Peritoneum, 15, 87, 655–656
Peritonitis, 686
Perivascular space, 502
Permanent cell, 64
Permanent dentition, 662
Pernicious anemia, 345
Peroxisomes, 42
  function of, 46E
Perspiration, 103
Phagocytic vesicle, 34
Phagocytosis, 34
Phalanges
  of the finger, 167
  of the foot, 176
Phantom pain, 566–568
Pharmacology, 57
Pharyngeal tonsil, 432
Pharynx, 629, 632–634
  muscles of, 256E
Pheochromocytoma, 617
Phlebitis, 423
Phosphate group, 48
Phospholipid bilayer, 30
Photophobia, 597
Physiology, 2
Pia mater, 472, 499
Piezoelectric effect, 120
Piles. *See* Hemorrhoids
Pimples, 103
Pineal gland, 621–622
Pinealocyte, 622
Pinkeye. *See* Conjunctivitis
Pinocytosis, 34
Pituitary dwarfism, 606–607
Pituitary gland, 604, 605–608, 609E
  adenohypophysis, 606
  development of, 622–623
  disorders of, 606–607, 609E
  hormones of, 606, 609E
  neurohypophysis, 607–608
Pivot joint, 192, 193E
Placenta, 417
Placenta previa, 777
Plane, 11–12
Plantar aponeurosis, 311E
Plantar flexion, 188, 190E
Plasma, 86, 336, 339
  chemical composition and description of substances in, 336E
Plasma cell, 342
Plasma membrane, 30–34
  chemistry and structure of, 30
  functions of, 30, 31, 46E
  movement of materials across, 32–34
Plasma membrane protein (PMP), 30
Plasmapheresis, 339
Plasma protein, 336
Platelet concentration, 348
Platelets. *See* Thrombocytes
Pleural cavity, 15, 15E, 639
Pleural membrane, 639
Pleurisy, 639
Plexus, 483–485
  brachial, 484–485, 487E
  cervical, 483, 484E
  lumbar, 486, 489, 489E
  pulmonary, 553
  renal, 704
  sacral, 489, 491E
Pneumocystis carinii pneumonia (PCP), 447
Pneumonectomy, 648
Pneumonia, 647
Pneumothorax, 639
Poliomyelitis (polio), 537
Polycystic disease, 713
Polycythemia, 346
Polyp, 107
Polyuria, 714
Pons, 508–509, 524E
Popliteal fossa, 303
Porphyria, 348
Portal triad, 675
Positron, 24E
Positron emission tomography (PET), 24E
Postcentral gyrus, 516
Posterior spinocerebellar tract, 478E, 570
Postganglionic fiber, 547
Pott's fracture, 123, 175
Preeclampsia, 786
Preganglionic fiber, 547
Pregnancy
  definition, 769
  development during, 769–773
  diagnosis of, 748
  ectopic, 770–771
Premenstrual syndrome (PMS), 761
Prenatal diagnostic techniques, 782–784
Preoptic region, 512
Prepuce, 733, 747
Presbyopia, 598
Pressure, 466
Pressure sensations, 565
Prevertebral ganglion, 550
Primary follicle, 738, 744
Primary hypertension, 422
Primary lysosome, 42
Primary ossification center, 116
Primary somesthetic area, 519

Primary tumor, 54
Prime mover muscle, 239–240
Primitive gut, 684
Primitive heart tube, 367
Primordial germ cell, 723
Proctodeum, 684
Proctology, 680
Proerythroblasts, 339, 340–341
Progeny, 57
Progesterone (PROG), 621, 621E, 624, 723E, 744
Projection, 562
Prolactin (PRL), 606, 609E
Pronation, 188, 190E
Pronator muscle, 241
Pronephros, 711
Prone position, 11
Prophase, 50, 51E
Proprioceptive sensation, 567–569
Proprioceptor, 563
Prosencephalon, 531
Prostatectomy, 760
Prostate gland, 730
  enlarged, 760
Prostatic urethra, 730
Prostatitis
  acute, 760
  chronic, 760
Proto-oncogene, 55
Protoplasmic astrocyte, 458E
Protraction, 188, 190E
Pruritus, 107
Pseudopodia, 34
Pseudostratified columnar epithelium, 64, 70E, 73
Psoriasis, 106
Pterygoid process, 139
Pterygopalantine ganglion, 553
Ptosis, 598
Pudendum. *See* Vulva
Puerperal fever, 786
Puerperium, 786
Pulmonary circulation, 417
Pulmonary edema, 648
Pulmonary embolism (PE), 648
Pulmonary plexus, 553
Pulmonary trunk, 358, 417
Pulmonary ventilation, 629
Pulsing electromagnetic fields (PEMFs), 125
Punch biopsy, 741
Punctate distribution, 563
Pupil, 318E, 578–579
Purkinje fiber. *See* Conduction myofiber
P wave, 364
Pyelitis, 713
Pyelonephritis, 713
Pyloric sphincter, 665
Pyloric stenosis, 665–666
Pylorospasm, 665–666
Pylorus, 665
Pyorrhea, 686
Pyramidal pathway, 572
Pyramid, 503

QRS wave, 364
Quadriplegia, 490

Radial keratotomy (RK), 578
Radial nerve damage, 485
Radical mastectomy, 751
Radiocarpal joint. *See* Wrist joint
Radiograph, 22E
Radiographic anatomy, 2E, 22E
Radioisotopic scanning, 24E
Rales, 648
Rami, 142, 483
Raphe, 718
Rapidly adapting receptor, 562
Raynaud's disease, 423
Receptor, 480–481
  of the autonomic nervous system, 554–556, 557
  as sense organs, 562, 567–569, 573
Receptor-mediated endocytosis, 34
Reciprocal transfusion, 384
Recruitment, 219
Rectum, 680
Red blood cells. *See* Erythrocytes
Red muscle fiber, 223
Red nucleus, 509, 517
Referred pain, 566
Reflex arc, 480
Reflexes, 479–481
Regeneration of epithelial tissue, 64
Regeneration of muscle tissue, 228–229
Regeneration of nervous tissue, 466–467
Regional anatomy, 2E
Regulating hormone, 606
Relaxin, 621, 621E, 624, 723E, 744
Remodeling, 119
Renal calculi, 711
Renal capsule, 695
Renal failure, 713
  acute, 713
  chronic, 713
Renal fascia, 698
Renal pelvis, 698
Renal plexus, 704
Renal pyramid, 698
Renal sinus, 695
Rennin, 669
Reproduction, 6
  definition, 718
Reproductive cell division, 47, 51–53
  meiosis, 47, 52–53
Reproductive system, 9E, 717–767
  aging and, 754
  birth control, 752–754, 755E
  development of, 754–757
  disorders of, 757–763
  female organs, 733–751
  gamete production, 718, 723
  key medical terms, 763
  male organs, 718–733
Resistance, 238
Respiration, 629

Respirator, 648
Respiratory area, 629–645
Respiratory bronchiole, 641
Respiratory distress syndrome (RDS) of the newborn, 647
Respiratory failure, 647
Respiratory system, 7E, 628–651
  aging and, 645
  conducting portion, 629
  development of, 645–646
  disorders of, 646–648
  key medical terms, 648–649
  organs of, 629, 630–645
  respiratory portion, 629
Responsiveness, 5
Retention, 710
Rete testis, 726
Reticular connective tissue, 79E, 85
Reticular fiber, 77
Reticular lamina, 64
Reticular region, 98
Reticulocyte count, 341
Retina (nervous tunic), 579–581
  detachment of, 581
Retinoblastoma, 55, 598
Retraction, 188, 190E
Retrograde degeneration, 467
Reverberating circuit, 467
Reye's syndrome, 540
Rh blood grouping system, 341
Rheumatism, 207
Rheumatoid arthritis (RA), 207–208
Rheumatology, 209
Rhinitis, 648
Rhinoplasty, 630
Rhombencephalon, 531
Rhythm method, 753
Rib fractures, 156
Ribonuclease, 671, 679
Ribosomes, 39
  function of, 46E
Ribs, 155–156
Rickets, 122
Right iliac region, 20E
Right lumbar region, 20E
Right lymphatic duct, 436–437
Rigidity, 538
Rigor mortis, 220–221
Rima glottidis, 635
Rima vestibuli, 635
Rod, 581
Root canal, 662
Root canal therapy, 662
Root of the hair, 102
Rotation, 187–188, 190E
Rotator cuff, 275E
Rotator muscle, 241
Rubber band ligation, 681
Rubrospinal tract, 479E
Runner's knee, 204
Rupture of the spleen, 434
Sacral parasympathetic outflow, 552
Sacral plexus, 489
  nerves of, 491E
Sacral vertebrae, 152
Sacrum, 152
Saddle joint, 192, 193E
Sagittal suture, 131
Saliva, 661
Salivary amylase, 661
Salivary gland, 555E, 658–661
Salpingectomy, 763
SA node. *See* Sinoatrial node
Sarcolemma, 87, 217
Sarcoma, 54
  osteogenic, 54
Sarcomere, 217
Sarcoplasm, 87, 217
Sarcoplasmic reticulum, 217
Satellite cell, 458E
Satiety center, 513
Scala vestibuli, 588
Scapula, 162
Scapular muscle, 275E
Schwann cell. *See* Neurolemmocyte
Sciatica, 491–492
Sclera (''white of the eye''), 578
Scleral venous sinus, 578
Scleroses, 538
Sclerotome, 229
Scoliosis, 156–157
Scotoma, 583
Scrotum, 718
Sebaceous ciliary gland, 576
Sebaceous gland, 102–103
  development of, 105
Sebum, 103
Secondary follicle, 744
Secondary hypertension, 422
Secondary lysosome, 42
Secondary ossification center, 118
Secondary somesthetic area, 519
Secondary spermatocyte, 724
Secondary tumor, 55
Second-class lever, 239
Secretin, 624
Secretion, 72
Secretory granule, 39
Secretory vesicle, 39
Section, 11–12
Segmental muscle, 292E
Segmentation, 669
Selective permeability, 32
Semen (seminal fluid), 731–732
Semen analysis, 731
Semicircular canal, 588
Semilunar ganglion, 527
Semilunar valve, 358
Seminalplasmin, 731
Seminal vesicle, 730–731
Senescence, 57
Senile macular degeneration (SMD), 581

Sensations, 561–563
characteristics of, 562–563
complexity of, 563
cutaneous, 563–567
definition, 562
gustatory, 574–576
levels of, 569
location of, 563
olfactory, 573–574
pain, 565–567
proprioceptive, 567–569
receptors, 567–569
simplicity of, 563
stimulus detected, 563
tactile, 563–565
thermoreceptive, 565
Sense organ, 562
Sensory areas of the cerebral cortex, 519–520
Sensory pathways, 569–571
cerebellar tract, 570
somatosensory cortex, 569–570
Septal defect, 370–372
Septicemia, 348
Serosa, 654–655
Serous membrane, 87
Serous pericardium, 354
Serum hepatitis. *See* Hepatitis B
Severe combined immunodeficiency disease (SCID), 449–450
Sex chromosome, 51, 723
Sex hormone, 616
Sex organs, effect of sympathetic and parasympathetic stimulation on, 555E
Sexually transmitted diseases (STDs), 757–759
Shaft of hair, 100
Shingles, 493
Shinsplints, 175
Shoulder, 195E
bones of, 162
muscles that move, 272E
surface anatomy of, 323, 323E
Shoulder blade. *See* Scapula
Shoulder joint, 195E
Shunt, 423
Siamese twins. *See* Conjoined twins
Sickle-cell anemia, 346
Simple epithelium, 64, 66E, 67E, 68E, 72
Simple gland, 73, 74
Simple goiter, 612
Simple-series circuit, 467
Sinoatrial (SA) node, 362
Sinusitis, 141
Sinusoids, 380
Skeletal muscles, 4E
abductor digiti minimi, 289E, 311E
abductor hallucis, 311E
abductor pollicis brevis, 241E, 288E, 303E
abductor pollicis longus, 283E
adductor brevis, 297E
adductor hallucis, 311E
adductor longus, 241E, 297E, 303E
adductor magnus, 297E, 303E
adductor pollicis, 288E
anconeus, 297E
anterior scalene, 293E
arytenoid, 260E
biceps brachii, 279E
biceps femoris, 303E
brachialis, 279E
brachioradialis, 279E
buccinator, 244E
bulbocavernosus, 270E
coccygeus, 268E
coracobrachialis, 275E
corrugator supercilii, 245E
cricothyroid, 260E
deep transverse perineus, 270E
deltoid, 275E
depressor labii inferioris, 241E, 244E
diagastric, 259E
diaphragm, 266E
dorsal interossei, 289E, 312E
effect of sympathetic and parasympathetic stimulation on, 555E
epicranius, 244E
external anal sphincter, 241E, 270E
extensor carpi radialis brevis, 283E
extensor carpi radialis longus, 283E
extensor carpi ulnarius, 241E, 283E
extensor digiti minimi, 283E
extensor digitorum, 283E
extensor digitorum brevis, 311E
extensor digitorum longus, 305E
extensor hallucis longus, 305E
extensor indicis, 283E
extensor pollicis brevis, 283E
extensor pollicis longus, 283E
external intercostal, 266E
external oblique, 263E
facial expression muscles, 244E, 245E
fasciculi arrangement, 239
flexor carpi radialis, 241E, 282E
flexor carpi ulnaris, 282E
flexor digiti mimini brevis, 289E, 311E
flexor digitorum brevis, 311E
flexor digitorum longus, 306E
flexor digitorum profundus, 282E
flexor digitorum superficialis, 282E
flexor hallucis brevis, 311E
flexor hallucis longus, 306E
flexor pollicis brevis, 288E
flexor pollicis longus, 282E
frontalis, 244E
gastrocnemius, 305E
genioglossus, 253E
geniohyoid, 259E
gluteus maximus, 296E
gluteus medius, 296E
gluteus minimus, 296E
gracilis, 303E
group actions, 239–240
hypoglossus, 253E
iliacus, 296E
iliococcygeus, 268E

iliocostalis lumborum, 292E
iliocostalis thoracis, 292E
inferior constrictor, 256E, 260E
inferior gemellus, 296E
inferior oblique, 251E
inferior rectus, 251E
infraspinatus, 275E
internal intercostals, 266E
internal oblique, 263E
interspinales, 293E
intertransversarii, 293E
ischiocavernosus, 270E
lateral cricoarytenoid, 260E
lateral pterygoid, 248E
lateral rectus, 251E
latissimus dorsi, 275E
levator ani, 268E
levator labii superioris, 244E
levator palpebrae superioris, 245E
levator scapulae, 241E, 272E
levator veli palatini, 254E
lever systems and leverage, 238–239
longissimus capitis, 262E, 292E
longissimus cervicis, 292E
longissimus thoracis, 292E
lumbricales, 289E, 311E
masseter, 248E
medial pterygoid, 248E
medial rectus, 251E
mentalis, 244E
middle constrictor, 256E, 260E
middle scalene, 293E
multifidus, 293E
musculus uvulae, 254E
mylohyoid, 259E
naming, 240
obturator externus, 296E
obturator internus, 241E, 296E
occipitalis, 244E
omohyoid, 260E
opponens digiti minimi, 289E
opponens pollicis, 288E
orbicularis oculi, 244E
orbicularis oris, 244E
origin and insertion of, 237–238
palatoglossus, 253E, 254E
palatopharyngeus, 254E, 256E, 260E
palmar interossei, 289E
palmaris brevis, 288E
palmaris longus, 282E
pectineus, 297E, 303E
pectoralis major, 272E, 275E
pectoralis minor, 275E
peroneus brevis, 305E
peroneus longus, 305E
peroneus tertius, 305E
piriformis, 296E
plantar interossei, 312E
plantaris, 305E
platysma, 244E
popliteus, 305E
posterior cricoarytenoid, 260E
posterior scalene, 293E
principal actions of, 241E
pronator quadratus, 279E
pronator teres, 241E, 279E
psoas major, 296E
pubococcygeus, 268E
quadratus femoris, 296E
quadratus lumborum, 263E
quadratus plantae, 311E
quadriceps femoris, 303E
rectus abdominis, 263E
rectus femoris, 303E
rhomboideus major, 272E
rhomboideus minor, 272E
risorius, 244E
rotatores, 293E
salpingopharyngeus, 256E
sartorius, 303E
semimembranosus, 303E
semispinalis capitis, 262E, 293E
semispinalis cervicis, 293E
semispinalis thoracis, 293E
semitendinosus, 303E
serratus anterior, 272E
soleus, 305E
spinalis capitis, 293E
spinalis cervicis, 293E
spinalis thoracis, 293E
splenius capitis, 262E, 292E
splenius cervicis, 292E
sternocleidomastoid, 262E
sternohyoid, 260E
sternothyroid, 260E
styloglossus, 253E
stylohyoid, 259E
stylopharyngeus, 256E, 260E
subclavius, 272E
subscapularis, 275E
superficial transverse perineus, 270E
superior constrictor, 256E
superior gemellus, 296E
superior oblique, 251E
superior rectus, 251E
supinator, 241E, 279E
supraspinatus, 275E
temporalis, 248E
tensor fasciae latae, 241E, 296E
tensor veli palatini, 254E
teres major, 275E
teres minor, 275E
thyroarytenoid, 260E
thyrohyoid, 260E
tibialis anterior, 305E
tibialis posterior, 306E
transversus abdominis, 263E
trapezius, 272E
triceps brachii, 279E
urethral sphincter, 270E
vastus intermedius, 303E
vastus lateralis, 303E

Skeletal muscles (*Continued*)
vastus medialis, 303E
zygomaticus major, 244E
Skeletal muscle tissue, 87, 89E, 213, 218, 229E
blood supply of, 215
connective tissue components, 213–215
contraction of, 222E
definition, 213
fascia, 213
histology, 217–218
nerve supply of, 215
relaxation of, 222E
types of, 223–228
Skeletal system, 4E
aging and, 120
appendicular skeleton, 129, 130, 131E, 162–179
axial skeleton, 129–157
definition, 111
development of, 120, 122
divisions of, 129–130, 131E
functions of, 111
histology, 111–114
ossification, 115–118
Skeleton of the heart, 358
Skin, 95–106
blood supply of, 100
color of, 98–99
dermis, 97–98
effect of sympathetic and parasympathetic stimulation on, 555E
epidermal ridges and grooves, 99–100
epidermis, 95–97
functions of, 98
products eliminated from, 695E
structure, 95
Skin cancer, 105
risk factors for, 105
Skull, 131–144
cranial bones, 131, 133–139
cranial fossae, 139
facial bones, 131, 133–139
fontanels, 130E, 131–132
foramina, 144, 145E
sutures, 131
Sliding-filament mechanism, 218
Slipped disc, 156
Slowly adapting receptor, 562
Small intestine, 675–679
blood supply of, 679
nerve supply of, 679
Smegma, 763
Smoke inhalation injury, 648
Smooth (visceral) muscle tissue, 87, 89E, 213, 225, 228, 229E
Soft palate, 657
muscles of, 254E
Solar plexus, 552
Somatic cell division, 47–50
cytokinesis, 50, 51E
mitosis, 48–50
Somatic nervous system (SNS), 456
autonomic nervous system compared with, 547, 548E
Somatic pain, 565
Somatic reflex, 481
Somatosensory cortex, 569–570
Somatostatin, 617
Somatotroph cell, 606
Somites, 229
mesoderms of, 229
Sonogram, 23E, 783–784
Spasm, 231
Spastic, 541
Spastic paralysis, 223, 573
Special senses, 573–597
auditory sensations, 583–595
equilibrium, 593–595
gustatory sensations, 574–576
olfactory sensations, 573–574
visual sensations, 576–583
S period, 48
activity of, 51E
Spermatic cord, 728
Spermatid, 724
Spermatogenesis, 52, 718, 719, 723–725
Spermatogonia, 719, 723–724
Spermatozoon, 719
Spermiation, 725
Spermiogenesis, 725
Sphincter
external, 708
of the hepatopancreatic ampulla, 678
ileocecal, 676, 679
internal, 708
lower esophageal, 665
precapillary, 380
pyloric, 665
upper esophageal, 665
Sphincter muscle, 241
Spider burst veins, 382
Spina bifida, 157
Spinal cord, 471–481
ascending tract, 472, 478E
coverings, 472–474
descending tract, 472, 478E
functions of, 478
general features, 474
injury to, 490–491
protection, 472–474
structure, 474–478
Spinal meninges, 472
Spinal nerves, 481–489
branches, 483–485
coverings, 481–483
dermatomes, 489–490
intercostal, 489
names, 481
plexus, 483–485
Spinal (lumbar) puncture (tap), 474
Spinal reflex, 481
Spinal segment, 474
Spinal shock, 490
Spine. *See* Vertebral column
Spiral fracture, 123
Spiral ganglion, 518

Spleen, 428, 432–434
effect of sympathetic and parasympathetic stimulation on, 553E
rupture of, 434
Splenectomy, 434
Splenius muscles, 292E
Splenomegaly, 451
Split-brain concept, 521
Sprain, 209
Spurs, 208
Squamosal suture, 131
Squamous cell, 64
Squamous cell carcinoma (SCC), 105
Squamous suture, 131
Stable cell, 64
Staging, 54
Stapedius muscle, 588
Static equilibrium, 593–594
Stereocilia, 726
Sterility. *See* Infertility
Sterilization, 752, 755E
Sternal puncture, 154–155
Sternocleidomastoid muscle, 320
Sternum, 154–155, 321E
Steroid, 223
Steroid hormone, 42
Stimulus, 463, 562
Stokes-Adams syndrome, 373
Stomach, 665–669
blood supply of, 669
effect of sympathetic and parasympathetic stimulation on, 555E
nerve supply of, 669
Stomach gastrin, 624
Strabismus, 598
Strain, 209
Stratified epithelium, 64, 68E, 69E, 72–73
Stratum basale, 95
Stratum corneum, 97
Stratum granulosum, 97
Stratum lucidum, 97
Stratum spinosum, 95
Stress fracture, 124
Stress-relaxation, 228
Striae, 98
Striated muscle tissue, 87, 213
Stricture, 714
Stroke, 536
Stroma, 77, 734–738
Structural organization, levels of, 2–5
Stupor, 541
Stylomastoid foramen, 138
Subcutaneous, 107
Subcutaneous (SC) layer, 84, 95
Subluxation, 209
Submandibular duct, 659
Submucosa, 654
Subneural cleft, 219
Substantia nigra, 509, 517
Subthalamic nucleus, 517
Subthalamus, 513, 524E
Subthreshold stimulus, 221
Sucrase, 679
Suction lipectomy, 85
Sudden infant death syndrome (SIDS), 647
Sudoriferous gland, 102
development of, 105
Sunburn, 106
Superficial fascia, 84, 95, 213
Superior vena cava (SVC), 358, 384, 402E
Supination, 188, 190E
Supinator muscle, 241
Supine position, 11
Suprahyoid muscle, 259
Supraoptic region, 512
Suprarenal gland. *See* Adrenal gland
Surface anatomy, 317–332
of the abdomen, 322E, 323
of the ankle, 328E, 331
of the arm, 323E, 326
of the buttocks, 328, 328E
of the chest, 320–323, 321E–322E
definition, 318
of the ear, 319E
of the elbow, 323E, 326
of the eye, 318E–319E
of the face, 319, 318E–319E
of the foot, 328E, 331
of the forearm, 325E, 326–327
of the hand, 326E, 327–328
of the head, 318, 318E
of the knee, 328E, 331
of the leg, 328E, 331
of the lips, 319E
of the lower extremity, 328, 328E, 331
of the neck, 319–320, 320E
of the nose, 319E
of the pelvis, 322E, 323
of the shoulder, 323, 323E
of the thigh, 320, 331, 328E
of the thorax, 320
of the trunk, 320–323, 321E–322E
of the upper extremity, 323, 323E, 325E–326E, 326–328
of the wrist, 325E, 327
Surfactant, 643
Suture, 131, 182, 193E
Swallowing, 665
Sweat, 103
Sweat gland, 555E
Sympathetic nervous system, 551–552
structural features of, 553E
Sympathetic trunk ganglia, 548
Sympathomimetic hormone, 617
Symphysis, 183, 193E
Sympto-thermal method of birth control, 753
Synapse
chemical, 464–465
electrical, 464–465
Synapsis, 52, 724
Synaptic cleft, 219, 465
Synaptic conduction, 465–466
Synaptic end-bulb, 218, 460
Synaptic gutter, 219

Synaptic vesicle, 218, 460, 465
Synarthroses, 182, 193
Synchondrosis, 183, 193E
Syncytiotrophoblast, 771
Syndesmosis, 182, 193E
Synergist, 240
Syngamy, 769
Synostoses, 182
Synovial fluid, 87, 183
Synovial joints, 182, 183–192, 193E
  ball-and-socket, 192, 193E
  ellipsoidal, 192, 193E
  gliding, 190, 193E
  hinge, 190, 193E
  movements of, 185–190, 190E
  pivot, 192, 193E
  saddle, 192, 193E
  structure, 183–185
  types, 190
Synovial membranes, 87, 183
Synovitis, 209
Syphilis, 757–759
System, 4, 95
Systemic anatomy, 4
Systemic circulation of blood, 384, 385E, 388E, 390E, 394E, 398E, 399E, 402E, 404E, 406E, 407E, 410E, 413E–414E
Systemic lupus erythematosus (SLE), 106
System level of structural organization, 3
Systole, 364

T_3 (triiodothyronine), 609, 612E
T_4 (thyroxine), 609, 612E
Tachypnea, 648
Tactile disc, 95
Tactile sensation, 563–565
Taeniae coli, 681
Talocrural joint. *See* Ankle joint
Target cell, 605
Tarsal gland, 576
Tarsals, 175
Taste pore, 574
Tay-Sachs disease, 539–540
T cell, 343, 622
Tectorial membrane, 588
Tectospinal tract, 479E
Tectum, 508
Teeth, 661–662
  blood supply of, 662
  nerve supply of, 662
Telencephalon, 536
Telophase, 50
Temporomandibular joint (TMJ), 143
TMJ syndrome, 143
Tendon, 85
Tendon organ, 568
Tendon sheath, 215
Tennis elbow, 198
Tenosynovitis, 215
Tension line, 98
Tensor muscle, 241
Terminal cistern, 217
Terminal ganglion, 550
Terminal web, 65
Tertiary bronchus, 638
Testes (testicles), 621, 621E, 718–725
  ducts of, 725–730
Testicular cancer, 759–760
Testosterone, 621, 621E, 723E
Tetany, 614
Tetralogy of Fallot, 372
Thalamus, 509–511, 524E
  anterior nucleus of, 511
Thalassemia, 346
Therapeutic plasma exchange (TPE), 339
Thermoreceptive sensation, 565
Thermoreceptor, 563
Thigh
  muscles that move, 296E, 297E
  surface anatomy of, 328, 328E, 331
Third-class lever, 239
Thirst center, 513
Thoracic aorta, 358
Thoracic cavity, 15E
Thoracic duct, 435–436
Thoracic vertebrae, 151, 152
Thoracolumbar outflow, 548
Thorax, 154–155
  lymph nodes of, 444E
  surface anatomy of, 320
  veins of, 410E
Threshold stimulus, 221
Throat. *See* Pharynx
Thrombectomy, 423
Thrombocytes (platelets), 86, 336, 344–345, 345E
Thrombocytopenia, 348
Thrombophlebitis, 434
Thymic corpuscle, 622
Thymic factor (TF), 622
Thymic humoral factor (THF), 622
Thymine nucleotide, 48
Thymopoietin, 622
Thymosin, 622
Thymus gland, 428, 434, 622, 622E
Thyroglobulin (TGB), 610
Thyroid cartilage. *See* Adam's apple
Thyroid gland, 608–612, 612E
  development of, 623
  disorders of, 611–612, 612E
  hormone of, 609, 612E
Thyroid-stimulating hormone (THS), 606, 609E
Thyroid storm, 624
Thyrotroph cell, 606
Thyroxine (T_4), 609, 612E
Tibia, 174–175
  muscles that act on, 303E
Tibia stress syndrome, 175
Tibiofemoral joint. *See* Knee joint
Tic, 231
Tic douloureux. *See* Trigeminal neuralgia
Tight junction, 64–65
Tinnitus, 598

Tissue, 63–90
blood, 83E, 86
bone, 83E, 86
connective, 63, 75–85
definition, 2, 63
epithelial, 63–73
muscular, 87–89
nervous, 63, 89
types, 63
Tissue level of structural organization, 2
Tissue rejection, 450
Toes, muscles that move, 305E, 306E
Tongue, 657–658
muscles that move, 253E
Tonofilament, 65
Tonsils, 428, 432
Topical, 107
Torn cartilage, 185
Torpor, 541
Torticollis, 231
Touch sensations, 563–564
Toxic shock syndrome (TSS), 761
Trabecula, 114, 116
Trabeculae carneae, 355
Trachea, 629, 635–638
Trachealis muscle, 636
Tracheostomy, 636–637
Trachoma, 596
Tracts, 462, 477
definition, 472
Transcendental meditation (TM), 558
Transducer, 23E
Transfusion, 348
direct, 347
exchange, 348
indirect, 348
reciprocal, 348
Transient ischemic attack (TIA), 537
Transitional cell, 64
Transplantation (transplant), 450
bone marrow, 344
corneal, 578
heart, 365–366
knee, 205
Transvaginal oocyte retrieval, 773
Transverospinalis muscle, 292E
Transverse arch of the foot, 178
Transverse fracture, 123
Transverse tubule, 217
Trapezius muscle, 320
Tremor, 517, 538
Triaxial joint, 192
Trichinosis, 231
Trichomoniasis, 759
Trigeminal neuralgia, 540
Trigone, 708
Triiodothyronine (T_3), 609, 609E
Trochanter, 130E
Trophoblast, 771
Tropic hormone, 606
Tropomyosin, 218
Tropomyosin-troponin complex, 218
Troponin, 218
True capillary, 380
True labor, 784
True rib, 155
Trunk, surface anatomy of, 320–323, 321E–322E
Trypsin, 670
Tubal ligation, 752
Tuberal region, 512
Tuber cinereum, 512, 605
Tubercle, 130E, 156
Tuberculosis (TB), 647
Tubular gland, 73, 74
Tubulin, 44
Tubuloacinar gland, 73, 74
Tumor, 53
brain, 537
of the GI tract, 686
grading, 54
of the male reproductive system, 760
malignant, 53
primary, 54
secondary, 55
Tumor angiogenesis factor (TAF), 55
Tumorigenesis, 55
Tunica externa, 378
Tunica interna, 378
Tunica media, 378
T wave, 364
Tympanic membrane. *See* Eardrum
Type I alveolar cell, 643
Type I cutaneous mechanoreceptor, 565
Type I diabetes, 620
Type I muscle fiber, 224
Type II alveolar cell, 643
Type II A muscle fiber, 224
Type II B muscle fiber, 224
Type II diabetes, 620
Type 2 herpes simplex virus, 55
Tyrosinase, 99

Ulcer
decubitus, 106
duodenal, 686
peptic, 686
Ulna, 164, 166
Ultrasound (US), 23E, 751
fetal, 783–784
Umbilical cord, 777
Umbilical cord accident, 777–778
Umbilical region, 20E
Umbilicus. *See* Navel
Unconsciousness, 504
Unicellular gland, 73
Unipennate muscle, 239
Upper extremities, 130, 162, 323–328
lymph nodes of, 438E
surface anatomy of, 323, 323E, 325E–326E, 326–328
veins of, 406E–407E
Upper respiratory system, 629
Uremia, 714

Ureter, 707
  blood supply of, 707
  effect of sympathetic and parasympathetic stimulation on, 555E
  nerve supply of, 707
Ureteric bud, 711
Urethra, 710–711, 729–730
Urethritis, 714
Urinary bladder, 707–710
  blood supply of, 710
  effect of sympathetic and parasympathetic stimulation on, 555E
  nerve supply of, 710
Urinary system, 8E, 694–716
  aging and, 711
  development of, 711
  disorders of, 711–713
  key medical terms, 713–714
  organs of, 695–711
Urinary tract infection (UTI), 713
Urine, 704
Urogenital diaphragm, 270
Urogenital sinus, 711
Urogenital triangle, 270
Urology, 695
Uterine tubes, 739–742
Uterus, 740–742
  cancer of, 740–741
  effect of sympathetic and parasympathetic stimulation on, 555E
  removal of, 763
Uvula, 657

Vacuole, 42
Vagina, 746–747
  lubrication of, 748
Vaginal orifice, 746, 747
Vaginitis, 763
Valves of the heart, 358–362
Valvular stenosis, 372
Varicose veins (VVs), 381–382
Vasa recta, 704
Vasa vasorum, 378
Vascular tissue. *See* Blood tissue
Vascular tunic, 578–579
Vasectomy, 728, 752
Vasoconstriction, 378
Vasodilation, 378
Vasomotor center, 504
Veins, 378, 381–382
  accessory hemiazygos, 410E
  anterior tibial, 414E
  arcuate, 704
  axillary, 407E
  azygos, 210E
  basilic, 407E
  as blood reservoirs, 382–383
  brachial, 407E
  brachiocephalic, 410E
  central retinal, 581
  cephalic, 407E
  common iliac, 410E
  definition, 378
  external iliac, 410E
  external jugular, 320, 404E
  femoral, 414E
  gastric, 417
  gastroepiploic, 417
  gonadal, 410E
  great cardiac, 381
  great saphenous, 413E
  hemiazygos, 410E
  hepatic, 410E, 417
  hepatic portal, 417
  inferior phrenic, 410E
  inferior vena cava, 384, 402E, 410E
  interlobar, 704
  interlobular, 704
  internal iliac, 410E
  internal jugular, 404E
  lumbar, 410E
  median antebrachial, 407E
  middle cardiac, 361
  pancreatic, 417
  popliteal, 414E
  posterior tibial, 414E
  pulmonary, 358, 417
  pyloric, 417
  radial, 407E
  renal, 410E, 704
  small saphenous, 413E
  spider burst, 382
  splenic, 417
  subclavian, 407E
  superior mesenteric, 417
  superior vena cava, 358, 384, 402E
  suprarenal, 410E
  of systemic circulation, 402E
  ulnar, 407E
  umbilical, 418, 420
  uterine, 742
  varicose, 381–382
  vascular sinus, 382
Venereal disease (VD), 757–759
Venesection, 348
Venipuncture, 334
Ventral body cavity, 15
Ventricle, 355, 499–502
Ventricular fibrillation (VF), 372
Ventricular premature contraction (VPC), 372
Venules, 378
  definition, 378
Vermiform appendix, 679
Vermilion, 656–657
Vermis, 522
Vertebral canal, 15
Vertebral column, 8, 145–154
  cervical region, 148, 150, 151
  disorders of, 156–157
  divisions of, 145–146
  fracture of, 157
  lumbar region, 152
  muscles that move, 292E, 293E
  sacrum, 152

structure, 146–147
thoracic region, 151, 152
Vertigo, 597
Vesicle
brain, 531–536
phagocytic, 34
seminal, 730–731
synaptic, 218, 460
Vestibular ganglion, 528
Vestibular nuclear complex, 508
Vestibule
of nasal cavity, 630
of oral cavity, 657
Vestibulospinal tract, 479E
Vibration sensation, 565
Viral encephalitis, 541
Virilism, 616
Virilizing adenoma, 616
Viscera, 15
Visceral autonomic reflex, 556–557
Visceral effector, 547, 557
Visceral efferent fiber, 547
Visceral efferent neuron, 547
Visceral peritoneum, 655
Visceral reflex, 481
Visceroceptor, 563
Visual sensation, 576–583
accessory structures of the eye, 576–578
structure of the eyeball, 578
visual pathway, 583
Vitamin D, 98, 624
deficiency of, 122
Vitiligo, 99
Vitreous body, 582
Vitreous chamber, 582
Vocal fold (vocal cord), 635
Volkmann's contracture, 231
Voluntary muscle tissue, 87, 213
Vomiting, 669
Vulva, 747–748
Vulvovaginal candidiasis, 763

Wallerian degeneration, 767
Wandering macrophage, 342
Wart, 107
''Watery'' eye, 577–578
Whiplash injury, 150
White blood cell (WBC). *See* Leukocyte
White coat hypertension, 423
White commissure, 474
White matter, 472, 474–477, 516
White muscle fiber, 223
White rami communicantes, 551
Whole blood, 348
Windpipe. *See* Trachea
Word blindness, 521
Word deafness, 521
Wrist, 199E
muscles that move, 282E, 283E
surface anatomy of, 325E, 327
Wrist crease, 325E, 327
Wrist joint, 199E, 208–209
Wryneck, 231

X chromosome, 51, 723
Xenograft, 451
Xeromammography, 750–751
Xiphoid process, 154

Y chromosome, 51, 723
Yellow marrow, 112
Yoga, 558
Yolk sac, 774–775

Z disc, 217
Zona fasciculata, 614
Zona glomerulosa, 614
Zona pellucida, 730, 769
Zona reticularis, 616
Zone of calcified matrix, 118
Zone of hypertrophic cartilage, 118
Zone of proliferating cartilage, 118
Zone of reserved cartilage, 118
Zygomatic arch, 142
Zygomaticofacial foramen, 141–142
Zygote, 51, 723
Zymogenic (peptic) cell, 666

MEASUREMENTS

METRIC UNITS OF LENGTH AND SOME U.S. EQUIVALENTS

Metric Unit	Meaning of Prefix	Metric Equivalent	U.S. Equivalent
1 kilometer (km)	kilo = 1,000	1,000m = 10^3 m	3,280.84 ft = 0.62 mi 1 mi = 1.61 km
1 hectometer (hm)	hecto = 100	100 m = 10^2 m	328 ft
1 dekameter (dam)	deka = 10	10 m = 10^1 m	32.8 ft
1 meter (m)		Standard unit of length	39.37 in. = 3.28 ft = 1.09 yd
1 decimeter (dm)	deci = $\frac{1}{10}$	0.1 m = 10^{-1} m	3.94 in.
1 centimeter (cm)	centi = $\frac{1}{100}$	0.01 m = 10^{-2} m	0.394 in. 1 in. = 2.54 cm
1 millimeter (mm)	milli = $\frac{1}{1,000}$	0.001 m = $\frac{1}{10}$ cm = 10^{-3} m	0.0394 in.
1 micrometer (μm) [formerly micron (μ)]	micro = $\frac{1}{1,000,000}$	0.000,001 m = $\frac{1}{10,000}$ cm = 10^{-6} m	3.94 × 10^{-5} in.
1 nanometer (nm) [formerly millimicron (mμ)]	nano = $\frac{1}{1,000,000,000}$	0.000,000,001 m = $\frac{1}{10,000,000}$ cm = 10^{-9} m	3.94 × 10^{-8} in.

METRIC UNITS OF MASS AND SOME U.S. EQUIVALENTS

Metric Unit	Metric Equivalent	U.S. Equivalent
1 kilogram (kg)	1,000 g	2.205 lb 1 ton = 907 kg
1 hectogram (hg)	100 g	
1 dekagram (dag)	10 g	0.0353 oz
1 gram (g)	1 g	1 lb = 453.6 g 1 oz = 28.35 g
1 decigram (dg)	0.1 g	
1 centigram (cg)	0.01 g	
1 milligram (mg)	0.001 g	0.015 g
1 microgram (μg)	0.000,001 g	
1 nanogram (ng)	0.000,000,001 g	
1 picogram (pg)	0.000,000,000,001 g	

METRIC UNITS OF VOLUME AND SOME U.S. EQUIVALENTS

Metric Unit	Metric Equivalent	U.S. Equivalent
1 liter (1)	1,000 ml	33.81 fl oz = 1.057 qt 946 ml = 1 qt
1 milliliter (ml)	0.001 liter	0.0338 fl oz 30 ml = 1 fl oz 5 ml = 1 teaspoon
1 cubic centimeter (cm^3)	0.999972 ml	0.0338 fl oz